JACARANDA
MATHS QUEST 10+10A

AUSTRALIAN CURRICULUM | FOURTH EDITION

CATHERINE SMITH

JAMES SMART

GEETHA JAMES

CAITLIN MAHONY

BEVERLY LANGSFORD WILLING

CONTRIBUTING AUTHORS

Michael Sheedy | Kahni Burrows | Paul Menta

jacaranda
A Wiley Brand

Fourth edition published 2022 by
John Wiley & Sons Australia, Ltd
42 McDougall Street, Milton, Qld 4064

First edition published 2012
Second edition published 2015
Third edition published 2018

Typeset in 10.5/14 pt Times LT Std

ISBN: 978-0-7303-9275-0

Front cover image: © Perepadia Y/Shutterstock

Illustrated by diacriTech and Wiley Composition Services

Typeset in India by diacriTech

A catalogue record for this book is available from the National Library of Australia

Printed in Singapore
M116291R1_130522

Contents

Everything you need (and *want*) at your fingertips

A full lesson on one screen in your online course. Trusted, curriculum-aligned theory. Engaging, rich multimedia. All the teacher support resources you need. Deep insights into progress. Immediate feedback for students. Create custom assignments in just a few clicks.

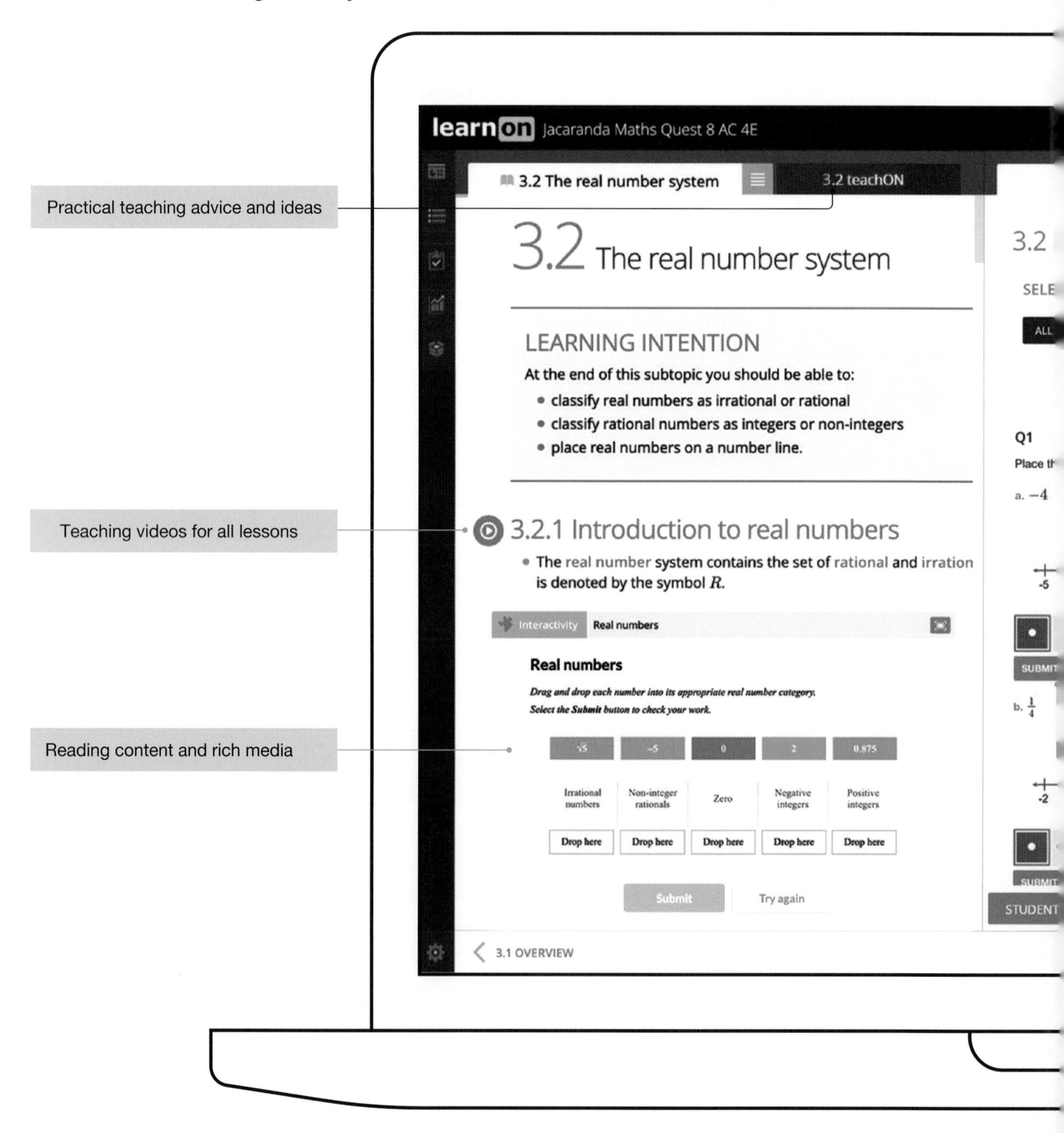

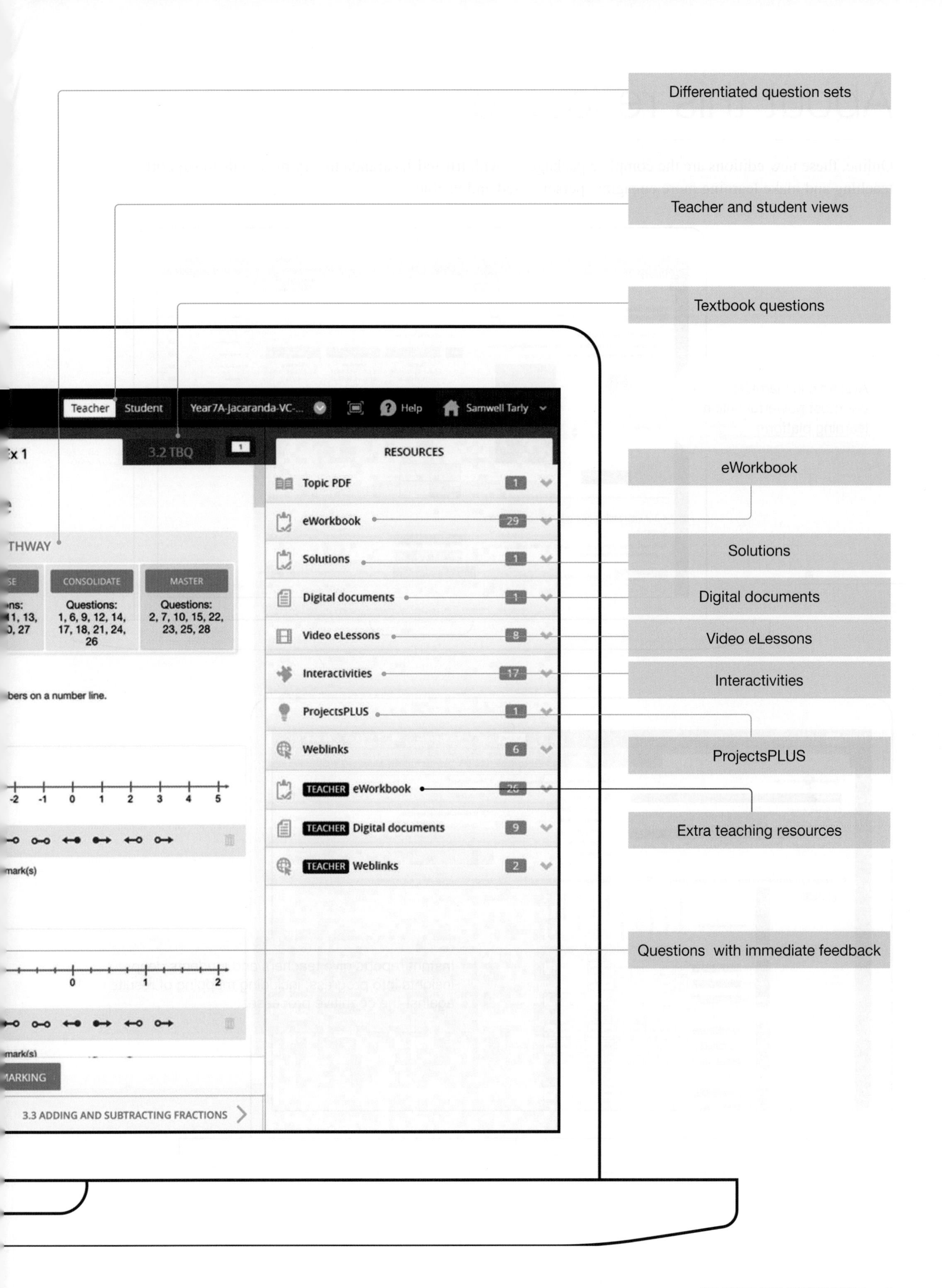

Differentiated question sets
Teacher and student views
Textbook questions
eWorkbook
Solutions
Digital documents
Video eLessons
Interactivities
ProjectsPLUS
Extra teaching resources
Questions with immediate feedback
Teacher
Student
Year7A-Jacaranda-VC-...
Help
Samwell Tarly
3.2 TBQ
RESOURCES
Topic PDF
eWorkbook
Solutions
Digital documents
Video eLessons
Interactivities
ProjectsPLUS
Weblinks
TEACHER eWorkbook
TEACHER Digital documents
TEACHER Weblinks
CONSOLIDATE
Questions: 1, 6, 9, 12, 14, 17, 18, 21, 24, 26
MASTER
Questions: 2, 7, 10, 15, 22, 23, 25, 28
3.3 ADDING AND SUBTRACTING FRACTIONS

About this resource

Online, these new editions are the complete package — with trusted Jacaranda theory plus tools to support teaching and make learning more engaging, personalised and visible.

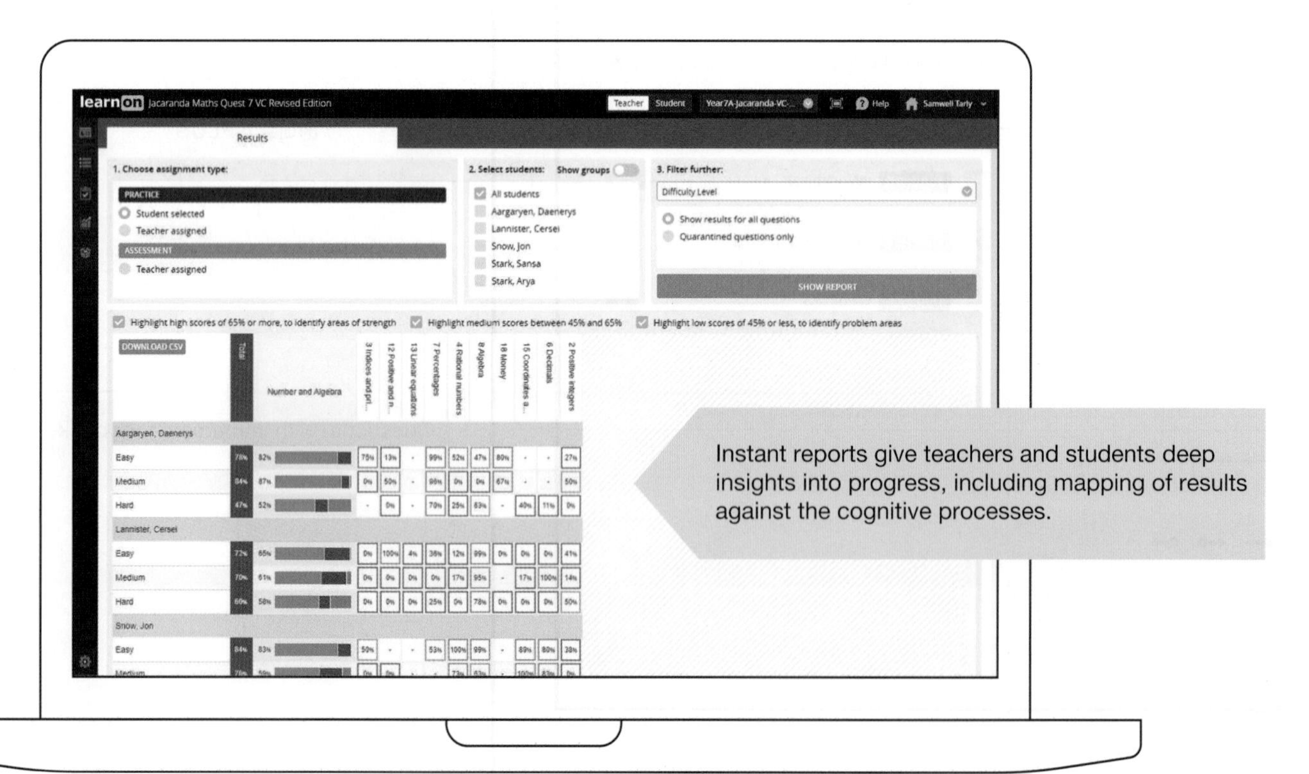

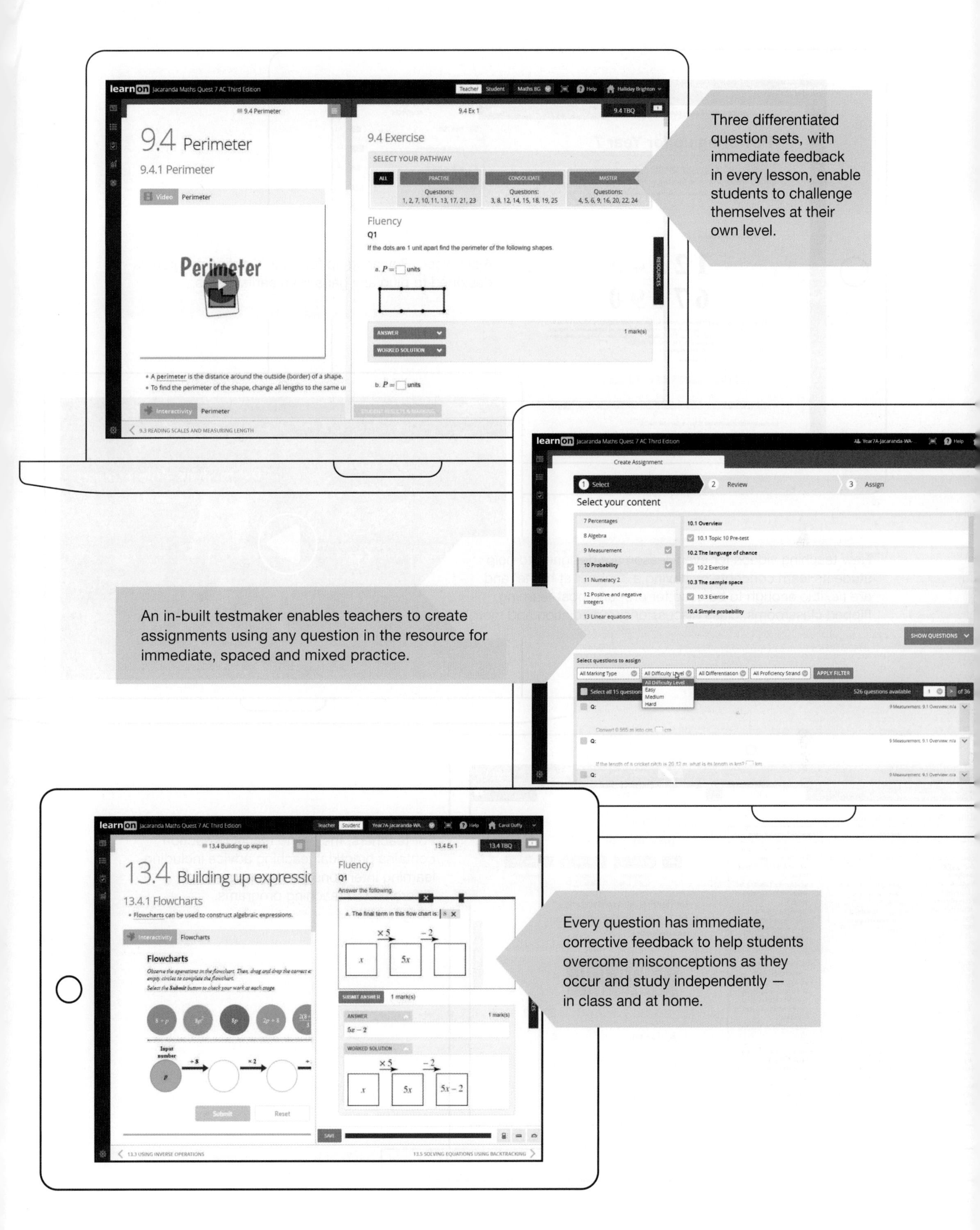
Three differentiated question sets, with immediate feedback in every lesson, enable students to challenge themselves at their own level.
An in-built testmaker enables teachers to create assignments using any question in the resource for immediate, spaced and mixed practice.
Every question has immediate, corrective feedback to help students overcome misconceptions as they occur and study independently — in class and at home.
9.4 Perimeter
9.4.1 Perimeter
9.4 Exercise
SELECT YOUR PATHWAY
Fluency
Select your content
13.4 Building up expressions
13.4.1 Flowcharts
Flowcharts

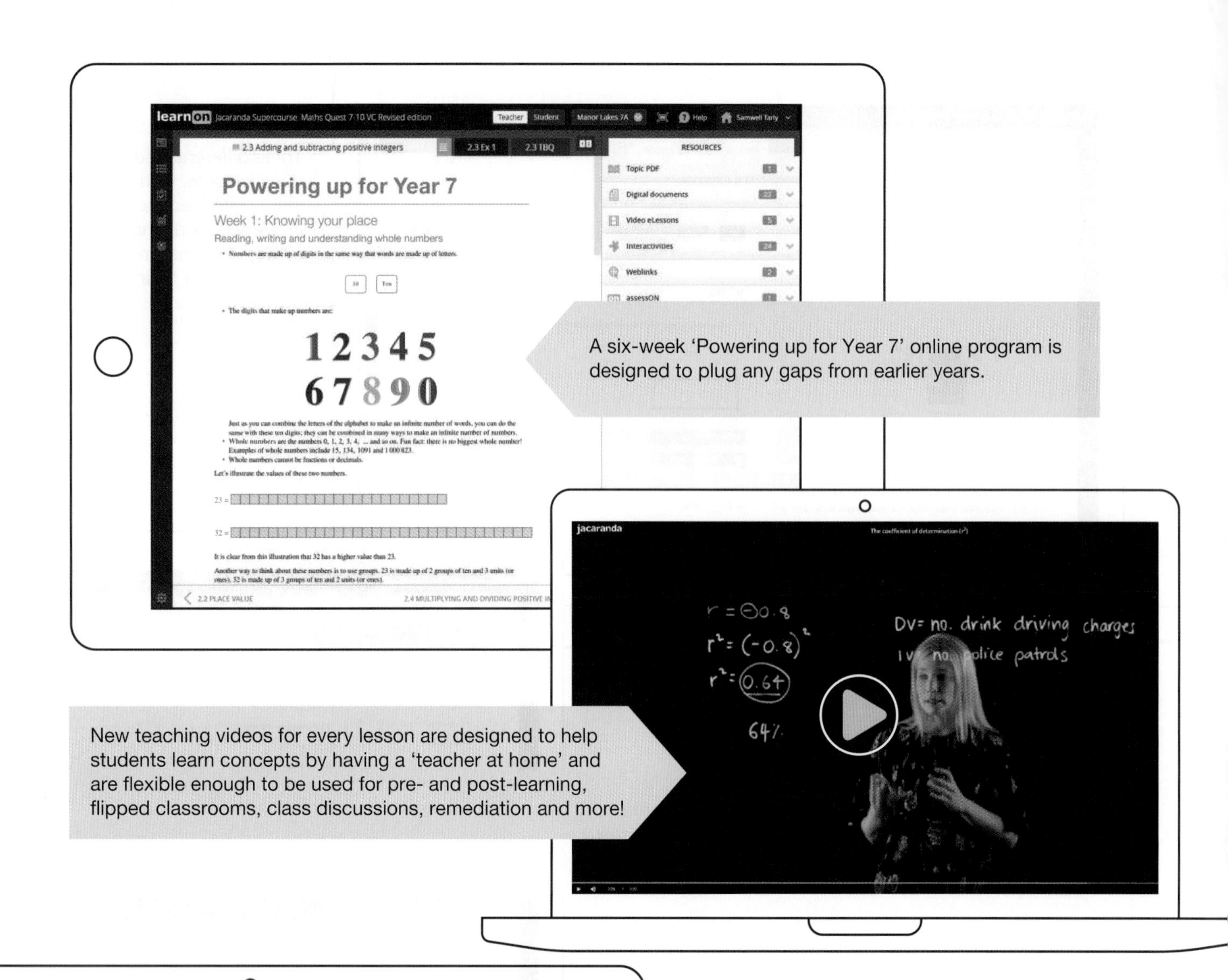

A six-week ‘Powering up for Year 7’ online program is designed to plug any gaps from earlier years.

New teaching videos for every lesson are designed to help students learn concepts by having a ‘teacher at home’ and are flexible enough to be used for pre- and post-learning, flipped classrooms, class discussions, remediation and more!

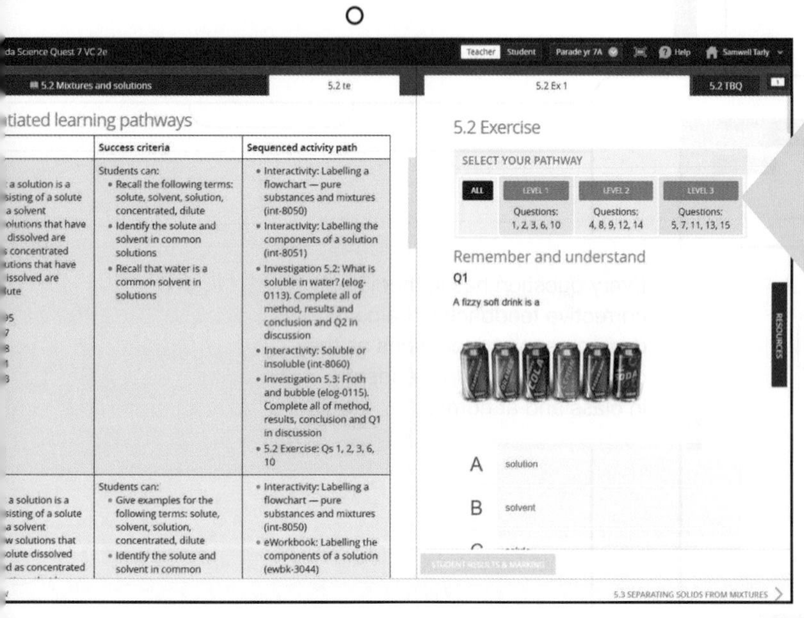

For teachers, the online teachON section contains practical teaching advice including learning intentions and three levels of differentiated teaching programs.

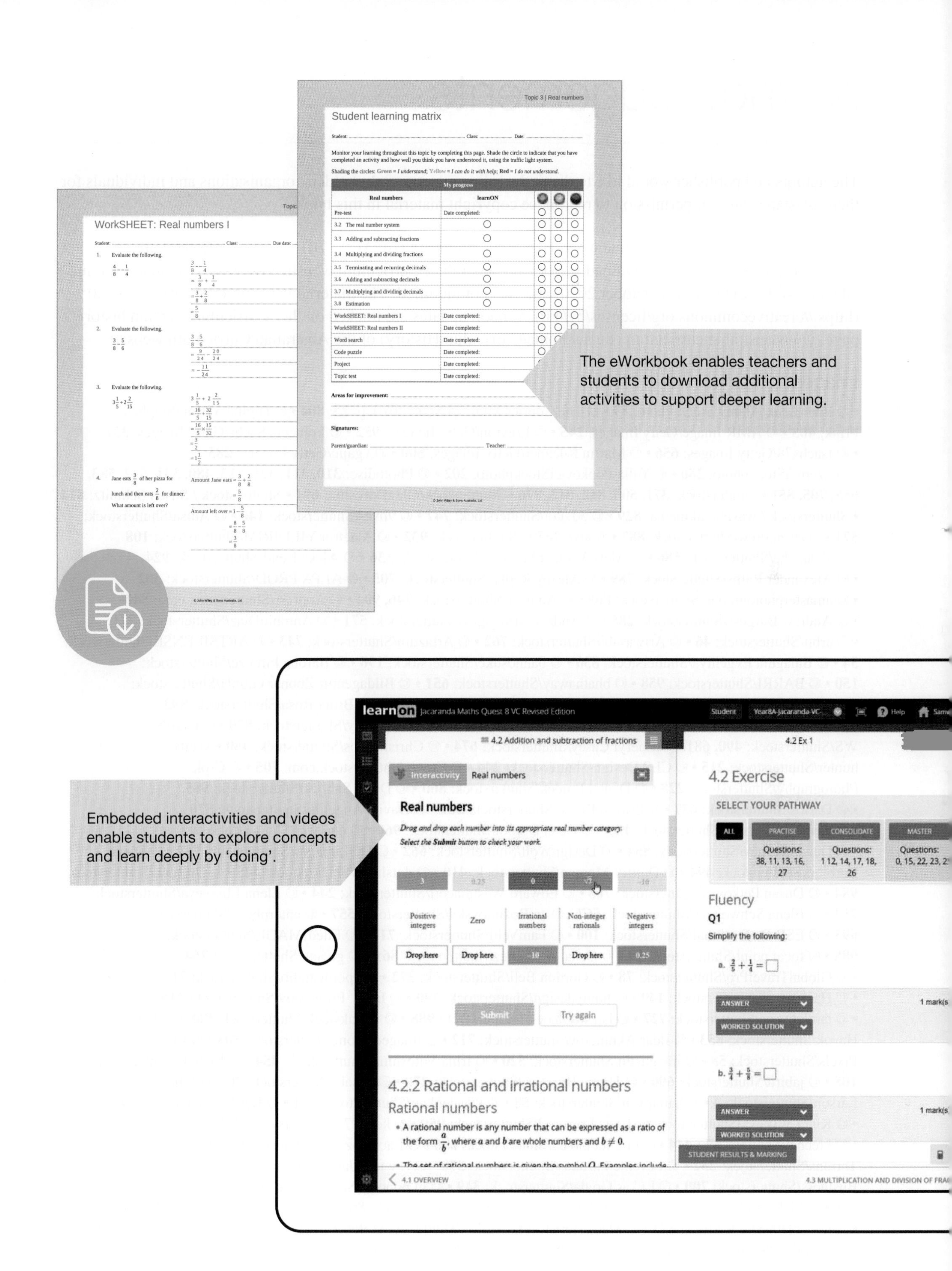
Topic 3 | Real numbers
Student learning matrix
Monitor your learning throughout this topic by completing this page. Shade the circle to indicate that you have completed an activity and how well you think you have understood it, using the traffic light system.
Shading the circles: Green = I understand; Yellow = I can do it with help; Red = I do not understand.
My progress
Real numbers
learnON
Pre-test
3.2 The real number system
3.3 Adding and subtracting fractions
3.4 Multiplying and dividing fractions
3.5 Terminating and recurring decimals
3.6 Adding and subtracting decimals
3.7 Multiplying and dividing decimals
3.8 Estimation
WorkSHEET: Real numbers I
WorkSHEET: Real numbers II
Word search
Code puzzle
Project
Topic test
Date completed:
Areas for improvement:
Signatures:
Parent/guardian:
Teacher:
WorkSHEET: Real numbers I
1. Evaluate the following.
2. Evaluate the following.
3. Evaluate the following.
4. Jane eats 3/8 of her pizza for lunch and then eats 2/8 for dinner. What amount is left over?
The eWorkbook enables teachers and students to download additional activities to support deeper learning.
Embedded interactivities and videos enable students to explore concepts and learn deeply by ‘doing’.
learnON Jacaranda Maths Quest 8 VC Revised Edition
4.2 Addition and subtraction of fractions
Interactivity Real numbers
Real numbers
Drag and drop each number into its appropriate real number category. Select the Submit button to check your work.
Positive integers
Zero
Irrational numbers
Non-integer rationals
Negative integers
Drop here
Submit
Try again
4.2.2 Rational and irrational numbers
Rational numbers
A rational number is any number that can be expressed as a ratio of the form a/b, where a and b are whole numbers and b ≠ 0.
4.2 Exercise
SELECT YOUR PATHWAY
ALL
PRACTISE
CONSOLIDATE
MASTER
Fluency
Q1
Simplify the following:
ANSWER
WORKED SOLUTION
STUDENT RESULTS & MARKING
4.1 OVERVIEW
4.3 MULTIPLICATION AND DIVISION OF FRA

Acknowledgements

The authors and publisher would like to thank the following copyright holders, organisations and individuals for their assistance and for permission to reproduce copyright material in this book.

© Australian Curriculum, Assessment and Reporting Authority (ACARA) 2010 to present, unless otherwise indicated. This material was downloaded from the Australian Curriculum website (www.australiancurriculum.edu.au) (Website) (accessed October 2020) and was not modified. The material is licensed under CC BY 4.0 (https://creativecommons.org/licenses/by/4.0). Version updates are tracked on the 'Curriculum version history' page (www.australiancurriculum.edu.au/Home/CurriculumHistory) of the Australian Curriculum website.

Images

• © Piter Lenk/Alamy Stock Photo: **89** • © The Courier Mail, 14 Sept. 2010, p. 25: **804** • © Digital Vision/Stephen Frink: **403** • © AMR Image/Getty Images: **245** • © Eraxion/Getty Images: **952** • © Francois Sachs/Getty Images: **357** • © fsachs78/Getty Images: **656** • © Martin Ruegner/Getty Images: **500** • © vgajic/Getty Images: **285** • © Szepy/iStockphoto: **280** • © Yulia Popkova/iStockphoto: **202** • © Photodisc: **310, 311, 391, 413, 480, 531, 532, 583, 658, 705, 851** • Shutterstock: **371, 505, 812, 813, 876** • Shutterstock/OlegDoroshin: **691** • Shutterstock / Susan Schmitz: **814** • Shutterstock / wavebreakmedia: **829** • © 3d_kot/Shutterstock: **747** • © 9lives/Shutterstock: **140** • © Adisa/Shutterstock: **571** • © agsandrew/Shutterstock: **887** • © aiyoshi597/Shutterstock: **932** • © Alaettin YILDIRIM/Shutterstock: **108** • © alean che/Shutterstock: **586** • © Aleix Ventayol Farrés/Shutterstock: **336** • © Aleks Kend/Shutterstock: **924** • © Alexander Raths/Shutterstock: **789** • © Alexey Repka/Shutterstock: **705** • © ALPA PROD/Shutterstock: **302** • © amasterphotographe/Shutterstock: **869** • © Andresr/Shutterstock: **746, 904** • © Andresr/Shutterstock.com: **840** • © Andrew Burgess/Shutterstock: **285** • © Andrey Armyagov/Shutterstock: **571** • © AntonioDiaz/Shutterstock: **771** • © arbit/Shutterstock: **46** • © Ariwasabi/Shutterstock: **762** • © Artazum/Shutterstock: **743** • © ARTSILENSE/Shutterstock: **34** • © Balaguta Evgeniya/Shutterstock: **830** • © bannosuke/Shutterstock: **196** • © Barone Firenze/Shutterstock: **150** • © BARRI/Shutterstock: **958** • © bhathaway/Shutterstock: **651** • © Bildagentur Zoonar GmbH/Shutterstock: **785** • © BlueSkyImage/Shutterstock: **171** • © Bobex-73/Shutterstock: **704** • © BrunoRosa/Shutterstock: **893** • © BSG_1974/Shutterstock: **1000** • © CC7/Shutterstock: **568** • © Cemil Aksoy/Shutterstock: **870** • © CHEN WS/Shutterstock: **490, 681** • © Cheryl Casey/Shutterstock: **674** • © Chrislofotos/Shutterstock: **460** • © city hunter/Shutterstock: **215** • © CLS Design/Shutterstock: **244** • © Crevis/Shutterstock.com: **705** • © Crok Photography/Shutterstock: **228** • © Daniel Prudek/Shutterstock: **860** • © Dario Sabljak/Shutterstock: **985** • © Darios/Shutterstock: **657** • © Darren Brode/Shutterstock: **261** • © DavideAngelini/Shutterstock: **570** • © davidundderriese/Shutterstock: **139** • © Dean Drobot/Shutterstock: **960** • © del-Mar/Shutterstock: **2** • © Denis Kuvaev/Shutterstock: **854** • © DesignWolf/Shutterstock: **662** • © DGLimages/Shutterstock: **792** • © Diego Barbieri/Shutterstock: **494** • © Dmitry Morgan/Shutterstock: **410** • © dotshock/Shutterstock: **445** • © DrHitch/Shutterstock: **984** • © Dusan Petkovic/Shutterstock: **763** • © Edward Westmacott/Shutterstock: **244** • © Elena Elisseeva/Shutterstock: **753** • © Elena Schweitzer/Shutterstock: **567** • © en Rozhnovsky/Shutterstock: **557** • © enterphoto/Shutterstock: **695** • © ESB Professional/Shutterstock: **100** • © FamVeld/Shutterstock: **710** • © FiledIMAGE/Shutterstock: **988** • © focal point/Shutterstock: **99** • © Gena Melendrez/Shutterstock: **861** • © givaga/Shutterstock: **764** • © GlobalTravelPro/Shutterstock: **78** • © Gordon Bell/Shutterstock: **372** • © gpointstudio/Shutterstock: **71** • © Hannari_eli/Shutterstock: **149** • © homydesign/Shutterstock: **790** • © Iakov Filimonov/Shutterstock: **216** • © iamlukyeee/Shutterstock: **727** • © ian woolcock/Shutterstock: **988** • © ibreakstock/Shutterstock: **868** • © Igor Havok/Shutterstock: **853** • © Ildar Akhmerov/Shutterstock: **712** • © imagedb.com/Shutterstock: **808** • © Inked Pixels/Shutterstock: **58** • © iravgustin/Shutterstock: **520** • © Irina Voloshina/Shutterstock: **454** • © Iscatel/Shutterstock: **108** • © jabiru/Shutterstock: **690** • © Jayme Burrows/Shutterstock: **270** • © Jirsak/Shutterstock: **26** • © Johan Larson/Shutterstock: **4** • © karamysh/Shutterstock: **58** • © kavalenkau/Shutterstock: **571** • © Keo/Shutterstock: **921** • © Klara Viskova/Shutterstock: **661** • © Kotruro2/Shutterstock: **959** • © koya979/Shutterstock: **407** • © krechet/Shutterstock: **444** • © l i g h t p o e t/Shutterstock: **286** • © LacoKozyna/Shutterstock: **125** • © Lee Torrens/Shutterstock: **232** • © Levente Fazakas/Shutterstock: **570** • © Liv friis-larsen/Shutterstock: **131** • © Lucky Business/Shutterstock: **700** • © Lukas Gojda/Shutterstock: **349** • © m.bonotto/Shutterstock: **413** • © Madlen/Shutterstock: **760** • © Maksym Dykha/Shutterstock: **168** • © Maria Maarbes/Shutterstock: **802** • © Maridav/Shutterstock: **349** • © max blain/Shutterstock: **843** • © Meawstory15 Production/Shutterstock: **960** • © Meder Lorant/Shutterstock: **714** • © Megapixel/Shutterstock: **113** • © mezzotint/Shutterstock: **775** • © mfauzisaim/Shutterstock: **782**

• © Micha Klootwijk/Shutterstock: **503** • © Mikhail Kolesnikov/Shutterstock: **557** • © Monkey Business Images/Shutterstock: **711, 781** • © mw2st/Shutterstock: **191** • © Natali Glado/Shutterstock: **803** • © Neale Cousland/Shutterstock: **476, 780** • © Nic Vilceanu/Shutterstock: **888** • © NigelSpiers/Shutterstock: **62** • © nito/Shutterstock: **504, 551** • © nyker/Shutterstock: **469** • © OlegDoroshin/Shutterstock: **691** • © Oleksandr Khoma/Shutterstock: **396** • © Oleksiy Mark/Shutterstock: **241, 508** • © Olga Danylenko/Shutterstock: **312** • © Omer N Raja/Shutterstock: **742** • © Orla/Shutterstock: **543** • © P Meybruck/Shutterstock: **340** • © PabloBenii/Shutterstock: **485** • © Palatinate Stock/Shutterstock: **618** • © Paulo M.F. Pires/Shutterstock: **77** • © PavleMarjanovic/Shutterstock: **791** • © Pavlo Baliukh/Shutterstock: **892** • © peacefoo/Shutterstock: **617** • © Peterfz30/Shutterstock: **87** • © peterschreiber.media/Shutterstock: **519** • © Photo Melon/Shutterstock: **652** • © photoiconix/Shutterstock: **407** • © Piotr Zajc/Shutterstock: **752** • © Popartic/Shutterstock: **406** • © Poprotskiy Alexey/Shutterstock: **412** • © Poznyakov/Shutterstock: **409** • © PrinceOfLove/Shutterstock: **1** • © psynovec/Shutterstock: **763** • © R. Gino Santa Maria/Shutterstock: **4** • © Rashevskyi Viacheslav/Shutterstock: **107** • © Repina Valeriya/Shutterstock: **348** • © Robyn Mackenzie/Shutterstock: **130** • © royaltystockphoto.com/Shutterstock: **566** • © Ruth Peterkin/Shutterstock: **480** • © science photo/Shutterstock: **957** • © Serg64/Shutterstock: **384** • © sirtravelalot/Shutterstock: **470** • © Soloviova Liudmyla/Shutterstock: **850** • © SpeedKingz/Shutterstock: **842** • © spotmatik/Shutterstock: **745** • © Stanislav Komogorov/Shutterstock.com: **450** • © Stefan Schurr/Shutterstock: **862** • © Steve Tritton/Shutterstock: **423** • © Stocker_team/Shutterstock: **457** • © Suzanne Tucker/Shutterstock: **710** • © Syda Productions/Shutterstock: **277, 916** • © Sylverarts Vectors/Shutterstock: **408** • © szefei/Shutterstock: **337** • © Tepikina Nastya/Shutterstock: **262** • © Tetiana Yurchenko/Shutterstock: **728** • © theskaman306/Shutterstock: **245** • © tickcharoen04/Shutterstock: **673** • © tj-rabbit/Shutterstock: **931** • © TnT Designs/Shutterstock: **81** • © topae/Shutterstock: **756** • © travelview/Shutterstock: **418** • © trekandshoot/Shutterstock: **239** • © Valdis Skudre/Shutterstock: **999** • © VectorMine/Shutterstock: **1033, 1034** • © Viaceslav/Shutterstock: **692** • © vikky/Shutterstock: **706** • © VILevi/Shutterstock: **200** • © Vladi333/Shutterstock: **424** • © wonlopcolors/Shutterstock: **313** • © XiXinXing/Shutterstock: **543** • © Yellowj/Shutterstock: **783** • © Yuriy Rudyy/Shutterstock: **277** • © zhu difeng/Shutterstock: **558** • © Zurijeta/Shutterstock: **853** • © Viewfinder Australia Photo Library: **52** • © Amy Johansson/Shutterstock: **525** • © John Carnemolla/Shutterstock: **525** • © Nadezda Cruzova/Shutterstock: **751** • © Paul D Smith/Shutterstock: **525**

Every effort has been made to trace the ownership of copyright material. Information that will enable the publisher to rectify any error or omission in subsequent reprints will be welcome. In such cases, please contact the Permissions Section of John Wiley & Sons Australia, Ltd.

[illegible]

• © sirtravelalot/Shutterstock, 428 • © Solovyova Lyudmyla/Shutterstock, 850 • © SpeedKingz/Shutterstock, 842
• © spotmatik/Shutterstock, 745 • © Stanislav Komogorov/Shutterstock.com, 450 • © Stefan Schurr/Shutterstock,
862 • © Steve [illegible]/Shutterstock, 423 • © Stocker_team/Shutterstock, 457 • © Suzanne Tucker/Shutterstock, 710
• © Syda Productions/Shutterstock, 277, 916 • [illegible]

[illegible]

Every effort has been made to trace the ownership of copyright material. Information that will enable the publisher to rectify any error or omission in subsequent reprints will be welcome. In such cases, please contact the Permissions Section of John Wiley & Sons Australia, Ltd.

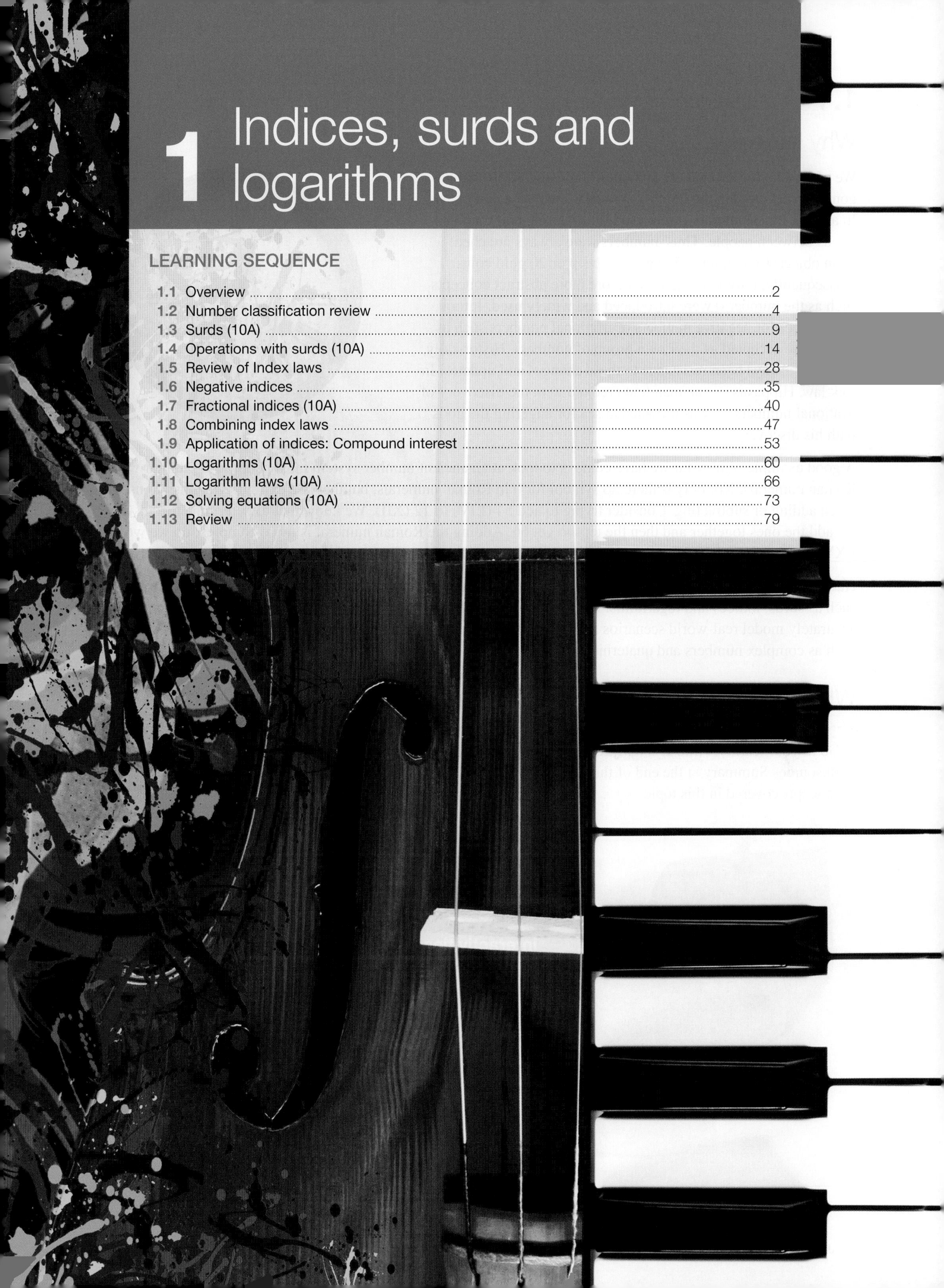

1 Indices, surds and logarithms

LEARNING SEQUENCE

1.1 Overview

Why learn this?

We often take for granted the amount of time and effort that has gone into developing the number system we use on a daily basis. In ancient times, numbers were used for bartering and trading goods between people. Thus, numbers were always attached to an object; for example, 5 cows, 13 sheep or 20 gold coins. Consequently, it took a long time before more abstract concepts such as the number 0 were introduced and widely used. It took even longer for negative numbers or irrational numbers such as surds to be accepted as their own group of numbers. Historically, there has always been resistance to these changes and updates. In folk law, Hippasus — the man first credited with the discovery of irrational numbers — was drowned at sea for angering the gods with his discovery.

A good example of how far we have come is to look at an ancient number system most people are familiar with: Roman numerals. Not only is there no symbol for 0 in Roman numerals, but they are extremely clumsy to use when adding or subtracting. Consider trying to add 54 (LIV) to 12 (XII). We know that to determine the answer we add the ones together and then the tens to get 66. Adding the Roman numeral is more complex; do we write LXVIII or LIVXII or LVXI or LXVI?

Having a better understanding of our number system makes it easier to understand how to work with concepts such as surds, indices and logarithms. By building our understanding of these concepts, it is possible to more accurately model real-world scenarios and extend our understanding of number systems to more complex sets, such as complex numbers and quaternions.

Where to get help

Go to your learnON title at **www.jacplus.com.au** to access the following digital resources. The Online Resources Summary at the end of this topic provides a full list of what's available to help you learn the concepts covered in this topic.

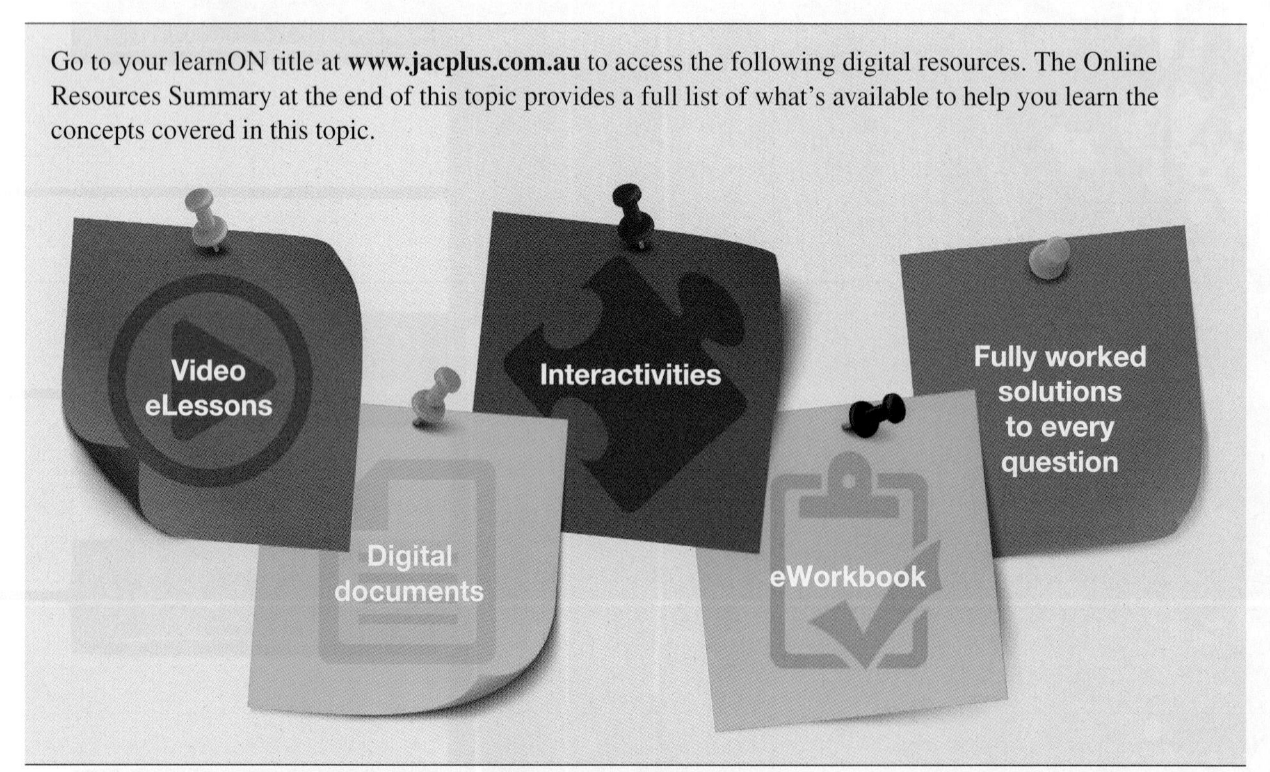

Exercise 1.1 Pre-test

Complete this pre-test in your learnON title at www.jacplus.com.au and receive **automatic marks, immediate corrective feedback** and **fully worked solutions.**

1. Positive numbers are also known as natural numbers. Is this statement true or false?
2. State whether $\sqrt{36}$ is a rational or irrational number.
3. Simplify the following: $3n^{\frac{1}{5}} \times 5n^{\frac{1}{3}}$.
4. Simplify the following: $\sqrt[5]{32p^{10}q^{15}}$.
5. Determine the exact value of $81^{-\frac{3}{4}}$.
6. MC Select which of the numbers of the set $\left\{\sqrt{0.25},\ \pi,\ 0.\overline{261},\ -5,\ \frac{2}{3}\right\}$ are rational.

   A. $\left\{\sqrt{0.25},\ \pi,\ 0.\overline{261}\right\}$
   B. $\left\{0.261,\ -5,\ \frac{2}{3}\right\}$
   C. $\left\{\pi, 0.\overline{261}\right\}$
   D. $\left\{\sqrt{0.25},\ 0.\overline{261},\ -5,\ \frac{2}{3}\right\}$
   E. $\left\{\sqrt{0.25},\ \pi,\ 0.\overline{261},\ -5\right\}$
7. MC $\dfrac{12x^8 \times 3x^7}{9x^{10} \times x^3}$ simplifies to:

   A. $\dfrac{5x^2}{3}$
   B. $4x^2$
   C. $4x^{26}$
   D. $\dfrac{5x^{26}}{3}$
   E. $\dfrac{x^2}{4}$
8. Simplify the following expression: $3\sqrt{2} \times \sqrt{10}$.
9. Simplify the following expression: $5\sqrt{2} + 12\sqrt{2} - 3\sqrt{2}$.
10. MC Choose the most simplified form of the following expression: $\sqrt{8a^3} + \sqrt{18a} + \sqrt{a^5}$

    A. $5\sqrt{2a} + a\sqrt{a}$
    B. $2a\sqrt{2a^2} + 3\sqrt{2a} + a^4\sqrt{a}$
    C. $2a^2\sqrt{2a} + 2\sqrt{3a} + a^4\sqrt{a}$
    D. $2a^2\sqrt{2a} + 2\sqrt{3a} + a^2\sqrt{a}$
    E. $2a\sqrt{2a} + 3\sqrt{2a} + a^2\sqrt{a}$
11. Solve the following equation for y: $\dfrac{1}{125} = 5^{y+2}$.
12. Solve the following equation for x: $x = \log_{\frac{1}{4}} 16$.
13. Calculate the amount of interest earned on an investment of \$3000 compounding annually at 3% p. a. for 3 years, correct to the nearest cent.
14. Simplify the following expression. $\log_2\left(\frac{1}{4}\right) + \log_2(32) - \log_2(8)$.
15. MC Choose the correct value for x in $3 + \log_2 3 = \log_2 x$.

    A. $x = 0$
    B. $x = 3$
    C. $x = 9$
    D. $x = 24$
    E. $x = 27$

1.2 Number classification review

LEARNING INTENTION

At the end of this subtopic you should be able to:
- define the real, rational, irrational, integer and natural numbers
- determine whether a number is rational or irrational.

1.2.1 The real number system

eles-4661

- The number systems used today evolved from a basic and practical need of primitive people to count and measure magnitudes and quantities such as livestock, people, possessions, time and so on.
- As societies grew and architecture and engineering developed, number systems became more sophisticated. Number use developed from solely whole numbers to fractions, decimals and irrational numbers.

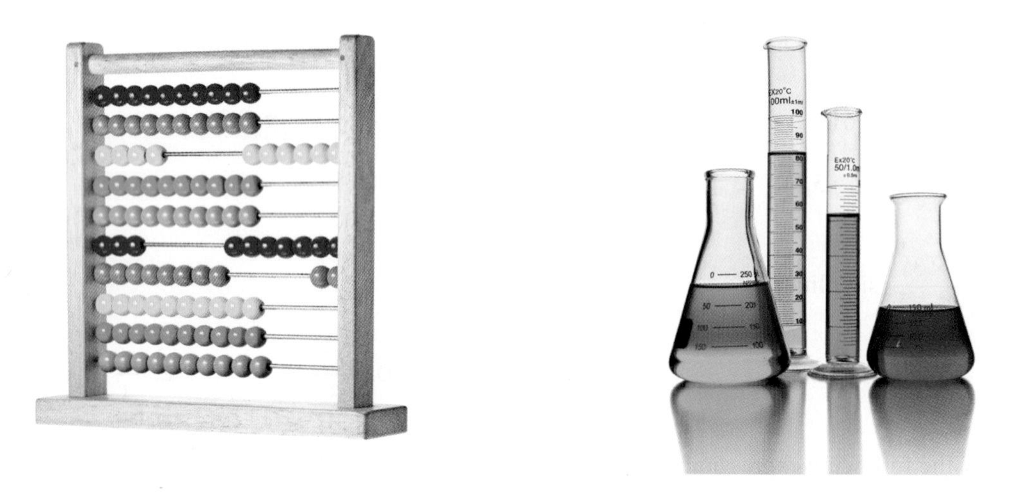

- The real number system contains the set of rational and irrational numbers. It is denoted by the symbol R. The set of real numbers contains a number of subsets which can be classified as shown in the chart below.

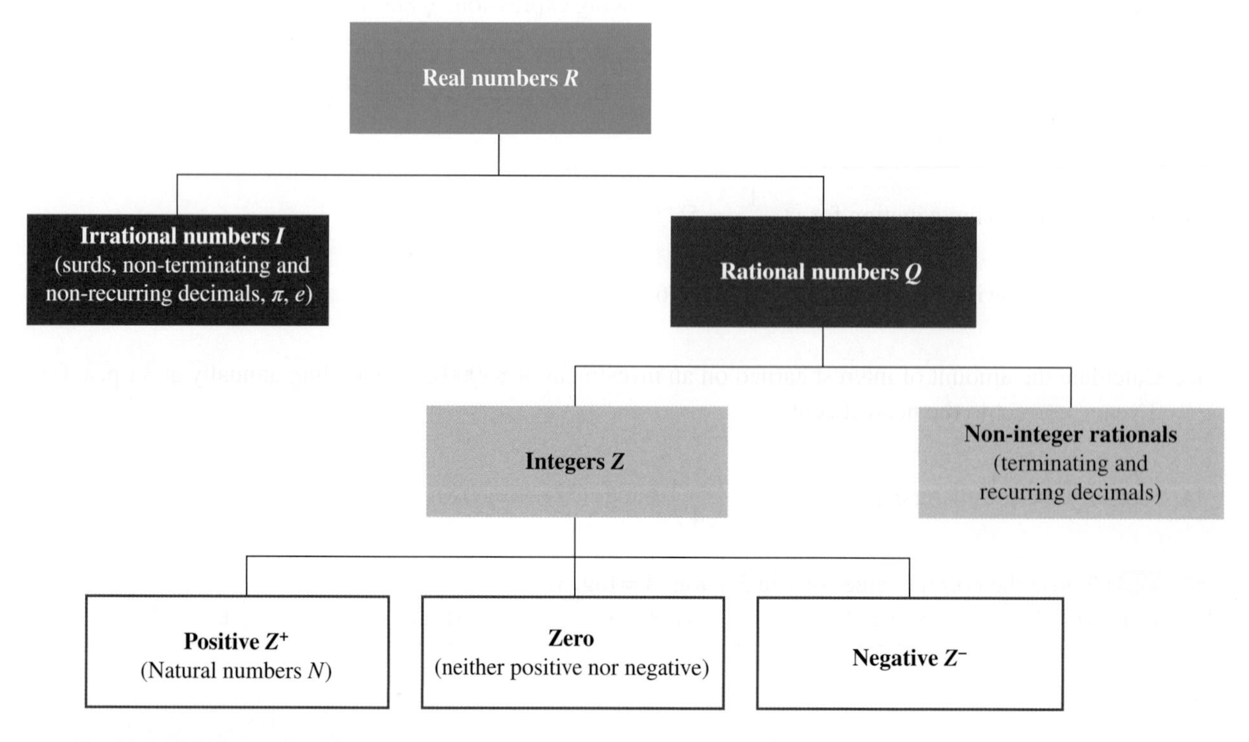

Integers (Z)

- The set of **integers** consists of whole positive and negative numbers and 0 (which is neither positive nor negative).
- The set of integers is denoted by the symbol Z and can be visualised as:

$$Z = \{\ldots, -3, -2, -1, 0, 1, 2, 3, \ldots\}$$

- The set of positive integers are known as the **natural numbers** (or counting numbers) and is denoted Z^+ or N. That is:

$$Z^+ = N = \{1, 2, 3, 4, 5, 6, \ldots\}$$

- The set of negative integers is denoted Z^-.

$$Z^- = \{\ldots -6, -5, -4, -3, -2, -1\}$$

- Integers may be represented on the number line as illustrated below.

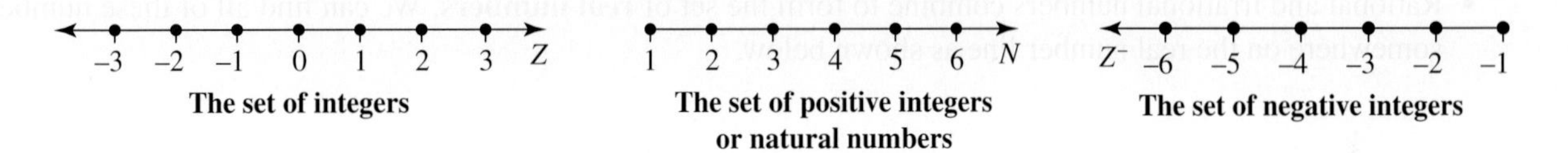

The set of integers **The set of positive integers or natural numbers** **The set of negative integers**

Rational numbers (Q)

- A **rational number** is a number that can be expressed as a ratio of two integers in the form $\frac{a}{b}$, where $b \neq 0$.
- The set of rational numbers are denoted by the symbol Q.
- Rational numbers include all whole numbers, fractions and all terminating and recurring decimals.
- **Terminating decimals** are decimal numbers which terminate after a specific number of digits.
  Examples are:

$$\frac{1}{4} = 0.25, \frac{5}{8} = 0.625, \frac{9}{5} = 1.8.$$

- **Recurring decimals** do not terminate but have a specific digit (or number of digits) repeated in a pattern.
  Examples are:

$$\frac{1}{3} = 0.333\,333\ldots = 0.\dot{3} \text{ or } 0.\overline{3}$$

$$\frac{133}{666} = 0.199\,699\,699\,6\ldots = 0.1\dot{9}9\dot{6} \text{ or } 0.1\overline{996}$$

- Recurring decimals are represented by placing a dot or line above the repeating digit/s.
- Using set notations, we can represent the set of rational numbers as:

$$Q = \left\{\frac{a}{b} : a, b \in Z, b \neq 0\right\}$$

- This can be read as 'Q is all numbers of the form $\frac{a}{b}$ given a and b are integers and b is not equal to 0'.

Irrational numbers (I)

- An **irrational number** is a number that cannot be expressed as a ratio of two integers in the form $\frac{a}{b}$, where $b \neq 0$.
- All irrational numbers have a decimal representation that is non-terminating and non-recurring. This means the decimals do not terminate and do not repeat in any particular pattern or order.
For example:

$$\sqrt{5} = 2.236\,067\,997\,5\,...$$
$$\pi = 3.141\,592\,653\,5\,...$$
$$e = 2.718\,281\,828\,4\,...$$

- The set of irrational numbers is denoted by the symbol I. Some common irrational numbers that you may be familiar with are $\sqrt{2}, \pi, e, \sqrt{5}$.
- The symbol π **(pi)** is used for a particular number that is the circumference of a circle whose diameter is 1 unit.
- In decimal form, π has been calculated to more than 29 million decimal places with the aid of a computer.

Rational or irrational

- Rational and irrational numbers combine to form the set of **real numbers**. We can find all of these number somewhere on the real number line as shown below.

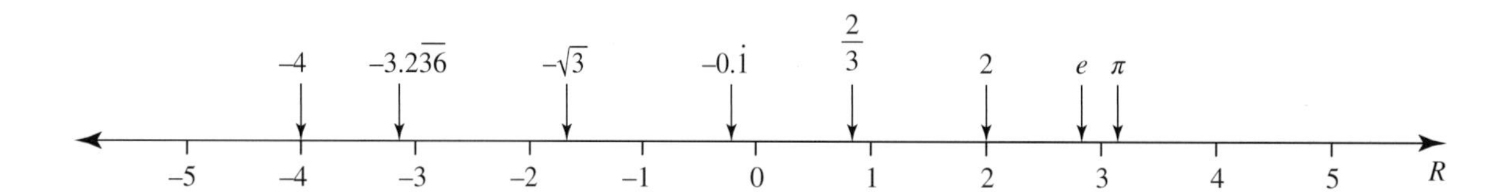

- To classify a number as either rational or irrational:
  1. Determine whether it can be expressed as a whole number, a fraction, or a terminating or recurring decimal.
  2. If the answer is yes, the number is rational. If no, the number is irrational.

WORKED EXAMPLE 1 Classifying numbers as rational or irrational

Classify whether the following numbers are rational or irrational.

a. $\frac{1}{5}$ b. $\sqrt{25}$ c. $\sqrt{13}$ d. 3π

e. 0.54 f. $\sqrt[3]{64}$ g. $\sqrt[3]{32}$ h. $\sqrt[3]{\frac{1}{27}}$

THINK	WRITE
a. $\frac{1}{5}$ is already a rational number.	a. $\frac{1}{5}$ is rational.
b. 1. Evaluate $\sqrt{25}$.	b. $\sqrt{25} = 5$
2. The answer is an integer, so classify $\sqrt{25}$.	$\sqrt{25}$ is rational.
c. 1. Evaluate $\sqrt{13}$.	c. $\sqrt{13} = 3.605\,551\,275\,46\,...$
2. The answer is a non-terminating and non-recurring decimal; classify $\sqrt{13}$.	$\sqrt{13}$ is irrational.

d. 1. Use your calculator to find the value of 3π.	**d.** $3\pi = 9.424\,777\,960\,77\ \ldots$
2. The answer is a non-terminating and non-recurring decimal; classify 3π.	3π is irrational.
e. 0.54 is a terminating decimal; classify it accordingly.	**e.** 0.54 is rational.
f. 1. Evaluate $\sqrt[3]{64}$.	**f.** $\sqrt[3]{64} = 4$
2. The answer is a whole number, so classify $\sqrt[3]{64}$.	$\sqrt[3]{64}$ is rational.
g. 1. Evaluate $\sqrt[3]{32}$.	**g.** $\sqrt[3]{32} = 3.17480210394\ldots$
2. The result is a non-terminating and non-recurring decimal; classify $\sqrt[3]{32}$.	$\sqrt[3]{32}$ is irrational.
h. 1. Evaluate $\sqrt[3]{\frac{1}{27}}$.	**h.** $\sqrt[3]{\frac{1}{27}} = \frac{1}{3}$.
2. The result is a number in a rational form.	$\sqrt[3]{\frac{1}{27}}$ is rational.

on Resources

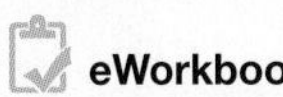

eWorkbook Topic 1 Workbook (worksheets, code puzzle and project) (ewbk-2027)

Interactivities Individual pathway interactivity: Number classification review (int-8332)
The number system (int-6027)
Recurring decimals (int-6189)

Exercise 1.2 Number classification review

learnon

Individual pathways

■ PRACTISE	■ CONSOLIDATE	■ MASTER
1, 4, 7, 10, 13, 14, 17, 20, 23	2, 5, 8, 11, 15, 18, 21, 24	3, 6, 9, 12, 16, 19, 22, 25

To answer questions online and to receive **immediate corrective feedback** and **fully worked solutions** for all questions, go to your learnON title at www.jacplus.com.au.

Fluency

For questions 1 to 6, classify whether the following numbers are rational (Q) or irrational (I).

1. **WE1**

a. $\sqrt{4}$ b. $\frac{4}{5}$ c. $\frac{7}{9}$ d. $\sqrt{2}$

2. a. $\sqrt{7}$ b. $\sqrt{0.04}$ c. $2\frac{1}{2}$ d. $\sqrt{5}$

3. a. $\frac{9}{4}$ b. 0.15 c. -2.4 d. $\sqrt{100}$

4. a. $\sqrt{14.4}$ b. $\sqrt{1.44}$ c. π d. $\sqrt{\frac{25}{9}}$

5. a. 7.32 b. $-\sqrt{21}$ c. $\sqrt{1000}$ d. 7.216 349 157 ...

6. a. $-\sqrt{81}$ b. 3π c. $\sqrt[3]{62}$ d. $\sqrt{\frac{1}{16}}$

For questions 7 to 12, classify the following numbers as rational (Q), irrational (I) or neither.

7. a. $\frac{1}{8}$ b. $\sqrt{625}$ c. $\frac{11}{4}$ d. $\frac{0}{8}$

8. a. $-6\frac{1}{7}$ b. $\sqrt[3]{81}$ c. $-\sqrt{11}$ d. $\sqrt{\frac{1.44}{4}}$

9. a. $\sqrt{\pi}$ b. $\frac{8}{0}$ c. $\sqrt[3]{21}$ d. $\frac{\pi}{7}$

10. a. $\sqrt[3]{(-5)^2}$ b. $-\frac{3}{11}$ c. $\sqrt{\frac{1}{100}}$ d. $\frac{64}{16}$

11. a. $\sqrt{\frac{2}{25}}$ b. $\frac{\sqrt{6}}{2}$ c. $\sqrt[3]{27}$ d. $\frac{1}{\sqrt{4}}$

12. a. $\frac{22\pi}{7}$ b. $\sqrt[3]{-1.728}$ c. $6\sqrt{4}$ d. $4\sqrt{6}$

13. **MC** Identify a rational number from the following.

A. π B. $\sqrt{\frac{4}{9}}$ C. $\sqrt{\frac{9}{12}}$ D. $\sqrt[3]{3}$ E. $\sqrt{5}$

14. **MC** Identify which of the following best represents an irrational number from the following numbers.

A. $-\sqrt{81}$ B. $\frac{6}{5}$ C. $\sqrt[3]{343}$ D. $\sqrt{22}$ E. $\sqrt{144}$

15. **MC** Select which one of the following statements regarding the numbers $-0.69, \sqrt{7}, \frac{\pi}{3}, \sqrt{49}$ is correct.

A. $\frac{\pi}{3}$ is the only rational number.
B. $\sqrt{7}$ and $\sqrt{49}$ are both irrational numbers.
C. -0.69 and $\sqrt{49}$ are the only rational numbers.
D. -0.69 is the only rational number.
E. $\sqrt{7}$ is the only rational number.

16. **MC** Select which one of the following statements regarding the numbers $2\frac{1}{2}, -\frac{11}{3}, \sqrt{624}, \sqrt[3]{99}$ is correct.

A. $-\frac{11}{3}$ and $\sqrt{624}$ are both irrational numbers.
B. $\sqrt{624}$ is an irrational number and $\sqrt[3]{99}$ is a rational number.
C. $\sqrt{624}$ and $\sqrt[3]{99}$ are both irrational numbers.
D. $2\frac{1}{2}$ is a rational number and $-\frac{11}{3}$ is an irrational number.
E. $\sqrt[3]{99}$ is the only rational number.

Understanding

17. Simplify $\sqrt{\dfrac{a^2}{b^2}}$.

18. **MC** If $p < 0$, then $\sqrt{p}$ is:

    A. positive B. negative C. rational D. irrational E. none of these

19. **MC** If $p < 0$, then $\sqrt{p^2}$ must be:

    A. positive B. negative C. rational D. irrational E. any of these

Reasoning

20. Simplify $(\sqrt{p} - \sqrt{q}) \times (\sqrt{p} + \sqrt{q})$. Show full working.

21. Prove that if $c^2 = a^2 + b^2$, it does not follow that $a = b + c$.

22. Assuming that x is a rational number, for what values of k will the expression $\sqrt{x^2 + kx + 16}$ always be rational? Justify your response.

Problem solving

23. Determine the value of m and n if $\dfrac{36}{11}$ is written as:

    a. $3 + \dfrac{1}{\frac{m}{n}}$ b. $3 + \dfrac{1}{3 + \frac{m}{n}}$ c. $3 + \dfrac{1}{3 + \dfrac{1}{3 + \frac{m}{n}}}$ d. $3 + \dfrac{1}{3 + \dfrac{1}{1 + \frac{m}{n}}}$

24. If x^{-1} means $\dfrac{1}{x}$, determine the value of $\dfrac{3^{-1} - 4^{-1}}{3^{-1} + 4^{-1}}$.

25. If $x^{-n} = \dfrac{1}{x^n}$, evaluate $\dfrac{3^{-n} - 4^{-n}}{3^{-n} + 4^{-n}}$ when $n = 3$.

1.3 Surds (10A)

LEARNING INTENTION

At the end of this subtopic you should be able to:

- determine whether a number under a root or radical sign is a surd
- prove that a surd is irrational by contradiction.

1.3.1 Identifying surds

eles-4662

- A **surd** is an irrational number that is represented by a root sign or a radical sign, for example: $\sqrt{}, \sqrt[3]{}, \sqrt[4]{}$.

  Examples of surds include: $\sqrt{7}, \sqrt{5}, \sqrt[3]{11}, \sqrt[4]{15}$.
- The numbers $\sqrt{9}, \sqrt{16}, \sqrt[3]{125}$, and $\sqrt[4]{81}$ are not surds as they can be simplified to rational numbers, that is: $\sqrt{9} = 3$, $\sqrt{16} = 4$, $\sqrt[3]{125} = 5$, $\sqrt[4]{81} = 3$.

WORKED EXAMPLE 2 Identifying surds

Determine which of the following numbers are surds.

a. $\sqrt{16}$ b. $\sqrt{13}$ c. $\sqrt{\frac{1}{16}}$ d. $\sqrt[3]{17}$ e. $\sqrt[4]{63}$ f. $\sqrt[3]{1728}$

THINK	WRITE
a. 1. Evaluate $\sqrt{16}$.	a. $\sqrt{16} = 4$
2. The answer is rational (since it is a whole number), so state your conclusion.	$\sqrt{16}$ is not a surd.
b. 1. Evaluate $\sqrt{13}$.	b. $\sqrt{13} = 3.605\,551\,275\,46\ldots$
2. The answer is irrational (since it is a non-recurring and non-terminating decimal), so state your conclusion.	$\sqrt{13}$ is a surd.
c. 1. Evaluate $\sqrt{\frac{1}{16}}$.	c. $\sqrt{\frac{1}{16}} = \frac{1}{4}$
2. The answer is rational (a fraction); state your conclusion.	$\sqrt{\frac{1}{16}}$ is not a surd.
d. 1. Evaluate $\sqrt[3]{17}$.	d. $\sqrt[3]{17} = 2.571\,281\,590\,66\ldots$
2. The answer is irrational (a non-terminating and non-recurring decimal), so state your conclusion.	$\sqrt[3]{17}$ is a surd.
e. 1. Evaluate $\sqrt[4]{63}$.	e. $\sqrt[4]{63} = 2.817\,313\,247\,26\ldots$
2. The answer is irrational, so classify $\sqrt[4]{63}$ accordingly.	$\sqrt[4]{63}$ is a surd.
f. 1. Evaluate $\sqrt[3]{1728}$.	f. $\sqrt[3]{1728} = 12$
2. The answer is rational; state your conclusion.	$\sqrt[3]{1728}$ is not a surd. So **b**, **d** and **e** are surds.

1.3.2 Proof that a number is irrational

- In Mathematics you are required to study a variety of types of proofs. One such method is called proof by contradiction.
- This proof is so named because the logical argument of the proof is based on an assumption that leads to contradiction within the proof. Therefore the original assumption must be false.
- An irrational number is one that cannot be expressed in the form $\frac{a}{b}$ (where a and b are integers). The next worked example sets out to prove that $\sqrt{2}$ is irrational.

WORKED EXAMPLE 3 Proving the irrationality of $\sqrt{2}$

Prove that $\sqrt{2}$ is irrational.

THINK	WRITE
1. Assume that $\sqrt{2}$ is rational; that is, it can be written as $\frac{a}{b}$ in simplest form. We need to show that a and b have no common factors.	Let $\sqrt{2} = \frac{a}{b}$, where a and b are integers that have no common factors and $b \neq 0$.
2. Square both sides of the equation.	$2 = \frac{a^2}{b^2}$
3. Rearrange the equation to make a^2 the subject of the formula.	$a^2 = 2b^2$ [1]
4. $2b^2$ is an even number and $2b^2 = a^2$.	$\therefore a^2$ is an even number and a must also be even; that is, a has a factor of 2.
5. Since a is even it can be written as $a = 2r$.	$\therefore a = 2r$
6. Square both sides.	$a^2 = 4r^2$ [2] But $a^2 = 2b^2$ from [1]
7. Equate [1] and [2].	$\therefore 2b^2 = 4r^2$ $b^2 = \frac{4r^2}{2}$ $= 2r^2$ $\therefore b^2$ is an even number and b must also be even; that is, b has a factor of 2.
8. Use reasoning to deduce that $\sqrt{2} = \frac{a}{b}$ where a and b have no common factor.	Both a and b have a common factor of 2. This contradicts the original assumption that $\sqrt{2} = \frac{a}{b}$, where a and b have no common factors. $\therefore \sqrt{2}$ is not rational. $\therefore$ It must be irrational.

- *Note:* An irrational number written in surd form gives an exact value of the number; whereas the same number written in decimal form (for example, to 4 decimal places) gives an approximate value.

DISCUSSION

How can you be certain that root $\sqrt{a}$ is a surd?

Resources

eWorkbook	Topic 1 Workbook (worksheets, code puzzle and project) (ewbk-2027)
Digital document	SkillSHEET Identifying surds (doc-5354)
Interactivity	Surds on the number line (int-6029)

Exercise 1.3 Surds (10A)

learnON

Individual pathways

■ PRACTISE	■ CONSOLIDATE	■ MASTER
1, 4, 7, 8, 11, 14, 17	2, 5, 9, 12, 15, 18	3, 6, 10, 13, 16, 19

To answer questions online and to receive **immediate corrective feedback** and **fully worked solutions** for all questions, go to your learnON title at www.jacplus.com.au.

Fluency

WE2 For questions 1 to 6, determine which of the following numbers are surds.

1. a. $\sqrt{81}$ b. $\sqrt{48}$ c. $\sqrt{16}$ d. $\sqrt{1.6}$

2. a. $\sqrt{0.16}$ b. $\sqrt{11}$ c. $\sqrt{\frac{3}{4}}$ d. $\sqrt[3]{\frac{3}{27}}$

3. a. $\sqrt{1000}$ b. $\sqrt{1.44}$ c. $4\sqrt{100}$ d. $2+\sqrt{10}$

4. a. $\sqrt[3]{32}$ b. $\sqrt{361}$ c. $\sqrt[3]{100}$ d. $\sqrt[3]{125}$

5. a. $\sqrt{6}+\sqrt{6}$ b. 2π c. $\sqrt[3]{169}$ d. $\sqrt{\frac{7}{8}}$

6. a. $\sqrt[4]{16}$ b. $\left(\sqrt{7}\right)^2$ c. $\sqrt[3]{33}$ d. $\sqrt{0.0001}$
   e. $\sqrt[5]{32}$ f. $\sqrt{80}$

7. MC The correct statement regarding the set of numbers $\left\{\sqrt{\frac{6}{9}}, \sqrt{20}, \sqrt{54}, \sqrt[3]{27}, \sqrt{9}\right\}$ is:
   A. $\sqrt[3]{27}$ and $\sqrt{9}$ are the only rational numbers of the set.
   B. $\sqrt{\frac{6}{9}}$ is the only surd of the set.
   C. $\sqrt{\frac{6}{9}}$ and $\sqrt{20}$ are the only surds of the set.
   D. $\sqrt{20}$ and $\sqrt{54}$ are the only surds of the set.
   E. $\sqrt{9}$ and $\sqrt{20}$ are the only surds of the set.

8. MC Identify the numbers from the set $\left\{\sqrt{\frac{1}{4}}, \sqrt[3]{\frac{1}{27}}, \sqrt{\frac{1}{8}}, \sqrt{21}, \sqrt[3]{8}\right\}$ that are surds.
   A. $\sqrt{21}$ only
   B. $\sqrt{\frac{1}{8}}$ only
   C. $\sqrt{\frac{1}{8}}$ and $\sqrt[3]{8}$
   D. $\sqrt{\frac{1}{8}}$ and $\sqrt{21}$ only
   E. $\sqrt{\frac{1}{4}}$ and $\sqrt{21}$ only

9. **MC** Select a statement regarding the set of numbers $\left\{\pi, \sqrt{\frac{1}{49}}, \sqrt{12}, \sqrt{16}, \sqrt{3}, +1\right\}$ that is *not* true.

A. $\sqrt{12}$ is a surd.

B. $\sqrt{12}$ and $\sqrt{16}$ are surds.

C. π is irrational but not a surd.

D. $\sqrt{12}$ and $\sqrt{3}+1$ are not rational.

E. π is not a surd.

10. **MC** Select a statement regarding the set of numbers $\left\{6\sqrt{7}, \sqrt{\frac{144}{16}}, 7\sqrt{6}, 9\sqrt{2}, \sqrt{18}, \sqrt{25}\right\}$ that is *not* true.

A. $\sqrt{\frac{144}{16}}$ when simplified is an integer.

B. $\sqrt{\frac{144}{16}}$ and $\sqrt{25}$ are not surds.

C. $7\sqrt{6}$ is smaller than $9\sqrt{2}$.

D. $9\sqrt{2}$ is smaller than $6\sqrt{7}$.

E. $\sqrt{18}$ is a surd.

Understanding

11. Complete the following statement by selecting appropriate words, suggested in brackets: $\sqrt{a}$ is definitely not a surd, if a is... (any multiple of 4; a perfect square; cube).

12. Determine the smallest value of m, where m is a positive integer, so that $\sqrt[3]{16m}$ is not a surd.

13. **a.** Determine any combination of m and n, where m and n are positive integers with $m < n$, so that $\sqrt[4]{(m+4)(16-n)}$ is not a surd.

b. If the condition that $m < n$ is removed, how many possible combinations are there?

Reasoning

14. Determine whether the following are rational or irrational.

a. $\sqrt{5}+\sqrt{2}$

b. $\sqrt{5}-\sqrt{2}$

c. $\left(\sqrt{5}+\sqrt{2}\right)\left(\sqrt{5}-\sqrt{2}\right)$

15. **WE3** Prove that the following numbers are irrational, using a proof by contradiction:

a. $\sqrt{3}$

b. $\sqrt{5}$

c. $\sqrt{7}$.

16. π is an irrational number and so is $\sqrt{3}$. Therefore, determine whether $\left(\pi-\sqrt{3}\right)\left(\pi+\sqrt{3}\right)$ is an irrational number.

Problem solving

17. Many composite numbers have a variety of factor pairs. For example, factor pairs of 24 are 1 and 24, 2 and 12, 3 and 8, 4 and 6.

a. Use each pair of possible factors to simplify the following surds.

i. $\sqrt{48}$

ii. $\sqrt{72}$

b. Explain if the factor pair chosen when simplifying a surd affect the way the surd is written in simplified form.

c. Explain if the factor pair chosen when simplifying a surd affect the value of the surd when it is written in simplified form.

18. Consider the expression $(\sqrt{p}+\sqrt{q})(\sqrt{m}-\sqrt{n})$. Determine under what conditions will the expression produce a rational number.

19. Solve $\sqrt{3}x-\sqrt{12}=\sqrt{3}$ and indicate whether the result is rational or irrational.

1.4 Operations with surds (10A)

LEARNING INTENTION

At the end of this subtopic you should be able to:
- multiply and simplify surds
- add and subtract like surds
- divide surds
- rationalise the denominator of a fraction.

1.4.1 Multiplying and simplifying surds

eles-4664

Multiplication of surds

- To multiply surds, multiply the expressions under the radical sign.
  For example: $\sqrt{8} \times \sqrt{3} = \sqrt{8 \times 3} = \sqrt{24}$
- If there are coefficients in front of the surds that are being multiplied, multiply the coefficients and then multiply the expressions under the radical signs.
  For example: $2\sqrt{3} \times 5\sqrt{7} = (2 \times 5)\sqrt{3 \times 7} = 10\sqrt{21}$

Multiplication of surds

In order to multiply two or more surds, use the following:

- $\sqrt{a} \times \sqrt{b} = \sqrt{a \times b}$
- $m\sqrt{a} \times n\sqrt{b} = mn\sqrt{a \times b}$

where a and b are positive real numbers.

Simplification of surds

- To simplify a surd means to make the number under the radical sign as small as possible.
- Surds can only be simplified if the number under the radical sign has a factor which is a perfect square $(4, 9, 16, 25, 36, ...)$.
- Simplification of a surd uses the method of multiplying surds in reverse.
- The process is summarised in the following steps:
  1. Split the number under the radical into the product of two factors, one of which is a perfect square.
  2. Write the surd as the product of two surds multiplied together. The two surds must correspond to the factors identified in step 1.
  3. Simplify the surd of the perfect square and write the surd in the form $a\sqrt{b}$.
- The example below shows the how the surd $\sqrt{45}$ can be simplified by following the steps 1 to 3.

$$\begin{aligned} \sqrt{45} &= \sqrt{9 \times 5} && \text{(Step 1)} \\ &= \sqrt{9} \times \sqrt{5} && \text{(Step 2)} \\ &= 3 \times \sqrt{5} = 3\sqrt{5} && \text{(Step 3)} \end{aligned}$$

- If possible, try to factorise the number under the radical sign so that the largest possible perfect square is used. This will ensure the surd is simplified in 1 step.

Simplification of surds

$$\begin{aligned}\sqrt{n} &= \sqrt{a^2 \times b}\\ &= \sqrt{a^2} \times \sqrt{b}\\ &= a \times \sqrt{b}\\ &= a\sqrt{b}\end{aligned}$$

WORKED EXAMPLE 4 Simplifying surds

Simplify the following surds. Assume that x and y are positive real numbers.

a. $\sqrt{384}$ b. $3\sqrt{405}$ c. $-\frac{1}{8}\sqrt{175}$ d. $5\sqrt{180x^3y^5}$

THINK	WRITE
a. 1. Express 384 as a product of two factors where one factor is the largest possible perfect square.	a. $\sqrt{384} = \sqrt{64 \times 6}$
2. Express $\sqrt{64 \times 6}$ as the product of two surds.	$= \sqrt{64} \times \sqrt{6}$
3. Simplify the square root from the perfect square (that is, $\sqrt{64} = 8$).	$= 8\sqrt{6}$
b. 1. Express 405 as a product of two factors, one of which is the largest possible perfect square.	b. $3\sqrt{405} = 3\sqrt{81 \times 5}$
2. Express $\sqrt{81 \times 5}$ as a product of two surds.	$= 3\sqrt{81} \times \sqrt{5}$
3. Simplify $\sqrt{81}$.	$= 3 \times 9\sqrt{5}$
4. Multiply together the whole numbers outside the square root sign (3 and 9).	$= 27\sqrt{5}$
c. 1. Express 175 as a product of two factors in which one factor is the largest possible perfect square.	c. $-\frac{1}{8}\sqrt{175} = -\frac{1}{8}\sqrt{25 \times 7}$
2. Express $\sqrt{25 \times 7}$ as a product of 2 surds.	$= -\frac{1}{8} \times \sqrt{25} \times \sqrt{7}$
3. Simplify $\sqrt{25}$.	$= -\frac{1}{8} \times 5\sqrt{7}$
4. Multiply together the numbers outside the square root sign.	$= -\frac{5}{8}\sqrt{7}$
d. 1. Express each of 180, x^3 and y^5 as a product of two factors where one factor is the largest possible perfect square.	d. $5\sqrt{180x^3y^5} = 5\sqrt{36 \times 5 \times x^2 \times x \times y^4 \times y}$
2. Separate all perfect squares into one surd and all other factors into the other surd.	$= 5 \times \sqrt{36x^2y^4} \times \sqrt{5xy}$

3. Simplify $\sqrt{36x^2y^4}$. $= 5 \times 6 \times x \times y^2 \times \sqrt{5xy}$

4. Multiply together the numbers and the pronumerals outside the square root sign. $= 30xy^2\sqrt{5xy}$

WORKED EXAMPLE 5 Multiplying surds

Multiply the following surds, expressing answers in the simplest form. Assume that x and y are positive real numbers.

a. $\sqrt{11} \times \sqrt{7}$ b. $5\sqrt{3} \times 8\sqrt{5}$ c. $6\sqrt{12} \times 2\sqrt{6}$ d. $\sqrt{15x^5y^2} \times \sqrt{12x^2y}$

THINK	WRITE
a. Multiply the surds together, using $\sqrt{a} \times \sqrt{b} = \sqrt{ab}$ (that is, multiply expressions under the square root sign). *Note:* This expression cannot be simplified any further.	a. $\sqrt{11} \times \sqrt{7} = \sqrt{11 \times 7}$ $= \sqrt{77}$
b. Multiply the coefficients together and then multiply the surds together.	b. $5\sqrt{3} \times 8\sqrt{5} = 5 \times 8 \times \sqrt{3} \times \sqrt{5}$ $= 40 \times \sqrt{3 \times 5}$ $= 40\sqrt{15}$
c. 1. Simplify $\sqrt{12}$.	c. $6\sqrt{12} \times 2\sqrt{6} = 6\sqrt{4 \times 3} \times 2\sqrt{6}$ $= 6 \times 2\sqrt{3} \times 2\sqrt{6}$ $= 12\sqrt{3} \times 2\sqrt{6}$
2. Multiply the coefficients together and multiply the surds together.	$= 24\sqrt{18}$
3. Simplify the surd.	$= 24\sqrt{9 \times 2}$ $= 24 \times 3\sqrt{2}$ $= 72\sqrt{2}$
d. 1. Simplify each of the surds.	d. $\sqrt{15x^5y^2} \times \sqrt{12x^2y}$ $= \sqrt{15 \times x^4 \times x \times y^2} \times \sqrt{4 \times 3 \times x^2 \times y}$ $= x^2 \times y \times \sqrt{15 \times x} \times 2 \times x \times \sqrt{3 \times y}$ $= x^2y\sqrt{15x} \times 2x\sqrt{3y}$
2. Multiply the coefficients together and the surds together.	$= x^2y \times 2x\sqrt{15x \times 3y}$ $= 2x^3y\sqrt{45xy}$ $= 2x^3y\sqrt{9 \times 5xy}$
3. Simplify the surd.	$= 2x^3y \times 3\sqrt{5xy}$ $= 6x^3y\sqrt{5xy}$

- When working with surds, it is sometimes necessary to multiply surds by themselves; that is, square them. Consider the following examples:

$$\left(\sqrt{2}\right)^2 = \sqrt{2} \times \sqrt{2} = \sqrt{4} = 2$$
$$\left(\sqrt{5}\right)^2 = \sqrt{5} \times \sqrt{5} = \sqrt{25} = 5$$

- Observe that squaring a surd produces the number under the radical sign. This is not surprising, because squaring and taking the square root are *inverse operations* and, when applied together, leave the original unchanged.

Squaring surds

When a surd is squared, the result is the expression under the radical sign; that is:

$$\left(\sqrt{a}\right)^2 = a$$

where a is a positive real number.

WORKED EXAMPLE 6 Squaring surds

Simplify each of the following.

a. $\left(\sqrt{6}\right)^2$

b. $\left(3\sqrt{5}\right)^2$

THINK	WRITE
a. Use $\left(\sqrt{a}\right)^2 = a$, where $a = 6$.	a. $\left(\sqrt{6}\right)^2 = 6$
b. 1. Square 3 and apply $\left(\sqrt{a}\right)^2 = a$ to square $\sqrt{5}$.	b. $\left(3\sqrt{5}\right)^2 = 3^2 \times \left(\sqrt{5}\right)^2$ $= 9 \times 5$
2. Simplify.	$= 45$

1.4.2 Addition and subtraction of surds

eles-4665

- Surds may be added or subtracted only if they are *alike*.
  Examples of *like* surds include $\sqrt{7}, 3\sqrt{7}$ and $-5\sqrt{7}$.
  Examples of *unlike* surds include $\sqrt{11}$, $\sqrt{5}, 2\sqrt{13}$ and $-2\sqrt{3}$.
- In some cases surds will need to be simplified before you decide whether they are like or unlike, and then addition and subtraction can take place. The concept of adding and subtracting surds is similar to adding and subtracting like terms in algebra.

WORKED EXAMPLE 7 Adding and subtracting surds

Simplify each of the following expressions containing surds. Assume that a and b are positive real numbers.

a. $3\sqrt{6}+17\sqrt{6}-2\sqrt{6}$

b. $5\sqrt{3}+2\sqrt{12}-5\sqrt{2}+3\sqrt{8}$

c. $\frac{1}{2}\sqrt{100a^3b^2}+ab\sqrt{36a}-5\sqrt{4a^2b}$

THINK	WRITE
a. All 3 terms are alike because they contain the same surd ($\sqrt{6}$). Simplify.	a. $3\sqrt{6}+17\sqrt{6}-2\sqrt{6}=(3+17-2)\sqrt{6}$ $=18\sqrt{6}$
b. 1. Simplify surds where possible.	b. $5\sqrt{3}+2\sqrt{12}-5\sqrt{2}+3\sqrt{8}$ $=5\sqrt{3}+2\sqrt{4\times 3}-5\sqrt{2}+3\sqrt{4\times 2}$ $=5\sqrt{3}+2\times 2\sqrt{3}-5\sqrt{2}+3\times 2\sqrt{2}$
2. Add like terms to obtain the simplified answer.	$=5\sqrt{3}+4\sqrt{3}-5\sqrt{2}+6\sqrt{2}$ $=9\sqrt{3}+\sqrt{2}$
c. 1. Simplify surds where possible.	c. $\frac{1}{2}\sqrt{100a^3b^2}+ab\sqrt{36a}-5\sqrt{4a^2b}$ $=\frac{1}{2}\times 10\sqrt{a^2\times a\times b^2}+ab\times 6\sqrt{a}-5\times 2\times a\sqrt{b}$ $=\frac{1}{2}\times 10\times a\times b\sqrt{a}+ab\times 6\sqrt{a}-5\times 2\times a\sqrt{b}$
2. Add like terms to obtain the simplified answer.	$=5ab\sqrt{a}+6ab\sqrt{a}-10a\sqrt{b}$ $=11ab\sqrt{a}-10a\sqrt{b}$

TI | THINK	DISPLAY/WRITE
a–c. In a new document, on a Calculator page, complete the entry lines as: $3\sqrt{6}+17\sqrt{6}-2\sqrt{6}$ $5\sqrt{3}+2\sqrt{12}-5\sqrt{2}$ $+3\sqrt{8}$ $\frac{1}{2}\sqrt{100a^3b^2}+ab\sqrt{36a}-$ $5\sqrt{4a^2b}\|a>0$ and $b>0$ Press ENTER after each entry.	a–c. 1.1 *Real numbers RAD $3\cdot\sqrt{6}+17\cdot\sqrt{6}-2\cdot\sqrt{6}$ $18\cdot\sqrt{6}$ $5\cdot\sqrt{3}+2\cdot\sqrt{12}-5\cdot\sqrt{2}+3\cdot\sqrt{8}$ $9\cdot\sqrt{3}+\sqrt{2}$ $\frac{1}{2}\cdot\sqrt{100\cdot a^3\cdot b^2}+a\cdot b\cdot\sqrt{36\cdot a}-5\cdot\sqrt{4\cdot a^2\cdot b}$ $11\cdot a^{\frac{3}{2}}\cdot b-10\cdot a\cdot\sqrt{b}$ $3\sqrt{6}+17\sqrt{6}-2\sqrt{6}=18\sqrt{6}$ $5\sqrt{3}+2\sqrt{12}-5\sqrt{2}+3\sqrt{8}$ $=9\sqrt{3}+\sqrt{2}$ $\frac{1}{2}\sqrt{100a^3b^2}+ab\sqrt{36a}-$ $5\sqrt{4a^2b}=11a^{\frac{2}{3}}b-10a\sqrt{b}$

CASIO | THINK	DISPLAY/WRITE
a–c. On the Main screen, complete the entry lines as: $3\sqrt{6}+17\sqrt{6}-2\sqrt{6}$ $5\sqrt{3}+2\sqrt{12}-5\sqrt{2}$ $+3\sqrt{8}$ simplify $\left(\frac{1}{2}\sqrt{100a^3b^2}+\right.$ $a\times b\times\sqrt{36a}-$ $5\sqrt{4a^2b}\|a>0\|b>0)$ Press EXE after each entry.	a–c. 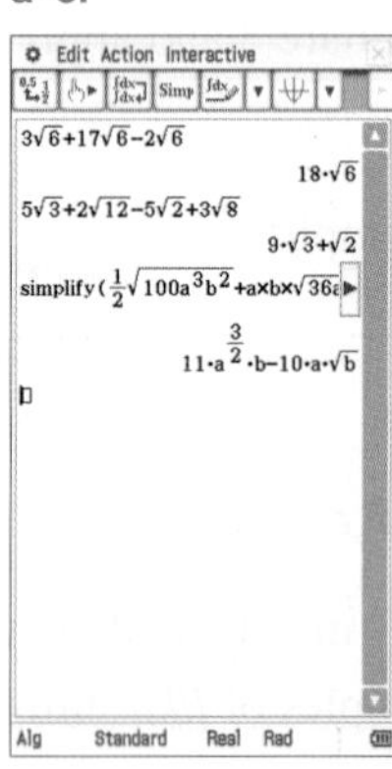 $3\sqrt{6}+17\sqrt{6}-2\sqrt{6}=18\sqrt{6}$ $5\sqrt{3}+2\sqrt{12}-5\sqrt{2}+3\sqrt{8}$ $=9\sqrt{3}+\sqrt{2}$ $\frac{1}{2}\sqrt{100a^3b^2}+ab\sqrt{36a}-$ $5\sqrt{4a^2b}=11a^{\frac{2}{3}}b-10a\sqrt{b}$

1.4.3 Dividing surds

eles-4666

- To divide surds, divide the expressions under the radical signs.

Dividing surds

$$\frac{\sqrt{a}}{\sqrt{b}}=\sqrt{\frac{a}{b}}$$

where a and b are positive real numbers.

- When dividing surds it is best to simplify them (if possible) first. Once this has been done, the coefficients are divided next and then the surds are divided.

$$\frac{m\sqrt{a}}{n\sqrt{b}}=\frac{m}{n}\sqrt{\frac{a}{b}}$$

WORKED EXAMPLE 8 Dividing surds

Divide the following surds, expressing answers in the simplest form. Assume that x and y are positive real numbers.

a. $\frac{\sqrt{55}}{\sqrt{5}}$ b. $\frac{\sqrt{48}}{\sqrt{3}}$ c. $\frac{9\sqrt{88}}{6\sqrt{99}}$ d. $\frac{\sqrt{36xy}}{\sqrt{25x^9y^{11}}}$

THINK	WRITE
a. 1. Rewrite the fraction, using $\frac{\sqrt{a}}{\sqrt{b}}=\sqrt{\frac{a}{b}}$.	a. $\frac{\sqrt{55}}{\sqrt{5}}=\sqrt{\frac{55}{5}}$
2. Divide the numerator by the denominator (that is, 55 by 5). Check if the surd can be simplified any further.	$=\sqrt{11}$
b. 1. Rewrite the fraction, using $\frac{\sqrt{a}}{\sqrt{b}}=\sqrt{\frac{a}{b}}$.	b. $\frac{\sqrt{48}}{\sqrt{3}}=\sqrt{\frac{48}{3}}$
2. Divide 48 by 3.	$=\sqrt{16}$
3. Evaluate $\sqrt{16}$.	$=4$
c. 1. Rewrite surds, using $\frac{\sqrt{a}}{\sqrt{b}}=\sqrt{\frac{a}{b}}$.	c. $\frac{9\sqrt{88}}{6\sqrt{99}}=\frac{9}{6}\sqrt{\frac{88}{99}}$
2. Simplify the fraction under the radical by dividing both numerator and denominator by 11.	$=\frac{9}{6}\sqrt{\frac{8}{9}}$

3. Simplify surds. $= \dfrac{9 \times 2\sqrt{2}}{6 \times 3}$

4. Multiply the whole numbers in the numerator together and those in the denominator together. $= \dfrac{18\sqrt{2}}{18}$

5. Cancel the common factor of 18. $= \sqrt{2}$

d. 1. Simplify each surd.

d. $\dfrac{\sqrt{36xy}}{\sqrt{25x^9y^{11}}} = \dfrac{6\sqrt{xy}}{5\sqrt{x^8 \times x \times y^{10} \times y}}$

$= \dfrac{6\sqrt{xy}}{5x^4y^5\sqrt{xy}}$

2. Cancel any common factors — in this case $\sqrt{xy}$. $= \dfrac{6}{5x^4y^5}$

1.4.4 Rationalising denominators

eles-4667

- If the **denominator** of a fraction is a surd, it can be changed into a rational number through multiplication. In other words, it can be rationalised.
- As discussed earlier in this chapter, squaring a simple surd (that is, multiplying it by itself) results in a rational number. This fact can be used to rationalise denominators as follows.

Rationalising the denominator

$$\frac{\sqrt{a}}{\sqrt{b}} = \frac{\sqrt{a}}{\sqrt{b}} \times \frac{\sqrt{b}}{\sqrt{b}} = \frac{\sqrt{ab}}{b}$$

- If both numerator and denominator of a fraction are multiplied by the surd contained in the denominator, the denominator becomes a rational number. The fraction takes on a different appearance, but its numerical value is unchanged, because multiplying the numerator and denominator by the same number is equivalent to multiplying by 1.

WORKED EXAMPLE 9 Rationalising the denominator

Express the following in their simplest form with a rational denominator.

a. $\dfrac{\sqrt{6}}{\sqrt{13}}$

b. $\dfrac{2\sqrt{12}}{3\sqrt{54}}$

c. $\dfrac{\sqrt{17} - 3\sqrt{14}}{\sqrt{7}}$

THINK | **WRITE**

a. 1. Write the fraction.

a. $\dfrac{\sqrt{6}}{\sqrt{13}}$

2. Multiply both the numerator and denominator by the surd contained in the denominator (in this case $\sqrt{13}$). This has the same effect as multiplying the fraction by 1, because $\dfrac{\sqrt{13}}{\sqrt{13}} = 1$.

$$= \frac{\sqrt{6}}{\sqrt{13}} \times \frac{\sqrt{13}}{\sqrt{13}}$$

$$= \frac{\sqrt{78}}{13}$$

b. 1. Write the fraction.

b. $\dfrac{2\sqrt{12}}{3\sqrt{54}}$

2. Simplify the surds. (This avoids dealing with large numbers.)

$$\frac{2\sqrt{12}}{3\sqrt{54}} = \frac{2\sqrt{4 \times 3}}{3\sqrt{9 \times 6}}$$

$$= \frac{2 \times 2\sqrt{3}}{3 \times 3\sqrt{6}}$$

$$= \frac{4\sqrt{3}}{9\sqrt{6}}$$

3. Multiply both the numerator and denominator by $\sqrt{6}$. This has the same effect as multiplying the fraction by 1, because $\dfrac{\sqrt{6}}{\sqrt{6}} = 1$.

Note: We need to multiply only by the surd part of the denominator (that is, by $\sqrt{6}$ rather than by $9\sqrt{6}$.)

$$= \frac{4\sqrt{3}}{9\sqrt{6}} \times \frac{\sqrt{6}}{\sqrt{6}}$$

$$= \frac{4\sqrt{18}}{9 \times 6}$$

4. Simplify $\sqrt{18}$.

$$= \frac{4\sqrt{9 \times 2}}{9 \times 6}$$

$$= \frac{4 \times 3\sqrt{2}}{54}$$

$$= \frac{12\sqrt{2}}{54}$$

5. Divide both the numerator and denominator by 6 (cancel down).

$$= \frac{2\sqrt{2}}{9}$$

c. 1. Write the fraction.

c. $\dfrac{\sqrt{17} - 3\sqrt{14}}{\sqrt{7}}$

2. Multiply both the numerator and denominator by $\sqrt{7}$. Use grouping symbols (brackets) to make it clear that the whole numerator must be multiplied by $\sqrt{7}$.

$$= \frac{(\sqrt{17} - 3\sqrt{14})}{\sqrt{7}} \times \frac{\sqrt{7}}{\sqrt{7}}$$

3. Apply the Distributive Law in the numerator.
$a(b+c)=ab+ac$

$$=\frac{\sqrt{17}\times\sqrt{7}-3\sqrt{14}\times\sqrt{7}}{\sqrt{7}\times\sqrt{7}}$$
$$=\frac{\sqrt{119}-3\sqrt{98}}{7}$$

4. Simplify $\sqrt{98}$.

$$=\frac{\sqrt{119}-3\sqrt{49\times 2}}{7}$$
$$=\frac{\sqrt{119}-3\times 7\sqrt{2}}{7}$$
$$=\frac{\sqrt{119}-21\sqrt{2}}{7}$$

1.4.5 Rationalising denominators using conjugate surds

eles-4668

- The product of pairs of **conjugate surds** results in a rational number.
- Examples of pairs of conjugate surds include $\sqrt{6}+11$ and $\sqrt{6}-11$, $\sqrt{a}+b$ and $\sqrt{a}-b$, $2\sqrt{5}-\sqrt{7}$ and $2\sqrt{5}+\sqrt{7}$.
This fact is used to rationalise denominators containing a sum or a difference of surds.

Using conjugates to rationalise the denominator

- To rationalise the denominator that contains a sum or a difference of surds, multiply both numerator and denominator by the conjugate of the denominator.
Two examples are given below:

1. To rationalise the denominator of the fraction $\frac{1}{\sqrt{a}+\sqrt{b}}$, multiply it by $\frac{\sqrt{a}-\sqrt{b}}{\sqrt{a}-\sqrt{b}}$.
2. To rationalise the denominator of the fraction $\frac{1}{\sqrt{a}-\sqrt{b}}$, multiply it by $\frac{\sqrt{a}+\sqrt{b}}{\sqrt{a}+\sqrt{b}}$.

- A quick way to simplify the denominator is to use the difference of two squares identity:

$$\left(\sqrt{a}-\sqrt{b}\right)\left(\sqrt{a}+\sqrt{b}\right)=\left(\sqrt{a}\right)^2-\left(\sqrt{b}\right)^2$$
$$=a-b$$

WORKED EXAMPLE 10 Using conjugates to rationalise the denominator

Rationalise the denominator and simplify the following.

a. $\frac{1}{4-\sqrt{3}}$

b. $\frac{\sqrt{6}+3\sqrt{2}}{3+\sqrt{3}}$

THINK	WRITE
a. **1.** Write the fraction.	**a.** $\dfrac{1}{4-\sqrt{3}}$
2. Multiply the numerator and denominator by the conjugate of the denominator. (Note that $\dfrac{(4+\sqrt{3})}{(4+\sqrt{3})}=1$).	$=\dfrac{1}{(4-\sqrt{3})}\times\dfrac{(4+\sqrt{3})}{(4+\sqrt{3})}$
3. Apply the Distributive Law in the numerator and the difference of two squares identity in the denominator.	$=\dfrac{4+\sqrt{3}}{(4)^2-(\sqrt{3})^2}$
4. Simplify.	$=\dfrac{4+\sqrt{3}}{16-3}$ $=\dfrac{4+\sqrt{3}}{13}$
b. **1.** Write the fraction.	**b.** $\dfrac{\sqrt{6}+3\sqrt{2}}{3+\sqrt{3}}$
2. Multiply the numerator and denominator by the conjugate of the denominator. (Note that $\dfrac{(3-\sqrt{3})}{(3-\sqrt{3})}=1$.)	$=\dfrac{(\sqrt{6}+3\sqrt{2})}{(3+\sqrt{3})}\times\dfrac{(3-\sqrt{3})}{(3-\sqrt{3})}$
3. Multiply the expressions in grouping symbols in the numerator, and apply the difference of two squares identity in the denominator.	$=\dfrac{\sqrt{6}\times 3+\sqrt{6}\times(-\sqrt{3})+3\sqrt{2}\times 3+3\sqrt{2}\times-(\sqrt{3})}{(3)^2-(\sqrt{3})^2}$
4. Simplify.	$=\dfrac{3\sqrt{6}-\sqrt{18}+9\sqrt{2}-3\sqrt{6}}{9-3}$ $=\dfrac{-\sqrt{18}+9\sqrt{2}}{6}$ $=\dfrac{-\sqrt{9\times 2}+9\sqrt{2}}{6}$ $=\dfrac{-3\sqrt{2}+9\sqrt{2}}{6}$ $=\dfrac{6\sqrt{2}}{6}$ $=\sqrt{2}$

TI | THINK	DISPLAY/WRITE
a-b. On a Calculator page, complete the entry lines as: $\frac{1}{4-\sqrt{3}}$ $\frac{\sqrt{6}+3\sqrt{2}}{3+\sqrt{3}}$ Press ENTER after each entry.	**a-b.** 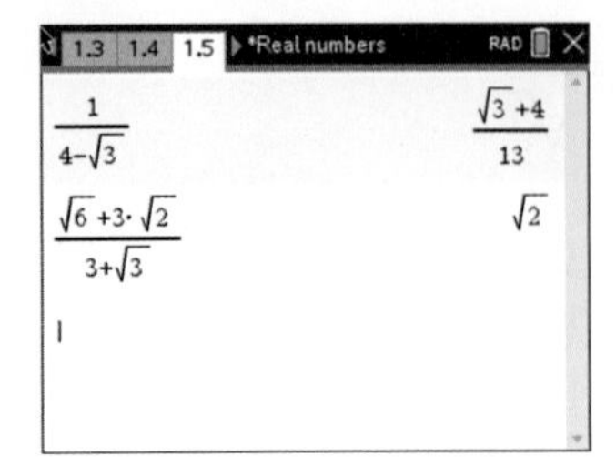 $\frac{1}{4-\sqrt{3}}=\frac{4+\sqrt{3}}{13}$ $\frac{\sqrt{6}+3\sqrt{2}}{3+\sqrt{3}}=\sqrt{2}$

CASIO | THINK	DISPLAY/WRITE
a-b. On the Main screen, complete the entry lines as: simplify $\left(\frac{1}{4-\sqrt{3}}\right)$ simplify $\left(\frac{\sqrt{6}+3\sqrt{2}}{3+\sqrt{3}}\right)$ Press EXE after each entry.	**a-b.** 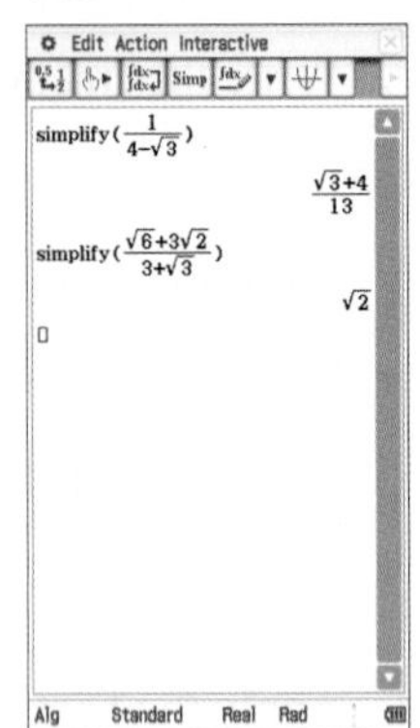 $\frac{1}{4-\sqrt{3}}=\frac{4+\sqrt{3}}{13}$ $\frac{\sqrt{6}+3\sqrt{2}}{3+\sqrt{3}}=\sqrt{2}$

on Resources

eWorkbook	Topic 1 Workbook (worksheets, code puzzle and project) (ewbk-2027)
Digital documents	SkillSHEET Simplifying surds (doc-5355) SkillSHEET Adding and subtracting surds (doc-5356) SkillSHEET Multiplying and dividing surds (doc-5357) SkillSHEET Rationalising denominators (doc-5360) SkillSHEET Conjugate pairs (doc-5361) SkillSHEET Applying the difference of two squares rule to surds (doc-5362)
Video eLessons	Surds (eles-1906) Rationalisation of surds (eles-1948)
Interactivities	Addition and subtraction of surds (int-6190) Multiplying surds (int-6191) Dividing surds (int-6192) Simplifying surds (int-6028) Conjugate surds (int-6193)

Exercise 1.4 Operations with surds (10A)

learnon

Individual pathways

■ PRACTISE	■ CONSOLIDATE	■ MASTER
1, 4, 7, 10, 12, 15, 18, 21, 24, 27, 30, 33, 36, 39	2, 5, 8, 11, 13, 16, 19, 22, 25, 28, 31, 34, 37, 40	3, 6, 9, 14, 17, 20, 23, 26, 29, 32, 35, 38, 41

To answer questions online and to receive **immediate corrective feedback** and **fully worked solutions** for all questions, go to your learnON title at www.jacplus.com.au.

Fluency

WE4a For questions 1 to 3, simplify the following surds.

1. a. $\sqrt{12}$ b. $\sqrt{24}$ c. $\sqrt{27}$ d. $\sqrt{125}$

2. a. $\sqrt{54}$ b. $\sqrt{112}$ c. $\sqrt{68}$ d. $\sqrt{180}$

3. a. $\sqrt{88}$ b. $\sqrt{162}$ c. $\sqrt{245}$ d. $\sqrt{448}$

WE4b,c For questions 4 to 6, simplify the following surds.

4. a. $2\sqrt{8}$ b. $8\sqrt{90}$ c. $9\sqrt{80}$ d. $7\sqrt{54}$

5. a. $-6\sqrt{75}$ b. $-7\sqrt{80}$ c. $16\sqrt{48}$ d. $\frac{1}{7}\sqrt{392}$

6. a. $\frac{1}{9}\sqrt{162}$ b. $\frac{1}{4}\sqrt{192}$ c. $\frac{1}{9}\sqrt{135}$ d. $\frac{3}{10}\sqrt{175}$

WE4d For questions 7 to 9, simplify the following surds. Assume that a, b, c, d, e, f, x and y are positive real numbers.

7. a. $\sqrt{16a^2}$ b. $\sqrt{72a^2}$ c. $\sqrt{90a^2b}$ d. $\sqrt{338a^4}$

8. a. $\sqrt{338a^3b^3}$ b. $\sqrt{68a^3b^5}$ c. $\sqrt{125x^6y^4}$ d. $5\sqrt{80x^3y^2}$

9. a. $6\sqrt{162c^7d^5}$ b. $2\sqrt{405c^7d^9}$ c. $\frac{1}{2}\sqrt{88ef}$ d. $\frac{1}{2}\sqrt{392e^{11}f^{11}}$

10. **WE5a** Simplify the following expressions containing surds. Assume that x and y are positive real numbers.
  a. $3\sqrt{5}+4\sqrt{5}$ b. $2\sqrt{3}+5\sqrt{3}+\sqrt{3}$
  c. $8\sqrt{5}+3\sqrt{3}+7\sqrt{5}+2\sqrt{3}$ d. $6\sqrt{11}-2\sqrt{11}$

11. Simplify the following expressions containing surds. Assume that x and y are positive real numbers.
  a. $7\sqrt{2}+9\sqrt{2}-3\sqrt{2}$ b. $9\sqrt{6}+12\sqrt{6}-17\sqrt{6}-7\sqrt{6}$
  c. $12\sqrt{3}-8\sqrt{7}+5\sqrt{3}-10\sqrt{7}$ d. $2\sqrt{x}+5\sqrt{y}+6\sqrt{x}-2\sqrt{y}$

WE5b For questions 12 to 14, simplify the following expressions containing surds. Assume that a and b are positive real numbers.

12. a. $\sqrt{200}-\sqrt{300}$ b. $\sqrt{125}-\sqrt{150}+\sqrt{600}$
  c. $\sqrt{27}-\sqrt{3}+\sqrt{75}$ d. $2\sqrt{20}-3\sqrt{5}+\sqrt{45}$

13. a. $6\sqrt{12}+3\sqrt{27}-7\sqrt{3}+\sqrt{18}$ b. $\sqrt{150}+\sqrt{24}-\sqrt{96}+\sqrt{108}$
  c. $3\sqrt{90}-5\sqrt{60}+3\sqrt{40}+\sqrt{100}$ d. $5\sqrt{11}+7\sqrt{44}-9\sqrt{99}+2\sqrt{121}$

14. a. $2\sqrt{30}+5\sqrt{120}+\sqrt{60}-6\sqrt{135}$ b. $6\sqrt{ab}-\sqrt{12ab}+2\sqrt{9ab}+3\sqrt{27ab}$
  c. $\frac{1}{2}\sqrt{98}+\frac{1}{3}\sqrt{48}+\frac{1}{3}\sqrt{12}$ d. $\frac{1}{8}\sqrt{32}-\frac{7}{6}\sqrt{18}+3\sqrt{72}$

WE5c For questions 15 to 17, simplify the following expressions containing surds. Assume that a and b are positive real numbers.

15. a. $7\sqrt{a}-\sqrt{8a}+8\sqrt{9a}-\sqrt{32a}$ b. $10\sqrt{a}-15\sqrt{27a}+8\sqrt{12a}+14\sqrt{9a}$
  c. $\sqrt{150ab}+\sqrt{96ab}-\sqrt{54ab}$ d. $16\sqrt{4a^2}-\sqrt{24a}+4\sqrt{8a^2}+\sqrt{96a}$

16. a. $\sqrt{8a^3}+\sqrt{72a^3}-\sqrt{98a^3}$ b. $\frac{1}{2}\sqrt{36a}+\frac{1}{4}\sqrt{128a}-\frac{1}{6}\sqrt{144a}$
  c. $\sqrt{9a^3}+\sqrt{3a^5}$ d. $6\sqrt{a^5b}+\sqrt{a^3b}-5\sqrt{a^5b}$

17. **a.** $ab\sqrt{ab}+3ab\sqrt{a^2b}+\sqrt{9a^3b^3}$
b. $\sqrt{a^3b}+5\sqrt{ab}-2\sqrt{ab}+5\sqrt{a^3b}$
c. $\sqrt{32a^3b^2}-5ab\sqrt{8a}+\sqrt{48a^5b^6}$
d. $\sqrt{4a^2b}+5\sqrt{a^2b}-3\sqrt{9a^2b}$

WE6 For questions 18 to **20**, multiply the following surds, expressing answers in the simplest form. Assume that a, b, x and y are positive real numbers.

18. a. $\sqrt{2}\times\sqrt{7}$
b. $\sqrt{6}\times\sqrt{7}$
c. $\sqrt{8}\times\sqrt{6}$
d. $\sqrt{10}\times\sqrt{10}$
e. $\sqrt{21}\times\sqrt{3}$
f. $\sqrt{27}\times 3\sqrt{3}$

19. a. $5\sqrt{3}\times 2\sqrt{11}$
b. $10\sqrt{15}\times 6\sqrt{3}$
c. $4\sqrt{20}\times 3\sqrt{5}$
d. $10\sqrt{6}\times 3\sqrt{8}$
e. $\frac{1}{4}\sqrt{48}\times 2\sqrt{2}$
f. $\frac{1}{9}\sqrt{48}\times 2\sqrt{3}$

20. **a.** $\frac{1}{10}\sqrt{60}\times\frac{1}{5}\sqrt{40}$
b. $\sqrt{xy}\times\sqrt{x^3y^2}$
c. $\sqrt{3a^4b^2}\times\sqrt{6a^5b^3}$
d. $\sqrt{12a^7b}\times\sqrt{6a^3b^4}$
e. $\sqrt{15x^3y^2}\times\sqrt{6x^2y^3}$
f. $\frac{1}{2}\sqrt{15a^3b^3}\times 3\sqrt{3a^2b^6}$

WE7 For questions 21 to **23**, simplify each of the following.

21. a. $\left(\sqrt{2}\right)^2$
b. $\left(\sqrt{5}\right)^2$
c. $\left(\sqrt{12}\right)^2$

22. a. $\left(\sqrt{15}\right)^2$
b. $\left(3\sqrt{2}\right)^2$
c. $\left(4\sqrt{5}\right)^2$

23. a. $\left(2\sqrt{7}\right)^2$
b. $\left(5\sqrt{8}\right)^2$

WE8 For questions 24 to **26**, divide the following surds, expressing answers in the simplest form. Assume that a, b, x and y are positive real numbers.

24. a. $\frac{\sqrt{15}}{\sqrt{3}}$
b. $\frac{\sqrt{8}}{\sqrt{2}}$
c. $\frac{\sqrt{60}}{\sqrt{10}}$
d. $\frac{\sqrt{128}}{\sqrt{8}}$

25. a. $\frac{\sqrt{18}}{4\sqrt{6}}$
b. $\frac{\sqrt{65}}{2\sqrt{13}}$
c. $\frac{\sqrt{96}}{\sqrt{8}}$
d. $\frac{7\sqrt{44}}{14\sqrt{11}}$

26. **a.** $\frac{9\sqrt{63}}{15\sqrt{7}}$
b. $\frac{\sqrt{2040}}{\sqrt{30}}$
c. $\frac{\sqrt{x^4y^3}}{\sqrt{x^2y^5}}$
d. $\frac{\sqrt{16xy}}{\sqrt{8x^7y^9}}$
e. $\frac{\sqrt{xy}}{\sqrt{x^5y^7}}\times\frac{\sqrt{12x^8y^{12}}}{\sqrt{x^2y^3}}$
f. $\frac{2\sqrt{2a^2b^4}}{\sqrt{5a^3b^6}}\times\frac{\sqrt{10a^9b^3}}{3\sqrt{a^7b}}$

WE9a,b For questions 27 to **29**, express the following in their simplest form with a rational denominator.

27. a. $\frac{5}{\sqrt{2}}$
b. $\frac{7}{\sqrt{3}}$
c. $\frac{4}{\sqrt{11}}$
d. $\frac{8}{\sqrt{6}}$
e. $\frac{\sqrt{12}}{\sqrt{7}}$

28. a. $\frac{\sqrt{15}}{\sqrt{6}}$ b. $\frac{2\sqrt{3}}{\sqrt{5}}$ c. $\frac{3\sqrt{7}}{\sqrt{5}}$ d. $\frac{5\sqrt{2}}{2\sqrt{3}}$ e. $\frac{4\sqrt{3}}{3\sqrt{5}}$

Understanding

29. a. $\frac{5\sqrt{14}}{7\sqrt{8}}$ b. $\frac{16\sqrt{3}}{6\sqrt{5}}$ c. $\frac{8\sqrt{3}}{7\sqrt{7}}$ d. $\frac{8\sqrt{60}}{\sqrt{28}}$ e. $\frac{2\sqrt{35}}{3\sqrt{14}}$

WE9c For questions 30 to 32, express the following in their simplest form with a rational denominator.

30. a. $\frac{\sqrt{6}+\sqrt{12}}{\sqrt{3}}$ b. $\frac{\sqrt{15}-\sqrt{22}}{\sqrt{6}}$ c. $\frac{6\sqrt{2}-\sqrt{15}}{\sqrt{10}}$ d. $\frac{2\sqrt{18}+3\sqrt{2}}{\sqrt{5}}$

31. a. $\frac{3\sqrt{5}+6\sqrt{7}}{\sqrt{8}}$ b. $\frac{4\sqrt{2}+3\sqrt{8}}{2\sqrt{3}}$ c. $\frac{3\sqrt{11}-4\sqrt{5}}{\sqrt{18}}$ d. $\frac{2\sqrt{7}-2\sqrt{5}}{\sqrt{12}}$

32. a. $\frac{7\sqrt{12}-5\sqrt{6}}{6\sqrt{3}}$ b. $\frac{6\sqrt{2}-\sqrt{5}}{4\sqrt{8}}$ c. $\frac{6\sqrt{3}-5\sqrt{5}}{7\sqrt{20}}$ d. $\frac{3\sqrt{5}+7\sqrt{3}}{5\sqrt{24}}$

WE10 For questions 33 to 35, rationalise the denominator and simplify.

33. a. $\frac{1}{\sqrt{5}+2}$ b. $\frac{1}{\sqrt{8}-\sqrt{5}}$ c. $\frac{4}{2\sqrt{11}-\sqrt{13}}$

34. a. $\frac{5\sqrt{3}}{3\sqrt{5}+4\sqrt{2}}$ b. $\frac{\sqrt{8}-3}{\sqrt{8}+3}$ c. $\frac{\sqrt{12}-\sqrt{7}}{\sqrt{12}+\sqrt{7}}$

35. a. $\frac{\sqrt{3}-1}{\sqrt{5}+1}$ b. $\frac{3\sqrt{6}-\sqrt{15}}{\sqrt{6}+2\sqrt{3}}$ c. $\frac{\sqrt{5}-\sqrt{3}}{4\sqrt{2}-\sqrt{3}}$

Reasoning

36. Calculate the area of a triangle with base length $\frac{3}{\sqrt{2}+2}$ and perpendicular height $\frac{5}{\sqrt{8}-1}$. Express your answer with a rational denominator. Show full working.

37. Determine the average of $\frac{1}{2\sqrt{x}}$ and $\frac{1}{3-2\sqrt{x}}$, writing your answer with a rational denominator. Show full working.

38. a. Show that $\left(\sqrt{a}+\sqrt{b}\right)^2=a+b+2\sqrt{ab}$.
    b. Use this result to evaluate:
       i. $\sqrt{8+2\sqrt{15}}$ ii. $\sqrt{8-2\sqrt{15}}$ iii. $\sqrt{7+4\sqrt{3}}$.

Problem solving

39. Simplify $\frac{\sqrt{5}+\sqrt{3}}{\sqrt{3}+\sqrt{3}+\sqrt{5}}-\frac{\sqrt{5}-\sqrt{3}}{\sqrt{3}+\sqrt{3}-\sqrt{5}}$.

40. Solve for x.

a. $\sqrt{9+x}-\sqrt{x}=\dfrac{5}{\sqrt{9+x}}$

b. $\dfrac{9\sqrt{x}-7}{3\sqrt{x}}=\dfrac{3\sqrt{x}+1}{\sqrt{x}+5}$

41. Solve the following for x: $2-\sqrt{2-\sqrt{2-\sqrt{2-\ldots}}}=x$

1.5 Review of index laws

LEARNING INTENTION

At the end of this subtopic you should be able to:

- recall and apply the index or exponent laws
- simplify expressions involving multiplication and division of terms with the same base
- evaluate expressions involving powers of zero
- simplify expressions involving raising a power to another power.

1.5.1 Review of index laws

eles-4669

Index notation

- When a number or pronumeral is repeatedly multiplied by itself, it can be written in a shorter form called **index form**.
- A number written in index form has two parts, the **base** and the **index**, and is written as:

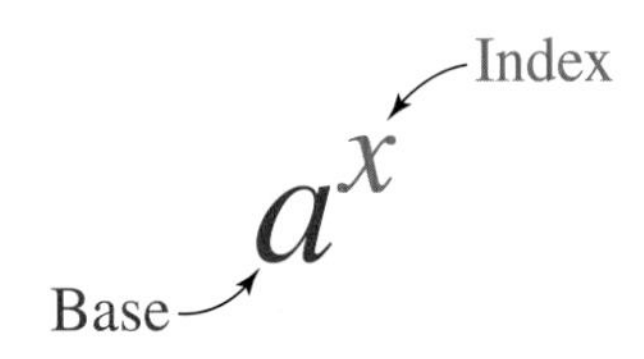

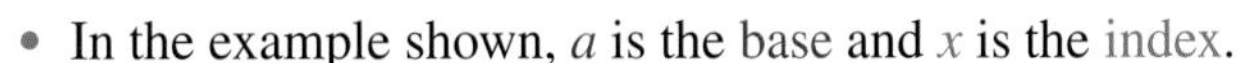

- In the example shown, a is the base and x is the index.
- Another name for an index is *exponent* or *power*.

Index laws

- Performing operations on numbers or pronumerals written in index form requires application of the index laws. There are six index laws.

First Index Law

When terms with the same base are multiplied, the indices are added.

$$a^m \times a^n = a^{m+n}$$

Second Index Law

When terms with the same base are divided, the indices are subtracted.

$$a^m \div a^n = a^{m-n}$$

WORKED EXAMPLE 11 Simplifying using the first two index laws

Simplify each of the following.

a. $m^4n^3p \times m^2n^5p^3$

b. $2a^2b^3 \times 3ab^4$

c. $\dfrac{2x^5y^4}{10x^2y^3}$

THINK	WRITE
a. 1. Write the expression.	**a.** $m^4n^3p \times m^2n^5p^3$
2. Multiply the terms with the same base by adding the indices. *Note:* $p = p^1$.	$= m^{4+2}n^{3+5}p^{1+3}$ $= m^6n^8p^4$
b. 1. Write the expression.	**b.** $2a^2b^3 \times 3ab^4$
2. Simplify by multiplying the coefficients, then multiply the terms with the same base by adding the indices.	$= 2 \times 3 \times a^{2+1} \times b^{3+4}$ $= 6a^3b^7$
c. 1. Write the expression.	**c.** $\dfrac{2x^5y^4}{10x^2y^3}$
2. Simplify by dividing both of the coefficients by the same factor, then divide terms with the same base by subtracting the indices.	$= \dfrac{1x^{5-2}y^{4-3}}{5}$ $= \dfrac{x^3y}{5}$

TI | THINK	DISPLAY/WRITE	CASIO | THINK	DISPLAY/WRITE
a–c. In a new document on a calculator page, complete the entry lines as: $m^4 \times n^3 \times p \times m^2 \times n^5 \times p^3$ $2 \times a^2 \times b^3 \times 3 \times a \times b^4$ $\dfrac{2 \times x^5 \times y^4}{10 \times x^2 \times y^3}$ Press ENTER after each entry. Be sure to include the multiplication sign between each variable.	**a–c.** 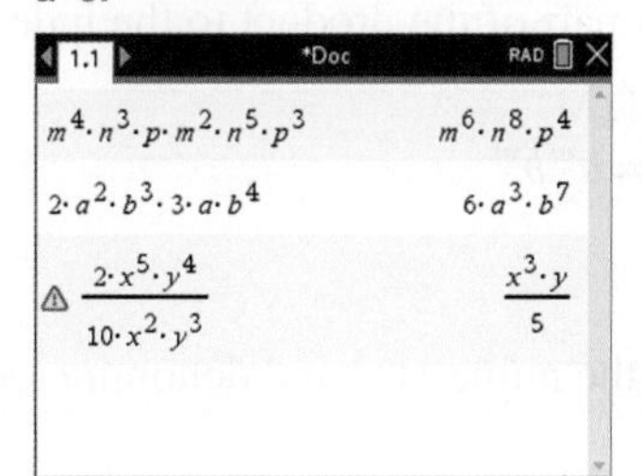 $m^4n^3p \times m^2n^5p^3 = m^6n^8p^4$ $2a^2b^3 \times 3ab^4 = 6a^3b^7$ $\dfrac{2x^5y^4}{10x^2y^3} = \dfrac{x^3y}{5}$	**a–c.** On the main screen, using the Var tab, complete the entry lines as: $m^4n^3p \times m^2n^5p^3$ $2a^2b^3 \times 3ab^4$ $\dfrac{2x^5y^4}{10x^2y^3}$ Press EXE after each entry.	**a–c.** 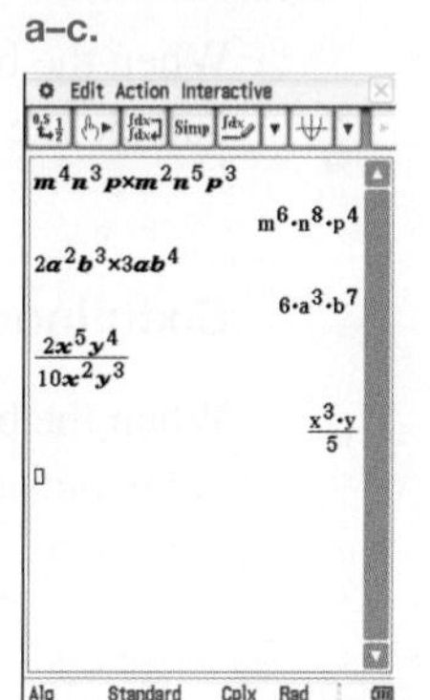 $m^4n^3p \times m^2n^5p^3 = m^6n^8p^4$ $2a^2b^3 \times 3ab^4 = 6a^3b^7$ $\dfrac{2x^5y^4}{10x^2y^3} = \dfrac{x^3y}{5}$

Third Index Law

Any term (excluding 0) with an index of 0 is equal to 1.

$$a^0 = 1,\ a \neq 0$$

WORKED EXAMPLE 12 Simplifying terms with indices of zero

Simplify each of the following.

a. $(2b^3)^0$

b. $-4(a^2b^5)^0$

THINK	WRITE
a. 1. Write the expression.	a. $(2b^3)^0$
2. Apply the Third Index Law, which states that any term (excluding 0) with an index of 0 is equal to 1.	$= 1$
b. 1. Write the expression.	b. $-4(a^2b^5)^0$
2. The entire term inside the brackets has an index of 0, so the bracket is equal to 1.	$= -4 \times 1$
3. Simplify.	$= -4$

Fourth Index Law

When a power (a^m) is raised to a power, the indices are multiplied.

$$(a^m)^n = a^{mn}$$

Fifth Index Law

When the base is a product, raise every part of the product to the index outside the brackets.

$$(ab)^m = a^m b^m$$

Sixth Index Law

When the base is a fraction, raise both the numerator and denominator to the index outside the brackets.

$$\left(\frac{a}{b}\right)^m = \frac{a^m}{b^m}$$

WORKED EXAMPLE 13 Simplifying terms in index form raised to a power

Simplify each of the following.

a. $(2n^4)^3$ b. $(3a^2b^7)^3$ c. $\left(\frac{2x^3}{y^4}\right)^4$ d. $(-4)^3$

THINK	WRITE
a. 1. Write the term.	a. $(2n^4)^3$
2. Apply the Fourth Index Law and simplify.	$= 2^{1\times3} \times n^{4\times3}$ $= 2^3n^{12}$ $= 8n^{12}$
b. 1. Write the expression.	b. $(3a^2b^7)^3$
2. Apply the Fifth Index Law and simplify.	$= 3^{1\times3} \times a^{2\times3} \times b^{7\times3}$ $= 3^3a^6b^{21}$ $= 27a^6b^{21}$

c. 1. Write the expression.

c. $\left(\dfrac{2x^3}{y^4}\right)^4$

2. Apply the Sixth Index Law and simplify.

$= \dfrac{2^{1\times4}\times x^{3\times4}}{y^{4\times4}}$

$= \dfrac{16x^{12}}{y^{16}}$

d. 1. Write the expression.

d. $(-4)^3$

2. Write in expanded form.

$= -4\times-4\times-4$

3. Simplify, taking careful note of the negative sign.

$= -64$

TI | THINK

a–d.

On a Calculator page, use the brackets and complete the entry line as:

$\left(2n^4\right)^3$

$\left(3a^2b^7\right)^3$

$\left(\dfrac{2x^3}{y^4}\right)^4$

$(-4)^3$

Press Enter after each entry.

DISPLAY/WRITE

a–d.

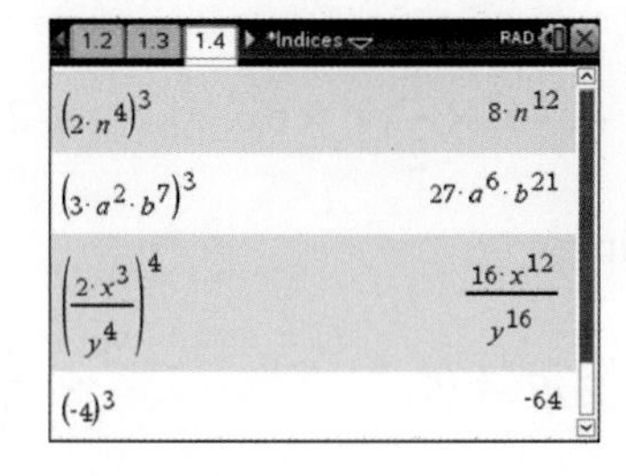

$\left(2n^4\right)^3 = 8n^{12}$

$\left(3a^2b^7\right)^3 = 27a^6b^{21}$

$\left(\dfrac{2x^3}{y^4}\right)^4 = \dfrac{16x^{12}}{y^{16}}$

$(-4)^3 = -64$

CASIO | THINK

a–d.

On the Main screen, use the brackets and complete the entry lines as:

$\left(2n^4\right)^3$

$\left(3a^2b^7\right)^3$

$\left(\dfrac{2x^3}{y^4}\right)^4$

$(-4)^3$

Press EXE after each entry.

DISPLAY/WRITE

a–d.

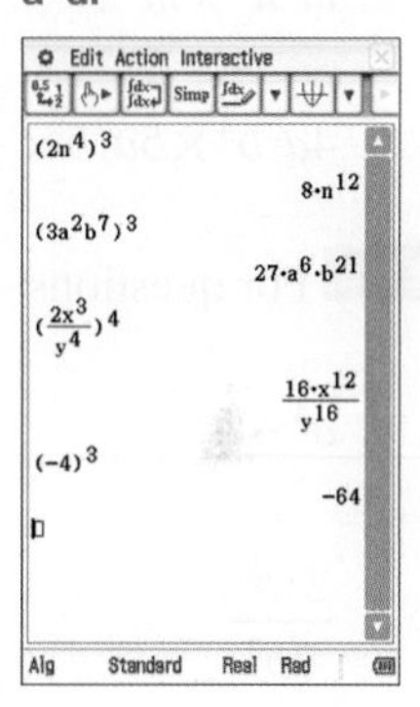

$\left(2n^4\right)^3 = 8n^{12}$

$\left(3a^2b^7\right)^3 = 27a^6b^{21}$

$\left(\dfrac{2x^3}{y^4}\right)^4 = \dfrac{16x^{12}}{y^{16}}$

$(-4)^3 = -64$

on Resources

eWorkbook Topic 1 Workbook (worksheets, code puzzle and project) (ewbk-2027)

Digital documents SkillSHEET Index form (doc-5168)

SkillSHEET Using a calculator to evaluate numbers given in index form (doc-5169)

Video eLesson Index laws (eles-1903)

Interactivities Individual pathway interactivity: Review of index laws (int-4652)

First Index Law (int-3709)

Second Index Law (int-3711)

Third Index Law (int-3713)

Fourth Index Law — Multiplication (int-3716)

Fifth and sixth index laws (int-6063)

Exercise 1.5 Review of index laws

learnON

Individual pathways

■ PRACTISE	■ CONSOLIDATE	■ MASTER
1, 4, 7, 10, 13, 15, 18, 21, 22, 26	2, 5, 8, 11, 14, 16, 19, 23, 24, 27	3, 6, 9, 12, 17, 20, 25, 28

To answer questions online and to receive **immediate corrective feedback** and **fully worked solutions** for all questions, go to your learnON title at www.jacplus.com.au.

Fluency

WE11a,b For questions 1 to **3**, simplify each of the following.

1. a. $a^3 \times a^4$ b. $a^2 \times a^3 \times a$ c. $b \times b^5 \times b^2$ d. $ab^2 \times a^3b^5$

2. a. $m^2n^6 \times m^3n^7$ b. $a^2b^5c \times a^3b^2c^2$ c. $mnp \times m^5n^3p^4$ d. $2a \times 3ab$

3. a. $4a^2b^3 \times 5a^2b \times \frac{1}{2}b^5$ b. $3m^3 \times 2mn^2 \times 6m^4n^5$ c. $4x^2 \times \frac{1}{2}xy^3 \times 6x^3y^3$ d. $2x^3y^2 \times 4x \times \frac{1}{2}x^4y^4$

WE11c For questions 4 to **6**, simplify each of the following.

4. a. $a^4 \div a^3$ b. $a^7 \div a^2$ c. $b^6 \div b^3$ d. $\frac{4a^7}{3a^3}$

5. a. $\frac{21b^6}{7b^2}$ b. $\frac{48m^8}{12m^3}$ c. $\frac{m^7n^3}{m^4n^2}$ d. $\frac{2x^4y^3}{4x^4y}$

6. a. $7ab^5c^4 \div ab^2c^4$ b. $\frac{20m^5n^3p^4}{16m^3n^3p^2}$ c. $\frac{14x^3y^4z^2}{28x^2y^2z^2}$

WE12 For questions 7 to **9**, simplify each of the following.

7. a. a^0 b. $(2b)^0$ c. $(3m^2)^0$

8. a. $3x^0$ b. $4b^0$ c. $-3 \times (2n)^0$

9. a. $4a^0 - \left(\frac{a}{4}\right)^0$ b. $5y^0 - 12$ c. $5x^0 - (5xy^2)^0$

WE13 For questions 10 to **12**, simplify each of the following.

10. a. $(a^2)^3$ b. $(2a^5)^4$ c. $\left(\frac{m^2}{3}\right)^4$ d. $\left(\frac{2n^4}{3}\right)^2$ e. $(-7)^2$

11. a. $(a^2b)^3$ b. $(3a^3b^2)^2$ c. $(2m^3n^5)^4$ d. $\left(\frac{3m^2n}{4}\right)^3$ e. $\left(\frac{a^2}{b^3}\right)^2$

12. a. $\left(\frac{5m^3}{n^2}\right)^4$ b. $\left(\frac{7x}{2y^5}\right)^3$ c. $\left(\frac{3a}{5b^3}\right)^4$ d. $(-3)^5$ e. $(-2)^5$

13. **MC** a. $2m^{10}n^5$ is the simplified form of:

A. $m^5n^3 \times 2m^4n^2$ B. $\frac{6m^{10}n^4}{3n}$ C. $(2m^5n^2)^2$ D. $2n(m^5)^2 \times n^4$ E. $\left(\frac{2m^5}{n^3}\right)^2$

b. The value of $4-(5a)^0$ is:

A. -1 B. 9 C. 1 D. 3 E. 5

14. **MC** a. $4a^3b \times b^4 \times 5a^2b^3$ simplifies to:

A. $9a^5b^8$ B. $20a^5b^7$ C. $20a^5b^8$ D. $9a^5b^7$ E. $21a^5b^8$

b. $\frac{15x^9 \times 3x^6}{9x^{10} \times x^4}$ simplifies to:

A. $5x^9$ B. $9x$ C. $5x^{29}$ D. $9x^9$ E. $5x$

c. $\frac{3p^7 \times 8q^9}{12p^3 \times 4q^5}$ simplifies to:

A. $2q^4$ B. $\frac{p^4q^4}{2}$ C. $\frac{q^4}{2}$ D. $\frac{p^4q^4}{24}$ E. $\frac{q^4}{24}$

d. $\frac{7a^5b^3}{5a^6b^2} \div \frac{7b^3a^2}{5b^5a^4}$ simplifies to:

A. $\frac{49a^3b}{25}$ B. $\frac{25a^3b}{49}$ C. a^3b D. ab^3 E. $\frac{25ab^3}{49}$

Understanding

For questions 15 to 17, evaluate each of the following.

15. a. $2^3 \times 2^2 \times 2$ b. $2 \times 3^2 \times 2^2$ c. $(5^2)^2$

16. a. $\frac{3^5 \times 4^6}{3^4 \times 4^4}$ b. $(2^3 \times 5)^2$ c. $\left(\frac{3}{5}\right)^3$

17. a. $\frac{4^4 \times 5^6}{4^3 \times 5^5}$ b. $(3^3 \times 2^4)^0$ c. $4(5^2 \times 3^5)^0$

For questions 18 to 20, simplify each of the following.

18. a. $(x^y)^{3z}$ b. $a^b \times (p^q)^0$

19. a. $m^a \times n^b \times (mn)^0$ b. $\left(\frac{a^2}{b^3}\right)^x$

20. a. $\frac{n^3m^2}{n^pm^q}$ b. $(a^{m+n})^p$

Reasoning

21. Explain why $a^3 \times a^2 = a^5$ and not a^6.

22. Is $2x$ ever the same as x^2? Explain your reasoning using examples.

23. Explain the difference between $3x^0$ and $(3x)^0$.

24. a. Complete the table for $a = 0$, 1, 2 and 3.

a	0	1	2	3
$3a^2$				
$5a$				
$3a^2 + 5a$				
$3a^2 \times 5a$				

b. Analyse what would happen as a becomes very large.

25. Evaluate algebraically the exact value of x if $4^{x+4} = 2^{x^2}$. Justify your answer.

Problem solving

26. Binary numbers (base 2 numbers) are used in computer operations. As the name implies, binary uses only two types of numbers, 0 and 1, to express all numbers. A binary number such as 101 (read one, zero, one) means $(1 \times 2^2) + (0 \times 2^1) + (1 \times 2^0) = 4 + 0 + 1 = 5$ (in base 10, the base we are most familiar with).
The number 1010 (read one, zero, one, zero) means

$$(1 \times 2^3) + (0 \times 2^2) + (1 \times 2^1) + (0 \times 2^0) = 8 + 0 + 2 + 0 = 10.$$

If we read the binary number from right to left, the index of 2 increases by one each time, beginning with a power of zero. Using this information, write out the numbers 1 to 10 in binary (base 2) form.

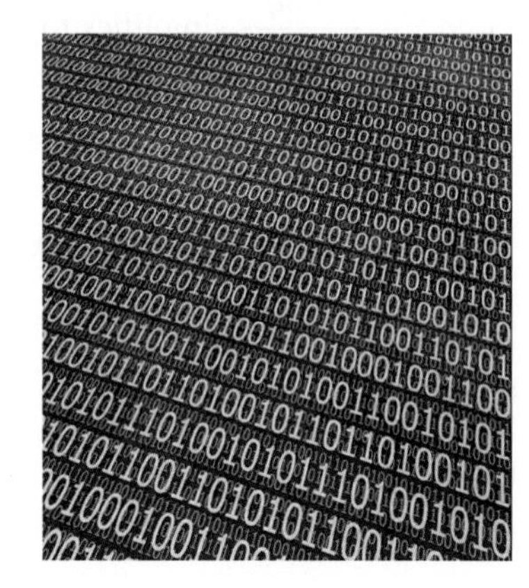

27. Solve for x:

a. $\dfrac{7^x \times 7^{1+2x}}{(7^x)^2} = 16\,807$

b. $2^{2x} - 5(2^x) = -4$

28. For the following:

a. determine the correct answer

b. identify the error in the solution.

$$\begin{aligned}\left(\frac{a^2b^3c}{a^2b^2}\right)^3 \times \left(\frac{a^3b^2c^2}{a^2b^3}\right)^2 &= \left(\frac{b^3c}{b^2}\right)^3 \times \left(\frac{ab^2c^2}{b^2}\right)^2 \\ &= \left(\frac{bc}{1}\right)^3 \times \left(\frac{ac^2}{b}\right)^2 \\ &= \left(\frac{abc^3}{b}\right)^6 \\ &= \left(\frac{ac^3}{1}\right)^6 \\ &= a^6c^{18}\end{aligned}$$

1.6 Negative indices

LEARNING INTENTION

At the end of this subtopic you should be able to:
- evaluate expressions involving negative indices
- simplify expressions involving negative indices and re-write expressions so that all indices are positive.

1.6.1 Negative indices and the Seventh Index Law

eles-4670

- Consider the expression. $\frac{a^3}{a^5}$. This expression can be simplified in two different ways.
  1. Written in expanded form: $\frac{a^3}{a^5} = \frac{a \times a \times a}{a \times a \times a \times a \times a}$
  $= \frac{1}{a \times a}$
  $= \frac{1}{a^2}$
  2. Using the Second Index Law: $\frac{a^3}{a^5} = a^{3-5}$
  $= a^{-2}$
- Equating the results of both of these simplifications we get $a^{-2} = \frac{1}{a^2}$.
- In general, $\frac{1}{a^n} = \frac{a^0}{a^n} \left(1 = a^0\right)$
  $= a^{0-n}$ (using the Second Index Law)
  $= a^{-n}$
  This statement is the Seventh Index Law.

Seventh Index Law

A term raised to a negative index is equivalent to 1 over the original term with a positive index.

$$\boldsymbol{a^{-n} = \frac{1}{a^n}}$$

- The converse of this law can be used to rewrite terms with positive indices only.

$$\frac{1}{a^{-n}} = a^n$$

- It is also worth noting that applying a negative index to a fraction has the effect of swapping the numerator and denominator.

$$\left(\frac{a}{b}\right)^{-n} = \frac{b^n}{a^n}$$

Note: It is proper mathematical convention for an algebraic term to be written with each variable in alphabetical order with positive indices only. For example: $\frac{b^3a^2c^{-4}}{y^6x^{-5}}$ should be written as $\frac{a^2b^3x^5}{c^4y^6}$.

WORKED EXAMPLE 14 Writing terms with positive indices only

Express each of the following with positive indices.

a. x^{-3}

b. $2m^{-4}n^2$

c. $\dfrac{4}{a^{-3}}$

THINK	WRITE
a. 1. Write the expression.	a. x^{-3}
2. Apply the Seventh Index Law.	$= \dfrac{1}{x^3}$
b. 1. Write the expression.	b. $2m^{-4}n^2$
2. Apply the Seventh Index Law to write the expression with positive indices.	$= \dfrac{2n^2}{m^4}$
c. 1. Write the expression and rewrite the fraction, using a division sign.	c. $\dfrac{4}{a^{-3}} = 4 \div a^{-3}$
2. Apply the Seventh Index Law to write the expression with positive indices.	$= 4 \div \dfrac{1}{a^3}$
3. To divide the fraction, change fraction division into multiplication.	$= 4 \times \dfrac{a^3}{1}$ $= 4a^3$

WORKED EXAMPLE 15 Simplifying expressions with negative indices

Simplify each of the following, expressing the answers with positive indices.

a. $a^2b^{-3} \times a^{-5}b$

b. $\dfrac{2x^4y^2}{3xy^5}$

c. $\left(\dfrac{2m^3}{n^{-2}}\right)^{-2}$

THINK	WRITE
a. 1. Write the expression.	a. $a^2b^{-3} \times a^{-5}b$
2. Apply the First Index Law. Multiply terms with the same base by adding the indices.	$= a^{2+-5}b^{-3+1}$ $= a^{-3}b^{-2}$
3. Apply the Seventh Index Law to write the answer with positive indices.	$= \dfrac{1}{a^3b^2}$
b. 1. Write the expression.	b. $\dfrac{2x^4y^2}{3xy^5}$
2. Apply the Second Index Law. Divide terms with the same base by subtracting the indices.	$= \dfrac{2x^{4-1}y^{2-5}}{3}$ $= \dfrac{2x^3y^{-3}}{3}$
3. Apply the Seventh Index Law to write the answer with positive indices.	$= \dfrac{2x^3}{3y^3}$

c. 1. Write the expression.

c. $\left(\frac{2m^3}{n^{-2}}\right)^{-2}$

2. Apply the Sixth Index Law. Multiply the indices of both the numerator and denominator by the index outside the brackets.

$= \frac{2^{-2}m^{-6}}{n^4}$

3. Apply the Seventh Index Law to express all terms with positive indices.

$= \frac{1}{2^2m^6n^4}$

4. Simplify.

$= \frac{1}{4m^6n^4}$

WORKED EXAMPLE 16 Evaluating expressions containing negative indices

Evaluate 6×3^{-3} without using a calculator.

THINK	WRITE
1. Write the multiplication.	6×3^{-3}
2. Apply the Seventh Index Law to write 3^{-3} with a positive index.	$= 6 \times \frac{1}{3^3}$
3. Multiply the numerator of the fraction by the whole number.	$= \frac{6}{3^3}$
4. Evaluate the denominator.	$= \frac{6}{27}$
5. Cancel by dividing both the numerator and denominator by the highest common factor (3).	$= \frac{2}{9}$

on Resources

eWorkbook Topic 1 Workbook (worksheets, code puzzle and project) (ewbk-2027)

Video eLesson Negative indices (eles-1910)

Interactivities Individual pathway interactivity: Negative indices (int-4563)

Negative indices (int-6064)

Exercise 1.6 Negative indices

learnON

Individual pathways

■ PRACTISE	■ CONSOLIDATE	■ MASTER
1, 4, 7, 10, 13, 15, 17, 18, 28, 31	2, 5, 8, 11, 14, 16, 19, 20, 23, 26, 29, 32	3, 6, 9, 12, 21, 22, 24, 25, 27, 30, 33

To answer questions online and to receive **immediate corrective feedback** and **fully worked solutions** for all questions, go to your learnON title at www.jacplus.com.au.

Fluency

WE14 For questions 1 to 3, express each of the following with positive indices.

1. a. x^{-5} b. y^{-4} c. $2a^{-9}$ d. $\frac{4}{5}a^{-3}$

2. a. $3x^2y^{-3}$ b. $2^{-2}m^{-3}n^{-4}$ c. $6a^3b^{-1}c^{-5}$ d. $\frac{1}{a^{-6}}$

3. a. $\frac{2}{3a^{-4}}$ b. $\frac{6a}{3b^{-2}}$ c. $\frac{7a^{-4}}{2b^{-3}}$ d. $\frac{2m^3n^{-5}}{3a^{-2}b^4}$

WE15 For questions 4 to 6, simplify each of the following, expressing the answers with positive indices.

4. a. $a^3b^{-2}\times a^{-5}b^{-1}$ b. $2x^{-2}y\times 3x^{-4}y^{-2}$ c. $3m^2n^{-5}\times m^{-2}n^{-3}$
   d. $4a^3b^2\div a^5b^7$ e. $2xy^6\div 3x^2y^5$

5. a. $5x^{-2}y^3\div 6xy^2$ b. $\frac{6m^4n}{2n^3m^6}$ c. $\frac{4x^2y^9}{x^7y^{-3}}$ d. $\frac{2m^2n^{-4}}{6m^5n^{-1}}$ e. $\left(2a^3m^4\right)^{-5}$

6. a. $4\left(p^7q^{-4}\right)^{-2}$ b. $3\left(a^{-2}b^{-3}\right)^4$ c. $\left(\frac{2p^2}{3q^3}\right)^{-3}$ d. $\left(\frac{a^{-4}}{2b^{-3}}\right)^2$ e. $\left(\frac{6a^2}{3b^{-2}}\right)^{-3}$

WE16 For questions 7 to 9, evaluate each of the following without using a calculator.

7. a. 2^{-3} b. 6^{-2} c. 3^{-4} d. $3^{-2}\times 2^3$

8. a. $4^{-3}\times 2^2$ b. 5×6^{-2} c. $\frac{6}{2^{-3}}$ d. $\frac{4\times 3^{-3}}{2^{-3}}$

9. a. $\frac{1}{3}\times 5^{-2}\times 3^4$ b. $\frac{16^0\times 2^4}{8^2\times 2^{-4}}$ c. $\frac{5^3\times 25^0}{25^2\times 5^{-4}}$ d. $\frac{3^4\times 4^2}{12^3\times 15^0}$

10. Write each of these numbers as a power of 2.
    a. 8 b. $\frac{1}{8}$ c. 32 d. $\frac{1}{64}$

11. Solve each of the following for x.
    a. $125=5^x$ b. $\frac{1}{16}=4^x$ c. $\frac{1}{7}=7^x$ d. $216=6^x$ e. $0.01=10^x$

12. Solve each of the following for x.
    a. $1=8^x$ b. $64=4^x$ c. $\frac{1}{64}=4^x$ d. $\frac{1}{64}=2^x$ e. $\frac{1}{64}=8^x$

13. Evaluate the following expressions.

a. $\left(\frac{2}{3}\right)^{-1}$ b. $\left(\frac{5}{4}\right)^{-1}$ c. $\left(3\frac{1}{2}\right)^{-1}$ d. $\left(\frac{1}{5}\right)^{-1}$

14. Write the following expressions with positive indices.

a. $\left(\frac{a}{b}\right)^{-1}$ b. $\left(\frac{a^2}{b^3}\right)^{-1}$ c. $\left(\frac{a^{-2}}{b^{-3}}\right)^{-1}$ d. $\left(\frac{m^3}{n^{-2}}\right)^{-1}$

15. Evaluate each of the following, using a calculator.

a. 3^{-6} b. 12^{-4} c. 7^{-5}

16. Evaluate each of the following, using a calculator.

a. $\left(\frac{1}{2}\right)^{-8}$ b. $\left(\frac{3}{4}\right)^{-7}$ c. $(0.04)^{-5}$

Understanding

17. **MC** $\frac{1}{a^{-4}}$ is the same as:

A. $4a$ B. $-4a$ C. a^4 D. $\frac{1}{a^4}$ E. $-a^4$

18. **MC** $\frac{1}{8}$ is the same as:

A. 2^3 B. 2^{-3} C. 3^2 D. 3^{-2} E. $\frac{1}{2^{-3}}$

19. **MC** Select which of the following, when simplified, gives $\frac{3m^4}{4n^2}$.

A. $\frac{3m^{-4}n^{-2}}{4}$ B. $3\times 2^{-2}\times m^4\times n^{-2}$ C. $\frac{3n^{-2}}{2^{-2}m^{-4}}$
D. $\frac{2^2n^{-2}}{3^{-1}m^{-4}}$ E. $3m^4\times 2^2n^{-2}$

20. **MC** When simplified, $3a^{-2}b^{-7}\div\left(\frac{3}{4}a^{-4}b^6\right)$ is equal to:

A. $\frac{4}{a^6b^{13}}$ B. $\frac{9b}{4a^6}$ C. $\frac{9a^2}{4b}$ D. $\frac{4a^2}{b^{13}}$ E. $\frac{4a^2}{b}$

21. **MC** When $(2x^6y^{-4})^{-3}$ is simplified, it is equal to:

A. $\frac{2x^{18}}{y^{12}}$ B. $\frac{x^{18}}{8y^{12}}$ C. $\frac{y^{12}}{8x^{18}}$ D. $\frac{8y^{12}}{x^{18}}$ E. $\frac{x^{18}}{6y^{12}}$

22. **MC** If $\left(\frac{2a^x}{b^y}\right)^3$ is equal to $\frac{8b^9}{a^6}$, then x and y (in that order) are:

A. −3 and −6 B. −6 and −3 C. −3 and 2 D. −3 and −2 E. −2 and −3

23. Simplify, expressing your answer with positive indices.

a. $\frac{m^{-3}n^{-2}}{m^{-5}n^6}$ b. $\frac{(m^3n^{-2})^{-7}}{(m^{-5}n^3)^4}$ c. $\frac{5(a^3b^{-3})^2}{(ab^{-4})^{-1}}\div\frac{(5a^{-2}b)^{-1}}{(a^{-4}b)^3}$

24. Simplify, expanding any expressions in brackets.

a. $(r^3+s^3)(r^3-s^3)$

b. $(m^5+n^5)^2$

c. $\dfrac{(x^{a+1})^b \times x^{a+b}}{x^{a(b+1)} \times x^{2b}}$

d. $\left(\dfrac{p^{x+1}}{p^{x-1}}\right)^{-4} \times \dfrac{p^{8(x+1)}}{(p^{2x})^4} \times \dfrac{p^2}{(p^{12x})^0}$

25. Write $\left(\dfrac{2^r \times 8^r}{2^{2r} \times 16}\right)$ in the form 2^{ar+b}.

26. Write $2^{-m} \times 3^{-m} \times 6^{2m} \times 3^{2m} \times 2^{2m}$ as a power of 6.

27. Solve for x if $4^x - 4^{x-1} = 48$.

Reasoning

28. Consider the equation $y = \dfrac{6}{x}$. Clearly $x \neq 0$ as $\dfrac{6}{x}$ would be undefined.

Explain what happens to the value of y as x gets closer to zero coming from:

a. the positive direction

b. the negative direction.

29. Consider the expression 2^{-n}. Explain what happens to the value of this expression as n increases.

30. Explain why each of these statements is false. Illustrate each answer by substituting a value for the pronumeral

a. $5x^0 = 1$

b. $9x^5 \div (3x^5) = 3x$

c. $a^5 \div a^7 = a^2$

d. $2c^{-4} = \dfrac{1}{2c^4}$

Problem solving

31. Solve the following pair of simultaneous equations.

$3^{y+1} = \dfrac{1}{9}$ and $\dfrac{5^y}{125^x} = 125$

32. Simplify $\dfrac{x^{n+2} + x^{n-2}}{x^{n-4} + x^n}$.

33. Solve for x and y if $5^{x-y} = 625$ and $3^{2x} \times 3^y = 243$.

Hence, evaluate $\dfrac{35^x}{7^{-2y} \times 5^{-3y}}$.

1.7 Fractional indices (10A)

LEARNING INTENTION

At the end of this subtopic you should be able to:

- evaluate expressions involving fractional indices
- simplify expressions involving fractional indices.

1.7.1 Fractional indices and the Eighth Index Law

eles-4671

- Consider the expression $a^{\frac{1}{2}}$. Now consider what happens if we square that expression.

$$\left(a^{\frac{1}{2}}\right)^2 = a \text{ (Using the Fourth Index Law, } (a^m)^n = a^{m \times n})$$

- From our work on surds, we know that $\left(\sqrt{a}\right)^2 = a$.

- Equating the two facts above, $\left(a^{\frac{1}{2}}\right)^2 = \left(\sqrt{a}\right)^2$. Therefore, $a^{\frac{1}{2}} = \sqrt{a}$.
- Similarly, $b^{\frac{1}{3}} \times b^{\frac{1}{3}} \times b^{\frac{1}{3}} = \left(b^{\frac{1}{3}}\right)^3 = b$ implying that $b^{\frac{1}{3}} = \sqrt[3]{b}$.
- This pattern can be continued and generalised to produce $a^{\frac{1}{n}} = \sqrt[n]{a}$.
- Now consider: $a^{\frac{m}{n}} = a^{m \times \frac{1}{n}}$ or $a^{\frac{m}{n}} = a^{\frac{1}{n} \times m}$

$$a^{\frac{m}{n}} = a^{m\times\frac{1}{n}} = (a^m)^{\frac{1}{n}} = \sqrt[n]{a^m}$$

$$a^{\frac{m}{n}} = a^{\frac{1}{n}\times m} = \left(a^{\frac{1}{n}}\right)^m = \left(\sqrt[n]{a}\right)^m$$

Eighth Index Law

A term raised to a fractional index $\frac{m}{n}$ is equivalent to the nth root of the term raised to the power m.

$$a^{\frac{m}{n}} = \sqrt[n]{a^m} = \left(\sqrt[n]{a}\right)^m$$

WORKED EXAMPLE 17 Converting fractional indices to surd form

Write each of the following expressions in simplest surd form.

a. $10^{\frac{1}{2}}$ **b.** $5^{\frac{3}{2}}$

THINK	WRITE
a. Since an index of $\frac{1}{2}$ is equivalent to taking the square root, this term can be written as the square root of 10.	a. $10^{\frac{1}{2}} = \sqrt{10}$
b. 1. A power of $\frac{3}{2}$ means the square root of the number cubed.	b. $5^{\frac{3}{2}} = \sqrt{5^3}$
2. Evaluate 5^3.	$= \sqrt{125}$
3. Simplify $\sqrt{125}$.	$= 5\sqrt{5}$

WORKED EXAMPLE 18 Evaluating fractional indices without a calculator

Evaluate each of the following without using a calculator.

a. $9^{\frac{1}{2}}$ **b.** $16^{\frac{3}{2}}$

THINK	WRITE
a. 1. Rewrite the number using Eighth Index Law.	a. $9^{\frac{1}{2}} = \sqrt{9}$
2. Evaluate.	$= 3$

b. 1. Rewrite the number using $a^{\frac{m}{n}} = \left(\sqrt[n]{a}\right)^m$.

2. Simplify and evaluate the result.

b. $16^{\frac{3}{2}} = \left(\sqrt{16}\right)^3$
$= 4^3$
$= 64$

WORKED EXAMPLE 19 Evaluating fractional indices with a calculator

Use a calculator to determine the value of the following, correct to 1 decimal place.

a. $10^{\frac{1}{4}}$ **b.** $200^{\frac{1}{5}}$

THINK	WRITE
a. Use a calculator to produce the answer.	a. $10^{\frac{1}{4}} = 1.77827941$ ≈ 1.8
b. Use a calculator to produce the answer.	b. $200^{\frac{1}{5}} = 2.885399812$ ≈ 2.9

TI | THINK

a.
In a new document on a Calculator page, complete the entry line as:
$10^{\frac{1}{4}}$
Then press ENTER.
To convert the answer to decimal press:
- MENU
- 2: Number
- 1: Convert to Decimal

Then press ENTER.

DISPLAY/WRITE

a.

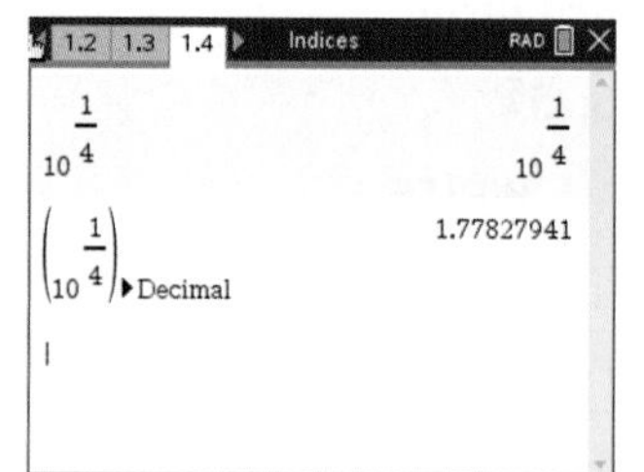

$10^{\frac{1}{4}} = 1.77827941$
≈ 1.8

b.
In a new document on a Calculator page, complete the entry line as:
$200^{\frac{1}{5}}$
Then press ENTER.
To convert the answer to decimal press:
- MENU
- 2: Number
- 1: Convert to Decimal

Then press ENTER.

b.

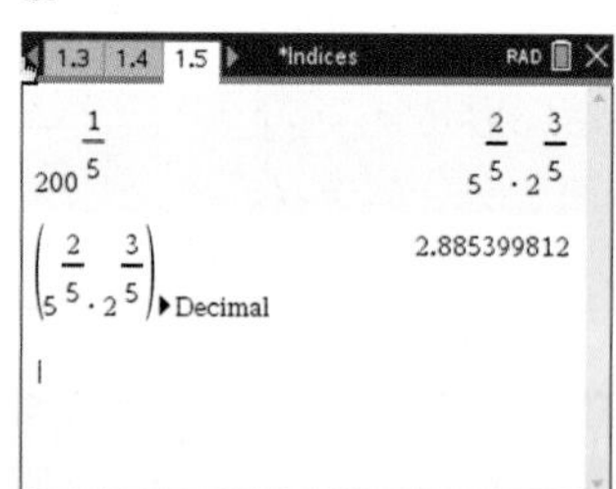

$200^{\frac{1}{5}} = 2.885399812$
≈ 2.9

CASIO | THINK

a–b.
On the Main screen, complete the entry lines as:
$10^{\frac{1}{4}}$
$200^{\frac{1}{5}}$
Press EXE after each entry.

Note: Change Standard to Decimal.

DISPLAY/WRITE

a–b.

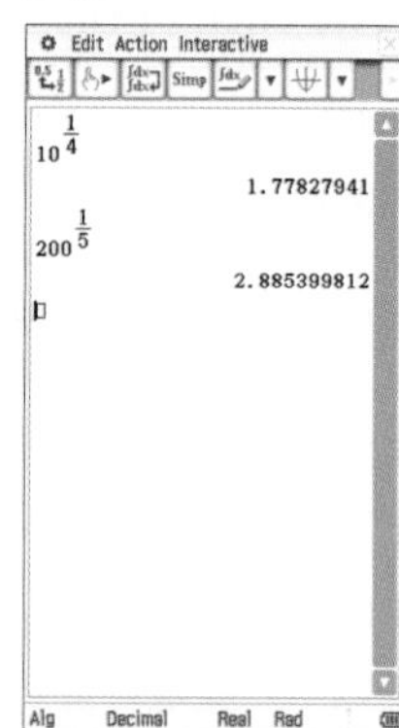

$10^{\frac{1}{4}} = 1.77827941$
≈ 1.8

$200^{\frac{1}{5}} = 2.885399812$
≈ 2.9

WORKED EXAMPLE 20 Simplifying expressions with fractional indices

Simplify each of the following.

a. $m^{\frac{1}{5}} \times m^{\frac{2}{5}}$

b. $(a^2b^3)^{\frac{1}{6}}$

c. $\left(\dfrac{x^{\frac{2}{3}}}{y^{\frac{3}{4}}}\right)^{\frac{1}{2}}$

THINK	WRITE
a. 1. Write the expression.	a. $m^{\frac{1}{5}} \times m^{\frac{2}{5}}$
2. Multiply numbers with the same base by adding the indices.	$= m^{\frac{3}{5}}$
b. 1. Write the expression.	b. $(a^2b^3)^{\frac{1}{6}}$
2. Multiply each index inside the grouping symbols (brackets) by the index on the outside.	$= a^{\frac{2}{6}}b^{\frac{3}{6}}$
3. Simplify the fractions.	$= a^{\frac{1}{3}}b^{\frac{1}{2}}$
c. 1. Write the expression.	c. $\left(\dfrac{x^{\frac{2}{3}}}{y^{\frac{3}{4}}}\right)^{\frac{1}{2}}$
2. Multiply the index in both the numerator and denominator by the index outside the grouping symbols.	$= \dfrac{x^{\frac{1}{3}}}{y^{\frac{3}{8}}}$

Resources

- **eWorkbook** Topic 1 Workbook (worksheets, code puzzle and project) (ewbk-2027)
- **Digital documents** SkillSHEET Addition of fractions (doc-5176)
  SkillSHEET Subtraction of fractions (doc-5177)
  SkillSHEET Multiplication of fractions (doc-5178)
  SkillSHEET Writing roots as fractional indices (doc-5179)
- **Video eLesson** Fractional indices (eles-1950)
- **Interactivities** Individual pathway interactivity: Fractional indices (int-4564)
  Fractional indices (int-6107)

Exercise 1.7 Fractional indices (10A)

learnon

Individual pathways

■ PRACTISE	■ CONSOLIDATE	■ MASTER
1, 3, 7, 10, 13, 15, 17, 20, 23, 25, 28, 31	2, 4, 5, 8, 11, 14, 18, 21, 26, 29, 32	6, 9, 12, 16, 19, 22, 24, 27, 30, 33, 34

To answer questions online and to receive **immediate corrective feedback** and **fully worked solutions** for all questions, go to your learnON title at www.jacplus.com.au.

Fluency

WE17 For questions 1 to 4, write the following in surd form.

1. a. $15^{\frac{1}{2}}$ b. $m^{\frac{1}{4}}$ c. $7^{\frac{2}{5}}$ d. $7^{\frac{5}{2}}$
2. a. $w^{\frac{3}{8}}$ b. $w^{1.25}$ c. $5^{3\frac{1}{3}}$ d. $a^{0.3}$
3. a. $\sqrt{t}$ b. $\sqrt[4]{5^7}$ c. $\sqrt[6]{6^{11}}$ d. $\sqrt[7]{x^6}$
4. a. $\sqrt[6]{x^7}$ b. $\sqrt[5]{w^{10}}$ c. $^{10}\sqrt{w^5}$ d. $\sqrt[x]{11^n}$
5. **WE18** Evaluate each of the following without using a calculator.
   a. $16^{\frac{1}{2}}$ b. $25^{\frac{1}{2}}$ c. $81^{\frac{1}{2}}$
6. Evaluate each of the following without using a calculator.
   a. $8^{\frac{1}{3}}$ b. $64^{\frac{1}{3}}$ c. $81^{\frac{1}{4}}$

WE19 For questions 7 to 9, use a calculator to evaluate each of the following, correct to 1 decimal place.

7. a. $5^{\frac{1}{2}}$ b. $7^{\frac{1}{5}}$ c. $8^{\frac{1}{9}}$
8. a. $12^{\frac{3}{8}}$ b. $100^{\frac{5}{9}}$ c. $50^{\frac{2}{3}}$
9. a. $(0.6)^{\frac{4}{5}}$ b. $\left(\frac{3}{4}\right)^{\frac{3}{4}}$ c. $\left(\frac{4}{5}\right)^{\frac{2}{3}}$

For questions 10 to 19, simplify each of the expressions.

10. a. $4^{\frac{3}{5}} \times 4^{\frac{1}{5}}$ b. $2^{\frac{1}{8}} \times 2^{\frac{3}{8}}$ c. $a^{\frac{1}{2}} \times a^{\frac{1}{3}}$
11. a. $x^{\frac{3}{4}} \times x^{\frac{2}{5}}$ b. $5m^{\frac{1}{3}} \times 2m^{\frac{1}{5}}$ c. $\frac{1}{2}b^{\frac{3}{7}} \times 4b^{\frac{2}{7}}$
12. a. $-4y^2 \times y^{\frac{2}{9}}$ b. $\frac{2}{5}a^{\frac{3}{8}} \times 0.05a^{\frac{3}{4}}$ c. $5x^3 \times x^{\frac{1}{2}}$
13. a. $a^{\frac{2}{3}}b^{\frac{3}{4}} \times a^{\frac{1}{3}}b^{\frac{3}{4}}$ b. $x^{\frac{3}{5}}y^{\frac{2}{9}} \times x^{\frac{1}{5}}y^{\frac{1}{3}}$ c. $2ab^{\frac{1}{3}} \times 3a^{\frac{3}{5}}b^{\frac{4}{5}}$
14. a. $6m^{\frac{3}{7}} \times \frac{1}{3}m^{\frac{1}{4}}n^{\frac{2}{5}}$ b. $x^3y^{\frac{1}{2}}z^{\frac{1}{3}} \times x^{\frac{1}{6}}y^{\frac{1}{3}}z^{\frac{1}{2}}$ c. $2a^{\frac{2}{5}}b^{\frac{3}{8}}c^{\frac{1}{4}} \times 4b^{\frac{3}{4}}c^{\frac{3}{4}}$
15. a. $x^3y^2 \div x^{\frac{4}{3}}y^{\frac{3}{5}}$ b. $a^{\frac{5}{9}}b^{\frac{2}{3}} \div a^{\frac{2}{5}}b^{\frac{2}{5}}$ c. $m^{\frac{3}{8}}n^{\frac{4}{7}} \div 3n^{\frac{3}{8}}$
16. a. $10x^{\frac{4}{5}}y \div 5x^{\frac{2}{3}}y^{\frac{1}{4}}$ b. $\dfrac{5a^{\frac{3}{4}}b^{\frac{3}{5}}}{20a^{\frac{1}{5}}b^{\frac{1}{4}}}$ c. $\dfrac{p^{\frac{7}{8}}q^{\frac{1}{4}}}{7p^{\frac{2}{3}}q^{\frac{1}{6}}}$
17. a. $\left(2^{\frac{3}{4}}\right)^{\frac{3}{5}}$ b. $\left(5^{\frac{2}{3}}\right)^{\frac{1}{4}}$ c. $\left(7^{\frac{1}{5}}\right)^{6}$

18. a. $(a^3)^{\frac{1}{10}}$ b. $\left(m^{\frac{4}{9}}\right)^{\frac{3}{8}}$ c. $\left(2b^{\frac{1}{2}}\right)^{\frac{1}{3}}$

19. a. $4\left(p^{\frac{3}{7}}\right)^{\frac{14}{15}}$ b. $\left(x^{\frac{m}{n}}\right)^{\frac{n}{p}}$ c. $\left(3m^{\frac{a}{b}}\right)^{\frac{b}{c}}$

Understanding

WE20 For questions **20** to **22**, simplify each of the following.

20. a. $\left(a^{\frac{1}{2}}b^{\frac{1}{3}}\right)^{\frac{1}{2}}$ b. $(a^4b)^{\frac{3}{4}}$ c. $\left(x^{\frac{3}{5}}y^{\frac{7}{8}}\right)^2$

21. a. $\left(3a^{\frac{1}{3}}b^{\frac{3}{5}}c^{\frac{3}{4}}\right)^{\frac{1}{3}}$ b. $5\left(x^{\frac{1}{2}}y^{\frac{2}{3}}z^{\frac{2}{5}}\right)^{\frac{1}{2}}$ c. $\left(\frac{a^{\frac{3}{4}}}{b}\right)^{\frac{2}{3}}$

22. a. $\left(\frac{m^{\frac{4}{5}}}{n^{\frac{7}{8}}}\right)^2$ b. $\left(\frac{b^{\frac{3}{5}}}{c^{\frac{4}{9}}}\right)^{\frac{2}{3}}$ c. $\left(\frac{4x^7}{2y^{\frac{3}{4}}}\right)^{\frac{1}{2}}$

23. **MC** a. $y^{\frac{2}{5}}$ is equal to:

A. $\left(y^{\frac{1}{2}}\right)^5$ B. $y\times\frac{2}{5}$ C. $(y^5)^{\frac{1}{2}}$ D. $2\sqrt[5]{y}$ E. $\left(y^{\frac{1}{5}}\right)^2$

b. $k^{\frac{2}{3}}$ is not equal to:

A. $\left(k^{\frac{1}{3}}\right)^2$ B. $\sqrt[3]{k^2}$ C. $\left(k^{\frac{1}{2}}\right)^3$ D. $\left(\sqrt[3]{k}\right)^2$ E. $(k^2)^{\frac{1}{3}}$

c. $\frac{1}{\sqrt[5]{g^{-2}}}$ is equal to:

A. $g^{\frac{2}{5}}$ B. $g^{-\frac{2}{5}}$ C. $g^{\frac{5}{2}}$ D. $g^{-\frac{5}{2}}$ E. $2g^{\frac{1}{5}}$

24. **MC** a. If $\left(a^{\frac{3}{4}}\right)^{\frac{m}{n}}$ is equal to $a^{\frac{1}{4}}$, then m and n could not be:

A. 1 and 3 B. 2 and 6 C. 3 and 8 D. 4 and 9 E. both **C** and **D**

b. When simplified, $\left(\frac{a^{\frac{m}{n}}}{b^{\frac{n}{p}}}\right)^{\frac{p}{m}}$ is equal to:

A. $\frac{a^{\frac{m}{p}}}{b^{\frac{n}{m}}}$ B. $\frac{a^{\frac{p}{n}}}{b^{\frac{n}{m}}}$ C. $\frac{a^{\frac{mp}{n}}}{b^{\frac{n}{m}}}$ D. $\frac{a^p}{b^m}$ E. $\frac{a^{\frac{m^2}{np}}}{b^{\frac{nm}{p^2}}}$

25. Simplify each of the following.

a. $\sqrt{a^8}$ b. $\sqrt[3]{b^9}$ c. $\sqrt[4]{m^{16}}$

26. Simplify each of the following.

a. $\sqrt{16x^4}$ b. $\sqrt[3]{8y^9}$ c. $\sqrt[4]{16x^8y^{12}}$

27. Simplify each of the following.

a. $\sqrt[3]{27m^9n^{15}}$

b. $\sqrt[5]{32p^5q^{10}}$

c. $\sqrt[3]{216a^6b^{18}}$

Reasoning

28. The relationship between the length of a pendulum (L) in a grandfather clock and the time it takes to complete one swing (T) in seconds is given by the following rule. Note that g is the acceleration due to gravity and will be taken as 9.8.

$$T=2\pi\left(\frac{L}{g}\right)^{\frac{1}{2}}$$

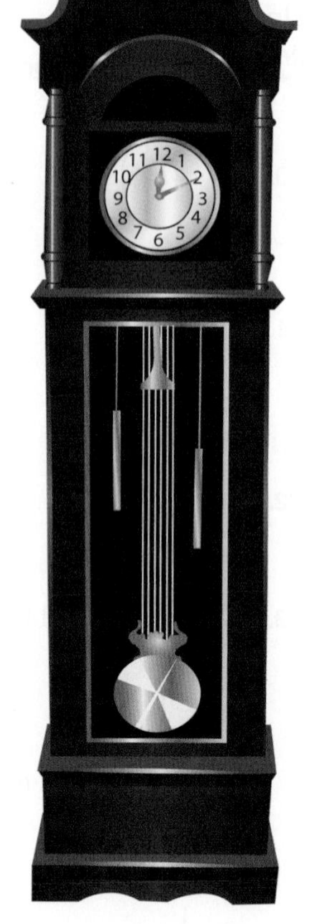

a. Calculate the time it takes a 1 m long pendulum to complete one swing.
b. Determine the time it takes the pendulum to complete 10 swings.
c. Determine how many swings will be completed after 10 seconds.

29. Using the index laws, show that $\sqrt[5]{32a^5b^{10}}=2ab^2$.

30. To rationalise a fraction means to remove all non-rational numbers from the denominator of the fraction. Rationalise $\dfrac{a^2}{3+\sqrt{b^3}}$ by multiplying the numerator and denominator by $3-\sqrt{b^3}$, and then evaluate if $b=a^2$ and $a=2$. Show all of your working.

Problem solving

31. Simplify:

a. $\dfrac{x+2x^{\frac{1}{2}}y^{\frac{1}{2}}+y-z}{\left(x^{\frac{1}{2}}+y^{\frac{1}{2}}+y^{\frac{1}{2}}\right)}$

b. $\sqrt[5]{\dfrac{t^2}{\sqrt{t^3}}}$

32. Expand $\left(m^{\frac{3}{4}}+m^{\frac{1}{2}}n^{\frac{1}{2}}+m^{\frac{1}{4}}n+n^{\frac{3}{2}}\right)\left(m^{\frac{1}{4}}-n^{\frac{1}{2}}\right)$.

33. Simplify $\dfrac{m^{\frac{2}{5}}-2m^{\frac{1}{5}}n^{\frac{1}{5}}+n^{\frac{2}{5}}-p^{\frac{2}{5}}}{m^{\frac{1}{5}}-n^{\frac{1}{5}}-p^{\frac{1}{5}}}$

34. A scientist has discovered a piece of paper with a complex formula written on it. She thinks that someone has tried to disguise a simpler formula. The formula is:

$$\frac{\sqrt[4]{a^{13}}a^2\sqrt{b^3}}{\sqrt{a^1b}}\times b^3\times\left(\frac{\sqrt{a^3b}}{ab^2}\right)^2\times\left(\frac{b^2}{a^2\sqrt{b}}\right)^3$$

a. Simplify the formula using index laws so that it can be worked with.
b. From your simplified formula, can a take a negative value? Explain.
c. Evaluate the smallest value for a for which the expression will give a rational answer. Consider only integers.

1.8 Combining index laws

LEARNING INTENTION

At the end of this subtopic you should be able to:
- simplify algebraic expressions involving brackets, fractions, multiplication and division using appropriate index laws.

1.8.1 Combining index laws

eles-4672

- When it is clear that multiple steps are required to simplify an expression, expand brackets first.
- When fractions are involved, it is usually easier to carry out all multiplications first, leaving one division as the final process.
- Make sure to simplify terms to a common base, before attempting to apply the index laws.
  For example: $5^{2x} \times 25^3 = 5^{2x} \times \left(5^2\right)^3 = 5^{2x} \times 5^6 = 5^{2x+6}$.
- Finally, write the answer with positive indices and variables in alphabetical order, as is convention.

WORKED EXAMPLE 21 Simplifying expressions in multiple steps

Simplify each of the following.

a. $\dfrac{(2a)^4 b^4}{6a^3 b^2}$

b. $\dfrac{3^{n-2} \times 9^{n+1}}{81^{n-1}}$

THINK	WRITE
a. 1. Write the expression.	a. $\dfrac{(2a)^4 b^4}{6a^3 b^2}$
2. Apply the Fourth Index Law to remove the bracket.	$= \dfrac{16a^4 b^4}{6a^3 b^2}$
3. Apply the Second Index Law for each number and pronumeral to simplify.	$= \dfrac{8a^{4-3} b^{4-2}}{3}$
4. Write the answer.	$= \dfrac{8ab^2}{3}$
b. 1. Write the expression.	b. $\dfrac{3^{n-2} \times 9^{n+1}}{81^{n-1}}$
2. Rewrite each term in the expression so that it has a base of 3.	$= \dfrac{3^{n-2} \times (3^2)^{n+1}}{(3^4)^{n-1}}$
3. Apply the Fourth Index Law to expand the brackets.	$= \dfrac{3^{n-2} \times 3^{2n+2}}{3^{4n-4}}$
4. Apply the First and Second Index Laws to simplify and write your answer.	$= \dfrac{3^{3n}}{3^{4n-4}}$ $= \dfrac{1}{3^{n-4}}$

WORKED EXAMPLE 22 Simplifying complex expressions involving multiple steps

Simplify each of the following.

a. $(2a^3b)^4 \times 4a^2b^3$

b. $\dfrac{7xy^3}{(3x^3y^2)^2}$

c. $\dfrac{2m^5n \times 3m^7n^4}{7m^3n^3 \times mn^2}$

THINK	WRITE
a. 1. Write the expression.	a. $(2a^3b)^4 \times 4a^2b^3$
2. Apply the Fourth Index Law. Multiply each index inside the brackets by the index outside the brackets.	$= 2^4a^{12}b^4 \times 4a^2b^3$
3. Evaluate the number.	$= 16a^{12}b^4 \times 4a^2b^3$
4. Multiply coefficients and multiply pronumerals. Apply the First Index Law to multiply terms with the same base by adding the indices.	$= 16 \times 4 \times a^{12+2}b^{4+3}$ $= 64a^{14}b^7$
b. 1. Write the expression.	b. $\dfrac{7xy^3}{(3x^3y^2)^2}$
2. Apply the Fourth Index Law in the denominator. Multiply each index inside the brackets by the index outside the brackets.	$= \dfrac{7xy^3}{9x^6y^4}$
3. Apply the Second Index Law. Divide terms with the same base by subtracting the indices.	$= \dfrac{7x^{-5}y^{-1}}{9}$
4. Use $a^{-m} = \dfrac{1}{a^m}$ to express the answer with positive indices.	$= \dfrac{7}{9x^5y}$
c. 1. Write the expression.	c. $\dfrac{2m^5n \times 3m^7n^4}{7m^3n^3 \times mn^2}$
2. Simplify each numerator and denominator by multiplying coefficients and then terms with the same base.	$= \dfrac{6m^{12}n^5}{7m^4n^5}$
3. Apply the Second Index Law. Divide terms with the same base by subtracting the indices.	$= \dfrac{6m^8n^0}{7}$
4. Simplify the numerator using $a^0 = 1$.	$= \dfrac{6m^8 \times 1}{7}$ $= \dfrac{6m^8}{7}$

WORKED EXAMPLE 23 Simplifying expressions with multiple fractions

Simplify each of the following.

a. $\dfrac{(5a^2b^3)^2}{a^{10}} \times \dfrac{a^2b^5}{(a^3b)^7}$

b. $\dfrac{8m^3n^{-4}}{(6mn^2)^3} \div \dfrac{4m^{-2}n^{-4}}{6m^{-5}n}$

THINK	WRITE
a. 1. Write the expression.	a. $\dfrac{(5a^2b^3)^2}{a^{10}} \times \dfrac{a^2b^5}{(a^3b)^7}$
2. Remove the brackets in the numerator of the first fraction and in the denominator of the second fraction.	$= \dfrac{25a^4b^6}{a^{10}} \times \dfrac{a^2b^5}{a^{21}b^7}$
3. Multiply the numerators and then multiply the denominators of the fractions. (Simplify across.)	$= \dfrac{25a^6b^{11}}{a^{31}b^7}$
4. Divide terms with the same base by subtracting the indices. (Simplify down.)	$= 25a^{-25}b^4$
5. Express the answer with positive indices.	$= \dfrac{25b^4}{a^{25}}$
b. 1. Write the expression.	b. $\dfrac{8m^3n^{-4}}{(6mn^2)^3} \div \dfrac{4m^{-2}n^{-4}}{6m^{-5}n}$
2. Remove the brackets.	$= \dfrac{8m^3n^{-4}}{216m^3n^6} \div \dfrac{4m^{-2}n^{-4}}{6m^{-5}n}$
3. Multiply by the reciprocal.	$= \dfrac{8m^3n^{-4}}{216m^3n^6} \times \dfrac{6m^{-5}n}{4m^{-2}n^{-4}}$
4. Multiply the numerators and then multiply the denominators. (Simplify across.)	$= \dfrac{48m^{-2}n^{-3}}{864mn^2}$
5. Cancel common factors and divide pronumerals with the same base. (Simplify down.)	$= \dfrac{m^{-3}n^{-5}}{18}$
6. Simplify and express the answer with positive indices.	$= \dfrac{1}{18m^3n^5}$

on Resources

eWorkbook Topic 1 Workbook (worksheets, code puzzle and project) (ewbk-2027)

Interactivities Individual pathway interactivity: Combining index laws (int-4565)

Combining index laws (int-6108)

Exercise 1.8 Combining index laws

learnON

Individual pathways

■ PRACTISE	■ CONSOLIDATE	■ MASTER
1, 4, 7, 10, 13, 16, 17, 23, 26	2, 5, 8, 11, 14, 18, 19, 21, 24, 27	3, 6, 9, 12, 15, 20, 22, 25, 28

To answer questions online and to receive **immediate corrective feedback** and **fully worked solutions** for all questions, go to your learnON title at www.jacplus.com.au.

Fluency

WE22 For questions 1 to 3, simplify each of the following.

1. a. $(3a^2b^2)^3 \times 2a^4b^3$ b. $(4ab^5)^2 \times 3a^3b^6$ c. $2m^3n^{-5} \times (m^2n^{-3})^{-6}$

2. a. $(2pq^3)^2 \times (5p^2q^4)^3$ b. $(2a^7b^2)^2 \times (3a^3b^3)^2$ c. $5(b^2c^{-2})^3 \times 3(bc^5)^{-4}$

3. a. $6x^{\frac{1}{2}}y^{\frac{1}{3}} \times \left(4x^{\frac{3}{4}}y^{\frac{4}{5}}\right)^{\frac{1}{2}}$ b. $(16m^3n^4)^{\frac{3}{4}} \times \left(m^{\frac{1}{2}}n^{\frac{1}{4}}\right)^3$

c. $2\left(p^{\frac{2}{3}}q^{\frac{1}{3}}\right)^{-\frac{3}{4}} \times 3\left(p^{\frac{1}{4}}q^{-\frac{3}{4}}\right)^{-\frac{1}{3}}$ d. $\left(8p^{\frac{1}{5}}q^{\frac{2}{3}}\right)^{-\frac{1}{3}} \times \left(64p^{\frac{1}{3}}q^{\frac{3}{4}}\right)^{\frac{2}{3}}$

WE21 For questions 4 to 6, simplify each of the following.

4. a. $\dfrac{5a^2b^3}{(2a^3b)^3}$ b. $\dfrac{4x^5y^6}{(2xy^3)^4}$ c. $\dfrac{(3m^2n^3)^3}{(2m^5n^5)^7}$

5. a. $\left(\dfrac{4x^3y^{10}}{2x^7y^4}\right)^6$ b. $\dfrac{3a^3b^{-5}}{(2a^7b^4)^{-3}}$ c. $\left(\dfrac{3g^2h^5}{2g^4h}\right)^3$

6. a. $\dfrac{\left(5p^6q^{\frac{1}{3}}\right)^2}{25\left(p^{\frac{1}{2}}q^{\frac{1}{4}}\right)^{\frac{2}{3}}}$ b. $\left(\dfrac{3b^2c^3}{5b^{-3}c^{-4}}\right)^{-4}$ c. $\dfrac{\left(x^{\frac{1}{2}}y^{\frac{1}{4}}z^{\frac{1}{2}}\right)^2}{\left(x^{\frac{2}{3}}y^{-\frac{1}{4}}z^{\frac{1}{3}}\right)^{-\frac{3}{2}}}$

WE22c For questions 7 to 9, simplify each of the following.

7. a. $\dfrac{2a^2b \times 3a^3b^4}{4a^3b^5}$ b. $\dfrac{4m^6n^3 \times 12mn^5}{6m^7n^6}$ c. $\dfrac{10m^6n^5 \times 2m^2n^3}{12m^4n \times 5m^2n^3}$

8. a. $\dfrac{6x^3y^2 \times 4x^6y}{9xy^5 \times 2x^3y^6}$ b. $\dfrac{(6x^3y^2)^4}{9x^5y^2 \times 4xy^7}$ c. $\dfrac{5x^2y^3 \times 2xy^5}{10x^3y^4 \times x^4y^2}$

9. a. $\dfrac{a^3b^2 \times 2(ab^5)^3}{6(a^2b^3)^3 \times a^4b}$ b. $\dfrac{(p^6q^2)^{-3} \times 3pq}{2p^{-4}q^{-2} \times (5pq^4)^{-2}}$ c. $\dfrac{6x^{\frac{3}{2}}y^{\frac{1}{2}} \times x^{\frac{4}{5}}y^{\frac{3}{5}}}{2\left(x^{\frac{1}{2}}y\right)^{\frac{1}{5}} \times 3x^{\frac{1}{2}}y^{\frac{1}{5}}}$

WE23a For questions **10** to **12**, simplify each of the following.

10. a. $\dfrac{a^3b^2}{5a^4b^7} \times \dfrac{2a^6b}{a^9b^3}$ b. $\dfrac{(2a^6)^2}{10a^7b^3} \times \dfrac{4ab^6}{6a^3}$ c. $\dfrac{(m^4n^3)^2}{(m^6n)^4} \times \dfrac{(m^3n^3)^3}{(2mn)^2}$

11. a. $\left(\dfrac{2m^3n^2}{3mn^5}\right)^3 \times \dfrac{6m^2n^4}{4m^3n^{10}}$ b. $\left(\dfrac{2xy^2}{3x^3y^5}\right)^4 \times \left(\dfrac{x^3y^9}{2y^{10}}\right)^2$ c. $\dfrac{4x^{-5}y^{-3}}{(x^2y^2)^{-2}} \times \dfrac{3x^5y^6}{2^{-2}x^{-7}y}$

12. a. $\dfrac{5p^6q^{-5}}{3q^{-4}} \times \left(\dfrac{5p^6q^4}{3p^5}\right)^{-2}$ b. $\dfrac{2a^{\frac{1}{2}}b^{\frac{1}{3}}}{6a^{\frac{1}{3}}b^{\frac{1}{2}}} \times \dfrac{\left(4a^{\frac{1}{4}}b\right)^{\frac{1}{2}}}{b^{\frac{1}{4}}a}$ c. $\dfrac{3x^{\frac{2}{3}}y^{\frac{1}{5}}}{9x^{\frac{1}{3}}y^{\frac{1}{4}}} \times \dfrac{4x^{\frac{1}{2}}}{x^{\frac{3}{4}}y}$

WE23b For questions **13** to **15**, simplify each of the following.

13. a. $\dfrac{5a^2b^3}{6a^7b^5} \div \dfrac{a^9b^4}{3ab^6}$ b. $\dfrac{7a^2b^4}{3a^6b^7} \div \left(\dfrac{3ab}{2a^6b^4}\right)^3$

14. a. $\left(\dfrac{4a^9}{b^6}\right)^3 \div \left(\dfrac{3a^7}{2b^5}\right)^4$ b. $\dfrac{5x^2y^6}{(2x^4y^5)^2} \div \dfrac{(4x^6y)^3}{10xy^3}$ c. $\left(\dfrac{x^5y^{-3}}{2xy^5}\right)^{-4} \div \dfrac{4x^6y^{-10}}{(3x^{-2}y^2)^{-3}}$

15. a. $\dfrac{3m^3n^4}{2m^{-6}n^{-5}} \div \left(\dfrac{2m^4n^6}{m^{-1}n}\right)^{-2}$ b. $4m^{\frac{1}{2}}n^{\frac{3}{4}} \div \dfrac{6m^{\frac{1}{3}}n^{\frac{1}{4}}}{8m^{\frac{3}{4}}n^{\frac{1}{2}}}$ c. $\left(\dfrac{4b^3c^{\frac{1}{3}}}{6c^{\frac{1}{5}}b}\right)^{\frac{1}{2}} \div (2b^3c^{-\frac{1}{5}})^{-\frac{3}{2}}$

Understanding

16. Evaluate each of the following.

a. $(5^2 \times 2)^0 \times (5^{-3} \times 2^0)^5 \div (5^6 \times 2^{-1})^{-3}$ b. $(2^3 \times 3^3)^{-2} \div \dfrac{(2^6 \times 3^9)^0}{2^6 \times (3^{-2})^{-3}}$

17. Evaluate the following for $x = 8$. (*Hint:* Simplify first.)

$$(2x)^{-3} \times \left(\frac{x}{2}\right)^2 \div \frac{2x}{(2^3)^4}$$

18. a. Simplify the following fraction: $\dfrac{a^{2y} \times 9b^y \times (5ab)^y}{(a^y)^3 \times 5(3b^y)^2}$

b. Determine the value of y if the fraction is equal to 125.

19. **MC** Select which of the following is not the same as $(4xy)^{\frac{3}{2}}$.

A. $8x^{\frac{3}{2}}y^{\frac{3}{2}}$ B. $(\sqrt{4xy})^3$ C. $\sqrt{64x^3y^3}$ D. $\dfrac{(2x^3y^3)^{\frac{1}{2}}}{\left(\sqrt{32}\right)^{-1}}$ E. $4xy^{\frac{1}{2}} \times (2xy^2)^{\frac{1}{2}}$

20. **MC** The expression $\dfrac{x^2y}{(2xy^2)^3} \div \dfrac{xy}{16x^0}$ is equal to:

A. $\dfrac{2}{x^2y^6}$ B. $\dfrac{2x^2}{b^6}$ C. $2x^2y^6$ D. $\dfrac{2}{xy^6}$ E. $\dfrac{1}{128xy^5}$

21. Simplify the following.

a. $\sqrt[3]{m^2n} \div \sqrt{mn^3}$

b. $(g^{-2}h)^3 \times \left(\dfrac{1}{n^{-3}}\right)^{\frac{1}{2}}$

c. $\dfrac{45^{\frac{1}{3}}}{9^{\frac{3}{4}} \times 15^{\frac{3}{2}}}$

22. Simplify the following.

a. $2^{\frac{3}{2}} \times 4^{-\frac{1}{4}} \times 16^{-\frac{3}{4}}$

b. $\left(\dfrac{a^3b^{-2}}{3^{-3}b^{-3}}\right)^{-2} \div \left(\dfrac{3^{-3}a^{-2}b}{a^4b^{-2}}\right)^2$

c. $\left(\sqrt[5]{d^2}\right)^{\frac{3}{2}} \times \left(\sqrt[3]{d^5}\right)^{\frac{1}{5}}$

Reasoning

23. The population of the number of bacteria on a petri dish is modelled by $N = 6 \times 2^{t+1}$, where N is the number of bacteria after t days.

a. Determine the initial number of bacteria.
b. Determine the number of bacteria after one week.
c. Calculate when the number of bacteria will first exceed 100 000.

24. In a controlled breeding program at the Melbourne Zoo, the population (P) of koalas at t years is modelled by $P = P_0 \times 10^{kt}$. Given $P_0 = 20$ and $k = 0.3$:

a. Evaluate the number of koalas after 2 years.
b. Determine when the population will be equal to 1000. Show full working.

25. The decay of uranium is modelled by $D = D_0 \times 2^{-kt}$. It takes 6 years for the mass of uranium to halve. Giving your answers to the nearest whole number, determine the percentage remaining after:

a. 2 years
b. 5 years
c. 10 years.

Problem solving

26. Solve the following for x: $2^{2x+2} - 2^{2x-1} - 28 = 0$.

27. Simplify $\dfrac{7^{2x+1} - 7^{2x-1} - 48}{36 \times 7^{2x} - 252}$.

28. Simplify $\dfrac{z^4 + z^{-4} - 3}{z^2 + z^{-2} - 5^{\frac{1}{2}}}$.

1.9 Application of indices: Compound interest

LEARNING INTENTION

At the end of this subtopic you should be able to:
- calculate the future value of an investment earning compound interest
- calculate the amount of interest earned after a period of time on an investment with compound interest.

1.9.1 Application of indices: compound interest

eles-4673

- One practical application of indices is compound interest.
- **Compound interest** is the type of interest that is applied to savings in a bank account, term deposits and bank loans.
- Unlike simple interest, which has a fixed amount of interest added at each payment, compound interest depends on the balance (or principal) of the account. This means that the amount of interest increases with each successive payment. It is often calculated per year (per annum or p.a.)
- The following graph shows how compound interest increases over time. Each interest amount is 20% of the previous balance. As the balance grows, the interest increases and the balance growth accelerates.

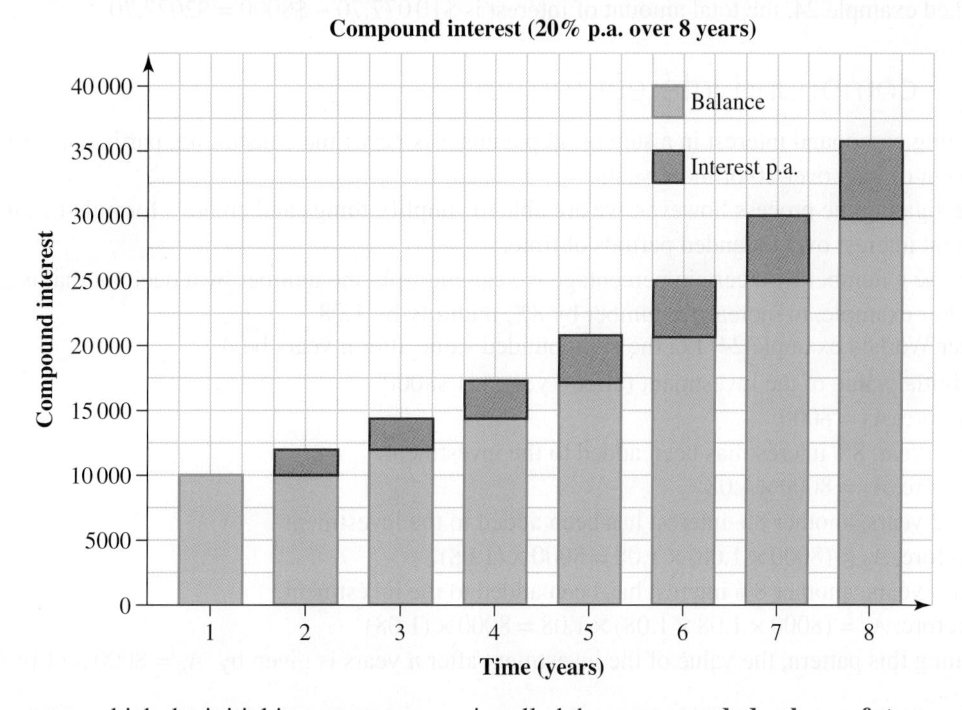

- The amount to which the initial investment grows is called the **compounded value** or **future value**.
- Compound interest can be calculated by methodically calculating the amount of interest earned at each time, and adding it to the value of the investment.

WORKED EXAMPLE 24 Calculating compound interest step by step

Kyna invests \$8000 at 8% p.a. for 3 years with interest paid at the end of each year. Determine the compounded value of the investment by calculating the simple interest on each year separately.

THINK	WRITE
1. Write the initial (first year) principal.	Initial principal = \$8000
2. Calculate the interest for the first year.	Interest for year 1 = 8% of \$8000 = \$640

3. Calculate the principal for the second year by adding the first year's interest to the initial principal.	Principal for year 2 = $8000 + $640 = $8640
4. Calculate the interest for the second year.	Interest for year 2 = 8% of $8640 = $691.20
5. Calculate the principal for the third year by adding the second year's interest to the second year's principal.	Principal for year 3 = $8640 + $691.20 = $9331.20
6. Calculate the interest for the third year.	Interest for year 3 = 8% of $9331.20 = $746.50
7. Calculate the future value of the investment by adding the third year's interest to the third year's principal.	Compounded value after 3 years = $9331.20 + $746.50 = $10 077.70

- To calculate the total amount of interest received, subtract the initial value from the future value.
- In Worked example 24, the total amount of interest is $10 077.70 − $8000 = $2077.70.

1.9.2 The compound interest formula

eles-4674

- Calculating compound interest in a step by step manner is very time consuming, particularly for an investment or loan over 20 or more years.
- By investigating the process however, we are able to simplify things and create a formula to calculate compound interest over extended periods of time.
- To increase a number by a certain percentage we can multiply the number by a decimal that is greater than 1. For example, to increase a number by 8%, multiply by 1.08.
- Consider Worked example 24. Let the compounded value after n years be A_n.
  - The initial value of the investment (after 0 years) is $8000. Therefore, $A_0 = 8000$.
  - After 1 year, 8% interest has been added to the investment. Therefore, $A_1 = 8000 \times 1.08$
  - After 2 years, another 8% interest has been added to the investment. Therefore, $A_2 = (8000 \times 1.08) \times 1.08 = 8000 \times (1.08)^2$
  - After 3 years, another 8% interest has been added to the investment. Therefore, $A_3 = (8000 \times 1.08 \times 1.08) \times 1.08 = 8000 \times (1.08)^3$
- Continuing this pattern, the value of the investment after n years is given by $A_n = 8000 \times (1.08)^n$.

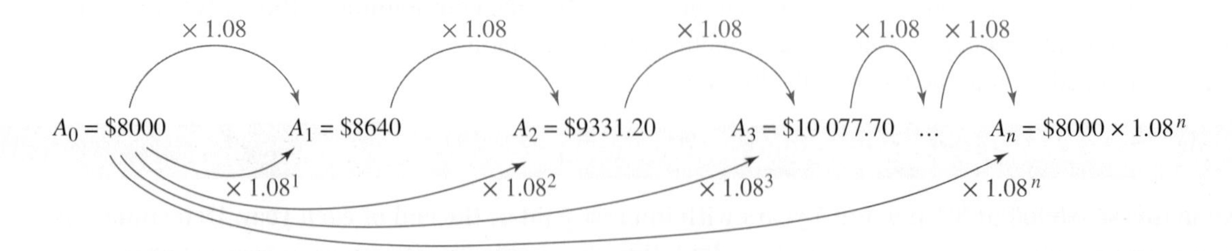

- We can simplify this so that we skip all of the values in the middle and focus on the initial value (principal) and the final (future) value.

Compound interest formula

For any investment or loan, the balance after n compounding periods is given by:

$$A = P(1+i)^n$$

where:

- A is the future value of the investment in \$
- P is the principal (initial value of the investment) in \$
- i is the interest rate per compounding period as a decimal. (*Note:* The pronumerals r or R can also be used in the formula to represent the interest rate.)
- n is the number of compounding periods.

- The interest, (I), can then be calculated using the formula:

$$I = A - P$$

WORKED EXAMPLE 25 Using the compound interest formula

William has \$14 000 to invest. He invests the money at 9% p.a. for 5 years with interest compounded annually.

a. Use the formula $A = P(1+i)^n$ to calculate the amount to which this investment will grow.

b. Calculate the compound interest earned on the investment.

THINK	WRITE
a. 1. Write the compound interest formula.	a. $A = P(1 + i)^n$
2. Write down the values of P, i and n.	$P = \$14\,000,\ i = 0.09,\ n = 5$
3. Substitute the values into the formula.	$A = \$14\,000 \times 1.09^5$
4. Calculate.	$= \$21\,540.74$ The investment will grow to \$21 540.74.
b. Calculate the compound interest earned.	b. $I = A - P$ $= \$21\,540.74 - \$14\,000$ $= \$7540.74$ The compound interest earned is \$7540.74.

TI | THINK

a–b.

On a Calculator page, store the value of p. To do this, complete the entry line as:

$p{:}= 14000$

Then press:

- MENU
- 3: Algebra
- 1: Solve

Complete the entry line as:

solve $(a = p \times (1+i)^n, a)$ $|i = 0.09$ and $n = 5$

Then press ENTER and complete as shown.

DISPLAY/WRITE

a–b.

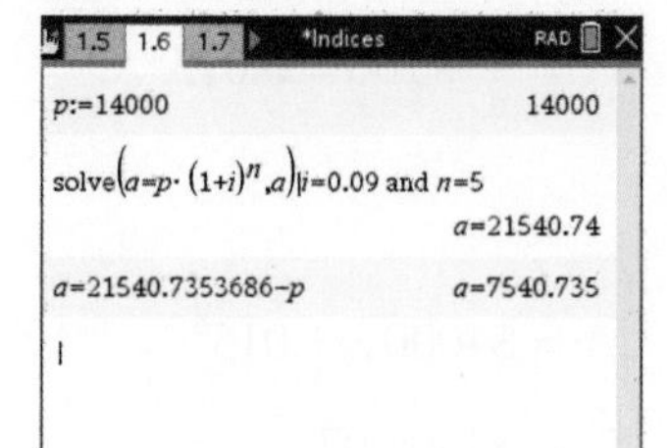

The investment will grow to \$21 540.74.

The compound interest earned is \$7540.74.

CASIO | THINK

a–b.

On the Main screen, complete the entry line as:

solve $(A = P \times (1+i)^n, A)|i = 0.09|n = 5|P = 14000$

Then press EXE and complete as shown.

DISPLAY/WRITE

a–b.

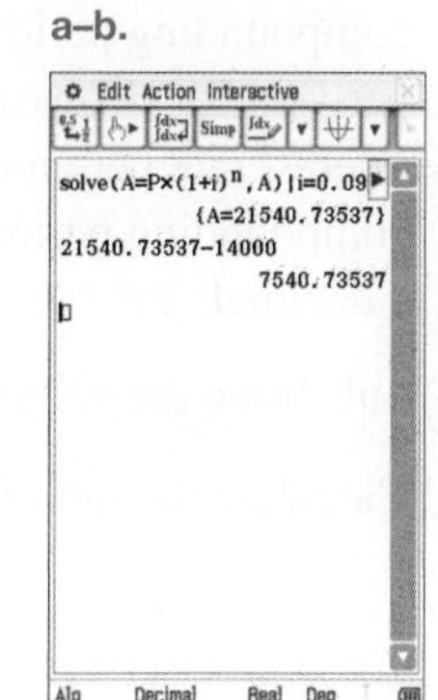

The investment will grow to \$21 540.74

The compound interest earned is \$7540.74.

1.9.3 Compounding period

eles-4675

- In Worked example 25, interest is paid annually. Interest can be paid more regularly — it may be paid six-monthly (twice a year), quarterly (4 times a year), monthly or even daily.
- The frequency of interest payment is called the **compounding period**.
- In general, the time period of a loan will be stated in years and the interest rate will be quoted as % p.a. (per annum). If the compounding period is not annual, these values must be adjusted in the compound interest formula.
- For example, an investment over 5 years at 6% p.a. compounding quarterly will have:

$$n = 5 \times 4 = 20 \text{ compounding periods}$$

$$i = \frac{6}{4}\% = 1.5\% = 0.015$$

Calculating n and i for different compounding time periods

- n is the total number of compounding time periods:

$$n = \textbf{number of years} \times \textbf{number of compounding periods per year}$$

- i is the interest rate per compounding time period:

$$i = \frac{\textbf{interest rate per annum}}{\textbf{number of compounding periods per year}}$$

WORKED EXAMPLE 26 Calculating the future value of an investment

Calculate the future value of an investment of $4000 at 6% p.a. for 2 years with interest compounded quarterly.

THINK	WRITE
1. Write the compound interest formula.	$A = P(1 + i)^n$
2. Write the values of P, n and i. The number of compounding periods, n, is 4 compounding periods per year for two years. The interest rate, i, is the interest rate per annum divided by the number of compounding periods per year, expressed as a decimal.	$P = \$4000$, $n = 2 \times 4 = 8$, $i = \frac{6}{4} \div 100 = 0.015$
3. Substitute the values into the formula.	$A = \$4000 \times 1.015^8$
4. Calculate the future value of an investment.	$= \$4505.97$ The future value of the investment is $4505.97.

TI | THINK	DISPLAY/WRITE
Use the finance functions available on the calculator for this question. On a Calculator page, press: • MENU • 8: Finance • 2: TVM Functions • 5: Future Value Complete the entry line as: tvmFV(8, 1.5, 4000, 0) Press ENTER. Note that the number of compounding periods is 8, that is 4 times a year for 2 years, and the interest is $\frac{6}{4} = 1.5\%$ quarterly.	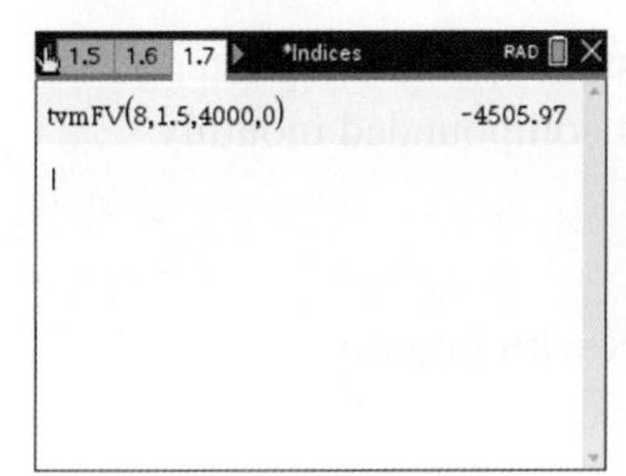 The future value is \$4505.97.

CASIO | THINK	DISPLAY/WRITE
On the Financial screen, press: • Compound Interest Enter the values as shown in the screenshot. The FV is left blank. Tap it and it will calculate the value. Note that the number of compounding periods is 8, that is 4 times a year for 2 years, and the interest is $\frac{6}{4} = 1.5\%$ quarterly.	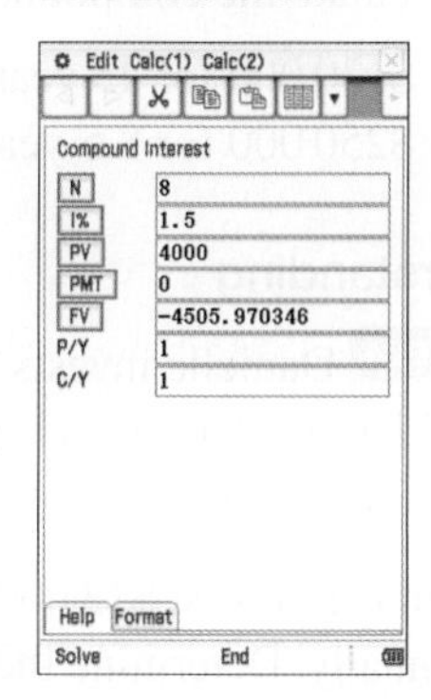 The future value is \$4505.97.

on Resources

 **eWorkbook** Topic 1 Workbook (worksheets, code puzzle and project) (ewbk-2027)

Interactivities Individual pathway interactivity: Compound interest (int-4636)
Compound interest (int-6075)
Compounding periods (int-6186)

Exercise 1.9 Application of indices: Compound interest

learnon

Individual pathways

■ PRACTISE	■ CONSOLIDATE	■ MASTER
1, 4, 7, 9, 10, 14, 19, 22	2, 5, 8, 11, 15, 17, 20, 23	3, 6, 12, 13, 16, 18, 21, 24

To answer questions online and to receive **immediate corrective feedback** and **fully worked solutions** for all questions, go to your learnON title at www.jacplus.com.au.

Fluency

For questions 1 to 3, use the formula $A = P(1 + i)^n$ to calculate the amount to which each of the following investments will grow with interest compounded annually.

1. a. \$3000 at 4% p.a. for 2 years b. \$9000 at 5% p.a. for 4 years
2. a. \$16000 at 9% p.a. for 5 years b. \$12500 at 5.5% p.a. for 3 years
3. a. \$9750 at 7.25% p.a. for 6 years b. \$100 000 at 3.75% p.a. for 7 years

For questions 4 and 5, calculate the compounded value of each of the following investments.

4. a. \$870 for 2 years at 3.50% p.a. with interest compounded six-monthly
   b. \$9500 for $2\frac{1}{2}$ years at 4.6% p.a. with interest compounded quarterly
5. a. \$148 000 $3\frac{1}{2}$ for years at 9.2% p.a. with interest compounded six-monthly
   b. \$16 000 for 6 years at 8 p.a. with interest compounded monthly

6. Calculate the compounded value of each of the following investments.
   a. \$130 000 for 25 years at 12.95% p.a. with interest compounded quarterly
   b. \$250 000 for 8.5 years at 6.75% p.a. with interest compounded monthly

Understanding

7. **WE24** Danielle invests \$6000 at 10% p.a. for 4 years with interest paid at the end of each year. Determine the compounded value of the investment by calculating the simple interest on each year separately.
8. Ben is to invest \$13 000 for 3 years at 8% p.a. with interest paid annually. Determine the amount of interest earned by calculating the simple interest for each year separately.

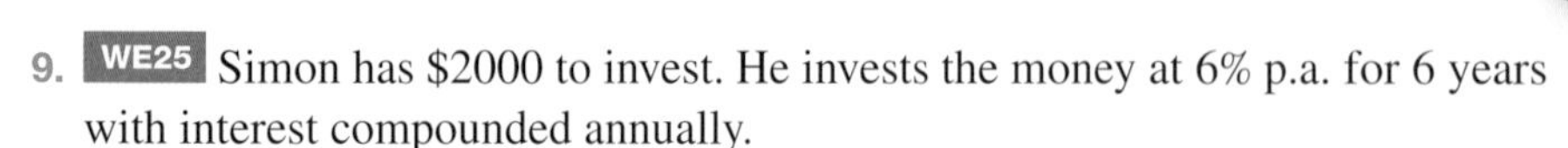

9. **WE25** Simon has \$2000 to invest. He invests the money at 6% p.a. for 6 years with interest compounded annually.
   a. Use the formula $A = P(1 + i)^n$ to calculate the amount to which this investment will grow.
   b. Calculate the compound interest earned on the investment.
10. **WE26** Calculate the future value of an investment of \$14 000 at 7% p.a. for 3 years with interest compounded quarterly.
11. A passbook savings account pays interest of 0.3% p.a. Jill has \$600 in such an account. Calculate the amount in Jill's account after 3 years, if interest is compounded quarterly.
12. Damien is to invest \$35 000 at 7.2% p.a. for 6 years with interest compounded six-monthly. Calculate the compound interest earned on the investment.
13. Sam invests \$40 000 in a one-year fixed deposit at an interest rate of 7% p.a. with interest compounding monthly.
    a. Convert the interest rate of 7% p.a. to a rate per month.
    b. Calculate the value of the investment upon maturity.
14. **MC** A sum of \$7000 is invested for 3 years at the rate of 5.75% p.a., compounded quarterly. The interest paid on this investment, to the nearest dollar, is:
    A. \$1208 B. \$1308 C. \$8208 D. \$8308 E. \$8508.
15. **MC** After selling their house and paying off their mortgage, Mr and Mrs Fong have \$73 600. They plan to invest it at 7% p.a. with interest compounded annually. The value of their investment will first exceed \$110 000 after:
    A. 5 years
    B. 6 years
    C. 8 years
    D. 10 years
    E. 15 years
16. **MC** Maureen wishes to invest \$150 000 for a period of 7 years. The following investment alternatives are suggested to her. The best investment would be:
    A. simple interest at 8% p.a.
    B. compound interest at 6.7% p.a. with interest compounded annually
    C. compound interest at 6.6% p.a. with interest compounded six-monthly
    D. compound interest at 6.5% p.a. with interest compounded quarterly
    E. compound interest at 6.4% p.a. with interest compounded monthly

17. **MC** An amount is to be invested for 5 years and compounded semi-annually at 7% p.a. Select which of the following investments will have a future value closest to $10 000.

A. $700 **B.** $6500 **C.** $7400 **D.** $9000 **E.** $9900

18. Jake invests $120 000 at 9% p.a. for a 1-year term. For such large investments interest is compounded daily.
    a. Calculate the daily percentage interest rate, correct to 4 decimal places. Use 1 year = 365 days.
    b. Hence, calculate the compounded value of Jake's investment on maturity.
    c. Calculate the amount of interest paid on this investment.
    d. Calculate the extra amount of interest earned compared with the case where the interest is calculated only at the end of the year.

Reasoning

19. Daniel has $15 500 to invest. An investment over a 2-year term will pay interest of 7% p.a.
    a. Calculate the compounded value of Daniel's investment if the compounding period is:
       i. 1 year ii. 6 months iii. 3 months iv. monthly.
    b. Explain why it is advantageous to have interest compounded on a more frequent basis.
20. Jasmine invests $6000 for 4 years at 8% p.a. simple interest. David also invests $6000 for 4 years, but his interest rate is 7.6% p.a. with interest compounded quarterly.
    a. Calculate the value of Jasmine's investment on maturity.
    b. Show that the compounded value of David's investment is greater than Jasmine's investment.
    c. Explain why David's investment is worth more than Jasmine's investment despite receiving a lower rate of interest.
21. Quan has $20 000 to invest over the next 3 years. He has the choice of investing his money at 6.25% p.a. simple interest or 6% p.a. compound interest.
    a. Calculate the amount of interest that Quan will earn if he selects the simple interest option.
    b. Calculate the amount of interest that Quan will earn if the interest is compounded:
       i. annually ii. six monthly iii. quarterly.
    c. Clearly Quan's decision will depend on the compounding period. Explain the conditions under which Quan should accept the lower interest rate on the compound interest investment.
    d. Consider an investment of $10000 at 8% p.a. simple interest over 5 years. Use a trial-and-error method to determine an equivalent rate of compound interest over the same period.
    e. State whether this equivalent rate be the same if we change:
       i. the amount of the investment ii. the period of the investment.

Problem solving

22. A building society advertises investment accounts at the following rates:
    i. 3.875% p.a. compounding daily
    ii. 3.895% p.a. compounding monthly
    iii. 3.9% p.a. compounding quarterly.

    Peter thinks the first account is the best one because the interest is calculated more frequently. Paul thinks the last account is the best one because it has the highest interest rate. Explain whether either is correct.
23. Two banks offer the following investment packages.
    Bank west: 7.5% p. a. compounded annually fixed for 7 years.
    Bank east: 5.8% p. a. compounded annually fixed for 9 years.
    a. Determine which bank's package will yield the greatest interest.
    b. If a customer invests $20 000 with Bank west, determine how much she would have to invest with Bank east to produce the same amount as Bank west at the end of the investment period.

24. **a.** Consider an investment of $1 invested at 100% interest for 1 year. Calculate the value of the investment if it is compounded:
  i. quarterly **ii.** monthly **iii.** daily **iv.** once every hour.
  b. Comment on the pattern you notice as the compounding period become more frequent. Do you notice any connection to an important mathematical constant?

1.10 Logarithms (10A)

LEARNING INTENTION

At the end of this subtopic you should be able to:
- convert between index form and logarithmic form
- evaluate logarithms and use logarithms in scale measurement.

1.10.1 Logarithms

eles-4676

- The index, power or exponent in the statement $y = a^x$ is also known as a **logarithm** (or log for short).

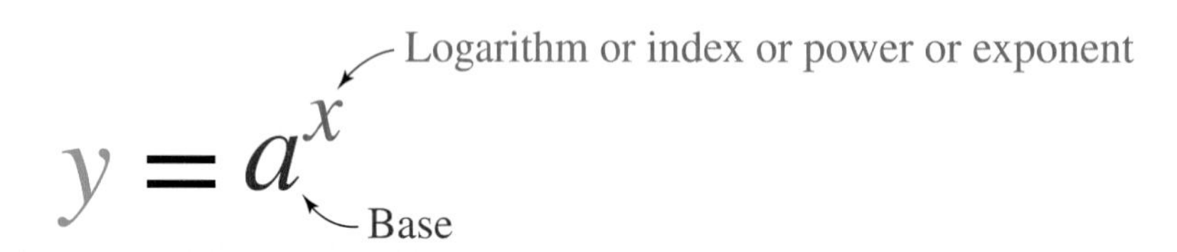

- This statement $y = a^x$ can be written in an alternative form as $\log_a y = x$, which is read as 'the logarithm of y to the base a is equal to x'. These two statements are equivalent.

Index form		Logarithmic form
$a^x = y$	$\Leftrightarrow$	$\log_a (y) = x$

- For example, $3^2 = 9$ can be written as $\log_3 9 = 2$. The log form would be read as 'the logarithm of 9, to the base of 3, is 2'. In both forms, the base is 3 and the logarithm is 2.
- It helps to remember that the output of a logarithm would be the power of the equivalent expression in index form. Logarithms take in large numbers and output small numbers (powers).

WORKED EXAMPLE 27 Converting to logarithmic form

Write the following in logarithmic form.

a. $\mathbf{10^4 = 10\,000}$ **b.** $\mathbf{6^x = 216}$

THINK	WRITE
a. 1. Write the given statement.	**a.** $10^4 = 10\,000$

2. Identify the base (10) and the logarithm (4) and write the equivalent statement in logarithmic form. (Use $a^x = y \Leftrightarrow \log_a y = x$, where the base is a and the log is x.)	$\log_{10}(10\,000) = 4$
b. 1. Write the given statement.	b. $6^x = 216$
2. Identify the base (6) and the logarithm (x) and write the equivalent statement in logarithmic form.	$\log_6(216) = x$

WORKED EXAMPLE 28 Converting to index form

Write the following in index form.

a. $\log_2 (8) = 3$

b. $\log_{25}(5) = \frac{1}{2}$

THINK	WRITE
a. 1. Write the statement.	a. $\log_2(8) = 3$
2. Identify the base (2) and the log (3), and write the equivalent statement in index form. Remember that the log is the same as the index.	$2^3 = 8$
b. 1. Write the statement.	b. $\log_{25}(5) = \frac{1}{2}$
2. Identify the base (25) and the log $\left(\frac{1}{2}\right)$, and write the equivalent statement in index form.	$25^{\frac{1}{2}} = 5$

- In the previous examples, we found that:

$$\log_2(8) = 3 \Leftrightarrow 2^3 = 8 \text{ and } \log_{10}(10\,000) = 4 \Leftrightarrow 10^4 = 10\,000.$$

We could also write $\log_2(8) = 3$ as $\log_2(2^3) = 3$ and $\log_{10}(10\,000) = 4$ as $\log_2\left(10^4\right) = 4$.

- Can this pattern be used to work out the value of $\log_3(81)$? We need to find the power when the base of 3 is raised to that power to give 81.

WORKED EXAMPLE 29 Evaluating a logarithm

Evaluate $\log_3 (81)$.

THINK	WRITE
1. Write the log expression.	$\log_3(81)$
2. Express 81 in index form with a base of 3.	$= \log_3(3^4)$
3. Write the value of the logarithm.	$= 4$

1.10.2 Using logarithmic scales in measurement

eles-4677

- Logarithms can also be used to display data sets that cover a range of values which vary greatly in size. For example, when measuring the amplitude of earthquake waves, some earthquakes will have amplitudes of around 10 000, whereas other earthquakes may have amplitudes of around 10 000 000 (1000 times greater). Rather than trying to display this data on a linear scale, we can take the logarithm of the amplitude, which gives us the magnitude of each earthquake.
- The Richter scale uses the magnitudes of earthquakes to display the difference in their power.
- The logarithm that is used in these scales is the logarithm with base 10, which means that an increase by 1 on the scale, is an increase of 10 in the actual value.
- The logarithm with base 10 is often written simply as $\log(x)$, with the base omitted.

WORKED EXAMPLE 30 Real world application of logarithms

Convert the following amplitudes of earthquakes into values on the Richter scale, correct to 1 decimal place.

a. 1989 Newcastle earthquake: amplitude 398 000
b. 2010 Canterbury earthquake: amplitude 12 600 000
c. 2010 Chile earthquake: amplitude 631 000 000

THINK	WRITE
a. Use a calculator to calculate the logarithmic value of the amplitude. Round the answer to 1 decimal place. Write the answer in words.	a. $\log(398\,000) = 5.599\ldots$ $= 5.6$ The 1989 Newcastle earthquake rated 5.6 on the Richter scale.
b. Use a calculator to calculate the logarithmic value of the amplitude. Round the answer to 1 decimal place. Write the answer in words.	b. $\log(12\,600\,000) = 7.100\ldots$ $= 7.1$ The 2010 Canterbury earthquake rated 7.1 on the Richter scale.
c. Use a calculator to calculate the logarithmic value of the amplitude. Round the answer to 1 decimal place. Write the answer in words.	c. $\log(631\,000\,000) = 8.800\ldots$ $= 8.8$ The 2010 Chile earthquake rated 8.8 on the Richter scale.

Displaying logarithmic data in histograms

- If we are given a data set in which the data vary greatly in size, we can use logarithms to transform the data into more manageable figures, and then group the data into intervals to provide an indication of the spread of the data.

WORKED EXAMPLE 31 Creating a histogram with a log scale

The following table displays the population of 10 different towns and cities in Victoria (using data from the 2011 census).

a. Convert the populations into logarithmic form, correct to 2 decimal places.

b. Group the data into a frequency table.

c. Draw a histogram to represent the data.

Town or city	Population
Benalla	9328
Bendigo	76 051
Castlemaine	9124
Echuca	12 613
Geelong	143 921
Kilmore	6 142
Melbourne	3 707 530
Stawell	5734
Wangaratta	17 377
Warrnambool	29 284

THINK

a. Use a calculator to calculate the logarithmic values of all of the populations. Round the answers to 2 decimal places.

WRITE

a.

Town or city	log(population)
Benalla	3.97
Bendigo	4.88
Castlemaine	3.96
Echuca	4.10
Geelong	5.16
Kilmore	3.79
Melbourne	6.57
Stawell	3.76
Wangaratta	4.24
Warrnambool	4.67

b. Group the logarithmic values into class intervals and create a frequency table.

b.

log(population)	Frequency
3– <4	4
4– <5	4
5– <6	1
6– <7	1

c. Construct a histogram of the data set.

c.

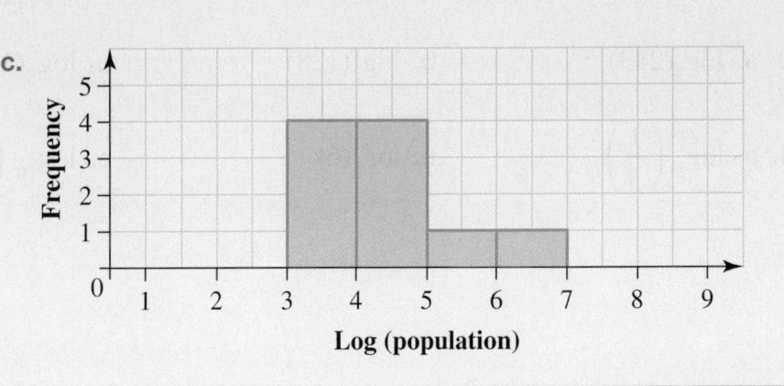

on Resources

eWorkbook Topic 1 Workbook (worksheets, code puzzle and project) (ewbk-2027)

Interactivity Logarithms (int-6194)

Exercise 1.10 Logarithms (10A)

learnon

Individual pathways

PRACTISE	CONSOLIDATE	MASTER
1, 4, 5, 9, 12, 14, 17, 20	2, 6, 8, 10, 13, 15, 18, 21	3, 7, 11, 16, 19, 22

To answer questions online and to receive **immediate corrective feedback** and **fully worked solutions** for all questions, go to your learnON title at www.jacplus.com.au.

Fluency

WE27 For questions 1 to 3, write the following in logarithmic form.

1. a. $4^2 = 16$ b. $2^5 = 32$ c. $3^4 = 81$ d. $6^2 = 36$ e. $1000 = 10^3$
2. a. $25 = 5^2$ b. $4^3 = x$ c. $5^x = 125$ d. $7^x = 49$ e. $p^4 = 16$
3. a. $9^{\frac{1}{2}} = 3$ b. $0.1 = 10^{-1}$ c. $2 = 8^{\frac{1}{3}}$ d. $2^{-1} = \frac{1}{2}$ e. $4^{\frac{3}{2}} = 8$
4. **MC** The statement $w = h^t$ is equivalent to:
   A. $w = \log_t(h)$ B. $h = \log_t(w)$ C. $t = \log_w(h)$ D. $t = \log_h(w)$ E. $h = \log_w(t)$

WE28 For questions 5 to 7, write the following in index form.

5. a. $\log_2(16) = 4$ b. $\log_3(27) = 3$ c. $\log_{10}(1\,000\,000) = 6$ d. $\log_5(125) = 3$
6. a. $\log_{16}(4) = \frac{1}{2}$ b. $\log_4(64) = x$ c. $\frac{1}{2} = \log_{49}(7)$ d. $\log_3(x) = 5$
7. a. $\log_{81}(9) = \frac{1}{2}$ b. $\log_{10}(0.01) = -2$ c. $\log_8(8) = 1$ d. $\log_{64}(4) = \frac{1}{3}$
8. **MC** The statement $q = \log_r(p)$ is equivalent to:
   A. $q = r^p$ B. $p = r^q$ C. $r = p^q$ D. $r = q^p$ E. $p = q^r$

WE29 For questions 9 to 11, evaluate the following logarithms.

9. a. $\log_2(16)$ b. $\log_4(16)$ c. $\log_{11}(121)$ d. $\log_{10}(100\,000)$
10. a. $\log_3(243)$ b. $\log_2(128)$ c. $\log_5(1)$ d. $\log_9(3)$
11. a. $\log_3\left(\frac{1}{3}\right)$ b. $\log_6(6)$ c. $\log_{10}\left(\frac{1}{100}\right)$ d. $\log_{125}(5)$

12. Write the value of each of the following.

a. $\log_{10}(1)$ b. $\log_{10}(10)$ c. $\log_{10}(100)$

13. Write the value of each of the following.

a. $\log_{10}(1000)$ b. $\log_{10}(10\,000)$ c. $\log_{10}(100\,000)$

Understanding

14. Use your results to question 12 and 13 to answer the following.

a. Between which two whole numbers would $\log_{10}(7)$ lie?
b. Between which two whole numbers would $\log_{10}(4600)$ lie?
c. Between which two whole numbers would $\log_{10}(85)$ lie?

15. a. Between which two whole numbers would $\log_{10}(12\,750)$ lie?
b. Between which two whole numbers would $\log_{10}(110)$ lie?
c. Between which two whole numbers would $\log_{10}(81\,000)$ lie?

16. WE30 Convert the following amplitudes of earthquakes into values on the Richter scale, correct to 1 decimal place.

a. 2016 Northern Territory earthquake: amplitude 1 260 000.
b. 2011 Christchurch earthquake: amplitude 2 000 000.
c. 1979 Tumaco earthquake: amplitude 158 000 000.

Reasoning

17. a. If $\log_{10}(g) = k$, determine the value of $\log_{10}\left(g^2\right)$. Justify your answer.
b. If $\log_x(y) = 2$, determine the value of $\log_y(x)$. Justify your answer.
c. By referring to the equivalent index statement, explain why x must be a positive number given $\log_4(x) = y$, for all values of y.

18. Calculate each of the following logarithms.

a. $\log_2(64)$ b. $\log_3\left(\frac{1}{81}\right)$ c. $\log_{10}(0.00001)$

19. Calculate each of the following logarithms.

a. $\log_3(243)$ b. $\log_4\left(\frac{1}{64}\right)$ c. $\log_5\left(\sqrt{125}\right)$

Problem solving

20. For each of the following, determine the value of x.

a. $\log_x\left(\frac{1}{243}\right) = -5$ b. $\log_x(343) = 3$ c. $\log_{64}(x) = -\frac{1}{2}$

21. Simplify $10^{\log_{10}(x)}$.

22. Simplify the expression $3^{2-\log_3(x)}$.

1.11 Logarithm laws (10A)

LEARNING INTENTION

At the end of this subtopic you should be able to:
- simplify expressions using logarithm laws.

1.11.1 Logarithm laws

eles-4678

- Recall the index laws:

Index Law 1: $a^m \times a^n = a^{m+n}$	Index Law 2: $\frac{a^m}{a^n} = a^{m-n}$
Index Law 3: $a^0 = 1$	Index Law 4: $(a^m)^n = a^{mn}$
Index Law 5: $(ab)^m = a^m b^m$	Index Law 6: $\left(\frac{a}{b}\right)^m = \frac{a^m}{b^m}$
Index Law 7: $a^{-m} = \frac{1}{a^m}$	Index Law 8: $a^{\frac{m}{n}} = \sqrt[n]{a^m}$

- The index laws can be used to produce a set of equivalent logarithm laws.

Logarithm Law 1

- If $x = a^m$ and $y = a^n$, then $\log_a x = m$ and $\log_a y = n$ (equivalent log form).

Now	$xy = a^m \times a^n$	
or	$xy = a^{m+n}$	(First Index Law).
So	$\log_a(xy) = m + n$	(equivalent log form)
or	$\log_a(xy) = \log_a x + \log_a y$	(substituting for m and n).

Logarithm Law 1

$$\mathbf{\log_a(x) + \log_a(y) = \log_a(xy)}$$

- This means that the sum of two logarithms with the same base is equal to the logarithm of the product of the numbers.

WORKED EXAMPLE 32 Adding logarithms

Evaluate $\log_{10}(20) + \log_{10}(5)$.

THINK	WRITE
1. Since the same base of 10 used in each log term, use $\log_a(x) + \log_a(y) = \log_a(xy)$ and simplify.	$\log_{10}(20) + \log_{10}(5) = \log_{10}(20 \times 5)$ $= \log_{10}(100)$
2. Evaluate. (Remember that $100 = 10^2$.)	$= 2$

Logarithm Law 2

- If $x = a^m$ and $y = a^n$, then $\log_a(x) = m$ and $\log_a(y) = n$ (equivalent log form).

Now $\dfrac{x}{y} = \dfrac{a^m}{a^n}$

or $\dfrac{x}{y} = a^{m-n}$ (Second Index Law).

So $\log_a\left(\dfrac{x}{y}\right) = m - n$ (equivalent log form)

or $\log_a\left(\dfrac{x}{y}\right) = \log_a(x) - \log_a(y)$ (substituting for m and n).

> **Logarithm Law 2**
>
> $$\mathbf{\log_a(x) - \log_a(y) = \log_a\left(\frac{x}{y}\right)}$$

- This means that the difference of two logarithms with the same base is equal to the logarithm of the quotient of the numbers.

WORKED EXAMPLE 33 Subtracting logarithms

Evaluate $\log_4(20) - \log_4(5)$.

THINK	WRITE
1. Since the same base of 4 is used in each log term, use $\log_a(x) - \log_a(y) = \log_a\left(\frac{x}{y}\right)$ and simplify.	$\log_4(20) - \log_4(5) = \log_4\left(\frac{20}{5}\right)$ $= \log_4(4)$
2. Evaluate. (Remember that $4 = 4^1$.)	$= 1$

WORKED EXAMPLE 34 Simplifying multiple logarithm terms

Evaluate $\log_5(35) + \log_5(15) - \log_5(21)$.

THINK	WRITE
1. Since the first two log terms are being added, use $\log_a(x) + \log_a(y) = \log_a(xy)$ and simplify.	$\log_5(35) + \log_5(15) - \log_5(21)$ $= \log_5(35 \times 15) - \log_5(21)$ $= \log_5(525) - \log_5(21)$
2. To find the difference between the two remaining log terms, use $\log_a(x) - \log_a(y) = \log_a\left(\frac{x}{y}\right)$ and simplify.	$= \log_5\left(\frac{525}{21}\right)$ $= \log_5(25)$
3. Evaluate. (Remember that $25 = 5^2$.)	$= 2$

TI | THINK	DISPLAY/WRITE	CASIO | THINK	DISPLAY/WRITE
On a Calculator page, press CTRL log (above 10^x) and complete the entry line as: $\log_5(35) + \log_5(15) - \log_5(21)$ Then press ENTER.	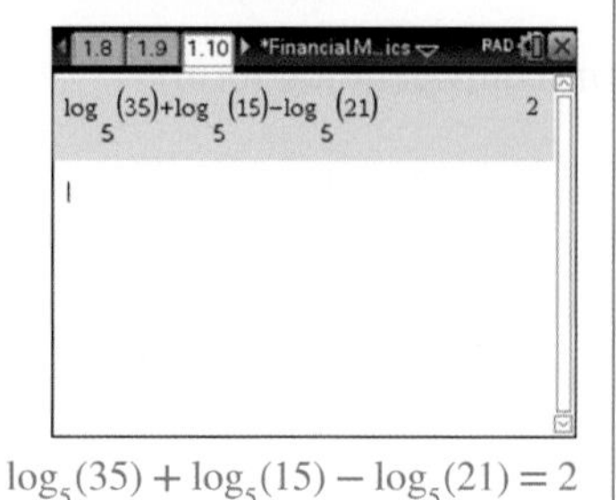 $\log_5(35) + \log_5(15) - \log_5(21) = 2$	On the Math1 keyboard screen, tap: • $\log_{\blacksquare}\square$ Complete the entry line as: $\log_5(35) + \log_5(15) - \log_5(21)$ Then press EXE.	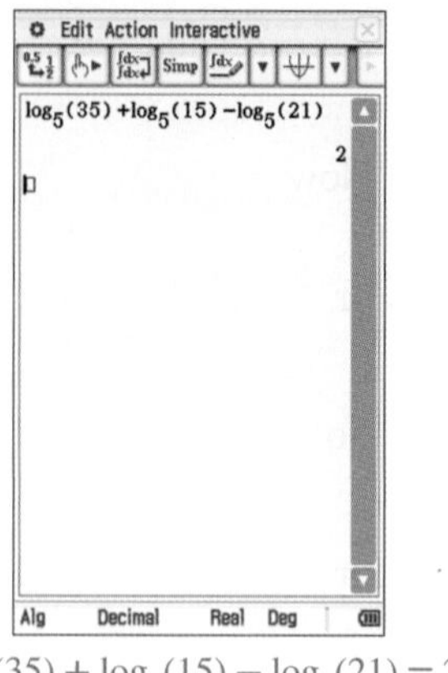 $\log_5(35) + \log_5(15) - \log_5(21) = 2$

- Once you have gained confidence in using the first two laws, you can reduce the number of steps of working by combining the application of the laws. In Worked example 34, we could write:

$$\begin{aligned}\log_5(35) + \log_5(15) - \log_5(21) &= \log_5\left(\frac{35 \times 15}{21}\right) \\ &= \log_5(25) \\ &= 2\end{aligned}$$

Logarithm Law 3

- If $x = a^m$, then $\log_a(x) = m$ (equivalent log form).

Now	$x^n = (a^m)^n$	
or	$x^n = a^{mn}$	(Fourth Index Law)
So	$\log_a(x^n) = mn$	(equivalent log form)
or	$\log_a(x^n) = \left(\log_a(x)\right) \times n$	(substituting for m)
or	$\log_a(x^n) = n\log_a(x)$	

> **Logarithm Law 3**
>
> $$\boldsymbol{\log_a(x^n) = n\log_a(x)}$$

- This means that the logarithm of a number raised to a power is equal to the product of the power and the logarithm of the number.

WORKED EXAMPLE 35 Simplifying a logarithm of a number raised to a power

Evaluate $2\log_6(3) + \log_6(4)$.

THINK	WRITE
1. The first log term is not in the required form to use the log law relating to sums. Use $\log_a(x^n) = n\log_a(x)$ to rewrite the first term in preparation for applying the first log law.	$2\log_6(3) + \log_6(4) = \log_6\left(3^2\right) + \log_6(4)$ $= \log_6(9) + \log_6(4)$
2. Use $\log_a(x) + \log_a(y) = \log_a(xy)$ to simplify the two log terms to one.	$= \log_6(9 \times 4)$ $= \log_6(36)$
3. Evaluate. (Remember that $36 = 6^2$.)	$= 2$

Logarithm Law 4

- As $a^0 = 1$ (Third Index Law)

$\log_a(1) = 0$ (equivalent log form)

> **Logarithm Law 4**
>
> $$\mathbf{\log_a(1) = 0}$$

- This means that the logarithm of 1 with any base is equal to 0.

Logarithm Law 5

- As $a^1 = a$

$\log_a(a) = 1$ (equivalent log form)

> **Logarithm Law 5**
>
> $$\mathbf{\log_a(a) = 1}$$

- This means that the logarithm of any number a with base a is equal to 1.

Logarithm Law 6

- Now $\log_a\left(\frac{1}{x}\right) = \log_a\left(x^{-1}\right)$ (Seventh Index law)

or $\log_a\left(\frac{1}{x}\right) = -1 \times \log_a(x)$ (using the fourth log law)

or $\log_a\left(\frac{1}{x}\right) = -\log_a(x)$.

> **Logarithm Law 6**
>
> $$\mathbf{\log_a\left(\frac{1}{x}\right) = -\log_a(x)}$$

Logarithm Law 7

- Now $\log_a(a^x) = x\log_a(a)$ (using the third log law)

or $\log_a(a^x) = x \times 1$ (using the fifth log law)

or $\log_a(a^x) = x$.

> **Logarithm Law 7**
>
> $$\mathbf{\log_a(a^x) = x}$$

on Resources

eWorkbook Topic 1 Workbook (worksheets, code puzzle and project) (ewbk-2027)

Interactivities The first law of logarithms (int-6195)
The second law of logarithms (int-6196)
The third law of logarithms (int-6197)
The fourth law of logarithms (int-6198)
The fifth law of logarithms (int-6199)
The sixth law of logarithms (int-6200)
The seventh law of logarithms (int-6201)

Exercise 1.11 Logarithm laws (10A)

learn**on**

Individual pathways

■ PRACTISE	■ CONSOLIDATE	■ MASTER
1, 2, 3, 6, 9, 12, 13, 15, 18, 22, 26, 29	4, 7, 10, 14, 16, 19, 21, 23, 27, 30	5, 8, 11, 17, 20, 24, 25, 28, 31

To answer questions online and to receive **immediate corrective feedback** and **fully worked solutions** for all questions, go to your learnON title at www.jacplus.com.au.

Fluency

1. Use a calculator to evaluate the following, correct to 5 decimal places.
   a. $\log_{10}(50)$ b. $\log_{10}(25)$ c. $\log_{10}(5)$ d. $\log_{10}(2)$
2. Use your answers to question 1 to show that each of the following statements is true.
   a. $\log_{10}(25)+\log_{10}(2)=\log_{10}(50)$
   b. $\log_{10}(50)-\log_{10}(2)=\log_{10}(25)$
   c. $\log_{10}(25)=2\log_{10}(5)$
   d. $\log_{10}(50)-\log_{10}(25)-\log_{10}(2)=\log_{10}(1)$

WE32 For questions 3 to 5, evaluate the following.

3. a. $\log_6(3)+\log_6(2)$ b. $\log_4(8)+\log_4(8)$
4. a. $\log_{10}(25)+\log_{10}(4)$ b. $\log_8(32)+\log_8(16)$
5. a. $\log_6(108)+\log_6(12)$ b. $\log_{14}(2)+\log_{14}(7)$

WE33 For questions 6 to 8, evaluate the following.

6. a. $\log_2(20)-\log_2(5)$ b. $\log_3(54)-\log_3(2)$
7. a. $\log_4(24)-\log_4(6)$ b. $\log_{10}(30\,000)-\log_{10}(3)$
8. a. $\log_6(648)-\log_6(3)$ b. $\log_2(224)-\log_2(7)$

WE34 For questions 9 to 11, evaluate the following.

9. a. $\log_3(27)+\log_3(2)-\log_3(6)$ b. $\log_4(24)-\log_4(2)-\log_4(6)$
10. a. $\log_6(78)-\log_6(13)+\log_6(1)$ b. $\log_2(120)-\log_2(3)-\log_2(5)$

11. a. $\log_7(15) + \log_7(3) - \log_7(315)$ b. $\log_9(80) - \log_9(8) - \log_9(30)$

12. Evaluate $2\log_4(8)$.

WE35 For questions 13 to 17, evaluate the following.

13. a. $2\log_{10}(5) + \log_{10}(4)$ b. $\log_3(648) - 3\log_3(2)$

14. a. $4\log_5(10) - \log_5(80)$ b. $\log_2(50) + \frac{1}{2}\log_2(16) - 2\log_2(5)$

15. a. $\log_8(8)$ b. $\log_5(1)$ c. $\log_2\left(\frac{1}{2}\right)$ d. $\log_4\left(4^5\right)$

16. a. $\log_6\left(6^{-2}\right)$ b. $\log_{20}(20)$ c. $\log_2(1)$ d. $\log_3\left(\frac{1}{9}\right)$

17. a. $\log_4\left(\frac{1}{2}\right)$ b. $\log_5\left(\sqrt{5}\right)$ c. $\log_3\left(\frac{1}{\sqrt{3}}\right)$ d. $\log_2\left(8\sqrt{2}\right)$

Understanding

For questions 18 to 20, use the logarithm laws to simplify each of the following.

18. a. $\log_a(5) + \log_a(8)$ b. $\log_a(12) + \log_a(3) - \log_a(2)$
c. $4\log_x(2) + \log_x(3)$ d. $\log_x(100) - 2\log_x(5)$

19. a. $3\log_a(x) - \log_a\left(x^2\right)$ b. $5\log_a(a) - \log_a\left(a^4\right)$
c. $\log_x(6) - \log_x(6x)$ d. $\log_a\left(a^7\right) + \log_a(1)$

20. a. $\log_p\left(\sqrt{p}\right)$ b. $\log_k\left(k\sqrt{k}\right)$ c. $6\log_a\left(\frac{1}{a}\right)$ d. $\log_a\left(\frac{1}{\sqrt[3]{a}}\right)$

21. **MC** *Note:* There may be more than one correct answer.

a. The equation $y = 10^x$ is equivalent to:
A. $x = 10^y$ B. $x = \log_{10}(y)$ C. $x = \log_x(10)$
D. $x = \log_y(10)$ E. $x = \log_{10}(10)$

b. The equation $y = 10^{4x}$ is equivalent to:
A. $x = \log_{10}\left(\sqrt{4y}\right)$ B. $x = \log_{10}\left(\sqrt[4]{y}\right)$
C. $x = 10^{\frac{1}{4}y}$ D. $x = \frac{1}{4}\log_{10}(y)$
E. $x = 4\log_{10}(y)$

c. The equation $y = 10^{3x}$ is equivalent to:
A. $x = \frac{1}{3}\log_{10}(y)$ B. $x = \log_{10}\left(y^{\frac{1}{3}}\right)$ C. $x = \log_{10}(y) - 3$
D. $x = 10^{y-3}$ E. $x = 3\log_{10}(y)$

d. The equation $y = ma^{nx}$ is equivalent to:
A. $x = \frac{1}{n}a^{my}$ B. $x = \log_a\left(\frac{m}{y}\right)^n$ C. $x = \frac{1}{n}\left(\log_a(y) - \log_a(m)\right)$
D. $x = \frac{1}{n}\log_a\left(\frac{y}{m}\right)$ E. $x = n\log_a\left(\frac{y}{m}\right)$

For questions 22 to 24, simplify, and evaluate where possible, each of the following without a calculator.

22. a. $\log_2(8)+\log_2(10)$ b. $\log_3(7)+\log_3(15)$ c. $\log_{10}(20)+\log_{10}(5)$ d. $\log_6(8)+\log_6(7)$

23. a. $\log_2(20)-\log_2(5)$ b. $\log_3(36)-\log_3(12)$ c. $\log_5(100)-\log_5(8)$ d. $\log_2\left(\frac{1}{3}\right)+\log_2(9)$

24. a. $\log_4(25)+\log_4\left(\frac{1}{5}\right)$ b. $\log_{10}(5)-\log_{10}(20)$ c. $\log_3\left(\frac{4}{5}\right)-\log_3\left(\frac{1}{5}\right)$
d. $\log_2(9)+\log_2(4)-\log_2(12)$ e. $\log_3(8)-\log_3(2)+\log_3(5)$ f. $\log_4(24)-\log_4(2)-\log_4(6)$

25. MC a. The expression $\log_{10}(xy)$ is equal to:
A. $\log_{10}(x)\times\log_{10}(y)$ B. $\log_{10}(x)-\log_{10}(y)$ C. $\log_{10}(x)+\log_{10}(y)$
D. $y\log_{10}(x)$ E. $x\log_{10}(y)$
b. The expression $\log_{10}(x^y)$ is equal to:
A. $x\log_{10}(y)$ B. $y\log_{10}(x)$ C. $10\log_x(y)$
D. $\log_{10}(x)+\log_{10}(y)$ E. $10\log_y(x)$
c. The expression $\frac{1}{3}\log_2(64)+\log_2(10)$ is equal to:
A. $\log_2(40)$ B. $\log_2(80)$ C. $\log_2\left(\frac{64}{10}\right)$
D. 1 E. 2

Reasoning

26. For each of the following, write the possible strategy you intend to use.
a. Evaluate $\left(\log_3(81)\right)\left(\log_3(27)\right)$.
b. Evaluate $\frac{\log_a(81)}{\log_a(3)}$.
c. Evaluate $5^{\log_5(7)}$.
In each case, explain how you obtained your final answer.

27. Simplify $\log_5(10)+2\log_5(2)-3\log_5(10)$.

28. Simplify $\log_2\left(\frac{8}{125}\right)-3\log_2\left(\frac{3}{5}\right)-4\log_2\left(\frac{1}{2}\right)$.

Problem solving

29. Simplify $\log_a\left(a^5+a^3\right)-\log_a\left(a^4+a^2\right)$.

30. If $2\log_a(x)=1+\log_a(8x-15a)$, determine the value of x in terms of a where a is a positive constant and x is positive.

31. Solve the following for x:
$$\log_3(x+2)+\log_3(x-4)=3$$

1.12 Solving equations (10A)

LEARNING INTENTION

At the end of this subtopic you should be able to:
- simplify and solve equations involving logarithms using the logarithm laws and index laws.

1.12.1 Solving equations with logarithms

eles-4679

- The equation $\log_a(y) = x$ is an example of a general **logarithmic equation**. Laws of logarithms and indices are used to solve these equations.

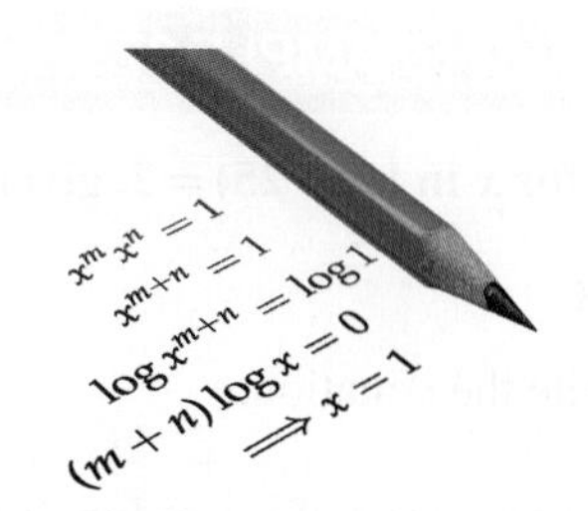

WORKED EXAMPLE 36 Solving by converting to index form

Solve for x in the following equations.

a. $\log_2(x) = 3$ b. $\log_6(x) = -2$ c. $\log_3(x^4) = -16$ d. $\log_5(x-1) = 2$

THINK	WRITE
a. 1. Write the equation.	a. $\log_2(x) = 3$
2. Rewrite using $a^x = y \Leftrightarrow \log_a(y) = x$.	$2^3 = x$
3. Rearrange and simplify.	$x = 8$
b. 1. Write the equation.	b. $\log_6(x) = -2$
2. Rewrite using $a^x = y \Leftrightarrow \log_a(y) = x$.	$6^{-2} = x$
3. Rearrange and simplify.	$x = \frac{1}{6^2}$ $= \frac{1}{36}$
c. 1. Write the equation.	c. $\log_3\left(x^4\right) = -16$
2. Rewrite using $\log_a(x^n) = n\log_a(x)$.	$4\log_3(x) = -16$
3. Divide both sides by 4.	$\log_3(x) = -4$
4. Rewrite using $a^x = y \Leftrightarrow \log_a(y) = x$.	$3^{-4} = x$
5. Rearrange and simplify.	$x = \frac{1}{3^4}$ $= \frac{1}{81}$

d. 1. Write the equation. — d. $\log_5(x-1)=2$

2. Rewrite using $a^x = y \Leftrightarrow \log_a(y) = x$. — $5^2 = x-1$

3. Solve for x. — $x-1=25$

$x=26$

WORKED EXAMPLE 37 Solving for the base of a logarithm

Solve for x in $\log_x(25) = 2$, given that $x > 0$.

THINK	WRITE
1. Write the equation.	$\log_x(25) = 2$
2. Rewrite using $a^x = y \Leftrightarrow \log_a(y) = x$.	$x^2 = 25$
3. Solve for x. *Note:* $x = -5$ is rejected as a solution because $x > 0$.	$x = 5$ (because $x > 0$)

WORKED EXAMPLE 38 Evaluating logarithms

Solve for x in the following.

a. $\log_2(16) = x$ b. $\log_3\left(\frac{1}{3}\right) = x$ c. $\log_9(3) = x$

THINK	WRITE
a. 1. Write the equation.	a. $\log_2(16) = x$
2. Rewrite using $a^x = y \Leftrightarrow \log_a(y) = x$.	$2^x = 16$
3. Write 16 with base 2.	$= 2^4$
4. Equate the indices.	$x = 4$
b. 1. Write the equation.	b. $\log_3\left(\frac{1}{3}\right) = x$
2. Rewrite using $a^x = y \Leftrightarrow \log_a(y) = x$.	$3^x = \frac{1}{3}$ $= \frac{1}{3^1}$
3. Write $\frac{1}{3}$ with base 3.	$3^x = 3^{-1}$
4. Equate the indices.	$x = -1$
c. 1. Write the equation.	c. $\log_9(3) = x$
2. Rewrite using $a^x = y \Leftrightarrow \log_a(y) = x$.	$9^x = 3$
3. Write 9 with base 3.	$(3^2)^x = 3$

4. Remove the grouping symbols.	$3^{2x} = 3^1$
5. Equate the indices.	$2x = 1$
6. Solve for x.	$x = \frac{1}{2}$

WORKED EXAMPLE 39 Solving equations with multiple logarithm terms

Solve for x in the equation $\log_2(4) + \log_2(x) - \log_2(8) = 3$.

THINK	WRITE
1. Write the equation.	$\log_2(4) + \log_2(x) - \log_2(8) = 3$
2. Simplify the left-hand side. Use $\log_a(x) + \log_a(y) = \log_a(xy)$ and $\log_a(x) - \log_a(y) = \log_a\left(\frac{x}{y}\right)$.	$\log_2\left(\frac{4 \times x}{8}\right) = 3$
3. Simplify.	$\log_2\left(\frac{x}{2}\right) = 3$
4. Rewrite using $a^x = y \Leftrightarrow \log_a(y) = x$.	$2^3 = \frac{x}{2}$
5. Solve for x.	$x = 2 \times 2^3$ $= 2 \times 8$ $= 16$

on Resources

eWorkbook Topic 1 Workbook (worksheets, code puzzle and project) (ewbk-2027)

Interactivity Solving logarithmic equations (int-6202)

Exercise 1.12 Solving equations (10A)

learnon

Individual pathways

■ PRACTISE	■ CONSOLIDATE	■ MASTER
1, 4, 7, 10, 14, 17, 20	2, 5, 8, 11, 15, 18, 21	3, 6, 9, 12, 13, 16, 19, 22

To answer questions online and to receive **immediate corrective feedback** and **fully worked solutions** for all questions, go to your learnON title at www.jacplus.com.au.

Fluency

WE38 For questions 1 to 3, solve for x in the following equations.

1. a. $\log_5(x) = 2$ b. $\log_3(x) = 4$ c. $\log_2(x) = -3$ d. $\log_4(x) = -2$ e. $\log_{10}\left(x^2\right) = 4$

2. a. $\log_2(x^3) = 12$ b. $\log_3(x+1) = 3$ c. $\log_5(x-2) = 3$ d. $\log_4(2x-3) = 0$ e. $\log_{10}(2x+1) = 0$

3. **a.** $\log_2(-x) = -5$ **b.** $\log_3(-x) = -2$ **c.** $\log_5(1-x) = 4$ **d.** $\log_{10}(5-2x) = 1$

WE39 For questions **4** to **6**, solve for x in the following equations, given that $x > 0$.

4. a. $\log_x(9) = 2$ b. $\log_x(16) = 4$ c. $\log_x(25) = \frac{2}{3}$

5. a. $\log_x(125) = \frac{3}{4}$ b. $\log_x\left(\frac{1}{8}\right) = -3$ c. $\log_x\left(\frac{1}{64}\right) = -2$

6. **a.** $\log_x(6^2) = 2$ **b.** $\log_x(4^3) = 3$

WE40 For questions **7** to **9**, solve for x in the following equations.

7. a. $\log_2(8) = x$ b. $\log_3(9) = x$ c. $\log_5\left(\frac{1}{5}\right) = x$

8. a. $\log_4\left(\frac{1}{16}\right) = x$ b. $\log_4(2) = x$ c. $\log_8(2) = x$

9. **a.** $\log_6(1) = x$ **b.** $\log_8(1) = x$ **c.** $\log_{\frac{1}{2}}(2) = x$ **d.** $\log_{\frac{1}{3}}(9) = x$

WE41 For questions **10** to **12**, solve for x in the following.

10. a. $\log_2(x) + \log_2(4) = \log_2(20)$ b. $\log_5(3) + \log_5(x) = \log_5(18)$
c. $\log_3(x) - \log_3(2) = \log_3(5)$ d. $\log_{10}(x) - \log_{10}(4) = \log_{10}(2)$

11. a. $\log_4(8) - \log_4(x) = \log_4(2)$ b. $\log_3(10) - \log_3(x) = \log_3(5)$
c. $\log_6(4) + \log_6(x) = 2$ d. $\log_2(x) + \log_2(5) = 1$

12. **a.** $3 - \log_{10}(x) = \log_{10}(2)$ **b.** $5 - \log_4(8) = \log_4(x)$
c. $\log_2(x) + \log_2(6) - \log_2(3) = \log_2(10)$ **d.** $\log_2(x) + \log_2(5) - \log_2(10) = \log_2(3)$
e. $\log_3(5) - \log_3(x) + \log_3(2) = \log_3(10)$ **f.** $\log_5(4) - \log_5(x) + \log_5(3) = \log_5(6)$

13. **MC** **a.** The solution to the equation $\log_7(343) = x$ is:

A. $x = 2$ B. $x = 3$ C. $x = 1$
D. $x = 0$ E. $x = 4$

b. If $\log_8(x) = 4$, then x is equal to:

A. 4096 B. 512 C. 64
D. 2 E. 16

c. Given that $\log_x(3) = \frac{1}{2}$, x must be equal to:

A. 3 B. 6 C. 81
D. 9 E. 18

d. If $\log_a(x) = 0.7$, then $\log_a(x^2)$ is equal to:

A. 0.49 B. 1.4 C. 0.35
D. 0.837 E. 0.28

Understanding

For questions **14** to **16**, solve for x in the following equations.

14. a. $2^x = 128$ **b.** $3^x = 9$ **c.** $7^x = \frac{1}{49}$ **d.** $9^x = 1$ **e.** $5^x = 625$

15. a. $64^x = 8$ **b.** $6^x = \sqrt{6}$ **c.** $2^x = 2\sqrt{2}$ **d.** $3^x = \frac{1}{\sqrt{3}}$ **e.** $4^x = 8$

16. a. $9^x = 3\sqrt{3}$ **b.** $2^x = \frac{1}{4\sqrt{2}}$ **c.** $3^{x+1} = 27\sqrt{3}$ **d.** $2^{x-1} = \frac{1}{32\sqrt{2}}$ **e.** $4^{x+1} = \frac{1}{8\sqrt{2}}$

Reasoning

17. The apparent brightness of stars is measured on a logarithmic scale called magnitude, in which lower numbers mean brighter stars. The relationship between the ratio of apparent brightness of two objects and the difference in their magnitudes is given by the formula:

$$m_2 - m_1 = -2.5 \log_{10}\left(\frac{b_2}{b_1}\right)$$

where m is the magnitude and b is the apparent brightness. Determine how many times brighter a magnitude 2.0 star is than a magnitude 3.0 star.

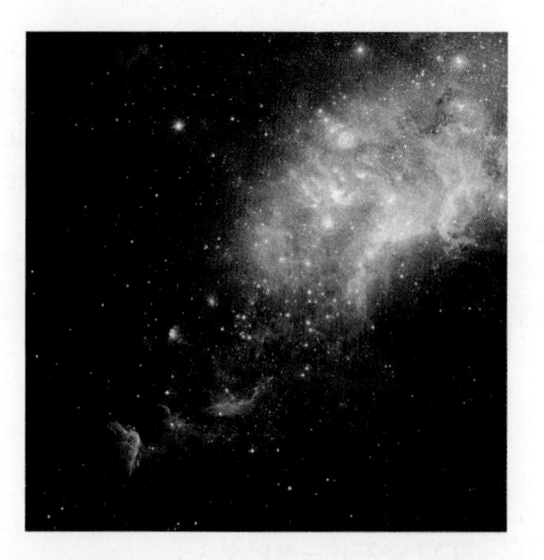

18. The decibel (dB) scale for measuring loudness, d, is given by the formula $d = 10\log_{10}(I \times 10^{12})$, where I is the intensity of sound in watts per square metre.

a. Determine the number of decibels of sound if the intensity is 1.
b. Evaluate the number of decibels of sound produced by a jet engine at a distance of 50 metres if the intensity is 10 watts per square metre.
c. Determine the intensity of sound if the sound level of a pneumatic drill 10 metres away is 90 decibels.
d. Determine how the value of d changes if the intensity is doubled. Give your answer to the nearest decibel.
e. Evaluate how the value of d changes if the intensity is 10 times as great.
f. Determine by what factor does the intensity of sound have to be multiplied in order to add 20 decibels to the sound level.

19. The Richter scale is used to describe the energy of earthquakes. A formula for the Richter scale is: $R = \frac{2}{3}\log_{10}(K) - 0.9$, where R is the Richter scale value for an earthquake that releases K kilojoules (kJ) of energy.

a. Determine the Richter scale value for an earthquake that releases the following amounts of energy:

i. 1000 kJ ii. 2000 kJ iii. 3000 kJ iv. 10 000 kJ v. 100 000 kJ vi. 1 000 000 kJ

b. Does doubling the energy released double the Richter scale value? Justify your answer.

c. Determine the energy released by an earthquake of:

i. magnitude 4 on the Richter scale
ii. magnitude 5 on the Richter scale
iii. magnitude 6 on the Richter scale.

d. Explain the effect (on the amount of energy released) of increasing the Richter scale value by 1.

e. Explain why an earthquake measuring 8 on the Richter scale so much more devastating than one that measures 5.

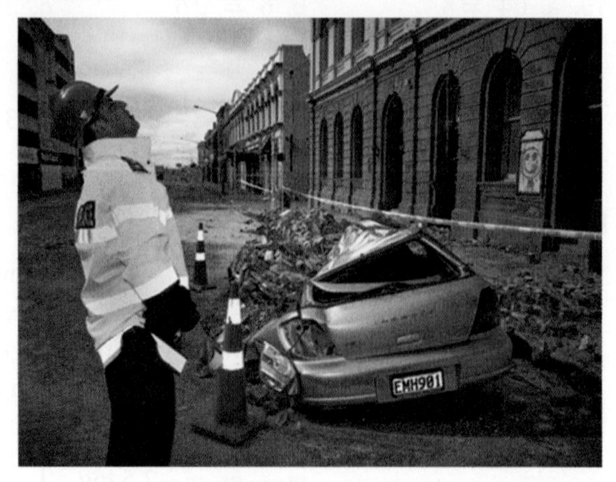

Problem solving

20. Solve for x.

a. $3^{x+1} = 7$

b. $3^{x+1} = 7^x$

21. Solve the following for x.
$(27 \times 3^x)^3 = 81^x \times 3^2$

22. Solve $\left\{x : (3^x)^2 = 30 \times 3^x - 81\right\}$.

1.13 Review

1.13.1 Topic summary

Number sets

- Natural numbers: $N = \{1, 2, 3, 4, 5 \ldots\}$
- Integers: $Z = \{\ldots, -2, -1, 0, 1, 2 \ldots\}$
- Rational numbers: $Q = \left\{\frac{1}{4}, -0.36, 2, \frac{9}{7}, 0.\dot{3}\dot{6}\right\}$
- Irrational numbers: $I = \{\sqrt{2}, \pi, e\}$
- Real numbers: $R = Q + I$

Index notation

- Index notation is a short way of writing a repeated multiplication.
  e.g. $2 \times 2 \times 2 \times 2 \times 2 \times 2$ can be written as 2^6, which is read '2 to the power of 6'.
- The **base** is the number that is being repeatedly multiplied and the **index** is the number of times it is multiplied.
  e.g. $2^6 = 2 \times 2 \times 2 \times 2 \times 2 \times 2 = 64$

Surds

- A surd is any number that requires a $\sqrt{x}$ or $\sqrt[n]{x}$ symbol and does not simplify to a whole number.
  e.g. $\sqrt{2}$ is a surd, but $\sqrt{9} = 3$ is not.
- To simplify a surd look for the highest square factor.
  e.g. $\sqrt{48} = \sqrt{16 \times 3} = 4\sqrt{3}$
- Only like surds can be added and subtracted.
  e.g. $\sqrt{5}$, $3\sqrt{5}$ and $-6\sqrt{5}$ are like surds whereas $\sqrt{7}$ and $2\sqrt{11}$ are not like surds.
- Surds are added and subtracted the same way like terms are combined in algebra. You may need to simplify first.
  e.g. $3\sqrt{2} + 7\sqrt{2} - 2\sqrt{2} = 10\sqrt{2} - 2\sqrt{2} = 8\sqrt{2}$
  e.g. $\sqrt{12} + \sqrt{75} = 2\sqrt{3} + 5\sqrt{3} = 7\sqrt{3}$

INDICES, SURDS AND LOGARITHMS

Index laws

- 1st law: $a^m \times a^n = a^{m+n}$
- 2nd law: $a^m \div a^n = a^{m-n}$
- 3rd law: $a^0 = 1, a \neq 0$
- 4th law: $(a^m)^n = a^{m \times n} = a^{mn}$
- 5th law: $(ab)^n = a^n b^n$
- 6th law: $\left(\frac{a}{b}\right)^n = \frac{a^n}{b^n}$
- 7th law: $a^{-n} = \frac{1}{a^n}$
- 8th law: $a^{\frac{1}{n}} = \sqrt[n]{a}$

Multiplying and dividing surds

- $\sqrt{a} \times \sqrt{b} = \sqrt{ab}$
- $m\sqrt{a} \times n\sqrt{b} = mn\sqrt{ab}$
- $\sqrt{a} \div \sqrt{b} = \frac{\sqrt{a}}{\sqrt{b}} = \sqrt{\frac{a}{b}}$
- $m\sqrt{a} \div (n\sqrt{b}) = \frac{m\sqrt{a}}{n\sqrt{b}} = \frac{m}{n}\sqrt{\frac{a}{b}}$

Rationalising the denominator

- Involves re-writing a fraction with a rational denominator.
  e.g. $\frac{2}{\sqrt{5}} = \frac{2}{\sqrt{5}} \times \frac{\sqrt{5}}{\sqrt{5}} = \frac{2\sqrt{5}}{5}$
- It may be necessary to multiply by the conjugate in order to rationalise.
  e.g. $\frac{1}{\sqrt{6}+2} \times \frac{\sqrt{6}-2}{\sqrt{6}-2} = \frac{\sqrt{6}-2}{2}$

Compound interest

- To calculate the value of an investment earning compound interest:
  $$\boldsymbol{A = P(1+i)^n}$$
  - A = future value
  - P = principal (starting value)
  - i = interest rate as a decimal (e.g. 7.5% p.a. is equal to 0.075 p.a.)
  - n = number of periods
- To calculate the amount of interest earned: **Interest** $\boldsymbol{= I = A - P}$

Logarithms

- Index form: $y = a^x$
- Logarithmic form: $\log_a(y) = x$
- Log laws:
  - $\log_a(x) + \log_a(y) = \log_a(xy)$
  - $\log_a(x) - \log_a(y) = \log_a\left(\frac{x}{y}\right)$
  - $\log_a(x)^n = n\log_a(x)$
  - $\log_a(1) = 0$
  - $\log_a(a) = 1$
  - $\log_a\left(\frac{1}{x}\right) = -\log_a(x)$
  - $\log_a(a^x) = x$
- Each log law is equivalent to one of the index laws.

Solving logarithmic equations

- Simplify both sides of the equation so there is at most a single logarithm on each side.
- Switch to index form or log form as required.
  e.g.
  $\log_3(x) = 4 \Longrightarrow x = 3^4$
  $5^x = 7 \Longrightarrow \log_5(7) = x$

Compounding multiple times per year

- The more frequently interest is compounded per year, the faster the value of an investment will grow.
- To account for this, adjust the rule for compound interest by:
  n = periods per year × years
  i = interest rate ÷ periods per year

1.13.2 Success criteria

Tick a column to indicate that you have completed the subtopic and how well you think you have understood it using the traffic light system.

(**Green:** I understand; **Yellow:** I can do it with help; **Red:** I do not understand)

Subtopic	Success criteria	Green	Yellow	Red
1.2	I can define the real, rational, irrational, integer and natural numbers.			
	I can determine whether a number is rational or irrational.			
1.3	I can determine whether the number under a root or radical sign is a surd.			
	I can prove that a surd is irrational by contradiction.			
1.4	I can multiply and simplify surds.			
	I can add and subtract like surds.			
	I can divide surds.			
	I can rationalise the denominator of a fraction.			
1.5	I can recall and apply the index laws.			
	I can simplify expressions involving multiplication and division of terms with the same base.			
	I evaluate expressions involving powers of zero.			
	I simplify expressions involving raising a power to another power.			
1.6	I can evaluate expressions involving negative indices.			
	I can simplify expressions involving negative indices and re-write expressions so that all indices are positive.			
1.7	I can evaluate expressions involving fractional indices.			
	I can simplify expressions involving fractional indices.			
1.8	I can simplify algebraic expressions involving brackets, fractions, multiplication and division using appropriate index laws.			
1.9	I can calculate the future value of an investment earning compound interest.			
	I can calculate the amount of interest earned after a period of time on an investment with compound interest.			
1.10	I can convert between index form and logarithmic form.			
	I can evaluate logarithms and use logarithms in scale measurement.			
1.11	I can simplify expressions using logarithm laws.			
1.12	I can simplify and solve equations involving logarithms using the logarithm laws and index laws.			

1.13.3 Project

Other number systems

Throughout history, different systems have been used to aid with counting. Ancient tribes are known to have used stones, bones and knots in rope to help keep count. The counting system that is used around the world today is called the Hindu-Arabic system. This system had its origin in India around 300–200BC . The Arabs brought this method of counting to Europe in the Middle Ages.

The Hindu–Arabic method is known as the decimal or base 10 system, as it is based on counting in lots of ten. This system uses the digits 0, 1, 2, 3, 4, 5, 6, 7, 8 and 9. Notice that the largest digit is one less than the base number, that is, the largest digit in base 10 is 9. To make larger numbers, digits are grouped together. The position of the digit tells us about its value. We call this *place value*. For example, in the number 325 , the 3 has a value of 'three lots of a hundred', the 2 has a value of 'two lots of ten' and the 5 has a value of 'five lots of units'. Another way to write this is:

$$3 \times 100 + 2 \times 10 + 5 \times 1 \text{ or } 3 \times 10^2 + 2 \times 10^1 + 5 \times 10^0$$

In a decimal system, every place value is based on the number 10 raised to a power. The smallest place value (units) is described by 10^0, the tens place value by 10^1, the hundreds place value by 10^2, the thousands by 10^3 and so on.

Computers do not use a decimal system. The system for computer languages is based on the number 2 and is known as the binary system. The only digits needed in the binary system are the digits 0 and 1. Can you see why?

Decimal number	0	1	2	3	4	5	6	7	8	9	10	11	12	13
Binary number	0	1	10	11	100	101	110	111	1000	1001	1010	1011	1100	1101

Consider the decimal number 7. From the table above, you can see that its binary equivalent is 111. How can you be sure this is correct?

$$111 = 1 \times 2^2 + 1 \times 2^1 + 1 \times 2^0 = 4 + 2 + 1 = 7$$

Notice that this time each place value is based on the number 2 raised to a power. You can use this technique to change any binary number into a decimal number. (The same pattern applies to other bases, for example, in base 6 the place values are based on the number 6 raised to a power.)

Binary operations

When adding in the decimal system, each time the addition is greater than 9, we need to 'carry over' into the next place value. In the example below, the units column adds to more than 9, so we need to carry over into the next place value.

$$\begin{array}{r} {}^{1}17 \\ +\ 13 \\ \hline 30 \end{array}$$

The same is true when adding in binary, except we need to 'carry over' every time the addition is greater than 1.

$$\begin{array}{r} {}^{1}01 \\ \hline +\ 01 \\ 10 \end{array}$$

1. Perform the following binary additions.

a. $\begin{array}{r} 11_2 \\ +01_2 \\ \hline \end{array}$ b. $\begin{array}{r} 111_2 \\ +110_2 \\ \hline \end{array}$ c. $\begin{array}{r} 1011_2 \\ +\ 101_2 \\ \hline \end{array}$

2. Perform the following binary subtractions. Remember that if you need to borrow a number from a column on the left-hand side, you will actually be borrowing a 2 (not a 10).

a. $\begin{array}{r} 11_2 \\ -01_2 \\ \hline \end{array}$ b. $\begin{array}{r} 111_2 \\ -110_2 \\ \hline \end{array}$ c. $\begin{array}{r} 1011_2 \\ -\ 101_2 \\ \hline \end{array}$

3. Try some multiplication. Remllember to carry over lots of 2.

a. $\begin{array}{r} 11_2 \\ \times 01_2 \\ \hline \end{array}$ b. $\begin{array}{r} 111_2 \\ \times 110_2 \\ \hline \end{array}$ c. $\begin{array}{r} 1011_2 \\ \times\ 101_2 \\ \hline \end{array}$

4. What if our number system had an 8 as its basis (that is, we counted in lots of 8)? The only digits available for use would be 0, 1, 2, 3, 4, 5, 6 and 7. (Remember the maximum digit is 1 less than the base value.) Give examples to show how numbers would be added, subtracted and multiplied using this base system. Remember that you would 'carry over' or 'borrow' lots of 8.
5. The hexadecimal system has 16 as its basis. Investigate this system. Explain how it would be possible to have 15, for example, in a single place position. Give examples to show how the system would add, subtract and multiply.

on Resources

eWorkbook Topic 1 Workbook (worksheets, code puzzle and project) (ewbk-2027)

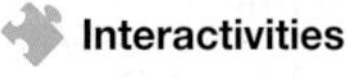
Interactivities Crossword (int-2872)
Sudoku puzzle (int-3891)

Exercise 1.13 Review questions

learnON

To answer questions online and to receive **immediate corrective feedback** and **fully worked solutions** for all questions, go to your learnON title at www.jacplus.com.au.

Fluency

1. **MC** Identify which of the given numbers are rational.

$$\sqrt{\frac{6}{12}}, \sqrt{0.81}, 5, -3.26, 0.5, \frac{\pi}{5}, \sqrt{\frac{3}{12}}$$

A. $\sqrt{0.81}$, 5, −3.26, 0.5 and $\sqrt{\frac{3}{12}}$

B. $\sqrt{\frac{6}{12}}$ and $\frac{\pi}{5}$

C. $\sqrt{\frac{6}{12}}$, $\sqrt{0.81}$ and $\sqrt{\frac{3}{12}}$

D. 5, −3.26 and $\sqrt{\frac{6}{12}}$

E. $\sqrt{0.81}$ and $\sqrt{\frac{3}{12}}$

2. For each of the following, state whether the number is rational or irrational and give the reason for your answer:

a. $\sqrt{12}$
b. $\sqrt{121}$
c. $\frac{2}{9}$
d. $0.\dot{6}$
e. $\sqrt[3]{0.08}$

3. **MC** Identify which of the numbers of the given set are surds.

$$\{3\sqrt{2}, 5\sqrt{7}, 9\sqrt{4}, 6\sqrt{10}, 7\sqrt{12}, 12\sqrt{64}\}$$

A. $9\sqrt{4}, 12\sqrt{64}$

B. $3\sqrt{2}$ and $7\sqrt{12}$ only

C. $3\sqrt{2}, 5\sqrt{7}$ and $6\sqrt{10}$ only

D. $3\sqrt{2}, 5\sqrt{7}, 6\sqrt{10}$ and $7\sqrt{12}$

E. $5\sqrt{7}$ only

4. Identify which of $\sqrt{2m}$, $\sqrt{25m}$, $\sqrt{\frac{m}{16}}$, $\sqrt{\frac{m}{20}}$, $\sqrt[3]{m}$, $\sqrt[3]{8m}$ are surds:

a. if $m = 4$
b. if $m = 8$

5. Simplify each of the following.

a. $\sqrt{50}$
b. $\sqrt{180}$
c. $2\sqrt{32}$
d. $5\sqrt{80}$

6. **MC** The expression $\sqrt{392x^8y^7}$ may be simplified to:

A. $196x^4y^3\sqrt{2y}$
B. $2x^4y^3\sqrt{14y}$
C. $14x^4y^3\sqrt{2y}$
D. $14x^4y^3\sqrt{2}$
E. $14x^8y^7\sqrt{2}$

7. Simplify the following surds. Give the answers in the simplest form.

a. $4\sqrt{648x^7y^9}$
b. $-\frac{2}{5}\sqrt{\frac{25}{64}x^5y^{11}}$

8. Simplify the following, giving answers in the simplest form.

a. $7\sqrt{12}+8\sqrt{147}-15\sqrt{27}$

b. $\frac{1}{2}\sqrt{64a^3b^3}-\frac{3}{4}ab\sqrt{16ab}+\frac{1}{5ab}\sqrt{100a^5b^5}$

9. Simplify each of the following.

a. $\sqrt{3}\times\sqrt{5}$ b. $2\sqrt{6}\times3\sqrt{7}$ c. $3\sqrt{10}\times5\sqrt{6}$ d. $\left(\sqrt{5}\right)^2$

10. Simplify the following, giving answers in the simplest form.

a. $\frac{1}{5}\sqrt{675}\times\sqrt{27}$ b. $10\sqrt{24}\times6\sqrt{12}$

11. Simplify the following.

a. $\dfrac{\sqrt{30}}{\sqrt{10}}$ b. $\dfrac{6\sqrt{45}}{3\sqrt{5}}$ c. $\dfrac{3\sqrt{20}}{12\sqrt{6}}$ d. $\dfrac{\left(\sqrt{7}\right)^2}{14}$

12. Rationalise the denominator of each of the following.

a. $\dfrac{2}{\sqrt{6}}$ b. $\dfrac{\sqrt{3}}{2\sqrt{6}}$ c. $\dfrac{2}{\sqrt{5}-2}$ d. $\dfrac{\sqrt{3}-1}{\sqrt{3}+1}$

13. Evaluate each of the following, correct to 1 decimal place if necessary.

a. $64^{\frac{1}{3}}$ b. $20^{\frac{1}{2}}$ c. $10^{\frac{1}{3}}$ d. $50^{\frac{1}{4}}$

14. Evaluate each of the following, correct to 1 decimal place.

a. $20^{\frac{2}{3}}$ b. $2^{\frac{3}{4}}$ c. $(0.7)^{\frac{3}{5}}$ d. $\left(\frac{2}{3}\right)^{\frac{2}{3}}$

15. Write each of the following in simplest surd form.

a. $2^{\frac{1}{2}}$ b. $18^{\frac{1}{2}}$ c. $5^{\frac{3}{2}}$ d. $8^{\frac{4}{3}}$

16. Evaluate each of the following, without using a calculator. Show all working.

a. $\dfrac{16^{\frac{3}{4}}\times81^{\frac{1}{4}}}{6\times16^{\frac{1}{2}}}$ b. $\left(125^{\frac{2}{3}}-27^{\frac{2}{3}}\right)^{\frac{1}{2}}$

17. Evaluate each of the following, giving your answer as a fraction.

a. 4^{-1} b. 9^{-1} c. 4^{-2} d. 10^{-3}

18. Determine the value of each of the following, correct to 3 significant figures.

a. 12^{-1} b. 7^{-2} c. $(1.25)^{-1}$ d. $(0.2)^{-4}$

19. Write down the value of each of the following.

a. $\left(\frac{2}{3}\right)^{-1}$ b. $\left(\frac{7}{10}\right)^{-1}$ c. $\left(\frac{1}{5}\right)^{-1}$ d. $\left(3\frac{1}{4}\right)^{-1}$

20. **MC** a. The expression $\sqrt{250}$ may be simplified to:

A. $25\sqrt{10}$ B. $5\sqrt{10}$ C. $10\sqrt{5}$ D. $5\sqrt{50}$ E. 25

b. When expressed in its simplest form, $2\sqrt{98}-3\sqrt{72}$ is equal to:

A. $-4\sqrt{2}$ B. -4 C. $-2\sqrt{4}$
D. $4\sqrt{2}$ E. None of these options.

c. When expressed in its simplest form, $\sqrt{\frac{8x^3}{32}}$ is equal to:

A. $\frac{x\sqrt{x}}{2}$ B. $\frac{\sqrt{x^3}}{4}$ C. $\frac{\sqrt{x^3}}{2}$
D. $\frac{x\sqrt{x}}{4}$ E. None of these options.

21. Determine the value of the following, giving your answer in fraction form.

a. $\left(\frac{2}{5}\right)^{-1}$ b. $\left(\frac{2}{3}\right)^{-2}$

22. Determine the value of each of the following, leaving your answer in fraction form.

a. 2^{-1} b. 3^{-2} c. 4^{-3} d. $\left(\frac{1}{2}\right)^{-1}$

23. **MC** $3d^{10}e^4$ is the simplified form of:

A. $d^6e^2 \times 3d^4e^3$ B. $\frac{6d^{10}e^5}{2e^2}$ C. $\left(3d^5e^2\right)^2$ D. $3e\left(d^5\right)^2 \times e^3$ E. $3\left(\frac{d^5}{e^2}\right)^2$

24. **MC** $8m^3n \times n^4 \times 2m^2n^3$ simplifies to:

A. $10m^5n^8$ B. $16m^5n^7$ C. $16m^5n^8$ D. $10m^5n^7$ E. $17m^5n^8$

25. **MC** $8x^3 \div 4x^{-3}$ is equal to:

A. 2 B. $2x^0$ C. $2x^6$ D. $2x^{-1}$ E. $\frac{2}{x^9}$

26. **MC** $\frac{12x^8 \times 2x^7}{6x^9 \times x^5}$ simplifies to:

A. $4x^5$ B. $8x$ C. $4x$ D. $8x^5$ E. $4x^{29}$

27. **MC** The expression $\frac{(a^2b^3)^5}{(2a^2b)^2}$ is equal to:

A. $\frac{a^6b^{13}}{4}$ B. $2a^6b^{13}$ C. $\frac{a^3b^6}{2}$ D. $\frac{a^6b^{13}}{2}$ E. $\frac{a^3b^6}{4}$

28. **MC** $\frac{\left(p^2q\right)^4}{\left(2p^5q^2\right)^3} \div \frac{\left(p^5q^2\right)^2}{2pq^5}$ can be simplified to:

A. $\frac{1}{4p^{16}q}$ B. $\frac{2^2}{p^{16}q}$ C. $\frac{1}{4p^8}$ D. $\frac{1}{2p^{16}q}$ E. $2^2p^{16}q$

29. **MC** $16^{-\frac{3}{4}} \div 9^{\frac{3}{2}}$ can be simplified to:

A. 2 **B.** $\frac{1}{216}$ **C.** $\frac{8}{27}$ **D.** $3\frac{3}{8}$ **E.** $\frac{1}{2}$

30. **MC** $\dfrac{\left(2l^{\frac{2}{9}}m^{-1}\right)^{-3}}{8\left(\frac{1}{16}lm^{-2}\right)^{2}}$ can be simplified to:

A. $\dfrac{8m^7}{l^{\frac{11}{3}}}$ **B.** $\dfrac{2m^7}{l^{\frac{7}{3}}}$ **C.** $\dfrac{4m^7}{l^{\frac{8}{3}}}$ **D.** $\dfrac{16m^7}{l^{\frac{5}{3}}}$ **E.** $\dfrac{m^7}{2l^{\frac{7}{3}}}$

31. **MC** $\sqrt[5]{32i^{\frac{10}{7}}j^{\frac{5}{11}}k^2}$ can be simplified to:

A. $\dfrac{32i^{\frac{2}{7}}j^{\frac{1}{11}}k^{\frac{2}{5}}}{5}$ **B.** $2i^{\frac{2}{7}}j^{\frac{1}{11}}k^{\frac{2}{5}}$ **C.** $\dfrac{32i^{\frac{10}{7}}j^{\frac{5}{11}}k^2}{5}$ **D.** $2i^{\frac{50}{7}}j^{\frac{25}{11}}k^{10}$ **E.** $\dfrac{2i^{\frac{2}{7}}j^{\frac{1}{11}}k^{\frac{2}{5}}}{5}$

32. Simplify each of the following.

a. $5x^3 \times 3x^5y^4 \times \frac{3}{5}x^2y^6$

b. $\dfrac{26a^4b^6c^5}{12a^3b^3c^3}$

c. $\left(\dfrac{20m^5n^2}{6}\right)^3$

d. $\left(\dfrac{14p^7}{21q^3}\right)^4$

33. Evaluate each of the following.

a. $5a^0 - \left(\dfrac{2a}{3}\right)^0 + 12$

b. $-(3b)^0 - \dfrac{(4b)^0}{2}$

34. Simplify each of the following and express your answer with positive indices.

a. $2a^{-5}b^2 \times 4a^{-6}b^{-4}$ **b.** $4x^{-5}y^{-3} \div 20x^{12}y^{-5}$ **c.** $\left(2m^{-3}n^2\right)^{-4}$

35. Evaluate each of the following without using a calculator.

a. $\left(\dfrac{1}{2}\right)^{-3}$ **b.** $2 \times (3)^{-3} \times \left(\dfrac{9}{2}\right)^2$ **c.** $4^{-3} \times \dfrac{5}{8^{-2}} - 5$

36. Simplify each of the following.

a. $2a^{\frac{4}{5}}b^{\frac{1}{2}} \times 3a^{\frac{1}{2}}b^{\frac{3}{4}} \times 5a^{\frac{3}{4}}b^{\frac{2}{5}}$ **b.** $\dfrac{4^3x^{\frac{3}{4}}y^{\frac{1}{9}}}{16x^{\frac{4}{5}}y^{\frac{1}{3}}}$ **c.** $\left(\dfrac{4a^{\frac{1}{3}}}{b^3}\right)^{\frac{1}{2}}$

37. Evaluate each of the following without using a calculator. Show all working.

a. $\dfrac{16^{\frac{3}{4}} \times 81^{\frac{1}{4}}}{6 \times 16^{\frac{1}{2}}}$

b. $\left(125^{\frac{2}{3}} - 27^{\frac{2}{3}}\right)^{\frac{1}{2}}$

38. Simplify:

a. $\sqrt[3]{a^9} + \sqrt[4]{16a^8b^2} - 3\left(\sqrt[5]{a}\right)^{15}$

b. $\sqrt[5]{32x^5y^{10}} + \sqrt[3]{64x^3y^6}$

39. Simplify each of the following.

a. $\dfrac{\left(5a^{-2}b\right)^{-3} \times 4a^{6}b^{-2}}{2a^{2}b^{3} \times 5^{-2}a^{-3}b^{-6}}$

b. $\dfrac{2x^{4}y^{-5}}{3y^{6}x^{-2}} \times \left(\dfrac{4xy^{-2}}{3x^{-6}y^{3}}\right)^{-3}$

c. $\left(\dfrac{2m^{3}n^{4}}{5m^{\frac{1}{2}}n}\right)^{\frac{1}{3}} \div \left(\dfrac{4m^{\frac{1}{3}}n^{-2}}{5^{-\frac{2}{3}}}\right)^{-\frac{1}{2}}$

40. Simplify each of the following and then evaluate.

a. $\left(3 \times 5^{6}\right)^{\frac{1}{2}} \times 3^{\frac{3}{2}} \times 5^{-2} + \left(3^{6} \times 5^{-\frac{1}{2}}\right)^{0}$

b. $\left(6 \times 3^{-2}\right)^{-1} \div \dfrac{\left(3^{\frac{1}{2}} \times 6^{\frac{1}{3}}\right)^{6}}{-6^{2} \times \left(3^{-3}\right)^{0}}$

41. Ryan invests \$12 500 for 3 years at 8% p.a. with interest paid annually. By calculating the amount of simple interest earned each year separately, determine the amount to which the investment will grow.

42. Calculate the compound interest earned on \$45 000 at 12% p.a. over 4 years if interest is compounded:

a. annually
b. six-monthly
c. quarterly
d. monthly.

43. Evaluate the following.

a. $\log_{12}(18) + \log_{12}(8)$
b. $\log_{4}(60) - \log_{4}(15)$
c. $\log_{9}\left(9^{8}\right)$
d. $2\log_{3}(6) - \log_{3}(4)$

44. Use the logarithm laws to simplify each of the following.

a. $\log_{a}(16) + \log_{a}(3) - \log_{a}(2)$
b. $\log_{x}\left(x\sqrt{x}\right)$
c. $4\log_{a}(x) - \log_{a}\left(x^{2}\right)$
d. $5\log_{x}\left(\dfrac{1}{x}\right)$

45. Solve for x in the following, given that $x > 0$.

a. $\log_{2}(x) = 9$
b. $\log_{5}(x) = -2$
c. $\log_{x}(25) = 2$
d. $\log_{x}\left(2^{6}\right) = 6$
e. $\log_{3}(729) = x$
f. $\log_{7}(1) = x$

46. Solve for x in the following.

a. $\log_{5}(4) + \log_{5}(x) = \log_{5}(24)$
b. $\log_{3}(x) - \log_{3}(5) = \log_{3}(7)$

47. Solve for x in the following equations.

a. $6^{x} = \dfrac{1}{36}$
b. $7^{x} = \dfrac{1}{\sqrt{7}}$
c. $2^{x+1} = 8\sqrt{2}$

48. Solve for x in the following equations, correct to 3 decimal places.

a. $2^{x} = 25$
b. $0.6^{x} = 7$
c. $9^{-x} = 0.84$

Problem solving

49. Answer the following. Explain how you reached your answer.

a. What is the hundred's digit in 3^{3^3}?
b. What is the one's digit in 6^{704}?
c. What is the thousand's digit in 9^{1000}?

50. a. Plot a graph of $y = 4^x$ by first producing a table of values. Label the y-intercept and the equation of any asymptotes.
b. Draw the line $y = x$ on the same set of axes.
c. Use the property of inverse graphs to draw the graph of $y = \log_4(x)$. Label any intercepts and the equation of any asymptotes. Use a graphics calculator or graphing software to check your graphs.

51. Solve for x: $\left(\frac{6}{x}\right)^{-1} + \frac{1}{6} = x^{-1}$

52. Simplify $\left(\left(\frac{(a^2)^{-1}}{b^{\frac{1}{2}}}\right)^{-1}\right)^{-1}$

53. If $m = 2$, determine the value of:

$$\frac{6a^{3m} \times 2b^{2m} \times (3ab)^{-m}}{(4b)^m \times (9a^{4m})^{\frac{1}{2}}}$$

54. Answer the following and explain your reasoning.

a. Identify the digit in the tens of 3^{3^3}.
b. Identify the digit in the ones of 6^{309}.
c. Identify the digit in the ones of 8^{1007}.

55. For the work shown below:

a. calculate the correct answer
b. identify where the student has made mistakes.

$$\left(\frac{3a^3b^5c^3}{5a^2b}\right)^2 \div \left(\frac{2ab}{c}\right) = \frac{3a^6b^{10}c^6}{10a^4b^2} \div \frac{2ab}{c}$$

$$= \frac{3a^6b^{10}c^6}{10a^4b^2} \times \frac{c}{2ab}$$

$$= \frac{3a^6b^{10}c^7}{20a^5b^3}$$

$$= \frac{3ab^7c^7}{20}$$

56. A friend is trying to calculate the volume of water in a reservoir amid fears there may be a severe water shortage. She comes up with the following expression:

$$W = \frac{r^4u^2}{r^{\frac{3}{2}}d^2\sqrt{u}} \times \frac{ru \times d^2}{dr^3u^4},$$

where r is the amount of rain, d is how dry the area is, u is the usage of water by the townsfolk, and W is the volume of water in kL.

a. Help your friend simplify the expression by simplifying each pronumeral one at a time.
b. Explain whether the final expression contain any potential surds.
c. Express the fraction with a rational denominator.
d. List the requirements for the possible values of r, u and d to give a rational answer.
e. Calculate the volume of water in the reservoir when $r = 4$, $d = 60$ and $u = 9$. Write your answer in:
i. kL
ii. L
iii. mL.
f. Does a high value for d mean the area is dry? Explain using working.

57. The speed of a toy plane can be modelled by the equation $S = \dfrac{p^2}{2 + \sqrt{w^3}}$, where:

w = wind resistance
p = battery power (from 0 (empty) to 10 (full)).

a. Rationalise the denominator of the expression.
b. Using your knowledge of perfect squares, estimate the speed of a toy plane with its battery half full and a wind resistance of 2. Check your answer with a calculator.
c. How does the speed of the toy plane change with increasing wind resistance? Explain providing supportive calculations.

on To test your understanding and knowledge of this topic, go to your learnON title at www.jacplus.com.au and complete the **post-test**.

Online Resources

Below is a full list of **rich resources** available online for this topic. These resources are designed to bring ideas to life, to promote deep and lasting learning and to support the different learning needs of each individual.

eWorkbook

Download the workbook for this topic, which includes worksheets, a code puzzle and a project (ewbk-2027) ☐

Solutions

Download a copy of the fully worked solutions to every question in this topic (sol-0735) ☐

Digital documents

- 1.3 SkillSHEET Identifying surds (doc-5354) ☐
- 1.4 SkillSHEET Simplifying surds (doc-5355) ☐
  - SkillSHEET Adding and subtracting surds (doc-5356) ☐
  - SkillSHEET Multiplying and dividing surds (doc-5357) ☐
  - SkillSHEET Rationalising denominators (doc-5360) ☐
  - SkillSHEET Conjugate pairs (doc-5361) ☐
  - SkillSHEET Applying the difference of two squares rule to surds (doc-5362) ☐
- 1.5 SkillSHEET Index form (doc-5168) ☐
  - SkillSHEET Using a calculator to evaluate numbers given in index form (doc-5169) ☐
- 1.7 SkillSHEET Addition of fractions (doc-5176) ☐
  - SkillSHEET Subtraction of fractions (doc-5177) ☐
  - SkillSHEET Multiplication of fractions (doc-5178) ☐
  - SkillSHEET Writing roots as fractional indices (doc-5179) ☐

Video eLessons

- 1.2 The real number system (eles-4661) ☐
- 1.3 Identifying surds (eles-4662) ☐
  - Proof that a number is irrational (eles-4663) ☐
- 1.4 Multiplying and simplifying surds (eles-4664) ☐
  - Addition and subtraction of surds (eles-4665) ☐
  - Dividing surds (eles-4666) ☐
  - Rationalising denominators (eles-4667) ☐
  - Rationalising denominators using conjugate surds (eles-4668) ☐
  - Surds (eles-1906) ☐
  - Rationalisation of surds (eles-1948) ☐
- 1.5 Review of index laws (eles-4669) ☐
  - Index laws (eles-1903) ☐
- 1.6 Negative indices and the Seventh Index Law (eles-4670) ☐
  - Negative indices (eles-1905) ☐
- 1.7 Fractional indices and the Eighth Index Law (eles-4671) ☐
  - Fractional indicies (eles-1950) ☐
- 1.8 Combining index laws (eles-4672) ☐
- 1.9 Application of indices: compound interest (eles-4673) ☐
  - The compound interest formula (eles-4674) ☐
  - Compounding period (eles-4675) ☐
- 1.10 Logarithms (eles-4676) ☐
  - Using logarithmic scales in measurement (eles-4677) ☐
- 1.11 Logarithm laws (eles-4678) ☐
- 1.12 Solving equations with logarithms (eles-4679) ☐

Interactivities

- 1.2 Individual pathway interactivity: Number classification review (int-8332) ☐
  - The number system (int-6027) ☐
  - Recurring decimals (int-6189) ☐
- 1.3 Surds on the number line (int-6029) ☐
- 1.4 Addition and subtraction of surds (int-6190) ☐
  - Multiplying surds (int-6191) ☐
  - Dividing surds (int-6192) ☐
  - Simplifying surds (int-6028) ☐
  - Conjugate surds (int-6193) ☐
- 1.5 Individual pathway interactivity: Review of index laws (int-4652) ☐
  - First Index Law (int-3709) ☐
  - Second Index Law (int-3711) ☐
  - Third Index Law (int-3713) ☐
  - Fourth Index Law — Multiplication (int-3716) ☐
  - Fifth and sixth index laws (int-6063) ☐
- 1.6 Individual pathway interactivity: Negative indices (int-4563) ☐
  - Negative indices (int-6064) ☐
- 1.7 Individual pathway interactivity: Fractional indices (int-4564) ☐
  - Fractional indices (int-6107) ☐
- 1.8 Individual pathway interactivity: Combining index laws (int-4565) ☐
  - Combining index laws (int-6108) ☐
- 1.9 Individual pathway interactivity: Compound interest (int-4636) ☐
  - Compound interest (int-6075) ☐
  - Compounding periods (int-6186) ☐
- 1.10 Logarithms (int-6194) ☐
- 1.11 The first law of logarithms (int-6195) ☐
  - The second law of logarithms (int-6196) ☐
  - The third law of logarithms (int-6197) ☐
  - The fourth law of logarithms (int-6198) ☐
  - The fifth law of logarithms (int-6199) ☐
  - The sixth law of logarithms (int-6200) ☐
  - The seventh law of logarithms (int-6201) ☐
- 1.12 Solving logarithmic equations (int-6202) ☐
- 1.13 Crossword (int-2872) ☐
  - Sudoku puzzle (int-3891) ☐

Teacher resources

There are many resources available exclusively for teachers online.

To access these online resources, log on to **www.jacplus.com.au**.

Answers

Topic 1 Indices, surds and logarithms

Exercise 1.1 Pre-test

1. True
2. Rational
3. $15n^{\frac{8}{15}}$
4. $2p^2q^3$
5. $\frac{1}{27}$
6. D
7. B
8. $6\sqrt{5}$
9. $14\sqrt{2}$
10. E
11. $y = -5$
12. $x = -2$
13. $278.18
14. 0
15. D

Exercise 1.2 Number classification review

1. a. Q b. Q c. Q d. I
2. a. I b. Q c. Q d. I
3. a. Q b. Q c. Q d. Q
4. a. I b. Q c. I d. Q
5. a. Q b. I c. I d. I
6. a. Q b. I c. I d. Q
7. a. Q b. Q c. Q d. Q
8. a. Q b. I c. I d. Q
9. a. I b. Undefined c. I d. I
10. a. I b. Q c. Q d. Q
11. a. I b. I c. Q d. Q
12. a. I b. Q c. Q d. I
13. B
14. D
15. C
16. C
17. $\frac{a}{b}$
18. E
19. A
20. $p - q$
21. Sample responses can be found in the worked solutions in the online resources.
22. 8, -8
23. a. $m = 11, n = 3$ b. $m = 2, n = 3$ c. $m = 3, n = 2$ d. $m = 1, n = 2$
24. $\frac{1}{7}$ or 7^{-1}
25. $\frac{37}{91}$

Exercise 1.3 Surds

1. b and d
2. b, c and d
3. a and d
4. a and c
5. a, c and d
6. c and f
7. A
8. D
9. B
10. C
11. Any perfect square.
12. $m = 4$
13. a. $m = 5$, $n = 7$ and $m = 4, n = 14$
    b. 15
14. a. Irrational b. Irrational c. Rational
15. Sample responses can be found in the worked solutions in the online resources.
16. Irrational
17. a. i. $4\sqrt{3}$ ii. $6\sqrt{2}$
    b. Yes. If you don't choose the largest perfect square, then you will need to simplify again.
    c. No
18. $p = m$ and $q = n$
19. $x = 3$, rational

Exercise 1.4 Operations with surds

1. a. $2\sqrt{3}$ b. $2\sqrt{6}$ c. $3\sqrt{3}$ d. $5\sqrt{5}$
2. a. $3\sqrt{6}$ b. $4\sqrt{7}$ c. $2\sqrt{17}$ d. $6\sqrt{5}$
3. a. $2\sqrt{22}$ b. $9\sqrt{2}$ c. $7\sqrt{5}$ d. $8\sqrt{7}$
4. a. $4\sqrt{2}$ b. $24\sqrt{10}$ c. $36\sqrt{5}$ d. $21\sqrt{6}$
5. a. $-30\sqrt{3}$ b. $-28\sqrt{5}$ c. $64\sqrt{3}$ d. $2\sqrt{2}$
6. a. $\sqrt{2}$ b. $2\sqrt{3}$ c. $\frac{1}{3}\sqrt{15}$ d. $\frac{3}{2}\sqrt{7}$
7. a. $4a$ b. $6a\sqrt{2}$ c. $3a\sqrt{10b}$ d. $13a^2\sqrt{2}$
8. a. $13ab\sqrt{2ab}$ b. $2ab^2\sqrt{17ab}$ c. $5x^3y^2\sqrt{5}$ d. $20xy\sqrt{5x}$
9. a. $54c^3d^2\sqrt{2cd}$ b. $18c^3d^4\sqrt{5cd}$ c. $\sqrt{22ef}$ d. $7e^5f^5\sqrt{2ef}$

10. a. $7\sqrt{5}$ b. $8\sqrt{3}$
c. $15\sqrt{5}+5\sqrt{3}$ d. $4\sqrt{11}$

11. a. $13\sqrt{2}$ b. $-3\sqrt{6}$
c. $17\sqrt{3}-18\sqrt{7}$ d. $8\sqrt{x}+3\sqrt{y}$

12. a. $10\left(\sqrt{2}-\sqrt{3}\right)$ b. $5\left(\sqrt{5}+\sqrt{6}\right)$
c. $7\sqrt{3}$ d. $4\sqrt{5}$

13. a. $14\sqrt{3}+3\sqrt{2}$ b. $3\sqrt{6}+6\sqrt{3}$
c. $15\sqrt{10}-10\sqrt{15}+10$ d. $-8\sqrt{11}+22$

14. a. $12\sqrt{30}-16\sqrt{15}$ b. $12\sqrt{ab}+7\sqrt{3ab}$
c. $\frac{7}{2}\sqrt{2}+2\sqrt{3}$ d. $15\sqrt{2}$

15. a. $31\sqrt{a}-6\sqrt{2a}$ b. $52\sqrt{a}-29\sqrt{3a}$
c. $6\sqrt{6ab}$ d. $32a+2\sqrt{6a}+8a\sqrt{2}$

16. a. $a\sqrt{2a}$ b. $\sqrt{a}+2\sqrt{2a}$
c. $3a\sqrt{a}+a^2\sqrt{3a}$ d. $(a^2+a)\sqrt{ab}$

17. a. $4ab\sqrt{ab}+3a^2b\sqrt{b}$
b. $3\sqrt{ab}(2a+1)$
c. $-6ab\sqrt{2a}+4a^2b^3\sqrt{3a}$
d. $-2a\sqrt{b}$

18. a. $\sqrt{14}$ b. $\sqrt{42}$ c. $4\sqrt{3}$
d. 10 e. $3\sqrt{7}$ f. 27

19. a. $10\sqrt{33}$ b. $180\sqrt{5}$ c. 120
d. $120\sqrt{3}$ e. $2\sqrt{6}$ f. $2\frac{2}{3}$

20. a. $\frac{2}{5}\sqrt{6}$ b. $x^2y\sqrt{y}$ c. $3a^4b^2\sqrt{2ab}$
d. $6a^5b^2\sqrt{2b}$ e. $3x^2y^2\sqrt{10xy}$ f. $\frac{9}{2}a^2b^4\sqrt{5ab}$

21. a. 2 b. 5 c. 12

22. a. 15 b. 18 c. 80

23. a. 28 b. 200

24. a. $\sqrt{5}$ b. 2
c. $\sqrt{6}$ d. 4

25. a. $\frac{\sqrt{3}}{4}$ b. $\frac{\sqrt{5}}{2}$
c. $2\sqrt{3}$ d. 1

26. a. $1\frac{4}{5}$ b. $2\sqrt{17}$ c. $\frac{x}{y}$
d. $\frac{\sqrt{2}}{x^3y^4}$ e. $2xy\sqrt{3y}$ f. $\frac{4\sqrt{a}}{3}$

27. a. $\frac{5\sqrt{2}}{2}$ b. $\frac{7\sqrt{3}}{3}$ c. $\frac{4\sqrt{11}}{11}$
d. $\frac{4\sqrt{6}}{3}$ e. $\frac{2\sqrt{21}}{7}$

28. a. $\frac{\sqrt{10}}{2}$ b. $\frac{2\sqrt{15}}{5}$ c. $\frac{3\sqrt{35}}{5}$
d. $\frac{5\sqrt{6}}{6}$ e. $\frac{4\sqrt{15}}{15}$

29. a. $\frac{5\sqrt{7}}{14}$ b. $\frac{8\sqrt{15}}{15}$ c. $\frac{8\sqrt{21}}{49}$
d. $\frac{8\sqrt{105}}{7}$ e. $\frac{\sqrt{10}}{3}$

30. a. $\sqrt{2}+2$ b. $\frac{3\sqrt{10}-2\sqrt{33}}{6}$
c. $\frac{12\sqrt{5}-5\sqrt{6}}{10}$ d. $\frac{9\sqrt{10}}{5}$

31. a. $\frac{3\sqrt{10}+6\sqrt{14}}{4}$ b. $\frac{5\sqrt{6}}{3}$
c. $\frac{3\sqrt{22}-4\sqrt{10}}{6}$ d. $\frac{\sqrt{21}-\sqrt{15}}{3}$

32. a. $\frac{14-5\sqrt{2}}{6}$ b. $\frac{12-\sqrt{10}}{16}$ c. $\frac{6\sqrt{15}-25}{70}$
d. $\frac{\sqrt{30}+7\sqrt{2}}{20}$

33. a. $\sqrt{5}-2$ b. $\frac{2\sqrt{2}+\sqrt{5}}{3}$ c. $\frac{8\sqrt{11}+4\sqrt{13}}{31}$

34. a. $\frac{15\sqrt{15}-20\sqrt{6}}{13}$
b. $12\sqrt{2}-17$
c. $\frac{19-4\sqrt{21}}{5}$

35. a. $\frac{\sqrt{15}-\sqrt{3}-\sqrt{5}+1}{4}$
b. $\frac{-6+6\sqrt{2}+\sqrt{10}-2\sqrt{5}}{2}$
c. $\frac{4\sqrt{10}+\sqrt{15}-4\sqrt{6}-3}{29}$

36. $\frac{15\left(3\sqrt{2}-2\right)}{28}$

37. $\frac{9\sqrt{x}+6x}{36x-16x^2}$

38. a. Sample responses can be found in the worked solutions in the online resources.
b. i. $\sqrt{5}+\sqrt{3}$ ii. $\sqrt{5}+\sqrt{3}$ iii. $\sqrt{3}+2$

39. $\frac{2}{7}$

40. a. $x=16$ b. $x=1$

41. 1

Exercise 1.5 Review of index laws

1. a. a^7 b. a^6 c. b^8 d. a^4b^7

2. a. m^5n^{13} b. $a^5b^7c^3$ c. $m^6n^4p^5$ d. $6a^2b$

3. a. $10a^4b^9$ b. $36m^8n^7$ c. $12x^6y^6$ d. $4x^8y^6$

4. a. a b. a^5 c. b^3 d. $\frac{4}{3}a^4$

5. a. $3b^4$ b. $4m^5$ c. m^3n d. $\frac{1}{2}y^2$

6. a. $7b^3$ b. $\frac{5}{4}m^2p^2$ c. $\frac{1}{2}xy^2$

7. a. 1 b. 1 c. 1

8. a. 3 b. 4 c. −3

9. a. 3 b. −7 c. 4

10. a. a^6 b. $16a^{20}$
c. $\frac{1}{81}m^8$ d. $\frac{4}{9}n^8$
e. 49

11. a. a^6b^3 b. $9a^6b^4$ c. $16m^{12}n^{20}$
d. $\frac{27}{64}m^6n^3$ e. $\frac{a^4}{b^6}$

12. a. $\frac{625m^{12}}{n^8}$ b. $\frac{343x^3}{8y^{15}}$ c. $\frac{81a^4}{625b^{12}}$
d. −243 e. −32

13. a. D b. D

14. a. C b. E
c. B d. D

15. a. 64 b. 72 c. 625

16. a. 48 b. 1600 c. $\frac{27}{125}$

17. a. 20 b. 1 c. 4

18. a. x^{3yz} b. a^b

19. a. m^an^b b. $\frac{a^{2x}}{b^{3x}}$

20. a. $n^{3-p}m^{2-q}$ b. a^{mp+np}

21.
$$a^3 = a \times a \times a$$
$$a^2 = a \times a$$
$$a^3 \times a^2 = a \times a \times a \times a \times a$$
$$= a^5, \text{ not } a^6$$
Explanations will vary.

22. They are equal when $x = 2$. Explanations will vary.

23. $3x^0 = 3$ and $(3x)^0 = 1$. Explanations will vary.

24. a.

a	0	1	2	3
$3a^2$	0	3	12	27
$5a$	0	5	10	15
$3a^2 + 5a$	0	8	22	42
$3a^2 \times 5a$	0	15	120	405

b. $3a^2 \times 5a$ will become much larger than $3a^2 + 5a$.

25. $x = -2$ or 4

26. $1 \equiv 1$
$2 \equiv 10$
$3 \equiv 11$
$4 \equiv 100$
$5 \equiv 101$
$6 \equiv 110$
$7 \equiv 111$
$8 \equiv 1000$
$9 \equiv 1001$
$10 \equiv 1010$

27. a. $x = 4$ b. $x = 0, 2$

28. a. a^2bc^7
b. The student made a mistake when multiplying the two brackets in line 3. Individual brackets should be expanded first.

Exercise 1.6 Negative indices

1. a. $\frac{1}{x^5}$ b. $\frac{1}{y^4}$ c. $\frac{2}{a^9}$ d. $\frac{4}{5a^3}$

2. a. $\frac{3x^2}{y^3}$ b. $\frac{1}{4m^3n^4}$ c. $\frac{6a^3}{bc^5}$ d. a^6

3. a. $\frac{2a^4}{3}$ b. $2ab^2$ c. $\frac{7b^3}{2a^4}$ d. $\frac{2m^3a^2}{3b^4n^5}$

4. a. $\frac{1}{a^2b^3}$ b. $\frac{6}{x^6y}$ c. $\frac{3}{n^8}$
d. $\frac{4}{a^2b^5}$ e. $\frac{2y}{3x}$

5. a. $\frac{5y}{6x^3}$ b. $\frac{3}{m^2n^2}$ c. $\frac{4y^{12}}{x^5}$
d. $\frac{1}{3m^3n^3}$ e. $\frac{1}{32a^{15}m^{20}}$

6. a. $\frac{4q^8}{p^{14}}$ b. $\frac{3}{a^8b^{12}}$ c. $\frac{27q^9}{8p^6}$
d. $\frac{b^6}{4a^8}$ e. $\frac{1}{8a^6b^6}$

7. a. $\frac{1}{8}$ b. $\frac{1}{36}$ c. $\frac{1}{81}$ d. $\frac{8}{9}$

8. a. $\frac{1}{16}$ b. $\frac{5}{36}$ c. 48 d. $\frac{32}{27}$

9. a. $\frac{27}{25} = 1\frac{2}{25}$ b. 4
c. 125 d. $\frac{3}{4}$

10. a. 2^3 b. 2^{-3} c. 2^5 d. 2^{-6}

11. a. $x = 3$ b. $x = -2$ c. $x = -1$
d. $x = 3$ e. $x = -2$

12. a. $x = 0$ b. $x = 3$ c. $x = -3$
d. $x = -6$ e. $x = -2$

13. a. $\frac{3}{2}$ b. $\frac{4}{5}$ c. $\frac{2}{7}$ d. 5

14. a. $\frac{b}{a}$ b. $\frac{b^3}{a^2}$ c. $\frac{a^2}{b^3}$ d. $\frac{1}{m^3n^2}$

15. a. $\frac{1}{729}$
b. $\frac{1}{20\,736}$
c. 0.000059499 or $\frac{1}{16807}$

16. a. 256 b. $\frac{16\,384}{2187}$ c. 9 765 625

17. C

18. B

19. B

20. D

21. C

22. E

23. a. $\frac{m^2}{n^8}$ b. $\frac{n^2}{m}$ c. $\frac{25}{a^7b^6}$

24. a. $r^6 - s^6$
 b. $m^{10} + 2m^5n^5 + n^{10}$
 c. 1
 d. p^2

25. 2^{2r-4}

26. 6^{3m}

27. $x = 3$

28. a. As x gets closer to 0 coming from the positive direction, y gets more and more positive, approaching ∞.
 b. As x gets closer to 0 coming from the negative direction, y gets more and more negative, approaching $-\infty$.

29. $2^{-n} = \frac{1}{2^n}$
 A n increases, the value of 2^n increases, so the value of 2^{-n} gets closer to 0.

30. Sample responses can be found in the worked solutions of your online resources.

31. $x = -2, y = -3$

32. x^2

33. $x = 3, y = -1; 7$

Exercise 1.7 Fractional indices

1. a. $\sqrt{15}$ b. $\sqrt[4]{m}$ c. $\sqrt[5]{7^2}$ d. $\sqrt{7^5}$

2. a. $\sqrt[8]{w^3}$ b. $\sqrt[4]{w^5}$ c. $\sqrt[3]{5^{10}}$ d. $\sqrt[10]{a^3}$

3. a. $t^{\frac{1}{2}}$ b. $5^{\frac{7}{4}}$ c. $6^{\frac{11}{6}}$ d. $x^{\frac{6}{7}}$

4. a. $x^{\frac{7}{6}}$ b. w^2 c. $w^{\frac{1}{2}}$ d. $11^{\frac{n}{x}}$

5. a. 4 b. 5 c. 9

6. a. 2 b. 4 c. 3

7. a. 2.2 b. 1.5 c. 1.3

8. a. 2.5 b. 12.9 c. 13.6

9. a. 0.7 b. 0.8 c. 0.9

10. a. $4^{\frac{4}{5}}$ b. $2^{\frac{1}{2}}$ c. $a^{\frac{5}{6}}$

11. a. $x^{\frac{23}{20}}$ b. $10m^{\frac{8}{15}}$ c. $2b^{\frac{5}{7}}$

12. a. $-4y^{\frac{20}{9}}$ b. $0.02a^{\frac{9}{8}}$ c. $5x^{\frac{7}{2}}$

13. a. $ab^{\frac{3}{2}}$ b. $x^{\frac{4}{5}}y^{\frac{5}{9}}$ c. $6a^{\frac{8}{5}}b^{\frac{17}{15}}$

14. a. $2m^{\frac{19}{28}}n^{\frac{2}{5}}$ b. $x^{\frac{19}{6}}y^{\frac{5}{6}}z^{\frac{5}{6}}$ c. $8a^{\frac{2}{5}}b^{\frac{9}{8}}c$

15. a. $x^{\frac{5}{3}}y^{\frac{7}{5}}$ b. $a^{\frac{7}{45}}b^{\frac{4}{15}}$ c. $\frac{1}{3}m^{\frac{3}{8}}n^{\frac{11}{56}}$

16. a. $2x^{\frac{2}{15}}y^{\frac{3}{4}}$ b. $\frac{1}{4}a^{\frac{11}{20}}b^{\frac{7}{20}}$ c. $\frac{1}{7}p^{\frac{5}{24}}q^{\frac{1}{12}}$

17. a. $2^{\frac{9}{20}}$ b. $5^{\frac{1}{6}}$ c. $7^{\frac{6}{5}}$

18. a. $a^{\frac{3}{10}}$ b. $m^{\frac{1}{6}}$ c. $2^{\frac{1}{3}}b^{\frac{1}{6}}$

19. a. $4p^{\frac{2}{5}}$ b. $x^{\frac{m}{p}}$ c. $3^{\frac{b}{c}}m^{\frac{a}{c}}$

20. a. $a^{\frac{1}{4}}b^{\frac{1}{6}}$ b. $a^3b^{\frac{3}{4}}$ c. $x^{\frac{6}{5}}y^{\frac{7}{4}}$

21. a. $3^{\frac{1}{3}}a^{\frac{1}{9}}b^{\frac{1}{5}}c^{\frac{1}{4}}$ b. $5x^{\frac{1}{4}}y^{\frac{1}{3}}z^{\frac{1}{5}}$ c. $\frac{a^{\frac{1}{2}}}{b^{\frac{2}{3}}}$

22. a. $\frac{m^{\frac{8}{5}}}{n^{\frac{7}{4}}}$ b. $\frac{b^{\frac{2}{5}}}{c^{\frac{8}{27}}}$ c. $\frac{2^{\frac{1}{2}}x^{\frac{7}{2}}}{y^{\frac{3}{8}}}$

23. a. E b. C c. B

24. a. E b. B

25. a. a^4 b. b^3 c. m^4

26. a. $4x^2$ b. $2y^3$ c. $2x^2y^3$

27. a. $3m^3n^5$ b. $2pq^2$ c. $6a^2b^6$

28. a. 2.007 s b. 20.07 s c. 4.98 swings

29. $\left(2^5a^5b^{10}\right)^{\frac{1}{5}} = 2ab^2$

30. $\frac{a^2\left(3 - \sqrt{b^3}\right)}{9 - b^3}; \frac{4}{11}$

31. a. $x^{\frac{1}{2}} + y^{\frac{1}{2}} - z^{\frac{1}{2}}$ b. $t^{\frac{1}{10}}$

32. $m - n^2$

33. $m^{\frac{1}{5}} - n^{\frac{1}{5}} + p^{\frac{1}{5}}$

34. a. $a^{-\frac{1}{4}} \times b^{\frac{13}{2}}$
 b. No, because you can't take the fourth root of a negative number.
 c. $a = 1$

Exercise 1.8 Combining index laws

1. a. $54a^{10}b^9$ b. $48a^5b^{16}$ c. $\frac{2n^{13}}{m^9}$

2. a. $500p^8q^{18}$ b. $36a^{20}b^{10}$ c. $\frac{15b^2}{c^{26}}$

3. a. $12x^{\frac{7}{8}}y^{\frac{11}{15}}$ b. $8m^{\frac{15}{4}}n^{\frac{15}{4}}$ c. $\frac{6}{p^{\frac{7}{12}}}$
 d. $8p^{\frac{7}{45}}q^{\frac{5}{18}}$

4. a. $\frac{5}{8a^7}$ b. $\frac{x}{4y^6}$ c. $\frac{27}{128m^{29}n^{26}}$

5. a. $\frac{64y^{36}}{x^{24}}$ b. $24a^{24}b^7$ c. $\frac{27h^{12}}{8g^6}$

6. a. $p^{\frac{35}{3}}q^{\frac{1}{2}}$ b. $\frac{625}{81b^{20}c^{28}}$ c. $x^{\frac{5}{3}}y^{\frac{1}{8}}z^{\frac{3}{2}}$

7. a. $\frac{3a^2}{2}$ b. $8n^2$ c. $\frac{m^2n^4}{3}$

8. a. $\frac{4x^5}{3y^8}$ b. $\frac{36x^6}{y}$ c. $\frac{y^2}{x^4}$

9. a. $\frac{b^7}{3a^4}$ b. $\frac{75q^5}{2p^{11}}$ c. $x^{\frac{17}{10}}y^{\frac{7}{10}}$

10. a. $\frac{2}{5a^4b^7}$ b. $\frac{4a^3b^3}{15}$ c. $\frac{n^9}{4m^9}$

11. a. $\frac{4m^5}{9n^{15}}$ b. $\frac{4}{81x^2y^{14}}$ c. $48x^{11}y^6$

12. a. $\frac{3p^4}{5q^9}$ b. $\frac{2b^{\frac{1}{12}}}{3a^{\frac{17}{24}}}$ c. $\frac{4x^{\frac{1}{12}}}{3y^{\frac{21}{20}}}$

13. a. $\frac{5}{2a^{13}}$ b. $\frac{56a^{11}b^6}{81}$

14. a. $\frac{1024b^2}{81a}$ b. $\frac{25}{128x^{23}y^4}$ c. $\frac{4y^{36}}{27x^{16}}$

15. a. $6m^{19}n^{19}$ b. $\frac{16m^{\frac{11}{12}}n}{3}$ c. $\frac{4b^{\frac{11}{2}}}{3^{\frac{1}{2}}c^{\frac{7}{30}}}$

16. a. $\frac{125}{8}$ b. 1

17. 1

18. a. 5^{y-1} b. $y = 4$

19. E

20. A

21. a. $m^{\frac{1}{6}}n^{-\frac{7}{6}}$ or $\sqrt[6]{\frac{m}{n^7}}$

b. $g^{-6}h^3n^{\frac{3}{2}}$

c. $3^{-\frac{7}{3}} \times 5^{-\frac{7}{6}}$

22. a. 2^{-2} or $\frac{1}{4}$ b. a^6b^{-8} or $\frac{a^6}{b^8}$ c. $d^{\frac{14}{15}}$ or $\sqrt[15]{d^{14}}$

23. a. 12 b. 1536 c. 14 days

24. a. 80 koalas

b. During the 6th year.

25. a. 79% b. 56% c. 31%

26. $\frac{3}{2}$

27. $\frac{4}{21}$

28. $z^2 + z^{-2} + \sqrt{5}$

Exercise 1.9 Compound interest

1. a. \$3244.80 b. \$10 939.56

2. a. \$24 617.98 b. \$14 678.02

3. a. \$14 838.45 b. \$129 394.77

4. a. \$932.52 b. \$10 650.81

5. a. \$202 760.57

b. \$25 816.04

6. a. \$3 145 511.41

b. \$443 014.84

7. \$8784.60

8. \$3376.26

9. a. \$2837.04 b. \$837.04

10. \$17 240.15

11. \$605.42

12. \$18 503.86

13. a. 0.5833% b. \$42 891.60

14. B

15. B

16. C

17. C

18. a. 0.0247%

b. ≈ \$131 295.85

c. ≈ \$11 295.85

d. ≈ \$495.85

19. a. i. \$17 745.95

ii. \$17 786.61

iii. \$17 807.67

iv. \$17 821.99

b. The interest added to the principal also earns interest.

20. a. \$7920

b. David's investment = \$8108.46

c. Because David's interest is compounded, the interest is added to the principal each quarter and earns itself interest.

21. a. \$3750 interest

b. i. \$3820.32 interest

ii. \$3881.05

iii. \$3912.36

c. Compound quarterly gives the best return.

d. If we assume that interest is compounded annually, an equivalent return of $I = 7\%$ would be achieved.

e. i. Yes

ii. No

22. Neither is correct. The best option is to choose 3.895% p.a. compounding monthly.

23. a. Bank east b. \$19 976.45

24. a. i. \$2.44 ii. \$2.61

iii. \$2.71 iv. \$2.71

b. Compounding more frequently increases the final value, but the amount of increase becomes less and less. The final value of iv. is 2.7181 which is almost equal to iii.

Exercise 1.10 Logarithms

1. a. $\log_4(16) = 2$ b. $\log_2(32) = 5$

c. $\log_3(81) = 4$ d. $\log_6(36) = 2$

e. $\log_{10}(1000) = 3$

2. a. $\log_5(25) = 2$ b. $\log_4(x) = 3$

c. $\log_5(125) = x$ d. $\log_7(49) = x$

e. $\log_p(16) = 4$

3. a. $\log_9(3) = \frac{1}{2}$ b. $\log_{10}(0.1) = -1$
   c. $\log_8(2) = \frac{1}{3}$ d. $\log_2\left(\frac{1}{2}\right) = -1$
   e. $\log_4(8) = \frac{3}{2}$
4. D
5. a. $2^4 = 16$ b. $3^3 = 27$
   c. $10^6 = 1\,000\,000$ d. $5^3 = 125$
6. a. $16^{\frac{1}{2}} = 4$ b. $4^x = 64$
   c. $49^{\frac{1}{2}} = 7$ d. $3^5 = x$
7. a. $81^{\frac{1}{2}} = 9$ b. $10^{-2} = 0.01$
   c. $8^1 = 8$ d. $64^{\frac{1}{3}} = 4$
8. B
9. a. 4 b. 2
   c. 2 d. 5
10. a. 5 b. 7
    c. 0 d. $\frac{1}{2}$
11. a. -1 b. 1
    c. -2 d. $\frac{1}{3}$
12. a. 0 b. 1 c. 2
13. a. 3 b. 4 c. 5
14. a. 0 and 1 b. 3 and 4 c. 1 and 2
15. a. 4 and 5 b. 2 and 3 c. 4 and 5
16. a. 6.1 b. 6.3 c. 8.2
17. a. $\log_{10}(g) = k$ implies that $g = k$ so $g^2 = \left(10^k\right)^2$. That is, $g^2 = 10^{2k}$, therefore, $\log_{10}\left(g^2\right) = 2k$.
    b. $\log_x(y) = 2$ implies that $y = x^2$, so $x = y^{\frac{1}{2}}$ and therefore $\log_y(x) = \frac{1}{2}$.
    c. The equivalent exponential statement is $x = 4^y$, and we know that 4^y is greater than zero for all values of y. Therefore, x is a positive number.
18. a. 6 b. -4 c. -5
19. a. 5 b. -3 c. $\frac{3}{2}$
20. a. 3 b. 7 c. $\frac{1}{8}$
21. x
22. $\frac{9}{x}$

Exercise 1.11 Logarithm laws

1. a. 1.698 97 b. 1.397 94
   c. 0.698 97 d. 0.301 03
2. Sample responses can be found in the worked solutions in the online resources.
3. a. 1 b. 3
4. a. 2 b. 3
5. a. 4 b. 1
6. a. 2 b. 3
7. a. 1 b. 4
8. a. 3 b. 5
9. a. 2 b. $\frac{1}{2}$
10. a. 1 b. 3
11. a. -1 b. $-\frac{1}{2}$
12. 3
13. a. 2 b. 4
14. a. 3 b. 3
15. a. 1 b. 0
    c. -1 d. 5
16. a. -2 b. 1
    c. 0 d. -2
17. a. $-\frac{1}{2}$ b. $\frac{1}{2}$
    c. $-\frac{1}{2}$ d. $\frac{7}{2}$
18. a. $\log_a(40)$ b. $\log_a(18)$
    c. $\log_x(48)$ d. $\log_x(4)$
19. a. $\log_a(x)$ b. 1
    c. -1 d. 7
20. a. $\frac{1}{2}$ b. $\frac{3}{2}$
    c. -6 d. $-\frac{1}{3}$
21. a. B b. B, D
    c. A, B d. C, D
22. a. $\log_2(80)$ b. $\log_3(105)$
    c. $\log_{10}(100) = 2$ d. $\log_6(56)$
23. a. $\log_2(4) = 2$ b. $\log_3(3) = 1$ c. $\log_5(12.5)$
    d. $\log_2(3)$
24. a. $\log_4(5)$ b. $\log_{10}\left(\frac{1}{4}\right)$
    c. $\log_3(4)$ d. $\log_2(3)$
    e. $\log_3(20)$ f. $\log_4(2) = \frac{1}{2}$
25. a. C b. B c. A
26. a. 12 (Evaluate each logarithm separately and then find the product.)
    b. 4 (First simplify the numerator by expressing 81 as a power of 3.)
    c. 7 (Let $y = 5^{\log_5(7)}$ and write an equivalent statement in logarithmic form.)
27. -2
28. $7 - 3\log_2(3)$
29. 1
30. $x = 3a, 5a$
31. 7

Exercise 1.12 Solving equations

1. a. 25 b. 81 c. $\frac{1}{8}$
   d. $\frac{1}{16}$ e. 100, −100
2. a. 16 b. 26 c. 127
   d. 2 e. 0
3. a. $-\frac{1}{32}$ b. $-\frac{1}{9}$ c. −624
   d. −2.5
4. a. 3 b. 2 c. 125
5. a. 625 b. 2 c. 8
6. a. 6 b. 4
7. a. 3 b. 2 c. −1
8. a. −2 b. $\frac{1}{2}$ c. $\frac{1}{3}$
9. a. 0 b. 0 c. −1
   d. −2
10. a. 5 b. 6 c. 10
    d. 8
11. a. 4 b. 2 c. 9
    d. $\frac{2}{5}$
12. a. 500 b. 128 c. 5
    d. 6 e. 1 f. 2
13. a. B b. A c. D
    d. B
14. a. 7 b. 2 c. −2
    d. 0 e. 4
15. a. $\frac{1}{2}$ b. $\frac{1}{2}$ c. $\frac{3}{2}$
    d. $-\frac{1}{2}$ e. $\frac{3}{2}$
16. a. $\frac{3}{4}$ b. $-\frac{5}{2}$ c. $\frac{5}{2}$
    d. $-\frac{9}{2}$ e. $-\frac{11}{4}$
17. Approximately 2.5 times brighter.
18. a. 120 b. 130
    c. 0.001 d. 3 dB are added.
    e. 10 dB are added. f. 100
19. a. i. 1.1 ii. 1.3 iii. 1.418
    iv. 1.77 v. 2.43 vi. 3.1
    b. No; see answers to 19a i and ii above.
    c. i. 22 387 211 KJ
    ii. 707 945 784 KJ
    iii. 22 387 211 386 KJ.
    d. The energy is increased by a factor of 31.62.
    e. It releases 31.62^3 times more energy.
20. a. $x = 0.7712$ b. $x = 1.2966$
21. $x = 7$
22. $x = 1, 3$

Project

1. a. 100_2 b. 1101_2 c. 10000_2
2. a. 10_2 b. 1_2 c. 110_2
3. a. 11_2 b. 101010_2 c. 110111_2
4. Sample responses can be found in the worked solutions in the online resources. The digits in octal math are 0, 1, 2, 3, 4, 5, 6, and 7. The value "eight" is written as "1 eight and 0 ones", or 108.
5. Sample responses can be found in the worked solutions in the online resources. The numbers 10, 11, 12, 13, 14 and 15 are allocated the letters A, B,C, D, E and F respectively.

Exercise 1.13 Review questions

1. A
2. a. Irrational, since equal to non-recurring and non-terminating decimal
   b. Rational, since can be expressed as a whole number
   c. Rational, since given in a rational form
   d. Rational, since it is a recurring decimal
   e. Irrational, since equal to non-recurring and non-terminating decimal
3. D
4. a. $\sqrt{2m}, \sqrt{\frac{20}{m}}, \sqrt[3]{m}, \sqrt[3]{8m}$
   b. $\sqrt{25m}, \sqrt{\frac{m}{16}}, \sqrt{\frac{20}{m}}$
5. a. $5\sqrt{2}$ b. $6\sqrt{5}$ c. $8\sqrt{2}$ d. $20\sqrt{5}$
6. C
7. a. $72x^3y^4\sqrt{2xy}$ b. $-\frac{1}{4}x^2y^5\sqrt{xy}$
8. a. $25\sqrt{3}$ b. $3ab\sqrt{ab}$
9. a. $\sqrt{15}$ b. $6\sqrt{42}$
   c. $30\sqrt{15}$ d. 5
10. a. 27 b. $720\sqrt{2}$
11. a. $\sqrt{3}$ b. 6
    c. $\frac{\sqrt{10}}{4\sqrt{3}}$ or $\frac{\sqrt{30}}{12}$ d. $\frac{1}{2}$
12. a. $\frac{\sqrt{6}}{3}$ b. $\frac{\sqrt{2}}{4}$
    c. $2\sqrt{5}+4$ d. $2-\sqrt{3}$
13. a. 4 b. 4.5
    c. 2.2 d. 2.7
14. a. 7.4 b. 1.7
    c. 0.8 d. 0.8
15. a. $\sqrt{2}$ b. $3\sqrt{2}$
    c. $5\sqrt{5}$ d. 16
16. a. 1 b. 4
17. a. $\frac{1}{4}$ b. $\frac{1}{9}$
    c. $\frac{1}{16}$ d. $\frac{1}{1000}$

18. a. 0.0833   b. 0.0204
c. 0.800   d. 625

19. a. $1\frac{1}{2}$ b. $1\frac{3}{7}$
c. 5   d. $\frac{4}{13}$

20. a. B   b. A   c. A

21. a. $2\frac{1}{2}$ b. $2\frac{1}{4}$

22. a. $\frac{1}{2}$ b. $\frac{1}{9}$
c. $\frac{1}{64}$ d. $\frac{2}{1}$

23. D
24. C
25. C
26. C
27. A
28. A
29. B
30. C
31. B

32. a. $9x^{10}y^{10}$ b. $\frac{13ab^3c^2}{6}$
c. $\frac{1000m^{15}n^6}{27}$ d. $\frac{16p^{28}}{81q^{12}}$

33. a. 16   b. $-\frac{3}{2}$

34. a. $\frac{8}{a^{11}b^2}$ b. $\frac{y^2}{5x^{17}}$ c. $\frac{m^{12}}{16n^8}$

35. a. 8   b. $\frac{3}{2}$ c. 0

36. a. $30a^{\frac{41}{20}}b^{\frac{33}{20}}$ b. $\frac{4}{x^{\frac{1}{20}}y^{\frac{2}{9}}}$ c. $\frac{2a^{\frac{1}{6}}}{b^{\frac{3}{2}}}$

37. a. 1   b. 4

38. a. $-2a^3 + 2a^2b^{\frac{1}{2}}$
b. $6xy^2$

39. a. $\frac{2a^{13}}{5b^2}$ b. $\frac{9y^4}{32x^{15}}$ c. $2^{\frac{4}{3}}m$

40. a. 46   b. $-\frac{1}{18}$

41. $15 746.40

42. a. $25 808.37   b. $26 723.16
c. $27 211.79   d. $27 550.17

43. a. 2   b. 1
c. 8   d. 2

44. a. $\log_a(24)$ b. $\frac{3}{2}$
c. $\log_a\left(x^2\right)$ or $2\log_a(x)$ d. -5

45. a. 512   b. $\frac{1}{25}$ c. 5
d. 2   e. 6   f. 0

46. a. 6   b. 35

47. a. -2 b. $-\frac{1}{2}$ c. $\frac{5}{2}$

48. a. 4.644   b. -3.809 c. 0.079

49. a. 9   b. 6   c. 0

50. a, b, c

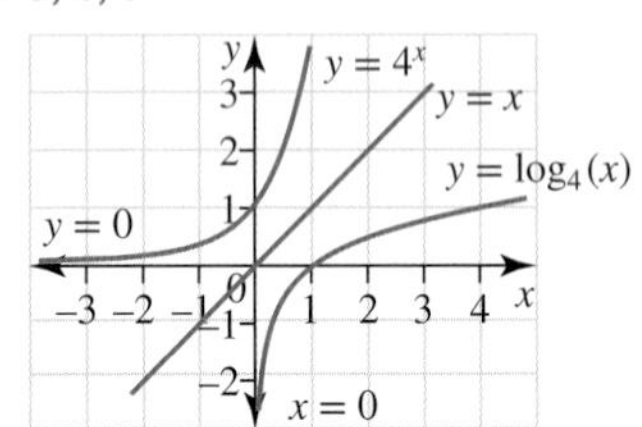

51. $x = 2, -3$

52. $\frac{1}{a^2b^{\frac{1}{2}}}$

53. $\frac{1}{36}$

54. a. 8   b. 6   c. 2

55. a. $\frac{9ab^7c^7}{50}$
b. The student has made two mistakes when squaring the left-hand bracket in line 1 : $3^2 = 9, 5^2 = 25$.

56. a. $\frac{\sqrt{r}}{d\sqrt{u^3}}$
b. Yes, $\sqrt{r}$, $\sqrt{u^3}$
c. $\frac{\sqrt{ru^3}}{du^3}$
d. r should be a perfect square, u should be a perfect cube and d should be a rational number.
e. i. 0.0012346 kL
ii. 1.2346 L
iii. 1234.6 mL
f. A high value for d causes the expression to be smaller, as d only appears on the denominator of the fraction. This means that when d is high there is less water in the reservoir and the area is dry.

57. a. $\frac{p^2\left(2 - \sqrt{w^3}\right)}{4 - w^3}$
b. Sample responses can be found in the worked solutions in the online resources; approximately 5.
c. Speed decreases as wind resistance increases.

2 Algebra and equations

LEARNING SEQUENCE

2.1 Overview

Why learn this?

Algebra is like the language of maths; it holds the key to understanding the rules, formulae and relationships that summarise much of our understanding of the universe. Every maths student needs this set of skills in order to process mathematical information and move on to more challenging concepts.

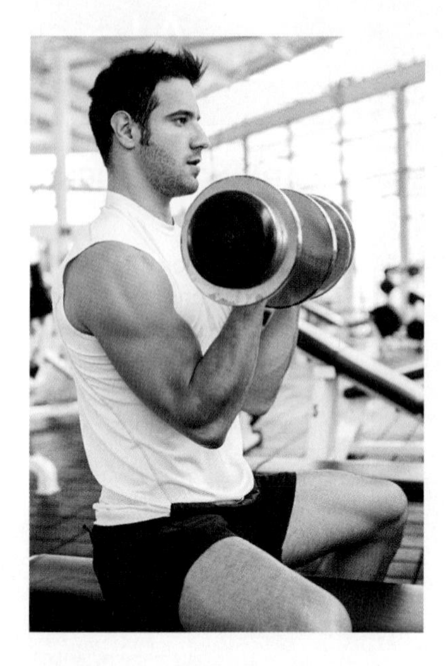

To some extent, this explains why those who want to pursue a career in maths need algebra. Every maths teacher is faced with the question 'Why do *I* need to study algebra, I'm never going to use it?' and yet no one asks why a professional footballer would lift weights when they don't lift any weight in their sport. The obvious answer for the footballer is that they are training their muscles to be fitter and stronger for upcoming matches. Learning algebra is no different, in that you are training your mind to better handle abstract concepts. Abstraction is the ability to consider concepts beyond what we observe. Spatial reasoning, complex reasoning, understanding verbal and non-verbal ideas, recognising patterns, analysing ideas and solving problems all involve abstract thinking to some degree. If some food were to fall on the ground, an adult would think about how long the food has been there, whether the ground is clean, whether the food surface can be washed; whereas a young child would just pick up the food and eat it off the ground, because they lack the ability to think abstractly. Being able to think about all these considerations is just a simple example of abstract thinking. We use abstract thinking every day, and develop this skill over our life. Those who have strong abstract reasoning skills tend to perform highly on intelligence tests and are more likely to be successful in later life. Algebra helps us develop our abstract reasoning skills and thus is of use to all students!

Where to get help

Go to your learnON title at **www.jacplus.com.au** to access the following digital resources. The Online Resources Summary at the end of this topic provides a full list of what's available to help you learn the concepts covered in this topic.

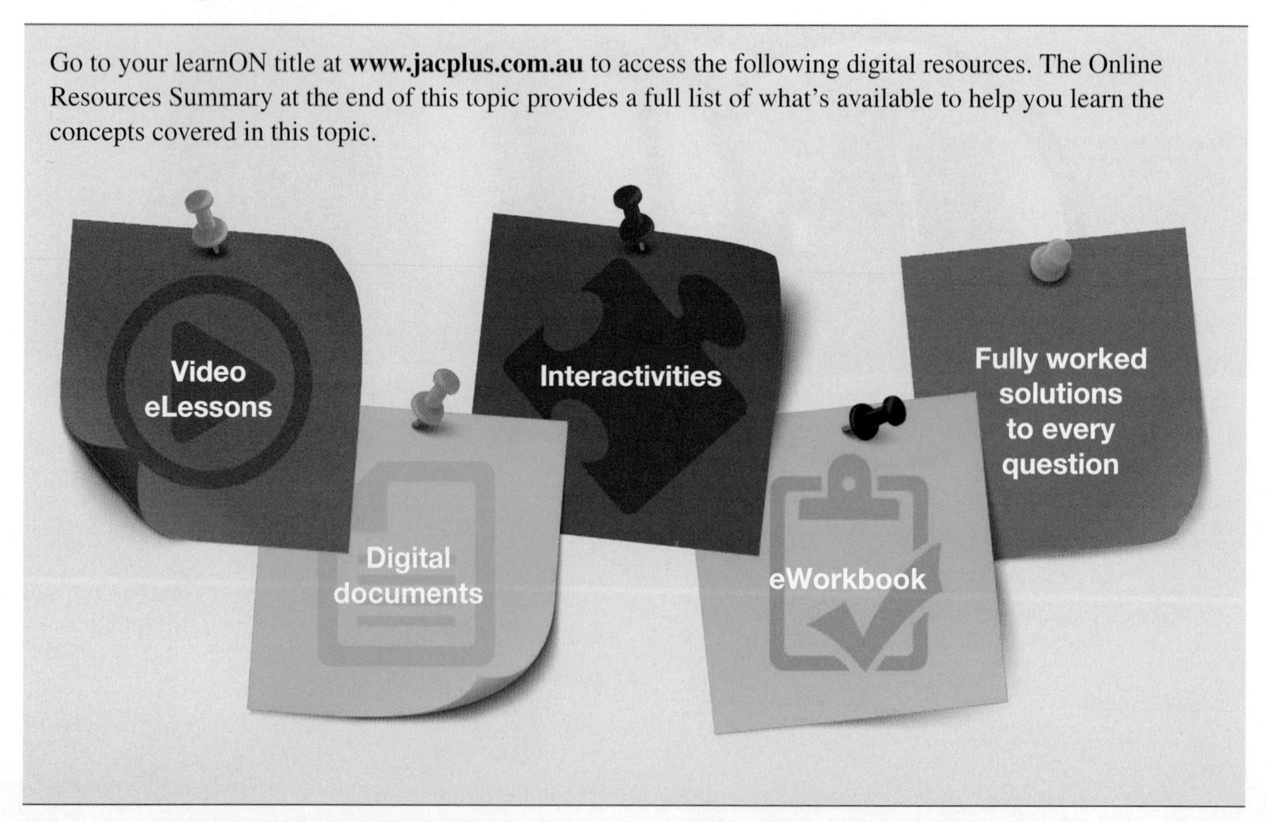

Exercise 2.1 Pre-test

Complete this pre-test in your learnON title at www.jacplus.com.au and receive **automatic marks**, **immediate corrective feedback** and **fully worked solutions**.

1. Evaluate $\frac{(-d)^2}{9c}$, if $c = \frac{1}{3}$ and $d = -6$.

2. If $c = \sqrt{a^2 + b^2}$, calculate c, if $a = 4$ and $b = 3$.

3. **MC** Given the integer values $x = 3$ and $y = -2$, state whether the Closure Law holds for $3y \div x$.
   A. Yes, the answer obtained is an integer value.
   B. No, the answer obtained is a negative integer.
   C. Yes, the answer obtained is a natural number.
   D. No, the answer obtained is irrational.
   E. No, the answer obtained is a terminating decimal.

4. Simplify the following:

$$\frac{y}{5} - \frac{y}{6}$$

5. **MC** The expression $\frac{2}{x+1} - \frac{1}{(x+1)^2}$ can be simplified to:

   A. $\frac{1}{x+1}$ B. $\frac{1}{(x+1)^2}$ C. $\frac{x-1}{(x+1)^2}$ D. $\frac{2x}{(x+1)^2}$ E. $\frac{2x+1}{(x+1)^2}$

6. **MC** The expression $\frac{5}{2x} + \frac{1}{3x}$ simplified is:

   A. $\frac{6}{5x}$ B. $\frac{17}{6x}$ C. $\frac{18}{6x}$ D. $\frac{5}{6x^2}$ E. $\frac{17}{6x^2}$

7. Simplify the expression:

$$\frac{x}{5} \times \frac{-10}{3y}$$

8. If the side length of a cube is x cm, then the cube's volume, V, is given by $V = x^3$. Calculate the side length, in cm, of a cube that has a volume of 1 m^3.

9. Solve the equation $\frac{2(4r+3)}{5} = \frac{3(2r+5)}{4}$.

10. Solve the equation $\frac{8x+3}{5} - \frac{3(x-1)}{2} = \frac{1}{2}$.

11. Solve the equation $\frac{\sqrt[3]{a}}{4} = -2$.

12. At a charity fundraising event, three-eighths of the profit came from sales of tickets, one-fifth came from donations. A third of the profit came from the major raffle and a pop up stall raised $2200. Determine the amount of money raised at the event.

13. **MC** If $\dfrac{x+4}{(x+1)(x-2)} = \dfrac{a}{x+1} + \dfrac{b}{x-2}$, the values of a and b respectively are:

A. $a = x$ and $b = 4$
B. $a = 1$ and $b = 2$
C. $a = -1$ and $b = 2$
D. $a = 1$ and $b = -2$
E. $a = -1$ and $b = -2$

14. **MC** Solve the literal equation $\dfrac{1}{a} + \dfrac{1}{b} = \dfrac{1}{c}$ for a.

A. $a = \dfrac{bc}{b-c}$
B. $a = \dfrac{1}{b-c}$
C. $a = \dfrac{bc}{b+c}$
D. $a = c - b$
E. $a = b + c$

15. **MC** Rearrange the literal equation $m = \dfrac{pa+qb}{p-q}$ to make p the subject.

A. $p = \dfrac{qb}{-a}$
B. $p = \dfrac{q(m-b)}{m+a}$
C. $p = \dfrac{q(m+b)}{m+a}$
D. $p = \dfrac{q(m+b)}{m-a}$
E. $p = \dfrac{a+b}{m+q}$

2.2 Substitution

LEARNING INTENTION

At the end of this subtopic you should be able to:
- evaluate expressions by substituting the numeric values of pronumerals
- understand and apply the Commutative, Associative, Identity and Inverse laws.

2.2.1 Substituting values into expressions

eles-4696

- An expression can be evaluated by substituting the numerical value of pronumerals into an algebraic expression.
- The substituted values are placed in brackets when evaluating an expression.

WORKED EXAMPLE 1 Substituting values into an expression

If $a = 4$, $b = 2$ and $c = -7$, evaluate the following expressions.

a. $a - b$

b. $a^3 + 9b - c$

THINK	WRITE
a. 1. Write the expression.	a. $a - b$
2. Substitute $a = 4$ and $b = 2$ into the expression.	$= 4 - 2$
3. Simplify and write the answer.	$= 2$
b. 1. Write the expression.	b. $a^3 + 9b - c$
2. Substitute $a = 4$, $b = 2$ and $c = -7$ into the expression.	$= (4)^3 + 9(2) - (-7)$
3. Simplify and write the answer.	$= 64 + 18 + 7$ $= 89$

WORKED EXAMPLE 2 Substituting into the Pythagoras theorem

If $c = \sqrt{a^2 + b^2}$, calculate c if $a = 12$ and $b = -5$.

THINK	WRITE
1. Write the expression.	$c = \sqrt{a^2 + b^2}$
2. Substitute $a = 12$ and $b = -5$ into the expression.	$= \sqrt{(12)^2 + (-5)^2}$
3. Simplify.	$= \sqrt{144 + 25}$
	$= \sqrt{169}$
4. Write the answer.	$= 13$

TI | THINK

In a new document, open a calculator page. To substitute values, use the symbol |. Press CTRL and then = to bring up the palette; use the Touchpad to select the | symbol. Then type 'and' or find it in the CATALOG.
Complete the entry line as:
$c = \sqrt{a^2 + b^2} \mid a = 12$ and $b = -5$
Then press ENTER.

DISPLAY/WRITE

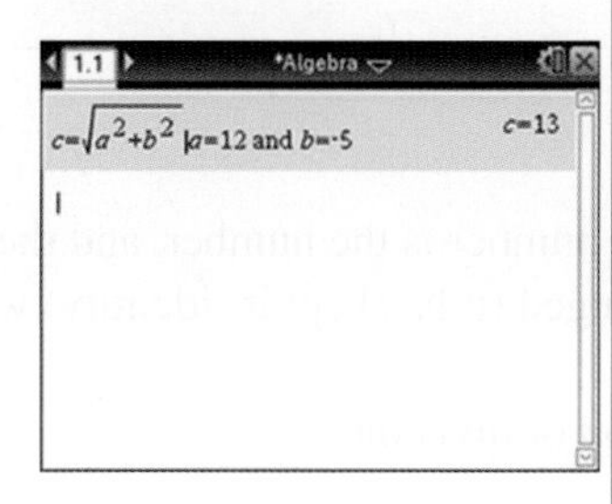

If $a = 12$ and $b = -5$, then $c = \sqrt{a^2 + b^2} = 13$.

CASIO | THINK

To type the equation, $\sqrt{\ }$ is on the Keyboard Math1 screen. The vertical line | is on the Keyboard Math3 screen.
Complete the entry line as:
$c = \sqrt{a^2 + b^2} \mid a = 12$ and $b = -5$
Then press EXE.

DISPLAY/WRITE

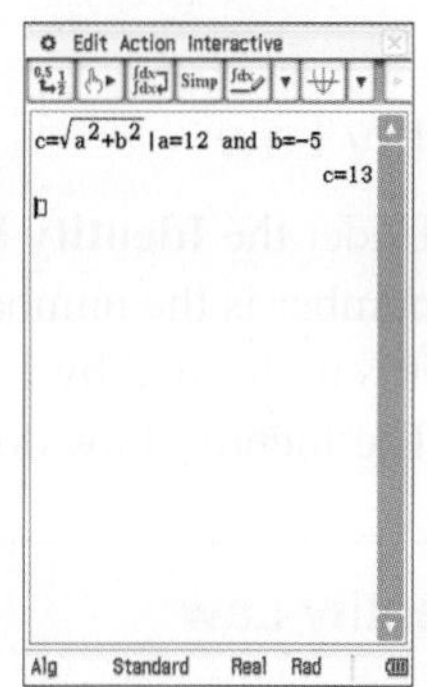

If $a = 12$ and $b = -5$ then $c = \sqrt{a^2 + b^2} = 13$.

2.2.2 Number laws

eles-4697

- Recall from previous studies that when dealing with numbers and pronumerals, particular rules must be obeyed. Before progressing further, let us briefly review the Commutative, Associative, Identity and Inverse Laws.
- Consider any three pronumerals, y and z, where x, y and z are elements of the set of real numbers.

Commutative Law

- The **Commutative Law** holds true for addition and multiplication. That is, you can add or multiply in any order, since the order in which two numbers or pronumerals are added or multiplied does not affect the result.
- The Commutative Law does not hold true for subtraction or division.

Commutative Law

$x + y = y + x$	For example: $3 + 2 = 5$ and $2 + 3 = 5$
$x - y \neq y - x$	For example: $3 - 2 = 1$ but $2 - 3 = -1$
$x \times y = y \times x$	For example: $3 \times 2 = 6$ and $2 \times 3 = 6$
$x \div y \neq y \div x$	For example: $3 \div 2 = \frac{3}{2}$, but $2 \div 3 = \frac{2}{3}$

Associative Law

- The **Associative Law** holds true for addition and multiplication since grouping two or more numbers or pronumerals and calculating them in a different order does not affect the result.
- The Associative Law does not hold true for subtraction or division.

Associative Law

$x+(y+z)=(x+y)+z$ For example: $2+(3+4)=2+7=9$ and $(2+3)+4=5+4=9$

$x-(y-z)\neq(x-y)-z$ For example: $2-(3-4)=2--1=3$ but $(2-3)-4=-1-4=-5$

$x\times(y\times z)=(x\times y)\times z$ For example: $2\times(3\times4)=2\times12=24$ and $(2\times3)\times4=6\times4=24$

$x\div(y\div z)\neq(x\div y)\div z$ For example: $2\div(3\div4)=2\div\frac{3}{4}=2\times\frac{4}{3}=\frac{8}{3}$ but

$(2\div3)\div4=\frac{2}{3}\div4=\frac{2}{3}\times\frac{1}{4}=\frac{2}{12}=\frac{1}{6}$

Identity Law

- Under the **Identity Law**, the sum of zero and any number is the number, and the product of 1 and any number is the number. That is, x has not been changed (it has kept its identity) when zero is added to it or it is multiplied by 1.
- The Identity Law does not hold true for subtraction or division.

Identity Law

$\boldsymbol{x+0=0+x=x}$ For example: $5+0=0+5=5$

$\boldsymbol{x-0\neq0-x}$ For example: $5-0=5$ and $0-5=-5$

$\boldsymbol{x\times1=1\times x=x}$ For example: $7\times1=1\times7=7$

$\boldsymbol{x\div1\neq1\div x}$ For example: $8\div1=8$ and $1\div8=\frac{1}{8}$

Inverse Law

- The inverse of a real number x under addition is $-x$.
- The inverse of a real number x under multiplication is its reciprocal, $\frac{1}{x}$.
- The **Inverse Law** states that in general:
  - when the additive inverse of a number or pronumeral is added to itself, it equals 0.
  - when the multiplicative inverse of a number or pronumeral is multiplied by itself, it equals 1.

Inverse Law

$\boldsymbol{x+-x=-x+x=0}$ For example: $5+-5=-5+5=0$

$\boldsymbol{x\times\frac{1}{x}=\frac{1}{x}\times x=1}$ For example: $7\times\frac{1}{7}=\frac{1}{7}\times7=1$

- It is worth noting that the subtraction $(5-2=3)$ is equivalent to adding an inverse $(5+(-2)=3)$, and that the division $(10\div2=5)$ is equivalent to multiplication by an inverse $(10\times\frac{1}{2}=5)$.

Closure Law

- The **Closure Law** states that, when an operation is performed on an element (or elements) of a set, the result produced must also be an element of that set.
  For example, addition is closed on natural numbers (that is, positive integers: $1, 2, 3, \ldots$) since adding a pair of natural numbers produces a natural number.
- Subtraction is not closed on natural numbers.
  For example, 5 and 7 are natural numbers and the result of adding them is 12, a natural number. However, the result of subtracting 7 from 5 is -2, which is not a natural number.

WORKED EXAMPLE 3 Determining which operations with the integers are closed

Determine the value of the following expressions, given the integer values $x = 4$ and $y = -12$. Comment on whether the Closure Law for integers holds for each of the expressions when these values are substituted.

a. $x + y$ b. $x - y$ c. $x \times y$ d. $x \div y$

THINK	WRITE
a. 1. Substitute each pronumeral into the expression.	a. $x + y = 4 + (-12)$
2. Evaluate and write the answer.	$= -8$
3. Determine whether the Closure Law holds; that is, is the result an integer?	The Closure Law holds for these substituted values.
b. Repeat steps 1–3 of part a.	b. $x - y = 4 - (-12)$ $= 16$ The Closure Law holds for these substituted values.
c. Repeat steps 1–3 of part a.	c. $x \times y = 4 \times (-12)$ $= -48$ The Closure Law holds for these substituted values.
d. Repeat steps 1–3 of part a.	d. $x \div y = 4 \div (-12)$ $= \frac{4}{-12}$ $= -\frac{1}{3}$ The Closure Law does not hold for these substituted values since the answer obtained is a fraction, not an integer.

- It is important to note that, although a particular set of numbers may be closed under a given operation, for example multiplication, another set of numbers may not be closed under that same operation.
  For example, in part c of Worked example 3, integers were closed under multiplication.
- In some cases, however, the set of *irrational numbers* is not closed under multiplication, since $\sqrt{3} \times \sqrt{3} = \sqrt{9} = 3$. In this example, two irrational numbers produced a rational number under multiplication.

eWorkbook	Topic 2 Workbook (worksheets, code puzzle and project) (ewbk-2028)
Digital documents	SkillSHEET Like terms (doc-5183) SkillSHEET Collecting like terms (doc-5184) SkillSHEET Finding the highest common factor (doc-5185) SkillSHEET Order of operations (doc-5189)
Video eLesson	Substitution (eles-1892)
Interactivities	Individual pathway interactivity: Substitution (int-4566) Substituting positive and negative numbers (int-3765) Commutative Law (int-6109) Associative Law (int-6110) Identity Law (int-6111) Inverse Law (int-6112)

Exercise 2.2 Substitution

learnon

Individual pathways

■ PRACTISE	■ CONSOLIDATE	■ MASTER
1, 4, 8, 10, 13, 14, 17, 22	2, 5, 9, 11, 15, 18, 19, 23	3, 6, 7, 12, 16, 20, 21, 24

To answer questions online and to receive **immediate corrective feedback** and **fully worked solutions** for all questions, go to your learnON title at www.jacplus.com.au.

Fluency

WE1 For questions 1 to 3, if $a = 2$, $b = 3$ and $c = 5$, evaluate the following expressions.

1. a. $a + b$ b. $c - b$ c. $c - a - b$ d. $c - (a - b)$

2. a. $7a + 8b - 11c$ b. $\frac{a}{2} + \frac{b}{3} + \frac{c}{5}$ c. abc d. $ab(c - b)$

3. a. $a^2 + b^2 - c^2$ b. $c^2 + a$ c. $-a \times b \times -c$ d. $2.3a - 3.2b$

For questions 4 to 6, if $d = -6$ and $k = -5$, evaluate the following.

4. a. $d + k$ b. $d - k$ c. $k - d$

5. a. kd b. $-d(k + 1)$ c. d^2

6. a. k^3 b. $\frac{k - 1}{d}$ c. $3k - 5d$

7. If $x = \frac{1}{3}$ and $y = \frac{1}{4}$, evaluate the following.

   a. $x + y$ b. $y - x$ c. xy

   d. $\frac{x}{y}$ e. x^2y^3 f. $\frac{9x}{y^2}$

8. If $x = 3$, determine the value of the following.

   a. x^2 b. $-x^2$ c. $(-x)^2$

   d. $2x^2$ e. $-2x^2$ f. $(-2x)^2$

9. If $x = -3$, determine the value of the following.

a. x^2
b. $-x^2$
c. $(-x)^2$
d. $2x^2$
e. $-2x^2$
f. $(-2x)^2$

WE2 For questions 10 to 12, calculate the unknown variable in the following real-life mathematical formulas.

10. a. If $c = \sqrt{a^2 + b^2}$, calculate c if $a = 8$ and $b = 15$.

b. If $A = \frac{1}{2}bh$, determine the value of A if $b = 12$ and $h = 5$.

c. The perimeter, P, of a rectangle is given by $P = 2L + 2W$. Calculate the perimeter, P, of a rectangle, given $L = 1.6$ and $W = 2.4$.

11. a. If $T = \frac{C}{L}$, determine the value of T if $C = 20.4$ and $L = 5.1$.

b. If $K = \frac{n+1}{n-1}$, determine the value of K if $n = 5$.

c. Given $F = \frac{9C}{5} + 32$, calculate F if $C = 20$.

12. a. If $v = u + at$, evaluate v if $u = 16$, $a = 5$, $t = 6$.

b. The area, A, of a circle is given by the formula $A = \pi r^2$. Calculate the area of a circle, correct to 1 decimal place, if $r = 6$.

c. If $E = \frac{1}{2}mv^2$, calculate m if $E = 40$, $v = 4$.

d. Given $r = \sqrt{\frac{A}{\pi}}$, evaluate A to 1 decimal place if $r = 14.1$.

13. MC a. If $p = -5$ and $q = 4$, then pq is equal to:

A. 20
B. 1
C. −1
D. −20
E. $-\frac{5}{4}$

b. If $c^2 = a^2 + b^2$, and $a = 6$ and $b = 8$, then c is equal to:

A. 28
B. 100
C. 10
D. 14
E. 44

c. Given $h = 6$ and $k = 7$, then kh^2 is equal to:

A. 294
B. 252
C. 1764
D. 5776
E. 85

Understanding

14. Knowing the length of two sides of a right-angled triangle, the third side can be calculated using Pythagoras' theorem. If the two shorter sides have lengths of 1.5 cm and 3.6 cm, calculate the length of the hypotenuse.

15. The volume of a sphere can be calculated using the formula $\frac{4}{3}\pi r^3$. What is the volume of a sphere with a radius of 2.5 cm? Give your answer correct to 2 decimal places.

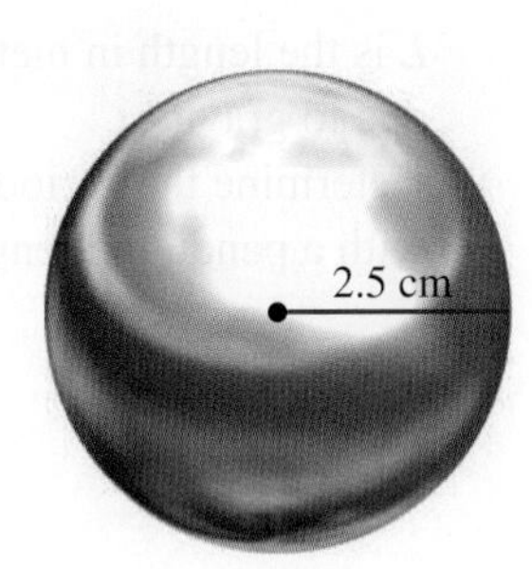

16. A rectangular park is 200 m by 300 m. If Blake runs along the diagonal of the park, calculate how far he will run. Give your answer to the nearest metre.

Reasoning

17. WE3 Determine the value of the following expressions, given the integer values $x = 1$, $y = -2$ and $z = -1$. Comment on whether the Closure Law for integers holds true for each of the expressions when these values are substituted.

a. $x + y$
b. $y - z$
c. $y \times z$

18. Determine the value of the following expressions, given the integer values $x = 1$, $y = -2$ and $z = -1$. Comment on whether the Closure Law for integers holds true for each of the expressions when these values are substituted.

a. $x \div z$ b. $z - x$ c. $x \div y$

19. For each of the following, complete the relationship to illustrate the stated law. Justify your reasoning.

a. $(a + 2b) + 4c =$ ____________ Associative Law
b. $(x \times 3y) \times 5c =$ ____________ Associative Law
c. $2p \div q \neq$ ____________ Commutative Law
d. $5d + q =$ ____________ Commutative Law

20. Calculate the value of the following expressions, given the natural number values $x = 8$, $y = 2$ and $z = 6$. Comment on whether the Closure Law for natural numbers holds true for each of the expressions.

a. $x + y$ b. $y - z$ c. $y \times z$
d. $x \div z$ e. $z - x$ f. $x \div y$

21. For each of the following, complete the relationship to illustrate the stated law. Justify your reasoning.

a. $3z + 0 =$ ____________ Identity Law
b. $2x \times$ ________ = ________ Inverse Law
c. $(4x \div 3y) \div 5z \neq$ ____________ Associative Law
d. $3d - 4y \neq$ ____________ Commutative Law

Problem solving

22. $s = ut + \frac{1}{2}at^2$ where t is the time in seconds, s is the displacement in metres, u is the initial velocity and a is the acceleration due to gravity.

a. Calculate s when $u = 16.5$ m/s, $t = 2.5$ seconds and $a = 9.8$ m/s^2.
b. A body has an initial velocity of 14.7 m/s and after t seconds has a displacement of 137.2 metres. Determine the value of t if $a = 9.8$ m/s^2.

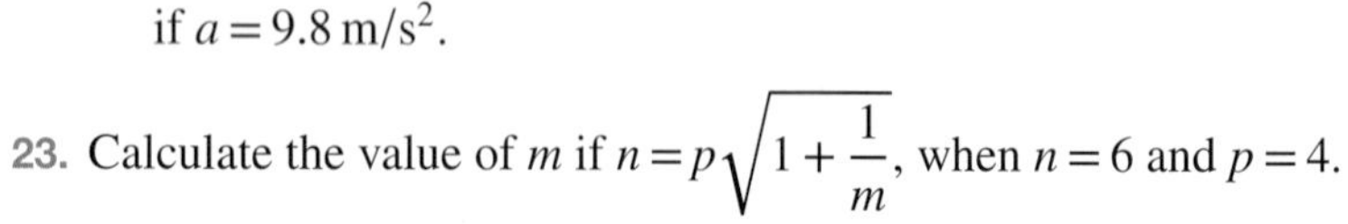

23. Calculate the value of m if $n = p\sqrt{1 + \frac{1}{m}}$, when $n = 6$ and $p = 4$.

24. The formula for the period (T) of a pendulum in seconds is $T = 2\pi\sqrt{\frac{L}{g}}$, where L is the length in metres of the pendulum and $g = 9.81$ m/s^2 is the acceleration due to gravity.
Determine the period of a pendulum, to 1 decimal place, in a grandfather clock with a pendulum length of 154 cm.

2.3 Adding and subtracting algebraic fractions

LEARNING INTENTION

At the end of this subtopic you should be able to:
- determine the lowest common denominator of two or more fractions with pronumerals in the denominator
- add and subtract fractions involving algebraic expressions.

2.3.1 Algebraic fractions

eles-4698

- In an algebraic fraction, the denominator, the numerator or both are algebraic expressions.
  For example, $\frac{x}{2}$, $\frac{3x+1}{2x-5}$ and $\frac{1}{x^2+5}$ are all **algebraic fractions**.
- As with all fractions, algebraic fractions must have a common denominator if they are to be added or subtracted, so an important step is to determine the lowest common denominator (LCD).

WORKED EXAMPLE 4 Simplifying fractions with algebraic numerators

Simplify the following expressions.

a. $\frac{2x}{3}-\frac{x}{2}$

b. $\frac{x+1}{6}+\frac{x+4}{4}$

THINK	WRITE
a. 1. Write the expression.	a. $\frac{2x}{3}-\frac{x}{2}$
2. Rewrite each fraction as an equivalent fraction using the LCD of 3 and 2, which is 6.	$=\frac{2x}{3}\times\frac{2}{2}-\frac{x}{2}\times\frac{3}{3}$ $=\frac{4x}{6}-\frac{3x}{6}$
3. Express as a single fraction.	$=\frac{4x-3x}{6}$
4. Simplify the numerator and write the answer.	$=\frac{x}{6}$
b. 1. Write the expression.	b. $\frac{x+1}{6}+\frac{x+4}{4}$
2. Rewrite each fraction as an equivalent fraction using the LCD of 6 and 4, which is 12.	$=\frac{x+1}{6}\times\frac{2}{2}+\frac{x+4}{4}\times\frac{3}{3}$ $=\frac{2(x+1)}{12}+\frac{3(x+4)}{12}$
3. Express as a single fraction.	$=\frac{2(x+1)+3(x+4)}{12}$
4. Simplify the numerator by expanding brackets and collecting like terms.	$=\frac{2x+2+3x+12}{12}$
5. Write the answer.	$=\frac{5x+14}{12}$

2.3.2 Pronumerals in the denominator

eles-4699

- If pronumerals appear in the denominator, the process involved in adding and subtracting the fractions is to determine a lowest common denominator as usual.
- When there is an algebraic expression in the denominator of each fraction, a common denominator can be obtained by writing the product of the denominators. For example, if $x+3$ and $2x-5$ are in the denominator of each fraction, then a common denominator of the two fractions will be $(x+3)(2x-5)$.

WORKED EXAMPLE 5 Simplifying fractions with algebraic denominators

Simplify $\frac{2}{3x}-\frac{1}{4x}$.

THINK	WRITE
1. Write the expression.	$\frac{2}{3x}-\frac{1}{4x}$
2. Rewrite each fraction as an equivalent fraction using the LCD of $3x$ and $4x$, which is $12x$. *Note:* $12x^2$ is not the lowest LCD.	$=\frac{2}{3x}\times\frac{4}{4}-\frac{1}{4x}\times\frac{3}{3}$ $=\frac{8}{12x}-\frac{3}{12x}$
3. Express as a single fraction.	$=\frac{8-3}{12x}$
4. Simplify the numerator and write the answer.	$=\frac{5}{12x}$

WORKED EXAMPLE 6 Simplifying by finding the LCD of two algebriac expressions

Simplify $\frac{x+1}{x+3}+\frac{2x-1}{x+2}$ by writing it first as a single fraction.

THINK	WRITE
1. Write the expression.	$\frac{x+1}{x+3}+\frac{2x-1}{x+2}$
2. Rewrite each fraction as an equivalent fraction using the LCD of $x+3$ and $x+2$, which is the product $(x+3)(x+2)$.	$=\frac{(x+1)}{(x+3)}\times\frac{(x+2)}{(x+2)}+\frac{(2x-1)}{(x+2)}\times\frac{(x+3)}{(x+3)}$ $=\frac{(x+1)(x+2)}{(x+3)(x+2)}+\frac{(2x-1)(x+3)}{(x+3)(x+2)}$
3. Express as a single fraction.	$=\frac{(x+1)(x+2)+(2x-1)(x+3)}{(x+3)(x+2)}$
4. Simplify the numerator by expanding brackets and collecting like terms. *Note:* The denominator is generally kept in factorised form. That is, it is not expanded.	$=\frac{\left(x^2+2x+x+2\right)+\left(2x^2+6x-x-3\right)}{(x+3)(x+2)}$ $=\frac{\left(x^2+3x+2+2x^2+5x-3\right)}{(x+3)(x+2)}$
5. Write the answer.	$=\frac{3x^2+8x-1}{(x+3)(x+2)}$

WORKED EXAMPLE 7 Simplification involving repeated linear factors

Simplify $\dfrac{x+2}{x-3}+\dfrac{x-1}{(x-3)^2}$ by writing it first as a single fraction.

THINK	WRITE
1. Write the expression.	$\dfrac{x+2}{x-3}+\dfrac{x-1}{(x-3)^2}$
2. Rewrite each fraction as an equivalent fraction using the LCD of $x-3$ and $(x-3)^2$, which is $(x-3)^2$.	$=\dfrac{x+2}{x-3}\times\dfrac{x-3}{x-3}+\dfrac{x-1}{(x-3)^2}$ $=\dfrac{(x+2)(x-3)}{(x-3)^2}+\dfrac{x-1}{(x-3)^2}$ $=\dfrac{x^2-x-6}{(x-3)^2}+\dfrac{x-1}{(x-3)^2}$
3. Express as a single fraction.	$=\dfrac{x^2-x-6+x-1}{(x-3)^2}$
4. Simplify the numerator and write the answer.	$=\dfrac{x^2-7}{(x-3)^2}$

TI | THINK	DISPLAY/WRITE
On a Calculator page, press CTRL and ÷ to get the fraction template, and then complete the entry line as: $\dfrac{x+2}{x-3}+\dfrac{x-1}{(x-3)^2}$ Then press ENTER.	$\dfrac{x+2}{x-3}+\dfrac{x-1}{(x-3)^2}=\dfrac{x^2-7}{(x-3)^2}$

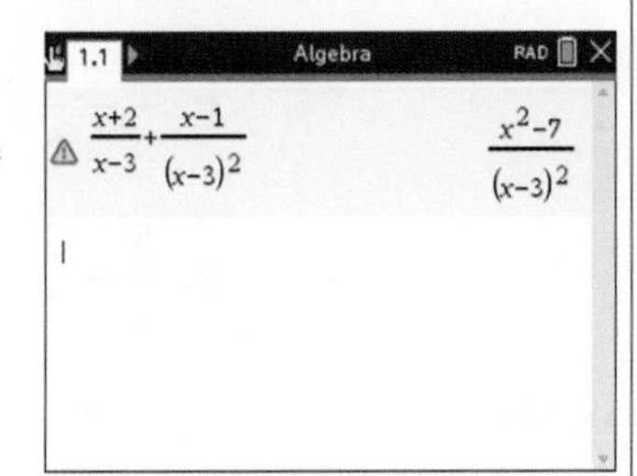

CASIO | THINK	DISPLAY/WRITE
On the Main screen, complete the entry line as: combine$\left(\dfrac{x+2}{x-3}+\dfrac{x-1}{(x-3)^2}\right)$ Then press EXE.	$\dfrac{x+2}{x-3}+\dfrac{x-1}{(x-3)^2}=\dfrac{x^2-7}{(x-3)^2}$

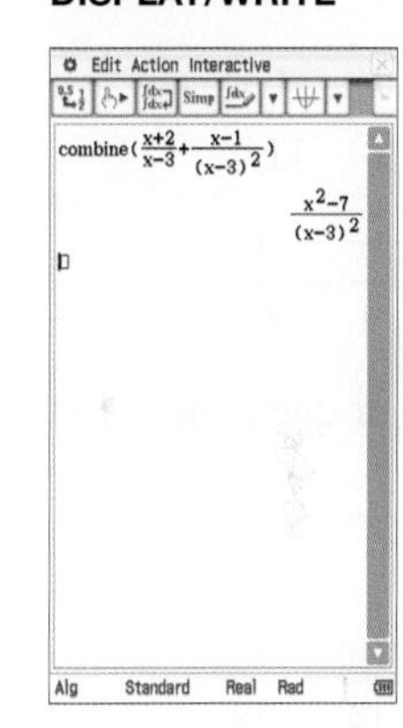

DISCUSSION

Explain why we can't just add the numerators and the denominators of fractions, as shown in the incorrect expression below:

$$\frac{a}{b}+\frac{c}{d}=\frac{a+c}{b+d}$$

Resources

eWorkbook — Topic 2 Workbook (worksheets, code puzzle and project) (ewbk-2028)

Digital documents — SkillSHEET Addition and subtraction of fractions (doc-5186)
SkillSHEET Writing equivalent algebraic fractions with the lowest common denominator (doc-5190)

Interactivities — Individual pathway interactivity: Adding and subtracting algebraic fractions (int-4567)
Adding and subtracting algebraic fractions (int-6113)
Lowest common denominators with pronumerals (int-6114)

Exercise 2.3 Adding and subtracting algebraic fractions

learnON

Individual pathways

PRACTISE	CONSOLIDATE	MASTER
1, 4, 7, 10, 13, 16, 19	2, 5, 8, 11, 14, 17, 20	3, 6, 9, 12, 15, 18, 21

To answer questions online and to receive **immediate corrective feedback** and **fully worked solutions** for all questions, go to your learnON title at www.jacplus.com.au.

Fluency

For questions 1 to 3, simplify each of the following.

1. a. $\frac{4}{7}+\frac{2}{3}$ b. $\frac{1}{8}+\frac{5}{9}$ c. $\frac{3}{5}+\frac{6}{15}$

2. a. $\frac{4}{9}-\frac{3}{11}$ b. $\frac{3}{7}-\frac{2}{5}$ c. $\frac{1}{5}-\frac{x}{6}$

3. a. $\frac{5x}{9}-\frac{4}{27}$ b. $\frac{3}{8}-\frac{2x}{5}$ c. $\frac{5}{x}-\frac{2}{3}$

WE4 For questions 4 to 6, simplify the following expressions.

4. a. $\frac{2y}{3}-\frac{y}{4}$ b. $\frac{y}{8}-\frac{y}{5}$ c. $\frac{4x}{3}-\frac{x}{4}$ d. $\frac{8x}{9}+\frac{2x}{3}$

5. a. $\frac{2w}{14}-\frac{w}{28}$ b. $\frac{y}{20}-\frac{y}{4}$ c. $\frac{12y}{5}+\frac{y}{7}$ d. $\frac{10x}{5}+\frac{2x}{15}$

6. a. $\frac{x+1}{5}+\frac{x+3}{2}$ b. $\frac{x+2}{4}+\frac{x+6}{3}$ c. $\frac{2x-1}{5}-\frac{2x+1}{6}$ d. $\frac{3x+1}{2}+\frac{5x+2}{3}$

WE5 For questions 7 to 9, simplify the following.

7. a. $\frac{2}{4x}+\frac{1}{8x}$ b. $\frac{3}{4x}-\frac{1}{3x}$ c. $\frac{5}{3x}+\frac{1}{7x}$

8. a. $\frac{12}{5x}+\frac{4}{15x}$ b. $\frac{1}{6x}+\frac{1}{8x}$ c. $\frac{9}{4x}-\frac{9}{5x}$

9. a. $\frac{2}{100x}+\frac{7}{20x}$ b. $\frac{1}{10x}+\frac{5}{x}$ c. $\frac{4}{3x}-\frac{3}{2x}$

WE6,7 For questions 10 to 12, simplify the following by writing as single fractions.

10. a. $\frac{2}{x+4}+\frac{3x}{x-2}$ b. $\frac{2x}{x+5}+\frac{5}{x-1}$ c. $\frac{5}{2x+1}+\frac{x}{x-2}$ d. $\frac{2x}{x+1}-\frac{3}{2x-7}$

11. a. $\dfrac{4x}{x+7}+\dfrac{3x}{x-5}$ b. $\dfrac{x+2}{x+1}+\dfrac{x-1}{x+4}$ c. $\dfrac{x+8}{x+1}-\dfrac{2x+1}{x+2}$ d. $\dfrac{x+5}{x+3}-\dfrac{x-1}{x-2}$

12. a. $\dfrac{x+1}{x+2}-\dfrac{2x-5}{3x-1}$ b. $\dfrac{2}{x-1}-\dfrac{3}{1-x}$ c. $\dfrac{4}{(x+1)^2}+\dfrac{3}{x+1}$ d. $\dfrac{3}{x-1}-\dfrac{1}{(x-1)^2}$

Understanding

13. A classmate attempted to complete an algebraic fraction subtraction problem.

$$\begin{aligned}\frac{x}{x-1}-\frac{3}{x-2}&=\frac{x}{x-1}\times\frac{(x-2)}{(x-2)}-\frac{3}{x-2}\times\frac{(x-1)}{(x-1)}\\&=\frac{x(x-2)-3(x-1)}{(x-1)(x+2)}\\&=\frac{x^2-2x-3x-1}{(x-1)(x+2)}\\&=\frac{x^2-5x-1}{(x-1)(x+2)}\end{aligned}$$

a. Identify the mistake she made.
b. Determine the correct answer.

14. Simplify the following.

a. $\dfrac{y-x}{x-y}$ b. $\dfrac{3}{x-2}+\dfrac{4}{2-x}$

15. Simplify the following.

a. $\dfrac{3}{3-x}+\dfrac{3x}{(x-3)^2}$ b. $\dfrac{1}{x-2}-\dfrac{2x}{(2-x)^2}+\dfrac{x^2}{(x-2)^3}$

Reasoning

16. Simplify the following.

a. $\dfrac{1}{x+2}+\dfrac{2}{x+1}+\dfrac{1}{x+3}$

b. $\dfrac{1}{x-1}+\dfrac{4}{x+2}+\dfrac{2}{x-4}$

17. Simplify the following.

a. $\dfrac{3}{x+1}+\dfrac{2}{x+3}-\dfrac{1}{x+2}$

b. $\dfrac{2}{x-4}-\dfrac{3}{x-1}+\dfrac{5}{x+3}$

c. Explain why the process that involves determining the lowest common denominator is important in parts **a** and **b**.

18. The reverse process of adding or subtracting algebraic fractions is quite complex. Use trial and error, or technology, to determine the value of a if $\dfrac{7x-4}{(x-8)(x+5)}=\dfrac{a}{x-8}+\dfrac{3}{x+5}$.

Problem solving

19. Simplify $\dfrac{3}{x^2+7x+12}-\dfrac{1}{x^2+x-6}+\dfrac{2}{x^2+2x-8}$.

20. Simplify $\dfrac{x^2+3x-18}{x^2-x-42}-\dfrac{x^2-3x+2}{x^2-5x+4}$.

21. Simplify $\dfrac{x^2-25}{x^2-2x-15}+\dfrac{x^2+12x+32}{x^2+4x-32}-\dfrac{2x^2}{x^2-x-12}$.

2.4 Multiplying and dividing algebraic fractions

LEARNING INTENTION

At the end of this subtopic you should be able to:
- cancel factors, including algebraic expressions, that are common to the numerator and denominator of fractions
- multiply and divide fractions involving algebraic expression and simplify the result.

2.4.1 Multiplying algebraic fractions

eles-4700

- Algebraic fractions can be simplified using the index laws and by cancelling factors common to the numerator and denominator.
- A fraction can only be simplified if:
  - there is a common factor in the numerator and the denominator
  - the numerator and denominator are both written in factorised form, that is, as the *product* of two or more factors.

$$\frac{3ab}{12a}=\frac{{}^{1}\cancel{3}\times{}^{1}\cancel{a}\times b}{{}^{4}\cancel{12}\times{}^{1}\cancel{a}}\begin{matrix}\leftarrow \text{product of factors}\\ \leftarrow \text{product of factors}\end{matrix}$$
$$=\frac{b}{4}$$

$$\frac{3a+b}{12a}=\frac{3\times a+b}{12\times a}\begin{matrix}\leftarrow \text{not a product of factors}\\ \leftarrow \text{product of factors}\end{matrix}$$

Cannot be simplified

- Multiplication of algebraic fractions follows the same rules as multiplication of numerical fractions: multiply the numerators, then multiply the denominators.

WORKED EXAMPLE 8 Multiplying algebraic fractions and simplifying the result

Simplify each of the following.

a. $\dfrac{5y}{3x}\times\dfrac{6z}{7y}$

b. $\dfrac{2x}{(x+1)(2x-3)}\times\dfrac{x+1}{x}$

THINK	WRITE
a. 1. Write the expression.	a. $\dfrac{5y}{3x}\times\dfrac{6z}{7y}$

2. Cancel common factors in the numerator and denominator. The y can be cancelled in the denominator and the numerator. Also, the 3 in the denominator can divide into the 6 in the numerator.

$$= \frac{5y^1}{_1 3x} \times \frac{6^2 z}{7y^1}$$
$$= \frac{5}{x} \times \frac{2z}{7}$$

3. Multiply the numerators, then multiply the denominators and write the answer.

$$= \frac{10z}{7x}$$

b. 1. Write the expression.

b. $$\frac{2x}{(x+1)(2x-3)} \times \frac{x+1}{x}$$

2. Cancel common factors in the numerator and the denominator. $(x + 1)$ and the x are both common in the numerator and the denominator and can therefore be cancelled.

$$= \frac{2x^1}{^1(x+1)(2x-3)} \times \frac{x+1^1}{x^1}$$
$$= \frac{2}{2x-3} \times \frac{1}{1}$$

3. Multiply the numerators, then multiply the denominators and write the answer.

$$= \frac{2}{2x-3}$$

2.4.2 Dividing algebraic fractions

eles-4701

- When dividing algebraic fractions, follow the same rules as for division of numerical fractions: write the division as a multiplication and invert the second fraction.
- This process is sometimes known as multiplying by the **reciprocal**.

WORKED EXAMPLE 9 Dividing algebraic fractions

Simplify the following expressions.

a. $\frac{3xy}{2} \div \frac{4x}{9y}$ **b.** $\frac{4}{(x+1)(3x-5)} \div \frac{x-7}{x+1}$

THINK / **WRITE**

a. 1. Write the expression.

a. $$\frac{3xy}{2} \div \frac{4x}{9y}$$

2. Change the division sign to a multiplication sign and write the second fraction as its reciprocal.

$$= \frac{3xy}{2} \times \frac{9y}{4x}$$

3. Cancel common factors in the numerator and denominator. The pronumeral x is common to both the numerator and denominator and can therefore be cancelled.

$$= \frac{3y}{2} \times \frac{9y}{4}$$

4. Multiply the numerators, then multiply the denominators and write the answer.

$$= \frac{27y^2}{8}$$

b. 1. Write the expression.

b. $$\frac{4}{(x+1)(3x-5)} \div \frac{x-7}{x+1}$$

2. Change the division sign to a multiplication sign and write the second fraction as its reciprocal.

$$= \frac{4}{(x+1)(3x-5)} \times \frac{x+1}{x-7}$$

3. Cancel common factors in the numerator and denominator. $(x+1)$ is common to both the numerator and denominator and can therefore be cancelled.

$$= \frac{4}{3x-5} \times \frac{1}{x-7}$$

4. Multiply the numerators, then multiply the denominators and write the answer.

$$= \frac{4}{(3x-5)(x-7)}$$

TI | THINK

a.

On a Calculator page the fraction template, twice to complete the entry line as:

$$\frac{3xy}{2} \div \frac{4x}{9y}$$

Then press ENTER.

DISPLAY/WRITE

a.

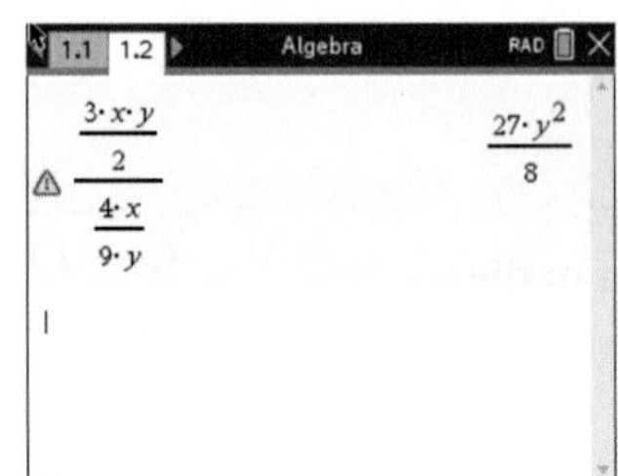

$$\frac{3xy}{2} \div \frac{4x}{9y} = \frac{27y^2}{8}$$

b.

On a Calculator page, use the fraction template twice to complete the entry line as:

$$\frac{4}{(x+1)(3x-5)} \div \frac{x-7}{x+1}$$

Then press ENTER.

b.

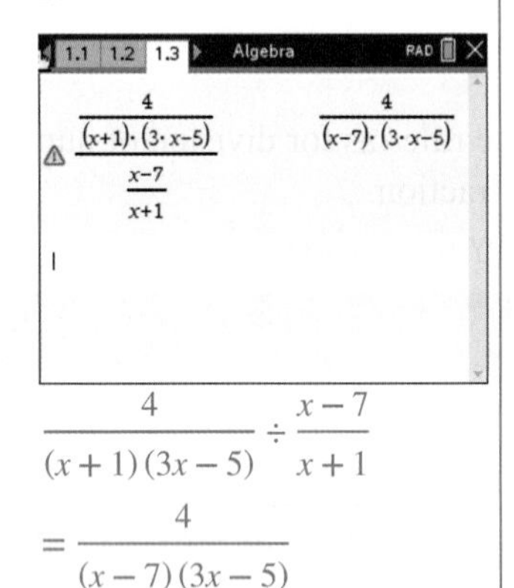

$$\frac{4}{(x+1)(3x-5)} \div \frac{x-7}{x+1}$$
$$= \frac{4}{(x-7)(3x-5)}$$

CASIO | THINK

a-b.

On the Main screen used the fraction template, twice to complete the entry line as:

$$\frac{3 \times xy}{2} \div \frac{4x}{9y}$$
$$\frac{4}{(x+1)(3x-5)} \div \frac{x-7}{x+1}$$

Press EXE after each entry.

DISPLAY/WRITE

a-b.

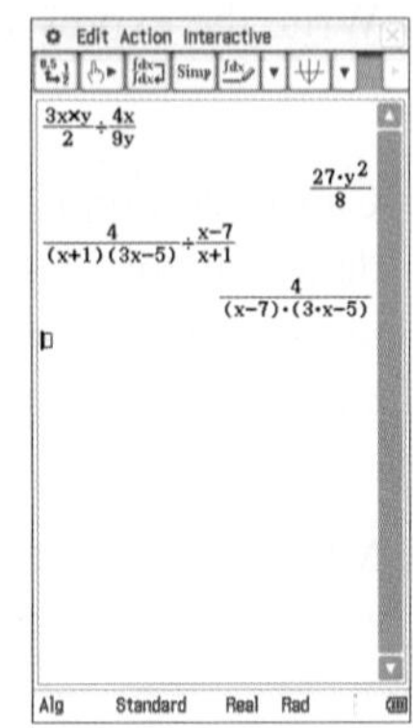

$$\frac{3xy}{2} \div \frac{4x}{9y} = \frac{27y^2}{8}$$
$$\frac{4}{(x+1)(3x-5)} \div \frac{x-7}{x+1}$$
$$= \frac{4}{(x-7)(3x-5)}$$

DISCUSSION

Explain how multiplying and dividing algebraic fractions is different to adding and subtracting them.

on Resources

eWorkbook Topic 2 Workbook (worksheets, code puzzle and project) (ewbk-2028)

Digital documents SkillSHEET Multiplication of fractions (doc-5187)
SkillSHEET Division of fractions (doc-5188)
SkillSHEET Simplification of algebraic fractions (doc-5191)

Interactivities Individual pathway interactivity: Multiplying and dividing algebraic fractions (int-4568)
Simplifying algebraic fractions (int-6115)
Multiplying algebraic fractions (int-6116)
Dividing algebraic fractions (int-6117)

Exercise 2.4 Multiplying and dividing algebraic fractions

learnON

Individual pathways

■ PRACTISE	■ CONSOLIDATE	■ MASTER
1, 4, 7, 10, 13, 16, 19	2, 5, 8, 11, 14, 17, 20	3, 6, 9, 12, 15, 18, 21

To answer questions online and to receive **immediate corrective feedback** and **fully worked solutions** for all questions, go to your learnON title at www.jacplus.com.au.

Fluency

WE8a For questions **1** to **3**, simplify each of the following.

1. a. $\frac{x}{5} \times \frac{20}{y}$ b. $\frac{x}{4} \times \frac{12}{y}$ c. $\frac{y}{4} \times \frac{16}{x}$ d. $\frac{x}{2} \times \frac{9}{2y}$

2. a. $\frac{x}{10} \times \frac{-25}{2y}$ b. $\frac{3w}{-14} \times \frac{-7}{x}$ c. $\frac{3y}{4x} \times \frac{8z}{7y}$ d. $\frac{-y}{3x} \times \frac{6z}{-7y}$

3. a. $\frac{x}{3z} \times \frac{-9z}{2y}$ b. $\frac{5y}{3x} \times \frac{x}{8y}$ c. $\frac{-20y}{7x} \times \frac{-21z}{5y}$ d. $\frac{y}{-3w} \times \frac{x}{2y}$

WE8b For questions **4** to **6**, simplify the following expressions.

4. a. $\frac{2x}{(x-1)(3x-2)} \times \frac{x-1}{x}$ b. $\frac{5x}{(x-3)(4x+7)} \times \frac{4x+7}{x}$
   c. $\frac{9x}{(5x+1)(x-6)} \times \frac{5x+1}{2x}$ d. $\frac{(x+4)}{(x+1)(x+3)} \times \frac{x+1}{x+4}$

5. a. $\frac{2x}{x+1} \times \frac{x-1}{(x+1)(x-1)}$ b. $\frac{2}{x(2x-3)} \times \frac{x(x+1)}{4}$
   c. $\frac{2x}{4(a+3)} \times \frac{3a}{15x}$ d. $\frac{15c}{12(d-3)} \times \frac{21d}{6c}$

6. a. $\frac{6x^2}{20(x-2)^2} \times \frac{15(x-2)}{16x^4}$ b. $\frac{7x^2(x-3)}{5x(x+1)} \times \frac{3(x-3)(x+1)}{14(x-3)^2(x-1)}$

WE9a For questions **7** to **9**, simplify the following expressions.

7. a. $\frac{3}{x} \div \frac{5}{x}$ b. $\frac{2}{x} \div \frac{9}{x}$ c. $\frac{4}{x} \div \frac{12}{x}$ d. $\frac{20}{y} \div \frac{20}{3y}$

8. a. $\frac{1}{5w} \div \frac{5}{w}$ b. $\frac{7}{2x} \div \frac{3}{5x}$ c. $\frac{3xy}{7} \div \frac{3x}{4y}$ d. $\frac{2xy}{5} \div \frac{5x}{y}$

9. a. $\frac{6y}{9} \div \frac{3x}{4xy}$ b. $\frac{8wx}{5} \div \frac{3w}{4y}$ c. $\frac{2xy}{5} \div \frac{3xy}{5}$ d. $\frac{10xy}{7} \div \frac{20x}{14y}$

WE9b For questions **10** to **12**, simplify the following expressions.

10. a. $\frac{9}{(x-1)(3x-7)} \div \frac{x+3}{x-1}$ b. $\frac{1}{(x+2)(2x-5)} \div \frac{x-9}{2x-5}$

11. a. $\frac{12(x-3)^2}{(x+5)(x-9)} \div \frac{4(x-3)}{7(x-9)}$ b. $\frac{13}{6(x-4)^2(x-1)} \div \frac{3(x+1)}{2(x-4)(x-1)}$

12. a. $\dfrac{16(x+5)(x-4)^2}{(x+3)(x+4)} \div \dfrac{8(x+3)(x+5)}{(x-4)(x+4)}$

b. $\dfrac{(x+2)(x+3)(x+4)}{(x-2)(x+3)^2} \div \dfrac{x^2+x-12}{x^2-4}$

Understanding

For questions **13** to **15**, determine the missing fraction.

13. a. $\dfrac{x+2}{3} \times \square = 5$

b. $\dfrac{3}{x^2} \div \square = \dfrac{1}{4}$

14. a. $\dfrac{(x+3)(x+2)}{(x-4)} \times \square = \dfrac{x-5}{x+2}$

b. $\dfrac{x^2(x-3)}{(x+4)(x-5)} \div \square = \dfrac{3x}{2(x+4)}$

15. a. $\dfrac{x^2+8x+15}{x^2-4x-21} \times \square = \dfrac{x^2-25}{x^2-11x+28}$

b. $\square \div \dfrac{x^2-2x-24}{x^2-36} = \dfrac{x^2+12x+36}{x^2}$

Reasoning

16. Explain whether $\dfrac{3}{x+2}$ is the same as $\dfrac{1}{x+2} + \dfrac{1}{x+2} + \dfrac{1}{x+2}$.

17. Does $\dfrac{12xy+16yz^2}{20xyz}$ simplify to $\dfrac{3+4z}{5}$? Explain your reasoning.

18. a. Simplify $\dfrac{(x-4)(x+3)}{4x-x^2} \times \dfrac{x^2-x}{(x+3)(x-1)}$.

b. Identify and explain the error in the following reasoning.

$$\begin{aligned} &\frac{(x-4)(x+3)}{4x-x^2} \times \frac{x^2-x}{(x+3)(x-1)} \\ &= \frac{(x-4)(x+3)}{x(4-x)} \times \frac{x(x-1)}{(x+3)(x-1)} = 1 \end{aligned}$$

Problem solving

19. Simplify $\dfrac{x^2-2x-3}{x^4-1} \times \dfrac{x^2+4x-5}{x^2-5x+6} \div \dfrac{x^2+7x+10}{x^4-3x^2-4}$.

20. Simplify $\dfrac{x+1}{x-\dfrac{x}{\dfrac{x}{a}}}$ where $a = \dfrac{x-1}{x+1}$.

21. Simplify $\left(\dfrac{\dfrac{x^2+1}{x-1}-x}{\dfrac{x^2-1}{x+1}+1}\right) \times \left(1-\dfrac{2}{1+\dfrac{1}{x}}\right)$.

2.5 Solving simple equations

LEARNING INTENTION

At the end of this subtopic you should be able to:
- solve one- and two-step equations using inverse operations.
- solve equations with pronumerals on both sides of the equals sign.

2.5.1 Solving equations using inverse operations

eles-4702

- **Equations** show the equivalence of two expressions.
- Equations can be solved using inverse operations.
- Determining the solution of an equation involves calculating the value or values of a variable that, when substituted into that equation, produces a true statement.
- When solving equations, the last operation performed on the pronumeral when building the equation is the first operation undone by applying inverse operations to both sides of the equation.
  For example, the equation $2x + 3 = 5$ is built from x by:
  First operation: multiplying by 2 to give $2x$
  Second operation: adding 3 to give $2x + 3$.
- In order to solve the equation, undo the second operation of adding 3 by subtracting 3, then undo the first operation of multiplying by 2 by dividing by 2.

Inverse operations

+ and − are inverse operations

× and ÷ are inverse operations

^{2} and $\sqrt{}$ are inverse operations

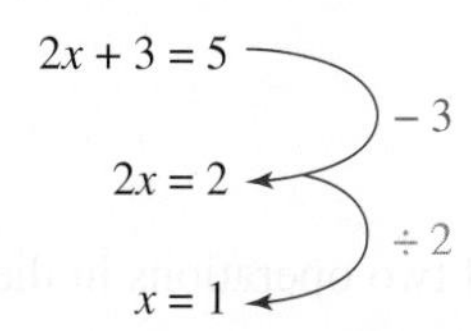

- Equations that require one step to solve are called one-step equations.

WORKED EXAMPLE 10 Solving equations using inverse operations

Solve the following equations.

a. $a + 27 = 71$ **b.** $\frac{d}{16} = 3\frac{1}{4}$ **c.** $\sqrt{e} = 0.87$ **d.** $f^2 = \frac{4}{25}$

THINK	WRITE
a. 1. Write the equation.	a. $a + 27 = 71$
2. 27 has been added to a resulting in 71. The addition of 27 has to be reversed by subtracting 27 from both sides of the equation to obtain the solution.	$a + 27 - 27 = 71 - 27$ $a = 44$
b. 1. Write the equation.	b. $\frac{d}{16} = 3\frac{1}{4}$
2. Express $3\frac{1}{4}$ as an improper fraction.	$\frac{d}{16} = \frac{13}{4}$

3. The pronumeral d has been divided by 16 resulting in $\frac{13}{4}$. Therefore the division has to be reversed by multiplying both sides of the equation by 16 to obtain d.

$$\frac{d}{\cancel{16}} \times \cancel{16} = \frac{13}{\cancel{4}_1} \times \cancel{16}^4$$

$$d = 52$$

c. 1. Write the equation.

c. $\sqrt{e} = 0.87$

2. The square root of e has been taken to result in 0.87. Therefore, the square root has to be reversed by squaring both sides of the equation to obtain e.

$$\left(\sqrt{e}\right)^2 = 0.87^2$$

$$e = 0.7569$$

d. 1. Write the equation.

d. $f^2 = \frac{4}{25}$

2. The pronumeral f has been squared, resulting in $\frac{4}{25}$. Therefore the squaring has to be reversed by taking the square root of both sides of the equation to obtain f. Note that there are two possible solutions, one positive and one negative, since two negative numbers can also be multiplied together to produce a positive result.

$$f = \pm\sqrt{\frac{4}{25}}$$

$$f = \pm\frac{2}{5}$$

2.5.2 Two-step equations

eles-4704

- Two-step equations involve the inverse of two operations in their solutions.

WORKED EXAMPLE 11 Solving two-step equations

Solve the following equations.

a. $\mathbf{5y - 6 = 79}$

b. $\mathbf{\frac{4x}{9} = 5}$

THINK / **WRITE**

a. 1. Write the equation.

a. $5y - 6 = 79$

2. Step 1: Add 6 to both sides of the equation.

$$5y - 6 + 6 = 79 + 6$$

$$5y = 85$$

3. Step 2: Divide both sides of the equation by 5 to obtain y.

$$\frac{5y}{5} = \frac{85}{5}$$

4. Write the answer.

$$y = 17$$

b. 1. Write the equation.

b. $\frac{4x}{9} = 5$

2. Step 1: Multiply both sides of the equation by 9.

$$\frac{4x}{9} \times 9 = 5 \times 9$$

$$4x = 45$$

3. Step 2: Divide both sides of the equation by 4 to obtain x.	$\frac{4x}{4} = \frac{45}{4}$ $x = \frac{45}{4}$
4. Express the answer as a mixed number.	$x = 11\frac{1}{4}$

TI | THINK	DISPLAY/WRITE	CASIO | THINK	DISPLAY/WRITE
a. On a Calculator page to solve equation press: • MENU • 3: Algebra • 1: Solve Then complete the line as: solve($5y - 6 = 79, y$) The 'comma y' (,y) instructs the calculator to solve for the variable y. The press ENTER.	**a.** 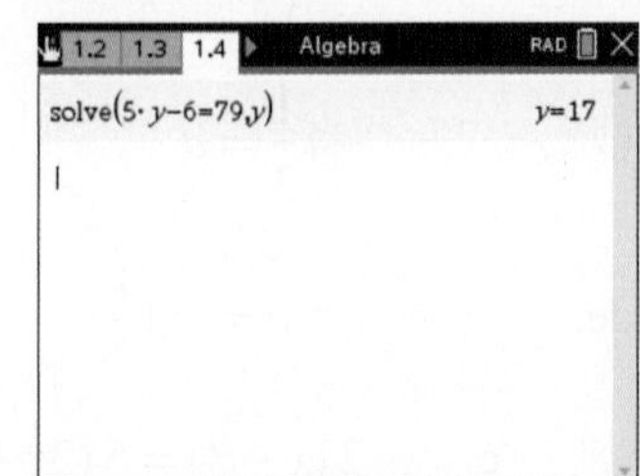 $5y - 6 = 79$ $\Rightarrow y = 17$	**a-b.** On the Main screen, to solve the equation tap: • Action • Advanced • slove Then complete the entry line as: solve($5y - 6 = 79, y$) Then press EXE. The 'comma y' (,y) instructs the calculator to solve for the variable y. Then complete the entry line as: solve $\left(\frac{4x}{9} = 5\right)$ The result is given as an improper fraction. If required, to change to a proper fraction, tap: • Action • Transformation • Fraction • propFrac Then complete as shown and press EXE. If x is the only pronumeral, it is not necessary to include x at the end of the entry line.	**a-b.** 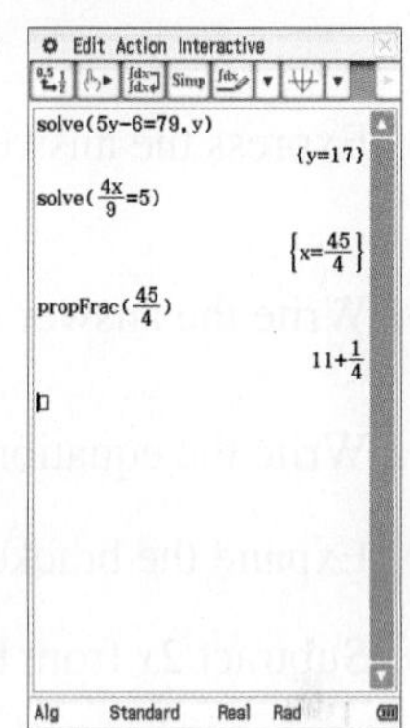 $5y - 6 = 79$ $\Rightarrow y = 17$ $\frac{4x}{y} = 5$ $\Rightarrow x = 11\frac{1}{4}$
b. On a Calculator page, complete the entry line as: solve $\left(\frac{4x}{9} = 5, x\right)$ The result is given as an improper fraction. To change to a proper fraction, press: • MENU • 2: Number • 7: Fraction Tools • 1: Proper Fraction Then complete as shown and press ENTER.	**b.** 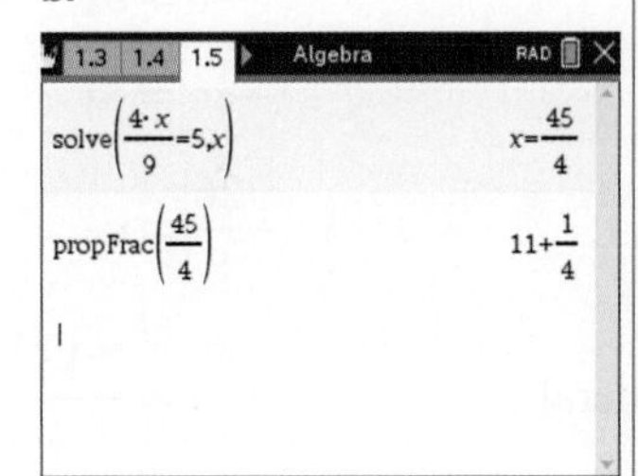 $\frac{4x}{y} = 5$ $\Rightarrow x = 11\frac{1}{4}$		

2.5.3 Equations where the pronumeral appears on both sides

eles-4705

- In solving equations where the pronumeral appears on both sides, subtract the smaller pronumeral term so that it is eliminated from both sides of the equation.

WORKED EXAMPLE 12 Solving equations with multiple pronumeral terms

Solve the following equations.

a. $5h + 13 = 2h - 2$ **b.** $14 - 4d = 27 - d$ **c.** $2(x - 3) = 5(2x + 4)$

THINK	WRITE
a. 1. Write the equation.	**a.** $5h + 13 = 2h - 2$
2. Eliminate the pronumeral from the right-hand side by subtracting $2h$ from both sides of the equation.	$3h + 13 = -2$

3. Subtract 13 from both sides of the equation.	$3h = -15$
4. Divide both sides of the equation by 3 and write the answer.	$h = -5$
b. 1. Write the equation.	**b.** $14 - 4d = 27 - d$
2. Add $4d$ to both sides of the equation.	$14 = 27 + 3d$
3. Subtract 27 from both sides of the equation.	$-13 = 3d$
4. Divide both sides of the equation by 3.	$-\frac{13}{3} = d$
5. Express the answer as a mixed number.	$-4\frac{1}{3} = d$
6. Write the answer so that d is on the left-hand side.	$d = -4\frac{1}{3}$
c. 1. Write the equation.	**c.** $2(x-3) = 5(2x+4)$
2. Expand the brackets on both sides of the equation.	$2x - 6 = 10x + 20$
3. Subtract $2x$ from both sides of the equation.	$-6 = 8x + 20$
4. Subtract 20 from both sides of the equation.	$-26 = 8x$
5. Divide both sides of the equation by 8.	$-\frac{26}{8} = x$
6. Simplify and write the answer with the pronumeral on the left-hand side.	$x = -\frac{13}{4}$

DISCUSSION

Describe in one sentence what it means to solve linear equations.

on Resources

eWorkbook Topic 2 Workbook (worksheets, code puzzle and project) (ewbk-2028)

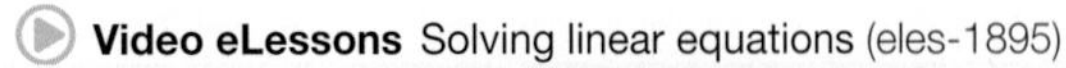

Video eLessons Solving linear equations (eles-1895)
Solving linear equations with pronumerals on both sides (eles-1901)

Interactivity Individual pathway interactivity: Solving simple equations (int-4569)
Using algebra to solve problems (int-3805)
One-step equations (int-6118)
Two-step equations (int-6119)

Exercise 2.5 Solving simple equations

learnON

Individual pathways

■ PRACTISE	■ CONSOLIDATE	■ MASTER
1, 4, 7, 10, 13, 16, 18, 21, 25, 26, 30, 33, 36, 40, 43, 46	2, 5, 8, 11, 14, 17, 19, 22, 23, 27, 28, 31, 34, 37, 41, 44, 47	3, 6, 9, 12, 15, 20, 24, 29, 32, 35, 38, 39, 42, 45, 48

To answer questions online and to receive **immediate corrective feedback** and **fully worked solutions** for all questions, go to your learnON title at www.jacplus.com.au.

Fluency

WE10a For questions **1** to **3**, solve the following equations.

1. a. $a+61=85$ b. $k-75=46$ c. $g+9.3=12.2$

2. a. $r-2.3=0.7$ b. $h+0.84=1.1$ c. $i+5=3$

3. a. $t-12=-7$ b. $q+\frac{1}{3}=\frac{1}{2}$ c. $x-2=-2$

WE10b For questions **4** to **6**, solve the following equations.

4. a. $\frac{f}{4}=3$ b. $\frac{i}{10}=-6$ c. $6z=-42$

5. a. $9v=63$ b. $6w=-32$ c. $\frac{k}{12}=\frac{5}{6}$

6. a. $4a=1.7$ b. $\frac{m}{19}=\frac{7}{8}$ c. $\frac{y}{4}=5\frac{3}{8}$

WE10c, d For questions **7** to **12**, solve the following equations.

7. a. $\sqrt{t}=10$ b. $y^2=289$ c. $\sqrt{q}=2.5$

8. a. $f^2=1.44$ b. $\sqrt{h}=\frac{4}{7}$ c. $p^2=\frac{9}{64}$

9. a. $\sqrt{g}=\frac{15}{22}$ b. $j^2=\frac{196}{961}$ c. $a^2=2\frac{7}{9}$

10. a. $\sqrt{t}-3=2$ b. $5x^2=180$ c. $3\sqrt{m}=12$
    d. $-2t^2=-18$ e. $t^2+11=111$ f. $\sqrt{m}-5=0$

11. a. $\sqrt[3]{x}=2$ b. $x^3=-27$ c. $\sqrt[3]{m}=\frac{1}{2}$
    d. $x^3=\frac{27}{64}$ e. $\sqrt[3]{m}=0.2$ f. $w^3=15\frac{5}{8}$

12. a. $x^3+1=0$ b. $3x^3=-24$ c. $\sqrt[3]{m}+5=6$
    d. $-2\times\sqrt[3]{w}=16$ e. $\sqrt[3]{t}-13=-8$ f. $2x^3-14=2$

WE11a For questions **13** to **20**, solve the following.

13. a. $5a+6=26$ b. $6b+8=44$ c. $8i-9=15$

14. a. $7f-18=45$ b. $8q+17=26$ c. $10r-21=33$

15. a. $6s+46=75$ b. $5t-28=21$ c. $8a+88=28$

16. a. $\frac{f}{4}+6=16$ b. $\frac{g}{6}+4=9$ c. $\frac{r}{10}+6=5$

17. a. $\frac{m}{9}-12=-10$ b. $\frac{n}{8}+5=8.5$ c. $\frac{p}{12}-1.8=3.4$

18. a. $6(x+8)=56$ b. $7(y-4)=35$ c. $5(m-3)=7$

19. a. $3(2k+5)=24$ b. $5(3n-1)=80$ c. $6(2c+7)=58$

20. a. $2(x-5)+3(x-7)=19$ b. $3(x+5)-5(x-1)=12$ c. $3(2x-7)-(x+3)=-60$

WE11b For questions 21 to 24, solve the following.

21. a. $\frac{3k}{5}=15$ b. $\frac{9m}{8}=18$ c. $\frac{7p}{10}=-8$

22. a. $\frac{8u}{11}=-3$ b. $\frac{11x}{4}=2$ c. $\frac{4v}{15}=0.8$

23. a. $\frac{x-5}{3}=7$ b. $\frac{2m+1}{3}=-3$ c. $\frac{3w-1}{4}=6$

24. a. $\frac{t-5}{2}=0$ b. $\frac{6-x}{3}=-1$ c. $\frac{3n-5}{4}=-6$

25. **MC** a. The solution to the equation $\frac{p}{5}+2=7$ is:

A. $p=5$ B. $p=25$ C. $p=45$ D. $p=10$ E. $p=1$

b. If $5h+8=53$, then h is equal to:

A. $\frac{1}{5}$ B. 12.2 C. 225 D. 10 E. 9

c. The exact solution to the equation $14x=75$ is:

A. $x=5.357\,142\,857$

B. $x=5.357$ (to 3 decimal places)

C. $x=5\frac{5}{14}$

D. $x=5.4$

E. $x=5.5$

For questions 26 to 29, solve the following equations.

26. a. $-5h=10$ b. $2-d=3$ c. $5-p=-2$ d. $-7-x=4$

27. a. $-6t=-30$ b. $-\frac{v}{5}=4$ c. $-\frac{r}{12}=\frac{1}{4}$ d. $-4g=3.2$

28. a. $6-2x=8$ b. $10-3v=7$ c. $9-6l=-3$ d. $-3-2g=1$

29. a. $-5-4t=-17$ b. $-\frac{3e}{5}=14$ c. $-\frac{k}{4}-3=6$ d. $-\frac{4f}{7}+1=8$

WE12a For questions 30 to 32, solve the following equations.

30. a. $6x+5=5x+7$ b. $7b+9=6b+14$ c. $11w+17=6w+27$

31. a. $8f-2=7f+5$ b. $10t-11=5t+4$ c. $12r-16=3r+5$

32. a. $12g - 19 = 3g - 31$ **b.** $7h + 5 = 2h - 6$ **c.** $5a - 2 = 3a - 2$

WE12b For questions **33** to **35**, solve the following equations.

33. a. $5 - 2x = 6 - x$ **b.** $10 - 3c = 8 - 2c$ **c.** $3r + 13 = 9r - 3$

34. a. $k - 5 = 2k - 6$ **b.** $5y + 8 = 13y + 17$ **c.** $17 - 3g = 3 - g$

35. a. $14 - 5w = w + 8$ **b.** $4m + 7 = 8 - m$ **c.** $14 - 5p = 9 - 2p$

WE12c For questions **36** to **38**, solve the following equations.

36. a. $3(x + 5) = 2x$ **b.** $8(y + 3) = 3y$ **c.** $6(t - 5) = 4(t + 3)$

37. a. $10(u + 1) = 3(u - 3)$ **b.** $12(f - 10) = 4(f - 5)$ **c.** $2(4r + 3) = 3(2r + 7)$

38. a. $5(2d + 9) = 3(3d + 13)$ **b.** $5(h - 3) = 3(2h - 1)$ **c.** $2(4x + 1) = 5(3 - x)$

39. **MC** **a.** The solution to $8 - 4k = -2$ is:

A. $k = 2\frac{1}{2}$ B. $k = -2\frac{1}{2}$ C. $k = 1\frac{1}{2}$ D. $k = -1\frac{1}{2}$ E. $k = \frac{2}{5}$

b. The solution to $-\frac{6n}{5} + 3 = -7$ is:

A. $n = 3\frac{1}{3}$ B. $n = -3\frac{1}{3}$ C. $n = \frac{1}{3}$ D. $n = 8\frac{1}{3}$ E. $n = -8\frac{1}{3}$

c. The solution to $p - 6 = 8 - 4p$ is:

A. $p = \frac{2}{5}$ B. $p = 2\frac{4}{5}$ C. $p = 4\frac{2}{3}$ D. $p = \frac{2}{3}$ E. $p = \frac{4}{5}$

Understanding

40. If the side length of a cube is x cm, then its volume V is given by $V = x^3$. Calculate the side length (correct to the nearest cm) of a cube that has a volume of:

a. 216 cm^3 b. 2 m^3.

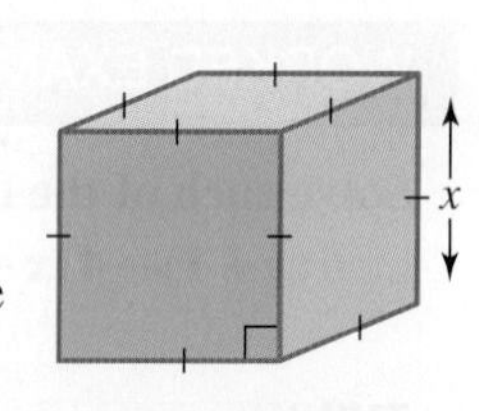

41. The surface area of a cube with side length x cm is given by $A = 6x^2$. Determine the side length (correct to the nearest cm) of a cube that has a surface area of:

a. 37.5 cm^2 b. 1 m^2.

42. A pebble is dropped down a well. In time t seconds it falls a distance of d metres, given by $d = 5t^2$.

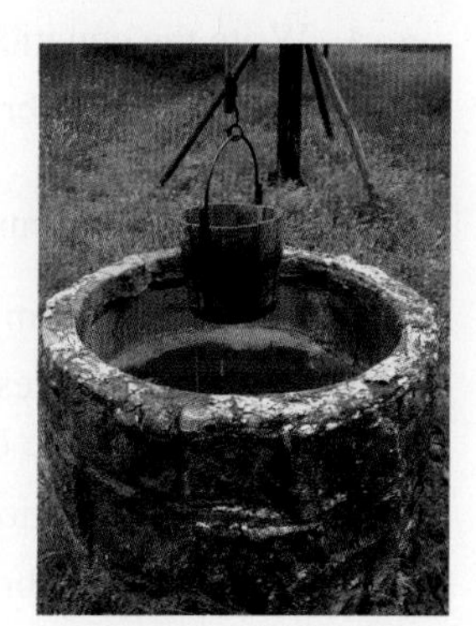

a. Calculate the distance the pebble falls in 1 second.

b. Calculate the time the pebble will take to fall 40 m. (Answer in seconds correct to 1 decimal place.)

Reasoning

43. The surface area of a sphere is given by the formula $A = 4\pi r^2$, where r is the radius of the sphere.

a. Determine the surface area of a sphere that has a radius of 5 cm. Show your working.

b. Evaluate the radius of a sphere that has a surface area equal to 500 cm^2. (Answer correct to the nearest mm.)

44. Determine the radius of a circle of area 10 cm^2. Show your working.

45. The volume of a sphere is given by the formula $V = \frac{4}{3}\pi r^3$, where r is the radius of the sphere. If the sphere can hold 1 litre of water, determine its radius correct to the nearest mm. Show your working.

Problem solving

46. The width of a room is three-fifths of its length. When the width is increased by 2 metres and the length is decreased by 2 metres, the resultant shape is a square. Determine the dimensions of the room.

47. Four years ago, Leon was one third of James' age. In six years' time, the sum of their ages will be 60. Determine their current ages.

48. A target board for a dart game has been designed as three concentric circles where each coloured region is the same area. If the radius of the blue circle is r cm and the radius of the outer circle is 10 cm, determine the value of r.

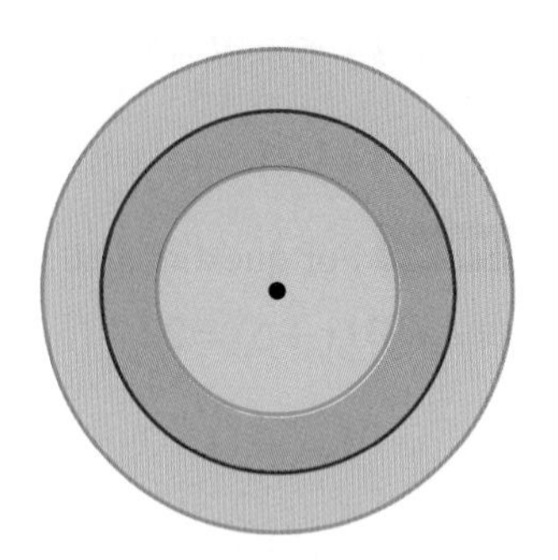

2.6 Solving multi-step equations

LEARNING INTENTION

At the end of this subtopic you should be able to:
- expand brackets and collect like terms in order to solve a multi-step equation
- solve equations involving algebraic fractions by determining the LCM of the denominators.

2.6.1 Equations with multiple brackets

eles-4706

- Equations can be simplified by expanding brackets and collecting like terms before they are solved.

WORKED EXAMPLE 13 Solving equations with brackets

Solve each of the following linear equations.

a. $6(x+1) - 4(x-2) = 0$ **b.** $7(5-x) = 3(x+1) - 10$

THINK	WRITE
a. 1. Write the equation.	a. $6(x+1) - 4(x-2) = 0$
2. Expand all the brackets. (Be careful with the −4.)	$6x + 6 - 4x + 8 = 0$
3. Collect like terms.	$2x + 14 = 0$
4. Subtract 14 from both sides of the equation.	$2x = -14$
5. Divide both sides of the equation by 2 to obtain the value of x.	$x = -7$
b. 1. Write the equation.	b. $7(5-x) = 3(x+1) - 10$
2. Expand all the brackets.	$35 - 7x = 3x + 3 - 10$
3. Collect like terms.	$35 - 7x = 3x - 7$
4. Create a single pronumeral term by adding $7x$ to both sides of the equation.	$35 = 10x - 7$
5. Add 7 to both sides of the equation.	$42 = 10x$

6. Divide both sides of the equation by 10 to solve for x and simplify.

$$\frac{42}{10} = x$$

$$\frac{21}{5} = x$$

7. Express the improper fraction as a mixed number fraction.

$$4\frac{1}{5} = x$$

8. Rewrite the equation so that x is on the left-hand side.

$$x = 4\frac{1}{5}$$

TI | THINK

a–b.

On a Calculator page, complete the entry lines as:
solve
$(6(x+1) - 4(x-2) = 0, x)$
solve
$(7(5-x) = 3(x+1) - 10, x)$
Press ENTER after each entry.
Convert **b** to a proper fraction.

DISPLAY/WRITE

a–b.

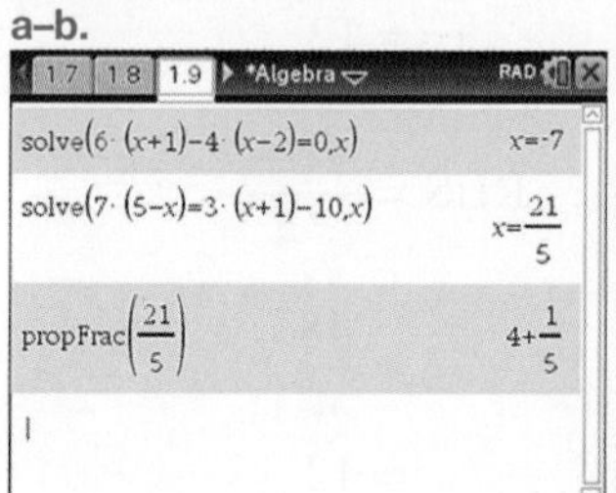

$6(x+1) - 4(x-2) = 0$
$\Rightarrow x = -7$
$7(5-x) = 3(x+1) - 10$
$\Rightarrow x = 4\frac{1}{5}$

CASIO | THINK

a–b.

On the Main screen, complete the entry lines as:
solve
$(6(x+1) - 4(x-2) = 0, x)$
solve
$(7(5-x) = 3(x+1) - 10, x)$
Press EXE after each entry.
Convert **b** to a proper fraction.

DISPLAY/WRITE

a–b.

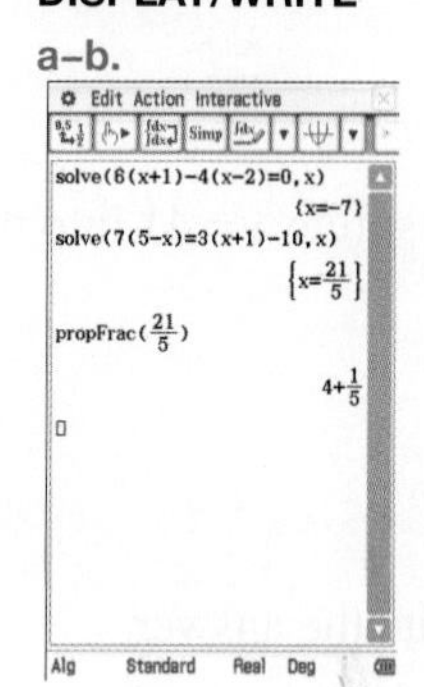

$6(x+1) - 4(x-2) = 0$
$\Rightarrow x = -7$
$7(5-x) = 3(x+1) - 10$
$\Rightarrow = 4\frac{1}{5}$

2.6.2 Equations involving algebraic fractions

eles-4707

- To solve an equation containing algebraic fractions, multiply both sides of the equation by the lowest common multiple (LCM) of the denominators. This gives an equivalent form of the equation without fractions.

WORKED EXAMPLE 14 Solving equations with algebraic fractions

Solve the equation $\frac{x-5}{3} = \frac{x+7}{4}$ and verify the solution.

THINK / **WRITE**

1. Write the equation.

$$\frac{x-5}{3} = \frac{x+7}{4}$$

2. The LCM is $3 \times 4 = 12$. Multiply both sides of the equation by 12.

$$\frac{{}^{4}\cancel{12}(x-5)}{{}^{1}\cancel{3}} = \frac{{}^{3}\cancel{12}(x+7)}{{}^{1}\cancel{4}}$$

3. Simplify the fractions.

$$4(x-5) = 3(x+7)$$

4. Expand the brackets.

$$4x - 20 = 3x + 21$$

5. Subtract $3x$ from both sides of the equation.

$$x - 20 = 21$$

6. Add 20 to both sides of the equation and write the answer. $x = 41$

7. To verify, check that the answer $x = 41$ is true for both the left-hand side (LHS) and the right-hand side (RHS) of the equation by substitution.

Substitute $x = 41$ into the LHS.

$$\text{LHS} = \frac{41-5}{3} = \frac{36}{3} = 12$$

Substitute $x = 41$ into the RHS.

$$\text{RHS} = \frac{41+7}{4} = \frac{48}{4} = 12$$

8. Write the answer. Because the LHS = RHS, the solution $x = 41$ is correct.

WORKED EXAMPLE 15 Solving involving algebraic fractions

Solve each of the following equations.

a. $\dfrac{5(x+3)}{6} = 4 + \dfrac{3(x-1)}{5}$ b. $\dfrac{4}{3(x-1)} = \dfrac{1}{x+1}$

THINK	WRITE
a. 1. Write the equation.	a. $\dfrac{5(x+3)}{6} = 4 + \dfrac{3(x-1)}{5}$
2. The lowest common denominator of 5 and 6 is 30. Write each term as an equivalent fraction with a denominator of 30.	$\dfrac{25(x+3)}{30} = \dfrac{120}{30} + \dfrac{18(x-1)}{30}$
3. Multiply each term by 30. This effectively removes the denominator.	$25(x+3) = 120 + 18(x-1)$
4. Expand the brackets and collect like terms.	$25x + 75 = 120 + 18x - 18$ $25x + 75 = 102 + 18x$
5. Subtract $18x$ from both sides of the equation.	$7x + 75 = 102$
6. Subtract 75 from both sides of the equation.	$7x = 27$
7. Divide both sides of the equation by 7 to solve for x.	$x = \dfrac{27}{7}$
8. Express the answer as a mixed number.	$x = 3\dfrac{6}{7}$

b. 1. Write the equation.

b. $\frac{4}{3(x-1)} = \frac{1}{x+1}$

2. The lowest common denominator of 3, $x+1$ and $x-1$ is $3(x-1)(x+1)$. Write each term as an equivalent fraction with a common denominator of $3(x-1)(x+1)$.

$\frac{4(x+1)}{3(x-1)(x+1)} = \frac{3(x-1)}{3(x-1)(x+1)}$

3. Multiply each term by the common denominator.

$4(x+1) = 3(x-1)$

4. Expand the brackets.

$4x + 4 = 3x - 3$

5. Subtract $3x$ from both sides of the equation.

$x + 4 = -3$

6. Subtract 4 from both sides of the equation to solve for x.

$x + 4 - 4 = -3 - 4$

7. Write the answer.

$x = -7$

DISCUSSION

Do the rules for the order of operations apply to algebraic fractions? Explain.

on Resources

eWorkbook Topic 2 Workbook (worksheets, code puzzle and project) (ewbk-2028)

Video eLesson Solving linear equations with algebraic fractions (eles-1857)

Interactivities Individual pathway interactivity: Solving multi-step equations (int-4570)
Expanding brackets: Distributive Law (int-3774)

Exercise 2.6 Solving multi-step equations

learnon

Individual pathways

■ PRACTISE	■ CONSOLIDATE	■ MASTER
1, 4, 7, 10, 13, 17, 20	2, 5, 8, 11, 14, 18, 21	3, 6, 9, 12, 15, 16, 19, 22

To answer questions online and to receive **immediate corrective feedback** and **fully worked solutions** for all questions, go to your learnON title at www.jacplus.com.au.

Fluency

WE13 1 to 3, solve each of the following linear equations.

1. a. $6(4x-3) + 7(x+1) = 9$
   b. $9(3-2x) + 2(5x+1) = 0$

2. a. $8(5-3x) - 4(2+3x) = 3$
   b. $9(1+x) - 8(x+2) = 2x$

3. a. $6(4+3x) = 7(x-1) + 1$
   b. $10(4x+2) = 3(8-x) + 6$

WE14 For questions 4 to 6, solve each of the following equations and verify the solutions.

4. a. $\frac{x+1}{2}=\frac{x+3}{3}$ b. $\frac{x-7}{5}=\frac{x-8}{4}$ c. $\frac{x-6}{4}=\frac{x-2}{2}$

5. a. $\frac{8x+3}{5}=2x$ b. $\frac{2x-1}{5}=\frac{x-3}{4}$ c. $\frac{4x+1}{3}=\frac{x+2}{4}$

6. a. $\frac{6-x}{3}=\frac{2x-1}{5}$ b. $\frac{8-x}{9}=\frac{2x+1}{3}$ c. $\frac{2(x+1)}{5}=\frac{3-2x}{4}$

For questions 7 to 9, solve each of the following linear equations.

7. a. $\frac{x}{3}+\frac{4x}{5}=\frac{1}{3}$ b. $\frac{x}{4}-\frac{x}{5}=\frac{3}{4}$ c. $\frac{x}{4}-\frac{4x}{7}=2$ d. $\frac{-3x}{5}+\frac{x}{8}=\frac{1}{4}$

8. a. $\frac{2x}{3}-\frac{x}{6}=-\frac{3}{4}$ b. $\frac{5x}{8}-8=\frac{2x}{3}$ c. $\frac{2}{7}-\frac{x}{8}=\frac{3x}{8}$ d. $\frac{4}{x}-\frac{1}{6}=\frac{2}{x}$

9. a. $\frac{15}{x}-4=\frac{2}{x}$ b. $\frac{1}{3}+\frac{4}{x}=\frac{5}{x}$ c. $\frac{2x-4}{5}+6=\frac{x}{2}$ d. $\frac{4x-1}{2}-\frac{2x+5}{3}=0$

WE15 For questions 10 to 12, solve each of the following linear equations.

10. a. $\frac{3(x+1)}{2}+\frac{5(x+1)}{3}=4$ b. $\frac{2(x+1)}{7}+\frac{3(2x-5)}{8}=0$
c. $\frac{2(4x+3)}{5}-\frac{6(x-2)}{2}=\frac{1}{2}$ d. $\frac{8(x+3)}{5}=\frac{3(x+2)}{4}$

11. a. $\frac{5(7-x)}{2}=\frac{2(2x-1)}{7}+1$ b. $\frac{2(6-x)}{3}=\frac{9(x+5)}{6}+\frac{1}{3}$
c. $\frac{-5(x-2)}{3}-\frac{6(2x-1)}{5}=\frac{1}{3}$ d. $\frac{9(2x-1)}{7}=\frac{4(x-5)}{3}$

12. a. $\frac{1}{x-1}+\frac{3}{x+1}=\frac{8}{x+1}$ b. $\frac{3}{x+1}+\frac{5}{x-4}=\frac{5}{x+1}$
c. $\frac{1}{x-1}-\frac{3}{x}=\frac{-1}{x-1}$ d. $\frac{4}{2x-1}-\frac{5}{x}=\frac{-1}{x}$

Understanding

13. Last week Maya broke into her money box. She spent one-quarter of the money on a birthday present for her brother and one-third of the money on an evening out with her friends, leaving her with \$75.
Determine the amount of money in her money box.

14. At work Keith spends one-fifth of his time in planning and buying merchandise. He spends seven-twelfths of his time in customer service and one-twentieth of his time training the staff. This leaves him ten hours to deal with the accounts.
Determine the number of hours he works each week.

15. Last week's school fete was a great success, raising a good deal of money. Three-eighths of the profit came from sales of food and drink, and the market stalls recorded one-fifth of the total. A third of the profit came from the major raffle, and the jumping castle raised \$1100.
Determine the amount of money raised at the fete.

16. Lucy had half as much money as Mel, but since Grandma gave them each \$20 she now has three-fifths as much. Determine the amount of money Lucy has.

Reasoning

17. Answer the following question and justify your answer:
    a. Determine numbers smaller than 100 that have exactly 3 factors (including 1 and the number itself).
    b. Determine the two numbers smaller than 100 that have exactly 5 factors.
    c. Determine a number smaller than 100 that has exactly 7 factors.

18. To raise money for a charity, a Year 10 class has decided to organise a school lunch. Tickets will cost \$6 each. The students have negotiated a special deal for delivery of drinks and pizzas, and they have budgeted \$200 for drinks and \$250 for pizzas. If they raise \$1000 or more, they qualify for a special award.

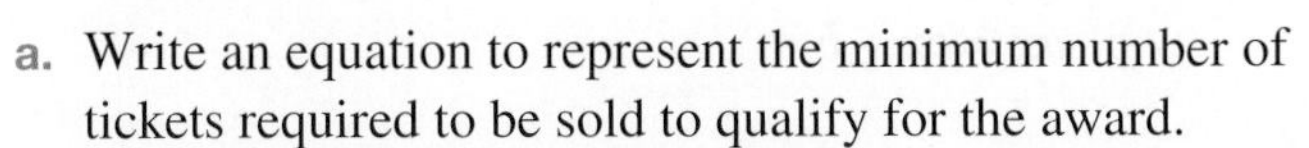

    a. Write an equation to represent the minimum number of tickets required to be sold to qualify for the award.
    b. Solve the equation to find the number of tickets they must sell to qualify for the award. Explain your answer.

19. If $\dfrac{x+7}{(x+2)(x+3)} \equiv \dfrac{a}{x+2} - \dfrac{4}{x+3}$, explain why a must be equal to 5.
(*Note:* '$\equiv$' means identically equal to.)

Problem solving

20. Solve for x:
$\dfrac{2}{9}(x-1) - \dfrac{5}{8}(x-2) = \dfrac{2}{5}(x-4) - \dfrac{7}{12}$

21. If $\dfrac{2(4x+3)}{(x-3)(x+7)} \equiv \dfrac{a}{x-3} + \dfrac{b}{x+7}$, determine the values of a and b.

22. If $\dfrac{7x+20}{x^2+7x+12} = \dfrac{a}{x+3} + \dfrac{b}{x+4} + \dfrac{a+b}{x^2+7x+12}$, determine the values of a and b.

2.7 Literal equations

LEARNING INTENTION

At the end of this subtopic you should be able to:

- solve literal equations, which include multiple variables, by changing the subject of an equation to a particular pronumeral
- determine any restriction on a variable in an equation due to limitations imposed by the equation or context of the question.

2.7.1 Literal equations

eles-4708

- **Literal equations** are equations that include several pronumerals or variables. Solving literal equations involves changing the subject of the equation to a particular pronumeral.
- A variable is the subject of an equation if it expressed in terms of the other variables. In $v = u + at$, the subject of the equation is v as it is written in terms of the variables u, a and t.
- A formula is a literal equation that records an interesting or important real-life relationship.

WORKED EXAMPLE 16 Solving literal equations

Solve the following literal equations for x.

a. $ax^2 + bd = c$

b. $ax = cx + b$

THINK	WRITE
a. 1. Write the equation.	**a.** $ax^2 + bd = c$
2. Subtract bd from both sides of the equation.	$ax^2 = c - bd$
3. Divide both sides by a.	$x^2 = \dfrac{c - bd}{a}$
4. To solve for x, take the square root of both sides. This gives both a positive and negative result for x.	$x = \pm\sqrt{\dfrac{c - bd}{a}}$
b. 1. Write the equation.	**b.** $ax = cx + b$
2. Subtract cx from both sides.	$ax - cx = b$
3. Factorise by taking x as a common factor.	$x(a - c) = b$
4. To solve for x, divide both sides by $a - c$.	$x = \dfrac{b}{a - c}$

TI | THINK

a–b.

In a new problem on a Calculator page, complete the entry lines as:
solve $(a \times x^2 + b \times d = c, x)$
solve $(a \times x = c \times x + b, x)$
Press ENTER after each entry.

DISPLAY/WRITE

a–b.

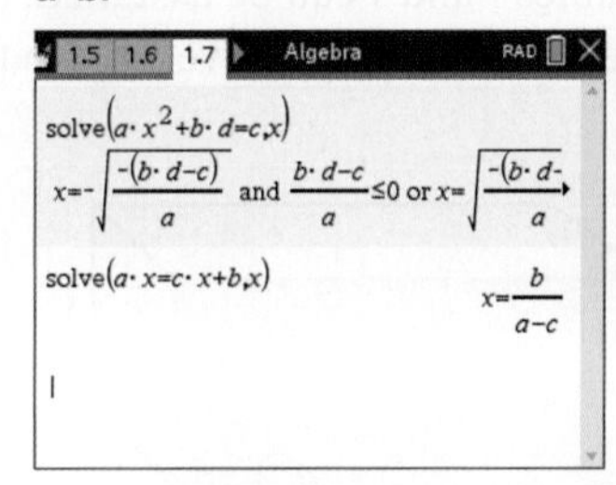

$$x = \pm\sqrt{\frac{c - bd}{a}}$$

$$x = \frac{b}{a - c}$$

CASIO | THINK

a–b.

On the Main screen, complete the entry lines as:
solve $(a \times x^2 + b \times d = c, x)$
solve $(a \times x = c \times x + b, x)$
Press EXE after each entry.

DISPLAY/WRITE

a–b.

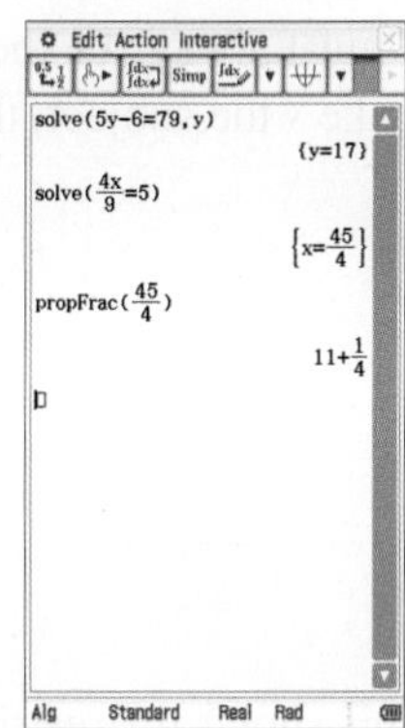

$$x = \pm\sqrt{\frac{c - bd}{a}}$$

$$x = \frac{b}{a - c}$$

WORKED EXAMPLE 17 Rearranging to make a variable the subject of an equation

Make b the subject of the formula $D = \sqrt{b^2 - 4ac}$.

THINK	WRITE
1. Write the formula.	$D = \sqrt{b^2 - 4ac}$
2. Square both sides.	$D^2 = b^2 - 4ac$
3. Add $4ac$ to both sides of the equation.	$D^2 + 4ac = b^2$
4. Take the square root of both sides.	$\pm\sqrt{D^2 + 4ac} = b$
5. Make b the subject of the formula by solving for b.	$b = \pm\sqrt{D^2 + 4ac}$

2.7.2 Restrictions on variables

eles-4709

- Some variables may have implicit restrictions on the values that they may be assigned in an equation or formula.
  For example:
  - if $V = \frac{d}{t}$, then t cannot equal zero, otherwise the value of V would be undefined.
  - if $d = \sqrt{x - 9}$, then:
    - the value of d will be restricted to positive values or 0
    - the value of $x - 9$ must be greater than or equal to zero because the square root of a negative number cannot be found.

$$x - 9 \geq 0$$
$$x \geq 9 \quad \text{(Hence } x \text{ must be greater than or equal to 9)}$$

- Other restrictions may arise once a formula is rearranged. For example, if we look at the formula $V = ls^2$, there are no restrictions on the values that the variables l and s can be assigned. (However, the sign of V must always be the same as the sign of l because s^2 is always positive.) If the formula is transposed to make s the subject, then:

$$V = ls^2$$

$$\frac{V}{l} = s^2$$

$$\text{or } s = \pm\sqrt{\frac{V}{l}}$$

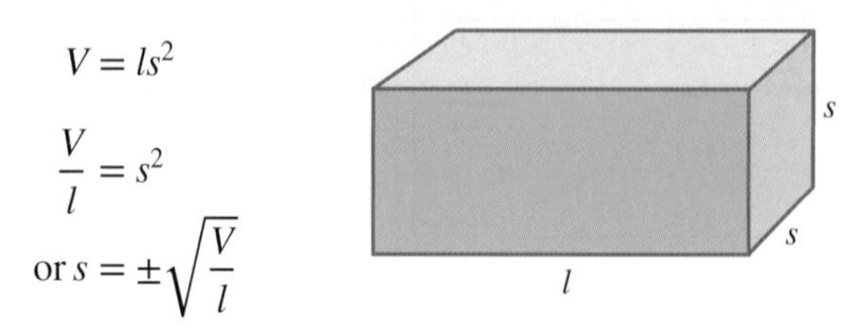

This shows the restrictions that $l \neq 0$ and $\frac{V}{l} \geq 0$.

- If the formula $V = ls^2$ represents the volume of the rectangular prism shown, additional restrictions become evident: the variables l and s represent a length and must be positive numbers. Hence, when we make s the subject we get $s = \sqrt{\frac{V}{l}}$.

WORKED EXAMPLE 18 Identifying restrictions on variables

List any restrictions on the variables in the equations below.

a. The literal equation: $x = \dfrac{100\sqrt{y+4}}{z-10}$

b. The area of a triangle: $A = \dfrac{bh}{2}$ **where** b **= base length and** h **= height**

THINK	WRITE
a. We cannot substitute a negative value into a square root. This affects the possible values for y.	a. $y + 4 \geq 0$ $y \geq -4$
A fraction is undefined if the denominator is equal to 0. This affects the possible values for z.	$z - 10 \neq 0$ $z \neq 10$
b. In this case the restrictions do not come from the equation, but from the context of the equation. Since b and h represent lengths of a shape, they must have positive values or else the shape would not exist.	b. $b > 0$ and $h > 0$ This also implies that $A > 0$.

DISCUSSION

Why is it important to consider restrictions on variables when solving literal equations?

Resources

 **eWorkbook** Topic 2 Workbook (worksheets, code puzzle and project) (ewbk-2028)

 **Interactivities** Individual pathway interactivity: Literal equations (int-4571)
Restrictions on variables (int-6120)

Exercise 2.7 Literal equations

learnON

Individual pathways

■ PRACTISE	■ CONSOLIDATE	■ MASTER
1, 4, 7, 10, 13, 16	2, 5, 8, 11, 14, 17	3, 6, 9, 12, 15, 18

To answer questions online and to receive **immediate corrective feedback** and **fully worked solutions** for all questions, go to your learnON title at www.jacplus.com.au.

Fluency

WE16 For questions 1 to 3, solve the following literal equations for x.

1. a. $\dfrac{ax}{bc}=d$ b. $\dfrac{x}{a}-bc=d$ c. $\sqrt{x+n}=m$

2. a. $acx^2=w$ b. $\dfrac{a}{x}=\dfrac{b}{y}$ c. $\dfrac{x+m}{n}=w$

3. a. $ab(x+b)=c$ b. $\dfrac{a}{x}=\dfrac{b}{c}+m$
   c. $mx=ay-bx$ d. $\dfrac{x}{m}+a=\dfrac{c}{d}$

WE17 For questions 4 to 6, rearrange each of the following literal equations to make the variable in brackets the subject.

4. a. $V=lbh\ [l]$ b. $P=2l+2b\ [b]$
   c. $A=\dfrac{1}{2}bh\ [h]$ d. $c=\sqrt{a^2+b^2}\ [a]$

5. a. $F=\dfrac{9C}{5}+32\ [C]$ b. $A=\pi r^2\ [r]$ c. $v=u+at\ [a]$
   d. $I=\dfrac{PRN}{100}\ [N]$ e. $E=\dfrac{1}{2}mv^2\ [m]$

6. a. $E=\dfrac{1}{2}mv^2\ [v]$ b. $v^2=u^2+2as\ [a]$ c. $v^2=u^2+2as\ [u]$
   d. $\dfrac{1}{x}=\dfrac{1}{a}+\dfrac{1}{b}\ [a]$ e. $x=\dfrac{nx_1+mx_2}{m+n}\ [x_1]$

For questions 7 to 9, complete the following.

7. a. If $c=\sqrt{a^2+b^2}$, calculate a if $c=13$ and $b=5$.
   b. If $A=\dfrac{1}{2}bh$, calculate the value of h if $A=56$ and $b=16$.
   c. If $F=\dfrac{9C}{5}+32$, calculate the value of C if $F=86$.

8. a. If $v = u + at$, calculate the value of a if $v = 83.6$, $u = 15$ and $t = 7$.
   b. If $V = ls^2$, calculate the value of s if $V = 2028$ and $l = 12$.
   c. If $v^2 = u^2 + 2as$, calculate the value of u if $v = 16$, $a = 10$ and $s = 6.75$.

9. a. If $A = \frac{1}{2}h(a + b)$, calculate the value of a if $A = 360$, $b = 15$ and $h = 18$.
   b. If $x = \frac{nx_1 + mx_2}{m + n}$, calculate the value of x_2 if $x = 10$, $m = 2$, $n = 1$ and $x_1 = 4$.

Understanding

10. For the following equations:
   i. **WE18** list any restrictions on the variables in the equation.
   ii. rearrange the equation to make the variable in brackets the subject.
   iii. list any new restrictions on the variables in the equation formed in part ii.

   a. $y = x^2 + 4 \quad [x]$
   b. $y = \frac{2}{x - 3} \quad [x]$
   c. $v = u + at \quad [t]$

11. For the following equations:
   i. list any restrictions on the variables in the equation.
   ii. rearrange the equation to make the variable in brackets the subject.
   iii. list any new restrictions on the variables in the equation formed in part ii.

   a. $c = \sqrt{a^2 + b^2} \quad [b]$
   b. $s = \frac{a}{1 - r} \quad [r]$
   c. $m = \frac{pb + qa}{p + q} \quad [b]$

12. For the following equations:
   i. list any restrictions on the variables in the equation.
   ii. rearrange the equation to make the variable in brackets the subject.
   iii. list any new restrictions on the variables in the equation formed in part ii.

   a. $x = \frac{-b \pm \sqrt{b^2 - 4ac}}{2a} \quad [c]$
   b. $m = \frac{pb + qa}{p + q} \quad [p]$
   c. $E^2 = (pc)^2 + \left(mc^2\right)^2 \quad [m]$

Reasoning

13. The area of a trapezium is given by $A = \frac{1}{2}(a + b)h$, where a and b are the lengths of the top and the base and h is the height of the trapezium.
   a. State any restrictions on the variables in the formula. Justify your response.
   b. Make b the subject of the equation.
   c. Determine the length of the base of a trapezium with a height of 4 cm and top of 5 cm and a total area of 32 cm^2. Show your working.

14. The volume of a cylinder is given by $V = \pi r^2 h$, where r is the radius and h is the height of the cylinder.
   a. State any restrictions on the values of the variables in this formula. Justify your response.
   b. Make r the subject of the formula.
   c. List any new restrictions on the variables in the formula. Justify your response.

15. T is the period of a pendulum whose length is l and g is the acceleration due to gravity. The formula relating these variables is $T = 2\pi\sqrt{\frac{l}{g}}$.
   a. State what restrictions are applied to the variables T and l. Justify your response.
   b. Make l the subject of the equation.
   c. Justify if the restrictions stated in part **a** still apply.
   d. Determine the length of a pendulum that has a period of 3 seconds, given that $g = 9.8$ m/s^2. Give your answer correct to 1 decimal place.

Problem solving

16. $F = 32 + \frac{9}{5}C$ is the formula relating degrees Celsius (C) to degrees Fahrenheit (F).
    a. Transform the equation to make C the subject.
    b. Determine the temperature when degrees Celsius is equal to degrees Fahrenheit.

17. Newton's law of universal gravitation, $F = G\frac{m_1 m_2}{r^2}$, tells us the gravitational force acting between two objects with masses m_1 and m_2, at a distance r metres apart. In this equation, G is the gravitational constant and has a fixed value of 6.67×10^{-11}.

    a. Transform the equation to make m_1 the subject.
    b. Evaluate the mass of the Moon, to 2 decimal places, if the value of F between Earth and the Moon is 2.0×10^{20} N and the distance between Earth and the Moon is assumed to be 3.84×10^{8} m. Take the mass of Earth to be approximately 5.97×10^{24} kg.

18. Jing Jing and Pieter live on the same main road but Jing Jing lives a kilometres to the east of Pieter. Both Jing Jing and Pieter set off on their bicycles at exactly the same time and both ride in a westerly direction.
    Jing Jing rides at j kilometres per hour and Pieter rides at p kilometres per hour. It is known that $j > p$. Determine an equation in terms of a, j and p for the distance Jing Jing has ridden in order to catch up with Pieter.

2.8 Review

2.8.1 Topic summary

Algebraic basics

- We can only add and subtract like terms: $3x + 6y - 7x + 2z = 6y + 2z - 4x$
- Multiplying algebraic terms: $10x^3y^2 \times 4x^2z = 40x^5y^2z$
- Cancelling down fractions: only cancel what is common to all terms in both the numerator and denominator.
  $\frac{3ac + 5ab}{10abc} = \frac{3\cancel{a}c + 5\cancel{a}b}{10\cancel{a}bc} = \frac{3c + 5b}{10bc}$
- Expanding brackets: $x(a + b) = ax + bx$

Substitution

- When the numeric value of a pronumeral is known, it can be substituted into an expression to evaluate the expression.
- It can be helpful to place substituted values inside brackets.
  e.g. Evaluate the expression $b^2 - 4ac$ when $a = -3$, $b = -2$ and $c = 4$:
  $$\begin{aligned} b^2 - 4ac &= (-2)^2 - 4 \times (-3) \times (4) \\ &= 4 + 48 \\ &= 52 \end{aligned}$$

Number laws

- **Commutative law:** the order in which an operation is carried out does not affect the result. It holds true for:
  - Addition: $x + y = y + x$
  - Multiplication: $x \times y = y \times x$
- **Associative law:** when calculating two or more numerals, how they are grouped does not affect the result. It holds true for:
  - Addition: $x + (y + z) = (x + y) + z$
  - Multiplication: $x \times (y \times z) = (x \times y) \times z$
- **Identity law:** an identity is any number the when applied to another number under a specific operation does not change the result.
  - For Addition the identity is 0
  - For Multiplication the identity is 1
- **Inverse law:** an inverse is any number that when applied to another number produce 0 for addition and 1 for multiplication.
  - Under addition, the inverse of x is $-x$ as $x + (-x) = 0$
  - Under multiplication, the inverse of x is $\frac{1}{x}$ as $x \times \frac{1}{x} = 1$

Solving equations

- Inverse operations are used to solve equations.
  - Add (+) and subtract (−) are inverses
  - Multiply (×) and divide (÷) are inverses
  - Squares (x^2) and square roots ($\sqrt{x}$) are inverses
- One-step equations can be solved using one inverse operation:
  e.g. $$\begin{aligned} x + 5 &= 12 \\ x + 5 - 5 &= 12 - 5 \\ x &= 7 \end{aligned}$$

Solving complex equations

- Solving two-step and multi-step equations will involving the following.
  - Using inverse operations
  - Expanding brackets
  - Collecting like terms
  - Finding the LCM of algebraic fractions, then multiplying by the LCM to remove all denominators

Literal equations

- Literal equations are equations that involve multiple pronumerals or variables.
- The same processes (inverse operations etc.) are used to solve literal equations.
- Solving a literal equation is the same as making one variable the subject of the equation. This means it is expressed in terms of the other variables.
  e.g. P is the subject of the equation: $P = \frac{nRT}{V}$
  To make T the subject, transpose to get: $T = \frac{PV}{nR}$

ALGEBRA & EQUATIONS

Closure

- A set of numbers is closed if under a specific operation, the result produced is also an element of the set.
  e.g. For integers, multiplication is closed as the product of two integers is always an integer. (i.e. $3 \times (-5) = -15$)
  For integers, division is not closed as the quotient of two numbers is often not an integer.
  (i.e. $3 \div (-5) = -\frac{3}{5}$)

Algebraic fractions: + and −

- Fractions can be added and subtracted if they have the same common denominator.
  e.g.
  $$\frac{5}{2x} + \frac{4}{3y} = \frac{15y}{6xy} + \frac{8x}{6xy} = \frac{15y + 8x}{6xy}$$
  Or
  $$\begin{aligned} \frac{3}{(x+2)} - \frac{2}{(x-2)} &= \frac{3(x-2)}{x^2-4} - \frac{2(x+2)}{x^2-4} \\ &= \frac{3x - 6 - 2x - 4}{x^2 - 4} \\ &= \frac{x - 10}{x^2 - 4} \end{aligned}$$

Algebraic fractions: ×

- When multiplying fractions, multiply the numerators together and the denominators together.
- Cancel any common factors in the numerator and denominator.
  e.g.
  $$\frac{5y}{12x} \times \frac{7x^2}{15z} = \frac{(1)\cancel{5}y}{12\cancel{x}} \times \frac{7x^{\cancel{2}}}{(3)\cancel{15}z} = \frac{7xy}{36z}$$
  Cancel common factors from top and bottom.
  Write variables in alphabetical order.

Algebraic fractions: ÷

- When dividing two fractions, multiply the first fraction by the reciprocal of the second.
- The reciprocal of $\frac{a}{b}$ is $\frac{b}{a}$.
  e.g.
  $$\frac{10x^2}{33z} \div \frac{6x^2}{11y} = \frac{(5)\cancel{10x^2}}{(3)\cancel{33}z} \times \frac{(1)\cancel{11}y}{(3)\cancel{6x^2}} = \frac{5y}{9z}$$

2.8.2 Success criteria

Tick the column to indicate that you have completed the subtopic and how well you have understood it using the traffic light system.

(**Green:** I understand; **Yellow:** I can do it with help; **Red:** I do not understand)

Subtopic	Success criteria	Green	Yellow	Red
2.2	I can evaluate an expression by substituting in values for each pronumeral.			
	I understand the Commutative, Associative, Identity and Inverse laws and determine in which situations they apply.			
2.3	I can determine the lowest common denominator of two or more fractions with pronumerals in the denominator.			
	I can add and subtract fractions involving algebraic expressions.			
2.4	I can cancel factors, including algebraic expressions, that are common to the numerator and denominator of fractions.			
	I can multiply and divide fractions involving algebraic expression and simplify the result.			
2.5	I can solve one and two-step equations using inverse operations.			
	I can solve equations with pronumerals on both sides of the equals sign.			
2.6	I can expand brackets and collect like terms in order to solve a multi-step equation.			
	I can solve equations involving algebraic fractions by determining the LCM of the denominators.			
2.7	I can solve a linear literal equation, which include multiple variables, by changing the subject of an equation to a particular pronumeral.			
	I can determine any restrictions on a variable in an equation due to limitations imposed by the equation or context of the question.			

2.8.3 Project

Checking for data entry errors

When entering numbers into an electronic device, or even writing numbers down, errors frequently occur. A common type of error is a transposition error, which occurs when two digits are written in the reverse order. Take the number 2869, for example. With this type of error, it could be written as 8269, 2689 or 2896. A common rule for checking these errors is as follows.

If the difference between the correct number and the recorded number is a multiple of 9, a transposition error has occurred.

We can use algebraic expressions to check this rule. Let the digit in the thousands position be represented by a, the digit in the hundreds position by b, the digit in the tens position by c and the digit in the ones position by d. So the real number can be represented as $1000a + 100b + 10c + d$.

1. If the digits in the ones position and the tens position were written in the reverse order, the number would be $1000a + 100b + 10d + c$. The difference between the correct number and the incorrect one would then be: $1000a + 100b + 10c + d - (1000a + 100b + 10d + c)$.
   a. Simplify this expression.
   b. Is the expression a multiple of 9? Explain.
2. If a transposition error had occurred in the tens and hundreds position, the incorrect number would be $1000a + 100c + 10b + d$. Perform the procedure shown in question 1 to determine whether the difference between the correct number and the incorrect one is a multiple of 9.
3. Consider, lastly, a transposition error in the thousands and hundreds positions. Is the difference between the two numbers a multiple of 9?
4. Comment on the checking rule for transposition errors.

on Resources

 **eWorkbook** Topic 2 Workbook (worksheets, code puzzle and project) (ewbk-2028)

 **Interactivities** Crossword (int-2830)
Sudoku puzzle (int-3589)

Exercise 2.8 Review questions

learnon

To answer questions online and to receive **immediate corrective feedback** and **fully worked solutions** for all questions, go to your learnON title at www.jacplus.com.au.

Fluency

1. **MC** Given $E = \frac{1}{2}mv^2$ where $m = 0.2$ and $v = 0.5$, the value of E is:
   A. 0.000625 B. 0.1 C. 0.005 D. 0.025 E. 0.0025
2. **MC** The expression $-6d + 3r - 4d - r$ simplifies to:
   A. $2d + 2r$ B. $-10d + 2r$ C. $-10d - 4r$ D. $2d + 4r$ E. $-8dr$
3. **MC** The expression $5(2f + 3) + 6(4f - 7)$ simplifies to:
   A. $34f + 2$ B. $34f - 4$ C. $34f - 27$ D. $34f + 14$ E. $116f - 14$
4. **MC** The expression $7(b - 1) - (8 - b)$ simplifies to:
   A. $8b - 9$ B. $8b - 15$ C. $6b - 9$ D. $6b - 15$ E. $8b + 1$
5. **MC** If $14p - 23 = 6p - 7$ then p equals:
   A. −3 B. −1 C. 1 D. 2 E. 4
6. Simplify the following by collecting like terms.
   a. $3c - 5 + 4c - 8$
   b. $-3k + 12m - 4k - 9m$
   c. $-d + 3c - 8c - 4d$
   d. $6y^2 + 2y + y^2 - 7y$
7. If $A = \frac{1}{2}bh$, determine the value of A if $b = 10$ and $h = 7$.

8. For each of the following, complete the relationship to illustrate the stated law.

a. $(a+3b)+6c=$ _________ Associative Law
b. $12a-3b\neq$ _____________ Commutative Law
c. $7p\times$ _____ $=$ ___________ Inverse Law
d. $(x\times 5y)\times 7z=$ __________ Associative Law
e. $12p+0=$ ______________ Identity Law
f. $(3p\div 5q)\div 7r=$ __________ Associative Law
g. $9d+11e=$ ______________ Commutative Law
h. $4a\div b\neq$ _______________ Commutative Law

9. Determine the value of the following expressions given the natural number values $x=12$, $y=8$ and $z=4$. Comment on whether the Closure Law holds for each of the expressions when the values are substituted.

a. $x\times y$
b. $z\div x$
c. $y-x$

10. Simplify the following.

a. $\dfrac{5y}{3}-\dfrac{y}{2}$
b. $\dfrac{x+4}{5}+\dfrac{x+2}{2}$
c. $\dfrac{5}{3x}-\dfrac{1}{5x}$
d. $\dfrac{x-1}{x+3}+\dfrac{2x-5}{x+2}$

11. Simplify the following.

a. $\dfrac{y}{4}\times\dfrac{32}{x}$
b. $\dfrac{20y}{7x}\times\dfrac{35z}{16y}$
c. $\dfrac{x+6}{(x+1)(x+3)}\times\dfrac{5(x+1)}{x+6}$
d. $\dfrac{25}{x}\div\dfrac{30}{x}$
e. $\dfrac{xy}{5}\div\dfrac{10x}{y}$
f. $\dfrac{2x}{(x+8)(x-1)}\div\dfrac{9x+1}{x+8}$

12. Solve the following equations.

a. $p-20=68$
b. $s-0.56=2.45$
c. $3b=48$
d. $\dfrac{r}{7}=-5$
e. $\sqrt{x}=12$
f. $2(x+5)=-3$
g. $\dfrac{y}{4}-3=12$
h. $a^2=36$
i. $5-k=-7$

13. Solve the following.

a. $42-7b=14$
b. $12t-11=4t+5$
c. $2(4p-3)=2(3p-5)$

14. Solve each of the following linear equations.

a. $5(x-2)+3(x+2)=0$
b. $7(5-2x)-3(1-3x)=1$
c. $5(x+1)-6(2x-1)=7(x+2)$
d. $8(3x-2)+(4x-5)=7x$
e. $7(2x-5)-4(x+20)=x-5$
f. $3(x+1)+6(x+5)=3x+40$

15. Solve each of the following equations.

a. $\dfrac{x}{2}+\dfrac{x}{5}=\dfrac{3}{5}$
b. $\dfrac{x}{3}-\dfrac{x}{5}=3$
c. $-\dfrac{1}{21}=\dfrac{x}{7}-\dfrac{x}{6}$
d. $\dfrac{3}{x}+\dfrac{2}{5}=\dfrac{5}{x}$
e. $\dfrac{2x-3}{2}-\dfrac{3}{5}=\dfrac{x+3}{5}$
f. $\dfrac{2(x+2)}{3}=\dfrac{3}{7}+\dfrac{5(x+1)}{3}$

16. a. Make x the subject of $bx+cx=\dfrac{d}{2}$.

b. Make r the subject of $V=\dfrac{4}{3}\pi r^3$.

Problem solving

17. A production is in town and many parents are taking their children. An adult ticket costs \$15 and a child's ticket costs \$8. Every child must be accompanied by an adult and each adult can have no more than 4 children with them. It costs the company \$12 per adult and \$3 per child to run the production. There is a seating limit of 300 people and all tickets are sold.

a. Determine how much profit the company makes on each adult ticket and on each child's ticket.

b. To maximise profit, the company should sell as many children's tickets as possible. Of the 300 available seats, determine how many should be allocated to children if there is a maximum of 4 children per adult.

c. Using your answer to part **b**, determine how many adults would make up the remaining seats.

d. Construct an equation to represent the profit that the company can make depending on the number of children and adults attending the production.

e. Substitute your values to calculate the maximum profit the company can make.

18. You are investigating prices for having business cards printed for your new games store. A local printing company charges a flat rate of \$250 for the materials used and \$40 per hour for labour.

a. If h is the number of hours of labour required to print the cards, construct an equation for the cost of the cards, C.

b. You have budgeted \$1000 for the printing job. Determine the number of hours of labour you can afford. Give your answer to the nearest minute.

c. The printer estimates that it can print 1000 cards per hour of labour. Evaluate the number of cards that will be printed with your current budget.

d. An alternative to printing is photocopying. The company charges 15 cents per side for the first 10 000 cards and then 10 cents per side for the remaining cards. Justify which is the cheaper option for 18 750 single-sided cards and by how much.

19. A scientist tried to use a mathematical formula to predict people's moods based on the number of hours of sleep they had the previous night. One formula that he used was what he called the 'grumpy formula', $g = 0.16(h-8)^2$, which was valid on a 'grumpy scale' from 0 to 10 (least grumpy to most grumpy).

a. Calculate the number of hours needed to not be grumpy.

b. Evaluate the grumpy factor for somebody who has had:

i. 4 hours of sleep

ii. 6 hours of sleep

iii. 10 hours of sleep.

c. Determine the number of hours of sleep required to be most grumpy.

Another scientist already had his own grumpy formula and claims that the scientist above stole his idea and has just simplified it. The second scientist's grumpy formula was

$$g = \frac{0.16(h-8)}{8-h} \times \frac{2(8-h)}{3(h-8)} \div \frac{2h}{3(h-8)^2}$$

d. Write the second scientist's formula in simplified form.

e. Are the second scientist's claims justified? Explain.

on To test your understanding and knowledge of this topic, go to your learnON title at www.jacplus.com.au and complete the **post-test**.

Online Resources

Below is a full list of **rich resources** available online for this topic. These resources are designed to bring ideas to life, to promote deep and lasting learning and to support the different learning needs of each individual.

eWorkbook

Download the workbook for this topic, which includes worksheets, a code puzzle and a project (ewbk-2028) ☐

Solutions

Download a copy of the fully worked solutions to every question in this topic (sol-0736) ☐

Digital documents

2.2 SkillSHEET Like terms (doc-5183) ☐
SkillSHEET Collecting like terms (doc-5184) ☐
SkillSHEET Finding the highest common factor (doc-5185) ☐
SkillSHEET Order of operations (doc-5189) ☐
2.3 SkillSHEET Addition and subtraction of fractions (doc-5186) ☐
SkillSHEET Writing equivalent algebraic fractions with the lowest common denominator (doc-5190) ☐
2.4 SkillSHEET Multiplication of fractions (doc-5187) ☐
SkillSHEET Division of fractions (doc-5188) ☐
SkillSHEET Simplification of algebraic fractions (doc-5191) ☐

Video eLessons

2.2 Substituting values into expressions (eles-4696) ☐
Number laws (eles-4697) ☐
Substitution (eles-1892) ☐
2.3 Algebraic fractions (eles-4698) ☐
Pronumerals in the denominator (eles-4699) ☐
2.4 Multiplying algebraic fractions (eles-4700) ☐
Dividing algebraic fractions (eles-4701) ☐
2.5 Solving equations using inverse operations (eles-4702) ☐
Two-step equations (eles-4704) ☐
Equations where the pronumeral appears on both sides (eles-4705) ☐
Solving linear equations (eles-1895) ☐
Solving linear equations with pronumerals on both sides (eles-1901) ☐
2.6 Equations with multiple brackets (eles-4706) ☐
Equations involving algebraic fractions (eles-4707) ☐
Solving linear equations with algebraic fractions (eles-1857) ☐
2.7 Literal equations (eles-4708) ☐
Restrictions on variables (eles-4709) ☐

Interactivities

2.2 Individual pathway interactivity: Substitution (int-4566) ☐
Substituting positive and negative numbers (int-3765) ☐
Commutative Law (int-6109) ☐
Associative Law (int-6110) ☐
Identity Law (int-6111) ☐
Inverse Law (int-6112) ☐
2.3 Individual pathway interactivity: Adding and subtracting algebraic fractions (int-4567) ☐
Adding and subtracting algebraic fractions (int-6113) ☐
Lowest common denominators with pronumerals (int-6114) ☐
2.4 Individual pathway interactivity: Multiplying and dividing algebraic fractions (int-4568) ☐
Simplifying algebraic fractions (int-6115) ☐
Multiplying algebraic fractions (int-6116) ☐
Dividing algebraic fractions (int-6117) ☐
2.5 Individual pathway interactivity: Solving simple equations (int-4569) ☐
Using algebra to solve problems (int-3805) ☐
One-step equations (int-6118) ☐
Two-step equations (int-6119) ☐
2.6 Individual pathway interactivity: Solving multi-step equations (int-4570) ☐
Expanding brackets: Distributive Law (int-3774) ☐
2.7 Individual pathway interactivity: Literal equations (int-4571) ☐
Restrictions on variables (int-6120) ☐
2.8 Crossword (int-2830) ☐
Sudoku puzzle (int-3589) ☐

Teacher resources

There are many resources available exclusively for teachers online.

To access these online resources, log on to **www.jacplus.com.au**.

Answers

Topic 2 Algebra and equations

Exercise 2.1 Pre-test

1. 12
2. $c = \pm 5$
3. A
4. $\frac{y}{30}$
5. E
6. B
7. $\frac{-2x}{3y}$
8. 100 cm
9. $25\frac{1}{2}$
10. $x = -16$
11. $a = -512$
12. $24 000
13. C
14. A
15. D

Exercise 2.2 Substitution

1. a. 5 b. 2 c. 0 d. 6
2. a. −17 b. 3 c. 30 d. 12
3. a. −12 b. 27 c. 30 d. −5
4. a. −11 b. −1 c. 1
5. a. 30 b. −24 c. 36
6. a. −125 b. 1 c. 15
7. a. $\frac{7}{12}$ b. $-\frac{1}{12}$ c. $\frac{1}{12}$ d. $1\frac{1}{3}$ e. $\frac{1}{576}$ f. 48
8. a. 9 b. −9 c. 9 d. 18 e. −18 f. 36
9. a. 9 b. −9 c. 9 d. 18 e. −18 f. 36
10. a. 17 b. 30 c. 8
11. a. 4 b. 1.5 c. 68
12. a. 46 b. 113.1 c. 5 d. 624.6
13. a. D b. C c. B
14. 3.9 cm
15. 65.45 cm^3
16. 361 m
17. a. −1; in this case, addition is closed on integers.
    b. −1; in this case, subtraction is closed on integers.
    c. 2; in this case, multiplication is closed on integers.
18. a. −1; in this case, division is closed on integers.
    b. −2; in this case, subtraction is closed on integers.
    c. $-\frac{1}{2}$; in this case, division is not closed on integers.
19. a. $(a + 2b) + 4c = a + (2b + 4c)$
    b. $(x \times 3y) \times 5c = x \times (3y \times 5c)$
    c. $2p \div q \neq q \div 2p$
    d. $5d + q = q + 5d$
20. a. 10; in this case, addition is closed on natural numbers.
    b. −4; in this case, subtraction is not closed on natural numbers.
    c. 12; in this case, multiplication is closed on natural numbers.
    d. $\frac{4}{3}$; in this case, division is not closed on natural numbers.
    e. −2; in this case, subtraction is not closed on natural numbers.
    f. 4; in this case, division is closed on natural numbers.
21. a. $3z + 0 = 0 + 3z = 3z$
    b. $2x \times \frac{1}{2x} = \frac{1}{2x} \times 2x = 1$
    c. $(4x \div 3y) \div 5z \neq 4x \div (3y \div 5z)$
    d. $3d - 4y \neq 4y - 3d$
22. a. $s = 71.875$ metres b. $t = 4$ seconds
23. $m = \frac{4}{5}$
24. 2.5 seconds

Exercise 2.3 Adding and subtracting algebraic fractions

1. a. $\frac{26}{21}$ or $1\frac{5}{21}$ b. $\frac{49}{72}$ c. 1
2. a. $\frac{17}{99}$ b. $\frac{1}{35}$ c. $\frac{6-5x}{30}$
3. a. $\frac{15x-4}{27}$ b. $\frac{15-16x}{40}$ c. $\frac{15-2x}{3x}$
4. a. $\frac{5y}{12}$ b. $-\frac{3y}{40}$ c. $\frac{13x}{12}$ d. $\frac{14x}{9}$
5. a. $\frac{3w}{28}$ b. $-\frac{y}{5}$ c. $\frac{89y}{35}$ d. $\frac{32x}{15}$
6. a. $\frac{7x+17}{10}$ b. $\frac{7x+30}{12}$ c. $\frac{2x-11}{30}$ d. $\frac{19x+7}{6}$
7. a. $\frac{5}{8x}$ b. $\frac{5}{12x}$ c. $\frac{38}{21x}$
8. a. $\frac{8}{3x}$ b. $\frac{7}{24x}$ c. $\frac{9}{20x}$
9. a. $\frac{37}{100x}$ b. $\frac{51}{10x}$ c. $-\frac{1}{6x}$
10. a. $\frac{3x^2+14x-4}{(x+4)(x-2)}$ b. $\frac{2x^2+3x+25}{(x+5)(x-1)}$ c. $\frac{2x^2+6x-10}{(2x+1)(x-2)}$ d. $\frac{4x^2-17x-3}{(x+1)(2x-7)}$
11. a. $\frac{7x^2+x}{(x+7)(x-5)}$ b. $\frac{2x^2+6x+7}{(x+1)(x+4)}$ c. $\frac{-x^2+7x+15}{(x+1)(x+2)}$ d. $\frac{x-7}{(x+3)(x-2)}$

12. a. $\dfrac{x^2+3x+9}{(x+2)(3x-1)}$ b. $\dfrac{5-5x}{(x-1)(1-x)}=\dfrac{5}{x-1}$
c. $\dfrac{3x+7}{(x+1)^2}$ d. $\dfrac{3x-4}{(x-1)^2}$

13. a. The student transcribed the denominator incorrectly and wrote $(x+2)$ instead of $(x-2)$ in line 2.
Also, the student forgot that multiplying a negative number by a negative number gives a positive number. Line 3 should have $+3$ in the numerator, not -1. They didn't multiply.
b. $\dfrac{x^2-5x+3}{(x-1)(x-2)}$

14. a. -1 b. $\dfrac{-1}{(x-2)}$

15. a. $\dfrac{9}{(x-3)^2}$ b. $\dfrac{4}{(x-2)^3}$

16. a. $\dfrac{4x^2+17x+17}{(x+2)(x+1)(x+3)}$ b. $\dfrac{7x^2-20x+4}{(x-1)(x+2)(x-4)}$

17. a. $\dfrac{4x^2+17x+19}{(x+1)(x+3)(x+2)}$ b. $\dfrac{2\left(2x^2-9x+25\right)}{(x-4)(x-1)(x+3)}$
c. The lowest common denominator may not always be the product of the denominators. Each fraction must be multiplied by the correct multiple.

18. $a=4$

19. $\dfrac{4(x-1)}{(x+3)(x+4)(x-2)}$

20. $\dfrac{2(x-1)}{(x-7)(x-4)}$

21. $\dfrac{8(x-1)}{(x-4)(x+3)}$

Exercise 2.4 Multiplying and dividing algebraic fractions

1. a. $\dfrac{4x}{y}$ b. $\dfrac{3x}{y}$ c. $\dfrac{4y}{x}$ d. $\dfrac{9x}{4y}$

2. a. $\dfrac{-5x}{4y}$ b. $\dfrac{3w}{2x}$ c. $\dfrac{6z}{7x}$ d. $\dfrac{2z}{7x}$

3. a. $\dfrac{-3x}{2y}$ b. $\dfrac{5}{24}$ c. $\dfrac{12z}{x}$ d. $\dfrac{-x}{6w}$

4. a. $\dfrac{2}{3x-2}$ b. $\dfrac{5}{x-3}$ c. $\dfrac{9}{2(x-6)}$ d. $\dfrac{1}{x+3}$

5. a. $\dfrac{2x}{(x+1)^2}$ b. $\dfrac{x+1}{2(2x-3)}$
c. $\dfrac{a}{10(a+3)}$ d. $\dfrac{35d}{8(d-3)}$

6. a. $\dfrac{9}{32x^2(x-2)}$ b. $\dfrac{3x}{10(x-1)}$

7. a. $\dfrac{3}{5}$ b. $\dfrac{2}{9}$ c. $\dfrac{1}{3}$ d. 3

8. a. $\dfrac{1}{25}$ b. $\dfrac{35}{6}$ or $5\dfrac{5}{6}$
c. $\dfrac{4y^2}{7}$ d. $\dfrac{2y^2}{25}$

9. a. $\dfrac{8y^2}{9}$ b. $\dfrac{32xy}{15}$ c. $\dfrac{2}{3}$ d. y^2

10. a. $\dfrac{9}{(3x-7)(x+3)}$ b. $\dfrac{1}{(x+2)(x-9)}$

11. a. $\dfrac{21(x-3)}{x+5}$ b. $\dfrac{13}{9(x-4)(x+1)}$

12. a. $\dfrac{2(x-4)^3}{(x+3)^2}$ b. $\dfrac{(x+2)^2}{(x+3)(x-3)}$

13. a. $\dfrac{15}{(x+2)}$ b. $\dfrac{12}{x^2}$

14. a. $\dfrac{(x-4)(x-5)}{(x+3)(x+2)^2}$ b. $\dfrac{2x(x-3)}{3(x-5)}$

15. a. $\dfrac{(x-5)}{(x-4)}$ b. $\dfrac{(x+4)(x+6)}{x^2}$

16. Yes, because all of the fractions have the same denominator and therefore can be added together.

17. No, x and z are not common to all terms so cannot be cancelled down.

18. a. -1
b. $4-x$ considered to be the same as $x-4$.

19. 1

20. $\dfrac{(x+1)^2}{x^2+1}$

21. $\dfrac{-1}{x}$

Exercise 2.5 Solving simple equations

1. a. $a=24$ b. $k=121$ c. $g=2.9$

2. a. $r=3$ b. $h=0.26$ c. $i=-2$

3. a. $t=5$ b. $q=\dfrac{1}{6}$ c. $x=0$

4. a. $f=12$ b. $i=-60$ c. $z=-7$

5. a. $v=7$ b. $w=-5\dfrac{1}{3}$ c. $k=10$

6. a. $a=0.425$ b. $m=16\dfrac{5}{8}$ c. $y=21\dfrac{1}{2}$

7. a. $t=100$ b. $y=\pm17$ c. $q=6.25$

8. a. $f=\pm1.2$ b. $h=\dfrac{16}{49}$ c. $p=\pm\dfrac{3}{8}$

9. a. $g=\dfrac{225}{484}$ b. $j=\pm\dfrac{14}{31}$ c. $a=\pm1\dfrac{2}{3}$

10. a. $t=25$ b. $x=\pm6$ c. $m=16$
d. $t=\pm3$ e. $t=\pm10$ f. $m=25$

11. a. $x=8$ b. $x=-3$ c. $m=\dfrac{1}{8}$
d. $x=\dfrac{3}{4}$ e. $m=0.008$ f. $w=2\dfrac{1}{2}$

12. a. $x=-1$ b. $x=-2$ c. $m=1$
d. $w=-512$ e. $t=125$ f. $x=2$

13. a. $a=4$ b. $b=6$ c. $i=3$

14. a. $f=9$ b. $q=1\dfrac{1}{8}$ c. $r=5\dfrac{2}{5}$

15. a. $s=4\frac{5}{6}$ b. $t=9\frac{4}{5}$ c. $a=-7\frac{1}{2}$
16. a. $f=40$ b. $g=30$ c. $r=-10$
17. a. $m=18$ b. $n=28$ c. $p=62.4$
18. a. $x=1\frac{1}{3}$ b. $y=9$ c. $m=4\frac{2}{5}$
19. a. $k=1\frac{1}{2}$ b. $n=5\frac{2}{3}$ c. $c=1\frac{1}{3}$
20. a. $x=10$ b. $x=4$ c. $x=-7\frac{1}{5}$
21. a. $k=25$ b. $m=16$ c. $p=-11\frac{3}{7}$
22. a. $u=-4\frac{1}{8}$ b. $x=\frac{8}{11}$ c. $v=3$
23. a. $x=26$ b. $m=-5$ c. $w=\frac{25}{3}$
24. a. $t=5$ b. $x=9$ c. $n=-\frac{19}{3}$
25. a. B b. E c. C
26. a. $h=-2$ b. $d=-1$ c. $p=7$
d. $x=-11$
27. a. $t=5$ b. $v=-20$ c. $r=-3$
d. $g=-0.8$
28. a. $x=-1$ b. $v=1$ c. $l=2$
d. $g=-2$
29. a. $t=3$ b. $e=-23\frac{1}{3}$ c. $k=-36$
d. $f=-12\frac{1}{4}$
30. a. $x=2$ b. $b=5$ c. $w=2$
31. a. $f=7$ b. $t=3$ c. $r=2\frac{1}{3}$
32. a. $g=-1\frac{1}{3}$ b. $h=-2\frac{1}{5}$ c. $a=0$
33. a. $x=-1$ b. $c=2$ c. $r=2\frac{2}{3}$
34. a. $k=1$ b. $y=-1\frac{1}{8}$ c. $g=7$
35. a. $w=1$ b. $m=\frac{1}{5}$ c. $p=1\frac{2}{3}$
36. a. $x=-15$ b. $y=-4\frac{4}{5}$ c. $t=21$
37. a. $u=-2\frac{5}{7}$ b. $f=12\frac{1}{2}$ c. $r=7\frac{1}{2}$
38. a. $d=-6$ b. $h=-12$ c. $x=1$
39. a. A b. D c. B
40. a. 6 cm b. 1.26 m
41. a. 2.5 cm b. 41 cm
42. a. 5 m b. 2.8 s
43. a. 314 cm^2 b. 6.3 cm
44. 1.8 cm
45. 6.2 cm
46. Dimensions are 10 m by 6 m.
47. Leon is 14 and James is 34.
48. $\frac{10\sqrt{3}}{3}$ cm

Exercise 2.6 Solving multi-step equations

1. a. $x=\frac{20}{31}$ b. $x=3\frac{5}{8}$
2. a. $x=\frac{29}{36}$ b. $x=-7$
3. a. $x=-2\frac{8}{11}$ b. $x=\frac{10}{43}$
4. a. $x=3$ b. $x=12$ c. $x=-2$
5. a. $x=\frac{3}{2}$ b. $x=-\frac{11}{3}$ or $x=-3\frac{2}{3}$
c. $x=\frac{2}{13}$
6. a. $x=3$ b. $x=\frac{5}{7}$ c. $x=\frac{7}{18}$
7. a. $x=\frac{5}{17}$ b. $x=15$
c. $x=-6\frac{2}{9}$ d. $x=-\frac{10}{19}$
8. a. $x=-1\frac{1}{2}$ b. $x=-192$
c. $x=\frac{4}{7}$ d. $x=12$
9. a. $x=3\frac{1}{4}$ b. $x=3$
c. $x=52$ d. $x=1\frac{5}{8}$
10. a. $x=\frac{5}{19}$ b. $x=1\frac{31}{58}$
c. $x=4\frac{11}{14}$ d. $x=-3\frac{15}{17}$
11. a. $x=5\frac{20}{43}$ b. $x=-1\frac{10}{13}$
c. $x=1\frac{2}{61}$ d. $x=-4\frac{9}{26}$
12. a. $x=1.5$ b. $x=-4\frac{1}{3}$
c. $x=3$ d. $x=1$
13. $180
14. 60 hours
15. $12000
16. $60
17. a. 4, 9, 25, 49 b. 16, 81 c. 64
18. a. $6x-450=1000$
b. $241\frac{1}{3}$ tickets. This means they need to sell 242 tickets to qualify, as the number of tickets must be a whole number.
19. Sample responses can be found in the worked solutions in the online resources.
20. 4
21. $a=3, b=5$
22. $a=-8$ and $b=15$

Exercise 2.7 Literal equations

1. a. $x=\frac{bcd}{a}$ b. $x=a(d+bc)$ c. $x=(m-n)^2$

2. a. $x=\pm\sqrt{\frac{w}{ac}}$ b. $x=\frac{ay}{b}$ c. $x=nw-m$

3. a. $x=\frac{c}{ab}-b$ b. $x=\frac{ac}{b+mc}$
   c. $x=\frac{ay}{m+b}$ d. $x=\frac{mc-amd}{d}$

4. a. $l=\frac{V}{bh}$ b. $b=\frac{P-2l}{2}$
   c. $h=\frac{2A}{b}$ d. $a=\pm\sqrt{c^2-b^2}$

5. a. $C=\frac{5}{9}(F-32)$ b. $r=\pm\sqrt{\frac{A}{\pi}}$
   c. $a=\frac{v-u}{t}$ d. $N=\frac{100I}{PR}$
   e. $m=\frac{2E}{v^2}$

6. a. $v=\pm\sqrt{\frac{2E}{m}}$ b. $a=\frac{v^2-u^2}{2s}$
   c. $u=\pm\sqrt{v^2-2as}$ d. $a=\frac{xb}{b-x}$
   e. $x_1=\frac{x(m+n)-mx_2}{n}$

7. a. $a=\pm12$ b. $h=7$ c. $C=30$

8. a. $a=9.8$ b. $s=\pm13$ c. $u=\pm11$

9. a. $a=25$ b. $x_2=13$

10. a. i. No restrictions on x
    ii. $x=\pm\sqrt{y-4}$
    iii. $y\geq4$
    b. i. $x\neq3$
    ii. $x=\frac{2}{y}+3$
    iii. $y\neq0$
    c. i. No restrictions
    ii. $t=\frac{v-u}{a}$
    iii. $a\neq0$

11. a. i. $c\geq0$
    ii. $b=\pm\sqrt{c^2-a^2}$
    iii. $|c|\geq|a|$
    b. i. $r\neq1$
    ii. $r=\frac{s-a}{s}$
    iii. $s\neq0$
    c. i. $p\neq-q$
    ii. $b=\frac{m(p+q)-qa}{p}$
    iii. $p\neq0$

12. a. i. $a\neq0, b^2\geq4ac$
    ii. $c=\frac{b^2-(2ax+b)^2}{4a}$ or $c=-ax^2-bx$
    iii. No new restrictions
    b. i. $p\neq-q$
    ii. $p=\frac{q(a-m)}{m-b}$
    iii. $m\neq b$
    c. i. $E>(pc)$
    ii. $m=\frac{\sqrt{E^2-(pc)^2}}{c^2}$
    iii. $c\neq0$

13. a. No restriction, but all values must be positive for the trapezium to exist.
    b. $b=\frac{2A}{h}-a$
    c. $b=11$ cm

14. a. No restrictions, all values must be positive for a cylinder to exist.
    b. $r=\sqrt{\frac{V}{\pi h}}$
    c. $h\neq0$, no new restrictions

15. a. T and l must be greater than zero.
    b. $l=\frac{T^2g}{4\pi^2}$
    c. The restrictions still hold.
    d. 2.2 m

16. a. $C=\frac{5}{9}(F-32)$
    b. $-40°$

17. a. $m_1=\frac{Fr^2}{Gm_2}$ b. 7.41×10^{22} kg

18. Distance Jing Jing has ridden is $\frac{ja}{j-p}$ kilometres.

Project

1. a. $9(c-d)$
   b. Yes, this is a multiple of 9 as the number that multiples the brackets is 9.
2. $90(b-c)$; 90 is a multiple of 9 so the difference between the correct and incorrect one is a multiple of 9.
3. $900(a-b)$; again 900 is a multiple of 9.
4. If two adjacent digits are transposed, the difference between the correct number and the transposed number is a multiple of 9.

Exercise 2.8 Review questions

1. D
2. B
3. C
4. B
5. D
6. a. $7c-13$ b. $-7k+3m$
   c. $-5c-5d$ d. $7y^2-5y$
7. 35
8. a. $(a+3b)+6c=a+(3b+6c)$
   b. $12a-3b\neq3b-12a$
   c. $7p\times\frac{1}{7p}=\frac{1}{7p}\times7p=1$
   d. $(x\times5y)\times7z=x\times(5y\times7z)$

e. $12p + 0 = 0 + 12p = 12p$

f. $(3p \div 5q) \div 7r \neq 3p \div (5q \div 7r)$

g. $9d + 11e = 11e + 9d$

h. $4a \div b \neq b \div 4a$

9. a. 96; in this case, multiplication is closed on natural numbers.

b. $\frac{1}{3}$; in this case, division is not closed on natural numbers.

c. -4; in this case, subtraction is not closed on natural numbers.

10. a. $\frac{7y}{6}$ b. $\frac{7x + 18}{10}$

c. $\frac{22}{15x}$ d. $\frac{3x^2 + 2x - 17}{(x + 3)(x + 2)}$

11. a. $\frac{8y}{x}$ b. $\frac{25z}{4x}$ c. $\frac{5}{x + 3}$

d. $\frac{5}{6}$ e. $\frac{y^2}{50}$ f. $\frac{2x}{(x - 1)(9x + 1)}$

12. a. $p = 88$ b. $s = 3.01$ c. $b = 16$

d. $r = -35$ e. $x = 144$ f. $x = -\frac{13}{2}$

g. $y = 60$ h. $a = \pm 6$ i. $k = 12$

13. a. $b = 4$ b. $t = 2$ c. $p = -2$

14. a. $x = \frac{1}{2}$ b. $x = 6\frac{1}{5}$ c. $x = -\frac{3}{14}$

d. $x = 1$ e. $x = 12\frac{2}{9}$ f. $x = 1\frac{1}{6}$

15. a. $x = \frac{6}{7}$ b. $x = 22\frac{1}{2}$ c. $x = 2$

d. $x = 5$ e. $x = 3\frac{3}{8}$ f. $x = -\frac{16}{21}$

16. a. $x = \frac{d}{2(b + c)}$ b. $r = \sqrt[3]{\frac{3V}{4\pi}}$

17. a. \$3 per adult ticket; \$5 per child's ticket.

b. 240

c. 60

d. $P = 3a + 5c$, where $a =$ number of adults and $c =$ number of children.

e. \$1380

18. a. $C = 250 + 40h$

b. 18 hours 45 minutes

c. 18 750

d. Printing is the cheaper option by \$1375.

19. a. 8 hours

b. i. 2.56

ii. 0.64

iii. 0.64

c. 0.094 hours or 15.9 hours

d. $g = \frac{0.16(h - 8)^2}{h}$

e. No, the formula is not the same.

3 Coordinate geometry

LEARNING SEQUENCE

3.1 Overview

Why learn this?

Coordinate geometry in many ways represents the foundation upon which your understanding in maths will be built upon over the final years of your secondary schooling. The principles you learn in this topic will be applied to a variety of contexts you encounter as you learn about higher order polynomial functions and conic sections. Indeed, skills presented in this subject, such as determining the midpoint and length of a line segment, are regularly applicable to the study of differential calculus, which forms a large part of your study in the final years of high-school mathematics.

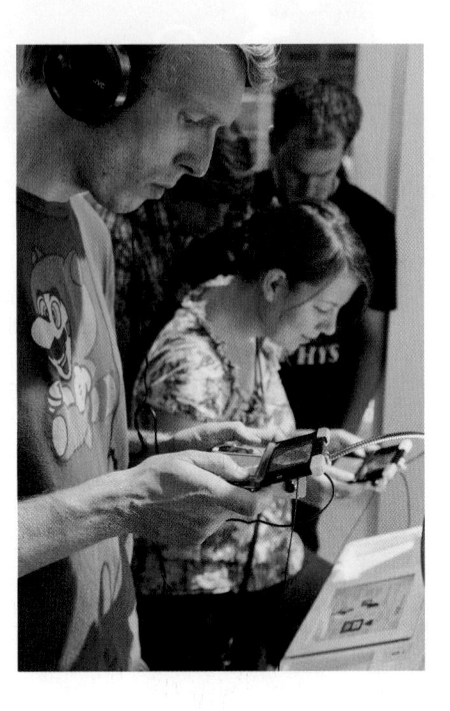

In the world beyond education, understanding the principles of coordinate geometry will help you model real-world data and behaviour, interpret the nature of market trends and population trends, and determine points of market equilibrium in the finance sector. A knowledge of algebra, linear quadratic and simultaneous equations is used to create the computer games. Establishing a relationship between variables is also fundamental to the study of science, and the principles learned in this topic will help inform your understanding of the world around us!

Where to get help

Go to your learnON title at **www.jacplus.com.au** to access the following digital resources. The Online Resources Summary at the end of this topic provides a full list of what's available to help you learn the concepts covered in this topic.

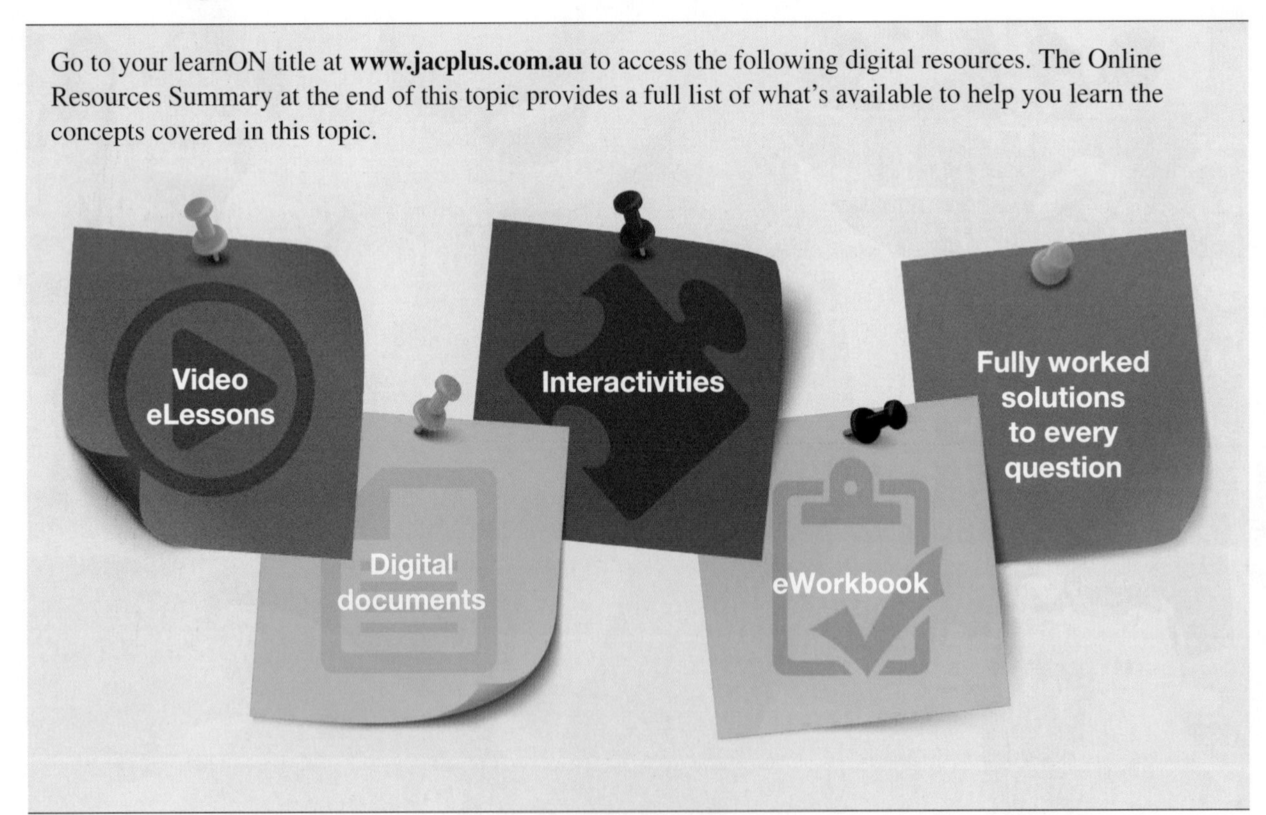

Exercise 3.1 Pre-test

learnon

Complete this pre-test in your learnON title at www.jacplus.com.au and receive **automatic marks, immediate corrective feedback** and **fully worked solutions**.

1. **MC** Lines that have the same gradient are:
   A. parallel
   B. collinear
   C. perpendicular
   D. of same lengths
   E. of different lengths

2. Determine the x-intercept of the line $6x + y - 3 = 0$.

3. Sammy has \$35 credit from an App Store. She only buys apps that cost \$2.50 each. Calculate the number of apps Sammy can buy and still have \$27.50 credit.

4. Determine the equation of the line, in the form $y = mx + c$.

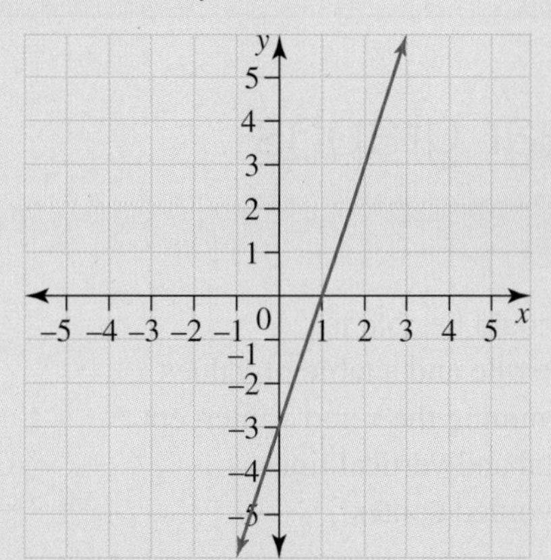

5. **MC** The distance between the points $(-3a, 6b)$ and $(a, 2b)$ is:
   A. $\sqrt{4a^2 + 16b^2}$
   B. $\sqrt{2a^2 + 4b^2}$
   C. $\sqrt{2a^2 + 8b^2}$
   D. $4\sqrt{a^2 + b^2}$
   E. $16\sqrt{a^2 + b^2}$

6. **MC** Identify the equation of the vertical line passing through the point $(-2, 3)$.
   A. $y = -2$
   B. $x = -2$
   C. $y = 3$
   D. $x = 3$
   E. $x = 2$

7. The distance between the points $(-3, 10)$ and $(6, a)$ is 15 units. Determine the possible values of a. Write the lowest value first.

8. **MC** The midpoint of a line segment AB is $(3, -2)$. If the coordinates of A are $(10, 7)$, the coordinates of B are:
   A. $\left(\frac{13}{2}, \frac{5}{2}\right)$
   B. $\left(\frac{7}{2}, \frac{9}{2}\right)$
   C. $\left(\frac{7}{2}, \frac{5}{2}\right)$
   D. $(5, 9)$
   E. $(-4, 11)$

9. **MC** The equation of the straight line, in the form $y = mx + c$, passing through the point $(3, -1)$ with a gradient of -2 is:
   A. $y = -2x + 2$
   B. $y = -2x + 5$
   C. $y = -2x + 3$
   D. $y = -2x - 1$
   E. $y = 3x - 1$

10. **MC** The equation of the straight line, in the form $by + ax = k$, that passes through $\left(2, -\frac{1}{2}\right)$ and $\left(-6, \frac{3}{2}\right)$ is:
    A. $2y - 4x = 15$
    B. $2y + 4x = 7$
    C. $2y + 8x = 7$
    D. $2y + 8x = 15$
    E. $2y - 8x = 15$

11. ABCD is a parallelogram. The coordinates are A(3, 8), B(6, 1), C(4, −1) and D(1, a). Calculate the value of a.

12. Determine the equation of the straight line, in the form $y = mx + c$, that passes through the midpoint of A(0.5, −3) and B(−2.5, 7) and has a gradient of −2.

13. If $2y + 5x = 7$ is perpendicular to $3y + 12 = nx$, determine the value of n.

14. **MC** A is the point (−3, 2) and B is the point (7, −4).
The equation of the perpendicular bisector of AB, in the form $y = mx + c$, is:
A. $y = \frac{5}{3}x - \frac{13}{3}$ **B.** $y = \frac{5}{3}x + \frac{11}{3}$ **C.** $y = -\frac{3}{5}x + \frac{1}{5}$ **D.** $y = -\frac{3}{5}x + \frac{7}{5}$ **E.** $y = -\frac{3}{5}x + \frac{11}{5}$

15. Write the equation of the straight line $8(y - 2) = -2(x + 3)$ in the standard form $y = mx + c$.

3.2 Sketching linear graphs

LEARNING INTENTION

At the end of this subtopic you should be able to:
- plot points on a graph using a rule and a table of values
- sketch linear graphs by determining the x and y intercept
- sketch the graphs of horizontal and vertical lines
- model linear graphs from a worded context.

3.2.1 Plotting linear graphs

eles-4736

- If a series of points (x, y) is plotted using the rule $y = mx + c$, then the points always lie in a straight line whose gradient equals m and whose y-intercept equals c.
- The rule $y = mx + c$ is called the equation of a straight line written in 'gradient–intercept' form.
- To plot a **linear graph**, complete a table of values to determine the points.

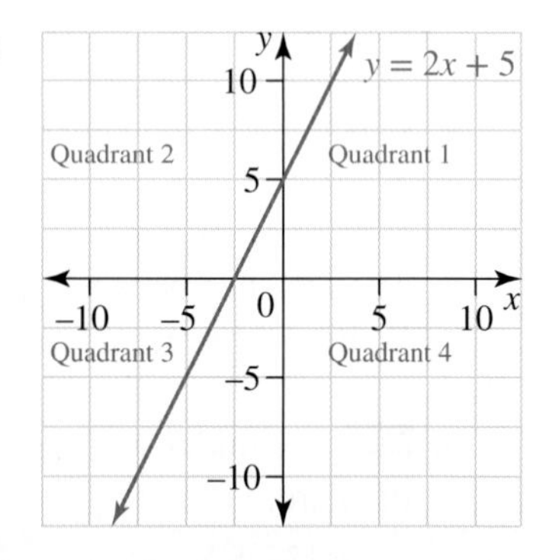

WORKED EXAMPLE 1 Plotting linear graphs

Plot the linear graph defined by the rule $y = 2x - 5$ for the x-values −3, −2, −1, 0, 1, 2 and 3.

THINK

1. Create a table of values using the given x-values.

WRITE/DRAW

x	−3	−2	−1	0	1	2	3
y							

2. Determine the corresponding y-values by substituting each x-value into the rule.

x	−3	−2	−1	0	1	2	3
y	−11	−9	−7	−5	−3	−1	1

3. Plot the points on a Cartesian plane and rule a straight line through them. Since the x-values have been specified, the line should only be drawn between the x-values of −3 and 3.
4. Label the graph.

TI | THINK

1. In a new document, on a Lists & Spreadsheet page, label column A as x and label column B as y. Enter the x-values into column A. Then in cell B1, complete the entry line as: $= 2a1 - 5$ Then press ENTER.

2. Highlight cell B1, then press CTRL then click (the button in the middle of the direction arrows). Press the down arrow until you reach cell B7 then press ENTER.

CASIO | THINK

1. On the Spreadsheet screen, enter the x-values into column A. Then in cell B1, complete the entry line as: $= 2A1 - 5$ Then press EXE.

DISPLAY/WRITE

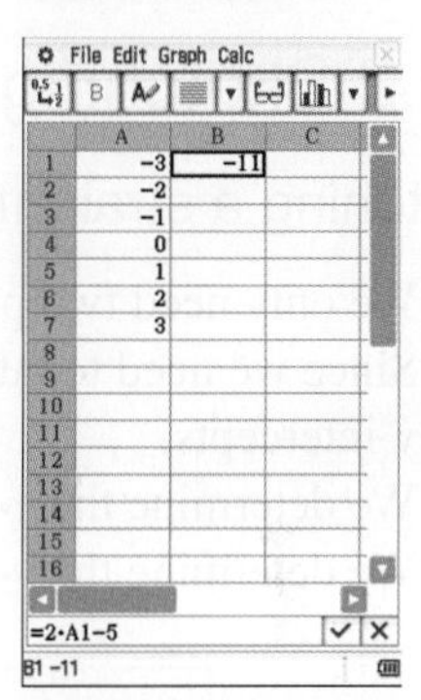

2. Highlight cell B1 to B7, then tap:
   - Edit
   - Fill
   - Fill Range
   - OK

DISPLAY/WRITE

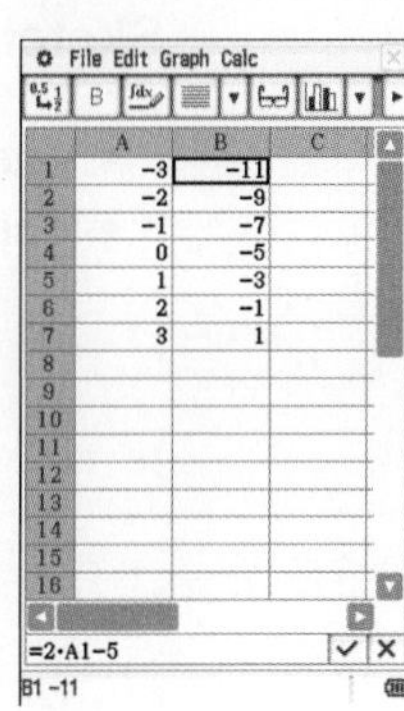

3. Open a Data & Statistics page.
Press TAB to locate the label of the horizontal axis and select the variable x.
Press TAB again to locate the label of the vertical axis and select the variable y. The graph will be plotted as shown.

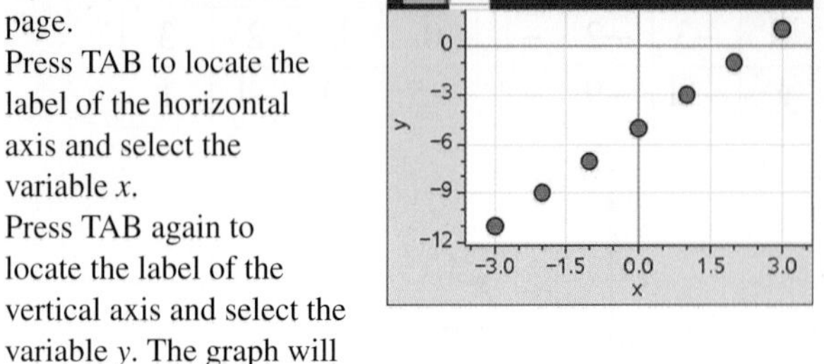

4. To join the dots with a line, press:
- MENU
- 2: Plot Properties
- 1: Connect Data Points

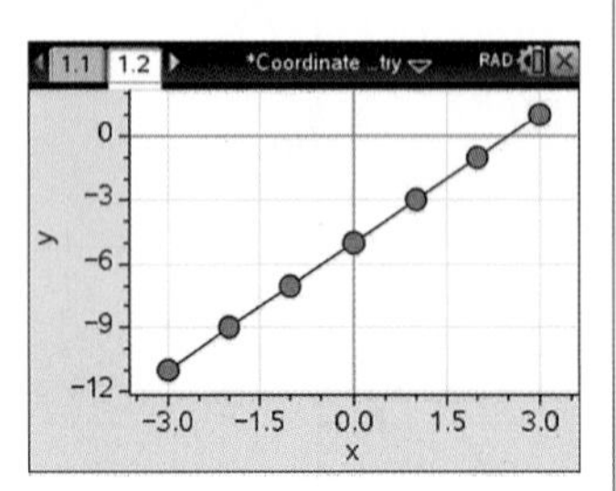

3. Highlight cells A1 to B7, then tap:
- Graph
- Scatter

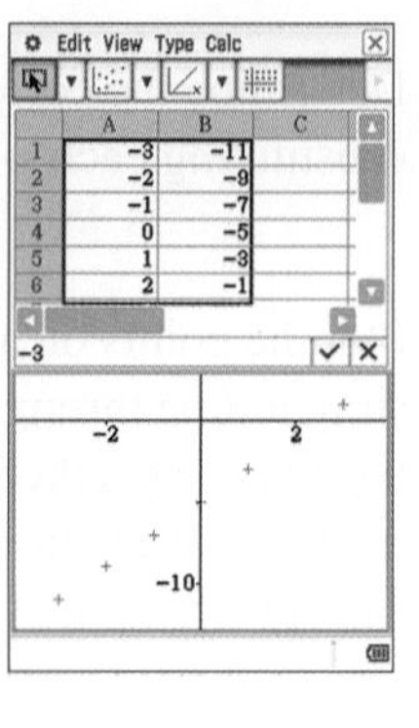

4. To join the dots with a line, tap:
- Calc
- Regression
- Linear Reg

Note that the equation is given, if required.

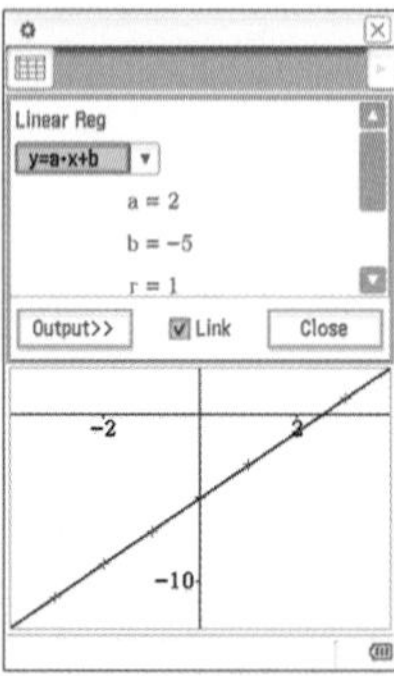

3.2.2 Sketching linear graphs

eles-4737

Sketching a straight line using the x- and y-intercepts

- We only need two points in order to sketch a straight-line (linear) graph.
- Since we need to label all critical points, it is most efficient to plot these graphs by determining the x- and y-intercepts.
- We determine the x-**intercept** by substituting $y = 0$.
- We determine the y-**intercept** by substituting $x = 0$.

Sketching a straight-line graph

- The x- and y-intercepts need to be labelled.
- The equation needs to be labelled.

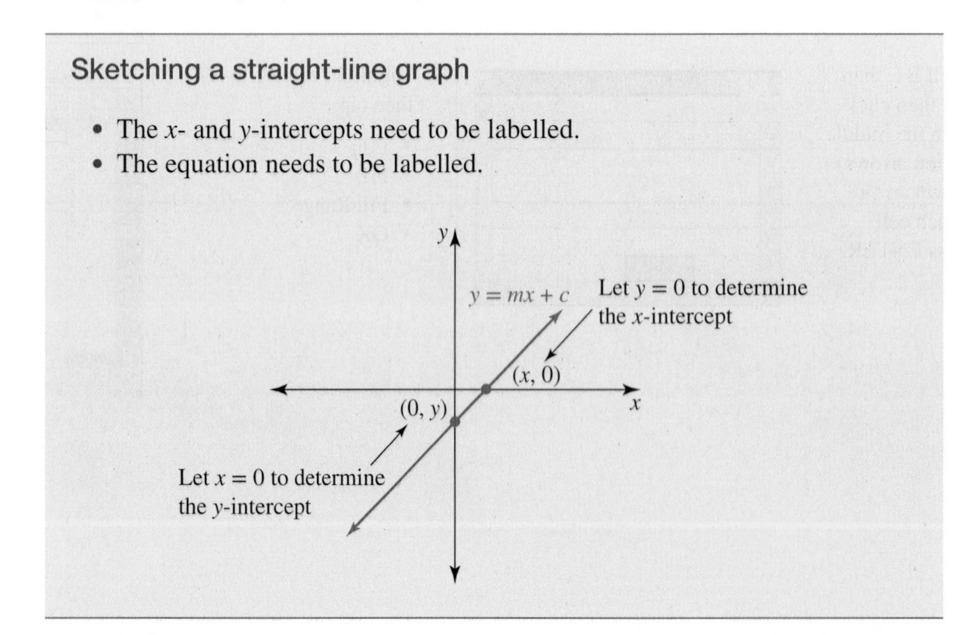

WORKED EXAMPLE 2 Sketching linear graphs

Sketch graphs of the following linear equations.

a. $2x + y = 6$ **b.** $y = -3x - 12$

THINK	WRITE/DRAW
a. 1. Write the equation.	**a.** $2x + y = 6$
2. Determine the x-intercept by substituting $y = 0$.	x-intercept: when $y = 0$, $2x + 0 = 6$ $2x = 6$ $x = 3$ x-intercept is (3, 0).
3. Determine the y-intercept by substituting $x = 0$.	y-intercept: when $x = 0$, $2(0) + y = 6$ $y = 6$ y-intercept is (0, 6).
4. Plot both points and rule the line. **5.** Label the graph.	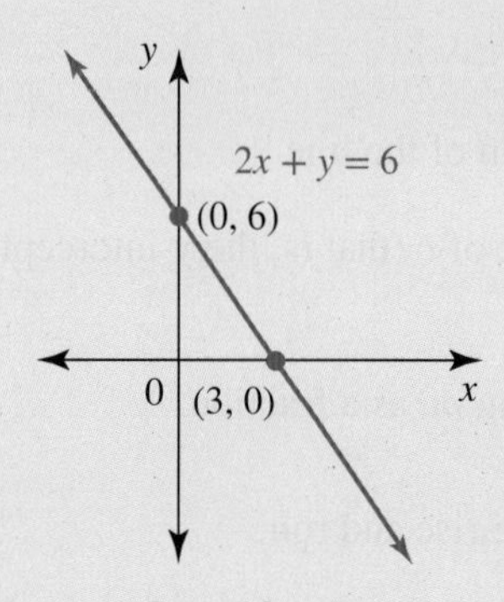
b. 1. Write the equation.	**b.** $y = -3x - 12$
2. Determine the x-intercept by substituting $y = 0$ i. Add 12 to both sides of the equation. ii. Divide both sides of the equation by -3.	x-intercept: when $y = 0$, $-3x - 12 = 0$ $-3x = 12$ $x = -4$ x-intercept is (−4, 0).
3. Determine the y-intercept. The equation is in the form $y = mx + c$, so compare this with our equation to determine the y-intercept, c.	$c = -12$ y-intercept is (0, −12).
4. Plot both points and rule the line. **5.** Label the graph.	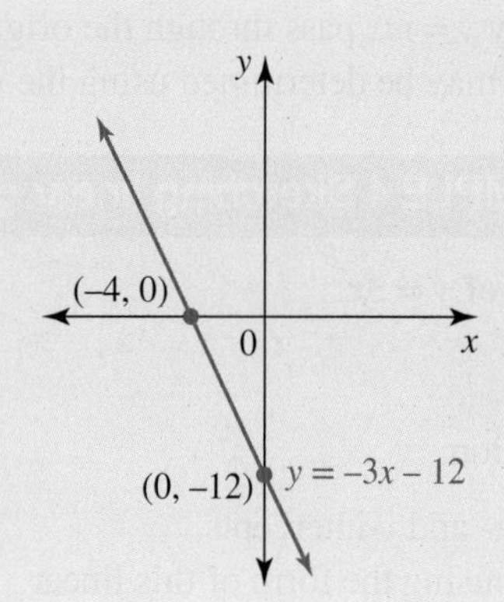

Sketching a straight line using the gradient–intercept method

- This method is often used if the equation is in the form $y = mx + c$, where m represents the gradient (slope) of the straight line, and c represents the y-intercept.
- The steps below outline how to use the gradient–intercept method to sketch a linear graph.

Step 1: Plot a point at the y-intercept.

Step 2: Write the gradient in the form $m = \frac{\text{rise}}{\text{run}}$. (To write a whole number as a fraction, place it over a denominator of 1.)

Step 3: Starting from the y-intercept, move up the number of units suggested by the rise (move down if the gradient is negative).

Step 4: Move to the right the number of units suggested by the run and plot the second point.

Step 5: Rule a straight line through the two points.

WORKED EXAMPLE 3 Sketching more linear graphs

Sketch the graph of $y = \frac{2}{5}x - 3$ using the gradient–intercept method.

THINK	WRITE
1. Write the equation of the line.	$y = \frac{2}{5}x - 3$
2. Identify the value of c (that is, the y-intercept) and plot this point.	$c = -3$, so y-intercept: $(0, -3)$
3. Write the gradient, m, as a fraction.	$m = \frac{2}{5}$
4. $m = \frac{\text{rise}}{\text{run}}$, note the rise and run.	So rise = 2; run = 5.
5. Starting from the y-intercept at $(0, -3)$, move 2 units up and 5 units to the right to find the second point $(5, -1)$. We have still not found the x-intercept.	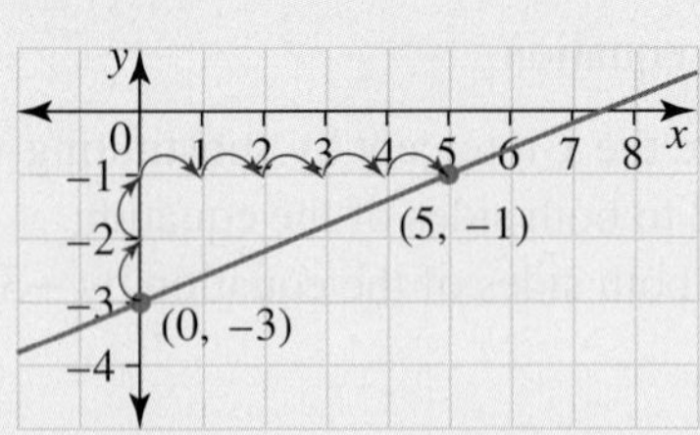

Sketching linear graphs of the form $y = mx$

eles-4738

- Graphs given by $y = mx$ pass through the origin $(0, 0)$, since $c = 0$.
- A second point may be determined using the rule $y = mx$ by substituting a value for x to determine y.

WORKED EXAMPLE 4 Sketching linear graphs of the form $y = mx$

Sketch the graph of $y = 3x$.

THINK	WRITE/DRAW
1. Write the equation.	$y = 3x$
2. Determine the x- and y-intercepts. *Note:* By recognising the form of this linear equation, $y = mx$ you can simply state that the graph passes through the origin, $(0, 0)$.	x-intercept: when $y = 0$, $0 = 3x$ $x = 0$ y-intercept: $(0, 0)$ Both the x- and y-intercepts are at $(0, 0)$.

3. Determine another point to plot by calculating the y-value when $x = 1$.	When $x = 1$, $\quad y = 3 \times 1$ $\qquad\qquad\quad = 3$ Another point on the line is (1, 3).
4. Plot the two points (0, 0) and (1, 3) and rule a straight line through them. **5.** Label the graph.	

3.2.3 Sketching linear graphs of the form $y = c$ and $x = a$

eles-4739

- The line $y = c$ is parallel to the x-axis, having a gradient of zero and a y-intercept of c.
- The line $x = a$ is parallel to the y-axis and has an undefined (infinite) gradient.

Horizontal and vertical lines

- Horizontal lines are in the form $y = c$.
- Vertical lines are in the form $x = a$.

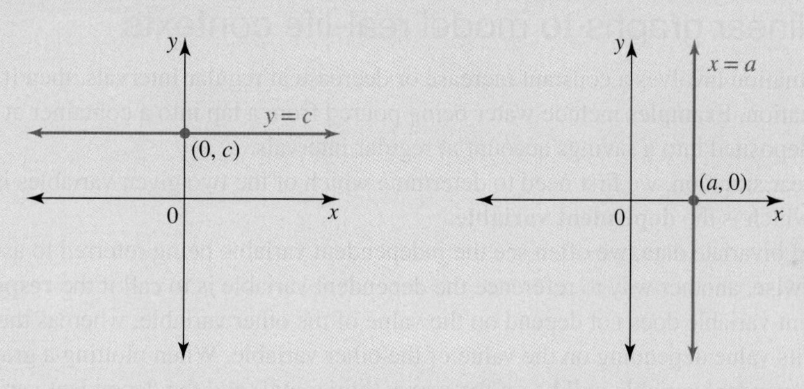

WORKED EXAMPLE 5 Sketching graphs of the form $y = c$ and $x = a$

Sketch graphs of the following linear equations.

a. $y = -3$ **b.** $x = 4$

THINK	WRITE/DRAW
a. 1. Write the equation.	**a.** $y = -3$
2. The y-intercept is -3. As x does not appear in the equation, the line is parallel to the x-axis, such that all points on the line have a y-coordinate equal to -3. That is, this line is the set of points $(x, -3)$ where x is an element of the set of real numbers.	y-intercept $= -3$, $(0, -3)$

3. Sketch a horizontal line through $(0, -3)$.

4. Label the graph.

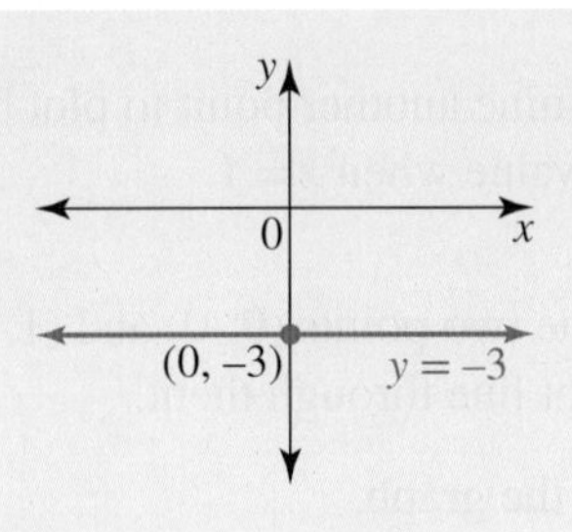

b. 1. Write the equation.

b. $x = 4$

2. The x-intercept is 4. As y does not appear in the equation, the line is parallel to the y-axis, such that all points on the line have an x-coordinate equal to 4. That is, this line is the set of points $(4, y)$ where y is an element of the set of real numbers.

x-intercept $= 4$, $(4, 0)$

3. Sketch a vertical line through $(4, 0)$.

4. Label the graph.

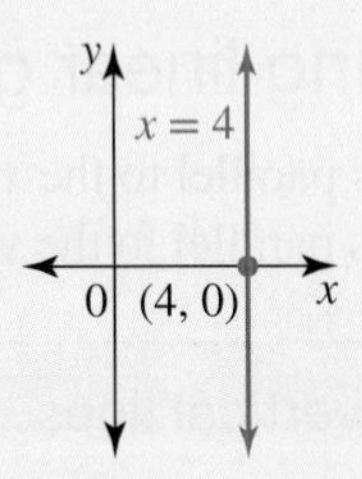

3.2.4 Using linear graphs to model real-life contexts

eles-4740

- If a real-life situation involves a constant increase or decrease at regular intervals, then it can be modelled by a linear equation. Examples include water being poured from a tap into a container at a constant rate, or money being deposited into a savings account at regular intervals.
- To model a linear situation, we first need to determine which of the two given variables is the **independent variable** and which is the **dependent variable**.
- With numerical bivariate data, we often see the independent variable being referred to as the **explanatory variable**. Likewise, another way to reference the dependent variable is to call it the **response variable**.
- The independent variable does not depend on the value of the other variable, whereas the dependent variable takes its value depending on the value of the other variable. When plotting a graph of a linear model, the independent variable will be on the x-axis (horizontal) and the dependent variable will be on the y-axis (vertical).
- Real-life examples identifying the variables are shown in the following table.

Situation	Independent variable (explanatory variable)	Dependent variable (response variable)
Money being deposited into a savings account at regular intervals	Time	Money in account
The age of a person in years and their height in cm	Age in years	Height in cm
The temperature at a snow resort and the depth of the snow	Temperature	Depth of snow
The length of Pinocchio's nose and the number of lies he told	Number of lies Pinocchio told	Length of Pinocchio's nose
The number of workers building a house and the time taken to complete the project	Number of workers	Time

- Note that if time is one of the variables, it will usually be the independent variable. The final example above is a rare case of time being the dependent variable. Also, some of the above cases can't be modelled by linear graphs, as the increases or decreases aren't necessarily happening at constant rates.

WORKED EXAMPLE 6 Using linear graphs to model real-life situations

Water is leaking from a bucket at a constant rate. After 1 minute there is 45 litres in the bucket; after 3 minutes there is 35 litres in the bucket; after 5 minutes there is 25 litres in the bucket; and after 7 minutes there is 15 litres in the bucket.

a. **Define two variables to represent the given information.**
b. **Determine which variable is the independent variable and which is the dependent variable.**
c. **Represent the given information in a table of values.**
d. **Plot a graph to represent how the amount of water in the bucket is changing.**
e. **Use your graph to determine how much water was in the bucket at the start and how long it will take for the bucket to be empty.**

THINK

a. Determine which two values change in the relationship given.

WRITE/DRAW

a. The two variables are 'time' and 'amount of water in bucket'.

THINK

b. The dependent variable takes its value depending on the value of the independent variable.
In this situation the amount of water depends on the amount of time elapsed, not the other way round.

WRITE/DRAW

b. Independent variable = time
Dependent variable = amount of water in bucket

THINK

c. The independent variable should appear in the top row of the table of values, with the dependent variable appearing in the second row.

WRITE/DRAW

c.

Time (minutes)	1	3	5	7
Amount of water in bucket (litres)	45	35	25	15

THINK

d. The values in the top row of the table represent the values on the horizontal axis, and the values in the bottom row of the table represent the values on the vertical axis. As the value for time can't be negative and there can't be a negative amount of water in the bucket, only the first quadrant needs to be drawn for the graph. Plot the 4 points and rule a straight line through them. Extend the graph to meet the vertical and horizontal axes.

WRITE/DRAW

d.

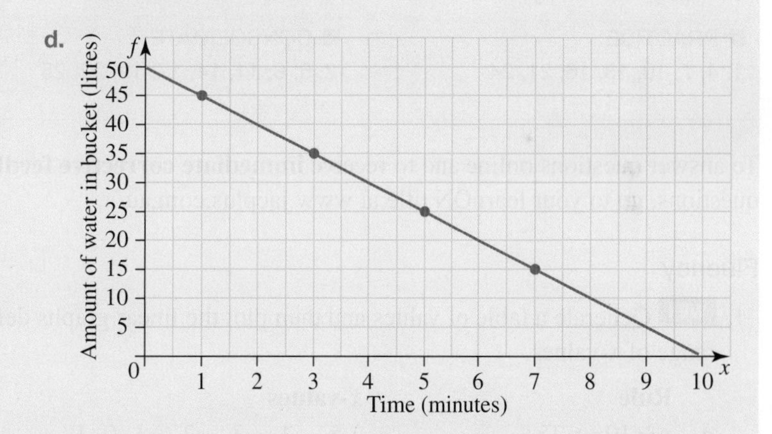

THINK

e. The amount of water in the bucket at the start is the value at which the line meets the vertical axis, and the time taken for the bucket to be empty is the value at which the line meets the horizontal axis.
Note: Determining the time when the bucket will be empty is an example of *extrapolation* as this time is determined by extending the graph beyond the known data points.

WRITE/DRAW

e. There was 50 litres of water in the bucket at the start, and it will take 10 minutes for the bucket to be empty.

DISCUSSION

What types of straight lines have an x- and y-intercept of the same value?

Resources

eWorkbook Topic 3 Workbook (worksheets, code puzzle and project) (ewbk-2029)

Digital documents SkillSHEET Describing the gradient of a line (doc-5197)
SkillSHEET Plotting a line using a table of values (doc-5198)
SkillSHEET Stating the y-intercept from a graph (doc-5199)
SkillSHEET Solving linear equations that arise when determining x- and y-intercepts (doc-5200)
SkillSHEET Using Pythagoras' theorem (doc-5201)
SkillSHEET Substitution into a linear rule (doc-5202)
SkillSHEET Transposing linear equations to standard form (doc-5203)

Video eLessons Sketching linear graphs (eles-1919)
Sketching linear graphs using the gradient-intercept method (eles-1920)

Interactivities Individual pathway interactivity: Sketching graphs (int-4572)
Plottling linear graphs (int-3834)
The gradient-intercept method (int-3839)
The intercept method (int-3840)
Equations of straight lines (int-6485)

Exercise 3.2 Sketching linear graphs

learnon

Individual pathways

PRACTISE	CONSOLIDATE	MASTER
1, 4, 7, 10, 13, 16, 21, 24	2, 5, 8, 11, 14, 17, 19, 22, 25	3, 6, 9, 12, 15, 18, 20, 23, 26

To answer questions online and to receive **immediate corrective feedback** and **fully worked solutions** for all questions, go to your learnON title at www.jacplus.com.au.

Fluency

1. WE1 Generate a table of values and then plot the linear graphs defined by the following rules for the given range of x-values.

	Rule	x-values
a.	$y = 10x + 25$	$-5, -4, -3, -2, -1, 0, 1$
b.	$y = 5x - 12$	$-1, 0, 1, 2, 3, 4$
c.	$y = -0.5x + 10$	$-6, -4, -2, 0, 2, 4$

2. Generate a table of values and then plot the linear graphs defined by the following rules for the given range of x-values.

	Rule	x-values
a.	$y = 100x - 240$	$0, 1, 2, 3, 4, 5$
b.	$y = -5x + 3$	$-3, -2, -1, 0, 1, 2$
c.	$y = 7 - 4x$	$-3, -2, -1, 0, 1, 2$

3. Plot the linear graphs defined by the following rules for the given range of x-values.

Rule | ***x*-values**

a. $y = -3x + 2$

x	−6	−4	−2	0	2	4	6
y							

b. $y = -x + 3$

x	−3	−2	−1	0	1	2	3
y							

c. $y = -2x + 3$

x	−6	−4	−2	0	2	4	6
y							

WE2 For questions 4 to 6, sketch graphs of the following linear equations by determining the x- and y-intercepts.

4. a. $5x - 3y = 10$ b. $5x + 3y = 10$ c. $-5x + 3y = 10$ d. $-5x - 3y = 10$ e. $2x - 8y = 20$

5. a. $4x + 4y = 40$ b. $-x + 6y = 120$ c. $-2x + 8y = -20$
d. $10x + 30y = -150$ e. $5x + 30y = -150$

6. a. $-9x + 4y = 36$ b. $6x - 4y = -24$ c. $y = 2x - 10$ d. $y = -5x + 20$ e. $y = -\frac{1}{2}x - 4$

WE3 For questions 7 to 9, sketch graphs of the following using the gradient–intercept method.

7. a. $y = 4x + 1$ b. $y = 3x - 7$ c. $y = -2x + 3$

8. a. $y = -5x - 4$ b. $y = \frac{1}{2}x - 2$ c. $y = -\frac{2}{7}x + 3$

9. a. $y = 0.6x + 0.5$ b. $y = 8x$ c. $y = x - 7$

WE4 For questions 10 to 12, sketch the graphs of the following linear equations on the same set of axes.

10. a. $y = 2x$ b. $y = \frac{1}{2}x$ c. $y = -2x$

11. a. $y = 5x$ b. $y = \frac{1}{3}x$ c. $y = -\frac{5}{2}x$

12. a. $y = \frac{2}{3}x$ b. $y = -3x$ c. $y = -\frac{3}{2}x$

WE5 For questions 13 to 15, sketch the graphs of the following linear equations.

13. a. $y = 10$ b. $x = -10$ c. $x = 0$

14. a. $y = -10$ b. $y = 100$ c. $x = -100$

15. a. $x = 10$ b. $y = 0$ c. $y = -12$

Understanding

For questions 16 to 18, transpose each of the equations to standard form (that is, $y = mx + c$). State the x- and y-intercept for each.

16. a. $5(y + 2) = 4(x + 3)$ b. $5(y - 2) = 4(x - 3)$ c. $2(y + 3) = 3(x + 2)$

17. a. $10(y - 20) = 40(x - 2)$ b. $4(y + 2) = -4(x + 2)$ c. $2(y - 2) = -(x + 5)$

18. a. $-5(y + 1) = 4(x - 4)$ b. $5(y + 2.5) = 2(x - 3.5)$ c. $2.5(y - 2) = -6.5(x - 1)$

19. Determine the x- and y-intercepts of the following lines.

a. $-y = 8 - 4x$ b. $6x - y + 3 = 0$ c. $2y - 10x = 50$

20. Explain why the gradient of a horizontal line is equal to zero and the gradient of a vertical line is undefined.

Reasoning

21. **WE6** Your friend loves to download music. She earns \$50 and spends some of it buying music online at \$1.75 per song. She saves the remainder. Her saving is given by the function $y = 50 - 1.75x$.

a. Determine which variable is the independent variable and which is the dependent variable.
b. Sketch the function.
c. Determine the number of songs your friend can buy and still save \$25.

22. Determine whether $\frac{x}{3} - \frac{y}{2} = \frac{7}{6}$ is the equation of a straight line by rearranging into an appropriate form and hence sketch the graph, showing all relevant features.

23. Nikita works a part-time job and is interested in sketching a graph of her weekly earnings. She knows that in a week where she does not work any hours, she will still earn \$25.00 for being 'on call'. On top of this initial payment, Nikita earns \$20.00 per hour for her regular work. Nikita can work a maximum of 8 hours per day as her employer is unwilling to pay her overtime.

a. Write a linear equation that represents the amount of money Nikita could earn in a week. (*Hint:* You might want to consider the 'on call' amount as an amount of money earned for zero hours worked.)
b. Sketch a graph of Nikita's weekly potential earnings.
c. Determine the maximum amount of money that Nikita can earn in a single week.

Problem solving

24. The temperature in a room is rising at a constant rate. Initially (when time equals zero), the temperature of the room is 15 °C. After 1 hour, the temperature of the room has risen to 18 °C. After 3 hours, the temperature has risen to 24 °C.

a. Using the variables t to represent the time in hours and T to represent the temperature of the room, identify the dependent and the independent variable in this linear relationship.
b. i. Construct a table of values to represent this information.
   ii. Plot this relationship on a suitable axis.
c. If the maximum temperature of the room was recorded to be 30 °C, evaluate after how many hours was this recording taken.

25. Water is flowing from a tank at a constant rate. The equation relating the volume of water in the tank, V litres, to the time the water has been flowing from the tank, t minutes, is given by $V = 80 - 4t,\ t \geq 0$.

a. Determine which variable is the independent variable and which is the dependent variable.
b. Calculate how much water is in the tank initially.
c. Explain why it is important that $t \geq 0$.
d. Determine the rate the water is flowing from the tank.
e. Determine how long it takes for the tank to empty.
f. Sketch the graph of V versus t.

26. A straight line has a general equation defined by $y = mx + c$. This line intersects the lines defined by the rules $y = 7$ and $x = 3$. The lines $y = mx + c$ and $y = 7$ have the same y-intercept while $y = mx + c$ and $x = 3$ have the same x-intercept.

a. On the one set of axes, sketch all three graphs.
b. Determine the y-axis intercept for $y = mx + c$.
c. Determine the gradient for $y = mx + c$.
d. **MC** The equation of the line defined by $y = mx + c$ is:

A. $x + y = 3$ B. $7x + 3y = 21$ C. $3x + 7y = 21$ D. $x + y = 7$ E. $7x + 3y = 7$

3.3 Determining linear equations

LEARNING INTENTION

At the end of this subtopic you should be able to:

- determine the equation of a straight line when given its graph
- determine the equation of a straight line when given the gradient and the y-intercept
- determine the equation of a straight line passing through two points
- formulate the equation of a straight line from a written context.

3.3.1 Determining a linear equation given two points

eles-4741

- The gradient of a straight line can be calculated from the coordinates of two points (x_1, y_1) and (x_2, y_2) that lie on the line.
- The equation of the straight line can then be found in the form $y = mx + c$, where c is the y-intercept.

Gradient of a straight line

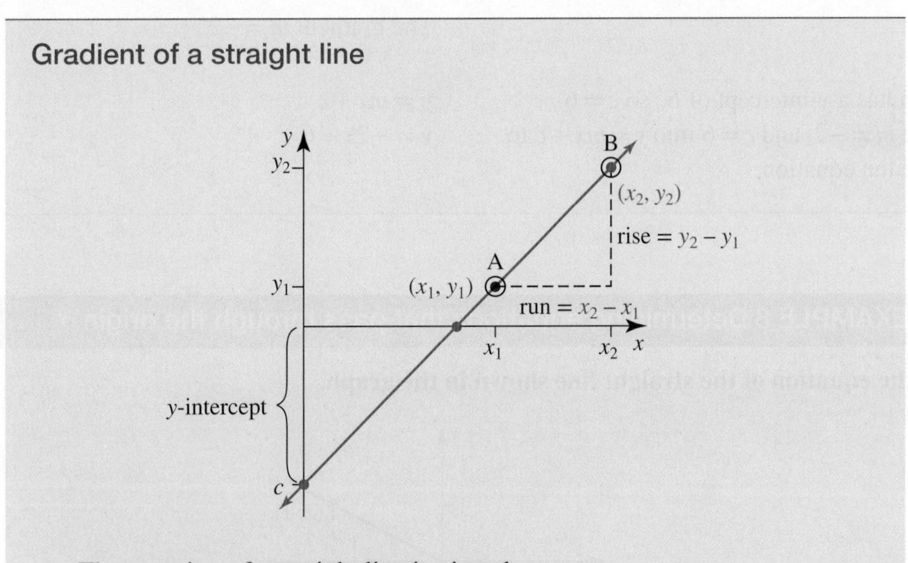

- The equation of a straight line is given by $y = mx + c$.
- m is the value of the gradient and c is the value of the y-intercept.

$$\textbf{Gradient} = m = \frac{\textbf{rise}}{\textbf{run}} = \frac{y_2 - y_1}{x_2 - x_1}$$

WORKED EXAMPLE 7 Determining equations with a known y-intercept

Determine the equation of the straight line shown in the graph.

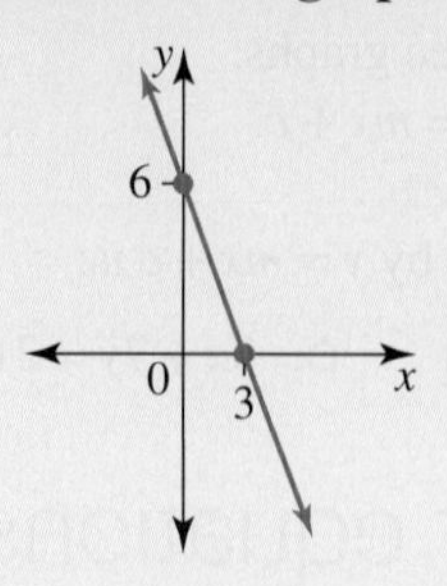

THINK	WRITE
1. There are two points given on the straight line: the x-intercept $(3, 0)$ and the y-intercept $(0, 6)$.	$(3,\ 0), (0,\ 6)$
2. Calculate the gradient of the line by applying the formula $m = \frac{\text{rise}}{\text{run}} = \frac{y_2 - y_1}{x_2 - x_1}$, where $(x_1, y_1) = (3,\ 0)$ and $(x_2, y_2) = (0,\ 6)$.	$m = \frac{\text{rise}}{\text{run}}$ $= \frac{y_2 - y_1}{x_2 - x_1}$ $= \frac{6 - 0}{0 - 3}$ $= \frac{6}{-3}$ $= -2$ The gradient $m = -2$.
3. The graph has a y-intercept of 6, so $c = 6$. Substitute $m = -2$, and $c = 6$ into $y = mx + c$ to determine the equation.	$y = mx + c$ $y = -2x + 6$

WORKED EXAMPLE 8 Determining equations that pass through the origin

Determine the equation of the straight line shown in the graph.

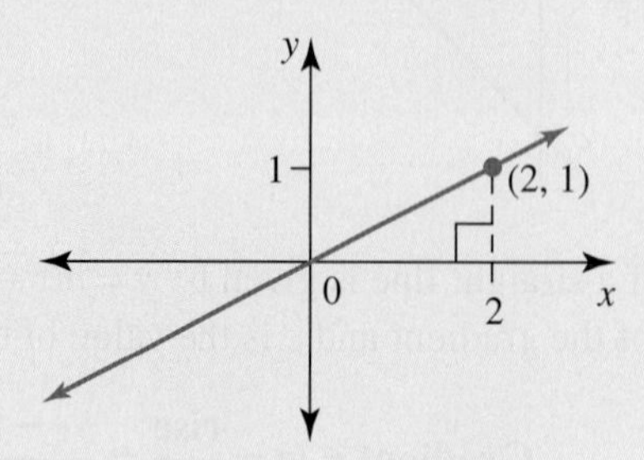

THINK	WRITE
1. There are two points given on the straight line: the x- and y-intercept $(0,\ 0)$ and another point $(2,\ 1)$.	$(0,\ 0), (2,\ 1)$

2. Calculate the gradient of the line by applying the formula $m = \frac{\text{rise}}{\text{run}} = \frac{y_2 - y_1}{x_2 - x_1}$, where $(x_1, y_1) = (0, 0)$ and $(x_2, y_2) = (2, 1)$.

$$m = \frac{\text{rise}}{\text{run}}$$
$$= \frac{y_2 - y_1}{x_2 - x_1}$$
$$= \frac{1 - 0}{2 - 0}$$
$$= \frac{1}{2}$$

The gradient $m = \frac{1}{2}$.

3. The y-intercept is 0, so $c = 0$. Substitute $m = \frac{1}{2}$ and $c = 0$ into $y = mx + c$ to determine the equation.

$$y = mx + c$$
$$y = \frac{1}{2}x + 0$$
$$y = \frac{1}{2}x$$

3.3.2 A simple formula

eles-4742

- The diagram shows a line of gradient m passing through the point (x_1, y_1).
- If (x, y) is any other point on the line, then:

$$m = \frac{\text{rise}}{\text{run}}$$
$$m = \frac{y - y_1}{x - x_1}$$
$$m(x - x_1) = y - y_1$$
$$y - y_1 = m(x - x_1)$$

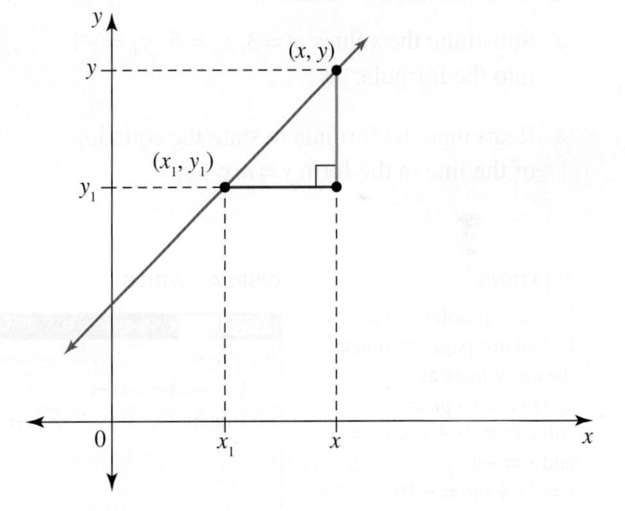

- The formula $y - y_1 = m(x - x_1)$ can be used to write down the equation of a line, given the gradient and the coordinates of one point.

The equation of a straight line

- Determining the equation of a straight line with coordinates of one point (x_1, y_1) and the gradient (m):

$$\boldsymbol{y - y_1 = m(x - x_1)}$$

WORKED EXAMPLE 9 Determining the equation using the gradient and the y-intercept

Determine the equation of the straight line with a gradient of 2 and a y-intercept of -5.

THINK	WRITE
1. Write the gradient formula.	$y - y_1 = m(x - x_1)$
2. State the known variables.	$m = 2,\ (x_1, y_1) = (0, -5)$
3. Substitute the values into the formula.	$y - (-5) = 2(x - 0)$ $y + 5 = 2x$
4. Rearrange the formula. *Note:* You could also solve this by using the equation $y = mx + c$ and substituting directly for m and c.	$y = 2x - 5$

WORKED EXAMPLE 10 Determining the equation using the gradient and another point

Determine the equation of the straight line with a gradient of 3 and passing through the point (5, -1).

THINK	WRITE
1. Write out the gradient formula.	$y - y_1 = m(x - x_1)$
2. State the known variables.	$m = 3,\ x_1 = 5,\ y_1 = -1$
3. Substitute the values $m = 3, x_1 = 5, y_1 = -1$ into the formula.	$y - (-1) = 3(x - 5)$ $y + 1 = 3x - 15$
4. Rearrange the formula to state the equation of the line in the form $y = mx + c$.	$y = 3x - 16$

TI | THINK

In a new problem on a Calculator page, complete the entry lines as:
$y = m \times x + c | m = 3$
solve $(y = 3x + c,\ c) | x = 5$ and $y = -1$
$y = 3x + c | c = -16$
Press ENTER after each entry.

DISPLAY/WRITE

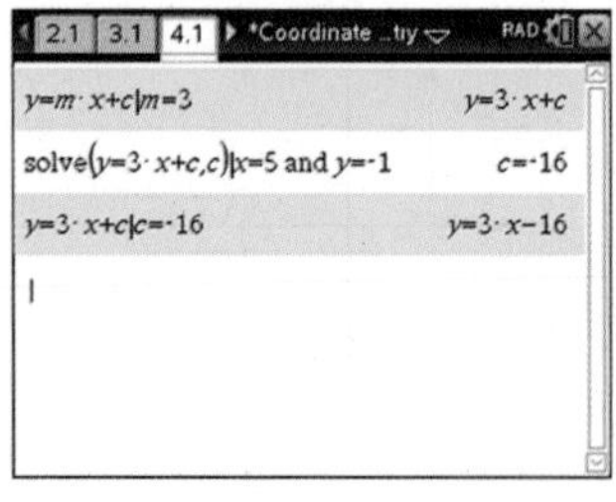

The equation is $y = 3x - 16$.

CASIO | THINK

On the Main screen, complete the entry lines as:
solve $(y = 3x + c,\ c)$
$|x = 5|y = -1$
Press EXE.

DISPLAY/WRITE

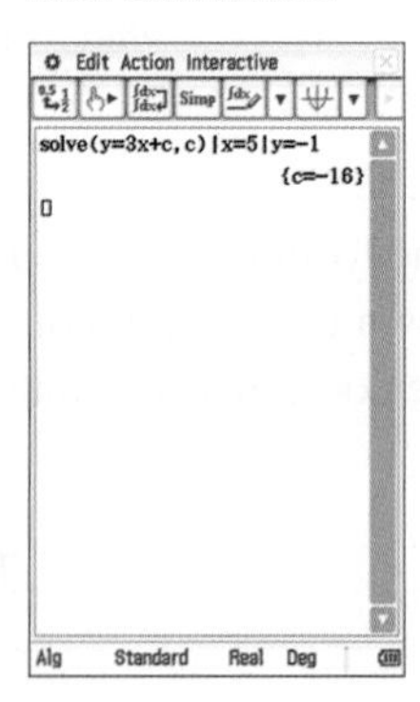

The equation is $y = 3x - 16$.

WORKED EXAMPLE 11 Determining the equation of a line using two points

Determine the equation of the straight line passing through the points (−2, 5) and (1, −1).

THINK	WRITE
1. Write out the gradient formula.	$y - y_1 = m(x - x_1)$
2. State the known variables.	$(x_1, y_1) = (-2, 5)$ $(x_2, y_2) = (1, -1)$
3. Substitute the values $(x_1, y_1) = (-2, 5)$ and $(x_2, y_2) = (1, -1)$ to calculate the gradient from the given points.	$m = \frac{y_2 - y_1}{x_2 - x_1}$ $m = \frac{-1 - 5}{1 - -2}$ $m = \frac{-6}{3}$ $= -2$
4. Substitute the values $m = -2$, $(x_1, y_1) = (-2, 5)$ into the formula for the equation of a straight line.	$y - y_1 = m(x - x_1)$ $y - 5 = -2(x - -2)$
5. Rearrange the formula to state the equation of the line in the form $y = mx + c$.	$y - 5 = -2(x + 2)$ $y = -2x - 4 + 5$ $y = -2x + 1$

WORKED EXAMPLE 12 Writing an equation in the form $ax + by + c = 0$

Determine the equation of the line with a gradient of −2 which passes through the point (3, −4). Write the equation in general form, that is in the form $ax + by + c = 0$.

THINK	WRITE
1. Use the formula $y - y_1 = m(x - x_1)$. Write the values of x_1, y_1, and m.	$m = -2, \quad x_1 = 3, \quad y_1 = -4$ $y - y_1 = m(x - x_1)$
2. Substitute for x_1, y_1, and m into the equation.	$y - (-4) = -2(x - 3)$ $y + 4 = -2x + 6$
3. Transpose the equation into the form $ax + by + c = 0$.	$y + 4 + 2x - 6 = 0$ $2x + y - 2 = 0$

WORKED EXAMPLE 13 Applying equations of lines to model real-life situations

A printer prints pages at a constant rate. It can print 165 pages in 3 minutes and 275 pages in 5 minutes.

a. **Identify which variable is the independent variable (x) and which is the dependent variable (y).**
b. **Calculate the gradient of the equation and explain what this means in the context of the question.**
c. **Write an equation, in algebraic form, linking the independent and dependent variables.**
d. **Rewrite your equation in words.**
e. **Using the equation, determine how many pages can be printed in 11 minutes.**

THINK	WRITE/DRAW
a. The dependent variable takes its value depending on the value of the independent variable. In this situation the number of pages depends on the time elapsed, not the other way round.	a. Independent variable = time Dependent variable = number of pages
b. 1. Determine the two points given by the information in the question.	b. $(x_1, y_1) = (3, 165)$ $(x_2, y_2) = (5, 275)$
2. Substitute the values of these two points into the formula to calculate the gradient.	$m = \frac{y_2 - y_1}{x_2 - x_1}$ $= \frac{275 - 165}{5 - 3}$ $= \frac{110}{2}$ $= 55$
3. The gradient states how much the dependent variable increases for each increase of 1 unit in the independent variable.	In the context of the question, this means that each minute 55 pages are printed.
c. The graph travels through the origin, as the time elapsed for the printer to print 0 pages is 0 seconds. Therefore, the equation will be in the form $y = mx$. Substitute in the value of m.	c. $y = mx$ $y = 55x$
d. Replace x and y in the equation with the independent and dependent variables.	d. Number of pages = 55 × time
e. 1. Substitute $x = 11$ into the equation.	e. $y = 55x$ $= 55 \times 11$ $= 605$
2. Write the answer in words.	The printer can print 605 pages in 11 minutes.

DISCUSSION

What problems might you encounter when calculating the equation of a line whose graph is actually parallel to one of the axes?

on Resources

eWorkbook	Topic 3 Workbook (worksheets, code puzzle and project) (ewbk-2029)
Digital documents	SkillSHEET Measuring the rise and the run (doc-5196) SkillSHEET Determining the gradient given two points (doc-5204)
Video eLesson	The equation of a straight line (eles-2313)
Interactivities	Individual pathway interactivity: Determining the equation (int-4573) Linear graphs (int-6484)

Exercise 3.3 Determining linear equations

learnON

Individual pathways

■ PRACTISE	■ CONSOLIDATE	■ MASTER
1, 4, 7, 10, 13, 16	2, 5, 8, 11, 14, 17	3, 6, 9, 12, 15, 18

To answer questions online and to receive **immediate corrective feedback** and **fully worked solutions** for all questions, go to your learnON title at www.jacplus.com.au.

Fluency

1. WE7 Determine the equation for each of the straight lines shown.

a. 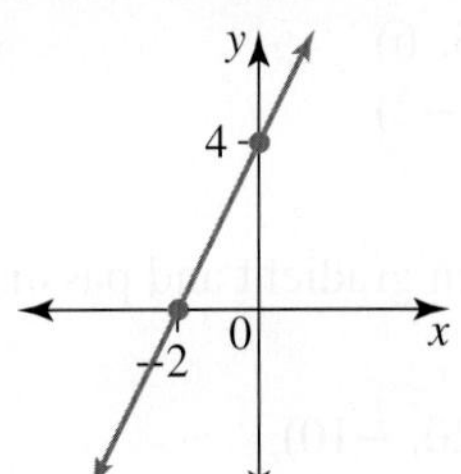

b. 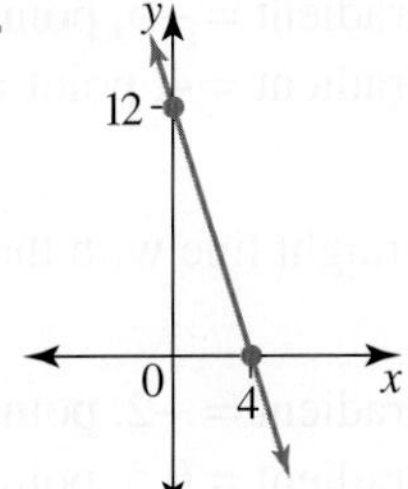

c. 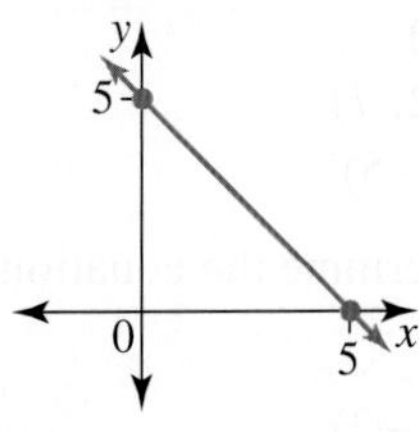

d. 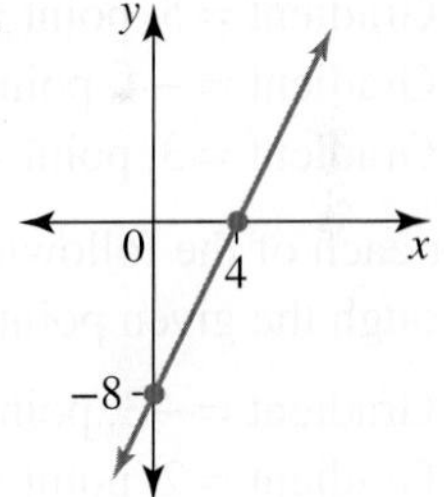

2. Determine the equation for each of the straight lines shown.

a. 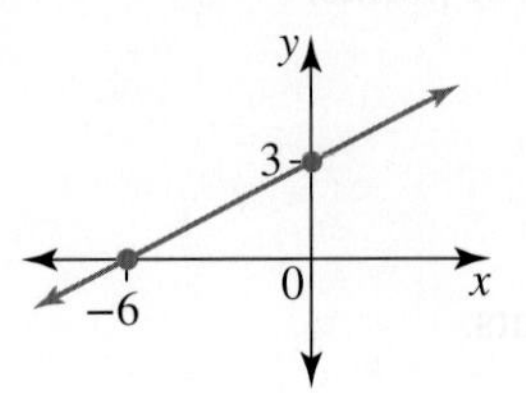

b. 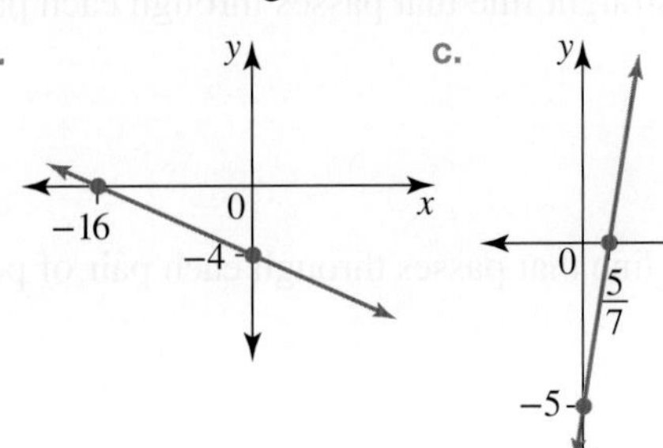

c. 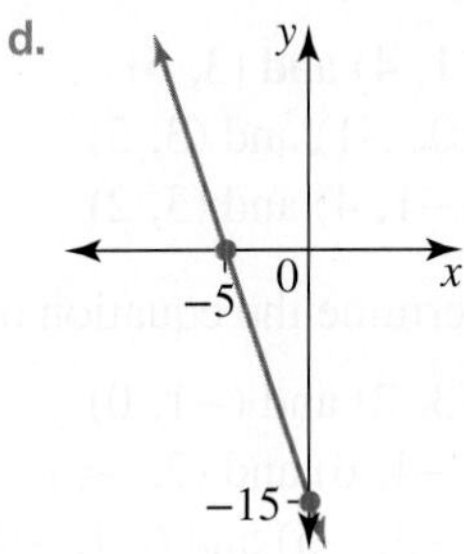

d.

3. WE8 Determine the equation of each of the straight lines shown.

a.

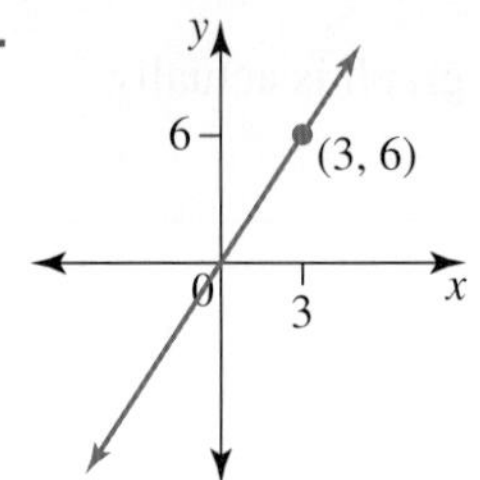

b.

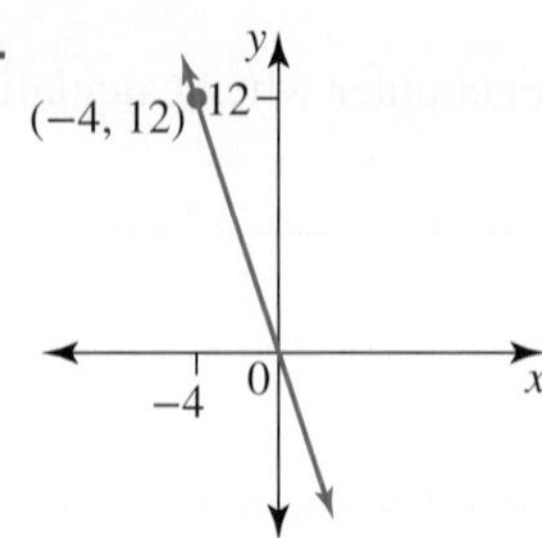

c.

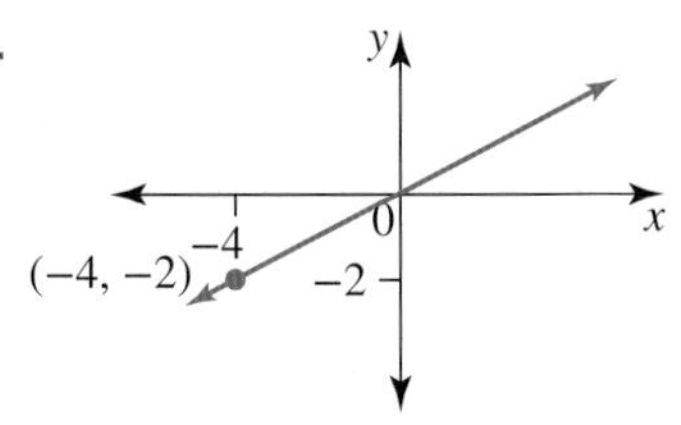

d.

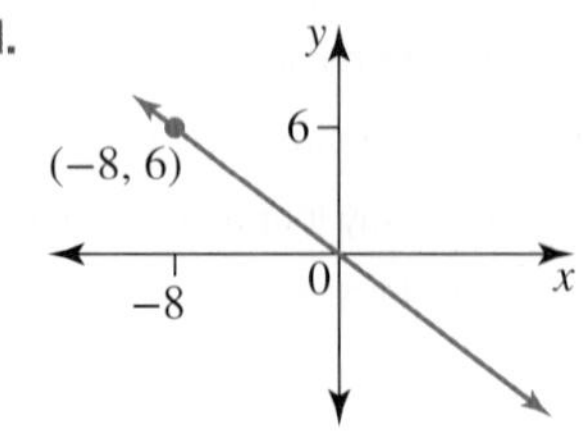

4. WE9 Determine the linear equation given the information in each case below.

a. Gradient = 3, y-intercept = 3
b. Gradient = −3, y-intercept = 4
c. Gradient = − 4, y-intercept = 2
d. Gradient = 4, y-intercept = 2
e. Gradient = −1, y-intercept = 4

5. Determine the linear equation given the information in each case below.

a. Gradient = 0.5, y-intercept = −4
b. Gradient = 5, y-intercept = 2.5
c. Gradient = −6, y-intercept = 3
d. Gradient = −2.5, y-intercept = 1.5
e. Gradient = 3.5, y-intercept = 6.5

6. WE10 For each of the following, determine the equation of the straight line with the given gradient and passing through the given point.

a. Gradient = 5, point = (5, 6)
b. Gradient = −5, point = (5, 6)
c. Gradient = −4, point = (−2, 7)
d. Gradient = 4, point = (8, −2)
e. Gradient = 3, point = (10, −5)

7. For each of the following, determine the equation of the straight line with the given gradient and passing through the given point.

a. Gradient = −3, point = (3, −3)
b. Gradient = −2, point = (20, −10)
c. Gradient = 2, point = (2, −0.5)
d. Gradient = 0.5, point = (6, −16)
e. Gradient = −0.5, point = (5, 3)

8. WE11 Determine the equation of the straight line that passes through each pair of points.

a. (1, 4) and (3, 6)
b. (0, −1) and (3, 5)
c. (−1, 4) and (3, 2)

9. Determine the equation of the straight line that passes through each pair of points.

a. (3, 2) and (−1, 0)
b. (−4, 6) and (2, −6)
c. (−3, −5) and (−1, −7)

Understanding

10. **WE13** a. Determine which variable (time or cost) is the independent variable and which is the dependent variable in the Supa-Bowl advertisement on the right.

> Save $$$ with Supa-Bowl!!!
> NEW Ten-Pin Bowling Alley
> *Shoe rental just $2 (fixed fee)*
> Rent a lane for ONLY $6/hour!

   b. If t represents the time in hours and C represents cost ($), construct a table of values for 0–3 hours for the cost of playing ten-pin bowling at the new alley.
   c. Use your table of values to plot a graph of time versus cost. (*Hint:* Ensure your time axis (horizontal axis) extends to 6 hours and your cost axis (vertical axis) extends to $40.)
   d. i. Identify the y-intercept.
      ii. Describe what the y-intercept represents in terms of the cost.
   e. Calculate the gradient and explain what this means in the context of the question.
   f. Write a linear equation to describe the relationship between cost and time.
   g. Use your linear equation from part **f** to calculate the cost of a 5-hour tournament.
   h. Use your graph to check your answer to part **g**.

11. A local store has started renting out scooters to tour groups who pass through the city. Groups are charged based on the number of people hiring the equipment. There is a flat charge of $10.00 any time you book a day of rentals and it is known that the cost for 20 people to hire scooters is $310.00. The cost for 40 people to hire scooters is $610.00.

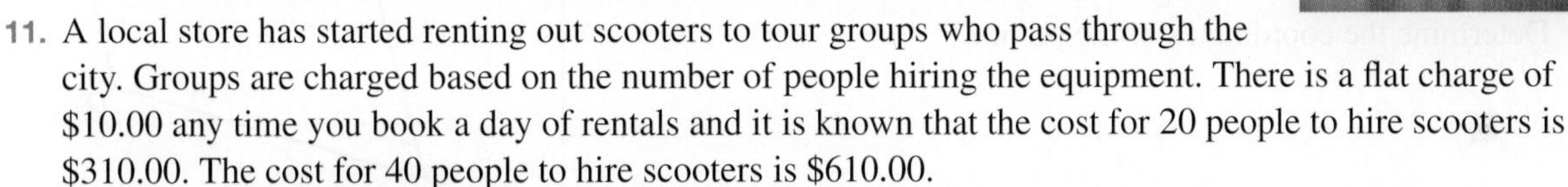

   a. Label the cost in dollars for hiring scooters for a day as the variable C. Let the number of people hiring scooters be the variable n. Identify which is the dependent variable and which is the independent variable.
   b. Formulate a linear equation that models the cost of hiring scooters for a day.
   c. Calculate how much it will cost to hire 30 scooters.
   d. Sketch a graph of the cost function you created in part **b**.

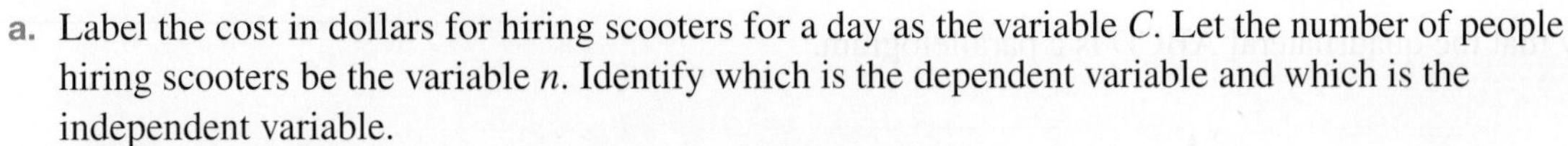

12. The Robinsons' water tank sprang a leak and has been losing water at a steady rate. Four days after the leak occurred, the tank contained 552 L of water, and ten days later it held only 312 L.
   a. Determine the rule linking the amount of water in the tank (w) and the number of days (t) since the leak occurred.
   b. Calculate how much water was in the tank initially.
   c. If water loss continues at the same rate, determine when the tank will be empty.

Reasoning

13. When using the gradient to draw a line, does it matter if you rise before you run or run before you rise? Explain your answer.

14. a. Using the graph shown, write a general formula for the gradient m in terms of x, y and c.
   b. Transpose your formula to make y the subject. Explain what you notice.

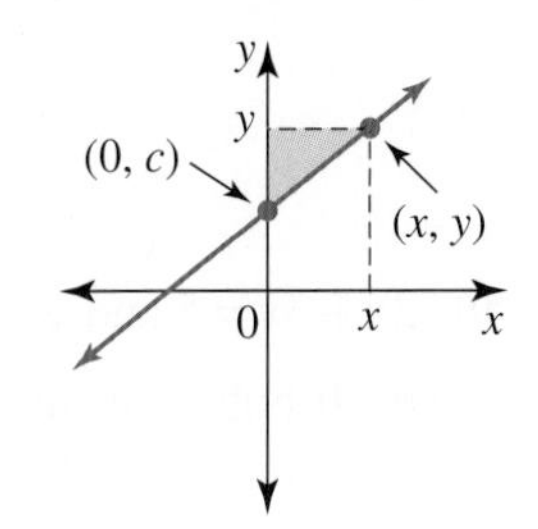

15. The points $A(x_1, y_1)$, $B(x_2, y_2)$ and $P(x, y)$ all lie on the same line. P is a general point that lies anywhere on the line. Given that the gradient from A to P must be equal to the gradient from P to B, show that an equation relating these three points is given by:

$$y - y_1 = \frac{y_2 - y_1}{x_2 - x_1}(x - x_1)$$

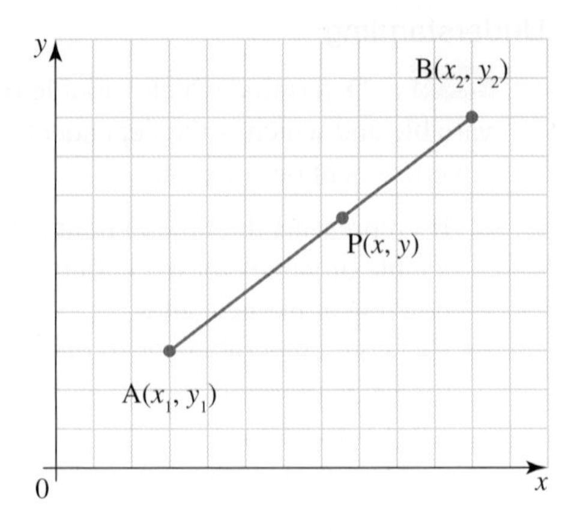

Problem solving

16. *ABCD* is a parallelogram with coordinates *A*(2, 1), *B*(3, 6) and *C*(7, 10).
    a. Calculate the value of the gradient of the line AB.
    b. Determine the equation of the line AB.
    c. Calculate the value of the gradient of the line CD.
    d. Determine the coordinates of the point *D*.

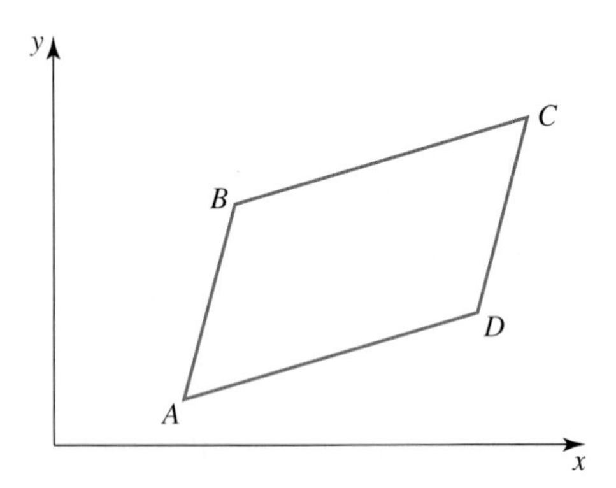

17. Show that the quadrilateral ABCD is a parallelogram.

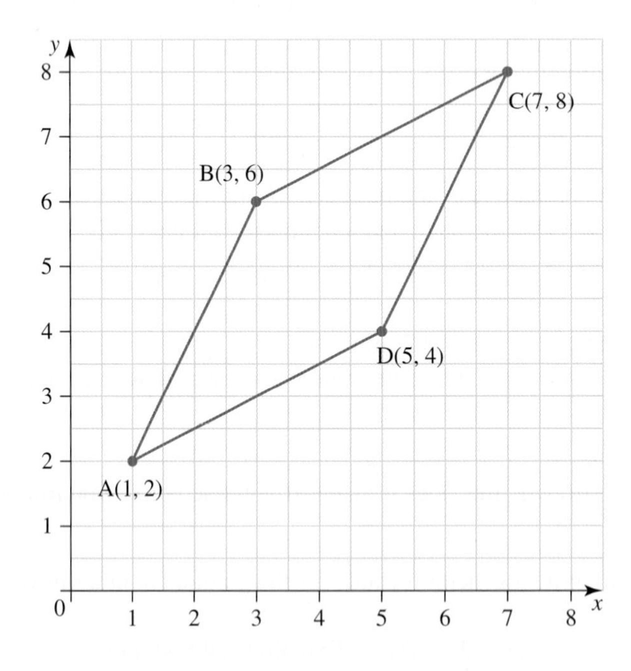

18. $2x + 3y = 5$ and $ax - 6y = b$ are the equations of two lines.
    a. If both lines have the same *y* intercept, determine the value of *b*.
    b. If both lines have the same gradient (but a different *y* intercept), determine the value of *a*.

3.4 Parallel and perpendicular lines

LEARNING INTENTION

At the end of this subtopic you should be able to:
- determine whether two lines are parallel, perpendicular or neither
- determine the equation of a straight line that is parallel to a given line
- determine the equation of a straight line that is perpendicular to a given line.

3.4.1 Parallel lines

eles-4743

- Lines that have the same gradient are **parallel** lines.
- The three lines (pink, green and blue) on the graph shown all have a gradient of 1 and are parallel to each other.
- Parallel lines will never intersect with on another.

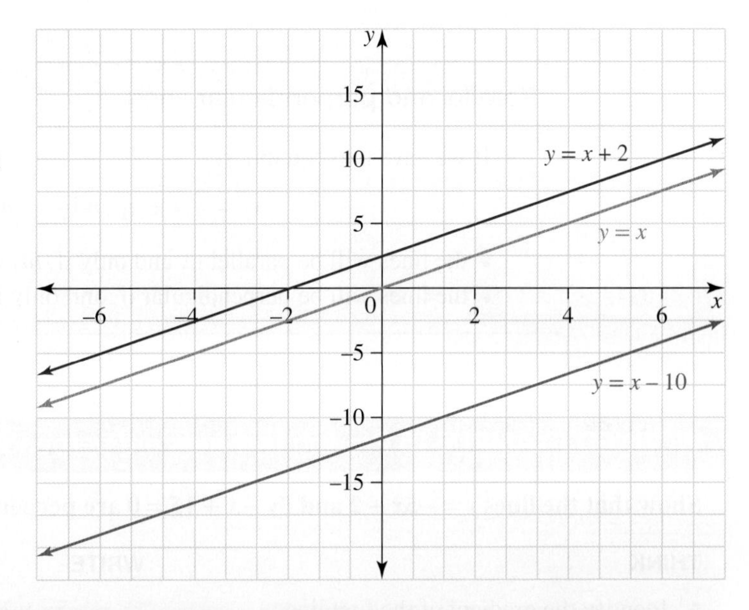

WORKED EXAMPLE 14 Proving that two lines are parallel

Show that AB is parallel to CD given that A has coordinates (−1, −5), B has coordinates (5, 7), C has coordinates (−3, 1) and D has coordinates (4, 15).

THINK	WRITE
1. Calculate the value of the gradient of AB by applying the formula $m = \dfrac{y_2 - y_1}{x_2 - x_1}$.	Let $A(-1, -5) = (x_1, y_1)$ and $B(5, 7) = (x_2, y_2)$ Since $m = \dfrac{y_2 - y_1}{x_2 - x_1}$ $m_{AB} = \dfrac{7-(-5)}{5-(-1)}$ $= \dfrac{12}{6}$ $= 2$
2. Calculate the value of the gradient of CD.	Let $C(-3, 1) = (x_1, y_1)$ and $D(4, 15) = (x_2, y_2)$ $m_{CD} = \dfrac{15-1}{4-(-3)}$ $= \dfrac{14}{7}$ $= 2$
3. Draw a conclusion. (*Note*: ǁ means 'is parallel to'.)	Since $m_{AB} = m_{CD} = 2$, then AB ǁ CD.

3.4.2 Perpendicular lines

eles-4744

- Perpendicular lines are lines that intersect at *right angles* as seen in the diagram. We can think of these lines as each having opposite gradients.
- More formally, we would state that perpendicular lines have gradients in which one gradient is the *negative reciprocal* of the other.
- Mathematically, we denote this as $m_1 \times m_2 = -1$.
- The gradient properties of parallel and perpendicular lines can be used to solve many problems.

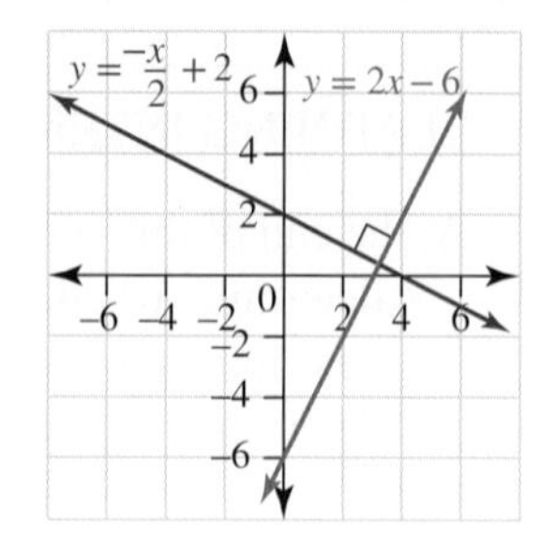

Parallel and perpendicular lines

- If we have two equations:

$$y_1 = m_1x + c_1 \text{ and } y_2 = m_2x + c_2$$

  - the lines will be parallel if, and only if, $\boldsymbol{m_1 = m_2}$
  - the lines will be perpendicular if, and only if, $\boldsymbol{m_1 \times m_2 = -1}$.

WORKED EXAMPLE 15 Proving two lines are perpendicular

Show that the lines $y = -5x + 2$ and $5y - x + 15 = 0$ are perpendicular.

THINK	WRITE
1. Identify the gradient of the first line $y_1 = -5x + 2$.	$y_1 = -5x + 2$ $m_1 = -5$
2. Identify the gradient of the second line $5y - x + 15 = 0$ by rearranging the equation in the form $y = mx + c$.	$5y - x + 15 = 0$ $5y = x - 15$ $y = \frac{x}{5} - 3$ $m_2 = \frac{1}{5}$
3. Test if the two points are perpendicular by checking whether the product of the two gradients is equal to -1.	$m_1 \times m_2 = -5 \times \frac{1}{5}$ $= -1$
4. Write the answer in a sentence.	As the product of the two gradients is equal to -1, therefore these two lines are perpendicular.

WORKED EXAMPLE 16 Determining the equation of a parallel line

Determine the equation of the line that passes through the point (3, 1) and is parallel to the line with equation $y - 2x + 1 = 0$.

THINK	WRITE
1. To determine the equation of a line, we need both a gradient and a point. State the known values.	$(x_1, y_1) = (3, 1)$
2. Identify the gradient of the line $y - 2x + 1 = 0$ to which another line is parallel.	$y - 2x + 1 = 0$ $y = 2x - 1$ $m = 2$
3. Determine the line equation by using the gradient = 2 and the point (3, 1). Rearrange and write the answer.	$y - y_1 = m(x - x_1)$ $y - 1 = 2(x - 3)$ $y - 1 = 2x - 6$ $y = 2x - 5$

WORKED EXAMPLE 17 Determining the equation of a perpendicular line

Determine the equation of the line that passes through the point (2, 1) and is perpendicular to the line with a gradient of 5.

THINK	WRITE
1. To determine the equation of a line, we need both a gradient and a point. State the known values.	$(x_1, y_1) = (2, 1)$ $m_1 = 5$
2. As the lines are perpendicular so $m_1 \times m_2 = -1$. Calculate the value of the gradient of the other line using this formula.	$m_1 \times m_2 = -1$ $5 \times m_2 = -1$ $m_2 = \frac{-1}{5}$
3. Determine the line equation by using the gradient = $\frac{-1}{5}$ and the point (2, 1). Rearrange and write the answer.	$y - y_1 = m(x - x_1)$ $y - 1 = \frac{-1}{5}(x - 2)$ $y - 1 = \frac{-x}{5} + \frac{2}{5}$ $y = \frac{-x}{5} + \frac{7}{5}$ or $5y + x = 7$

on Resources

eWorkbook Topic 3 Workbook (worksheets, code puzzle and project) (ewbk-2029)

Interactivities Individual pathway interactivity: Parallel and perpendicular lines (int-4576)
Parallel lines (int-3841)
Perpendicular lines (int-6124)

Exercise 3.4 Parallel and perpendicular lines

learnon

Individual pathways

■ PRACTISE	■ CONSOLIDATE	■ MASTER
1, 3, 7, 8, 13, 14, 18, 19, 22	2, 4, 9, 12, 17, 20, 23	5, 6, 10, 11, 15, 16, 21, 24

To answer questions online and to receive **immediate corrective feedback** and **fully worked solutions** for all questions, go to your learnON title at www.jacplus.com.au.

Fluency

WE14 For questions 1 to 4, determine whether AB is parallel to CD given the following sets of points.

1. a. A(4, 13), B(2, 9), C(0, −10), D(15, 0)
   b. A(2, 4), B(8, 1), C(−6, −2), D(2, −6)
   c. A(−3, −10), B(1, 2), C(1, 10), D(8, 16)

2. a. A(1, −1), B(4, 11), C(2, 10), D(−1, −5)
   b. A(1, 0), B(2, 5), C(3, 15), D(7, 35)
   c. A(1, −6), B(−5, 0), C(0, 0), D(5, −4)

3. a. A(1, 6), B(3, 8), C(4, −6), D(−3, 1)
   b. A(2, 12), B(−1, −9), C(0, 2), D(7, 1)
   c. A(1, 3), B(4, 18), C(−5, 4), D(5, 0)

4. a. A(1, −5), B(0, 0), C(5, 11), D(−10, 8)
   b. A(−4, 9), B(2, −6), C(−5, 8), D(10, 14)
   c. A(4, 4), B(−8, 5), C(−6, 2), D(3, 11)

5. Determine which pairs of the following straight lines are parallel.
   a. $2x+y+1=0$
   b. $y=3x-1$
   c. $2y-x=3$
   d. $y=4x+3$
   e. $y=\frac{x}{2}-1$
   f. $6x-2y=0$
   g. $3y=x+4$
   h. $2y=5-x$

6. **WE15** Show that the lines $y=6x-3$ and $x+6y-6=0$ are perpendicular to one another.

7. **WE16** Determine the equation of the line that passes through the point (4, −1) and is parallel to the line with equation $y=2x-5$.

8. **WE17** Determine the equation of the line that passes through the point (−2, 7) and is perpendicular to a line with a gradient of $\frac{2}{3}$.

9. Determine the equations of the following lines.
   a. Gradient 3 and passing through the point (1, 5)
   b. Gradient −4 and passing through the point (2, 1)
   c. Passing through the points (2, −1) and (4, 2)
   d. Passing through the points (1, −3) and (6, −5)
   e. Passing through the point (5, −2) and parallel to $x+5y+15=0$
   f. Passing through the point (1, 6) and parallel to $x-3y-2=0$
   g. Passing through the point (−1, −5) and perpendicular to $3x+y+2=0$

10. Determine the equation of the line that passes through the point (−2, 1) and is:
    a. parallel to the line with equation $2x-y-3=0$
    b. perpendicular to the line with equation $2x-y-3=0$.

11. Determine the equation of the line that contains the point (1, 1) and is:

a. parallel to the line with equation $3x - 5y = 0$
b. perpendicular to the line with equation $3x - 5y = 0$.

Understanding

12. **MC** a. The vertical line passing through the point (3, −4) is given by:

A. $y = -4$ B. $x = 3$ C. $y = 3x - 4$ D. $y = -4x + 3$ E. $x = -4$

b. Select the point which passes through the horizontal line given by the equation $y = -5$.

A. (−5, 4) B. (4, 5) C. (3, −5) D. (5, −4) E. (5, 5)

c. Select which of the following statements is true.

A. Vertical lines have a gradient of zero.
B. The y-coordinates of all points on a vertical line are the same.
C. Horizontal lines have an undefined gradient.
D. The x-coordinates of all points on a vertical line are the same.
E. A horizontal line has the general equation $x = a$.

d. Select which of the following statements is false.

A. Horizontal lines have a gradient of zero.
B. The line joining the points (1, −1) and (−7, −1) is vertical.
C. Vertical lines have an undefined gradient.
D. The line joining the points (1, 1) and (−7, 1) is horizontal.
E. A horizontal line has the general equation $y = c$.

13. **MC** The point (−1, 5) lies on a line parallel to $4x + y + 5 = 0$. Another point on the same line as (−1, 5) is:

A. (2, 9) B. (4, 2) C. (4, 0) D. (−2, 3) E. (3, −11)

14. Determine the equation of the straight line given the following conditions.

a. Passes through the point (−1, 3) and parallel to $y = -2x + 5$
b. Passes through the point (4, −3) and parallel to $3y + 2x = -3$

15. Determine which pairs of the following lines are perpendicular.

a. $x + 3y - 5 = 0$ b. $y = 4x - 7$ c. $y = x$ d. $2y = x + 1$
e. $y = 3x + 2$ f. $x + 4y - 9 = 0$ g. $2x + y = 6$ h. $x + y = 0$

16. Determine the equation of the straight line that cuts the x-axis at 3 and is perpendicular to the line with equation $3y - 6x = 12$.

17. Calculate the value of m for which lines with the following pairs of equations are perpendicular to each other.

a. $2y - 5x = 7$ and $4y + 12 = mx$
b. $5x - 6y = -27$ and $15 + mx = -3y$

18. **MC** The gradient of the line perpendicular to the line with equation $3x - 6y = 2$ is:

A. 3 B. −6 C. 2 D. $\frac{1}{2}$ E. −2

Reasoning

19. Determine the equation of a line, in the form of $ax + by + c = 0$, that is perpendicular to the line with equation $2x - y = 3$ and passes through the point (2, 3).

20. Form the equation of the line, in the form of $ax + by + c = 0$, that is perpendicular to the line with equation $-4x - 3y = 3$ and passes through the point (−1, 4).

21. **MC** Triangle ABC has a right angle at B. The vertices are A(−2, 9), B(2, 8) and C(1, z). The value of z is:

A. $8\frac{1}{4}$ B. 4 C. 12 D. $7\frac{3}{4}$ E. −4

Problem solving

22. a. Sketch the graph of the equation $y = 2x - 4$.
   b. On the same set of axes, sketch the graph of the line parallel to $y = 2x - 4$ that has a y-intercept of −2.
   c. Sketch the graph of the line that is perpendicular to the lines found in part a and b that also passes through the origin.

23. Determine the value(s) of a such that there would be no point of intersection between the lines $ay + 3x = 4a$ and $2x - y = 5$.

24. A family of parallel lines has the equation $3x - 2y = k$ where k is a real number.
   a. Determine the gradient of each member of this family of lines.
   b. Show that all lines in the family contain the point (k, k).

3.5 The distance between two points

LEARNING INTENTION

At the end of this subtopic you should be able to:
- calculate the straight-line distance between two points
- determine the value of an unknown coordinate given the distance between two points.

3.5.1 The distance between two points

eles-4745

- The distance between two points can be calculated using Pythagoras' theorem.
- Consider two points $A(x_1, y_1)$ and $B(x_2, y_2)$ on the Cartesian plane as shown.
- If point C is placed as shown, ABC is a right-angled triangle and AB is the hypotenuse.

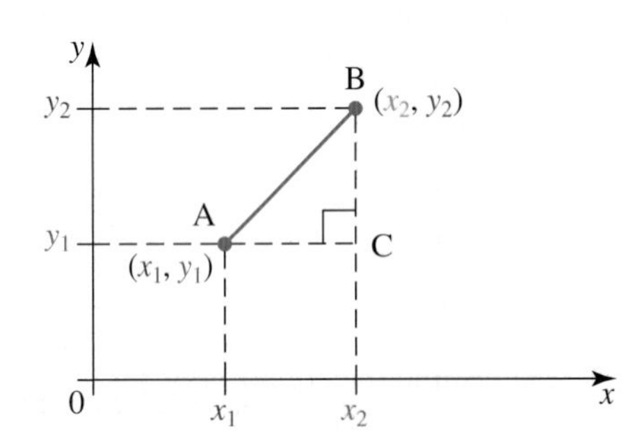

$$AC = x_2 - x_1$$
$$BC = y_2 - y_1$$

By Pythagoras' theorem:

$$AB^2 = AC^2 + BC^2$$
$$= (x_2 - x_1)^2 + (y_2 - y_1)^2$$

$$\text{Hence } AB = \sqrt{(x_2 - x_1)^2 + (y_2 - y_1)^2}$$

The distance between two points

The distance between two points $A(x_1, y_1)$ and $B(x_2, y_2)$ is:

$$\mathbf{AB} = \sqrt{(x_2 - x_1)^2 + (y_2 - y_1)^2}$$

- This distance formula can be used to calculate the distance between any two points on the Cartesian plane.
- The distance formula has many geometric applications.

Note: If the coordinates were named in the reverse order, the formula would still give the same answer. Check this for yourself using $(x_1, y_1) = (3, 4)$ and $(x_2, y_2) = (-3, 1)$.

WORKED EXAMPLE 18 Determining the distance between two points on a graph

Determine the distance between the points A and B in the figure.

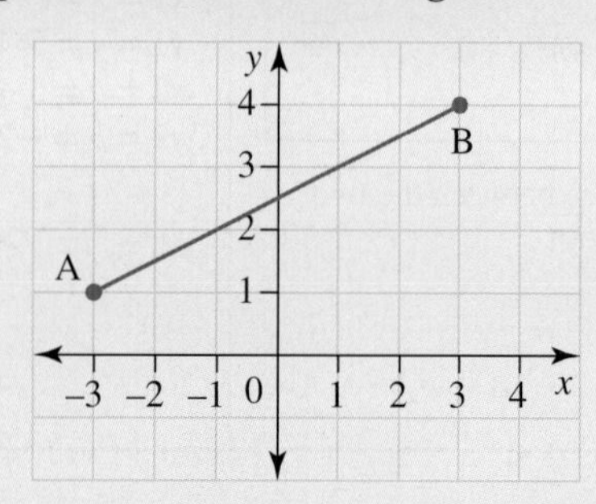

THINK	WRITE
1. From the graph, locate points A and B.	A(−3, 1) and B(3, 4)
2. Let A have coordinates (x_1, y_1).	Let $(x_1, y_1) = (-3, 1)$
3. Let B have coordinates (x_2, y_2).	Let $(x_2, y_2) = (3, 4)$
4. Calculate the length AB by applying the formula for calculating the distance between two points.	$\text{AB} = \sqrt{(x_2 - x_1)^2 + (y_2 - y_1)^2}$ $= \sqrt{(3 - (-3))^2 + (4 - 1)^2}$ $= \sqrt{(6)^2 + (3)^2}$ $= \sqrt{36 + 9}$ $= \sqrt{45}$ $= 3\sqrt{5}$

WORKED EXAMPLE 19 Calculating the distance between two points

Calculate the distance between the points P(−1, 5) and Q(3, −2).

THINK	WRITE
1. Let P have coordinates (x_1, y_1).	Let $(x_1, y_1) = (-1, 5)$
2. Let Q have coordinates (x_2, y_2).	Let $(x_2, y_2) = (3, -2)$
3. Calculate the length PQ by applying the formula for the distance between two points.	$\text{PQ} = \sqrt{(x_2 - x_1)^2 + (y_2 - y_1)^2}$ $= \sqrt{(3 - (-1))^2 + (-2 - 5)^2}$ $= \sqrt{(4)^2 + (-7)^2}$ $= \sqrt{16 + 49}$ $= \sqrt{65}$

TI | THINK	DISPLAY/WRITE
On a Calculator page, complete the entry lines as: $x1: = -1$ $y1: = 5$ $x2: = 3$ $y2: = -2$ $\sqrt{(x2-x1)^2+(y2-y1)^2}$ Press ENTER after each entry.	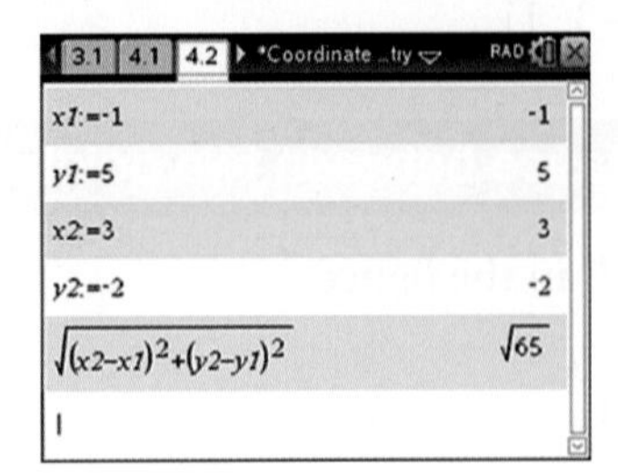 The distance between the two points is $\sqrt{65}$.

CASIO | THINK	DISPLAY/WRITE
On the Main screen, complete the entry lines as: $\sqrt{(x-a)^2+(y-b)^2}\|x=-1$ $\|y=5\|a=3\|b=-2$ $x1=x=-1$ $y1=y=5$ $x2=a=3$ $y2=b=-2$	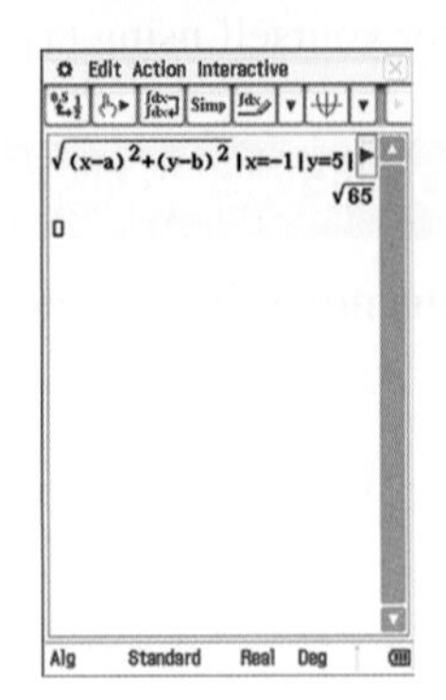 The distance between the two points is $\sqrt{65}$.

WORKED EXAMPLE 20 Applying the distance formula

Prove that the points A(1, 1), B(3, −1) and C(−1, −3) are the vertices of an isosceles triangle.

THINK	WRITE/DRAW
1. Plot the points and draw the triangle. *Note*: For triangle ABC to be isosceles, two sides must have the same magnitude.	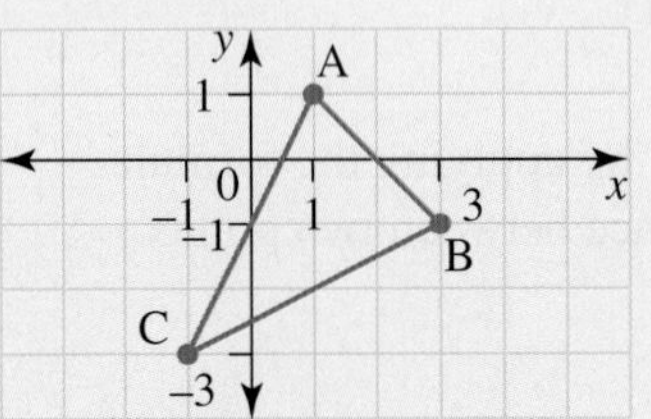
2. AC and BC seem to be equal. Calculate the length AC. $A(1, 1) = (x_2, y_2)$ $C(-1, -3) = (x_1, y_1)$	$AC = \sqrt{[1-(-1)]^2+[1-(-3)]^2}$ $= \sqrt{(2)^2+(4)^2}$ $= \sqrt{20}$ $= 2\sqrt{5}$
3. Calculate the length BC. $B(3, -1) = (x_2, y_2)$ $C(-1, -3) = (x_1, y_1)$	$BC = \sqrt{[3-(-1)]^2+[-1-(-3)]^2}$ $= \sqrt{(4)^2+(2)^2}$ $= \sqrt{20}$ $= 2\sqrt{5}$
4. Calculate the length AB. $A(1, 1) = (x_1, y_1)$ $B(3, -1) = (x_2, y_2)$	$AB = \sqrt{[3-(1)]^2+[-1-(1)]^2}$ $= \sqrt{(2)^2+(-2)^2}$ $= \sqrt{4+4}$ $= 2\sqrt{2}$
5. Write your conclusion.	Since $AC = BC \neq AB$, triangle ABC is an isosceles triangle.

DISCUSSION

How could you use the distance formula to show that a series of points lay on the circumference of a circle with centre C?

Resources

eWorkbook Topic 3 Workbook (worksheets, code puzzle and project) (ewbk-2029)

Interactivities Individual pathway interactivity: The distance between two points (int-4574)

Distance between two points (int-6051)

Exercise 3.5 The distance between two points

learnon

Individual pathways

PRACTISE	CONSOLIDATE	MASTER
1, 4, 7, 11	2, 5, 8, 9, 12	3, 6, 10, 13

To answer questions online and to receive **immediate corrective feedback** and **fully worked solutions** for all questions, go to your learnON title at www.jacplus.com.au.

Fluency

1. WE18 Determine the distance between each pair of points shown in the graph.

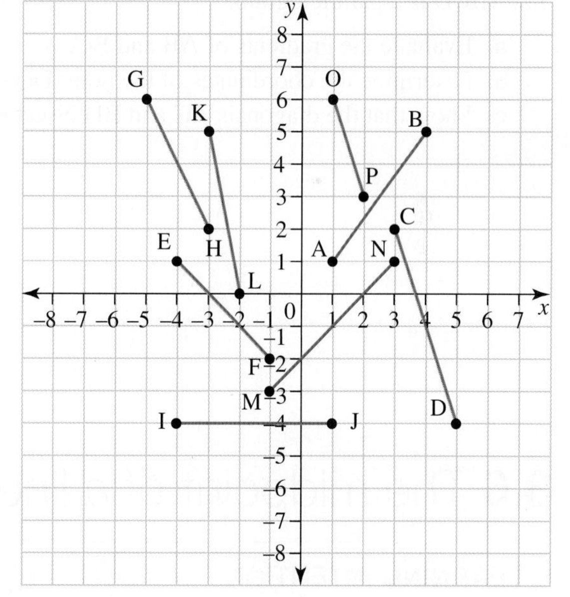

2. WE19 Calculate the distance between the following pairs of points.
   a. (2, 5), (6, 8)
   b. (−1, 2), (4, 14)
   c. (−1, 3), (−7, −5)
   d. (5, −1), (10, 4)
   e. (4, −5), (1, 1)
3. Calculate the distance between the following pairs of points.
   a. (−3, 1), (5, 13)
   b. (5, 0), (−8, 0)
   c. (1, 7), (1, −6)
   d. (a, b), $(2a, -b)$
   e. $(-a, 2b)$, $(2a, -b)$
4. The vertices of a quadrilateral are A(1, 4), B(−1, 8), C(1, 9) and D(3, 5).
   a. Determine the lengths of the sides.
   b. Determine the lengths of the diagonals.
   c. State the type of quadrilateral.

Understanding

5. MC If the distance between the points $(3, b)$ and $(-5, 2)$ is 10 units, then the value of b is:

   A. −8 B. −4 C. 4 D. 0 E. 2

6. MC A rhombus has vertices A(1, 6), B(6, 6), C(−2, 2) and D(x, y). The coordinates of D are:

   A. (2, −3) B. (2, 3) C. (−2, 3) D. (3, 2) E. (3, −2)

Reasoning

7. **WE20** Prove that the points A(0, −3), B(−2, −1) and C(4, 3) are the vertices of an isosceles triangle.
8. The points P(2, −1), Q(−4, −1) and R$(-1, 3\sqrt{3} - 1)$ are joined to form a triangle. Prove that triangle PQR is equilateral.
9. Prove that the triangle with vertices D(5, 6), E(9, 3) and F(5, 3) is a right-angled triangle.
10. A rectangle has vertices A(1, 5), B(10.6, z), C(7.6, −6.2) and D(−2, 1). Determine:
    a. the length of CD
    b. the length of AD
    c. the length of the diagonal AC
    d. the value of z.

Problem solving

11. Triangle ABC is an isosceles triangle where AB = AC, B is the point (−1, 2), C is the point (6, 3) and A is the point (a, $3a$). Determine the value of the integer constant a.

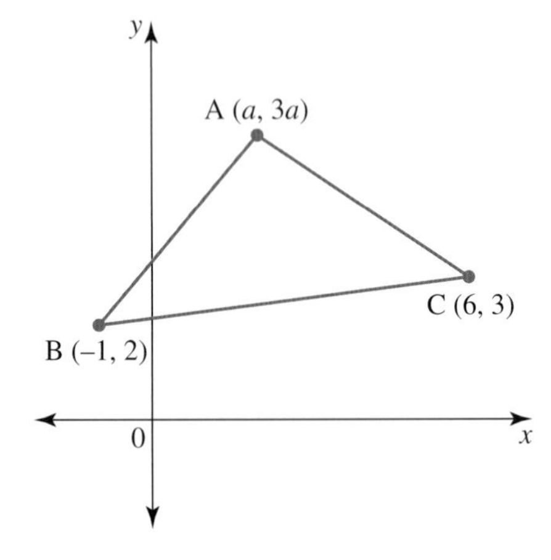

12. Show that the triangle ABC with coordinates A(a, a), B(m, $-a$) and C($-a$, m) is isosceles.
13. ABCD is a parallelogram.
    a. Evaluate the gradients of AB and BC.
    b. Determine the coordinates of the point D(x, y).
    c. Show that the diagonals AC and BD bisect each other.

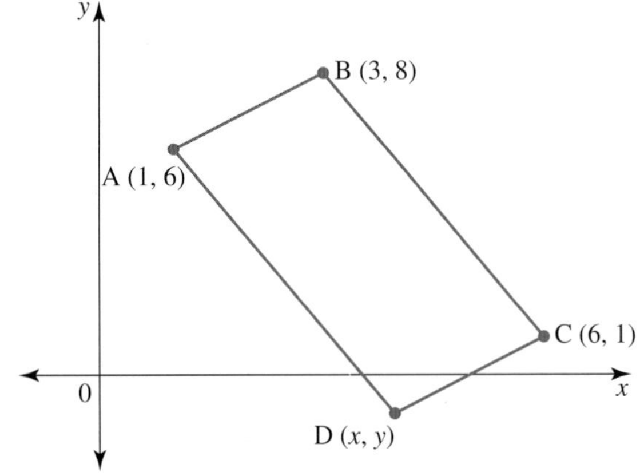

3.6 The midpoint of a line segment

LEARNING INTENTION

At the end of this subtopic you should be able to:
- calculate the midpoint between two points
- determine the value of an unknown coordinate given the midpoint between two points.

3.6.1 Midpoint of a line segment

eles-4746

- The **midpoint** of a **line segment** is the halfway point.
- The x- and y-coordinates of the midpoint are halfway between those of the coordinates of the end points.
- The following diagram shows the line interval AB joining points A(x_1, y_1) and B(x_2, y_2).

The midpoint of AB is P, so AP = PB.

Points C(x, y_1) and D(x_2, y) are added to the diagram and are used to make the two right-angled triangles ΔABC and ΔPBD.

The two triangles are congruent:

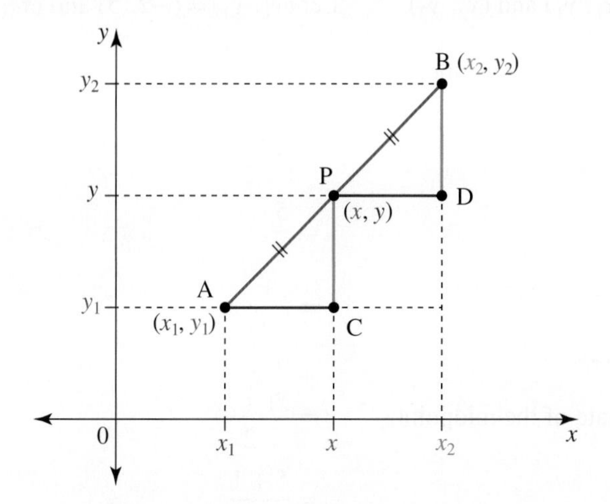

$$\begin{aligned} AP &= PB && \text{(given)} \\ \angle APC &= \angle PBD && \text{(corresponding angles)} \\ \angle CAP &= \angle DPB && \text{(corresponding angles)} \\ \text{So } \Delta APC &= \Delta PBD && \text{(ASA)} \end{aligned}$$

This means that AC = PD;

$$\text{i.e. } x - x_1 = x_2 - x \quad \text{(solve for } x\text{)}$$

$$\text{i.e. } 2x = x_1 + x_2$$

$$x = \frac{x_1 + x_2}{2}$$

In other words, x is simply the average x_1 and x_2.

Similarly, $y = \dfrac{y_1 + y_2}{2}$.

The midpoint formula

To calculate the midpoint (x, y) of the two points A (x_1, y_1) and B (x_2, y_2):

- The x-value is the average of x_1 and x_2.
- The y-value is the average of y_1 and y_2.

$$\text{Midpoint} = \left(\frac{x_1 + x_2}{2}, \frac{y_1 + y_2}{2}\right)$$

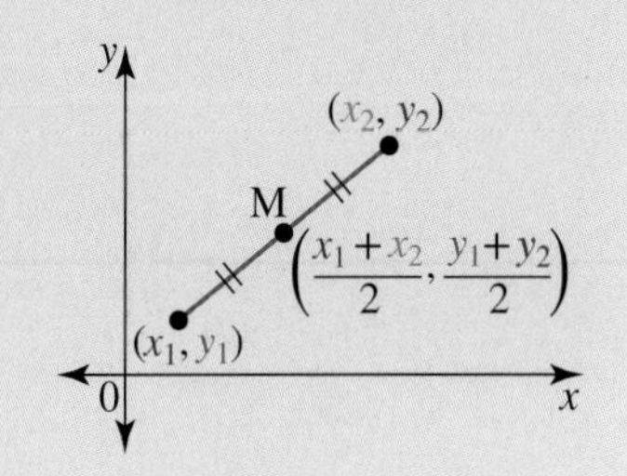

WORKED EXAMPLE 21 Calculating the midpoint

Calculate the coordinates of the midpoint of the line segment joining (−2, 5) and (7, 1).

THINK	WRITE
1. Label the given points (x_1, y_1) and (x_2, y_2).	Let $(x_1, y_1) = (-2, 5)$ and $(x_2, y_2) = (7, 1)$
2. Determine the x-coordinate of the midpoint.	$x = \frac{x_1 + x_2}{2}$ $= \frac{-2+7}{2}$ $= \frac{5}{2}$ $= 2\frac{1}{2}$
3. Determine the y-coordinate of the midpoint.	$y = \frac{y_1 + y_2}{2}$ $= \frac{5+1}{2}$ $= \frac{6}{2}$ $= 3$
4. Write the coordinates of the midpoint.	The midpoint is $\left(2\frac{1}{2}, 3\right)$.

TI | THINK	DISPLAY/WRITE	CASIO | THINK	DISPLAY/WRITE
On a Calculator page, complete the entry lines as: $\frac{x1 + x2}{2} \vert x1 = -2 \text{ and } x2 = 7$ $\frac{y1 + y2}{2} \vert y1 = 5 \text{ and } y2 = 1$ Press ENTER after each entry.	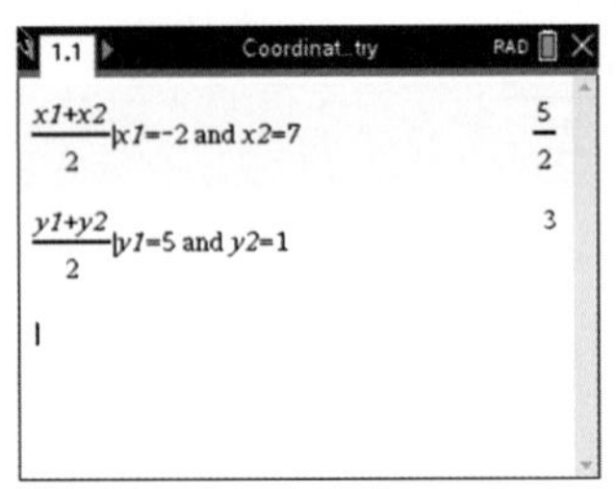 The midpoint is $\left(2\frac{1}{2}, 3\right)$.	On the Main screen, complete the entry lines as: $\frac{x + a}{2} \vert x = -2 \vert a = 7$ $\frac{y + b}{2} \vert y = 5 \vert b = 1$ Press EXE after each entry.	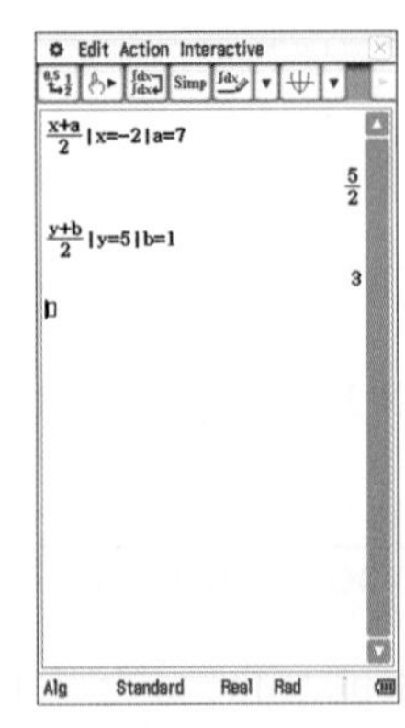 The midpoint is $\left(2\frac{1}{2}, 3\right)$.

WORKED EXAMPLE 22 Determining an unknown coordinate

The coordinates of the midpoint, M, of the line segment AB are (7, 2). If the coordinates of A are (1, −4), determine the coordinates of B.

THINK	WRITE/DRAW
1. Let the start of the line segment be (x_1, y_1) and the midpoint be (x, y).	Let $(x_1, y_1) = (1, -4)$ and $(x, y) = (7, 2)$
2. The average of the x-coordinates is 7. Determine the x-coordinate of the end point.	$x = \frac{x_1 + x_2}{2}$ $7 = \frac{1 + x_2}{2}$ $14 = 1 + x_2$ $x_2 = 13$
3. The average of the y-coordinates is 2. Determine the y-coordinate of the end point.	$y = \frac{y_1 + y_2}{2}$ $2 = \frac{-4 + y_2}{2}$ $4 = -4 + y_2$ $y_2 = 8$
4. Write the coordinates of the end point.	The coordinates of the point B are (13, 8).
5. Check that the coordinates are feasible by drawing a diagram.	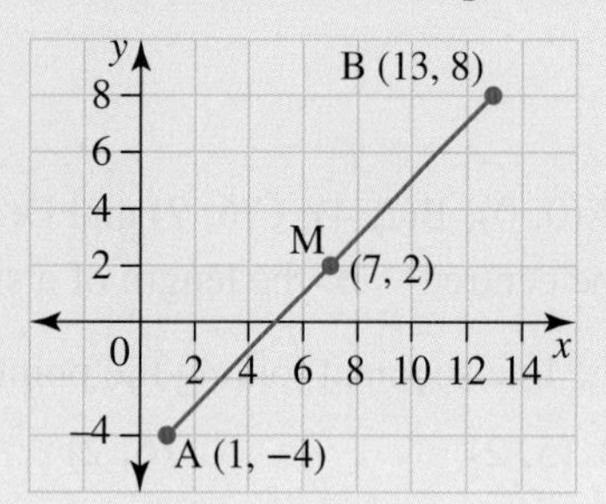

DISCUSSION

If the midpoint of a line segment is the origin, what are the possible values of the x- and y- coordinates of the end points?

on Resources

 **eWorkbook** Topic 3 Workbook (worksheets, code puzzle and project) (ewbk-2029)

 **Interactivities** Individual pathway interactivity: The midpoint of a line segment (int-4575)

Midpoints (int-6052)

Exercise 3.6 The midpoint of a line segment

learnON

Individual pathways

PRACTISE	CONSOLIDATE	MASTER
1, 3, 6, 7, 12, 15	4, 5, 8, 9, 13, 16	2, 10, 11, 14, 17

To answer questions online and to receive **immediate corrective feedback** and **fully worked solutions** for all questions, go to your learnON title at www.jacplus.com.au.

Fluency

1. WE21 Calculate the coordinates of the midpoint of the line segment joining the following pairs of points.
   a. $(-5, 1)$, $(-1, -8)$ b. $(4, 2)$, $(11, -2)$ c. $(0, 4)$, $(-2, -2)$
2. Calculate the coordinates of the midpoint of the line segment joining the following pairs of points.
   a. $(3, 4)$, $(-3, -1)$ b. $(a, 2b)$, $(3a, -b)$ c. $(a+3b, b)$, $(a-b, a-b)$
3. WE 22 The coordinates of the midpoint, M, of the line segment AB are $(2, -3)$. If the coordinates of A are $(7, 4)$, determine the coordinates of B.
4. Determine the midpoint of the following sets of coordinates.
   a. $(1, 2)$ and $(3, -4)$ b. $(7, -2)$ and $(-4, 13)$ c. $(3, a)$ and $(1, 4a)$
5. If $M(2, -2)$ is the midpoint of the line segment joining the points $X(4, y)$ and $Y(x, -1)$, then calculate the value of $x+y$.

Understanding

6. A square has vertices $A(0, 0)$, $B(2, 4)$, $C(6, 2)$ and $D(4, -2)$. Determine:
   a. the coordinates of the centre b. the length of a side c. the length of a diagonal.
7. MC The midpoint of the line segment joining the points $(-2, 1)$ and $(8, -3)$ is:
   A. $(6, -2)$ B. $(5, 2)$ C. $(6, 2)$ D. $(3, -1)$ E. $(5, -2)$
8. MC If the midpoint of AB is $(-1, 5)$ and the coordinates of B are $(3, 8)$, then A has coordinates:
   A. $(1, 6.5)$ B. $(2, 13)$ C. $(-5, 2)$ D. $(4, 3)$ E. $(7, 11)$
9. a. The vertices of a triangle are $A(2, 5)$, $B(1, -3)$ and $C(-4, 3)$. Determine:
      i. the coordinates of P, the midpoint of AC
      ii. the coordinates of Q, the midpoint of AB
      iii. the length of PQ.
   b. Show that $BC = 2PQ$.
10. a. A quadrilateral has vertices $A(6, 2)$, $B(4, -3)$, $C(-4, -3)$ and $D(-2, 2)$. Determine:
      i. the midpoint of the diagonal AC
      ii. the midpoint of the diagonal BD.
    b. State what you can infer about the quadrilateral.
11. a. The points $A(-5, 3.5)$, $B(1, 0.5)$ and $C(-6, -6)$ are the vertices of a triangle. Determine:
      i. the midpoint, P, of AB
      ii. the length of PC
      iii. the length of AC
      iv. the length of BC.
    b. Describe the triangle. State what PC represents.

Reasoning

12. a. Plot the following points on a Cartesian plane: A(−1, −4), B(2, 3), C(−3, 8) and D(4, −5).
    b. Show that the midpoint of the interval AC is (−2, 2).
    c. Calculate the exact distance between the points A and C.
    d. If B is the midpoint of an interval CM, determine the coordinates of point M.
    e. Show that the gradient of the line segment AB is $\frac{7}{3}$.
    f. Determine the equation of the line that passes through the points B and D.

13. Write down the coordinates of the midpoint of the line joining the points $(3k - 1, 4 - 5k)$ and $(5k - 1, 3 - 5k)$. Show that this point lies on the line with equation $5x + 4y = 9$.

14. The points A $(2m, 3m)$, B $(5m, -2m)$ and C $(-3m, 0)$ are the vertices of a triangle. Show that this is a right-angled triangle.

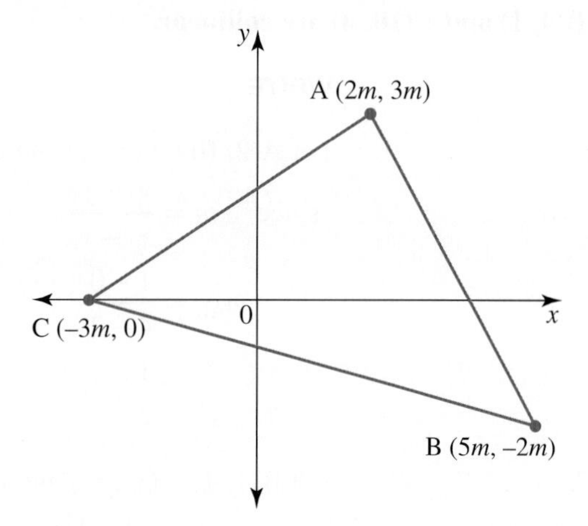

Problem solving

15. Determine the equation of the straight line that passes through the midpoint of A(−2, 5) and B(−2, 3), and has a gradient of −3.

16. Determine the equation of the straight line that passes through the midpoint of A(−1, −3) and B(3, −5), and has a gradient of $\frac{2}{3}$.

17. Determine the equation of the straight line passing through the midpoint of (3, 2) and (5, −2) that is also perpendicular to the line $3x - 2y = 7$.

3.7 Applications and collinearity

LEARNING INTENTION

At the end of this subtopic you should be able to:
- determine whether a set of coordinates are collinear
- determine the equation of a perpendicular bisector
- apply your understanding from multiple skills learned in this topic to a single problem.

3.7.1 Collinear points

eles-4747

- **Collinear points** are points that all lie on the same straight line.
- If A, B and C are collinear, then $m_{AB} = m_{BC}$.

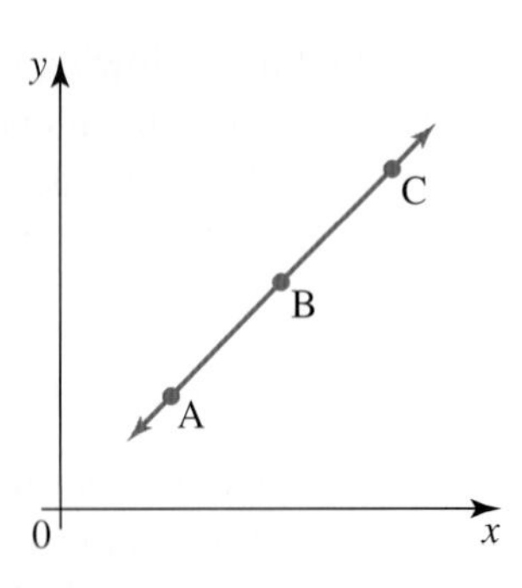

WORKED EXAMPLE 23 Proving points are collinear

Show that the points A(2, 0), B(4, 1) and C(10, 4) are collinear.

THINK	WRITE
1. Calculate the gradient of AB.	Let A(2, 0) = (x_1, y_1) and B(4, 1) = (x_2, y_2) since $m = \frac{y_2 - y_1}{x_2 - x_1}$ $m_{AB} = \frac{1-0}{4-2}$ $= \frac{1}{2}$
2. Calculate the gradient of BC.	Let B(4, 1) = (x_1, y_1) and C(10, 4) = (x_2, y_2) $m_{BC} = \frac{4-1}{10-4}$ $= \frac{3}{6}$ $= \frac{1}{2}$
3. Show that A, B and C are collinear.	Since $m_{AB} = m_{BC} = \frac{1}{2}$ and B is common to both line segments, A, B and C are collinear.

3.7.2 Equations of horizontal and vertical lines

eles-4748

- Horizontal lines are parallel to the x-axis, have a gradient of zero, are expressed in the form $y = c$ and have no x-intercept.
- Vertical lines are parallel to the y-axis, have an undefined (infinite) gradient, are expressed in the form $x = a$ and have no y-intercept.

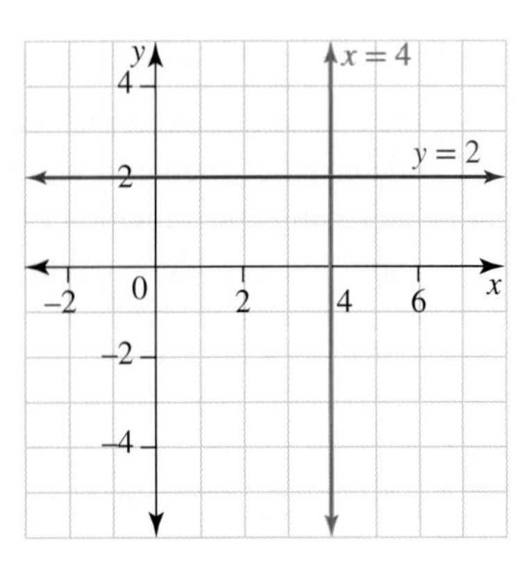

WORKED EXAMPLE 24 Determining the equation of vertical and horizontal lines

Determine the equation of:
a. the vertical line that passes through the point (2, −3)
b. the horizontal line that passes through the point (−2, 6).

THINK	WRITE
a. The equation of a vertical line is $x = a$. The x-coordinate of the given point is 2.	a. $x = 2$
b. The equation of a horizontal line is $y = c$. The y-coordinate of the given point is 6.	b. $y = 6$

3.7.3 Perpendicular bisectors

eles-4749

- A perpendicular bisector is a line that intersects another line at a right angle and cuts it into two equal lengths.
- A perpendicular bisector passes through the midpoint of a line segment.

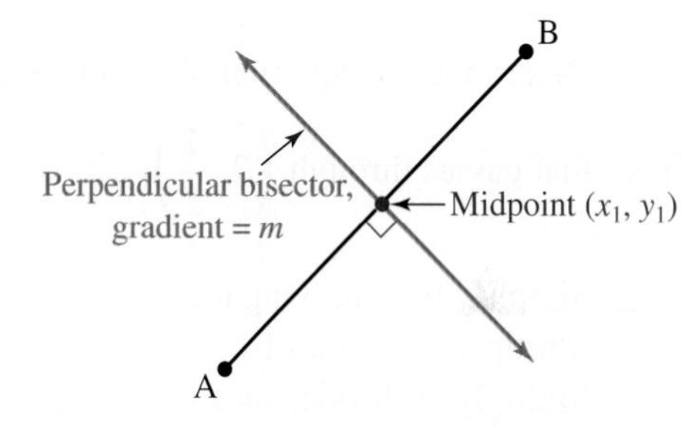

The perpendicular bisector

- The equation of a perpendicular bisector can be found by using the formula:

$$y - y_1 = m(x - x_1)$$

where (x_1, y_1) represents the midpoint of a given set of coordinates and m is the gradient perpendicular to the gradient of the given line (i.e. the gradient of the perpendicular bisector).

WORKED EXAMPLE 25 Determining the equation of a perpendicular bisector

Determine the equation of the perpendicular bisector of the line joining the points (0, −4) and (6, 5). (A bisector is a line that crosses another line at right angles and cuts it into two equal lengths.)

THINK	WRITE
1. Determine the gradient of the line joining the given points by applying the formula. $m = \dfrac{y_2 - y_1}{x_2 - x_1}$.	Let $(0, -4) = (x_1, y_1)$. Let $(6, 5) = (x_2, y_2)$. $m_1 = \dfrac{y_2 - y_1}{x_2 - x_1}$ $m_1 = \dfrac{5 - (-4)}{6 - 0}$ $= \dfrac{9}{6}$ $= \dfrac{3}{2}$

2. Calculate the gradient of the perpendicular line.
$m_1 \times m_2 = -1$

$$m_1 = \frac{3}{2}$$
$$m_2 = -\frac{2}{3}$$

3. Determine the midpoint of the line joining the given points.
$M = \left(\frac{x_1 + x_2}{2}, \frac{y_1 + y_2}{2}\right)$ where $(x_1, y_1) = (0, -4)$ and $(x_2, y_2) = (6, 5)$.

$$x = \frac{x_1 + x_2}{2} = \frac{0+6}{2} = 3 \qquad y = \frac{y_1 + y_2}{2} = \frac{-4+5}{2} = \frac{1}{2}$$

Hence $\left(3, \frac{1}{2}\right)$ are the coordinates of the midpoint.

4. Determine the equation of the line with gradient $-\frac{2}{3}$ that passes through $\left(3, \frac{1}{2}\right)$.

Since $y - y_1 = m(x - x_1)$,
then $y - \frac{1}{2} = -\frac{2}{3}(x - 3)$

5. Simplify by removing the fractions.
Multiply both sides by 3.
Multiply both sides by 2.

$$3\left(y - \frac{1}{2}\right) = -2(x - 3)$$
$$3y - \frac{3}{2} = -2x + 6$$
$$6y - 3 = -4x + 12$$
$$4x + 6y - 15 = 0$$

DISCUSSION

How could you use coordinate geometry to design a logo for an organisation?

on Resources

eWorkbook Topic 3 Workbook (worksheets, code puzzle and project) (ewbk-2029)

Interactivities Individual pathway interactivity: Applications and collinearity (int-8469)
Vertical and horizontal lines (int-6049)

Exercise 3.7 Applications and collinearity

learnON

Individual pathways

■ PRACTISE	■ CONSOLIDATE	■ MASTER
1, 3, 4, 11, 12, 13, 18, 19	2, 7, 9, 14, 15, 20	5, 6, 8, 10, 16, 17, 21

To answer questions online and to receive **immediate corrective feedback** and **fully worked solutions** for all questions, go to your learnON title at www.jacplus.com.au.

Fluency

1. WE23 Show that the points A(0, −2), B(5, 1) and C(−5, −5) are collinear.
2. Show that the line that passes through the points (−4, 9) and (0, 3) also passes through the point (6, −6).
3. WE24 Determine the equation of:
   a. the vertical line that passes through the point (1, −8)
   b. the horizontal line that passes through the point (−5, −7).
4. WE25 Determine the equation of the perpendicular bisector of the line joining the points (1, 2) and (−5, −4).
5. a. Show that the following three points are collinear: $\left(1, \frac{7}{5}\right)$, $\left(\frac{5}{2}, 2\right)$ and (5, 3).
   b. Determine the equation of the perpendicular bisector of the line joining the points $\left(1, \frac{7}{5}\right)$ and (5, 3).
6. The triangle ABC has vertices A(9, − 2), B(3, 6), and C(1, 4).
   a. Determine the midpoint, M, of BC.
   b. Determine the gradient of BC.
   c. Show that AM is the perpendicular bisector of BC.
   d. Describe triangle ABC.

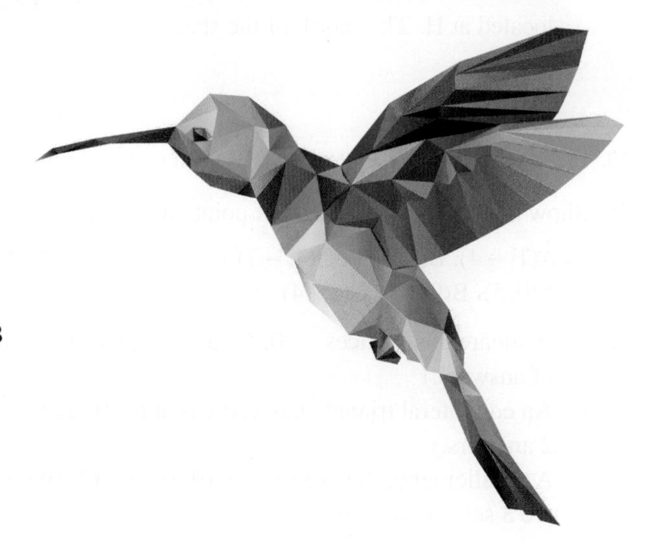

7. Determine the equation of the perpendicular bisector of the line joining the points (−2, 9) and (4, 0).
8. ABCD is a parallelogram. The coordinates of A, B and C are (4, 1), (1, −2) and (−2, 1) respectively. Determine:
   a. the equation of AD
   b. the equation of DC
   c. the coordinates of D.

Understanding

9. In each of the following, show that ABCD is a parallelogram.

   a. $A(2, 0), B(4, -3), C(2, -4), D(0, -1)$
   b. $A(2, 2), B(0, -2), C(-2, -3), D(0, 1)$
   c. $A(2.5, 3.5), B(10, -4), C(2.5, -2.5), D(-5, 5)$

10. In each of the following, show that ABCD is a trapezium.

   a. $A(0, 6), B(2, 2), C(0, -4), D(-5, -9)$
   b. $A(26, 32), B(18, 16), C(1, -1), D(-3, 3)$
   c. $A(2, 7), B(1, -1), C(-0.6, -2.6), D(-2, 3)$

11. **MC** The line that passes through the points (0, −6) and (7, 8) also passes through:

   A. (4, 3) **B.** (5, 4) **C.** (−2, 10) **D.** (1, −8) **E.** (1, 4)

Reasoning

12. The map shows the proposed course for a yacht race. Buoys have been positioned at $A(1, 5)$, $B(8, 8)$, $C(12, 6)$, and $D(10, w)$.

   a. Calculate how far it is from the start, O, to buoy A.
   b. The race marshall boat, M, is situated halfway between buoys A and C. Determine the coordinates of the boat's position.
   c. Stage 4 of the race (from C to D) is perpendicular to stage 3 (from B to C). Evaluate the gradient of CD.
   d. Determine the linear equation that describes stage 4.
   e. Hence determine the exact position of buoy D.
   f. An emergency boat is to be placed at point E, (7, 3). Determine how far the emergency boat is from the hospital, located at H, 2 km north of the start.

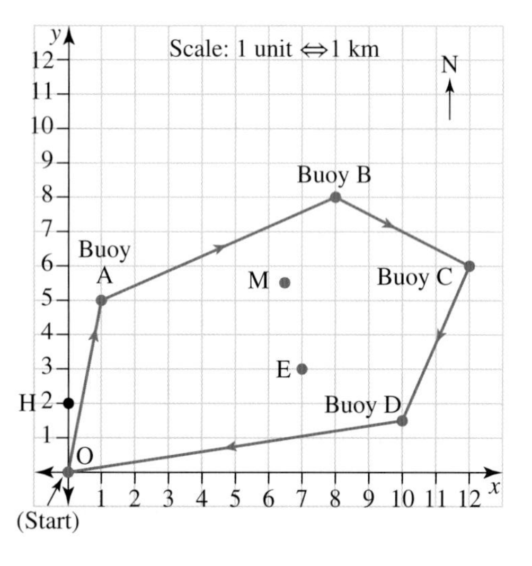

13. Prove that the quadrilateral ABCD is a rectangle with A(2, 5), B(6, 1), C(3, −2) and D(−1, 2).

14. Show that the following sets of points form the vertices of a right-angled triangle.

   a. $A(1, -4), B(2, -3), C(4, -7)$
   b. $A(3, 13), B(1, 3), C(-4, 4)$
   c. $A(0, 5), B(9, 12), C(3, 14)$

15. a. A square has vertices at (0, 0) and (2, 0). Determine where the other 2 vertices are. (There are 3 sets of answers.)
   b. An equilateral triangle has vertices at (0, 0) and (2, 0). Determine where the other vertex is. (There are 2 answers.)
   c. A parallelogram has vertices at (0, 0) and (2, 0) and (1, 1). Determine where the other vertex is. (There are 3 sets of answers.)

16. Prove that the quadrilateral ABCD is a rhombus, given $A(2, 3), B(3, 5), C(5, 6)$ and $D(4, 4)$.
   Hint: A rhombus is a parallelogram with diagonals that intersect at right angles.

17. A is the point (0, 0) and B is the point (0, 2).

   a. Determine the perpendicular bisector of AB.
   b. Show that any point on this line is equidistant from A and B.

Problem solving

Questions 18 and 19 relate to the diagram.

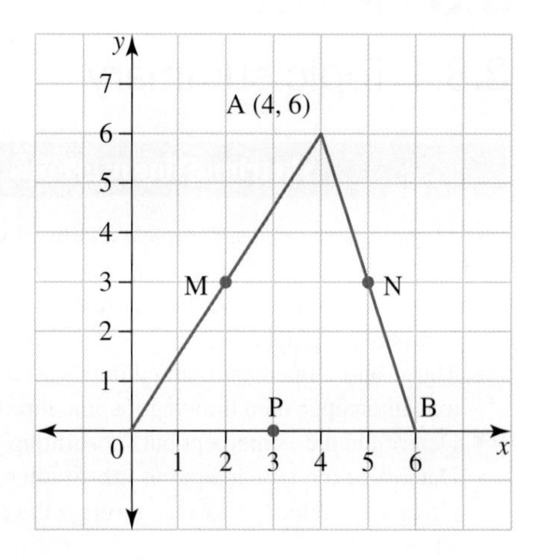

M is the midpoint of OA.
N is the midpoint of AB.
P is the midpoint of OB.

18. A simple investigation:
    a. Show that MN is parallel to OB.
    b. Is PN parallel to OA? Explain.
    c. Is PM parallel to AB? Explain.

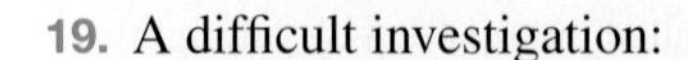

19. A difficult investigation:
    a. Determine the perpendicular bisectors of OA and OB.
    b. Determine the point W where the two bisectors intersect.
    c. Show that the perpendicular bisector of AB also passes through W.
    d. Explain why W is equidistant from O, A and B.
    e. W is called the circumcentre of triangle OAB. Using W as the centre, draw a circle through O, A, and B.

20. Line A is parallel to the line with equation $2x - y = 7$ and passes through the point (2, 3). Line B is perpendicular to the line with equation $4x - 3y + 3 = 0$ and also passes through the point (2, 3). Line C intersects with line A where it cuts the y-axis and intersects with line B where it cuts the x-axis.
    a. Determine the equations for all three lines. Give answers in the form $ax + by + c = 0$.
    b. Sketch all three lines on the one set of axes.
    c. Determine whether the triangle formed by the three lines is scalene, isosceles or equilateral.

21. The lines l_1 and l_2 are at right angles to each other. The line l_1 has the equation $px + py + r = 0$. Show that the distance from M to the origin is given by $\dfrac{r}{\sqrt{p^2 + p^2}}$.

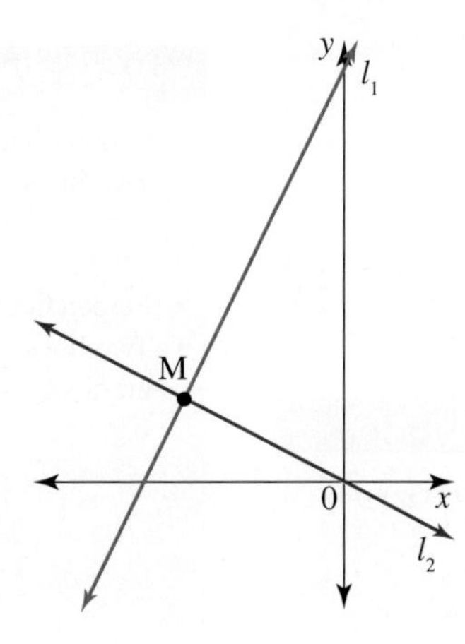

3.8 Review

3.8.1 Topic summary

Sketching linear graphs

- To plot linear graphs, complete a table of values to determine the points and use a rule.
- Only two points are needed in order to sketch a straight-line graph.
- The x- and y-intercept method involves calculating both axis intercepts, then drawing the line through them.
- Determine the x-intercept but substituting $y = 0$.
- Determine the y-intercept but substituting $x = 0$.
- Graphs given by $y = mx$ pass through the origin (0, 0), since $c = 0$.
- The line $y = c$ is parallel to the x-axis, having a gradient of zero and a y-intercept of c.
- The line $x = a$ is parallel to the y-axis, having a undefined (infinite) gradient and a x-intercept of a.

Equation of a straight line

- The equation of a straight line is:
  $$y = mx + c$$
  Where: m is the gradient and c is the y-intercept
  e.g. $y = 2x + 5$ → 5: y-intercept; 2: gradient
- The rule $y = mx + c$ is called the equation of a straight line in the **gradient-intercept** form.
- The gradient of a straight line can be determined by the formulas:
  $$m = \frac{\text{rise}}{\text{run}} = \frac{y_2 - y_1}{x_2 - x_1}$$

COORDINATE GEOMETRY

The midpoint of a line segment

- The midpoint of two points, (x_1, y_1) and (x_2, y_2) is:
  $$M = \left(\frac{x_1 + x_2}{2}, \frac{y_1 + y_2}{2}\right)$$

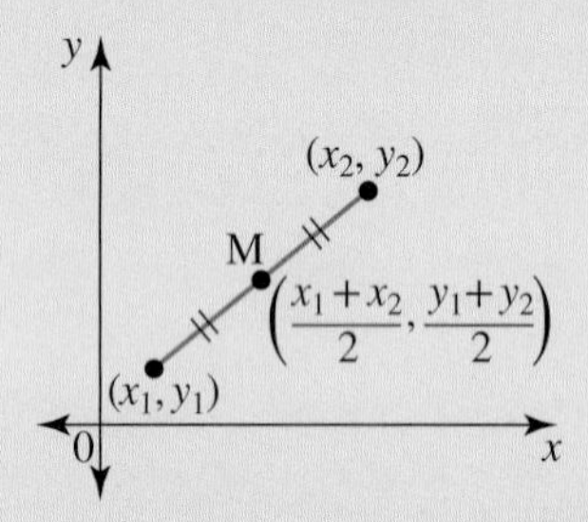

Determining linear equations

- The formula:
  $$y - y_1 = m(x - x_1)$$
  can be used to write the equation of a line, given the gradient and the coordinates of one point.

Parallel and perpendicular lines

- Parallel lines will never intersect with each other.
- Two lines are parallel if they have the same gradient.
  e.g. $y = 3x - 6$
  $y = 3x + 1$
- Perpendicular lines are lines that intersect at *right angles*.
- Two lines are perpendicular if the product of their gradients is −1.
  e.g. $y = 2x + 3$
  $y = -\frac{x}{2} - 4$
  $m_1 \times m_2 = 2 \times -\frac{1}{2} = -1$

The distance between two points

- The distance between two points, (x_1, y_1) and (x_2, y_2) is:
  $$d = \sqrt{(x_2 - x_1)^2 + (y_2 - y_1)^2}$$

Applications and collinearity

- **Collinear points** are points that all lie on the same straight line.
- Horizontal lines are parallel to the x-axis, have a gradient of zero, are expressed in the form $y = c$ and have no x-intercept.
- Vertical lines are parallel to the y-axis, have an undefined (infinite) gradient, are expressed in the form $x = a$ and have no y-intercept.

Perpendicular bisector

- A perpendicular bisector is a line that intersects another line at a right angle and cuts it into two equal lengths.
- A perpendicular bisector passes through the midpoint of a line segment.

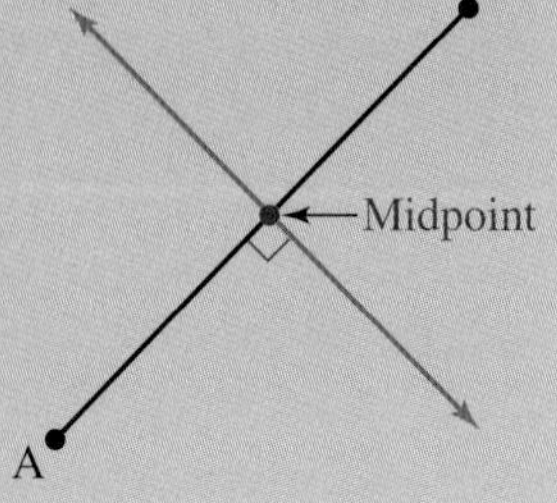

3.8.2 Success criteria

Tick the column to indicate that you have completed the subtopic and how well you have understood it using the traffic light system.

(**Green:** I understand; **Yellow:** I can do it with help; **Red:** I do not understand)

Subtopic	Success criteria	Green	Yellow	Red
3.2	I can plot points on a graph using a rule and a table of values.			
	I can sketch linear graphs by determining the x- and y-intercepts.			
	I can sketch the graphs of horizontal and vertical lines.			
	I can model linear graphs from a worded context.			
3.3	I can determine the equation of a straight line when given its graph.			
	I can determine the equation of a straight line when given the gradient and the y-intercept.			
	I can determine the equation of a straight line passing through two points.			
	I can formulate the equation of a straight line from a written context.			
3.4	I can determine whether two lines are parallel, perpendicular or neither.			
	I can determine the equation of a straight line that is parallel to a given line.			
	I can determine the equation of a straight line that is perpendicular to a given line.			
3.5	I can calculate the straight-line distance between two points.			
	I can determine the value of an unknown coordinate given the distance between two points.			
3.6	I can calculate the midpoint between two points.			
	I can determine the value of an unknown coordinate given the midpoint between two points.			
3.7	I can determine whether a set of coordinates are collinear.			
	I can determine the equation of a perpendicular bisector.			
	I can apply my understanding from multiple skills learned in this topic to a single problem.			

3.8.3 Project

What common computer symbol is this?

On computer hardware, and on many different software applications, a broad range of symbols is used. These symbols help us to identify where things need to be plugged into, what buttons we need to push, or what option needs to be selected. The main focus of this task involves constructing a common symbol found on the computer. The instructions are given below. Use grid paper to construct the symbol.

The construction part of this task requires you to graph nine lines to reveal a common computer symbol. Draw the scale of your graph to accommodate x- and y-values in the following ranges: $-10 \le x \le 16$ and $-10 \le y \le 16$. Centre the axes on the grid lines.

- Line 1 has an equation $y = x - 1$. Graph this line in the range $-7 \le x \le -2$.
- Line 2 is perpendicular to line 1 and has a y-intercept of -5. Determine the equation of this line, and then draw the line in the range $-5 \le x \le -1$.
- Line 3 is parallel to line 1, with a y-intercept of 3. Determine the equation of the line, and then graph the line in the range $-9 \le x \le -4$.
- Line 4 is parallel to line 1, with a y-intercept of -3. Determine the equation of the line, and then graph the line in the range $-1 \le x \le 2$.
- Line 5 has the same length as line 4 and is parallel to it. The point $(-2, 3)$ is the starting point of the line, which decreases in both x- and y-values from there.
- Line 6 commences at the same starting point as line 5, and then runs at right angles to line 5. It has an x-intercept of 1 and is the same length as line 2.
- Line 7 commences at the same starting point as both lines 5 and 6. Its equation is $y = 6x + 15$. The point $(-1, 9)$ lies at the midpoint.
- Line 8 has the equation $y = -x + 15$. Its midpoint is the point $(7, 8)$ and its extremities are the points where the line meets line 7 and line 9.
- Line 9 has the equation $6y - x + 8 = 0$. It runs from the intersection of lines 4 and 6 until it meets line 8.

1. Determine what common computer symbol you have drawn.
2. The top section of your figure is a familiar geometric shape. Use the coordinates on your graph, together with the distance formula to determine the necessary lengths to calculate the area of this figure.
3. Using any symbol of interest to you, draw your symbol on grid lines and provide instructions for your design. Ensure that your design involves aspects of coordinate geometry that have been used throughout this task.

Resources

eWorkbook Topic 3 Workbook (worksheets, code puzzle and project) (ewbk-2029)

Interactivities Crossword (int-2833)

Sudoku puzzle (int-3590)

Exercise 3.8 Review questions

learnon

To answer questions online and to receive **immediate corrective feedback** and **fully worked solutions** for all questions, go to your learnON title at www.jacplus.com.au.

Fluency

1. **MC** The equation of the following line is:

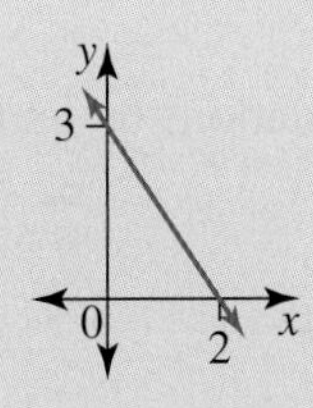

A. $3x+2y=6$ B. $3x-2y=6$ C. $2x+3y=6$
D. $2x-3y=6$ E. $2x-3y=-6$

2. **MC** The equation of a linear graph with gradient -3 and x-intercept of 4 is:
A. $y=-3x-12$ B. $y=-3x+4$ C. $y=-3x-4$
D. $y=-3x+12$ E. $y=4x-3$

3. **MC** The equation of a linear graph which passes through $(2, -7)$ and $(-2, -2)$ is:
A. $4x-5y+18=0$ B. $5x+4y+18=0$ C. $5x+4y-18=0$
D. $5x-4y-18=0$ E. $4x+5y+18=0$

4. **MC** The distance between the points $(1, 5)$ and $(6, -7)$ is:
A. $\sqrt{53}$ B. $\sqrt{29}$ C. 13 D. $\sqrt{193}$ E. 12

5. **MC** The midpoint of the line segment joining the points $(-4, 3)$ and $(2, 7)$ is:
A. $(-1, 5)$ B. $(-2, 10)$ C. $(-6, 4)$ D. $(-2, 4)$ E. $(-1, 2)$

6. **MC** If the midpoint of the line segment joining the points A$(3, 7)$ and B(x, y) has coordinates $(6, 2)$, then the coordinates of B are:
A. $(15, 3)$ B. $(0, -6)$ C. $(9, -3)$ D. $(4.5, 4.5)$ E. $(-9, 3)$

7. **MC** If the points $(-6, -11)$, $(2, 1)$ and $(x, 4)$ are collinear, then the value of x is:
A. 4 B. 3.2 C. $\frac{1}{4}$ D. $\frac{5}{16}$ E. 3

8. **MC** The gradient of the line perpendicular to $3x-4y+7=0$ is:
A. $\frac{3}{4}$ B. $\frac{4}{3}$ C. $-\frac{4}{3}$ D. 3 E. -4

9. **MC** The equation of the line perpendicular to $2x + y - 1 = 0$ and passing through the point (1, 4) is:
   A. $2x + y - 6 = 0$ B. $2x + y - 2 = 0$ C. $x - 2y + 7 = 0$ D. $x + 2y + 9 = 0$ E. $x - 2y = 0$

10. Produce a table of values, and sketch the graph of the equation $y = -5x + 15$ for values of x between -10 and $+10$.

11. Sketch the graph of the following linear equations, labelling the x- and y-intercepts.
   a. $y = 3x - 2$ b. $y = -5x + 15$ c. $y = -\frac{2}{3}x + 1$ d. $y = \frac{7}{5}x - 3$

12. Determine the x- and y-intercepts of the following straight lines.
   a. $y = -7x + 6$ b. $y = \frac{3}{8}x - 5$ c. $y = \frac{4}{7}x - \frac{3}{4}$ d. $y = 0.5x + 2.8$

13. Sketch graphs of the following linear equations by finding the x- and y-intercepts.
   a. $2x - 3y = 6$ b. $3x + y = 0$ c. $5x + y = -3$ d. $x + y + 3 = 0$

14. Sketch the graph of each of the following.
   a. $y = \frac{1}{2}x$ b. $y = -4x$ c. $x = -2$ d. $y = 7$

15. Sketch the graph of the equation $3(y - 5) = 6(x + 1)$.

16. Determine the equations of the straight lines in the following graphs.

a.
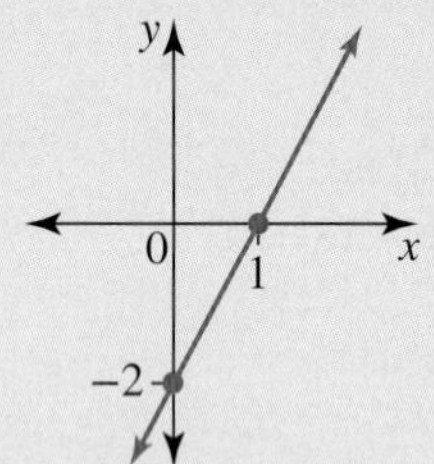

b.
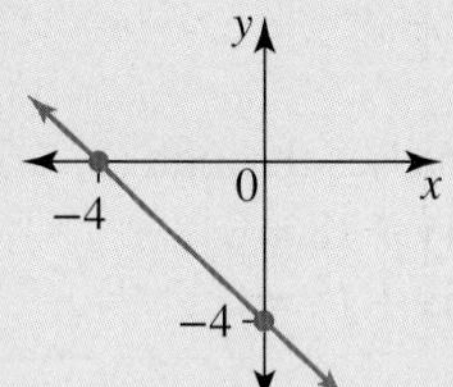

c.
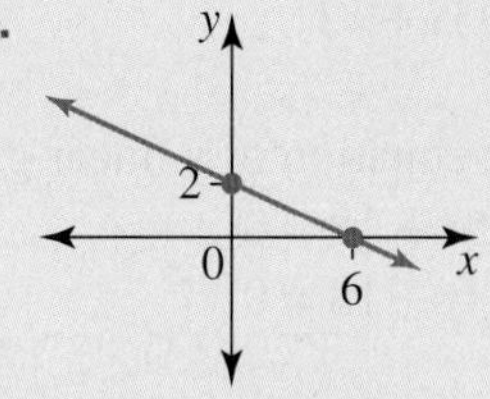

d.
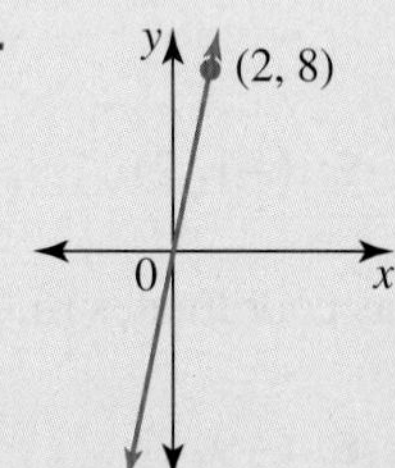

e.
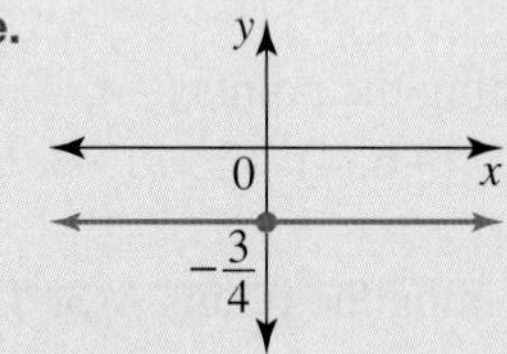

f.
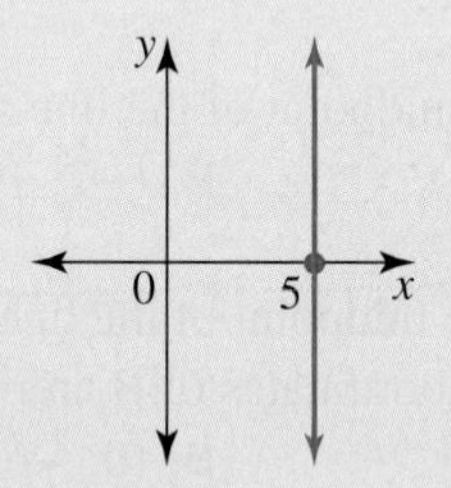

17. Determine the linear equation given the information in each case below.
   a. Gradient = 3, y-intercept = −4 b. Gradient = −2, y-intercept = −5
   c. Gradient = $\frac{1}{2}$, y-intercept = 5 d. Gradient = 0, y-intercept = 6

18. For each of the following, determine the equation of the straight line with the given gradient and passing through the given point.
    a. Gradient = 7, point (2, 1)
    b. Gradient = −3, point (1, 1)
    c. Gradient = $\frac{1}{2}$, point (−2, 5)
    d. Gradient = $\frac{3}{5}$, point (1, −3)
19. Determine the distance between the points (1, 3) and (7, −2) in exact form.
20. Prove that triangle ABC is isosceles given A(3, 1), B(−3, 7) and C(−1, 3).
21. Show that the points A(1, 1), B(2, 3) and C(8, 0) are the vertices of a right-angled triangle.
22. The midpoint of the line segment AB is (6, −4). If B has coordinates (12, 10), determine the coordinates of A.
23. Show that the points A(3, 1), B(5, 2) and C(11, 5) are collinear.
24. Show that the lines $y = 2x - 4$ and $x + 2y - 10 = 0$ are perpendicular to one another.
25. Determine the equation of the straight line passing through the point (6, −2) and parallel to the line $x + 2y - 1 = 0$.
26. Determine the equation of the line perpendicular to $3x - 2y + 6 = 0$ and having the same y-intercept.
27. Determine the equation of the perpendicular bisector of the line joining the points (−2, 7) and (4, 11).
28. Determine the equation of the straight line joining the point (−2, 5) and the point of intersection of the straight lines with equations $y = 3x - 1$ and $y = 2x + 5$.
29. Use the information given in the diagram to complete the following.

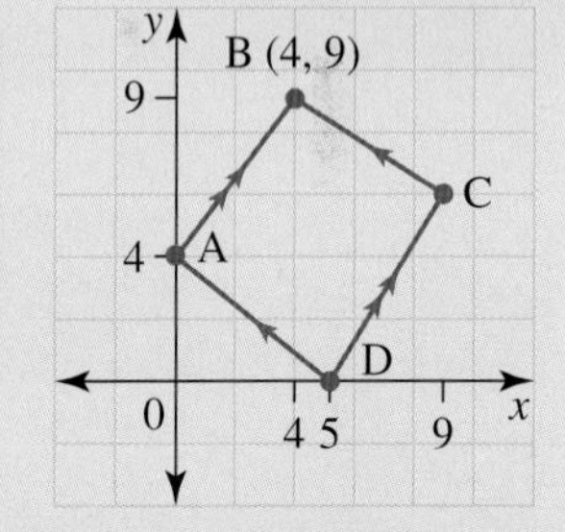

    a. Determine:
        i. the gradient of AD
        ii. the gradient of AB
        iii. the equation of BC
        iv. the equation of DC
        v. the coordinates of C.
    b. Describe quadrilateral ABCD.
30. In triangle ABC, A is (1, 5), B is (−2, −3) and C is (8, −2).
    a. Determine:
        i. the gradient of BC
        ii. the midpoint, P, of AB
        iii. the midpoint, Q, of AC.
    b. Hence show that:
        i. PQ is parallel to BC
        ii. PQ is half the length of BC.

Problem solving

31. John has a part-time job working as a gardener and is paid \$13.50 per hour.

a. Complete the following table of values relating the amount of money received to the number of hours worked.

Number of hours	0	2	4	6	8	10
Pay \$						

b. Determine a linear equation relating the amount of money received to the number of hours worked.

c. Sketch the linear equation on a Cartesian plane over a suitable domain.

d. Using algebra, calculate the pay that John will receive if he works for $6\frac{3}{4}$ hours.

32. A fun park charges a \$12.50 entry fee and an additional \$2.50 per ride.

a. Complete the following table of values relating the total cost to the number of rides.

Number of rides	0	2	4	6	8	10
Cost \$						

b. Determine a linear equation relating total cost to the number of rides.

c. Sketch the linear equation on a Cartesian plane over a suitable domain.

d. Using algebra, calculate the cost for 7 rides.

33. The cost of hiring a boat is \$160 plus \$22.50 per hour.

a. Sketch a graph showing the total cost for between 0 and 12 hours.

b. State the equation relating cost to time rented.

c. Predict the cost of hiring a boat for 12 hours and 15 minutes.

34. ABCD is a quadrilateral with vertices A(4, 9), B(7, 4), C(1, 2) and D(a, 10).
Given that the diagonals are perpendicular to each other, determine:

a. the equation of the diagonal AC

b. the equation of the diagonal BD

c. the value of a.

35. An architect decides to design a building with a 14-metre-square base such that the external walls are initially vertical to a height of 50 metres, but taper so that their separation is 8 metres at its peak height of 90 metres. A profile of the building is shown with the point (0, 0) marked as a reference at the centre of the base.

a. Write the equation of the vertical line connecting A and B.

b. Write the coordinates of B and C.

c. Determine the length of the tapered section of wall from B to C.

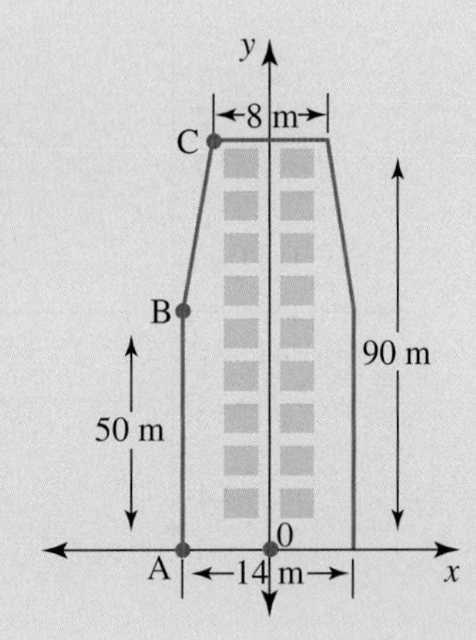

36. In a game of lawn bowls, the object is to bowl a biased ball so that it gets as close as possible to a smaller white ball called a jack. During a game, a player will sometimes bowl a ball quite quickly so that it travels in a straight line in order to displace an opponents 'guard balls'. In a particular game, player x has 2 guard balls close to the jack. The coordinates of the jack are (0, 0) and the coordinates of the guard balls are A $\left(-1, \frac{4}{5}\right)$ and B $\left(-\frac{1}{2}, \frac{57}{40}\right)$. Player Y bowls a ball so that it travels in a straight line toward the jack. The ball is bowled from the position S, with the coordinates (−30, 24).

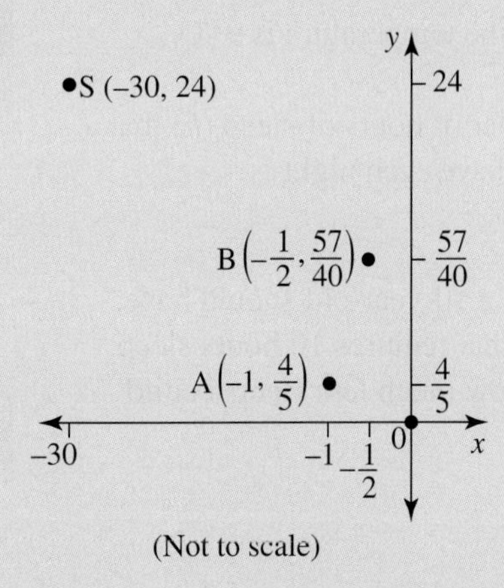

(Not to scale)

a. Will player Y displace one of the guard balls? If so, which one? Explain your answer.

b. Due to bias, the displaced guard ball is knocked so that it begins to travel in a straight line (at right angles to the path found in part **a**. Determine the equation of the line of the guard ball.

c. Show that guard ball A is initially heading directly toward guard ball B.

d. Given its initial velocity, guard ball A can travel in a straight line for 1 metre before its bias affects it path. Calculate and explain whether guard ball A will collide with guard ball B.

37. The graph shows the line p passing through the points A(−1, 1) and B(5, 5). Given that C is the point (4, 1), determine:

a. the gradient of p

b. the equation of p

c. the area of ΔABC

d. the length BC, giving your answer correct to 2 decimal places.

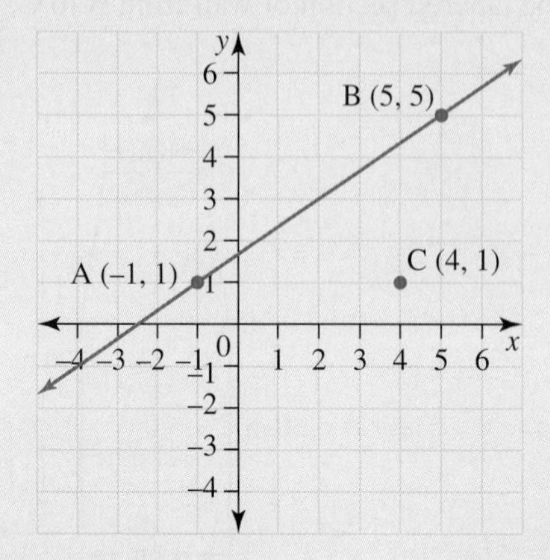

38. The temperature of the air (T °C) is related to the height above sea level (h metres) by the formula $T = 18 - 0.005h$.

a. Evaluate the temperature at the heights of:

i. 600 m

ii. 1000 m

iii. 3000 m

b. Draw a graph using the results from part **a**.

c. Use the graph to determine the temperature at 1200 m and 2500 m.

d. Predict the height at which the temperature is 9 °C.

39. An old theory is that the number of hours of sleep (h) that a child of c years of age should have each night is $h = 8 + \frac{18 - c}{2}$.

a. Determine how many hours a 10-year-old should have.

b. Evaluate the age of a child that requires 10 hours sleep.

c. For every year, determine how much less sleep a child requires.

on To test your understanding and knowledge of this topic, go to your learnON title at www.jacplus.com.au and complete the **post-test**.

Online Resources

Below is a full list of **rich resources** available online for this topic. These resources are designed to bring ideas to life, to promote deep and lasting learning and to support the different learning needs of each individual.

eWorkbook

Download the workbook for this topic, which includes worksheets, a code puzzle and a project (ewbk-2029) ☐

Solutions

Download a copy of the fully worked solutions to every question in this topic (sol-0737) ☐

Digital documents

3.2 SkillSHEET Describing the gradient of a line (doc-5197) ☐
SkillSHEET Plotting a line using a table of values (doc-5198) ☐
SkillSHEET Stating the y-intercept from a graph (doc-5199) ☐
SkillSHEET Solving linear equations that arise when finding x- and y-intercepts (doc-5200) ☐
SkillSHEET Using Pythagoras' theorem (doc-5201) ☐
SkillSHEET Substitution into a linear rule (doc-5202) ☐
SkillSHEET Transposing linear equations to standard form (doc-5203) ☐
3.3 SkillSHEET Measuring the rise and the run (doc-5196) ☐
SkillSHEET Determining the gradient given two points (doc-5204) ☐

Video eLessons

3.2 Plotting linear graphs (eles-4736) ☐
Sketching linear graphs (eles-4737) ☐
Sketching linear graphs of the form $y = mx$ (eles-4738) ☐
Sketching linear graphs of the form $y = c$ and $x = a$ (eles-4739) ☐
Using linear graphs to model real-life contexts (eles-4740) ☐
Sketching linear graphs (eles-1919) ☐
Sketching linear graphs using the gradient–intercept method (eles-1920) ☐
3.3 Determining a linear equation given two points (eles-4741) ☐
A simple formula (eles-4742) ☐
The equation of a straight line (eles-2313) ☐
3.4 Parallel lines (eles-4743) ☐
Perpendicular lines (eles-4744) ☐
3.5 The distance between two points (eles-4745) ☐
3.6 Midpoint of a line segment (eles-4746) ☐
3.7 Collinear points (eles-4747) ☐
Equations of horizontal and vertical lines (eles-4748) ☐
Perpendicular bisectors (eles-4749) ☐

Interactivities

3.2 Individual pathway interactivity: Sketching graphs (int-4572) ☐
Plotting linear graphs (int-3834) ☐
The gradient–intercept method (int-3839) ☐
The intercept method (int-3840) ☐
Equations of straight lines (int-6485) ☐
3.3 Individual pathway interactivity: Determining the equation (int-4573) ☐
Linear graphs (int-6484) ☐
3.4 Individual pathway interactivity: Parallel and perpendicular lines (int-4576) ☐
Parallel lines (int-3841) ☐
Perpendicular lines (int-6124) ☐
3.5 Individual pathway interactivity: The distance between two points (int-4574) ☐
Distance between two points (int-6051) ☐
3.6 Individual pathway interactivity: The midpoint of a line segment (int-4575) ☐
Midpoints (int-6052) ☐
3.7 Individual pathway interactivity: Applications and collinearity (int-8469) ☐
Vertical and horizontal lines (int-6049) ☐
3.8 Crossword (int-2833) ☐
Sudoku puzzle (int-3590) ☐

Teacher resources

There are many resources available exclusively for teachers online.

To access these online resources, log on to **www.jacplus.com.au**.

Answers

Topic 3 Coordinate geometry

Exercise 3.1 Pre-test

1. A
2. $x = \frac{1}{2}$
3. 3 apps
4. $y = 3x - 3$
5. D
6. B
7. $a = -2$ or $a = 22$
8. E
9. B
10. D
11. $a = 6$
12. $y = -2x$
13. $n = \frac{6}{5}$
14. A
15. $y = -\frac{1}{4}x + \frac{5}{4}$

Exercise 3.2 Sketching linear graphs

1. a.

x	y
−5	−25
−4	−15
−3	−5
−2	5
−1	15
0	25
1	35

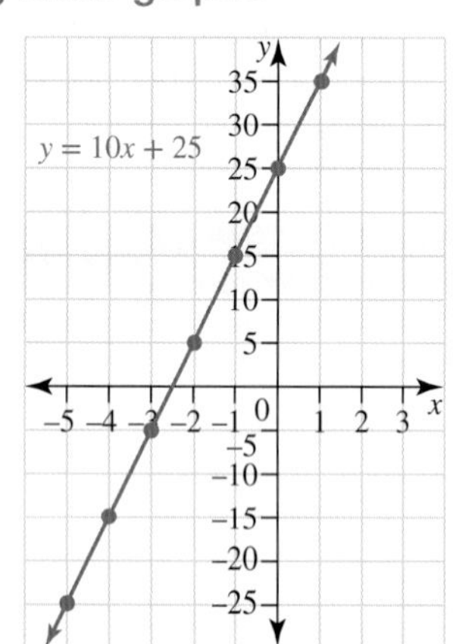

b.

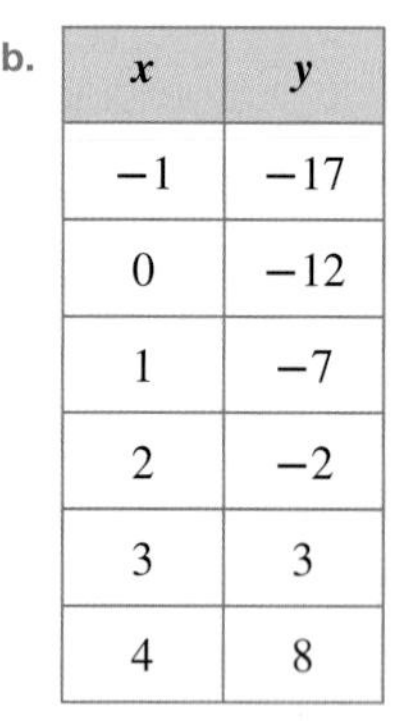

x	y
−1	−17
0	−12
1	−7
2	−2
3	3
4	8

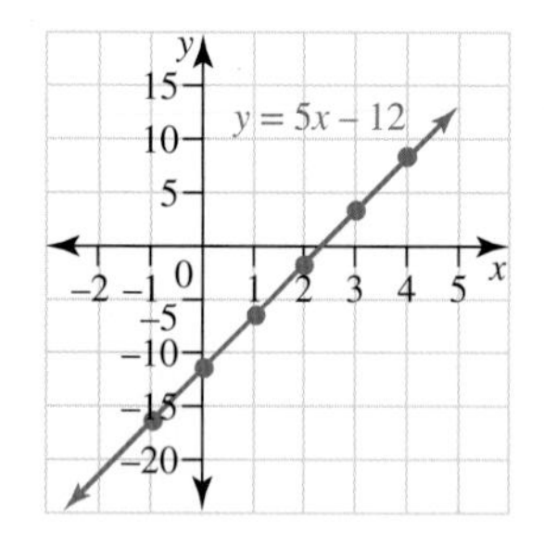

c.

x	y
−6	13
−4	12
−2	11
0	10
2	9
4	8

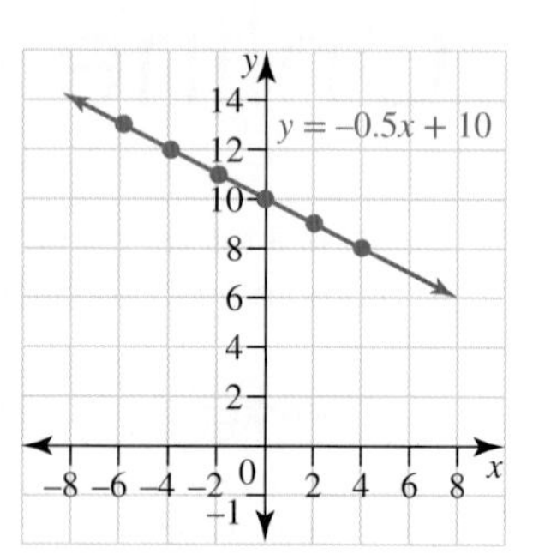

2. a.

x	y
0	−240
1	−140
2	−40
3	60
4	160
5	260

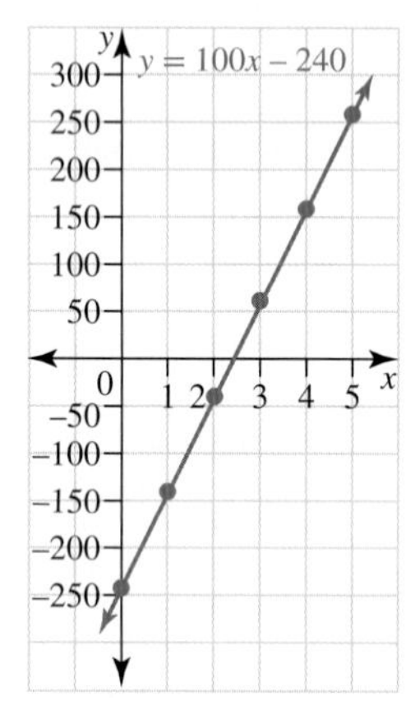

b.

x	y
−3	18
−2	13
−1	8
0	3
1	−2
2	−7

c.

x	y
−3	19
−2	15
−1	11
0	7
1	3
2	−1

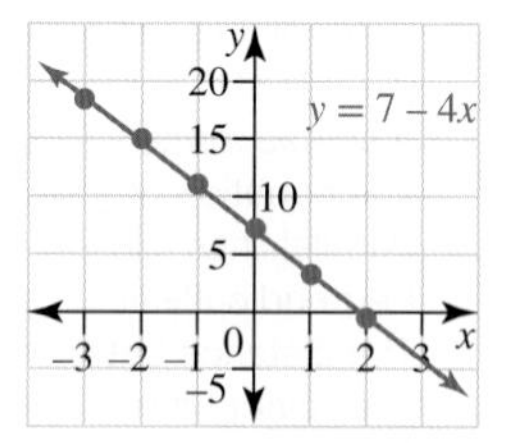

3. a.

x	y
−6	20
−4	14
−2	8
0	2
2	−4
4	−10
6	−16

$y = -3x + 2$

b.

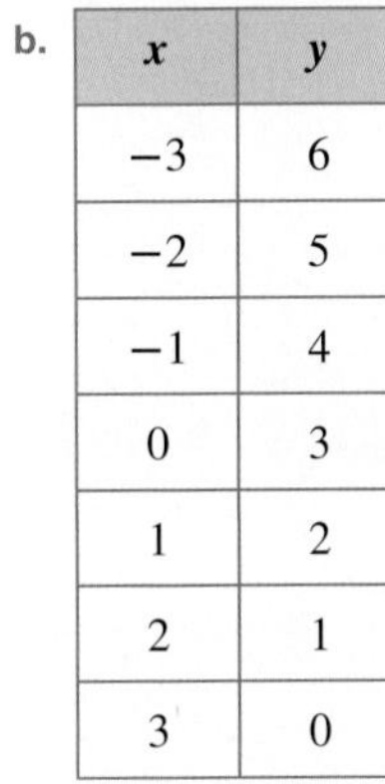

x	y
−3	6
−2	5
−1	4
0	3
1	2
2	1
3	0

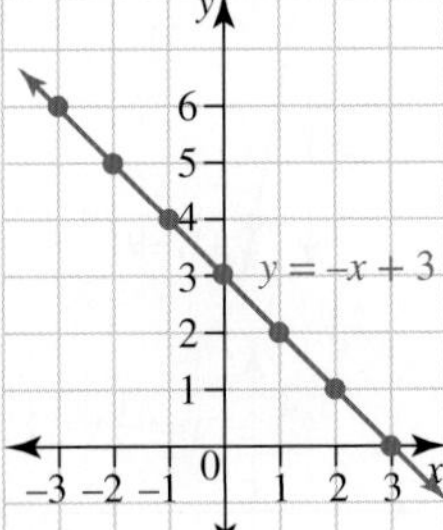

c.

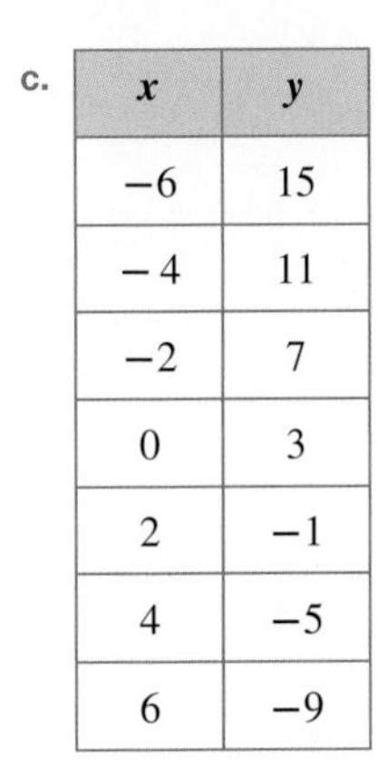

x	y
−6	15
−4	11
−2	7
0	3
2	−1
4	−5
6	−9

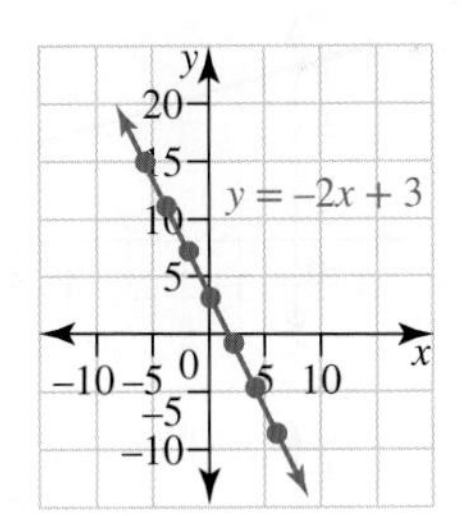

4. a.

$5x - 3y = 10$

b.

c.

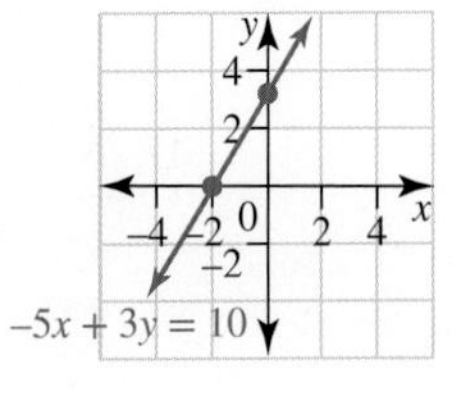

d.

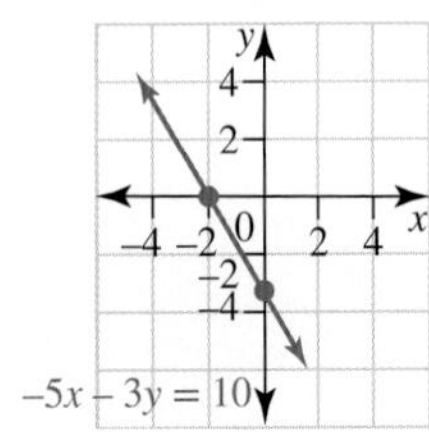

e.

$2x - 8y = 20$

5. a.

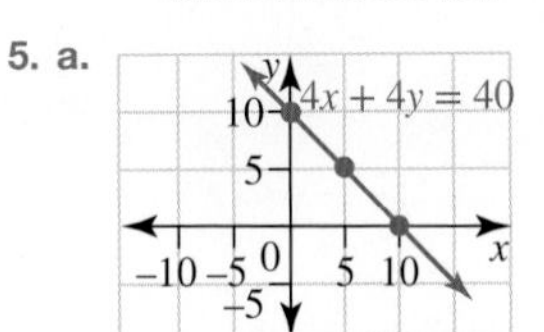

b.

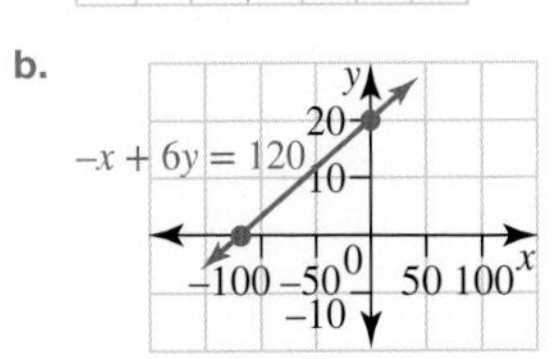

c.

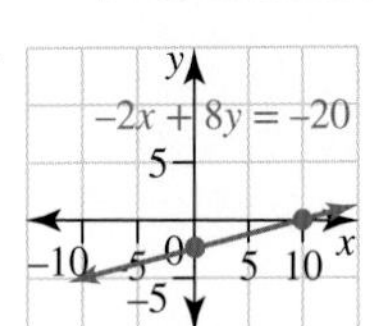

d.

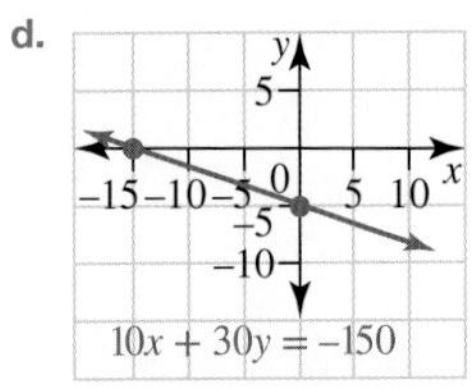

e.

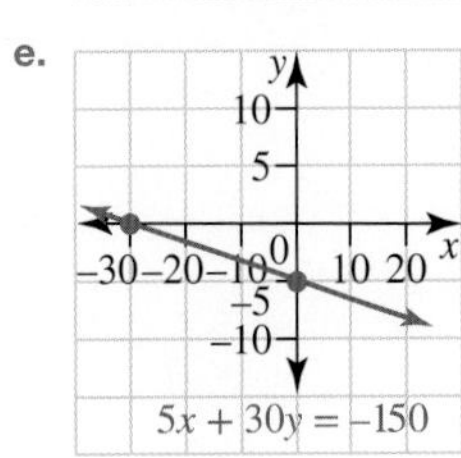

6. a.

b.

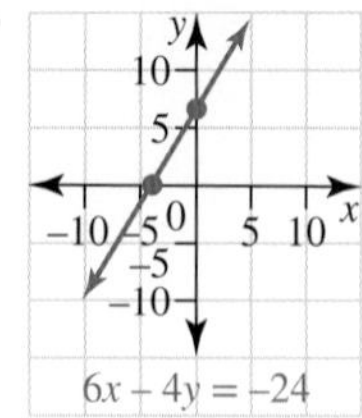

c.

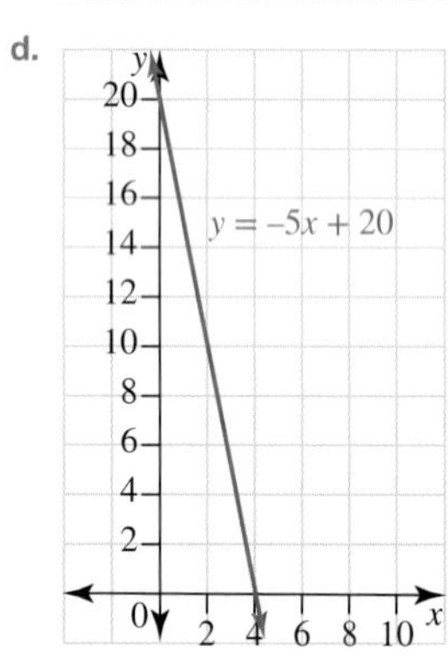

d.

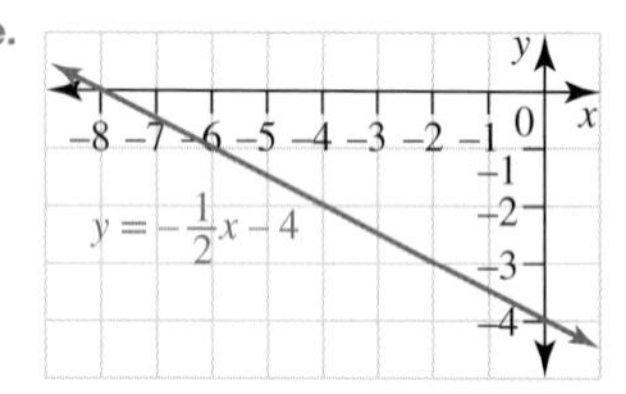

e.

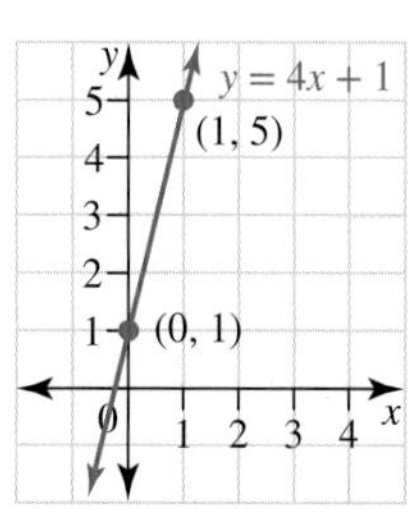

7. a.

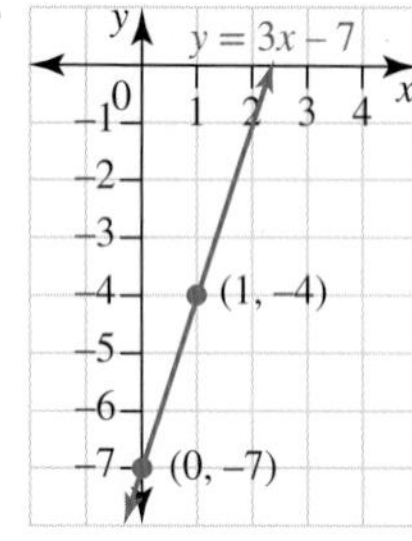

b.

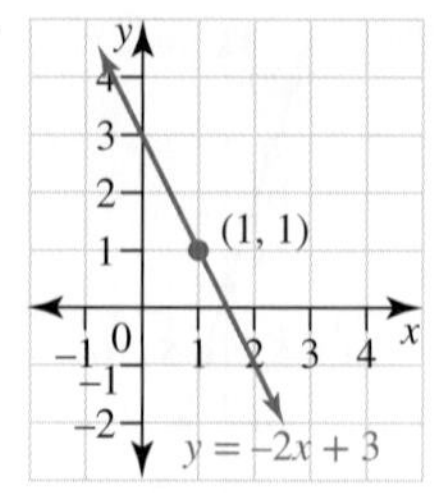

c.

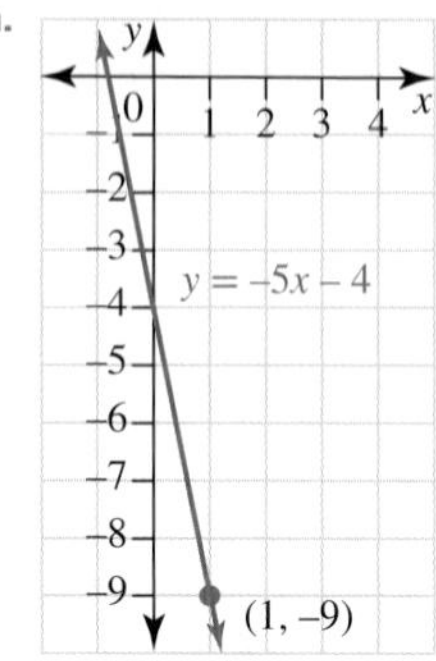

8. a.

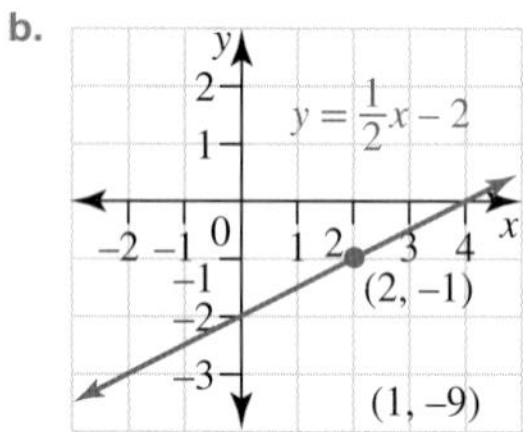

b.

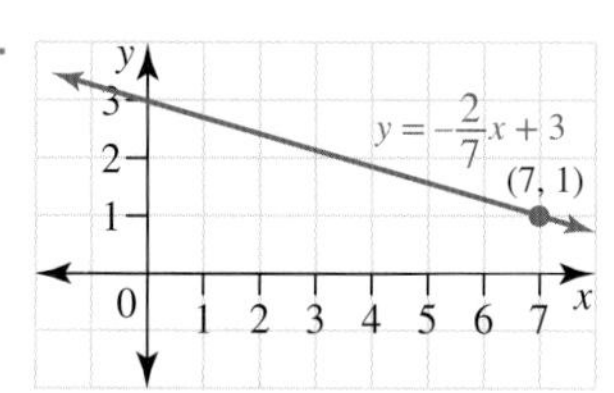

c.

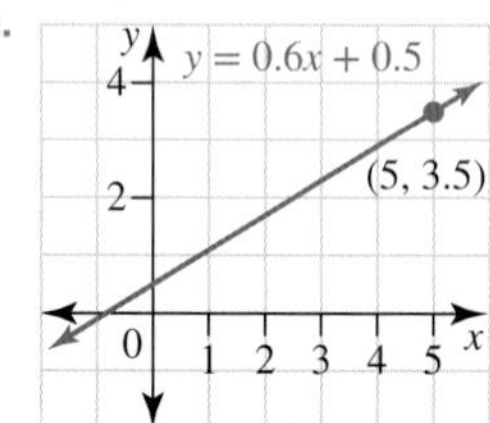

9. a.

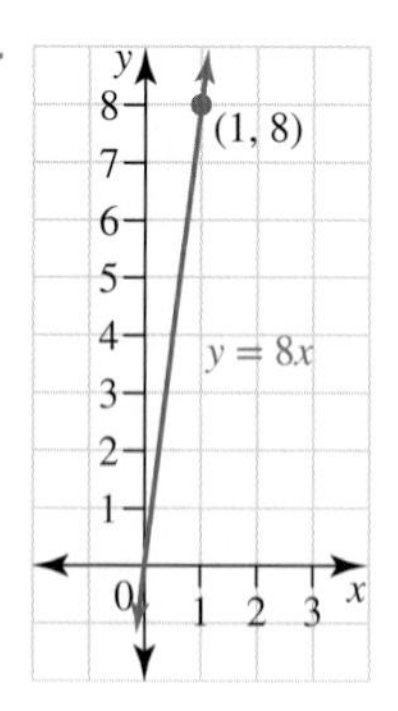

b.

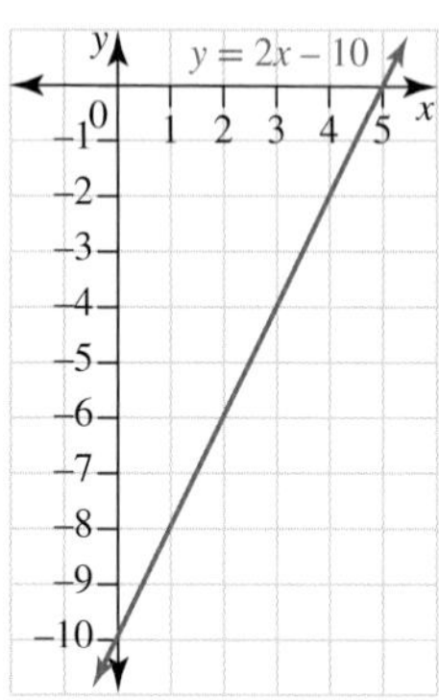

c. 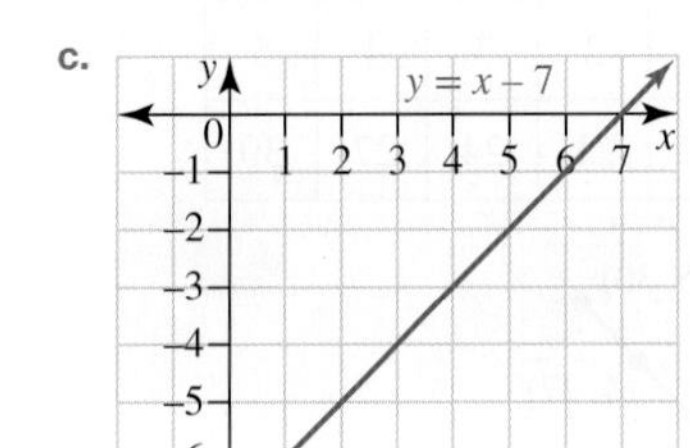

10. 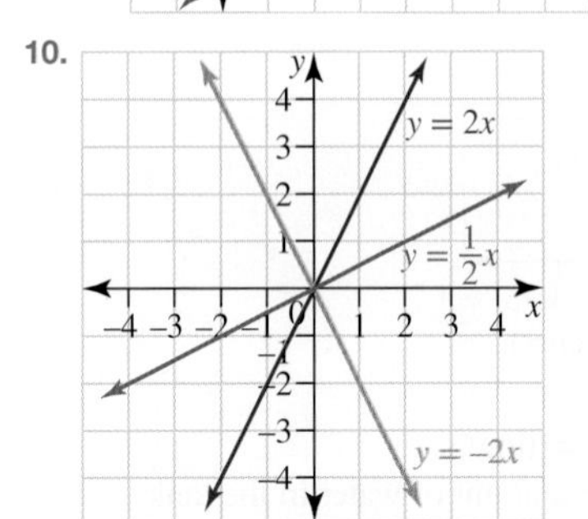

11. 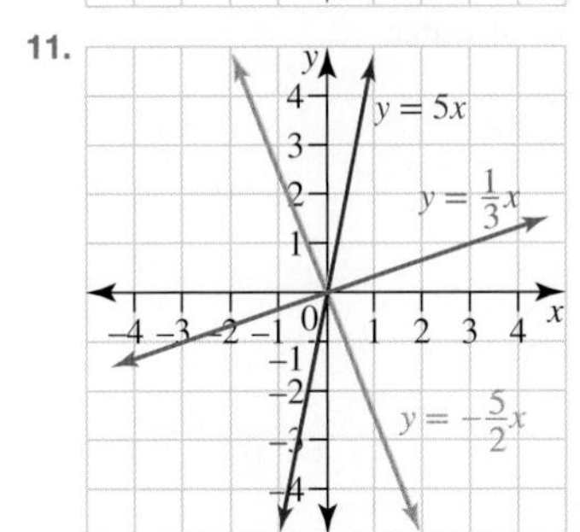

12. 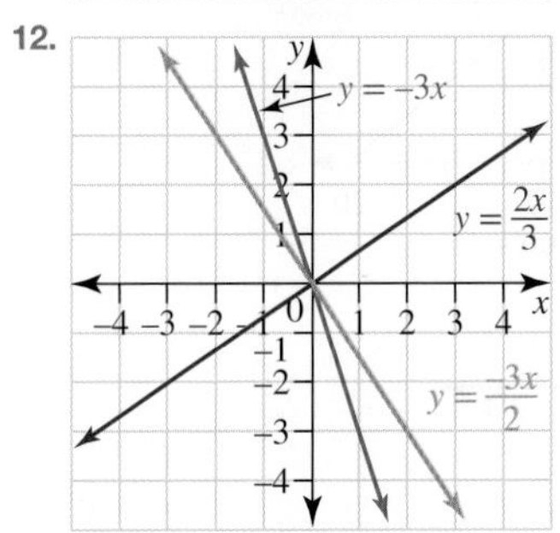

13. a. 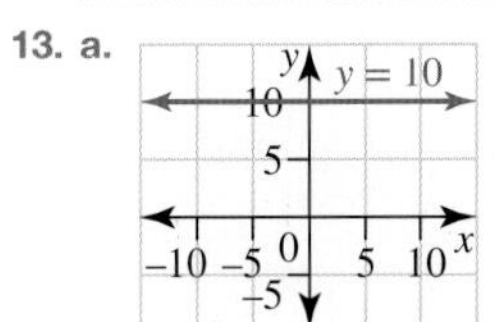

b. 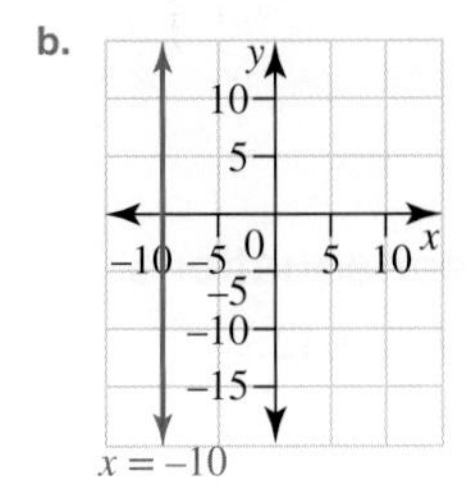

c. 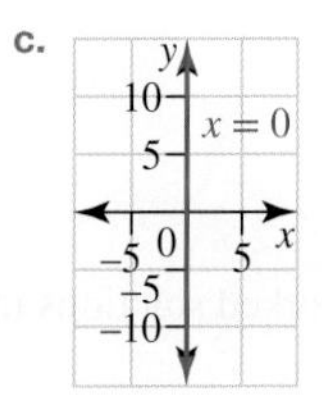

14. a. 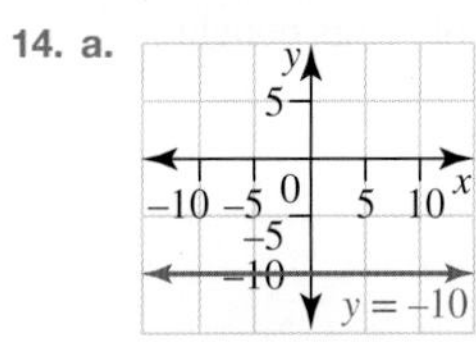

b. 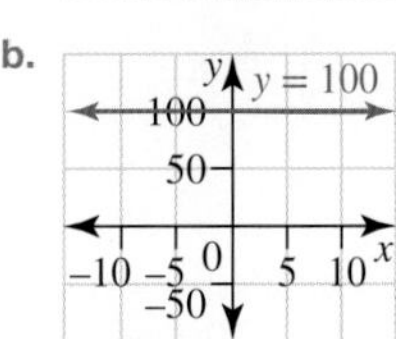

c. 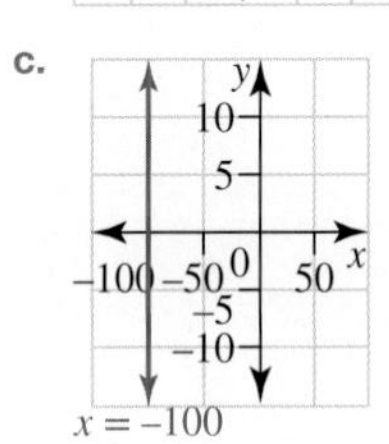

15. a. 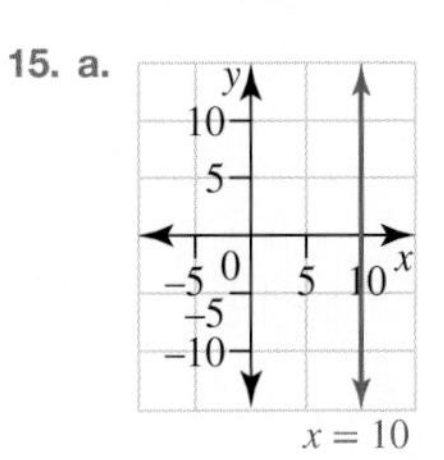

b. 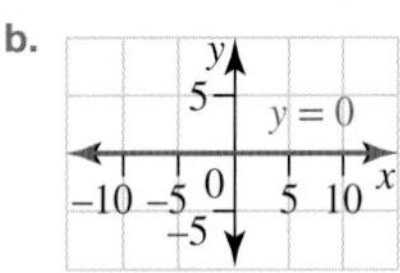

c. 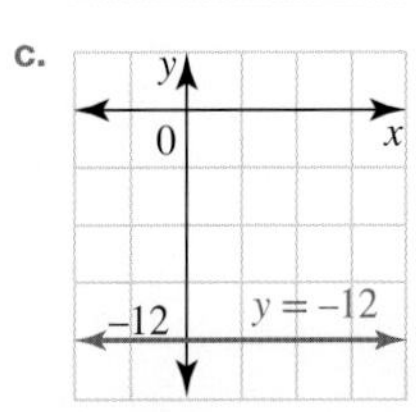

16. a. x-intercept: -0.5; y-intercept: 0.4

b. x-intercept: 0.5; y-intercept: -0.4

c. x-intercept: 0; y-intercept: 0

17. a. x-intercept: -3; y-intercept: 12

b. x-intercept: -4; y-intercept: -4

c. x-intercept: -1; y-intercept: -0.5

18. a. x-intercept: 2.75; y-intercept: 2.2

b. x-intercept: 9.75; y-intercept: -3.9

c. x-intercept: $-\frac{23}{13} \approx 1.77$; y-intercept: 4.6

19. a. $(2,\ 0),\ (0,\ -8)$

b. $\left(-\frac{1}{2},\ 0\right),\ (0,\ 3)$

c. $(-5,\ 0),\ (0,\ 25)$

20. Sample responses can be found in the worked solutions in the online resources.

21. a. Independent variable = number of songs bought, dependent variable = amount of money saved

b.

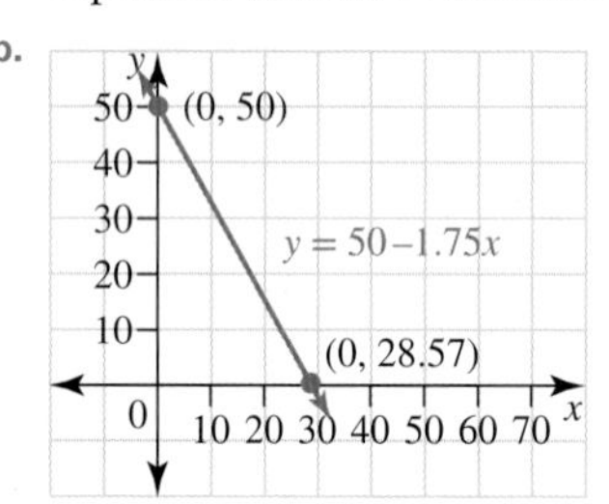

c. 14 songs

22. $y = \frac{2}{3}x - \frac{7}{3}$

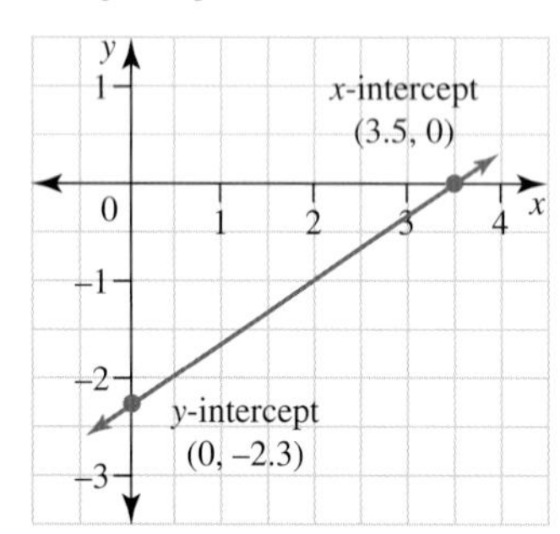

23. a. $y = 20x + 25$

b.

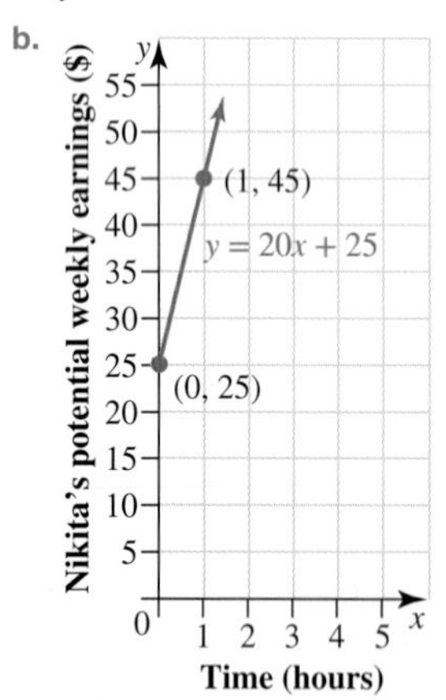

c. Nikita can earn a maximum of $1145.00 in a single week.

24. a. T is the dependent variable (temperature) and t is the independent variable (time).

b. i.

t	0	1	2	3	4	5
T	15	18	21	24	27	30

ii.

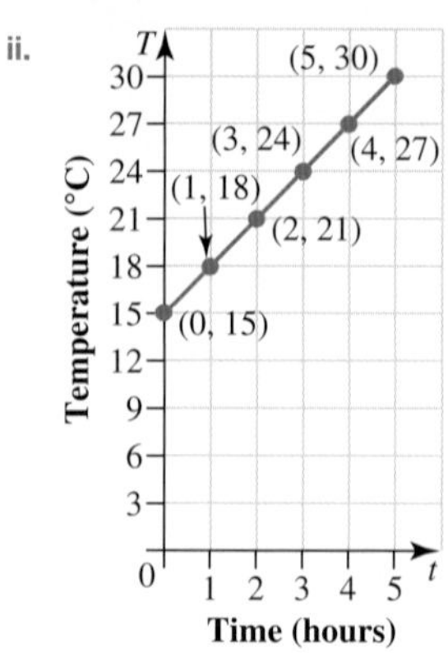

c. 5 hours

25. a. Independent variable = time, dependent variable = amount of water in the tank

b. Initially there are 80 litres of water.

c. Time cannot be negative.

d. 4 litres per minute

e. 20 minutes

f.

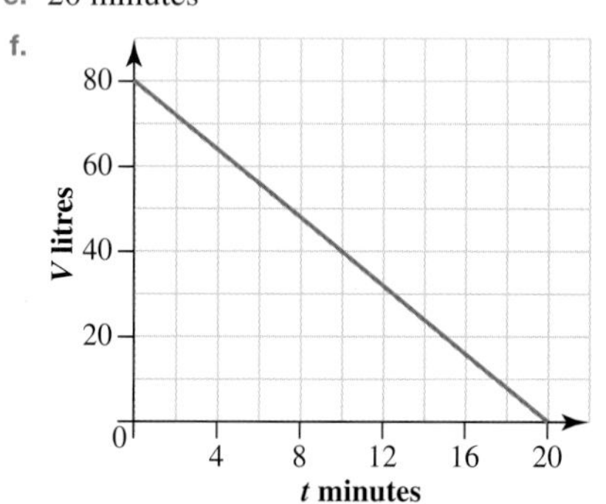

26. a.

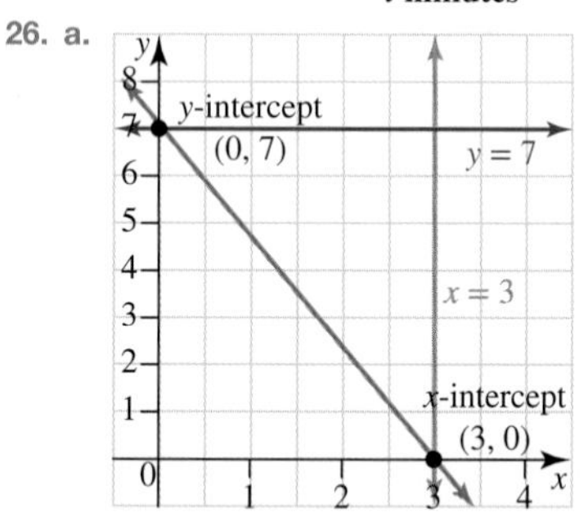

b. 7

c. $-\frac{7}{3}$

d. B

Exercise 3.3 Determining linear equations

1. a. $y = 2x + 4$ b. $y = -3x + 12$
c. $y = -x + 5$ d. $y = 2x - 8$

2. a. $y = \frac{1}{2}x + 3$ b. $y = -\frac{1}{4}x - 4$
c. $y = 7x - 5$ d. $y = -3x - 15$

3. a. $y = 2x$ b. $y = -3x$
c. $y = \frac{1}{2}x$ d. $y = -\frac{3}{4}x$

4. a. $y = 3x + 3$ b. $y = -3x + 4$ c. $y = -4x + 2$
d. $y = 4x + 2$ e. $y = -x - 4$

5. a. $y = 0.5x - 4$ b. $y = 5x + 2.5$
c. $y = -6x + 3$ d. $y = -2.5x + 1.5$
e. $y = 3.5x + 6.5$

6. a. $y = 5x - 19$ b. $y = -5x + 31$
c. $y = -4x - 1$ d. $y = 4x - 34$
e. $y = 3x - 35$

7. a. $y = -3x + 6$ b. $y = -2x + 30$
c. $y = 2x - 4.5$ d. $y = 0.5x - 19$
e. $y = -0.5x + 5.5$

8. a. $y = x + 3$ b. $y = 2x - 1$ c. $y = -\frac{1}{2}x + \frac{7}{2}$

9. a. $y = \frac{1}{2}x + \frac{1}{2}$ b. $y = -2x - 2$ c. $y = -x - 8$

10. a. Independent variable = time (in hours), dependent variable = cost (in \$)

b.

t	0	1	2	3
c	2	8	14	20

c.

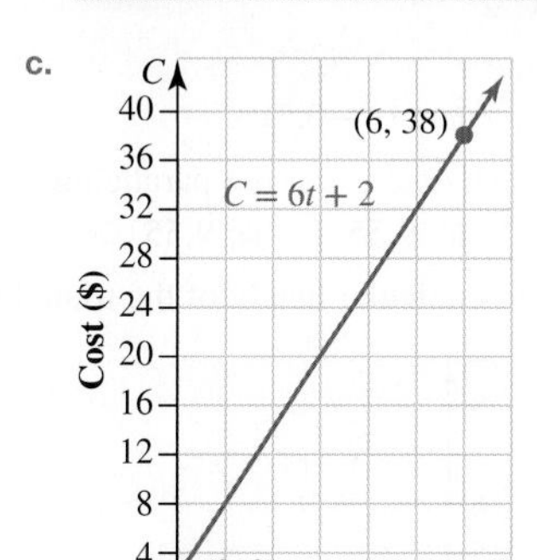

d. i. (0, 2)

ii. The y-intercept represents the initial cost of bowling at the alley, which is the shoe rental.

e. $m = 6$, which represents the cost to hire a lane for an additional hour.

f. $C = 6t + 2$

g. \$32

h. Sample responses can be found in the worked solutions in the online resources.

11. a. C = dependent variable, n = independent variable

b. $C = 15n + 10$

c. \$460.00

d.

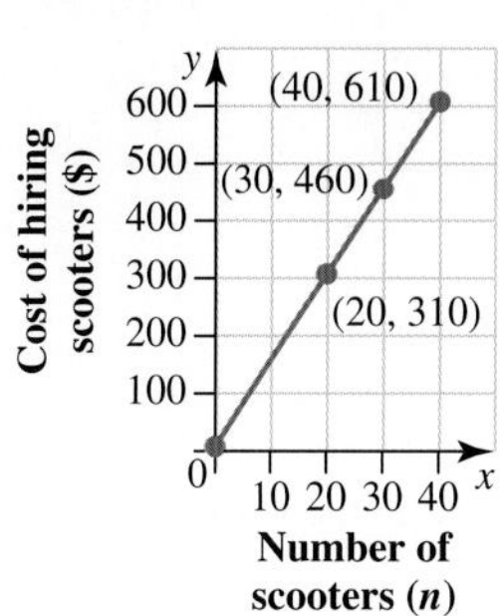

12. a. $W = -40t + 712$

b. 712 L

c. 18 days

13. It does not matter if you rise before you run or run before you rise, as long as you take into account whether the rise or run is negative.

14. a. $m = \frac{y - c}{x}$ b. $y = mx + c$

15. Sample responses can be found in the worked solutions in the online resources.

16. a. $m_{AB} = 5$ b. $y = 5x - 3$
c. $m_{CD} = 5$ d. $D = (6, 5)$

17. $m_{AB} = m_{CD} = 2$ and $m_{BC} = m_{AD} = \frac{1}{2}$. As opposite sides have the same gradients, this quadrilateral is a parallelogram.

18. a. $b = -10$ b. $a = -4$

Exercise 3.4 Parallel and perpendicular lines

1. a. No b. Yes c. No

2. a. No b. Yes c. No

3. a. No b. No c. No

4. a. No b. No c. No

5. b and f are parallel. c and e are parallel.

6. Sample responses can be found in the worked solutions in the online resources.

7. $y = 2x - 9$

8. $3x + 2y - 8 = 0$

9. a. $y = 3x + 2$ b. $y = -4x + 9$
c. $y = \frac{3x}{2} - 4$ d. $y = \frac{-2x}{5} - \frac{13}{5}$
e. $y = \frac{x}{5} - 1$ f. $y = \frac{x}{3} + \frac{17}{3}$
g. $y = \frac{x}{3} - \frac{14}{3}$

10. a. $2x - y + 5 = 0$ b. $x + 2y = 0$

11. a. $3x - 5y + 2 = 0$ b. $5x + 3y - 8 = 0$

12. a. B b. C c. D d. B

13. E

14. a. $y = -2x + 1$ b. $y = \frac{-2x}{3} - \frac{1}{3}$

15. a and e are perpendicular; b and f are perpendicular; c and h are perpendicular; d and g are perpendicular.

16. $y = \frac{-x}{2} + \frac{3}{2}$

17. a. $m = \frac{-8}{5}$ b. $m = \frac{18}{5}$

18. E

19. $2y + x - 8 = 0$

20. $4y - 3x + 15 = 0$

21. B

22. a.

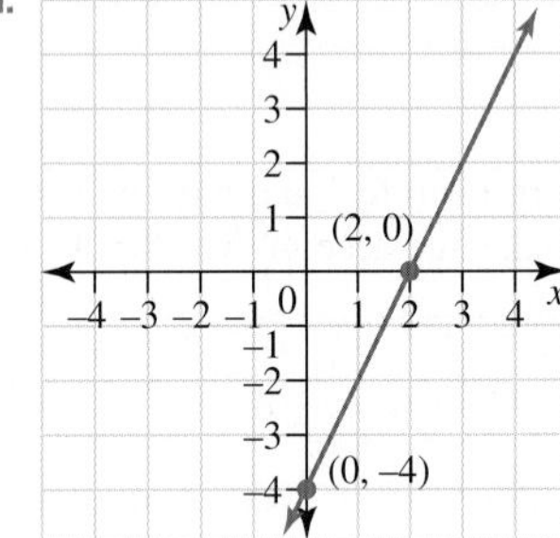

b.

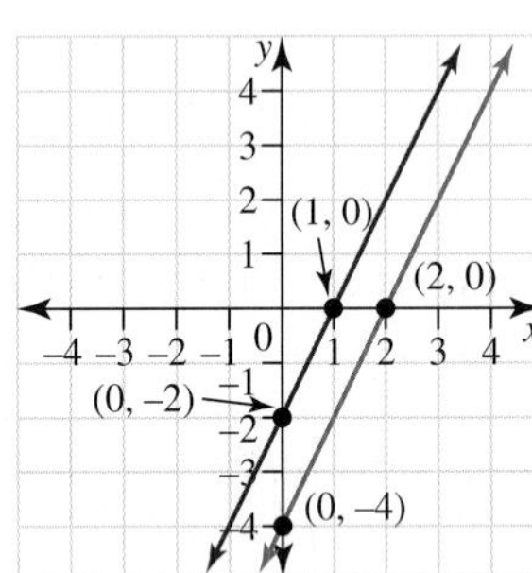

c.

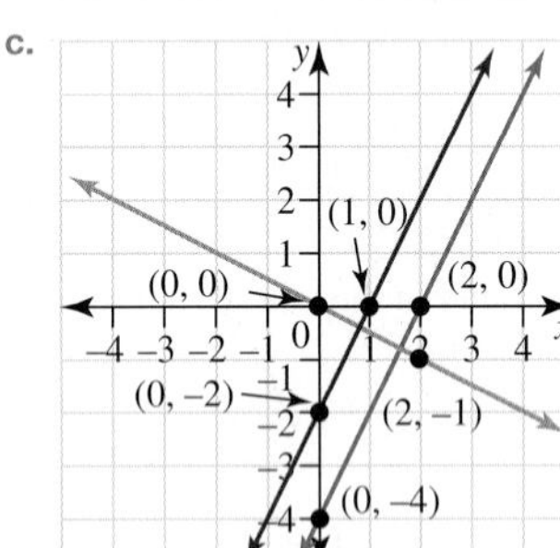

23. $a = \frac{-3}{2}$

24. a. $\frac{3}{2}$

b. Sample responses can be found in the worked solutions in the online resources.

Exercise 3.5 The distance between two points

1. $AB = 5$, $CD = 2\sqrt{10}$ or 6.32, $EF = 3\sqrt{2}$ or 4.24, $GH = 2\sqrt{5}$ or 4.47, $IJ = 5$, $KL = \sqrt{26}$ or 5.10, $MN = 4\sqrt{2}$ or 5.66, $OP = \sqrt{10}$ or 3.16
2. a. 5 b. 13 c. 10 d. 7.07 e. 6.71
3. a. 14.42 b. 13 c. 13
   d. $\sqrt{a^2 + 4b^2}$ e. $3\sqrt{a^2 + b^2}$
4. a. $AB = 4.47$, $BC = 2.24$, $CD = 4.47$, $DA = 2.24$
   b. $AC = 5$, $BD = 5$
   c. Rectangle
5. B
6. D
7. 8. and 9. Sample responses can be found in the worked solutions in the online resources.
10. a. 12 b. 5 c. 13 d. -2.2
11. $a = 2$
12. Sample responses can be found in the worked solutions in the online resources.

13. a. $m_{AB} = 1$ and $m_{BC} = -\frac{7}{3}$
    b. $D(4, -1)$
    c. Sample responses can be found in the worked solutions in the online resources.

Exercise 3.6 The midpoint of a line segment

1. a. $\left(-3, -3\frac{1}{2}\right)$ b. $\left(7\frac{1}{2}, 0\right)$ c. $(-1, 1)$
2. a. $\left(0, 1\frac{1}{2}\right)$ b. $\left(2a, \frac{1}{2}b\right)$ c. $\left(a + b, \frac{1}{2}a\right)$
3. $(-3, -10)$
4. a. $(2, -1)$ b. $\left(\frac{3}{2}, \frac{11}{2}\right)$ c. $\left(2, \frac{5a}{2}\right)$
5. $x + y = -3$
6. a. $(3, 1)$ b. 4.47 c. 6.32
7. D
8. C
9. a. i. $(-1, 4)$ ii. $\left(1\frac{1}{2}, 1\right)$ iii. 3.91
   b. $BC = 7.8 = 2PQ$
10. a. i. $(1, -0.5)$ ii. $(1, -0.5)$
    b. The diagonals bisect each other, so it is a parallelogram.
11. a. i. $(-2, 2)$ ii. 8.94 iii. 9.55 iv. 9.55
    b. Isosceles. PC is the perpendicular height of the triangle.
12. a.

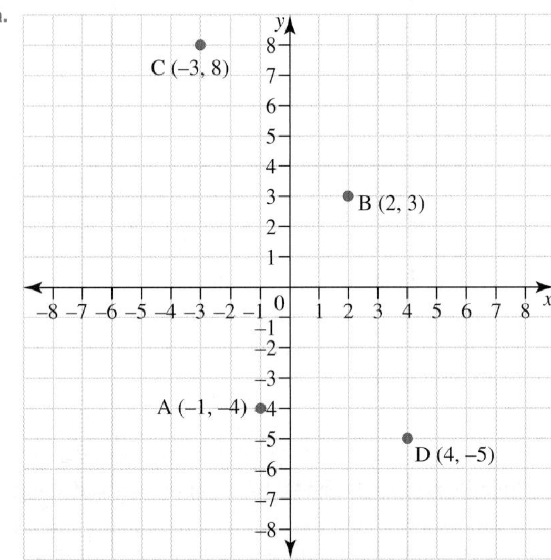

b. $M = \left(\frac{-1 + -3}{2}, \frac{-4 + 8}{2}\right)$
   $= (-2, 2)$

c. $2\sqrt{37}$

d. $(7, -2)$

e. $\frac{3 - (-4)}{2 - (-1)} = \frac{7}{3}$

f. $y = -4x + 11$

13. $(4k - 1, 3.5 - 5k)$
14. Sample responses can be found in the worked solutions in the online resources.
15. $y = -3x - 2$

16. $3y - 2x + 14 = 0$

17. $y = \frac{-2x}{3} + \frac{8}{3}$

Exercise 3.7 Applications and collinearity

1. and 2. Sample responses can be found in the worked solutions in the online resources.

3. a. $x = 1$ b. $y = -7$

4. $x + y + 3 = 0$

5. a. Since $m_1 = m_2 = \frac{2}{5}$ and $\left(\frac{5}{2}, 2\right)$ is common to both line segments, these three points are collinear.

   b. $y = \frac{-5}{2}x + \frac{97}{10}$

6. a. (2, 5)

   b. 1

   c. Sample responses can be found in the worked solutions in the online resources.

   d. Isosceles triangle

7. $4x - 6y + 23 = 0$

8. a. $y = -x + 5$ b. $y = x + 3$ c. (1, 4)

9. and 10. Sample responses can be found in the worked solutions in the online resources.

11. B

12. a. 5.10 km b. (6.5, 5.5) c. 2
    d. $y = 2x - 18$ e. (10, 2) f. 7.07 km

13. and 14. Sample responses can be found in the worked solutions in the online solutions.

15. a. (0, 2), (2, 2) or (0, −2), (−2, −2) or (1, 1), (1, −1)

    b. $\left(1, \sqrt{3}\right)$ or $\left(1, -\sqrt{3}\right)$

    c. (3, 1), (−1, 1) or (1, −1)

16. Sample responses can be found in the worked solutions in the online resources.

17. Sample responses can be found in the worked solutions in the online resources.

    a. $y = 1$

    b. Sample responses can be found in the worked solutions in the online resources.

18. a. Sample responses can be found in the worked solutions in the online resources.

    b. Yes

    c. Yes

19. a. OA: $2x + 3y - 13 = 0$; OB: $x = 3$

    b. $\left(3, \frac{7}{3}\right)$

    c. d. and e. Sample responses can be found in the worked solutions in the online resources.

20. a. Line A: $2x - y - 1 = 0$, Line B: $3x + 4y - 18 = 0$, Line C: $x - 6y - 6 = 0$

    b.

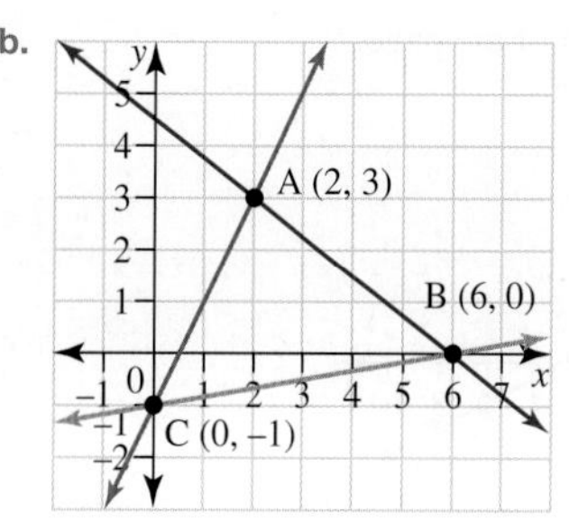

    c. Scalene

21. Sample responses can be found in the worked solutions in the online resources.

Project

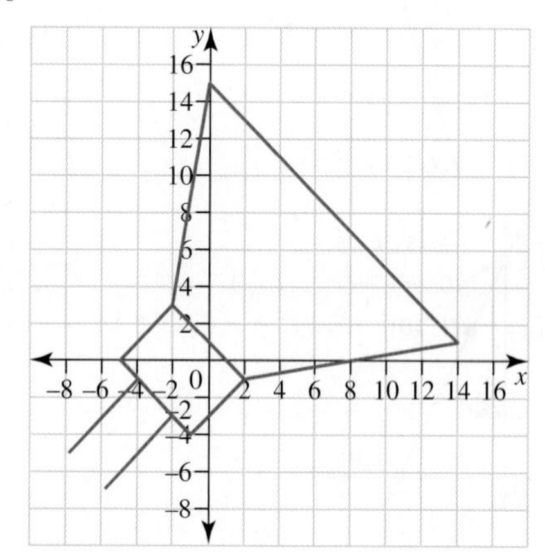

1. The symbol is the one used to represent a speaker.

2. The shape is a trapezium.

$$\text{Area} = \frac{1}{2}\left(\text{length line 6} + \text{length line 8}\right)$$

$$\times \text{ Perpendicular distance between these lines.}$$

$$= \frac{1}{2}\left(4\sqrt{2} + 14\sqrt{2}\right) \times 7\sqrt{2}$$

$$= 126 \text{ units}^2$$

3. Sample responses can be found in the worked solutions in the online resources. You could use any symbol of interest and provide instructions for your design. Ensure that your design involves aspects of coordinate geometry that have been used throughout this task.

Exercise 3.8 Review questions

1. A
2. D
3. B
4. C
5. A
6. C
7. A
8. C
9. C

10. See table at the bottom of the page.*

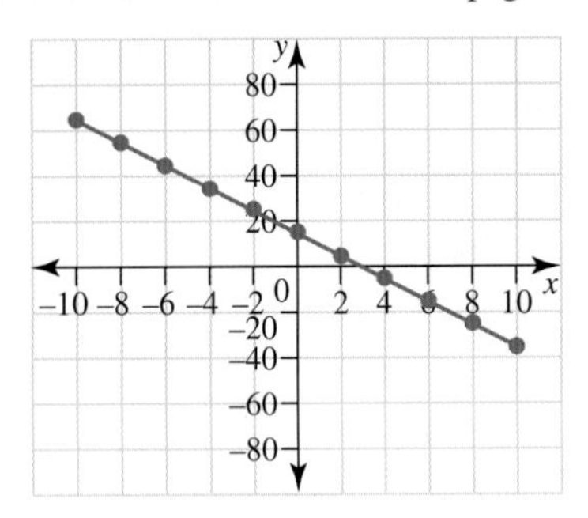

11. a.

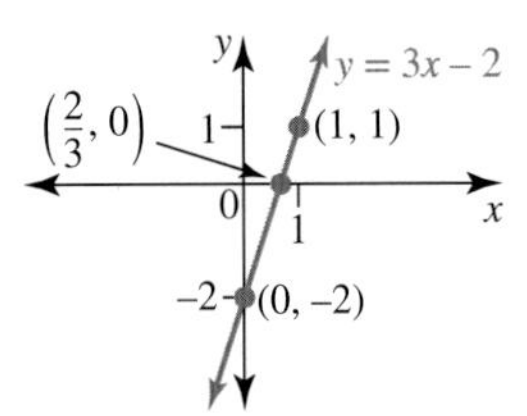

b.

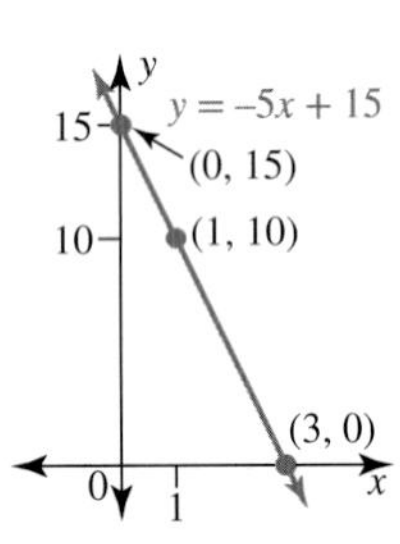

c.

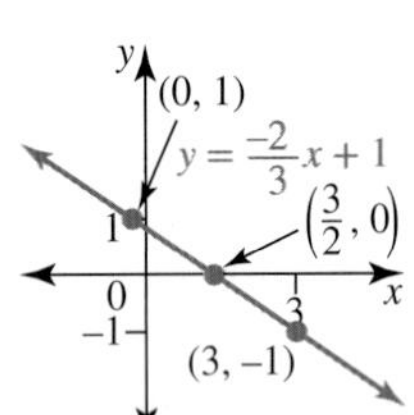

d.

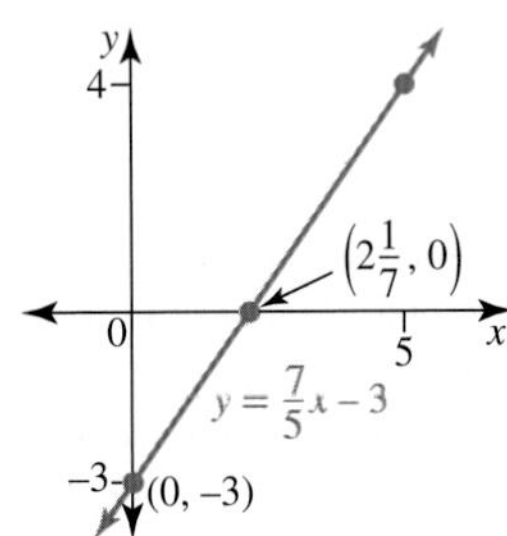

12. a. x-intercept: $= \frac{6}{7}$; y-intercept: $b = 6$

b. x-intercept: $= \frac{40}{3}, \left(= 13\frac{1}{3}\right)$, y-intercept: $= -5$

c. x-intercept: $= \frac{21}{16}, \left(= 1\frac{5}{16}\right)$, y-intercept: $= -\frac{3}{4}$

d. x-intercept: -5.6 y-intercept: $= 2.8$

13. a.

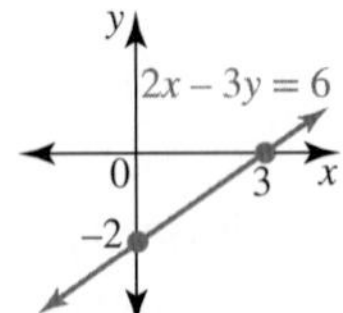

b.

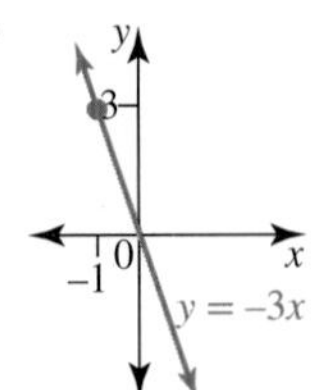

c.

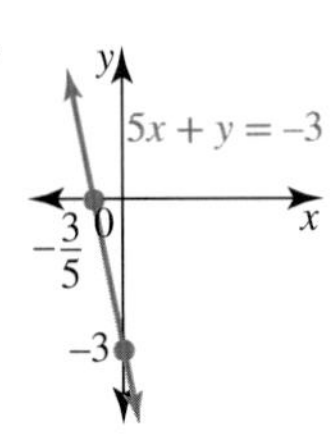

d.

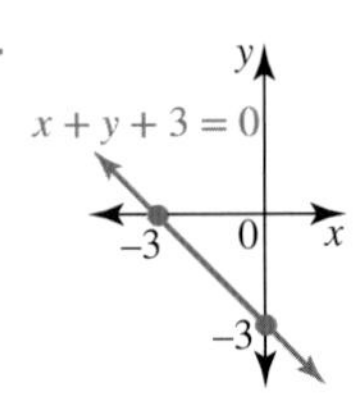

14. a.

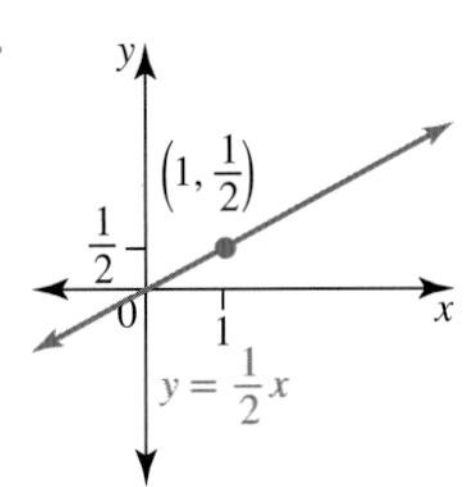

b.

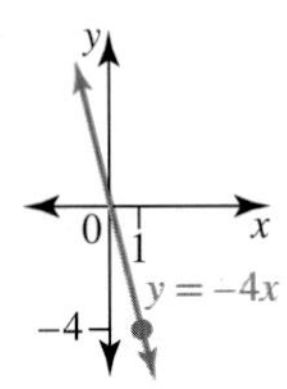

c.

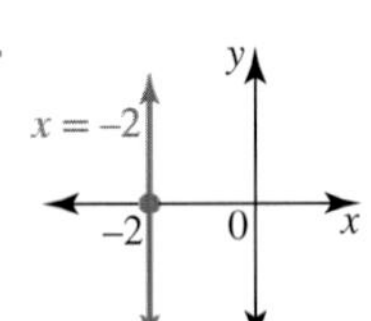

d.

y

7 y = 7

0 x

*10.	x	−10	−8	−6	−4	−2	0	2	4	6	8	10
	y	65	55	45	35	25	15	5	−5	−15	−25	−35

15.

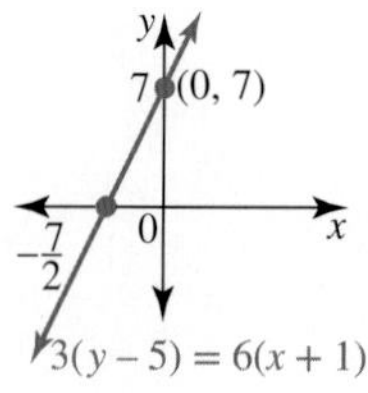

16. a. $y = 2x - 2$ b. $y = -x - 4$ c. $y = -\frac{1}{3}x + 2$
d. $y = 4x$ e. $y = -\frac{3}{4}$ f. $x = 5$

17. a. $y = 3x - 4$ b. $y = -2x - 5$
c. $y = \frac{1}{2}x + 5$ d. $y = 6$

18. a. $y = 7x - 13$ b. $y = -3x + 4$
c. $y = \frac{1}{2}x + 6$ d. $y = \frac{3}{5}x - \frac{18}{5}$

19. $\sqrt{61}$

20. and 21. Sample responses can be found in the worked solutions in the online resources.

22. $(0, -18)$

23. and 24. Sample responses can be found in the worked solutions in the online resources.

25. $x + 2y - 2 = 0$

26. $2x + 3y - 9 = 0$

27. $3x + 2y - 21 = 0$

28. $3x - 2y + 16 = 0$

29. a. i. $-\frac{4}{5}$ ii. $\frac{5}{4}$
iii. $4x + 5y - 61 = 0$ iv. $5x - 4y - 25 = 0$
v. $(9, 5)$
b. Square

30. a. i. $\frac{1}{10}$ ii. $\left(-\frac{1}{2}, 1\right)$ iii. $\left(4\frac{1}{2}, 1\frac{1}{2}\right)$
b. Sample responses can be found in the worked solutions in the online resources.

31. a. See table at the bottom of the page.*
b. Pay = \$13.50 × (number of hours worked)
c.

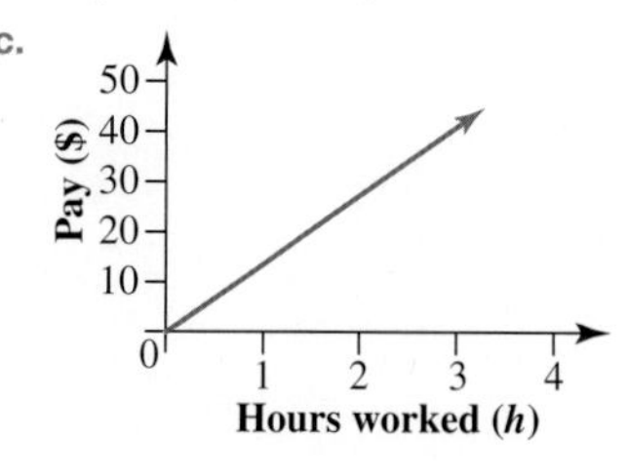

d. \$91.13

32. a. See table at the bottom of the page.*
b. Cost = \$2.50 × number of rides + \$12.50
c.

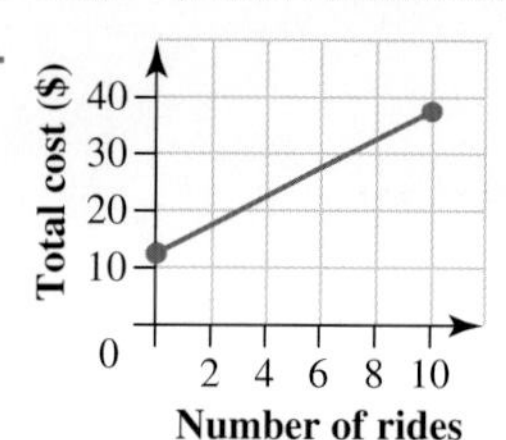

d. \$30

33. a.

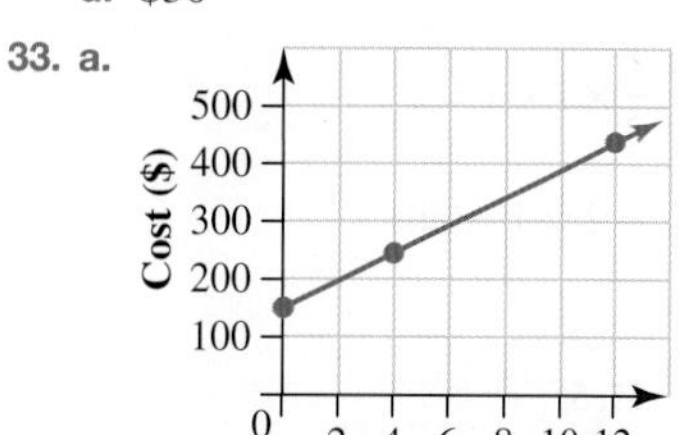

b. $C = 22.50h + 160$
c. Approximately \$436

34. a. $7x - 3y - 1 = 0$
b. $3x + 7y - 49 = 0$
c. -7

35. a. $x = -7$
b. B $(-7, 50)$, C $(-4, 90)$
c. 40.11 metres

36. a. Since the gradient of SA equals the gradient of SO = −0.8, the points S, A and O are collinear. Player Y will displace guard ball A.
b. $y = \frac{5}{4}x + \frac{41}{20}$ or $25x - 20y + 41 = 0$
c. Since the gradient of the path AB is $\frac{5}{4}$, which is the same as the gradient of the known path of travel from the common point A, the direction of travel is toward B.
d. $d_{AB} = 0.80$ m. Yes, guard ball A will collide with guard ball B as it will not be deviated from its linear path under 1 metre of travel.

37. a. Gradient $= m = \frac{5-1}{5--1} = \frac{4}{6} = \frac{2}{3}$

*31a.

Number of hours	0	2	4	6	8	10
Pay (\$)	0	27	54	81	108	135

*32a.

Number of rides	0	2	4	6	8	10
Cost(\$)	12.50	17.50	22.50	27.50	32.50	37.50

b. $y = mx + b$, $y = \frac{2}{3}x + b$

If $x = -1$ and $y = 1$, substitute in the question:

$1 = \frac{2}{3}(-1) + b$

$b = 1\frac{2}{3}$

$y = \frac{2}{3}x + 1\frac{2}{3}$

c. Plot the point (5, 1).

Area of large $\Delta = \frac{1}{2} \times 6 \times 4 = 12$

Area of small $\Delta = \frac{1}{2} \times 1 \times 4 = 2$

Area of ΔABC $= 12 - 2 = 10$ units^2

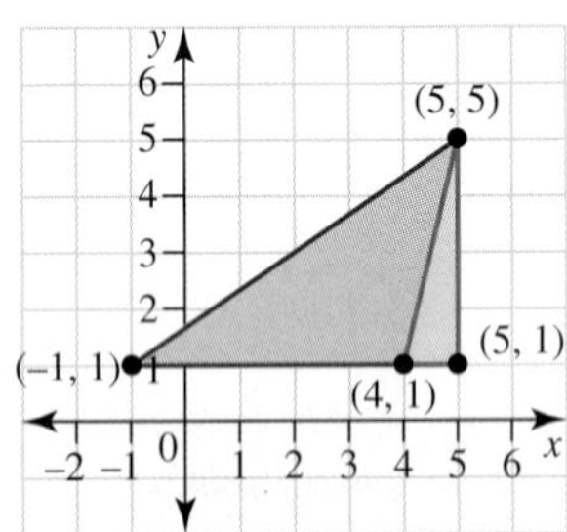

d. $BC^2 = 4^2 + 1^2$

$BC^2 = 16 + 1$

$BC^2 = 17$

$BC^2 = \sqrt{17} \approx 4.12$ units

38. a. i. $T = 18 - 0.005\,(600) = 15\,°C$

ii. $T = 18 - 0.005\,(1000) = 13\,°C$

iii. $T = 18 - 0.005\,(3000) = 3\,°C$

b.

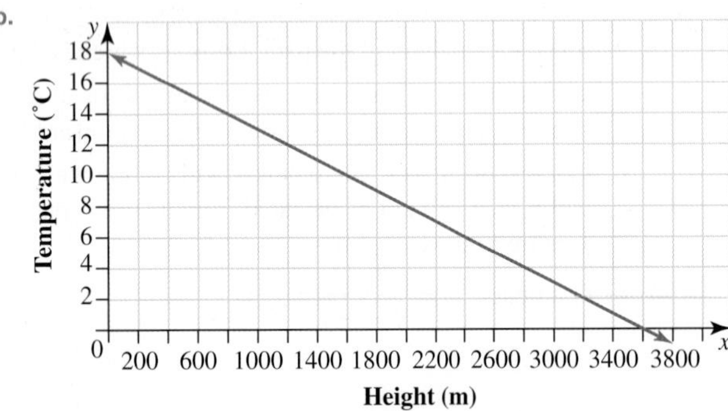

c. 1200 m = 12 °C, 2500 m = 5.5 °C

d. 1800 m

39. a. 12 hours

b. 14 years old

c. $h = 8 + \frac{18 - c}{2}$

$2h = 16 + 18 - c$

$3h = -c + 34$

$h = \frac{1}{2}c + 17$

For every year, the child requires half an hour less sleep.

4 Simultaneous linear equations and inequalities

LEARNING SEQUENCE

4.1 Overview

Why learn this?

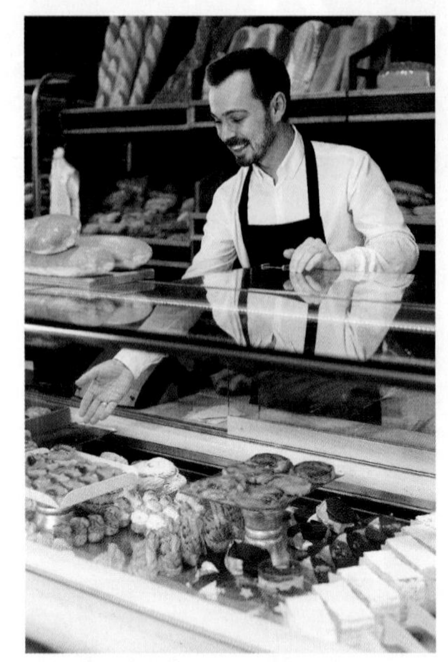

Often in life, we will be faced with a trade-off situation. This means that you are presented with multiple options and must decide on a combination of outcomes that provides you with the best result. Imagine a race with both swimming and running components, in which athletes start from a boat, swim to shore, and then run along the beach to the finish line. Each athlete would have the following options:

- swim directly to shore and run a longer distance along the beach
- swim a longer distance diagonally through the ocean and reduce the distance required to run to reach the finish line
- swim directly through the ocean to the finish, covering the shortest possible distance.

Which option should an athlete take? This would depend on how far the athlete can swim or run, because reducing the swimming distance increases the running distance. To find the best combination of swimming and running, an athlete could form equations based on speed, time and distance and solve simultaneously to find the best combination.

Just like the athletes in the scenario above, businesses face trade-offs like these every day, where they have to decide how much of each product they should produce in order to produce the highest possible profit. As an example, a baker might make the most per-item profit from selling cakes; but if they don't produce muffins, bread and a range of other products then they will attract fewer customers and miss out on sales, reducing overall profit. Thus, a baker could use simultaneous equations to find the best combination of baked goods to produce in order to maximise profit.

Where to get help

Go to your learnON at **www.jacplus.com.au** to access the following digital resources. The Online Resources Summary at the end of this topic provides a full list of what's available to help you learn the concepts covered in this topic.

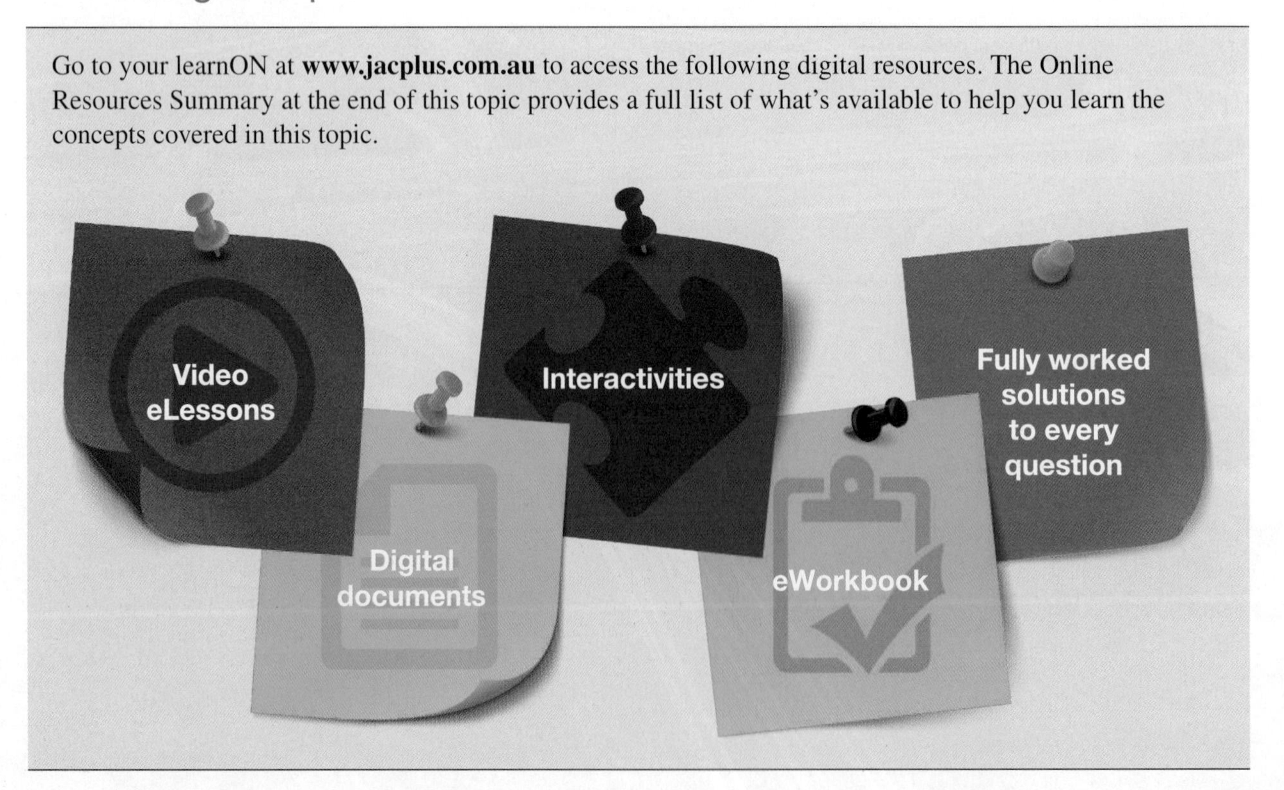

Exercise 4.1 Pre-test

learnon

Complete this pre-test in your learnON title at www.jacplus.com.au and receive **automatic marks**, **immediate corrective feedback** and **fully worked solutions**.

1. State whether the following is True or False. The point $\left(\frac{10}{3}, 5\right)$ is the solution to the simultaneous equations $3x + 2y = 10$ and $-x - 4y = 5$.

2. Identify the solution to the simultaneous equations shown in the graph.

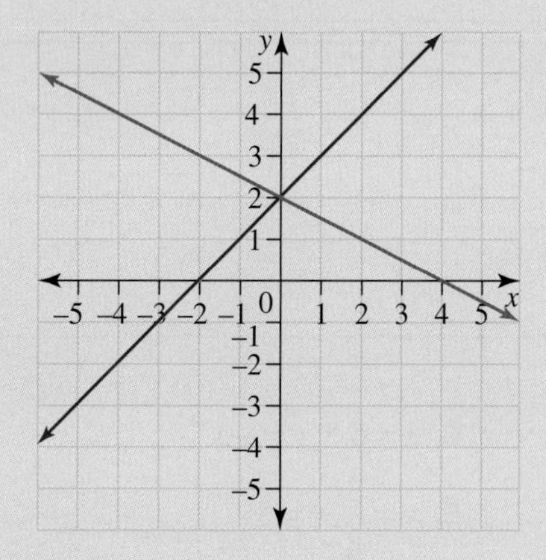

3. State the number of solutions to the pair of simultaneous equations $2x - y = 1$ and $-6x + 3y = -3$.

4. Use substitution to solve the simultaneous equations $y = 0.2x$ and $y = -0.3x + 0.5$.
   Give your answer as a coordinate pair.

5. **MC** Solve for x, the inequality $2x + 3 > 5x - 6$.
   A. $x < -1$ **B.** $x < 1$ **C.** $x > 1$ **D.** $x < 3$ **E.** $x > 3$

6. **MC** Solve for m, the inequality $\frac{5 - 2m}{3} \le 2$.
   A. $m \le -\frac{1}{2}$ **B.** $m \ge -\frac{1}{2}$ **C.** $m \ge -\frac{11}{2}$ **D.** $m \le -\frac{11}{2}$ **E.** $m \le -2$

7. Dylan received a better result for his Maths test than for his English test. If the sum of his two test results is 159 and the difference is 25, determine Dylan's maths test result.

8. **MC** Solve the pair of simultaneous equations $mx + ny = m$ and $x = y + n$ for x and y in terms of m and n.

   A. $x = \frac{m + n^2}{m + n}$ and $y = \frac{m(1 - n)}{m + n}$

   B. $x = 1$ and $y = \frac{m(1 - n)}{m + n}$

   C. $x = \frac{1}{n}$ and $y = \frac{m(1 - n)}{m + n}$

   D. $x = \frac{m + n^2}{m + n}$ and $y = \frac{1 - n}{n}$

   E. $x = m - mn$ and $y = m + n$

9. **MC** Solve the pair of simultaneous equations $\frac{x}{3} - \frac{y}{2} = \frac{1}{6}$ and $\frac{x}{4} + \frac{y}{3} = \frac{1}{2}$.

A. $x = \frac{221}{6}$ and $y = \frac{9}{2}$
B. $x = \frac{13}{102}$ and $y = \frac{1}{34}$
C. $x = \frac{81}{17}$ and $y = \frac{17}{9}$
D. $x = \frac{17}{22}$ and $y = \frac{9}{22}$
E. $x = \frac{22}{17}$ and $y = \frac{9}{17}$

10. **MC** If the perimeter of a rectangle is 22 cm and the area is 24 cm^2. Select all possible values of x and y.

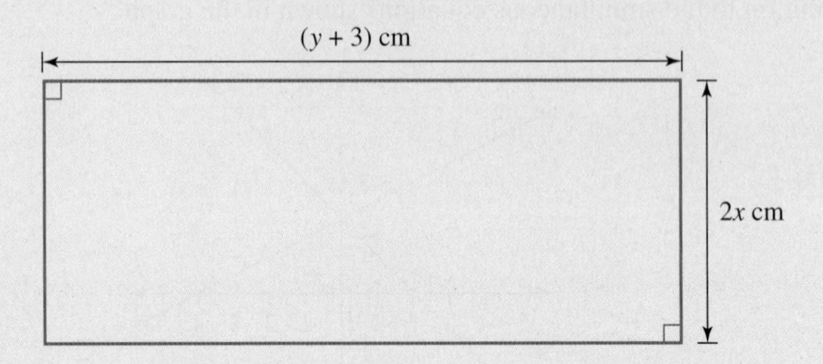

A. $x = 4, y = 0$
B. $x = 6.5, y = -5$
C. $x = 1\frac{1}{2}, y = 5$
D. $x = -\frac{3}{2}, y = -5$
E. $x = \frac{3}{2}, y = 5$
F. $x = -4, y = 0$
G. $x = 1.5, y = 5$

11. **MC** Identify the points of intersection between the line $y = x + 4$ and the hyperbola $y = \frac{5}{x}$.

A. $(-4, 4)$ and $(1, 5)$
B. $(-1, -5)$ and $(5, 1)$
C. $(-5, -1)$ and $(1, 5)$
D. $(5, 9)$ and $(-1, 3)$
E. $(0, 9)$ and $(-1, 5)$

12. **MC** Select all the point(s) of intersection between the circle $x^2 + y^2 = 8$ and the line $y = x$.

A. $(4, 4)$
B. $(2, 2)$
C. $(-4, -4)$
D. $(-2, -2)$
E. $(0, 0)$
F. $(-2, 2)$
G. $(2, -2)$

13. State at how many points the line $y = 2$ intersects with parabola $y = x^2 - 4$.

14. **MC** Identify the three inequalities that define the shaded region in the diagram.

A. $x > 5, y < 1, y < x + 2$
B. $x > 5, y < 1, y > -x + 2$
C. $x > 5, y < 1, y < -x + 2$
D. $x > 1, y < 5, y < x + 2$
E. $x > 1, y < 5, y > x + 2$

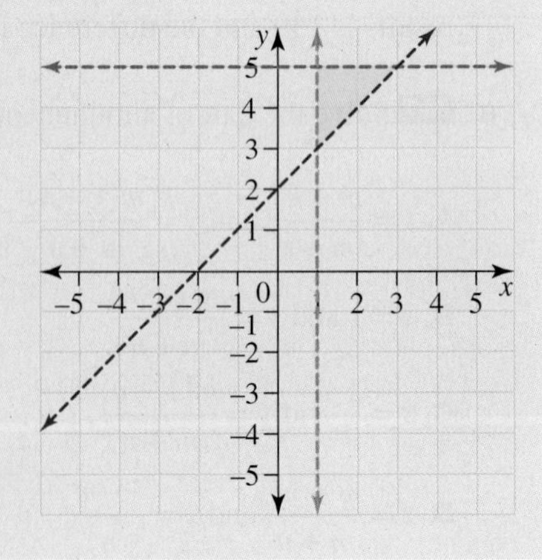

15. **MC** Identify the region satisfying the systems of inequalities $2y - 3x > 1$ and $y + x < -2$.

A.

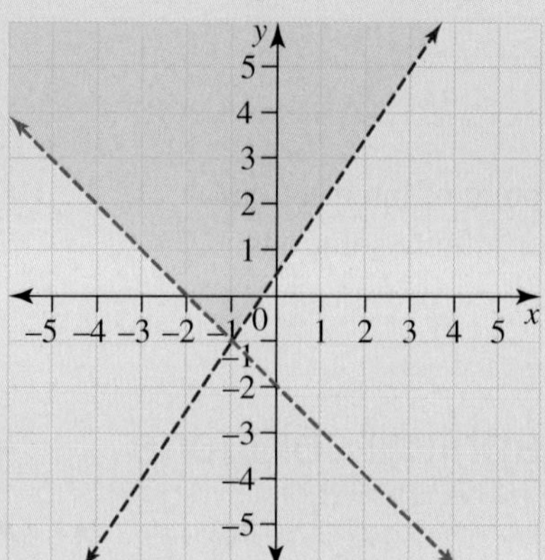

B.

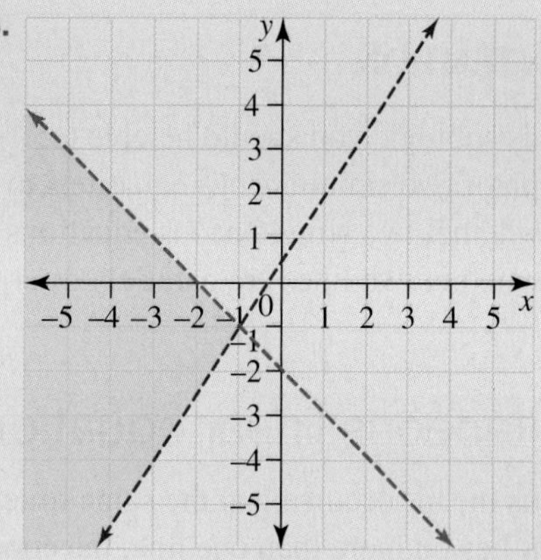

C.

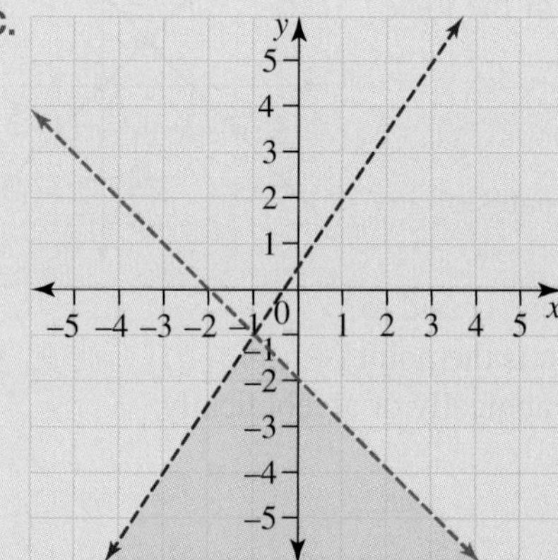

D.

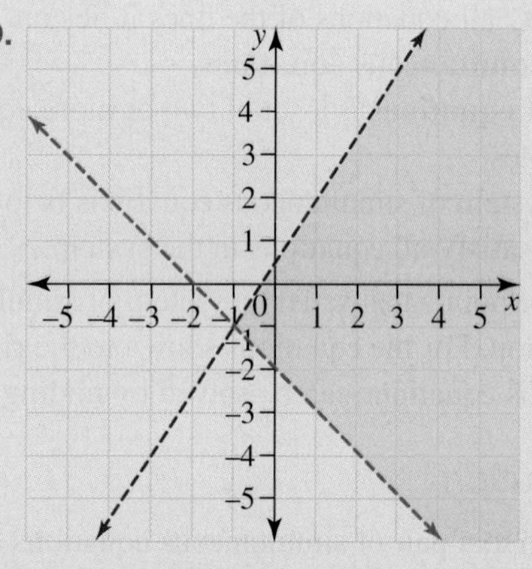

E.

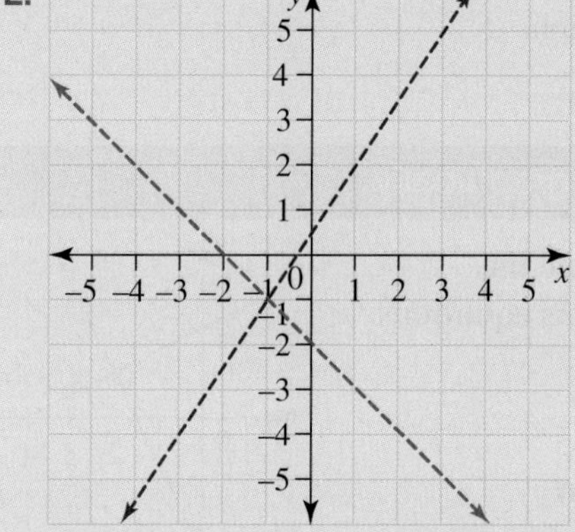

4.2 Graphical solution of simultaneous linear equations

LEARNING INTENTION

At the end of this subtopic you should be able to:
- use the graph of two simultaneous equations to determine the point of intersection
- determine whether two simultaneous equations will have 0, 1 or infinite solutions
- determine whether two lines are parallel or perpendicular.

4.2.1 Simultaneous linear equations and graphical solutions

eles-4763

- **Simultaneous** means occurring at the same time.
- When a point lies on more than one line, the coordinates of that point are said to satisfy all equations of the lines it lies on. The equations of the lines are called **simultaneous equations**.
- A **system of equations** is a set of two or more equations with the same variables.
- Solving a system of simultaneous equations is to find the coordinates of any point/s that satisfy all equations in the system.

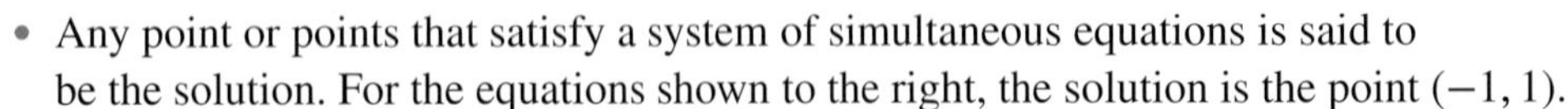

- Any point or points that satisfy a system of simultaneous equations is said to be the solution. For the equations shown to the right, the solution is the point $(-1, 1)$.
- Simultaneous equations can be solved by finding these points graphically or algebraically.

Graphical solution

- The solution to a pair of simultaneous equations can be found by graphing the two equations and identifying the coordinates of the point of intersection.
- The accuracy of the solution depends on having an accurate graph.

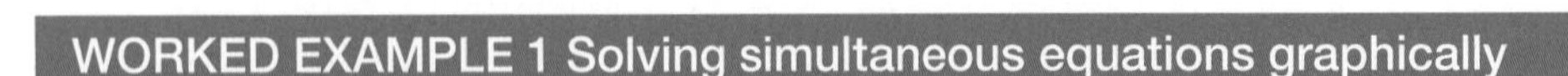

WORKED EXAMPLE 1 Solving simultaneous equations graphically

Use the graphs of the given simultaneous equations to determine the point of intersection and, hence, the solution of the simultaneous equations.

$$x + 2y = 4$$
$$y = 2x - 3$$

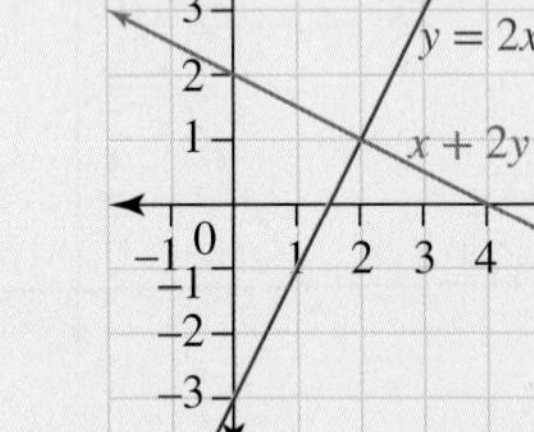

THINK	WRITE/DRAW
1. Write the equations, one under the other and number them.	$x + 2y = 4$ [1] $y = 2x - 3$ [2]
2. Locate the point of intersection of the two lines. This gives the solution.	Point of intersection $(2, 1)$ Solution: $x = 2$ and $y = 1$

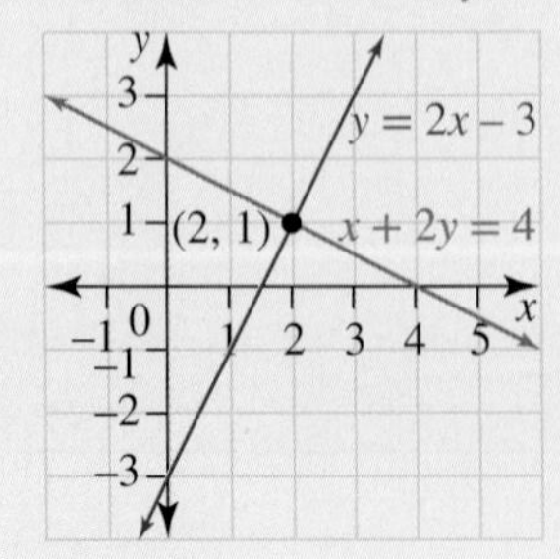

3. Check the solution by substituting $x = 2$ and $y = 1$ into the given equations. Comment on the results obtained.	Check equation [1]: LHS $= x + 2y$ RHS $= 4$ $= 2 + 2(1)$ $= 4$ LHS = RHS Check equation [2]: LHS $= y$ RHS $= 2x - 3$ $= 1$ $= 2(2) - 3$ $= 4 - 3$ $= 1$ LHS = RHS
4. State the solution.	In both cases LHS = RHS, therefore the solution set (2, 1) is correct.

WORKED EXAMPLE 2 Verifying a solution using substitution

Verify whether the given pair of coordinates, (5, −2), is the solution to the following pair of simultaneous equations.

$$\mathbf{3x - 2y = 19}$$
$$\mathbf{4y + x = -3}$$

THINK	WRITE
1. Write the equations and number them.	$3x - 2y = 19$ [1] $4y + x = -3$ [2]
2. Substitute $x = 5$ and $y = -2$ into equation [1].	Check equation [1]: LHS $= 3x - 2y$ RHS $= 19$ $= 3(5) - 2(-2)$ $= 15 + 4$ $= 19$ LHS = RHS
3. Substitute $x = 5$ and $y = -2$ into equation [2].	Check equation [2]: LHS $= 4y + x$ RHS $= -3$ $= 4(-2) + 5$ $= -3$ LHS = RHS
4. State the solution.	Therefore, the solution set (5, −2) is a solution to both equations.

WORKED EXAMPLE 3 Using a graphical method to solve simultaneously

Solve the following pair of simultaneous equations using a graphical method.

$$\mathbf{x + y = 6}$$
$$\mathbf{2x + 4y = 20}$$

THINK	WRITE/DRAW
1. Write the equations, one under the other and number them.	$x + y = 6$ [1] $2x + 4y = 20$ [2]

2. Calculate the x- and y-intercepts for equation [1]. For the x-intercept, substitute $y = 0$ into equation [1].	Equation [1] x-intercept: when $y = 0$, $x + 0 = 6$ $x = 6$ The x-intercept is at $(6, 0)$.
For the y-intercept, substitute $x = 0$ into equation [1].	y-intercept: when $x = 0$, $0 + y = 6$ $y = 6$ The y-intercept is at $(0, 6)$.
3. Calculate the x- and y-intercepts for equation [2]. For the x-intercept, substitute $y = 0$ into equation [2]. Divide both sides by 2.	Equation [2] x-intercept: when $y = 0$, $2x + 0 = 20$ $2x = 20$ $x = 10$ The x-intercept is at $(10, 0)$.
For the y-intercept, substitute $y = 0$ into equation [2]. Divide both sides by 4.	y-intercept: when $x = 0$, $0 + 4y = 20$ $4y = 20$ $y = 5$ The y-intercept is at $(0, 5)$.
4. Use graph paper to rule up a set of axes and label the x-axis from 0 to 10 and the y-axis from 0 to 6. **5.** Plot the x- and y-intercepts for each equation. **6.** Produce a graph of each equation by ruling a straight line through its intercepts. **7.** Label each graph.	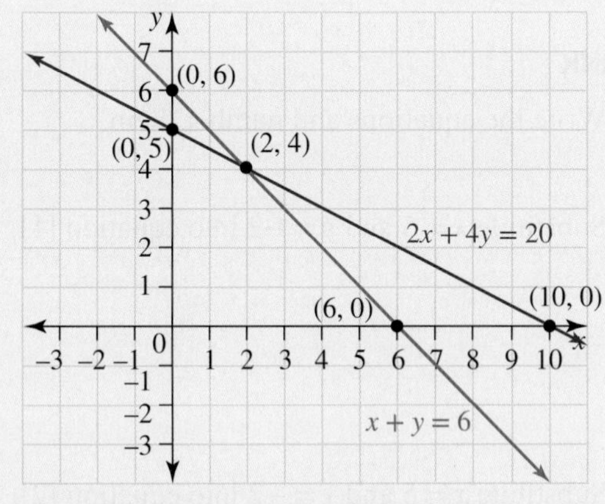
8. Locate the point of intersection of the lines.	The point of intersection is $(2, 4)$.
9. Check the solution by substituting $x = 2$ and $y = 4$ into each equation.	Check [1]: $\text{LHS} = x + y$ $\quad$ $\text{RHS} = 6$ $= 2 + 4$ $= 6$ LHS = RHS Check [2]: $\text{LHS} = 2x + 4y$ $\quad$ $\text{RHS} = 20$ $= 2(2) + 4(4)$ $= 4 + 16$ $= 20$ LHS = RHS
10. State the solution.	In both cases, LHS = RHS. Therefore, the solution set $(2, 4)$ is correct. The solution is $x = 2, y = 4$.

TI | THINK	DISPLAY/WRITE	CASIO | THINK	DISPLAY/WRITE
1. In a new problem on a Graphs page, complete the function entry lines as: $f1(x) = 6 - x$ $f2(x) = 5 - \frac{x}{2}$ Press the down arrow between entering the functions. The graphs will be displayed.	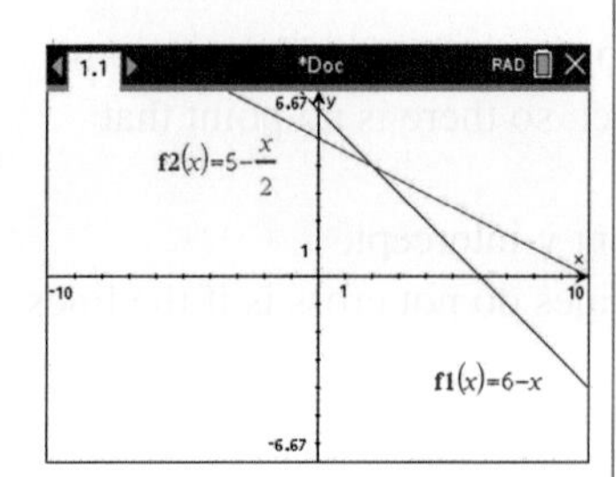	1. On a Graph & Table screen, complete the function entry lines as: $y1 = 6 - x$ $y2 = 5 - \frac{x}{2}$ Then tap the graphing icon. The graphs will be displayed.	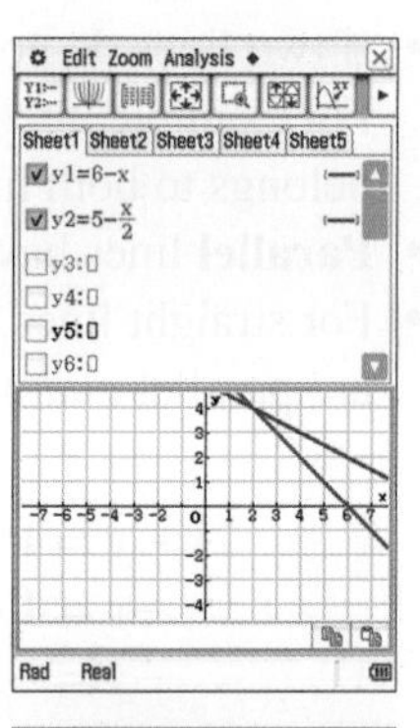
2. To locate the point of intersection, press: • MENU • 6: Analyze Graph • 4: Intersection Drag the dotted line to the left of the point of intersection (the lower bound), press ENTER and then drag the dotted line to the right of the point of intersection (the upper bound) and press ENTER. The point of intersection will be shown.	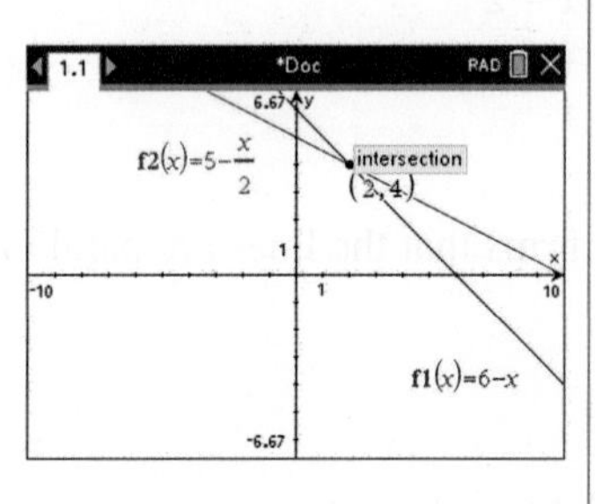	2. To locate the point of intersection, tap: • Analysis • G-Solve • Intersection The point of intersection will be shown.	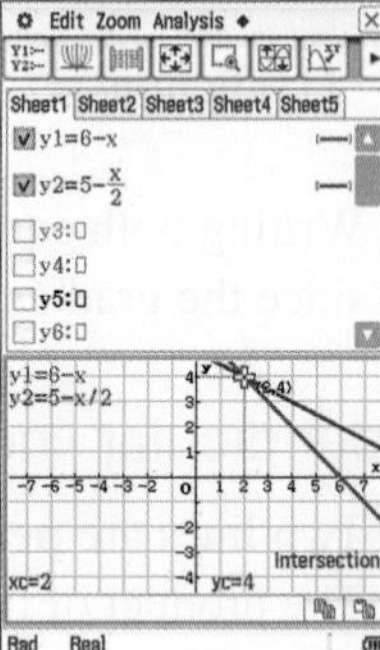
3. State the point of intersection.	The point of intersection (the solution) is (2, 4). That is, $x = 2, y = 4$.	3. State the point of intersection.	The point of intersection (the solution) is (2, 4). That is, $x = 2, y = 4$.

4.2.2 Solutions to coincident, parallel and perpendicular lines

eles-4764

- Two lines are **coincident** if they lie one on top of the other. For example, in the graph shown, the line in blue and line segment in pink are coincident.

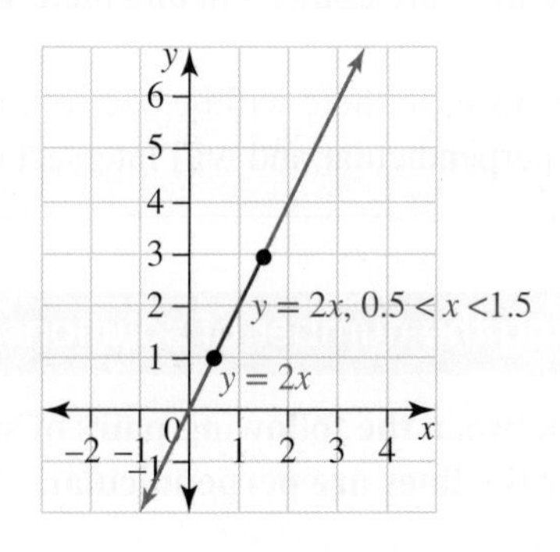

- There are an infinite number of solutions to coincident equations. Every point where the lines coincide satisfies both equations and hence is a solution to the simultaneous equations.
- Coincident equations have the same equation, although the equations may have been transposed so they look different. For example, $y = 2x + 3$ and $2y - 4x = 6$ are coincident equations.

Parallel lines

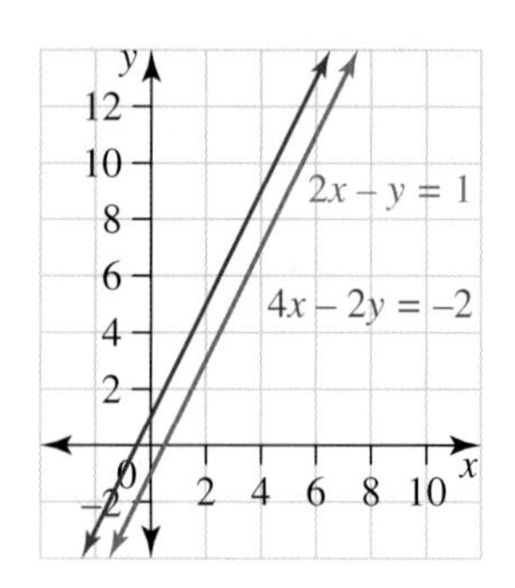

- If two lines do not intersect, there is no simultaneous solution to the equations. For example, the graph lines shown do not intersect, so there is no point that belongs to both lines.
- **Parallel** lines have the same gradient but a different y-intercept.
- For straight lines, the only situation in which the lines do not cross is if the lines are parallel *and* not coincident.

$$\begin{aligned} 2x - y &= 1 \quad [1] \\ -y &= 1 - 2x \\ -y &= -2x - 1 \\ y &= 2x - 1 \\ \text{Gradient } m &= 2 \end{aligned} \qquad \begin{aligned} 4x - 2y &= -2 \quad [2] \\ -2y &= -2 - 4x \\ -2y &= -4x - 2 \\ y &= 2x + 1 \\ \text{Gradient } m &= 2 \end{aligned}$$

- Writing both equations in the form $y = mx + c$ confirms that the lines are parallel since the gradients are equal.

Perpendicular lines

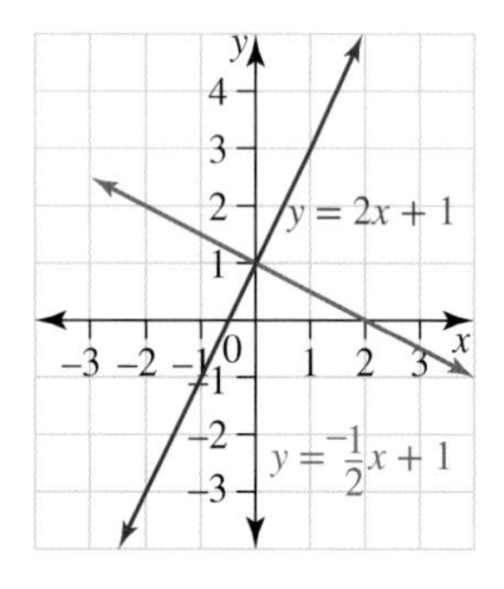

- Two lines are **perpendicular** if they intersect at right angles (90°).
- The product of the gradients of two perpendicular lines is equal to −1:

$$m_1 \times m_2 = -1 \text{ or } m_1 = -\frac{1}{m_2}$$

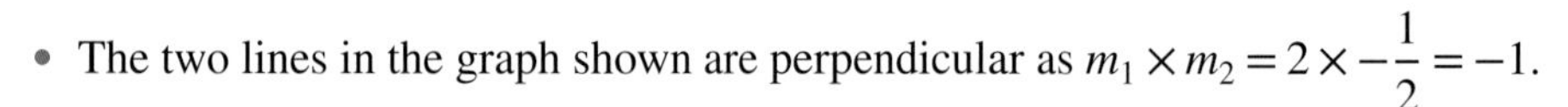

- The two lines in the graph shown are perpendicular as $m_1 \times m_2 = 2 \times -\frac{1}{2} = -1$.

Number of solutions for a pair of simultaneous linear equations

For two linear equations given by $y_1 = m_1x + c_1$ and $y_2 = m_2x + c_2$:

- If $m_1 = m_2$ and $c_1 \neq c_2$ then the two lines are parallel and there will be no solutions between the two lines.
- If $m_1 = m_2$ and $c_1 = c_2$ then the two lines are coincident and there will be infinite solutions between the two lines.
- If $m_1 \neq m_2$ then the lines will cross once, so there will be one solution.
- If $m_1 \times m_2 = -1$ then the lines are perpendicular and will intersect (once) at right angles (90°).

WORKED EXAMPLE 4 Determining the number of solutions between two lines

Determine the number of solutions between the following pairs of simultaneous equations. If there is only one solution, determine whether the lines are perpendicular.

a. $\mathbf{2y = 4x + 6}$ **and** $\mathbf{-3y = -6x - 12}$
b. $\mathbf{y = -3x + 2}$ **and** $\mathbf{-3y = x + 15}$
c. $\mathbf{5y = 25x - 30}$ **and** $\mathbf{2y - 10x + 12 = 0}$

THINK	WRITE
a. 1. Re-write both equations in the form $y = mx + c$.	a. $\begin{aligned} 2y &= 4x + 6 \\ y &= 2x + 3 \; [1] \\ -3y &= -6x - 12 \\ y &= 2x + 4 \; [2] \end{aligned}$

2. Determine the gradient of both lines.	$m_1 = 2$ and $m_2 = 2$
3. Check if the lines are parallel, coincident or perpendicular.	The gradients are the same and the y-intercepts different. So, the two lines are parallel.
4. Write the answer.	There will be *no* solutions between this pair of simultaneous equations as the lines are parallel.
b. 1. Re-write both equations in the form $y = mx + c$.	**b.** $y = -3x + 2$ [1] $-3y = x + 15$ $y = \frac{x}{-3} - 5$ [2]
2. Determine the gradient of both lines.	$m_1 = -3$ and $m_2 = -\frac{1}{3}$
3. Check if the lines are parallel, coincident or perpendicular and comment on the number of solutions.	The gradients are different so there will be *one* solution.
4. Determine if the lines are perpendicular by calculating the product of the gradients.	$m_1 \times m_2 = -3 \times -\frac{1}{3} = 1$
5. Write the answer.	The lines have one solution but they are not perpendicular.
c. 1. Re-write both equations in the form $y = mx + c$.	**c.** $5y = 25x - 30$ $y = 5x - 6$ [1] $2y - 10x + 12 = 0$ $2y = 10x - 12$ $y = 5x - 6$ [2]
2. Determine the gradient of both lines.	$m_1 = 5$ and $m_2 = 5$
3. Check if the lines are parallel, coincident, or perpendicular.	The gradients are the same and the y-intercepts are also the same. So, the two lines are coincident.
4. Write the answer.	The lines are coincident so there are *infinite* solutions between the two lines.

DISCUSSION

What do you think is the major error made when solving simultaneous equations graphically?

on Resources

eWorkbook	Topic 4 Workbook (worksheets, code puzzle and a project) (ewbk-2030)
Digital document	SkillSHEET Graphing linear equations using the x- and y-intercept method (doc-5217)
Interactivities	Individual pathway interactivity: Graphical solution of simultaneous linear equations (int-4577) Solving simultaneous equations graphically (int-6452) Parallel lines (int-3841) Perpendicular lines (int-6124)

Exercise 4.2 Graphical solution of simultaneous linear equations

learnON

Individual pathways

■ PRACTISE	■ CONSOLIDATE	■ MASTER
1, 4, 7, 10, 14, 15, 18	2, 5, 8, 11, 12, 16, 19	3, 6, 9, 13, 17, 20, 21

To answer questions online and to receive **immediate corrective feedback** and **fully worked solutions** for all questions, go to your learnON title at www.jacplus.com.au.

Fluency

WE1 For questions 1 to 3, use the graphs to determine the point of intersection and hence the solution of the simultaneous equations.

1. a. $x+y=3$
   $x-y=1$

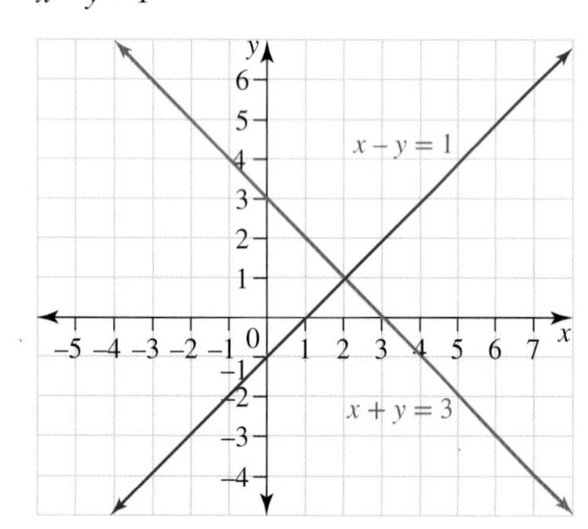

b. $x+y=2$
   $3x-y=2$

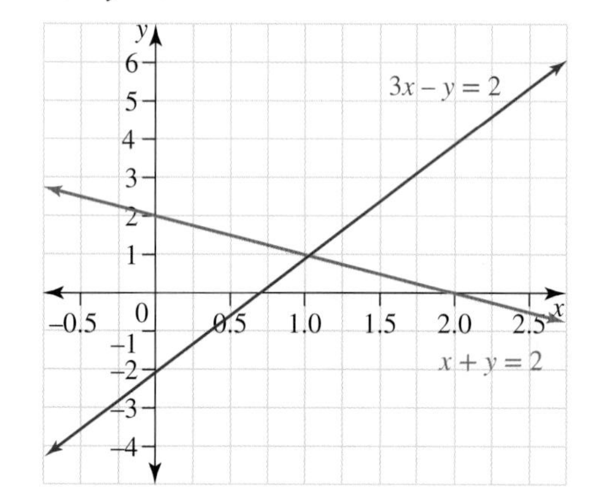

2. a. $y-x=4$
   $3x+2y=8$

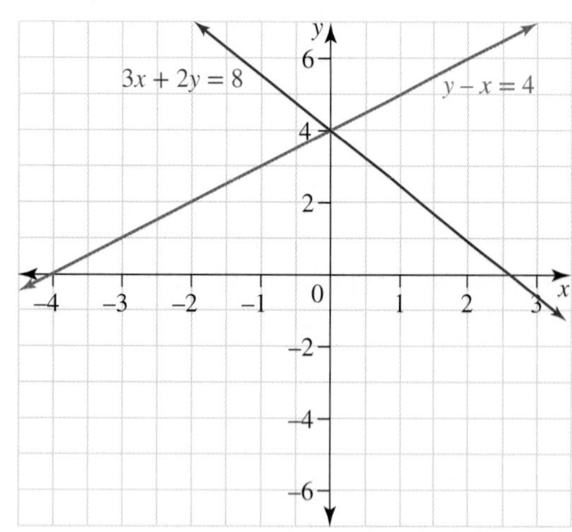

b. $y+2x=3$
   $2y+x=0$

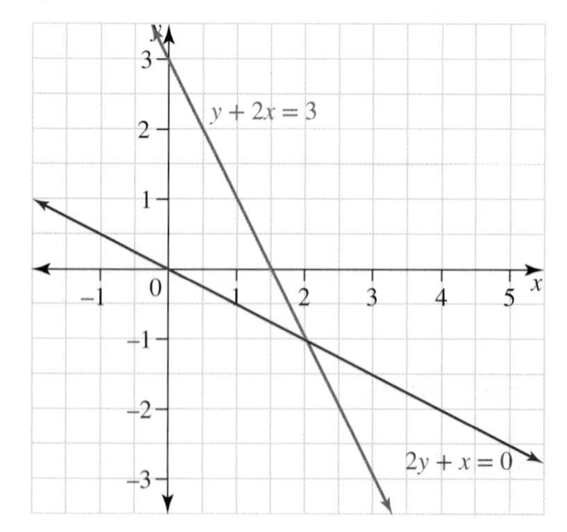

3. a. $y - 3x = 2$
$x - y = 2$

b. $2y - 4x = 5$
$4y + 2x = 5$

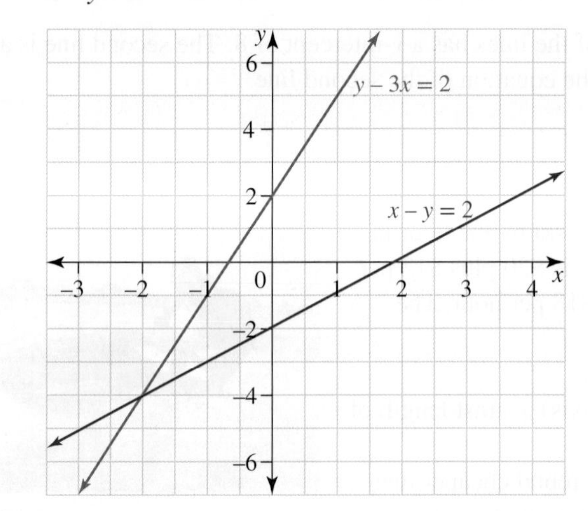

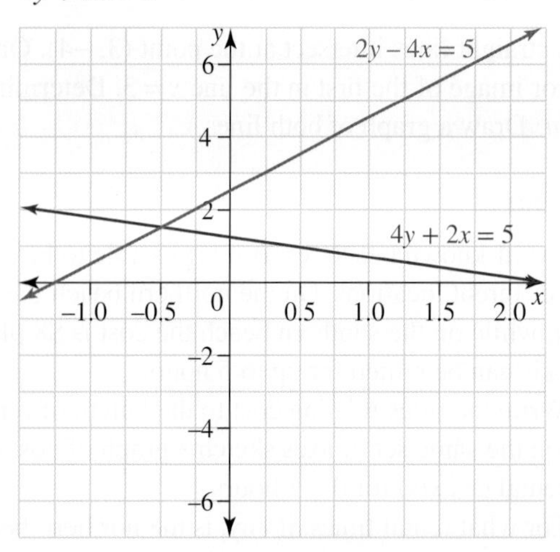

WE2 For questions 4 to 6, use substitution to check if the given pair of coordinates is a solution.

4. a. (7, 5) $3x + 2y = 31$, $2x + 3y = 28$
b. (3, 7) $y - x = 4$, $2y + x = 17$
c. (9, 1) $x + 3y = 12$, $5x - 2y = 143$
d. (2, 5) $x - y = 7$, $2x + 3y = 18$

5. a. (4, −3) $y = 3x - 15$, $4x + 7y = -5$
b. (6, −2) $x - y = 7$, $3x + y = 16$
c. (4, −2) $2x + y = 6$, $x - 3y = 8$
d. (5, 1) $y - 5x = -24$, $3y + 4x = 23$

6. a. (−2, −5) $3x - 2y = -4$, $2x - 3y = 11$
b. (−3, −1) $y - x = 2$, $2y - 3x = 7$
c. $\left(-\frac{1}{2}, 2\right)$ $6x + 4y = 5$, $20x - 5y = 0$
d. $\left(\frac{3}{2}, \frac{5}{3}\right)$ $8x + 6x = 22$, $10x - 9y = 0$

WE3 For questions 7 to 9, solve each of the following pairs of simultaneous equations using a graphical method.

7. a. $x + y = 5$, $2x + y = 8$
b. $x + 2y = 10$, $3x + y = 15$
c. $2x + 3y = 6$, $2x - y = -10$
d. $x - 3y = -8$, $2x + y = -2$

8. a. $6x + 5y = 12$, $5x + 3y = 10$
b. $y + 2x = 6$, $2y + 3x = 9$
c. $y = 3x + 10$, $y = 2x + 8$
d. $y = 8$, $3x + y = 17$

9. a. $4x - 2y = -5$, $x + 3y = 4$
b. $3x + y = 11$, $4x - y = 3$
c. $3x + 4y = 27$, $x + 2y = 11$
d. $3y + 3x = 8$, $3y + 2x = 6$

Understanding

For questions 10 to 12, using technology, determine which of the following pairs of simultaneous equations have no solutions. Confirm by finding the gradient of each line.

10. a. $y = 2x - 4$, $3y - 6x = 10$
b. $5x - 3y = 13$, $4x - 2y = 10$
c. $x + 2y = 8$, $5x + 10y = 45$
d. $y = 4x + 5$, $2y - 10x = 8$

11. a. $3y + 2x = 9$, $6x + 4y = 22$
b. $y = 5 - 3x$, $3y = -9x + 18$
c. $4y + 3x = 7$, $12y + 9x = 22$
d. $2y - x = 0$, $14y - 6x = 2$

12. a. $y = 3x - 4$
$5y = 12 + 15x$

b. $4x - 6y = 12$
$6x - 4y = 12$

c. $3y = 5x - 22$
$5x = 3y + 26$

d. $3x = 12 - 4y$
$8y + 6x = 14$

13. Two straight lines intersect at the point $(3, -4)$. One of the lines has a y-intercept of 8. The second line is a mirror image of the first in the line $x = 3$. Determine the equation of the second line.
(*Hint:* Draw a graph of both lines.)

Reasoning

14. At a well-known beach resort it is possible to hire a jet-ski by the hour in two different locations. On the northern beach the cost is \$20 plus \$12 per hour, while on the southern beach the cost is \$8 plus \$18 per hour. The jet-skis can be rented for up to 5 hours.

a. Write the rules relating cost to the length of rental.
b. On the same set of axes sketch a graph of cost (y-axis) against length of rental (x-axis) for $0 - 5$ hours.
c. For what rental times, if any, is the northern beach rental cheaper than the southern beach rental? Use your graph to justify your answer.
d. For what length of rental time are the two rental schemes identical? Use the graph and your rules to justify your answer.

15. For each of the pairs of simultaneous equations below, determine whether they are the same line, parallel lines, perpendicular lines or intersecting lines. Show your working.

a. $2x - y = -9$
$-4x - 18 = -2y$

b. $x - y = 7$
$x + y = 7$

c. $x + 6 = y$
$2x + y = 6$

d. $x + y = -2$
$x + y = 7$

16. For each of the following, explain if the equations have one solution, an infinite number of solutions or no solution.

a. $x - y = 1$
$2x - 3y = 2$

b. $2x - y = 5$
$4x - 2y = -6$

c. $x - 2y = -8$
$4x - 8y = -16$

17. Determine whether the following pairs of equations will have one, infinite or no solutions. If there is only one solution, determine whether the lines are perpendicular.

a. $3x + 4y = 14$
$4x - 3y = 2$

b. $2x + y = 5$
$3y + 6x = 15$

c. $3x - 5y = -6$
$5x - 3y = 24$

d. $2y - 4x = 6$
$2x - y = -10$

Problem solving

18. Use the information below to determine the value of a in each of the following equations:

a. $y = ax + 3$, which is parallel to $y = 3x - 2$
b. $y = ax - 2$, which is perpendicular to $y = -4x + 6$
c. $y = ax - 4$, which intersects the line $y = 3x + 6$ when $x = 2$.

19. Line A is parallel to the line with equation $y - 3x - 3 = 0$ and passes through the point $(1, 9)$. Line B is perpendicular to the line with equation $2y - x + 6 = 0$ and passes through the point $(2, -3)$.

a. Determine the equation of line A.
b. Determine the equation of line B.
c. Sketch both lines on the one set of axes to find where they intersect.

20. Solve the system of three simultaneous equations graphically.

$$3x - y = 2$$
$$y + 3x = 4$$
$$2y - x = 1$$

21. A line with equation $4x + 5y = 4$ intersects a second line when $x = -4$. Determine the equation of the second line if it is perpendicular to the first line.

4.3 Solving simultaneous linear equations using substitution

LEARNING INTENTION

At the end of this subtopic you should be able to:
- identify when it is appropriate to solve using the substitution method
- solve a system of two linear simultaneous equations using the substitution method.

4.3.1 Solving simultaneous equations using the substitution method

eles-4766

- A variable is considered the **subject of an equation** if it is expressed in terms of the other variables. In the equation $y = 3x + 4$, the variable y is the subject.
- The **substitution method** is used when one (or both) of the equations is presented in a form where one of the two variables is the subject of the equation.
- When solving two linear simultaneous equation, the substitution method involves replacing a variable in one equation with the other equation. This produces a new third equation expressed in terms of a single variable.
- Consider the pair of simultaneous equations:

$$y = 2x - 4$$
$$3x + 2y = 6$$

- In the first equation, y is written as the subject and is equal to $(2x - 4)$. In this case, substitution is performed by replacing y in the second equation with the expression $(2x - 4)$.

$$y = 2x - 4$$

$$3x + 2(y) = 10$$

$$3x + 2(2x - 4) = 6$$

- This produces a third equation, all in terms of x, so that the value of x can be found.
- Once a value for one variable is found, it can be substituted back into either equation to find the value of the other variable.
- It is often helpful to use brackets when substituting an expression into another equation.

WORKED EXAMPLE 5 Solving using the substitution method

Solve the simultaneous equations $y = 2x - 1$ and $3x + 4y = 29$ using the substitution method.

THINK	WRITE
1. Write the equations, one under the other and number them.	$y = 2x - 1$ [1] $3x + 4y = 29$ [2]
2. y and $2x - 1$ are equal so substitute the expression $(2x - 1)$ for y into equation [2].	Substituting $(2x - 1)$ into [2]: $3x + 4(2x - 1) = 29$
3. Solve for x.	
i. Expand the brackets on the LHS of the equation.	$3x + 8x - 4 = 29$
ii. Collect like terms.	$11x - 4 = 29$

iii. Add 4 to both sides of the equation.	$11x = 33$
iv. Divide both sides by 11.	$x = 3$
4. Substitute $x = 3$ into either of the equations, say [1], to find the value of y.	Substituting $x = 3$ into [1]: $y = 2(3) - 1$ $= 6 - 1$ $= 5$
5. Write your answer.	Solution: $x = 3, y = 5$ or $(3, 5)$
6. Check the solution by substituting $(3, 5)$ into equation [2].	Check: Substitute $(3, 5)$ into $3x + 4y = 29$. LHS $= 3(3) + 4(5)$ RHS $= 29$ $= 9 + 20$ $= 29$ As LHS = RHS, the solution is correct.

4.3.2 Equating equations

eles-4767

- To **equate** in mathematics is to take two expressions that have the same value and make them equal to each other.
- When both linear equations are written with the same variable as the subject, we can equate the equations to solve for the other variable. Consider the following simultaneous equations:

$$y = 4x - 3$$
$$y = 2x + 9$$

- In the first equation y is equal to $(4x - 3)$ and in the second equation y is equal to $(2x + 9)$. Since both expressions are equal to the same thing (y), they must also be equal to each other. Thus, equating the equations gives:

$$y = 4x - 3$$
$$(y) = 2x + 9$$
$$4x - 3 = 2x + 9$$

- As can be seen above, equating equations is still a form of substitution. A third equation is produced, all in terms of x, allowing for a value of x to be solved.

WORKED EXAMPLE 6 Substitution by equating two equations

Solve the pair of simultaneous equations $y = 5x - 8$ and $y = -3x + 16$ by equating the equations.

THINK	WRITE
1. Write the equations, one under the other and number them.	$y = 5x - 8$ [1] $y = -3x + 16$ [2]
2. Both equations are written with y as the subject, so equate them.	$5x - 8 = -3x + 16$

3. Solve for x. i. Add $3x$ to both sides of the equation. ii. Add 8 to both sides of the equation. iii. Divide both sides of the equation by 8.	$8x-8=16$ $8x=24$ $x=3$
4. Substitute the value of x into either of the original equations, say [1], and solve for y.	Substituting $x=3$ into [1]: $y=5(3)-8$ $=15-8$ $=3$
5. Write your answer.	Solution: $x=3, y=7$ or $(3, 7)$
6. Check the answer by substituting the point of intersection into equation [2].	Check: Substitute into $y=-3x+16$. LHS $= y$ $=7$ RHS $=-3x+16$ $=-3(3)+16$ $=-9+16$ $=7$ As LHS = RHS, the solution is correct.

DISCUSSION

When would you choose the substitution method in solving simultaneous equations?

on Resources

eWorkbook Topic 4 Workbook (worksheets, code puzzle and a project) (ewbk-2030)

Video eLesson Solving simultaneous equations using substitution (eles-1932)

Interactivities Individual pathway interactivity: Solving simultaneous linear equations using substitution (int-4578)
Solving simultaneous equations using substitution (int-6453)

Exercise 4.3 Solving simultaneous linear equations using substitution

learnon

Individual pathways

■ PRACTISE	■ CONSOLIDATE	■ MASTER
1, 4, 7, 8, 14, 15, 18	2, 5, 9, 12, 16, 19	3, 6, 10, 11, 13, 17, 20

To answer questions online and to receive **immediate corrective feedback** and **fully worked solutions** for all questions, go to your learnON title at www.jacplus.com.au.

Fluency

WE5 For questions 1 to 3, solve the following simultaneous equations using the substitution method. Check your solutions using technology.

1. a. $x=-10+4y$
$3x+5y=21$
b. $3x+4y=2$
$x=7+5y$
c. $3x+y=7$
$x=-3-3y$
d. $3x+2y=33$
$y=41-5x$

2. a. $y=3x-3$
$-5x+3y=3$

b. $4x+y=9$
$y=11-5x$

c. $x=-5-2y$
$5y+x=-11$

d. $x=-4-3y$
$-3x-4y=12$

3. a. $x=7+4y$
$2x+y=-4$

b. $x=14+4y$
$-2x+3y=-18$

c. $3x+2y=12$
$x=9-4y$

d. $y=2x+1$
$-5x-4y=35$

WE6 For questions 4 to **6**, solve the following pairs of simultaneous equations by equating the equations. Check your solutions using technology.

4. a. $y=2x-11$ and $y=4x+1$
b. $y=3x+8$ and $y=7x-12$
c. $y=2x-10$ and $y=-3x$
d. $y=x-9$ and $y=-5x$

5. a. $y=-4x-3$ and $y=x-8$
b. $y=-2x-5$ and $y=10x+1$
c. $y=-x-2$ and $y=x+1$
d. $y=6x+2$ and $y=-4x$

6. a. $y=0.5x$ and $y=0.8x+0.9$
b. $y=0.3x$ and $y=0.2x+0.1$
c. $y=-x$ and $y=-\frac{2}{7}x+\frac{4}{7}$
d. $y=-x$ and $y=-\frac{3}{4}x-\frac{1}{4}$

Understanding

7. A small farm has sheep and chickens. There are twice as many chicken as sheep, and there are 104 legs between the sheep and the chickens. Calculate the total number of chickens.

For questions 8 to **10**, use substitution to solve each of the following pairs of simultaneous equations.

8. a. $5x+2y=17$
$y=\dfrac{3x-7}{2}$

b. $2x+7y=17$
$x=\dfrac{1-3y}{4}$

9. a. $2x+3y=13$
$y=\dfrac{4x-15}{5}$

b. $-2x-3y=-14$
$x=\dfrac{2+5y}{3}$

10. a. $3x+2y=6$
$y=3-\dfrac{5x}{3}$

b. $-3x-2y=-12$
$y=\dfrac{5x-20}{3}$

11. Use substitution to solve each of the following pairs of simultaneous equations for x and y in terms of m and n.

a. $mx+y=n$
$y=mx$

b. $x+ny=m$
$y=nx$

c. $mx-ny=n$
$y=x$

d. $mx-ny=n$
$y=x$

e. $mx-ny=-m$
$x=y-n$

f. $mx+y=m$
$x=\frac{y+m}{n}$

12. Determine the values of a and b so that the pair of equations $ax+by=17$ and $2ax-by=-11$ has a unique solution of $(-2, 3)$.

13. The earliest record of magic squares is from China in about 2200BC. In magic squares the sums of the numbers of each row, column and diagonal are all equal to a magic number. Let z be the magic number. By creating a set of equations, solve to find the magic number and the missing values in the magic square.

m	11	7
9		
n	5	10

Reasoning

14. a. Consider the pair of simultaneous equations:

$$8x-7y=9$$
$$x+2y=4$$

Identify which equation is the logical choice to make x the subject.

b. Use the substitution method to solve the system of equations. Show all your working.

15. A particular chemistry book costs \$6 less than a particular physics book, while two such chemistry books and three such physics books cost a total of \$123. Construct two simultaneous equations and solve them using the substitution method. Show your working.

16. The two shorter sides of a right triangle are 1 cm and 8 cm shorter than the hypotenuse. If the area of the triangle is 30 cm^2, determine the perimeter of the triangle.

17. Andrew is currently ten years older than his sister Prue. In four years time he will be twice as old as Prue. Determine how old Andrew and Prue are now.

Problem solving

18. Use the substitution method to solve the following.

$$2x+y-9=0$$
$$4x+5y+3=0$$

19. Use the substitution method to solve the following.

$$\frac{y-x}{2}-\frac{x+y}{3}=\frac{1}{6}$$
$$\frac{x}{5}+\frac{y}{2}=\frac{1}{2}$$

20. Consider the following pair of equations:

$$kx-\frac{y}{k}=2$$
$$27x-3y=12k-18$$

Determine the values of k when they will have:

a. one solution

b. no solutions

c. infinite solutions.

4.4 Solving simultaneous linear equations using elimination

LEARNING INTENTION

At the end of this subtopic you should be able to:
- solve two simultaneous linear equations using the elimination method.

4.4.1 Solving simultaneous equations using the elimination method

eles-4768

- The **elimination method** is an algebraic method to solve simultaneous linear equations. It involves adding or subtracting equations in order to eliminate one of the variables.
- In order to eliminate a variable, the variable must be on the same side of the equal sign in both equations and must have the same coefficient.
- If the coefficients of the variable have the same sign, we subtract one equation from the other to eliminate the variable.
- If the coefficients of the variables have the opposite sign, we add the two equations together to eliminate the variable.

$$\begin{array}{cc} 3x+4y=14 & 6x-2y=12 \\ 5x-4y=2 & 6x+3y=27 \\ \text{(add equations to eliminate } y) & \text{(subtract equations to eliminate } x) \end{array}$$

- The process of elimination is carried out by adding (or subtracting) the left-hand sides and the right-hand sides of each equation together. Consider the equations $2x+y=5$ and $x+y=3$. The process of subtracting each side of the equation from each other is visualised on the scales to the right.
- To represent this process algebraically, the setting out would look like:

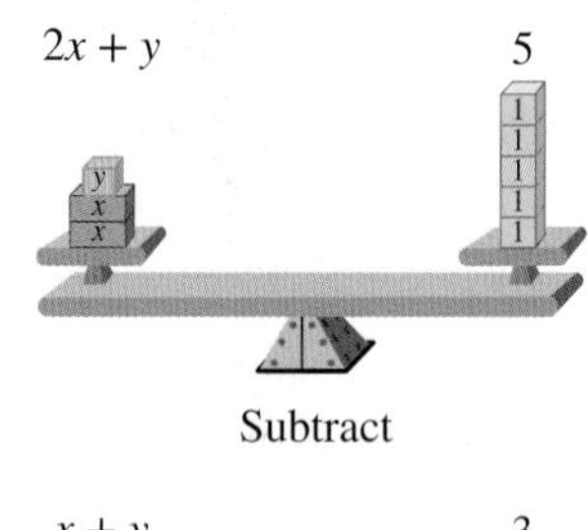

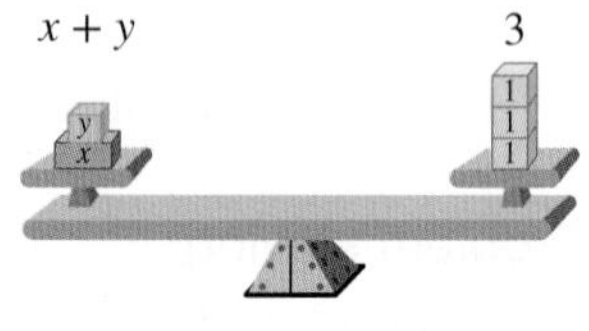

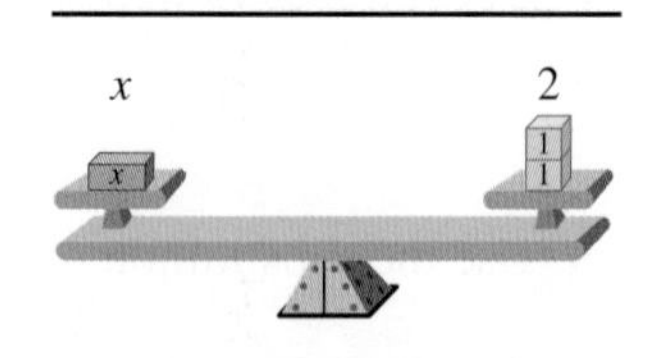

$$\begin{array}{r} 2x+y=5 \\ -(x+y=3) \\ \hline x=2 \end{array}$$

- Once the value of x has been found, it can be substituted into either original equation to find y.

$$2(2)+y=5 \Rightarrow y=1$$

WORKED EXAMPLE 7 Solving using the elimination method

Solve the following pair of simultaneous equations using the elimination method.

$$\begin{aligned} -2x-3y &= -9 \\ 2x+y &= 7 \end{aligned}$$

THINK	WRITE
1. Write the equations, one under the other and number them.	$-2x-3y=-9$ [1] $2x+y=7$ [2]

2. Look for an addition or subtraction that will eliminate either x or y. *Note:* Adding equations [1] and [2] in order will eliminate x.	[1] + [2]: $-2x - 3y + (2x + y) = -9 + 7$ $-2x - 3y + 2x + y = -2$ $-2y = -2$
3. Solve for y by dividing both sides of the equation by -2.	$y = 1$
4. Substitute the value of y into equation [2]. *Note:* $y = 1$ may be substituted into either equation.	Substituting $y = 1$ into [2]: $2x + 1 = 7$
5. Solve for x.	
i. Subtract 1 from both sides of the equation.	$2x = 6$
ii. Divide both sides of the equation by 2.	$x = 3$
6. Write the solution.	Solution: $x = 3$, $y = 1$ or $(3, 1)$
7. Check the solution by substituting $(3, 1)$ into equation [1] since equation [2] was used to find the value of x.	Check: Substitute into $-2x - 3y = -9$. LHS $= -2(3) - 3(1)$ $= -6 - 3$ $= -9$ RHS $= -9$ LHS = RHS, so the solution is correct.

4.4.2 Solving simultaneous equations by multiplying by a constant

eles-4769

- If neither variable in the two equations have the same coefficient, it will be necessary to multiply one or both equations by a constant so that a variable can be eliminated.
- The equals sign in an equation acts like a balance, so as long as both sides of equation are correctly multiplied by the same value, the new statement is still a valid equation.

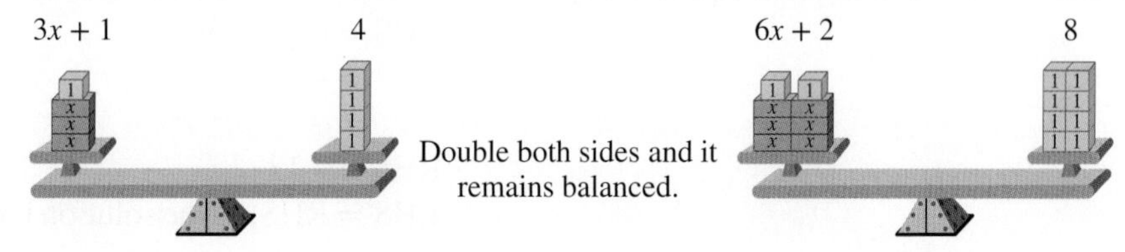

- Consider the following pairs of equations:

$$3x + 7y = 23 \quad 4x + 5y = 22$$
$$6x + 2y = 22 \quad 3x - 4y = -6$$

- For the first pair: the easiest starting point is to work towards eliminating x. This is done by first multiplying the top equation by 2 so that both equations have the same coefficient of x.

$$2(3x + 7y = 23) \Rightarrow 6x + 14y = 46$$

- For the second pair: in this case, both equations will need to be multiplied by a constant. Choosing to eliminate x would require the top equation to be multiplied by 3 and the bottom equation by 4 in order to produce two new equations with the same coefficient of x.

$$3(4x + 5y = 23) \Rightarrow 12x + 15y = 69$$

$$4(3x - 4y = -6) \Rightarrow 12x - 16y = -24$$

- Once the coefficient of one of the variables is the same, you can begin the elimination method.

WORKED EXAMPLE 8 Multiplying one equation by a constant to eliminate

Solve the following pair of simultaneous equations using the elimination method.

$$x - 5y = -17$$
$$2x + 3y = 5$$

THINK	WRITE
1. Write the equations, one under the other and number them.	$x - 5y = -17$ [1] $2x + 3y = 5$ [2]
2. Look for a single multiplication that will create the same coefficient of either x or y. Multiply equation [1] by 2 and call the new equation [3].	[1] × 2 : $2x - 10y = -34$ [3]
3. Subtract equation [2] from [3] in order to eliminate x.	[3] − [2]: $2x - 10y - (2x + 3y) = -34 - 5$ $2x - 10y - 2x - 3y = -39$ $-13y = -39$
4. Solve for y by dividing both sides of the equation by −13.	$y = 3$
5. Substitute the value of y into equation [2].	Substituting $y = 3$ into [2]: $2x + 3(3) = 5$ $2x + 9 = 5$
6. Solve for x. i. Subtract 9 from both sides of the equation. ii. Divide both sides of the equation by 2.	 $2x = -4$ $x = -2$
7. Write the solution.	Solution: $x = -2, y = 3$ or $(-2, 3)$
8. Check the solution by substituting into equation [1].	Check: Substitute into $x - 5y = -17$. LHS $= (-2) - 5(3)$ $= -2 - 15$ $= -17$ RHS $= -17$ LHS = RHS, so the solution is correct.

Note: In this example, equation [1] could have been multiplied by −2 (instead of by 2), then the two equations added (instead of subtracted) to eliminate x.

WORKED EXAMPLE 9 Multiplying both equations by a constant to eliminate

Solve the following pair of simultaneous equations using the elimination method.

$$6x + 5y = 3$$
$$5x + 4y = 2$$

THINK	WRITE
1. Write the equations, one under the other and number them.	$6x + 5y = 3$ [1] $5x + 4y = 2$ [2]
2. Decide which variable to eliminate, say y. Multiply equation [1] by 4 and call the new equation [3]. Multiply equation [2] by 5 and call the new equation [4].	Eliminate y. [1] × 4: $24x + 20y = 12$ [3] [1] × 5: $25x + 20y = 10$ [4]
3. Subtract equation [3] from [4] in order to eliminate y.	[4] − [3]: $25x + 20y - (24x + 20y) = 10 - 12$ $25x + 20y - 24x - 20y = -2$ $x = -2$
4. Substitute the value of x into equation [1].	Substituting $x = -2$ into [1]: $6(-2) + 5y = 3$ $-12 + 5y = 3$
5. Solve for y. i. Add 12 to both sides of the equation. ii. Divide both sides of the equation by 5.	 $5y = 15$ $y = 3$
6. Write your answer.	Solution $x = -2$, $y = 3$ or $(-2, 3)$
7. Check the answer by substituting the solution into equation [2].	Check: Substitute into $5x + 4y = 2$. LHS $= 5(-2) + 4(3)$ $= -10 + 12$ $= 2$ RHS $= 2$ LHS = RHS, so the solution is correct.

Note: Equation [1] could have been multiplied by −4 (instead of by 4), then the two equations added (instead of subtracted) to eliminate y.

on Resources

eWorkbook Topic 4 Workbook (worksheets, code puzzle and a project) (ewbk-2030)

Video eLesson Solving simultaneous equations using elimination (eles-1931)

Interactivities Individual pathway interactivity: Solving simultaneous linear equations using elimination (int-4579)
Solving simultaneous equations using elimination (int-6127)

Exercise 4.4 Solving simultaneous linear equations using elimination

learnon

Individual pathways

■ PRACTISE	■ CONSOLIDATE	■ MASTER
1, 3, 5, 10, 13, 18	2, 6, 8, 11, 14, 15, 19	4, 7, 9, 12, 16, 17, 20

To answer questions online and to receive **immediate corrective feedback** and **fully worked solutions** for all questions, go to your learnON title at www.jacplus.com.au.

Fluency

1. WE7 Solve the following pairs of simultaneous equations by adding equations to eliminate either x or y.

a. $x+2y=5$
$-x+4y=1$

b. $5x+4y=2$
$5x-4y=-22$

c. $-2x+y=10$
$2x+3y=14$

2. Solve the following pairs of equations by subtracting equations to eliminate either x or y.

a. $3x+2y=13$
$5x+2y=23$

b. $2x-5y=-11$
$2x+y=7$

c. $-3x-y=8$
$-3x+4y=13$

3. Solve each of the following equations using the elimination method.

a. $6x-5y=-43$
$6x-y=-23$

b. $x-4y=27$
$3x-4y=17$

c. $-4x+y=-10$
$4x-3y=14$

4. Solve each of the following equations using the elimination method.

a. $-5x+3y=3$
$-5x+y=-4$

b. $5x-5y=1$
$2x-5y=-5$

c. $4x-3y-1=0$
$4x+7y-11=0$

WE8 For questions 5 to 7, solve the following pairs of simultaneous equations.

5. a. $6x+y=9$
$-3x+2y=3$

b. $x+3y=14$
$3x+y=10$

c. $5x+y=27$
$4x+3y=26$

6. a. $-6x+5y=-14$
$-2x+y=-6$

b. $2x+5y=14$
$3x+y=-5$

c. $-3x+2y=6$
$x+4y=-9$

7. a. $3x-5y=7$
$x+y=-11$

b. $2x+3y=9$
$4x+y=-7$

c. $-x+5y=7$
$5x+5y=19$

8. WE9 Solve the following pairs of simultaneous equations.

a. $-4x+5y=-9$
$2x+3y=21$

b. $2x+5y=-6$
$3x+2y=2$

c. $2x-2y=-4$
$5x+4y=17$

9. Solve the following pairs of simultaneous equations.

a. $2x-3y=6$
$4x-5y=9$

b. $\frac{x}{2}+\frac{y}{3}=2$
$\frac{x}{4}+\frac{y}{3}=4$

c. $\frac{x}{3}+\frac{y}{2}=\frac{3}{2}$
$\frac{x}{2}+\frac{y}{5}=-\frac{1}{2}$

Understanding

For questions **10** to **12**, solve the following simultaneous equations using an appropriate method. Check your answer using technology.

10. a. $7x + 3y = 16$
$y = 4x - 1$

b. $2x + y = 8$
$4x + 3y = 16$

c. $-3x + 2y = 19$
$4x + 5y = 13$

11. a. $-3x + 7y = 9$
$4x - 3y = 7$

b. $-4x + 5y = -7$
$x = 23 - 3y$

c. $y = -x$
$y = -\frac{2}{5}x - \frac{1}{5}$

12. a. $4x + 5y = 41$
$y = \frac{3x}{2} - 1$

b. $3x - 2y = 9$
$2x + 5y = -13$

c. $\frac{x}{3} + \frac{y}{4} = 7$
$3y - 2x = 12$

Reasoning

13. The cost of a cup of coffee and croissant is \$8.50 from a local bakery, and an order of 5 coffees and three croissants costs \$35.70. Determine the cost of one croissant.

14. Celine notices that she only has 5 cent and 10 cent coins in her coin purse. She counts up how much she has and finds that from the 34 coins in the purse the total value is \$2.80. Determine how many of each type of coin she has.

15. Abena, Bashir and Cecily wanted to weigh themselves, but the scales they had were broken and would only give readings over 100 kg. They decided to weigh themselves in pairs and calculate their weights from the results.
- Abena and Bashir weighed 119 kg
- Bashir and Cecily weighed 112 kg
- Cecily and Abena weighed 115 kg

Determine the weight of each student.

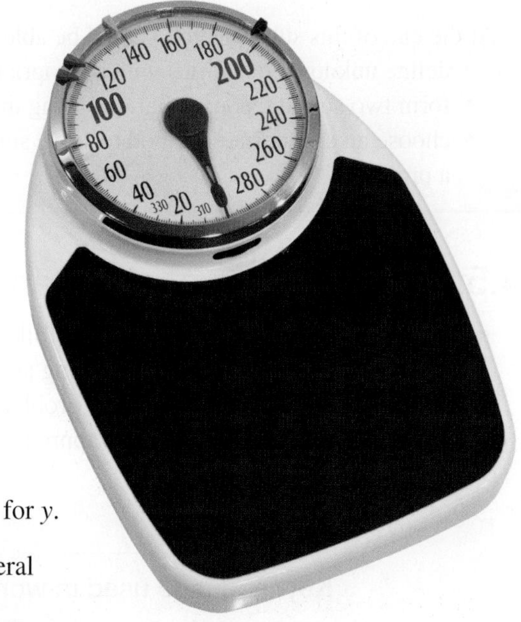

16. a. For the general case $ax + by = e$ [1]
$cx + dy = f$ [2]
y can be found by eliminating x.
i. Multiply equation [1] by c to create equation 3.
ii. Multiply equation [2] by a to create equation 4.
iii. Use the elimination method to find a general solution for y.

b. Use a similar process to that outlined above to find a general solution for x.

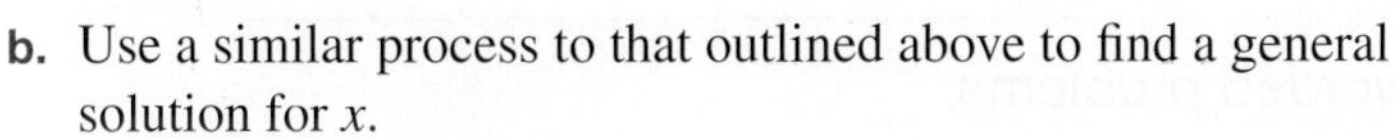

c. Use the general solution for x and y to solve each of the following.
i. $2x + 5y = 7$
$7x + 2y = 24$
ii. $3x - 5y = 4$
$x + 3y = 5$

Choose another method to check that your solutions are correct in each part.

d. For y to exist, it is necessary to state that $bc - ad \neq 0$. Explain.

e. Is there a necessary condition for x to exist? Explain.

17. A family of two parents and four children go to the movies and spend \$95 on the tickets. Another family of one parent and two children go to see the same movie and spend \$47.50 on the tickets. Determine if it possible to work out the cost of an adult's ticket and child's ticket from this information.

Problem solving

18. The sum of two numbers is equal to k. The difference of the two numbers is given by $k-20$. Determine the possible solutions for the two numbers.

19. Use the method of elimination to solve:

$$\frac{x-4}{3}+y=-2$$
$$\frac{2y-1}{7}+x=6$$

20. Use an appropriate method to solve:

$$\begin{aligned}2x+3y+3z&=-1\\3x-2y+z&=0\\z+2y&=0\end{aligned}$$

4.5 Applications of simultaneous linear equations

LEARNING INTENTION

At the end of this subtopic you should be able to:
- define unknown quantities with appropriate variables
- form two simultaneous equations using the information presented in a problem
- choose an appropriate method to solve simultaneous equations in order to find the solution to a problem.

4.5.1 Applications of simultaneous linear equations

eles-4770

- When solving practical problems, the following steps can be useful:
  1. Define the unknown quantities using appropriate pronumerals.
  2. Use the information given in the problem to form two equations in terms of these pronumerals.
  3. Solve these equations using an appropriate method.
  4. Write the solution in words.
  5. Check the solution.

Key language used in worded problems

To help set up equations from the information presented in a problem question, make sure you look out for the following key terms:
- **Addition:** sum, altogether, add, more than, and, in total
- **Subtraction:** difference, less than, take away, take off, fewer than
- **Multiplication:** product, groups of, times, of, for each, double, triple
- **Division:** quotient, split into, halve, thirds
- **Equals:** gives, is

WORKED EXAMPLE 10 Applying the elimination method

Ashley received better results for his Mathematics test than for his English test. If the sum of the two marks is 164 and the difference is 22, calculate the mark he received for each subject.

THINK	WRITE
1. Define the two variables.	Let $x =$ the Mathematics mark. Let $y =$ the English mark.
2. Formulate two equations from the information given and number them. The sum of the two marks is $x+y$. The difference of the two marks is $x-y$.	$x+y=164$ [1] $x-y=22$ [2]
3. Use the elimination method by adding equations [1] and [2] to eliminate y.	$[1]+[2]: 2x=186$
4. Solve for x by dividing both sides of the equation by 2.	$x=93$
5. Substitute the value of x into equation [1].	Substituting $x=93$ into [1]: $x+y=164$ $93+y=164$
6. Solve for y by subtracting 93 from both sides of the equation.	$y=71$
7. Write the solution.	Solution: Mathematics mark $(x)=93$ English mark $(y)=71$
8. Check the solution by substituting $x=93$ and $y=71$ into equation [1].	Check: Substitute into $x+y=164$ LHS $=93+71$ RHS $=164$ $=164$ As LHS = RHS, the solution is correct.

WORKED EXAMPLE 11 Applying the substitution method

To finish a project, Genevieve bought a total of 25 nuts and bolts from a hardware store. If each nut costs 12 cents, each bolt costs 25 cents and the total purchase price is \$4.30, calculate how many nuts and how many bolts Genevieve bought.

THINK	WRITE
1. Define the two variables.	Let $x =$ the number of nuts. Let $y =$ the number of bolts.
2. Formulate two equations from the information given and number them. *Note:* The total number of nuts and bolts is 25. Each nut cost 12 cents, each bolt cost 25 cents and the total cost is 430 cents (\$4.30).	$x+y=25$ [1] $12x+25y=430$ [2]

3. Solve simultaneously using the substitution method, since equation [1] is easy to rearrange. Rearrange equation [1] to make x the subject by subtracting y from both sides of equation [1].

Rearrange equation [1]:
$$x + y = 25$$
$$x = 25 - y$$

4. Substitute the expression $(25 - y)$ for x into equation [2].

Substituting $(25 - y)$ into [2]:
$$12\,(25 - y) + 25y = 430$$

5. Solve for y.

$$\begin{aligned} 300 - 12x + 25y &= 430 \\ 300 + 13y &= 430 \\ 13y + 300 &= 430 \\ 13y &= 130 \\ y &= 10 \end{aligned}$$

6. Substitute the value of y into the rearranged equation $x = 25 - y$ from step 3.

Substituting $y = 10$ into $x = 25 - y$
$$x = 25 - 10$$
$$x = 15$$

7. Write the solution.

Solution:
The number of nuts $(x) = 15$.
The number of bolts $(y) = 10$.

8. Check the solution by substituting $x = 15$ and $y = 10$ into equation [1].

Check: Substitute into $x + y = 25$.
$$\text{LHS} = 15 + 10 \qquad \text{RHS} = 25$$
$$= 25$$
As LHS = RHS, the solution is correct.

- It is also possible to determine solutions to worded problems using the graphical method by forming and then graphing equations.

WORKED EXAMPLE 12 Applying the graphical method

Cecilia buys 2 pairs of shorts and 3 T-shirts for $160. Ida buys 1 pair of shorts and 2 T-shirts for $90. Develop two equations to describe the situation and solve them graphically to determine the cost of one pair of shorts and one T-shirt.

THINK / **WRITE**

1. Define the two variables.

Let x = cost of a pair of shorts.
Let y = cost of a T-shirt.

2. Formulate two equations from the information given and number them.

$$2x + 3y = 160 \quad [1]$$
$$x + 2y = 90 \quad [2]$$

3. Calculate the x- and y-intercepts for both graphs.

Equation [1]
$$2x + 3y = 160$$
x-intercept, $y = 0$
$$\begin{aligned} 2x + 3 \times 0 &= 160 \\ 2x &= 160 \\ x &= 80 \end{aligned}$$
y-intercept, $x = 0$
$$\begin{aligned} 2 \times 0 + 3y &= 160 \\ 3y &= 160 \\ y &= 53\frac{1}{3} \end{aligned}$$

Equation [2]
$$x + 2y = 90$$
x-intercept, $y = 0$
$$\begin{aligned} x + 2 \times 0 &= 90 \\ x &= 90 \end{aligned}$$
y-intercept, $x = 0$
$$\begin{aligned} 0 + 2y &= 90 \\ 2y &= 90 \\ y &= 45 \end{aligned}$$

4. Graph the two lines either by hand or using technology. Only the first quadrant of the graph is required, as cost cannot be negative.

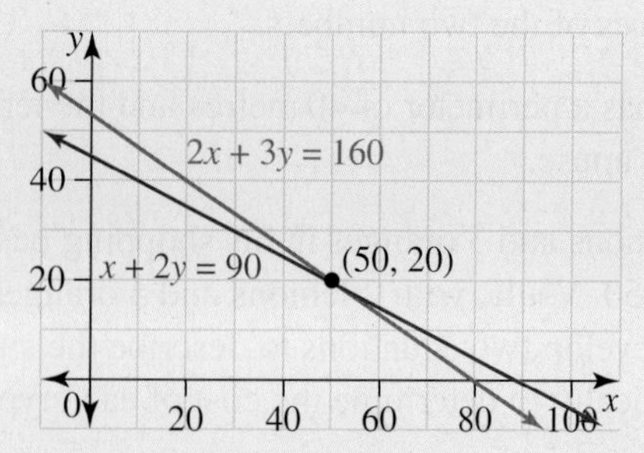

5. Identify the point of intersection to solve the simultaneous equations. — The point of intersection is (50, 20).

6. Write the answer as a sentence. — The cost of one pair of shorts is \$50 and the cost of one T-shirt is \$20.

DISCUSSION

How do you decide which method to use when solving problems using simultaneous linear equations?

on Resources

eWorkbook Topic 4 Workbook (worksheets, code puzzle and a project) (ewbk-2030)

Interactivity Individual pathway interactivity: Applications of simultaneous linear equations (int-4580)

Exercise 4.5 Applications of simultaneous linear equations

learnon

Individual pathways

■ PRACTISE	■ CONSOLIDATE	■ MASTER
1, 7, 8, 12, 13, 15, 17, 22	2, 5, 9, 10, 14, 18, 19, 23	3, 4, 6, 11, 16, 20, 21, 24, 25

To answer questions online and to receive **immediate corrective feedback** and **fully worked solutions** for all questions, go to your learnON title at www.jacplus.com.au.

Fluency

1. **WE10** Rick received better results for his Maths test than for his English test. If the sum of his two marks is 163 and the difference is 31, calculate the mark recieved for each subject.
2. **WE11** Rachael buys 30 nuts and bolts to finish a project. If each nut costs 10 cents, each bolt costs 20 cents and the total purchase price is \$4.20, how many nuts and how many bolts does she buy?
3. Eloise has a farm that raises chicken and sheep. Altogether there are 1200 animals on the farm. If the total number of legs from all the animals is 4000, calculate how many of each type of animal there is on the farm.

Understanding

4. Determine the two numbers whose difference is 5 and whose sum is 11.
5. The difference between two numbers is 2. If three times the larger number minus twice the smaller number is 13, determine the values of the two numbers.

6. One number is 9 less than three times a second number. If the first number plus twice the second number is 16, determine the values of the two numbers.

7. A rectangular house has a perimeter of 40 metres and the length is 4 metres more than the width. Calculate the dimensions of the house.

8. **WE12** Mike has 5 lemons and 3 oranges in his shopping basket. The cost of the fruit is \$3.50. Voula, with 2 lemons and 4 oranges, pays \$2.10 for her fruit. Develop two equations to describe the situation and solve them graphically to determine the cost of each type of fruit.

9. A surveyor measuring the dimensions of a block of land finds that the length of the block is three times the width. If the perimeter is 160 metres, calculate the dimensions of the block.

10. Julie has \$3.10 in change in her pocket. If she has only 50 cent and 20 cent pieces and the total number of coins is 11, calculate how many coins of each type she has.

11. Mr Yang's son has a total of twenty-one \$1 and \$2 coins in his moneybox. When he counts his money, he finds that its total value is \$30. Determine how many coins of each type he has.

12. If three Magnums and two Paddlepops cost \$8.70 and the difference in price between a Magnum and a Paddlepop is 90 cents, calculate how much each type of ice-cream costs.

13. If one Red Frog and four Killer Pythons cost \$1.65, whereas two Red Frogs and three Killer Pythons cost \$1.55, calculate how much each type of lolly costs.

14. A catering firm charges a fixed cost for overheads and a price per person. It is known that a party for 20 people costs \$557, whereas a party for 35 people costs \$909.50. Determine the fixed cost and the cost per person charged by the company.

15. The difference between Sally's PE mark and Science mark is 12, and the sum of the marks is 154. If the PE mark is the higher mark, calculate what mark Sally got for each subject.

16. Mozza's Cheese Supplies sells six Mozzarella cheeses and eight Swiss cheeses to Munga's deli for \$83.60, and four Mozzarella cheeses and four Swiss cheeses to Mina's deli for \$48. Calculate how much each type of cheese costs.

Reasoning

17. If the perimeter of the triangle in the diagram is 12 cm and the length of the rectangle is 1 cm more than the width, determine the value of x and y.

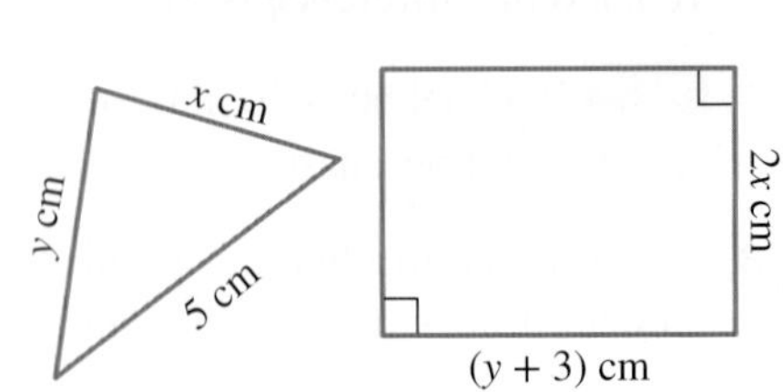

18. Mr and Mrs Waugh want to use a caterer for a birthday party for their twin sons. The manager says the cost for a family of four would be \$160. However, the sons want to invite 8 friends, making 12 people in all. The cost for this would be \$360. If the total cost in each case is made up of the same cost per person and the same fixed cost, calculate the cost per person and the fixed cost. Show your working.

19. Joel needs to buy some blank DVDs and USB sticks to back up a large amount of data that has been generated by an accounting firm. He buys 6 DVDs and 3 USB sticks for \$96. He later realises these are not sufficient and so buys another 5 DVDs and 4 USB sticks for \$116. Determine how much each DVD and each USB stick cost. (Assume the same rate per item was charged for each visit.) Show your working.

20. Four years ago Tim was 4 times older than his brother Matthew. In six years' time Tim will only be double his brother's age. Calculate how old the two brothers currently are.

21. A local cinema has different prices for movie tickets for children (under 12), adults and seniors (over 60). Consider the following scenarios:
    - For a senior couple (over 60) and their four grandchildren, the total cost is \$80.
    - For two families with four adults and seven children, the total cost is \$160.50.
    - For a son (under 12), his father and his grandfather (over 60), the total cost is \$45.75.

    Determine the cost of each type of ticket.

Problem solving

22. Reika completes a biathlon (swimming and running) that has a total distance of 37 km. Reika knows that her swimming speed is 3.2 km per hour and her running speed is 12.4 km per hour. If her total time for the race was 6 hours and 39 minutes, calculate the length of the swimming component of the race.

23. At the football hot chips are twice as popular as meat pies and three times as popular as hot dogs. Over the period of half an hour during half time, a fast-food outlet serves 121 people who each bought one item. Determine how many serves of each of the foods were sold during this half-hour period.

24. Three jet-skis in a 300 kilometres handicap race leave at two hour intervals. Jet-ski 1 leaves first and has an average speed of 25 kilometres per hour for the entire race. Jet-ski 2 leaves two hours later and has an average speed of 30 kilometres per hour for the entire race. Jet-ski 3 leaves last, two hours after jet-ski 2 and has an average speed of 40 kilometres per hour for the entire race.
    a. Sketch a graph to show each jet-ski's journey on the one set of axes.
    b. Determine who wins the race.
    c. Check your findings algebraically and describe what happened to each jet-ski during the course of the race.

25. Alice is competing in a cycling race on an extremely windy day. The race is an 'out and back again' course, so the wind is against Alice in one direction and assisting her in the other. For the first half of the race the wind is blowing against Alice, slowing her down by 4 km per hour. Given that on a normal day Alice could maintain a pace of 36 km per hour and that this race took her 4 hours and 57 minutes, calculate the total distance of the course.

4.6 Solving simultaneous linear and non-linear equations

LEARNING INTENTION

At the end of this subtopic you should be able to:
- determine the point or points of intersection between a linear equation and various non-linear equations using various techniques
- use digital technology to find the points of intersection between a linear equation and a non-linear equation.

4.6.1 Solving simultaneous linear and quadratic equations

eles-4771

- The graph of a quadratic function is called a parabola.
- A parabola and a straight line may:
  - intersect at only one point

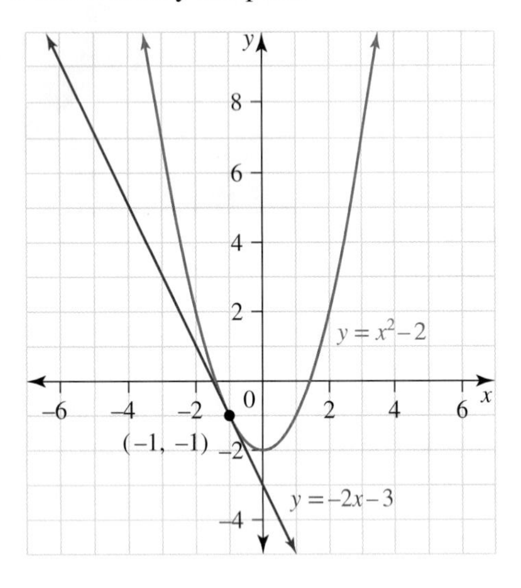

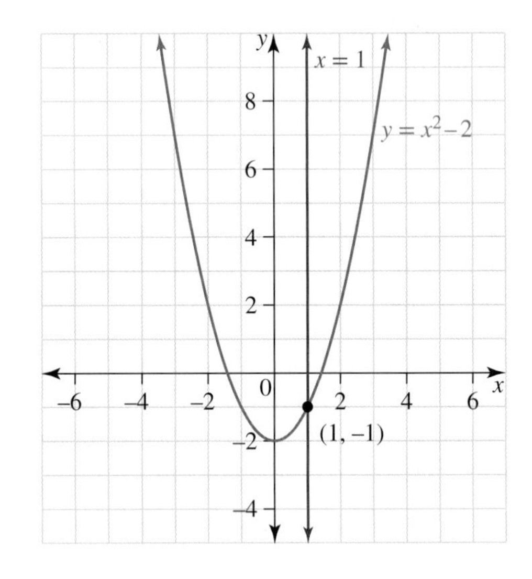

  - intersect at two points

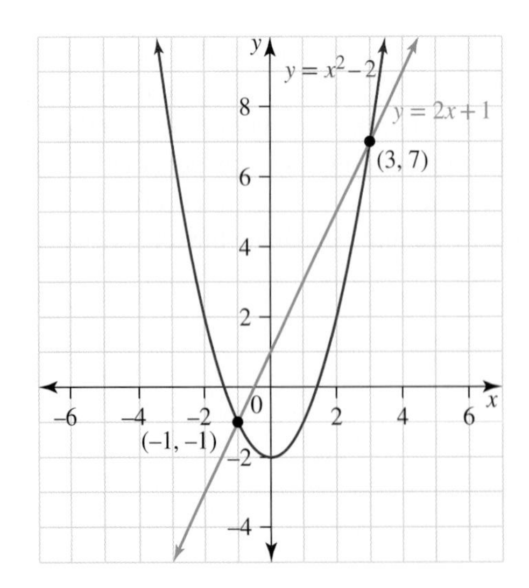

- not intersect at all.

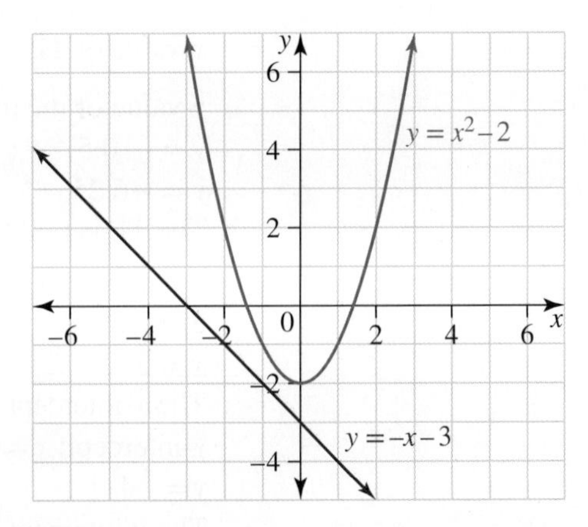

WORKED EXAMPLE 13 Solving linear and quadratic simultaneous equations

Determine the points of intersection of $y = x^2 + x - 6$ and $y = 2x - 4$:

a. algebraically

b. graphically.

THINK	WRITE/DRAW
a. 1. Number the equations. Equate [1] and [2].	**a.** $y = x^2 + x - 6$ [1] $y = 2x - 4$ [2] $x^2 + x - 6 = 2x - 4$
2. Collect all the terms on one side and simplify.	$x^2 + x - 6 - 2x + 4 = 2x - 4 - 2x + 4$ $x^2 + x - 6 - 2x + 4 = 0$ $x^2 - x - 2 = 0$
3. Factorise and solve the quadratic equation, using the Null Factor Law.	$(x - 2)(x + 1) = 0$ $x - 2 = 0$ or $x + 1 = 0$ $x = 2$ $x = -1$
4. Identify the y-coordinate for each point of intersection by substituting each x-value into one of the equations.	When $x = 2$, $y = 2(2) - 4$ $= 4 - 4$ $= 0$ Intersection point $(2, 0)$ When $x = -1$ $y = 2(-1) - 4$ $= -2 - 4$ $= -6$
5. Write the solution.	Intersection point $(-1, -6)$
b. 1. To sketch the graph of $y = x^2 + x - 6$, determine the x- and y-intercepts and the turning point (TP). The x-value of the TP is the average of the x-axis intercepts. The y-value of the TP is calculated by substituting the x-value into the equation of the parabola.	**b.** x-intercepts: $y = 0$ $0 = x^2 + x - 6$ $0 = (x + 3)(x - 2)$ $x = -3, x = 2$ The x-intercepts are $(-3, 0)$ and $(2, 0)$. y-intercept: $x = 0$ $y = -6$ The y-intercept is $(0, -6)$

x-value of TP: $\dfrac{-3+2}{2} = -0.5$

y-value of the turning point; when $x = -0.5$:

$y = (-0.5)^2 + (-0.5) - 6$

$y = -6.25$

The TP is $(-0.5, -6.25)$

2. To sketch the graph of $y = 2x - 4$, find the x- and y-intercepts.

x-intercept: $y = 0$

$0 = 2x - 4$

$x = 2$

The x-intercept is $(2, 0)$

y-intercept: $x = 0$

$y = -4$

The y-intercept is $(0, -4)$

3. On the same set of axes, sketch the graphs of $y = x^2 + x - 6$ and $y = 2x - 4$, labelling both.

$y = 2x - 4$

$(2, 0)$

$y = x^2 + x - 6$

$(-1, -6)$

4. On the graph, locate the points of intersection and write the solutions.

The points of intersection are $(2, 0)$ and $(-1, -6)$.

TI | THINK

a.

1. On a Calculator page, press:
   - MENU
   - 3: Algebra
   - 1: Solve

   Complete the entry line as:
   solve($y = x^2 + x - 6$ and $y = 2x - 4, \{x, y\}$)
   Then press ENTER.

DISPLAY/WRITE

a.

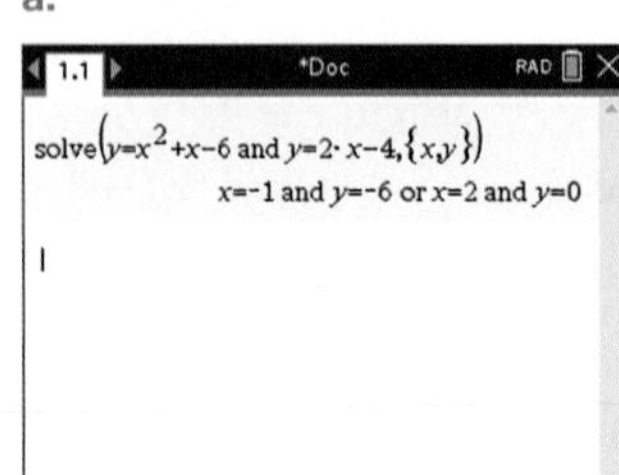

2. Write the solutions.

The points of intersection are $(-1, -6)$ and $(2, 0)$.

CASIO | THINK

a.

1. On a Main screen, complete the entry line as:
   solve $(x^2 + x - 6 = 2x - 4, x)$
   The x-values of the solutions will be shown.
   To determine the corresponding y-values, complete the entry lines as:
   $2x - 4|x = -1$
   $2x - 4|x = 2$
   Press EXE after each entry.

DISPLAY/WRITE

a.

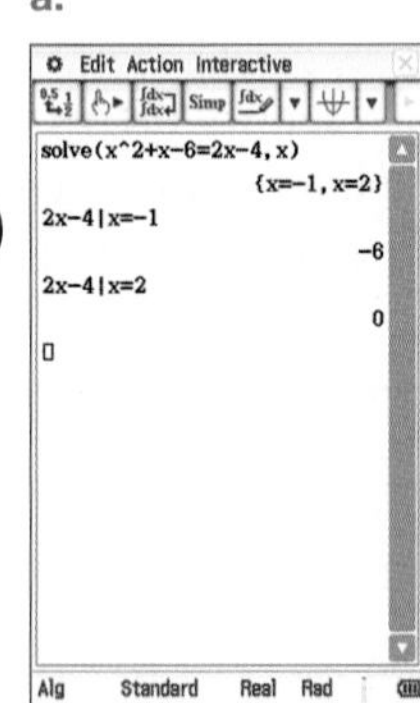

2. Write the solutions.

The points of intersection are $(-1, -6)$ and $(2, 0)$.

b.

1. On a Graphs page, complete the function entry lines as:
$f1\,(x) = x^2 + x - 6$
$f2\,(x) = 2x - 4$
Press the down arrow between entering the functions. The graphs will be displayed.

b.

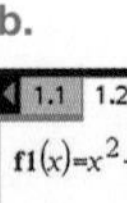

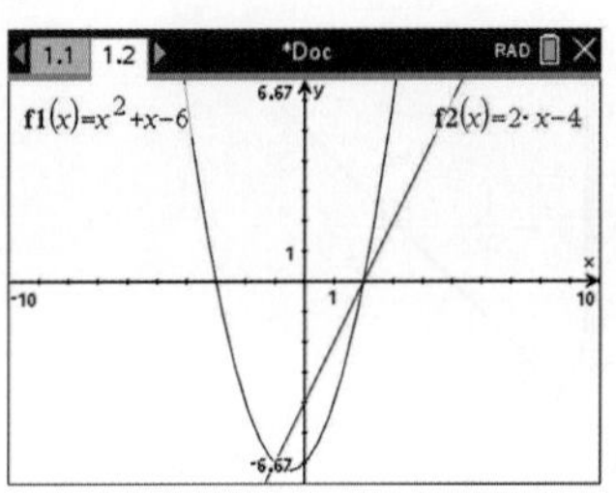

2. To locate the point of intersection, press:
   - MENU
   - 6: Analyze Graph
   - 4: Intersection

   Drag the dotted line to the left of the first point of intersection (the lower bound), press ENTER and then drag the dotted line to the right of the point of intersection (the upper bound) and press ENTER. Repeat for the second point.
   The points of intersection will be shown.

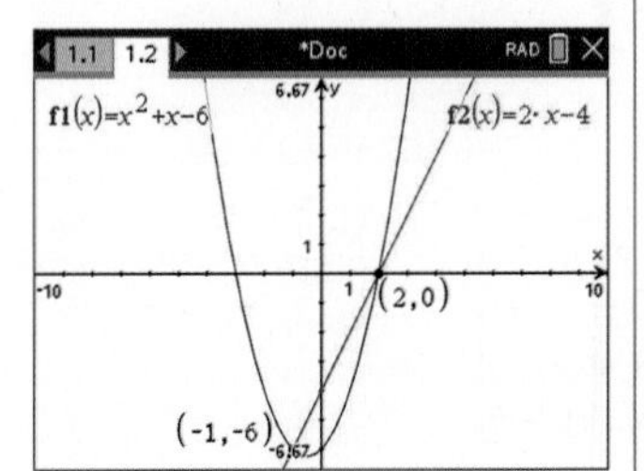

3. State the points of intersection.

The points of intersection are $(-1, -6)$ and $(2, 0)$

b.

1. On a Graph & Table screen, complete the function entry lines as:
$y1 = x^2 + x - 6$
$y2 = 2x - 4$
Then tap the graphing icon. The graphs will be displayed.

b.

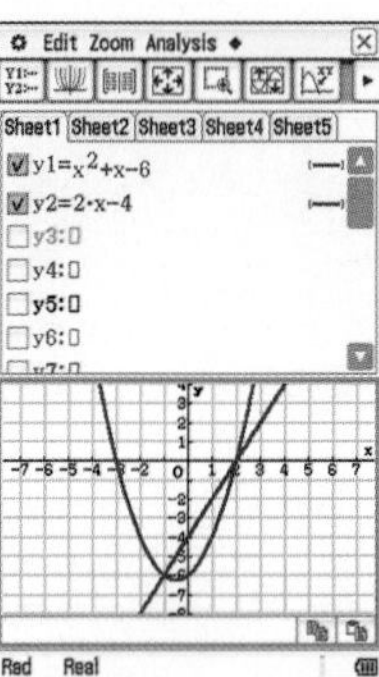

2. To locate the first point of intersection, press:
   - Analysis
   - G-Solve
   - Intersection

   The point of intersection will be shown. Tap the right arrow for the second point.

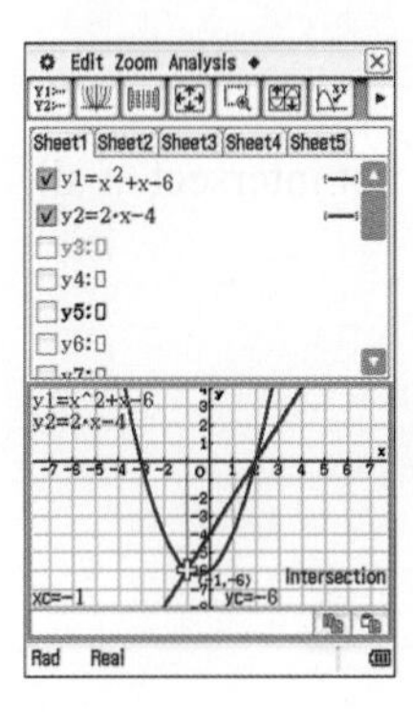

3. State the points of intersection.

The points of intersection are $(-1, -6)$ and $(2, 0)$

4.6.2 Solving simultaneous linear and hyperbolic equations

eles-4772

- A hyperbola and a straight line may:
  - intersect at only one point. In the first case, the line is a tangent to the curve.

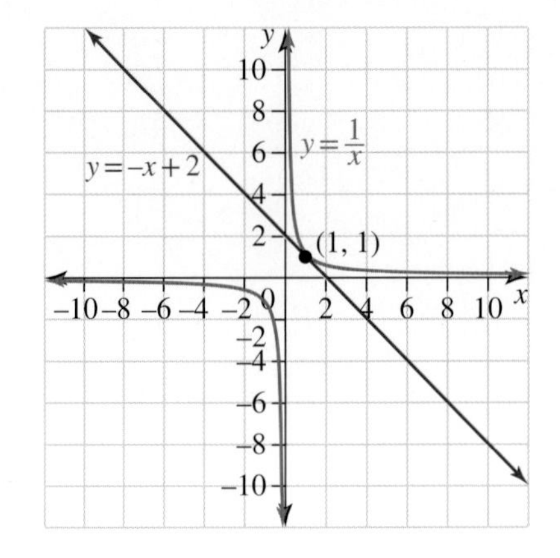

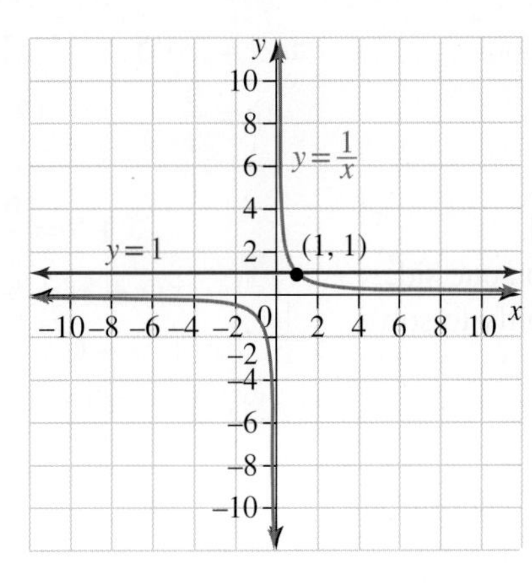

- intersect at two points

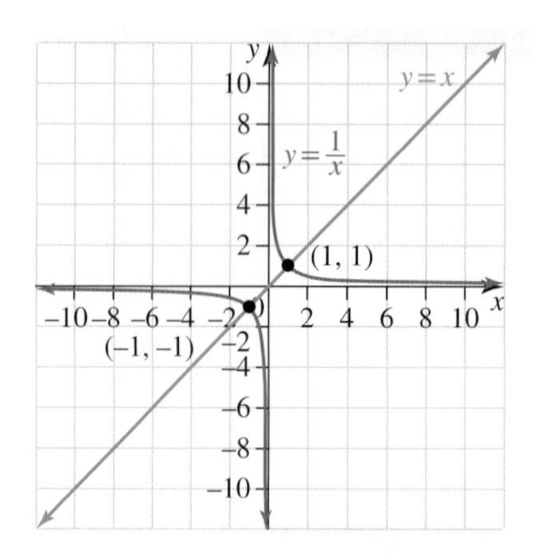

- not intersect at all.

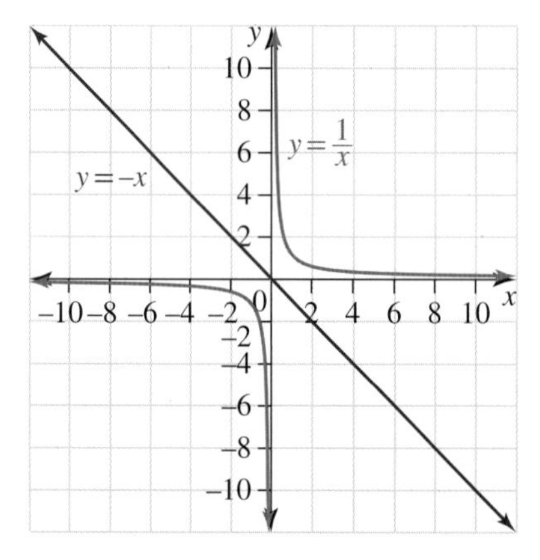

WORKED EXAMPLE 14 Solving linear and hyperbolic simultaneous equations

Determine the point(s) of intersection between $y = x + 5$ and $y = \frac{6}{x}$:

a. algebraically **b. graphically.**

THINK	WRITE/DRAW
a. 1. Number the equations.	**a.** $y = x + 5 \quad [1]$ $y = \frac{6}{x} \quad [2]$
2. Equate [1] and [2]. Collect all terms on one side, factorise and simplify to solve for x.	$x + 5 = \frac{6}{x}$ $x(x+5) = 6$ $x^2 + 5x - 6 = 0$ $(x+6)(x-1) = 0$ $x = -6, x = 1$
3. To determine the y-coordinates of the points of intersection, substitute the values of x into [1].	$x = -6 \qquad x = 1$ $y = -6 + 5 \qquad y = 1 + 5$ $y = -1 \qquad y = 6$

4. Write the solutions. | The points of intersection are $(-6, -1)$ and $(1, 6)$.

b. 1. To sketch the graph of $y = \frac{6}{x}$, draw a table of values.

b.

x	-6	-5	-4	-3	-2	-1	-0	1	2
y	-1	$-1\frac{1}{5}$	$-1\frac{1}{2}$	-2	-3	-6	Undef.	6	3

2. To sketch the graph of $y = x + 5$, find the x- and y-intercepts.

x-intercept: $y = 0$

$$0 = x + 5$$
$$x = -5$$

The x-intercept is $(-5, 0)$.

y-intercept: $x = 0$

$$y = 5$$

The y-intercept is $(0, 5)$.

3. On the same set of axes, sketch the graphs of $y = x + 5$ and $y = \frac{6}{x}$, labelling both.

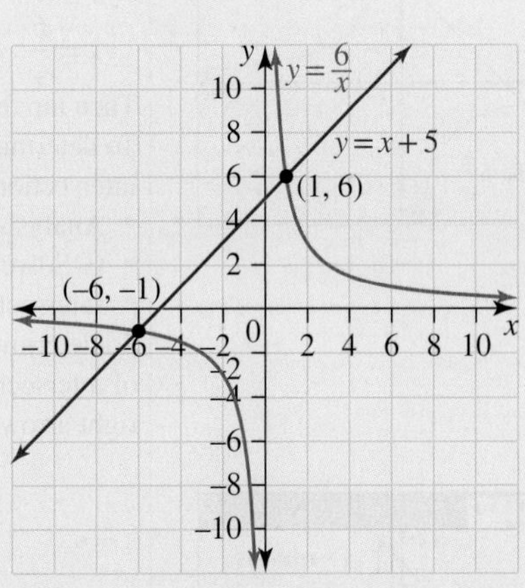

4. On the graph, locate the points of intersection and write the solutions.

The points of intersection are $(1, 6)$ and $(-6, -1)$.

TI | THINK

a.

1. In a new problem, on a Calculator page, press:
   - MENU
   - 1: Actions
   - 1: Define

   Complete the entry line as:
   Define $f1(x) = x + 5$
   Repeat for the second function:
   Define $f2(x) = \frac{6}{x}$
   Press ENTER after each entry.

DISLPAY/WRITE

a.

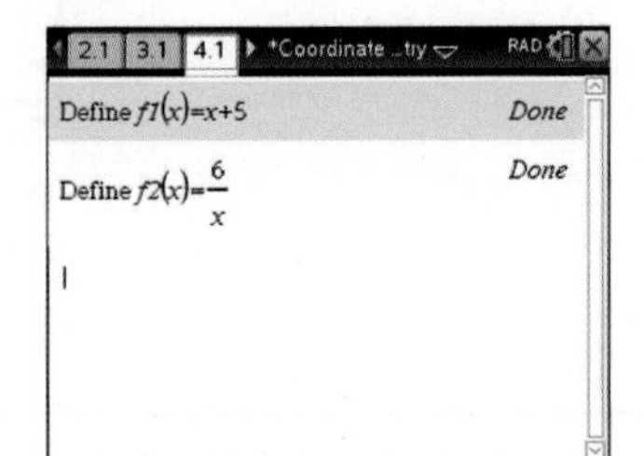

CASIO | THINK

a.

On the Main screen, tap:
- Action
- Advanced
- solve

Complete the entry lines as:

$\text{solve}\left(x + 5 = \frac{6}{x}, x\right)$

$x + 5|x = -6$

$x + 5|x = 1$

Press EXE after each entry.

DISPLAY/WRITE

a.

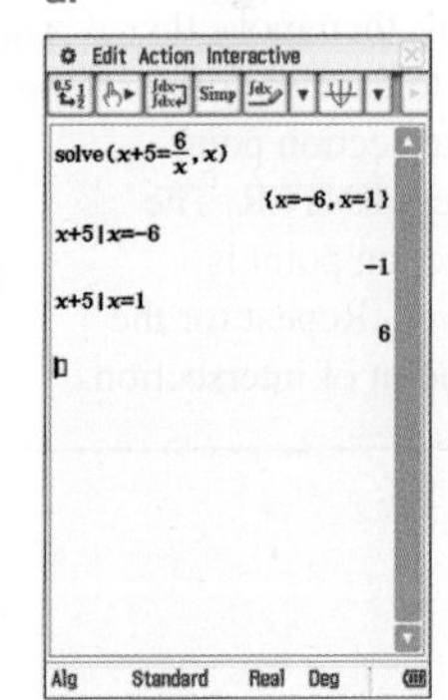

The points of intersection are $(-6, -1)$ and $(1, 6)$.

2. To determine the intersection points algebraically, press:
   - MENU
   - 3: Algebra
   - 1: Solve

   Complete the entry line as:
   solve $(f1(x) = f2(x), x)$
   $f1(-6)$
   $f2(1)$

Define $f1(x)=x+5$	Done
Define $f2(x)=\frac{6}{x}$	Done
solve$(f1(x)=f2(x),x)$	$x=-6$ or $x=1$
$f1(-6)$	-1
$f2(1)$	6

The points $(-6, -1)$ and $(1, 6)$ are the points of intersection.

b.

1. On a Graphs page, press the up arrow ▲ to select the function $f2(x)$, then press ENTER. The graph will be displayed. Now press TAB, select the function $f1(x)$ and press ENTER to draw the function. Apply colour if you would like to.

b.

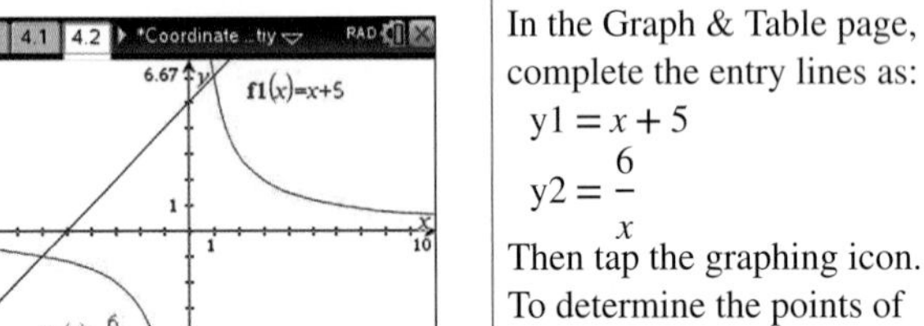

2. To determine the points of intersection between the two graphs, press:
   - MENU
   - 6: Analyze Graph
   - 4: Intersection

   Move the cursor to the left of one of the intersection points, press ENTER, then move the cursor to the right of this intersection point and press ENTER. The intersection point is displayed. Repeat for the other point of intersection.

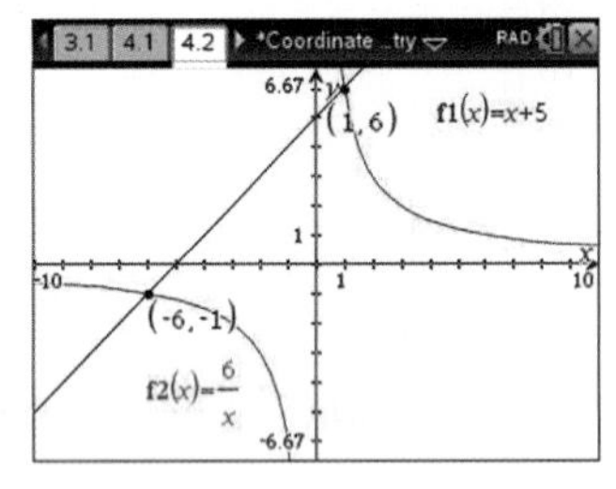

The points $(-6, -1)$ and $(1, 6)$ are the points of intersection.

b.

In the Graph & Table page, complete the entry lines as:

$y1 = x + 5$

$y2 = \frac{6}{x}$

Then tap the graphing icon. To determine the points of intersection, tap:
- Analysis
- G-Solve
- Intersection

To determine the next point of intersection, press the right arrow.

b.

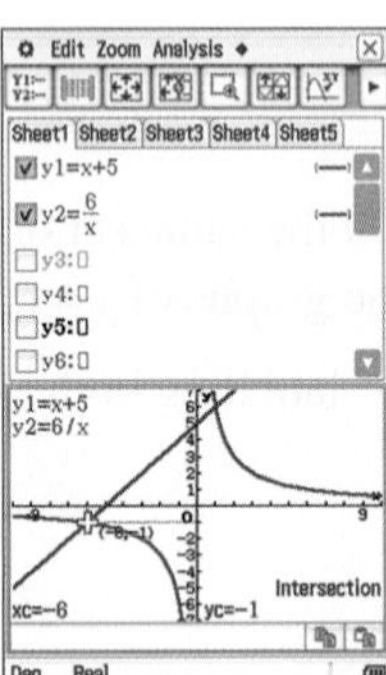

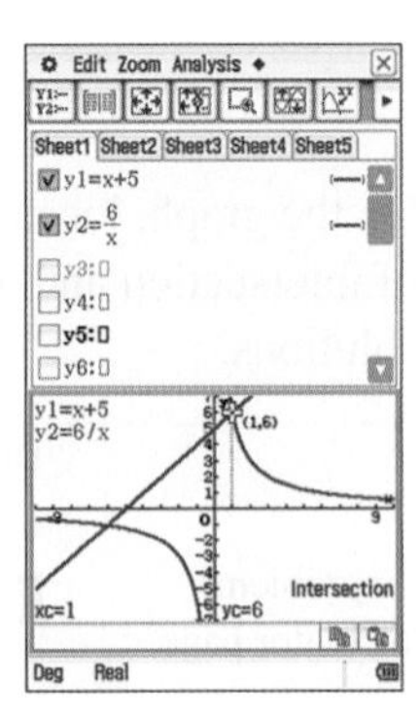

The points of intersection are $(-6, -1)$ and $(1, 6)$.

Solving simultaneous linear equations and circles

- A circle and a straight line may:
  - intersect at only one point. Here, the line is a tangent to the curve.
  - intersect at two points
  - not intersect at all.

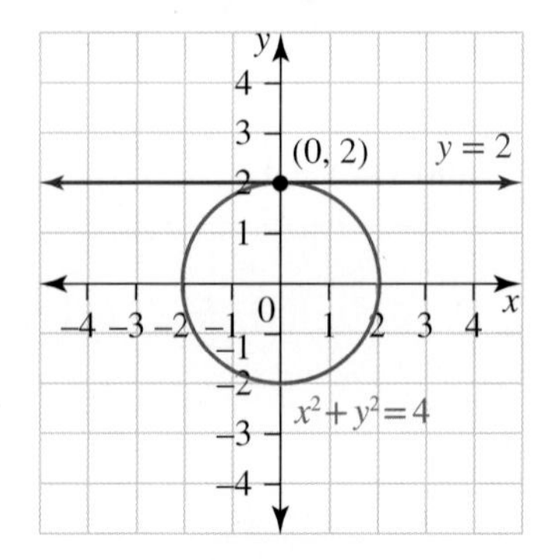

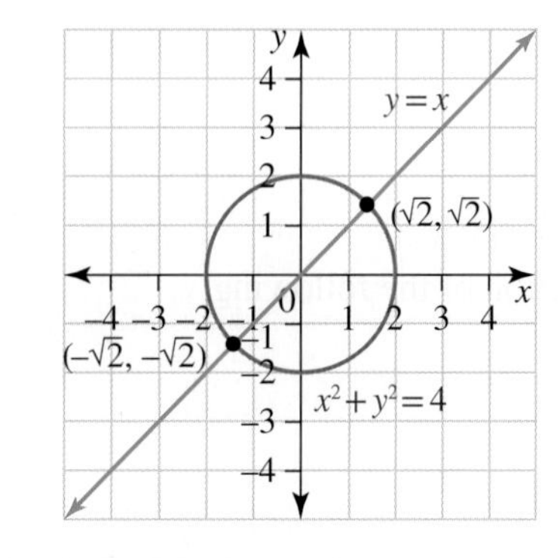

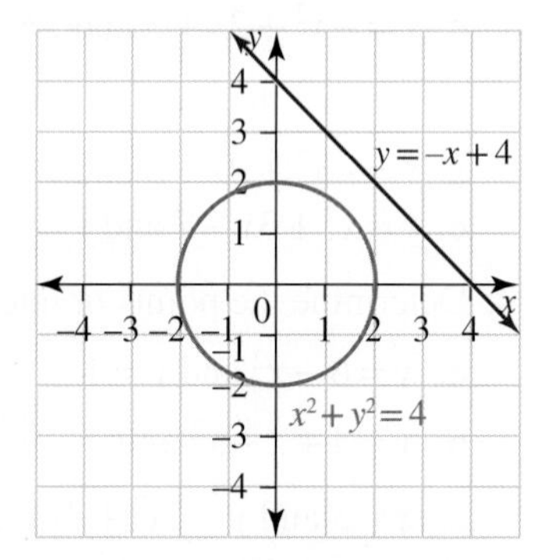

Solutions of a linear and non-linear equation

Depending on the equations, a linear equation and a non-linear equation can have a different number of solutions. For a linear equation and any of the following:

- quadratic equations
- hyperbolic equations
- circles

the number of possible solutions (points of intersections) is 0, 1 or 2.

DISCUSSION

What does it mean if a straight line touches a curve only once?

on Resources

eWorkbook Topic 4 Workbook (worksheets, code puzzle and a project) (ewbk-2030)

Interactivities Individual pathway interactivity: Solving simultaneous linear and non-linear equations (int-4581)
Solving simultaneous linear and non-linear equations (int-6128)

Exercise 4.6 Solving simultaneous linear and non-linear equations

learnon

Individual pathways

■ PRACTISE	■ CONSOLIDATE	■ MASTER
1, 2, 5, 8, 9, 12, 15	3, 6, 10, 13, 16, 17	4, 7, 11, 14, 18, 19, 20

To answer questions online and to receive **immediate corrective feedback** and **fully worked solutions** for all questions, go to your learnON title at www.jacplus.com.au.

Fluency

1. Describe how a parabola and straight line may intersect. Use diagrams to illustrate your explanation.

2. **WE13** Determine the points of intersection of the following:
   i. algebraically
   ii. algebraically using a calculator
   iii. graphically using a calculator.

   a. $y = x^2 + 5x + 4$ and $y = -x - 1$
   b. $y = -x^2 + 2x + 3$ and $y = -2x + 7$
   c. $y = -x^2 + 2x + 3$ and $y = -6$

3. Determine the points of intersection of the following.

   a. $y = -x^2 + 2x + 3$ and $y = 3x - 8$
   b. $y = -(x - 1)^2 + 2$ and $y = x - 1$
   c. $y = x^2 + 3x - 7$ and $y = 4x + 2$

4. Determine the points of intersection of the following.

   a. $y = 6 - x^2$ and $y = 4$
   b. $y = 4 + x - x^2$ and $y = \dfrac{3 - x}{2}$
   c. $x = 3$ and $y = 2x^2 + 7x - 2$

5. **MC** Identify which of the following graphs shows the parabola $y = x^2 + 3x + 2$, $x \in R$, and the straight line $y = x + 3$.

A.
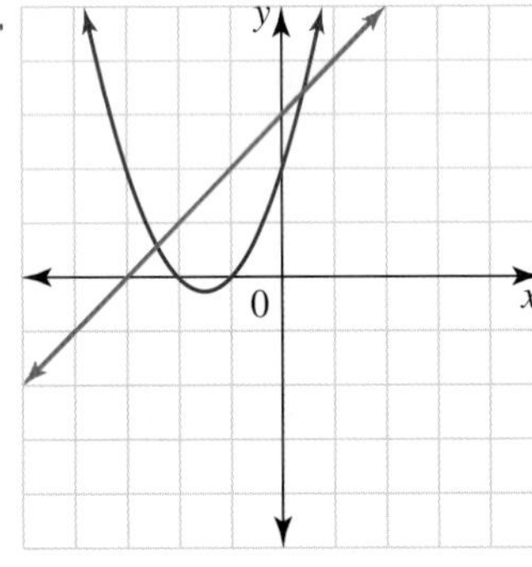

B.
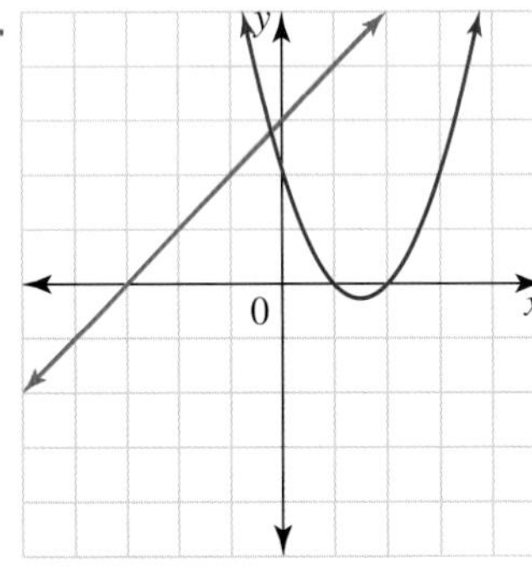

C.
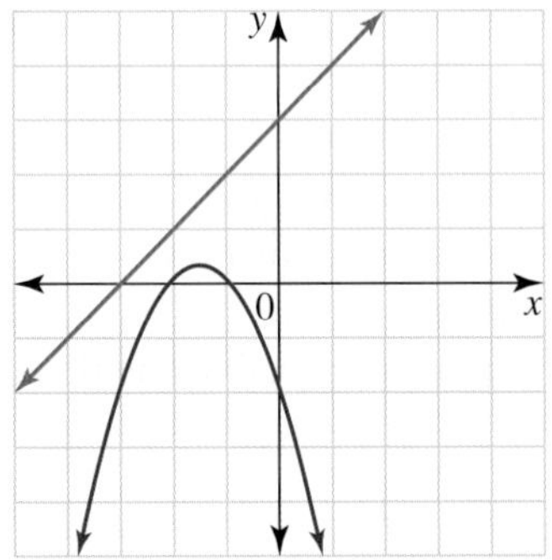

D.
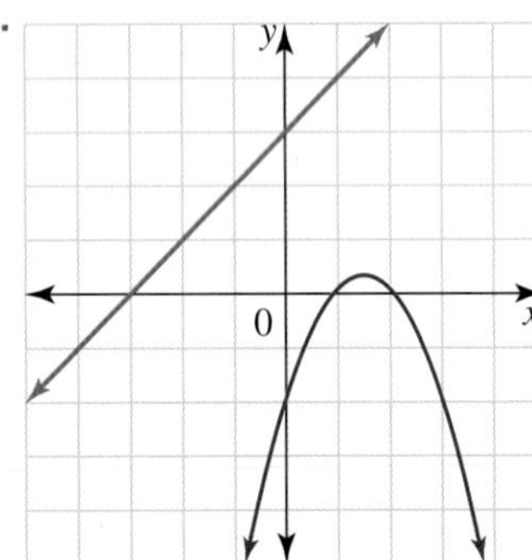

E.
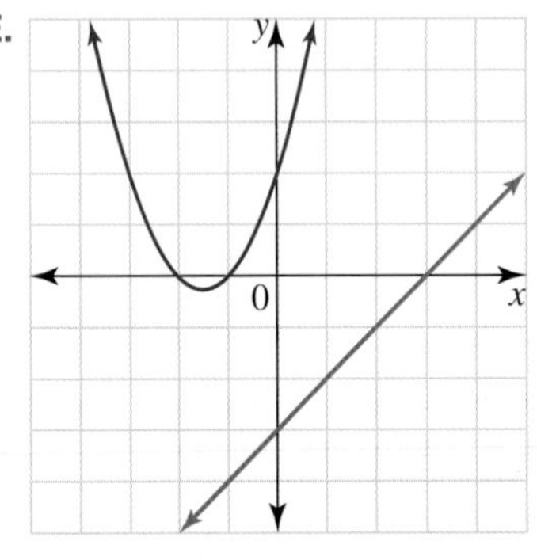

6. **MC** Identify which of the following equations are represented by the graph shown.

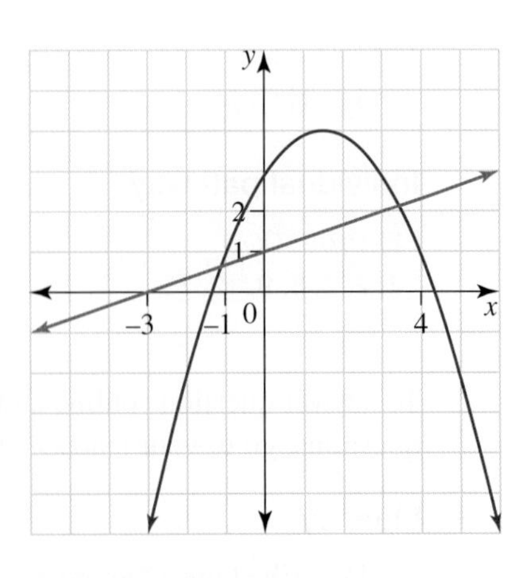

   A. $y = 0.5(x + 1.5)^2 + 4$ and $y = -\dfrac{1}{3}x + 1$
   B. $y = -0.5(x + 1.5)^2 - 4$ and $y = -\dfrac{1}{3}x + 1$
   C. $y = -0.5(x - 1.5)^2 + 4$ and $y = \dfrac{1}{3}x + 1$
   D. $y = 0.5(x - 1.5)^2 + 4$ and $y = -\dfrac{1}{3}x + 1$
   E. $y = 0.5(x - 1.5)^2 + 4$ and $y = -\dfrac{1}{3}x + 1$

7. Determine whether the following graphs intersect.

   a. $y = -x^2 + 3x + 4$ and $y = x - 4$
   b. $y = -x^2 + 3x + 4$ and $y = 2x + 5$
   c. $y = -(x + 1)^2 + 3$ and $y = -4x - 1$
   d. $y = (x - 1)^2 + 5$ and $y = -4x - 1$

Understanding

WE14 For questions 8 to 11, determine the point(s) of intersection between the following.

8. a. $y = x$, $y = \frac{1}{x}$
   b. $y = x - 2$, $y = \frac{1}{x}$
   c. $y = 3x$, $y = \frac{5}{x}$
   d. $y = \frac{6}{x}$, $y = \frac{x}{2} + 2$

9. a. $y = 3x$, $x^2 + y^2 = 10$
   b. $x^2 + y^2 = 25$, $3x + 4y = 0$
   c. $x^2 + y^2 = 50$, $y = 5 - 2x$
   d. $x^2 + y^2 = 9$, $y = 2 - x$

10. a. $y = \frac{1}{x}$, $y = 4x$
    b. $x^2 + y^2 = 25$, $y = -2x + 5$
    c. $y = 2x + 3$, $y = -4x^2 + 3$
    d. $3x + 4y = 7$, $y = \frac{10}{x} - 4$

11. a. $y = x^2$, $y = 2x - 1$
    b. $x^2 + (y + 1)^2 = 25$, $y = 3$
    c. $y = -4x - 5$, $y = x^2 + 2x + 3$
    d. $\frac{x}{3} + \frac{y}{4} = 7$, $y = \frac{x^2}{16} + 3$

Reasoning

12. Consider the following equations: $y = \frac{1}{2}(x - 3)^2 + \frac{5}{2}$ and $y = x + k$.
    Identify for what values of k the two lines would have:
    a. no points of intersection
    b. one point of intersection
    c. two points of intersection.

13. Show that there is at least one point of intersection between the parabola $y = -2(x + 1)^2 - 5$, where $y = f(x)$, and the straight line $y = mx - 7$, where $y = f(x)$.

14. a. Using technology, sketch the following graphs and state how many ways a straight line could intersect with the equation.
       i. $y = x^3 - 4x$.
       ii. $y = x^4 - 8x^2 + 16$.
       iii. $y = x^5 - 8x^3 + 16x$.
    b. Comment on the connection between the highest power of x and the number of possible points of intersection.

Problem solving

15. If two consecutive numbers have a product of 306, calculate the possible values for these numbers.

16. The perimeter of a rectangular paddock is 200 m and the area is 1275 m^2. Determine the length and width of the paddock.

17. a. Determine the point(s) of intersection between the circle $x^2 + y^2 = 50$ and the linear equation $y = 2x - 5$.
    b. Confirm your solution to part a by plotting the equation of the circle and the linear equation on the same graph.

18. The sum of two positive numbers is 21. Twice the square of the larger number minus three times the square of the smaller number is 45. Determine the value of the two numbers.

19. a. Omar is running laps around a circular park with equation $x^2 + y^2 = 32$. Chae-won is running along another track where the path is given by $y = \left(\sqrt{2} - 1\right)x + \left(8 - 4\sqrt{2}\right)$. Determine the point(s) where the two paths intersect.
b. Omar and Chae-won both start from the same point. If Chae-won gets between the two points in two hours, calculate the possible speeds Omar could run at along his circle in order to collide with Chae-won at the other point of intersection. Assume all distances are in kilometres and give your answer to 2 decimal places.

20. Adam and Eve are trying to model the temperature of a cup of coffee as it cools.
- Adam's model: Temperature (°C) $= 100 - 5 \times \text{time}$
- Eve's model: Temperature (°C) $= \dfrac{800}{10 + \text{time}} + 20$

Time is measured in minutes.
a. Using either model, identify the initial temperature of the cup of coffee.
b. Determine at what times the two models predict the same temperature for the cup of coffee.
c. Evaluate whose model is more realistic. Justify your answer.

4.7 Solving linear inequalities

LEARNING INTENTION

At the end of this subtopic you should be able to:
- solve an inequality and represent the solution on a number line
- convert a worded statement to an inequality in order to solve a problem.

4.7.1 Inequalities between two expressions

eles-4774

- An equation, such as $y = 2x$, is a statement of *equality* as both sides are equal to each other.
- An **inequation**, such as $y < x + 3$, is a statement of **inequality** between two expressions.
- A linear equation such as $3x = 6$ will have a unique solution ($x = 2$), whereas an inequation such as $3x < 6$ will have an infinite number of solutions ($x = 1, 0, -1, -2 \ldots$ are all solutions).
- We use a number line to represent all possible solutions to a linear inequation. When representing an inequality on a number line, an open circle is used to represent that a value is not included, while a closed circle is used to indicate that a number is included.
- The table below shows four basic inequalities and their representation on a number line.

Mathematical statement	Worded statement	Number line diagram
$x > 2$	x is greater than 2	−10 −8 −6 −4 −2 0 2 4 6 8 10
$x \geq 2$	x is greater than or equal to 2	−10 −8 −6 −4 −2 0 2 4 6 8 10
$x < 2$	x is less than 2	−10 −8 −6 −4 −2 0 2 4 6 8 10
$x \leq 2$	x is less than or equal to 2	−10 −8 −6 −4 −2 0 2 4 6 8 10

Solving inequalities

- The following operations may be done to both sides of an inequality without affecting its truth.
  - A number can be added or subtracted from both sides of the inequality.

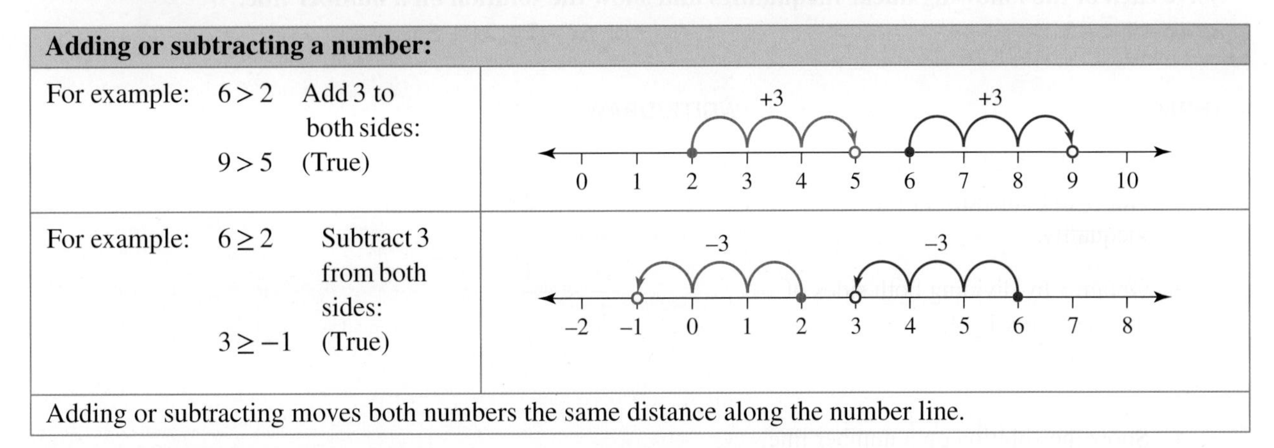

Adding or subtracting a number:	
For example: $6 > 2$ Add 3 to both sides: $9 > 5$ (True)	
For example: $6 \geq 2$ Subtract 3 from both sides: $3 \geq -1$ (True)	
Adding or subtracting moves both numbers the same distance along the number line.	

- A number can be multiplied or divided by a positive number.

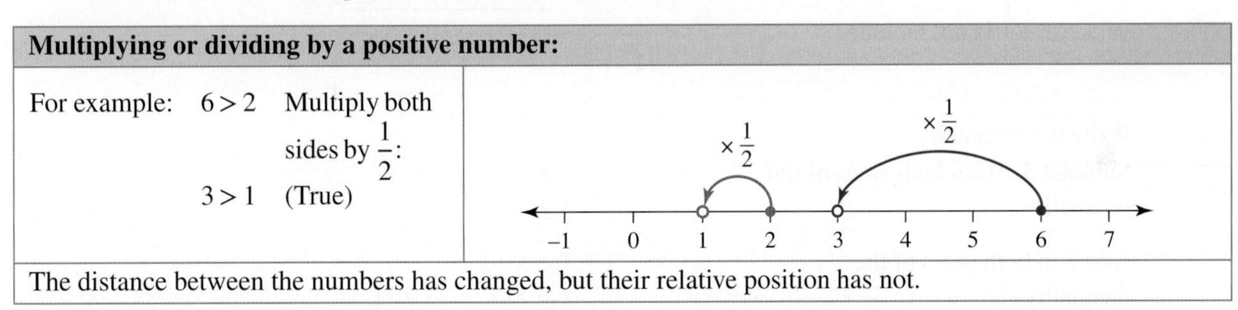

Multiplying or dividing by a positive number:	
For example: $6 > 2$ Multiply both sides by $\frac{1}{2}$: $3 > 1$ (True)	
The distance between the numbers has changed, but their relative position has not.	

- Care must be taken when multiplying or dividing by a negative number.

Multiplying or dividing by a negative number:	
For example: $6 > 2$ Multiply both sides by -1: $-6 > 2$ (False)	$\times -1$ $\times -1$ –6 –5 –4 –3 –2 –1 0 1 2 3 4 5 6
Multiplying or dividing by a negative number reflects numbers about $x = 0$. Their relative positions are reversed.	

- When solving inequalities, if both sides are multiplied or divided by a negative number, then the inequality sign must be reversed.
  For example, $6 > 2$ implies that $-6 < -2$.

Solving a linear inequality

Solving a linear inequality is a similar process to solving a standard linear equation. We can perform the following inverse operations as normal:

- a number or term can be added to or subtracted from each side of the inequality
- each side of an inequation can be multiplied or divided by a positive number.

We must take care to **change the direction of the inequality sign** when:

- each side of an inequation is to be multiplied or divided by a **negative number**.

WORKED EXAMPLE 15 Solving linear inequalities

Solve each of the following linear inequalities and show the solution on a number line.

a. $4x - 1 < -2$

b. $6x - 7 \geq 3x + 5$

THINK / **WRITE/DRAW**

a. 1. Write the inequality.
a. $4x - 1 < -2$

2. Add 1 to both sides of the inequality.
$4x - 1 + 1 < -2 + 1$
$4x < -1$

3. Obtain x by dividing both sides of the inequality by 4.
$\frac{4x}{4} < -\frac{1}{4}$
$x < -\frac{1}{4}$

4. Show the solution on a number line. Use an open circle to show that the value of $-\frac{1}{4}$ is not included.

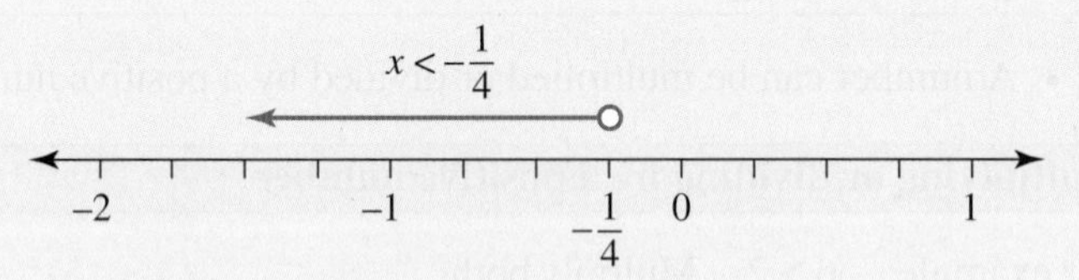

b. 1. Write the inequality.
b. $6x - 7 \geq 3x + 5$

2. Subtract $3x$ from both sides of the inequality.
$6x - 7 - 3x \geq 3x + 5 - 3x$
$3x - 7 \geq 5$

3. Add 7 to both sides of the inequality.
$3x - 7 + 7 \geq 5 + 7$
$3x \geq 12$

4. Obtain x by dividing both sides of the inequality by 3.
$3x \geq 12$
$\frac{3x}{3} \geq \frac{12}{3}$

5. Show the solution on a number line. Use a closed circle to show that the value of 4 is included.

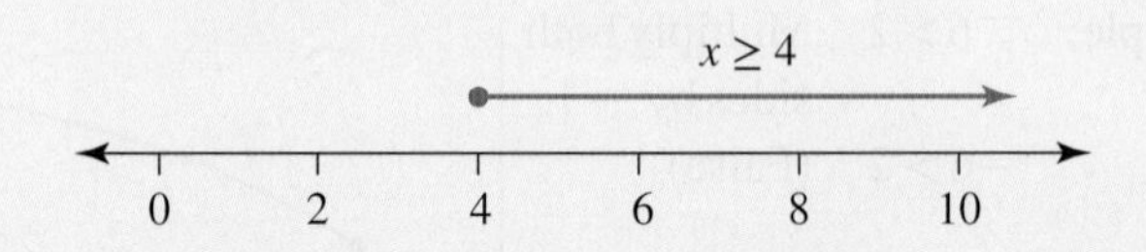

WORKED EXAMPLE 16 Solving complex linear inequalities

Solve each of the following linear inequalities.

a. $-3m + 5 < -7$

b. $5(x - 2) \geq 7(x + 3)$

THINK / **WRITE**

a. 1. Write the inequality.
a. $-3m + 5 < -7$

2. Subtract 5 from both sides of the inequality. (No change to the inequality sign.)
$-3m + 5 - 5 < -7 - 5$
$-3m < -12$

3. Obtain m by dividing both sides of the inequation by –3. Reverse the inequality sign, since we are dividing by a negative number.
$\frac{-3m}{-3} > \frac{-12}{-3}$
$m > 4$

b. 1. Write the inequality.	**b.** $5(x-2) \geq 7(x+3)$
2. Expand both brackets.	$5x - 10 \geq 7x + 21$
3. Subtract $7x$ from both sides of the inequality.	$5x - 10 - 7x \geq 7x + 21 - 7x$ $-2x - 10 \geq 21$
4. Add 10 to both sides of the inequation.	$-2x - 10 + 10 \geq 21 + 10$ $-2x \geq 31$
5. Divide both sides of the inequality by –2. Reverse the direction of the inequality sign as we are dividing by a negative number.	$\frac{-2x}{-2} \leq \frac{31}{-2}$ $x \leq \frac{-31}{2}$ $x \leq -15\frac{1}{2}$

DISCUSSION

What is are the similarities and differences when solving linear inequations compared to linear equations?

on Resources

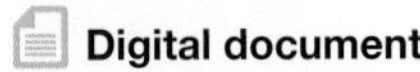

eWorkbook Topic 4 Workbook (worksheets, code puzzle and a project) (ewbk-2030)

Digital documents SkillSHEET Checking whether a given point makes the inequation a true statement (doc-5218)
SkillSHEET Writing equations from worded statements (doc-5219)

Interactivities Individual pathway interactivity: Solving linear inequalities (int-4582)
Inequalities on the number line (int-6129)

Exercise 4.7 Solving linear inequalities

learnon

Individual pathways

■ PRACTISE	■ CONSOLIDATE	■ MASTER
1, 4, 7, 10, 15, 18, 19, 22, 25, 28, 31	2, 5, 8, 11, 13, 16, 20, 23, 26, 29, 32	3, 6, 9, 12, 14, 17, 21, 24, 27, 30, 33

To answer questions online and to receive **immediate corrective feedback** and **fully worked solutions** for all questions, go to your learnON title at www.jacplus.com.au.

Fluency

WE15a For questions 1 to 3, solve each of the following inequalities and show the solution on a number line.

1. a. $x + 1 > 3$ b. $a + 2 > 1$ c. $y - 3 \geq 4$ d. $m - 1 \geq 3$
2. a. $p + 4 < 5$ b. $x + 2 < 9$ c. $m - 5 \leq 4$ d. $a - 2 \leq 5$
3. a. $x - 4 > -1$ b. $5 + m \geq 7$ c. $6 + q \geq 2$ d. $5 + a > -3$

For questions 4 to 6, solve each of the following inequalities. Check your solutions by substitution.

4. a. $3m > 9$ b. $5p \leq 10$ c. $2a < 8$ d. $4x \geq 20$

5. a. $5p > -25$ b. $3x \leq -21$ c. $2m \geq -1$ d. $4b > -2$

6. a. $\frac{m}{3} > 6$ b. $\frac{x}{2} < 4$ c. $\frac{a}{7} \leq -2$ d. $\frac{m}{5} \geq 5$

For questions 7 to 9, solve each of the following inequalities.

7. a. $2m + 3 < 12$ b. $3x + 4 \geq 13$ c. $5p - 9 > 11$ d. $4n - 1 \leq 7$

8. a. $2b - 6 < 4$ b. $8y - 2 > 14$ c. $10m + 4 \leq -6$ d. $2a + 5 \geq -5$

9. a. $3b + 2 < -11$ b. $6c + 7 \leq 1$ c. $4p - 2 > -10$ d. $3a - 7 \geq -28$

WE15b For questions 10 to 14, solve each of the following linear inequalities and show the solution on a number line.

10. a. $2m + 1 > m + 4$ b. $2a - 3 \geq a - 1$ c. $5a - 3 < a - 7$ d. $3a + 4 \leq a - 2$

11. a. $5x - 2 > 40 - 2x$ b. $7x - 5 \leq 11 - x$ c. $7b + 5 < 2b + 25$ d. $2(a + 4) > a + 13$

12. a. $3(m - 1) < m + 1$ b. $5(2m - 3) \leq 3m + 6$ c. $3(5b + 2) \leq -10 + 4b$ d. $5(3m + 1) \geq 2(m + 9)$

13. a. $\frac{x+1}{2} \leq 4$ b. $\frac{x-2}{5} \geq -4$ c. $\frac{x+7}{3} < -1$

14. a. $\frac{2x+3}{4} > 6$ b. $\frac{3x-1}{7} \geq 2$ c. $\frac{5x+9}{6} < 0$

WE16 For questions 15 to 17, solve each of the following inequalities.

15. a. $-2m > 4$ b. $-5p \leq 15$ c. $-2a \geq -10$ d. $-p - 3 \leq 2$ e. $10 - y \geq 13$

16. a. $14 - x < 7$ b. $1 - 6p > 1$ c. $2 - 10a \leq 0$ d. $2(3 - x) < 12$ e. $-4(a + 9) \geq 8$

17. a. $-15 \leq -3(2 + b)$ b. $2x - 3 > 5x + 6$ c. $k + 5 < 2k - 3$ d. $3(x - 4) < 5(x + 5)$ e. $7(a + 4) \geq 4(2a - 3)$

18. **MC** When solving the inequality $-2x > -7$ we need to:

A. change the sign to $\geq$
B. change the sign to $<$
C. change the sign to $=$
D. change the sign to $\leq$
E. keep the sign unchanged

For questions 19 to 24, solve each of the following inequalities.

19. a. $\frac{2-x}{3} > 1$ b. $\frac{5-m}{4} \geq 2$

20. a. $\frac{-3-x}{5} < -4$ b. $\frac{3-8a}{2} < -1$

21. a. $\frac{4-3m}{2} \leq 0$ b. $\frac{-2m+6}{10} \leq 3$

22. a. $3k > 6$ b. $-a - 7 < -2$ c. $5 - 3m \geq 0$ d. $x + 4 > 9$

23. a. $10 - y \leq 3$ b. $5 + 3d < -1$ c. $\frac{7p}{3} \geq -2$ d. $\frac{1 - x}{3} \leq 2$

24. a. $\frac{-4 - 2m}{5} > 0$ b. $5a - 2 < 4a + 7$ c. $6p + 2 \leq 7p - 1$ d. $2(3x + 1) > 2x - 16$

Understanding

25. Write linear inequalities for the following statements, using x to represent the unknown. (Do not attempt to solve the equations.)
   a. The product of 5 and a certain number is greater than 10.
   b. When three is subtracted from a certain number the result is less than or equal to 5.
   c. The sum of seven and three times a certain number is less than 42.

26. Write linear inequalities for the following statements. Choose an appropriate letter to represent the unknown.
   a. Four more than triple a number is more than 19.
   b. Double the sum of six and a number is less than 10.
   c. Seven less the half the difference between a number and 8 is at least 9.

27. Write linear inequalities for the following situations. Choose an appropriate letter to represent the unknown.
   a. John makes \$50 profit for each television he sells. Determine how many televisions John needs to sell to make at least \$650 in profit.
   b. Determine what distances a person can travel with \$60 if the cost of a taxi ride is \$2.50 per km with a flagfall cost of \$5.

Reasoning

28. Tom is the youngest of 5 children. The five children were all born 1 year apart. If the sum of their ages is at most 150, set up an inequality and solve it to find the possible ages of Tom.

29. Given the positive numbers a, b, c and d and the variable x, there is the following relationship:

$$-c < ax + b < -d.$$

   a. Determine the possible range of values of x if $a = 2$, $b = 3$, $c = 10$ and $d = 1$.
   b. Rewrite the original relationship in terms of x only (x by itself between the $<$ signs), using a, b, c and d.

30. Two speed boats are racing along a section of Lake Quikalong. The speed limit along this section of the lake is 50 km/h. Ella is travelling 6 km/h faster than Steven and the sum of the speeds at which they are travelling is greater than 100 km/h.

   a. Write an inequation and solve it to describe all possible speeds that Steven could be travelling at.
   b. At Steven's lowest possible speed, is he over the speed limit?
   c. The water police issue a warning to Ella for exceeding the speed limit on the lake. Show that the police were justified in issuing a warning to Ella.

Problem solving

31. Mick the painter has fixed costs (e.g. insurance, equipment, etc) of \$3400 per year. His running cost to travel to jobs is based on \$0.75 per kilometre. Last year Mick had costs that were less than \$16 000.
    a. Write an inequality and solve it to find how many kilometres Mick travelled for the year.
    b. Explain the information you have found.

32. A coffee store produces doughnuts and croissants to sell alongside its coffee. Each morning the bakery has to decide how many of each it will produce. The store has 240 minutes to produce food in the morning.
    It takes 20 minutes to make a batch of doughnuts and 10 minutes to make a batch of croissants. The store also has 36 kg of flour to use each day. A batch of doughnuts uses 2 kg of flour and a batch of croissants require 2 kg of flour.
    a. Set up an inequality around the amount of time available to produce doughnuts and croissants.
    b. Set up an inequality around the amount of flour available to produce doughnuts and croissants.
    c. Use technology to work out the possible number of each that can be made, taking into account both inequalities.

33. I have \$40 000 to invest. Part of this I intend to invest in a stable 5% simple interest account. The remainder will be invested in my friend's IT business. She has said that she will pay me 7.5% interest on any money I give to her. I am saving for a European trip so want the best return for my money. Calculate the least amount of money I should invest with my friend so that I receive at least \$2500 interest per year from my investments.

4.8 Inequalities on the Cartesian plane

LEARNING INTENTION

At the end of this subtopic you should be able to:
- sketch the graph of a half plane: the region represented by an inequality
- sketch inequalities using digital technology.

4.8.1 Inequalities on the Cartesian plane

eles-4776

- A solution to a linear inequality is any ordered pair (coordinate) that makes the inequality true.
- There is an infinite number of points that can satisfy an inequality. If we consider the inequality $x + y < 10$, the following points $(1, 7)$, $(5, 2)$ and $(4, 3)$ are all solutions, whereas $(6, 8)$ is not as it does not satisfy the inequality ($6 + 8$ is not less than 10).
- These points that satisfy an inequality are represented by a region that is found on one side of a line and is called a **half plane**.
- To indicate whether the points on a line satisfy the inequality, a specific type of **boundary line** is used.

Points on the line	Symbol	Type of boundary line used
Do not satisfy the inequality	$<$ or $>$	Dashed - - - - - - - -
Satisfy the inequality	$\le$ or $\ge$	Solid ————

- The **required region** is the region that contains the points that satisfy the inequality.
- Shading or no shading is used to indicate which side of the line is the required region, and a key is shown to indicate the region.

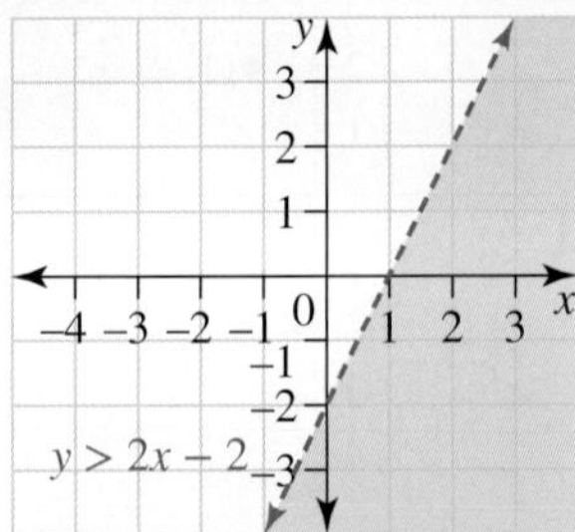

The required region is □.

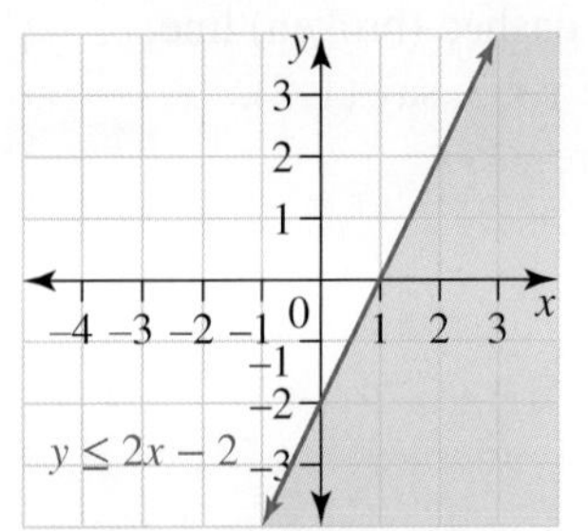

The required region is ■.

- Consider the line $x = 2$. It divides the **Cartesian plane** into two distinct regions or half-planes.

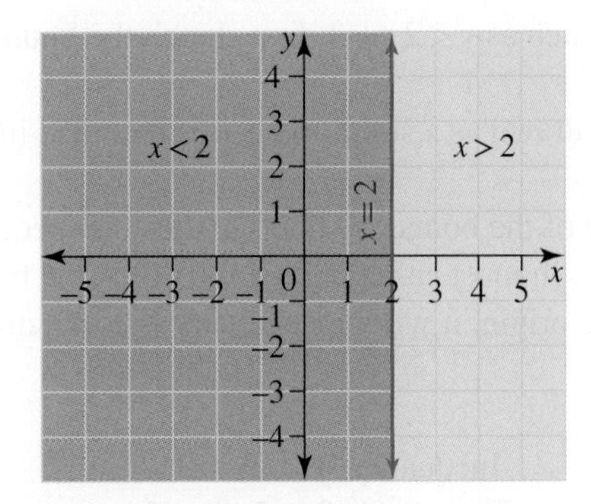

- The region on the left (shaded pink) contains all the points whose x -coordinate is less than 2, for example $(1, 3)$, so this region is given the name $x < 2$.
- The region on the right (shaded blue) contains all the points whose x -coordinate is greater than 2, for example $(3, -2)$, so this region is given the name $x > 2$.
- There are three distinct parts to the graph:
  - the boundary line, where $x = 2$
  - the pink region, where $x < 2$
  - the blue region, where $x > 2$.

WORKED EXAMPLE 17 Sketching simple inequalities

Sketch a graph of each of the following regions.

a. $x \geq -1$

b. $y < 3$

THINK

a. 1. $x \geq -1$ includes the line $x = -1$ and the region $x > -1$.

2. On a neat Cartesian plane sketch the line $x = -1$. Because the line is required, it will be drawn as a continuous (unbroken) line.

3. Identify a point where $x > -1$, say (2, 1).

4. Shade the region that includes this point. Label the region $x \geq -1$.

DRAW

a.

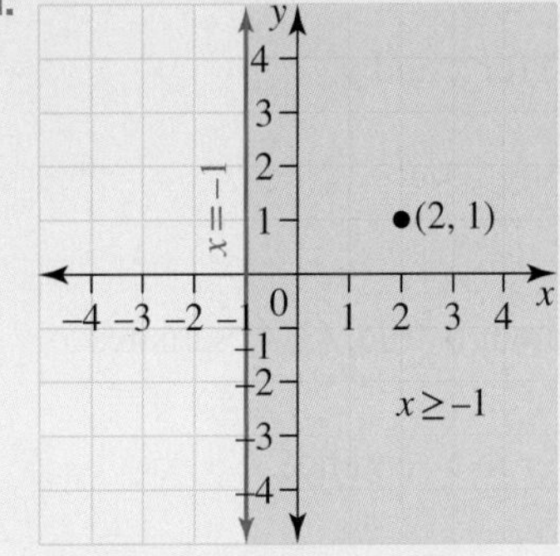

b. 1. The line $y = 3$ is not included.
2. Sketch the line $y = 3$. Because the line is not included, show it as a dashed (broken) line.
3. Identify a point where $y < 3$, say $(1, 2)$.
4. Shade the region where $y < 3$.
5. Label the region.

b.
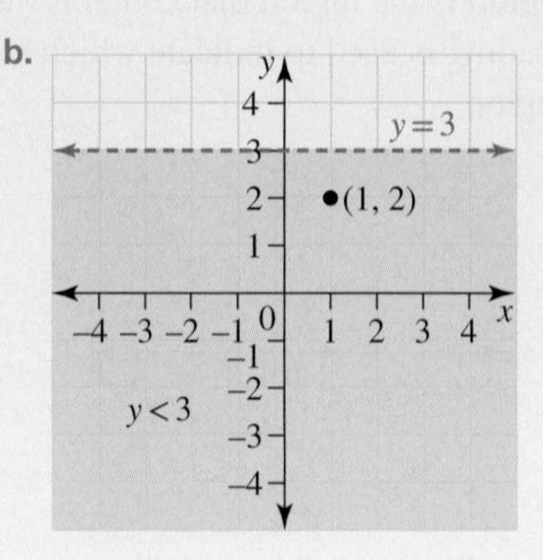

4.8.2 Determining the required region on the Cartesian plane

eles-4777

- For a more complex inequality, such as $y < 2x + 3$, first sketch the boundary line which is given by the equation $y = 2x + 3$.
  Note: The boundary line will be drawn as a solid line if it is included in the inequality ($y \leq x$) or as a broken line if it is not included ($y < x$).
- In order to determine which side of the boundary line satisfies the inequality, choose a point and test whether it satisfies the inequality. In most cases the point (0, 0) is the best point to choose, but if the boundary line passes through the origin, it will be necessary to test a different point such as (0, 1).
  For example:

$$\begin{aligned} \text{Inequality: } y &< 2x + 3: \\ \text{Test } (0, 0)\text{: } 0 &< 2(0) + 3 \\ 0 &< 3 \qquad \text{True} \end{aligned}$$

- Since 0 is less than 3, the point $(0, 0)$ does satisfy the inequality. Thus, the half plane containing $(0, 0)$ is the required region.

WORKED EXAMPLE 18 Verifying inequalities at points on the Cartesian plane

Determine whether the points $(0, 0)$ and $(3, 4)$ satisfy either of the following inequalities.

a. $x - 2y < 3$

b. $y > 2x - 3$

THINK	WRITE
a. 1. Substitute $(0, 0)$ for x and y.	a. $x - 2y < 3$ Substitute $(0, 0)$:
2. Since the statement is true, $(0, 0)$ satisfies the inequality.	$0 - 0 < 3$ $0 < 3$ True
3. Substitute $(3, 4)$ for x and y.	$x - 2y < 3$ Substitute $(3, 4)$: $3 - 2(4) < 3$ $3 - 8 < 3$
4. Since the statement is true, $(3, 4)$ satisfies the inequality.	$-5 < 3$ True
5. Write the answer in a sentence.	The points $(0, 0)$ and $(3, 4)$ both satisfy the inequality

b. 1. Substitute $(0, 0)$ for x and y.	**b.** $y > 2x - 3$ Substitute $(0, 0)$: $0 > 0 - 3$
2. Since the statement is true, $(0, 0)$ satisfies the inequality.	$0 > -3$ True
3. Substitute $(3, 4)$ for x and y.	$y > 2x - 3$ Substitute $(3, 4)$: $4 > 2(3) - 3$ $4 > 6 - 3$
4. Since the statement is true, $(3, 4)$ satisfies the inequality.	$4 > 3$ True
5. Write the answer in a sentence.	The points $(0, 0)$ and $(3, 4)$ both satisfy the inequality

WORKED EXAMPLE 19 Sketching a linear inequality

Sketch a graph of the region $2x + 3y < 6$.

THINK	**WRITE/DRAW**
1. Locate the boundary line $2x + 3y < 6$ by finding the x- and y-intercepts.	$x = 0$: $0 + 3y = 6$ $y = 2$ $y = 0$: $2x + 0 = 6$ $x = 3$
2. The line is not required due to the < inequality, so rule a broken line.	
3. Test with the point $(0, 0)$. Does $(0, 0)$ satisfy $2x + 3y < 6$?	Test $(0, 0)$: $2(0) + 3(0) = 0$ As $0 < 6$, $(0, 0)$ is in the required region.
4. Shade the region that includes $(0, 0)$.	
5. Label the region.	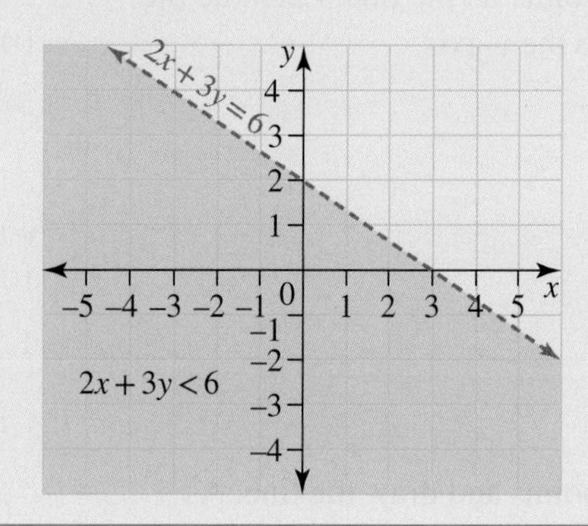

TI | THINK	DISPLAY/WRITE	CASIO | THINK	DISPLAY/WRITE
Rearrange the inequality as: $y < 2 - \frac{2x}{3}$ On a Graphs page at the function entry line, delete the = symbol, then select: • 2: $y <$ Complete the function entry line as: $y < 2 - \frac{2x}{3}$ Press ENTER. The shaded region will be displayed.	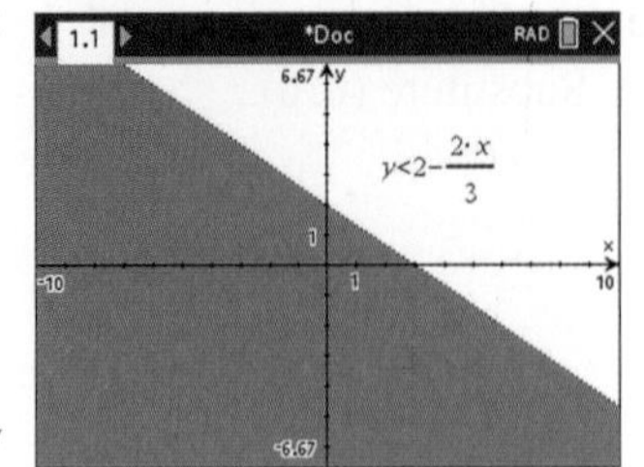	Rearrange the inequality as: $y < 2 - \frac{2x}{3}$ On a Graphs & Table screen, tap: • Type • Inequality • $y <$ Type Complete the function entry line as: $y < 2 - \frac{2x}{3}$ Then tap the graphing icon. The shaded region will be displayed.	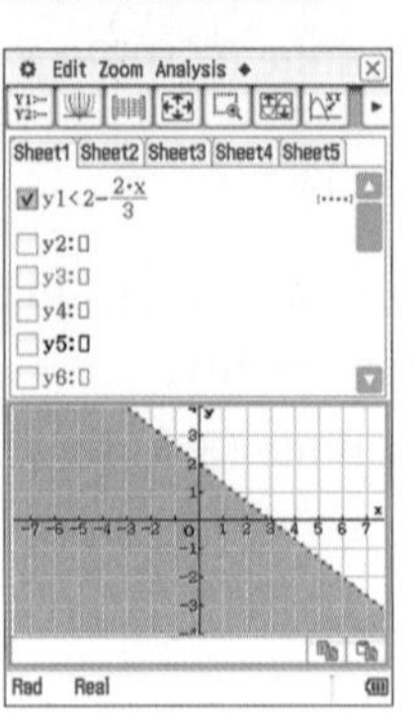

WORKED EXAMPLE 20 Modelling real-life situations

In the school holidays you have been given \$160 to arrange some activities for your family. A ticket to the movies costs \$10 and a ticket for the trampoline park costs \$16.

a. If m represents the movie tickets and t represents the trampoline park tickets, write an inequality in terms of m and t that represents your entertainment budget.

b. Sketch the inequality from part a on the Cartesian plane.

c. Using the graph from part b explore the maximum number of movie and trampoline park tickets you can buy to use the maximum amount of your holiday budget.

THINK	WRITE
a. Each movie ticket, m, costs \$12, and each trampoline ticket, t, costs \$15. The maximum amount you have to spend is \$160.	a. $10m + 16t \leq 160$
b. 1. To draw the boundary line $10m + 15t \leq 160$, identify two points on the line. Let m be the x-axis and t be the y-axis.	b. For the line $10m + 16t = 160$ x-intercept; let $t = 0$ $10m + 16 \times 0 = 160$ $10m = 160$ $m = 16$ x-intercept is (16, 0) y-intercept; let $m = 0$ $10 \times 0 + 16t = 160$ $16t = 160$ $t = 10$ y-intercept is (0, 10)
2. Plot the two points and draw the line. As you can spend up to and including \$160, the boundary line is solid. Only the first quadrant of the graph is required, as the number of tickets cannot be negative.	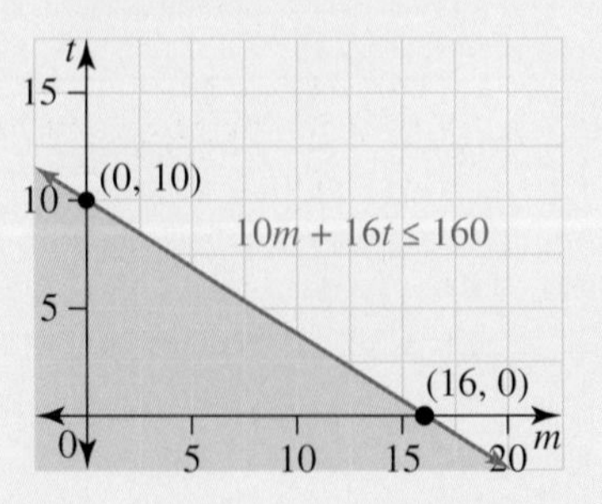

c. 1. To determine the maximum number of movie and trampoline park tickets, identify the nearest whole numbers of each to the graph line. These must be whole numbers as you cannot buy part of a ticket.

c. To spend the entire $160, only 16 movie tickets or 10 trampoline park tickets can be purchased. If less than $160 was spent you could purchase any whole number combinations, such as 6 movie tickets and 6 trampoline park tickets for $156.

DISCUSSION

Think of some real-life situations where inequalities could be used to help solve a problem.

Resources

eWorkbook Topic 4 Workbook (worksheets, code puzzle and a project) (ewbk-2030)

Interactivities Individual pathway interactivity: Inequalities on the Cartesian plane (int-4583)
Linear inequalities in two variables (int-6488)

Exercise 4.8 Inequalities on the Cartesian plane

learnon

Individual pathways

■ PRACTISE	■ CONSOLIDATE	■ MASTER
1, 4, 7, 10, 13, 16, 19	2, 5, 8, 11, 14, 17, 20	3, 6, 9, 12, 15, 18, 21

To answer questions online and to receive **immediate corrective feedback** and **fully worked solutions** for all questions, go to your learnON title at www.jacplus.com.au.

Fluency

WE18 For questions 1 to 3, sketch a graph of each of the following regions.

1. a. $x < 1$ b. $y \geq -2$ c. $x \geq 0$ d. $y < 0$

2. a. $x > 2$ b. $x \leq -6$ c. $y \geq 3$ d. $y \leq 2$

3. a. $x < \frac{1}{2}$ b. $y < \frac{3}{2}$ c. $y \geq -4$ d. $x \leq \frac{3}{2}$

WE17 For questions 4 to 6, determine which of the points A (0, 0), B (1, −2) and C (4, 3) satisfy each of the following inequalities.

4. a. $x + y > 6$ b. $x - 3y < 2$

5. a. $y > 2x - 5$ b. $y < x + 3$

6. a. $3x + 2y < 0$ b. $x \geq 2y - 2$

WE19 For questions 7 to 9, sketch the graphs for the regions given by each of the following inequations. Verify your solutions using technology.

7. a. $y \geq x + 1$ b. $y < x - 6$ c. $y > -x - 2$ d. $y < 3 - x$

8. a. $y > x - 2$ b. $y < 4$ c. $2x - y < 6$ d. $y \leq x - 7$

9. a. $x - y > 3$ b. $y < x + 7$ c. $x + 2y \leq 5$ d. $y \leq 3x$

10. **MC** The shaded region satisfying the inequality $y > 2x - 1$ is:

A.
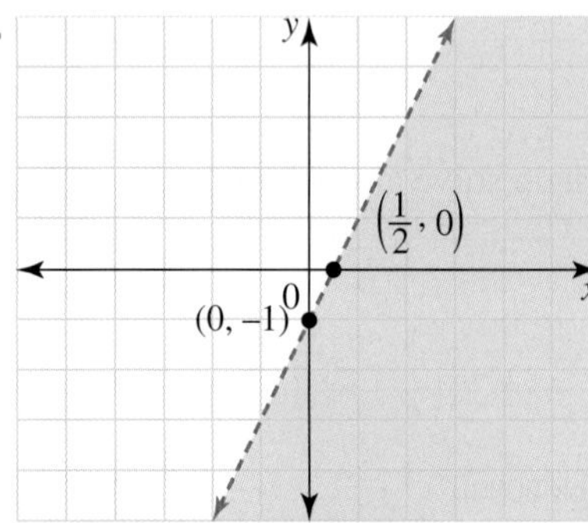

B.
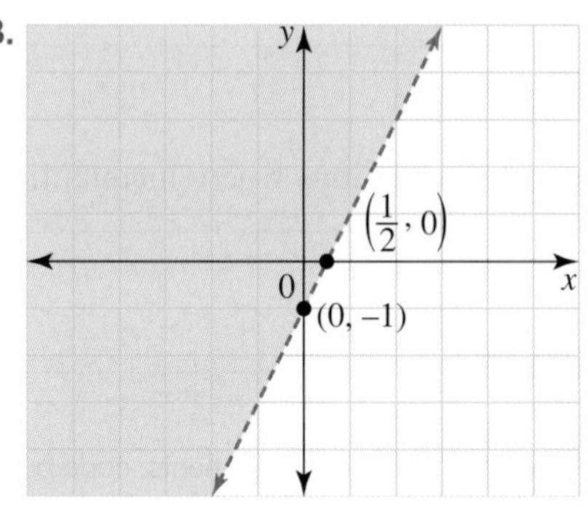

C.
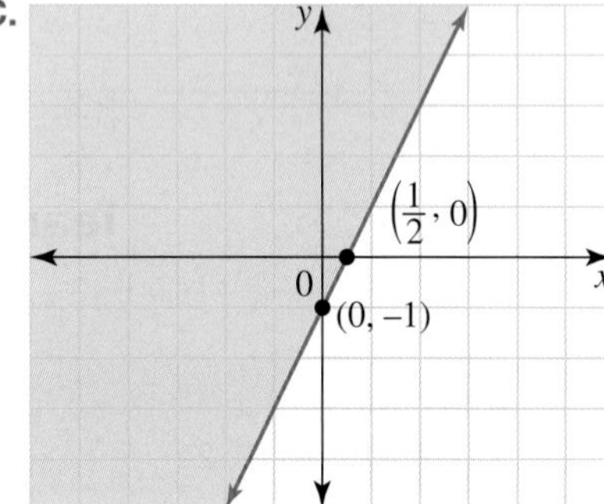

D.
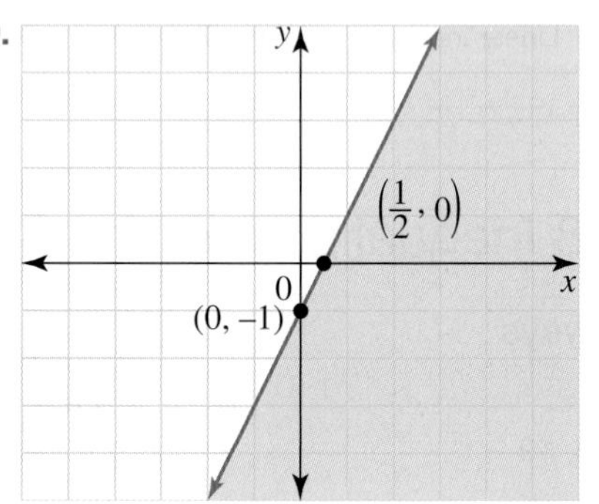

E.
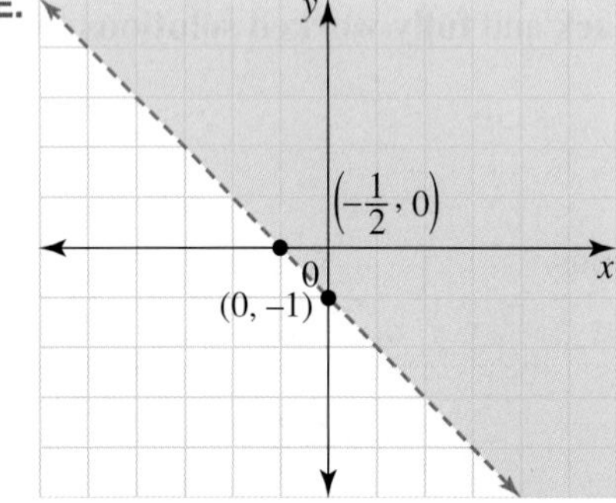

11. **MC** The shaded region satisfying the inequality $y \leq x + 4$ is:

A.

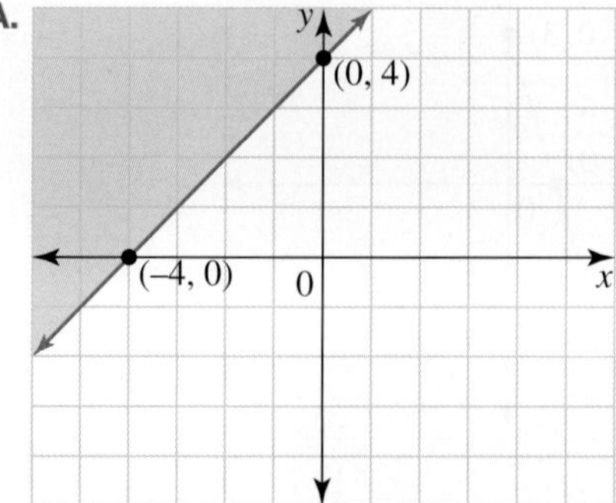

B.

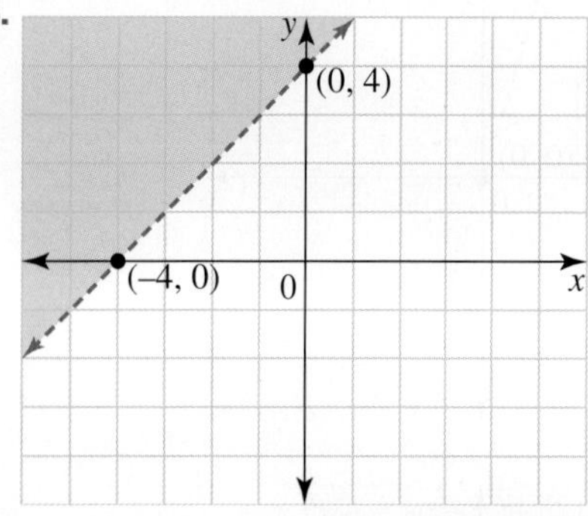

C.

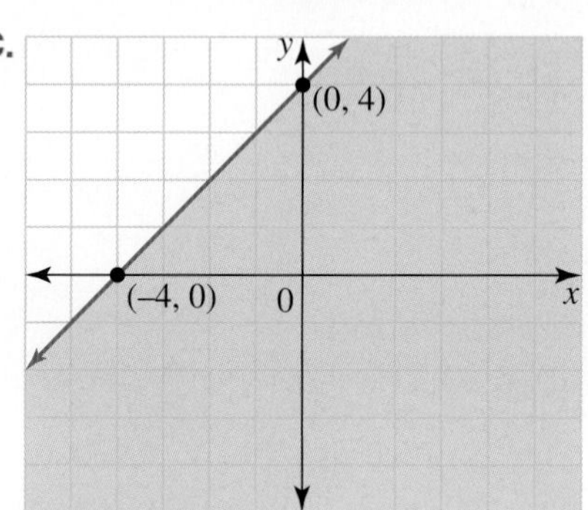

D.

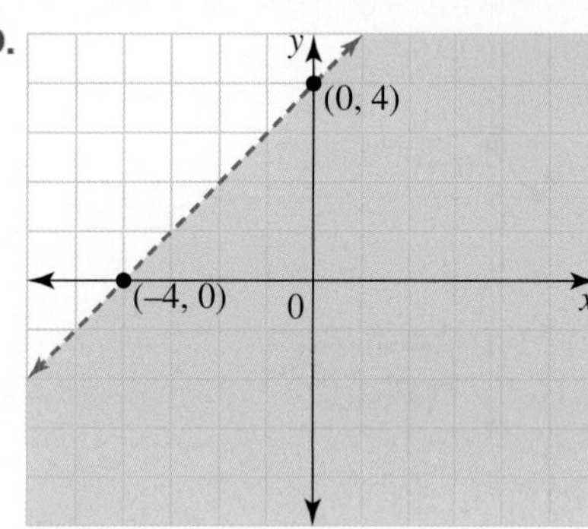

E.

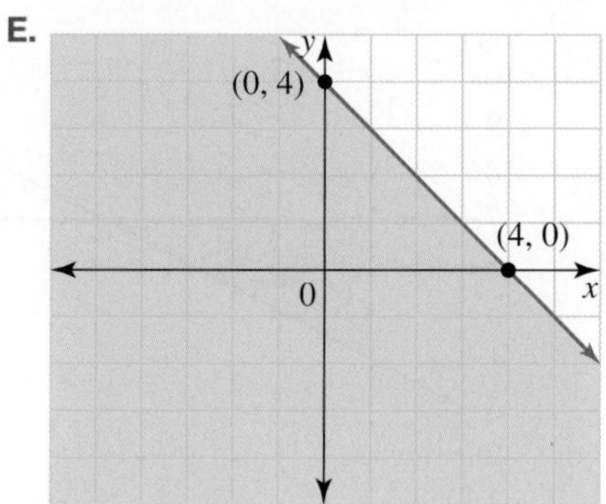

12. **MC** The region satisfying the inequality $y < 3x$ is:

A.

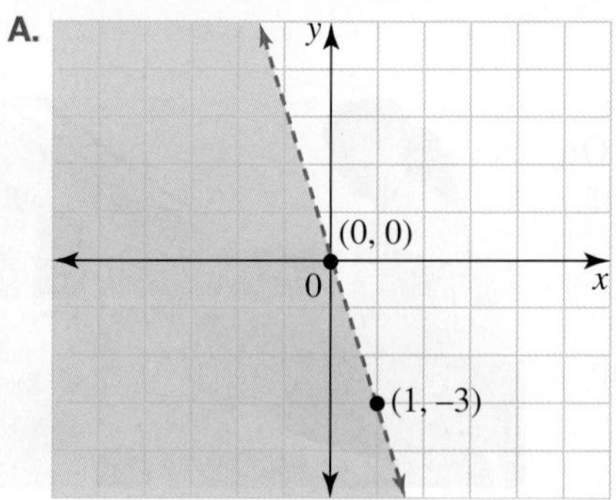

B.

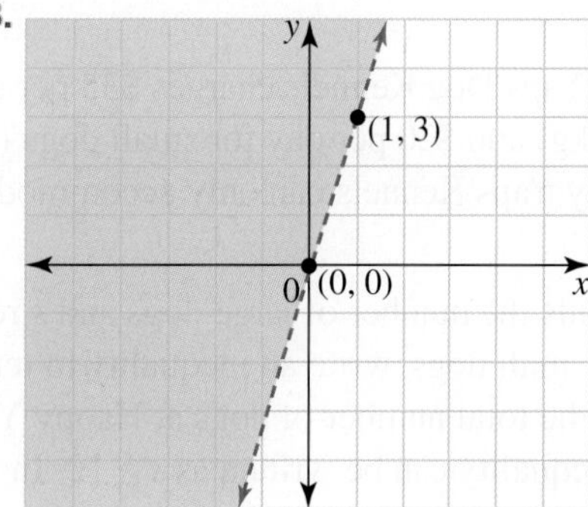

C.

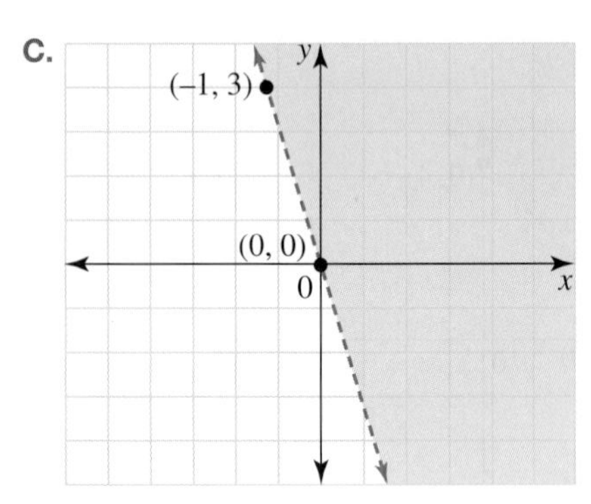

D.

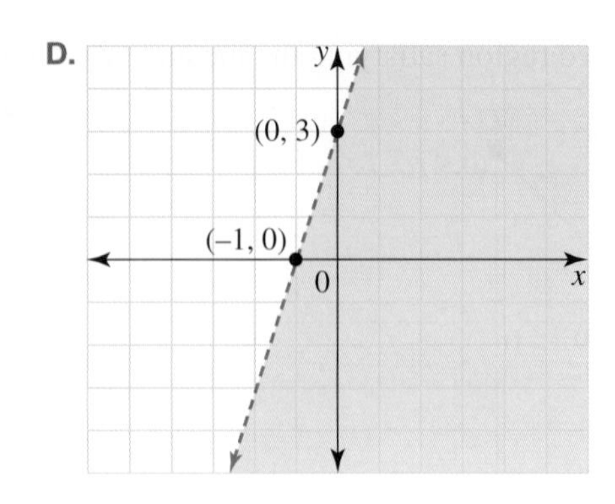

E.

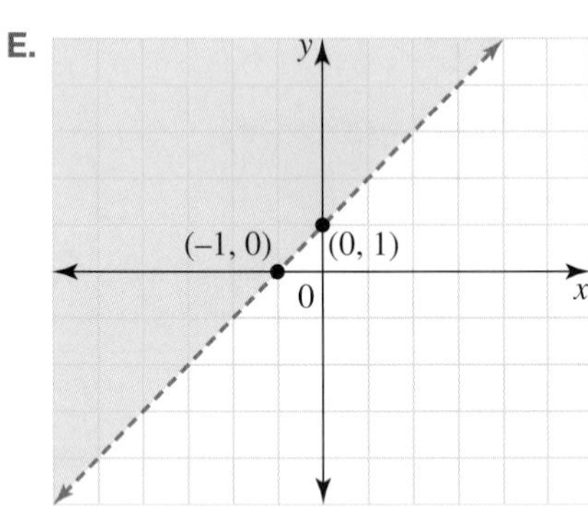

Understanding

13. a. Determine the equation of the line l shown in the diagram.
    b. Write down three inequalities that define the region R.

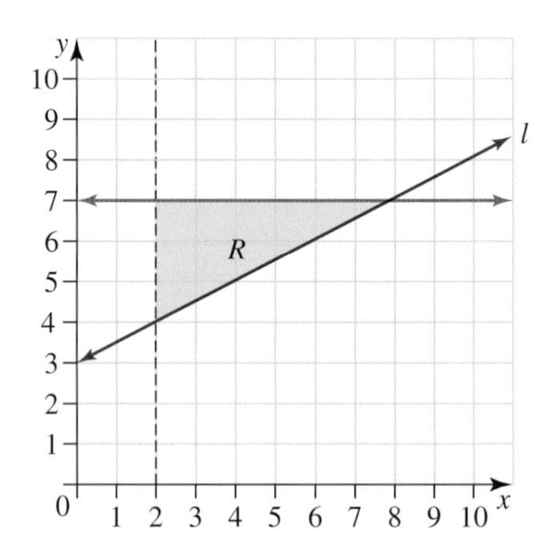

14. Identify all points with integer coordinates that satisfy the following inequalities:

$$x \geq 3$$

$$y > 2$$

$$3x + 2y \leq 19$$

15. **WE20** Happy Yaps Dog Kennels charges \$35 per day for large dogs (dogs over 20 kg) and \$20 per day for small dogs (less than 20 kg). On any day, Happy Yaps Kennels can only accommodate a maximum of 30 dogs.
    a. If l represents the number of large dogs and s represents the number of small dogs, write an inequality in terms of l and s that represents the total number of dogs at Happy Yaps.
    b. Another inequality can be written as $s \geq 12$. In the context of this problem, write down what this inequality represents.
    c. The inequality $l \leq 15$ represents the number of large dogs that Happy Yaps can accommodate on any day. Draw a graph that represents this situation.
    d. Explore the maximum number of small and large dogs Happy Yaps Kennels can accommodate to receive the maximum amount in fees.

Reasoning

16. Use technology to sketch and then find the area of the region formed by the following inequalities.

$$y \geq -4$$
$$y < 2x - 4$$
$$2y + x \leq 2$$

17. Answer the following questions.

   a. Given the following graph, state the inequality it represents.
   b. Choose a point from each half plane and show how this point confirms your answer to part a.

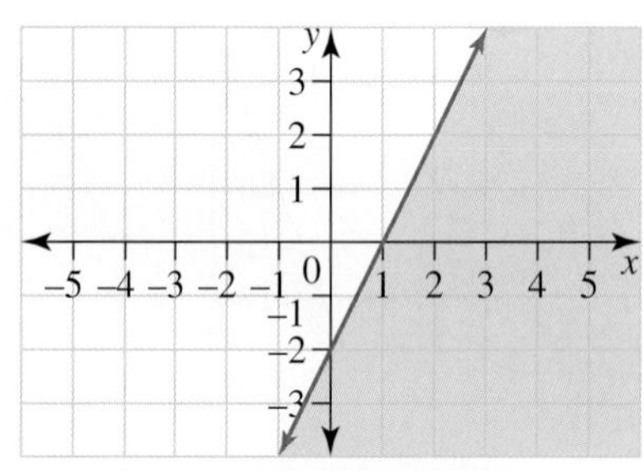

The required region is □.

18. Answer the following questions.

   a. Determine the equation of the line, l.
   b. Write an inequation to represent the unshaded region.
   c. Write an inequation to represent the shaded region.
   d. Rewrite the answer for part **b** if the line was not broken.

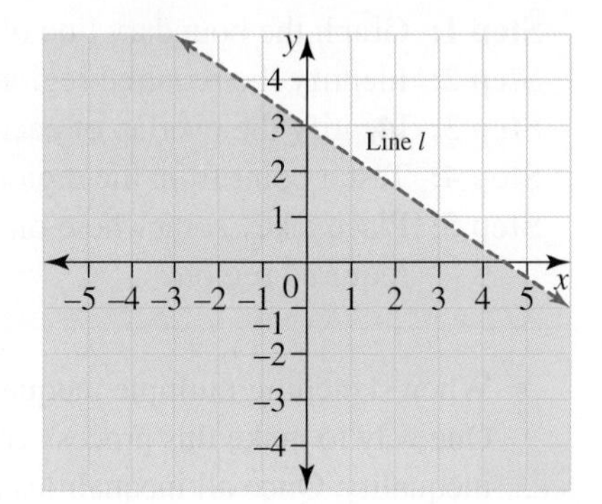

Problem solving

19. a. Sketch the graph of:

$$\frac{x+1}{2} - \frac{x+1}{3} = 2 - y$$

   b. Shade the region that represents:

$$\frac{x+1}{2} - \frac{x+1}{3} \leq 2 - y$$

20. Use your knowledge about linear inequations to sketch the regions defined by:

   a. $x^2 + y^2 < 16$
   b. $x^2 + y^2 > 36$

21. Use your knowledge about linear inequations to sketch the region defined by $y \geq x^2 + 4x + 3$.

4.9 Solving simultaneous linear inequalities

LEARNING INTENTION

At the end of this subtopic you should be able to:

- sketch multiple linear inequalities on the same Cartesian plane and determine the required region that satisfies both inequalities.

4.9.1 Multiple inequalities on the Cartesian plane

eles-4778

- The graph of an inequality represents a region of the Cartesian plane.
- When sketching multiple inequalities on the same set of axes, the required region is the overlap of each inequality being sketched.

- The required region given when placing $y < 3x$ and $y > x$ is shown below:

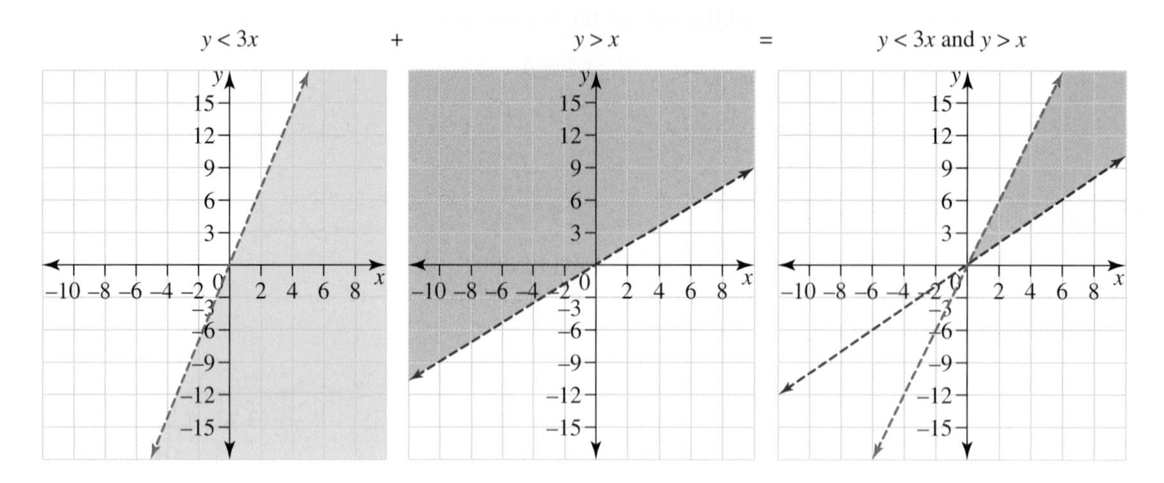

Graphing simultaneous linear inequalities

Step 1: Graph the boundary line of all linear inequalities.
Step 2: Identify the required region for each individual inequality by testing a point.
Step 3: Identify the overlap of each required region and shade this section or sections.
Step 4: Test a point from the region found in step 3 and make sure it satisfies all inequalities.
Step 5: Place a key somewhere on or below the Cartesian plane to indicate which section is the required region.

- When sketching multiple inequalities, finding the required region can get fairly tricky (and messy). One way to make this process easier is to shade the region for each inequality that **does not satisfy** the inequality. Once all inequalities have been sketched, the only section not shaded in is the solution to the simultaneous inequalities.

WORKED EXAMPLE 21 Solving simultaneous linear inequalities

Identify the required region in the following pair of linear inequalities:

$$2x + 3y \geq 6 \text{ and } y < 2x - 3$$

THINK	WRITE/DRAW	
1. To sketch each inequality, the boundary line needs to be drawn first. • To draw each line, identify two points on each line. • Use the intercepts method for $2x + 3y \geq 6$. • Use substitution of values for $y < 2x - 3$. • Write the coordinates. *Note:* The intercepts method could also have been used for the second equation.	$2x + 3y \geq 6$ For the line $2x + 3y = 6$, x-intercept: let $y = 0$ $2x + 0 = 6$ $x = 3$ y-intercept: let $x = 0$ $0 + 3y = 6$ $y = 2$ $(3, 0)$, $(0, 2)$	$y < 2x - 3$ For the line $y = 2x - 3$, let $x = 0$ $y = 2(0) - 3$ $y = -3$ let $x = 2$ $y = 2(2) - 3$ $y = 4 - 3$ $y = 1$ $(0, -3)$, $(2, 1)$

2. Plot the two points for each line.
   - Plot the x- and y-intercepts for $2x + 3y = 6$, as shown in blue.
   - Plot the two points for $y = 2x - 3$, as shown in pink.

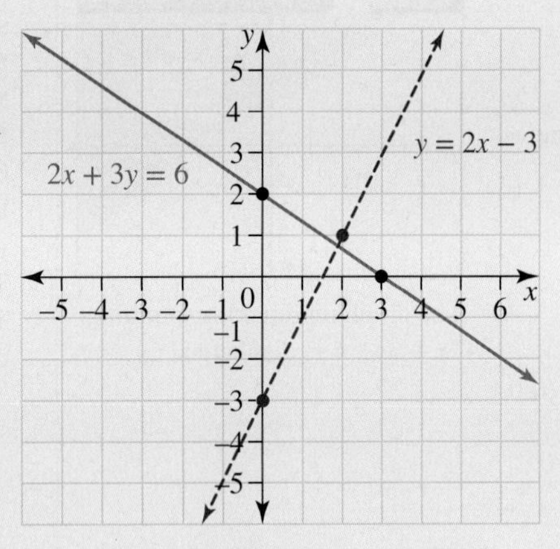

3. Draw the boundary lines.
   - For $2x + 3y \geq 6$, the points on the line are included. The boundary line is solid, as shown in blue.
   - For $y < 2x - 3$, the points on the line are not included. The boundary line is dashed, as shown in pink.

4. To determine which side of the line is the required region, select a point on one side of the line and check to see whether the point satisfies the equation. Choose the point (3, 1) to substitute into the equation.

Check the point (3, 1):

$x = 3, y = 1$

$2x + 3y \geq 6$

$\text{LHS} = 2x + 3y$
$= 2(3) + 3(1)$
$= 6 + 3$
$= 9$

$\text{RHS} = 6$

$\text{LHS} > \text{RHS}$

The point (3, 1) satisfies the inequality and is in the required region for $2x + 3y \geq 6$

$y < 2x - 3$

$\text{LHS} = y$
$= 1$

$\text{RHS} = 2(3) - 3$
$= 6 - 3$
$= 3$

$\text{LHS} < \text{RHS}$

The point (3, 1) satisfies the inequality and is in the required region for $y < 2x - 3$

5. The region *not* required for:
   $2x + 3y \geq 6$ is shaded pink.
   $y < 2x - 3$ is shaded green.
   Since the point (3, 1) satisfies both inequalities, it is in the required region. The required region is the unshaded section of the graph. Write a key.

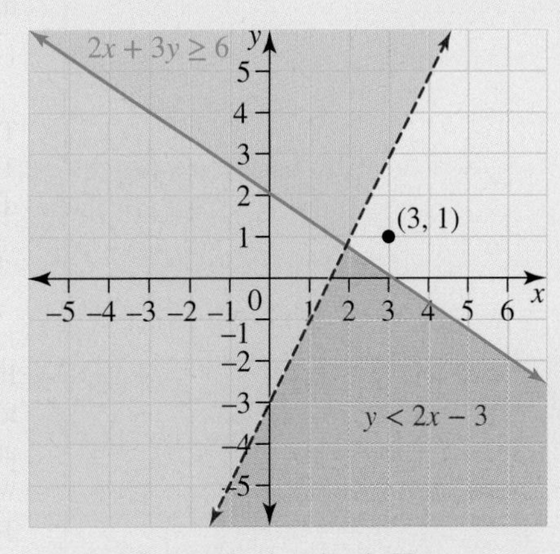

The required region is □.

TI | THINK	DISPLAY/WRITE	CASIO | THINK	DISPLAY/WRITE
1. In a new problem, on a Graphs page at the function entry line delete the = symbol, then complete entry line as $y \geq 2 - \frac{2x}{3}$. Then press ENTER.	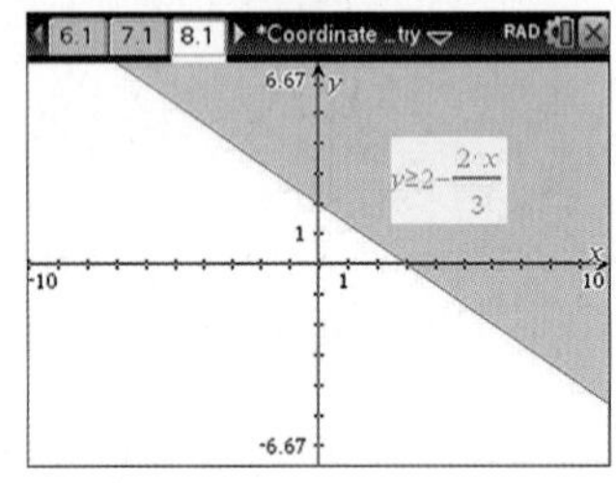 The graph region corresponding $2x + 3y \geq 6$ is displayed.	1. On the Graph & Table screen tap: • Type • Inequality • $y <$ Type. Complete the funtion entry line as: $y < 2x - 3$. Then tap the graphing icon. The shaded region will be displayed.	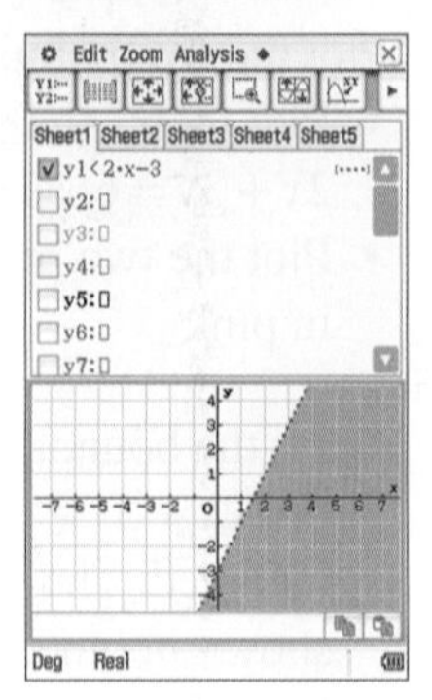 The graph region corresponding to $y < 2x - 3$ is displayed.
2. Press TAB. Complete the entry line as $y < 2x - 3$. Then press ENTER. You may need to change the Line Colour and Fill Colour of this inequality to green to see the shaded region in dark green as shown.	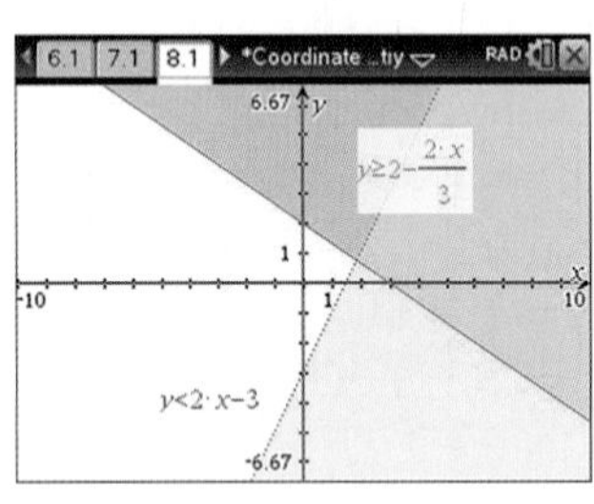 The shaded region indicated is the area corresponding to $2x + 3y \geq 6$ and $y < 2x - 3$.	2. On the Main screen, complete the entry line as: solve $(2x + 3y \geq 6, y)$. Highlight the previous answer and drag it to complete the entry line as: $\text{simplify}\left(y \geq \frac{-(2x-6)}{3}\right)$. Press EXE after each entry line.	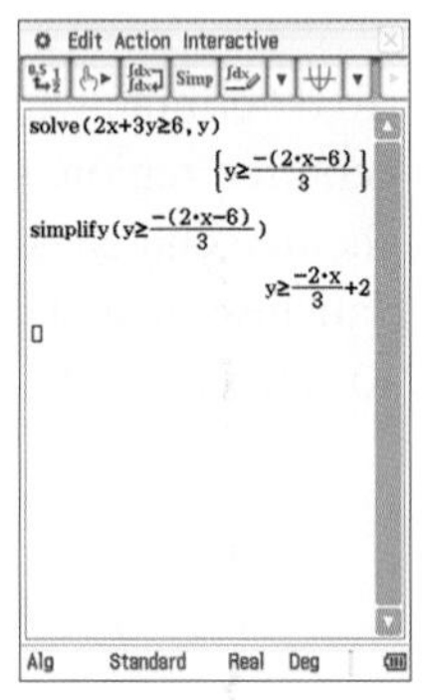 The inequality is given by $y \geq -\frac{2x}{3} + 2$
		3. Go back to the Graph & Table screen and complete the function entry line as: $y \geq -\frac{2x}{3} + 2$. Then tap the graphing icon. The shaded region will be displayed.	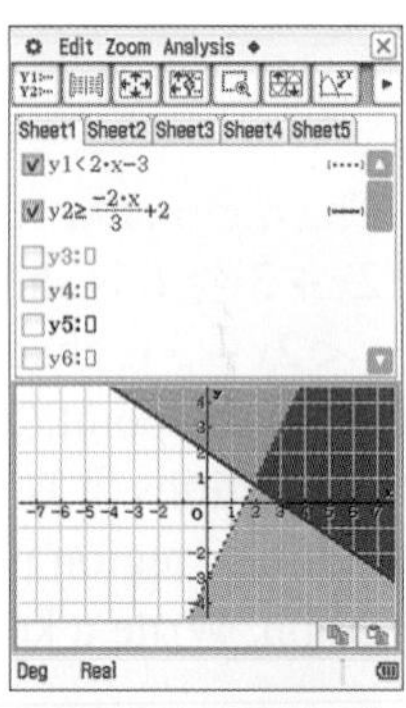
		4. If the solution region is hard to see, fix this by setting an appropriate viewing window. To do this, tap [icon]. Select the values as shown in the screenshot and tap OK. 5. The darker shaded region to the top right is the area corresponding to $y < 2x - 3$ and $2x + 3y \geq 6$.	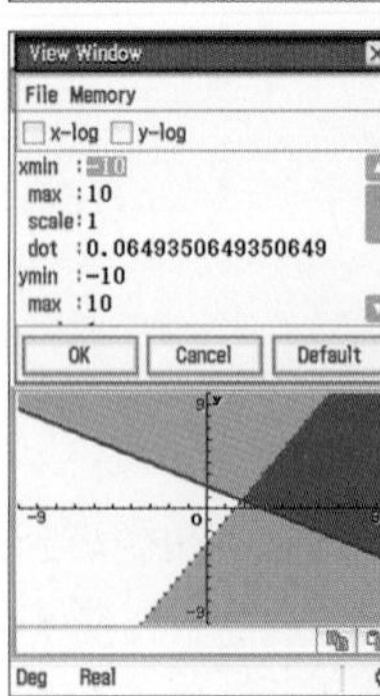

Resources

eWorkbook Topic 4 Workbook (worksheets, code puzzle and a project) (ewbk-2030)

Interactivities Individual pathway interactivity: Solving simultaneous linear inequalities (int-4584)
Graphing simultaneous linear inequalities (int-6283)

Exercise 4.9 Solving simultaneous linear inequalities

learnON

Individual pathways

■ PRACTISE	■ CONSOLIDATE	■ MASTER
1, 3, 6, 9, 12	4, 7, 10, 13	2, 5, 8, 11, 14

To answer questions online and to receive **immediate corrective feedback** and **fully worked solutions** for all questions, go to your learnON title at www.jacplus.com.au.

Fluency

1. WE21 Identify the required region in the following pair of inequalities.

$$4x + 7y \geq 21$$
$$10x - 2y \geq 16$$

2. Given the graph shown, determine the inequalities that represent the shaded region.

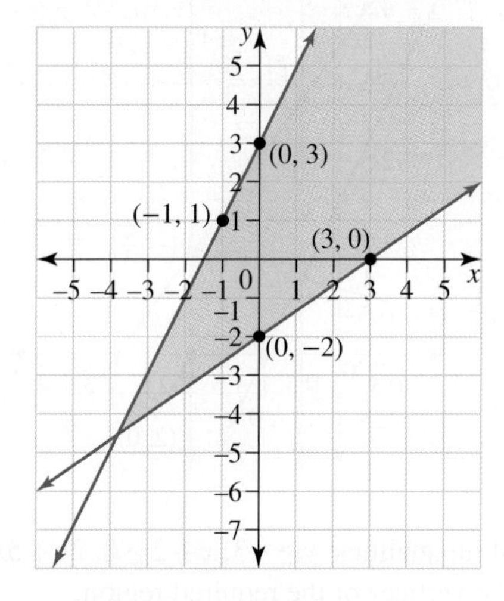

For questions 3 to 5, sketch the following pairs of inequalities.

3. a. $y < 4$
$y \leq -x$

b. $y + 3x > 6$
$y - 2x < 9$

4. a. $5y - 3x \geq -10$
$6y + 4x \geq 12$

b. $\frac{1}{3}y + 2x \leq 4$
$y - 4x \geq -8$

5. a. $3x + 4y < 24$
$y > 2x - 5$

b. $6x - 5y > 30$
$x + y < 16$

Understanding

6. **MC** Identify which system of inequalities represents the required region on the graph.

   A. $y \le x - 2$
      $y > -3x - 6$

   B. $y \ge x - 2$
      $y \ge -3x - 6$

   C. $y \le x - 2$
      $y \le -3x - 6$

   D. $y \ge x + 3$
      $y \le -3x - 6$

   E. $y > x + 2$
      $y < -3x + 6$

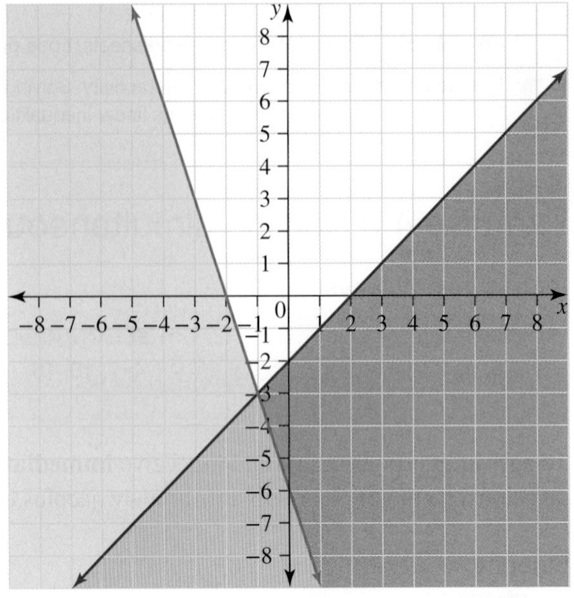

The required region is $\square$.

7. Given the diagram, write the inequalities that created the shaded region.

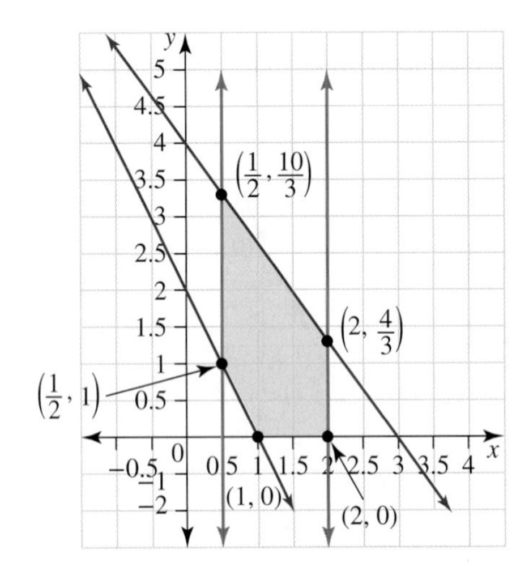

8. a. Graph the following system of inequalities: $y \ge -3$, $x + 2 \ge 0$, $2y + 5x \le 7$
   b. Calculate the coordinates of the vertices of the required region.

Reasoning

9. The sum of the lengths of any two sides of a triangle must be greater than the third side.
   a. Given a triangle with sides x, 9 and 4, draw diagrams to show the possible triangles, using the above statement to establish inequalities.
   b. Determine the possible solutions for x and explain how you determined this.

10. Create a triangle with the points $(0, 0)$, $(0, 8)$ and $(6, 0)$.
    a. Calculate the equations of the lines for the three sides.
    b. If you shade the interior of the triangle (including the boundary lines), determine the inequalities that would create the shaded region.
    c. Calculate the side lengths of this triangle.

11. A rectangle must have a length that is a least 4 cm longer than its width. The area of the rectangle must be less than 25 cm^2.

a. Write three inequalities that represent this scenario.
b. Determine how many possible rectangles could be formed, with integer side lengths, under these conditions.

Problem solving

12. a. Determine the equations of the two lines in the diagram shown.
b. Determine the coordinates of the point A.
c. Write a system of inequations to represent the shaded region.

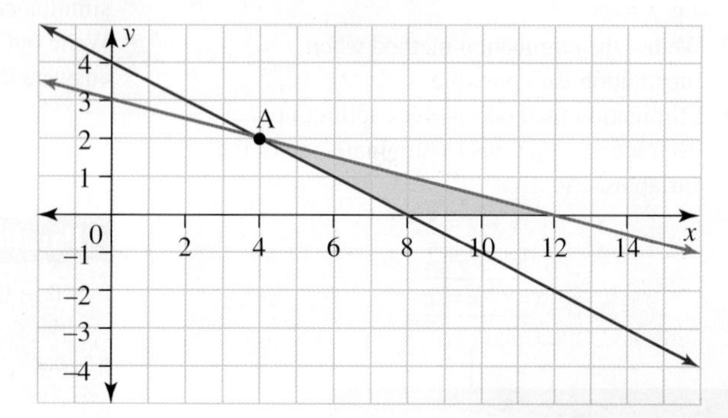

13. The Ecofriendly company manufactures two different detergents. *Shine* is specifically for dishwashers while *Motherearth* is a washing machine detergent. For the first week of June, the production manager has specified that the total amount of the two products produced should be at least 400 litres as one client has already pre-ordered 125 litres of *Shine* for that week. The time that is required to process one litre of *Shine* is 30 minutes while one litre of *Motherearth* requires 15 minutes. During the week mentioned, the factory can process the detergents for up to 175 hours.

a. If x represents the number of litres of *Shine* produced and y represents the number of litres of *Motherearth* produced, formulate the constraints as linear inequations.
b. Show the feasible region.
c. State the coordinates of the vertices of the region.

14. Ethan is a bodybuilder who maintains a strict diet. To supplement his current diet, he wants to mix two different products, *Proteinplus* and *Carboload*, in order to produce a desired balance composed of 100 g of protein, 160 g of carbs and 70 g of fat. Each product is sold in 50 g sachets that contain the following:

Product	Protein (per 50 g)	Carbohydrates (per 50 g)	Fats (per 50 g)
Proteinplus	24 g	14 g	5 g
Carboload	10 g	32 g	20 g

a. Set up three inequalities to represent this situation.
b. Sketch the feasible region.
c. Determine what combination of the two products requires the fewest number of sachets to be used.

4.10 Review

4.10.1 Topic summary

SIMULTANEOUS LINEAR EQUATIONS AND INEQUALITIES

Simultaneous equations

- Solving simultaneous equations involves finding the point (or points) of intersection between two lines.
- We can determine these points by accurately sketching both equations, or using technology to find these points.

Substitution and elimination

- The substitution and elimination methods are two algebraic techniques used to solve simultaneous equations.
- We can use substitution when one (or both) of the equations have a variable as the subject. e.g. $y = 3x - 4$
- We use the elimination method when substitution isn't possible.
- Elimination method involves adding or subtracting equations to eliminate one of the variables. e.g.

$$\begin{aligned} 3x + y &= 5 \\ 4x - y &= 2 \\ \hline 7x &= 7 \end{aligned}$$

Applications

- Choose appropriate variables to define the unknown quantities.
- Use the information in the question to form two or more equations.
- Pick an appropriate technique to solve simultaneously.
- Write out a statement that explicitly answers the question.

Parallel and perpendicular lines

- Two lines are parallel if they have the same gradient.
  e.g. $y = \mathbf{3}x - 6$
  $y = \mathbf{3}x + 1$
- Two lines are perpendicular if the product of their gradients is -1.
  e.g. $y = 2x + 3$
  $y = -\frac{x}{2} - 4$
  $m_1 \times m_2 = 2 \times -\frac{1}{2} = -1$

Inequations

- An equation has a = sign.
- An inequation will have one of the following:
  $>, \geq, <, \leq$
- An inequation, such as $x < 5$, will have an infinite number of solutions: $x = \{4, 3, 2, 1, \ldots\}$
- We can represent an inequality on a number line.
  e.g. $x > 2$

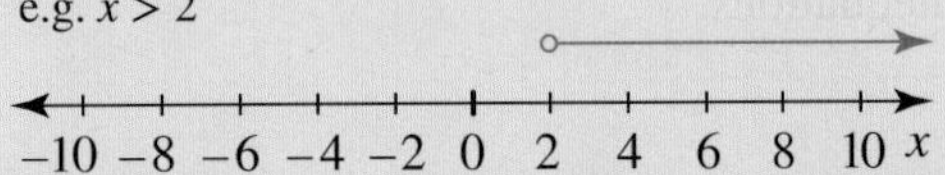

- An open circle mean that a value **is not included** as a solution. A closed circle means that the value **is included** in the solution.

Simultaneous linear and non-linear equations

- A system of equations which contains a linear equation and a non-linear equation can have 0, 1 or 2 solutions (points of intersection).
- The number of solutions will depend on the equations of both lines.
- Non-linear equations include:
  - quadratic equations (parabolas)
  - hyperbolic equations
  - circles.

Number of solutions

- Parallel lines with *different* y-intercepts will never intersect.
- Parallel lines with the *same* y-intercept are called **coincident lines** and will intersect an infinite number of times.
- Perpendicular lines will intersect once and cross at right angles to each other.

Inequalities and half planes

- The graph of a linear inequality is called a **half plane** and is the region above or below a boundary line.

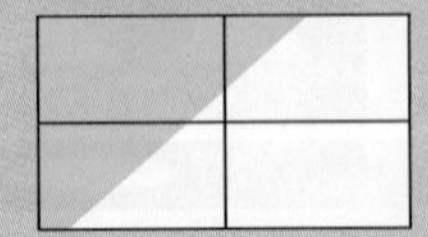
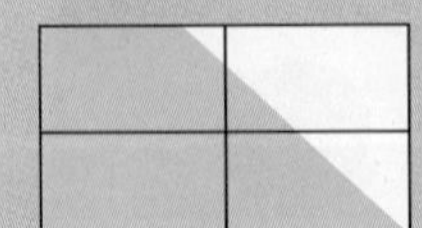

- If the inequality has < or > the boundary line is dotted as it is not included in the solution.
- If the inequality has ≤ or ≥ the boundary line is solid as it is included in the solution.

Sketching inequalities

- When sketching a linear inequality such as $y < 3x + 4$, sketch the boundary line first which is given by $y = 3x + 4$.
- To determine which side of the boundary line is the required region test the point (0, 0) and see if it satisfies the inequality.
  Test $(0, 0) = 0 < 3(0) + 4$
  $0 < 4$ which is true
- In this case, the region required is the region with the point (0, 0).
- When sketching simultaneous inequalities, the required region is the overlap region of each individual inequality.

4.10.2 Success criteria

Tick the column to indicate that you have completed the subtopic and how well you have understood it using the traffic light system.

(**Green:** I understand; **Yellow:** I can do it with help; **Red:** I do not understand)

Subtopic	Success criteria	Green	Yellow	Red
4.2	I can use the graph of two simultaneous equations to determine the point of intersection			
	I can determine whether two simultaneous equations will have 0, 1 or infinite solutions			
	I can determine whether two lines are parallel or perpendicular.			
4.3	I can identify when it is appropriate to solve using the substitution method.			
	I can solve a system of two linear simultaneous equations using the substitution method.			
4.4	I can solve two linear simultaneous equations using the elimination method.			
4.5	I can define unknown quantities with appropriate variables.			
	I can form two simultaneous equations using the information presented in a problem.			
	I can choose an appropriate method to solve simultaneous equations in order to find the solution to a problem.			
4.6	I can determine the point or points of intersection between a linear equation and various non-linear equations using various techniques.			
	I can use digital technology to find the points of intersection between a linear equation and a non-linear equation.			
4.7	I can solve an inequality and represent the solution on a number line.			
	I can convert a worded statement to an inequality in order to solve a problem.			
4.8	I can sketch the graph of a half plane: the region represented by an inequality.			
	I can sketch inequalities using digital technology.			
4.9	I can sketch multiple linear inequalities on the same Cartesian plane and determine the required region that satisfies both inequalities.			

4.10.3 Project

Documenting business expenses

In business, expenses can be represented graphically, so that relevant features are clearly visible. The graph compares the costs of hiring cars from two different car rental companies. It will be cheaper to use Plan A when travelling distances less than 250 kilometres, and Plan B when travelling more than 250 kilometres. Both plans cost the same when you are travelling exactly 250 kilometres.

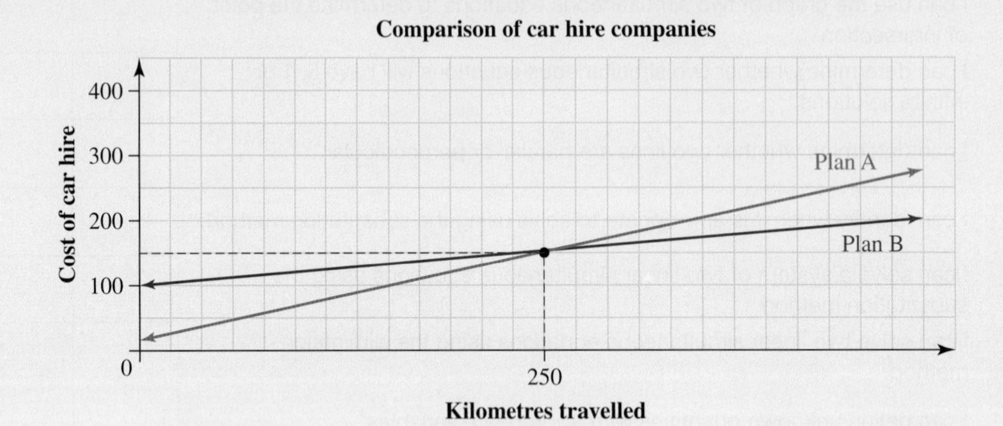

Andrea works as a travelling sales representative. She needs to plan her next business trip to Port Hedland, which she anticipates will take her away from the office for 3 or 4 days. Due to other work commitments, she is not sure whether she can make the trip by the end of this month or early next month.

She plans to fly to Port Hedland and use a hire car to travel when she arrives. Andrea's boss has asked her to supply documentation detailing the anticipated costs for the hire car, based on the following quotes received.

A1 Rentals	\$35 per day plus 28c per kilometre of travel
Cut Price Rentals	\$28 per day plus 30c per kilometre of travel

Andrea is aware that, although the Cut Price Rentals deal looks cheaper, it could work out more expensive in the long run, because of the higher cost per kilometre of travel; she intends to travel a considerable distance.

Andrea is advised by both rental companies that their daily hire charges are due to rise by \$2 per day from the first day of next month.

Assuming that Andrea is able to travel this month and her trip will last 3 days, use the information given to answer questions 1 to 4.

1. Write equations to represent the costs of hiring a car from A1 Rentals and Cut Price Rentals. Use the pronumeral C to represent the cost (in dollars) and d to represent the distance travelled (in kilometres).

2. Copy the following set of axes to plot the two equations from question 1 to show how the costs compare over 1500 km.

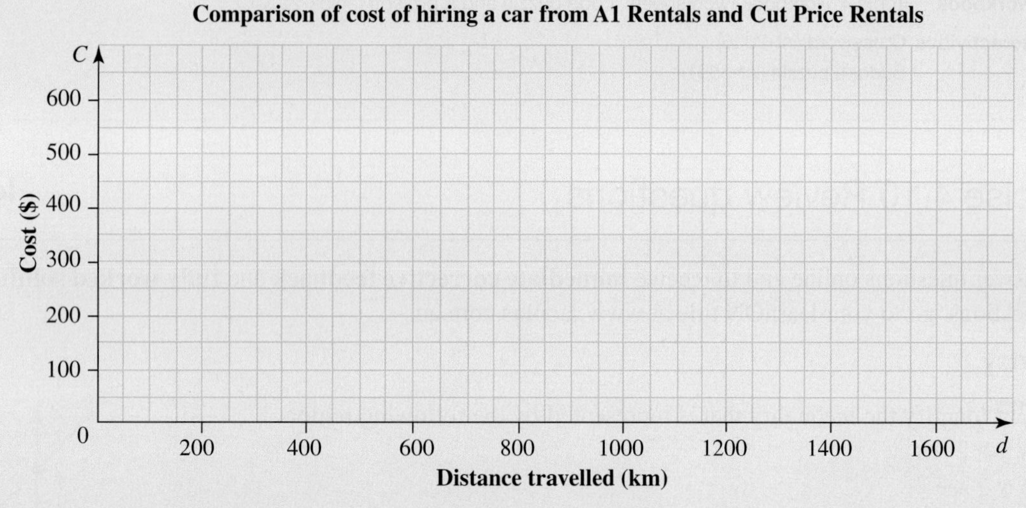

3. Use the graph to determine how many kilometres Andrea would have to travel to make the hire costs the same for both rental companies.
4. Assume Andrea's trip is extended to four days. Use an appropriate method to show how this changes the answer found in question 3.

For questions 5 to 7, assume that Andrea has delayed her trip until next month when the hire charges have increased.

5. Write equations to show the cost of hiring a car from both car rental companies for a trip lasting:
   a. 3 days
   b. 4 days.
6. Copy the following set of axes to plot the four equations from question 5 to show how the costs compare over 1500 km.

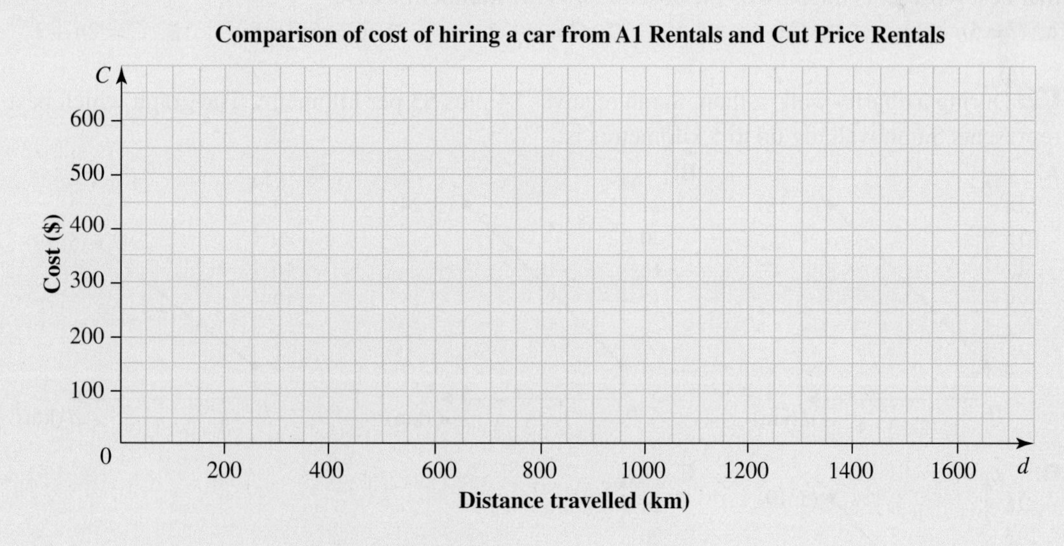

7. Comment on the results displayed in your graph.
8. Andrea needs to provide her boss with documentation of the hire car costs, catering for all options. Prepare a document for Andrea to hand to her boss.

Resources

eWorkbook Topic 4 Workbook (worksheets, code puzzle and a project) (ewbk-2030)

Interactivities Crossword (int-2836)

Sudoku puzzle (int-3591)

Exercise 4.10 Review questions

learnON

To answer questions online and to receive **immediate corrective feedback** and **fully worked solutions** for all questions, go to your learnON title at www.jacplus.com.au.

Fluency

1. **MC** Identify the inequality that is represented by the following region.

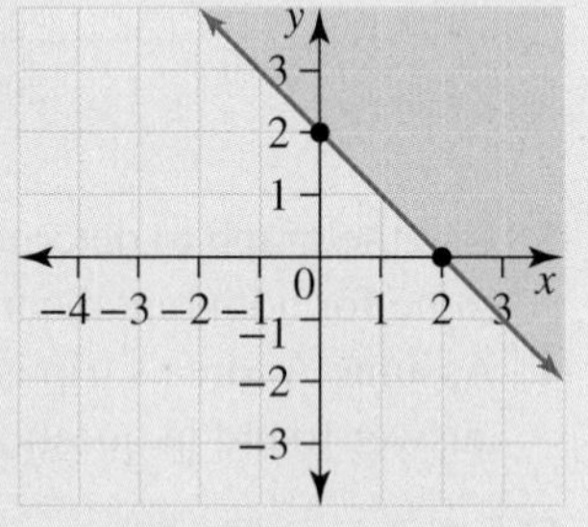

A. $y \geq 2 - x$
B. $y \geq x - 2$
C. $y \leq 2 - x$
D. $y \leq x - 2$
E. $y \geq 2x$

2. **MC** Identify the equation of a linear graph which passes through the origin with gradient -3.
A. $y = -3$ B. $x = -3$ C. $y = -3x$ D. $y = 3 - 3x$ E. $y = 3x - 3$

3. **MC** A music shop charges a flat rate of \$5 postage for 2 CDs and \$11 for 5 CDs. Identify the equation that best represents this, if C is the cost and n is the number of CDs.
A. $C = 5n + 11$ B. $C = 6n + 5$ C. $C = n + 2$ D. $C = 5n + 1$ E. $C = 2n + 1$

4. **MC** During a charity walk-a-thon, Sarah receives \$4 plus \$3 per kilometre. The graph which best represents Sarah walking up to 5 kilometres is:

A.
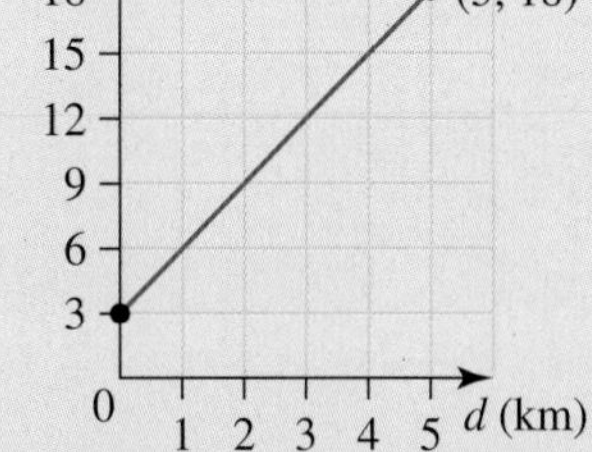

B.
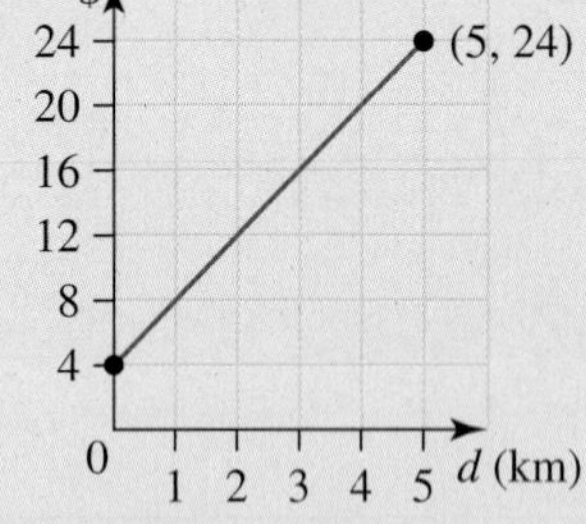

C.
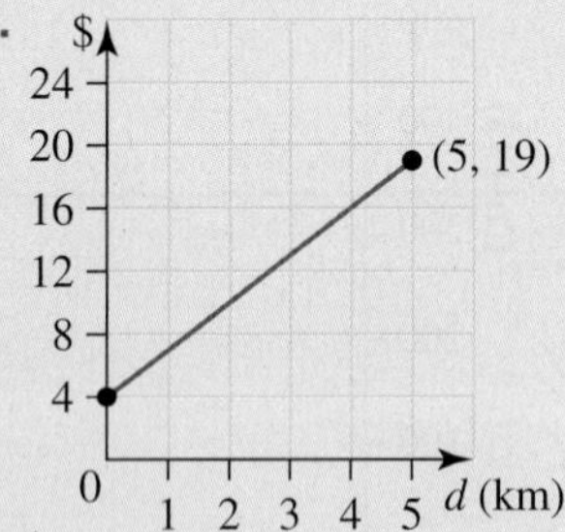

D. (5, 19)

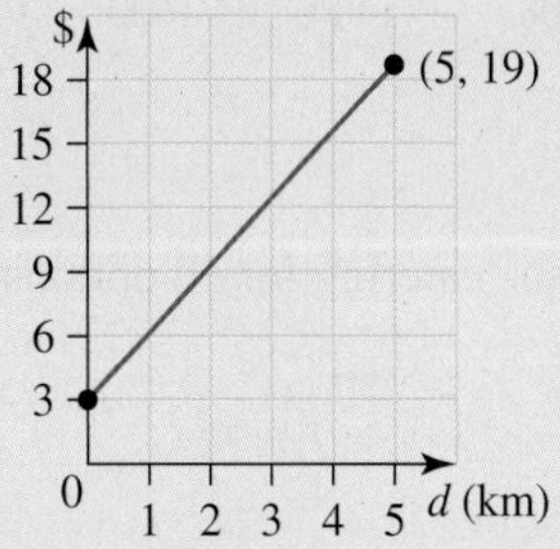

E.
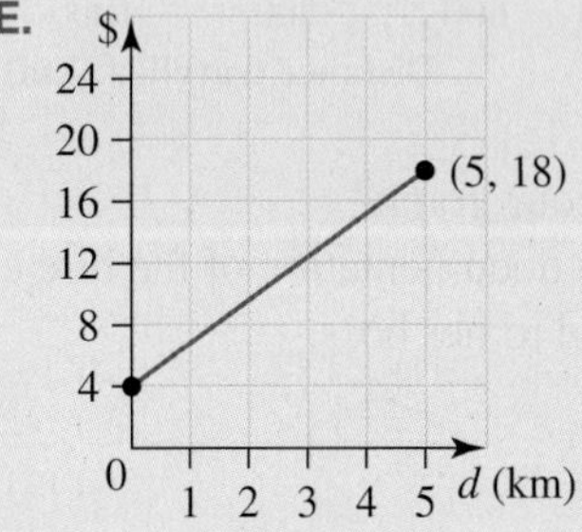

5. **MC** Identify which of the following pairs of coordinates is the solution to the simultaneous equations:

$$2x + 3y = 18$$
$$5x - y = 11$$

A. $(6, 2)$ **B.** $(3, -4)$ **C.** $(3, 9)$ **D.** $(3, 4)$ **E.** $(5, 11)$

6. **MC** Identify the graphical solution to the following pair of simultaneous equations:

$$y = 5 - 2x$$
$$y = 3x - 10$$

A.

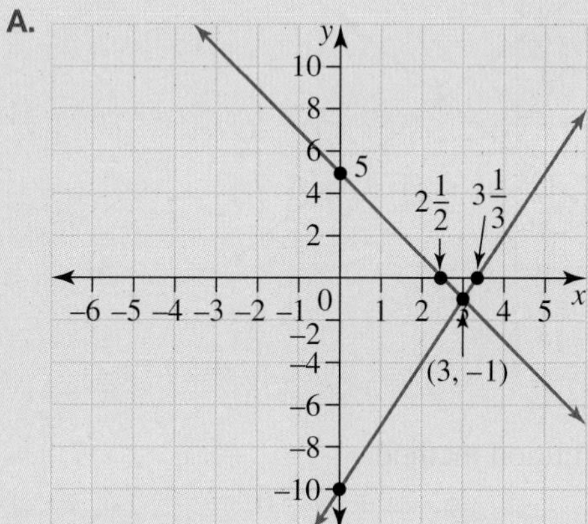

B.

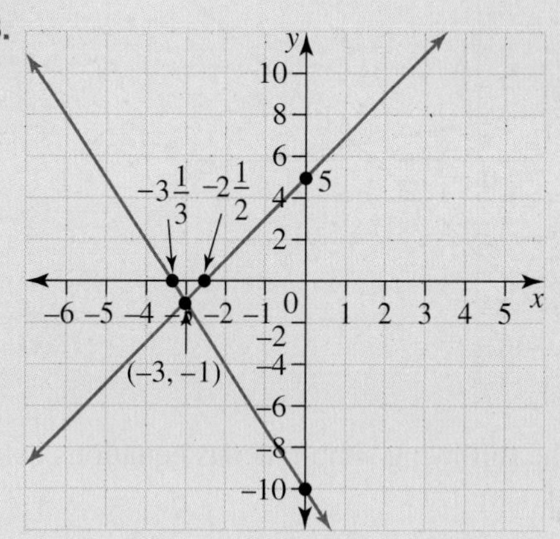

C.

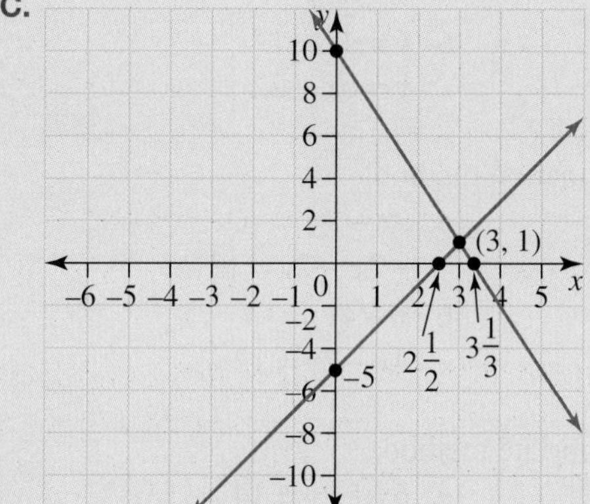

D.

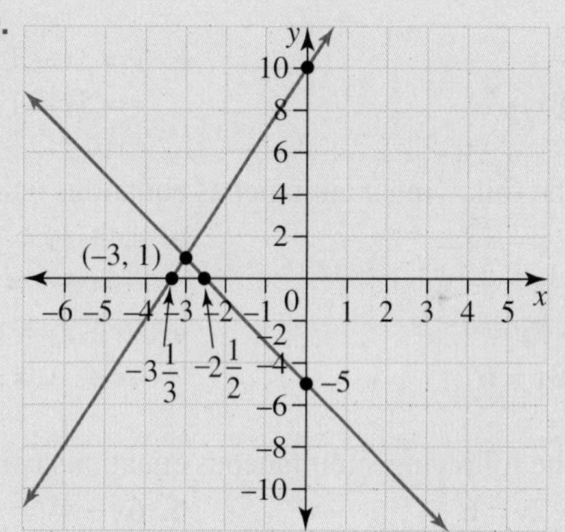

E. None of the above

7. Sketch the half plane given by each of the following inequalities.

a. $y \leq x + 1$ **b.** $y \geq 2x + 10$ **c.** $y > 3x - 12$ **d.** $y < 5x$

e. $x \geq 7$ **f.** $y \leq \frac{1}{2}x + 1$ **g.** $2x + y \geq 9$ **h.** $y > -12$

8. Use substitution to check if the given pair of coordinates is a solution to the given simultaneous equations.

a. $(7, 1)$ $x - 2y = 5$
$5y + 2x = 18$

b. $(4, 3)$ $y = 7 - 3$
$5y - 2x = 7$

9. Solve each of the following pairs of simultaneous equations using a graphical method.

a. $4y - 2x = 8$
$x + 2y = 0$

b. $y = 2x - 2$
$x - 4y = 8$

c. $2x + 5y = 20$
$y = 7$

10. Use the graphs below, showing the given simultaneous equations, to write the point of intersection of the graphs and, hence, the solution of the simultaneous equations.

a. $x + 3y = 6$
$y = 2x - 5$

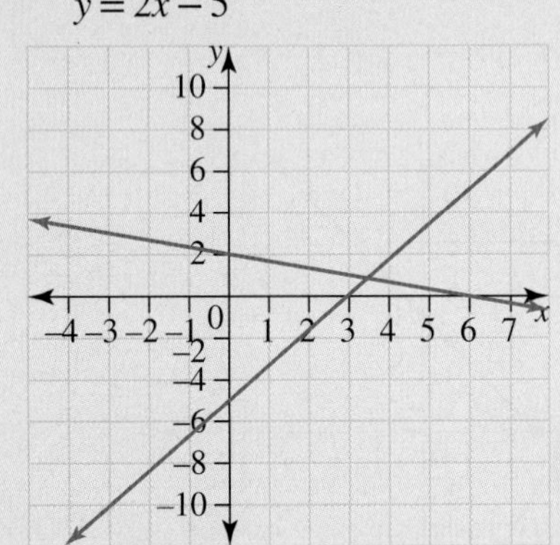

b. $3x + 2y = 12$
$2y = 3x$

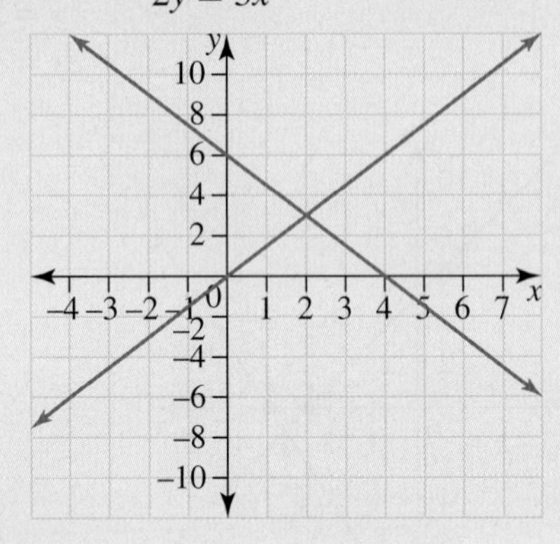

11. Solve the following simultaneous equations using the substitution method.

a. $y = 3x + 1$
$x + 2y = 16$

b. $y = 2x + 7$
$3y - 4x = 11$

c. $2x + 5y = 6$
$y = \frac{3}{2}x + 5$

d. $y = -x$
$y = 8x + 21$

e. $y = 3x - 11$
$y = 5x + 17$

f. $y = 4x - 17$
$y = 6x - 22$

12. Solve the following simultaneous equations using the elimination method.

a. $3x + y = 17$
$7x - y = 33$

b. $4x + 3y = 1$
$-4x + y = 11$

c. $3x - 7y = -2$
$-2x - 7y = 13$

d. $4y - 3x = 9$
$y + 3x = 6$

e. $5x + 2y = 6$
$4x + 3y = 2$

f. $x - 4y = -4$
$4x - 2y = 12$

13. Solve the following simultaneous equations using an appropriate method.

a. $3x + 2y = 6$
$3y + 5x = 9$

b. $6x - 4y = -6$
$7x + 3y = -30$

c. $6x + 2y = 14$
$x = -3 + 5y$

14. Sketch the following pairs of inequalities.

a. $y \leq x + 4$
$y \geq 3$

b. $2y - 3x \geq 12$
$y + 3x > 0$

c. $5x + y < 10$
$x + 2y < 11$

15. Determine the point(s) of intersection for each of the following pairs of lines.

a. $y = x^2 - 6$
$y = 5x - 3$

b. $y = \frac{2}{x}$
$y = 5x - 3$

c. $x^2 + y^2 = 2$
$y = 5x - 3$

Problem solving

16. Write the following as a pair of simultaneous equations and solve.

a. Determine which two numbers have a difference of 5, and their sum is 23.

b. A rectangular house has a total perimeter of 34 metres and the width is 5 metres less than the length. Calculate the dimensions of the house.

c. If two Chupa Chups and three Wizz Fizzes cost \$2.55, but five Chupa Chups and seven Wizz Fizzes cost \$6.10, determine the price of each type of lolly.

17. Laurie buys milk and bread for his family on the way home from school each day, paying with a \$10 note. If he buys three cartons of milk and two loaves of bread, he receives 5 cents in change. If he buys two cartons of milk and one loaf of bread, he receives \$4.15 in change. Calculate how much each item costs.

18. A paddock contains some cockatoos (2-legged) and kangaroos (4-legged). The total number of animals is 21 and they have 68 legs in total. Using simultaneous equations, determine how many cockatoos and kangaroos there are in the paddock.

19. Warwick was solving a pair of simultaneous equations using the elimination method and reached the result that $0 = -5$. Suggest a solution to the problem, giving a reason for your answer.

20. There are two sections to a concert hall. Seats in the 'Dress circle' are arranged in rows of 40 and cost \$140 each. Seats in the 'Bleachers' are arranged in rows of 70 and cost \$60 each. There are 10 more rows in the 'Dress circle' than in the 'Bleachers' and the capacity of the hall is 7000.

a. If d represents the number of rows in the 'Dress circle' and b represents the number of rows in the 'Bleachers' then write an equation in terms of these two variables based on the fact that there are 10 more rows in the 'Dress circle' than in the 'Bleachers'.

b. Write an equation in terms of these two variables based on the fact that the capacity of the hall is 7000 seats.

c. Solve the two equations from **a** and **b** simultaneously using the method of your choice to find the number of rows in each section.

d. Now that you have the number of rows in each section, calculate the number of seats in each section.

e. Hence, calculate the total receipts for a concert where all tickets are sold.

21. John is comparing two car rental companies, Golden Ace Rental Company and Silver Diamond Rental Company.
Golden Ace Rental Company charges a flat rate of \$38 per day and \$0.20 per kilometre. The Silver Diamond Rental Company charges a flat rate of \$30 per day plus \$0.32 per kilometre.
  a. Write an algebraic equation for the cost of renting a car for three days from the Golden Ace Rental Company in terms of the number of kilometres travelled, k.
  b. Write an algebraic equation for the cost of renting a car for three days from the Silver Diamond Rental Company in terms of the number of kilometres travelled, k.
  c. Determine how many kilometres John would have to travel so that the cost of hiring from each company for three days is the same.
  d. Write an inequation that, when solved, will tell you the number of kilometres for which it is cheaper to use Golden Ace Rental Company when renting for three days.
  e. Determine the number of kilometres for which it is cheaper to use Silver Diamond Rental Company for three days' hire.

22. Frederika has \$24 000 saved for a holiday and a new stereo. Her travel expenses are \$5400 and her daily expenses are \$260.
  a. Write down an equation for the cost of her holiday if she stays for d days. Upon her return from holidays Frederika wants to purchase a new stereo system that will cost her \$2500.
  b. Calculate how many days can she spend on her holiday if she wishes to purchase a new stereo upon her return.

on To test your understanding and knowledge of this topic, go to your learnON title at www.jacplus.com.au and complete the **post-test**.

Online Resources

Below is a full list of **rich resources** available online for this topic. These resources are designed to bring ideas to life, to promote deep and lasting learning and to support the different learning needs of each individual.

eWorkbook

Download the workbook for this topic, which includes worksheets, a code puzzle and a project (ewbk-2030) ☐

Solutions

Download a copy of the fully worked solutions to every question in this topic (sol-0738) ☐

Digital documents

4.2 SkillSHEET Graphing linear equations using the x- and y-intercept method (doc-5217) ☐
4.7 SkillSHEET Checking whether a given point makes the inequation a true statement (doc-5218) ☐
SkillSHEET Writing equations from worded statements (doc-5219) ☐

Video eLessons

4.2 Simultaneous linear equations and graphical solutions (eles-4763) ☐
Solutions to coincident, parallel and perpendicular lines (eles-4764) ☐
4.3 Solving simultaneous equations using the substitution method (eles-4766) ☐
Equating equations (eles-4767) ☐
Solving simultaneous equations using substitution (eles-1932) ☐
4.4 Solving simultaneous equations using the elimination method (eles-4768) ☐
Solving simultaneous equations by multiplying by a constant (eles-4769) ☐
Solving simultaneous equations using elimination (eles-1931) ☐
4.5 Applications of simultaneous linear equations (eles-4770) ☐
4.6 Solving simultaneous linear and quadratic equations (eles-4771) ☐
Solving simultaneous linear and hyperbolic equations (eles-4772) ☐
4.7 Inequalities between two expressions (eles-4774) ☐
4.8 Inequalities on the Cartesian plane (eles-4776) ☐
Determining the required region on the Cartesian plane (eles-4777) ☐
4.9 Multiple inequalities on the Cartesian plane (eles-4778) ☐

Interactivities

4.2 Individual pathway interactivity: Graphical solution of simultaneous linear equations (int-4577) ☐
Solving simultaneous equations graphically (int-6452) ☐
Parallel lines (int-3841) ☐
Perpendicular lines (int-6124) ☐
4.3 Individual pathway interactivity: Solving simultaneous linear equations using substitution (int-4578) ☐
Solving simultaneous equations using substitution (int-6453) ☐
4.4 Individual pathway interactivity: Solving simultaneous linear equations using elimination (int-4520) ☐
Solving simultaneous equations using elimination (int-6127) ☐
4.5 Individual pathway interactivity: Applications of simultaneous linear equations (int-4580) ☐
4.6 Individual pathway interactivity: Solving simultaneous linear and non-linear equations (int-4581) ☐
Solving simultaneous linear and non-linear equations (int-6128) ☐
4.7 Individual pathway interactivity: Solving linear inequalities (int-4582) ☐
Inequalities on the number line (int-6129) ☐
4.8 Individual pathway interactivity: Inequalities on the Cartesian plane (int-4583) ☐
Linear inequalities in two variables (int-6488) ☐
4.9 Individual pathway interactivity: Solving simultaneous linear inequalities (int-4584) ☐
Graphing simultaneous linear inequalities (int-6283) ☐
4.10 Crossword (int-2836) ☐
Sudoku puzzle (int-3591) ☐

Teacher resources

There are many resources available exclusively for teachers online.

To access these online resources, log on to **www.jacplus.com.au**.

Answers

Topic 4 Simultaneous linear equations and inequalities

Exercise 4.1 Pre-test

1. False
2. $(0, 2)$
3. An infinite number of solutions
4. $(1, 0.2)$
5. D
6. B
7. 92
8. A
9. E
10. A, C, E, G
11. C
12. B, D
13. At two points
14. E
15. B

Exercise 4.2 Graphical solution of simultaneous linear equations

1. a. $(2, 1)$ b. $(1, 1)$
2. a. $(0, 4)$ b. $(2, -1)$
3. a. $(-2, -4)$ b. $(-0.5, 1.5)$
4. a. No b. Yes c. Yes d. No
5. a. Yes b. No c. No d. Yes
6. a. No b. Yes c. No d. Yes
7. a. $(3, 2)$ b. $(4, 3)$ c. $(-3, 4)$ d. $(-2, 2)$
8. a. $(2, 0)$ b. $(3, 0)$ c. $(-2, 4)$ d. $(3, 8)$
9. a. $\left(-\frac{1}{2}, 1\frac{1}{2}\right)$ b. $(2, 5)$ c. $(5, 3)$ d. $\left(2, \frac{2}{3}\right)$
10. a. No solution b. $(2, -1)$ c. No solution d. $(1, 9)$
11. a. $(3, 1)$ b. No solution c. No solution d. $(2, 1)$
12. a. No solution b. $\left(\frac{6}{5}, -\frac{6}{5}\right)$ c. $(2, -4)$ d. No solution
13. $y = 4x - 16$
14. a. Northern beach
   $C = 20 + 12t, 0 \le t \le 5$
   Southern beach
   $D = 8 + 18t, 0 \le t \le 5$
   b. Northern beaches in red, southern beaches in blue
   c. Time $>$ 2 hours
   d. Time = 2 hours, cost = \$44

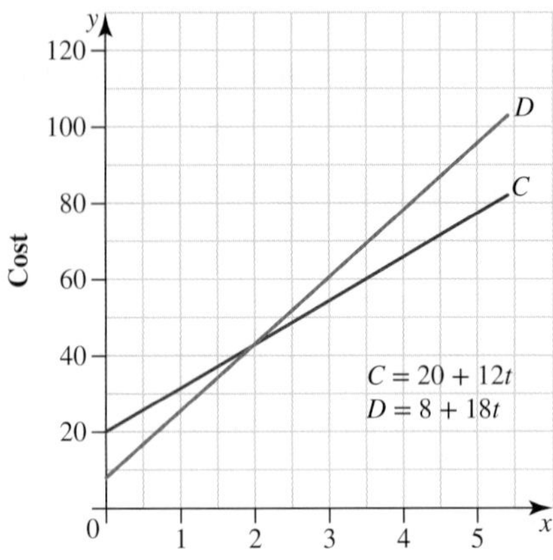

15. a. Same line b. Perpendicular c. Intersecting d. Parallel
16. a. 1 solution
   b. No solution (parallel lines)
   c. No solution (parallel lines)
17. a. 1 solution (perpendicular lines)
   b. Infinite solutions (coincident)
   c. 1 solution
   d. No solution (parallel lines)
18. a. $a = 3$ b. $a = \frac{1}{4}$ c. $a = 8$
19. a. $y = 3x + 6$
   b. $y = -2x + 1$
   c.

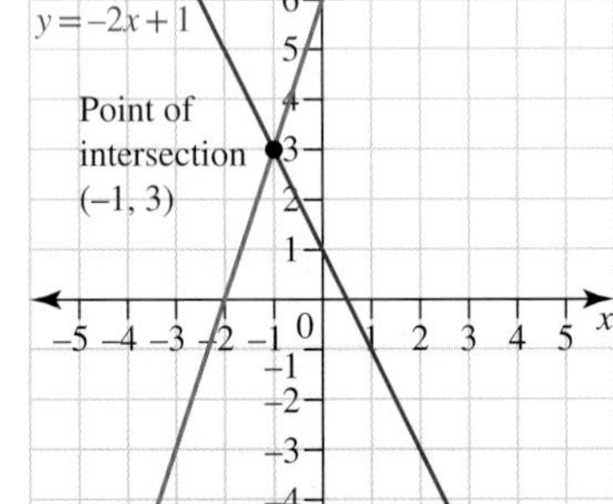

20.

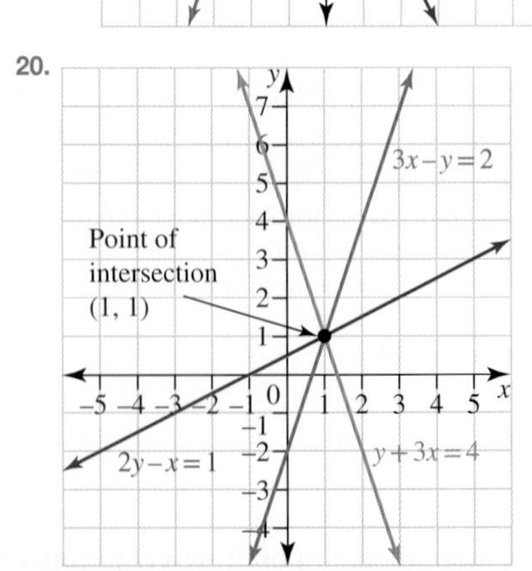

21. $y = \frac{5x}{4} + 9$

Exercise 4.3 Solving simultaneous linear equations using substitution

1. a. $(2, 3)$ b. $(2, -1)$
 c. $(3, -2)$ d. $(7, 6)$

2. a. $(3, 6)$ b. $(2, 1)$
 c. $(-1, -2)$ d. $(-4, 0)$

3. a. $(-1, -2)$ b. $(6, -2)$
 c. $\left(3, 1\frac{1}{2}\right)$ d. $(-3, -5)$

4. a. $(-6, -23)$ b. $(5, 23)$
 c. $(2, -6)$ d. $\left(\frac{3}{2}, -\frac{15}{2}\right)$

5. a. $(1, -7)$ b. $\left(-\frac{1}{2}, -4\right)$
 c. $\left(-\frac{3}{2}, -\frac{1}{2}\right)$ d. $\left(-\frac{1}{5}, \frac{4}{5}\right)$

6. a. $(-3, -1.5)$ b. $(1, 0.3)$
 c. $\left(-\frac{4}{5}, \frac{4}{5}\right)$ d. $(1, -1)$

7. 26 chickens

8. a. $(3, 1)$ b. $(-2, 3)$

9. a. $(5, 1)$ b. $(4, 2)$

10. a. $(0, 3)$ b. $(4, 0)$

11. a. $x = \frac{n}{2m}, y = \frac{n}{2}$
 b. $x = \frac{m}{n^2 + 1}, y = \frac{mn}{n^2 + 1}$
 c. $x = \frac{n}{m - n}, y = \frac{n^2}{m - n}$
 d. $x = \frac{n}{m - n}, y = \frac{n}{m - n}$
 e. $x = \frac{n^2 - m}{m - n}, y = \frac{m(n - 1)}{m - n}$
 f. $x = \frac{2m}{m + n}, y = \frac{m(n - m)}{m + n}$

12. $a = -1, b = 5$

13. $z = 24, m = 6, n = 9$

m	11	7
9	8	7
n	5	10

14. a. $x + 2y = 4$ b. $x = 2, y = 1$

15. Chemistry $21, physics $27.

16. 30 cm

17. Andrew is 16, Prue is 6.

18. $x = 8, y = -7$

19. $x = 0, y = 1$

20. a. $k \neq \pm 3$ b. $k = 3$ c. $k = -3$

Exercise 4.4 Solving simultaneous linear equations using elimination

1. a. $(3, 1)$ b. $(-2, 3)$ c. $(-2, 6)$

2. a. $(5, -1)$ b. $(2, 3)$ c. $(-3, 1)$

3. a. $(-3, 5)$ b. $(-5, -8)$ c. $(2, -2)$

4. a. $\left(1\frac{1}{2}, 3\frac{1}{2}\right)$ b. $\left(2, 1\frac{4}{5}\right)$ c. $(1, 1)$

5. a. $(1, 3)$ b. $(2, 4)$ c. $(5, 2)$

6. a. $(4, 2)$ b. $(-3, 4)$ c. $\left(-3, -1\frac{1}{2}\right)$

7. a. $(-6, -5)$ b. $(-3, 5)$ c. $(2, 1.8)$

8. a. $(6, 3)$ b. $(2, -2)$ c. $(1, 3)$

9. a. $(-1.5, -3)$ b. $(-8, 18)$ c. $(-3, 5)$

10. a. $(1, 3)$ b. $(4, 0)$ c. $(-3, 5)$

11. a. $(4, 3)$ b. $(8, 5)$ c. $\left(\frac{1}{3}, -\frac{1}{3}\right)$

12. a. $(4, 5)$ b. $(1, -3)$ c. $(12, 12)$

13. $3.40 (coffee is $5.10)

14. 12 five cent coins and 22 ten cent coins.

15. Abena 61 kg, Bashir 58 kg, Cecily 54 kg.

16. a. i. $acx + bcy = ce$ (3)
 ii. $acx + ady = af$ (4)
 iii. $y = \frac{ce - af}{bc - ad}$
 b. $x = \frac{de - bf}{ad - bc}$
 c. i. $\left(\frac{106}{31}, \frac{1}{31}\right)$ ii. $\left(\frac{37}{14}, \frac{11}{14}\right)$
 d. Because you cannot divide by 0.
 e. $ad - bc \neq 0$

17. It is not possible. When the two equations are set up it is impossible to eliminate one variable without eliminating the other.

18. $k - 10, 10$

19. $x = 7, y = -3$

20. $x = 4, y = 3, z = -6$

Exercise 4.5 Applications of simultaneous linear equations

1. Maths mark = 97, English mark = 66.
2. 18 nuts, 12 bolts.
3. 800 sheep, 400 chickens.
4. 8 and 3
5. 9 and 7
6. 6 and 5
7. Length = 12 m and width = 8 m.
8. Lemons cost 55 cents and oranges cost 25 cents.
9. Length 60 m and width 20 m.
10. Eight 20-cent coins and three 50-cent coins.
11. Twelve $1 coins and nine $2 coins.
12. Paddlepops cost $1.20 and a Magnum costs $2.10.

13. Cost of the Killer Python = 35 cents and cost of the Red frog = 25 cents.
14. Fixed costs = \$87, cost per person = \$23.50.
15. PE mark is 83 and Science mark is 71.
16. Mozzarella costs \$6.20, Swiss cheese costs \$5.80.
17. $x = 3$ and $y = 4$.
18. Fixed costs = \$60, cost per person = \$25.
19. \$4 each for DVDs and \$24 each for zip disks.
20. 9 and 24 years old.
21. Child \$12.50, Adult \$18.25, Elderly \$15.
22. 6.5 km
23. 66 cups of hot chips, 33 meat pies and 22 hot dogs were sold during the half-hour period.
24. a. See graph at the bottom of the page.*
    b. Jet-ski 3 wins the race.
    c. Jet-ski 1 and 2 reach the destination at the same time although jet-ski 2 started two hours after jet-ski 1. Jet-ski 3 overtakes jet-ski 1 6 hours and 40 minutes after its race begins or 10 hours and 40 minutes after jet-ski 1 starts the race. Jet-ski 3 overtakes jet-ski 2 6 hours after it starts the race or 8 hours after jet-ski 2 started the race.
25. 176 km

Exercise 4.6 Solving simultaneous linear and non-linear equations

1. A parabola may intersect with a straight line twice, once or not at all.
2. a. $(-5, 4)$ and $(-1, 0)$
   b. $(2, 3)$
   c. $\left(1-\sqrt{10}, -6\right)$ and $\left(1+\sqrt{10}, -6\right)$
3. a. $\left(\frac{-1}{2}-\frac{3\sqrt{5}}{2}, \frac{-19}{2}-\frac{9\sqrt{5}}{2}\right)$ and $\left(\frac{-1}{2}+\frac{3\sqrt{5}}{2}, \frac{-19}{2}+\frac{9\sqrt{5}}{2}\right)$
   b. $(-1, -2)$ and $(2, 1)$
   c. $(-2.54, -8.17)$ and $(3.54, 16.17)$
4. a. $(-1.41, 4)$ and $(1.41, 4)$
   b. $(-1, 2)$ and $\left(\frac{5}{2}, \frac{1}{4}\right)$
   c. $(3, 37)$
5. B
6. C
7. a. Yes b. No c. Yes d. No
8. a. $(1, 1), (-1, -1)$
   b. $\left(1+\sqrt{2}, -1+\sqrt{2}\right), \left(1-\sqrt{2}, -1-\sqrt{2}\right)$
   c. $\left(\frac{-\sqrt{15}}{3}, -\sqrt{15}\right), \left(\frac{\sqrt{15}}{3}, \sqrt{15}\right)$
   d. $(-6, -1), (2, 3)$
9. a. $(-1, -3), (1, 3)$
   b. $(-4, 3), (4, -3)$
   c. $(-1, 7), (5, -5)$
   d. $\left(\frac{\left(-1\sqrt{4}-2\right)}{2}, \frac{\left(\sqrt{4}+2\right)}{2}\right), \left(\frac{\left(\sqrt{4}+2\right)}{2}, \frac{\left(-1\sqrt{4}-2\right)}{2}\right)$
10. a. $\left(-\frac{1}{2}, -2\right), \left(\frac{1}{2}, 2\right)$ b. $(0, 5), (4, -3)$
    c. $\left(-\frac{1}{2}, 2\right), (0, 3)$ d. $\left(\frac{8}{3}, -\frac{1}{4}\right), (5, -2)$
11. a. $(1, 1)$ b. $(-3, 3), (3, 3)$
    c. $(-4, 11), (-2, 3)$ d. $\left(-\frac{100}{3}, \frac{652}{9}\right), (12, 12)$
12. a. $k < 0$ b. $k = 0$ c. $k > 0$
13. The straight line crosses the parabola at $(0, -7)$ so no matter what value m takes, there will be at least one intersection point.
14. a. i. 1, 2, 3
       ii. 0, 1, 2, 4
       iii. 1, 2, 3, 4, 5
    b. The number of possible intersections between an equation and a straight line is equal to the highest power of x.

*24a.

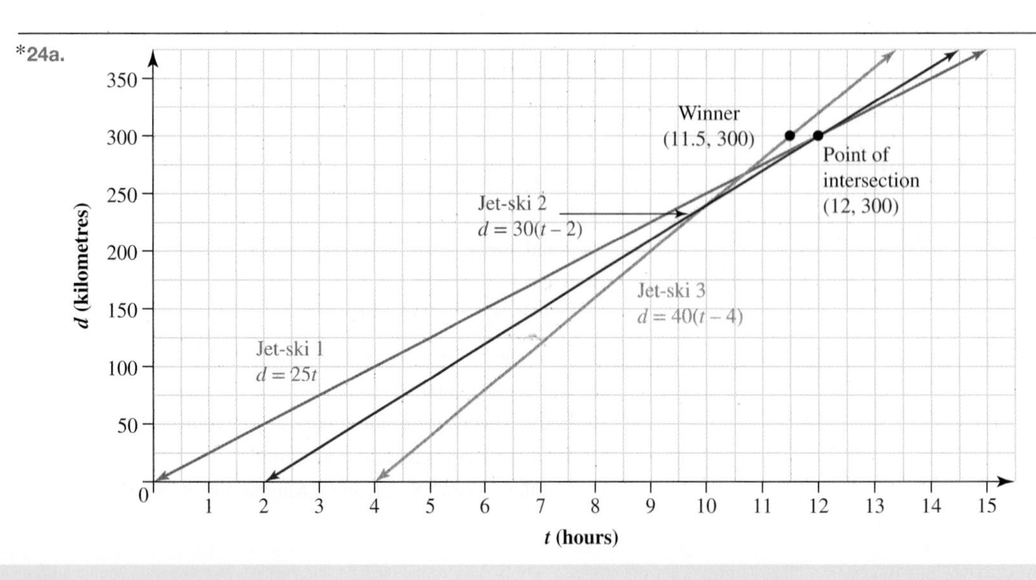

15. 17, 18 and −17, −18

16. Length 15 m, width 85 m.

17. a. $(5,5),(-1,-7)$

b. 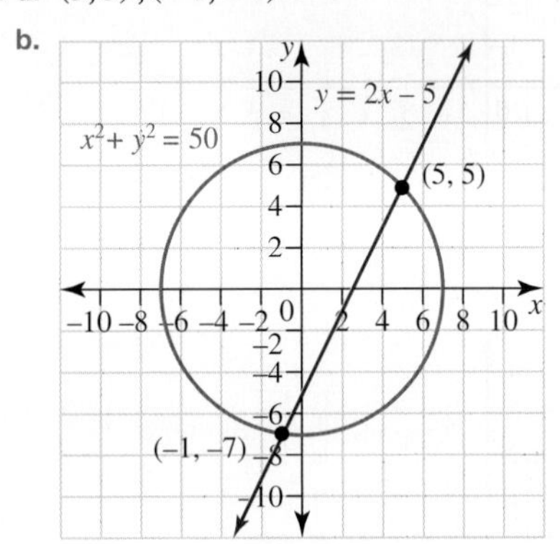

18. 9, 12

19. a. $(4,4), \left(-\sqrt{32}, 0\right)$

b. 6.66 km per hour or 11.11 km per hour

20. a. 100 °C

b. 0 minutes, 6 minutes

c. Eve's model. This model flattens out at 20 °C, whereas Adam's becomes negative which would not occur.

Exercise 4.7 Solving linear inequalities

1. a. $x > 2$

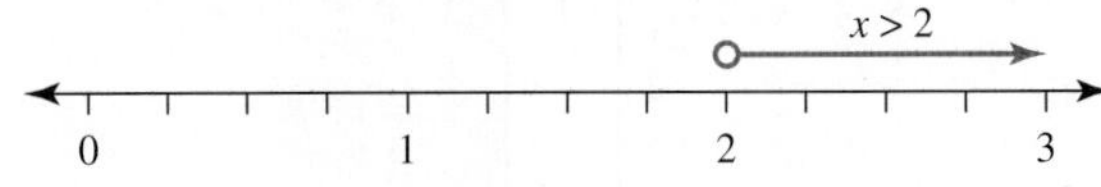

b. $a > -1$

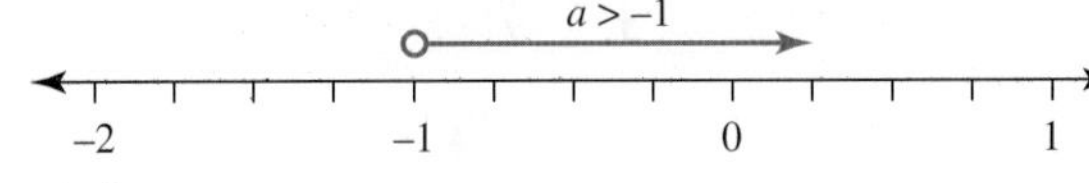

c. $y \geq 7$

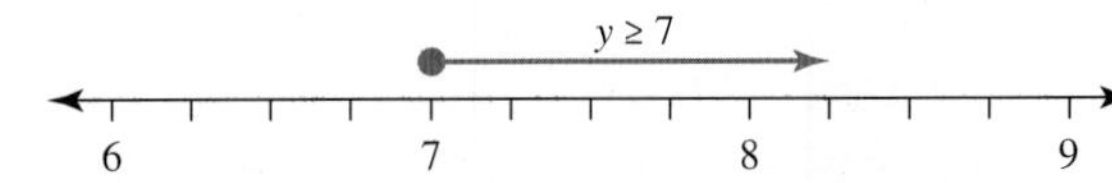

d. $m \geq 4$

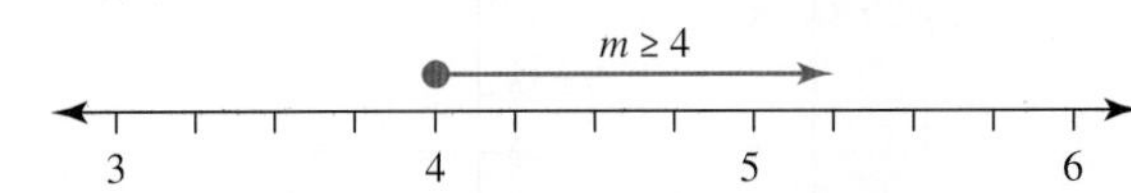

2. a. $p < 1$

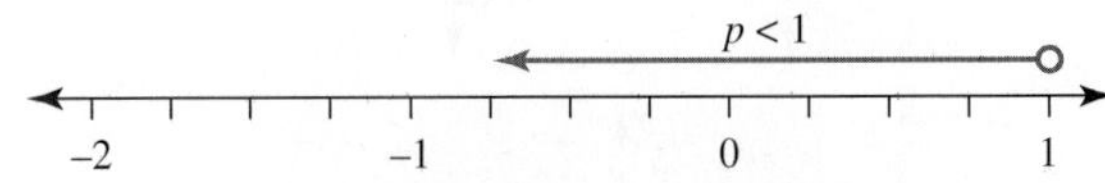

b. $x < 7$

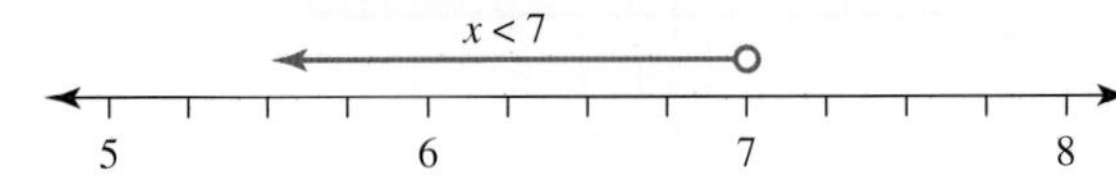

c. $m \leq 9$

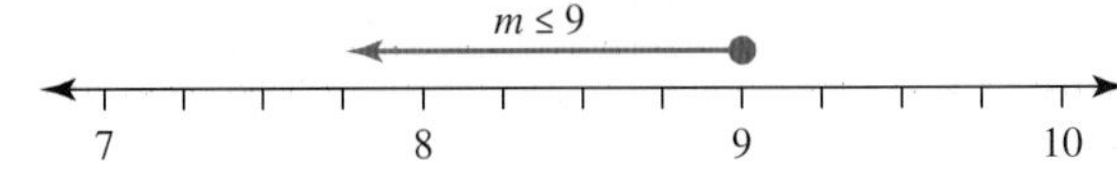

d. $a \leq 7$

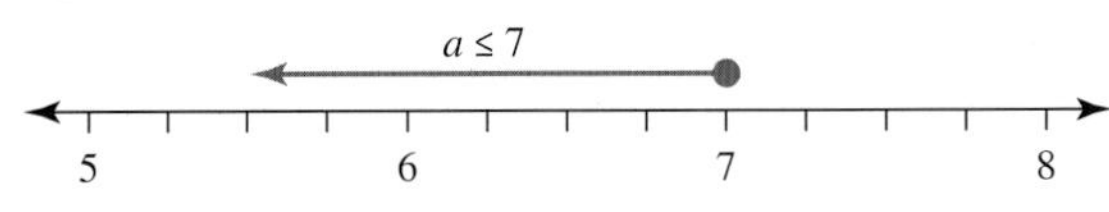

3. a. $x > 3$

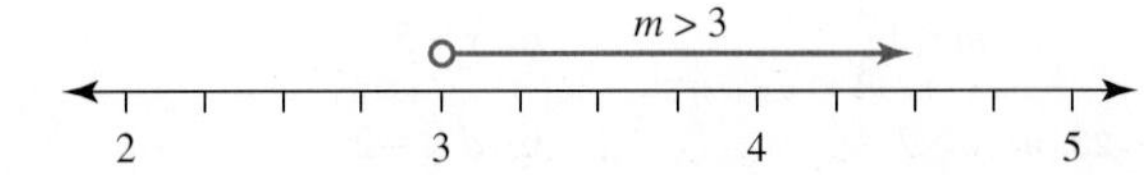

b. $m \geq 2$

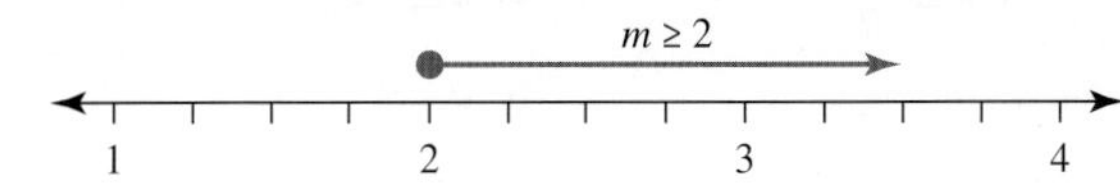

c. $q \geq -4$

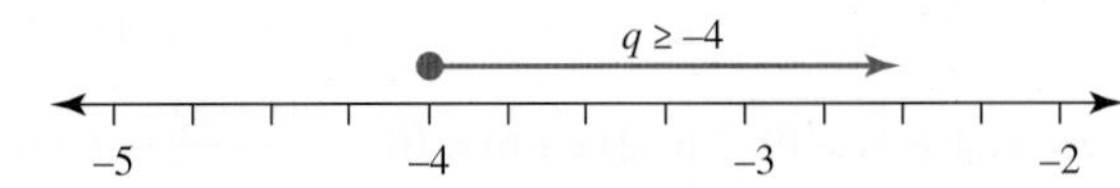

d. $a > -8$

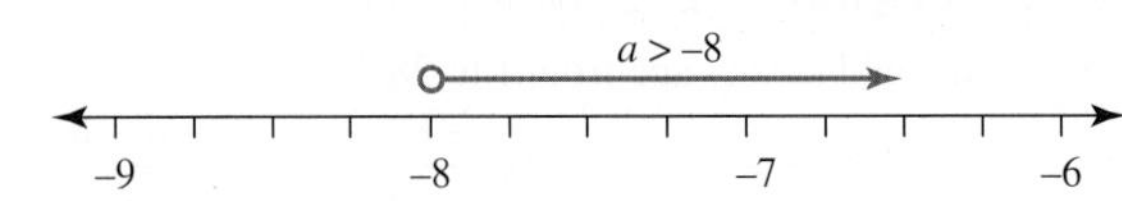

4. a. $m > 3$ b. $p \leq 2$ c. $a < 4$ d. $x \geq 5$

5. a. $p > -5$ b. $x \leq -7$ c. $m \geq -0.5$ d. $b > -0.5$

6. a. $m > 18$ b. $x < 8$ c. $a \leq -14$ d. $m \geq 25$

7. a. $m < 4.5$ b. $x \geq 3$ c. $p > 4$ d. $n \leq 2$

8. a. $b < 5$ b. $y > 2$ c. $m \leq -1$ d. $a \geq -5$

9. a. $b < -4\frac{1}{3}$ b. $c \leq -1$ c. $p > -2$ d. $a \geq -7$

10. a. $m > 3$ b. $a \geq 2$ c. $a < -1$ d. $a \leq -3$

11. a. $x > 6$ b. $x \leq 2$ c. $b < 4$ d. $a > 5$

12. a. $m < 2$ b. $m \leq 3$ c. $b \leq -\frac{16}{11}$ d. $m \geq 1$

13. a. $x \leq 7$ b. $x \geq -18$ c. $x < -10$

14. a. $x > 10\frac{1}{2}$ b. $x \geq 5$ c. $x < -1\frac{4}{5}$

15. a. $m < -2$ b. $p \geq -3$ c. $a \leq 5$ d. $p \geq -5$ e. $y \leq -3$

16. a. $x > 7$ b. $p < 0$ c. $a \geq \frac{1}{5}$ d. $x > -3$ e. $a \leq -11$

17. a. $b \leq 3$ b. $x < -3$ c. $k > 8$ d. $x > -18\frac{1}{2}$ e. $a \leq 40$

18. B

19. a. $x < -1$ b. $m \leq -3$

20. a. $x > 17$ b. $a > \frac{5}{8}$

21. a. $m \geq 1\frac{1}{3}$ b. $m \geq -12$

22. a. $k > 2$ b. $a > -5$
c. $m \leq 1\frac{2}{3}$ d. $x > 5$

23. a. $y \geq 7$ b. $d < -2$
c. $p \geq \frac{-6}{7}$ d. $x \geq -5$

24. a. $m < -2$ b. $a < 9$
c. $p \geq 3$ d. $x > -4\frac{1}{2}$

25. a. $5x > 10$ b. $x - 3 \leq 5$ c. $7 + 3x < 42$

26. a. $4 + 3x > 19$ b. $2(x + 6) < 10$ c. $\frac{(x - 8)}{2} - 7 \geq 9$

27. a. $50x \geq 650$ b. $2.50d + 5 \leq 60$

28. Tom could be any age from 1 to 28.

29. a. $-6.5 < x < -2$ b. $\frac{-c - b}{a} < x < \frac{-d - b}{a}$

30. a. $S > 47$
b. No
c. Sample responses can be found in the worked solutions in the online resources.

31. a. $n < 16\,800$ km
b. Mick travelled less than 16 800 km for the year and his costs stayed below \$16 000.

32. a. $20d + 10c \leq 240$
b. $2d + 2c \leq 36$
c. $0 \leq d \leq 12$ and $0 \leq C \leq 18$

33. \$20 000

Exercise 4.8 Inequalities on the Cartesian plane

1. a.

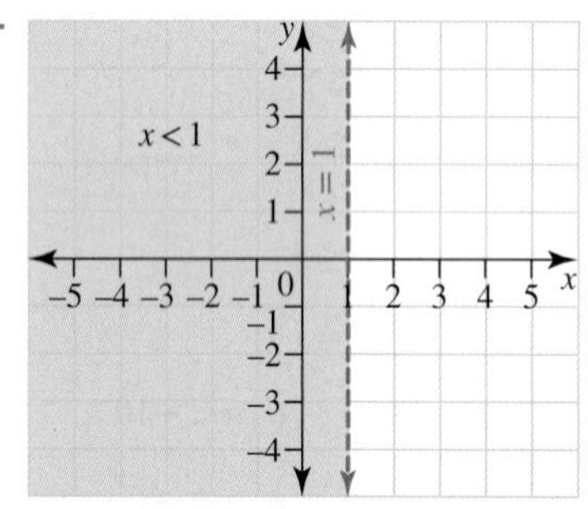

b.

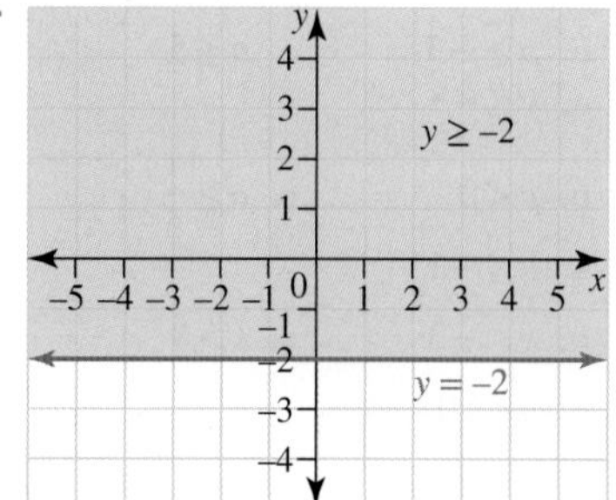

c.

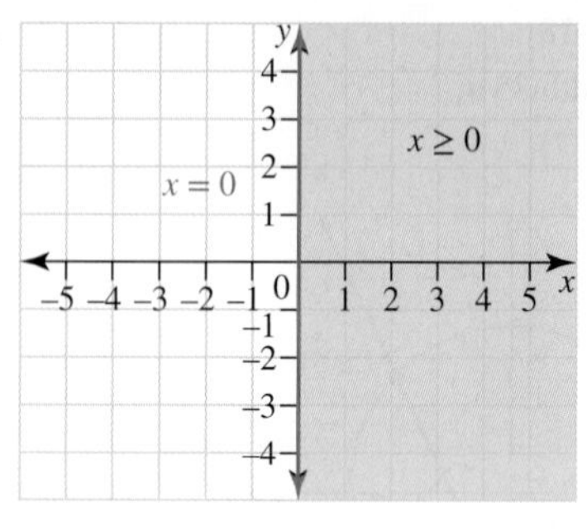

d.

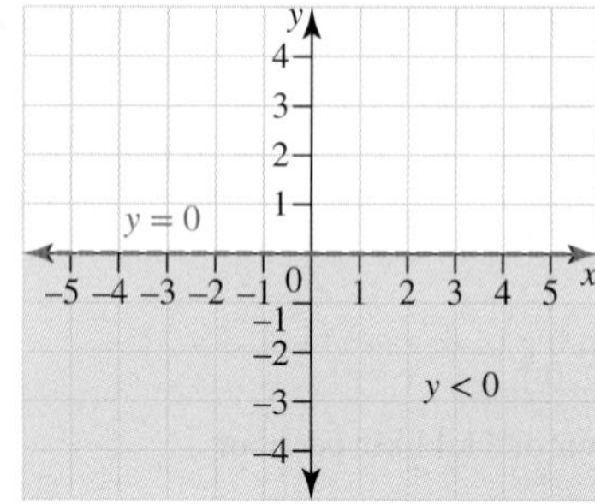

2. a.

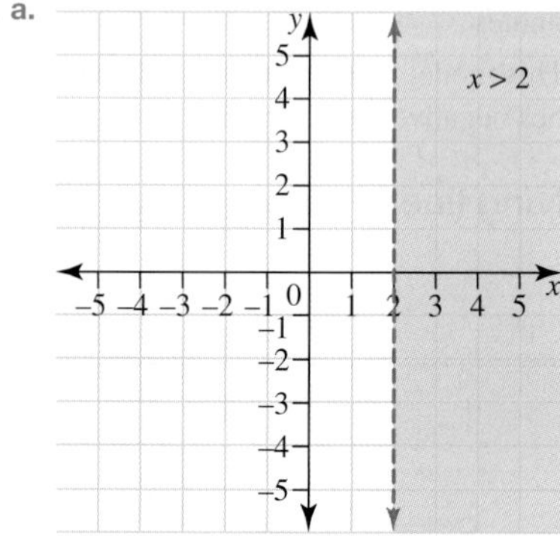

b.

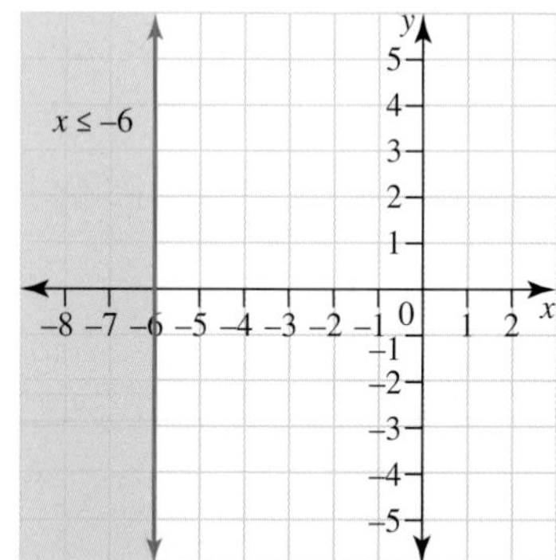

c.

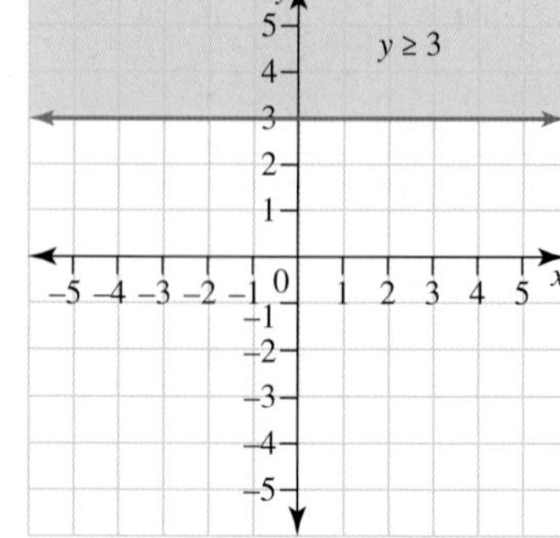

d.

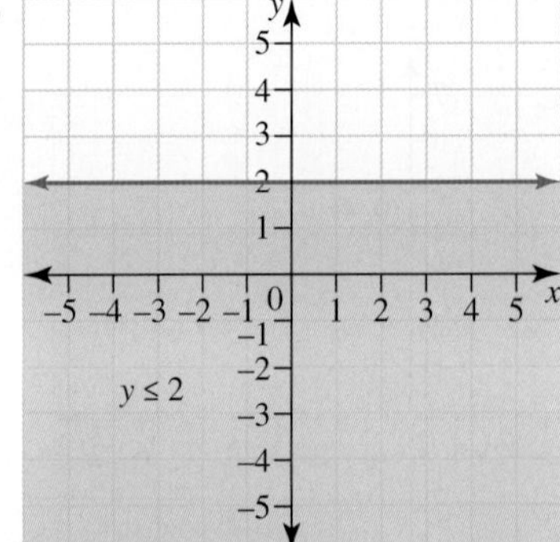

3. a.

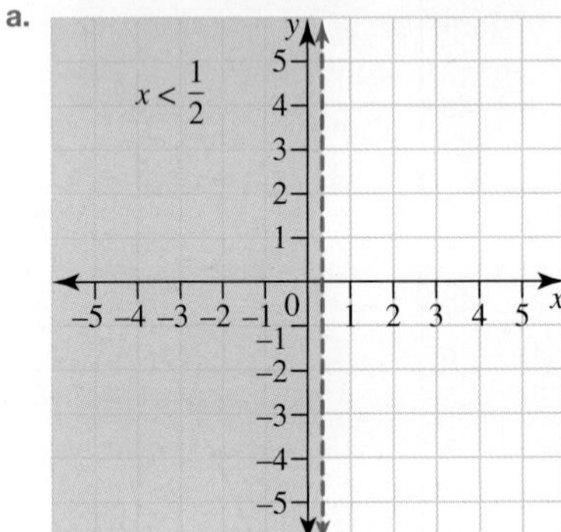

b.

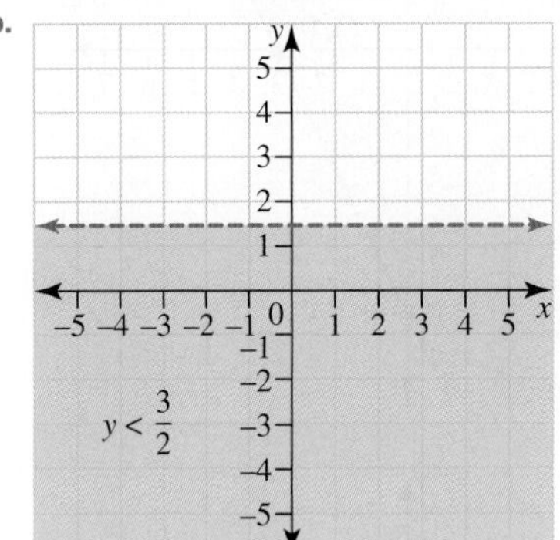

c.

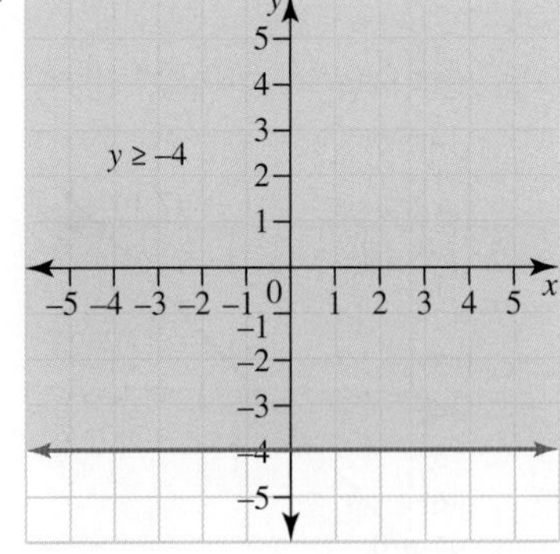

d.

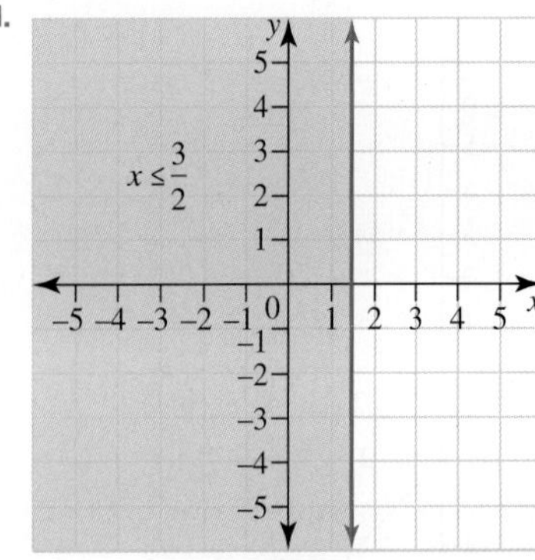

4. a. C b. A, C

5. a. A, B b. A, B, C

6. a. B b. A, B, C

7. a. $y \geq x + 1$

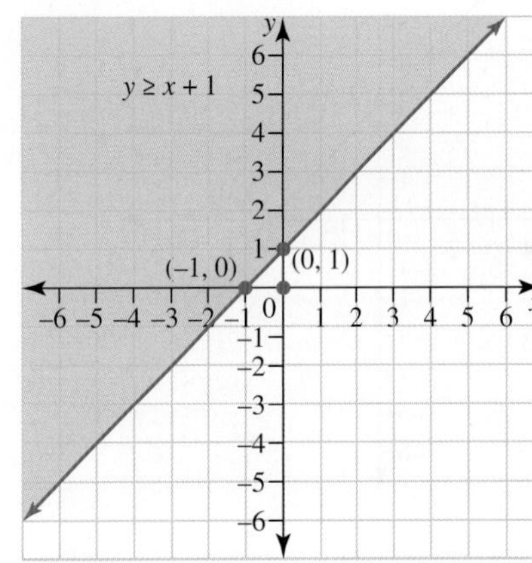

b. $y < x - 6$

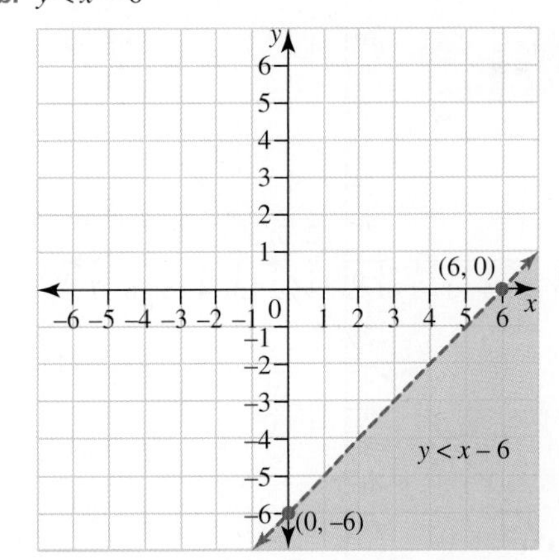

c. $y > -x - 2$

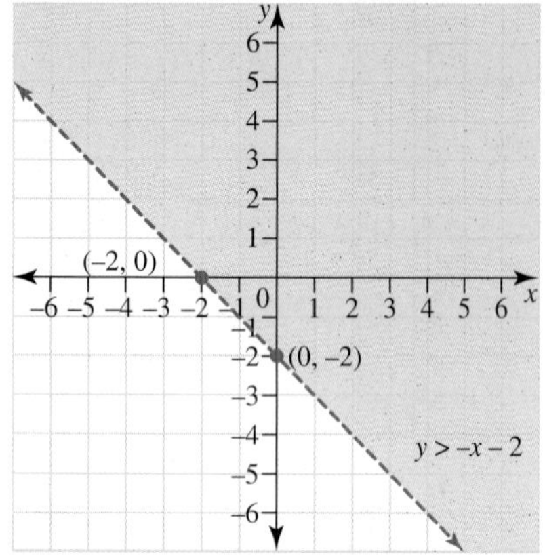

d. $y < 3 - x$

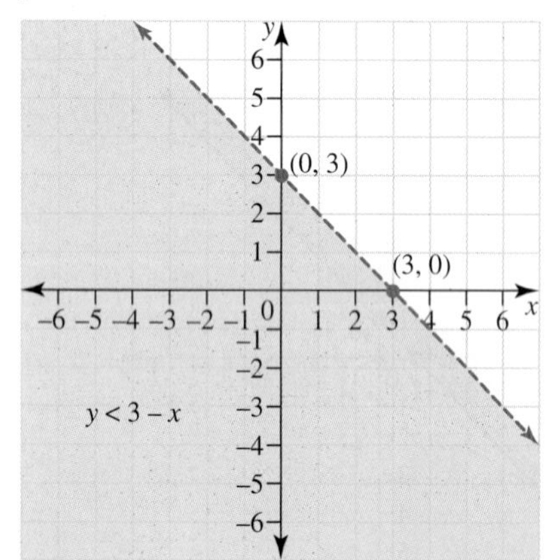

8. a. $y > x - 2$

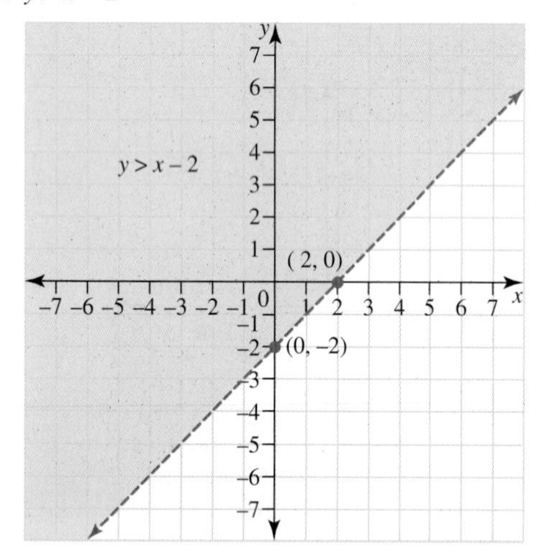

b. $y < 4$

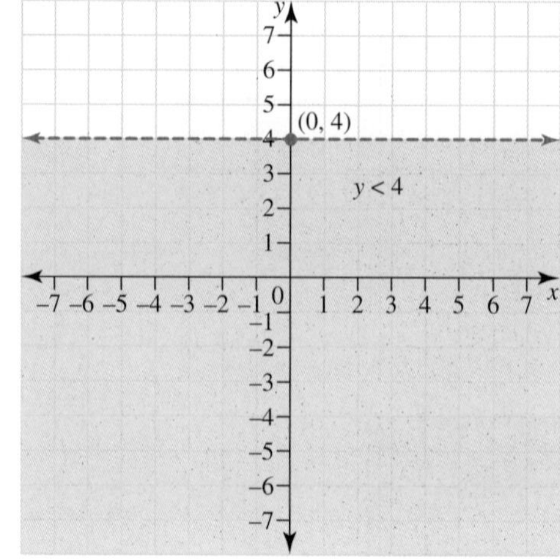

c. $2x - y < 6$

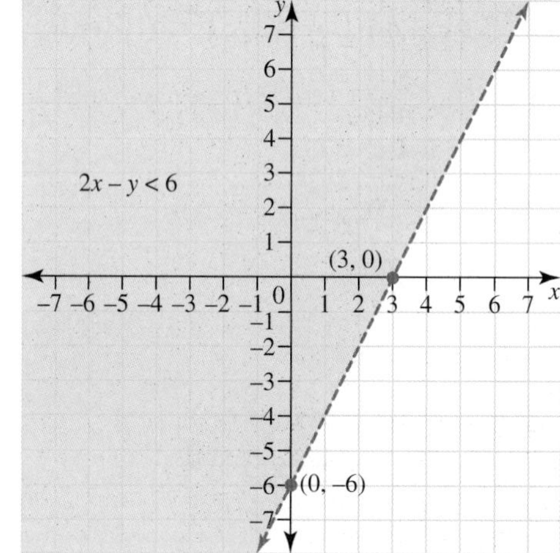

d. $y \le x - 7$

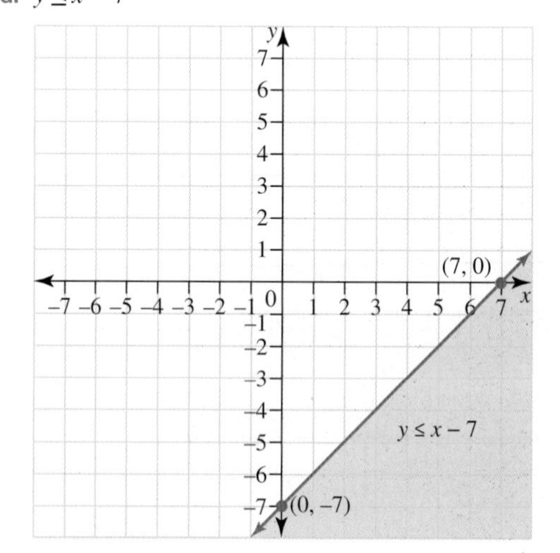

9. a. $x - y > 3$

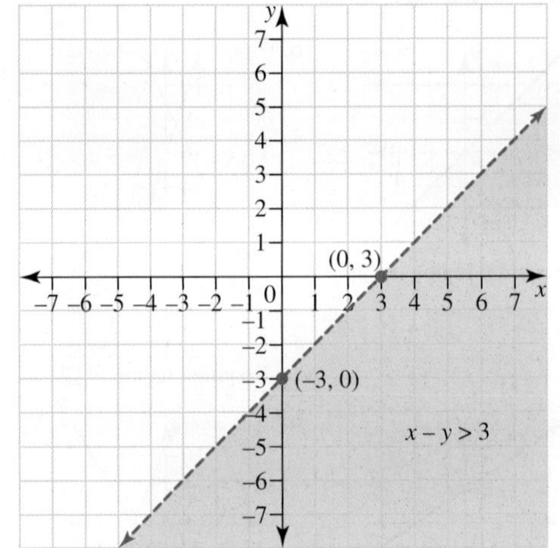

b. $y < x + 7$

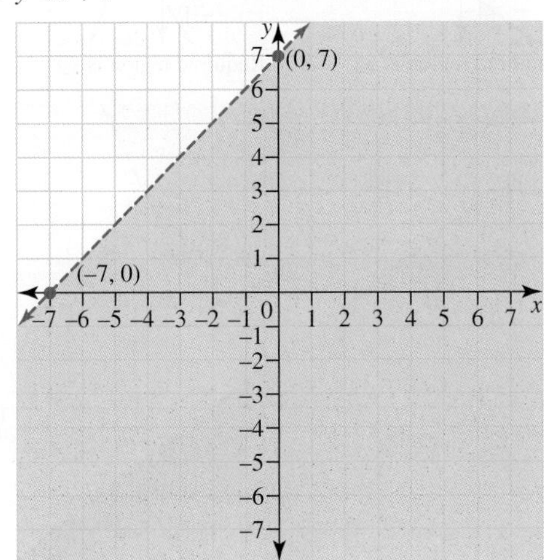

c. $x + 2y \leq 5$

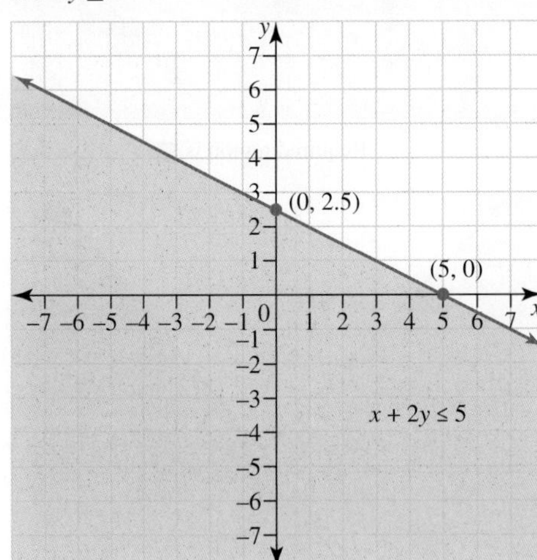

d. $y \leq 3x$

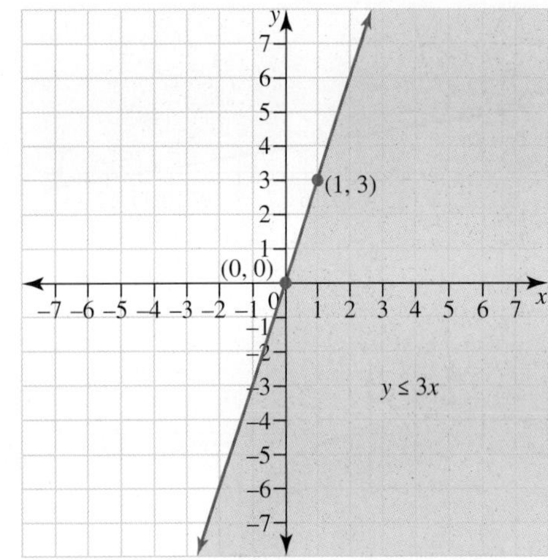

10. B

11. D

12. A

13. a. $y = \frac{1}{2}x + 3$

b. $y \geq \frac{1}{2}x + 3, x > 2, y \leq 7$

14. $(3, 3), (3, 4), (3, 5), (4, 3)$

15. a. $l + s \leq 30$

b. At least 12 small dogs

c.

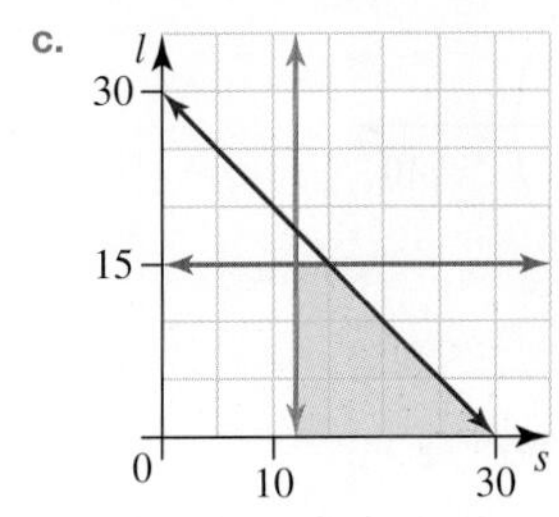

d. 15 large and 15 small dogs

16. 20 units^2

17. a. $y \leq 2x - 2$

b. Sample responses can be found in the worked solutions in the online resources.

18. a. $y = -\frac{2}{3}x + 3$

b. $y > -\frac{2}{3}x + 3$

c. $y < -\frac{2}{3}x + 3$

d. $y \geq -\frac{2}{3}x + 3$

19. a. $y = \frac{11}{6} - \frac{1}{6}x$

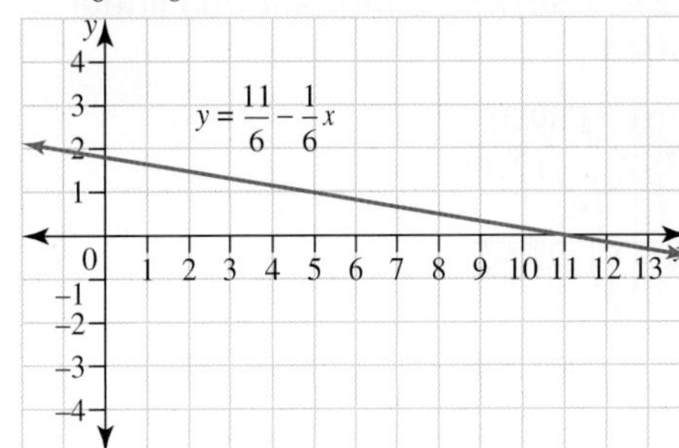

b. The unshaded region is the required region.

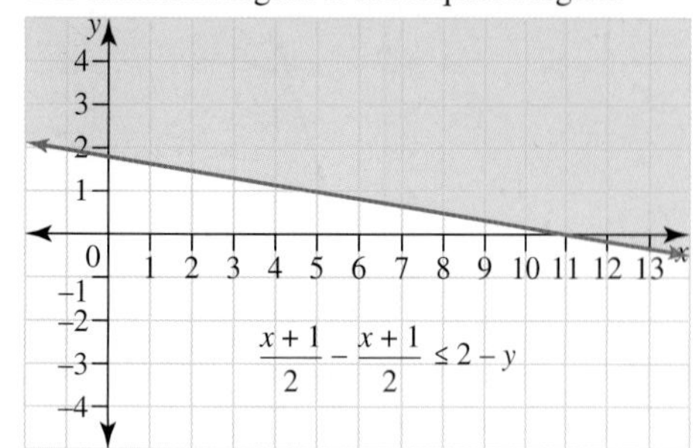

20. a.

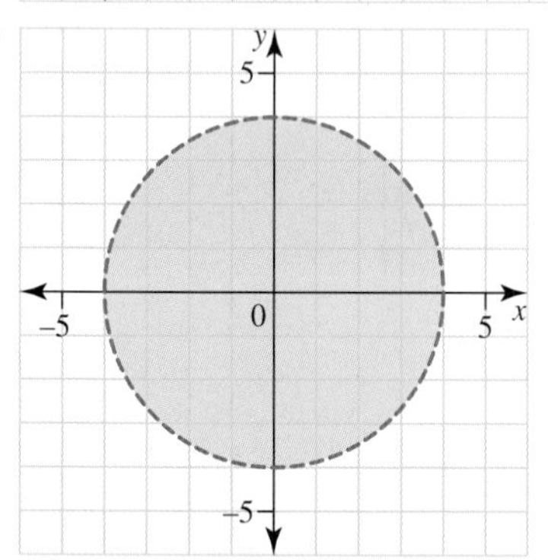

b.

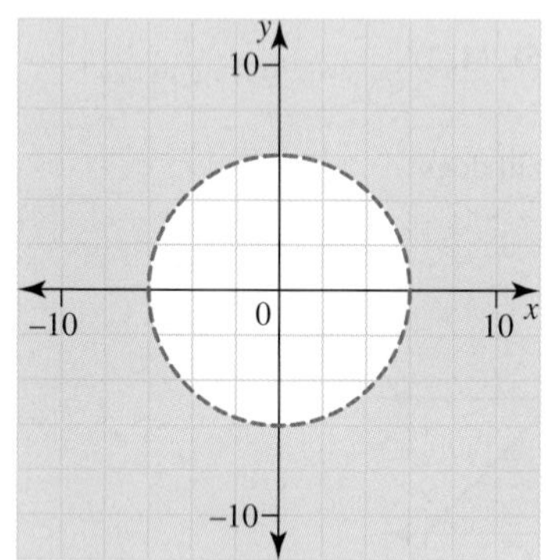

21. The unshaded region is the required region.

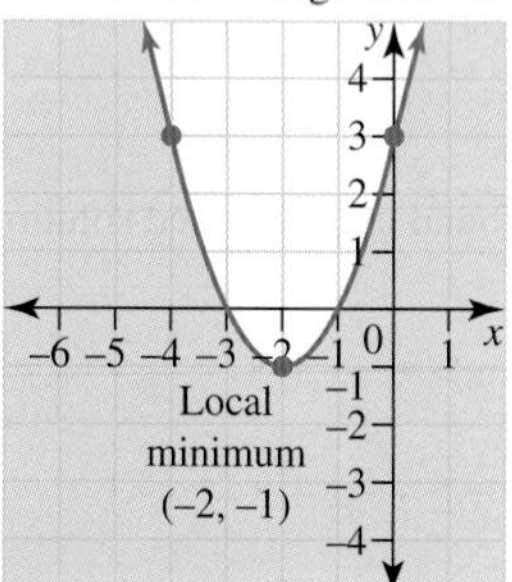

Exercise 4.9 Solving simultaneous linear inequalities

1.

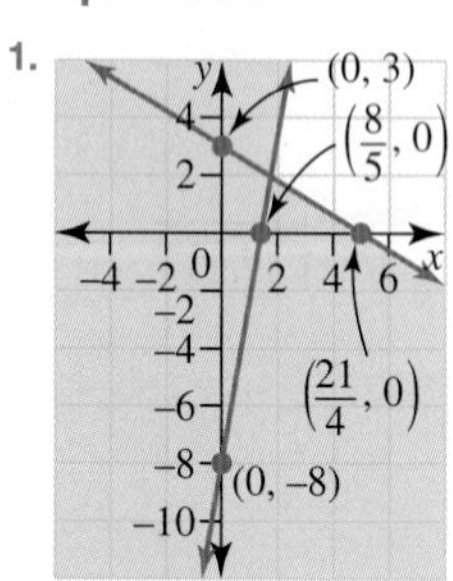

Required region is □

2. $y \leq 2x + 3$ and $y \geq \frac{2}{3}x - 2$

3. a.

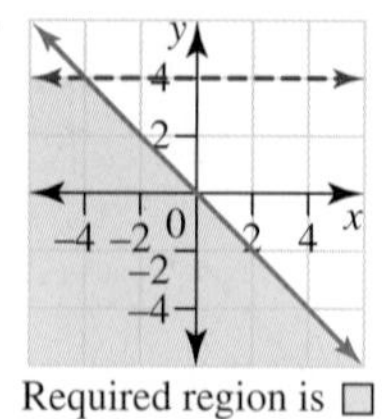

Required region is □

b.

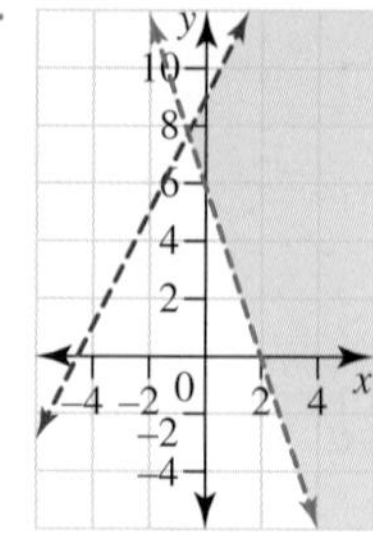

Required region is □

4. a.

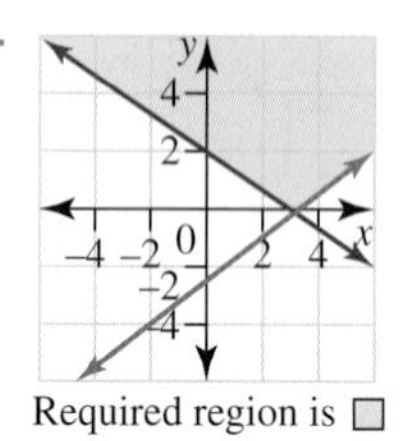

Required region is □

b.

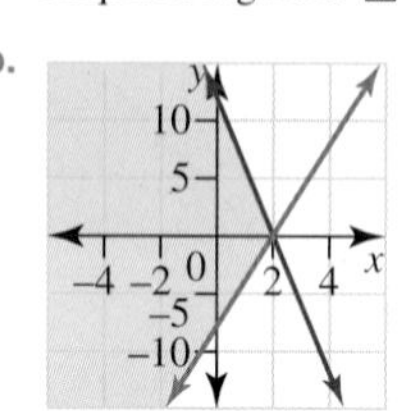

Required region is □

5. a.

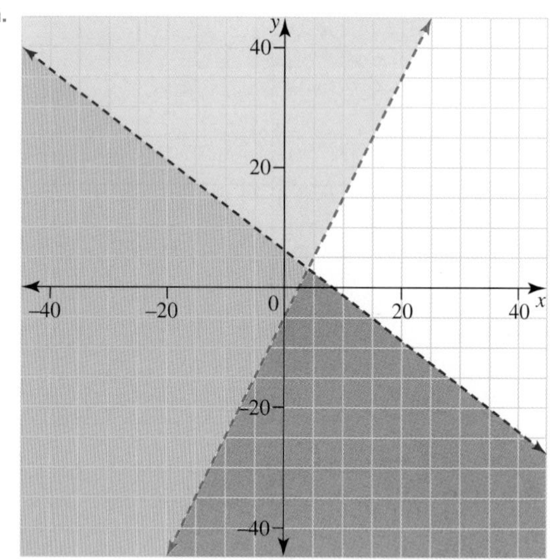

Required region is ▨

b.

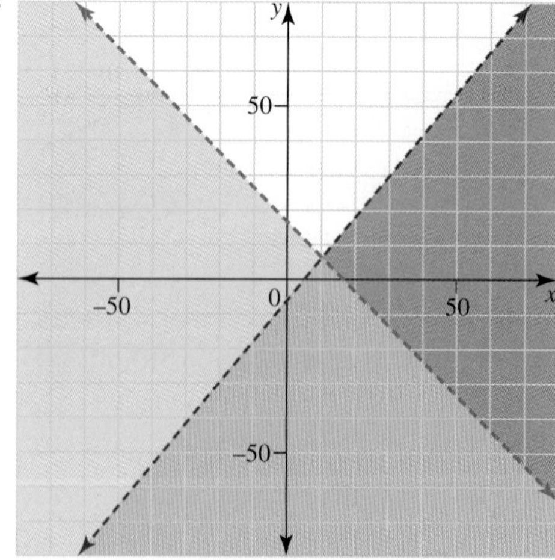

Required region is ▨

6. B

7. $\frac{1}{2} \le x \le 2, y \ge 0, 2x + y \ge 2, 4x + 3y \le 12$

8. a.

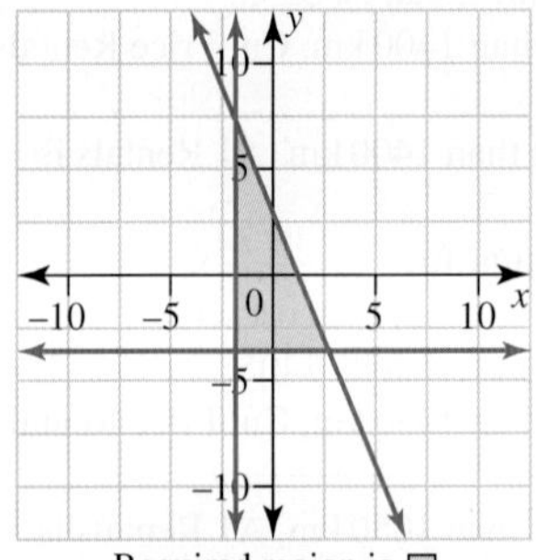

Required region is ▢

b. $(-2, 8.5), (-2, -3), (2.6, -3)$

9. a.

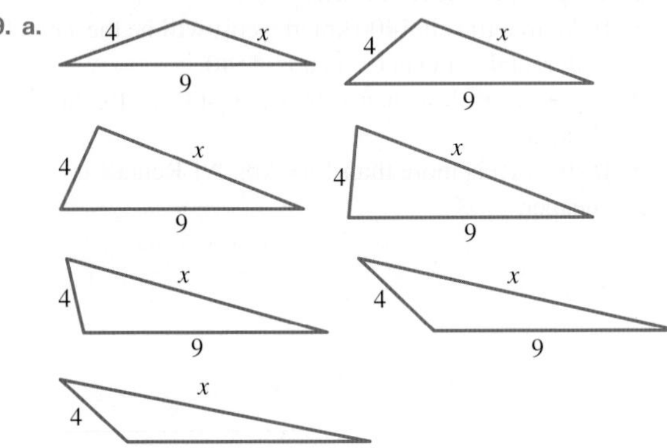

b. $5 < x < 13$

10.

a. $AB\colon x = 0, AC\colon y = 0, BC\colon 3y + 4x = 24$

b. $x \ge 0, y \ge 0, 3y + 4x \ge 24$

c. $a = 10$ units, $b = 6$ units, $c = 8$ units

11. a. $L > 0, W > 4, L \times W < 25$

b. 29 possible rectangles

12. a. $y = \frac{1}{4}x + 3$ or $x + 4y = 12$ and $y = \frac{1}{2}x + 4$ or $x + 2y = 8$

b. A$(4, 2)$

c. $y \le -\frac{1}{4}x + 3$ or $x + 4y \le 12$

$y \ge -\frac{1}{2}x + 4$ or $x + 2y \ge 8$

$x \ge 4$

$y \ge 0$

13. a. $y \ge 0$

$x + y \ge 400$

$x \ge 125$

$\frac{1}{2}x + \frac{1}{4}y \le 175$ or $2x + y \le 700$

b. See graph at the bottom of page.*

c. Vertices are $(125, 275), (125, 450)$ and $(300, 100)$.

14. a. $24p + 10c \ge 100, 14p + 32c \ge 160, 5p + 20c \ge 70$

b.

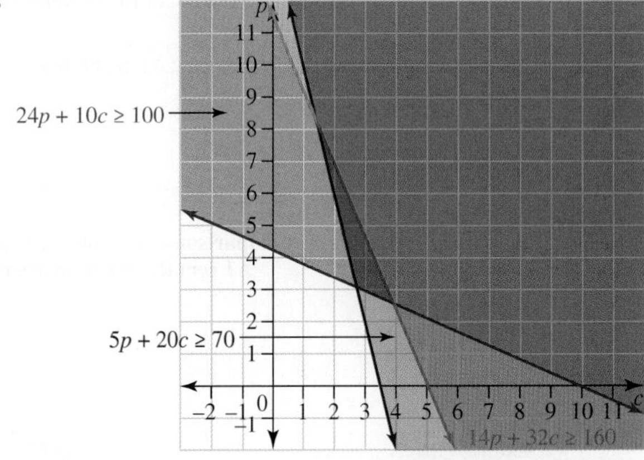

Required region is ▢

c. The minimum number of sachets required is 7 ($p = 3$ and $c = 4$ satisfies all conditions).

*13. b.

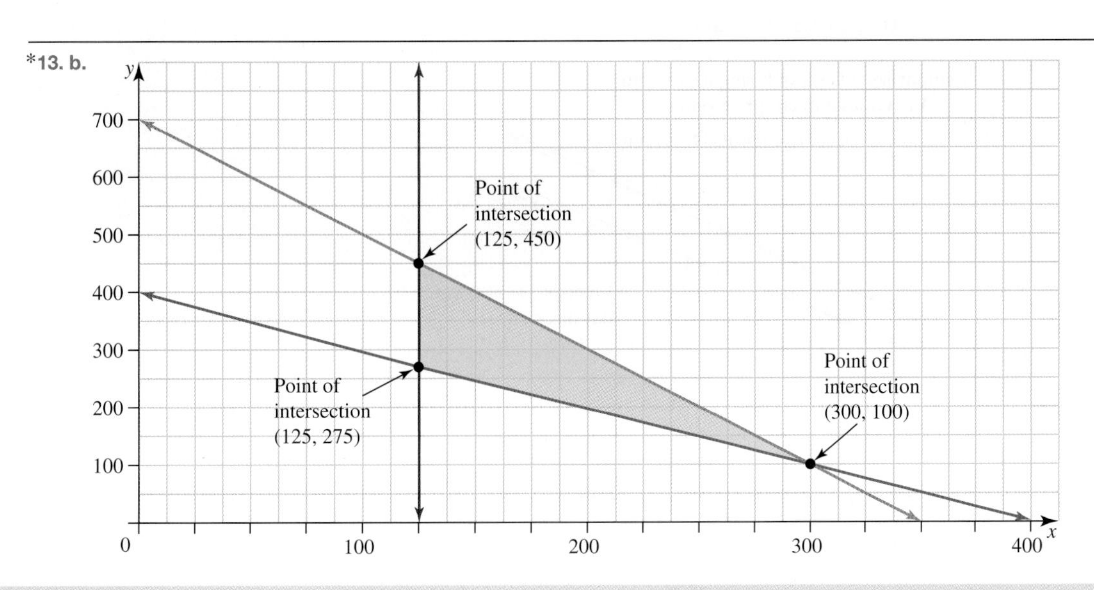

Project

1. A1 Rentals: $C = \$35 \times 3 + 0.28d$
   Cut Price Rentals: $C = \$28 \times 3 + 0.3d$
2. See graph at the bottom of the page.*
3. 1050 km
4. 1400 km
5. a. A1 Rentals: $C = \$37 \times 3 + 0.28d$
   Cut Price Rentals: $C = \$30 \times 3 + 0.3d$
   b. A1 Rentals: $C = \$37 \times 4 + 0.28d$
   Cut Price Rentals: $C = \$30 \times 4 + 0.3d$
6. See graph at the bottom of the page.*
7. The extra cost of \$2 per day for both rental companies has not affected the charges they make for the distances travelled. However, the overall costs have increased.
8. Presentation of the answers will vary. Answers will include:
   Travelling 3 days this month:
   - If Andrea travels 1050 km, the cost will be the same for both rental companies; that is, \$399.
   - If she travels less than 1050 km, Cut Price Rentals is cheaper.
   - If she travels more than 1050 km, A1 Rentals is cheaper.

   Travelling 4 days this month:
   - If Andrea travels 1400 km, the cost will be the same for both rental companies; that is, \$532.
   - If she travels less than 1400 km, Cut Price Rentals is cheaper.
   - If she travels more than 1400 km, A1 Rentals is cheaper.

   Travelling 3 days next month:
   - If Andrea travels 1050 km, the cost will be the same for both rental companies; that is, \$405.
   - If she travels less than 1050 km, Cut Price Rentals is cheaper.
   - If she travels more than 1050 km, A1 Rentals is cheaper.

   Travelling 4 days next month:
   - If Andrea travels 1400 km, the cost will be the same for both rental companies; that is, \$540.
   - If she travels less than 1400 km, Cut Price Rentals is cheaper.
   - If she travels more than 1400 km, A1 Rentals is cheaper.

*2.

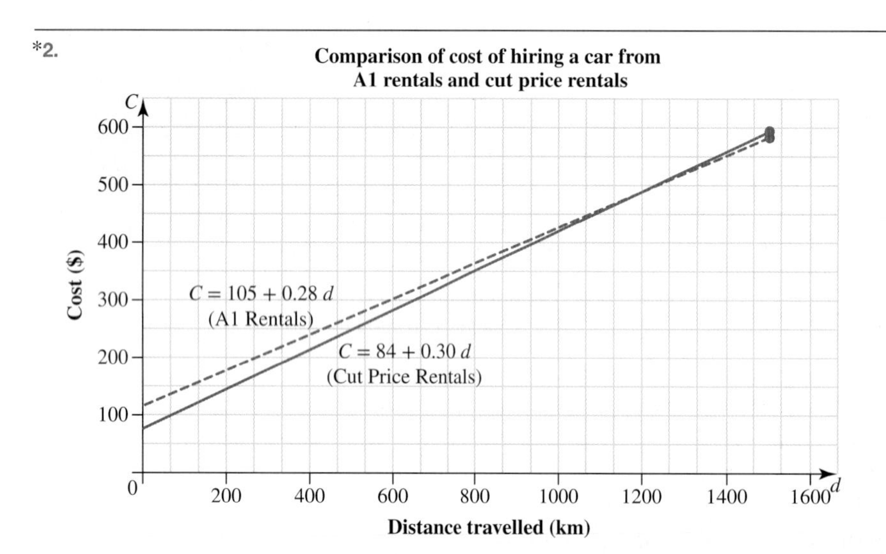

*6.

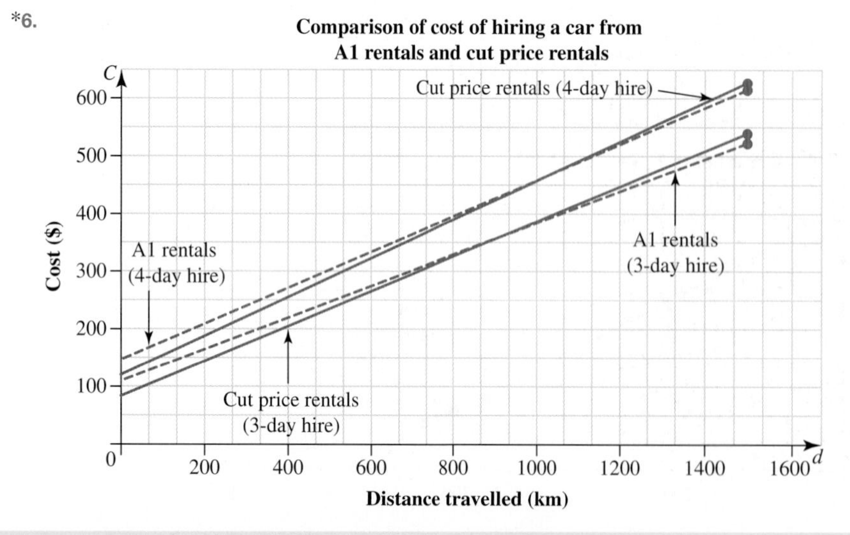

Exercise 4.10 Review questions

1. A
2. C
3. E
4. C
5. D
6. A
7. *Note:* The shaded region is the region required.

a. 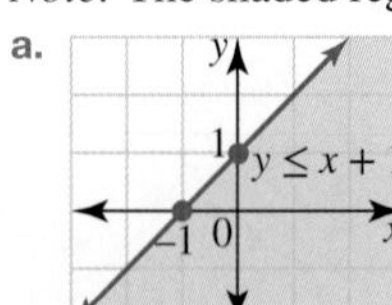

b. 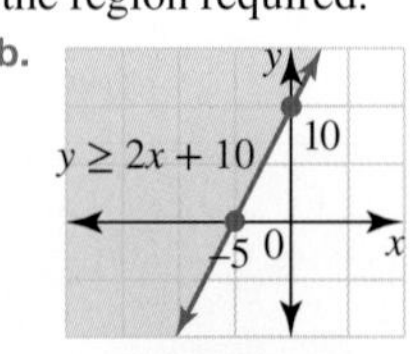

c. 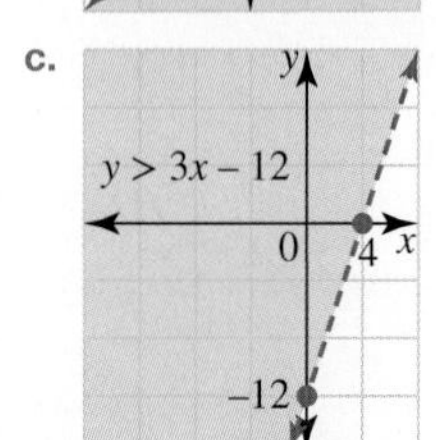

d. 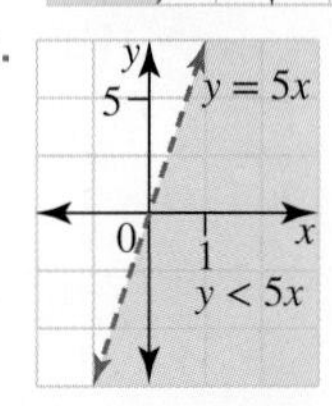

e. 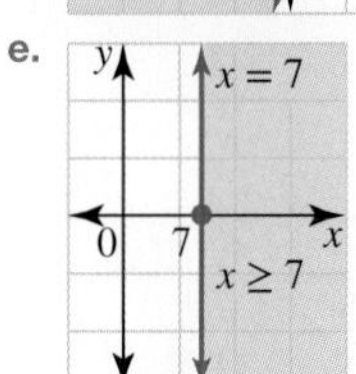

f. 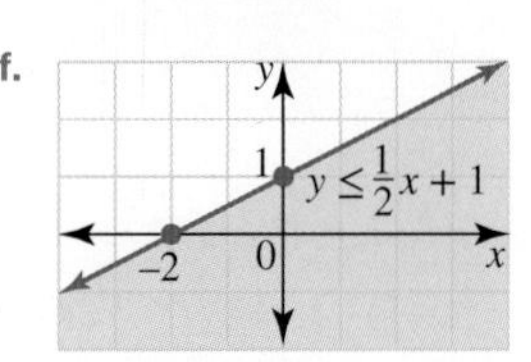

g. 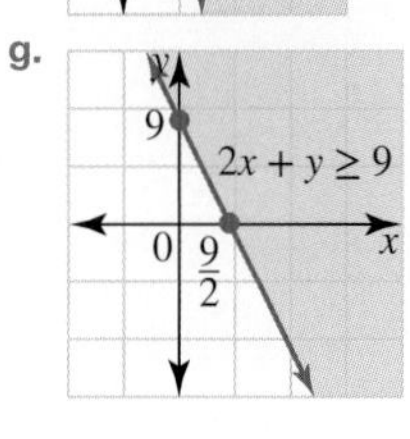

h. 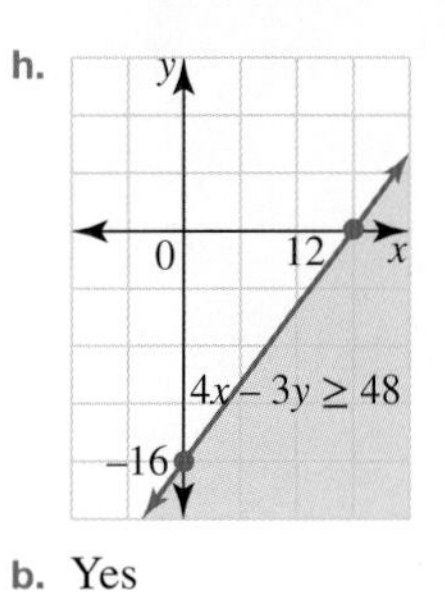

8. a. No b. Yes
9. a. $(-2, 1)$ b. $(0, -2)$
   c. $(-7.5, 7)$
10. a. $(3, 1)$ b. $(2, 3)$
11. a. $(2, 7)$ b. $(-5, -3)$
    c. $(-2, 2)$ d. $\left(-\frac{7}{3}, \frac{7}{3}\right)$
    e. $(-14, -53)$ f. $\left(\frac{5}{2}, -7\right)$
12. a. $(5, 2)$ b. $(-2, 3)$
    c. $(-3, -1)$ d. $(1, 3)$
    e. $(2, -2)$ f. $(4, 2)$
13. a. $(0, 3)$ b. $(-3, -3)$
    c. $(2, 1)$

14. *Note:* The shaded region is the region required.

a. 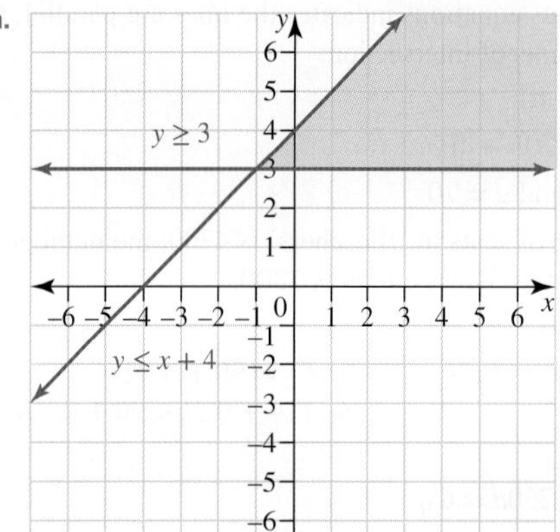

b. 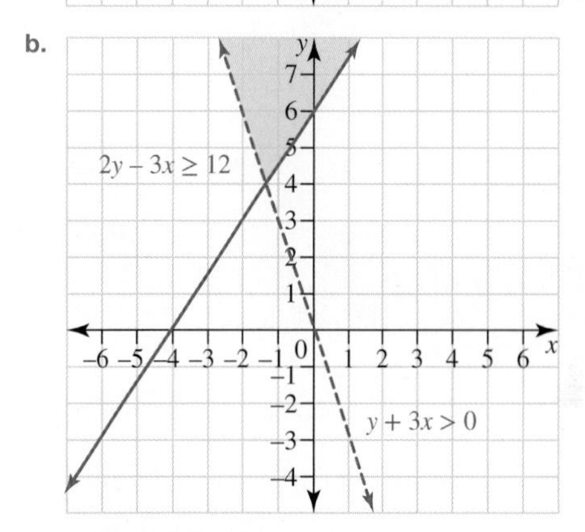

c. 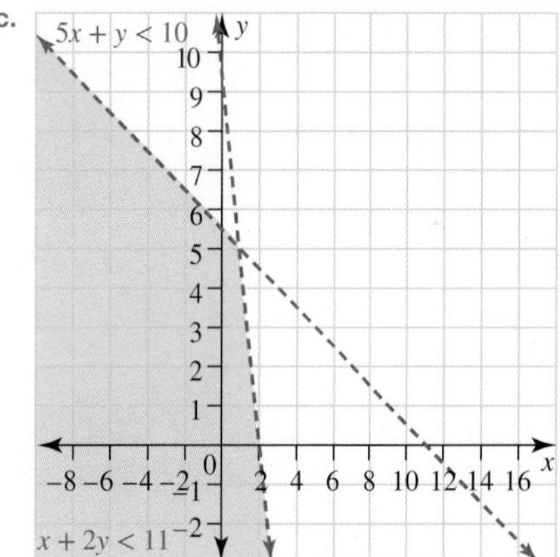

15. a. 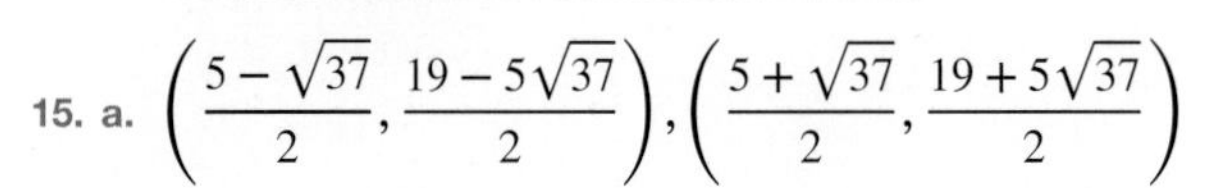
$$\left(\frac{5-\sqrt{37}}{2}, \frac{19-5\sqrt{37}}{2}\right), \left(\frac{5+\sqrt{37}}{2}, \frac{19+5\sqrt{37}}{2}\right)$$

b. $\left(-\frac{2}{5}, -5\right), (1, 2)$

c. $\left(\frac{15+\sqrt{43}}{26}, \frac{-3+5\sqrt{43}}{26}\right), \left(\frac{15-\sqrt{43}}{26}, \frac{-3-5\sqrt{43}}{26}\right)$

16. a. Numbers are 9 and 14.
    b. Length = 11 meters, width = 6 meters.
    c. Chupa-chups cost 45 cents and Whizz fizzes cost 55 cents.
17. Milk \$1.75, bread \$2.35.
18. 13 kangaroos and 8 cockatoos.

19. Any false statement that occurs during the solving of simultaneous equations indicates the lines are parallel, and have no points of intersection.

20. a. $d = b + 10$
   b. $7000 = 70b + 40d$
   c. $b = 60$ and $d = 70$
   d. Number of seats in 'Bleachers' is 4200; the number of seats in the 'Dress circle' is 2800.
   e. $644 000

21. a. $C_G = 114 + 0.2k$
   b. $C_S = 90 + 0.32k$
   c. 200 km
   d. $114 + 0.2k < 90 + 0.32k$
   e. $k < 200$

22. a. $5400 + 260d = C_H$
   b. 61 days

5 Trigonometry I

LEARNING SEQUENCE

5.1 Overview

Why learn this?

Nearly 2000 years ago Ptolemy of Alexandria published the first book of trigonometric tables, which he used to chart the heavens and plot the courses of the Moon, stars and planets. He also created geographical charts and provided instructions on how to create maps.

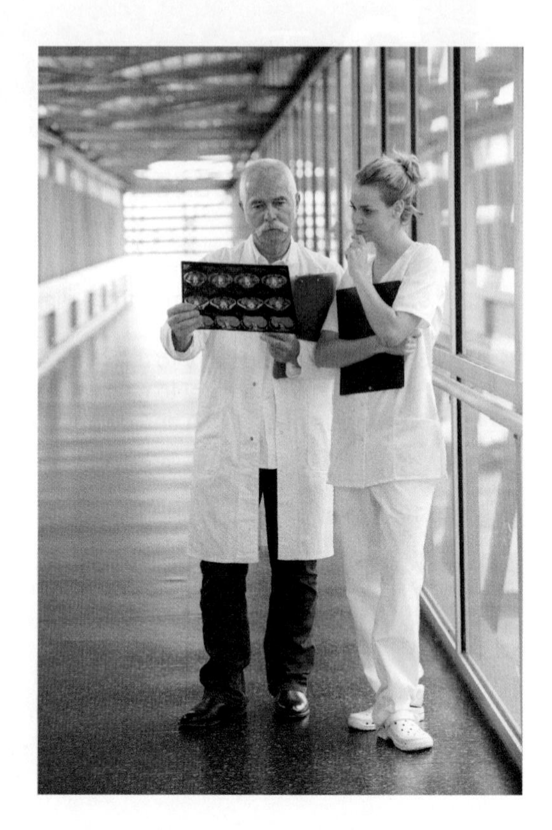

The word trigonometry is derived from Greek words 'trigonon' and 'metron' meaning *triangles* and *measure* respectively, and reportedly has been studied since the third century BCE. This field of mathematics was studied across the world, with major discoveries made in India, China, Greece and Persia, to name a few. The works ranged from developing relationships, axioms and proofs to its application to everyday use and life.

Trigonometry is the branch of mathematics that makes the whole universe more easily understood. The role and use of trigonometry in navigation in the early years were crucial and its application and study grew from there. Today, it is used in architecture, surveying, astronomy and, as previously mentioned, navigation. It also provides the foundation of the study of sound and light waves, resulting in its application in the areas of music manufacturing and composition, study of tides, radiology and many other fields.

Where to get help

Go to your learnON title at **www.jacplus.com.au** to access the following digital resources. The Online Resources Summary at the end of this topic provides a full list of what's available to help you learn the concepts covered in this topic.

Video eLessons

Interactivities

Fully worked solutions to every question

Digital documents

eWorkbook

Exercise 5.1 Pre-test

learnon

Complete this pre-test in your learnON title at www.jacplus.com.au and receive **automatic marks, immediate corrective feedback** and **fully worked solutions**.

1. Determine the value of the pronumeral w, correct to two decimal places.

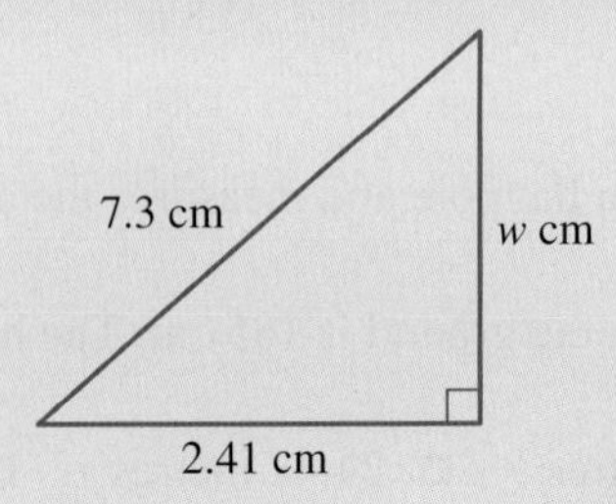

2. Determine the value of x, correct to two decimal places.

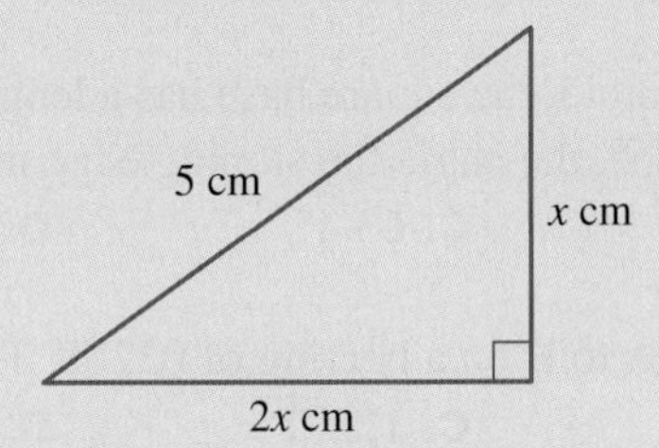

3. A square-based pyramid is 16 cm high. Each sloping edge is 20 cm long. Calculate the length of the sides of the base, in cm correct to two decimal places.

4. A cork is in the shape of a truncated cone; both the top and the base of the cork are circular.

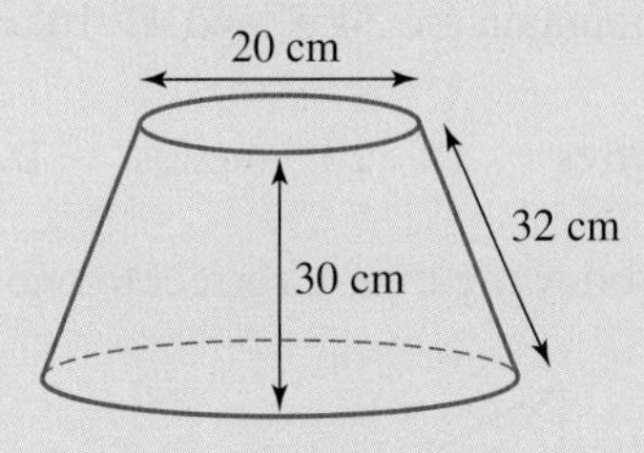

Calculate the sum of diameters of the top and the base. Give your answer in cm to 2 decimal places.

5. Evaluate $\sin(20°37')$ correct to four decimal places.

6. Calculate the size of the angle θ, correct to the nearest minute, given that $\cos(\theta) = 0.5712$. Give your answer in degrees and minutes.

7. Determine the size of the angle θ, correct to the nearest second.

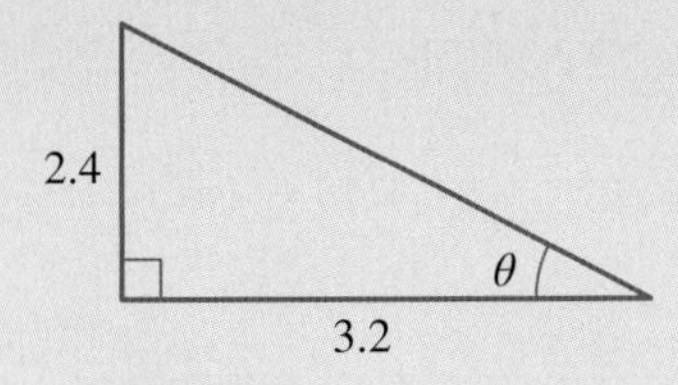

8. Calculate y, correct to one decimal place.

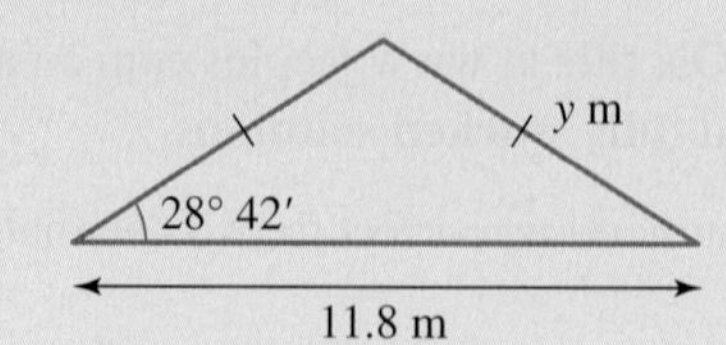

9. **MC** Tyler is standing 12 m from a flagpole and measures the angle of elevation from his eye level to the top of the pole as 62°.
The distance from Tyler's eyes to the ground is 185 cm. The height of the flagpole correct to two decimal places is:
A. 20.71 metres **B.** 22.56 metres **C.** 24.42 metres **D.** 207.56 metres **E.** 209.42 metres

10. Change each of the following compass bearings to true bearings.
a. N20°E **b.** S47°W **c.** N33°W **d.** S17°E

11. **MC** In a right square-based pyramid, the square base has a length of 7.2 cm. If the angle between the triangular face and the base is 55°, the angle the sloping edge makes with the base is:
A. 45.3° **B.** 55° **C.** 63.4° **D.** 47.7° **E.** 26.6°

12. **MC** A boat travels 15 km from A to B on a bearing of 032°T. The bearing from B to A is:
A. 032°T **B.** 058°T **C.** 122°T **D.** 212°T **E.** 328°T

13. A bushwalker travels *N*50°*W* for 300 m and then changes direction 220°*T* for 0.5 km. Determine how many metres west the bushwalker is from his starting point.
Give your answer in km correct to one decimal place.

14. **MC** P and Q are two points on a horizontal line that are 120 metres apart. If the angles of elevation from P and Q to the top of the mountain are 34°5′ and 41°16′ respectively, the height of the mountain correct to one decimal place is:
A. 81.2 metres **B.** 105.3 metres **C.** 120.5 metres **D.** 253.8 metres **E.** 354.7 metres

15. Determine the value of x in the following figure, correct to one decimal place.

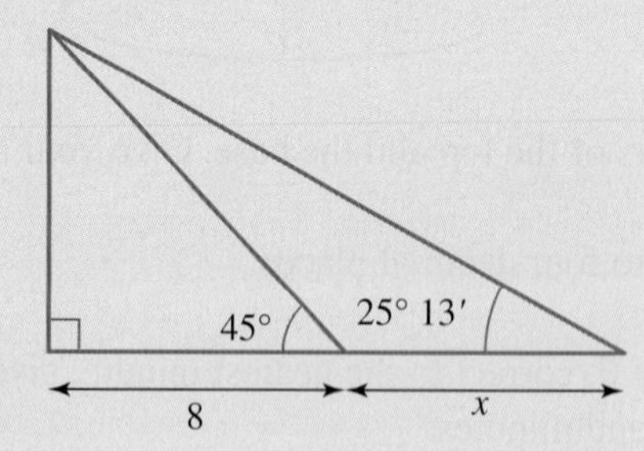

5.2 Pythagoras' theorem

LEARNING INTENTION

At the end of this subtopic you should be able to:
- identify similar right-angled triangles when corresponding sides are in the same ratio and corresponding angles are congruent
- apply Pythagoras' theorem to calculate the third side of a right-angled triangle when two other sides are known.

5.2.1 Similar right-angled triangles

eles-4799

- Two similar right-angled triangles have the same angles when the corresponding sides are in the same ratio.
- The **hypotenuse** is the longest side of a right-angled triangle and is always the side that is opposite the right angle.
- The corresponding sides are in the same ratio.

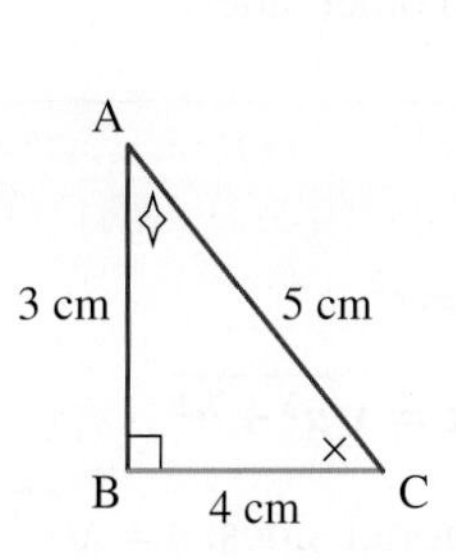

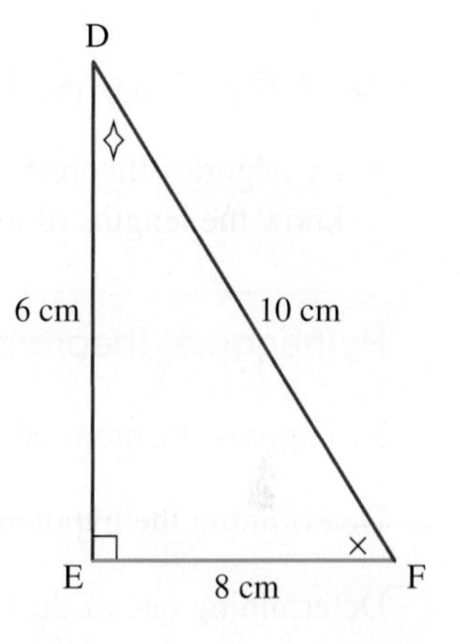

$$\frac{AB}{DE} = \frac{AC}{DF} = \frac{BC}{EF}$$

- To write this using the side lengths of the triangles gives:

$$\frac{AB}{DE} = \frac{3}{6} = \frac{1}{2}$$
$$\frac{AC}{DF} = \frac{5}{10} = \frac{1}{2}$$
$$\frac{BC}{EF} = \frac{4}{8} = \frac{1}{2}$$

This means that for right-angled triangles, when the angles are fixed, the ratios of the sides in the triangle are constant.

- We can examine this idea further by completing the following activity.
Using a protractor and ruler, draw an angle of 70° measuring horizontal distances of 3 cm, 7 cm and 10 cm as demonstrated in the diagram below.

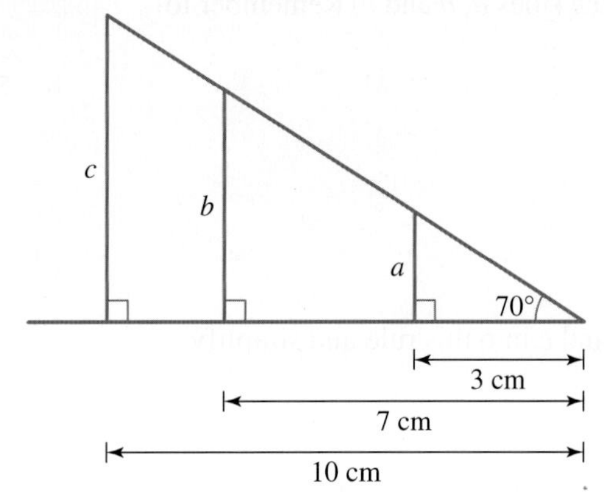

Note: Diagram not drawn to scale.

Measure the perpendicular heights a, b and c.

$$a \approx 8.24\text{ cm}, b \approx 19.23\text{ cm}, c \approx 27.47\text{ cm}$$

- To test if the theory for right-angled triangles, that when the angles are fixed the ratios of the sides in the triangle are constant, is correct, calculate the ratios of the side lengths.

$$\frac{a}{3} \approx \frac{8.24}{3} \approx 2.75, \quad \frac{b}{7} \approx \frac{19.23}{7} \approx 2.75, \quad \frac{c}{10} \approx \frac{27.47}{10} \approx 2.75$$

The ratios are the same because the triangles are similar. This important concept forms the basis of trigonometry.

5.2.2 Review of Pythagoras' theorem

eles-4800

- Pythagoras' theorem gives us a way of finding the length of the third side in a right angle triangle, if we know the lengths of the two other sides.

Pythagoras' theorem

Pythagoras' theorem: $a^2 + b^2 = c^2$

Determining the hypotenuse: $c = \sqrt{a^2 + b^2}$

Determining one of the two shorter sides: $a = \sqrt{c^2 - b^2}$ or $b = \sqrt{c^2 - a^2}$

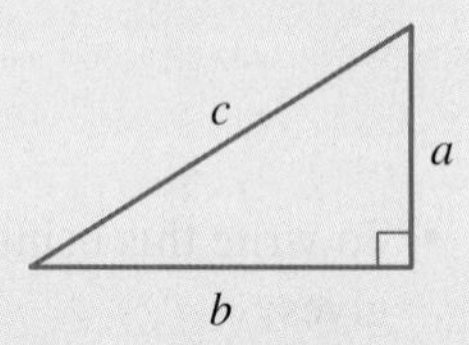

WORKED EXAMPLE 1 Calculating the hypotenuse

For the following triangle, calculate the length of the hypotenuse x, correct to 1 decimal place.

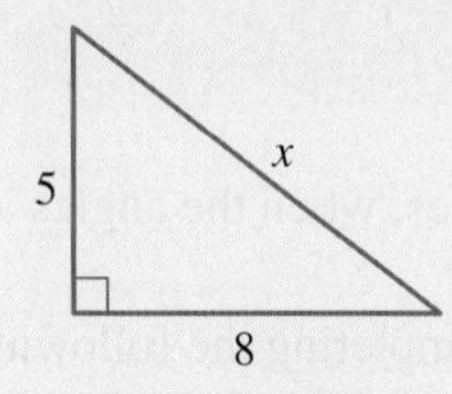

THINK

1. Copy the diagram and label the sides a, b and c. Remember to label the hypotenuse as c.

WRITE/DRAW

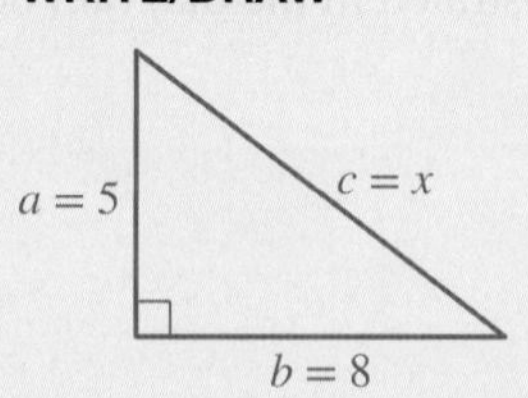

2. Write Pythagoras' theorem.

$c^2 = a^2 + b^2$

3. Substitute the values of a, b and c into this rule and simplify.

$$\begin{aligned} x^2 &= 5^2 + 8^2 \\ &= 25 + 64 \\ &= 89 \end{aligned}$$

4. Take the square root of both sides. Round the positive answer correct to 1 decimal place, since $x > 0$.

$$\begin{aligned} x &= \pm\sqrt{89} \\ x &\approx 9.4 \end{aligned}$$

WORKED EXAMPLE 2 Calculating the length of the shorter side

Calculate the length, correct to 1 decimal place, of the unmarked side of the following triangle.

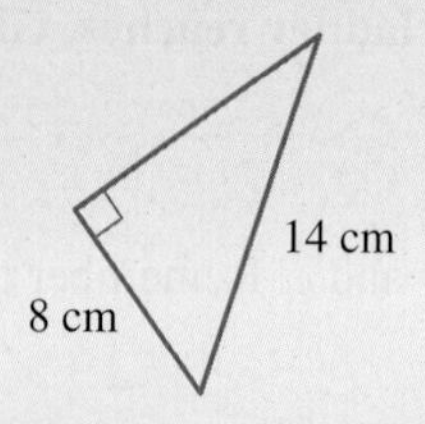

THINK

1. Copy the diagram and label the sides a, b and c. Remember to label the hypotenuse as c; it does not matter which side is a and which side is b.

WRITE/DRAW

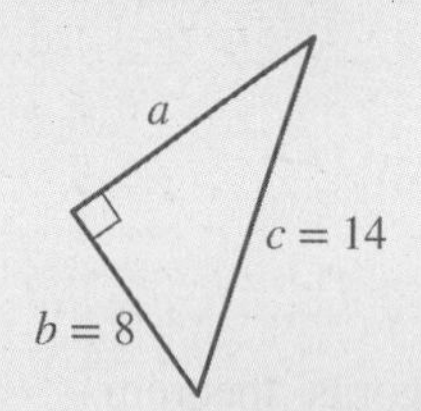

2. Write Pythagoras' theorem.

$c^2 = a^2 + b^2$

3. Substitute the values of a, b and c into this rule and solve for a.

$$14^2 = a^2 + 8^2$$
$$196 = a^2 + 64$$
$$a^2 = 196 - 64$$
$$= 132$$

4. Evaluate a by taking the square root of both sides and round to 1 decimal place ($a > 0$).

$$a = \pm\sqrt{132}$$
$$\approx 11.5\,\text{cm}$$

TI | THINK

In a new document, on a Calculator page, to solve equations press:
- MENU
- 3: Algebra
- 1: Solve

Complete the entry line as:
solve $(c^2 = a^2 + b^2, a)$
$|b = 8$ and $c = 14$ and $a > 0$
Then press ENTER.
Press CTRL ENTER to get a decimal approximation.

DISPLAY/WRITE

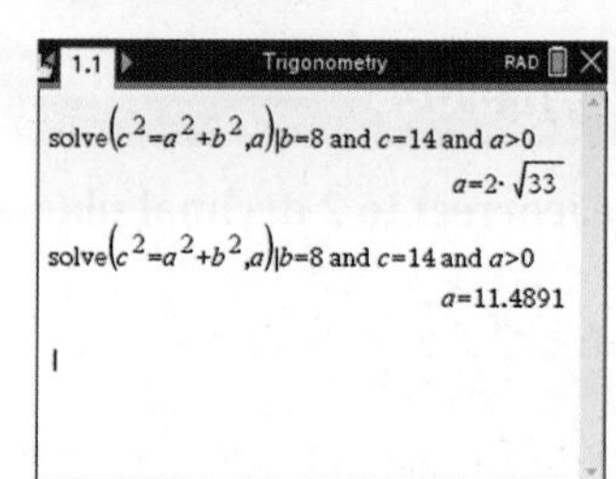

The length of the unmarked side is $a = 2\sqrt{33} = 11.5$ correct to 1 decimal place.

CASIO | THINK

On the Main screen, complete the entry line as:
solve $(c^2 = a^2 + b^2, a)$
$|b = 8|\, c = 14$
Then press EXE.
To convert to decimals, highlight the answer which is greater than 0 and drag it to a new line. Change the mode to decimal then press EXE.

DISPLAY/WRITE

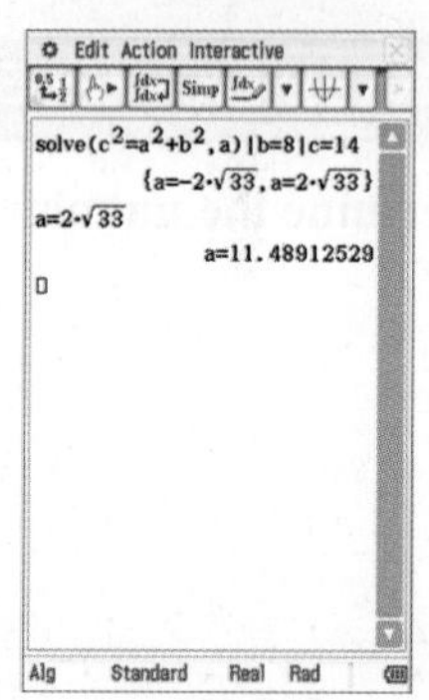

The length of the unmarked side is $a = 2\sqrt{33} = 11.5$ correct to 1 decimal place.

WORKED EXAMPLE 3 Solving a practical problem using Pythagoras' theorem

A ladder that is 5.5 m long leans up against a vertical wall. The foot of the ladder is 1.5 m from the wall. Determine how far up the wall the ladder reaches. Give your answer in metres correct to 1 decimal place.

THINK	WRITE/DRAW
1. Draw a diagram and label the sides a, b and c. Remember to label the hypotenuse as c.	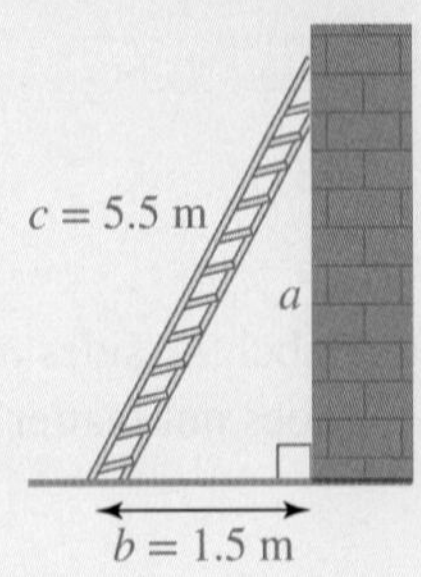
2. Write Pythagoras' theorem.	$c^2 = a^2 + b^2$
3. Substitute the values of a, b and c into this rule and simplify.	$5.5^2 = a^2 + 1.5^2$ $30.25 = a^2 + 2.25$ $a^2 = 30.25 - 2.25$ $= 28$
4. Evaluate a by taking the square root of 28. Round to 1 decimal place, $a > 0$.	$a = \pm\sqrt{28}$ ≈ 5.3
5. Write the answer in a sentence.	The ladder reaches 5.3 m up the wall.

WORKED EXAMPLE 4 Determining the unknown sides

Determine the unknown side lengths of the triangle, correct to 2 decimal places.

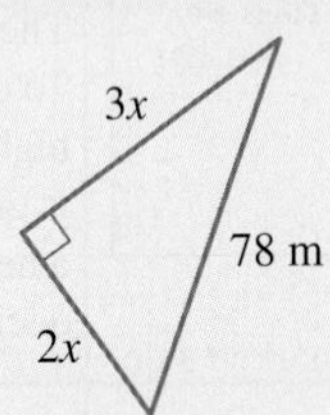

THINK	WRITE/DRAW
1. Copy the diagram and label the sides a, b and c.	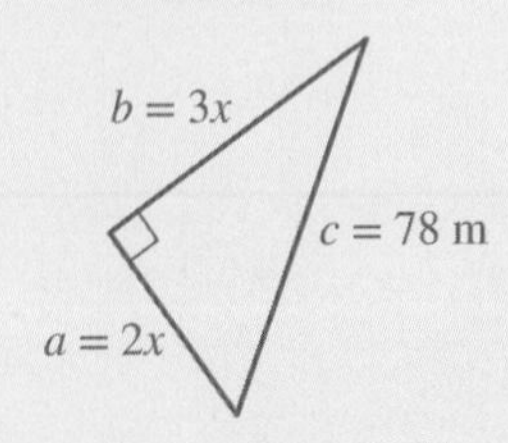
2. Write Pythagoras' theorem.	$c^2 = a^2 + b^2$
3. Substitute the values of a, b and c into this rule and simplify.	$78^2 = (3x)^2 + (2x)^2$ $6084 = 9x^2 + 4x^2$ $6084 = 13x^2$

4. Rearrange the equation so that the pronumeral is on the left-hand side of the equation. $13x^2 = 6084$

5. Divide both sides of the equation by 13. $\frac{13x^2}{13} = \frac{6084}{13}$
$x^2 = 468$

6. Evaluate x by taking the square root of both sides. Round the answer correct to 2 decimal places. $x = \pm\sqrt{468}$
≈ 21.6333

7. Substitute the value of x into $2x$ and $3x$ to determine the lengths of the unknown sides. $2x \approx 43.27\text{m}$
$3x \approx 64.90\text{ m}$

DISCUSSION

Pythagoras' theorem was known about before the age of Pythagoras. Research which other civilisations knew about the theory and construct a timeline for its history.

Resources

eWorkbook Topic 5 Workbook (worksheets, code puzzle and project) (ewbk-2031)

Digital document SkillSHEET Rounding to a given number of decimal places (doc-5224)

Interactivities Individual pathway interactivity: Pythagoras' theorem (int-4585)
Finding a shorter side (int-3845)
Finding the hypotenuse (int-3844)

Exercise 5.2 Pythagoras' theorem

learnON

Individual pathways

■ PRACTISE	■ CONSOLIDATE	■ MASTER
1, 5, 7, 10, 11, 14, 17, 21, 22, 25	2, 3, 8, 12, 15, 19, 20, 23, 26	4, 6, 9, 13, 16, 18, 24, 27

To answer questions online and to receive **immediate corrective feedback** and **fully worked solutions** for all questions, go to your learnON title at www.jacplus.com.au.

Fluency

1. WE1 For each of the following triangles, calculate the length of the hypotenuse, giving answers correct to 2 decimal places.

a. 4.7, 6.3

b. 19.3, 27.1

c. 804, 562

2. For each of the following triangles, calculate the length of the hypotenuse, giving answers correct to 2 decimal places.

a. 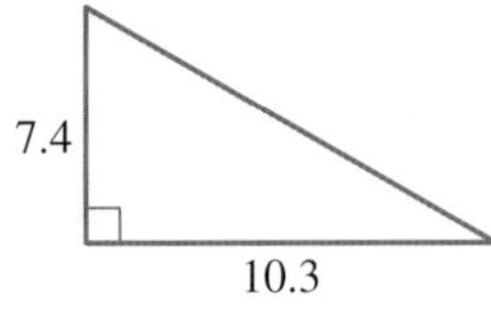

b. 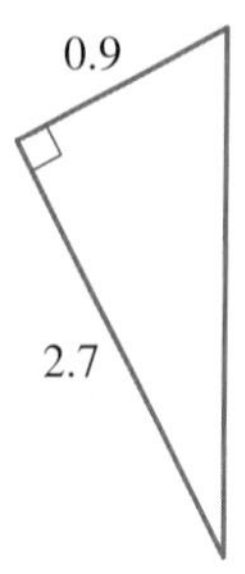

c. 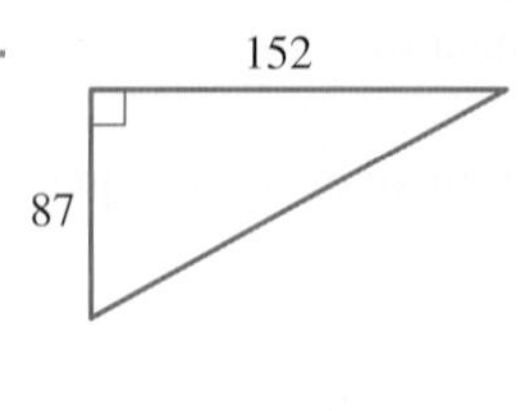

3. **WE2** Determine the value of the pronumeral, correct to 2 decimal places.

a. 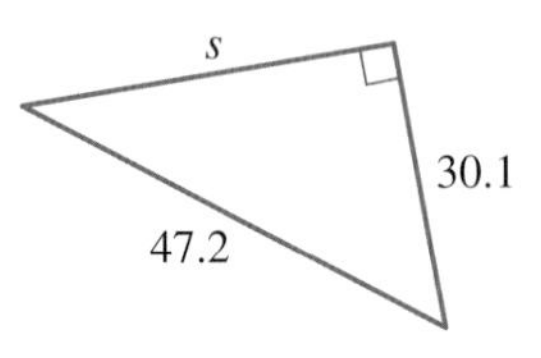

b. 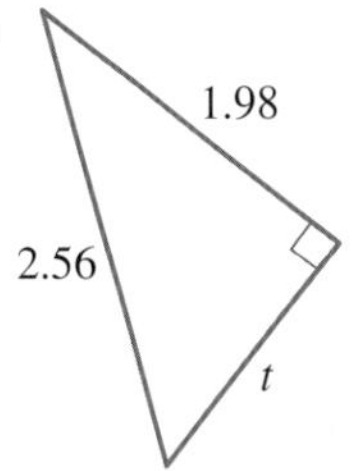

c. 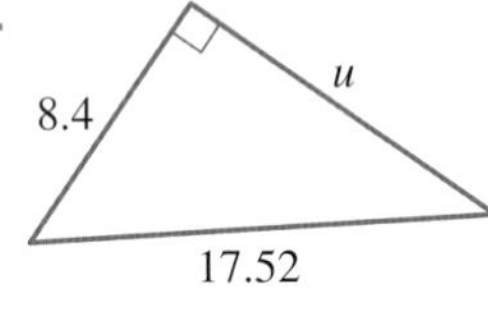

4. Determine the value of the pronumeral, correct to 2 decimal places.

a. 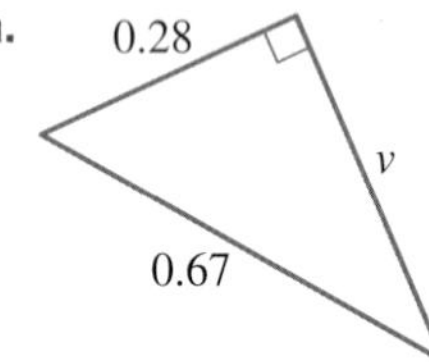

b. 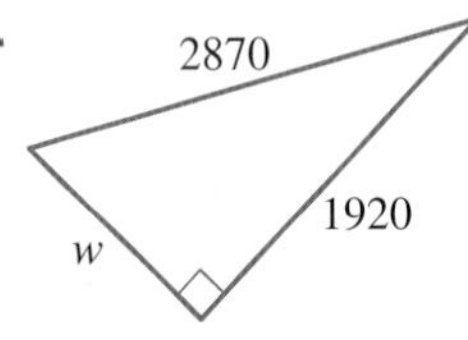

c. 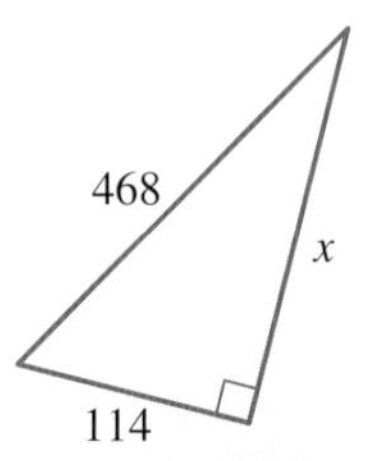

5. **WE3** The diagonal of the rectangular NO SMOKING sign is 34 cm. If the height of this sign is 25 cm, calculate the width of the sign, in cm correct to 2 decimal places.

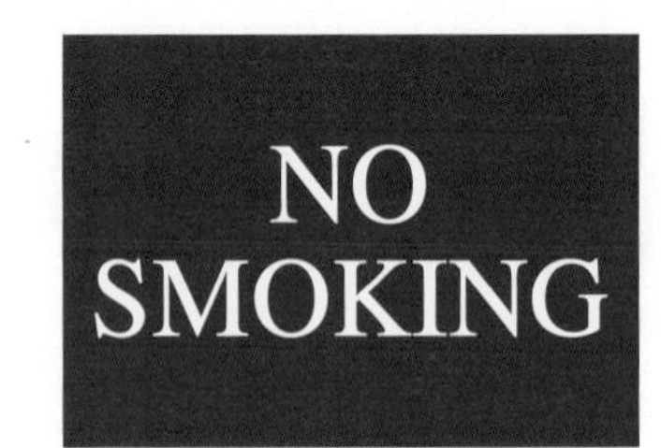

6. A right-angled triangle has a base of 4 cm and a height of 12 cm. Calculate the length of the hypotenuse in cm correct to 2 decimal places.
7. Calculate the lengths of the diagonals (in cm to 2 decimal places) of squares that have side lengths of:

a. 10 cm b. 17 cm c. 3.2 cm.

8. The diagonal of a rectangle is 90 cm. One side has a length of 50 cm. Determine, correct to 2 decimal places:

a. the length of the other side
b. the perimeter of the rectangle
c. the area of the rectangle.

9. **WE4** Determine the value of the pronumeral, correct to 2 decimal places for each of the following.

a. 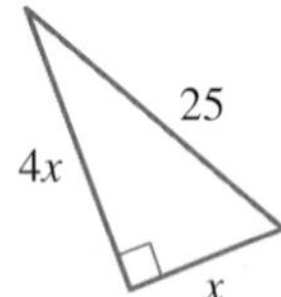

b. 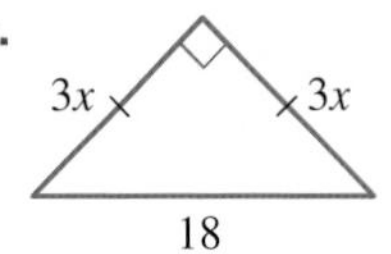

c. 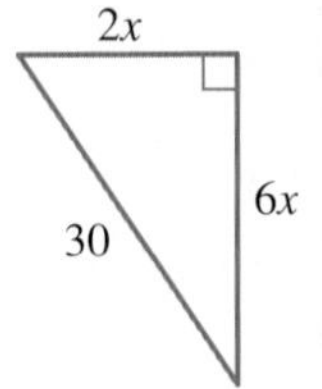

Understanding

10. An isosceles triangle has a base of 25 cm and a height of 8 cm. Calculate the length of the two equal sides, in cm correct to 2 decimal places.
11. An equilateral triangle has sides of length 18 cm. Determine the height of the triangle, in cm correct to 2 decimal places.
12. A right-angled triangle has a height of 17.2 cm, and a base that is half the height. Calculate the length of the hypotenuse, in cm correct to 2 decimal places.
13. The road sign shown is based on an equilateral triangle. Determine the height of the sign and, hence, calculate its area. Round your answers to 2 decimal places.

14. A flagpole, 12 m high, is supported by three wires, attached from the top of the pole to the ground. Each wire is pegged into the ground 5 m from the pole. Determine how much wire is needed to support the pole, correct to the nearest metre.
15. Sarah goes canoeing in a large lake. She paddles 2.1 km to the north, then 3.8 km to the west. Use the triangle shown to determine how far she must then paddle to get back to her starting point in the shortest possible way, in km correct to 2 decimal places.

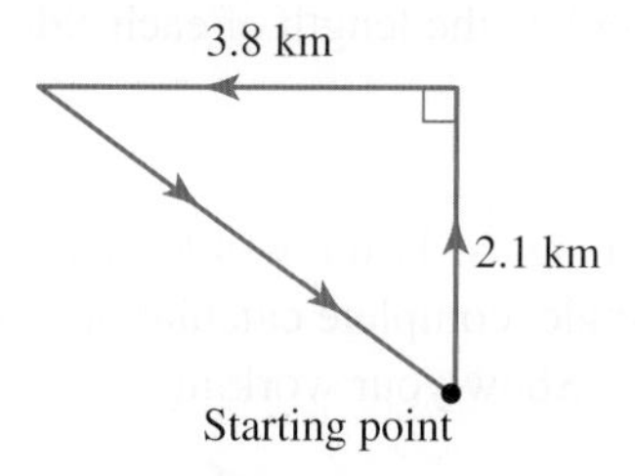

16. A baseball diamond is a square of side length 27 m. When a runner on first base tries to steal second base, the catcher has to throw the ball from home base to second base. Calculate the distance of the throw, in metres correct to 1 decimal place.

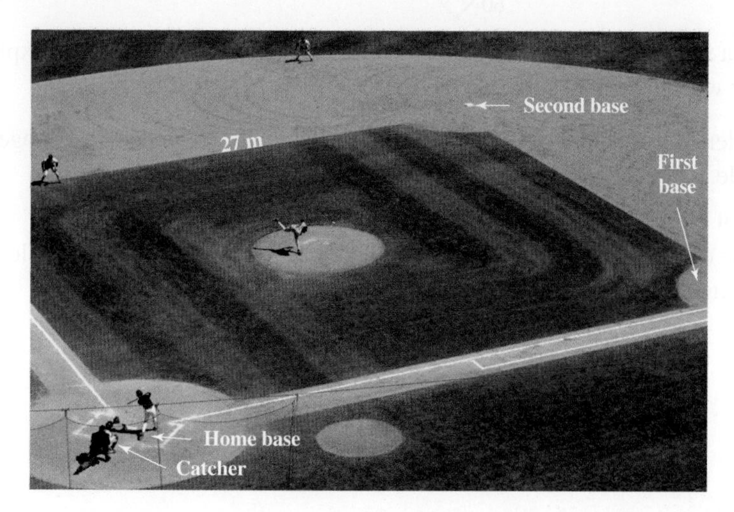

17. A rectangle measures 56 mm by 2.9 cm. Calculate the length of its diagonal in millimetres correct to 2 decimal places.

18. A rectangular envelope has a length of 24 cm and a diagonal measuring 40 cm. Calculate:

    a. the width of the envelope, correct to the nearest cm
    b. the area of the envelope, correct to the nearest cm^2.

19. A swimming pool is 50 m by 25 m. Peter is bored by his usual training routine, and decides to swim the diagonal of the pool. Determine how many diagonals he must swim to complete his normal distance of 1500 m.

20. A hiker walks 2.9 km north, then 3.7 km east. Determine how far in metres she is from her starting point. Give your answer in metres to 2 decimal places.

21. A square has a diagonal of 14 cm. Calculate the length of each side, in cm correct to 2 decimal places.

Reasoning

22. The triangles below are right-angled triangles. Two possible measurements have been suggested for the hypotenuse in each case. For each triangle, complete calculations to determine which of the lengths is correct for the hypotenuse in each case. Show your working.

a.
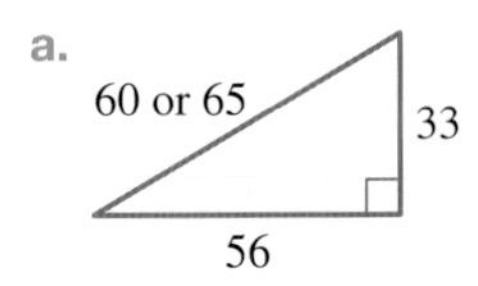

b.
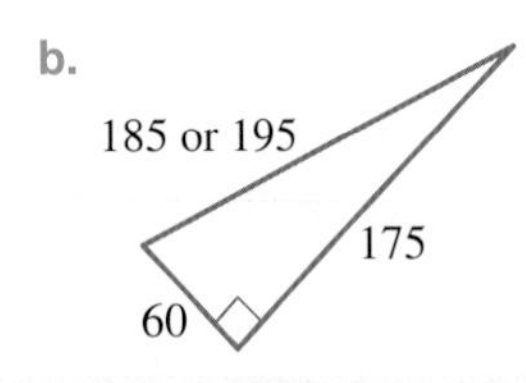

c.
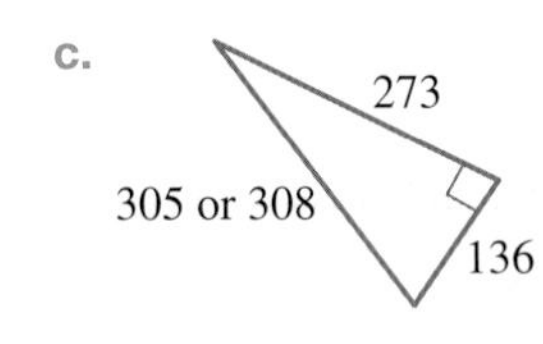

23. The square root of a number usually gives us both a positive and negative answer. Explain why we take only the positive answer when using Pythagoras' theorem.

24. Four possible side length measurements are 105, 208, 230 and 233. Three of them together produce a right-angled triangle.

    a. Explain which of the measurements could not be the hypotenuse of the triangle.
    b. Complete as few calculations as possible to calculate which combination of side lengths will produce a right-angled triangle.

Problem solving

25. The area of the rectangle MNPQ is 588 cm^2. Angles MRQ and NSP are right angles.

a. Determine the integer value of x.
b. Determine the length of MP.
c. Calculate the value of y and hence determine the length of RS, in cm correct to 1 decimal place.

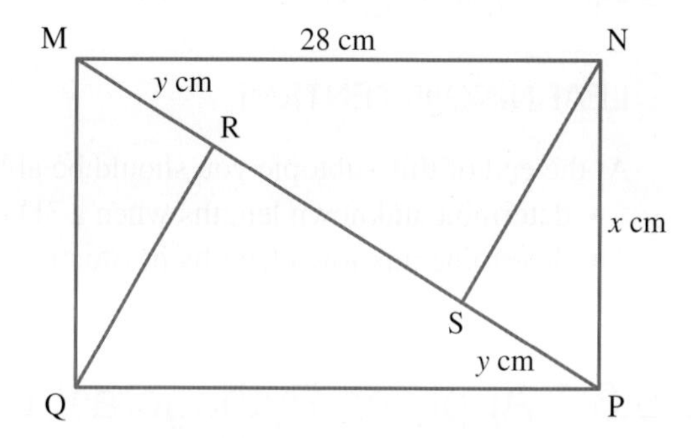

26. Triangle ABC is an equilateral triangle of side length x cm. Angles ADB and DBE are right angles. Determine the value of x in cm, correct to 2 decimal places.

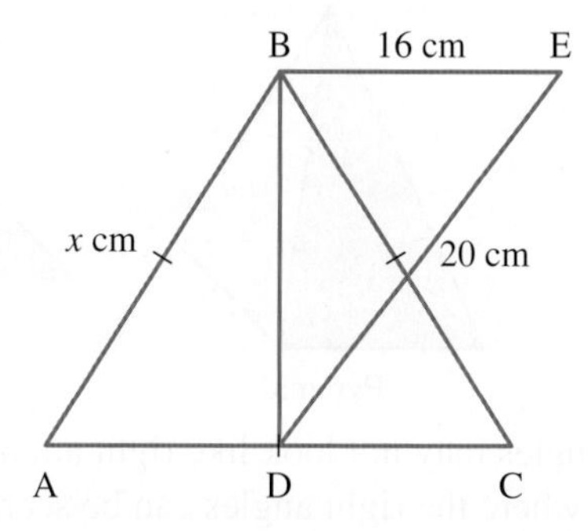

27. The distance from Earth to the Moon is approximately 385 000 km and the distance from Earth to the Sun is approximately 147 million kilometres.
In a total eclipse of the Sun, the moon moves between the Sun and Earth, thus blocking the light of the Sun from reaching Earth and causing a total eclipse of the Sun.
If the diameter of the Moon is approximately 3474 km, evaluate the diameter of the Sun. Express your answer to the nearest 10 000 km.

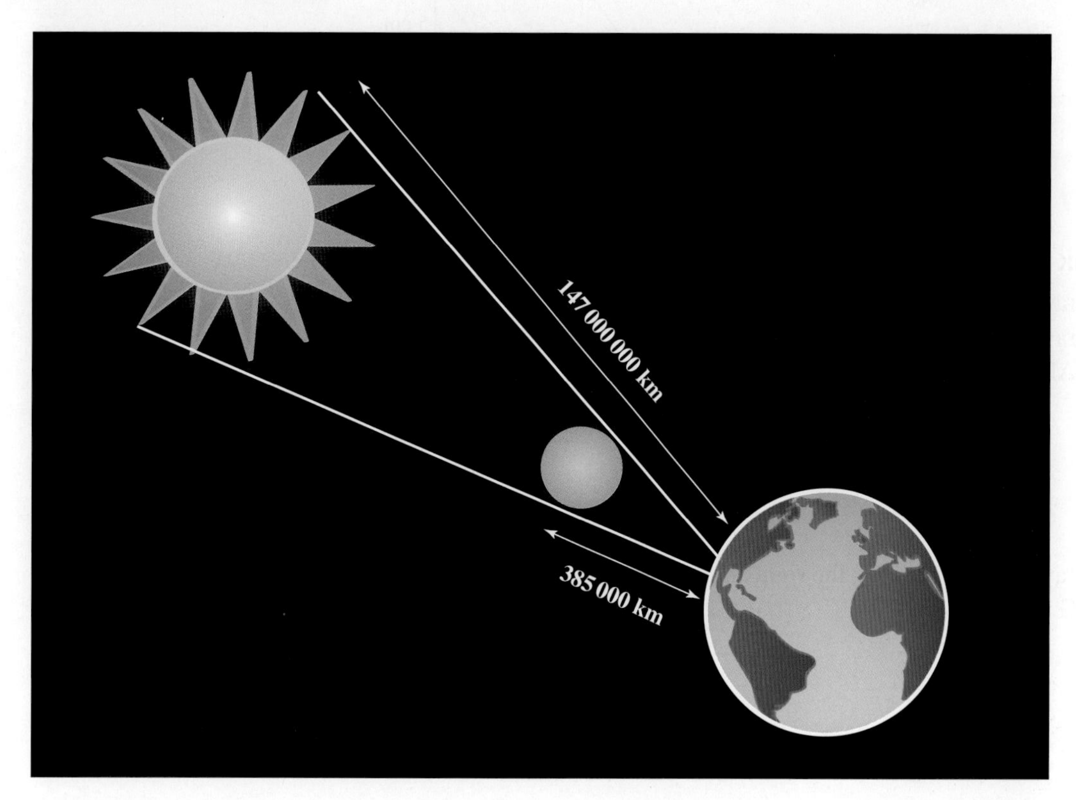

5.3 Pythagoras' theorem in three dimensions (10A)

LEARNING INTENTION

At the end of this subtopic you should be able to apply Pythagoras' theorem to:
- determine unknown lengths when a 3D diagram is given
- determine unknown lengths in situations by first drawing a diagram.

5.3.1 Applying Pythagoras' theorem in three dimensions

eles-4801

- Many real-life situations involve **3-dimensional** (3-D) objects: objects with length, width and height. Some common 3-D objects used in this section include cuboids, pyramids and right-angled wedges.

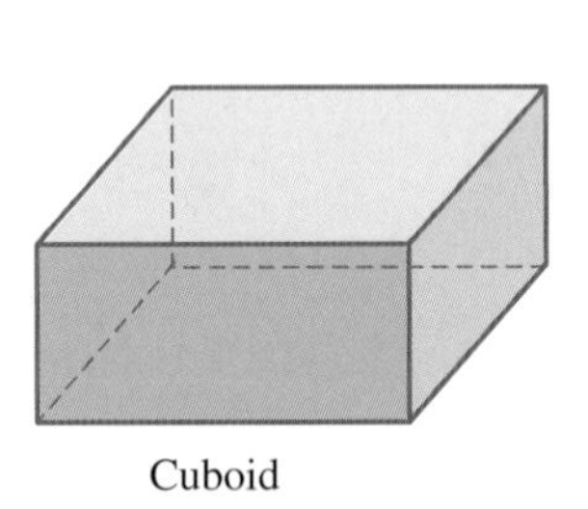
Cuboid

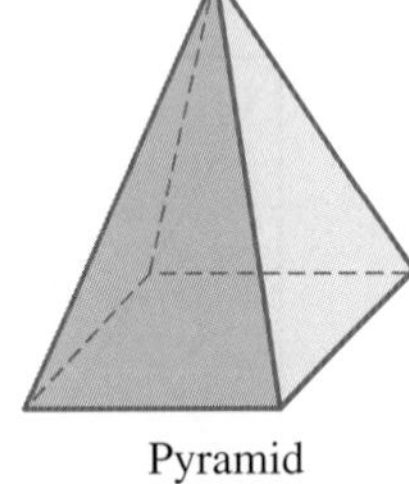
Pyramid

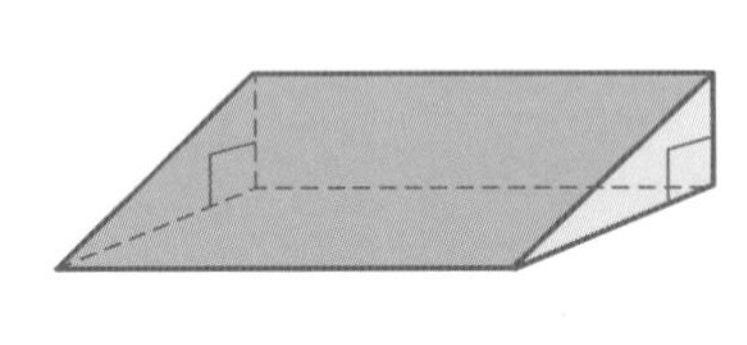
Right-angled wedge

- In diagrams of 3-D objects, right angles may not look like right angles, so it is important to redraw sections of the diagram in two dimensions, where the right angles can be seen accurately.

WORKED EXAMPLE 5 Applying Pythagoras' theorem to 3D objects

Determine the length AG in this rectangular prism (cuboid), in cm correct to two decimal places.

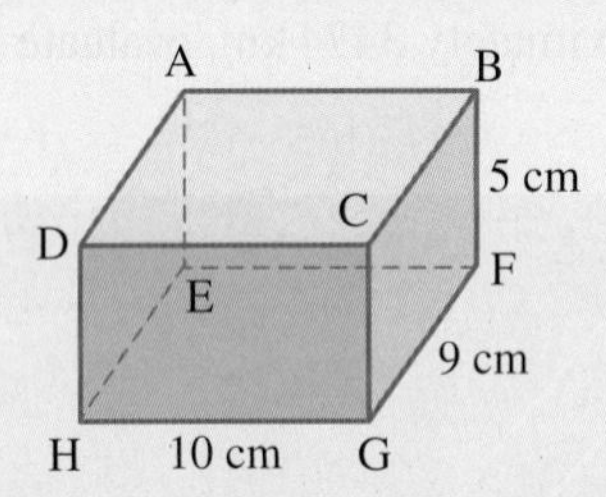

THINK

1. Draw the diagram in three dimensions.
Draw the lines AG and EG.
∠AEG is a right angle.

WRITE/DRAW

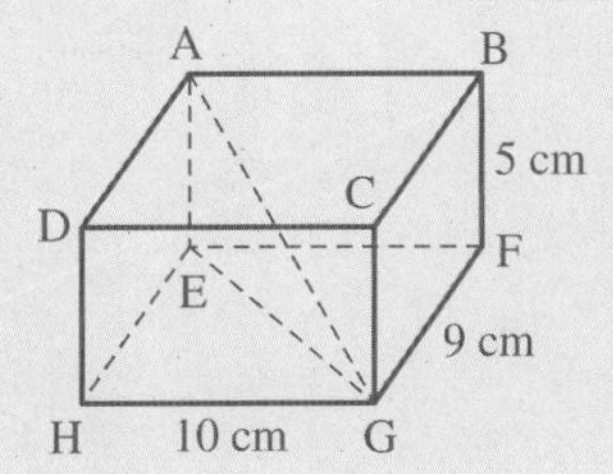

2. Draw ΔAEG, showing the right angle. Only 1 side is known, so EG must be found.

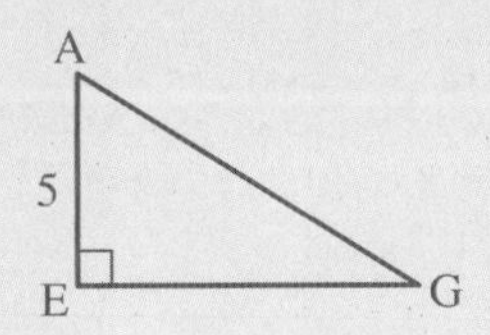

3. Draw EFGH in two dimensions and label the diagonal EG as x.

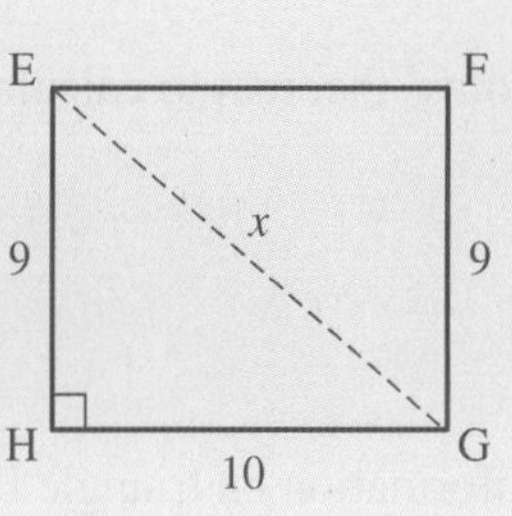

4. Use Pythagoras' theorem to calculate x. $\left(c^2 = a^2 + b^2\right)$

$$\begin{aligned} x^2 &= 9^2 + 10^2 \\ &= 81 + 100 \\ &= 181 \\ x &= \sqrt{181} \end{aligned}$$

5. Place this information on triangle AEG. Label the side AG as y.

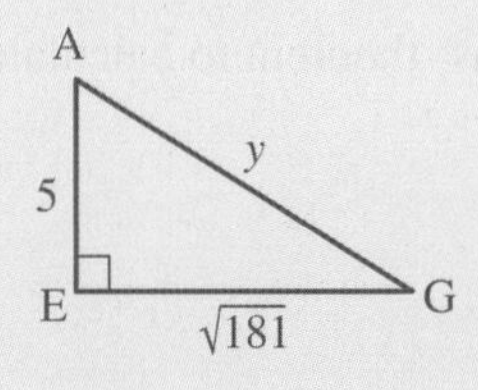

6. Use Pythagoras' theorem to calculate y. $\left(c^2 = a^2 + b^2\right)$

$$\begin{aligned} y^2 &= 5^2 + \left(\sqrt{181}\right)^2 \\ &= 25 + 181 \\ &= 206 \\ y &= \sqrt{206} \\ &\approx 14.35 \end{aligned}$$

7. Write the answer in a sentence. The length of AG is 14.35 cm.

WORKED EXAMPLE 6 Drawing a diagram to solve problems

A piece of cheese in the shape of a right-angled wedge sits on a table. It has a rectangular base measuring 14 cm by 8 cm, and is 4 cm high at the thickest point. An ant crawls diagonally across the sloping face. Determine how far, to the nearest millimetre, the ant walks.

THINK	WRITE/DRAW
1. Draw a diagram in three dimensions and label the vertices. Mark BD, the path taken by the ant, with a dotted line. ∠BED is a right angle.	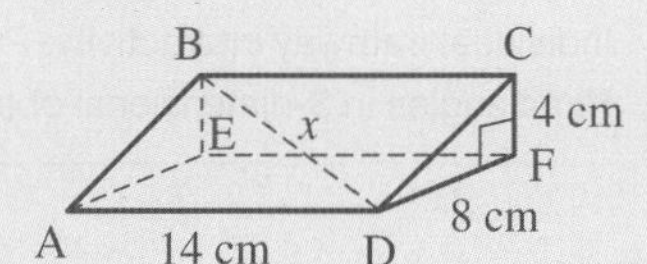
2. Draw ΔBED, showing the right angle. Only one side is known, so ED must be found.	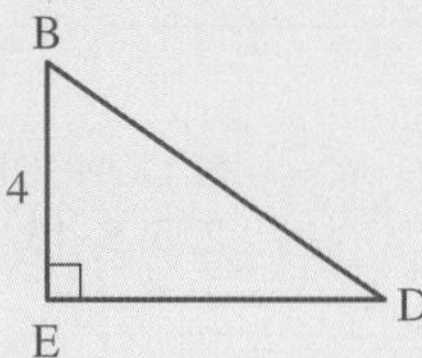
3. Draw EFDA in two dimensions, and label the diagonal ED. Label the side ED as x.	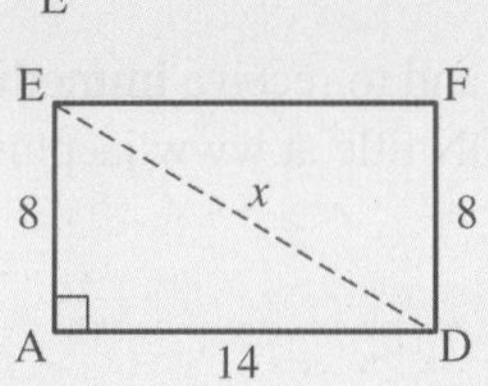

4. Use Pythagoras' theorem to calculate x.

$$c^2 = a^2 + b^2$$
$$x^2 = 8^2 + 14^2$$
$$= 64 + 196$$
$$= 260$$
$$x = \sqrt{260}$$

5. Place this information on triangle BED. Label the side BD as y.

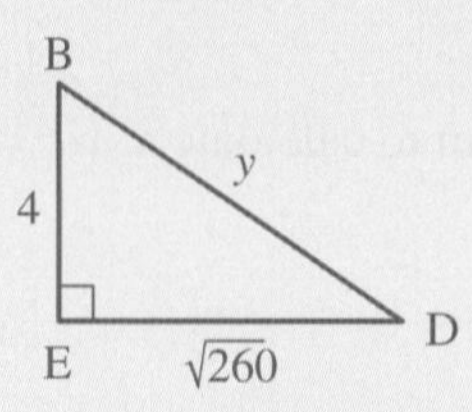

6. Use Pythagoras' theorem to calculate y.

$$y^2 = 4^2 + \left(\sqrt{260}\right)^2$$
$$= 16 + 260$$
$$= 276$$
$$y = \sqrt{276}$$
$$\approx 16.61 \text{ cm}$$
$$\approx 166.1 \text{ mm}$$

7. Write the answer in a sentence.

The ant walks 166 mm, correct to the nearest millimetre.

DISCUSSION

Look around the room you are in. How many right angles you can spot in three-dimensional objects? Make a list of them and compare your list to that of another student.

on Resources

eWorkbook Topic 5 Workbook (worksheets, code puzzle and project) (ewbk-2031)

Digital document SkillSHEET Drawing 3-D shapes (doc-5229)

Video eLesson Pythagoras' theorem in three dimensions (eles-1913)

Interactivities Individual pathway interactivity: Pythagoras' theorem in three dimensions (int-4586)
Right angles in 3-dimensional objects (int-6132)

Exercise 5.3 Pythagoras' theorem in three dimensions (10A) learnON

Individual pathways

■ PRACTISE	■ CONSOLIDATE	■ MASTER
1, 4, 7, 11, 12, 15	2, 5, 8, 13, 16	3, 6, 9, 10, 14, 17

To answer questions online and to receive **immediate corrective feedback** and **fully worked solutions** for all questions, go to your learnON title at www.jacplus.com.au.

Where appropriate in this exercise, give answers correct to 2 decimal places.

Fluency

1. **WE5** Calculate the length of AG in each of the following figures.

a.

b.

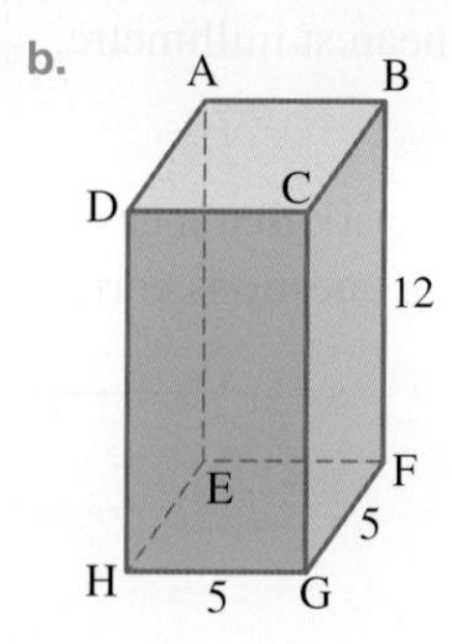

c.

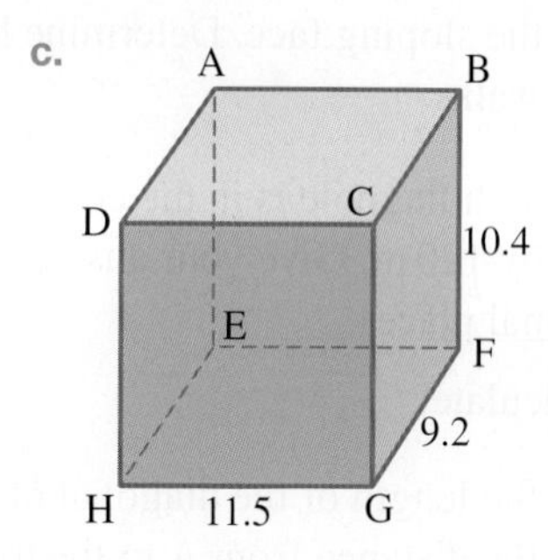

2. Consider the wedge shown. Calculate the length of CE in the wedge and, hence, obtain the length of AC .

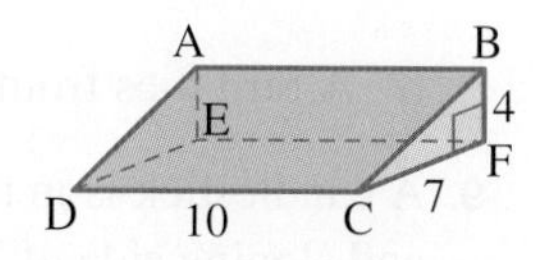

3. If DC = 3.2 m, AC = 5.8 m, and CF = 4.5 m in the figure, calculate the length of AD and BF.

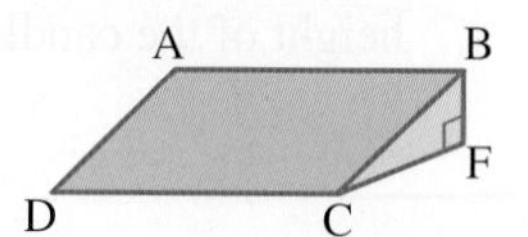

4. Consider the pyramid shown. Calculate the length of BD and, hence, the height of the pyramid.

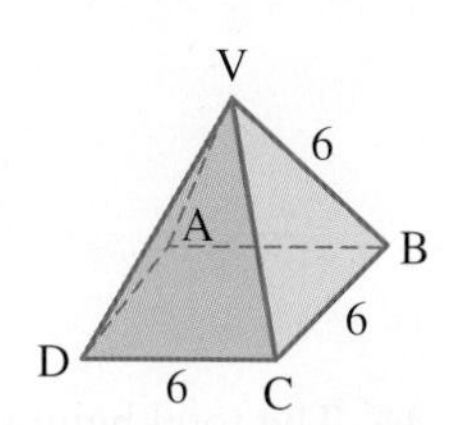

5. The pyramid ABCDE has a square base. The pyramid is 20 cm high. Each sloping edge measures 30 cm. Calculate the length of the sides of the base.

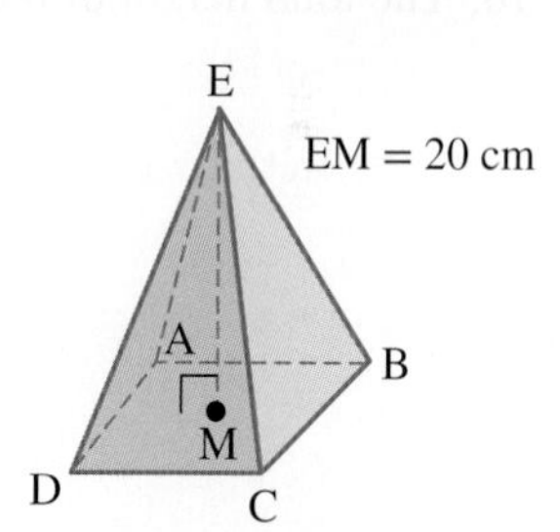

6. The sloping side of a cone is 16 cm and the height is 12 cm. Determine the length of the radius of the base.

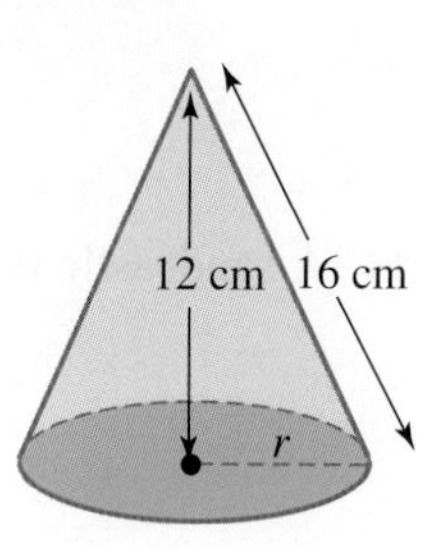

Understanding

7. WE6 A piece of cheese in the shape of a right-angled wedge sits on a table. It has a base measuring 20 mm by 10 mm, and is 4 mm high at the thickest point, as shown in the figure. A fly crawls diagonally across the sloping face. Determine how far, to the nearest millimetre, the fly walks.

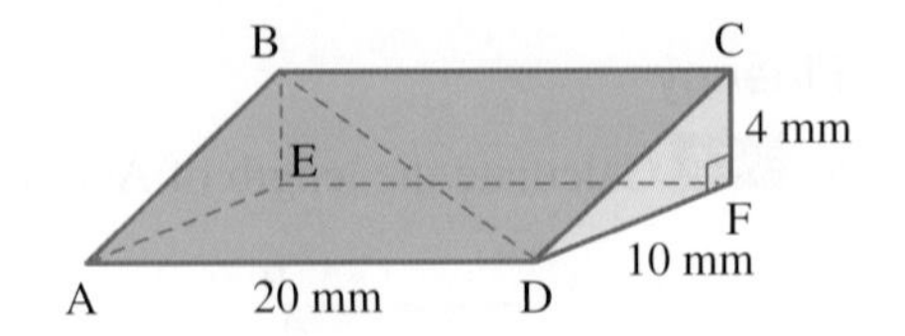

8. A 7 m high flagpole is in the corner of a rectangular park that measures 200 m by 120 m. Give your answers to the following questions correct to 2 decimal places.

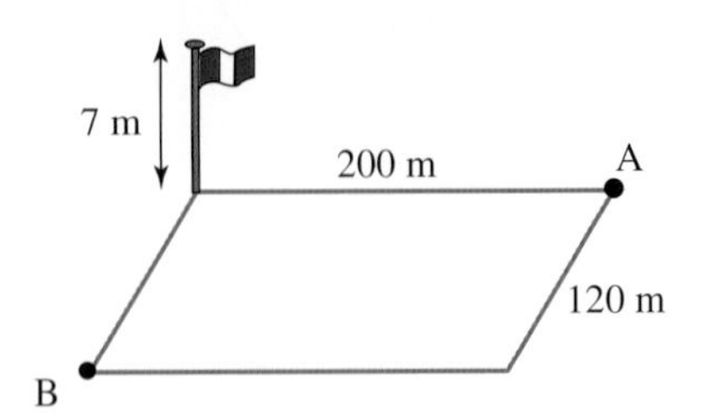

   a. Calculate:
      i. the length of the diagonal of the park
      ii. the distance from A to the top of the pole
      iii. the distance from B to the top of the pole.
   b. A bird flies from the top of the pole to the centre of the park. Calculate how far it flies.

9. A candlestick is in the shape of two cones, joined at the vertices as shown. The smaller cone has a diameter and sloping side of 7 cm, and the larger one has a diameter and sloping side of 10 cm. Calculate the total height of the candlestick.

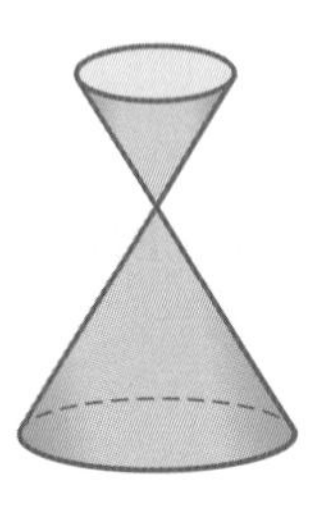

10. The total height of the shape below is 15 cm. Calculate the length of the sloping side of the pyramid.

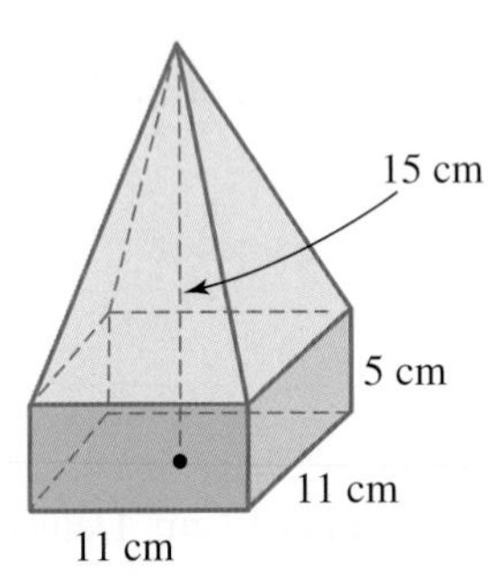

11. A sandcastle is in the shape of a truncated cone as shown. Calculate the length of the diameter of the base.

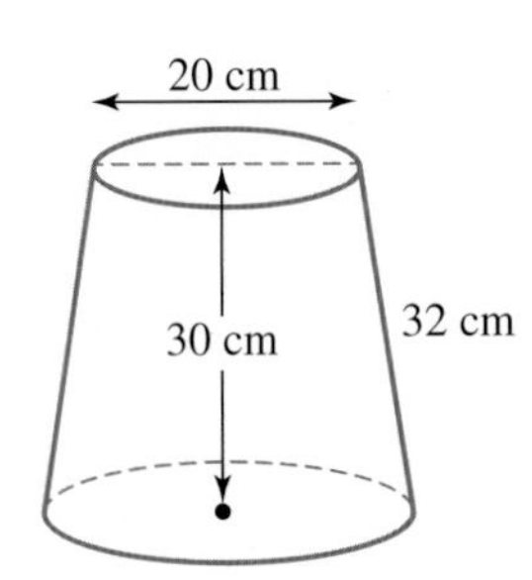

Reasoning

12. Stephano is renovating his apartment, which he accesses through two corridors. The corridors of the apartment building are 2 m wide with 2 m high ceilings, and the first corridor is at right angles to the second. Show that he can carry lengths of timber up to 6 m long to his apartment.

13. The Great Pyramid in Egypt is a square-based pyramid. The square base has a side length of 230.35 metres and the perpendicular height is 146.71 metres. Determine the slant height, s, of the great pyramid. Give your answer correct to 1 decimal place.

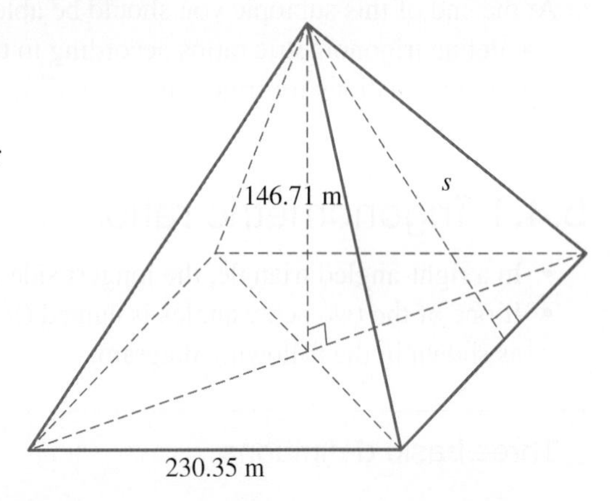

14. A tent is in the shape of a triangular prism, with a height of 140 cm as shown in the diagram. The width across the base of the door is 1 m and the tent is 2.5 m long.
    a. Calculate the length of each sloping side, in metres.
    b. Using your answer from part **a** calculate the area of fabric used in the construction of the sloping rectangles which form the sides. Show full working.

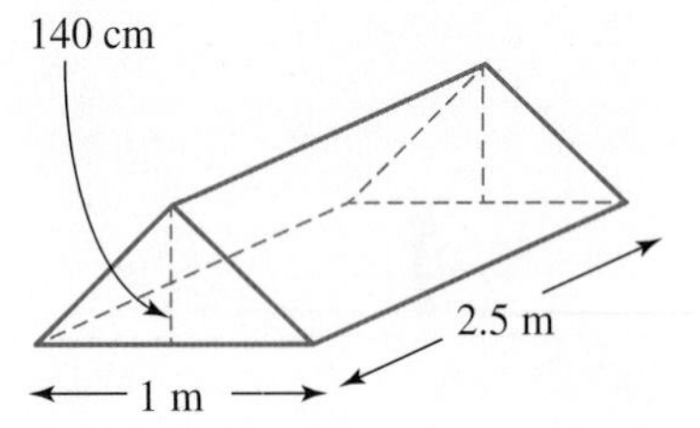

Problem solving

15. Determine the exact length of the longest steel rod that can sit inside a cuboid with dimensions $32\text{ cm} \times 15\text{ cm} \times 4\text{ cm}$. Ignore the thickness of the steel rod.

16. Angles ABD, CBD and ABC are right angles. Determine the value of h, correct to 3 decimal places.

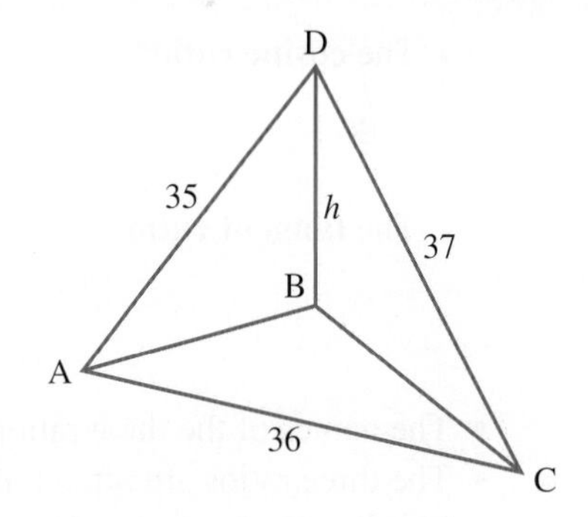

17. The roof of a squash centre is constructed to allow for maximum use of sunlight. Determine the value of h, giving your answer correct to 1 decimal place.

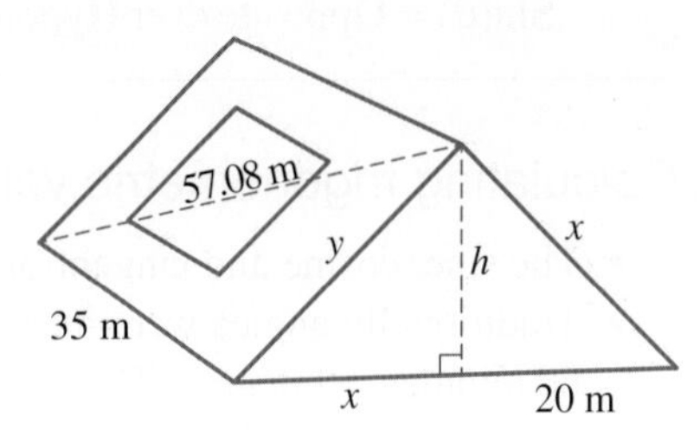

5.4 Trigonometric ratios

LEARNING INTENTION

At the end of this subtopic you should be able to:
- define trigonometric ratios according to the lengths of the relevant sides
- write equations for trigonometric ratios.

5.4.1 Trigonometric ratios

eles-4802

- In a right-angled triangle, the longest side is called the hypotenuse.
- If one of the two acute angles is named (for example, θ), then the other two sides can also be given names, as shown in the following diagram.

Three basic definitions

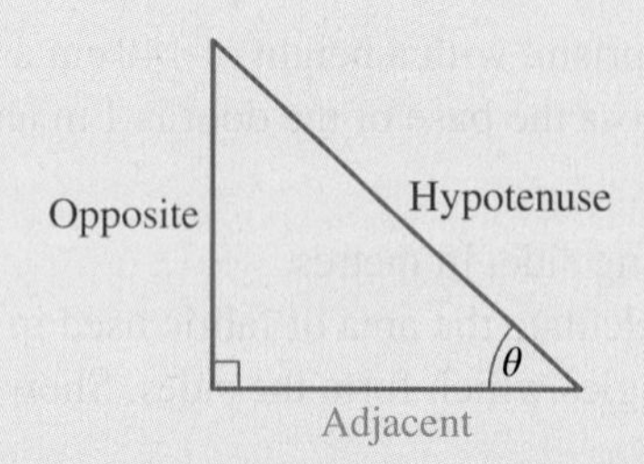

- Using the diagram, the following three **trigonometric ratios** can be defined.
  - The **sine ratio**:

$$\textbf{sine}(\theta) = \frac{\text{length of Opposite side}}{\text{length of Hypotenuse}}$$

  - The **cosine ratio**:

$$\textbf{cosine}(\theta) = \frac{\text{length of Adjacent side}}{\text{length of Hypotenuse}}$$

  - The **tangent ratio**:

$$\textbf{tangent}(\theta) = \frac{\text{length of Opposite side}}{\text{length of Adjacent side}}$$

- The names of the three ratios are usually shortened to **sin(θ)**, **cos(θ)** and **tan(θ)**.
- The three ratios are often remembered using the mnemonic **SOHCAHTOA**, where SOH means **S**in(θ) = **O**pposite over **H**ypotenuse and so on.

Calculating trigonometric values using a calculator

- The sine, cosine and tangent of an angle have numerical values that can be found using a calculator.
- Traditionally angles were measured in **degrees**, minutes and seconds, where 60 seconds = 1 minute and 60 minutes = 1 degree. This is known as a sexagesimal system as the division are based on 60. For example, $50°33'48''$ means 50 degrees, 33 minutes and 48 seconds.

WORKED EXAMPLE 7 Calculating values (ratios) from angles

Calculate the value of each of the following, correct to 4 decimal places, using a calculator. (Remember to first work to 5 decimal places before rounding.)

a. $\cos(65°57')$

b. $\tan(56°45'30'')$

THINK

a. Write your answer to the required number of decimal places.

WRITE

a. $\cos(65°57') \approx 0.40753$
≈ 0.4075

THINK

b. Write your answer to the correct number of decimal places.

WRITE

b. $\tan(56°45'30'') \approx 1.52573$
≈ 1.5257

TI | THINK

a-b.

1. To ensure your calculator is set to degree and approximate mode, press:
   - HOME
   - 5: Settings
   - 2: Document Settings

   In the Display Digits, select Fix 4. Tab to Angle and select Degree; tab to Calculation Mode and select Approximate.
   Tab to OK and press ENTER

DISPLAY/WRITE

a-b.

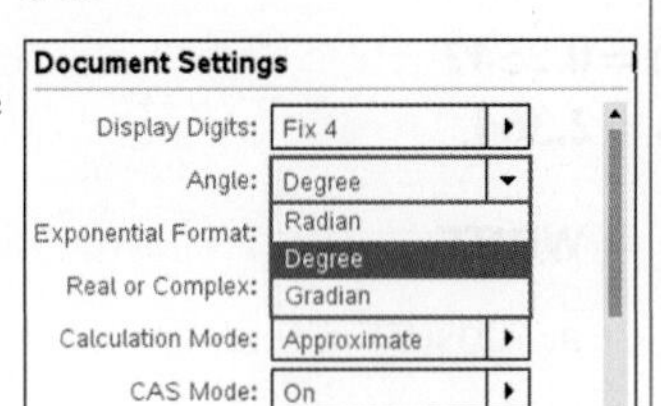

2. On a Calculator page, press TRIG to access and select the appropriate trigonometric ratio. Then press the template key and choose the template for degrees, minutes and seconds as shown.

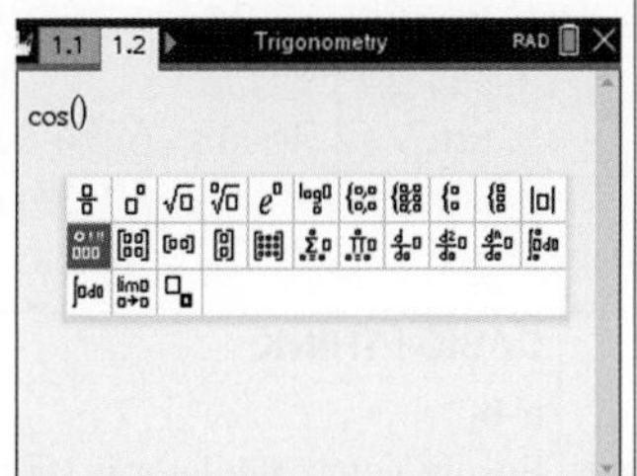

3. Complete the entry lines as:
   $\cos(65°57')$
   $\tan(56°45'30'')$
   Press ENTER after each entry.
   Since the Calculation Mode is set to Approximate and Fix 4, the answer are shown correct to 4 decimal places.

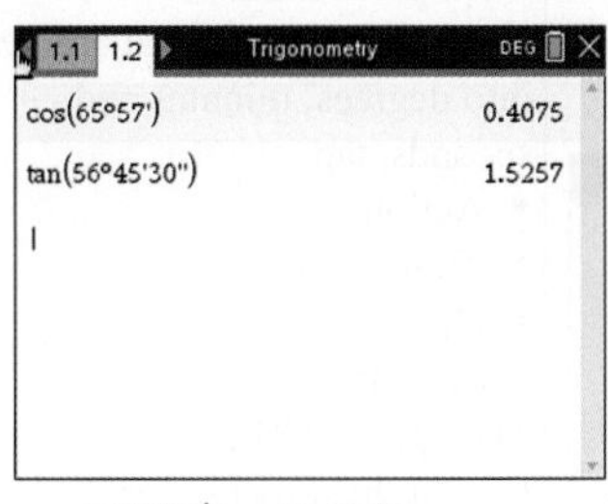

$\cos(65°57') = 0.4075$
$\tan(56°45'30'') = 1.5257$

CASIO | THINK

a-b.

1. Ensure your calculator is set to degrees at the bottom right. If not, tap 'Rad' or 'Gra'.
   Tap unit Deg appears.
   Set to Decimal.
   Set the keyboard to Trig.

DISPLAY/WRITE

a-b.

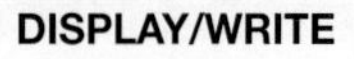

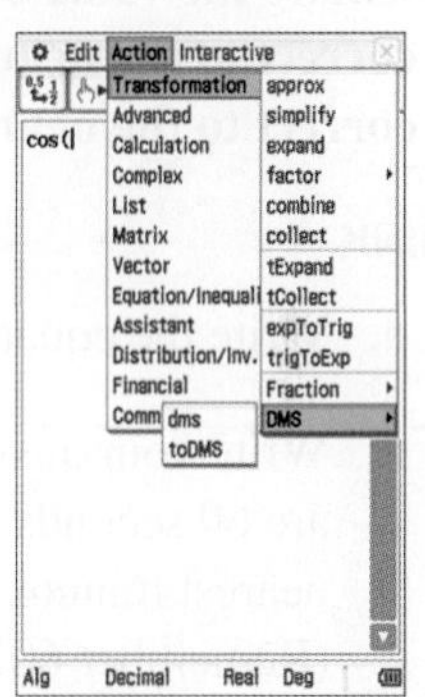

2. Complete the entry line as:
   cos(dms(65, 57))
   tan(dms(56, 45, 30))
   To get dms, tap
   - Action
   - Transformation
   - DMS
   - dms

   Then press EXE after each entry.

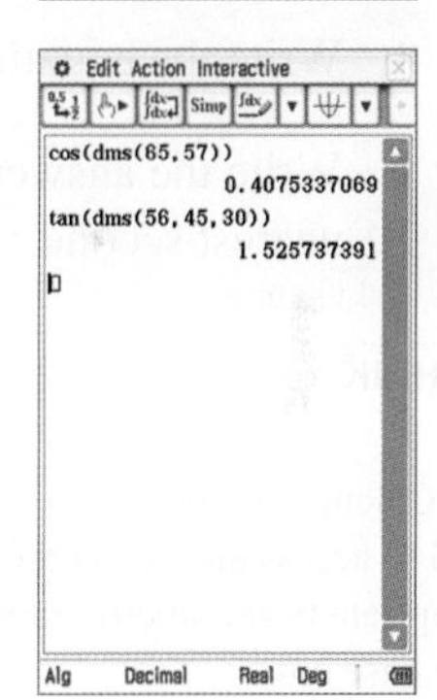

$\cos(65°57') = 0.4075$
$\tan(56°45'30'') = 1.5257$

WORKED EXAMPLE 8 Calculating angles from ratios

Calculate the size of angle θ, correct to the nearest degree, given $\sin(\theta) = 0.7854$.

THINK	WRITE
1. Write the given equation.	$\sin(\theta) = 0.7854$
2. To calculate the size of the angle, we need to undo sine with its inverse, $\sin^{-1}$. (Ensure your calculator is in degrees mode.)	$\theta = \sin^{-1}(0.7854)$ $\approx 51.8°$
3. Write your answer to the nearest degree.	$\theta \approx 52°$

WORKED EXAMPLE 9 Expressing angles in degrees, minutes and seconds

Calculate the value of θ:

a. correct to the nearest minute, given that $\cos(\theta) = 0.2547$

b. correct to the nearest second, given that $\tan(\theta) = 2.364$.

THINK	WRITE
a. 1. Write the equation.	a. $\cos(\theta) = 0.2547$
2. Write your answer, including seconds. There are 60 seconds in 1 minute. Round to the nearest minute. (Remember $60'' = 1'$, so $39''$ is rounded up.)	$\cos^{-1}(0.2547) \approx 75°14'39''$ $\approx 75°15'$
b. 1. Write the equation.	b. $\tan(\theta) = 2.364$
2. Write the answer, rounding to the nearest second.	$\tan^{-1}(2.364) \approx 67°4'15.8''$ $\approx 67°4'16''$

TI | THINK	DISPLAY/WRITE
a-b. On a Calculator page, press TRIG to access and select the appropriate trigonometric ratio, in the case $\cos^{-1}$. Complete the entry lines as: $\cos^{-1}(0.2547)$ To convert the decimal degree into degrees, minutes and seconds, press: • CATALOG • 1 • D Scroll and select ▶ DMS Then press ENTER. Repeat this process for $\tan^{-1}(2.364)$.	a-b. 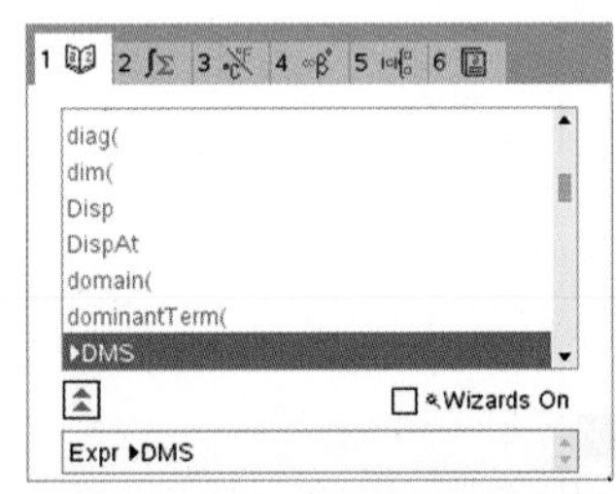 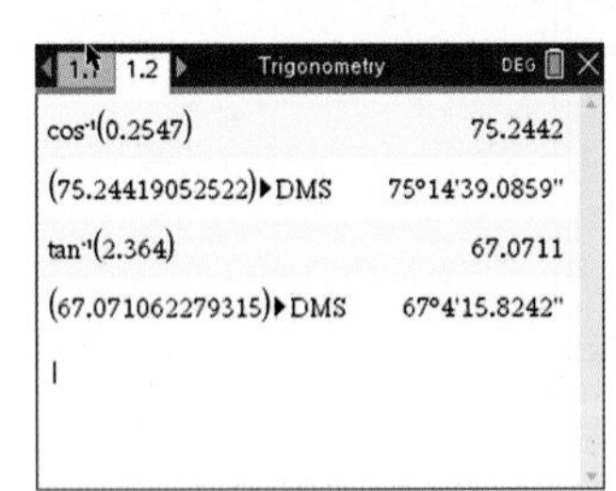 $\cos^{-1}(0.2547) = 75°15'$ rounding to the nearest minute. $\tan^{-1}(2.364) = 64°4'16''$ rounding to the nearest second.

CASIO | THINK	DISPLAY/WRITE
a-b. Ensure your calculator is set to degrees at the bottom right. On the Main screen in decimal and degree mode, complete the entry line as: $\cos^{-1}(0.2547)$ Then press EXE. To convert the decimal answer into degrees, minutes and seconds, tap: • Action • Transformation • DMS • to DMS Then press EXE. Repeat this process for $\tan^{-1}(2.364)$.	a-b. 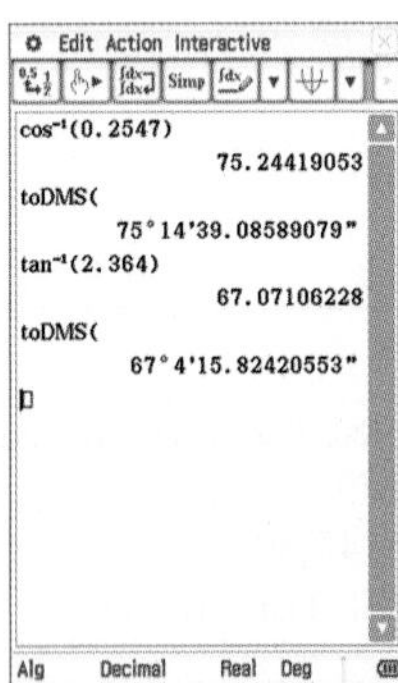 $\cos^{-1}(0.2547) = 75°15'$ rounding to the nearest minute. $\tan^{-1}(2.364) = 64°4'16''$ rounding to the nearest second.

WORKED EXAMPLE 10 Expressing trigonometric ratios as equations

Write the equation that relates the two marked sides and the marked angle.

a.

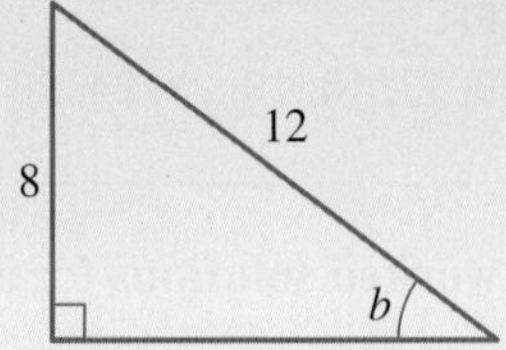

b.

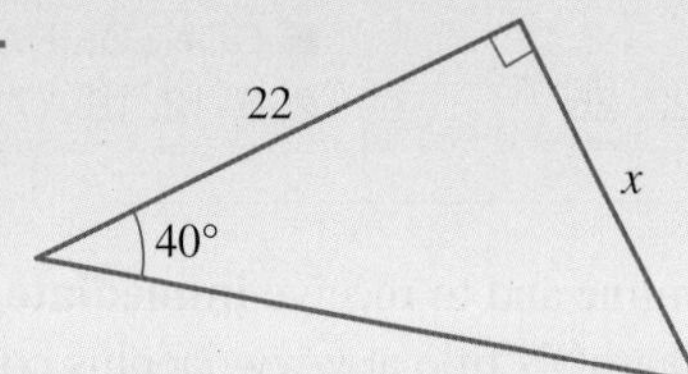

THINK	WRITE/DRAW
a. 1. Label the given sides of the triangle.	a. 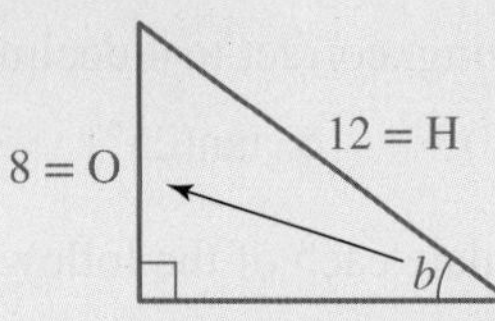
2. Write the ratio that contains O and H.	$\sin(\theta) = \dfrac{O}{H}$
3. Identify the values of the pronumerals.	$O = 8, H = 12$
4. Substitute the values of the pronumerals into the ratio and simplify the fraction. (Since the given angle is denoted with the letter b, replace θ with b.)	$\sin(b) = \dfrac{8}{12} = \dfrac{2}{3}$
b. 1. Label the given sides of the triangle.	b. 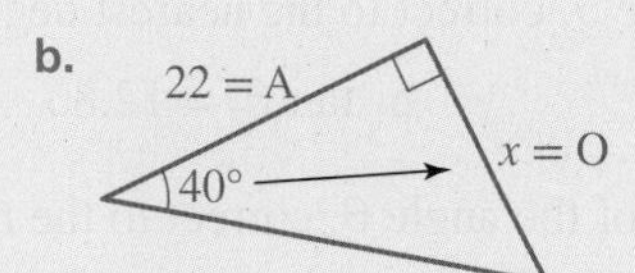
2. Write the ratio that contains O and A.	$\tan(\theta) = \dfrac{O}{A}$
3. Identify the values of the pronumerals.	$O = x, A = 22, \theta = 40°$
4. Substitute the values of the pronumerals into the ratio.	$\tan(40°) = \dfrac{x}{22}$

DISCUSSION

Do you know of any other mnemonics that you can use to help you remember important information?

on Resources

eWorkbook Topic 5 Workbook (worksheets, code puzzle and project) (ewbk-2031)

Digital documents SkillSHEET Labelling the sides of a right-angled triangle (doc-5226)
SkillSHEET Selecting an appropriate trigonometric ratio based on the given information (doc-5231)

Interactivities Individual pathway interactivity: Trigonometric ratios (int-4587)
Trigonometric ratios (int-2577)

Exercise 5.4 Trigonometric ratios

learnON

Individual pathways

PRACTISE	CONSOLIDATE	MASTER
1, 5, 7, 9, 11, 14, 15, 22, 25	2, 6, 12, 16, 17, 20, 23, 26, 27	3, 4, 8, 10, 13, 18, 19, 21, 24, 28

To answer questions online and to receive **immediate corrective feedback** and **fully worked solutions** for all questions, go to your learnON title at www.jacplus.com.au.

Fluency

1. Calculate each of the following, correct to 4 decimal places.
   a. $\sin(30°)$ b. $\cos(45°)$ c. $\tan(25°)$ d. $\sin(57°)$ e. $\tan(83°)$ f. $\cos(44°)$

WE7 For questions 2 to 4, calculate each of the following, correct to 4 decimal places.

2. a. $\sin(40°30')$ b. $\cos(53°57')$ c. $\tan(27°34')$ d. $\tan(123°40')$ e. $\sin(92°32')$ f. $\sin(42°8')$

3. a. $\cos(35°42'35'')$ b. $\tan(27°42'50'')$ c. $\cos(143°25'23'')$ d. $\sin(23°58'21'')$ e. $\cos(8°54'2'')$

4. a. $\sin(286)°$ b. $\tan(420°)$ c. $\cos(845°)$ d. $\sin(367°35')$

5. WE8 Calculate the size of angle θ, correct to the nearest degree, for each of the following.
   a. $\sin(\theta)=0.763$ b. $\cos(\theta)=0.912$ c. $\tan(\theta)=1.351$

6. Calculate the size of angle θ, correct to the nearest degree, for each of the following.
   a. $\cos(\theta)=0.321$ b. $\tan(\theta)=12.86$ c. $\cos(\theta)=0.756$

7. WE9a Calculate the size of the angle θ, correct to the nearest minute.
   a. $\sin(\theta)=0.814$ b. $\sin(\theta)=0.110$ c. $\tan(\theta)=0.015$

8. Calculate the size of the angle θ, correct to the nearest minute.
   a. $\cos(\theta)=0.296$ b. $\tan(\theta)=0.993$ c. $\sin(\theta)=0.450$

9. WE9b Calculate the size of the angle θ, correct to the nearest second.
   a. $\tan(\theta)=0.5$ b. $\cos(\theta)=0.438$ c. $\sin(\theta)=0.9047$

10. Calculate the size of the angle θ, correct to the nearest second.
    a. $\tan(\theta)=1.1141$ b. $\cos(\theta)=0.8$ c. $\tan(\theta)=43.76.$

For questions 11 to 13, calculate the value of each expression, correct to 3 decimal places.

11. a. $3.8\cos(42°)$ b. $118\sin(37°)$ c. $2.5\tan(83°)$ d. $\dfrac{2}{\sin(45°)}$

12. a. $\dfrac{220}{\cos(14°)}$ b. $\dfrac{2\cos(23°)}{5\sin(18°)}$ c. $\dfrac{12.8}{\tan(60°32')}$ d. $\dfrac{18.7}{\sin(35°25'42'')}$

13. a. $\dfrac{55.7}{\cos(89°21')}$ b. $\dfrac{3.8\tan(1°51'44'')}{4.5\sin(25°45')}$ c. $\dfrac{2.5\sin(27°8')}{10.4\cos(83°2')}$ d. $\dfrac{3.2\cos(34°52')}{0.8\sin(12°48')}$

For questions **14** to **19**, write an expression for:

a. sine **b.** cosine **c.** tangent.

14.

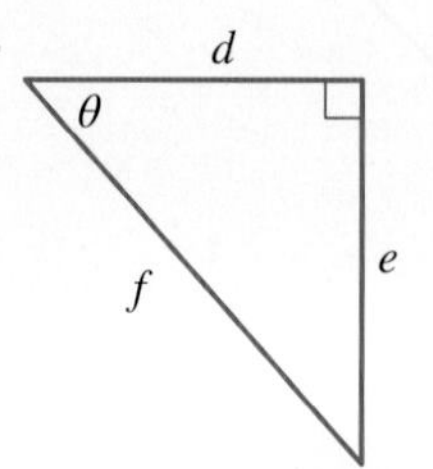

15.

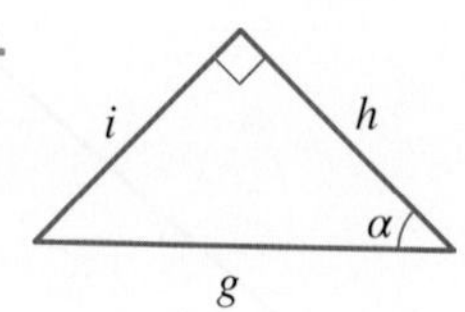

16.

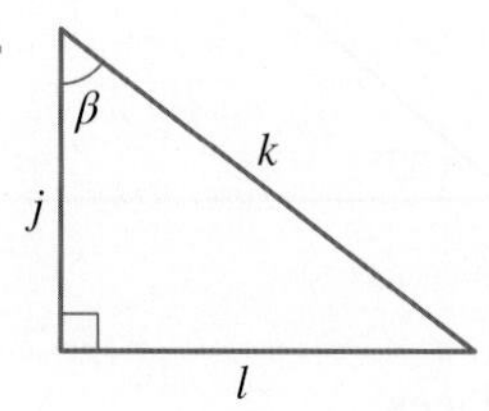

17.

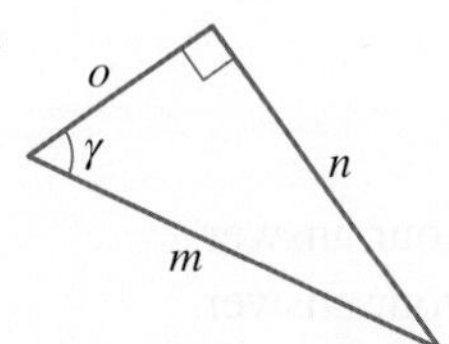

18.

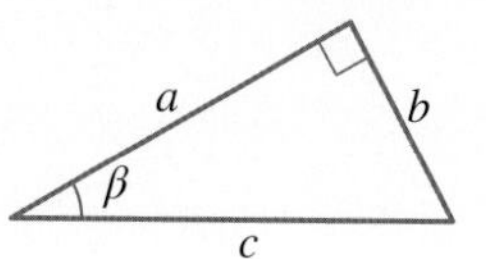

19.

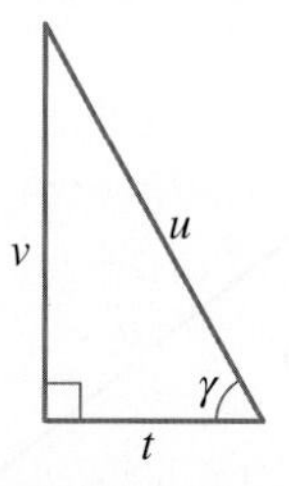

Understanding

20. **WE10** Write the equation that relates the two marked sides and the marked angle in each of the following triangles.

a.

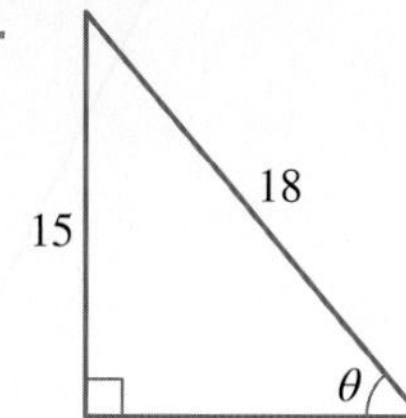

b.

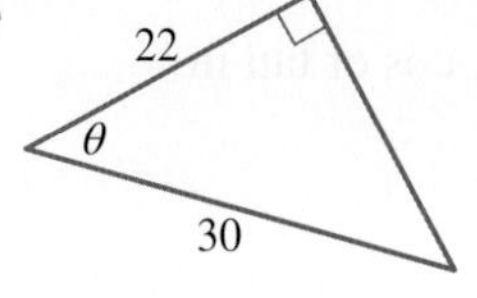

c.

9
θ
7

21. Write the equation that relates the two marked sides and the marked angle in each of the following triangles.

a.

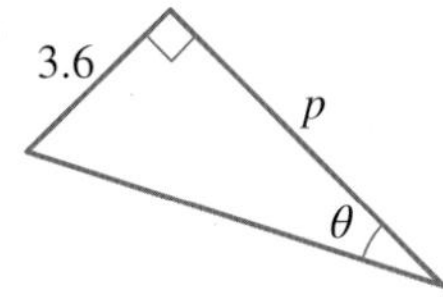

b.

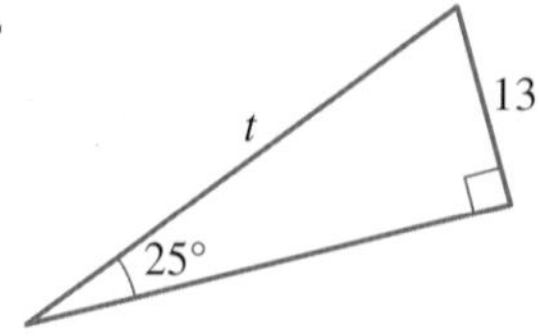

c.

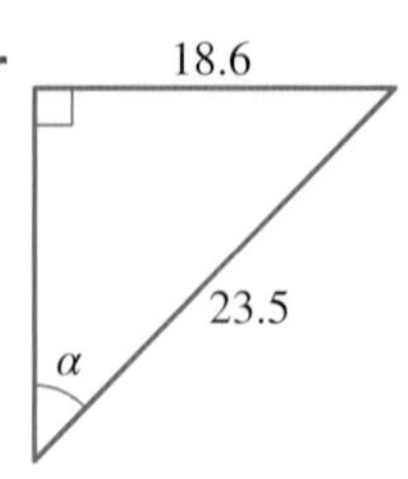

Reasoning

22. Consider the right-angled triangle shown.

a. Label each of the sides using the letters O, A and H with respect to the 37° angle.

b. Determine the value of each trigonometric ratio. (Where applicable, answers should be given correct to 2 decimal places.)

i. $\sin(37^\circ)$ ii. $\cos(37^\circ)$ iii. $\tan(37^\circ)$

c. Determine the value of the unknown angle, α.

α

37°

23. Consider the right-angled triangle shown in 22.

a. Determine the value of each of these trigonometric ratios, correct to 2 decimal places.

i. $\sin(\alpha)$ ii. $\cos(\alpha)$ iii. $\tan(\alpha)$

(*Hint:* First relabel the sides of the triangle with respect to angle α)

b. What do you notice about the relationship between $\sin(37^\circ)$ and $\cos(\alpha)$? Explain your answer.

c. What do you notice about the relationship between $\sin(\alpha)$ and $\cos(37^\circ)$? Explain your answer.

d. Make a general statement about the two angles.

24. Using a triangle labelled with a, h and o, algebraically show that $\tan(\theta) = \dfrac{\sin(\theta)}{\cos(\theta)}$.

(*Hint:* Write all the sides in terms of the hypotenuse.)

Problem solving

25. ABC is a scalene triangle with side lengths a, b and c as shown. Angles BDA and BDC are right angles.

a. Express h^2 in terms of a and x.

b. Express h^2 in terms of b, c and x.

c. Equate the two equations for h^2 to show that $c^2 = a^2 + b^2 - 2bx$.

d. Use your knowledge of trigonometry to produce the equation $c^2 = a^2 + b^2 - 2ab\cos(C)$, which is known as the cosine rule for non-right-angled triangles.

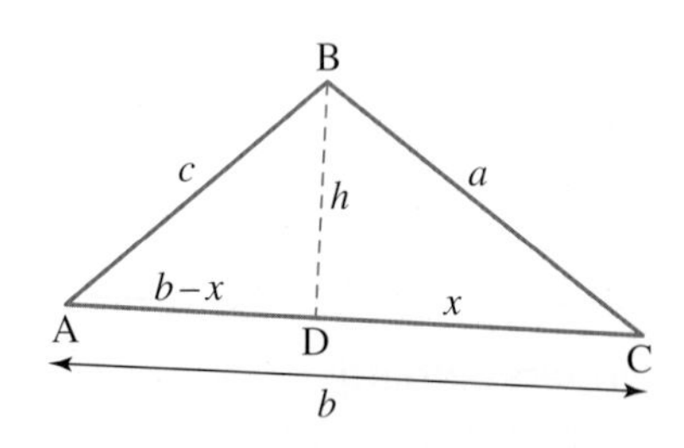

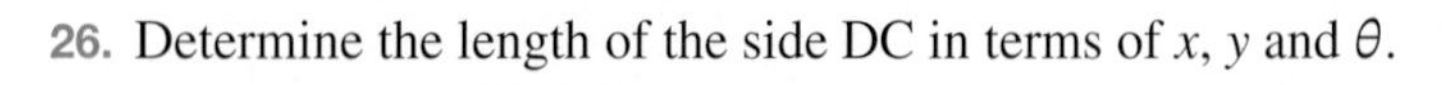

26. Determine the length of the side DC in terms of x, y and θ.

27. Explain how we determine whether to use sin, cos or tan in trigonometry questions.

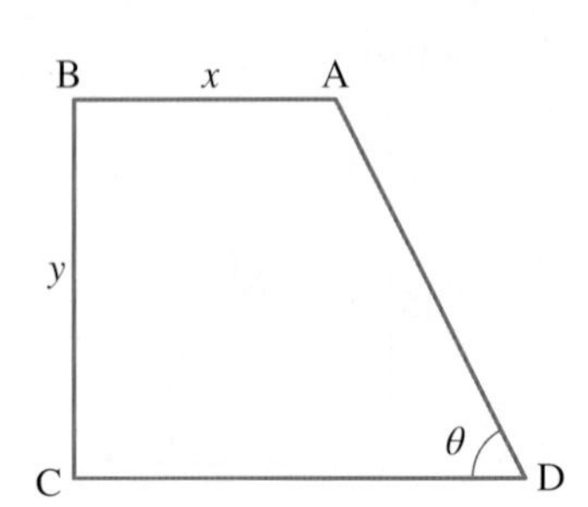

28. From an observer on a boat 110 m away from a vertical cliff with height c, the angle from the base of the cliff to the top of the cliff is $\alpha°$. There is a lighthouse with height t on the cliff. From the observer, the angle from the base of the lighthouse to the top of the lighthouse is another $\theta°$ more than $\alpha°$.
Express the height of the lighthouse, t, in terms of $\theta°$ and $\alpha°$.

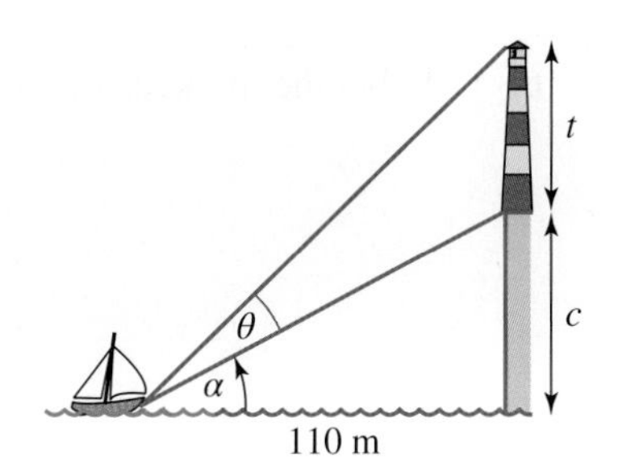

5.5 Using trigonometry to calculate side lengths

LEARNING INTENTION

At the end of this subtopic you should be able to:
- apply trigonometric ratios to find the length of an unknown side when the length of one other side and an acute angle is known.

5.5.1 Using trigonometry to calculate side lengths

eles-4804

- When one acute angle and one side length are known in a right-angled triangle, this information can be used to find all other unknown sides or angles.

WORKED EXAMPLE 11 Using trigonometry to calculate side lengths

Calculate the value of each pronumeral, giving answers correct to 3 decimal places.

a.

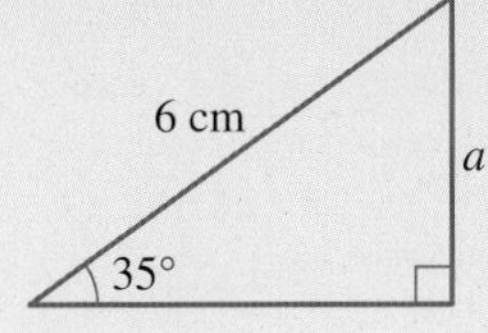

b.

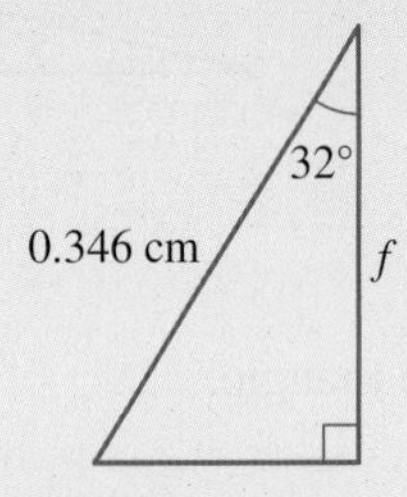

THINK

WRITE/DRAW

a. 1. Label the marked sides of the triangle.

a.

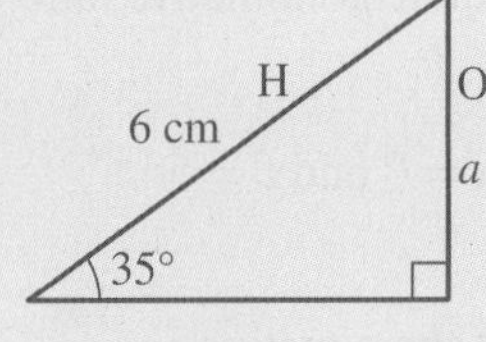

2. Identify the appropriate trigonometric ratio to use.

$\sin(\theta) = \frac{\text{O}}{\text{H}}$

3. Substitute O = a, H = 6 and $\theta = 35^{\circ}$.

$\sin(35°) = \frac{a}{6}$

4. Make a the subject of the equation.

$6\sin(35°) = a$

$x = 6\sin(35°)$

5. Calculate and round the answer, correct to 3 decimal places.

$a \approx 3.441$ cm

b. 1. Label the marked sides of the triangle.

b.

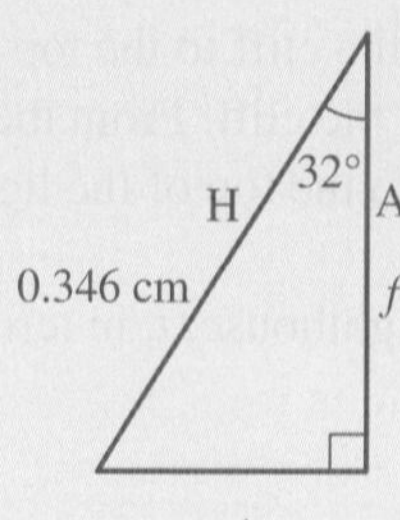

2. Identify the appropriate trigonometric ratio to use.

$\cos(\theta) = \dfrac{A}{H}$

3. Substitute $A = f$, $H = 0.346$ and $\theta = 32°$.

$\cos(32°) = \dfrac{f}{0.346}$

4. Make f the subject of the equation.

$0.346\cos(32°) = f$

$f = 0.346\cos(32°)$

5. Calculate and round the answer, correct to 3 decimal places.

$f \approx 0.293$ cm

WORKED EXAMPLE 12 Using trigonometry to calculate side lengths

Calculate the value of the pronumeral in the triangle shown. Give the answer correct to 2 decimal places.

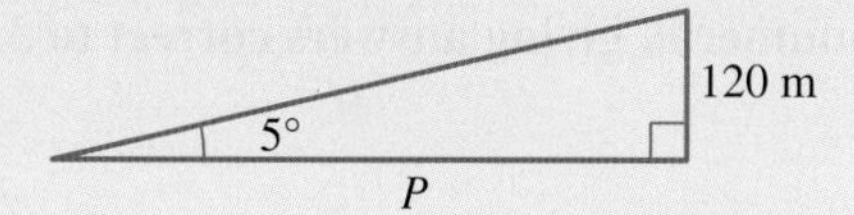

THINK

WRITE/DRAW

1. Label the marked sides of the triangle.

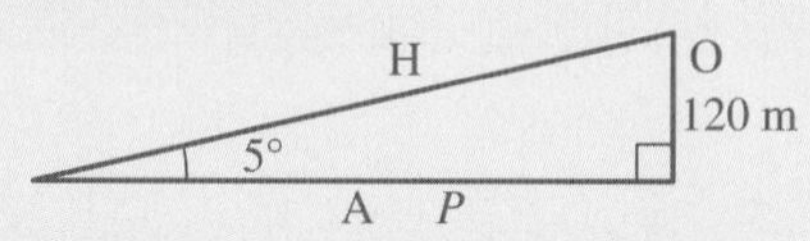

2. Identify the appropriate trigonometric ratio to use.

$\tan(\theta) = \dfrac{O}{A}$

3. Substitute $O = 120$, $A = P$ and $\theta = 5°$.

$\tan(5°) = \dfrac{120}{p}$

4. Make P the subject of the equation.
   i. Multiply both sides of the equation by P.
   ii. Divide both sides of the equation by tan (5°).

$P \times \tan(5°) = 120$

$P = \dfrac{120}{\tan(5°)}$

5. Calculate and round the answer, correct to 2 decimal places.

$P \approx 1371.61$ m

DISCUSSION

How does solving a trigonometric equation differ when we are finding the length of the hypotenuse side compared to when finding the length of a shorter side?

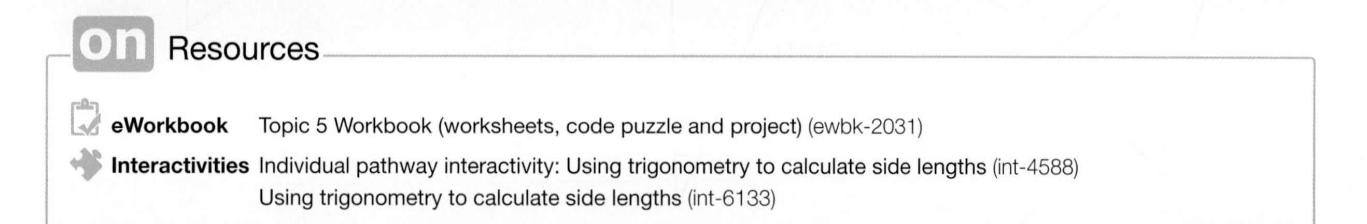

on Resources

eWorkbook Topic 5 Workbook (worksheets, code puzzle and project) (ewbk-2031)

Interactivities Individual pathway interactivity: Using trigonometry to calculate side lengths (int-4588)
Using trigonometry to calculate side lengths (int-6133)

Exercise 5.5 Using trigonometry to calculate side lengths **learnon**

Individual pathways

■ PRACTISE	■ CONSOLIDATE	■ MASTER
1, 3, 7, 9, 12	2, 5, 8, 10, 13	4, 6, 11, 14, 15

To answer questions online and to receive **immediate corrective feedback** and **fully worked solutions** for all questions, go to your learnON title at www.jacplus.com.au.

Fluency

1. **WE11** Calculate the value of each pronumeral in each of the following, correct to 3 decimal places.

a. 10 cm, a, 60°

b. 8, 25°, a

c. x, 31°, 14

2. **WE12** Calculate the value of each pronumeral in each of the following, correct to 2 decimal places.

a. 71°, m, 2.3 m

b. 4.6 m, 13°, n

c.

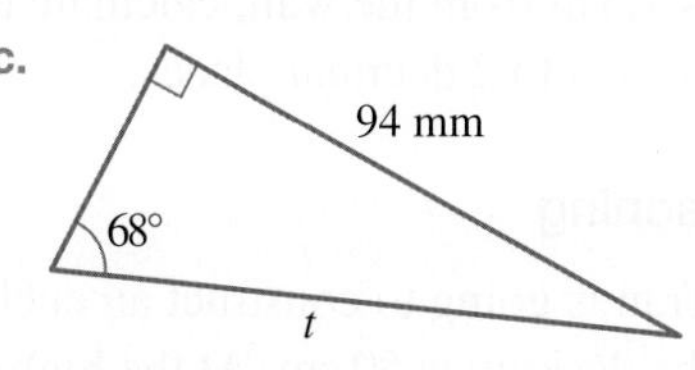

3. Determine the length of the unknown side in each of the following, correct to 2 decimal places.

a. 43.95 m, 40°26′, x

b.

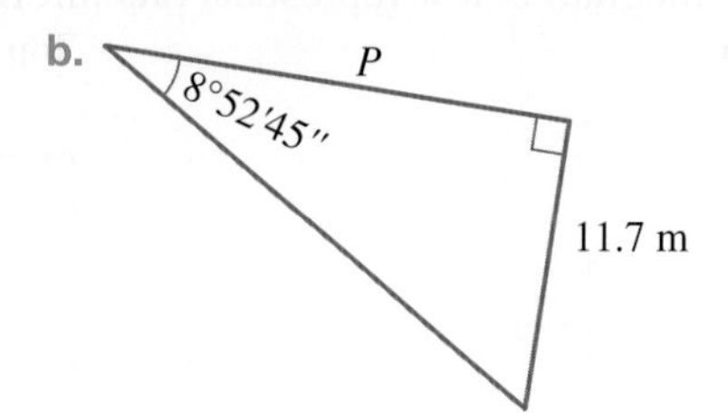

c.

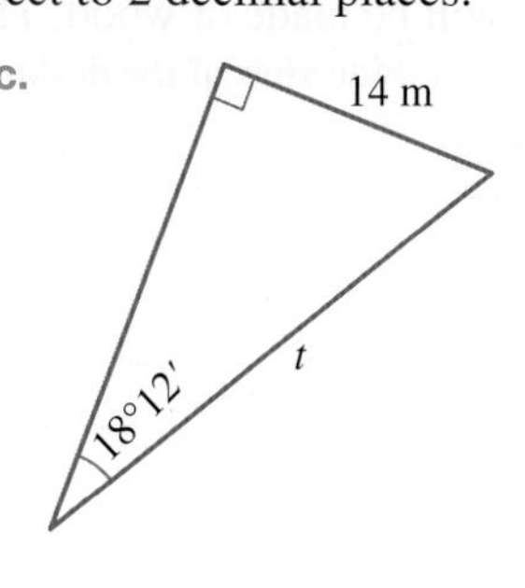

4. Determine the length of the unknown side in each of the following, correct to 2 decimal places.

a. 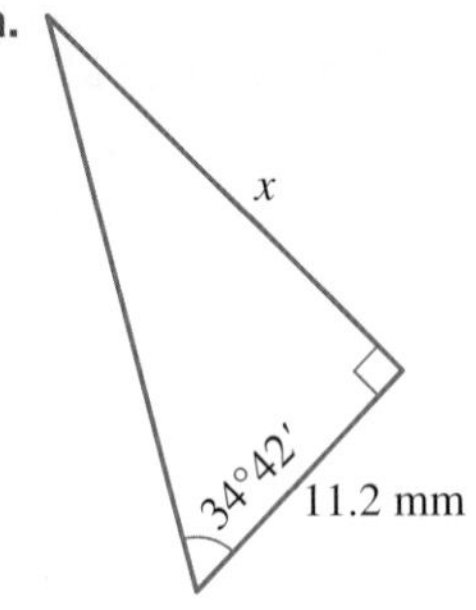

b. 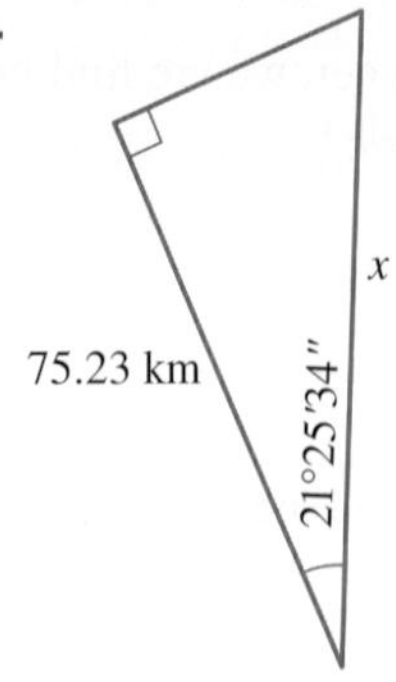

c. 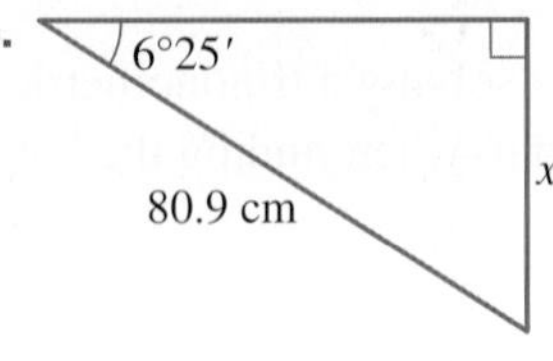

5. Calculate the value of the pronumeral in each of the following, correct to 2 decimal places.

a. 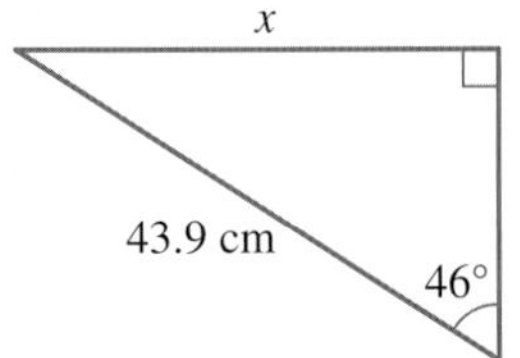

b. 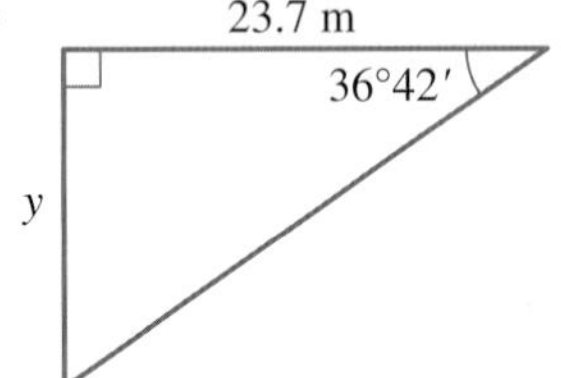

c. 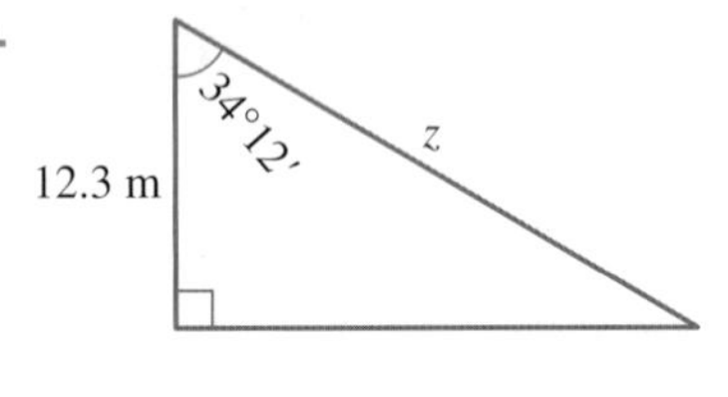

6. Calculate the value of the pronumeral in each of the following, correct to 2 decimal places.

a. 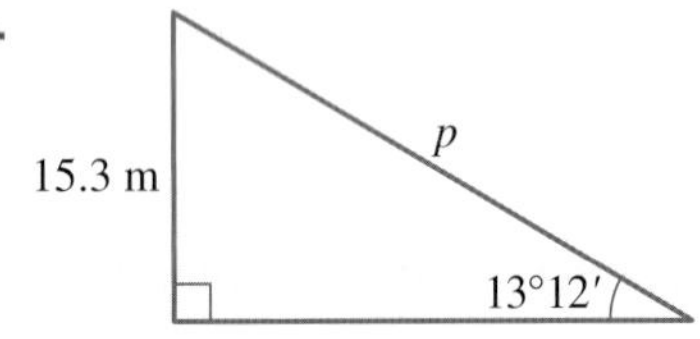

b. 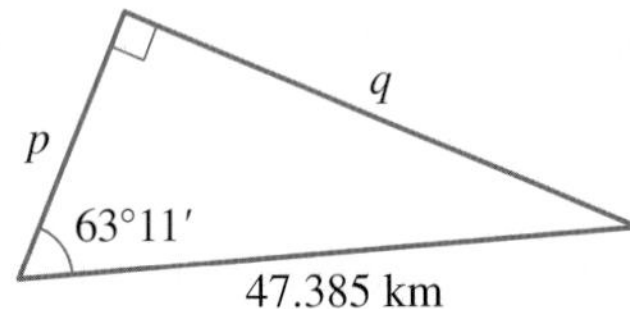

c. 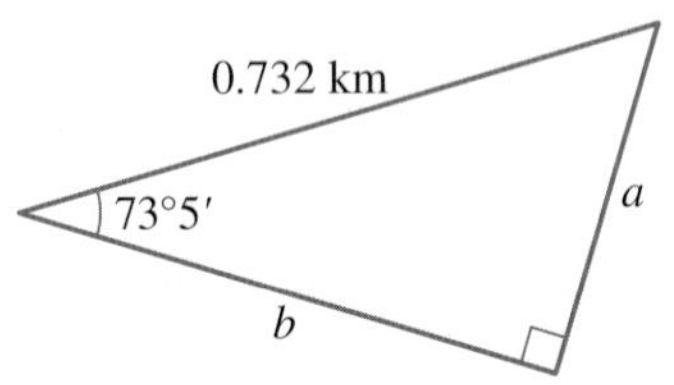

Understanding

7. Given that the angle θ is 42° and the length of the hypotenuse is 8.95 m in a right-angled triangle, calculate the length of:

a. the opposite side

b. the adjacent side.

Give each answer correct to 1 decimal place.

8. A ladder rests against a wall. If the angle between the ladder and the ground is 35° and the foot of the ladder is 1.5 m from the wall, calculate how high up the wall the ladder reaches. Write your answer in metres correct to 2 decimal places.

Reasoning

9. Tran is going to construct an enclosed rectangular desktop that is at an incline of 15°. The diagonal length of the desktop is 50 cm. At the high end, the desktop, including top, bottom and sides, will be raised 8 cm. The desktop will be made of wood. The diagram below represents this information.

Side view of the desktop

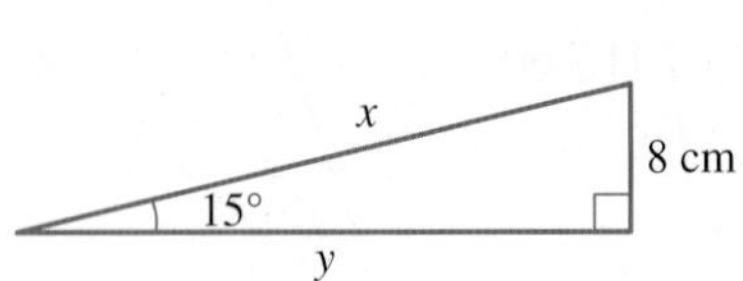

Top view of the desktop

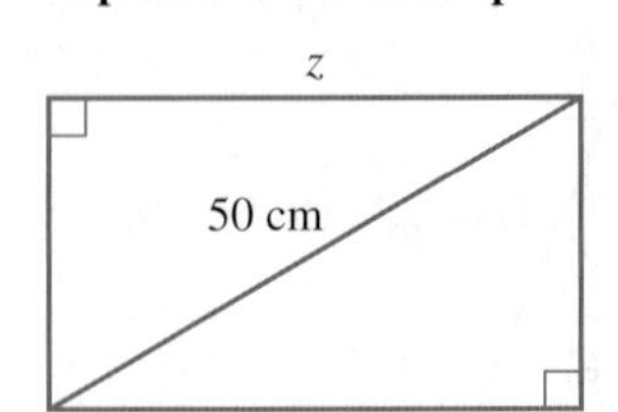

a. Determine the values (in centimetres) of x, y and z of the desktop. Write your answers correct to 2 decimal places.

b. Using your answer from part **a** determine the minimum area of wood, in cm^2, Tran needs to construct his desktop including top, bottom and sides. Write your answer correct to 2 decimal places.

10. a. In a right-angled triangle, under what circumstances will the opposite side and the adjacent side have the same length?
b. In a right-angled triangle, for what values of θ (the reference angle) will the adjacent side be longer than the opposite side?

11. In triangle ABC shown, the length of x correct to two decimal places is 8.41 cm. AM is perpendicular to BC. Jack found the length of x to be 5.11 cm. Below is his working. Identify his error and what he should have done instead.

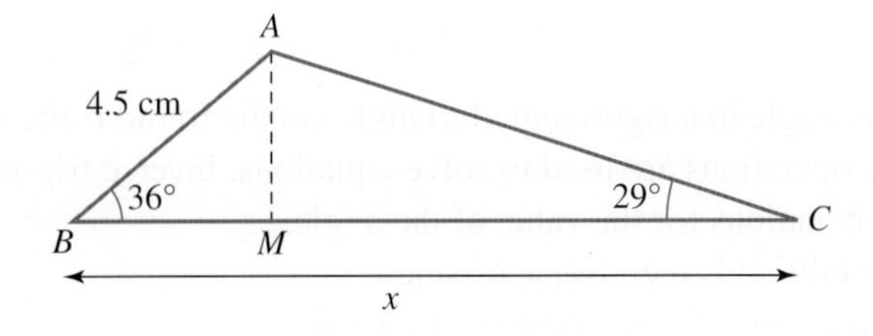

$$\cos(36°) = \frac{BM}{4.5}$$
$$BM = 3.641$$
$$\sin(36°) = \frac{AM}{4.5}$$
$$AM = 2.645$$
$$\tan(29°) = \frac{2.645}{MC}$$
$$MC = 1.466$$
$$BC = 3.641 + 1.466 = 5.107 \text{ cm}$$
$$= 5.11 \text{ cm (rounded to 2 decimal places)}$$

Problem solving

12. A surveyor needs to determine the height of a building. She measures the angle of elevation of the top of the building from two points, 64 m apart. The surveyor's eye level is 195 cm above the ground.
a. Determine the expressions for the height of the building, h, in terms of x using the two angles.
b. Solve for x by equating the two expressions obtained in part a. Give your answer to 2 decimal places.
c. Determine the height of the building correct to 2 decimal places.

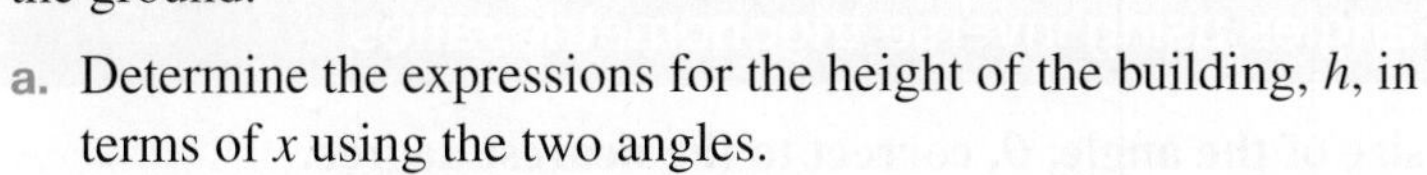

13. Building A and Building B are 110 m apart. From the base of Building A to the top of Building B, the angle is 15°. From the top of Building A looking down to the top of Building B, the angle is 22°. Evaluate the heights of each of the two buildings correct to one decimal place.

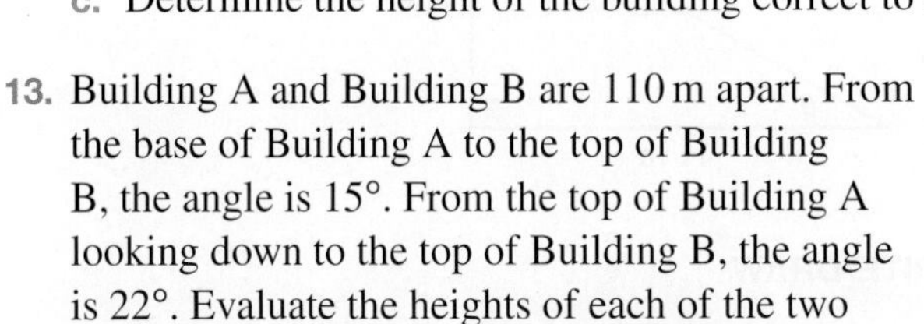

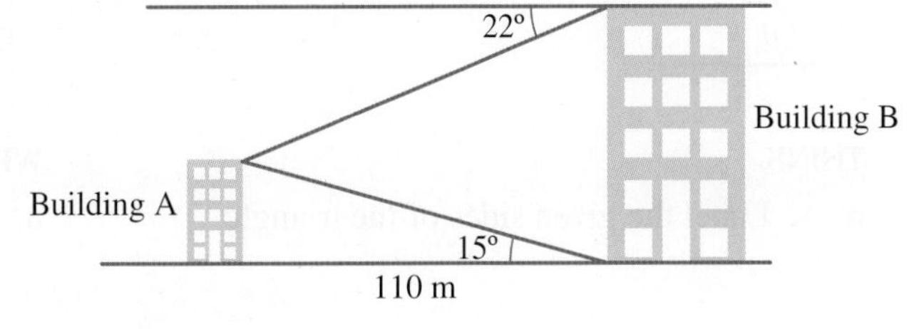

14. If angles QNM, QNP and MNP are right angles, determine the length of NQ.

15. Determine how solving a trigonometric equation differs when we are calculating the length of the hypotenuse side compared to when determining the length of a shorter side.

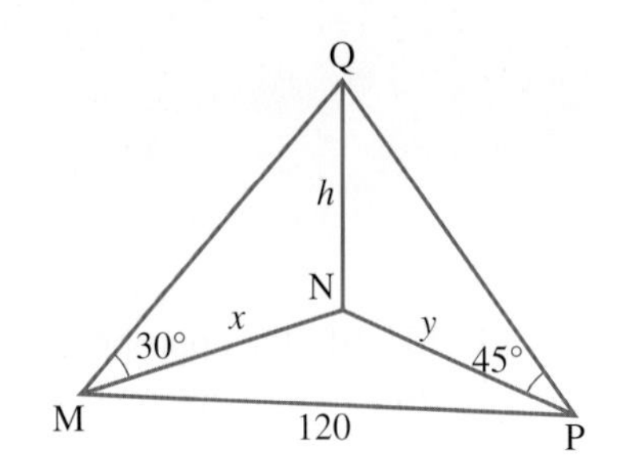

5.6 Using trigonometry to calculate angle size

LEARNING INTENTION

At the end of this subtopic you should be able to:
- apply inverse operations to calculate a known acute angle when two sides are given.

5.6.1 Using trigonometry to calculate angle size

eles-4805

- The size of any angle in a right-angled triangle can be found if the lengths of any two sides are known.
- Just as inverse operations are used to solve equations, inverse trigonometric ratios are used to solve trigonometric equations for the value of the angle.
  - Inverse sine ($\sin^{-1}$) is the inverse of sine.
  - Inverse cosine ($\cos^{-1}$) is the inverse of cosine.
  - Inverse tangent ($\tan^{-1}$) is the inverse of tangent.

Inverse operations

$$\text{If } \sin(\theta) = a, \text{ then } \sin^{-1}(a) = \theta.$$
$$\text{If } \cos(\theta) = a, \text{ then } \cos^{-1}(a) = \theta.$$
$$\text{If } \tan(\theta) = a, \text{ then } \tan^{-1}(a) = \theta.$$

For example, since $\sin(30°) = 0.5$, then $\sin^{-1}(0.5) = 30°$; this is read as 'inverse sine of 0.5 is 30 degrees'.
- A calculator can be used to calculate the values of inverse trigonometric ratios.

WORKED EXAMPLE 13 Evaluating angles using inverse trigonometric ratios

For each of the following, calculate the size of the angle, θ, correct to the nearest degree.

a.

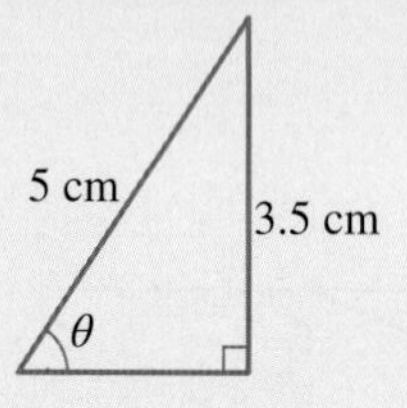

b.

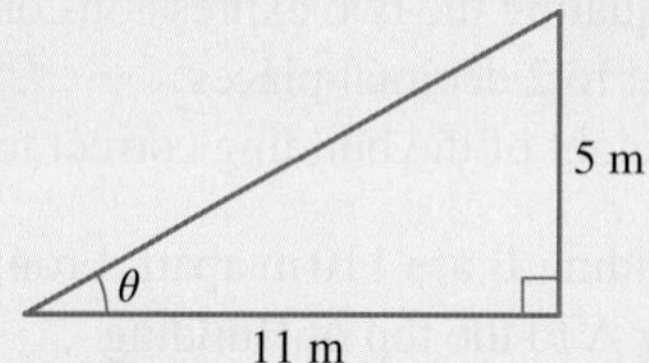

THINK	WRITE/DRAW
a. 1. Label the given sides of the triangle.	a. 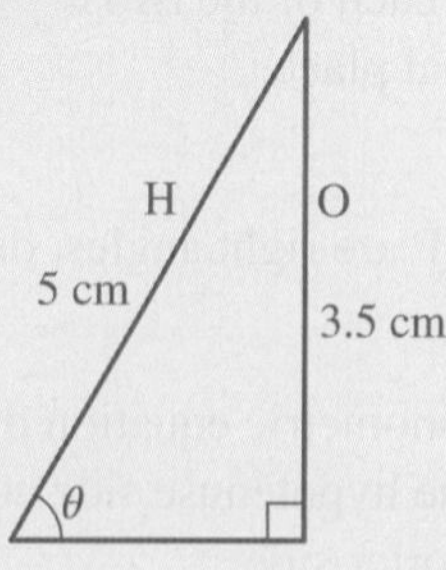
2. Identify the appropriate trigonometric ratio to use. We are given O and H.	$\sin(\theta) = \frac{O}{H}$

3. Substitute O = 3.5 and H = 5 and evaluate the expression. $\sin(\theta) = \dfrac{3.5}{5}$
$= 0.7$

4. Make θ the subject of the equation using inverse sine. $\theta = \sin^{-1}(0.7)$
$= 44.427\,004°$

5. Evaluate θ and round the answer, correct to the nearest degree. $\theta \approx 44°$

b. 1. Label the given sides of the triangle. **b.**

2. Identify the appropriate trigonometric ratio to use. Given O and A. $\tan(\theta) = \dfrac{O}{A}$

3. Substitute O = 5 and A = 11. $\tan(\theta) = \dfrac{5}{11}$

4. Make θ the subject of the equation using inverse tangent. $\theta = \tan^{-1}\left(\dfrac{5}{11}\right)$
$= 24.443\,954\,78°$

5. Evaluate θ and round the answer, correct to the nearest degree. $\theta \approx 24°$

WORKED EXAMPLE 14 Evaluating angles in minutes and seconds

Calculate the size of angle θ:
a. correct to the nearest second
b. correct to the nearest minute.

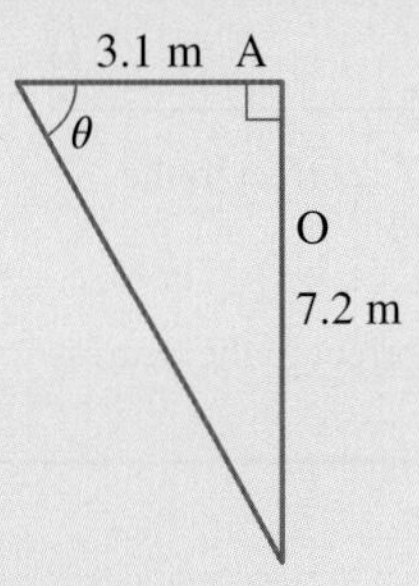

THINK | **WRITE/DRAW**

a. 1. Label the given sides of the triangle. **a.**

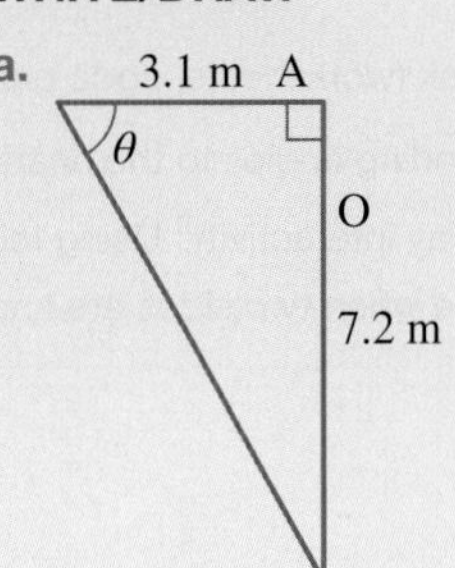

2. Identify the appropriate trigonometric ratio to use. $\tan(\theta) = \frac{O}{A}$

3. Substitute O = 7.2 and A = 3.1. $\tan(\theta) = \frac{7.2}{3.1}$

4. Make θ the subject of the equation using inverse tangent. $\theta = \tan^{-1}\left(\frac{17.2}{3.1}\right)$

5. Evaluate θ and write the calculator display. $\theta = 66.70543675°$

6. Use the calculator to convert the answer to degrees, minutes and seconds. $\theta = 66°42'19.572''$

7. Round the answer to the nearest second. $\theta \approx 66°42'20''$

b. Round the answer to the nearest minute. **b.** $\theta \approx 66°42'$

TI | THINK

a.

On a Calculator page, in degree mode, complete the entry line as:

$\tan^{-1}\left(\frac{7.2}{3.1}\right)$

To convert the decimal degree into degrees, minutes and seconds, press:

- CATALOG
- 1
- D

Scroll and select ▶ DMS. Then press ENTER.

DISPLAY/WRITE

a.

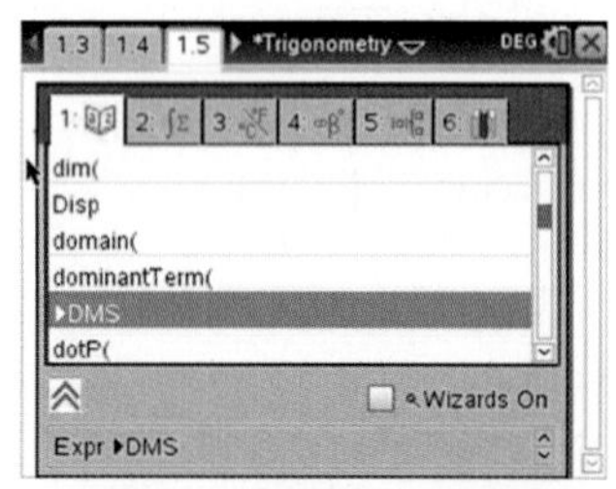

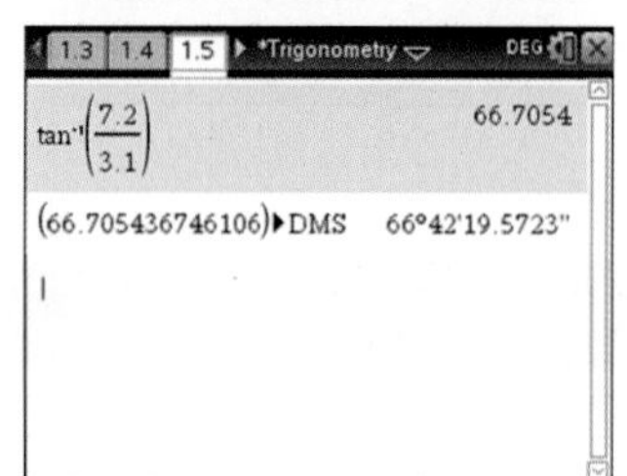

$\theta = 66°42'20''$ correct to the nearest second.

b.

Using the same screen, round to the nearest minute.

b.

$\theta = 66°42'$ correct to the nearest minute.

CASIO | THINK

a.

On the Main screen in decimal and degree mode, complete the entry line as:

$\tan^{-1}\left(\frac{7.2}{3.1}\right)$

To convert the decimal answer into degrees, minutes and seconds, tap:

- Action
- Transformation
- DMS
- toDMS

Highlight the decimal answer and drag it into this line.

Then press EXE.

DISPLAY/WRITE

a.

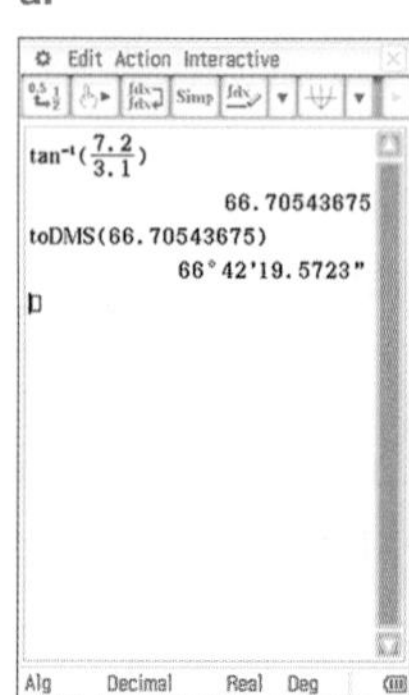

$\theta = 66°42'20''$ correct to the nearest second.

b.

Using the same screen, round to the nearest minute.

b.

$\theta = 66°42'$ correct to the nearest minute.

Resources

eWorkbook Topic 5 Workbook (worksheets, code puzzle and project) (ewbk-2031)

Digital document SkillSHEET Rounding angles to the nearest degree (doc-5232)

Interactivities Individual pathway interactivity: Using trigonometry to calculate angle size (int-4589)

Finding the angle when two sides are known (int-6046)

Exercise 5.6 Using trigonometry to calculate angle size

learnON

Individual pathways

■ PRACTISE	■ CONSOLIDATE	■ MASTER
1, 4, 7, 9, 12	2, 5, 8, 10, 13	3, 6, 11, 14

To answer questions online and to receive **immediate corrective feedback** and **fully worked solutions** for all questions, go to your learnON title at www.jacplus.com.au.

Fluency

1. **WE13** Calculate the size of the angle, θ, in each of the following. Give your answer correct to the nearest degree.

a. 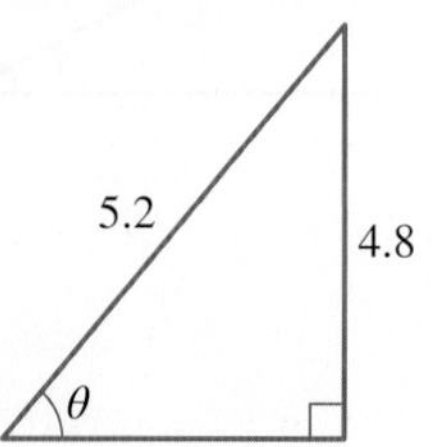

b. 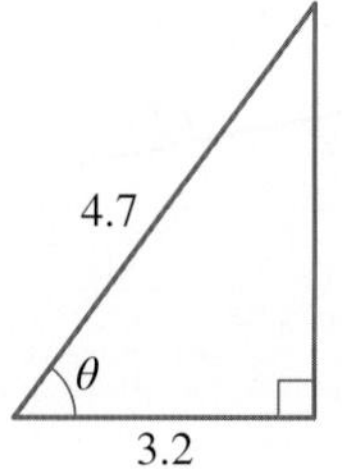

c. 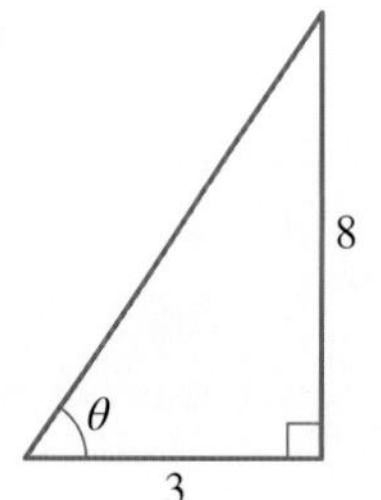

2. **WE14b** Calculate the size of the angle marked with the pronumeral in each of the following. Give your answer correct to the nearest minute.

a. 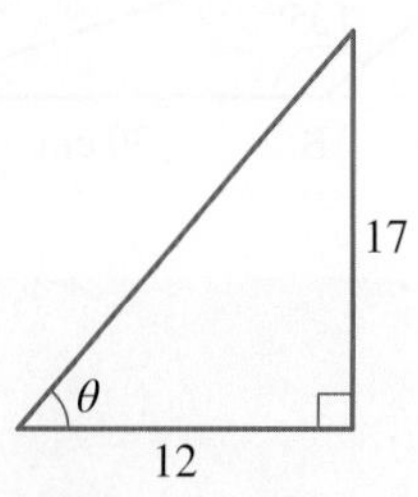

b. 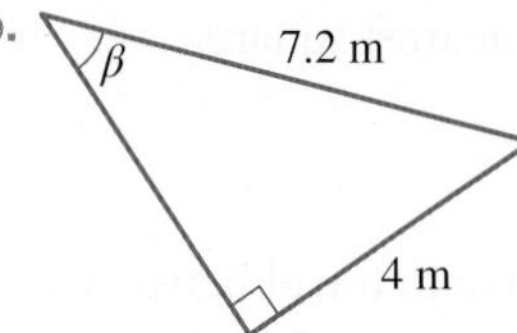

c. 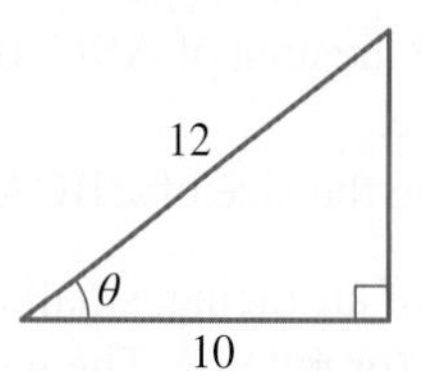

3. **WE14a** Calculate the size of the angle marked with the pronumeral in each of the following. Give your answer correct to the nearest second.

a. 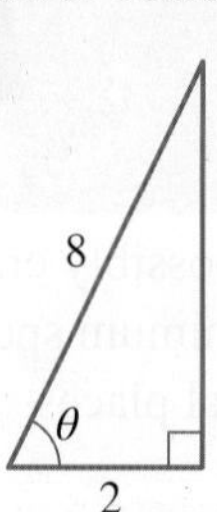

b. 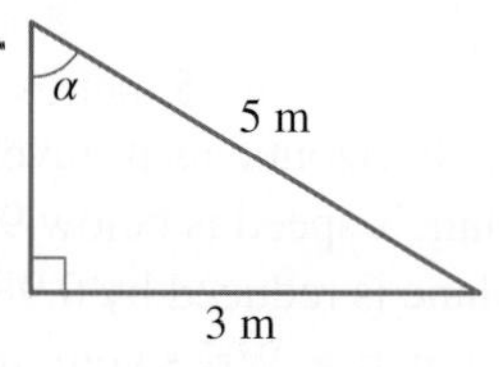

c. 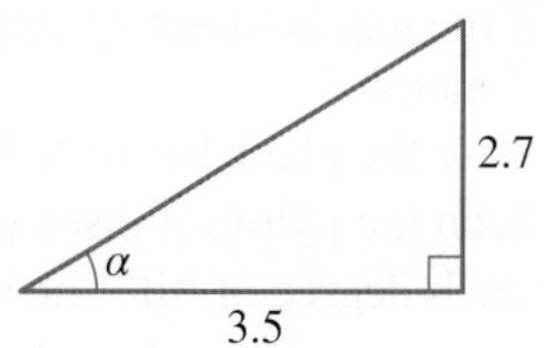

4. Calculate the size of the angle marked with the pronumeral in each of the following, giving your answer correct to the nearest degree.

a. 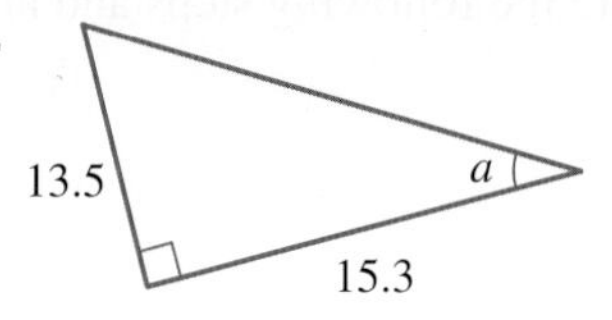

b. 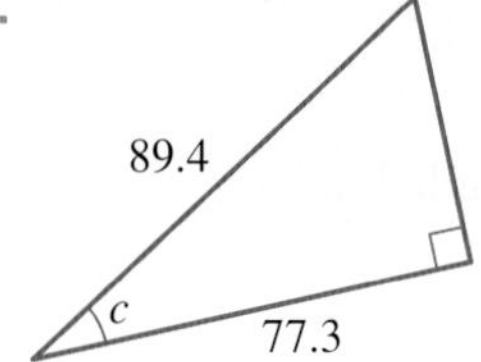

c. 106.4

92.7

b

5. Calculate the size of the angle marked with the pronumeral in each of the following, giving your answer correct to the nearest degree.

a. 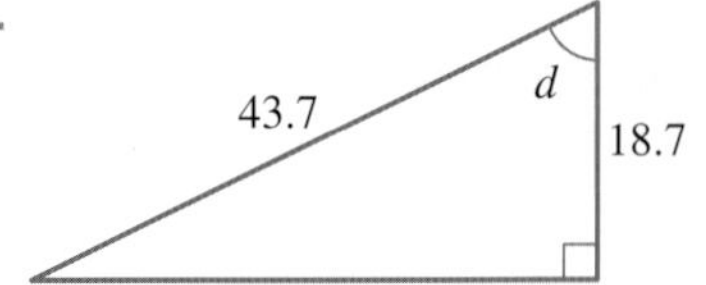

b. 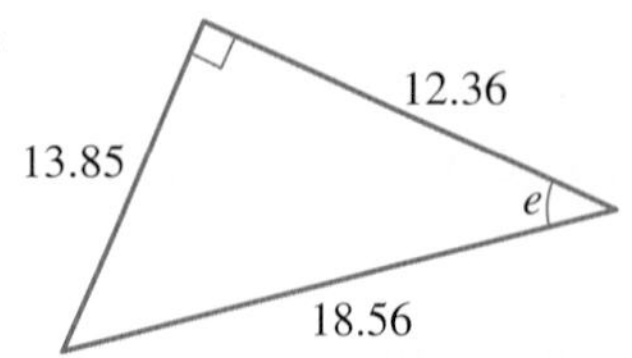

c. 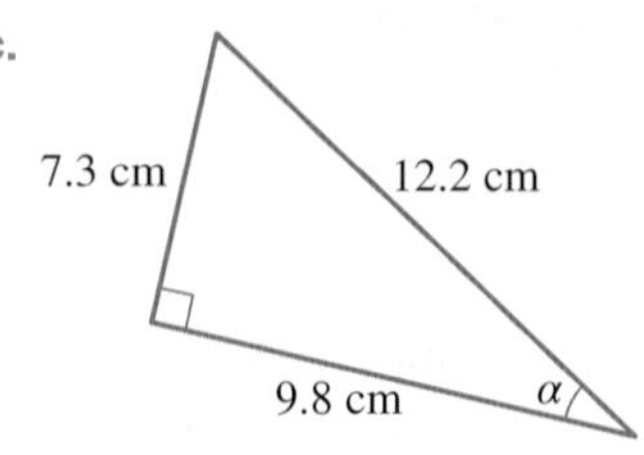

6. Calculate the size of each of the angles in the following, giving your answers correct to the nearest minute.

a. 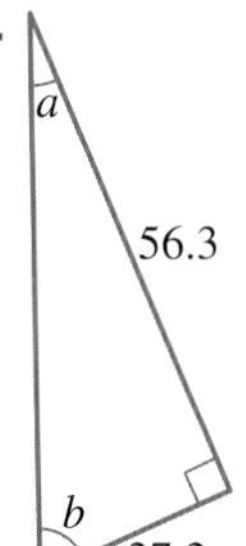

b. 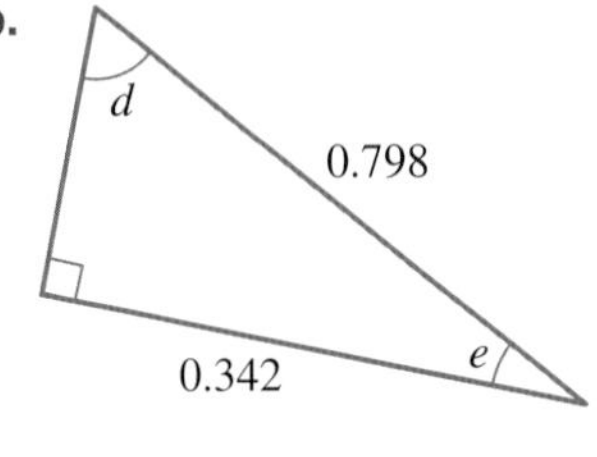

c. 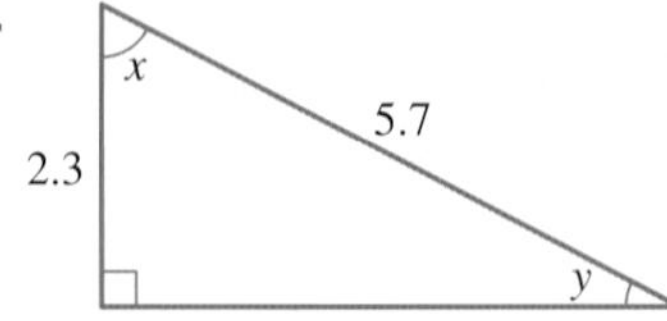

Understanding

7. Answer the following questions for the triangle shown.

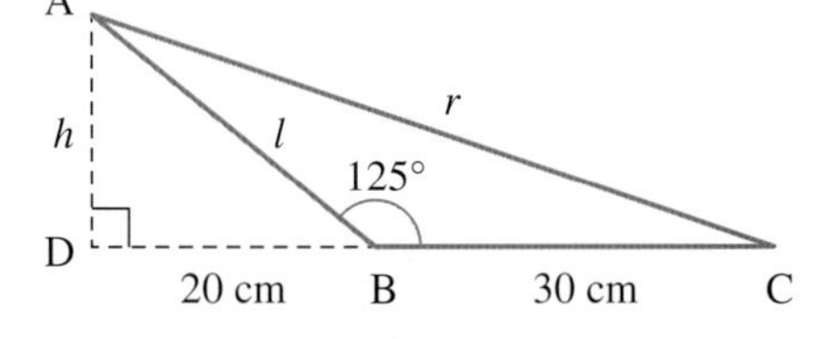

   a. Calculate the length of the sides r, l and h. Write your answers correct to 2 decimal places.
   b. Calculate the area of ABC, correct to the nearest square centimetre.
   c. Determine the size of ∠BCA.

8. In the sport of air racing, small aeroplanes have to travel between two large towers (or pylons). The gap between a pair of pylons is smaller than the wing-span of the plane, so the plane has to go through on an angle with one wing higher than the other. The wing-span of a competition plane is 8 metres.
   a. Determine the angle, correct to 1 decimal place, that the plane has to tilt if the gap between pylons is:
      i. 7 metres ii. 6 metres iii. 5 metres.
   b. Because the plane has rolled away from the horizontal as it travels between the pylons it loses speed. If the plane's speed is below 96 km/h it will stall and possibly crash. For each degree of 'tilt' the speed of the plane is reduced by 0.98 km/h. Calculate the minimum speed the plane must go through each of the pylons in part a. Write your answer correct to 2 decimal places.

Reasoning

9. Explain how calculating the angle of a right-angled triangle is different to calculating a side length.

10. There are two important triangles commonly used in trigonometry. Complete the following steps and answer the questions to create these triangles.

Triangle 1

- Sketch an equilateral triangle with side length 2 units.
- Calculate the size of the internal angles.
- Bisect the triangle to form two right-angled triangles.
- Redraw one of the triangles formed.
- Calculate the side lengths of this right-angled triangle as exact values.
- Fully label your diagram showing all side lengths and angles.

Triangle 2

- Draw a right-angled isosceles triangle.
- Calculate the sizes of the internal angles.
- Let the sides of equal length be 1 unit long each.
- Calculate the length of the third side as an exact value.
- Fully label your diagram showing all side lengths and angles.

11. a. Use the triangles formed in question **10** to calculate exact values for sin(30°), cos(30°) and tan(30°). Justify your answers.

b. Use the exact values for sin(30°), cos(30°) and tan(30°) to show that $\tan(30°) = \dfrac{\sin(30°)}{\cos(30°)}$.

c. Use the formulas $\sin(\theta) = \dfrac{O}{H}$ and $\cos(\theta) = \dfrac{A}{H}$ to prove that $\tan(\theta) = \dfrac{\sin(\theta)}{\cos(\theta)}$.

Problem solving

12. During a Science excursion, a class visited an underground cave to observe rock formations. They were required to walk along a series of paths and steps as shown in the diagram below.

a. Calculate the angle of the incline (slope) required to travel down between each site. Give your answers to the nearest whole number.

b. Determine which path would have been the most challenging; that is, which path had the steepest slope.

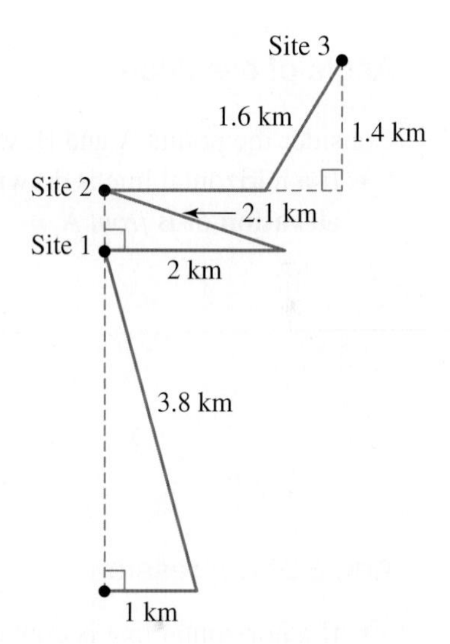

13. Determine the angle θ in degrees and minutes.

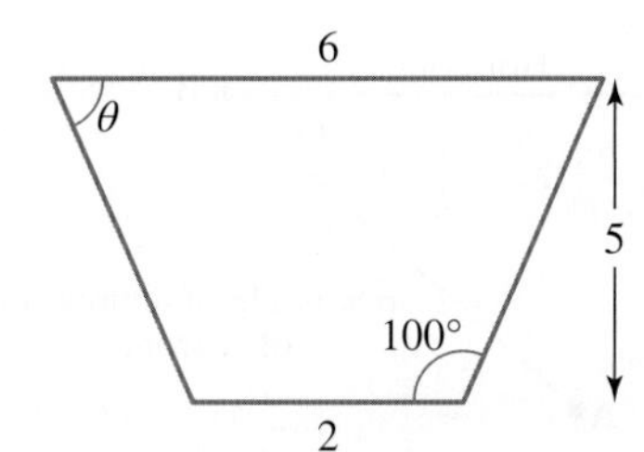

14. At midday, the hour hand and the minute hand on a standard clock are both pointing at the twelve. Calculate the angles the minute hand and the hour hand have moved 24.5 minutes later. Express both answers in degrees and minutes.

5.7 Angles of elevation and depression

LEARNING INTENTION

At the end of this subtopic you should be able to:
- identify angles of elevation and depression and solve for unknown side lengths and angles.

5.7.1 Angles of elevation and depression

eles-4806

- Solving real-life problems usually involves the person measuring angles or lengths from their position using trigonometry.
- They may have to either look up at the object or look down to it; hence the terms 'angle of elevation' and 'angle of depression' respectively.

Angle of elevation

Consider the points A and B, where B is at a higher elevation than A.
- If a horizontal line is drawn from A as shown, forming the angle θ, then θ is called the **angle of elevation** of B *from* A.

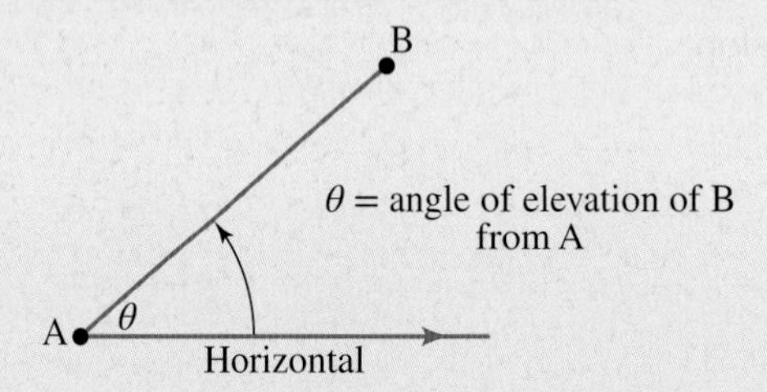

Angle of depression

- If a horizontal line is drawn from B, forming the angle α, then α is called the **angle of depression** of A *from* B.

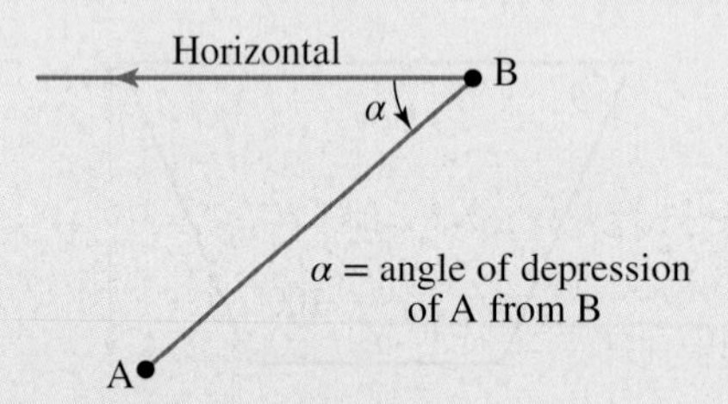

Alternate angle rule

- Because the horizontal lines are parallel, θ and α have the same size (alternate angles).

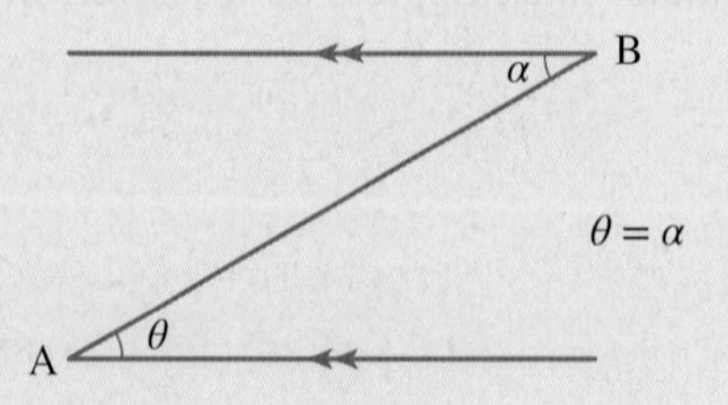

WORKED EXAMPLE 15 Applying angles of elevation to solve problems

From a point P, on the ground, the angle of elevation of the top of a tree is 50°. If P is 8 metres from the tree, determine the height of the tree correct to 2 decimal places.

THINK	WRITE/DRAW
1. Let the height of the tree be h. Sketch a diagram and show the relevant information.	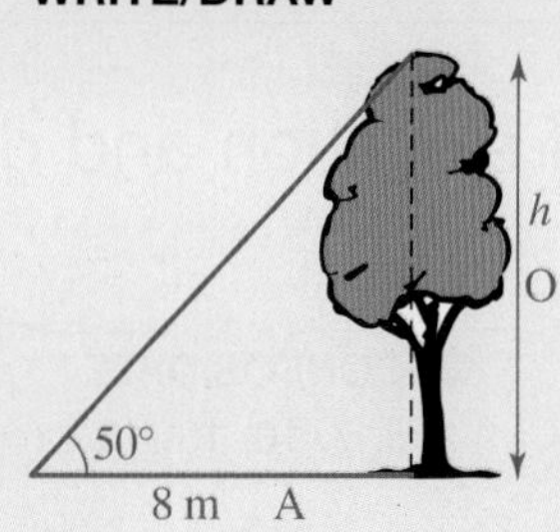
2. Identify the appropriate trigonometric ratio.	$\tan(\theta) = \frac{O}{A}$
3. Substitute O = h, A = 8 and $\theta = 50°$.	$\tan(50°) = \frac{h}{8}$
4. Rearrange to make h the subject.	$h = 8\tan(50°)$
5. Calculate and round the answer to 2 decimal places.	≈ 9.53
6. Write the answer in a sentence.	The height of the tree is 9.53 m.

WORKED EXAMPLE 16 Applying angles of depression to solve problems

The angle of depression from a helicopter, at point H, to a swimmer in distress in the water is 60°. If the helicopter is hovering 800 m above sea level, determine how far horizontally the swimmer is from the helicopter. Write your answer in metres correct to 2 decimal places.

THINK	WRITE/DRAW
1. Let the horizontal distance between the swimmer and the helicopter be d. Sketch a diagram and show the relevant information.	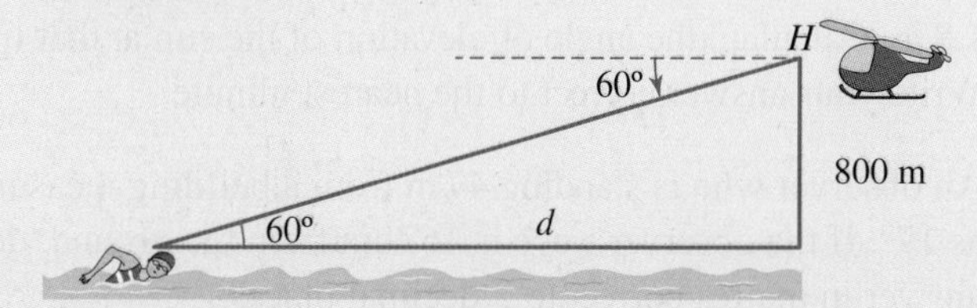
2. Identify the appropriate trigonometric ratio.	$\tan(\theta) = \frac{O}{A}$
3. Substitute $O = 800$, $Q = 60°$ and $A = d$.	$\tan(60°) = \frac{800}{d}$
4. Rearrange to make d the subject.	$d = \frac{800}{\tan(60°)}$
5. Calculate and round to 2 decimal places.	$d \approx 461.88$ m
6. Write the answer in a sentence.	The horizontal distance between the swimmer and the helicopter is 461.88 m.

on Resources

eWorkbook Topic 5 Workbook (worksheets, code puzzle and project) (ewbk-2031)

Interactivities Individual pathway interactivity: Angles of elevation and depression (int-4590)
Finding the angle of elevation and angle of depression (int-6047)

Exercise 5.7 Angles of elevation and depression

learnon

Individual pathways

■ PRACTISE	■ CONSOLIDATE	■ MASTER
1, 4, 6, 9, 12, 15	2, 5, 7, 10, 13, 16	3, 8, 11, 14, 17

To answer questions online and to receive **immediate corrective feedback** and **fully worked solutions** for all questions, go to your learnON title at www.jacplus.com.au.

Fluency

1. **WE15** From a point P on the ground the angle of elevation from an observer to the top of a tree is 54°22′. If the tree is known to be 12.19 m high, determine how far P is from the tree (measured horizontally). Write your answer in metres correct to 2 decimal places.
2. **WE16** From the top of a cliff 112 m high, the angle of depression to a boat is 9°15′. Determine how far the boat is from the foot of the cliff. Write your answer in metres correct to 1 decimal place.
3. A person on a ship observes a lighthouse on the cliff, which is 830 metres away from the ship. The angle of elevation of the top of the lighthouse is 12°.
   a. Determine how far above sea level the top of the lighthouse is, correct to 2 decimal places.
   b. If the height of the lighthouse is 24 m, calculate the height of the cliff, correct to 2 decimal places.

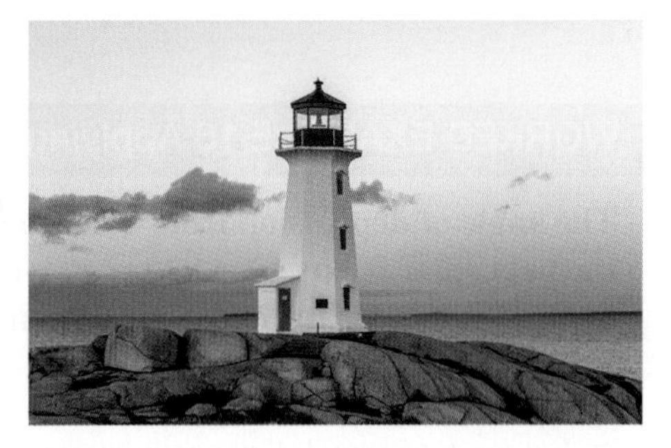

4. At a certain time of the day a post, 4 m tall, casts a shadow of 1.8 m. Calculate the angle of elevation of the sun at that time. Write your answer correct to the nearest minute.
5. An observer who is standing 47 m from a building measures the angle of elevation of the top of the building as 17°. If the observer's eye is 167 cm from the ground, determine the height of the building. Write your answer in metres correct to 2 decimal places.

Understanding

6. A surveyor needs to determine the height of a building. She measures the angle of elevation of the top of the building from two points, 38 m apart. The surveyor's eye level is 180 cm above the ground.
   a. Determine two expressions for the height of the building, h, in terms of x using the two angles.
   b. Solve for x by equating the two expressions obtained in a. Write your answer in metres correct to 2 decimal places
   c. Determine the height of the building, in metres correct to 2 decimal places.

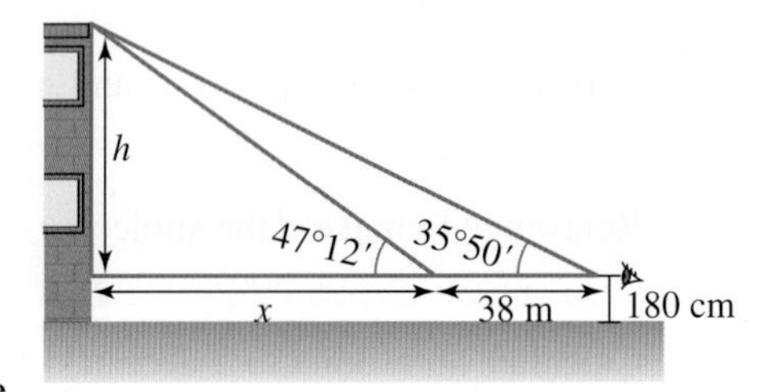

7. The height of another building needs to be determined but cannot be found directly. The surveyor decides to measure the angle of elevation of the top of the building from different sites, which are 75 m apart. The surveyor's eye level is 189 cm above the ground.
   a. Determine two expressions for the height of the building above the surveyor's eye level, h, in terms of x using the two angles.
   b. Solve for x. Write your answer in metres correct to 2 decimal places.
   c. Determine the height of the building, in metres correct to 2 decimal places.

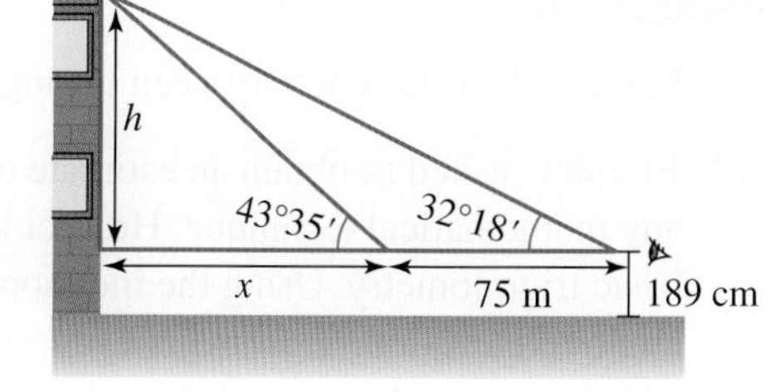

8. A lookout tower has been erected on top of a cliff. At a distance of 5.8 km from the foot of the cliff, the angle of elevation to the base of the tower is 15.7° and to the observation deck at the top of the tower is 16° respectively, as shown in the figure below. Determine how high from the top of the cliff the observation deck is, to the nearest metre.

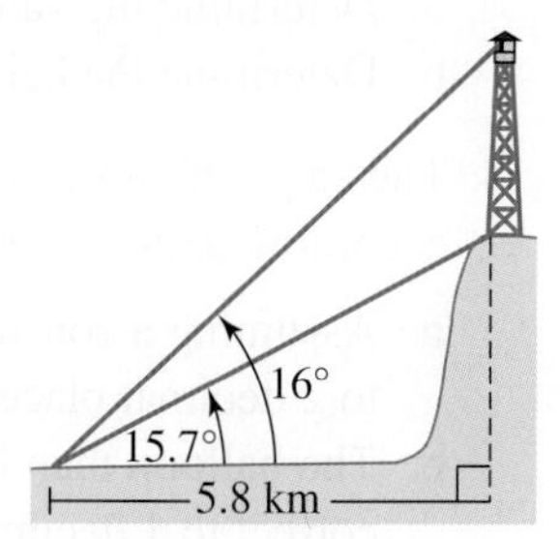

9. Elena and Sonja were on a camping trip to the Grampians, where they spent their first day hiking. They first walked 1.5 km along a path inclined at an angle of 10° to the horizontal. Then they had to follow another path, which was at an angle of 20° to the horizontal. They walked along this path for 1.3 km, which brought them to the edge of the cliff. Here Elena spotted a large gum tree 1.4 km away. If the gum tree is 150 m high, calculate the angle of depression from the top of the cliff to the top of the gum tree. Express your answer in degrees correct to the nearest degree.

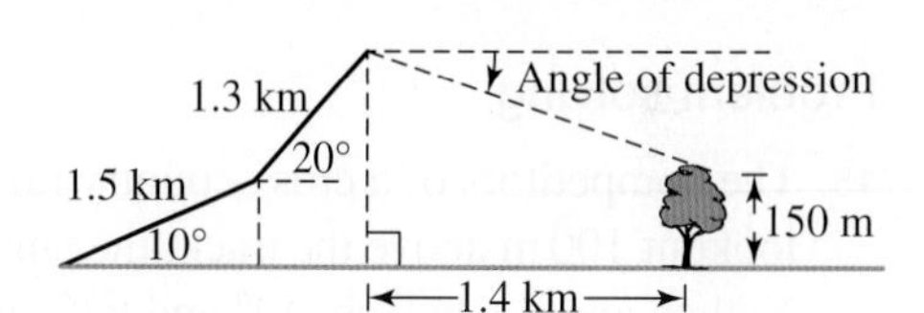

10. From a point on top of a cliff, two boats are observed. If the angles of depression are 58° and 32° and the cliff is 46 m above sea level, determine how far apart the boats are, in metres correct to 2 decimal places.

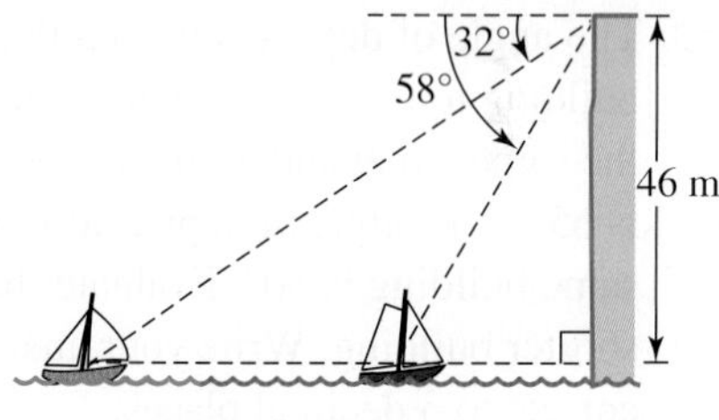

11. A 2.05 m tall man, standing in front of a street light 3.08 m high, casts a 1.5 m shadow.
   a. Calculate the angle of elevation, to the nearest degree, from the ground to the source of light.
   b. Determine how far the man is from the bottom of the light pole, in metres correct to 2 decimal places.

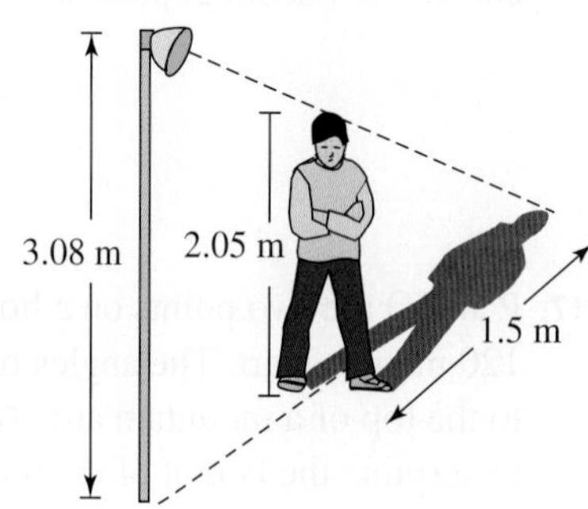

Reasoning

12. Explain the difference between an angle of elevation and an angle of depression.

13. Joseph is asked to obtain an estimate of the height of his house using any mathematical technique. He decides to use an inclinometer and basic trigonometry. Using the inclinometer, Joseph determines the angle of elevation, θ, from his eye level to the top of his house to be 42°. The point from which Joseph measures the angle of elevation is 15 m away from his house and the distance from Joseph's eyes to the ground is 1.76 m.

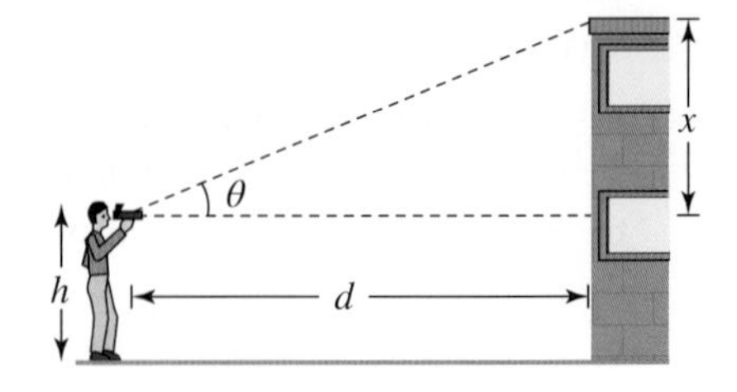

   a. Determine the values for the pronumerals h, d and θ.
   b. Determine the height of Joseph's house, in metres correct to 2 decimal places.

14. The angle of elevation of a vertically rising hot air balloon changes from 27° at 7:00 am to 61° at 7:03 am, according to an observer who is 300 m away from the take-off point.
   a. Assuming a constant speed, calculate that speed (in m/s and km/h) at which the balloon is rising, correct to 2 decimal places.
   b. The balloon then falls 120 metres. Determine the angle of elevation now. Write your answer in degrees correct to 1 decimal place.

Problem solving

15. The competitors of a cross-country run are nearing the finish line. From a lookout 100 m above the track, the angles of depression to the two leaders, Nathan and Rachel, are 40° and 62° respectively. Evaluate how far apart, to the nearest metre, the two competitors are.

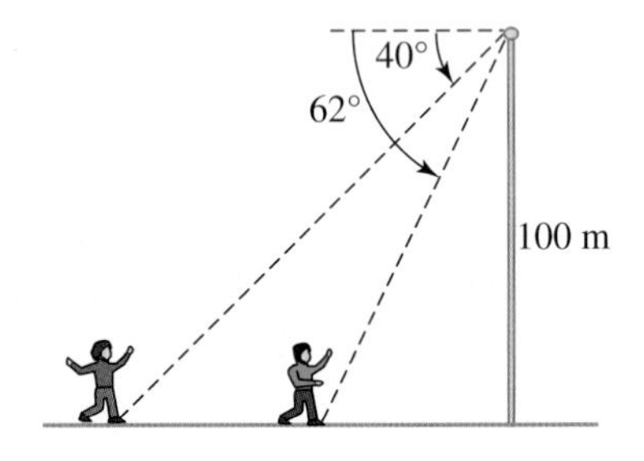

16. The angle of depression from the top of one building to the foot of another building across the same street and 45 metres horizontally away is 65°. The angle of depression to the roof of the same building is 30°. Evaluate the height of the shorter building. Write your answer in metres correct to 3 decimal places.

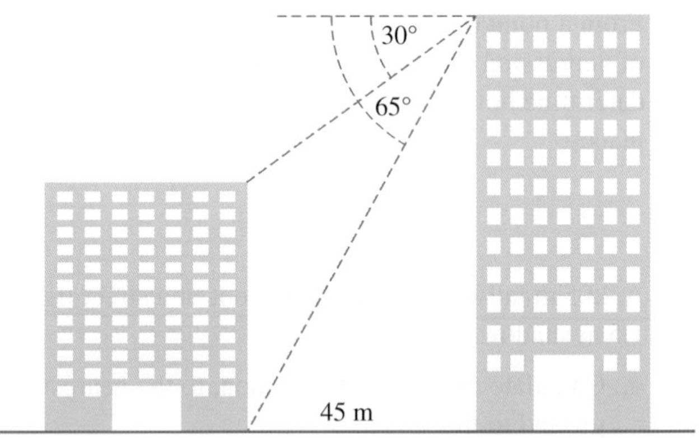

17. P and Q are two points on a horizontal line that are 120 metres apart. The angles of elevation from P and Q to the top of a mountain are 36° and 42° respectively. Determine the height of the mountain, in metres, correct to 1 decimal place.

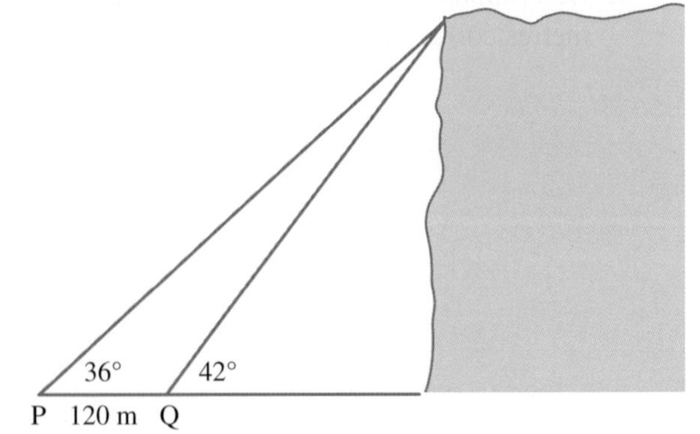

5.8 Bearings

LEARNING INTENTION

At the end of this subtopic you should be able to:
- draw diagrams with correct angles to represent information to help solve triangles
- apply trigonometry to solve bearing problems involving compass and true bearings.

5.8.1 Using bearings

eles-4807

- A bearing gives the direction of travel from one point or object to another.
- The bearing of B from A tells how to get to B *from* A. A compass rose would be drawn at A.

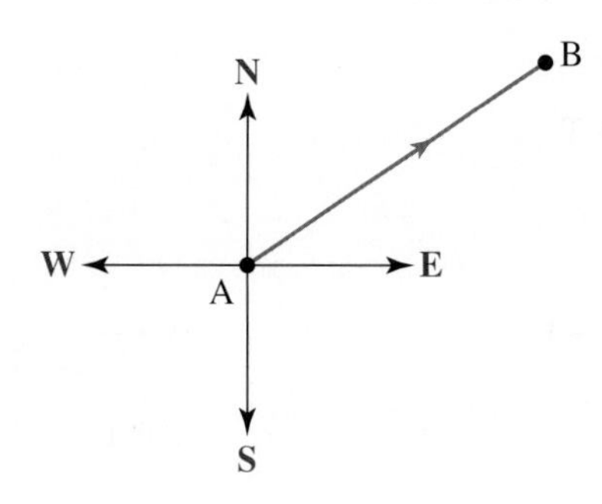

To illustrate the bearing of A *from* B, a compass rose would be drawn at B.

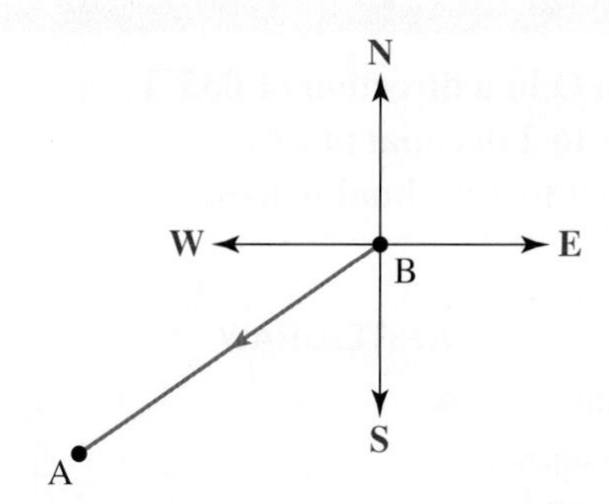

- There are two ways in which bearings are commonly written. They are compass bearings and true bearings.

Compass bearings

- A **compass bearing** (for example N40°E or S72°W) has three parts.
  - The first part is either N or S (for north or south).
  - The second part is an acute angle.
  - The third part is either E or W (for east or west).
- For example, the compass bearing S20°E means start by facing south and then turn 20° towards the east. This is the direction of travel.
  N40°W means start by facing north and then turn 40° towards the west.

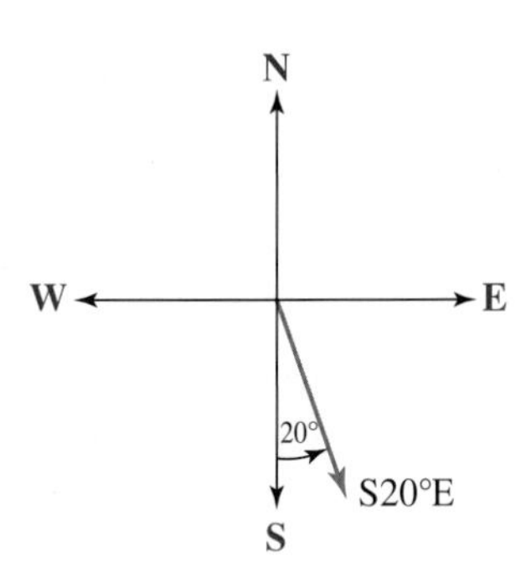

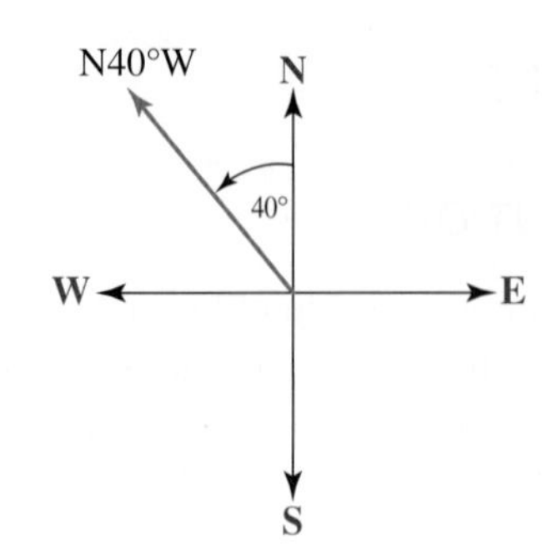

True bearings

- **True bearings** are measured from north in a clockwise direction and are expressed in 3 digits.
- The diagrams below show the bearings of 025° true and 250° true respectively. (These true bearings are more commonly written as 025°T and 250°T.)

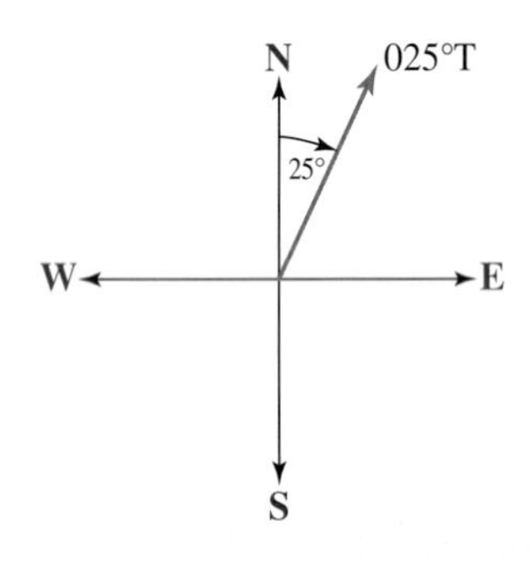

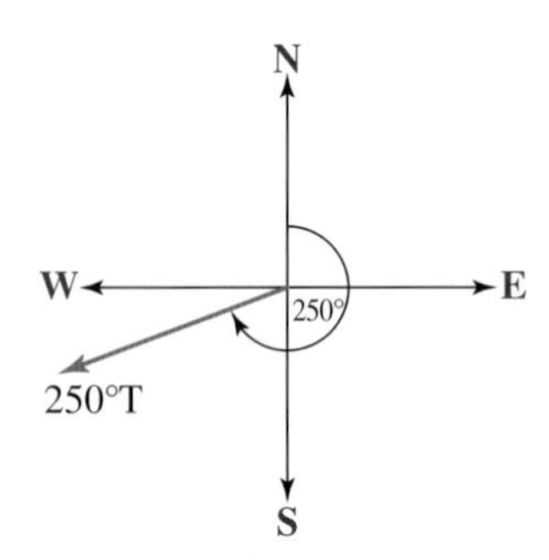

WORKED EXAMPLE 17 Solving trigonometric problems involving bearings

A boat travels a distance of 5 km from P to Q in a direction of 035°T.

a. **Calculate how far east of P is Q, correct to 2 decimal places.**

b. **Calculate how far north of P is Q, correct to 2 decimal places.**

c. **Calculate the true bearing of P from Q.**

THINK	WRITE/DRAW
a. 1. Draw a diagram showing the distance and bearing of Q from P. Complete a right-angled triangle travelling x km due east from P and then y km due north to Q.	a. 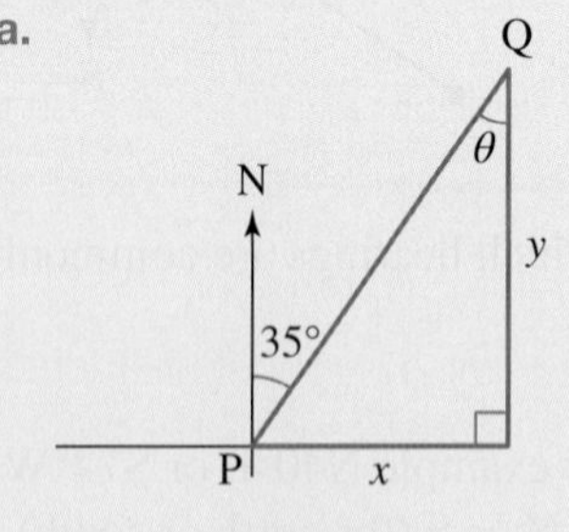
2. To determine how far Q is east of P, we need to determine the value of x. We are given the length of the hypotenuse (H) and need to find the length of the opposite side (O). Write the sine ratio.	$\sin(\theta) = \frac{\text{O}}{\text{H}}$
3. Substitute O = x, H = 5 and $\theta = 35°$.	$\sin(35°) = \frac{x}{5}$

4. Make x the subject of the equation. | $x = 5\sin(35°)$

5. Evaluate and round the answer, correct to 2 decimal places. | ≈ 2.87

6. Write the answer in a sentence. | Point Q is 2.87 km east of P.

b. 1. To determine how far Q is north of P, we need to find the value of y. This can be done in several ways, namely: using the cosine ratio, the tangent ratio, or Pythagoras' theorem. Write the cosine ratio. | **b.** $\cos(\theta) = \frac{A}{H}$

2. Substitute $A = y$, $H = 5$ and $\theta = 35°$. | $\cos(35°) = \frac{y}{5}$

3. Make y the subject of the equation. | $y = 5\cos(35°)$

4. Evaluate and round the answer, correct to 2 decimal places. | ≈ 4.10

5. Write the answer in a sentence. | Point B is 4.10 km north of A.

c. 1. To determine the bearing of P from Q, draw a compass rose at Q. The true bearing is given by $\angle\theta$. | **c.**

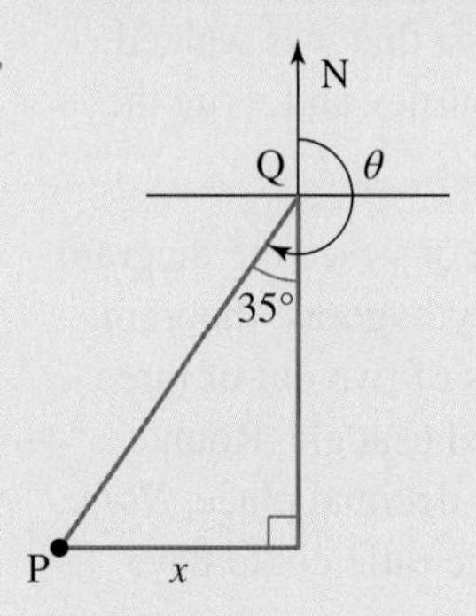

2. The value of θ is the sum of 180° (from north to south) and 35°. Write the value of θ. |

$$\text{True bearing} = 180° + \alpha$$
$$\alpha = 35°$$
$$\text{True bearing} = 180° + 35°$$
$$= 215°$$

3. Write the answer in a sentence. | The bearing of P from Q is 215°T.

- Sometimes a journey includes a change in directions. In such cases, each section of the journey should be dealt with separately.

WORKED EXAMPLE 18 Solving bearings problems with 2 stages

A boy walks 2 km on a true bearing of 090° and then 3 km on a true bearing of 130°.

a. Calculate how far east of the starting point the boy is at the completion of his walk, correct to 1 decimal place.

b. Calculate how far south of the starting point the boy is at the completion of his walk, correct to 1 decimal place.

c. To return directly to his starting point, calculate how far the boy must walk and on what bearing. Write your answers in km correct to 2 decimal places and in degrees and minutes correct to the nearest minute.

THINK | **WRITE/DRAW**

a. 1. Draw a diagram of the boy's journey. The first leg of the journey is due east. Label the easterly component x and the southerly component y.

a.

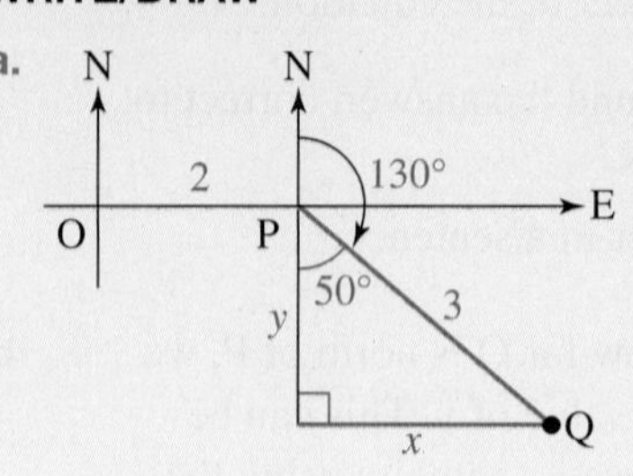

2. Write the ratio to determine the value of x.

$$\sin(\theta) = \frac{\text{O}}{\text{H}}$$

3. Substitute $\text{O} = x$, $\text{H} = 3$ and $\theta = 50°$.

$$\sin(50°) = \frac{x}{3}$$

4. Make x the subject of the equation.

$$x = 3\sin(50°)$$

5. Evaluate and round correct to 1 decimal place.

$$\approx 2.3 \text{ km}$$

6. Add to this the 2 km east that was walked in the first leg of the journey and write the answer in a sentence.

$$\text{Total distance east} = 2 + 2.3$$
$$= 4.3 \text{ km}$$

The boy is 4.3 km east of the starting point.

b. 1. To determine the value of y (see the diagram in part **a**) we can use Pythagoras' theorem, as we know the lengths of two out of three sides in the right-angled triangle. Round the answer correct to 1 decimal place. *Note:* Alternatively, the cosine ratio could have been used.

b. Distance south $= y$ km

$$a^2 = c^2 - b^2$$
$$y^2 = 3^2 - 2.3^2$$
$$= 9 - 5.29$$
$$= 3.71$$
$$y = \sqrt{3.71}$$
$$= 1.9 \text{ km}$$

2. Write the answer in a sentence.

The boy is 1.9 km south of the starting point.

c. 1. Draw a diagram of the journey and write in the results found in parts **a** and **b**. Draw a compass rose at Q.

c.

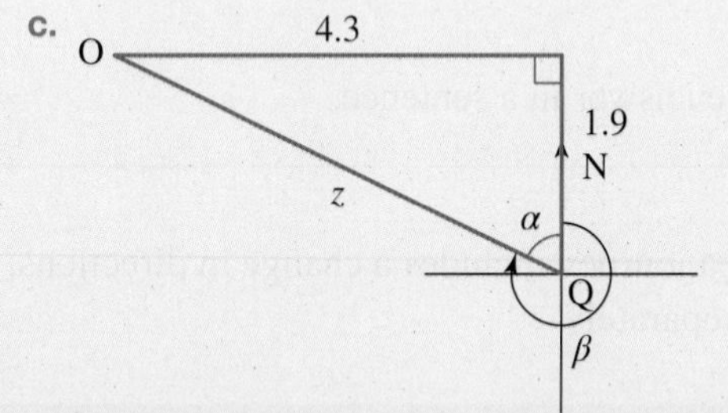

2. Determine the value of z using Pythagoras' theorem.

$$z^2 = 1.9^2 + 4.3^2$$
$$= 22.1$$
$$z = 22.1$$
$$\approx 4.70$$

3. Determine the value of α using trigonometry.

$$\tan(\alpha) = \frac{4.3}{1.9}$$

4. Make α the subject of the equation using the inverse tangent function.

$$\alpha = \tan^{-1}\left(\frac{4.3}{1.9}\right)$$

5. Evaluate and round to the nearest minute. $= 66.161\,259\,82°$ $= 66°9'40.535''$ $= 66°10'$

6. The angle β gives the bearing. $\beta = 360° - 66°10'$ $= 293°50'$

7. Write the answer in a sentence. The boy travels 4.70 km on a bearing of 293°50 T.

DISCUSSION

Explain the difference between true bearings and compass directions.

on Resources

eWorkbook	Topic 5 Workbook (worksheets, code puzzle and project) (ewbk-2031)
Digital document	SkillSHEET Drawing a diagram from given directions (doc-5228)
Video eLesson	Bearings (eles-1935)
Interactivities	Individual pathway interactivity: Bearings (int-4591) Bearings (int-6481)

Exercise 5.8 Bearings

learnon

Individual pathways

■ PRACTISE	■ CONSOLIDATE	■ MASTER
1, 3, 7, 10, 14, 17	2, 5, 8, 11, 12, 15, 18	4, 6, 9, 13, 16, 19

To answer questions online and to receive **immediate corrective feedback** and **fully worked solutions** for all questions, go to your learnON title at www.jacplus.com.au.

Fluency

1. Change each of the following compass bearings to true bearings.
   a. N20°E b. N20°W c. S35°W
2. Change each of the following compass bearings to true bearings.
   a. S28°E b. N34°E c. S42°W
3. Change each of the following true bearings to compass bearings.
   a. 049°T b. 132°T c. 267°T
4. Change each of the following true bearings to compass bearings.
   a. 330°T b. 086°T c. 234°T

5. Describe the following paths using true bearings.

a. 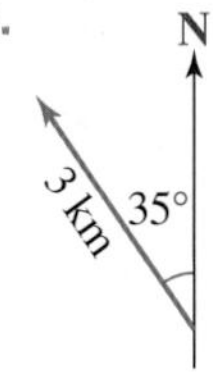

b. 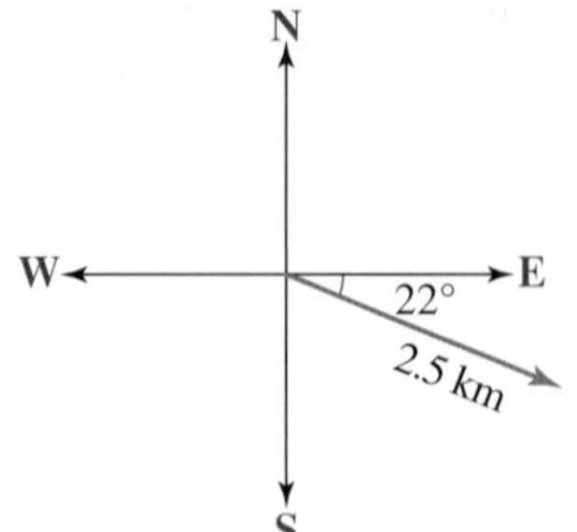

c. 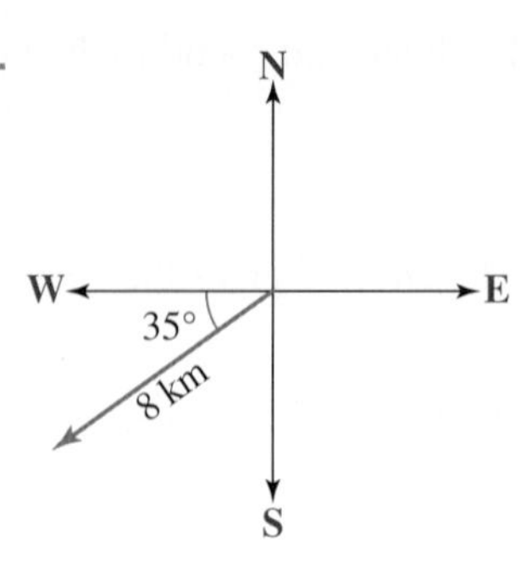

6. Describe the following paths using true bearings.

a. 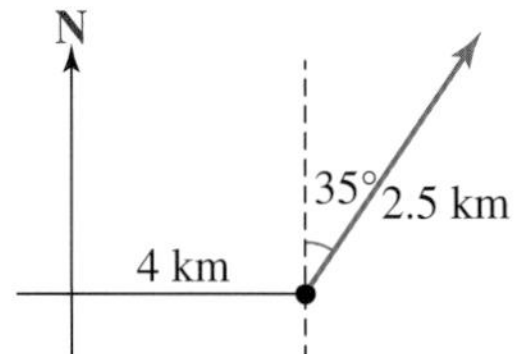

b. 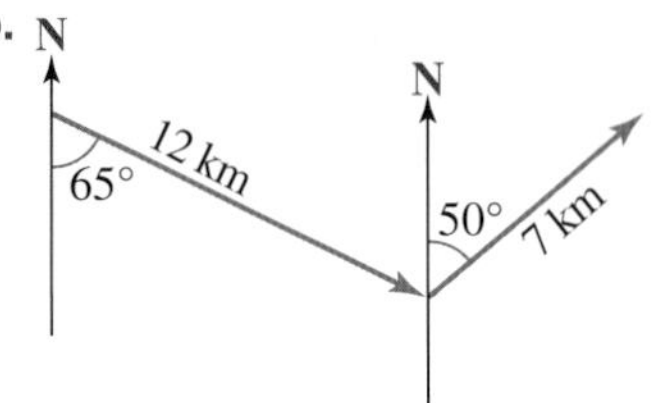

c. 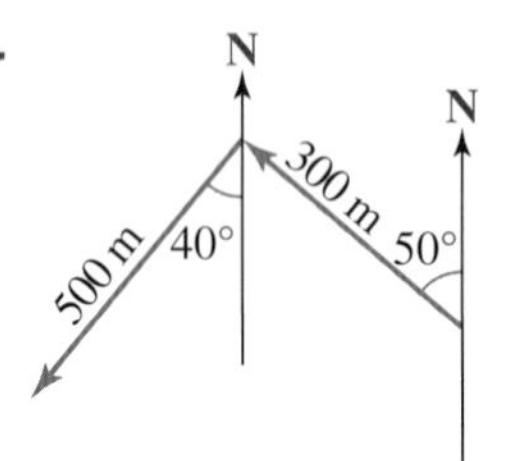

7. Show each of the following journeys as a diagram.
   a. A ship travels 040°T for 40 km and then 100°T for 30 km.
   b. A plane flies for 230 km in a direction 135°T and a further 140 km in a direction 240°T.

8. Show each of the following journeys as a diagram.
   a. A bushwalker travels in a direction 260°T for 0.8 km, then changes direction to 120°T for 1.3 km, and finally travels in a direction of 32° for 2.1 km.
   b. A boat travels N40°W for 8 km, then changes direction to S30°W for 5 km and then S50°E for 7 km.
   c. A plane travels N20°E for 320 km, N70°E for 180 km and S30°E for 220 km.

9. **WE17** A yacht travels 20 km from *A* to *B* on a bearing of 042°T.
   a. Calculate how far east of A is B, in km correct to 2 decimal places.
   b. Calculate how far north of A is B, in km correct to 2 decimal places.
   c. Calculate the bearing of A from B.

   The yacht then sails 80 km from B to C on a bearing of 130°T.
   d. Show the journey using a diagram.
   e. Calculate how far south of B is C, in km correct to 2 decimal places.
   f. Calculate how far east of B is C, in km correct to 2 decimal places.
   g. Calculate the bearing of B from C.

10. If a farmhouse is situated 220 m N35°E from a shed, calculate the true bearing of the shed from the house.

Understanding

11. A pair of hikers travel 0.7 km on a true bearing of 240° and then 1.3 km on a true bearing of 300°. Calculate how far west have they travelled from their starting point, in km correct to 3 decimal places.

12. **WE18** A boat travels 6 km on a true bearing of 120° and then 4 km on a true bearing of 080°
   a. Calculate how far east the boat is from the starting point on the completion of its journey, in km correct to 3 decimal places.
   b. Calculate how far south the boat is from the starting point on the completion of its journey, in km correct to 3 decimal places.
   c. Calculate the bearing of the boat from the starting point on the completion of its journey, correct to the nearest minute.

13. A plane flies on a true bearing of 320° for 450 km. It then flies on a true bearing of 350° for 130 km and finally on a true bearing of 050° for 330 km. Calculate how far north of its starting point the plane is. Write your answer in km correct to 2 decimal places.

Reasoning

14. A bushwalker leaves her tent and walks due east for 4.12 km, then walks a further 3.31 km on a bearing of N20°E. If she wishes to return directly to her tent, determine how far she must walk and what bearing she should take. Write your answers in km correct to 2 decimal places and to the nearest degree.

15. A car travels due south for 3 km and then due east for 8 km. Determine the bearing of the car from its starting point, to the nearest degree. Show full working.

16. If the bearing of A from O is θ°T, then (in terms of theta) determine the bearing of O from A:

    a. if $0° < \theta° < 180°$

    b. if $180° < \theta° < 360°$.

Problem solving

17. A boat sails on a compass direction of E12°S for 10 km then changes direction to S27°E for another 20 km. The boat then decides to return to its starting point.

    a. Determine how far, correct to 2 decimal places, the boat is from its starting point.

    b. Determine on what bearing should the boat travel to return to its starting point. Write the angle correct to the nearest degree.

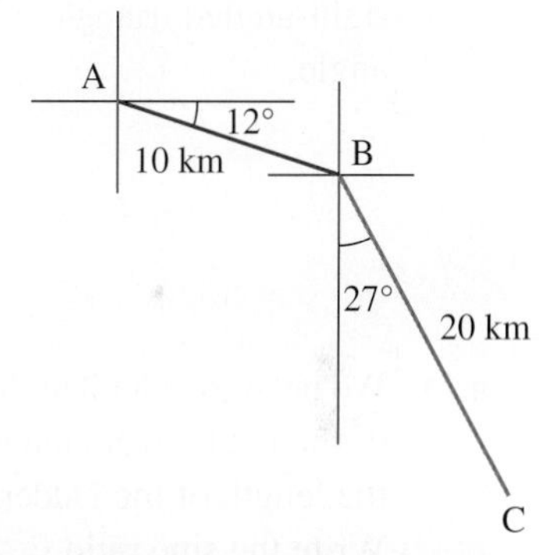

18. Samira and Tim set off early from the car park of a national park to hike for the day. Initially they walk N60°E for 12 km to see a spectacular waterfall. They then change direction and walk in a south-easterly direction for 6 km, then stop for lunch. Give all answers correct to 2 decimal places.

    a. Make a scale diagram of the hiking path they completed.

    b. Determine how far north of the car park they are at the lunch stop.

    c. Determine how far east of the car park they are at the lunch stop.

    d. Determine the bearing of the lunch stop from the car park.

    e. If Samira and Tim then walk directly back to the car park, calculate the distance they have covered after lunch.

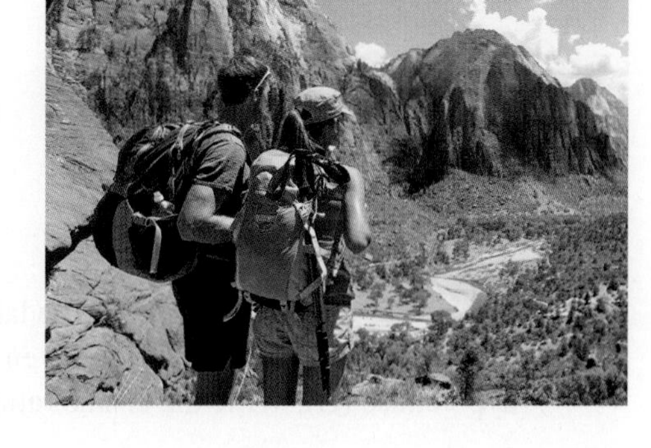

19. Starting from their base in the national park, a group of bushwalkers travel 1.5 km at a true bearing of 030°, then 3.5 km at a true bearing of 160°, and then 6.25 km at a true bearing of 300°. Evaluate how far, and at what true bearing, the group should walk to return to its base. Write your answers in km correct to 2 decimal places and to the nearest degree.

5.9 Applications

LEARNING INTENTION

At the end of this subtopic you should be able to:
- draw well-labelled diagrams to represent information
- apply trigonometry to solve various problems involving triangles.

5.9.1 Applications of trigonometry

eles-4808

- When applying trigonometry to practical situations, it is essential to draw good mathematical diagrams using points, lines and angles.
- Several diagrams may be required to show all the necessary right-angled triangles.

WORKED EXAMPLE 19 Applying trigonometry to solve problems

A ladder of length 3 m makes an angle of 32° with the wall.
a. Calculate how far the foot of the ladder is from the wall, in metres, correct to 2 decimal places.
b. Calculate how far up the wall the ladder reaches, in metres, correct to 2 decimal places.
c. Calculate the value of the angle the ladder makes with the ground.

THINK	WRITE/DRAW
Sketch a diagram and label the sides of the right-angled triangle with respect to the given angle.	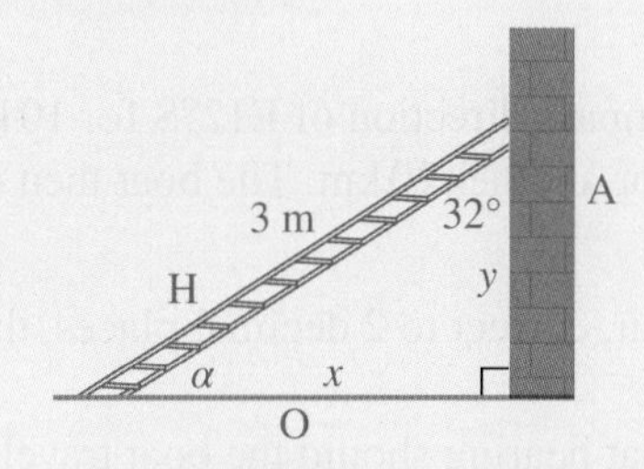
a. 1. We need to calculate the distance of the foot of the ladder from the wall (O) and are given the length of the ladder (H). Write the sine ratio.	**a.** $\sin(\theta) = \frac{\text{O}}{\text{H}}$
2. Substitute $\text{O} = x$, $\text{H} = 3$ and $\theta = 32°$.	$\sin(32°) = \frac{x}{3}$
3. Make x the subject of the equation.	$x = 3\sin(32°)$
4. Evaluate and round the answer to 2 decimal places.	$\approx 1.59\text{ m}$
5. Write the answer in a sentence.	The foot of the ladder is 1.59 m from the wall.
b. 1. We need to calculate the height the ladder reaches up the wall (A) and are given the hypotenuse (H). Write the cosine ratio.	**b.** $\cos(\theta) = \frac{\text{A}}{\text{H}}$
2. Substitute $\text{A} = y$, $\text{H} = 3$ and $\theta = 32°$.	$\cos(32°) = \frac{y}{3}$
3. Make y the subject of the equation.	$y = 3\cos(32°)$
4. Evaluate and round the answer to 2 decimal places.	$y \approx 2.54\text{ m}$
5. Write the answer in a sentence.	The ladder reaches 2.54 m up the wall.

c. 1. To calculate the angle that the ladder makes with the ground, we could use any of the trigonometric ratios, as the lengths of all three sides are known. However, it is quicker to use the angle sum of a triangle.	**c.** $\alpha + 90° + 32° = 180°$ $\alpha + 122° = 180°$ $\alpha = 180° - 122°$ $\alpha = 58°$
2. Write the answer in a sentence.	The ladder makes a 58° angle with the ground.

TI | THINK	**DISPLAY/WRITE**	**CASIO | THINK**	**DISPLAY/WRITE**
On a Calculator page, complete the entry lines as: solve $\left(\sin(32) = \frac{x}{3}, x\right)$ solve $\left(\cos(32) = \frac{y}{3}, y\right)$ $180 - (90 + 32)$ Press ENTER after each entry.	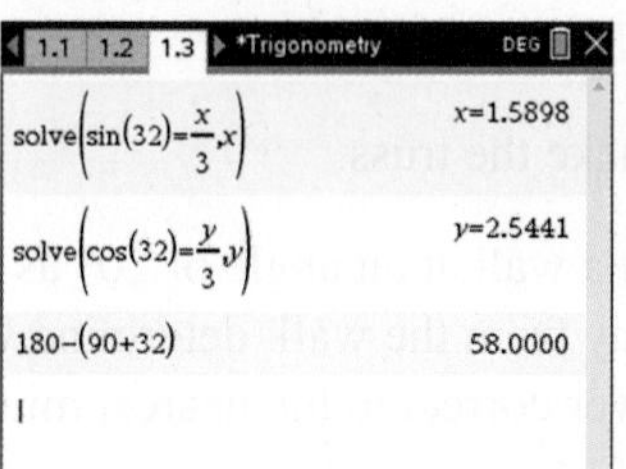 $x = 1.59$ m correct to 2 decimal places. $y = 2.54$ mcorrect to 2 decimal places. $\alpha = 58°$	On the Main screen, complete the entry lines as: solve $\left(\sin(32) = \frac{x}{3}, x\right)$ solve $\left(\cos(32) = \frac{y}{3}, y\right)$ $180 - (90 + 32)$ Press EXE after each entry.	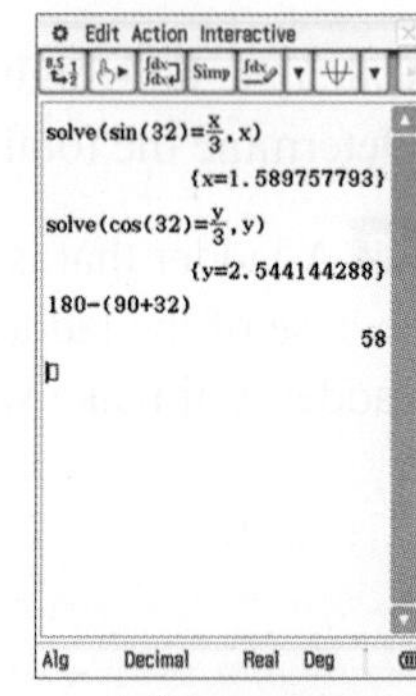 $x = 1.59$ m correct to 2 decimal places. $y = 2.54$ mcorrect to 2 decimal places. $\alpha = 58°$

DISCUSSION

What are some real-life applications of trigonometry?

Resources

eWorkbook Topic 5 Workbook (worksheets, code puzzle and project) (ewbk-2031)

Interactivity Individual pathway interactivity: Applications (int-4592)

Exercise 5.9 Applications

learnon

Individual pathways

■ PRACTISE	■ CONSOLIDATE	■ MASTER
1, 3, 4, 10, 11, 17	2, 5, 8, 12, 14, 16, 18	6, 7, 9, 13, 15, 19

To answer questions online and to receive **immediate corrective feedback** and **fully worked solutions** for all questions, go to your learnON title at www.jacplus.com.au.

Fluency

1. A carpenter wants to make a roof pitched at 29°30′, as shown in the diagram. Calculate how long, in metres correct to 2 decimal places, he should cut the beam PR.

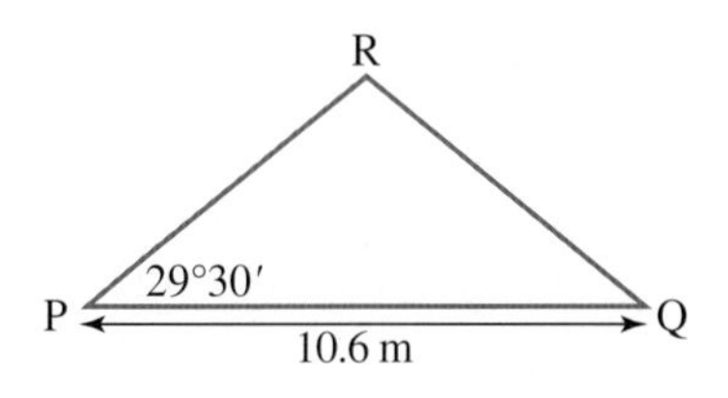

2. The mast of a boat is 7.7 m high. A guy wire from the top of the mast is fixed to the deck 4 m from the base of the mast. Determine the angle, correct to the nearest minute, the wire makes with the horizontal.

Understanding

3. A steel roof truss is to be made to the following design. Write your answers in metres correct to 2 decimal places.

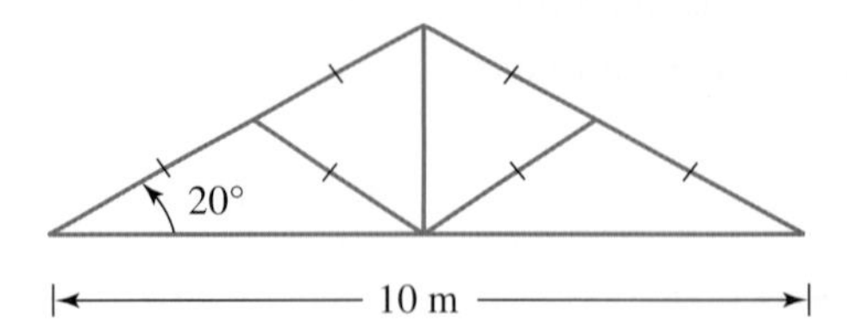

   a. Calculate how high the truss is.
   b. Determine the total length of steel required to make the truss.

4. **WE19** A ladder that is 2.7 m long is leaning against a wall at an angle of 20° as shown. If the base of the ladder is moved 50 cm further away from the wall, determine what angle the ladder will make with the wall. Write your answer correct to the nearest minute.

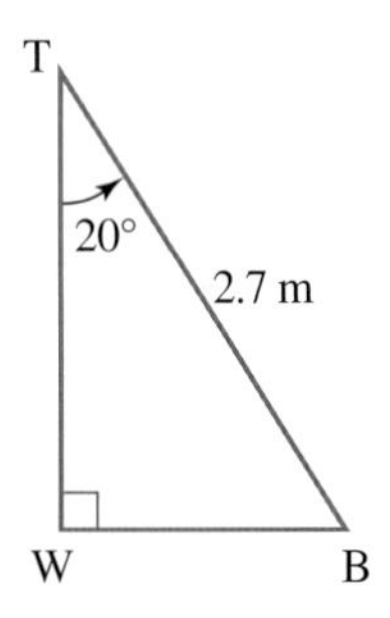

5. A wooden framework is built as shown.
   Bella plans to reinforce the framework by adding a strut from C to the midpoint of AB. Calculate the length of the strut, in metres correct to the 2 decimal places.

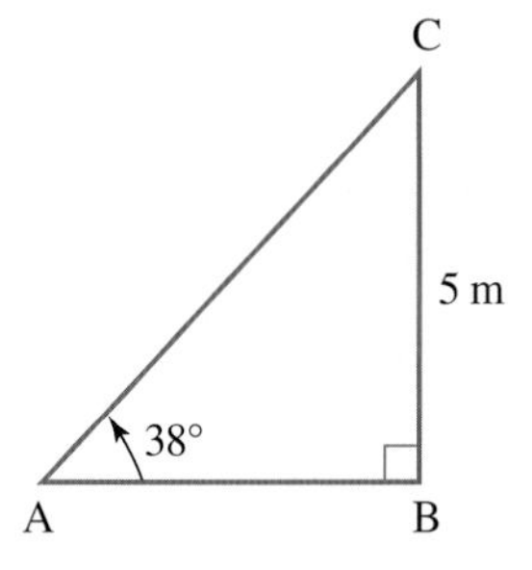

6. Atlanta is standing due south of a 20 m flagpole at a point where the angle of elevation of the top of the pole is 35°. Ginger is standing due east of the flagpole at a point where the angle of elevation of the top of the pole is 27°. Calculate how far, to the nearest metre, Ginger is from Atlanta.

7. From a point at ground level, Henry measures the angle of elevation of the top of a tall building to be 41°. After walking directly towards the building, he finds the angle of elevation to be 75°. If the building is 220 m tall, determine how far Henry walked between measurements. Write your answer correct to the nearest metre.

8. Sailing in the direction of a mountain peak of height 893 m, Imogen measured the angle of elevation to be 14°. A short time later the angle of elevation was 27°. Calculate how far, in km correct to 3 decimal places, Imogen had sailed in that time.

9. A desk top of length 1.2 m and width 0.5 m rises to 10 cm.

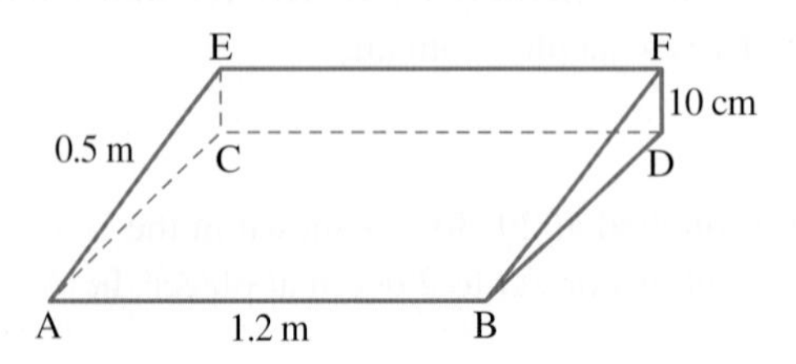

   Calculate, correct to the nearest minute:

   a. $\angle DBF$ **b.** $\angle CBE$.

10. A cuboid has a square end. If the length of the cuboid is 45 cm and its height and width are 25 cm each, calculate:

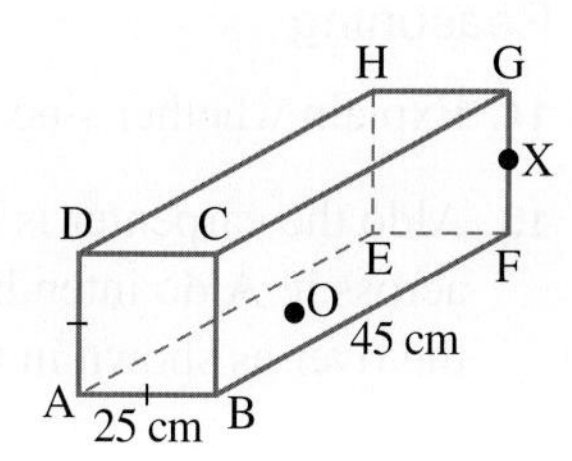

a. the length of BD, correct to 2 decimal places
b. the length of BG, correct to 2 decimal places
c. the length of BE, correct to 2 decimal places
d. the length of BH, correct to 2 decimal places
e. ∠FBG, correct to the nearest minute
f. ∠EBH, correct to the nearest minute.

If the midpoint of FG is X and the centre of the rectangle ABFE is O calculate:

g. the length OF, correct to 2 decimal places
h. the length FX, correct to 1 decimal place
i. ∠FOX, correct to the nearest minute
j. the length OX, correct to 2 decimal places.

11. In a right square-based pyramid, the length of the side of the base is 12 cm and the height is 26 cm.

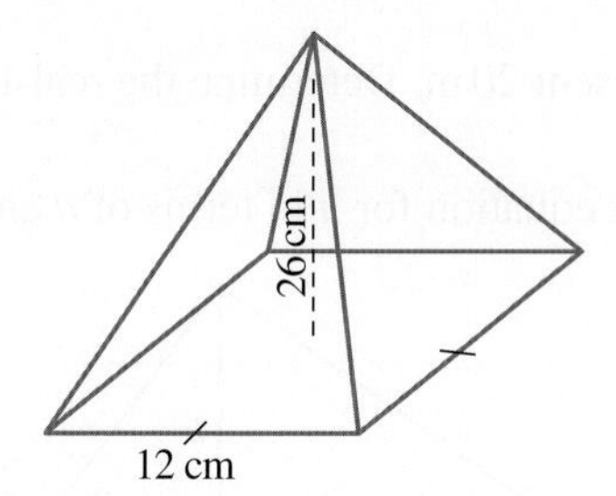

Determine:

a. the angle the triangular face makes with the base, correct to the nearest degree
b. the angle the sloping edge makes with the base, correct to the nearest minute
c. the length of the sloping edge, in cm correct to 2 decimal places.

12. In a right square-based pyramid, the length of the side of the square base is 5.7 cm.

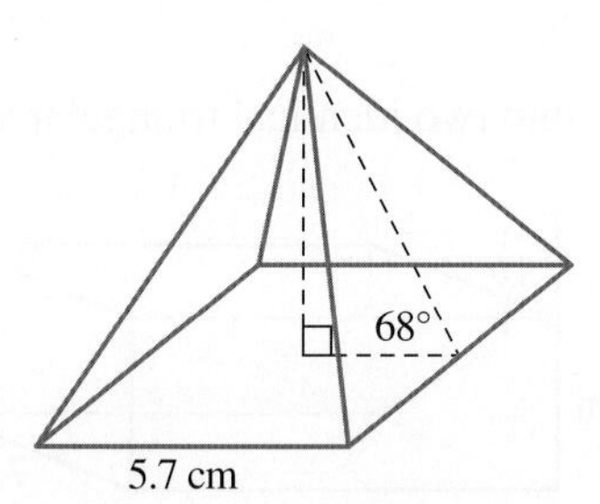

If the angle between the triangular face and the base is 68°, calculate:

a. the height of the pyramid, in cm correct to 2 decimal places
b. the angle the sloping edge makes with the base, correct to the nearest minute
c. the length of the sloping edge, in cm correct to 2 decimal places.

13. In a right square-based pyramid, the height is 47 cm. If the angle between a triangular face and the base is 73°, calculate:

a. the length of the side of the square base, in cm correct to 2 decimal places
b. the length of the diagonal of the base, in cm correct to 2 decimal places
c. the angle the sloping edge makes with the base, correct to the nearest minute.

Reasoning

14. Explain whether sine of an acute angle can be 1 or greater.

15. Aldo the carpenter is lost in a rainforest. He comes across a large river and he knows that he can not swim across it. Aldo intends to build a bridge across the river. He draws some plans to calculate the distance across the river as shown in the diagram below.

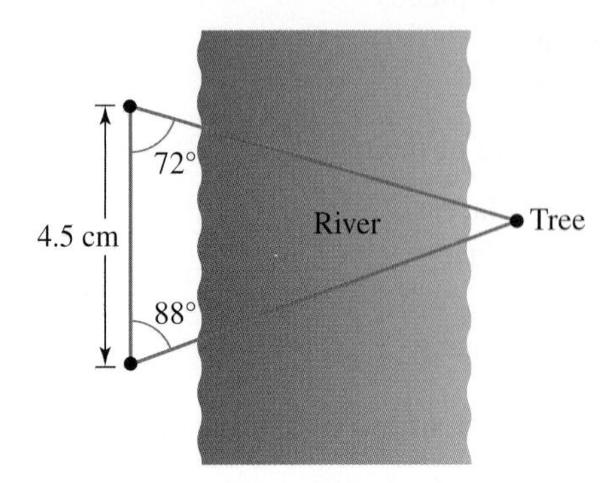

a. Aldo used a scale of 1 cm to represent 20 m. Determine the real-life distance represented by 4.5 cm in Aldo's plans.

b. Use the diagram below to write an equation for h in terms of d and the two angles.

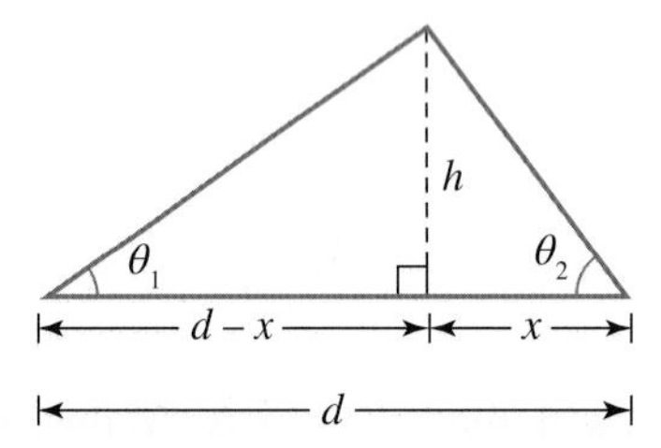

c. Use your equation from part **b** to find the distance across the river, correct to the nearest metre.

16. A block of cheese is in the shape of a rectangular prism as shown. The cheese is to be sliced with a wide blade that can slice it in one go. Calculate the angle (to the vertical correct to 2 decimal places) that the blade must be inclined if:

a. the block is to be sliced diagonally into two identical triangular wedges

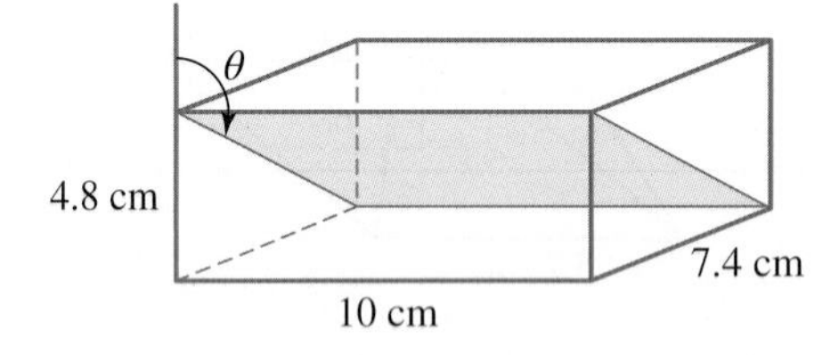

b. the blade is to be placed in the middle of the block and sliced through to the bottom corner, as shown.

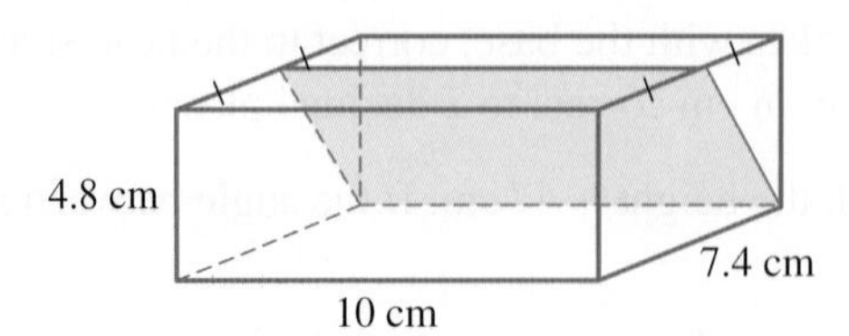

Problem solving

17. A ship travels north for 7 km, then on a true bearing of 140° for another 13 km.
    a. Draw a sketch of the situation.
    b. Determine how far south the ship is from its starting point, in km correct to 2 decimal places.
    c. Evaluate the bearing, correct to the nearest degree, the ship is now from its starting point.
18. The ninth hole on a municipal golf course is 630 m from the tee. A golfer drives a ball from the tee a distance of 315 m at a 10° angle off the direct line as shown.

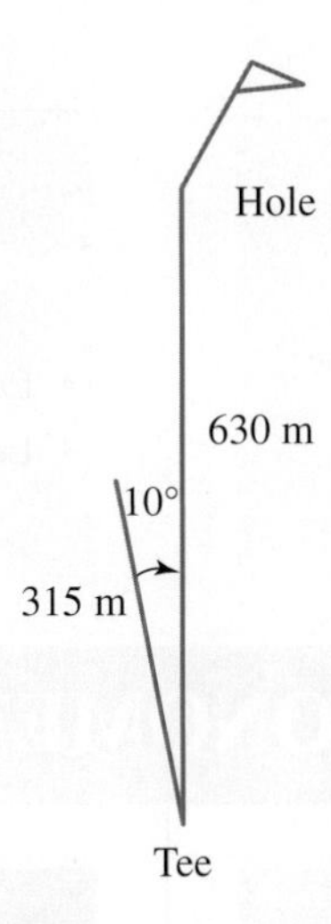

Determine how far the ball is from the hole and state the angle of the direct line that the ball must be hit along to go directly to the hole. Give your answers correct to 1 decimal place.

19. A sphere of radius length 2.5 cm rests in a hollow inverted cone as shown. The height of the cone is 12.5 cm and its vertical angle is equal to 36°.

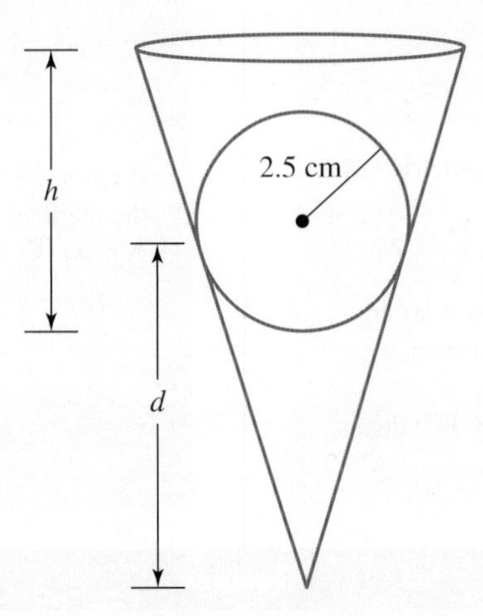

   a. Evaluate the distance, d, from the tip of the cone to the point of contact with the sphere, correct to 2 decimal places.
   b. Determine the distance, h, from the open end of the cone to the bottom of the ball, correct to 2 decimal places.

5.10 Review

5.10.1 Topic summary

TRIGONOMETRY I

Similar triangles

- When triangles have common angles, they are said to be similar.
  e.g. Triangles OAH, OBG and ODE are similar.

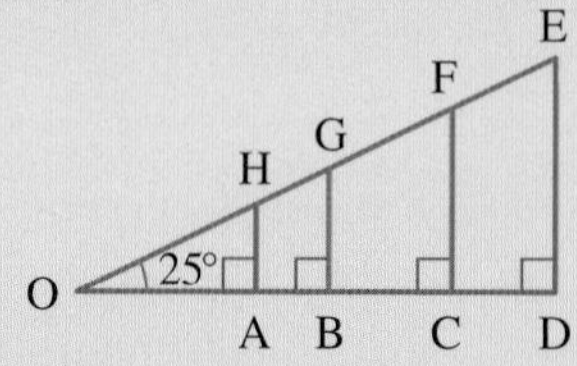

- Corresponding sides of similar triangles will have the same ratio.
  e.g. $\frac{FC}{OC} = \frac{ED}{OD} = \frac{HA}{OA}$

Pythagoras' theorem

- When the length of two sides are known in a right-angled triangle, the third side can be found using the rule $a^2 + b^2 = c^2$.

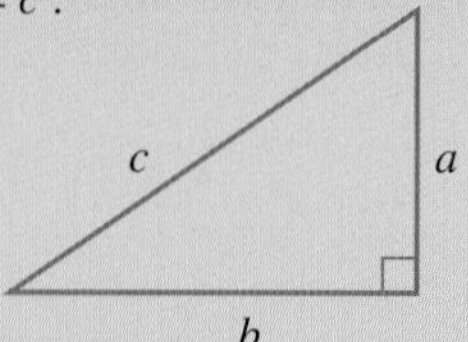

- Length of the longest side $c = \sqrt{a^2 + b^2}$
- Length of the shorter sides $a = \sqrt{c^2 - b^2}$ or $b = \sqrt{c^2 - a^2}$

Trigonometric ratios (SOHCAHTOA)

- In a right-angled triangle, the longest side is called the hypotenuse.

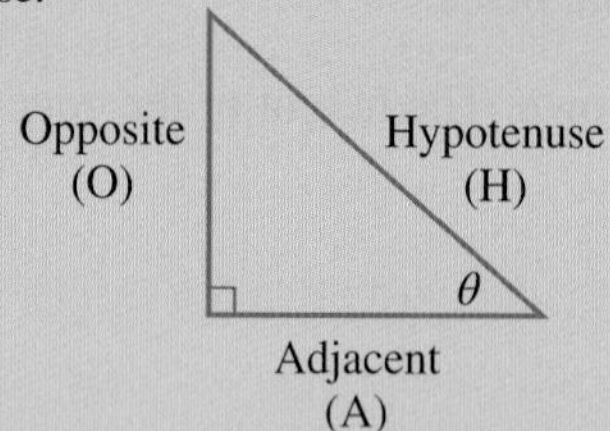

- If an acute angle is known, then the trigonometric ratios can be defined as:

$$\sin\theta = \frac{O}{H}, \cos\theta = \frac{A}{H}, \tan\theta = \frac{O}{A}$$

- An acute angle can be calculated when two sides are known using the inverse operation of the correct trigonometric ratio.
  e.g. Since sin (30°) = 0.5, then $\sin^{-1}$ (0.5) = 30°; this is read as 'inverse sine of 0.5 is 30°'.

Angles of elevation and depression

- If a horizontal line is drawn from A as shown, forming the angle θ, then θ is called the **angle of elevation** of B *from* A.

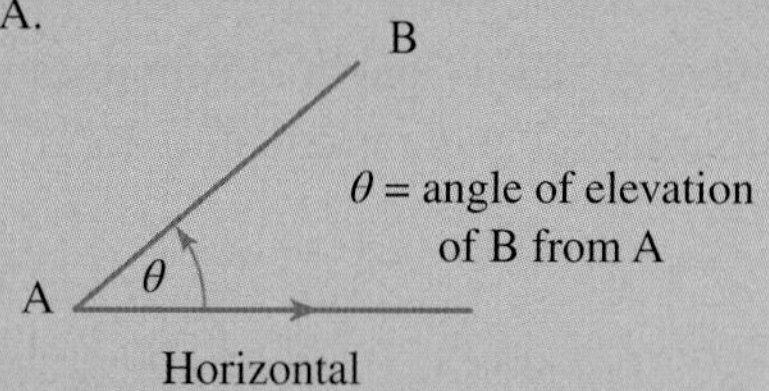

- If a horizontal line is drawn from B as shown, forming the angle α, then α is called the **angle of depression** of A *from* B.

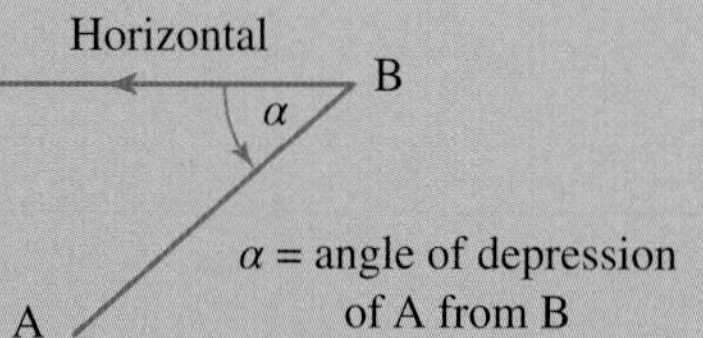

Bearings

There are two ways in which bearings can be written:

- **Compass bearings** have 3 parts:
  - First part is either N or S (for north or south).
  - Second part is an acute angle.
  - Third part is either E or W (for east or west).

  e.g. S20°E means start by facing south and then turn 20° towards the east.

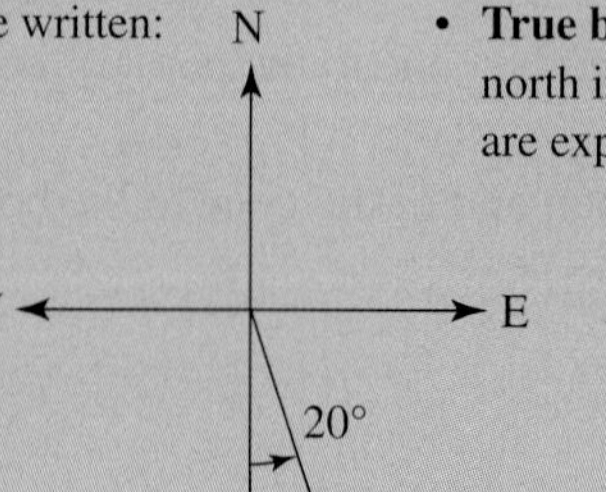

- **True bearings** are measured from north in a clockwise direction and are expressed in 3 digits.

N
025°T
25°
W
E
S

5.10.2 Success criteria

Tick the column to indicate that you have completed the subtopic and how well you have understood it using the traffic light system.

(**Green:** I understand; **Yellow:** I can do it with help; **Red:** I do not understand)

Subtopic	Success criteria	Green	Yellow	Red
5.2	I can identify similar right-angled triangles when corresponding sides are in the same ratio and corresponding angles are congruent.			
	I can apply Pythagoras' theorem to calculate the third side of a right-angled triangle when two other sides are known.			
5.3	I can apply Pythagoras' theorem to determine unknown lengths when a 3D diagram is given.			
	I can apply Pythagoras' theorem to determine unknown lengths in situations by first drawing a diagram.			
5.4	I can define trigonometric ratios according to the lengths of the relevant sides.			
	I can write equations for trigonometric ratios.			
5.5	I can apply trigonometric ratios to find the length of an unknown side when the length of one other side and an acute angle is known.			
5.6	I can apply inverse operations to calculate a known acute angle when two sides are given.			
5.7	I can identify angles of elevation and depression and solve for unknown side lengths and angles.			
5.8	I can draw diagrams with correct angles to represent information to help solve triangles.			
	I can apply trigonometry to solve bearing problems involving compass and true bearings.			
5.9	I can draw well-labelled diagrams to represent information.			
	I can apply trigonometry to solve various problems involving triangles.			

5.10.3 Project

How steep is the land?

When buying a block of land on which to build a house, the slope of the land is often not very obvious. The slab of a house built on the ground must be level, so it is frequently necessary to remove or build up soil to obtain a flat area. The gradient of the land can be determined from a contour map of the area.

Consider the building block shown. The contour lines join points having the same height above sea level. Their measurements are in metres. The plan clearly shows that the land rises from A to B. The task is to determine the angle of this slope.

1. A cross-section shows a profile of the surface of the ground. Let us look at the cross-section of the ground between A and B. The technique used is as follows.
   - Place the edge of a piece of paper on the line joining A and B.
   - Mark the edge of the paper at the points where the contour lines intersect the paper.
   - Transfer this paper edge to the horizontal scale of the profile and mark these points.
   - Choose a vertical scale within the range of the heights of the contour lines.
   - Plot the height at each point where a contour line crosses the paper.
   - Join the points with a smooth curve.

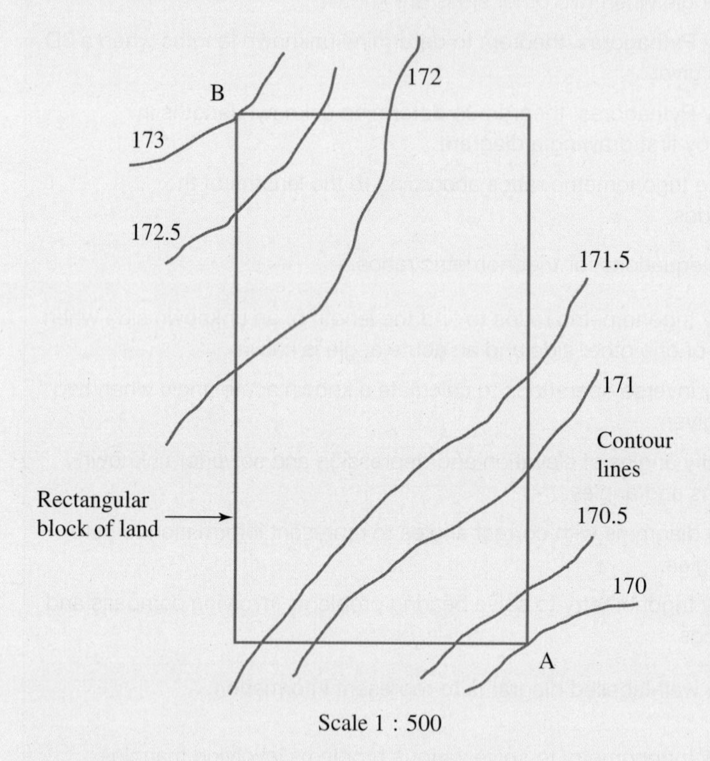

The cross-section has been started for you. Complete the profile of the line A B. You can now see a visual picture of the profile of the soil between A and B.

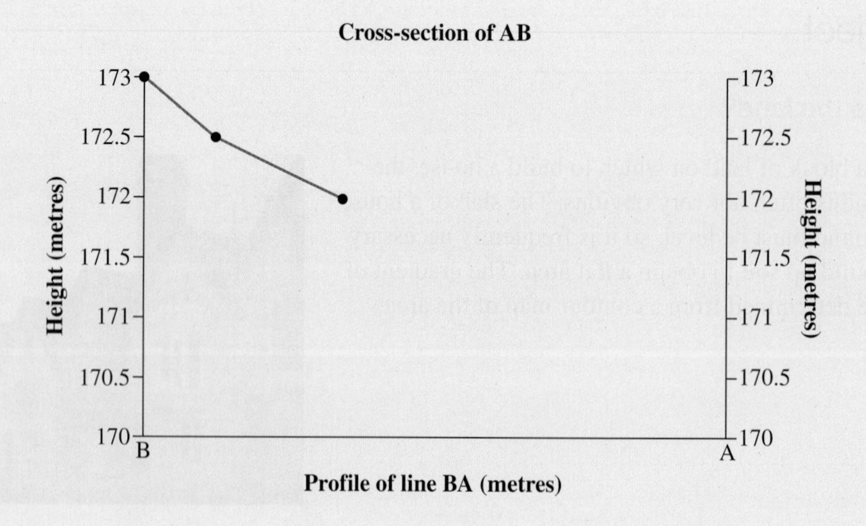

2. We now need to determine the horizontal distance between A and B.
   a. Measure the map distance between A and B using a ruler. What is the map length?
   b. Using the scale of 1 : 500, calculate the actual horizontal distance AB (in metres).
3. The vertical difference in height between A and B is indicated by the contour lines. Calculate this vertical distance.
4. Complete the measurements on this diagram.

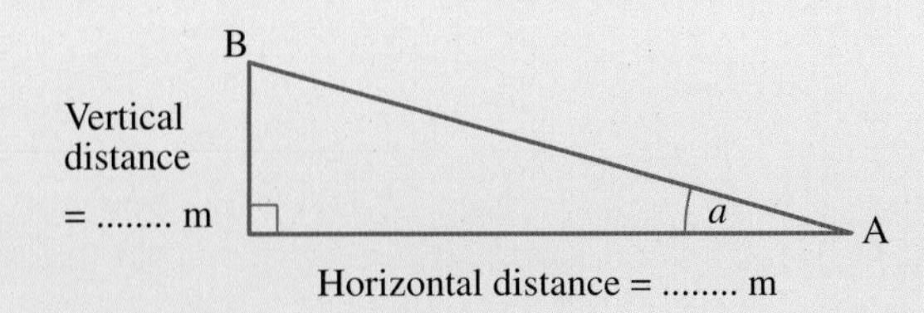

5. The angle a represents the angle of the average slope of the land from A to B. Use the tangent ratio to calculate this angle (to the nearest minute).
6. In general terms, an angle less than 5° can be considered a gradual to moderate rise. An angle between 5° and 15° is regarded as moderate to steep while more than 15° is a steep rise. How would you describe this block of land?
7. Imagine that you are going on a bushwalk this weekend with a group of friends. A contour map of the area is shown. Starting at X, the plan is to walk directly to the hut.
   Draw a cross-section profile of the walk and calculate the average slope of the land. How would you describe the walk?

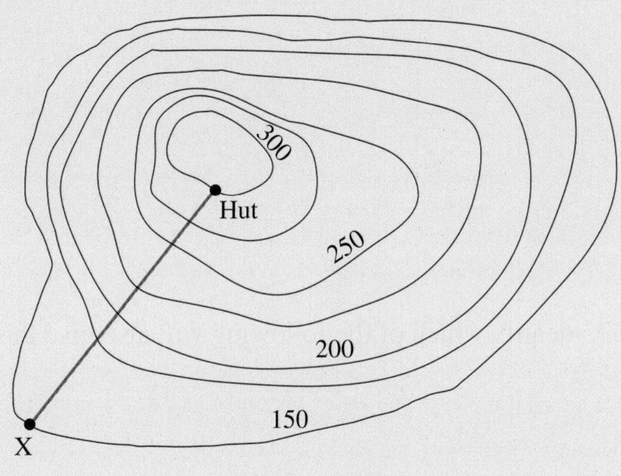

Scale 1 : 20 000

Resources

eWorkbook Topic 5 Workbook (worksheets, code puzzle and project) (ewbk-2031)

Interactivities Crossword (int-2869)
Sudoku puzzle (int-3592)

Exercise 5.10 Review questions

learnON

To answer questions online and to receive **immediate corrective feedback** and **fully worked solutions** for all questions, go to your learnON title at www.jacplus.com.au.

Fluency

1. **MC** The most accurate measurement for the length of the third side in the triangle is:

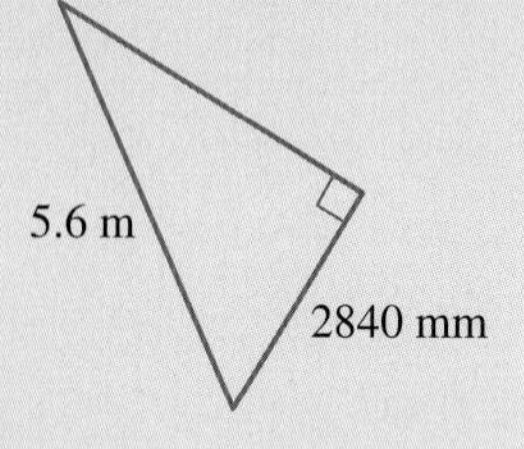

   A. 483 m
   B. 23.3 cm
   C. 3.94 m
   D. 2330 mm
   E. 4826 mm

2. **MC** The value of x in this figure is:

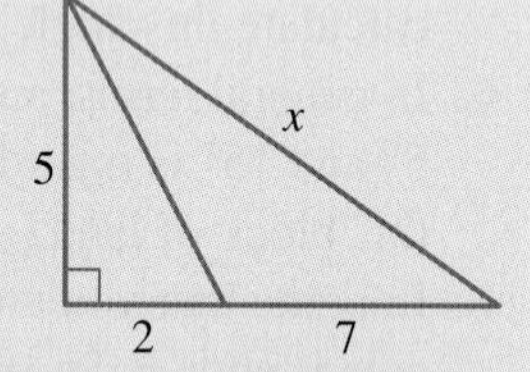

   A. 5.4
   B. 7.5
   C. 10.1
   D. 10.3
   E. 4

3. **MC** Select the closest length of AG of the cube.

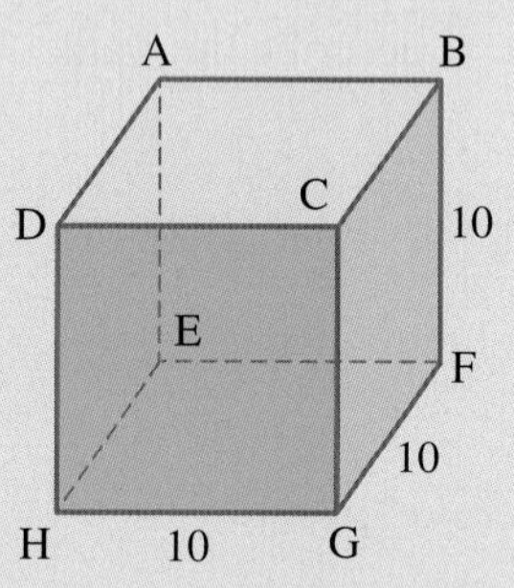

   A. 10
   B. 30
   C. 20
   D. 14
   E. 17

4. **MC** If $\sin(38°) = 0.6157$, identify which of the following will also give this result.
   A. $\sin(218°)$
   B. $\sin(322°)$
   C. $\sin(578°)$
   D. $\sin(682°)$
   E. $\sin(142°)$

5. **MC** The angle $118°52'34''$ is also equal to:
   A. $118.5234°$
   B. $118\frac{52}{34}°$
   C. $118.861°$
   D. $118.876°$
   E. $118.786°$

6. **MC** Identify which trigonometric ratio for the triangle shown below is incorrect.

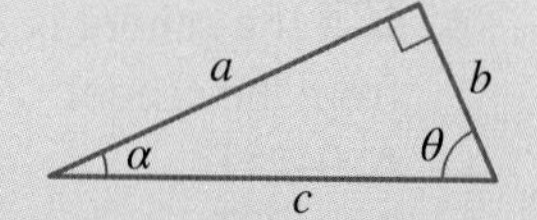

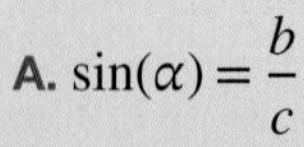

A. $\sin(\alpha) = \dfrac{b}{c}$

B. $\sin(\alpha) = \dfrac{a}{c}$

C. $\cos(\alpha) = \dfrac{a}{c}$

D. $\tan(\alpha) = \dfrac{b}{a}$

E. $\tan(\theta) = \dfrac{a}{b}$

7. **MC** Identify which of the following statements is correct.

A. $\sin(55°) = \cos(55°)$
B. $\sin(45°) = \cos(35°)$
C. $\cos(15°) = \sin(85°)$
D. $\sin(30°) = \sin(60°)$
E. $\sin(42°) = \cos(48°)$

8. **MC** Identify which of the following can be used to determine the value of x in the diagram.

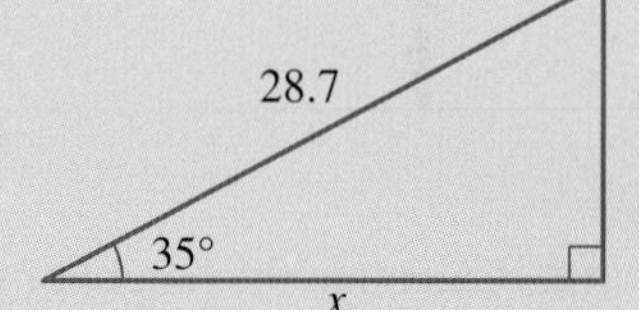

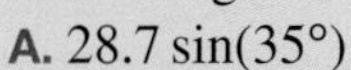

A. $28.7\sin(35°)$
B. $28.7\cos(35°)$
C. $28.7\tan(35°)$
D. $\dfrac{28.7}{\sin(35°)}$
E. $\dfrac{28.7}{\cos(35°)}$

9. **MC** Identify which of the following expressions can be used to determine the value of a in the triangle shown.

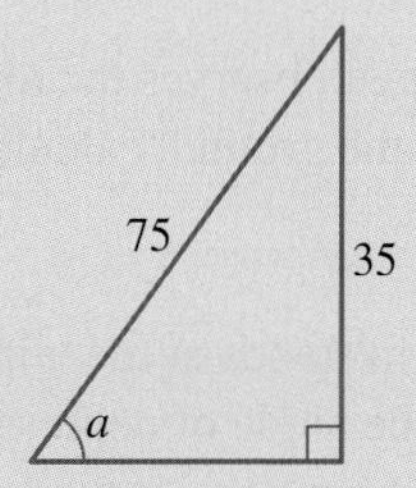

A. $35\sin(75°)$

B. $\sin^{-1}\left(\dfrac{35}{75}\right)$

C. $\sin^{-1}\left(\dfrac{75}{35}\right)$

D. $\cos^{-1}\left(\dfrac{35}{75}\right)$

E. $\cos^{-1}\left(\dfrac{75}{35}\right)$

10. **MC** If a school is 320 m S42°W from the police station, calculate the true bearing of the police station from the school.
A. 042°T
B. 048°T
C. 222°T
D. 228°T
E. 312°T

11. Calculate x, correct to 2 decimal places.

a.
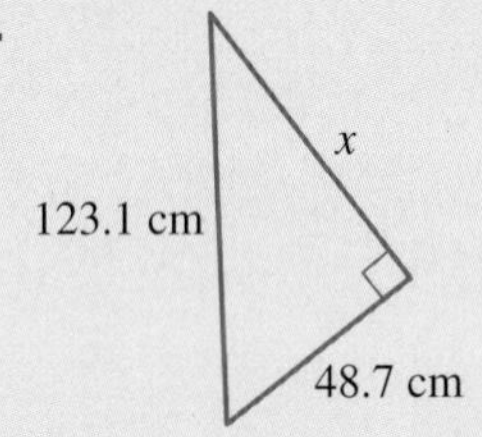

b.
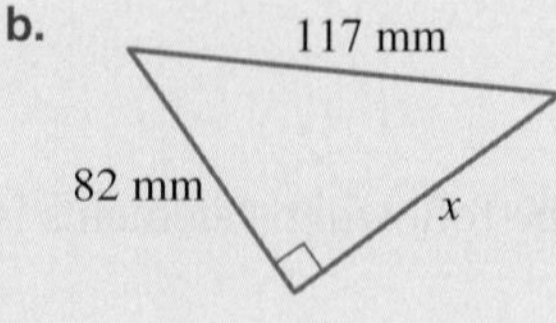

12. Calculate the value of the pronumeral, correct to 2 decimal places.

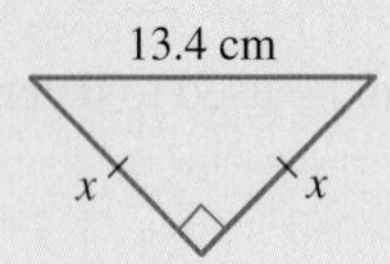

13. Calculate the height of this pyramid, in mm correct to 2 decimal places.

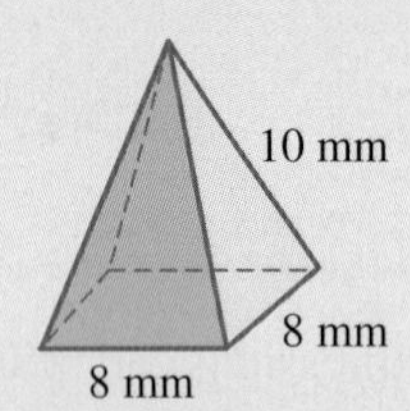

14. A person standing 23 m away from a tree observes the top of the tree at an angle of elevation of 35°. If the person's eye level is 1.5 m from the ground, calculate the height of the tree, in metres correct to 1 decimal place.

15. A man with an eye level height of 1.8 m stands at the window of a tall building. He observes his young daughter in the playground below. If the angle of depression from the man to the girl is 47° and the floor on which the man stands is 27 m above the ground, determine how far from the bottom of the building the child is, in metres correct to 2 decimal places.

16. A plane flies 780 km in a direction of 185°T. Evaluate how far west it has travelled from the starting point, in km correct to 2 decimal places.

17. A hiker travels 3.2 km on a bearing of 250°T and then 1.8 km on a bearing of 320°T. Calculate far west she has travelled from the starting point, in km correct to 2 decimal places.

18. If a 4 m ladder is placed against a wall and the foot of the ladder is 2.6 m from the wall, determine the angle (in degrees and minutes, correct to the nearest minute) the ladder makes with the wall.

Problem solving

19. The height of a right square-based pyramid is 13 cm. If the angle the face makes with the base is 67°, determine:

a. the length of the edge of the square base, in cm correct to 2 decimal places
b. the length of the diagonal of the base, in cm correct to 2 decimal places
c. the angle the slanted edge makes with the base in degrees and minutes, correct to the nearest minute.

20. A car is travelling northwards on an elevated expressway 6 m above ground at a speed of 72 km/h. At noon another car passes under the expressway, at ground level, travelling west, at a speed of 90 km/h.

a. Determine how far apart, in metres, the two cars are 40 seconds after noon, in metres correct to 2 decimal places.
b. At this time the first car stops, while the second car keeps going. Determine the time when they will be 3.5 km apart. Write your answer correct to the nearest tenth of a second.

21. Two towers face each other separated by a distance, d, of 20 metres. As seen from the top of the first tower, the angle of depression of the second tower's base is 59° and that of the top is 31°. Calculate the height, in metres correct to 2 decimal places, of each of the towers.

22. A piece of flat pastry is cut in the shape of a right-angled triangle. The longest side is $6b$ cm and the shortest is $2b$ cm.

a. Determine the length of the third side. Give your answer in exact form.
b. Determine the sizes of the angles in the triangle.
c. Show that the area of the triangle is equal to $4\sqrt{2}b^2$ cm^2.

23. A yacht is anchored off an island. It is 2.3 km from the yacht club and 4.6 km from a weather station. The three points form a right angled triangle at the yacht club.

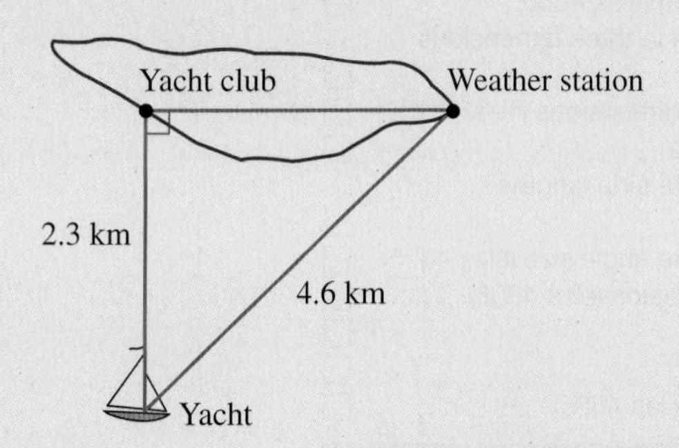

a. Calculate the angle at the yacht between the yacht club and the weather station.
b. Evaluate the distance between the yacht club and the weather station, in km correct to 2 decimal places.

The next day the yacht travels directly towards the yacht club, but is prevented from reaching the club because of dense fog. The weather station notifies the yacht that it is now 4.2 km from the station.

c. Calculate the new angle at the yacht between the yacht club and the weather station, in degrees correct to 1 decimal place.
d. Determine how far the yacht is now from the yacht club, correct to 2 decimal places.

on To test your understanding and knowledge of this topic, go to your learnON title at www.jacplus.com.au and complete the **post-test**.

Online Resources

Below is a full list of **rich resources** available online for this topic. These resources are designed to bring ideas to life, to promote deep and lasting learning and to support the different learning needs of each individual.

eWorkbook

Download the workbook for this topic, which includes worksheets, a code puzzle and a project (ewbk-2031) ☐

Solutions

Download a copy of the fully worked solutions to every question in this topic (sol-0739) ☐

Digital documents

5.2 SkillSHEET Rounding to a given number of decimal places (doc-5224) ☐
5.3 SkillSHEET Drawing 3-D shapes (doc-5229) ☐
5.4 SkillSHEET Labelling the sides of a right-angled triangle (doc-5226) ☐
SkillSHEET Selecting an appropriate trigonometric ratio based on the given information (doc-5231) ☐
5.6 SkillSHEET Rounding angles to the nearest degree (doc-5232) ☐
5.8 SkillSHEET Drawing a diagram from given directions (doc-5228) ☐

Video eLessons

5.2 Similar right-angled triangles (eles-4799) ☐
Review of Pythagoras' theorem (eles-4800) ☐
5.3 Applying Pythagoras' theorem in three dimensions (eles-4801) ☐
Pythagoras' theorem in three dimensions (eles-1913) ☐
5.4 Trigonometric ratios (eles-4802) ☐
5.5 Using trigonometry to calculate side lengths (eles-4804) ☐
5.6 Using trigonometry to calculate angle size (eles-4805) ☐
5.7 Angles of elevation and depression (eles-4806) ☐
5.8 Using bearings (eles-4807) ☐
Bearings (eles-1935) ☐
5.9 Applications of trigonometry (eles-4808) ☐

Interactivities

5.2 Individual pathway interactivity: Pythagoras' theorem (int-4585) ☐
Finding a shorter side (int-3845) ☐
Finding the hypotenuse (int-3844) ☐
5.3 Individual pathway interactivity: Pythagoras' theorem in three dimensions (int-4586) ☐
Right angles in 3-dimensional objects (int-6132) ☐
5.4 Individual pathway interactivity: Trigonometric ratios (int-4587) ☐
Trigonometric ratios (int-2577) ☐
5.5 Individual pathway interactivity: Using trigonometry to calculate side lengths (int-4588) ☐
Using trigonometry to calculate side lengths (int-6133) ☐
5.6 Individual pathway interactivity: Using trigonometry to calculate angle size (int-4589) ☐
Finding the angle when two sides are known (int-6046) ☐
5.7 Individual pathway interactivity: Angles of elevation and depression (int-4590) ☐
Finding the angle of elevation and angle of depression (int-6047) ☐
5.8 Individual pathway interactivity: Bearings (int-4591) ☐
Bearings (int-6481) ☐
5.9 Individual pathway interactivity: Applications (int-4592) ☐
5.10 Crossword (int-2869) ☐
Sudoku puzzle (int-3592) ☐

Teacher resources

There are many resources available exclusively for teachers online.

To access these online resources, log on to **www.jacplus.com.au**.

Answers

Topic 5 Trigonometry I

Exercise 5.1 Pre-test

1. $w = 6.89$ cm
2. $x = 2.24$ cm
3. 16.97 cm
4. 62.28 cm
5. 0.3521
6. $\theta = 55°10'$
7. $\theta = 36°52'12''$
8. $y = 6.7$ m
9. C
10. a. 020°T b. 227°T c. 327°T d. 163°T
11. A
12. D
13. 551.2 m
14. E
15. $x = 9.0$

Exercise 5.2 Pythagoras' theorem

1. a. 7.86 b. 33.27 c. 980.95
2. a. 12.68 b. 2.85 c. 175.14
3. a. 36.36 b. 1.62 c. 15.37
4. a. 0.61 b. 2133.19 c. 453.90
5. 23.04 cm
6. 12.65 cm
7. a. 14.14 cm b. 24.04 cm c. 4.53 cm
8. a. 74.83 cm b. 249.67 cm c. 3741.66 cm^2
9. a. 6.06 b. 4.24 c. 4.74
10. 14.84 cm
11. 15.59 cm
12. 19.23 cm
13. 72.75 cm; 3055.34 cm^2
14. 39 m
15. 4.34 km
16. 38.2 m
17. 63.06 mm
18. a. 32 cm b. 768 cm^2
19. 26.83 diagonals, so would need to complete 27
20. 4701.06 m
21. 9.90 cm
22. a. 65 b. 185 c. 305
23. The value found using Pythagoras' theorem represents length and therefore can't be negative.
24. a. Neither 105 nor 208 can be the hypotenuse of the triangle, because they are the two smallest values. The other two values could be the hypotenuse if they enable the creation of a right-angled triangle.
    b. 105, 208, 233
25. a. 21 cm
    b. 35 cm
    c. $y = 12.6$ cm and RS = 9.8 cm
26. 13.86 cm
27. 1.33 million km

Exercise 5.3 Pythagoras' theorem in three dimensions

1. a. 13.86 b. 13.93 c. 18.03
2. 12.21, 12.85
3. 4.84 m, 1.77 m
4. 8.49, 4.24
5. 31.62 cm
6. 10.58 cm
7. 23 mm
8. a. i. 233.24 m ii. 200.12 m iii. 120.20 m
   b. 116.83 m
9. 14.72 cm
10. 12.67 cm
11. 42.27 cm
12. Sample responses can be found in the worked solutions in the online resources.
13. 186.5 m
14. a. 1.49 m b. 7.43 m^2
15. $\sqrt{1265}$ cm
16. 25.475
17. 28.6 m

Exercise 5.4 Trigonometric ratios

1. a. 0.5000 b. 0.7071 c. 0.4663
   d. 0.8387 e. 8.1443 f. 0.7193
2. a. 0.6494 b. 0.5885 c. 0.5220
   d. −1.5013 e. 0.9990 f. 0.6709
3. a. 0.8120 b. 0.5253 c. −0.8031
   d. 0.4063 e. 0.9880
4. a. −0.9613 b. 1.7321 c. −0.5736 d. 0.1320
5. a. 50° b. 24° c. 53°
6. a. 71° b. 86° c. 41°
7. a. 54°29′ b. 6°19′ c. 0°52′
8. a. 72°47′ b. 44°48′ c. 26°45′
9. a. 26°33′54″ b. 64°1′25″ c. 64°46′59″
10. a. 48°5′22″ b. 36°52′12″ c. 88°41′27″
11. a. 2.824 b. 71.014 c. 20.361 d. 2.828
12. a. 226.735 b. 1.192 c. 7.232 d. 32.259
13. a. 4909.913 b. 0.063 c. 0.904 d. 14.814
14. a. $\sin(\theta) = \frac{e}{f}$ b. $\cos(\theta) = \frac{d}{f}$ c. $\tan(\theta) = \frac{e}{d}$
15. a. $\sin(\alpha) = \frac{i}{g}$ b. $\cos(\alpha) = \frac{h}{g}$ c. $\tan(\alpha) = \frac{i}{h}$

16. a. $\sin(\beta) = \frac{l}{k}$ b. $\cos(\beta) = \frac{j}{k}$ c. $\tan(\beta) = \frac{l}{j}$

17. a. $\sin(\gamma) = \frac{n}{m}$ b. $\cos(\gamma) = \frac{o}{m}$ c. $\tan(\gamma) = \frac{n}{o}$

18. a. $\sin(\beta) = \frac{b}{c}$ b. $\cos(\beta) = \frac{a}{c}$ c. $\tan(\beta) = \frac{b}{a}$

19. a. $\sin(\gamma) = \frac{v}{u}$ b. $\cos(\gamma) = \frac{t}{u}$ c. $\tan(\gamma) = \frac{v}{t}$

20. a. $\sin(\theta) = \frac{15}{18}$ b. $\cos(\theta) = \frac{22}{30}$ c. $\tan(\theta) = \frac{7}{9}$

21. a. $\tan(\theta) = \frac{3.6}{p}$ b. $\sin(25°) = \frac{13}{t}$ c. $\sin(\alpha) = \frac{18.6}{23.5}$

22. a.

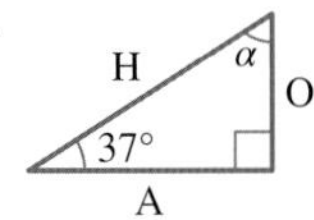

b. i. $\sin(37°) = 0.60$

ii. $\cos(37°) = 0.80$

iii. $\tan(37°) = 0.75$

c. $\alpha = 53°$

23. a. i. $\sin(53°) = 0.80$

ii. $\cos(53°) = 0.60$

iii. $\tan(53°) = 1.33$

b. They are equal.

c. They are equal.

d. The sin of an angle is equal to the cos of its complement angle.

24. $\sin(\theta) = \frac{\text{opp}}{\text{hyp}}, \cos(\theta) = \frac{\text{adj}}{\text{hyp}} \Rightarrow \frac{\sin(\theta)}{\cos(\theta)} = \frac{\text{opp}}{\text{adj}} = \tan(\theta)$

25. a. $h^2 = a^2 - x^2$

b. $h^2 = c^2 - b^2 + 2bx - x^2$

c. Sample responses can be found in the worked solutions in the online resources.

d. Sample responses can be found in the worked solutions in the online resources.

26. $\text{DC} = x + \frac{y}{\tan(\theta)}$

27. To determine which trigonometric ratio to apply, the sides in relation to the angle relevant in the question need to be identified and named.

28. $110\tan(\theta + \alpha) - 110\tan(\alpha)$

Exercise 5.5 Using trigonometry to calculate side lengths

1. a. 8.660 b. 7.250 c. 8.412

2. a. 0.79 b. 4.72 c. 101.38

3. a. 33.45 m b. 74.89 m c. 44.82 m

4. a. 7.76 mm b. 80.82 km c. 9.04 cm

5. a. $x = 31.58$ cm b. $y = 17.67$ m c. $z = 14.87$ m

6. a. $p = 67.00$ m

b. $p = 21.38$ km, $q = 42.29$ km

c. $a = 0.70$ km, $b = 0.21$ km

7. a. 6.0 m b. 6.7 m

8. 1.05 m

9. a. $x = 30.91$ cm, $y = 29.86$ cm, $z = 39.30$ cm

b. 2941.54 cm^2

10. a. In an isosceles right-angled triangle

b. $\theta < 45°$

11. Sample responses can be found in the worked solutions in the online resources.

12. a. $h = \tan(47°48')x$ m

$h = \tan(36°24')(x + 64)$ m

b. 129.07 m

c. 144.29 m

13. Building A is 73.9 m and Building B is 29.5 m.

14. 60

15. Sample responses can be found in the worked solutions in the online resources.

Exercise 5.6 Using trigonometry to calculate angle size

1. a. 67° b. 47° c. 69°

2. a. 54°47′ b. 33°45′ c. 33°33′

3. a. 75°31′21″ b. 36°52′12″ c. 37°38′51″

4. a. 41° b. 30° c. 49°

5. a. 65° b. 48° c. 37°

6. a. $a = 25°47'$, $b = 64°13'$

b. $d = 25°23'$, $e = 64°37'$

c. $x = 66°12'$, $y = 23°48'$

7. a. $r = 57.58$, $l = 34.87$, $h = 28.56$

b. 428 cm^2

c. 29.7°

8. a. i. 29.0° ii. 41.4° iii. 51.3°

b. i. 124.42 km/h ii. 136.57 km/h iii. 146.27 km/h

9. To find the size of acute angles, use inverse operations.

10. Sample responses can be found in the worked solutions in the online resources.

11. a. $\sin(30°) = \frac{1}{2}$, $\cos(30°) = \frac{\sqrt{3}}{2}$, $\tan(30°) = \frac{\sqrt{3}}{3}$

b. Sample responses can be found in the worked solutions in the online resources.

c. Sample responses can be found in the worked solutions in the online resources.

12. a. Between site 3 and site 2: 61°
Between site 2 and site 1: 18°
Between site 1 and bottom: 75°

b. Between site 1 and bottom: 75° slope

13. 58°3′

14. 147°0′; 12°15′

Exercise 5.7 Angles of elevation and depression

1. 8.74 m

2. 687.7 m

3. a. 176.42 m b. 152.42 m

4. 65°46′

5. 16.04 m

6. a. $h = x\tan(47°12')$ m; $h = (x + 38)\tan(35°50')$ m
   b. $x = 76.69$ m
   c. 84.62 m

7. a. $h = x\tan(43°35')$ m; $h = (x + 75)\tan(32°18')$ m
   b. 148.37 m
   c. 143.10 m

8. 0.033 km or 33 m

9. 21°

10. 44.88 m

11. a. 54° b. 0.75 m

12. Angle of elevation is an angle measured upwards from the horizontal. Angle of depression is measured from the horizontal downwards.

13. a.

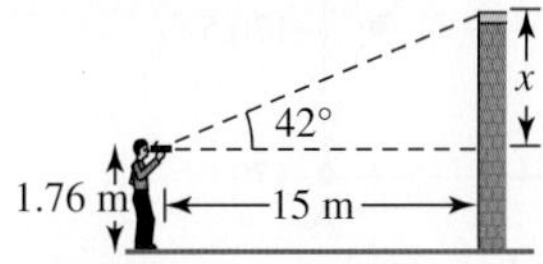

   b. 15.27 m

14. a. 2.16 m/s, 7.77 km/h
   b. 54.5°

15. 66 m

16. 70.522 m

17. 451.5 m

Exercise 5.8 Bearings

1. a. 020°T b. 340°T c. 215°T

2. a. 152°T b. 034°T c. 222°T

3. a. N49°E b. S48°E c. S87°W

4. a. N30°W b. N86°E c. S54°W

5. a. 3 km 325°T b. 2.5 km 112°T c. 8 km 235°T

6. a. 4 km 090°T, then 2.5 km 035°T
   b. 12 km 115°T, then 7 km 050°T
   c. 300 m 310°T, then 500 m 220°T

7. a.

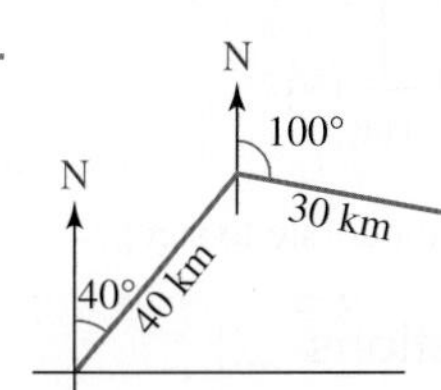

   b.

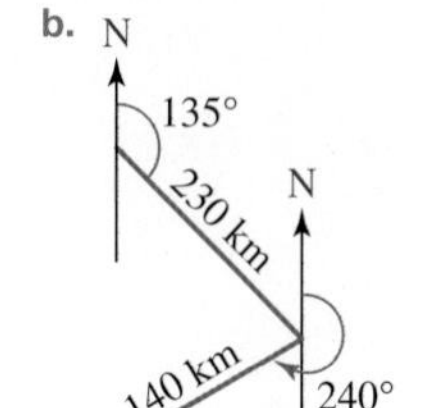

8. a.

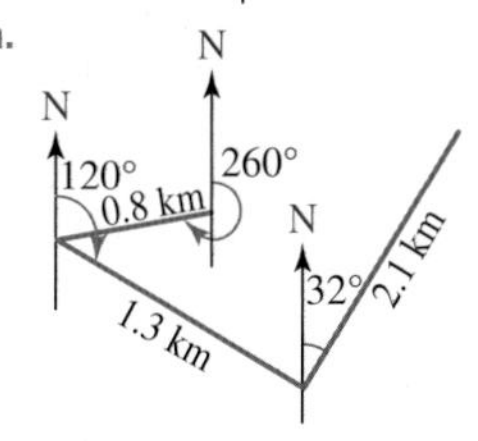

   b.

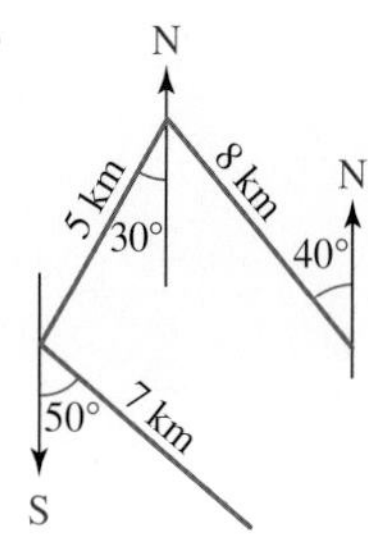

   c.

N
70°
180 km
220 km
N
30°
20°
S
320 km

9. a. 13.38 km
   b. 14.86 km
   c. 222°T
   d.

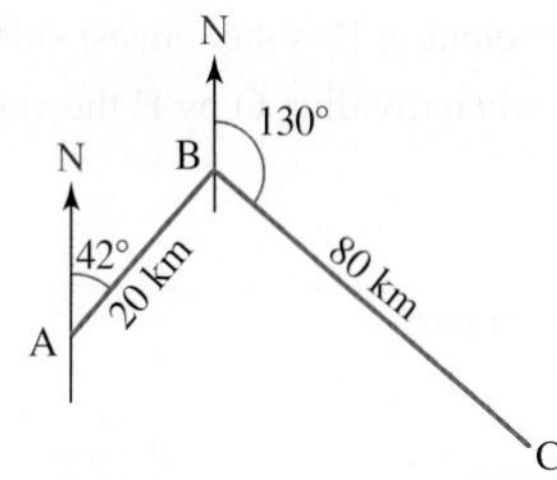

   e. 51.42 km
   f. 61.28 km
   g. 310°T

10. 215°T

11. 1.732 km

12. a. 9.135 km b. 2.305 km c. 104°10′T

13. 684.86 km

14. 6.10 km and 239°T

15. 111°T

16. a. $(180 - \theta)$ °T b. $(\theta - 180)$ °T

17. a. 27.42 km b. N43°W or 317°T

18. a.

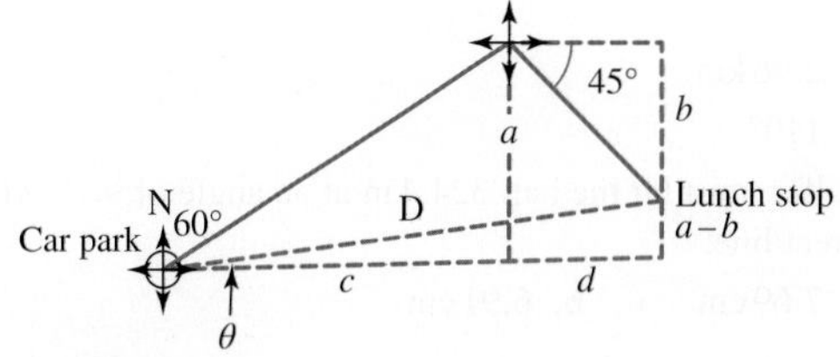

   b. 1.76 km North
   c. 14.63 km East
   d. N83.15°E
   e. D = 14.74 km

19. 3.65 km on a bearing of 108°T

Exercise 5.9 Applications

1. 6.09 m
2. $62°33'$
3. a. 1.82 m b. 27.78 m
4. $31°49'$
5. 5.94 m
6. 49 m
7. 194 m
8. 1.829 km
9. a. $11°32'$ b. $4°25'$
10. a. 35.36 cm b. 51.48 cm c. 51.48 cm
    d. 57.23 cm e. $29°3'$ f. $25°54'$
    g. 25.74 cm h. 12.5 cm i. $25°54'$
    j. 28.61 cm
11. a. 77° b. $71°56'$ c. 27.35 cm
12. a. 7.05 cm b. $60°15'$ c. 8.12 cm
13. a. 28.74 cm b. 40.64 cm c. $66°37'$
14. $\sin(\theta) = \frac{O}{H}$. Since the hypotenuse H is the longest side in the right-angled triangle, when dividing O by H the value will be between 0 and 1.
15. a. 90 m
    b. $h = \dfrac{d\tan(\theta_1)}{\tan(\theta_1) + \tan(\theta_2)} \times \tan(\theta_2)$
    c. 250 m
16. a. 122.97° b. 142.37°
17. a.

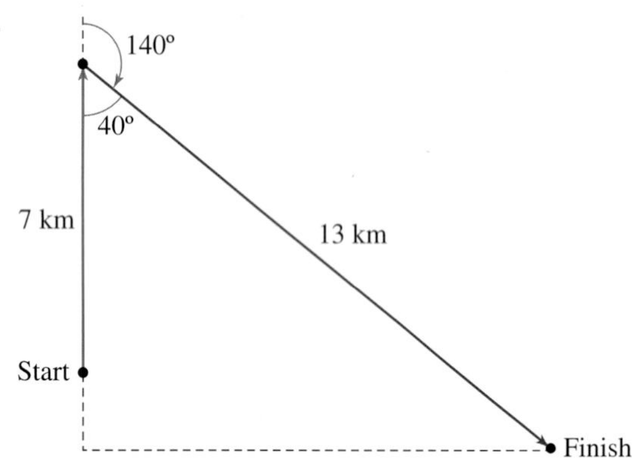

    b. 2.96 km
    c. 110°
18. Golfer must hit the ball 324.4 m at an angle of 9.7° off the direct line.
19. a. 7.69 cm b. 6.91 cm

Project

1.

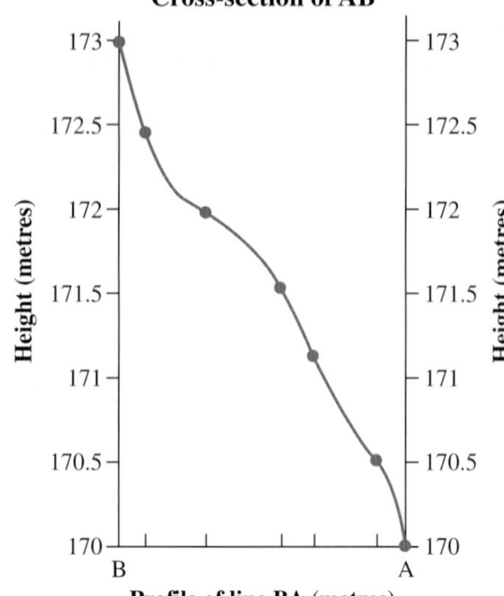

2. a. 8 cm b. 40 m
3. 3 m
4.

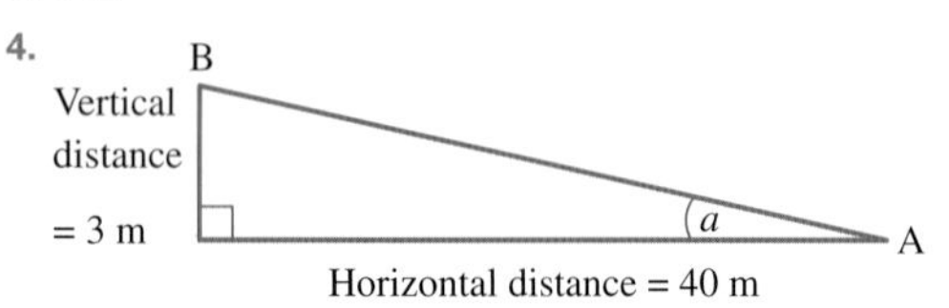

5. $a = 4°17'$
6. Gradual to moderate
7.

Cross-section X to hut

Height (metres): 150, 200, 250, 300

X — Hut

Profile of X to hut

The average slope is 11.46° — moderate to steep.

Exercise 5.10 Review questions

1. E
2. D
3. E
4. E
5. D
6. B
7. E
8. B
9. B
10. A
11. a. $x = 113.06$ cm b. $x = 83.46$ mm
12. 9.48 cm
13. 8.25 mm

14. 17.6 m

15. 26.86 m

16. 67.98 km

17. 4.16 km

18. $40°32'$

19. a. 11.04 cm b. 15.61 cm c. $59°1'$

20. a. 1280.64 m b. 12 : 02 : 16.3 pm

21. 33.29 m, 21.27 m

22. a. $4\sqrt{2}b$

b. 19.5°, 70.5°, 90°.

c. $\text{Area} = \frac{1}{2}\text{base} \times \text{height}$

$= \frac{1}{2} \times 2b \times 4\sqrt{2}b$

$= 4\sqrt{2}b^2 \text{cm}^2.$

23. a. 60° b. 3.98 km c. 71.5° d. 1.33 km

6 Surface area and volume

LEARNING SEQUENCE

6.1 Overview

Why learn this?

People must measure! How much paint or carpet will you need to redecorate your bedroom? How many litres of water will it take to fill the new pool? How many tiles do you need to order to retile the bathroom walls? How far is it from the North Pole to the South Pole? These are just a few examples where measurement skills are needed.

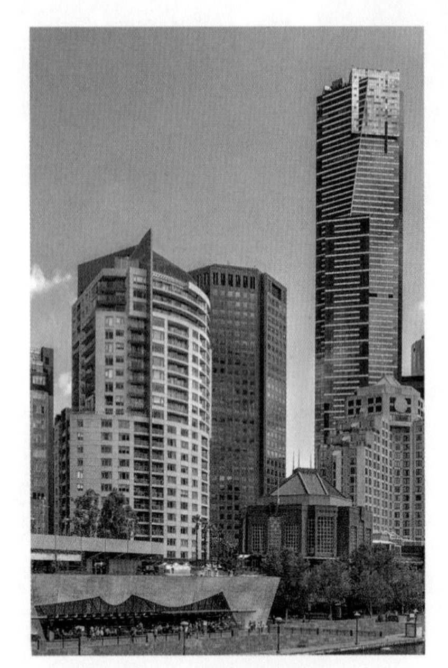

Measuring tools have advanced significantly in their capability to measure extremely small and extremely large amounts and objects, leading to many breakthroughs in medicine, engineering, science, architecture and astronomy.

In architecture, not all buildings are simple rectangular prisms. In our cities and towns, you will see buildings that are cylindrical in shape, buildings with domes and even buildings that are hexagonal or octagonal in shape. Architects, engineers and builders all understand the relationships between these various shapes and how they are connected. Industrial and interior designers use the properties of plane figures, prisms, pyramids and spheres in various aspects of their work.

Have you ever wondered why tennis balls are sold in cylindrical containers? This is an example of manufacturers wanting to minimise the amount of waste in packaging. Understanding the concepts involved in calculating the surface area and volume of common shapes we see around us is beneficial in many real-life situations.

Where to get help

Go to your learnON title at **www.jacplus.com.au** to access the following digital resources. The Online Resources Summary at the end of this topic provides a full list of what's available to help you learn the concepts covered in this topic.

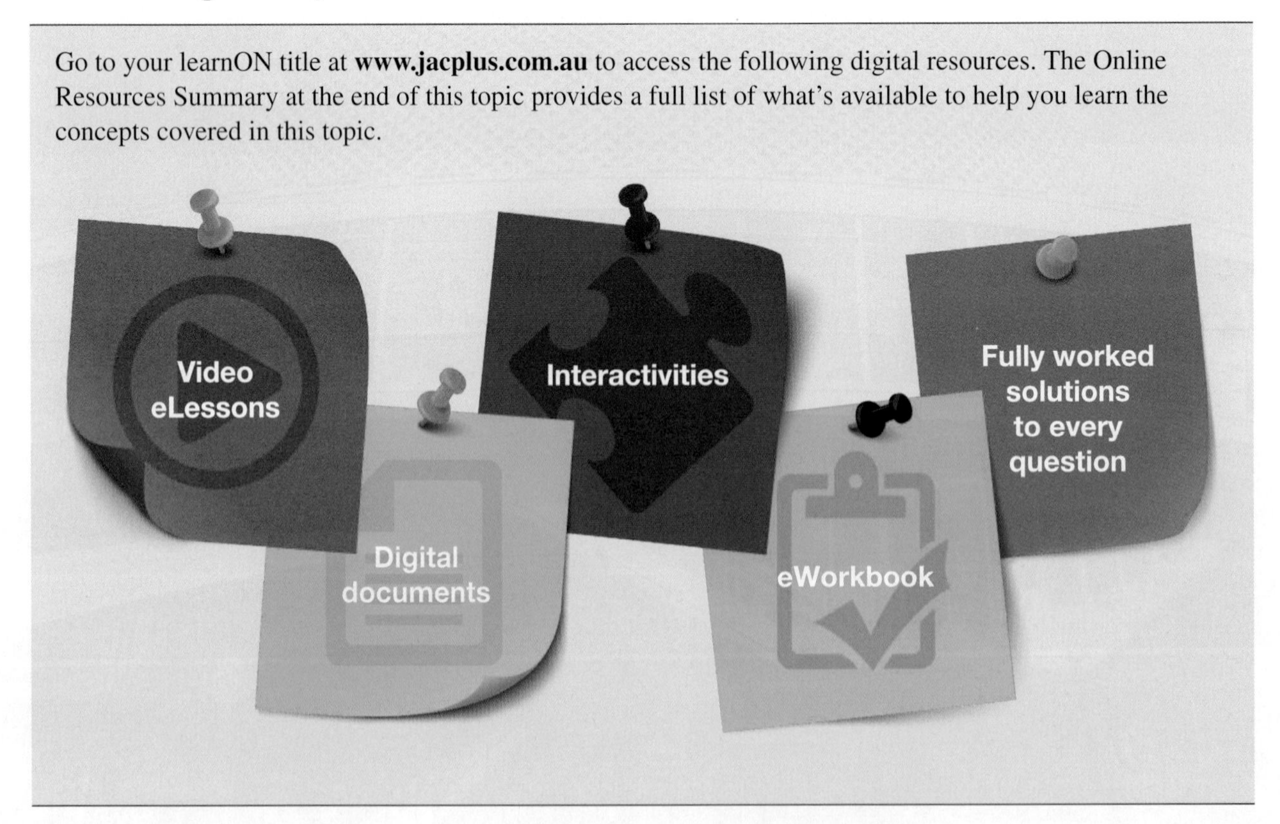

Exercise 6.1 Pre-test

learnON

Complete this pre-test in your learnON title at www.jacplus.com.au and receive **automatic marks**, **immediate corrective feedback** and **fully worked solutions**.

1. Calculate the area of the shape, correct to 2 decimal places.

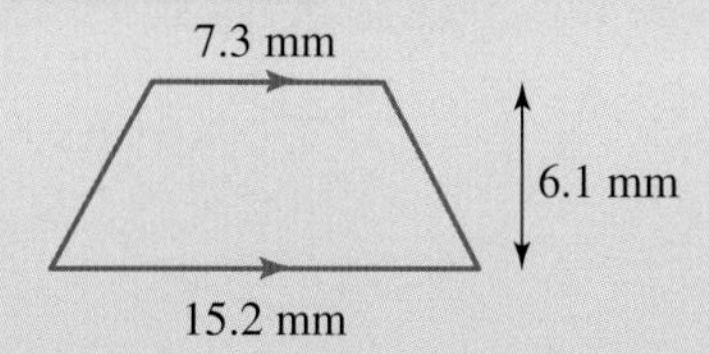

2. Calculate the area of the ellipse, correct to 1 decimal place.

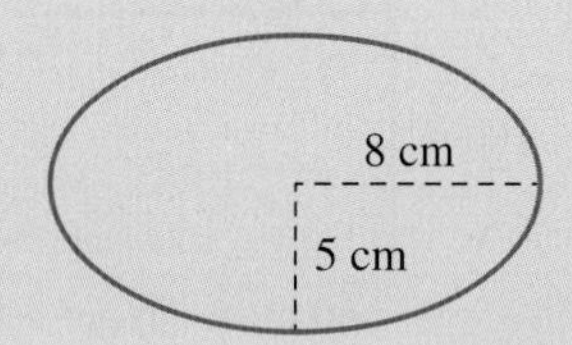

3. **MC** Select the total surface area of the rectangular prism from the following.

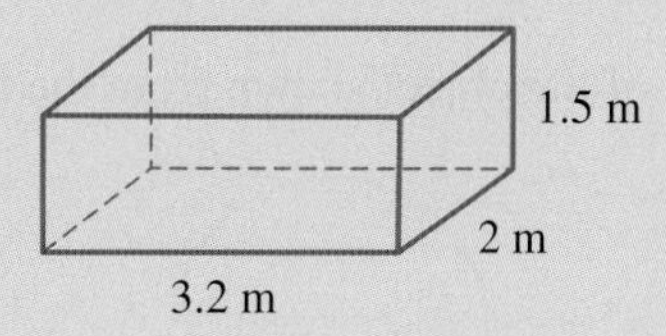

A. $9.6\,m^2$ **B.** $14.2\,m^2$ **C.** $22.0\,m^2$ **D.** $25.4\,m^2$ **E.** $28.4\,m^2$

4. Calculate the total surface area of the sphere, correct to 1 decimal place.

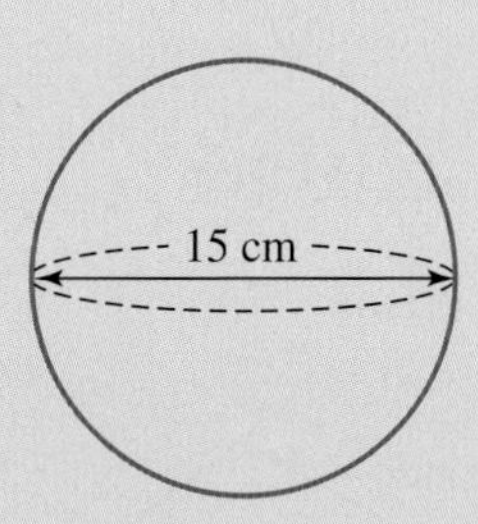

5. Calculate the volume of the solid.

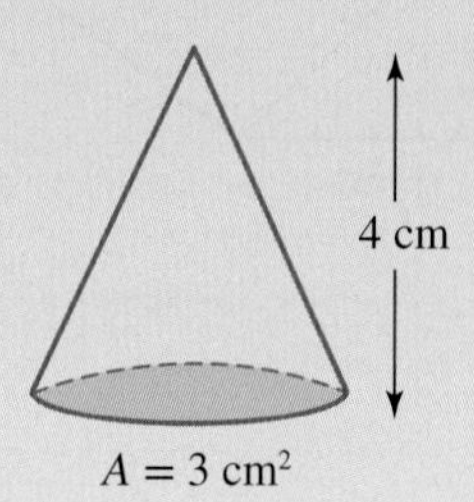

6. Calculate the area of the shape, correct to 1 decimal place.

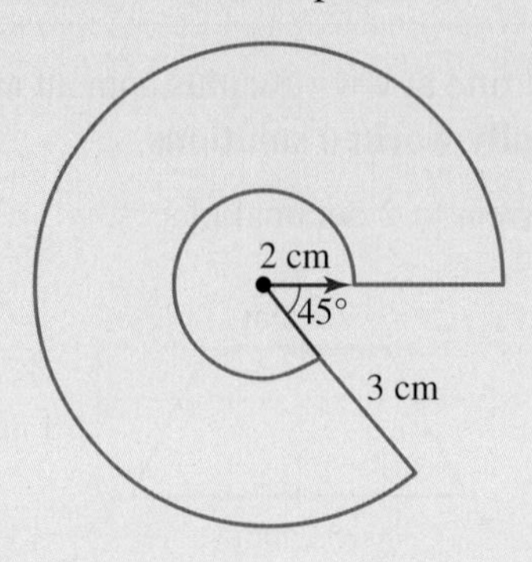

7. A council park is shown below.

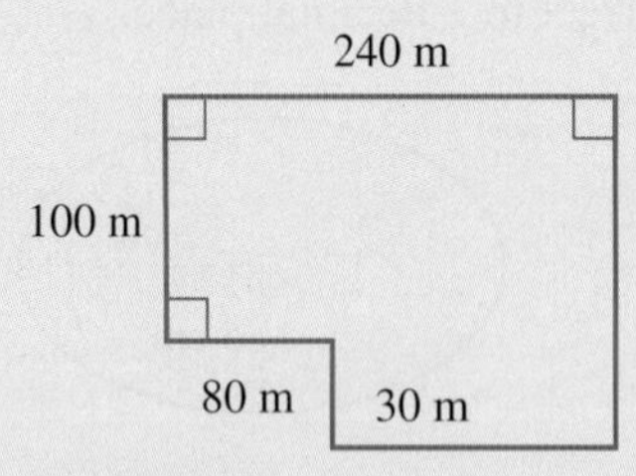

A worker charges \$30 per 1000 m^2 to mow the grass. Determine how much it will cost the council to have the grass mown.

8. **MC** Select the total surface area of the object shown from the following.

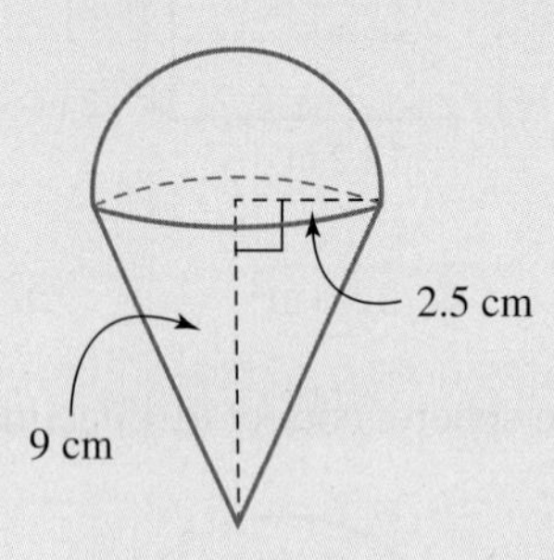

A. 109.96 cm^2 **B.** 112.63 cm^2 **C.** 151.9 cm^2 **D.** 124.36 cm^2 **E.** 91.63 cm^2

9. Determine the volume of the triangular prism.

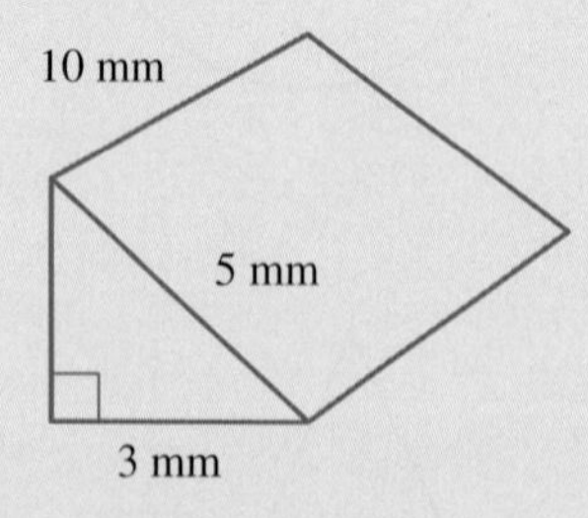

10. **MC** Select the volume of the object from the following.

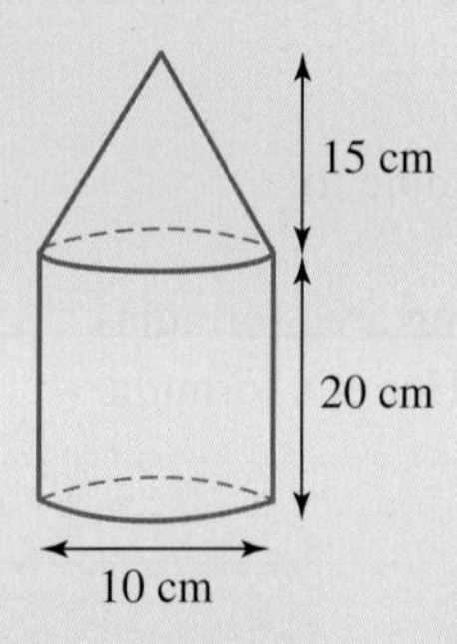

A. 2748.9 cm^3 **B.** 1701.7 cm^3 **C.** 1963.5 cm^3 **D.** 7854 cm^3 **E.** 6806.8 cm^3

11. **MC** The volume of the frustum of a square-based pyramid is given by:

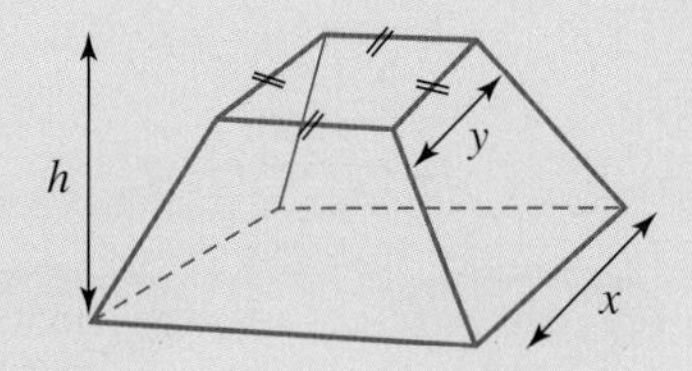

A. $V = \frac{1}{3}\pi\ (x^2 + y^2)$ **B.** $V = \frac{1}{3}\pi\ (x^2 + xy + y^2)$ **C.** $V = \frac{1}{3}h\pi\ (x^2 + xy + y^2)$

D. $V = \frac{1}{3}h\ (x^2 + 2xy + y^2)$ **E.** $V = \frac{1}{3}h\ (x^2 + xy + y^2)$

12. The volume of a ball is given by the formula $V = \frac{4}{3}\pi r^3$. Evaluate the radius of a ball with a volume of 384.66 cm^3. Give your answer correct to 1 decimal place.

13. **MC** Determine what effect doubling the radius and halving the height of a cone will have on its volume.
A. The volume will be the same.
B. The volume will be halved.
C. The volume will be doubled.
D. The volume will be quadrupled.
E. The volume will be divided by a quarter.

14. Using Heron's formula, evaluate the area of the triangle correct to 1 decimal place.

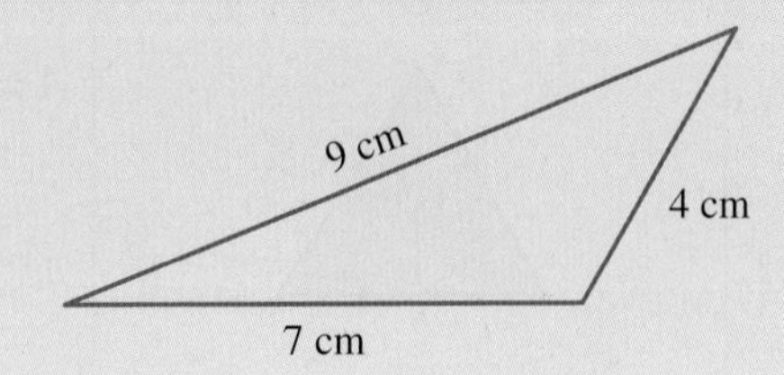

15. A cylindrical soft drink can has a diameter of 6.4 cm and a height of 14.3 cm.
If the can is only half full, determine what capacity of soft drink remains, to the nearest millilitre.

6.2 Area

LEARNING INTENTION

At the end of this subtopic you should be able to:
- convert between units of area
- calculate the area of plane figures using area formulas
- calculate the area of a triangle using Heron's formula.

6.2.1 Area

eles-4809

- The **area** of a figure is the amount of surface covered by the figure.
- The units used for area are mm^2, cm^2, m^2, km^2 and ha (hectares).
- One unit that is often used when measuring land is the hectare. It is equal to $10\,000\ m^2$.
- The following diagram can be used to convert between units of area.

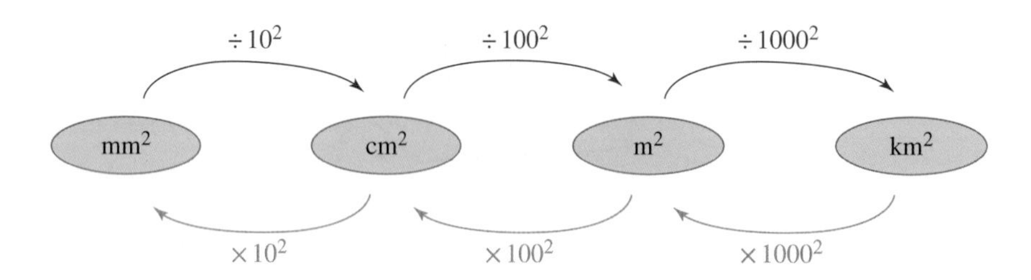

Area formulas

- The table below shows the formula for the area of some common shapes.

Shape	Diagram	Formula
Square		$A = l^2$
Rectangle		$A = lw$
Triangle		$A = \frac{1}{2}bh$
Parallelogram		$A = bh$

Shape	Diagram	Formula
Trapezium		$A = \frac{1}{2}(a+b)h$
Kite (including rhombus)		$A = \frac{1}{2}xy$
Circle		$A = \pi r^2$
Sector		$A = \frac{\theta°}{360°} \times \pi r^2$
Ellipse		$A = \pi ab$

Heron's formula

- The area of a triangle can be calculated if the lengths of all three sides are known.

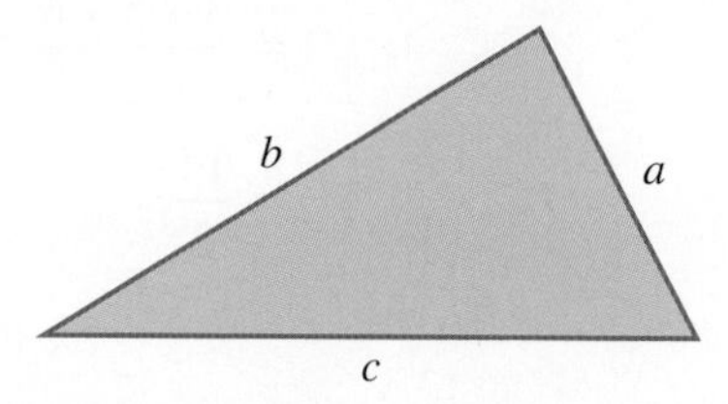

- The area, A, of a triangle given the lengths of the three sides a, b and c is:

$$A = \sqrt{s(s-a)(s-b)(s-c)}$$

where $s = \frac{a+b+c}{2}$, the semi-perimeter.

Digital technology

When using the number π in calculations, it is best to use a calculator. Calculators use the exact value of π, which will ensure your answer is exactly correct.

The π button on the TI-*n*spire CX CAS calculator is found near the bottom left of the calculator, as can be seen in the image at right.

If you do not have a calculator to hand, you can use the approximations $\pi \approx \frac{22}{7} \approx 3.14$; however, your answer may differ from the exact answer by a small amount.

WORKED EXAMPLE 1 Calculating areas of plane figures

Calculate the areas of the following plane figures, correct to 2 decimal places.

a. 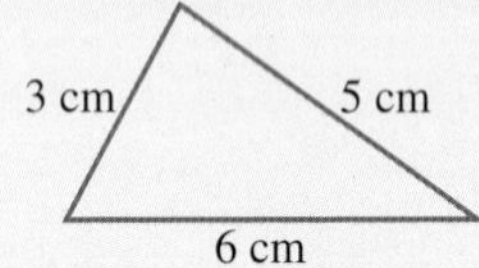

b. 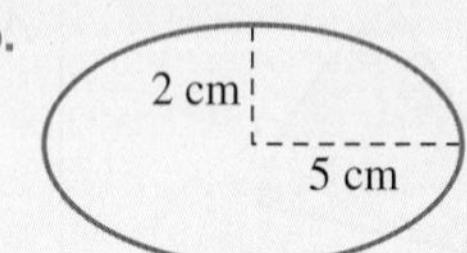

c. 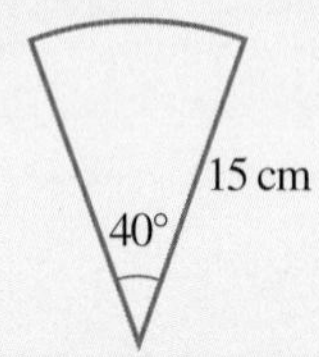

THINK	WRITE
a. 1. Three side lengths are known; apply Heron's formula.	a. $A = \sqrt{s(s-a)(s-b)(s-c)}$
2. Identify the values of a, b and c.	$a = 3, b = 5, c = 6$
3. Calculate the value of s, the semi-perimeter of the triangle.	$s = \frac{a+b+c}{2}$ $= \frac{3+5+6}{2}$ $= \frac{14}{2}$ $= 7$
4. Substitute the values of a, b, c and s into Heron's formula and evaluate, correct to 2 decimal places.	$A = \sqrt{7(7-3)(7-5)(7-6)}$ $= \sqrt{7 \times 4 \times 2 \times 1}$ $= \sqrt{56}$ $= 7.48\ \text{cm}^2$
b. 1. The shape shown is an ellipse. Write the appropriate area formula.	b. $A = \pi ab$
2. Identify the values of a and b (the semi-major and semi-minor axes).	$a = 5,\ b = 2$

3. Substitute the values of a and b into the formula and evaluate, correct to 2 decimal places.

$A = \pi \times 5 \times 2$
$= 31.42\text{ cm}^2$

c. 1. The shape shown is a sector. Write the formula for finding the area of a sector.

c. $A = \dfrac{\theta}{360°} \times \pi r^2$

2. Write the value of θ and r.

$\theta = 40°,\ r = 15$

3. Substitute and evaluate the expression, correct to 2 decimal places.

$A = \dfrac{40°}{360°} \times \pi \times 15^2$
$= 78.54\text{ cm}^2$

6.2.2 Areas of composite figures

eles-4810

- A **composite figure** is a figure made up of a combination of simple figures.
- The area of a composite figure can be calculated by:
  - calculating the sum of the areas of the simple figures that make up the composite figure
  - calculating the area of a larger shape and then subtracting the extra area involved.

WORKED EXAMPLE 2 Calculating areas of composite shapes

Calculate the area of each of the following composite shapes.

a.

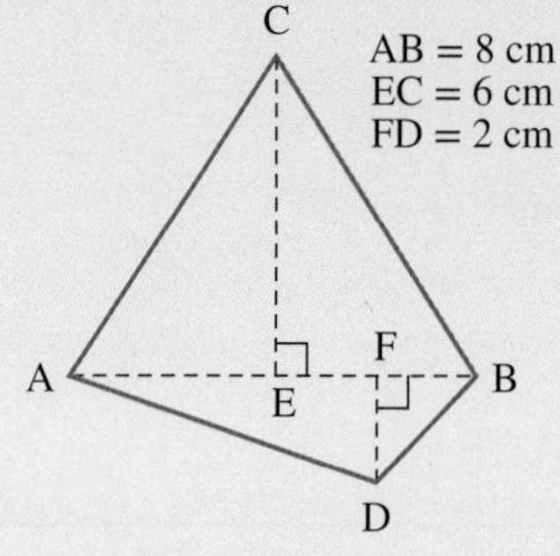

b.

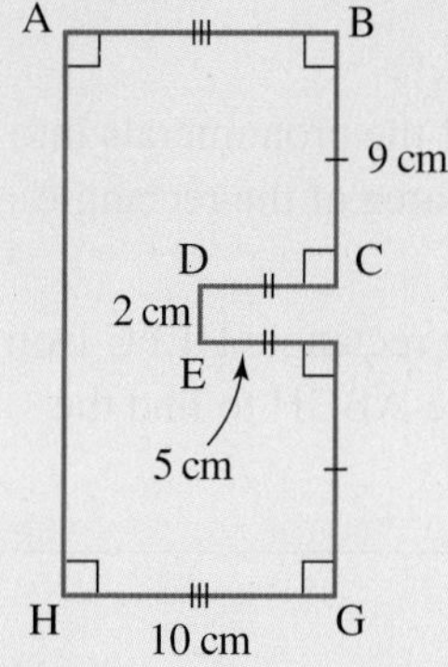

THINK

WRITE

a. 1. ACBD is a quadrilateral that can be split into two triangles: ΔABC and ΔABD.

a. Area ACBD = Area ΔABC + Area ΔABD

2. Write the formula for the area of a triangle containing base and height.

$A_{\text{triangle}} = \dfrac{1}{2}bh$

3. Identify the values of b and h for ΔABC.

ΔABC: $b = AB = 8,\ h = EC = 6$

4. Substitute the values of the pronumerals into the formula and calculate the area of ΔABC.

Area of $\Delta ABC = \dfrac{1}{2} \times AB \times EC$
$= \dfrac{1}{2} \times 8 \times 6$
$= 24\text{ cm}^2$

5. Identify the values of b and h for ΔABD.

ΔABD: $b = AB = 8,\ h = FD = 2$

6. Calculate the area of ΔABD.

$$\text{Area of } \Delta ABD = \frac{1}{2}AB \times FD$$
$$= \frac{1}{2} \times 8 \times 2$$
$$= 8\text{ cm}^2$$

7. Add the areas of the two triangles together to find the area of the quadrilateral ACBD.

$$\text{Area of ACBD} = 24\text{ cm}^2 + 8\text{ cm}^2$$
$$= 32\text{ cm}^2$$

b. 1. One way to find the area of the shape shown is to find the total area of the rectangle ABGH and then subtract the area of the smaller rectangle DEFC.

b. Area = Area ABGH − Area DEFC

2. Write the formula for the area of a rectangle.

$$A_{\text{rectangle}} = l \times w$$

3. Identify the values of the pronumerals for the rectangle ABGH.

Rectangle ABGH: $l = 9 + 2 + 9$
$= 20$
$w = 10$

4. Substitute the values of the pronumerals into the formula to find the area of the rectangle ABGH.

$$\text{Area of ABGH} = 20 \times 10$$
$$= 200\text{ cm}^2$$

5. Identify the values of the pronumerals for the rectangle DEFC.

Rectangle DEFC: $l = 5,\ w = 2$

6. Substitute the values of the pronumerals into the formula to find the area of the rectangle DEFC.

$$\text{Area of DEFC} = 5 \times 2$$
$$= 10\text{ cm}^2$$

7. Subtract the area of the rectangle DEFC from the area of the rectangle ABGH to find the area of the given shape.

$$\text{Area} = 200 - 10$$
$$= 190\text{ cm}^2$$

on Resources

eWorkbook Topic 6 workbook (worksheets, code puzzle and a project) (ewbk-2032)

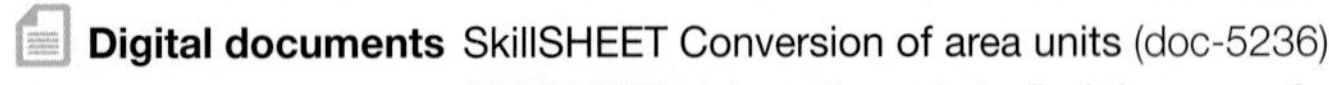
Digital documents SkillSHEET Conversion of area units (doc-5236)
SkillSHEET Using a formula to find the area of a common shape (doc-5237)

Video eLesson Composite area (eles-1886)

Interactivities Individual pathway interactivity: Area (int-4593)
Conversion chart for area (int-3783)
Area of rectangles (int-3784)
Area of parallelograms (int-3786)
Area of trapeziums (int-3790)
Area of circles (int-3788)
Area of a sector (int-6076)
Area of a kite (int-6136)
Area of an ellipse (int-6137)
Using Heron's formula to find the area of a triangle (int-6475)

Exercise 6.2 Area

learnon

Individual pathways

■ PRACTISE	■ CONSOLIDATE	■ MASTER
1, 4, 7, 9, 11, 14, 15, 19, 22	2, 5, 8, 12, 16, 17, 20, 23	3, 6, 10, 13, 18, 21, 24

To answer questions online and to receive **immediate feedback** and **sample responses** for every question, go to your learnON title at www.jacplus.com.au.

Unless told otherwise, where appropriate, give answers correct to 2 decimal places.

Fluency

1. Calculate the areas of the following shapes.

a.

4 cm

b.

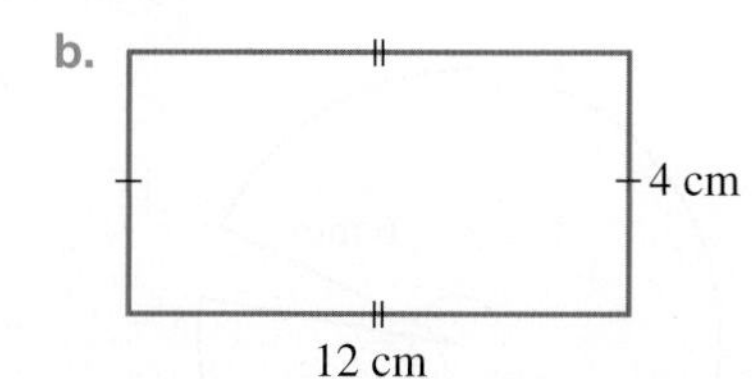

c.

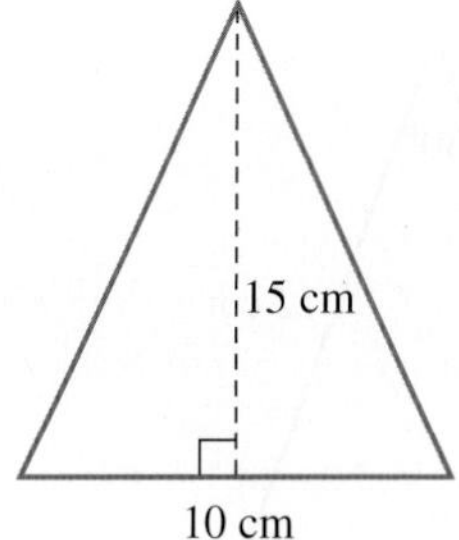

2. Calculate the areas of the following shapes.

a.

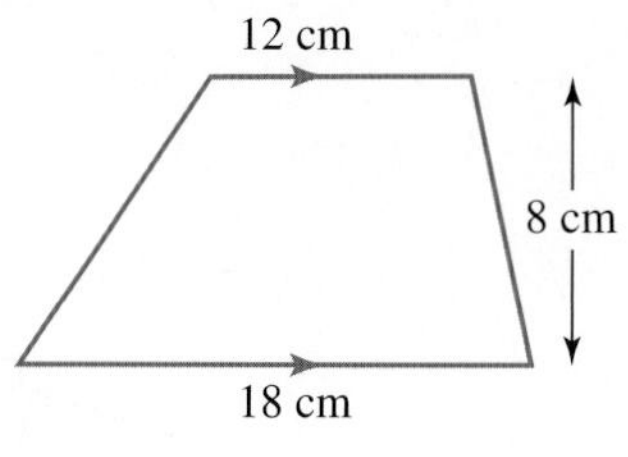

b.

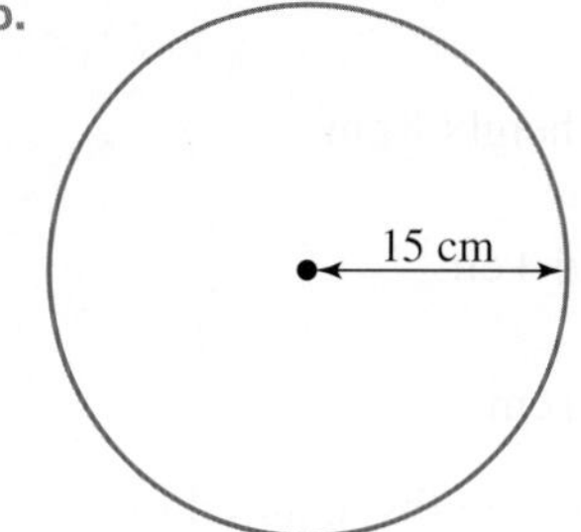

c.

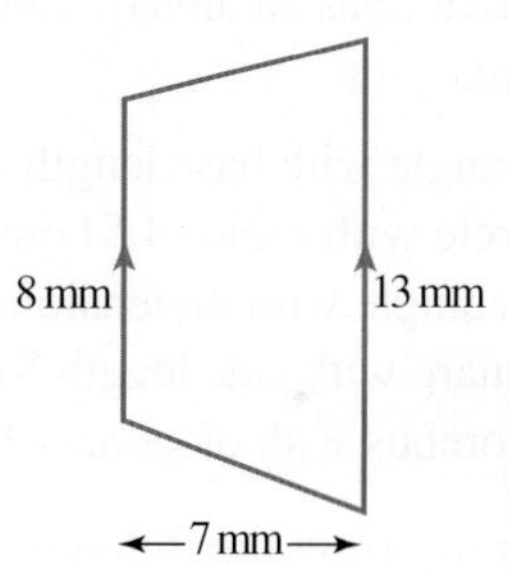

3. Calculate the areas of the following shapes.

a.

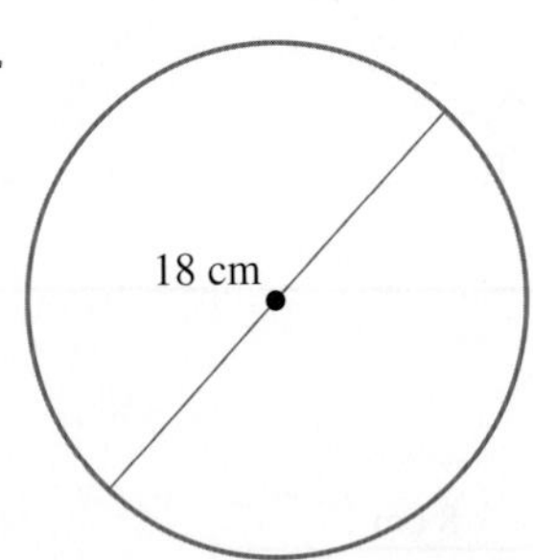

b.

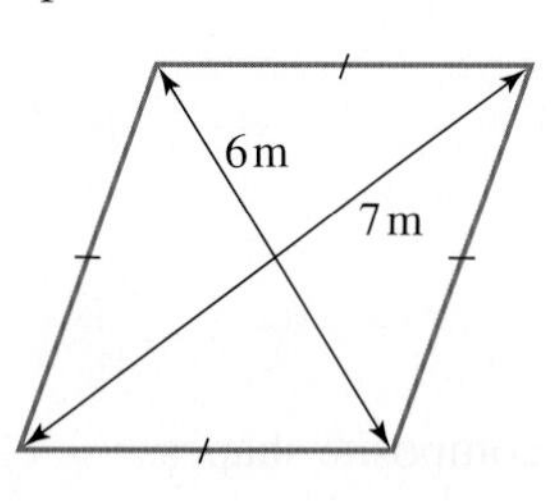

c.

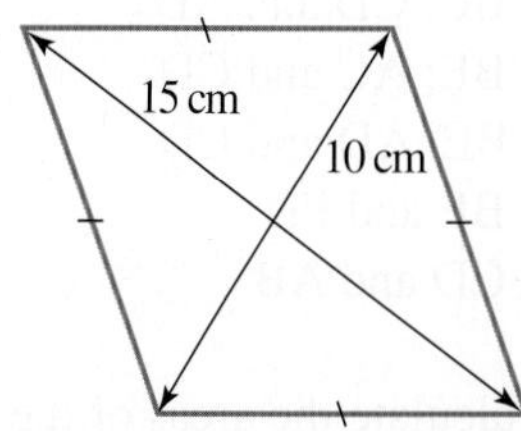

4. **WE1a** Use Heron's formula to calculate the area of the following triangles correct to 2 decimal places.

a.

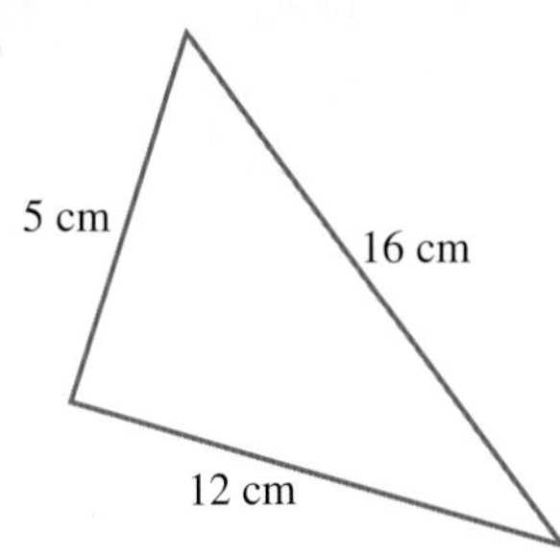

b.

8 cm

3 cm

6 cm

5. **WE1b** Calculate the areas of the following ellipses. Answer correct to 1 decimal place.

a.

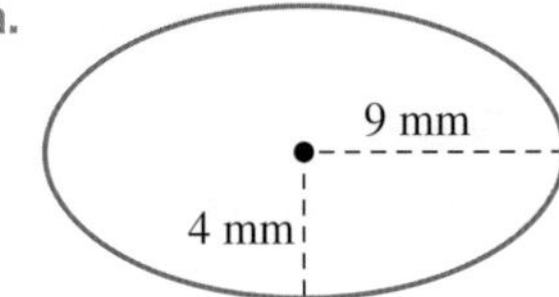

b.

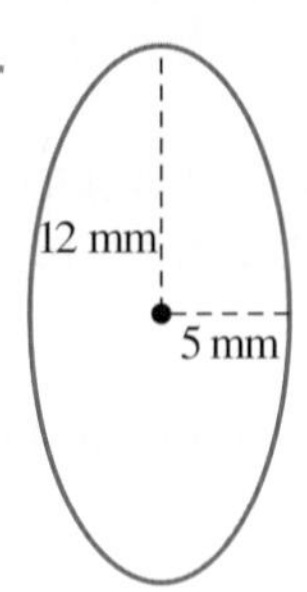

6. **WE1c** Calculate the area of each of the following shapes:

i. stating the answer exactly, that is, in terms of π
ii. correct to 2 decimal places.

a.

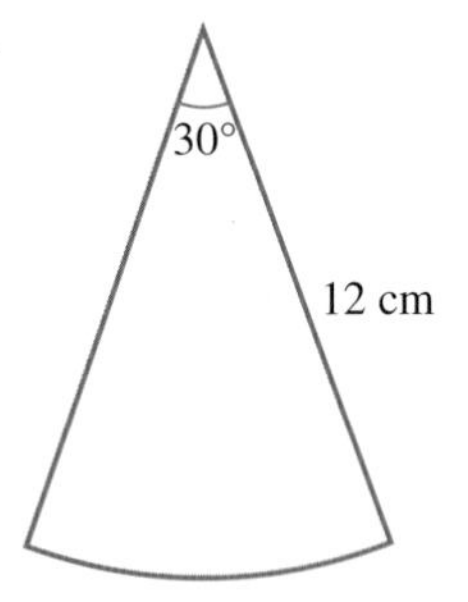

b.

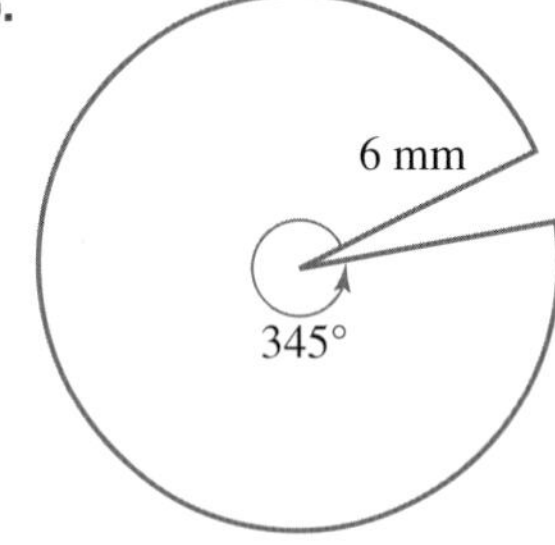

c.

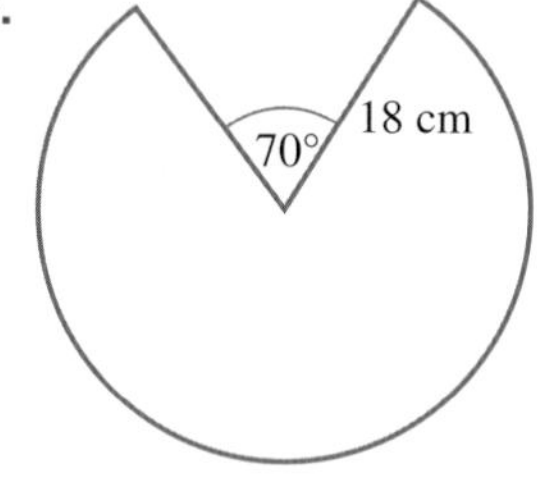

7. **MC** A figure has an area of about 64 cm^2. Identify which of the following *cannot* possibly represent the figure.

A. A triangle with base length 16 cm and height 8 cm
B. A circle with radius 4.51 cm
C. A rectangle with dimensions 16 cm and 4 cm
D. A square with side length 8 cm
E. A rhombus with diagonals 16 cm and 4 cm

8. **MC** Identify from the following list, all the lengths required to calculate the area of the quadrilateral shown.

A. AB, BC, CD and AD
B. AB, BE, AC and CD
C. BC, BE, AD and CD
D. AC, BE and FD
E. AC, CD and AB

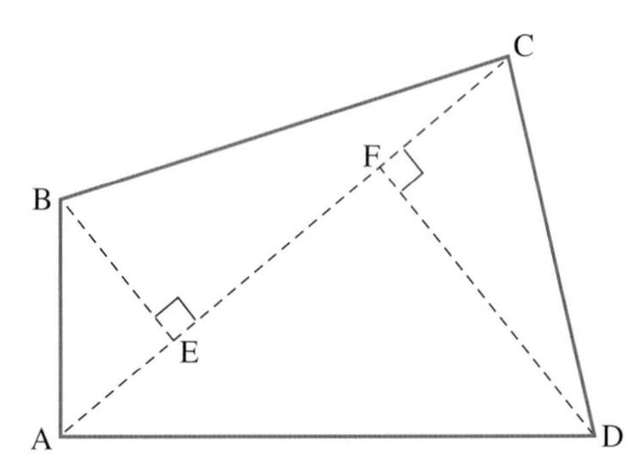

9. **WE2** Calculate the areas of the following composite shapes.

a.

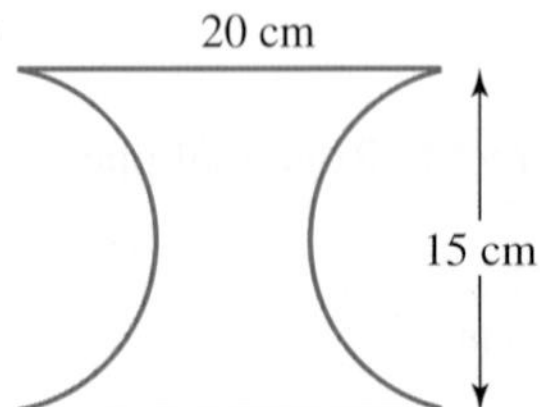

b.

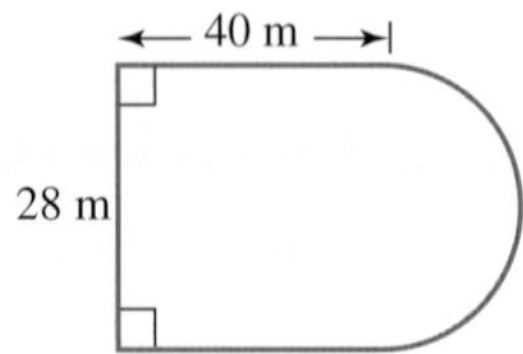

c.

8 cm
3 cm
2 cm
4 cm

10. Calculate the areas of the following composite shapes.

a. 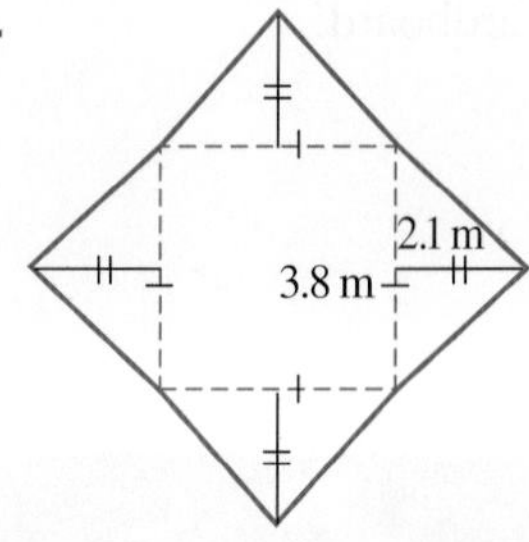

b. 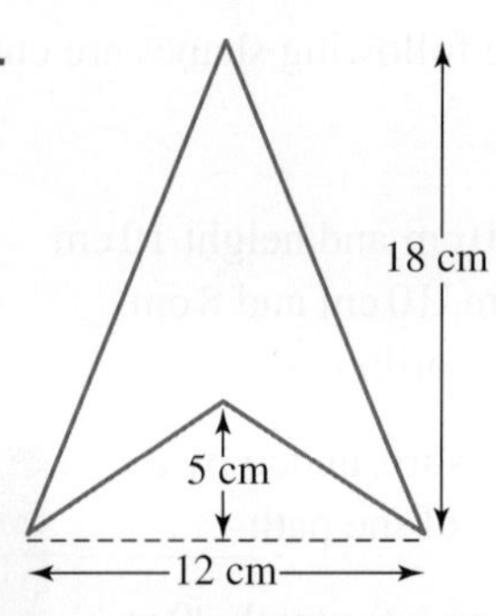

c. 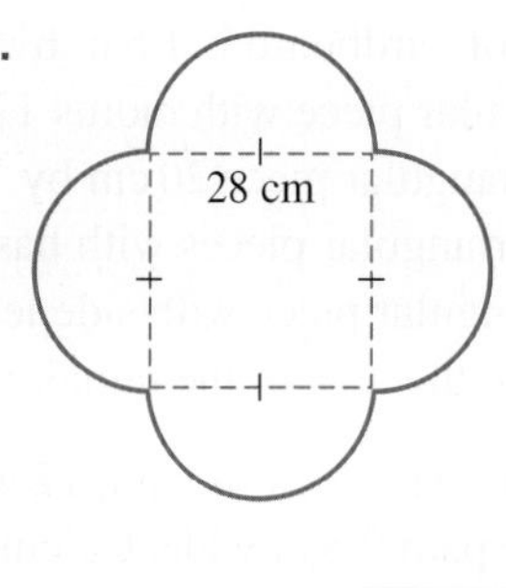

11. Calculate the shaded area in each of the following.

a. 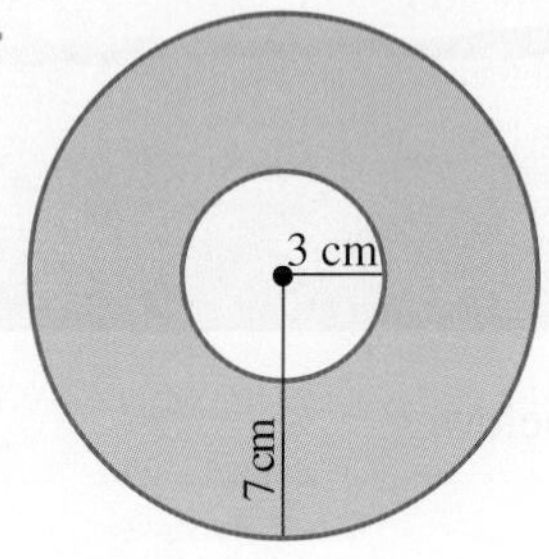

b. 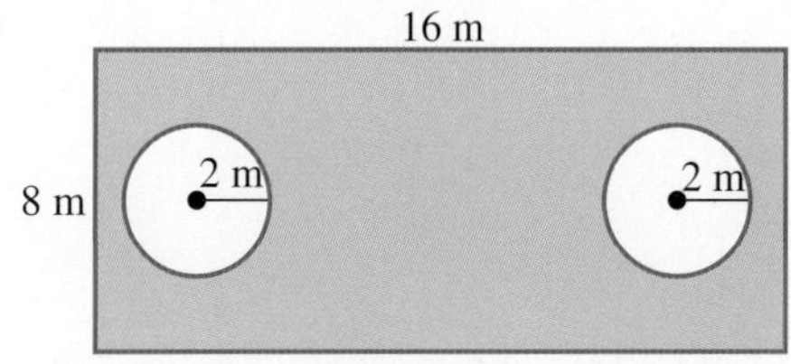

12. Calculate the shaded area in each of the following.

a. 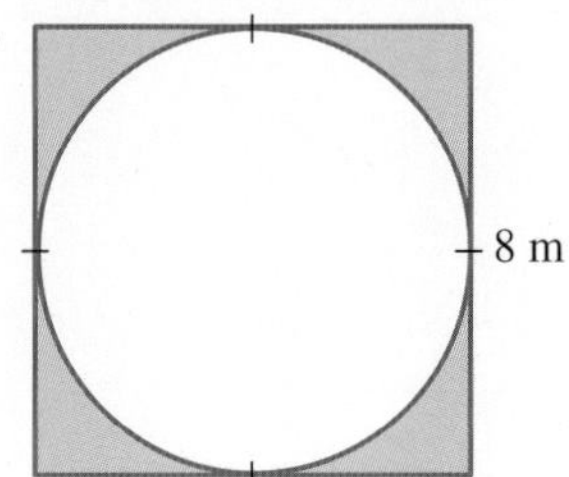

b. 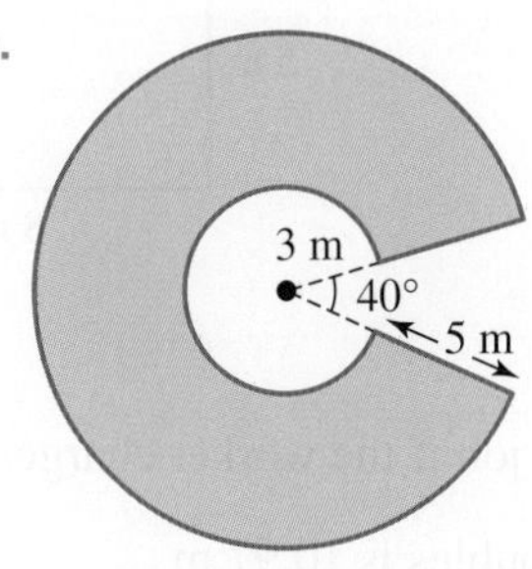

13. Calculate the shaded area in each of the following.

a. 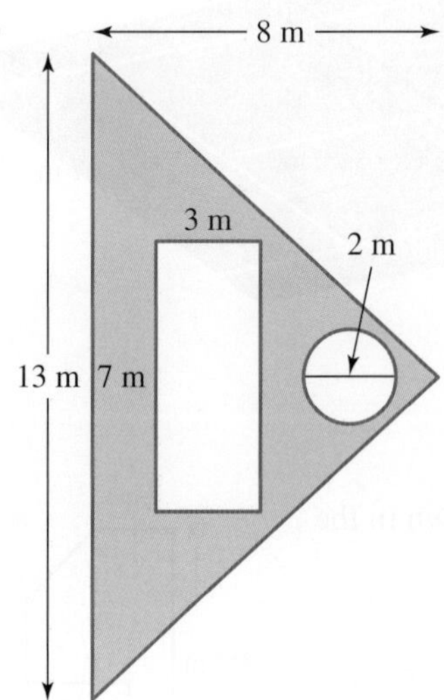

b. 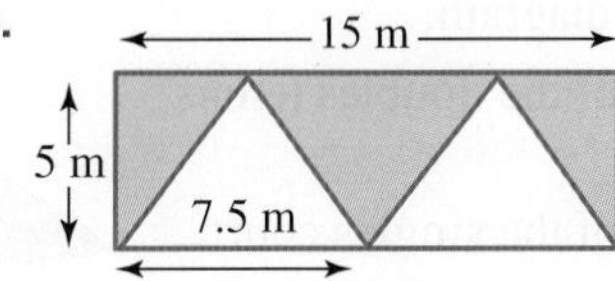

Understanding

14. A sheet of cardboard is 1.6 m by 0.8 m. The following shapes are cut from the cardboard:
    - a circular piece with radius 12 cm
    - a rectangular piece 20 cm by 15 cm
    - two triangular pieces with base length 30 cm and height 10 cm
    - a triangular piece with side lengths 12 cm, 10 cm and 8 cm.

    Calculate the area of the remaining piece of cardboard.

15. A rectangular block of land, 12 m by 8 m, is surrounded by a concrete path 0.5 m wide. Calculate the area of the path.

16. Concrete slabs 1 m by 0.5 m are used to cover a footpath 20 m by 1.5 m. Determine how many slabs are needed.

17. A city council builds a 0.5 m wide concrete path around the garden as shown below.

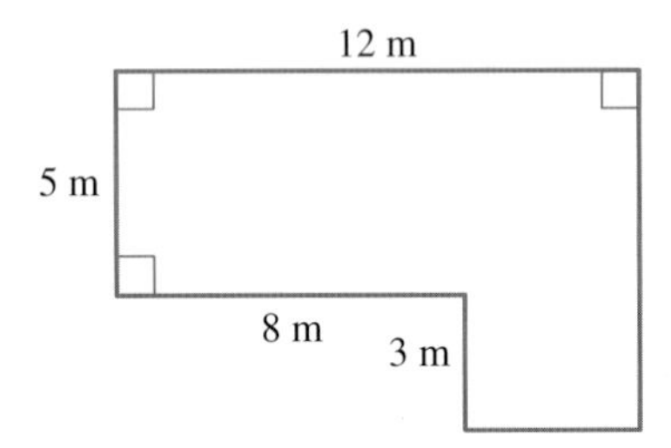

Determine the cost of the job if the worker charges \$40.00 per m^2.

18. A tennis court used for doubles is 10.97 m wide, but a singles court is only 8.23 m wide, as shown in the diagram.
    a. Calculate the area of the doubles tennis court.
    b. Calculate the area of the singles court.
    c. Determine the percentage of the doubles court that is used for singles. Give your answer to the nearest whole number.

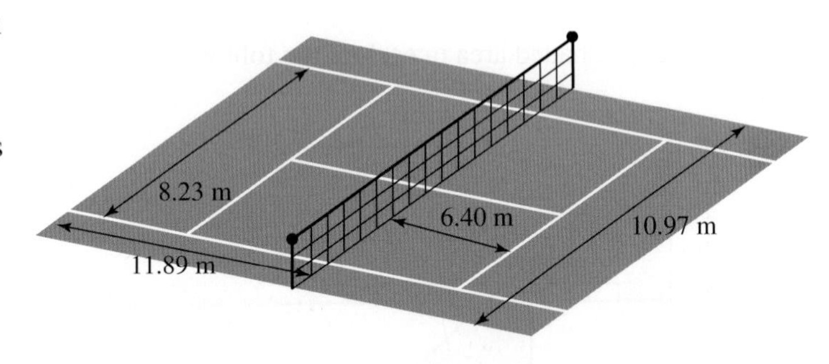

Reasoning

19. Dan has purchased a country property with layout and dimensions as shown in the diagram.
    a. Show that the property has a total area of 987.5 ha.
    b. Dan wants to split the property in half (in terms of area) by building a straight-lined fence running either north–south or east–west through the property. Assuming the cost of the fencing is a fixed amount per linear metre, justify where the fence should be built (that is, how many metres from the top left-hand corner and in which direction) to minimise the cost.

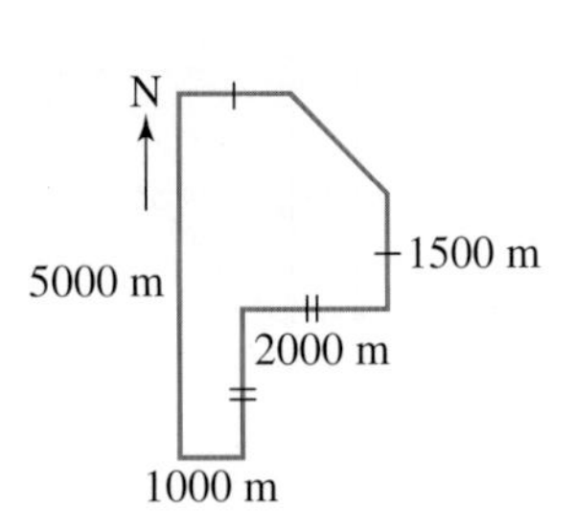

20. Ron the excavator operator has 100 metres of barricade mesh and needs to enclose an area to safely work in. He chooses to make a rectangular region with dimensions x and y. Show your working when required.

a. Write an equation that connects x, y and the perimeter.
b. Write y in terms of x.
c. Write an equation for the area of the region in terms of x.
d. Fill in the table for different values of x.

x	0	5	10	15	20	25	30	35	40	45	50
Area $(\mathbf{m}^2)$											

e. Can x have a value more than 50? Why?
f. Sketch a graph of area against x.
g. Determine the value of x that makes the area a maximum.
h. Determine the value of y for maximum area.
i. Determine the shape that encloses the maximum area.
j. Calculate the maximum area.

Ron decides to choose to make a circular area with the barricade mesh.

k. Calculate the radius of this circular region.
l. Calculate the area that is enclosed in this circular region.
m. Determine how much extra area Ron now has compared to his rectangular region.

21. In question 20, Ron the excavator operator could choose to enclose a rectangular or circular area with 150 m of barricade mesh. In this case, the circular region resulted in a larger safe work area.

a. Show that for 150 m of barricade mesh, a circular region again results in a larger safe work area as opposed to a rectangular region.
b. Show that for n metres of barricade mesh, a circular region will result in a larger safe work area as opposed to a rectangular region.

Problem solving

22. A vegetable gardener is going to build four new rectangular garden beds side by side. Each garden bed measures 12.5 metres long and 3.2 metres wide. To access the garden beds, the gardener requires a path 1 metre wide between each garden bed and around the outside of the beds.

a. Evaluate the total area the vegetable gardener would need for the garden beds and paths.
b. The garden beds need to be mulched. Bags of mulch, costing $29.50 each, cover an area of 25 square metres. Determine how many bags of mulch the gardener will need to purchase.
c. The path is to be resurfaced at a cost of $39.50 per 50 square metres. Evaluate the cost of resurfacing the path.
d. The gardener needs to spend a further $150 on plants. Determine the total cost of building these new garden beds and paths.

23. The diagram shows one smaller square drawn inside a larger square on grid paper.

a. Determine what fraction of the area of the larger square is the area of the smaller square.

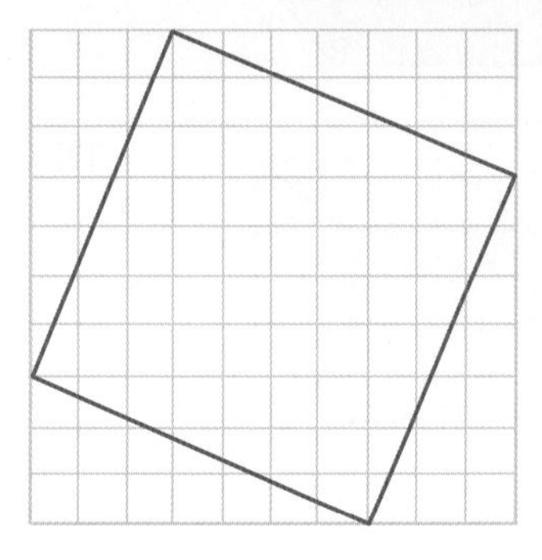

b. Another square with side lengths of 10 cm has a smaller square drawn inside. Determine the values of x and y if the smaller square is half the larger square.

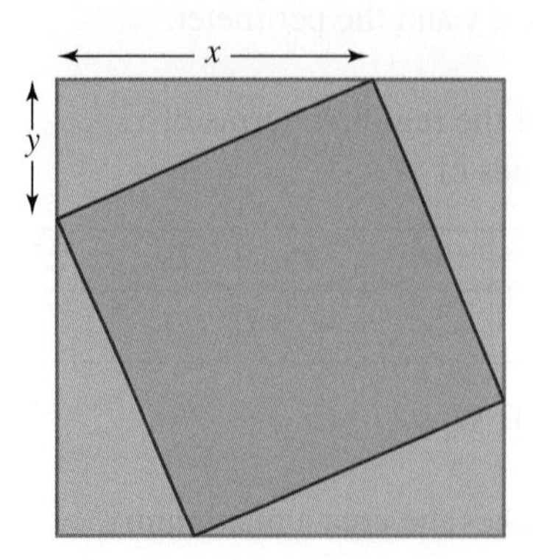

24. The shaded area in the diagram is called a segment of a circle.
A circle with a radius of 10 cm has ∠AOB equal to 90°.
A second circle, also with a radius of 10 cm, has ∠AOB equal to 120°.
Evaluate the difference in the areas of the segments of these two circles, correct to 2 decimal places.

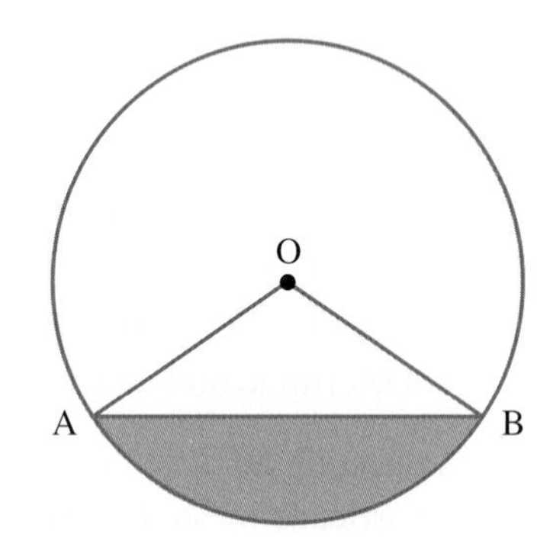

6.3 Total surface area

LEARNING INTENTION

At the end of this subtopic you should be able to:
- calculate the total surface area of rectangular prisms and pyramids
- calculate the total surface area of cylinders and spheres
- calculate the total surface area of cones

6.3.1 Total surface area of solids

eles-4811

- The **total surface area (TSA)** of a solid is the sum of the areas of all the faces of that solid.

TSA of rectangular prisms and cubes

Shape	Diagram	Formula
Rectangular prism (cuboid)	*h*, *w*, *l*	$\text{TSA} = 2(lh + lw + wh)$
Cube	*h*, *w*, *l*	$\text{TSA} = 6l^2$

TSA of spheres and cylinders

Shape	Diagram	Formula
Sphere		$\text{TSA} = 4\pi r^2$
Cylinder		$\text{TSA} = A_{\text{curved surface}} + A_{\text{circular ends}}$ $= 2\pi rh + 2\pi r^2$ $= 2\pi r(h + r)$

WORKED EXAMPLE 3 Calculating TSA of solids

Calculate the total surface area of the solids, correct to the nearest cm^2.

a.

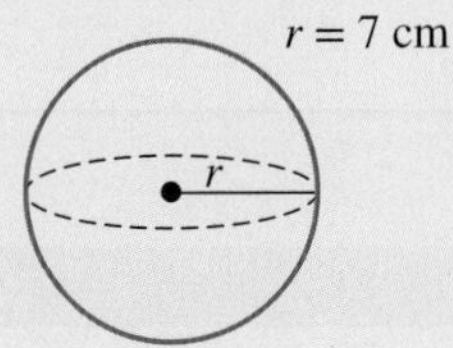

b.

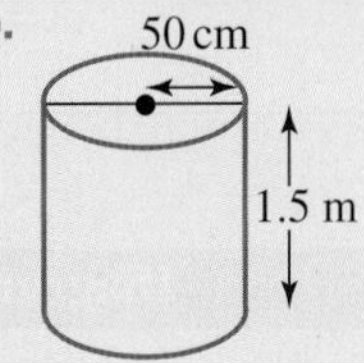

THINK	WRITE
a. 1. Write the formula for the TSA of a sphere.	a. $\text{TSA} = 4\pi r^2$
2. Identify the value for r.	$r = 7$
3. Substitute and evaluate.	$\text{TSA} = 4 \times \pi \times 7^2$ $\approx 615.8\,\text{cm}^2$
4. Write the answer to correct to the nearest cm^2.	$\approx 616\,\text{cm}^2$
b. 1. Write the formula for the TSA of a cylinder.	b. $\text{TSA} = 2\pi r(r + h)$
2. Identify the values for r and h. Note that the units will need to be the same.	$r = 50\,\text{cm},\ h = 1.5\,\text{m}$ $= 150\,\text{cm}$
3. Substitute and evaluate.	$\text{TSA} = 2 \times \pi \times 50 \times (50 + 150)$ $= 62\,831.9\,\text{cm}^2$
4. Write the answer to correct to the nearest cm^2.	$\approx 62\,832\,\text{cm}^2$

6.3.2 Total surface area of cones

eles-4812

- The total surface area of a cone can be found by considering its net, which is comprised of a small circle and a sector of a larger circle.

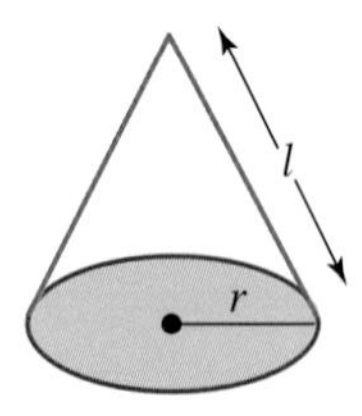

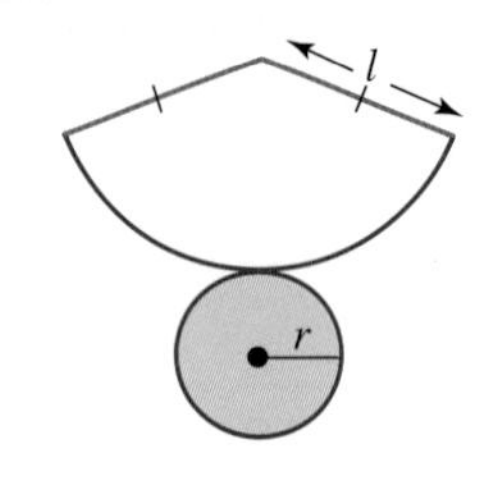

r = radius of the cone
l = slant height of the cone

- The sector is a fraction of the full circle of radius l with circumference $2\pi l$.
- The sector has an arc length equivalent to the circumference of the base of the cone, $2\pi r$.
- The fraction of the full circle represented by the sector can be found by writing the arc length as a fraction of the circumference of the full circle, $\frac{2\pi r}{2\pi l} = \frac{r}{l}$.

$$\begin{aligned}\text{Area of a sector} &= \text{fraction of the circle} \times \pi l^2 \\ &= \frac{r}{l} \times \pi l^2 \\ &= \pi r l\end{aligned}$$

Total surface area of a cone

Shape	Diagram	Formula
Cone	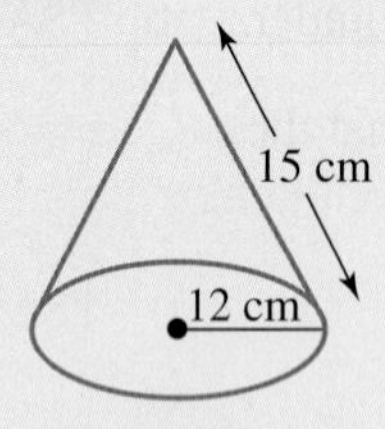	$\begin{aligned}\text{TSA} &= A_{\text{curved surface}} + A_{\text{circular end}} \\ &= \pi r^2 + \pi r l \\ &= \pi r(r + l)\end{aligned}$

WORKED EXAMPLE 4 Calculating the TSA of a cone

Calculate the total surface area of the cone shown.

THINK	WRITE
1. Write the formula for the TSA of a cone.	$\text{TSA} = \pi r(r + l)$
2. State the values of r and l.	$r = 12, l = 15$
3. Substitute and evaluate to obtain the answer.	$\text{TSA} = \pi \times 12 \times (12 + 15)$ $= 1017.9\ \text{cm}^2$

TI | THINK	DISPLAY/WRITE	CASIO | THINK	DISPLAY/WRITE
On the Calculator page, complete the entry line as: $\pi(r+s) \mid r=12$ and $s=15$ Press CTRL ENTER to get a decimal approximation.	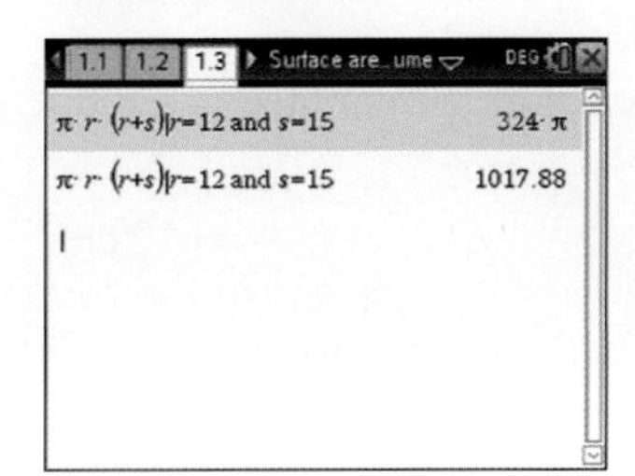 The total surface area of the cone is 1017.9 cm^2 correct to 1 decimal place.	On the Main screen in decimal mode, complete the entry lines as: $\pi r(r+s) \mid r=12 \mid s=15$ Then press EXE.	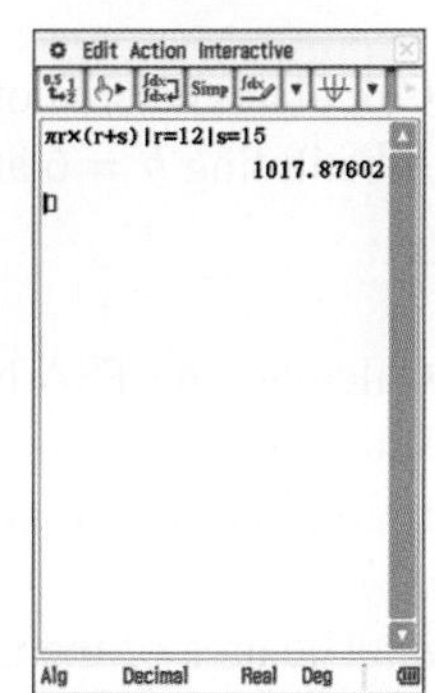 The total surface area of the cone is 1017.9 cm^2 correct to1 decimal place.

6.3.3 Total surface area of other solids

eles-4813

- TSA can be found by summing the areas of each face.
- Check the total number of faces to ensure that none are left out.

WORKED EXAMPLE 5 Calculating the TSA of a pyramid

Calculate the total surface area of the square-based pyramid shown.

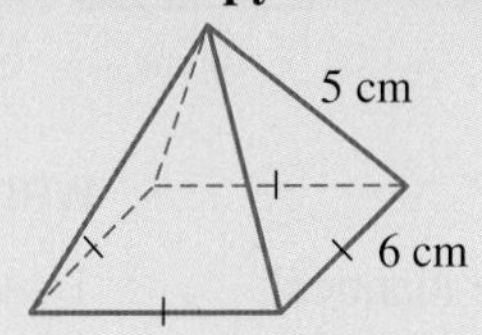

THINK	WRITE/DRAW
1. There are five faces: The square base and four identical triangles.	TSA = Area of square base + area of four triangular faces
2. Calculate the area of the square base.	Area of base $= l^2$, where $l = 6$ Area of base $= 6^2$ $= 36\text{ cm}^2$
3. Draw and label one triangular face and write the formula for determining its area.	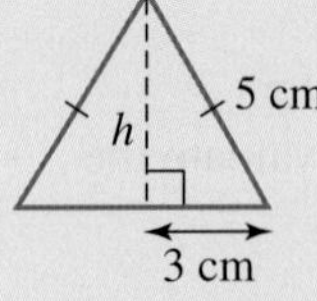 Area of a triangular face $= \frac{1}{2}bh$; $b = 6$
4. Calculate the height of the triangle, h, using Pythagoras' theorem.	$a^2 = c^2 - b^2$, where $a = h$, $b = 3$, $c = 5$ $h^2 = 5^2 - 3^2$ $h^2 = 25 - 9$ $h^2 = 16$ $h = 4$ cm

5. Calculate the area of the triangular face by substituting $b = 6$ and $h = 4$.

$$\text{Area of triangular face} = \frac{1}{2} \times 6 \times 4$$
$$= 12\text{ cm}^2$$

6. Calculate the TSA by adding the area of the square base and the area of four identical triangular faces together.

$$\text{TSA} = 36 + 4 \times 12$$
$$= 36 + 48$$
$$= 84\text{ cm}^2$$

Note: The area of the triangular faces can be found using Heron's formula. This method is demonstrated in the following worked example.

WORKED EXAMPLE 6 Calculating the TSA of a solid

Calculate the total surface area of the solid shown correct to 1 decimal place.

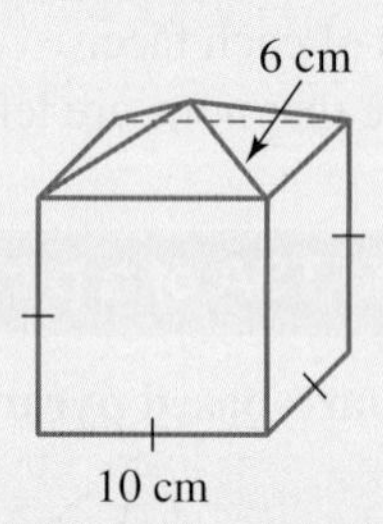

THINK

1. The solid shown has nine faces — five identical squares and four identical triangles.
2. Calculate the area of one square face with the side length 10 cm.
3. Draw a triangular face and label the three sides. Use Heron's formula to calculate the area.
4. State the formula for s, the semi-perimeter. Substitute the values of a, b and c and evaluate the value of s.
5. State Heron's formula for the area of one triangle. Substitute and evaluate.

WRITE/DRAW

$$\text{TSA} = 5 \times \text{area of a square} + 4 \times \text{area of a triangle}$$

$$A_{\text{square}} = l^2,\ \text{where } l = 10$$
$$A = 10^2$$
$$A = 100\text{ cm}^2$$

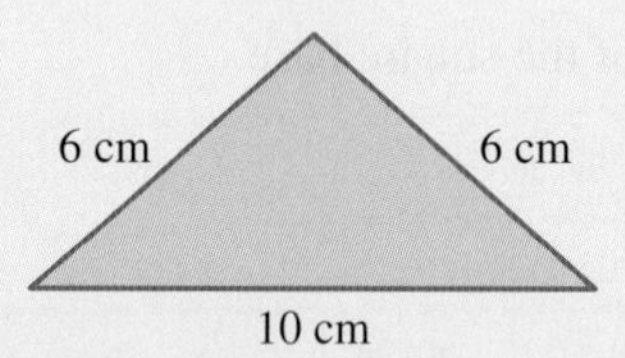

$$s = \frac{a+b+c}{2}$$
$$s = \frac{6+6+10}{2}$$
$$s = 11$$

$$A = \sqrt{s(s-a)(s-b)(s-c)}$$
$$A = \sqrt{11(11-6)(11-6)(11-10)}$$
$$A = \sqrt{275}$$
$$A = 16.583\,124\ldots\text{ cm}^2$$

6. Determine the TSA of the solid by adding the area of the five squares and four triangles.

$$\begin{aligned} \text{TSA} &= 5 \times 100 + 4 \times 16.583\,124\ldots \\ &= 566.3325\ldots \\ &= 566.3\ \text{cm}^2 \text{ (to 1 decimal place)} \end{aligned}$$

Note: Rounding is not done until the final step. It is important to realise that rounding too early can affect the accuracy of results.

WORKED EXAMPLE 7 Applying surface area in worded problems

The silo shown is to be built from metal. The top portion of the silo is a cylinder of diameter 4 m and height 8 m. The bottom part of the silo is a cone of slant height 3 m. The silo has a circular opening of radius 30 cm on the top.

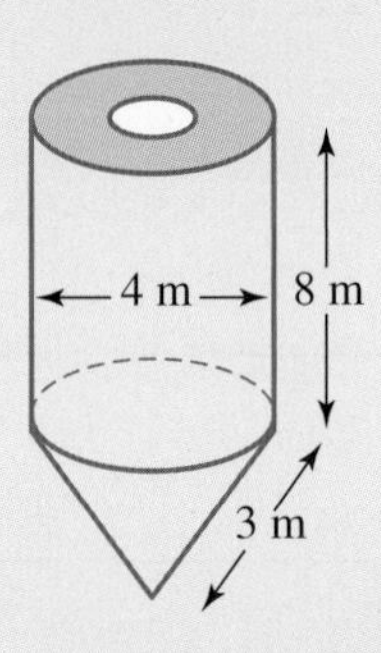

a. **Calculate the area of metal (to the nearest m^2) that is required to build the silo.**
b. **If it costs \$12.50 per m^2 to cover the surface with an anti-rust material, determine how much will it cost to cover the silo completely.**

THINK	WRITE
a. 1. The surface area of the silo consists of an annulus, the curved part of the cylinder and the curved section of the cone.	a. TSA = area of annulus + area of curved section of a cylinder + area of curved section of a cone
2. To calculate the area of the annulus, subtract the area of the small circle from the area of the larger circle. Let R = radius of small circle. Remember to convert all measurements to the same units.	Area of annulus $= A_{\text{large circle}} - A_{\text{small circle}}$ $= \pi r^2 - \pi R^2$ where $r = \frac{4}{2} = 2$ m and $R = 30$ cm $= 0.3$ m. Area of annulus $= \pi \times 2^2 - \pi \times 0.3^2$ $= 12.28$ m
3. The middle part of the silo is the curved part of a cylinder. Determine its area. (Note that in the formula $\text{TSA}_{\text{cylinder}} = 2\pi r^2 + 2\pi rh$, the curved part is represented by $2\pi rh$.)	Area of curved section of cylinder $= 2\pi rh$ where $r = 2$, $h = 8$. Area of curved section of cylinder $= 2 \times \pi \times 2 \times 8$ $= 100.53$ m^2
4. The bottom part of the silo is the curved section of a cone. Determine its area. (Note that in the formula $\text{TSA}_{\text{cone}} = \pi r^2 + \pi rl$, the curved part is given by πrl.)	Area of curved section of cone $= \pi rl$ where $r = 2$, $l = 3$. Area of curved section of cone $= \pi \times 2 \times 3$ $= 18.85$ m^2

5. Calculate the total surface area of the silo by finding the sum of the surface areas calculated above.	$\text{TSA} = 12.28 + 100.53 + 18.85$ $= 131.66\text{ m}^2$
6. Write the answer in words.	The area of metal required is 132 m^2, correct to the nearest square metre.
b. To determine the total cost, multiply the total surface area of the silo by the cost of the anti-rust material per m^2 (\$12.50).	**b.** $\text{Cost} = 132 \times \$12.50$ $= \$1650.00$

Resources

 **eWorkbook** Topic 6 workbook (worksheets, code puzzle and a project) (ewbk-2032)

 **Digital document** SkillSHEET Total surface area of cubes and rectangular prisms (doc-5238)

 **Video eLesson** Total surface area of prisms (eles-1909)

Interactivities Individual pathway interactivity: Total surface area (int-4594)
Surface area of a prism (int-6079)
Surface area of a cylinder (int-6080)
Surface area (int-6477)

Exercise 6.3 Total surface area

learnON

Individual pathways

■ PRACTISE	■ CONSOLIDATE	■ MASTER
1, 5, 7, 10, 12, 17, 20	2, 6, 9, 11, 14, 15, 18, 21	3, 4, 8, 13, 16, 19, 22

To answer questions online and to receive **immediate feedback** and **sample responses** for every question, go to your learnON title at www.jacplus.com.au.

Unless told otherwise, where appropriate, give answers correct to 1 decimal place.

Fluency

1. Calculate the total surface areas of the solids shown.

a. 10 cm

b.

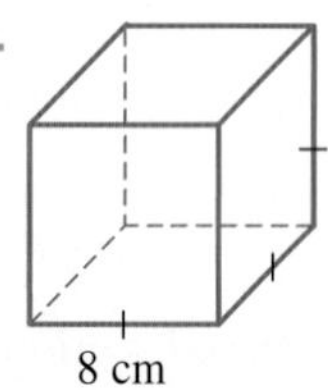

c.

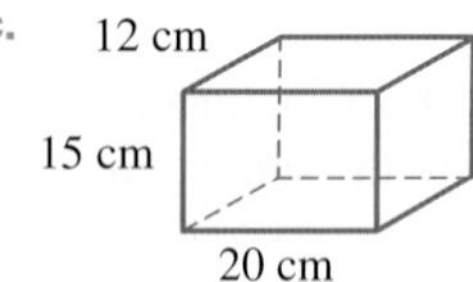

d.

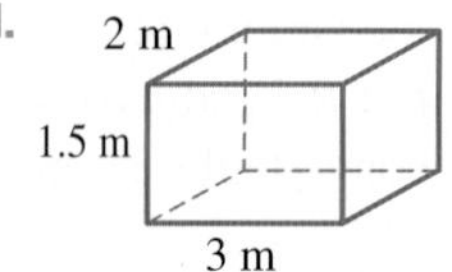

2. WE3 Calculate the total surface area of the solids shown below.

a. $r = 3$ m

b.

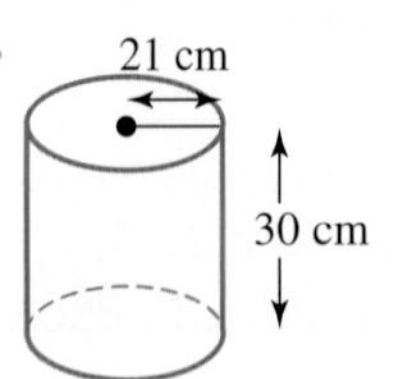

c.

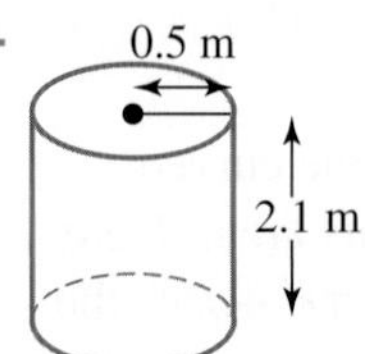

d. 12 cm

3. WE4 Calculate the total surface area of the cones below.

a.

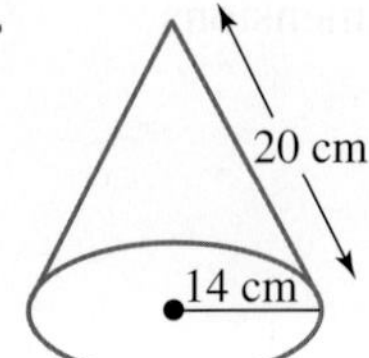

b.

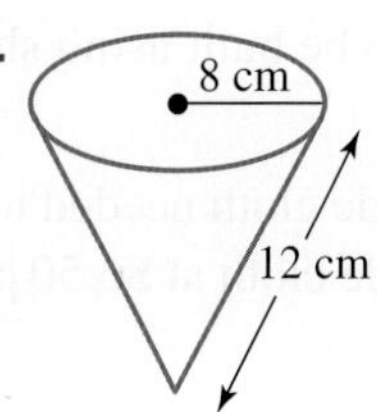

4. WE5 Calculate the total surface area of the solids below.

a.

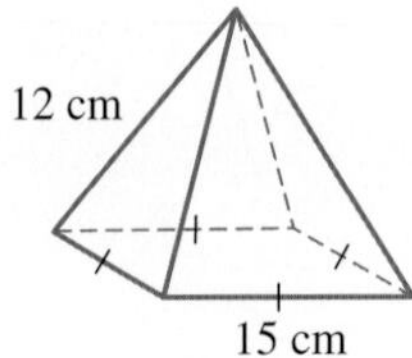

b.

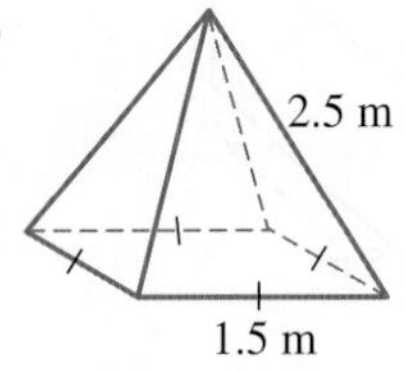

c.

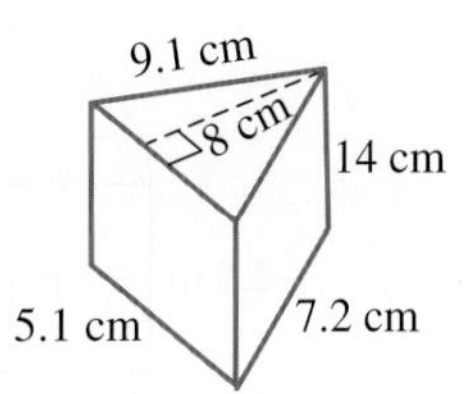

d.

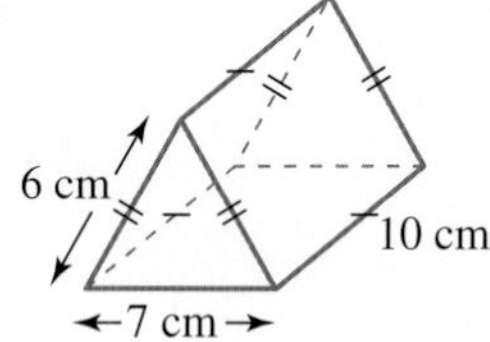

5. Calculate the surface areas of the following.
   a. A cube of side length 1.5 m
   b. A rectangular prism $6\,\text{m} \times 4\,\text{m} \times 2.1\,\text{m}$
   c. A cylinder of radius 30 cm and height 45 cm, open at one end
6. Calculate the surface areas of the following.
   a. A sphere of radius 28 mm
   b. An open cone of radius 4 cm and slant height 10 cm
   c. A square pyramid of base length 20 cm and slant edge 30 cm
7. WE6 Calculate the total surface area of the objects shown.

a.

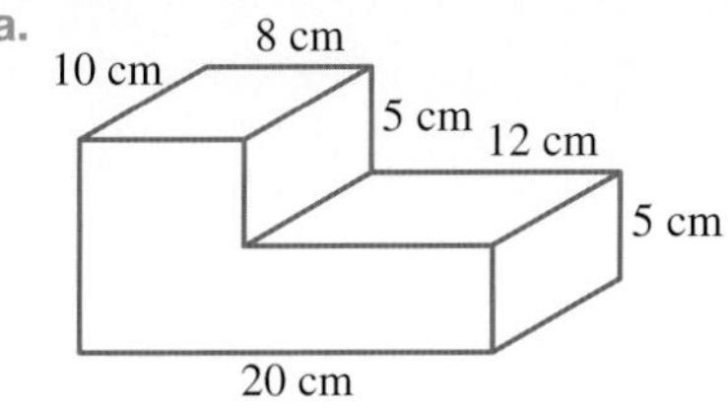

b.

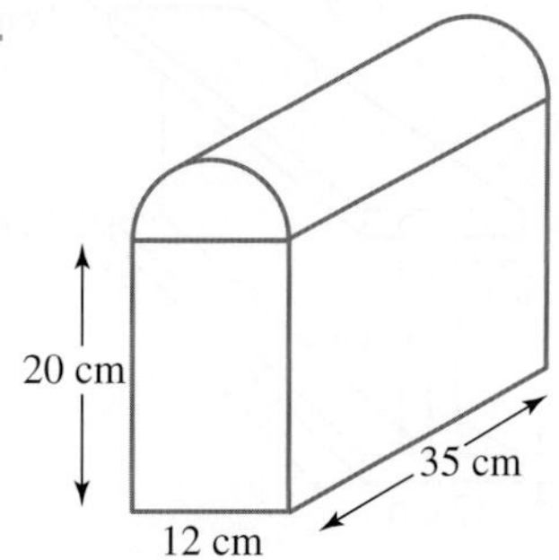

c.

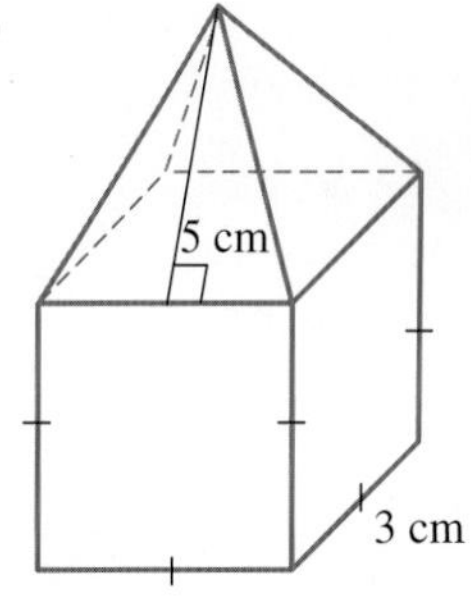

8. Calculate the total surface area of the objects shown.

a.

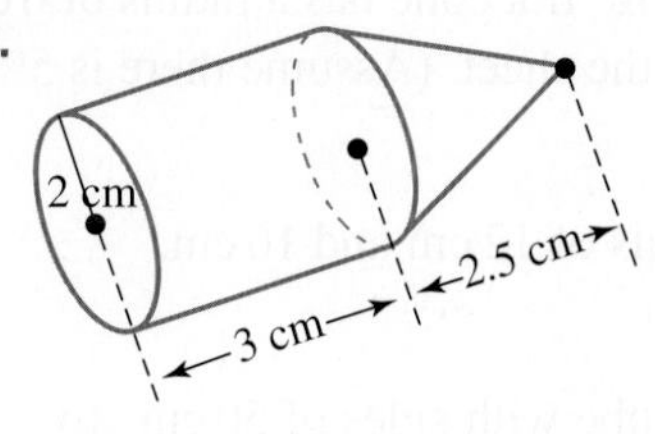

b.

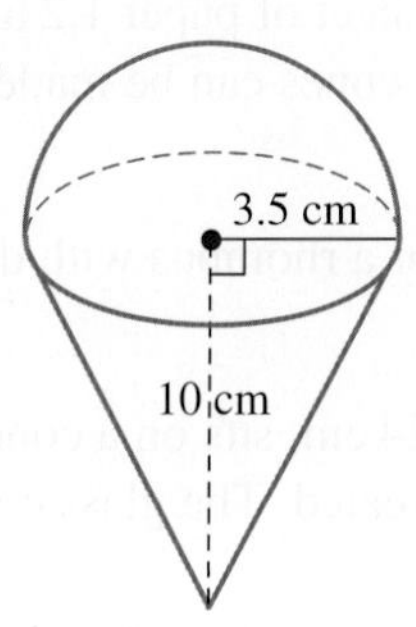

c.

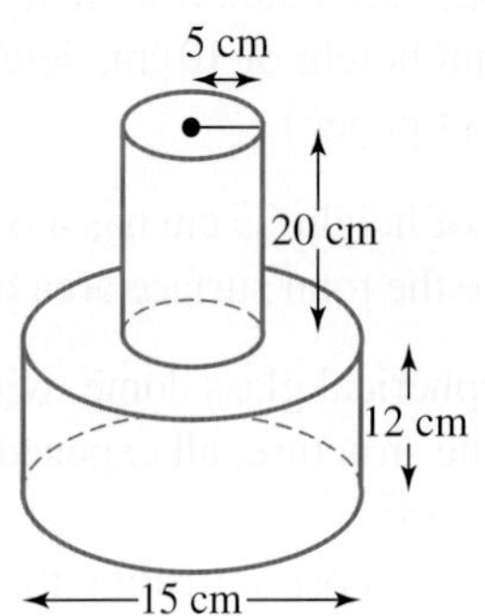

9. MC A cube has a total surface area of $384\,\text{cm}^2$. Calculate the length of the edge of the cube.

   A. 9 cm B. 8 cm C. 7 cm D. 6 cm E. 5 cm

Understanding

10. **WE7** The greenhouse shown is to be built using shade cloth. It has a wooden door of dimensions $1.2\,\text{m} \times 0.5\,\text{m}$.
   a. Calculate the total area of shade cloth needed to complete the greenhouse.
   b. Determine the cost of the shade cloth at \$6.50 per m^2.

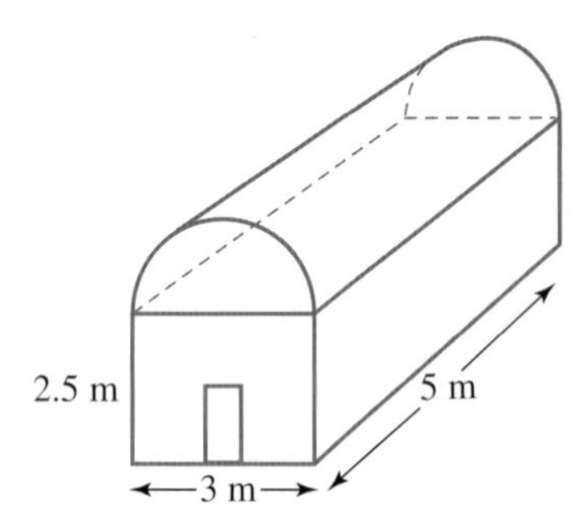

11. A cylinder is joined to a hemisphere to make a cake holder, as shown. The surface of the cake holder is to be chromed at 5.5 cents per cm^2.
   a. Calculate the total surface area to be chromed.
   b. Determine the cost of chroming the cake holder.

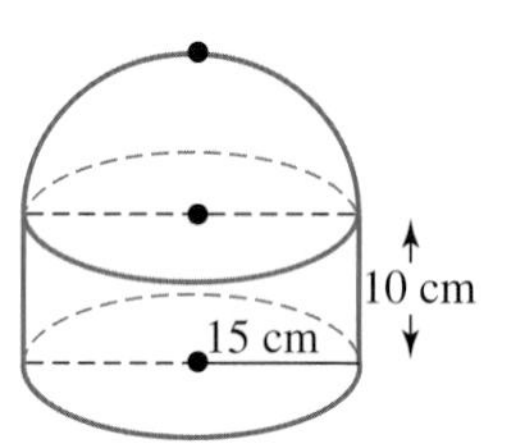

12. A steel girder is to be painted. Calculate the area of the surface to be painted.

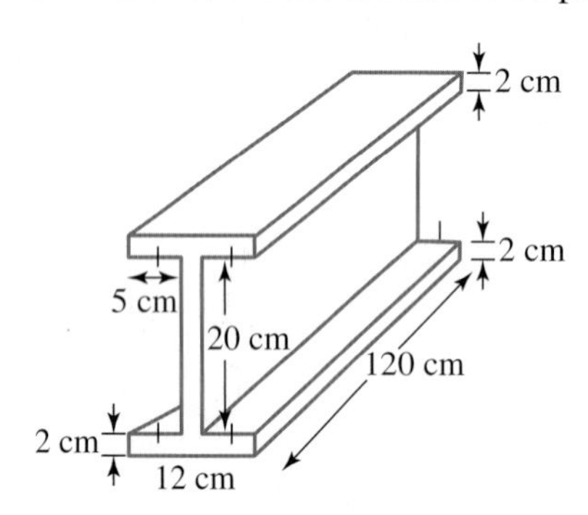

13. Open cones are made from nets cut from a large sheet of paper $1.2\,\text{m} \times 1.0\,\text{m}$. If a cone has a radius of 6 cm and a slant height of 10 cm, determine how many cones can be made from the sheet. (Assume there is 5% wastage of paper.)

14. A prism of height 25 cm has a base in the shape of a rhombus with diagonals of 12 cm and 16 cm. Calculate the total surface area of the prism.

15. A hemispherical glass dome, with a diameter of 24 cm, sits on a concrete cube with sides of 50 cm. To protect the structure, all exposed sides are to be treated. The glass costs \$1.50/cm^2 to treat and the concrete costs 5 c/cm^2.
   Calculate the cost in treating the structure if the base of the cube is already fixed to the ground. Give your answer to the nearest dollar.

16. An inverted cone with side length 4 metres is placed on top of a sphere such that the centre of the cone's base is 0.5 metres above the centre of the sphere. The radius of the sphere is $\sqrt{2}$ metres.

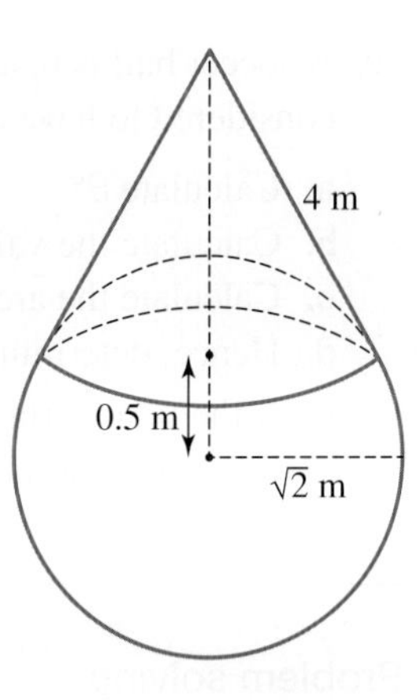

a. Calculate the exact total surface area of the sphere.

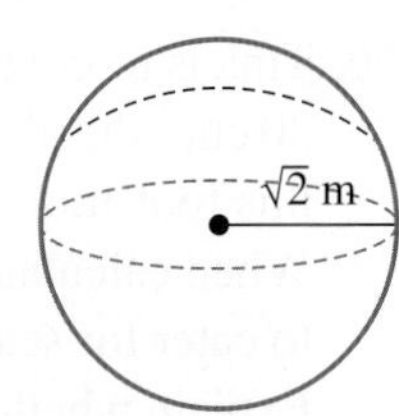

b. Calculate the radius of the cone exactly.
c. Calculate the area of the curved surface of the cone exactly.

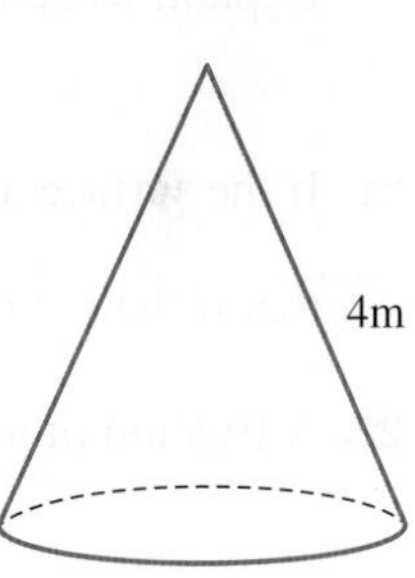

Reasoning

17. A shower recess with dimensions 1500 mm (back wall) by 900 mm (side wall) needs to have the back and two side walls tiled to a height of 2 m.
   a. Calculate the area to be tiled in m^2.
   b. Justify that 180 tiles (including those that need to be cut) of dimension 20 cm by 20 cm will be required. Disregard the grout and assume that once a tile is cut, only one piece of the tile can be used.
   c. Evaluate the cheapest option of tiling; \$1.50/tile or \$39.50/box, where a box covers 1 m^2, or tiles of dimension 30 cm by 30 cm costing \$3.50/tile.
18. The table shown below is to be varnished (including the base of each leg). The tabletop has a thickness of 180 mm and the cross-sectional dimensions of the legs are 50 mm by 50 mm.
A friend completes the calculation without a calculator as shown. Assume there are no simple calculating errors. Analyse the working presented and justify if the TSA calculated is correct.

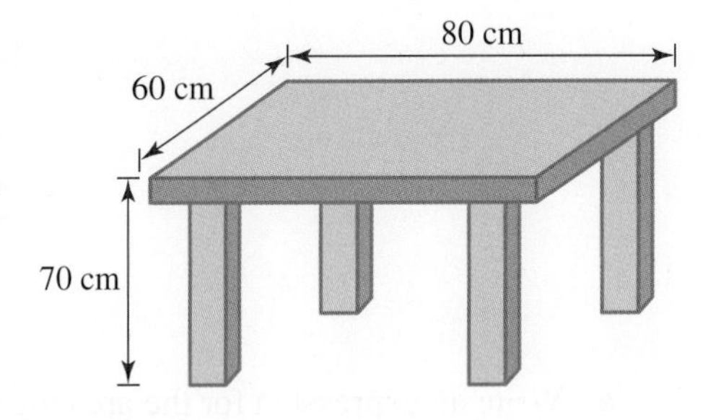

Tabletop (inc. leg bases)	0.96	$2 \times (0.8 \times 0.6)$
Legs	0.416	$16 \times (0.52 \times 0.05)$
Tabletop edging	0.504	$0.18 \times (2\,(0.8 + 0.6))$
TSA	1.88 m^2	

19. A soccer ball is made up of a number of hexagons sewn together on its surface. Each hexagon can be considered to have dimensions as shown in the diagram.

   a. Calculate $\theta°$.
   b. Calculate the values of x and y exactly.
   c. Calculate the area of the trapezium in the diagram.
   d. Hence, determine the area of the hexagon.
   e. If the total surface area of the soccer ball is $192\sqrt{3}\,\text{cm}^2$, determine how many hexagons are on its surface.

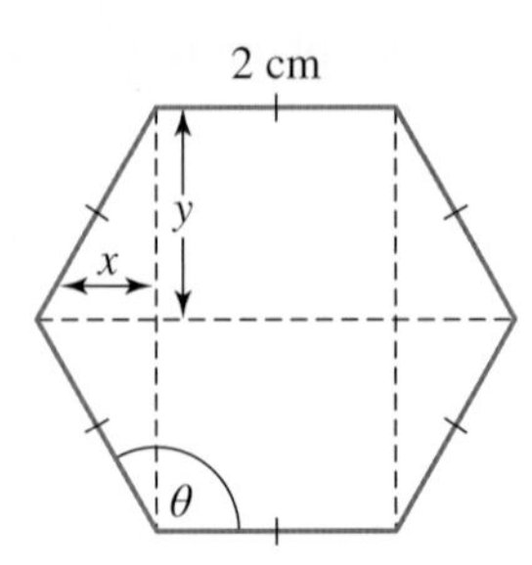

Problem solving

20. Tina is re-covering a footstool in the shape of a cylinder with diameter 50 cm and height 30 cm. She also intends to cover the base of the cushion. She has 1 m^2 of fabric to make this footstool.
When calculating the area of fabric required, allow an extra 20% of the total surface area to cater for seams and pattern placings.
Explain whether Tina has enough material to cover the footstool.

21. If the surface area of a sphere to that of a cylinder is in the ratio 4 : 3 and the sphere has a radius of $3a$, show that if the radius of the cylinder is equal to its height, then the radius of the cylinder is $\dfrac{3\sqrt{3}a}{2}$.

22. A frustum of a cone is a cone with the top sliced off, as shown.

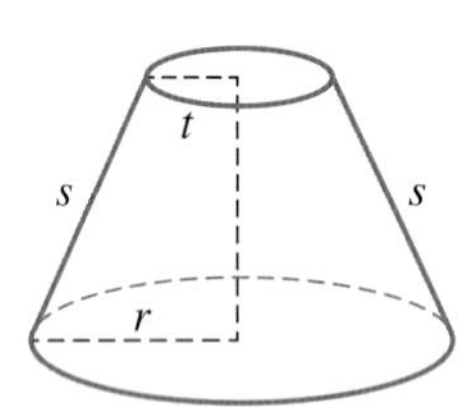

When the curved side is 'opened up', it creates a shape, ABYX, as shown in the diagram.

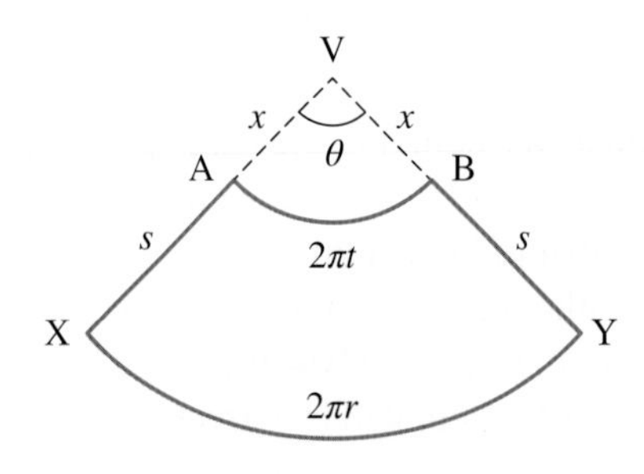

   a. Write an expression for the arc length XY in terms of the angle θ. Write another expression for the arc length AB in terms of the same angle θ. Show that, in radians, $\theta = \dfrac{2\pi(r-t)}{s}$.
   b. i. Using the above formula for θ, show that $x = \dfrac{st}{(r-t)}$.
      ii. Use similar triangles to confirm this formula.
   c. Determine the area of sectors AVB and XVY and hence determine the area of ABYX. Add the areas of the 2 circles to the area of ABYX to determine the TSA of a frustum.

6.4 Volume

LEARNING INTENTION

At the end of this subtopic you should be able to:
- calculate the volume of prisms, including cylinders
- calculate the volume of spheres
- calculate the volume of pyramids.

6.4.1 Volume

eles-4814

- The **volume** of a 3-dimensional object is the amount of space it takes up.
- Volume is measured in units of mm^3, cm^3 and m^3.
- The following diagram can be used to convert between units of volume.

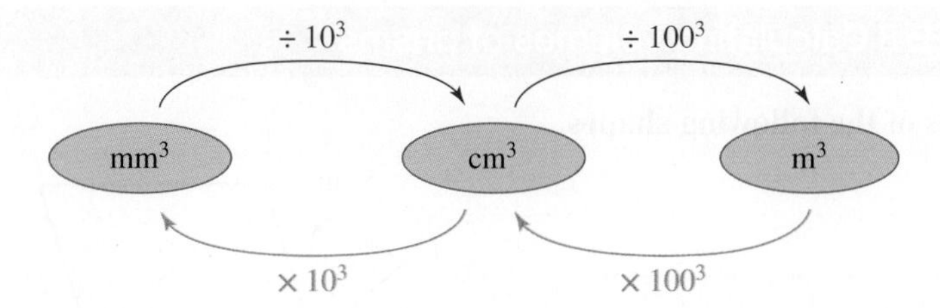

Volume of a prism

- The volume of any solid with a uniform cross-sectional area is given by the formula shown below.

Volume of a solid with uniform cross-sectional area

$$V = AH$$

where A is the area of the cross-section and H is the height of the solid.

Shape	Diagram	Formula
Cube		Volume $= AH$ $=$ area of a square $\times$ height $= l^2 \times l$ $= l^3$
Rectangular prism		Volume $= AH$ $=$ area of a rectangle $\times$ height $= lwh$

(continued)

Shape	Diagram	Formula
Cylinder		Volume $= AH$ $=$ area of a circle $\times$ height $= \pi r^2 h$
Triangular prism		Volume $= AH$ $=$ area of a triangle $\times$ height $= \frac{1}{2}bh \times H$

WORKED EXAMPLE 8 Calculating volumes of prisms

Calculate the volumes of the following shapes.

a.

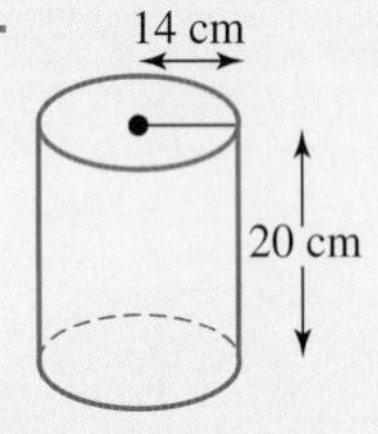

b.

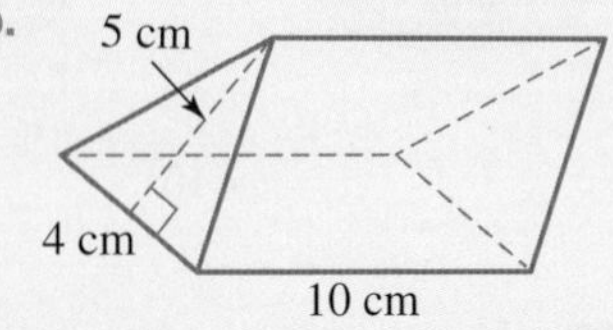

THINK

a. 1. Write the formula for the volume of the cylinder (prism).

2. Identify the value of the pronumerals.

3. Substitute and evaluate the answer.

b. 1. Write the formula for the volume of a triangular prism.

2. Identify the value of the pronumerals. (*Note: h* is the height of the triangle and *H* is the depth of the prism.)

3. Substitute and evaluate the answer.

WRITE

a. $V = AH$
$= \pi r^2 h$

$r = 14,\ h = 20$

$V = \pi \times 14^2 \times 20$
$\approx 12\,315.04\ \text{cm}^3$

b. $V = \frac{1}{2}bh \times H$

$b = 4,\ h = 5,\ H = 10$

$V = \frac{1}{2} \times 4 \times 5 \times 10$
$= 100\ \text{cm}^3$

WORKED EXAMPLE 9 Changing the dimensions of a prism

a. If each of the side lengths of a cube are doubled, then determine the effect on its volume.

b. If the radius is halved and the height of a cylinder is doubled, then determine the effect on its volume.

THINK	WRITE
a. 1. Write the formula for the volume of the cube.	**a.** $V = l^3$
2. Identify the value of the pronumeral. *Note:* Doubling is the same as multiplying by 2.	$l_{new} = 2l$
3. Substitute and evaluate.	$V_{new} = (2l)^3$
4. Compare the answer obtained in step 3 with the volume of the original shape.	$= 8l^3$
5. Write your answer.	Doubling each side length of a cube increases the volume by a factor of 8; that is, the new volume will be 8 times as large as the original volume.
b. 1. Write the formula for the volume of the cylinder.	**b.** $V = \pi r^2 h$
2. Identify the value of the pronumerals. *Note:* Halving is the same as dividing by 2.	$r_{new} = \frac{r}{2},\ h_{new} = 2h$
3. Substitute and evaluate.	$V_{new} = \pi\left(\frac{r}{2}\right)^2 2h$ $= \pi \times \frac{r^2}{\cancel{4}_2} \times \cancel{2}h$ $= \frac{\pi r^2 h}{2}$
4. Compare the answer obtained in step 3 with the volume of the original shape.	$= \frac{1}{2}\pi r^2 h$
5. Write your answer.	Halving the radius and doubling the height of a cylinder decreases the volume by a factor of 2; that is, the new volume will be half the original volume.

6.4.2 Volumes of common shapes

eles-4815

Volume of a sphere

- The volume of a sphere of radius r is given by the following formula.

Volume of a sphere

Shape	Diagram	Formula
Sphere	r	$V = \frac{4}{3}\pi r^3$

WORKED EXAMPLE 10 Calculating the volume of a sphere

Find the volume of a sphere of radius 9 cm. Answer correct to 1 decimal place.

THINK	WRITE
1. Write the formula for the volume of a sphere.	$V = \frac{4}{3}\pi r^3$
2. Identify the value of r.	$r = 9$
3. Substitute and evaluate.	$V = \frac{4}{3} \times \pi \times 9^3$ $= 3053.6\ \text{cm}^3$

Volume of a pyramid

- Pyramids are not prisms, as the cross-section changes from the base upwards.
- The volume of a pyramid is one-third the volume of the prism with the same base and height.

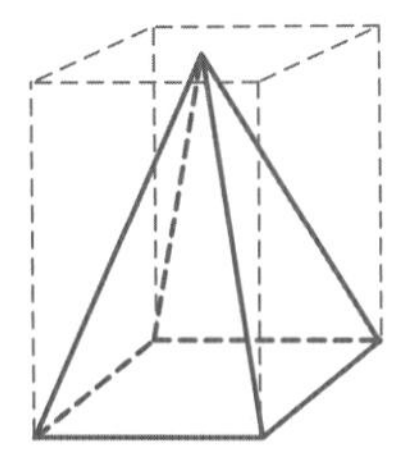

Volume of a pyramid

Shape	Diagram	Formula
Pyramid	H Area of base = A Base	$V_{\text{pyramid}} = \frac{1}{3}AH$

Volume of a cone

- The cone is a pyramid with a circular base.

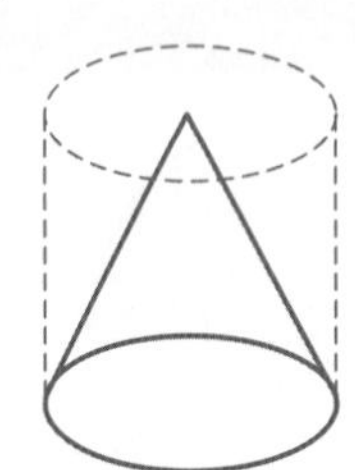

Volume of a cone

Shape	Diagram	Formula
Cone		$V_{\text{cone}} = \frac{1}{3}\pi r^2 h$

WORKED EXAMPLE 11 Calculating the volume of pyramids and cones

Calculate the volume of each of the following solids.

a.

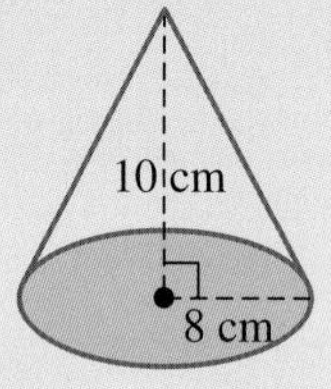

b.

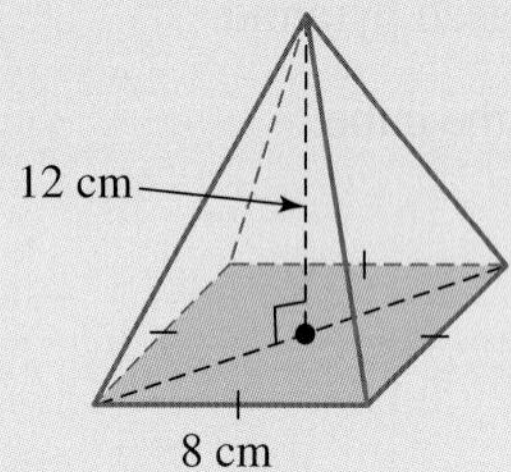

THINK	WRITE
a. 1. Write the formula for the volume of a cone.	a. $V = \frac{1}{3}\pi r^2 h$
2. Identify the values of r and h.	$r = 8,\ h = 10$
3. Substitute and evaluate.	$V = \frac{1}{3} \times \pi \times 8^2 \times 10$ $= 670.21\ \text{cm}^3$
b. 1. Write the formula for the volume of a pyramid.	b. $V = \frac{1}{3}AH$
2. Calculate the area of the square base.	$A = l^2$ where $l = 8$ $A = 8^2$ $= 64\ \text{cm}^2$
3. Identify the value of H.	$H = 12$
4. Substitute and evaluate.	$V = \frac{1}{3} \times 64 \times 12$ $= 256\ \text{cm}^3$

6.4.3 Volume of composite solids

eles-4816

- A composite solid is a combination of a number of solids.
- Calculate the volume of each solid separately.
- Sum these volumes to give the volume of the composite solid.

WORKED EXAMPLE 12 Calculating the volume of a composite solid

Calculate the volume of the composite solid shown.

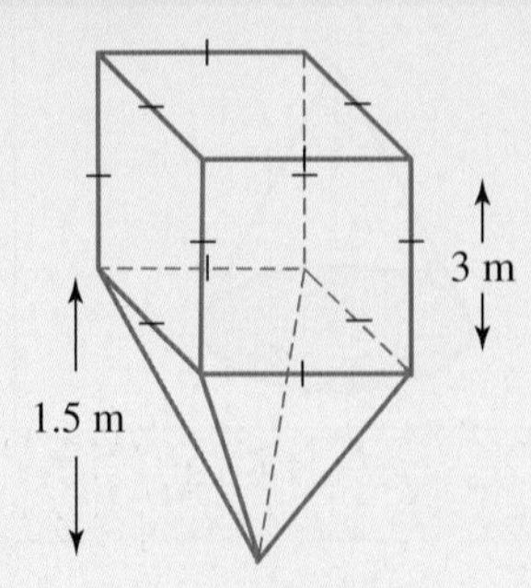

THINK	WRITE
1. The given solid is a composite figure, made up of a cube and a square-based pyramid.	$V =$ Volume of cube + Volume of pyramid
2. Calculate the volume of the cube.	$V_{cube} = l^3$ where $l = 3$ $V_{cube} = 3^3$ $= 27\text{ m}^3$
3. Write the formula for the volume of a square-based pyramid.	$V_{\text{square-based pyramid}} = \frac{1}{3}AH$
4. Calculate the area of the square base.	$A = l^2$ $= 3^2$ $= 9\text{ m}^2$
5. Identify the value of H.	$H = 1.5$
6. Substitute and evaluate the volume of the pyramid.	$V_{\text{square-based pyramid}} = \frac{1}{3} \times 9 \times 1.5$ $= 4.5\text{ m}^3$
7. Calculate the total volume by adding the volume of the cube and pyramid.	$V = 27 + 4.5$ $= 31.5\text{ m}^3$

6.4.4 Capacity

eles-4817

- Some 3-dimensional objects are hollow and can be filled with liquid or some other substance.
- The amount of substance that a container can hold is called its capacity.
- **Capacity** is essentially the same as volume but is usually measured in mL, L, kL and ML (megalitres) where $1\text{ mL} = 1\text{ cm}^3$

$$1\text{ L} = 1000\text{ cm}^3$$
$$1\text{ kL} = 1\text{ m}^3.$$

- The following diagram can be used to convert between units of capacity.

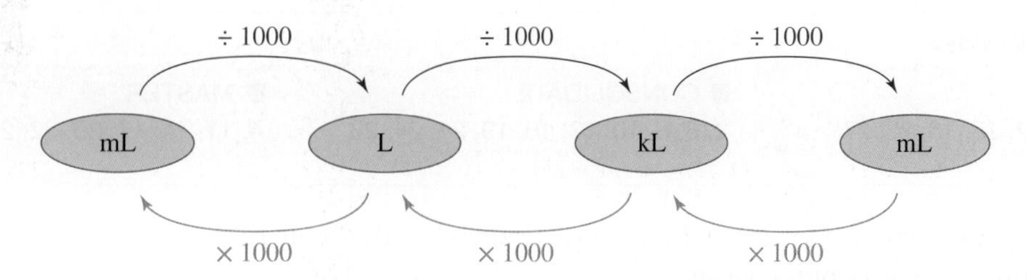

WORKED EXAMPLE 13 Calculating the capacity of a prism

Determine the capacity (in litres) of a cuboidal aquarium that is 50 cm long, 30 cm wide and 40 cm high.

THINK	WRITE
1. Write the formula for the volume of a rectangular prism.	$V = lwh$
2. Identify the values of the pronumerals.	$l = 50,\ w = 30,\ h = 40$
3. Substitute and evaluate.	$V = 50 \times 30 \times 40$ $= 60\,000\text{ cm}^3$
4. State the capacity of the container in millilitres, using $1\text{ cm}^3 = 1\text{ mL}$.	$= 60\,000\text{ mL}$
5. Since $1\text{ L} = 1000\text{ mL}$, to convert millilitres to litres divide by 1000.	$= 60\text{ L}$
6. Write the answer in a sentence.	The capacity of the fish tank is 60 L.

on Resources

 **eWorkbook** Topic 6 workbook (worksheets, code puzzle and a project) (ewbk-2032)

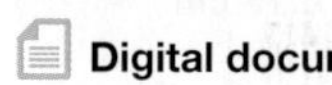 **Digital documents** SkillSHEET Conversion of volume units (doc-5239)
SkillSHEET Volume of cubes and rectangular prisms (doc-5240)

 **Interactivities** Individual pathway interactivity: Volume (int-4595)
Volume 1 (int-3791)
Volume 2 (int-6476)
Volume of solids (int-3794)

Exercise 6.4 Volume

learnon

Individual pathways

■ PRACTISE	■ CONSOLIDATE	■ MASTER
1, 3, 5, 7, 9, 12, 15, 18, 23, 27	2, 6, 8, 10, 13, 16, 19, 21, 24, 28	4, 11, 14, 17, 20, 22, 25, 26, 29, 30

To answer questions online and to receive **immediate feedback** and **sample responses** for every question, go to your learnON title at www.jacplus.com.au.

Fluency

1. Calculate the volumes of the following prisms.

a. 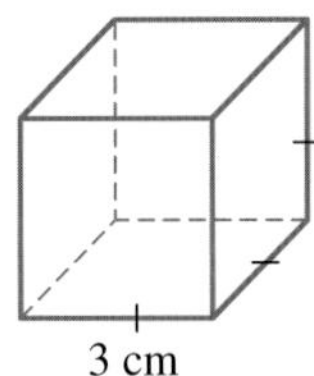

b. 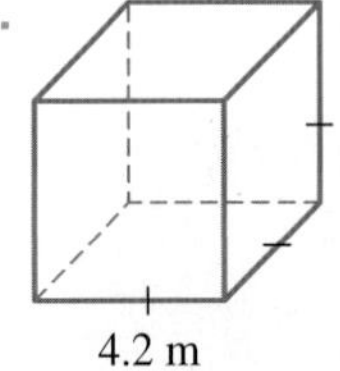

c. 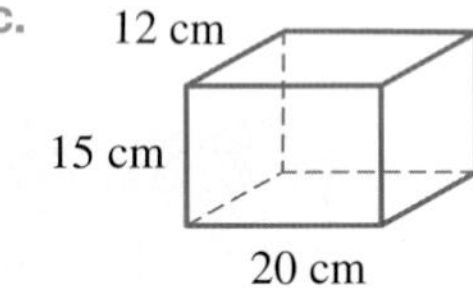

d. 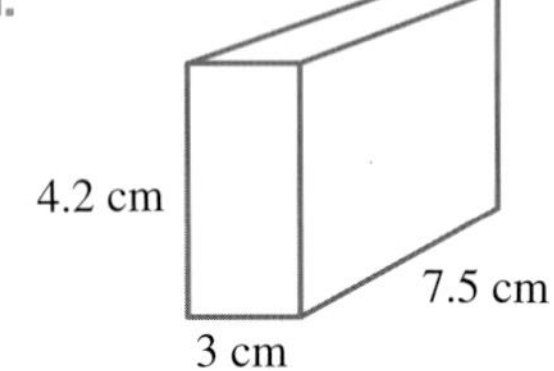

2. Calculate the volume of each of these solids.

a. 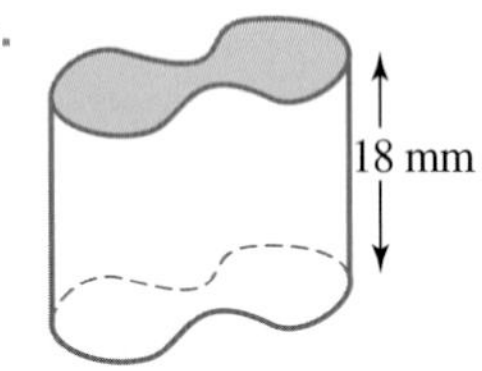

[Base area: 25 mm^2]

b. 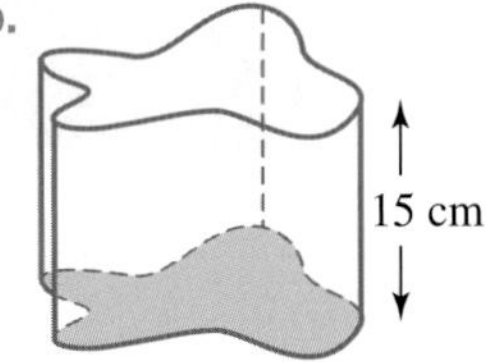

[Base area: 24 cm^2]

3. **WE8** Calculate the volume of each of the following. Give each answer correct to 1 decimal place where appropriate.

a. 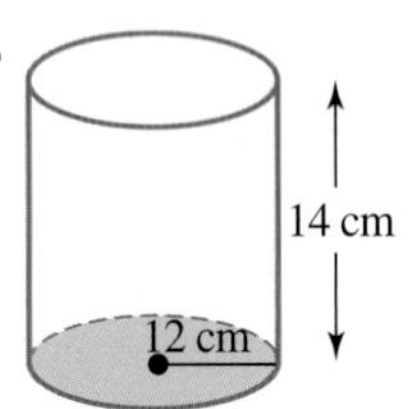

b. 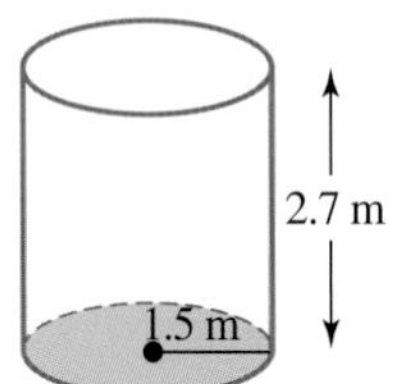

c. 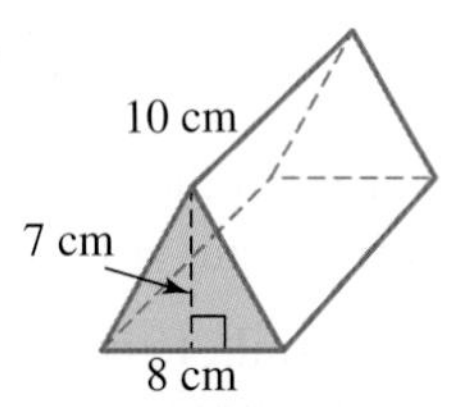

4. Calculate the volume of each of the following. Give each answer correct to 1 decimal place where appropriate.

a. 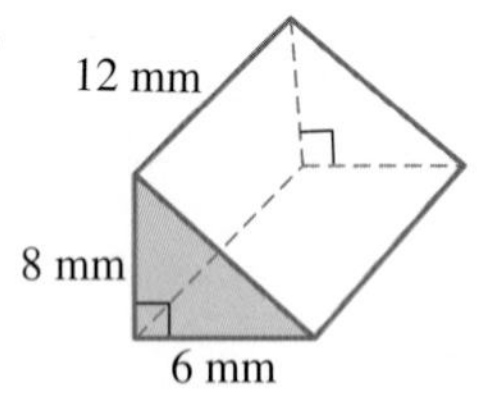

b. 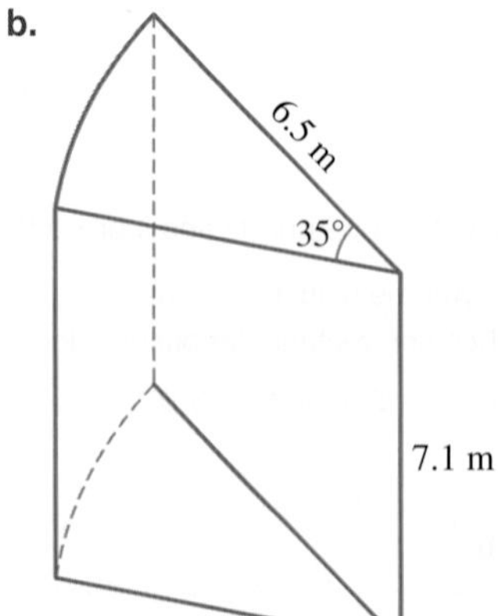

c. 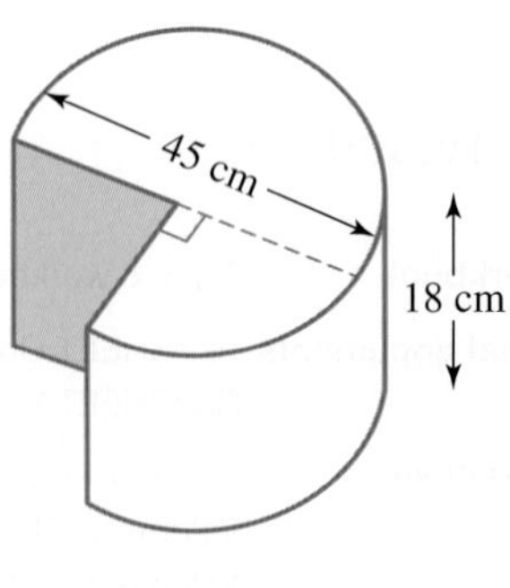

5. **WE10** Determine the volume of a sphere (correct to 1 decimal place) with a radius of:

a. 1.2 m
b. 15 cm
c. 7 mm
d. 50 cm

6. Calculate the volume of each of these figures, correct to 2 decimal places.

a.

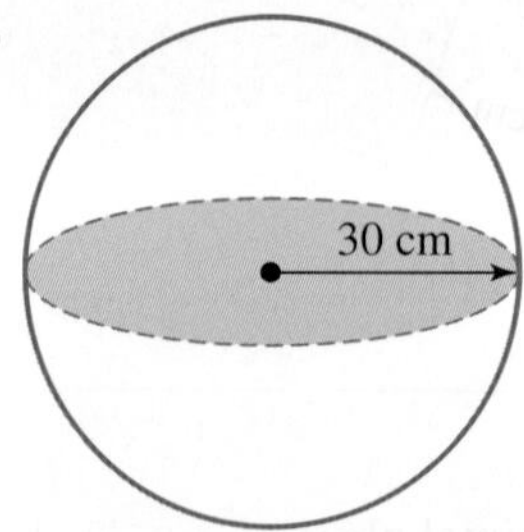

b.

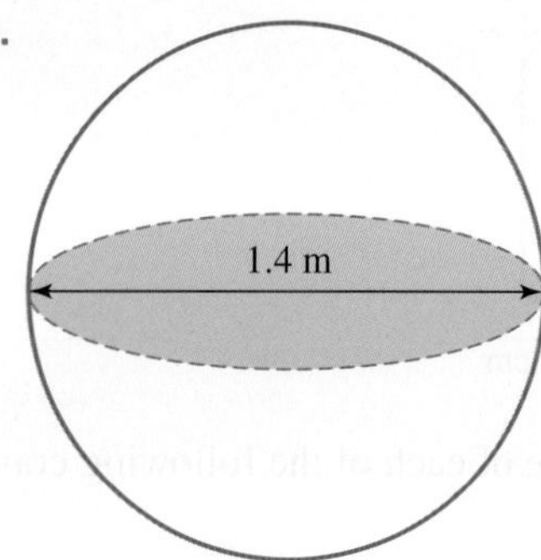

c.

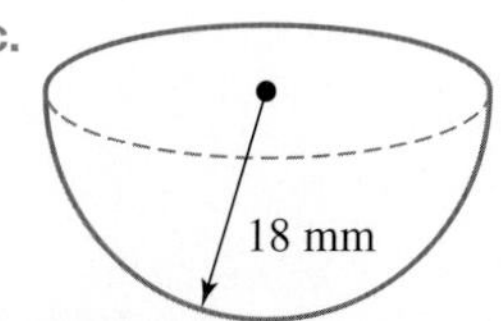

d.

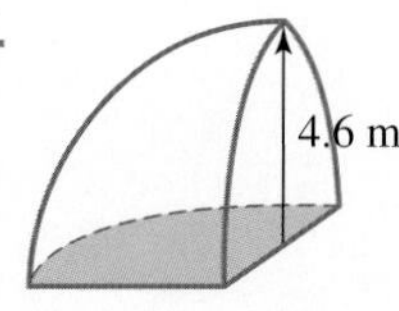

7. **WE11a** Determine the volume of each of the following cones, correct to 1 decimal place.

a.

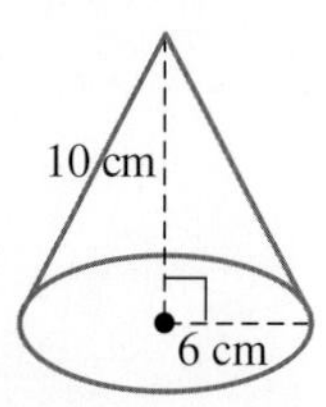

b.

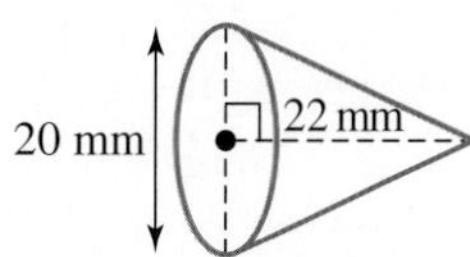

8. **WE11b** Calculate the volume of each of the following pyramids.

a.

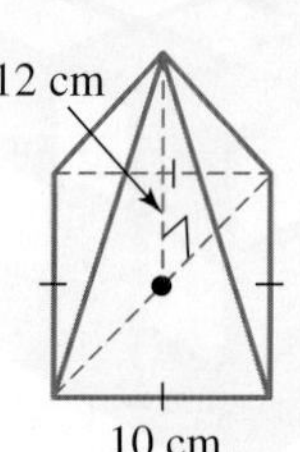

b.

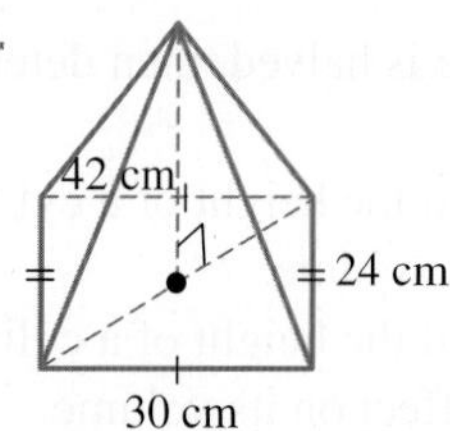

9. **WE12** Calculate the volume of each of the following composite solids correct to 2 decimal places where appropriate.

a.

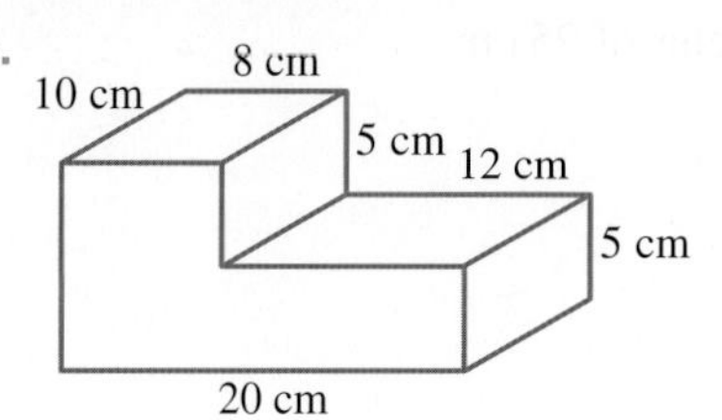

b.

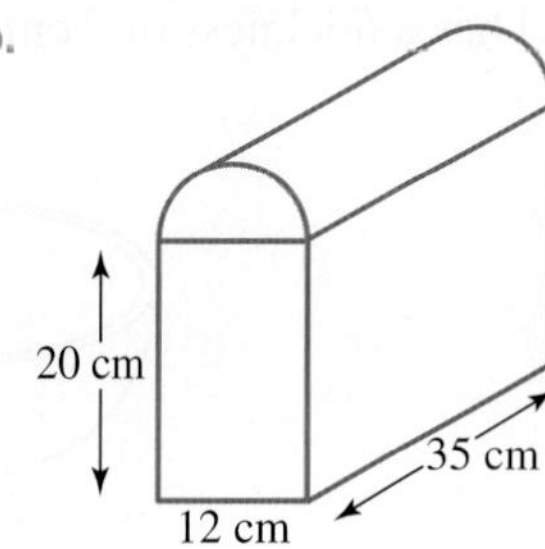

10. Calculate the volume of each of the following composite solids correct to 2 decimal places where appropriate.

a.

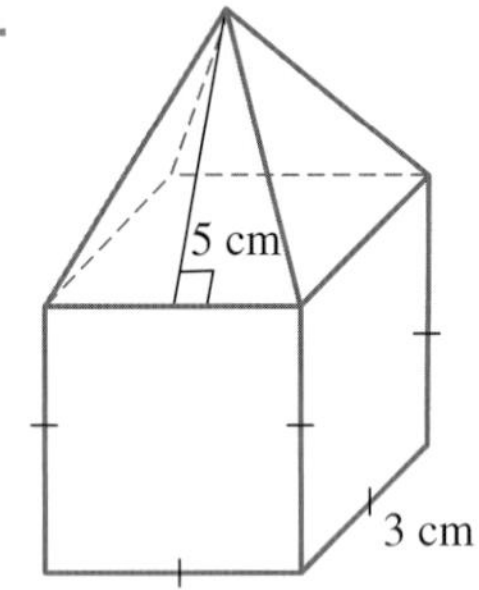

b.

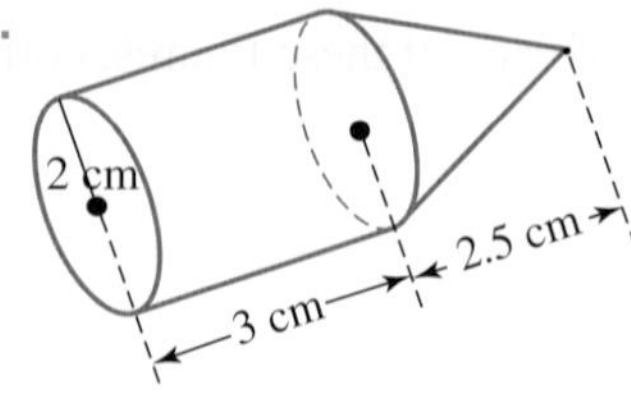

11. Calculate the volume of each of the following composite solids correct to 2 decimal places where appropriate.

a.

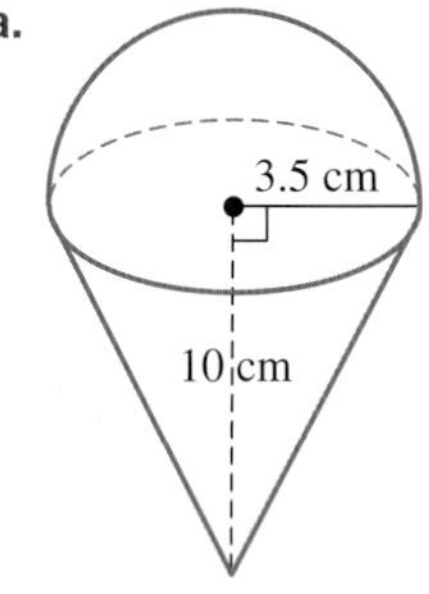

b.

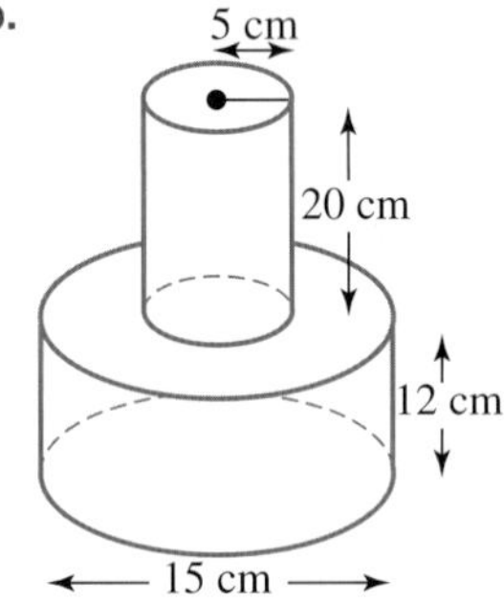

Understanding

12. WE9 Answer the following questions.

a. If the side length of a cube is tripled, then determine the effect on its volume.
b. If the side length of a cube is halved, then determine the effect on its volume.
c. If the radius is doubled and the height of a cylinder is halved, then determine the effect on its volume.
d. If the radius is doubled and the height of a cylinder is divided by four, then determine the effect on its volume.
e. If the length is doubled, the width is halved and the height of a rectangular prism is tripled, then determine the effect on its volume.

13. MC A hemispherical bowl has a thickness of 2 cm and an outer diameter of 25 cm.

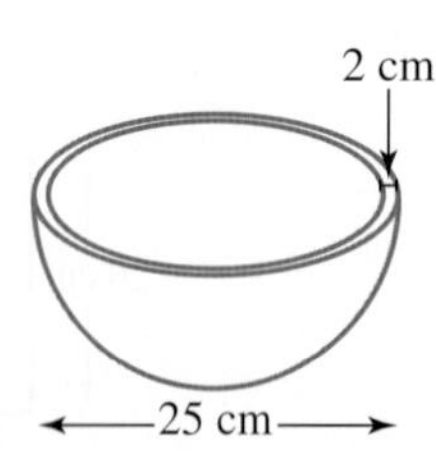

If the bowl is filled with water, the capacity of the water will be closest to:

A. 1.526 L B. 1.30833 L C. 3.05208 L D. 2.61666 L E. 2.42452 L

14. Tennis balls of diameter 8 cm are packed in a box 40 cm × 32 cm × 10 cm, as shown. Determine, correct to 2 decimal places, how much space is left unfilled.

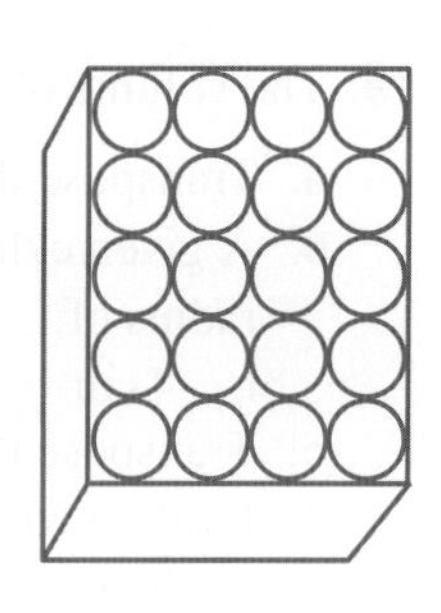

15. **WE13** A cylindrical water tank has a diameter of 1.5 m and a height of 2.5 m. Determine the capacity (in litres) of the tank, correct to 1 decimal place.

16. A monument in the shape of a rectangular pyramid (base length of 10 cm, base width of 6 cm, height of 8 cm), a spherical glass ball (diameter of 17 cm) and conical glassware (radius of 14 cm, height of 10 cm) are packed in a rectangular prism of dimensions 30 cm by 25 cm by 20 cm. The extra space in the box is filled up by a packing material. Determine, correct to 2 decimal places, the volume of packing material that is required.

17. A swimming pool is being constructed so that it is the upper part of an inverted square-based pyramid.

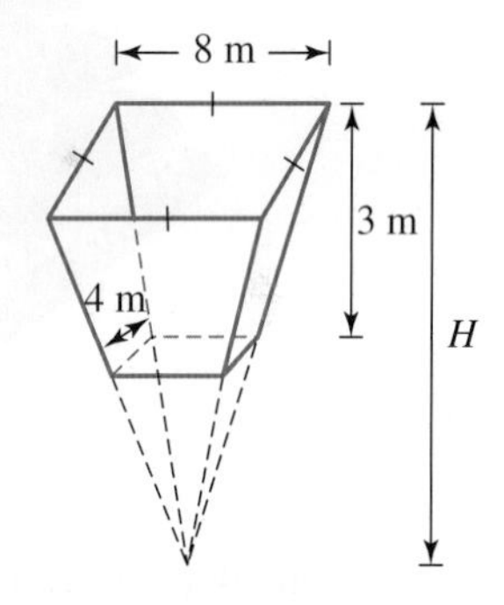

a. Calculate H.
b. Calculate the volume of the pool.
c. Determine how many 6 m^3 bins will be required to take the dirt away.
d. Determine how many litres of water are required to fill this pool.
e. Determine how deep the pool is when it is half-filled.

18. A soft drink manufacturer is looking to repackage cans of soft drink to minimise the cost of packaging while keeping the volume constant. Consider a can of soft drink with a capacity of 400 mL.

a. If the soft drink was packaged in a spherical can:
   i. calculate the radius of the sphere, correct to 2 decimal places
   ii. determine the total surface area of this can, correct to 1 decimal place.

b. If the soft drink was packaged in a cylindrical can with a radius of 3 cm:
   i. calculate the height of the cylinder, correct to 2 decimal places
   ii. determine the total surface area of this can, correct to 2 decimal places.

c. If the soft drink was packaged in a square-based pyramid with a base side length of 6 cm:
   i. calculate the height of the pyramid, correct to 2 decimal places
   ii. determine the total surface area of this can, correct to 2 decimal places.

d. Explain which can you would recommend the soft drink manufacturer use for its repackaging.

19. The volume of a cylinder is given by the formula $V = \pi r^2 h$.
   a. Transpose the formula to make h the subject.
   b. A given cylinder has a volume of 1600 cm^3. Calculate its height, correct to 1 decimal place, if it has a radius of:
      i. 4 cm
      ii. 8 cm.
   c. Transpose the formula to make r the subject.
   d. Explain what restrictions must be placed on r.
   e. A given cylinder has a volume of 1800 cm^3. Determine its radius, correct to 1 decimal place, if it has a height of:
      i. 10 cm
      ii. 15 cm.

20. A toy maker has enough rubber to make one super-ball of radius 30 cm. Determine how many balls of radius 3 cm he can make from this rubber.

21. A manufacturer plans to make a cylindrical water tank to hold 2000 L of water.
   a. Calculate the height, correct to 2 decimal places, if he uses a radius of 500 cm.
   b. Calculate the radius, correct to 2 decimal places if he uses a height of 500 cm.
   c. Determine the surface area of each of the two tanks. Assume the tank is a closed cylinder and give your answer in square metres correct to 2 decimal places.

22. The ancient Egyptians knew that the volume of the frustum of a square-based pyramid was given by the formula $V = \frac{1}{3}h\left(x^2 + xy + y^2\right)$, although how they discovered this is unclear. (A frustum is the part of a cone or pyramid that is left when the top is cut off.)

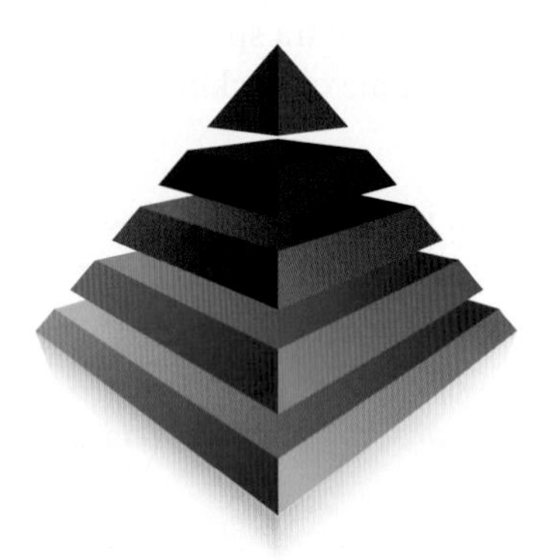

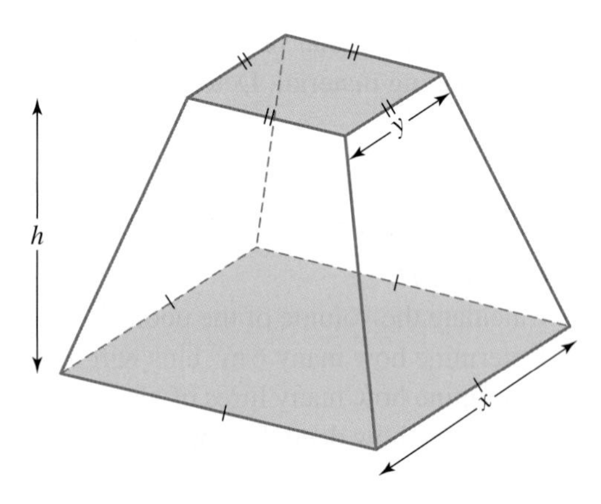

   a. Calculate the volume of the frustum below, correct to 2 decimal places.
   b. Determine the volume of the missing portion of the square-based pyramid shown, correct to 2 decimal places.

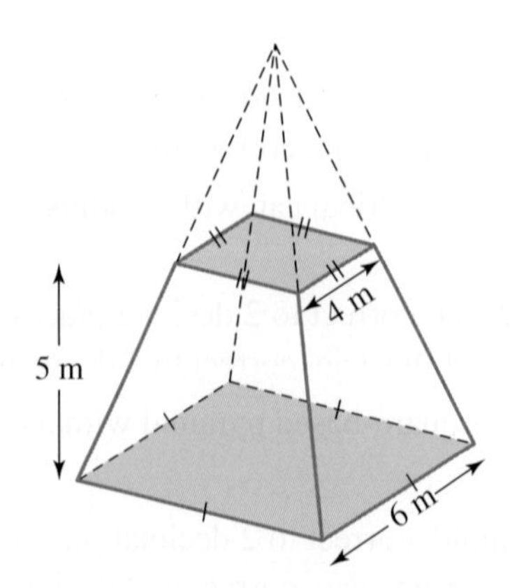

Reasoning

23. The Hastings' family house has a rectangular roof with dimensions $17\text{ m} \times 10\text{ m}$ providing water to three cylindrical water tanks, each with a radius of 1.25 m and a height of 2.1 m. Show that approximately 182 millimetres of rain must fall on the roof to fill the tanks.

24. Archimedes is considered to be one of the greatest mathematicians of all time. He discovered several of the formulas used in this chapter. Inscribed on his tombstone was a diagram of his proudest discovery. It shows a sphere inscribed (fitting exactly) into a cylinder. Show that:

$$\frac{\text{volume of the cylinder}}{\text{volume of the sphere}} = \frac{\text{surface area of the cylinder}}{\text{surface area of the sphere}}$$

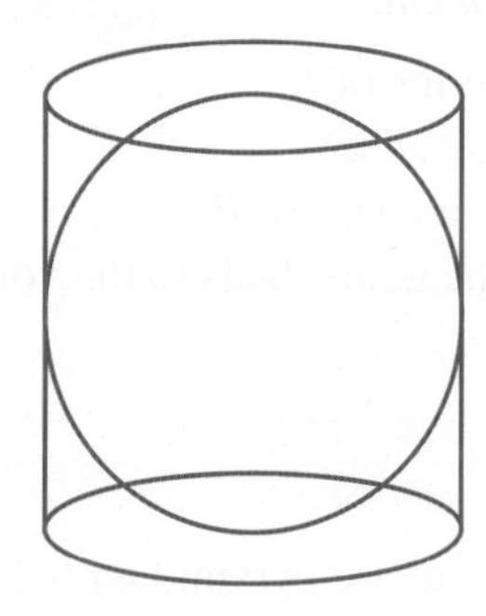

25. Marion has mixed together ingredients for a cake. The recipe requires a baking tin that is cylindrical in shape with a diameter of 20 cm and a height of 5 cm. Marion only has a tin in the shape of a trapezoidal prism and a muffin tray consisting of 24 muffin cups. Each of the muffin cups in the tray is a portion of a cone. Both the tin and muffin cup are shown in the diagrams. Explain whether Marion should use the tin or the muffin tray.

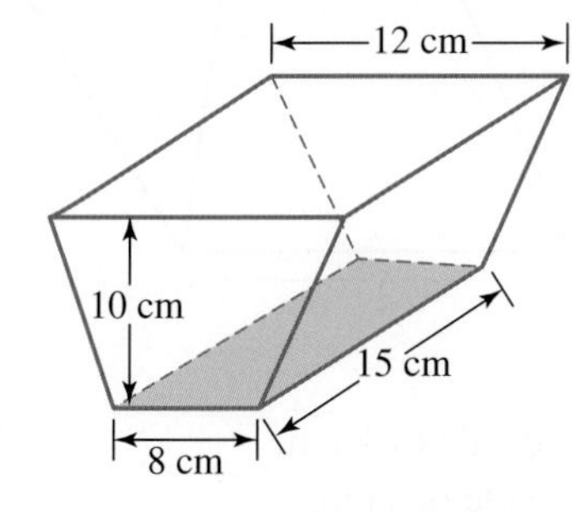

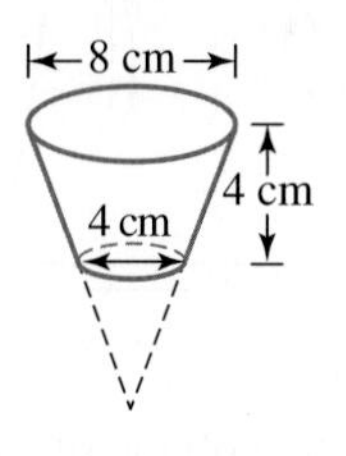

26. Sam is having his 16th birthday party and wants to make an ice trough to keep drinks cold. He has found a square piece of sheet metal with a side length of 2 metres. He cuts squares of side length x metres from each corner, then bends the sides of the remaining sheet.
When four squares of the appropriate side length are cut from the corners, the capacity of the trough can be maximised at 588 litres. Explain how Sam should proceed to maximise the capacity of the trough.

Problem solving

27. Nathaniel and Annie are going to the snow for survival camp. They plan to construct an igloo, consisting of an entrance and a hemispherical living section as shown. Nathaniel and Annie are asked to redraw their plans and increase the height of the liveable region (hemispherical structure) so that the total volume (including entrance) is doubled. Determine what must the new height of the hemisphere be to achieve this so that the total volume (including entrance) is doubled. Write your answer in metres correct to 2 decimal places.

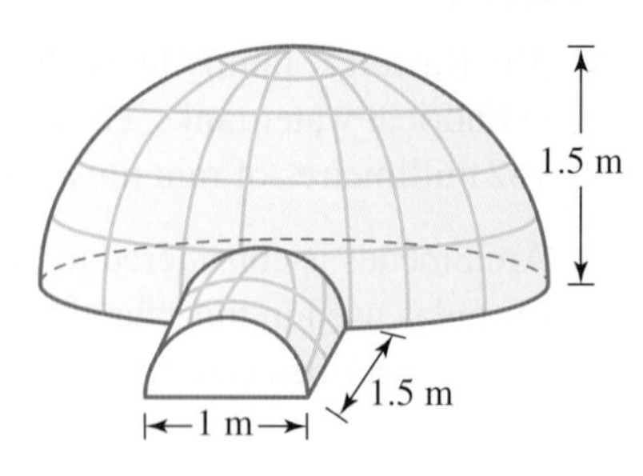

28. Six tennis balls are just contained in a cylinder as the balls touch the sides and the end sections of the cylinder. Each tennis ball has a radius of R cm.
    a. Express the height of the cylinder in terms of R.
    b. Evaluate the total volume of the tennis balls.
    c. Determine the volume of the cylinder in terms of R.
    d. Show that the ratio of the volume of the tennis balls to the volume of the cylinder is 2 : 3.

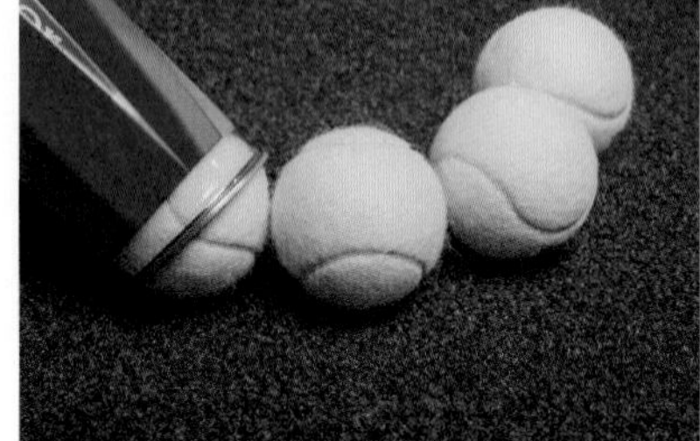

29. A frustum of a square-based pyramid is a square pyramid with the top sliced off. H is the height of the full pyramid and h is the height of the frustum.

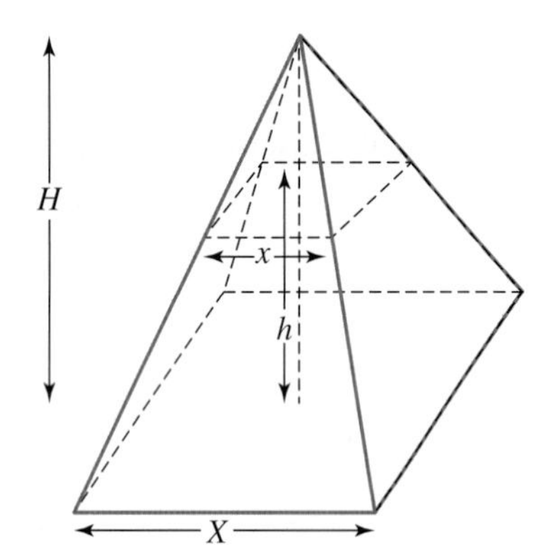

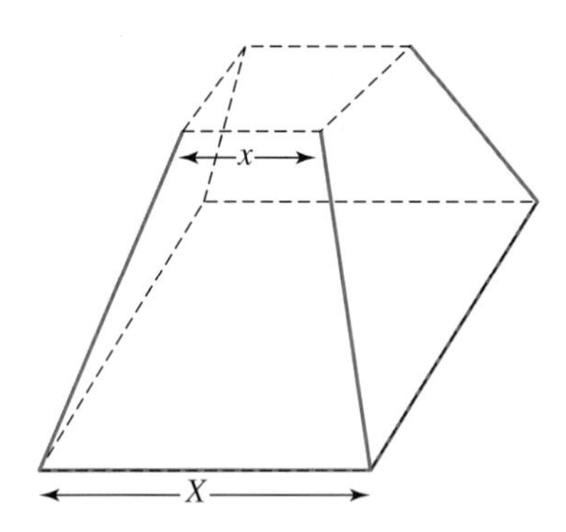

   a. Determine the volume of the large pyramid that has a square base side of X cm.
   b. Evaluate the volume of the small pyramid that has a square base side of x cm.
   c. Show that the relationship between H and h is given by $H = \frac{Xh}{X - x}$.
   d. Show that the volume of the frustum is given by $\frac{1}{3}h\left(X^2 + x^2 + Xx\right)$.

30. A large container is five-eighths full of ice-cream. After removing 27 identical scoops, it is one-quarter full. Determine how many scoops of ice-cream are left in the container.

6.5 Review

6.5.1 Topic summary

Prisms and cylinders

- Prisms are 3D objects that have a uniform cross section and all flat surfaces.
- The surface area of a prism is calculated by adding the areas of its faces.
- The volume of a prism is $V = AH$, where A is the cross-sectional area of the prism, and H is the perpendicular height.
- A cylinder is a 3D object that has a circular cross-section.
- The curved surface area of a cylinder is $2\pi rh$, and the total surface area is $2\pi r^2 + 2\pi rh = 2\pi rh(r + h)$.
- The volume of a cylinder is $V = \pi r^2 h$.

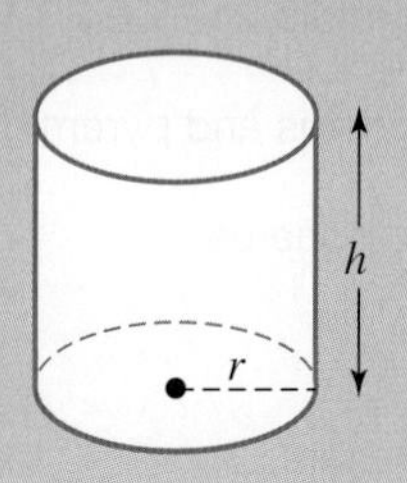

Units of area, volume and capacity

Area:

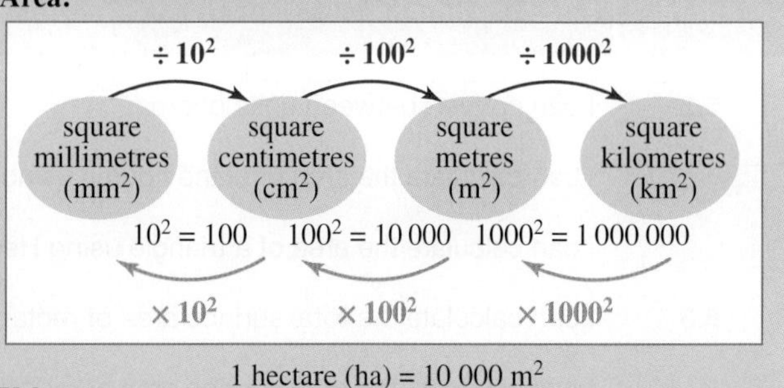

1 hectare (ha) = 10 000 m²

Volume:

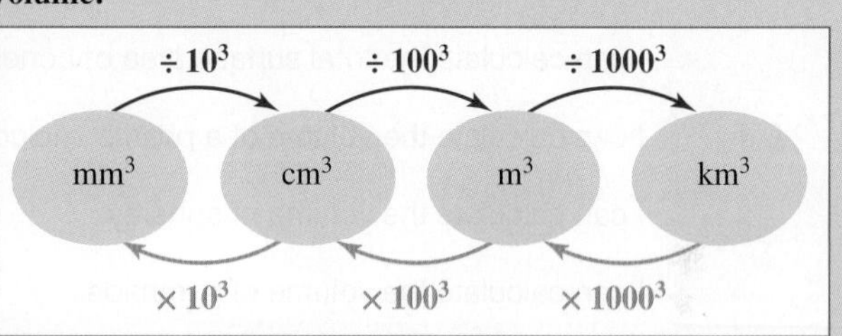

Capacity:

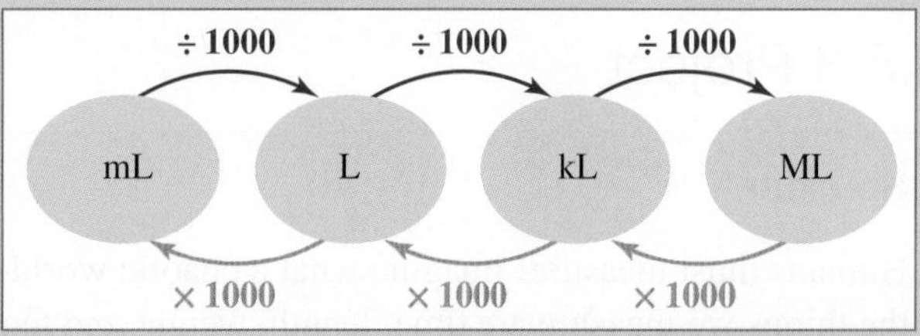

1 cm³ = 1 mL
1 L = 1 000 cm³

SURFACE AREA AND VOLUME

Pyramids

- The surface area of a pyramid can be calculated by adding the surface areas of its faces.
- The volume of a pyramid is $V = \frac{1}{3}AH$, where A is the area of the base and H is the height.

Cones

- The curved surface area of a cone is $SA_{\text{curved}} = \pi rl$, where l is the slant height.
- The total surface area is $SA = \pi rl + \pi r^2 = \pi r(l + r)$.
- The volume of a cone is $V = \frac{1}{3}\pi r^2 h$.

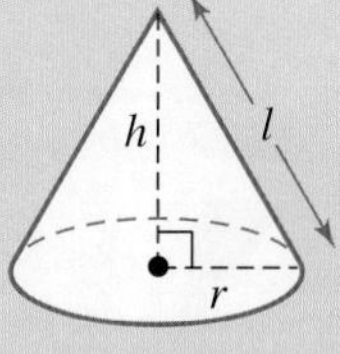

Spheres

- The surface area of a sphere is $A = 4\pi r^2$.
- The volume of a sphere is $V = \frac{4}{3}\pi r^3$.

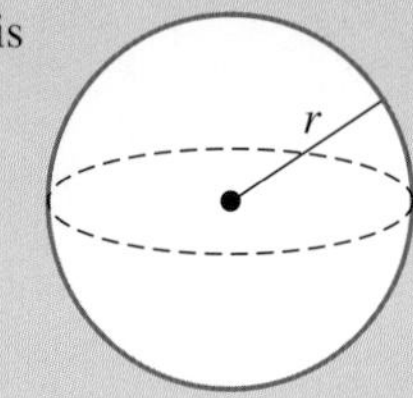

Area formulas

- Square: $A = l^2$
- Rectangle: $A = lw$
- Triangle: $A = \frac{1}{2}bh$
- Parallelogram: $A = bh$
- Trapezium: $A = \frac{1}{2}(a + b)h$
- Kite: $A = \frac{1}{2}xy$
- Circle: $A = \pi r^2$
- Sector: $A = \frac{\theta}{360} \times \pi r^2$
- Ellipse: $A = \pi ab$

Heron's formula

- Heron's formula is an alternate method to calculate the area of a triangle.
- For a triangle with sides of length a, b and c the area is $A = \sqrt{s(s-a)(s-b)(s-c)}$, where $s = \frac{a+b+c}{2}$.

6.5.2 Success criteria

Tick the column to indicate that you have completed the subtopic and how well you have understood it using the traffic light system.

(**Green:** I understand; **Yellow:** I can do it with help; **Red:** I do not understand)

Subtopic	Success criteria	Green	Yellow	Red
6.2	I can convert between units of area.			
	I can calculate the area of plane figures using area formulas.			
	I can calculate the area of a triangle using Heron's formula.			
6.3	I can calculate the total surface area of rectangular prisms and pyramids.			
	I can calculate the total surface area of cylinders and spheres.			
	I can calculate the total surface area of cones.			
6.4	I can calculate the volume of a prisms, including cylinders.			
	I can calculate the volume of spheres.			
	I can calculate the volume of pyramids.			

6.5.3 Project

So close!

Humans must measure! Imagine what a chaotic world it would be if we didn't measure anything. Some of the things we measure are time, length, weight and temperature; we also use other measures derived from these such as area, volume, speed.

Accurate measurement is important. The accuracy of a measurement depends on the instrument being used to measure and the interpretation of the measurement. There is no such thing as a perfectly accurate measurement. The best we can do is learn how to make meaningful use of the numbers we read off our devices. It is also important to use appropriate units of measurement.

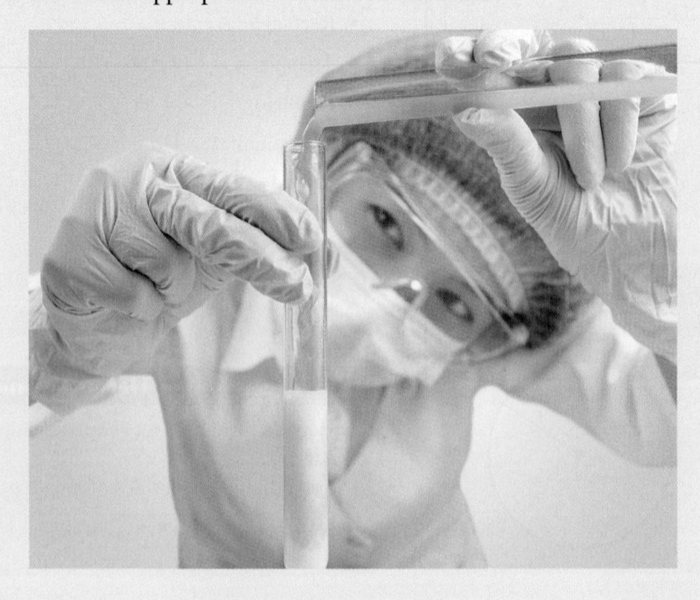

Measurement errors

When we measure a quantity by using a scale, the accuracy of our measurement depends on the markings on the scale. For example, the ruler shown can measure both in centimetres and millimetres.

Measurements made with this ruler would have ± 0.5 mm added to the measurement. The quantity ± 0.5 mm is called the tolerance of measurement or measurement error.

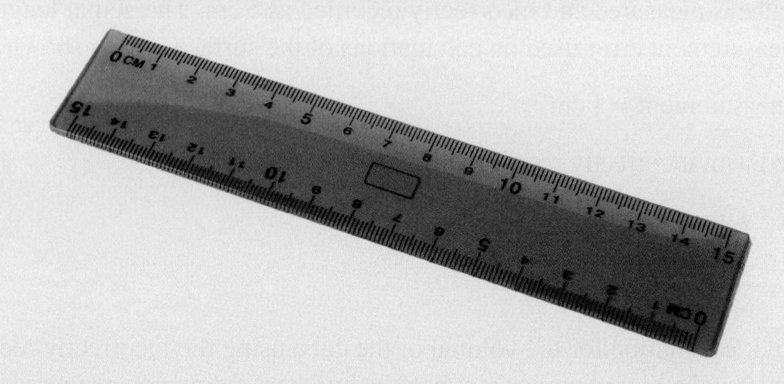

$$\text{Tolerance of measurement} = \frac{1}{2} \times \text{size of smallest marked unit}$$

For a measurement of 5.6 ± 0.5 mm, the largest possible value is 5.6 cm + 0.5 mm = 5.65 cm, and the smallest value is 5.6 cm − 0.5 mm = 5.55 cm.

1. For the thermometer scale shown:
   a. identify the temperature
   b. state the measurement with its tolerance
   c. calculate the largest and smallest possible values.
2. Calculate the largest and smallest values for:
   a. $(56.2 \pm 0.1) - (19.07 \pm 0.05)$
   b. $(78.4 \pm 0.25) \times (34 \pm 0.1)$.

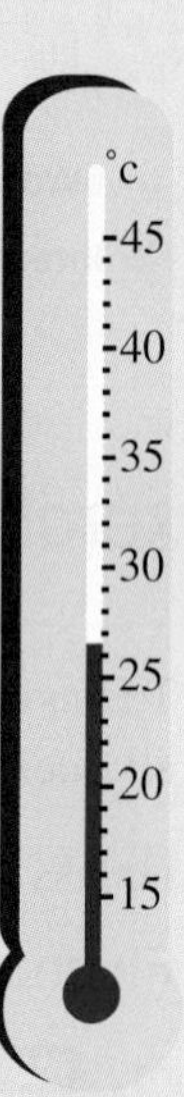

Significant figures in measurement

A significant figure is any non zero-digit, any zero appearing between two non-zero digits, any trailing zeros in a number containing a decimal point, and any digits in the decimal places. For example, the number 345.6054 has 7 significant figures, whereas 300 has 1 significant figure.

The number of significant figures is an expression of the accuracy of a measurement. The greater the number of significant figures, the more accurate the measurement. For example, a fast food chain claims it has sold 6 000 000 000 hamburgers, not 6 453 456 102. The first measurement has only 1 significant figure and is a very rough approximation of the actual number sold, which has 10 significant figures.

Reducing the number of significant figures is a process that is similar to rounding.

Rounding and measurement error in calculations

When you perform calculations, it is important to keep as many significant digits as practical and to perform any rounding as the *final* step. For example, calculating 5.34×341 by rounding to 2 significant figures *before* multiplying gives $5.30 \times 340 = 1802$, compared with 1820 if the rounding is carried out after the multiplication.

Calculations that involve numbers from measurements containing errors can result in answers with even larger errors. The smaller the tolerances, the more accurate the answers will be.

3. a. Calculate $45\,943.4503 \times 86.765\,303$ by:
   i. first rounding each number to 2 significant figures
   ii. rounding only the answer to 2 significant figures.
   b. Compare the two results.

Error in area and volume resulting from an error in a length measurement

The side length of a cube is measured and incorrectly recorded as 5 cm. The actual length is 6 cm. The effect of the length measurement error used on calculations of the surface area is shown below.

Error used in length measurement = 1 cm

Surface area calculated with incorrectly recorded value $= 5^2 \times 6 = 150\text{ cm}^2$

Surface area calculated with actual value $= 6^2 \times 6 = 216\text{ cm}^2$

Percentage error $= \dfrac{216 - 150}{6} \times 100\% \approx 30.5\%$

4. a. Complete a similar calculation for the volume of the cube using the incorrectly recorded length. What conclusion can you make regarding errors when the number of dimensions increase?
   b. Give three examples of a practical situation where an error in measuring or recording would have a potentially disastrous impact.

 Resources

 **eWorkbook** Topic 6 workbook (worksheets, code puzzle and a project) (ewbk-2032)

 **Interactivities** Crossword (int-2842)
Sudoku puzzle (int-3593)

Exercise 6.5 Review questions

To answer questions online and to receive **immediate corrective feedback** and **fully worked solutions** for all questions, go to your learnON title at www.jacplus.com.au.

Unless told otherwise, where appropriate, give answers correct to 2 decimal places.

Fluency

1. **MC** If all measurements are in cm, the area of the figure is:

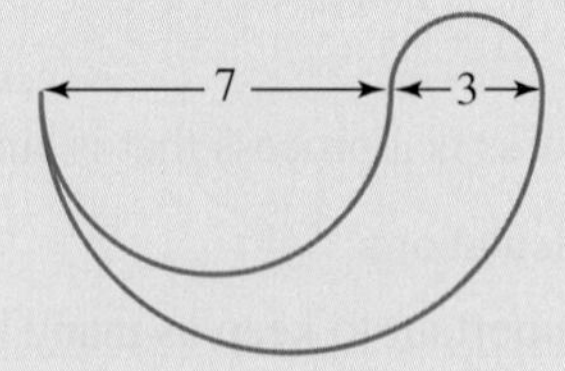

A. 16.49 cm^2 B. 39.25 cm^2 C. 9.81 cm^2 D. 23.56 cm^2 E. 30 cm^2

2. **MC** If all measurements are in centimetres, the area of the figure is:

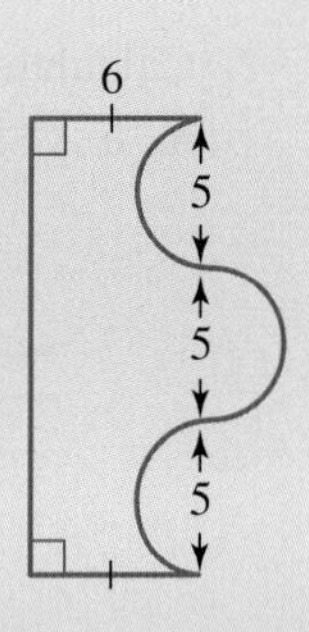

A. $50.73\,\text{cm}^2$
B. $99.82\,\text{cm}^2$
C. $80.18\,\text{cm}^2$
D. $90\,\text{cm}^2$
E. $119.45\,\text{cm}^2$

3. **MC** If all measurements are in centimetres, the shaded area of the figure is:

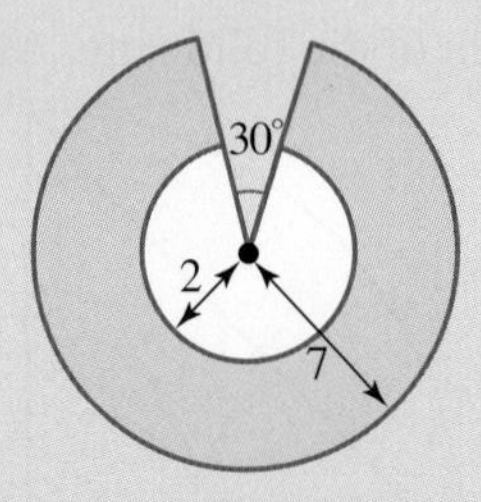

A. $3.93\,\text{cm}^2$ B. $11.52\,\text{cm}^2$ C. $388.77\,\text{cm}^2$ D. $141.11\,\text{cm}^2$ E. $129.59\,\text{cm}^2$

4. **MC** The total surface area of the solid is:

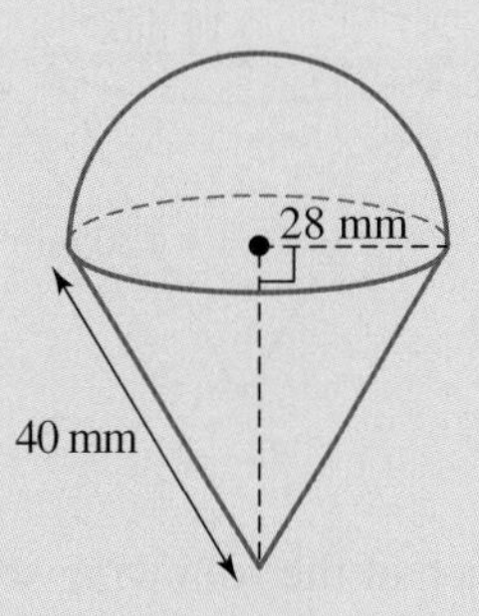

A. $8444.6\,\text{mm}^2$ B. $9221\,\text{mm}^2$ C. $14\,146.5\,\text{mm}^2$ D. $50\,271.1\,\text{mm}^2$ E. $16\,609.5\,\text{mm}^2$

5. Calculate the areas of the following plane figures. All measurements are in cm.

a.

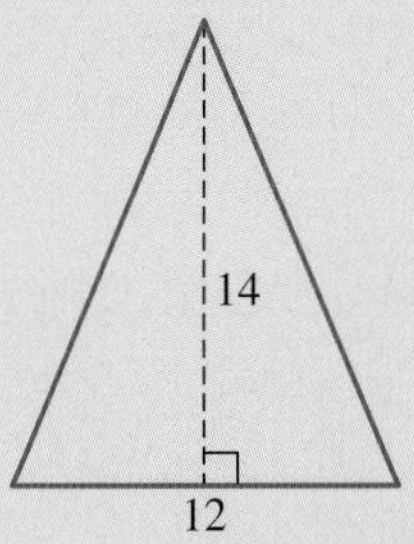

b.

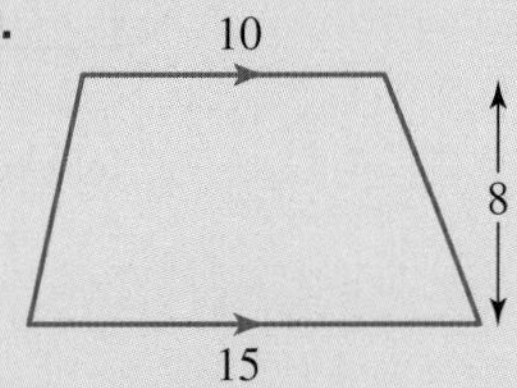

c.

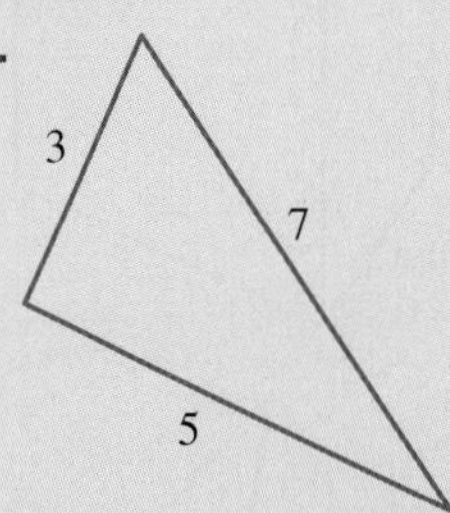

6. Calculate the areas of the following plane figures. All measurements are in cm.

a.

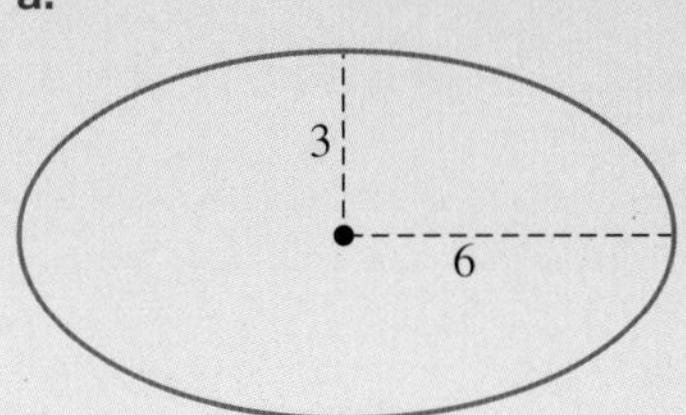

b.

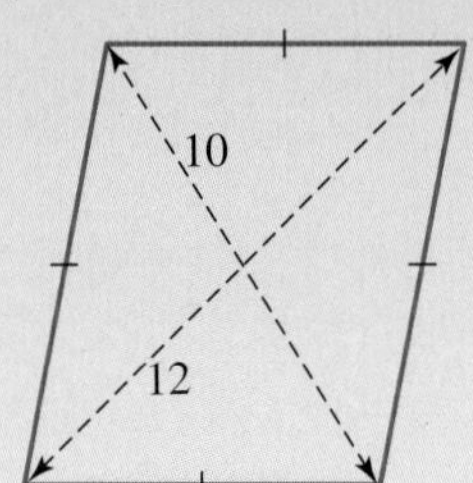

c.

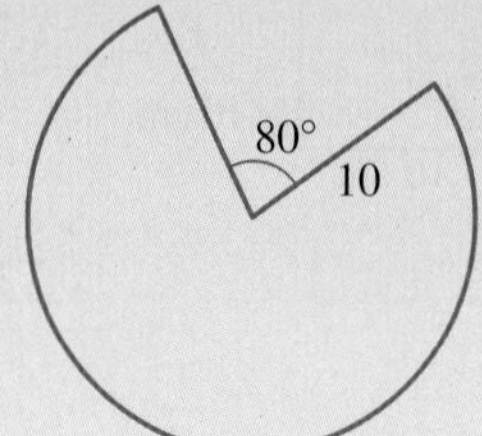

7. Calculate the areas of the following figures. All measurements are in cm.

a.

b.

c.

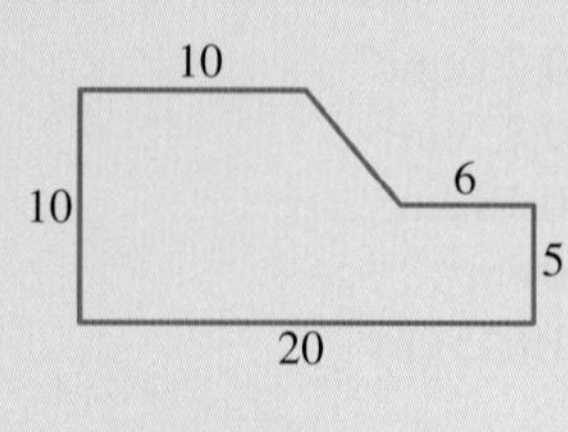

8. Calculate the blue shaded area in each of the following. All measurements are in cm.

a.

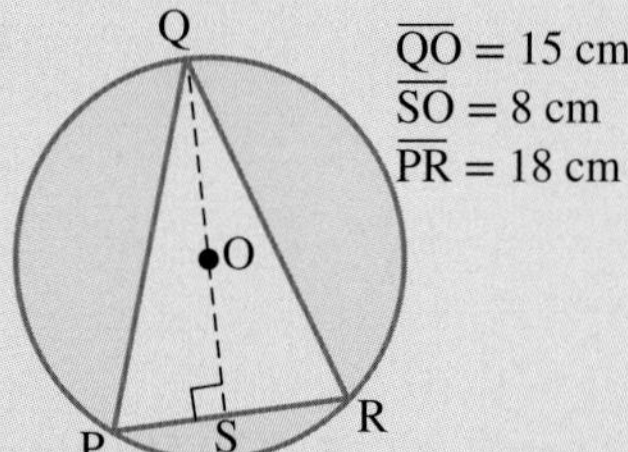

$\overline{QO} = 15$ cm
$\overline{SO} = 8$ cm
$\overline{PR} = 18$ cm

b.

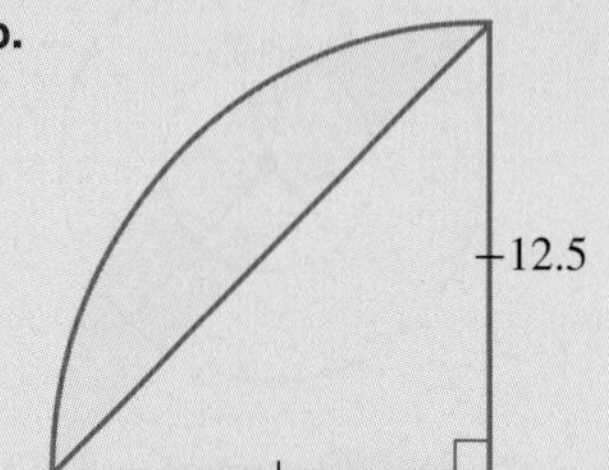

c.

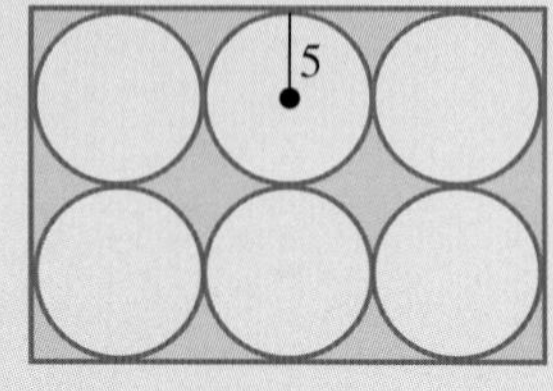

9. Calculate the total surface area of each of the following solids.

a.

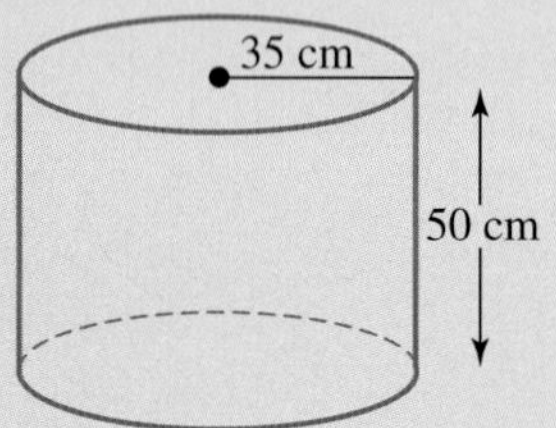

b.

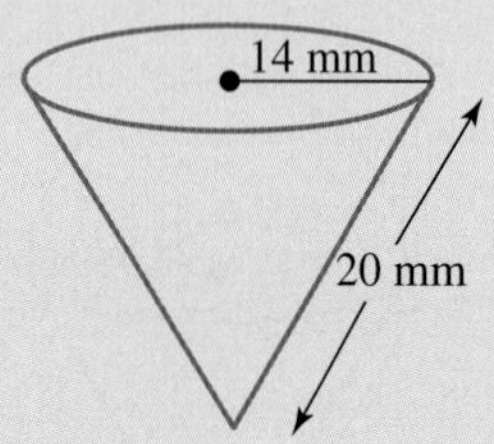

c.

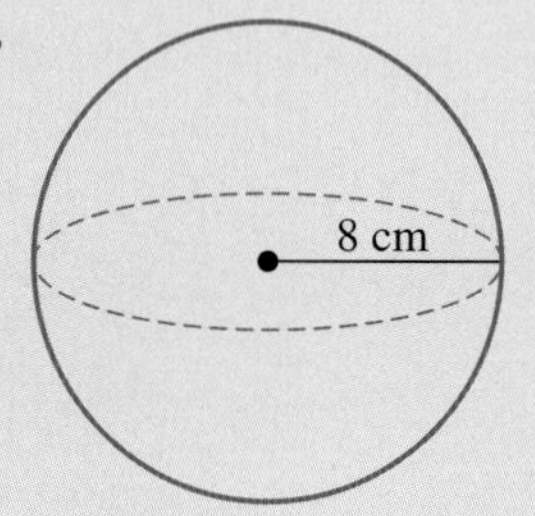

10. Calculate the total surface area of each of the following solids.

a.

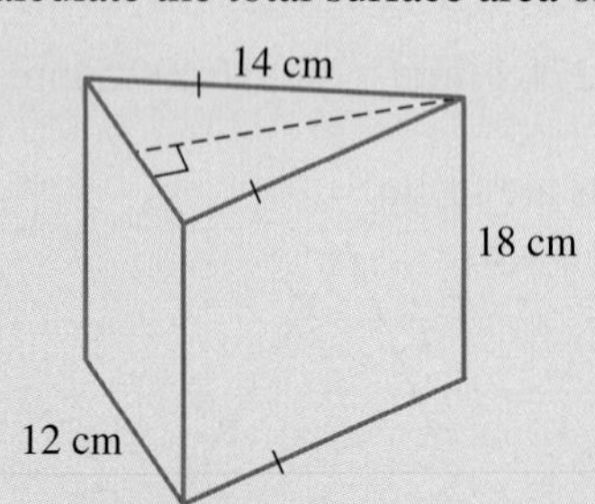

b.

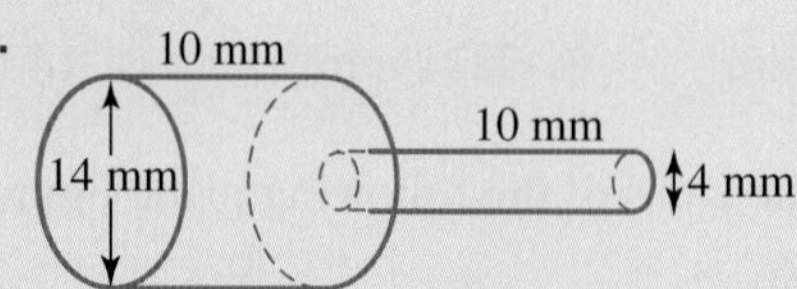

[closed at both ends]

c.

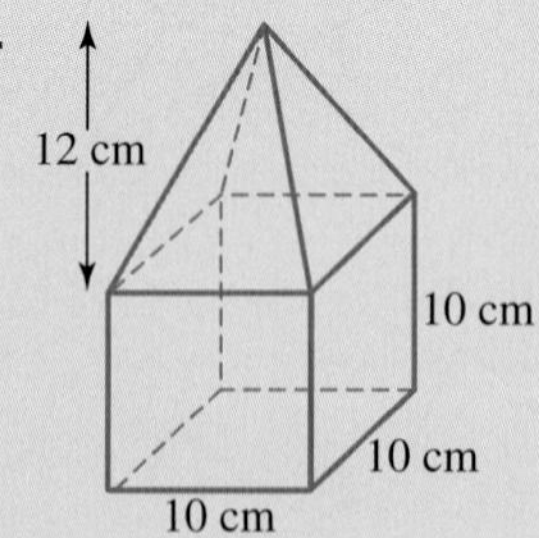

11. Calculate the volume of each of the following.

a.

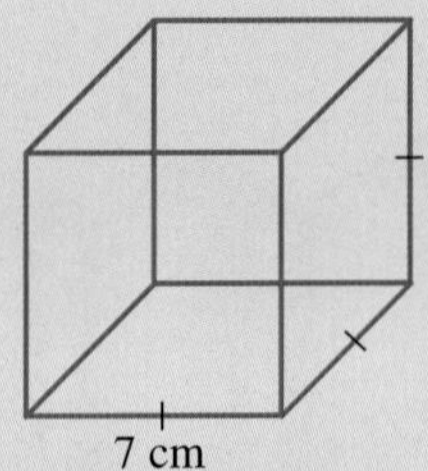

b.

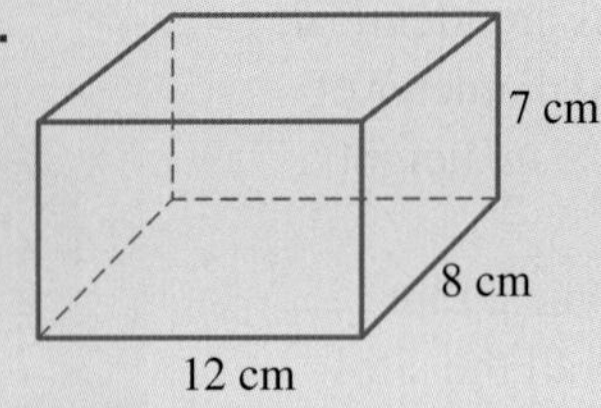

c.

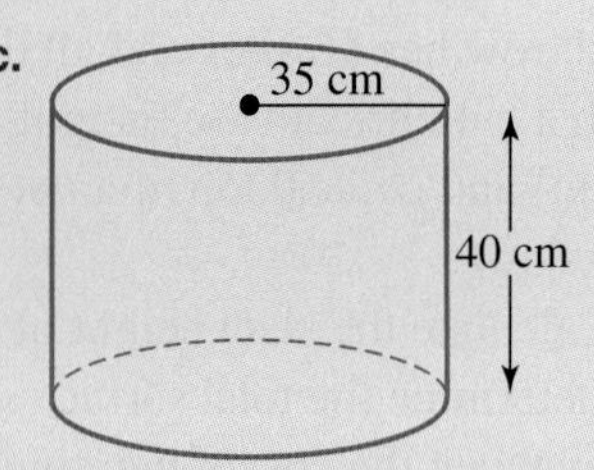

12. Determine the volume of each of the following.

a.

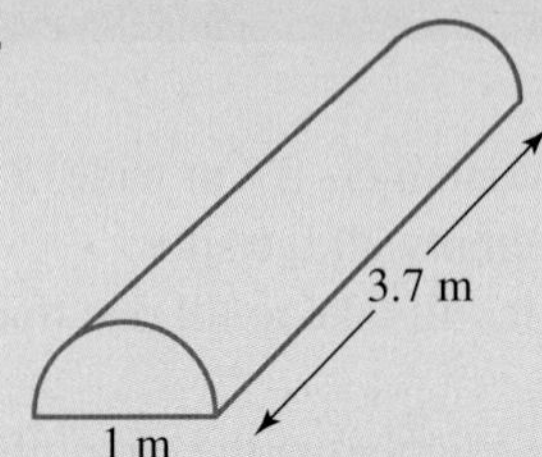

b.

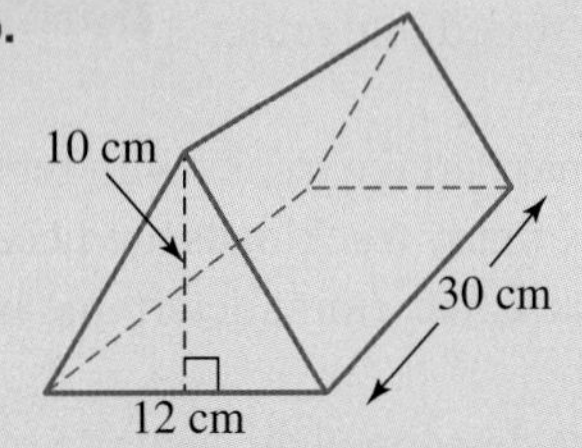

c.

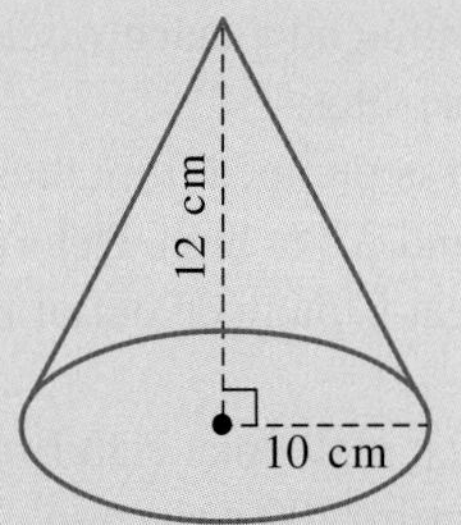

13. Determine the volume of each of the following.

a.

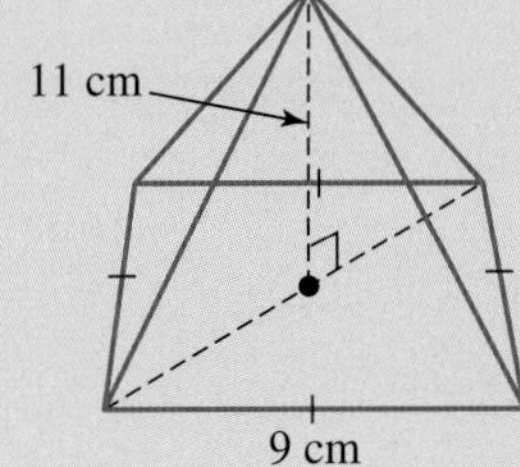

b.

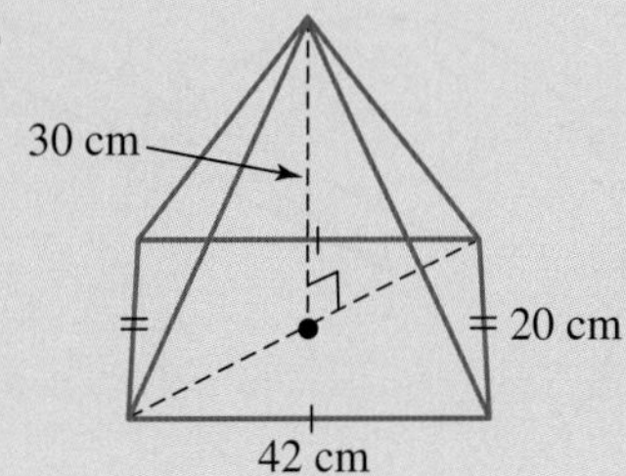

c.

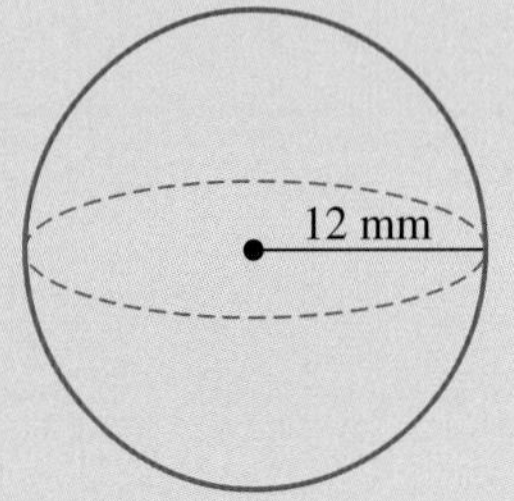

Problem solving

14. A rectangular block of land $4\text{ m} \times 25\text{ m}$ is surrounded by a concrete path 1 m wide.

a. Calculate the area of the path.

b. Determine the cost of concreting at \$45 per square metre.

15. If the radius is tripled and the height of a cylinder is divided by six, then determine the effect on its volume (in comparison with the original shape).

16. If the length is halved, the width is tripled and the height of a rectangular prism is doubled, then determine the effect on its volume (in comparison with the original shape).

17. A cylinder of radius 14 cm and height 20 cm is joined to a hemisphere of radius 14 cm to form a bread holder.

a. Calculate the total surface area.

b. Determine the cost of chroming the bread holder on the outside at \$0.05 per cm^2.

c. Calculate the storage volume of the bread holder.

d. Determine how much more space is in this new bread holder than the one it is replacing, which had a quarter circle end with a radius of 18 cm and a length of 35 cm.

18. Bella Silos has two rows of silos for storing wheat. Each row has 16 silos and all the silos are identical, with a cylindrical base (height of 5 m, diameter of 1.5 m) and conical top (diameter of 1.5 m, height of 1.1 m).
   a. Calculate the slant height of the conical tops.
   b. Determine the total surface area of all the silos.
   c. Evaluate the cost of painting the silos if one litre of paint covers 40 m^2 at a bulk order price of $28.95 per litre.
   d. Determine how much wheat can be stored altogether in these silos.

   e. Wheat is pumped from these silos into cartage trucks with rectangular containers 2.4 m wide, 5 m long and 2.5 m high. Determine how many truckloads are necessary to empty all the silos.
   f. If wheat is pumped out of the silos at 2.5 m^3/min, determine how long it will take to fill one truck.

19. The Greek mathematician Eratosthenes developed an accurate method for calculating the circumference of the Earth 2200 years ago! The figure illustrates how he did this.

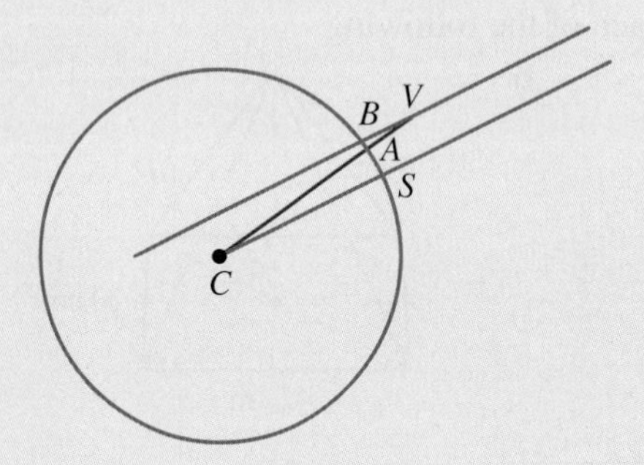

In this figure, A is the town of Alexandria and S is the town of Syene, exactly 787 km due south. When the sun's rays (blue lines) were vertical at Syene, they formed an angle of 7.2° at Alexandria ($\angle BVA = 7.2°$), obtained by placing a stick at A and measuring the angle formed by the sun's shadow with the stick.
   a. Assuming that the sun's rays are parallel, evaluate the angle $\angle SCA$, correct to 1 decimal place.
   b. Given that the arc $AS = 787$ km, determine the radius of the Earth, SC. Write your answer correct to the nearest kilometre.
   c. Given that the true radius is 6380 km, determine Eratosthenes' percentage error, correct to 1 decimal place.

To test your understanding and knowledge of this topic, go to your learnON title at www.jacplus.com.au and complete the **post-test**.

Online Resources

Below is a full list of **rich resources** available online for this topic. These resources are designed to bring ideas to life, to promote deep and lasting learning and to support the different learning needs of each individual.

eWorkbook

Download the workbook for this topic, which includes worksheets, a code puzzle and a project (ewbk-2032) ☐

Solutions

Download a copy of the fully worked solutions to every question in this topic (sol-0740) ☐

Digital documents

6.2 SkillSHEET Conversion of area units (doc-5236) ☐
SkillSHEET Using a formula to find the area of a common shape (doc-5237) ☐
6.3 SkillSHEET Total surface area of cubes and rectangular prisms (doc-5238) ☐
6.4 SkillSHEET Conversion of volume units (doc-5239) ☐
SkillSHEET Volume of cubes and rectangular prisms (doc-5240) ☐

Video eLessons

6.2 Area (eles-4809) ☐
Areas of composite figures (eles-4810) ☐
Composite area (eles-1886) ☐
6.3 Total surface area of solids (eles-4811) ☐
Total surface area of cones (eles-4812) ☐
Total surface area of other solids (eles-4813) ☐
Total surface area of prisms (eles-1909) ☐
6.4 Volume (eles-4814) ☐
Volumes of common shapes (eles-4815) ☐
Volume of composite solids (eles-4816) ☐
Capacity (eles-4817) ☐

Interactivities

6.2 Individual pathway interactivity: Area (int-4593) ☐
Conversion chart for area (int-3783) ☐
Area of rectangles (int-3784) ☐
Area of parallelograms (int-3786) ☐
Area of trapeziums (int-3790) ☐
Area of circles (int-3788) ☐
Area of a sector (int-6076) ☐
Area of a kite (int-6136) ☐
Area of an ellipse (int-6137) ☐
Using Heron's formula to find the area of a triangle (int-6475) ☐
6.3 Individual pathway interactivity: Total surface area (int-4594) ☐
Surface area of a prism (int-6079) ☐
Surface area of a cylinder (int-6080) ☐
Surface area (int-6477) ☐
6.4 Individual pathway interactivity: Volume (int-4595) ☐
Volume 1 (int-3791) ☐
Volume 2 (int-6476) ☐
Volume of solids (int-3794) ☐
6.5 Crossword (int-2842) ☐
Sudoku puzzle (int-3593) ☐

Teacher resources

There are many resources available exclusively for teachers online.

To access these online resources, log on to **www.jacplus.com.au**.

Answers

Topic 6 Surface area and volume

Exercise 6.1 Pre-test

1. $68.63\ mm^2$
2. $125.7\ cm^2$
3. E
4. $706.9\ cm^2$
5. $4\ cm^3$
6. $57.7\ cm^2$
7. $864
8. B
9. $60\ mm^3$
10. C
11. E
12. 4.5 cm
13. C
14. $13.4\ cm^2$
15. 230 mL

Exercise 6.2 Area

1. a. $16\ cm^2$ b. $48\ cm^2$ c. $75\ cm^2$
2. a. $120\ cm^2$ b. $706.86\ cm^2$ c. $73.5\ mm^2$
3. a. $254.47\ cm^2$ b. $21\ m^2$ c. $75\ cm^2$
4. a. $20.66\ cm^2$ b. $7.64\ cm^2$
5. a. $113.1\ mm^2$ b. $188.5\ mm^2$
6. a. i. $12\pi\ cm^2$ ii. $37.70\ cm^2$
   b. i. $\frac{69\pi}{2}\ mm^2$ ii. $108.38\ mm^2$
   c. i. $261\pi\ cm^2$ ii. $819.96\ cm^2$
7. E
8. D
9. a. $123.29\ cm^2$ b. $1427.88\ m^2$ c. $52\ cm^2$
10. a. $30.4\ m^2$ b. $78\ cm^2$ c. $2015.50\ cm^2$
11. a. $125.66\ cm^2$ b. $102.87\ m^2$
12. a. $13.73\ m^2$ b. $153.59\ m^2$
13. a. $27.86\ m^2$ b. $37.5\ m^2$
14. $11\,707.92\ cm^2$
15. $21\ m^2$
16. 60
17. $840
18. a. $260.87\ m^2$ b. $195.71\ m^2$ c. 75%
19. a. Sample responses can be found in the worked solutions in the online resources.
    b. 2020.83 m; horizontal. If vertical split 987.5 m.
20. a. $50 = x + y$
    b. $y = 50 - x$
    c. $\text{Area} = 50x - x^2$
    d. See the table at bottom of the page.*
    e. No, impossible to make a rectangle.
    f. 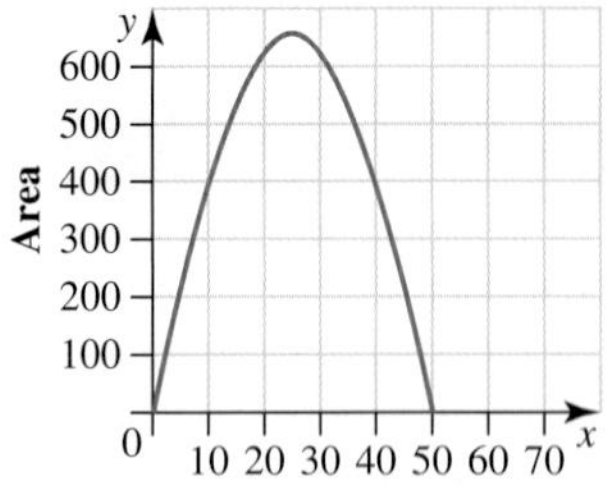

    g. $x = 25$
    h. $y = 25$
    i. Square
    j. $625\ m^2$
    k. $r = 15.92\ m$
    l. $795.77\ m^2$
    m. $170.77\ m^2$
21. a. Circular area, $1790.49\ m^2$; rectangular area, $1406.25\ m^2$
    b. Circular area, $\left(\frac{1}{4\pi}n^2\right)\ m^2$; rectangular (square) area, $\left(\frac{1}{16}n^2\right)\ m^2$. Circular area is always $\frac{4}{\pi}$ or 1.27 times larger.
22. a. $258.1\ m^2$ b. 7 bags c. $79
    d. $435.50
23. a. $\frac{29}{50}$ b. $x = 5, y = 5$
24. $32.88\ cm^2$

Exercise 6.3 Total surface area

1. a. $600\ cm^2$ b. $384\ cm^2$ c. $1440\ cm^2$
   d. $27\ m^2$
2. a. $113.1\ m^2$ b. $6729.3\ cm^2$ c. $8.2\ m^2$
   d. $452.4\ cm^2$
3. a. $1495.4\ cm^2$ b. $502.7\ cm^2$
4. a. $506.0\ cm^2$ b. $9.4\ m^2$ c. $340.4\ cm^2$
   d. $224.1\ cm^2$
5. a. $13.5\ m^2$ b. $90\ m^2$ c. $11\,309.7\ cm^2$
6. a. $9852.0\ mm^2$ b. $125.7\ cm^2$ c. $1531.4\ cm^2$
7. a. $880\ cm^2$ b. $3072.8\ cm^2$ c. $75\ cm^2$
8. a. $70.4\ cm^2$ b. $193.5\ cm^2$ c. $1547.2\ cm^2$
9. B
10. a. $70.0\ m^2$ b. $455
11. a. $3063.1\ cm^2$ b. $168.47
12. $11\,216\ cm^2$

*20. d.

x	0	5	10	15	20	25	30	35	40	45	50
Area(m^2)	0	225	400	525	600	625	600	525	400	225	0

13. 60

14. $1592\,\text{cm}^2$

15. $1960

16. a. $8\pi\ \text{m}^2$ b. $\dfrac{\sqrt{7}}{2}\text{m}$ c. $4\sqrt{2\pi}\ \text{m}^2$

17. a. $6.6\,\text{m}^2$
    b. Back wall = 80 tiles
    Side wall = 50 tiles
    80 + 50 + 50 = 180 tiles
    c. Cheapest: 30 cm by 30 cm, $269.50; 20 cm by 20 cm (individually) $270; 20 cm by 20 cm (boxed) $276.50

18. The calculation is correct.

19. a. $\theta = 120°$ b. $x = 1;\ y = \sqrt{3}$
    c. $3\sqrt{3}\,\text{cm}^2$ d. $6\sqrt{3}\,\text{cm}^2$
    e. 32

20. The area of material required is $1.04\,\text{m}^2$. If Tina is careful in placing the pattern pieces, she may be able to cover the footstool.

21. $r = \dfrac{3\sqrt{3}\text{a}}{2}$

22. a. Arc length XY $= (x + s)\theta$
    Arc length AB $= x\theta$
    b. i. $x = \dfrac{2\pi t}{\theta} = \dfrac{st}{r - t}$
    ii. $\dfrac{x}{x + s} = \dfrac{t}{r}$
    c. Area of sector AVB $= \dfrac{x^2\theta}{2}$
    Area of sector XVY $= \dfrac{(s + x)^2\theta}{2}$
    Area of ABYX $= \dfrac{s\theta\,(s + 2x)}{2}$
    TSA of frustum $= \pi\left(t^2 + r^2\right) + \dfrac{s\theta\,(s + 2x)}{2}$

Exercise 6.4 Volume

1. a. $27\,\text{cm}^3$ b. $74.088\,\text{m}^3$ c. $3600\,\text{cm}^3$
   d. $94.5\,\text{cm}^3$

2. a. $450\,\text{mm}^3$ b. $360\,\text{cm}^2$

3. a. $6333.5\,\text{cm}^3$ b. $19.1\,\text{m}^3$ c. $280\,\text{cm}^3$

4. a. $288\,\text{mm}^3$ b. $91.6\,\text{m}^3$ c. $21\,470.8\,\text{cm}^3$

5. a. $7.2\,\text{m}^3$ b. $14\,137.2\,\text{cm}^3$ c. $1436.8\,\text{mm}^3$
   d. $523\,598.8\,\text{cm}^3$

6. a. $113\,097.34\,\text{cm}^3$ b. $1.44\,\text{m}^3$
   c. $12\,214.51\,\text{mm}^3$ d. 101.93

7. a. $377.0\,\text{cm}^3$ b. $2303.8\,\text{mm}^3$

8. a. $400\,\text{cm}^3$ b. $10\,080\,\text{cm}^3$

9. a. $1400\,\text{cm}^3$ b. $10\,379.20\,\text{cm}^3$

10. a. $41.31\,\text{cm}^3$ b. $48.17\,\text{cm}^3$

11. a. $218.08\,\text{cm}^3$ b. $3691.37\,\text{cm}^3$

12. a. $V_{\text{new}} = 27l^3$, the volume will be 27 times as large as the original volume.
    b. $V_{\text{new}} = \dfrac{1}{8}l^2$, the volume will be $\dfrac{1}{8}$ of the original volume.
    c. $V_{\text{new}} = 2\pi r^2 h$, the volume will be twice as large as the original volume.
    d. $V_{\text{new}} = \pi r^2 h$, the volume will remain the same.
    e. $V_{\text{new}} = 3lwh$, the volume will be 3 times as large as the original value.

13. E

14. $7438.35\,\text{cm}^3$

15. 4417.9 L

16. $10\,215.05\,\text{cm}^3$

17. a. $H = 6\,\text{m}$ b. $112\,\text{m}^3$ c. 19 bins
    d. 112 000 L e. 1.95 m from floor

18. a. i. 4.57 cm
    ii. $262.5\,\text{cm}^2$
    b. i. 14.15 cm
    ii. $323.27\,\text{cm}^2$
    c. i. 33.33 cm
    ii. $437.62\,\text{cm}^2$
    d. Sphere. Costs less for a smaller surface area.

19. a. $h = \dfrac{V}{\pi r^2}$
    b. i. 31.8 cm
    ii. 8.0 cm
    c. $\sqrt{\dfrac{V}{\pi h}}$
    d. $r \geq 0$, since r is a length
    e. i. 7.6 cm
    ii. 6.2 cm

20. 1000

21. a. 2.55 cm
    b. 35.68 cm
    c. $A_a = 157.88\,\text{m}^2$, $A_b = 12.01\,\text{m}^2$

22. a. $126.67\,\text{m}^3$ b. $53.33\,\text{m}^3$

23. Volume of water needed; $30.9\,\text{m}^3$.

24. Sample responses can be found in the worked solutions in the online resources.

25. Required volume = $1570.80\,\text{cm}^3$; tin volume = $1500\,\text{cm}^3$; muffin tray volume = $2814.72\,\text{cm}^3$. Marion could fill the tin and have a small amount of mixture left over, or she could almost fill 14 of the muffin cups and leave the remaining cups empty.

26. Cut squares of side length $s = 0.3$ m or 0.368 m from the corners.

27. 1.94 m.

28. a. $H = 12R$ b. $8\pi R^3$ c. $12\pi R^3$
    d. 8 : 12 = 2 : 3

29. a. $\dfrac{1}{3}X^2H$ b. $\dfrac{1}{3}x^2(H - h)$

30. 18 scoops

Project

1. a. The temperature reading is 26.5 °C.
   b. The smallest unit mark is 1°C, so the tolerance is 0.5.
   c. Largest possible value = 27 °C,
      smallest possible value = 26 °C
2. a. Largest value = 37.28, smallest value = 36.98
   b. Largest value = 2681.965, smallest value = 2649.285
3. a. i. 4 002 000
      ii. 4 000 000
   b. The result for i has 4 significant figures, whereas ii has only 1 significant figure after rounding. However, ii is closer to the actual value (3 986 297.386 144 940 9).
4. Volume using the incorrectly recorded value = 125 cm^3
   Volume using the actual value = 216 cm^3
   The percentage error is 42.1%, which shows that the error compounds as the number of dimensions increases.

Exercise 6.5 Review questions

1. D
2. C
3. E
4. A
5. a. 84 cm^2 b. 100 cm^2 c. 6.50 cm^2
6. a. 56.52 cm^2 b. 60 cm^2 c. 244.35 cm^2
7. a. 300 cm^2 b. 224.55 cm^2 c. 160 cm^2
8. a. 499.86 cm^2 b. 44.59 cm^2 c. 128.76 cm^2
9. a. 18 692.48 cm^2 b. 1495.40 mm^2 c. 804.25 cm^2
10. a. 871.79 cm^2 b. 873.36 mm^2 c. 760 cm^2
11. a. 343 cm^3 b. 672 cm^3 c. 153 938.04 cm^3
12. a. 1.45 m^3 b. 1800 cm^3 c. 1256.64 cm^3
13. a. 297 cm^3 b. 8400 cm^3 c. 7238.23 mm^3
14. a. 62 m^2 b. $2790
15. $V = \frac{3}{2}\pi r^2 h$, the volume will be 1.5 times as large as the original volume.
16. $V = 3lwh$, the volume will be 3 times as large as (or triple) the original volume.
17. a. 3606.55 cm^2 b. $180.33 c. 18 062.06 cm^3
    d. 9155.65 cm^3
18. a. 1.33 m
    b. 910.91 m^2
    c. $618.35 or $636.90 assuming you have to buy full litres (i.e not 0.7 of a litre)
    d. 303.48 m^3
    e. 11 trucks
    f. 12 minutes
19. a. 7.2° b. 6263 km c. 1.8% error

7 Quadratic expressions

LEARNING SEQUENCE

7.1 Overview

Why learn this?

How is your algebraic tool kit? Is there some room to expand your skills? As expressions become more complex, more power will be needed to manipulate them and to carry out basic algebraic skills such as adding, multiplying, expanding and factorising.

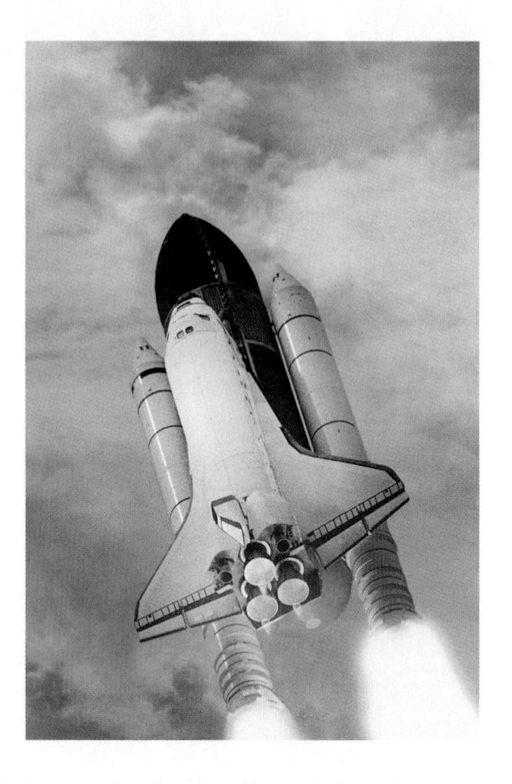

We have sent humans into space to live for months at a time on the International Space Station and even landed people on the moon. We have satellites circling our globe which enable us to communicate with friends and family around the world. Satellites also send out weather information, allowing meteorologists to study the patterns and changes and predict the upcoming weather conditions.

Technology connects us with others via our mobile phones or computers, often using a digital social media platform. All of these things that have become such important parts of our modern lives are only possible due to the application of mathematical techniques that we will begin to explore in this topic.

Many careers require a strong understanding of algebraic techniques. An example is a structural engineer. When designing structures, bridges, high rise buildings or domestic homes, they need to make decisions based on mathematics to ensure the structure is strong, durable and passes all regulations. This topic will help you develop your knowledge of algebraic expressions, so that you can apply them to real-life situations.

Where to get help

Go to your learnON title at www.jacplus.com.au to access the following digital resources. The Online Resources Summary at the end of this topic provides a full list of what's available to help you learn the concepts covered in this topic.

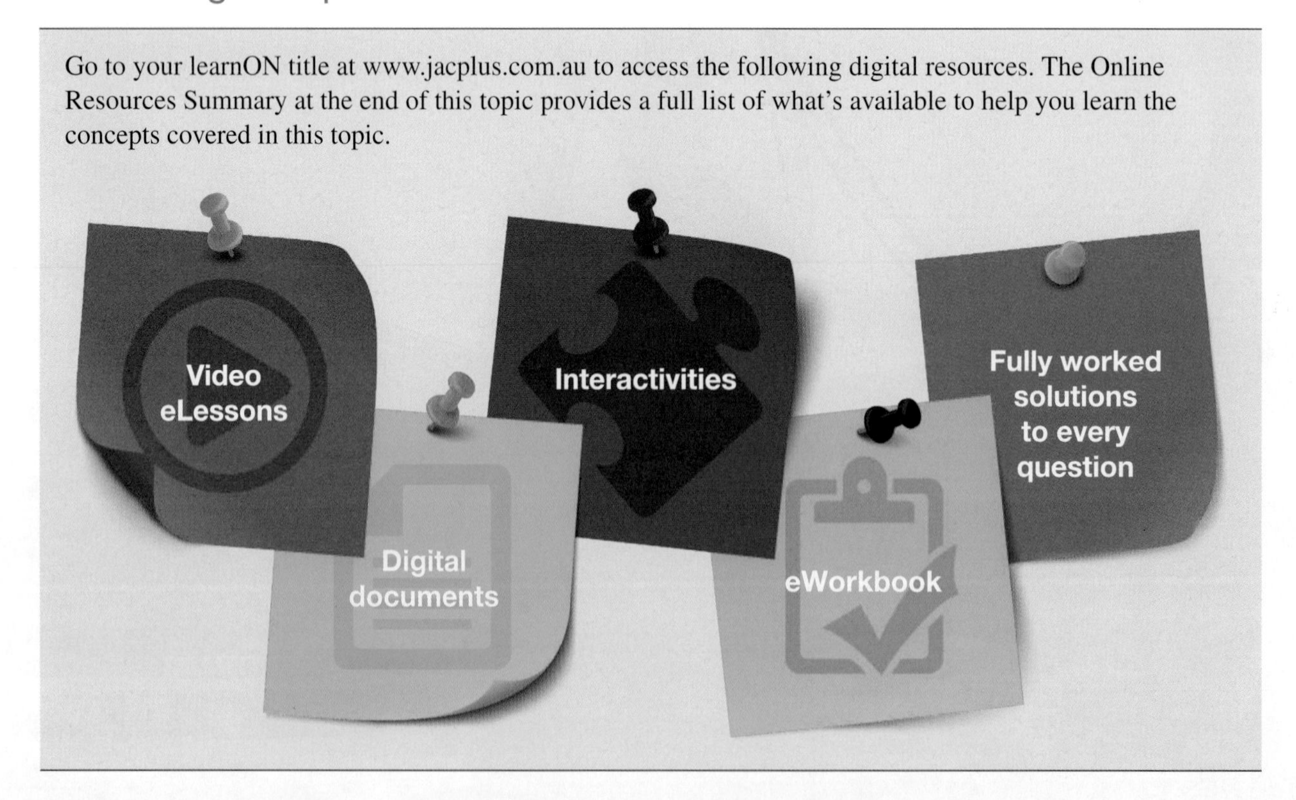

Exercise 7.1 Pre-test

learnON

Complete this pre-test in your learnON title at www.jacplus.com.au and receive **automatic marks**, **immediate corrective feedback** and **fully worked solutions**.

1. Expand $-2(x+3)$.
2. Factorise $4m-20m^2$.
3. Factorise $4a^2+a^2b+4ac+abc$.
4. Factorise $x^2+3x-10$.
5. Expand and simplify $-(3x+1)^2$. Write your answer in descending powers.
6. **MC** $\left(\sqrt{2}+5x\right)\left(\sqrt{5}-2x\right)+\sqrt{2}x$ when expanded and simplified is:
   A. $\sqrt{10}-\sqrt{2}x+5\sqrt{5}x-10x^2$
   B. $\sqrt{10}-\sqrt{2}x+5\sqrt{5}x-10x$
   C. $10-\sqrt{2}x+5\sqrt{5}x-10x^2$
   D. $10-\sqrt{2}x+5\sqrt{5}x-10x$
   E. $\sqrt{10}-10\sqrt{5}x-10x^2$
7. **MC** $7x^2-700$ when factorised is:
   A. $\left(\sqrt{7}x-100\right)\left(\sqrt{7}x+100\right)$
   B. $\left(\sqrt{7}x-10\right)\left(\sqrt{7}x+10\right)$
   C. $(7x-100)(7x+100)$
   D. $7(x-100)(x+100)$
   E. $7(x-10)(x+10)$
8. **MC** $4(2x+1)^2-9(y-4)^2$ when factorised is:
   A. $(4x+3y-11)(4x-3y-11)$
   B. $(8x-9y+1)(8x+9y+16)$
   C. $(4x+3y-10)(4x-3y+14)$
   D. $(2x+y-3)(2x-y+5)$
   E. $6(2x+y-4)(2x-y+4)$
9. Complete the square to factorise x^2+4x-6.
10. Expand $(x-2)(x+1)(2x+3)$. Write your answer in descending powers.
11. Factorise fully the algebraical expression $2x^2-6x-10$.
12. **MC** $-2x^2+10x+6$ when factorised is:
    A. $-2\left(x-\frac{5}{2}-\frac{\sqrt{17}}{2}\right)\left(x-\frac{5}{2}+\frac{\sqrt{17}}{2}\right)$
    B. $-2\left(x-5-\sqrt{37}\right)\left(x-5+\sqrt{37}\right)$
    C. $2\left(x+\frac{5}{2}-\frac{\sqrt{37}}{2}\right)\left(x+\frac{5}{2}+\frac{\sqrt{37}}{2}\right)$
    D. $-2\left(x-\frac{5}{2}-\frac{\sqrt{37}}{2}\right)\left(x-\frac{5}{2}+\frac{\sqrt{37}}{2}\right)$
    E. $-2\left(x-\frac{5}{2}-\frac{\sqrt{13}}{2}\right)\left(x-\frac{5}{2}+\frac{\sqrt{13}}{2}\right)$
13. Simplify $\dfrac{x^2-1}{x^2+2x-3}\div\dfrac{x+1}{2x+6}$.

14. Factorise $4(x+6)^2 + 11(x+6) - 3$.

15. MC $\dfrac{6x^2+11x-2}{6x^2-x} \times \dfrac{x^2-4x+4}{x^2-4}$ can be simplified to:

A. $\dfrac{-15x-2}{-x}$ **B.** $\dfrac{x-2}{x}$ **C.** -2 **D.** $\dfrac{-44x-2}{x}$ **E.** $\dfrac{x+2}{x}$

7.2 Expanding algebraic expressions

LEARNING INTENTION

At the end of this subtopic you should be able to:

- expand binomial expressions using FOIL
- expand squares of binomial expressions
- expand binomial expressions using the difference of two squares (DOTS).

7.2.1 Binomial expansion

eles-4825

- Consider the rectangle of length $a+b$ and width $c+d$ shown below. Its area is equal to $(a+b)(c+d)$.

	a	b
c	ac	bc
d	ad	bd

The diagram shows that $\underbrace{(a+b)(c+d)}_{\text{factorised form}} = \underbrace{ac+ad+bc+bd}_{\text{expanded form}}$.

- Expansion of the binomial expression $(x+3)(x+2)$ can be shown by this area model.

	x	3
x	$x \times x = x^2$	$3 \times x = 3x$
2	$2 \times x = 2x$	$3 \times 2 = 6$

Expressed mathematically this is:

$$\begin{aligned}(x+3)(x+2) &= x^2+2x+3x+6 \\ &= x^2+5x+6\end{aligned}$$

factorised form — expanded form

- There are several methods that can be used to expand binomial factors.

FOIL method

- The word **FOIL** provides us with an acronym for the expansion of a binomial product.
  - **First:** multiply the first terms in each bracket

F
$$(x + a)(x - b)$$

  - **Outer:** multiply the two outer terms

O
$$(x + a)(x - b)$$

  - **Inner:** multiply the two inner terms

I
$$(x + a)(x - b)$$

  - **Last:** multiply the last terms in each bracket

L
$$(x + a)(x - b)$$

WORKED EXAMPLE 1 Expanding binomial expressions using FOIL

Expand each of the following.

a. $(x + 3)(x + 2)$

b. $(x - 7)(6 - x)$

THINK	WRITE
a. 1. Write the expression.	**a.** $(x + 3)(x + 2)$
2. Use FOIL to expand the pair of brackets.	$(x \times x) + (x \times 2) + (3 \times x) + (3 \times 2)$
3. Simplify and then collect like terms.	$= x^2 + 2x + 3x + 6$ $= x^2 + 5x + 6$
b. 1. Write the expression.	**b.** $(x - 7)(6 - x)$
2. Use FOIL to expand the pair of brackets. Remember to include the negative signs where appropriate.	$(x \times 6) + (x \times -x) + (-7 \times 6) + (-7 \times -x)$
3. Simplify and then collect like terms.	$= 6x - x^2 - 42 + 7x$ $= -x^2 + 13x - 42$

- If there is a term outside the pair of brackets, expand the brackets and then multiply each term of the expansion by that term.

WORKED EXAMPLE 2 Expanding binomial expressions

Expand $3(x+8)(x+2)$.

THINK	WRITE
1. Write the expression.	$3(x+8)(x+2)$
2. Use FOIL to expand the pair of brackets.	$=3(x^2+2x+8x+16)$
3. Collect like terms within the brackets.	$=3(x^2+10x+16)$
4. Multiply each of the terms inside the brackets by the term outside the brackets.	$=3x^2+30x+48$

7.2.2 The square of a binomial expression

eles-4827

- The expansion of $(a+b)^2$ can be represented by this area model.

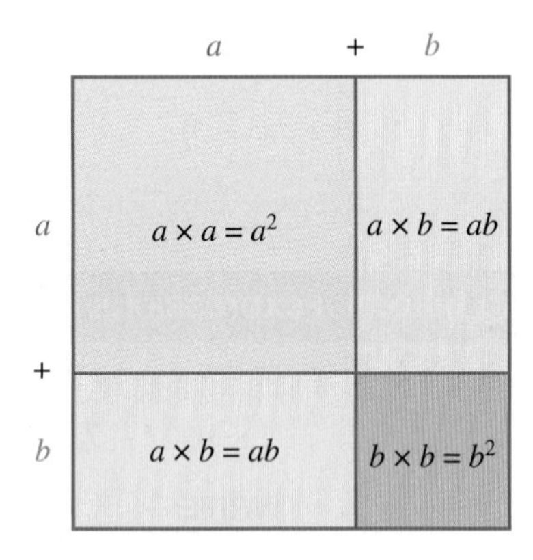

- This result provides a shortcut method for squaring binomials.

Squaring binomial expressions

$$(a+b)^2 = a^2 + 2ab + b^2$$

Similarly,

$$(a-b)^2 = a^2 - 2ab + b^2$$

- This expansion is often memorised. To determine the square of a binomial:
  - square the first term
  - multiply the two terms together and then double the product
  - square the last term.

WORKED EXAMPLE 3 Expanding and simplifying binomial expressions

Expand and simplify each of the following.

a. $(2x-5)^2$

b. $-3(2x+7)^2$

THINK	WRITE
a. 1. Write the expression.	a. $(2x-5)^2$
2. Expand using the rule $(a-b)^2 = a^2 - 2ab + b^2$.	$= (2x)^2 - 2 \times 2x \times 5 + (5)^2$ $= 4x^2 - 20x + 25$
b. 1. Write the expression.	b. $-3(2x+7)^2$
2. Expand the brackets using the rule $(a+b)^2 = a^2 + 2ab + b^2$.	$= -3[(2x)^2 + 2 \times 2x \times 7 + (7)^2]$ $= -3(4x^2 + 28x + 49)$
3. Multiply every term inside the brackets by the term outside the brackets.	$= -12x^2 - 84x - 147$

TI | THINK

a–b.

In a new problem, on a Calculator page, press:

- CATALOG
- 1
- E

Then scroll down to select expand(
Complete the entry lines as:
expand $((2x-5)^2)$
expand $(-3(2x+7)^2)$
Press ENTER after each entry.

DISPLAY/WRITE

a–b.

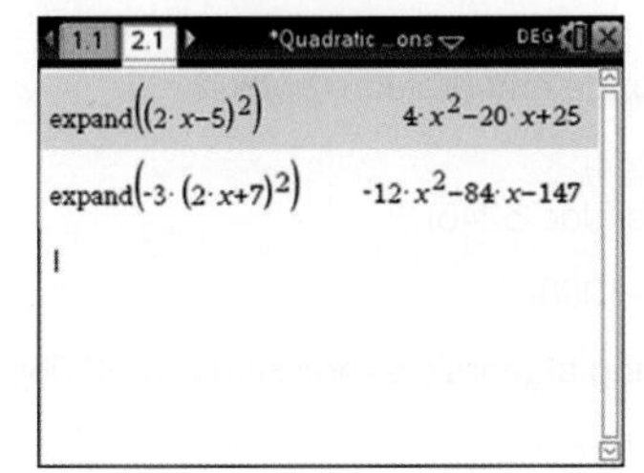

$(2x-5)^2 = 4x^2 - 20x + 25$

$-3(2x+7)^2 = -12x^2 - 84x - 147$

CASIO | THINK

a–b.

On the Main screen, tap:

- Action
- Transformation
- expand

Complete the entry lines as:
expand $(2x-5)^2$
expand $(-3(2x+7)^2)$
Press EXE after each entry.

DISPLAY/WRITE

a–b.

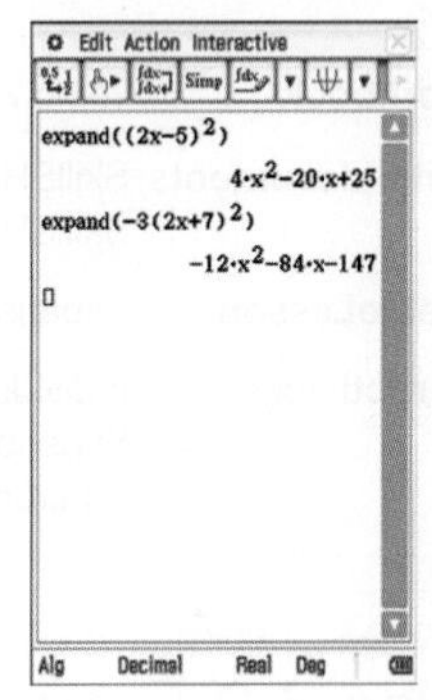

$(2x-5)^2 = 4x^2 - 20x + 25$

$-3(2x+7)^2 = -12x^2 - 84x - 147$

7.2.3 The difference of two squares

eles-4828

- When $(a+b)$ is multiplied by $(a-b)$ (or vice-versa),

$$\begin{aligned}(a+b)(a-b) &= a^2 - ab + ab - b^2 \\ &= a^2 - b^2\end{aligned}$$

The expression is called the difference of two squares and is often referred to as DOTS.

D Difference
O Of
T Two
S Squares

Difference of two squares

$$(a+b)(a-b) = a^2 - b^2$$

WORKED EXAMPLE 4 Expanding using DOTS

Expand and simplify each of the following.

a. $(3x+1)(3x-1)$

b. $4(2x-7)(2x+7)$

THINK	WRITE
a. 1. Write the expression.	a. $(3x+1)(3x-1)$
2. Expand using the rule $(a+b)(a-b)=a^2-b^2$.	$=(3x)^2-(1)^2$ $=9x^2-1$
b. 1. Write the expression.	b. $4(2x-7)(2x+7)$
2. Expand using the difference of two squares rule.	$=4[(2x)^2-(7)^2]$ $=4(4x^2-49)$
3. Multiply by 4.	$=16x^2-196$

Resources

eWorkbook Topic 7 Workbook (worksheets, code puzzle and project) (ewbk-2033)

Digital documents SkillSHEET Expanding brackets (doc-5244)
SkillSHEET Expanding a pair of brackets (doc-5245)

Video eLesson Expansion of binomial expressions (eles-1908)

Interactivities Individual pathway interactivity: Expanding algebraic expressions (int-4596)
Expanding binomial factors (int-6033)
Difference of two squares (int-6036)

Exercise 7.2 Expanding algebraic expressions

learnon

Individual pathways

■ PRACTISE	■ CONSOLIDATE	■ MASTER
1, 4, 7, 10, 14, 16, 17, 20, 23, 28, 29, 30, 34	2, 5, 8, 11, 12, 15, 18, 21, 24, 26, 31, 35	3, 6, 9, 13, 19, 22, 25, 27, 32, 33, 36

To answer questions online and to receive **immediate feedback** and **sample responses** for every question, go to your learnON title at www.jacplus.com.au.

Fluency

For questions 1–3, expand the expressions.

1. a. $2(x+3)$ b. $4(x-5)$ c. $3(7-x)$ d. $-(x+3)$
2. a. $x(x+2)$ b. $2x(x-4)$ c. $3x(5x-2)$ d. $5x(2-3x)$
3. a. $2x(4x+1)$ b. $2x^2(2x-3)$ c. $3x^2(2x-1)$ d. $5x^2(3x+4)$

WE1 For questions 4–6, expand the expressions.

4. a. $(x+3)(x-4)$ b. $(x+1)(x-3)$ c. $(x-7)(x+2)$ d. $(x-1)(x-5)$

5. a. $(2-x)(x+3)$ b. $(x-4)(x-2)$ c. $(2x-3)(x-7)$ d. $(x-1)(3x+2)$

6. a. $(3x-1)(2x-5)$ b. $(3-2x)(7-x)$ c. $(5-2x)(3+4x)$ d. $(11-3x)(10+7x)$

WE2 For questions 7–11, expand the expressions.

7. a. $2(x+1)(x-3)$ b. $4(2x+1)(x-4)$ c. $-2(x+1)(x-7)$

8. a. $2x(x-1)(x+1)$ b. $3x(x-5)(x+5)$ c. $6x(x-3)(x+3)$

9. a. $-2x(3-x)(x-3)$ b. $-5x(2-x)(x-4)$ c. $6x(x+5)(4-x)$

10. a. $(x-1)(x+1)(x+2)$ b. $(x-3)(x-1)(x+2)$ c. $(x-5)(x+1)(x-1)$

11. a. $(x-1)(x-2)(x-3)$ b. $(2x-1)(x+1)(x-4)$ c. $(3x+1)(2x-1)(x-1)$

For questions 12 and 13, expand and simplify the expressions.

12. a. $(x+2)(x-1)-2x$ b. $3x-(2x-5)(x+2)$
c. $(2x-3)(x+1)+(3x+1)(x-2)$ d. $(3-2x)(2x-1)+(4x-5)(x+4)$

13. a. $(x+1)(x-7)-(x+2)(x-3)$ b. $(x-2)(x-5)-(x-1)(x-4)$
c. $(x-3)(x+1)+\sqrt{3}x$ d. $\left(\sqrt{2}-3x\right)\left(\sqrt{3}+2x\right)-\sqrt{5}x$

14. MC $(3x-1)(2x+4)$ expands to:

A. $6x^2+10x-4$ B. $5x^2-24x+3$ C. $3x^2+2x-4$
D. $6x^2-10x-4$ E. $6x^2-4$

15. MC $-2x(x-1)(x+3)$ expands to:

A. x^2+2x-3 B. $-2x^2-4x+6$ C. $-2x^3-4x^2+6x$
D. $-2x^3+4x^2-6x$ E. $-2x^3-3$

16. MC The expression $(x-1)(x-3)(x+2)$ is *not* the same as:

A. $(x-3)(x-1)(x+2)$ B. $(x+3)(x-1)(x-2)$ C. $(x-1)(x+2)(x-3)$
D. $(x+2)(x-1)(x-3)$ E. $(x-3)(x+2)(x-1)$

WE3a For questions 17–19, expand and simplify the expressions.

17. a. $(x-1)^2$ b. $(x+2)^2$ c. $(x+5)^2$ d. $(4+x)^2$

18. a. $(7-x)^2$ b. $(12-x)^2$ c. $(3x-1)^2$ d. $(12x-3)^2$

19. a. $(5x+2)^2$ b. $(2-3x)^2$ c. $(5-4x)^2$ d. $(1-5x)^2$

Understanding

WE3b For questions 20–22, expand and simplify the expressions.

20. a. $2(x-3)^2$ b. $4(x-7)^2$ c. $3(x+1)^2$

21. a. $-(2x+3)^2$ b. $-(7x-1)^2$ c. $2(2x-3)^2$

22. a. $-3(2-9x)^2$ b. $-5(3-11x)^2$ c. $-4(2x+1)^2$

WE4 For questions 23–25, expand and simplify the expressions.

23. a. $(x+7)(x-7)$ **b.** $(x+9)(x-9)$ **c.** $(x-5)(x+5)$

24. a. $(x-1)(x+1)$ **b.** $(2x-3)(2x+3)$ **c.** $(3x-1)(3x+1)$

25. a. $(7-x)(7+x)$ **b.** $(8+x)(8-x)$ **c.** $(3-2x)(3+2x)$

26. The length of the side of a rectangle is $(x+1)$ cm and the width is $(x-3)$ cm.

a. Determine an expression for the area of the rectangle.
b. Simplify the expression by expanding.
c. If $x = 5$ cm, calculate the dimensions of the rectangle and, hence, its area.

27. Chickens are kept in a square enclosure with sides measuring x m. The number of chickens is increasing and so the size of the enclosure is to have 1 metre added to one side and 2 metres to the adjacent side.

a. Draw a diagram of the original enclosure.
b. Add to the first diagram or draw another one to show the new enclosure. Mark the lengths on each side on your diagram.
c. Write an expression for the area of the new enclosure in factorised form.
d. Expand and simplify the expression by removing the brackets.
e. If the original enclosure had sides of 2 metres, calculate the area of the original square and then the area of the new enclosure.

28. Write an expression in factorised and expanded form that is:

a. a quadratic trinomial
b. the square of a binomial
c. the difference of two squares
d. both **a** and **b**.

Reasoning

29. Shown below are three students' attempts at expanding $(3x+4)(2x+5)$.

STUDENT A

$(3x+4)(2x+5)$
$= 3x \times 2x + 3x \times 5 + 4 \times 2x + 4 \times 5$
$= 6x + 15x + 8x + 20$
$= 29x + 20$
$= 49x$

STUDENT B

$(3x+4)(2x+5)$
$= 3x \times 2x + 4 \times 2x + 4 \times 5$
$= 6x^2 + 8x + 20$

STUDENT C

$(3x+4)(2x+5)$
$= 3x \times 2x + 3x \times 5 + 4 \times 2x + 4 \times 5$
$= 6x^2 + 15x + 8x + 20$
$= 6x^2 + 23x + 20$

a. Identify which student's work was correct.
b. Correct the mistakes in each wrong case as though you were the teacher of these students. Explain your answer.

30. If $a = 5$ and $b = 3$, show that $(a - b)(a + b) = a^2 - b^2$ by evaluating both expressions.

31. If $a = 5$ and $b = 3$, show that $(a + b)^2 = a^2 + 2ab + b^2$ by evaluating both expressions.

32. Show that $(a + b)(c + d) = (c + d)(a + b)$.

33. Explain the difference between 'the square of a binomial' and 'the difference between two squares'.

Problem solving

34. Determine an expanded expression for the volume of the cuboid shown.

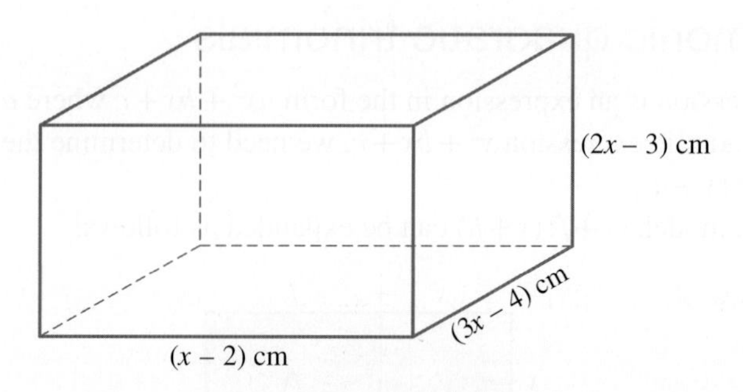

35. Determine an expanded expression for the total surface area of the square-based pyramid.

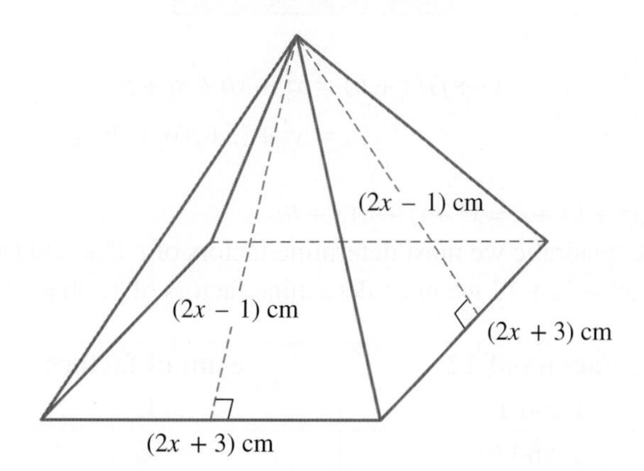

36. Expand the following.

a. $(2x + 3y - 5z)^2$

b. $\left(\left(1 + \frac{1}{2x}\right) - 2x\right)^2$

7.3 Factorising expressions with three terms

LEARNING INTENTION

At the end of this subtopic you should be able to:
- recognise monic and non-monic quadratic expressions
- factorise monic quadratic trinomials
- factorise non-monic quadratic trinomials.

7.3.1 Factorising monic quadratic trinomials

eles-4829

- A monic quadratic expression is an expression in the form $ax^2 + bx + c$ where $a = 1$.
- To factorise a monic quadratic expression $x^2 + bx + c$, we need to determine the numbers f and h such that $x^2 + bx + c = (x + f)(x + h)$.
- Using FOIL, or the area model, $(x + f)(x + h)$ can be expanded as follows:

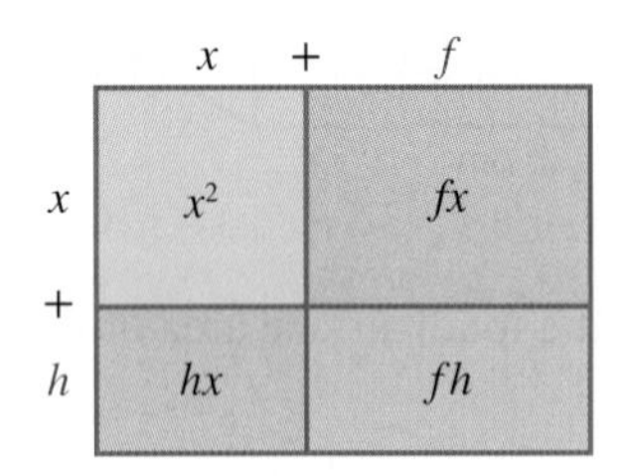

$$\begin{aligned}(x + f)(x + h) &= x^2 + xh + xf + fh \\ &= x^2 + (f + h)x + fh\end{aligned}$$

- Equating the quadratics: $x^2 + bx + c = x^2 + (f + h)x + fh$.
- Therefore, to factorise the quadratic we must determine factors of c that add to b.
- For example, to factorise $x^2 + 7x + 12$ we must determine factors of 12 that add to 7.

Factors of 12	Sum of factors
1 and 12	13
2 and 6	8
3 and 4	7

The factors of 12 that add to 7 are 3 and 4. Therefore, $x^2 + 7x + 12 = (x + 3)(x + 4)$.

- This result can be checked by expanding using FOIL.

$$\begin{aligned}(x + 3)(x + 4) &= x^2 + 4x + 3x + 12 \\ &= x^2 + 7x + 12\end{aligned}$$

Factorising a monic quadratic

$$\mathbf{x^2 + bx + c = (x + f)(x + h)}$$

where f and h are factors of c that sum to b.

WORKED EXAMPLE 5 Factorising monic quadratic trinomials

Factorise the following quadratic expressions.

a. $x^2 + 5x + 6$

b. $x^2 + 10x + 24$

THINK

a. 1. List all of the factors of 6 and determine the sum of each pair of factors. Highlight the pair of factors which add to 5.

WRITE

Factors of 6	Sum of factors
1 and 6	7
2 and 3	**5**

2. Factorise the quadratic using the factor pair highlighted in step 1.

$x^2 + 5x + 6 = (x + 2)(x + 3)$

b. 1. List all of the factors of 24 and determine the sum of each pair of factors. Highlight the pair of factors which add to 10.

Factors of 24	Sum of factors
1 and 24	25
2 and 12	14
3 and 8	11
4 and 6	**10**

2. Factorise the quadratic using the factor pair highlighted in step 1.

$x^2 + 10x + 24 = (x + 4)(x + 6)$

7.3.2 Factorising non-monic quadratic trinomials

eles-4830

- A non-monic quadratic expression is an expression in the form $ax^2 + bx + c$, where $a \neq 1$.
- Quadratic expressions of the form $ax^2 + bx + c$ can be factorised by following the process shown below. The example in the right column will help you to understand each step.

Process	Example: $2x^2 + 11x + 12$
Step 1: Determine factors of ac that add up to b.	$a = 2,\ b = 11, c = 12,\ ac = 24$ The factors of 24 that add up to 11 are 8 and 3.
Step 2: Rewrite the quadratic expression as $ax^2 + mx + nx + c$, where m and n are the factors of ac that add up to b (as found in step 1).	$2x^2 + 11x + 12 = 2x^2 + 8x + 3x + 12$
Step 3: Order the terms in $ax^2 + mx + nx + c$ so that the two terms on the left share a common factor and the two terms on the right also share a common factor.	$= \left(2x^2 + 8x\right) + (3x + 12)$
Step 4: Factorise the two left terms and two right terms independently by taking out their common factor.	$= 2x(x + 4) + 3(x + 4)$
Step 5: Factorise the expression by taking out the binomial factor.	$= (x + 4)(2x + 3)$

- The reasoning behind step 1 is not given in the table above. The working below shows why we use factors of ac that add to b. You do not need to be able to complete this working yourself, but try to read along and understand it.

$$(dx+e)(fx+g) = dfx^2 + dgx + efx + eg$$
$$= dfx^2 + (dg+ef)x + eg$$

$$m+n = dg+ef \quad \text{and} \quad m \times n = dg \times ef$$
$$= b \qquad\qquad = dgef$$
$$= dfeg$$
$$= ac$$

Factorising a non-monic quadratic

To factorise a general quadratic of the form ax^2+bx+c, find factors of ac that sum to b. Then rewrite the expression as four terms that can then be regrouped and factorised.

$$ax^2 + bx + c = ax^2 + mx + nx + c$$

Factors of ac that sum to b

WORKED EXAMPLE 6 Factorising a non-monic quadratic trinomial

Factorise $6x^2 - 11x - 10$.

THINK	WRITE
1. Write the expression and look for common factors and special patterns. The expression is a general quadratic with $a=6$, $b=-11$ and $c=-10$.	$6x^2 - 11x - 10$
2. List factors of −60 and determine the sum of each pair of factors. Highlight the pair of factors which add to −11.	(see table below)
3. Rewrite the quadratic expression: $ax^2+bx+c = ax^2+mx+nx+c$ with $m=4$ and $n=-15$.	$6x^2 - 11x - 10 = 6x^2 + 4x + -15x - 10$
4. Factorise using the grouping method: $6x^2+4x = 2x(3x+2)$ and $-15x-10 = -5(3x+2)$. Write the answer.	$6x^2 - 11x - 10 = 2x(3x+2) + -5(3x+2)$ $= (3x+2)(2x-5)$

Factors of −60 (6 × −10)	Sum of factors
−60, 1	−59
−20, 3	−17
−30, 2	−28
15, − 4	11
−15, 4	**−11**

Resources

eWorkbook Topic 7 Workbook (worksheets, code puzzle and project) (ewbk-2033)

Digital document SkillSHEET Finding a factor pair that adds to a given number (doc-5250)

Video eLesson Factorisation of trinomials (eles-1921)

Interactivities Individual pathway interactivity: Factorising expressions with three terms (int-4597)
Factorising monic quadratic trinomials (int-6143)
Factorising trinomials by grouping (int-6144)

Exercise 7.3 Factorising expressions with three terms

learnON

Individual pathways

PRACTISE	CONSOLIDATE	MASTER
1, 4, 7, 10, 12, 13, 16, 19, 22, 25	2, 5, 8, 11, 14, 17, 20, 23, 26, 27	3, 6, 9, 15, 18, 21, 24, 28, 29

To answer questions online and to receive **immediate feedback** and **sample responses** for every question, go to your learnON title at www.jacplus.com.au.

Fluency

WE5 For questions 1–3, factorise the expressions.

1. a. x^2+3x+2 b. x^2+4x+3 c. $x^2+10x+16$
   d. $x^2+8x+16$ e. x^2-2x-3

2. a. x^2-3x-4 b. $x^2-11x-12$ c. $x^2-4x-12$
   d. x^2+3x-4 e. x^2+4x-5

3. a. x^2+6x-7 b. $x^2+3x-10$ c. x^2-4x+3
   d. $x^2-9x+20$ e. $x^2+9x-70$

WE6 For questions 4–9, factorise the expressions.

4. a. $-2x^2-20x-18$ b. $-3x^2-9x-6$ c. $-x^2-3x-2$ d. $-x^2-11x-10$

5. a. $-x^2-7x-10$ b. $-x^2-13x-12$ c. $-x^2-7x-12$ d. $-x^2-8x-12$

6. a. $2x^2+14x+20$ b. $3x^2+33x+30$ c. $5x^2+105x+100$ d. $5x^2+45x+100$

7. a. a^2-6a-7 b. t^2-6t+8 c. b^2+5b+4 d. $m^2+2m-15$

8. a. $p^2-13p-48$ b. $c^2+13c-48$ c. $k^2+22k+57$ d. $s^2-16s-57$

9. a. g^2-g-72 b. $v^2-28v+75$ c. $x^2+14x-32$ d. $x^2-19x+60$

10. MC To factorise $-14x^2-49x+21$, the first step is to:
    A. Determine factors of 14 and 21 that will add to −49.
    B. Take out 14 as a common factor.
    C. Take out −7 as a common factor.
    D. Determine factors of 14 and −49 that will add to make 21.
    E. Take out −14 as a common factor.

11. MC The expression $42x^2-9x-6$ can be completely factorised to:
    A. $(6x-3)(7x+2)$ B. $3(2x-1)(7x+2)$ C. $(2x-1)(21x+6)$
    D. $3(2x+1)(7x-2)$ E. $42(x-3)(x+2)$

12. MC When factorised, $(x+2)^2-(y+3)^2$ equals:
    A. $(x+y-2)(x+y+2)$ B. $(x-y-1)(x+y-1)$ C. $(x-y-1)(x+y+5)$
    D. $(x-y+1)(x+y+5)$ E. $(x+y-1)(x+y+2)$

Understanding

For questions 13–15, factorise the expressions using an appropriate method.

13. a. $2x^2 + 5x + 2$ b. $2x^2 - 3x + 1$ c. $4x^2 - 17x - 15$ d. $4x^2 + 4x - 3$

14. a. $2x^2 - 9x - 35$ b. $3x^2 + 10x + 3$ c. $6x^2 - 17x + 7$ d. $12x^2 - 13x - 14$

15. a. $10x^2 - 9x - 9$ b. $20x^2 + 3x - 2$ c. $12x^2 + 5x - 2$ d. $15x^2 + x - 2$

For questions 16–18, factorise the expressions. Remember to look for a common factor first.

16. a. $4x^2 + 2x - 6$ b. $9x^2 - 60x - 21$ c. $72x^2 + 12x - 12$ d. $-18x^2 + 3x + 3$

17. a. $-60x^2 + 150x + 90$ b. $24ax^2 + 18ax - 105a$ c. $-8x^2 + 22x - 12$ d. $-10x^2 + 31x + 14$

18. a. $-24x^2 + 35x - 4$ b. $-12x^2 - 2xy + 2y^2$
c. $-30x^2 + 85xy + 70y^2$ d. $-600x^2 - 780xy - 252y^2$

19. Consider the expression $(x - 1)^2 + 5(x - 1) - 6$.
   a. Substitute $w = x - 1$ in this expression.
   b. Factorise the resulting quadratic.
   c. Replace w with $x - 1$ and simplify each factor. This is the factorised form of the original expression.

For questions 20 and 21, use the method outlined in question 19 to factorise the expressions.

20. a. $(x + 1)^2 + 3(x + 1) - 4$ b. $(x + 2)^2 + (x + 2) - 6$ c. $(x - 3)^2 + 4(x - 3) + 4$

21. a. $(x + 3)^2 + 8(x + 3) + 12$ b. $(x - 7)^2 - 7(x - 7) - 8$ c. $(x - 5)^2 - 3(x - 5) - 10$

Reasoning

22. Fabric pieces comprising yellow squares, white squares and black rectangles are sewn together to make larger squares (patches) as shown in the diagram. The length of each black rectangle is twice its width. These patches are then sewn together to make a patchwork quilt. A finished square quilt, made from 100 patches, has an area of $1.44\ m^2$.

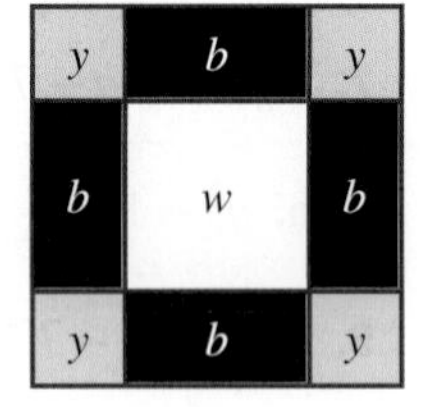

   a. Determine the size of each yellow, black and white section in one fabric piece. Show your working.
   b. Determine how much $\left(\text{in } m^2\right)$ of each of the coloured fabrics would be needed to construct the quilt. (Ignore seam allowances.)
   c. Sketch a section of the finished product.

23. Each factorisation below contains an error. Identify the error in each statement.
   a. $x^2 - 7x + 12 = (x + 3)(x - 4)$ b. $x^2 - x - 12 = (x - 3)(x + 4)$
   c. $x^2 - x - 2 = (x - 1)(x + 2)$ d. $x^2 - 4x - 21 = (x - 3)(x - 7)$

24. Each factorisation below contains an error. Identify the error in each statement.
   a. $x^2 + 4x - 21 = (x + 3)(x - 7)$ b. $x^2 - x - 30 = (x - 5)(x + 6)$
   c. $x^2 + 7x - 8 = (x + 1)(x - 8)$ d. $x^2 - 11x + 30 = (x - 5)(x + 6)$

Problem solving

25. Cameron wants to build an in-ground 'endless' pool. Basic models have a depth of 2 metres and a length triple the width. A spa will also be attached to the end of the pool.
   a. The pool needs to be tiled. Write an expression for the surface area of the empty pool (that is, the floor and walls only).
   b. The spa needs an additional $16\ m^2$ of tiles. Write an expression for the total area of tiles needed for both the pool and the spa.
   c. Factorise this expression.

d. Cameron decides to use tiles that are selling at a discount price, but there are only 280 m^2 of the tile available. Determine the maximum dimensions of the pool he can build if the width is in whole metres. Assume the spa is to be included in the tiling.
e. Evaluate the area of tiles that is actually needed to construct the spa and pool.
f. Determine the volume of water the pool can hold.

26. Factorise $x^2 + x - 0.75$.

27. The area of a rectangular playground is given by the general expression $\left(6x^2 + 11x + 3\right)$ m^2 where x is a positive whole number.

a. Determine the length and width of the playground in terms of x.
b. Write an expression for the perimeter of the playground.
c. If the perimeter of a particular playground is 88 metres, evaluate x.

28. Factorise:

a. $6(3a-1)^2 - 13(3a-1) - 5$
b. $3m^4 - 19m^2 - 14$
c. $2\sin^2(x) - 3\sin(x) + 1$.

29. Students decide to make Science Day invitation cards.
The total area of each card is equal to $(x^2 - 4x - 5)$ cm^2.

a. Factorise the expression to determine the dimensions of the cards in terms of x.
b. Write down the length of the shorter side in terms of x.
c. If the shorter side of a card is 10 cm in length and the longer side is 16 cm in length, determine the value of x.
d. Evaluate the area of the card proposed in part **c**.
e. If the students want to make 3000 Science Day invitation cards, determine how much cardboard will be required. Give your answer in terms of x.

7.4 Factorising expressions with two or four terms

LEARNING INTENTION

At the end of this subtopic you should be able to:
- factorise expressions by taking out the highest common factor
- recognise and factorise expressions that can be written as DOTS
- factorise expressions with four terms by grouping.

7.4.1 Factorising expressions with two terms

eles-4831

- To factorise an expression with two terms, follow the following steps:
  Step 1: look for the highest common factor (HCF) of the terms
  Step 2: take out the HCF
  Step 3: check to see if the remaining expression is a DOTS.
  For example:

$$\begin{aligned} 4x^2 - 36 &= 4\left(x^2 - 9\right) && \textbf{taking out the HCF of 4} \\ &= 4(x-3)(x+3) && \textbf{using DOTS} \end{aligned}$$

Remember, the formula for the difference of two squares (DOTS) is:

$$a^2 - b^2 = (a-b)(a+b)$$

WORKED EXAMPLE 7 Factorising expressions with two terms

Factorise the following.

a. $12k^2 + 18$

b. $16a^2 - 25b^4$

THINK	WRITE
a. 1. Write the expression and look for common factors. The terms have a highest common factor of 6. Write the 6 in front of a set of brackets, then determine what must go inside the brackets. $12k^2 = 6 \times 2k^2$, $18 = 6 \times 3$	a. $12k^2 + 18 = 6\left(2k^2 + 3\right)$
2. Look for patterns in the expression inside the brackets to factorise further. The expression inside the brackets cannot be factorised further.	
b. 1. Write the expression and look for common factors. The expression has no common factors.	b. $16a^2 - 25b^4$
2. Look for the DOTS pattern in the expression. Write the equation showing squares.	$= 4^2a^2 - 5^2\left(b^2\right)^2$ $= (4a)^2 - \left(5b^2\right)^2$
3. Use the pattern for DOTS to write the factors. $a^2 - b^2 = (a+b)(a-b)$	$= \left(4a + 5b^2\right)\left(4a - 5b^2\right)$

7.4.2 Factorising expressions with four terms

eles-4832

- To factorise an expression with four terms, follow the following steps:
  Step 1: look for a common factor
  Step 2: group in pairs that have a common factor
  Step 3: factorise each pair by taking out the common factor
  Step 4: factorise fully by taking out the common binomial factor
- This process is known as grouping 'two and two'.
  For example,

$10a - 30b + 15ac - 45bc$	**terms have a common factor of 5**
$= 5\,(2a - 6b + 3ac - 9bc)$	**grouping in pairs**
$= 5\,(2(a - 3b) + 3c(a - 3b))$	**factorising each pair**
$= 5\,(a - 3b)(2 + 3c)$	**common binomial factor**

WORKED EXAMPLE 8 Factorising expressions with four terms by grouping 'two and two'

Factorise each of the following.

a. $x - 4y + mx - 4my$

b. $x^2 + 3x - y^2 + 3y$

THINK	WRITE
a. 1. Write the expression and look for a common factor. (There isn't one.)	a. $x - 4y + mx - 4my$
2. Group the terms so that those with common factors are next to each other.	$= (x - 4y) + (mx - 4my)$

3. Take out a common factor from each group (it may be 1).	$= 1(x - 4y) + m(x - 4y)$
4. Factorise by taking out a common binomial factor. The factor $(x - 4y)$ is common to both groups.	$= (x - 4y)(1 + m)$
b. 1. Write the expression and look for a common factor.	**b.** $x^2 + 3x - y^2 + 3y$
2. Group the terms so that those with common factors are next to each other.	$= \left(x^2 - y^2\right) + (3x + 3y)$
3. Factorise each group.	$= (x + y)(x - y) + 3(x + y)$
4. Factorise by taking out a common binomial factor. The factor $(x + y)$ is common to both groups.	$= (x + y)(x - y + 3)$

- Sometimes an expression containing four terms can be factorised by grouping three terms together and then factorising. This method is known as grouping "three and one" and it is demonstrated in the Worked example 9.

WORKED EXAMPLE 9 Factorising by grouping 'three and one'

Factorise the following expression: $x^2 + 12x + 36 - y^2$.

THINK	WRITE
1. Write the expression and look for a common factor.	$x^2 + 12x + 36 - y^2$
2. Group the terms so that those that can be factorised are next to each other.	$= \left(x^2 + 12x + 36\right) - y^2$
3. Factorise the quadratic trinomial. This is the form of a perfect square.	$= (x + 6)(x + 6) - y^2$ $= (x + 6)^2 - y^2$
4. Factorise the expression using $a^2 - b^2 = (a + b)(a - b)$.	$= (x + 6 + y)(x + 6 - y)$

on Resources

eWorkbook Topic 7 Workbook (worksheets, code puzzle and project) (ewbk-2033)

Digital documents SkillSHEET Factorising by taking out the highest common factor (doc-5246)
SkillSHEET Factorising by taking out a common binomial factor (doc-5247)

Interactivities Individual pathway interactivity: Factorising expressions with two or four terms (int-4598)
Factorising expressions with four terms (int-6145)

Exercise 7.4 Factorising expressions with two or four terms

learnON

Individual pathways

PRACTISE	CONSOLIDATE	MASTER
1, 4, 6, 9, 12, 15, 16, 19, 22, 25, 31, 32, 35, 38	2, 5, 7, 10, 13, 17, 20, 23, 26, 27, 29, 33, 36, 39	3, 8, 11, 14, 18, 21, 24, 28, 30, 34, 37, 40

To answer questions online and to receive **immediate feedback** and **sample responses** for every question, go to your learnON title at www.jacplus.com.au.

Fluency

For questions 1–3, factorise the expressions by taking out a common factor.

1. a. $x^2 + 3x$ b. $x^2 - 4x$ c. $3x^2 - 6x$
2. a. $4x^2 + 16x$ b. $9x^2 - 3x$ c. $8x - 8x^2$
3. a. $12x - 3x^2$ b. $8x - 12x^2$ c. $8x^2 - 11x$

For questions 4 and 5, factorise the expressions by taking out a common binomial factor.

4. a. $3x(x-2) + 2(x-2)$ b. $5(x+3) - 2x(x+3)$ c. $(x-1)^2 + 6(x-1)$
5. a. $(x+1)^2 - 2(x+1)$ b. $(x+4)(x-4) + 2(x+4)$ c. $7(x-3) - (x+3)(x-3)$

WE7 For questions 6–11, factorise the expressions.

6. a. $x^2 - 1$ b. $x^2 - 9$ c. $x^2 - 25$
7. a. $x^2 - 100$ b. $y^2 - k^2$ c. $4x^2 - 9y^2$
8. a. $16a^2 - 49$ b. $25p^2 - 36q^2$ c. $1 - 100d^2$
9. a. $4x^2 - 4$ b. $5x^2 - 80$ c. $ax^2 - 9a$
10. a. $2b^2 - 8d^2$ b. $100x^2 - 1600$ c. $3ax^2 - 147a$
11. a. $4px^2 - 256p$ b. $36x^2 - 16$ c. $108 - 3x^2$
12. MC If the factorised expression is $(x+7)(x-7)$, then the expanded expression must have been:
    A. $x^2 - 7$ B. $x^2 + 7$ C. $x^2 - 49$ D. $x^2 + 49$ E. $x^2 - 14x + 49$
13. MC If the factorised expression is $\left(\frac{x}{4} - \frac{3}{5}\right)\left(\frac{x}{4} + \frac{3}{5}\right)$ then the original expression must have been:
    A. $\frac{x^2}{4} - \frac{3}{5}$
    B. $\frac{x^2}{16} - \frac{9}{25}$
    C. $\frac{x^2}{4} - \frac{\left(\sqrt{3}\right)^2}{\left(\sqrt{5}\right)^2}$
    D. $\frac{x^2}{4} - \frac{9}{25}$
    E. $\frac{x^2}{16} - \frac{\left(\sqrt{3}\right)^2}{\left(\sqrt{5}\right)^2}$
14. MC The factorised form of $64x^2 - 9y^2$ is:
    A. $(64x + 9y)(64x - 9y)$ B. $(8x + 3y)(8x - 3y)$ C. $(8x - 3y)(8x - 3y)$
    D. $(8x + 3y)(8x + 3y)$ E. $(16x + 3y)(16x - 3y)$

15. MC Which of the following expressions would be factorised by grouping 'two and two'?

A. $x^2 - a^2 + 12a - 36$
B. $x^2 - 7x - 10$
C. $2x^2 - 6x - xy + 3y$
D. $(s-5)^2 - 25(s+3)^2$
E. $(r+5) - (r+3)(r+5)$

For questions 16–18, factorise the expressions over the set of real numbers.

16. a. $x^2 - 11$ b. $x^2 - 7$ c. $x^2 - 15$

17. a. $4x^2 - 13$ b. $9x^2 - 19$ c. $3x^2 - 66$

18. a. $5x^2 - 15$ b. $2x^2 - 4$ c. $12x^2 - 36$

Understanding

For questions 19–21, factorise the expressions.

19. a. $(x-1)^2 - 4$ b. $(x+1)^2 - 25$ c. $(x-2)^2 - 9$

20. a. $(x+3)^2 - 16$ b. $49 - (x+1)^2$ c. $36 - (x-4)^2$

21. a. $(x-1)^2 - (x-5)^2$ b. $4(x+2)^2 - 9(x-1)^2$ c. $25(x-2)^2 - 16(x+3)^2$

WE8a For questions 22–26, factorise the expressions.

22. a. $x - 2y + ax - 2ay$ b. $2x + ax + 2y + ay$ c. $ax - ay + bx - by$ d. $4x + 4y + xz + yz$

23. a. $ef - 2e + 3f - 6$ b. $mn - 7m + n - 7$
c. $6rt - 3st + 6ru - 3su$ d. $7mn - 21n + 35m - 105$

24. a. $64 - 8j + 16k - 2jk$ b. $3a^2 - a^2b + 3ac - abc$
c. $5x^2 + 10x + x^2y + 2xy$ d. $2m^2 - m^2n + 2mn - mn^2$

25. a. $xy + 7x - 2y - 14$ b. $mn + 2n - 3m - 6$ c. $pq + 5p - 3q - 15$

26. a. $s^2 + 3s - 4st - 12t$ b. $a^2b - cd - bc + a^2d$ c. $xy - z - 5z^2 + 5xyz$

WE8b For questions 27 and 28, factorise the expressions.

27. a. $a^2 - b^2 + 4a - 4b$ b. $p^2 - q^2 - 3p + 3q$ c. $m^2 - n^2 + lm + ln$

28. a. $7x + 7y + x^2 - y^2$ b. $5p - 10pq + 1 - 4q^2$ c. $49g^2 - 36h^2 - 28g - 24h$

WE9 For questions 29 and 30, factorise the expressions.

29. a. $x^2 + 14x + 49 - y^2$ b. $x^2 + 20x + 100 - y^2$ c. $a^2 - 22a + 121 - b^2$

30. a. $9a^2 + 12a + 4 - b^2$ b. $25p^2 - 40p + 16 - 9t^2$ c. $36t^2 - 12t + 1 - 5v^2$

31. MC In the expression $3(x-2) + 4y(x-2)$, the common binomial factor is:

A. $3 + 4y$ B. $3 - 4y$ C. x D. $-x + 2$ E. $x - 2$

32. MC Identify which of the following terms is a perfect square.

A. 9 B. $(x+1)(x-1)$ C. $3x^2$ D. $5(a+b)^2$ E. $25x$

33. MC Identify which of the following expressions can be factorised using grouping.

A. $x^2 - y^2$
B. $1 + 4y - 2xy + 4x^2$
C. $3a^2 + 8a + 4$
D. $x^2 + x + y - y^2$
E. $2a + 4b - 6ab + 18$

34. **MC** When factorised, $6(a+b)-x(a+b)$ equals:

A. $6-x(a+b)$
B. $(6-x)(a+b)$
C. $6(a+b-x)$
D. $(6+x)(a-b)$
E. $(6+x)(a+b)$

Reasoning

35. Jack and Jill both attempt to factorise the expression $4x^2-12x+8$.

Jack's solution	Jill's solution
$4x^2-12x+8$ $=4(x^2-3x+2)$ $=4(x-1)(x+3)$	$4x^2-12x+8$ $=4(x^2-3x+2)$ $=4(x-2)(x-1)$

a. State who was correct.
b. Explain what error was made in the incorrect solution.

36. Jack and Jill attempted to factorise another expression. Their solutions are given below. Explain why they both had the same correct answer but their working was different.

Jack's solution	Jill's solution
$6x^2+7x-20$ $=6x^2-8x+15x-20$ $=2x(3x-4)+5(3x-4)$ $=(3x-4)(2x+5)$	$6x^2+7x-20$ $=6x^2+15x-8x-20$ $=3x(2x+5)-4(2x+5)$ $=(2x+5)(3x-4)$

37. Factorise the following expressions. Explain your answer.

a. $x^2-4xy+4y^2-a^2+6ab-b^2$
b. $12x^2-75y^2-9(4x-3)$

Problem solving

38. The area of a rectangle is $\left(x^2-25\right)$ cm^2.

a. Factorise the expression.
b. Determine the length of the rectangle if the width is $(x+5)$ cm.
c. If $x=7$ cm, calculate the dimensions of the rectangle.
d. Hence, evaluate the area of the rectangle.
e. If $x=13$ cm, determine how much bigger the area of this rectangle would be.

39. A circular garden of diameter $2r$ m is to have a gravel path laid around it. The path is to be 1 m wide.

a. Determine the area of the garden in terms of r.
b. Determine the area of the garden and path together in terms of r, using the formula for the area of a circle.
c. Write an expression for the area of the path in fully factorised form.
d. If the radius of the garden is 5 m, determine the area of the path, correct to 2 decimal places.

40. A roll of material is $(x+2)$ metres wide. Annie buys $(x+3)$ metres of the material and Bronwyn buys 5 metres of the material.
   a. Write an expression, in terms of x, for the area of each piece of material purchased.
   b. If Annie has bought more material than Bronwyn, write an expression for how much more she has than Bronwyn.
   c. Factorise and simplify this expression.
   d. Evaluate the width of the material if Annie has 5 m^2 more than Bronwyn.
   e. Determine how much material each person has.

7.5 Factorising by completing the square

LEARNING INTENTION

At the end of this subtopic you should be able to:
- complete the square for a given quadratic expression
- factorise a quadratic expression by completing the square.

7.5.1 Completing the square

eles-4833

- **Completing the square** is the process of writing a general quadratic expression in turning point form.

$$\underbrace{ax^2+bx+c}_{\text{General form}} = \underbrace{a(x-h)^2+k}_{\text{Turning point form}}$$

- The expression x^2+8x can be modelled as a square with a smaller square missing from the corner, as shown below.

$$x^2+8x = (x+4)^2-(4)^2$$

- In 'completing the square', the general equation is written as the area of the large square minus the area of the small square.
- In general, to complete the square for $x^2 + bx$, the small square has a side length equal to half of the coefficient of x; that is, the area of the small square is $\left(\frac{b}{2}\right)^2$.

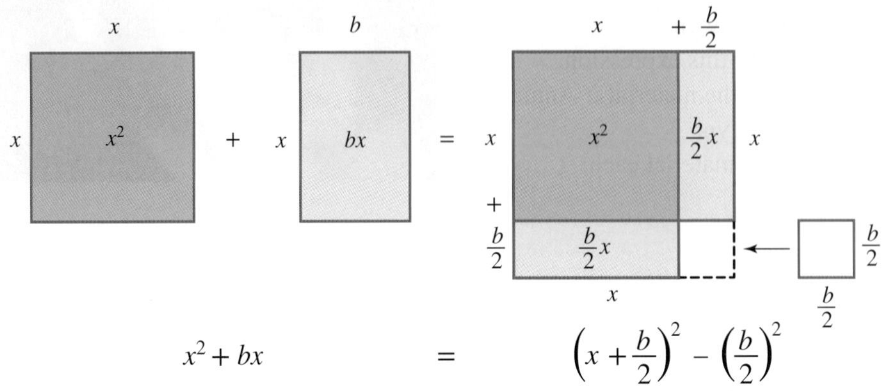

$$x^2 + bx = \left(x + \frac{b}{2}\right)^2 - \left(\frac{b}{2}\right)^2$$

- The process of completing the square is sometimes described as the process of adding the square of half of the coefficient of x then subtracting it, as shown in purple below. The result of this process is a perfect square that is then factorised, as shown in blue.

$$\begin{aligned} x^2 + bx &= x^2 + bx + \left(\frac{b}{2}\right)^2 - \left(\frac{b}{2}\right)^2 \\ &= x^2 + bx + \left(\frac{b}{2}\right)^2 - \left(\frac{b}{2}\right)^2 \\ &= \left(x + \frac{b}{2}\right)^2 - \left(\frac{b}{2}\right)^2 \end{aligned}$$

WORKED EXAMPLE 10 Converting expressions into turning point form

Write the following in turning point form by completing the square.

a. $\mathbf{x^2 + 4x}$

b. $\mathbf{x^2 + 7x + 1}$

THINK	WRITE
a. • The square will consist of a square that has an area of x^2 and two identical rectangles with a total area of $4x$. • The length of the large square is $(x + 2)$ so its area is $(x + 2)^2$. • The area of the smaller square is $(2)^2$. • Write $x^2 + 4x$ in turning point form.	a. 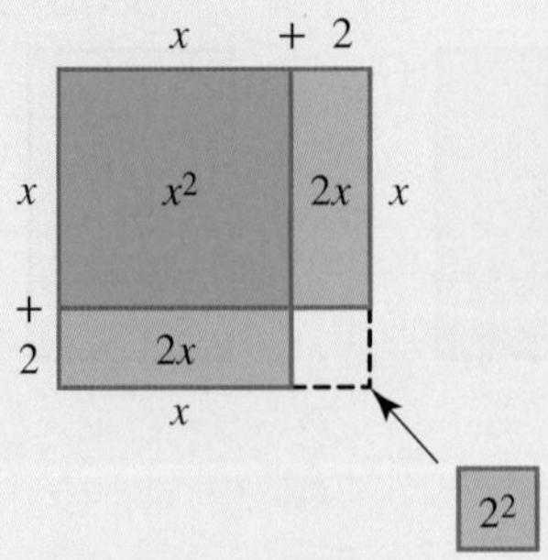 $x^2 + 4x = (x + 2)^2 - (2)^2$ $= (x + 2)^2 - 4$

b. 1. • Complete the square with the terms containing x.
• The square will consist of a square that has an area of x^2 and two identical rectangles with a total area of $7x$.
• The length of the large square is $\left(x+\frac{7}{2}\right)$ so its area is $\left(x+\frac{7}{2}\right)^2$.
• The area of the smaller square is $\left(\frac{7}{2}\right)^2$.
• Write x^2+7x+1 in turning point form.

b.

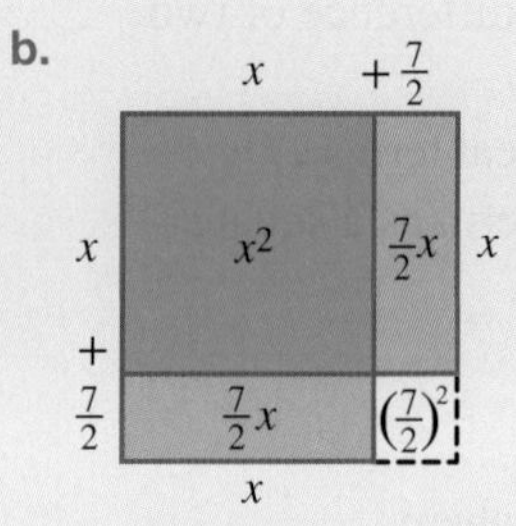

$$x^2+7x+1=\left(x+\frac{7}{2}\right)^2-\left(\frac{7}{2}\right)^2+1$$

2. Simplify the last two terms.

$$=\left(x+\frac{7}{2}\right)^2-\frac{49}{4}+\frac{4}{4}$$

$$=\left(x+\frac{7}{2}\right)^2-\frac{45}{4}$$

7.5.2 Factorising by completing the square

eles-4834

- When an equation is written in turning point form, it can be factorised as a difference of two squares. For example, factorise x^2+8x+2 by completing the square.

$$x^2+8x+\left(\frac{8}{2}\right)^2-\left(\frac{8}{2}\right)^2+2$$

$$=x^2+8x+(4)^2-(4)^2+2$$
$$=x^2+8x+16-16+2$$
$$=(x+4)^2-14$$
$$=(x+4)^2-14$$
$$=\left(x+4-\sqrt{14}\right)\left(x+4+\sqrt{14}\right)$$

WORKED EXAMPLE 11 Factorising by completing the square

Factorise the following by completing the square.

a. $\boldsymbol{x^2+4x+2}$

b. $\boldsymbol{x^2-9x+1}$

THINK

a. 1. To complete the square, add the square of half of the coefficient of x and then subtract it.

2. Write the perfect square created in its factorised form.

WRITE

a. $x^2+4x+2=x^2+4x+\left(\frac{4}{2}\right)^2-\left(\frac{4}{2}\right)^2+2$

$$=x^2+4x+(2)^2-(2)^2+2$$
$$=(x+2)^2-(2)^2+2$$

THINK	WRITE
3. Write the expression as a difference of two squares by: • simplifying the numerical terms • writing the numerical term as a square $\left(2=\left(\sqrt{2}\right)^2\right)$.	$=(x+2)^2-4+2$ $=(x+2)^2-2$ $=(x+2)^2-\left(\sqrt{2}\right)^2$
4. Use the pattern for DOTS, $a^2-b^2=(a-b)(a+b)$, where $a=(x+2)$ and $b=\sqrt{2}$.	$=\left(x+2+\sqrt{2}\right)\left(x+2-\sqrt{2}\right)$
b. 1. To complete the square, add the square of half of the coefficient of x, then subtract it.	**b.** $x^2-9x+1=x^2-9x+\left(\frac{9}{2}\right)^2-\left(\frac{9}{2}\right)^2+1$
2. Write the perfect square created in its factorised form.	$=x^2-9x+\left(\frac{9}{2}\right)^2-\left(\frac{9}{2}\right)^2+1$ $=\left(x-\frac{9}{2}\right)^2-\left(\frac{9}{2}\right)^2+1$
3. Write the expression as a difference of two squares by: • simplifying the numerical terms • writing the numerical term as a square. $\frac{77}{4}=\left(\sqrt{\frac{77}{4}}\right)^2=\left(\frac{\sqrt{77}}{2}\right)^2$	$=\left(x-\frac{9}{2}\right)^2-\frac{81}{4}+1$ $=\left(x-\frac{9}{2}\right)^2-\frac{77}{4}$ $=\left(x-\frac{9}{2}\right)^2-\left(\frac{\sqrt{77}}{2}\right)^2$
4. Use the pattern for DOTS: $a^2-b^2=(a-b)(a+b)$, where $a=\left(x+-\frac{9}{2}\right)$ and $b=\frac{\sqrt{77}}{2}$.	$=\left(x-\frac{9}{2}+\frac{\sqrt{77}}{2}\right)\left(x-\frac{9}{2}-\frac{\sqrt{77}}{2}\right)$

TI | THINK

a–b.

In a new problem, on a Calculator page, press:
- CATALOG
- 1
- F

Then scroll down to select factor(

Complete the entry lines as:

factor $\left(x^2+4x+2, x\right)$

factor $\left(x^2-9x+1, x\right)$

Press EXE after each entry.

DISPLAY/WRITE

a–b.

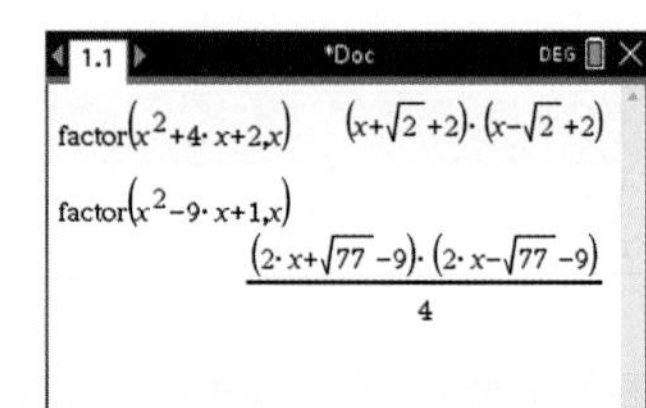

$$x^2+4x+2=\left(x+\sqrt{2}+2\right)\left(x-\sqrt{2}+2\right)$$

$$x^2-9x+1=\frac{\left(2x+\sqrt{77}-9\right)\left(2x-\sqrt{77}-9\right)}{4}$$

CASIO | THINK

a–b.

On the Main screen, tap:
- Action
- Transformation
- factor
- rFactor

Complete the entry lines as:

rfactor $\left(x^2+4x+2\right)$

rfactor $\left(x^2-9x+1\right)$

Press EXE after each entry.

DISPLAY/WRITE

a–b.

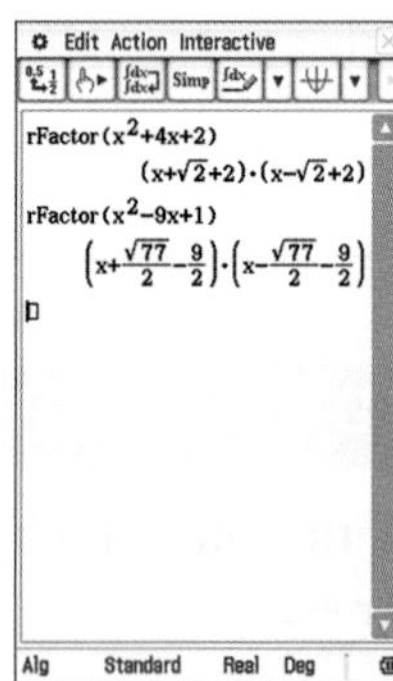

$$x^2+4x+2=\left(x+\sqrt{2}+2\right)\left(x-\sqrt{2}+2\right)$$

$$x^2-9x+1=\left(x+\frac{\sqrt{77}}{2}-\frac{9}{2}\right)\left(x-\frac{\sqrt{77}}{2}-\frac{9}{2}\right)$$

- Remember that you can expand the brackets to check your answer.
- If the coefficient of $x^2 \neq 1$, factorise the expression before completing the square.
- For example, $2x^2 - 8x + 2$

$$
\begin{aligned}
&= 2(x^2 - 4x + 1) \\
&= 2\left(x^2 - 4x + \left(\frac{-4}{2}\right)^2 - \left(\frac{-4}{2}\right)^2 + 1\right) \\
&= 2((x-2)^2 - 4 + 1) \\
&= 2((x-2)^2 - 3) \\
&= 2\left(x - 2 - \sqrt{3}\right)\left(x - 2 + \sqrt{3}\right)
\end{aligned}
$$

on Resources

eWorkbook Topic 7 Workbook (worksheets, code puzzle and project) (ewbk-2033)

Video eLesson Factorisation by completing the square (eles-1939)

Interactivities Individual pathway interactivity: Factorising by completing the square (int-4599)
Completing the square (int-2559)

Exercise 7.5 Factorising by completing the square

learnon

Individual pathways

PRACTISE	CONSOLIDATE	MASTER
1, 3, 4, 7, 10, 13, 16	2, 5, 8, 11, 14, 17	6, 9, 12, 15, 18

To answer questions online and to receive **immediate feedback** and **sample responses** for every question, go to your learnON title at www.jacplus.com.au.

Fluency

WE10 For questions 1 and 2, write the expressions in turning point form by completing the square.

1. a. $x^2 + 10x$ b. $x^2 + 6x$ c. $x^2 - 4x$ d. $x^2 + 16x$ e. $x^2 - 20x$

2. a. $x^2 + 8x$ b. $x^2 - 14x$ c. $x^2 + 50x$ d. $x^2 + 7x$ e. $x^2 - x$

WE11 For questions 3–6, factorise the expressions by completing the square.

3. a. $x^2 - 4x - 7$ b. $x^2 + 2x - 2$ c. $x^2 - 10x + 12$ d. $x^2 + 6x - 10$ e. $x^2 + 16x - 1$

4. a. $x^2 - 14x + 43$ b. $x^2 + 8x + 9$ c. $x^2 - 4x - 13$ d. $x^2 - 12x + 25$ e. $x^2 - 6x + 4$

5. a. $x^2 - x - 1$ b. $x^2 - 3x - 3$ c. $x^2 + x - 5$ d. $x^2 + 3x - 1$ e. $x^2 + 5x + 2$

6. a. $x^2 + 5x - 2$ b. $x^2 - 7x - 1$ c. $x^2 - 9x + 13$ d. $x^2 - x - 3$ e. $x^2 - x - 1$

For questions 7 and 8, factorise the expressions by first looking for a common factor and then completing the square.

7. a. $2x^2 + 4x - 4$
b. $4x^2 - 8x - 20$
c. $5x^2 + 30x + 5$
d. $3x^2 - 12x - 39$
e. $5x^2 - 30x + 10$

8. a. $6x^2 + 24x - 6$
b. $3x^2 + 30x + 39$
c. $2x^2 - 8x - 14$
d. $6x^2 + 36x - 30$
e. $4x^2 - 8x - 16$

Understanding

9. From the following list, determine which method of factorising is the most appropriate for each of the expressions given.
a. Factorising using common factors
b. Factorising using the difference of two squares rule
c. Factorising by grouping
d. Factorising quadratic trinomials
e. Completing the square
i. $3x^2 - 8x - 3$
ii. $49m^2 - 16n^2$
iii. $x^2 + 8x + 4 - y^2$
iv. $7x^2 - 28x$
v. $6a - 6b + a^2 - b^2$
vi. $x^2 + x - 5$
vii. $(x - 3)^2 + 3(x - 3) - 10$
viii. $x^2 - 7x - 1$

10. **MC** To complete the square, the term which should be added to $x^2 + 4x$ is:
A. 16
B. 4
C. $4x$
D. 2
E. $2x$

11. **MC** To factorise the expression $x^2 - 3x + 1$, the term that must be both added and subtracted is:
A. 9
B. 3
C. $3x$
D. $\frac{3}{2}$
E. $\frac{9}{4}$

12. **MC** The factorised form of $x^2 - 6x + 2$ is:
A. $\left(x + 3 - \sqrt{7}\right)\left(x + 3 + \sqrt{7}\right)$
B. $\left(x + 3 - \sqrt{7}\right)\left(x - 3 + \sqrt{7}\right)$
C. $\left(x - 3 - \sqrt{7}\right)\left(x - 3 - \sqrt{7}\right)$
D. $\left(x - 3 - \sqrt{7}\right)\left(x + 3 + \sqrt{7}\right)$
E. $\left(x - 3 + \sqrt{7}\right)\left(x - 3 - \sqrt{7}\right)$

Reasoning

13. Show that $x^2 + 4x + 6$ cannot be factorised by completing the square.

14. A square measuring x cm in side length has a cm added to its length and b cm added to its width. The resulting rectangle has an area of $\left(x^2 + 6x + 3\right)$ cm^2. Evaluate a and b, correct to 2 decimal places.

15. Show that $2x^2 + 3x + 4$ cannot be factorised by completing the square.

Problem solving

16. Students were asked to choose one quadratic expression from a given list.
Peter chose $x^2 - 4x + 9$ and Annabelle chose $x^2 - 4x - 9$.
Using the technique of completing the square to write the expression in turning point form, determine which student was able to factorise their expression.

17. Use the technique of completion of the square to factorise $x^2 + 2(1 - p)x + p(p - 2)$.

18. For each of the following, complete the square to factorise the expression.
a. $2x^2 + 8x + 1$
b. $3x^2 - 7x + 5$

7.6 Mixed factorisation

LEARNING INTENTION

At the end of this subtopic you should be able to:
- factorise by grouping
- factorise by completing the square
- factorise using DOTS.

7.6.1 Mixed factorisation

eles-4835

- To factorise quadratic trinomials and other expressions, remember:

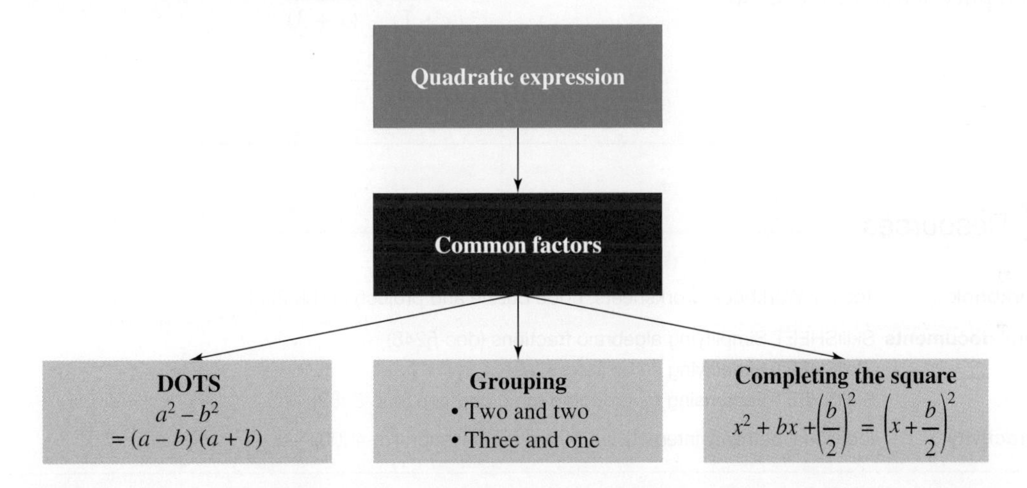

WORKED EXAMPLE 12 Applying mixed factorisation to expressions

Factorise and simplify each of the following expressions.

a. $\dfrac{6x-24}{x^2-16} \times \dfrac{5x+10}{8(x+2)}$

b. $\left(\dfrac{x^2+6x+8}{x^2+2x+1}\right) \div \left(\dfrac{x^2+5x+6}{(x+1)}\right)$

THINK	WRITE
a. 1. Write the expression.	a. $\dfrac{6x-24}{x^2-16} \times \dfrac{5x+10}{8(x+2)}$
2. Look for common factors.	$= \dfrac{6(x-4)}{x^2-16} \times \dfrac{5(x+2)}{8(x+2)}$
3. Look for the DOTS pattern in the expression. Use $a^2 - b^2 = (a+b)(a-b)$ to write the factors.	$= \dfrac{6(x-4)}{(x-4)(x+4)} \times \dfrac{5(x+2)}{8(x+2)}$
4. Cancel the common factors.	$= \dfrac{{}^{3}\cancel{6}\cancel{(x-4)}^{1}}{{}_{1}\cancel{(x-4)}(x+4)} \times \dfrac{5\cancel{(x+2)}^{1}}{{}_{4}\cancel{8}_{1}\cancel{(x+2)}}$
5. Simplify and write the answer.	$= \dfrac{3}{(x+4)} \times \dfrac{5}{4}$ $= \dfrac{15}{4(x+4)}$

b. 1. Write the expression.

$$\text{b. } \frac{x^2+6x+8}{x^2+2x+1} \div \frac{x^2+5x+6}{(x+1)}$$

2. Change the division to multiplication. Flip the second expression.

$$= \frac{x^2+6x+8}{x^2+2x+1} \times \frac{x+1}{x^2+5x+6}$$

3. Factorise and look for common factors.

$$= \frac{(x+2)(x+4)}{(x+1)^2} \times \frac{x+1}{(x+3)(x+2)}$$

4. Cancel the common factors.

$$= \frac{{}^{1}\cancel{(x+2)}(x+4)}{(x+1)^{\cancel{2}1}} \times \frac{\cancel{(x+1)}^{1}}{(x+3)_{1}\cancel{(x+2)}}$$

5. Simplify and write the answer.

$$= \frac{(x+4)}{(x+1)} \times \frac{1}{(x+3)}$$

$$= \frac{(x+4)}{(x+1)(x+3)}$$

on Resources

 **eWorkbook** Topic 7 Workbook (worksheets, code puzzle and project) (ewbk-2033)

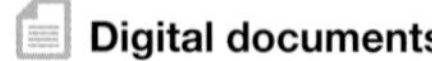 **Digital documents** SkillSHEET Simplifying algebraic fractions (doc-5248)
SkillSHEET Simplifying surds (doc-5249)
SkillSHEET Factorising by grouping three and one (doc-5252)

 **Interactivity** Individual pathway interactivity: Mixed factorisation (int-4600)

Exercise 7.6 Mixed factorisation

learnon

Individual pathways

■ PRACTISE	■ CONSOLIDATE	■ MASTER
1, 4, 7, 10, 11, 14, 17	2, 5, 8, 12, 15, 18	3, 6, 9, 13, 16, 19

To answer questions online and to receive **immediate feedback** and **sample responses** for every question, go to your learnON title at www.jacplus.com.au.

Fluency

For questions 1–9, factorise each of the following expressions.

1. a. $3x+9$ b. $x^2+4x+4-9y^2$ c. x^2-36
   d. x^2-49 e. $5x^2-9x-2$

2. a. $15x-20y$ b. $5c+de+dc+5e$ c. $5x^2-80$
   d. $-x^2-6x-5$ e. x^2+x-12

3. a. $mn+1+m+n$ b. x^2-7 c. $16x^2-4x$
   d. $5x^2+60x+100$ e. $18+9x-6y-3xy$

4. a. $x^2-8x+16-y^2$ b. $4x^2+8$ c. $fg+2h+2g+fh$
   d. x^2-5 e. $10mn-5n+10m-5$

5. a. x^2+6x+5
b. $x^2-10x-11$
c. x^2-4
d. $-5a+bc+ac-5b$
e. $xy-1+x-y$

6. a. $3x^2+5x+2$
b. $7x^2-28$
c. $-4x^2-28x-24$
d. $2p-rs+pr-2s$
e. $3x^2-27$

7. a. $-3u+tv+ut-3v$
b. x^2-11
c. $12x^2-7x+1$
d. $(x-1)^2-4$
e. $(x+2)^2-16$

8. a. $(2x+3)^2-25$
b. $3(x+5)^2-27$
c. $25-(x-2)^2$
d. $4(3-x)^2-16y^2$
e. $(x+2y)^2-(2x+y)^2$

9. a. $(x+3)^2-(x+1)^2$
b. $(2x-3y)^2-(x-y)^2$
c. $(x+3)^2+5(x+3)+4$
d. $(x-3)^2+3(x-3)-10$
e. $2(x+1)^2+5(x+1)+2$

Understanding

10. Consider the following product of algebraic fractions.

$$\frac{x^2+3x-10}{x^2-4}\times\frac{x^2+4x+4}{x^2-2x-8}$$

a. Factorise the expression in each numerator and denominator.
b. Cancel factors common to both the numerator and the denominator.
c. Simplify the expression as a single fraction.

WE12 For questions 11–13, factorise and simplify each of the following expressions.
Note: You may choose to follow the procedure in question 10.

11. a. $\dfrac{x^2-4x+3}{x^2-4x-12}\times\dfrac{x^2+5x+6}{x^2-9}$
b. $\dfrac{3x^2-17x+10}{6x^2+5x-6}\times\dfrac{x^2-1}{x^2-6x+5}$
c. $\dfrac{6x-12}{x^2-4}\times\dfrac{3x+6}{x(x-5)}$

12. a. $\dfrac{6x^2-x-2}{2x^2+3x+1}\times\dfrac{2x^2+x-1}{3x^2+10x-8}$
b. $\dfrac{x^2+4x-5}{x^2+x-2}\div\dfrac{x^2+10x+25}{x^2+4x+4}$
c. $\dfrac{x^2-7x+6}{x^2+x-2}\div\dfrac{x^2-x-12}{x^2-2x-8}$

13. a. $\dfrac{4ab+8a}{(c-3)}\div\dfrac{5ac+5a}{c2-2c-3}$
b. $\dfrac{p^2-7p}{p^2-49}\div\dfrac{p^2+p-6}{p^2+14p+49}$
c. $\dfrac{m^2+4m+4-n^2}{4m^2-4m-15}\div\dfrac{2m^2+4m-2mn}{10m^2+15m}$

Reasoning

14. a. Determine the original expression if the factorised expression is:

i. $\left(\dfrac{x}{4}+\dfrac{3}{5}\right)\left(\dfrac{x}{4}-\dfrac{3}{5}\right)$
ii. $\left(\dfrac{-x}{4}+\dfrac{3}{5}\right)\left(\dfrac{-x}{4}-\dfrac{3}{5}\right)$

b. Explain why your answers are the same.

15. Factorise the following using grouping 'three and one' and DOTS.

a. $x^2 - 18x + 81 - y^2$

b. $4x^2 + 12x - 16y^2 + 9$

16. **The expansion of perfect squares**
$(a+b)^2 = a^2 + 2ab + b^2$ and $(a-b)^2 = a^2 - 2ab + b^2$ can be used to simplify some arithmetic calculations. For example:

$$\begin{aligned} 97^2 &= (100-3)^2 \\ &= 100^2 - 2 \times 100 \times 3 + 3^2 \\ &= 9409 \end{aligned}$$

Use this method to calculate the following.

a. 103^2 b. 62^2 c. 997^2 d. 1012^2 e. 53^2 f. 98^2

Problem solving

17. Use grouping 'two and two' and DOTS to factorise the following. Show your working.

a. $x^2 + 3x - y^2 + 3y$

b. $7x + 7y + x^2 - y^2$

c. $5p - 10pq + 1 - 4q^2$

18. The expansions for the sum and difference of two cubes are given below.

$$a^3 + b^3 = (a+b)\left(a^2 - ab + b^2\right)$$

$$a^3 - b^3 = (a-b)\left(a^2 + ab + b^2\right)$$

Using these expansions, simplify:

$$\frac{2a^2 - 7a + 6}{a^3 + 8} \times \frac{5a^2 + 11a + 2}{a^3 - 8} \div \frac{10a^2 - 13a - 3}{a^2 - 2a + 4}$$

19. Factorise:

a. $x^2 + 12x + 40 - 4(x^2y^2 + 1)$

b. $225x^4y^2 - 169x^2y^6$

7.7 Review

7.7.1 Topic summary

Expanding using FOIL

- FOIL is a process that can be applied to expand brackets.
  It is an acronym that stands for:
  First
  Outer
  Inner
  Last
  e.g.
  $(x+2)(x-5)$
  $= (x \times x) + (x \times -5) + (2 \times x) + (2 \times -5)$
  $= x^2 - 5x + 2x - 10$
  $= x^2 - 3x - 10$

F: $(x+a)(x-b)$
O: $(x+a)(x-b)$
I: $(x+a)(x-b)$
L: $(x+a)(x-b)$

Factorising monic quadratics

- To factorise monic quadratics, $x^2 + bx + c$, determine factors of c that add to b.
- $x^2 + bx + c = (x+m)(x+n)$ where m and n are the factors of c that add to b.
  e.g.
  $x^2 - 2x - 15$
  The factors of -15 that add to -2 are -5 and 3.
  $x^2 - 2x - 15 = (x-5)(x+3)$.

Factorising non-monic quadratics

- To factorise non-monic quadratics, $ax^2 + bx + c$, determine factors of ac that add to b.
- Split the bx term into two terms based on the factors of ac that add to b.
- Pair the four terms so that the terms on the left share a common factor and the terms on the right share a common factor.
- Take the common factors out of the pairs of terms.
- Take out the binomial factor.
  e.g.
  $5x^2 + 8x + 3$
  The factors of 15 that add to 8 are 5 and 3.
  $5x^2 + 8x + 3$
  $= 5x^2 + 5x + 3x + 3$
  $= 5x(x+1) + 3(x+1)$
  $= (x+1)(5x+3)$

QUADRATIC EXPRESSIONS

Binomial expansion (perfect squares)

- When a binomial expression is squared the expansion can be determined using the following formulas.
  - $(a+b)^2 = a^2 + 2ab + b^2$
  - $(a-b)^2 = a^2 - 2ab + b^2$

  e.g.
  $(3x-5)^2$
  $= (3x)^2 - 2 \times 3x \times 5 + (5)^2$
  $= 9x^2 - 30x + 25$

Difference of two squares (DOTS)

- $a^2 - b^2 = (a+b)(a-b)$
- This formula can be used to expand or factorise quadratic expressions.
  e.g.
  $4x^2 - 9$
  $= (2x)^2 - (3)^2$
  $= (2x+3)(2x-3)$

Completing the square

- Completing the square is a method to convert a quadratic expression into turning point form.
- $x^2 + bx + \left(\frac{b}{2}\right)^2 = \left(x + \frac{b}{2}\right)^2$
- By adding $\left(\frac{b}{2}\right)^2$ to the expression a perfect square is formed. This part of the expression can then be factorised.
  e.g.
  $x^2 + 6x = x^2 + 6x + \left(\frac{6}{2}\right)^2 - \left(\frac{6}{2}\right)^2$
  $= x^2 + 6x + 9 - 9$
  $= (x+3)^2 - 9$
- Once the square has been completed the quadratic can often be factorised further using DOTS.
  e.g.
  $(x+3)^2 - 9 = (x+3)^2 - (3)^2$
  $= (x+3-3)(x+3+3)$
  $= x(x+6)$

7.7.2 Success criteria

Tick the column to indicate that you have completed the subtopic and how well you have understood it using the traffic light system.

(**Green:** I understand; **Yellow:** I can do it with help; **Red:** I do not understand)

Subtopic	Success criteria	Green	Yellow	Red
7.2	I can expand binomial expressions using FOIL.			
	I can expand squares of binomial expressions.			
	I can expand binomial expressions using the difference of two squares (DOTS).			
7.3	I can recognise monic and non-monic quadratic expressions.			
	I can factorise monic quadratic trinomials.			
	I can factorise non-monic quadratic trinomials.			
7.4	I can factorise expressions by taking out the HCF.			
	I can recognise and factorise expressions that can be written as DOTS.			
	I can factorise expressions with four terms by grouping.			
7.5	I can complete the square for a given quadratic expression.			
	I can factorise a quadratic expression by completing the square.			
7.6	I can factorise by grouping.			
	I can factorise by completing the square.			
	I can factorise using DOTS.			

7.7.3 Project

Celebrity squares and doubles

In small groups or as a class, use the process of elimination to find your 'square and double pair' by playing 'Celebrity squares and doubles' as outlined below.

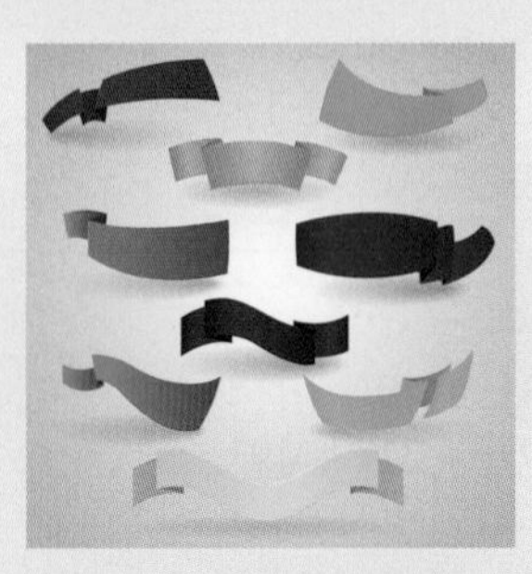

Equipment: roll of calculator paper, scissors, sticky tape, marker pen

1. **Set-up**
   - Make a class set of headbands. Each headband will be part of a matching pair made by a number being squared and that same original number being doubled (16 and 8 would be a pair, because $4^2 = 16$ and $4 \times 2 = 8$). Your teacher will direct the class as to what number should be written on each headband.
   - Place the headbands randomly on a table.
2. **Beginning the game**
   - There is to be no communication between players at this time.
   - Your teacher will randomly allocate a headband to each player by placing a headband on their head without the player seeing the number on their headband.
3. **Playing the game**

   The object of the game
   - The object of the game is to use the process of elimination for you to find your pair. A possible train of thought is illustrated at right.

   Starting the game
   - Once all headbands have been allocated, stand in a circle or walk around freely.
   - Without speaking, determine who is a match; then, by a process of elimination, determine who might be your match.

   Making a match
   - When you think you have found your match, approach that person and say 'I think I am your match.'
   - The other player should now check to see if you have a match elsewhere and can reply by saying one of two things: 'Yes, I think I am your match,' or 'I know your match is still out there.'
   - If a match is agreed upon, the players should sit out for the remainder of the game. If a match is not agreed upon, players should continue looking.

   Ending the game
   - The class should continue until everyone is in a pair, at which time the class can check their results.
   - The class should now discuss the different trains of thought they used to find their pair and how this relates to factorising quadratic trinomials.

on Resources

eWorkbook Topic 7 Workbook (worksheets, code puzzle and project) (ewbk-2033)

Interactivities Crossword (int-2845)

Sudoku puzzle (int-3594)

Exercise 7.7 Review questions

learnON

To answer questions online and to receive **immediate corrective feedback** and **fully worked solutions** for all questions, go to your learnON title at www.jacplus.com.au.

Fluency

1. **MC** When expanded, $-3x(x+4)(5-x)$ becomes:
   A. $-3x^3-3x^2-27x$
   B. $-3x^3+3x-27x$
   C. $3x^3+3x^2-60x$
   D. $-3x^3+3x^2-60x$
   E. $3x^3-3x^2-60x$

2. **MC** When expanded, $(3x+7)^2$ becomes:
   A. $9x^2+49$
   B. $3x^2+49$
   C. $3x^2+21x+49$
   D. $9x^2+42x+49$
   E. $9x^2+21x+49$

3. **MC** The factorised form of $-3d^2-9d+30$ is:
   A. $-3(d-5)(d-2)$
   B. $-3(d+5)(d-6)$
   C. $-(3d+5)(d-2)$
   D. $-(3d+5)(d-6)$
   E. $-3(d+5)(d-2)$

4. **MC** If the factorised expression is $(2x-5)(2x+5)$, then the original expression must have been:
   A. $2x^2-5$
   B. $4x^2-5$
   C. $4x^2-25$
   D. $4x^2-20x+25$
   E. $2x^2+25$

5. **MC** To factorise $-5x^2-45x+100$, the first step is to:
   A. determine factors of 5 and 100 that sum to -45.
   B. take out 5 as a common factor.
   C. take out -5 as a common factor.
   D. determine factors of 5 and -45 that will add to make 100.
   E. take out $-5x$ as a common factor.

6. **MC** To complete the square, the term which should be added to x^2-12x is:
   A. 36
   B. -12
   C. $-12x$
   D. -6
   E. $-6x$

7. **MC** Identify which of the following is equivalent to $5x^2-20x-5$.
   A. $5(x-2)^2$
   B. $5(x-2)^2-3$
   C. $5(x-2)^2-15$
   D. $5(x-2)^2-20$
   E. $5(x-2)^2-25$

8. **MC** In the expanded form of $(x-3)(x+5)$, determine which of the following is incorrect.
   A. The value of the constant is -15.
   B. The coefficient of the x term is 2.
   C. The coefficient of the x term is -8.
   D. The coefficient of the x^2 term is 1.
   E. The expansion shows this to be a trinomial expression.

For questions **9** and **10**, expand each of the following and simplify where necessary.

9. a. $3x(x-4)$
   b. $-7x(3x+1)$
   c. $(x-7)(x+1)$
   d. $(2x-5)(x-3)$
   e. $(4x-1)(3x-5)$

10. a. $3(x-4)(2x+7)$
    b. $(2x-5)(x+3)(x+7)$
    c. $(x+5)(x+7)+(2x-5)(x-6)$
    d. $(x+3)(5x-1)-2x$

For questions **11–13**, expand and simplify each of the following.

11. a. $(x-7)^2$ **b.** $(2-x)^2$ **c.** $(3x+1)^2$

12. a. $-2(3x-2)^2$ **b.** $-7(2x+5)^2$ **c.** $-10(4x-5)^2$

13. a. $(x+9)(x-9)$ **b.** $(3x-1)(3x+1)$ **c.** $(5+2x)(5-2x)$

For questions **14–17**, factorise each of the following.

14. a. $2x^2-8x$ **b.** $-4x^2+12x$ **c.** $3ax-2ax^2$

15. a. $(x+1)^2+(x+1)$ **b.** $3(2x-5)-(2x-5)^2$ **c.** $(x-4)(x+2)-(x-4)$

16. a. x^2-16 **b.** x^2-25 **c.** $2x^2-72$

17. a. $3x^2-27y^2$ **b.** $4ax^2-16ay^2$ **c.** $(x-4)^2-9$

For questions **18–21**, factorise each of the following by grouping.

18. a. $ax-ay+bx-by$ **b.** $7x+ay+ax+7y$ **c.** $xy+2y+5x+10$

19. a. $mn-q-2q^2+2mnq$ **b.** $pq-5r^2-r+5pqr$ **c.** $uv-u+9v-9$

20. a. $a^2-b^2+5a-5b$ **b.** $d^2-4c^2-3d+6c$ **c.** $2+2m+1-m^2$

21. a. $4x^2+12x+9-y^2$ **b.** $49a^2-28a+4-4b^2$ **c.** $64s^2-16s+1-3t$

For questions **22–24**, factorise each of the following.

22. a. $x^2+10x+9$ **b.** $x^2-11x+18$ **c.** $x^2-4x-21$
d. $x^2+3x-28$ **e.** $-x^2+6x-9$

23. a. $3x^2+33x-78$ **b.** $-2x^2+8x+10$ **c.** $-3x^2+24x-36$
d. $8x^2+2x-1$ **e.** $6x^2+x-1$

24. a. $8x^2+4x-12$ **b.** $105x^2-10x-15$ **c.** $-12x^2+62x-70$
d. $-45x^2-3x+6$ **e.** $-60x^2-270x-270$

For questions **25** and **26**, factorise each of the following by completing the square.

25. a. x^2+6x+1 **b.** $x^2-10x-3$ **c.** x^2+4x-2

26. a. x^2-5x+2 **b.** x^2+7x-1 **c.** $2x^2+18x-2$

For questions **27** and **28**, factorise each of the following using the most appropriate method.

27. a. $3x^2-12x$ **b.** x^2+6x+2 **c.** $4x^2-25$

28. a. $2x^2+9x+10$ **b.** $2ax+4x+3a+6$ **c.** $-3x^2-3x+18$

29. First factorise then simplify each of the following.

a. $\dfrac{x+4}{5x-30}\times\dfrac{2x-12}{x+1}$ **b.** $\dfrac{3x+6}{4x-24}\times\dfrac{7x-42}{6x+12}$ **c.** $\dfrac{x^2-4}{x^2+5x}\times\dfrac{x^2+4x-5}{x^2-2x-8}$

Problem solving

30. A large storage box has a square base with sides measuring $(x + 2)$ cm and is 32 cm high.

a. Write an expression for the area of the base of the box.
b. Write an expression for the volume of the box ($V =$ area of base $\times$ height).
c. Simplify the expression by expanding the brackets.
d. If $x = 30$ cm, calculate the volume of the box in cm^3.

31. A section of garden is to have a circular pond of radius $2r$ with a 2 m path around its edge.

a. State the diameter of the pond.
b. State the radius of the pond and path.
c. Calculate the area of the pond.
d. Calculate the area of the pond and path.
e. Write an expression to determine the area of the path only and write it in factorised form.
f. If the radius of the pond is 3 metres, calculate the area of the path.

32. In order to make the most of the space available for headlines and stories, the front page of a newspaper is given an area of $(x^2 - 5x - 14)\,\text{cm}^2$.

a. If the length is $(x + 2)$ cm, calculate the width.
b. Write down the length of the shorter side in terms of x.
c. If the shorter side of the front page is 28 cm, determine the value of x.
d. Evaluate the area of this particular paper.

33. Here is a well-known puzzle. Let $a = b = 1$.

Step 1: Write $a = b$.	$a = b$
Step 2: Multiply both sides by a.	$a^2 = ab$
Step 3: Subtract b^2 from both sides.	$a^2 - b^2 = ab - b^2$
Step 4: Factorise.	$(a + b)(a - b) = b(a - b)$
Step 5: Simplify by dividing by $(a - b)$.	$(a + b) = b$
Step 6: Substitute $a = b = 1$.	$1 + 1 = 1$

Explain where the error is. Show your thinking.

To test your understanding and knowledge of this topic, go to your learnON title at www.jacplus.com.au and complete the **post-test**.

Online Resources

Resources

Below is a full list of **rich resources** available online for this topic. These resources are designed to bring ideas to life, to promote deep and lasting learning and to support the different learning needs of each individual.

eWorkbook

Download the workbook for this topic, which includes worksheets, a code puzzle and a project (ewbk-2033) ☐

Solutions

Download a copy of the fully worked solutions to every question in this topic (sol-0741) ☐

Digital documents

7.2 SkillSHEET Expanding brackets (doc-5244) ☐
SkillSHEET Expanding a pair of brackets (doc-5245) ☐
7.3 SkillSHEET Finding a factor pair that adds to a given number (doc-5250) ☐
7.4 SkillSHEET Factorising by taking out the highest common factor (doc-5246) ☐
SkillSHEET Factorising by taking out a common binomial factor (doc-5247) ☐
7.6 SkillSHEET Simplifying algebraic fractions (doc-5248) ☐
SkillSHEET Simplifying surds (doc-5249) ☐
SkillSHEET Factorising by grouping three and one (doc-5252) ☐

Video eLessons

7.2 Binomial expansion (eles-4825) ☐
The square of a binomial expression (eles-4827) ☐
The difference of two squares (eles-4828) ☐
Expansion of binomial expressions (eles-1908) ☐
7.3 Factorising monic quadratic trinomials (eles-4829) ☐
Factorising non-monic quadratic trinomials (eles-4830) ☐
Factorisation of trinomials (eles-1921) ☐
7.4 Factorising expressions with two terms (eles-4831)
Factorising expressions with four terms (eles-4832) ☐
7.5 Completing the square (eles-4833) ☐
Factorising by completing the square (eles-4834) ☐
Factorising by completing the square (int-1939) ☐
7.6 Mixed factorisation (eles-4835) ☐

Interactivities

7.2 Individual pathway interactivity: Expanding algebraic expressions (int-4596) ☐
Expanding binomial factors (int-6033) ☐
Difference of two squares (int-6036) ☐
7.3 Individual pathway interactivity: Factorising expressions with three terms (int-4597) ☐
Factorising monic quadratic trinomials (int-6143) ☐
Factorising trinomials by grouping (int-6144) ☐
7.4 Individual pathway interactivity: Factorising expressions with two or four terms (int-4598)
Factorising expressions with four terms (int-6145) ☐
7.5 Individual pathway interactivity: Factorising by completing the square (int-4599) ☐
Completing the square (int-2559) ☐
7.6 Individual pathway interactivity: Mixed factorisation (int-4600) ☐
7.7 Crossword (int-2845) ☐
Sudoku puzzle (int-3594) ☐

Teacher resources

There are many resources available exclusively for teachers online.

To access these online resources, log on to **www.jacplus.com.au**.

Answers

Topic 7 Quadratic expressions

Exercise 7.1 Pre-test

1. $-2x-6$
2. $4m(1-5m)$
3. $a(a+c)(4+b)$
4. $(x+5)(x-2)$
5. $-9x^2-6x-1$
6. A
7. E
8. C
9. $\left(x+2-\sqrt{10}\right)\left(x+2+\sqrt{10}\right)$
10. $2x^3+x^2-7x-6$
11. $2\left(x-\frac{3}{2}-\frac{\sqrt{29}}{2}\right)\left(x-\frac{3}{2}+\frac{\sqrt{29}}{2}\right)$
12. D
13. 2
14. $(4x+23)(x+9)$
15. B

Exercise 7.2 Expanding algebraic expressions

1. a. $2x+6$ b. $4x-20$
   c. $21-3x$ d. $-x-3$
2. a. x^2+2x b. $2x^2-8x$
   c. $15x^2-6x$ d. $10x-15x^2$
3. a. $8x^2+2x$ b. $4x^3-6x^2$
   c. $6x^3-3x^2$ d. $15x^3+20x^2$
4. a. x^2-x-12 b. x^2-2x-3
   c. $x^2-5x-14$ d. x^2-6x+5
5. a. $-x^2-x+6$ b. x^2-6x+8
   c. $2x^2-17x+21$ d. $3x^2-x-2$
6. a. $6x^2-17x+5$ b. $21-17x+2x^2$
   c. $15+14x-8x^2$ d. $110+47x-21x^2$
7. a. $2x^2-4x-6$ b. $8x^2-28x-16$
   c. $-2x^2+12x+14$
8. a. $2x^3-2x$ b. $3x^3-75x$ c. $6x^3-54x$
9. a. $2x^3-12x^2+18x$ b. $5x^3-30x^2+40x$
   c. $-6x^3-6x^2+120x$
10. a. x^3+2x^2-x-2 b. x^3-2x^2-5x+6
    c. x^3-5x^2-x+5
11. a. $x^3-6x^2+11x-6$ b. $2x^3-7x^2-5x+4$
    c. $6x^3-7x^2+1$
12. a. x^2-x-2 b. $-2x^2+4x+10$
    c. $5x^2-6x-5$ d. $19x-23$
13. a. $-5x-1$
    b. $-2x+6$
    c. $x^2-2x-3+\sqrt{3}x$
    d. $\sqrt{6}+2\sqrt{2}x-3\sqrt{3}x-6x^2-\sqrt{5}x$
14. A
15. C
16. B
17. a. x^2-2x+1 b. x^2+4x+4
    c. $x^2+10x+25$ d. $16+8x+x^2$
18. a. $49-14x+x^2$ b. $144-24x+x^2$
    c. $9x^2-6x+1$ d. $144x^2-72x+9$
19. a. $25x^2+20x+4$ b. $4-12x+9x^2$
    c. $25-40x+16x^2$ d. $1-10x+25x^2$
20. a. $2x^2-12x+18$ b. $4x^2-56x+196$
    c. $3x^2+6x+3$
21. a. $-4x^2-12x-9$ b. $-49x^2+14x-1$
    c. $8x^2-24x+18$
22. a. $-12+108x-243x^2$ b. $-45+330x-605x^2$
    c. $-16x^2-16x-4$
23. a. x^2-49 b. x^2-81 c. x^2-25
24. a. x^2-1 b. $4x^2-9$ c. $9x^2-1$
25. a. $49-x^2$ b. $64-x^2$ c. $9-4x^2$
26. a. $(x+1)(x-3)$ b. x^2-2x-3
    c. 6 cm, 2 cm, 12 cm^2
27. a. 

    b. 

    c. $(x+1)(x+2)$ d. x^2+3x+2
    e. 4 m^2, 12 m^2
28. Sample responses can be found in the worked solutions in the online resources. Examples are shown.
    a. $(x+4)(x+3)=x^2+7x+12$
    b. $(x+4)^2=x^2+8x+16$
    c. $(x+4)(x-4)=x^2-16$
    d. $(x+4)^2=x^2+8x+16$
29. a. Student C
    b. Student B
    $$(3x+4)(2x+5)$$
    $$=3x\times 2x+3x\times 5+4\times 2x+4\times 5$$
    $$=6x^2+23x+20$$
    Student A
    $$(3x+4)(2x+5)$$
    $$=3x\times 2x+3x\times 5+4\times 2x+4\times 5$$
    $$=6x^2+15x+8x+20$$
    $$=6x^2+23x+20$$
30. $(a-b)(a+b)=a^2-b^2$
    LHS
    $(5-3)(5+3)$
    $=2\times 8$
    $=16$
    RHS :
    5^2-3^2
    $=25-9$
    $=16$
    LHS = RHS ⇒ True

31. $(a+b)^2 = a^2 + 2ab + b^2$
LHS :
$(5+3)^2$
$= 8^2$
$= 64$
RHS :
$5^2 + 2 \times 5 \times 3 + 3^2$
$= 25 + 30 + 9$
$= 64$
LHS = RHS ⇒ True

32. $(a+b)(c+d) = (c+d)(a+b)$
LHS :
$(a+b)(c+d) = ac + ad + bc + bd$
RHS :
$(c+d)(a+b) = ca + cb + da + db$
LHS = RHS ⇒ True

33. The square of a binomial is a trinomial; the difference of two squares has two terms.

34. $V = 6x^3 - 29x^2 + 46x - 24$

35. $TSA = 12x^2 + 20x + 3$

36. a. $4x^2 + 12xy - 20xz + 9y^2 - 30yz + 25z^2$
b. $\frac{1}{x} + \frac{1}{4x^2} - 4x + 4x^2 - 1$

Exercise 7.3 Factorising expressions with three terms

1. a. $(x+2)(x+1)$ b. $(x+3)(x+1)$
c. $(x+8)(x+2)$ d. $(x+4)^2$
e. $(x-3)(x+1)$

2. a. $(x-4)(x+1)$ b. $(x-12)(x+1)$
c. $(x-6)(x+2)$ d. $(x+4)(x-1)$
e. $(x+5)(x-1)$

3. a. $(x+7)(x-1)$ b. $(x+5)(x-2)$
c. $(x-3)(x-1)$ d. $(x-4)(x-5)$
e. $(x+14)(x-5)$

4. a. $-2(x+9)(x+1)$ b. $-3(x+2)(x+1)$
c. $-(x+2)(x+1)$ d. $-(x+10)(x+1)$

5. a. $-(x+2)(x+5)$ b. $-(x+12)(x+1)$
c. $-(x+3)(x+4)$ d. $-(x+2)(x+6)$

6. a. $2(x+2)(x+5)$ b. $3(x+1)(x+10)$
c. $5(x+20)(x+1)$ d. $5(x+4)(x+5)$

7. a. $(a-7)(a+1)$ b. $(t-4)(t-2)$
c. $(b+4)(b+1)$ d. $(m+5)(m-3)$

8. a. $(p-16)(p+3)$ b. $(c+16)(c-3)$
c. $(k+19)(k+3)$ d. $(s-19)(s+3)$

9. a. $(g+8)(g-9)$ b. $(v-25)(v-3)$
c. $(x+16)(x-2)$ d. $(x-15)(x-4)$

10. C

11. B

12. C

13. a. $(2x+1)(x+2)$ b. $(2x-1)(x-1)$
c. $(4x+3)(x-5)$ d. $(2x-1)(2x+3)$

14. a. $(x-7)(2x+5)$ b. $(3x+1)(x+3)$
c. $(3x-7)(2x-1)$ d. $(4x-7)(3x+2)$

15. a. $(5x+3)(2x-3)$ b. $(4x-1)(5x+2)$
c. $(3x+2)(4x-1)$ d. $(3x-1)(5x+2)$

16. a. $2(x-1)(2x+3)$ b. $3(3x+1)(x-7)$
c. $12(2x+1)(3x-1)$ d. $-3(3x+1)(2x-1)$

17. a. $-30(2x+1)(x-3)$ b. $3a(4x-7)(2x+5)$
c. $-2(4x-3)(x-2)$ d. $-(2x-7)(5x+2)$

18. a. $-(8x-1)(3x-4)$
b. $-2(3x-y)(2x+y)$
c. $-5(2x-7y)(3x+2y)$
d. $-12(5x+3y)(10x+7y)$

19. a. $w^2 + 5w - 6$ b. $(w+6)(w-1)$
c. $(x+5)(x-2)$

20. a. $x(x+5)$
b. $x(x+5)$
c. $(x-1)^2$

21. a. $(x+9)(x+5)$ b. $(x-15)(x-6)$
c. $(x-10)(x-3)$

22. a. Yellow = 3 cm × 3 cm
Black = 3 cm × 6 cm
White = 6 cm × 6 cm
b. Yellow = 0.36 m^2
Black = 0.72 m^2
White = 0.36 m^2
c. 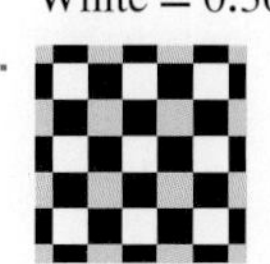

23. a. $x^2 - 7x + 12 = (x-3)(x-4)$
b. $x^2 + 7x - 12 = (x-3)(x-4)$
c. $x^2 - x - 2 = (x-2)(x+1)$
d. $x^2 - 4x - 21 = (x+3)(x-7)$

24. a. $x^2 + 4x - 21 = (x-3)(x+7)$
b. $x^2 - x - 30 = (x+5)(x-6)$
c. $x^2 + 7x - 8 = (x-1)(x+8)$
d. $x^2 - 11x + 30 = (x-5)(x-6)$

25. a. SA = $3x^2 + 16x$
b. Total area = $3x^2 + 16x + 16$
c. $(3x+4)(x+4)$
d. $l = 21$ m; $w = 7$ m; $d = 2$ m
e. 275 m^2
f. 294 m^3

26. $(x-0.5)(x+1.5)$

27. a. $(2x+3)(3x+1)$ b. $P = 10x + 8$
c. $x = 8$ metres

28. a. $(9a-2)(6a-7)$
b. $(3m^2+2)\left(m-\sqrt{7}\right)\left(m+\sqrt{7}\right)$
c. $(2\sin(x)-1)(\sin(x)-1)$

29. a. $(x-5)(x+1)$
b. $(x-5)$ cm
c. $x = 15$ m
d. 160 cm^2
e. $3000(x-5)(x+1)$ cm^2 or $(3000x^2 - 12000x - 15000)$ cm^2

Exercise 7.4 Factorising expressions with two or four terms

1. a. $x(x+3)$ b. $x(x-4)$ c. $3x(x-2)$
2. a. $4x(x+4)$ b. $3x(3x-1)$ c. $8x(1-x)$
3. a. $3x(4-x)$ b. $4x(2-3x)$ c. $x(8x-11)$
4. a. $(x-2)(3x+2)$ b. $(x+3)(5-2x)$
 c. $(x-1)(x+5)$
5. a. $(x+1)(x-1)$ b. $(x+4)(x-2)$ c. $(x-3)(4-x)$
6. a. $(x+1)(x-1)$ b. $(x+3)(x-3)$ c. $(x+5)(x-5)$
7. a. $(x+10)(x-10)$ b. $(y+k)(y-k)$
 c. $(2x+3y)(2x-3y)$
8. a. $(4a+7)(4a-7)$ b. $(5p+6q)(5p-6q)$
 c. $(1+10d)(1-10d)$
9. a. $4(x+1)(x-1)$ b. $5(x+4)(x-4)$
 c. $a(x+3)(x-3)$
10. a. $2(b+2d)(b-2d)$ b. $100(x+4)(x-4)$
 c. $3a(x+7)(x-7)$
11. a. $4p(x+8)(x-8)$ b. $4(3x+2)(3x-2)$
 c. $3(6+x)(6-x)$
12. C
13. B
14. B
15. C
16. a. $\left(x+\sqrt{11}\right)\left(x-\sqrt{11}\right)$
 b. $\left(x+\sqrt{7}\right)\left(x-\sqrt{7}\right)$
 c. $\left(x+\sqrt{15}\right)\left(x-\sqrt{15}\right)$
17. a. $\left(2x+\sqrt{13}\right)\left(2x-\sqrt{13}\right)$
 b. $\left(3x+\sqrt{19}\right)\left(3x-\sqrt{19}\right)$
 c. $3\left(x+\sqrt{22}\right)\left(x-\sqrt{22}\right)$
18. a. $5\left(x+\sqrt{3}\right)\left(x-\sqrt{3}\right)$
 b. $2\left(x+\sqrt{2}\right)\left(x-\sqrt{2}\right)$
 c. $12\left(x+\sqrt{3}\right)\left(x-\sqrt{3}\right)$
19. a. $(x-3)(x+1)$ b. $(x-4)(x+6)$ c. $(x-5)(x+1)$
20. a. $(x-1)(x+7)$ b. $(6-x)(x+8)$ c. $(10-x)(x+2)$
21. a. $8(x-3)$ b. $(7-x)(5x+1)$
 c. $(x-22)(9x+2)$
22. a. $(x-2y)(1+a)$ b. $(x+y)(2+a)$
 c. $(x-y)(a+b)$ d. $(x+y)(4+z)$
23. a. $(f-2)(e+3)$ b. $(n-7)(m+1)$
 c. $3(2r-s)(t+u)$ d. $7(m-3)(n+5)$
24. a. $2(8-j)(4+k)$ b. $a(3-b)(a+c)$
 c. $x(5+y)(x+2)$ d. $m(m+n)(2-n)$
25. a. $(y+7)(x-2)$ b. $(m+2)(n-3)$
 c. $(q+5)(p-3)$
26. a. $(s+3)(s-4t)$ b. $(b+d)(a^2-c)$
 c. $(1+5z)(xy-z)$
27. a. $(a-b)(a+b+4)$ b. $(p-q)(p+q-3)$
 c. $(m+n)(m-n+l)$
28. a. $(x+y)(7+x-y)$
 b. $(1-2q)(5p+1+2q)$
 c. $(7g+6h)(7g-6h-4)$
29. a. $(x+7+y)(x+7-y)$
 b. $(x+10+y)(x+10-y)$
 c. $(a-11+b)(a-11-b)$
30. a. $(3a+2+b)(3a+2-b)$
 b. $(5p-4+3t)(5p-4-3t)$
 c. $\left(6t-1+\sqrt{5}v\right)\left(6t-1-\sqrt{5}v\right)$
31. E
32. A
33. D
34. B
35. a. Jill
 b. The factors have to add to the -3 and multiple to 2. Jack's pair added to 2 and multiplied to -3, the wrong way around.
36. When splitting the middle term, the order is not important. The order will change the first binomial factor but gives the same answer.
37. a. $(x-2y-a+3b)(x-2y+a-3b)$
 b. $3(2x-5y-3)(2x+5y-3)$
38. a. $(x-5)(x+5)$ b. $(x-5)$ cm, $(x+5)$ cm
 c. 2 cm, 12 cm d. $24\,\text{cm}^2$
 e. $120\,\text{cm}^2$ or 6 times bigger
39. a. $A_1=\pi r^2\ \text{m}^2$
 b. $A_2=\pi(r+1)^2\ \text{m}^2$
 c. $A=\pi(r+1)^2-\pi r^2=\pi(2r+1)\,\text{m}^2$
 d. $34.56\,\text{m}^2$
40. a. Annie $=(x+3)(x+2)\ \text{m}^2$ Bronwyn $=5(x+2)\ \text{m}^2$
 b. $(x+3)(x+2)-5(x+2)$
 c. $(x+2)(x-2)=x^2-4$
 d. Width $=5$ m
 e. Annie has $30\,\text{m}^2$ and Bronwyn has $25\,\text{m}^2$.

Exercise 7.5 Factorising by completing the square

1. a. 25 b. 9 c. 4 d. 64 e. 100
2. a. 16 b. 49 c. 625
 d. $\frac{49}{4}$ e. $\frac{1}{4}$
3. a. $\left(x-2+\sqrt{11}\right)\left(x-2-\sqrt{11}\right)$
 b. $\left(x+1+\sqrt{3}\right)\left(x+1-\sqrt{3}\right)$
 c. $\left(x-5+\sqrt{13}\right)\left(x-5-\sqrt{13}\right)$

d. $\left(x+3+\sqrt{19}\right)\left(x+3-\sqrt{19}\right)$
e. $\left(x+8+\sqrt{65}\right)\left(x+8-\sqrt{65}\right)$

4. a. $\left(x-7+\sqrt{6}\right)\left(x-7-\sqrt{6}\right)$
b. $\left(x+4+\sqrt{7}\right)\left(x+4-\sqrt{7}\right)$
c. $\left(x-2+\sqrt{17}\right)\left(x-2-\sqrt{17}\right)$
d. $\left(x-6+\sqrt{11}\right)\left(x-6-\sqrt{11}\right)$
e. $\left(x-3-\sqrt{5}\right)\left(x-3+\sqrt{5}\right)$

5. a. $\left(x-\frac{1}{2}+\frac{\sqrt{5}}{2}\right)\left(x-\frac{1}{2}-\frac{\sqrt{5}}{2}\right)$
b. $\left(x-\frac{3}{2}+\frac{\sqrt{21}}{2}\right)\left(x-\frac{3}{2}-\frac{\sqrt{21}}{2}\right)$
c. $\left(x+\frac{1}{2}+\frac{\sqrt{21}}{2}\right)\left(x+\frac{1}{2}-\frac{\sqrt{21}}{2}\right)$
d. $\left(x+\frac{3}{2}+\frac{\sqrt{13}}{2}\right)\left(x+\frac{3}{2}-\frac{\sqrt{13}}{2}\right)$
e. $\left(x+\frac{5}{2}+\frac{\sqrt{17}}{2}\right)\left(x+\frac{5}{2}-\frac{\sqrt{17}}{2}\right)$

6. a. $\left(x+\frac{5}{2}+\frac{\sqrt{33}}{2}\right)\left(x+\frac{5}{2}-\frac{\sqrt{33}}{2}\right)$
b. $\left(x-\frac{7}{2}+\frac{\sqrt{53}}{2}\right)\left(x-\frac{7}{2}-\frac{\sqrt{53}}{2}\right)$
c. $\left(x-\frac{9}{2}+\frac{\sqrt{29}}{2}\right)\left(x-\frac{9}{2}-\frac{\sqrt{29}}{2}\right)$
d. $\left(x-\frac{1}{2}+\frac{\sqrt{13}}{2}\right)\left(x-\frac{1}{2}-\frac{\sqrt{13}}{2}\right)$
e. $\left(x-\frac{1}{2}-\frac{\sqrt{5}}{2}\right)\left(x-\frac{1}{2}+\frac{\sqrt{5}}{2}\right)$

7. a. $2\left(x+1+\sqrt{3}\right)\left(x+1-\sqrt{3}\right)$
b. $4\left(x-1+\sqrt{6}\right)\left(x-1-\sqrt{6}\right)$
c. $5\left(x+3+2\sqrt{2}\right)\left(x+3-2\sqrt{2}\right)$
d. $3\left(x-2+\sqrt{17}\right)\left(x-2-\sqrt{17}\right)$
e. $5\left(x-3+\sqrt{7}\right)\left(x-3-\sqrt{7}\right)$

8. a. $6\left(x+2+\sqrt{5}\right)\left(x+2-\sqrt{5}\right)$
b. $3\left(x+5+2\sqrt{3}\right)\left(x+5-2\sqrt{3}\right)$
c. $2\left(x-2+\sqrt{11}\right)\left(x-2-\sqrt{11}\right)$
d. $6\left(x+3+\sqrt{14}\right)\left(x+3-\sqrt{14}\right)$
e. $4\left(x-1-\sqrt{5}\right)\left(x-1+\sqrt{5}\right)$

9. i. **d** ii. **b** iii. **c** iv. **a**
v. **c** vi. **d** vii. **d** viii. **e**

10. B
11. E
12. E
13. This expression cannot be factorised as there is no difference of two squares after completing the square.
14. $a = 0.55$; $b = 5.45$
15. This expression cannot be factorised as it does not become DOTS.
16. Annabelle: $\left(x-2-\sqrt{13}\right)\left(x-2+\sqrt{13}\right)$
17. $(x-p)(x-p+2)$
18. a. $2\left(x+2-\frac{\sqrt{14}}{2}\right)\left(x+2+\frac{\sqrt{14}}{2}\right)$
b. This expression cannot be factorised as there is no difference of two squares.

Exercise 7.6 Mixed factorisation

1. a. $3(x+3)$
b. $(x+2+3y)(x+2-3y)$
c. $(x+6)(x-6)$
d. $(x+7)(x-7)$
e. $(5x+1)(x-2)$

2. a. $5(3x-4y)$
b. $(c+e)(5+d)$
c. $5(x+4)(x-4)$
d. $-(x+5)(x+1)$
e. $(x+4)(x-3)$

3. a. $(m+1)(n+1)$
b. $\left(x+\sqrt{7}\right)\left(x-\sqrt{7}\right)$
c. $4x(4x-1)$
d. $5(x+10)(x+2)$
e. $3(3-y)(x+2)$

4. a. $(x-4+y)(x-4-y)$
b. $4(x^2+2)$
c. $(g+h)(f+2)$
d. $\left(x+\sqrt{5}\right)\left(x-\sqrt{5}\right)$
e. $5(n+1)(2m-1)$

5. a. $(x+5)(x+1)$
b. $(x+1)(x-11)$
c. $(x+2)(x-2)$
d. $(a+b)(c-5)$
e. $(y+1)(x-1)$

6. a. $(3x+2)(x+1)$
b. $7(x+2)(x-2)$
c. $-4(x+6)(x+1)$
d. $(2+r)(p-s)$
e. $3(x+3)(x-3)$

7. a. $(u+v)(t-3)$
   b. $\left(x+\sqrt{11}\right)\left(x-\sqrt{11}\right)$
   c. $(4x-1)(3x-1)$
   d. $(x+1)(x-3)$
   e. $(x+6)(x-2)$

8. a. $4(x-1)(x+4)$
   b. $3(x+2)(x+8)$
   c. $(3+x)(7-x)$
   d. $4(3-x+2y)(3-x-2y)$
   e. $3(y+x)(y-x)$

9. a. $(3x-4y)(x-2y)$
   b. $(x+7)(x+4)$
   c. $(x+2)(x-5)$
   d. $(2x+3)(x+3)$
   e. $4(x+2)$

10. a. $\dfrac{(x+5)(x-2)}{(x+2)(x-2)}\times\dfrac{(x+2)(x+2)}{(x-4)(x+2)}$
   b. $\dfrac{(x+5)\cancel{(x+2)}}{\cancel{(x+2)}\cancel{(x-2)}}\times\dfrac{\cancel{(x+2)}\cancel{(x+2)}}{(x-4)\cancel{(x+2)}}$
   c. $\dfrac{x+5}{x-4}$

11. a. $\dfrac{x-1}{x-6}$ b. $\dfrac{x+1}{2x+3}$ c. $\dfrac{18}{x(x-5)}$

12. a. $\dfrac{2x-1}{x+4}$ b. $\dfrac{x+2}{x+5}$ c. $\dfrac{x-6}{x+3}$

13. a. $\dfrac{4(b+2)}{5}$
   b. $\dfrac{p(p+7)}{(p+3)(p-2)}$
   c. $\dfrac{5(m+2+n)}{2(2m-5)}$

14. a. i. $\dfrac{x^2}{16}-\dfrac{9}{25}$
   ii. $\dfrac{x^2}{16}-\dfrac{9}{25}$
   b. $\left(\dfrac{-x}{4}\right)^2=\left(\dfrac{x}{4}\right)^2=\dfrac{x^2}{16}$

15. a. $(x-9-y)(x-9+y)$
   b. $(2x+3-4y)(2x+3+4y)$

16. a. 10 609 b. 3844 c. 99 409
   d. 1 024 144 e. 2809 f. 9604

17. a. $(x+y)(x-y+3)$
   b. $(x+y)(7+x-y)$
   c. $(1-2q)(5p+1+2q)$

18. $\dfrac{1}{a^2+2a+4}$

19. a. $(x+6-2xy)(x+6+2xy)$
   b. $x^2y^2(15x-13y^2)(15x+13y^2)$ or $x^2y^2(-15x-13y^2)(-15x+13y^2)$

Project

Students will apply their skills of elimination to find 'square and double pair' by playing the game with the given instructions.

Exercise 7.7 Review questions

1. E
2. D
3. E
4. C
5. C
6. A
7. E
8. C

9. a. $3x^2-12x$ b. $-21x^2-7x$
   c. x^2-6x-7 d. $2x^2-11x+15$
   e. $12x^2-23x+5$

10. a. $6x^2-3x-84$ b. $2x^3+15x^2-8x-105$
   c. $3x^2-5x+65$ d. $5x^2+12x-3$

11. a. $x^2-14x+49$ b. $4-4x+x^2$ c. $9x^2+6x+1$

12. a. $-18x^2+24x-8$
   b. $-28x^2-140x-175$
   c. $-160x^2+400x-250$

13. a. x^2-81 b. $9x^2-1$ c. $25-4x^2$

14. a. $2x(x-4)$ b. $-4x(x-3)$ c. $ax(3-2x)$

15. a. $(x+1)(x+2)$
   b. $2(2x-5)(4-x)$
   c. $(x-4)(x+1)$

16. a. $(x+4)(x-4)$
   b. $(x+5)(x-5)$
   c. $2(x+6)(x-6)$

17. a. $3(x+3y)(x-3y)$
   b. $4a(x+2y)(x-2y)$
   c. $(x-1)(x-7)$

18. a. $(x-y)(a+b)$ b. $(x+y)(7+a)$
   c. $(x+2)(y+5)$

19. a. $(1+2q)(mn-q)$ b. $(5r+1)(pq-r)$
   c. $(v-1)(u+9)$

20. a. $(a-b)(a+b+5)$ b. $(d-2c)(d+2c-3)$
   c. $(1+m)(3-m)$

21. a. $(2x+3+y)(2x+3-y)$
   b. $(7a-2+2b)(7a-2-2b)$
   c. $\left(8s-1+\sqrt{3t}\right)\left(8s-1-\sqrt{3t}\right)$

22. a. $(x+9)(x+1)$ b. $(x-9)(x-2)$
   c. $(x-7)(x+3)$ d. $(x+7)(x-4)$
   e. $-(x-3)^2$

23. a. $3(x+13)(x-2)$ b. $-2(x-5)(x+1)$
   c. $-3(x-6)(x-2)$ d. $(4x-1)(2x+1)$
   e. $(3x-1)(2x+1)$

24. a. $4(2x+3)(x-1)$ b. $5(7x-3)(3x+1)$
   c. $-2(3x-5)(2x-7)$ d. $-3(3x-1)(5x+2)$
   e. $-30(2x+3)(x+3)$

25. a. $\left(x+3+2\sqrt{2}\right)\left(x+3-2\sqrt{2}\right)$

b. $\left(x-5+2\sqrt{7}\right)\left(x-5-2\sqrt{7}\right)$

c. $\left(x+2+\sqrt{6}\right)\left(x+2-\sqrt{6}\right)$

26. a. $\left(x-\frac{5}{2}+\frac{\sqrt{17}}{2}\right)\left(x-\frac{5}{2}-\frac{\sqrt{17}}{2}\right)$

b. $\left(x+\frac{7}{2}+\frac{\sqrt{53}}{2}\right)\left(x+\frac{7}{2}-\frac{\sqrt{53}}{2}\right)$

c. $2\left(x+\frac{9}{2}+\frac{\sqrt{85}}{2}\right)\left(x+\frac{9}{2}-\frac{\sqrt{85}}{2}\right)$

27. a. $3x(x-4)$
b. $\left(x+3+\sqrt{7}\right)\left(x+3-\sqrt{7}\right)$
c. $(2x+5)(2x-5)$

28. a. $(2x+5)(x+2)$
b. $(a+2)(2x+3)$
c. $-3(x-2)(x+3)$

29. a. $\frac{2(x+4)}{5(x+1)}$
b. $\frac{7}{8}$
c. $\frac{(x-2)(x-1)}{x(x-4)}$

30. a. $(x+2)^2$
b. $32(x+2)^2$
c. $32x^2+128x+128$
d. 32768 cm^3

31. a. $4r$
b. $2r+2$
c. $4\pi r^2$
d. $(4r^2+8r+4)\pi$
e. $4\pi(2r+1)$
f. $28\pi\text{ m}^2$

32. a. $(x-7)$
b. $x-7$ cm
c. 35
d. 1036 cm^2

33. Dividing by zero in step 5

8 Quadratic equations

LEARNING SEQUENCE

8.1 Overview

Why learn this?

Have you ever thought about the shape a ball makes as it flies through the air? Or have you noticed the shape that the stream of water from a drinking fountain makes? These are both examples of quadratic equations in the real world. When you learn about quadratic equations, you learn about the mathematics of these real-world shapes, but quadratic equations are so much more than interesting shapes.

Being able to use and understand quadratic equations lets you unlock incredible problem-solving skills. It allows you to quickly understand situations and solve complicated problems that would be close to impossible without these skills. Many professionals rely on their understanding of quadratic equations to make important decisions, designs and discoveries. Some examples include: sports people finding the perfect spot to intercept a ball, architects designing complex buildings such as the Sydney Opera House and scientists listening to sound waves coming from the furthest regions of space using satellite dishes. All of these professionals use the principles of quadratic equations.

By learning about quadratic equations you are also embarking on your first step on the path to learning about non-linear equations. When you learn about non-linear equations, a whole world of different shaped curves and different problem-solving skills will open up.

Where to get help

Go to your learnON title at **www.jacplus.com.au** to access the following digital resources. The Online Resources Summary at the end of this topic provides a full list of what's available to help you learn the concepts covered in this topic.

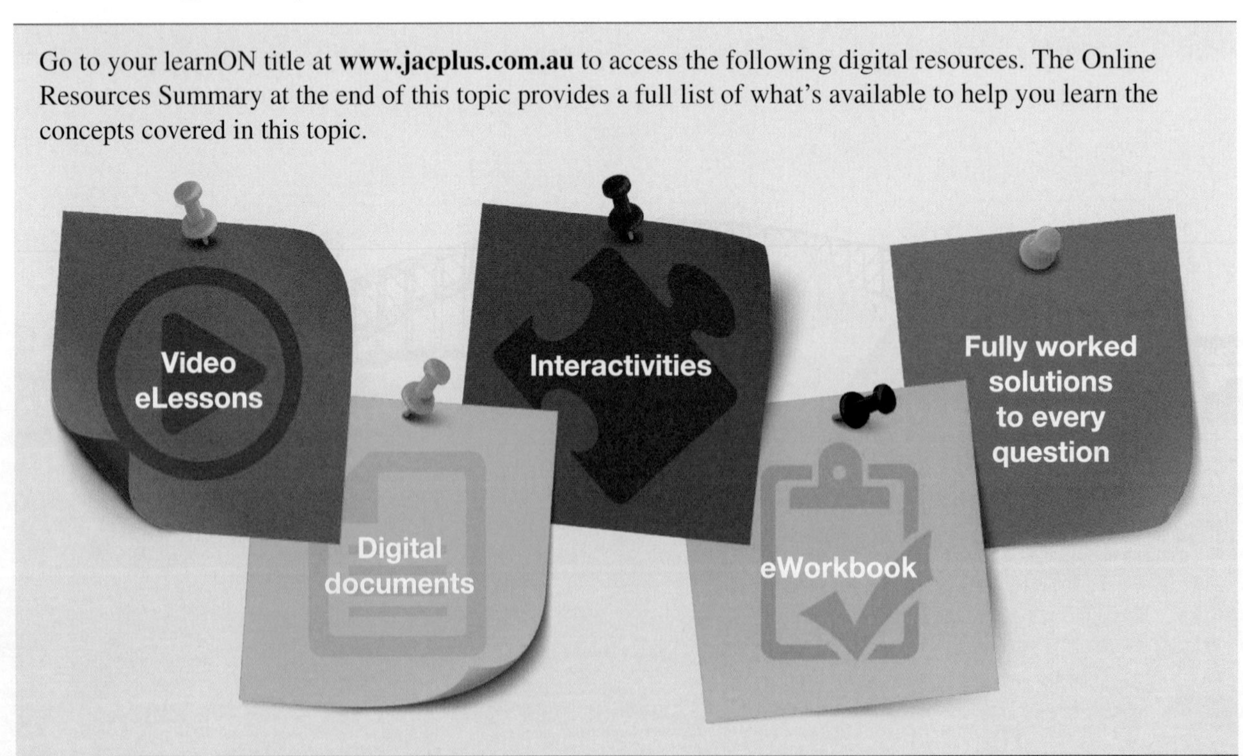

Exercise 8.1 Pre-test

learnON

Complete this pre-test in your learnON title at www.jacplus.com.au and receive **automatic marks**, **immediate corrective feedback** and **fully worked solutions**.

1. State whether the following statement is true or false.
   The Null Factor Law states that if the product of two numbers is zero then one or both of the numbers must equal zero.

2. Calculate the two solutions for the equation $(2x - 1)(x + 5) = 0$.

3. Solve $16x^2 - 9 = 0$ for x.

4. **MC** The solutions for $x(x - 3)(x + 2) = 0$ are:
   A. $x = 3$ or $x = -2$
   B. $x = 3$ or $x = -2$
   C. $x = 0$ or $x = -3$ or $x = 2$
   D. $x = 0$ or $x = 3$ or $x = -2$
   E. $x = 1$ or $x = 3$ or $x = -2$

5. **MC** The solutions to the equation $x^2 + x - 6 = 0$ are:
   A. $x = 3$ or $x = -2$
   B. $x = -3$ or $x = 2$
   C. $x = 3$ or $x = 2$
   D. $x = -6$ or $x = 1$
   E. $x = -1$ or $x = 6$

6. **MC** Four times a number is subtracted from 3 times its square. If the result is 4, the possible numbers are:
   A. $x = -4$
   B. $x = 0$ or $x = -4$
   C. $x = 0$ or $x = 4$
   D. $x = -\frac{2}{3}$ or $x = 2$
   E. $x = -\frac{3}{2}$ or $x = 2$

7. **MC** An exact solution to the equation $x^2 - 3x - 1 = 0$ is $x =$
   A. $\frac{3 \pm \sqrt{13}}{2}$
   B. $\frac{-3 \pm \sqrt{13}}{2}$
   C. $\frac{3 \pm 2\sqrt{2}}{2}$
   D. $\frac{-3 \pm 2\sqrt{2}}{2}$
   E. -3 or -1

8. Calculate the discriminant for $5x^2 - 8x + 2 = 0$.

9. Match the discriminant value to the number of solutions for a quadratic equation.

Discriminant value	Number of solutions
a. $\Delta = 0$	No real solutions
b. $\Delta < 0$	Two solutions
c. $\Delta > 0$	Only one solution

10. **MC** If $\Delta = 0$, the graph of a quadratic:
    A. does not cross or touch the x-axis
    B. touches the x-axis
    C. intersects the x-axis twice
    D. intersects the x-axis three times
    E. There is not enough information to know whether the graph touches the x-axis.

11. **MC** The equation $3x^2 - 27 = 0$ has:
   - A. two rational solutions
   - B. two irrational solutions
   - C. one solution
   - D. one rational and one irrational solution
   - E. no solutions.

12. A rectangle's length is 4 cm more than its width. Determine the dimensions of the rectangle if its area is $(4x + 9)$ cm^2.

13. Calculate the value of m for which $mx^2 - 12x + 9 = 0$ has one solution.

14. **MC** Identify for what values of a the straight line $y = ax - 12$ intersects once with the parabola $y = x^2 - 2x - 8$.
   - A. $a = -6$ or $a = 2$
   - B. $a = 6$ or $a = -2$
   - C. $a = 4$ or $a = -2$
   - D. $a = -4$ or $a = 2$
   - E. $a = 3$ or $a = 4$

15. **MC** Solve for x, $\frac{x}{x-2} - \frac{x+1}{x+4} = 1$.
   - A. $x = -1$ or $x = -4$
   - B. $x = 0$ or $x = 2$
   - C. $x = 0$ or $x = -1$
   - D. $x = 2$ or $x = -4$
   - E. $x = 5$ or $x = -2$

8.2 Solving quadratic equations algebraically

LEARNING INTENTION

At the end of this subtopic you should be able to:
- use the Null Factor Law to solve quadratic equations
- use the completing the square technique to solve quadratic equations
- solve worded questions algebraically.

8.2.1 Quadratic equations

eles-4843

- **Quadratic equations** are equations in the form:

$$ax^2 + bx + c = 0, \text{ where } a, b \text{ and } c \text{ are numbers.}$$

- For example, the equation $3x^2 + 5x - 7 = 0$ has $a = 3, b = 5$ and $c = -7$.

The Null Factor Law

- We know that any number multiplied by zero will equal zero.

Null Factor Law

- If the product of two numbers is zero, then one or both numbers must be zero.

$$\text{a number} \times \text{another number} = 0$$

- In the above equation, we know that a number = 0 and/or another number = 0.

- To apply the Null Factor Law, quadratic equations must be in a factorised form.
- To review factorising quadratic expressions, see topic 7.

WORKED EXAMPLE 1 Applying the Null Factor Law to solve quadratic equations

Solve the equation $(x-7)(x+11)=0$.

THINK	WRITE
1. Write the equation and check that the right-hand side equals zero. (The product of the two numbers is zero.)	$(x-7)(x+11)=0$
2. The left-hand side is factorised, so apply the Null Factor Law.	$x-7=0$ or $x+11=0$
3. Solve for x.	$x=7$ or $x=-11$

WORKED EXAMPLE 2 Factorising then applying the Null Factor Law

Solve each of the following equations.

a. $x^2-3x=0$

b. $3x^2-27=0$

c. $x^2-13x+42=0$

d. $36x^2-21x=2$

THINK	WRITE
a. 1. Write the equation. Check that the right-hand side equals zero.	a. $x^2-3x=0$
2. Factorise by taking out the common factor of x^2 and $3x$, which is x.	$x(x-3)=0$
3. Apply the Null Factor Law.	$x=0$ or $x-3=0$
4. Solve for x.	$x=0$ or $x=3$
b. 1. Write the equation. Check that the right-hand side equals zero.	b. $3x^2-27=0$
2. Factorise by taking out the common factor of $3x^2$ and 27, which is 3.	$3(x^2-9)=0$
3. Factorise using the difference of two squares rule.	$3(x^2-3^2)=0$ $3(x+3)(x-3)=0$
4. Apply the Null Factor Law.	$x+3=0$ or $x-3=0$
5. Solve for x.	$x=-3$ or $x=3$ (Alternatively, $x=\pm3$.)
c. 1. Write the equation. Check that the right-hand side equals zero.	c. $x^2-13x+42=0$
2. Factorise by identifying a factor pair of 42 that adds to -13.	
3. Use the Null Factor Law to write two linear equations.	$(x-6)(x-7)=0$ $x-6=0$ or $x-7=0$
4. Solve for x.	$x=6$ or $x=7$

Factors of 42	Sum of factors
-6 and -7	-13

d. 1. Write the equation. Check that the right-hand side equals zero. (It does not.)	**d.** $36x^2 - 21x = 2$
2. Rearrange the equation so the right-hand side of the equation equals zero as a quadratic trinomial in the form $ax^2 + bx + c = 0$.	$36x^2 - 21x - 2 = 0$
3. Recognise that the expression to factorise is a quadratic trinomial. Calculate $ac = 36 \times -2 = -72$. Factorise by identifying a factor pair of -72 that adds to -21.	(see table below) $36x^2 - 24x + 3x - 2 = 0$
4. Factorise the expression.	$12x(3x-2) + (3x-2) = 0$ $(3x-2)(12x+1) = 0$
5. Use the Null Factor Law to write two linear equations.	$3x - 2 = 0$ or $12x + 1 = 0$ $3x = 2$ or $12x = -1$
6. Solve for x.	$x = \frac{2}{3}$ or $x = -\frac{1}{12}$

Factors of −72	Sum of factors
3 and − 24	−21

8.2.2 Solving quadratic equations by completing the square

eles-4844

- Completing the square is another technique that can be used for factorising quadratics.
- Use completing the square when other techniques will not work.
- To complete the square, the value of a must be 1. If a is not 1, divide everything by the coefficient of x^2 so that $a = 1$.
- The completing the square technique was introduced in topic 9.

Steps to complete the square

Factorise $x^2 + 6x + 2 = 0$.

Step 1: $ax^2 + bx + c = 0$. Since $a = 1$ in this example, we can complete the square.

Step 2: Add and subtract $\left(\frac{1}{2}b\right)^2$. In this example, $b = 6$.

$$x^2 + 6x + \left(\frac{1}{2} \times 6\right)^2 + 2 - \left(\frac{1}{2} \times 6\right)^2 = 0$$

$$x^2 + 6x + 9 + 2 - 9 = 0$$

Step 3: Factorise the first three terms.

$$x^2 + 6x + 9 - 7 = 0$$

$$(x+3)^2 - 7 = 0$$

Step 4: Factorise the quadratic by using difference of two squares

$$(x+3)^2 - \left(\sqrt{7}\right)^2 = 0 \text{ because } \left(\sqrt{7}\right)^2 = 7$$

$$\left(x + 3 + \sqrt{7}\right)\left(x + 3 - \sqrt{7}\right) = 0$$

When $a \neq 1$:

- factorise the expression if possible before completing the square
- if a is not a factor of each term, divide each term by a to ensure the coefficient of x^2 is 1.

WORKED EXAMPLE 3 Solving quadratic equations by completing the square

Solve the equation $x^2 + 2x - 4 = 0$ by completing the square. Give exact answers.

THINK	WRITE
1. Write the equation.	$x^2 + 2x - 4 = 0$
2. Identify the coefficient of x, halve it and square the result.	$\left(\frac{1}{2} \times 2\right)^2$
3. Add the result of step 2 to the equation, placing it after the x-term. To balance the equation, we need to subtract the same amount as we have added.	$x^2 + 2x + \left(\frac{1}{2} \times 2\right)^2 - 4 - \left(\frac{1}{2} \times 2\right)^2 = 0$ $x^2 + 2x + (1)^2 - 4 - (1)^2 = 0$ $x^2 + 2x + 1 - 4 - 1 = 0$
4. Insert brackets around the first three terms to group them and then simplify the remaining terms.	$(x^2 + 2x + 1) - 5 = 0$
5. Factorise the first three terms to produce a perfect square.	$(x + 1)^2 - 5 = 0$
6. Express as the difference of two squares and then factorise.	$(x + 1)^2 - \left(\sqrt{5}\right)^2 = 0$ $\left(x + 1 + \sqrt{5}\right)\left(x + 1 - \sqrt{5}\right) = 0$
7. Apply the Null Factor Law to identify linear equations.	$x + 1 + \sqrt{5} = 0$ or $x + 1 - \sqrt{5} = 0$
8. Solve for x. Keep the answer in surd form to provide an exact answer.	$x = -1 - \sqrt{5}$ or $x = -1 + \sqrt{5}$ (Alternatively, $x = -1 \pm \sqrt{5}$.)

TI | THINK

In a new document, on a Calculator page, press:

- Menu
- 3: Algebra
- 1: Solve

Complete the entry line as:
solve $(x^2 + 2x - 4 = 0, x)$
Then press ENTER.

DISPLAY/WRITE

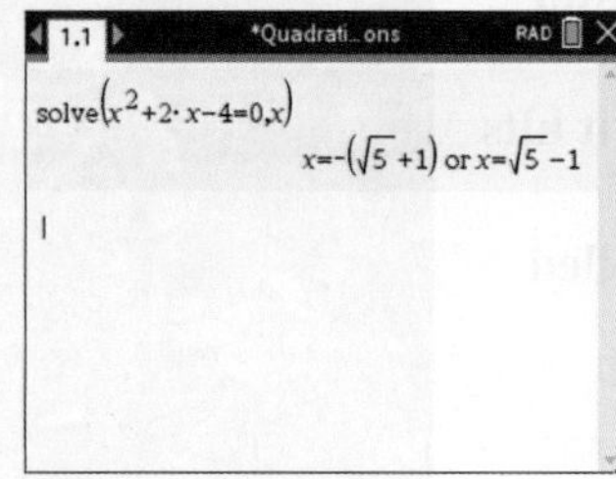

$x^2 + 2x - 4 = 0$
$\Rightarrow x = -1 + \sqrt{5}$ or $-1 - \sqrt{5}$

CASIO | THINK

On the Main screen in standard, tap:

- Action
- Advanced
- solve

Then press EXE.

DISPLAY/WRITE

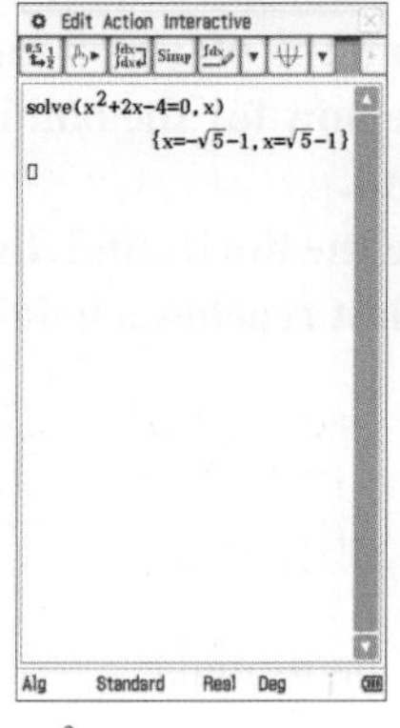

$x^2 + 2x - 4 = 0$
$\Rightarrow x = -1 + \sqrt{5}$ or $-1 - \sqrt{5}$

8.2.3 Solving worded questions

eles-4845

- For worded questions:
  - start by identifying the unknowns
  - then write the equation and solve it
  - give answers in a full sentence and include units if required.

WORKED EXAMPLE 4 Solving worded questions

When two consecutive numbers are multiplied together, the result is 20. Determine the numbers.

THINK	WRITE
1. Define the unknowns. First number $= x$, second number $= x + 1$.	Let the two numbers be x and $(x + 1)$.
2. Write an equation using the information given in the question.	$x(x + 1) = 20$
3. Transpose the equation so that the right-hand side equals zero.	$x(x + 1) - 20 = 0$
4. Expand to remove the brackets.	$x^2 + x - 20 = 0$
5. Factorise.	$(x + 5)(x - 4) = 0$
6. Apply the Null Factor Law to solve for x.	$x + 5 = 0$ or $x - 4 = 0$ $x = -5$ or $x = 4$
7. Use the answer to determine the second number.	If $x = -5, x + 1 = -4$. If $x = 4, x + 1 = 5$.
8. Check the solutions.	Check: $4 \times 5 = 20 \quad (-5) \times (-4) = 20$
9. Write the answer in a sentence.	The numbers are 4 and 5 or −5 and −4.

WORKED EXAMPLE 5 Solving worded application questions

The height of a football after being kicked is determined by the formula $h = -0.1d^2 + 3d$, where d is the horizontal distance from the player in metres.

a. Calculate how far the ball is from the player when it hits the ground.

b. Calculate the horizontal distance the ball has travelled when it first reaches a height of 20 m.

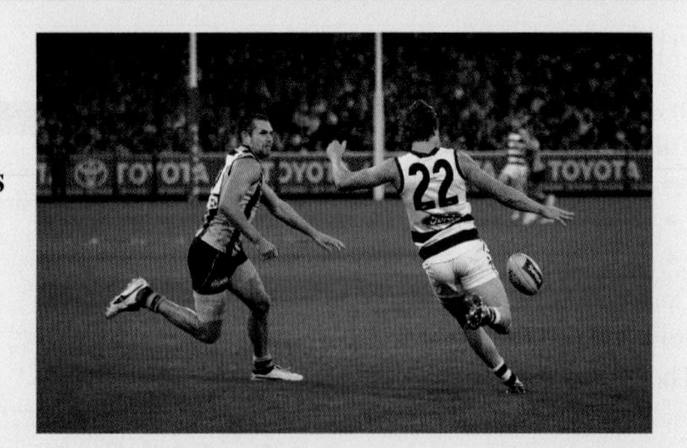

THINK	WRITE
a. 1. Write the formula.	a. $h = -0.1d^2 + 3d$
2. The ball hits the ground when $h = 0$. Substitute $h = 0$ into the formula.	$-0.1d^2 + 3d = 0$
3. Factorise.	$-0.1d^2 + 3d = 0$ $d(-0.1d + 3) = 0$

4. Apply the Null Factor Law and simplify.

$$d = 0 \text{ or } -0.1d + 3 = 0$$
$$-0.1d = -3$$
$$d = \frac{-3}{-0.1}$$
$$= 30$$

5. Interpret the solutions.

$d = 0$ is the origin of the kick.
$d = 30$ is the distance from the origin that the ball has travelled when it lands.

6. Write the answer in a sentence.

The ball is 30 m from the player when it hits the ground.

b. 1. The height of the ball is 20 m, so, substitute $h = 20$ into the formula.

b. $h = -0.1d^2 + 3d$

$20 = -0.1d^2 + 3d$

2. Transpose the equation so that zero is on the right-hand side.

$0.1d^2 - 3d + 20 = 0$

3. Multiply both sides of the equation by 10 to remove the decimal from the coefficient.

$d^2 - 30d + 200 = 0$

4. Factorise.

$(d - 20)(d - 10) = 0$

5. Apply the Null Factor Law.

$d - 20 = 0$ or $d - 10 = 0$

6. Solve.

$d = 20$ or $d = 10$

7. Interpret the solution. The ball reaches a height of 20 m on the way up and on the way down. The *first* time the ball reaches a height of 20 m is the smaller value of d. Write the answer in a sentence.

The ball first reaches a height of 20 m after it has travelled a distance of 10 m.

DISCUSSION

What does the Null Factor Law mean?

on Resources

eWorkbook Topic 8 Workbook (worksheets, code puzzle and a project) (ewbk-2034)

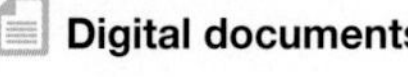

Digital documents SkillSHEET Factorising by taking out the highest common factor (doc-5256)
SkillSHEET Finding a factor pair that adds to a given number (doc-5257)
SkillSHEET Simplifying surds (doc-5258)
SkillSHEET Substituting into quadratic equations (doc-5259)
SkillSHEET Equation of a vertical line (doc-5260)

Video eLesson The Null Factor Law (eles-2312)

Interactivities Individual pathway interactivity: Solving quadratic equations algebraically (int-4601)
The Null Factor Law (int-6095)

Exercise 8.2 Solving quadratic equations algebraically

learnON

Individual pathways

■ PRACTISE	■ CONSOLIDATE	■ MASTER
1, 4, 7, 10, 13, 16, 17, 18, 21, 27, 30, 31, 35, 39, 44	2, 5, 8, 11, 14, 19, 22, 24, 25, 28, 32, 33, 36, 40, 41, 45	3, 6, 9, 12, 15, 20, 23, 26, 29, 34, 37, 38, 42, 43, 46

To answer questions online and to receive **immediate corrective feedback** and **fully worked solutions** for all questions, go to your learnON title at www.jacplus.com.au.

Fluency

WE1 For questions 1–6 solve each of the following equations.

1. a. $(x+7)(x-9)=0$ b. $(x-3)(x+2)=0$ c. $(x-2)(x-3)=0$ d. $x(x-3)=0$

2. a. $x(x-1)=0$ b. $x(x+5)=0$ c. $2x(x-3)=0$ d. $9x(x+2)=0$

3. a. $\left(x-\frac{1}{2}\right)\left(x+\frac{1}{2}\right)=0$ b. $-(x+1.2)(x+0.5)=0$ c. $2(x-0.1)(2x-1.5)=0$ d. $\left(x+\sqrt{2}\right)\left(x-\sqrt{3}\right)=0$

4. a. $(2x-1)(x-1)=0$ b. $(3x+2)(x+2)=0$ c. $(4x-1)(x-7)=0$

5. a. $(7x+6)(2x-3)=0$ b. $(5x-3)(3x-2)=0$ c. $(8x+5)(3x-2)=0$

6. a. $x(x-3)(2x-1)=0$ b. $x(2x-1)(5x+2)=0$ c. $x(x+3)(5x-2)=0$

WE2a For questions 7–9 solve each of the following equations.

7. a. $x^2-2x=0$ b. $x^2+5x=0$ c. $x^2=7x$

8. a. $3x^2=-2x$ b. $4x^2-6x=0$ c. $6x^2-2x=0$

9. a. $4x^2-2\sqrt{7}x=0$ b. $3x^2+\sqrt{3}x=0$ c. $15x-12x^2=0$

WE2b For questions 10–12 solve each of the following equations.

10. a. $x^2-4=0$ b. $x^2-25=0$ c. $3x^2-12=0$ d. $4x^2-196=0$

11. a. $9x^2-16=0$ b. $4x^2-25=0$ c. $9x^2=4$ d. $36x^2=9$

12. a. $x^2-\frac{1}{25}=0$ b. $\frac{1}{36}x^2-\frac{4}{9}=0$ c. $x^2-5=0$ d. $9x^2-11=0$

WE2c For questions 13–15 solve each of the following equations.

13. a. $x^2-x-6=0$ b. $x^2+6x+8=0$ c. $x^2-6x-7=0$ d. $x^2-8x+15=0$

14. a. $x^2 - 3x - 4 = 0$ b. $x^2 - 10x + 25 = 0$ c. $x^2 - 3x - 10 = 0$
d. $x^2 - 8x + 12 = 0$ e. $x^2 - 4x - 21 = 0$

15. a. $x^2 - x - 30 = 0$ b. $x^2 - 7x + 12 = 0$ c. $x^2 - 8x + 16 = 0$
d. $x^2 + 10x + 25 = 0$ e. $x^2 - 20x + 100 = 0$

16. MC The solutions to the equation $x^2 + 9x - 10 = 0$ are:
A. $x = 1$ and $x = 10$ B. $x = 1$ and $x = -10$ C. $x = -1$ and $x = 10$
D. $x = -1$ and $x = -10$ E. $x = 1$ and $x = 9$

17. MC The solutions to the equation $x^2 - 100 = 0$ are:
A. $x = 0$ and $x = 10$ B. $x = 0$ and $x = -10$ C. $x = -10$ and $x = 10$
D. $x = 0$ and $x = 100$ E. $x = -100$ and $x = 100$

WE2d For questions 18–20 solve each of the following equations.

18. a. $2x^2 - 5x = 3$ b. $3x^2 + x - 2 = 0$
c. $5x^2 + 9x = 2$ d. $6x^2 - 11x + 3 = 0$

19. a. $14x^2 - 11x = 3$ b. $12x^2 - 7x + 1 = 0$
c. $6x^2 - 7x = 20$ d. $12x^2 + 37x + 28 = 0$

20. a. $10x^2 - x = 2$ b. $6x^2 - 25x + 24 = 0$
c. $30x^2 + 7x - 2 = 0$ d. $3x^2 - 21x = -36$

WE3 For questions 21–26 solve the following equations by completing the square. Give exact answers.

21. a. $x^2 - 4x + 2 = 0$ b. $x^2 + 2x - 2 = 0$ c. $x^2 + 6x - 1 = 0$

22. a. $x^2 - 8x + 4 = 0$ b. $x^2 - 10x + 1 = 0$ c. $x^2 - 2x - 2 = 0$

23. a. $x^2 + 2x - 5 = 0$ b. $x^2 + 4x - 6 = 0$ c. $x^2 + 4x - 11 = 0$

24. a. $x^2 - 3x + 1 = 0$ b. $x^2 + 5x - 1 = 0$ c. $x^2 - 7x + 4 = 0$

25. a. $x^2 - 5 = x$ b. $x^2 - 11x + 1 = 0$ c. $x^2 + x = 1$

26. a. $x^2 + 3x - 7 = 0$ b. $x^2 - 3 = 5x$ c. $x^2 - 9x + 4 = 0$

For questions 27–29 solve each of the following equations, rounding answers to 2 decimal places.

27. a. $2x^2 + 4x - 6 = 0$ b. $3x^2 + 12x - 3 = 0$ c. $5x^2 - 10x - 15 = 0$

28. a. $4x^2 - 8x - 8 = 0$ b. $2x^2 - 6x + 2 = 0$ c. $3x^2 - 9x - 3 = 0$

29. a. $5x^2 - 15x - 25 = 0$ b. $7x^2 + 7x - 21 = 0$ c. $4x^2 + 8x - 2 = 0$

Understanding

30. WE4 When two consecutive numbers are multiplied, the result is 72. Determine the numbers.

31. When two consecutive even numbers are multiplied, the result is 48. Determine the numbers.

32. When a number is added to its square the result is 90. Determine the number.

33. Twice a number is added to three times its square. If the result is 16, determine the number.

34. Five times a number is added to two times its square. If the result is 168, determine the number.

35. WE5 A soccer ball is kicked. The height, h, in metres, of the soccer ball t seconds after it is kicked can be represented by the equation $h = -t(t-6)$. Calculate how long it takes for the soccer ball to hit the ground again.

36. The length of an Australian flag is twice its width and the diagonal length is 45 cm.
    a. If x cm is the width of the flag, express its length in terms of x.
    b. Draw a diagram of the flag marking in the diagonal. Label the length and the width in terms of x.
    c. Use Pythagoras' theorem to write an equation relating the lengths of the sides to the length of the diagonal.
    d. Solve the equation to calculate the dimensions of the Australian flag. Round your answer to the nearest cm.

37. If the length of a paddock is 2 m more than its width and the area is 48 m^2, calculate the length and width of the paddock.

38. Solve for x.
    a. $x+5=\frac{6}{x}$
    b. $x=\frac{24}{x-5}$
    c. $x=\frac{1}{x}$

Reasoning

39. The sum of the first n numbers $1, 2, 3, 4 \ldots n$ is given by the formula $S=\frac{n(n+1)}{2}$.
    a. Use the formula to calculate the sum of the first 6 counting numbers.
    b. Determine how many numbers are added to give a sum of 153.

40. If these two rectangles have the same area, determine the value of x.

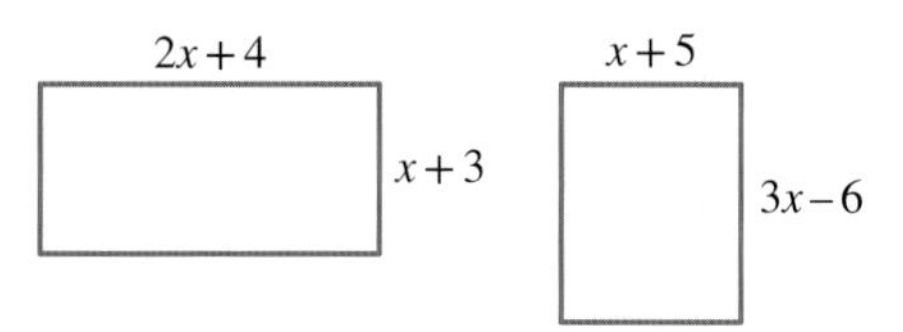

41. Henrietta is a pet rabbit who lives in an enclosure that is 2 m wide and 4 m long. Her human family has decided to purchase some more rabbits to keep her company and so the size of the enclosure must be increased.
    a. Draw a diagram of Henrietta's enclosure, clearly marking the lengths of the sides.
    b. If the length and width of the enclosure are increased by x m, express the new dimensions in terms of x.
    c. If the new area is to be 24 m^2, write an equation relating the sides and the area of the enclosure (area = length × width).
    d. Use the equation to calculate the value of x and, hence, the length of the sides of the new enclosure. Justify your answer.

42. The cost per hour, C, in thousands of dollars of running two cruise ships, *Annabel* and *Betty*, travelling at a speed of s knots is given by the following relationships.
    $C_{Annabel} = 0.3s^2 + 4.2s + 12$ and $C_{Betty} = 0.4s^2 + 3.6s + 8$
    a. Determine the cost per hour for each ship if they are both travelling at 28 knots.
    b. Calculate the speed in knots at which both ships must travel for them to have the same cost.
    c. Explain why only one of the solutions obtained in your working for part **b** is valid.

43. Explain why the equation $x^2 + 4x + 10 = 0$ has no real solutions.

Problem solving

44. Solve $(x^2 - x)^2 - 32(x^2 - x) + 240 = 0$ for x.

45. Solve $\frac{3z^2 - 35}{16} - z = 0$ for z.

46. A garden measuring 12 metres by 16 metres is to have a pedestrian pathway installed all around it, increasing the total area to 285 square metres. Determine the width of the pathway.

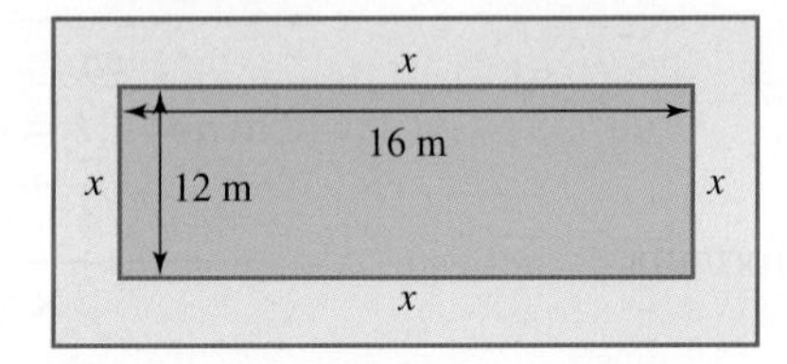

8.3 The quadratic formula

LEARNING INTENTION

At the end of this subtopic you should be able to:
- use the quadratic formula to solve equations
- use the quadratic formula to solve worded questions.

8.3.1 Using the quadratic formula

eles-4846

- The **quadratic formula** can be used to solve any equations in the form:

$$ax^2 + bx + c = 0, \text{ where } a, b \text{ and } c \text{ are numbers.}$$

- To use the quadratic formula:
  - identify the values for a, b and c
  - substitute the values into the quadratic formula.
- Note that the derivation of this formula is beyond the scope of this course.

The quadratic formula

The quadratic formula is:

$$x = \frac{-b \pm \sqrt{b^2 - 4ac}}{2a}$$

Note that the $\pm$ symbol is called the plus–minus sign. To use it, identify the first solution by treating it as a plus, then identify the second solution by treating it as a minus.

- There will be no real solutions if the value in the square root is negative; that is, there will be no solutions if $b^2 - 4ac < 0$.

WORKED EXAMPLE 6 Using the quadratic formula to solve equations

Use the quadratic formula to solve each of the following equations.

a. $3x^2 + 4x + 1 = 0$ **(exact answer)**

b. $-3x^2 - 6x - 1 = 0$ **(round to 2 decimal places)**

THINK	WRITE
a. 1. Write the equation.	**a.** $3x^2 + 4x + 1 = 0$
2. Write the quadratic formula.	$x = \dfrac{-b \pm \sqrt{b^2 - 4ac}}{2a}$
3. State the values for a, b and c.	where $a = 3, b = 4, c = 1$
4. Substitute the values into the formula.	$x = \dfrac{-4 \pm \sqrt{4^2 - 4 \times 3 \times 1}}{2 \times 3}$
5. Simplify and solve for x.	$= \dfrac{-4 \pm \sqrt{4}}{6}$ $= \dfrac{-4 \pm 2}{6}$ $x = \dfrac{-4 + 2}{6}$ or $x = \dfrac{-4 - 2}{6}$
6. Write the two solutions.	$x = -\dfrac{1}{3}$ or $x = -1$
b. 1. Write the equation.	**b.** $-3x^2 - 6x - 1 = 0$
2. Write the quadratic formula.	$x = \dfrac{-b \pm \sqrt{b^2 - 4ac}}{2a}$
3. State the values for a, b and c.	where $a = -3, b = -6, c = -1$
4. Substitute the values into the formula.	$x = \dfrac{-(-6) \pm \sqrt{(-6)^2 - 4 \times (-3) \times (-1)}}{2 \times -3}$
5. Simplify the fraction.	$= \dfrac{6 \pm \sqrt{24}}{-6}$ $= \dfrac{6 \pm 2\sqrt{6}}{-6}$ $= \dfrac{3 \pm \sqrt{6}}{-3}$ $x = \dfrac{3 + \sqrt{6}}{-3}$ or $\dfrac{3 - \sqrt{6}}{-3}$
6. Write the two solutions correct to 2 decimal places.	$x \approx -1.82$ or $x \approx -0.18$

Note: When asked to give an answer in exact form, you should simplify any surds as necessary.

TI | THINK

a-b.

In a new problem, on a Calculator page, complete the entry lines as:
solve $(3x^2 + 4x + 1 = 0)$
solve $(-3x^2 - 6x - 1 = 0)$
Then press ENTER after each entry.
Press CTRL ENTER to get a decimal approximation for **b**.

DISPLAY/WRITE

a-b.

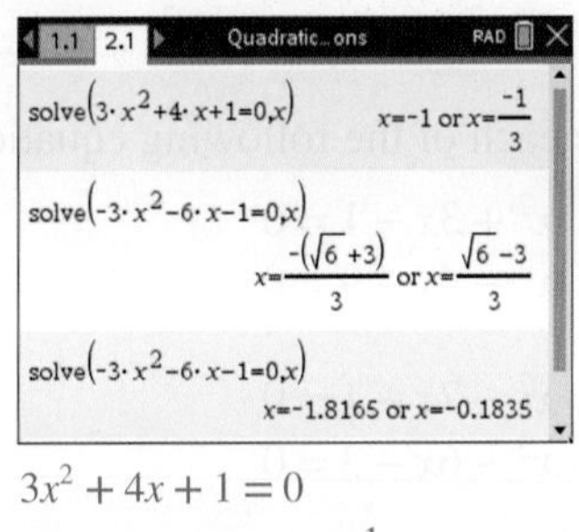

$3x^2 + 4x + 1 = 0$

$\Rightarrow x = -1 \quad \text{or} \quad -\dfrac{1}{3}$

$-3x^2 - 6x - 1 = 0$

$\Rightarrow x = \dfrac{-\left(\sqrt{6}+3\right)}{3} \quad \text{or} \quad \dfrac{\sqrt{6}-3}{3}$

$x = -1.82$ or -0.18 rounding to 2 decimal places.

CASIO | THINK

a-b.

On the Main screen in standard, complete the entry lines as:
solve $(3x^2 + 4x + 1 = 0)$
solve $(-3x^2 - 6x - 1 = 0)$
Press EXE after each entry.
To change the final answer to a decimal, change from standard to decimal mode and press EXE again.

DISPLAY/WRITE

a-b.

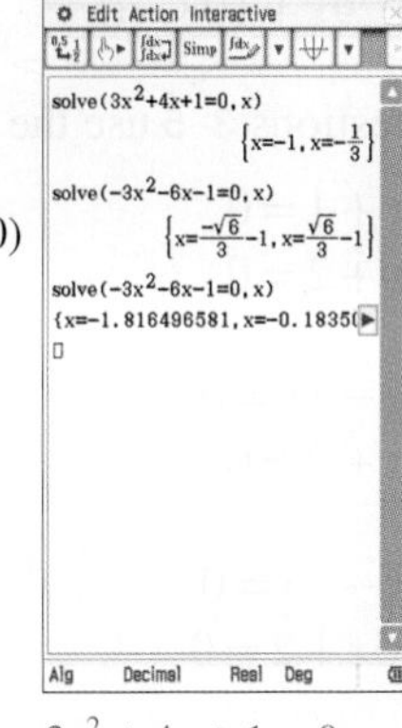

$3x^2 + 4x + 1 = 0$

$\Rightarrow x = -1 \quad \text{or} \quad -\dfrac{1}{3}$

$-3x^2 - 6x - 1 = 0$

$\Rightarrow x = \dfrac{-\left(\sqrt{6}+3\right)}{3} \quad \text{or} \quad \dfrac{\sqrt{6}-3}{3}$

$x = -1.82$ or -0.18 rounding to 2 decimal places.

DISCUSSION

What kind of answer will you get if the value inside the square root sign in the quadratic formula is zero?

on Resources

 **eWorkbook** Topic 8 Workbook (worksheets, code puzzle and a project) (ewbk-2034)

 **Digital document** SkillSHEET Substituting into the quadratic formula (doc-5262)

 **Video eLesson** The quadratic formula (eles-2314)

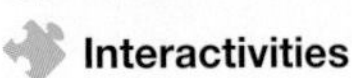 **Interactivities** Individual pathway interactivity: The quadratic formula (int-4602)
The quadratic formula (int-2561)

Exercise 8.3 The quadratic formula

learnon

Individual pathways

■ PRACTISE	■ CONSOLIDATE	■ MASTER
1, 3, 6, 9, 13, 18, 19, 22	2, 4, 7, 10, 11, 14, 17, 20, 23, 24	5, 8, 12, 15, 16, 21, 25

To answer questions online and to receive **immediate corrective feedback** and **fully worked solutions** for all questions, go to your learnON title at www.jacplus.com.au.

Fluency

1. State the values for a, b and c in each of the following equations of the form $ax^2 + bx + c = 0$.

a. $3x^2 - 4x + 1 = 0$

b. $7x^2 - 12x + 2 = 0$

c. $8x^2 - x - 3 = 0$

d. $x^2 - 5x + 7 = 0$

2. State the values for a, b and c in each of the following equations of the form $ax^2 + bx + c = 0$.

a. $5x^2 - 5x - 1 = 0$
b. $4x^2 - 9x - 3 = 0$
c. $12x^2 - 29x + 103 = 0$
d. $43x^2 - 81x - 24 = 0$

WE6a For questions 3–5 use the quadratic formula to solve each of the following equations. Give exact answers.

3. a. $x^2 + 5x + 1 = 0$
b. $x^2 + 3x - 1 = 0$
c. $x^2 - 5x + 2 = 0$
d. $x^2 - 4x - 9 = 0$

4. a. $x^2 + 2x - 11 = 0$
b. $x^2 - 7x + 1 = 0$
c. $x^2 - 9x + 2 = 0$
d. $x^2 - 6x - 3 = 0$

5. a. $x^2 + 8x - 15 = 0$
b. $-x^2 + x + 5 = 0$
c. $-x^2 + 5x + 2 = 0$
d. $-x^2 - 2x + 7 = 0$

WE6b For questions 6–8 use the quadratic formula to solve each of the following equations. Give approximate answers rounded to 2 decimal places.

6. a. $3x^2 - 4x - 3 = 0$
b. $4x^2 - x - 7 = 0$
c. $2x^2 + 7x - 5 = 0$
d. $7x^2 + x - 2 = 0$
e. $5x^2 - 8x + 1 = 0$

7. a. $2x^2 - 13x + 2 = 0$
b. $-3x^2 + 2x + 7 = 0$
c. $-7x^2 + x + 8 = 0$
d. $-12x^2 + x + 9 = 0$
e. $-6x^2 + 4x + 5 = 0$

8. a. $-11x^2 - x + 1 = 0$
b. $-4x^2 - x + 7 = 0$
c. $-2x^2 + 12x - 1 = 0$
d. $-5x^2 + x + 3 = 0$

9. **MC** The solutions of the equation $3x^2 - 7x - 2 = 0$ are:

A. $1, 2$
B. $1, -2$
C. $-0.257, 2.59$
D. $-0.772, 7.772$
E. $-1.544, 15.544$

10. **MC** In the expansion of $(6x - 5)(3x + 4)$, the coefficient of x is:

A. 18
B. −15
C. 9
D. 6
E. −2

11. **MC** In the expanded form of $(x - 2)(x + 4)$, identify which of the following statements is incorrect.

A. The value of the constant is −8.
B. The coefficient of the x term is −6.
C. The coefficient of the x term is 2.
D. The coefficient of the x^2 term is 1.
E. The expansion shows this to be a trinomial expression.

12. **MC** Identify an exact solution to the equation $x^2 + 2x - 5 = 0$.

A. -3.449
B. $-1 + \sqrt{24}$
C. $-1 + \sqrt{6}$
D. $\dfrac{2 + \sqrt{-16}}{2}$
E. $\dfrac{2 + \sqrt{24}}{2}$

Understanding

For questions 13–15 use each of the following equations using any suitable method. Round to 3 decimal places where appropriate.

13. a. $2x^2 - 7x + 3 = 0$
b. $x^2 - 5x = 0$
c. $x^2 - 2x - 3 = 0$
d. $x^2 - 3x + 1 = 0$

14. a. $x^2 - 6x + 8 = 0$
b. $x^2 - 5x + 8 = 0$
c. $x^2 - 7x - 8 = 0$
d. $x^2 + 2x - 9 = 0$

15. a. $2x^2 + 11x - 21 = 0$
b. $7x^2 - 2x + 1 = 0$
c. $-x^2 + 9x - 14 = 0$
d. $-6x^2 - x + 1 = 0$

16. The surface area of a closed cylinder is given by the formula $\text{SA} = 2\pi r(r + h)$, where r cm is the radius of the can and h cm is the height.
The height of a can of wood finish is 7 cm and its surface area is 231 cm^2.
   a. Substitute values into the formula to form a quadratic equation using the pronumeral, r.
   b. Use the quadratic formula to solve the equation and, hence, determine the radius of the can correct to 1 decimal place.
   c. Calculate the area of the curved surface of the can, correct to the nearest square centimetre.

17. To satisfy lighting requirements, the window shown must have an area of 1500 cm^2.
   a. Write an expression for the area of the window in terms of x.
   b. Write an equation so that the window satisfies the lighting requirements.
   c. Use the quadratic formula to solve the equation and calculate x to the nearest mm.

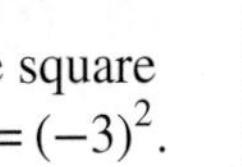

18. When using the quadratic formula, you are required to calculate b^2 (inside the square root sign). When $b = -3$, Breanne says that $b^2 = -(3)^2$ but Kelly says that $b^2 = (-3)^2$.
   a. What answer did Breanne calculate for b^2?
   b. What answer did Kelly calculate for b^2?
   c. Identify who is correct and explain why.

Reasoning

19. There is one solution to the quadratic equation if $b^2 - 4ac = 0$. The following have one solution only. Determine the missing value.
   a. $a = 1$ and $b = 4$, $c = ?$
   b. $a = 2$ and $c = 8$, $b = ?$
   c. $2x^2 + 12x + c = 0$, $c = ?$

20. Two competitive neighbours build rectangular pools that cover the same area but are different shapes. Pool A has a width of $(x + 3)$ m and a length that is 3 m longer than its width. Pool B has a length that is double the width of Pool A. The width of Pool B is 4 m shorter than its length.
   a. Determine the exact dimensions of each pool if their areas are the same.
   b. Verify that the areas are the same.

21. A block of land is in the shape of a right-angled triangle with a perimeter of 150 m and a hypotenuse of 65 m. Determine the lengths of the other two sides. Show your working.

Problem solving

22. Solve $\left(x + \dfrac{1}{x}\right)^2 - 14\left(x + \dfrac{1}{x}\right) = 72$ for x.

23. Gunoor's tennis serve can be modelled using $y = -\dfrac{1}{16}x^2 + \dfrac{7}{8}x + 2$. The landing position of his serve will be one of the solutions to this equation when $y = 0$.
   a. For $-\dfrac{1}{16}x^2 + \dfrac{7}{8}x + 2 = 0$, determine the values of a, b and c in $ax^2 + bx + c = 0$.
   b. Using a calculator, calculate the solutions to this equation.
   c. Gunoor's serve will be 'in' if one solution is between 12 and 18, and a 'fault' if no solutions are between those values. Interpret whether Gunoor's serve was 'in' or if it was a 'fault'.

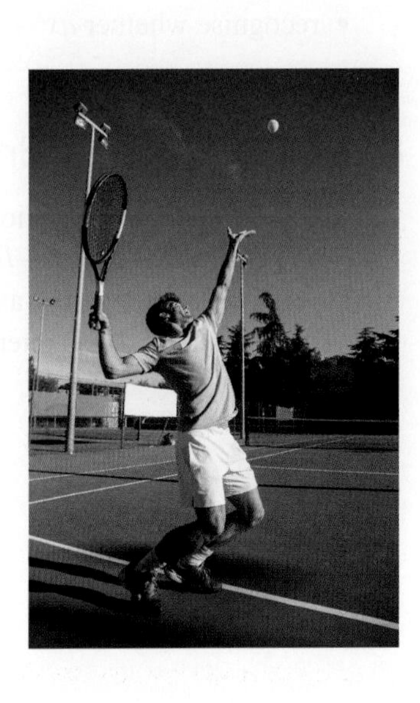

24. Triangle MNP is an isosceles triangle with sides MN = MP = 3 cm. Angle MPN is equal to 72°. The line NQ bisects the angle MNP.

a. Prove that triangles MNP and NPQ are similar.
b. If NP = m cm and $PQ = 3 - m$ cm, show that $m^2 + 3m - 9 = 0$.
c. Solve the equation $m^2 + 3m - 9 = 0$ and determine the side length of NP, giving your answer correct to 2 decimal places.

25. The equation $ax^4 + bx^2 + c = 0$ can be solved by applying substitution and the rules used to solve quadratics.
For example, $x^4 - 5x^2 + 4 = 0$ is solved for x as follows.
Notice that $x^4 - 5x^2 + 4 = (x^2)^2 - 5(x)^2 + 4$. Now let $x^2 = u$ and substitute.

$$(x^2)^2 - 5(x)^2 + 4 = u^2 - 5u + 4$$

Solve for u. That is,

$$\begin{aligned} u^2 - 5u + 4 &= 0 \\ (u-4)(u-1) &= 0 \\ u - 4 = 0 \text{ or } u - 1 &= 0 \\ u = 4 \text{ or } u &= 1 \end{aligned}$$

Since $x^2 = u$, that implies that

$$\begin{aligned} x^2 &= 4 \text{ or } x^2 = 1 \\ x &= \pm 2 \text{ or } x = \pm 1 \end{aligned}$$

Using this or another method, solve the following for x.

a. $x^4 - 13x^2 + 36 = 0$
b. $4x^4 - 17x^2 = -4$

8.4 Solving quadratic equations graphically

LEARNING INTENTION

At the end of this subtopic you should be able to:
- identify the solutions to $ax^2 + bx + c = 0$ by inspecting the graph of $y = ax^2 + bx + c$
- recognise whether $ax^2 + bx + c = 0$ has two, one or no solutions by looking at the graph.

8.4.1 Solving quadratic equations graphically

eles-4847

- To solve quadratic equations, look at the graph of $y = ax^2 + bx + c$.
  - The solutions to $ax^2 + bx + c = 0$ are the x-axis intercepts, where the graph 'cuts' the x-axis.
- The number of x-axis intercepts will indicate the number of solutions.

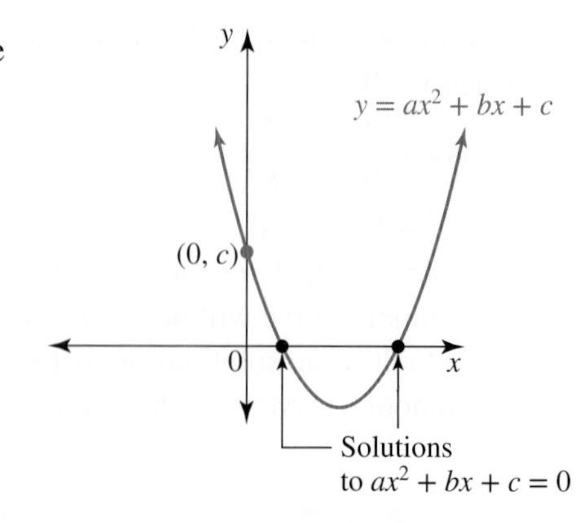

Number of solutions

Two solutions	One solution	No real solutions
The graph 'cuts' the x-axis twice. For example: 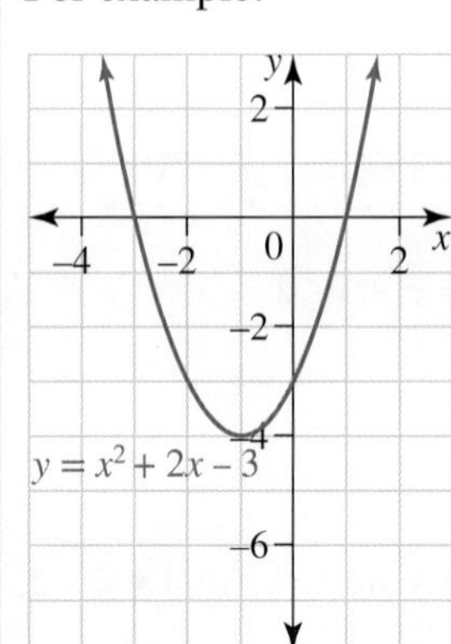 	The graph 'touches' the x-axis. For example: 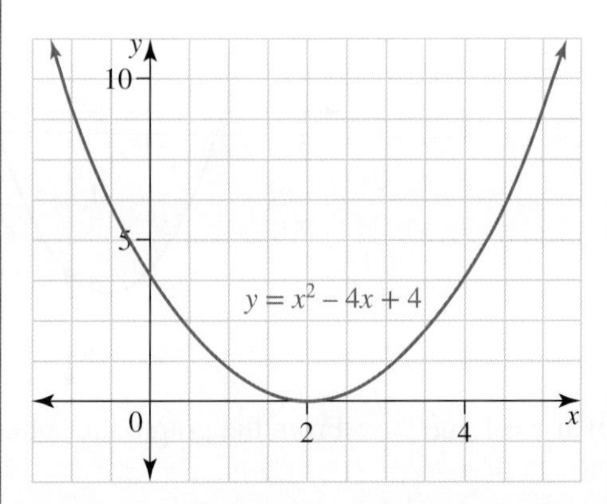 	The graph does not reach the x-axis. For example: 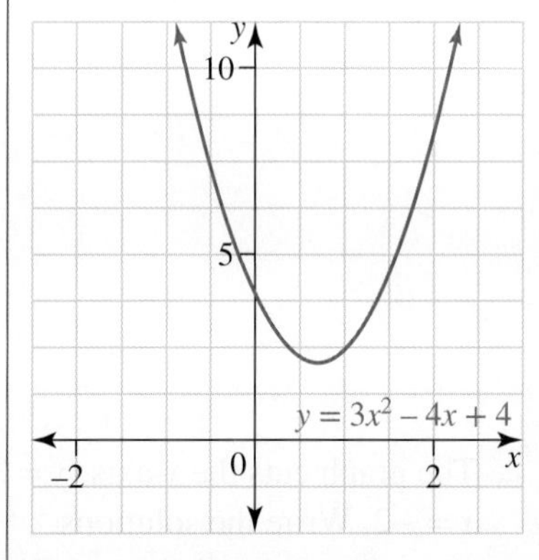

WORKED EXAMPLE 7 Determining solutions from the graph

Determine the solutions of each of the following quadratic equations by inspecting their corresponding graphs. Give answers to 1 decimal place where appropriate.

a. $x^2 + x - 2 = 0$

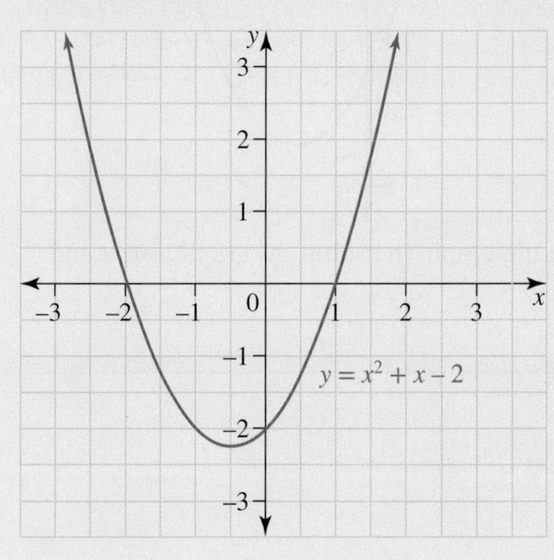

b. $2x^2 + 4x - 5 = 0$

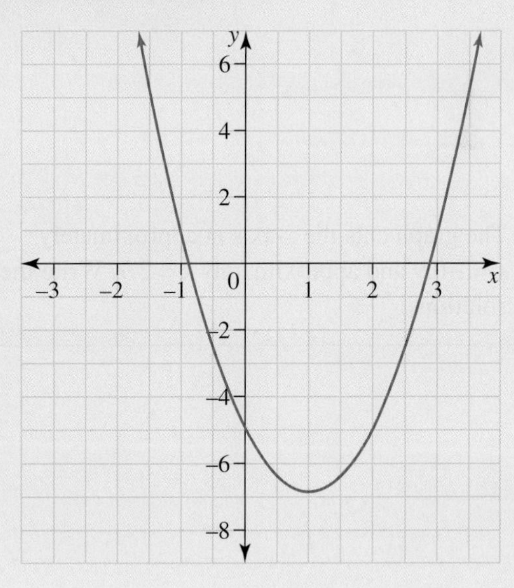

THINK	WRITE/DRAW
a. 1. Examine the graph of $y = x^2 + x - 2$ and locate the points where $y = 0$, that is, where the graph intersects the x-axis.	**a.** 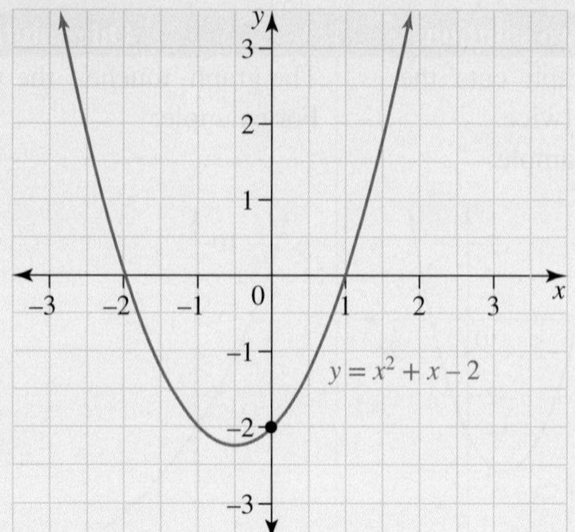
2. The graph cuts the x-axis ($y = 0$) at $x = 1$ and $x = -2$. Write the solutions.	From the graph, the solutions are $x = 1$ and $x = -2$.
b. 1. The graph of $y = 2x^2 - 4x - 5$ is equal to zero when $y = 0$. Look at the graph to see where $y = 0$, that is, where it intersects the x-axis. By sight, we can only give estimates of the solutions.	**b.** 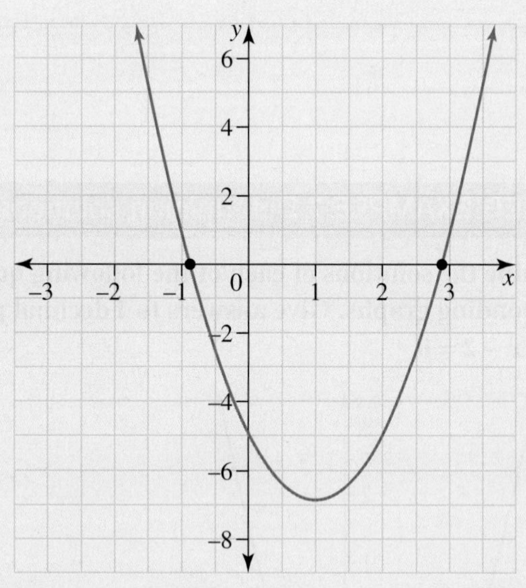
2. The graph cuts the x-axis at approximately $x = -0.9$ and approximately $x = 2.9$. Write the solutions.	From the graph, the solutions are $x \approx -0.9$ and $x \approx 2.9$.

TI | THINK

a.

In a new problem, on a Graphs page, complete the function entry line as:
$f1(x) = x^2 + x - 2$
Then press ENTER. The graph will be displayed.
To locate the x-intercepts, press:
- MENU
- 6: Analyze Graph
- 1: Zero

Move the cursor to the left of the zero, press ENTER, then move the cursor to the right of the zero and press ENTER. The coordinates of the x-intercept are displayed. Repeat for the other x-intercept.

DISPLAY/WRITE

a.

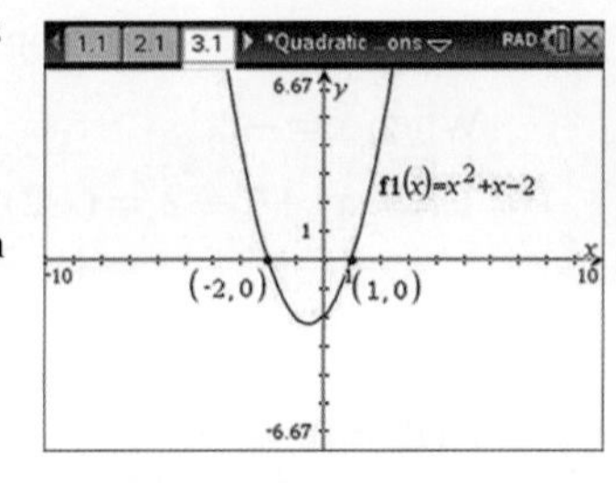

$x^2 + x - 2 = 0$
$\Rightarrow x = 1$ or -2

TI | THINK

b.

On a Graphs page, complete the function entry line as:
$f1(x) = 2x^2 - 4x - 5$
Then press ENTER. The graph will be displayed.
To locate the x-intercepts, press:
- MENU
- 6: Analyze Graph
- 1: Zero

Move the cursor to the left of the zero, press ENTER, then move the cursor to the right of the zero and press ENTER. The coordinates of the x-intercept are displayed. Repeat for the other x-intercept.

DISPLAY/WRITE

b.

$2x^2 - 4x - 5 = 0$
$\Rightarrow x \approx -0.9$ or 2.91
correct to 1 decimal place.

CASIO | THINK

a.

On the Graph & Table screen, complete the function entry line as:
$y1 = x^2 + x - 2$
Press EXE.
Then tap the graphing icon.
The graph will be displayed. To locate the x-intercepts, tap the $Y = 0$ icon.
To locate the second root, tap the right arrow.

DISPLAY/WRITE

a.

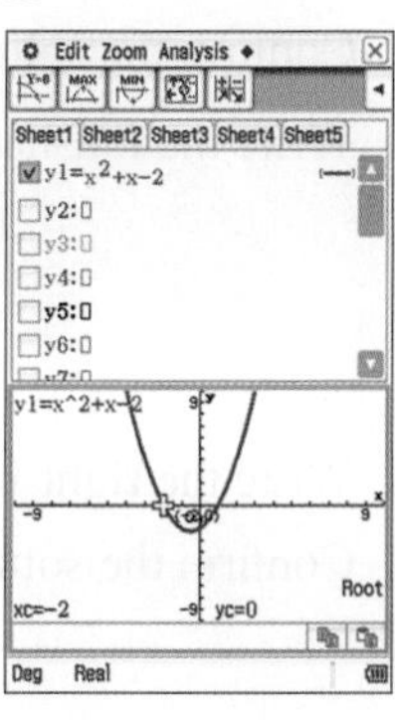

$x^2 + x - 2 = 0$
$\Rightarrow x = 1$ or -2

CASIO | THINK

b.

On the Graph & Table screen, complete the function entry line as:
$y1 = 2x^2 - 4x - 5$
Press EXE.
Then tap the graphing icon.
The graph will be displayed. To locate the x-intercepts, tap the $Y = 0$ icon.
To locate the second root, tap the right arrow.

DISPLAY/WRITE

b.

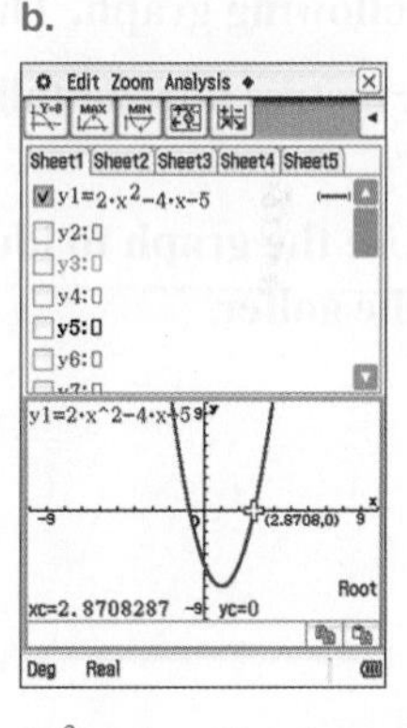

$2x^2 - 4x - 5 = 0$
$\Rightarrow x \approx -0.9$ or 2.9
correct to 1 decimal place.

8.4.2 Confirming solutions of quadratic equations

eles-4848

- To confirm a solution is correct:
  - substitute one solution for x into the quadratic equation
  - the solution is confirmed if the left-hand side (LHS) equals the right-hand side (RHS)
  - repeat for the second solution if applicable.

WORKED EXAMPLE 8 Confirming solutions by substitution

Confirm, by substitution, the solutions obtained in Worked example 7a: $x^2 + x - 2 = 0$; solutions: $x = 1$ and $x = -2$.

THINK	WRITE
1. Write the left-hand side of the equation and substitute $x = 1$ into the expression.	When $x = 1$, LHS: $x^2 + x - 2 = 1^2 + 1 - 2$ $= 0$

2. Write the right-hand side.	RHS $= 0$
3. Confirm the solution.	LHS = RHS $\Rightarrow$ Solution is confirmed.
4. Write the left-hand side and substitute $x = -2$.	When $x = -2$, LHS: $x^2 + x - 2 = (-2)^2 + -2 - 2$ $= 4 - 2 - 2$ $= 0$
5. Write the right-hand side.	RHS $= 0$
6. Confirm the solution.	LHS = RHS $\Rightarrow$ Solution is confirmed

WORKED EXAMPLE 9 Solving application questions using a graph

A golf ball hit along a fairway follows the path shown in the following graph. The height, *h* metres after it has travelled *x* metres horizontally, follows the rule $h = \frac{1}{270}\left(x^2 - 180x\right)$.

Use the graph to identify how far the ball landed from the golfer.

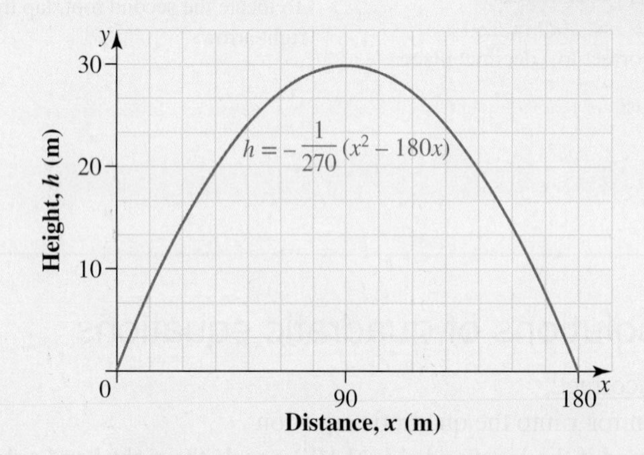

THINK	WRITE
On the graph, the ground is represented by the x-axis since this is where $h = 0$. The golf ball lands when the graph intersects the x-axis.	The golf ball lands 180 m from the golfer.

DISCUSSION

What does 'the solution of a graph' mean?

on Resources

eWorkbook Topic 8 Workbook (worksheets, code puzzle and a project) (ewbk-2034)

Interactivities Individual pathway interactivity: Solving quadratic equations graphically (int-4603)

Solving quadratic equations graphically (int-6148)

Exercise 8.4 Solving quadratic equations graphically

learn**on**

Individual pathways

■ PRACTISE	■ CONSOLIDATE	■ MASTER
1, 4, 7, 10	2, 5, 8, 11	3, 6, 9, 12

To answer questions online and to receive **immediate corrective feedback** and **fully worked solutions** for all questions, go to your learnON title at www.jacplus.com.au.

Fluency

1. a. **WE7** Determine the solutions of each of the following quadratic equations by inspecting the corresponding graphs.

i. $x^2 - x - 6 = 0$

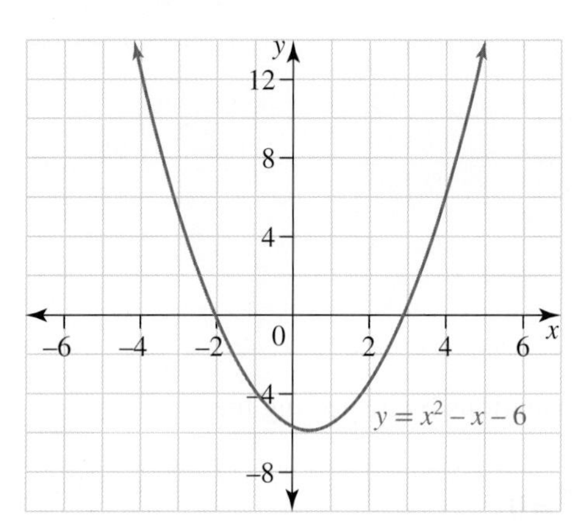

ii. $x^2 - 11x + 10 = 0$

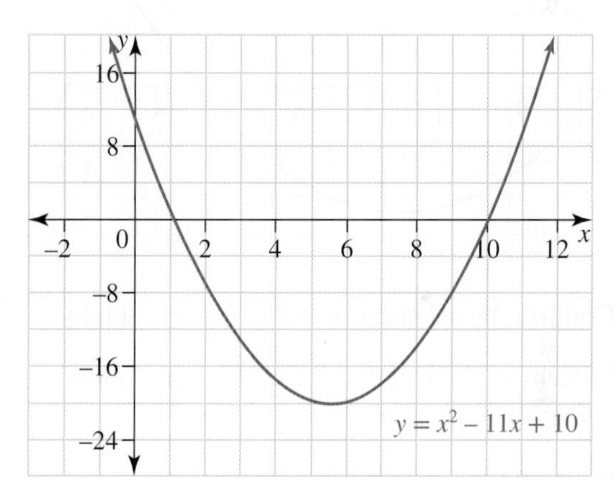

iii. $-x^2 + 25 = 0$

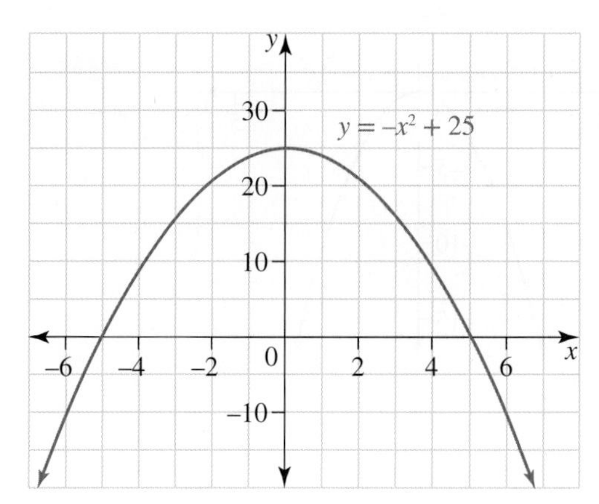

b. **WE8** Confirm, by substitution, the solutions obtained in part a.

2. a. Determine the solutions of each of the following quadratic equations by inspecting the corresponding graphs. Give answers correct to 1 decimal place where appropriate.

i. $2x^2 - 8x + 8 = 0$

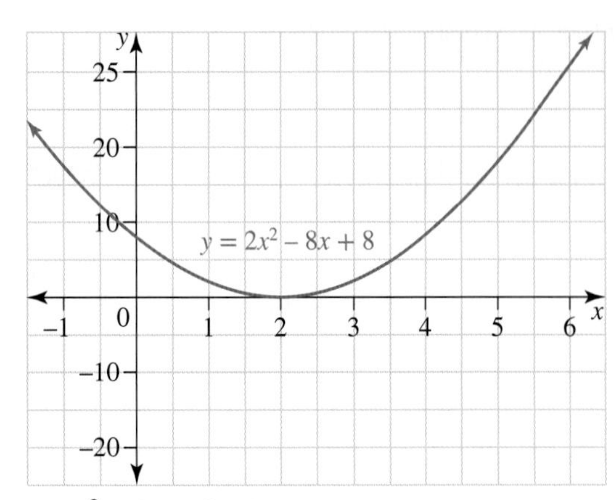

ii. $x^2 - 3x - 4 = 0$

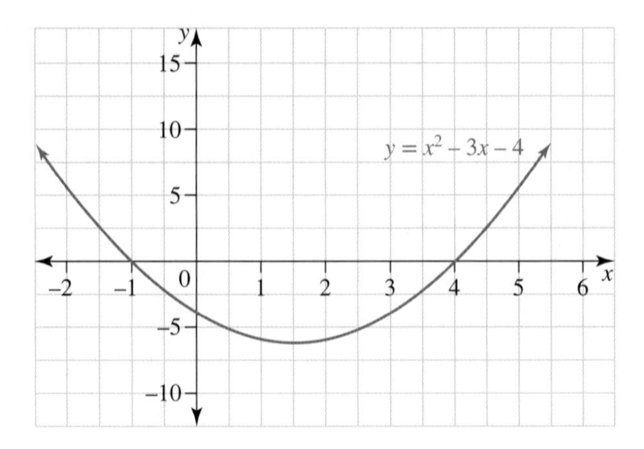

iii. $x^2 - 3x - 6 = 0$

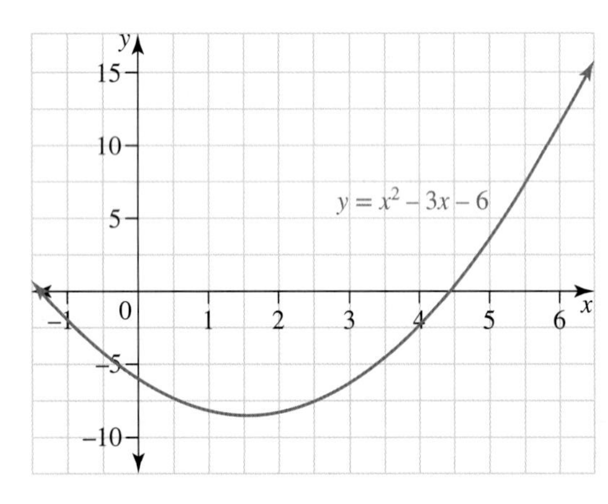

b. Confirm, by substitution, the solutions obtained in part a.

3. a. Determine the solutions of each of the following quadratic equations by inspecting the corresponding graphs.

i. $x^2 + 15x - 250 = 0$

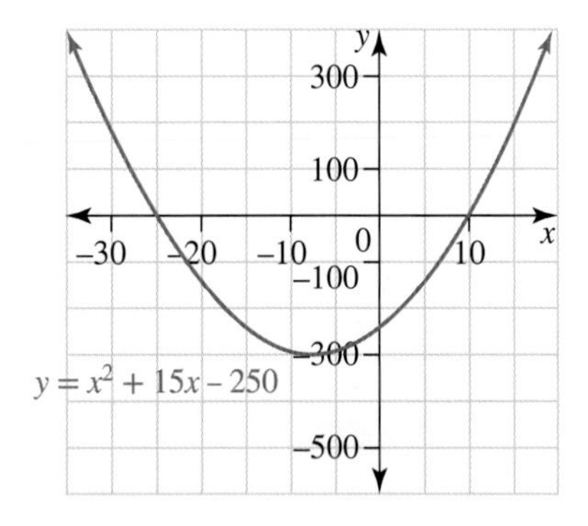

ii. $-x^2 = 0$

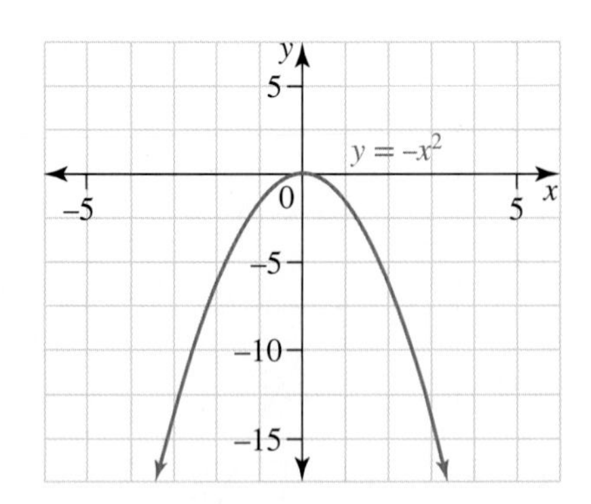

iii. $x^2 + x - 3 = 0$

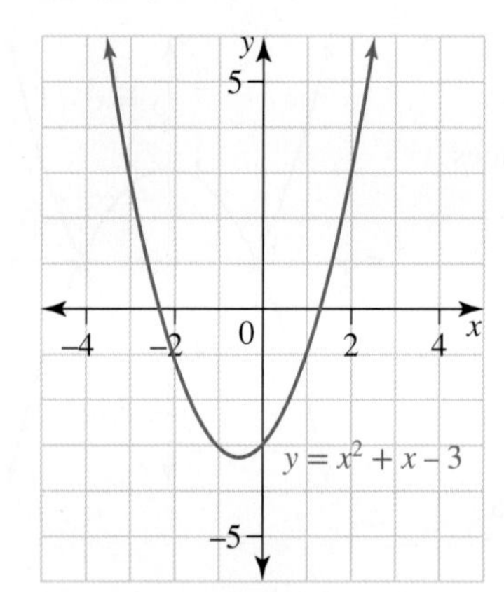

iv. $2x^2 + x - 3 = 0$

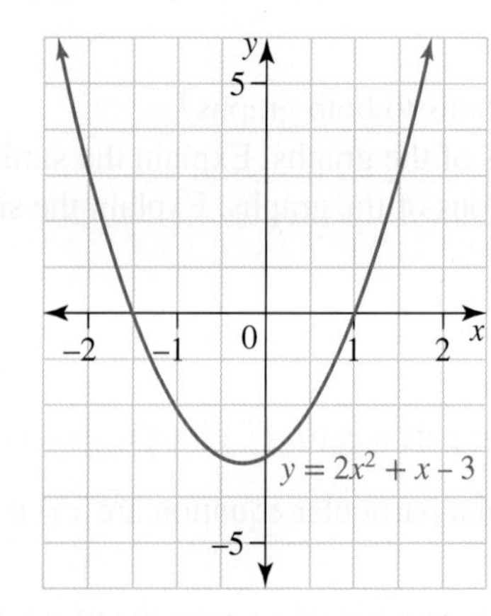

b. Confirm, by substitution, the solutions obtained in part **a**.

Understanding

4. **WE9** A golf ball hit along a fairway follows the path shown in the graph.
The height, h metres after it has travelled x metres horizontally, follows the rule $h = -\frac{1}{200}(x^2 - 150x)$. Use the graph to identify how far the ball lands from the golfer.

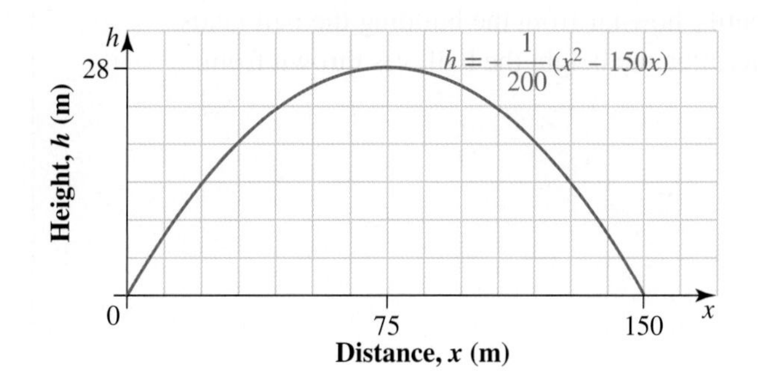

5. **MC** Use the graph to determine how many solutions the equation $x^2 + 3 = 0$ has.

A. No real solutions
B. One solution
C. Two solutions
D. Three solutions
E. Four solutions

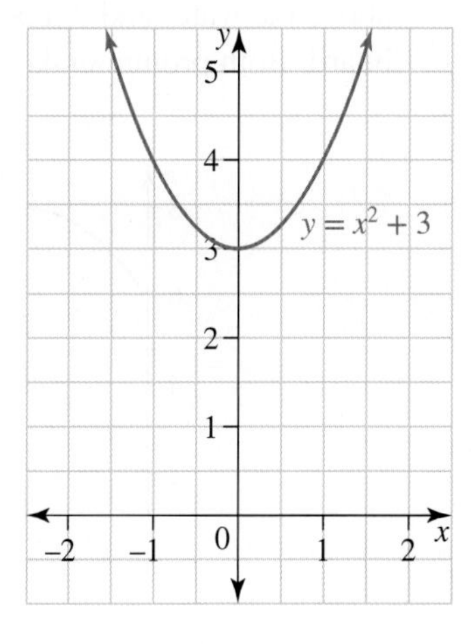

6. **MC** For the equation $x^2 + c = 0$, what values of c will give two solutions, one solution and no real solutions?

A. Two solutions $c > 0$; one solution $c = 0$; no solutions $c < 0$
B. Two solutions $c = 2$; one solution $c = 0$; no solutions $c = -2$
C. Two solutions $c < 0$; one solution $c = 0$; no solutions $c > 0$
D. Two solutions $c = 2$; one solution $c = 1$; no solutions $c = 0$
E. Two solutions $c < 0$; one solution $c = 0$; one solutions $c = 1$

Reasoning

7. Two graphs are shown.

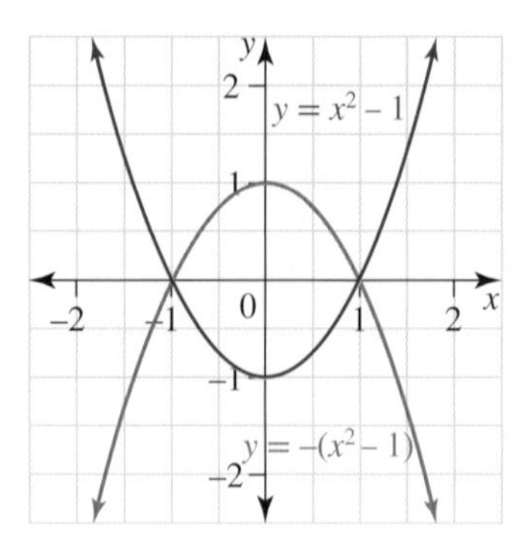

   a. What are the solutions to both graphs?
   b. Look at the shapes of the graphs. Explain the similarities and differences.
   c. Look at the equations of the graphs. Explain the similarities and differences.

8. a. The x-intercepts of a particular equation are $x = 2$ and $x = 5$. Suggest a possible equation.
   b. If the y-intercept in part **a** is $(0, 4)$, give the exact equation.

9. a. The x-intercepts of a particular equation are $x = p$ and $x = q$. Suggest a possible equation.
   b. If the y-intercept in part **a** is $(0, r)$, give the exact equation.

Problem solving

10. A ball is thrown upwards from a building and follows the path shown in the graph until it lands on the ground.
The ball is h metres above the ground when it is a horizontal distance of x metres from the building.
The path of the ball follows the rule $h = -x^2 + 4x + 21$.

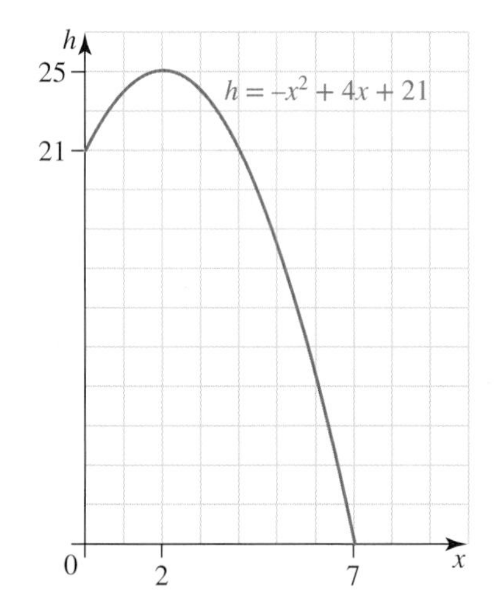

   a. Use the graph to identify how far from the building the ball lands.
   b. Use the graph to determine the height the ball was thrown from.

11. A platform diver follows a path determined by the equation $h = -0.5d^2 + 2d + 6$, where h represents the height of the diver above the water and d represents the distance from the diving board. Both pronumerals are measured in metres.

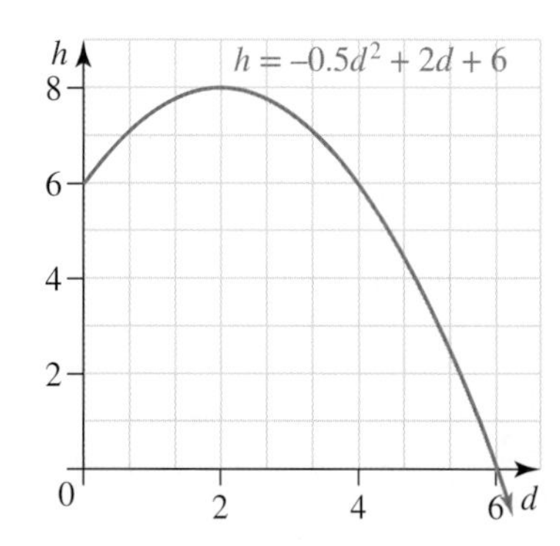

Use the graph to determine:

   a. how far the diver landed from the edge of the diving board
   b. how high the diving board is above the water
   c. the maximum height reached by the diver when they are in the air.

12. Determine the equation of the given parabola. Give your answer in the form $y = ax^2 + bx + c$.

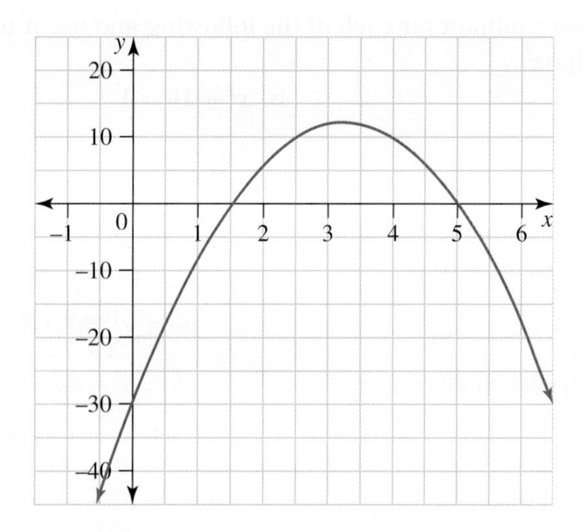

8.5 The discriminant

LEARNING INTENTION

At the end of this subtopic you should be able to:
- calculate the value of the discriminant
- use the value of the discriminant to determine the number of solutions and type of solutions
- calculate the discriminant for simultaneous equations to determine whether two graphs intersect.

8.5.1 Using the discriminant

eles-4849

- **The discriminant** is the value inside the square root sign in the quadratic formula.
- The discriminant can be found for equations in the form $ax^2 + bx + c = 0$.

The discriminant

$$\Delta = b^2 - 4ac$$

- The symbol Δ represents the *discriminant*. This symbol is the Greek capital letter delta.
- The discriminant is found from:

$$x = \frac{-b \pm \sqrt{b^2 - 4ac}}{2a}$$

$$x = \frac{-b \pm \sqrt{\Delta}}{2a}$$

- The value of the discriminant indicates the number of solutions for a quadratic equation.
  - $\Delta < 0$: if the discriminant is negative there are *no real solutions*.
  - $\Delta = 0$: if the discriminant is zero there is *one solution*.
  - $\Delta > 0$: if the discriminant is positive there are *two solutions*.
    - A positive discriminant that is a perfect square (e.g. 16, 25, 144) gives two *rational* solutions.
    - A positive discriminant that is not a perfect square (e.g. 7, 11, 15) gives two *irrational (surd)* solutions.

WORKED EXAMPLE 10 Using the discriminant to determine the number of solutions

Calculate the value of the discriminant for each of the following and use it to determine how many solutions the equation will have.

a. $2x^2 + 9x - 5 = 0$

b. $x^2 + 10 = 0$

THINK	WRITE
a. 1. Write the expression and determine the values of a, b and c given $ax^2 + bx + c = 0$.	a. $2x^2 + 9x - 5 = 0$ $2x^2 + 9x + -5 = 0$ $a = 2, b = 9, c = -5$
2. Write the formula for the discriminant and substitute values of a, b and c.	$\Delta = b^2 - 4ac$ $= 9^2 - 4 \times 2 \times -5$
3. Simplify the equation and solve.	$= 81 - (-40)$ $= 121$
4. State the number of solutions. In this case $\Delta > 0$, which means there are two solutions.	$\Delta > 0$, so there will be two solutions to the equation $2x^2 + 9x - 5 = 0$.
b. 1. Write the expression and determine the values of a, b and c given $ax^2 + bx + c = 0$.	b. $x^2 + 10 = 0$ $1x^2 + 0x + 10 = 0$ $a = 1, b = 0, c = 10$
2. Write the formula for the discriminant and substitute the values of a, b and c.	$\Delta = b^2 - 4ac$ $= 0^2 - 4 \times 1 \times 10$ $= 0 - 40$ $= -40$
3. State the number of solutions. In this case $\Delta < 0$, which means there are no solutions.	$\Delta < 0$, so there will be no solutions to the equation $x^2 + 10 = 0$.

8.5.2 Using the discriminant to determine if graphs intersect

eles-4850

- The table below summarises the number of points of intersection by the graph, indicated by the discriminant.

	$\Delta < 0$ (negative)	$\Delta = 0$ (zero)	$\Delta > 0$ (positive)	
			Perfect square	Not a perfect square
Number of solutions	No real solutions	1 rational solution	2 rational solutions	2 irrational (surd) solutions
Description	Graph does not cross or touch the x-axis	Graph touches the x-axis	Graph intersects the x-axis twice	
Graph	y, x	y, a, x	y, −a, b, x	

WORKED EXAMPLE 11 Using the discriminant to determine the number and type of solutions

By using the discriminant, determine whether the following equations have:

i. two rational solutions
ii. two irrational solutions
iii. one solution
iv. no real solutions.

a. $x^2 - 9x - 10 = 0$
b. $x^2 - 2x - 14 = 0$
c. $x^2 - 2x + 14 = 0$
d. $x^2 - 14x = -49$

THINK	WRITE
a. 1. Write the equation.	**a.** $x^2 - 9x - 10 = 0$
2. Identify the coefficients a, b and c.	$a = 1, b = -9, c = -10$
3. Calculate the discriminant.	$\Delta = b^2 - 4ac$ $= (-9)^2 - 4 \times 1 \times (-10)$ $= 121$
4. Identify the number and type of solutions when $\Delta > 0$ and is a perfect square.	The equation has two rational solutions.
b. 1. Write the equation.	**b.** $x^2 - 2x - 14 = 0$
2. Identify the coefficients a, b and c.	$a = 1, b = -2, c = -14$
3. Calculate the discriminant.	$\Delta = b^2 - 4ac$ $= (-2)^2 - 4 \times 1 \times (-14)$ $= 60$
4. Identify the number and type of solutions when $\Delta > 0$ but not a perfect square.	The equation has two irrational solutions.
c. 1. Write the equation.	**c.** $x^2 - 2x + 14 = 0$
2. Identify the coefficients a, b and c.	$a = 1, b = -2, c = 14$
3. Calculate the discriminant.	$\Delta = b^2 - 4ac$ $= (-2)^2 - 4 \times 1 \times 14$ $= -52$
4. Identify the number and type of solutions when $\Delta < 0$.	The equation has no real solutions.
d. 1. Write the equation, then rewrite it so the right side equals zero.	**d.** $x^2 + 14x = -49$ $x^2 + 14x + 49 = 0$
2. Identify the coefficients a, b and c.	$a = 1, b = 14, c = 49$
3. Calculate the discriminant.	$\Delta = b^2 - 4ac$ $= 14^2 - 4 \times 1 \times 49$ $= 0$
4. Identify the number and types of solutions when $\Delta = 0$.	The equation has one solution.

TI | THINK

a-d.

On a Calculator page, complete the entry lines as:

$b^2 - 4ac \mid a = 1$ and $b = -9$ and $c = -10$

$b^2 - 4ac \mid a = 1$ and $b = -2$ and $c = -14$

$b^2 - 4ac \mid a = 1$ and $b = -2$ and $c = 14$

$b^2 - 4ac \mid a = 1$ and $b = 14$ and $c = 49$

Press ENTER after each entry.

DISPLAY/WRITE

a-d.

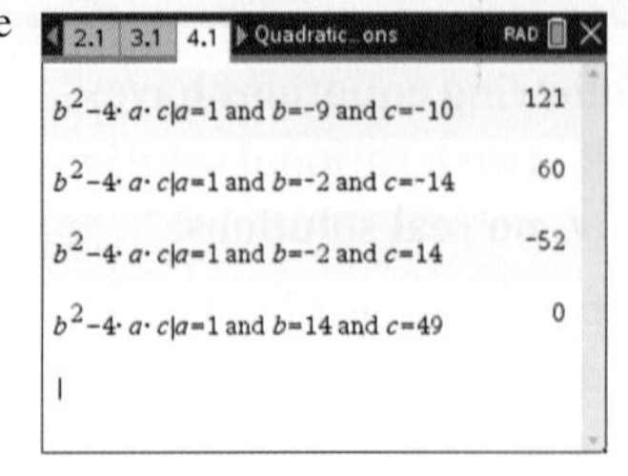

If $\Delta = 121$, the equation has two rational solutions.
If $\Delta = 60$, the equation has two irrational solutions.
If $\Delta = -52$,the equation has no real solutions.
If $\Delta = 0$, the equation has one solution.

CASIO | THINK

a-d.

On the Main screen, complete the entry lines as:

$b^2 - 4ac \mid a = 1 \mid b = -9 \mid c = -10$

$b^2 - 4ac \mid a = 1 \mid b = -2 \mid c = -14$

$b^2 - 4ac \mid a = 1 \mid b = -2 \mid c = 14$

$b^2 - 4ac \mid a = 1 \mid b = 14 \mid c = 49$

Press EXE after each entry.

DISPLAY/WRITE

a-d.

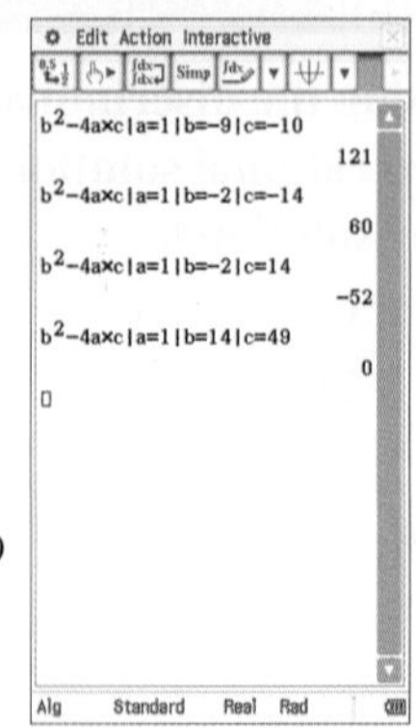

If $\Delta = 121$, the equation has two rational solutions.
If $\Delta = 60$, the equation has two irrational solutions.
If $\Delta = -52$,the equation has no real solutions.
If $\Delta = 0$, the equation has one solution.

- In topic 4 we saw that simultaneous equations can be solved graphically, where the intersection of the two graphs is the solution.
- The discriminant can be used to determine whether a solution exists for two equations and, hence, whether the graphs intersect.

WORKED EXAMPLE 12 Using the discriminant to determine if graphs intersect

Determine whether the parabola $y = x^2 - 2$ and the line $y = x - 3$ intersect.

THINK	WRITE
1. If the parabola and the line intersect, there will be at least one solution to the simultaneous equations: let $y_1 = y_2$.	$y_1 = x_2 - 2$ $y_2 = x - 3$ $y_1 = y_2$ $x^2 - 2 = x - 3$
2. Collect all terms on one side and simplify.	$x^2 - 2 - x + 3 = x - 3 - x + 3$ $x^2 - 2 - x + 3 = 0$ $x^2 - x + 1 = 0$
3. Use the discriminant to check if any solutions exist. If $\Delta < 0$, then no solutions exist.	$\Delta = b^2 - 4ac$ $a = 1, b = -1, c = 1$ $\Delta = (-1)^2 - 4 \times 1 \times 1$ $= 1 - 4$ $= -3$ $\Delta < 0, \therefore$ no solutions exist
4. Write the answer in a sentence.	The parabola and the line do not intersect.

DISCUSSION

What does the discriminant tell us?

on Resources

eWorkbook Topic 8 Workbook (worksheets, code puzzle and a project) (ewbk-2034)

Video eLesson The discriminant (eles-1946)

Interactivities Individual pathway interactivity: The discriminant (int-4604)

The discriminant (int-2560)

Exercise 8.5 The discriminant

learnon

Individual pathways

■ PRACTISE	■ CONSOLIDATE	■ MASTER
1, 4, 9, 11, 15, 19, 22	2, 5, 7, 12, 14, 18, 20, 23	3, 6, 8, 10, 13, 16, 17, 21, 24

To answer questions online and to receive **immediate corrective feedback** and **fully worked solutions** for all questions, go to your learnON title at www.jacplus.com.au.

Fluency

WE10 For questions 1–3 calculate the value of the discriminant for each of the following and use it to determine how many solutions the equation will have.

1. a. $6x^2 + 13x - 5 = 0$
   b. $x^2 + 9x - 90 = 0$
   c. $x^2 + 4x - 2 = 0$
   d. $36x^2 - 1 = 0$
   e. $x^2 + 2x + 8 = 0$

2. a. $x^2 - 5x - 14 = 0$
   b. $36x^2 + 24x + 4 = 0$
   c. $x^2 - 19x + 88 = 0$
   d. $x^2 - 10x + 17 = 0$
   e. $30x^2 + 17x - 21 = 0$

3. a. $x^2 + 16x + 62 = 0$
   b. $9x^2 - 36x + 36 = 0$
   c. $2x^2 - 16x = 0$
   d. $x^2 - 64 = 0$

4. WE11 By using the discriminant, determine whether the equations below have:
   i. two rational solutions
   ii. two irrational solutions
   iii. one solution
   iv. no real solutions.

   a. $x^2 - 3x + 5$
   b. $4x^2 - 20x + 25 = 0$
   c. $x^2 + 9x - 22 = 0$
   d. $9x^2 + 12x + 4$

5. By using the discriminant, determine whether the equations below have:
   i. two rational solutions
   ii. two irrational solutions
   iii. one solution
   iv. no real solutions.

   a. $x^2 + 3x - 7 = 0$
   b. $25x^2 - 10x + 1 = 0$
   c. $3x^2 - 2x - 4 = 0$
   d. $2x^2 - 5x + 4 = 0$

6. By using the discriminant, determine whether the equations below have:
   i. two rational solutions
   ii. two irrational solutions
   iii. one solution
   iv. no real solutions.

   a. $x^2 - 10x + 26 = 0$
   b. $3x^2 + 5x - 7 = 0$
   c. $2x^2 + 7x - 10 = 0$
   d. $x^2 - 11x + 30 = 0$

7. **WE12** Determine whether the following graphs intersect.

a. $y = -x^2 + 3x + 4$ and $y = x - 4$
b. $y = -x^2 + 3x + 4$ and $y = 2x + 5$
c. $y = -(x+1)^2 + 3$ and $y = -4x - 1$
d. $y = (x-1)^2 + 5$ and $y = -4x - 1$

8. Consider the equation $3x^2 + 2x + 7 = 0$.

a. Identify the values of a, b and c.
b. Calculate the value of $b^2 - 4ac$.
c. State how many real solutions, and hence x-intercepts, are there for this equation.

9. Consider the equation $-6x^2 + x + 3 = 0$.

a. Identify the values of a, b and c.
b. Calculate the value of $b^2 - 4ac$.
c. State how many real solutions, and hence x-intercepts, there are for this equation.

10. Consider the equation $-6x^2 + x + 3 = 0$. With the information gained from the discriminant, use the most efficient method to solve the equation. Give an exact answer.

11. **MC** Identify the discriminant of the equation $x^2 - 4x - 5 = 0$.

A. 36
B. 11
C. 4
D. 0
E. −4

12. **MC** Identify which of the following quadratic equations has two irrational solutions.

A. $x^2 - 8x + 16 = 0$
B. $2x^2 - 7x = 0$
C. $x^2 + 8x + 9 = 0$
D. $x^2 - 4 = 0$
E. $x^2 - 6x + 15 = 0$

13. **MC** The equation $x^2 = 2x - 3$ has:

A. two rational solutions
B. exactly one solution
C. no solutions
D. two irrational solutions
E. one rational and one irrational solution

Understanding

14. Determine the value of k if $x^2 - 2x - k = 0$ has one solution.

15. Determine the value of m for which $mx^2 - 6x + 5 = 0$ has one solution.

16. Determine the values of n when $x^2 - 3x - n = 0$ has two solutions.

17. The path of a dolphin as it leaps out of the water can be modelled by the equation $h = -0.4d^2 + d$, where h is the dolphin's height above water and d is the horizontal distance from its starting point. Both h and d are in metres.

a. Calculate how high above the water the dolphin is when it has travelled 2 m horizontally from its starting point.
b. Determine the horizontal distance the dolphin has covered when it first reaches a height of 25 cm.
c. Determine the horizontal distance the dolphin has covered when it next reaches a height of 25 cm. Explain your answer.
d. Determine the horizontal distance the dolphin covers in one leap. (*Hint:* What is the value of h when the dolphin has completed its leap?)
e. During a leap, determine whether this dolphin reaches a height of:

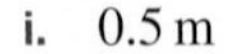

i. 0.5 m
ii. 1 m.

Explain how you can determine this without actually solving the equation.

f. Determine the greatest height the dolphin reaches during a leap.

18. a. Determine how many times the parabolas $y = x^2 - 4$ and $y = 4 - x^2$ intersect.
b. Determine the coordinates of their points of intersection.

Reasoning

19. Show that $3x^2 + px - 2 = 0$ will have real solutions for all values of p.

20. Answer the following questions.

a. Determine the values of a for which the straight line $y = ax + 1$ will have one intersection with the parabola $y = -x^2 - x - 8$.
b. Determine the values of b for which the straight line $y = 2x + b$ will *not* intersect with the parabola $y = x^2 + 3x - 5$.

21. Answer the following questions.

a. Identify how many points of intersection exist between the parabola $y = -2(x+1)^2 - 5$, where $y = f(x), x \in R$, and the straight line $y = mx - 7$, where $y = f(x), x \in R$.
b. Determine the value of m (where $m < 0$) such that $y = mx - 7$ has one intersection point with $y = -m(x+1)^2 - 5$.

Problem solving

22. Answer the following questions.

a. If $\Delta = 9$, $a = 1$ and $b = 5$, use the quadratic formula to determine the two solutions of x.
b. If $\Delta = 9$, $a = 1$ and $b = 5$, calculate the value of c.

23. The parabola with the general equation $y = ax^2 + bx + 9$ where $0 < a < 0$ and $0 < b < 20$ touches the x-axis at one point only. The graph passes through the point $(1, 25)$. Determine the values of a and b.

24. The line with equation $kx + y = 3$ is a tangent to the curve with equation $y = kx^2 + kx - 1$. Determine the value of k.

8.6 Review

8.6.1 Topic summary

Standard quadratic equation

- Quadratic equations are in the form: $ax^2 + bx + c = 0$ where a, b and c are numbers. e.g. In the equation $-5x^2 + 2x - 4 = 0$, $a = -5$, $b = 2$ and $c = -4$.

Solving quadratic equations graphically

- Inspect the graph of $y = ax^2 + bx + c$.
- The solutions to $ax^2 + bx + c = 0$ are the x-axis intercepts.

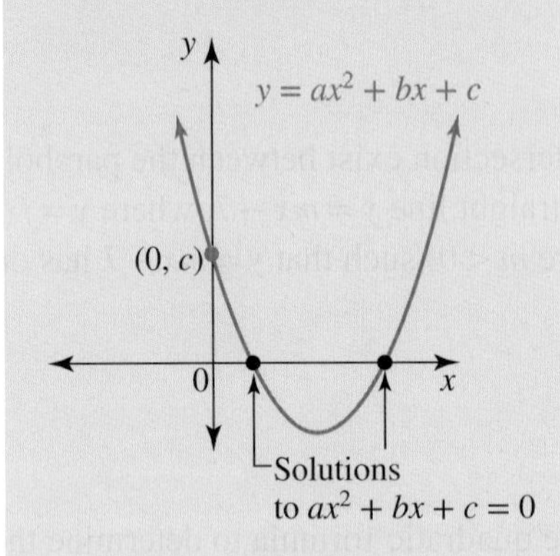

The discriminant

- The discriminant indicates the number and type of solutions.

$$\Delta = b^2 - 4ac$$

- $\Delta < 0$: no real solutions
- $\Delta = 0$: one solution
- $\Delta > 0$: two solutions (if perfect square then two rational solutions, if not a perfect square then two irrational solutions)
- A positive discriminant that is a perfect square (e.g. 16, 25, 144) gives two *rational* solutions.
- A positive discriminant that is not a perfect square (e.g. 7, 11, 15) gives two *irrational* (surd) solutions.

QUADRATIC EQUATIONS

Solving quadratic equations algebraically

- Use algebra to determine the x-value solutions to $ax^2 + bc + c = 0$.
- There will be two, one or no solutions.

Factorising

Factorise using:

- common factor
- difference of two squares
- factor pairs
- completing the square.

Null Factor Law

- Once an equation is factorised, then apply the Null Factor Law.
  - Set the products equal to zero.
  - Solve for x.
- If the product of two numbers is zero, then one or both numbers must be zero.

Quadratic formula

To use the quadratic formula:

- identify the values for a, b and c
- substitute the values into the quadratic formula.

$$x = \frac{-b \pm \sqrt{b^2 - 4ac}}{2a}$$

Note: The ± symbol is called the plus–minus sign.

8.6.2 Success criteria

Tick the column to indicate that you have completed the subtopic and how well you have understood it using the traffic light system.

(**Green:** I understand; **Yellow:** I can do it with help; **Red:** I do not understand)

Subtopic	Success criteria	Green	Yellow	Red
8.2	I can use the Null Factor Law to solve quadratic equations.			
	I can use the completing the square technique to solve quadratic equations.			
	I can solve worded questions algebraically.			
8.3	I can use the quadratic formula to solve equations.			
	I can use the quadratic formula to solve worded questions.			
8.4	I can identify the solutions to $ax^2 + bx + c = 0$ by inspecting the graph of $y = ax^2 + bx + c$.			
	I can recognise whether $ax^2 + bx + c = 0$ has two, one or no solutions by looking at the graph.			
8.5	I can calculate the value of the discriminant.			
	I can use the value of the discriminant to determine the number of solutions and type of solutions.			
	I can calculate the discriminant for simultaneous equations to determine whether two graphs intersect.			

8.6.3 Project

Weaving

Many articles of clothing are sewn from materials that show designs and patterns made by weaving together threads of different colours. Intricate and complex designs can result. Let's investigate some very simple repetitive patterns. Knowledge of quadratic equations and the quadratic formula is helpful in creating these designs.

We need to understand the process of weaving. Weaving machines have parts called *warps*. Each warp is divided into a number of *blocks*. Consider a pattern that is made up of a series of blocks, where the first block is all one colour except for the last thread, which is a different colour.

Let's say our pattern is red and blue. The first block contains all red threads, except for the last one, which is blue. The next block has all red threads, except for the last two threads, which are blue. The pattern continues in this manner. The last block has the first thread as red and the remainder as blue. The warp consists of a particular number of threads, let's say 42 threads.

How many blocks and threads per block would be necessary to create a pattern of this type?

To produce this pattern, we need to divide the warp into equally sized blocks, if possible. What size block and how many threads per block would give us the 42-thread warp? We will need to look for a mathematical pattern. Look at the table (below), where we consider the smallest block consisting of 2 threads through to a block consisting of 7 threads.

Block 1

Block 2

Block n

Pattern	Number of threads per block	Number of blocks	Total threads in warp
RB	2	1	2
RRB RBB	3	2	6
RRRB RRBB RBBB	4		
	5		
	6		
	7		

1. Complete the entries in the table.
2. Consider a block consisting of n threads.
   a. How many blocks would be needed?
   b. What would be the total number of threads in the warp?

The 42-thread warp was chosen as a simple example to show the procedure involved in determining the number of blocks required and the number of threads per block. In this particular case, 6 blocks of 7 threads per block would give us our design for a 42-thread warp. In practice, you would not approach the problem by drawing up a table to determine the number of blocks and the size of each block.

3. Take your expression in question 2b and let it equal 42. This should form a quadratic equation. Solve this equation to verify that you would need 6 blocks with 7 threads per block to fulfil the size of a 42-thread warp.
4. In reality, the size of each block is not always clearly defined. Also, the thread warp sizes are generally much larger, about 250. Let's determine the number of threads per block and the number of blocks required for a 250-thread warp.
   a. Form your quadratic equation with the thread warp size equal to 250.
   b. A solution to this equation can be found using the quadratic formula. Use the quadratic formula to determine a solution.
   c. The number of threads per block is represented by n and this obviously must be a whole number. Round your solution down to the nearest whole number.
   d. How many whole blocks are needed?
   e. Use your solutions to c and d to determine the total number of threads used for the pattern.
   f. How many more threads do you need to make the warp size equal to 250 threads?
   g. Distribute these threads by including them at the beginning of the first block and the end of the last block. Describe your overall pattern.
5. Investigate the number of blocks required and threads per block required for a 400-thread warp.
6. Investigate changing the pattern. Let the first block be all red. In the next block, change the colour of the first and last threads to blue. With each progressive block, change the colour of an extra thread at the top and bottom to blue until the last block is all blue. On a separate sheet of paper, draw a table to determine the thread warp size for a block size of n threads. Draw the pattern and describe the result for a particular warp size.

Resources

eWorkbook Topic 8 workbook (worksheets, code puzzle and a project) (ewbk-2034)

Interactivities Crossword (int-2848)

Sudoku puzzle (int-3595)

Exercise 8.6 Review questions

learnON

To answer questions online and to receive **immediate corrective feedback** and **fully worked solutions** for all questions, go to your learnON title at www.jacplus.com.au.

Fluency

1. **MC** Identify the solutions to the equation $x^2 + 10x - 11 = 0$.
   A. $x = 1$ and $x = 11$
   B. $x = 1$ and $x = -11$
   C. $x = -1$ and $x = 11$
   D. $x = -1$ and $x = -11$
   E. $x = 1$ and $x = 10$

2. **MC** Identify the solutions to the equation $-5x^2 + x + 3 = 0$.
   A. $x = 1$ and $x = \frac{3}{5}$
   B. $x = -0.68$ and $x = 0.88$
   C. $x = 3$ and $x = -5$
   D. $x = 0.68$ and $x = -0.88$
   E. $x = 1$ and $x = -\frac{3}{5}$

3. **MC** Identify the discriminant of the equation $x^2 - 11x + 30 = 0$.
   A. 1
   B. 241
   C. 91
   D. 19
   E. −11

4. **MC** Choose from the following equations which has two irrational solutions.
   A. $x^2 - 6x + 9 = 0$
   B. $4x^2 - 11x = 0$
   C. $x^2 - 25 = 0$
   D. $x^2 + 8x + 2 = 0$
   E. $x^2 - 4x + 10 = 0$

5. The area of a pool is $(6x^2 + 11x + 4)\text{ m}^2$. Determine the length of the rectangular pool if its width is $(2x + 1)\text{ m}$.

6. Determine the solutions of the following equations, by first factorising the left-hand side.
   a. $x^2 + 8x + 15 = 0$
   b. $x^2 + 7x + 6 = 0$
   c. $x^2 + 11x + 24 = 0$
   d. $x^2 + 4x - 12 = 0$
   e. $x^2 - 3x - 10 = 0$

7. Determine the solutions of the following equations, by first factorising the left-hand side.
   a. $x^2 + 3x - 28 = 0$
   b. $x^2 - 4x + 3 = 0$
   c. $x^2 - 11x + 30 = 0$
   d. $x^2 - 2x - 35 = 0$

8. Determine the solutions of the following equations, by first factorising the left-hand side.
   a. $2x^2 + 16x + 24 = 0$
   b. $3x^2 + 9x + 6 = 0$
   c. $4x^2 + 10x - 6 = 0$
   d. $5x^2 + 25x - 70 = 0$
   e. $2x^2 - 7x - 4 = 0$

9. Determine the solutions of the following equations, by first factorising the left-hand side.
   a. $6x^2 - 8x - 8 = 0$
   b. $2x^2 - 6x + 4 = 0$
   c. $6x^2 - 25x + 25 = 0$
   d. $2x^2 + 13x - 7 = 0$

10. Determine the solutions to the following equations by completing the square.
    a. $x^2 + 8x - 1 = 0$
    b. $3x^2 + 6x - 15 = 0$
    c. $-4x^2 - 3x + 1 = 0$

11. Ten times an integer is added to seven times its square. If the result is 152, calculate the original number.

12. By using the quadratic formula, determine the solutions to the following equations. Give your answers correct to 3 decimal places.
    a. $4x^2 - 2x - 3 = 0$
    b. $7x^2 + 4x - 1 = 0$
    c. $-8x^2 - x + 2 = 0$

13. By using the quadratic formula, determine the solutions to the following equations. Give your answers correct to 3 decimal places.
    a. $18x^2 - 2x - 7 = 0$
    b. $29x^2 - 105x - 24 = 0$
    c. $-5x^2 + 2 = 0$

14. The graph of $y = x^2 - 4x - 21$ is shown.

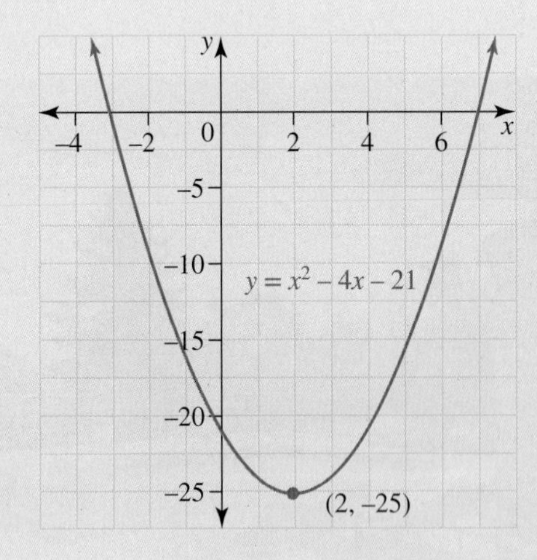

Use the graph to identify the solutions to the quadratic equation $x^2 - 4x - 21 = 0$.

15. Determine the solutions to the equation $-2x^2 - 4x + 6 = 0$.

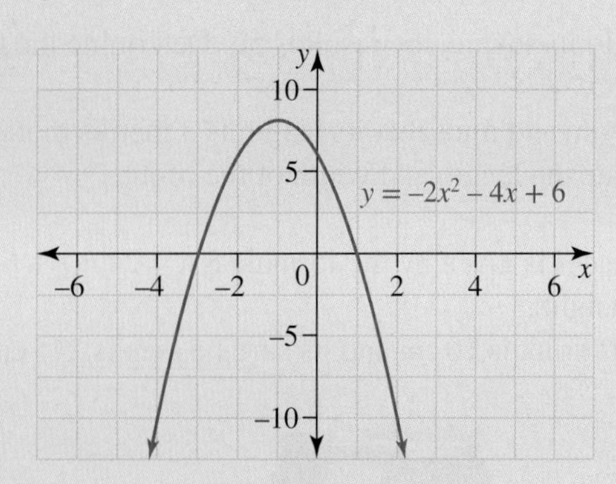

16. By using the discriminant, determine the number and nature of the solutions for the following equations.
 a. $x^2 + 11x + 9 = 0$
 b. $3x^2 + 2x - 5 = 0$
 c. $x^2 - 3x + 4 = 0$

17. What are the solutions to the pair of simultaneous equations shown below?

$$y = x^2 + 4x - 10$$
$$y = 6 - 2x$$

18. Solve the following pair of simultaneous equations to determine the point(s) of intersection.

$$y = x^2 - 7x + 20$$
$$y = 3x - 5$$

19. Determine the solutions to the pair of simultaneous equations shown to determine the point(s) of intersection.

$$y = x^2 + 7x + 11$$
$$y = x$$

20. For each of the following pairs of equations:
 i. solve simultaneously to determine the points of intersection
 ii. illustrate the solution (or lack of solution) by sketching a graph.
 a. $y = x^2 + 6x + 5$ and $y = 11x - 1$
 b. $y = x^2 + 5x - 6$ and $y = 8x - 8$
 c. $y = x^2 + 9x + 14$ and $y = 3x + 5$

21. For each of the following pairs of equations:
 i. solve simultaneously to determine the points of intersection
 ii. illustrate the solution (or lack of solution) by sketching a graph.
 a. $y = x^2 - 7x + 10$ and $y = -11x + 6$
 b. $y = -x^2 + 14x - 48$ and $y = 13x - 54$
 c. $y = -x^2 + 4x + 12$ and $y = 9x + 16$

Problem solving

22. When a number is added to its square, the result is 56. Determine the number.

23. Leroy measures his bedroom and finds that its length is 3 metres more than its width. If the area of the bedroom is 18 m^2, calculate the length and width of the room.

24. The surface area of a cylinder is given by the formula $\text{SA} = 2\pi r(r + h)$, where r cm is the radius of the cylinder and h cm is the height.
The height of a can of soft drink is 10 cm and its surface area is 245 cm^2.

a. Substitute values into the formula to form a quadratic equation using the pronumeral r.
b. Use the quadratic formula to solve the equation and, hence, determine the radius of the can. Round your answer to 1 decimal place.
c. Calculate the area of the label on the can. The label covers the entire curved surface. Round the answer to the nearest square centimetre.

25. Determine the value of d when $2x^2 - 5x - d = 0$ has one solution.

26. Determine the values of k where $(k - 1)x^2 - (k - 1)x + 2 = 0$ has two distinct solutions.

27. Let m and n be the solutions to the quadratic equation $x^2 - 2\sqrt{5}x - 2 = 0$. Determine the value of $m^2 + n^2$.

28. Although it requires a minimum of two points to determine the graph of a line, it requires a minimum of three points to determine the shape of a parabola. The general equation of a parabola is $y = ax^2 + bx + c$, where a, b and c are the constants to be determined.
a. Determine the equation of the parabola that has a y-intercept of $(0, -2)$, and passes though the points $(1, -5)$ and $(-2, 16)$.
b. Determine the equation of a parabola that goes through the points $(0, 0)$, $(2, 2)$ and $(5, 5)$. Show full working to justify your answer.

29. When the radius of a circle increases by 6 cm, its area increases by 25%. Use the quadratic formula to calculate the exact radius of the original circle.

30. A football player received a hand pass and ran directly towards goal. Right on the 50-metre line he kicked the ball and scored a goal. The graph shown represents the path of the ball. Using the graph, answer the following questions.

a. State the height of the ball from the ground when it was kicked.

b. Identify the greatest height the ball reached.

c. Identify the length of the kick.

d. If there were defenders in the goal square, explain if it would have been possible for one of them to mark the ball right on the goal line to prevent a goal. (*Hint:* What was the height of the ball when it crossed the goal line?)

e. As the footballer kicked the ball, a defender rushed at him to try to smother the kick. If the defender can reach a height of 3 m when he jumps, determine how close to the player kicking the ball he must be to just touch the football as it passes over his outstretched hands.

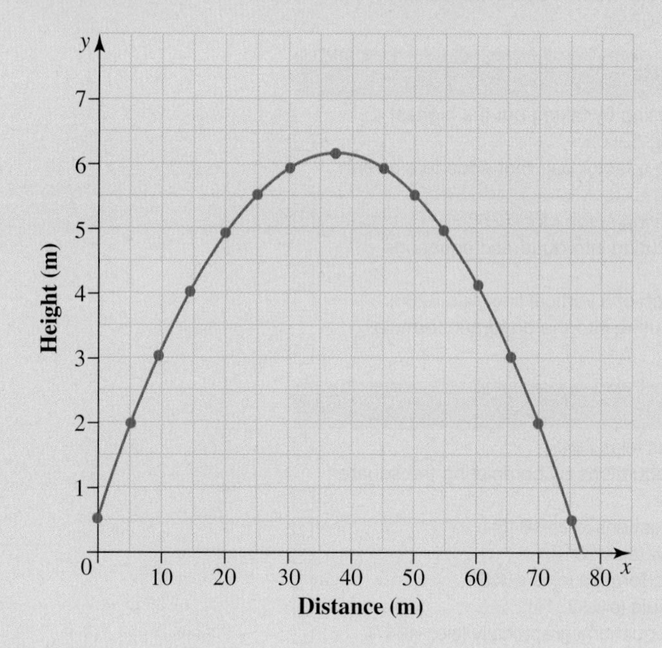

31. The quadratic formula is $x = \dfrac{-b \pm \sqrt{b^2 - 4ac}}{2a}$

An alternative form of the quadratic formula is $x = \dfrac{2c}{-b \pm \sqrt{b^2 - 4ac}}$.

Choose a quadratic equation and show that the two formulas give the same answers.

on To test your understanding and knowledge of this topic, go to your learnON title at www.jacplus.com.au and complete the **post-test**.

Online Resources

Below is a full list of **rich resources** available online for this topic. These resources are designed to bring ideas to life, to promote deep and lasting learning and to support the different learning needs of each individual.

eWorkbook

Download the workbook for this topic, which includes worksheets, a code puzzle and a project (ewbk-2034) ☐

Solutions

Download a copy of the fully worked solutions to every question in this topic (sol-0742) ☐

Digital documents

- 8.2 SkillSHEET: Factorising by taking out the highest common factor (doc-5256) ☐
- SkillSHEET: Finding a factor pair that adds to a given number (doc-5257) ☐
- SkillSHEET: Simplifying surds (doc-5258) ☐
- SkillSHEET: Substituting into quadratic equations (doc-5259) ☐
- SkillSHEET: Equation of a vertical line (doc-5260) ☐
- 8.3 SkillSHEET Substituting into the quadratic formula (doc-5262) ☐

Video eLessons

- 8.2 Quadratic equations (eles-4843) ☐
- Solving quadratic equations by completing the square (eles-4844) ☐
- Solving worded questions (eles-4845) ☐
- The Null Factor Law (eles-2312) ☐
- 8.3 Using the quadratic formula (eles-4846) ☐
- The quadratic formula (eles-2314) ☐
- 8.4 Solving quadratic equations graphically (eles-4847) ☐
- Confirming solutions of quadratic equations (eles-4848) ☐
- 8.5 Using the discriminant (eles-4849) ☐
- Using the discriminant to determine if graphs intersect (eles-4850) ☐
- The discriminant (eles-1946) ☐

Interactivities

- 8.2 Individual pathway interactivity: Solving quadratic equations algebraically (int-4601) ☐
- The Null Factor Law (int-6095) ☐
- 8.3 Individual pathway interactivity: The quadratic formula (int-4602) ☐
- The quadratic formula (int-2561) ☐
- 8.4 Individual pathway interactivity: Solving quadratic equations graphically (int-4603) ☐
- Solving quadratic equations graphically (int-6148) ☐
- 8.5 Individual pathway interactivity: The discriminant (int-4604) ☐
- The discriminant (int-2560) ☐
- 8.6 Crossword (int-2848) ☐
- Sudoku puzzle (int-3595) ☐

Teacher resources

There are many resources available exclusively for teachers online.

To access these online resources, log on to **www.jacplus.com.au**.

Answers

Topic 8 Quadratic equations

Exercise 8.1 Pre-test

1. True
2. $x = -5$ or $x = \frac{1}{2}$
3. $x = \pm\frac{3}{4}$
4. D
5. B
6. D
7. A
8. 24
9. a. Only one solution
   b. No real solutions
   c. Two solutions
10. B
11. A
12. Width = 3 cm, length = 7 cm
13. $m = 4$
14. A
15. E

Exercise 8.2 Solving quadratic equations algebraically

1. a. $-7, 9$ b. $-2, 3$ c. $2, 3$ d. $0, 3$
2. a. $0, 1$ b. $-5, 0$ c. $0, 3$ d. $-2, 0$
3. a. $-\frac{1}{2}, \frac{1}{2}$ b. $-1.2, -0.5$ c. $0.1, 0.75$ d. $-\sqrt{2}, \sqrt{3}$
4. a. $\frac{1}{2}, 1$ b. $-2, -\frac{2}{3}$ c. $\frac{1}{4}, 7$
5. a. $-\frac{6}{7}, 1\frac{1}{2}$ b. $\frac{3}{5}, \frac{2}{3}$ c. $-\frac{5}{8}, \frac{2}{3}$
6. a. $0, \frac{1}{2}, 3$ b. $0, \frac{1}{2}, -\frac{2}{5}$ c. $0, -3, \frac{2}{5}$
7. a. $0, 2$ b. $-5, 0$ c. $0, 7$
8. a. $-\frac{2}{3}, 0$ b. $0, 1\frac{1}{2}$ c. $0, \frac{1}{3}$
9. a. $0, \frac{\sqrt{7}}{2}$ b. $-\frac{\sqrt{3}}{3}, 0$ c. $0, 1\frac{1}{4}$
10. a. $-2, 2$ b. $-5, 5$ c. $-2, 2$ d. $-7, 7$
11. a. $-1\frac{1}{3}, 1\frac{1}{3}$ b. $-2\frac{1}{2}, 2\frac{1}{2}$ c. $-\frac{2}{3}, \frac{2}{3}$ d. $-\frac{1}{2}, \frac{1}{2}$
12. a. $-\frac{1}{5}, \frac{1}{5}$ b. $-4, 4$ c. $-\sqrt{5}, \sqrt{5}$ d. $-\frac{\sqrt{11}}{3}, \frac{\sqrt{11}}{3}$
13. a. $-2, 3$ b. $-4, -2$ c. $-1, 7$ d. $3, 5$
14. a. $-1, 4$ b. 5 c. $-2, 5$ d. $2, 6$ e. $-3, 7$
15. a. $-5, 6$ b. $3, 4$ c. 4 d. -5 e. 10
16. B
17. C
18. a. $-\frac{1}{2}, 3$ b. $\frac{2}{3}, -1$ c. $-2, \frac{1}{5}$ d. $\frac{1}{3}, 1\frac{1}{2}$
19. a. $-\frac{3}{14}, 1$ b. $\frac{1}{4}, \frac{1}{3}$ c. $-1\frac{1}{3}, 2\frac{1}{2}$ d. $-1\frac{3}{4}, -1\frac{1}{3}$
20. a. $-\frac{2}{5}, \frac{1}{2}$ b. $1\frac{1}{2}, 2\frac{2}{3}$ c. $-\frac{2}{5}, \frac{1}{6}$ d. $3, 4$
21. a. $2+\sqrt{2}, 2-\sqrt{2}$
    b. $-1+\sqrt{3}, -1-\sqrt{3}$
    c. $-3+\sqrt{10}, -3-\sqrt{10}$
22. a. $4+2\sqrt{3}, 4-2\sqrt{3}$
    b. $5+2\sqrt{6}, 5-2\sqrt{6}$
    c. $1+\sqrt{3}, 1-\sqrt{3}$
23. a. $-1+\sqrt{6}, -1-\sqrt{6}$
    b. $-2+\sqrt{10}, -2-\sqrt{10}$
    c. $-2+\sqrt{15}, -2-\sqrt{15}$
24. a. $\frac{2}{3}+\frac{\sqrt{5}}{2}, \frac{3}{2}-\frac{\sqrt{5}}{2}$
    b. $-\frac{5}{2}+\frac{\sqrt{29}}{2}, -\frac{5}{2}-\frac{\sqrt{29}}{2}$
    c. $\frac{7}{2}+\frac{\sqrt{33}}{2}, \frac{7}{2}-\frac{\sqrt{33}}{2}$
25. a. $\frac{1}{2}+\frac{\sqrt{21}}{2}, \frac{1}{2}-\frac{\sqrt{21}}{2}$
    b. $\frac{11}{2}+\frac{\sqrt{117}}{2}, \frac{11}{2}-\frac{\sqrt{117}}{2}$
    c. $-\frac{1}{2}+\frac{\sqrt{5}}{2}, -\frac{1}{2}-\frac{\sqrt{5}}{2}$
26. a. $\frac{3}{2}+\frac{\sqrt{37}}{2}, -\frac{3}{2}-\frac{\sqrt{37}}{2}$
    b. $\frac{5}{2}+\frac{\sqrt{37}}{2}, \frac{5}{2}-\frac{\sqrt{37}}{2}$
    c. $\frac{9}{2}+\frac{\sqrt{65}}{2}, \frac{9}{2}-\frac{\sqrt{65}}{2}$
27. a. $-3, 1$ b. $-4.24, 0.24$ c. $-1, 3$
28. a. $-0.73, 2.73$ b. $0.38, 2.62$ c. $-0.30, 3.30$
29. a. $-1.19, 4.19$ b. $-2.30, 1.30$ c. $-2.22, 0.22$
30. 8 and 9 or -8 and -9
31. 6 and 8, -6 and -8

32. 9 or -10

33. 2 or $-2\frac{2}{3}$

34. 8 or $-10\frac{1}{2}$

35. 6 seconds

36. a. $l = 2x$

b. 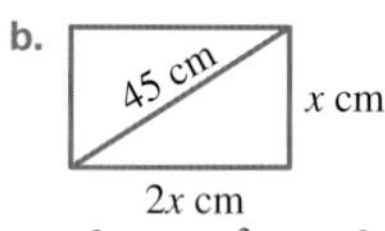

c. $x^2 + (2x)^2 = 45^2, 5x^2 = 2025$

d. Length 40 cm, width 20 cm

37. 8 m, 6 m

38. a. $-6, 1$ b. $8, -3$ c. $x = \pm 1$

39. a. 21 b. 17

40. a. 7

41. a. 2 m, 4 m

b. $(2 + x)$ m, $(4 + x)$ m

c. $(2 + x)(4 + x) = 24$

d. $x = 2$, 4 m wide, 6 m long

42. a. $C_{\text{Annabel}}(28) = \$364\,800$, $C_{\text{Betty}}(28) = \$422\,400$

b. 10 knots

c. Speed can only be a positive quantity, so the negative solution is not valid.

43. No real solutions — when we complete the square we get the sum of two squares, not the difference of two squares and we cannot factorise the expression.

44. $x = 5, -4, 4, -3$

45. $z = -\frac{5}{3}, 7$

46. The width of the pathway is 1.5 m.

Exercise 8.3 The quadratic formula

1. a. $a = 3, b = -4, c = 1$
b. $a = 7, b = -12, c = 2$
c. $a = 8, b = -1, c = -3$
d. $a = 1, b = -5, c = 7$

2. a. $a = 5, b = -5, c = -1$
b. $a = 4, b = -9, c = -3$
c. $a = 12, b = -29, c = 103$
d. $a = 43, b = -81, c = -24$

3. a. $\frac{-5 \pm \sqrt{21}}{2}$ b. $\frac{-3 \pm \sqrt{13}}{2}$
c. $\frac{5 \pm \sqrt{17}}{2}$ d. $2 \pm \sqrt{13}$

4. a. $-1 \pm 2\sqrt{3}$ b. $\frac{7 \pm 3\sqrt{5}}{2}$
c. $\frac{9 \pm \sqrt{73}}{2}$ d. $3 \pm 2\sqrt{3}$

5. a. $-4 \pm \sqrt{31}$ b. $\frac{1 \pm \sqrt{21}}{2}$
c. $\frac{5 \pm \sqrt{33}}{2}$ d. $-1 \pm 2\sqrt{2}$

6. a. $-0.54, 1.87$ b. $-1.20, 1.45$
c. $-4.11, 0.61$ d. $-0.61, 0.47$
e. 0.14, 1.46

7. a. 0.16, 6.34 b. $-1.23, 1.90$ c. $-1.00, 1.14$
d. $-0.83, 0.91$ e. $-0.64, 1.31$

8. a. $-0.35, 0.26$ b. $-1.45, 1.20$ c. 0.08, 5.92
d. $-0.68, 0.88$

9. C

10. C

11. B

12. C

13. a. 0.5, 3 b. 0, 5
c. $-1, 3$ d. 0.382, 2.618

14. a. 2, 4 b. No real solution
c. $-1, 8$ d. $-4.162, 2.162$

15. a. $-7, 1.5$ b. No real solution
c. 2, 7 d. $-\frac{1}{2}, \frac{1}{3}$

16. a. $2\pi r^2 + 14\pi r - 231 = 0$
b. 3.5 cm
c. 154 cm^2

17. a. $x(x + 30)$
b. $x(x + 30) = 1500$
c. 265 mm

18. a. -9 b. 9 c. Kelly

19. a. 4 b. 8 c. 18

20. a. Pool A: $3\frac{2}{3}$ m by $6\frac{2}{3}$ m; Pool B: $3\frac{1}{3}$ m by $7\frac{1}{3}$ m

b. The area of each is $24\frac{4}{9}$ m^2.

21. 25 m, 60 m

22. a. $a = -\frac{1}{16}$ $b = \frac{7}{8}$ $c = 2$

b. $x = -2$ and $x = 16$

c. One solution is between 12 and 18 ($x = 16$), which means his serve was 'in'.

23. $-2 \pm 3, 9 \pm 45$

24. a. Sample responses can be found in the worked solutions in the online resources.

b. Sample responses can be found in the worked solutions in the online resources.

c. $m = 1.85$ so NP is 1.85 cm.

25. a. $x = \pm 2$ or $x = \pm 3$

b. $x = \pm \frac{1}{2}$ or $x = \pm 2$

Exercise 8.4 Solving quadratic equations graphically

1. a. i. $x = -2, x = 3$ ii. $x = 1, x = 10$
iii. $x = -5, x = 5$

b. Sample responses can be found in the worked solutions in the online resources.

2. a. i. $x = 2$ ii. $x = -1, x = 4$
iii. $x \approx -1.4, x \approx 4.4$

b. Sample responses can be found in the worked solutions in the online resources.

3. a. i. $x = -25, x = 10$ ii. $x = 0$
iii. $x \approx -2.3, x \approx 1.3$ iv. $x \approx -1.5, x = 1$
b. Sample responses can be found in the worked solutions in the online resources.
4. 150 m
5. A
6. C
7. a. $x = -1$ and $x = 1$
b. Similarity: shapes; Difference: one is inverted
c. Similarity: both have $x^2 - 1$; Difference: one has a negative and brackets
8. a. $y = a(x-2)(x-5)$
b. $y = \frac{2}{5}(x-2)(x-5)$
9. a. $y = a(x-p)(x-q)$
b. $y = \frac{r}{pq}(x-p)(x-q)$
10. a. 7 m b. 21m
11. a. 6 m b. 6 m c. 8 m
12. $y = -4x^2 + 26x - 30$

Exercise 8.5 The discriminant

1. a. $\Delta = 289$, 2 solutions b. $\Delta = 441$, 2 solutions
c. $\Delta = 24$, 2 solutions d. $\Delta = 144$, 2 solutions
e. $\Delta = -28$, 0 solutions
2. a. $\Delta = 81$, 2 solutions b. $\Delta = 0$, 1 solution
c. $\Delta = 9$, 2 solutions d. $\Delta = 32$, 2 solutions
e. $\Delta = 2809$, 2 solutions
3. a. $\Delta = 8$, 2 solutions b. $\Delta = 0$, 1 solution
c. $\Delta = 256$, 2 solutions d. $\Delta = 256$, 2 solutions
4. a. No real solutions b. 1 solution
c. 2 rational solutions d. 1 solution
5. a. 2 irrational solutions b. 1 solution
c. 2 irrational solutions d. No real solutions
6. a. No real solutions b. 2 irrational solutions
c. 2 irrational solutions d. 2 rational solutions
7. a. Yes b. No c. Yes d. No
8. a. $a = 3, b = 2, c = 7$ b. -80
c. No real solutions
9. a. $a = -6, b = 1, c = 3$ b. 73
c. 2 real solutions
10. $\frac{1 \pm \sqrt{73}}{12}$
11. A
12. C
13. C
14. $k = -1$
15. $m = 1.8$
16. $n > -\frac{9}{4}$
17. a. 0.4 m
b. 0.28 m
c. 2.22 m
d. 2.5 m
e. i. Yes
ii. No
Identify the halfway point between the beginning and the end of the leap, and substitute this value into the equation to determine the maximum height.
f. 0.625 m
18. a. Two times b. $(-2, 0), (2, 0)$
19. p^2 can only give a positive number, which, when added to 24, is always a positive solution.
20. a. $a = -7$ or 5 will give one intersection point.
b. For values of $< -\frac{21}{4}$, there will be no intersection points.
21. a. The straight line crosses the parabola at $(0, -7)$, so no matter what value m takes, there will be at least one intersection point and a maximum of two.
b. $m = -\frac{8}{5}$
22. a. $x = -4$ and $x = -1$
b. 4
23. $a = 4, b = 12$
24. $k = -4$

Project

1. See table at the bottom of the page.*

*1.

Pattern	Number of threads per block	Number of blocks	Total threads in warp
RB	2	1	2
RRB RRB	3	2	6
RRRB RRBB RBBB	4	3	12
RRRRB RRRBB RRBBB RBBBB	5	4	20
RRRRRB RRRRBB RRRBBB RRBBBB RBBBBB	6	5	30
RRRRRRB RRRRRBB RRRRBBB RRRBBBB RRBBBBB RBBBBBB	7	6	42

2. a. $n - 1$ b. $n^2 - n$

3. Answers will vary. Students should form a quadratic equation and let it equal 42. Then solve this equation.

4. a. $n^2 - n = 250$

b. $n = \dfrac{\sqrt{1001} + 1}{2}$

c. $n = 16$

d. 15

e. 240

f. 10

g. Answers will vary. Students should show overall pattern.

5. Answers will vary. Students should investigate the number of blocks required and threads per block required for a 400-thread warp.

6. Answers will vary. Students should investigate changing the pattern. Students should draw a table to determine the thread warp size for a block size of n threads and also draw the pattern and describe the result for a particular warp size.

Exercise 8.6 Review questions

1. B
2. B
3. A
4. D
5. $(3x + 4)$ m

6. a. $-5, -3$ b. $-6, -1$ c. $-8, -3$
d. $2, -6$ e. $5, -2$

7. a. $4, -7$ b. $3, 1$ c. $5, 6$
d. $7, -5$

8. a. $-2, -6$ b. $-2, -1$ c. $\frac{1}{2}, -3$
d. $2, -7$ e. $-\frac{1}{2}, 4$

9. a. $-\frac{2}{3}, 2$ b. $2, 1$ c. $\frac{5}{3}, \frac{5}{2}$
d. $-7, \frac{1}{2}$

10. a. $-4 \pm \sqrt{17}$ b. $-1 \pm \sqrt{6}$ c. $-1, \frac{1}{4}$

11. 4

12. a. $-0.651, 1.151$ b. $-0.760, 0.188$
c. $0.441, -0.566$

13. a. $-0.571, 0.682$ b. $-0.216, 3.836$
c. $-0.632, 0.632$

14. $-3, 7$

15. $-3, 1$

16. a. 2 irrational solutions
b. 2 rational solutions
c. No real solutions

17. a. $(-8, 22)$ and $(2, 2)$

18. $(5, 10)$

19. No solution

20. a.

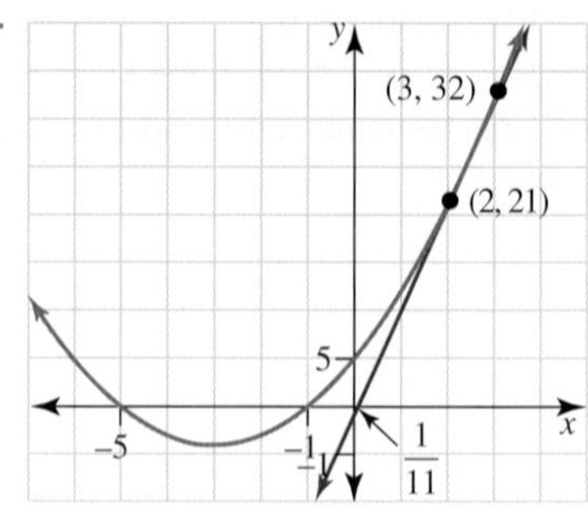

b.

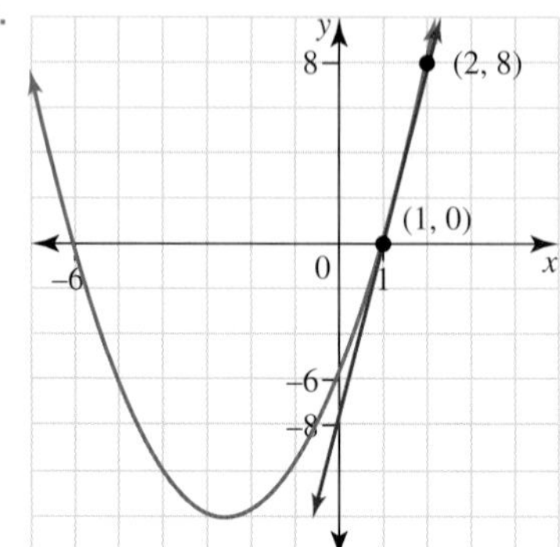

c.

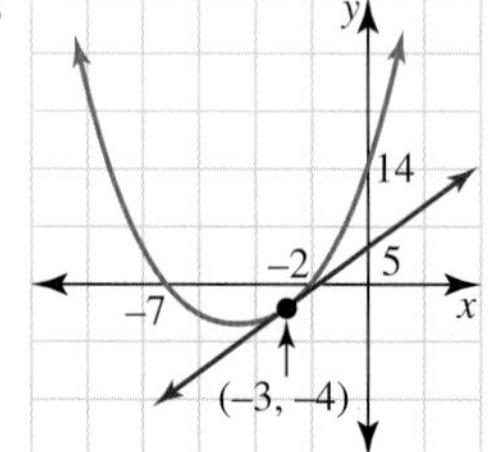

21. a.

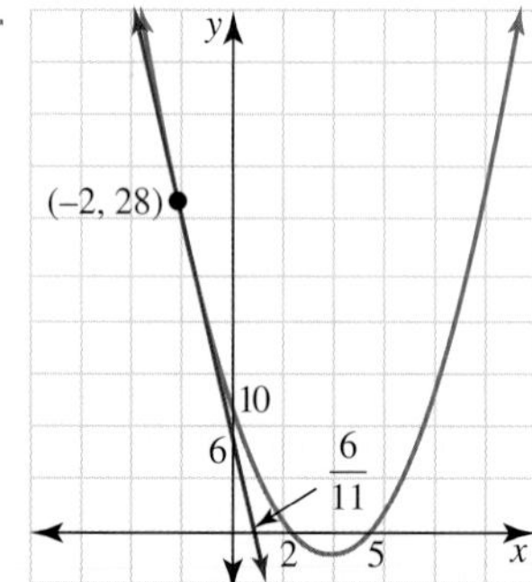

b.

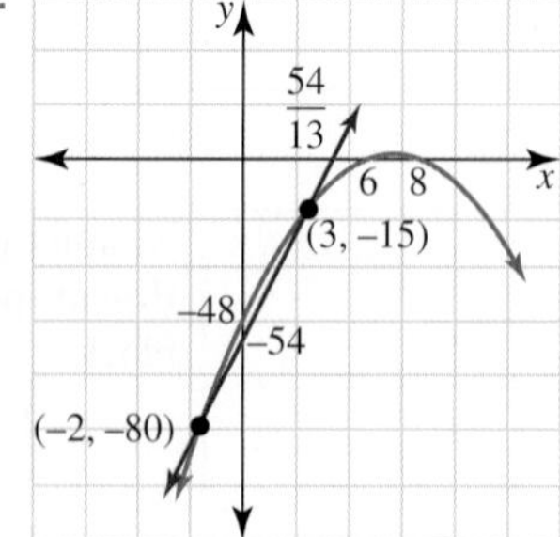

c. 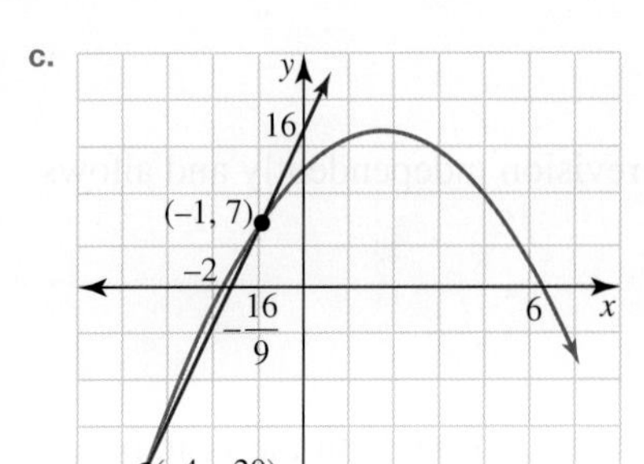

22. -8 and 7
23. Length $= 6$ m, width $= 3$ m
24. a. $2\pi r(r + 10) = 245$
    b. 3.0 cm
    c. 188 cm^2
25. $-\frac{25}{8}$
26. $k > 9$ and $k < 1$
27. 24
28. a. $y = 2x^2 - 5x - 2$
    b. No parabola is possible. The points are on the same straight line.
29. $12\left(\sqrt{5} + 2\right)$ cm
30. a. 0.5 m
    b. 6.1 m
    c. 76.5 m
    d. No, the ball is 5.5 m off the ground and nobody can reach it.
    e. 9.5 m away
31. Sample responses can be found in the worked solutions in the online resources.

Semester review 1

The learnON platform is a powerful tool that enables students to complete revision independently and allows teachers to set mixed and spaced practice with ease.

Student self-study

Review the **Course Content** to determine which topics and subtopics you studied throughout the year. Notice the green bubbles showing which elements were covered.

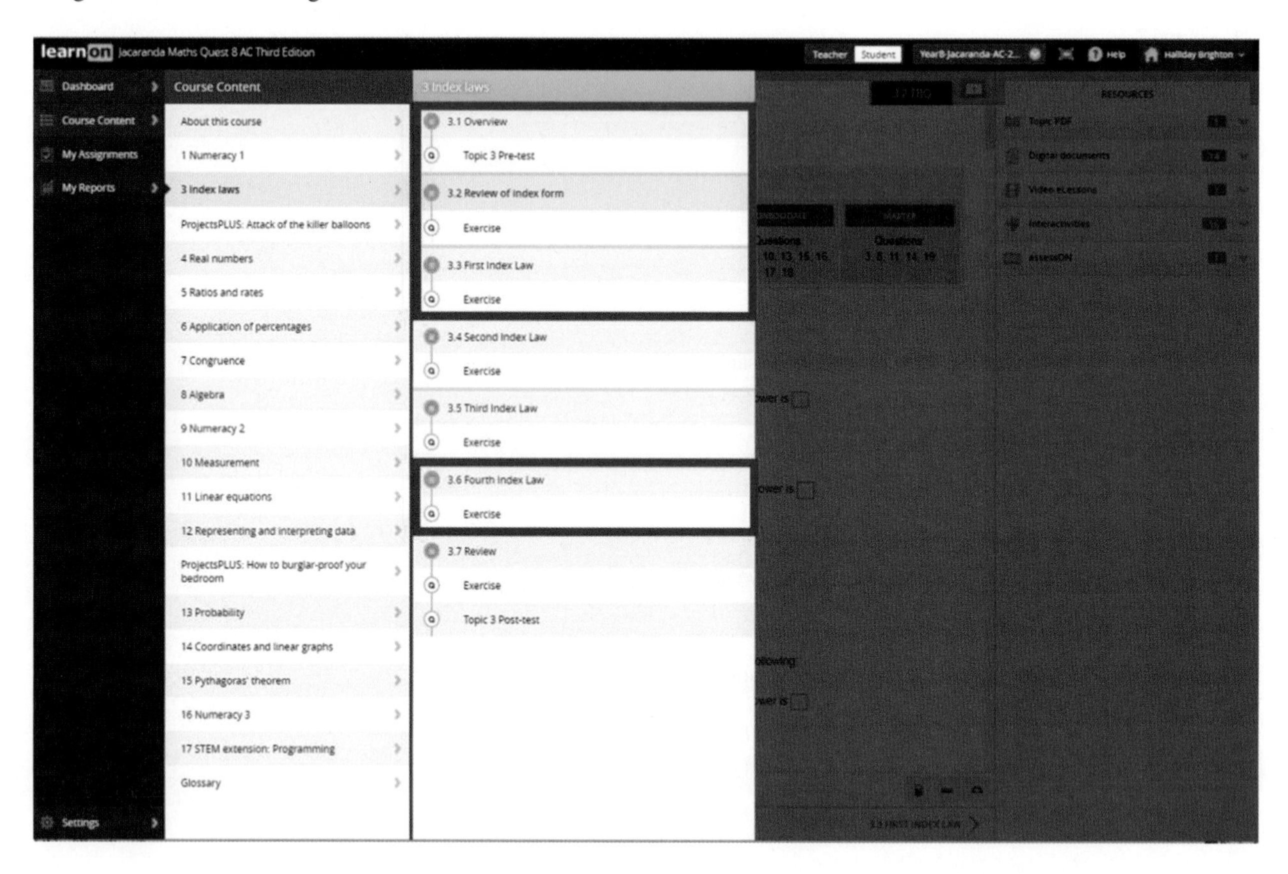

Review your results in **My Reports** and highlight the areas where you may need additional practice.

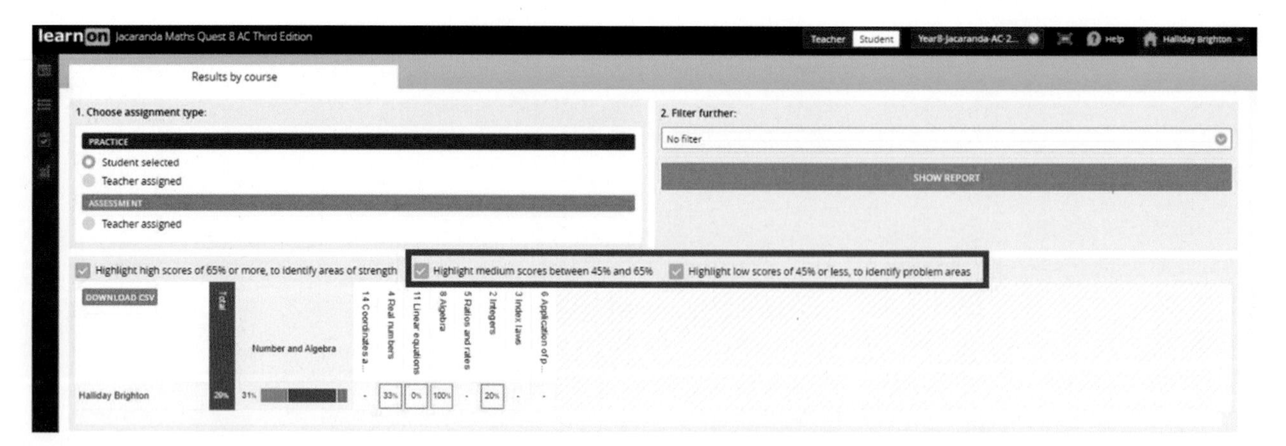

Use these and other tools to help identify areas of strengths and weakness and target those areas for improvement.

Teachers

It is possible to set questions that span multiple topics. These assignments can be given to individual students, to groups or to the whole class in a few easy steps.

Go to **Menu** and select **Assignments** and then **Create Assignment**. You can select questions from one or many topics simply by ticking the boxes as shown below.

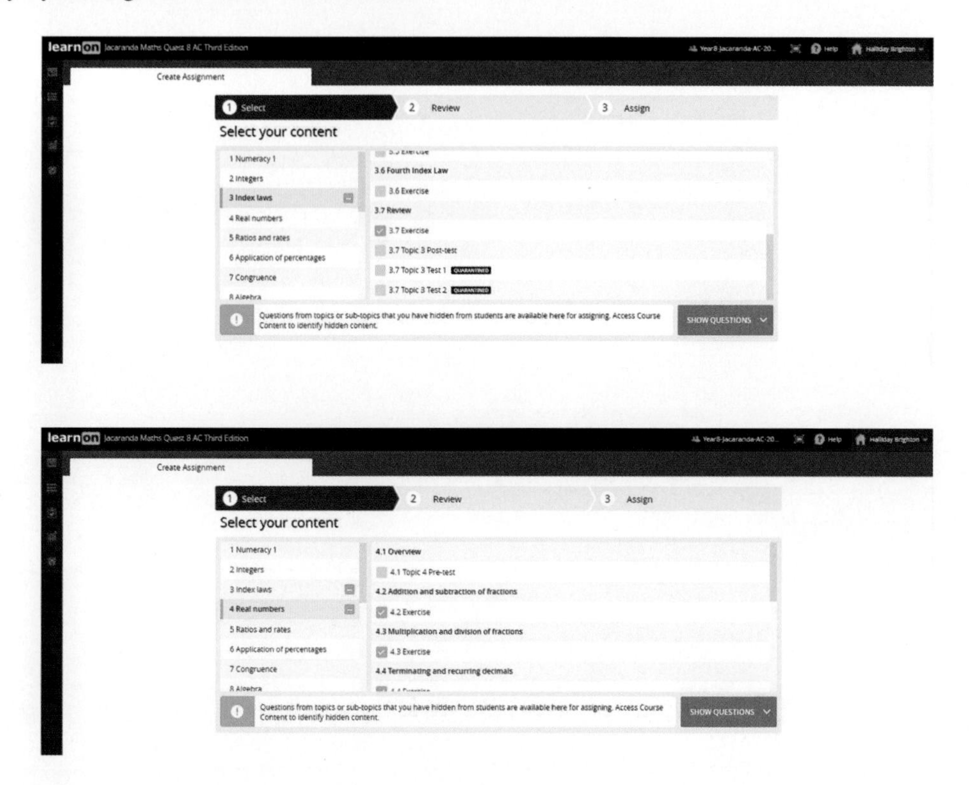

Once your selections are made, you can assign to your whole class or subsets of your class, with individualised start and finish times. You can also share with other teachers.

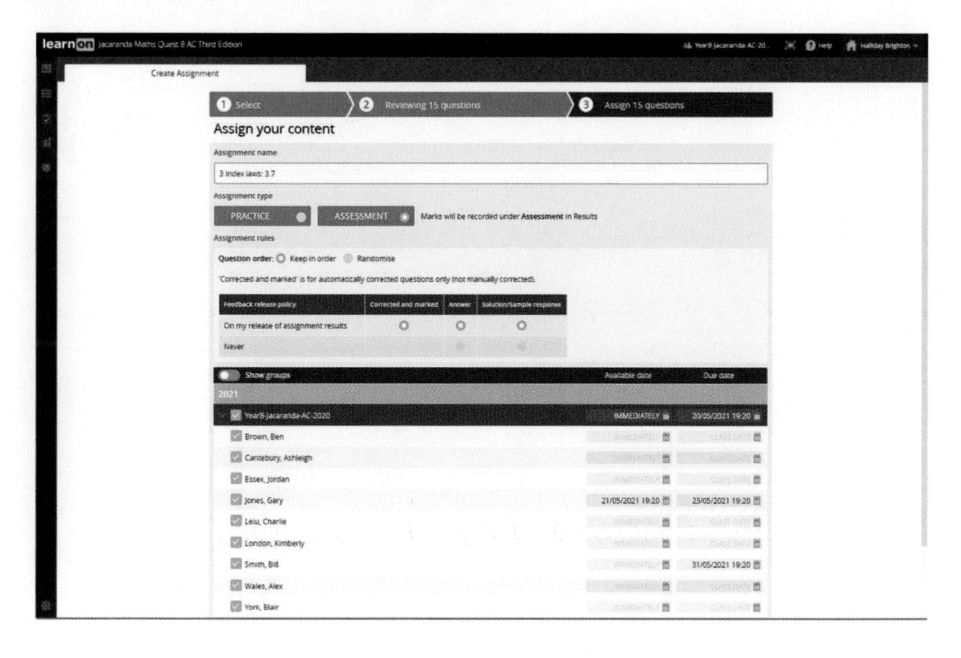

More instructions and helpful hints are available at www.jacplus.com.au.

Teachers

It is possible to set questions that span multiple topics. These assignments can be given to individual students, to groups or to the whole class in a few easy steps.

Go to Menu and select Assignments and then Create Assignment. You [illegible] topics by ticking the boxes as shown below.

Once your students have completed an assignment you can view their results, and share their results and their work with other teachers.

More instructions and helpful hints are available at www.jacplus.com.au.

9 Non-linear relationships

LEARNING SEQUENCE

9.1 Overview

Why learn this?

So far, throughout high school, much of the focus of algebraic sketching has been on linear graphs. A linear graph is the graph of a straight line; therefore, a non-linear graph is any graph of a curve that is not straight. This means that non-linear graphing is a huge field of mathematics, encompassing many topics and areas of study.

If we think of the purpose of sketching graphs, to some extent it is to model relationships between real-life variables. Yet how often do we actually come up across a linear relationship? Throwing a ball through the air, the speed of a car as it accelerates from rest, the path of a runner around a track, the temperature of coffee as it cools — none of these relationships are linear. Even the path of light from a star in the night sky may not be a straight line through space to Earth, due to the curvature of space itself based on gravity!

Non-linear graphs, whether they be parabolas, hyperbolas, exponentials or circles, are some of the more common types of graphs that are used to model phenomena in everyday life. Thus, it is important to study these relationships and their graphs so we can use them to help us model concepts such as exponential growth of a colony of bacteria, or an inversely proportional relationship such as the decay of radioactive material over time.

Where to get help

Go to your learnON title at **www.jacplus.com.au** to access the following digital resources. The Online Resources Summary at the end of this topic provides a full list of what's available to help you learn the concepts covered in this topic.

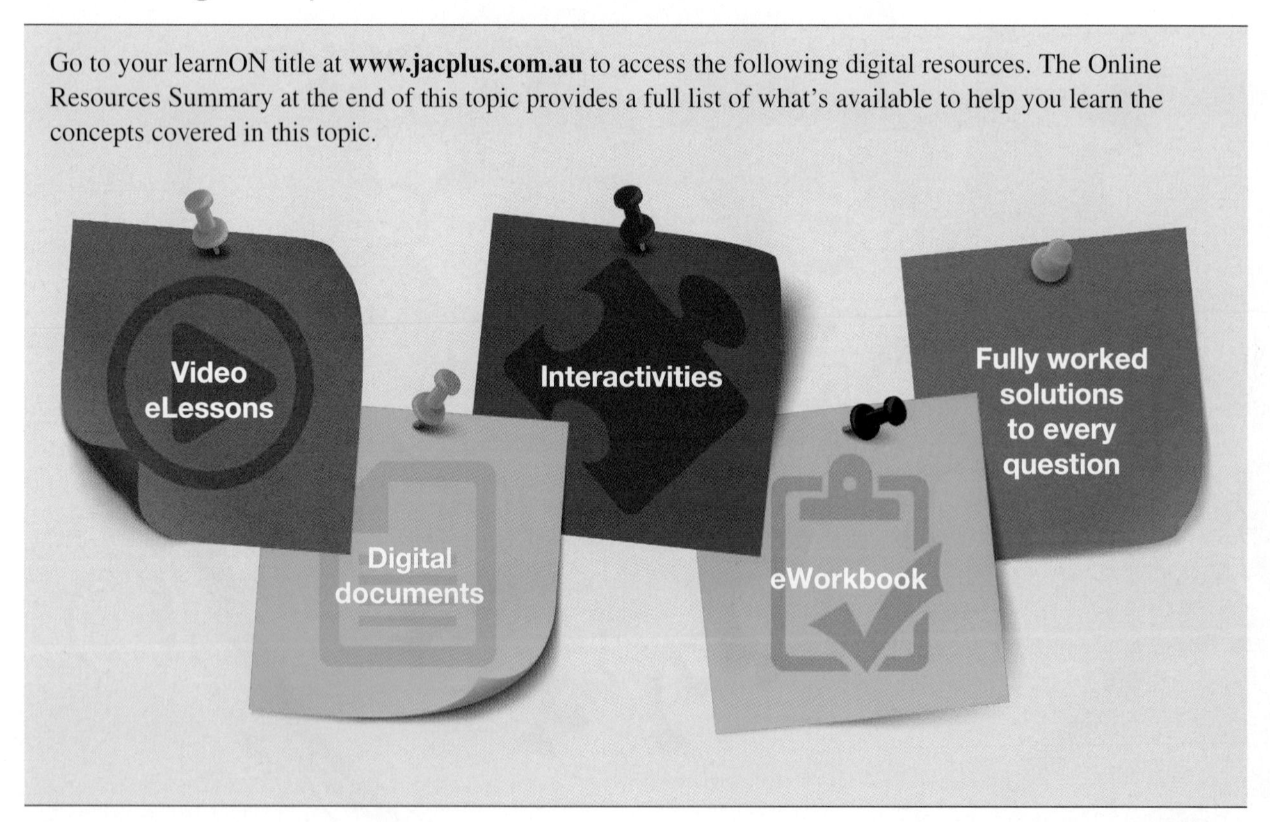

Exercise 9.1 Pre-test

learnON

Complete this pre-test in your learnON title at www.jacplus.com.au and receive **automatic marks**, **immediate corrective feedback** and **fully worked solutions**.

1. Complete the table of values for the equation $y = \frac{1}{2}(x-2)^2 - 3$.

x	-3	-2	-1	0
y				

2. For the graph of $y = -3(x+1)^2 - 2$, state the equation of the axis of symmetry.

3. MC For the graph of $y = 2(x+5)^2 + 8$, the turning point is:
   A. $(-5, -8)$ B. $(-5, 8)$ C. $(5, 8)$ D. $(8, 5)$ E. $(2, 8)$

4. MC The graph of $y = -(x-1)^2 - 3$ has:
   A. a maximum turning point
   B. a minimum turning point
   C. no turning point
   D. two maximum turning points
   E. two minimum turning points

5. MC The x-intercepts of $y = 2(x-2)^2 - 8$ are:
   A. $x = 2$ or $x = 8$
   B. $x = 0$ or $x = 8$
   C. $x = 10$ or $x = -6$
   D. $x = 0$ or $x = 4$
   E. $x = -8$ or $x = -4$

6. Calculate the coordinates of the turning point for the graph of $y = (x+3)(x+5)$. Write your answer in the form (a, b).

7. The equation $y = 2x^2 + bx - 1200$ has x-intercepts of $(-30, 0)$ and $(20, 0)$. Determine the value of b.

8. MC The radius r, of the circle $4x^2 + 4y^2 = 16$ is:
   A. $r = 16$ B. $r = 8$ C. $r = 4$ D. $r = 2$ E. $r = 1$

9. MC The center of the circle with equation $x^2 + 4x + y^2 - 6y + 9 = 0$ is:
   A. $(-4, -3)$ B. $(4, -6)$ C. $(-4, 6)$ D. $(2, -3)$ E. $(-2, 3)$

10. Calculate the points of intersection between the parabola $y = x^2$ and the circle $x^2 + y^2 = 1$, correct to two decimal places.

11. MC The horizontal asymptote for $y = 3^{-x} + 1$ is:
   A. $y = 0$ B. $y = 1$ C. $x = -1$ D. $x = 0$ E. $x = 1$

12. MC For the equation of a hyperbola $y = \frac{2}{x+3} - 1$, the vertical and horizontal asymptotes are:
   A. $x = \frac{2}{3}$ and $y = -1$
   B. $x = 3$ and $y = -1$
   C. $x = -3$ and $y = -1$
   D. $x = -1$ and $y = -3$
   E. $x = 1$ and $y = 3$

13. Match each graph with its correct equation.

Equation	Graph
a. $y = 2^x$	**1.**
b. -2^x	**2.**

14. MC From the graph of the hyperbola, the equation is:

A. $y = \dfrac{1}{x+2} - 1$

B. $y = \dfrac{-1}{x+2} - 1$

C. $y = \dfrac{2}{x+2} - 1$

D. $y = \dfrac{-2}{x+2} - 1$

E. $y = \dfrac{-1}{x-1} - 1$

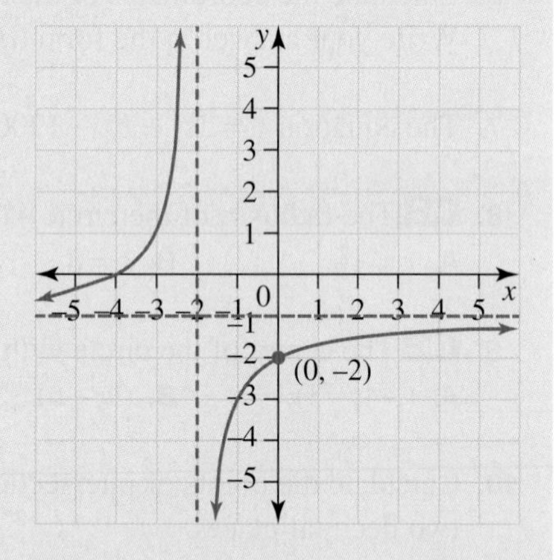

15. **MC** The graph of the truncus with equation $y = \dfrac{1}{x^2}$ reflected in the x-axis looks like:

A.

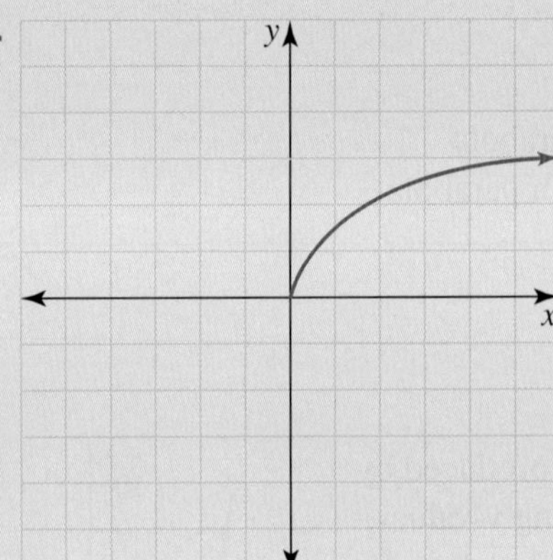

B.

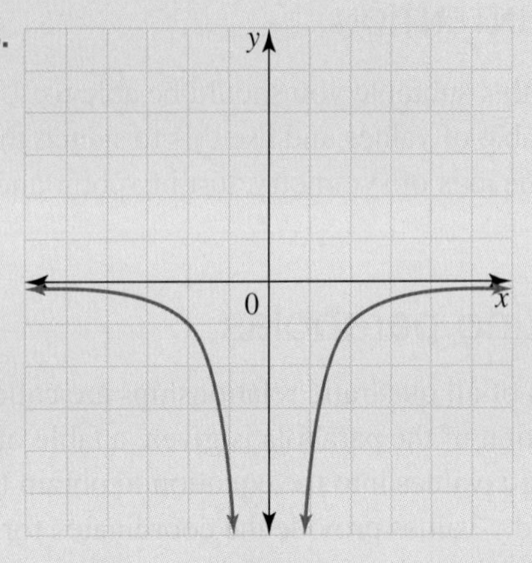

C.

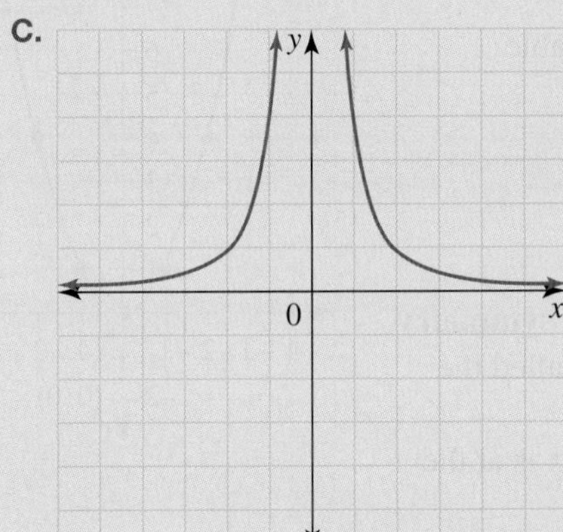

D.

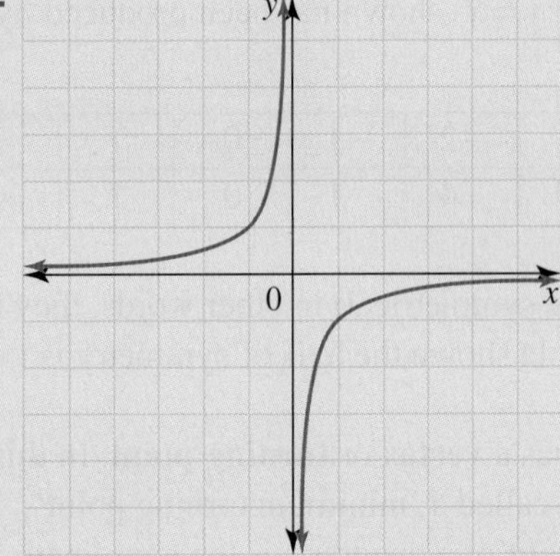

E.

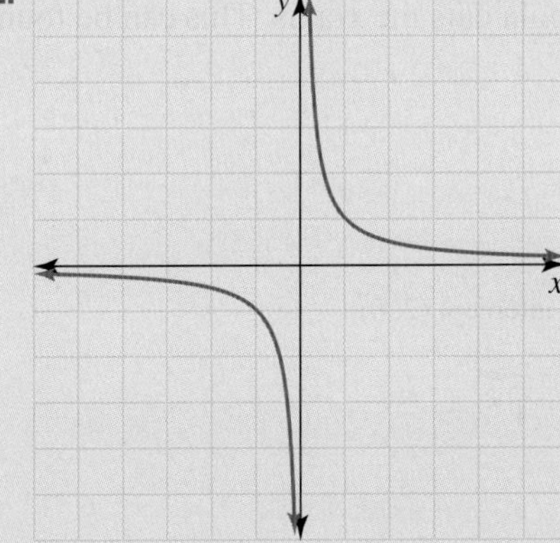

9.2 Plotting parabolas from a table of values

LEARNING INTENTION

At the end of this subtopic you should be able to:
- create a table of values and use this to sketch the graph of a parabola
- identify the axis of symmetry, turning point and y-intercept of a parabola.

9.2.1 Plotting parabolas

eles-4864

- The graphs of all quadratic relationships are called **parabolas**.
- If the equation of the parabola is given, a table of values can be produced by substituting x-values into the equation to obtain the corresponding y-values. These x-and y-values provide the coordinates for points that can be plotted and joined to form the shape of the graph.
- The graph of $y = x^2$ shown has been produced by generating a table of values.

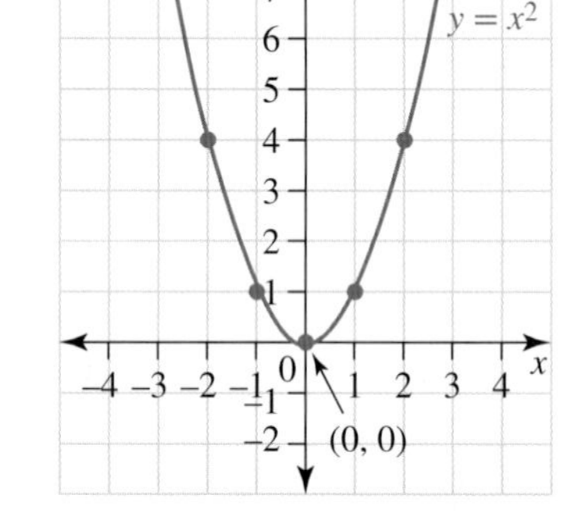

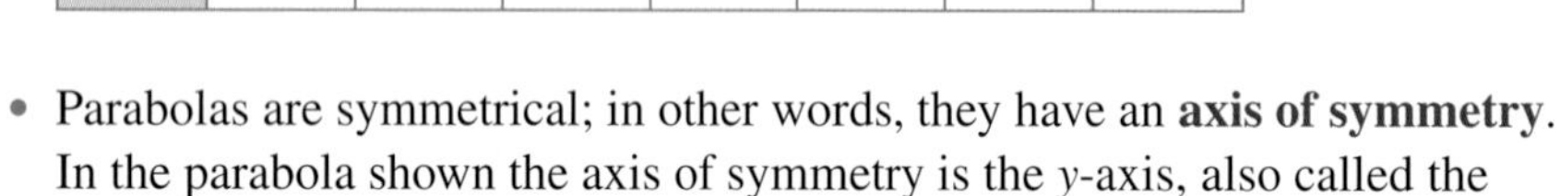

x	−3	−2	−1	0	1	2	3
y	9	4	1	0	1	4	9

- Parabolas are symmetrical; in other words, they have an **axis of symmetry**. In the parabola shown the axis of symmetry is the y-axis, also called the line $x = 0$.
- A parabola has a **vertex** or **turning point**. In this case the vertex is at the origin and is called a 'minimum turning point'.
- The **y-intercept** of a quadratic is the coordinate where the parabola cuts the y-axis. This can be found from a table of value by looking for the point where $x = 0$.
- The **x-intercept** of a quadratic is the coordinate where the parabola cuts the x-axis. This can be found from a table of value by looking for the point where $y = 0$.
- Consider the key features of the equation $y = x^2 + 2x + 8$.

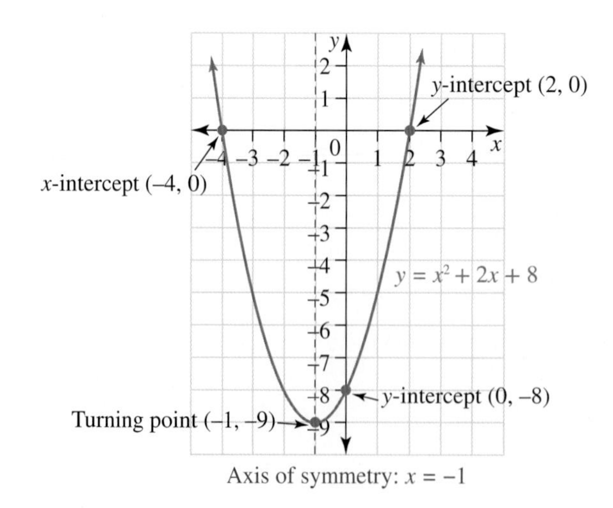

Axis of symmetry: $x = -1$

Shapes of parabolas

- Parabolas with the shape ∪ are said to be 'concave up' and have a minimum turning point.
- Parabolas with the shape ∩ are said to be 'concave down' and have a maximum turning point.

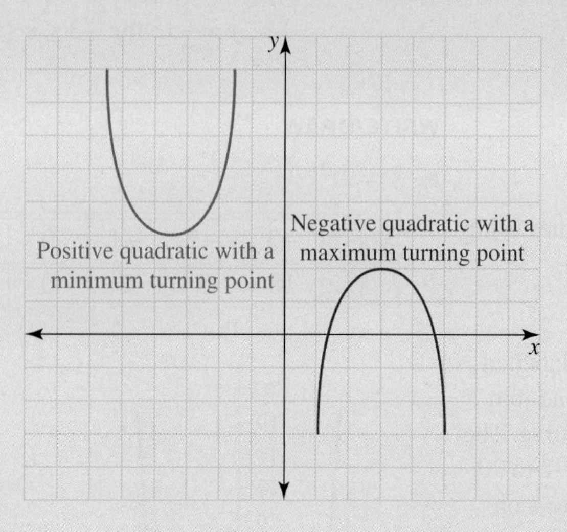

Parabolas in the world around us

- Parabolas abound in the world around us. Here are some examples.

Satellite dishes

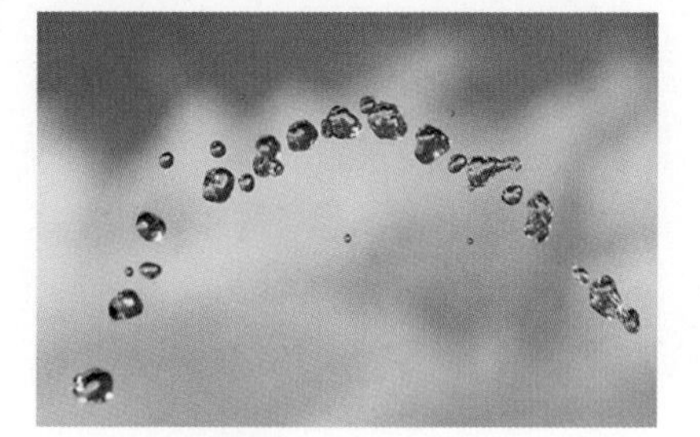

Water droplets from a hose

The cables from a suspension bridge

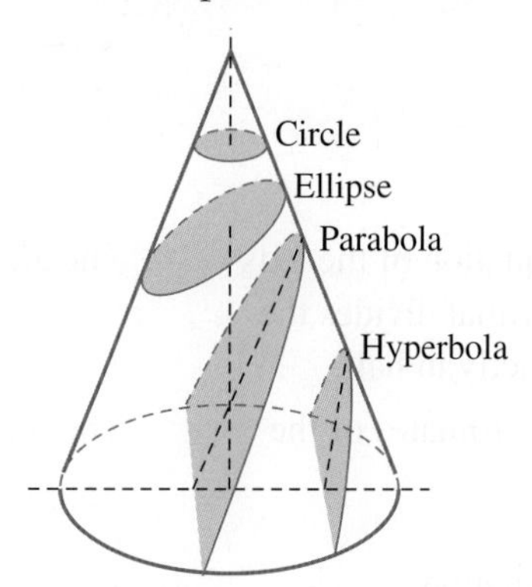

A cone when sliced parallel to its edge reveals a parabola.

WORKED EXAMPLE 1 Plotting parabolas using a table of values

Plot the graph of each of the following equations. In each case, use the values of x shown as the values in your table. State the equation of the axis of symmetry and the coordinates of the turning point.

a. $y = 2x^2$ for $-3 \le x \le 3$

b. $y = \frac{1}{2}x^2$ for $-3 \le x \le 3$

THINK | **WRITE/DRAW**

a. 1. Write the equation.

a. $y = 2x^2$

2. Produce a table of values using x-values from −3 to 3.

x	−3	−2	−1	0	1	2	3
y	18	8	2	0	2	8	18

3. Draw a set of clearly labelled axes, plot the points and join them with a smooth curve. The scale would be from −2 to 20 on the y-axis and −4 to 4 on the x-axis.

4. Label the graph.

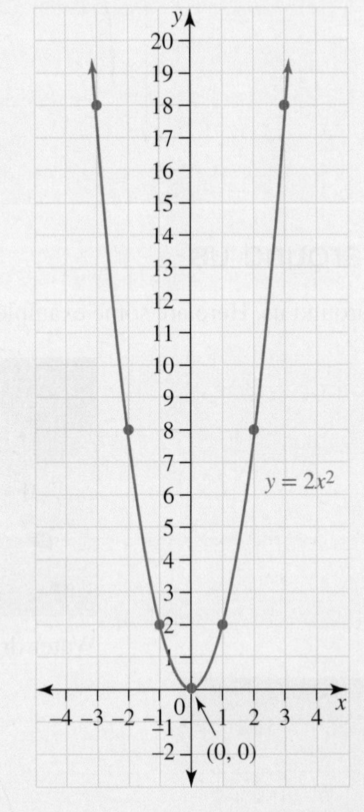

5. Write the equation of the axis of symmetry that divides the parabola exactly in half.

The equation of the axis of symmetry is $x = 0$.

6. Write the coordinates of the turning point.

The turning point is $(0, 0)$.

b. 1. Write the equation.

b. $y = \frac{1}{2}x^2$

2. Produce a table of values using x-values from −3 to 3.

x	−3	−2	−1	0	1	2	3
y	4.5	2	0.5	0	0.5	2	4.5

3. Draw a set of clearly labelled axes, plot the points and join them with a smooth curve. The scale would be from −2 to 6 on the y-axis and −4 to 4 on the x-axis.
4. Label the graph.

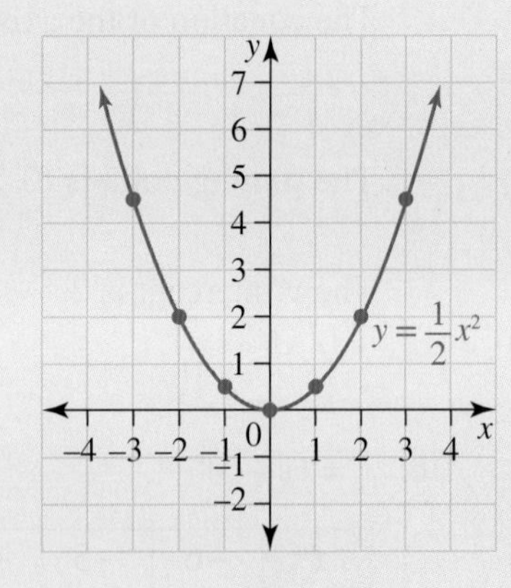

5. Write the equation of the line that divides the parabola exactly in half.

The equation of the axis of symmetry is $x = 0$.

6. Write the coordinates of the turning point.

The turning point is $(0, 0)$.

WORKED EXAMPLE 2 Determining the key features of a quadratic equation

Plot the graph of each of the following equations. In each case, use the values of x shown as the values in your table. State the equation of the axis of symmetry, the coordinates of the turning point and the y-intercept for each one.

a. $y = x^2 + 2$ **for** $-3 \le x \le 3$

b. $y = (x + 3)^2$ **for** $-6 \le x \le 0$

c. $y = -x^2$ **for** $-3 \le x \le 3$

THINK

WRITE/DRAW

a. 1. Write the equation.

a. $y = x^2 + 2$

2. Produce a table of values.

x	−3	−2	−1	0	1	2	3
y	11	6	3	2	3	6	11

3. Draw a set of clearly labelled axes, plot the points and join them with a smooth curve. The scale on the y-axis would be from −2 to 12 and −4 to 4 on the x-axis.
4. Label the graph.

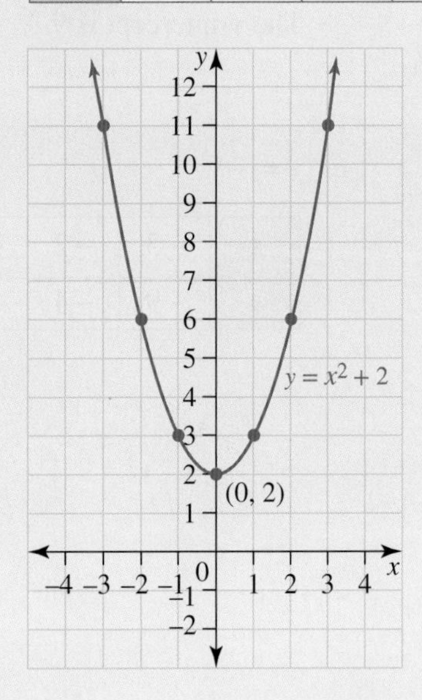

5. Write the equation of the line that divides the parabola exactly in half.

The equation of the axis of symmetry is $x = 0$.

6. Write the coordinates of the turning point.

The turning point is $(0, 2)$.

7. Determine the y-coordinate of the point where the graph crosses the y-axis.

The y-intercept is 2.

b. 1. Write the equation.

b. $y = (x + 3)^2$

2. Produce a table of values.

x	−6	−5	−4	−3	−2	−1	0
y	9	4	1	0	1	4	9

3. Draw a set of clearly labelled axes, plot the points and join them with a smooth curve. The scale on the y-axis would be from −2 to 10 and −7 to 1 on the x-axis.
4. Label the graph.

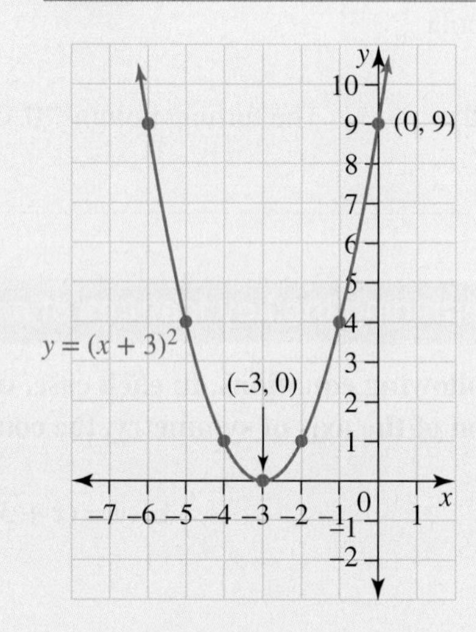

5. Write the equation of the line that divides the parabola exactly in half.

The equation of the axis of symmetry is $x = -3$.

6. Write the coordinates of the turning point.

The turning point is $(-3, 0)$.

7. Determine the y-coordinate of the point where the graph crosses the y-axis.

The y-intercept is 9.

c. 1. Write the equation.

c. $y = -x^2$

2. Produce a table of values.

x	−3	−2	−1	0	1	2	3
y	−9	−4	−1	0	−1	−4	−9

3. Draw a set of clearly labelled axes, plot the points and join them with a smooth curve. The scale on the y-axis would be from -10 to 1 and from -4 to 4 on the x-axis.
4. Label the graph.

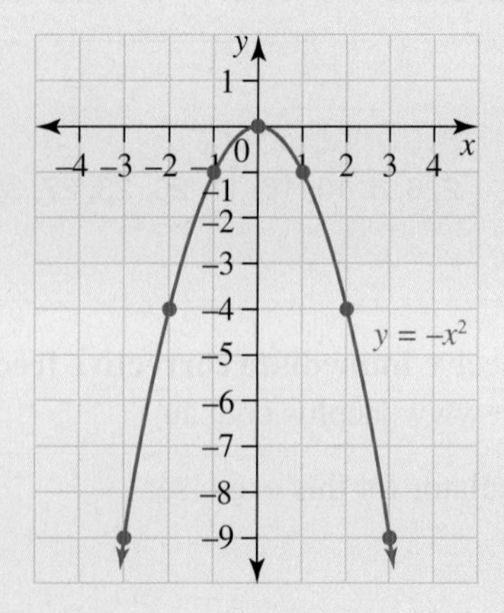

5. Write the equation of the line that divides the parabola exactly in half.

The equation of the axis of symmetry is $x = 0$.

6. Write the coordinates of the turning point.

The turning point is $(0, 0)$.

7. Determine the y-coordinate of the point where the graph crosses the y-axis.

The y-intercept is 0.

DISCUSSION

What x-values can a parabola have? What y-values can a parabola have?

Resources

eWorkbook Topic 9 Workbook (worksheets, code puzzle and a project) (ewbk-2035)

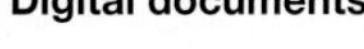

Digital documents SkillSHEET Substitution into quadratic equations (doc-5266)
SkillSHEET Equation of a vertical line (doc-5267)

Interactivities Individual pathway interactivity: Plotting parabolas (int-4605)
Plotting quadratic graphs (int-6150)
Parabolas in the world around us (int-7539)

Exercise 9.2 Plotting parabolas from a table of values

learnON

Individual pathways

PRACTISE	CONSOLIDATE	MASTER
1, 3, 5, 11, 13, 15, 19, 22, 26, 29	2, 6, 8, 10, 16, 17, 20, 23, 27, 30	4, 7, 9, 12, 14, 18, 21, 24, 25, 28, 31

To answer questions online and to receive **immediate corrective feedback** and **fully worked solutions** for all questions, go to your learnON title at www.jacplus.com.au.

You may wish to use a graphing calculator for this exercise.

Fluency

1. WE1 Plot the graph of each of the following equations. In each case, use the values of x shown as the values in your table. State the equation of the axis of symmetry and the coordinates of the turning point.

   a. $y = 3x^2$ for $-3 \leq x \leq 3$

   b. $y = \frac{1}{4}x^2$ for $-3 \leq x \leq 3$

2. Compare the graphs drawn for question 1 with that of $y = x^2$. Explain how placing a number in front of x^2 affects the graph obtained.

3. WE2a Plot the graph of each of the following for values of x between -3 and 3. State the equation of the axis of symmetry, the coordinates of the turning point and the y-intercept for each one.

   a. $y = x^2 + 1$

   b. $y = x^2 + 3$

   c. $y = x^2 - 3$

   d. $y = x^2 - 1$

4. Compare the graphs drawn for question 3 with the graph of $y = x^2$. Explain how adding to or subtracting from x^2 affects the graph obtained.

WE2b For questions 5 to 8, plot the graph of each of the following equations. In each case, use the values of x shown as the values in your table. State the equation of the axis of symmetry and the coordinates of the turning point and the y-intercept for each one.

5. $y = (x + 1)^2 \quad -5 \leq x \leq 3$

6. $y = (x - 2)^2 \quad -1 \leq x \leq 5$

7. $y = (x - 1)^2 \quad -2 \leq x \leq 4$

8. $y = (x + 2)^2 \quad -6 \leq x \leq 2$

9. Compare the graphs drawn for questions 5 to 8 with that for $y = x^2$. Explain how adding to or subtracting from x before squaring affects the graph obtained.

WE2c For questions 10 to 13, plot the graph of each of the following equations. In each case, use the values of x shown as the values in your table. State the equation of the axis of symmetry, the coordinates of the turning point and the y-intercept for each one.

10. $y = -x^2 + 1 \quad -3 \leq x \leq 3$

11. $y = -(x + 2)^2 \quad -5 \leq x \leq 1$

12. $y = -x^2 - 3 \quad -3 \leq x \leq 3$

13. $y = -(x - 1)^2 \quad -2 \leq x \leq 4$

14. Compare the graphs drawn for questions 10 to 13 with that for $y = x^2$. Explain how a negative sign in front of x^2 affects the graph obtained. Also compare the graphs obtained in questions 10 to 13 with those in questions 3 and 5 to 8. State which graphs have the same turning point. Describe how are they different.

Understanding

For questions 15 to 20:

a. plot the graph
b. state the equation of the axis of symmetry
c. state the coordinates of the turning point and whether it is a maximum or a minimum
d. state the y-intercept.

15. $y = (x-5)^2 + 1 \qquad 0 \le x \le 6$
16. $y = 2(x+2)^2 - 3 \qquad -5 \le x \le 1$
17. $y = -(x-3)^2 + 4 \qquad 0 \le x \le 6$
18. $y = -3(x-1)^2 + 2 \qquad -2 \le x \le 4$
19. $y = x^2 + 4x - 5 \qquad -6 \le x \le 2$
20. $y = -3x^2 - 6x + 24 \qquad -5 \le x \le 3$
21. Use the equation $y = a(x-b)^2 + c$ to answer the following.
    a. Explain how you can determine whether a parabola has a minimum or maximum turning point by looking only at its equation.
    b. Explain how you can determine the coordinates of the turning point of a parabola by looking only at the equation.
    c. Explain how you can obtain the equation of the axis of symmetry by looking only at the equation of the parabola.
22. **MC** For the graph of $y = (x-2)^2 + 5$, the turning point is:
    A. $(5, 2)$ B. $(2, -5)$ C. $(2, 5)$ D. $(-2, -5)$ E. $(-2, 5)$
23. **MC** For the graph of $y = 3(x-1)^2 + 12$, the turning point is:
    A. $(3, 12)$ B. $(1, 12)$ C. $(-1, 12)$ D. $(-3, 12)$ E. $(-1, -12)$
24. **MC** For the graph of $y = (x+2)^2 - 7$, the y-intercept is:
    A. -2 B. -7 C. -3 D. -11 E. 7
25. **MC** Select which of the following is true for the graph of $y = -(x-3)^2 + 4$.
    A. Turning point $(3, 4)$, y-intercept -5
    B. Turning point $(3, 4)$, y-intercept 5
    C. Turning point $(-3, 4)$, y-intercept -5
    D. Turning point $(-3, 4)$, y-intercept 5
    E. Turning point $(3, -4)$, y-intercept 13

Reasoning

26. A ball is thrown into the air. The height, h metres, of the ball at any time, t seconds, can be found by using the equation $h = -(t-4)^2 + 16$.
    a. Plot the graph for values of t between 0 and 8.
    b. Use the graph to determine:
        i. the maximum height of the ball
        ii. how long it takes for the ball to fall back to the ground from the moment it is thrown.

27. From a crouching position in a ditch, an archer wants to fire an arrow over a horizontal tree branch, which is 15 metres above the ground. The height, in metres (h), of the arrow t seconds after it has been fired is given by the equation $h = -8t(t - 3)$.

a. Plot the graph for $t = 0, 1, 1.5, 2, 3$.
b. From the graph, determine:
  i. the maximum height the arrow reaches
  ii. whether the arrow clears the branch and the distance by which it clears or falls short of the branch
  iii. the time it takes to reach maximum height
  iv. how long it takes for the arrow to hit the ground after it has been fired.

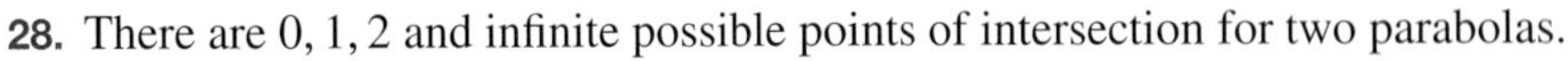

28. There are 0, 1, 2 and infinite possible points of intersection for two parabolas.
a. Illustrate these on separate graphs.
b. Explain why infinite points of intersection are possible. Give an example.
c. Determine how many points of intersection are possible for a parabola and a straight line. Illustrate these.

Problem solving

29. The area of a rectangle in cm^2 is given by the equation $A = \frac{-8w}{5}(w - 6)$, where w is the width of the rectangle in centimetres.
a. Complete a table of values for $-1 \leq w \leq 7$.
b. Explain which of the values for w from part a should be discarded and why.
c. Sketch the graph of A for suitable values of w.
d. Evaluate the maximum possible area of the rectangle. Show your working.
e. Determine the dimensions of the rectangle that produce the maximum area found in part d.

30. The path taken by a netball thrown by a rising Australian player is given by the quadratic equation $y = -x^2 + 3.2x + 1.8$, where y is the height of the ball and x is the horizontal distance from the player's upstretched hand.
a. Complete a table of values for $-1 \leq x \leq 4$.
b. Plot the graph.
c. Explain what values of x are 'not reasonable'.
d. Evaluate the maximum height reached by the netball.
e. Assuming that nothing hits the netball, determine how far away from the player the netball will strike the ground.

31. The values of a, b and c in the equation $y = ax^2 + bx + c$ can be calculated using three points that lie on the parabola. This requires solving triple simultaneous equations by algebra. This can also be done using a CAS calculator. If the points $(0, 1)$, $(1, 0)$ and $(2, 3)$ all lie on one parabola, determine the equation of the parabola.

9.3 Sketching parabolas using transformations

LEARNING INTENTION

At the end of this subtopic you should be able to:

- be familiar with the key features of the basic graph of a quadratic, $y = x^2$
- determine whether a dilation has made the graph of quadratic narrower or wider
- determine whether a translation has moved the graph of a quadratic function left/right or up/down
- sketch the graph of a quadratic equation that has undergone any of the following transformations: dilations, a reflection, or translations.

9.3.1 Sketching parabolas

eles-4865

- A sketch graph of a parabola does not show a series of plotted points, but it does accurately locate important features such as x- and y-intercepts and turning points.
- The basic quadratic graph has the equation $y = x^2$. Transformations or changes in the features of the graph can be observed when the equation changes. These transformations include:
  - dilation
  - translation
  - reflection.

Dilation

- A dilation stretches a graph away from an axis. A dilation of factor 3 from the x-axis triples the distance of each point from the x-axis. This means the point $(2, 4)$ would become $(2, 12)$.
- Compare the graph of $y = 2x^2$ with that of $y = x^2$. This graph is thinner or closer to the y-axis and has a dilation factor of 2 from the x-axis. As the magnitude (or size) of the coefficient of x^2 increases, the graph becomes **narrower** and closer to the y-axis.

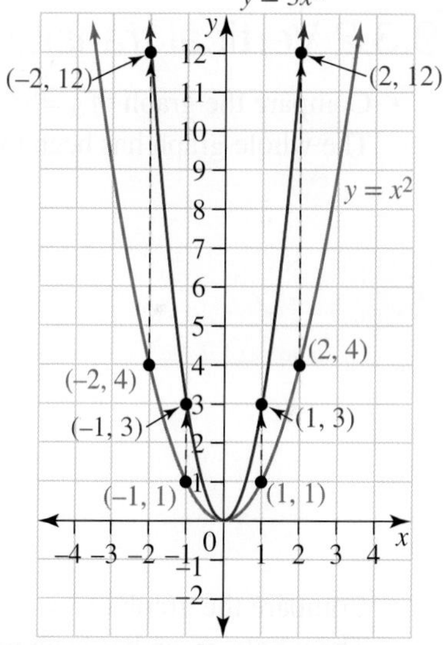

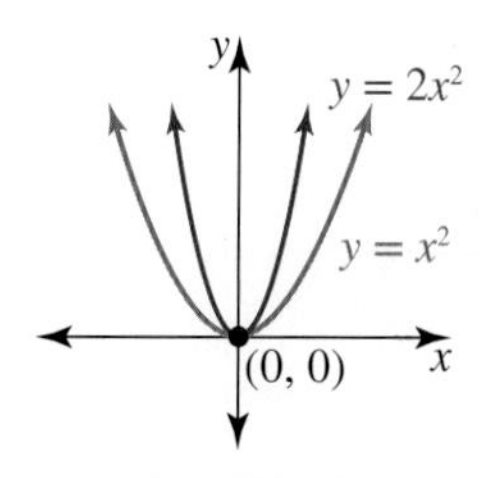

- The turning point has not changed under the transformation and is still $(0, 0)$.
- Compare the graph $y = \frac{1}{4}x^2$ with that of $y = x^2$.

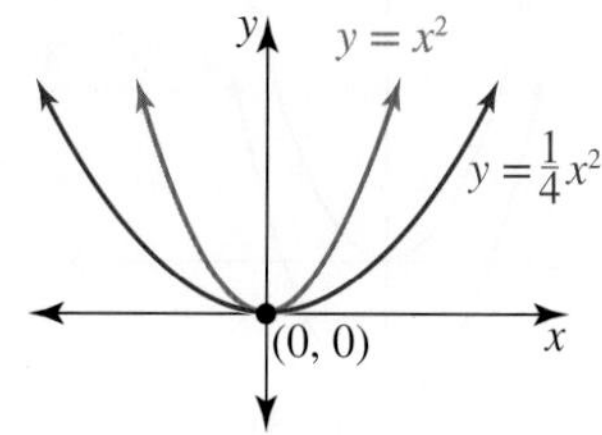

The graph is **wider** or closer to the x-axis and has a dilation factor of factor $\frac{1}{4}$.

The turning point has not changed and is still $(0, 0)$. As the coefficient of x^2 decreases (but remains positive), the graph becomes wider or closer to the x-axis.

WORKED EXAMPLE 3 Determining dilation and the turning point

State whether each of the following graphs is wider or narrower than the graph of $y = x^2$ and state the coordinates of the turning point of each one.

a. $y = \frac{1}{5}x^2$ **b.** $y = 4x^2$

THINK	WRITE
a. 1. Write the equation.	**a.** $y = \frac{1}{5}x^2$
2. Look at the coefficient of x^2 and decide whether it is greater than or less than 1.	$\frac{1}{5} < 1$, so the graph is wider than that of $y = x^2$.
3. The dilation doesn't change the turning point.	The turning point is (0, 0).
b. 1. Write the equation.	**b.** $y = 4x^2$
2. Look at the coefficient of x^2 and decide whether it is greater than or less than 1.	$4 > 1$, so the graph is narrower than that of $y = x^2$.
3. The dilation doesn't change the turning point.	The turning point is (0, 0).

9.3.2 Vertical translation

eles-4866

- Compare the graph of $y = x^2 + 2$ with that of $y = x^2$.
  The whole graph has been moved or translated 2 units upwards. The turning point has become (0, 2).

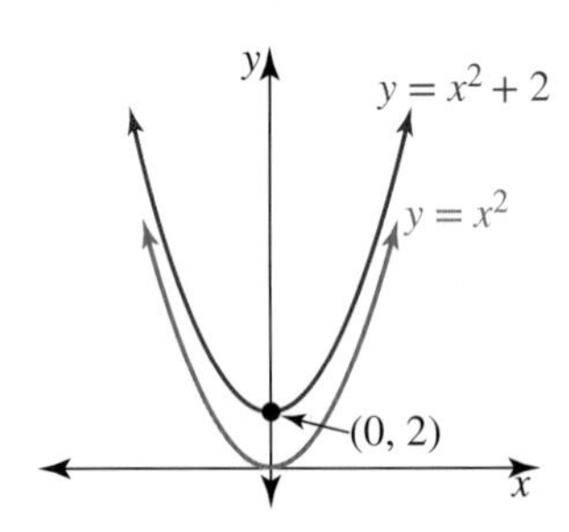

- Compare the graph of $y = x^2 - 3$ with that of $y = x^2$.
  The whole graph has been moved or translated 3 units downwards. The turning point has become (0, −3).

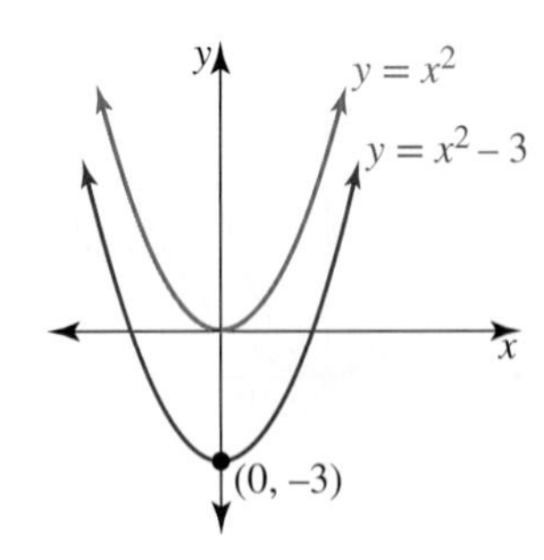

WORKED EXAMPLE 4 Determining the vertical translation and the turning point

State the vertical translation and the coordinates of the turning point for the graphs of the following equations when compared to the graph of $y = x^2$.

a. $y = x^2 + 5$

b. $y = x^2 - 4$

THINK	WRITE
a. 1. Write the equation.	a. $y = x^2 + 5$
2. +5 means the graph is translated upwards 5 units.	Vertical translation of 5 units up.
3. Translate the turning point of $y = x^2$ which is $(0, 0)$. The x-coordinate of the turning point remains 0, and the y-coordinate has 5 added to it.	The turning point becomes $(0, 5)$.
b. 1. Write the equation.	b. $y = x^2 - 4$
2. −4 means the graph is translated downwards 4 units.	Vertical translation of 4 units down.
3. Translate the turning point of $y = x^2$, which is $(0, 0)$. The x-coordinate of the turning point remains 0, and the y-coordinate has 4 subtracted from it.	The turning point becomes $(0, -4)$.

9.3.3 Horizontal translation

eles-4867

- Compare the graph of $y = (x - 2)^2$ with that of $y = x^2$.
  The whole graph has been moved or translated 2 units to the right. The turning point has become $(2, 0)$.

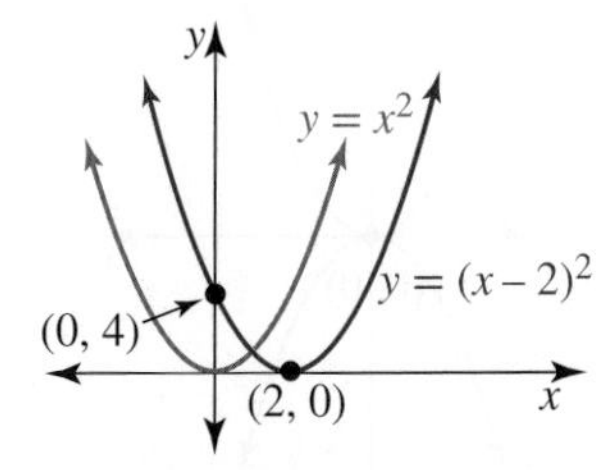

- Compare the graph of $y = (x + 1)^2$ with that of $y = x^2$.
  The whole graph has been moved or translated 1 unit left. The turning point has become $(-1, 0)$.

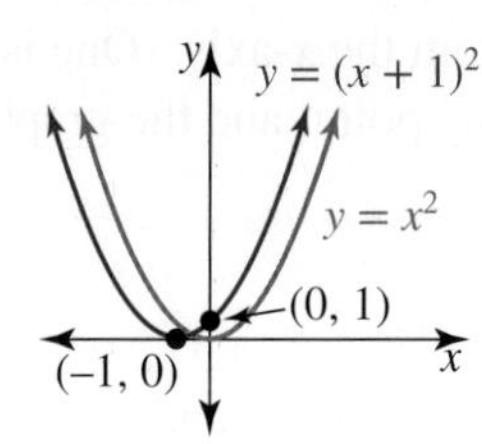

- *Note:* Horizontal translations appear to cause the graph to move in the opposite direction to the sign inside the brackets.

WORKED EXAMPLE 5 Determining the horizontal translation and the turning point

State the horizontal translation and the coordinates of the turning point for the graphs of the following equations when compared to the graph of $y = x^2$.

a. $y = (x - 3)^2$

b. $y = (x + 2)^2$

THINK	WRITE
a. 1. Write the equation.	a. $y = (x - 3)^2$
2. -3 means the graph is translated to the right 3 units.	Horizontal translation of 3 units to the right
3. Translate the turning point of $y = x^2$, which is $(0, 0)$. The y-coordinate of the turning point remains 0, and the x-coordinate has 3 added to it.	The turning point becomes $(3, 0)$.
b. 1. Write the equation.	b. $y = (x + 2)^2$
2. $+2$ means the graph is translated to the left 2 units.	Horizontal translation of 2 units to the left
3. Translate the turning point of $y = x^2$, which is $(0, 0)$. The y-coordinate of the turning point remains 0, and the x-coordinate has 2 subtracted from it.	The turning point becomes $(-2, 0)$.

9.3.4 Reflection

eles-4868

- Compare the graph of $y = -x^2$ with that of $y = x^2$.

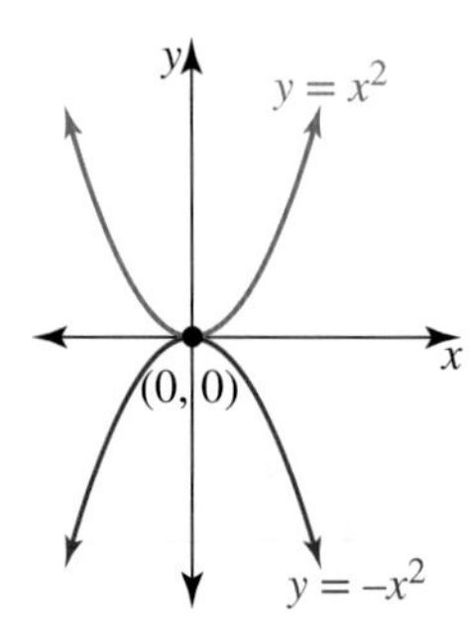

In each case the axis of symmetry is the line $x = 0$ and the turning point is $(0, 0)$. The only difference between the equations is the negative sign in $y = -x^2$, and the difference between the graphs is that $y = x^2$ 'sits' on the x-axis and $y = -x^2$ 'hangs' from the x-axis. (One is a reflection or mirror image of the other.) The graph of $y = x^2$ has a minimum turning point, and the graph of $y = -x^2$ has a maximum turning point.

Shapes of quadratic graphs

Any quadratic graph where x^2 is positive has a ∪ shape and is said to be upright. Conversely, if x^2 is negative the graph has a ∩ shape and is said to be inverted.

WORKED EXAMPLE 6 Identifying the coordinates of the turning point

For each of the following graphs, identify the coordinates of the turning point and state whether it is a maximum or a minimum.

a. $y = -(x-7)^2$ **b.** $y = 5 - x^2$

THINK	WRITE
a. 1. Write the equation.	**a.** $y = -(x-7)^2$
2. It is a horizontal translation of 7 units to the right, so 7 units is added to the x-coordinate of $(0, 0)$.	The turning point is $(7, 0)$.
3. The sign in front of the x^2 term is negative, so it is inverted.	Maximum turning point.
b. 1. Write the equation.	**b.** $y = 5 - x^2$
2. Rewrite the equation so that the x^2 term is first.	$y = -x^2 + 5$
3. The vertical translation is 5 units up, so 5 units is added to the y-coordinate of $(0, 0)$.	The turning point is $(0, 5)$.
4. The sign in front of the x^2 term is negative, so the graph is inverted.	Maximum turning point.

TI | THINK

a.

In a new problem, on a Graphs page, complete the function entry line as:
$f1(x) = -(x-7)^2$
Then press ENTER.
To locate the turning point, press:
- MENU
- 6: Analyze Graph
- 3: Maximum

Drag the dotted line to the left of the turning point (the lower bound), click ENTER and then drag the dotted line to the right of the turning point (the upper bound) and press ENTER.
The turning point will be shown.

DISPLAY/WRITE

a.

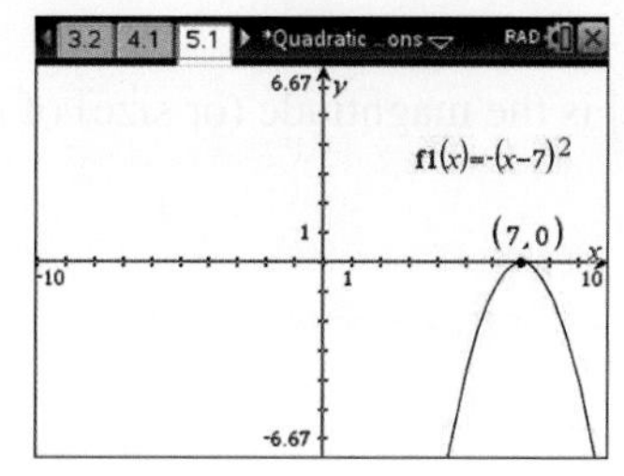

The turning point (7, 0) is a maximum.

CASIO | THINK

a.

On the Graph & Table screen, complete the function entry line as:
$y1 = -(x-7)^2$
Tap the graphing icon and the graph will be displayed.
To locate the turning point, tap:
- Analysis
- G-Solve
- Max

The turning point will be shown.

DISPLAY/WRITE

a.

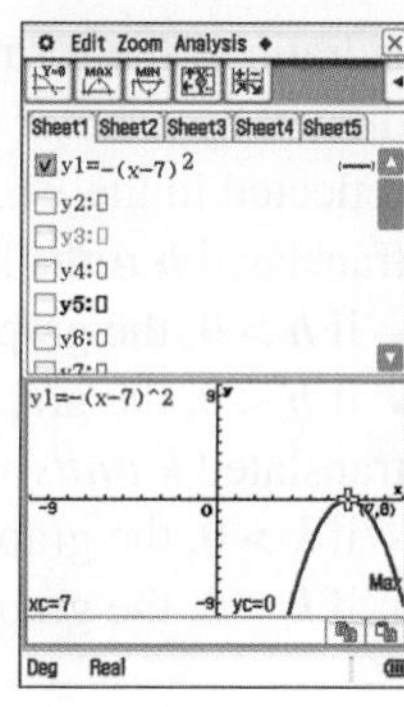

The turning point $(7, 0)$ is a maximum.

b. In a new problem, on a new Graphs page, complete the function entry line as: $f1(x) = 5 - x^2$ Then press ENTER. To locate the turning point, press: • MENU • 6: Analyze Graph • 3: Maximum Drag the dotted line to the left of the turning point (the lower bound), click ENTER and then drag the dotted line to the right of the turning point (the upper bound) and press ENTER. The turning point will be shown.	**b.** 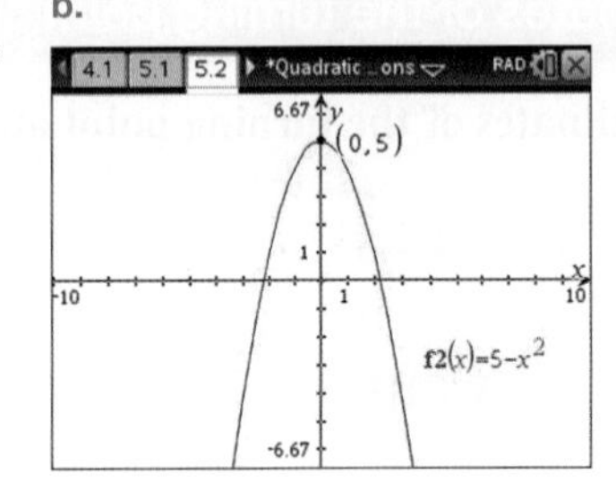 The turning point (0, 5) is a maximum.	**b.** On the Graph & Table screen, complete the function entry line as: $y1 = 5 - x^2$ Tap the graphing icon and the graph will be displayed. To locate the turning point, press: • Analysis • G-Solve • Max The turning point will be shown.	**b.** 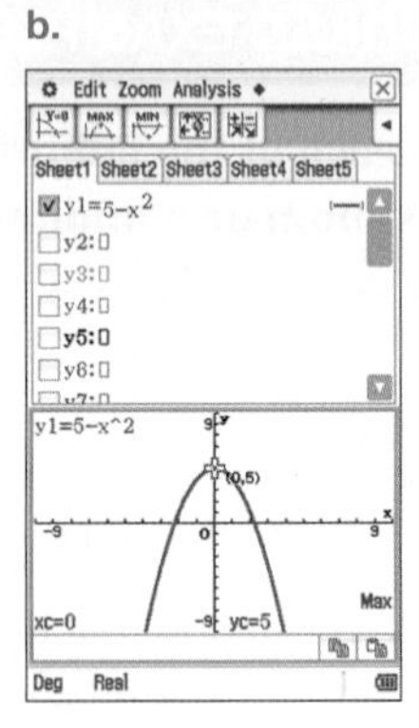 The turning point (0, 5) is a maximum.

9.3.5 Combining transformations

eles-4869

- Often, multiple transformations will be applied to the equation $y = x^2$ to produce a new graph.
- We can determine the transformations applied by looking at the equation of the resulting quadratic.

Combining transformations

A quadratic of the form $y = a(x - h)^2 + k$ has been:

- dilated by a factor of $|a|$ from the *x-axis*, where $|a|$ is the magntiude (or size) of a.
- reflected in the *x-axis* if $a < 0$
- translated ***h* units** horizontally:
  - if $h > 0$, the graph is translated to the *right*
  - if $h < 0$, the graph is translated to the *left*
- translated ***k* units** vertically:
  - if $k > 0$, the graph is translated *upwards*
  - if $k < 0$, the graph is translated *downwards*.

WORKED EXAMPLE 7 Determining transformations

For each of the following quadratic equations:

i. state the appropriate dilation, reflection and translation of the graph of $y = x^2$ needed to obtain the graph

ii. state the coordinates of the turning point

iii. hence, sketch the graph.

a. $y = (x + 3)^2$ **b.** $y = -2x^2$

THINK	WRITE/DRAW
a. 1. Write the quadratic equation.	**a.** $y = (x + 3)^2$
2. Identify the transformation needed — horizontal translation only, no dilation or reflection.	**i.** Horizontal translation of 3 units to the left

3. State the turning point.	ii. The turning point is $(-3, 0)$.
4. Sketch the graph of $y=(x+3)^2$. You may find it helpful to lightly sketch the graph of $y=x^2$ on the same set of axes first.	iii. 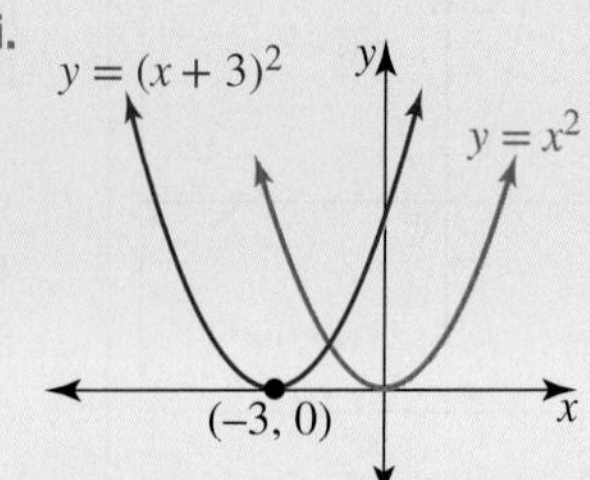
b. 1. Write the quadratic equation.	b. $y=-2x^2$
2. Identify the transformations needed — dilation (2 in front of x^2) and reflection (negative in front of x^2), no translation.	i. This is a reflection, so the graph is inverted. As $2>1$, the graph is narrower than that of $y=x^2$.
3. The turning point remains the same as there is no translation.	ii. The turning point is $(0, 0)$.
4. Sketch the graph of $y=-2x^2$. You may find it helpful to lightly sketch the graph of $y=x^2$ on the same set of axes first.	iii. 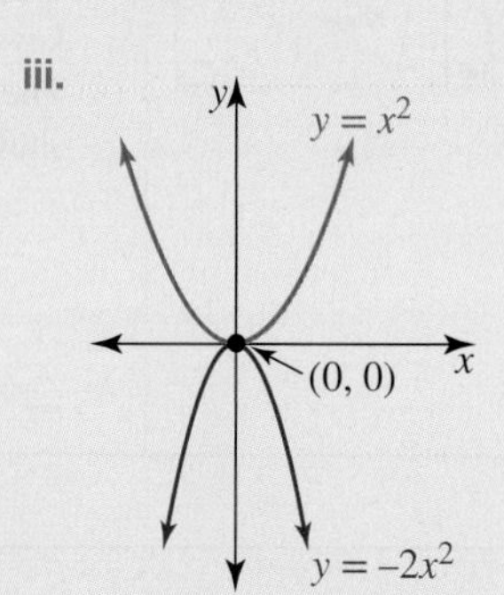

TI | THINK	DISPLAY/WRITE	CASIO | THINK	DISPLAY/WRITE
a.	a.	a.	a.
1. On a Graphs page, complete the function entry lines as: $f1(x)=x^2$ $f2(x)=(x+3)^2$ The graphs will be displayed.	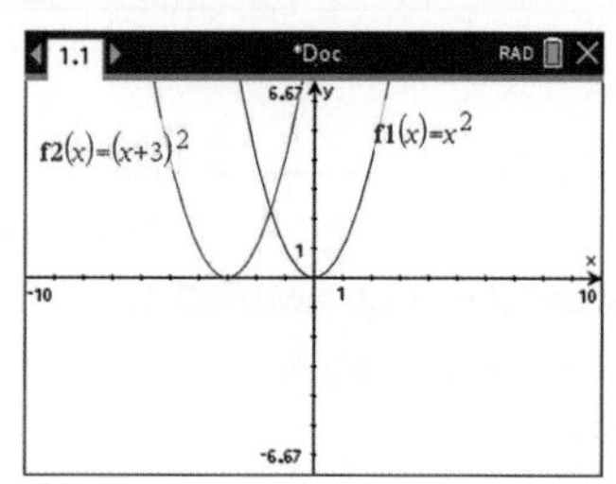	1. On a Graph & Table screen, complete the function entry lines as: $y1=x^2$ $y2=(x+3)^2$ Tap the graphing icon and the graphs will be displayed.	 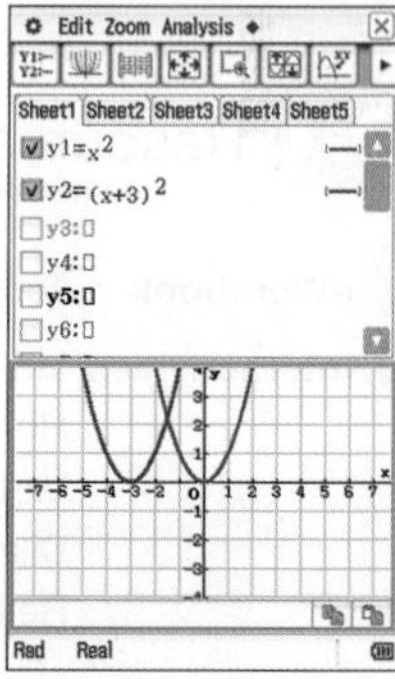
2. To locate the turning point, press: • MENU • 6: Analyze Graph • 2: Minimum Drag the dotted line to the left of the turning point (the lower bound), click ENTER and then drag the dotted line to the right of the turning point (the upper bound) and press ENTER. The turning point will be shown.	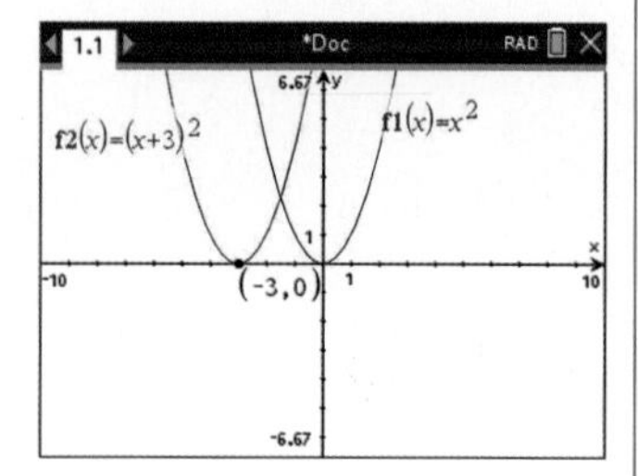	2. To locate the turning point, press: • Analysis • G-Solve • Min Select the graph you want using the up and down arrow keys and then press EXE. The turning point will be shown.	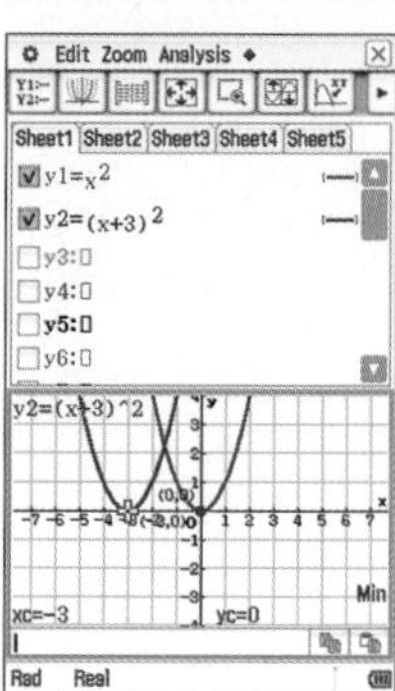

b.

1. On a Graphs page, complete the function entry lines as:

   $f1(x) = x^2$
   $f2(x) = -2x^2$

   The graphs will be displayed.

b.

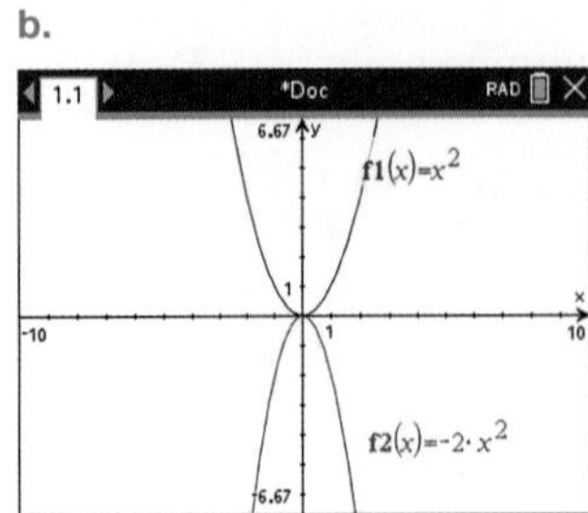
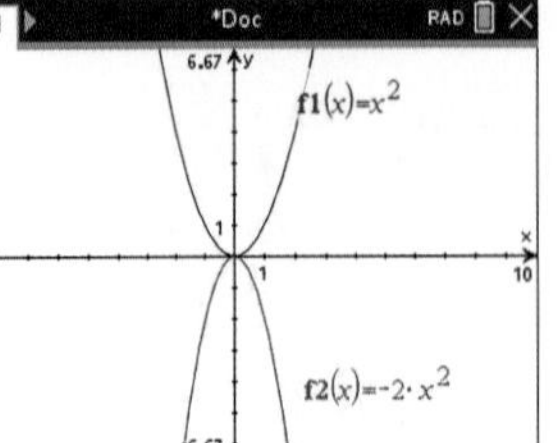

2. To locate the turning point, press:
   - MENU
   - 6: Analyze Graph
   - 2: Minimum

   Drag the dotted line to the left of the turning point (the lower bound), click ENTER and then drag the dotted line to the right of the turning point (the upper bound) and press ENTER.
   The turning point will be shown.

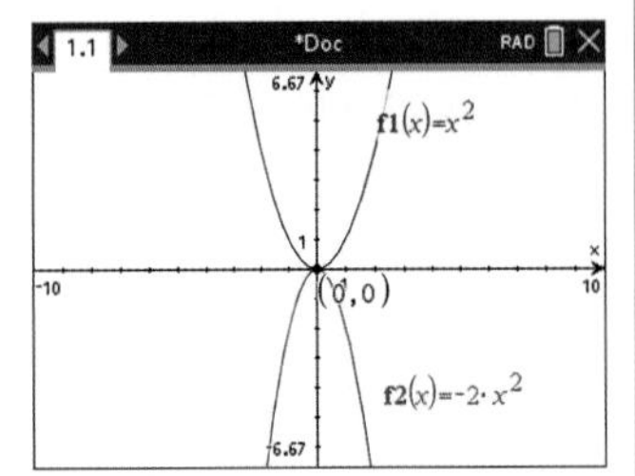

b.

1. On a Graph & Table screen, complete the function entry lines as:

   $y1 = x^2$
   $y2 = -2x^2$

   Tap the graphing icon and the graphs will be displayed.

b.

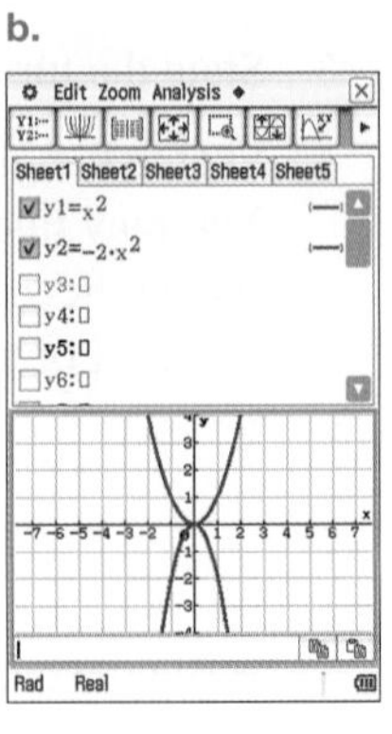

2. To locate the point of intersection, press:
   - Analysis
   - G-Solve
   - Max

   Select the graph you want using the up and down arrow keys and then press EXE.
   The turning point will be shown.

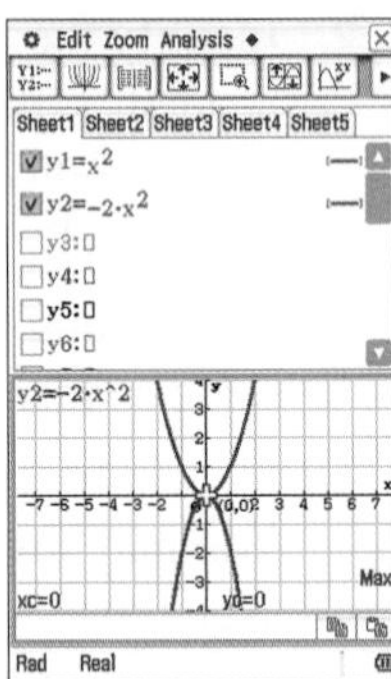

DISCUSSION

Determine the turning points of the graphs $y = x^2 + k$ and $y = (x - h)^2$.

 Resources

 **eWorkbook** Topic 9 Workbook (worksheets, code puzzle and a project) (ewbk-2035)

 **Interactivities** Individual pathway interactivity: Sketching parabolas (int-4606)
Horizontal translations of parabolas (int-6054)
Vertical translations of parabolas (int-6055)
Dilation of parabolas (int-6096)
Reflection of parabolas (int-6151)

Exercise 9.3 Sketching parabolas using transformations

learnON

Individual pathways

PRACTISE	CONSOLIDATE	MASTER
1, 4, 7, 10, 13, 16, 17, 22, 25	2, 5, 8, 11, 14, 18, 19, 23, 26	3, 6, 9, 12, 15, 20, 21, 24, 27, 28

To answer questions online and to receive **immediate corrective feedback** and **fully worked solutions** for all questions, go to your learnON title at www.jacplus.com.au.

Fluency

WE3 For questions 1 to 3, state whether each of the following graphs is wider or narrower than the graph of $y = x^2$ and state the coordinates of the turning point of each one.

1. a. $y = 5x^2$ b. $y = \frac{1}{3}x^2$

2. a. $y = 7x^2$ b. $y = 10x^2$ c. $y = \frac{2}{5}x^2$

3. a. $y = 0.25x^2$ b. $y = 1.3x^2$ c. $y = \sqrt{3}x^2$

WE4 For questions 4 to 6, state the vertical translation and the coordinates of the turning point for the graphs of each of the following equations when compared to the graph of $y = x^2$.

4. a. $y = x^2 + 3$ b. $y = x^2 - 1$

5. a. $y = x^2 - 7$ b. $y = x^2 + \frac{1}{4}$ c. $y = x^2 - \frac{1}{2}$

6. a. $y = x^2 - 0.14$ b. $y = x^2 + 2.37$ c. $y = x^2 + \sqrt{3}$

WE5 For questions 7 to 9, state the horizontal translation and the coordinates of the turning point for the graphs of the following equations when compared to the graph of $y = x^2$.

7. a. $y = (x - 1)^2$ b. $y = (x - 2)^2$

8. a. $y = (x + 10)^2$ b. $y = (x + 4)^2$ c. $y = \left(x - \frac{1}{2}\right)^2$

9. a. $y = \left(x + \frac{1}{5}\right)^2$ b. $y = (x + 0.25)^2$ c. $y = (x + \sqrt{3})^2$

WE6 For questions 10 to 12, for each of the following graphs identify the coordinates of the turning point and state whether it is a maximum or a minimum.

10. a. $y = -x^2 + 1$ b. $y = x^2 - 3$

11. a. $y = -(x + 2)^2$ b. $y = 3x^2$ c. $y = 4 - x^2$

12. a. $y = -2x^2$ b. $y = (x - 5)^2$ c. $y = 1 + x^2$

For questions 13 to **15**, in each of the following state whether the graph is wider or narrower than $y = x^2$ and whether it has a maximum or a minimum turning point.

13. a. $y = 3x^2$ b. $y = -3x^2$

14. a. $y = \frac{1}{2}x^2$ b. $y = -\frac{1}{5}x^2$ c. $y = -\frac{4}{3}x^2$

15. a. $y = 0.25x^2$ **b.** $y = \sqrt{3}x^2$ **c.** $y = -0.16x^2$

Understanding

WE7 For questions 16 to **21**:

i. state the appropriate dilation, reflection and translation of the graph of $y = x^2$ needed to obtain the graph

ii. state the coordinates of the turning point

iii. hence, sketch the graph.

16. a. $y = (x + 1)^2$ b. $y = -3x^2$ c. $y = x^2 + 1$

17. a. $y = \frac{1}{3}x^2$ b. $y = x^2 - 3$

18. a. $y = (x - 4)^2$ b. $y = -\frac{2}{5}x^2$ c. $y = 5x^2$

19. a. $y = -x^2 + 2$ b. $y = -(x - 6)^2$

20. a. $y = -x^2 - 4$ **b.** $y = 2(x + 1)^2 - 4$ **c.** $y = \frac{1}{2}(x - 3)^2 + 2$

21. a. $y = -\frac{1}{3}(x + 2)^2 + \frac{1}{4}$ **b.** $y = -\frac{7}{4}(x - 1)^2 - \frac{3}{2}$

Reasoning

22. A vase 25 cm tall is positioned on a bench near a wall as shown. The shape of the vase follows the curve $y = (x - 10)^2$, where y cm is the height of the vase and x cm is the distance of the vase from the wall.

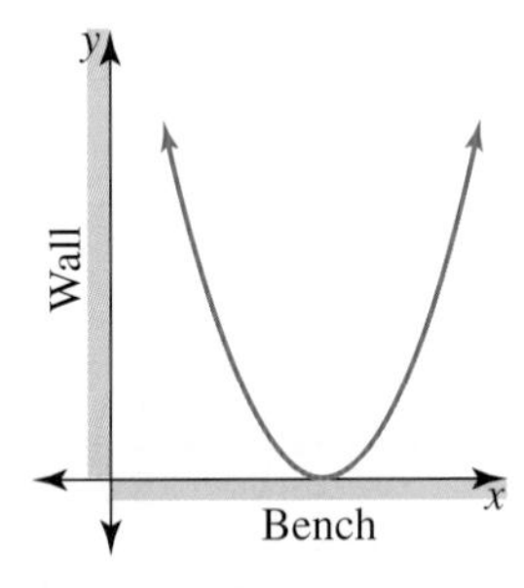

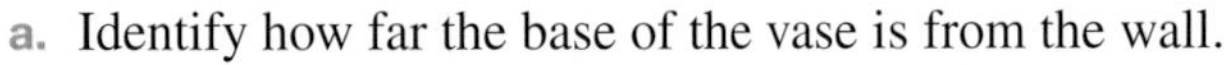

a. Identify how far the base of the vase is from the wall.

b. Determine the shortest distance from the top of the vase to the wall.

c. If the vase is moved so that the top just touches the wall, determine the new distance from the wall to the base.

d. Determine the new equation that follows the shape of the vase.

23. Tom is standing at the start of a footpath at (0, 0) that leads to the base of a hill. The height of the hill is modelled by the equation $h = -\frac{1}{10}(d - 25)^2 + 40$, where h is the height of the hill in metres and d is the horizontal distance from the start of the path.

a. Calculate how tall the hill is.

b. Determine how far the base of the hill is from the beginning of the footpath.

c. If the footpath is to be extended so the lead in to the hill is 50 m, determine the new equation that models the height of the hill.

d. The height of the hill has been incorrectly measured and should actually be 120 m. Adjust the equation from part **c** to correct this error and state the transformation applied.

24. A ball is thrown vertically upwards. Its height in metres after t seconds is given by $h = 7t - t^2$.

a. Sketch the graph of the height of the ball against time.
b. Evaluate the highest point reached by the ball. Show your full working.

A second ball is thrown vertically upwards. Its height in metres after t seconds is given by $h = 10t - t^2$.

c. On the same set of axes used for part **a**, sketch the graph of the height of the second ball against time.
d. State the difference in the highest point reached by the two balls.

Problem solving

25. Consider the quadratic equation $y = x^2 - 4x + 7$.
   a. Determine the equivalent inverted equation of the quadratic that just touches the one above at the turning point.
   b. Confirm your result graphically.

26. Consider the equation $y = 3(x-2)^2 - 7$.
   a. State the coordinates of the turning point and y-intercept.
   b. State a sequence of transformations that when applied to the graph of $y = \frac{-3}{2}(x-2)^2 - 7$ will produce the graph of $y = x^2$.

27. A parabola has the equation $y = -\frac{1}{2}(x-3)^2 + 4$. A second parabola has an equation defined by $Y = 2(y-1) - 3$.
   a. Determine the equation relating Y to x.
   b. State the appropriate dilation, reflection and translation of the graph of $Y = x^2$ required to obtain the graph of $Y = 2(y-1) - 3$.
   c. State the coordinates of the turning point $Y = 2(y-1) - 3$.
   d. Sketch the graph of $Y = 2(y-1) - 3$.

28. A ball shot at a certain angle to the horizontal follows a parabolic path. It reaches a maximum height of 200 m when its horizontal distance from its starting point is 10 m. When the ball's horizontal distance from the starting point was 1 m, the ball had reached a height of 38 m.
   Determine an equation to model the ball's flight, clearly defining your chosen pronumerals.

9.4 Sketching parabolas using turning point form

LEARNING INTENTION

At the end of this subtopic you should be able to:
- determine the axis of symmetry and turning point of a quadratic in turning point form
- calculate the y-intercept and any x-intercepts of a quadratic in turning point form.

9.4.1 Turning point form

eles-4870

- When a quadratic equation is expressed in the form $y = a(x - h)^2 + k$:
  - the turning point is the point (h, k)
  - the axis of symmetry is $x = h$
  - the x-intercepts are calculated by solving $a(x - h)^2 + k = 0$.
- Changing the values of a, h and k in the equation transforms the shape and position of the parabola when compared with the parabola $y = x^2$.

$y = a(x - h)^2 + k$

Reflects and dilates (a) — Translates left and right (h) — Translates up and down (k)

Turning point form

A quadratic of the form $\boldsymbol{y = a(x - h)^2 + k}$ has:
- a turning point at the coordinate $\boldsymbol{(h, k)}$
  - the turning point will be a minimum if $\boldsymbol{a > 0}$
  - the turning point will be a maximum if $\boldsymbol{a < 0}$
- an axis of symmetry of $\boldsymbol{x = h}$
- a y-intercept of $\boldsymbol{\left(0, ah^2 + k\right)}$.

The number of x-intercepts depends on the values of a, h and k. Changing the value of a does not change the position of the turning point, only h and k.

WORKED EXAMPLE 8 Determining the turning point from turning point form

For each of the following equations, state the coordinates of the turning point of the graph and whether it is a maximum or a minimum.

a. $\boldsymbol{y = (x - 6)^2 - 4}$ **b.** $\boldsymbol{y = -(x + 3)^2 + 2}$

THINK	WRITE
a. 1. Write the equation.	**a.** $y = (x - 6)^2 - 4$
2. Identify the transformations — horizontal translation of 6 units to the right and a vertical translation of 4 units down. State the turning point.	The turning point is $(6, -4)$.
3. As a is positive ($a = 1$), the graph is upright with a minimum turning point.	Minimum turning point.
b. 1. Write the equation	**b.** $y = -(x + 3)^2 + 2$
2. Identify the transformations — horizontal translation of 3 units to the left and a vertical translation of 2 units up. State the turning point.	The turning point is $(-3, 2)$.
3. As a is negative ($a = -1$), the graph is inverted with a maximum turning point.	Maximum turning point.

9.4.2 x- and y-intercepts of quadratic graphs

eles-4871

- Other key features such as the x- and y-intercepts can also be determined from the equation of a parabola.
- The point(s) where the graph cuts or touches the x-axis are called the x-intercept(s). At these points, $y = 0$.
- The point where the graph cuts the x-axis is called the y-intercept. At this point, $x = 0$.

WORKED EXAMPLE 9 Determining the axial intercepts from turning point form

For the parabolas with the following equations:

i. determine the y-intercept

ii. determine the x-intercepts (where they exist).

a. $y = (x+3)^2 - 4$ **b.** $y = 2(x-1)^2$ **c.** $y = -(x+2)^2 - 1$

THINK	WRITE
a. 1. Write the equation.	**a.** $y = (x+3)^2 - 4$
2. Calculate the y-intercept by substituting $x = 0$ into the equation.	y-intercept: when $x = 0$, $y = (0+3)^2 - 4$ $= 9 - 4$ $= 5$ The y-intercept is 5.
3. Calculate the x-intercepts by substituting $y = 0$ into the equation and solving for x. Add 4 to both sides of the equation. Take the square root of both sides of the equation. Subtract 3 from both sides of the equation. Solve for x.	x-intercepts: when $y = 0$, $(x+3)^2 - 4 = 0$ $(x+3)^2 = 4$ $(x+3) = +2$ or -2 $x = 2 - 3$ or $x = -2 - 3$ $x = -1 \quad x = -5$ The x-intercepts are -5 and -1.
b. 1. Write the equation.	**b.** $y = 2(x-1)^2$
2. Calculate the y-intercept by substituting $x = 0$ into the equation.	y-intercept: when $x = 0$, $y = 2(0-1)^2$ $= 2 \times 1$ $= 2$ The y-intercept is 2.
3. Calculate the x-intercepts by substituting $y = 0$ into the equation and solving for x. Note that there is only one solution for x and so there is only one x-intercept. (The graph touches the x-axis.)	x-intercepts: when, $y = 0$, $2(x-1)^2 = 0$ $(x-1)^2 = 0$ $x - 1 = 0$ $x = 0 + 1$ $x = 1$ The x-intercept is 1.
c. 1. Write the equation.	**c.** $y = -(x+2)^2 - 1$
2. Calculate the y-intercept by substituting $x = 0$ into the equation.	y-intercept: when $x = 0$, $y = -(0+2)^2 - 1$ $= -4 - 1$ $= -5$ The y-intercept is -5.

3. Calculate the x-intercepts by substituting $y = 0$ into the equation and solving for x. We cannot take the square root of -1 to obtain real solutions; therefore, there are no x-intercepts.

x-intercepts: when $y = 0$,
$-(x+2)^2 - 1 = 0$
$(x+2)^2 = -1$
There are no real solutions, so there are no x-intercepts.

WORKED EXAMPLE 10 Sketching a quadratic in turning point form

For each of the following:

i. **write the coordinates of the turning point**
ii. **state whether the graph has a maximum or a minimum turning point**
iii. **state whether the graph is wider, narrower or the same width as the graph of $y = x^2$**
iv. **calculate the y-intercept**
v. **calculate the x-intercepts**
vi. **sketch the graph.**

a. $y = (x-2)^2 + 3$

b. $y = -2(x+1)^2 + 6$

THINK / **WRITE/DRAW**

a. 1. Write the equation.

a. $y = (x-2)^2 + 3$

2. State the coordinates of the turning point from the equation. Use (h, k) as the equation is in the turning point form of $y = a(x-h)^2 + k$ where $a = 1, h = 2$ and $k = 3$.

The turning point is $(2, 3)$.

3. State the nature of the turning point by considering the sign of a.

The graph has a minimum turning point as the sign of a is positive.

4. Specify the width of the graph by considering the magnitude of a.

The graph has the same width as $y = x^2$ since $a = 1$.

5. Calculate the y-intercept by substituting $x = 0$ into the equation.

y-intercept: when $x = 0$,
$y = (0-2)^2 + 3$
$= 4 + 3$
$= 7$
y-intercept is 7.

6. Calculate the x-intercepts by substituting $y = 0$ into the equation and solving for x. As we have to take the square root of a negative number, we cannot solve for x.

x-intercepts: when $y = 0$,
$(x-2)^2 + 3 = 0$
$(x-2)^2 = -3$
There are no real solutions, and hence no x-intercepts.

7. Sketch the graph, clearly showing the turning point and the y-intercept.

8. Label the graph.

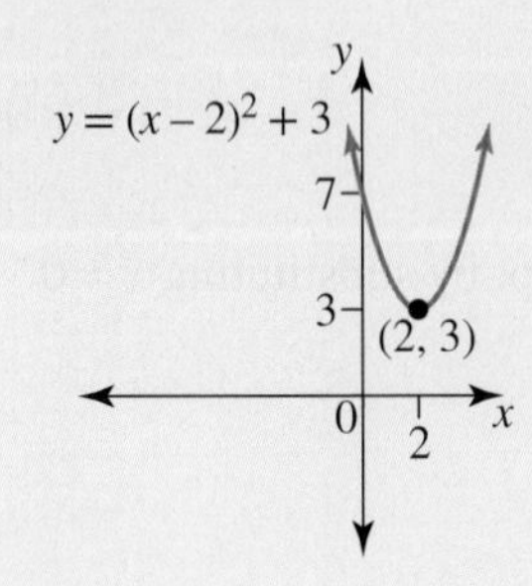

b. 1. Write the equation.	**b.** $y = -2(x+1)^2 + 6$
2. State the coordinates of the turning point from the equation. Use (h, k) as the equation is in the turning point form of $y = a(x-h)^2 + k$ where $a = -2, h = -1$ and $k = 6$.	The turning point is $(-1, 6)$.
3. State the nature of the turning point by considering the sign of a.	The graph has a maximum turning point as the sign of a is negative.
4. Specify the width of the graph by considering the magnitude of a.	The graph is narrower than $y = x^2$ since $\|a\| > 1$.
5. Calculate the y-intercept by substituting $x = 0$ into the equation.	y-intercept: when $x = 0$, $y = -2(0+1)^2 + 6$ $= -2 \times 1 + 6$ $= 4$ The y-intercept is 4.
6. Calculate the x-intercepts by substituting $y = 0$ into the equation and solving for x.	x-intercepts: when $y = 0$, $-2(x+1)^2 + 6 = 0$ $2(x+1)^2 = 6$ $(x+1)^2 = 3$ $x + 1 = \sqrt{3}$ or $x + 1 = -\sqrt{3}$ $x = -1 + \sqrt{3}$ or $x = -1 - \sqrt{3}$ The x-intercepts are $-1 - \sqrt{3}$ and $-1 + \sqrt{3}$ (or approximately -2.73 and 0.73).
7. Sketch the graph, clearly showing the turning point and the x- and y-intercepts. **8.** Label the graph.	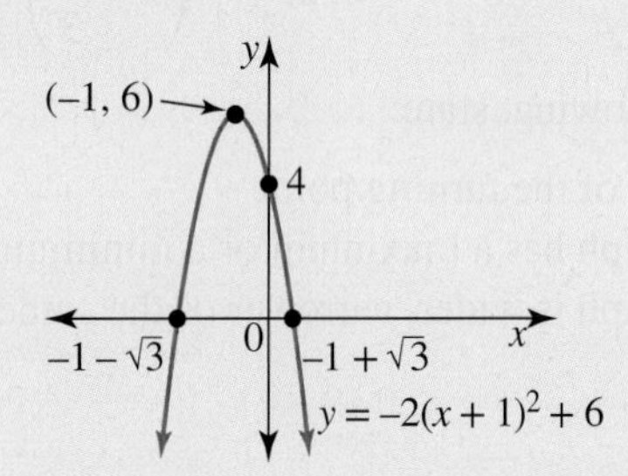

Note: Unless otherwise stated, exact values for the intercepts should be shown on sketch graphs.

DISCUSSION

Does a in the equation $y = a(x-h)^2 + k$ have any impact on the turning point?

Resources

eWorkbook Topic 9 Workbook (worksheets, code puzzle and a project) (ewbk-2035)

Video eLessons Sketching quadratics in turning point form (eles-1926)
Solving quadratics in turning point form (eles-1941)

Interactivities Individual pathway interactivity: Sketching parabolas in turning point form (int-4607)
Quadratic functions (int-2562)

Exercise 9.4 Sketching parabolas using turning point form

learnon

Individual pathways

■ PRACTISE	■ CONSOLIDATE	■ MASTER
1, 4, 7, 9, 12, 14, 17, 20, 23, 27	2, 5, 8, 10, 13, 15, 18, 21, 24, 28	3, 6, 11, 16, 19, 22, 25, 26, 29

To answer questions online and to receive **immediate corrective feedback** and **fully worked solutions** for all questions, go to your learnON title at www.jacplus.com.au.

Fluency

WE8 For questions 1 to 3, for each of the following equations, state the coordinates of the turning point of the graph and whether it is a maximum or a minimum.

1. a. $y=(x-1)^2+2$ b. $y=(x+2)^2-1$ c. $y=(x+1)^2+1$
2. a. $y=-(x-2)^2+3$ b. $y=-(x-5)^2+3$ c. $y=(x+2)^2-6$
3. a. $y=\left(x-\frac{1}{2}\right)^2-\frac{3}{4}$ b. $y=\left(x-\frac{1}{3}\right)^2+\frac{2}{3}$ c. $y=(x+0.3)^2-0.4$
4. For each of the following, state:
   i. the coordinates of the turning point
   ii. whether the graph has a maximum or a minimum turning point
   iii. whether the graph is wider, narrower or the same width as that of $y=x^2$.

   a. $y=2(x+3)^2-5$ b. $y=-(x-1)^2+1$
5. For each of the following, state:
   i. the coordinates of the turning point
   ii. whether the graph has a maximum or a minimum turning point
   iii. whether the graph is wider, narrower or the same width as that of $y=x^2$.

   a. $y=-5(x+2)^2-4$ b. $y=\frac{1}{4}(x-3)^2+2$
6. For each of the following, state:
   i. the coordinates of the turning point
   ii. whether the graph has a maximum or a minimum turning point
   iii. whether the graph is wider, narrower or the same width as that of $y=x^2$.

   a. $y=-\frac{1}{2}(x+1)^2+7$ b. $y=0.2\left(x-\frac{1}{5}\right)^2-\frac{1}{2}$

7. Select the equation that best suits each of the following graphs.

i.

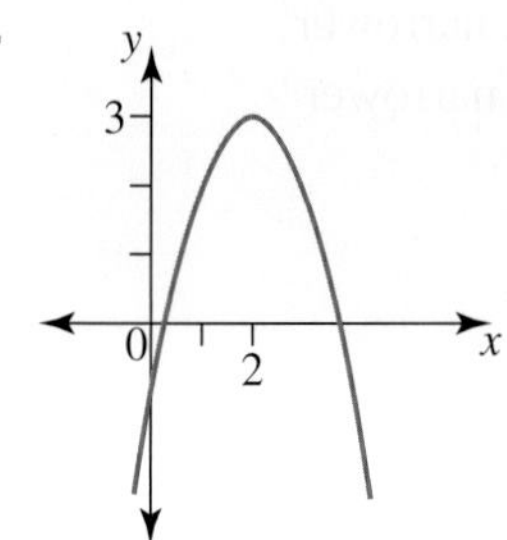

ii.

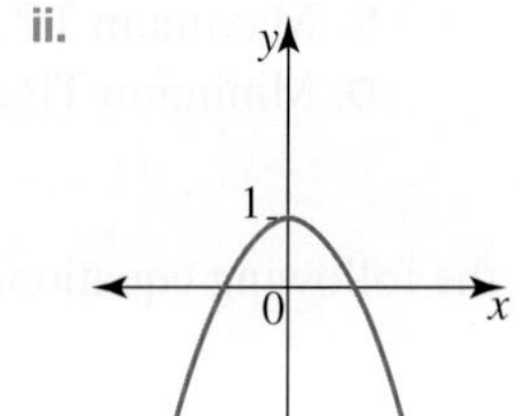

iii.

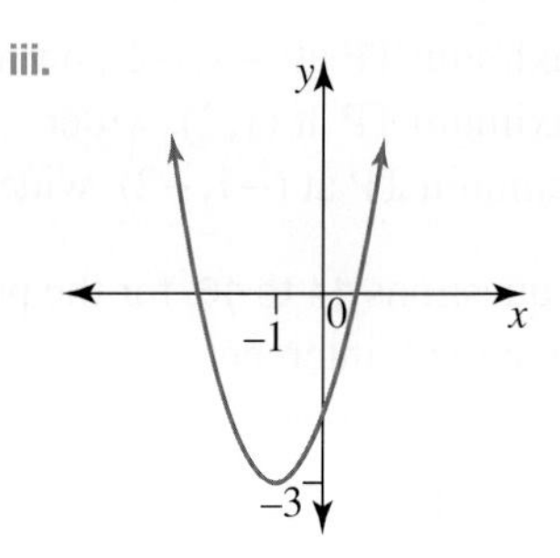

a. $y=(x+1)^2-3$

b. $y=-(x-2)^2+3$

c. $y=-x^2+1$

8. Select the equation that best suits each of the following graphs.

i.

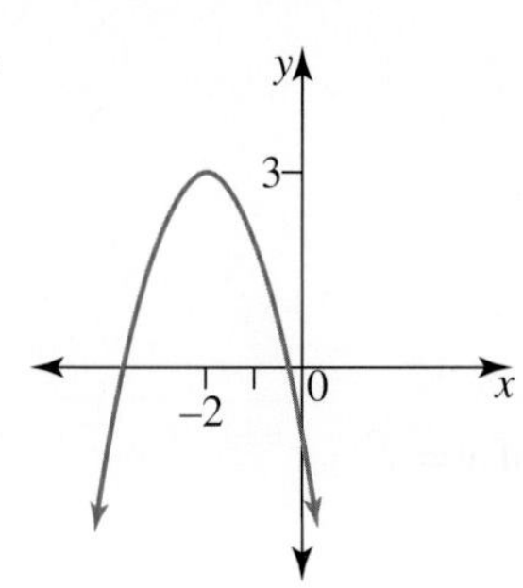

ii.

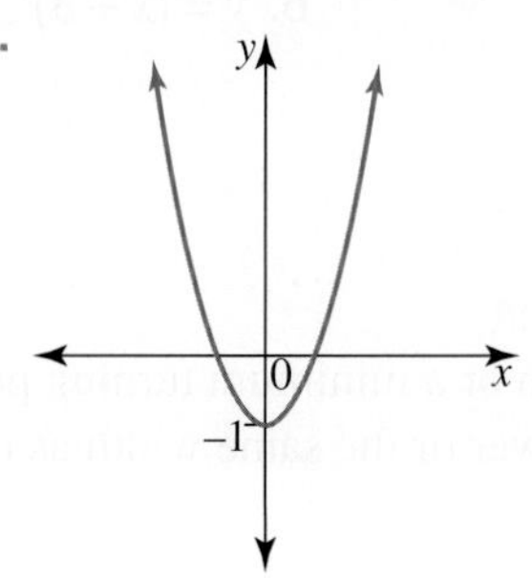

iii.

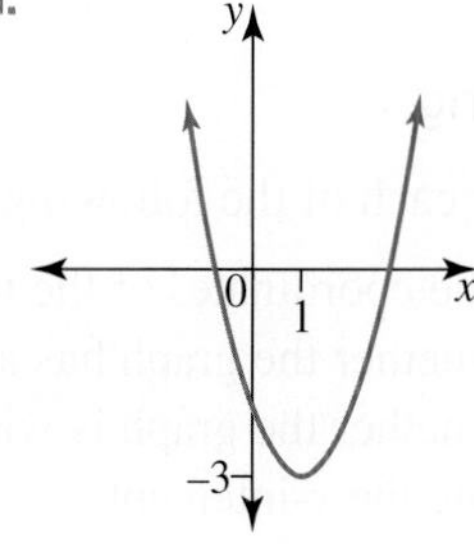

a. $y=(x-1)^2-3$

b. $y=-(x+2)^2+3$

c. $y=x^2-1$

9. **MC** The translations required to change $y=x^2$ into $y=\left(x-\frac{1}{2}\right)^2+\frac{1}{3}$ are:

A. right $\frac{1}{2}$, up $\frac{1}{3}$

B. left $\frac{1}{2}$, down $\frac{1}{3}$

C. right $\frac{1}{2}$, down $\frac{1}{3}$

D. left $\frac{1}{2}$, up $\frac{1}{3}$

E. right $\frac{1}{3}$, up $\frac{1}{2}$

10. **MC** For the graph $\frac{1}{4}\left(x-\frac{1}{2}\right)^2+\frac{1}{3}$, the effect of the $\frac{1}{4}$ on the graph is:

A. no effect

B. to make the graph narrower

C. to make the graph wider

D. to invert the graph

E. to translate the graph up $\frac{1}{4}$ of a unit

11. **MC** Compared to the graph of $y=x^2$, $y=-2(x+1)^2-4$ is:

A. inverted and wider

B. inverted and narrower

C. upright and wider

D. upright and narrower

E. inverted and the same width

12. **MC** A graph that has a minimum turning point $(1, 5)$ and that is narrower than the graph of $y=x^2$ is:

A. $y=(x-1)^2+5$

B. $y=\frac{1}{2}(x+1)^2+5$

C. $y=2(x-1)^2+5$

D. $y=2(x+1)^2+5$

E. $y=\frac{1}{2}(x-1)^2+5$

13. **MC** Compared to the graph of $y=x^2$, the graph of $y=-3(x-1)^2-2$ has the following features.

A. Maximum TP at $(-1,-2)$, narrower
B. Maximum TP at $(1,-2)$, narrower
C. Maximum TP at $(1,2)$, wider
D. Minimum TP at $(1,-2)$, narrower
E. Minimum TP at $(-1,-2)$, wider

WE9 For questions **14** to **16**, for the parabolas with the following equations:

i. determine the y-intercept
ii. determine the x-intercepts (where they exist).

14. a. $y=(x+1)^2-4$
b. $y=3(x-2)^2$

15. a. $y=-(x+4)^2-2$
b. $y=(x-2)^2-9$

16. a. $y=2x^2+4$
b. $y=(x+3)^2-5$

Understanding

17. **WE10** For each of the following:

i. write the coordinates of the turning point
ii. state whether the graph has a maximum or a minimum turning point
iii. state whether the graph is wider, narrower or the same width as the graph of $y=x^2$
iv. calculate the y-intercept
v. calculate the x-intercepts
vi. sketch the graph.

a. $y=(x-4)^2+2$
b. $y=(x-3)^2-4$
c. $y=(x+1)^2+2$

18. For each of the following:

i. write the coordinates of the turning point
ii. state whether the graph has a maximum or a minimum turning point
iii. state whether the graph is wider, narrower or the same width as the graph of $y=x^2$
iv. calculate the y-intercept
v. calculate the x-intercepts
vi. sketch the graph.

a. $y=(x+5)^2-3$
b. $y=-(x-1)^2+2$
c. $y=-(x+2)^2-3$

19. For each of the following:

i. write the coordinates of the turning point
ii. state whether the graph has a maximum or a minimum turning point
iii. state whether the graph is wider, narrower or the same width as the graph of $y=x^2$
iv. calculate the y-intercept
v. calculate the x-intercepts
vi. sketch the graph.

a. $y=-(x+3)^2-2$
b. $y=2(x-1)^2+3$
c. $y=-3(x+2)^2+1$

20. Consider the equation $2x^2-3x-8=0$.

a. Complete the square.
b. Use the result to determine the exact solutions to the original equation.
c. Determine the turning point of $y=2x^2-3x-8$ and indicate its type.

21. Answer the following questions.

a. Determine the equation of a quadratic that has a turning point of $(-4,6)$ and has an x-intercept at $(-1,0)$.
b. State the other x-intercept (if any).

22. Write the new equation for the parabola $y = x^2$ that has been:

a. reflected in the x-axis
b. dilated by a factor of 7 away from the x-axis
c. translated 3 units in the negative direction of the x-axis
d. translated 6 units in the positive direction of the y-axis
e. dilated by a factor of $\frac{1}{4}$ from the x-axis, reflected in the x-axis, and translated 5 units in the positive direction of the x-axis and 3 units in the negative direction of the y-axis.

Reasoning

23. The price of shares in fledgling company 'Lollies'r'us' plunged dramatically one afternoon, following the breakout of a small fire on the premises. However, Ms Sarah Sayva of Lollies Anonymous agreed to back the company, and share prices began to rise.
Sarah noted at the close of trade that afternoon that the company's share price followed the curve:
$P = 0.1(t - 3)^2 + 1$ where $\$P$ is the price of shares t hours after noon.

a. Sketch a graph of the relationship between time and share price to represent the situation.
b. Determine the initial share price.
c. Determine the lowest price of shares that afternoon.
d. Evaluate the time when the price was at its lowest.
e. Determine the final price of 'Lollies'r'us' shares as trade closed at 5 pm.

24. Rocky is practising for a football kicking competition. After being kicked, the path that the ball follows can be modelled by the quadratic relationship:

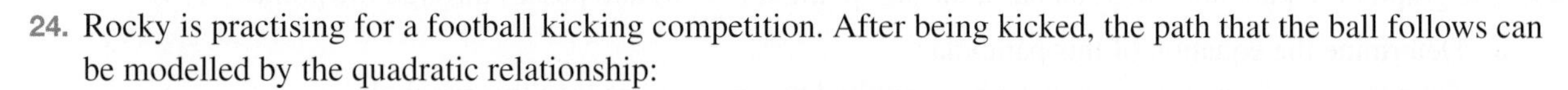

$$h = -\frac{1}{30}(d - 15)^2 + 8$$

where h is the vertical distance the ball reaches (in metres), and d is the horizontal distance (in metres).

a. Determine the initial vertical height of the ball.
b. Determine the exact maximum horizontal distance the ball travels.
c. Write down both the maximum height and the horizontal distance when the maximum height is reached.

25. Answer the following questions.

a. If the turning point of a particular parabola is $(2, 6)$, suggest a possible equation for the parabola.
b. If the y-intercept in part **a** is $(0, 4)$, determine the exact equation for the parabola.

26. Answer the following questions.

a. If the turning point of a particular parabola is (p, q), suggest a possible equation for the parabola.
b. If the y-intercept in part **a** is $(0, r)$, determine the exact equation for the parabola.

Problem solving

27. Use the completing the square method to write each of the following in turning point form and sketch the parabola for each part.

a. $y = x^2 - 8x + 1$
b. $y = x^2 + 4x - 5$
c. $y = x^2 + 3x + 2$

28. Use the information given in the graph shown to answer the following questions.

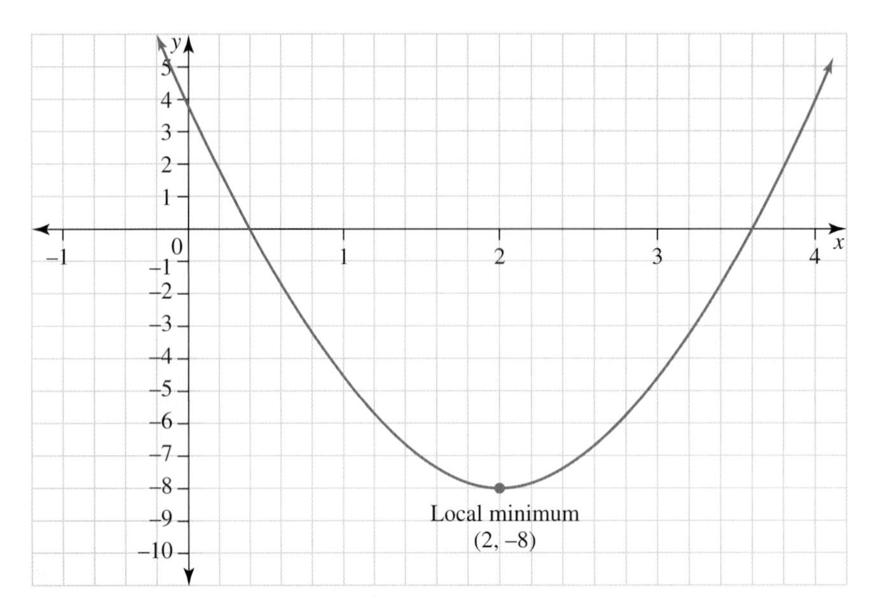

a. Determine the equation of the parabola shown.
b. State the dilation and translation transformations that have been applied to $y = x^2$ to achieve this parabola.
c. This graph is reflected in the x-axis. Determine the equation of the reflected graph.
d. Sketch the graph of the reflected parabola.

29. The graph of a quadratic equation has a turning point at $(-3, 8)$ and passes through the point $(-1, 6)$.

a. Determine the equation of this parabola.
b. State the transformations that have been applied to $y = x^2$ to produce this parabola.
c. Calculate the x and y-intercepts of this parabola.
d. The graph is reflected in the x-axis, dilated by a factor of 2 from the x-axis, and then reflected in the y-axis. Sketch the graph of this new parabola.

9.5 Sketching parabolas in expanded form

LEARNING INTENTION

At the end of this subtopic you should be able to:
- determine the x-intercept/s, y-intercept and turning point of a quadratic equation and sketch its graph by first factorising the equation
- determine the x-intercept/s, y-intercept and turning point of a quadratic equation and sketch its graph using the quadratic formula.

9.5.1 Parabolas of the form $y = ax^2 + bx + c$

eles-4872

- The general form of a quadratic equation is $y = ax^2 + bx + c$ where a, b and c are constants.
- The x-intercepts can be found by letting $y = 0$, factorising and using the Null Factor Law to solve for x.
- The y-intercept can be found by letting $x = 0$ and solving for y.
- The x-coordinate of the turning point lies midway between the x-intercepts.
- Once the midpoint of the x-intercepts is found, this value can be substituted into the original equation to find the y-coordinate of the turning point.
- Once these intercepts and turning point have been found, it is possible to sketch the parabola.
- If an equation is not written in turning point form, and cannot be readily factorised, then we will need to use the quadratic formula to help find all key points.

Quadratic formula

A quadratic of the form $y = ax^2 + bx + c$ has:
- a y-intercept at the coordinate $(0, c)$.
- x-intercepts that can be found using the quadratic formula; that is, when the equation $ax^2 + bx + c = 0$
$$x = \frac{-b \pm \sqrt{b^2 - 4ac}}{2a}$$
- a turning point at the coordinate $\left(-\frac{b}{2a}, c - \frac{b^2}{4a}\right)$.

For example, given the equation $y = x^2 + 4x - 6$:
- In this example we have $a = 1, b = 4$ and $c = -6$.
- The y-intercept is $(0, c) = (0, -6)$
- The x-intercepts are given by:

$$x = \frac{-b \pm \sqrt{b^2 - 4ac}}{2a} = \frac{-(4) \pm \sqrt{(4)^2 - 4 \times (1) \times (-6)}}{2 \times (1)} = \frac{-4 \pm \sqrt{40}}{2} = \frac{-4 \pm 2\sqrt{10}}{2} = -2 \pm \sqrt{10}$$

- Therefore, the x-intercepts are $\left(-2 + \sqrt{10}, 0\right)$ and $\left(-2 - \sqrt{10}, 0\right)$.
- The turning point is given by $\left(-\frac{b}{2a}, c - \frac{b^2}{4a}\right) = \left(-\frac{(4)}{2 \times (1)}, -6 - \frac{(4)^2}{4 \times (1)}\right) = (-2, -10)$.

Note: Do not convert answer to decimals unless specified by the question. It is always best practice to leave coordinates in exact form.

WORKED EXAMPLE 11 Sketching a factorised quadratic equation

Sketch the graph of $y = (x - 3)(x + 2)$.

THINK	WRITE/DRAW
1. The equation is in factorised form. To calculate the x-intercepts, let $y = 0$ and use the Null Factor Law.	$y = (x - 3)(x + 2)$ $0 = (x - 3)(x + 2)$ $x - 3 = 0$ or $x + 2 = 0$ (NFL) $x = 3$ or $x = -2$ x-intercepts: $(3, 0)\ (-2, 0)$
2. The x-coordinate of the turning point is midway between the x-intercepts. Calculate the average of the two x-intercepts to determine the midpoint between them.	$x_{TP} = \frac{3 + (-2)}{2}$ $= 0.5$
3. • To calculate the y-coordinate of the turning point, substitute x_{TP} into the equation.	$y = (x - 3)(x + 2)$ $y_{TP} = (0.5 - 3)(0.5 + 2)$ $= -6.25$
• State the turning point.	Turning point: $(0.5, -6.25)$
4. • To calculate the y-intercept, let $x = 0$ and substitute.	$y = (0 - 3)(0 + 2)$ $= -6$
• State the y-intercept.	y-intercept: $(0, -6)$

5. • Sketch the graph, showing all the important features.
• Label the graph.

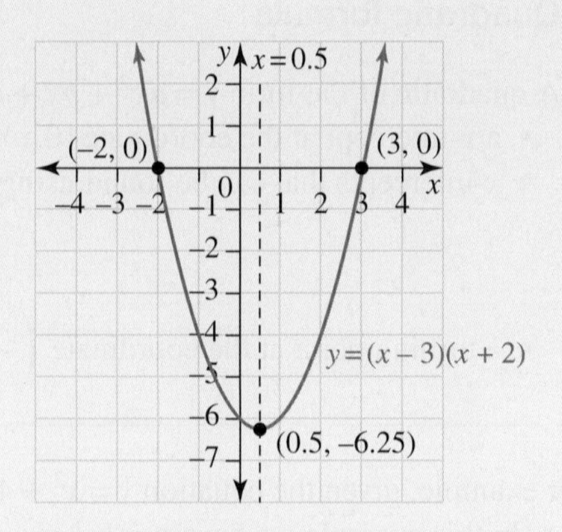

WORKED EXAMPLE 12 Sketching a quadratic in expanded form

Sketch the graph of $y = 2x^2 - 6x - 6$.

THINK	WRITE/DRAW
1. The equation is not in factorised form, but there is a common factor of 2. Take out the common factor of 2.	$y = 2x^2 - 6x - 6$ $= 2(x^2 - 3x - 3)$
2. The equation cannot be factorised (no factors of −3 add to −3), so use completing the square to write the equation in turning point form. • Halve and then square the coefficient of x. • Add this and then subtract it from the equation. • Collect the terms for and write the perfect square. • Simplify the brackets to write the equation in turning point form. • Identify the coordinates of the turning point (h, k).	$y = 2\left(x^2 - 3x + \left(\frac{3}{2}\right)^2 - \left(\frac{3}{2}\right)^2 - 3\right)$ $= 2\left(\left[x - \frac{3}{2}\right]^2 - \left(\frac{3}{2}\right)^2 - 3\right)$ $= 2\left(\left[x - \frac{3}{2}\right]^2 - \frac{9}{4} - 3\right)$ $= 2\left(\left[x - \frac{3}{2}\right]^2 - \frac{21}{4}\right)$ $= 2\left(x - \frac{3}{2}\right)^2 - 2 \times \frac{21}{4}$ $= 2\left(x - \frac{3}{2}\right)^2 - \frac{21}{2}$ Turning point : $\left(\frac{3}{2}, \frac{-21}{2}\right)$
3. • To calculate the x-intercepts, let $y = 0$. No factors of −3 add to −3, so use the quadratic formula to calculate the x-intercepts.	x-intercepts: let $y = 0$. $0 = 2x^2 - 6x - 6$ $x = \frac{-b \pm \sqrt{b^2 - 4ac}}{2a}$ where $a = 2, b = -6, c = -6$

$$x = \frac{-(-6) \pm \sqrt{(-6)^2 - 4 \times 2 \times (-6)}}{2(2)}$$

$$= \frac{6 \pm \sqrt{36 + 48}}{4}$$

$$= \frac{6 \pm \sqrt{84}}{4} = \frac{6 \pm 2\sqrt{21}}{4}$$

- State the x-intercepts.

The x-intercepts are:

$x = \dfrac{3 + \sqrt{21}}{2}$ and $x = \dfrac{3 - \sqrt{21}}{2}$

$x \approx 3.79$ $\quad x \approx -0.79$

4. • To calculate the y-intercepts, let $x = 0$ and substitute.
   • State the y-intercept.

$y = 2x^2 - 6x - 6$

$y = 2(0)^2 - 6(0) - 6$

$= -6$

y-intercept: $(0, -6)$

5. • Sketch the graph, showing all the important features.
   • Label the graph and show the exact values of the x-intercepts.

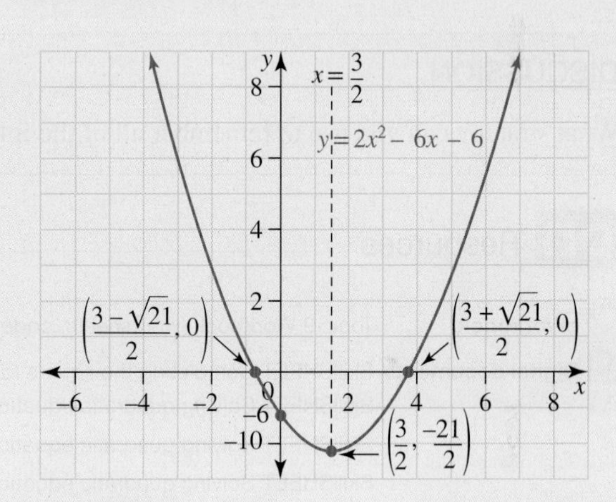

TI | THINK	DISPLAY/WRITE	CASIO | THINK	DISPLAY/WRITE
1. On a Graphs page, complete the function entry line as: $f1(x) = 2x^2 - 6x - 6$ The graph will be displayed.	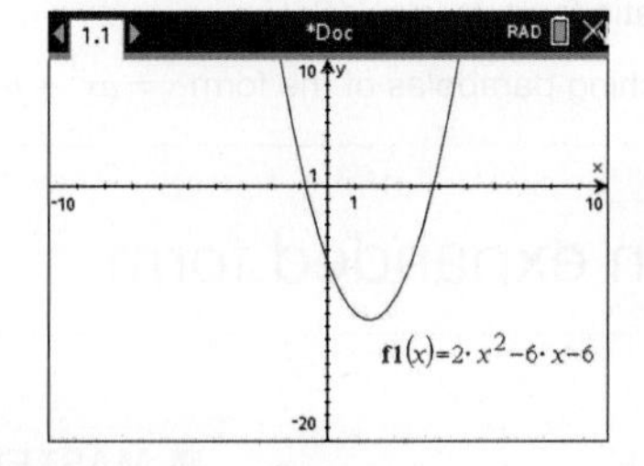	1. On a Graph & Table screen, complete the function entry line as: $y1 = 2x^2 - 6x - 6$ Tap the graphing icon and the graph will be displayed.	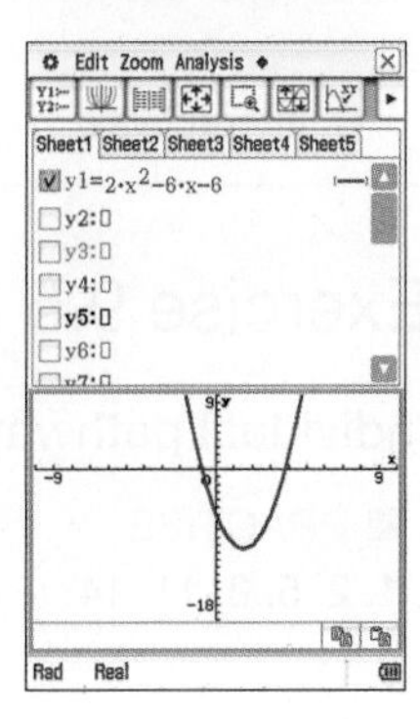

2. To locate the turning point, press:
   - MENU
   - 6: Analyze Graph
   - 2: Minimum

   Drag the dotted line to the left of the turning point (the lower bound), click ENTER and then drag the dotted line to the right of the turning point (the upper bound) and press ENTER.

   To locate the intercepts, press:
   - MENU
   - 6: Analyze Graph
   - 1: Zero

   Locate the points as described above.

   The points of interest will be shown.

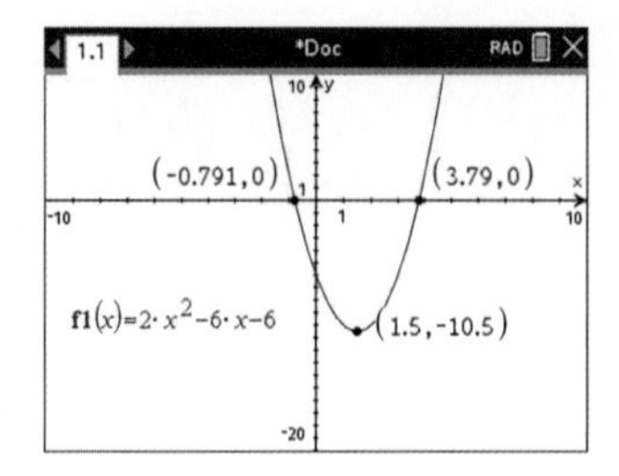

The graph is shown, along with the critical points.

2. To locate the turning point, press:
   - Analysis
   - G-Solve
   - Min

   To locate the turning point, press:
   - Analysis
   - G-Solve
   - Root

   To locate the second root, tap the right arrow.

   The points of interest will be shown.

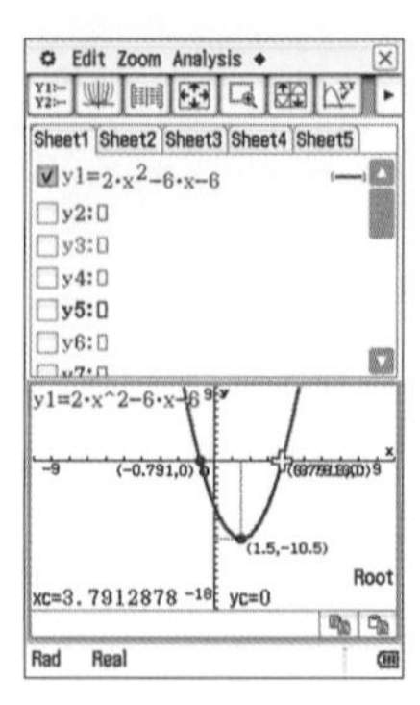

The turning point is $(1.5, 10.5)$.
The x-intercepts are $(-0.791, 0)$ and $(3.791, 0)$.

DISCUSSION

What strategy can you use to remember all of the information necessary to sketch a parabola?

Resources

eWorkbook Topic 9 Workbook (worksheets, code puzzle and a project) (ewbk-2035)

Digital documents SkillSHEET Completing the square (doc-5268)
SkillSHEET Solving quadratic equations using the quadratic formula (doc-5269)
SkillSHEET Solving quadratic equations of the type $ax^2 + bx + c = 0$ where $a = 1$ (doc-5270)
SkillSHEET Solving quadratic equations of the type $ax^2 + bx + c = 0$ where $a \neq 1$ (doc-5271)

Video eLessons Sketching quadratics in factorised form (eles-1927)
Sketching parabolas using the quadratic formula (eles-1945)

Interactivity Individual pathway interactivity: Sketching parabolas of the form $y = ax^2 + bx + c$ (int-4608)

Exercise 9.5 Sketching parabolas in expanded form

learnON

Individual pathways

■ PRACTISE	■ CONSOLIDATE	■ MASTER
1, 2, 5, 8, 11, 14	3, 6, 9, 12, 15	4, 7, 10, 13, 16

To answer questions online and to receive **immediate corrective feedback** and **fully worked solutions** for all questions, go to your learnON title at www.jacplus.com.au.

Fluency

1. What information is necessary to be able to sketch a parabola?

WE11 For questions 2 to 4, sketch the graph of each of the following.

2. a. $y = (x-5)(x-2)$ b. $y = (x+4)(x-7)$

3. a. $y = (x + 3)(x + 5)$

b. $y = (2x + 3)(x + 5)$

4. a. $y = (4 - x)(x + 2)$

b. $y = \left(\frac{x}{2} + 3\right)(5 - x)$

WE12 For questions 5 to 7, sketch the graph of each of the following.

5. a. $y = x^2 + 4x + 2$

b. $y = x^2 - 4x - 5$

c. $y = 2x^2 - 4x - 3$

6. a. $y = -2x^2 + 11x + 5$

b. $y = -2x^2 + 12x$

c. $y = 3x^2 + 6x + 1$

7. a. $y = -3x^2 - 5x + 2$

b. $y = 2x^2 + 8x - 10$

c. $y = -3x^2 + 7x + 3$

Understanding

8. The path of a soccer ball kicked by the goal keeper can be modelled by the equation $y = -\frac{1}{144}(x^2 - 24x)$ where y is the height of the soccer ball and x is the horizontal distance from the goalie, both in metres.

   a. Sketch the graph.
   b. Calculate how far away from the player does the ball first bounce.
   c. Calculate the maximum height of the ball.

9. The monthly profit or loss, p, (in thousands of dollars) for a new brand of chicken loaf is given by $p = 3x^2 - 15x - 18$, where x is the number of months after its introduction (when $x = 0$).

   a. Sketch the graph.
   b. Determine during which month a profit was first made.
   c. Calculate the month in which the profit is \$54 000.

10. The height, h metres, of a model rocket above the ground t seconds after launch is given by the equation $h = 4t(50 - t)$, where $0 \le t \le 50$.

    a. Sketch the graph of the rocket's flight.
    b. State the height of the rocket above the ground when it is launched.
    c. Calculate the greatest height reached by the rocket.
    d. Determine how long the rocket takes to reach its greatest height.
    e. Determine how long the rocket is in the air.

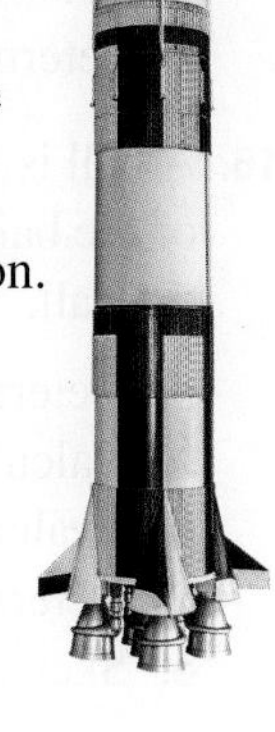

Reasoning

11. The equation $y = x^2 + bx + 7500$ has x-intercepts of $(-150, 0)$ and $(-50, 0)$. Determine the value of b in the equation. Justify your answer.

12. The equation $y = x^2 + bx + c$ has x-intercepts of m and n. Determine the value of b in the equation. Justify your answer.

13. A ball thrown from a cliff follows a parabolic path of the form $y = ax^2 + bx + c$. The ball is released at the point (0, 9), reaches a maximum height at (2, 11) and passes through the point (6, 3) on its descent.
    Determine the equation of the ball's path. Show full working.

Problem solving

14. A ball is thrown upwards from a building and follows the path given by the formula $h = -x^2 + 4x + 21$. The ball is h metres above the ground when it is a horizontal distance of x metres from the building.
    a. Sketch the graph of the path of the ball.
    b. Determine the maximum height of the ball.
    c. Determine how far the ball is from the wall when it reaches the maximum height.
    d. Determine how far from the building the ball lands.

15. During an 8-hour period, an experiment is done in which the temperature of a room follows the relationship $T = h^2 - 8h + 21$, where T is the temperature in degrees Celsius h hours after starting the experiment.

    a. Sketch the graph of this quadratic.
    b. Identify the initial temperature.
    c. Determine if the temperature is increasing or decreasing after 3 hours.
    d. Determine if the temperature is increasing or decreasing after 5 hours.
    e. Determine the minimum temperature and when it occurred.
    f. Determine the temperature after 8 hours.

16. A ball is thrown out of a window, passing through the points $(0, 7)$, $(6, 18)$ and $(18, 28)$. A rule for the height of the ball in metres is given by $h = ax^2 + bx + c$ where x is the horizontal distance, in metres, covered by the ball.
    a. Determine the values of a, b and c in the rule for the height of the ball.
    b. Calculate the height that the ball was thrown from.
    c. Evaluate the maximum height reached by the ball.
    d. Determine horizontal distance covered by the ball when it hits the ground.
    e. Sketch the flight path of the ball, making sure you show all key points.

9.6 Exponential graphs

LEARNING INTENTION

At the end of this subtopic you should be able to:
- determine the asymptote of an exponential equation and sketch a graph of the equation
- sketch the graph of an exponential graph after it has undergone a number of transformations that may include dilations, a reflection in the x- or y-axis, or translations.

9.6.1 Exponential functions

eles-4873

- Relationships of the form $y = a^x$ are called **exponential functions** with base a, where a is a real number not equal to 1, and x is the index power or exponent.
- The term 'exponential' is used, as x is an exponent (or index).
  For example, the graph of the exponential function $y = 2^x$ can be plotted by completing a table of values.

$$\text{Remember that } 2^{-3} = \frac{1}{2^3}$$
$$= \frac{1}{8}, \text{ and so on.}$$

x	−4	−3	−2	−1	0	1	2	3	4
y	$\frac{1}{16}$	$\frac{1}{8}$	$\frac{1}{4}$	$\frac{1}{2}$	1	2	4	8	16

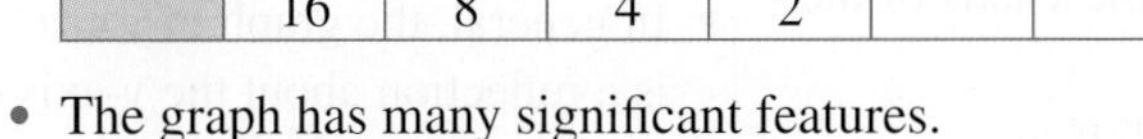

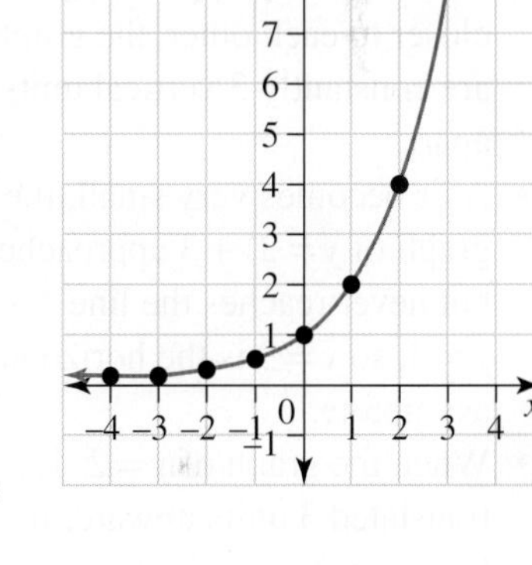

- The graph has many significant features.
  - The y-intercept is 1.
  - The value of y is always greater than zero.
  - As x decreases, y gets closer to but never reaches zero. So the graph gets closer to but never reaches the x-axis. The x-axis (or the line $y = 0$) is called an **asymptote.**
  - As x increases, y becomes very large.

Comparing exponential graphs

- The diagram at right shows the graphs of $y = 2^x$ and $y = 3^x$.
- The graphs both pass through the point (0, 1).
- The graph of $y = 3^x$ climbs more steeply than the graph of $y = 2^x$.
- $y = 0$ is an asymptote for both graphs.

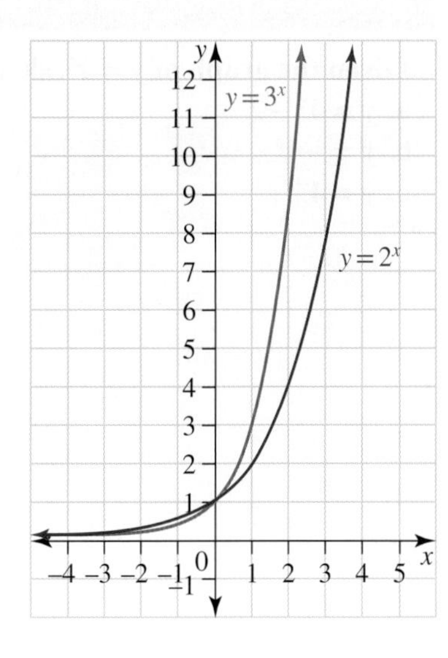

Translation and reflection

Vertical translation

- The diagram shows the graphs of $y = 2^x$ and $y = 2^x + 3$.
- The graphs have identical shape.
- Although they appear to get closer to each other, the graphs are constantly 3 vertical units apart.
- As x becomes very small, the graph of $y = 2^x + 3$ approaches but never reaches the line $y = 3$, so $y = 3$ is the horizontal asymptote.
- When the graph of $y = 2^x$ is translated 3 units upward, it becomes the graph of $y = 2^x + 3$.

Reflection about the x-axis

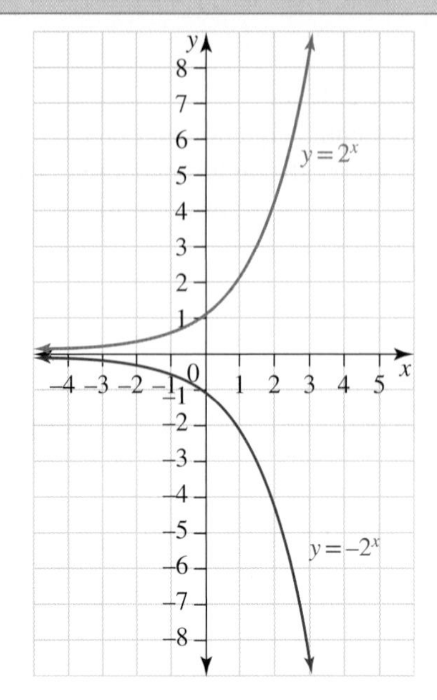

- The diagram shows the graphs of $y = 2^x$ and $y = -2^x$.
- The graphs have identical shape.
- The graph of $y = -2^x$ is a reflection about the x-axis of the graph of $y = 2^x$.
- The x-axis ($y = 0$) is an asymptote for both graphs.
- In general, the graph of $y = -a^x$ is a reflection about the x-axis of the graph of $y = a^x$.

Reflection about the y-axis

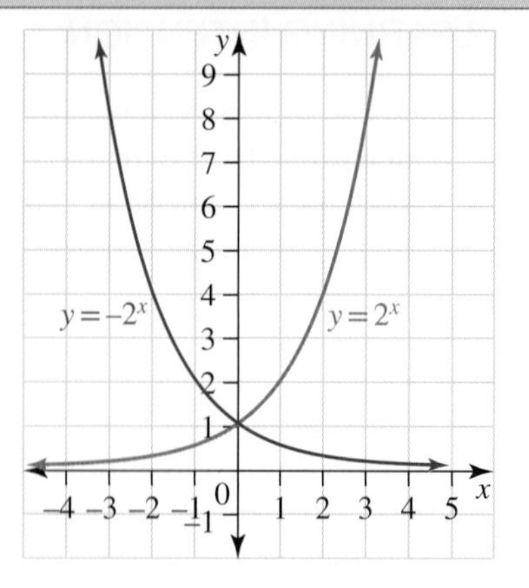

- The diagram shows the graphs of $y = 2^x$ and $y = 2^{-x}$.
- The graphs have identical shape.
- The graph of $y = 2^{-x}$ is a reflection about the y-axis of the graph of $y = 2^x$.
- Both graphs pass through the point (0, 1).
- The x-axis ($y = 0$) is an asymptote for both graphs.
- In general, the graph of $y = a^{-x}$ is a reflection about the y-axis of the graph of $y = a^x$.

WORKED EXAMPLE 13 Sketching exponential graphs

Given the graph of $y = 4^x$, sketch on the same axes the graphs of:

a. $y = 4^x - 2$

b. $y = -4^x$

c. $y = 4^{-x}$.

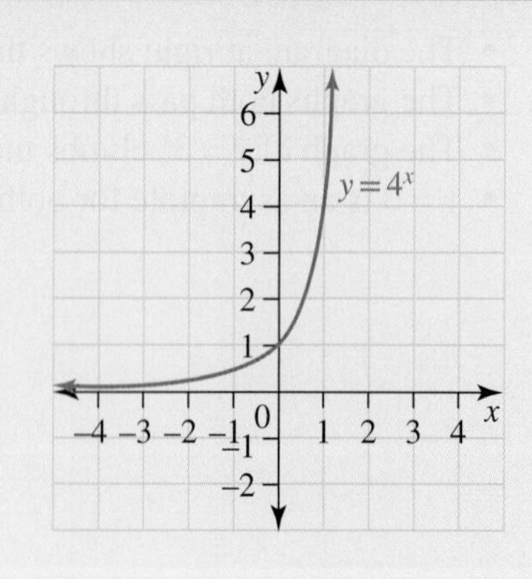

THINK

DRAW

a. The graph of $y = 4^x$ has already been drawn. It has a y-intercept of 1 and a horizontal asymptote at $y = 0$. The graph of $y = 4^x - 2$ has the same shape as $y = 4^x$ but is translated 2 units vertically down. It has a y-intercept of -1 and a horizontal asymptote at $y = -2$.

a.

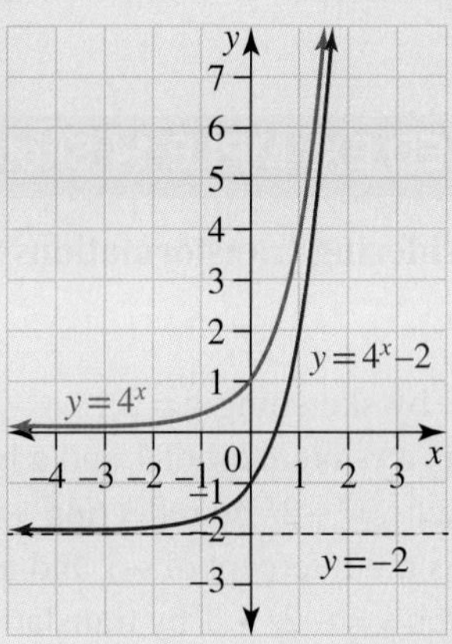

b. $y = -4^x$ has the same shape as $y = 4^x$ but is reflected about the x-axis. It has a y-intercept of -1 and a horizontal asymptote at $y = 0$.

b.

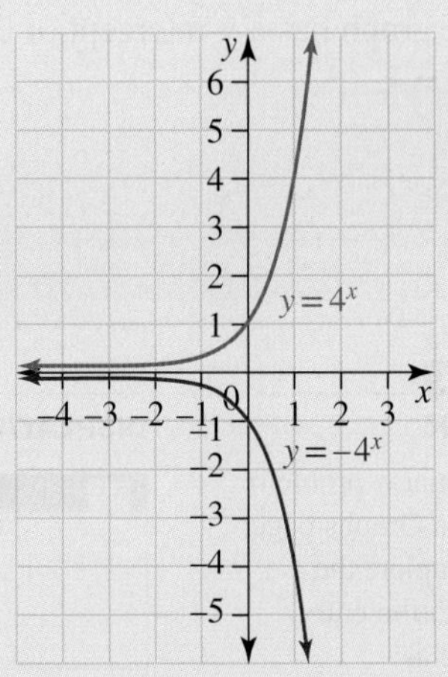

c. $y = 4^{-x}$ has the same shape as $y = 4^x$ but is reflected about the y-axis. The graphs have the same y-intercept and the same horizontal asymptote ($y = 0$).

c.

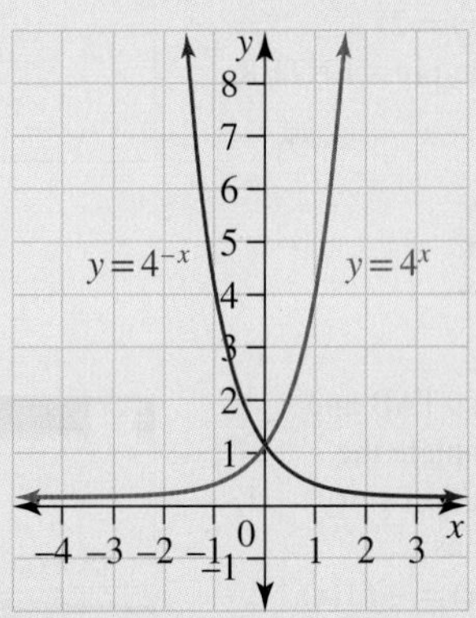

9.6.2 Combining transformations of exponential graphs

eles-4874

- It is possible to combine translations, dilations and reflections in one graph.

WORKED EXAMPLE 14 Sketching exponentials with multiple transformations

By considering transformations to the graph of $y = 2^x$, sketch the graph of $y = -2^x + 1$.

THINK

1. Start by sketching $y = 2^x$.
   It has a y-intercept of 1 and a horizontal asymptote at $y = 0$.
2. Sketch $y = -2^x$ by reflecting $y = 2^x$ about the x-axis.
   It has a y-intercept of -1 and a horizontal asymptote at $y = 0$.
3. Sketch $y = -2^x + 1$ by translating $y = -2^x$ upwards by 1 unit.
   The graph has a y-intercept of 0 and a horizontal asymptote at $y = 1$.

DRAW

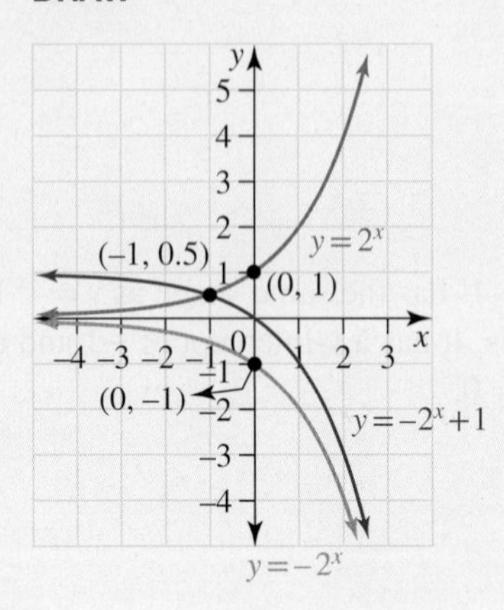

TI | THINK

1. In a new problem, on a Graphs page, complete the function entry line as:
   $f1(x) = 2^x$
   Then press ENTER.

DISPLAY/WRITE

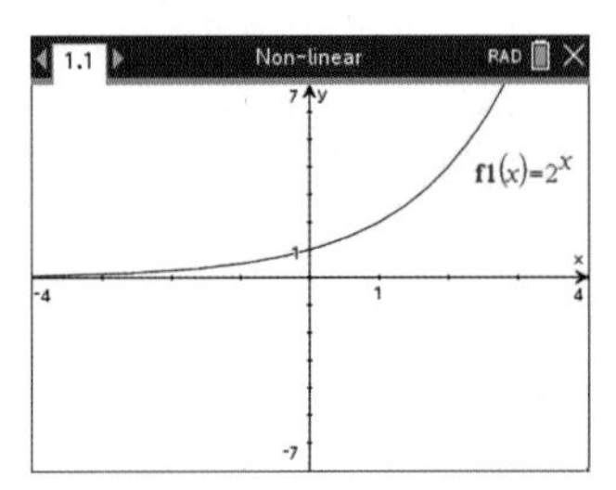

2. Press TAB and complete the function entry line as:
   $f2(x) = -f1(x)$
   Then press ENTER.
   Press TAB and complete the function entry line as:
   $f3(x) = f2(x) + 1$
   Then press ENTER.

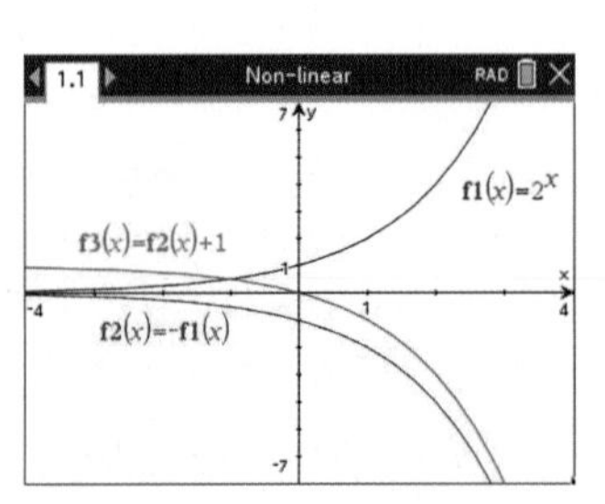

The graph of $y = -2^x$ is the reflection of the graph of $y = 2^x$ in the x-axis.
The graph of $y = -2^x + 1$ is the graph of $y = -2^x$ translated upwards by 1 unit.
The graph of $y = -2^x + 1$ passes through the origin and has a horizontal asymptote at $y = 1$.

CASIO | THINK

Open the Graph & Table screen and complete the function entry line as:
$y1 = 2^x$
$y2 = -2^x$
$y3 = -2^x + 1$
Then tap the graphing icon and the graphs will be displayed.

DISPLAY/WRITE

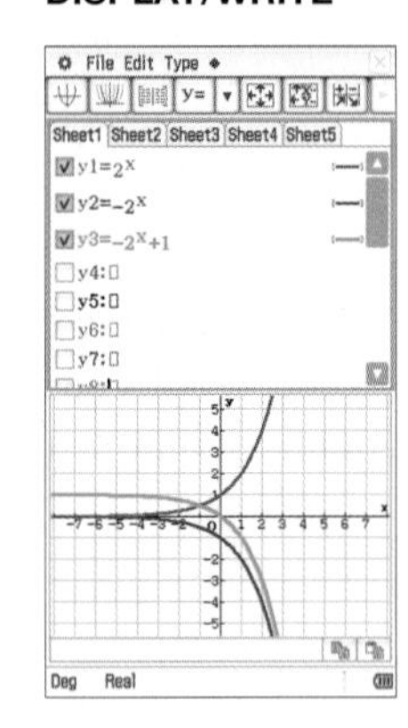

From $y1$ to $y2$ the graph has undergone a reflection about the x-axis. From $y2$ to $y3$ the graph has undergone a translation upwards by 1 unit.

DISCUSSION

Explain whether the graph of an exponential function will always have a horizontal asymptote.

on Resources

eWorkbook Topic 9 Workbook (worksheets, code puzzle and a project) (ewbk-2035)

Interactivities Individual pathway interactivity: Exponential functions and graphs (int-4609)

Exponential functions (int-5959)

Exercise 9.6 Exponential graphs

learnon

Individual pathways

■ PRACTISE	■ CONSOLIDATE	■ MASTER
1, 6, 8, 10, 13, 15, 18	2, 4, 7, 11, 14, 16, 19	3, 5, 9, 12, 17, 20

To answer questions online and to receive **immediate corrective feedback** and **fully worked solutions** for all questions, go to your learnON title at www.jacplus.com.au.

Fluency

1. Complete the following table and use it to plot the graph of $y = 3^x$, for $-3 \le x \le 3$.

x	-3	-2	-1	0	1	2	3
y							

2. If $x = 1$, calculate the value of y when:

a. $y = 2^x$ b. $y = 3^x$ c. $y = 4^x$

3. If $x = 1$, calculate the value of y when:

a. $y = 10^x$ b. $y = a^x$.

4. Using a calculator or graphing program, sketch the graphs of $y = 2^x, y = 3^x$ and $y = 4^x$ on the same set of axes.

a. Describe the common features among the graphs.
b. Describe how the value of the base $(2, 3, 4)$ affects the graph.
c. Predict where the graph $y = 8^x$ would lie and sketch it in.

5. Using graphing technology, sketch the following graphs on one set of axes.
$y = 3^x, y = 3^x + 2, y = 3^x + 5$ and $y = 3^x - 3$

a. State what remains the same in all of these graphs.
b. State what is changed.
c. For the graph of $y = 3^x + 10$, write down:
   i. the y-intercept
   ii. the equation of the horizontal asymptote.

6. Using graphing technology, sketch the graphs of:

a. $y = 2^x$ and $y = -2^x$
b. $y = 3^x$ and $y = -3^x$
c. $y = 6^x$ and $y = -6^x$.
d. State the relationship between these pairs of graphs.

7. Using graphing technology, sketch the graphs of:

a. $y = 2^x$ and $y = 2^{-x}$
b. $y = 3^x$ and $y = 3^{-x}$
c. $y = 6^x$ and $y = 6^{-x}$
d. State the relationship between these pairs of graphs.

8. WE13 Given the graph of $y = 2^x$, sketch on the same axes the graphs of:

a. $y = 2^x + 6$
b. $y = -2^x$
c. $y = 2^{-x}$.

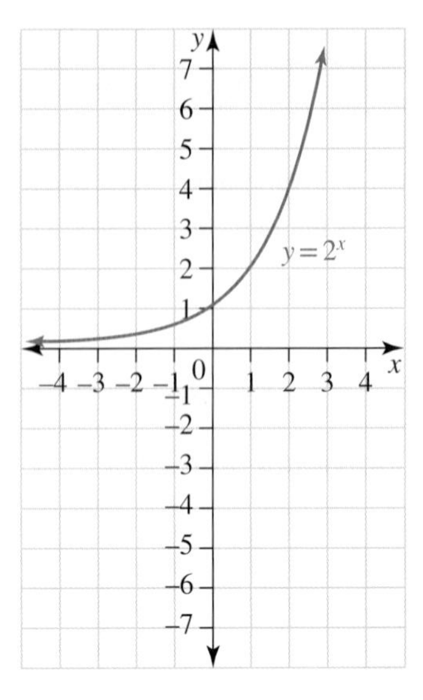

9. Given the graph of $y = 3^x$, sketch on the same axes the graphs of:

a. $y = 3^x + 2$
b. $y = -3^x$.

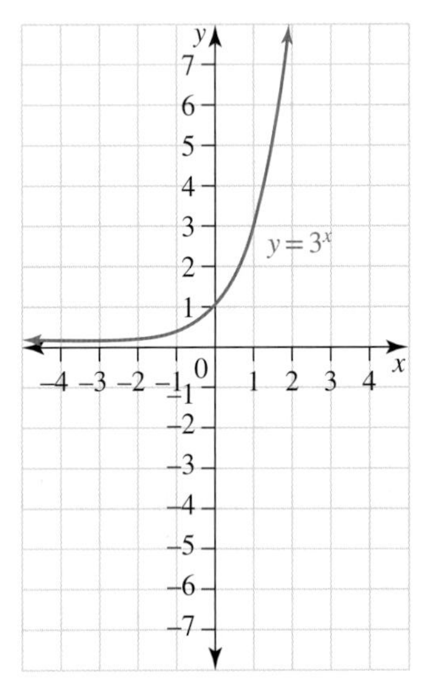

10. Given the graph of $y = 4^x$, sketch on the same axes the graphs of:

a. $y = 4^x - 3$
b. $y = 4^{-x}$.

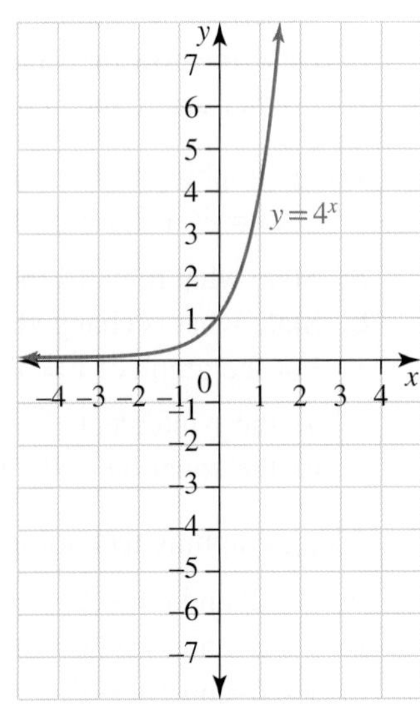

Understanding

11. WE14 By considering transformations of the graph of $y = 2^x$, sketch the following graphs on the same set of axes.

a. $y = 2^{-x} + 2$
b. $y = -2^x + 3$

12. By considering transformations of the graph of $y = 5^x$, sketch the following graphs on the same set of axes.

a. $y = -5^x + 10$
b. $y = 5^{-x} + 10$

13. Match each equation with its correct graph.

A. $y = 2^x$ B. $y = 3^x$ C. $y = -4^x$ D. $y = 5^{-x}$

a.

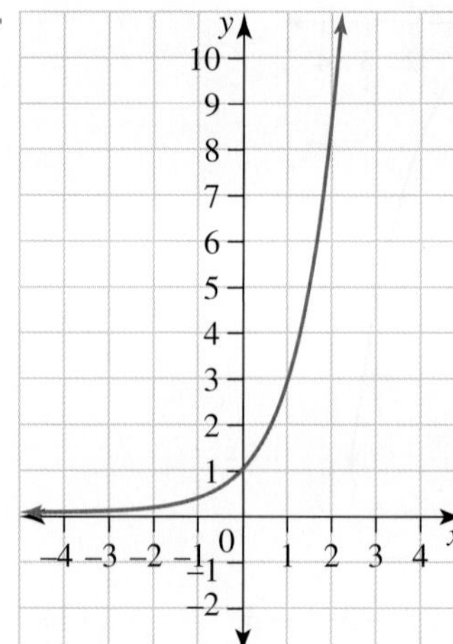

b.

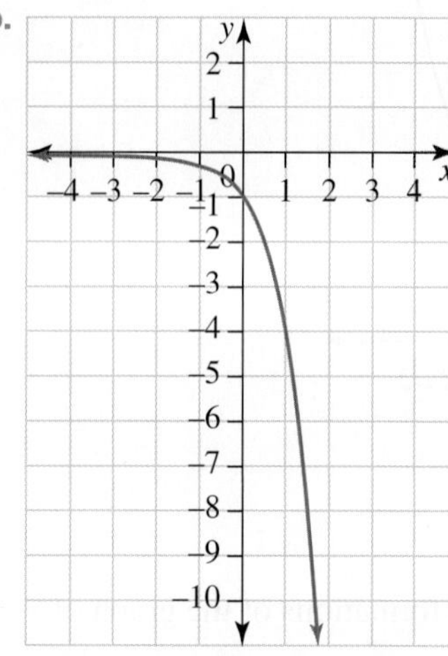

c.

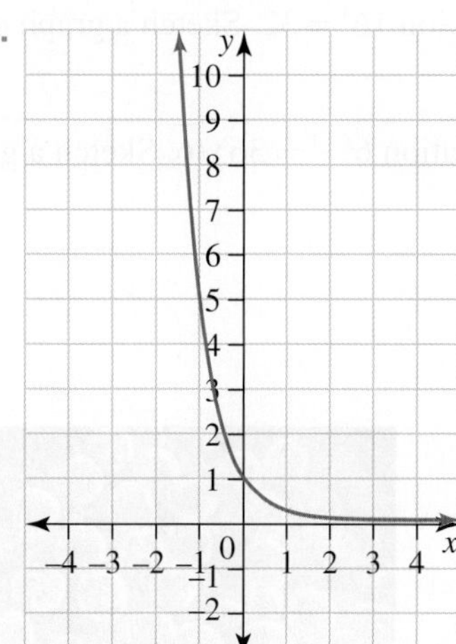

d.

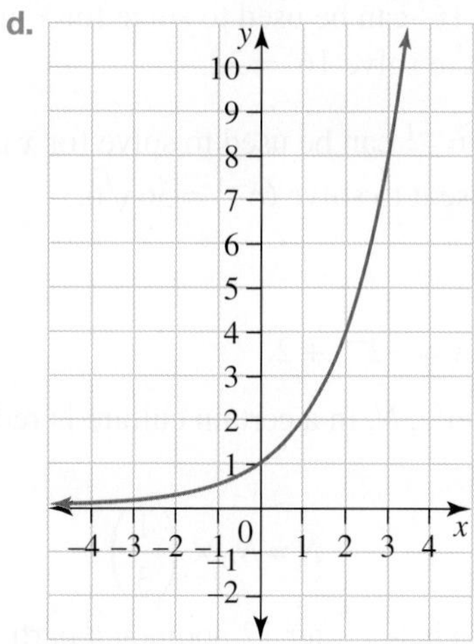

14. Match each equation with its correct graph. Explain your answer.

A. $y = 2^x + 1$ B. $y = 3^x + 1$ C. $y = -2^x + 1$ D. $y = 2^{-x} + 1$

a.

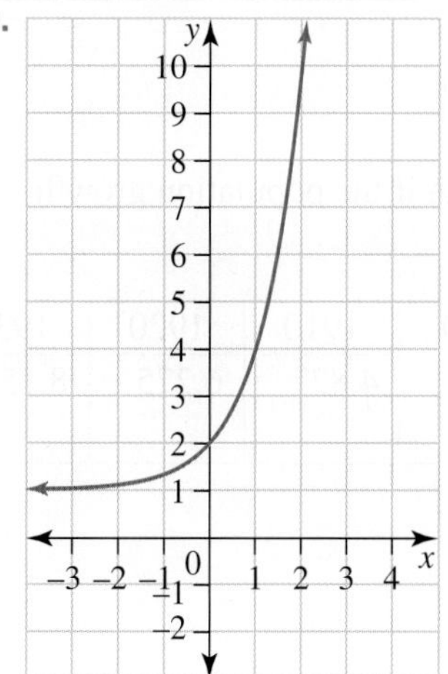

b.

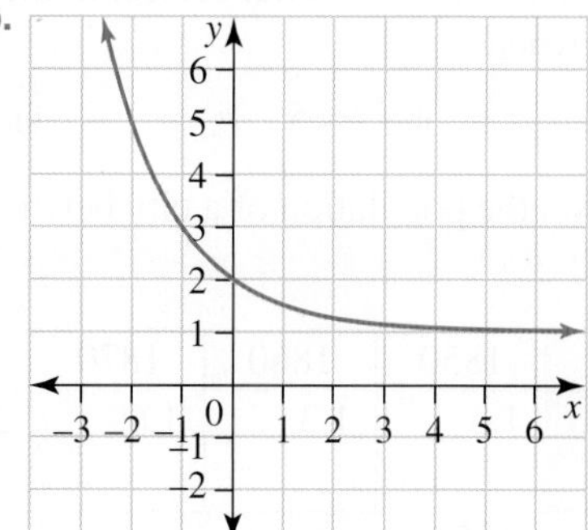

c.

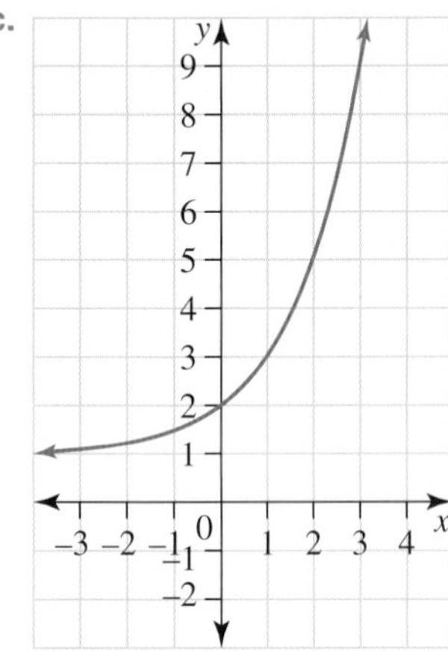

d.

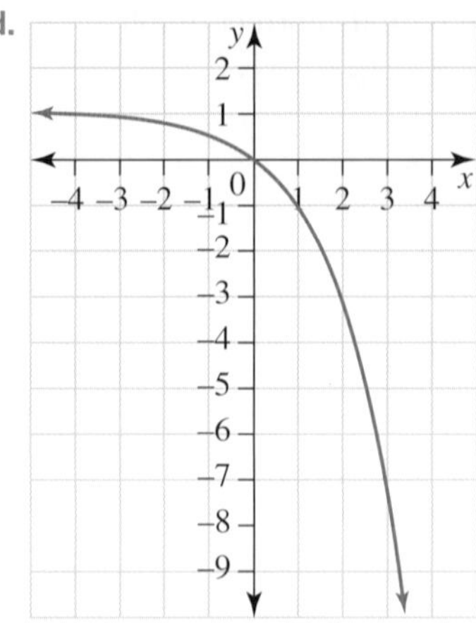

Reasoning

15. By considering transformations of the graph of $y = 3^x$, sketch the graph of $y = -3^{-x} - 3$.

16. The graph of $f(x) = 16^x$ can be used to solve for x in the exponential equation $16^x = 32$. Sketch a graph of $f(x) = 16^x$ and use it to solve $16^x = 32$.

17. The graph of $f(x) = 6^{x-1}$ can be used to solve for x in the exponential equation $6^{x-1} = 36\sqrt{6}$. Sketch a graph of $f(x) = 6^{x-1}$ and use it to solve $6^{x-1} = 36\sqrt{6}$.

Problem solving

18. Sketch the graph of $y = -2^{-x} + 2$.

19. The number of bacteria, N, in a certain culture is reduced by a third every hour so

$$N = N_0 \times \left(\frac{1}{3}\right)^t$$

where t is the time in hours after 12 noon on a particular day. Initially there are 10 000 bacteria present.

   a. State the value of N_0.
   b. Calculate the number of bacteria, correct to the nearest whole number, in the culture when:
      i. $t = 2$ ii. $t = 5$ iii. $t = 10$.

20. a. The table shows the population of a city between 1850 and 1930. Explain if the population growth is exponential.

Year	1850	1860	1870	1880	1890	1900	1910	1920	1930
Population (million)	1.0	1.3	1.69	2.197	2.856	3.713	4.827	6.275	8.157

   b. Determine the common ratio in part **a**.
   c. Evaluate the percentage increase every ten years.
   d. Estimate the population in 1895.
   e. Estimate the population in 1980.

9.7 Inverse proportion

LEARNING INTENTION

At the end of this subtopic you should be able to:

- calculate the constant of proportionality, k, between two variables that are inversely proportional to each other
- determine the rule between two inversely proportional variables and plot the graph of their relationship.

9.7.1 Inverse proportion

eles-4875

- If 24 sweets are shared between 4 children, then each child will receive 6 sweets. If the sweets are shared by 3 children, then each will receive 8 sweets.
- The relationship between the number of children (C) and the number sweets for each child (n) can be given in a table.

C	1	2	3	4	6	8	12
n	24	12	8	6	4	3	2

- As the number of children (C) increases, the number of sweets for each child (n) decreases. This is an example of **inverse proportion** or inverse variation.
- We say that 'n is inversely proportional to C' or 'n varies inversely as C'.
- This is written as $C \propto \frac{1}{n}$, or $C = \frac{k}{n}$, where k is a constant (the **constant of proportionality**). This formula can be rearranged to $Cn = k$. Note that multiplying any pair of values in the table (3×8, 12×2) gives the same result.
- The relationship has some important characteristics:
  - As C increases, n decreases, and vice versa.
  - The graph of the relationship is a **hyperbola**.

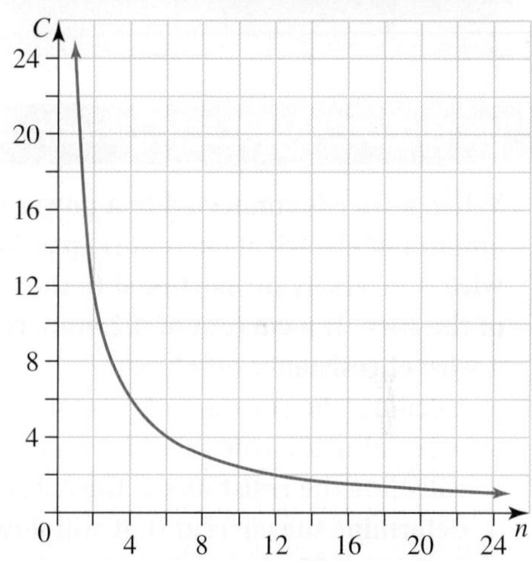

WORKED EXAMPLE 15 Calculating the constant of proportionality

***y* is inversely proportional to *x* and *y* = 10 when *x* = 2.**

a. **Calculate the constant of proportionality, *k*, and hence the rule relating *x* and *y*.**

b. **Plot a graph of the relationship between *x* and *y*, for values of *x* from 2 to 10.**

THINK	WRITE/DRAW
a. 1. Write the relationship between the variables.	a. $y \propto \frac{1}{x}$
2. Rewrite as an equation using k, the constant of proportionality	$y = \frac{k}{x}$
3. Substitute $y = 10, x = 2$, into $y = \frac{k}{x}$ and solve for k	$10 = \frac{k}{2}$ $k = 20$

4. Write the rule by substituting $k=20$ into $y=\frac{k}{x}$. $\quad y=\frac{20}{x}$

b. 1. Use the rule $y=\frac{20}{x}$ to set up a table of values for x and y, taking values for x which are positive factors of k so that only whole number values of y are obtained. For example, $x=4, y=\frac{20}{4}=5$.

b.

x	2	4	6	8	10
y	10	5	3.3	2.5	2

2. Plot the points on a clearly labelled set of axes and join the points with a smooth curve. Label the graph.

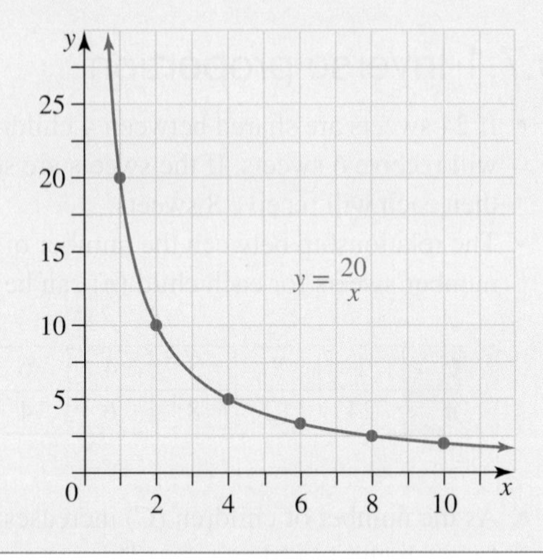

WORKED EXAMPLE 16 Determining the rule for an inversly proportional relationship

When a wire is connected to a power source, the amount of electrical current (I) passing through the wire is inversely proportional to the resistance (R) of the wire. If a current of 0.2 amperes flows through a wire of resistance 60 ohms:

a. calculate the constant of proportionality

b. determine the rule relating R and I

c. calculate the resistance if the current equals 5 amperes

d. determine the current that will flow through a wire of resistance 20 ohms.

THINK

Summarise the information in a table.

$I \propto \frac{1}{R}$

Write the rule.

a. 1. Substitute $R=60$, $I=0.2$, into $I=\frac{k}{R}$.

2. Solve for k.

b. Write the rule using $k=12$.

WRITE

R	60		20
I	0.2	5	

$I=\frac{k}{R}$

a. $0.2=\frac{k}{60}$

$0.2\times 60=k$

$k=12$

b. $I=\frac{12}{R}$

c. 1. Substitute $I = 5$ into $I = \frac{12}{R}$.

c. $5 = \frac{12}{R}$

$5R = 12$

$R = \frac{12}{5}$

2. Solve for R.

$= 2.4$

3. Write the answer in a sentence.

The resistance equals 2.4 ohms.

d. 1. Substitute $R = 20$ into $I = \frac{12}{R}$.

d. $I = \frac{12}{20}$

$= 0.6$

2. Write the answer in a sentence.

The current will be 0.6 amperes.

DISCUSSION

Explain what is meant by inverse proportion.

on Resources

eWorkbook Topic 9 Workbook (worksheets, code puzzle and a project) (ewbk-2035)

Interactivities Individual pathway interactivity: Inverse proportion (int-4512)

Inverse proportion (int-6058)

Exercise 9.7 Inverse proportion

learnon

Individual pathways

■ PRACTISE	■ CONSOLIDATE	■ MASTER
1, 3, 6, 10, 13	2, 4, 7, 11, 14	5, 8, 9, 12, 15

To answer questions online and to receive **immediate corrective feedback** and **fully worked solutions** for all questions, go to your learnON title at www.jacplus.com.au.

Fluency

1. Decide whether inverse proportion exists between each pair of variables. If it does exist, write an equation to describe the relationship.
   a. The speed of a car (s) and the time (t) it takes to complete one lap of a race circuit.
   b. The amount of money (D) that I have and the number (n) of cards that I can buy.
   c. The time (t) that it takes to make a pair of jeans and the number of pairs (p) that can be made in one day.
   d. The price (P) of petrol and the amount (L) that can be bought for \$80.
   e. The price (P) of petrol and the cost (C) of buying 80 L.
   f. The number of questions (n) in a test and the amount of time (t) available to answer each one.
2. List three examples of inverse proportion.

3. **WE15** y varies inversely as x and $y = 100$ when $x = 10$.

   a. Calculate the constant of proportionality, k, and hence the rule relating x and y.
   b. Plot a graph of the relationship between x and y, for values of x that are positive factors of k less than 21.

4. p is inversely proportional to q and $p = 12$ when $q = 4$.

   a. Calculate the constant of proportionality, k, and hence the rule relating p and q.
   b. Plot a graph of the relationship between q and p, for values of q that are positive factors of k less than 11.

5. y varies inversely as x and $y = 42$ when $x = 1$.

   a. Calculate the constant of proportionality, k, and hence the rule relating x and y.
   b. Plot a graph of the relationship between x and y, for values of x from 1 to 10.

Understanding

6. **WE16** When a constant force is applied to an object, its acceleration is inversely proportional to its mass. When the acceleration of an object is 40 m/s^2, the corresponding mass is 100 kg.

   a. Calculate the constant of proportionality.
   b. Determine the rule relating mass and acceleration.
   c. Determine the acceleration of a 200 kg object.
   d. Determine the acceleration of a 1000 kg object.

7. The number of colouring pencils sold is inversely proportional to the price of each pencil.
   Two thousand pencils are sold when the price is \$0.25 each.

   a. Calculate the constant of proportionality.
   b. Determine the number of pencils that could be sold for \$0.20 each.
   c. Determine the number of pencils that could be sold for \$0.50 each.

8. The time taken to complete a journey is inversely proportional to the speed travelled. A trip is completed in 4.5 hours travelling at 75 km per hour.

   a. Calculate the constant of proportionality.
   b. Determine how long, to the nearest minute, the trip would take if the speed was 85 km per hour.
   c. Determine the speed required to complete the journey in 3.5 hours, correct to 1 decimal place.
   d. Determine the distance travelled in each case.

9. The cost per person travelling in a charter plane is inversely proportional to the number of people in the charter group. It costs \$350 per person when 50 people are travelling.

   a. Calculate the constant of variation.
   b. Determine the cost per person, to the nearest cent, if there are 75 people travelling.
   c. Determine how many people are required to reduce the cost to \$250 per person.
   d. Determine the total cost of hiring the charter plane.

Reasoning

10. The electrical current in a wire is inversely proportional to the resistance of the wire to that current. There is a current of 10 amperes when the resistance of the wire is 20 ohms.
    a. Calculate the constant of proportionality.
    b. Determine the current possible when the resistance is 200 ohms.
    c. Determine the resistance of the wire when the current is 15 amperes.
    d. Justify your answer to parts **b** and **c** using a graph.

11. The pressure of an ideal gas is inversely proportional to the volume taken up by the gas. A balloon is filled with air so it takes up 3 L at a pressure of 5 atmospheres.

    a. Calculate the constant of proportionality.
    b. Determine the new volume of the balloon if the pressure was dropped to 0.75 atmospheres.
    c. Determine the pressure if the same amount of air took up a volume of 6 L.

12. Two equations relating the time of a trip, T, and the speed at which they travel, S, are given. For both cases the time is inversely proportional to the speed: $T_1 = \frac{5}{S_1}$ and $T_2 = \frac{7}{S_2}$. Explain what impact the different constants of proportionality have on the time of the trip.

Problem solving

13. The time it takes to pick a field of strawberries is inversely proportional to the number of pickers. It takes 2 people 5 hours to pick all of the strawberries in a field.

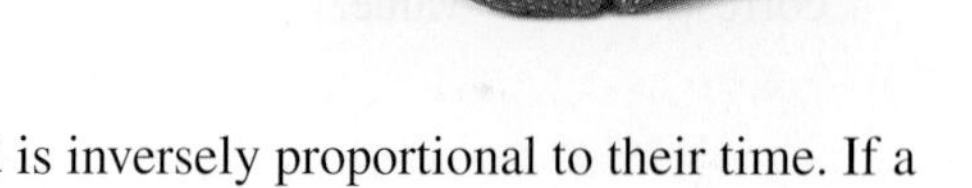

    a. Calculate the constant of proportionality.
    b. Determine the rule relating time (T) and the number of pickers (P).
    c. Determine the time spent if there are 6 pickers.

14. For a constant distance covered by a sprinter, the sprinter's speed is inversely proportional to their time. If a sprinter runs at a speed of 10.4 m/s, the corresponding time is 9.62 seconds.
    a. Calculate the constant of variation.
    b. Determine the rule relating speed (V) and time (T).
    c. Determine the time, correct to 2 decimal places, if they ran at a speed of 10.44 m/s.
    d. Determine the time, correct to 2 decimal places, if they ran at a speed of 6.67 m/s.

15. A holiday hostel is built to accommodate group bookings of up to 45 people. It is known that it would cost each of the individuals in a group of 20 people $67.50 per night to rent this venue. The cost of the venue will remain the same no matter how many people are part of the group booking.

    a. Write the rule for the cost per person (C) and the number of people in a group booking (n).
    b. If 6 people from the original group are no longer able to attend, determine the new cost per person.
    c. Calculate the cheapest possible cost per person.

9.8 Sketching the hyperbola

LEARNING INTENTION

At the end of this subtopic you should be able to:
- determine the equation of the vertical and horizontal asymptote of a hyperbola
- sketch the graph of a hyperbola, and understand the effect dilations and reflections have on the shape of the graph.

9.8.1 Hyperbolas

eles-4876

- A hyperbola is a function of the form $xy = k$ or $y = \frac{k}{x}$.

WORKED EXAMPLE 17 Creating a table of values in order to graph a hyperbola

Complete the table of values below and use it to plot the graph of $y = \frac{1}{x}$.

x	-3	-2	-1	$-\frac{1}{2}$	0	$\frac{1}{2}$	1	2	3
y									

THINK

1. Substitute each x-value into the function $y = \frac{1}{x}$ to obtain the corresponding y-value.

WRITE/DRAW

x	-3	-2	-1	$-\frac{1}{2}$	0	$\frac{1}{2}$	1	2	3
y	$-\frac{1}{3}$	$-\frac{1}{2}$	-1	-2	Undef.	2	1	$\frac{1}{2}$	$\frac{1}{3}$

2. Draw a set of axes and plot the points from the table. Join them with a smooth curve.

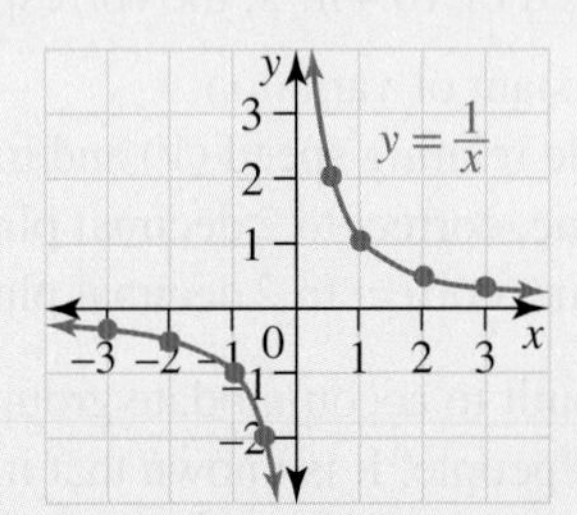

- The graph in Worked example 17 has several important features.
  1. There is no function value (y-value) when $x = 0$. At this point the hyperbola is undefined. When this occurs, the line that the graph approaches ($x = 0$) is called a vertical asymptote.
  2. As x becomes larger and larger, the graph gets very close to but will never touch the x-axis. The same is true as x becomes smaller and smaller. The hyperbola also has a horizontal asymptote at $y = 0$.
  3. The hyperbola has two separate branches. It cannot be drawn without lifting your pen from the page and is an example of a discontinuous graph.
- Graphs of the form $y = \frac{k}{x}$ are the same basic shape as $y = \frac{1}{x}$ with y-values dilated by a factor of k.

General form of a hyperbola

The general form of a hyperbola with multiple transformations is given by the equation

$$y = \frac{a}{x-h} + k.$$

- This graph has a horizontal asymptote with equation $y = k$. This occurs because the fraction $\frac{a}{x-h}$ cannot be made to equal 0, which means y can never be equal to k.
- The graph has a vertical asymptote with equation $x = h$. This is because the denominator of a fraction cannot be equal to 0, so x can never be equal to h.
- The graph will be reflected in the x-axis if $a < 0$.

Note: Not all exponential graphs will have an x- or y-intercept. This will depend on the equation of the asymptotes.

WORKED EXAMPLE 18 Identifying the asymptotes of a hyperbola

a. Plot the graph of $y = \frac{4}{x}$ for $-2 \le x \le 2$.

b. Write down the equation of each asymptote.

THINK

a. 1. Prepare a table of values taking x-values from −2 to 2. Fill in the table by substituting each x-value into the given equation to find the corresponding y-value.

WRITE/DRAW

a.

x	−2	−1	$-\frac{1}{2}$	0	$\frac{1}{2}$	1	2
y	−2	−4	−8	Undef.	8	4	2

2. Draw a set of axes and plot the points from the table. Join them with a smooth curve.

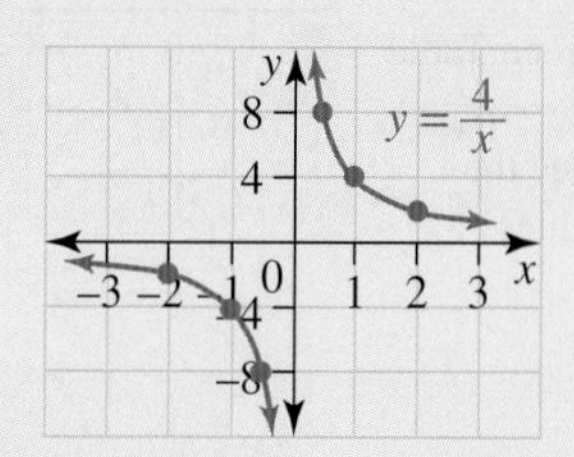

b. Consider any lines that the curve approaches but does not cross.

b. Vertical asymptote is $x = 0$.
Horizontal asymptote is $y = 0$.

WORKED EXAMPLE 19 Sketching a reflected hyperbola

Plot the graph of $y = \frac{-3}{x}$ for $-3 \le x \le 3$.

THINK

1. Draw a table of values and substitute each x-value into the given equation to find the corresponding y-value.

WRITE/DRAW

x	−3	−2	−1	$-\frac{1}{2}$	0	$\frac{1}{2}$	1	2	3
y	1	1.5	3	6	Undef.	−6	−3	−1.5	−1

2. Draw a set of axes and plot the points from the table. Join them with a smooth curve.

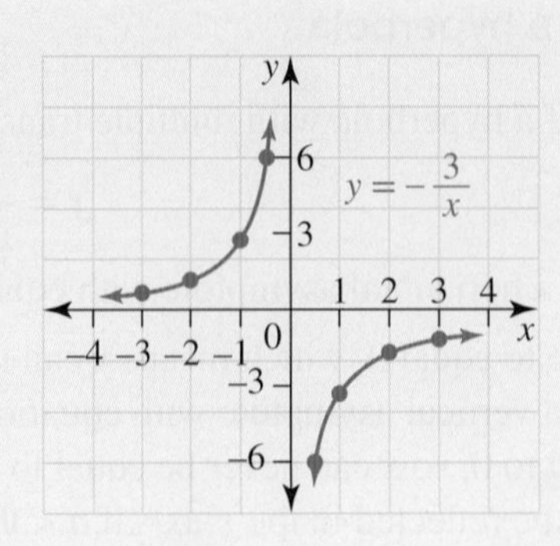

TI | THINK

1. In a new problem, open a Graphs page to sketch the graph, complete the entry line as:

$$f1(x) = \frac{-3}{x}$$

Then press ENTER.

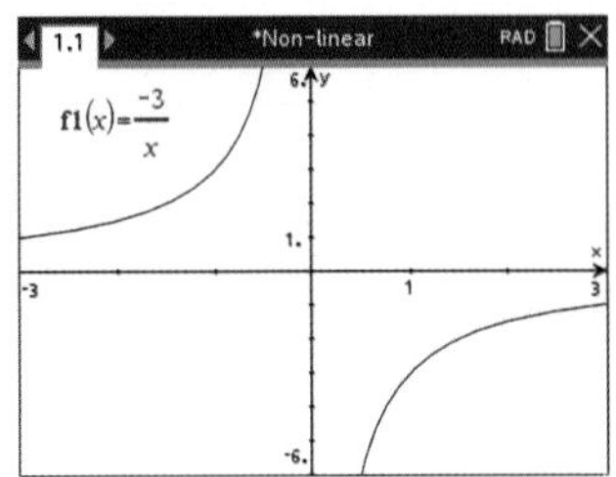

2. To see the table of values (for x and y values), press:
- Menu
- 7: Table
- 1: Split-screen Table (Ctrl+T)

Scroll up to see the negative x-values.

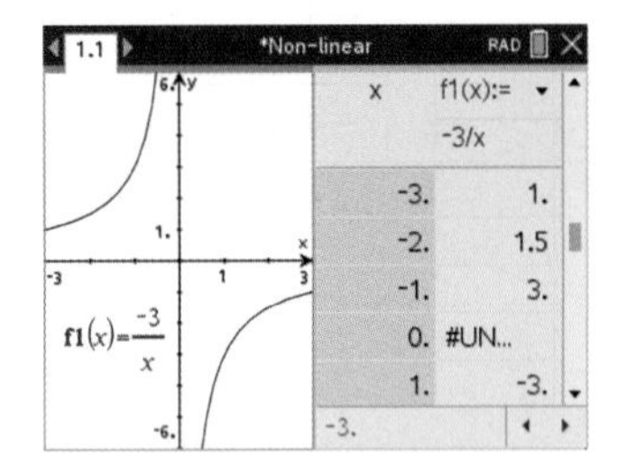

CASIO | THINK

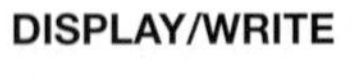

1. Open the Graph & Table screen and complete the function entry line as:

$$f1(x) = \frac{-3}{x}$$

Then tap the graphing icon and the graph will be displayed.

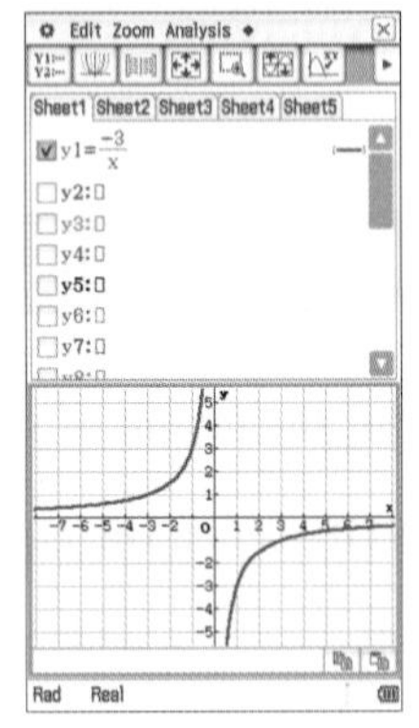

2. To see the table of values (for x and y values), tap on the table of values icon. The values displayed can be changed by editing the Table Input.

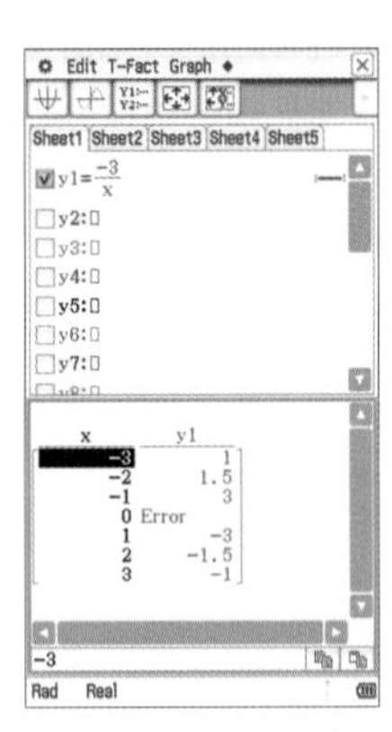

WORKED EXAMPLE 20 Determining the equation of the asymptotes of a hyperbola

Write the equation of the asymptotes of the following hyperbolas:

a. $y = \dfrac{3}{x+2}$ b. $y = -\dfrac{1}{x} + 4$ c. $y = \dfrac{5}{10-x} - 1$

THINK	WRITE
a. Consider what value of x would make the denominator equal to 0, as well as any value added to the fraction.	a. Vertical asymptote is $x = -2$ Horizontal asymptote is $y = 0$
b. Consider what value of x would make the denominator equal to 0, as well as any value added to the fraction.	b. Vertical asymptote is $x = 0$ Horizontal asymptote is $y = 4$
c. Consider what value of x would make the denominator equal to 0, as well as any value added to the fraction.	c. Vertical asymptote is $x = 10$ Horizontal asymptote is $y = -1$

WORKED EXAMPLE 21 Sketching a transformed hyperbola

Plot the graph of $y = -\dfrac{2}{x+1} + 3$.

THINK	WRITE
1. Consider what value of x would make the denominator equal to 0, as well as any value added to the fraction.	Vertical asymptote is $x = -1$ Horizontal asymptote is $y = 3$
2. Calculate the y-intercept (when $x = 0$).	$y = -\dfrac{2}{0+1} + 3 = 1$ y-intercept is (0, 1)
3. Calculate the x-intercept (when $y = 0$).	$0 = -\dfrac{2}{x+1} + 3$ $-3(x+1) = -2$ $x + 1 = \dfrac{-2}{-3}$ $x = \dfrac{2}{3} - 1 = -\dfrac{1}{3}$ The x-intercept is $\left(-\dfrac{1}{3}, 0\right)$ $x = -1$
4. Sketch the graph with coordinates of both intercepts and both asymptotes labelled.	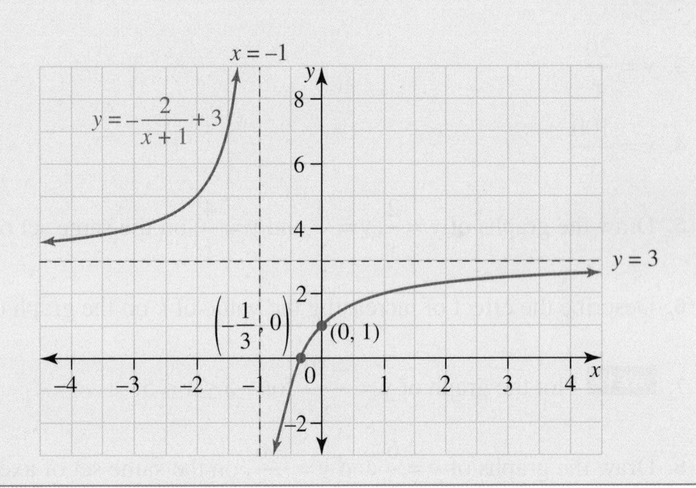

DISCUSSION

How could you summarise the effect of the transformations dealt with in this subtopic on the shape of the basic hyperbola $y = \dfrac{1}{x}$?

on Resources

eWorkbook Topic 9 Workbook (worksheets, code puzzle and a project) (ewbk-2035)

Interactivities Individual pathway interactivity: The hyperbola (int-4610)
Hyperbolas (int-6155)

Exercise 9.8 Sketching the hyperbola

learnon

Individual pathways

PRACTISE	CONSOLIDATE	MASTER
1, 4, 7, 10, 14, 17, 20	2, 5, 8, 11, 15, 18, 21	3, 6, 9, 12, 13, 16, 19, 22, 23

To answer questions online and to receive **immediate corrective feedback** and **fully worked solutions** for all questions, go to your learnON title at www.jacplus.com.au.

Fluency

1. **WE17** Complete the table of values below and use it to plot the graph of $y = \frac{10}{x}$.

x	−5	−4	−3	−2	−1	0	1	2	3	4	5
y											

WE18 For questions 2 to 4, answer the following.

a. Plot the graph of each hyperbola.

b. Write down the equation of each asymptote.

2. $y = \frac{5}{x}$

3. $y = \frac{20}{x}$

4. $y = \frac{100}{x}$

5. Draw the graphs of $y = \frac{2}{x}$, $y = \frac{3}{x}$ and $y = \frac{4}{x}$, on the same set of axes.

6. Describe the effect of increasing the value of k on the graph of $y = \frac{k}{x}$.

7. **WE19** Plot the graph of $y = \frac{-10}{x}$ for $-5 \le x \le 5$.

8. Draw the graphs of $y = \frac{6}{x}$ and $y = \frac{-6}{x}$, on the same set of axes.

9. Describe the effect of the negative in $y = \frac{-k}{x}$.

10. Complete the table of values below and use the points to plot $y = \frac{1}{x-1}$. State the equation of the vertical asymptote.

x	−3	−2	−1	0	1	2	3	4
y								

Understanding

11. Plot the graph of each hyperbola and label the vertical asymptote.

a. $y = \frac{1}{x-2}$

b. $y = \frac{1}{x-3}$

12. Plot the graph of $y=\frac{1}{x+1}$ and label the vertical asymptote.

13. Describe the effect of a in $y=\frac{1}{x-a}$.

For questions 14 to 16:

i. WE20 write the equations of the asymptotes of the following hyperbolas

ii. sketch the graphs of the following hyperbolas.

14. a. $y=\frac{-4}{x+1}$

b. $y=\frac{2}{x-1}$

15. a. $y=\frac{5}{x+2}$

b. $y=\frac{3}{x+2}-2$

16. a. $y=-\frac{4}{x+2}+1$

b. $y=\frac{7}{3-x}+5$

Reasoning

17. Give an example of the equation of a hyperbola that has a vertical asymptote of:

a. $x=3$

b. $x=-10$.

18. Give an example of a hyperbola that has the following key features.

a. Asymptotes of $x=2$ and $y=3$

b. Asymptotes of $x=-2$ and $y=4$ and a y-intercept of -3

19. The graph of $y=\frac{1}{x}$ is reflected in the x-axis, dilated by a factor of 2 parallel to the y-axis or from the x-axis and translated 3 units to the left and down 1 unit. Determine the equation of the resulting hyperbola and give the equations of any asymptotes.

Problem solving

20. a. Complete the following table in order to graph the hyperbola defined by $y=\frac{1}{x^2}$.

x	-2	-1	$-\frac{1}{2}$	$\frac{1}{2}$	1	2
y						

b. This hyperbola is also known as a truncus. Give the equations of any asymptotes.

c. Determine the equation of the truncus which results when $y=\frac{1}{x^2}$ is reflected in the x-axis.

d. Determine the equation of the truncus which results when $y=\frac{1}{x^2}$ is reflected in the y-axis.

21. The temperature of a cup of coffee as it cools is modelled by the equation $T=\frac{780}{t+10}+22$, where T represents the temperature in °C and t is the time in minutes since the coffee was first made.

a. State the initial temperature of the cup of coffee.

b. Calculate the temperature, to 1 decimal place, of the coffee after it has been left to cool for an hour.

c. A coffee will be too hot to drink unless its temperature has dropped below 50 °C. Determine how long someone would have to wait, to the nearest minute, before drinking the coffee.

d. Explain whether the coffee will ever cool to 0 °C. Justify your answer.

22. Consider the equation $y = \dfrac{6}{x^2 - 4} + 2$.
   a. Calculate the x- and y-intercepts of the graph of the equation.
   b. Use your knowledge of quadratics and hyperbolas to state the equation of any asymptotes of the above equation.
   c. Using a table of values, or otherwise, sketch the graph of the equation.

23. Consider the truncus defined by $y = \dfrac{1}{x^2}$. This hyperbola is reflected in the x-axis, dilated by a factor of 3 parallel to the y-axis or from the x-axis and translated 1 unit to the left and up 2 units. Determine the equation of the resulting hyperbola and give the equations of any asymptotes.

9.9 Sketching the circle

LEARNING INTENTION

At the end of this subtopic you should be able to:
- identify the centre and radius of the graph of a circle from its equation and then sketch its graph
- use completing the square to turn an equation in expanded form into the form of an equation of a circle.

9.9.1 Circles

eles-5352

- A circle is the path traced out by a point at a constant distance (the radius) from a fixed point (the centre).
- Consider the circles shown. The first circle has its centre at the origin and radius r.
  Let $P(x, y)$ be a point on the circle.
  By Pythagoras: $x^2 + y^2 = r^2$.
  This relationship is true for all points, P, on the circle.
  The equation of a circle, with centre $(0, 0)$ and radius r is:

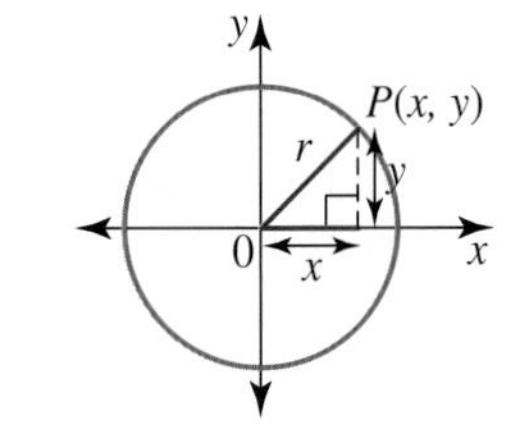

$$x^2 + y^2 = r^2$$

- If the circle is translated h units to the right, parallel to the x-axis, and k units upwards, parallel to the y-axis, then the centre of the circle will become (h, k). The radius will remain unchanged.

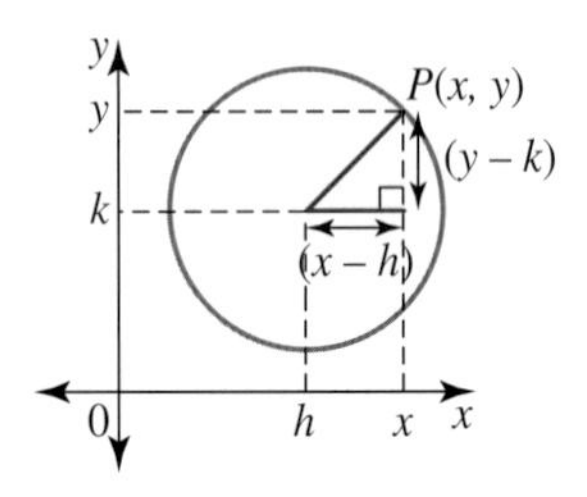

Equation of a circle

- The equation of a circle, with centre $(\mathbf{0}, \mathbf{0})$ and radius $\boldsymbol{r}$, is:

$$\boldsymbol{x^2 + y^2 = r^2}$$

- The equation of a circle, with centre $(\boldsymbol{h}, \boldsymbol{k})$ and radius $\boldsymbol{r}$, is:

$$\boldsymbol{(x - h)^2 + (y - k)^2 = r^2}$$

Note: We can produce an ellipse by dilating a circle from one or both of the axes.

WORKED EXAMPLE 22 Identifying the centre and radius of a circle from its equation

Sketch the graph of $4x^2 + 4y^2 = 25$, stating the centre and radius.

THINK	WRITE/DRAW
1. Express the equation in general form by dividing both sides by 4.	$x^2 + y^2 = r^2$ $4x^2 + 4y^2 = 25$ $x^2 + y^2 = \frac{25}{4}$
2. State the coordinates of the centre.	Centre $(0, 0)$
3. Calculate the length of the radius by taking the square root of both sides. (Ignore the negative results.)	$r^2 = \frac{25}{4}$ $r = \frac{5}{2}$ Radius $= 2.5$ units
4. Sketch the graph.	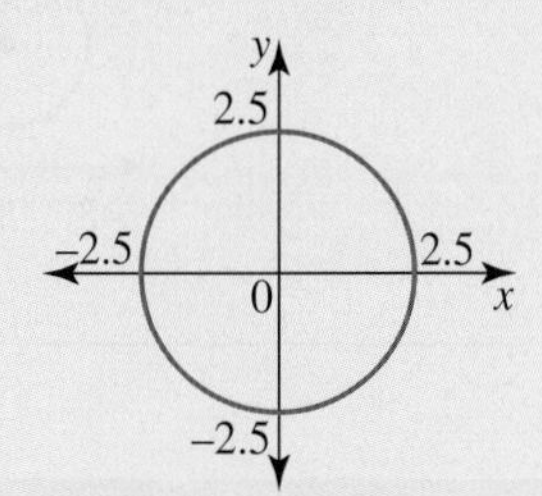

WORKED EXAMPLE 23 Sketching a circle from its equation

Sketch the graph of $(x - 2)^2 + (y + 3)^2 = 16$, clearly showing the centre and radius.

THINK	WRITE/DRAW
1. Express the equation in general form.	$(x - h)^2 + (y - k)^2 = r^2$ $(x - 2)^2 + (y + 3)^2 = 16$
2. State the coordinates of the centre.	Centre $(2, -3)$
3. State the length of the radius.	$r^2 = 16$ $r = 4$ Radius $= 4$ units
4. Sketch the graph.	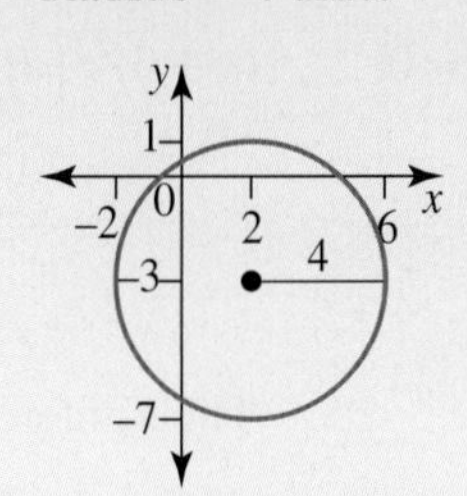

WORKED EXAMPLE 24 Completing the square to determine the equation of a circle

Sketch the graph of the circle $x^2 + 2x + y^2 - 6y + 6 = 0$.

THINK	WRITE/DRAW
1. Express the equation in general form by completing the square on the x terms and again on the y terms.	$(x-h)^2 + (y-k)^2 = r^2$ $x^2 + 2x + y^2 - 6y + 6 = 0$ $(x^2 + 2x + 1) - 1 + (y^2 - 6y + 9) - 9 + 6 = 0$ $(x+1)^2 + (y-3)^2 - 4 = 0$ $(x+1)^2 + (y-3)^2 = 4$
2. State the coordinates of the centre.	Centre $(-1, 3)$
3. State the length of the radius.	$r^2 = 4$ $r = 2$ Radius = 2 units
4. Sketch the graph.	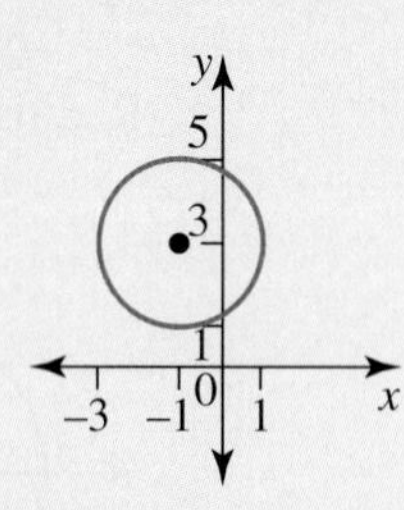

WORKED EXAMPLE 25 Sketching a circle including the intercepts

Sketch the graph of the circle $(x+3)^2 + (y-2)^2 = 25$. Make sure to show all axial intercepts.

THINK	WRITE
1. State the coordinate of the centre.	Centre $(-3, 2)$
2. State the length of the radius.	$r^2 = 25$ $r = 5$ Radius = 5 units
3. To determine the y-intercepts let $x = 0$.	$(0+3)^2 + (y-2)^2 = 25$ $(y-2)^2 = 25 - 9$ $(y-2)^2 = 16$ $y = \pm 4$ $y = 4 + 2$ or $4 + 2$ The y-intercepts are $(0, 6)$ and $(0, -2)$
4. To determine the x-intercepts let $y = 0$.	$(x+3)^2 + (0-2)^2 = 25$ $(x+3)^2 = 25 - 4$ $x + 3 = \pm\sqrt{21}$ $x = -3 \pm \sqrt{21}$ The x-intercepts are $\left(-3 + \sqrt{21}, 0\right)$ and $\left(-3 - \sqrt{21}, 0\right)$.

5. Sketch the graph.

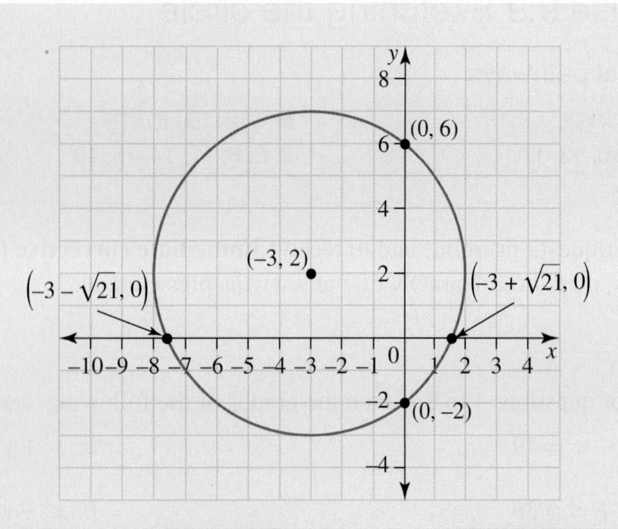

TI | THINK	DISPLAY/WRITE	CASIO | THINK	DISPLAY/WRITE
On a Graphing page, press: • MENU • 3: Graph Entry/Edit • 2: Relation Then type: $(x+3)^2+(y-2)^2=25$ Press ENTER and the graph will be displayed.	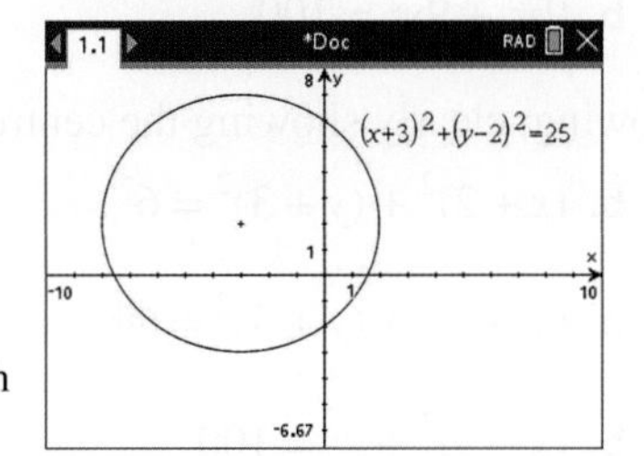	On a Conics screen, type: $(x+3)^2+(y-2)^2=25$ Press the graphing icon and the graph will be displayed.	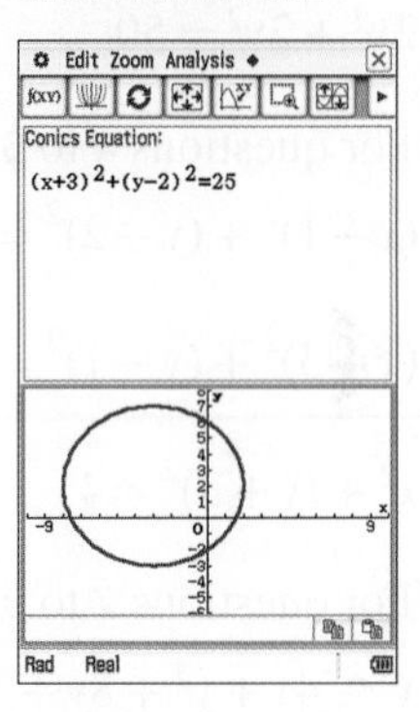

DISCUSSION

How could you write equations representing a set of concentric circles (circles with the same centre, but different radii)?

on Resources

eWorkbook Topic 9 Workbook (worksheets, code puzzle and a project) (ewbk-2035)

Interactivities Individual pathway interactivity: The circle (int-4611)

Graphs of circles (int-6156)

Exercise 9.9 Sketching the circle

learnON

Individual pathways

■ PRACTISE	■ CONSOLIDATE	■ MASTER
1, 4, 7, 10, 13, 17	2, 5, 8, 11, 14, 15, 18	3, 6, 9, 12, 16, 19, 20

To answer questions online and to receive **immediate corrective feedback** and **fully worked solutions** for all questions, go to your learnON title at www.jacplus.com.au.

Fluency

WE22 For questions 1 to 3, sketch the graphs of the following, stating the centre and radius of each.

1. a. $x^2 + y^2 = 49$ b. $x^2 + y^2 = 4^2$
2. a. $x^2 + y^2 = 36$ b. $x^2 + y^2 = 81$
3. a. $2x^2 + 2y^2 = 50$ b. $9x^2 + 9y^2 = 100$

WE23 For questions 4 to 6, sketch the graphs of the following, clearly showing the centre and the radius.

4. a. $(x - 1)^2 + (y - 2)^2 = 5^2$ b. $(x + 2)^2 + (y + 3)^2 = 6^2$
5. a. $(x + 3)^2 + (y - 1)^2 = 49$ b. $(x - 4)^2 + (y + 5)^2 = 64$
6. a. $x^2 + (y + 3)^2 = 4$ b. $(x - 5)^2 + y^2 = 100$

WE24 For questions 7 to 9, sketch the graphs of the following circles.

7. a. $x^2 + 4x + y^2 + 8y + 16 = 0$ b. $x^2 - 10x + y^2 - 2y + 10 = 0$
8. a. $x^2 - 14x + y^2 + 6y + 9 = 0$ b. $x^2 + 8x + y^2 - 12y - 12 = 0$
9. a. $x^2 + y^2 - 18y - 19 = 0$ b. $2x^2 - 4x + 2y^2 + 8y - 8 = 0$

Understanding

10. **MC** The graph of $(x - 2)^2 + (y + 5)^2 = 4$ is:

A.
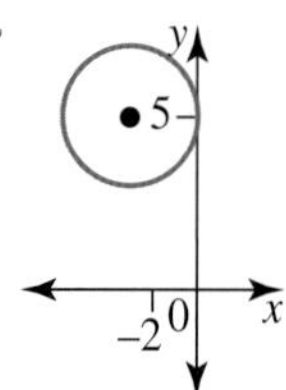

B.
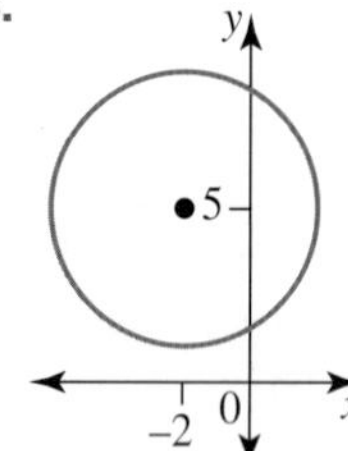

C.
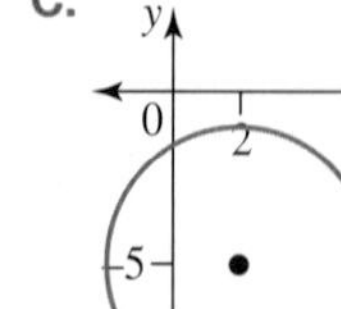

D.
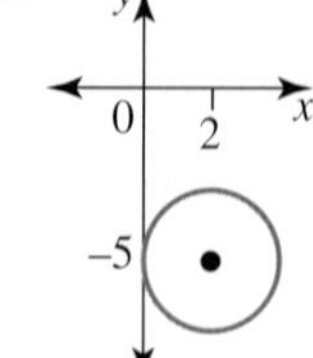

E. None of these

11. **MC** The centre and radius of the circle $(x+1)^2+(y-3)^2=4$ is:

A. $(1,-3), 4$
B. $(-1,3), 2$
C. $(3,-1), 4$
D. $(1,-3), 2$
E. $(-1,3), 4$

12. **MC** The centre and radius of the circle with equation $x^2+y^2+8x-10y=0$ is:

A. $(4,5), \sqrt{41}$
B. $(-4,5), 9$
C. $(4,-5), 3$
D. $(-4,5), \sqrt{41}$
E. $(-4,-5), \sqrt{41}$

Reasoning

13. Determine the equation representing the outer edge of the galaxy as shown in the photo below, using the astronomical units provided.

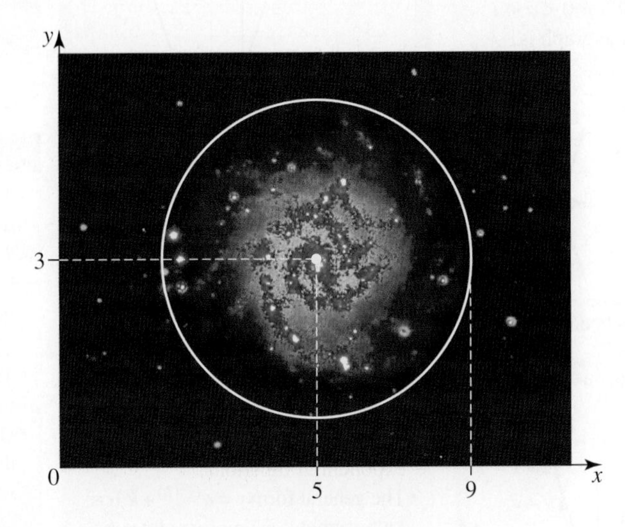

14. Circular ripples are formed when a water drop hits the surface of a pond.
If one ripple is represented by the equation $x^2+y^2=4$ and then 3 seconds later by $x^2+y^2=190$, where the length of measurements are in centimetres:
   a. identify the radius (in cm) of the ripple in each case
   b. calculate how fast the ripple is moving outwards.
   (State your answers to 1 decimal place.)

15. Two circles with equations $x^2+y^2=4$ and $(x-1)^2+y^2=9$ intersect. Determine the point(s) of intersection. Show your working.

16. a. Graph the line $y=x$, the parabola $y=x^2$ and the circle $x^2+y^2=1$ on the one set of axes.
   b. Evaluate algebraically the points of intersection of:
      i. the line and the circle
      ii. the line and the parabola
      iii. the parabola and the circle.

Problem solving

17. Sketch the graph of $(x+6)^2+(y-3)^2=100$ showing all axial intercepts.

18. Determine the points of intersection between the quadratic equation $y=x^2-5$ and the circle given by $x^2+y^2=25$.

19. Determine the point(s) of intersection of the circles $x^2+y^2-2x-2y-2=0$ and $x^2+y^2-8x-2y+16=0$ both algebraically and graphically.

20. The general equation of a circle is given by $x^2+y^2+ax+by+c=0$. Determine the equation of the circle which passes through the points $(4,5)$, $(2,3)$ and $(0,5)$. State the centre of the circle and its radius.

9.10 Review

9.10.1 Topic summary

NON-LINEAR RELATIONSHIPS

Transformations of the parabola

- The parabola with the equation $y = x^2$ has a turning point at (0,0).
- A **dilation** from the x-axis stretches the parabola.
- A **dilation** of factor a from the x-axis produces the equation $y = ax^2$ and:
  - will produce a narrow graph for $a > 0$
  - will produce a wider graph for $0 < a < 1$
  - For $a < 0$ (a is negative), the graph is **reflected** in the x-axis.

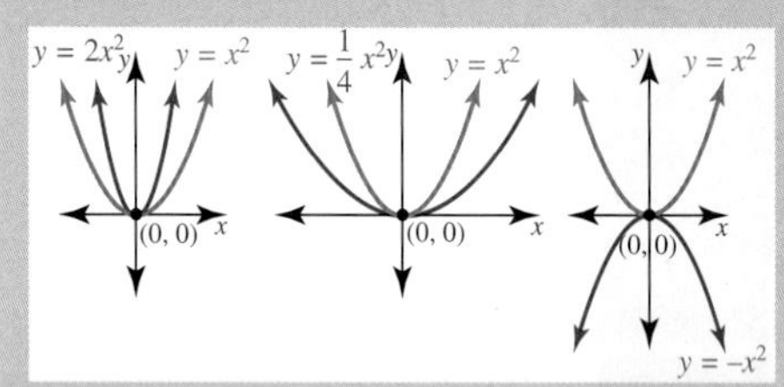

- The graph of $y = (x - h)^2$ has been translated left/right h units.
- The graph of $y = x^2 + k$ has been translated up/down k units.

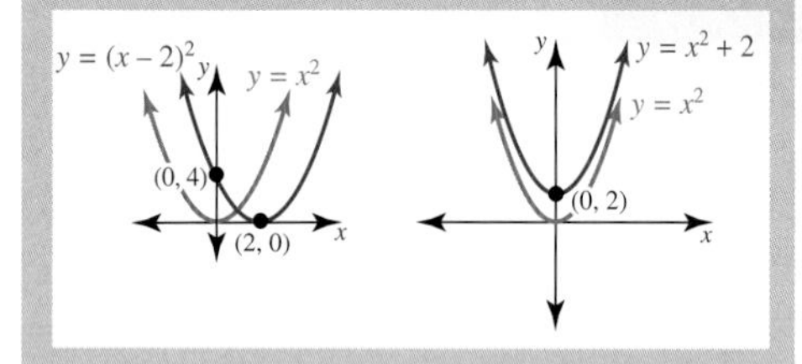

Inverse proportion and hyperbolas

- If y and x are inversely proportional then $y \propto \frac{1}{x}$ or $y = \frac{k}{x}$, where k is the constant of proportionality.
- The graph of an equation in the form $y = \frac{k}{x}$ is called a hyperbola.
- The general form of a hyperbola is $y = \frac{a}{x - h} + k$ and has:
  - vertical asymptote at $x = h$
  - horizontal asymptote at $y = k$
  - y-intercept at $\left(0, \frac{-a}{h + k}\right)$
  - x-intercept at $\left(\frac{-a}{k + h}, 0\right)$.
- If $a < 0$ the graph is reflected in the x-axis.

The parabola

- The graph of a quadratic equation is called a parabola.

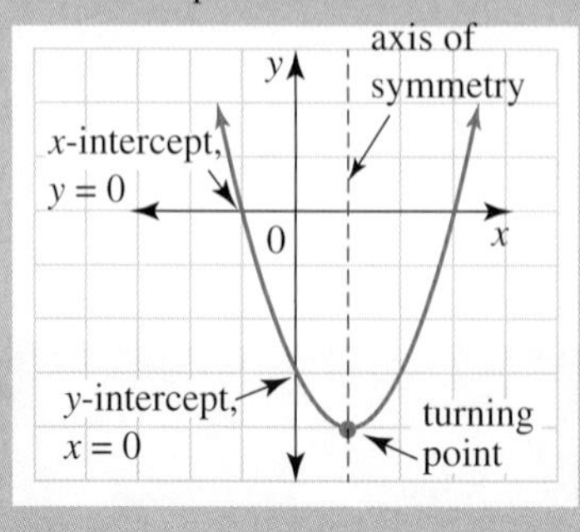

Exponential graphs

- Equations $y = a^x$ are called exponential functions.
- The general form $y = a^{(x-h)} + k$ has:
  - a horizontal asymptote at $y = k$
  - a y-intercept at $(0, a^{-h} + k)$
- A negative in front of a results in a reflection in the x-axis.
- A negative in front of x results in a reflection in the y-axis.

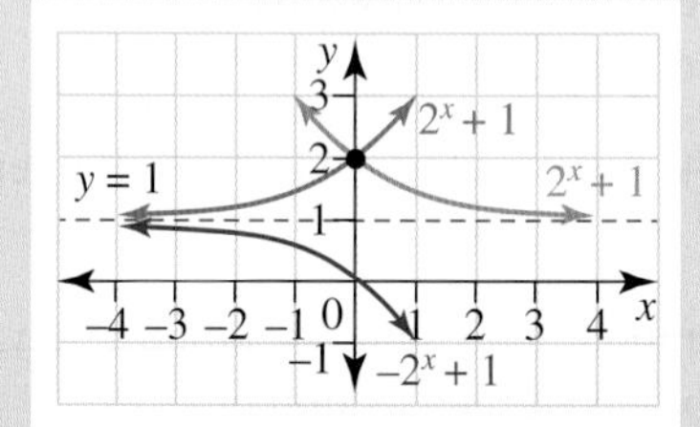

Circles

- $x^2 + y^2 = r^2$ is a circle with centre (0, 0) and radius r.
- $(x - h)^2 + (y - k)^2 = r^2$ is a circle with centre (h, k) and radius r.
- Use completing the square to determine the equation of a circle:

$x^2 + 6x + y^2 + 8x = 0$
$x^2 + 6x + 9 + y^2 + 8x + 16 = 9 + 16$
$(x + 3)^2 + (y + 4)^2 = 25$

Turning point form

- Completing the square allows us to write a quadratic in turning point form.
- Turning point form:
  $$y = a(x - h)^2 + k$$
  - Turning point at (h, k)
  - The y-intercept is determined by setting $x = 0$ and solving for y.
  - x-intercept/s is determined by setting $y = 0$ and solving for x.

Expanded form

- The expanded form of a quadratic is $y = ax^2 + bx + c$.
- There are two methods for sketching: factorising or using the quadratic formula.

e.g. Consider $y = x^2 - 2x - 3$.

- To sketch, first factorise: $y = (x - 3)(x + 1)$
- Use the null-factor law to determine the x-intercepts: when $y = 0$, $0 = (x - 3)(x + 1)$ so $x = 3$ or -1.
- Axis of symmetry: $x = \frac{3 + (-1)}{2} = 1$
- y-coordinate of the turning point: $x = 1$, $y = (1 - 3)(1 + 1) = -4$. Turning point is (1, –4). The y-intercept when $x = 0$ is (0, –3).

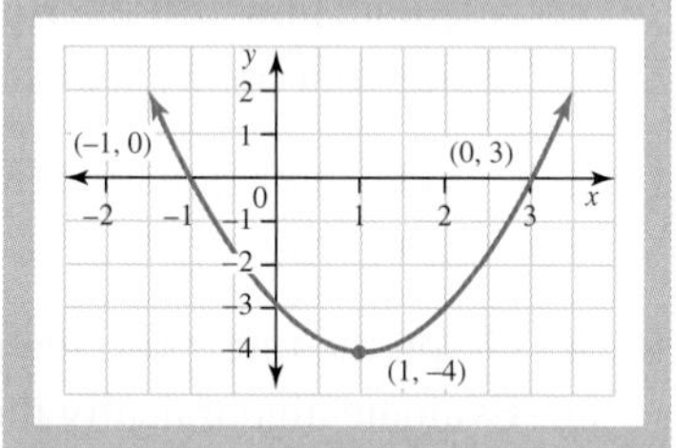

Quadratic formula

- For $y = ax^2 + bx + c$
  - axis of symmetry: $x = -\frac{b}{2a}$
  - turning point: $\left(-\frac{b}{2a}, c - \frac{b^2}{4a}\right)$
- x-intercepts can be calculated as: $x = \frac{-b \pm \sqrt{b^2 - 4ac}}{2a}$
- Discriminant: $\Delta = b^2 - 4ac$ is used to determine the number of x-intercepts.
  - $\Delta > 0$, two x-intercepts
  - $\Delta = 0$, one x-intercept
  - $\Delta < 0$, no x-intercept

9.10.2 Success criteria

Tick the column to indicate that you have completed the subtopic and how well you have understood it using the traffic light system.

(**Green:** I understand; **Yellow:** I can do it with help; **Red:** I do not understand)

Subtopic	Success criteria	Green	Yellow	Red
9.2	I can create a table of values and use this to sketch the graph of a parabola.			
	I can identify the axis of symmetry, turning point and y-intercept of a parabola.			
9.3	I am familiar with the key features of the basic quadratic $y = x^2$.			
	I can determine whether a dilation has made the graph of quadratic narrower or wider			
	I can determine whether a translation has moved the graph of a quadratic function left/right or up/down			
	I can sketch a quadratic equation that has undergone any of the following transformations: dilations, reflections or translations.			
9.4	I can determine the axis of symmetry and turning point of a quadratic expressed in turning point form.			
	I can calculate the y-intercept and any x-intercepts of a quadratic in turning point form.			
9.5	I can determine the x-intercept/s, y-intercept and turning point of a quadratic equation and sketch its graph by first factorising the equation.			
	I can determine the x-intercept/s, y-intercept and turning point of a quadratic equation and sketch its graph using the quadratic formula.			
9.6	I can determine the equation of the asymptote of an exponential equation and sketch a graph of the equation.			
	I can sketch the graph of an exponential equation after it has undergone any number of transformations that may include dilations, a reflection in the x- or y-axis, or translations.			
9.7	I can calculate the constant of proportionality, k, between two variables that are inversely proportional to each other.			
	I can determine the rule between two inversely proportional variables and plot the graph of their relationship.			
9.8	I can determine the equation of the vertical and horizontal asymptote of a hyperbola.			
	I can sketch the graph of a hyperbola, and understand the effect dilations and reflections have on the shape of the graph.			
9.9	I can identify the centre and radius of the graph of a circle from its equation and then sketch its graph.			
	I can use completing the square to turn an equation in expanded form into the form of an equation of a circle.			

9.10.3 Project

Parametric equations

You are familiar with the quadratic equation $y = x^2$ and its resulting graph. Let us consider an application of this equation by forming a relationship between x and y through a third variable, say, t.

$$x = t \text{ and } y = t^2$$

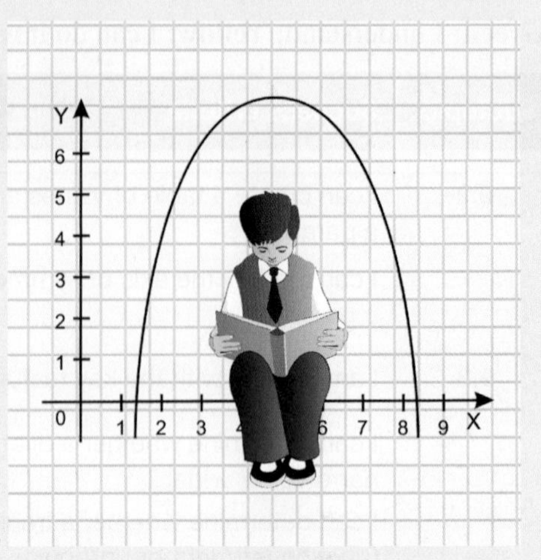

It is obvious that these two equations are equivalent to the equation $y = x^2$. This third variable t is known as a parameter, and the two equations are now called parametric equations. We cannot automatically assume that the resulting graph of these two parametric equations is the same as that of $y = x^2$ for all real values of x. It is dependent on the range of values of t.

Consider the parametric equations $x = t$ and $y = t^2$ for values of the parameter $t \geq 0$ for questions 1 to 3.

1. Complete the following table by calculating x- and y-values from the corresponding t-value.

t	x	y
0		
1		
2		
3		
4		
5		

2. Graph the x-values and corresponding y-values on this Cartesian plane. Join the points with a smooth curve and place an arrow on the curve to indicate the direction of increasing t-values.

Parametric equations

3. Is there any difference between this graph and that of $y = x^2$? Explain your answer.
4. Consider now the parametric equations $x = 1 - t$ and $y = (1 - t)^2$. These clearly are also equivalent to the equation $y = x^2$. Complete the table and draw the graph of these two equations for values of the parameter $t \geq 0$. Draw an arrow on the curve in the direction of increasing t-values.

t	x	y
0		
1		
2		
3		
4		
5		

Parametric equations

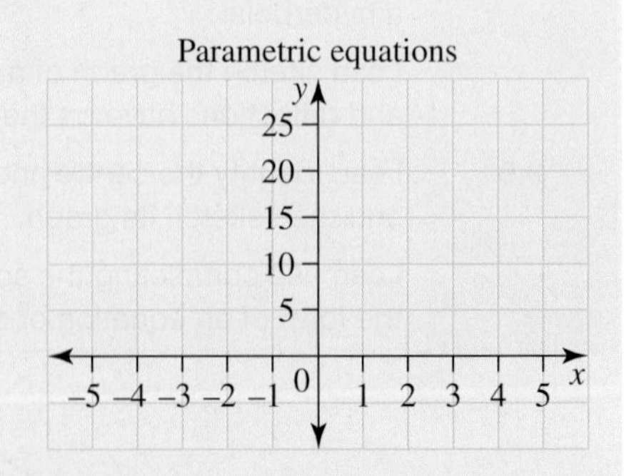

Describe the shape of your resulting graph. What values of the parameter t would produce the same curve as that obtained in question 2?

5. The graph of $y = -x^2$ is a reflection of $y = x^2$ in the x-axis. Construct a table and draw the graph of the parametric equations $x = t$ and $y = -t^2$ for parameter values $t \geq 0$. Remember to place an arrow on the curve in the direction of increasing t-values.
6. Without constructing a table, predict the shape of the graph of the parametric equations $x = 1 - t$ and $y = -(1 - t)^2$ for parameter values $t \geq 0$. Draw a sketch of the shape.

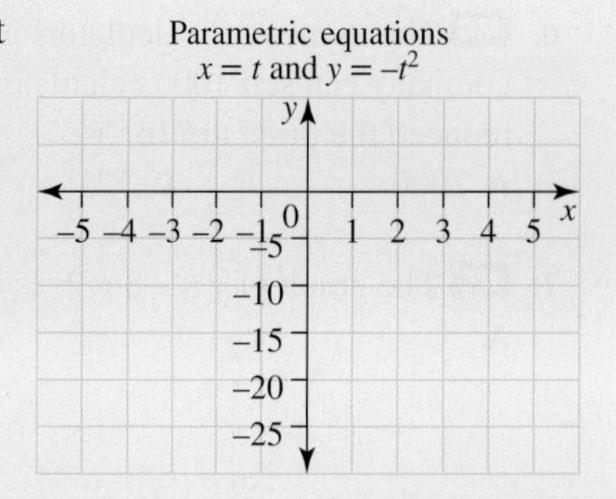

Resources

 **eWorkbook** Topic 9 Workbook (worksheets, code puzzle and a project) (ewbk-2035)

 **Interactivities** Crossword (int-2851)

Sudoku puzzle (int-3596)

Exercise 9.10 Review questions

To answer questions online and to receive **immediate corrective feedback** and **fully worked solutions** for all questions, go to your learnON title at www.jacplus.com.au.

Fluency

1. MC The turning point for the graph $y = 3x^2 - 4x + 9$ is:
   - A. $\left(\frac{1}{3}, 1\frac{2}{3}\right)$
   - B. $\left(\frac{1}{3}, \frac{2}{3}\right)$
   - C. $\left(\frac{1}{6}, 1\frac{1}{6}\right)$
   - D. $\left(\frac{2}{3}, 7\frac{2}{3}\right)$
   - E. $\left(\frac{2}{3}, 6\frac{2}{3}\right)$
2. MC Select which graph of the following equations has the x-intercepts closest together.
   - A. $y = x^2 + 3x + 2$
   - B. $y = x^2 + x - 2$
   - C. $y = 2x^2 + x - 15$
   - D. $y = 4x^2 + 27x - 7$
   - E. $y = x^2 - 2x - 8$
3. MC Select which graph of the equations below has the largest y-intercept.
   - A. $y = 3(x - 2)^2 + 9$
   - B. $y = 5(x - 1)^2 + 8$
   - C. $y = 2(x - 1)^2 + 19$
   - D. $y = 2(x - 5)^2 + 4$
   - E. $y = 12(x - 1)^2 + 10$
4. MC The translation required to change $y = x^2$ into $y = (x - 3)^2 + \frac{1}{4}$ is:
   - A. right 3, up $\frac{1}{4}$
   - B. right 3, down $\frac{1}{4}$
   - C. left 3, down $\frac{1}{4}$
   - D. left 3, up $\frac{1}{4}$
   - E. right $\frac{1}{4}$, up 3
5. MC If y is inversely proportional to x, then select which of the following statements is true.
   - A. $x + y$ is a constant value.
   - B. $y \div x$ is a constant value.
   - C. $y \times x$ is a constant value.
   - D. $y - x$ is a constant value.
   - E. $x \div y$ is a constant value.

6.  The number of calculators a company sells is inversely proportional to the selling price. If a company can sell 1000 calculators when the price is \$22, determine how many they could sell if they reduced the price to \$16.

A. 2000 B. 727 C. 6000 D. 1375 E. 137.5

7.  The graph of $y = -3 \times 2^x$ is best represented by:

A.
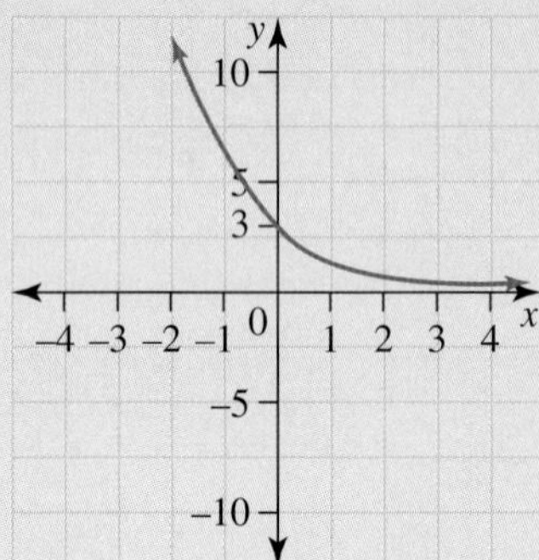

B.
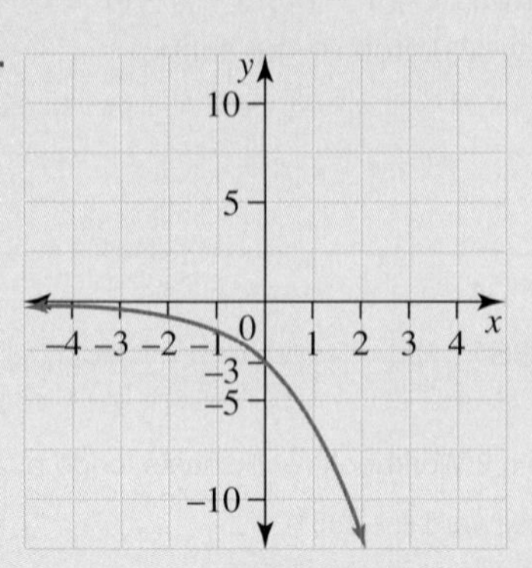

C.
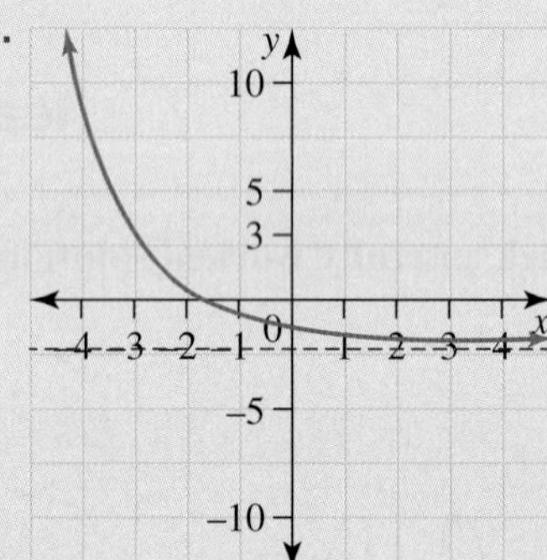

D.
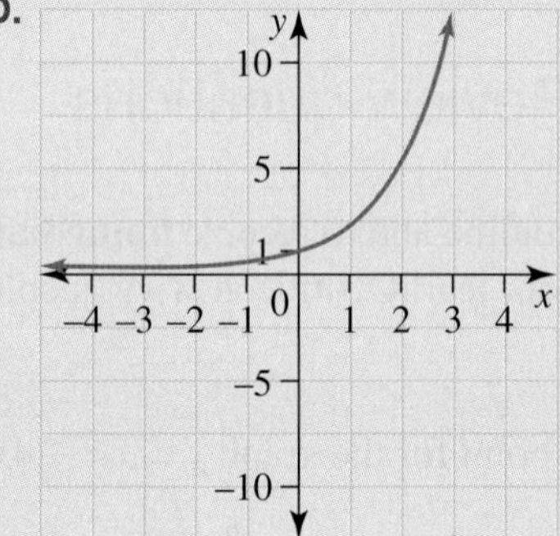

E.
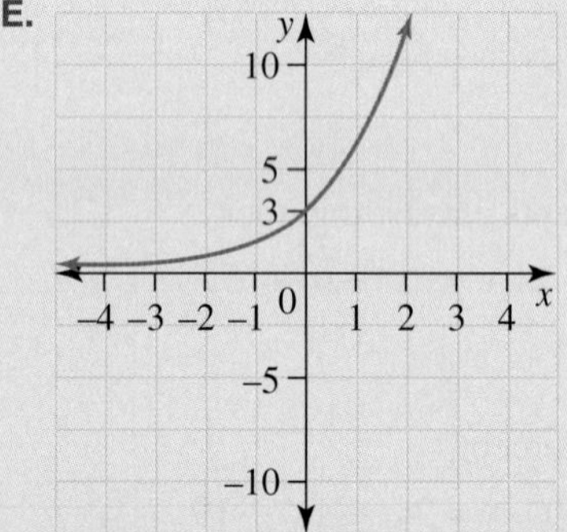

8. Use the completing the square method to determine the turning point for each of the following graphs.

a. $y = x^2 - 8x + 1$
b. $y = x^2 + 4x - 5$

9. For the graph of the equation $y = x^2 + 8x + 7$, produce a table of values for the x-values between -9 and 1, and then plot the graph. Show the y-intercept and turning point. From your graph, state the x-intercepts.

10. For each of the following, determine the coordinates of the turning point and the x- and y-intercepts and sketch the graph.

a. $y = (x - 3)^2 + 1$
b. $y = 2(x + 1)^2 - 5$

11. For the equation $y = -x^2 - 2x + 15$, sketch the graph and determine the x- and y-intercepts and the coordinates of the turning point.

12. Draw the graph of $y = 10 \times 3^x$ for $-2 \leq x \leq 4$.

13. Draw the graph of $y = 10^{-x}$ for $-2 \leq x \leq 2$.

14. For the exponential function $y = 5^x$:
    a. complete the table of values

x	y
−3	
−2	
−1	
0	
1	
2	
3	

    b. plot the graph.

15. a. On the same axes, draw the graphs of $y = (1.2)^x$ and $y = (1.5)^x$.
    b. Use your answer to part **a** to explain the effect of changing the value of a in the equation of $y = a^x$.

16. a. On one set of axes, draw the graphs of $y = 2 \times 3^x$, $y = 5 \times 3^x$ and $y = \frac{1}{2} \times 3^x$
    b. Use your answer to part **a** to explain the effect of changing the value of k in the equation of $y = ka^x$.

17. a. On the same set of axes, sketch the graphs of $y = (2.5)^x$ and $y = (2.5)^{-x}$.
    b. Use your answer to part **a** to explain the effect of a negative index on the equation $y = a^x$.

18. Sketch each of the following.
    a. $y = \frac{4}{x}$
    b. $y = -\frac{2}{x}$

19. Sketch $y = \frac{-3}{x - 2}$.

20. Give an example of an equation of a hyperbola that has a vertical asymptote at $x = -3$.

21. Sketch each of these circles. Clearly show the centre and the radius.
    a. $x^2 + y^2 = 16$
    b. $(x - 5)^2 + (y + 3)^2 = 64$

22. Sketch the following circles. Remember to first complete the square.
    a. $x^2 + 4x + y^2 - 2y = 4$
    b. $x^2 + 8x + y^2 + 8y = 32$

23. Determine the equation of this circle.

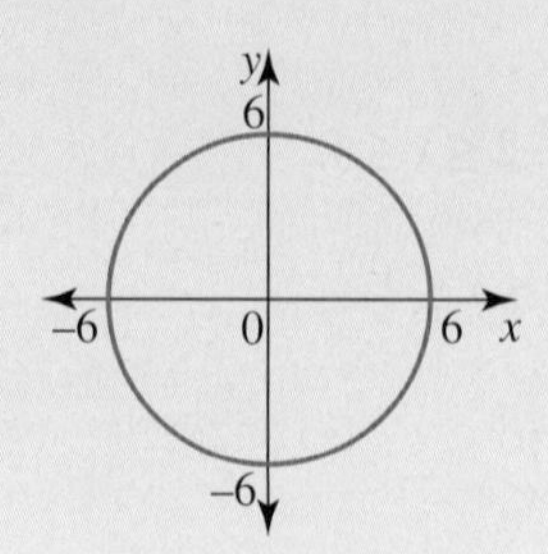

Problem solving

24. The height, h, in metres of a golf ball t seconds after it is hit is given by the formula $h = 4t - t^2$.
    a. Sketch the graph of the path of the ball.
    b. Calculate the maximum height the golf ball reaches.
    c. Determine how long it takes for the ball to reach the maximum height.
    d. Determine how long is it before the ball lands on the ground after it has been hit.

25. A soccer ball is kicked upwards in the air. The height, h, in metres, t seconds after the kick is modelled by the quadratic equation $h = -5t^2 + 20t$.
    a. Sketch the graph of this relationship.
    b. Determine how many seconds the ball is in the air for.
    c. Determine how many seconds the ball is above a height of 15 m. That is, solve the quadratic inequation $-5t^2 + 20t > 15$.
    d. Calculate how many seconds the ball is above a height of 20 m.

26. The height of the water level in a cave is determined by the tides. At any time, t, in hours after 9 am, the height, $h(t)$, in metres, can be modelled by the function $h(t) = t^2 - 12t + 32, 0 \le t \le 12$.
    a. Determine the values of t for which the model is valid. Write your answer in interval notation.
    b. State the initial height of the water.
    c. Bertha has dropped her keys onto a ledge which is 7 metres from the bottom of the cave. By using a graphics calculator, determine the times in which she would be able to climb down to retrieve her keys. Write your answers correct to the nearest minute.

27. When a drop of water hits the flat surface of a pool, circular ripples are made. One ripple is represented by the equation $x^2 + y^2 = 9$ and 5 seconds later, the ripple is represented by the equation $x^2 + y^2 = 225$, where the lengths of the radii are in cm.
    a. State the radius of each of the ripples.
    b. Sketch these graphs.
    c. Evaluate how fast the ripple are moving outwards.
    d. If the ripple continues to move at the same rate, determine when it will hit the edge of the pool which is 2 m from its centre.

28. Seventy grams of ammonium sulfate crystals are dissolved in 0.5 L of water.
    a. Determine the concentration of the solution in g/mL.
    b. Another 500 mL of water is added. Determine the concentration of the solution now.

29. A grassed area is planted in a courtyard that has a width of 5 metres and length of 7 metres. The shape of the grassed area is described by the function $P = -x^2 + 5x$, where P is the distance, in metres, from the house and x is the distance, in metres from the side wall. The diagram represents this information on a Cartesian plane.

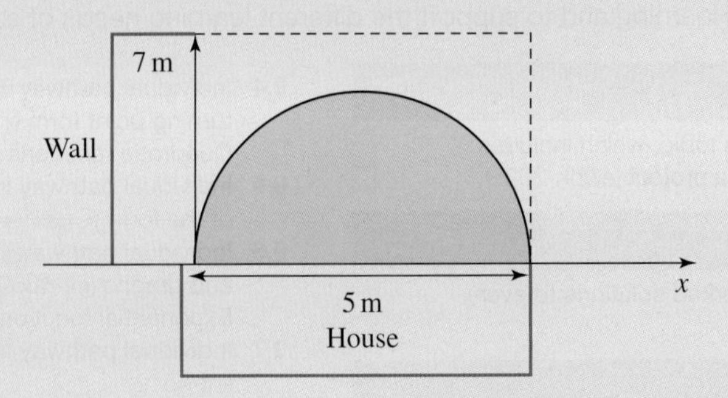

a. In terms of P, write down an inequality that describes the region where the grass has been planted.
b. Determine the maximum distance the grass area is planted from the house.
c. The owners of the house have decided that they would prefer all of the grass to be within a maximum distance of 3.5 metres from the house. The shape of the lawn following this design can be described by the equation $N(x) = ax^2 + bx + c$
i. Using algebra, show that this new design can be described by the function $N(x) = -0.56x(x - 5)$.
ii. Describe the transformation that maps $P(x)$ to $N(x)$.
d. If the owners decide on the first design, $P(x)$, the percentage of area within the courtyard without grass is 40.5%. By using any method, determine the approximate percentage of area of courtyard without lawn with the new design, $N(x)$.

30. A stone arch bridge has a span of 50 metres. The shape of the curve AB can be modelled using a quadratic equation.

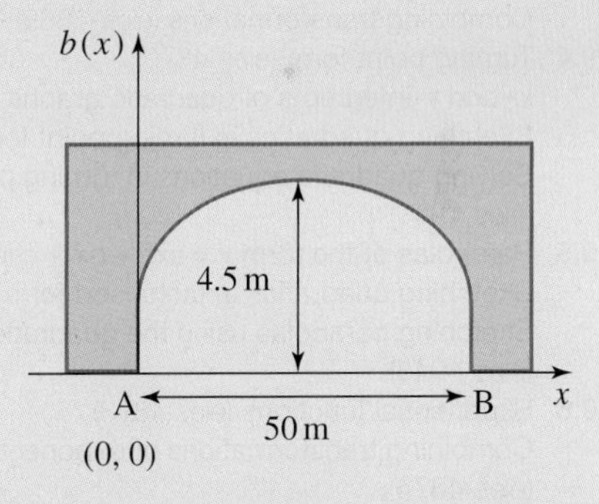

a. Taking A as the origin $(0, 0)$ and given that the maximum height of the arch above the water level is 4.5 metres, show using algebra, that the shape of the arch can be modelled using the equation $b(x) = -0.0072x^2 + 0.36x$, where $b(x)$ is the vertical height of the bridge, in metres, and x is the horizontal distance, in metres.
b. A floating platform p metres high is towed under the bridge. Given that the platform needs to have a clearance of at least 30 centimetres on each side, explain why the maximum value of p is 10.7 centimetres.

on To test your understanding and knowledge of this topic, go to your learnON title at www.jacplus.com.au and complete the **post-test**.

Online Resources

Below is a full list of **rich resources** available online for this topic. These resources are designed to bring ideas to life, to promote deep and lasting learning and to support the different learning needs of each individual.

eWorkbook

Download the workbook for this topic, which includes worksheets, a code puzzle and a project (ewbk-2035) ☐

Solutions

Download a copy of the fully worked solutions to every question in this topic (sol-0743) ☐

Digital documents

9.2 SkillSHEET Substitution into quadratic equations (doc-5266) ☐
SkillSHEET Equation of a vertical line (doc-5267)
9.5 SkillSHEET Completing the square (doc-5268) ☐
SkillSHEET Solving quadratic equations using the quadratic formula (doc-5269) ☐
SkillSHEET Solving quadratic equations of the type $ax^2 + bx + c = 0$ where $a = 1$ (doc-5270) ☐
SkillSHEET Solving quadratic equations of the type $ax^2 + bx + c = 0$ where $a \neq 1$ (doc-5271) ☐

Video eLessons

9.2 Plotting parabolas (eles-4864) ☐
9.3 Sketching parabolas (eles-4865) ☐
Vertical translation (eles-4866) ☐
Horizontal translation (eles-4867) ☐
Reflection (eles-4868) ☐
Combining transformations (eles-4869) ☐
9.4 Turning point form (eles-4870) ☐
x- and y-intercepts of quadratic graphs (eles-4871) ☐
Sketching quadratics in turning point form (eles-1926) ☐
Solving quadratic equations in turning point form (eles-1941) ☐
9.5 Parabolas of the form $y = ax^2 + bx + c$ (eles-4872) ☐
Sketching quadratics in factorised form (eles-1927) ☐
Sketching parabolas using the quadratic formula (eles-1945) ☐
9.6 Exponential functions (eles-4873) ☐
Combining transformations of exponential graphs (eles-4874) ☐
9.7 Inverse proportion (eles-4875) ☐
9.8 Hyperbolas (eles-4876) ☐
9.9 Circles (eles-5352) ☐

Interactivities

9.2 Individual pathway interactivity: Plotting parabolas (int-4605) ☐
Plotting quadratic graphs (int-6150) ☐
Parabolas in the world around us (int-7539) ☐
9.3 Individual pathway interactivity: Sketching parabolas (int-4606) ☐
Horizontal translations of parabolas (int-6054) ☐
Vertical translations of parabolas (int-6055) ☐
Dilation of parabolas (int-6096) ☐
Reflection of parabolas (int-6151) ☐
9.4 Individual pathway interactivity: Sketching parabolas in turning point form (int-4607) ☐
Quadratic functions (int-2562) ☐
9.5 Individual pathway interactivity: Sketching parabolas of the form $y = ax^2 + bx + c$ (int-4608) ☐
9.6 Individual pathway interactivity: Exponential functions and graphs (int-4609) ☐
Exponential functions (int-5959) ☐
9.7 Individual pathway interactivity: Inverse proportion (int-4512) ☐
Inverse proportion (int-6058) ☐
9.8 Individual pathway interactivity: The hyperbola (int-4610) ☐
Hyperbolas (int-6155) ☐
9.9 Individual pathway interactivity: The circle (int-4611) ☐
Graphs of circles (int-6156) ☐
9.10 Crossword (int-2851) ☐
Sudoku puzzle (int-3596) ☐

Teacher resources

There are many resources available exclusively for teachers online.

To access these online resources, log on to **www.jacplus.com.au**.

Answers

Topic 9 Non-linear relationships

Exercise 9.1 Pre-test

1.

x	-3	-2	-1	0
y	9.5	5	1.5	-1

2. $x = -1$
3. B
4. A
5. D
6. $(-4, -1)$
7. $b = 20$
8. D
9. E
10. $(-0.79, 0.62)$ and $(0.79, 0.62)$
11. B
12. C
13. a. 2 b. 1
14. D
15. B

Exercise 9.2 Plotting parabolas from a table of values

1. a.

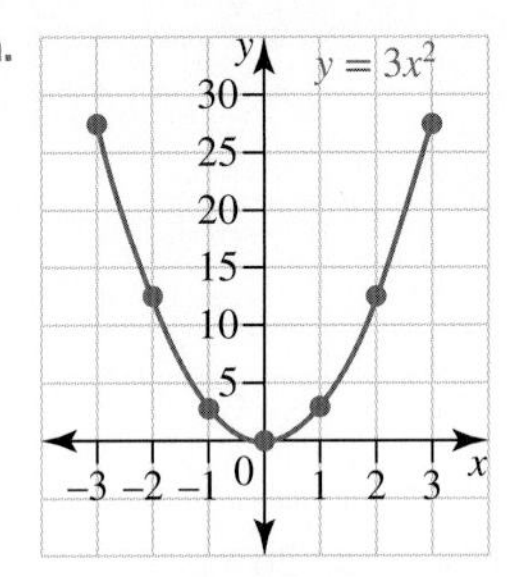

$x = 0, (0, 0)$

b.

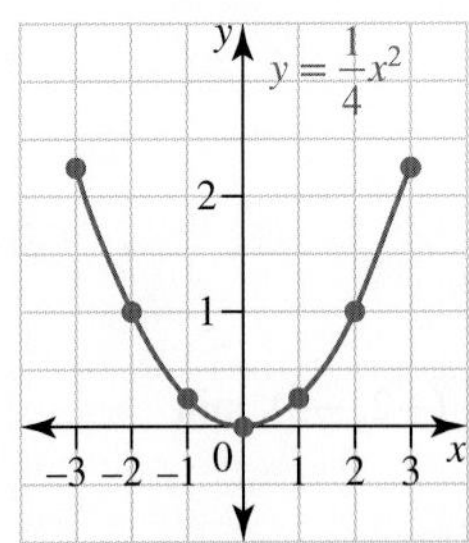

$x = 0, (0, 0)$

2. Placing a number greater than 1 in front of x^2 makes the graph thinner. Placing a number greater than 0 but less than 1 in front of x^2 makes the graph wider.

3. a.

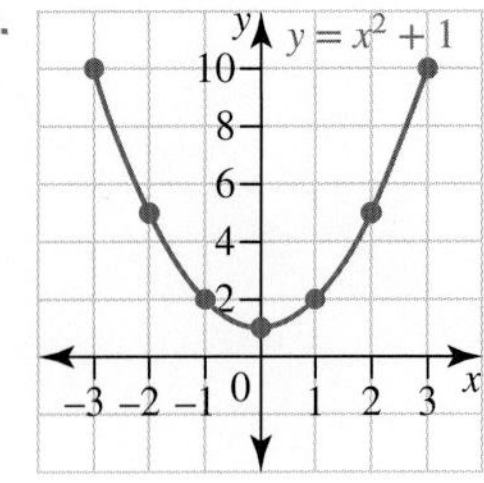

$x = 0, (0, 1), 1$

b.

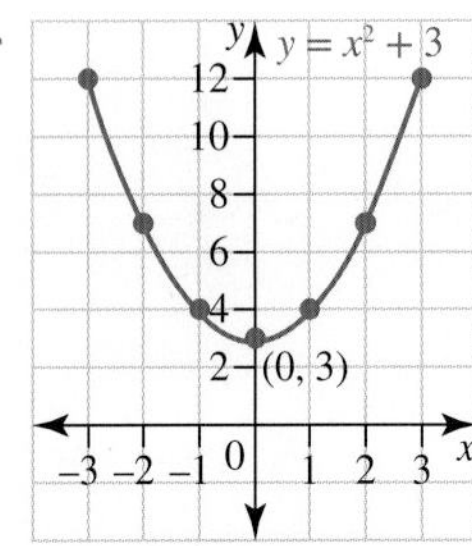

$x = 0, (0, 3), 3$

c.

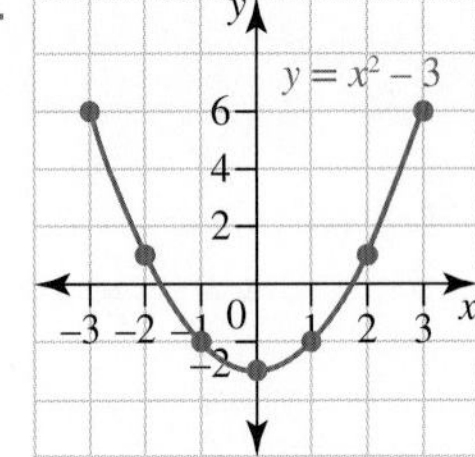

$x = 0, (0, -3), -3$

d.

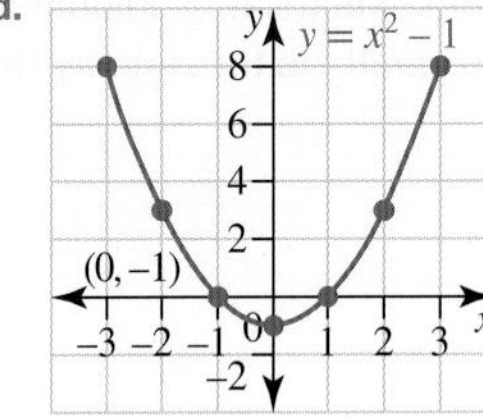

$x = 0, (0, -1), -1$

4. Adding a number raises the graph of $y = x^2$ vertically that number of units. Subtracting a number lowers the graph of $y = x^2$ vertically that number of units.

5.

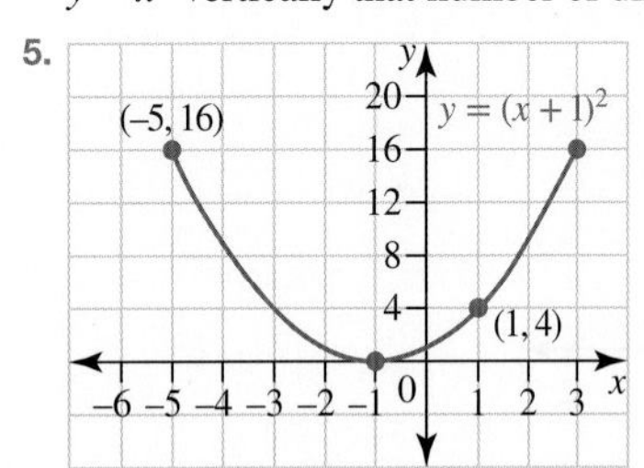

$x = -1, (-1, 0), 1$

6. 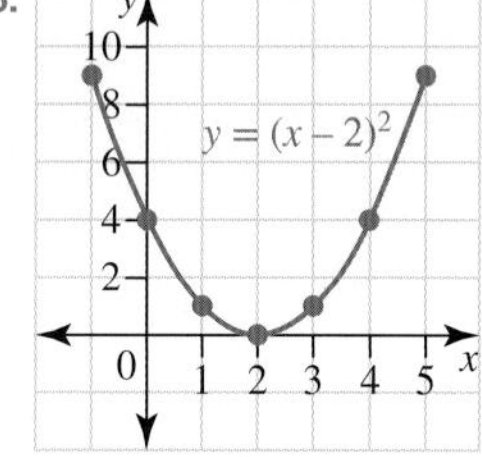

$x = 2, (2, 0), 4$

7. 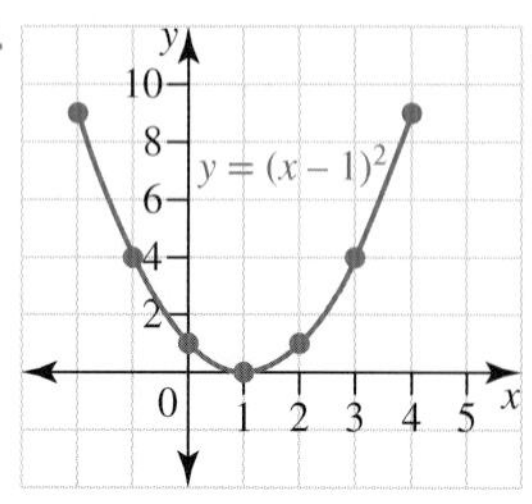

$x = 1, (1, 0), 1$

8. 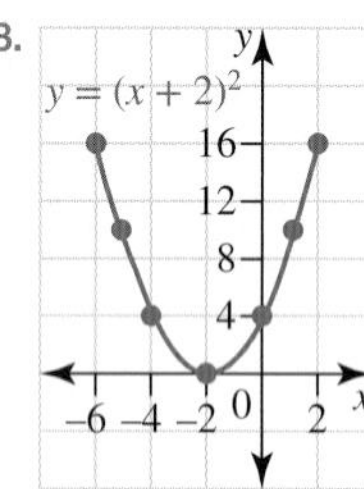

$x = -2, (-2, 0), 4$

9. Adding a number moves the graph of $y = x^2$ horizontally to the left by that number of units. Subtracting a number moves the graph of $y = x^2$ horizontally to the right by that number of units.

10. 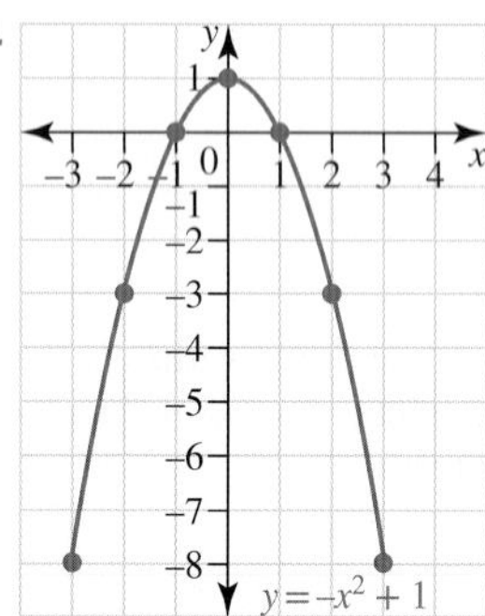

$x = 0, (0, 1), 1$

11. 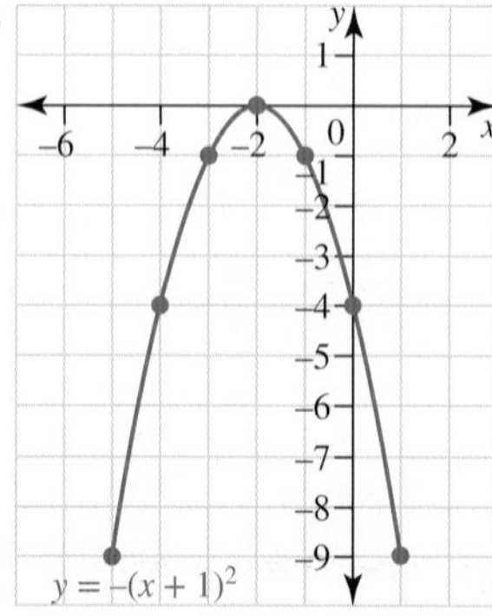

$x = -2, (-2, 0), -4$

12. 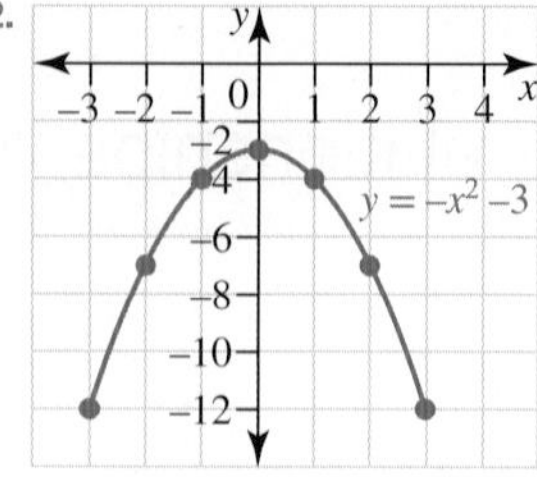

$x = 0, (0, -3), -3$

13. 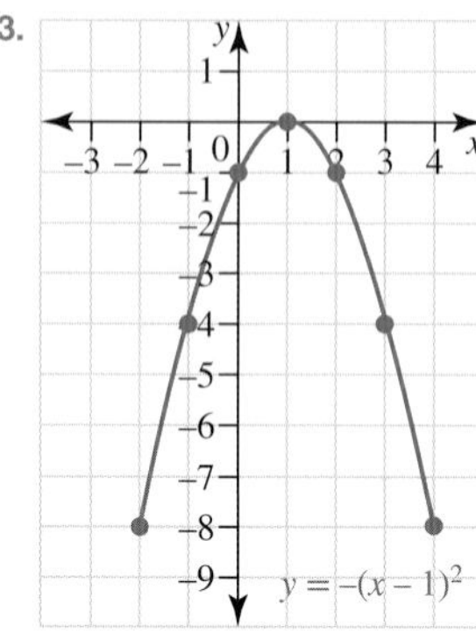

$x = 1, (1, 0), -1$

14. The negative sign inverts the graph of $y = x^2$. The graphs with the same turning points are: $y = x^2 + 1$ and $y = -x^2 + 1$; $y = (x - 1)^2$ and $y = -(x - 1)^2$; $y = (x + 2)$ and $y = -(x + 2)^2$; $y = x^2 - 3$ and $y = -x^2 - 3$. They differ in that the first graph is upright while the second graph is inverted.

15. a. 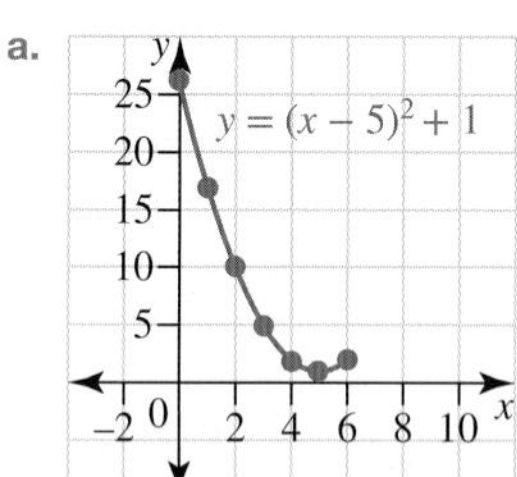

b. $x = 5$ c. (5, 1), min d. 26

16. a. 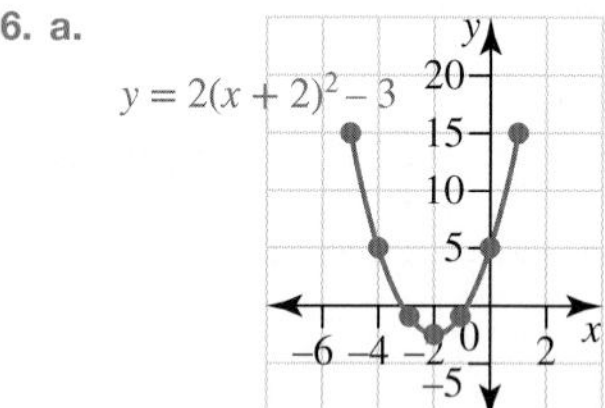

b. $x = -2$ c. $(-2, -3)$, min
d. 5

17. a. 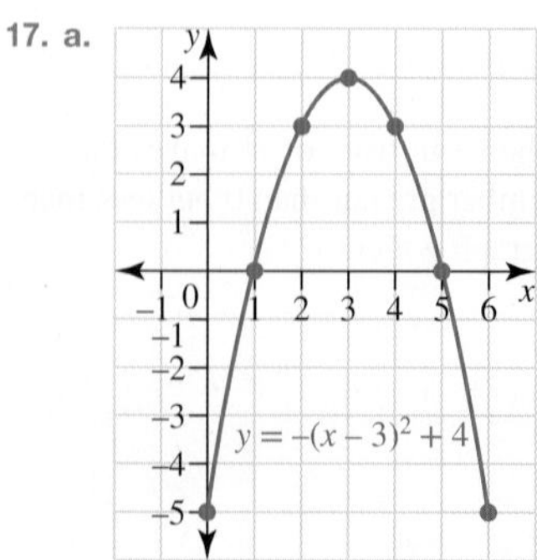

b. $x = 3$ c. (3, 4), max d. -5

18. a. 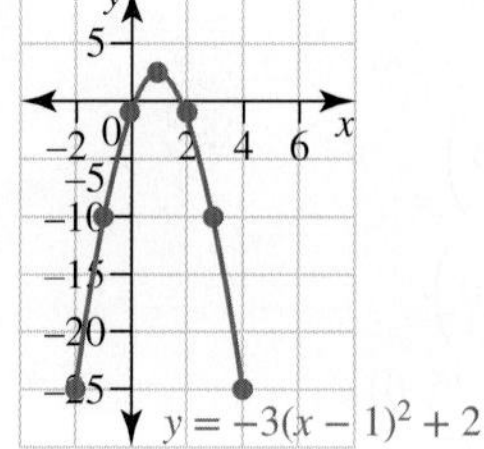

b. $x = 1$ c. $(1, 2)$, max d. -1

19. a. 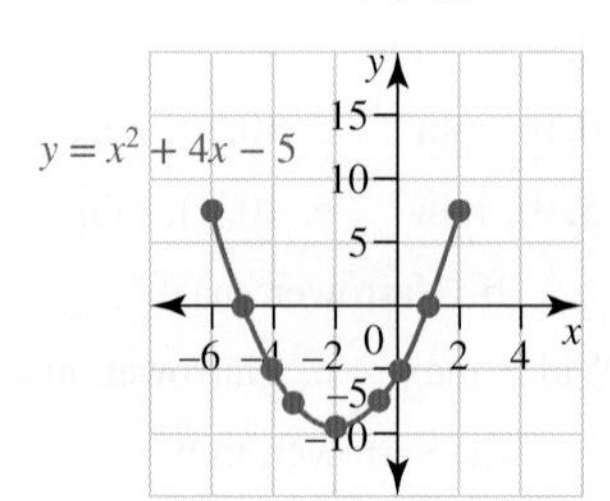

b. $x = -2$ c. $(-2, -9)$, min
d. -5

20. a. 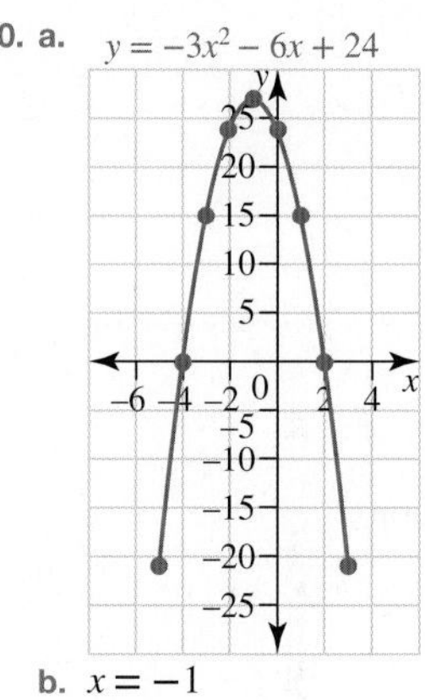

b. $x = -1$ c. $(-1, 27)$, max
d. 24

21. a. If the x^2 term is positive, the parabola has a minimum turning point. If the x^2 term is negative, the parabola has a maximum turning point.
b. If the equation is of the form $y = a(x - b)^2 + c$, the turning point has coordinates (b, c).
c. The equation of the axis of symmetry can be found from the x-coordinate of the turning point. That is, $x = b$.

22. C
23. B
24. C
25. A

26. a. 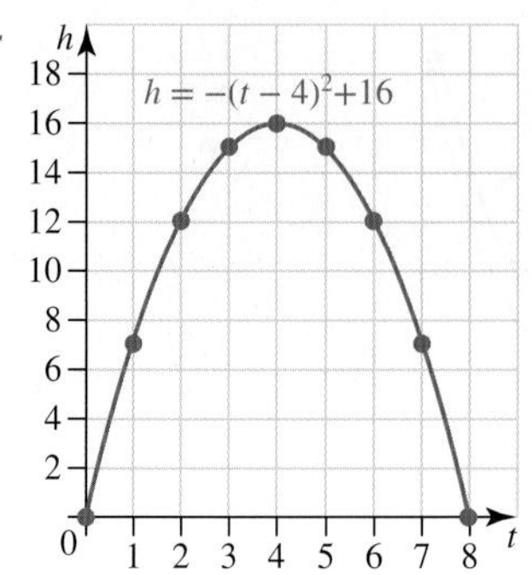

b. i. 16 m ii. 8 s

27. a. 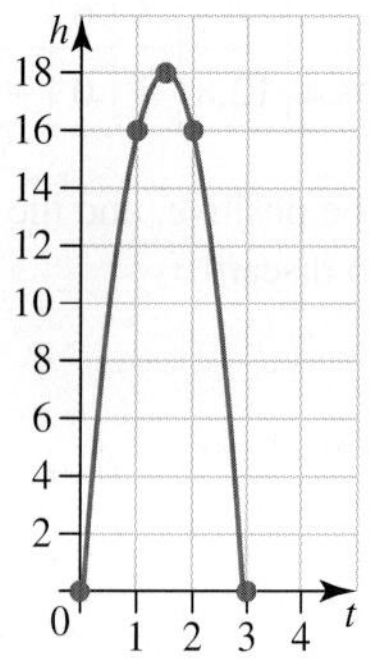

b. i. 18 m ii. Yes, by 3 m
iii. 1.5 s iv. 3 s

28. a. 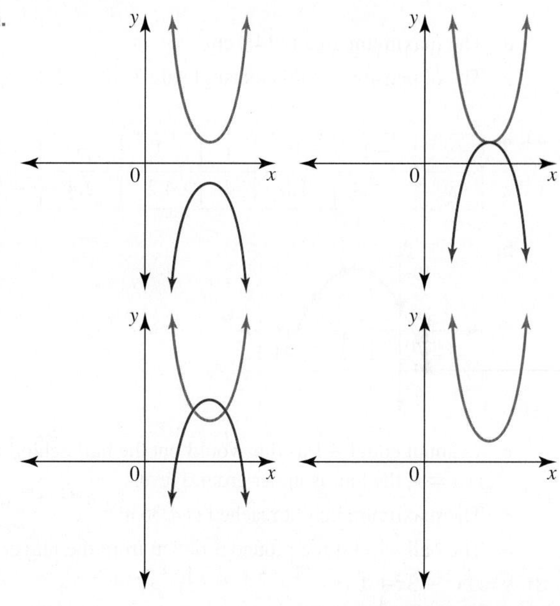

b. An infinite number of points of intersection occur when the two equations represent the same parabola, with the effect that the two parabolas superimpose. For example $y = x^2 + 4x + 3$ and $2y = 2x^2 + 8x + 6$.
c. It is possible to have 0, 1 or 2 points of intersection.

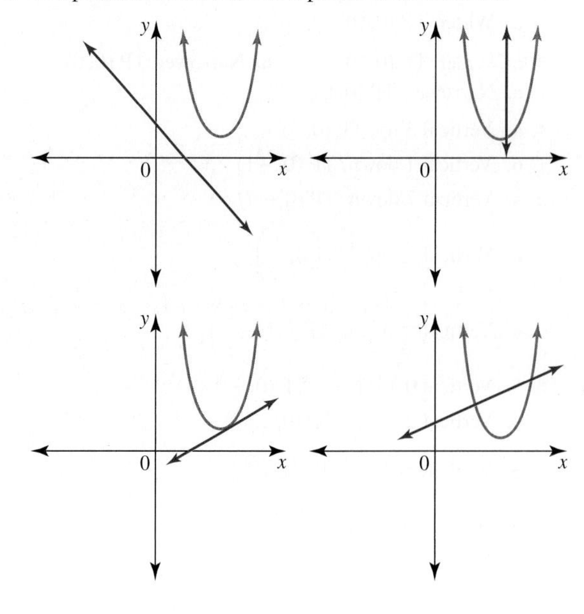

29. a.

w cm	−1	0	1	2	3	4	5	6	7
A cm^2	−11.2	0	8	12.8	14.4	12.8	8	0	−11.2

b. The length of a rectangle must be positive, and the area must also be positive, so we can discard $w = -1, 0, 6$ and 7.

c. 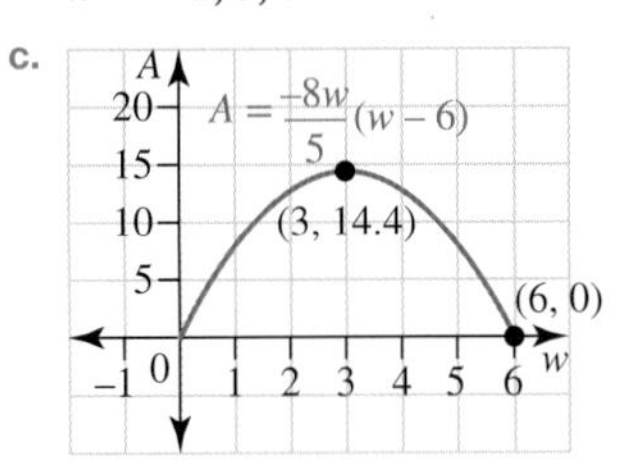

d. The maximum area is $14.4\,\text{cm}^2$.

e. The dimensions of this rectangle are $3\,\text{cm} \times 4.8\,\text{cm}$.

30. a.

x	−1	0	1	2	3	4
y	−2.4	1.8	4	4.2	2.4	−1.4

b. 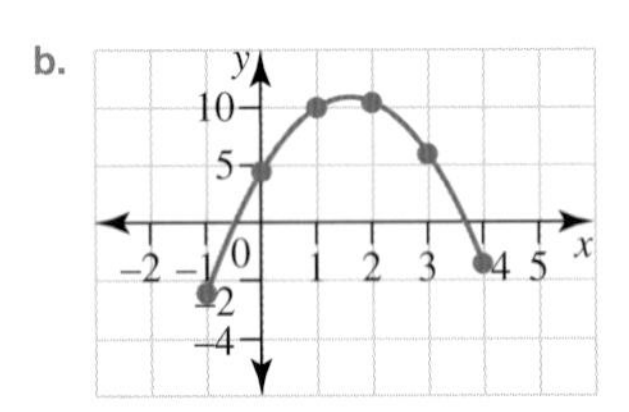

c. x cannot equal −1 as this would put the ball behind her; at $x = 4$, the ball is under ground level.

d. The maximum height reached is 4.36 m.

e. The ball will hit the ground 3.688 m from the player.

31. $y = 2x^2 - 3x + 1$

Exercise 9.3 Sketching parabolas using transformations

1. a. Narrower, TP $(0, 0)$ b. Wider, $(0, 0)$

2. a. Narrower, TP $(0, 0)$ b. Narrower, TP $(0, 0)$
   c. Wider, TP $(0, 0)$

3. a. Wider, TP $(0, 0)$ b. Narrower, TP $(0, 0)$
   c. Narrower, TP $(0, 0)$

4. a. Vertical 3 up, TP $(0, 3)$
   b. Vertical 1 down, TP $(0, -1)$

5. a. Vertical 7 down, TP $(0, -7)$
   b. Vertical $\frac{1}{4}$ up, TP $\left(0, \frac{1}{4}\right)$
   c. Vertical $\frac{1}{2}$ down, TP $\left(0, -\frac{1}{2}\right)$

6. a. Vertical 0.14 down, TP $(0, -0.14)$
   b. Vertical 2.37 up, TP $(0, 2.37)$
   c. Vertical $\sqrt{3}$ up, TP $\left(0, \sqrt{3}\right)$

7. a. Horizontal 1 right, $(1, 0)$
   b. Horizontal 2 right, $(2, 0)$

8. a. Horizontal 10 left, $(-10, 0)$
   b. Horizontal 4 left, $(-4, 0)$
   c. Horizontal $\frac{1}{2}$ right, $\left(\frac{1}{2}, 0\right)$

9. a. Horizontal $\frac{1}{5}$ left, $\left(-\frac{1}{5}, 0\right)$
   b. Horizontal 0.25 left, $(-0.25, 0)$
   c. Horizontal $\sqrt{3}$ left, $\left(-\sqrt{3}, 0\right)$

10. a. $(0, 1)$, max b. $(0, -3)$, min

11. a $(-2, 0)$, max b $(0, 0)$, min c $(0, 4)$, max

12. a. $(0, 0)$, max b. $(5, 0)$, min c. $(0, 1)$, min

13. a. Narrower, min b. Narrower, max

14. a. Wider, min b. Wider, max c. Narrower, max

15. a. Wider, min b. Narrower, min
    c. Wider, max

16. a. i. Horizontal translation 1 left
    ii. $(-1, 0)$
    iii. 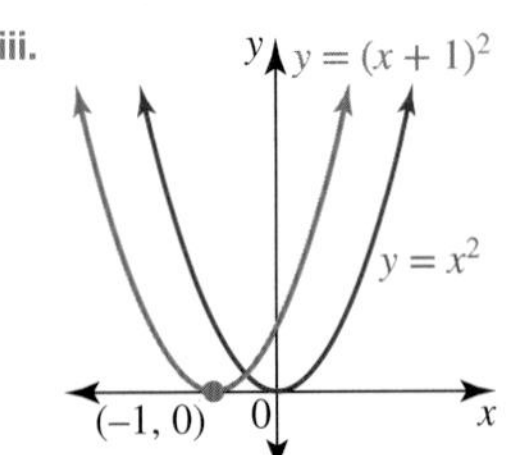

    b. i. Reflected, narrower (dilation)
    ii. $(0, 0)$
    iii. 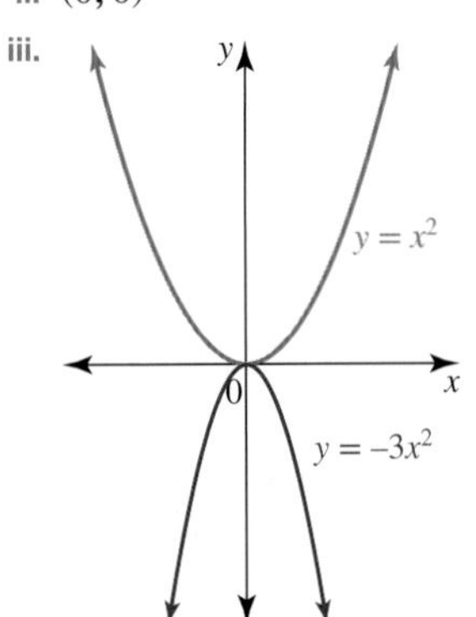

    c. i. Vertical translation 1 up
    ii. $(0, 1)$
    iii. 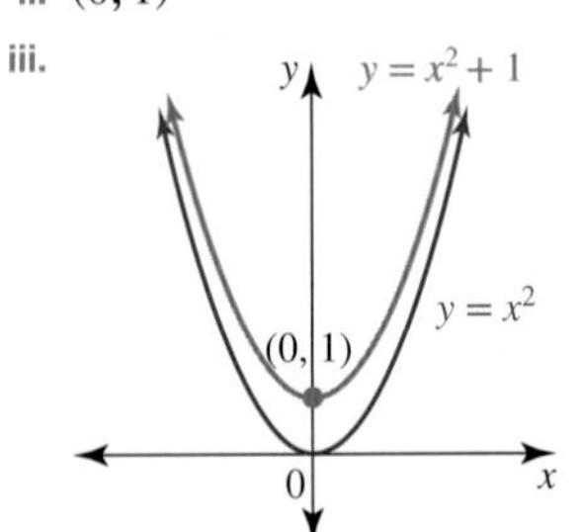

17. a. i. Wider (dilation)
    ii. $(0, 0)$

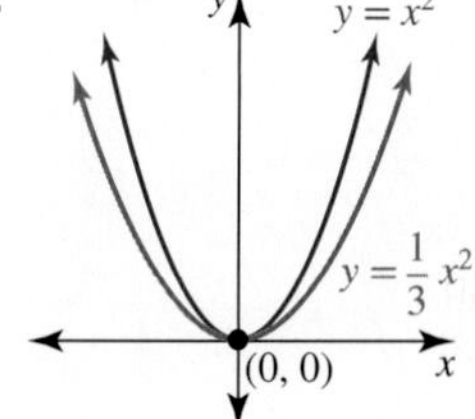

b. i. Vertical translation 3 down

ii. $(0, -3)$

iii.

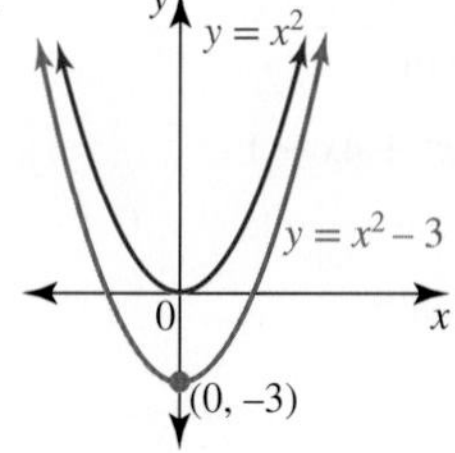

18. a. i. Horizontal translation 4 right

ii. $(4, 0)$

iii.

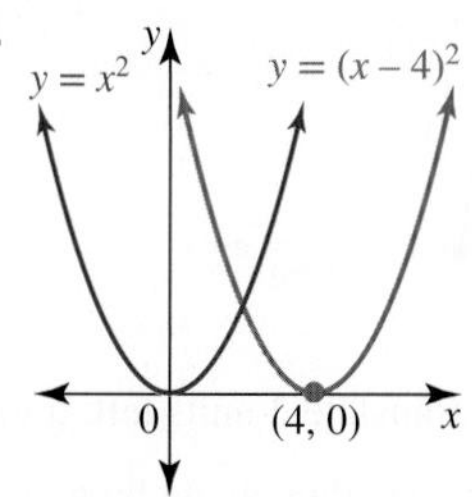

b. i. Reflected, wider (dilation)

ii. $(0, 0)$

iii.

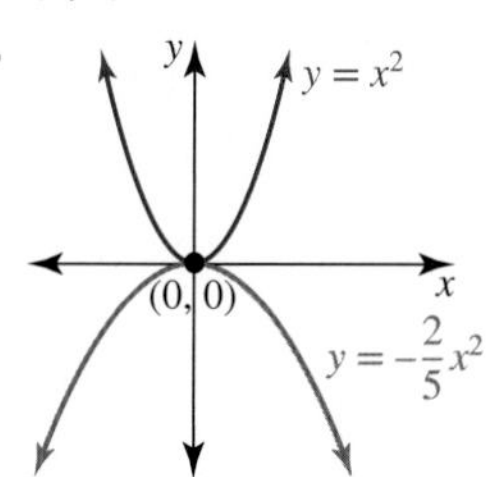

c. i. Narrower (dilation)

ii. $(0, 0)$

iii.

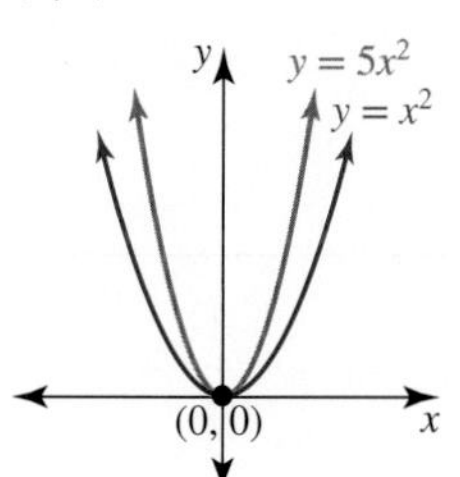

19. a. i. Reflected, vertical translation 2 up

ii. $(0, 2)$

iii.

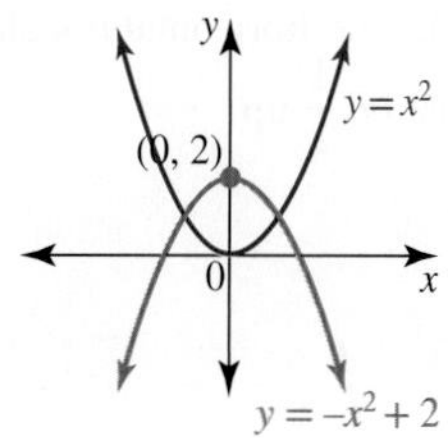

b. i. Reflected, horizontal translation 6 right

ii. $(6, 0)$

iii.

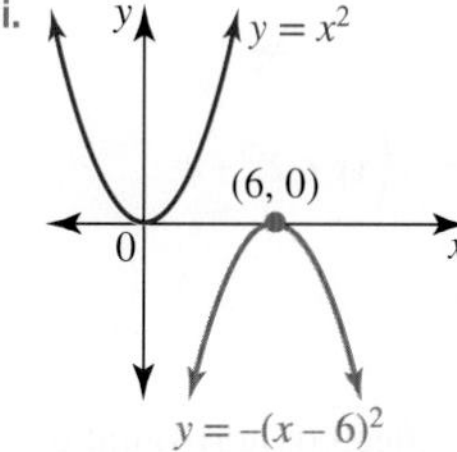

20. a. i. Reflected, vertical translation 4 down

ii. $(0, -4)$

iii.

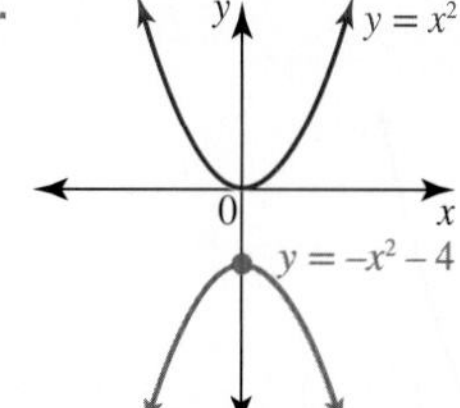

b. i. Narrower (dilation), horizontal translation 1 left, vertical translation 4 down

ii. $(-1, -4)$

iii.

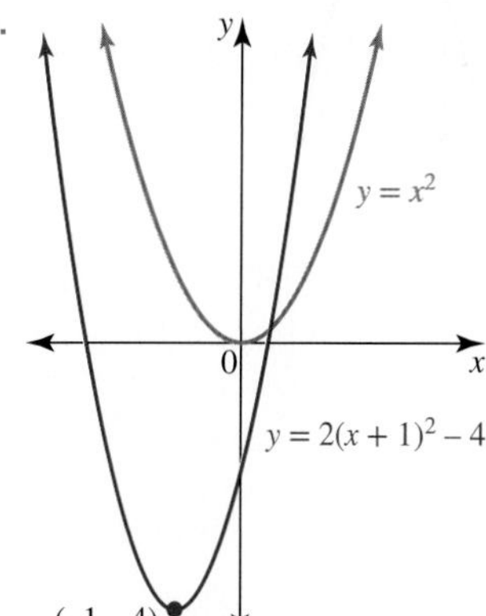

c. i. Wider (dilation), horizontal translation 3 right, vertical translation 2 up

ii. $(3, 2)$

iii.

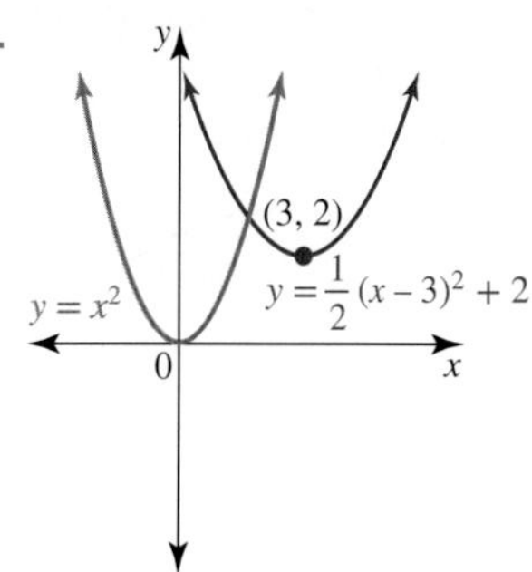

21. a. i. Wider (dilation), reflected, horizontal translation 2 left, vertical translation $\frac{1}{4}$ up

ii. $\left(-2, \frac{1}{4}\right)$

iii.

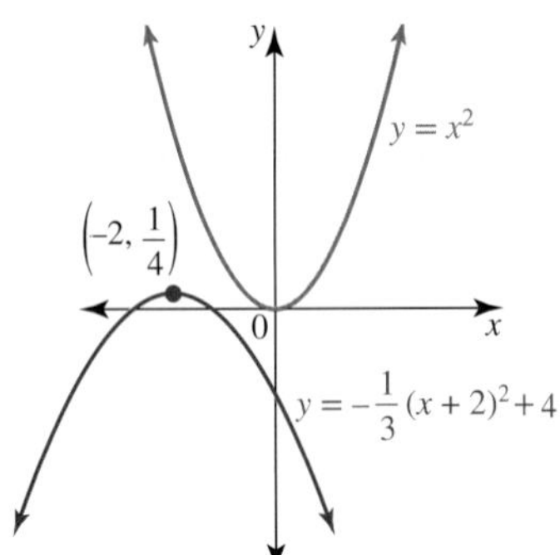

b. i. Narrower (dilation), reflected, horizontal translation 1 right, vertical translation $\frac{3}{2}$ down

ii. $\left(1, -\frac{3}{2}\right)$

iii.

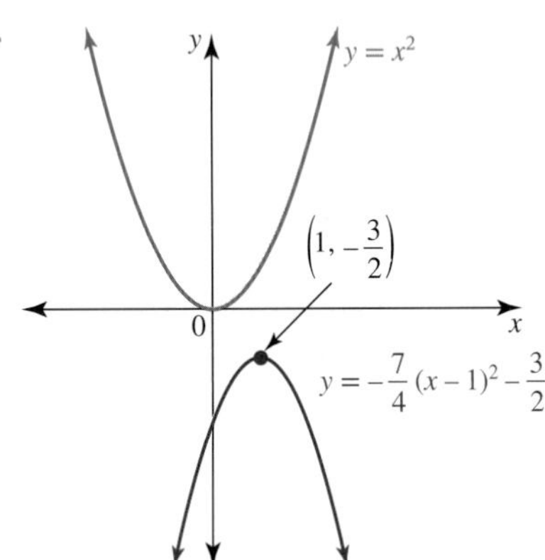

22. a. 10 cm b. 5 cm c. 5 cm
d. $y = (x - 5)^2$

23. a. 40 m
b. 5 m
c. $h = -\frac{1}{10}(d - 70)^2 + 40$
d. $h = -\frac{3}{10}(d - 70)^2 + 120$, this is a dilation by a factor of 3 from the d axis.

24. a. and c.
See figure at the bottom of the page.*
b. 12.25 m d. 12.75 m

25. a. $y = -(x - 2)^2 + 3 = -x^2 + 4x - 1$
b.

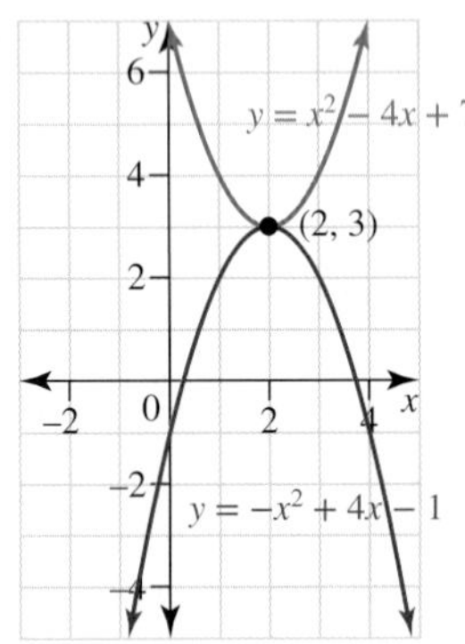

26. a. (2, −7) and (0, 5)
b. In the following order: translate 2 units left, translate 7 units up, reflect in the x-axis then dilate by a factor of $\frac{2}{3}$ from the x-axis.

*24. a. and c.

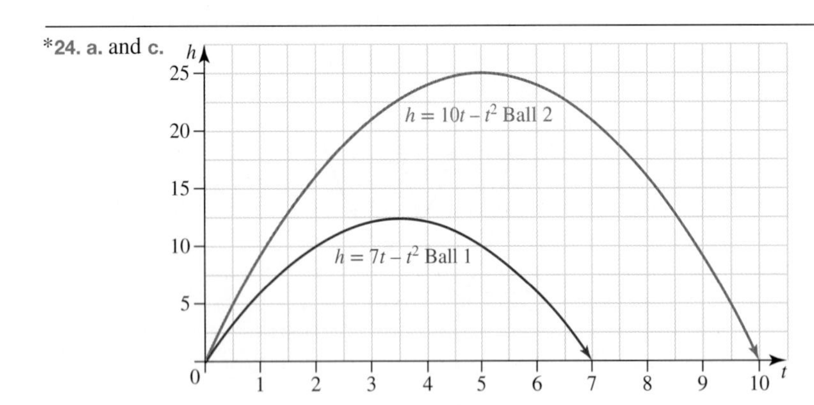

27. a. $Y = -(x-3)^2 + 3$

b. Reflected in x-axis, translated 3 units to the right and up 3 units. No dilation.

c. $(3, 3)$

d. See figure at the bottom of the page.*

28. $v = -2h^2 + 40h$, where v is the vertical distance h is the horizontal distance.

Exercise 9.4 Sketching parabolas using turning point form

1. a. $(1, 2)$, min

b. $(-2, -1)$, min

c. $(-1, 1)$, min

2. a. $(2, 3)$, max b. $(5, 3)$, max c. $(-2, -6)$, min

3. a. $\left(\frac{1}{2}, -\frac{3}{4}\right)$, min b. $\left(\frac{1}{3}, \frac{2}{3}\right)$, min

c. $(-0.3, -0.4)$, min

4. a. i. $(-3, -5)$ ii. Min iii. Narrower

b. i. $(1, 1)$ ii. Max iii. Same

5. a. i. $(-2, -4)$ ii. Max iii. Narrower

b. i. $(3, 2)$ ii. Min iii. Wider

6. a. i. $(-1, 7)$ ii. Max iii. Wider

b. i. $\left(\frac{1}{5}, -\frac{1}{2}\right)$ ii. Min iii. Wider

7. a. $y = (x+1)^2 - 3$, graph iii

b. $y = -(x-2)^2 + 3$, graph i

c. $y = -x^2 + 1$, graph ii

8. a. $y = (x-1)^2 - 3$, graph iii

b. $y = -(x+2)^2 + 3$, graph i

c. $y = x^2 - 1$, graph ii

9. A

10. C

11. B

12. C

13. B

14. a. i. -3 ii. $-3, 1$

b. i. 12 ii. 2

15. a. i. -18 ii. No x-intercepts

b. i. -5 ii. $-1, 5$

16. a. i. 4 ii. No x-intercepts

b. i. 4

ii. $-3 - \sqrt{5}, -3 + \sqrt{5}$ (approx. $-5.24, -0.76$)

17. a. i. $(4, 2)$

ii. Min

iii. Same width

iv. 18

v. No x-intercepts

vi.

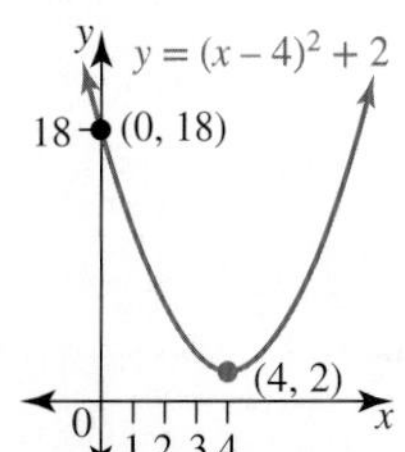

b. i. $(3, -4)$

ii. Min

iii. Same width

iv. 5

v. 1, 5

vi.

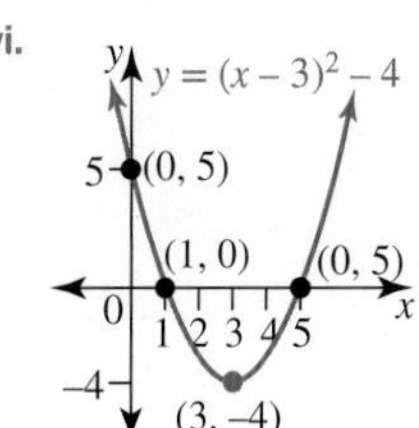

c. i. $(-1, 2)$

ii. Min

iii. Same width

iv. 3

v. No x-intercepts

*27. d.

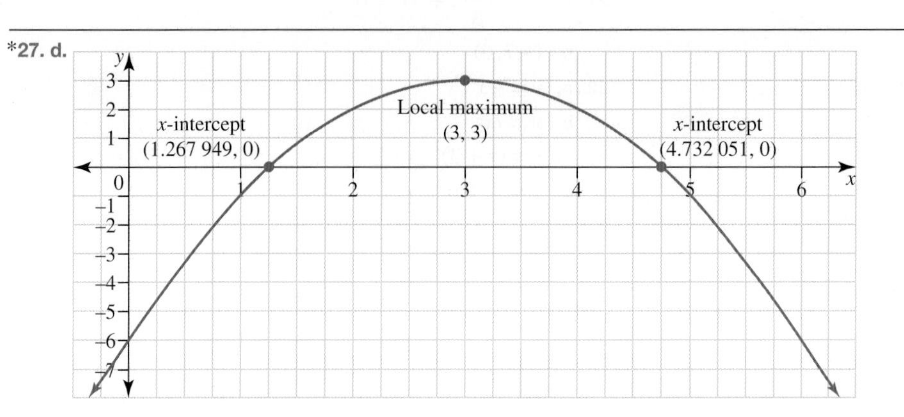

vi.

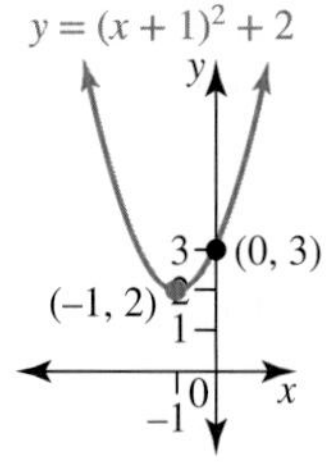

18. a. i. $(-5, -3)$

ii. Min

iii. Same width

iv. 22

v. $-5 - \sqrt{3}, -5 + \sqrt{3}$ (approx. $-6.73, -3.27$)

vi.

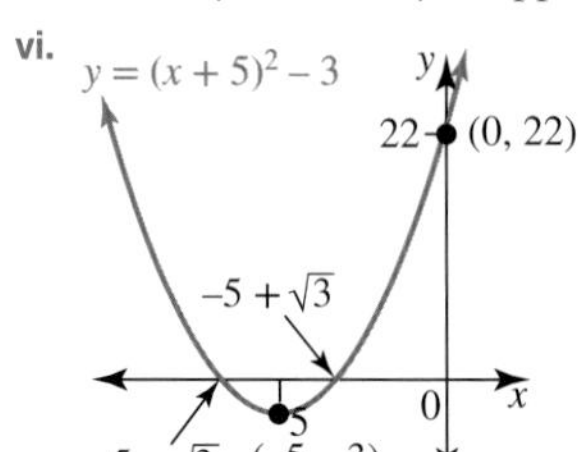

b. i. $(1, 2)$

ii. Max

iii. Same width

iv. 1

v. $1 - \sqrt{2}, 1 + \sqrt{2}$ (approx. $-0.41, 2.41$)

vi.

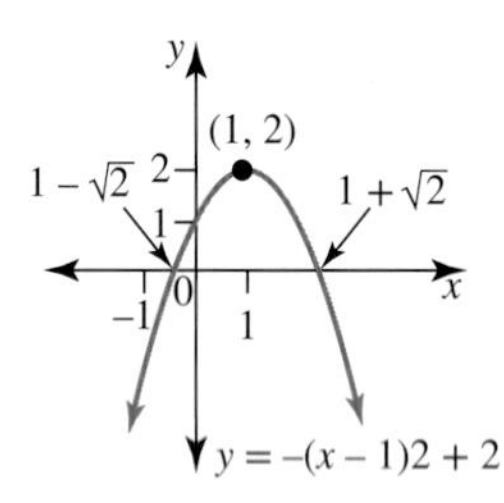

c. i. $(-2, -3)$

ii. Max

iii. Same width

iv. -7

v. No x-intercepts

vi.

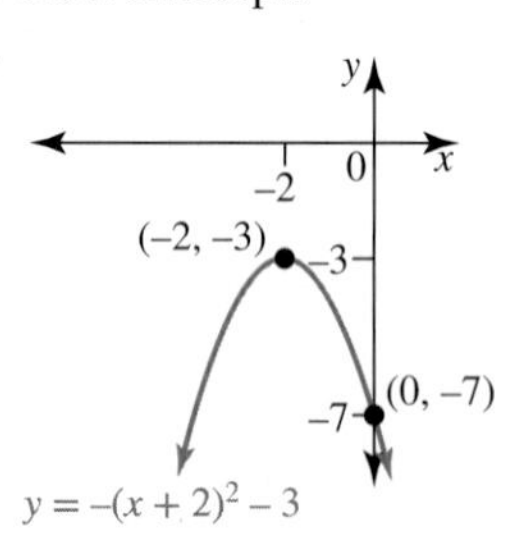

19. a. i. $(-3, -2)$

ii. Max

iii. Same width

iv. -11

v. No x-intercepts

vi.

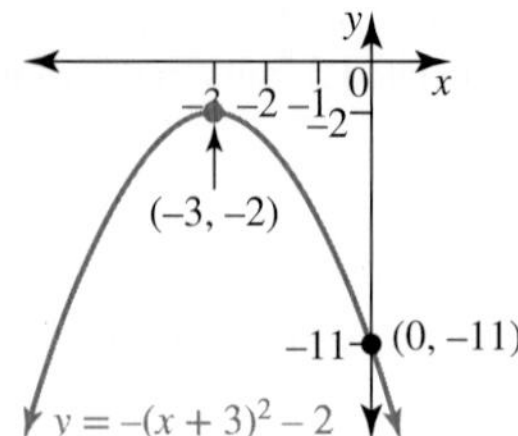

b. i. $(1, 3)$

ii. Min

iii. Narrower

iv. 5

v. No x-intercepts

vi.

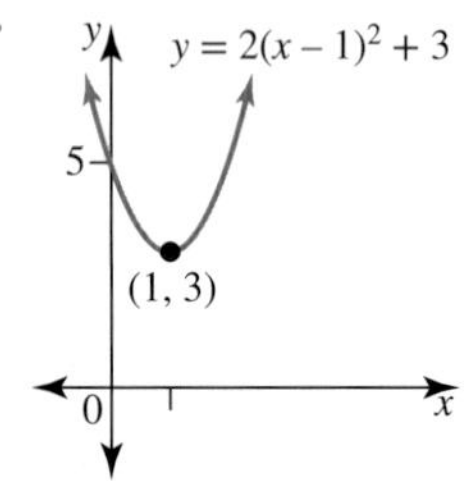

c. i. $(-2, 1)$

ii. Max

iii. Narrower

iv. -11

v. $-2 - \frac{1}{\sqrt{3}}, -2 + \frac{1}{\sqrt{3}}$ (approx. $-2.58, -1.42$)

vi.

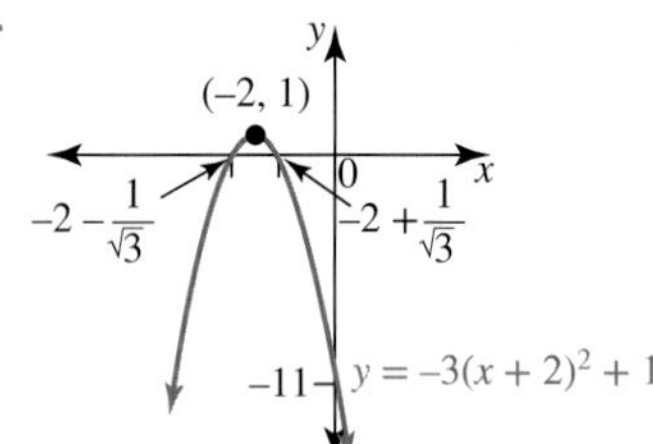

20. a. $2\left(x - \frac{3}{4}\right)^2 - \frac{73}{8} = 0$

b. $x = \frac{3}{4} \pm \frac{\sqrt{73}}{4}$

c. $\left(\frac{3}{4}, -\frac{73}{8}\right)$, minimum

21. a. $y = -\frac{2}{3}(x + 4)^2 + 6$

b. $(-7, 0)$

22. a. $y = -x^2$

b. $y = 7x^2$

c. $y = (x + 3)^2$

d. $y = x^2 + 6$

e. $y = -\frac{1}{4}(x - 5)^2 - 3$

23. a.

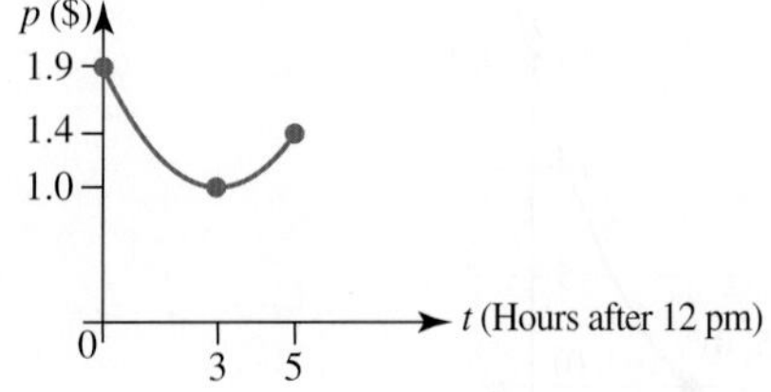

b. \$1.90 **c.** \$1 **d.** 3 pm **e.** \$1.40

24. a. 0.5 m

b. $\left(15 + 4\sqrt{15}\right)$ m

c. Maximum height is 8 meters when horizontal distance is 15 meters.

25. a. Sample responses can be found in the worked solutions in the online resources.
An example is $y = (x - 2)^2 + 6$.

b. $y = -\frac{1}{2}(x - 2)^2 + 6$

26. a. Sample responses can be found in the worked solutions in the online resources.
An example is $y = (x - p)^2 + q$.

b. $y = \left(\frac{r - q}{p^2}\right)(x - p)^2 + q$

27. a. $y = (x - 4)^2 - 15$

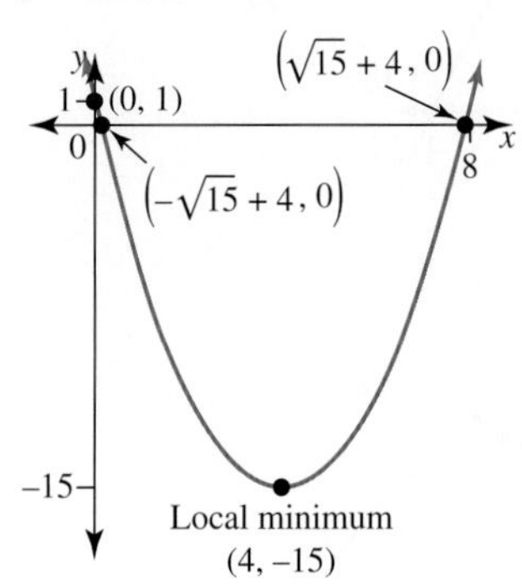

b. $y = (x + 2)^2 - 9$

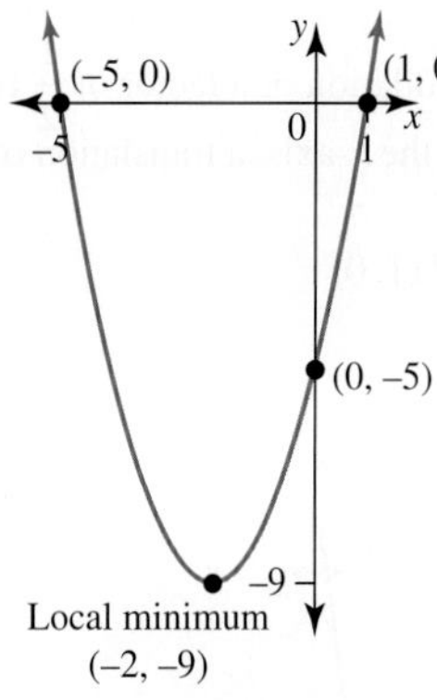

c. $y = \left(x + \frac{3}{2}\right)^2 - \frac{1}{4}$

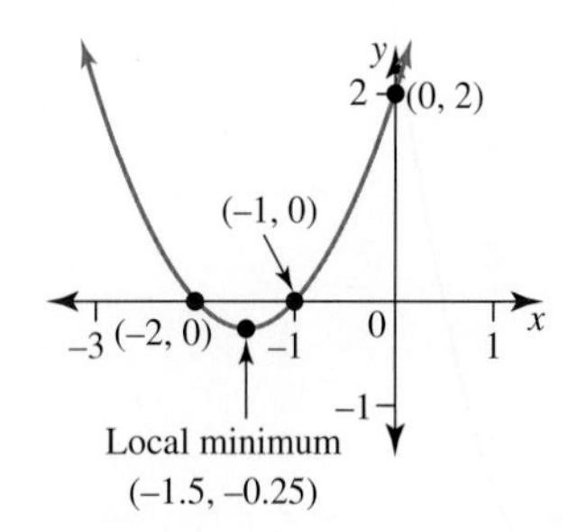

28. a. $y = 3(x - 2)^2 - 8$

b. Dilated by a factor of 3 parallel to the y-axis or from the x-axis as well as being translated 2 units to the right and down 8 units.

c. $y = -3(x - 2)^2 + 8$

d. See figure at the bottom of the page.*

***28. d.**

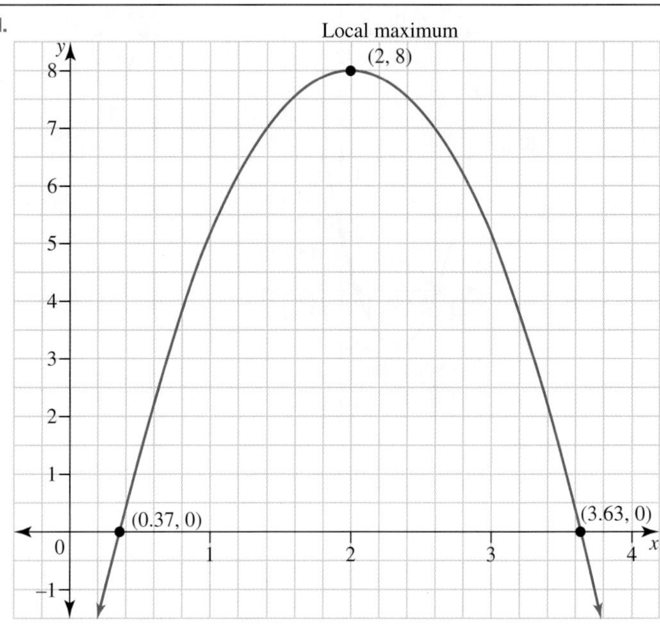

29. a. $y = -\frac{1}{2}(x+3)^2 + 8$

b. In the following order: a dilation of a factor of $\frac{1}{2}$ from the x-axis, a reflection in the x-axis, a translation of 3 units left and 8 units up.

c. x- intercepts: $(-7, 0)$ and $(1, 0)$

y- intercept: $\left(0, \frac{7}{2}\right)$

d. The new rule is $y = (x-3)^2 - 16$

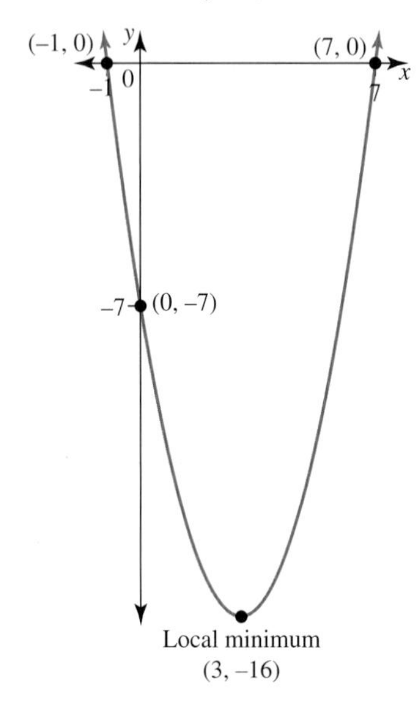

Exercise 9.5 Sketching parabolas in expanded form

1. You need the x-intercepts, the y-intercept and the turning point to sketch a parabola.

2. a.

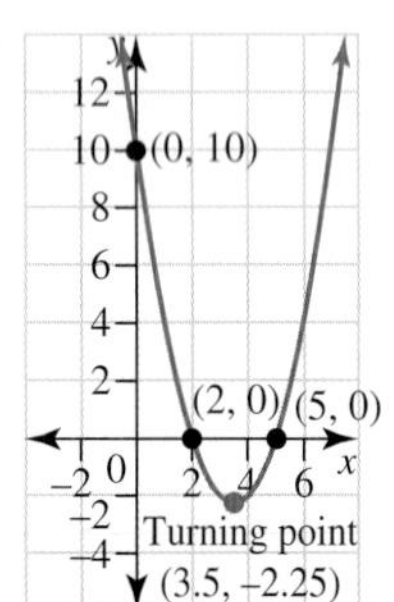

b.

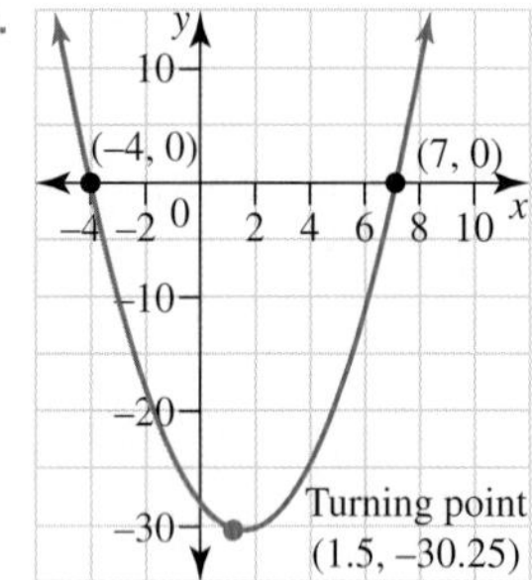

3. a.

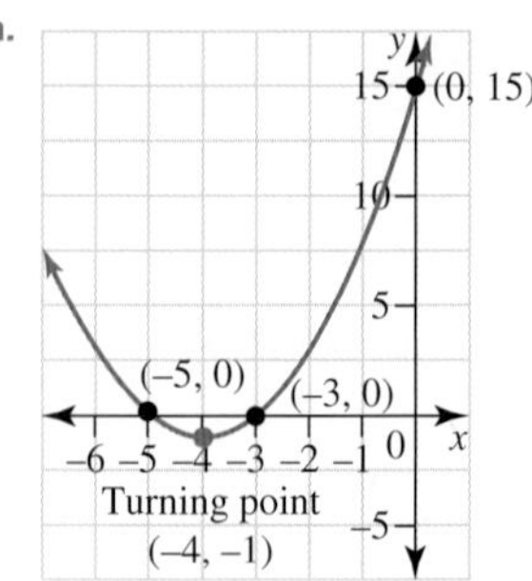

b.

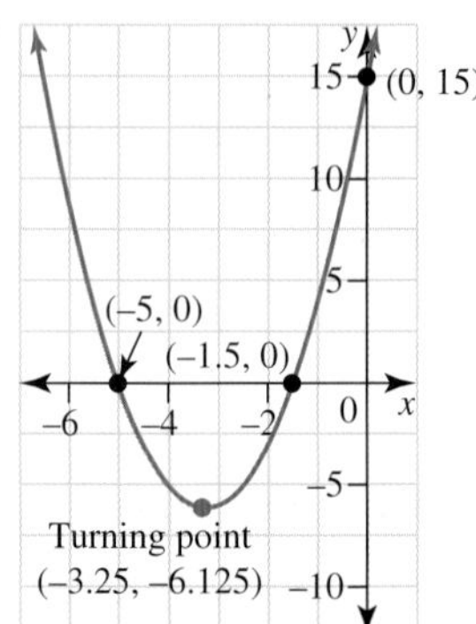

4. a.

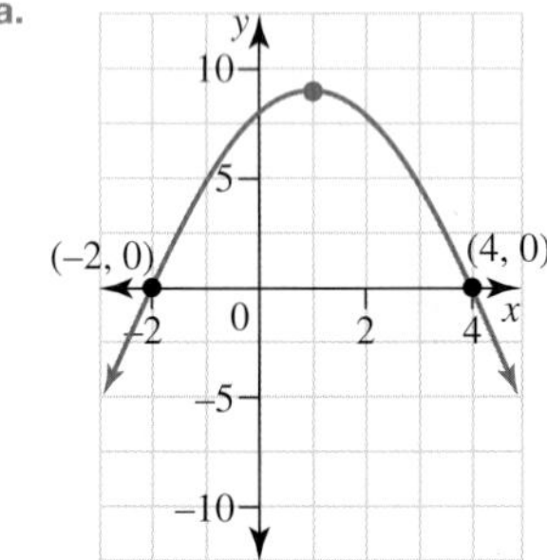

b.

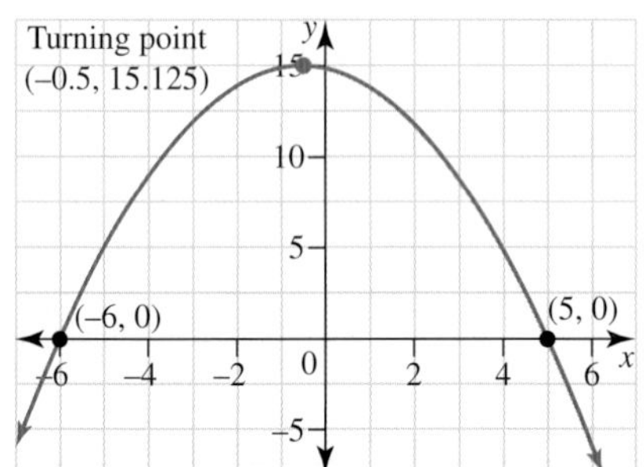

5. a.

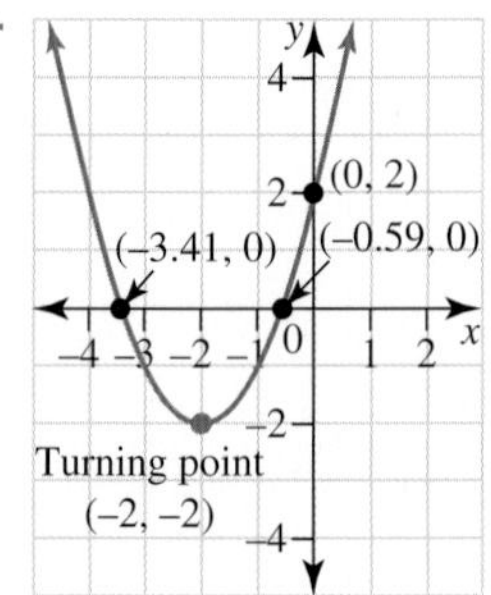

b.

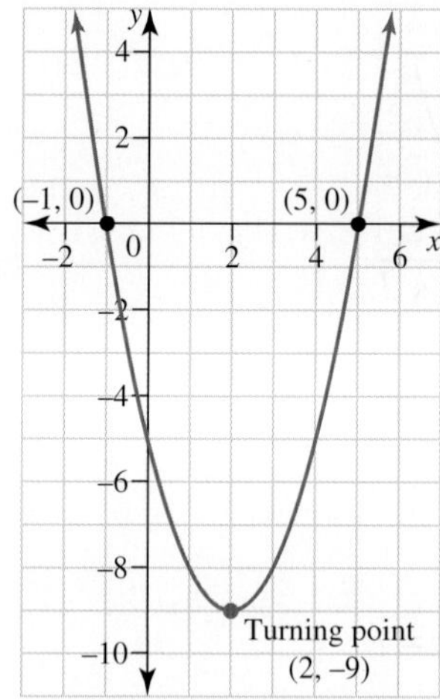

(–1, 0)
(5, 0)
Turning point
(2, –9)

c.

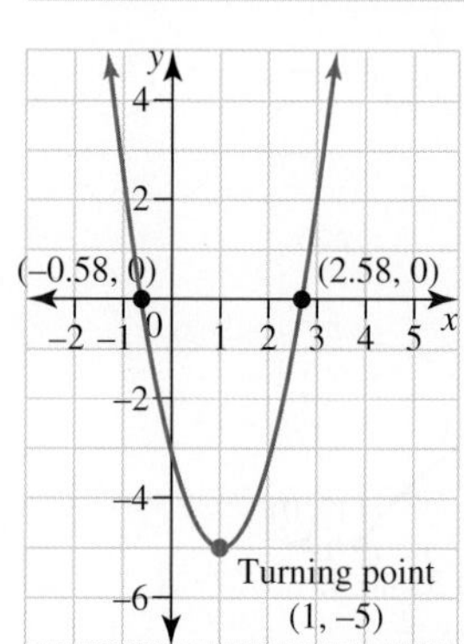

(–0.58, 0)
(2.58, 0)
Turning point
(1, –5)

6. a.

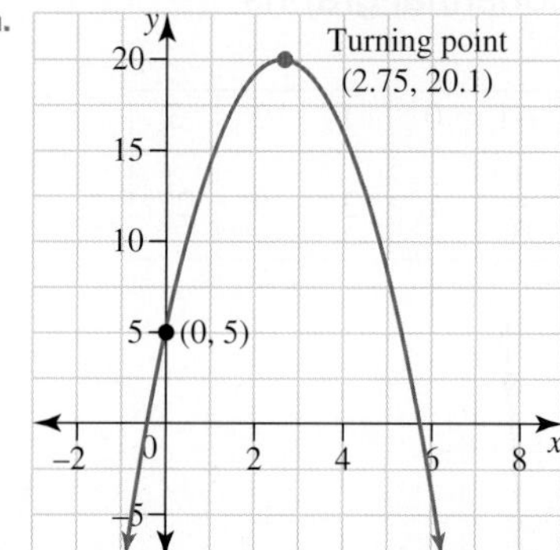

Turning point
(2.75, 20.1)
(0, 5)

b.

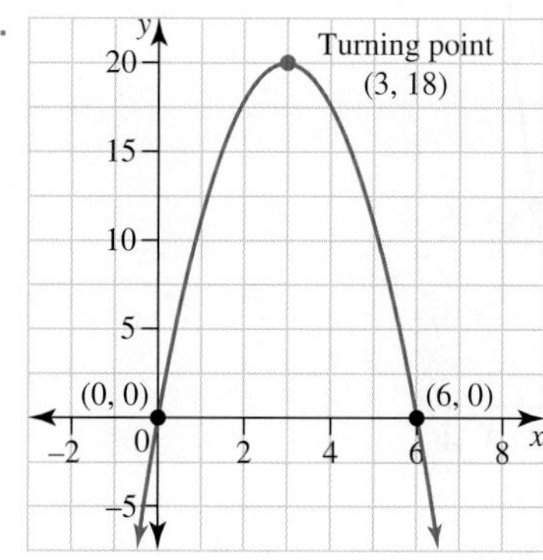

Turning point
(3, 18)
(0, 0)
(6, 0)

c.

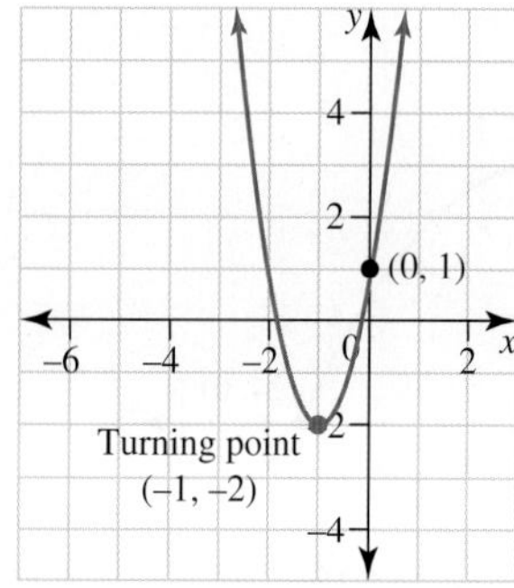

(0, 1)
Turning point
(–1, –2)

7. a.

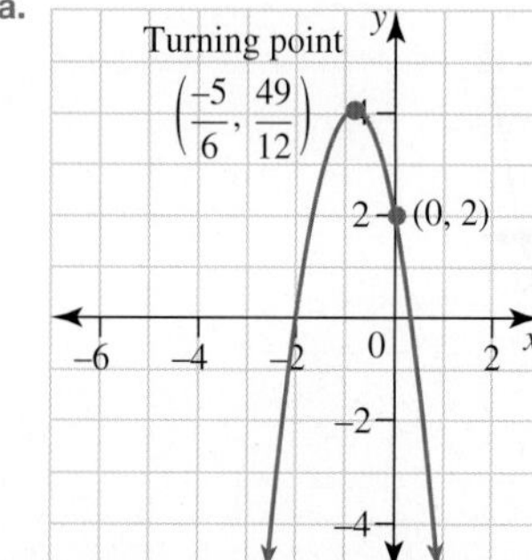

Turning point
$\left(\frac{-5}{6}, \frac{49}{12}\right)$
(0, 2)

b.

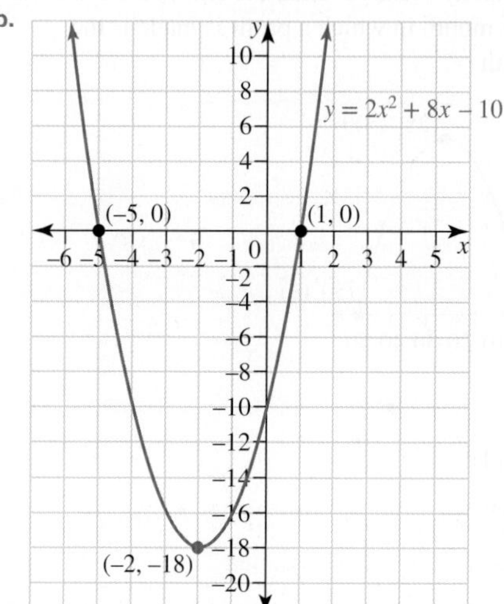

$y = 2x^2 + 8x - 10$
(–5, 0)
(1, 0)
(–2, –18)

c.

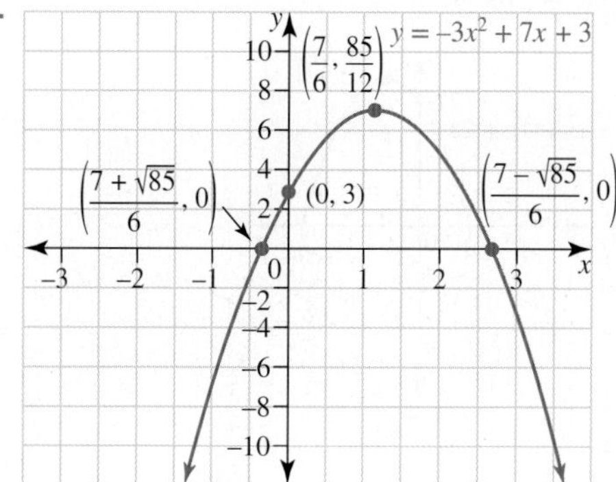

$\left(\frac{7}{6}, \frac{85}{12}\right)$
$y = -3x^2 + 7x + 3$
$\left(\frac{7+\sqrt{85}}{6}, 0\right)$
(0, 3)
$\left(\frac{7-\sqrt{85}}{6}, 0\right)$

8. a. 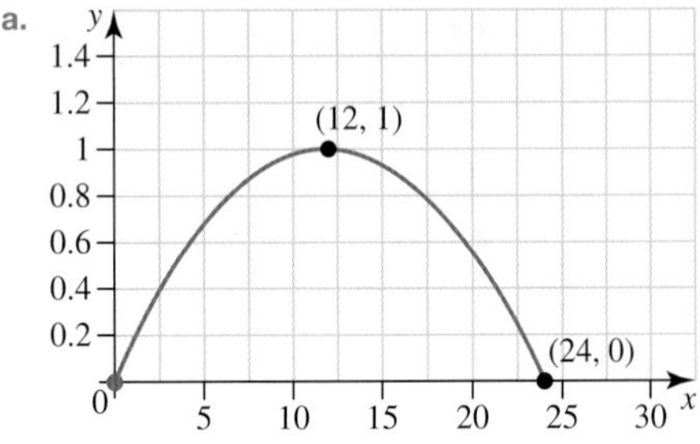

b. 24 m

c. 1 m

9. a. 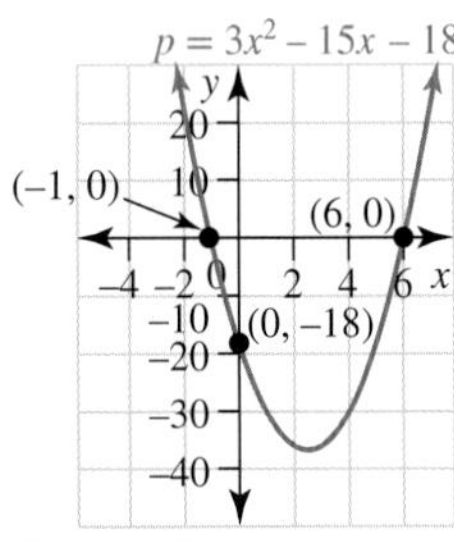

b. 6th month

c. The company breaks even at the end of the 6th month. The first month in which a profit is made is the 7th month.

10. a. 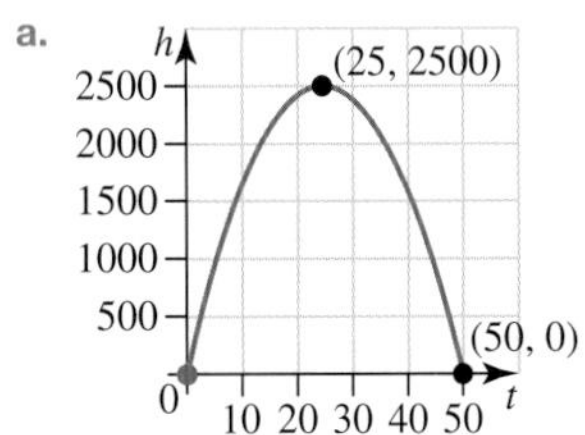

b. 0

c. 2500 m

d. 25 seconds

e. 50 seconds

11. 200

12. $-(m+n)$

13. $y = -\frac{1}{2}x^2 + 2x + 9$

14. a. 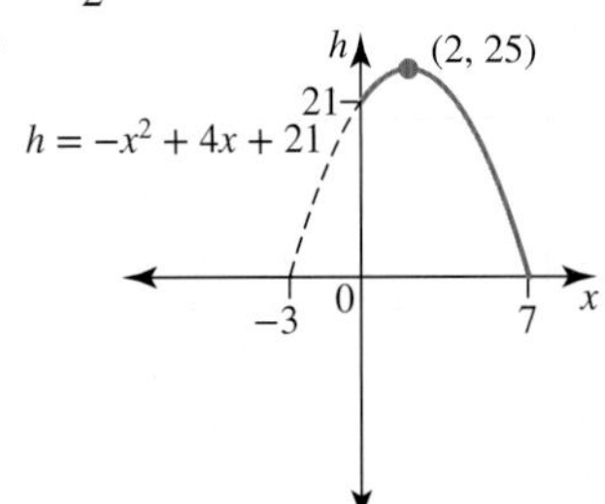

b. 25 m

c. 2 m

d. 7 m

15. a. 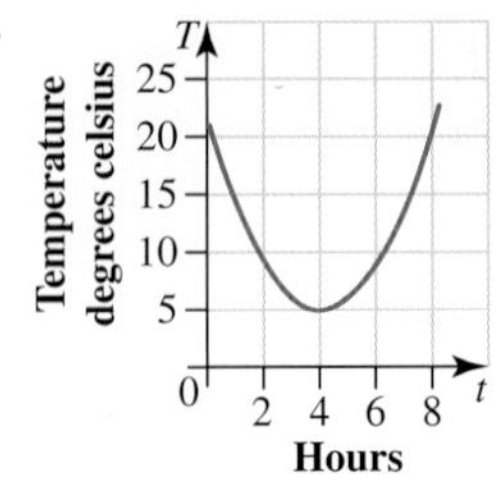

b. 21 °C

c. Decreasing

d. Increasing

e. 5 °C after 4 hours

f. 21 °C

16. a. $a = -\frac{1}{18}, b = \frac{13}{6}, c = 7$

b. 7 metres

c. $\frac{225}{8}$ metres

d. 42 metres

e. 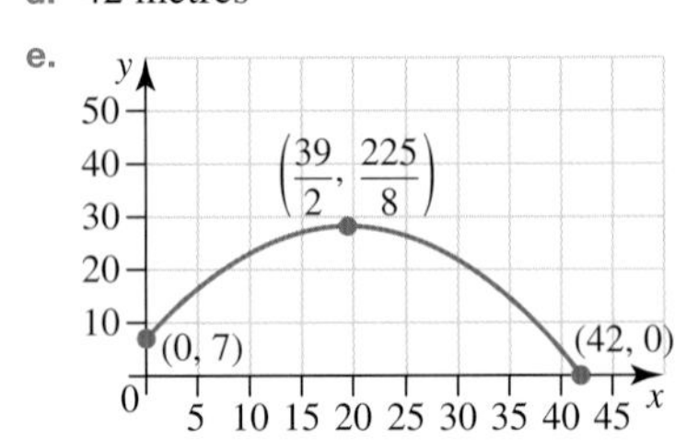

Exercise 9.6 Exponential graphs

1.

x	−3	−2	−1	0	1	2	3
y	$\frac{1}{27}$	$\frac{1}{9}$	$\frac{1}{3}$	1	3	9	27

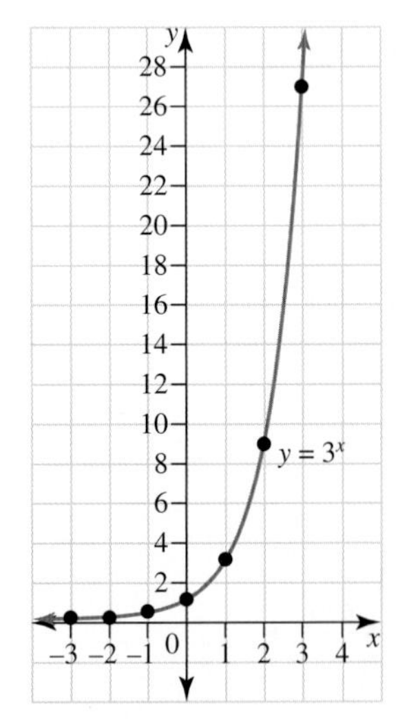

2. a. 2 b. 3 c. 4

3. a. 10 b. a

4.

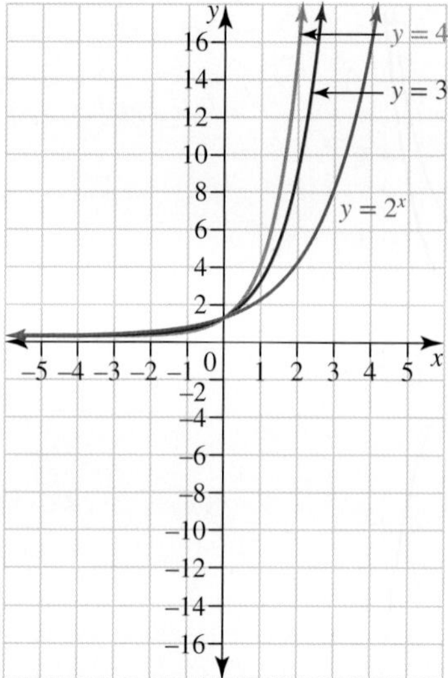

a. The graphs all pass through (0, 1). The graphs have the same horizontal asymptote, $(y = 0)$. The graphs are all very steep.

b. As the base grows larger, the graphs become steeper.

c.

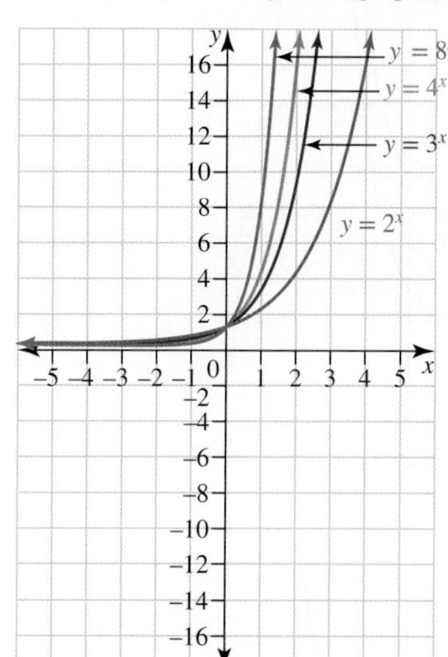

5.

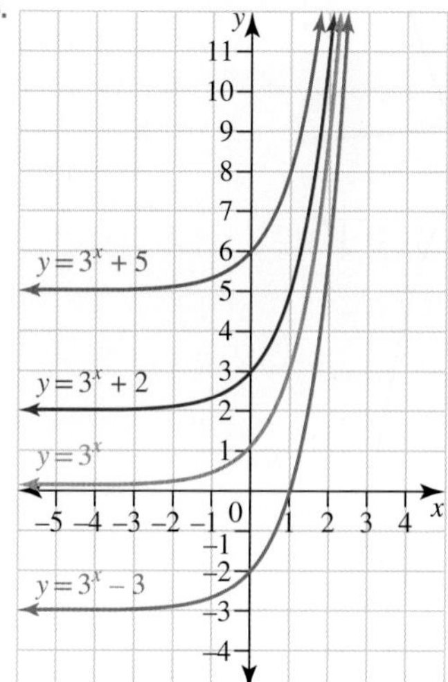

a. The shape of each graph is the same.

b. Each graph has a different y- intercept and a different horizontal asymptote.

c. i. $(0, 11)$

ii. $y = 10$

6. a.

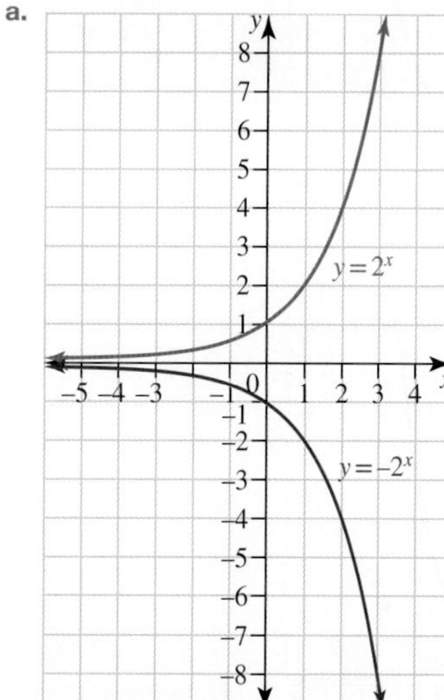

b.

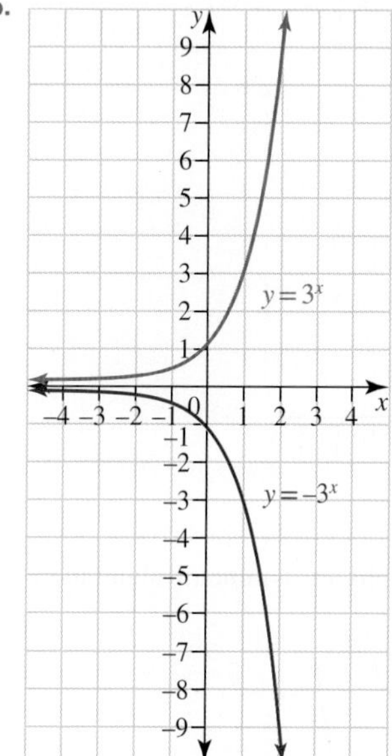

c.

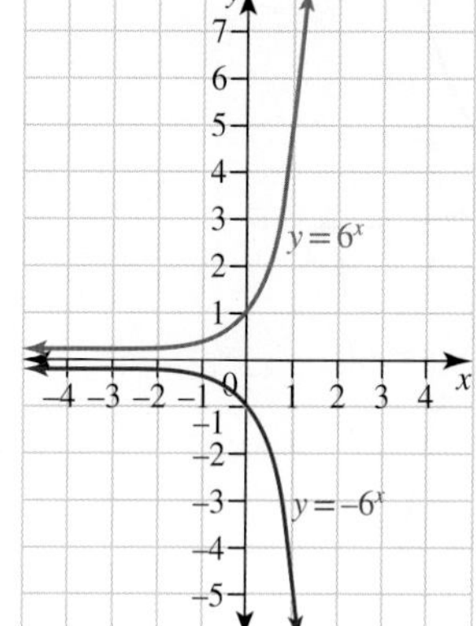

d. In each case the graphs are symmetric about the x-axis.

7. a.

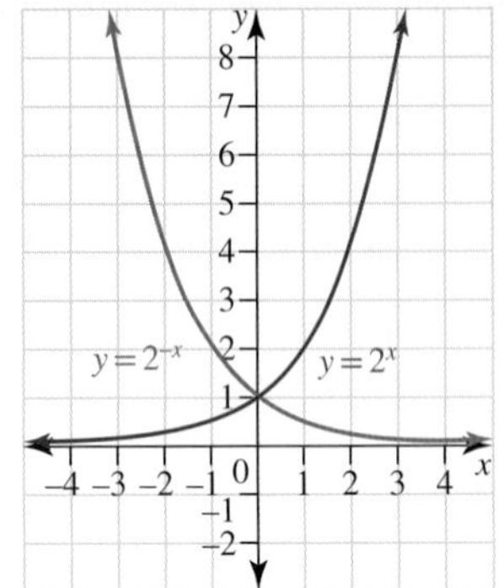

b.

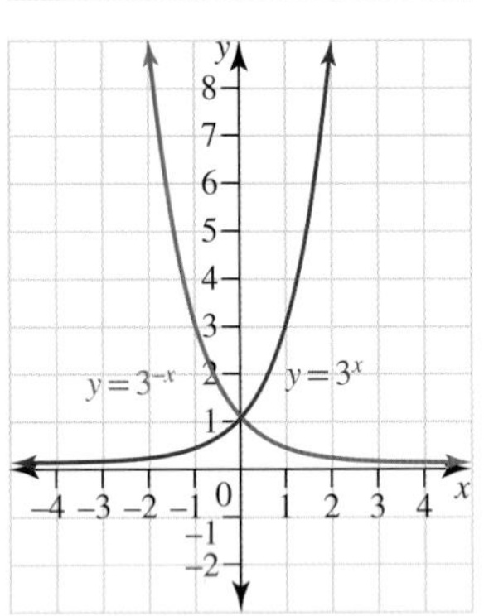

c.

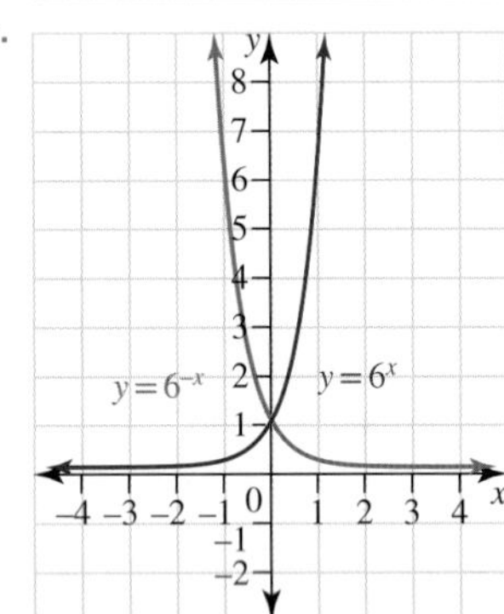

d. In each case the graphs are symmetric about the *y*-axis.

8. a–c.

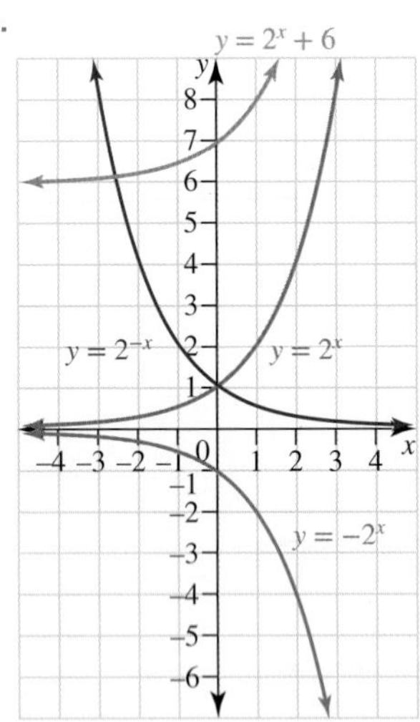

9. a, b

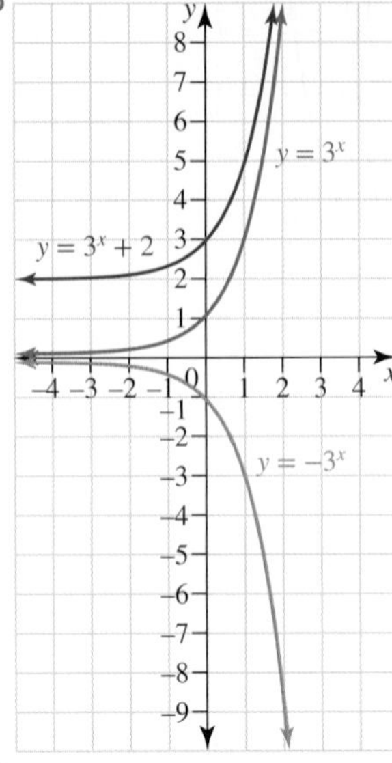

10. a, b

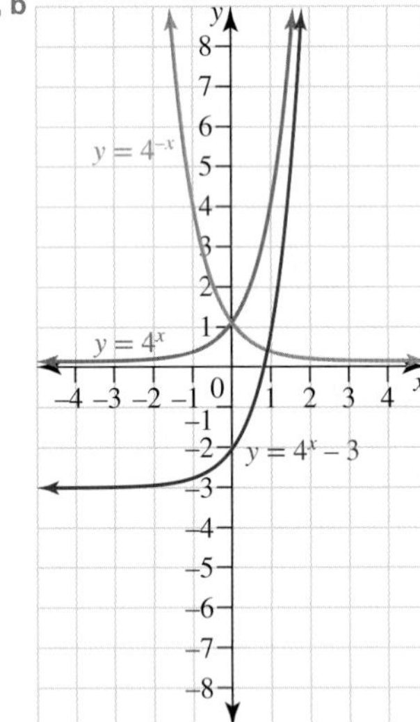

11. a, b

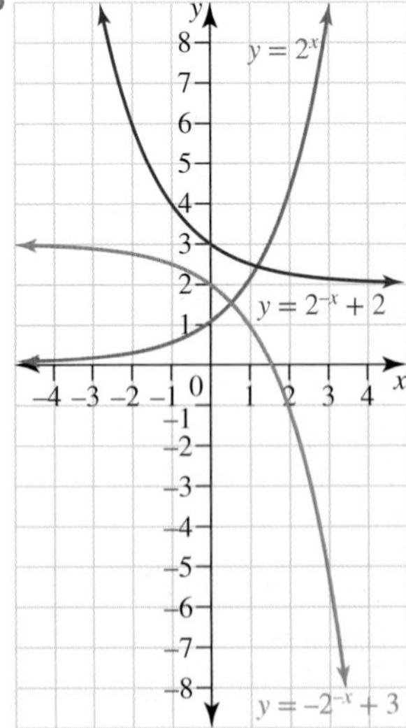

12. a, b

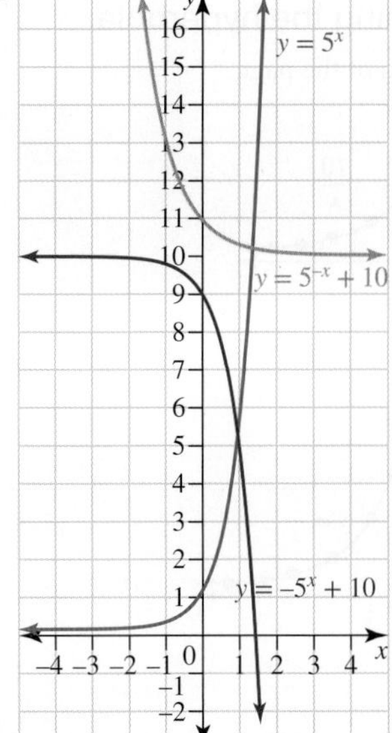

13. a. B b. C
c. D d. A

14. a. B b. D
c. A d. C

15.

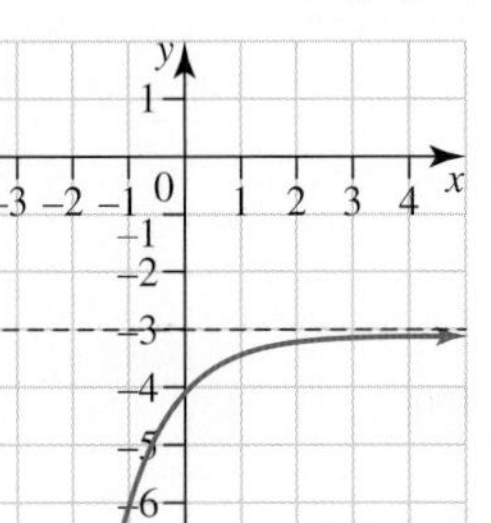

16.

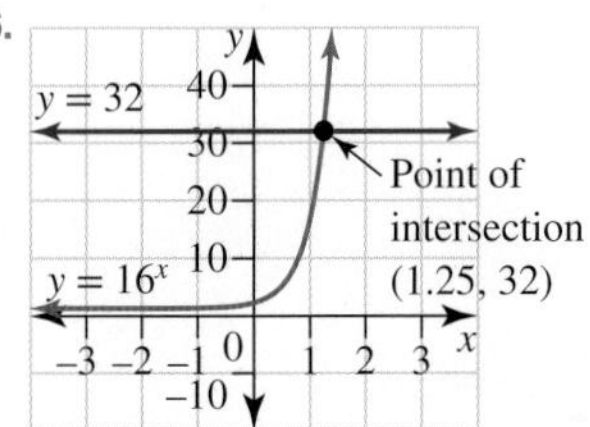

$x = 1.25$

17.

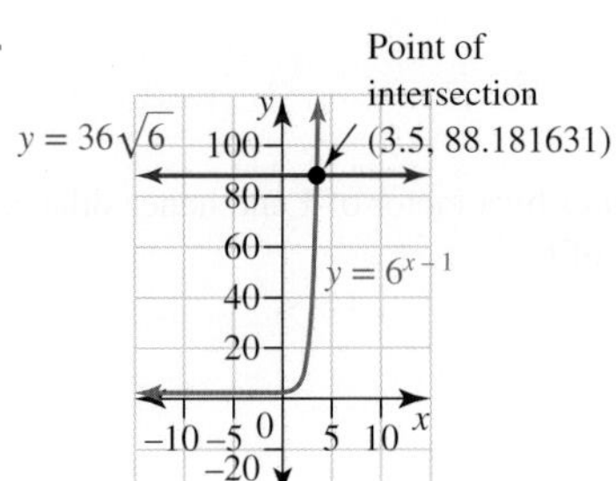

$x = 3.5$

18.

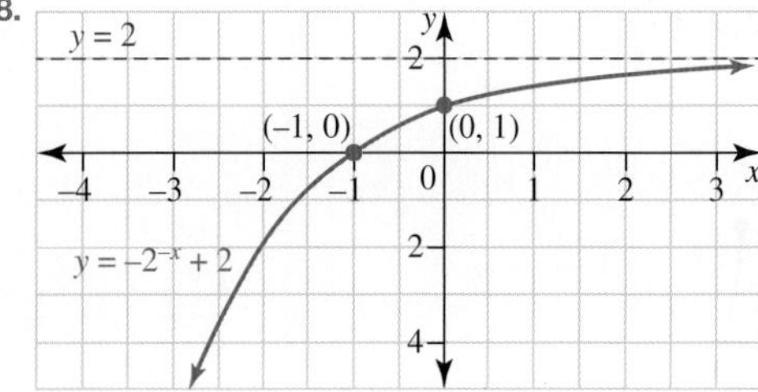

19. a. 10 000
b. i. 1111
ii. 41
iii. 0

20. a. Yes
b. There is a constant ratio of 1.3.
c. 30%
d. 3.26 million
e. 30.26 million

Exercise 9.7 Inverse proportion

1. a. $s = \frac{k}{t}$ or $t = \frac{k}{s}$ b. No
c. $t = \frac{k}{p}$ or $p = \frac{k}{t}$ d. $L = \frac{k}{p}$ or $p = \frac{k}{L}$
e. No f. $t = \frac{k}{n}$ or $n = \frac{k}{t}$

2. Sample responses can be found in the worked solutions in the online resources.

3. a. $k = 1000, y = \frac{1000}{x}$
b.

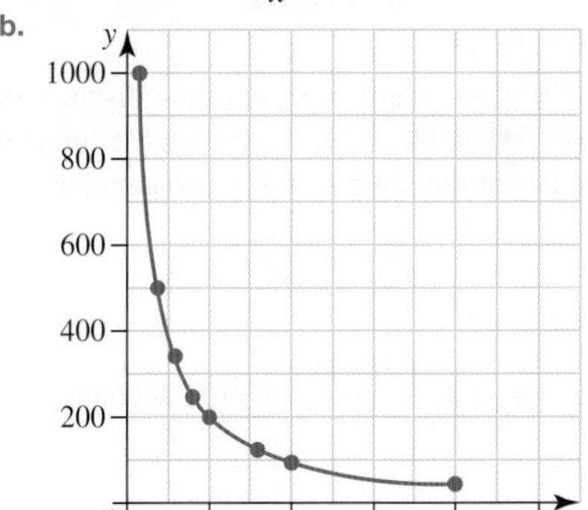

4. a. $k = 48, p = \frac{48}{q}$
b.

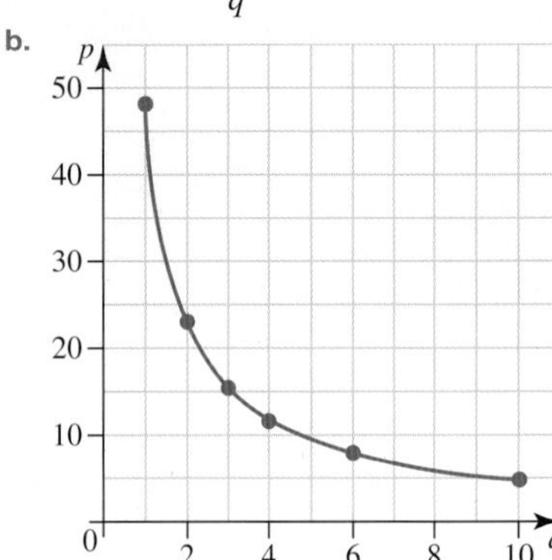

5. a. $k = 42, y = \frac{42}{x}$

b. 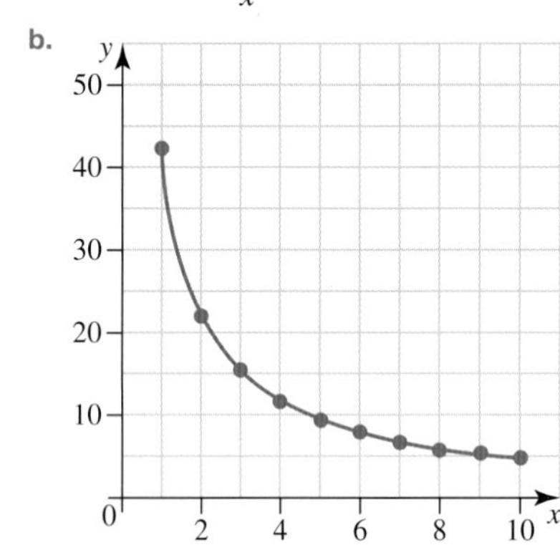

6. a. 4000 b. $a = \frac{4000}{m}$

c. $20\ \text{m/s}^2$ d. $4\ \text{m/s}^2$

7. a. 500 b. 2500 pencils

c. 1000 pencils

8. a. 337.5 b. 3.97 hours = 3 h 58 min

c. 96.4 km/h d. 337.5 km

9. a. 17 500 b. $233.33

c. 70 people d. $17 500

10. a. 200

b. 1 ampere

c. 13.3 ohms

d. Sample responses can be found in the worked solutions in the online resources.

11. a. k = 15 b. $V = 20\,L$

c. $P = 2.5$ atmospheres

12. The constant of proportionality represents the distance of the trip; therefore, when this value is smaller, the time taken to complete the trip at the same speed is also smaller.

13. a. 10 b. $T = \frac{10}{P}$

c. 1 hour 40 minutes

14. a. 100.048 b. $T = \frac{100.048}{V}$ c. 9.58 seconds

d. 15.00 seconds

15. a. $C = \frac{1350}{n}$ b. $96.43 per person

c. $30 per person

Exercise 9.8 Sketching the hyperbola

1. See table at the bottom of the page.*

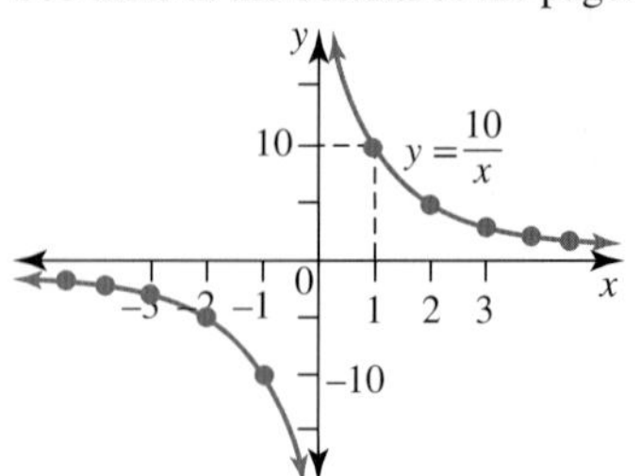

2. a. 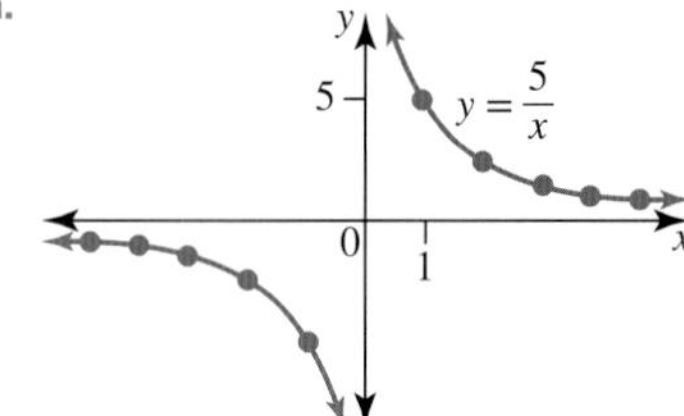

b. $x = 0, y = 0$

3. a. 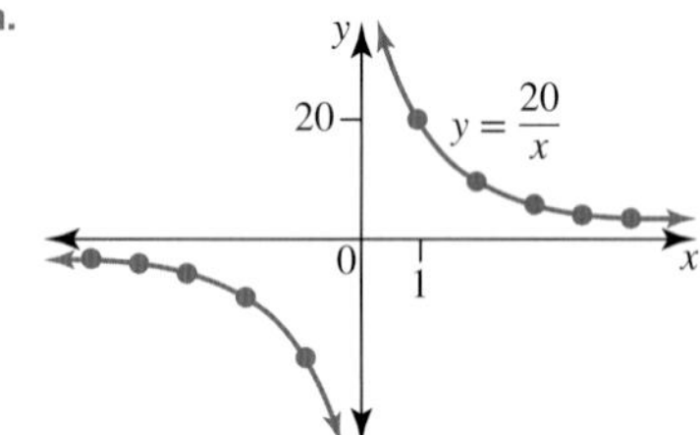

b. $x = 0, y = 0$

4. a. 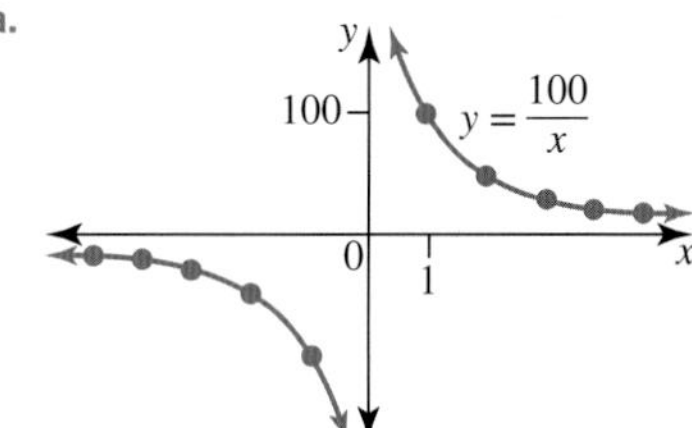

b. $x = 0, y = 0$

5. 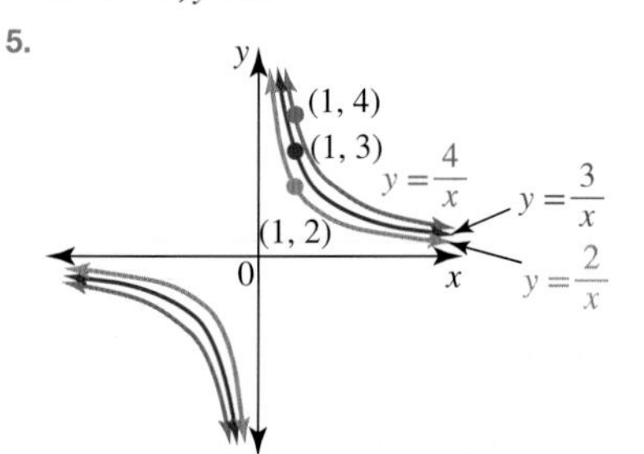

6. It increases the y-values by a factor of k and hence dilates the curve by a factor of k.

*1.

x	−5	−4	−3	−2	−1	0	1	2	3	4	5
y	−2	−2.5	−3.3	−5	−10	Undef.	10	5	3.3	2.5	2

7.

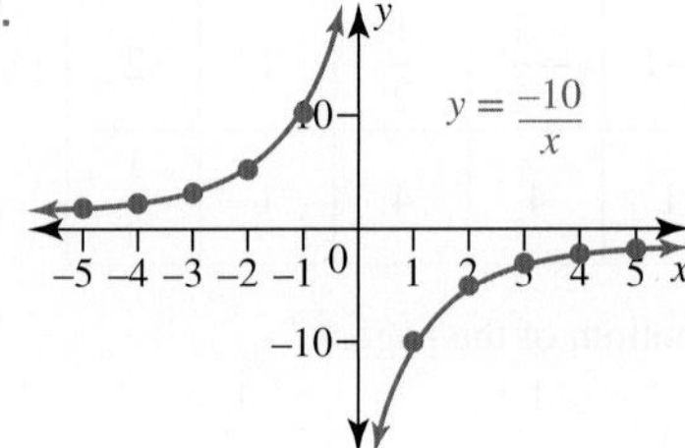

8.

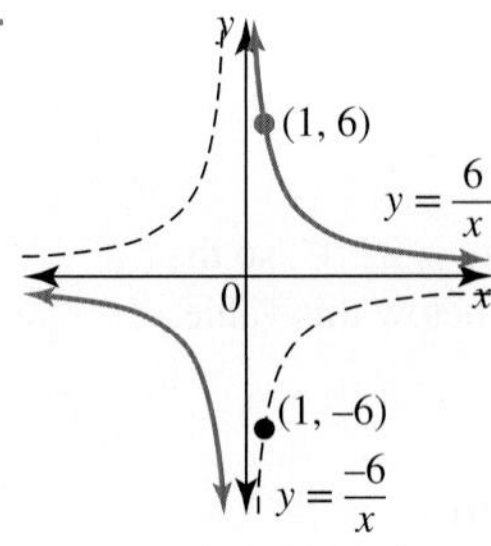

9. The negative reflects the curve $y = \frac{k}{x}$ in the x-axis.

10. See table bottom of the page.*

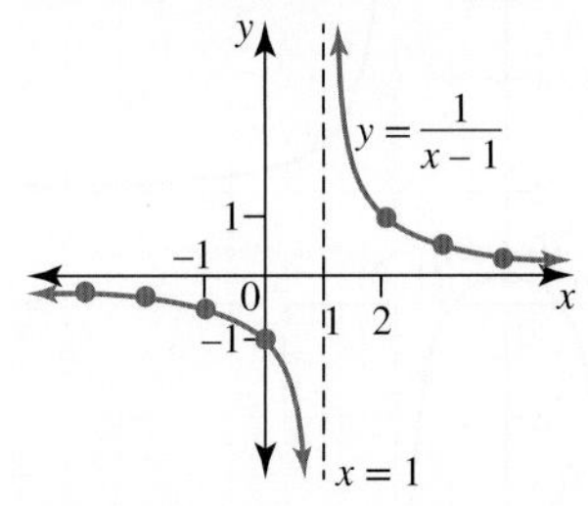

Equation of vertical asymptote is $x = 1$.

11. a.

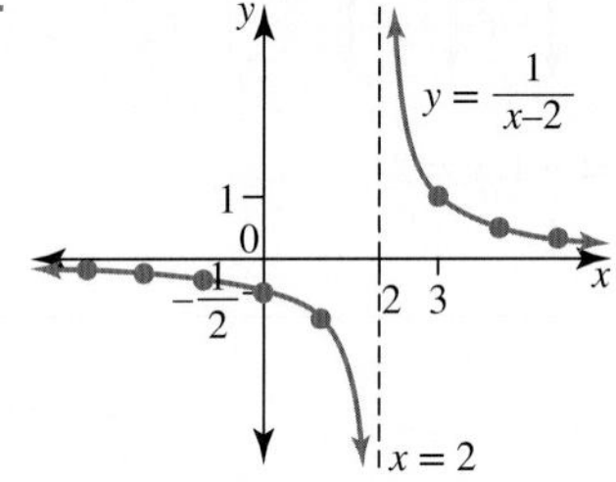

b.

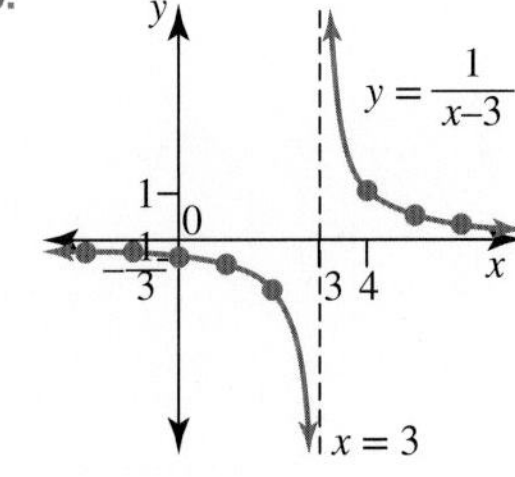

12.

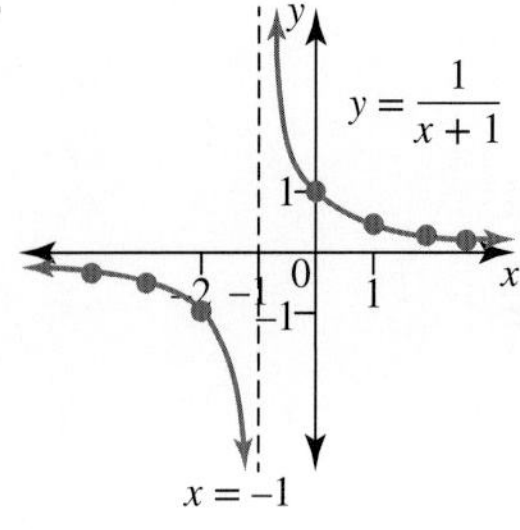

13. The a translates the graph left or right, and $x = a$ becomes the vertical asymptote.

14. a. i. $x = -1, y = 0$

ii.

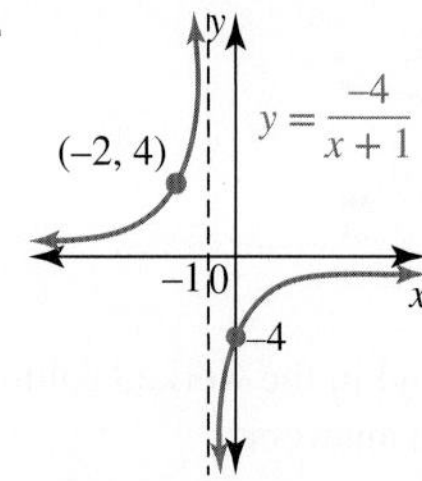

b. i. $x = 1, y = 0$

ii.

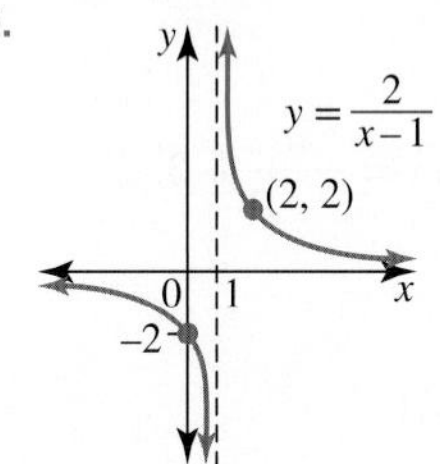

15. a. i. $x = -2, y = 0$

ii.

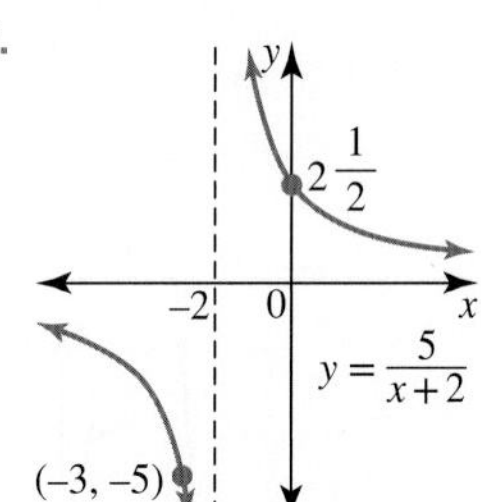

b. i. $x = -2, y = -2$

ii.

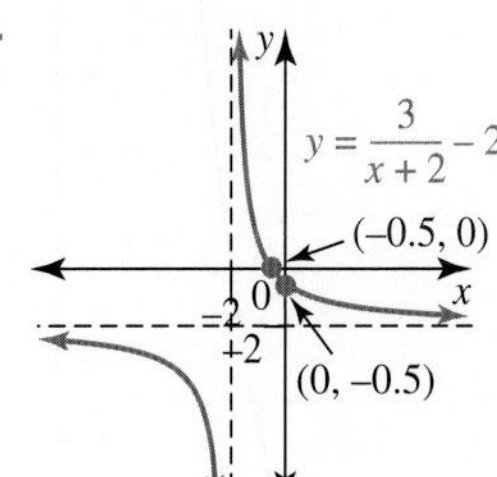

*10. a.

x	−3	−2	−1	0	1	2	3	4
y	−0.25	−0.33	−0.5	−1	Undef.	1	0.5	0.33

16. a. i. $x = -2, y = 1$

ii.

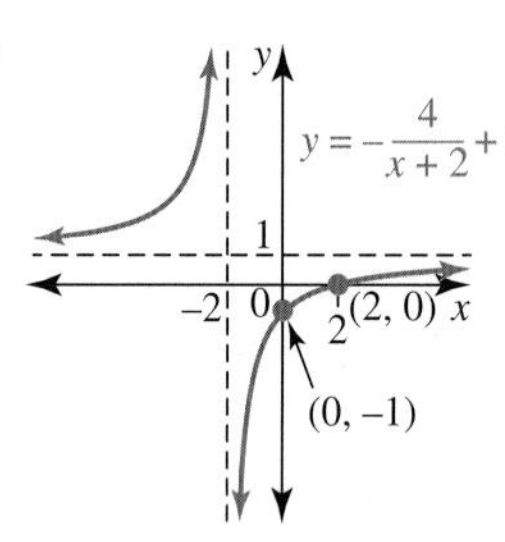

b. i. $x = 3, y = 5$

ii.

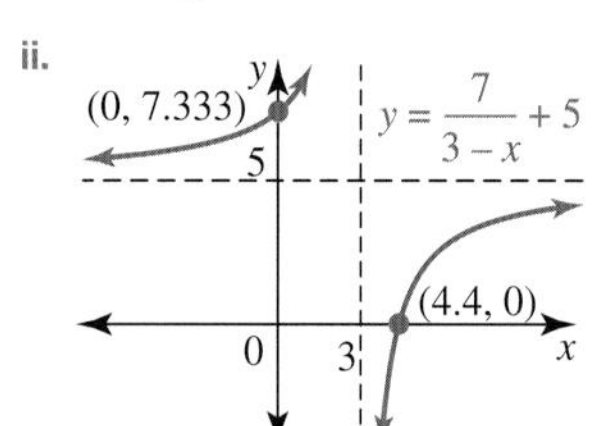

17. Sample responses can be found in the worked solutions in the online resources. Possible answers:

a. $y = \dfrac{1}{x-3}$

b. $y = \dfrac{1}{x+10}$

18. a. Any equation of the form $y = \dfrac{a}{x-2} + 3$.

b. $y = \dfrac{-14}{x+2} + 4$

19. $y = \dfrac{-2}{x+3} - 1, x = -3, y = -1$

20. a.

x	-2	-1	$-\frac{1}{2}$	$\frac{1}{2}$	1	2
y	$\frac{1}{4}$	1	4	4	1	$\frac{1}{4}$

See figure at the bottom of the page.*

b. $x = 0$ c. $y = -\dfrac{1}{x^2}$ d. $y = \dfrac{1}{x^2}$

21. a. 100 °C

b. 33.1 °C

c. 17.85 minutes $\approx$ 18 minutes

d. No. The horizontal asymptote is 22 °C, so the temperature will never drop below this value.

22. a. y-intercept: $\left(0, \dfrac{1}{2}\right)$

x-intercepts: $(1, 0)$ and $(-1, 0)$

b. $x = 2, x = -2$ and $y = 2$

c.

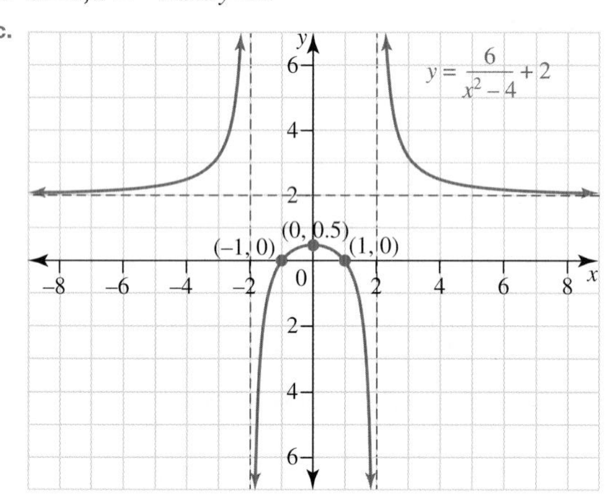

23. $y = -\dfrac{3}{(x+1)^2} + 2, x = -1, y = 2$

*20. a.

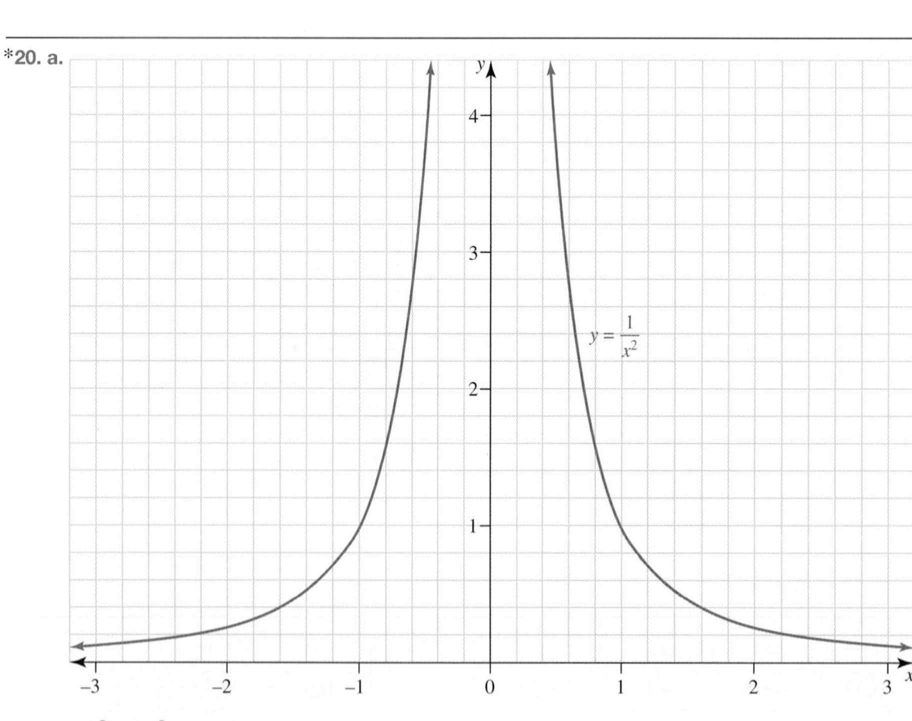

$x = 0, y = 0$

Exercise 9.9 Sketching the circle

1. a.

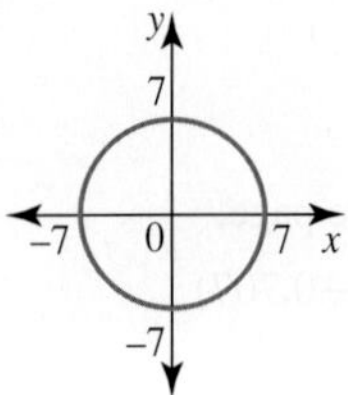

Centre $(0, 0)$, radius 7

b.

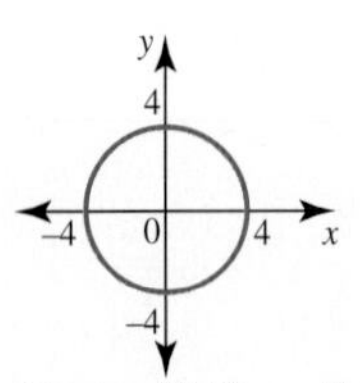

Centre $(0, 0)$, radius 4

2. a.

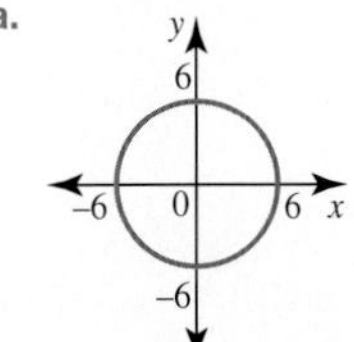

Centre $(0, 0)$, radius 6

b.

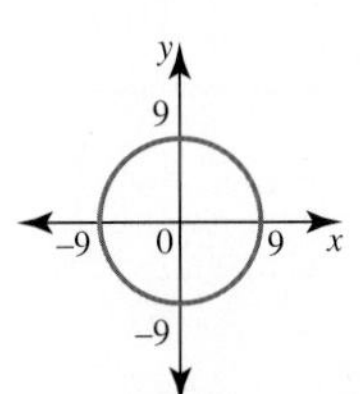

Centre $(0, 0)$, radius 9

3. a.

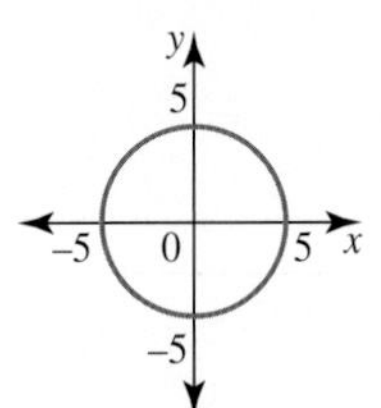

Centre $(0, 0)$, radius 5

b.

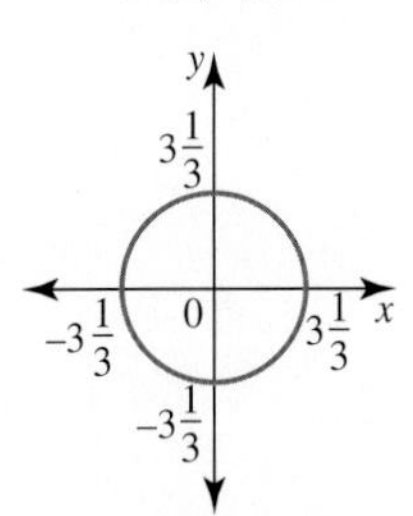

Centre $(0, 0)$, radius $\frac{10}{3}$

4. a.

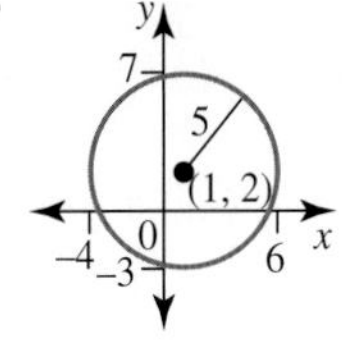

b.

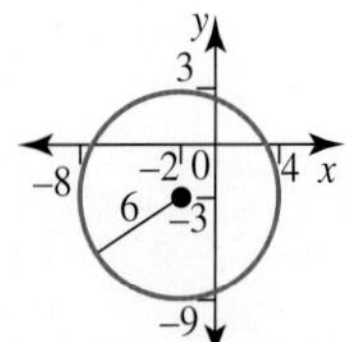

5. a.

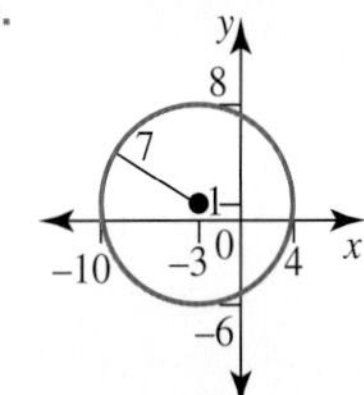

b.

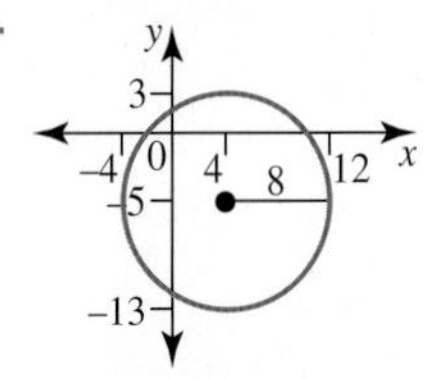

6. a.

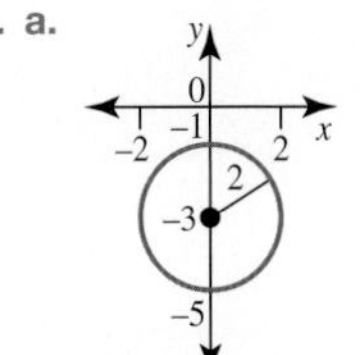

b.

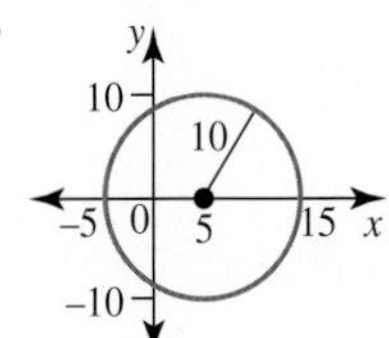

7. a. $(x + 2)^2 + (y + 4)^2 = 2^2$

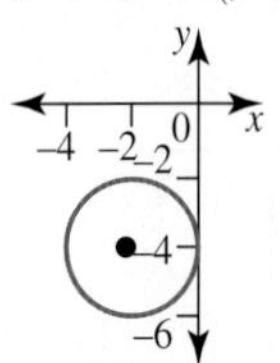

b. $(x - 5)^2 + (y - 1)^2 = 4^2$

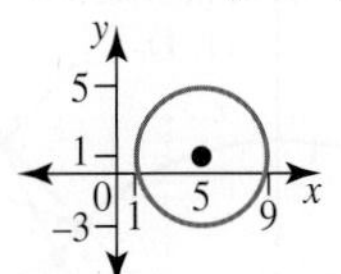

8. a. $(x - 7)^2 + (y + 3)^2 = 7^2$

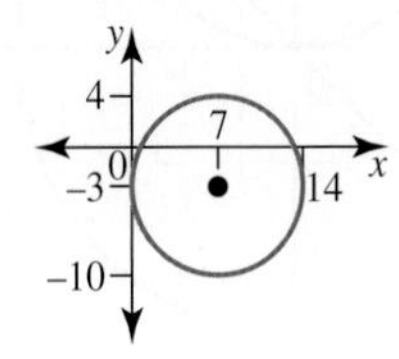

b. $(x + 4)^2 + (y - 6)^2 = 8^2$

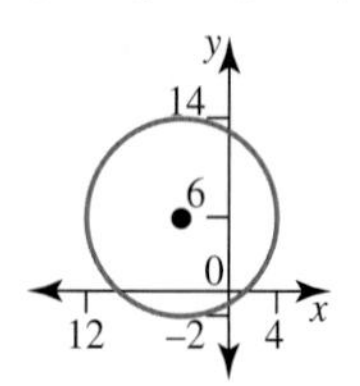

9. a. $x^2 + (y - 9)^2 = 10^2$

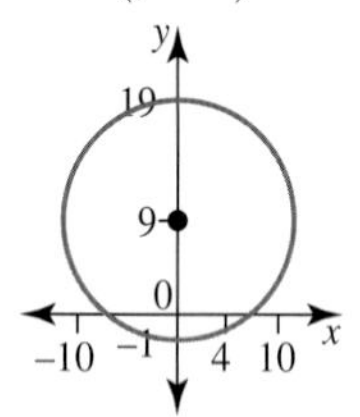

b. $(x - 1)^2 + (y + 2)^2 = 3^2$

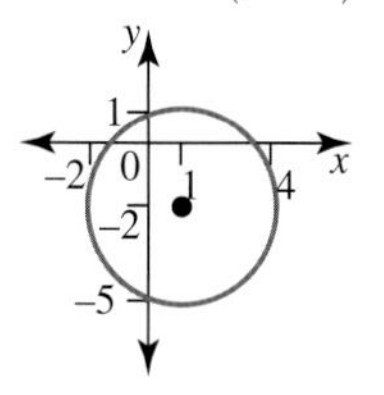

10. D

11. B

12. D

13. $(x - 5)^2 + (y - 3)^2 = 16$

14. a. 2 cm, 13.8 cm

b. 3.9 cm/s

15. $(-2, 0)$

16. a. See the figure at the bottom of the page.*

b. i. $(0.707, 0.707)$ and $(-0.707, -0.707)$

ii. $(0, 0)$ and $(1, 1)$

iii. $(0.786, 0.618)$ and $(-0.786, 0.618)$

17.

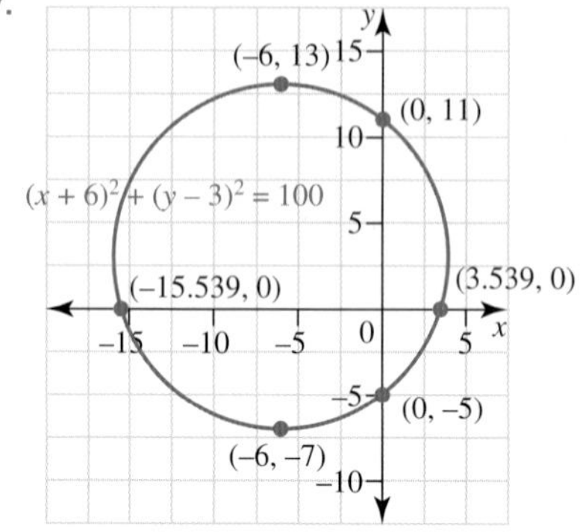

18. $(-3, 4), (3, 4), (0, 5)$ and $(0, -5)$

*16. a.

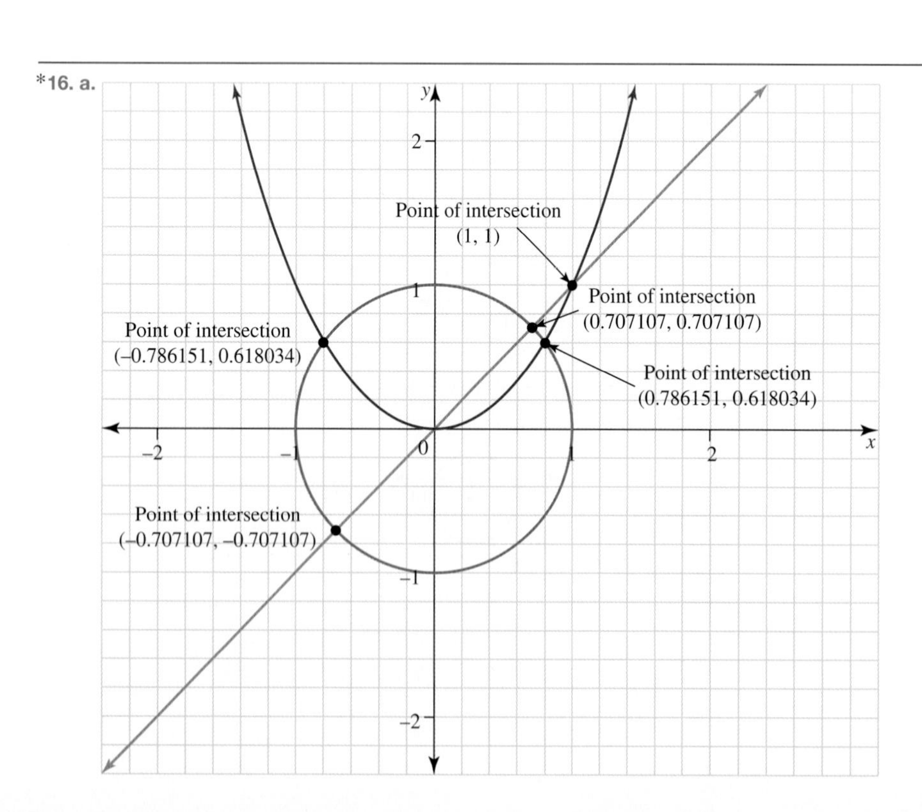

19. $(x-1)^2 + (y-1)^2 = 4$ centre at $(1, 1)$ and radius of 2 units.
$(x-4)^2 + (y-1)^2 = 1$ centre at $(4, 1)$ and radius of 1 unit.
The circles intersect (touch) at $(3, 1)$.
See figure at the bottom of the page.*

20. $(x-2)^2 + (y-5)^2 = 4$ centre at $(2, 5)$ and radius of 2 units.

Project

1.

t	x	y
0	0	0
1	1	1
2	2	4
3	3	9
4	4	16
5	5	25

2.

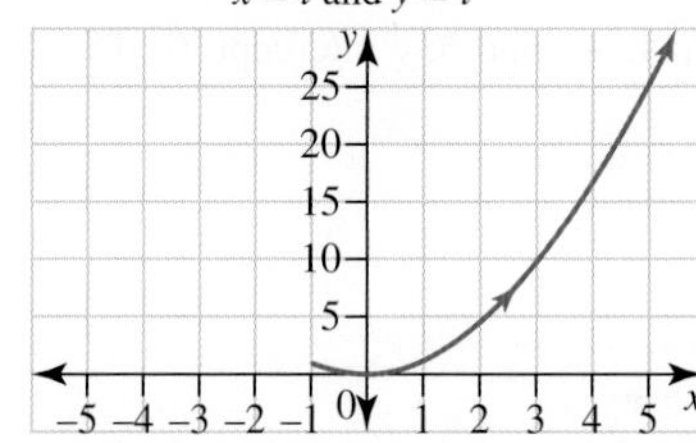

3. Sample response: Yes the graph is different. $Y = x^2$ is a parabola and therefore, the graph appears in the I and the II quadrants. Where as in Q2 the graph only appears in the I quadrant. Other sample responses can be found in the worked solutions in the online resources.

4.

t	x	y
0	1	1
1	0	0
2	−1	1
3	−2	4
4	−3	9
5	−4	16

Sample responses can be found in the worked solutions in the online resources.
$t = 1, 0, -1, -2, -3, -4$.

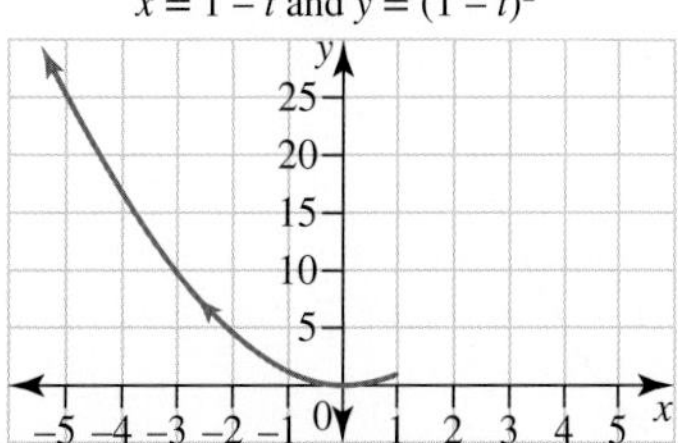

5.

t	x	y
0	0	0
1	1	−1
2	2	−4
3	3	−9
4	4	−16
5	5	−25

*19.

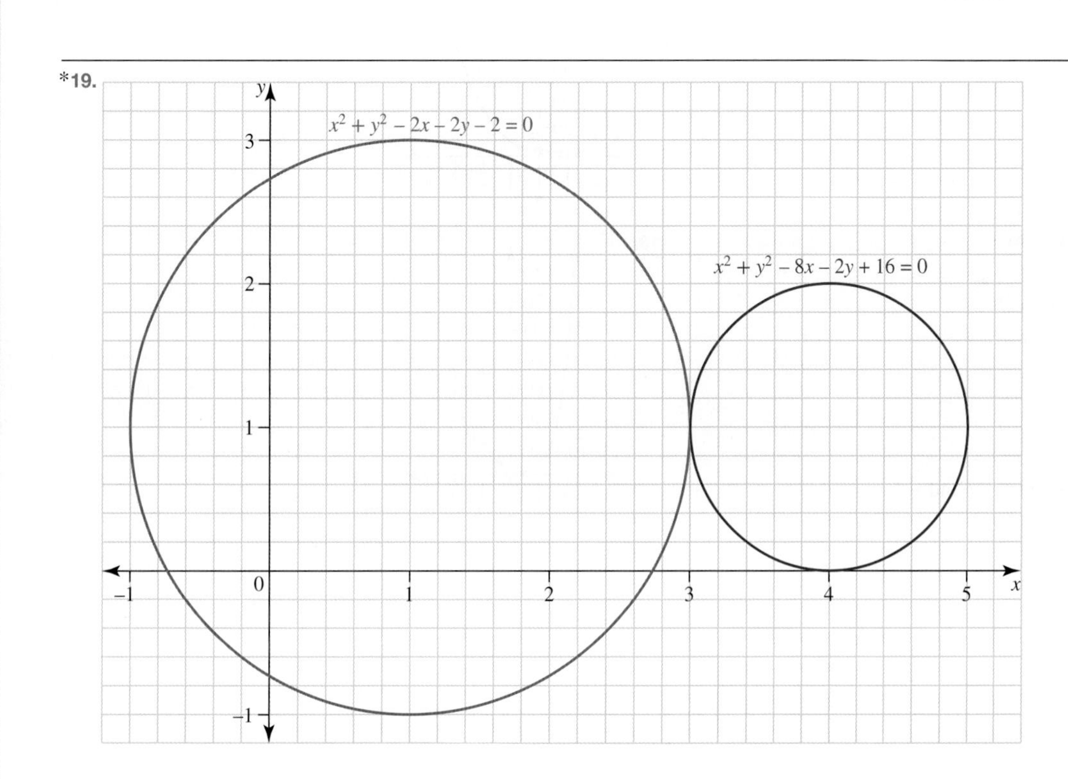

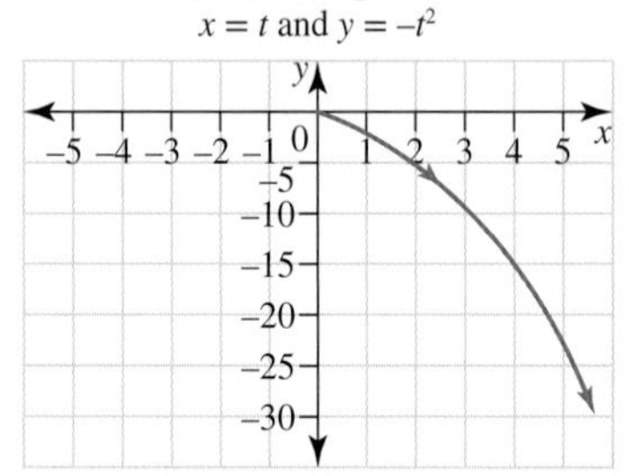

6. Sample response: The shape of the graph will be the same as shown in Q5 but it will be in quadrant III (reflection of the graph in Q5). Other sample responses can be found in the worked solutions in the online resources.

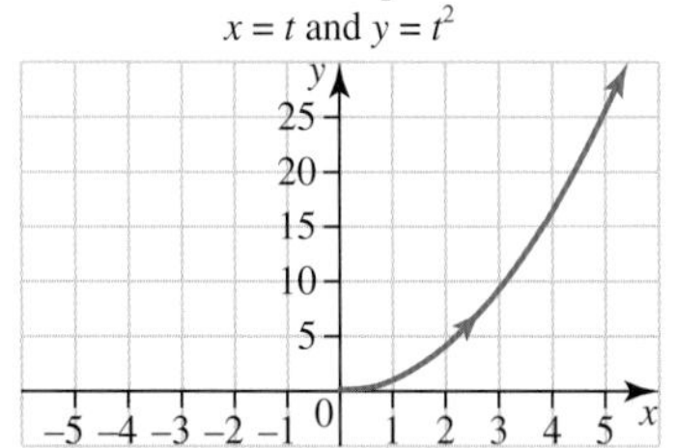

Exercise 9.10 Review questions

1. D
2. A
3. D
4. A
5. C
6. D
7. B
8. a. $(4, -15)$ b. $(-2, -9)$
9. See table at the bottom of the page.*

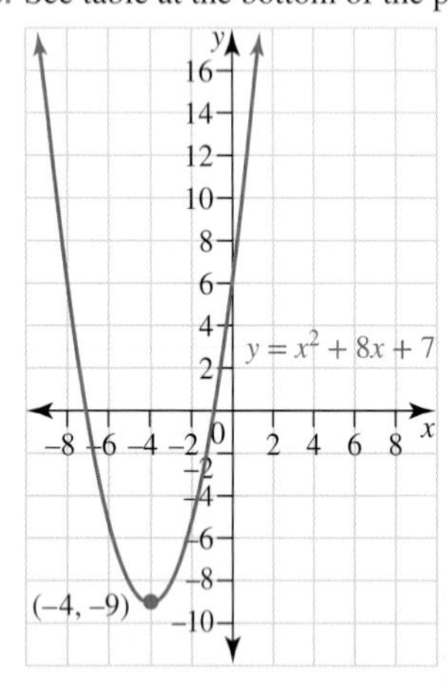

TP $(-4, -9)$; x-intercepts: -7 and -1

10. a. TP $(3, 1)$; no x-intercepts; y-intercept: $(0, 10)$

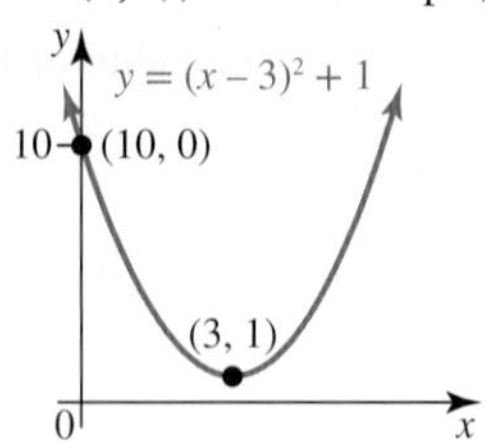

b. TP $(-1, -5)$; x-intercepts: $-1 - \sqrt{\frac{5}{2}}, -1 + \sqrt{\frac{5}{2}}$; y-intercept: $(0, -3)$

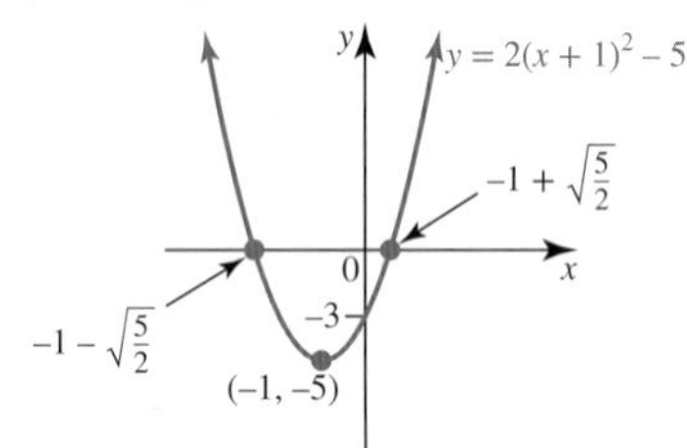

11. TP $(-1, 16)$; x-intercepts: -5 and 3; y-intercept: $(0, 15)$

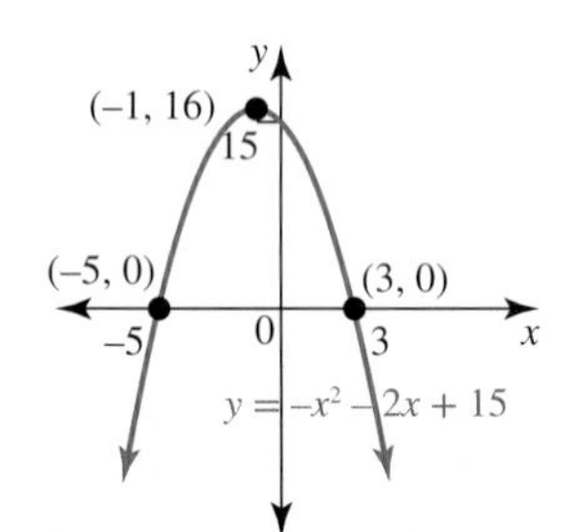

12.

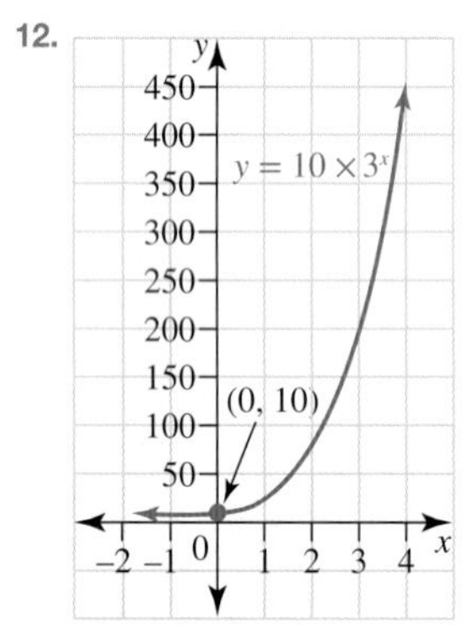

13.

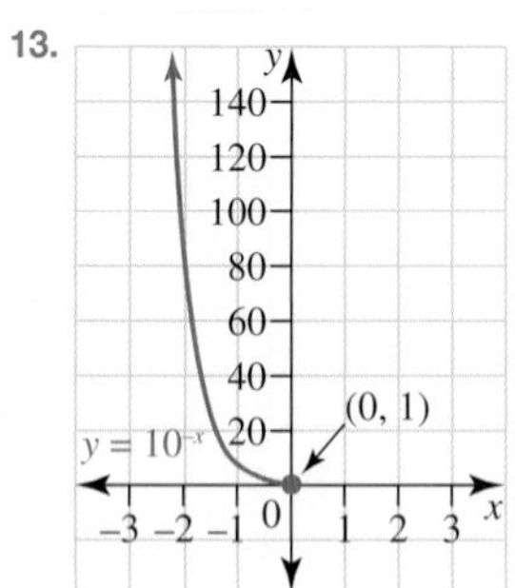

*9.

x	−9	−8	−7	−6	−5	−4	−3	−2	−1	0	1
y	16	7	0	−5	−8	−9	−8	−5	0	7	16

14. a. See table at the bottom of the page.*

b.

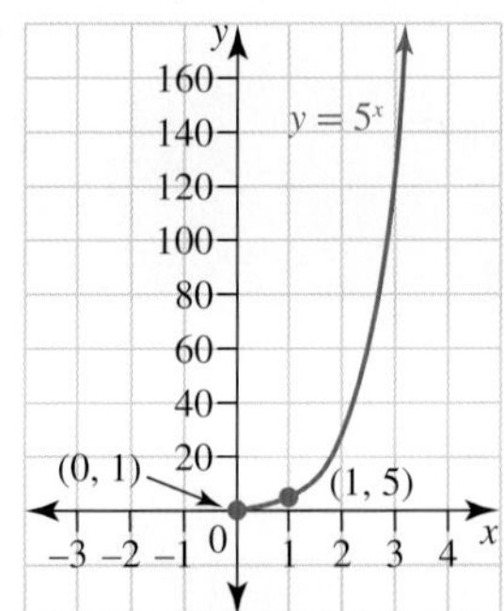

15. a.

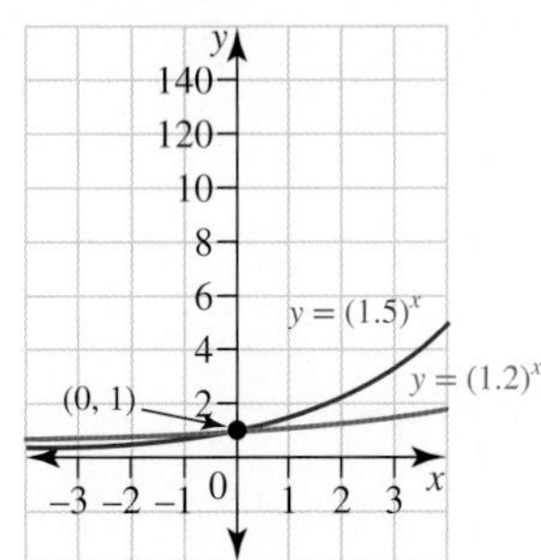

b. Increasing the value of a makes the graph steeper for positive x-values and flatter for negative x-values.

16. a.

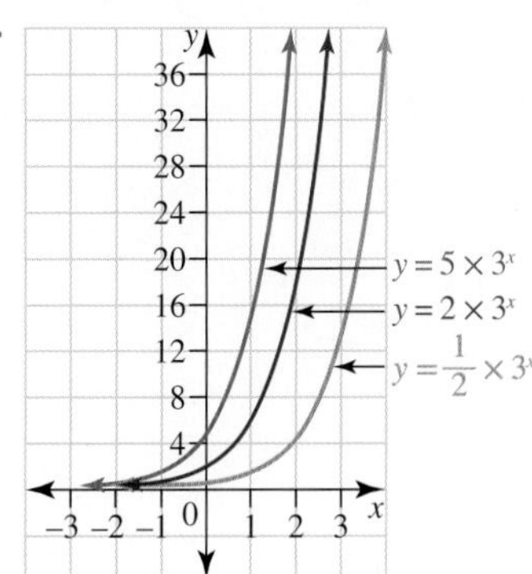

b. Increasing the value of k makes the graph steeper.

17. a.

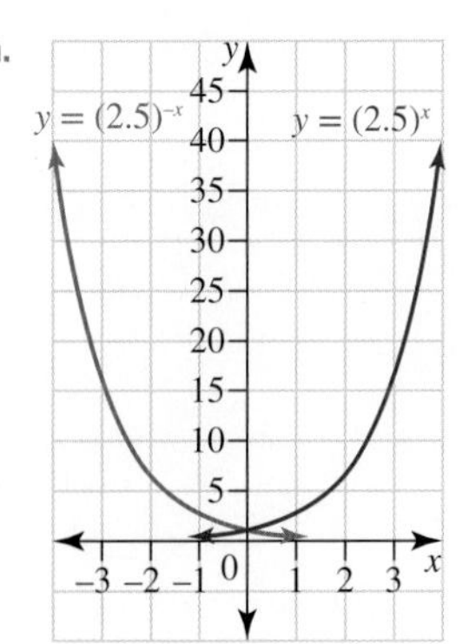

b. Changing the sign of the index reflects the graph in the y-axis.

18. a.

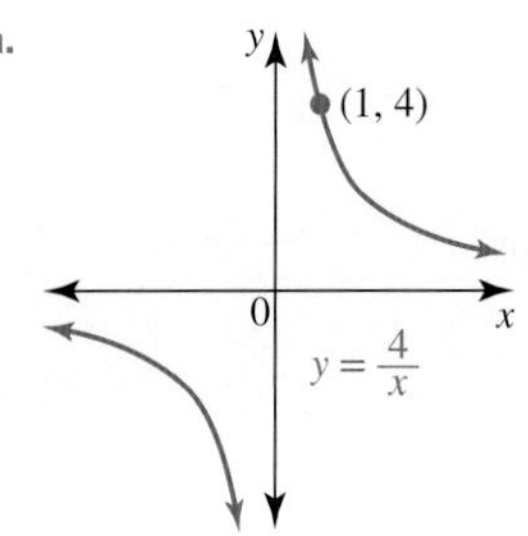

b.

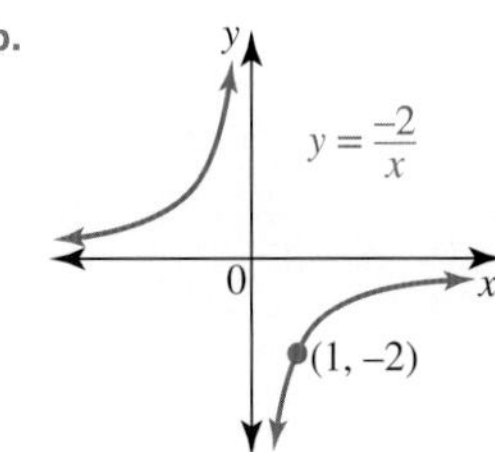

19.

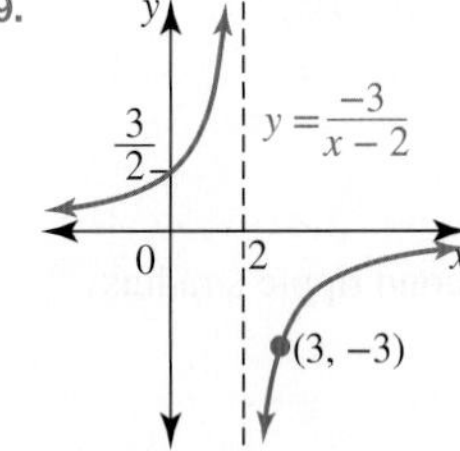

20. Sample responses can be found in the worked solutions in the online resources. Possible answer is $y = \dfrac{1}{x + 3}$.

21. a.

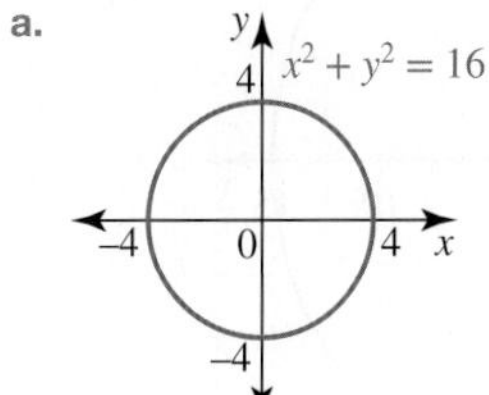

b.

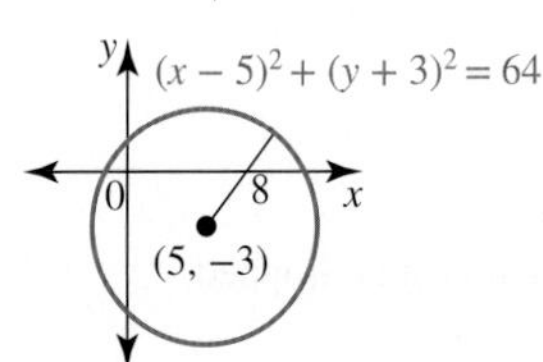

22. a. $x^2 + 4x + y^2 - 2y = 4$

(−2, 1)

b.

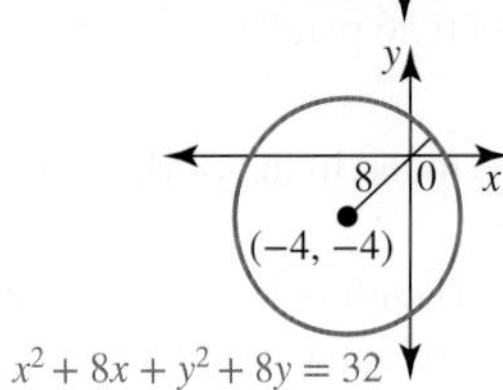

23. $x^2 + y^2 = 36$

*14. a.

x	−3	−2	−1	0	1	2	3
y	0.008	0.04	0.2	1	5	25	125

24. a. 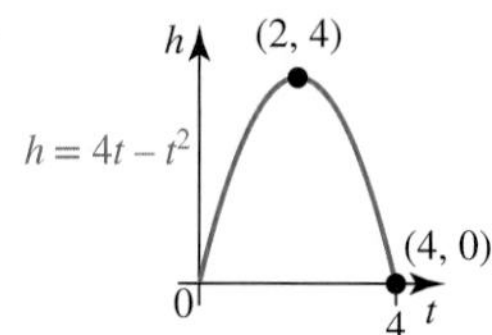

b. 4 m

c. 2 seconds

d. 4 seconds

25. a. 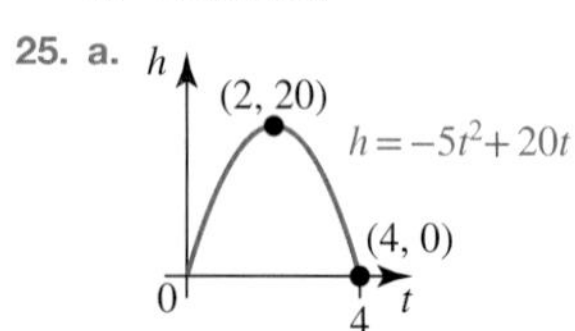

b. 4 seconds

c. 2 seconds ($1 < t < 3$)

d. The ball is never above a height of 20 m.

26. a. $[0, 12]$

b. 32 m

c. 11:41 am to 6:19 pm.

27. a. First ripple's radius is 3 cm, second ripple's radius is 15 cm.

b. 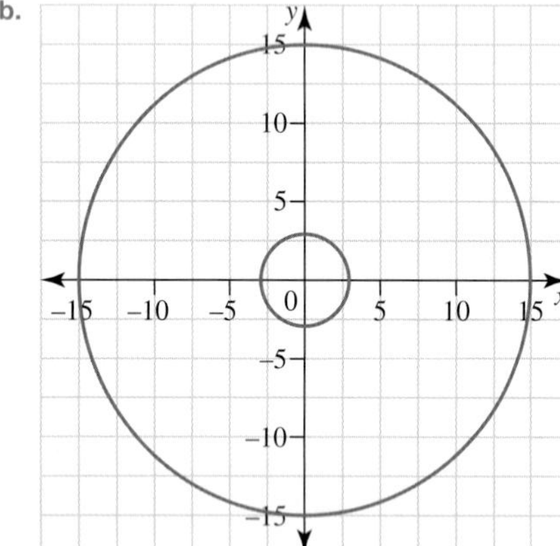

c. 2.4 cm/s

d. 1 minute 22.1 seconds after it is dropped.

28. a. 0.14 g/ml b. 0.7 g/ml

29. a. $P \leq -x^2 + 5x, 0 \leq x \leq 5$

b. 6.25 m

c. i. Sample responses can be found in the worked solutions in the online resources.

ii. Dilation by a factor of 0.56 parallel to the y-axis.

d. 66.7%

30. a. Sample responses can be found in the worked solutions in the online resources.

b. When $x = 0.3$, $b = 10.7$. Therefore if p is greater than 10.7 cm the platform would hit the bridge.

10 Deductive geometry

LEARNING SEQUENCE

10.1 Overview

Why learn this?

Geometry is an area of mathematics that has an abundance of real-life applications. The first important skill that geometry teaches is the ability to reason deductively and prove logically that certain mathematical statements are true.

Euclid (c. 300 BCE) was the mathematician who developed a systematic approach to geometry, now referred to as Euclidean geometry, that relied on mathematical proofs. Mathematicians and research scientists today spend a large part of their time trying to prove new theories, and they rely heavily on all the proofs that have gone before.

Geometry is used extensively in professions such as navigation and surveying. Planes circle our world, land needs to be surveyed before any construction can commence, and architects, designers and engineers all use geometry in their drawings and plans. Geometry is also used extensively in software and computing. Computer-aided design programs, computer imaging, animations, video games and 3D printers all rely greatly on built-in geometry packages.

Just about every sport involves geometry heavily. In cricket alone there are many examples; bowlers adjust the angle at which they release the ball to make the ball bounce towards the batsmen at different heights; fielders are positioned so they cover as much of the ground as efficiently as possible and batsmen angle their bat as they hit the ball to ensure the ball rolls along the ground instead of in the air. Netballers must consider the angle at which they shoot the ball to ensure it arcs into the ring and cyclists must consider the curved path of their turns that will allow them to corner in the quickest and most efficient way.

Where to get help

Go to your learnON title at **www.jacplus.com.au** to access the following digital resources. The Online Resources Summary at the end of this topic provides a full list of what's available to help you learn the concepts covered in this topic.

Video eLessons

Digital documents

Interactivities

eWorkbook

Fully worked solutions to every question

Exercise 10.1 Pre-test

learnON

Complete this pre-test in your learnON title at www.jacplus.com.au and receive **automatic marks**, **immediate corrective feedback** and **fully worked solutions**.

1. Calculate the value of the pronumeral x.

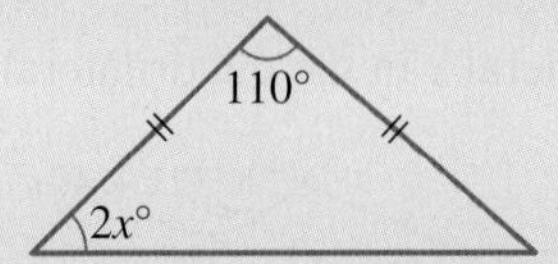

2. Calculate the value of the pronumeral x.

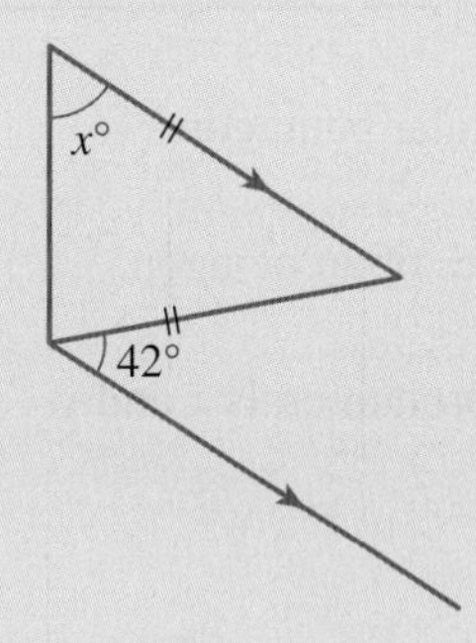

3. State what type of triangles have the same size and shape.

4. Determine the values of the pronumerals x and y.

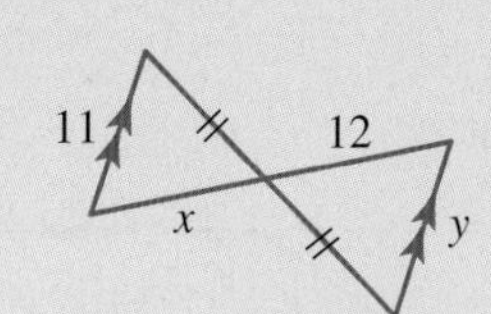

5. Triangles ABC and DEF are congruent; calculate the values of the pronumerals x, y and z.

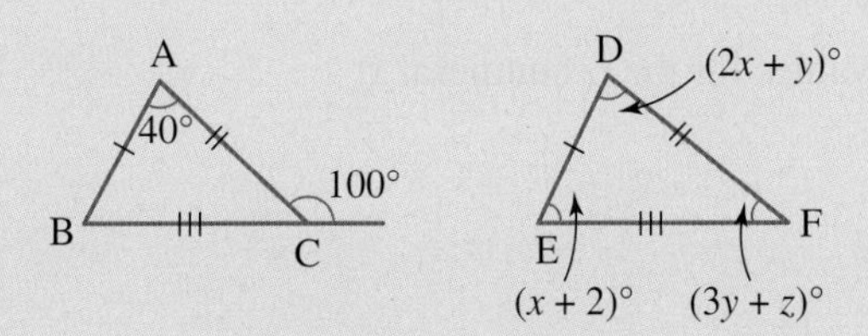

6. **MC** Choose which congruency test will prove these triangles are congruent.

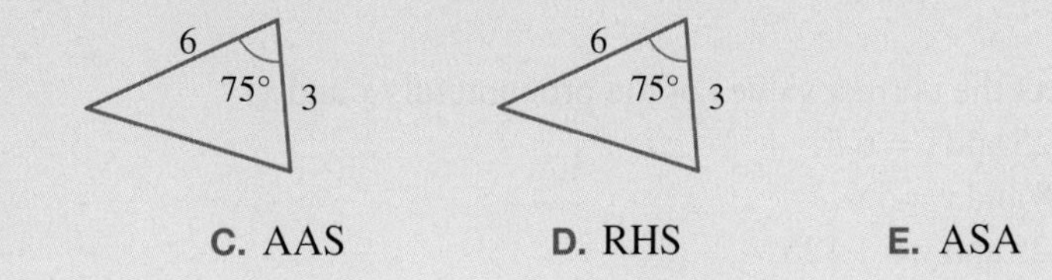

A. SAS **B.** SSS **C.** AAS **D.** RHS **E.** ASA

7. State what the AAA test checks about two triangles.

8. Calculate the value of the pronumeral x in the quadrilateral shown.

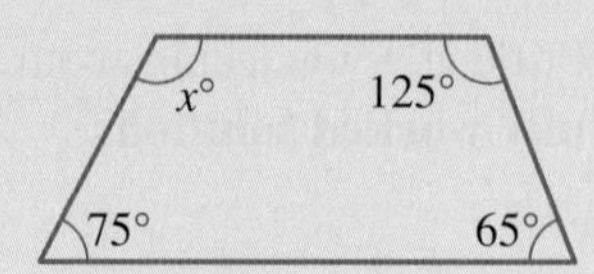

9. Determine the value of the pronumeral x in the quadrilateral shown.

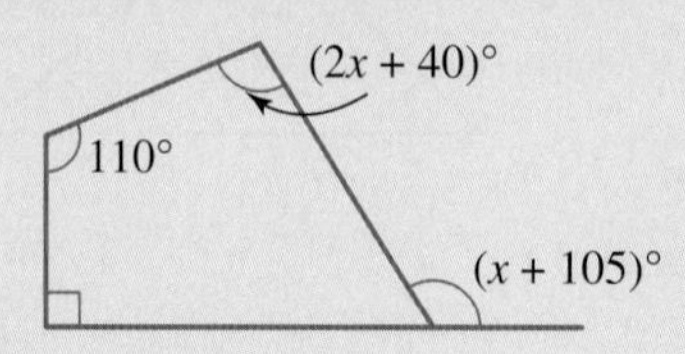

10. Determine the exterior angle of a regular pentagon.

11. Evaluate the sum of the interior angles of an octagon.

12. **MC** Select the correct values of the pronumerals x and y.

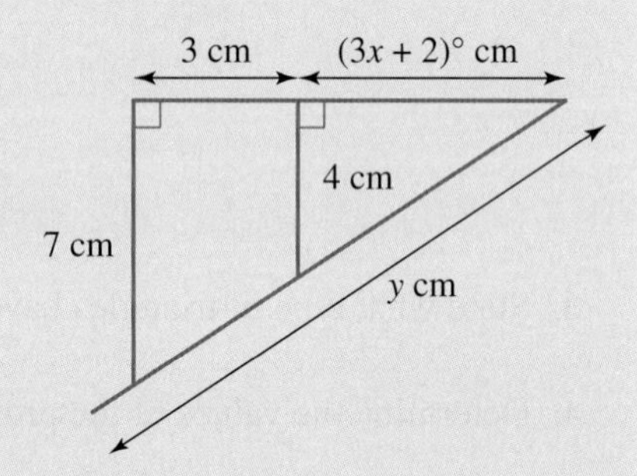

A. $x = \frac{1}{3}, y = 7\sqrt{2}$

B. $x = \frac{2}{3}, y = 5\sqrt{2}$

C. $x = \frac{2}{3}, y = 7\sqrt{2}$

D. $x = 1\frac{1}{3}, y = 7\sqrt{2}$

E. $x = \frac{2}{3}, y = \sqrt{78}$

13. Calculate the value of the pronumeral x.

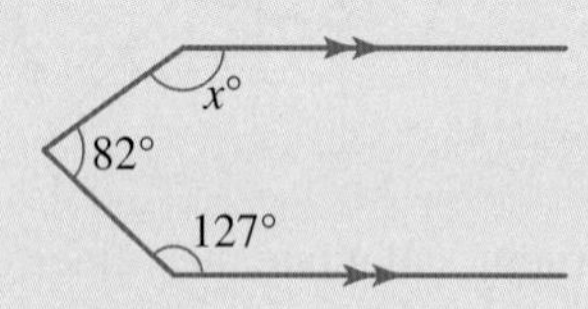

14. **MC** Choose the correct values for the pronumeral x.

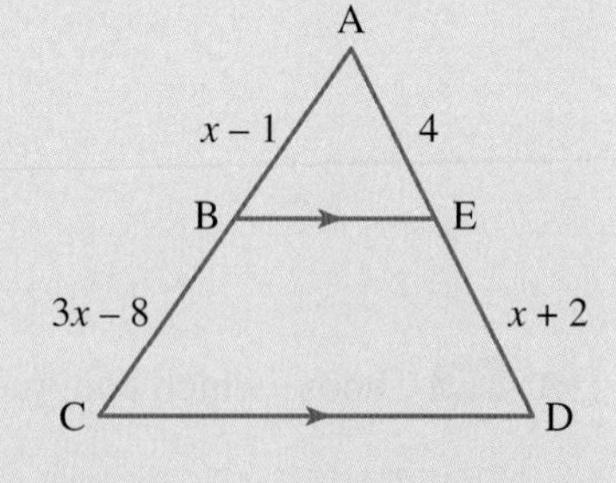

A. $x = 6$ or $x = 5$

B. $x = 4$ or $x = 4$

C. $x = 5$ or $x = 4$

D. $x = 6$ or $x = 4$

E. $x = 5$ or $x = 5$

15. **MC** Select the correct values of the pronumerals x and y.

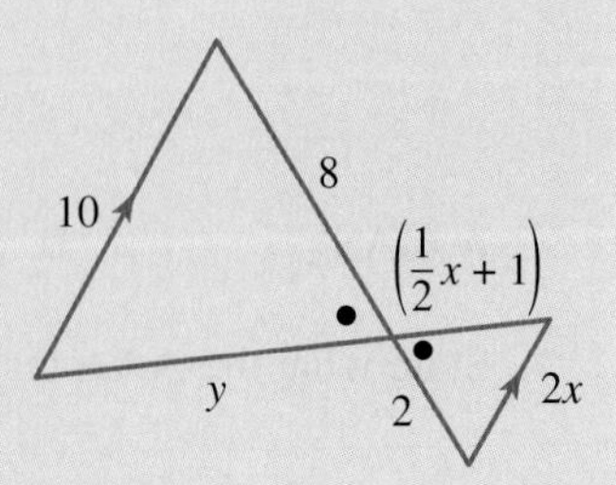

A. $x = 1.25$ and $y = 6.5$

B. $x = 0.8$ and $y = 5.6$

C. $x = 0.25$ and $y = 1.125$

D. $x = 5$ and $y = 3.5$

E. $x = 0.2$ and $y = 1.1$

10.2 Angles, triangles and congruence

LEARNING INTENTION

At the end of this subtopic you should be able to:
- apply properties of straight lines and triangles to determine the value of an unknown angle
- construct simple geometric proofs for angles in triangles or around intersecting lines
- prove that triangles are congruent by applying the appropriate congruency test.

10.2.1 Proofs and theorems of angles

eles-4892

- Euclid (c. 300 BC) was the mathematician who developed a systematic approach to geometry, now referred to as Euclidean geometry, that relied on mathematical proofs.
- A **proof** is an argument that shows why a statement is true.
- A **theorem** is a statement that can be demonstrated to be true. It is conventional to use the following structure when setting out a theorem.
  - **Given:** a summary of the information given
  - **To prove:** a statement that needs to be proven
  - **Construction:** a description of any additions to the diagram given
  - **Proof:** a sequence of steps that can be justified and form part of a formal mathematical proof.

Sums of angles

Angles at a point

- The sum of the angles at a point is 360°.

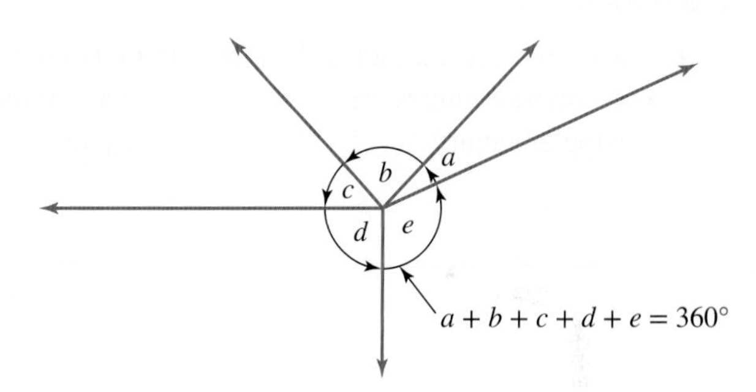

Supplementary angles

- The sum of the angles on a straight line is 180°.
- Angles that add up to 180° are called **supplementary angles**.
- In the diagram angles a, b and c are supplementary.

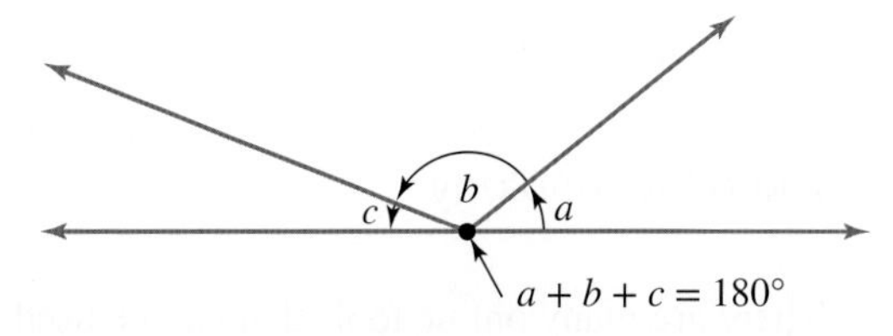

Complementary angles

- The sum of the angles in a right angle is 90°.
- Angles that add up to 90° are called **complementary angles**.
- In the diagram angles a, b and c are complementary.

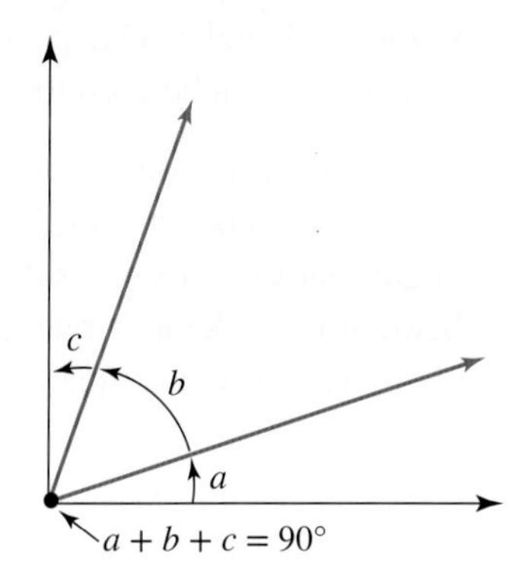

Vertically opposite angles

Theorem 1

- **Vertically opposite angles** are equal.

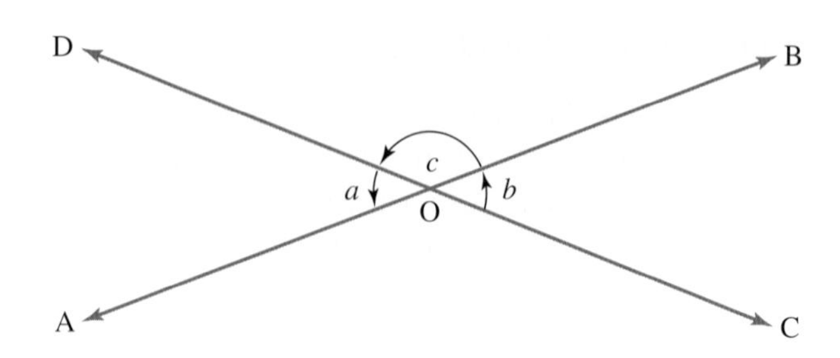

Given: Straight lines AB and CD, which intersect at O.

To prove: $\angle AOD = \angle BOC$ and $\angle BOD = \angle AOC$

Construction: Label $\angle AOD$ as a, $\angle BOC$ as b and $\angle BOD$ as c.

Proof: Let $\angle AOD = a°$, $\angle BOC = b°$ and $\angle BOD = c°$.

$a + c = 180°$ (supplementary angles)

$b + c = 180°$ (supplementary angles)

$\therefore a + c = b + c$

$\therefore a = b$

So, $\angle AOD = \angle BOC$.

Similarly, $\angle BOD = \angle AOC$.

Parallel lines

- If two lines are parallel and cut by a **transversal**, then:
  - co-interior angles are supplementary
  - corresponding angles are equal
  - alternate angles are equal.

$a + b = 180°$

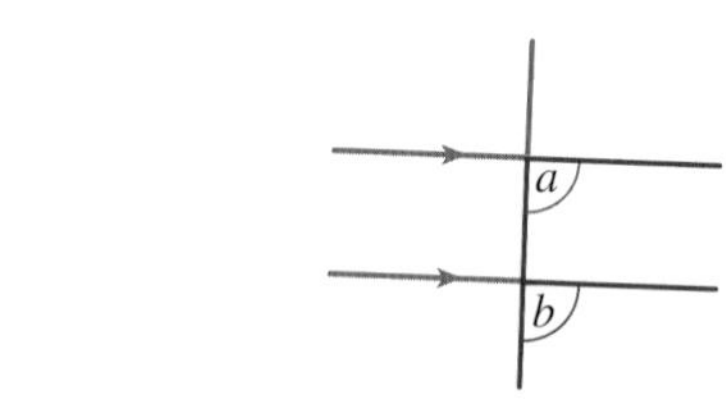

$a = b$

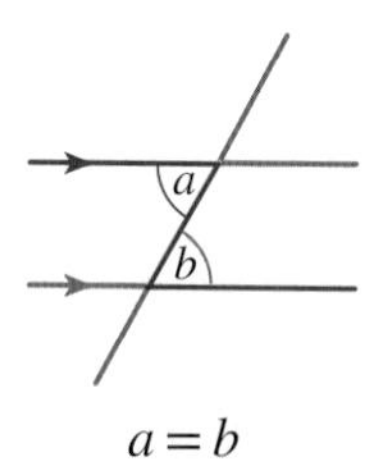

$a = b$

Digital technology

There are many online tools that can be used to play around with lines, shapes and angles. One good tool to explore is the Desmos geometry tool, which can be used for free at www.desmos.com/geometry.

In the Desmos geometry tool, you can draw lines, circles, polygons and all kinds of other shapes. You can then use the angle tool to explore the angles between lines or sides. The figure at right shows the angle tool being used to demonstrate that co-interior angles are supplementary and alternate angles are equal.

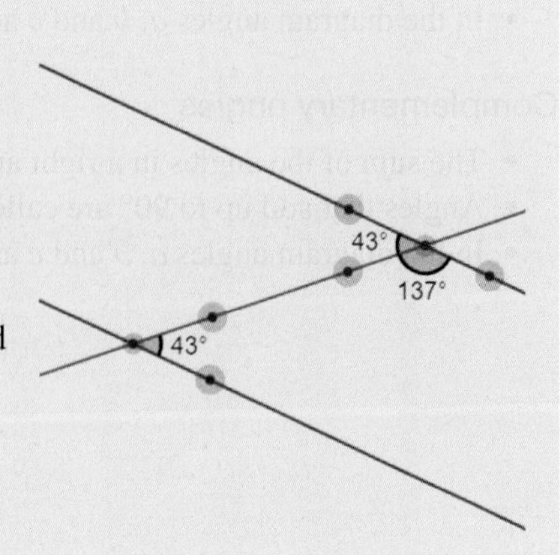

10.2.2 Angle properties of triangles

eles-5353

Theorem 2

- The sum of the interior angles of a triangle is 180°.

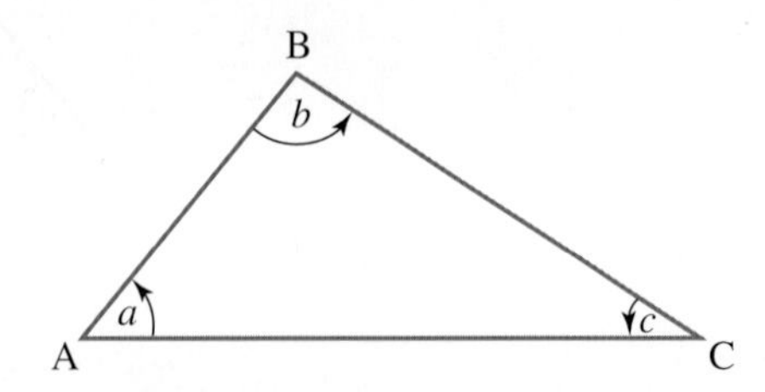

Given: ΔABC with interior angles a, b and c

To prove: $a + b + c = 180°$

Construction: Draw a line parallel to AC, passing through B and label it DE as shown.
Label $\angle$ABD as x and $\angle$CBE as y.

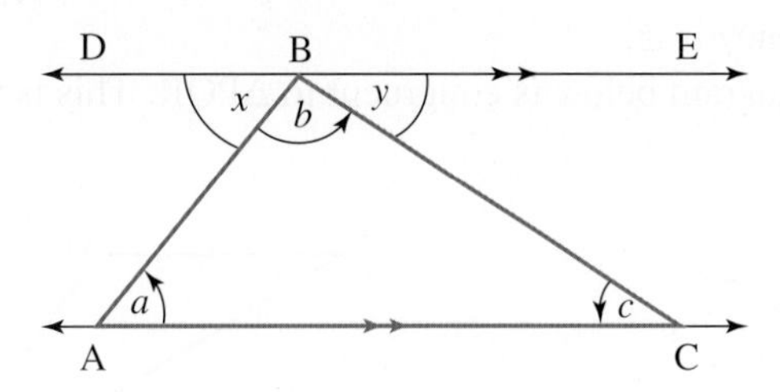

Proof:

$a = x$	(alternate angles)
$c = y$	(alternate angles)
$x + b + y = 180°$	(supplementary angles)
$\therefore a + b + c = 180°$	

Equilateral triangles

- It follows from Theorem 2 that each interior angle of an **equilateral triangle** is 60°, and, conversely, if the three angles of a triangle are equal, then the triangle is equiangular.

$$a + a + a = 180° \quad \text{(sum of interior angles in a triangle is 180°)}$$
$$3a = 180°$$
$$a = 60°$$

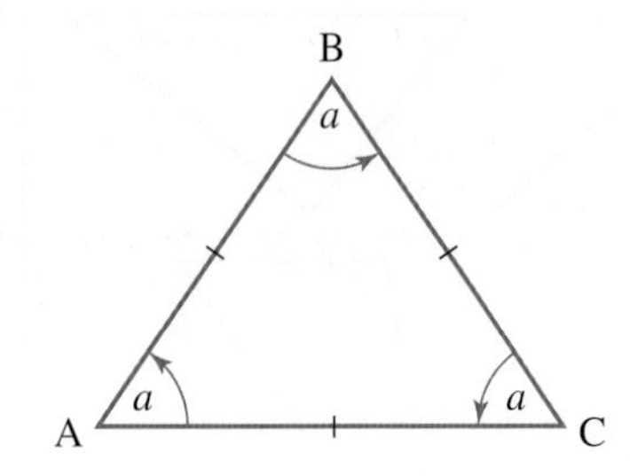

Theorem 3

- The exterior angle of a triangle is equal to the sum of the opposite interior angles.

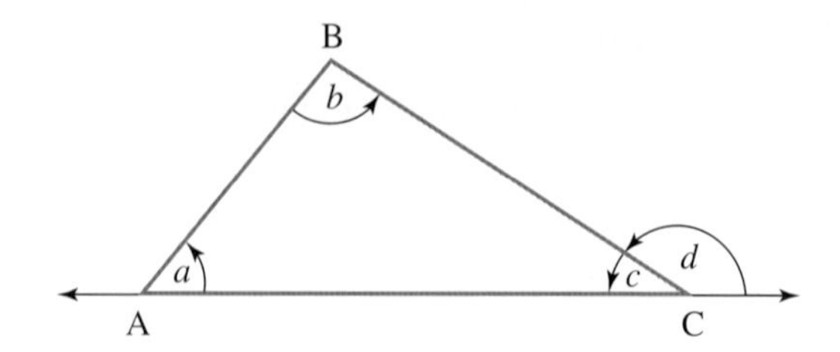

Given:	ΔABC with the exterior angle labelled d	
To prove:	$d = a + b$	
Proof:	$c + d = 180°$	(supplementary angles)
	$a + b + c = 180°$	(sum of interior angles in a triangle is 180°)
	$\therefore d = a + b$	

10.2.3 Congruent triangles

eles-4897

- **Congruent triangles** have the same size and the same shape; that is, they are identical in all respects.
- The symbol used for congruency is $\cong$.
- For example, ΔABC in the diagram below is congruent to ΔPQR. This is written as $\Delta ABC \cong \Delta PQR$.

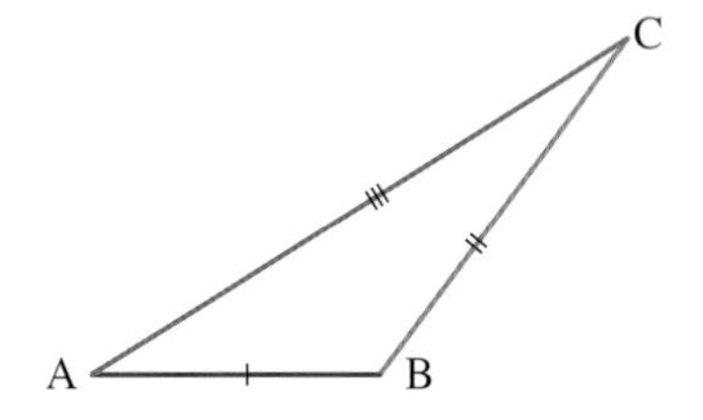

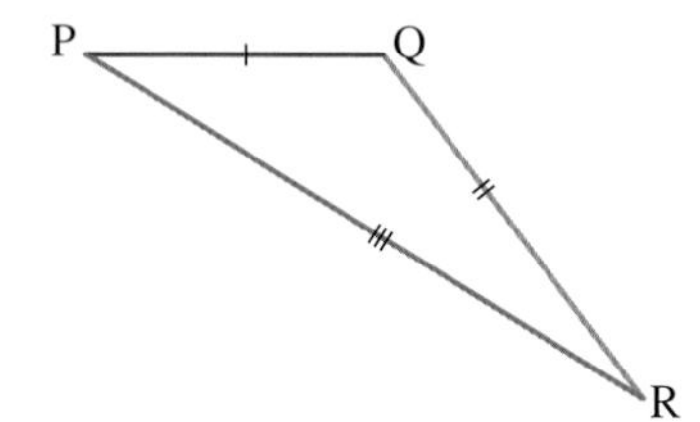

- Note that the vertices of the two triangles are written in corresponding order.
- There are four tests designed to check whether triangles are congruent. The tests are summarised in the table below.

Congruence test	Example	Description
Side-side-side (SSS)		The three corresponding sides are the same lengths.
Side-angle-side (SAS)		Two corresponding sides are the same length and the angle in between these sides is equal.
Angle-angle-side (AAS)		Two corresponding angles are equal and the side in between these angles is the same length.

Congruence test	Example	Description
		A pair of corresponding angles and a non-contained side are equal.
Right angle-hypotenuse-side (RHS)		The triangles are right-angled, and the hypotenuse and one other side of one triangle are equal to the hypotenuse and a side of the other triangle.

- In each of the tests we need to show three equal measurements about a pair of triangles in order to show they are congruent.

WORKED EXAMPLE 1 Determining pairs of congruent triangles

Select a pair of congruent triangles from the diagrams shown, giving a reason for your answer.

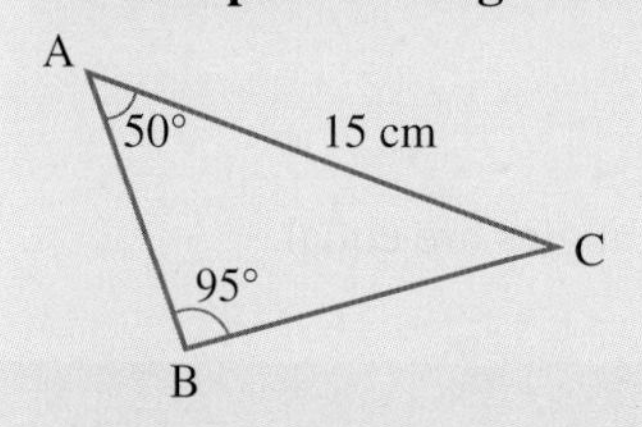

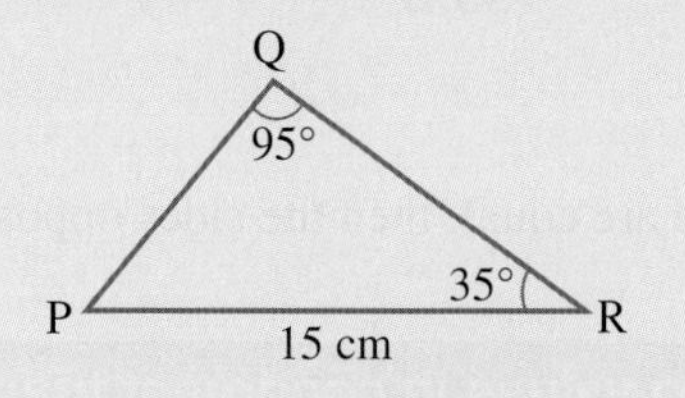

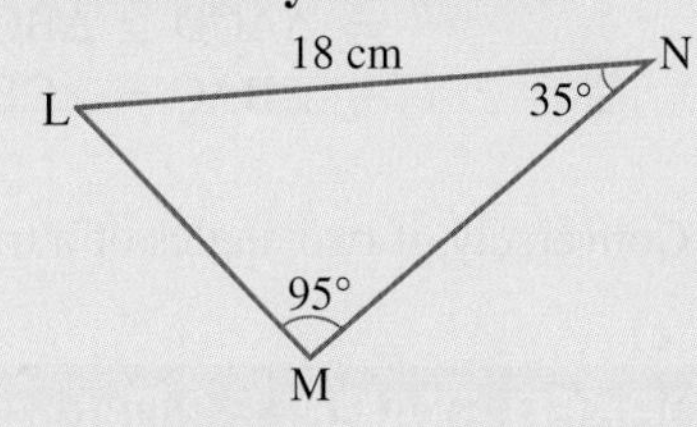

THINK	WRITE
1. In each triangle the length of the side opposite the 95° angle is given. If triangles are to be congruent, the sides opposite the angles of equal size must be equal in length. Draw your conclusion.	All three triangles have equal angles, but the sides opposite the angle 95° are not equal. $AC = PR = 15$ cm and $LN = 18$ cm
2. To test whether ΔABC is congruent to ΔPQR, first evaluate the angle C.	ΔABC: $\angle A = 50°$, $\angle B = 95°$, $\angle C = 180° - 50° - 95°$ $= 35°$
3. Apply a test for congruence. Triangles ABC and PQR have a pair of corresponding sides equal in length and 2 pairs of angles the same, so draw your conclusion.	A pair of corresponding angles ($\angle B = \angle Q$ and $\angle C = \angle R$) and a corresponding side ($AP = PR$) are equal. $\Delta ABC \cong \Delta PQR$ (AAS)

10.2.4 Isosceles triangles

eles-4898

- A triangle is isosceles if the lengths of two sides are equal but the third side is not equal.

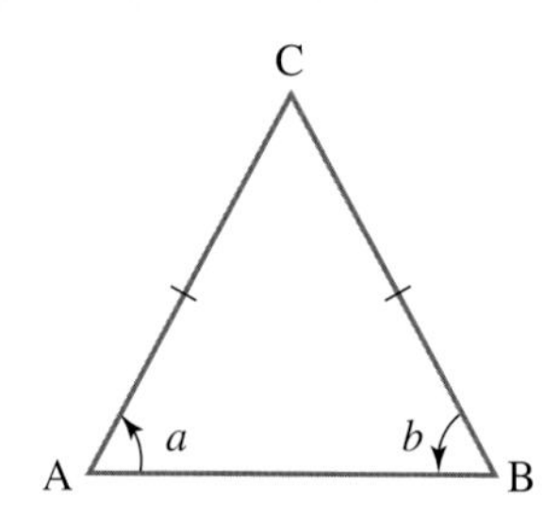

Theorem 4

- The angles at the base of an **isosceles triangle** are equal.

Given: AC = CB

To prove: ∠BAC = ∠CBA

Construction: Draw a line from the vertex C to the midpoint of the base AB and label the midpoint D. CD is the bisector of ∠ACB.

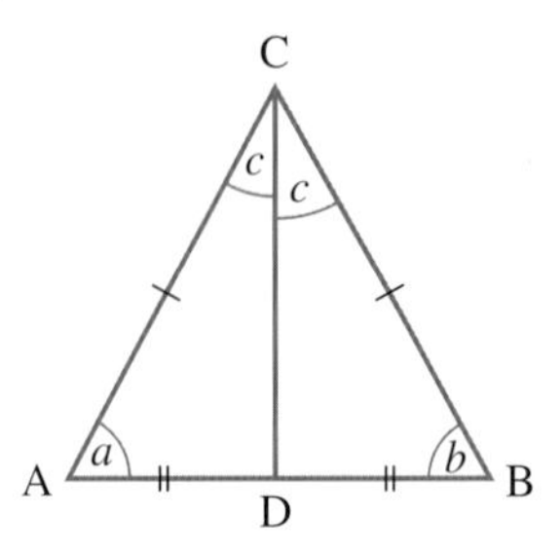

Proof: In ΔACD and ΔBCD,

CD = CD	(common side)
AD = DB	(construction, D is the midpoint of AB)
AC = CB	(given)
⇒ ΔACD ≅ ΔBCD	(SSS)
∴ ∠BAC = ∠CBA	

- Conversely, if two angles of a triangle are equal, then the sides opposite those angles are equal.

WORKED EXAMPLE 2 Determining values in congruent triangles

Given that ΔABD ≅ ΔCBD, determine the values of the pronumerals in the figure shown.

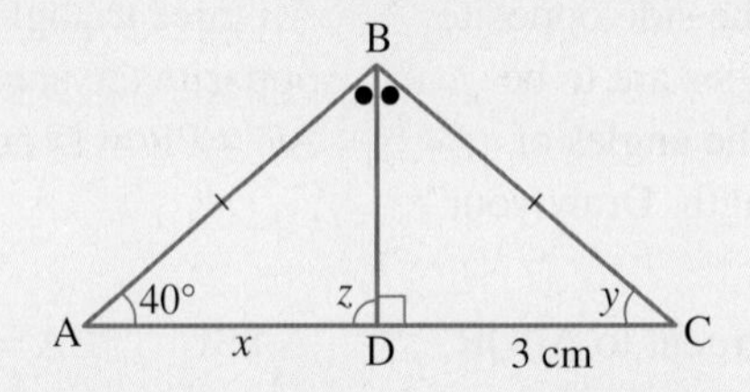

THINK	WRITE
1. In congruent triangles corresponding sides are equal in length. Side AD (marked x) corresponds to side DC, so state the value of x.	ΔABD ≅ ΔCBD AD = CD, AD = x, CD = 3 So x = 3 cm.
2. Since the triangles are congruent, corresponding angles are equal. State the angles corresponding to y and z and hence determine the values of these pronumerals.	∠BAD = ∠BCD ∠BAD = 40°, ∠BCD = y So y = 40° ∠BDA = ∠BDC ∠BDA = z, ∠BDC = 90° So z = 90°.

WORKED EXAMPLE 3 Proving two triangles are congruent

Prove that ΔPQS is congruent to ΔRSQ.

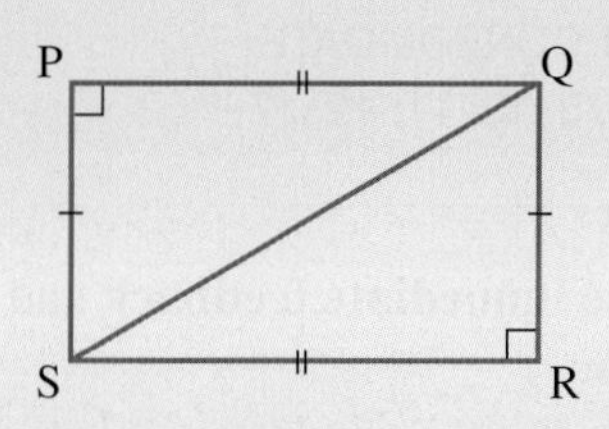

THINK

1. Write the information given.
2. Write what needs to be proved.
3. Select the appropriate congruency test for proof. (In this case, it is RHS because the triangles have an equal side, a right angle and a common hypotenuse.)

WRITE

Given: Rectangle PQRS with diagonal QS.

To prove: that ΔPQS is congruent to ΔRSQ.

QP = SR (opposite sides of a rectangle)

$\angle SPQ = \angle SRQ = 90°$ (given)

QS is common.

So ΔPQS ≅ ΔRSQ (RHS)

Resources

eWorkbook Topic 10 Workbook (worksheets, code puzzle and project) (ewbk-2036)

Digital documents SkillSHEET Naming angles, lines and figures (doc-5276)
SkillSHEET Corresponding sides and angles of congruent triangles (doc-5277)
SkillSHEET Angles and parallel lines (doc-5280)

Interactivities Individual pathway interactivity: Angles, triangles and congruence (int-4612)
Angles at a point (int-6157)
Supplementary angles (int-6158)
Angles in a triangle (int-3965)
Interior and exterior angles of a triangle (int-3966)
Vertically opposite and adjacent angles (int-3968)
Corresponding angles (int-3969)
Co-interior angles (int-3970)
Alternate angles (int-3971)
Congruency tests (int-3755)
Congruent triangles (int-3754)
Angles in an isosceles triangle (int-6159)

Exercise 10.2 Angles, triangles and congruence

learnON

Individual pathways

■ PRACTISE	■ CONSOLIDATE	■ MASTER
1, 4, 7, 8, 9, 16	2, 5, 10, 11, 12, 17	3, 6, 13, 14, 15, 18

To answer questions online and to receive **immediate feedback** and **sample responses** for every question, go to your learnON title at www.jacplus.com.au.

Fluency

1. Determine the values of the unknown in each of the following.

a. 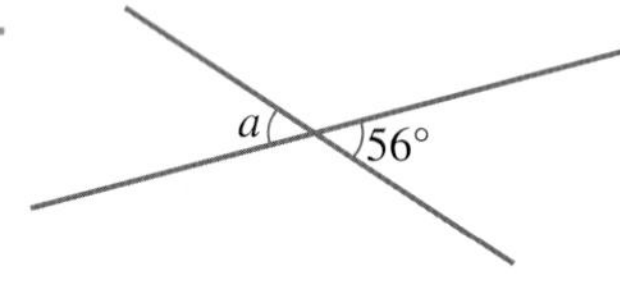

b. 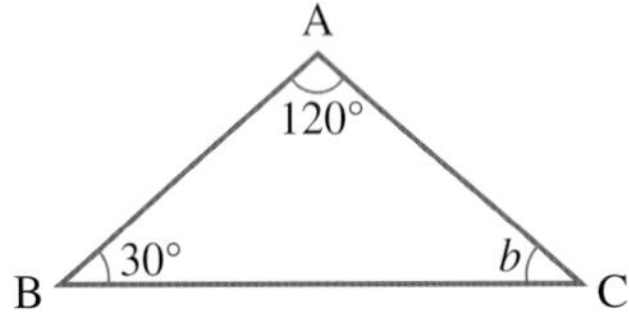

c. 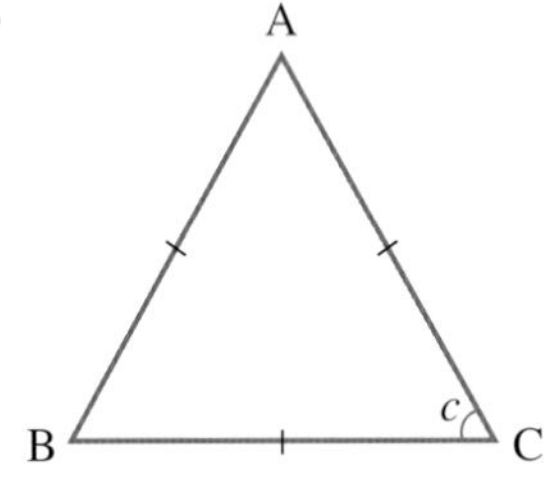

d. 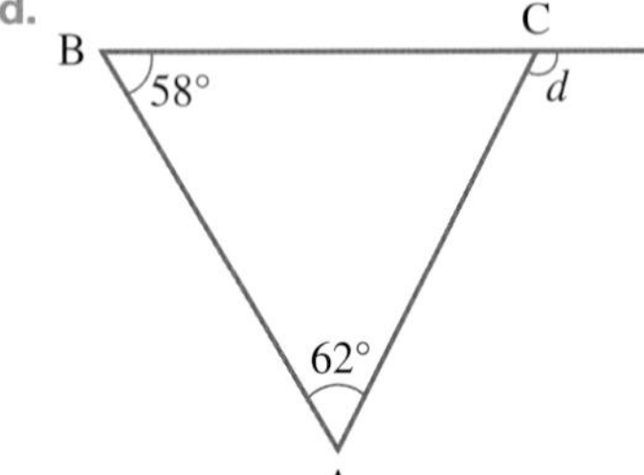

e. 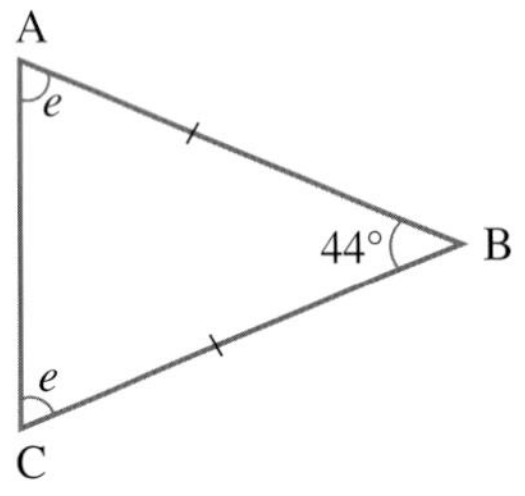

2. Determine the values of the pronumerals in the following diagrams.

a. 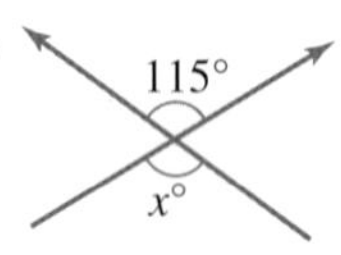

b. 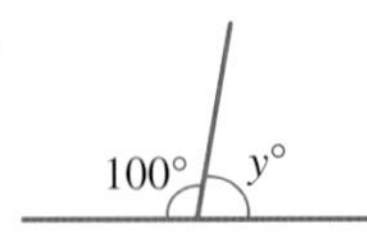

c. 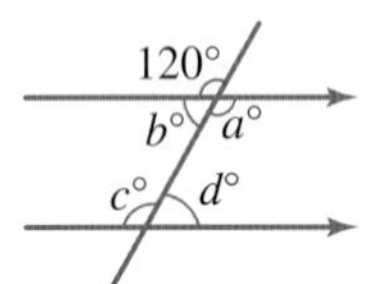

d. 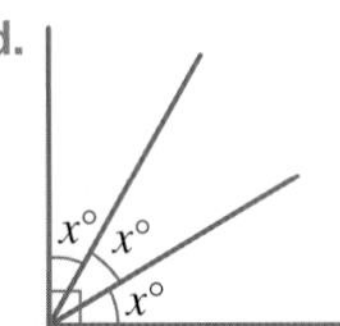

3. **WE1** Select a pair of congruent triangles in each of the following, giving a reason for your answer. All side lengths are in cm.

a. 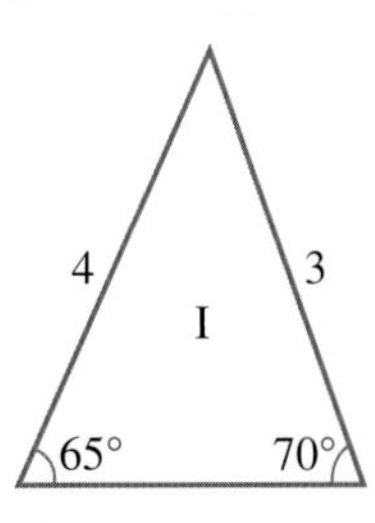

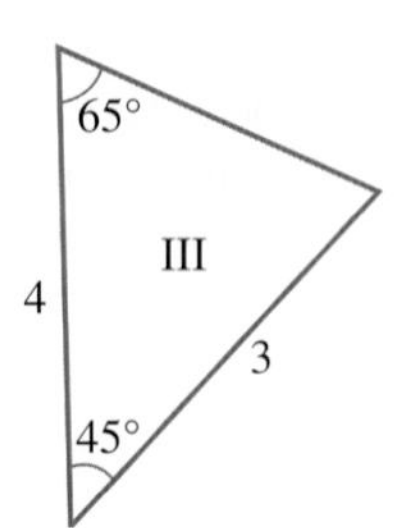

b. 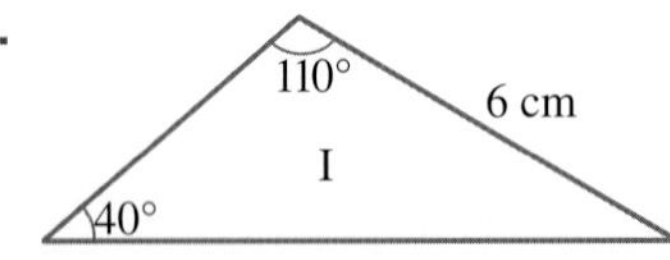

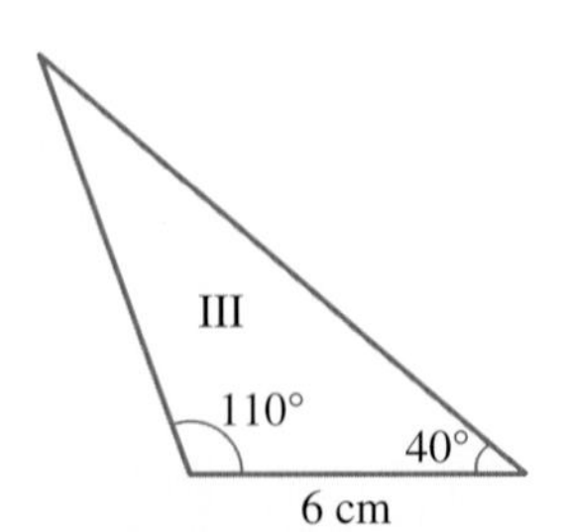

c.

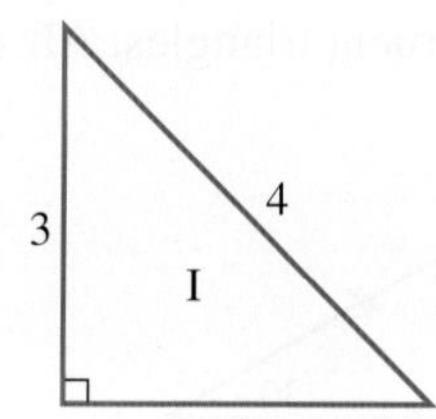

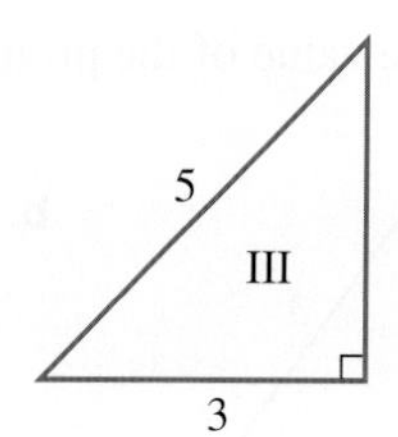

d.

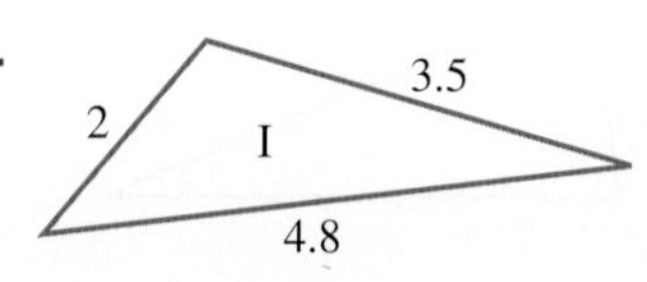

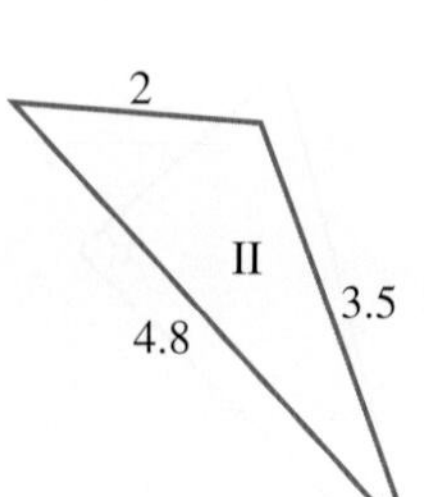

Understanding

4. Determine the missing values of x and y in each of the following diagrams. Give reasons for your answers.

a.

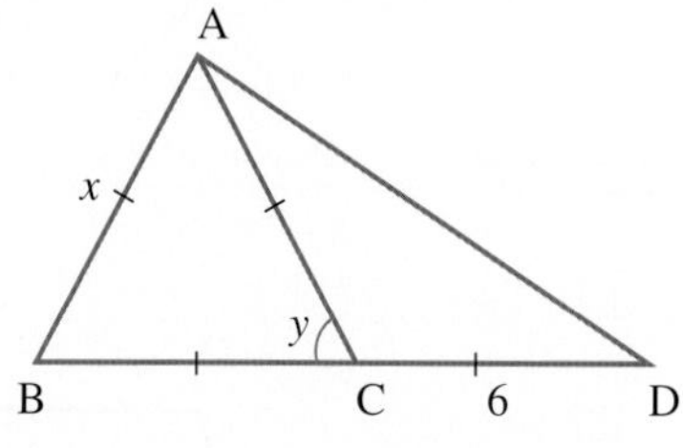

b.

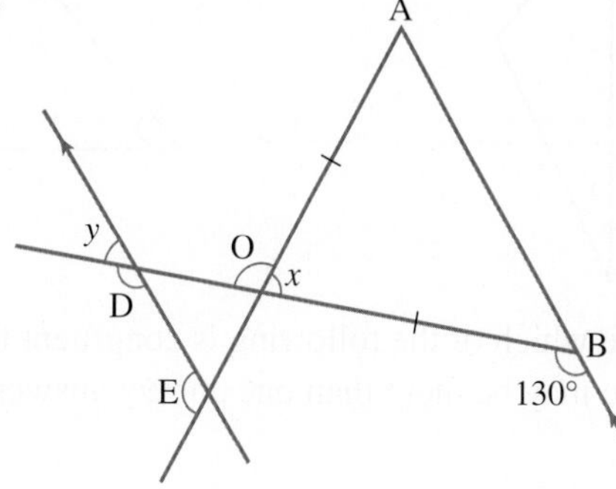

c.

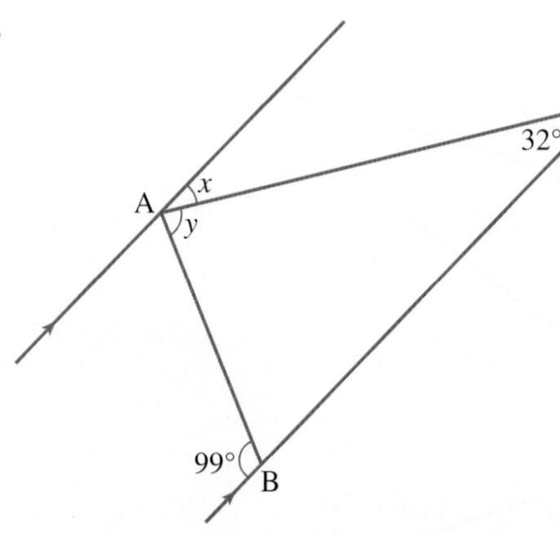

d.

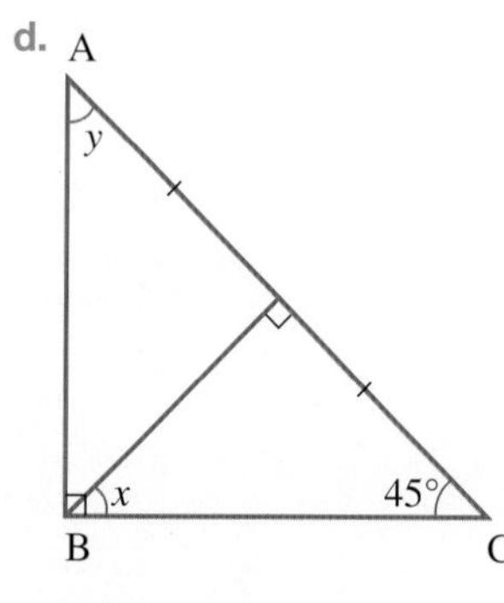

5. Determine the values of the pronumerals. Give reasons for your answers.

a.

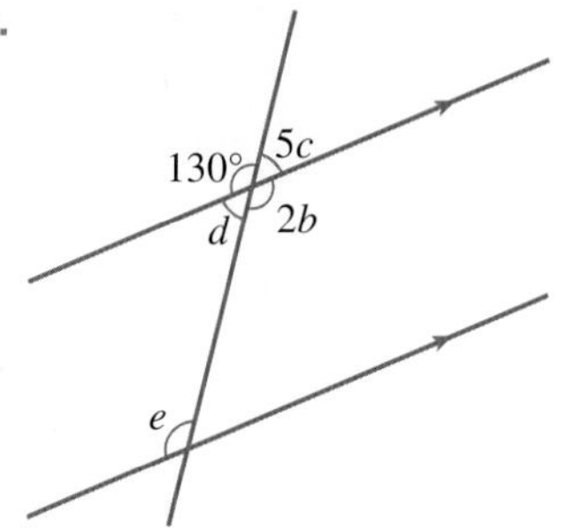

b.

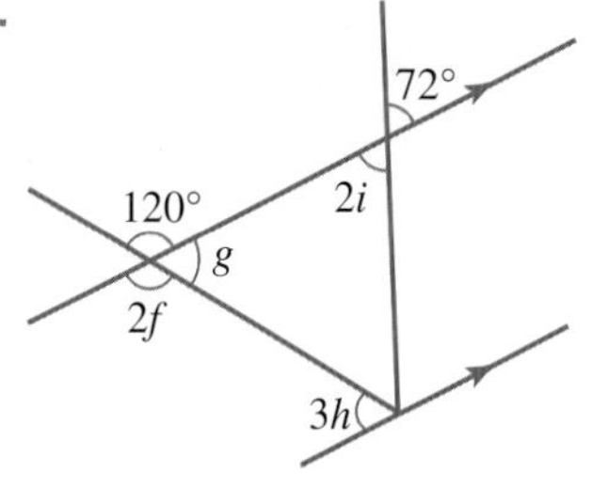

6. **WE2** Determine the value of the pronumeral in each of the following pairs of congruent triangles. All side lengths are in cm.

a.

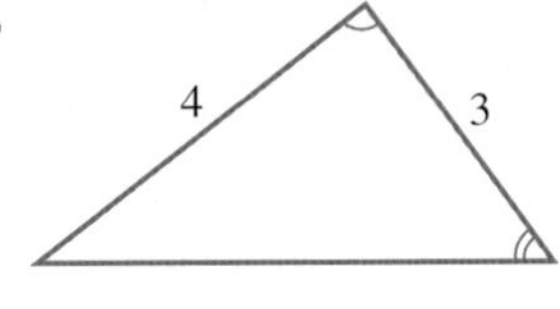

b.

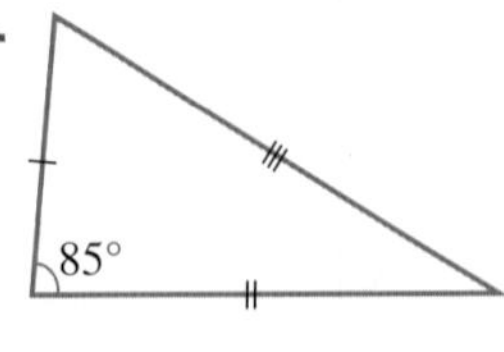

c.

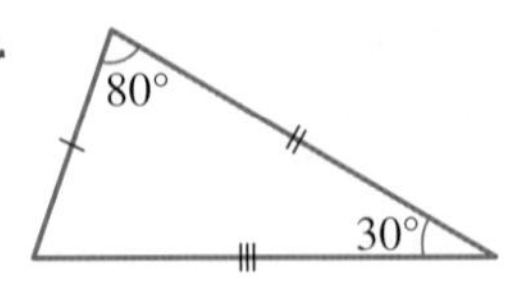

d.

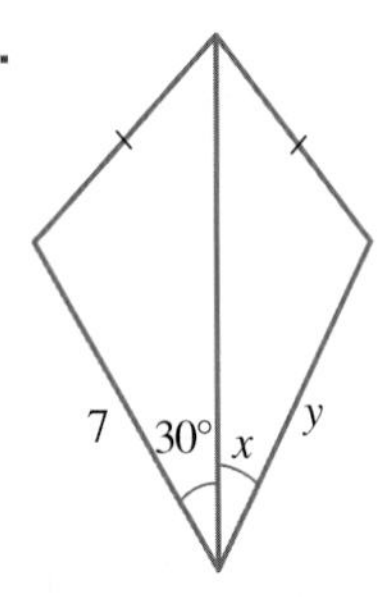

e.

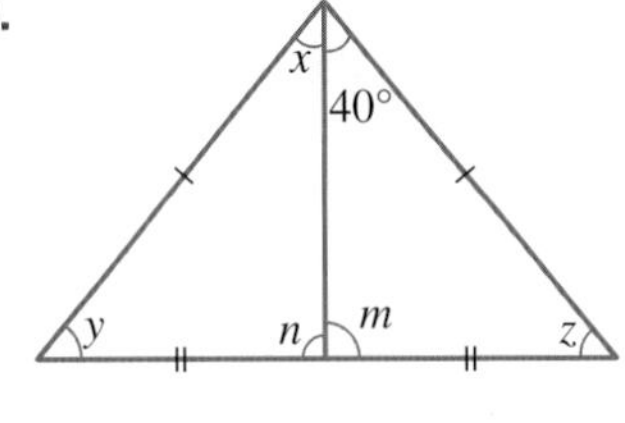

7. **MC** Choose which of the following is congruent to the triangle shown.
Note: There may be more than one correct answer.

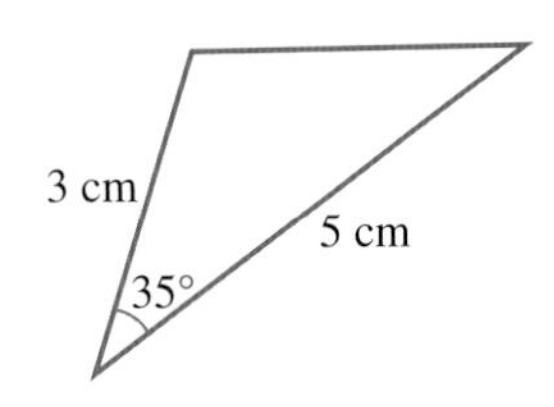

A.

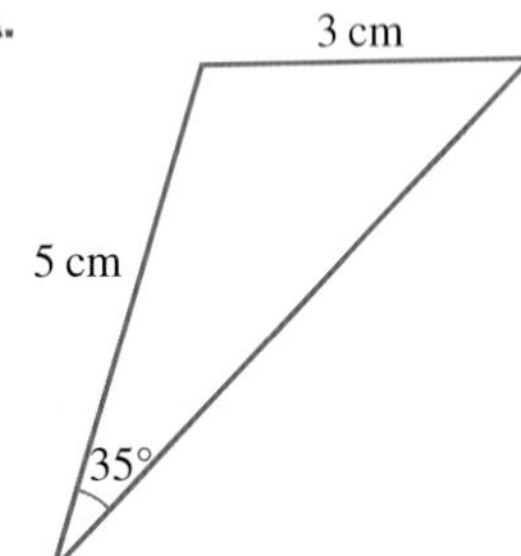

B.

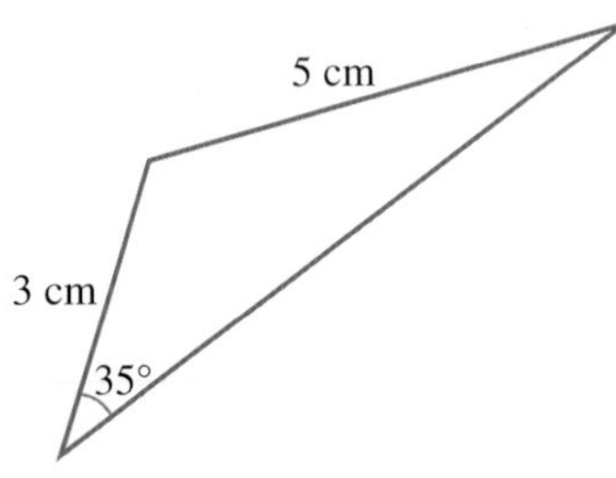

C.

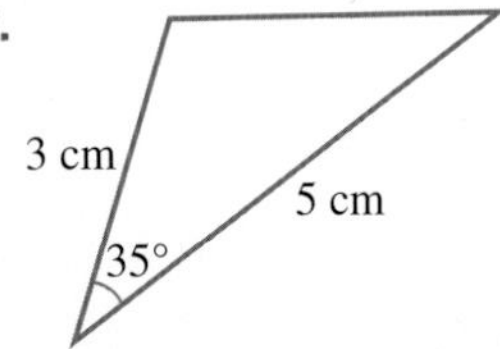

D.

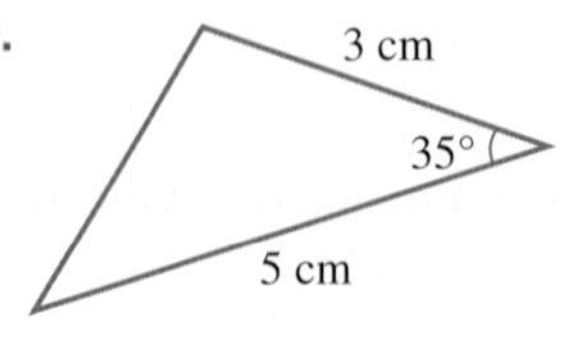

E. None of these

Reasoning

8. Prove that $\Delta ABC \cong \Delta ADC$ and hence determine the value of the pronumerals.

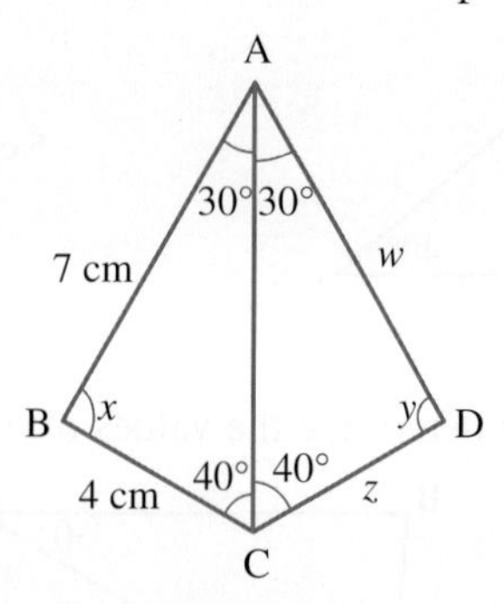

9. If $DA = DB = DC$, prove that $\angle ABC$ is a right angle.

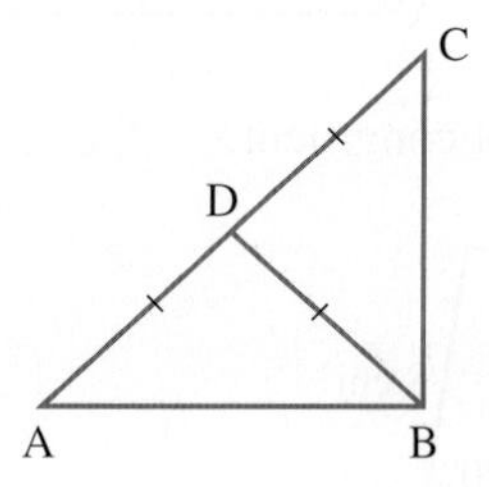

10. **WE3** Prove that each of the following pairs of triangles are congruent.

a.

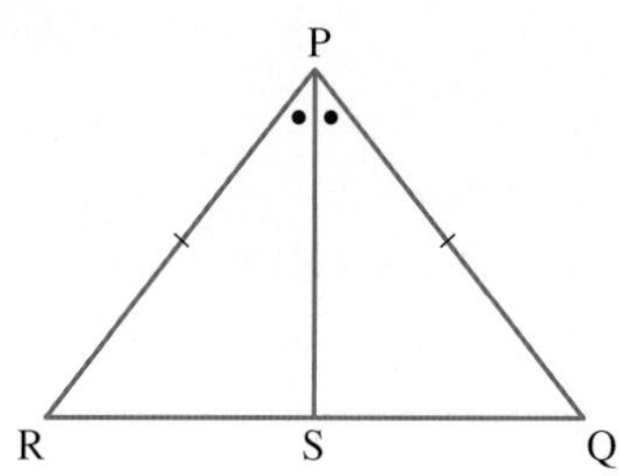

b.

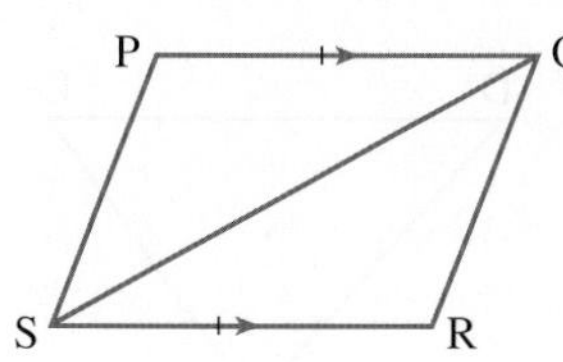

c.

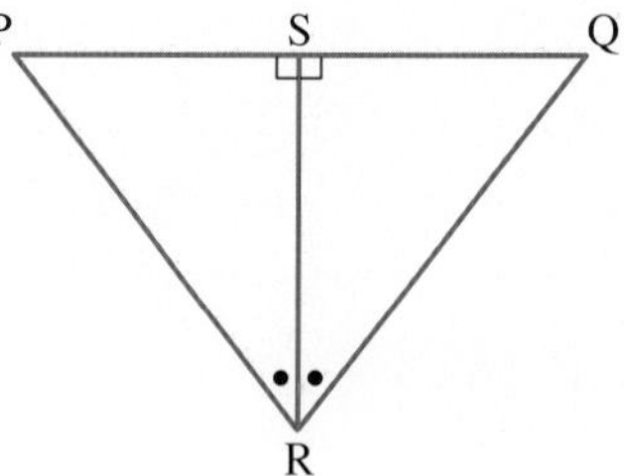

d.

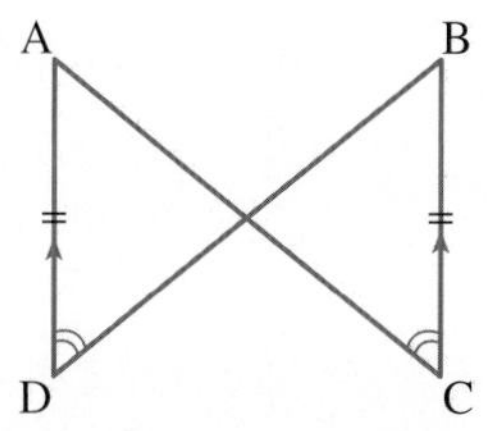

e.

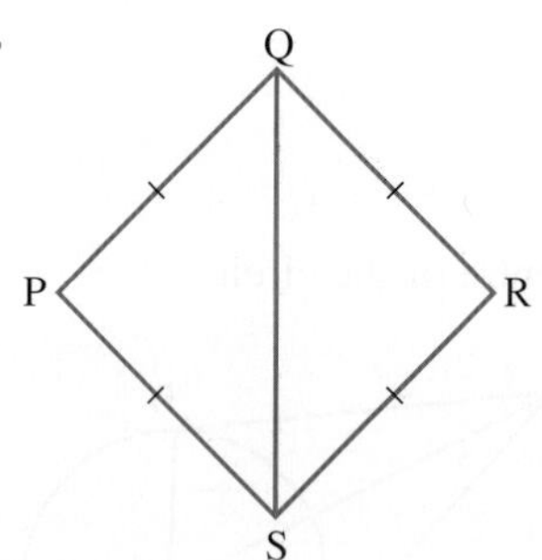

11. Prove that $\Delta ABC \cong \Delta ADC$ and hence determine the value of x.

B D

x 70°

A C

12. Explain why the triangles shown are not necessarily congruent.

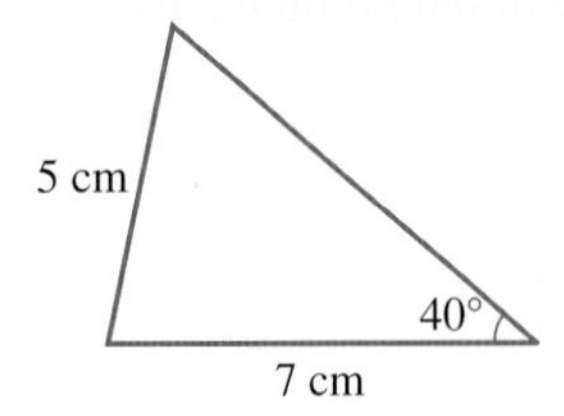

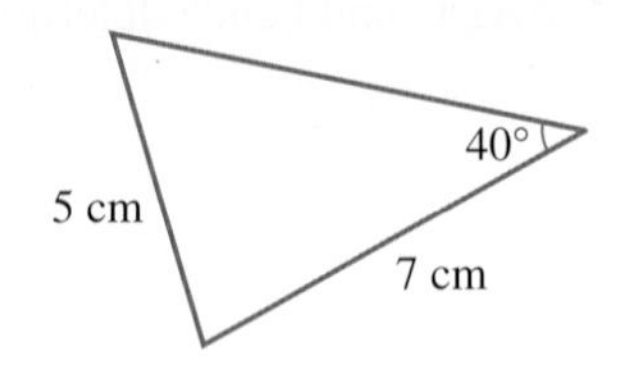

13. Prove that $\Delta ABC \cong \Delta ADC$ and hence determine the values of the pronumerals.

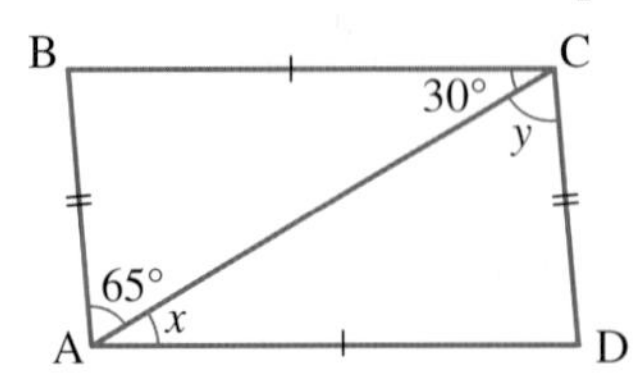

14. Explain why the triangles shown are not congruent.

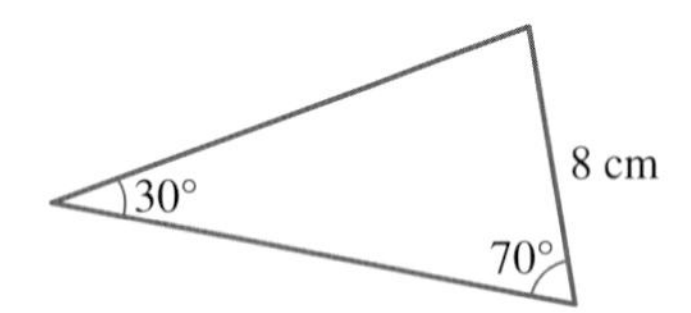

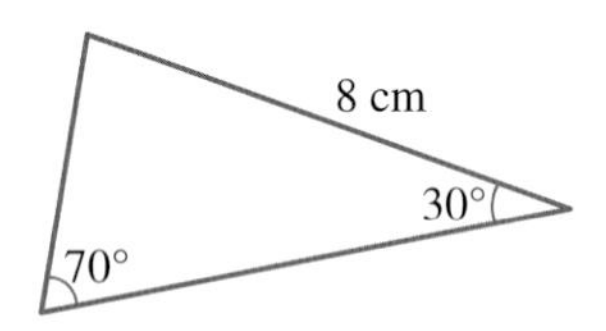

15. If $AC = CB$ and $DC = CE$ in the diagram shown, prove that $AB \parallel DE$.

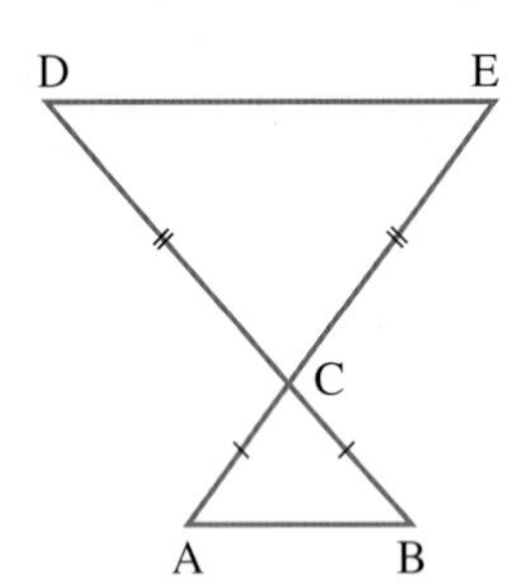

Problem solving

16. Show that $\Delta ABO \cong \Delta ACO$, if O is the centre of the circle.

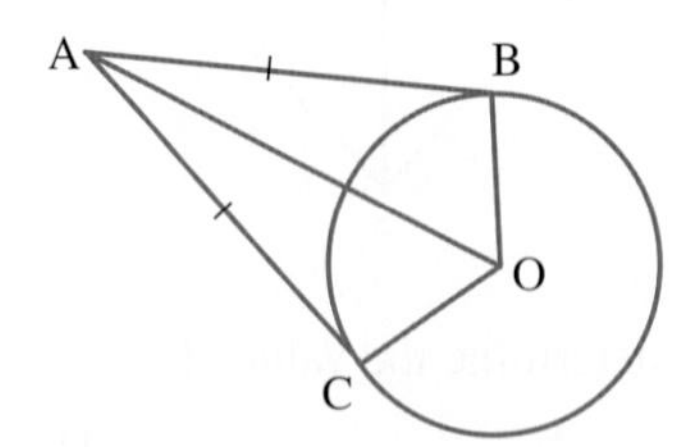

17. Triangles ABC and DEF are congruent.

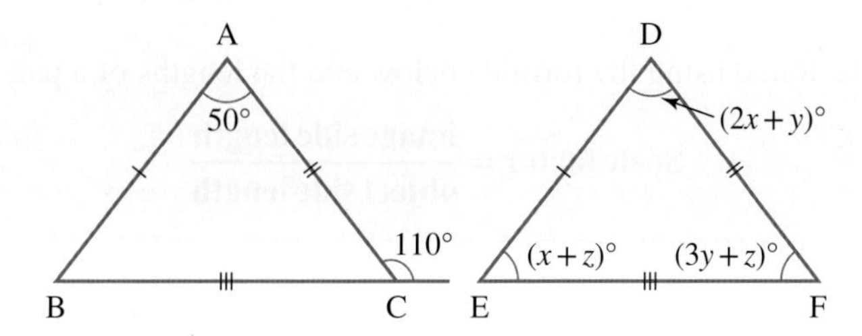

Determine the values of the pronumerals x, y and z.

18. ABC is an isosceles triangle in which AB and AC are equal in length. BDF is a right-angled triangle.

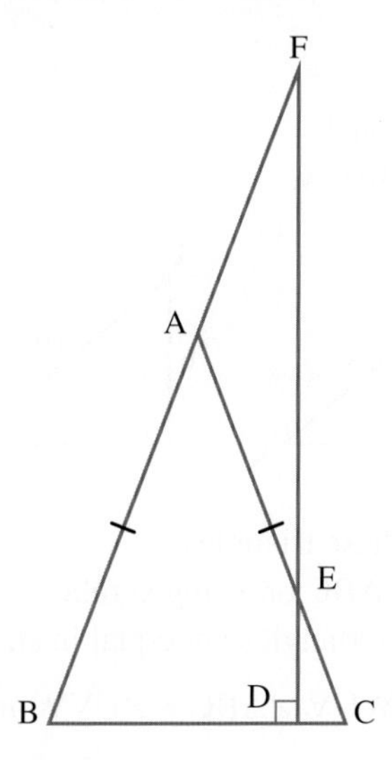

Show that triangle AEF is an isosceles triangle.

10.3 Similar triangles

LEARNING INTENTION

At the end of this subtopic you should be able to:
- identify similar figures
- calculate the scale factor in similar figures
- show that two triangles are similar using the appropriate similarity test.

10.3.1 Similar figures

eles-4899

- Two geometric shapes are **similar** when one is an **enlargement** or reduction of the other shape.
  - An enlargement increases the length of each side of a figure in all directions by the same factor. For example, in the diagram shown, triangle A′B′C′ is an enlargement of triangle ABC by a factor of 3 from its **centre of enlargement** at O.

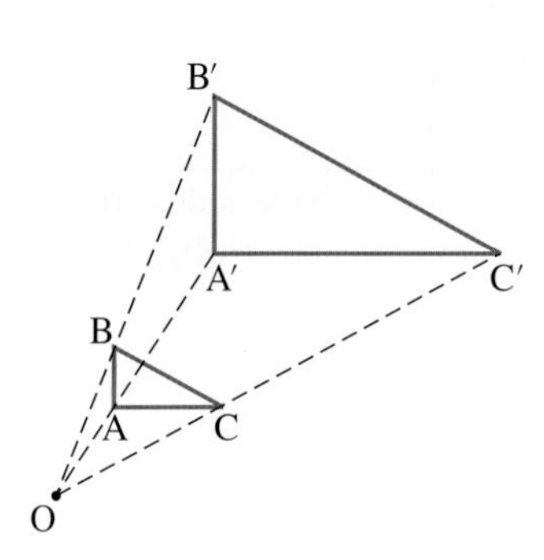

- Similar figures have the same shape. The corresponding angles are the same and each pair of corresponding sides is in the same ratio.
- The symbol for similarity is ~ and is read as 'is similar to'.
- The **image** of the original object is the enlarged or reduced shape.
- To create a similar shape, use a **scale factor** to enlarge or reduce the original shape called the object.

Calculating scale factor

- The scale factor can be found using the formula below and the lengths of a pair of corresponding sides.

$$\textbf{Scale factor} = \frac{\textbf{image side length}}{\textbf{object side length}}$$

- If the scale factor is less than 1, the image is a reduced version of the original shape. If the scale factor is greater than 1, the image is an enlarged version of the original shape.

Similar triangles

- Two triangles are similar if:
  - the angles are equal, or
  - the corresponding sides are proportional.
- Consider the pair of **similar triangles** below.

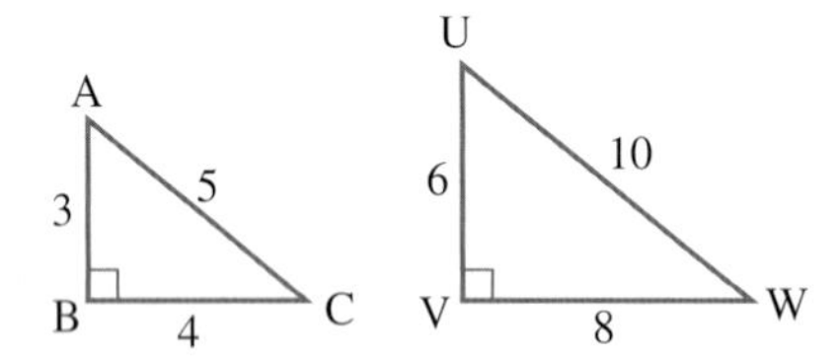

- The following statements are true for these triangles.
  - Triangle UVW is similar to triangle ABC or, using symbols, ΔUVW ~ ΔABC.
  - The corresponding angles of the two triangles are equal in size:

$$\angle CAB = \angle WUV, \angle ABC = \angle UVW \text{ and } \angle ACB = \angle UWV$$

  - The corresponding sides of the two triangles are in the same ratio. $\frac{UV}{AB} = \frac{VW}{BC} = \frac{UW}{AC} = 2$; that is, ΔUVW has each of its sides twice as long as the corresponding sides in ΔABC.
  - The scale factor is 2.

10.3.2 Testing triangles for similarity

eles-4900

- Triangles can be checked for similarity using one of the tests described in the table below.

Similarity test	Example	Description
Angle-angle-angle (AAA)		The three corresponding angles are equal.
Side-side-side (SSS)		The three sides of one triangle are proportional to the three sides of the other triangle.

Similarity test	Example	Description
Side-angle-side (SAS)		Two sides of one triangle are proportional to two sides of the other triangle, and the included angle is equal.

- *Note:* When using the equiangular test, only two corresponding angles have to be checked. Since the sum of the interior angles in any triangle is a constant number (180°), the third pair of corresponding angles will automatically be equal, provided that the first two pairs match exactly.

WORKED EXAMPLE 4 Determining pairs of similar triangles

Determine a pair of similar triangles among those shown. Give a reason for your answer.

a.
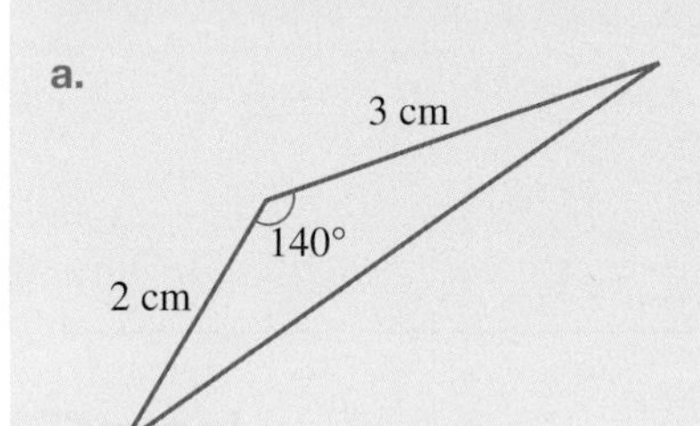

b.
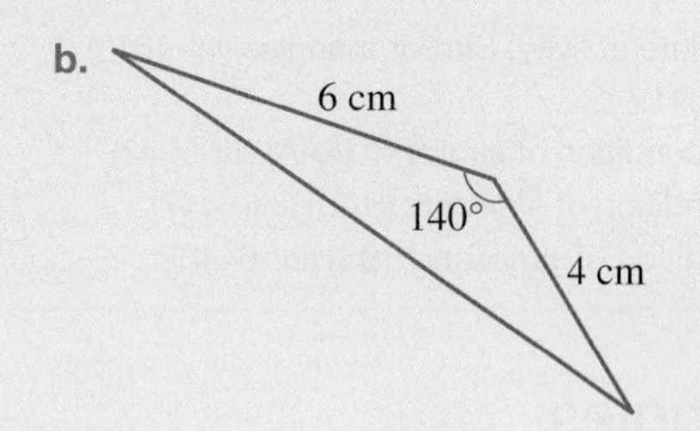

c.
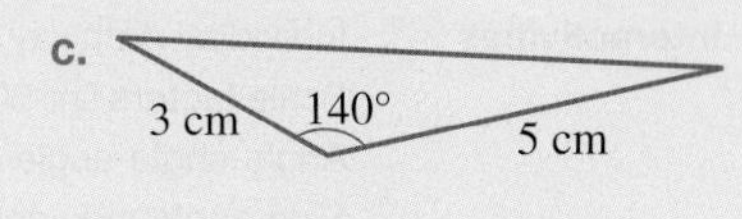

THINK	WRITE
1. In each triangle the lengths of two sides and the included angle are known, so the SAS test can be applied. Since all included angles are equal (140°), we need to the calculate ratios of corresponding sides, taking two triangles at a time.	For triangles **a** and **b**: $\frac{6}{3}=\frac{4}{2}=2$ For triangles **a** and **c**: $\frac{5}{3}=1.6, \frac{3}{2}=1.5$ For triangles **b** and **c**: $\frac{5}{6}=0.83, \frac{3}{4}=0.75$
2. Only triangles **a** and **b** have corresponding sides in the same ratio (and the included angle is equal). State your conclusion, specifying the similarity test you used.	Triangle **a** ~ triangle **b** (SAS)

WORKED EXAMPLE 5 Proving two triangles are similar

Prove that ΔABC is similar to ΔEDC.

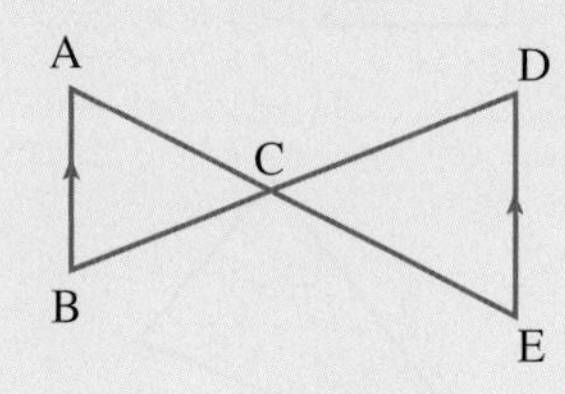

THINK	WRITE
1. Write the information given. AB is parallel to DE. Transversal BD forms two alternate angles: ∠ABC and ∠EDC.	Given: ΔABC and ΔDCE AB||DE C is common.
2. Write what is to be proved.	To prove: ΔABC ~ ΔEDC

3. Write the proof.	Proof: $\angle ABC = \angle EDC$ (alternate angles) $\angle BAC = \angle DEC$ (alternate angles) $\angle BCA = \angle DCE$ (vertically opposite angles) $\therefore \Delta ABC \sim \Delta EDC$ (equiangular, AAA)

on Resources

eWorkbook Topic 10 Workbook (worksheets, code puzzle and project) (ewbk-2036)

Digital documents SkillSHEET Writing similarity statements (doc-5278)
SkillSHEET Calculating unknown side lengths in a pair of similar triangles (doc-5281)

Video eLesson Similar triangles (eles-1925)

Interactivities Individual pathway interactivity: Similar triangles (int-4613)
Scale factors (int-6041)
Angle-angle-angle condition of similarity (AAA) (int-6042)
Side-angle-side condition of similarity (SAS) (int-6447)
Side-side-side condition of similarity (SSS) (int-6448)

Exercise 10.3 Similar triangles

learnON

Individual pathways

■ PRACTISE	■ CONSOLIDATE	■ MASTER
1, 3, 6, 11, 14	2, 4, 7, 9, 12, 15, 16	5, 8, 10, 13, 17, 18

To answer questions online and to receive **immediate feedback** and **sample responses** for every question, go to your learnON title at www.jacplus.com.au.

Fluency

1. WE4 Select a pair of similar triangles among those shown in each part. Give a reason for your answer.

a. i.

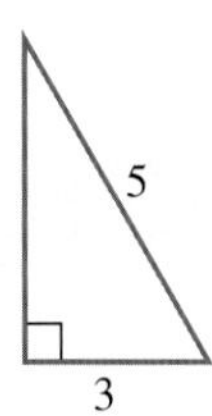

ii.

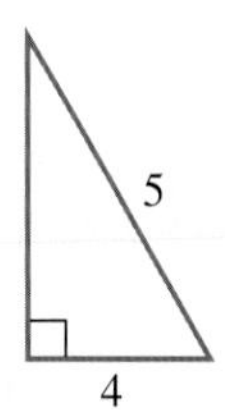

iii.

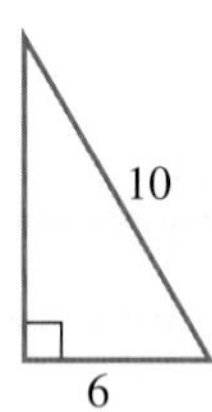

b. i.

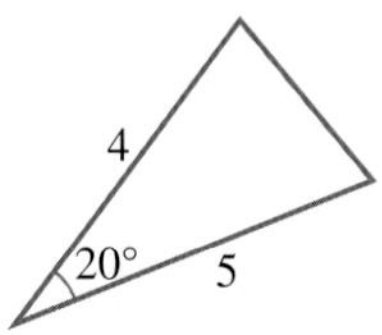

ii.

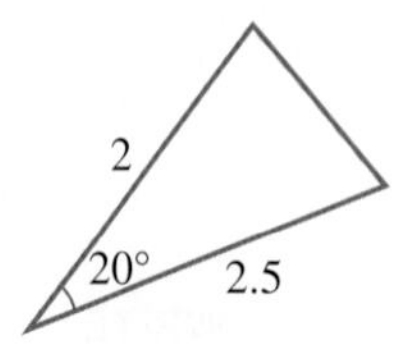

iii.

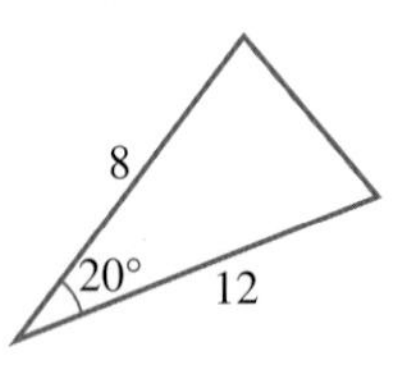

c. i.

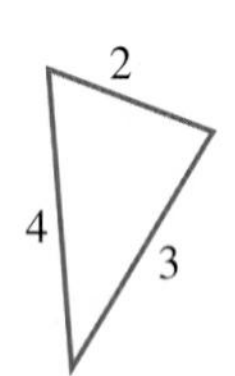

ii.

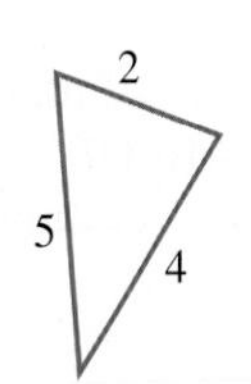

iii.

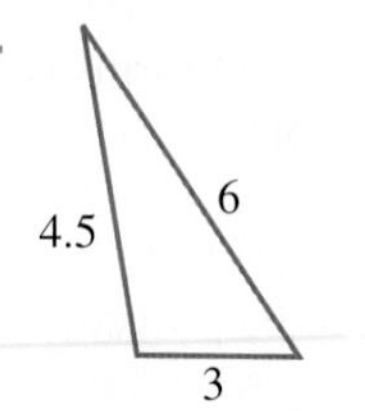

d. i. 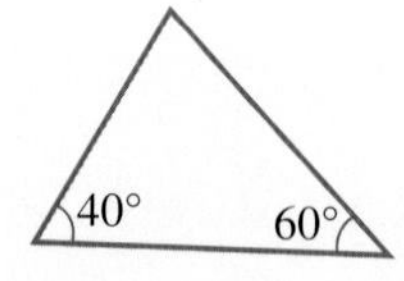

ii. 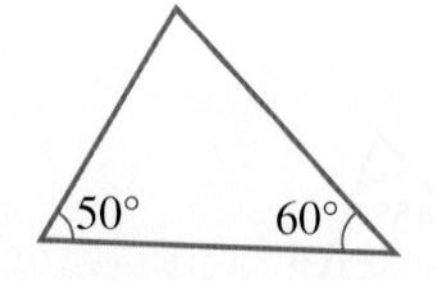

iii. 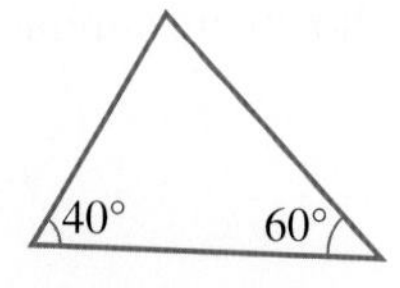

e. i. 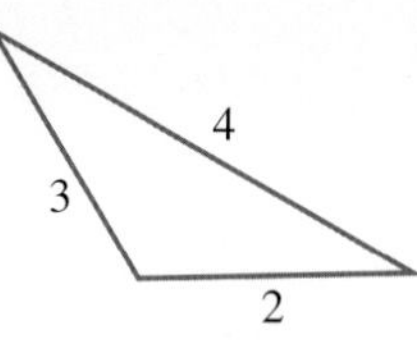

ii. 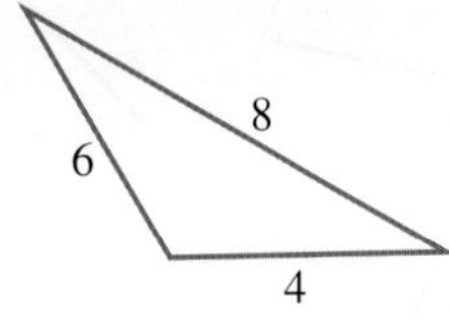

iii. 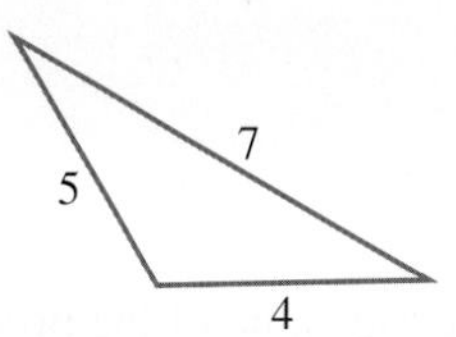

2. Name two similar triangles in each of the following figures.

a. 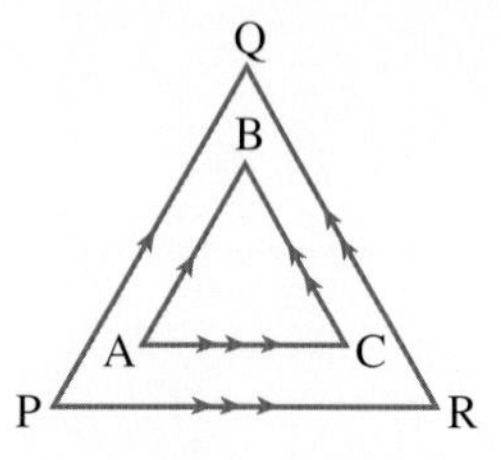

b. 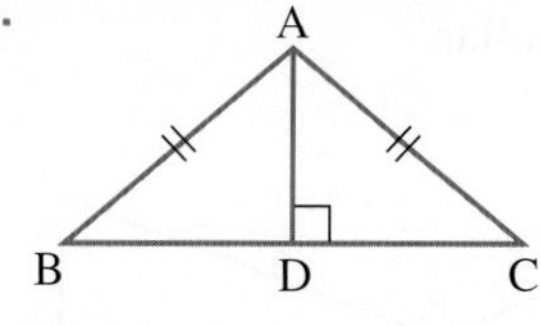

c. 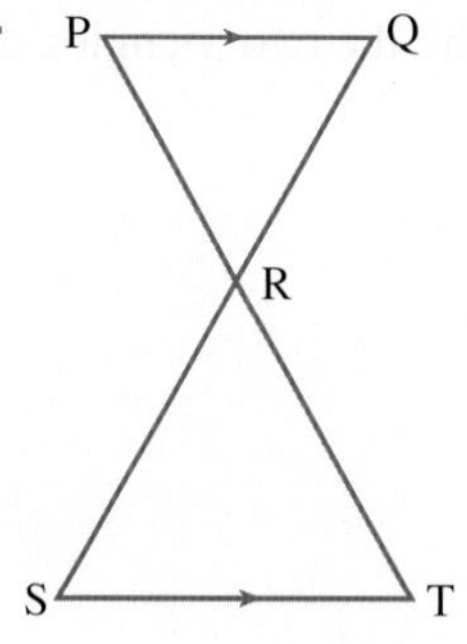

d. 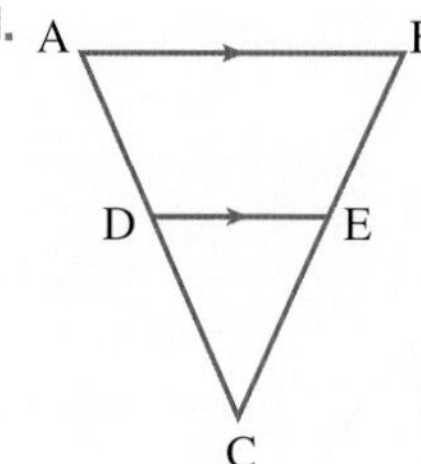

e. 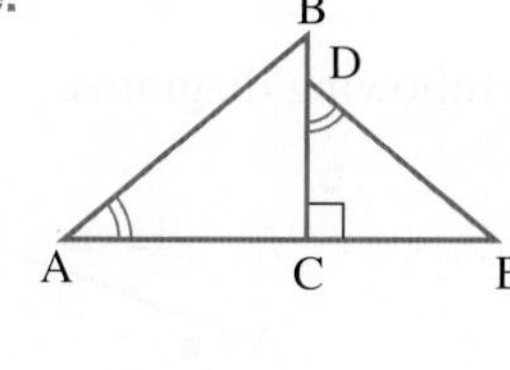

3. a. Complete this statement: $\dfrac{AB}{AD} = \dfrac{BC}{} = \dfrac{}{AE}$.

b. Determine the value of the pronumerals.

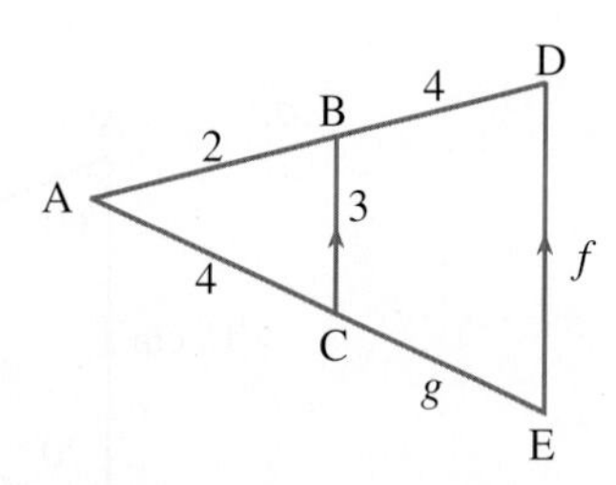

4. Determine the value of the pronumeral in the diagram shown.

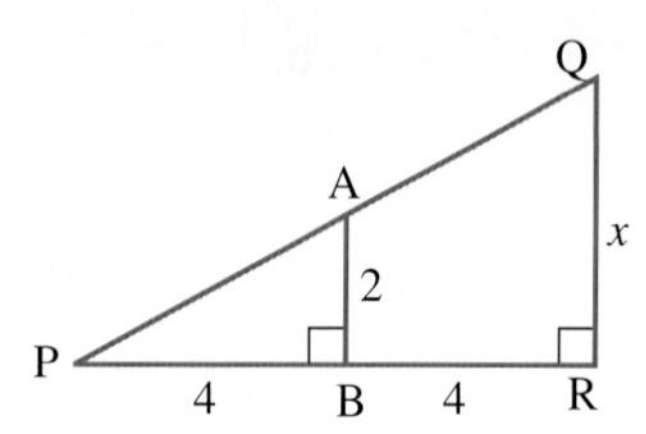

5. The triangles shown are similar.

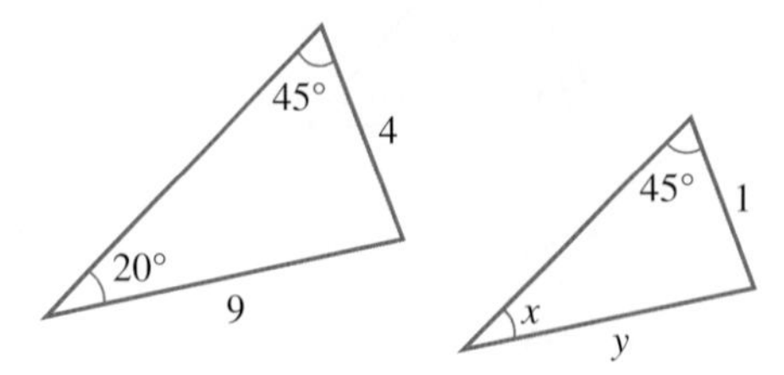

Determine the value of the pronumerals x and y.

Understanding

6. a. State why the two triangles shown are similar.

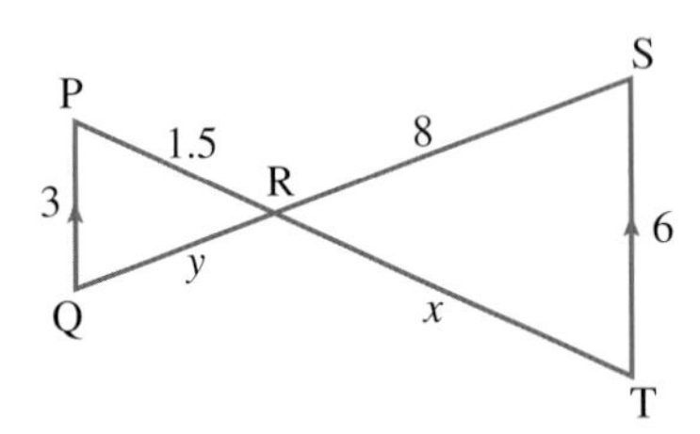

b. Determine the values of the pronumerals x and y in the diagram.

7. Calculate the values of the pronumerals in the following diagrams.

a.

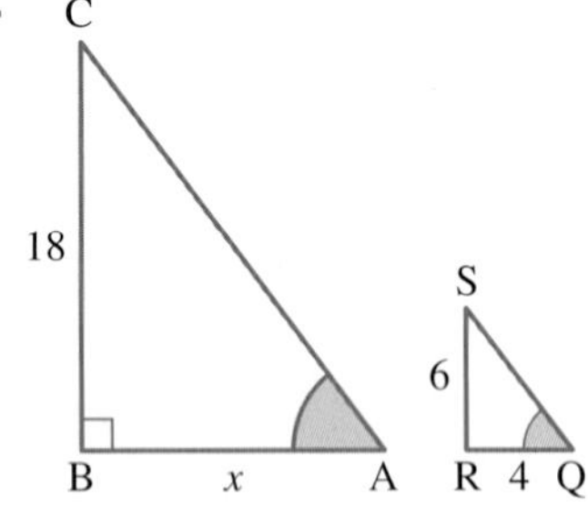

b.

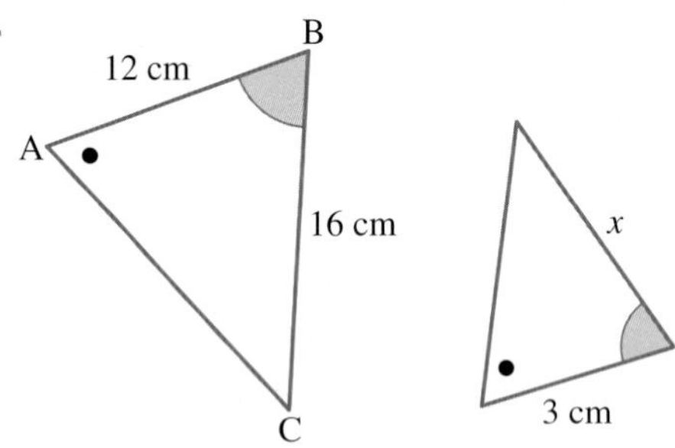

c.

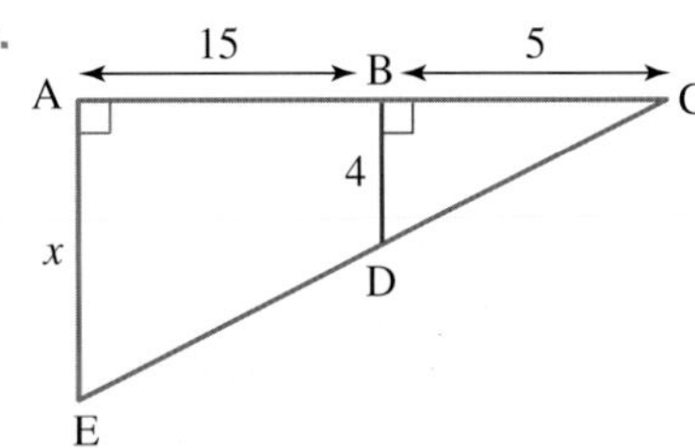

d.

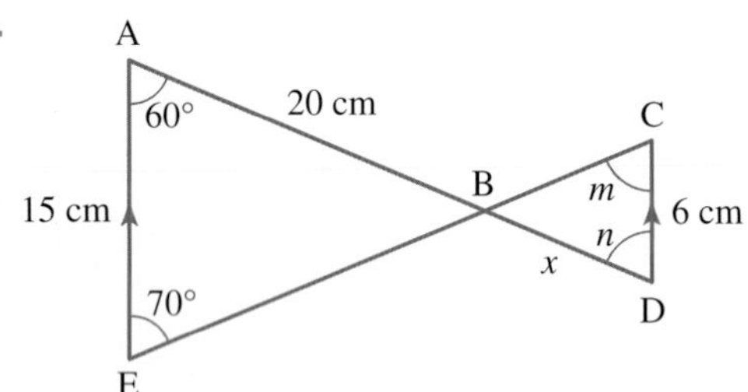

8. Calculate the values of the pronumerals in the following diagrams.

a.

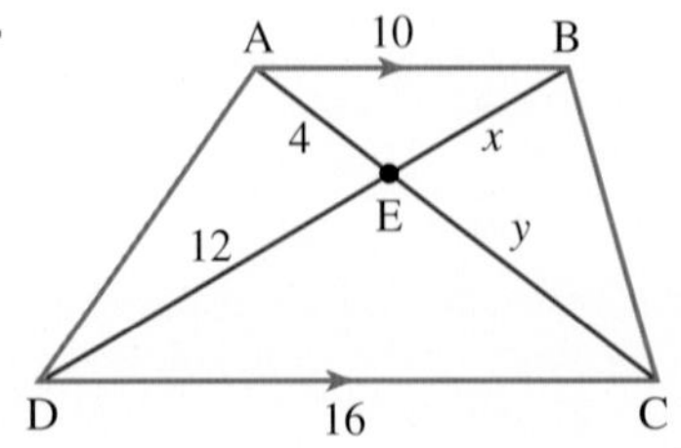

b.

9. Determine the value of x in the diagram.

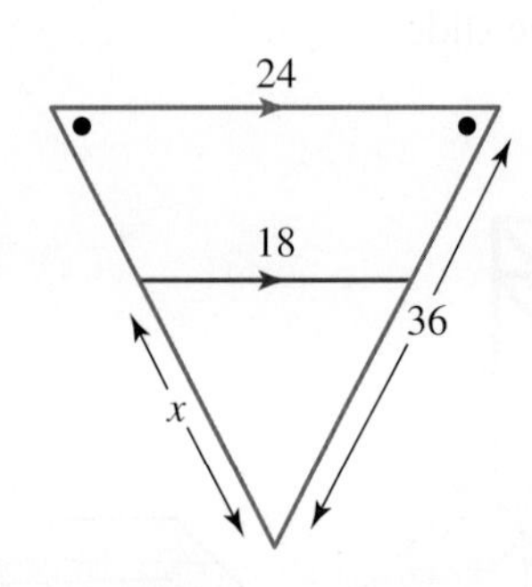

10. Calculate the values of the pronumerals.

a. 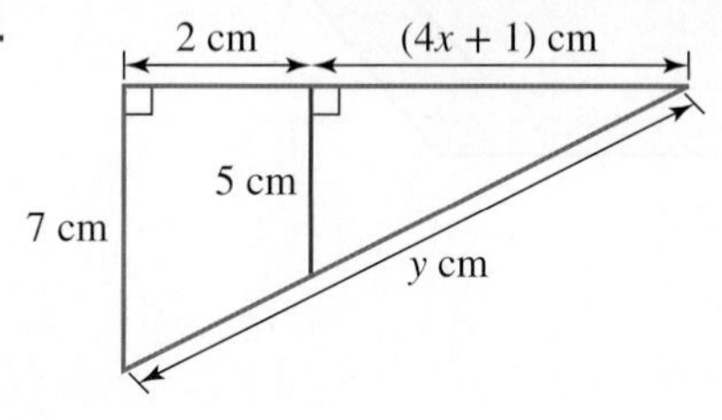

b. 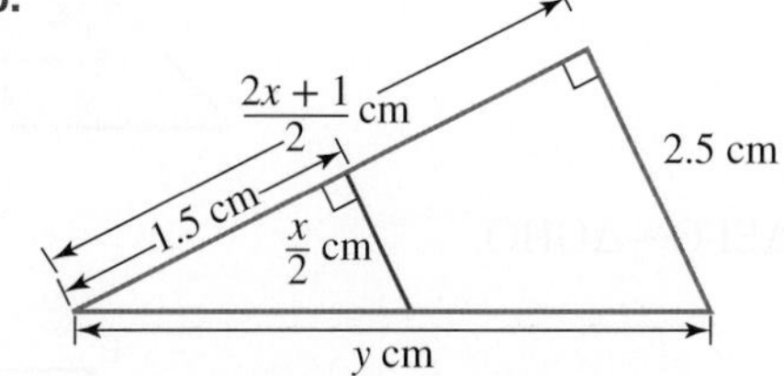

Reasoning

11. **WE5** Prove that ΔABC is similar to ΔEDC in each of the following.

a. 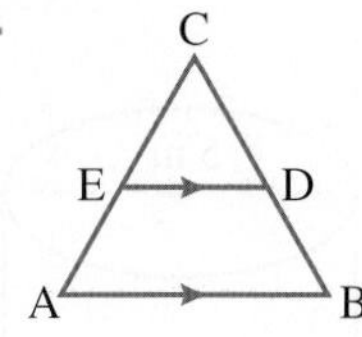

b. 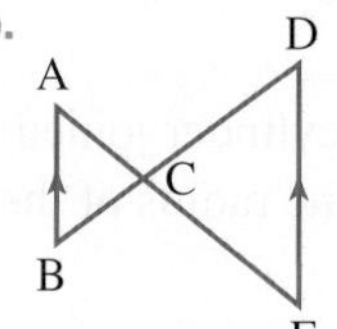

c. 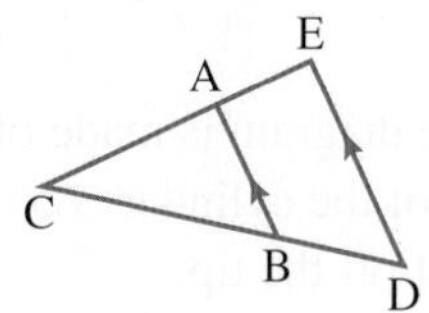

d. 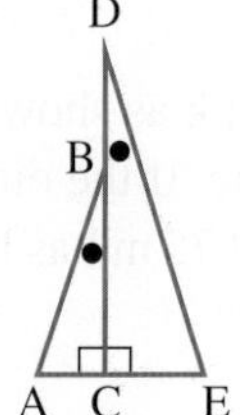

12. ΔABC is a right-angled triangle. A line is drawn from A to D as shown so that $AD \perp BC$.
Prove that:
a. $\Delta ABD \sim \Delta ACB$
b. $\Delta ACD \sim \Delta ACB$.

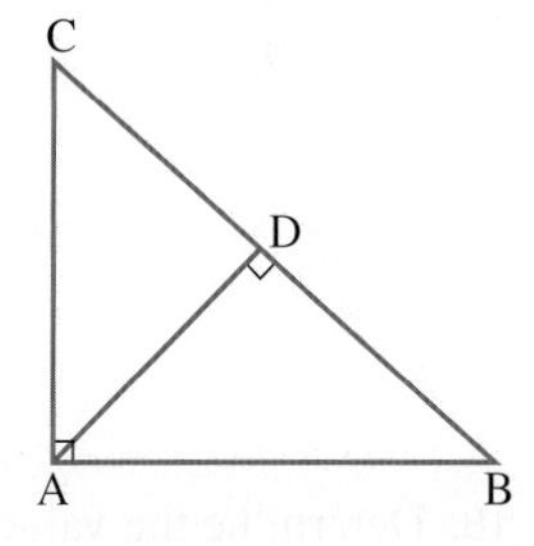

13. Explain why the AAA test cannot be used to prove congruence but can be used to prove similarity.

Problem solving

14. A student casts a shadow of 2.8 m. Another student, who is taller, stands in the same spot at the same time of day. If the diagrams are to the same scale, determine the length of the shadow cast by the taller student.

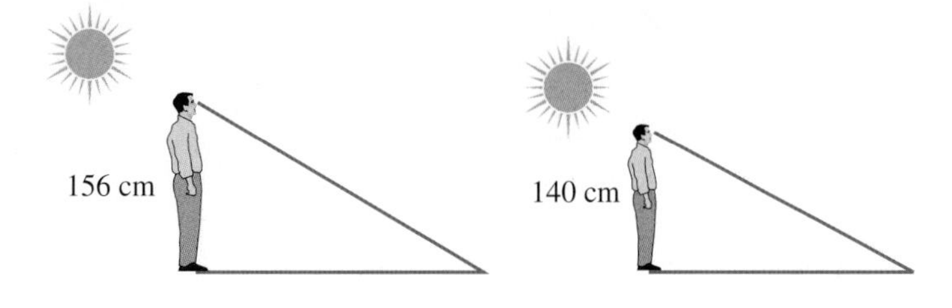

15. A waterslide is 4.2 m high and has a support 2.4 m tall. If a student reaches this support when she is 3.1 m down the slide, evaluate the length of the slide.

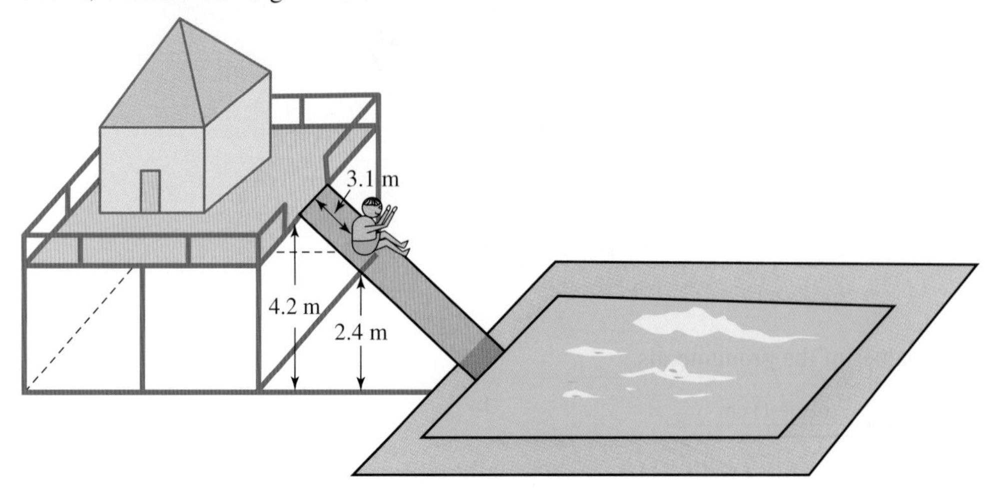

16. Prove that $\Delta EFO \sim \Delta GHO$.

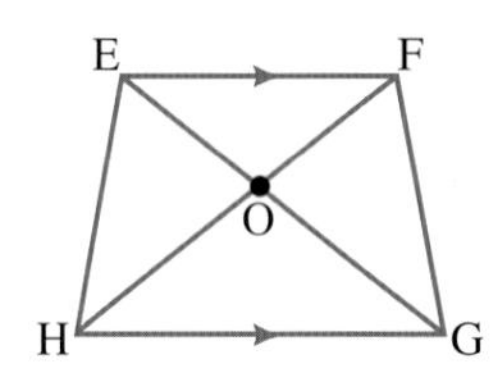

17. A storage tank as shown in the diagram is made of a 4-m-tall cylinder joined by a 3-m-tall cone. If the diameter of the cylinder is 5 m, evaluate the radius of the end of the cone if 0.75 m has been cut off the tip.

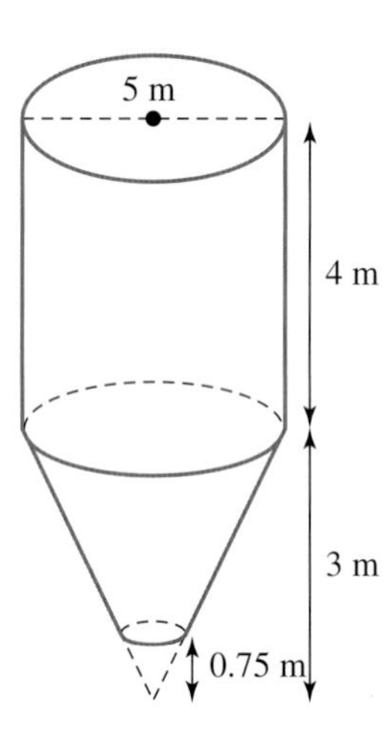

18. Determine the value of x in the diagram shown.

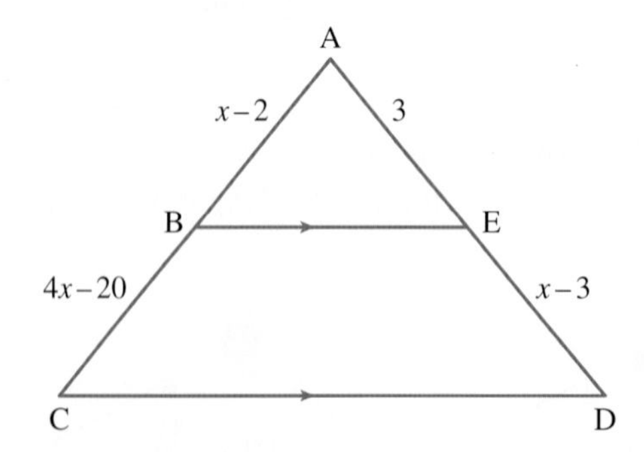

10.4 Quadrilaterals

LEARNING INTENTION

At the end of this subtopic you should be able to:
- identify the different types of quadrilaterals
- construct simple geometric proofs for angles, sides and diagonals in quadrilaterals.

10.4.1 Quadrilaterals

eles-4901

- Quadrilaterals are four-sided plane shapes whose interior angles sum to 360°.

Theorem 5

- The sum of the interior angles in a quadrilateral is 360°.

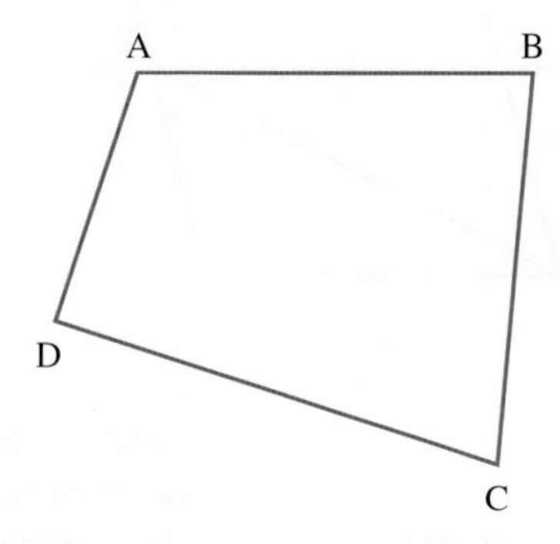

Given: A quadrilateral ABCD

To prove: $\angle ABC + \angle BCD + \angle ADC + \angle BAD = 360°$

Construction: Draw a line joining vertex A to vertex C. Label the interior angles of the triangles formed.

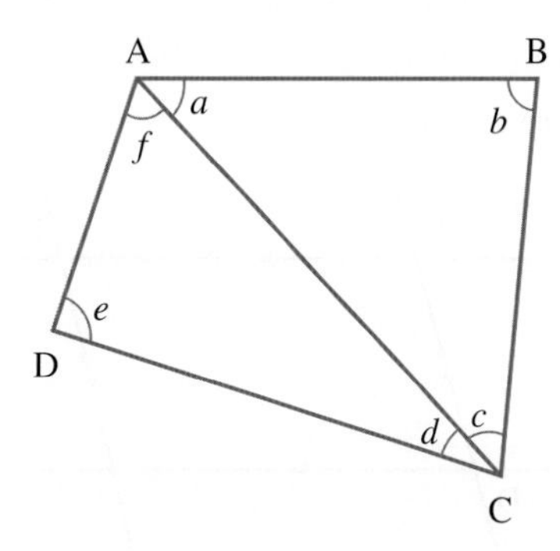

Proof:

$a + b + c = 180°$ (sum of interior angles in a triangle is 180°)

$d + e + f = 180°$ (sum of interior angles in a triangle is 180°)

$\Rightarrow a + b + c + d + e + f = 360°$

$\therefore \angle ABC + \angle BCD + \angle ADC + \angle BAD = 360°$

10.4.2 Parallelograms

eles-5354

- A **parallelogram** is a quadrilateral with two pairs of parallel sides.

Theorem 6

- Opposite angles of a parallelogram are equal.

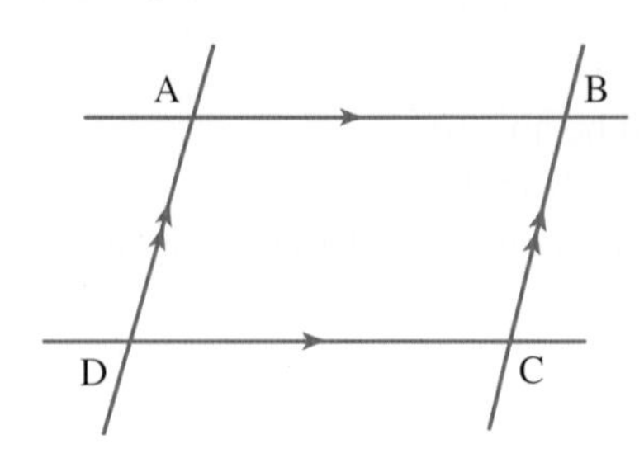

Given: AB || DC and AD || BC

To prove: $\angle ABC = \angle ADC$

Construction: Draw a diagonal from B to D.

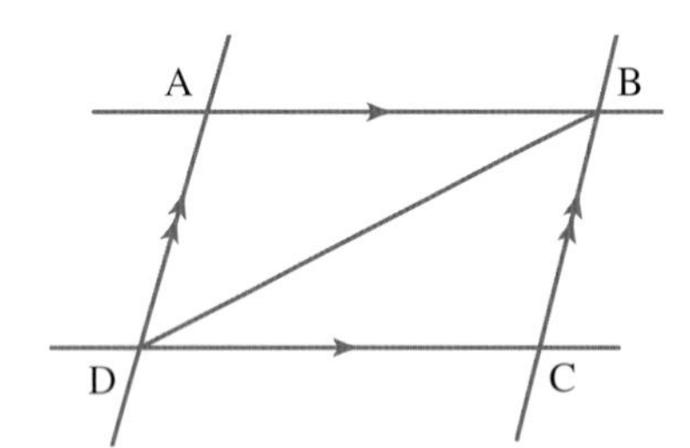

Proof:

$\angle ABD = \angle BDC$	(alternate angles)
$\angle ADB = \angle CBD$	(alternate angles)
$\angle ABC = \angle ABD + \angle CBD$	(by construction)
$\angle ADC = \angle BDC + \angle ADB$	(by construction)
$\therefore \angle ABC = \angle ADC$	

- Conversely, if each pair of opposite angles of a quadrilateral is equal, then it is a parallelogram.

Theorem 7

- Opposite sides of a parallelogram are equal.

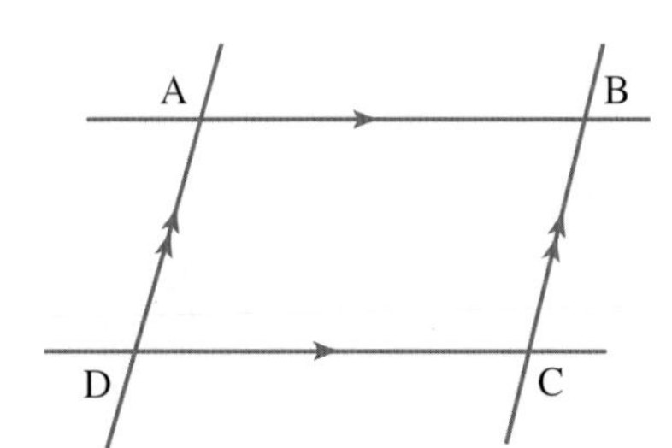

Given: AB || DC and AD || BC

To prove: AB = DC

Construction: Draw a diagonal from B to D.

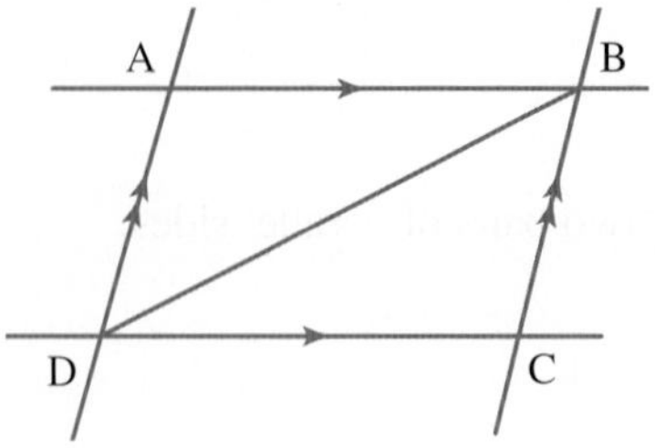

Proof: $\angle ABD = \angle BDC$ (alternate angles)
$\angle ADB = \angle CBD$ (alternate angles)
BD is common to ΔABD and ΔBCD.
$\Rightarrow \Delta ABD \cong \Delta BCD$ (ASA)
$\therefore AB = DC$

- Conversely, if each pair of opposite sides of a quadrilateral is equal, then it is a parallelogram.

Theorem 8

- The diagonals of a parallelogram bisect each other.

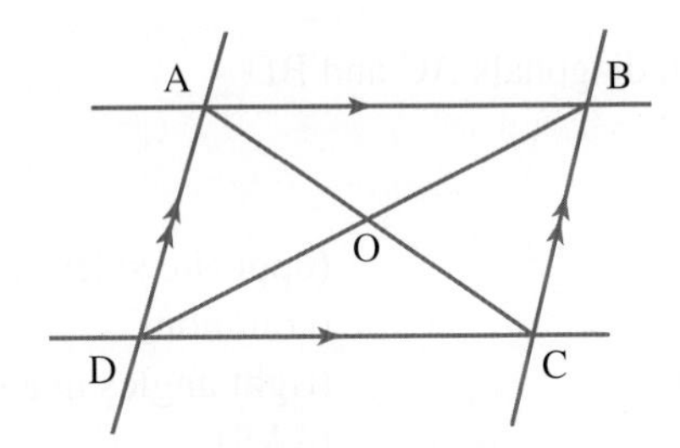

Given: AB || DC and AD || BC with diagonals AC and BD

To prove: AO = OC and BO = OD

Proof: In ΔAOB and ΔCOD,

$\angle OAB = \angle OCD$	(alternate angles)
$\angle OBA = \angle ODC$	(alternate angles)
$AB = CD$	(opposite sides of a parallelogram)
$\Rightarrow \Delta AOB \cong \Delta COD$	(ASA)
$\Rightarrow AO = OC$	(corresponding sides in congruent triangles)
and $BO = OD$	(corresponding sides in congruent triangles)

Rectangles

- A rectangle is a parallelogram with four right angles.

Theorem 9

- A parallelogram with a right angle is a rectangle.

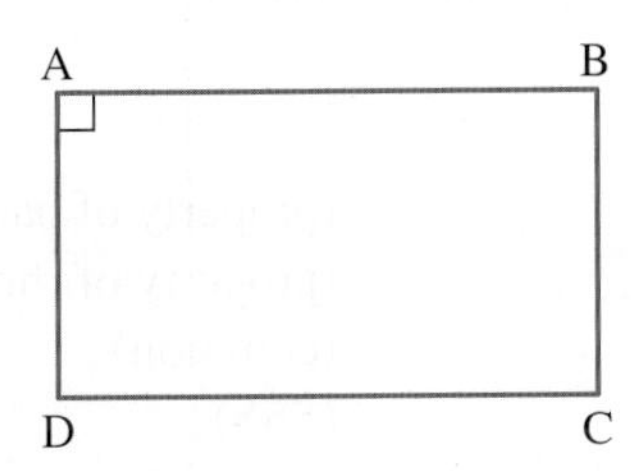

Given: Parallelogram ABCD with $\angle BAD = 90°$

To prove: $\angle BAD = \angle ABC = \angle BCD = \angle ADC = 90°$

Proof:

AB || CD	(properties of a parallelogram)
$\Rightarrow \angle BAD + \angle ADC = 180°$	(co-interior angles)
But $\angle BAD = 90°$	(given)
$\Rightarrow \angle ADC = 90°$	
Similarly, $\angle BCD = \angle ADC = 90°$	
$\therefore \angle BAD = \angle ABC = \angle BCD = \angle ADC = 90°$	

Theorem 10

- The diagonals of a rectangle are equal.

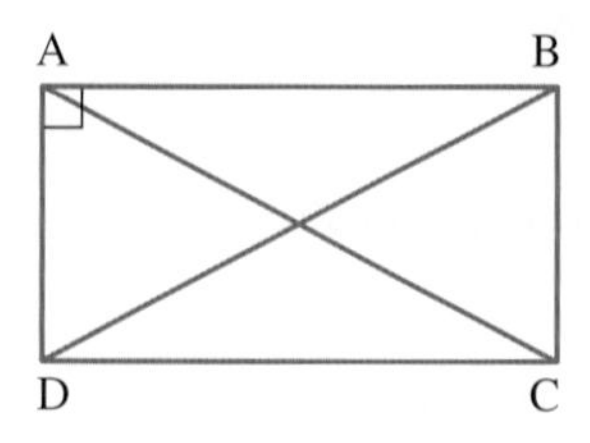

Given: Rectangle ABCD with diagonals AC and BD

To prove: $AC = BD$

Proof: In ΔADC and ΔBCD,

$AD = BC$	(opposite sides equal in a rectangle)
$DC = CD$	(common)
$\angle ADC = \angle BCD = 90°$	(right angles in a rectangle)
$\Rightarrow \Delta ADC \cong \Delta BCD$	(SAS)
$\therefore AC = BD$	

Rhombuses

eles-5356

- A **rhombus** is a parallelogram with four equal sides.

Theorem 11

- The diagonals of a rhombus are perpendicular.

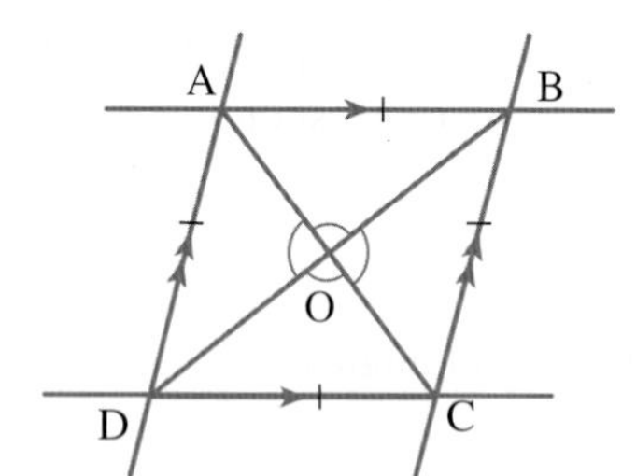

Given: Rhombus ABCD with diagonals AC and BD

To prove: $AC \perp BD$

Proof: In ΔAOB and ΔBOC,

$AO = OC$	(property of parallelogram)
$AB = BC$	(property of rhombus)
$BO = OB$	(common)
$\Rightarrow \Delta AOB \cong \Delta BOC$	(SSS)
$\Rightarrow \angle AOB = \angle BOC$	
But $\angle AOB + \angle BOC = 180°$	(supplementary angles)
$\Rightarrow \angle AOB = \angle BOC = 90°$	
Similarly, $\angle AOD = \angle DOC = 90°$.	
Hence, $AC \perp BD$	

WORKED EXAMPLE 6 Determining the values of the pronumerals of the quadrilaterals

a.

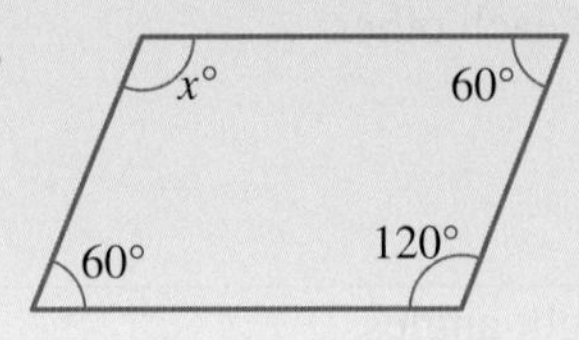

b.

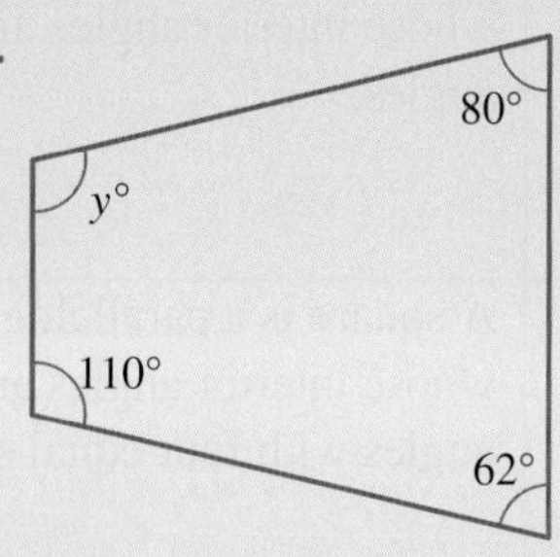

THINK	WRITE
a. 1. Identify the shape.	**a.** The shape is a parallelogram as the shape has two pairs of parallel sides.
2. To determine the values of $x°$, apply theorem 6, which states that opposite angles of a parallelogram are equal.	$x° = 120°$
b. 1. Identify the shape.	**b.** The shape is a trapezium, as one pair of opposite sides is parallel but not equal in length.
2. To determine the values of $y°$, apply theorem 5, which states the sum of interior angles in a quadrilateral is 360°.	The sum of all the angles = 360° $y° + 110° + 80° + 62° = 360°$
3. Simplify and solve for $y°$.	$y° + 110° + 80° + 62° = 360°$ $y° + 252° = 360°$ $y° = 360° - 252°$ $y° = 108°$

- A summary of the definitions and properties of quadrilaterals is shown in the following table.

Shape	Definition	Properties
Trapezium	A trapezium is a quadrilateral with one pair of opposite sides parallel.	• One pair of opposite sides is parallel but not equal in length.
Parallelogram	A parallelogram is a quadrilateral with both pairs of opposite sides parallel.	• Opposite angles are equal. • Opposite sides are equal. • Diagonals bisect each other.
Rhombus	A rhombus is a parallelogram with four equal sides.	• Diagonals bisect each other at right angles. • Diagonals bisect the angles at the vertex through which they pass.

(continued)

(continued)

Shape	Definition	Properties
Rectangle	A rectangle is a parallelogram whose interior angles are right angles.	• Diagonals are equal. • Diagonals bisect each other.
Square	A square is a parallelogram whose interior angles are right angles with four equal sides.	• All angles are right angles. • All side lengths are equal. • Diagonals are equal in length and bisect each other at right angles. • Diagonals bisect the vertex through which they pass (45°).

10.4.3 The midpoint theorem

eles-4905

- Now that the properties of quadrilaterals have been explored, the midpoint theorem can be tackled.

Theorem 12

- The interval joining the midpoints of two sides of a triangle is parallel to the third side and half its length.

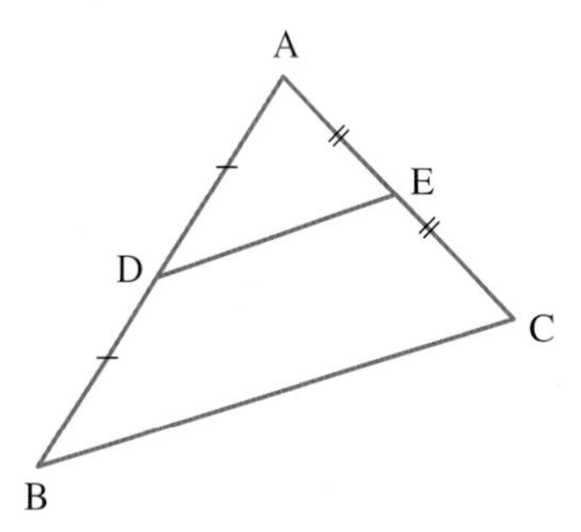

Given: ΔABC in which AD = DB and AE = EC

To prove: DE || BC and DE = $\frac{1}{2}$BC

Construction: Draw a line through C parallel to AB. Extend DE to F on the parallel line.

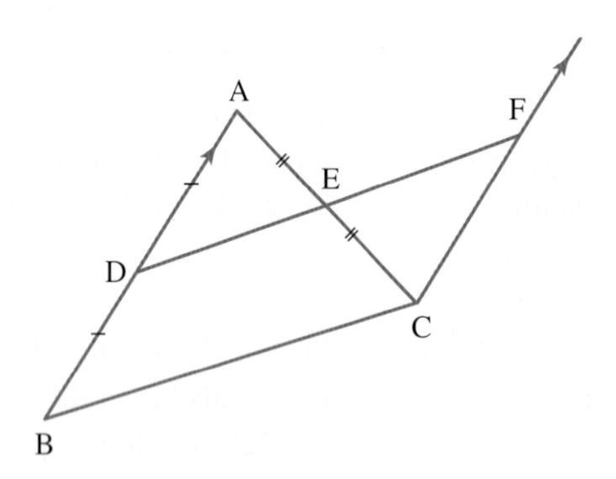

Proof:

In ΔADE and ΔCEF,

$AE = EC$ (E is the midpoint of AC, given)

$\angle AED = \angle CEF$ (vertically opposite angles)

$\angle EAD = \angle ECF$ (alternate angles)

$\Rightarrow \Delta ADE \cong \Delta CEF$ (ASA)

$\therefore AD = CF$ and $DE = EF$ (corresponding sides in congruent triangles)

So, $AD = DB = CF$.

We have $AB \parallel CF$ (by construction)

So BDFC is a parallelogram.

$\Rightarrow DE \parallel BC$ (opposite sides in parallelogram)

Also, $BC = DF$

But $DE = DF$ (sides in congruent triangles)

$\Rightarrow DE = \frac{1}{2}BC$

Therefore, $DE \parallel BC$ and $DE = \frac{1}{2}BC$.

- Conversely, if a line interval is drawn parallel to a side of a triangle and half the length of that side, then the line interval bisects each of the other two sides of the triangle.

WORKED EXAMPLE 7 Applying midpoint theorem to determine the unknown length

In triangle ABC, the midpoints of AC and AB are D and E respectively. Determine the value of DE, if BC = 18 cm.

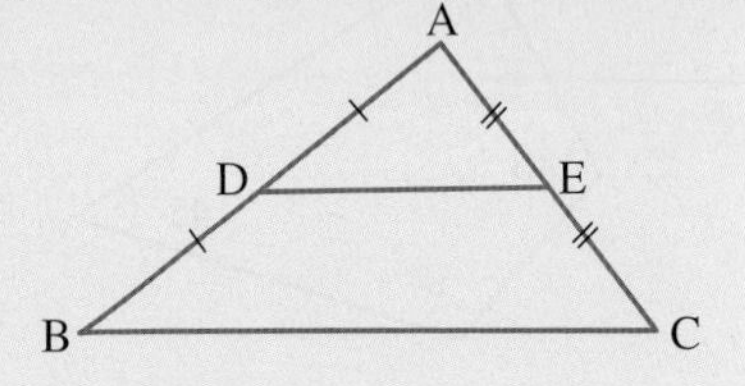

THINK	WRITE
1. Determine the midpoints on the line AB and AC.	D is midpoint of AB and E is midpoint of AC.
2. Apply the midpoint theorem to determine the length of DE.	$DE = \frac{1}{2} BC$
3. Substitute the value of BC = 18 cm into the formula.	$DE = \frac{1}{2} \times 18$
4. Simplify and determine the length of DE.	$DE = 9\text{ cm}$

on Resources

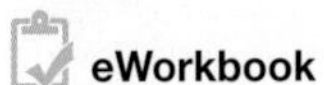
eWorkbook Topic 10 Workbook (worksheets, code puzzle and project) (ewbk-2036)

Digital document SkillSHEET Identifying quadrilaterals (doc-5279)

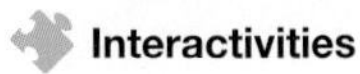
Interactivities Individual pathway interactivity: Quadrilaterals (int-4614)
Quadrilateral definitions (int-2786)
Angles in a quadrilateral (int-3967)
Opposite angles of a parallelogram (int-6160)
Opposite sides of a parallelogram (int-6161)
Diagonals of a parallelogram (int-6162)
Diagonals of a rectangle (int-6163)
Diagonals of a rhombus (int-6164)
The midpoint theorem (int-6165)
Quadrilaterals (int-3756)

Exercise 10.4 Quadrilaterals

Individual pathways

■ PRACTISE	■ CONSOLIDATE	■ MASTER
1, 3, 7, 9, 12, 13, 18	2, 4, 8, 10, 14, 15, 19, 20	5, 6, 11, 16, 17, 21, 22

To answer questions online and to receive **immediate feedback** and **sample responses** for every question, go to your learnON title at www.jacplus.com.au.

Fluency

1. Use the definitions of the five special quadrilaterals to decide if the following statements are true or false.
   a. A square is a rectangle.
   b. A rhombus is a parallelogram.
   c. A square is a rhombus.
   d. A rhombus is a square.
2. Use the definitions of the five special quadrilaterals to decide if the following statements are true or false.
   a. A square is a trapezium.
   b. A parallelogram is a rectangle.
   c. A trapezium is a rhombus.
   d. A rectangle is a square.
3. **WE6** Determine the value of the pronumeral in each of the following quadrilaterals.

a.
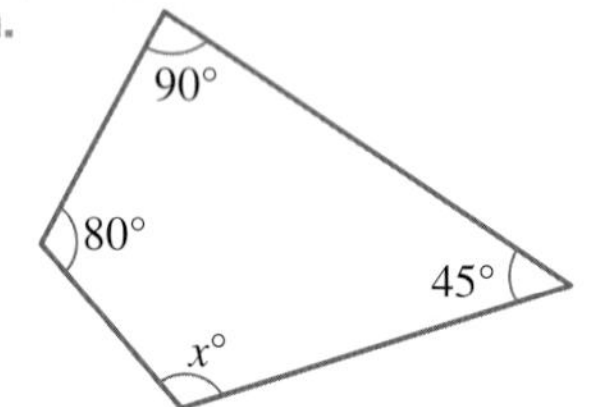

b.
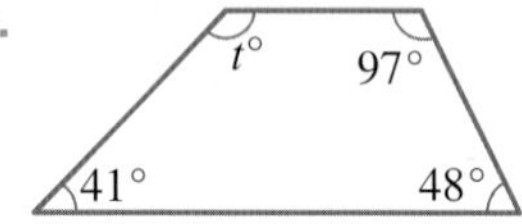

c.
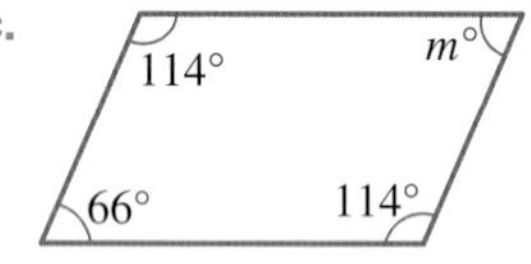

d.
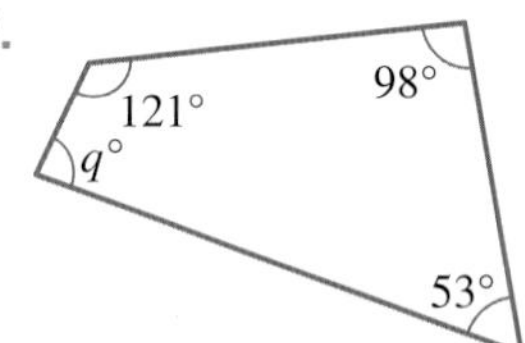

4. Determine the values of the pronumerals in the following diagrams.

a.
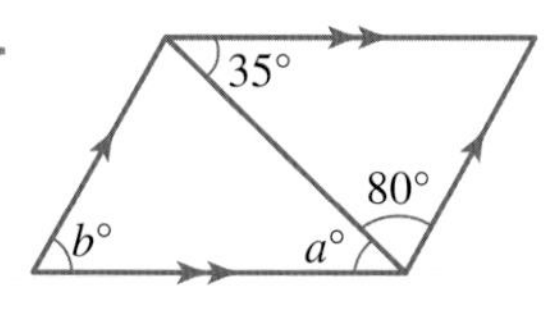

b.
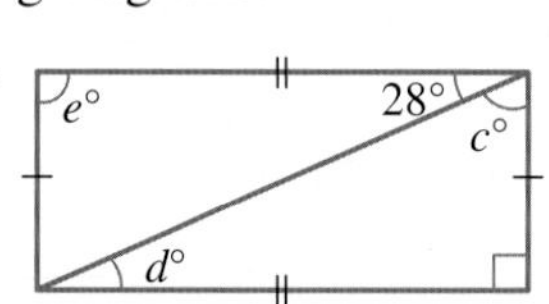

c.
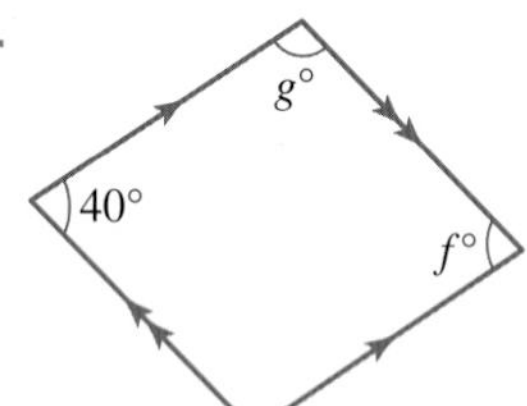

d.
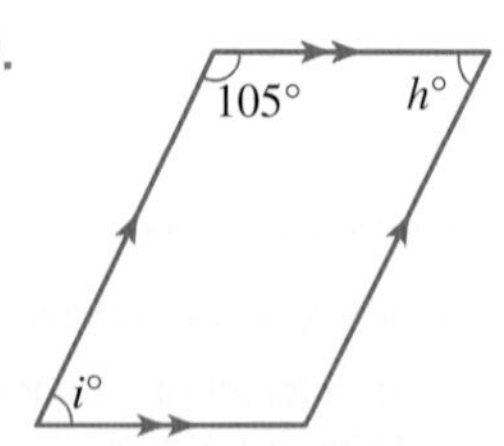

5. Determine the values of the pronumerals in each of the following figures.

a.
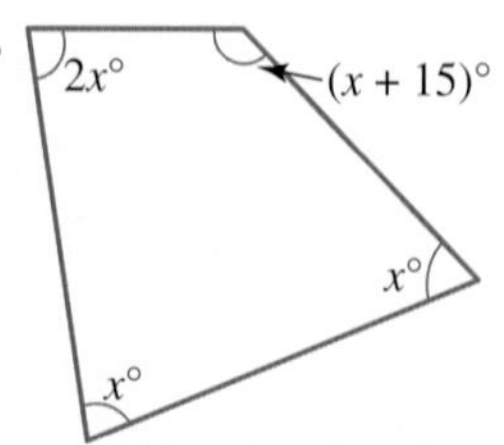

b. 80°, 150°, 3x°, 2x°, y°

6. Determine the values of x and y in each of the following figures.

a. 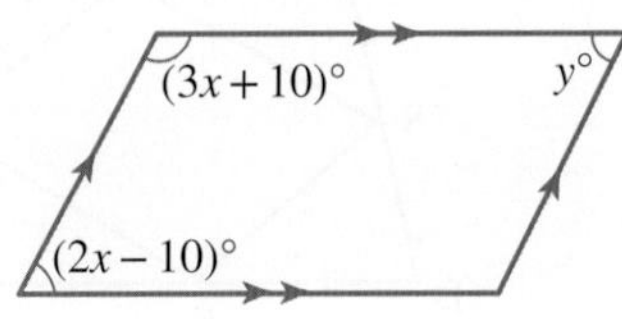

b. 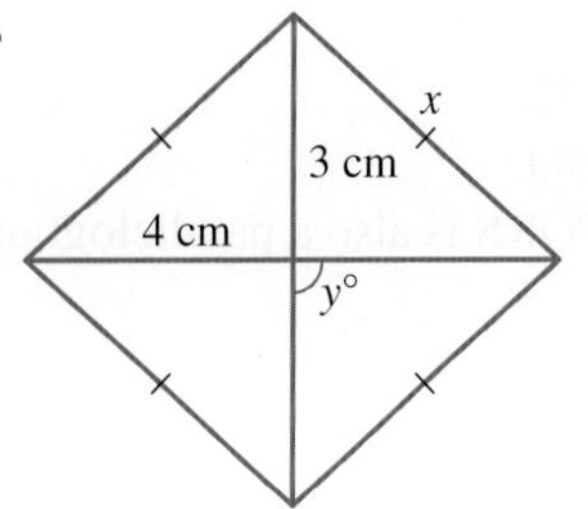

c. 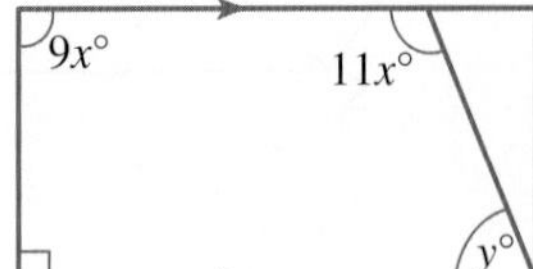

d. 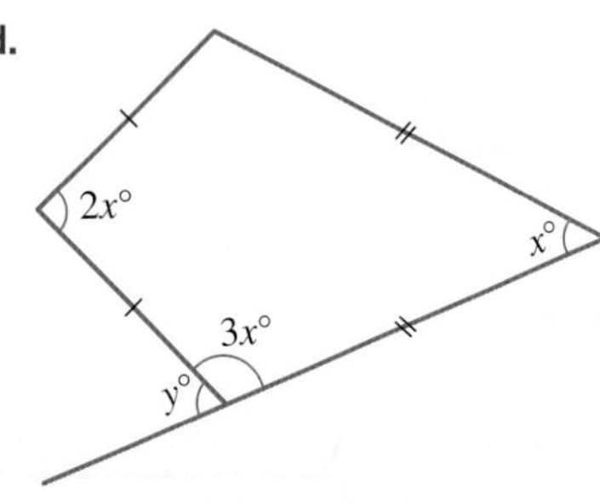

Understanding

7. Draw three different trapeziums. Using your ruler, compass and protractor, decide which of the following properties are true in a trapezium.
   a. Opposite sides are equal.
   b. All sides are equal.
   c. Opposite angles are equal.
   d. All angles are equal.
   e. Diagonals are equal in length.
   f. Diagonals bisect each other.
   g. Diagonals are perpendicular.
   h. Diagonals bisect the angles they pass through.

8. Draw three different parallelograms. Using your ruler and protractor to measure, decide which of the following properties are true in a parallelogram.
   a. Opposite sides are equal.
   b. All sides are equal.
   c. Opposite angles are equal.
   d. All angles are equal.
   e. Diagonals are equal in length.
   f. Diagonals bisect each other.
   g. Diagonals are perpendicular.
   h. Diagonals bisect the angles they pass through.

9. Choose which of the following properties are true in a rectangle.
   a. Opposite sides are equal.
   b. All sides are equal.
   c. Opposite angles are equal.
   d. All angles are equal.
   e. Diagonals are equal in length.
   f. Diagonals bisect each other.
   g. Diagonals are perpendicular.
   h. Diagonals bisect the angles they pass through.

10. Name four quadrilaterals that have at least one pair of opposite sides that are parallel and equal.

11. Name a quadrilateral that has equal diagonals that bisect each other and bisect the angles they pass through.

Reasoning

12. Prove that the diagonals of a rhombus bisect each other.

13. Give reasons why a square is a rhombus, but a rhombus is not necessarily a square.

14. **WE7** ABCD is a parallelogram. X is the midpoint of AB and Y is the midpoint of DC. Prove that AXYD is also a parallelogram.

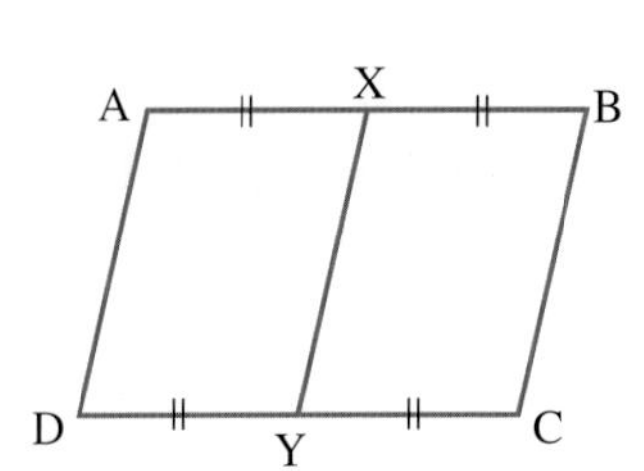

15. The diagonals of a parallelogram meet at right angles. Prove that the parallelogram is a rhombus.

16. ABCD is a parallelogram. P, Q, R and S are all midpoints of their respective sides of ABCD.

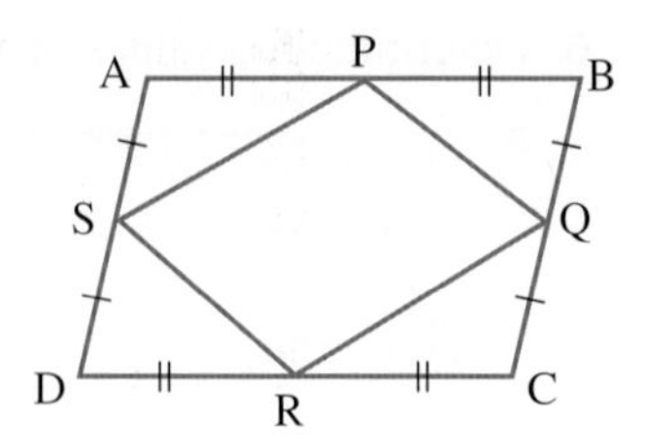

a. Prove $\Delta PAS \cong \Delta RCQ$.
b. Prove $\Delta SDR \cong \Delta PBQ$.
c. Hence, prove that PQRS is also a parallelogram.

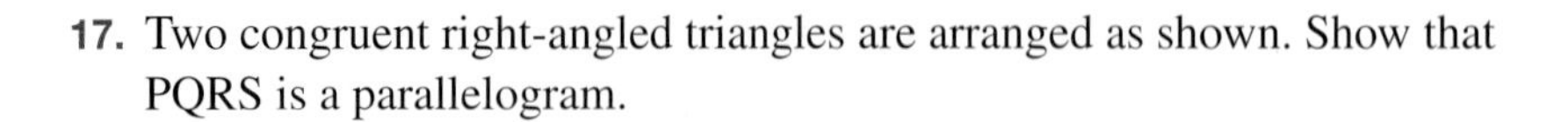

17. Two congruent right-angled triangles are arranged as shown. Show that PQRS is a parallelogram.

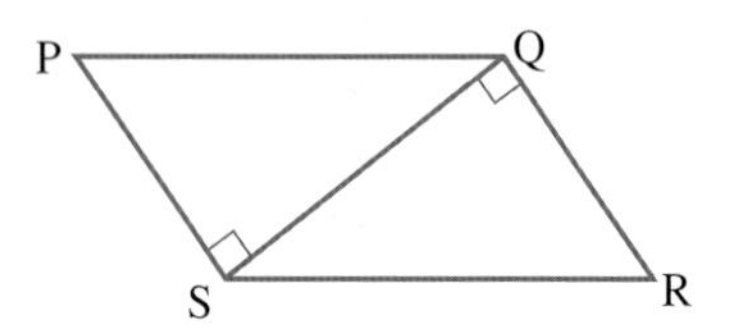

Problem solving

18. ABCD is a trapezium.

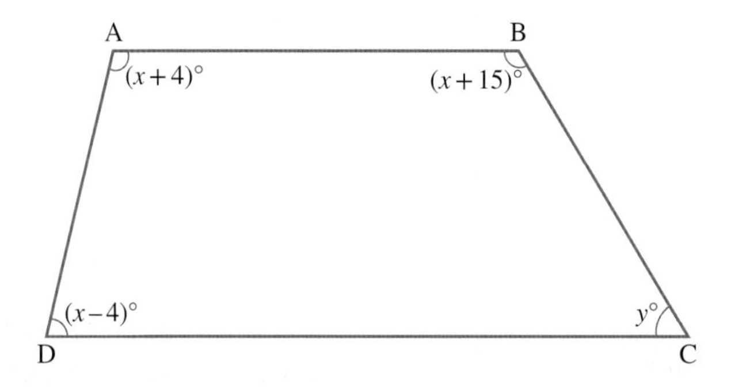

a. Describe a fact about a trapezium.
b. Determine the values of x and y.

19. ABCD is a kite where AC = 8 cm, BE = 5 cm and ED = 9 cm.

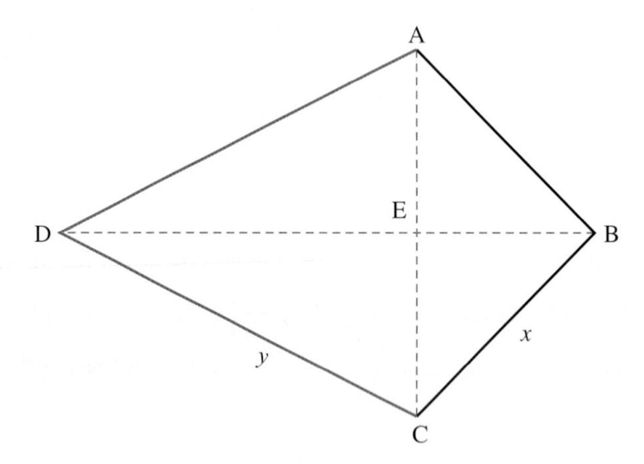

a. Determine the exact values of:
   i. x
   ii. y.
b. Evaluate angle BAD and hence angle BCD. Write your answer in degrees and minutes, correct to the nearest minute.

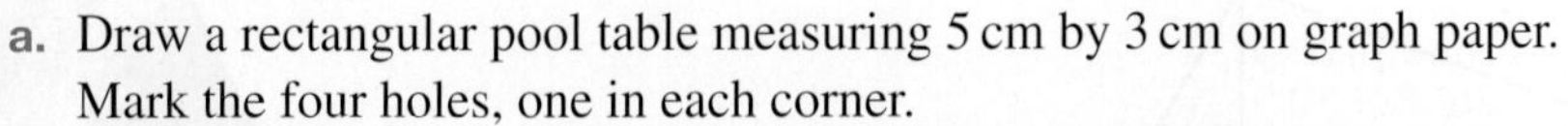

20. Pool is played on a rectangular table. Balls are hit with a cue and bounce off the sides of the table until they land in one of the holes or pockets.

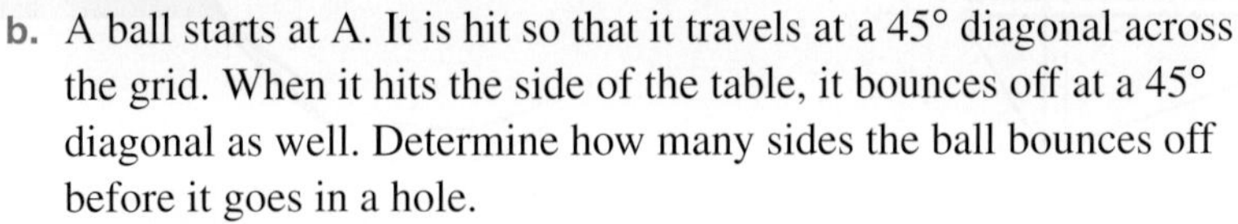

a. Draw a rectangular pool table measuring 5 cm by 3 cm on graph paper. Mark the four holes, one in each corner.

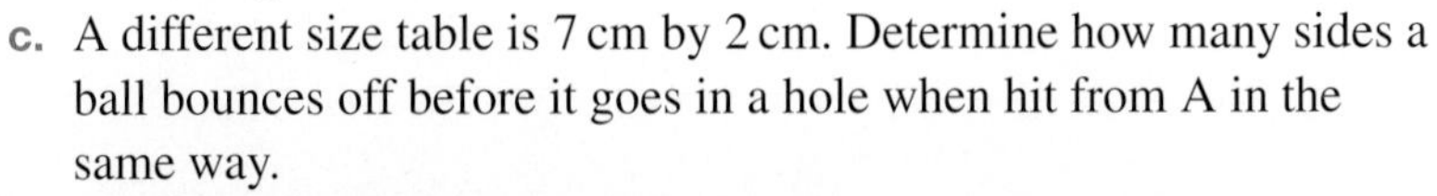

b. A ball starts at A. It is hit so that it travels at a 45° diagonal across the grid. When it hits the side of the table, it bounces off at a 45° diagonal as well. Determine how many sides the ball bounces off

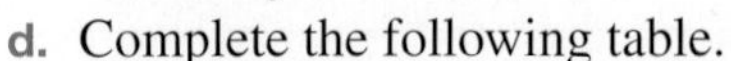

before it goes in a hole.

c. A different size table is 7 cm by 2 cm. Determine how many sides a ball bounces off before it goes in a hole when hit from A in the same way.

d. Complete the following table.

A

Table size	Number of sides hit
5 cm × 3 cm	
7 cm × 2 cm	
4 cm × 3 cm	
4 cm × 2 cm	
6 cm × 3 cm	
9 cm × 3 cm	
12 cm × 4 cm	

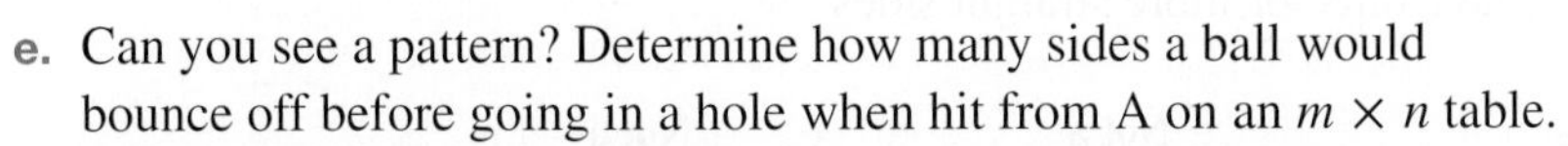

e. Can you see a pattern? Determine how many sides a ball would bounce off before going in a hole when hit from A on an $m \times n$ table.

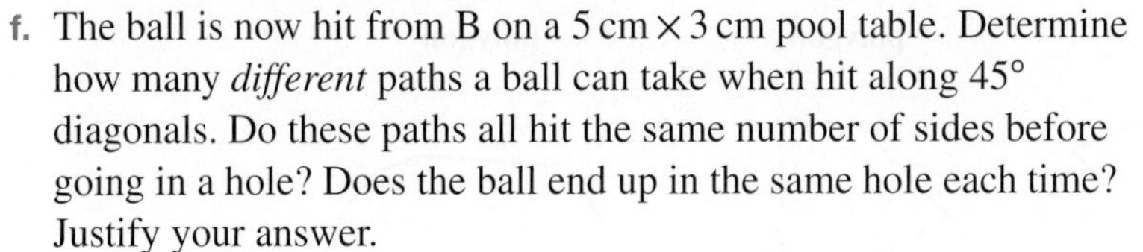

f. The ball is now hit from B on a 5 cm × 3 cm pool table. Determine how many *different* paths a ball can take when hit along 45° diagonals. Do these paths all hit the same number of sides before going in a hole? Does the ball end up in the same hole each time? Justify your answer.

B

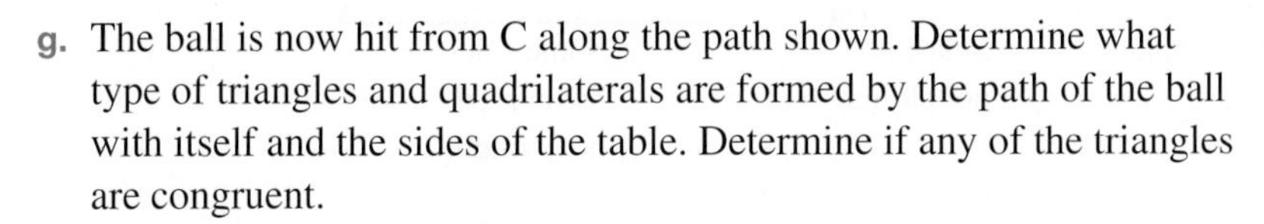

g. The ball is now hit from C along the path shown. Determine what type of triangles and quadrilaterals are formed by the path of the ball with itself and the sides of the table. Determine if any of the triangles are congruent.

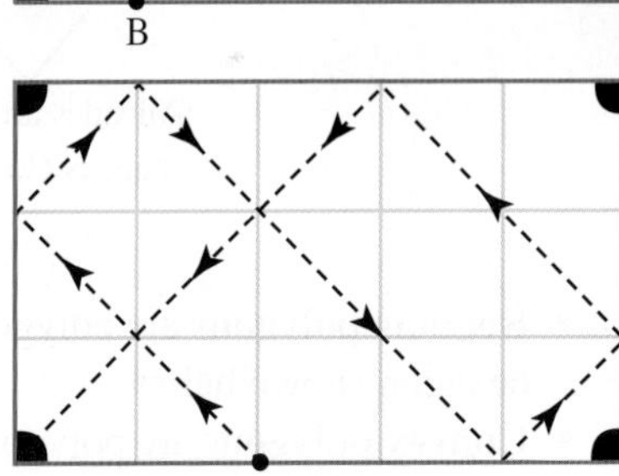

21. ABCD is called a cyclic quadrilateral because it is inscribed inside a circle.
A characteristic of a cyclic quadrilateral is that the opposite angles are supplementary.
Determine the value of x.

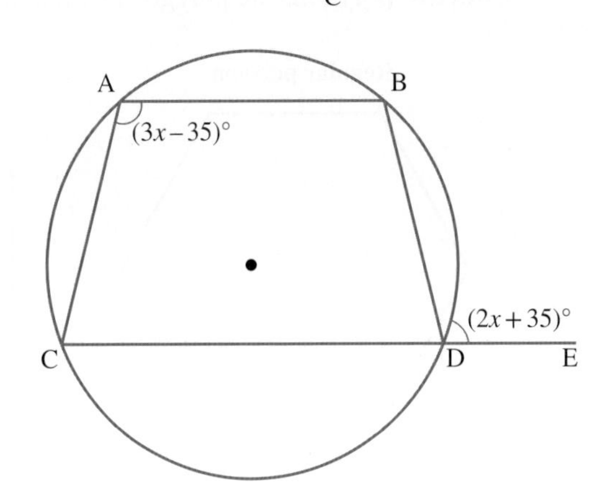

22. The perimeter of this kite is 80 cm. Determine the exact value of x.

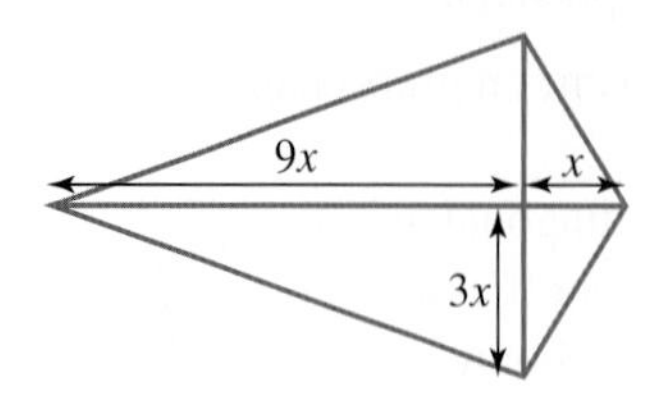

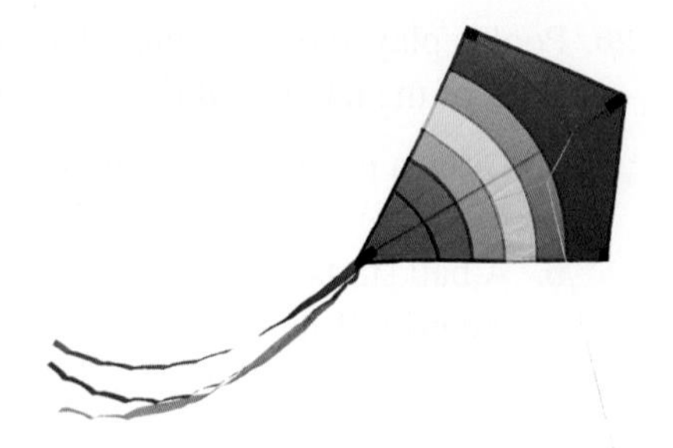

10.5 Polygons

LEARNING INTENTION

At the end of this subtopic you should be able to:
- identify regular and irregular polygons
- calculate the sum of the interior angles of a polygon
- determine the exterior angles of a regular polygon.

10.5.1 Polygons

eles-4906

- **Polygons** are closed shapes that have three or more straight sides.

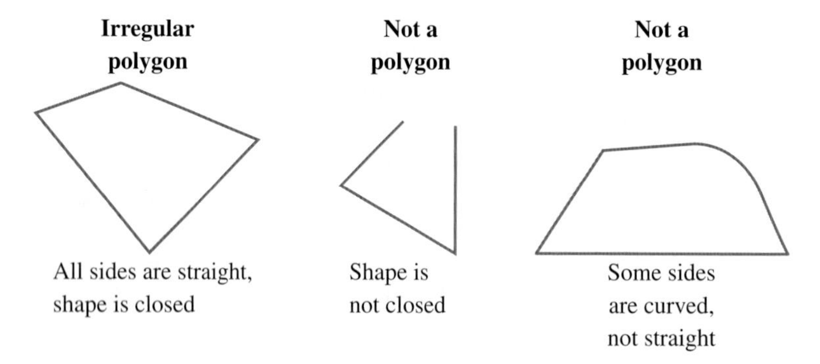

- **Regular polygons** are polygons with sides of the same length and interior angles of the same size, like the hexagon shown below.
- **Convex polygons** are polygons with no interior reflex angles.
- **Concave polygons** are polygons with at least one reflex interior angle.

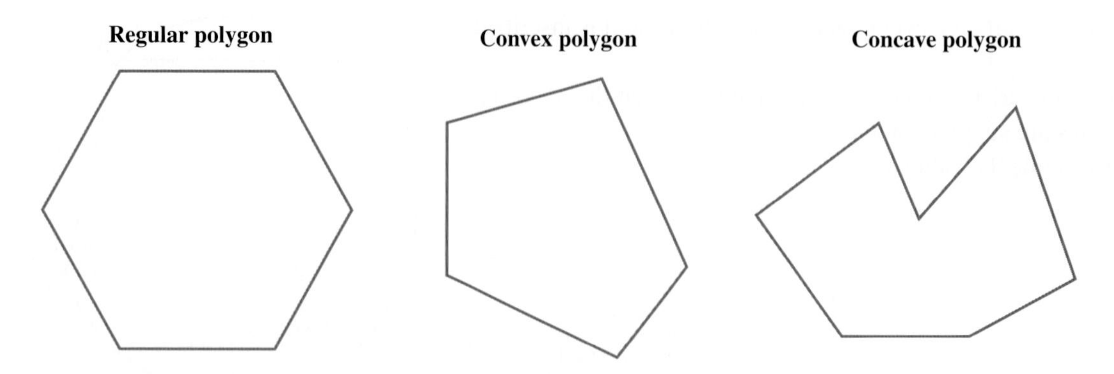

Interior angles of a polygon

- The sum of the interior angles of a polygon is given by the formula shown below.

> **Sum of interior angles of a polygon**
>
> $$\textbf{Sum of interior angles} = 180^\circ \times (n - 2)$$
>
> where n = the number of sides of the polygon.

WORKED EXAMPLE 8 Calculating the values of angles in a given diagram

Calculate the value of the pronumerals in the figure shown.

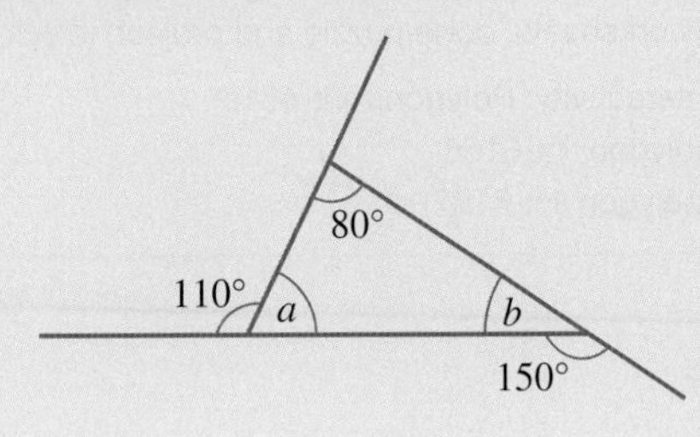

THINK	WRITE
1. Angles a and 110° form a straight line and so are supplementary (add to 180°).	$a + 110^\circ = 180^\circ$ $a + 110^\circ - 110^\circ = 180^\circ - 110^\circ$ $a = 70^\circ$
2. The interior angles of a triangle sum to 180°.	$b + a + 80^\circ = 180$
3. Substitute 70° for a and solve for b.	$b + 70^\circ + 80^\circ = 180^\circ$ $b + 150^\circ = 180^\circ$ $b = 30^\circ$
4. Write the value of the pronumerals.	$a = 70^\circ$, $b = 30^\circ$

Exterior angles of a polygon

- The exterior angles of a polygon are formed by the side of the polygon and an extension of its adjacent side. For example, x, y and z are exterior angles for the polygon (triangle) below and q, r, s and t are the exterior angles of the quadrilateral.

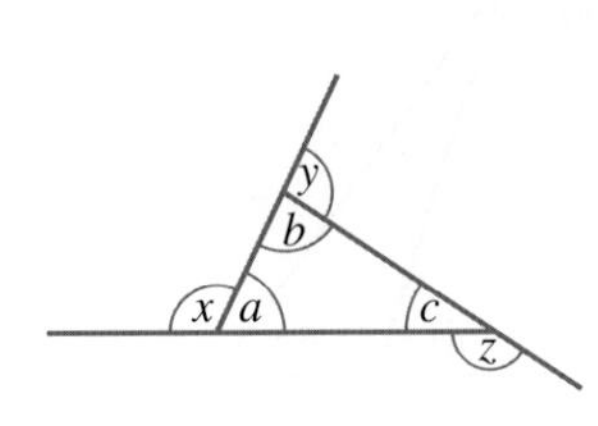

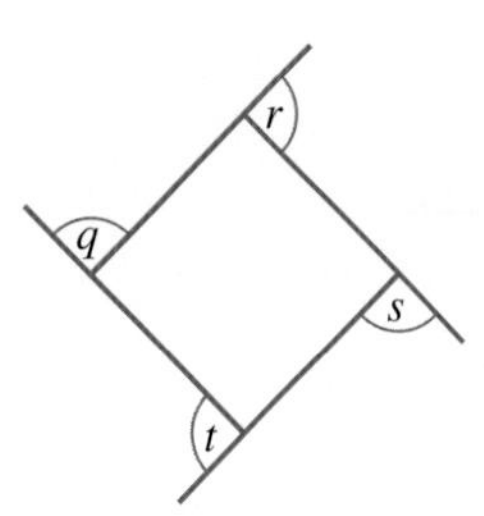

- The exterior angle and interior angle at that vertex are supplementary (add to 180°). For example, in the triangle above, $x + a = 180^\circ$.
- Exterior angles of polygons can be measured in a clockwise or anticlockwise direction.
- In a regular polygon, the size of the exterior angle can be found by dividing 360° by the number of sides.

> **Exterior angles of a regular polygon**
>
> $$\textbf{Exterior angles of a regular polygon} = \frac{360°}{n}$$
>
> where n = the number of sides of the regular polygon.

- The sum of the exterior angles of a polygon equals 360°.
- The exterior angle of a triangle is equal to the sum of the opposite interior angles.

Resources

 **eWorkbook** Topic 10 Workbook (worksheets, code puzzle and project) (ewbk-2036)

 **Interactivities** Individual pathway interactivity: Polygons (int-4615)
Interior angles of a polygon (int-6166)
Exterior angles of a polygon (int-6167)

Exercise 10.5 Polygons

learnON

Individual pathways

■ PRACTISE	■ CONSOLIDATE	■ MASTER
1, 3, 6, 7, 11, 14	2, 4, 8, 12, 15, 17	5, 9, 10, 13, 16, 18

To answer questions online and to receive **immediate feedback** and **sample responses** for every question, go to your learnON title at www.jacplus.com.au.

Fluency

1. WE8 Calculate the values of the pronumerals in the diagrams shown.

a.

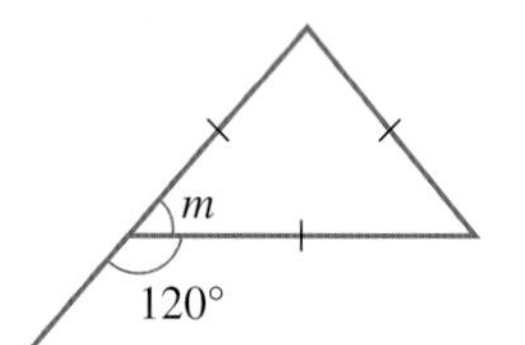

b.

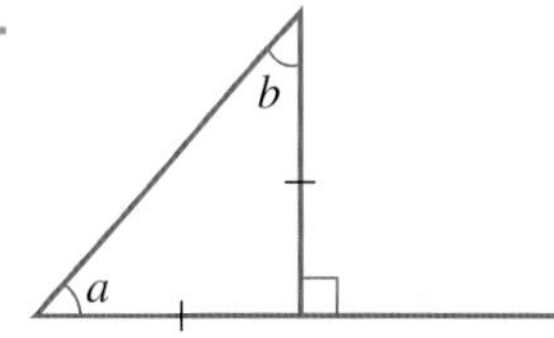

c.

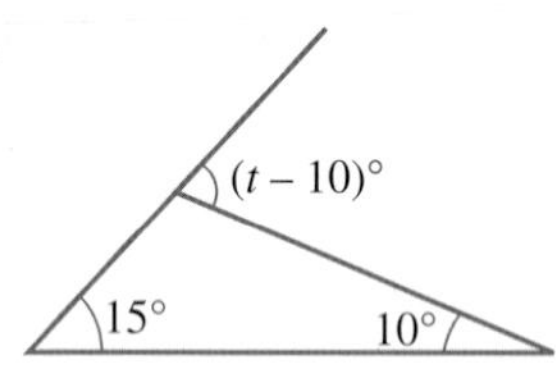

d.

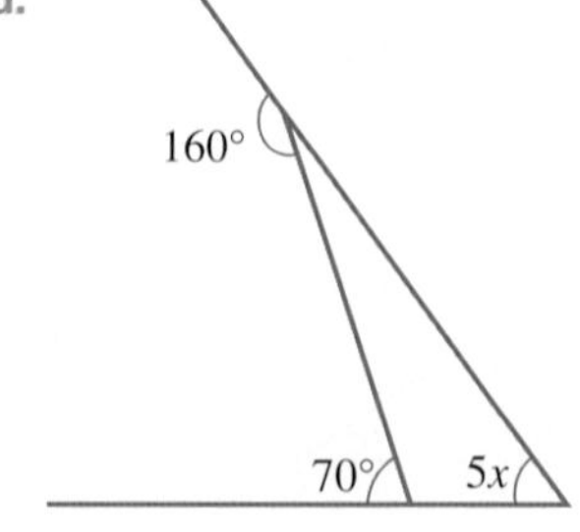

2. Determine the value of the pronumeral in each of the following polygons.

a. 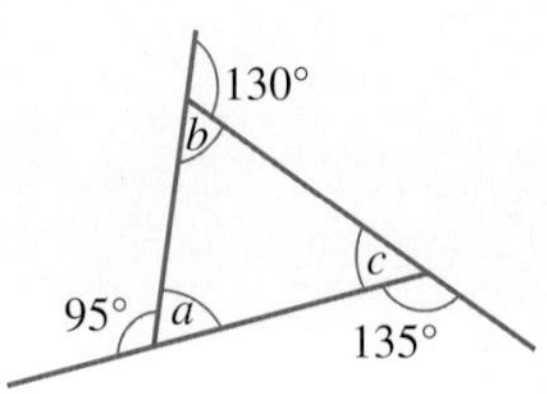

b. 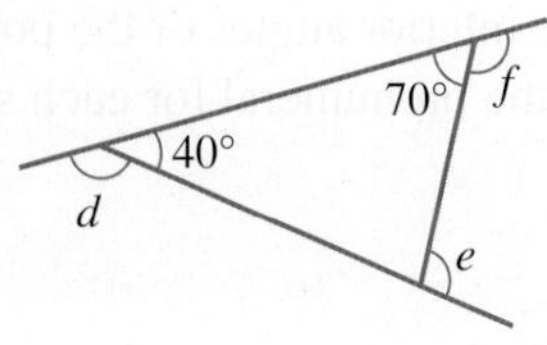

c. 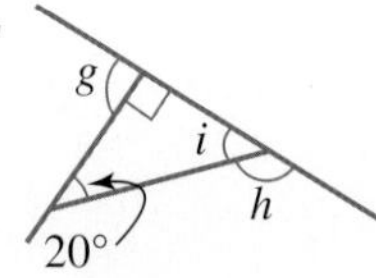

d. 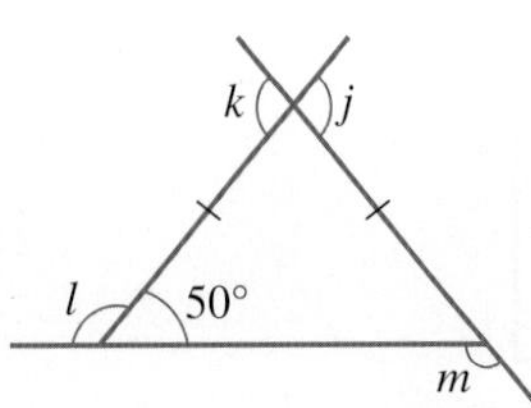

3. For the triangles shown, evaluate the pronumerals and determine the size of the interior angles.

a. 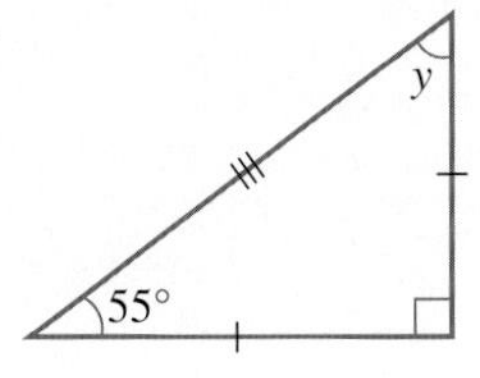

b. 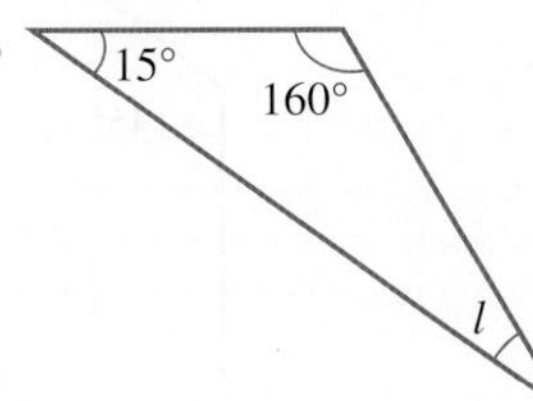

c. 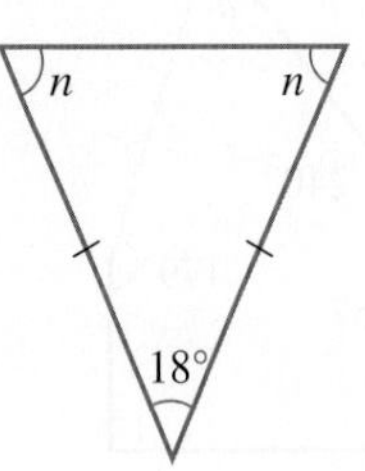

d. 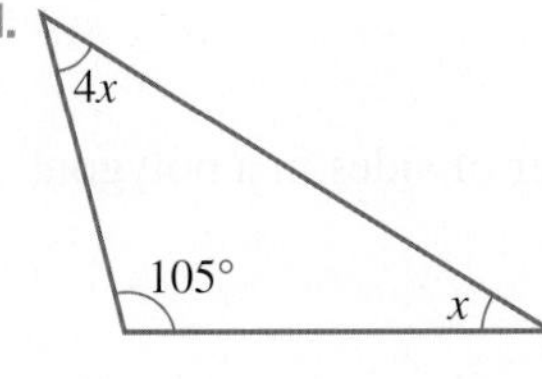

e. 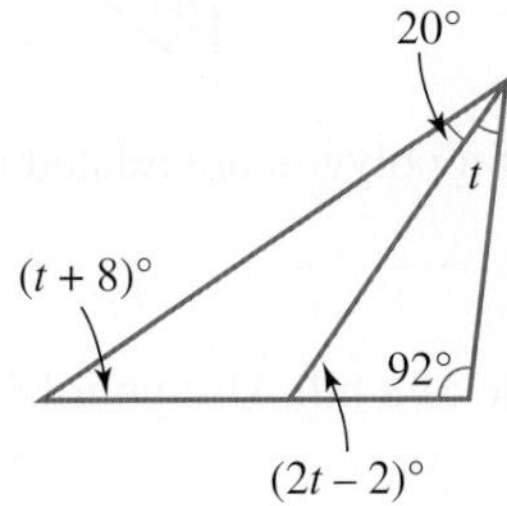

4. For the five quadrilaterals shown:

i. label the quadrilaterals as regular or irregular

ii. determine the value of the pronumeral for each shape.

a. 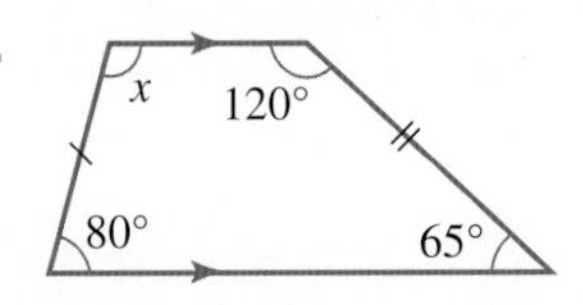

b. 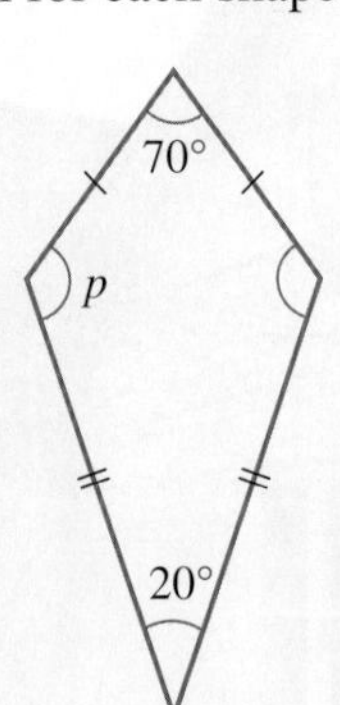

c. 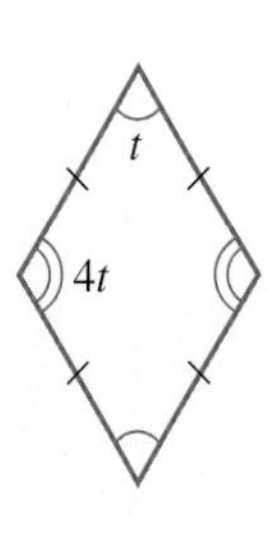

d. 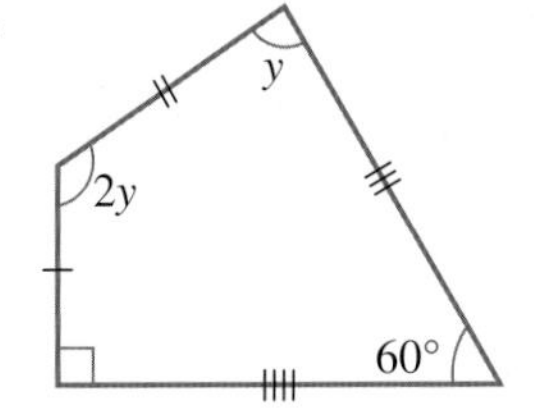

e. 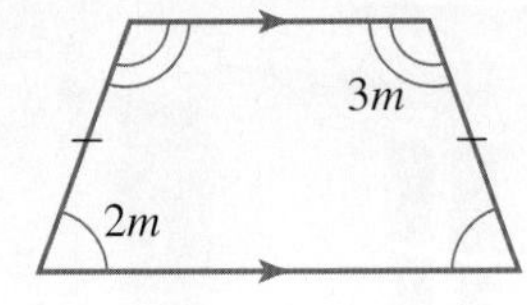

5. For the four polygons:
   i. calculate the sum of the interior angles of the polygon
   ii. determine the value of the pronumeral for each shape.

a.

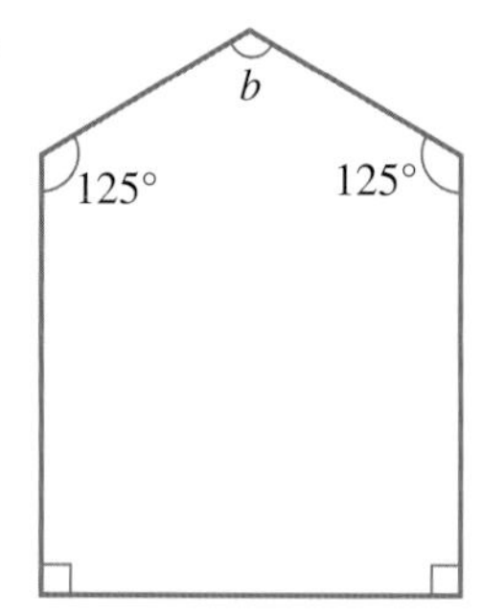

b.

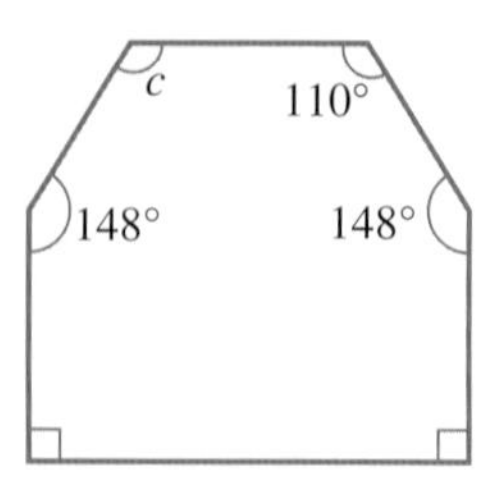

c.

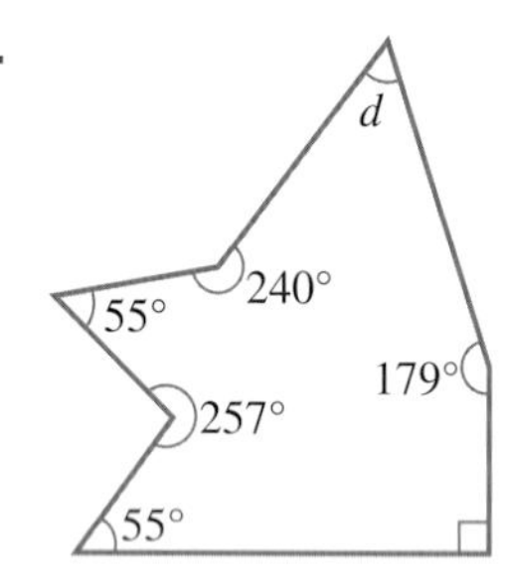

d.

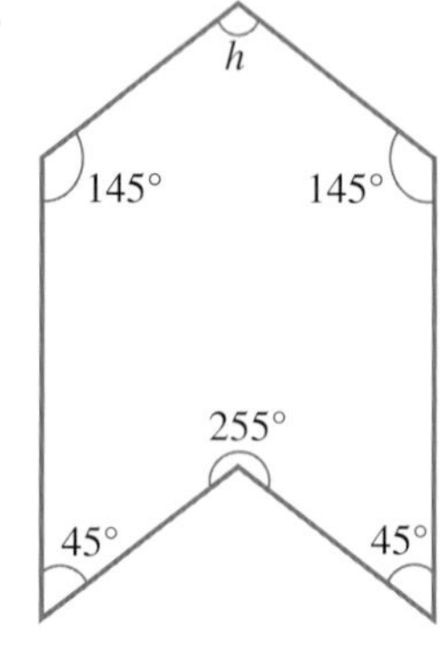

6. Explain how the interior and exterior angles of a polygon are related to the number of sides in a polygon.

Understanding

7. The photograph shows a house built on the side of a hill. Use your knowledge of angles to determine the values of the pronumerals. Show full working.

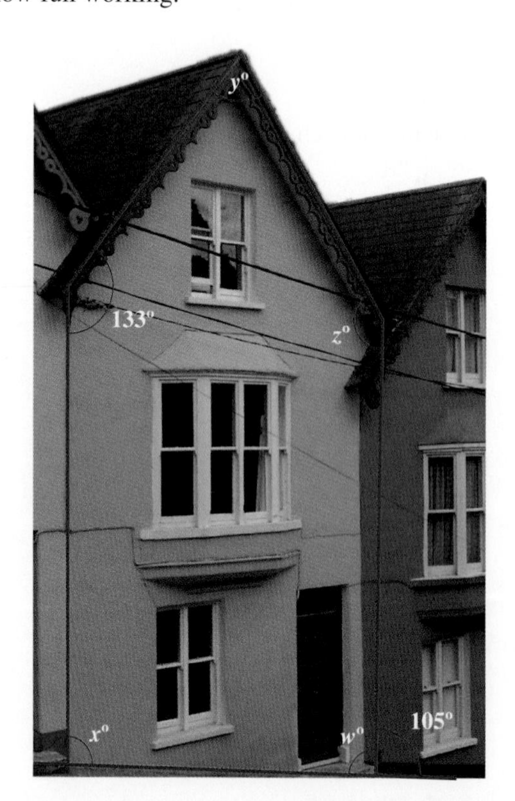

8. Determine the values of the four interior angles of the front face of the building in the photograph shown. Show full working.

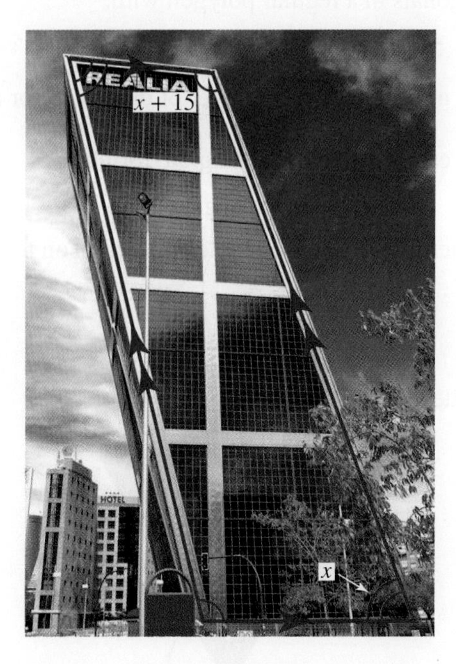

9. Determine the values of the pronumerals for the irregular polygons shown. Show full working.

a. 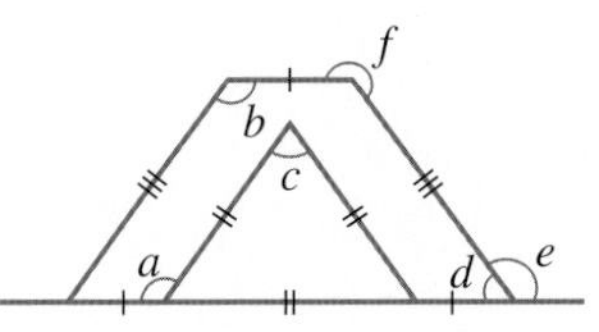

b. 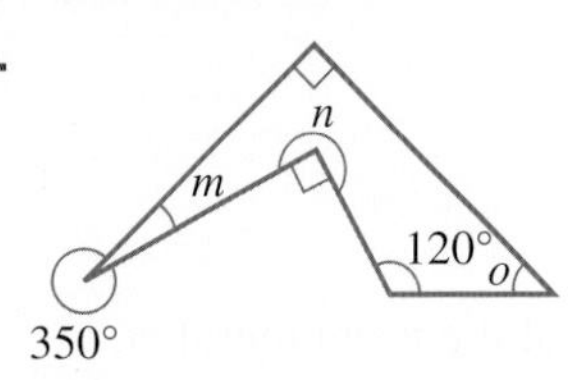

10. Calculate the size of the exterior angle of a regular hexagon (6 sides).

Reasoning

11. State whether the following polygons are regular or irregular. Give a reason for your answer.

a. 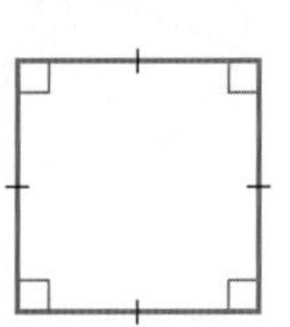

b. 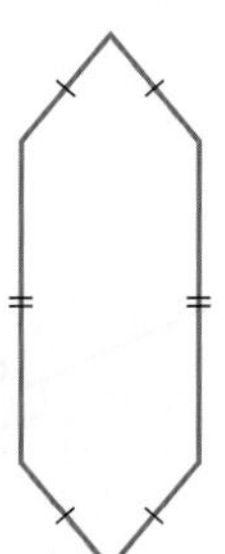

c. 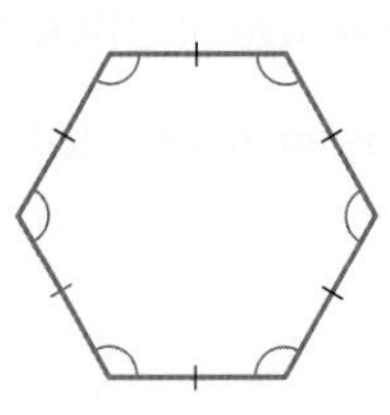

d. 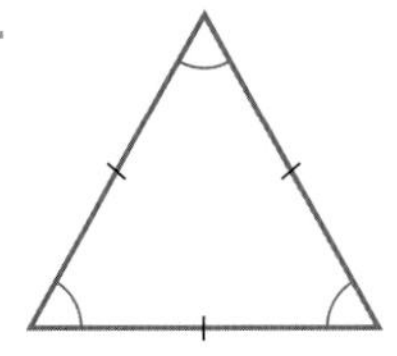

e. 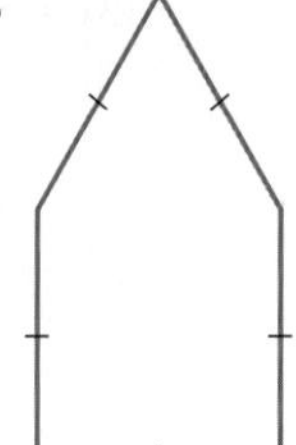

f. 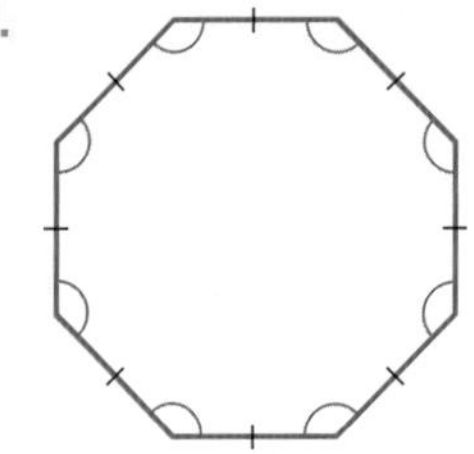

12. A diagonal of a polygon joins two vertices.

   a. Calculate the number of diagonals in a regular polygon with:

      i. 4 sides ii. 5 sides iii. 6 sides iv. 7 sides.

   b. Using your results from part a, show that the number of diagonals for an n-sided polygon is $\frac{1}{2}n(n-3)$.

13. The exterior angle of a polygon can be calculated using the formula:

$$\text{Exterior angle} = \frac{360°}{n}$$

Use the relationship between interior and exterior angles of a polygon to write a formula for the internal angle of a regular polygon.

Problem solving

14. a. Name the polygon that best describes the road sign shown.

   b. Determine the value of the pronumeral m.

15. The diagram shows a regular octagon with centre O.

   a. Calculate the size of ∠CBD.
   b. Calculate the size of ∠CBO.
   c. Calculate the size of the exterior angle of the octagon, ∠ABD.
   d. Calculate the size of ∠BOC.

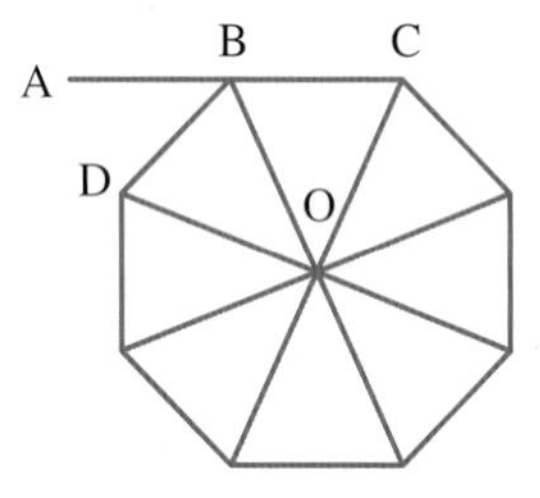

16. ABCDEFGH is an eight-sided polygon.

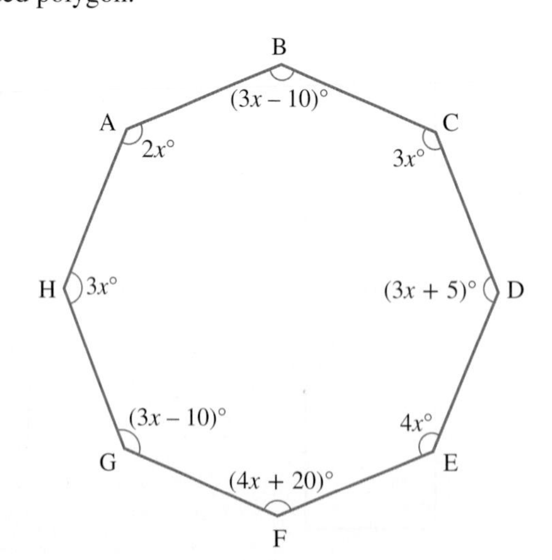

   a. Evaluate the sum of the interior angles of an eight-sided polygon.
   b. Determine the value of the pronumeral x.

17. Answer the following questions for the given shape.

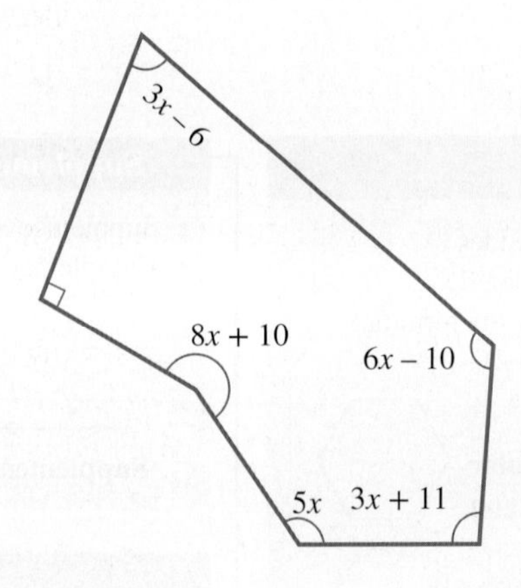

a. Evaluate the sum of the interior angles of this shape.
b. Determine the value of the pronumeral x.

18. Answer the following questions for the given shape.

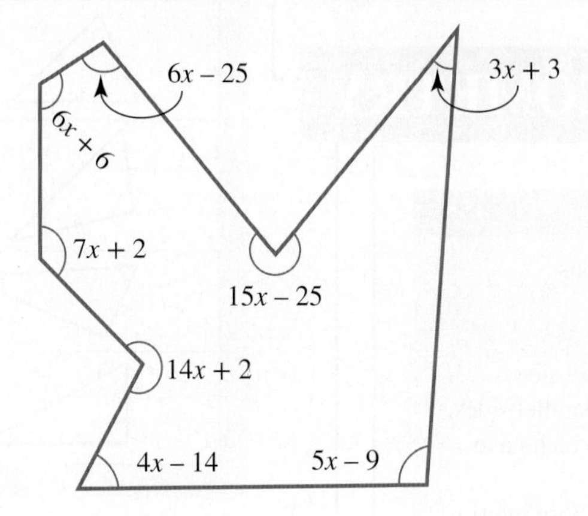

a. Evaluate the sum of the interior angles of this shape.
b. Determine the value of the pronumeral x.

10.6 Review

10.6.1 Topic summary

DEDUCTIVE GEOMETRY

Polygons

- Polygons are closed shapes with straight sides.
- The number of sides a polygon has is denoted n.
- The sum of the interior angles is given by the formula:
  Interior angle sum $= 180^\circ(n-2)$
- **Regular polygons** have all sides the same length and all interior angles equal.
- **Convex polygons** have no internal reflex angles.
- **Concave polygons** have at least one internal reflex angle.
- The exterior angles of a regular polygon are given by the formula:
  Exterior angles $= \dfrac{360^\circ}{n}$

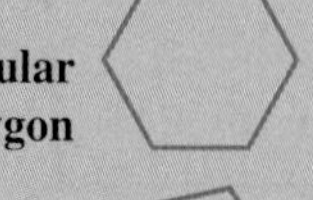

Regular polygon

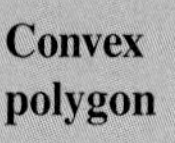

Convex polygon

Concave polygon

Quadrilaterals

- Quadrilaterals are four sided polygons.
- The interior angles sum to 360°.
- There are many types of quadrilaterals:
  - **Trapeziums** have 1 pair of parallel sides.
  - **Parallelograms** have 2 pairs of parallel sides.
  - **Rhombuses** are parallelograms which have 4 equal sides.
  - A **rectangle** is a parallelogram whose interior angles are right angles.
  - A **square** is a rectangle with 4 equal sides.

Parallel lines

- If parallel lines are cut by a transversal, then:

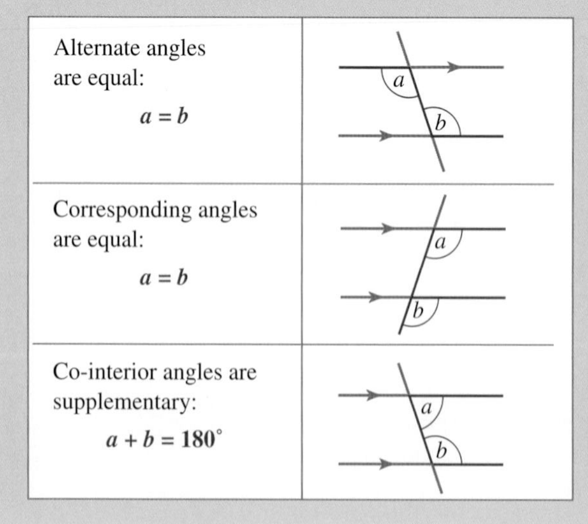

Supplementary and complementary angles

- Supplementary angles are angles that add up to 180°.
- Complementary angles are angles that add up to 90°.

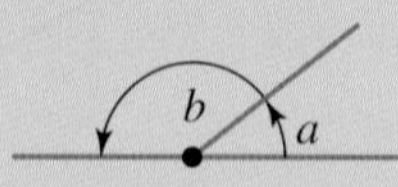

Supplementary angles

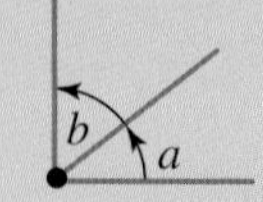

Complementary angles

Congruent figures

- Congruent figures have the same size and shape.
- The symbol for congruence is ≅.
- The following tests can be used to determine whether two triangles are congruent:

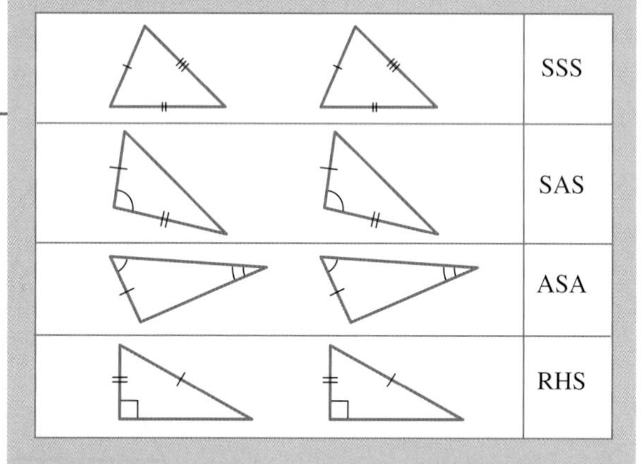

Similar figures

- Similar figures have the same shape but different size.
- The symbol for similarity is ~.
- Scale factor $= \dfrac{\text{image side length}}{\text{object side length}}$
- The following tests can be used to determine whether two triangles are similar:

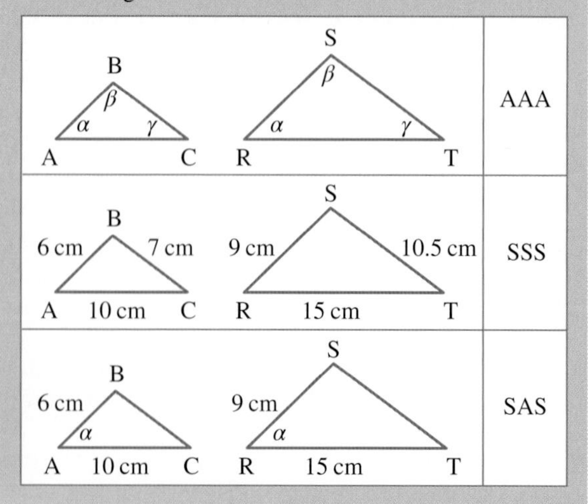

10.6.2 Success criteria

Tick the column to indicate that you have completed the subtopic and how well you have understood it using the traffic light system.

(**Green:** I understand; **Yellow:** I can do it with help; **Red:** I do not understand)

Subtopic	Success criteria	Green	Yellow	Red
10.2	I can apply properties of straight lines and triangles to determine the value of an unknown angle.			
	I can construct simple geometric proofs for angles in triangles or around intersecting lines.			
	I can prove that triangles are congruent by applying the appropriate congruency test.			
10.3	I can identify similar figures.			
	I can calculate the scale factor in similar figures.			
	I can show that two triangles are similar using the appropriate similarity test.			
10.4	I can identify the different types of quadrilaterals.			
	I can construct simple geometric proofs for angles, sides and diagonals in quadrilaterals.			
10.5	I can identify regular and irregular polygons.			
	I can calculate the sum of the interior angles of a polygon.			
	I can determine the exterior angles of a regular polygon.			

10.6.3 Project

Enlargement activity

Enlargement is the construction of a bigger picture from a small one. The picture is identical to the other except that it is bigger. The new picture is often called the image. This can also be called creating a similar figure.

The geometrical properties shared by a shape and its image under enlargement can be listed as:

- lines are enlarged as lines
- sides are enlarged to corresponding sides by the same factor
- matching angles on the two shapes are equal.

In this activity, we will start with a small cartoon character, and then 'blow it up' to almost life-size.

Equipment: ruler, pencil, cartoon print, butcher's paper or some other large piece of paper.

1. Do some research on the internet and select a cartoon character or any character of your choice.
2. Draw a grid of 2-cm squares over the small cartoon character.

   Example: The cat is 9 squares wide and 7 squares tall.
3. Label the grids with letters across the top row and numbers down the first column.
4. Get a large piece of paper and draw the same number of squares. You will have to work out the ratio of similitude (e.g. 2 cm : 8 cm).
5. If your small cartoon character stretches from one side of the 'small' paper (the paper the image is printed on) to the other, your 'large' cat must stretch from one side of the 'big' paper to the other. Your large grid squares may have to be 8 cm by 8 cm or larger, depending on the paper size.
6. Draw this enlarged grid on your large paper. Use a metre ruler or some other long straight-edged tool. Be sure to keep all of your squares the same size.
   - At this point, you are ready to draw. Remember, you do NOT have to be an artist to produce an impressive enlargement.
   - All you do is draw EXACTLY what you see in each small cell into its corresponding large cell.
   - For example, in cell B3 of the cat enlargement, you see the tip of his ear, so draw this in the big grid.
   - If you take your time and are very careful, you will produce an extremely impressive enlargement.
   - What you have used is called a 'ratio of similitude'. This ratio controls how large the new picture will be.

A 2 : 5 ratio will give you a smaller enlargement than a 2 : 7 ratio, because for every 2 units on the original you are generating only 5 units of enlargement instead of 7.

If the cat ratio is 1 : 4, it produces a figure that has a linear measure that is four times bigger.

The big cat's overall **area**, however, will be **16 times larger** than the small cat's. This is because area is found by taking length times width.

The length is 4 times longer and the width is 4 times longer. Thus the **area** is $4 \times 4 = 16$ times **larger** than the original cat.

The overall **volume** will be $4 \times 4 \times 4$ or **64 times larger!** This means that the big cat will weigh 64 times more than the small cat.

Resources

 **eWorkbook** Topic 10 Workbook (worksheets, code puzzle and project) (ewbk-2036)

 **Interactivities** Crossword (int-2854)

Sudoku puzzle (int-3597)

Exercise 10.6 Review questions

learnON

To answer questions online and to receive **immediate corrective feedback** and **fully worked solutions** for all questions, go to your learnON title at www.jacplus.com.au.

Fluency

1. Select a pair of congruent triangles in each of the following sets of triangles, giving a reason for your answer. All angles are in degrees and side lengths in cm. (The figures are not drawn to scale.)

   a.

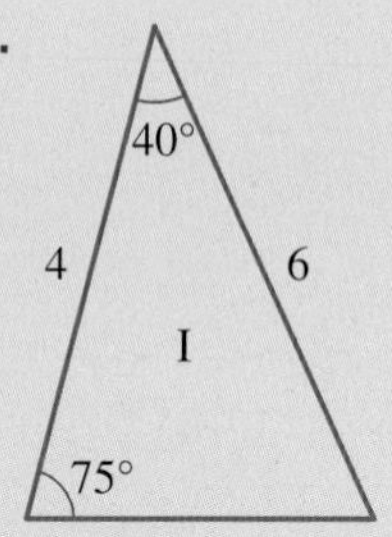

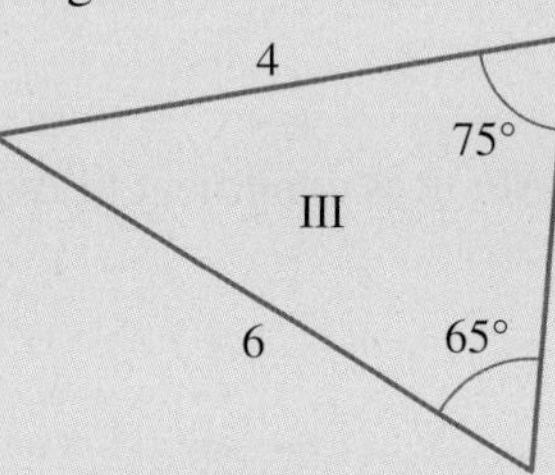

   b.

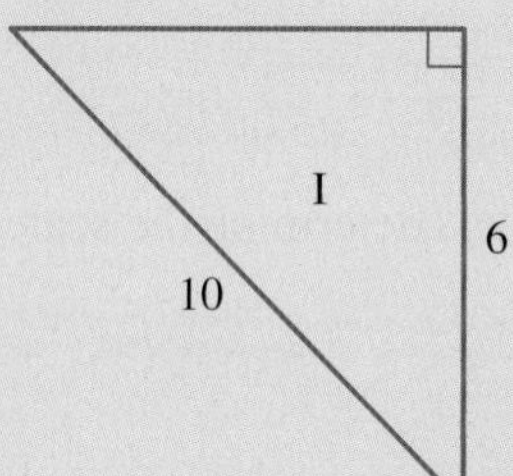

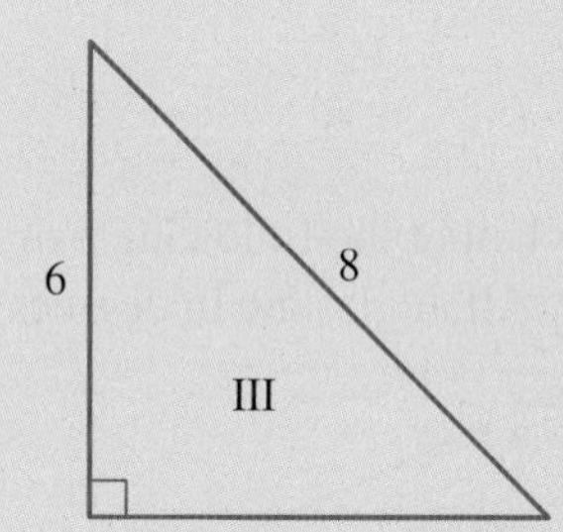

2. Determine the value of the pronumeral in each pair of congruent triangles. All angles are given in degrees and side lengths in cm.

   a.

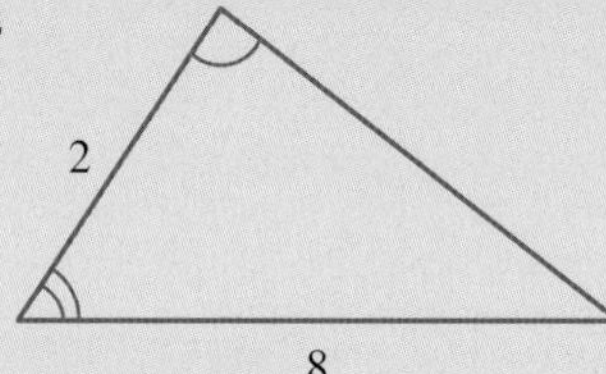

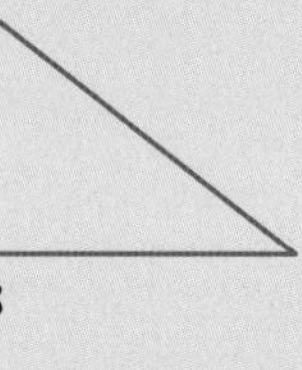

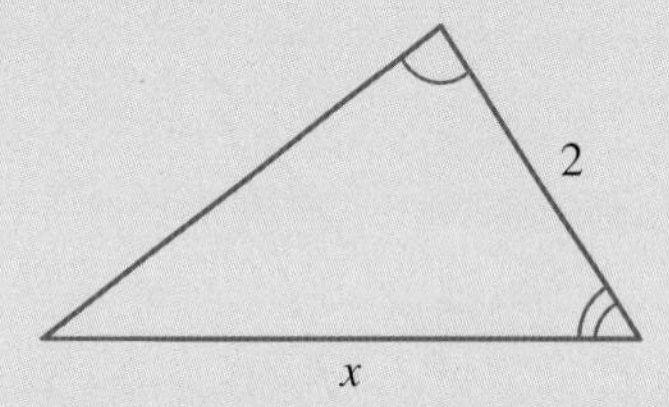

   b.

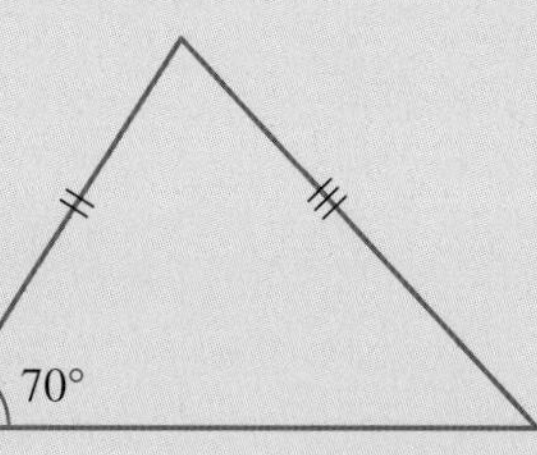

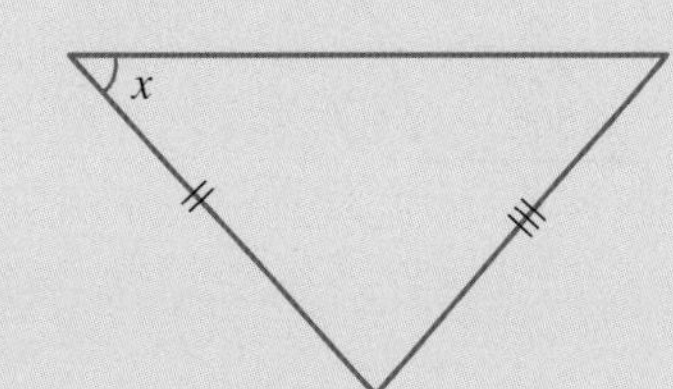

   c.

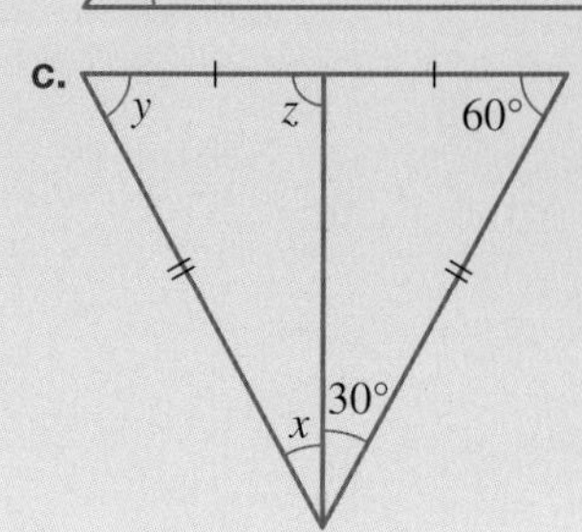

3. a. Prove that the two triangles shown in the diagram are congruent.

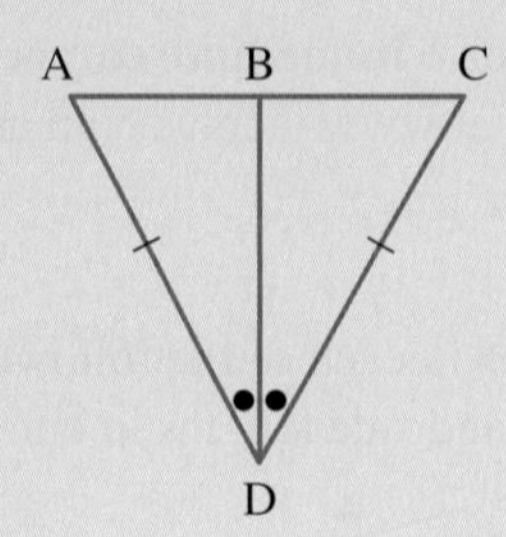

b. Prove that ΔPQR is congruent to ΔQPS.

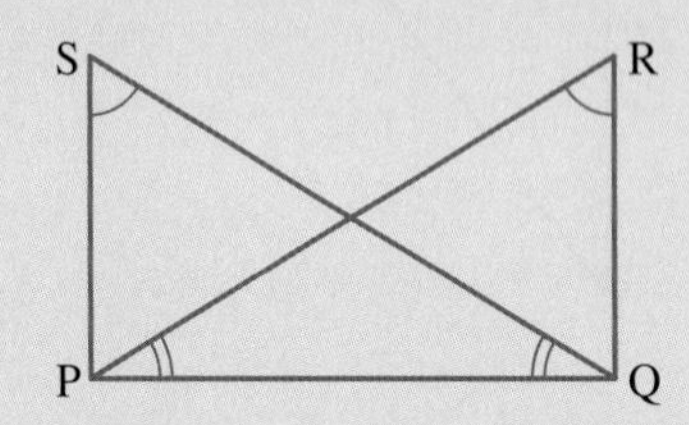

4. Test whether the following pairs of triangles are similar. For similar triangles, determine the scale factor. All angles are in degrees and side lengths in cm.

a.

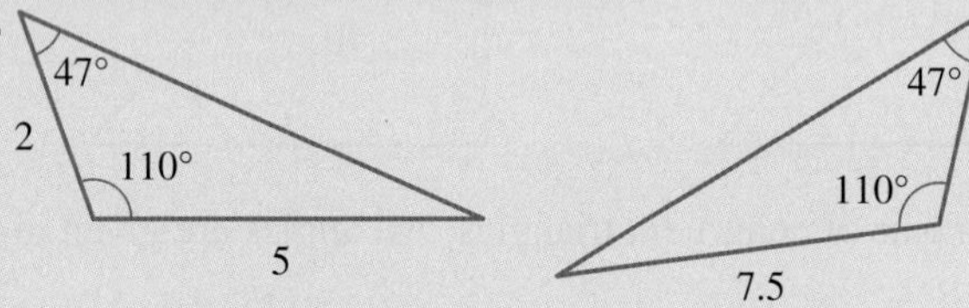

b.

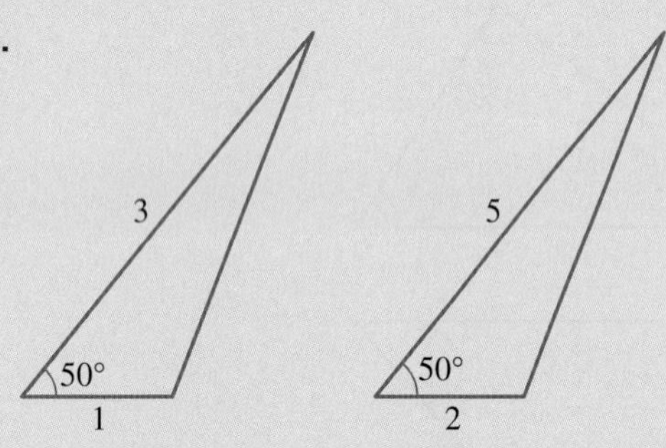

c.

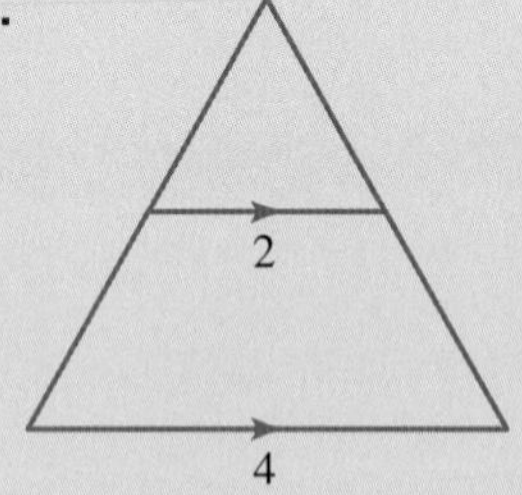

5. Determine the value of the pronumeral in each pair of similar triangles. All angles are given in degrees and side lengths in cm.

a. 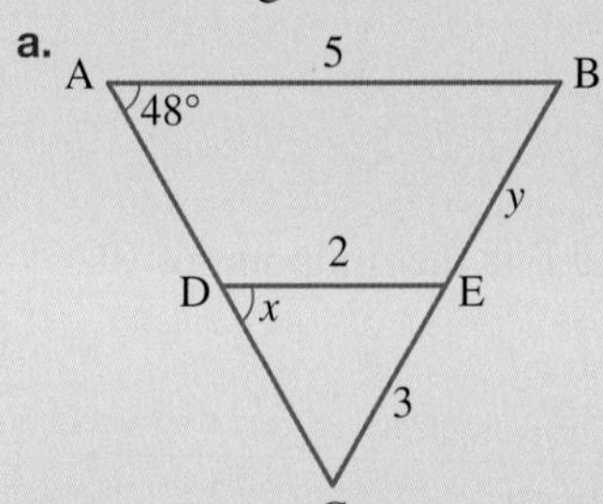

b. 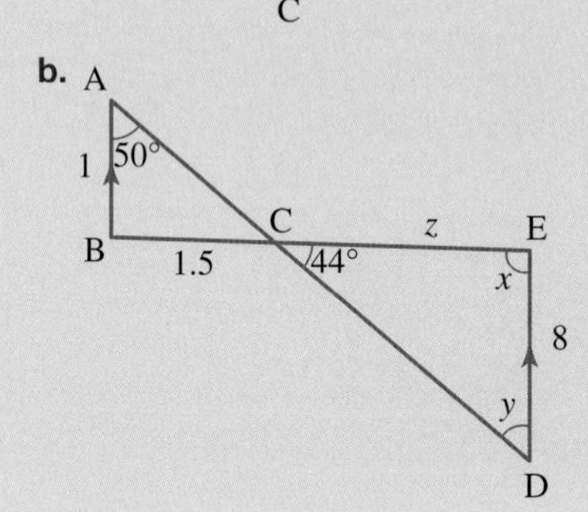

c. 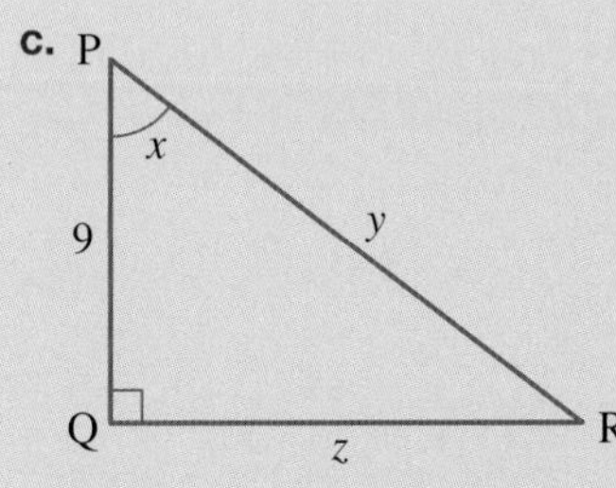

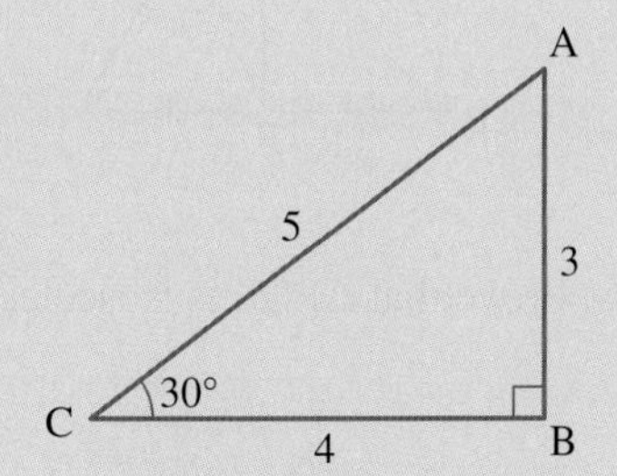

6. Prove that ΔABC ~ ΔEDC.

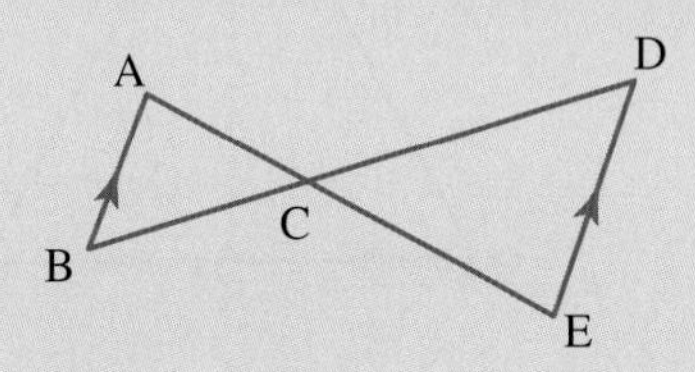

7. Prove that ΔPST ~ ΔPRQ.

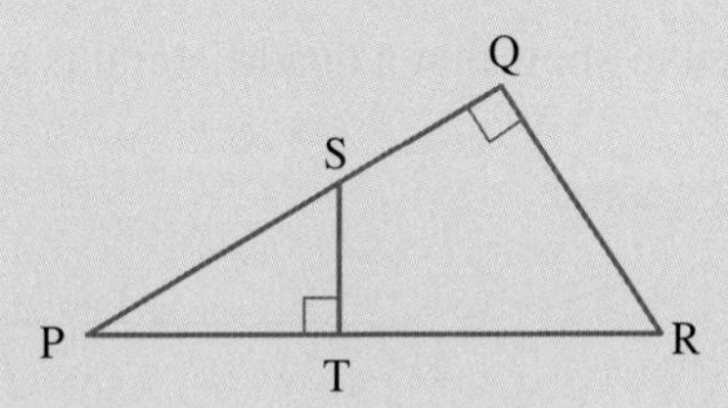

8. State the definition of a rhombus.

9. **MC** Two corresponding sides in a pair of similar octagons have lengths of 4 cm and 60 mm. The respective scale factor in length is:

A. 1 : 15 **B.** 3 : 20 **C.** 2 : 3 **D.** 3 : 2 **E.** 20 : 3

10. A regular nonagon has side length x cm. Use a scale factor of $\frac{x+1}{x}$ to calculate the side length of a similar nonagon.

Problem solving

11. ABC is a triangle. D is the midpoint of AB, E is the midpoint of AC and F is the midpoint of BC DG $\perp$ AB, EG $\perp$ AC and FG $\perp$ BC.
   a. Prove that ΔGDA $\cong \Delta$GDB.
   b. Prove that ΔGDE $\cong \Delta$GCE.
   c. Prove that ΔGBF $\cong \Delta$GCF.
   d. State what this means about AG, BG and CG.
   e. A circle centred at G is drawn through A.
   Determine what other points it must pass through.

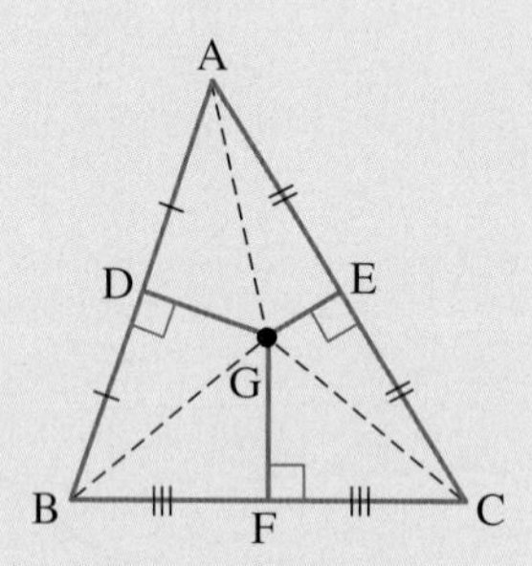

12. PR is the perpendicular bisector of QS. Prove that ΔPQS is isosceles.

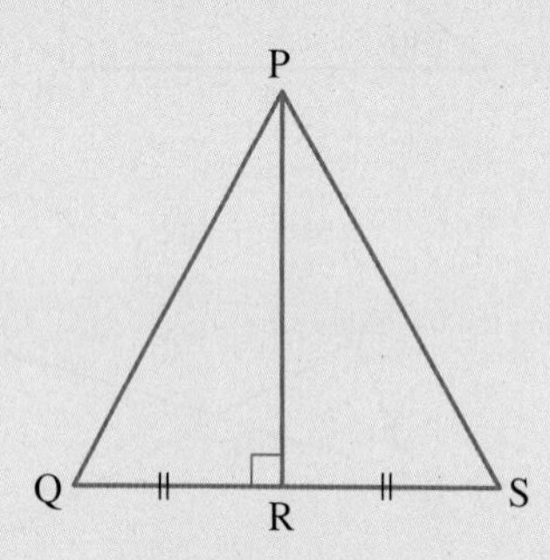

13. Name any quadrilaterals that have diagonals that bisect the angles they pass through.

14. State three tests that can be used to show that a quadrilateral is a rhombus.

15. Prove that WXYZ is a parallelogram.

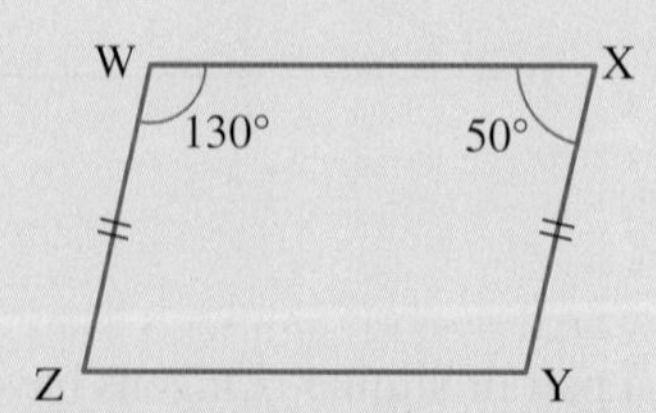

16. Prove that the diagonals in a rhombus bisect the angles they pass through.

17. Explain why the triangles shown are not congruent.

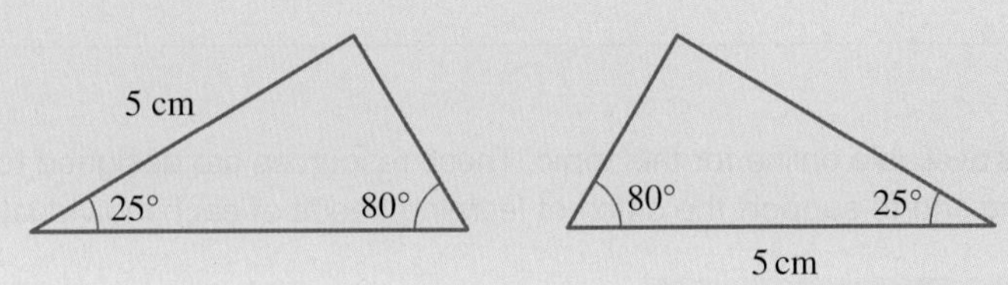

18. Prove that the angles opposite the equal sides in an isosceles triangle are equal.

19. Name any quadrilaterals that have equal diagonals.

20. This 8 cm by 12 cm rectangle is cut into two sections as shown.

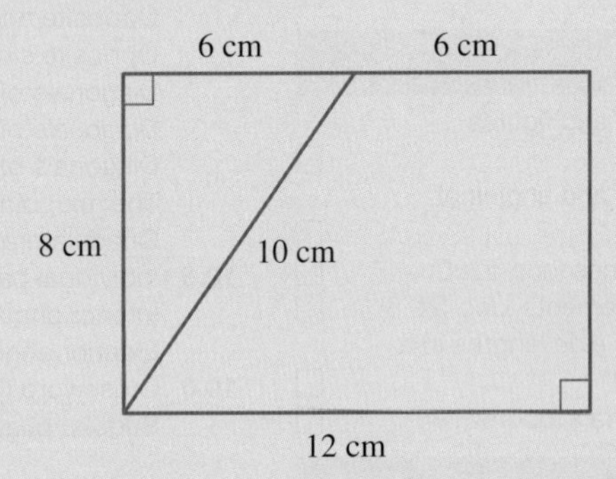

a. Draw labelled diagrams to show how the two sections can be rearranged to form:
   i. a parallelogram
   ii. a right-angled triangle
   iii. a trapezium.

b. Comment on the perimeters of the figures.

on To test your understanding and knowledge of this topic, go to your learnON title at www.jacplus.com.au and complete the **post-test**.

Online Resources

Below is a full list of **rich resources** available online for this topic. These resources are designed to bring ideas to life, to promote deep and lasting learning and to support the different learning needs of each individual.

eWorkbook

Download the workbook for this topic, which includes worksheets, a code puzzle and a project (ewbk-2036) ☐

Solutions

Download a copy of the fully worked solutions to every question in this topic (sol-0744) ☐

Digital documents

10.2 SkillSHEET Naming angles, lines and figures (doc-5276) ☐
SkillSHEET Corresponding sides and angles of congruent triangles (doc-5277) ☐
SkillSHEET Angles and parallel lines (doc-5280) ☐
10.3 SkillSHEET Writing similarity statements (doc-5278) ☐
SkillSHEET Calculating unknown side lengths in a pair of similar triangles (doc-5281) ☐
10.4 SkillSHEET Identifying quadrilaterals (doc-5279) ☐

Video eLessons

10.2 Proofs and theorems of angles (eles-4892) ☐
Angle properties of triangles (eles-5353) ☐
Congruent triangles (eles-4897) ☐
Isosceles triangles (eles-4898) ☐
10.3 Similar figures (eles-4899) ☐
Testing triangles for similarity (eles-4900) ☐
Similar triangles (eles-1925) ☐
10.4 Quadrilaterals (eles-4901) ☐
Parallelograms (eles-5354) ☐
Rectangles (eles-5355) ☐
Rhombuses (eles-5356) ☐
The midpoint theorem (eles-4905) ☐
10.5 Polygons (eles-4906) ☐

Interactivities

10.2 Individual pathway interactivity: Angles, triangles and congruence (int-4612) ☐
Angles at a point (int-6157) ☐
Supplementary angles (int-6158) ☐
Angles in a triangle (int-3965) ☐
Interior and exterior angles of a triangle (int-3966) ☐
Vertically opposite and adjacent angles (int-3968) ☐
Corresponding angles (int-3969) ☐
Co-interior angles (int-3970) ☐
Alternate angles (int-3971) ☐
Congruency tests (int-3755) ☐
Congruent triangles (int-3754) ☐
Angles in an isosceles triangle (int-6159) ☐
10.3 Individual pathway interactivity: Similar triangles (int-4613) ☐
Scale factors (int-6041) ☐
Angle-angle-angle condition of similarity (AAA) (int-6042) ☐
Side-angle-side condition of similarity (SAS) (int-6447) ☐
Side-side-side condition of similarity (SSS) (int-6448) ☐
10.4 Individual pathway interactivity: Quadrilaterals (int-4614) ☐
Quadrilateral definitions (int-2786) ☐
Angles in a quadrilateral (int-3967) ☐
Opposite angles of a parallelogram (int-6160) ☐
Opposite sides of a parallelogram (int-6161) ☐
Diagonals of a parallelogram (int-6162) ☐
Diagonals of a rectangle (int-6163) ☐
Diagonals of a rhombus (int-6164) ☐
The midpoint theorem (int-6165) ☐
Quadrilaterals (int-3756) ☐
10.5 Individual pathway interactivity: Polygons (int-4615) ☐
Interior angles of a polygon (int-6166) ☐
Exterior angles of a polygon (int-6167) ☐
10.6 Crossword (int-2854) ☐
Sudoku puzzle (int-3597) ☐

Teacher resources

There are many resources available exclusively for teachers online.

To access these online resources, log on to **www.jacplus.com.au**.

Answers

Topic 10 Deductive geometry

Exercise 10.1 Pre-test

1. 17.5°
2. 69°
3. Congruent triangles
4. $x = 12,\ y = 11$
5. $x = 58,\ y = -76,\ z = 308$
6. A
7. Whether they are similar or not
8. 95°
9. 45°
10. 72°
11. 1080°
12. C
13. 151°
14. A
15. A

Exercise 10.2 Angles, triangles and congruence

1. a. $a = 56°$ b. $b = 30°$ c. $c = 60°$
   d. $d = 120°$ e. $e = 68°$
2. a. $x = 115°$
   b. $y = 80°$
   c. $a = 120°, b = 60°, c = 120°, d = 60°$
   d. $x = 30°$
3. a. I and III, SAS b. I and II, AAS
   c. II and III, RHS d. I and II, SSS
4. a. $x = 6,\ y = 60°$ b. $x = 80°,\ y = 50°$
   c. $x = 32°,\ y = 67°$ d. $x = 45°,\ y = 45°$
5. a. $b = 65°, c = 10°, d = 50°, e = 130°$
   b. $f = 60°, g = 60°, h = 20°, i = 36°$
6. a. $x = 3$ cm
   b. $x = 85°$
   c. $x = 80°,\ y = 30°,\ z = 70°$
   d. $x = 30°, y = 7$ cm
   e. $x = 40°,\ y = 50°,\ z = 50°,\ m = 90°,\ n = 90°$
7. C, D
8. $x = 110°,\ y = 110°,\ z = 4$ cm, $w = 7$ cm
9. Sample responses can be found in the worked solutions in the online resources.
10. Sample responses can be found in the worked solutions in the online resources.
    a. Use SAS. b. Use SAS. c. Use ASA.
    d. Use ASA. e. Use SSS.
11. $x = 70°$
12. The third sides are not necessarily the same.
13. $x = 30°,\ y = 65°$
14. Corresponding sides are not the same.
15. Sample responses can be found in the worked solutions in the online resources.
16. Use SSS. Sample responses can be found in the worked solutions in the online resources.
17. $x = 20°,\ y = 10°$ and $z = 40°$
18. Sample responses can be found in the worked solutions in the online resources.

Exercise 10.3 Similar triangles

1. a. i and iii, RHS b. i and ii, SAS
   c. i and iii, SSS d. i and iii, AAA
   e. i and ii, SSS
2. a. Triangles PQR and ABC
   b. Triangles ADB and ADC.
   c. Triangles PQR and TSR.
   d. Triangles ABC and DEC.
   e. Triangles ABC and DEC.
3. a. $\frac{AB}{AD} = \frac{BC}{DE} = \frac{AC}{AE}$ b. $f = 9,\ g = 8$
4. $x = 4$
5. $x = 20°,\ y = 2\frac{1}{4}$
6. a. AAA b. $x = 3,\ y = 4$
7. a. $x = 12$
   b. $x = 4$ cm
   c. $x = 16$
   d. $x = 8$ cm, $n = 60°,\ m = 70°$
8. a. $x = 7.5,\ y = 6.4$ b. $x = 8$
9. $x = 27$
10. a. $x = 1,\ y = 7\sqrt{2}$ b. $x = 2.5,\ y = 3.91$
11. Sample responses can be found in the worked solutions in the online resources.
12. a. $\angle ABD = \angle ABC$ (common angle)
       $\angle ADB = \angle BAC = 90°$
       $\Delta ABD \sim \Delta ACB$ (AAA)
    b. $\angle ACD = \angle BCA$ (common angle)
       $\angle ADC = \angle CAB = 90°$
       $\Delta ACD \sim \Delta ACB$ (AAA)
13. Congruent triangles must be identical; that is, the angles must be equal and the side lengths must be equal. Therefore, it is not enough just to prove that the angles are equal.
14. The taller student's shadow is 3.12 metres long.
15. The slide is 7.23 m long.
16. $\angle FEO = \angle OGH$ (alternate angles equal as EF || HG)
    $\angle EFO = \angle OHG$ (alternate angles equal as EF || HG)
    $\angle EOF = \angle HOG$ (vertically opposite angles equal)
    $\therefore \Delta EFO \sim \Delta GHO$ (equiangular)
17. Radius = 0.625 m
18. $x = 6$ or $x = 11$

Exercise 10.4 Quadrilaterals

1. a. True b. True
   c. True d. False
2. a. False b. False
   c. False d. False
3. a. $x = 145°$ b. $t = 174°$
   c. $m = 66°$ d. $q = 88°$
4. a. $a = 35°, b = 65°$
   b. $c = 62°, d = 28°, e = 90°$

c. $f = 40°, g = 140°$
d. $h = 75°, i = 75°$

5. a. $x = 69°$ b. $x = 26°, y = 128°$

6. a. $x = 36°, y = 62°$ b. $x = 5\text{ cm}, y = 90°$
c. $x = 10°, y = 70°$ d. $x = 40°, y = 60°$

7. None are true, unless the trapezium is a regular trapezium, then **e** is true.

8. a, c, f

9. a, c, d, e, f

10. Parallelogram, rhombus, rectangle, square

11. Square

12. Use AAS. Sample responses can be found in the worked solutions in the online resources.

13. All the sides of a square are equal, so a square is a special rhombus. But the angles of a rhombus are not equal, so can't be a square.

14. AX || DY because ABCD is a parallelogram
AX = DY (given)
∴ AXYD is a parallelogram since opposite sides are equal and parallel.

15. Use SAS. Sample responses can be found in the worked solutions in the online resources.

16. a. Use SAS. Sample responses can be found in the worked solutions in the online resources.
b. Use SAS. Sample responses can be found in the worked solutions in the online resources.
c. Opposite sides are equal. Sample responses can be found in the worked solutions in the online resources.

17. PS = QR (corresponding sides in congruent triangles are equal)
PS || QR (alternate angles are equal)
∴ PQRS is a parallelogram since one pair of opposite sides are parallel and equal.

18. a. One pair of opposite sides are parallel.
b. $x = 90°, y = 75°$

19. a. i. $x = \sqrt{41}$ ii. $y = \sqrt{97}$
b. $\angle BAD = \angle BCD = 117°23'$

20. a.

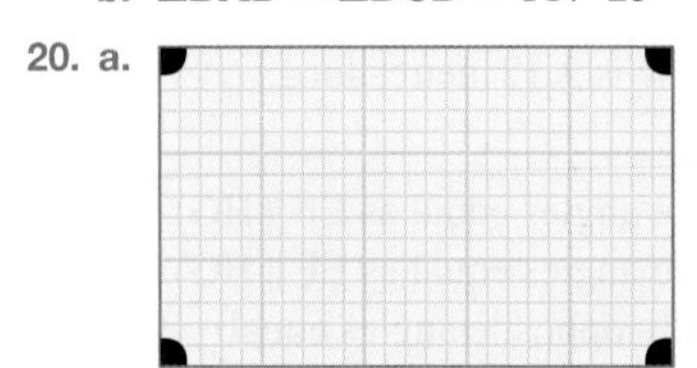

b. 6
c. 7
d.

Table size	Number of sides hit
5 cm × 3 cm	6
7 cm × 2 cm	7
4 cm × 3 cm	5
4 cm × 2 cm	1
6 cm × 3 cm	1
9 cm × 3 cm	2
12 cm × 4 cm	2

e. If the ratio of the sides is written in simplest form, then the pattern is $m + n - 2$.
f. There are two routes for the ball when hit from B. Either 2 or 3 sides are hit. The ball does not end up in the same hole each time.
A suitable justification would be a diagram — student to draw.
g. Isosceles triangles and parallelograms. The triangles are congruent.

21. 70°

22. $x = \sqrt{10}\text{ cm}$

Exercise 10.5 Polygons

1. a. $m = 60°$ b. $a = 45°, b = 45°$
c. $t = 35°$ d. $x = 10°$

2. a. $a = 85°, b = 50°, c = 45°$
b. $d = 140°, e = 110°, f = 110°$
c. $g = 90°, h = 110°, i = 70°$
d. $j = 100°, k = 100°, l = 130°, m = 130°$

3. a. $y = 35°$ b. $t = 5°$ c. $n = 81°$
d. $x = 15°$ e. $t = 30°$

4. a. i. Irregular ii. $x = 95°$
b. i. Irregular ii. $p = 135°$
c. i. Irregular ii. $t = 36°$
d. i. Irregular ii. $y = 70°$
e. i. Irregular ii. $p = 36°$

5. a. i. 540° ii. $b = 110°$
b. i. 720° ii. $c = 134°$
c. i. 900° ii. $d = 24°$
d. i. 720° ii. $h = 85°$

6. The sum of the interior angles is based on the number of sides of the polygon. The size of the exterior angle can be found by dividing 360° by the number of sides.

7. $w = 75°, x = 105°, y = 94°, z = 133°$

8. 82.5°, 82.5°, 97.5°, 97.5°

9. a. $a = 120°, b = 120°, c = 60°, d = 60°, e = 120°, f = 240°$
b. $m = 10°, n = 270°, o = 50°$

10. 60°

11. a. Regular: all sides and interior angles are equal.
b. Irregular: all sides and interior angles are not equal.
c. Regular: all sides and interior angles are equal.
d. Regular: all sides and interior angles are equal.
e. Irregular: the sides are all equal, but the interior angles are not equal.
f. Regular: all sides and interior angles are equal.

12. a. i. 2 ii. 5 iii. 9 iv. 14
b. Sample responses can be found in the worked solutions in the online resources.

13. Internal angle $= 180° - \dfrac{360°}{n}$

14. a. Equilateral triangle
b. $m = 150°$

15. a. 135° b. 67.5° c. 45° d. 45°

16. a. 1080° b. 43°

17. a. 720° b. $x = 25°$

18. a. 1080° b. $x = 17°$

Project

Students will apply the knowledge of deductive geometry to enlarge a cartoon character to almost life-size. Sample responses can be found in the worked solutions in the online resources.

Exercise 10.6 Review questions

1. a. I and III, ASA or SAS
   b. I and II, RHS
2. a. $x = 8$ cm
   b. $x = 70°$
   c. $x = 30°, y = 60°, z = 90°$
3. Sample responses can be found in the worked solutions in the online resources.
   a. Use SAS. b. Use ASA.
4. a. Similar, scale factor = 1.5
   b. Not similar
   c. Similar, scale factor = 2
5. a. $x = 48°$, $y = 4.5$ cm
   b. $x = 86°$, $y = 50°$, $z = 12$ cm
   c. $x = 60°$, $y = 15$ cm, $z = 12$ cm
6. Use the equiangular test. Sample responses can be found in the worked solutions in the online resources.
7. Use the equiangular test. Sample responses can be found in the worked solutions in the online resources.
8. A rhombus is a parallelogram with two adjacent sides equal in length.
9. C
10. $x + 1$
11. a. Use SAS. Sample responses can be found in the worked solutions in the online resources.
    b. Use SAS. Sample responses can be found in the worked solutions in the online resources.
    c. Use SAS. Sample responses can be found in the worked solutions in the online resources.
    d. They are all the same length.
    e. B and C
12. Use SAS. Sample responses can be found in the worked solutions in the online resources.
    PQ = PS (corresponding sides in congruent triangles are equal)
13. Rhombus, square.
14. A quadrilateral is a rhombus if:
    1 all sides are equal
    2 the diagonals bisect each other at right angles
    3 the diagonals bisect the angles they pass through.
15. WZ || XY (co-interior angles are supplementary) and WZ = XY (given)
    $\therefore$ WXYZ is a parallelogram since one pair of sides is parallel and equal.
16. Sample responses can be found in the worked solutions in the online resources.
17. Corresponding sides are not the same.
18.

Bisect ∠BAC
AB = AC (given)
∠BAD = ∠DAC
AD is common.
$\therefore \Delta ABD \cong \Delta ACD$ (SAS)
$\therefore \angle ABD = \angle ACD$ (corresponding sides in congruent triangles are equal)

19. Rectangle, square.
20. a. i.

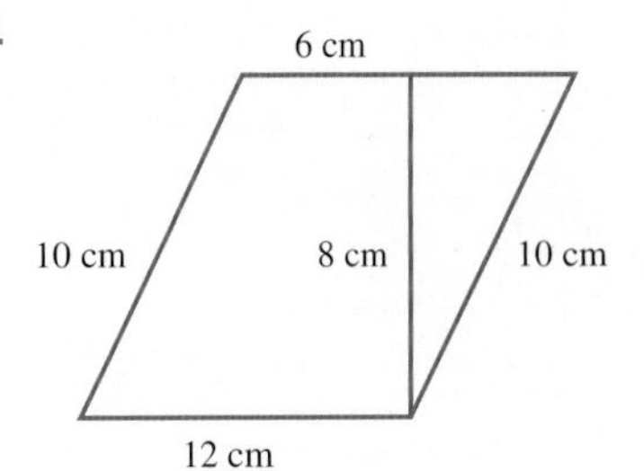

ii.

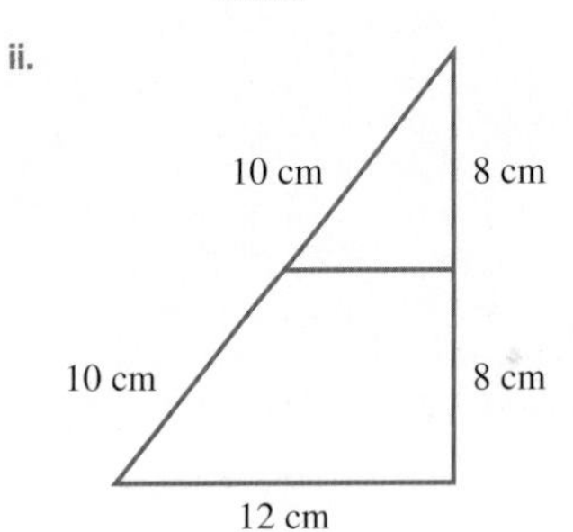

iii.

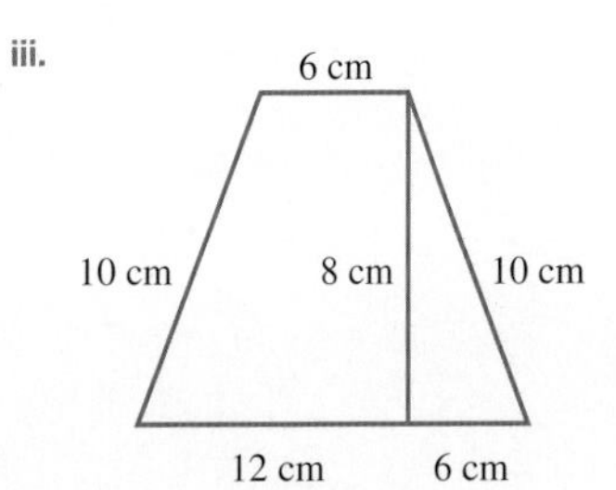

b. Perimeter of rectangle = 40 cm
Perimeter of parallelogram = 44 cm
Perimeter of triangle = 48 cm
Perimeter of trapezium = 44 cm
The triangle has the largest perimeter, while the rectangle has the smallest.

11 Probability

LEARNING SEQUENCE

11.1 Overview

Why learn this?

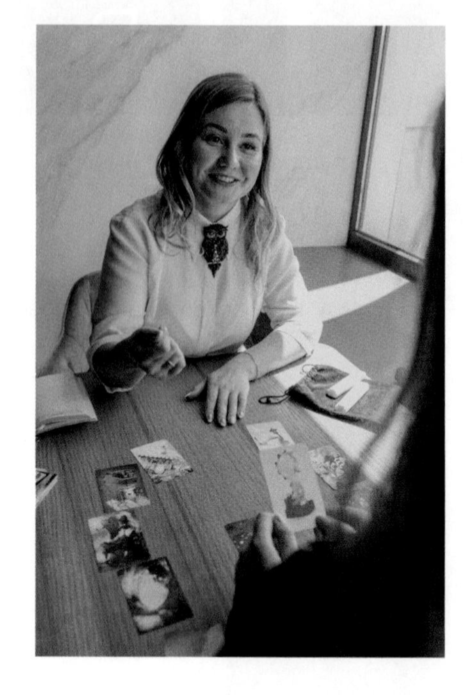

Probability is a broad and interesting area of mathematics that affects our day-to-day lives far more than we can imagine. Here is a fun fact: did you know there are so many possible arrangements of the 52 cards in a deck $(52! = 8.0658 \times 10^{67})$ that the probability of ever getting the same arrangement after shuffling is virtually zero? This means every time you shuffle a deck of cards, you are almost certainly producing an arrangement that has never been seen before. Probability is also a big part of computer and board games; letters X and Q in Scrabble are worth more points because you are less likely to be able to form a word using those letters. It goes without saying that probability is a big part of any casino game and of the odds and payouts when gambling on the outcome of racing or sports.

While it is handy to know probability factoids and understand gambling, this isn't the reason we spend time learning probability. Probability helps us build critical thinking skills, which are required for success in almost any career and even just for navigating our own lives. For example, if you were told your chance of catching a rare disease had doubled you probably wouldn't need to worry, as a 1-in-a-million-chance becoming a 2-in-a-million chance isn't a significant increase in the probability of you developing the disease. On the other hand, if a disease has a 1% mortality rate that may seem fairly low, but it means that if a billion people developed that disease, then 10 million would die. Using probability to understand risk helps us steer clear of manipulation by advertising, politicians and the media. Building on this understanding helps us as individuals make wise decisions in our day-to-day life, whether it be investing in the stock market, avoiding habits that increase our risk of sickness, or building our career.

Where to get help

Go to your learnON title at **www.jacplus.com.au** to access the following digital resources. The Online Resources Summary at the end of this topic provides a full list of what's available to help you learn the concepts covered in this topic.

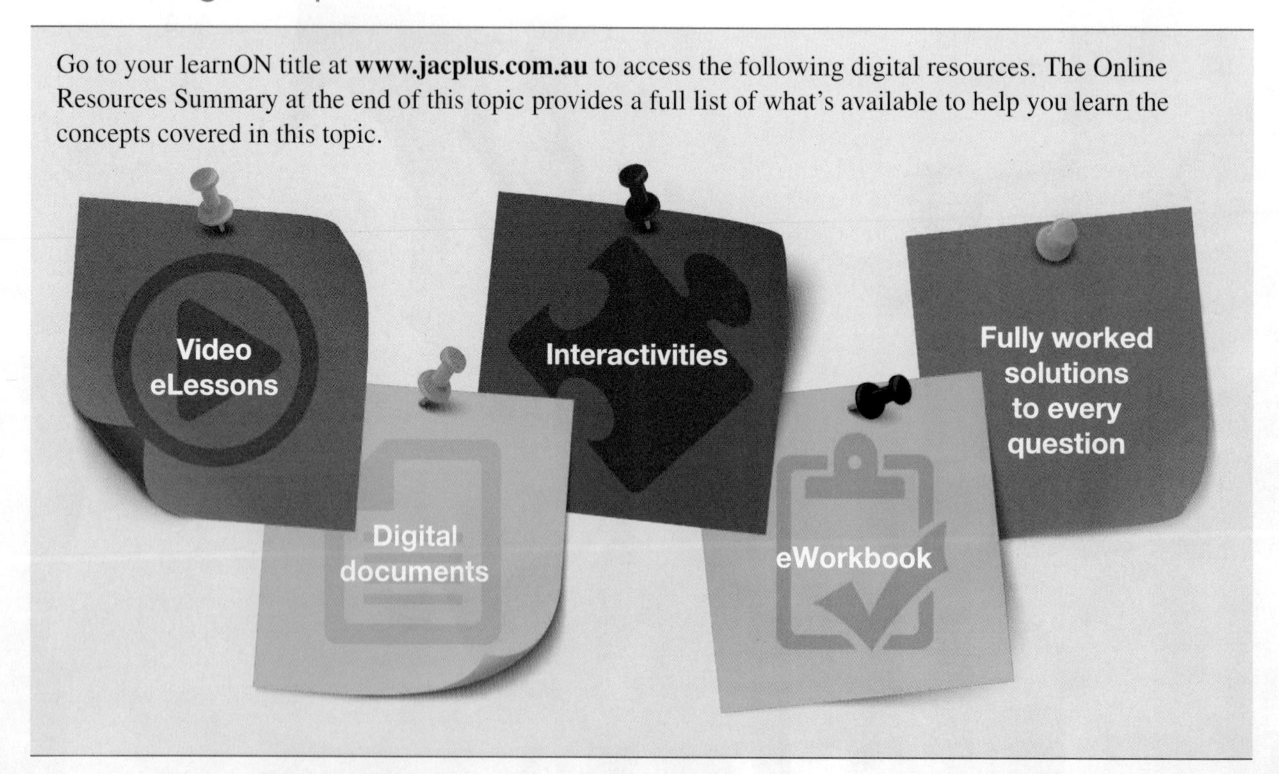

Exercise 11.1 Pre-test

learnON

Complete this pre-test in your learnON title at www.jacplus.com.au and receive **automatic marks**, **immediate corrective feedback** and **fully worked solutions**.

1. Calculate $\Pr(A \cap B)$ if $\Pr(A = 0.4)$, $\Pr(B = 0.3)$ and $\Pr(A \cup B) = 0.5$.

2. If events A and B are mutually exclusive, and $\Pr(B) = 0.38$ and $\Pr(A \cup B) = 0.89$, calculate $\Pr(A)$.

3. State whether the events $A = \{\text{drawing a red marble from a bag}\}$ and $B = \{\text{rolling a 1 on a die}\}$ are independent or dependent.

4. The Venn diagram shows the number of university students in a group of 25 who own a computer and/or tablet.
   In simplest form, calculate the probability that a university student selected at random will own only a tablet.

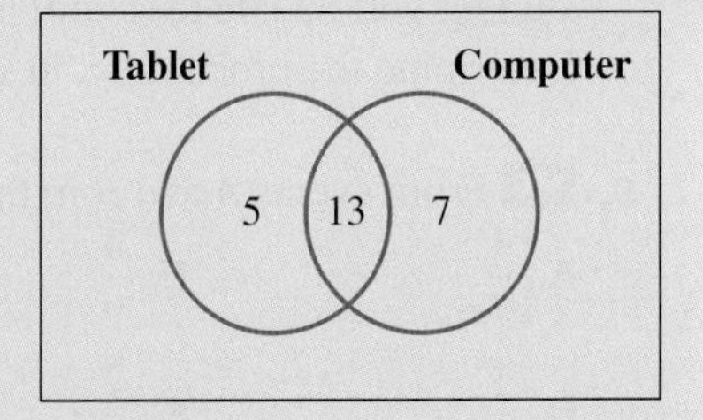

5. **MC** Two unbiased four-sided dice are rolled. Determine the probability that the total sum of two face-down numbers obtained is 6.

   A. $\frac{1}{8}$ B. $\frac{1}{16}$ C. $\frac{3}{16}$ D. $\frac{1}{4}$ E. $\frac{3}{8}$

6. **MC** Identify which Venn diagram best illustrates $\Pr(A \cup B)'$.

A.

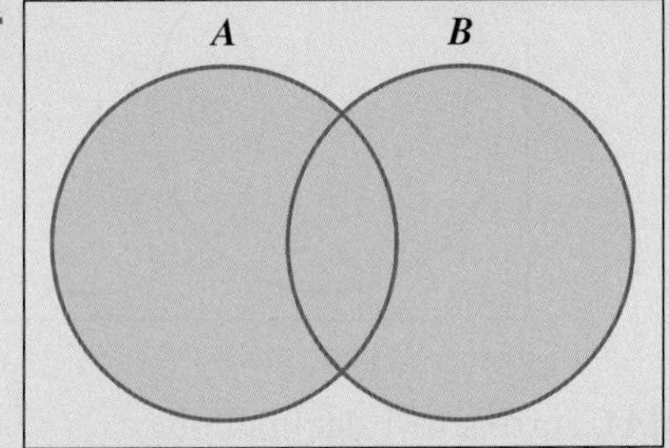

B.

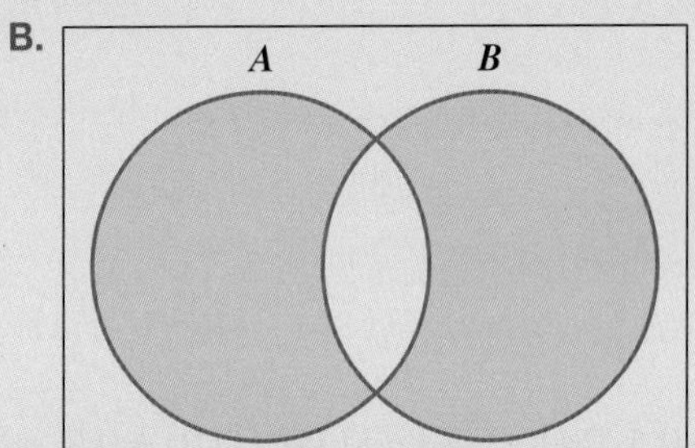

C.

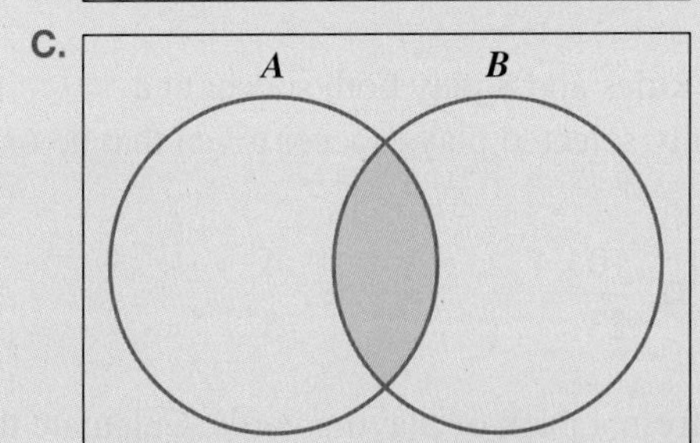

D.

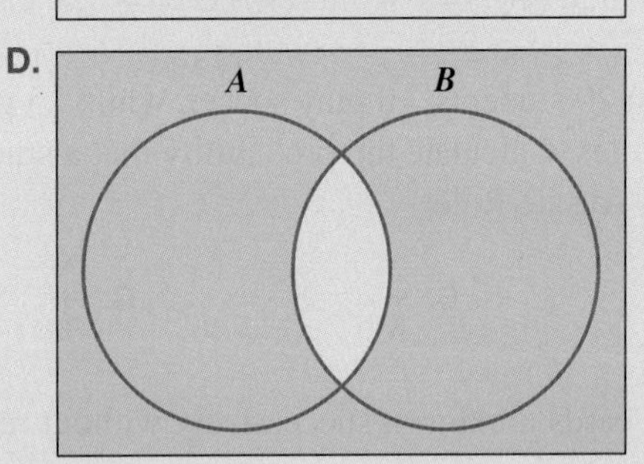

E.

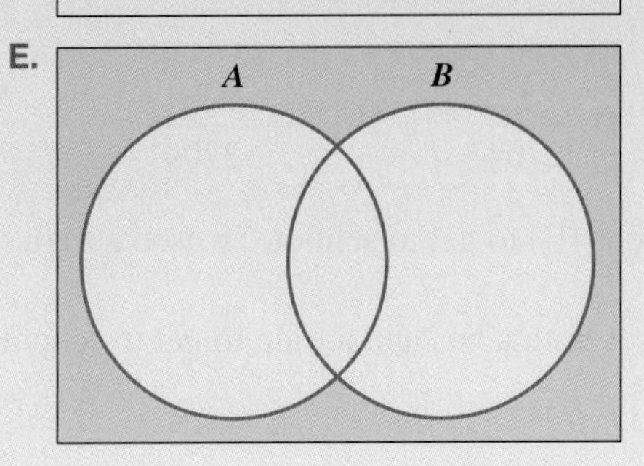

7. **MC** From a group of 25 people, 12 use Facebook (F), 14 use Snapchat (S), and 6 use both Facebook and Snapchat applications. Determine the probability that a person selected will use neither application.

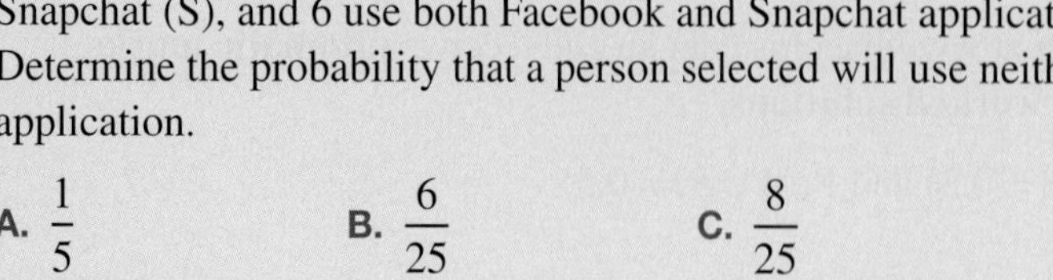

A. $\frac{1}{5}$ B. $\frac{6}{25}$ C. $\frac{8}{25}$

D. $\frac{12}{25}$ E. $\frac{14}{25}$

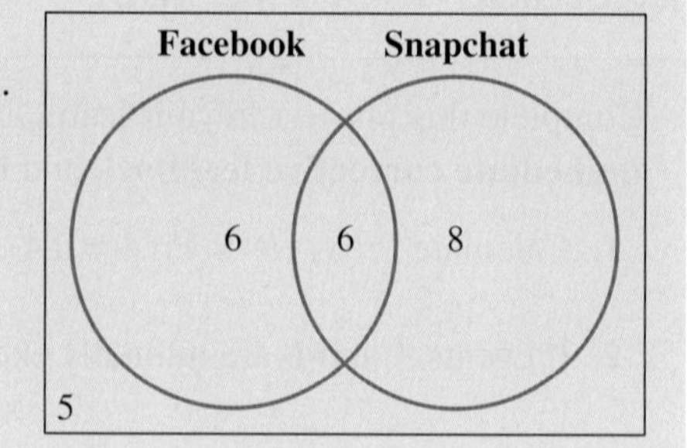

8. The probability that a student will catch a bus to school is 0.7 and the independent probability that a student will be late to school is 0.2.
Determine the probability, in simplest form, that a student catches a bus and is not late to school.

9. **MC** From events A and B in the Venn diagram, calculate $\Pr(A|B)$.

A. $\frac{4}{7}$ B. $\frac{4}{9}$ C. $\frac{2}{5}$

D. $\frac{1}{3}$ E. $\frac{7}{9}$

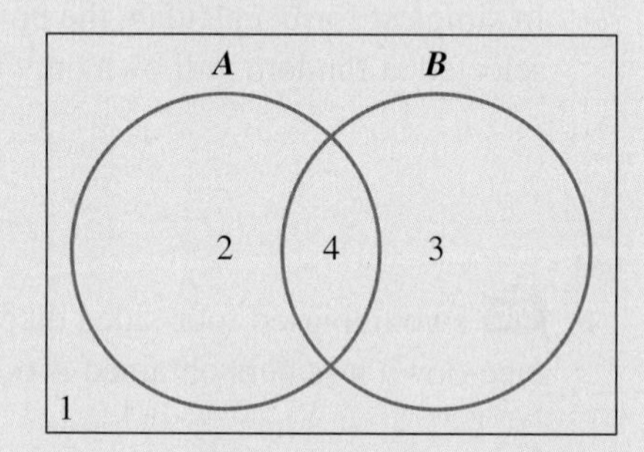

10. In the Venn diagram, events A and B are independent.
State whether this statement is true or false.

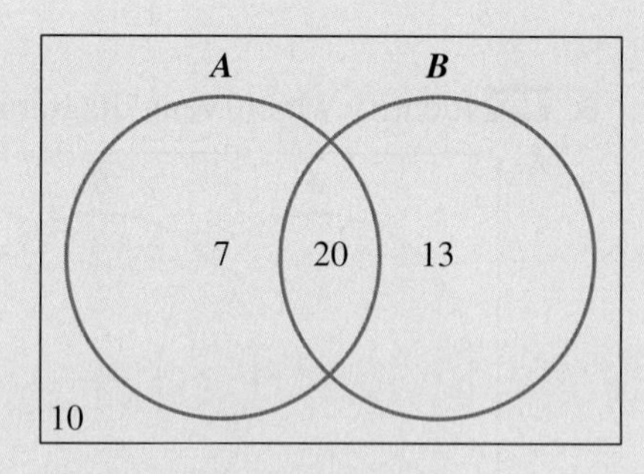

11. If $\Pr(A) = 0.5, \Pr(B) = 0.4$ and $\Pr(A \cup B) = 0.8$, calculate $\Pr(B|A)$, correct to 1 decimal place.

12. **MC** From 20 students, 10 play soccer, while 15 play Aussie Rules and 8 play both soccer and Aussie Rules. Calculate the probability that a student randomly selected plays soccer given that he or she plays Aussie Rules.

A. $\frac{8}{15}$ B. $\frac{2}{5}$ C. $\frac{2}{3}$ D. $\frac{10}{23}$ E. $\frac{4}{5}$

13. **MC** Two cards are drawn successively without replacement from a pack of playing cards. Calculate the probability of drawing 2 spades.

A. $\frac{1}{17}$ B. $\frac{2}{2652}$ C. $\frac{1}{2652}$ D. $\frac{1}{2704}$ E. $\frac{2}{2704}$

14. A survey of a school of 800 students found that 100 used a bus (B) to get to school, 75 used a train (T) and 650 used neither.
In simplest form, determine the probability that a student uses both a bus and a train to get to school.

15. **MC** On the first day at school, students are asked to tell the class about their holidays. There are 30 students in the class and all have spent part or all of their holidays at one of the following: a coastal resort, interstate, or overseas.
The teacher finds that:
- 5 students went to a coastal resort only
- 2 students went interstate only
- 2 students holidayed in all three ways
- 8 students went to a coastal resort and travelled overseas only
- 20 students went to a coastal resort
- no less than 4 students went overseas only
- no less than 13 students travelled interstate

Determine the probability that a student travelled overseas and interstate only.

A. $\frac{2}{15}$ B. $\frac{5}{6}$ C. $\frac{3}{15}$ D. $\frac{1}{3}$ E. $\frac{4}{15}$

11.2 Review of probability

LEARNING INTENTION

At the end of this subtopic you should be able to:
- use key probability terminology such as: trials, frequency, sample space, likely and unlikely events
- use two-way tables to represent sample spaces
- calculate experimental and theoretical probabilities
- determine the probability of complementary events and mutually exclusive events
- use the addition rule to calculate the probability of event 'A or B'.

11.2.1 The language of probability

eles-4922

- **Probability** measures the chance of an event taking place and ranges from 0 for an impossible event to 1 for a certain event.

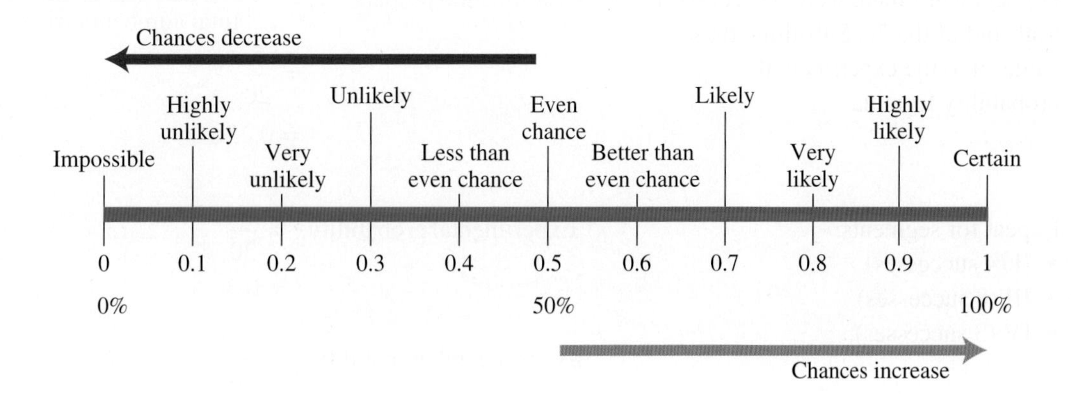

- The **experimental probability** of an event is based on the outcomes of experiments, simulations or surveys.
- A **trial** is a single experiment, for example, a single flip of a coin.

Experimental probability

$$\textbf{Experimental probability} = \frac{\textbf{number of successful trials}}{\textbf{total number of trials}}$$

- The experimental probability of an event is also known as the **relative frequency**.
- The list of all possible outcomes of an experiment is known as the **event space** or **sample space**. For example, when flipping a coin there are two possible outcomes: Heads or Tails. The event space can be written, using set notation, as {H, T} .

WORKED EXAMPLE 1 Sample space and calculating relative frequency

The spinner shown here is made up of 4 equal-sized segments. It is known that the probability that the spinner will land on any one of the four segments from one spin is $\frac{1}{4}$. To test if the spinner shown here is fair, a student spun the spinner 20 times and each time recorded the segment in which the spinner stopped. The spinner landed as follows.

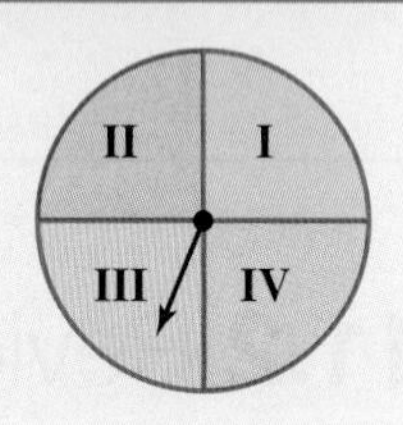

Segment	I	II	III	IV
Tally	5	4	8	3

a. **List the sample space.**
b. **Given the experimental results, determine the experimental probability of each segment.**
c. **Compare the experimental probabilities with the known probabilities and suggest how the experiment could be changed to ensure that the results give a better estimate of the true probability.**

THINK	WRITE
a. The sample space lists all possible outcomes from one spin of the spinner. There are four possible outcomes.	a. Sample space = {I, II, III, IV}
b. 1. For segment I there were 5 successful trials out of the 20. Substitute these values into the experimental probability formula.	b. $\text{Experimental probability}_{\text{I}} = \frac{\text{number of successful trials}}{\text{total number of trials}}$ $= \frac{5}{20}$ $= 0.25$
2. Repeat for segments: • II (4 successes) • III (8 successes) • IV (3 successes).	$\text{Experimental probability}_{\text{II}} = \frac{4}{20}$ $= 0.2$ $\text{Experimental probability}_{\text{III}} = \frac{8}{20}$ $= 0.4$ $\text{Experimental probability}_{\text{IV}} = \frac{3}{20}$ $= 0.15$

c. Compare the experimental frequency values with the known value of $\frac{1}{4}$ (0.25). Answer the question.	c. The experimental probability of segment I was the only segment that mirrored the known value. To ensure that experimental probability gives a better estimate of the true probability, the spinner should be spun many more times.

Two-way tables

- The sample space can be displayed using a **two-way table**.
- A two-way table represents two of the outcomes of events in a two-dimensional table. A two-way table for the experiment of tossing a coin and rolling a die simultaneously is shown below.

		Die outcomes					
		1	**2**	**3**	**4**	**5**	**6**
Coin outcomes	**H**	H, 1	H, 2	H, 3	H, 4	H, 5	H, 6
	T	T, 1	T, 2	T, 3	T, 4	T, 5	T, 6

WORKED EXAMPLE 2 Representing sample space with a two-way table

Two dice are rolled, and the values on the two uppermost faces are multiplied together. Draw a diagram to illustrate the sample space.

THINK

The sample space for rolling 1 die is {1, 2, 3, 4, 5, 6}. When two dice are rolled and the two uppermost faces are multiplied, the sample space is made up of 36 products. This is best represented with the use of a two-way table.

- Draw a 7×7 grid.
- In the first row and column list the outcomes of each die.
- At the intersection of a column and row, write the product of the relevant die outcomes.

WRITE

	First die						
Second die	×	1	2	3	4	5	6
	1	1	2	3	4	5	6
	2	2	4	6	8	10	12
	3	3	6	9	12	15	18
	4	4	8	12	16	20	24
	5	5	10	15	20	25	30
	6	6	12	18	24	30	36

Theoretical probability

- **Theoretical probability** is the probability of an event occurring, based on the number of possible favourable outcomes, $n(E)$, and the total number of possible outcomes, $n(\xi)$.

Theoretical probability

When all outcomes are equally likely, the theoretical probability of an event can be calculated using the formula:

$$\text{Pr(event)} = \frac{\text{number of favourable outcomes}}{\text{total number of possible outcomes}} \quad \text{or} \quad \text{Pr(event)} = \frac{n(E)}{n(\xi)}$$

where $n(E)$ is the number of favourable events and $n(\xi)$ is the total number of possible outcomes.

WORKED EXAMPLE 3 Calculating theoretical probability

A fair die is rolled and the value of the uppermost side is recorded. Calculate the theoretical probability that a 4 is uppermost.

THINK	WRITE
1. Write the number of favourable outcomes and the total number of possible outcomes. The number of 4s on a fair die is 1. There are 6 possible outcomes.	$n(E) = 1$ $n(\xi) = 6$
2. Substitute the values found in part 1 to calculate the probability of the event that a 4 is uppermost when a die is rolled.	$\Pr(\text{a } 4) = \frac{n(E)}{n(\xi)}$ $= \frac{1}{6}$
3. Write the answer in a sentence.	The probability that a 4 is uppermost when a fair die is rolled is $\frac{1}{6}$.

11.2.2 Properties of probability events

eles-4923

Complementary events

- The **complement** of the set A is the set of all elements that belong to the universal set (ξ) but that do *not* belong to A.
- The complement of A is written as A' and is read as 'A dashed' or 'A prime'.
- On a Venn diagram, **complementary events** appear as separate regions that together occupy the whole universal set.

Complementary events

Since complementary events fill the entire sample space:

$$\mathbf{Pr(A) + Pr(A') = 1}$$

- As an example, the complement of {drawing a diamond} from a deck of cards is {not drawing a diamond}, which can also be described as {drawing a heart, spade or club}. This is shown in the Venn diagram.

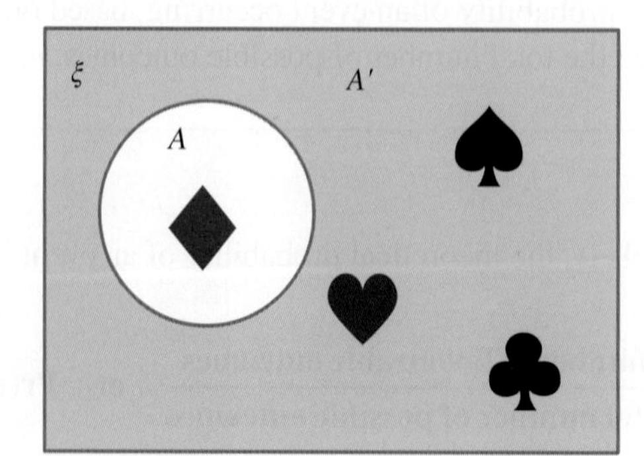

WORKED EXAMPLE 4 Determining complementary events

A player is chosen from a cricket team. Are the events 'selecting a batter' and 'selecting a bowler' complementary events if a player can have more than one role? Give a reason for your answer.

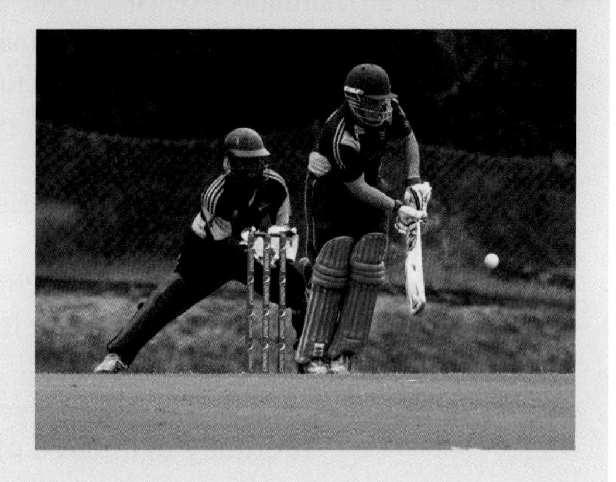

THINK	WRITE
Explain the composition of a cricket team. Players who can bat and bowl are not necessarily the only players in a cricket team. There is a wicket-keeper as well. Some players (all rounders) can bat and bowl.	No, the events 'selecting a batter' and 'selecting a bowler' are not complementary events. These events may have common elements, that is, the all rounders in the team who can bat and bowl. The cricket team also includes a wicket-keeper.

The intersection and union of A and B

- The intersection of two events A and B is written $A \cap B$. These are the outcomes that are in A 'and' in B and so the intersection is often referred to as the event 'A and B'.
- The union of two events A and B is written $A \cup B$. These are the outcomes that are in A 'or' in B and so the union of often referred to as the event 'A or B'.

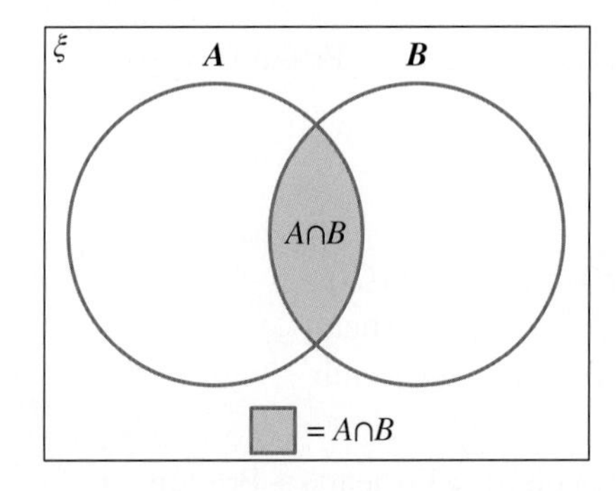

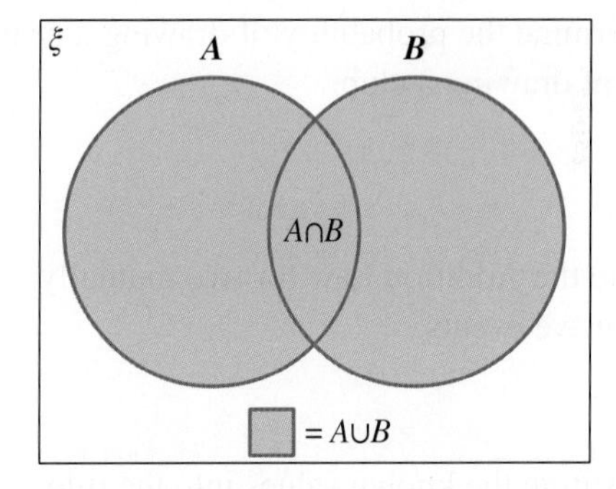

- When calculating the probability of $A \cup B$ we cannot simply add the probabilities of A and B, as $A \cap B$ would be counted twice.
- The formula for the probability of $A \cup B$ is therefore given by the following equation, which is known as the **Addition Law of probability**.

> **The Addition Law of probability**
>
> For intersecting events A and B:
>
> $$\Pr(A \cup B) = \Pr(A) + \Pr(B) - \Pr(A \cap B)$$

Mutually exclusive events

- Two events are **mutually exclusive** if one event happening excludes the other from happening. These events may not encompass all possible events. For example, when selecting a card from a deck of cards, selecting a black card excludes the possibility that the card is a heart.
- On a Venn diagram, mutually exclusive events appear as disjointed sets within the universal set.
- For mutually exclusive events A and B, $\Pr(A \cap B) = 0$. Therefore the formula for $\Pr(A \cup B)$ is simplified to the following.

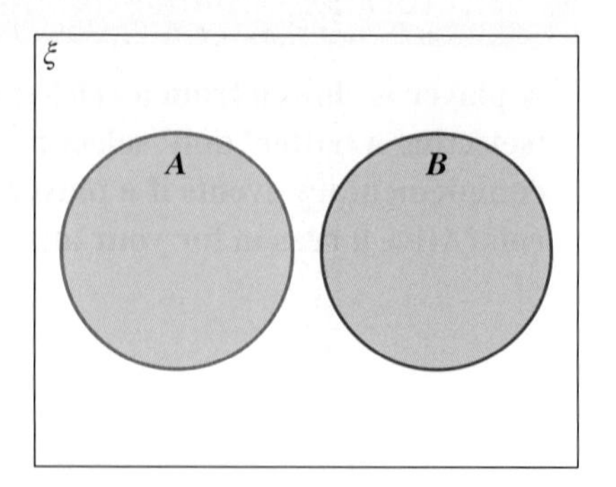

Mutually exclusive probabilities

For mutually exclusive events A and B:

$$\mathbf{Pr(A \cup B) = Pr(A) + Pr(B)}$$

WORKED EXAMPLE 5 Determining the probability of the union of two events

A card is drawn from a pack of 52 playing cards. Determine the probability that the card is a heart or a club.

THINK	WRITE
1. Determine whether the given events are mutually exclusive.	The two events are mutually exclusive as they have no common elements.
2. Determine the probability of drawing a heart and of drawing a club.	$\Pr(\text{heart}) = \frac{13}{52} = \frac{1}{4}$ $\quad \Pr(\text{club}) = \frac{13}{52} = \frac{1}{4}$
3. Write the Addition Law for two mutually exclusive events.	$\Pr(A \text{ or } B) = \Pr(A) + \Pr(B)$ where A = drawing a heart and B = drawing a club
4. Substitute the known values into the rule.	$\Pr(\text{heart or club}) = \Pr(\text{heart}) + \Pr(\text{club}) = \frac{1}{4} + \frac{1}{4} = \frac{2}{4}$
5. Evaluate and simplify.	$= \frac{1}{2}$
6. Write your answer.	The probability of drawing a heart or a club is $\frac{1}{2}$.

Note: Alternatively, we can use the formula for theoretical probability.	$\Pr(\text{heart or club}) = \dfrac{n(\text{heart or club})}{n(\xi)}$ $= \dfrac{26}{52}$ $= \dfrac{1}{2}$

WORKED EXAMPLE 6 Determining probabilities

A die is rolled. Determine:

a. Pr(an odd number)

b. Pr(a number less than 4)

c. Pr(an odd number or a number less than 4).

THINK	WRITE
a. 1. Determine the probability of obtaining an odd number, that is, $\{1, 3, 5\}$.	**a.** $\Pr(\text{odd}) = \dfrac{3}{6}$ $= \dfrac{1}{2}$
2. Write your answer.	The probability of obtaining an odd number is $\dfrac{1}{2}$.
b. 1. Determine the probability of obtaining a number less than 4, that is, $\{1, 2, 3\}$.	**b.** $\Pr(\text{less than } 4) = \dfrac{3}{6}$ $= \dfrac{1}{2}$
2. Write your answer.	The probability of obtaining a number less than 4 is $\dfrac{1}{2}$.
c. 1. Determine which numbers are odd or less than 4.	**c.** Less than $4 = \{1, 2, 3\}$ $\text{Odd} = \{1, 3, 5\}$ The numbers $\{1, 2, 3, 5\}$ are odd or less than 4.
2. Determine the probability of obtaining a number that is odd or less than 4.	$\Pr(\text{odd or less than } 4) = \dfrac{4}{6}$ $= \dfrac{2}{3}$
3. Write your answer.	The probability of obtaining an odd number or a number less than 4 is $\dfrac{2}{3}$.

WORKED EXAMPLE 7 Using the Addition Law to determine the intersection of two events

Given $\Pr(A) = 0.6$, $\Pr(B) = 0.4$ and $\Pr(A \cup B) = 0.9$:

a. use the Addition Law of probability to calculate the value of $\Pr(A \cap B)$

b. draw a Venn diagram to represent the universal set

c. calculate $\Pr(A \cap B')$.

THINK	WRITE
a. 1. Write the Addition Law of probability and substitute given values.	**a.** $\Pr(A \cup B) = \Pr(A) + \Pr(B) - \Pr(A \cap B)$ $0.9 = 0.6 + 0.4 - \Pr(A \cap B)$

2. Collect like terms and rearrange to make $\Pr(A \cap B)$ the subject. Solve the equation.

$$0.9 = 1.0 - \Pr(A \cap B)$$
$$\Pr(A \cap B) = 1.0 - 0.9$$
$$= 0.1$$

b. 1. Draw intersecting sets A and B within the universal set and write $\Pr(A \cap B) = 0.1$ inside the overlapping section, as shown.

b. 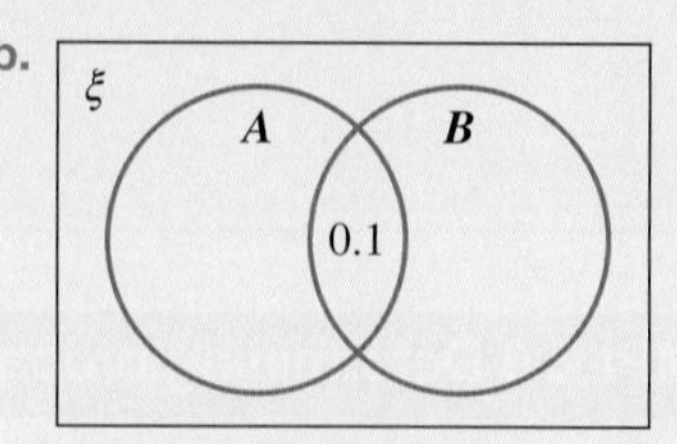

2. • As $\Pr(A) = 0.6$, 0.1 of this belongs in the overlap, the remainder of set A is 0.5 $(0.6 - 0.1)$.
• Since $\Pr(B) = 0.4$, 0.1 of this belongs in the overlap, the remainder of set B is 0.3 $(0.4 - 0.1)$.

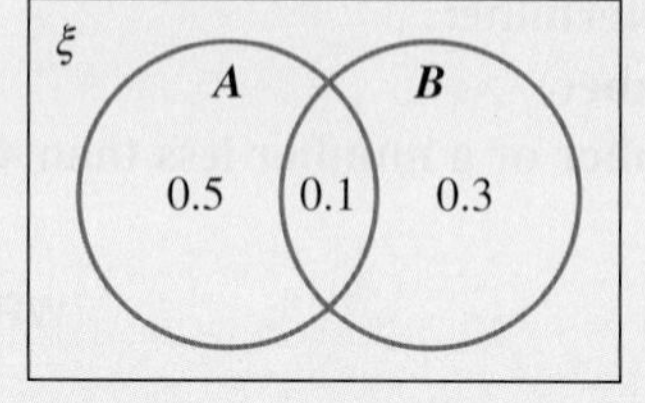

3. The total probability for the universal set is 1. That means $\Pr(A \cup B)' = 0.1$. Write 0.1 outside sets A and B to form the remainder of the universal set.

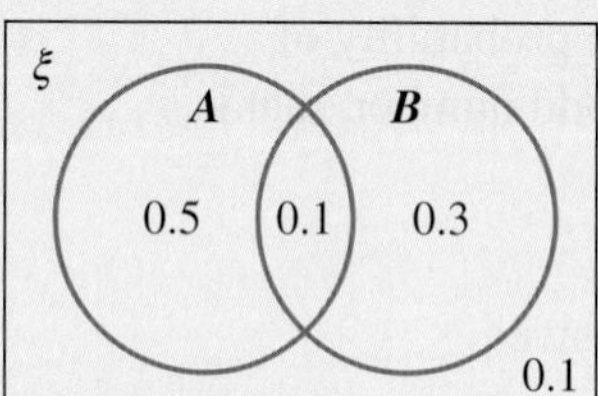

c. $\Pr(A \cap B')$ is the overlapping region of $\Pr(A)$ and $\Pr(B')$. Shade the region and write down the corresponding probability value for this area.

c. 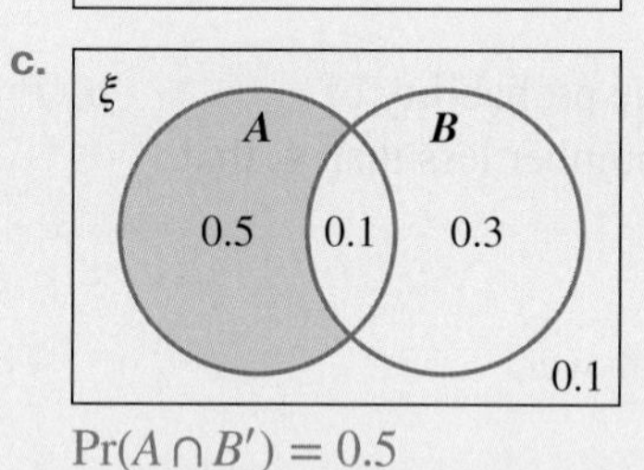

$\Pr(A \cap B') = 0.5$

WORKED EXAMPLE 8 Using a Venn diagram to represent sets and find probabilities

a. Draw a Venn diagram representing the relationship between the following sets. Show the position of all the elements in the Venn diagram.

$\xi = \{1, 2, 3, 4, 5, 6, 7, 8, 9, 10, 11, 12, 13, 14, 15, 16, 17, 18, 19, 20\}$
$A = \{3, 6, 9, 12, 15, 18\}$
$B = \{2, 4, 6, 8, 10, 12, 14, 16, 18, 20\}$

b. Determine:

i. $\Pr(A)$ ii. $\Pr(B)$ iii. $\Pr(A \cap B)$ iv. $\Pr(A \cup B)$ v. $\Pr(A' \cap B')$

THINK **WRITE/DRAW**

a. 1. Draw a rectangle with two partly intersecting circles labelled A and B.
2. Analyse sets A and B and place any common elements in the central overlap.
3. Place the remaining elements of set A in circle A.
4. Place the remaining elements of set B in circle B.
5. Place the remaining elements of the universal set ξ in the rectangle.

a. $n(\xi) = 20$

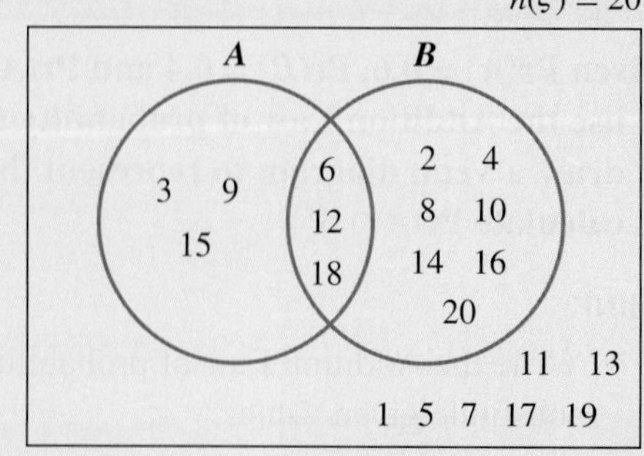

b. **i.** **1.** Write the number of elements that belong to set A and the total number of elements.

2. Write the rule for probability.

3. Substitute the known values into the rule.

4. Evaluate and simplify.

ii. **1.** Write the number of elements that belong to set B and the total number of elements.

2. Repeat steps 2 to 4 of part **b i.**

iii. **1.** Write the number of elements that belong to set $(A \cap B)$ and the total number of elements.

2. Repeat steps 2 to 4 of part **b i.**

iv. **1.** Write the number of elements that belong to set $(A \cup B)$ and the total number of elements.

2. Repeat steps 2 to 4 of part **b i.**

v. **1.** Write the number of elements that belong to set $A' \cap B'$ and the total number of elements.

2. Repeat steps 2 to 4 of part **b i.**

b. **i.** $n(A) = 6, n(\xi) = 20$

$$\Pr(A) = \frac{n(A)}{n(\xi)}$$

$$\Pr(A) = \frac{6}{20}$$

$$= \frac{3}{10}$$

ii. $n(B) = 10, n(\xi) = 20$

$$\Pr(B) = \frac{n(B)}{n(\xi)}$$

$$\Pr(B) = \frac{10}{20}$$

$$= \frac{1}{2}$$

iii. $n(A \cap B) = 3, n(\xi) = 20$

$$\Pr(A \cap B) = \frac{n(A \cap B)}{n(\xi)}$$

$$\Pr(A \cap B) = \frac{3}{20}$$

iv. $n(A \cup B) = 13, n(\xi) = 20$

$$n(A \cup B) = \frac{n(A \cup B)}{n(\xi)}$$

$$n(A \cup B) = \frac{13}{20}$$

v. $n(A' \cap B') = 7, n(\xi) = 20$

$$\Pr(A' \cap B') = \frac{n(A' \cap B')}{n(\xi)}$$

$$\Pr(A' \cap B') = \frac{7}{20}$$

WORKED EXAMPLE 9 Creating a Venn diagram for three intersecting events

In a class of 35 students, 6 students like all three subjects: PE, Science and Music. Eight of the students like PE and Science, 10 students like PE and Music, and 12 students like Science and Music. Also, 22 students like PE, 18 students like Science and 17 like Music. Two students don't like any of the subjects.

a. **Display this information on a Venn diagram.**

b. **Determine the probability of selecting a student who:**

 i. **likes PE only**

 ii. **does not like Music.**

c. **Find Pr[(Science ∪ Music) ∩ PE′].**

THINK	WRITE/DRAW
a. 1. Draw a rectangle with three partly intersecting circles, labelled PE, Science and Music.	a. 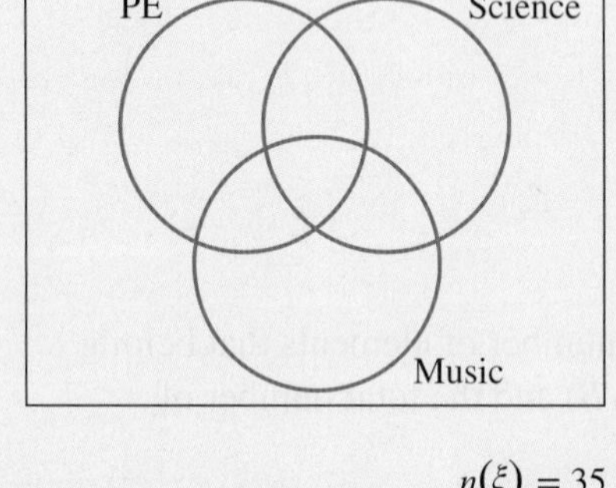
2. Extract the information relating to students liking all three subjects. *Note:* The central overlap is the key to solving these problems. Six students like all three subjects, so place the number 6 into the section corresponding to the intersection of the three circles.	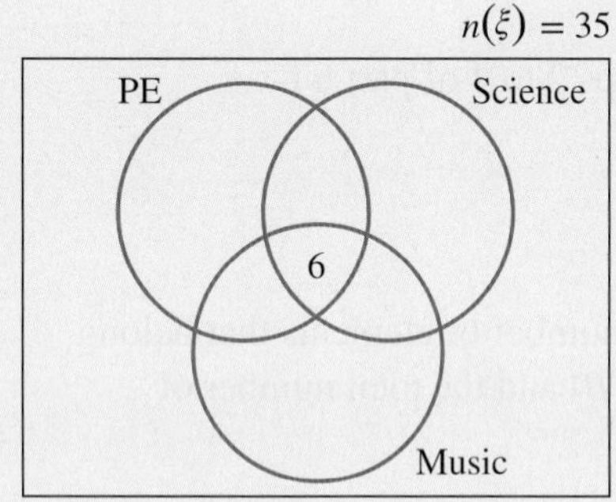
3. Extract the relevant information from the second sentence and place it into the appropriate position. *Note:* Eight students like PE and Science; however, 6 of these students have already been accounted for in step 2. Therefore, 2 will fill the intersection of only PE and Science. Similarly, 4 of the 10 who like PE and Music will fill the intersection of only PE and Music, and 6 of the 12 students will fill the intersection of only Science and Music.	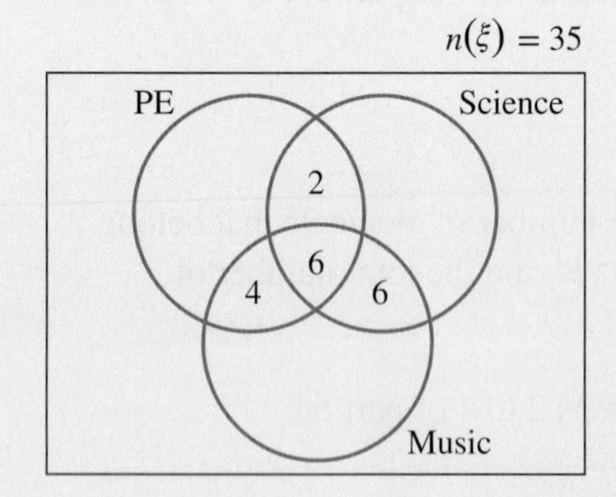

4. Extract the relevant information from the third sentence and place it into the appropriate position.
Note: Twenty-two students like PE and 12 have already been accounted for in the set. Therefore, 10 students are needed to fill the circle corresponding to PE only. Similarly, 4 students are needed to fill the circle corresponding to Science only to make a total of 18 for Science. One student is needed to fill the circle corresponding to Music only to make a total of 17 for Music.

$n(\xi) = 35$

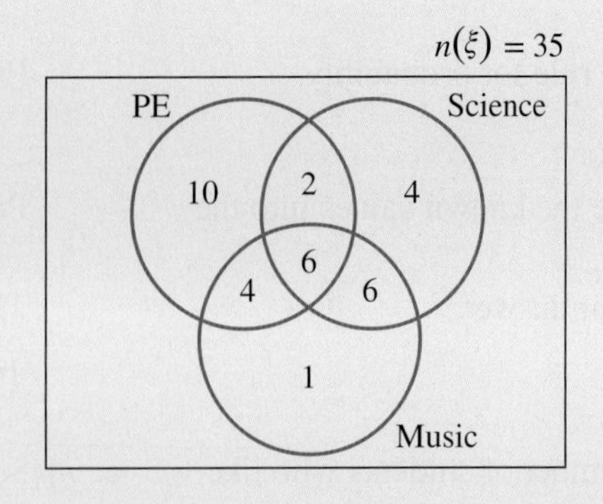

5. Extract the relevant information from the final sentence and place it into the appropriate position.
Note: Two students do not like any of the subjects, so they are placed in the rectangle outside the three circles.

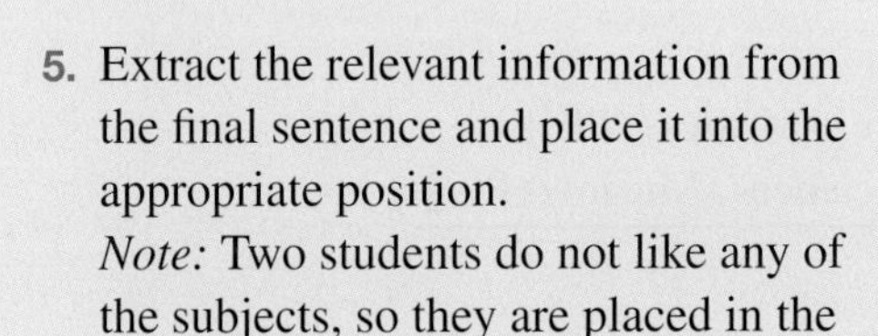

$n(\xi) = 35$

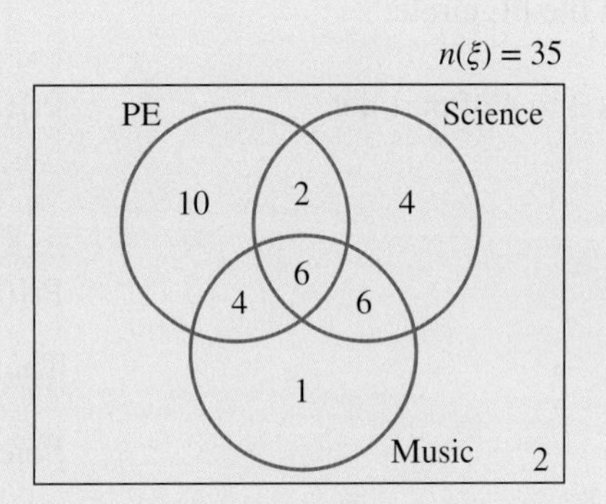

6. Check that the total number in all positions is equal to the number in the universal set.

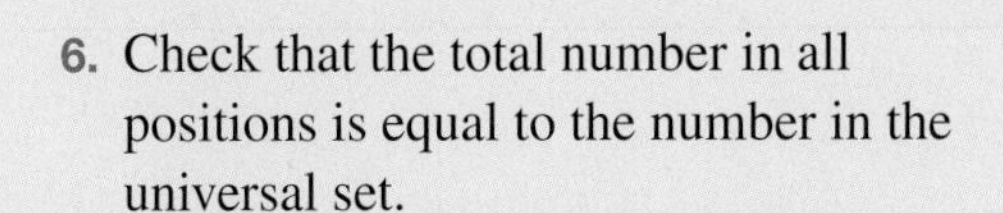

$10 + 2 + 4 + 4 + 6 + 6 + 1 + 2 = 35$

b. i. 1. Write the number of students who like PE only and the total number of students in the class.

b. i. $n(\text{students who like PE only}) = 10$
$n(\xi) = 35$

2. Write the rule for probability.

$$\Pr(\text{likes PE only}) = \frac{n(\text{likes PE only})}{n(\xi)}$$

3. Substitute the known values into the rule.

$$\Pr(\text{likes PE only}) = \frac{10}{35}$$

4. Evaluate and simplify.

$$= \frac{2}{7}$$

5. Write your answer.

The probability of selecting a student who likes PE only is $\frac{2}{7}$.

ii. 1. Write the number of students who do not like Music and the total number of students in the class.
Note: Add all the values that do not appear in the Music circle as well as the two that sit in the rectangle outside the circles.

ii. $n(\text{students who do not like Music}) = 18$
$n(\xi) = 35$

2. Write the rule for probability.

$$\Pr(\text{does not like Music}) = \frac{n(\text{does not like Music})}{n(\xi)}$$

3. Substitute the known values into the rule.

$$\Pr(\text{does not like Music}) = \frac{18}{35}$$

4. Write your answer.

The probability of selecting a student who does not like Music is $\frac{18}{35}$.

c. 1. Write the number of students who like Science and Music but not PE.
Note: Add the values that appear in the Science and Music circles but do not overlap with the PE circle.

c. $n[(\text{Science} \cup \text{Music}) \cap \text{PE}'] = 11$
$n(\xi) = 35$

2. Repeat steps 2 to 4 of part **b ii.**

$$\Pr[(\text{Science} \cup \text{Music}) \cap \text{PE}'] = \frac{n[(\text{Science} \cup \text{Music}) \cap \text{PE}']}{n(\xi)}$$

$$\Pr[(\text{Science} \cup \text{Music}) \cap \text{PE}'] = \frac{11}{35}.$$

The probability of selecting a student who likes Science or Music but not PE is $\frac{11}{35}$.

Resources

eWorkbook	Topic 11 Workbook (worksheets, code puzzle and project) (ewbk-2037)
Digital documents	SkillSHEET Set notation (doc-5286) SkillSHEET Simplifying fractions (doc-5287) SkillSHEET Determining complementary events (doc-5288) SkillSHEET Addition and subtraction of fractions (doc-5289) SkillSHEET Working with Venn diagrams (doc-5291) SkillSHEET Distinguishing between complementary and mutually exclusive events (doc-5294)
Video eLesson	Venn diagrams (eles-1934)
Interactivities	Individual pathway interactivity: Review of probability (int-4616) Experimental probability (int-3825) Two-way tables (int-6082) Theoretical probability (int-6081) Venn diagrams (int-3828) Addition Law of probability (int-6168)

Exercise 11.2 Review of probability

learnON

Individual pathways

■ PRACTISE	■ CONSOLIDATE	■ MASTER
1, 2, 4, 7, 11, 12, 16, 19	3, 5, 8, 13, 14, 17, 20	6, 9, 10, 15, 18, 21

To answer questions online and to receive **immediate corrective feedback** and **fully worked solutions** for all questions, go to your learnON title at www.jacplus.com.au.

Fluency

1. Explain the difference between experimental and theoretical probability.

2. **WE1** The spinner shown was spun 50 times and the outcome each time was recorded in the table.

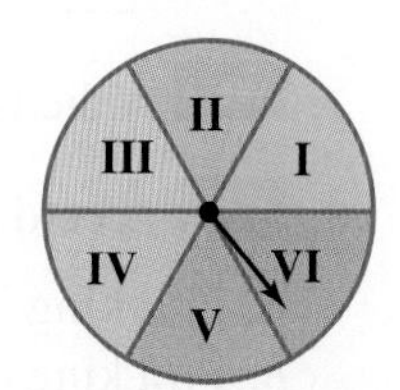

Segment	I	II	III	IV	V	VI
Tally	10	6	8	7	12	7

 a. List the event space.
 b. Given the experimental results, determine the relative frequency for each segment.
 c. The theoretical probability of the spinner landing on any particular segment with one spin is $\frac{1}{6}$. State how the experiment could be changed to give a better estimate of the true probabilities.

3. A laptop company conducted a survey to see what were the most appealing colours for laptop computers among 15–18-year-old students. The results were as follows:

Colour	Black Black	Sizzling Silver	Power Pink	Blazing Blue	Gooey Green	Glamour Gold
Number	102	80	52	140	56	70

 a. Calculate the number of students who were surveyed.
 b. Calculate the relative frequency of students who found silver the most appealing laptop colour.
 c. Calculate the relative frequency of students who found black and green to be their most appealing colours.
 d. State which colour was found to be most appealing.

4. **WE2** Two dice are rolled and the values on the two uppermost faces are added together.
 a. Construct a table to illustrate the sample space.
 b. Calculate the most likely outcome.
 c. Calculate the least likely outcome.

5. **WE7** Given $\Pr(A) = 0.5, \Pr(B) = 0.4$ and $\Pr(A \cup B) = 0.8$:
 a. use the Addition Law of probability to calculate the value of $\Pr(A \cap B)$
 b. draw a Venn diagram to represent the universal set
 c. calculate $\Pr(A \cap B')$.

6. Let $\Pr(A) = 0.25, \Pr(B) = 0.65$ and $\Pr(A \cap B) = 0.05$.
 a. Calculate:
  i. $\Pr(A \cup B)$
  ii. $\Pr(A \cap B)'$.

b. **MC** Choose which Venn diagram below best illustrates $\Pr(A \cap B)'$.

A. 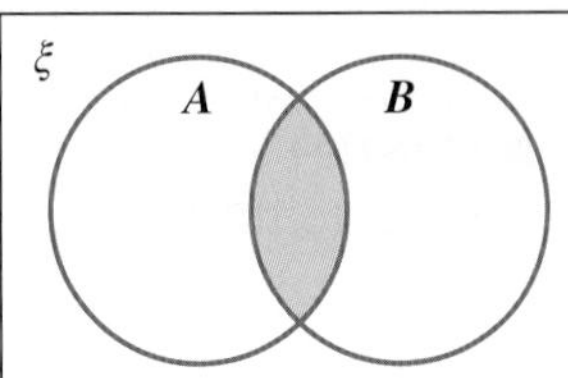

B. 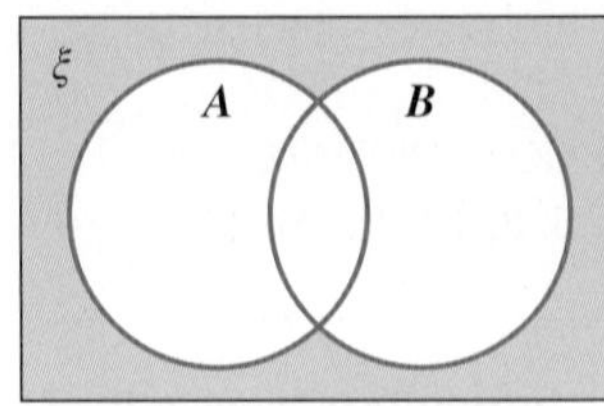

C. 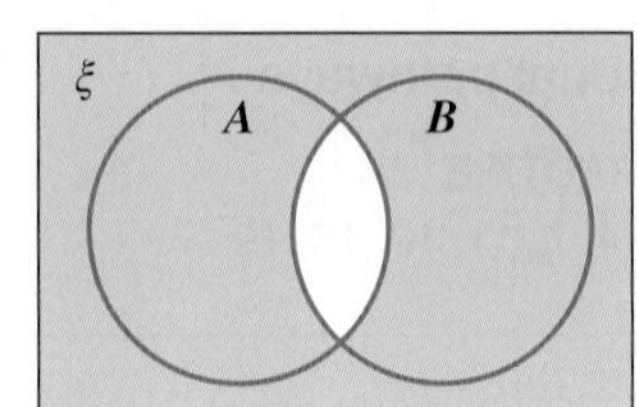

D. 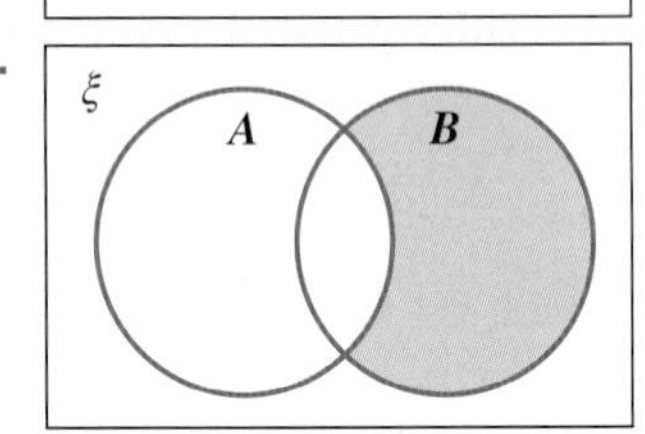

E. 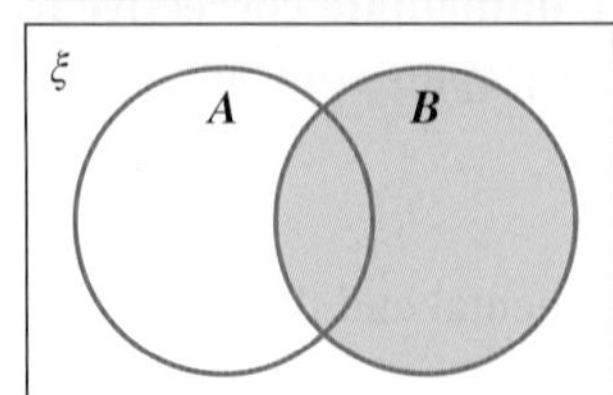

7. **WE3** A die is rolled. Calculate the probability that the outcome is an even number or a 5.

8. **WE6** A card is drawn from a well-shuffled pack of 52 playing cards. Calculate:
   a. Pr(a king is drawn)
   b. Pr(a heart is drawn)
   c. Pr(a king or a heart is drawn).

9. **WE4** For each of the following pairs of events:
   i. state, giving justification, if the pair are complementary events
   ii. alter the statements, where applicable, so that the events become complementary events.

   a. Having Weet Bix or having Strawberry Pops for breakfast
   b. Walking to a friend's place or driving there
   c. Watching TV or reading as a leisure activity
   d. Rolling a number less than 5 or rolling a number greater than 5 with a ten-sided die with faces numbered 1 to 10
   e. Passing a maths test or failing a maths test

10. a. **WE8** Draw a Venn diagram representing the relationship between the following sets. Show the position of all the elements in the Venn diagram.

$$\xi = \{1, 2, 3, 4, 5, 6, 7, 8, 9, 10, 11, 12, 13, 14, 15, 16, 17, 18, 19, 20\}$$
$$A = \{1, 3, 5, 7, 9, 11, 13, 15, 17, 19\}$$
$$B = \{1, 4, 9, 16\}$$

   b. Calculate:
      i. $\Pr(A)$
      ii. $\Pr(B)$
      iii. $\Pr(A \cap B)$
      iv. $\Pr(A \cup B)$
      v. $\Pr(A' \cap B')$.

Understanding

11. You and a friend are playing a dice game. You have an eight-sided die (with faces numbered 1 to 8 inclusive) and your friend has a six-sided die (with faces numbered 1 to 6 inclusive). You each roll your own die.
    a. The person who rolls the number 4 wins. Determine if this game is fair.
    b. The person who rolls an odd number wins. Determine if this game is fair.

12. A six-sided die has three faces numbered 5; the other faces are numbered 6. Determine if the events 'rolling a 5' and 'rolling a 6' are equally likely.

13. WE5 A card is drawn from a shuffled pack of 52 cards. Calculate the probability that the card drawn is:

a. an ace
b. a club
c. a red card
d. not a jack
e. a green card
f. not a red card.

14. A bag contains 4 blue marbles, 7 red marbles and 9 yellow marbles. All marbles are the same size. A marble is selected at random. Calculate the probability that the marble is:

a. blue
b. red
c. not yellow
d. black.

15. WE9 Thirty students were asked which lunchtime sports they enjoyed — volleyball, soccer or tennis. Five students chose all three sports. Six students chose volleyball and soccer, 7 students chose volleyball and tennis, and 9 chose soccer and tennis. Fifteen students chose volleyball, 14 students chose soccer and 18 students chose tennis.

a. Copy the Venn diagram shown and enter the given information.

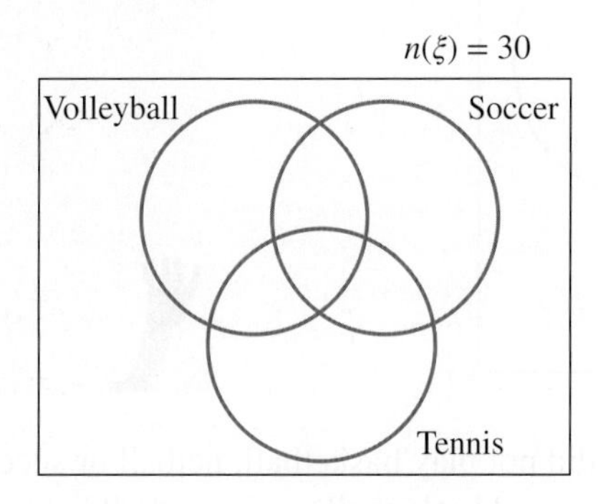

b. If a student is selected at random, determine the probability of selecting a student who:

i. chose volleyball
ii. chose all three sports
iii. chose both volleyball and soccer but not tennis
iv. did not choose tennis
v. chose soccer.

c. Determine:

i. $\Pr[(\text{soccer} \cup \text{tennis}) \cap \text{volleyball}']$
ii. $\Pr[(\text{volleyball} \cup \text{tennis}) \cap \text{soccer}']$.

Reasoning

16. A six-sided die has three faces numbered 1 and the other three faces numbered 2. Determine if the events 'rolling a 1' and 'rolling a 2' are equally likely.

17. With the use of diagrams, show that $\Pr(A' \cap B') = \Pr(A \cup B)'$.

18. A drawer contains purple socks and red socks. The chance of obtaining a red sock is 2 in 9. There are 10 red socks in the drawer.
Determine the smallest number of socks that need to be added to the drawer so that the probability of drawing a red sock increases to 3 in 7.

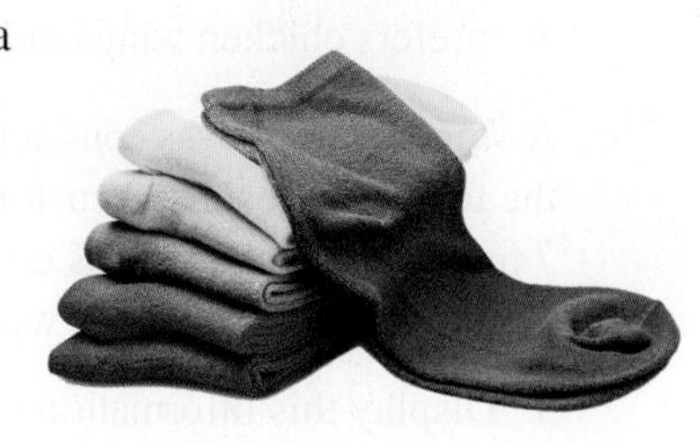

Problem solving

19. Ninety students were asked which lunchtime sports on offer, basketball, netball and soccer, they had participated in on at least one occasion in the last week. The results are shown in the following table.

Sport	Basketball	Netball	Soccer	Basketball and netball	Basketball and soccer	Netball and soccer	All three
Number of students	35	25	39	5	18	8	3

a. Copy and complete the Venn diagram shown below to illustrate the sample space.

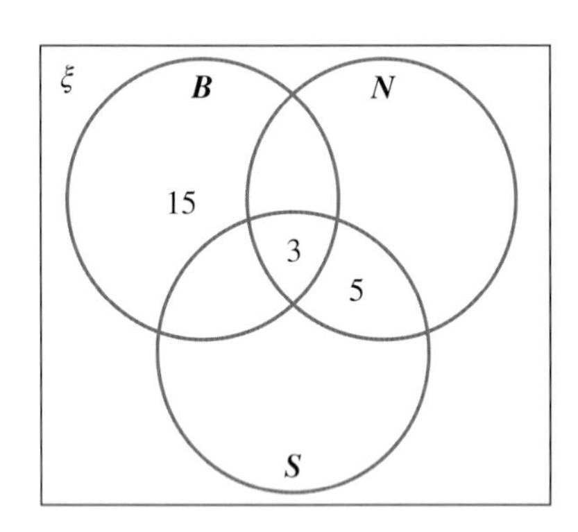

b. Determine how many students did not play basketball, netball or soccer at lunchtime.
c. Determine how many students played basketball and/or netball but not soccer.
d. Determine how many students are represented by the region (basketball ∩ not netball ∩ soccer).
e. Calculate the relative frequency of the region described in part d above.
f. Estimate the probability that a student will play three of the sports offered.

20. The Venn diagram shows the results of a survey completed by a Chinese restaurateur to find out the food preferences of his regular customers.

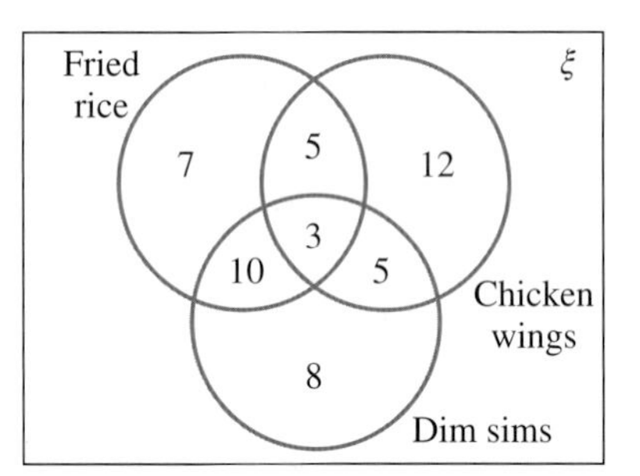

a. Determine the number of customers:
   i. surveyed
   ii. showing a preference for fried rice only
   iii. showing a preference for fried rice
   iv. showing a preference for chicken wings and dim sims.

b. A customer from this group won the draw for a lucky door prize. Determine the probability that this customer:
   i. likes fried rice
   ii. likes all three — fried rice, chicken wings and dim sims
   iii. prefers chicken wings only.

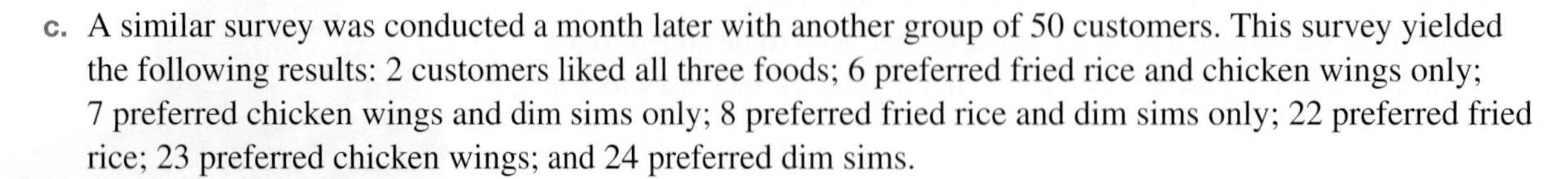

c. A similar survey was conducted a month later with another group of 50 customers. This survey yielded the following results: 2 customers liked all three foods; 6 preferred fried rice and chicken wings only; 7 preferred chicken wings and dim sims only; 8 preferred fried rice and dim sims only; 22 preferred fried rice; 23 preferred chicken wings; and 24 preferred dim sims.

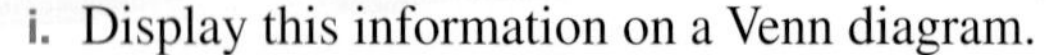

   i. Display this information on a Venn diagram.
   ii. Determine the probability of selecting a customer who prefers all three foods, if a random selection is made.

21. A pair of dice is rolled and the sum of the numbers shown is noted.
   a. Show the sample space in a two-way table.
   b. Determine how many different ways the sum of 7 can be obtained.
   c. Determine if all outcomes are equally likely.
   d. Complete the given table.

Sum	2	3	4	5	6	7	8	9	11	12
Frequency										

   e. Determine the relative frequencies of the following sums.
      i. 2 ii. 7 iii. 11
   f. Determine the probability of obtaining the following sums.
      i. 2 ii. 7 iii. 11
   g. If a pair of dice is rolled 300 times, calculate how many times you would expect the sum of 7.

11.3 Tree diagrams

LEARNING INTENTION

At the end of this subtopic you should be able to:
- create tree diagrams to represent two- and three-step chance experiments
- use tree diagrams to solve probability problems involving two or more trials or events.

11.3.1 Two-step chance experiments

eles-4924

- In **two-step chance experiments** the result is obtained after performing two trials. Two-step chance experiments are often represented using tree diagrams.
- **Tree diagrams** are used to list all possible outcomes of two or more events that are not necessarily equally likely.
- The probability of obtaining the result for a particular event is listed on the branches.
- The probability for each outcome in the sample space is the product of the probabilities associated with the respective branches. For example, the tree diagram shown here represents the sample space for flipping a coin, then choosing a marble from a bag containing three red marbles and one black marble.

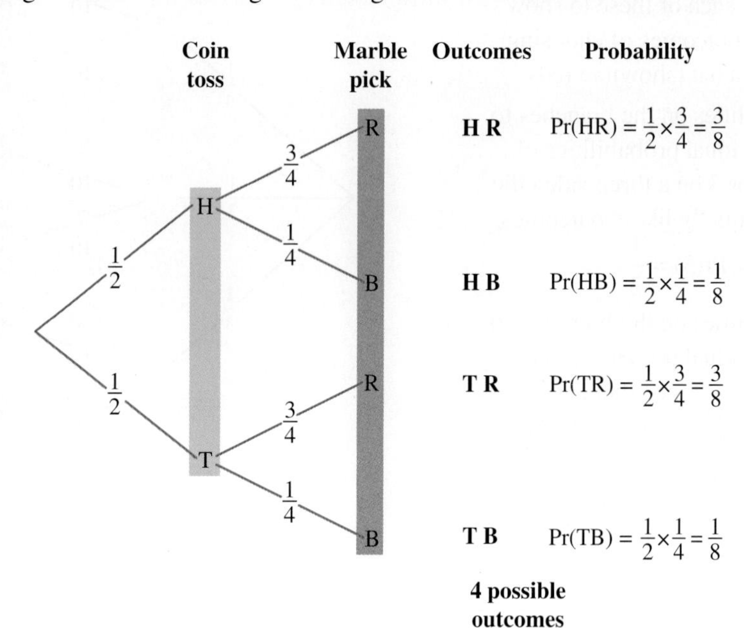

- When added together, all the probabilities for the outcomes should sum to 1. They are complementary events. For example,

$$\begin{aligned}\Pr(HR) + \Pr(HB) + \Pr(TR) + \Pr(TB) &= \frac{3}{8} + \frac{1}{8} + \frac{3}{8} + \frac{1}{8} \\ &= 1\end{aligned}$$

- Other probabilities can also be calculated from the tree diagram. For example, the probability of getting an outcome that contains a red marble can be calculated by summing the probabilities of each of the possible outcomes that include a red marble.
  Outcomes that contain a red marble are HR and TR, therefore:

$$\begin{aligned}\Pr(\text{red marble}) &= \Pr(HR) + \Pr(TR) \\ &= \frac{3}{8} + \frac{3}{8} \\ &= \frac{6}{8} \\ &= \frac{3}{4}\end{aligned}$$

WORKED EXAMPLE 10 Using a tree diagrams for a two-step chance experiment

A three-sided die is rolled and a name is picked out of a hat that contains 3 girls' names and 7 boys' names.

a. Construct a tree diagram to display the sample space.

b. Calculate the probability of:

i. rolling a 3, then choosing a boy's name

ii. choosing a boy's name after rolling an odd number.

THINK

a. 1. Draw 3 branches from the starting point to show the 3 possible outcomes of rolling a three-sided die (shown in blue), and then draw 2 branches off each of these to show the 2 possible outcomes of choosing a name out of a hat (shown in red).

2. Write probabilities on the branches to show the individual probabilities of rolling a 1, 2 or 3 on a three-sided die. As these are equally likely outcomes, $\Pr(1) = \Pr(2) = \Pr(3) = \frac{1}{3}$.

3. Write probabilities on the branches to show the individual probabilities of choosing a name. Since there are 3 girls' names and 7 boys' names in the hat, $\Pr(G) = \frac{3}{10}$ and $\Pr(B) = \frac{7}{10}$.

WRITE

a.

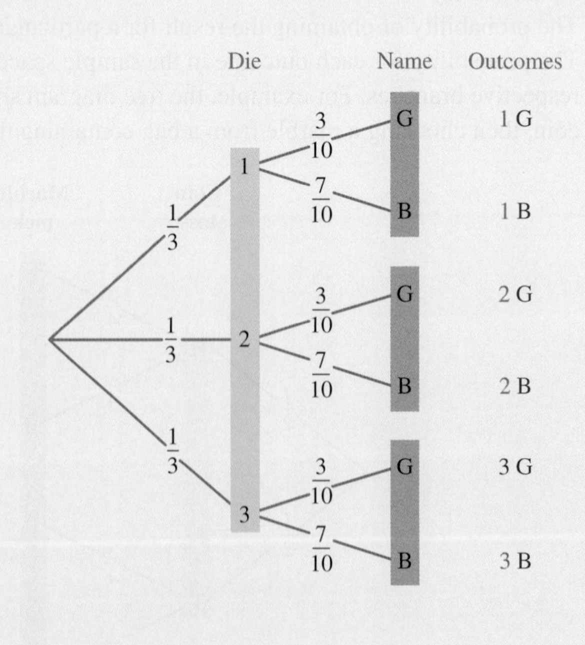

b. i. 1. Follow the pathway of rolling a 3 $\left[\Pr(3) = \frac{1}{3}\right]$ and choosing a boy's name $\left[\Pr(B) = \frac{7}{10}\right]$, and multiply the probabilities.	**b. i.** $\Pr(3B) = \Pr(3) \times \Pr(B)$ $= \frac{1}{3} \times \frac{7}{10}$ $= \frac{7}{30}$
2. Write the answer.	The probability of rolling a 3, then choosing a boy's name is $\frac{7}{30}$.
ii. 1. To roll an odd number (1 or 3) then choose a boy's name: • roll a 1, then choose a boy's name or • roll a 3, then choose a boy's name. Calculate the probability of each of these and add them together to calculate the total probability. Simplify the result if possible.	**ii.** $\Pr(\text{odd B}) = \Pr(1B) + \Pr(3B)$ $= \Pr(1) \times \Pr(B) + \Pr(3) \times \Pr(B)$ $= \frac{1}{3} \times \frac{7}{10} + \frac{1}{3} \times \frac{7}{10}$ $= \frac{7}{30} + \frac{7}{30}$ $= \frac{14}{30}$ $= \frac{7}{15}$
2. Write the answer.	The probability of choosing a boy's name after rolling an odd number is $\frac{7}{15}$.

11.3.2 Three-step chance experiments

eles-4925

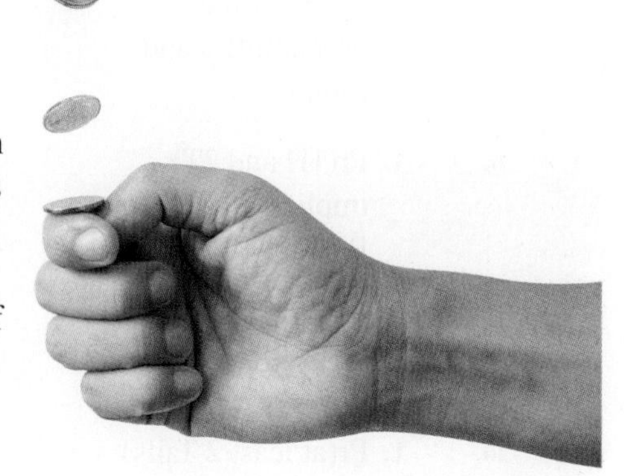

- Outcomes are often made up of combinations of events. For example, when a coin is flipped three times, three of the possible outcomes are HHT, HTH and THH. These outcomes all contain 2 Heads and 1 Tail.
- The probability of an outcome with a particular order is written such that the order required is shown. For example, Pr(HHT) is the probability of H on the first coin, H on the second coin and T on the third coin.
- The probability of an outcome with a particular combination of events in which the order is not important is written describing the particular combination required. For example, Pr(2 Heads and 1 Tail).

WORKED EXAMPLE 11 Using a tree diagram for a three-step chance experiment

A coin is biased so that the chance of it falling as a Head when flipped is 0.75.

a. **Construct a tree diagram to represent the coin being flipped three times.**

b. **Calculate the following probabilities, correct to 3 decimal places:**

i. **Pr(HTT)**

ii. **Pr(1H and 2T)**

iii. **Pr(at least 2 Tails).**

THINK	WRITE
a. 1. Tossing a coin has two outcomes. Draw 2 branches from the starting point to show the first toss, 2 branches off each of these to show the second toss and then 2 branches off each of these to show the third toss. 2. Write probabilities on the branches to show the individual probabilities of tossing a Head (0.75) and a Tail. Because tossing a Head and tossing a Tail are mutually exclusive, Pr(T) = 1 − Pr(H) = 0.25.	a. 1st toss, 2nd toss, 3rd toss, Outcomes H (0.75): H (0.75): H (0.75) HHH; T (0.25) HHT H (0.75): T (0.25): H (0.75) HTH; T (0.25) HTT T (0.25): H (0.75): H (0.75) THH; T (0.25) THT T (0.25): T (0.25): H (0.75) TTH; T (0.25) TTT
b. i. 1. Pr(HTT) implies the order: H(0.75), T (0.25), T (0.25). 2. Multiply the probabilities and round.	b. i. $\Pr(\text{HTT}) = \Pr(\text{H}) \times \Pr(\text{T}) \times \Pr(\text{T})$ $= (0.75) \times (0.25)^2$ $= 0.046875$ ≈ 0.047
ii. 1. Pr(1H and 2T) implies: Pr(HTT), Pr(THT), Pr(TTH). 2. Multiply the probabilities and round.	ii. $\Pr(\text{1H and 2T}) = \Pr(\text{HTT}) + \Pr(\text{THT}) + \Pr(\text{TTH})$ $= 3(0.75 \times 0.25^2)$ $= 0.140625$ ≈ 0.141
iii. 1. Pr(at least 2 Tails) implies: Pr(HTT), Pr(THT), Pr(TTH) and Pr(TTT). 2. Add these probabilities and round.	iii. $\Pr(\text{at least 2T}) = \Pr(\text{HTT}) + \Pr(\text{THT}) + \Pr(\text{TTH}) + \Pr(\text{TTT})$ $= 3(0.75 \times 0.25^2) + 0.25^3$ $= 0.15625$ ≈ 0.156

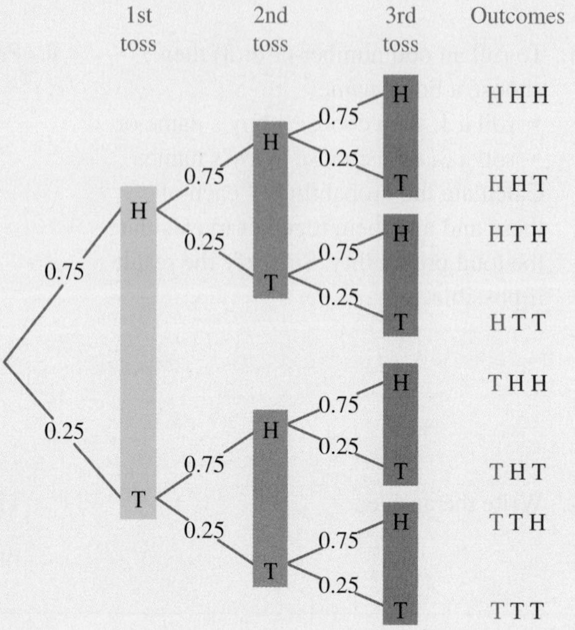

TI | THINK

a. i.

In a new document, on a Calculator page, to calculate the probability of an exact number of successes from a set number of trials press:

- MENU
- 5: Probability
- 5: Distributions
- A: BinomialPdf

Enter the values:

- Num Trials, n: 3
- Prob Success, p: 0.75
- X Value: 1

Then tap OK.

DISPLAY/WRITE

a. i.

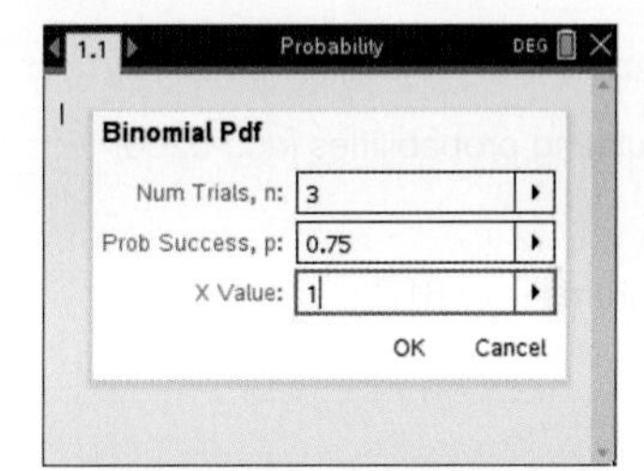

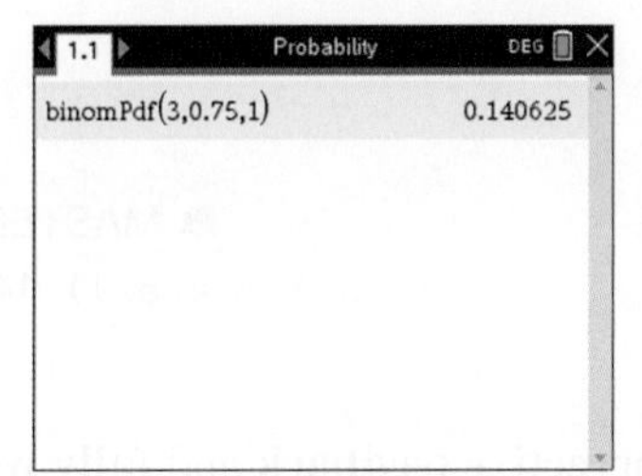

Pr(1H and 2T) = 0.141 (correct to 3 decimal places)

ii.

In a new document, on a Calculator page, to calculate the probability of at least x successes from a set number of trials press:

- MENU
- 5: Probability
- 5: Distributions
- B: BinomialCdf

Enter the values:

- Num Trials, n: 3
- Prob Success, p: 0.25
- Lower Bound: 2
- Upper Bound: 3

Then tap OK.

Note: In situations with repeated trials of independent events, such as flipping a coin, it becomes unfeasible to draw a tree diagram when the number of trials is 4 or more. BinomicalPdf and BinomialCdf are very helpful in these cases.

ii.

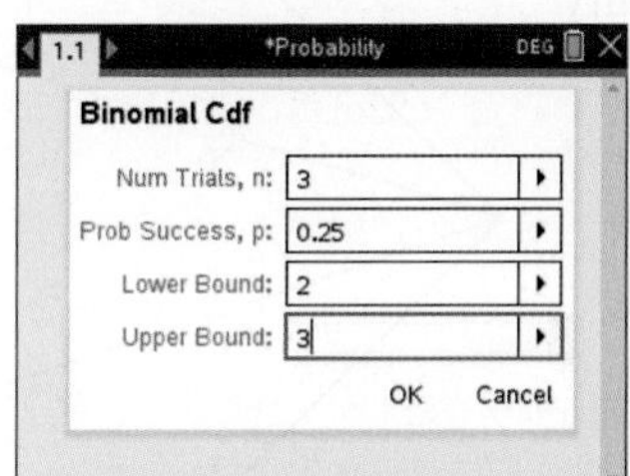

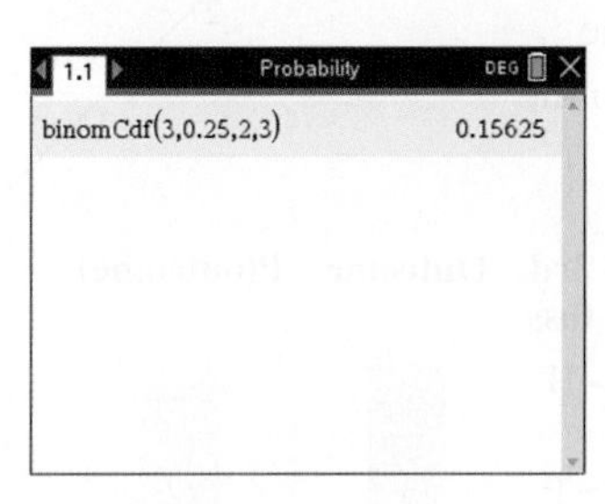

Pr(at least 2T) = 0.156 (correct to 3 decimal places)

CASIO | THINK

a. i.

On the Main screen, to calculate the probability of an exact number of successes from a set number of trials tap:

- Interactive
- Distribution/Inv. Dist
- Discrete
- binomialPDF

Enter the values:

- X: 1
- Numtrial: 3
- pos: 0.75

Then tap OK.

DISPLAY/WRITE

a. i.

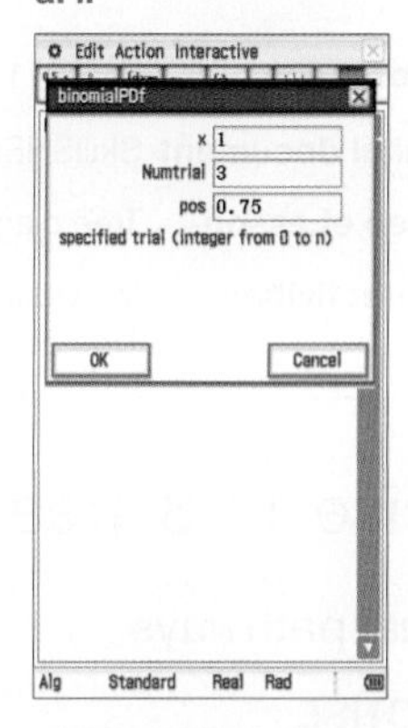

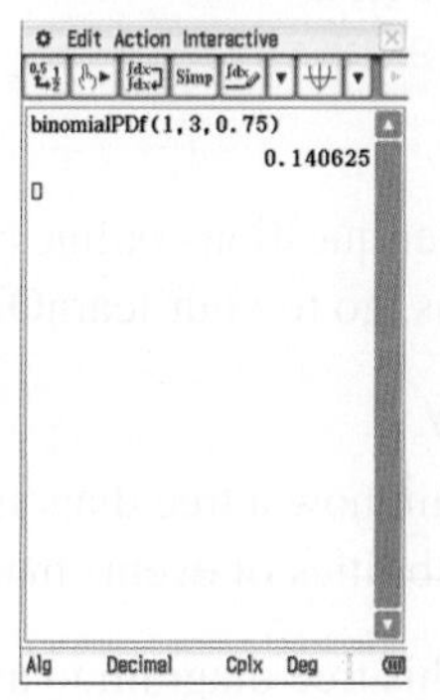

Pr(1H and 2T) = 0.141 (correct to 3 decimal places)

ii.

On the Main screen, to calculate the probability of at least x successes from a set number of trials tap:

- Interactive
- Distribution/Inv. Dist
- Discrete
- binomialCDF

Enter the values:

- Lower: 2
- Upper: 3
- Numtrial: 3
- pos: 0.25

Then tap OK.

Note: In situations with repeated trials of independent events, such as flipping a coin, it becomes unfeasible to draw a tree diagram when the number of trials is 4 or more. BinomicalPDF and BinomialCDF are very helpful in these cases.

ii.

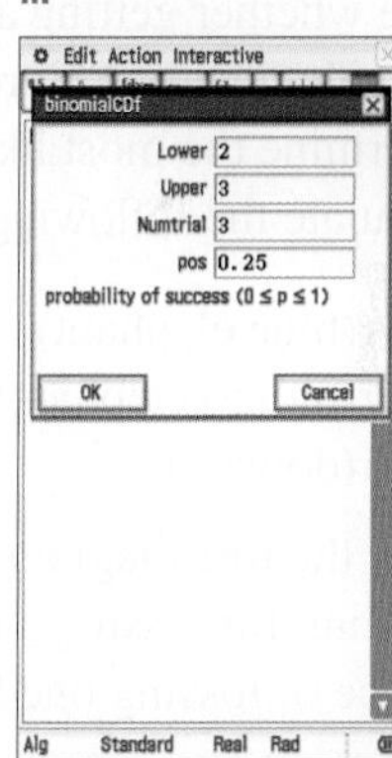

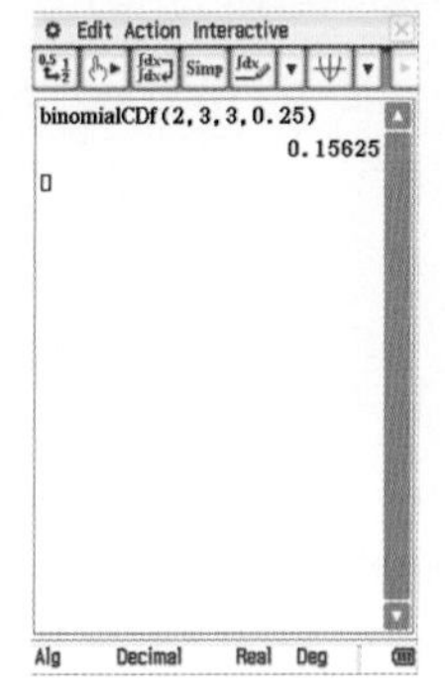

Pr(at least 2T) = 0.156 (correct to 3 decimal places)

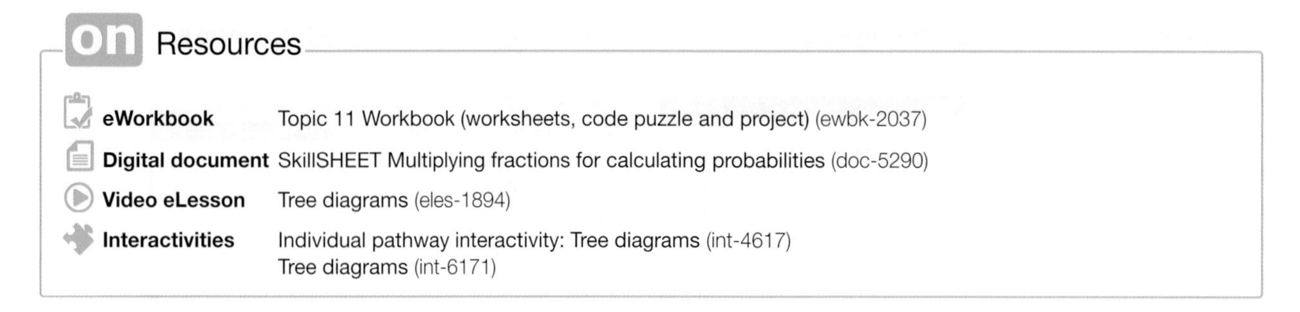

Resources

eWorkbook Topic 11 Workbook (worksheets, code puzzle and project) (ewbk-2037)

Digital document SkillSHEET Multiplying fractions for calculating probabilities (doc-5290)

Video eLesson Tree diagrams (eles-1894)

Interactivities Individual pathway interactivity: Tree diagrams (int-4617)
Tree diagrams (int-6171)

Exercise 11.3 Tree diagrams

learnON

Individual pathways

■ PRACTISE	■ CONSOLIDATE	■ MASTER
1, 2, 6, 9, 12	3, 5, 7, 10, 13	4, 8, 11, 14

To answer questions online and to receive **immediate corrective feedback** and **fully worked solutions** for all questions, go to your learnON title at www.jacplus.com.au.

Fluency

1. Explain how a tree diagram can be used to calculate probabilities of events that are not equally likely.
2. Use this tree diagram to answer the following questions.
   a. Identify how many different outcomes there are.
   b. Explain whether all outcomes are equally likely.
   c. State whether getting a red fish is more, less or equally likely than getting a green elephant.
   d. Determine the most likely outcome.
   e. Calculate the following probabilities.
      i. Pr(blue elephant)
      ii. Pr(indigo elephant)
      iii. Pr(donkey)

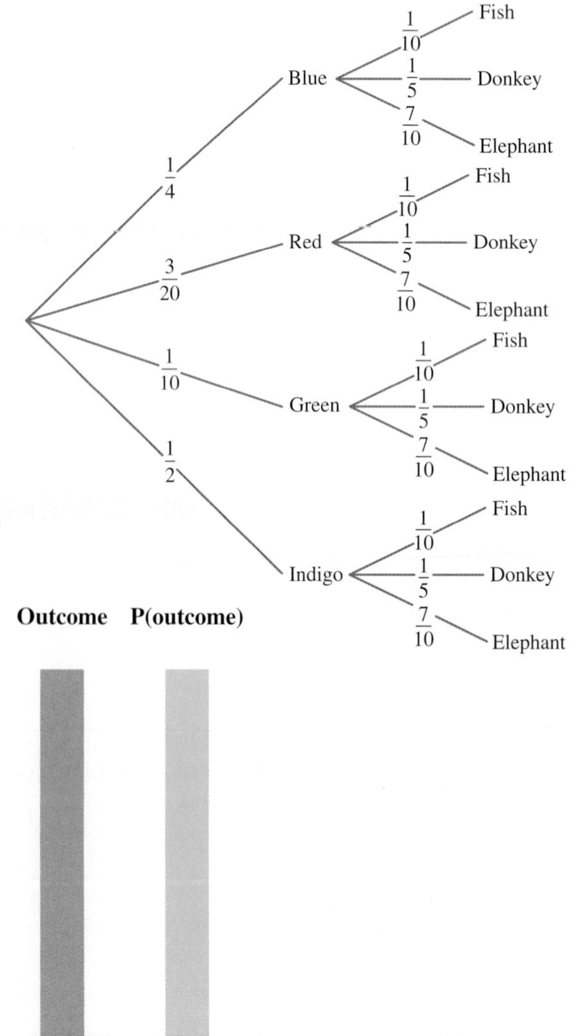

3. a. Copy the tree diagram shown here and complete the labelling for tossing a biased coin three times when the chance of tossing one Head in one toss is 0.7.

1st toss | 2nd toss | 3rd toss | Outcome | P(outcome)

H, T
H, T, H, T
H, T, H, T, H, T, H, T

b. Calculate the probability of tossing three Heads.
c. Determine the probability of getting at least one Tail.
d. Calculate the probability of getting exactly two Tails.

4. The questions below relate to rolling a fair die.
   a. Calculate the probability of each of the following outcomes from one roll of a die.
      i. Pr(rolling number < 4)
      ii. Pr(rolling a 4)
      iii. Pr(rolling a number other than a 6)

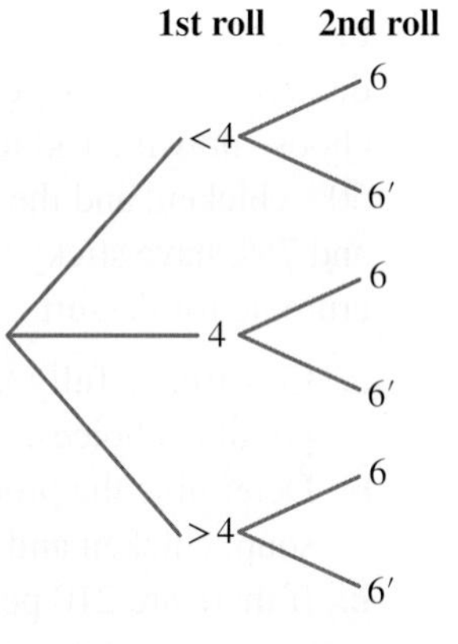

   b. The tree diagram shown has been condensed to depict rolling a die twice, noting the number relative to 4 on the first roll and 6 on the second. Complete a labelled tree diagram, showing probabilities on the branches and all outcomes, similar to that shown.
   c. Determine the probability of rolling the following with 2 rolls of the die.
      i. Pr(a 4 then a 6)
      ii. Pr(a number less than 4 then a 6)
      iii. Pr(a 4 then $6'$)
      iv. Pr(a number > 4 and then a number < 6)

5. **WE10** The spinner shown at right is divided into 3 equal-sized wedges labelled 1, 2 and 3. It is spun three times, and it is noted whether the spinner lands on a prime number, $\Pr = \{2, 3\} =$ 'prime', or not a prime number, $\Pr' = \{1\} =$ 'not prime'.
   a. Construct a labelled tree diagram for 3 spins of the spinner, showing probabilities on the branches and all possible outcomes.
   b. Calculate the following probabilities.
      i. Pr(3 prime numbers)
      ii. Pr(PPP' in this order)
      iii. Pr(PPP' in any order)

Understanding

6. **WE11** A coin is biased so that the chance of it falling as a Tail when tossed is 0.2.
   a. Construct a tree diagram to represent the coin being tossed 3 times.
   b. Determine the probability of getting the same outcome on each toss.

7. A die is tossed twice and each time it is recorded whether or not the number is a multiple of 3. If $M =$ the event of getting a multiple of 3 on any one toss and $M' =$ the event of not getting a multiple of 3 on any one toss:
   a. construct a tree diagram to represent the 2 tosses
   b. calculate the probability of getting two multiples of 3.

8. The biased spinner illustrated is spun three times.

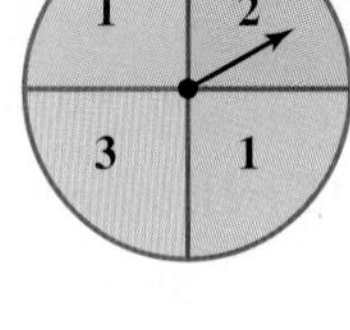

   a. Construct a completely labelled tree diagram for 3 spins of the spinner, showing probabilities on the branches and all possible outcomes and associated probabilities.
   b. Calculate the probability of getting exactly two 1s.
   c. Calculate the probability of getting at most two 1s.

Reasoning

9. Each morning when Ahmed gets dressed for work he has the following choices:
   - three suits that are grey, blue and white
   - four shirts that are white, blue, pink and grey
   - two ties that are grey and blue.

a. Construct a fully labelled tree diagram showing all possible clothing choices.
b. Calculate the probability of picking a grey suit, a pink shirt and a blue tie.
c. Calculate the probability of picking the same colour for all three options.
d. Ahmed works five days a week for 48 weeks of the year. Determine how many times each combination would get repeated over the course of one year at work.
e. Determine how many more combinations of clothing he would have if he bought another tie and another shirt.

10. A restaurant offers its customers a three-course dinner, where they choose between two entrées, three main meals and two desserts. The managers find that 30% choose soup and 70% choose prawn cocktail for the entrée; 20% choose vegetarian, 50% chicken, and the rest have beef for their main meal; and 75% have sticky date pudding while the rest have apple crumble for dessert.

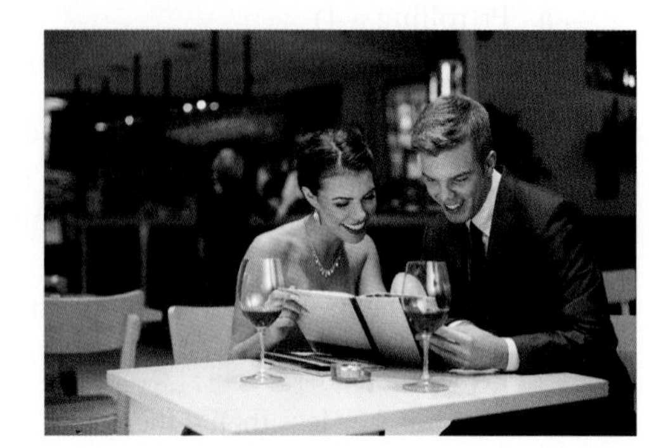

a. Construct a fully labelled tree diagram showing all possible choices.
b. Determine the probability that a customer will choose the soup, chicken and sticky date pudding.
c. If there are 210 people booked for the following week at the restaurant, determine how many you would expect to have the meal combination referred to in part b.

11. A bag contains 7 red and 3 white balls. A ball is taken at random, its colour noted and it is then placed back in the bag before a second ball is chosen at random and its colour noted.

a. i. Show the possible outcomes with a fully labelled tree diagram.
   ii. As the first ball was chosen, determine how many balls were in the bag.
   iii. As the second ball was chosen, determine how many balls were in the bag.
   iv. Explain whether the probability of choosing a red or white ball changes from the first selection to the second.
   v. Calculate the probability of choosing a red ball twice.

b. Suppose that after the first ball had been chosen it was not placed back in the bag.
   i. As the second ball is chosen, determine how many balls are in the bag.
   ii. Explain if the probability of choosing a red or white ball changes from the first selection to the second.
   iii. Construct a fully labelled tree diagram to show all possible outcomes.
   iv. Evaluate the probability of choosing two red balls.

Problem solving

12. An eight-sided die is rolled three times to see whether 5 occurs.

a. Construct a tree diagram to show the sample space.
b. Calculate:
   i. Pr(three 5s)
   ii. Pr(no 5s)
   iii. Pr(two 5s)
   iv. Pr(at least two 5s).

13. A tetrahedral die (four faces labelled 1, 2, 3 and 4) is rolled and a coin is tossed simultaneously.

a. Show all the outcomes on a two-way table.
b. Construct a tree diagram and list all outcomes and their respective probabilities.
c. Determine the probability of getting a Head on the coin and an even number on the die.

14. A biased coin that has an 80% chance of getting a Head is flipped four times. Use a tree diagram to answer to the following.
   a. Calculate the probability of getting 4 Heads.
   b. Determine the probability of getting 2 Heads then 2 Tails in that order.
   c. Calculate the probability of getting 2 Heads and 2 Tails in any order.
   d. Determine the probability of getting more Tails than Heads.

11.4 Independent and dependent events

LEARNING INTENTION

At the end of this subtopic you should be able to:
- determine whether two events are independent or dependent
- calculate the probability of outcomes from two-step chance experiments involving independent and dependent events.

11.4.1 Independent events

eles-4926

- **Independent events** are events that have no effect on each other. The outcome of the first event does not influence the outcome of the second.
- An example of independent events are successive coin tosses. The outcome of the first toss has no effect on the second coin toss. Similarly, the outcome of the roll on a die will not affect the outcome of the next roll.
- If events A and B are independent, then the **Multiplication Law of probability** states that:
  $\Pr(A \text{ and } B) = \Pr(A) \times \Pr(B)$ or $\Pr(A \cap B) = \Pr(A) \times \Pr(B)$
- The reverse is also true. If:
  $\Pr(A \text{ and } B) = \Pr(A) \times \Pr(B)$ or $\Pr(A \cap B) = \Pr(A) \times \Pr(B)$
  then event A and event B are independent events.

Independent events

For independent events A and B:

$$\mathbf{Pr}(A \cap B) = \mathbf{Pr}(A) \times \mathbf{Pr}(B)$$

WORKED EXAMPLE 12 Determining whether events are independent or dependent

Adam is one of the 10 young golfers to represent his state. Paz is one of the 12 netball players to represent her state. All the players in their respective teams have an equal chance of being nominated as captains.

a. Explain whether the events 'Adam is nominated as captain' and 'Paz is nominated as captain' are independent.

b. Calculate:

i. Pr(Adam is nominated as captain)

ii. Pr(Paz is nominated as captain).

c. Determine the probability that both Adam and Paz are nominated as captains of their respective teams.

THINK	WRITE
a. Determine whether the given events are independent and write your answer.	**a.** Adam's nomination has nothing to do with Paz's nomination and vice versa. Therefore, the events are independent.
b. i. 1. Determine the probability of Adam being nominated as captain. He is one of 10 players.	**b. i.** Pr(Adam is nominated) = Pr(A) $= \dfrac{n(\text{Adam is nominated})}{n(\xi)}$ Pr(Adam is nominated) $= \dfrac{1}{10}$
2. Write your answer.	The probability that Adam is nominated as captain is $\dfrac{1}{10}$.
ii. 1. Determine the probability of Paz being nominated as captain. She is one of 12 players.	**ii.** Pr(Paz is nominated) = Pr(P) $= \dfrac{n(\text{Paz is nominated})}{n(\xi)}$ Pr(Paz is nominated) $= \dfrac{1}{12}$
2. Write your answer.	The probability that Paz is nominated as captain is $\dfrac{1}{12}$.
c. 1. Write the Multiplication Law of probability for independent events.	**c.** Pr(A and P) $= \Pr(A \cap P)$ $= \Pr(A) \times \Pr(P)$ Pr(Adam and Paz are nominated) = Pr(Adam is nominated) × Pr(Paz is nominated)
2. Substitute the known values into the rule.	$= \dfrac{1}{10} \times \dfrac{1}{12}$
3. Evaluate.	$= \dfrac{1}{120}$
4. Write your answer.	The probability that both Adam and Paz are nominated as captains is $\dfrac{1}{120}$.

11.4.2 Dependent events

eles-4927

- **Dependent events** are events where the outcome of one event affects the outcome of the other event.
- An example of dependent events is drawing a card from a deck of playing cards. The probability that the first card drawn is a heart, Pr(hearts), is $\frac{13}{52}$ (or $\frac{1}{4}$). If this card is a heart and is not replaced, then this will affect the probability of subsequent draws. The probability that the second card drawn is a heart will be $\frac{12}{51}$, while the probability that the second card is not a heart will be $\frac{39}{51}$.
- If two events are dependent, then the probability of occurrence of one event affects that of the subsequent event.

WORKED EXAMPLE 13 Calculating the probability of outcomes of dependent events

A bag contains 5 blue, 6 green and 4 yellow marbles. The marbles are identical in all respects except their colours. Two marbles are picked in succession without replacement. Calculate the probability of picking 2 blue marbles.

THINK	WRITE
1. Determine the probability of picking the first blue marble.	$\Pr(\text{picking a blue marble}) = \frac{n(B)}{n(\xi)}$ $\Pr(\text{picking a blue marble}) = \frac{5}{15}$ $= \frac{1}{3}$
2. Determine the probability of picking the second blue marble. *Note:* The two events are dependent since marbles are not being replaced. Since we have picked a blue marble this leaves 4 blue marbles remaining out of a total of 14 marbles.	$\Pr(\text{picking second blue marble}) = \frac{n(B)}{n(\xi)}$ $\Pr(\text{picking second blue marble}) = \frac{4}{14}$ $= \frac{2}{7}$
3. Calculate the probability of obtaining 2 blue marbles.	$\Pr(2 \text{ blue marbles}) = \Pr(1\text{st blue}) \times \Pr(2\text{nd blue})$ $= \frac{1}{3} \times \frac{2}{7}$ $= \frac{2}{21}$
4. Write your answer.	The probability of obtaining 2 blue marbles is $\frac{2}{21}$.
Note: Alternatively, a tree diagram could be used to solve this question. The probability of selecting 2 blue marbles successively can be read directly from the first branch of the tree diagram.	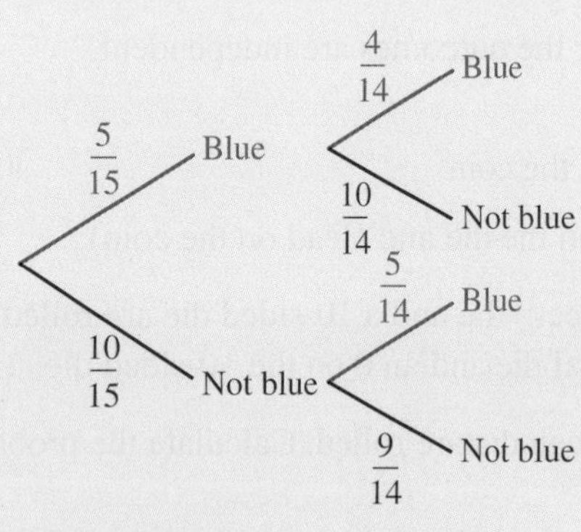 $\Pr(2 \text{ blue marbles}) = \frac{5}{15} \times \frac{4}{14}$ $= \frac{1}{3} \times \frac{2}{7}$ $= \frac{2}{21}$

on Resources

eWorkbook Toipc 11 Workbook (code puzzle and project) (ewbk-2037)

Interactivities Individual pathway interactivity: Independent and dependent events (int-4618)
Independent and dependent events (int-2787)
Multiplication Law of probability (int-6172)
Dependent events (int-6173)

Exercise 11.4 Independent and dependent events

learn**on**

Individual pathways

PRACTISE	CONSOLIDATE	MASTER
1, 4, 8, 11, 14, 16, 20	2, 5, 9, 12, 15, 17, 21	3, 6, 7, 10, 13, 18, 19, 22

To answer questions online and to receive **immediate corrective feedback** and **fully worked solutions** for all questions, go to your learnON title at www.jacplus.com.au.

Fluency

1. If A and B are independent events and $\Pr(A) = 0.7$ and $\Pr(B) = 0.4$, calculate:
   a. $\Pr(A \text{ and } B)$
   b. $\Pr(A' \text{ and } B)$ where A' is the complement of A
   c. $\Pr(A \text{ and } B')$ where B' is the complement of B
   d. $\Pr(A' \text{ and } B')$.
2. Determine whether two events A and B with $\Pr(A) = 0.6$, $\Pr(B') = 0.84$ and $\Pr(A \cap B) = 0.96$ are independent or dependent.
3. Determine whether two events A and B with $\Pr(A) = 0.25$, $\Pr(B) = 0.72$ and $\Pr(A \cup B) = 0.79$ are independent or dependent.

Understanding

4. **WE12** A die is rolled and a coin is tossed.
   a. Explain whether the outcomes are independent.
   b. Calculate:
      i. Pr(Head) on the coin
      ii. Pr(6) on the die.
   c. Calculate Pr(6 on the die and Head on the coin).
5. A tetrahedron (4-faced) die and a 10-sided die are rolled simultaneously. Calculate the probability of getting a 3 on the tetrahedral die and an 8 on the 10-sided die.
6. A blue die and a green die are rolled. Calculate the probability of getting a 5 on the blue die and not a 5 on the green die.
7. Dean is an archer. The experimental probability that Dean will hit the target is $\frac{4}{5}$.
   a. Calculate the probability that Dean will hit the target on two successive attempts.
   b. Calculate the probability that Dean will hit the target on three successive attempts.
   c. Calculate the probability that Dean will not hit the target on two successive attempts.
   d. Calculate the probability that Dean will hit the target on the first attempt but miss on the second attempt.

8. **MC** A bag contains 20 apples, of which 5 are bruised. Peter picks an apple and realises that it is bruised. He puts the apple back in the bag and picks another one.

a. The probability that Peter picks 2 bruised apples is:

A. $\frac{1}{4}$ B. $\frac{1}{2}$ C. $\frac{1}{16}$ D. $\frac{3}{4}$ E. $\frac{15}{16}$

b. The probability that Peter picks a bruised apple first but a good one on his second attempt is:

A. $\frac{1}{4}$ B. $\frac{1}{2}$ C. $\frac{3}{4}$ D. $\frac{3}{16}$ E. $\frac{1}{16}$

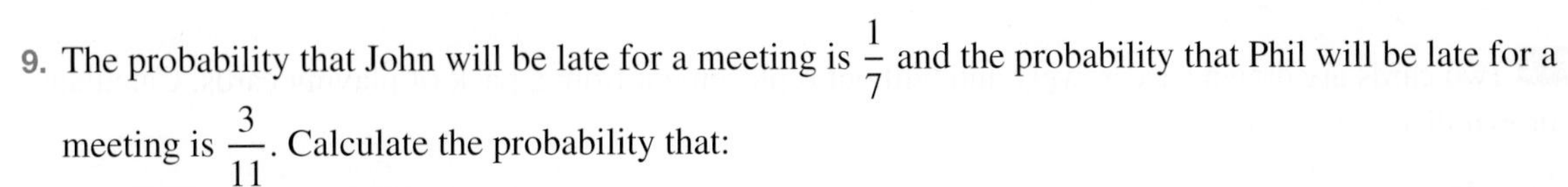

9. The probability that John will be late for a meeting is $\frac{1}{7}$ and the probability that Phil will be late for a meeting is $\frac{3}{11}$. Calculate the probability that:

a. John and Phil are both late
b. neither of them is late
c. John is late but Phil is not late
d. Phil is late but John is not late.

10. On the roulette wheel at the casino there are 37 numbers, 0 to 36 inclusive. Bidesi puts his chip on number 8 in game 20 and on number 13 in game 21.

a. Calculate the probability that he will win in game 20.
b. Calculate the probability that he will win in both games.
c. Calculate the probability that he wins at least one of the games.

11. Based on her progress through the year, Karen was given a probability of 0.8 of passing the Physics exam. If the probability of passing both Maths and Physics is 0.72, determine her probability of passing the Maths exam.

12. Suresh found that, on average, he is delayed 2 times out of 7 at Melbourne airport. Rakesh made similar observations at Brisbane airport, but found he was delayed 1 out of every 4 times. Determine the probability that both Suresh and Rakesh will be delayed if they are flying out of their respective airports.

13. Bronwyn has 3 pairs of Reebok and 2 pairs of Adidas running shoes. She has 2 pairs of Reebok, 3 pairs of Rio and a pair of Red Robin socks. Preparing for an early morning run, she grabs at random for a pair of socks and a pair of shoes. Calculate the probability that she chooses:

 a. Reebok shoes and Reebok socks
 b. Rio socks and Adidas shoes
 c. Reebok shoes and Red Robin socks
 d. Adidas shoes and socks that are not Red Robin.

14. **WE13** Two cards are drawn successively and without replacement from a pack of playing cards. Calculate the probability of drawing:

 a. 2 hearts
 b. 2 kings
 c. 2 red cards.

15. In a class of 30 students there are 17 students who study Music. Two students are picked randomly to represent the class in the Student Representative Council. Calculate the probability that:

 a. both students don't study Music
 b. both students do study Music
 c. one of the students doesn't study Music.

Reasoning

16. Greg has tossed a Tail on each of 9 successive coin tosses. He believes that his chances of tossing a Head on his next toss must be very high. Explain whether Greg is correct.

17. The Multiplication Law of probability relates to *independent events*. Tree diagrams can illustrate the sample space of successive *dependent events* and the probability of any one combination of events can be calculated by *multiplying* the stated probabilities along the branches. Explain whether this a contradiction to the Multiplication Law of probability.

18. Explain whether it is possible for two events, A and B, to be mutually exclusive and independent.

19. Consider the following sets:

$$\xi = \{1, 2, 3, 4, 5, 6, 7, 8, 9, 10, 11, 12, 13, 14, 15, 16, 17, 18, 19, 20\}$$

$$A = \{\text{evens in } \xi\}$$

$$B = \{\text{multiples of 3 in } \xi\}$$

 a. Explain whether the events A and B are independent.
 b. If ξ is changed to the integers from 1 to 10 only, explain whether the result from part **a** changes.

Problem solving

20. There are three coins in a box. One coin is a fair coin, one coin is biased with an 80% chance of landing Heads, and the third is a biased coin with a 40% chance of landing Heads. A coin is selected at random and flipped.
 If the result is a Head, determine the probability that the fair coin was selected.

21. A game at a carnival requires blindfolded contestants to throw balls at numbered ducks sitting on 3 shelves. The game ends when 3 ducks have been knocked off the shelves. Assume that the probability of hitting each duck is equal.
   a. Explain whether the events described in the game are dependent or independent.
   b. Determine the initial probabilities of hitting each number.
   c. Draw a labelled tree diagram to show the possible outcomes for a contestant.
   d. Calculate the probabilities of hitting the following:
      i. Pr(1, 1, 1) ii. Pr(2, 2, 2) iii. Pr(3, 3, 3) iv. Pr(at least one 3).

22. Question 21 described a game at a carnival. A contestant pays $3 to play and must make 3 direct hits to be eligible for a prize. The numbers on the ducks hit are then summed and the contestant wins a prize according to the winners' table.

Winners' table	
Total score	**Prize**
9	Major prize ($30 value)
7–8	Minor prize ($10 value)
5–6	$2 prize
3–4	No prize

   a. Determine the probability of winning each prize listed.
   b. Suppose 1000 games are played on an average show day. Evaluate the profit that could be expected to be made by the sideshow owner on any average show day.

11.5 Conditional probability

LEARNING INTENTION

At the end of this subtopic you should be able to:
- determine from the language in a question whether it involves conditional probability
- use the rule for conditional probability to calculate other probabilities.

11.5.1 Recognising conditional probability

eles-4928

- **Conditional probability** is when the probability of an event is conditional (depends) on another event occurring first.
- The effect of conditional probability is to reduce the sample space and thus increase the probability of the desired outcome.
- For two events, A and B, the conditional probability of event B, given that event A occurs, is denoted by $\Pr(B|A)$ and can be calculated using the following formula.

Probability of B given A

$$\Pr(B|A) = \frac{\Pr(A \cap B)}{\Pr(A)}, \ \Pr(A) \neq 0$$

- Conditional probability can be expressed using a variety of language. Some examples of conditional probability statements follow. The key words to look for in a conditional probability statement have been highlighted in each instance.
  - **If** a student receives a B+ or better in their first Maths test, **then** the chance of them receiving a B+ or better in their second Maths test is 75%.
  - **Given that** a red marble was picked out of the bag with the first pick, the probability of a blue marble being picked out with the second pick is 0.35.
  - **Knowing that** the favourite food of a student is hot chips, the probability of their favourite condiment being tomato sauce is 68%.

WORKED EXAMPLE 14 Using a Venn diagram to find conditional probabilities

A group of students was asked whether they like spaghetti (S) or lasagne (L). The results are illustrated in the Venn diagram shown. Use the Venn diagram to calculate the following probabilities relating to a student's preferred food.

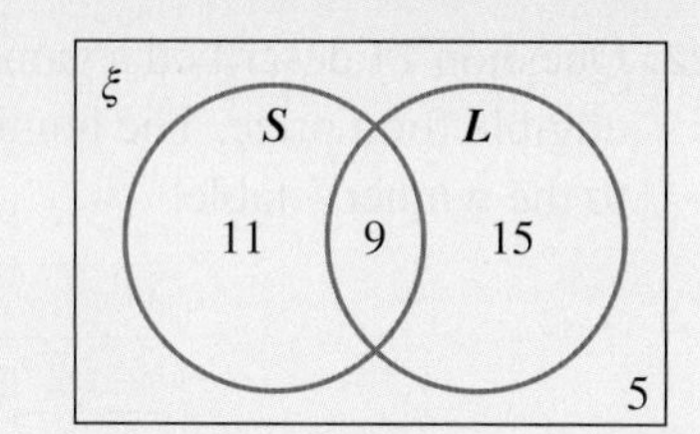

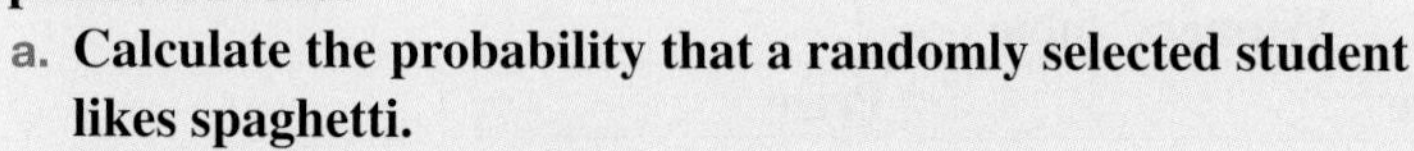

a. **Calculate the probability that a randomly selected student likes spaghetti.**

b. **Determine the probability that a randomly selected student likes lasagne given that they also like spaghetti.**

THINK	WRITE/DRAW
a. 1. Determine how many students were surveyed to identify the total number of possible outcomes. Add each of the numbers shown on the Venn diagram.	a. Total number of students = 11 + 9 + 15 + 5 = 40
2. There are 20 students that like 'spaghetti' or 'spaghetti and lasagne', as shown in pink.	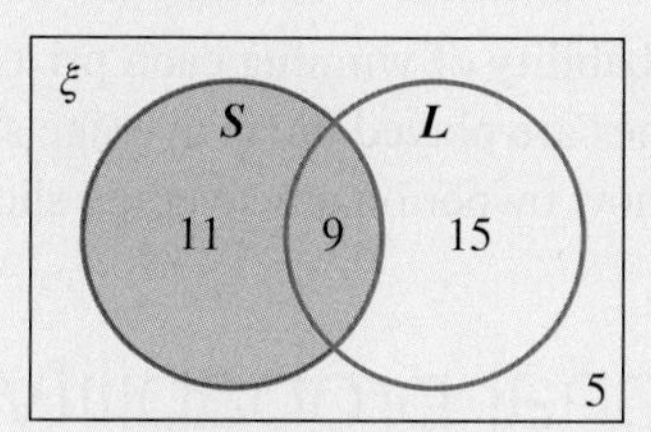
3. The probability that a randomly selected student likes spaghetti is found by substituting these values into the probability formula.	$\Pr(\text{event}) = \dfrac{\text{number of favourable outcomes}}{\text{total number of possible outcomes}}$ $\Pr(\text{spaghetti}) = \dfrac{20}{40}$ $= \dfrac{1}{2}$
b. 1. The condition imposed 'given they also like spaghetti' alters the sample space to the 20 students described in part **a**, as shaded in blue. Of these 20 students, 9 stated that they liked lasagne and spaghetti, as shown in pink.	b. 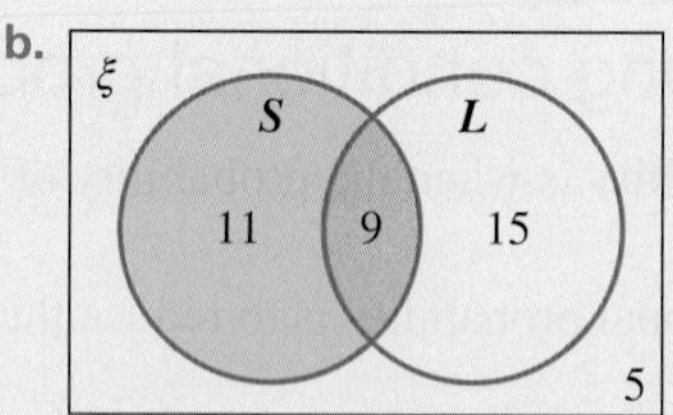
2. The probability that a randomly selected student likes lasagne, given that they like spaghetti, is found by substituting these values into the probability formula for conditional probability.	$\Pr(B\|A) = \dfrac{\Pr(A \cap B)}{\Pr(A)}$ $\Pr(L\|S) = \dfrac{\frac{9}{40}}{\frac{1}{2}}$ $= \dfrac{9}{20}$

WORKED EXAMPLE 15 Using the rule for conditional probability

If $\Pr(A) = 0.3, \Pr(B) = 0.5$ and $\Pr(A \cup B) = 0.6$, calculate:

a. $\Pr(A \cap B)$

b. $\Pr(B | A)$

THINK	WRITE
a. 1. State the Addition Law for probability to determine $\Pr(A \cup B)$.	a. $\Pr(A \cup B) = \Pr(A) + \Pr(B) - \Pr(A \cap B)$
2. Substitute the values given in the question into this formula and simplify.	$0.6 = 0.3 + 0.5 - \Pr(A \cap B)$ $\Pr(A \cap B) = 0.3 + 0.5 - 0.6$ $= 0.2$
b. 1. State the formula for conditional probability.	b. $\Pr(B \mid A) = \dfrac{\Pr(A \cap B)}{\Pr(A)}, \Pr(A) \neq 0$
2. Substitute the values given in the question into this formula and simplify.	$\Pr(B \mid A) = \dfrac{0.2}{0.3}$ $= \dfrac{2}{3}$

- It is possible to transpose the formula for conditional probability to calculate $\Pr(A \cap B)$:

$$\Pr(B|A) = \frac{\Pr(A \cap B)}{\Pr(A)}, \Pr(A) \neq 0$$

$$\Pr(A \cap B) = \Pr(A) \times \Pr(B|A)$$

on Resources

eWorkbook Topic 11 Workbook (worksheets, code puzzle and project) (ewbk-2037)

Video eLesson Conditional probability (eles-1928)

Interactivities Individual pathway interactivity: Conditional probability (int-4619)

Conditional probability (int-6085)

Exercise 11.5 Conditional probability

learn**on**

Individual pathways

■ PRACTISE	■ CONSOLIDATE	■ MASTER
1, 3, 5, 9, 12, 15	2, 6, 10, 13, 16	4, 7, 8, 11, 14, 17

To answer questions online and to receive **immediate corrective feedback** and **fully worked solutions** for all questions, go to your learnON title at www.jacplus.com.au.

Fluency

1. **WE14** A group of students was asked whether they liked the following forms of dance: hip hop (H) or jazz (J). The results are illustrated in the Venn diagram. Use the Venn diagram to calculate the following probabilities relating to a student's favourite form of dance.

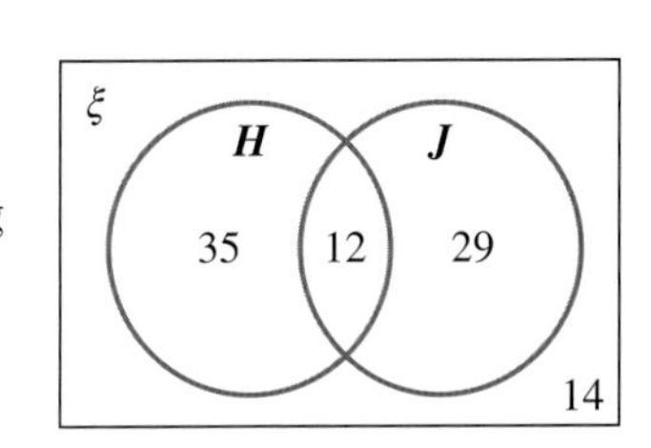

   a. Calculate the probability that a randomly selected student likes jazz.
   b. Determine the probability that a randomly selected student likes hip hop, given that they like jazz.

2. A group of students was asked which seats they liked: the seats in the computer lab or the science lab. The results are illustrated in the Venn diagram. Use the Venn diagram to calculate the following probabilities relating to the most comfortable seats.

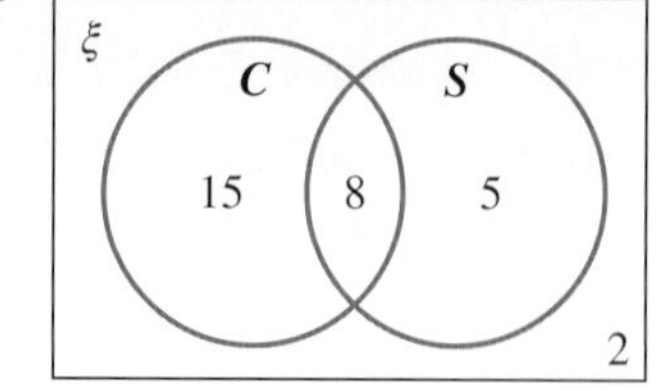

   a. Calculate the probability that a randomly selected student likes the seats in the science lab.
   b. Determine the probability that a randomly selected student likes the seats in the science lab, given that they like the seats in the computer lab or the science lab.

3. **WE15** If $\Pr(A) = 0.7$, $\Pr(B) = 0.5$ and $\Pr(A \cup B) = 0.9$, calculate:
   a. $\Pr(A \cap B)$
   b. $\Pr(B|A)$.

4. If $\Pr(A) = 0.65$, $\Pr(B) = 0.75$ and $\Pr(A \cap B) = 0.45$, calculate:
   a. $\Pr(B|A)$
   b. $\Pr(A|B)$.

Understanding

5. A medical degree requires applicants to participate in two tests: an aptitude test and an emotional maturity test. This year 52% passed the aptitude test and 30% passed both tests. Use the conditional probability formula to calculate the probability that a student who passed the aptitude test also passed the emotional maturity test.

6. At a school classified as a 'Music school for excellence', the probability that a student elects to study Music and Physics is 0.2. The probability that a student takes Music is 0.92. Determine the probability that a student takes Physics, given that the student is taking Music.

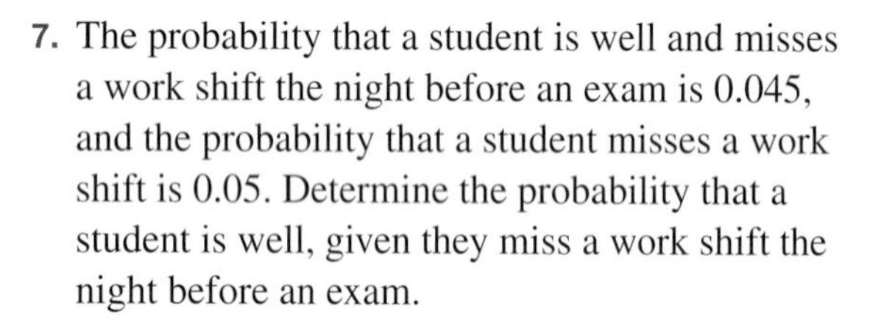

7. The probability that a student is well and misses a work shift the night before an exam is 0.045, and the probability that a student misses a work shift is 0.05. Determine the probability that a student is well, given they miss a work shift the night before an exam.

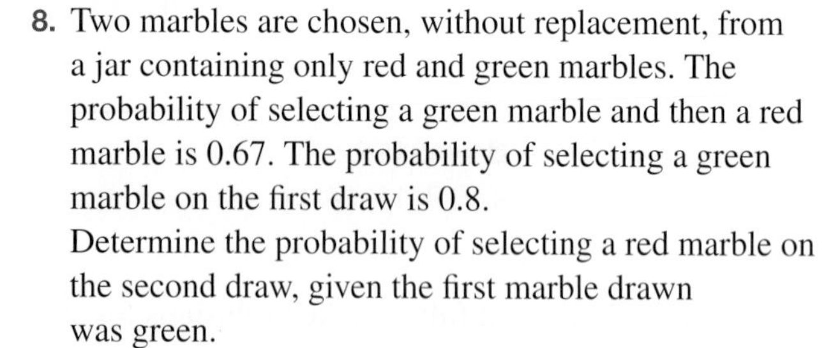

8. Two marbles are chosen, without replacement, from a jar containing only red and green marbles. The probability of selecting a green marble and then a red marble is 0.67. The probability of selecting a green marble on the first draw is 0.8.
Determine the probability of selecting a red marble on the second draw, given the first marble drawn was green.

9. Consider rolling a red and a black die and the probabilities of the following events:

Event A	the red die lands on 5
Event B	the black die lands on 2
Event C	the sum of the dice is 10

a. MC The initial probability of each event described is:

A. $\Pr(A) = \frac{1}{6}$, $\Pr(B) = \frac{1}{6}$, $\Pr(C) = \frac{1}{6}$

B. $\Pr(A) = \frac{5}{6}$, $\Pr(B) = \frac{2}{6}$, $\Pr(C) = \frac{7}{36}$

C. $\Pr(A) = \frac{5}{6}$, $\Pr(B) = \frac{2}{6}$, $\Pr(C) = \frac{5}{18}$

D. $\Pr(A) = \frac{1}{6}$, $\Pr(B) = \frac{1}{6}$, $\Pr(C) = \frac{1}{12}$

E. $\Pr(A) = \frac{1}{6}$, $\Pr(B) = \frac{2}{6}$, $\Pr(C) = \frac{1}{12}$

b. Calculate the following probabilities.

i. $\Pr(A|B)$ ii. $\Pr(B|A)$ iii. $\Pr(C|A)$ iv. $\Pr(C|B)$

10. MC A group of 80 schoolgirls consists of 54 dancers and 35 singers. Each member of the group is either a dancer, a singer, or both. The probability that a randomly selected student is a singer given that she is a dancer is:

A. 0.17 B. 0.44 C. 0.68 D. 0.11 E. 0.78

11. The following is the blood pressure data from 232 adult patients admitted to a hospital over a week. The results are displayed in a two-way frequency table.

Age	Blood pressure			Total
	Low	Medium	High	
Under 60 years	92	44	10	146
60 years or above	17	46	23	86
Total	109	90	33	**232**

a. Calculate the probability that a randomly chosen patient has low blood pressure.
b. Determine the probability that a randomly chosen patient is under 60 years of age.
c. Calculate the probability that a randomly chosen patient has high blood pressure, given they are aged 60 years or above.
d. Determine the probability that a randomly chosen patient is under the age of 60, given they have medium blood pressure.

Reasoning

12. Explain how imposing a condition alters probability calculations.

13. At a school, 65% of the students are male and 35% are female. Of the male students, 10% report that dancing is their favourite activity; of the female students, 25% report that dancing is their favourite activity.
Determine the probability that:
a. a student selected at random prefers dancing and is female
b. a student selected at random prefers dancing and is male.
c. Construct a tree diagram to present the information given, and use it to calculate:
   i. the probability that a student is male and does not prefer dancing
   ii. the overall percentage of students who prefer dancing.

14. Consider the following sets of numbers.

$$\xi = \{1, 2, 3, 4, 5, 6, 7, 8, 9, 10, 11, 12, 13, 14, 15, 16, 17, 18, 19, 20\}$$

$$A = \{1, 2, 3, 4, 5, 6, 7, 8, 9, 10\}$$

$$B = \{2, 3, 5, 7, 11, 13, 17, 19\}$$

a. Calculate the following:

i. $\Pr(A)$
ii. $\Pr(B)$
iii. $\Pr(A \cap B)$
iv. $\Pr(A|B)$
v. $\Pr(B|A)$

b. Explain whether the events A and B are independent.
c. Write a statement that connects $\Pr(A)$, $\Pr(A \cap B)$, $\Pr(A|B)$ and independent events.

Problem solving

15. The rapid test used to determine whether a person is infected with COVID-19 is not perfect. For one type of rapid test, the probability of a person with the disease returning a positive result is 0.98, while the probability of a person without the disease returning a positive result is 0.04.
At its peak in a certain country, the probability that a randomly selected person has COVID-19 is 0.05.
Determine the probability that a randomly selected person will return a positive result.

16. Two marbles are chosen, without replacement, from a jar containing only red and green marbles. The probability of selecting a green marble and then a red marble is 0.72. The probability of selecting a green marble on the first draw is 0.85.
Determine the probability of selecting a red marble on the second draw if the first marble drawn was green.

17. When walking home from school during the summer months, Harold buys either an ice-cream or a drink from the corner shop. If Harold bought an ice-cream the previous day, there is a 30% chance that he will buy a drink the next day.
If he bought a drink the previous day, there is a 40% chance that he will buy an ice-cream the next day. On Monday, Harold bought an ice-cream.
Determine the probability that he buys an ice-cream on Wednesday.

11.6 Review

11.6.1 Topic summary

PROBABILITY

Review of probability

- The sample space is the set of all possible outcomes. It is denoted ξ. For example, the sample space of rolling a die is: $\xi = \{1, 2, 3, 4, 5, 6\}$.
- An event is a set of favourable outcomes. For example, if A is the event {rolling an even number on a die}, $A = \{2, 4, 6\}$.
- The probability of an outcome or event is always between 0 and 1.
- The sum of all probabilities in a sample space is 1.
- An event that is **certain** has a probability of 1.
- An event that is **impossible** has a probability of 0.
- When all outcomes are equally likely the theoretical probability of an event is given by the following formula:

$$\Pr(\text{event}) = \frac{\text{number of favourable outcomes}}{\text{total number of outcomes}} = \frac{n(E)}{n(\xi)}$$

- The experimental probability is given by the following formula:

$$\text{Experimental Pr} = \frac{n(\text{Successful trials})}{n(\text{trials})}$$

- The complement of an event A is denoted A' and is the set of outcomes that are not in A.
  - $n(A) + n(A') = n(\xi)$
  - $\Pr(A) + \Pr(A') = 1$

Intersection and union

- The intersection of two events A and B is written $A \cap B$. These are the outcomes that are in A '**and**' in B.

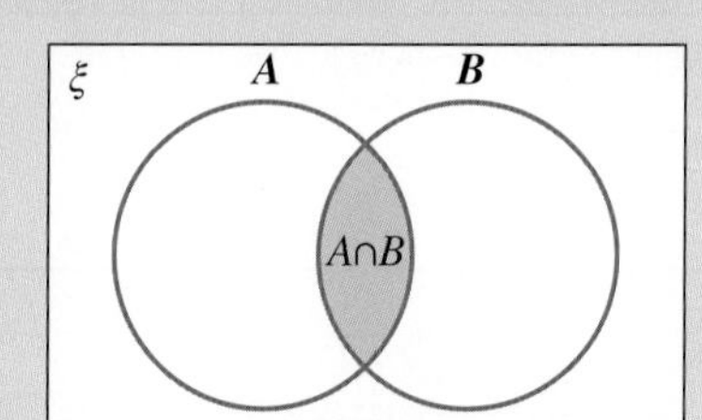

- The union of two events A and B is written $A \cup B$. These are the outcomes that are in A '**or**' in B.

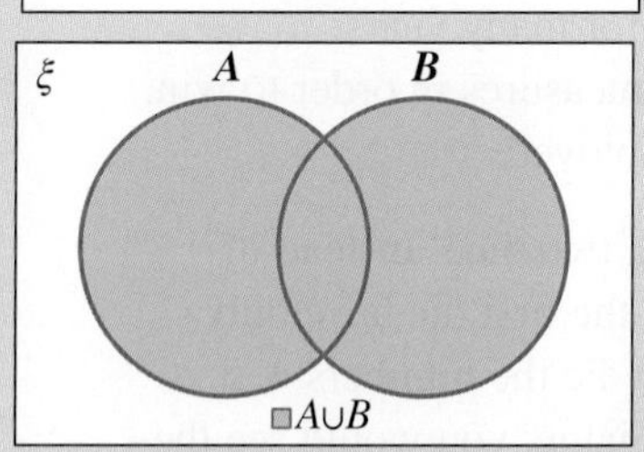

- A Venn diagram can be split into four distinct sections as shown at right.

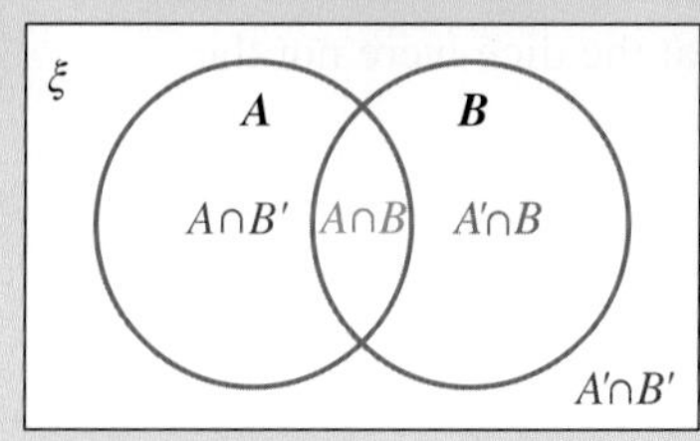

- The Addition Law of probability states that:

$$\Pr(A \cup B) = \Pr(A) + \Pr(B) - \Pr(A \cap B)$$

Independent, dependent events and conditional probability

- Two events are **independent** if the outcome of one event does not affect the outcome of the other event.
- If A and B are independent then:

$$\Pr(A \cap B) = \Pr(A) \times \Pr(B)$$

- Two events are **dependent** if the outcome of one event affects the outcome of the other event.
- Conditional probability applies to dependent events.
- The probability of A given that B has already occurred is given by:

$$\Pr(A \mid B) = \frac{\Pr(A \cap B)}{\Pr(B)}$$

- This can be rearranged into:

$$\Pr(A \cap B) = \Pr(A \mid B) \times \Pr(B)$$

Mutually exclusive events

- Mutually exclusive events have no elements in common.
- Venn diagrams show mutually exclusive events having no intersection. In the Venn diagram below A and B are mutually exclusive.

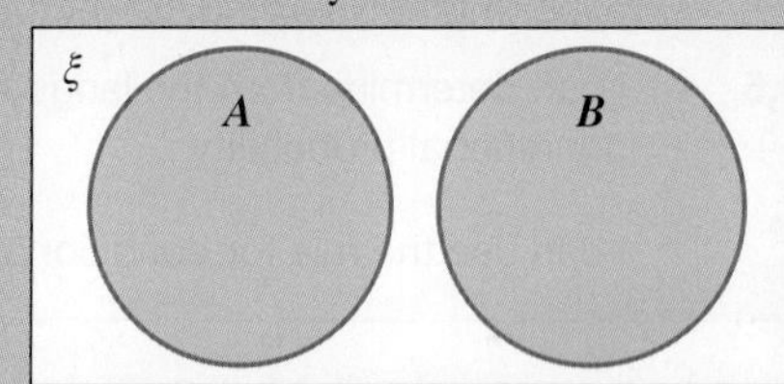

- For mutually exclusive events A and B:
  - $\Pr(A \cap B) = 0$
  - $\Pr(A \cup B) = \Pr(A) + \Pr(B)$

Multiple-step chance experiments

- Some chance experiments involve multiple trials performed separately.
- Two-way tables can be used to show the sample space of two-step experiments.
- Tree diagrams can be used to show the sample space and probabilities of multi-step experiments.

		Coin 1	
		H	T
Coin 2	H	HH	TH
	T	HT	TT

First selection	Second selection	Sample space
B	R	BR
	G	BG
	Y	BY
W	R	WR
	G	WG
	Y	WY

11.6.2 Success criteria

Tick the column to indicate that you have completed the subtopic and how well you have understood it using the traffic light system.

(**Green:** I understand; **Yellow:** I can do it with help; **Red:** I do not understand)

Subtopic	Success criteria	Green	Yellow	Red
11.2	I can use key probability terminology such as: trials, frequency, sample space, likely and unlikely events.			
	I can use two-way tables to represent sample spaces.			
	I can calculate experimental and theoretical probabilities.			
	I can determine the probability of complimentary events and mutually exclusive events.			
	I can use the addition rule to calculate the probability of event '*A* or *B*'.			
11.3	I can create a tree diagrams to represent two- and three-step chance experiments.			
	I can use a tree diagram to solve probability problems involving two or more trials or events.			
11.4	I can determine whether two events are dependent or independent.			
	I can calculate the probability of outcomes from two-step chance experiments involving independent or dependent events.			
11.5	I can determine from the language in a question whether it involves conditional probability.			
	I can use the rule for conditional probability to calculate other probabilities.			

11.6.3 Project

Tricky dice

Dice games have been played throughout the world for many years. Professional gamblers resort to all types of devious measures in order to win. Often the other players are unaware of the tricks employed.

Imagine you are playing a game that involves rolling two dice. Instead of having each die marked with the numbers 1 to 6, let the first die have only the numbers 1, 2 and 3 (two of each) and the second die the numbers 4, 5 and 6 (two of each). If you were an observer to this game, you would see the numbers 1 to 6 occurring and probably not realise that the dice were not the regular type.

1. Complete the grid below to show the sample space on rolling these two dice.
2. Identify how many different outcomes there are. Compare this with the number of different outcomes using two regular dice.

Die 1

Die 2	1	2	3	1	2	3
4						
5						
6						
4						
5						
6						

3. Calculate the chance of rolling a double using these dice.
4. The numbers on the two dice are added after rolling. Complete the table below to show the totals possible.

Die 1

Die 2	1	2	3	1	2	3
4						
5						
6						
4						
5						
6						

5. Identify how many different totals are possible and list them.
6. State which total you have the greatest chance of rolling. State which total you have the least chance of rolling.
7. If you played a game in which you had to bet on rolling a total of less than 7, equal to 7 or greater than 7, explain which option would you be best to take.
8. If you had to bet on an even-number outcome or an odd-number outcome, explain which would be the better option.
9. The rules are changed to subtracting the numbers on the two dice instead of adding them. Complete the following table to show the outcomes possible.

Die 1

Die 2	1	2	3	1	2	3
4						
5						
6						
4						
5						
6						

10. Identify how many different outcomes are possible in this case and list them.
11. State the most frequently occurring outcome and how many times it occurs.
12. Devise a game of your own using these dice. On a separate sheet of paper, write out rules for your game and provide a solution, indicating the best options for winning.

on Resources

eWorkbook Topic 11 Workbook (worksheets, code puzzle and project) (ewbk-2037)

Interactivities Crossword (int-2857)

Sudoku puzzle (int-3598)

Exercise 11.6 Review questions

learn**on**

To answer questions online and to receive **immediate corrective feedback** and **fully worked solutions** for all questions, go to your learnON title at www.jacplus.com.au.

Fluency

1. **MC** Choose which of the following is always true for an event, M, and its complementary event, M'.
   A. $\Pr(M) + \Pr(M') = 1$
   B. $\Pr(M) - \Pr(M') = 1$
   C. $\Pr(M) + \Pr(M') = 0$
   D. $\Pr(M) - \Pr(M') = 0$
   E. $\Pr(M) \times \Pr(M') = 1$

2. **MC** A number is chosen from the set $\{0, 1, 2, 3, 4, 5, 6, 7, 8, 9, 10\}$. Select which of the following pairs of events is mutually exclusive.
   A. $\{2, 4, 6\}$ and $\{4, 6, 7, 8\}$
   B. $\{1, 2, 3, 5\}$ and $\{4, 6, 7, 8\}$
   C. $\{0, 1, 2, 3\}$ and $\{3, 4, 5, 6\}$
   D. {multiples of 2} and {factors of 8}
   E. {even numbers} and {multiples of 3}

3. **MC** Choose which of the following states the Multiplication Law of probability correctly.
   A. $\Pr(A \cap B) = \Pr(A) + \Pr(B)$
   B. $\Pr(A \cap B) = \Pr(A) \times \Pr(B)$
   C. $\Pr(A \cup B) = \Pr(A) \times \Pr(B)$
   D. $\Pr(A \cup B) = \Pr(A) + \Pr(B)$
   E. $\Pr(A) = \Pr(A \cup B) + \Pr(B)$

4. **MC** Given $\xi = \{1, 2, 3, 4, 5, 6, 7, 8, 9, 10\}$ and $A = \{2, 3, 4\}$ and $B = \{3, 4, 5, 8\}$, $A \cap B$ is:
   A. $\{2, 3, 3, 4, 4, 5, 8\}$
   B. $\{3, 4\}$
   C. $\{2, 3, 4\}$
   D. $\{2, 3, 4, 5, 8\}$
   E. $\{2, 5, 8\}$

5. **MC** Given $\xi = \{1, 2, 3, 4, 5, 6, 7, 8, 9, 10\}$ and $A = \{2, 3, 4\}$ and $B = \{3, 4, 5, 8\}$, $A \cap B'$ is:
   A. $\{3, 4\}$
   B. $\{2\}$
   C. $\{2, 3, 4, 5, 8\}$
   D. $\{2, 3, 4\}$
   E. $\{1, 2, 6, 7, 9, 10\}$

6. Shade the region stated for each of the following Venn diagrams.
   a. $A' \cup B$
   b. $A' \cap B'$
   c. $A' \cap B' \cap C$

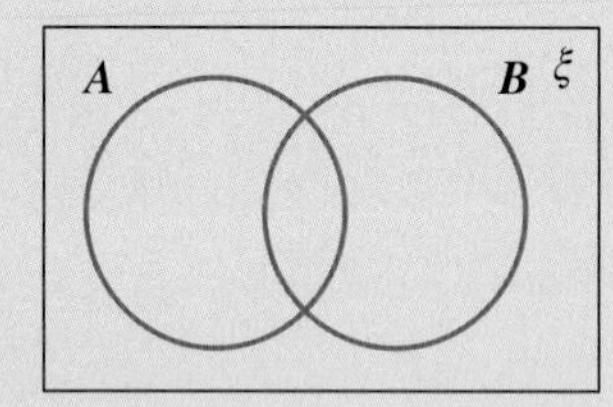

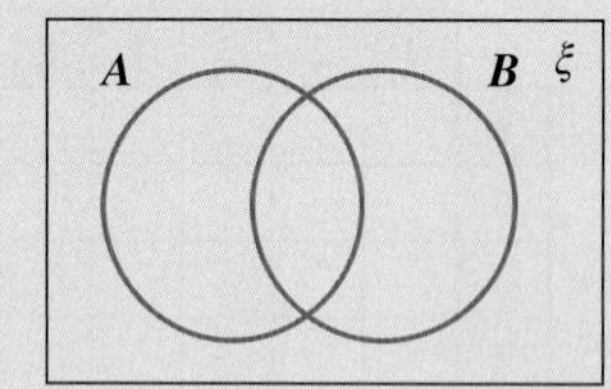

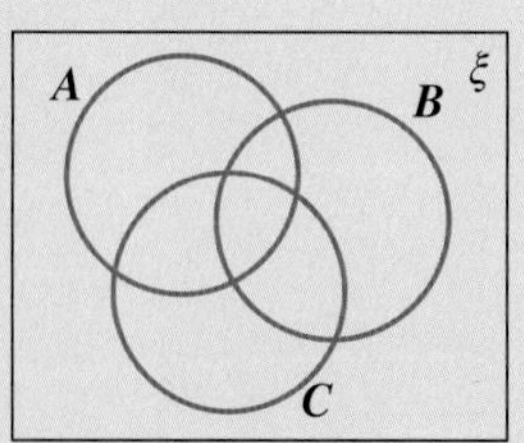

Problem solving

7. **MC** From past experience, it is concluded that there is a 99% probability that July will be a wet month in Launceston (it has an average rainfall of approximately 80 mm). The probability that July will not be a wet month next year in Launceston is:
   A. 99%
   B. 0.99
   C. $\frac{1}{100}$
   D. 1
   E. 0

8. **MC** A card is drawn from a well-shuffled deck of 52 cards. Select the theoretical probability of not selecting a red card.

A. $\frac{3}{4}$ B. $\frac{1}{4}$ C. $\frac{1}{13}$ D. $\frac{1}{2}$ E. 0

9. **MC** Choose which of the following events is not equally likely.

A. Obtaining a 5 or obtaining a 1 when a die is rolled
B. Obtaining a club or obtaining a diamond when a card is drawn from a pack of cards
C. Obtaining 2 Heads or obtaining 2 Tails when a coin is tossed twice
D. Obtaining 2 Heads or obtaining 1 Head when a coin is tossed twice
E. Obtaining a 3 or obtaining a 6 when a die is rolled

10. **MC** The Australian cricket team has won 12 of the last 15 Test matches. Select the experimental probability of Australia not winning its next Test match.

A. $\frac{4}{5}$ B. $\frac{1}{5}$ C. $\frac{1}{4}$ D. $\frac{3}{4}$ E. 1

11. A card is drawn from a well-shuffled pack of 52 cards. Calculate the theoretical probability of drawing:

a. an ace b. a spade c. a queen or a king d. not a heart.

12. A die is rolled five times.

a. Calculate the probability of rolling five 6s. b. Calculate the probability of not rolling five 6s.

13. Alan and Mary own 3 of the 8 dogs in a race. Evaluate the probability that:

a. one of Alan's or Mary's dogs will win b. none of Alan's or Mary's dogs will win.

14. A die is rolled. Event A is obtaining an even number. Event B is obtaining a 3.

a. Explain if events A and B are mutually exclusive.
b. Calculate $\Pr(A)$ and $\Pr(B)$.
c. Calculate $\Pr(A \cup B)$.

15. A card is drawn from a shuffled pack of 52 playing cards. Event A is drawing a club and event B is drawing an ace.

a. Explain if events A and B are mutually exclusive.
b. Calculate $\Pr(A)$, $\Pr(B)$ and $\Pr(A \cap B)$.
c. Calculate $\Pr(A \cup B)$.

16. A tetrahedral die is numbered 0, 1, 2 and 3. Two of these dice are rolled and the sum of the numbers (the number on the face that the die sits on) is taken.

a. Show the possible outcomes in a two-way table.
b. Determine if all the outcomes are equally likely.
c. Determine which total has the least chance of being rolled.
d. Determine which total has the best chance of being rolled.
e. Determine which sums have the same chance of being rolled.

17. A bag contains 20 pears, of which 5 are bad. Cathy picks 2 pears (without replacement) from the bag. Evaluate the probability that:

a. both pears are bad b. both pears are good c. one of the two pears is good.

18. Determine the probability of drawing 2 aces from a pack of cards if:

a. the first card is replaced before the second one is drawn
b. the first card drawn is not replaced.

19. On grandparents day at a school, a group of grandparents was asked where they most like to take their grandchildren — the beach (B) or shopping (S). The results are illustrated in the Venn diagram. Use the Venn diagram to calculate the following probabilities relating to the place grandparents most like to take their grandchildren.

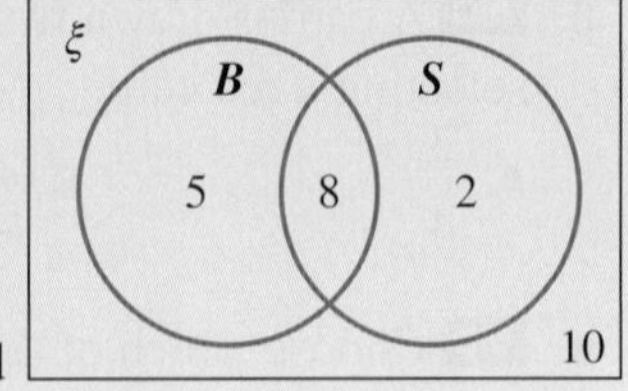

a. Determine the probability that a randomly selected grandparent preferred to take their grandchildren to the beach or shopping.

b. Determine the probability that a randomly selected grandparent liked to take their grandchildren to the beach, given that they liked to take their grandchildren shopping.

20. When all of Saphron's team players turn up for their twice weekly netball training the chance that they then win their Saturday game is 0.65. If not all players are at the training session, then the chance of winning their Saturday game is 0.40. Over a four-week period, Saphron's players all turn up for training three times.

a. Using a tree diagram, with T to represent all players training and W to represent a win, represent the winning chance of Saphron's netball team.

b. Using the tree diagram constructed in part **a**, determine the probability of Saphron's team winning their Saturday game. Write your answer correct to 4 decimal places.

c. Determine the exact probability that Saphron's team did not train given that they won their Saturday game.

21. Andrew does not know the answer to two questions on a multiple-choice exam. The first question has four choices and the second question he does not know has five choices.

a. Determine the probability that he will get both questions wrong.

b. If he is certain that one of the choices cannot be the answer in the first question, determine how this will change the probability that he will get both questions wrong.

22. Mariah the Mathematics teacher wanted to give her students a chance to win a reward at the end of the term. She placed 20 cards into a box, and wrote the word ON on 16 cards, and OFF on 4 cards. After a student chooses a card, that card is replaced into the box for the next student to draw. If a student chooses an OFF card, then they do not have to attend school on a specified day. If they choose an ON card, then they do not receive a day off.

a. Mick, a student, chose a random card from the box. Calculate the probability he received a day off.

b. Juanita, a student, chose a random card from the box after Mick. Calculate the probability that she did not receive a day off.

c. Determine the probability that Mick and Juanita both received a day off.

23. In the game of draw poker, a player is dealt 5 cards from a deck of 52. To obtain a flush, all 5 cards must be of the same suit.

a. Determine the probability of getting a diamond flush.

b. Determine the probability of getting any flush.

24. a. A Year 10 boy is talking with a Year 10 girl and asks her if she has any brothers or sisters. She says, 'Yes, I have one'. Determine the probability that she has at least one sister.

b. A Year 10 boy is talking with a Year 10 girl and asks her if she has any brothers or sisters. She says, 'Yes, I have three'. Determine the probability that she has at least one sister.

on To test your understanding and knowledge of this topic, go to your learnON title at www.jacplus.com.au and complete the **post-test**.

Online Resources

Below is a full list of **rich resources** available online for this topic. These resources are designed to bring ideas to life, to promote deep and lasting learning and to support the different learning needs of each individual.

eWorkbook

Download the workbook for this topic, which includes worksheets, a code puzzle and a project (ewbk-2037) ☐

Solutions

Download a copy of the fully worked solutions to every question in this topic (sol-0745) ☐

Digital documents

11.2 SkillSHEET Set notation (doc-5286) ☐
SkillSHEET Simplifying fractions (doc-5287) ☐
SkillSHEET Determining complementary events (doc-5288) ☐
SkillSHEET Addition and subtraction of fractions (doc-5289) ☐
SkillSHEET Working with Venn diagrams (doc-5291) ☐
SkillSHEET Distinguishing between complementary and mutually exclusive events (doc-5294) ☐
11.3 SkillSHEET Multiplying fractions for calculating probabilities (doc-5290) ☐

Video eLessons

11.2 The language of probability (eles-4922) ☐
Properties of probability events (eles-4923) ☐
Venn diagrams (eles-1934) ☐
11.3 Two-step chance experiments (eles-4924) ☐
Three-step chance experiments (eles-4925) ☐
Tree diagrams (eles-1894) ☐
11.4 Independent events (eles-4926) ☐
Dependent events (eles-4927) ☐
11.5 Recognising conditional probability (eles-4928) ☐
Conditional probability (eles-1928) ☐

Interactivities

11.2 Individual pathway interactivity: Review of probability (int-4616) ☐
Experimental probability (int-3825) ☐
Two-way tables (int-6082) ☐
Theoretical probability (int-6081) ☐
Venn diagrams (int-3828) ☐
Addition Law of probability (int-6168) ☐
11.3 Individual pathway interactivity: Tree diagrams (int-4617) ☐
Tree diagrams (int-6171) ☐
11.4 Individual pathway interactivity: Independent and dependent events (int-4618) ☐
Independent and dependent events (int-2787) ☐
Multiplication Law of probability (int-6172) ☐
Dependent events (int-6173) ☐
11.5 Individual pathway interactivity: Conditional probability (int-4619) ☐
Conditional probability (int-6085) ☐
11.6 Crossword (int-2857) ☐
Sudoku puzzle (int-3598) ☐

Teacher resources

There are many resources available exclusively for teachers online.

To access these online resources, log on to **www.jacplus.com.au**.

Answers

Topic 11 Probability

Exercise 11.1 Pre-test

1. $\Pr(A \cap B) = 0.2$
2. $\Pr(A) = 0.51$
3. Independent
4. $\frac{1}{5}$
5. C
6. E
7. A
8. 0.56
9. A
10. False
11. $\Pr(B|A) = 0.2$
12. A
13. A
14. 0.03125
15. A

Exercise 11.2 Review of probability

1. Experimental probability is based on the outcomes of experiments, simulations or surveys. Theoretical probability is based on the number of possible favourable outcomes and the total possible outcomes.
2. a. {I, II, III, IV, V, VI}
   b. Relative frequency for I = 0.2
   Relative frequency for II = 0.12
   Relative frequency for III = 0.16
   Relative frequency for IV = 0.14
   Relative frequency for V = 0.24
   Relative frequency for VI = 0.14
   c. The spinner should be spun a larger number of times.
3. a. 500 students
   b. Frequency for silver = 0.16
   c. Frequency for black and green = 0.316
   d. Blazing Blue
4. a.

+	1	2	3	4	5	6
1	2	3	4	5	6	7
2	3	4	5	6	7	8
3	4	5	6	7	8	9
4	5	6	7	8	9	10
5	6	7	8	9	10	11
6	7	8	9	10	11	12

   b. The sum of 7
   c. The sum of 2 or 12

5. a. $\Pr(A \cap B) = 0.1$
   b.

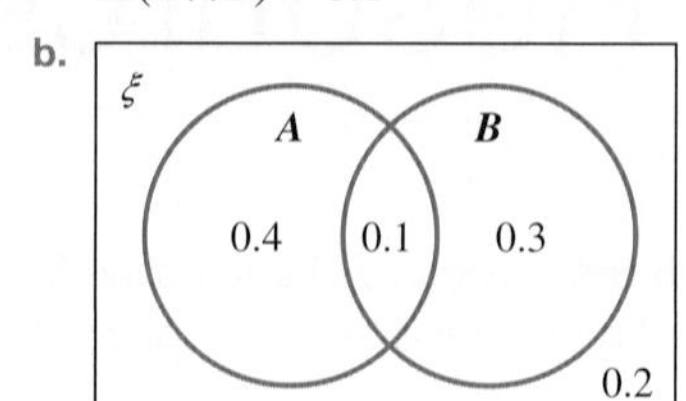

   c. $\Pr(A \cap B') = 0.4$
6. a. i. $\Pr(A \cup B) = 0.85$ ii. $\Pr(A \cap B)' = 0.95$
   b. C
7. $\frac{2}{3}$
8. a. $\frac{1}{13}$ b. $\frac{1}{4}$ c. $\frac{4}{13}$
9. Sample responses are given for part ii.
   a. i. No. There are many others foods one could have.
   ii. Having Weet Bix and not having Weet Bix
   b. i. No. There are other means of transport, for example, catching a bus.
   ii. Walking to a friend's place and not walking to a friend's place
   c. i. No. There are other possible leisure activities.
   ii. Watching TV and not watching TV
   d. i. No. The number 5 can be rolled too.
   ii. Rolling a number less than 5 and rolling a number 5 or greater
   e. i. Yes. There are only two possible outcomes: passing or failing.
   ii. No change is required to make these events complementary.
10. a.

A B ξ
3 5 11 7 13 15 17 19 | 1 9 | 4 16 | 20 14 12 2 6 8 10 18

   b. i. $\frac{10}{20} = \frac{1}{2}$ ii. $\frac{4}{20} = \frac{1}{5}$ iii. $\frac{2}{20} = \frac{1}{10}$
   iv. $\frac{12}{20} = \frac{3}{5}$ v. $\frac{8}{20} = \frac{2}{5}$
11. a. No. For a 6-sided die, $\Pr(4) = \frac{1}{6}$; for an 8-sided die, $\Pr(4) = \frac{1}{8}$.
   b. Yes; $\Pr(\text{odd}) = \frac{1}{2}$ for both 6-sided and 8-sided dice
12. Yes; $\Pr(5) = \frac{1}{2}, \Pr(6) = \frac{1}{2}$.
13. a. $\frac{1}{13}$ b. $\frac{1}{4}$ c. $\frac{1}{2}$ d. $\frac{12}{13}$
   e. 0 f. $\frac{1}{2}$

14. a. $\frac{1}{5}$ **b.** $\frac{7}{20}$ **c.** $\frac{11}{20}$ **d.** 0

15. a.

$\xi = 30$

Volleyball Soccer

7 1 4

5

2 4

7

Tennis

b. i. $\frac{1}{2}$ **ii.** $\frac{1}{6}$ **iii.** $\frac{1}{30}$ **iv.** $\frac{2}{5}$ **v.** $\frac{7}{15}$

c. i. $\frac{1}{2}$ **ii.** $\frac{8}{15}$

16. Yes. Both have a probability of $\frac{1}{2}$.

17. Sample responses can be found in the worked solutions in the online resources.

18. 17 red and 1 purple, i.e. 18 socks more

19. a.

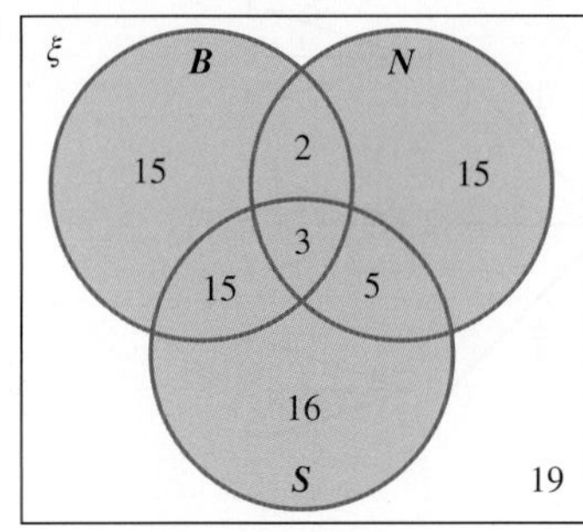

b. 19 students

c. 32 students

d. 15 students

e. Frequency $= \frac{15}{90}$ or $\frac{1}{6}$

f. Probability $= \frac{3}{90} = \frac{1}{30}$

20. a. i. 50 **ii.** 7 **iii.** 25 **iv.** 8

b. i. $\frac{1}{2}$ **ii.** $\frac{3}{50}$ **iii.** $\frac{6}{25}$

c. i.

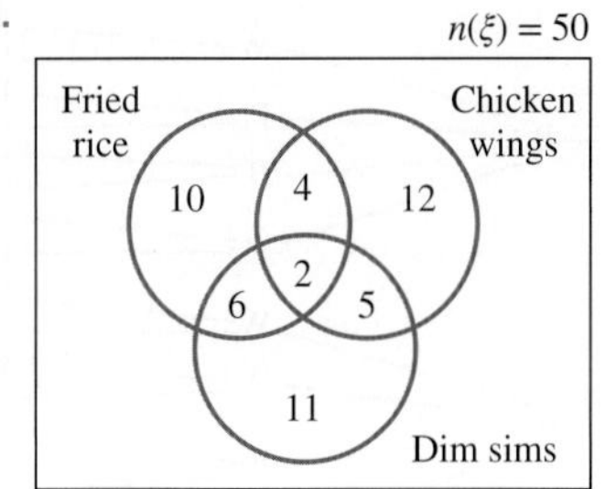

ii. $\frac{1}{25}$

21. a.

Die 2 outcomes (columns); Die 1 outcomes (rows)

	1	2	3	4	5	6
1	2	3	4	5	6	7
2	3	4	5	6	7	8
3	4	5	6	7	8	9
4	5	6	7	8	9	10
5	6	7	8	9	10	11
6	7	8	9	10	11	12

b. 6

c. No. The frequency of the numbers is different.

d.

Sum	2	3	4	5	6	7	8	9	10	11	12
Frequency	1	2	3	4	5	6	5	4	3	2	1

e. i. $\frac{1}{36}$ **ii.** $\frac{1}{6}$ **iii.** $\frac{1}{18}$

f. i. $\frac{1}{36}$ **ii.** $\frac{1}{6}$ **iii.** $\frac{1}{6}$

g. 50

Exercise 11.3 Tree diagrams

1. If the probabilities of two events are different, the first column of branches indicates the probabilities for the first event and the second column of branches indicates the probabilities for the second event. The product of each branch gives the probability. All probabilities add to 1.

2. a. 12 different outcomes

b. No. Each branch is a product of different probabilities.

c. Less likely

d. Indigo elephant

e. i. Pr(Blue elephant) $= \frac{7}{40}$

ii. Pr(Indigo elephant) $= \frac{7}{20}$

iii. Pr(Donkey) $= \frac{1}{5}$

3. a.

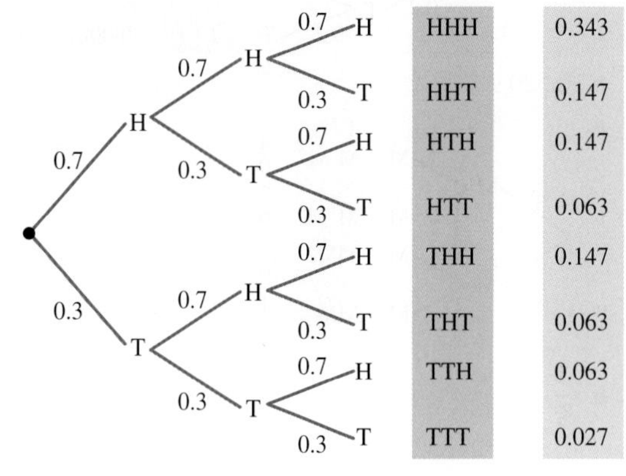

b. Pr(HHH) = 0.343

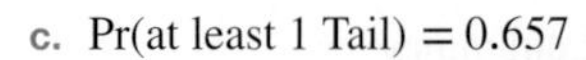

c. Pr(at least 1 Tail) = 0.657

d. Pr(exactly 2 Tails) = 0.189

4. a. i. $\frac{1}{2}$ ii. $\frac{1}{6}$ iii. $\frac{5}{6}$

b.

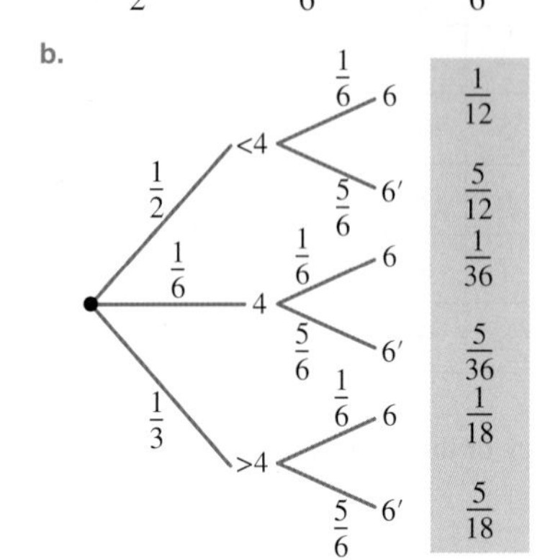

c. i. $\frac{1}{36}$ ii. $\frac{1}{12}$ iii. $\frac{5}{36}$ iv. $\frac{5}{18}$

5. a.

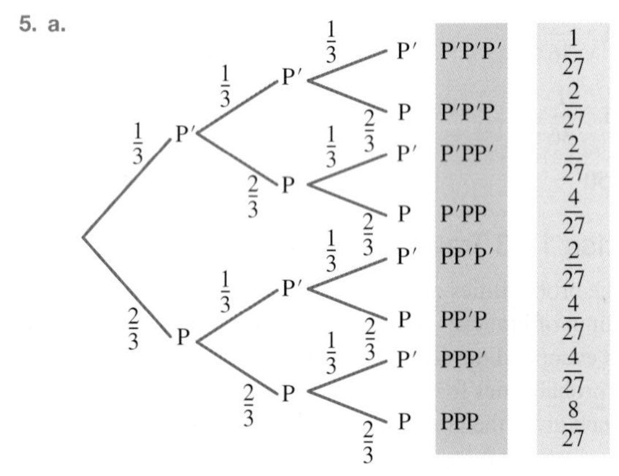

b. i. $\frac{8}{27}$ ii. $\frac{4}{27}$ iii. $\frac{12}{27}$

6. a.

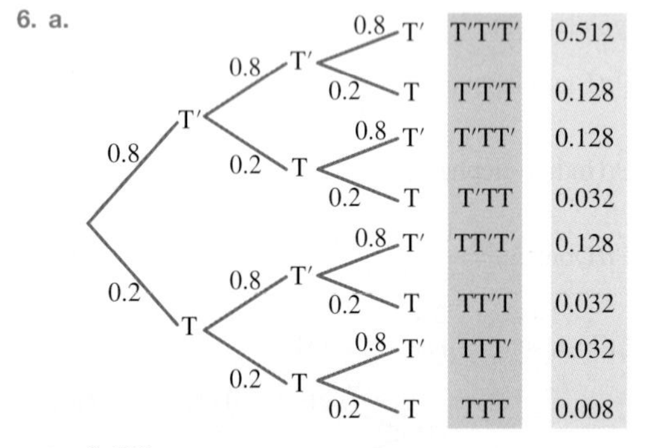

b. 0.520

7. a.

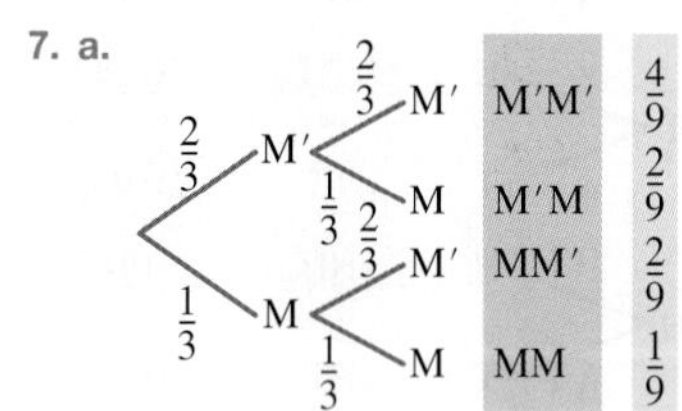

b. $\frac{1}{9}$

8. a.

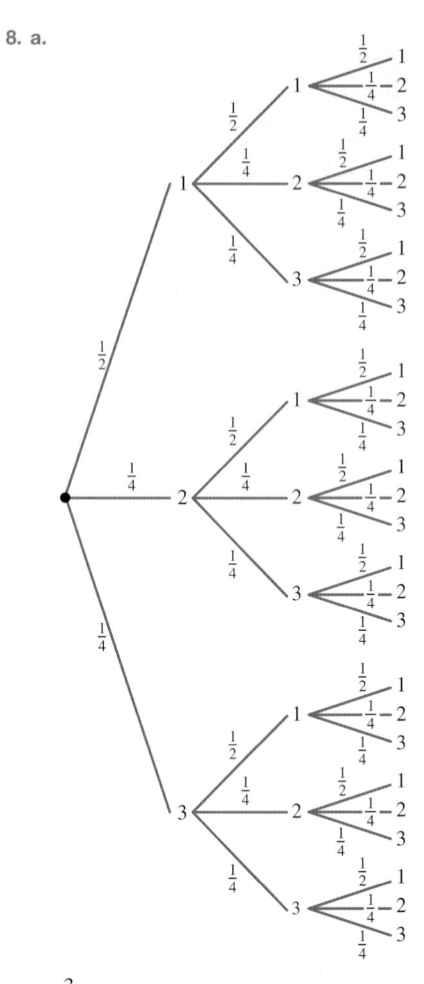

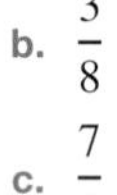

b. $\frac{3}{8}$

c. $\frac{7}{8}$

9. a.

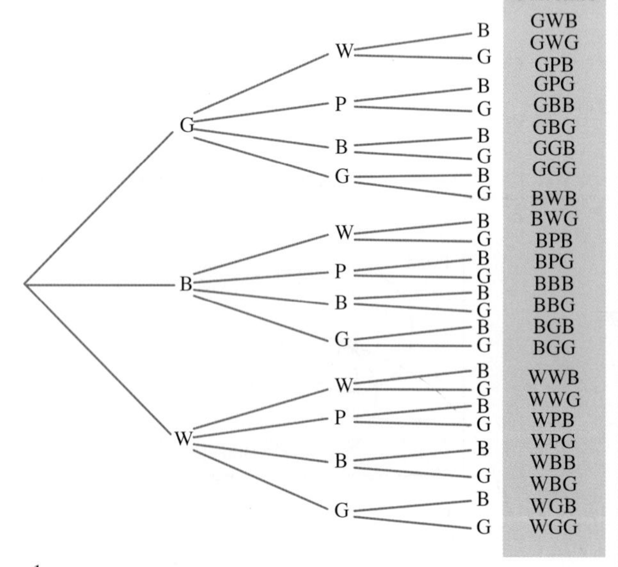

b. $\frac{1}{24}$

c. $\frac{1}{12}$

d. Each combination would be worn 10 times in a year.

e. 21 new combinations.

10. a.

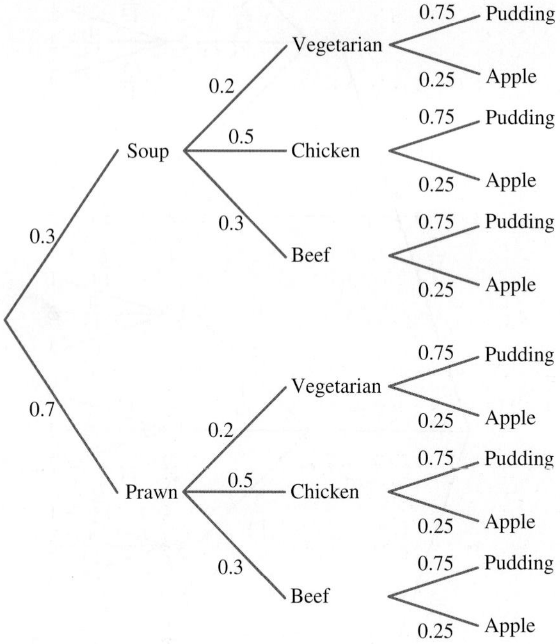

b. 0.1125

c. 24 people

11. a. i.

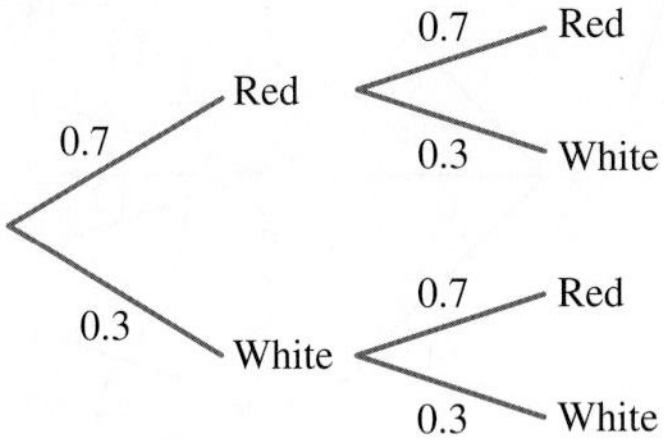

ii. 10 balls

iii. 10 balls

iv. No; the events are independent.

v. Pr(RR) = 0.49

b. i. 9 balls

ii. Yes. One ball has been removed from the bag.

iii.

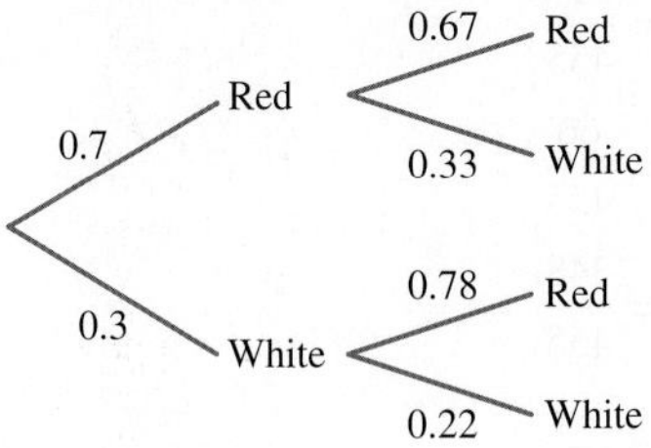

iv. Pr(RR) = $\frac{7}{12}$ or 0.469 using the rounded values from iii.

12. a.

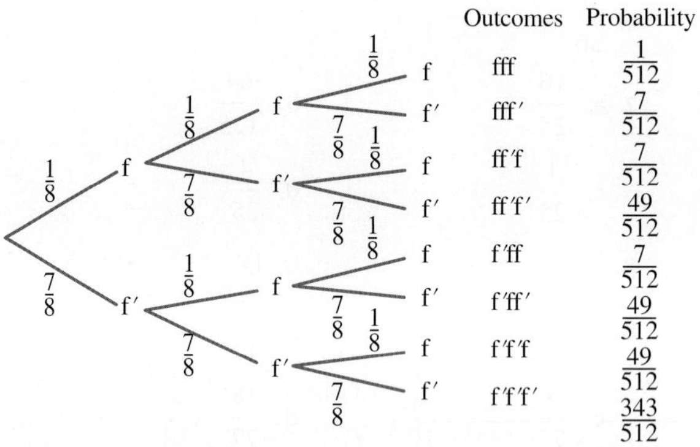

f = outcome of 5

b. i. $\frac{1}{512}$ ii. $\frac{343}{512}$ iii. $\frac{21}{512}$ iv. $\frac{11}{256}$

13. a.

Die outcomes

Coin outcomes	1	2	3	4
H	(H, 1)	(H, 2)	(H, 3)	(H, 4)
T	(T, 1)	(T, 2)	(T, 3)	(T, 4)

b.

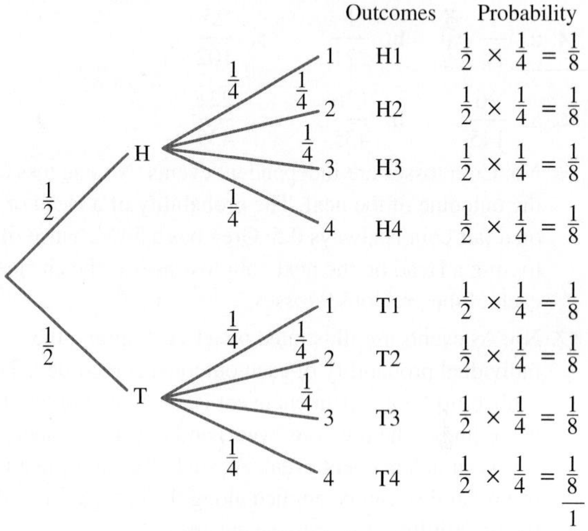

c. $\frac{1}{4}$

14. a. $\frac{256}{625}$ b. $\frac{16}{625}$ c. $\frac{96}{256}$ d. $\frac{17}{256}$

Exercise 11.4 Independent and dependent events

1. a. 0.28 b. 0.12 c. 0.42 d. 0.18

2. Dependent

3. Independent

4. a. Yes, the outcome is independent.

b. i. $\frac{1}{2}$ ii. $\frac{1}{6}$

c. $\frac{1}{12}$

5. $\frac{1}{40}$

6. $\frac{5}{36}$

7. a. $\frac{16}{25}$
b. $\frac{64}{125}$
c. $\frac{1}{25}$
d. $\frac{4}{25}$

8. a. C
b. D

9. a. $\frac{3}{77}$
b. $\frac{48}{77}$
c. $\frac{8}{77}$
d. $\frac{18}{77}$

10. a. $\frac{1}{37}$
b. $\frac{1}{1369}$
c. $\frac{73}{1369}$

11. 0.9

12. $\frac{1}{14}$

13. a. $\frac{1}{5}$
b. $\frac{1}{5}$
c. $\frac{1}{10}$
d. $\frac{1}{3}$

14. a. $\frac{1}{17}$
b. $\frac{1}{221}$
c. $\frac{25}{102}$

15. a. $\frac{26}{145}$
b. $\frac{136}{435}$
c. $\frac{221}{435}$

16. No. Coin tosses are independent events. No one toss affects the outcome of the next. The probability of a Head or Tail on a fair coin is always 0.5. Greg has a 50% chance of tossing a Head on the next coin toss as was the chance in each of the previous 9 tosses.

17. No. As events are illustrated on a tree diagram, the individual probability of each outcome is recorded. The probability of a dependent event is calculated (altered according to the previous event) and can be considered as if it was an independent event. As such, the multiplication law of probability can be applied along the branches to calculate the probability of successive events.

18. Only if $\Pr(A) = 0$ or $\Pr(B) = 0$ can two events be independent and mutually exclusive. For an event to have a probability of 0 means that it is impossible, so it is a trivial scenario.

19. a. A and B are independent.
b. A and B are not independent in this situation.

20. $\frac{5}{17}$

21. a. Dependent
b. $\Pr(1) = \frac{7}{15}$

$\Pr(2) = \frac{5}{15} = \frac{1}{3}$

$\Pr(3) = \frac{3}{15} = \frac{1}{5}$

c.

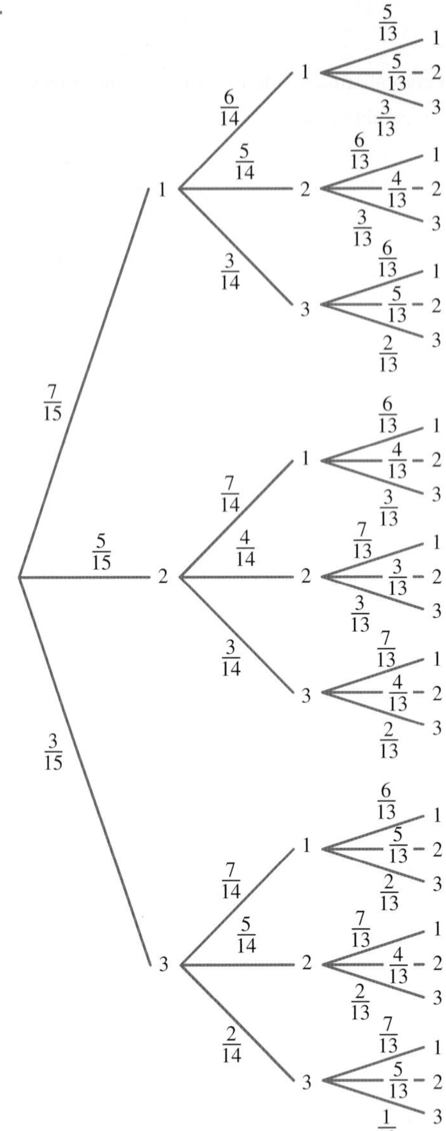

d. i. $\Pr(1, 1, 1) = \frac{1}{13}$
ii. $\Pr(2, 2, 2) = \frac{2}{91}$
iii. $\Pr(3, 3, 3) = \frac{1}{455}$
iv. $\Pr(\text{at least one } 3) = \frac{47}{91}$

22. a. $\Pr(9) = \frac{1}{455}$

$\Pr(7 - 8) = \frac{66}{455}$

$\Pr(5 - 6) = \frac{248}{455}$

$\Pr(3 - 4) = \frac{4}{13}$

b. $393.40

Exercise 11.5 Conditional probability

1. a. $\Pr(J) = \frac{41}{90}$ b. $\Pr(z|J) = \frac{12}{41}$
2. a. $\Pr(S) = \frac{13}{30}$ b. $\Pr(S|(C \cup S)) = \frac{13}{28}$
3. a. 0.3 b. $\frac{3}{7}$
4. a. $\frac{9}{13}$ b. $\frac{3}{5}$
5. 0.58 or $\frac{15}{26}$
6. 0.22 or $\frac{5}{23}$
7. 0.9
8. 0.8375
9. a. D
   b. i. $\Pr(A|B) = \frac{1}{6}$ ii. $\Pr(B|A) = \frac{1}{6}$
   iii. $\Pr(C|A) = \frac{1}{6}$ iv. $\Pr(C|B) = 0$
10. A
11. a. $\frac{109}{232}$ b. $\frac{73}{116}$ c. $\frac{23}{86}$ d. $\frac{22}{45}$
12. Conditional probability is when the probability of one event depends on the outcome of another event.
13. a. 0.0875
    b. 0.065
    c. i. 0.585 ii. 0.1525 or 15.25%.
14. a. i. $\frac{1}{2}$ ii. $\frac{2}{5}$ iii. $\frac{1}{5}$ iv. $\frac{1}{2}$ v. $\frac{2}{5}$
    b. Yes, A and B are independent.
    c. If $\Pr(A|B) = \frac{\Pr(A \cap B)}{\Pr(B)} = \Pr(A)$ then A and B will be independent events. This is because the probability of A given B occurs is the same as the probability of A, meaning the probability of A is independent of B occurring.
15. 0.087 or 8.7% chance.
16. 0.847
17. 0.61

Project

1.

Die 1

Die 2	1	2	3	1	2	3
4	(1, 4)	(2, 4)	(3, 4)	(1, 4)	(2, 4)	(3, 4)
5	(1, 5)	(2, 5)	(3, 5)	(1, 5)	(2, 5)	(3, 5)
6	(1, 6)	(2, 6)	(3, 6)	(1, 6)	(2, 6)	(3, 6)
4	(1, 4)	(2, 4)	(3, 4)	(1, 4)	(2, 4)	(3, 4)
5	(1, 5)	(2, 5)	(3, 5)	(1, 5)	(2, 5)	(3, 5)
6	(1, 6)	(2, 6)	(3, 6)	(1, 6)	(2, 6)	(3, 6)

2. 9
3. 0
4.

Die 1

Die 2	1	2	3	1	2	3
4	5	6	7	5	6	7
5	6	7	8	6	7	8
6	7	8	9	7	8	9
4	5	6	7	5	6	7
5	6	7	8	6	7	8
6	7	8	9	7	8	9

5. 5; 5, 6, 7, 8, 9
6. 7; 5, 9
7. Equal to 7; probability is the highest.
8. Odd-number outcome; probability is higher.

Die

Coin	1	2	3	4	5	6	7	8
Head	(H, 1)	(H, 2)	(H, 3)	(H, 4)	(H, 5)	(H, 6)	(H, 7)	(H, 8)
Tail	(T, 1)	(T, 2)	(T, 3)	(T, 4)	(T, 5)	(T, 6)	(T, 7)	(T, 8)

9.

Die 1

Die 2	1	2	3	1	2	3
4	3	2	1	3	2	1
5	4	3	2	4	3	2
6	5	4	3	5	4	3
4	3	2	1	3	2	1
5	4	3	2	4	3	2
6	5	4	3	5	4	3

10. 5; 1, 2, 3, 4, 5,
11. 3; 12
12. Students need to apply the knowledge of probability and create a new game using dice given. They also need to provide rules for the game and solution, indicating the best options for winning.

Exercise 11.6 Review questions

1. A
2. B
3. B
4. B
5. B
6. a.

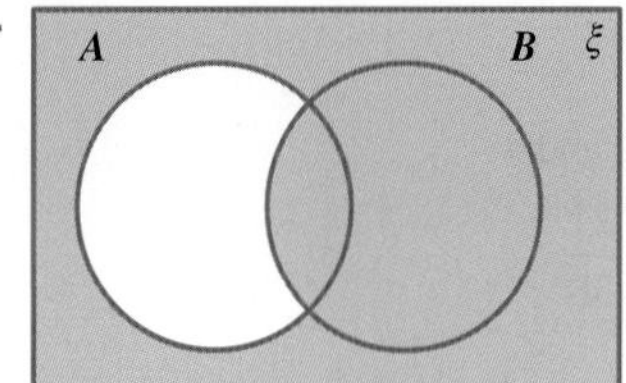

b. 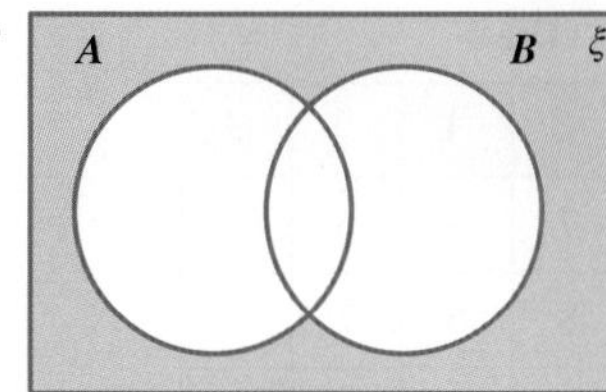

c. 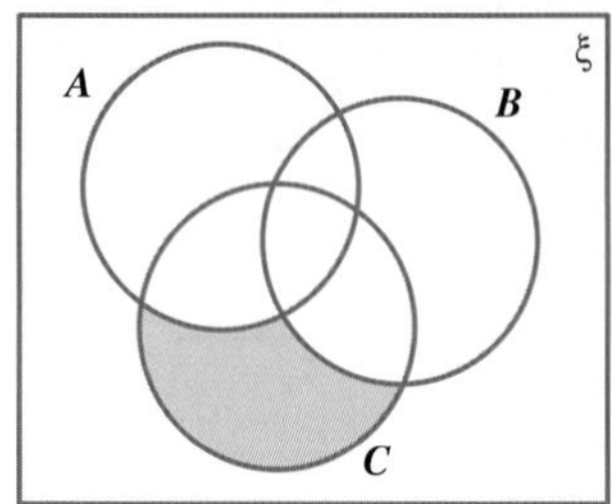

7. C

8. D

9. D

10. B

11. a. $\frac{1}{13}$ b. $\frac{1}{4}$ c. $\frac{2}{13}$ d. $\frac{3}{4}$

12. a. $\frac{1}{7776}$ b. $\frac{7775}{7776}$

13. a. $\frac{3}{8}$ b. $\frac{5}{8}$

14. a. Yes. It is not possible to roll an even number and for that number to be a 3.

b. $\Pr(A) = \frac{1}{2}$ and $\Pr(B) = \frac{1}{6}$

c. $\frac{2}{3}$

15. a. No. It is possible to draw a card that is a club and an ace.

b. $\Pr(A) = \frac{1}{4}$ and $\Pr(B) = \frac{1}{13}$, $\Pr(A \cap B) = \frac{1}{52}$

c. $\frac{4}{13}$

16. a.

Die 2 outcomes

Die 1 outcomes	0	1	2	3
0	0	1	2	3
1	1	2	3	4
2	2	3	4	5
3	3	4	5	6

b. No

c. 0 and 6

d. 3

e. 0 and 6, 1 and 5, 2 and 4

17. a. $\frac{1}{19}$ b. $\frac{21}{38}$ c. $\frac{15}{38}$

18. a. $\frac{1}{169}$ b. $\frac{1}{221}$

19. a. $\frac{15}{25} = \frac{3}{5}$ b. $\frac{8}{10} = \frac{4}{5}$

20. a. 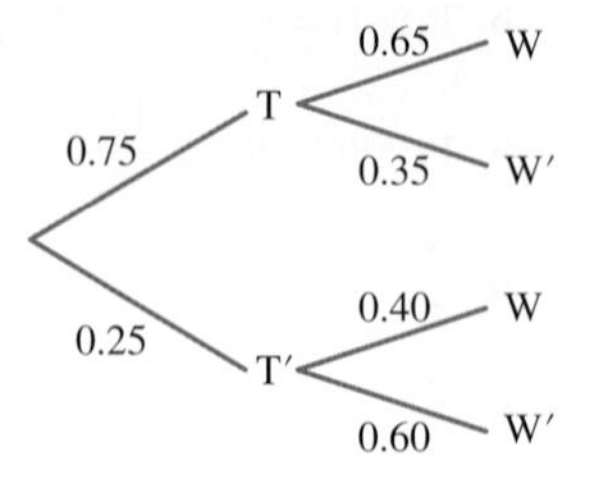

b. 0.5875

c. $\frac{8}{47}$

21. a. $\frac{3}{5}$ b. $\frac{8}{15}$

22. a. $\frac{1}{5}$ b. $\frac{4}{5}$ c. $\frac{1}{25}$

23. a. 0.000 495 b. 0.001 981

24. a. $\frac{1}{2}$ b. $\frac{7}{8}$

12 Univariate data

LEARNING SEQUENCE

12.1 Overview

Why learn this?

According to the novelist Mark Twain, 'There are three kinds of lies: lies, damned lies and statistics.' Statistics can easily be used to manipulate people unless they have an understanding of the basic concepts involved.

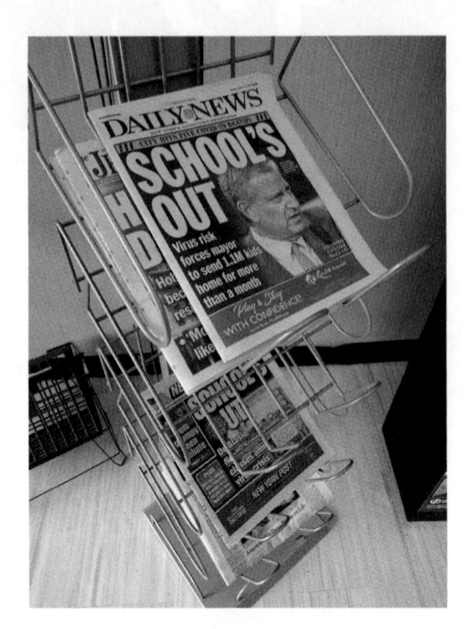

Statistics, when used properly, can be an invaluable aid to good decision-making. However, deliberate distortion of the data or meaningless pictures can be used to support almost any claim or point of view. Whenever you read an advertisement, hear a news report or are given some data by a friend, you need to have a healthy degree of scepticism about the reliability of the source and nature of the data presented. A solid understanding of statistics is crucially important, as it is very easy to fall prey to statistics that are designed to confuse and mislead.

In 2020 when the COVID19 pandemic hit, news and all forms of media were flooded with statistics. These statistics were used to inform governments worldwide about infection rates, recovery rates and all sorts of other important information. These statistics guided the decision-making process in determining the restrictions that were imposed or relaxed to maintain a safe community.

Statistics are also used to provide more information about a population in order to inform government policies. For example, the results of a census might indicate that the people in a particular city are fed up with traffic congestion. With this information now known, the government might prioritise works on public roads, or increase funding of public transport to try to create a more viable alternative to driving.

Where to get help

Go to your learnON title at **www.jacplus.com.au** to access the following digital resources. The Online Resources Summary at the end of this topic provides a full list of what's available to help you learn the concepts covered in this topic.

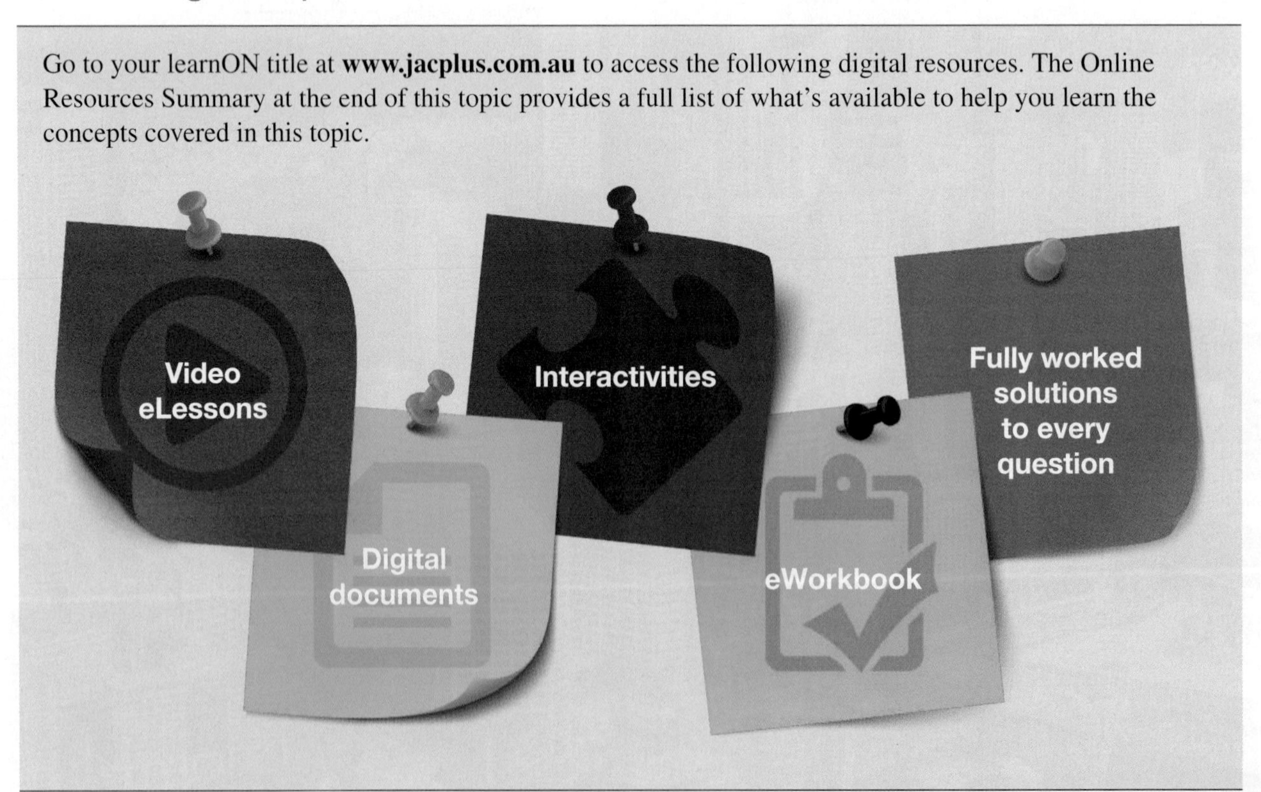

Exercise 12.1 Pre-test

learnON

Complete this pre-test in your learnON title at www.jacplus.com.au and receive **automatic marks, immediate corrective feedback** and **fully worked solutions**.

1. The following data show the number of cars in each of the 12 houses along a street.

$$2, 3, 3, 2, 2, 3, 2, 4, 3, 1, 1, 0$$

Calculate the median number of cars.

2. Calculate the range of the following data set: 5, 15, 23, 6, 31, 24, 26, 14, 12, 34, 18, 9, 17, 32.

3. The frequency table shows the scores obtained by 100 professional golfers in the final round of a tournament.

Score	Frequency
67	2
68	6
69	7
70	11
71	16
72	23
73	17
74	11
75	9

Identify the modal score.

4. A sample of 15 people was selected at random from those attending a local swimming pool. Their ages (in years) were recorded as follows:

$$19, 7, 83, 41, 17, 23, 62, 55, 15, 25, 32, 29, 11, 18, 10$$

Calculate the mean age of people attending the swimming pool, correct to one decimal place.

5. Complete the following sentence.
A sample is a _________ of a population.

6. At Einstein Secondary School a Year 10 mathematics class has 22 students. The following were the test scores for the class.

$$34, 47, 54, 59, 60, 63, 66, 69, 73, 77, 78, 78, 79, 80, 82, 83, 85, 86, 88, 89, 90, 91$$

Calculate the interquartile range (IQR).

7. The mean of a set of five scores is 11.8. If four of the scores are 17, 9, 14 and 6, calculate the fifth score.

▶

8. The box plot below shows the price of a meal for one person from ten fast-food shops.

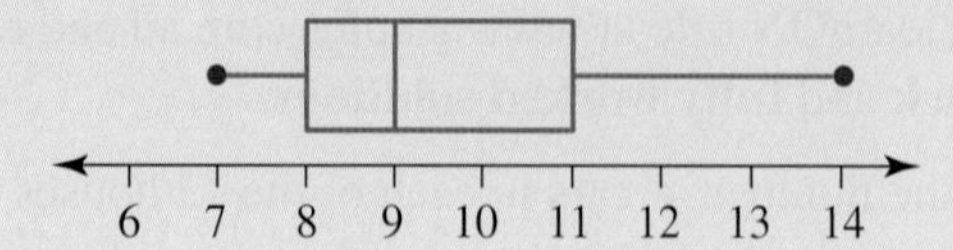

State whether the data is negatively skewed, positively skewed or symmetrical.

9. A frequency table for the time taken by 20 people to put together an item of flat-pack furniture is shown.

Time taken (min)	Frequency
0 – 40	1
5 – 95	3
10 – 14	5
15 – 19	2
20 – 24	4
25 – 29	2
30 – 34	2
35 – 39	1

Calculate the cumulative frequency to put together an item of flat-pack furniture in less than 20 minutes.

10. **MC** The frequency table below shows the scores obtained by 100 professional golfers in the final round of a tournament.

Score	Frequency
67	2
68	5
69	8
70	11
71	16
72	22
73	14
74	13
75	9

Select the median score.

A. 71 **B.** 71.75 **C.** 72 **D.** 72.5 **E.** 73

11. The heights of six basketball players (in cm) are:

178.1 185.6 173.3 193.4 183.1 193.0

Calculate the mean and standard deviation.

12. **MC** A group of 22 people recorded how many cans of soft drink they drank in a day. The table shows the number of cans drunk by each person.

0	2	2	2	1	1	3	4	4	2	1
2	4	1	6	3	3	5	4	1	2	5

Select the statement that is not true.

A. The maximum number of soft drinks cans drank is 6.
B. The minimum number of soft drink cans drank is 0.
C. The interquartile range is 3.
D. The median number of soft drink cans is 2.5.
E. The mean number of soft drink cans drank is 2.64.

13. **MC** Select the approximate median in the cumulative frequency percentage graph shown.

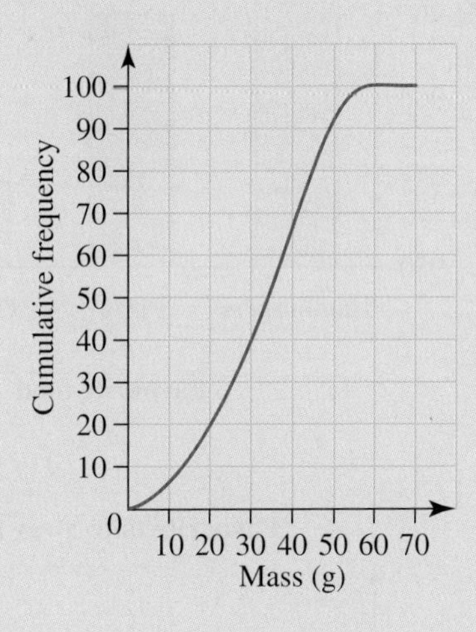

A. 30 B. 32 C. 40 D. 50 E. 92

14. **MC** The following back-to-back stem-and-leaf plot shows the typing speed in words per minute (wpm) of 30 Year 8 and Year 10 students.

Key: 2 | 6 = 26 wpm

Leaf: Year 8	*Stem*	*Leaf: Year 10*
9 2	0	
9 8 6 5 4 2	1	4 9
8 8 8 6 4 2 1 0 0	2	2 3 6 8 9
9 7 7 6 4 1 0	3	0 3 4 5 5 7 8 8
7 6 5 2 0	4	1 2 5 8 8 9 9
	5	0 3 5 7 8
	6	0 0 1

Calculate the mean typing speed and interquartile range for Year 8 and Year 10. Comment on your answers.

15. **MC** A survey was conducted on the favourite take-away foods for university students and the results were graphed using two bar charts.

Graph 1

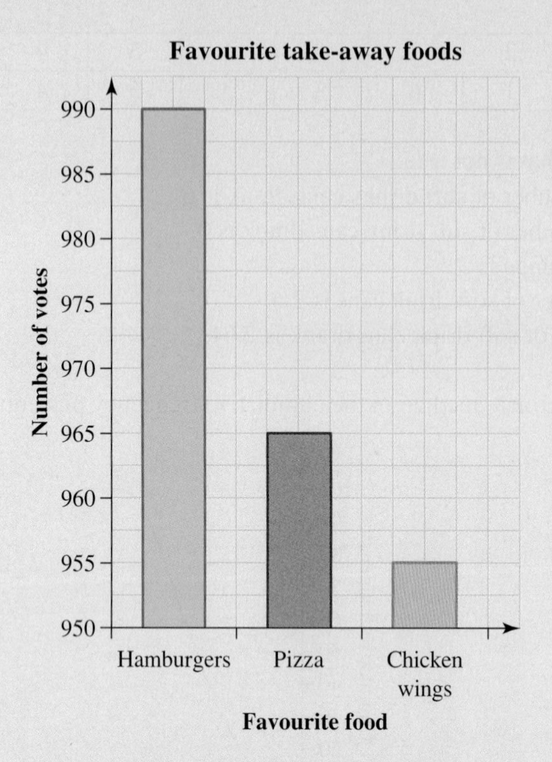

Graph 2

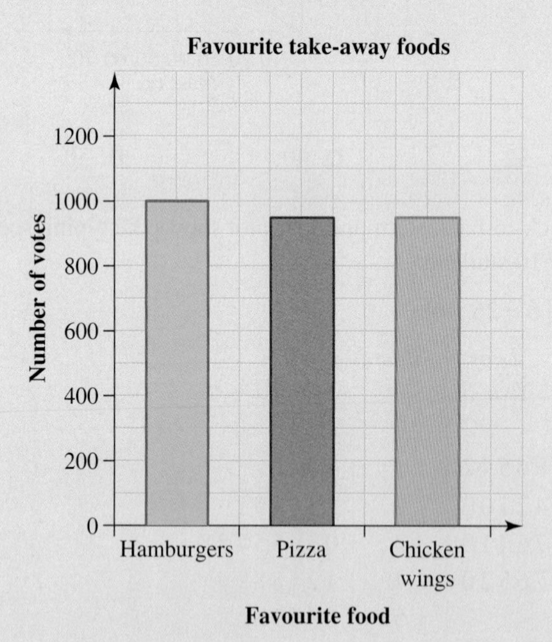

Select the statement that best describes these graphs.

A. Graph 1 is misleading as it suggests that students like hamburgers 7 times more than chicken wings.
B. Graph 1 is misleading as it suggests that students like pizza 3 times more than hamburgers.
C. Graph 1 is misleading as it suggests that students like pizza four times more than chicken wings.
D. Graph 2 is misleading as it suggests that students like all take-away foods evenly.
E. Graph 1 is misleading as the vertical axis does not start at zero.

12.2 Measures of central tendency

LEARNING INTENTION

At the end of this subtopic you should be able to:
- calculate the mean, median and mode of data presented as ungrouped data (in a single list), frequency distribution tables and grouped data.

12.2.1 Mean, median and mode of univariate data

eles-4949

- **Univariate data** are **data** with one variable; for example, the heights of Year 10 students.
- **Measures of central tendency** are summary statistics that measure the middle (or centre) of the data. These are known as the mean, median and mode.
  - The **mean** is the average of all observations in a set of data.
  - The **median** is the middle observation in an ordered set of data.
  - The **mode** is the most frequent observation in a data set.

The mean

- The *mean* of a set of data is what is referred to in everyday language as the *average*.
- The mean of a set of values is the sum of all the values divided by the number of values.
- The symbol we use to represent the mean is $\bar{x}$; that is, a lower-case x with a bar on top.

Calculating the mean

The formal definition of the mean is:

$$\textbf{Mean} = \frac{\textbf{sum of data values}}{\textbf{total number of data values}}$$

Using mathematical notation, this is written as:

$$\bar{x} = \frac{\sum x}{n}$$

The median

- The median represents the *middle* score when the data values are in ascending or descending order such that an equal number of data values will lie below the median and above it.

Calculating the median

When calculating the median:
1. Arrange the data values in order (usually in ascending order).
2. The *position* of the median is the $\left(\frac{n+1}{2}\right)$th data value, where n is the total number of data values.

Note: If there are an even number of data values then there will be two middle values. In this case the median is the average of those data values.

- When there are an odd number of data values, the median is the middle value.

1 1 3 (4) 6 7 8

median = 4

- When there are an even number of data values, the median is the average of the two middle values.

2 3 3 (5) (6) 6 7 9

$$\text{median} = \frac{5+6}{2} = 5.5$$

The mode

- The mode is the score that occurs most often.
- The data set can have no modes, one mode, two modes (bimodal) or more than two modes (multi modal).

Calculating the mode

When determining the mode:

1. Arrange the data values in ascending order (smallest to largest). This step is optional but does help.
2. Look for the number that occurs most often (has the highest frequency).

- If no value in a data set appears more than once then there is no mode.
- If a data set has multiple values that appear the most then it has multiple modes. All values that appear the most are modes.

For example, the set 1, 2, 2, 4, 5, 5, 7 has two modes, 2 and 5.

WORKED EXAMPLE 1 Calculating mean, median and mode

For the data 6, 2, 4, 3, 4, 5, 4, 5, calculate the:

a. mean **b. median** **c. mode.**

THINK	WRITE
a. 1. Calculate the sum of the scores; that is, $\sum x$.	a. $\sum x = 6+2+4+3+4+4+5+4+5$ $= 33$
2. Count the number of scores; that is, n.	$n = 8$
3. Write the rule for the mean.	$\bar{x} = \frac{\sum x}{n}$
4. Substitute the known values into the rule.	$= \frac{33}{8}$
5. Evaluate.	$= 4.125$
6. Write the answer.	The mean is 4.125.

b. 1. Write the scores in ascending numerical order.

b. 2 3 4 4 4 5 5 6

2. Locate the position of the median using the rule $\frac{n+1}{2}$, where $n = 8$. This places the median as the 4.5th score; that is, between the 4th and 5th score.

$$\text{Median} = \frac{n+1}{2}\text{th score}$$
$$= \frac{8+1}{2}\text{th score}$$
$$= 4.5\text{ th score}$$

2 3 4 **4 4** 5 5 6

3. Obtain the average of the two middle scores.

$$\text{Median} = \frac{4+4}{2}$$
$$= \frac{8}{2}$$
$$= 4$$

4. Write the answer

The median is 4.

c. 1. Systematically work through the set and make note of any repeated values (scores).

c.

$$\begin{array}{cccccccc} & & & & & \downarrow & \downarrow & \\ 2 & 3 & 4 & 4 & 4 & 5 & 5 & 6 \\ & & \uparrow & \uparrow & \uparrow & & & \end{array}$$

2. Write the answer.

The mode is 4.

TI | THINK	DISPLAY/WRITE	CASIO | THINK	DISPLAY/WRITE
1. In a new document, on a Lists & Spreadsheet page, label column A as 'xvalues', and enter the values in the data set. Press ENTER after entering each value.		1. On the Statistics screen, label list1 as 'x' and enter the values as shown in the table. Press EXE after entering each value.	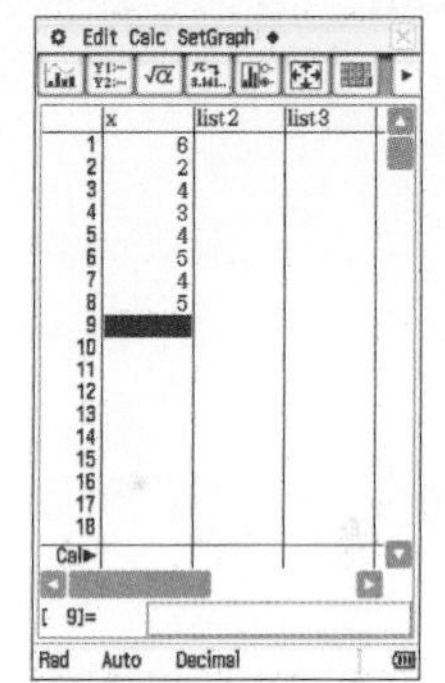
2. Although you can find many summary statistics, to find the mean only, open a Calculator page and press: • MENU • 6: Statistics • 3: List Math • 3: Mean Press VAR and select 'xvalues', then press ENTER. To calculate the median only, press: • MENU • 6: Statistics • 3: List Math • 4: Median Press VAR and select 'xvalues', then press ENTER.	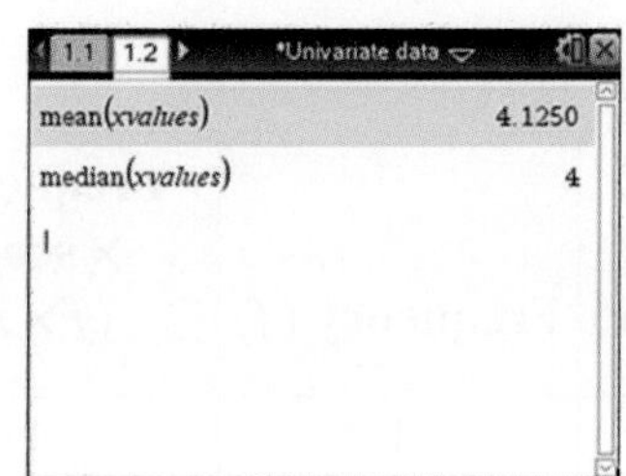 The mean is 4.125 and the median is 4. The mode is 4.	2. To calculate the mean, median and mode, tap: • Calc • One-Variable Set values as: • XList: main\x • Freq: 1 Tap OK. The mean is at the top $(\bar{x})$. Scroll down to find the median and mode.	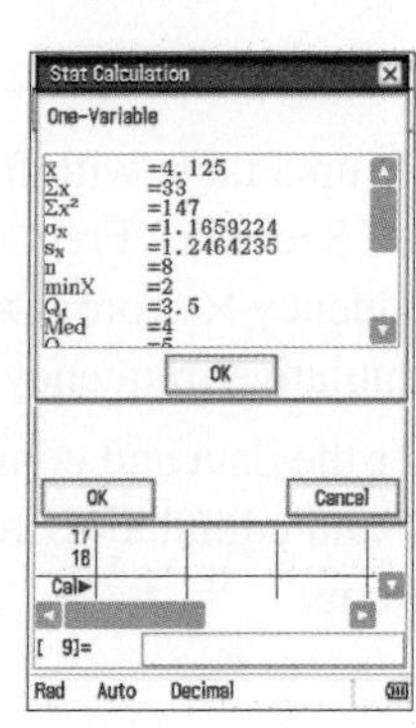 The mean is 4.125 and the median is 4. The mode is 4.

12.2.2 Calculating mean, median and mode from a frequency distribution table

eles-4950

- If data is provided in the form of a frequency distribution table we can determine the mean, median and mode using slightly different methods.
- The mode is the score with the highest **frequency**.
- To calculate the median, add a cumulative frequency column to the table and use it to determine the score that is the $\left(\frac{n+1}{2}\right)$ th data value.
- To calculate the mean, add a column that is the score multiplied by its frequency $f \times x$. The following formula can then be used to calculate the mean, where $\sum (f \times x)$ is the sum of the $(f \times x)$ column. $\sum$ is the uppercase Greek letter sigma.

Calculating the mean from a frequency table

$$\bar{x} = \frac{\sum (f \times x)}{n}$$

WORKED EXAMPLE 2 Calculations from a frequency distribution table

Using the frequency distribution table, calculate the:

a. **mean** b. **median** c. **mode.**

Score (x)	Frequency (f)
4	1
5	2
6	5
7	4
8	3
Total	15

THINK

1. Rule up a table with four columns titled Score (x), Frequency (f), Frequency × score ($f \times x$) and Cumulative frequency (cf).
2. Enter the data and complete both the $f \times x$ and cumulative frequency columns.

WRITE

Score (x)	Frequency (f)	Frequency × score ($f \times x$)	Cumulative frequency (cf)
4	1	4	1
5	2	10	$1+2=3$
6	5	30	$3+5=8$
7	4	28	$8+4=12$
8	3	24	$12+3=15$
	$n=15$	$\sum(f \times x)=96$	

a. 1. Write the rule for the mean.

a. $\bar{x} = \dfrac{\sum(f \times x)}{n}$

2. Substitute the known values into the rule and evaluate.

$\bar{x} = \dfrac{96}{15}$

$= 6.4$

3. Write the answer.

The mean of the data set is 6.4.

b. 1. Locate the position of the median using the rule $\dfrac{n+1}{2}$, where $n = 15$.
This places the median as the 8th score.

b. The median is the $\left(\dfrac{15+1}{2}\right)$th or 8th score.

2. Use the cumulative frequency column to find the 8th score and write the answer.

The median of the data set is 6.

c. 1. The mode is the score with the highest frequency.

c. The score with the highest frequency is 6.

2. Write the answer.

The mode of the data set is 6.

TI | THINK — **DISPLAY/WRITE**

1. In a new problem, on a Lists & Spreadsheet page, label column A as 'score' and column B as 'f'. Enter the values as shown in the table and press ENTER after entering each value.

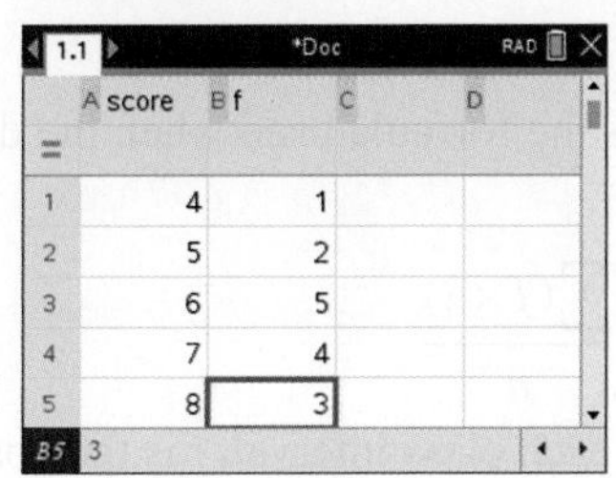

2. To find the summary statistics, press:
   - MENU
   - 4: Statistics
   - 1: Stat Calculations
   - 1: One-Variable Statistics...

   Select 1 as the number of lists. Then on the One-Variable Statistics page select 'score' as the X1 List and 'f' as the Frequency List. Leave the next fields empty, TAB to OK and press ENTER.

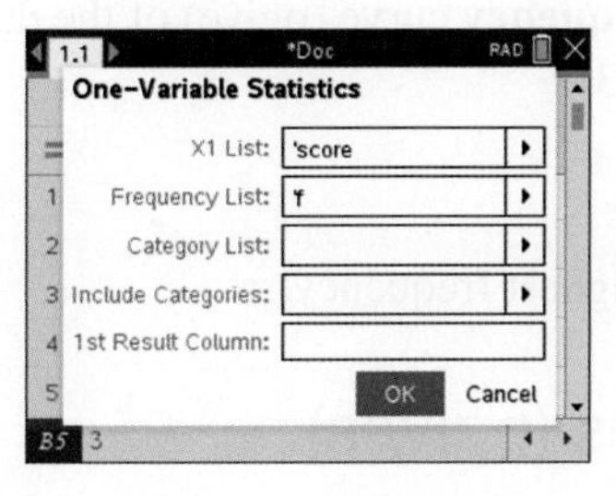

CASIO | THINK — **DISPLAY/WRITE**

1. On the Statistics screen, label list1 as 'score' and list2 as 'f', and enter the values as shown in the table.
Press EXE after entering each value.

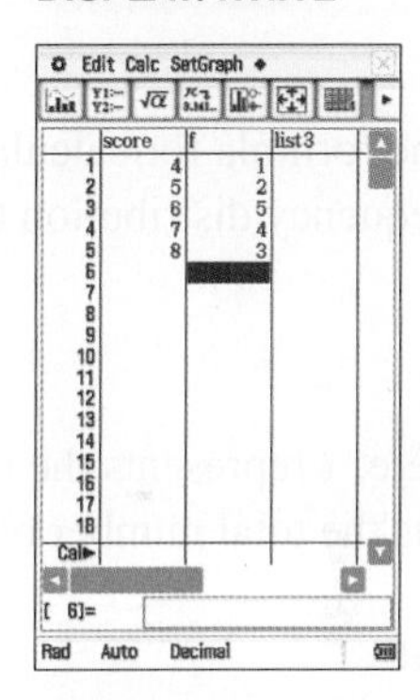

2. To find the summary statistics, tap:
   - Calc
   - One-Variable

   Set values as:
   - XList: main\score
   - Freq: main\f

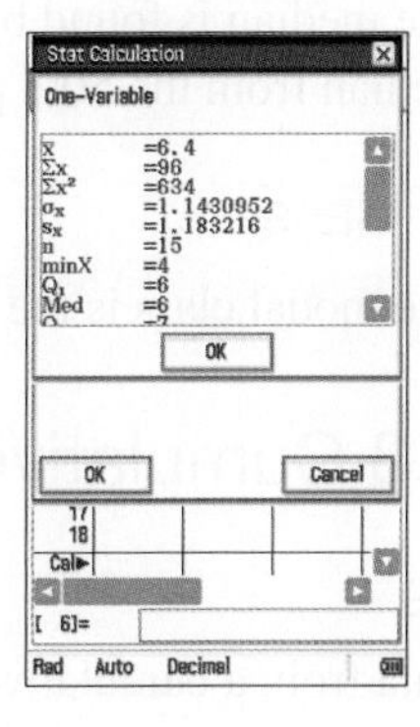

3. The results are displayed. The mean $\bar{x} = 6.4$ and the median is 6.
The mode is the data set with the highest frequency value, which in this case is 6.

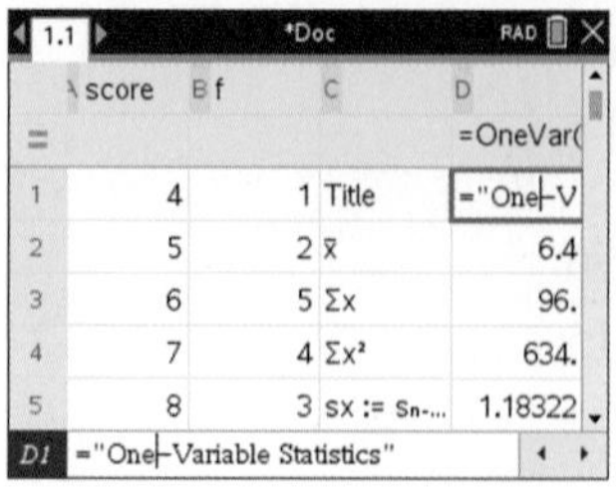

The mean is $\bar{x} = 6.4$ and the median is 6. The mode is the data set with the highest frequency value, which is 6.

3. Tap OK.
The mean is at the top $\bar{x}$.
Scroll down to find the median and mode.

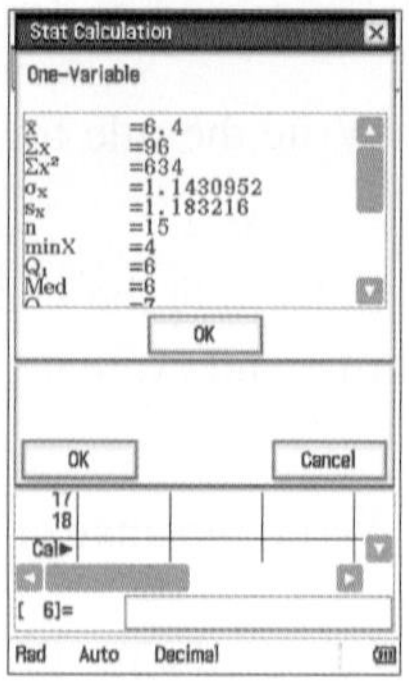

The mean is $\bar{x} = 6.4$ and the median is 6. The mode is the data set with the highest frequency value, which is 6.

Mean, median and mode of grouped data

- When the data are grouped into class intervals, the actual values (or data) are lost. In such cases we have to approximate the real values with the midpoints of the intervals into which these values fall.
  For example, if in a grouped frequency table showing the heights of different students, 4 students had a height between 180 and 185 cm, we have to assume that each of those 4 students is 182.5 cm tall.

Mean

- The formula for calculating the mean is the same as the formula used when the data is displayed in a frequency distribution table:

$$\bar{x} = \frac{\sum(f \times x)}{n}$$

  Here, x represents the midpoint (or class centre) of each class interval, f is the corresponding frequency and n is the total number of observations in a set.

Median

- The median is found by drawing a **cumulative frequency** curve (ogive) of the data and estimating the median from the 50th percentile (see section 12.2.3).

Modal class

- The modal class is the class interval that has the highest frequency.

12.2.3 Cumulative frequency curves (ogives)

eles-4951

Ogives

- Data from a cumulative frequency table can be plotted to form a **cumulative frequency curve** (sometimes referred to as cumulative frequency polgons), which is also called an **ogive** (pronounced '*oh-jive*').
- To plot an ogive for data that is in class intervals, the maximum value for the class interval is used as the value against which the cumulative frequency is plotted.

For example, the following table and graph show the mass of cartons of eggs ranging from 55 g to 65 g.

Mass (g)	Frequency (f)	Cumulative frequency (cf)	Percentage cumulative frequency ($\%cf$)
$55-<57$	2	2	6%
$57-<59$	6	$2+6=8$	22%
$59-<61$	12	$8+12=20$	56%
$61-<63$	11	$20+11=31$	86%
$63-<65$	5	$31+5=36$	100%

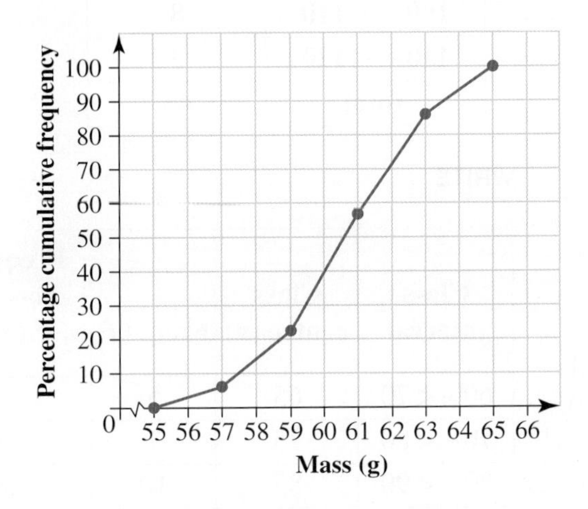

Quantiles

- An ogive can be used to divide the data into any given number of equal parts called **quantiles**.
- Quantiles are named after the number of parts that the data are divided into.
  - **Percentiles** divide the data into 100 equal-sized parts.
  - **Quartiles** divide the data into 4 equal-sized parts. For example, 25% of the data values lie at or below the first quartile.

Percentile	Quartile and symbol	Common name
25th percentile	First quartile, Q_1	Lower quartile
50th percentile	Second quartile, Q_2	Median
75th percentile	Third quartile, Q_3	Upper quartile
100th percentile	Fourth quartile, Q_4	Maximum

- A percentile is named after the percentage of data that lies at or below that value. For example, 60% of the data values lie at or below the 60th percentile.
- Percentiles can be read off a percentage cumulative frequency curve.
- A percentage cumulative frequency curve is created by:
  - writing the cumulative frequencies as a percentage of the total number of data values
  - plotting the percentage cumulative frequencies against the maximum value for each interval.

WORKED EXAMPLE 3 Estimating mean, median and modal class in grouped data

For the given data:

a. estimate the mean **b. estimate the median** **c. determine the modal class.**

Class interval	Frequency
60 – < 70	5
70 – < 80	7
80 – < 90	10
90 – < 100	12
100 – < 110	8
110 – < 120	3
Total	45

THINK

1. Draw up a table with 5 columns headed Class interval, Class centre (x), Frequency (f), Frequency × class centre ($f \times x$) and Cumulative frequency (cf).
2. Complete the x, $f \times x$ and cf columns.

WRITE

Class interval	Class centre (x)	Freq. (f)	Frequency × class centre ($f \times x$)	Cumulative frequency (cf)
60– < 70	65	5	325	5
70– < 80	75	7	525	12
80– < 90	85	10	850	22
90– < 100	95	12	1140	34
100– < 110	105	8	840	42
110– < 120	115	3	345	45
		$n = 45$	$\sum(f \times x) = 4025$	

a. 1. Write the rule for the mean.

a. $\bar{x} = \frac{\sum(f \times x)}{n}$

2. Substitute the known values into the rule and evaluate.

$\bar{x} = \frac{4025}{45}$

$\simeq 89.4$

3. Write the answer.

The mean for the given data is approximately 89.4.

b. 1. Draw a combined cumulative frequency histogram and ogive, labelling class centres on the horizontal axis and cumulative frequency on the vertical axis. Join the end-points of each class interval with a straight line to form the ogive.

b.

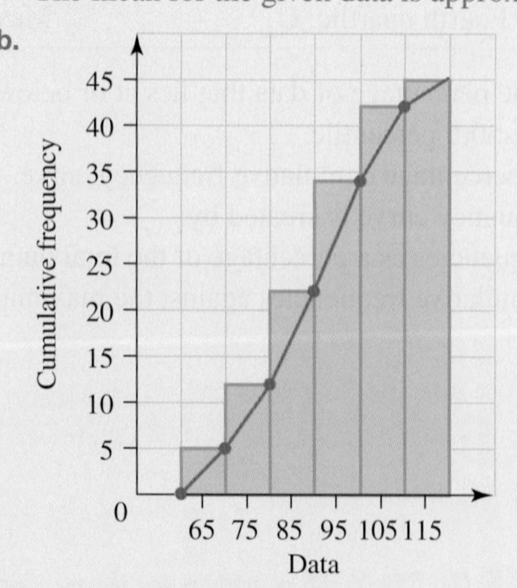

2. Locate the middle of the cumulative frequency axis, which is 22.5.
3. Draw a horizontal line from this point to the ogive and a vertical line to the horizontal axis.

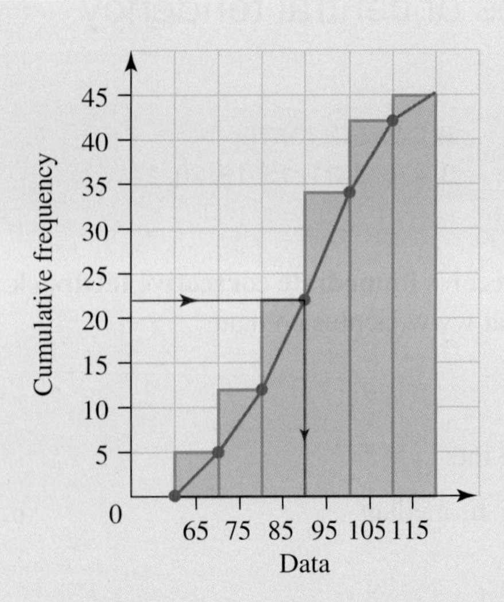

4. Read off the value of the median from the x-axis and write the answer.

The median for the given data is approximately 90.

c. 1. The modal class is the class interval with the highest frequency.

c. The class internal 90–100 occurs twelve times, which is the highest frequency.

2. Write the answer.

The modal class is the 90–100 class interval.

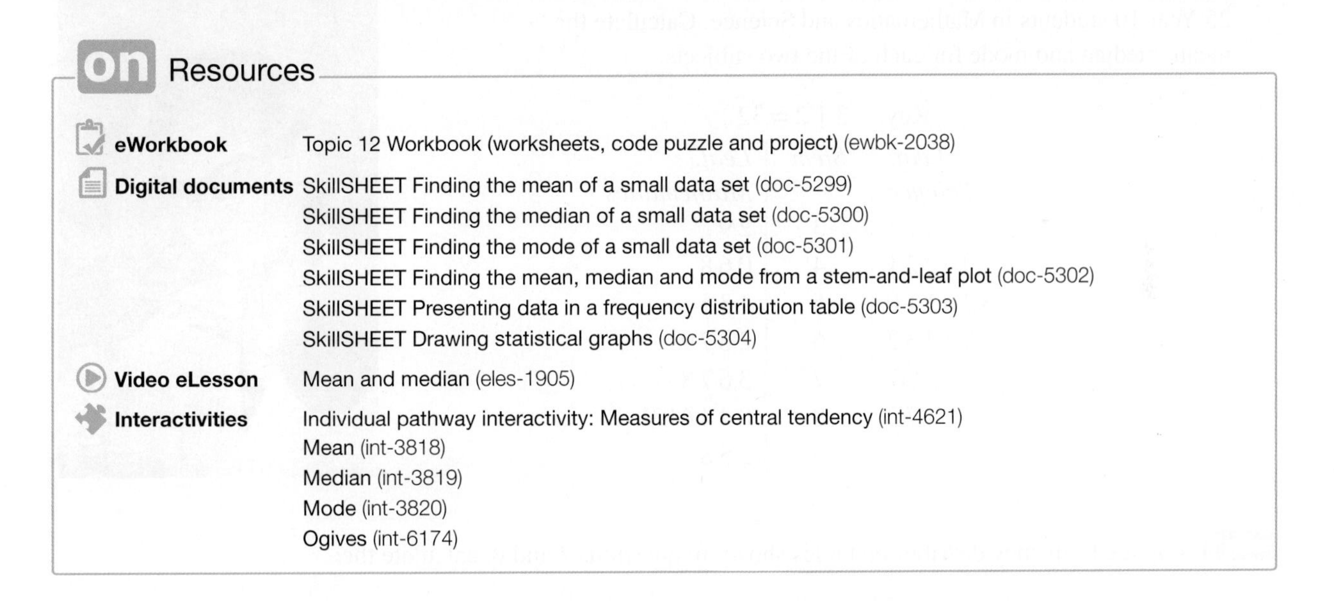

on Resources

eWorkbook	Topic 12 Workbook (worksheets, code puzzle and project) (ewbk-2038)
Digital documents	SkillSHEET Finding the mean of a small data set (doc-5299) SkillSHEET Finding the median of a small data set (doc-5300) SkillSHEET Finding the mode of a small data set (doc-5301) SkillSHEET Finding the mean, median and mode from a stem-and-leaf plot (doc-5302) SkillSHEET Presenting data in a frequency distribution table (doc-5303) SkillSHEET Drawing statistical graphs (doc-5304)
Video eLesson	Mean and median (eles-1905)
Interactivities	Individual pathway interactivity: Measures of central tendency (int-4621) Mean (int-3818) Median (int-3819) Mode (int-3820) Ogives (int-6174)

Exercise 12.2 Measures of central tendency

learnON

Individual pathways

■ PRACTISE	■ CONSOLIDATE	■ MASTER
1, 2, 7, 10, 14, 17, 18, 23	3, 4, 6, 8, 11, 15, 19, 20, 24	5, 9, 12, 13, 16, 21, 22, 25

To answer questions online and to receive **immediate corrective feedback** and **fully worked solutions** for all questions, go to your learnON title at www.jacplus.com.au.

Fluency

WE1 For questions 1 to 5, calculate the:

a. mean **b.** median **c.** mode.

1. 3, 5, 6, 8, 8, 9, 10
2. 4, 6, 7, 4, 8, 9, 7, 10
3. 17, 15, 48, 23, 41, 56, 61, 52
4. 4.5, 4.7, 4.8, 4.8, 4.9, 5.0, 5.3
5. $7\frac{1}{2}$, $10\frac{1}{4}$, 12, $12\frac{1}{4}$, 13, $13\frac{1}{2}$, $13\frac{1}{2}$, 14
6. The back-to-back stem-and-leaf plot below shows the test results of 25 Year 10 students in Mathematics and Science. Calculate the mean, median and mode for each of the two subjects.

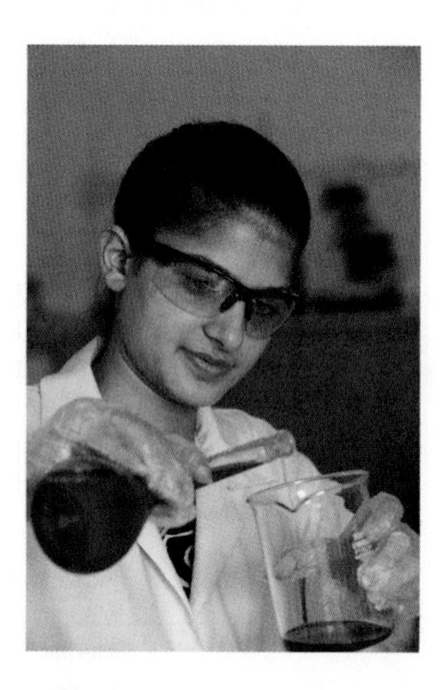

Key: $3 \mid 2 = 32$

Leaf: Science	*Stem*	*Leaf: Mathematics*
8 7 3	3	2 9
9 6 2 2 1	4	0 6 8
8 7 6 1 1 0	5	1 3 5
9 7 4 3 2	6	2 6 7 9
8 5 1 0	7	3 6 7 8
7 3	8	0 4 4 6 8 9
	9	2 5 8

WE2 Using the frequency distribution tables shown in questions 7 and 8, calculate the:

a. mean **b.** median **c.** mode.

7.

Score (x)	Frequency (f)
4	3
5	6
6	9
7	4
8	2
Total	24

8.

Score (x)	Frequency (f)
12	4
13	5
14	10
15	12
16	9
Total	40

9. The following data show the number of bedrooms in each of the 10 houses in a particular neighbourhood:
   2, 1, 3, 4, 2, 3, 2, 2, 3, 3.
   a. Calculate the mean and median number of bedrooms.
   b. A local motel contains 20 rooms. Add this observation to the set of data and recalculate the values of the mean and median.
   c. Compare the answers obtained in parts **a** and **b** and complete the following statement:
   When the set of data contains an unusually large value(s), called an outlier, the ________ (mean/median) is the better measure of central tendency, as it is less affected by this extreme value.

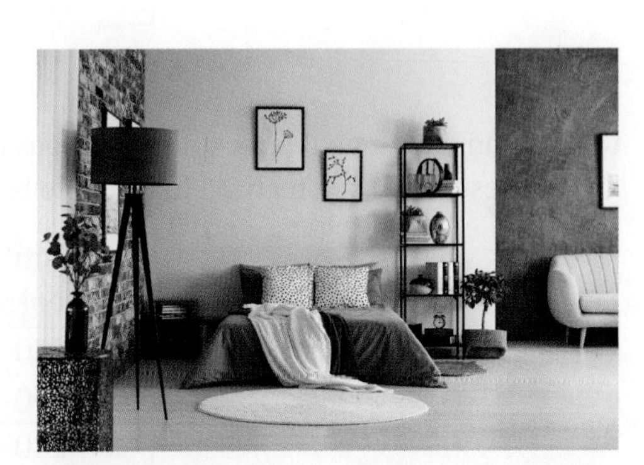

10. **WE3** For the given data:
    a. estimate the mean
    b. estimate the median
    c. determine the modal class.

Class interval	Frequency
$40-<50$	2
$50-<60$	4
$60-<70$	6
$70-<80$	9
$80-<90$	5
$90-<100$	4
Total	30

11. Calculate the mean of the grouped data shown in the table below.

Class interval	Frequency
100–109	3
110–119	7
120–129	10
130–139	6
140–149	4
Total	30

12. Determine the modal class of the data shown in the table below.

Class interval	Frequency
50 – < 55	1
55 – < 60	3
60 – < 65	4
65 – < 70	5
70 – < 75	3
75 – < 80	2
Total	18

13. The number of textbooks sold by various bookshops during the second week of December was recorded. The results are summarised in the table below.

Number of books sold	Frequency
220–229	2
230–239	2
240–249	3
250–259	5
260–269	4
270–279	4
Total	20

a. **MC** The modal class of the data is given by the class interval(s):

A. 220–229 and 230–239 **B.** 250–259 **C.** 260–269 and 270–279
D. of both A and **E.** none of these

b. **MC** The class centre of the first class interval is:

A. 224 **B.** 224.5 **C.** 224.75 **D.** 225 **E.** 227

c. **MC** The median of the data is in the interval:

A. 230–239 **B.** 240–249 **C.** 250–259 **D.** 260–269 **E.** 270–279

d. **MC** The estimated mean of the data is:

A. 251 **B.** 252 **C.** 253 **D.** 254 **E.** 255

Understanding

14. A random sample was taken, composed of 30 people shopping at a supermarket on a Tuesday night. The amount of money (to the nearest dollar) spent by each person was recorded as follows:

6 32 66 17 45 1 19 52 36 23 28 20 7 47 39
6 68 28 54 9 10 58 40 12 25 49 74 63 41 13

a. Calculate the mean and median amount of money spent at the checkout by the people in this sample.
b. Group the data into class intervals of 10 and complete the frequency distribution table. Use this table to estimate the mean amount of money spent.
c. Add the cumulative frequency column to your table and fill it in. Hence, construct the ogive. Use the ogive to estimate the median.
d. Compare the mean and the median of the original data from part **a** with the mean and the median obtained for grouped data in parts **b** and **c**. Explain if the estimates obtained in parts **b** and **c** were good enough.

15. Answer the following question and show your working.

a. Add one more number to the set of data 3, 4, 4, 6 so that the mean of a new set is equal to its median.
b. Design a set of five numbers so that mean = median = mode = 5.
c. In the set of numbers 2, 5, 8, 10, 15, change one number so that the median remains unchanged while the mean increases by 1.

16. Thirty men were asked to reveal the number of hours they spent doing housework each week. The results are detailed below.

1	5	2	12	2	6	2	8	14	18
0	1	1	8	20	25	3	0	1	2
7	10	12	1	5	1	18	0	2	2

a. Present the data in a frequency distribution table. (Use class intervals of 0–4, 5–9 etc.)
b. Use your table to estimate the mean number of hours that the men spent doing housework.
c. Determine the median class for hours spent by the men at housework.
d. Identify the modal class for hours spent by the men at housework.

Reasoning

17. The data shown give the age of 25 patients admitted to the emergency ward of a hospital.

18	16	6	75	24
23	82	75	25	21
43	19	84	76	31
78	24	20	63	79
80	20	23	17	19

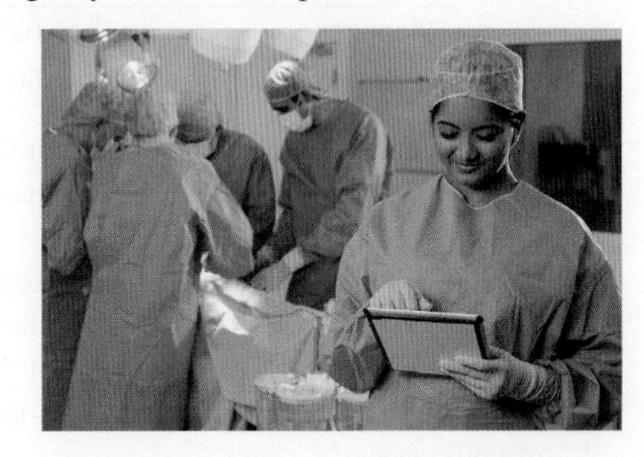

a. Present the data in a frequency distribution table. (Use class intervals of $0 - < 15$, $15 - < 30$ and so on.)
b. Draw a histogram of the data.
c. Suggest a word to describe the pattern of the data in this distribution.
d. Use your table to estimate the mean age of patients admitted.
e. Determine the median class for age of patients admitted.
f. Identify the modal class for age of patients admitted.
g. Draw an ogive of the data.
h. Use the ogive to determine the median age.
i. Explain if any of your statistics (mean, median or mode) give a clear representation of the typical age of an emergency ward patient.
j. Give some reasons which could explain the pattern of the distribution of data in this question.

18. **MC** In a set of data there is one score that is extremely small when compared to all the others. This outlying value is most likely to:

A. have greatest effect upon the mean of the data
B. have greatest effect upon the median of the data
C. have greatest effect upon the mode of the data
D. have very little effect on any of the statistics as we are told that the number is extremely small
E. none of these

19. The batting scores for two cricket players over 6 innings are as follows:

Player A 31, 34, 42, 28, 30, 41
Player B 0, 0, 1, 0, 250, 0

a. Calculate the mean score for each player.
b. State which player appears to be better, based upon mean result. Justify your answer.
c. Determine the median score for each player.
d. State which player appears to be better when the decision is based on the median result. Justify your answer.
e. State which player do you think would be the most useful to have in a cricket team. Justify your answer. Explain how can the mean result sometimes lead to a misleading conclusion.

20. The following frequency table gives the number of employees in different salary brackets for a small manufacturing plant.

Position	Salary ($)	Number of employees
Machine operator	18 000	50
Machine mechanic	20 000	15
Floor steward	24 000	10
Manager	62 000	4
Chief executive officer	80 000	1

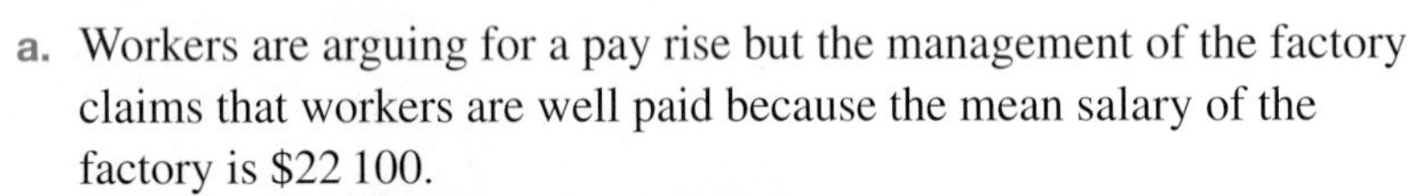

a. Workers are arguing for a pay rise but the management of the factory claims that workers are well paid because the mean salary of the factory is $22 100.
Explain whether the management is being honest.
b. Suppose that you were representing the factory workers and had to write a short submission in support of the pay rise. How could you explain the management's claim? Quote some other statistics in favour of your case.

21. The resting pulse rate of 20 female athletes was measured. The results are detailed below.

50	52	48	52	71	61	30	45	42	48
43	47	51	62	34	61	44	54	38	40

a. Construct a frequency distribution table. (Use class sizes of $1 - < 10$, $10 - < 20$ and so on.)
b. Use your table to estimate the mean of the data.
c. Determine the median class of the data.
d. Identify the modal class of the data.
e. Draw an ogive of the data. (You may like to use a graphics calculator for this.)
f. Use the ogive to determine the median pulse rate.

22. Design a set of five numbers with:

a. $\text{mean} = \text{median} = \text{mode}$
b. $\text{mean} > \text{median} > \text{mode}$
c. $\text{mean} < \text{median} = \text{mode}$.

Problem solving

23. The numbers 15, a, 17, b, 22, c, 10 and d have a mean of 14. Calculate the mean of a, b, c and d.

24. The numbers m, n, p, q, r, and s have a mean of a while x, y and z have a mean of b. Calculate the mean of all nine numbers.

25. The mean and median of six two-digit prime numbers is 39 and the mode is 31. The smallest number is 13. Determine the six numbers.

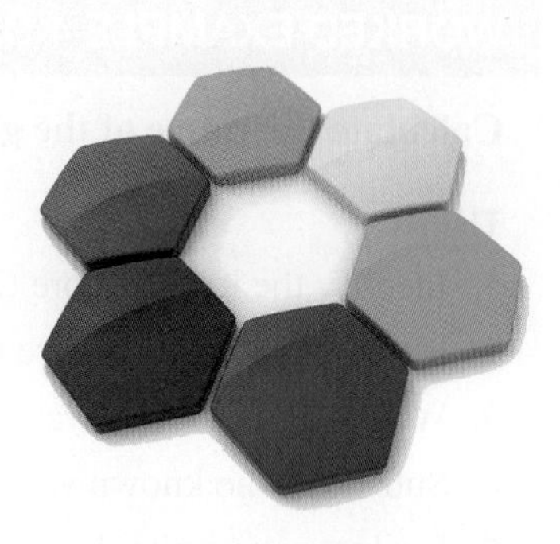

12.3 Measures of spread

LEARNING INTENTION

At the end of this subtopic you should be able to:
- calculate the range and interquartile range of a data set.

12.3.1 Measures of spread

eles-4952

- **Measures of spread** describe how far data values are spread from the centre or from each other.
- A shoe store proprietor has stores in Newcastle and Wollongong. The number of pairs of shoes sold each day over one week is recorded below.

$$\begin{array}{ll} \text{Newcastle:} & 45, 60, 50, 55, 48, 40, 52 \\ \text{Wollongong:} & 20, 85, 50, 15, 30, 60, 90 \end{array}$$

In each of these data sets consider the measures of central tendency.

Newcastle:	Mean = 50	Wollongong:	Mean = 50
	Median = 50		Median = 50
	No mode		No mode

With these measures being the same for both data sets we could come to the conclusion that both data sets are very similar; however, if we look at the data sets, they are very different. We can see that the data for Newcastle are very clustered around the mean, whereas the Wollongong data are spread out more.
- The data from Newcastle are between 40 and 60, whereas the Wollongong data are between 15 and 90.
- **Range** and **interquartile range (IQR)** are both measures of spread.

Range

- The most basic measure of spread is the range.
- The range is defined as the difference between the highest and the lowest values in the set of data.

Calculating the range of a data set

$$\textbf{Range} = \textbf{highest score} - \textbf{lowest score}$$
$$= X_{\max} - X_{\min}$$

WORKED EXAMPLE 4 Calculating the range of a data set

Calculate the range of the given data set: 2.1, 3.5, 3.9, 4.0, 4.7, 4.8, 5.2.

THINK	WRITE
1. Identify the lowest score (X_{min}) of the data set.	Lowest score = 2.1
2. Identify the highest score (X_{max}) of the data set.	Highest score = 5.2
3. Write the rule for the range.	Range $= X_{max} - X_{min}$
4. Substitute the known values into the rule.	$= 5.2 - 2.1$
5. Evaluate and write the answer.	$= 3.1$

Interquartile range

- The interquartile range (IQR) is the range of the middle 50% of all the scores in an ordered set. When calculating the interquartile range, the data are first organised into quartiles, each containing 25% of the data.
- The word 'quartile' comes from the word 'quarter'.

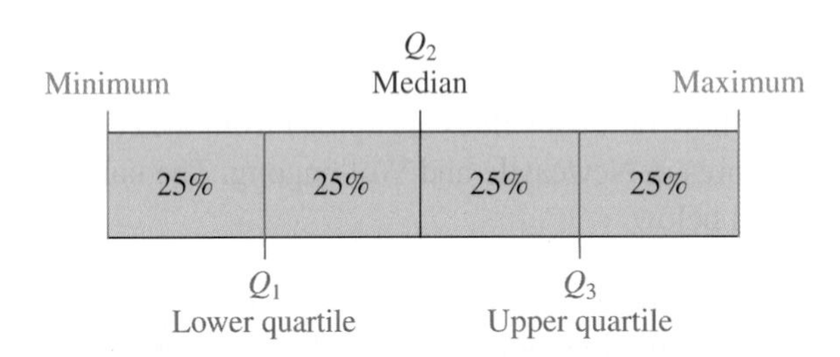

- The lower quartile (Q_1) is the median of the lower half of the data set.
- The upper quartile (Q_3) is the median of the upper half of the data set.

Calculating the IQR

$$\textbf{Interquartile range (IQR)} = \textbf{upper quartile} - \textbf{lower quartile}$$
$$= Q_{upper} - Q_{lower}$$
$$= Q_3 - Q_1$$

- The IQR is not affected by extremely large or extremely small data values (**outliers**), so in some circumstances the IQR is a better indicator of the spread of data than the range.

WORKED EXAMPLE 5 Calculating the IQR of a data set

Calculate the interquartile range (IQR) of the following set of data: 3, 2, 8, 6, 1, 5, 3, 7, 6.

THINK	WRITE
1. Arrange the scores in order.	1 2 3 3 5 6 6 7 8
2. Locate the median and use it to divide the data into two halves. *Note:* The median is the 5th score in this data set and should not be included in the lower or upper ends of the data.	1 2 3 3 5 6 6 7 8

3. Calculate Q_1, the median of the lower half of the data.

$$Q_1 = \frac{2+3}{2} = \frac{5}{2} = 2.5$$

4. Calculate Q_3, the median of the upper half of the data.

$$Q_3 = \frac{6+7}{2} = \frac{13}{2} = 6.5$$

5. Calculate the interquartile range.

$$\text{IQR} = Q_3 - Q_1 = 6.5 - 2.5$$

6. Write the answer.

$$= 4$$

TI | THINK / **DISPLAY/WRITE**

1. In a new problem, on a Lists & Spreadsheet page, label column A as 'xvalues'. Enter the values from the data set. Press ENTER after entering each value.

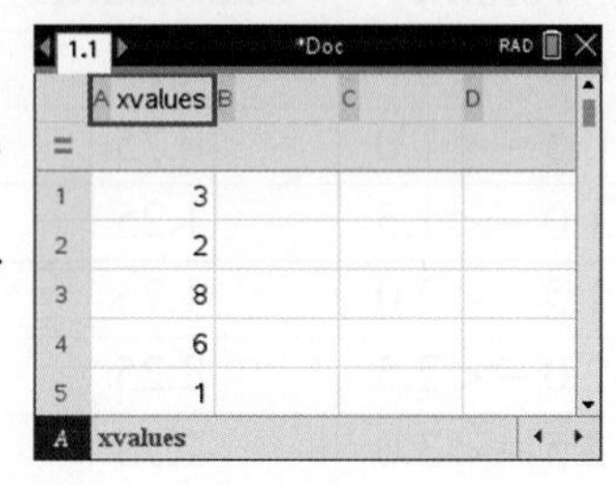

2. To find the summary statistics, open a Calculator page and press:
   - MENU
   - 6: Statistics
   - 1: Stat Calculations
   - 1: One-Variable Statistics

   Select 1 as the number of lists. Then on the One-Variable Statistics page select 'xvalues' as the X1 List and leave the Frequency as 1. Leave the remaining fields empty, TAB to OK, and then press ENTER. The summary statistics are shown.

"σx := σnx" 2.26623
"n" 9.
"MinX" 1.
"Q₁X" 2.5
"MedianX" 5.
"Q₃X" 6.5
"MaxX" 8.
SSX := Σ(x−x̄)² 46.2222

The IQR $= Q_3 - Q_1 = 6.5 - 2.5 = 4$

CASIO | THINK / **DISPLAY/WRITE**

1. On the Statistics screen, label list1 as 'x' and enter the values as shown in the table. Press EXE after entering each value.

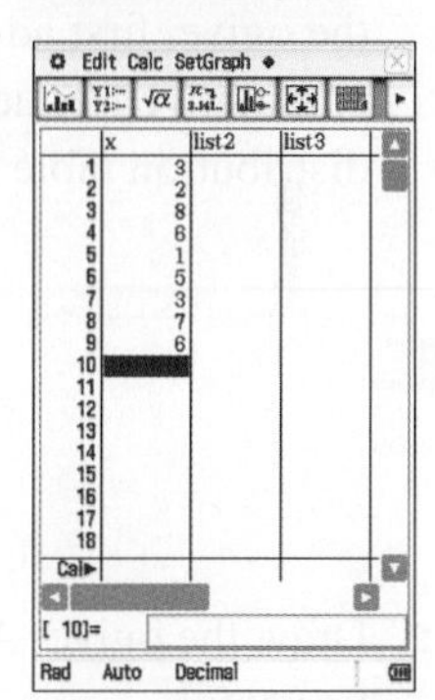

2. To find the summary statistics, tap:
   - Calc
   - One-Variable

   Set values as:
   - XList: main\x
   - Freq: 1

   Tap OK.
   Calculate the IQR $= Q_3 - Q_1$

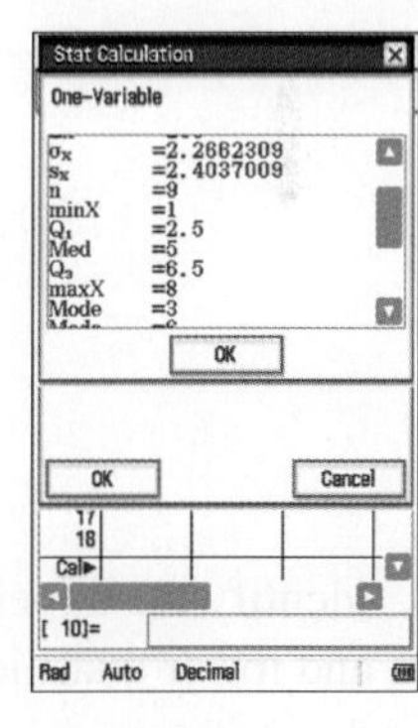

The IQR $= Q_3 - Q_1 = 6.5 - 2.5 = 4$

Determining the IQR from a graph

- When data are presented in a frequency distribution table, either ungrouped or grouped, the interquartile range is found by drawing an ogive.

WORKED EXAMPLE 6 Calculating the IQR from a graph

The following frequency distribution table gives the number of customers who order different volumes of concrete from a readymix concrete company during the course of a day. Calculate the interquartile range of the data.

Volume (m^3)	Frequency
0.0 – < 0.5	15
0.5 – < 1.0	12
1.0 – < 1.5	10

Volume (m^3)	Frequency
1.5 – < 2.0	8
2.0 – < 2.5	2
2.5 – < 3.0	4

THINK

WRITE/DRAW

1. To calculate the 25th and 75th percentiles from the ogive, first add a class centre column and a cumulative frequency column to the frequency distribution table and fill them in.

Volume	Class centre	f	cf
0.0 – < 0.5	0.25	15	15
0.5 – < 1.0	0.75	12	27
1.0 – < 1.5	1.25	10	37
1.5 – < 2.0	1.75	8	45
2.0 – < 2.5	2.25	2	47
2.5 – < 3.0	2.75	4	51

2. Draw the ogive. A percentage axis will be useful.

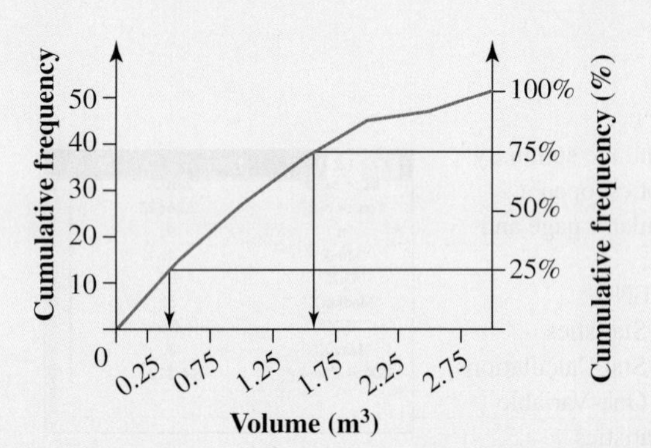

3. Identify the upper quartile (75th percentile) and lower quartile (25th percentile) from the ogive.

$Q_3 = 1.6\,\text{m}^3$

$Q_1 = 0.4\,\text{m}^3$

4. The interquartile range is the difference between the upper and lower quartiles.

$$\begin{aligned}\text{IQR} &= Q_3 - Q_1 \\ &= 1.6 - 0.4 \\ &= 1.2\,\text{m}^3\end{aligned}$$

on Resources

eWorkbook Topic 12 Workbook (worksheets, code puzzle and project) (ewbk-2038)

Interactivities Individual pathway interactivity: Measures of spread (int-4622)

Range (int-3822)

The interquartile range (int-4813)

Exercise 12.3 Measures of spread

learn**on**

Individual pathways

■ PRACTISE	■ CONSOLIDATE	■ MASTER
1, 4, 7, 10, 13	2, 6, 8, 11, 14	3, 5, 9, 12, 15

To answer questions online and to receive **immediate corrective feedback** and **fully worked solutions** for all questions, go to your learnON title at www.jacplus.com.au.

Fluency

1. **WE4** Calculate the range for each of the following sets of data.
   a. 4, 3, 9, 12, 8, 17, 2, 16
   b. 49.5, 13.7, 12.3, 36.5, 89.4, 27.8, 53.4, 66.8
   c. $7\frac{1}{2}, 12\frac{3}{4}, 5\frac{1}{4}, 8\frac{2}{3}, 9\frac{1}{6}, 3\frac{3}{4}$
2. **WE5** Calculate the interquartile range (IQR) for the following sets of data.
   a. 3, 5, 8, 9, 12, 14
   b. 7, 10, 11, 14, 17, 23
   c. 66, 68, 68, 70, 71, 74, 79, 80
   d. 19, 25, 72, 44, 68, 24, 51, 59, 36
3. The following stem-and-leaf plot shows the mass of newborn babies (rounded to the nearest 100g). Calculate the:
   a. range of the data
   b. IQR of the data.

Key: $1^* \mid 9 = 1.9\,kg$

Stem	*Leaf*
1*	9
2	2 4
2*	6 7 8 9
3	0 0 1 2 3 4
3*	5 5 6 7 8 8 8 9
4	0 1 3 4 4
4*	5 6 6 8 9
5	0 1 2 2

4. Use the ogive shown to calculate the interquartile range of the data.

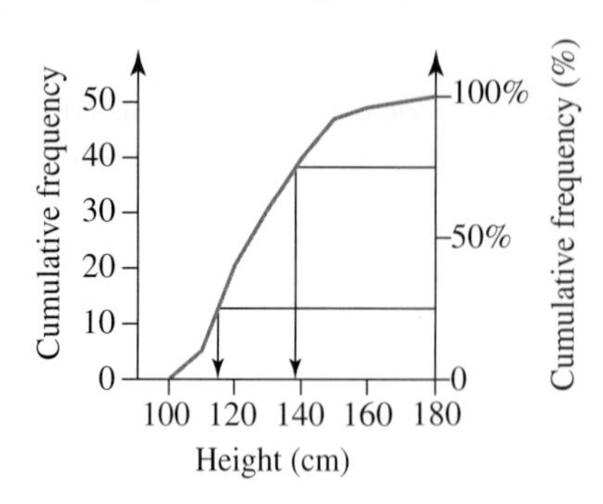

5. **WE6** The following frequency distribution table gives the amount of time spent by 50 people shopping for Christmas presents.

Time (h)	0 – < 0.5	0.5 – < 1	1 – < 1.5	1.5 – < 2	2 – < 2.5	2.5 – < 3	3 – < 3.5	3.5 – < 4
Frequency	1	2	7	15	13	8	2	2

Estimate the IQR of the data.

6. **MC** Calculate the interquartile range of the following data:

17, 18, 18, 19, 20, 21, 21, 23, 25

A. 8 **B.** 18 **C.** 4 **D.** 20 **E.** 25

Understanding

7. The following frequency distribution table shows the life expectancy in hours of 40 household batteries.

Life (h)	50 – < 55	55 – < 60	60 – < 65	65 – < 70	70 – < 75	75 – < 80
Frequency	4	10	12	8	5	1

a. Draw an ogive curve that represents the data in the table above.
b. Use the ogive to answer the following questions.
  i. Calculate the median score.
  ii. Determine the upper and lower quartiles.
  iii. Calculate the interquartile range.
  iv. Identify the number of batteries that lasted less than 60 hours.
  v. Identify the number of batteries that lasted 70 hours or more.

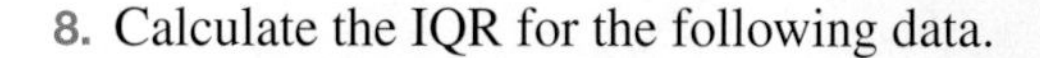

8. Calculate the IQR for the following data.

Class interval	**Frequency**
120 – < 130	2
130 – < 140	3
140 – < 150	9
150 – < 160	14
160 – < 170	10
170 – < 180	8
180 – < 190	6
190 – < 200	3

9. For each of the following sets of data, state:
   i. the range and
   ii. the IQR of each set.

   a. 6, 9, 12, 13, 20, 22, 26, 29
   b. 7, 15, 2, 26, 47, 19, 9, 33, 38
   c. 120, 99, 101, 136, 119, 87, 123, 115, 107, 100

Reasoning

10. Explain what the measures of spread tell us about a set of data.

11. As newly appointed coach of Terrorolo's Meteors netball team, Kate decided to record each player's statistics for the previous season. The number of goals scored by the leading goal shooter was:

1 3 8 18 19 23 25 25 25 26 27 28
28 28 28 29 29 30 30 33 35 36 37 40

   a. Calculate the mean of the data.
   b. Calculate the median of the data.
   c. Calculate the range of the data.
   d. Determine the interquartile range of the data.
   e. There are three scores that are much lower than most. Explain the effect these scores have on the summary statistics.

12. The following back-to-back stem-and-leaf plot shows the ages of 30 pairs of men and women when entering their first marriage.

Key: 1 | 6 = 16 years old

Leaf: Men	*Stem*	*Leaf: Women*
998	1	67789
99887644320	2	001234567789
9888655432	3	01223479
6300	4	1248
60	5	2

   a. Determine the mean, median, range and interquartile range of each set.
   b. Write a short paragraph comparing the two distributions.

Problem solving

13. Calculate the mean, median, mode, range and IQR of the following data collected when the temperature of the soil around 25 germinating seedlings was recorded:
28.9, 27.4, 23.6, 25.6, 21.1, 22.9, 29.6, 25.7, 27.4, 23.6, 22.4, 24.6, 21.8, 26.4, 24.9, 25.0, 23.5, 26.1, 23.6, 25.3, 29.5, 23.5, 22.0, 27.9, 23.6.

14. Four positive numbers a, b, c and d have a mean of 12, a median and mode of 9 and a range of 14. Determine the values of a, b, c and d.

15. A set of five positive integer scores have the following summary statistics:

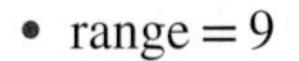

- range = 9
- median = 6
- $Q_1 = 3$ and $Q_3 = 9$.

   a. Explain whether the five scores could be 1, 4, 6, 9 and 10.
   b. A sixth score is added to the set. Determine whether there is a score that will maintain the summary statistics given above. Justify your answer.

12.4 Box plots

LEARNING INTENTION

At the end of this subtopic you should be able to:
- calculate the five-number summary for a set of data
- draw a box plot showing the five-number summary of a data set
- calculate outliers in a data set
- describe skewness of distributions
- compare box plots to dot plots or histograms
- draw parallel box plots and compare sets of data.

12.4.1 Five-number summary

eles-4953

- A five-number summary is a list consisting of the lowest score (X_{min}), lower quartile (Q_1), median (Q_2), upper quartile (Q_3) and greatest score (X_{max}) of a set of data.

WORKED EXAMPLE 7 Calculations using the five-number summary

From the following five-number summary, calculate:

a. the interquartile range

b. the range.

X_{min}	Q_1	Median (Q_2)	Q_3	X_{max}
29	37	39	44	48

THINK	WRITE
a. The interquartile range is the difference between the upper and lower quartiles.	a. $\text{IQR} = Q_3 - Q_1$ $= 44 - 37$ $= 7$
b. The range is the difference between the greatest score and the lowest score.	b. $\text{Range} = X_{max} - X_{min}$ $= 48 - 29$ $= 19$

12.4.2 Box plots

eles-4954

- A **box plot** is a graph of the five-number summary.
- Box plots consist of a central divided box with attached whiskers.
- The box spans the interquartile range.
- The median is marked by a vertical line drawn inside the box.
- The whiskers indicate the range of scores:

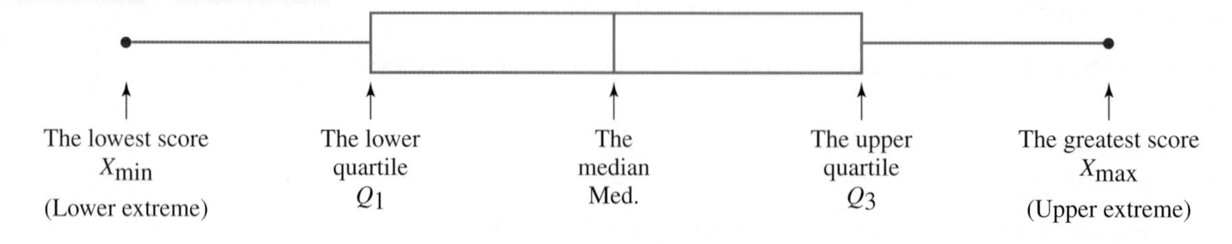

- Box plots are *always drawn to scale*.
- They are presented either with the five-number summary figures attached as labels (diagram at right) or with a scale presented alongside the box plot like the diagram below. They can also be drawn vertically.

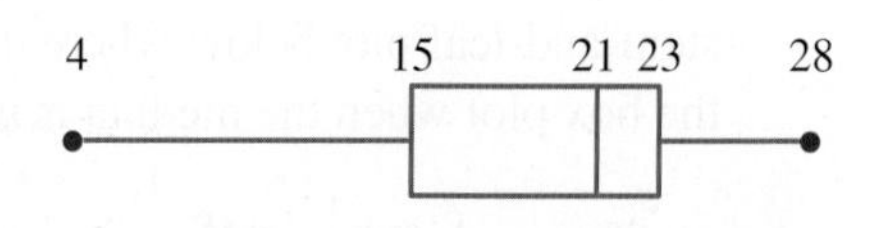

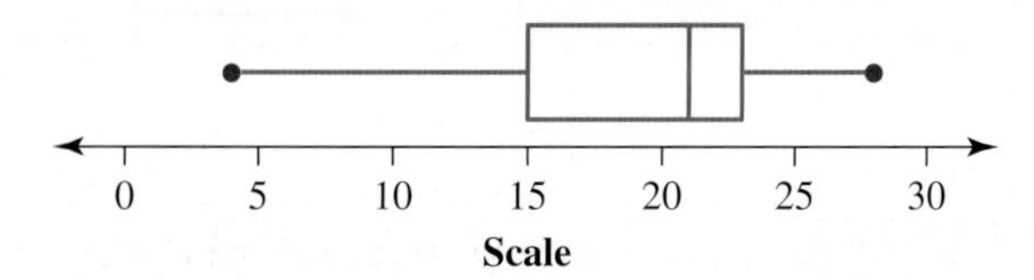

Identification of extreme values or outliers

- If an extreme value or outlier occurs in a set of data, it can be denoted by a small cross on the box plot. The whisker is then shortened to the next largest (or smallest) figure.
- The box plot below shows that the lowest score was 5. This was an extreme value as the rest of the scores were located within the range 15 to 42.
- Outliers are still included when calculating the range of the data.

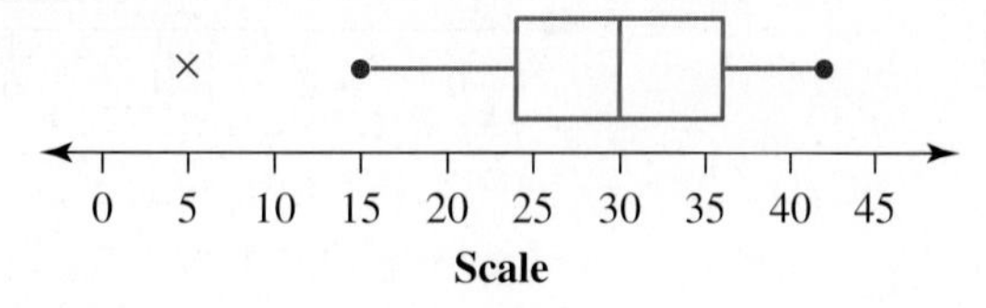

- Outliers sit $1.5 \times \text{IQR}$ or greater away from Q_1 or Q_3.

Identifying outliers

$$\textbf{Lower limit} = Q_1 - 1.5 \times \textbf{IQR}$$

$$\textbf{Upper limit} = Q_3 + 1.5 \times \textbf{IQR}$$

Any scores that sit outside these limits are considered outliers.

Symmetry and skewness in distributions

- A **symmetrical** plot has data that are evenly spaced around a central point. Examples of a stem-and-leaf plot and a symmetrical box plot are shown below.

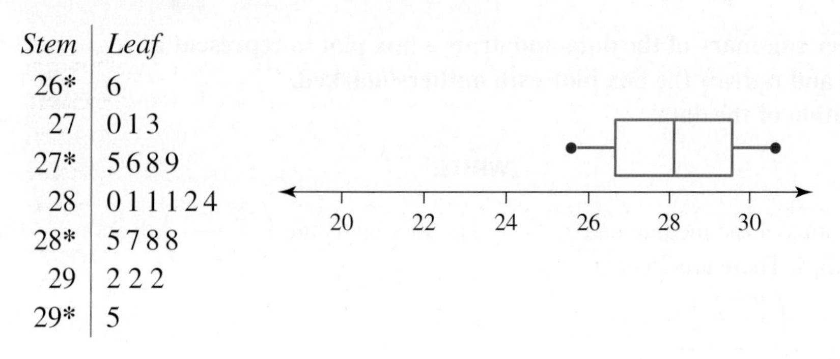

Stem	*Leaf*
26*	6
27	0 1 3
27*	5 6 8 9
28	0 1 1 1 2 4
28*	5 7 8 8
29	2 2 2
29*	5

- A **negatively skewed** plot has larger amounts of data at the higher end. This is illustrated by the stem-and-leaf plot below where the leaves increase in length as the data increase in value. It is illustrated on the box plot when the median is much closer to the maximum value than the minimum value.

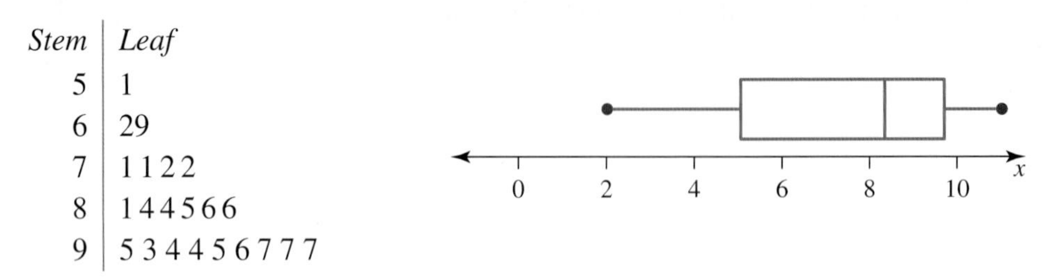

Stem	*Leaf*
5	1
6	29
7	1 1 2 2
8	1 4 4 5 6 6
9	5 3 4 4 5 6 7 7 7

- A **positively skewed** plot has larger amounts of data at the lower end. This is illustrated on the stem-and-leaf plot below where the leaves increase in length as the data decrease in value. It is illustrated on the box plot when the median is much closer to the minimum value than the maximum value.

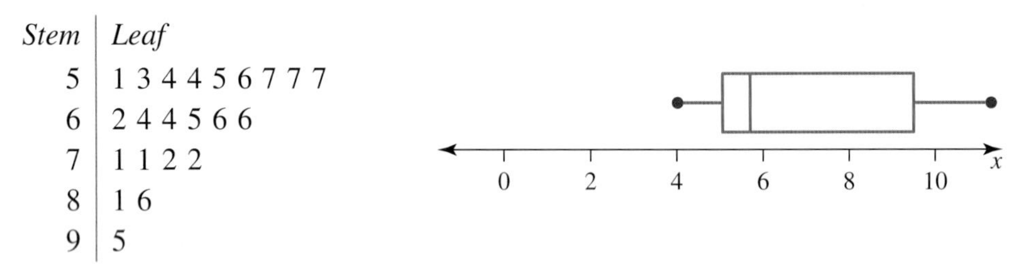

Stem	*Leaf*
5	1 3 4 4 5 6 7 7 7
6	2 4 4 5 6 6
7	1 1 2 2
8	1 6
9	5

WORKED EXAMPLE 8 Drawing a box plot

The following stem-and-leaf plot gives the speed of 25 cars caught by a roadside speed camera.

Key: 8 | 2 = 82 km/h, 8* | 6 = 86 km/h

Stem	***Leaf***
8	**2 2 4 4 4 4**
8*	**5 5 6 6 7 9 9 9**
9	**0 1 1 2 4**
9*	**5 6 9**
10	**0 2**
10*	
11	**4**

a. **Prepare a five-number summary of the data and draw a box plot to represent it.**
b. **Identify any outliers and redraw the box plot with outliers marked.**
c. **Describe the distribution of the data.**

THINK

1. First identify the positions of the median and upper and lower quartiles. There are 25 data values. The median is the $\left(\frac{n+1}{2}\right)$ th score.
The lower quartile is the median of the lower half of the data. The upper quartile is the median of the upper half of the data (each half contains 12 scores).

WRITE

The median is the $\left(\frac{25+1}{2}\right)$ th score — that is, the 13th score.
Q_1 is the $\left(\frac{12+1}{2}\right)$ th score in the lower half — that is, the 6.5th score. That is, halfway between the 6th and 7th scores.
Q_3 is halfway between the 6th and 7th scores in the upper half of the data.

2. Mark the positions of the median and upper and lower quartiles on the stem-and-leaf plot.

Key: 8 | 2 = 82 km/h
8* | 6 = 86 km/h

Stem	*Leaf*
8	2 2 4 4 4 4 | ← $Q1$
8*	5 5 6 6 7 9 ⑨ 9 ← Median
9	0 1 1 2 4 | ← $Q3$
9*	5 6 9
10	0 2
10*	
11	4

a. Write the five-number summary:
The lowest score is 82.
The lower quartile is between 84 and 85; that is, 84.5.
The median is 89.
The upper quartile is between 94 and 95; that is, 94.5.
The greatest score is 114.
Draw the box plot for this summary.

a. Five-number summary:

X_{min}	Q_1	Q_2	Q_3	X_{max}
82	84.5	89	94.5	114

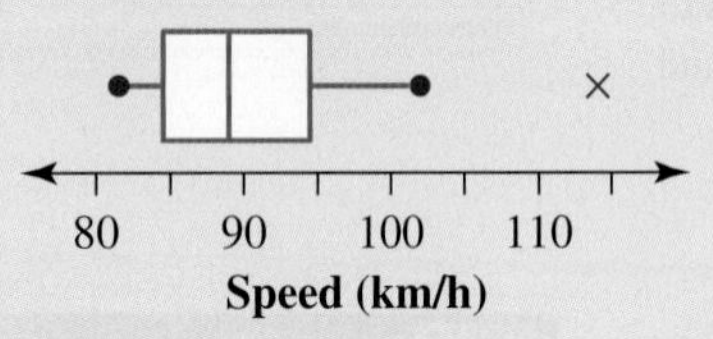

b. 1. Calculate the IQR.

b. $IQR = Q_3 - Q_1$
$= 94.5 - 84.5$
$= 10$

2. Calculate the lower and upper limits.

$\text{Lower limit} = 84.5 - 1.5 \times 10$
$= 69.5$

$\text{Upper limit} = 94.5 + 1.5 \times 10$
$= 109.5$

3. Identify the outliers.

114 is above the upper limit of 109.5, so it is an outlier.

4. Redraw the box plot, including the outlier marked as a cross. Draw the whisker to the next largest figure, 102.

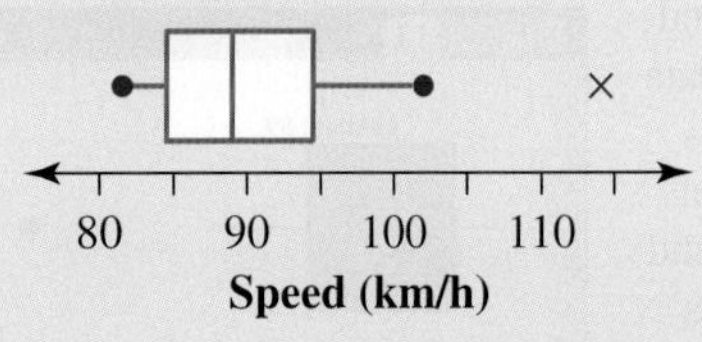

c. Describe the distribution.

c. The data is positively skewed.

TI | THINK

a.

1. In a new problem, on a Lists & Spreadsheet page, label column A as 'cars' and enter the values from the stem-and-leaf plot. Press ENTER after each value.

2. To find a five-point summary of the data, on a Calculator page press:
   - CATALOG
   - 1
   - F

   Then use the down arrow to scroll down to FiveNumSummary and press ENTER.

3. Press VAR and select 'cars' Complete the entry line as: FiveNumSummary cars and press ENTER. Then press VAR and select 'stat.results' and press ENTER.

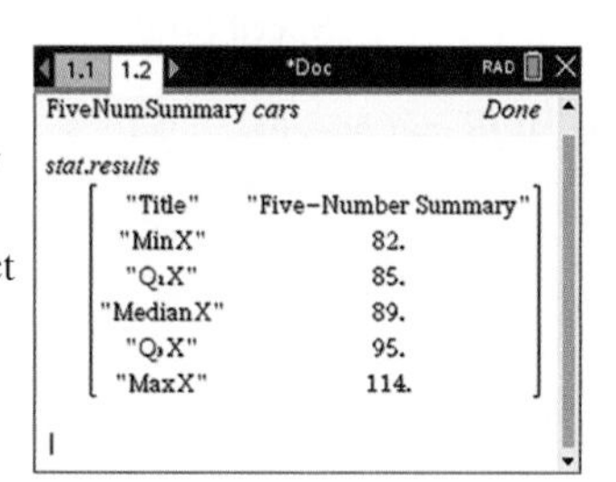

b.

To construct the box-and-whisker plot, open a Data & Statistics page. Press TAB to locate the label of the horizontal axis and select the variable 'cars'. Then press:
- MENU
- 1: Plot Type
- 2: Box Plot

To change the colour, place the pointer over one of the data points. Then press CTRL MENU. Then press:
- 6: Color
- 2: Fill Color.

Select whichever colour you like from the palette. Press ENTER.

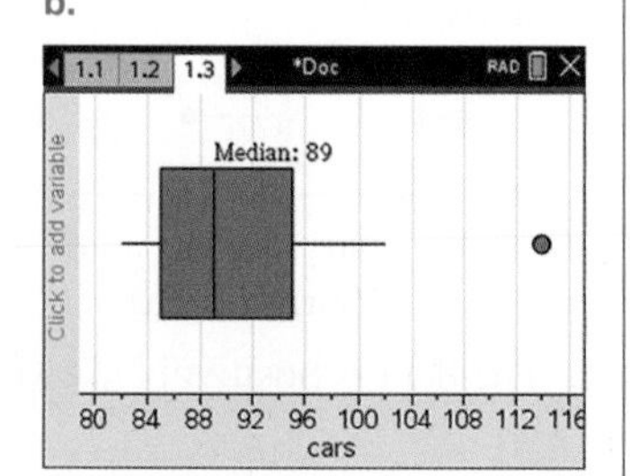

The box-and-whisker plot is displayed. As you scroll over the box-and-whisker plot, the values of the five-number summary statistics are displayed. The data are skewed (positively).

CASIO | THINK

a.

1. On the Statistics screen, label list1 as 'cars' and enter the values from the stem-and-leaf plot. Press EXE after entering each value.

2. To find the summary statistics, tap:
   - Calc
   - One-Variable

   Set values as:
   - XList: main\cars
   - Freq: 1

3. Tap OK. Scroll down to find more statistics.
   minX = 82
   Q1 = 84.5
   Med = 89
   Q3 = 95.5
   maxX = 114

b.

To construct the box-and-whisker plot, tap:
- SetGraph
- Setting...

Set:
- Type: MedBox
- XList: main\cars
- Freq: 1

  Then tap the graphing icon.

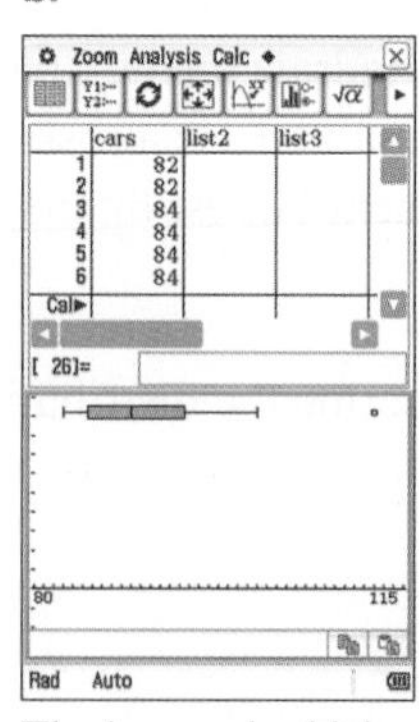

The box-and-whisker plot is displayed.

12.4.3 Comparing different graphical representations

eles-4956

Box plots and dot plots

- Box plots are a concise summary of data. A box plot can be directly related to a dot plot.
- **Dot plots** display each data value represented by a dot placed on a number line.
- The following data are the amount of money (in $) that a group of 27 five-year-olds had with them on a day visiting the zoo with their parents.

0	0.85	0	1.8	1.65	8.45	3.75	0.55	4.1	2.4	2.15
1.2	1.35	0.9	3.45	1	0	0	1.45	1.25	1.7	2.65
1.85	4.75	3.9	1.15							

- The dot plot below and its comparative box plot show the distribution of these data.

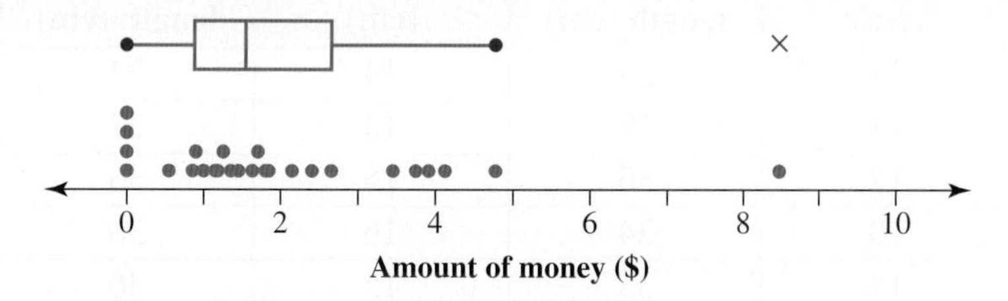

Both graphs indicate that the data is positively skewed and both graphs indicate the presence of the outlier. However, the box plot provides an excellent summary of the centre and spread of the distribution.

Box plots and histograms

- **Histograms** are graphs that display continuous numerical variables and do not retain all original data.
- The following data are the number of minutes, rounded to the nearest minute, that forty Year 10 students take to travel to their school on a particular day.

15	22	14	12	21	34	19	11	13	0	16
4	23	8	12	18	24	17	14	3	10	12
9	15	20	5	19	13	17	11	16	19	24
12	7	14	17	10	14	23				

The data are displayed in the histogram and box plot shown.

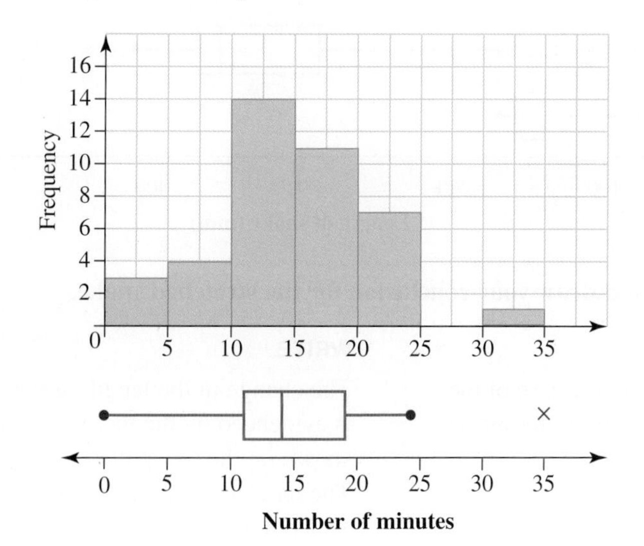

Both graphs indicate that the data is slightly positively skewed. The histogram clearly shows the frequencies of each class interval. Neither graph displays the original values. The histogram does not give precise information about the centre, but the distribution of the data is visible. However, the box plot shows the presence of an outlier and provides an excellent summary of the centre and spread of the distribution.

Parallel box plots

- A major reason for developing statistical skills is to be able to make comparisons between sets of data.

WORKED EXAMPLE 9 Comparing two sets of data

Each member of a class was given a jelly snake to stretch. They each measured the initial length of their snake to the nearest centimetre and then slowly stretched the snake to make it as long as possible. They then measured the maximum length of the snake by recording how far it had stretched at the time it broke. The results were recorded in the following table.

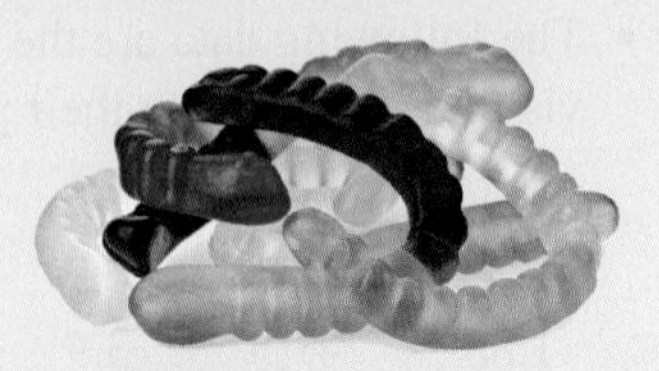

Initial length (cm)	Stretched length (cm)	Initial length (cm)	Stretched length (cm)
13	29	14	27
14	28	13	27
17	36	15	36
10	24	16	36
14	35	15	36
16	36	16	34
15	37	17	35
16	37	12	27
14	30	9	17
16	33	16	41
17	36	17	38
16	38	16	36
17	38	17	41
14	31	16	33
17	40	11	21

The above data was drawn on parallel box plots as shown below.

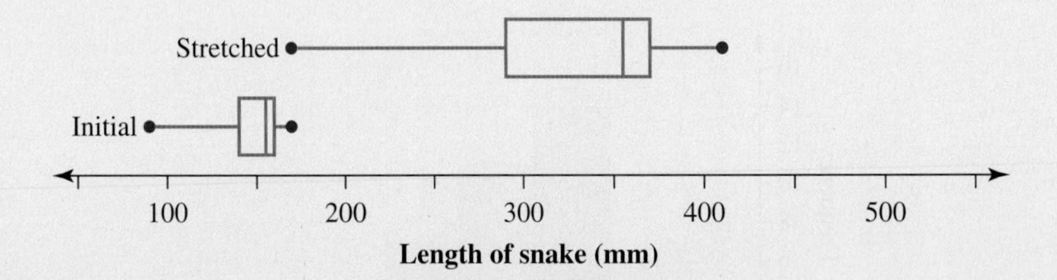

Compare the data sets and draw your conclusion for the stretched snake.

THINK	WRITE
1. Determine the median in the case of the initial and stretched length of the snake.	The change in the length of the snake when stretched is evidenced by the increased median and spread shown on the box plots. The median snake length before being stretched was 15.5 cm, but the median snake length after being stretched was 35 cm.
2. Draw your conclusion.	The range increased after stretching, as did the IQR.

Resources

eWorkbook Topic 12 Workbook (worksheets, code puzzle and project) (ewbk-2038)

Interactivities Individual pathway interactivity: Box-and-whisker plots (int-4623)
Skewness (int-3823)
Box plots (int-6245)
Parallel box plots (int-6248)

Exercise 12.4 Box plots

learnON

Individual pathways

■ PRACTISE	■ CONSOLIDATE	■ MASTER
1, 4, 6, 9, 12, 13, 16, 19	2, 5, 7, 10, 14, 17, 20	3, 8, 11, 15, 18, 21, 22

To answer questions online and to receive **immediate corrective feedback** and **fully worked solutions** for all questions, go to your learnON title at www.jacplus.com.au.

Fluency

1. WE7 From the following five-number summary calculate:

X_{min}	Q_1	Median	Q_3	X_{max}
6	11	13	16	32

a. the interquartile range
b. the range.

2. From the following five-number summary calculate:

X_{min}	Q_1	Median	Q_3	X_{max}
101	119	122	125	128

a. the interquartile range
b. the range.

3. From the following five-number summary calculate:

X_{min}	Q_1	Median	Q_3	X_{max}
39.2	46.5	49.0	52.3	57.8

a. the interquartile range
b. the range.

4. The box plot shows the distribution of final points scored by a football team over a season's roster.

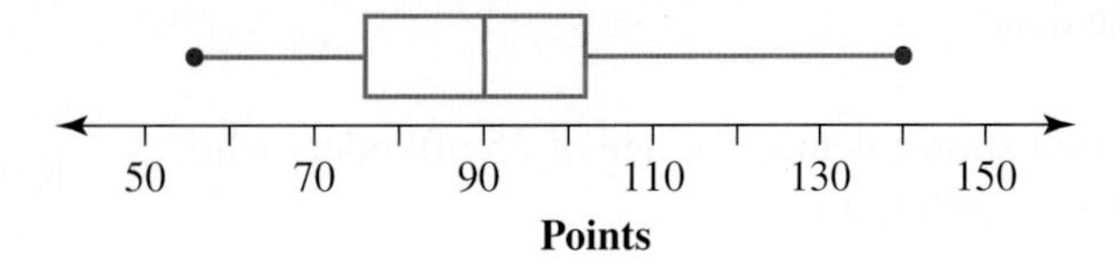

a. Identify the team's greatest points score.
b. Identify the team's least points score.
c. Calculate the team's median points score.
d. Calculate the range of points scored.
e. Calculate the interquartile range of points scored.

5. The box plot shows the distribution of data formed by counting the number of gummy bears in each of a large sample of packs.

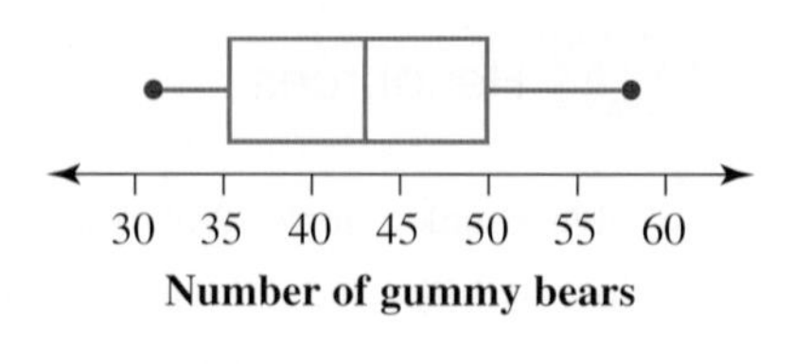

a. Identify the largest number of gummy bears in any pack.
b. Identify the smallest number of gummy bears in any pack.
c. Identify the median number of gummy bears in any pack.
d. Calculate the range of numbers of gummy bears per pack.
e. Calculate the interquartile range of gummy bears per pack.

Questions 6 to 8 refer to the following box plot:

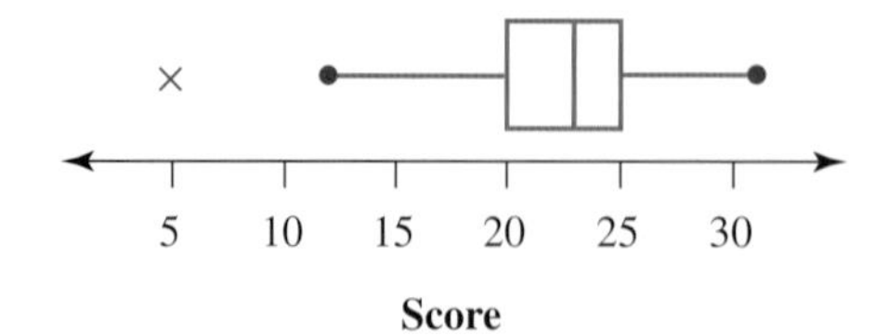

6. **MC** The median of the data is:

A. 20 B. 23 C. 25 D. 31 E. 5

7. **MC** The interquartile range of the data is:

A. 23 B. 26 C. 5 D. 20 to 25 E. 31

8. **MC** Select which of the following is not true of the data represented by the box plot.

A. One-quarter of the scores are between 5 and 20.
B. Half of the scores are between 20 and 25.
C. The lowest quarter of the data is spread over a wide range.
D. Most of the data are contained between the scores of 5 and 20.
E. One-third of the scores are between 5 and 20.

Understanding

9. The number of sales made each day by a salesperson is recorded over a 2-week period:

25, 31, 28, 43, 37, 43, 22, 45, 48, 33

a. Prepare a five-number summary of the data. (There is no need to draw a stem-and-leaf plot of the data. Just arrange them in order of size.)
b. Draw a box plot of the data.

10. The data below show monthly rainfall in millimetres.

J	F	M	A	M	J	J	A	S	O	N	D
10	12	21	23	39	22	15	11	22	37	45	30

a. Prepare a five-number summary of the data.
b. Draw a box plot of the data.

11. **WE8** The stem-and-leaf plot shown details the age of 25 offenders who were caught during random breath testing.

a. Prepare a five-number summary of the data.
b. Draw a box plot of the data.
c. Describe the distribution of the data.

Key: $1|8 = 18$ years

Stem	*Leaf*
1	8 8 9 9 9
2	0 0 0 1 1 3 4 6 9
3	0 1 2 7
4	2 5
5	3 6 8
6	6
7	4

12. The following stem-and-leaf plot details the price at which 30 blocks of land in a particular suburb sold for.

Key: 12|4 = \$124 000

Stem	*Leaf*
12	4 7 9
13	0 0 2 5 5
14	0 0 2 3 5 5 7 9 9
15	0 0 2 3 7 7 8
16	0 2 2 5 8
17	5

a. Prepare a five-number summary of the data.
b. Draw a box plot of the data.

13. Prepare comparative box plots for the following dot plots (using the same axis) and describe what each plot reveals about the data.

a. Number of sick days taken by workers last year at factory A

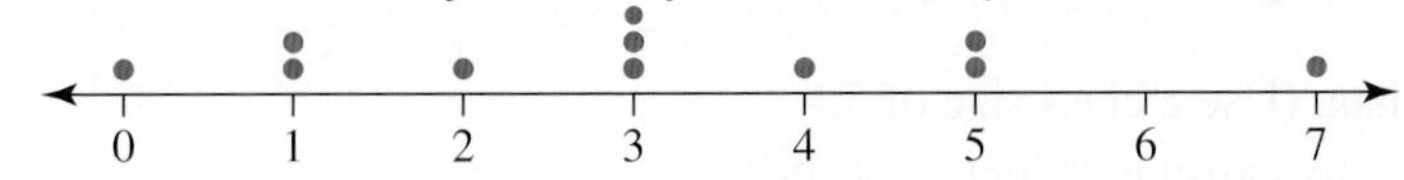

b. Number of sick days taken by workers last year at factory B

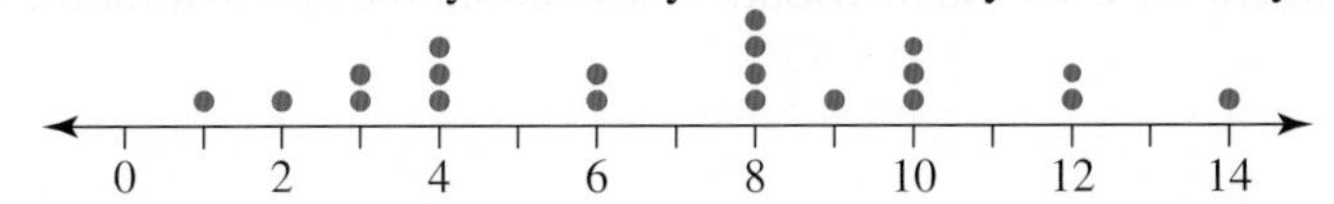

14. An investigation into the transport needs of an outer suburb community recorded the number of passengers boarding a bus during each of its journeys, as follows.

12 43 76 24 46 24 21 46 54 109 87 23 78
37 22 139 65 78 89 52 23 30 54 56 32 66 49

Display the data by constructing a histogram using class intervals of 20 and a comparative box plot on the same axis.

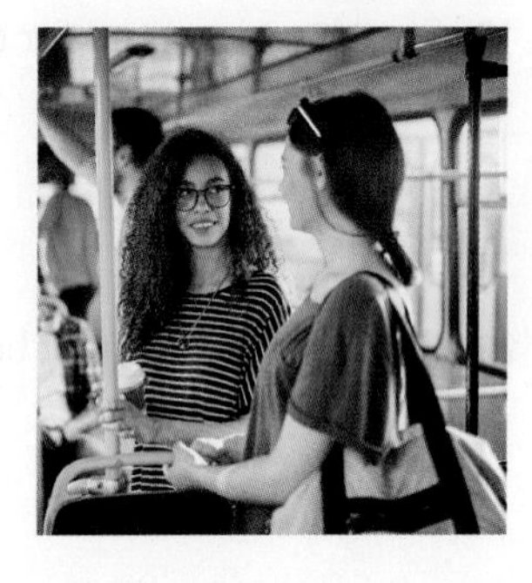

15. **WE12** At a weight-loss clinic, the following weights (in kilograms) were recorded before and after treatment.

Before	75	80	75	140	77	89	97	123	128	95	152	92
After	69	66	72	118	74	83	89	117	105	81	134	85

Before	85	90	95	132	87	109	87	129	135	85	137	102
After	79	84	90	124	83	102	84	115	125	81	123	94

a. Prepare a five-number summary for weight before and after treatment.
b. Draw parallel box plots for weight before and after treatment.
c. Comment on the comparison of weights before and after treatment.

Reasoning

16. Explain the advantages and disadvantages of box plots as a visual form of representing data.

17. The following data detail the number of hamburgers sold by a fast food outlet every day over a 4-week period.

M	T	W	T	F	S	S
125	144	132	148	187	172	181
134	157	152	126	155	183	188
131	121	165	129	143	182	181
152	163	150	148	152	179	181

a. Prepare a stem-and-leaf plot of the data. (Use a class size of 10.)
b. Draw a box plot of the data.
c. Comment on what these graphs tell you about hamburger sales.

18. The following data show the ages of 30 mothers upon the birth of their first baby.

22	21	18	33	17	23	22	24	24	20
25	29	32	18	19	22	23	24	28	20
31	22	19	17	23	48	25	18	23	20

a. Prepare a stem-and-leaf plot of the data. (Use a class size of 5.)
b. Draw a box plot of the data. Indicate any extreme values appropriately.
c. Describe the distribution in words. Comment on what the distribution says about the age that mothers have their first baby.

Problem solving

19. Sketch a histogram for the box plot shown.

20. Consider the box plot below which shows the number of weekly sales of houses by two real estate agencies.

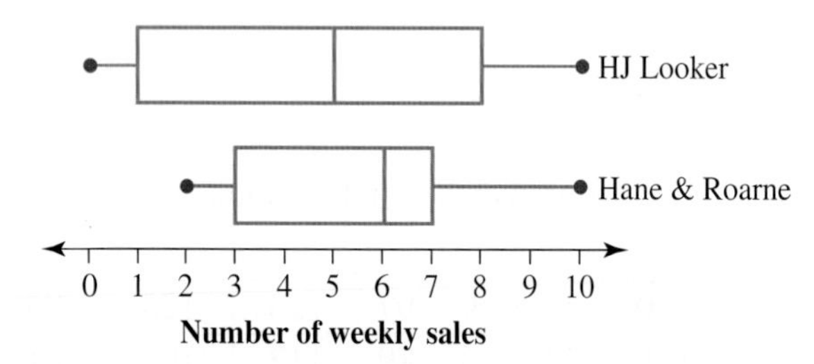

a. Determine the median number of weekly sales for each real estate agency.
b. State which agency had the greater range of sales. Justify your answer.
c. State which agency had the greater interquartile range of sales. Justify your answer.
d. State which agency performed better. Explain your answer.

21. Fifteen French restaurants were visited by three newspaper restaurant reviewers. The average price of a meal for a single person was investigated. The following box plot shows the results.

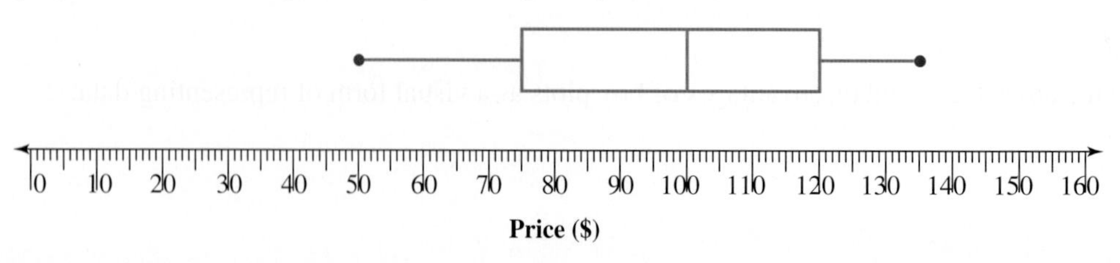

a. Identify the price of the cheapest meal.
b. Identify the price of the most expensive meal.
c. Identify the median cost of a meal.
d. Calculate the interquartile range for the price of a meal.
e. Determine the percentage of the prices that were below the median.

22. The following data give the box plots for three different age groups in a triathlon for under thirities.

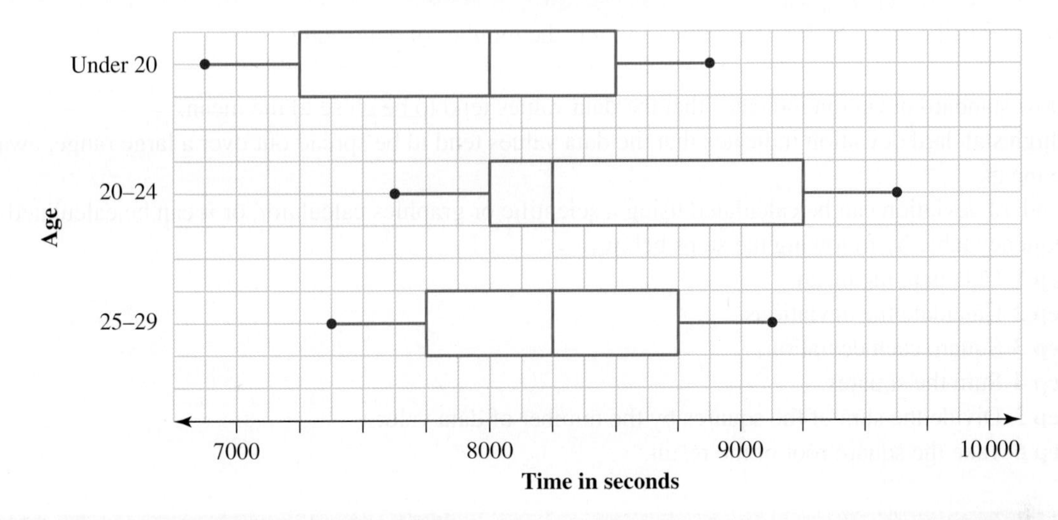

a. Identify the slowest time for the 20–24 year olds.
b. Estimate the difference in time between the fastest triathlete in:
   i. the under 20s and the 20–24 group
   ii. the under 20s and the 25–29 group
   iii. the under 20–24 group and the 25–29 group.
c. Comment on the overall performance of the three groups.

12.5 The standard deviation (10A)

LEARNING INTENTION

At the end of this subtopic you should be able to:
- calculate the standard deviation of a small data set by hand
- calculate the standard deviation using technology
- interpret the mean and standard deviation of data
- identify the effect of outliers on the standard deviation.

12.5.1 Standard deviation

eles-4958

- The **standard deviation** for a set of data is a measure of how far the data values are spread out (deviate) from the mean. The value of the standard deviation tells you the average deviation of the data from the mean.
- **Deviation** is the difference between each data value and the mean $(x - \bar{x})$. The standard deviation is calculated from the square of the deviations.

Standard deviation formula

- Standard deviation is denoted by the lowercase Greek letter sigma, σ, and can be calculated by using the following formula.

$$\sigma = \sqrt{\frac{\sum (x - \bar{x})^2}{n}}$$

where $\bar{x}$ is the mean of the data values and n is the number of data values.

- A low standard deviation indicates that the data values tend to be close to the mean.
- A high standard deviation indicates that the data values tend to be spread out over a large range, away from the mean.
- Standard deviation can be calculated using a scientific or graphics calculator, or it can be calculated from a frequency table by following the steps below.
  Step 1 Calculate the mean.
  Step 2 Calculate the deviations.
  Step 3 Square each deviation.
  Step 4 Sum the squares.
  Step 5 Divide the sum of the squares by the number of data values.
  Step 6 Take the square root of the result.

WORKED EXAMPLE 10 Calculating the standard deviation

The number of lollies in each of 8 packets is 11, 12, 13, 14, 16, 17, 18, 19. Calculate the mean and standard deviation correct to 2 decimal places.

THINK

1. Calculate the mean.

WRITE

$$\bar{x} = \frac{11 + 12 + 13 + 14 + 16 + 17 + 18 + 19}{8}$$

$$= \frac{120}{8}$$

$$= 15$$

2. To calculate the deviations $(x - \bar{x})$, set up a frequency table as shown and complete.

No. of lollies (x)	$(x - \bar{x})$
11	$11 - 15 = -4$
12	-3
13	-2
14	-1
16	1
17	2
18	3
19	4
Total	

3. Add another column to the table to calculate the square of the deviations, $(x-\bar{x})^2$. Then sum the results: $\Sigma(x-\bar{x})^2$.

No. of lollies (x)	$(x-\bar{x})$	$(x-\bar{x})^2$
11	$11-15=-4$	16
12	−3	9
13	−2	4
14	−1	1
16	1	1
17	2	4
18	3	9
19	4	16
Total		$\sum(x-\bar{x})^2=60$

4. To calculate the standard deviation, divide the sum of the squares by the number of data values, then take the square root of the result.

$$\sigma=\sqrt{\frac{\sum(x-\bar{x})^2}{n}}$$
$$=\sqrt{\frac{60}{8}}$$
$$\approx 2.74 \text{ (correct to 2 decimal places)}$$

5. Check the result using a calculator.

The calculator returns an answer of $\sigma_n=2.73861$. Answer confirmed.

6. Interpret the result.

The average (mean) number of lollies in each pack is 15 with a standard deviation of 2.74, which means that the number of lollies in each pack differs from the mean by an average of 2.74.

TI | THINK

1. In a new problem, on a Calculator page, complete the entry lines as shown. This stores the data values to the variable '*lollies*'.
lollies: = {11, 12, 13, 14, 15, 16, 17, 18, 19}
Although we can find many summary statistics, to calculate the mean only, open a Calculator page and press:
- MENU
- 6: Statistics
- 3: List Math
- 3: Mean

Press VAR and select '*lollies*', then press ENTER.

DISPLAY/WRITE

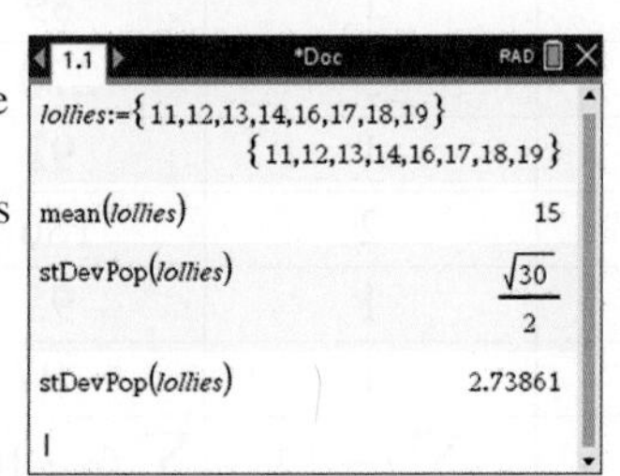

The mean number of lollies is 15 and the population standard deviation is $\sigma=2.74$.

CASIO | THINK

On the Statistics screen, label list1 as 'x' and enter the values from the question. Press EXE after entering each value. To find the summary statistics, tap:
- Calc
- One-Variable

Set values as:
- XList: main\x
- Freq: 1

Tap OK.
The standard deviation is shown as σ_x and the mean is shown as $\bar{x}$.

DISPLAY/WRITE

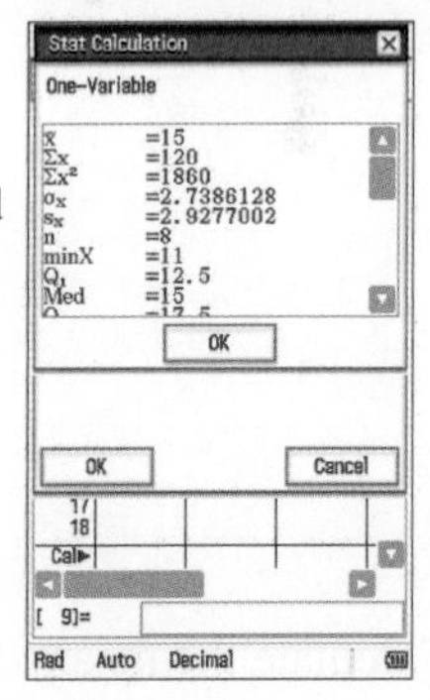

The mean number of lollies is 15 with a standard deviation of 2.74.

2. To calculate the population standard deviation only, press:
   - MENU
   - 6: Statistics
   - 3: List Math
   - 9: Population standard deviation

   Press VAR and select '*lollies*', then press ENTER. Press CTRL ENTER to get a decimal approximation.

- When calculating the standard deviation from a frequency table, the frequencies must be taken into account. Therefore, the following formula is used.

$$\sigma = \sqrt{\frac{\sum f(x-\bar{x})^2}{n}}$$

WORKED EXAMPLE 11 Calculating standard deviation from a frequency table

Lucy's scores in her last 12 games of golf were 87, 88, 88, 89, 90, 90, 90, 92, 93, 93, 95 and 97. Calculate the mean score and the standard deviation correct to 2 decimal places.

THINK

1. To calculate the mean, first set up a frequency table.

WRITE

Golf score (x)	Frequency (f)	fx
87	1	87
88	2	176
89	1	89
90	3	270
92	1	92
93	2	186
95	1	95
97	1	97
Total	$\sum f = 12$	$\sum fx = 1092$

2. Calculate the mean.

$$\bar{x} = \frac{\sum fx}{\sum f}$$

$$= \frac{1092}{12}$$

$$= 91$$

3. To calculate the deviations $(x - \bar{x})$, add another column to the frequency table and complete.

Golf score (x)	Frequency (f)	fx	$(x - \bar{x})$
87	1	87	$87 - 91 = -4$
88	8	176	−3
89	1	89	−2
90	3	270	−1
92	1	92	1
93	2	186	2
95	1	95	4
97	1	97	6
Total	$\sum f = 12$	$\sum fx = 1092$	

4. Add another column to the table and multiply the square of the deviations, $(x - \bar{x})^2$, by the frequency $f(x - \bar{x})^2$. Then sum the results: $\sum f(x - \bar{x})^2$.

Golf score (x)	Frequency (f)	fx	$(x - \bar{x})$	$f(x - \bar{x})^2$
87	1	87	$87 - 91 = -4$	$1 \times (-4)^2 = 16$
88	8	176	−3	18
89	1	89	−2	4
90	3	270	−1	3
92	1	92	1	1
93	2	186	2	8
95	1	95	4	16
97	1	97	6	36
Total	$\sum f = 12$	$\sum fx = 1092$		102

5. Calculate the standard deviation using the formula.

$$\sigma = \sqrt{\frac{\sum f(x - \bar{x})^2}{n}}$$

$$= \sqrt{\frac{102}{12}}$$

$$\approx 2.92$$

(correct to 2 decimal places)

6. Check the result using a calculator.

The calculator returns an answer of $\sigma_n = 2.91548$.
The answer is confirmed.

7. Interpret the result.

The average (mean) score for Lucy is 91 with a standard deviation of 2.92, which means that her score differs from the mean by an average of 2.92.

TI | THINK

1. In a new problem, on a Lists & Spreadsheet page, label column A as 'score' and label column B as 'f'. Enter the values and the frequency corresponding to each score as shown in the table. Press ENTER after each value.

DISPLAY/WRITE

	A score	B f
1	87	1
2	88	2
3	89	1
4	90	3
5	92	1

2. To find all the summary statistics, open a Calculator page and press:
   - MENU
   - 6: Statistics
   - 1: Stat Calculations
   - 1: One Variable Statistics...

   Select 1 as the number of lists, then on the One-Variable Statistics page, select 'score' as the X1 List and 'f' as the Frequency List. Leave the next two fields empty and TAB to OK, then press ENTER.

"Σx²"	99474.
"sx := sn-1x"	3.04512
"σx := σnx"	2.91548
"n"	12.
"MinX"	87.
"Q₁X"	88.5
"MedianX"	90.
"Q₃X"	93.
"MaxX"	97.
"SSX := Σ(x−x̄)²"	102.

The mean is $\bar{x} = 91$ and the population standard deviation is $\sigma = 2.92$ correct to 2 decimal places.

CASIO | THINK

1. On the Statistic screen, label list1 as 'score' and enter the values from the question. Press EXE after entering each value. To find the summary statistics, tap:
   - Calc
   - One-Variable

   Set values as:
   - XList: main\score
   - Freq: 1

DISPLAY/WRITE

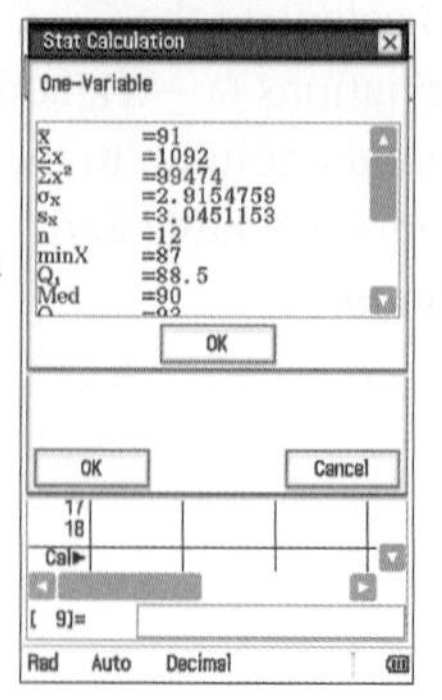

2. Tap OK. The standard deviation is shown as σ_x and the mean is shown as $\bar{x}$.

The mean is $\bar{x} = 91$ and the population standard deviation is $\sigma = 2.92$ correct to 2 decimal places.

Why the deviations are squared

- When you take an entire data set, the sum of the deviations from the mean is zero, that is, $\sum(x - \bar{x}) = 0$.
- When the data value is less than the mean ($x < \bar{x}$), the deviation is negative.
- When the data value is greater than the mean ($x > \bar{x}$), the deviation is positive.
- The negative and positive deviations cancel each other out; therefore, calculating the sum and average of the deviations is not useful.
- By squaring all of the deviations, each deviation becomes positive, so the average of the deviations becomes meaningful. This explains why the standard deviation is calculated using the squares of the deviations, $(x - \bar{x})^2$, for all data values.

Standard deviations of populations and samples

- So far we have calculated the standard deviation for a population of data, that is, for complete sets of data. There is another formula for calculating standard deviation for samples of data, that is, data that have been randomly selected from a larger population.
- The sample standard deviation is more commonly used in day-to-day life, as it is usually impossible to collect data from an entire population.
- For example, if you wanted to know how much time Year 10 students across the country spend on social media, you would not be able to collect data from every student in the country. You would have to take a sample instead.
- The sample standard deviation is denoted by the letter s, and can be calculated using the following formula.

Sample standard deviation formula

$$s = \sqrt{\frac{\sum f(x-\bar{x})^2}{n-1}}$$

- Calculators usually display both values for the standard deviation, so it is important to understand the difference between them.

12.5.2 Effects on standard deviation

eles-4959

- The standard deviation is affected by extreme values.

WORKED EXAMPLE 12 Interpreting the effects on standard deviation

On a particular day Lucy played golf brilliantly and scored 60.
The scores in her previous 12 games of golf were 87, 88, 88, 89, 90, 90, 90, 92, 93, 93, 95 and 97 (see Worked example 1).
Comment on the effect this latest score has on the standard deviation.

THINK	WRITE
1. Use a calculator to calculate the mean and the standard deviation.	$\bar{x} = 88.6154 \quad \sigma = 8.7225$ $\approx 88.62 \quad \approx 8.7225$
2. Interpret the result and compare it to the results found in Worked example 11.	In the first 12 games Lucy's mean score was 91 with a standard deviation of 2.92. This implied that Lucy's scores on average were 2.92 either side of her average of 91. Lucy's latest performance resulted in a mean score of 88.62 with a standard deviation of 8.72. This indicates a slightly lower mean score, but the much higher standard deviation indicates that the data are now much more spread out and that the extremely good score of 60 is an anomaly.

12.5.3 Properties of standard deviation

eles-4961

- If a constant c is added to all data values in a set, the deviations $(x-\bar{x})$ will remain unchanged and consequently the standard deviation remains unchanged.
- If all data values in a set are multiplied by a constant k, the deviations $(x-\bar{x})$ will be multiplied by k, that is $k(x-\bar{x})$; consequently the standard deviation is increased by a factor of k.
- The standard deviation can be used to measure consistency.
- When the standard deviation is low we are able to say that the scores in the data set are more consistent with each other.

WORKED EXAMPLE 13 Calculating numerical changes to the standard deviation

For the data 5, 9, 6, 11, 10, 7:

a. calculate the standard deviation

b. calculate the standard deviation if 4 is added to each data value. Comment on the effect.

c. calculate the standard deviation if all data values are multiplied by 2. Comment on the effect.

THINK	WRITE
a. 1. Calculate the mean.	a. $\bar{x} = \frac{5+9+6+11+10+7}{6}$ $= 8$
2. Set up a frequency table and enter the squares of the deviations.	(see table below)

(x)	$(x-\bar{x})$	$(x-\bar{x})^2$
5	$5-8=-3$	9
6	-2	4
7	-1	1
9	1	1
10	2	4
11	3	9
Total		$\sum(x-\bar{x})^2 = 28$

3. To calculate the standard deviation, apply the formula for standard deviation.

$$\sigma = \sqrt{\frac{\sum(x-\bar{x})^2}{n}}$$

$$= \sqrt{\frac{28}{6}}$$

≈ 2.16 (correct to 2 decimal places)

THINK	WRITE
b. 1. Add 4 to each data value in the set.	b. 9, 13, 10, 15, 14, 11
2. Calculate the mean.	$\bar{x} = \frac{9+13+10+15+14+11}{6}$ $= 12$
3. Set up a frequency table and enter the squares of the deviations.	(see table below)

(x)	$(x-\bar{x})$	$(x-\bar{x})^2$
9	$9-12=-3$	9
10	-2	4
11	-1	1
13	1	1
14	2	4
15	3	9
Total		$\sum(x-\bar{x})^2 = 28$

4. To calculate the standard deviation, apply the formula for standard deviation.

$$\sigma = \sqrt{\frac{\sum(x-\bar{x})^2}{n}}$$

$$= \sqrt{\frac{28}{6}}$$

≈ 2.16 (correct to 2 decimal places)

5. Comment on the effect of adding of 4 to each data value.

Adding 4 to each data value increased the mean but had no effect on the standard deviation, which remained at 2.16.

c. 1. Multiply each data value in the set by 2

c. 10, 18, 12, 22, 20, 14

2. Calculate the mean.

$$\bar{x} = \frac{10 + 18 + 12 + 22 + 20 + 14}{6}$$
$$= 16$$

3. Set up a frequency table and enter the squares of the deviations.

(x)	$(x - \bar{x})$	$(x - \bar{x})^2$
10	$10 - 16 = -6$	36
12	-4	16
14	-2	4
18	2	4
20	4	16
22	6	36
Total		$\sum (x - \bar{x})^2 = 112$

4. To calculate the standard deviation, apply the formula for standard deviation.

$$\sigma = \sqrt{\frac{\sum (x - \bar{x})^2}{n}} = \sqrt{\frac{112}{6}}$$
$$\approx 4.32 \text{ (correct to 2 decimal places)}$$

5. Comment on the effect of multiplying each data value by 2.

Multiplying each data value by 2 doubled the mean and doubled the standard deviation, which changed from 2.16 to 4.32.

Resources

eWorkbook Topic 12 Workbook (worksheets, code puzzle and project) (ewbk-2038)

Interactivities Individual pathway interactivity: The standard deviation (int-4624)

The standard deviation for a sample (int-4814)

Exercise 12.5 The standard deviation (10A)

learnON

Individual pathways

■ PRACTISE	■ CONSOLIDATE	■ MASTER
1, 2, 4, 8, 10, 13	3, 5, 7, 11, 14	6, 9, 12, 15

To answer questions online and to receive **immediate corrective feedback** and **fully worked solutions** for all questions, go to your learnON title at www.jacplus.com.au.

Fluency

1. **WE10** Calculate the standard deviation of each of the following data sets, correct to 2 decimal places.

a. 3, 5, 8, 2, 7, 1, 6, 5

b. 11, 8, 7, 12, 10, 11, 14

c. 25, 15, 78, 35, 56, 41, 17, 24

d. 5.2, 4.7, 5.1, 12.6, 4.8

2. Calculate the standard deviation of each of the following data sets, correct to 2 decimal places.

a.

Score (x)	Frequency (f)
1	1
2	5
3	9
4	7
5	3

b.

Score (x)	Frequency (f)
16	15
17	24
18	26
19	28
20	27

c.

Score (x)	Frequency (f)
8	15
10	19
12	18
14	7
16	6
18	2

d.

Score (x)	Frequency (f)
65	15
66	15
67	16
68	17
69	16
70	15
71	15
72	12

3. Complete the following frequency distribution table and use it to calculate the standard deviation of the data set, correct to 2 deciaml places.

Class	Class centre (x)	Frequency (f)
1 – 10		6
11 – 20		15
21 – 30		25
31 – 40		8
41 – 50		6

4. **WE11** First-quarter profit increases for 8 leading companies are given below as percentages.

2.3 0.8 1.6 2.1 1.7 1.3 1.4 1.9

Calculate the mean score and the standard deviation for this set of data and express your answer correct to 2 decimal places.

5. The heights in metres of a group of army recruits are given.

1.8 1.95 1.87 1.77 1.75 1.79 1.81 1.83 1.76 1.80 1.92 1.87 1.85 1.83

Calculate the mean score and the standard deviation for this set of data and express your answer correct to 2 decimal places.

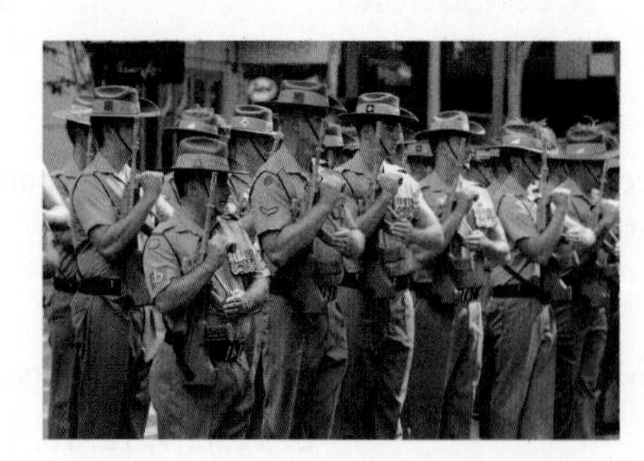

6. Times (to the nearest tenth of a second) for the heats in the open 100 m sprint at the school sports are given in the stem-and-leaf plot shown.
Calculate the standard deviation for this set of data and express your answer correct to 2 decimal places.

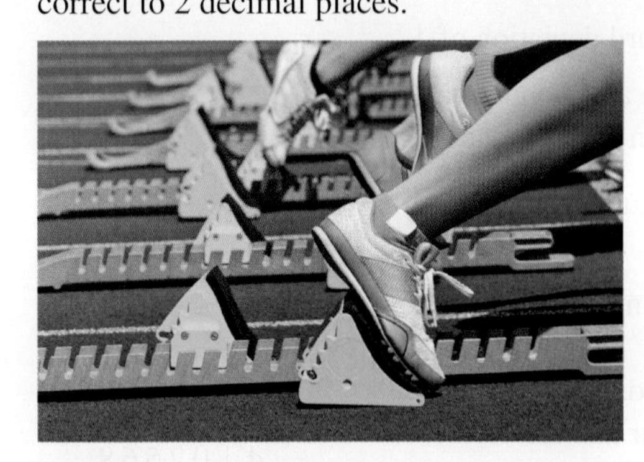

Key: 11|0 = 11.0 s

Stem	*Leaf*
11	0
11	2 3
11	4 4 5
11	6 6
11	8 8 9
12	0 1
12	2 2 3
12	4 4
12	6
12	9

7. The number of outgoing phone calls from an office each day over a 4-week period is shown in the stem-and-leaf plot.
Calculate the standard deviation for this set of data and express your answer correct to 2 decimal places.

Key: 1|3 = 13 calls

Stem	*Leaf*
0	8 9
1	3 4 7 9
2	0 1 3 7 7
3	3 4
4	1 5 6 7 8
5	3 8

8. **MC** A new legal aid service has been operational for only 5 weeks. The number of people who have made use of the service each day during this period is set out in the stem-and-leaf plot shown.
The standard deviation (to 2 decimal places) of these data is:
A. 6.00
B. 6.34
C. 6.47
D. 15.44
E. 9.37

Key: 1|6 = 16 people

Stem	*Leaf*
0	2 4
0	7 7 9
1	0 1 4 4 4 4
1	5 6 6 7 8 8 9
2	1 2 2 3 3 3
2	7

Understanding

9. **WE12** The speeds, in km/h, of the first 25 cars caught by a roadside speed camera on a particular day were:
82, 82, 84, 84, 84, 84, 85, 85, 85, 86, 86, 87, 89, 89, 89, 90, 91, 91, 92, 94, 95, 96, 99, 100, 102
The next car that passed the speed camera was travelling at 140 km/h.
Comment on the effect of the speed of this last car on the standard deviation for the data.

10. Explain what the standard deviation tells us about a set of data.

Reasoning

11. **WE13** For the data 1, 4, 5, 9, 11:
 a. calculate the standard deviation
 b. calculate the standard deviation if 7 is added to each data value. Comment on the effect.
 c. calculate the standard deviation if all data values are multiplied by 3. Comment on the effect.

12. Show using an example the effect, if any, on the standard deviation of adding a data value to a set of data that is equivalent to the mean.

Problem solving

13. If the mean for a set of data is 45 and the standard deviation is 6, determine how many standard deviations above the mean is a data value of 57.

14. Five numbers a, b, c, d and e have a mean of 12 and a standard deviation of 4.
    a. If each number is increased by 3, calculate the new mean and standard deviation.
    b. If each number is multiplied by 3, calculate the new mean and standard deviation.

15. Twenty-five students sat a test and the results for 24 of the students are given in the following stem-and-leaf plot.
    a. If the average mark for the test was 27.84, determine the mark obtained by the 25th student.
    b. Determine how many students scored higher than the median score.
    c. Calculate the standard deviation of the marks, giving your answer correct to 2 decimal places.

Stem	*Leaf*
0	8 9
1	1 2 3 7 8 9
2	2 3 5 6 8
3	0 1 2 4 6 8
4	0 2 5 6 8

12.6 Comparing data sets

LEARNING INTENTION

At the end of this subtopic you should be able to:
- choose an appropriate measure of centre and spread to analyse data
- interpret and make decisions based on measures of centre and spread.

12.6.1 Comparing data sets

eles-4962

- Besides locating the centre of the data (the mean, median or mode), any analysis of data must measure the extent of the spread of the data (range, interquartile range and standard deviation). Two data sets may have centres that are very similar but be quite differently distributed.
- Decisions need to be made about which measure of centre and which measure of spread to use when analysing and comparing data.
- The mean is calculated using every data value in the set. The median is the middle score of an ordered set of data, so it is a more useful measure of centre when a set of data contains outliers.
- The range is determined by calculating the difference between the maximum and minimum data values, so it includes outliers. It provides only a rough idea about the spread of the data and is inadequate in providing sufficient detail for analysis. It is useful, however, when we are interested in extreme values such as high and low tides or maximum and minimum temperatures.
- The interquartile range is the difference between the upper and lower quartiles, so it does not include every data value in its calculation, but it will overcome the problem of outliers skewing data.
- The standard deviation is calculated using every data value in the set.

WORKED EXAMPLE 14 Interpreting mean and standard deviation

For the two sets of data 6, 7, 8, 9, 10 and 12, 4, 10, 11, 3:
a. **calculate the mean**
b. **calculate the standard deviation**
c. **comment on the similarities and differences.**

THINK	WRITE
a. 1. Calculate the mean of the first set of data.	a. $\bar{x}_1 = \dfrac{6+7+8+9+10}{5}$ $= 8$
2. Calculate the mean of the second set of data.	$\bar{x}_2 = \dfrac{12+4+10+11+3}{5}$ $= 8$
b. 1. Calculate the standard deviation of the first set of data.	b. $\sigma_1 = \sqrt{\dfrac{(6-8)^2+(7-8)^2+(8-8)^2+(9-8)^2+(10-8)^2}{5}}$ ≈ 1.41
2. Calculate the standard deviation of the second set of data.	$\sigma_2 = \sqrt{\dfrac{(12-8)^2+(4-8)^2+(10-8)^2+(11-8)^2+(3-8)^2}{5}}$ ≈ 3.74
c. Comment on the findings.	c. For both sets of data the mean was the same, 8. However, the standard deviation for the second set (3.74) was much higher than the standard deviation of the first set (1.41), implying that the second set is more widely distributed than the first. This is confirmed by the range, which is $10-6=4$ for the first set and $12-3=9$ for the second.

- When multiple data displays are used to display similar sets of data, comparisons and conclusions can then be drawn about the data.
- We can use **back-to-back stem-and-leaf plots** and multiple or parallel box plots to help compare statistics such as the median, range and interquartile range.

Parallel box plots

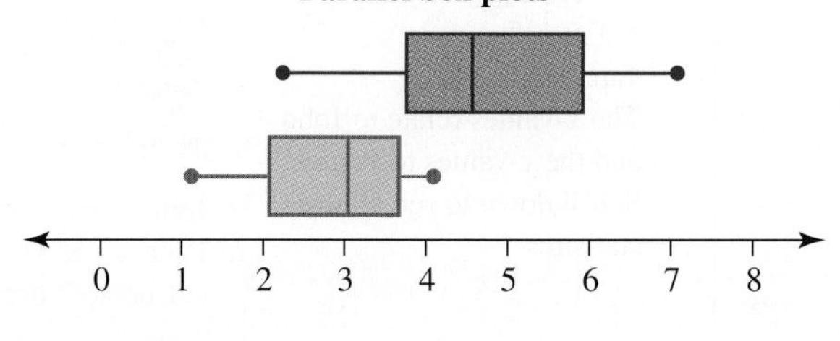

Back-to-back stem-and-leaf plots

Leaf	*Stem*	*Leaf*
5	0	8 9
6 6 3 2	1	4 5 6 8 9
8 4 1	2	0 4 5 7
9 8 5 3 0	3	1 6 9
7 1	4	1 3
	5	2 5

WORKED EXAMPLE 15 Comparing data sets

Below are the scores achieved by two students in eight Mathematics tests throughout the year.
John: 45, 62, 64, 55, 58, 51, 59, 62
Penny: 84, 37, 45, 80, 74, 44, 46, 50

a. **Determine the most appropriate measure of centre and measure of spread to compare the performance of the students.**
b. **Identify the student who performed better over the eight tests. Justify your answer.**
c. **Identify the student who was more consistent over the eight tests. Justify your answer.**

THINK	WRITE
a. In order to include all data values in the calculation of measures of centre and spread, calculate the mean and standard deviation.	**a.** John: $\bar{x} = 57$, $\sigma = 6$ Penny: $\bar{x} = 57.5$, $\sigma = 17.4$
b. Compare the mean for each student. The student with the higher mean performed better overall.	**b.** Penny performed slightly better on average as her mean mark was higher than John's.
c. Compare the standard deviation for each student. The student with the lower standard deviation performed more consistently.	**c.** John was the more consistent student because his standard deviation was much lower than Penny's. This means that his test results were closer to his mean score than Penny's were to hers.

TI | THINK — **DISPLAY/WRITE**

a.

1. In a new problem, on a Lists & Spreadsheet page, label column A as 'john' and column B as 'penny'. Enter the data sets from the question. Press ENTER after each value.

	A john	B penny	C	D
=				
1	45	84		
2	62	37		
3	64	45		
4	55	80		
5	58	74		

2. To calculate only the mean and standard deviation of each data set, open a Calculator page and complete the entry lines as:
mean(*john*)
stDevPop(*john*)
mean(*penny*)
stDevPop(*penny*)
Press CTRL ENTER after each entry to get a decimal approximation.

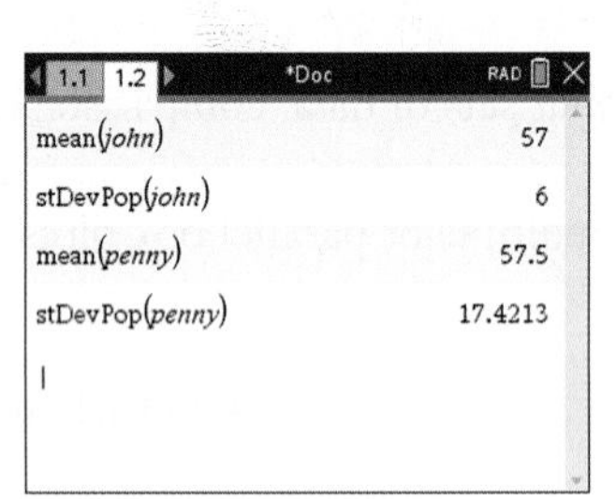

John: $\bar{x} = 57$, $\sigma = 6$
Penny: $\bar{x} = 57.5$, $\sigma = 17.4$
correct to 2 decimal places.

CASIO | THINK — **DISPLAY/WRITE**

a.

1. On the Statistics screen, label list1 as 'John' and list2 as 'Penny'. Enter the data set as shown. Press EXE after each value.

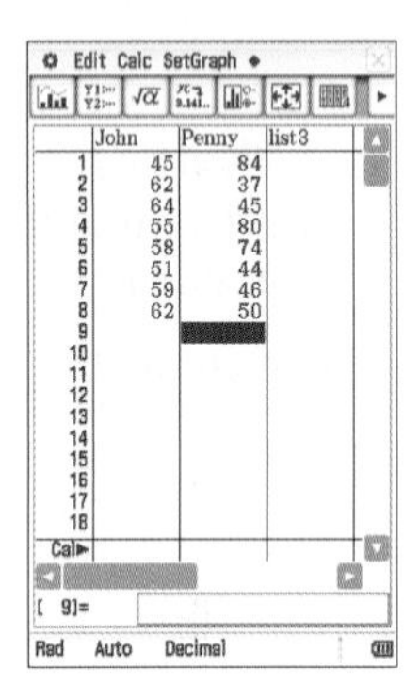

2. To calculate the mean and standard deviation of each data set, tap:
   - Calc
   - Two-Variable

   Set values as:
   - XList: main\John
   - YList: main\Penny
   - Freq: 1

   Tap OK.
   The x-values relate to John and the y-values to Penny. Scroll down to see all the statistics.

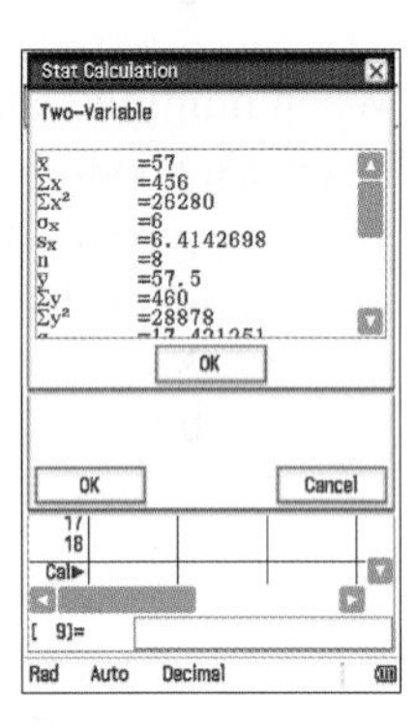

John: $\bar{x} = 57$, $\sigma = 6$
Penny: $\bar{x} = 57.5$, $\sigma = 17.4$
correct to 2 decimal places.

b–c.

To draw the two box plots on the same Data & Statistics page, press TAB to locate the label of the horizontal axis and select the variable 'john'.

Then press:

- MENU
- 1: Plot Type
- 2: Box Plot

Then press:

- MENU
- 2: Plot Properties
- 5: Add X-variable and select 'penny'.
  To change the colour, place the pointer over one of the data points. Then press CTRL MENU. Then press:
- 6: Color
- 2: Fill Color

Select whichever colour you like from the palette for each of the box plots.

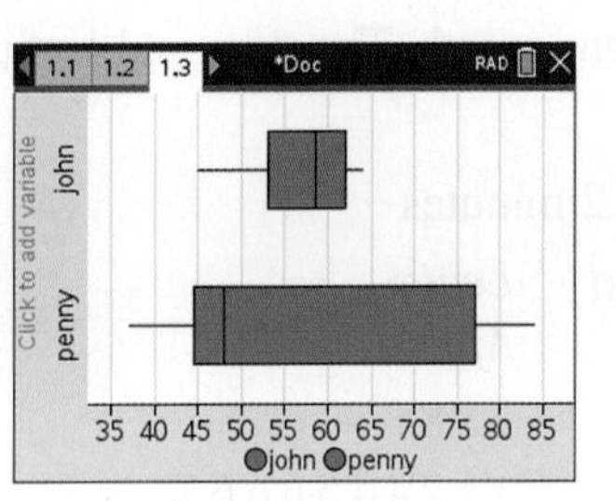

Penny performed slightly better overall as her mean mark was higher than John's; however, John was more consistent as his standard deviation was lower than Penny's.

b–c.

To draw the two box-and-whisker plots on the same Statistics screen, tap:

- SetGraph
- Setting...

Set values as:

- Type: MedBox
- XList: main\John
- Freq: 1

Tap 2 in the row of numbers at the top of the screen.

Set values as:

- Type: MedBox
- XList: main\Penny
- Freq: 1

Tap Set.

Tap SetGraph and tick StatGraph1 and StatGraph2.

Tap the graphing icon to display the graphs.

b–c.

Penny performed slightly better overall as her mean mark was higher than John's; however, John was more consistent as his standard deviation was lower than Penny's.

Resources

 **eWorkbook** Topic 12 Workbook (worksheets, code puzzle and project) (ewbk-2038)

 **Interactivities** Individual pathway interactivity: Comparing data sets (int-4625)

Back-to-back stem plots (int-6252)

Exercise 12.6 Comparing data sets

learnon

Individual pathways

■ PRACTISE	■ CONSOLIDATE	■ MASTER
1, 5, 8, 10, 11, 17	2, 4, 6, 9, 12, 13, 18	3, 7, 14, 15, 16, 19, 20

To answer questions online and to receive **immediate corrective feedback** and **fully worked solutions** for all questions, go to your learnON title at www.jacplus.com.au.

Fluency

1. WE14 For the two sets of data, 65, 67, 61, 63, 62, 60 and 56, 70, 65, 72, 60, 55:
   a. calculate the mean
   b. calculate the standard deviation
   c. comment on the similarities and differences.

2. A bank surveys the average morning and afternoon waiting times for customers. The figures were taken each Monday to Friday in the morning and afternoon for one month. The stem-and-leaf plot below shows the results.

Key: 1|2 = 1.2 minutes

Leaf: Morning	*Stem*	*Leaf: Afternoon*
7	0	7 8 8
8 6 3 1 1	1	1 1 2 4 4 5 6 6 6 7
9 6 6 6 5 5 4 3 3 1	2	2 5 5 8
9 5 2	3	1 6
5	4	
	5	7

a. Identify the median morning waiting time and the median afternoon waiting time.
b. Calculate the range for morning waiting times and the range for afternoon waiting times.
c. Use the information given in the display to comment about the average waiting time at the bank in the morning compared with the afternoon.

3. In a class of 30 students there are 15 boys and 15 girls. Their heights are measured in metres and are listed below.

Boys: 1.65, 1.71, 1.59, 1.74, 1.66, 1.69 1.72, 1.66, 1.65, 1.64, 1.68, 1.74, 1.57, 1.59, 1.60
Girls: 1.66, 1.69, 1.58, 1.55, 1.51, 1.56, 1.64, 1.69, 1.70, 1.57, 1.52, 1.58, 1.64, 1.68, 1.67

Display this information in a back-to-back stem-and-leaf plot and comment on their height distribution.

4. The stem-and-leaf plot at right is used to display the number of vehicles sold by the Ford and Hyundai dealerships in a Sydney suburb each week for a three-month period.

a. State the median of both distributions.
b. Calculate the range of both distributions.
c. Calculate the interquartile range of both distributions.
d. Show both distributions on a box plot.

Key: 1|5 = 15 vehicles

Leaf: Ford	*Stem*	*Leaf: Hyundai*
74	0	39
952210	1	111668
8544	2	2279
0	3	5

5. The box plot drawn below displays statistical data of two AFL teams over a season.

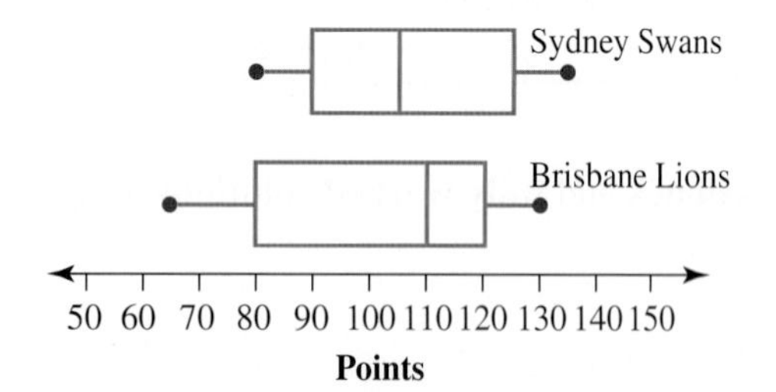

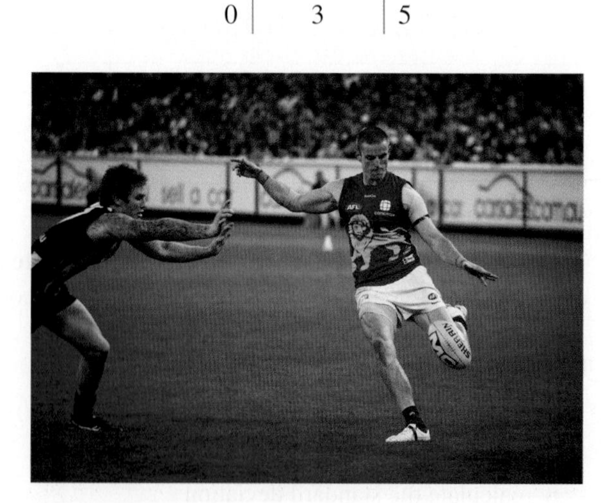

a. State the team that had the higher median score.
b. Determine the range of scores for each team.
c. For each team calculate the interquartile range.

Understanding

6. Tanya measures the heights (in m) of a group of Year 10 boys and girls and produces the following five-point summaries for each data set.

Boys:	1.45, 1.56, 1.62, 1.70, 1.81
Girls:	1.50, 1.55, 1.62, 1.66, 1.73

   a. Draw a box plot for both sets of data and display them on the same scale.
   b. Calculate the median of each distribution.
   c. Calculate the range of each distribution.
   d. Calculate the interquartile range for each distribution.
   e. Comment on the spread of the heights among the boys and the girls.

7. The box plots show the average daily sales of cold drinks at the school canteen in summer and winter.

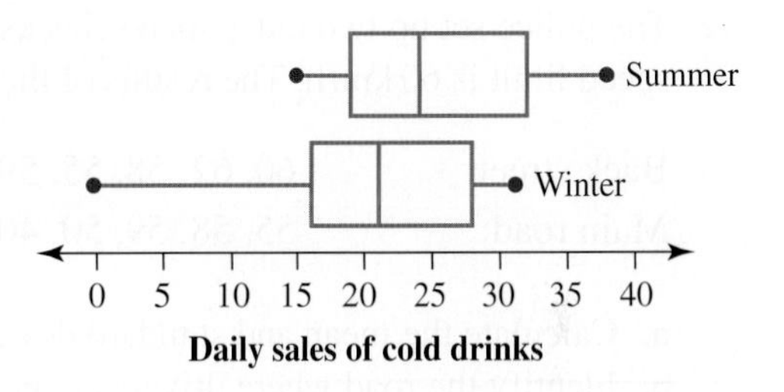

   a. Calculate the range of sales in both summer and winter.
   b. Calculate the interquartile range of the sales in both summer and winter.
   c. Comment on the relationship between the two data sets, both in terms of measures of centre and measures of spread.

8. MC Andrea surveys the age of people at two movies being shown at a local cinema. The box plot shows the results.

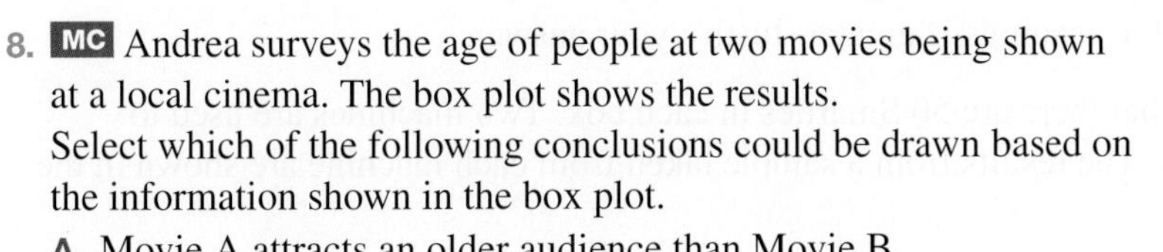

Select which of the following conclusions could be drawn based on the information shown in the box plot.

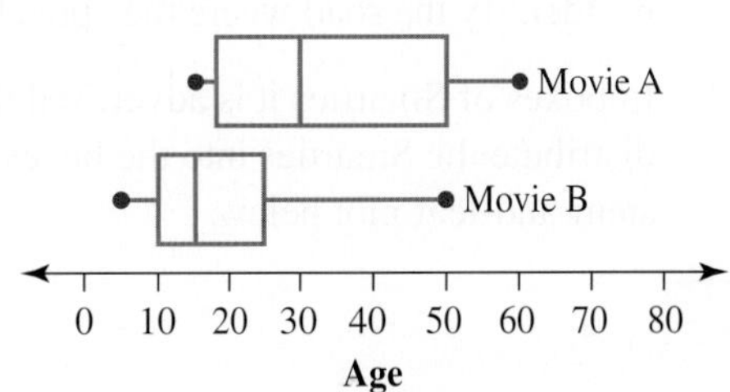

   A. Movie A attracts an older audience than Movie B
   B. Movie B attracts an older audience than Movie A.
   C. Movie A appeals to a wider age group than Movie B.
   D. Movie B appeals to a wider age group than Movie A.
   E. Both movies appeal equally to the same age groups.

9. MC *Note:* There may be more than one correct answer.

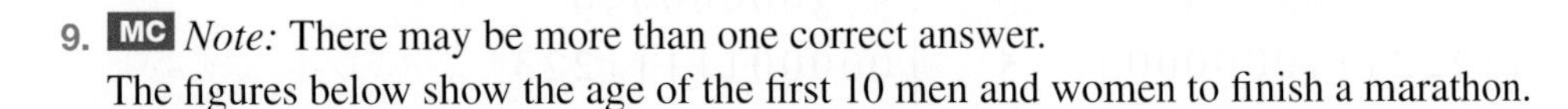

The figures below show the age of the first 10 men and women to finish a marathon.

Men:	28, 34, 25, 36, 25, 35, 22, 23, 40, 24
Women:	19, 27, 20, 26, 30, 18, 28, 25, 28, 22

Choose which of the following statements are correct.

   A. The mean age of the men is greater than the mean age of the women.
   B. The range is greater among the men than among the women.
   C. The interquartile range is greater among the men than among the women.
   D. The standard deviation is greater among the men than among the women.
   E. The standard deviation is less among the men than among the women.

Reasoning

10. **WE15** Cory recorded his marks for each test that he did in English and Science throughout the year.

English: 55, 64, 59, 56, 62, 54, 65, 50
Science: 35, 75, 81 32, 37, 62, 77, 75

a. Determine the most appropriate measure of centre and measure of spread to compare Cory's performance in the two subjects.
b. Identify the subject in which Cory received a better average. Justify your answer.
c. Identify the subject in which Cory performed more consistently. Justify your answer.

11. Draw an example of a graph that is:

a. symmetrical
b. positively skewed with one mode
c. negatively skewed with two modes.

12. The police set up two radar speed checks on a back street of Sydney and on a main road. In both places the speed limit is 60 km/h. The results of the first 10 cars that have their speed checked are given below.

Back street: 60, 62, 58, 55, 59, 56, 65, 70, 61, 64
Main road: 55, 58, 59, 50, 40, 90, 54, 62, 60, 60

a. Calculate the mean and standard deviation of the readings taken at each point.
b. Identify the road where drivers are generally driving faster. Justify your answer.
c. Identify the road where the spread of readings is greater. Justify your answer.

13. In boxes of Smarties it is advertised that there are 50 Smarties in each box. Two machines are used to distribute the Smarties into the boxes. The results from a sample taken from each machine are shown in the stem-and-leaf plot below.

Key: $5|1 = 51$ $5*|6 = 56$

Leaf: Machine A	*Stem*	*Leaf: Machine B*
4	4	
99877665	4*	57899999999
4322211110000000	5	0000011111223
55	5*	9

a. Display the data from both machines on parallel box plots.
b. Calculate the mean and standard deviation of the number of Smarties distributed from both machines.
c. State which machine is the more dependable. Justify your answer.

14. Nathan and Timana are wingers in their local rugby league team. The number of tries they have scored in each season are listed below.

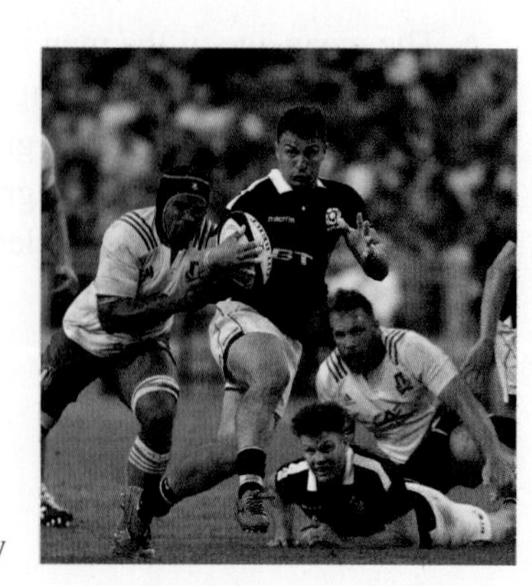

Nathan: 25, 23, 13, 36, 1, 8, 0, 9, 16, 20
Timana: 5, 10, 12, 14, 18, 11, 8, 14, 12, 19

a. Calculate the mean number of tries scored by each player.
b. Calculate the range of tries scored by each player. Justify your answer.
c. Calculate the interquartile range of tries scored by each player. Justify your answer.
d. State which player would you consider to be the more consistent. Justify your answer.

15. Year 10 students at Merrigong High School sit exams in Science and Maths. The results are shown in the table below.

Mark	Number of students in Science	Number of students in Maths
51 – 60	7	6
61 – 70	10	7
71 – 80	8	12
81 – 91	8	9
91 – 100	2	6

a. Determine if either distribution is symmetrical.
b. If either distribution is not symmetrical, state whether it is positively or negatively skewed.
c. Discuss the possible reasons for any skewness.
d. State the modal class of each distribution.
e. Determine which subject has the greater standard deviation greater. Explain your answer.

16. A new drug for the relief of cold symptoms has been developed. To test the drug, 40 people were exposed to a cold virus. Twenty patients were then given a dose of the drug while another 20 patients were given a placebo. (In medical tests a control group is often given a *placebo* drug. The subjects in this group believe that they have been given the real drug but in fact their dose contains no drug at all.)
All participants were then asked to indicate the time when they first felt relief of symptoms. The number of hours from the time the dose was administered to the time when the patients first felt relief of symptoms are detailed below.

Group A (drug)

25	29	32	45	18	21	37	42	62	13
42	38	44	42	35	47	62	17	34	32

Group B (placebo)

25	17	35	42	35	28	20	32	38	35
34	32	25	18	22	28	21	24	32	36

a. Display the data on a back-to-back stem-and-leaf plot.
b. Display the data for both groups on a parallel box plot.
c. Make comparisons of the data. Use statistics in your answer.
d. Explain if the drug works. Justify your answer.
e. Determine other considerations that should be taken into account when trying to draw conclusions from an experiment of this type.

Problem solving

17. The heights of Year 10 and Year 12 students (to the nearest centimetre) are being investigated. The results of some sample data are shown below.

Year 10	160	154	157	170	167	164	172	158	177	180	175	168	159	155	163	163	169	173	172	170
Year 12	160	172	185	163	177	190	183	181	176	188	168	167	166	177	173	172	179	175	174	108

a. Draw a back-to-back stem-and-leaf plot.
b. Draw a parallel box plot.
c. Comment on what the plots tell you about the heights of Year 10 and Year 12 students.

18. Kloe compares her English and Maths marks. The results of eight tests in each subject are shown below.

English:	76, 64, 90, 67, 83, 60, 85, 37
Maths:	80, 56, 92, 84, 65, 58, 55, 62

a. Calculate Kloe's mean mark in each subject.
b. Calculate the range of marks in each subject.
c. Calculate the standard deviation of marks in each subject.
d. Based on the above data, determine the subject that Kloe has performed more consistently in.

19. A sample of 50 students was surveyed on whether they owned an iPad or a mobile phone. The results showed that 38 per cent of the students owned both. Sixty per cent of the students owned a mobile phone and there were four students who had an iPad only. Evaluate the percentage of students that did not own a mobile phone or an iPad.

20. The life expectancy of non-Aboriginal and non–Torres Strait Islander people in Australian states and territories is shown on the boxplot below.

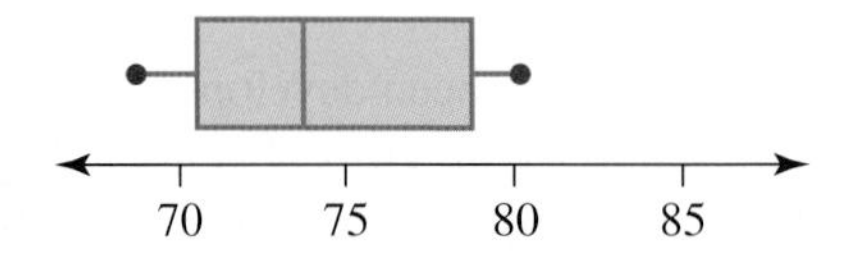

Life expectancy of non-Aboriginal and non–Torres Strait Islander people in Australian states and territories

The life expectancies of Aboriginal and Torres Strait Islander people in each of the Australian states and territories are 56, 58.4, 51.3, 57.8, 53.9, 55.4 and 61.0.

a. Draw parallel box plots on the same axes. Compare and comment on your results.
b. Comment on the advantage and disadvantage of using a box plot.

12.7 Populations and samples

LEARNING INTENTION

At the end of this subtopic you should be able to:
- describe the difference between populations and samples
- recognise the difference between a census and a survey and identify a preferred method in different circumstances.

12.7.1 Populations

eles-4946

- The term **population** refers to a complete set of individuals, objects or events belonging to some category.
- When data are collected from a *whole* population, the process is known as a **census**.
  - It is often not possible, nor cost-effective, to conduct a census.
  - or this reason, **samples** have to be selected carefully from the population. A sample is a subset of a population.

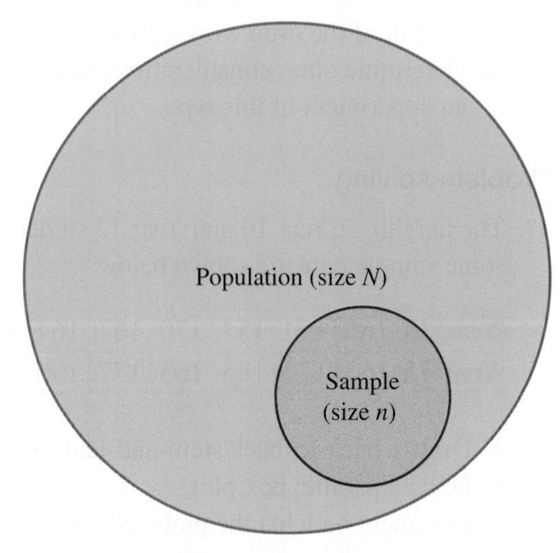

WORKED EXAMPLE 16 Identifying problems with collecting data on populations

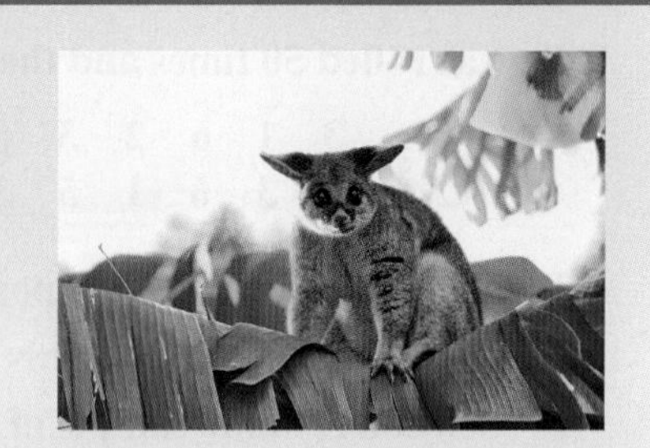

List some of the problems you might encounter in trying to collect data on the following populations.

a. The life of a mobile phone battery.

b. The number of possums in a local area.

c. The number of males in Australia.

d. The average cost of a loaf of white bread.

THINK	WRITE
For each of these scenarios, consider how the data might be collected, and the problems in obtaining these data.	
a. The life of a mobile phone battery.	a. The life of a mobile phone battery cannot be measured until it is dead. The battery life also depends on how the phone is used, and how many times it has been recharged.
b. The number of possums in a local area.	b. It would be almost impossible to find all the possums in a local area in order to count them. The possums also may stray into other areas.
c. The number of males in Australia.	c. The number of males in Australia is constantly changing. There are births and deaths every second.
d. The average cost of a loaf of white bread.	d. The price of one particular loaf of white bread varies widely from one location to another. Sometimes the bread is on 'Special' and this would affect the calculations.

12.7.2 Samples

eles-4947

- Surveys are conducted using samples. Ideally the sample should reveal generalisations about the population.
- The sample selected to be surveyed should be chosen without **bias**, as this may result in a sample that is not representative of the whole population. For example, the students conducting the investigation decide to choose a sample of 12 fellow students. While it would be simplest to choose 12 of their friends as the sample, this would introduce bias since they would not be representative of the population as a whole.
- A random sample is generally accepted as being an ideal representation of the population from which it was drawn. However, it must be remembered that different random samples from the same population can produce different results. This means that we must be cautious about making predictions about a population, as results of surveys conducted using random samples may vary.
- A sample size must be sufficiently large. As a general rule, the sample size should be about $\sqrt{N}$, where N is the size of the population. If the sample size is too small, the conclusions that are drawn from the sample data may not reflect the population as a whole.

WORKED EXAMPLE 17 Determining statistics from samples

A die was rolled 50 times and the following results were obtained.

6	**5**	**3**	**1**	**6**	**2**	**3**	**6**	**2**	**5**	**3**	**4**	**1**	**3**	**2**	**6**	**4**	**5**	**5**	**4**	**3**	**1**	**2**	**1**	**6**
4	**5**	**2**	**3**	**6**	**1**	**5**	**3**	**3**	**2**	**4**	**1**	**4**	**2**	**3**	**2**	**6**	**3**	**4**	**6**	**2**	**1**	**2**	**4**	**2**

a. **Determine the mean of the population (to 1 decimal place).**

b. **A suitable sample size for this population would be 7 $\left(\sqrt{50} \approx 7.1\right)$.**

   i. **Select a random sample of 7 scores, and determine the mean of these scores.**

   ii. **Select a second random sample of 7 scores, and determine the mean of these.**

   iii. **Select a third random sample of 20 scores, and determine the mean of these.**

c. **Comment on your answers to parts a and b.**

THINK

WRITE

a. Calculate the mean by first finding the sum of all the scores, then dividing by the number of scores (50).

a. Population mean $= \dfrac{\sum x}{n}$

$= \dfrac{169}{50}$

$= 3.4$

b. i. Use a calculator to randomly generate 7 scores from 1 to 50. Relate these numbers back to the scores, then calculate the mean.

b. i. The 7 scores randomly selected are numbers 17, 50, 40, 34, 48, 12, 19 in the set of 50 scores. These correspond to the scores: 4, 2, 3, 3, 2, 4, 5.

The mean of these scores $= \dfrac{23}{7} \approx 3.3$.

ii. Repeat b i to obtain a second set of 7 randomly selected scores. This second set of random numbers produced the number 1 twice. Try again. Another attempt produced the number 14 twice. Try again. A third attempt produced 7 different numbers. This set of 7 random numbers will then be used to, again, calculate the mean of the scores.

ii. Ignore the second and third attempts to select 7 random numbers because of repeated numbers. The second set of 7 scores randomly selected is numbers 16, 49, 2, 42, 31, 11, 50 of the set of 50. These correspond to the scores: 6, 4, 5, 6, 1, 3, 2.

The mean of these scores $= \dfrac{27}{7} \approx 3.9$.

iii. Repeat for a randomly selected 20 scores.

iii. The set of 20 randomly selected numbers produced a total of 68.

Mean of 20 random scores $= \dfrac{68}{20} = 3.4$

c. Comment on the results.

c. The population mean is 3.4.
The means of the two samples of 7 are 3.3 and 3.9.
This shows that, even though the samples are randomly selected, their calculated means may be different.
The mean of the sample of 20 scores is 3.4. This indicates that by using a bigger sample the result is more accurate than those obtained with the smaller samples.

TI | THINK

a.

1. In a new document, on a Lists & Spreadsheet page, label column A as 'die'. Enter the data from the question.

DISPLAY/WRITE

a.

2. Although you can find many summary statistics, to find the mean only, open a Calculator page and press:
   - MENU
   - 6: Statistics
   - 3: List Math
   - 3: Mean

   Press VAR and select 'die', then press CTRL ENTER to get a decimal approximation.

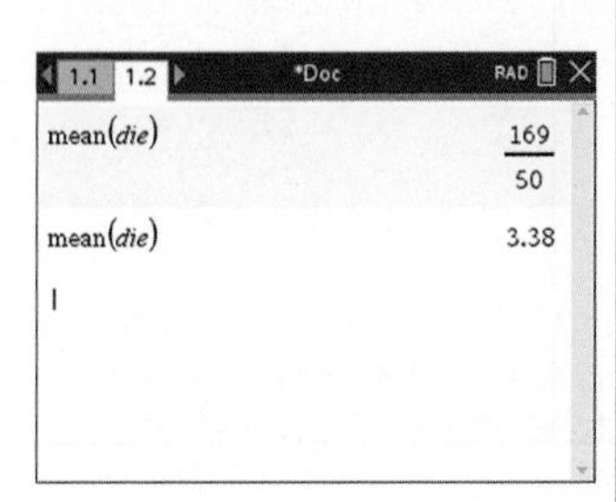

The mean of the 50 die rolls is 3.38.

b.

i-ii. To generate a random sample of 7 scores, on the Calculator page, press:
- MENU
- 5: Probability
- 4: Random
- 2: Integer

Type the entry line as: randInt (1,6,7)
Then press ENTER.
This randomly generates 7 numbers between 1 and 6.
Press:
- MENU
- 6: Statistics
- 3: List Math
- 3: Mean

Complete the entry line as: mean(ans)
Then press CTRL ENTER to get a decimal approximation of the mean.
Repeat this with another set of random numbers.

b.

i-ii.

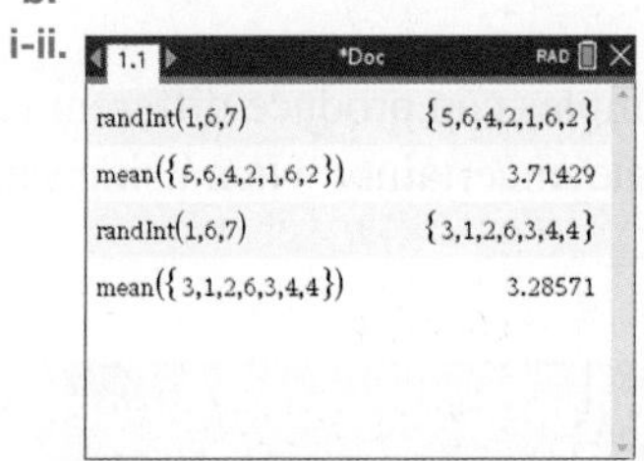

The mean of the first sample of 7 rolls is 3.71, and the mean of the second sample of 7 rolls is 3.29, correct to 2 decimal places.

CASIO | THINK

a.

1. On the Statistics screen, label list1 as 'die'. Enter the data from the question. Press EXE after each value.

DISPLAY/WRITE

a.

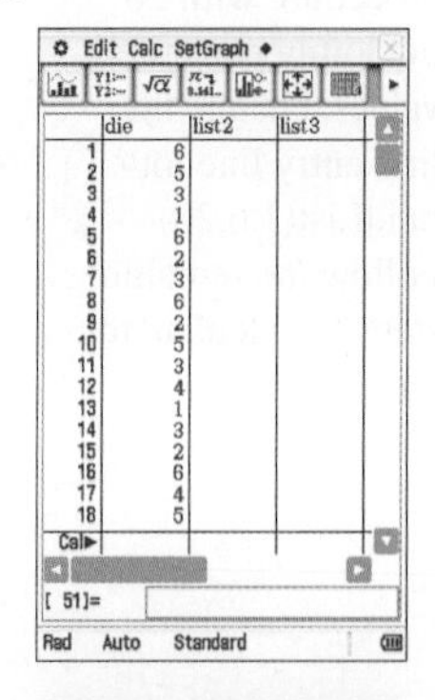

2. To find the statistics summary, tap:
   - Calc
   - One-Variable

   Set:
   - XList: main\die
   - Freq: 1

   Tap OK.

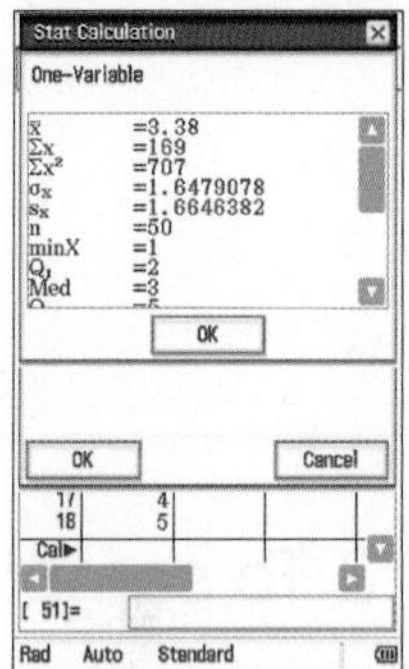

The mean of the 50 die rolls is 3.38.

b.

i-ii. To generate a random sample of 7 dice rolls, on the Main screen, type the entry line as: randList (7,1,6)
Then press EXE.
The randList can be found in the Catalog on the Keyboard.
This randomly generates 7 numbers between 1 and 6.

Tap:
- Action
- List
- Statistics
- Mean

Highlight the random list, including the brackets, and drag it to the 'mean('. Close the bracket and press EXE.
Repeat this with another set of random numbers.

b.

i-ii.

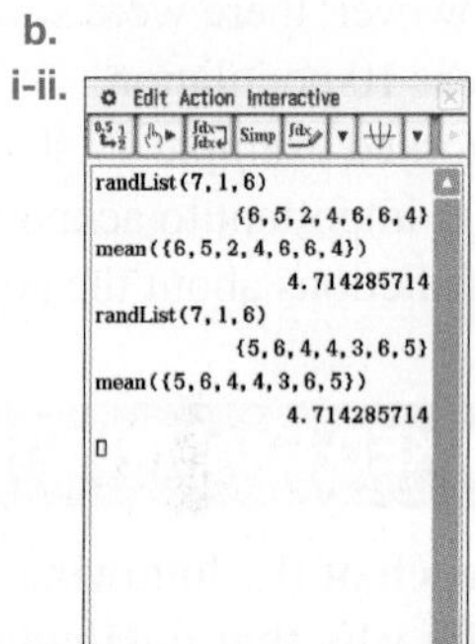

The mean of the first sample of 7 rolls is 4.71; the mean of the second sample of 7 rolls is also 4.71, correct to 2 decimal places.

iii. To repeat the procedure with 20 randomly generated values, change the first entry line to:
randList(1,6,20)
Follow the remaining steps to calculate the mean.

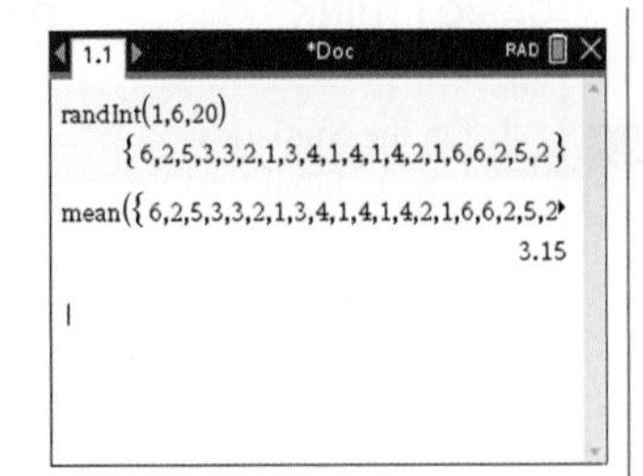

The mean of the third sample of 20 rolls is 3.15.

c.

c. This indicates that the results obtained from a bigger sample are more accurate than those from smaller samples.

iii. To repeat the procedure with 20 randomly generated values, change the first entry line to:
randList(20,1,6)
Follow the remaining steps to calculate the mean.

iii. 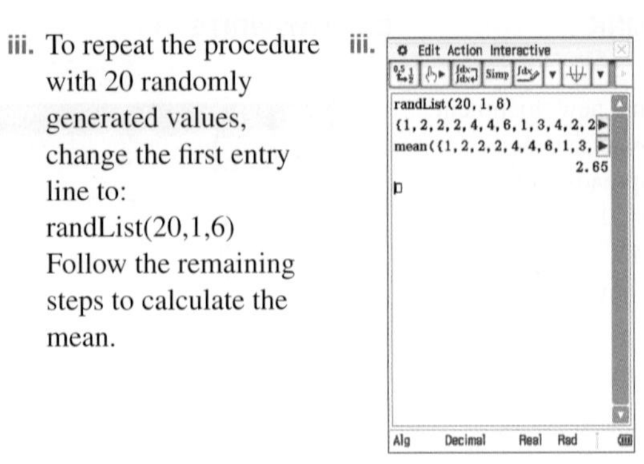

The mean of the third sample of 20 rolls is 2.65.

c.

c. This indicates that the results obtained from a bigger sample are more accurate than those from smaller samples.

12.7.3 To sample or to conduct a census?

eles-4948

- The particular circumstances determine whether data are collected from a population, or from a sample of the population. For example, suppose you collected data on the height of every Year 10 student in your class. If your class was the only Year 10 class in the school, your class would be the population. If, however, there were several Year 10 classes in your school, your class would be a sample of the Year 10 population.
- Worked example 17 showed that different random samples can produce different results. For this reason, it is important to acknowledge that there could be some uncertainty when using sample results to make predictions about the population.

WORKED EXAMPLE 18 Stating if the information was obtained by census or survey

For each of the following situations, state whether the information was obtained by census or survey. Justify why that particular method was used.

a. A roll call is conducted each morning at school to determine which students are absent.
b. TV ratings are collected from a selection of viewers to discover the popular TV shows.
c. Every hundredth light bulb off an assembly production line is tested to determine the life of that type of light bulb.
d. A teacher records the examination results of her class.

THINK	WRITE
a. Every student is recorded as being present or absent at the roll call.	a. This is a census. If the roll call only applied to a sample of the students, there would not be an accurate record of attendance at school. A census is essential in this case.
b. Only a selection of the TV audience contributed to these data.	b. This is a survey. To collect data from the whole viewer population would be time-consuming and expensive. For this reason, it is appropriate to select a sample to conduct the survey.

c. Only 1 bulb in every 100 is tested.

c. This is a survey. Light bulbs are tested to destruction (burn-out) to determine their life. If every bulb was tested in this way, there would be none left to sell! A survey on a sample is essential.

d. Every student's result is recorded.

d. This is a census. It is essential to record the result of every student.

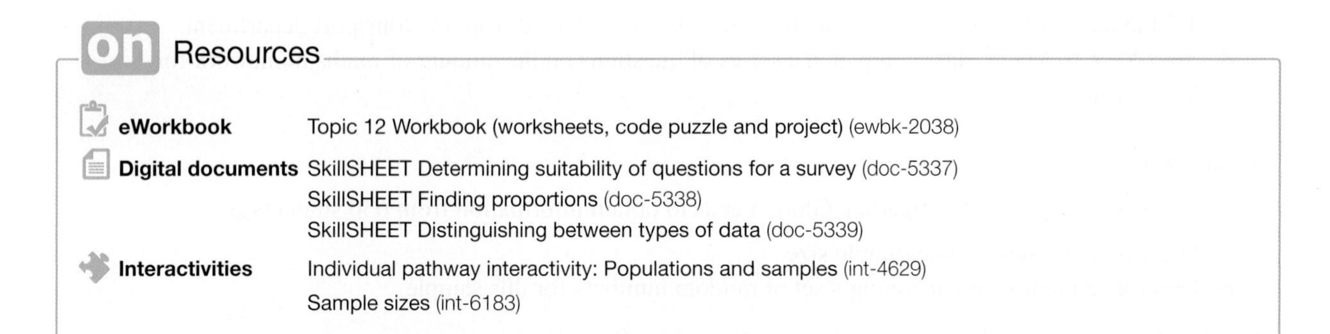

on Resources

eWorkbook Topic 12 Workbook (worksheets, code puzzle and project) (ewbk-2038)

Digital documents SkillSHEET Determining suitability of questions for a survey (doc-5337)
SkillSHEET Finding proportions (doc-5338)
SkillSHEET Distinguishing between types of data (doc-5339)

Interactivities Individual pathway interactivity: Populations and samples (int-4629)
Sample sizes (int-6183)

Exercise 12.7 Populations and samples

learnon

Individual pathways

■ PRACTISE	■ CONSOLIDATE	■ MASTER
1, 4, 5, 8, 11	2, 6, 9, 12	3, 7, 10, 13

To answer questions online and to receive **immediate feedback** and **sample responses** for every question, go to your learnON title at www.jacplus.com.au.

Fluency

1. **WE16** List some of the problems you might encounter in trying to collect data from the following populations.
   a. The life of a laptop computer battery.
   b. The number of dogs in your neighbourhood.
   c. The number of fish for sale at the fish markets.
   d. The average number of pieces of popcorn in a bag of popcorn.

2. **WE17** The data below show the results of the rolled die from Worked example 17.

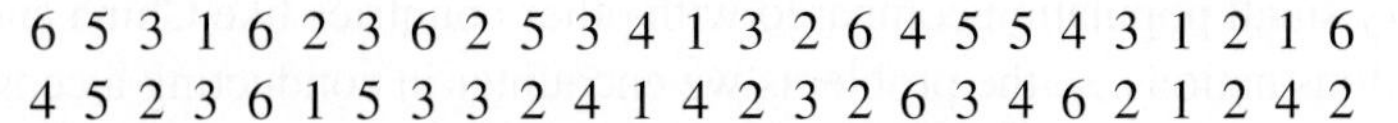

6 5 3 1 6 2 3 6 2 5 3 4 1 3 2 6 4 5 5 4 3 1 2 1 6
4 5 2 3 6 1 5 3 3 2 4 1 4 2 3 2 6 3 4 6 2 1 2 4 2

   The mean of the population is 3.4. Select your own samples for the following questions.
   a. Select a random sample of 7 scores, and determine the mean of these scores.
   b. Select a second random sample of 7 scores, and determine the mean of these.
   c. Select a third random sample of 20 scores, and determine the mean of these.
   d. Comment on your answers to parts a, b and c.

3. **WE18** In each of the following scenarios, state whether the information was obtained by census or survey. Justify why that particular method was used.
   a. Seating for all passengers is recorded for each aeroplane flight.
   b. Movie ratings are collected from a selection of viewers to discover the best movies for the week.

c. Every hundredth soft drink bottle off an assembly production line is measured to determine the volume of its contents.
d. A car driving instructor records the number of hours each learner driver has spent driving.

4. For each of the following, state whether a census or a survey has been used.
   a. Two hundred people in a shopping centre are asked to nominate the supermarket where they do most of their grocery shopping.
   b. To find the most popular new car on the road, new car buyers are asked what make and model they purchased.
   c. To find the most popular new car on the road, data are obtained from the transport department.
   d. Your Year 10 Maths class completed a series of questions on the amount of maths homework for Year 10 students.

Understanding

5. To conduct a statistical investigation, Gloria needs to obtain information from 630 students.
   a. Determine the appropriate sample size.
   b. Describe a method of generating a set of random numbers for this sample.
6. A local council wants the opinions of its residents regarding its endeavours to establish a new sporting facility for the community. It has specifically requested all residents over 10 years of age to respond to a set of on-line questions.
   a. State if this is a census or a survey.
   b. Determine what problems could be encountered when collecting data this way.

7. A poll was conducted at a school a few days before the election for Head Boy and Head Girl. After the election, it was discovered that the polls were completely misleading. Explain how this could have happened.

Reasoning

8. A sampling error is said to occur when results of a sample are different from those of the population from which the sample was drawn. Discuss some factors which could introduce sampling errors.
9. Since 1961, a census has been conducted in Australia every 5 years. Some people object to the census on the basis that their privacy is being invaded. Others say that the expense involved could be directed to a better cause. Others say that a sample could obtain statistics which are just as accurate. State are your views on this. Justify your statements.
10. Australia has a very small population compared with other countries like China and India. These are the world's most populous nations, so the problems we encounter in conducting a census in Australia would be insignificant compared with those encountered in those countries. Suggest what different problems authorities would come across when conducting a census in countries with large populations.

Problem solving

11. The game of Lotto involves picking the same 6 numbers in the range 1 to 45 as have been randomly selected by a machine containing 45 numbered balls. The balls are mixed thoroughly, then 8 balls are selected representing the 6 main numbers, plus 2 extra numbers, called supplementary numbers.
    Here is a list of the number of times each number had been drawn over a period of time, and also the number of weeks since each particular number has been drawn.

Number of weeks since each number drawn							
1 **1**	2 **5**	3 **2**	4 **1**	5 **1**	6 **7**	7 **-**	8 **4**
9 **3**	10 **3**	11 **1**	12 **5**	13 **5**	14 **7**	15 **-**	16 **4**
17 **9**	18 **-**	19 **9**	20 **2**	21 **2**	22 **12**	23 **10**	24 **8**
25 **5**	26 **11**	27 **17**	28 **2**	29 **3**	30 **3**	31 **-**	32 **22**
33 **4**	34 **3**	35 **-**	36 **1**	37 **12**	38 **-**	39 **6**	40 **-**
41 **6**	42 **1**	43 **7**	44 **-**	45 **31**			

Number of times each number drawn since draw 413							
1 **246**	2 **238**	3 **244**	4 **227**	5 **249**	6 **241**	7 **253**	8 **266**
9 **228**	10 **213**	11 **250**	12 **233**	13 **224**	14 **221**	15 **240**	16 **223**
17 **217**	18 **233**	19 **240**	20 **226**	21 **238**	22 **240**	23 **253**	24 **228**
25 **252**	26 **239**	27 **198**	28 **229**	29 **227**	30 **204**	31 **230**	32 **226**
33 **246**	34 **233**	35 **232**	36 **251**	37 **222**	38 **221**	39 **219**	40 **259**
41 **245**	42 **242**	43 **237**	44 **221**	45 **224**			

If these numbers are randomly chosen, explain the differences shown in the tables.

12. A sample of 30 people was selected at random from those attending a local swimming pool. Their ages (in years) were recorded as follows:

19, 7, 58, 41, 17, 23, 62, 55, 40, 37, 32, 29, 21, 18, 16,
10, 40, 36, 33, 59, 65, 68, 15, 9, 20, 29, 38, 24, 10, 30

a. Calculate the mean and the median age of the people in this sample.
b. Group the data into class intervals of (0–9 etc.) and complete the frequency distribution table.
c. Use the frequency distribution table to estimate the mean age.
d. Calculate the cumulative frequency and, hence, plot the ogive.
e. Estimate the median age from the ogive.
f. Compare the mean and median of the original data in part **a** with the estimates of the mean and the median obtained for the grouped data in parts **c** and **e**.
g. Were the estimates good enough? Explain your answer.

13. The typing speed (words per minute) was recorded for a group of Year 8 and Year 10 students. The results are displayed in this back-to-back stem plot.

Key: 2|6 = 26 wpm

Leaf: Year 8	*Stem*	*Leaf:* Year 10
9 9	0	
9 8 6 5 4 2 0	1	7 9
9 8 8 6 4 2 1 0 0	2	2 3 6 8 9
9 7 7 6 4 1 0	3	0 2 4 5 5 7 8 8
8 6 5 2 0	4	1 2 5 8 8 9 9
	5	0 3 5 7 8
	6	0 0 3

Write a report comparing the typing speeds of the two groups.

12.8 Evaluating inquiry methods and statistical reports

LEARNING INTENTION

At the end of this subtopic you should be able to:
- describe the difference between primary and secondary data collection
- identify misleading errors or graphical techniques in data
- evaluate the accuracy of a statistical report.

12.8.1 Data collection methods

eles-4963

- Data can be collected in different ways. The manner in which you collect data can affect the validity of the results you determine.
- **Primary data** is collected firsthand by observation, measurement, survey, experiment or simulation. and is owned by the data collector until it is published.
- **Secondary data** is obtained from external sources such as journals, newspapers, or any other previously collected data.

When collecting primary data through experiments, or using secondary data, considerations need to be given to:
- Validity. Results of valid investigations are supported by other investigations.
- Reliability. Data is reliable if the same data can be collected when the investigation is repeated. If the data cannot be repeated by other investigators, the data may not be valid or true.
- Accuracy and precision. When collecting primary data through experiments, the accuracy of the data is how close it is to a known value. Precise data is when multiple measurements of the same investigation are close to each other.
- Bias. When collecting primary data in surveys it is important to ensure your sample is representative of the population you are studying. When using secondary data you should consider who collected the data, and whether those researchers had any intentional or unintentional influence on the data.

WORKED EXAMPLE 19 Choosing appropriate collection methods

You have been given an assignment to investigate which year level uses the school library, after school, the most.

a. Explain whether it is more appropriate to use primary or secondary data in this case. Justify your choice.
b. Describe how the data could be collected. Discuss any problems which might be encountered.
c. Explain whether an alternative method would be just as appropriate.

THINK	WRITE
a. No records have been kept on library use.	a. Since records are not kept on the library use, secondary data is not an option. Primary data collection could be either sampling or census. A sufficiently large sample size could be chosen; this would take less time than conducting a census, although it would not be as accurate. Sampling would be considered appropriate in this case.

b. The data can be collected via a questionnaire or in person.

b. A questionnaire could be designed and distributed to a randomly-chosen sample. The problem here would be the non-return of the forms.
Observation could be used to personally interview students as they entered the library. This would take more time, but random interview times could be selected.

c. A census is the other option.

c. A census could be conducted, either by questionnaire or observation. This should yield a more accurate outcome.

WORKED EXAMPLE 20 Choosing appropriate collection methods

State which method would be the most appropriate to collect the following data. Suggest an alternative method in each case.

a. The number of cars parked in the staff car park each day.
b. The mass of books students carry to school each day.
c. The length a spring stretches when weights are added to it.
d. The cost of mobile phone plans with various network providers.

THINK	WRITE
a. Observation	a. The best way would probably be observation by visiting the staff car park to count the number of cars there. An alternative method would be to conduct a census of all workers to ask if they parked in the staff car park. This method may be prone to errors as it relies on accurate reporting by many people.
b. Measurement	b. The mass of the books could be measured by weighing each student's pack on scales. A random sample would probably yield a reasonably accurate result.
c. Experiment	c. Conduct an experiment and measure the extension of the spring with various weights. There are probably no alternatives to this method as results will depend upon the type of spring used.
d. Internet search	d. An internet search would enable data to be collected. Alternatively, a visit to mobile phone outlets would yield similar results.

12.8.2 Analysing the data

eles-4964

- Once the data have been collected and collated, a decision must be made with regard to the best methods for analysing the data.
  - a measure of central tendency — mean, median or mode
  - a measure of spread — range, interquartile range or standard deviation
  - an appropriate graph.

Statistical graphs

- Data can be graphed in a variety of ways — line graphs, bar graphs, histograms, stem plots, box plots, and so on.
- Because graphs give a quick visual impression, the temptation is to not look at them in great detail. Often graphs can be quite misleading.

- It is easy to manipulate a graph to give an impression which is supported by the creator of the graph. This is achieved by careful choice of scale on the horizontal and vertical axes.
  - Shortening the horizontal axis tends to highlight the increasing/decreasing nature of the trend of the graph. Lengthening the vertical axis tends to have the same effect.
  - Lengthening the horizontal and shortening the vertical axes tends to level out the trends.

WORKED EXAMPLE 21 Observing the effect of changing scales on bar graphs

The report shows the annual change in median house prices in the local government areas (LGA) of Queensland from 2019–20 to 2020–21.

a. Draw a bar graph which would give the impression that the percentage annual change was much the same throughout the whole state.

b. Construct a bar graph to give the impression that the percentage annual change in Brisbane was far greater than that in the other local government areas.

HOUSES Suburb/locality	Median house price 2020–21	2019–20	Annual change
Brisbane (LGA)	$700,000	$627,000	11.6%
Ipswich City (LGA)	$323,000	$310,000	4.2%
Redland City (LGA)	$467,500	$435,000	7.5%
Logan City (LGA)	$360,000	$340,000	5.9%
Moreton Bay (LGA)	$399,000	$372,000	7.3%
Gold Coast City (LGA)	$505,000	$465,000	8.6%
Toowoomba (LGA)	$360,000	$334,500	7.6%
Sunshine Coast (LGA)	$470,000	$445,000	5.6%
Fraser Coast (LGA)	$322,500	$312,500	3.2%
Bundaberg (LGA)	$282,000	$275,000	2.5%
Gladstone (LGA)	$286,000	$286,000	0.0%
Rockhampton (LGA)	$267,000	$254,000	5.1%
Mackay (LGA)	$398,000	$383,000	3.9%
Townsville City (LGA)	$375,000	$359,000	4.5%
Cairns (LGA)	$400,000	$389,000	2.8%

THINK

a. To flatten out trends, lengthen the horizontal axis and shorten the vertical axis.

WRITE/DRAW

a.

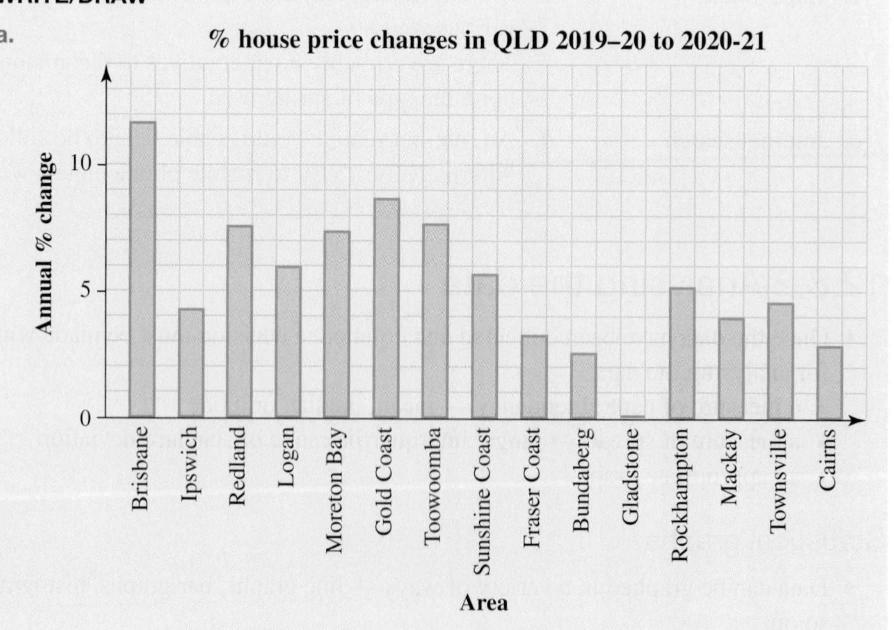

b. To accentuate trends, shorten the horizontal axis and lengthen the vertical axis.

b.

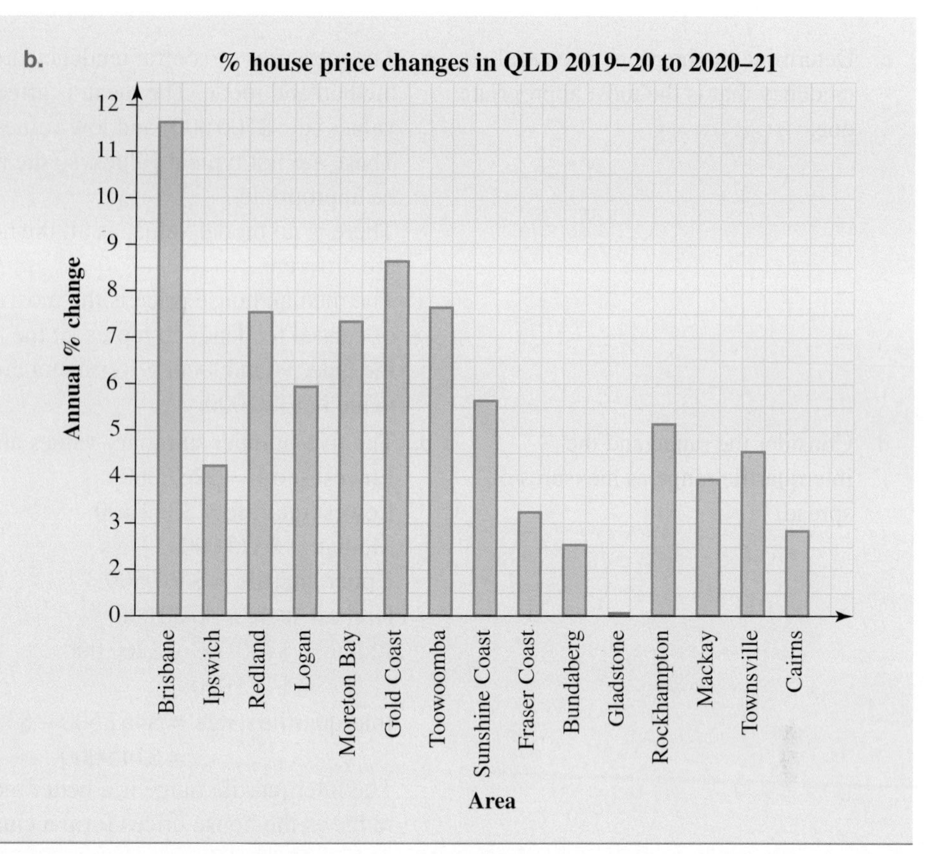

WORKED EXAMPLE 22 Comparing and choosing statistical measures

Consider the data displayed in the table of Worked example 21. Use the data collected for the median house prices in 2020–21.

a. **Explain whether the data would be classed as primary or secondary data.**

b. **Explain why the data shows median house prices rather than the mean or modal house price.**

c. **Calculate a measure of central tendency for the data. Explain the reason for this choice.**

d. **Give a measure of spread of the data, giving a reason for the particular choice.**

e. **Display the data in a graphical form, explaining why this particular form was chosen.**

THINK

a. These are data which have been collected by someone else.

WRITE

a. These are secondary data because they have been collected by someone else.

THINK

b. Median is the middle price, mean is the average price, and mode is the most frequently-occurring price.

WRITE

b. The median price is the middle value. It is not affected by outliers as the mean is. The modal house price may only occur for two house sales with the same value. On the other hand, there may not be any mode. The median price is the most appropriate in this case.

c. Determine the measure of central tendency that is the most appropriate one.

c. The measures of central tendency are the mean, median and mode. The mean is affected by high values (i.e. \$700 000) and low values (i.e. \$282 000). These are not typical values, so the mean would not be appropriate.
There is no modal value, as all the house prices are different.
The median house price is the most suitable measure of central tendency to represent the house prices in the Queensland local government areas. The median value is \$370 000.

d. Consider the range and the interquartile range as measures of spread.

d. The five-number summary values are:
Lowest score = \$267 000
Lowest quartile = \$322 500
Median = \$375 000
Upper quartile = \$467 500
Highest score = \$700 000
Range = \$700 000 − \$267 000
= \$433 000
Interquartile range = \$467 500 − \$322 500
= \$145 000
The interquartile range is a better measure for the range as the house prices form a cluster in this region.

e. Consider the graphing options.

e. Of all the graphing options, the box plot seems the most appropriate as it shows the spread of the prices as well as how they are grouped around the median price.

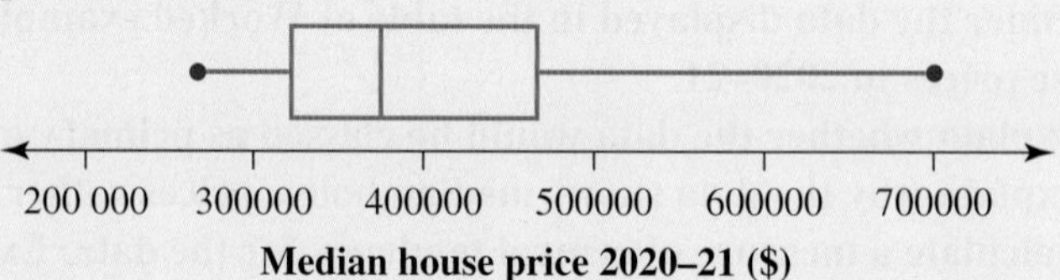

WORKED EXAMPLE 23 Analysing data and use of statistics to interpret results

The following data is the heights of the members of the Australian women's national basketball team (in metres):

1.73, 1.65, 1.8, 1.83, 1.96, 1.88, 1.63, 1.88, 1.83, 1.88, 1.8, 1.96

Provide calculations and explanations as evidence to verify or refute the following statements.

a. The mean height of the team is greater than their median height.
b. The range of the heights of the 12 players is almost 3 times their interquartile range.
c. Only 5 players are on the court at any one time. A team of 5 players can be chosen such that their mean, median and modal heights are all the same.

THINK	WRITE
a. 1. Calculate the mean height of the 12 players.	**a.** Mean $= \dfrac{\sum x}{n} = \dfrac{21.83}{12} = 1.82$ m

2. Order the heights to determine the median.

The heights of the players, in order, are: 1.63, 1.65, 1.73, 1.8, 1.8, 1.83, 1.83, 1.88, 1.88, 1.88, 1.96, 1.96. There are 12 scores, so the median is the average of the 6th and 7th scores.

$$\text{Median} = \frac{1.83 + 1.83}{2} = 1.83\text{ m}$$

3. Comment on the statement.

The mean is 1.82 m, while the median is 1.83 m. This means that the mean is less than the median, so the statement is not true.

b. 1. Determine the range and the interquartile range of the 12 heights.

b. Range = 1.96 − 1.63 = 0.33 m
Lower quartile is the average of 3rd and 4th scores.

$$\text{Lower quartile} = \frac{1.73 + 1.8}{2} = 1.765\text{ m}$$

Upper quartile is average of 3rd and 4th scores from the end.

$$\text{Upper quartile} = \frac{1.88 + 1.88}{2} = 1.88\text{ m}$$

Interquartile range = 1.88 − 1.765 = 0.115 m

2. Compare the two values.

Range = 0.33 m

Interquartile range = 0.115 m

$$\frac{\text{Range}}{\text{Interquartile range}} = \frac{0.33}{0.115} = 2.9$$

3. Comment on the statement.

Range = 2.9 × interquartile range
This is almost 3 times, so the statement is true.

c. 1. Choose 5 players whose mean, median and modal heights are all equal. Trial and error is appropriate here. There may be more than one answer.

c. Three players have a height of 1.88 m. If a player shorter and one taller are chosen, both the same measurement from 1.88 m, this would make the mean, median and mode all the same. Choose players with heights:

1.8, 1.88, 1.88, 1.88, 1.96

$$\text{Mean} = \frac{9.4}{5} = 1.88\text{ m}$$

Median = 3rd score = 1.88 m

Mode = Most frequent score = 1.88 m

2. Comment on the statement.

The 5 players with heights, 1.8 m, 1.88 m, 1.88 m, 1.88 m and 1.96 m have a mean, median and modal height of 1.88 m.
It is true that a team of 5 such players can be chosen.

TI | THINK | **DISPLAY/WRITE**

a.

1. In a new problem, on a Lists & Spreadsheet page, label column A as 'heights'. Enter the data from the question.

a.

	A heights	B	C	D
=				
1	1.73			
2	1.65			
3	1.8			
4	1.83			
5	1.96			

2. Open a Calculator page and complete the entry lines as:
mean(*heights*)
median(*heights*)
Press ENTER after each entry.

mean(*heights*) 1.81917
median(*heights*) 1.83

The mean heights are less than the median heights, so the statement is false.

b.

To find all the summary statistics, open the Calculator page and press:
- MENU
- 6: Statistics
- 1: Stat Calculations
- 1: One-Variable Statistics...

Select 1 as the number of lists. Then on the One-Variable Statistics page, select 'heights' as the X1 List and leave the frequency as 1. Leave the next two fields empty and TAB to OK and then press ENTER.

b.

"σx := σnx" 0.101691
"n" 12.
"MinX" 1.63
"Q₁X" 1.765
"MedianX" 1.83
"Q₃X" 1.88
"MaxX" 1.96
"SSX := Σ(x−x̄)²" 0.124092

The range is max − min = $1.96 - 1.63 = 0.33$.
$Q_1 = 1.765$ and $Q_3 = 1.88$.
$\text{IQR} = Q_3 - Q_1$
$= 1.88 - 1.765 = 0.115$.
Now $2.9 \times \text{IQR} \approx$ range so the statement is true.

CASIO | THINK | **DISPLAY/WRITE**

a.

1. On the Statistics screen, label list1 as 'heights'. Enter the data in the table as shown. Press EXE after each value.

a.

	heights	list2	list3
1	1.73		
2	1.65		
3	1.8		
4	1.83		
5	1.96		
6	1.88		
7	1.63		
8	1.88		
9	1.83		
10	1.88		
11	1.8		
12	1.96		

2. To find the statistics summary, tap:
- Calc
- One-Variable

Set values as:
- XList: main\heights
- Freq: 1

Tap OK.
The mean and median can be found in the list.

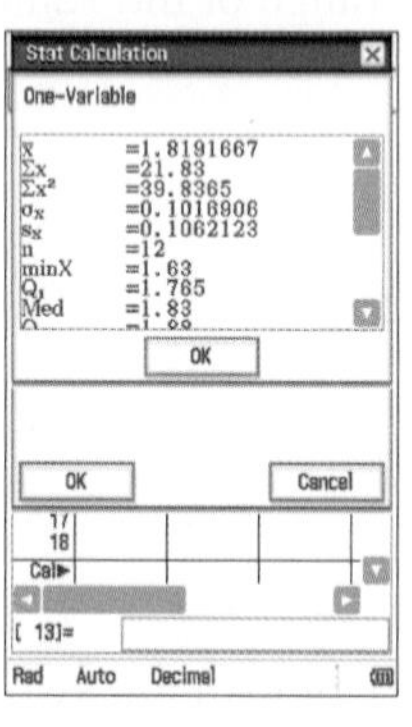

The mean heights are less than the median heights, so the statement is false.

b.

More statistics can be found from the statistics summary.

b.

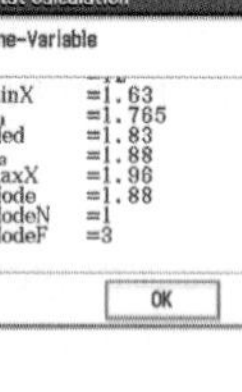

The range is max − min = $1.96 - 1.63 = 0.33$.
$Q_1 = 1.765$ and $Q_3 = 1.88$.
$\text{IQR} = Q_3 - Q_1$
$= 1.88 - 1.765 = 0.115$.
Now $2.9 \times \text{IQR} \approx$ range so the statement is true.

12.8.3 Statistical reports

eles-4955

- Reported data must not be simply taken at face value; all reports should be examined with a critical eye.

WORKED EXAMPLE 24 Analysing a statistical report

This is an excerpt from an article that appeared in a newspaper on Father's Day. It was reported to be a national survey findings of a *Gallup Poll* of data from 1255 fathers of children aged 17 and under.

THE GREAT AUSSIE DADS SURVEY

Thinking about all aspects of your life, how happy would you say you are?

	%
I am very happy	**26**
I am fairly happy	**49**
Totally happy	**75**
Some days I'm happy and some days I'm not	**21**
I am fairly unhappy	**3**
I am very unhappy	**1**
Totally unhappy	**4**

How often, if ever, do you regret having children?

Every day	**1**
Most days	**2**
Some days	**18**
Never	**79**

Which one of these best describes the impact of having children on your relationship with your partner?

We're closer than ever	**29**
We don't spend as much time together as we should	**40**
We're more like friends now than lovers	**21**
We have drifted apart	**6**
None of the above	**4**

Which one of these best describes the allocation of cooking and cleaning duties in your household?

My partner does nothing/I do everything	**1**
I do most of it	**11**
We share the cooking and cleaning	**42**
My partner does most of it	**41**
I do nothing/my partner does everything	**4**
None of the above	**1**

Which of these aspects of your children's future do you have concerns about?

	%
Their safety	**70**
Being exposed to drugs	**67**
Their health	**54**
Bullying or cyber-bullying	**50**
Teenage violence	**50**
Their ability to afford a home	**50**
Alcohol consumption and binge drinking	**47**
Achieving academic success	**47**
Achieving academic success	**47**
Feeling pressured into sex	**41**
Being able to afford the lifestyle they expect to have	**38**
Climate change	**23**
Having them living with you in their mid 20s	**14**
None of the above	**3**

What is the best thing about being a dad?

The simple pleasures of family life	**61**
Enjoying the successes of your kids	**24**
The unpredictability it brings	**9**
The comfort of knowing that you will be looked after in later life	**3**
None of the above	**3**

Key findings

75% of Aussie dads are totally happy
79% have never regretted having children
67% are worried about their children being exposed to drugs
57% would like more intimacy with their partner

"Work–life balance is definitely an issue for dads in 2010."

David Briggs
Galaxy principal

Source: *The Sunday Mail*, 5 Sept. 2010, pp. 14–15.

a. **Comment on the sample chosen.**
b. **Discuss the percentages displayed.**
c. **Comment on the claim that 57% of dads would like more intimacy with their partner.**

THINK	WRITE
a. How is the sample chosen? Is it truly representative of the population of Australian dads?	a. The results of a national survey such as this should reveal the outlook of the whole nation's dads. There is no indication of how the sample was chosen, so without further knowledge we tend to accept that it is representative of the population. A sample of 1255 is probably large enough.
b. Look at the percentages in each of the categories.	b. For the first question regarding happiness, the percentages total more than 100%. It seems logical that, in a question such as this, the respondents would tick only one box, but obviously this has not been the case. In the question regarding aspects of concern of 'your children's future', these percentages also total more than 100%. It seems appropriate here that dads would have more than one concerning area, so it is possible for the percentages to total more than 100%. In each of the other three questions, the percentages total 100%, which is appropriate.
c. Look at the tables to try to find the source of this figure.	c. Examining the reported percentages in the question regarding 'relationship with your partner', there is no indication how a figure of 57% was determined.

Note: Frequently media reports make claims where the reader has no hope of confirming their truth.

WORKED EXAMPLE 25 Analysing a statistical report

This article appeared in a newspaper. Read the article, then answer the following questions.

SPONGES ARE TOXIC

Washing dishes can pose a serious health risk, with more than half of all kitchen sponges containing high levels of dangerous bacteria, research shows.

A new survey dishing the dirt on washing up shows more than 50 per cent of kitchen sponges have high levels of *E.coli*, which can cause severe cramps and diarrhoea, and *staphylococcus aureus*, which releases toxins that can lead to food poisoning or toxic shock syndrome.

Microbiologist Craig Andrew-Kabilafkas of Australian Food Microbiology said the Westinghouse study of more than 1000 households revealed germs can spread easily to freshly washed dishes.

The only way to safeguard homes from sickness was to wash utensils at very high temperatures in a dishwasher.

Source: *The Sunday Mail*, 5 Sept. 2010, p. 36.

a. **Comment on the sample used in this survey.**
b. **Comment on the claims of the survey and identify any potential bias.**
c. **Is the heading of the article appropriate?**

THINK	WRITE
a. Look at sample size and selection of sample.	a. The report claims that the sample size was more than 1000. There is no indication how the sample was selected. The point to keep in mind is whether this sample is truly representative of the population consisting of all households. We have no way of knowing.
b. 1. Determine the results of the survey.	b. The survey claims that 50% of kitchen sponges have high levels of *E. coli* which can cause severe medical problems.
2. Identify any potential bias.	The study was conducted by Westinghouse, so it is not surprising they recommend using a dishwasher. There is no detail of how the sample was selected.
c. Examine the heading in the light of the contents of the article.	c. The heading is quite shocking, designed to catch the attention of readers.

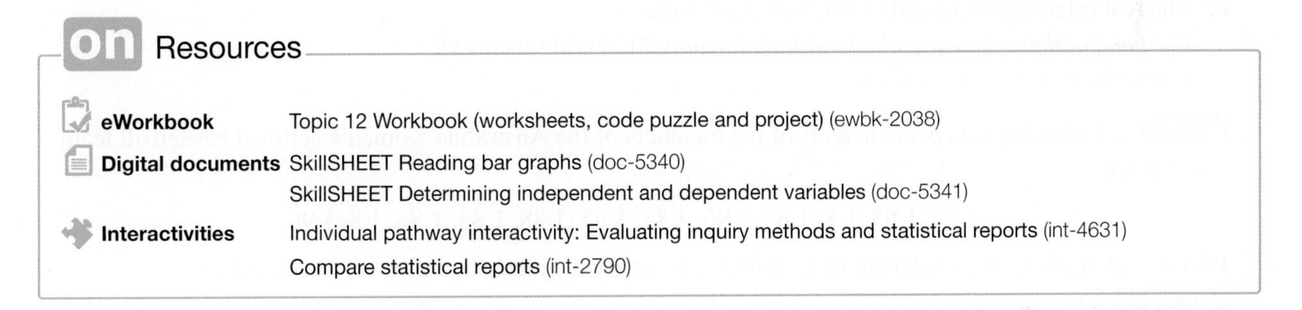

on Resources

eWorkbook Topic 12 Workbook (worksheets, code puzzle and project) (ewbk-2038)

Digital documents SkillSHEET Reading bar graphs (doc-5340)
SkillSHEET Determining independent and dependent variables (doc-5341)

Interactivities Individual pathway interactivity: Evaluating inquiry methods and statistical reports (int-4631)
Compare statistical reports (int-2790)

Exercise 12.8 Evaluating inquiry methods and statistical reports

learnON

Individual pathways

■ PRACTISE	■ CONSOLIDATE	■ MASTER
1, 4, 6, 9, 12	2, 7, 10, 13	3, 5, 8, 11, 14

To answer questions online and to receive **immediate feedback** and **sample responses** for every question, go to your learnON title at www.jacplus.com.au.

Fluency

1. WE19,20 You have been given an assignment to investigate which Year level has the greatest number of students who are driven to school each day by car.
   a. Explain whether it is more appropriate to use primary or secondary data in this case. Justify your choice.
   b. Describe how the data could be collected. Discuss any problems which might be encountered.
   c. Explain whether an alternative method would be just as appropriate.

2. WE21 You run a small company that is listed on the Australian Stock Exchange (ASX). During the past year you have given substantial rises in salary to all your staff. However, profits have not been as spectacular as in the year before. This table gives the figures for the salary and profits for each quarter.

	1st quarter	2nd quarter	3rd quarter	4th quarter
Profits ($′000 000)	6	5.9	6	6.5
Salaries ($′000 000)	4	5	6	7

Draw two graphs, one showing profits, the other showing salaries, which will show you in the best possible light to your shareholders.

3. WE22 The data below were collected from a real estate agent and show the sale prices of ten blocks of land in a new estate.
$150 000, $190 000, $175 000, $150 000, $650 000, $150 000, $165 000, $180 000, $160 000, $180 000
   a. Calculate a measure of central tendency for the data. Explain the reason for this choice.
   b. Give a measure of spread of the data, giving a reason for the particular choice.
   c. Display the data in a graphical form, explaining why this particular form was chosen.
   d. The real estate agent advertises the new estate land as:
   Own one of these amazing blocks of land for only $150 000 *(average)!*
   Comment on the agent's claims.

4. WE23 The following data is the heights of the members of the Australian women's national basketball team (in metres):

1.73, 1.65, 1.8, 1.83, 1.96, 1.88, 1.63, 1.88, 1.83, 1.88, 1.8, 1.96

Provide calculations and explanations as evidence to verify or refute the following statements.
   a. The mean height of the team is closer to the lower quartile than it is to the median.
   b. Half the players have a height within the interquartile range.
   c. Suggest which 5 players could be chosen to have the minimum range in heights.

5. The resting pulse of 20 female athletes was measured and is shown below.
50 62 48 52 71 61 30 45 42 48 43 47 51 52 34 61 44 54 38 40
   a. Represent the data in a distribution table using appropriate groupings.
   b. Calculate the mean, median and mode of the data.
   c. Comment on the similarities and differences between the three values.

6. The batting scores for two cricket players over six innings were recorded as follows.
Player A: 31, 34, 42, 28, 30, 41
Player B: 0, 0, 1, 0, 250, 0
Player B was hailed as a hero for his score of 250.
Comment on the performance of the two players.

Understanding

7. The table below shows the number of shoes of each size that were sold over a week at a shoe store.

Size	Number sold
4	5
5	7
6	19
7	24
8	16
9	8
10	7

a. Calculate the mean shoe size sold.
b. Determine the median shoe size sold.
c. Identify the modal shoe size sold.
d. Explain which measure of central tendency has the most meaning to the store proprietor.

8. WE24,25 This report from Woolworths appeared in a newspaper.

IT'S A RECORD

- Woolworths posted 10.1% gain in annual profit to $2.02b
- 11th consecutive year of double-digit growth
- Flags 8% to 11% growth in the current financial year
- Sales rose 4.8% to $51.2b
- Wants to increase its share of the fresh food market
- Announced $700m off-market share buyback
- Final fully franked dividend 62% a share

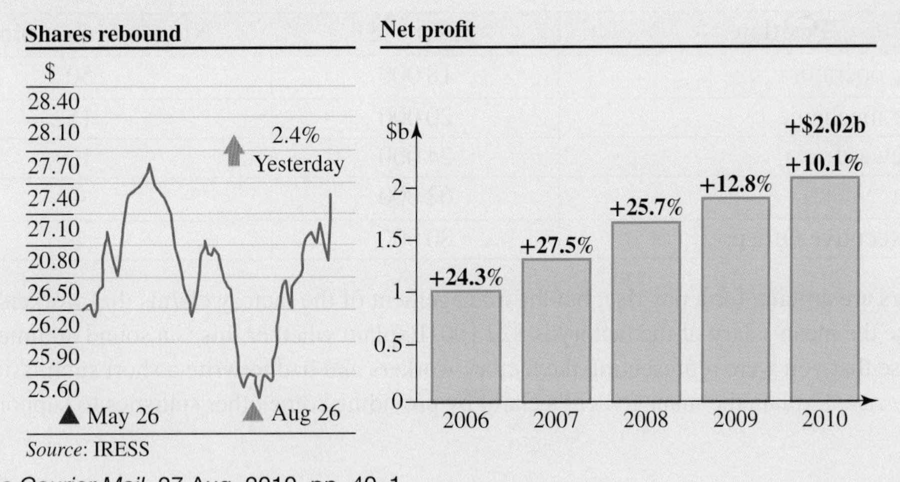

Source: *The Courier Mail*, 27 Aug. 2010, pp. 40–1.

Comment on the report.

Reasoning

9. Explain the point of drawing a misleading graph in a report.

10. The graph shows the fluctuation in the Australian dollar in terms of the US dollar during the period 13 July to 13 September 2010. The higher the Australian dollar, the cheaper it is for Australian companies to import goods from overseas, and the cheaper they should be able to sell their goods to the Australian public.
The manager of Company XYZ produced a graph to support his claim that, because there hasn't been much change in the Aussie dollar over that period, there hasn't been any change in the price he sells his imported goods to the Australian public.
Draw a graph that would support his claim. Explain how you were able to achieve this effect.

Source: *The Courier Mail*, 14 Sept. 2010, p. 25.

11. Two brands of light globes were tested by a consumer organisation. They obtained the following results.

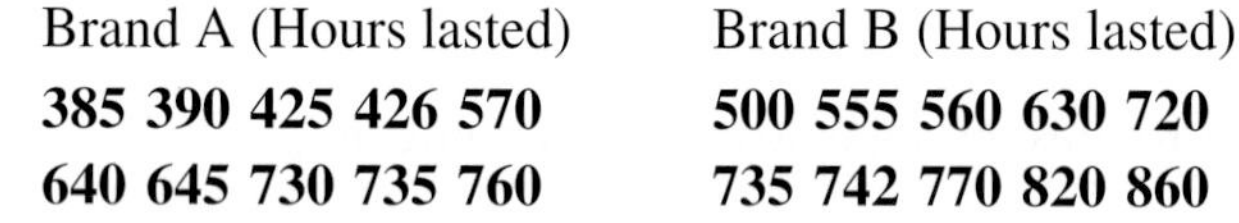

Brand A (Hours lasted)	Brand B (Hours lasted)
385 390 425 426 570	**500 555 560 630 720**
640 645 730 735 760	**735 742 770 820 860**

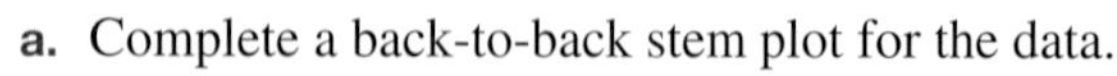

a. Complete a back-to-back stem plot for the data.
b. State the brand that had the shortest lifetime. Justify your answer.
c. State the brand that had the longest lifetime. Justify your answer.
d. If you wanted to be certain that a globe you bought would last at least 500 hours, determine which brand would you buy. Show your working.

Problem solving

12. A small manufacturing plant employs 80 workers. The table below shows the structure of the plant.

Position	Salary ($)	Number of employees
Machine operator	18 000	50
Machine mechanic	20 000	15
Floor steward	24 000	10
Manager	62 000	4
Chief Executive Officer	80 000	1

a. Workers are arguing for a pay rise, but the management of the factory claims that workers are well paid because the mean salary of the factory is $22 100. Explain whether this is a sound argument.
b. Suppose that you were representing the factory workers and had to write a short submission in support of the pay rise. Explain the management's claim by providing some other statistics to support your case.

13. Look at the following bar charts and discuss why the one on the left is misleading and what characteristics the one on the right possesses that makes it acceptable.

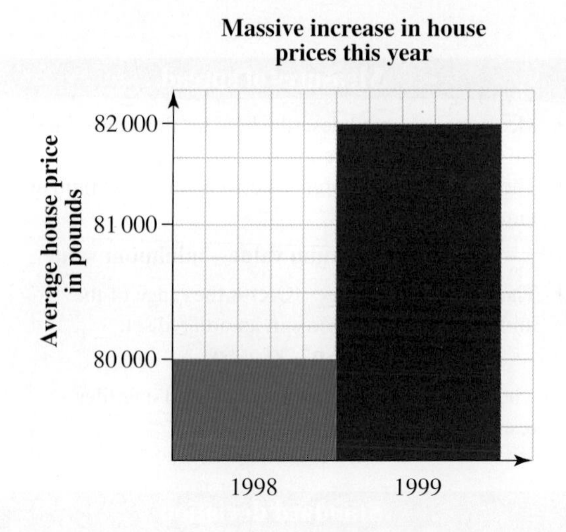

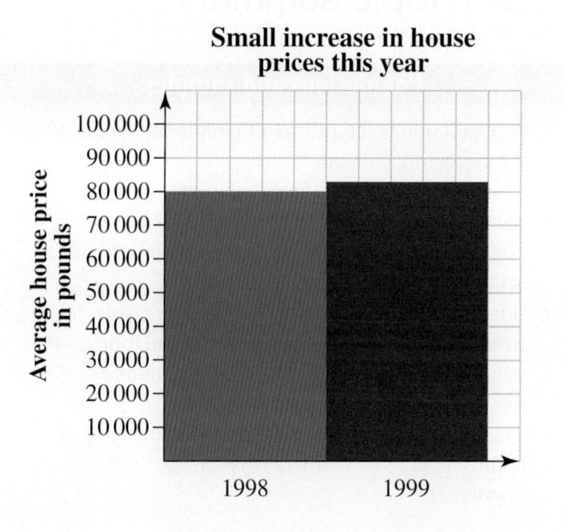

14. a. Determine what is wrong with this pie graph.

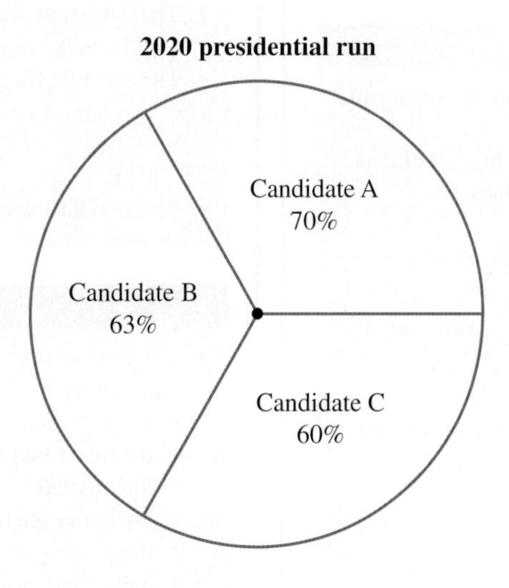

b. Explain why the following information is misleading.
Did scientists falsify research to support their own theories on global warming?
59% somewhat likely
35% very likely
26% not very likely

c. Discuss the implications of this falsification by statistics.

12.9 Review

12.9.1 Topic summary

UNIVARIATE DATA

Populations and samples

- A population is the full set of people/things that you are collecting data on.
- A sample is a subset of a population.
- Samples must be randomly selected from a population in order for the results of the sample to accurately reflect the population.
- It is important to acknowledge that the results taken from a sample may not reflect the population as a whole. This is particularly true if the sample size is too small.
- The minimum sample size that should be used is approximately $\sqrt{N}$, where N is the size of the population.

Measures of central tendency

- The three measures of central tendency are the mean, median and mode.
- The mean is the average of all values in a set of data. It is therefore affected by extreme values.

$$\bar{x} = \frac{\sum x}{n}$$

- The median is the middle value in an ordered set of data. It is located at the $\left(\frac{n+1}{2}\right)$th score.
- The mode is the most frequent value in a set of data.
- For the data set 2, 4, 5, 7, 7:

  The mean is $\frac{2+4+5+7+7}{5} = 5$.

  The median is the middle value, 5.

  The mode is 7.

Comparing data sets

- Measures of centre and measures of spread are used to compare data sets.
- It is important to consider which measure of centre or spread is the most relevant when comparing data sets.
- For example, if there are outliers in a data set, the median will likely be a better measure of centre than the mean, and the IQR would be a better measure of spread than the range.
- Back-to-back stem-and-leaf plots can be used to compare two data sets.
- Parallel boxplots can also be used to compare the spread of two or more data sets.

Measures of spread

- Measures of spread describe how far the data values are spread from the centre or from each other.
- The range is the difference between the maximum and minimum data values.

Range = maxiumum value – minimum value

- The interquartile range (IQR) is the range of the middle 50% of the scores in an ordered set:

$$\textbf{IQR} = Q_3 - Q_1$$

where Q_1 and Q_3 are the first and third quartiles respectively.

Standard deviation

- The standard deviation is a more sophisticated measure of spread.
- The standard deviation measures how far, on average, each data value is away from the mean.
- The deviation of a data value is the difference between it and the mean $(x - \bar{x})$.
- The formula for the standard deviation is:

$$\sigma = \sqrt{\frac{\sum (x-\bar{x})^2}{n}}$$

- The standard deviation is always a positive number.

Boxplots

- The five-number summary of a data set is a list containing:
  - the minimum value
  - the lower quartile, Q_1
  - the median
  - the upper quartile, Q_3
  - the maximum value.
- Boxplots are graphs of the five-number summary.

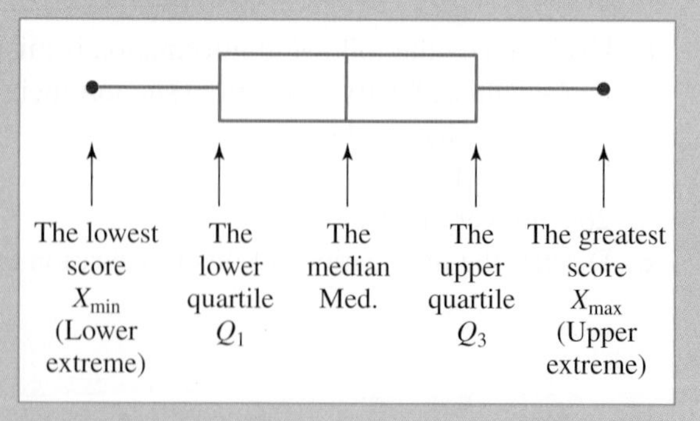

- Outliers are calculated and marked on the boxplot. A score is considered an outlier if it falls outside the upper or lower boundary.

  Lower boundary = $\mathbf{Q_1 - 1.5 \times IQR}$

  Upper boundary = $\mathbf{Q_3 + 1.5 \times IQR}$

12.9.2 Success criteria

Tick the column to indicate that you have completed the subtopic and how well you have understood it using the traffic light system.

(**Green:** I understand; **Yellow:** I can do it with help; **Red:** I do not understand.)

Subtopic	Success criteria	Green	Yellow	Red
12.2	I can calculate the mean, median and mode of data presented as ungrouped data (in a single list), frequency distribution tables and grouped data.			
12.3	I can calculate the range and interquartile range of a data set.			
12.4	I can calculate the five-number summary for a set of data.			
	I can draw a box plot showing the five-number summary of a data set.			
	I can calculate outliers in a data set.			
	I can describe skewness of distributions.			
	I can compare box plots to dot plots or histograms.			
	I can draw parallel box plots and compare sets of data.			
12.5	I can calculate the standard deviation of a small data set by hand.			
	I can calculate the standard deviation using technology.			
	I can interpret the mean and standard deviation of data.			
	I can identify the effect of outliers on the standard deviation.			
12.6	I can choose an appropriate measure of centre and spread to analyse data.			
	I can interpret and make decisions based on measures of centre and spread.			
12.7	I can describe the difference between populations and samples.			
	I can recognise the difference between a census and a survey and identify a preferred method in different circumstances.			
12.8	I can describe the difference between primary and secondary data collection.			
	I can identify misleading errors or graphical techniques in data.			
	I can evaluate the accuracy of a statistical report.			

12.9.3 Project

Cricket scores

Data are used to predict, analyse, compare and measure many aspects of the game of cricket. Attendance is tallied at every match. Players' scores are analysed to see if they should be kept on the team. Comparisons of bowling and batting averages are used to select winners for awards. Runs made, wickets taken, no-balls bowled, the number of ducks scored in a game as well as the number of 4s and 6s are all counted and analysed after the game. Data of all sorts are gathered and recorded, and measures of central tendency and spread are then calculated and interpreted.

Sets of data have been made available for you to analyse, and decisions based on the resultant measures can be made.

Batting averages

The following table shows the runs scored by four cricketers who are vying for selection to the state team.

Player	Runs in the last 25 matches	Mean	Median	Range	IQR
Will	13, 18, 23, 21, 9, 12, 31, 21, 20, 18, 14, 16, 28, 17, 10, 14, 9, 23, 12, 24, 0, 18, 14, 14, 20				
Rohit	2, 0, 112, 11, 0, 0, 8, 0, 10, 0, 56, 4, 8, 164, 6, 12, 2, 0, 5, 0, 0, 0, 8, 18, 0				
Marnus	12, 0, 45, 23, 0, 8, 21, 32, 6, 0, 8, 14, 1, 27, 23, 43, 7, 45, 2, 32, 0, 6, 11, 21, 32				
Ben	2, 0, 3, 12, 0, 2, 5, 8, 42, 0, 12, 8, 9, 17, 31, 28, 21, 42, 31, 24, 30, 22, 18, 20, 31				

1. Calculate the mean, median, range and IQR scored for each cricketer.
2. You need to recommend the selection of two of the four cricketers. For each player, write two points as to why you would or would not select them. Use statistics in your comments.

Bowling averages

The bowling average is the number of runs per wicket taken

$$\textbf{Bowling average} = \frac{\textbf{no. of runs scored}}{\textbf{no. of wicket taken}}$$

The smaller the average, the better the bowler has performed.

Josh and Ravi were competing for three bowling awards:
- Best in semifinal
- Best in final
- Best overall

The following table gives their scores.

	Semifinal		Final	
	Runs scored	Wickets taken	Runs scored	Wickets taken
Josh	12	5	28	6
Ravi	10	4	15	3

2. Calculate the bowling averages for the following and fill in the table below.
   - Semifinal
   - Final
   - Overall

	Semifinal average	Final average	Overall average
Josh			
Ravi			

3. Explain how Ravi can have the better overall average when Josh has the better average in both the semifinal and final.

 Resources

 **eWorkbook** Topic 12 Workbook (worksheets, code puzzle and project) (ewbk-2038)

 **Interactivities** Crossword (int-2860)

Sudoku puzzle (int-3599)

Exercise 12.9 Review questions

To answer questions online and to receive **immediate corrective feedback** and **fully worked solutions** for all questions, go to your learnON title at www.jacplus.com.au.

Fluency

1. List some problems you might encounter in trying to collect data from the following populations.
   a. The average number of mL in a can of soft drink.
   b. The number of fish in a dam.
   c. The number of workers who catch public transport to work each weekday morning.

2. For each of the following investigations, state whether a census or a survey has been used.
   a. The average price of petrol in Canberra was estimated by averaging the price at 30 petrol stations in the area.
   b. The performance of a cricketer is measured by looking at his performance in every match he has played.
   c. Public opinion on an issue is sought by a telephone poll of 2000 homes.

3. Calculate the mean, median and mode for each of the following sets of data:
   a. 7, 15, 8, 8, 20, 14, 8, 10, 12, 6, 19
   b. Key: $1|2 = 12$

Stem	*Leaf*
1	2 6
2	1 7 8
3	0 3 3 4 6 8
4	0 1 1 5 9
5	1 3 6

c.

Score (x)	Frequency (f)
70	2
71	6
72	9
73	7
74	4

4. For each of the following data sets, calculate the range.

a. 4, 3, 6, 7, 2, 5, 8, 4, 3

b.

x	13	14	15	16	17	18	19
f	3	6	7	12	6	7	8

c. Key: $1|8 = 18$

Stem	*Leaf*
1	7 8 8 9
2	1 2 4 4 5 7 7 7 8 9 9
3	0 0 0 1 3 4 7

5. For each of the following data sets, calculate the interquartile range.

a. 18, 14, 15, 19, 20, 11 16, 19, 18, 19

b. Key: $9|8 = 9.8$

Stem	*Leaf*
8	7 8 8 9
9	0 2 4 4 5 7 7 7 8 9 9
10	0 1 1 1 3

6. Consider the box plot shown.

a. Calculate the median.

b. Calculate the range.

c. Determine the interquartile range.

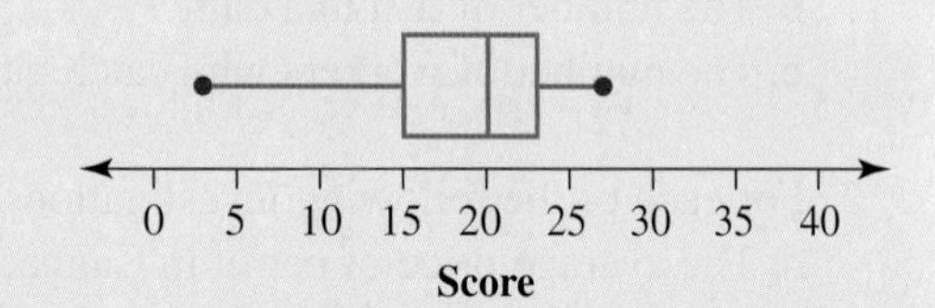

7. The following back-to-back stem-and-leaf plot shows the typing speed in words per minute (wpm) of 30 Year 8 and Year 10 students.

Key: $2|6 = 26$ wpm

Leaf *Year 8*	*Stem*	*Leaf* *Year 10*
9 9	0	
9 8 6 5 4 2 0	1	7 9
9 8 8 6 4	2	2 3 6 8 9
2 1 0 0		
9 7 7 6 4 1 0	3	0 2 5 5 7 8 8
8 6 5 2 0	4	1 2 5 8 8 9 9
	5	0 3 5 7 8
	6	0 0 3

a. Using a calculator or otherwise, construct a pair of parallel box-and-whisker plots to represent the two sets of data.
b. Calculate the mean, median, range, interquartile range and standard deviation of each set.
c. Compare the two distributions, using your answers to parts **a** and **b**.

8. The following data give the amount of cut meat (in kg) obtained from each of 20 lambs.

4.5	6.2	5.8	4.7	4.0	3.9	6.2	6.8	5.5	6.1
5.9	5.8	5.0	4.3	4.0	4.6	4.8	5.3	4.2	4.8

a. Detail the data on a stem-and-leaf plot. (Use a class size of 0.5 kg.)
b. Prepare a five-point summary of the data.
c. Draw a box plot of the data.

9. Calculate the standard deviation of each of the following data sets correct to one decimal place.
a. 58, 12, 98, 45, 60, 34, 42, 71, 90, 66
b.

x	1	2	3	4	5
f	2	6	12	8	5

c. Key: 1|4 = 14

Stem	*Leaf*
0	1 3 4 4 5 7 8
1	0 0 0 1 2 2 4 5 7 8 9
2	0 2 2 3 5 7

10. **MC** The Millers obtained a number of quotes on the price of having their home painted. The quotes, to the nearest hundred dollars, were:

4200 5100 4700 4600 4800 5000 4700 4900

The standard deviation for this set of data, to the nearest whole dollar, is:
A. 260 **B.** 278 **C.** 324 **D.** 325 **E.** 900

11. **MC** The number of Year 12 students who, during semester 2, spent all their spare periods studying in the resource centre is shown on the stem-and-leaf plot below.

Key: 2|5 = 25 students

Stem	*Leaf*
0	8
1	
2	5 6 6 7
3	0 2 3 6 9
4	7 9
5	6
6	1

The standard deviation for this set of data, to the nearest whole number is:
A. 12 **B.** 14 **C.** 17 **D.** 35 **E.** 53

12. Each week, varying amounts of a chemical are added to a filtering system. The amounts required (in mL) over the past 20 weeks are shown in the stem-and-leaf plot.

Key: 3|8 represents 0.38 ml

Stem	*Leaf*
2	1
2	2 2
2	4 4 4 5
2	6 6
2	8 8 99
3	0
3	2 2
3	4
3	6
3	8

Calculate to 2 decimal places the standard deviation of the amounts used.

13. Calculate the mean, median and mode of this data set: 2, 5, 6, 2, 5, 7, 8. Comment on the shape of the distribution.

14. The box plot shows the heights (in cm) of Year 12 students in a Maths class.

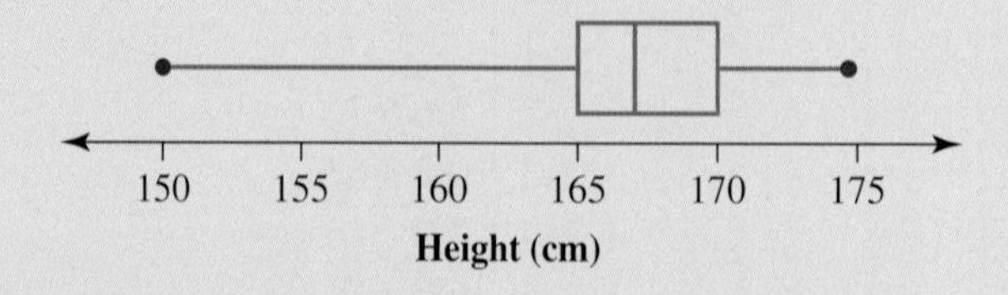

a. State the median class height.
b. Calculate the range of heights.
c. Calculate the interquartile range of the heights.

Problem solving

15. **MC** A data set has a mean of 75 and a standard deviation of 5. Another score of 50 is added to the data set. Choose which of the following will occur.
   A. The mean will increase and the standard deviation will increase.
   B. The mean will increase and the standard deviation will decrease.
   C. The mean will decrease and the standard deviation will increase.
   D. The mean will decrease and the standard deviation will decrease.
   E. The mean and the standard deviation will both remain unchanged.

16. **MC** A data set has a mean of 60 and a standard deviation of 10. A score of 100 is added to the data set. This score becomes the highest score in the data set. Choose which of the following will increase.
   Note: There may be more than one correct answer.
   A. Mean
   B. Standard deviation
   C. Range
   D. Interquartile range
   E. Median

17. A sample of 30 people was selected at random from those attending a local swimming pool. Their ages (in years) were recorded as follows:

19	7	58	41	17	23	62	55	40	37	32	29	21	18	16
10	40	36	33	59	65	68	15	9	20	29	38	24	10	30

   a. Calculate the mean and the median age of the people in this sample.
   b. Group the data into class intervals of 10 (0–9 etc.) and complete the frequency distribution table.
   c. Use the frequency distribution table to estimate the mean age.
   d. Calculate the cumulative frequency and, hence, plot the ogive.
   e. Estimate the median age from the ogive.
   f. Compare the mean and median of the original data in part **a** with the estimates of the mean and the median obtained for the grouped data in parts **c** and **e**.
   g. Determine if the estimates were good enough. Explain your answer.

18. The table below shows the number of cars that are garaged at each house in a certain street each night.

Number of cars	Frequency
1	9
2	6
3	2
4	1
5	1

a. Show these data in a frequency histogram.
b. State if the data is positively or negatively skewed. Justify your answer.

19. Consider the data set represented by the frequency histogram shown.
a. Explain if the data is symmetrical.
b. State if the mean and median of the data can be seen. If so, determine their values.
c. Evaluate the mode of the data.

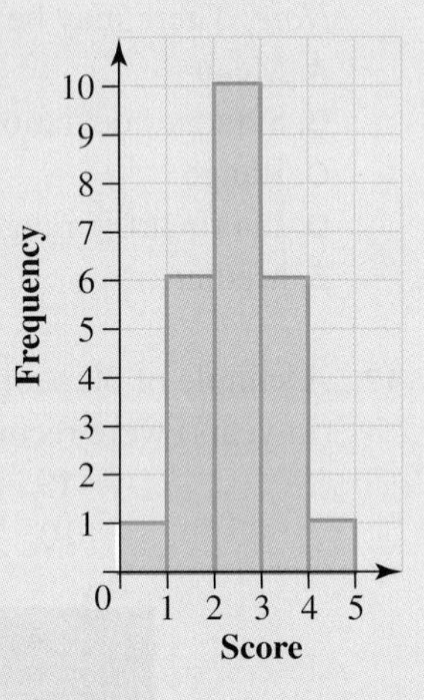

20. There are $3m$ values in a data set for which $\bar{x} = m$ and $\sigma = \dfrac{m}{2}$.
a. Comment on the changes to the mean and standard deviation if each value of the data set is multiplied by m.
b. An additional value is added to the original data set, giving a new mean of $m + 2$. Evaluate the additional value.

21. The following data show the number of pets in each of the 12 houses in Coral Avenue, Rosebud.

2, 3, 3, 2, 2, 3, 2, 4, 3, 1, 1, 0

a. Calculate the mean and median number of pets.
b. The empty block of land at the end of the street was bought by a Cattery and now houses 20 cats. Recalculate the mean and median.
c. Explain why the answers are so different, and which measure of central tendency is best used for certain data.

22. The number of Year 10 students in all the 40 schools in the Northern District of the Education Department was recorded as follows:

56, 134, 93, 67, 123, 107, 167, 124, 108, 78, 89, 99, 103, 107, 110, 45, 112, 127, 106, 111, 127, 145, 87, 75, 90, 123, 100, 87, 116, 128, 131, 106, 123, 87, 105, 112, 145, 115, 126, 92

a. Using an interval of 10, produce a table showing the frequency for each interval.
b. Use the table to estimate the mean.
c. Calculate the mean of the ungrouped data.
d. Compare the results from parts **b** and **c** and explain any differences.

23. The following back-to-back stem-and-leaf plot shows the ages of a group of 30 males and 30 females as they enter hospital for the first time.

Key: 1 | 7 = 17

Leaf: Male	*Stem*	*Leaf: Female*
9 8	0	5
9 9 8 8 8 6 3 2 1	1	7 7 8 9 9
8 7 7 6 4 3 2 0	2	0 0 1 2 4 5 5 6 7 9
8 6 3 1 0	3	0 1 3 3 5 8
7 5 2	4	2 3 6 8
5 3	5	1 3 4
	6	2
8	7	

a. Construct a pair of parallel box plots to represent the two sets of data, showing working out for the median and 1st and 3rd quartiles.
b. Calculate the mean, range and IQR for both sets of data.
c. Determine any outliers if they exist.
d. Write a short paragraph comparing the data.

24. The times, in seconds, of the duration of 20 TV advertisements shown in the 6–8 pm time slot are recorded below.

16 60 35 23 45 15 25 55 33 20 22 30 28 38 40 18 29 19 35 75

a. From the data, determine the:
i. mode
ii. median
iii. mean, write your answer correct to 2 decimal places
iv. range
v. lower quartile
vi. upper quartile
vii. interquartile range.

b. Using your results from part **a**, construct a box plot for the time, in seconds, for the 20 TV advertisements in the 6–8 pm time slot.

c. From your box plot, determine:
i. the percentage of advertisements that are more than 39 seconds in length
ii. the percentage of advertisements that last between 21 and 39 seconds
iii. the percentage of advertisements that are more than 21 seconds in length

The types of TV advertisements during the 6–8 pm time slot were categorised as Fast food, Supermarkets, Program information and Retail (clothing, sporting goods, furniture).
A frequency table for the frequency of these advertisements is shown below.

Type	Frequency
Fast food	7
Supermarkets	5
Program information	3
Retail	5

d. State the type of data that has been collected in the table.
e. Determine the percentage of advertisements that are advertisements for fast food outlets.
f. Suggest a good option for a graphical representation of this type of data.

25. The test scores, out of a total score of 50, for two classes A and B are shown in the back-to-back stem-and-leaf plot.

Key: 1 | 4 = 14

Leaf *Class* ***A***		*Leaf* *Class* ***B***
5	0	1 2 4
9 7 5 3	1	1 4 5
9 7 7 5 4	2	0 0 5
8 8 6 5 5 1	3	1 5 5
3 2 0	4	1 5 7 7 8 9
0	5	0 0

a. Ms Vinculum teaches both classes and made the statement that 'Class A's performance on the test showed that the students' ability was more closely matched than the students' ability in Class B'. By calculating the measure of centre, first and third quartiles, and the measure of spread for the test scores for each class, explain if Ms Vinculum's statement was correct.
b. Would it be correct to say that Class A performed better on the test than Class B? Justify your answer by comparing the quartiles and median for each class.

26. The speeds, in km/h, of 55 cars travelling along a major road are recorded below.

Speed	Frequency
60–64	1
65–69	1
70–74	10
75–79	13
80–84	9
85–89	8
90–94	6
95–99	3
100–104	2
105–109	1
110–114	1
Total	55

a. By calculating the midpoint for each class interval, determine the mean speed, in km/h, of the cars travelling along the road.
Write your answer correct to two decimal places.

b. The speed limit along the road is 75 km/h. A speed camera is set to photograph the license plates of cars travelling 7% more than the speed limit. A speeding fine is automatically sent to the owners of the cars photographed.
Based on the 55 cars recorded, determine the number of speeding fines that were issued.

c. Drivers of cars travelling 5 km/h up to 15 km/h over the speed limit are fined $135. Drivers of cars travelling more than 15 km/h and up to 25 km/h over the speed limit are fined $165 and drivers of cars recorded travelling more than 25 km/h and up to 35 km/h are fined $250. Drivers travelling more than 35 km/h pay a $250 fine in addition to having their driver's license suspended.
If it is assumed that this data is representative of the speeding habits of drivers along a major road and there are 30 000 cars travelling along this road on any given month.

i. Determine the amount, in dollars, collected in fines throughout the month. Write your answer correct to the nearest cent.

ii. Evaluate the number of drivers that would expect to have their licenses suspended throughout the month.

on To test your understanding and knowledge of this topic, go to your learnON title at www.jacplus.com.au and complete the **post-test**.

Online Resources

Below is a full list of **rich resources** available online for this topic. These resources are designed to bring ideas to life, to promote deep and lasting learning and to support the different learning needs of each individual.

eWorkbook

Download the workbook for this topic, which includes worksheets, a code puzzle and a project (ewbk-2038) ☐

Solutions

Download a copy of the fully worked solutions to every question in this topic (sol-0746) ☐

Digital documents

12.2 SkillSHEET Finding the mean of a small data set (doc-5299) ☐
SkillSHEET Finding the median of a small data set (doc-5300) ☐
SkillSHEET Finding the mode of a small data set (doc-5301) ☐
SkillSHEET Finding the mean, median and mode from a stem-and-leaf plot (doc-5302) ☐
SkillSHEET Presenting data in a frequency distribution table (doc-5303) ☐
SkillSHEET Drawing statistical graphs (doc-5304) ☐
12.7 SkillSHEET Determining suitability of questions for a survey (doc-5337) ☐
SkillSHEET Finding proportions (doc-5338) ☐
SkillSHEET Distinguishing between types of data (doc-5339) ☐
12.8 SkillSHEET Reading bar graphs (doc-5340) ☐
SkillSHEET Determining independent and dependent variables (doc-5341) ☐

Video eLessons

12.2 Mean, median and mode of univariate data (eles-4949) ☐
Calculating mean, median and mode from a frequency distribution table (eles-4950) ☐
Cumulative frequency curves (ogives) (eles-4951) ☐
Mean and median (eles-1905) ☐
12.3 Measures of spread (eles-4952) ☐
12.4 Five-number summary (eles-4953) ☐
Box plots (eles-4954) ☐
Comparing different graphical representations (eles-4956) ☐
12.5 Standard deviation (eles-4958) ☐
Effects on standard deviation (eles-4959) ☐
Properties of standard deviation (eles-4961) ☐
12.6 Comparing data sets (else-4962) ☐
12.7 Populations (eles-4946) ☐
Samples (eles-4947) ☐
To sample or to conduct a census? (eles-4948) ☐
12.8 Data collections methods (eles-4963) ☐
Analysing the data (eles-4964) ☐
Statistical reports (eles-4955) ☐

Interactivities

12.2 Individual pathway interactivity: Measures of central tendency (int-4621) ☐
Mean (int-3818) ☐
Median (int-3819) ☐
Mode (int-3820) ☐
Ogives (int-6174) ☐
12.3 Individual pathway interactivity: Measures of spread (int-4622) ☐
Range (int-3822) ☐
The interquartile range (int-4813) ☐
12.4 Individual pathway interactivity: Box-and-whisker plots (int-4623) ☐
Skewness (int-3823) ☐
Box plots (int-6245) ☐
Parallel box plots (int-6248) ☐
12.5 Individual pathway interactivity: The standard deviation (int-4624) ☐
The standard deviation for a sample (int-4814) ☐
12.6 Individual pathway interactivity: Comparing data sets (int-4625) ☐
Back-to-back stem plots (int-6252) ☐
12.7 Individual pathway interactivity: Populations and samples (int-4629) ☐
Sample sizes (int-6183) ☐
12.8 Individual pathway interactivity: Evaluating inquiry methods and statistical reports (int-4631) ☐
Compare statistical reports (int-2790) ☐
12.9 Crossword (int-2860) ☐
Sudoku puzzle (int-3599) ☐

Teacher resources

There are many resources available exclusively for teachers online.

To access these online resources, log on to **www.jacplus.com.au**.

Answers

Topic 12 Univariate data

Exercise 12.1 Pre-test

1. 2
2. 29
3. 72
4. 29.8
5. subset
6. 22
7. 13
8. Positively skewed
9. 11
10. C
11. $\bar{x} = 184.42$ and $\sigma = 7.31$
12. D
13. B
14. The mean typing speed is 26.53 and IQR is 19 for Year 8. The mean typing speed is 40.53 and IQR is 20 for Year 10. This suggests, that the mean typing speed for Year 10 is greater than the Year 8 students. The interquartile range is not the same for both Year 8 and Year 10.
15. E

Exercise 12.2 Measures of central tendency

1. a. 7 b. 8 c. 8
2. a. 6.875 b. 7 c. 4, 7
3. a. 39.125 b. 44.5 c. No mode
4. a. 4.857 b. 4.8 c. 4.8
5. a. 12 b. 12.625 c. 13.5
6. Science: mean = 57.6, median = 57, mode = 42, 51
   Maths: mean = 69.12, median = 73, mode = 84
7. a. 5.83 b. 6 c. 6
8. a. 14.425 b. 15 c. 15
9. a. Mean = 2.5, median = 2.5
   b. Mean = 4.09, median = 3
   c. Median
10. a. $72\frac{2}{3}$ b. 73 c. $70- < 80$
11. 124.83
12. $65- < 70$
13. a. B b. B
    c. C d. D
14. a. Mean = \$32.93, median = \$30

    b.

Class interval	Frequency	Cumulative frequency
0 – 9	5	5
10 – 19	5	10
20 – 29	5	15
30 – 39	3	18
40 – 49	5	23
50 – 59	3	26
60 – 69	3	29
70 – 79	1	30
Total	30	

Mean = \$32.50, median = \$30

c. 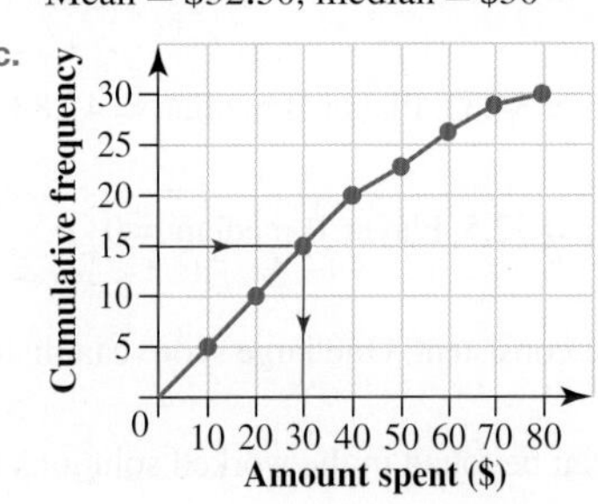

d. The mean is slightly underestimated; the median is exact. The estimate is good enough as it provides a guide only to the amount that may be spent by future customers.

15. a. 3
    b. 4, 5, 5, 5, 6 (one possible solution)
    c. One possible solution is to exchange 15 with 20.
16. a. Frequency column: 16, 6, 4, 2, 1, 1
    b. 6.8
    c. 0 – 4 hours
    d. 0 – 4 hours
17. a. Frequency column: 1, 13, 2, 0, 1, 8
    b. 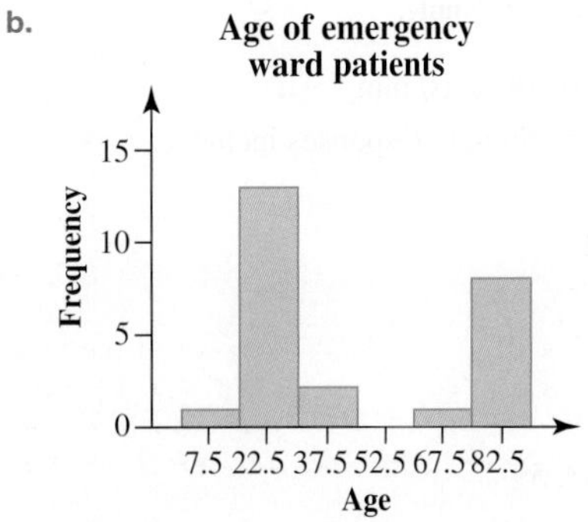

    c. Asymmetrical or bimodal (as if the data come from two separate graphs).
    d. 44.1
    e. 15 – <30
    f. 15 – <30

g. 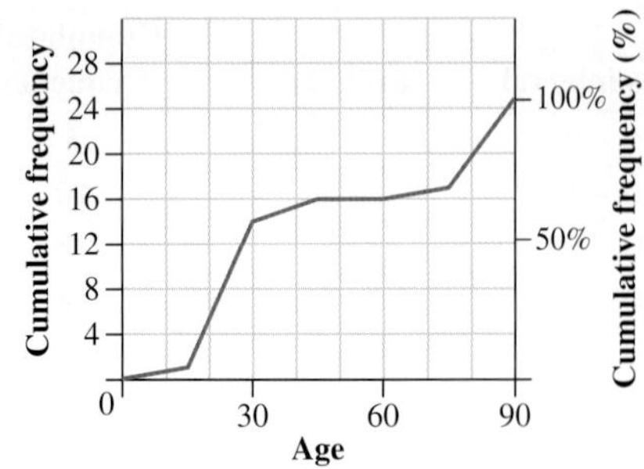

h. 28

i. No

j. Sample responses can be found in the worked solutions in the online resources.

18. A

19. a. Player A median = 34.33, Player B median = 41.83

b. Player B

c. Player A median = 32.5, Player B median = 0

d. Player A

e. Player A is more consistent. One large score can distort the mean.

20. Sample responses can be found in the worked solutions in the online resources.

21. a. Frequency column: 3, 8, 5, 3, 1

b. 50.5

c. 40– <50

d. 40– <50

e. 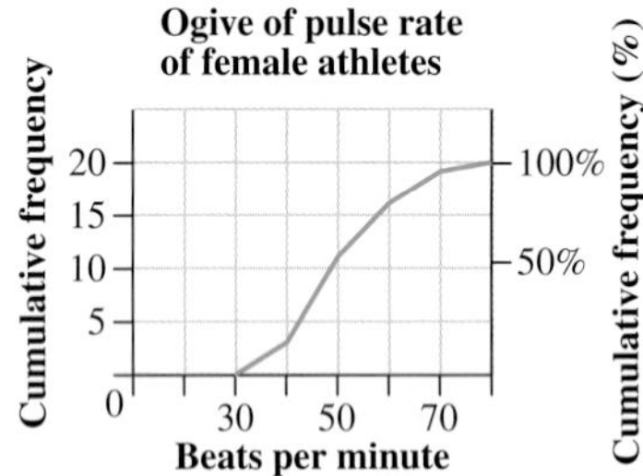

f. Approximately 48 beats/ min

22. Answers will vary. Sample responses include:

a. 3, 4, 5, 5, 8

b. 4, 4, 5, 10, 16

c. 2, 3, 6, 6, 12

23. 12

24. $\frac{2a+b}{3}$

25. 13, 31, 31, 47, 53, 59

Exercise 12.3 Measures of spread

1. a. 15 b. 77.1 c. 9

2. a. 7 b. 7 c. 8.5 d. 39

3. a. 3.3 kg b. 1.5 kg

4. 22 cm

5. 0.8

6. C

7. a. 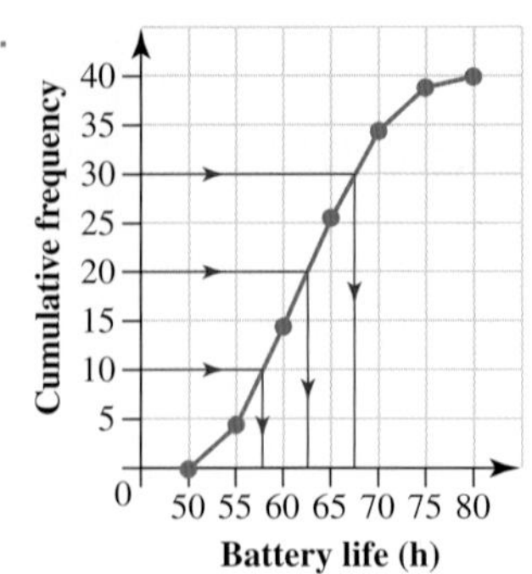

b. i. 62.5

ii. $Q_1 = 58$, $Q_3 = 67$

iii. 9

iv. 14

v. 6

8. IQR = 27

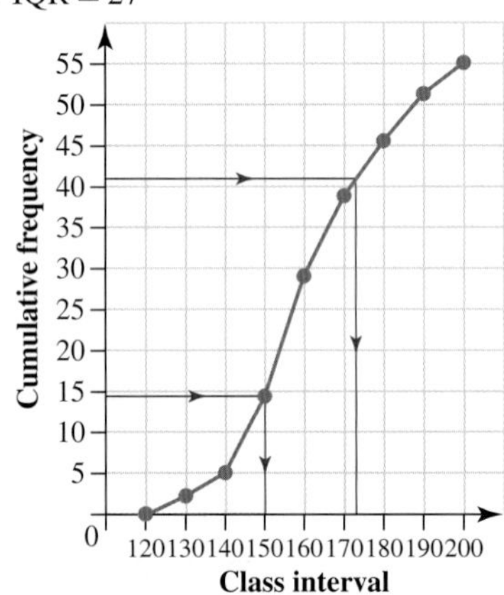

9. a. i. Range = 23

ii. IQR = 13.5

b. i. Range = 45

ii. IQR = 27.5

c. i. Range = 49

ii. IQR = 20

10. Measures of spread tell us how far apart the values (scores) are from one another.

11. a. 25.5

b. 28

c. 39

d. 6

e. The three lower scores affect the mean but not the median or mode.

12. a. Men: mean = 32.3; median = 32.5; range = 38; IQR = 14

Women: mean = 29.13; median = 27.5; range = 36; IQR = 13

b. Typically, women marry younger than men, although the spread of ages is similar.

13. Mean = 25.036, Median = 24.9, Mode = 23.6
Range = 8.5, IQR = 3.4

14. $a = 22$, $b = 9$, $c = 9$ and $d = 8$

15. a. Yes
Range = 10 − 1 = 9
Median (middle score) = 6

b. 6
To maintain range, min = 1 and max = 10.
If median = 6, then 3rd and 4th scores must be 6.
Therefore 6th score must be 6. This will maintain the range, $Q1$, $Q3$ and median.

Exercise 12.4 Box plots

1. a. 5 b. 26
2. a. 6 b. 27
3. a. 5.8 b. 18.6
4. a. 140 b. 56 c. 90 d. 84 e. 26
5. a. 58 b. 31 c. 43 d. 27 e. 7
6. B
7. C
8. D, E
9. a. 22, 28, 35, 43, 48

b.

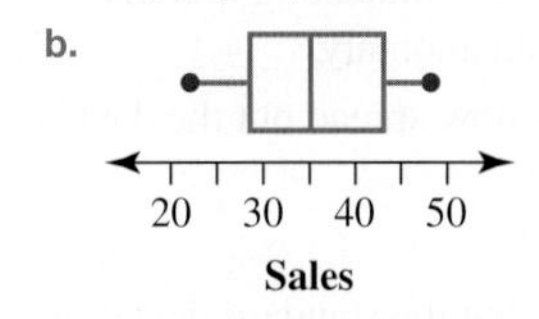

10. a. (10, 13.5, 22, 33.5, 45)

b.

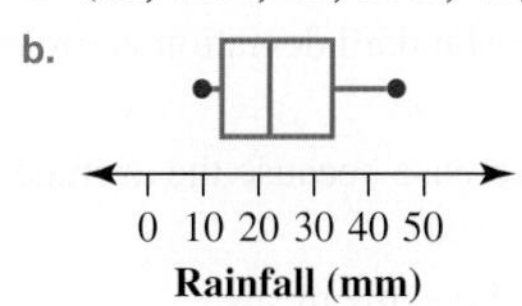

11. a. (18, 20, 26, 43.5, 74)

b.

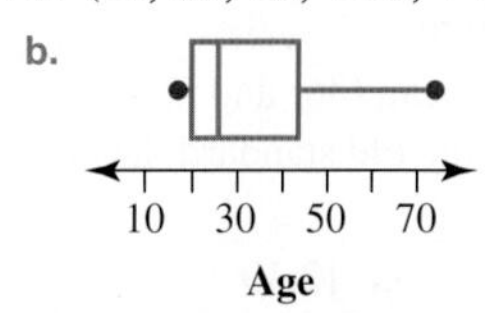

c. The distribution is positively skewed, with most of the offenders being young drivers.

12. a. (124 000, 135 000, 148 000, 157 000, 175 000)

b.

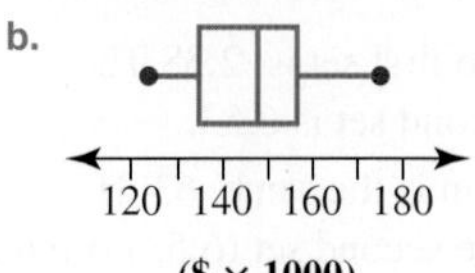

13. a.

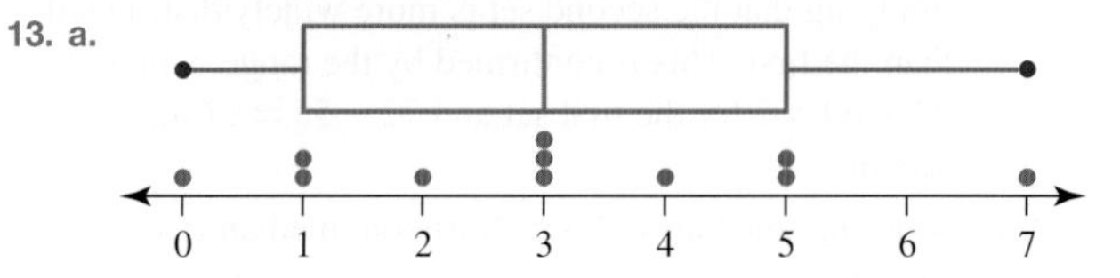

Both graphs indicate that the data is slightly positively skewed. However, the box plot provides an excellent summary of the centre and spread of the distribution.

b.

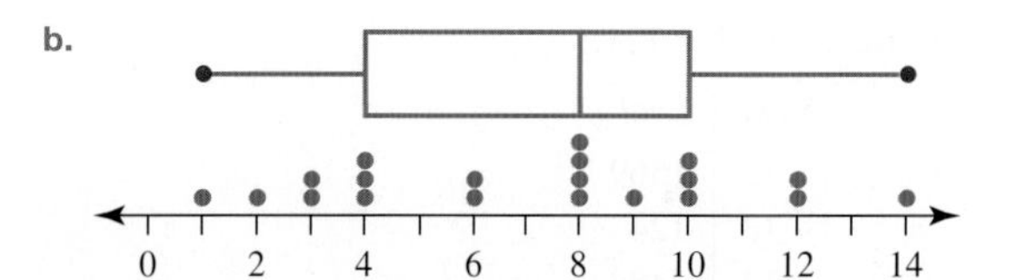

Both graphs indicate that the data is slightly negatively skewed. However, the box plot provides an excellent summary of the centre and spread of the distribution.

14.

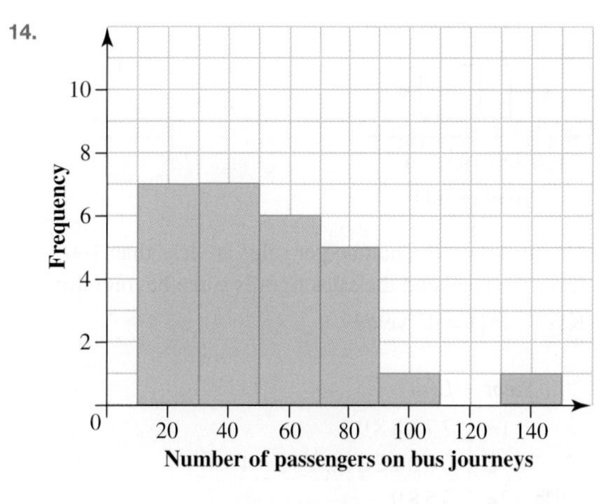

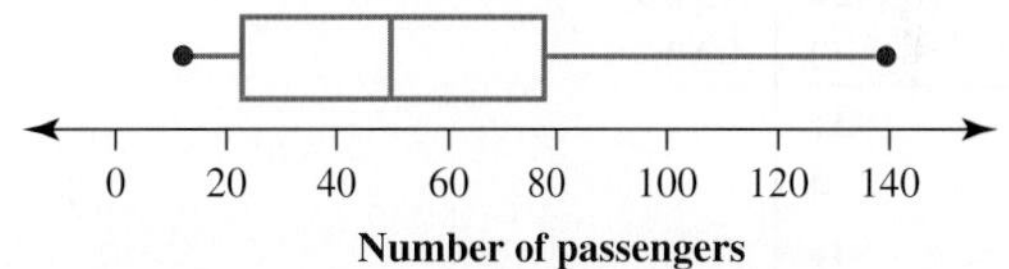

15.

	X_{min}	Q_1	Median	Q_3	X_{max}
Before	75	86	95	128.5	152
After	66	81	87	116	134

b.

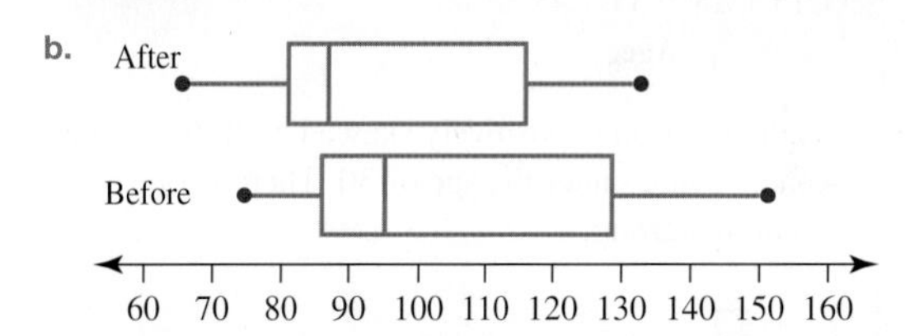

c. As a whole, the program was effective. The median weight dropped from 95 kg to 87 kg, a loss of 8 kg. A noticeable shift in the graph shows that after the program 50% of participants weighed between 66 and 87 kg, compared to 25% of participants weighing between 75 and 86 kg before they started. Before the program the range of weights was 77 kg (from 75 kg to 152 kg); after the program the range had decreased to 68 kg. The IQR also diminished from 42.5 kg to 35 kg.

16. The advantages of box plots is that they are clear visual representations of 5-number summary, display outliers and can handle a large volume of data. The disadvantage is that individual scores are lost.

17. a. Key: $12|1 = 121$.

Stem	*Leaf*
12	1569
13	124
14	3488
15	022257
16	35
17	29
18	1112378

b.

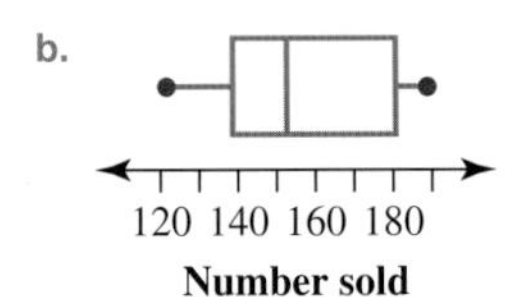

c. On most days the hamburger sales are less than 160. Over the weekend the sales figures spike beyond this.

18. a. Key: $1^*|7 = 17$ years

Stem	*Leaf*
1*	7 7 8 8 8 9 9
2	0 0 0 1 2 2 2 2 3 3 3 3 4 4 4
2*	5 5 8 9
3	1 2 3
3*	
4	
4*	8

b.

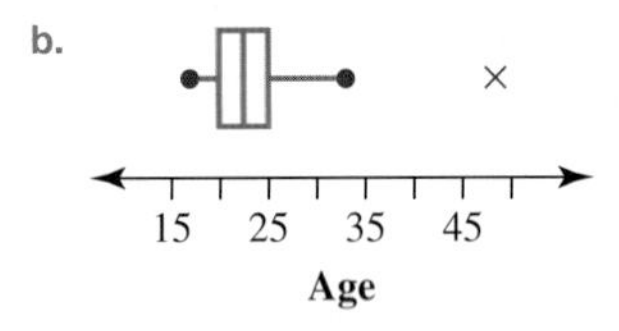

c. The distribution is positively skewed, with first-time mothers being under the age of 30. There is one outlier (48) in this group.

19.

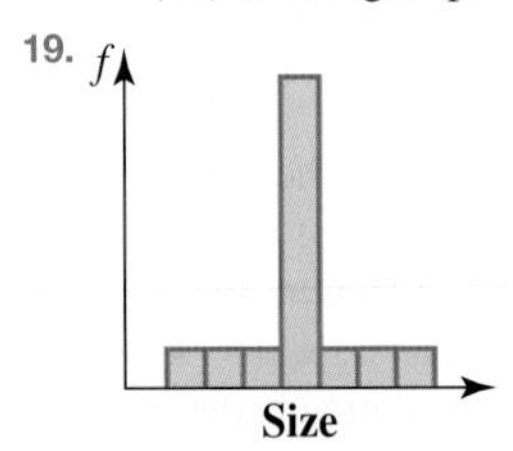

20. a. HJ Looker:median = 5; Hane and Roarne: median = 6

b. HJ Looker

c. HJ Looker

d. Hane and Roarne had a higher median and a lower spread and so they appear to have performed better.

21. a. \$50 b. \$135 c. \$100 d. \$45 e. 50%

22. a. 9625 seconds

b. i. Under 20 – (20–24): 750 seconds difference

ii. Under 20 – (25–29): 500 seconds difference

iii. (20–24) – (25–29): 250 seconds difference

c. The under-20s performed best of the three groups, with the fastest time for each metric (minimum time, first quartile, median, third quartile and maximum time). The next best performing group was the 25–29-year-olds. They had the same median as the 20–24-year-olds, but outperformed them in all of the other metrics. The 24–29-year-olds were the most consistent group, with a range of 1750 seconds compared to the range of 2000 seconds of the other groups.

Exercise 12.5 The standard deviation (10A)

1. a. 2.29 b. 2.19 c. 20.17 d. 3.07

2. a. 1.03 b. 1.33 c. 2.67 d. 2.22

3. 10.82

4. Mean = 1.64% Std dev. = 0.45%

5. Mean = 1.76 Std dev. = 0.06 m

6. 0.49 s

7. 15.10 calls

8. B

9. The mean of the first 25 cars is 89.24 km/h with a standard deviation of 5.60. The mean of the first 26 cars is 91.19 with a standard deviation of 11.20, indicating that the extreme speed of 140 km/h is an anomaly.

10. The standard deviation tells us how spread out the data is from the mean

11. a. $\sigma \approx 3.58$

b. The mean is increased by 7 but the standard deviation remains at $\sigma \approx 3.58$.

c. The mean is tripled and the standard deviation is tripled to $\sigma \approx 10.74$.

12. The standard deviation will decrease because the average distance to the mean has decreased.

13. 57 is two standard deviations above the mean.

14. a. New mean is the old mean increased by 3 (15) but no change to the standard deviation.

b. New mean is 3 times the old mean (36) and new standard deviation is 3 times the old standard deviation (12).

15. a. 43 b. 12 c. 12.19

Exercise 12.6 Comparing data sets

1. a. The mean of the first set is 63. The mean of the second set is 63.

b. The standard deviation of the first set is. 2.38 The standard deviation of the second set is 6.53.

c. For both sets of data the mean is the same, 63. However, the standard deviation for the second set (6.53) is much higher than the standard deviation of the first set (2.38), implying that the second set is more widely distributed than the first. This is confirmed by the range, which is $67 - 60 = 7$ for the first set and $72 - 55 = 17$ for the second.

2. a. Morning: median = 2.45; afternoon: median = 1.6

b. Morning: range = 3.8; afternoon: range = 5

c. The waiting time is generally shorter in the afternoon. One outlier in the afternoon data causes the range to be larger. Otherwise the afternoon data are far less spread out.

3. Key: 16|1 = 1.61 m

Leaf: Boys	*Stem*	*Leaf: Girls*
997	15	1256788
	16	4467899
98665540		
4421	17	0

4. a. Ford: median = 15; Hyundai: median = 16
   b. Ford: range = 26; Hyundai: range = 32
   c. Ford: IQR = 14; Hyundai: IQR = 13.5
   d.

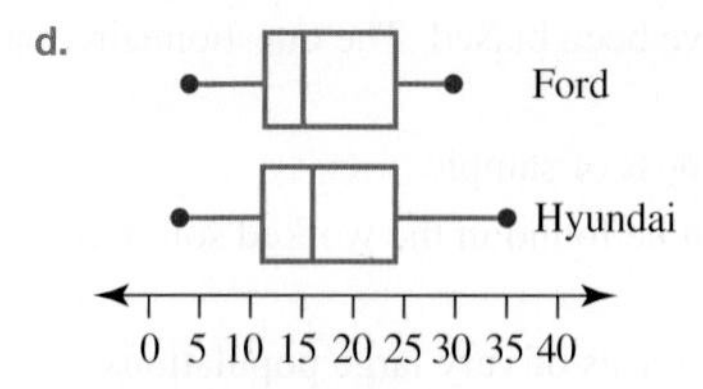

5. a. Brisbane Lions
   b. Brisbane Lions: range = 63; Sydney Swans: range = 55
   c. Brisbane Lions: IQR = 40; Sydney Swans: IQR = 35
6. a.

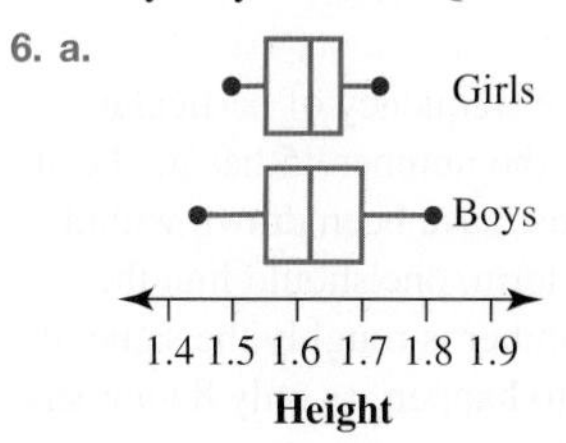

   b. Boys: median = 1.62; girls: median = 1.62
   c. Boys: range = 0.36; girls: range = 0.23
   d. Boys: IQR = 0.25; girls: IQR = 0.17
   e. Although boys and girls have the same median height, the spread of heights is greater among boys as shown by the greater range and interquartile range.
7. a. Summer: range = 23; winter: range = 31
   b. Summer: IQR = 14; winter: IQR = 11
   c. There are generally more cold drinks sold in summer as shown by the higher median. The spread of data is similar as shown by the IQR although the range in winter is greater.
8. A
9. A, B, C, D
10. a. In order to include all data values in the calculation of measures of centre and spread, calculate the mean and standard deviation.
    b. Cory achieved a better average mark in Science (59.25) than he did in English (58.125).
    c. Cory was more consistent in English ($\sigma = 4.9$) than he was in Science ($\sigma = 19.7$)
11. Sample responses can be found in the worked solutions in the online resources.
12. a. Back street: $\bar{x} = 61$, $\sigma = 4.3$; main road: $\bar{x} = 58.8$, $\sigma = 12.1$
    b. The drivers are generally driving faster on the back street.
    c. The spread of speeds is greater on the main road as indicated by the higher standard deviation.
13. a.

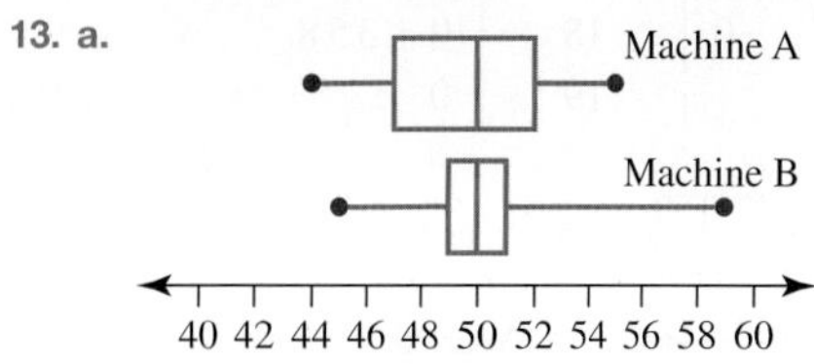

    b. Machine A: mean = 49.88, standared deviation = 2.87; Machine B: mean = 50.12, standared deviation = 2.44
    c. Machine B is more reliable, as shown by the lower standard deviation and IQR. The range is greater on machine B only because of a single outlier.
14. a. Nathan: mean = 15.1; Timana: mean = 12.3
    b. Nathan: range = 36; Timana: range = 14
    c. Nathan: IQR = 15; Timana: IQR = 4
    d. Timana's lower range and IQR shows that he is the more consistent player.
15. a. Yes — Maths
    b. Science: positively skewed
    c. The Science test may have been more difficult.
    d. Science: 61 – 70, Maths: 71 – 80
    e. Maths has a greater standard deviation (12.6) compared to Science (11.9).
16. a. Key: 2|3 = 2.3 hours

Leaf: Group A	*Stem*	*Leaf: Group B*
873	1	78
951	2	01245588
875422	3	222455568
754222	4	2
	5	
22	6	

    b. Five-point summary
       Group A: 13 27 36 43 62
       Group B: 17 23 30 35 42

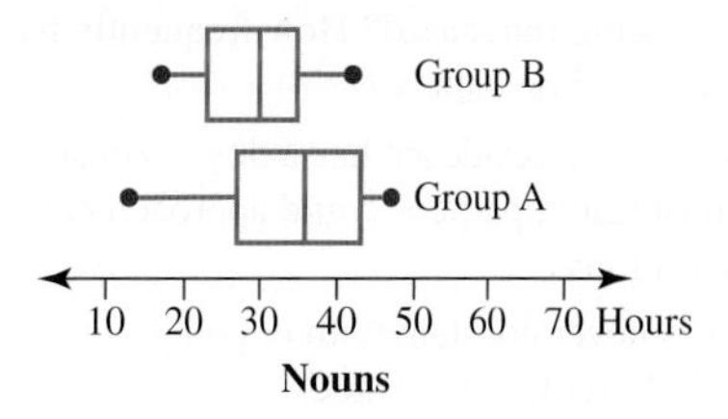

    c. Sample responses can be found in the worked solutions in the online resources.
    d. Sample responses can be found in the worked solutions in the online resources.
    e. Sample responses can be found in the worked solutions in the online resources.

17. a.

Leaf: Year 10	Stem	Leaf: Year 11
98754	15	
9874330	16	03678
7532200	17	223456779
0	18	01358
	19	0

b. 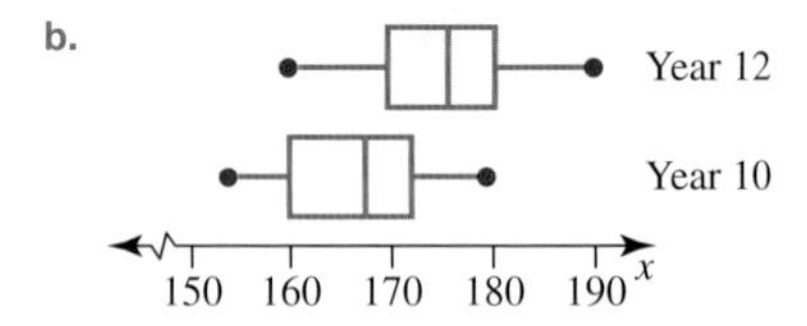

c. On average, the Year 12 students are about 6 – 10 cm taller than the Year 10 students. The heights of the majority of Year 12 students are between 170 cm and 180 cm, whereas the majority of the Year 10 students are between 160 and 172 cm in height.

18. a. English: mean = 70.25; Maths: mean = 69

b. English: range = 53; Maths: range = 37

c. English: $\sigma = 16.1$; Maths: $\sigma = 13.4$

d. Kloe has performed more consistently in Maths as the range and standard deviation are both lower.

19. 32%

20. a. 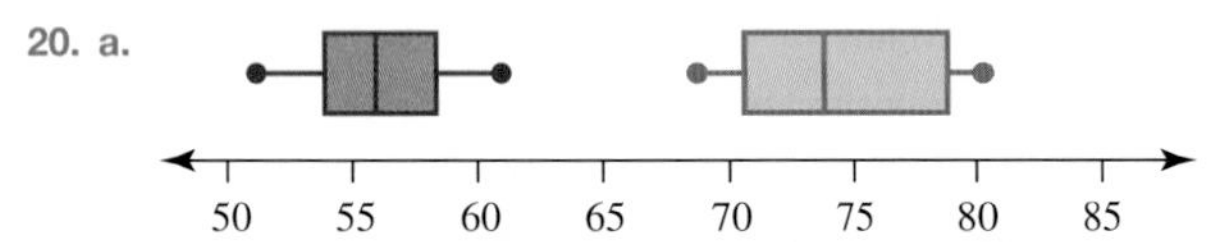

The parallel box plots show a significant gap between the life expectancy of Aboriginal and Torres Strait Islander people and that of non-Aboriginal and non–Torres Strait Islander people. Even the maximum median age of Aboriginal and Torres Strait Islander people is much lower than the minimum of non-Aboriginal and non–Torres Strait Islander people.

b. The advantage of box plots is that it gives a clear graphical representation of the results and in this case shows a significant difference between the median life expectancy of Aboriginal and Torres Strait Islander people and non-Aboriginal and non–Torres Strait Islander people. The disadvantage is that we lose the data for individual states and territories.

Exercise 12.7Populations and samples

1. a. When was it first put into the machine? How old was the battery before being purchased? How frequently has the computer been used on battery?

b. Can't always see if a residence has a dog; a census is very time-consuming; perhaps could approach council for dog registrations.

c. This number is never constant with ongoing purchases, and continuously replenishing stock.

d. Would have to sample in this case as a census would involve opening every packet.

2. Sample responses can be found in the worked solutions in the online resources.

3. a. Census. The airline must have a record of every passenger on every flight.

b. Survey. It would be impossible to interview everyone.

c. Survey. A census would involve opening every bottle.

d. Census. The instructor must have an accurate record of each learner driver's progress.

4. a. Survey b. Survey c. Census d. Survey

5. a. About 25

b. Drawing numbers from a hat, using a calculator.

6. a. The council is probably hoping it is a census, but it will probably be a survey because not all those over 10 will respond.

b. Residents may not all have internet access. Only those who are highly motivated are likely to respond.

7. The sample could have been biased. The questionnaire may have been unclear.

8. Sample size, randomness of sample

9. Sample responses can be found in the worked solutions in the online resources.

10. Sample response: A census of very large populations requires huge amounts of infrastructure and staff to collect the information for large numbers of people. These challenges could also be made harder because many people live in remote areas with poor transport access for census staff; forms may need to be created in multiple languages; and migrants who do not have residency permits may be unwilling to complete a census.

11. There is quite a variation in the frequency of particular numbers drawn. For example, the number 45 has not been drawn for 31 weeks, while most have been drawn within the last 10 weeks. In the long term, one should find the frequency of drawing each number is roughly the same. It may take a long time for this to happen, as only 8 numbers are drawn each week.

12. a. Mean = 32.03; median = 29.5

b.

Class interval	Frequency
0 – 9	2
10 – 19	7
20 – 29	6
30 – 39	6
40 – 49	3
50 – 59	3
60 – 69	3
Total	30

c. Mean = 31.83

d. 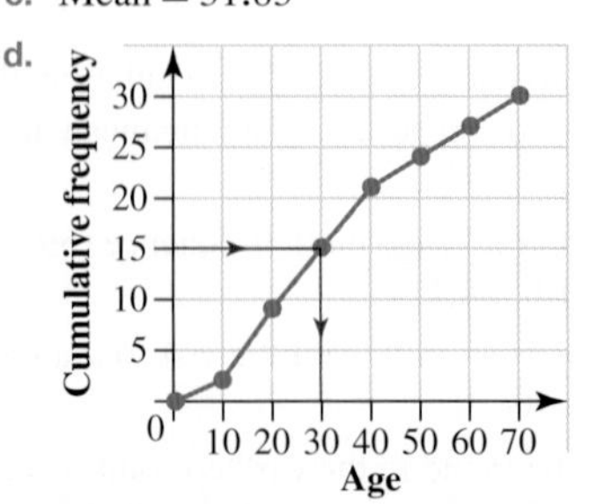

e. Median = 30

f. Estimates from parts c and e were fairly accurate.

g. Yes, they were fairly close to the mean and median of the raw data.

13. Year 8: mean = 26.83, median = 27, range = 39, IQR = 19
Year 10: mean = 40.7, median = 39.5 range = 46 IQR = 20
The typing speed of Year 10 students is about 13 to 14 wpm faster than that of Year students. The spread of data in Year 8 is slightly less than the spread in Year 10.

Exercise 12.8 Evaluating inquiry methods and statistical reports

1. a. Primary. There is probably no secondary data available.
   b. Sample responses can be found in the worked solutions in the online resources.
   c. Sample responses can be found in the worked solutions in the online resources.
2. Company profits

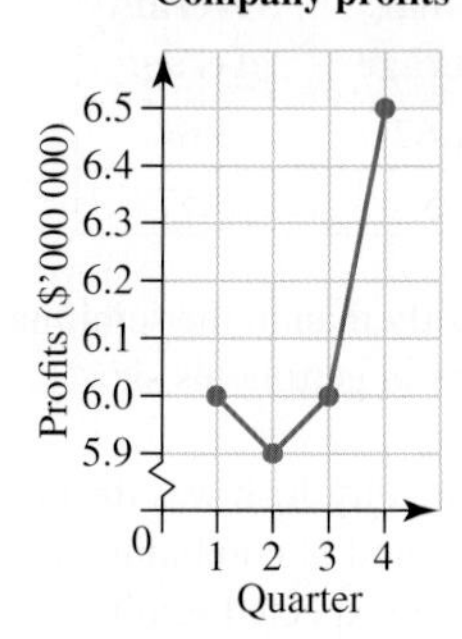

Mean salaries

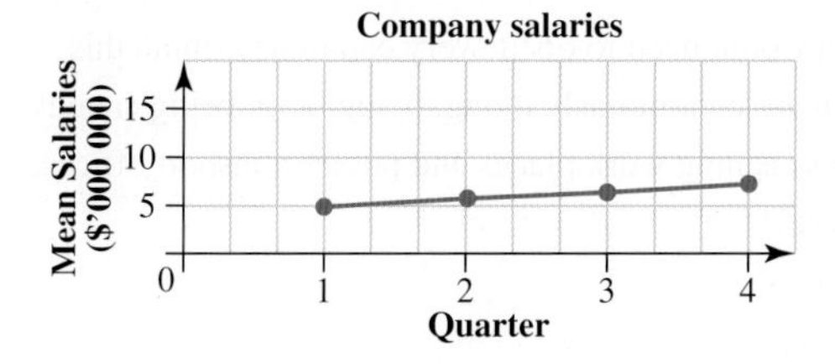

3. a. Mean = $21 5000, median = $170 000, mode = $150 000. The median best represents these land prices. The mean is inflated by one large score, and the mode is the lowest price.
   b. Range = $500 000, interquartile range = $30 000. The interquartile range is the better measure of spread.

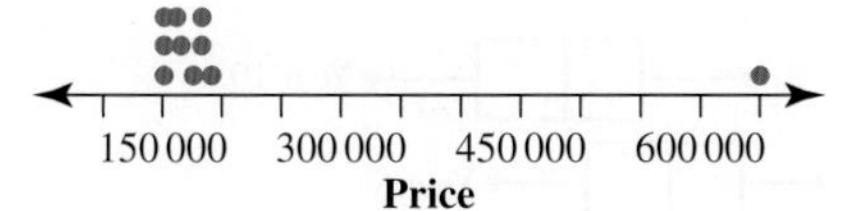

   c. This dot plot shows how 9 of the scores are grouped close together, while the score of $650 000 is an outlier.
   d. The agent is quoting the modal price, which is the lowest price. This is not a true reflection of the average price of these blocks of land.
4. a. False. Mean = 1.82 m, lower quartile = 1.765 m, median = 1.83 m
   b. True. This is the definition of interquartile range.
   c. Players with heights 1.83 m, 1.83 m, 1.88 m, 1.88 m
5. a.

Resting pulse	Frequency	Middle value, x	$x \times$ frequency	Cumulative frequency
30–39	3	34.5	103.5	3
40–49	8	44.5	356.0	11
50–59	5	54.5	272.5	16
60–69	3	64.5	193.5	19
70–79	1	74.5	74.5	20

   b. Mean = 48.65
   Median = 48
   Mode = 48
   c. From the table it can be clearly seen that the highest concentration of resting pulse readings was in the 40–49 and 50–59 groups. All three measures of central tendency fell within these two groupings.
   Because of the higher frequency in the 40–49 group, it was not surprising that the mode and median were contained there also. The mean was slightly higher, and this would have been influenced by the one reading in the 70–79 group.
6. Player B appears to be the better player if the mean result is used. However, Player A is the more consistent player.
7. a. 7.1 b. 7 c. 7
   d. The mode has the most meaning as this size sells the most.
8. Points which could be mentioned include:
10.1% is only just 'double digit' growth.
2006–08 showed mid to low 20% growth. Growth has been declining since 2008.
The share price has rebounded, but not to its previous high.
The share price scale is not consistent. Most increments are 30c, except for $27.70 to $28.10?(40c increment). Note also the figure of 20.80 — probably a typo instead of 26.80.
9. A misleading graph steers/convinces the reader towards a particular opinion. It can be biased and not accurate.
10. Shorten the y-axis and expand the x-axis.

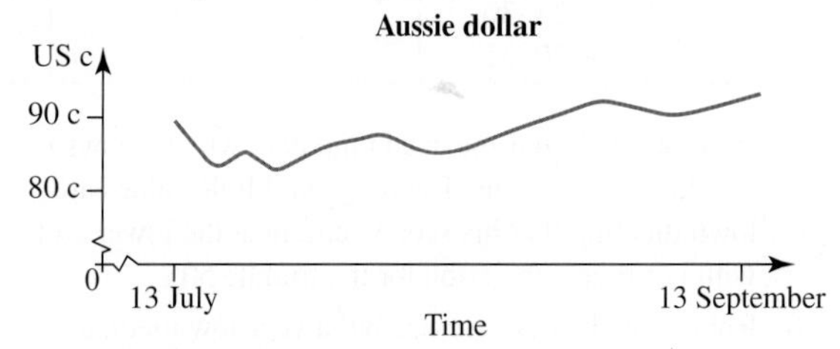

11. a. Key: 3|85 = 385 hours

Leaf: Brand B	*Stem*	*Leaf: Brand A*
	3	85 90
	4	25 26
60 55 00	5	70
30	6	40 45
70 42 35 20	7	30 35 60
60 20	8	

Brand A: mean = 570.6, median = 605
Brand B: mean = 644.2, median = 727.5
   b. Brand A had the shortest mean lifetime.

c. Brand B had the longest mean lifetime.

d. Brand B

12. a. The statement is true, but misleading as most of the employees earn \$18000.

b. The median and modal salary is \$18000 and only 15 out of 80 (less than 20%) earn more than the mean.

13. Left bar chart suggests prices have tripled in one year due to fact vertical axis does not start at zero. Bar chart on right is truly indicative of situation.

14. a. Percentages do not add to 100%.

b. Percentages do not add to 100%.

c. Such representation allows multiple choices to have closer percentages than really exist.

Project

1.

Player	Runs in the last 25 matches	Mean	Median	Range	IQR
Will	13, 18, 23, 21, 9, 12, 31, 21, 20, 18, 14, 16, 28, 17, 10, 14, 9, 23, 12, 24, 0, 18, 14, 14, 20	16.76	17	31	8.5
Rohit	2, 0, 112, 11, 0, 0, 8, 0, 10, 0, 56, 4, 8, 164, 6, 12, 2, 0, 5, 0, 0, 0, 8, 18, 0	17.04	4	164	10.5
Marnus	12, 0, 45, 23, 0, 8, 21, 32, 6, 0, 8, 14, 1, 27, 23, 43, 7, 45, 2, 32, 0, 6,11, 21, 32	16.76	12	45	25.5
Ben	2, 0, 3, 12, 0, 2, 5, 8, 42, 0, 12, 8, 9, 17, 31, 28, 21, 42, 31, 24, 30, 22, 18, 20, 31	16.72	17	42	25

2. a. Will: has a similar mean and median, which shows he was fairly consistent. The range and IQR values are lowindicating that his scores remain at the lower end with not much deviation for the middle 50% .

b. Rohit: has the best average but a very low median indicating his scores are not consistent. The range is extremely high and the IQR very low in comparison showing he can score very well at times but is not a consistent scorer.

c. Marnus: has a similar mean to Will and Ben but a lower median, indicating his scores are sometimes high but generally are lower than the average. The range and IQR show a consistent batting average and spread with only a few higher scores and some lower ones.

d. Ben: has a similar mean and median which shows he was a consistent player. The range and IQR show a consistent batting average and spread.

Players to be selected:
Would recommend **Will** if the team needs someone with very consistent batting scores every game but no outstanding runs.

Would recommend **Rohit** if the team needs someone who might score very high occasionally but in general fails to score many runs.

Would recommend **Marnus** if the team needs someone who is fairly consistent but can score quite well at times and the rest of the time does OK.

Would recommend **Ben** if the team needs someone who is fairly consistent but can score quite well at times and the rest of the time has a better median than Glenn.

3.

	Semifinal average	Final average	Overall average
Josh	2.4	4.67	3.64
Ravi	2.5	5	3.57

4. In the final, wickets were more costly than in the semifinal. Josh therefore conceded many runs in getting his six wickets.
This affected the overall mean. In reality Josh was the most valuable player overall, but this method of combining the data of the two matches led to this unexpected result.

Exercise 12.10 Review questions

1. a. You would need to open every can to determine this.

b. Fish are continuously dying, being born, being caught.

c. Approaching work places and public transport offices.

2. a. Survey

b. Census

c. Survey

3. a. Mean = 11.55; median = 10; mode = 8

b. Mean = 36; median = 36; mode = 33, 41

c. Mean = 72.18; median = 72; mode = 72

4. a. 6 b. 6 c. 20

5. a. 4 b. 0.85

6. a. 20 b. 24 c. 8

7. a.

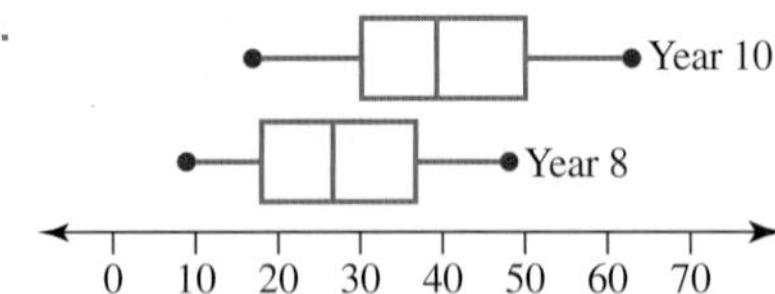

b. Year 8: mean = 26.83, median = 27, range = 39, IQR = 19, sd = 11.45
Year 10: mean = 40.7, median = 39.5, range = 46, IQR = 20, sd = 12.98

c. The typing speed of Year 10 students is about 13 to 14 wpm faster than that of Year 8 students. The spread of data in Year 8 is slightly less than in Year 10.

8. a. Key: $3^*|9 = 3.9$ kg

Stem	Leaf
3*	9
4	0 0 0 2 3
4*	5 6 7 8
5	0 3
5*	5 8 8 9
6	1 2 2
6*	8

b. (3.9, 4.4, 4.9, 5.85, 6.8)

c.

9. a. 24.4 b. 1.1 c. 7.3

10. A

11. B

12. 0.05 ml

13. Mean = 5, median = 5, mode = 2 and 5.
The distribution is positively skewed and bimodal.

14. a. Median height = 167 cm

b. Range = 25 cm

c. IQR = 5 cm

15. C

16. A, B and C

17. a. Mean = 32.03; median = 29.5

b.

Class interval	Frequency
0 – 9	2
10 – 19	7
20 – 29	6
30 – 39	6
40 – 49	3
50 – 59	3
60 – 69	3
Total	30

c. Mean = 31.83

d.

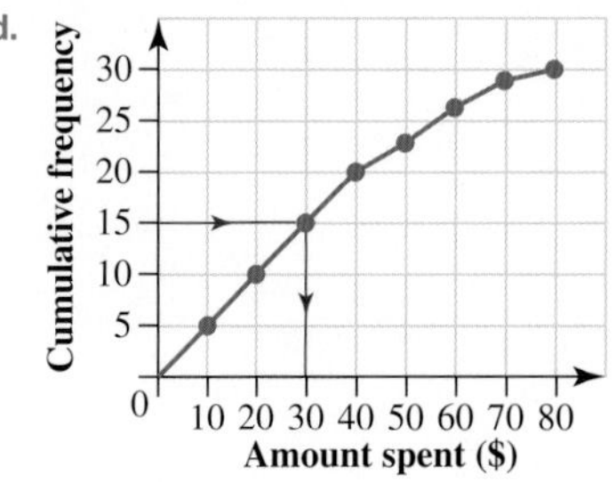

e. Median = 30

f. Estimates from parts c and e were fairly accurate.

g. Yes, they were fairly close to the mean and median of the raw data.

18. a.

b. Positively skewed — a greater number of scores is distributed at the lower end of the distribution.

19. a. Yes b. Yes. Both are 3

c. 3

20. a. $\bar{x} = m^2$ $\sigma = \dfrac{m^2}{2}$ b. $7m + 2$

21. a. Mean = 2.17, median = 2

b. Mean = 3.54, median = 2

c. The median relies on the middle value of the data and won't change much if an extra value is added. The mean however has increased because this large value will change the average of the numbers. The mean is used as a measure of central tendency if there are no outliers or if the data are symmetrical. The median is used as a measure of central tendency if there are outliers or the data are skewed.

22. a.

Interval	Frequency (f)	Midpoint × (f)
40 – 49	1	$44.5 \times 1 = 44.5$
50 – 59	1	$54.5 \times 1 = 54.5$
60 – 69	1	$64.5 \times 1 = 64.5$
70 – 79	2	$74.5 \times 2 = 149$
80 – 89	4	$84.5 \times 4 = 338$
90 – 99	4	$94.5 \times 4 = 378$
100 – 109	8	$104.5 \times 8 = 836$
110 – 119	6	$114.5 \times 6 = 687$
120 – 129	8	$124.5 \times 8 = 996$
130 – 139	2	$134.5 \times 2 = 269$
140 – 149	2	$144.5 \times 2 = 289$
150 – 159	0	$154.5 \times 0 = 0$
160 – 169	1	$164.5 \times 1 = 164.5$
Total	40	4270

b. 106.75

c. 107.15

d. The differences in this case were minimal; however, the grouped data mean is not based on the actual data but on the frequency in each interval and the interval midpoint. It is unlikely to yield an identical value to the actual mean. The spread of the scores within the class interval has a great effect on the grouped data mean.

23. a. 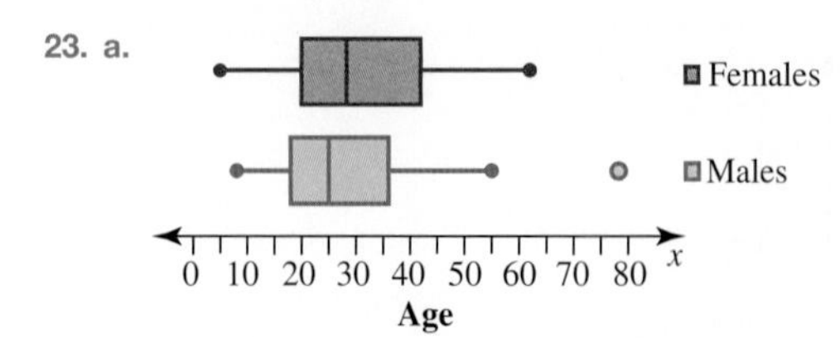

b.

	Males	Females
Mean	28.2	31.1
Range	70	57
IQR	18	22

c. There is one outlier — a male aged 78.

d. Typically males seem to enter hospital for the first time at a younger age than females.

24. a. i. 35 s

ii. 29.5 s

iii. 33.05 s

iv. 60 s

v. 21s

vi. 39 s

vii. 18 s

b. 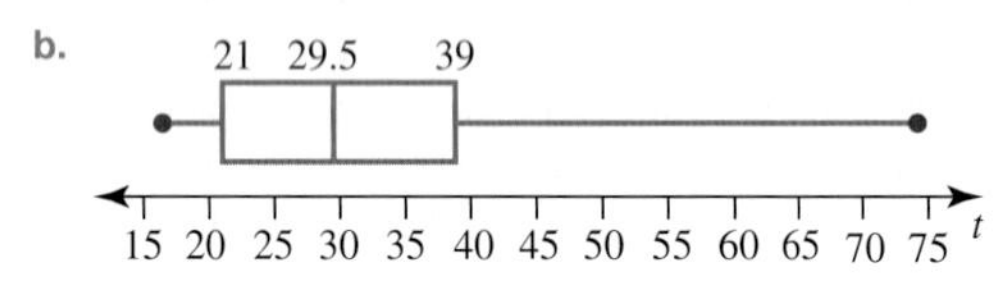

c. i. 25%

ii. 50%

iii. 75%

d. Categorical

e. 35%

f. Pictogram, pie chart or bar chart.

25. a. Class A: Q_1– 21.5, Median – 30, Q3 – 38, IQR – 16.5
Class B: Q_1 – 14.5, Median – 33, Q3 – 47, IQR – 32.5
Based on the comparison between Class A's IQR (16.5) and Class B's IQR (32.5), Ms Vinculum was correct in her statement.

b. No, Class B has a higher median and upper quartile score than Class A, while Class A has a higher lower quartile. You can't confidently say that either class did better in the test than the other.

26. a. 82.73 km/h

b. 30 cars

c. i. $2 607 272.73

ii. About 545

13 Bivariate data

LEARNING SEQUENCE

13.1 Overview

Why learn this?

Bivariate data can be collected from all kinds of place. This includes data about the weather, data about athletic performance and data about the profitability of a business. By learning the tools you need to analyse bivariate data, you will be gaining skills that help you turn numbers (data) into powerful information that can be used to make predictions (plots).

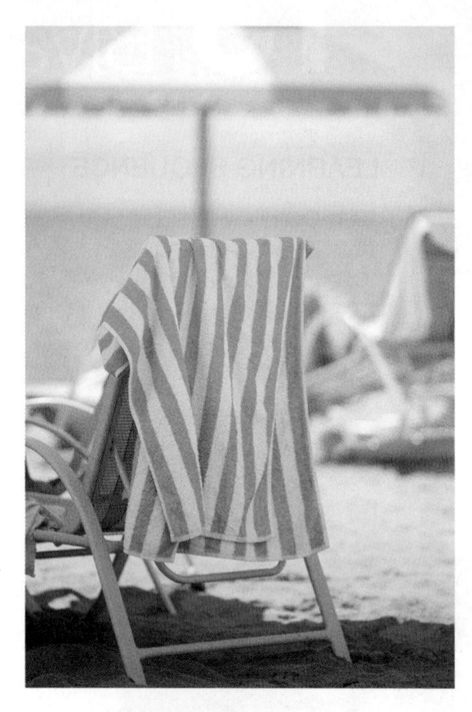

The use for bivariate data is not limited to the classroom; in fact, many professionals rely on bivariate data to help make decisions. Some examples of bivariate data in the real world are:

- when a new drug is created, scientists will run drug trials in which they collect bivariate data about how the drug works. When the drug is approved for use, the results of the scientific analysis help guide doctors, nurses, pharmacists and patients as to how much of the drug to use and how often.
- manufacturers of products can use bivariate data about sales to help make decisions about when to make products and how many to make. For example, a beach towel manufacturer would know that they need to produce more towels for summer, and analysis of the data would help them decide how many to make.

By studying bivariate data you can learn how to use data to make predictions. By studying and understanding how these predictions work, you will be able to understand the strengths and limitations of these types of predictions.

Where to get help

Go to your learnON title at **www.jacplus.com.au** to access the following digital resources. The Online Resources Summary at the end of this topic provides a full list of what's available to help you learn the concepts covered in this topic.

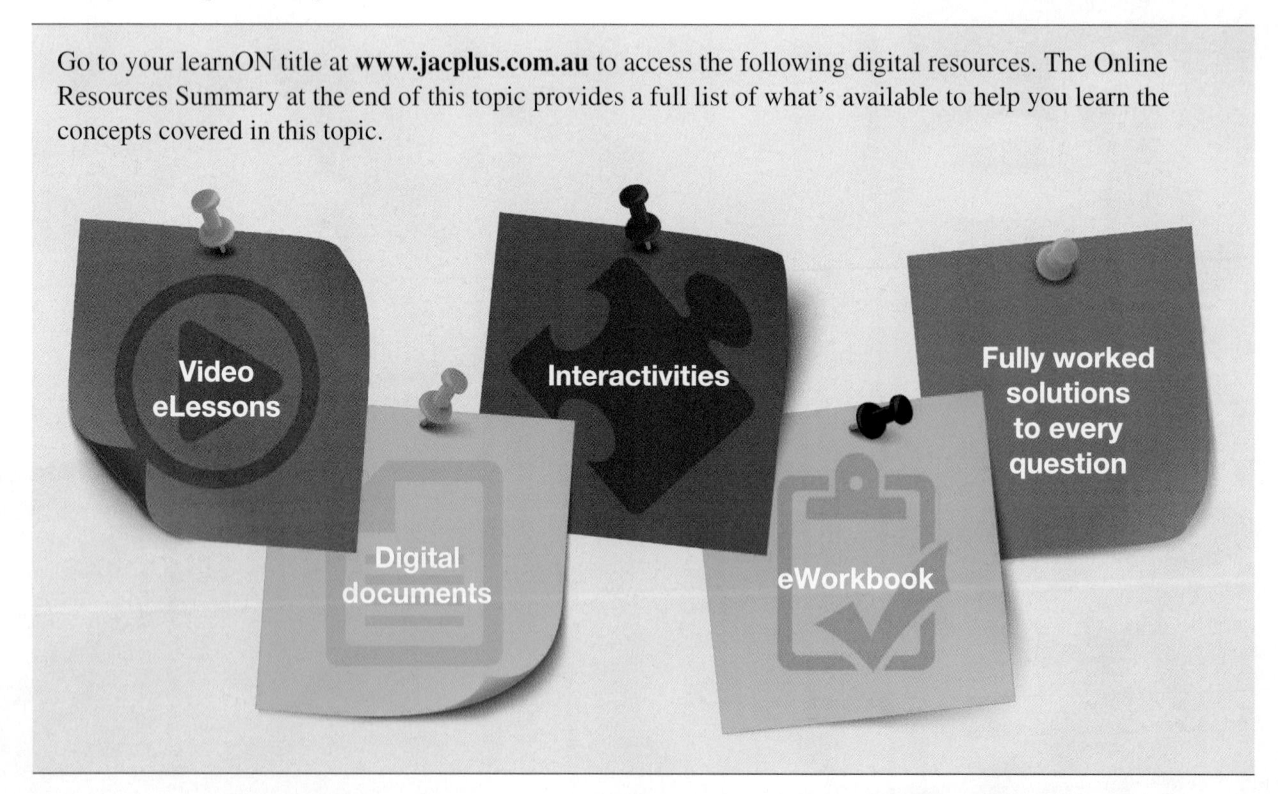

Exercise 13.1 Pre-test

learnON

Complete this pre-test in your learnON title at www.jacplus.com.au and receive **automatic marks, immediate corrective feedback** and **fully worked solutions**.

1. **MC** Choose the following graphs that shows whether there is a relationship between two variables and each data value is shown as a point on a Cartesian plane.
   A. Box plot B. Scatterplot C. Dot plot D. Ogive E. Histogram

2. **MC** Select which of the following statements is incorrect.
   A. Bivariate data are data with two variables.
   B. Correlation describes the strength, the direction and the form of the relationship between two variables.
   C. The independent variable is placed on the y-axis and the dependent variable on the x-axis.
   D. The dependent variable is the one whose value depends on the other variable.
   E. The independent variable takes on values that do not depend on the value of the other variable.

3. Data is compared from twenty students on the number of hours spent studying for an examination and the result of the examination. State if the number of hours spent studying is the independent or dependent variable.

4. Match the type of correlation with the data shown on the scatter plots.

Scatter plot	Type of correlation
a. 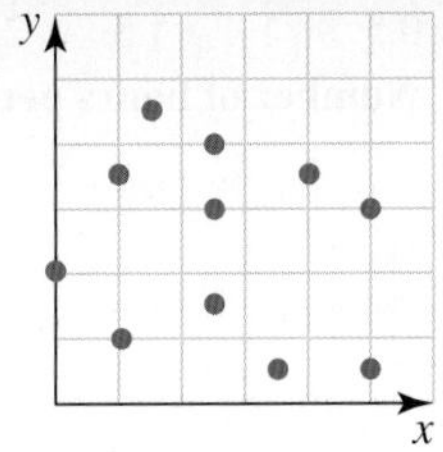	A. Strong negative linear correlation
b. 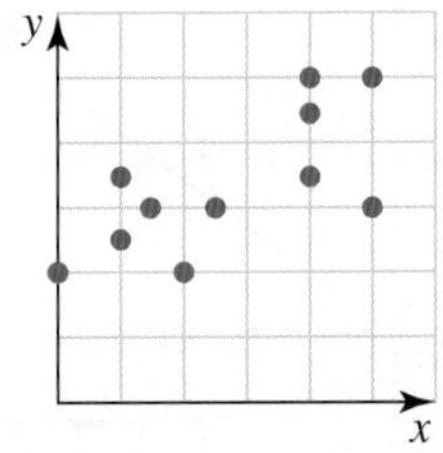	B. No correlation
c. 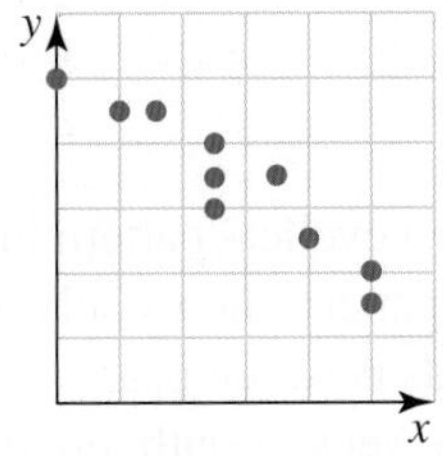	C. Weak positive linear correlation

5. **MC** The table below shows the number of hours spent doing a problem-solving task for a subject and the corresponding total score for task.

Number of hours spent on task	0	1.5	2	1	2	1.5	2.5	3	2	2.5
Task score %	20	50	60	45	80	70	75	97	85	20

Choose which data point is a possible outlier.

A. $(0, 20)$ B. $(1.5, 50)$ C. $(1.5, 70)$ D. $(2.5, 20)$ E. $(2.5, 70)$

6. **MC** Each point on the scatterplot shows the number of hours per week spent exercising by a person and their fitness level.

Choose the statement that best describes the scatterplot.

A. The more time exercising the worse the fitness level.
B. The number of hours per week spent exercising is the independent variable.
C. The correlation between the number of hours per week exercising and the fitness levels is a weak positive non-linear correlation.
D. There are six people's information collected.
E. There is an outlier.

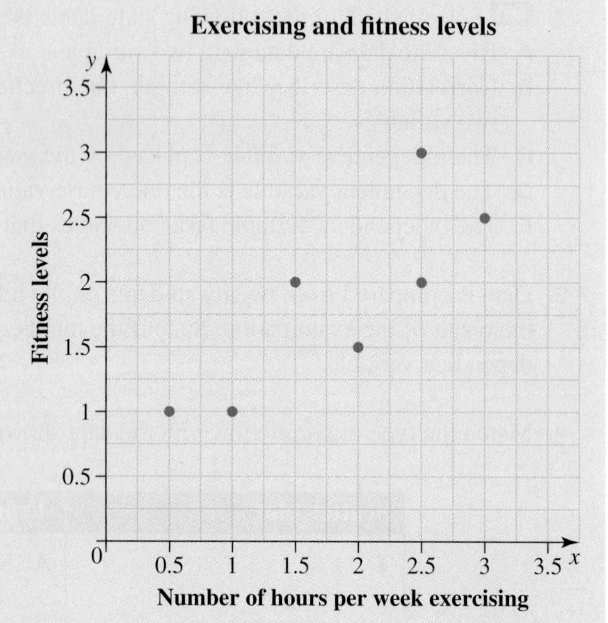

7. Select a term that describes a line of best fit being used to predict a value of a variable from within a given range from the following options: extrapolation, interpolation or regression.

8. Determine the gradient of the line of best fit shown in the scatter plot.

9. In time series data, explain whether time is the independent or dependent variable.

10. Select another term for an independent variable from the following options: a response variable or an explanatory variable.

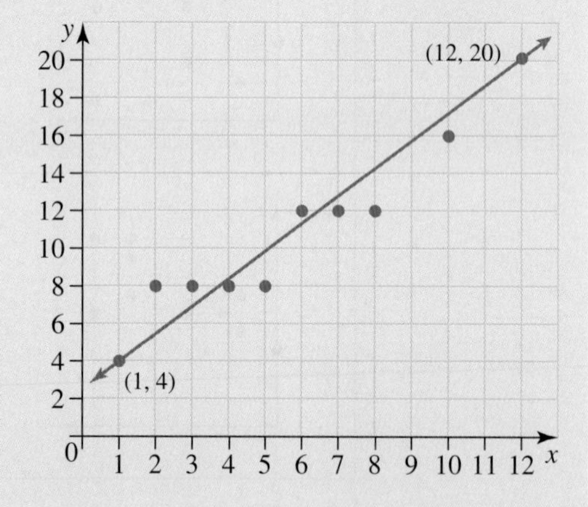

11. **MC** Select the correct difference between a seasonal pattern and a cyclical pattern in a time series plot.

A. A cyclical pattern shows upward trends, where as a seasonal pattern shows only downward trends.
B. A cyclical pattern displays fluctuations with no regular periods between peaks, where as a seasonal pattern displays fluctuations that repeat at the same time each week, month, quarter or year.
C. A cyclical pattern does not show any regular fluctuations, where as a seasonal pattern does.
D. A seasonal pattern displays fluctuations with no regular periods between peaks, where as a cyclical pattern displays fluctuations that repeat at the same time each week, month, quarter or year.
E. A seasonal pattern shows upward trends, where as a cyclical pattern shows only downward trends.

12. **MC** The time series plot shown can be classified as:
   A. upward trend
   B. downward trend
   C. seasonal pattern
   D. cyclical pattern
   E. random pattern

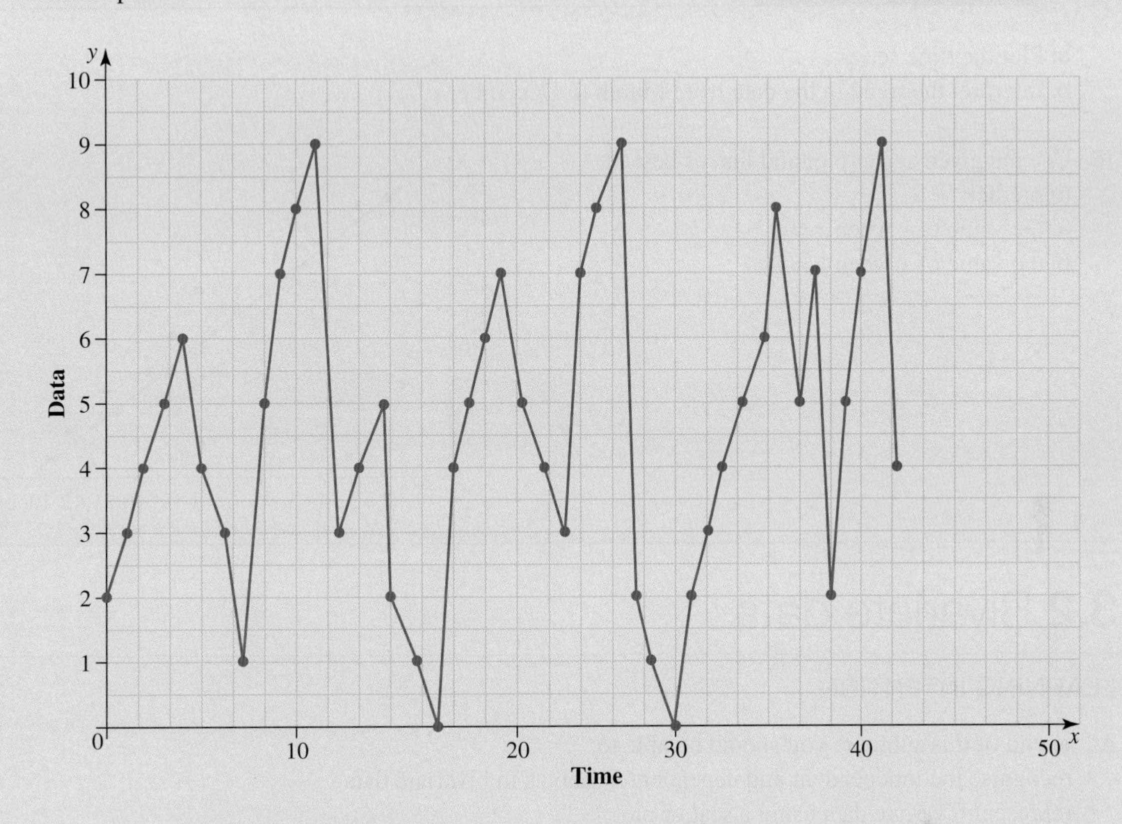

13. Use the given scatterplot and the line of best fit to determine the value of x when $y = 2$.

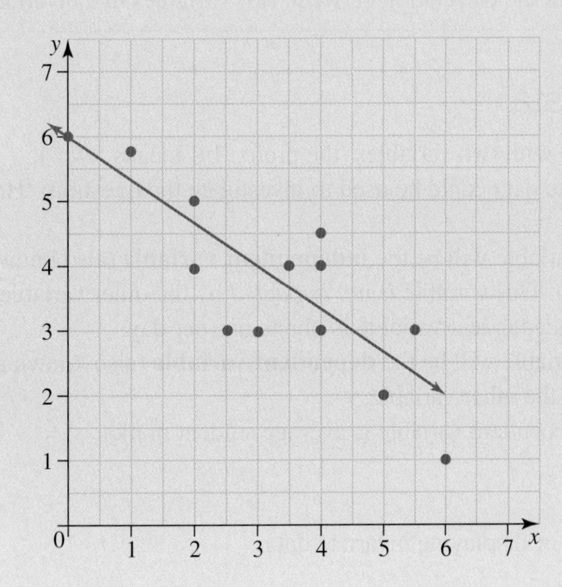

14. The table below shows the number of new COVID-19 cases per month reported in Australia in 2020.

Month	March	April	May	June	July
New COVID-19 cases	11	304	14	9	86
Month	August	September	October	November	December
New COVID-19 cases	377	73	18	5	8

a. Plot the time series.
b. Interpret the trend in the data from March to December.

15. Use the given scatterplot and line of best fit to predict:
a. the value of y when $x = 4$
b. the value of x when $y = 1$.

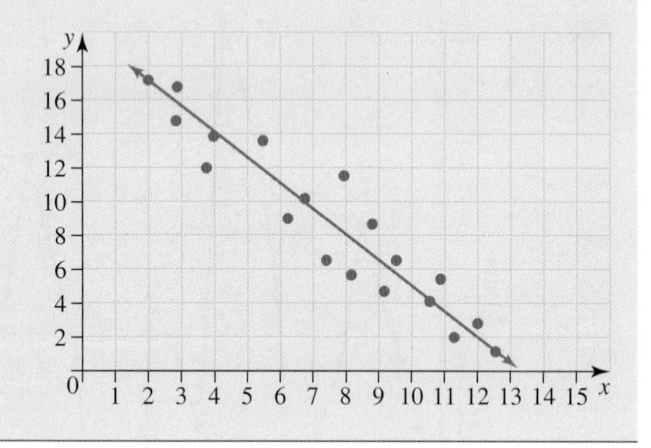

13.2 Bivariate data

LEARNING INTENTION

At the end of this subtopic you should be able to:
- recognise the independent and dependent variables in bivariate data
- represent bivariate data using a scatter plot
- describe the correlation between two variables in a bivariate data set
- draw conclusions about the correlation between two variables in a bivariate data set.

13.2.1 Bivariate data

eles-4965

- **Bivariate data** are data with two variables (the prefix 'bi' means 'two').
  - For example, bivariate data could be used to investigate the question: 'How are student marks affected by phone use?'
- In bivariate data, one variable will be the **independent variable** (also known as the experimental variable or explanatory variable). This variable *is not impacted* by the other variable.
  - In the example the independent variable is phone use per day.
- In bivariate data one variable will be the **dependent variable** (also known as the response variable). This variable *is impacted* by the other variable.
  - In the example the dependent variable is average student marks.

Scatter plots

- A **scatter plot** is a way of displaying bivariate data.
- A scatter plot will have:
  - the independent variable placed on the x-axis with a label and scale
  - the dependent variable placed on the y-axis with a label and scale
  - the data points shown on the plot.

Features of a scatter plot

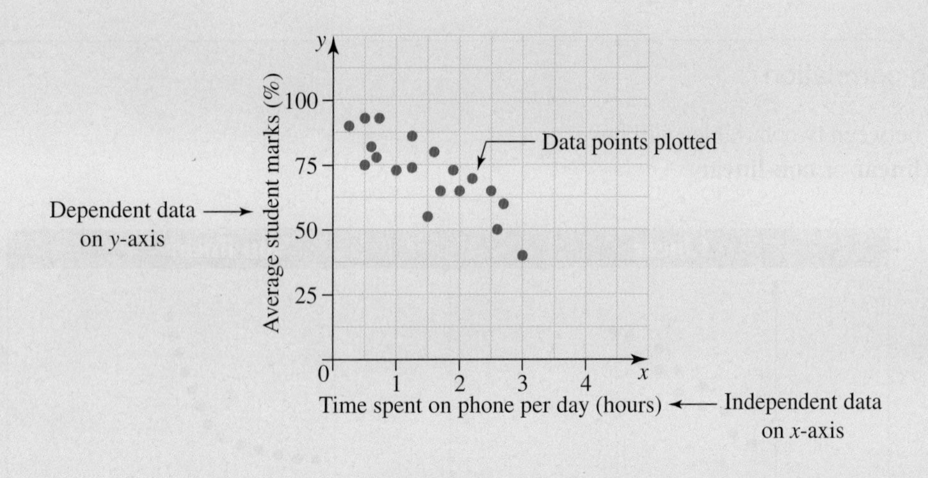

WORKED EXAMPLE 1 Representing bivariate data on a scatter plot

The table shows the total revenue from selling tickets for a number of different chamber music concerts. Represent the given data on a scatterplot.

Number of tickets sold	**400**	**200**	**450**	**350**	**250**	**300**	**500**	**400**	**350**	**250**
Total revenue ($)	**8000**	**3600**	**8500**	**7700**	**5800**	**6000**	**11 000**	**7500**	**6600**	**5600**

THINK

1. Determine which is the dependent variable and which is the independent variable.

WRITE/DRAW

The total revenue depends on the number of tickets being sold, so the total revenue is the dependent variable and the number of tickets in the independent variable.

2. Draw a set of axes. Label the title of the graph. Label the horizontal axis 'Number of tickets sold' and the vertical axis 'Total revenue ($)'.
3. Use an appropriate scale on the horizontal and vertical axes.
4. Plot the points on the scatterplot.

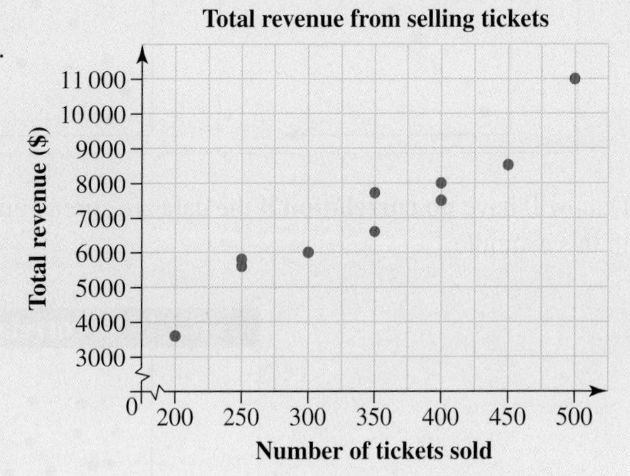

13.2.2 Correlation

eles-4966

- **Correlation** is a way of describing a connection between variables in a bivariate data set.

Describing correlation

Correlation between two variables will have:

- a **type** (**linear** or **non-linear**)

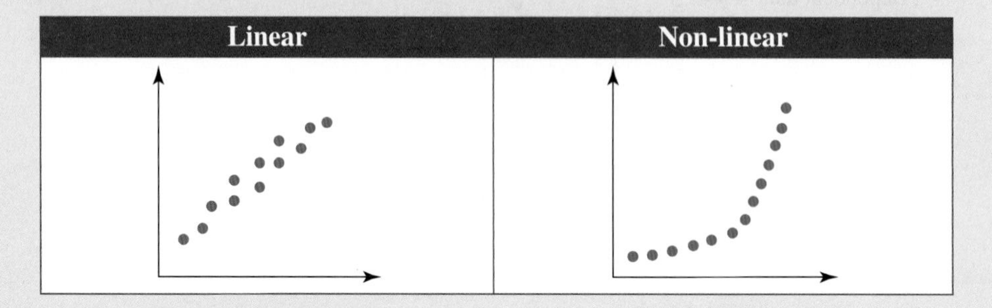

- a **direction** (**positive** or **negative**)

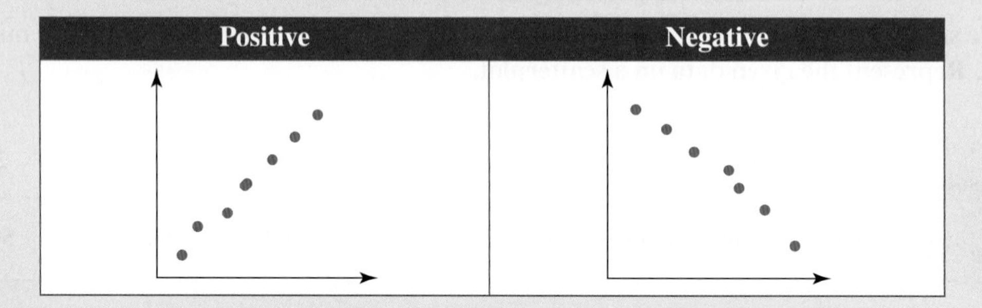

- a **strength** (**strong**, **moderate** or **weak**).

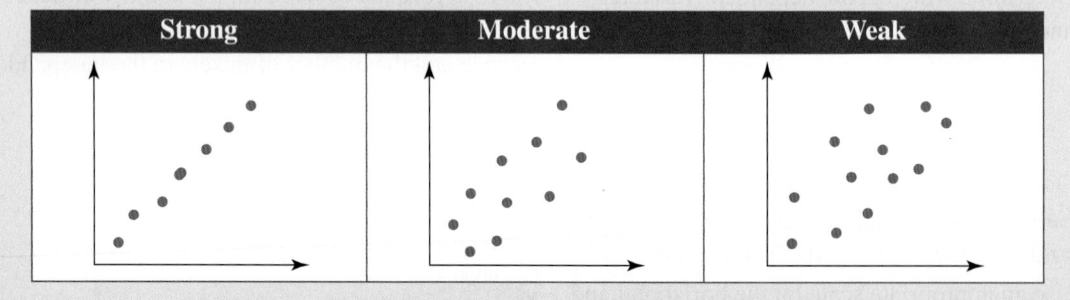

- Data will have **no correlation** if the data are spread out across the plot with no clear pattern, as shown in this example.

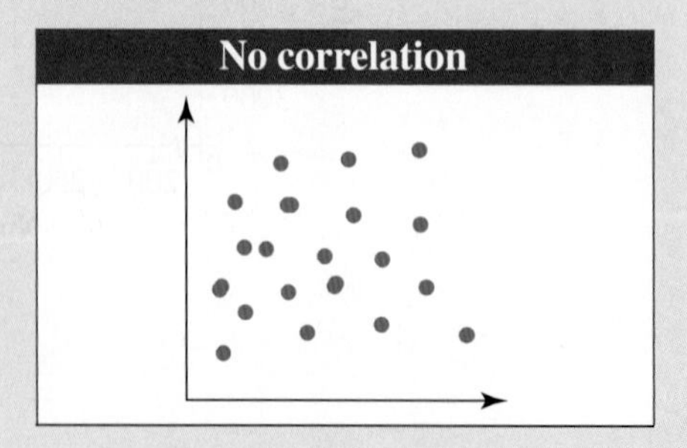

WORKED EXAMPLE 2 Describing the correlation

State the type of correlation between the variables x and y, shown on the scatterplot.

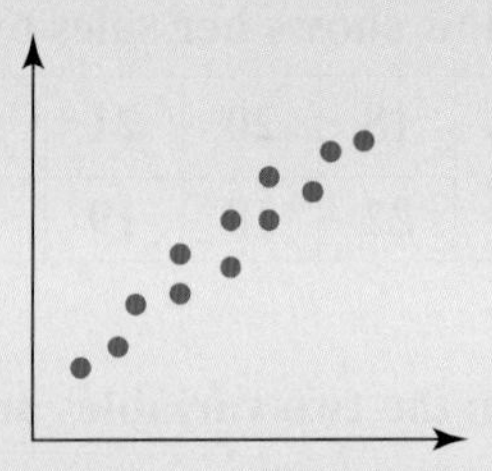

THINK	WRITE
Carefully analyse the scatterplot and comment on its form, direction and strength.	The points on the scatterplot are close together and constantly increasing therefore the relationship is linear. The path is directed from the bottom left corner to the top right corner and the value of y increases as x increases. Therefore the correlation is positive. The points are close together so the correlation can be classified as strong. There is a linear, positive and strong relationship between x and y.

13.2.3 Drawing conclusions from correlation

eles-4967

- When drawing a conclusion from a scatter plot, state how the independent variable appears to affect the dependent variable and explain what that means.
  - For the example of comparing time spent on phone to average marks, a good conclusion for the graph shown in section 13.2.1 would be:

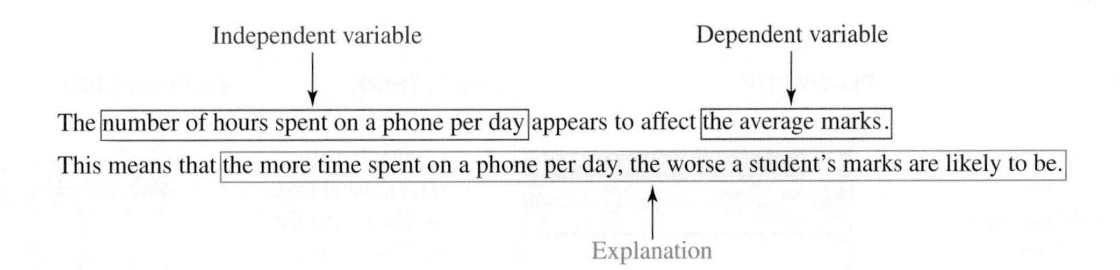

- Based on scatter plots, it is possible to draw conclusions about *correlation* but not *causation*.
  For example, for this graph:
  - it is correct to say that the amount of water drunk *appears to affect* the distance run
  - it is incorrect to say that drinking more water *causes* a person to run further, because someone might only be able to run 1 km and even if they drink 3 litres of water, they will still only be able to run 1 km.

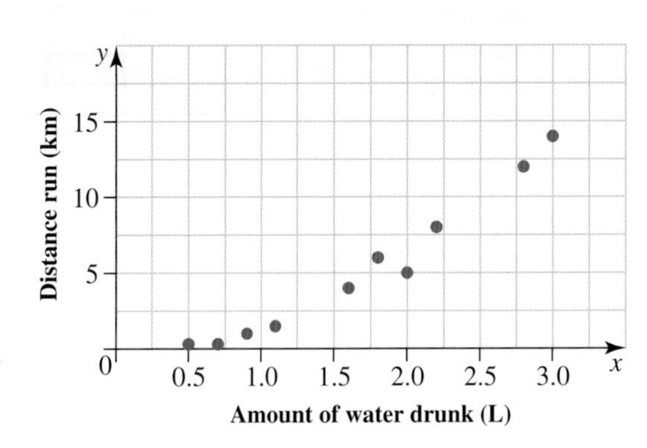

WORKED EXAMPLE 3 Stating conclusions from bivariate data

Mary sells business shirts in a department store. She always records the number of different styles of shirt sold during the day. The table below shows her sales over one week.

Price ($)	14	18	20	21	24	25	28	30	32	35
Number of shirts sold	21	22	18	19	17	17	15	16	14	11

a. Construct a scatterplot of the data.

b. State the type of correlation between the two variables and, hence, draw a corresponding conclusion.

THINK / **WRITE/DRAW**

a. Draw the scatterplot showing 'Price ($)' (independent variable) on the horizontal axis and 'Number of shirts sold' (dependent variable) on the vertical axis.

a.

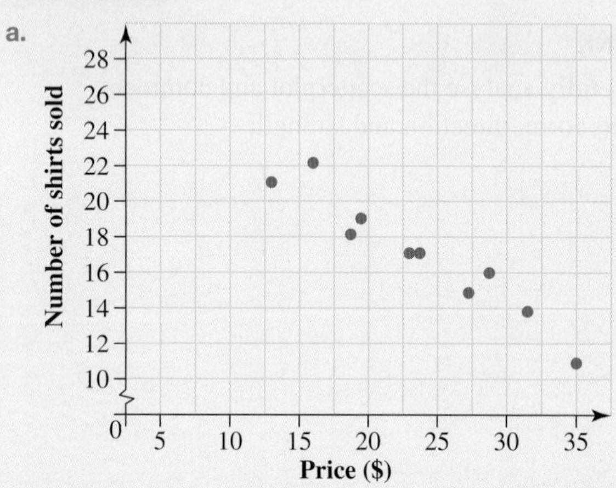

b. 1. Carefully analyse the scatterplot and comment on its form, direction and strength.

b. The points on the plot form a path that resembles a straight, narrow band, directed from the top left corner to the bottom right corner. The points are close to forming a straight line. There is a linear, negative and strong correlation between the two variables.

2. Draw a conclusion corresponding to the analysis of the scatterplot.

The price of the shirt appears to affect the number sold; that is, the more expensive the shirt the fewer sold.

TI | THINK / **DISPLAY/WRITE**

a–b.

1. In a new document, on a Lists & Spreadsheet page, label column A as 'price' and label column B as 'sold'. Enter the values from the question.

a–b.

1.1 *Bivariate data RAD

	A price	B sold	C	D
=				
6	25	17		
7	28	15		
8	30	16		
9	32	14		
10	35	11		

B10 11

CASIO | THINK / **DISPLAY/WRITE**

a–b.

1. On the Statistic screen, label list1 as 'Price' and list 2 as 'Shirts', then enter the values from the question. Press EXE after entering each value.

a–b.

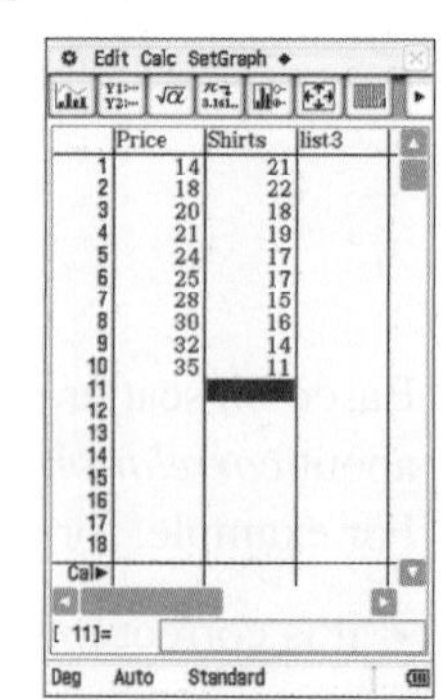

2. Open a Data & Statistics page. Press TAB to locate the label of the horizontal axis and select the variable 'price'. Press TAB again to locate the label of the vertical axis and select the variable 'sold'.

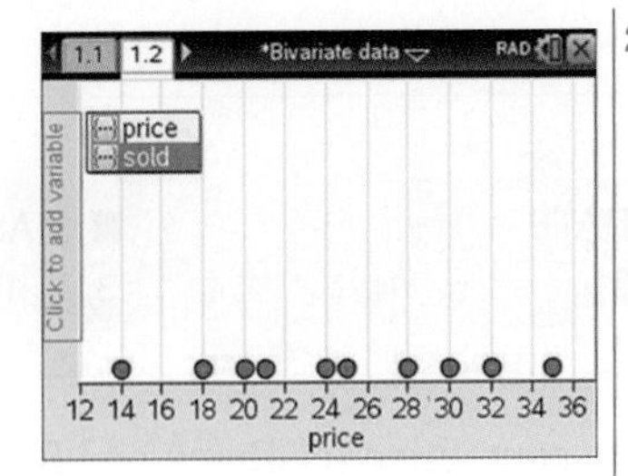

3. To change the colour of the scatterplot, place the pointer over one of the data points. Then press CTRL MENU. Press:
- Colour
- Fill Colour

Select a colour from the palette for the scatterplot. Press ENTER.

The scatterplot is shown, using a suitable scale for both axes. The points are close to forming a straight line. There is a strong negative, linear correlation between the two variables. The trend indicates that the price of a shirt appears to affect the number sold; that is, the more expensive the shirt, the fewer are sold.

2. Tap:
- SetGraph
- Setting...

Set values as shown in the screenshot, then tap Set.

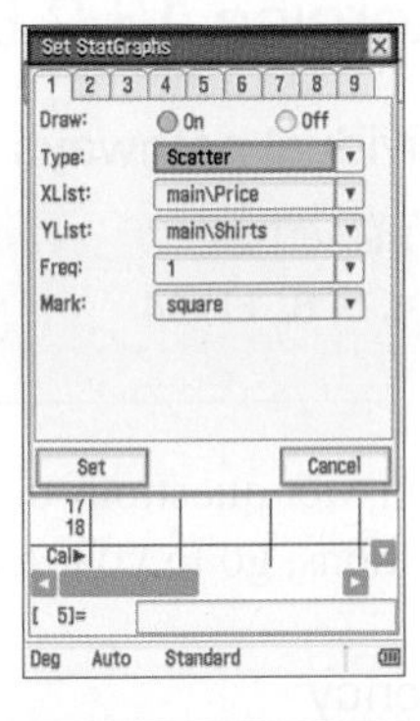

3. Tap the graphing icon and the scatterplot will appear.

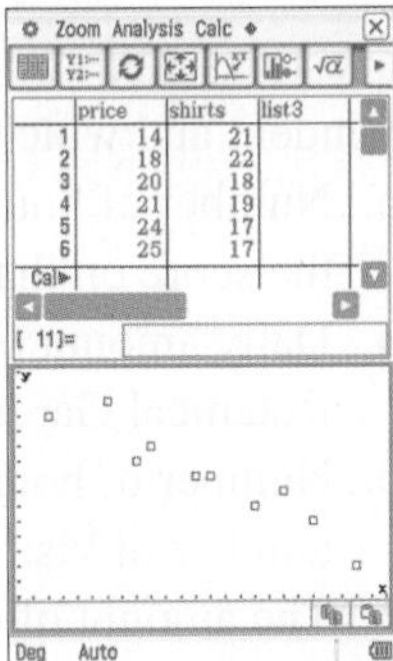

The scatterplot is shown, using a suitable scale for both axes. The points are close to forming a straight line. There is a strong negative linear correlation between the two variables. The trend indicates that the price of a shirt appears to affect the number sold; that is, the more expensive the shirt, the fewer are sold.

DISCUSSION

How could you determine whether the change in one variable causes the change in another variable?

Resources

eWorkbook Topic 13 Workbook (worksheets, code puzzle and project) (ewbk-2039)

Digital documents SkillSHEET Substitution into a linear rule (doc-5405)
SkillSHEET Solving linear equations that arise when finding x- and y-intercepts (doc-5406)
SkillSHEET Transposing linear equations to standard form (doc-5407)
SkillSHEET Measuring the rise and the run (doc-5408)
SkillSHEET Determining the gradient given two points (doc-5409)
SkillSHEET Graphing linear equations using the x- and y-intercept method (doc-5410)
SkillSHEET Determining independent and dependent variables (doc-5411)
SkillSHEET Determining the type of correlation (doc-5413)

Interactivity Individual pathway interactivity: Bivariate data (int-4626)

Exercise 13.2 Bivariate data

learnon

Individual pathways

■ PRACTISE	■ CONSOLIDATE	■ MASTER
1, 4, 5, 8, 11, 14	2, 6, 9, 12, 15	3, 7, 10, 13, 16

To answer questions online and to receive **immediate corrective feedback** and **fully worked solutions** for all questions, go to your learnON title at www.jacplus.com.au.

Fluency

For questions 1 and 2, decide which of the variables is independent and which is dependent.

1. a. Number of hours spent studying for a Mathematics test and the score on that test.
   b. Daily amount of rainfall (in mm) and daily attendance at the Botanical Gardens.
   c. Number of hours per week spent in a gym and the annual number of visits to the doctor.
   d. The amount of computer memory taken by an essay and the length of the essay (in words).
2. a. The cost of care in a childcare centre and attendance at the childcare centre.
   b. The cost of the property (real estate) and the age of the property.
   c. The entry requirements for a certain tertiary course and the number of applications for that course.
   d. The heart rate of a runner and the running speed.
3. **WE1** The following table shows the cost of a wedding reception at 10 different venues. Represent the data on a scatterplot.

No of guests	30	40	50	60	70	80	90	100	110	120
Total cost ($\times$ \$1000)	1.5	1.8	2.4	2.3	2.9	4	4.3	4.5	4.6	4.6

4. **WE2** State the type of relationship between x and y for each of the following scatterplots.

a. 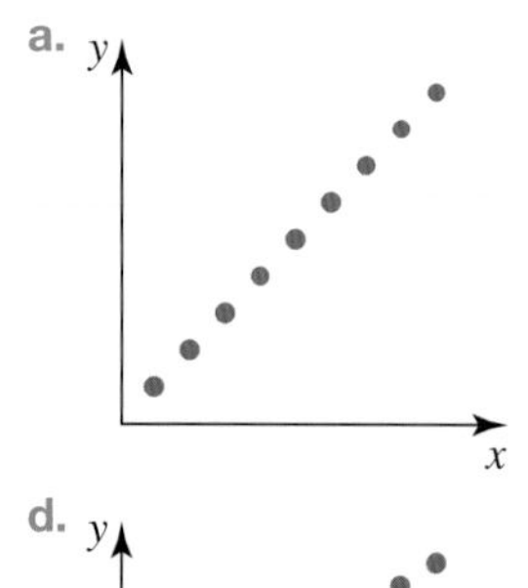

b.

c. 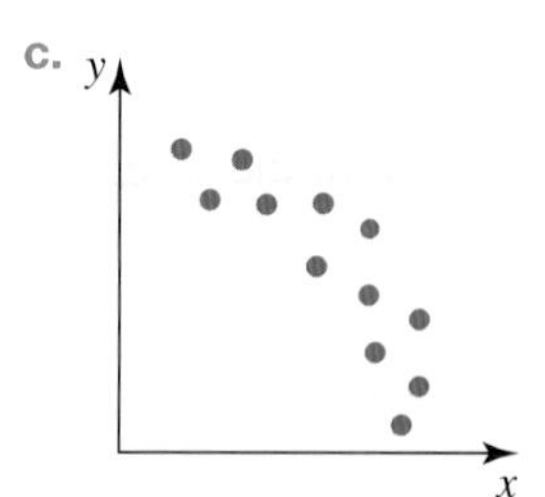

d. 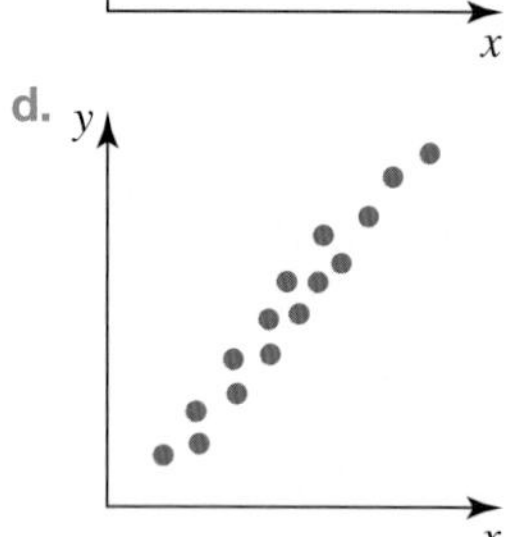

e. 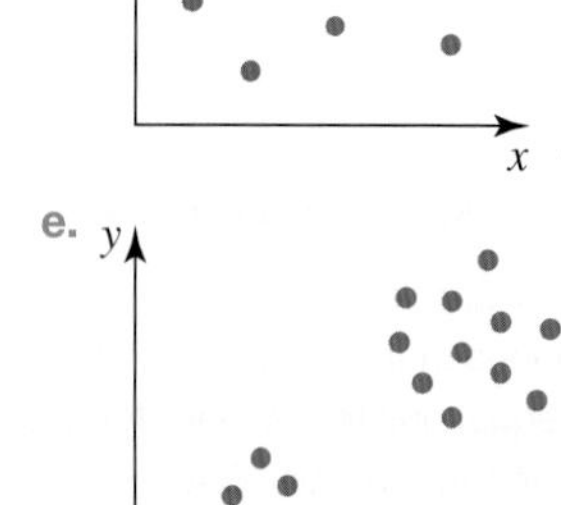

5. State the type of relationship between x and y for each of the following scatterplots.

a. 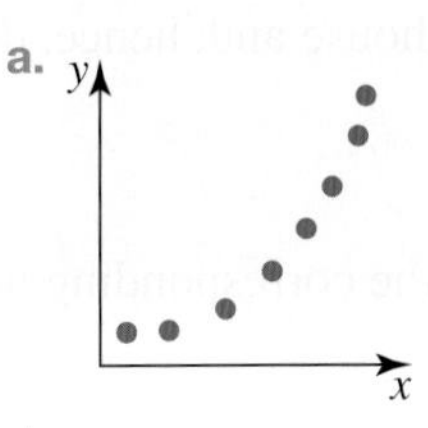

b. 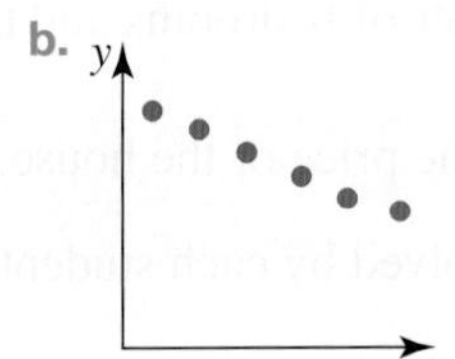

c. 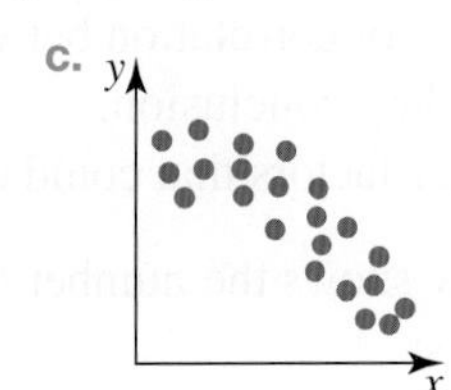

d. 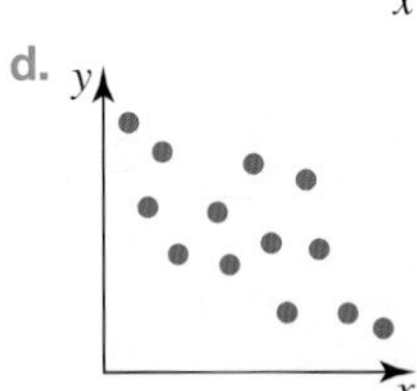

e. 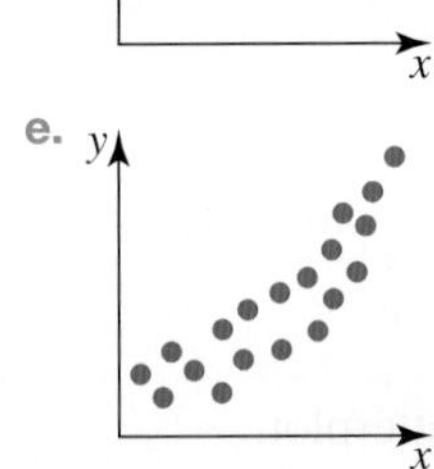

6. State the type of relationship between x and y for each of the following scatterplots.

a. 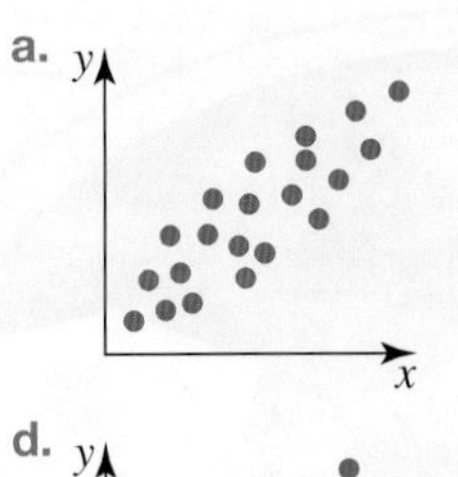

b. 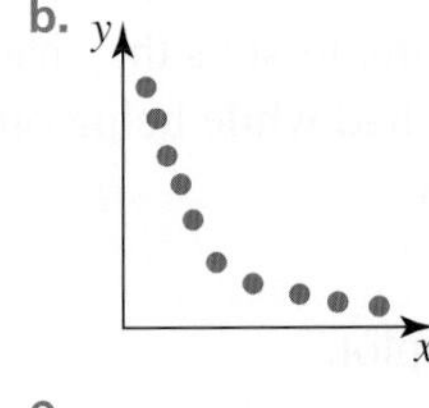

c. 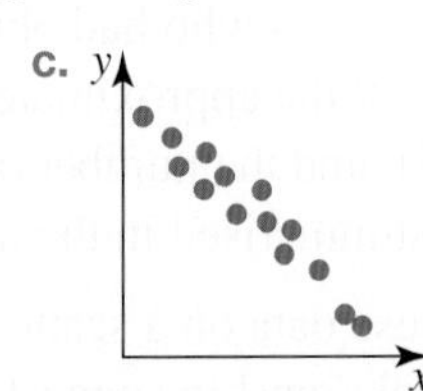

d. 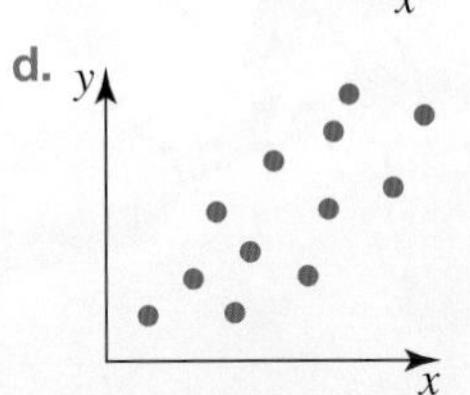

e. 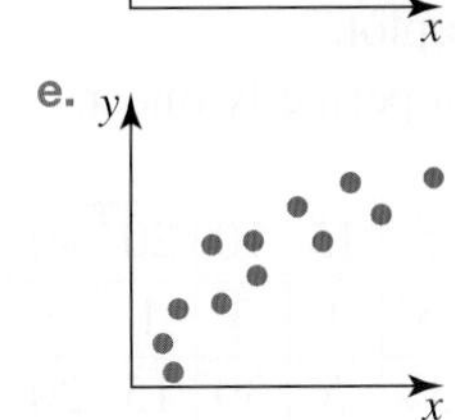

Understanding

7. **WE3** Eugene is selling leather bags at the local market. During the day he keeps records of his sales. The table below shows the number of bags sold over one weekend and their corresponding prices (to the nearest dollar).

Price ($) of a bag	30	35	40	45	50	55	60	65	70	75	80
Number of bags sold	10	12	8	6	4	3	4	2	2	1	1

a. Construct a scatterplot of the data.
b. State the type of correlation between the two variables and, hence, draw a corresponding conclusion.

8. The table below shows the number of bedrooms and the price of each of 30 houses.

Number of bedrooms	Price ($'000)	Number of bedrooms	Price ($'000)	Number of bedrooms	Price ($'000)
2	180	3	279	3	243
2	160	2	195	3	198
3	240	6	408	3	237
2	200	4	362	2	226
2	155	2	205	4	359
4	306	7	420	4	316
3	297	5	369	2	200
5	383	1	195	2	158
2	212	3	265	1	149
4	349	2	174	3	286

a. Construct a scatterplot of the data.
b. State the type of correlation between the number of bedrooms and the price of the house and, hence, draw a corresponding conclusion.
c. Suggest other factors that could contribute to the price of the house.

9. The table below shows the number of questions solved by each student on a test, and the corresponding total score on that test.

Number of questions	2	0	7	10	5	2	6	3	9	4	8	3	6
Total score (%)	22	39	69	100	56	18	60	36	87	45	84	32	63

a. Construct a scatterplot of the data.
b. Suggest the type of correlation shown in the scatterplot.
c. Give a possible explanation as to why the scatterplot is not perfectly linear.

10. A sample of 25 drivers who had obtained a full licence within the last month was asked to recall the approximate number of driving lessons they had taken (to the nearest 5), and the number of accidents they had while being on P plates. The results are summarised in the table that follows.

a. Represent these data on a scatterplot.
b. Specify the relationship suggested by the scatterplot.
c. Suggest some reasons why this scatterplot is not perfectly linear.

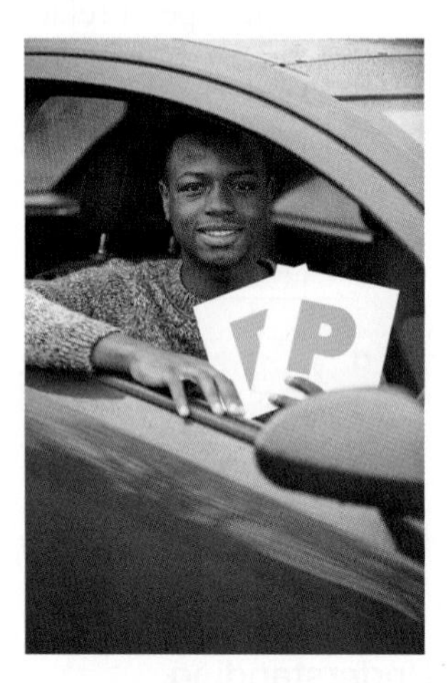

Number of lessons	5	20	15	25	10	35	5	15	10	20	40	25	10
Number of accidents	6	2	3	3	4	0	5	1	3	1	2	2	5
Number of lessons	5	20	40	25	30	15	35	5	30	15	20	10	
Number of accidents	5	3	0	4	1	4	1	4	0	2	3	4	

Reasoning

11. MC The scatterplot that best represents the relationship between the amount of water consumed daily by a certain household for a number of days in summer and the daily temperature is:

A. 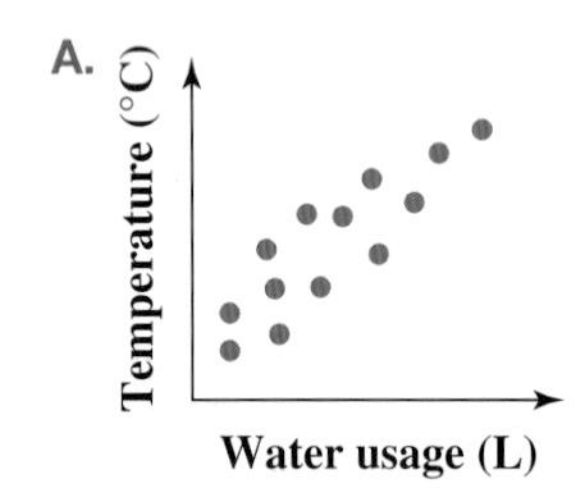

B. Water usage (L)
Temperature (°C)

C. 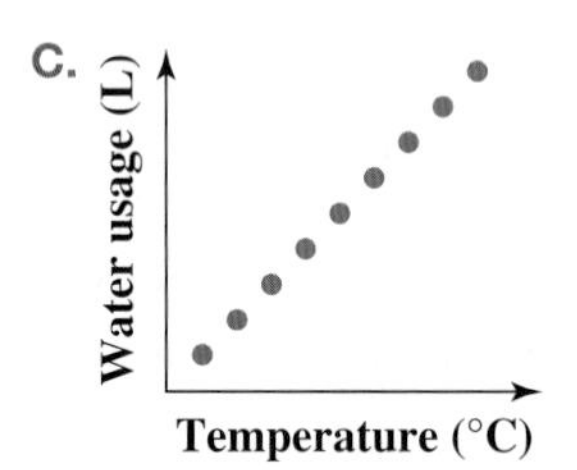

D. 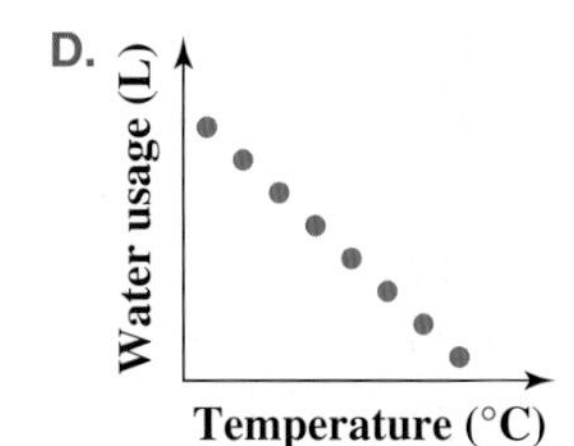

E. 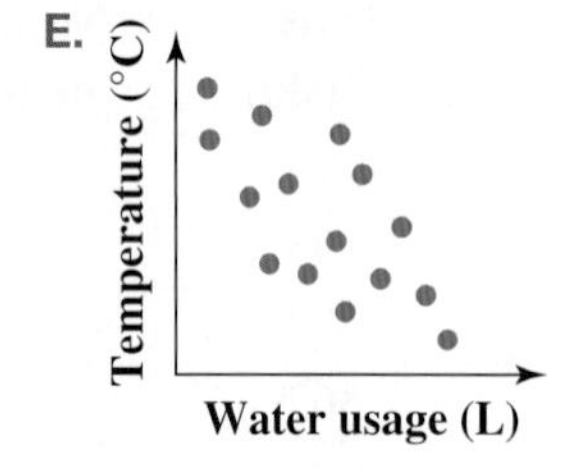

12. **MC** The scatterplot shows the number of sides and the sum of interior angles for a number of polygons.
Select the statement that is NOT true of the following statements.

A. The correlation between the number of sides and the angle sum of the polygon is perfectly linear.
B. The increase in the number of sides causes the increase in the size of the angle sum.
C. The number of sides depends on the sum of the angles.
D. The correlation between the two variables is positive.
E. There is a strong correlation between the two variables.

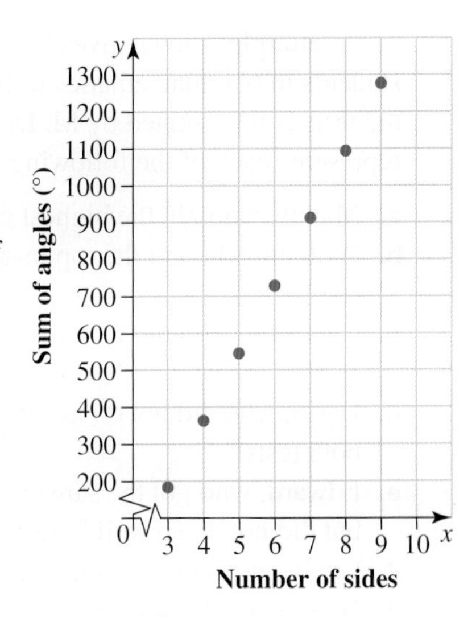

13. **MC** After studying a scatterplot, it was concluded that there was evidence that the greater the level of one variable, the smaller the level of the other variable. The scatterplot must have shown a:

A. strong, positive correlation
B. strong, negative correlation
C. moderate, positive correlation
D. moderate, negative correlation
E. weak, negative correlation

Problem solving

14. The table below gives the number of kicks and handballs obtained by the top 8 players in an AFL game.

Player	A	B	C	D	E	F	G	H
Number of kicks	20	27	21	19	17	18	21	22
Number of handballs	11	3	11	6	5	1	9	7

a. Represent this information on a scatterplot by using the x-axis as the number of kicks and the y-axis as the number of handballs.
b. State whether the scatterplot supports the claim that the more kicks a player obtains, the more handballs they give.

15. Each point on the scatterplot shows the time (in weeks) spent by a person on a healthy diet and the corresponding mass lost (in kg).
Study the scatterplot and state whether each of the following statements is true or false.

a. The number of weeks that the person stays on a diet is the independent variable.
b. The y-coordinates of the points represent the time spent by a person on a diet.
c. There is evidence to suggest that the longer the person stays on a diet, the greater the loss in mass.
d. The time spent on a diet is the only factor that contributes to the loss in mass.
e. The correlation between the number of weeks on a diet and the number of kilograms lost is positive.

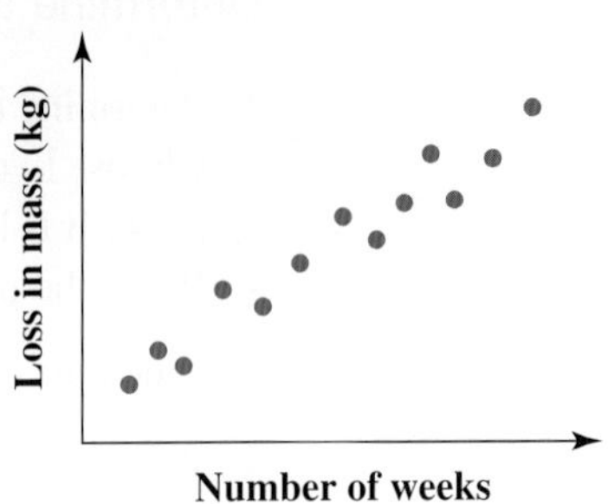

16. The scatterplot shown gives the marks obtained by students in two mathematics tests. Mardi's score in the tests is represented by M. Determine which point represents each of the following students.

a. Mandy, who got the highest mark in both tests.
b. William, who got the top mark in test 1 but not in test 2.
c. Charlotte, who did better on test 1 than Mardi but not as well in test 2.
d. Dario, who did not do as well as Charlotte in both tests.
e. Edward, who got the same mark as Mardi in test 2 but did not do so well in test 1.
f. Cindy, who got the same mark as Mardi for test 1 but did better than her for test 2.
g. Georgina, who was the lowest in test 1.
h. Harrison, who had the greatest discrepancy between his two marks.

13.3 Lines of best fit by eye

LEARNING INTENTION

At the end of this subtopic you should be able to:

- draw a line of best fit by eye
- determine the equation of the line of best fit
- use the line of best fit to make interpolation or extrapolation predictions.

13.3.1 Lines of best fit by eye

eles-4968

- A **line of best fit** is a line that follows the trend of the data in a scatter plot.
- A line of best fit is most appropriate for data with strong or moderate linear correlation.
- Drawing lines of best fit by eye is done by placing a line that:
  - represents the data trend
  - has an equal number of points above and below the line.

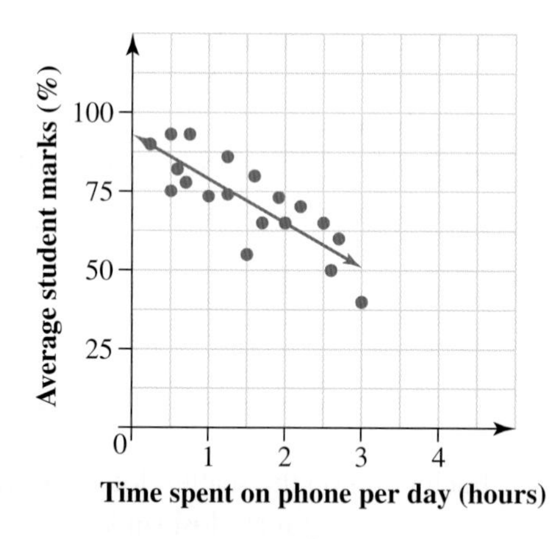

Determine the equation of a line of best fit by eye

To determine the equation of the line of best fit, follow these steps:

1. Choose two points on the line.
   Note: It is best to use two data points on the line if possible.
2. Write the points in the coordinate form (x_1, y_1) and (x_2, y_2).
3. Calculate the gradient using $m = \dfrac{y_2 - y_1}{x_2 - x_1}$.
4. Write the equation in the form $y = mx + c$ using the m found in step 3.
5. Substitute one coordinate into the equation and rearrange to find c.
6. Write the final equation, replacing x and y if needed.

WORKED EXAMPLE 4 Determining the equation for a line of best fit

The data in the table shows the cost of using the internet at a number of different internet cafes based on hours used per month.

Hours used per month	10	12	20	18	10	13	15	17	14	11
Total monthly cost ($)	15	18	30	32	18	20	22	23	22	18

a. **Construct a scatterplot of the data.**
b. **Draw the line of best fit by eye.**
c. **Determine the equation of the line of best fit in terms of the variables n (number of hours) and C (monthly cost).**

THINK

a. Draw the scatterplot placing the independent variable (hours used per month) on the horizontal axis and the dependent variable (total monthly cost) on the vertical axis. Label the axes.

WRITE/DRAW

a.

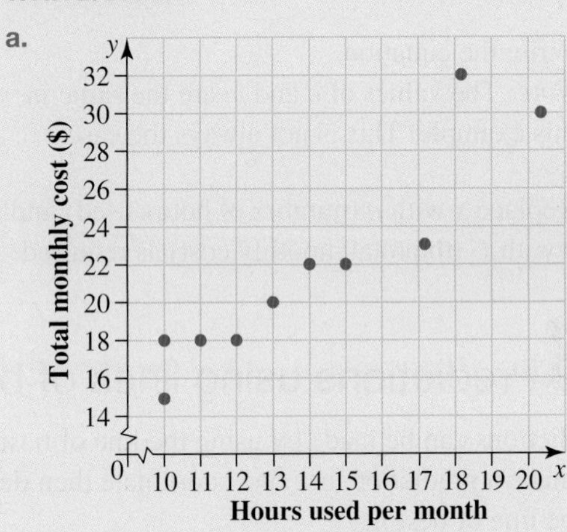

b. 1. Carefully analyse the scatterplot.

2. Position the line of best fit so there is approximately an equal number of data points on either side of the line and so that all points are close to the line.
Note: With the line of best fit, there is no single definite solution.

b.

y
32
30
28
26
24
22
20
18
16
14
0
10 11 12 13 14 15 16 17 18 19 20 21
x
(20, 30)
(13, 20)
Total monthly cost ($)
Hours used per month

c. 1. Select two points on the line that are not too close to each other.

c. Let $(x_1, y_1) = (13, 20)$ and $(x_2, y_2) = (20, 30)$.

2. Calculate the gradient of the line.

$$m = \frac{y_2 - y_1}{x_2 - x_1}$$

$$m = \frac{30 - 20}{20 - 13}$$

$$= \frac{10}{7}$$

3. Write the rule for the equation of a straight line.

$$y = mx + c$$

4. Substitute the known values into the equation.	$y = \frac{10}{7}x + c$
5. Substitute one pair of coordinates, say (13, 20) into the equation to evaluate c.	$20 = \frac{10}{7}(13) + c$ $c = 20 - \frac{130}{7}$ $= \frac{140 - 130}{7}$ $= \frac{10}{7}$
6. Write the equation. *Note:* The values of c and m are the same in this example. This is not always the case.	$y = \frac{10}{7}x + \frac{10}{7}$
7. Replace x with n (number of hours used) and y with C (the total monthly cost) as required.	$C = \frac{10}{7}n + \frac{10}{7}$

13.3.2 Predictions using lines of best fit

eles-4969

- Predictions can be made by using the line of best fit.
- To make a prediction, use one coordinate then determine the other coordinate by using:
  - the line of best fit
  - the equation of the line of best fit.
- Predictions will be made using:
  - **interpolation** if the prediction sits within the given data
  - **extrapolation** if the prediction sits outside the given data.

Interpolation vs extrapolation

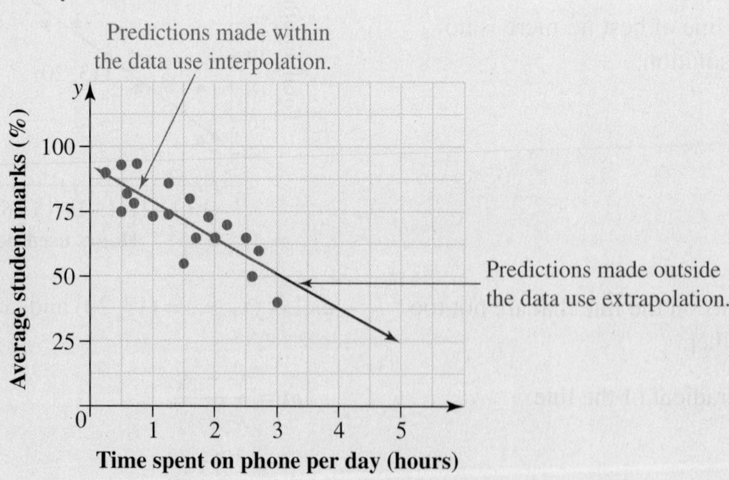

- Predictions will be reliable if they are made:
  - using interpolation
  - from data with a strong correlation
  - from a large number of data.

WORKED EXAMPLE 5 Making predictions using the line of best fit

Use the given scatterplot and line of best fit to predict:

a. the value of y when $x = 10$

b. the value of x when $y = 10$.

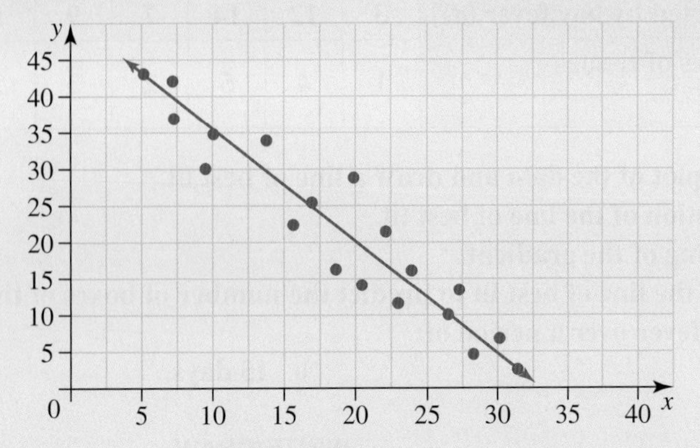

THINK | **WRITE/DRAW**

a. 1. Locate 10 on the x-axis and draw a vertical line until it meets with the line of best fit. From that point, draw a horizontal line to the y-axis. Read the value of y indicated by the horizontal line.

a.

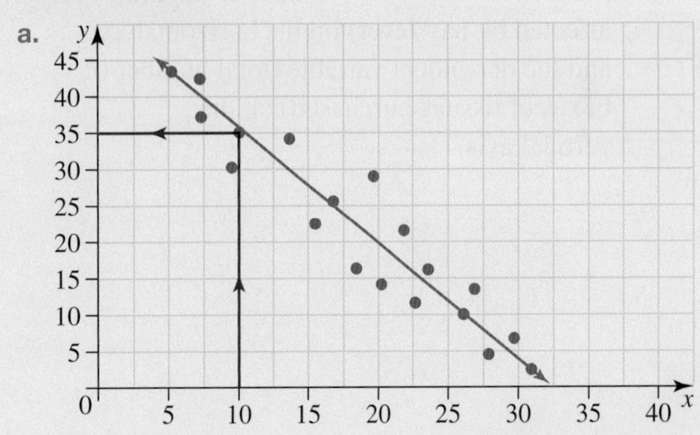

2. Write the answer.

When $x = 10$, y is predicted to be 35.

b. 1. Locate 10 on the y-axis and draw a horizontal line until it meets with the line of best fit. From that point draw a vertical line to the x-axis. Read the value of x indicated by the vertical line.

b.

2. Write the answer.

When $y = 10$, x is predicted to be 27.

WORKED EXAMPLE 6 Interpreting meaning and making predictions

The table below shows the number of boxes of tissues purchased by hay fever sufferers and the number of days affected by hay fever during the blooming season in spring.

Number of days affected by hay fever (d)	3	12	14	7	9	5	6	4	10	8
Total number of boxes of tissues purchased (T)	1	4	5	2	3	2	2	2	3	3

a. **Construct a scatterplot of the data and draw a line of best fit.**
b. **Determine the equation of the line of best fit.**
c. **Interpret the meaning of the gradient.**
d. **Use the equation of the line of best fit to predict the number of boxes of tissues purchased by people suffering from hay fever over a period of:**
 i. **11 days**
 ii. **15 days.**

THINK | **WRITE/DRAW**

a. 1. Draw the scatterplot showing the independent variable (number of days affected by hay fever) on the horizontal axis and the dependent variable (total number of boxes of tissues purchased) on the vertical axis.

a.

T
5
4
3
2
1
0
3 4 5 6 7 8 9 10 11 12 13 14 d
Total no. boxes of tissues purchased
No. days affected by hay fever

2. Position the line of best fit on the scatterplot so there is approximately an equal number of data points on either side of the line.

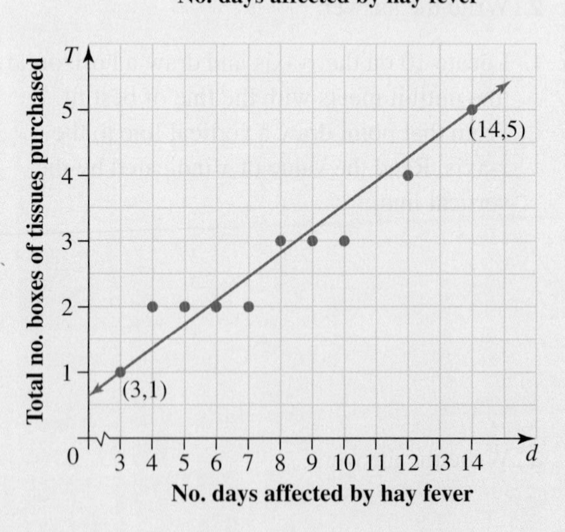

b. 1. Select two points on the line that are not too close to each other.

b. Let $(x_1, y_1) = (3, 1)$ and $(x_2, y_2) = (14, 5)$.

2. Calculate the gradient of the line.

$$m = \frac{y_2 - y_1}{x_2 - x_1}$$

$$m = \frac{5 - 1}{14 - 3} = \frac{4}{11}$$

3. Write the rule for the equation of a straight line.

$y = mx + c$

4. Substitute the known values into the equation, say (3, 1), into the equation to calculate c.

$$y = \frac{4}{11}x + c$$

$$1 = \frac{4}{11}(3) + c$$

$$c = 1 - \frac{12}{11}$$

$$= \frac{-1}{11}$$

5. Write the equation.

$$y = \frac{4}{11}x - \frac{1}{11}$$

6. Replace x with d (number of days with hay fever) and y with T (total number of boxes of tissues used) as required.

$$T = \frac{4}{11}d - \frac{1}{11}$$

c. Interpret the meaning of the gradient of the line of best fit.

c. The gradient indicates an increase in sales of tissues as the number of days affected by hay fever increases. A hay fever sufferer is using on average $\frac{4}{11}$ (or about 0.36) of a box of tissues per day.

d. i. 1. Substitute the value $d = 11$ into the equation and evaluate.

d. i. When $d = 11$,

$$T = \frac{4}{11} \times 11 - \frac{1}{11}$$

$$= 4 - \frac{1}{11}$$

$$= 3\frac{10}{11}$$

2. Interpret and write the answer.

In 11 days the hay fever sufferer will need 4 boxes of tissues.

ii. 1. Substitute the value $d = 15$ into the equation and evaluate.

ii. When $d = 15$,

$$T = \frac{4}{11} \times 15 - \frac{1}{11}$$

$$= \frac{60}{11} - \frac{1}{11}$$

$$= 5\frac{4}{11}$$

2. Interpret and write the answer.

In 15 days the hay fever sufferer will need about 6 boxes of tissues.

DISCUSSION

Why is extrapolation not considered to be reliable?

on Resources

eWorkbook Topic 13 Workbook (worksheets, code puzzle and project) (ewbk-2039)

Interactivities Individual pathway interactivity: Lines of best fit (int-4627)
Lines of best fit (int-6180)
Interpolation and extrapolation (int-6181)

Exercise 13.3 Lines of best fit by eye

learnON

Individual pathways

■ PRACTISE	■ CONSOLIDATE	■ MASTER
1, 2, 5, 8, 11	3, 6, 9, 12	4, 7, 10, 13

To answer questions online and to receive **immediate corrective feedback** and **fully worked solutions** for all questions, go to your learnON title at www.jacplus.com.au.

Fluency

1. WE4 The data in the table shows the distances travelled by 10 cars and the amount of petrol used for their journeys (to the nearest litre).

Distance travelled (km), d	52	36	83	12	44	67	74	23	56	95
Petrol used (L), P	7	5	9	2	7	9	12	3	8	14

a. Construct a scatterplot of the data and draw the line of best fit.
b. Determine the equation of the line of best fit in terms of the variables d (distance travelled) and P (petrol used).

2. WE5 Use the given scatterplot and line of best fit to predict:

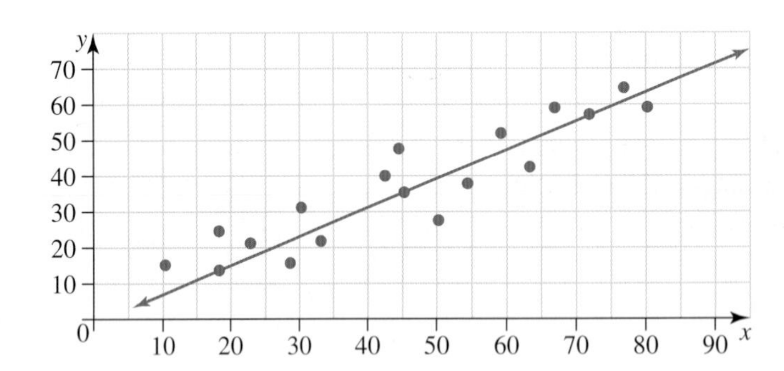

a. the value of y when $x = 45$
b. the value of x when $y = 15$.

3. Analyse the following graph.

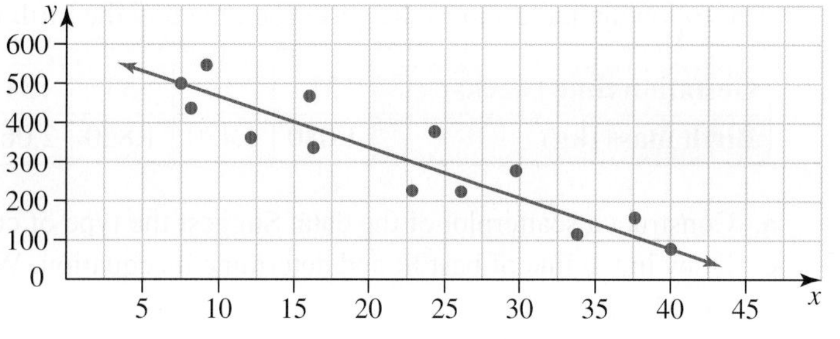

   a. Use the line of best fit to estimate the value of y when the value of x is:

      i. 7 ii. 22 iii. 36.

   b. Use the line of best fit to estimate the value of x when the value of y is:

      i. 120 ii. 260 iii. 480.

   c. Determine the equation of the line of best fit, if it is known that it passes through the points $(5, 530)$ and $(40, 75)$.

   d. Use the equation of the line to verify the values obtained from the graph in parts **a** and **b**.

4. A sample of ten Year 10 students who have part-time jobs was randomly selected. Each student was asked to state the average number of hours they work per week and their average weekly earnings (to the nearest dollar). The results are summarised in the table below.

Hours worked, *h*	4	8	15	18	10	5	12	16	14	6
Weekly earnings ($), *E*	23	47	93	122	56	33	74	110	78	35

   a. Construct a scatterplot of the data and draw the line of best fit.

   b. Write the equation of the line of best fit, in terms of variables h (hours worked) and E (weekly earnings).

   c. Interpret the meaning of the gradient.

Understanding

5. WE6 The following table shows the average weekly expenditure on food for households of various sizes.

Number of people in a household	1	2	4	7	5	4	3	5
Cost of food ($ per week)	70	100	150	165	150	140	120	155
Number of people in a household	2	4	6	5	3	1	4	
Cost of food ($ per week)	90	160	160	160	125	75	135	

   a. Construct a scatterplot of the data and draw in the line of best fit.

   b. Determine the equation of the line of best fit. Write it in terms of variables n (for the number of people in a household) and C (weekly cost of food).

   c. Interpret the meaning of the gradient.

   d. Use the equation of the line of best fit to predict the weekly food expenditure for a family of:

      i. 8 ii. 9 iii. 10.

6. The number of hours spent studying, and the percentage marks obtained by a group of students on a test are shown in this table.

Hours spent studying	45	30	90	60	105	65	90	80	55	75
Marks obtained	40	35	75	65	90	50	90	80	45	65

   a. State the values for marks obtained that can be used for interpolation.

   b. State the values for hours spent studying that can be used for interpolation.

7. The following table shows the gestation time and the birth mass of 10 babies.

Gestation time (weeks)	31	32	33	34	35	36	37	38	39	40
Birth mass (kg)	1.080	1.470	1.820	2.060	2.230	2.540	2.750	3.110	3.080	3.370

a. Construct a scatterplot of the data. Suggest the type of correlation shown by the scatterplot.
b. Draw in the line of best fit and determine its equation. Write it in terms of the variables t (gestation time) and M (birth mass).
c. Determine what the value of the gradient represents.
d. Although full term of gestation is considered to be 40 weeks, some pregnancies last longer. Use the equation obtained in part **b** to predict the birth mass of babies born after 41 and 42 weeks of gestation.
e. Many babies are born prematurely. Using the equation obtained in part **b**, predict the birth mass of a baby whose gestation time was 30 weeks.
f. Calculate their gestation time (to the nearest week), if the birth mass of the baby was 2.390 kg.

Reasoning

8. **MC** Consider the scatterplot shown.

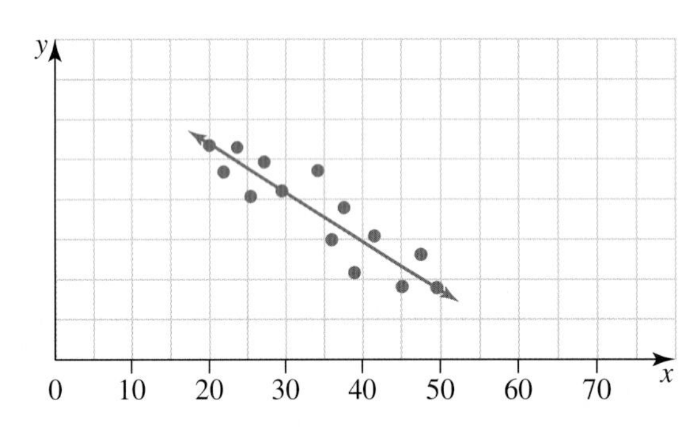

The line of best fit on the scatterplot is used to predict the values of y when $x = 15$, $x = 40$ and $x = 60$.

a. Interpolation would be used to predict the value of y when the value of x is:

A. 15 and 40 B. 15 and 60 C. 15 only D. 40 only E. 60 only

b. The prediction of the y-value(s) can be considered reliable when:

A. $x = 15$ and $x = 40$
B. $x = 15$, $x = 40$ and $x = 60$
C. $x = 40$
D. $x = 40$ and $x = 60$
E. $x = 60$

9. **MC** The scatterplot below is used to predict the value of y when $x = 300$. This prediction is:

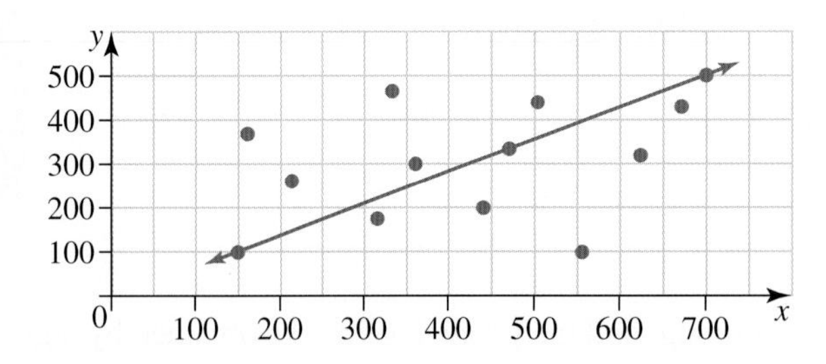

A. reliable, because it is obtained using interpolation
B. not reliable, because it is obtained using extrapolation
C. not reliable, because only x-values can be predicted with confidence
D. reliable because the scatterplot contains a large number of points
E. not reliable, because there is no correlation between x and y

10. As a part of her project, Rachel is growing a crystal. Every day she measures the crystal's mass using special laboratory scales and records it. The table below shows the results of her experiment.

Day number	1	2	3	4	5	8	9	10	11	12	15	16
Mass (g)	2.5	3.7	4.2	5.0	6.1	8.4	9.9	11.2	11.6	12.8	16.1	17.3

Measurements on days 6, 7, 13 and 14 are missing, since these were 2 consecutive weekends and, hence, Rachel did not have a chance to measure her crystal, which is kept in the school laboratory.

a. Construct the scatterplot of the data and draw in the line of best fit.
b. Determine the equation of the line of best fit. Write the equation, using variables d (day of the experiment) and M (mass of the crystal).
c. Interpret the meaning of the gradient.
d. For her report, Rachel would like to fill in the missing measurements (that is, the mass of the crystal on days 6, 7, 13 and 14). Use the equation of the line of best fit to help Rachel determine these measurements. Explain whether this is an example of interpolation or extrapolation.
e. Rachel needed to continue her experiment for 2 more days, but she fell ill and had to miss school. Help Rachel to predict the mass of the crystal on those two days (that is, days 17 and 18), using the equation of the line of best fit. Explain whether these predictions are reliable.

Problem solving

11. Ari was given a baby rabbit for his birthday. To monitor the rabbit's growth, Ari decided to measure it once a week.

The table below shows the length of the rabbit for various weeks.

Week number, n	1	2	3	4	6	8	10	13	14	17	20
Length (cm), l	20	21	23	24	25	30	32	35	36	37	39

a. Construct a scatterplot of the data.
b. Draw a line of best fit and determine its equation.
c. As can be seen from the table, Ari did not measure his rabbit on weeks 5, 7, 9, 11, 12, 15, 16, 18 and 19. Use the equation of the line of best fit to predict the length of the rabbit for those weeks.
d. Explain whether the predictions made in part c were an example of interpolation or extrapolation.
e. Predict the length of the rabbit in the next three weeks (that is, weeks 21–23), using the line of best fit from part c.
f. Explain whether the predictions that have been made in part e are reliable.

12. Laurie is training for the long jump, hoping to make the Australian Olympic team. His best jump each year is shown in the table below.

Age (a)	Best jump (B) (metres)
8	4.31
9	4.85
10	5.29
11	5.74
12	6.05
13	6.21
14	—
15	6.88
16	7.24
17	7.35
18	7.57

a. Plot the points generated by the table on a scatterplot.
b. Join the points generated with straight line segments.
c. Draw a line of best fit and determine its equation.
d. The next Olympic Games will occur when Laurie is 20 years old. Use the equation of the line of best fit to estimate Laurie's best jump that year and whether it will pass the qualifying mark of 8.1 metres.
e. Explain whether a line of best fit is a good way to predict future improvement in this situation. State the possible problems are there with using a line of best fit.
f. Olympic Games will also be held when Laurie is 24 years old and 28 years old. Using extrapolation, what length would you predict Laurie could jump at these two ages? Discuss whether this is realistic.
g. When Laurie was 14, he twisted a knee in training and did not compete for the whole season. In that year, a national junior championship was held. The winner of that championship jumped 6.5 metres. Use your line of best fit to predict whether Laurie would have won that championship.

13. Sam has a mean score of 88 per cent for his first nine tests of the semester. In order to receive an A^+ his score must be 90 per cent or higher. There is one test remaining for the semester. Explain whether it is possible for him to receive an A^+.

13.4 Linear regression using technology (10A)

LEARNING INTENTION

At the end of this subtopic you should be able to:
- display a scatter plot using technology
- determine the equation of the regression line using technology
- display a scatter plot with its regression line using technology
- use a regression line to make predictions.

13.4.1 Scatter plots using technology

eles-4970

- Scatter plots can be displayed using graphics calculators and spreadsheets.
- To display a scatter plot using technology:
  - first input the data with the independent variable in the first column and the dependent variable in the second column
  - then use the technology to draw the plot.

WORKED EXAMPLE 7 Displaying a scatter plot using technology

The following data shows the amount of time (hours) and the amount of distance walked (km) on a bushwalk. Display the data on a scatter plot using technology.

Time, hours (x)	1	2	3	4
Distance, km (y)	3.11	4.73	6.08	7.54

THINK

1. Determine which data will go in which column by identifying the independent and dependent variable.

WRITE

Time is the independent variable – it will go in the first column.
Distance is the dependent variable – it will go in the second column.

2. Input the data into the spreadsheet or calculator.

	A	B
1	Time	Distance
2	1	3.11
3	2	4.73
4	3	6.08
5	4	7.54
6		

3. Use the spreadsheet or calculator to create a scatter plot. Type your data into the spreadsheet and highlight all your data (including headings).
For Google Sheets:
Go to **Insert** and select **Chart**.
For Excel:
Go to **Insert**, select the scatter plot icon and choose the **Scatter** option.

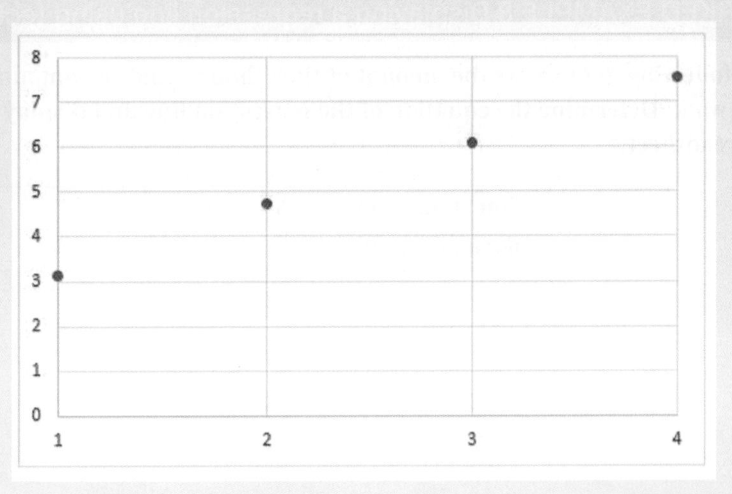

TI | THINK — **DISPLAY/WRITE**

1. In a new problem, on a Lists & Spreadsheet page, label column A as 'time' and B as 'distance'. Enter the data from the question.

2. Add a page by pressing CTRL then DOC and select:
   - 2: Add Graphs

   In the graphs page press:
   - MENU
   - 3: Graph Entry/Edit
   - 6: Scatter Plot

   On the first line (next to x ←) press VAR and select 'time' and on the second line (next to y ←) press VAR and select 'distance' then press ENTER.
   To adjust the screen press:
   - MENU
   - 4: Window / Zoom
   - 9: Zoom – Data

CASIO | THINK — **DISPLAY/WRITE**

1. From the menu, select Spreadsheet. Enter the data from the question in the columns A and B.

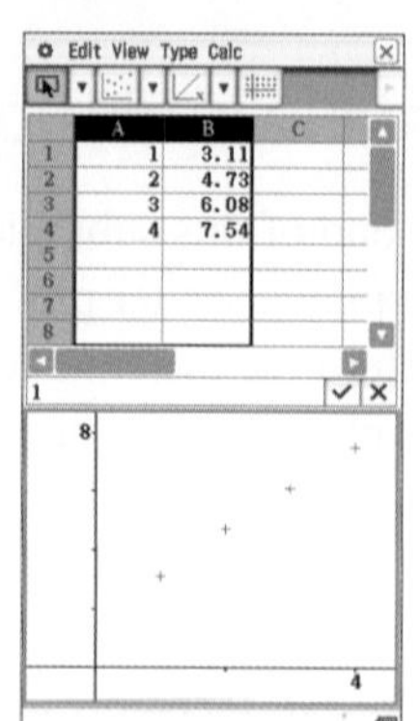

2. Highlight both columns A and B and tap:
   - Graph
   - Scatter

13.4.2 Regression lines using technology

eles-4971

- **Regression lines** are another name for lines of best fit.
- Using technology it is possible to calculate and sketch a regression line with more accuracy compared to using a line of best fit by eye.
- For regression lines using technology it is possible to:
  - draw the regression line
  - determine the equation of the line.
- Note that regression lines are **only valid** if the independent and dependent variables have a connection.

WORKED EXAMPLE 8 Displaying a regression line using technology

The following data shows the amount of time (hours) and the amount of distance walked (km) on a bushwalk. Determine the equation of the regression line and display the regression line using a spreadsheet.

Time, hours (x)	1	2	3	4
Distance, km (y)	3.11	4.73	6.08	7.54

THINK	WRITE
1. Use the scatter plot from Worked example 7. Start by sketching the regression line using the spreadsheet option called trendline.	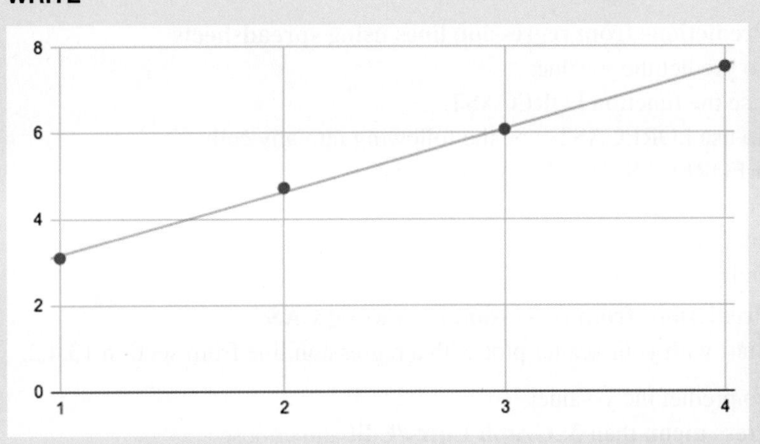
2. On the spreadsheet display the equation by using the spreadsheet option to show equations on the graph.	The equation is: $y = 1.46x + 1.71$

TI | THINK

1. Start with the scatter plot from Worked example 7. To determine the equation of the regression line, return to the spreadsheet page and click into the third column. Press:
   - MENU
   - 4: Statistics
   - 1: Stat Calculations
   - 3. Linear Regression (mx + b)

   In X List select 'time' and in Y List select 'distance', then click OK.

DISPLAY/WRITE

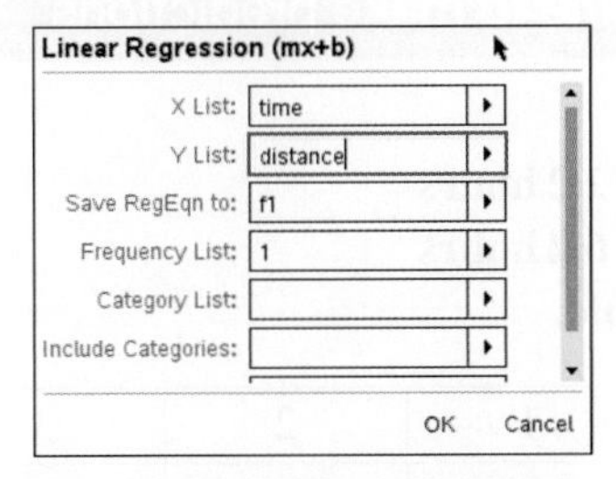

	distance	C	D	E
=				=LinRegM
1	3.11		Title	Linear R...
2	4.73		RegEqn	m*x+b
3	6.08		m	1.46
4	7.54		b	1.71
5			r²	0.999

E1 ="Linear Regression (mx+b)"

2. To plot the regression line, return to the graph page 1.2 and press:
   - MENU
   - 3: Graph Entry/Edit
   - 1: Function

   Press the up arrow to see 'f1 (x) = 1.464x + 1.705 Then press ENTER.

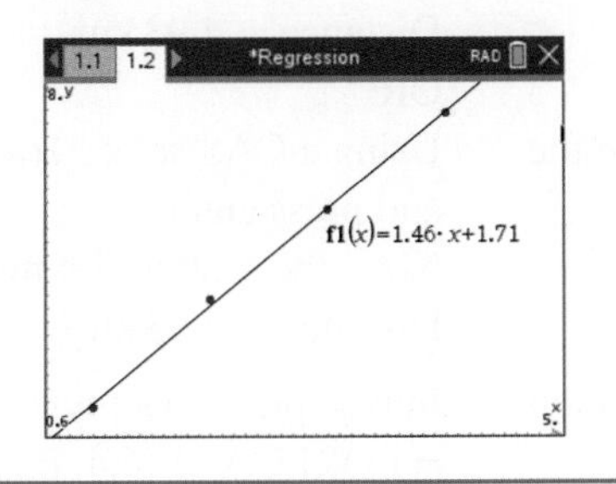

CASIO | THINK

Start with the scatter plot from Worked example 7. To determine the equation of the regression line, tap:
- Calc
- Regression
- Linear Reg

DISPLAY/WRITE

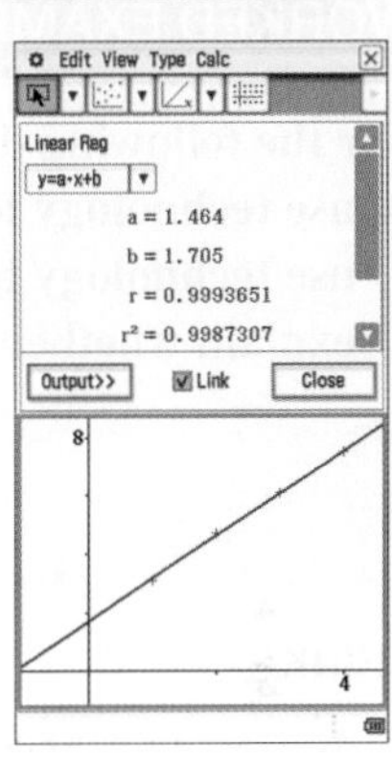

13.4.3 Using regression lines to make predictions

eles-4972

- The regression line can be used to make predictions by using technology.
- Predictions will be reliable if they are made: using interpolation, from data with a strong correlation and from a large number of data. (see section 13.3.2).

Digital technology

1. **Predictions from regression lines using spreadsheets**
   To predict the y-value:
   Use the function FORECAST.
   To use FORECAST type the following into any cell:
   = FORECAST(x-value, y-data, x-data)
   For example, for the x-value 3.5 type:
   = FORECAST (3.5, B2 : B6, A2 : A6)
   and find the y-value is 2.59.

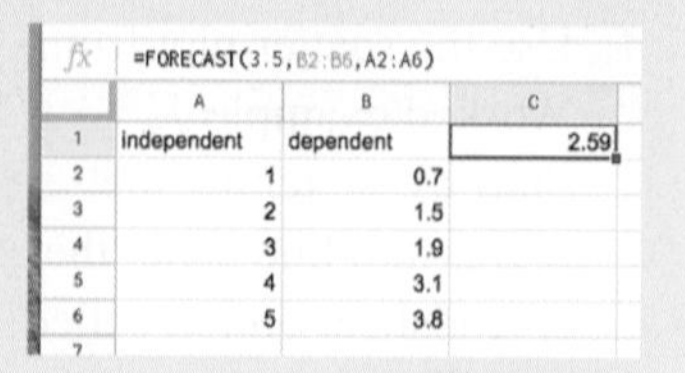

2. **Predictions from regression lines using CAS**
   Start with your scatter plot with a regression line from section 13.4.2.

   To predict the y-value:
   Press **menu** then **3: Graph Entry/Edit** and choose **6: Scatterplot** and press up to see s1. In s1 untick the blue box. Press **menu** then **5: Trace** then select **1: Graph Trace.** The type the x-value and press **enter.** The coordinate will appear.

   For the example, when x is 3.5, y is 2.59.

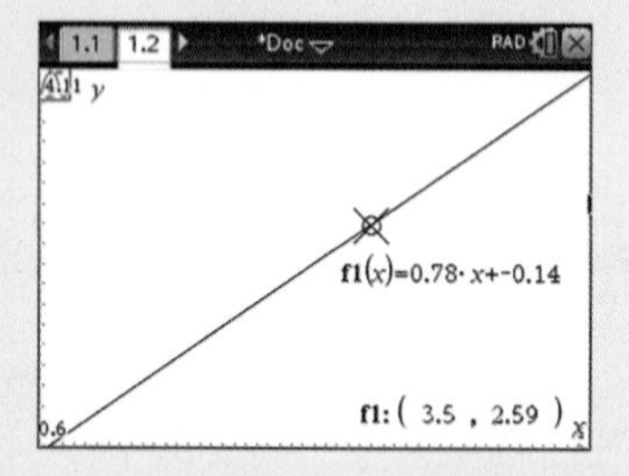

WORKED EXAMPLE 9 Using regression line to make predictions

For the following data:

a. use technology to predict the distance after 2.2 hours

b. use technology to predict the distance after 6.4 hours

c. explain whether these predictions are reliable.

Time, hours (x)	1	2	3	4
Distance, km (y)	3.11	4.73	6.08	7.54

THINK	WRITE
a. 1. Use the spreadsheet or CAS pages set up in Worked example 8.	See Worked example 8.
2. We need to calculate the distance value, which is the y-value. For a spreadsheet use the function FORECAST. OR For a CAS start on the graph page and use the Trace tool.	Into a spreadsheet type: = FORECAST (2.2, B2 : B5, A2 : A5) Note the order B2:B5 first and A2:A5 second. Distance = 4.93 km OR Using a CAS select Trace, type the value 2.2 and press enter. Note: the scatter plot must be turned off. Distance = 4.93 km
b. 1. We need to calculate the distance value which is the y-value. For a spreadsheet use the function FORECAST OR For a CAS start on the graph page and use the Trace tool.	Into a spreadsheet type: = FORECAST (6.4, B2 : B5, A2 : A5) Note the order B2:B5 first and A2:A5 second. Distance = 11.07 km OR Using a CAS select Trace, type the value 6.4 and press enter. Note: the scatter plot must be turned off. Distance = 11.07 km

c. Predictions that use interpolation are reliable and predictions that use extrapolation are not reliable.

Answer **a** is reliable because it uses interpolation.
Answer **b** is not reliable because it uses extrapolation.

on Resources

eWorkbook Topic 13 Workbook (worksheets, code puzzle and project) (ewbk-2039)

Exercise 13.4 Linear regression using technology (10A)

learnon

Individual pathways

■ PRACTISE	■ CONSOLIDATE	■ MASTER
1, 2, 3, 6, 11, 14	4, 7, 9, 12, 15	5, 8, 10, 13, 16

To answer questions online and to receive **immediate corrective feedback** and **fully worked solutions** for all questions, go to your learnON title at www.jacplus.com.au.

Fluency

1. WE 7 The following data shows the amount of time athletes spent training in preparation for a marathon and their finishing position in the race. Display the data on a scatter plot using technology.

Time, hours (x)	25	30	35
Finishing position (y)	15	11	8

2. WE 8 The following data shows the number of visitors to a store in a day and the profit of the store that day. Determine the equation of the regression line and display the regression line using either a spreadsheet or a CAS calculator.

Number of visitors (x)	80	85	94	101
Profit, dollars (y)	152	164	180	200

3. WE 9 Use the data from question 2 to answer the following questions.
   a. Use technology to predict the profit if there were 90 visitors.
   b. Use technology to predict the profit if there were 200 visitors. Round your answer to the nearest dollar.
   c. Explain if these predictions are reliable.

4. The following data shows how far away students live from school in kilometers and the hours those students spend in a car per week.

Distance from school, kilometres (x)	2	2.5	3	5
Hours spent in a car each week (y)	2.2	2.8	2.9	3.4

   a. Use technology to display the data on a scatter plot.
   b. Use technology to determine the equation of the regression line and display the regression line.
   c. Use technology to predict the hours spent in a car each week for a student that lives 6 km from school. Give your answer to one decimal place.

d. Use technology to predict the hours spent in a car each week for a student that lives 2.2 km from school. Give your answer to one decimal place.
e. Explain if these predictions are reliable.

5. The following data shows how many thousands of bees are in a hive and the amount of honey produced in that hive per year:

Number of bees ($\times 1000$) (x)	11	16	24	31
Honey produced, kilograms per year (y)	15	18	22	24

a. Use technology to display the data on a scatter plot.
b. Use technology to determine the equation of the regression line and display the regression line.
c. Use technology to predict the honey produced per year if a hive had 35 000 bees. Round the prediction to the nearest whole number.
d. Use technology to predict the honey produced per year if a hive had 18 000 bees. Round the prediction to the nearest whole number.
e. Explain if these predictions are reliable.

Understanding

6. The following data shows the temperature on certain days of the year. The days have been numbered like this: 1 January is 1, 2 January is 2 and so on for 365 days. Assume it is a non-leap year.

Day of the year (x)	1	5	12	20
Maximum temperature, °C (y)	32	33	38	42

a. Use technology to determine the equation of the regression line.
b. Use technology to predict the temperature on 15 March (day 75 of the year). Give your answer to one decimal place.
c. Explain if your answer to part b makes sense. Within your answer use the word *extrapolation*.

7. Chantal is a big fan of the Dugongs baseball team. The following data shows the number of games Chantal watched per year and the games won by the Dugongs per year.

Number of games watched per year (x)	8	12	16	20
Number of games won by the Dugongs (y)	10	11	15	16

a. Use technology to determine the equation of the regression line.
b. Use technology to predict the number of games won if Chantal watches 15 games. Give your answer to the nearest whole number.
c. Explain if your answer to part b makes sense. Explain your answer in terms of whether a fan watching a sports game has a connection to the outcome.

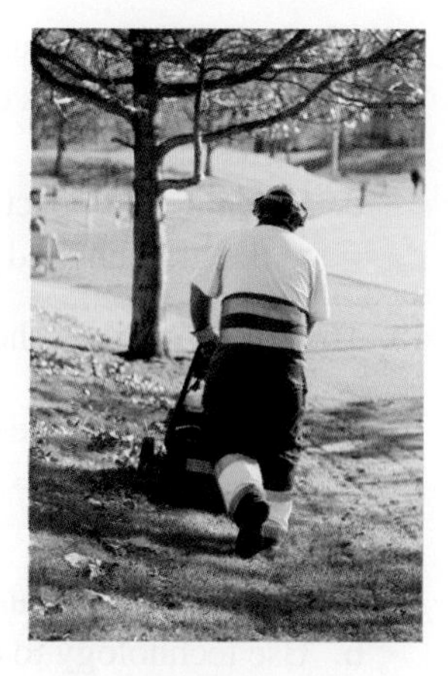

8. Below is data from four lawn mowing companies showing yard size and the cost of their most recent lawn mowing jobs:
Fred's mowing: 200 m^2 yard for \$80
Dial-a-gardner: 150 m^2 yard for \$75
Chopper chops limited: 50 m^2 yard for \$60
Landscapes-r-us: 400 m^2 yard for \$120
   a. Organise the data into a table and assign the x and y values.
   b. Explain how you have assigned the x and y values.
   c. Use technology to determine the equation of the regression line for this data.
   d. Based on your equation for the regression line from part **c**, estimate the call-out fee for lawn mowing.

9. MC Sally Miles is a world-famous pop star. By analysing Sally's tour data, the equation for a regression line is found that relates numbers of fans at a concert (x) to Sally's earning (y). The regression line equation is $y = 22x + 25144$. In the equation of the regression line, the number 22 represents:
   A. The amount Sally earns per fan at the concert.
   B. How much Sally earns per year.
   C. How much Sally would earn if she had 10 fans at her concert.
   D. The number of songs in Sally's playlist.
   E. How much Sally earns per month.

10. MC Regression lines are only valid for data where the independent variable has a connection to the dependent variable. Select which of the following would NOT have a valid regression line.
   A. Number of growing days and height of a sunflower.
   B. Average top speed of cars and years since cars were invented.
   C. Number of ice creams purchased and the price of ice cream.
   D. Amount of cheese eaten per capita and the number of injuries in an AFL season.
   E. Amount of cheese eaten per capita and the price of cheese.

Reasoning

11. Two friends, Yousef and Gavin, were having an eating competition. In the competition they both ate one, then two, then three apples and recorded their time. These were the results:
Yousef

Number of apples eaten (x)	1	2	3
Time taken, seconds (y)	46	67	124

Gavin

Number of apples eaten (x)	1	2	3
Time taken, seconds (y)	38	75	112

   a. Use technology to determine the equations of the regression lines for each set of data.
   b. Identify the gradients for each set of data.
   c. Compare the gradients. Explain what this information shows.
   d. Explain who won the apple-eating contest using the data.

12. The following data shows the time it took for five people to complete one lap of a BMX track.

Age, years (x)	13	14	15	16	25
Time, minutes (y)	2.5	2.2	1.9	1.8	1.2

a. Determine the equation of the regression line for this data.
b. The outlier in this data is the point $(25, 1.2)$. Remove this piece of data and find the equation of the new regression line.
c. Explain the impact on the equation of the regression line after removing an outlier. In your answer refer to the gradient and y-intercept.

13. The following data shows the time it took for packages of different weights to arrive in the post:

Weight, kilograms (x)	0.5	1.1	1.7	2.5	18
Time, days (y)	4	3	3	2	4

a. Determine the equation of the regression line for this data.
b. Use technology to draw the scatter plot for this data. Explain whether there is a correlation. Determine the type, direction and strength of the correlation.
c. There is an outlier in this data. Remove the outlier and determine the new regression line.
d. Use technology to draw the scatter plot for the data with the outlier removed. Explain whether there is a correlation; if so, describe the type, direction and strength.
e. Explain which regression line more accurately represents the data.

Problem solving

14. At the school athletics carnival Mr. Wall was in charge of recording the student year levels and jump heights for the winning jumps.

Year level (x)	7	8	9	10	11	12
Winning jump height, cm (y)	110	122	126		130	139

Mr Wall knows the regression line for this data is $y = 4.91x + 79.3$. Calculate the missing jump height.

15. Use the following data to answer these questions:

x	1	2	3	4
y	3	5	9	11

a. Determine the equation of the regression line for this data.
b. Determine what happens to the equation of the regression line if you double all the y values.
c. Determine the number that could be added to each y-value in the original data so that the regression line becomes $y = 2.8x + 5$. Show your working.

16. Answer the folllowing questions with full working.
a. Determine four data points that give a regression line equation of $y = 8x + 3$.
b. Now determine four different data points that still give a regression line equation of $y = 8x + 3$.

13.5 Time series

LEARNING INTENTION

At the end of this subtopic you should be able to:
- describe time series data using trends and patterns
- draw lines of best fit by eye and use it to make predictions for a time series

13.5.1 Describing time series

eles-4973

- **Time series** are a type of bivariate data with time as the independent variable. In other words, time series show time on the x-axis.
- To describe time series data use *trends* and *patterns*.
- Time series trends can be:
  - increasing or decreasing
  - linear or non-linear.

Increasing linear time series	Decreasing non-linear time series
An upwards slope from left to right with data that is approximately a straight line.	A downwards slope from left to right with data that is not a straight line.

- Time series patterns can be:
  - seasonal
  - cyclical
  - random.

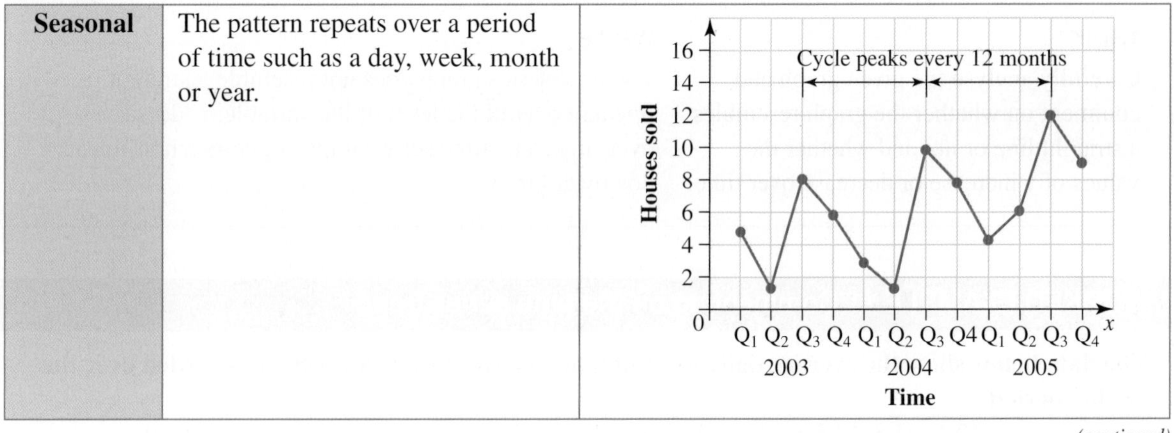

(continued)

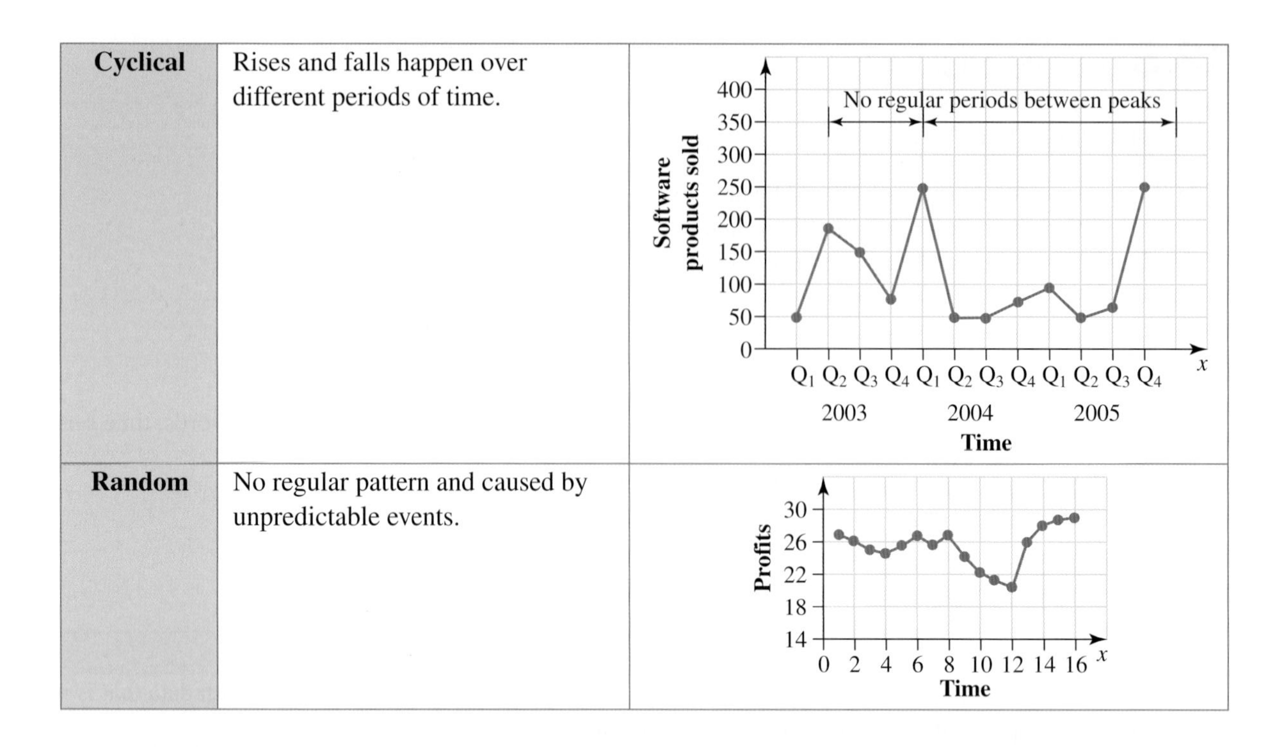

Cyclical	Rises and falls happen over different periods of time.	
Random	No regular pattern and caused by unpredictable events.	

WORKED EXAMPLE 10 Classifying the time series trend

Classify the trend suggested by the time series graph shown as being linear or non-linear, and upward, downward or no trend.

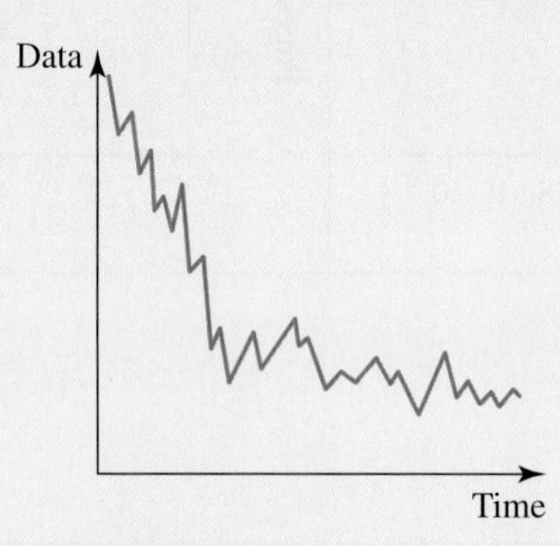

THINK

Carefully analyse the given graph and comment on whether the graph resembles a straight line or not and whether the values of y increase or decrease over time.

WRITE

The time series graph does not resemble a straight line and overall the level of the variable, y, decreases over time. The time series graph suggests a non-linear downward trend.

WORKED EXAMPLE 11 Commenting on the time series trend

The data below show the average daily mass of a person (to the nearest 100 g), recorded over the 28-day period.

63.6, 63.8, 63.5, 63.7, 63.2, 63.0, 62.8, 63.3, 63.1, 62.7, 62.6, 62.5, 62.9, 63.0, 63.1, 62.9, 62.6, 62.8, 63.0, 62.6, 62.5, 62.1, 61.8, 62.2, 62.0, 61.7, 61.5, 61.2

a. **Plot these masses as a time series graph.**

b. **Comment on the trend.**

THINK

a. 1. Draw the points on a scatterplot with time on the horizontal axis and mass on the vertical axis.

2. Join the points with straight line segments to create a time series plot.

WRITE/DRAW

a.

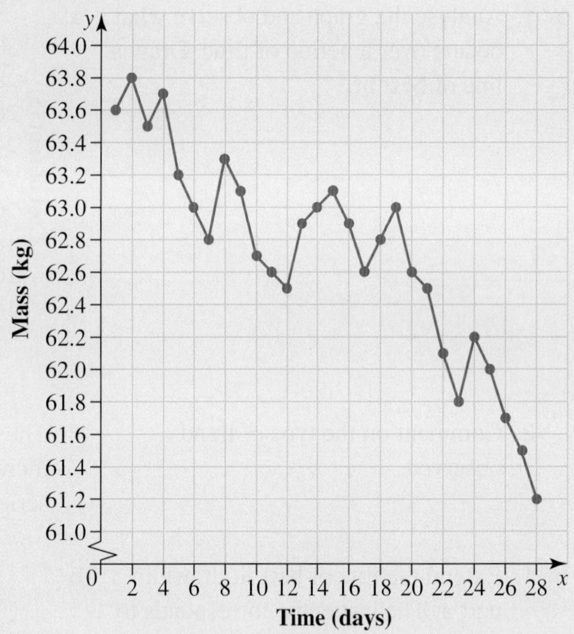

b. Carefully analyse the given graph and comment on whether the graph resembles a straight line or not and whether the values of y (in this case, mass) increase or decrease over time.

b. The graph resembles a straight line that slopes downwards from left to right (that is, mass decreases with increase in time). Although a person's mass fluctuates daily, the time series graph suggests a downward trend. That is, overall, the person's mass has decreased over the 28-day period.

13.5.2 Time series lines of best fit by eye

eles-4974

- It is possible to draw lines of best fit by eye for time series (see subtopic 13.3).
- Lines of best fit can be used to make predictions.
  - For interpolations, the predictions are reliable.
  - For extrapolations, the predictions are not very reliable since there is an assumption that the trend will continue.

WORKED EXAMPLE 12 Making predictions using a line of best fit

The graph at right shows the average cost of renting a one-bedroom flat, as recorded over a 10-year period.

a. If appropriate, draw in a line of best fit and comment on the type of the trend.

b. Assuming that the current trend will continue, use the line of best fit to predict the cost of rent in 5 years' time.

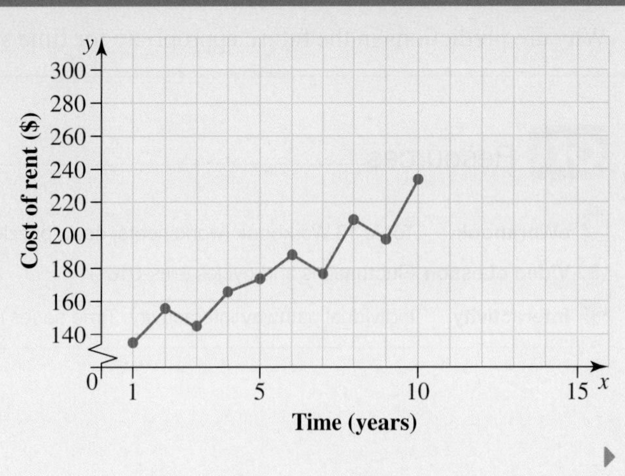

THINK	WRITE/DRAW
a. 1. Analyse the graph and observe what occurs over a period of time. Draw a line of best fit.	**a.**  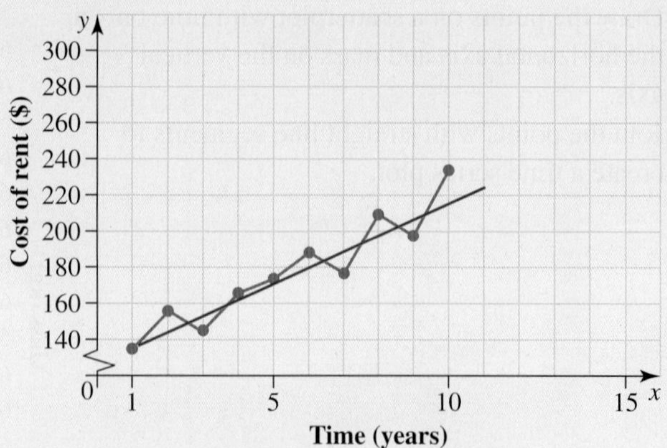
2. Comment on the type of trend observed.	The graph illustrates that the cost of rent increases steadily over the years. The time series graph indicates an upward linear trend.
b. 1. Extend the line of best fit drawn in part **a**. The last entry corresponds to the 10th year and we need to predict the cost of rent in 5 years' time; that is, in the 15th year. **2.** Locate the 15th year on the time axis and draw a vertical line until it meets with the line of best fit. From the trend line (line of best fit) draw a horizontal line to the cost axis.	**b.** 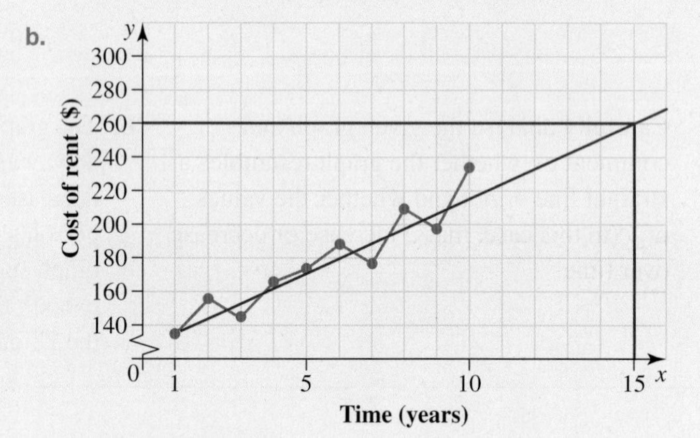
3. Read the cost from the vertical axis.	Cost of rent = \$260
4. Write the answer.	Assuming that the cost of rent will continue to increase at the present rate, in 5 years we can expect the cost of rent to reach \$260 per week.

DISCUSSION

Why are predictions in the future appropriate for time series even though they involve extrapolation?

 Resources

eWorkbook Topic 13 Workbook (worksheets, code puzzle and project) (ewbk-2039)

Video eLesson Fluctuations and cycles (eles-0181)

Interactivity Individual pathway interactivity: Time series (int-4628)

Exercise 13.5 Time series

learnON

Individual pathways

■ PRACTISE	■ CONSOLIDATE	■ MASTER
1, 4, 8, 11	2, 5, 7, 9, 12	3, 6, 10, 13

To answer questions online and to receive **immediate corrective feedback** and **fully worked solutions** for all questions, go to your learnON title at www.jacplus.com.au.

Fluency

1. **WE10** For questions 1 and 2, classify the trend suggested by each time series graph as being linear or non-linear, and upward, downward or stationary in the mean (no trend).

a.

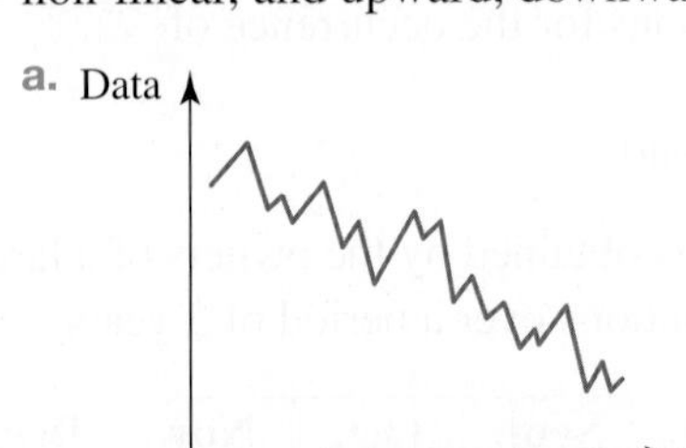

b.

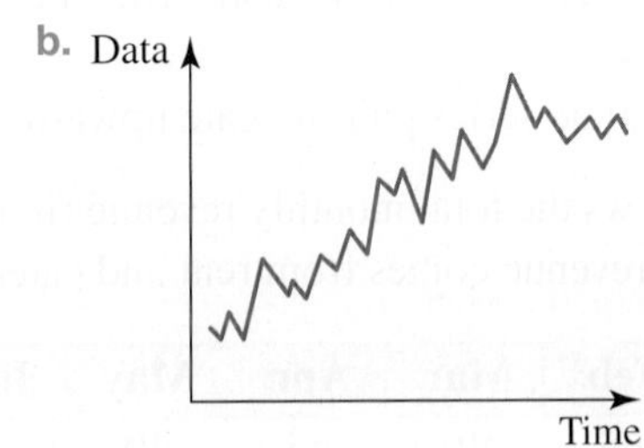

c.

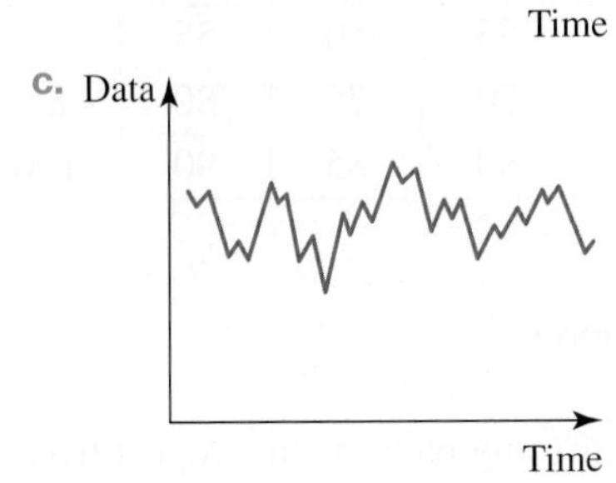

d.

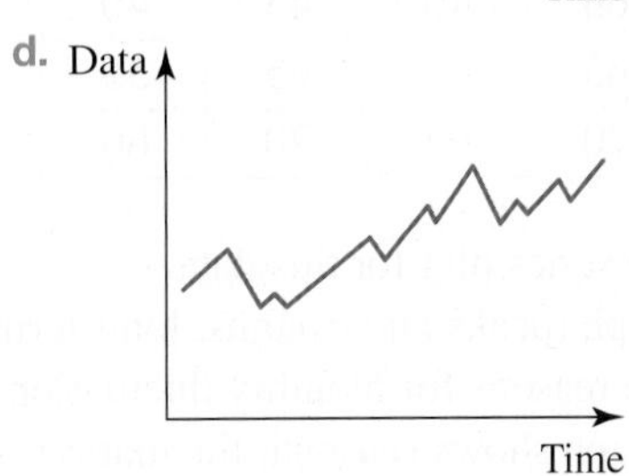

2. a.

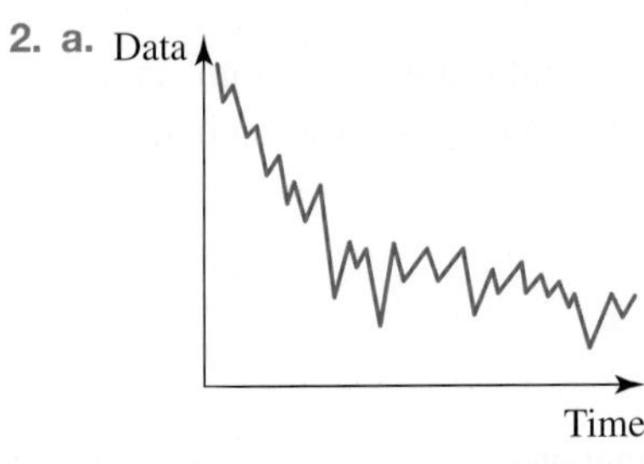

b.

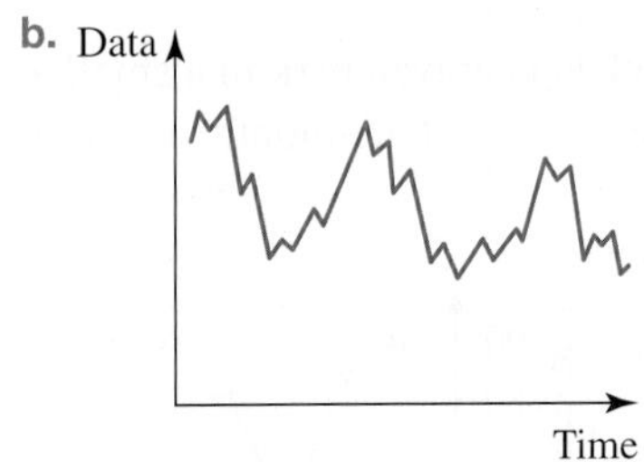

c.

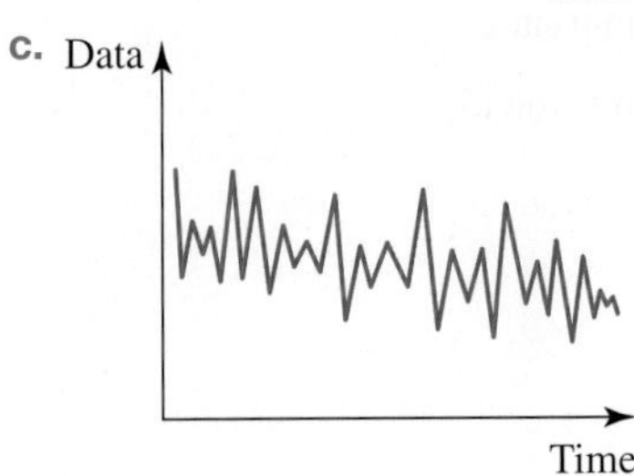

d.

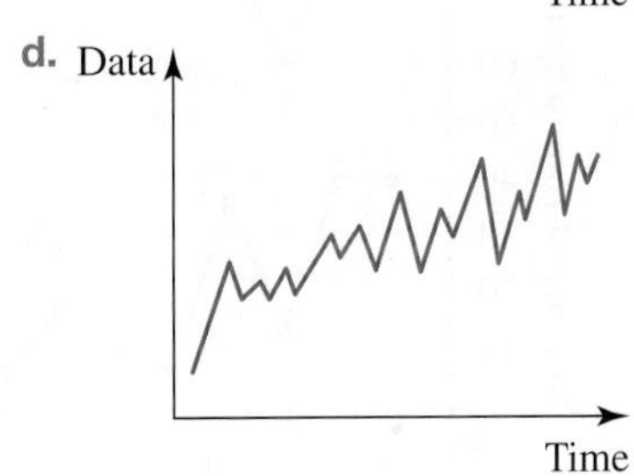

3. **WE11** The data below show the average daily temperatures recorded in June.

17.6, 17.4, 18.0, 17.2, 17.5, 16.9, 16.3, 17.1, 16.9, 16.2, 16.0, 16.6, 16.1, 15.4, 15.1,
15.5, 16.0, 16.0, 15.4, 15.2, 15.0, 15.5, 15.1, 14.8, 15.3, 14.9, 14.6, 14.4, 15.0, 14.2

a. Plot these temperatures as a time series graph.
b. Comment on the trend.

Understanding

4. The data below show the quarterly sales (in thousands of dollars) recorded by the owner of a sheepskin product store over a period of 4 years.

Quarter	2006	2007	2008	2009
1	57	59	50	52
2	100	102	98	100
3	125	127	120	124
4	74	70	72	73

a. Plot the time series.
b. The time series plot displays seasonal fluctuations of period 4 (since there are four quarters). Explain in your own words what this means. Also write one or two possible reasons for the occurrence of these fluctuations.
c. Determine if the time series plot indicate upward, downward or no trend.

5. The table below shows the total monthly revenue (in thousands of dollars) obtained by the owners of a large reception hall. The revenue comes from rent and catering for various functions over a period of 3 years.

	Jan.	Feb.	Mar.	Apr.	May	June	July	Aug.	Sept.	Oct.	Nov.	Dec.
2007	60	65	40	45	40	50	45	50	55	50	55	70
2008	70	65	60	65	55	60	60	65	70	75	80	85
2009	80	70	65	70	60	65	70	75	80	85	90	100

a. Construct a time series plot for this data.
b. Describe the graph (peaks and troughs, long-term trend, any other patterns).
c. Suggest possible reasons for monthly fluctuations.
d. Explain if the graph shows seasonal fluctuations over 12 months. Discuss any patterns that repeat from year to year.

6. The owner of a motel and caravan park in a small town keeps records of the total number of rooms and total number of camp sites occupied per month. The time series plots based on his records are shown below.

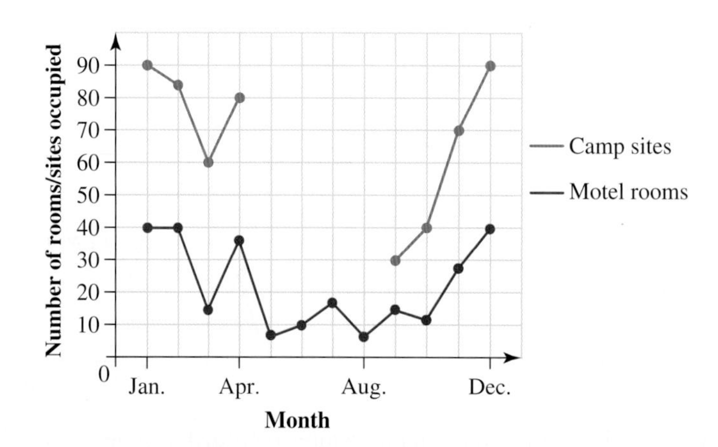

a. Describe each graph, discussing general trend, peaks and troughs and so on. Explain particular features of the graphs and give possible reasons.
b. Compare the two graphs and write a short paragraph commenting on any similarities and differences between them.

7. WE12 The graph shows enrolments in the Health and Nutrition course at a local college over a 10-year period.

a. If appropriate, draw in a line of best fit and comment on the type of the trend.
b. Assuming that the trend will continue, use the line of best fit to predict the enrolment for the course in 5 years' time; that is, in the 15th year.

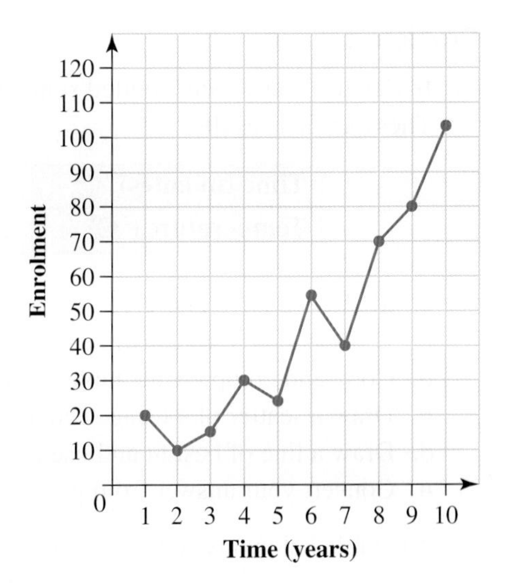

Reasoning

8. In June a new childcare centre was opened. The number of children attending full time (according to the enrolment at the beginning of each month) during the first year of operation is shown in the table.

June	July	Aug.	Sept.	Oct.	Nov.	Dec.	Jan.	Feb.	Mar.	Apr.	May
6	8	7	9	10	9	12	10	11	13	12	14

a. Plot this time series (*Hint*: Let June = 1, July = 2 etc.)
b. Justify if the childcare business is going well.
c. Draw a line of best fit.
d. Use your line of best fit to predict the enrolment in the centre during the second year of operation at the beginning of:
   i. August
   ii. January.
e. State any assumptions that you have made.

9. The graph shows the monthly sales of a certain book since its publication. Explain in your own words why linear trend forecasting of the future sales of this book is not appropriate.

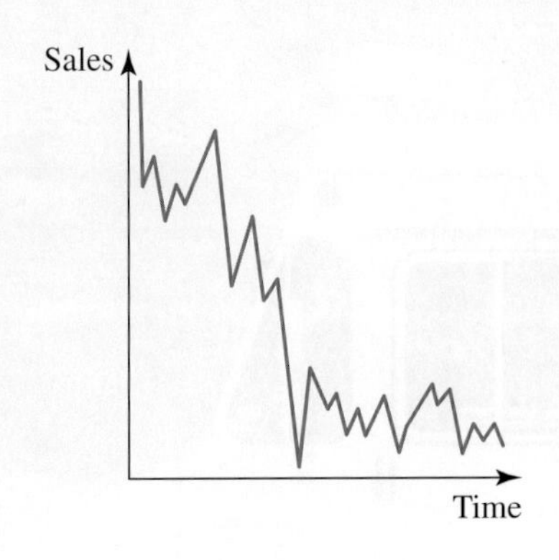

10. In the world of investing this phrase is commonly used when talking about investments: "Past performance is not an indicator of future returns."

a. Explain what this phrase means.
b. Explain why this phrase is true using the term *extrapolation*.

Problem solving

11. In Science class Melita boiled some water and then recorded the temperature of the water over ten minutes. These are her results:

Time (minutes)	0	1.5	3	4.5	6	7.5	10
Temperature (°C)	100	95	88	74	65	60	52

a. Melita wants to convert the time from minutes into seconds. She starts by converting 1.5 minutes to 150 seconds. Explain what she did wrong and find the correct number in seconds.
b. Copy and complete the table, changing the time in minutes to time in seconds.
c. Draw a scatter plot using seconds as the time scale.
d. Draw a line of best fit and use it to predict the time, in seconds, when the water will reach 20°C.
e. Convert your answer from part e back to minutes.

12. The table below gives the quarterly sales figures for a second-hand car dealer over a three-year period.

Year	Q1	Q2	Q3	Q4
2012	75	65	92	99
2013	91	79	115	114
2014	93	85	136	118

a. Represent this data on a time series plot.
b. Briefly describe how the car sales have altered over the time period.
c. Discuss if it appears that the car dealer can sell more cars in a particular period each year.

13. Jasper owns an ice-cream truck.
- In summer 2020/21 he sold 1536 ice-creams.
- In Autumn 2021 he sold one-quarter of that number.
- In Winter 2021 he sold one-eighth of that number
- In Spring 2021 he sold one-third of that number
- From Summer 2021/22 until Spring 2022 his sales doubled from the season in the previous year.

a. Represent his sales from Summer 2020/21 to Spring 2022 on a scatter plot.
b. Describe the trend and the patterns in the data.

13.6 Review

13.6.1 Topic summary

BIVARIATE DATA

Properties of bivariate data

- **Bivariate data** can be displayed, analysed and used to make predictions.
- **Types of variables:**
  - Independent (experimental or explanatory variable): not impacted by the other variable.
  - Dependent (response variable): impacted by the other variable.

Time series scatter plots

- A specific type of bivariate data where time is the independent variable.
- Describe time series by:
  - trends
  - patterns.
- Time series patterns can be:
  - seasonal
  - cyclical
  - random.

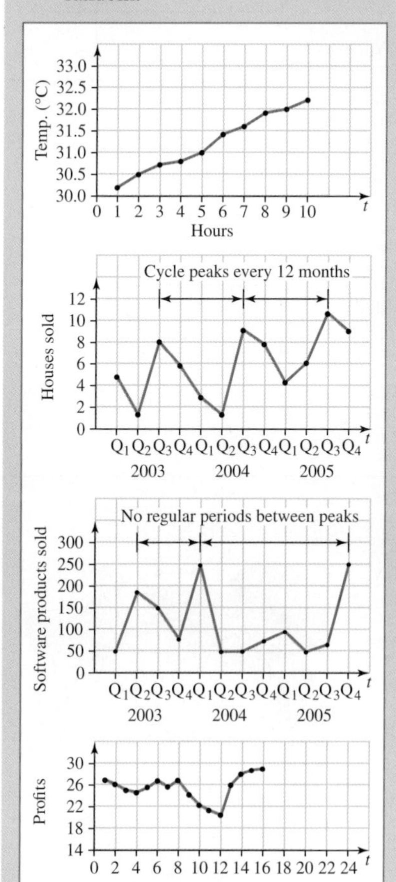

Representing the data

- **Scatter plots** can be created by hand or using technology (CAS or Excel).
- The independent variable is placed on the x-axis and the dependent variable on the y-axis.

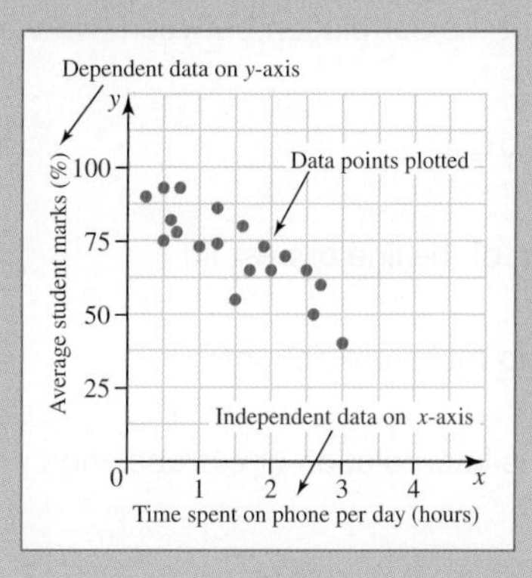

Line of best fit

- A line that follows the trend of the data in a scatter plot.
- It is most appropriate for data with strong or moderate linear correlation.
- Can be sketched as a line of best fit by eye or a regression line using technology.
  - The equation for the line can be found by using the gradient and equation of the straight line.
  - The line can be used to make predictions.
- Regression lines are only valid if the independent and dependent variables have a connection.

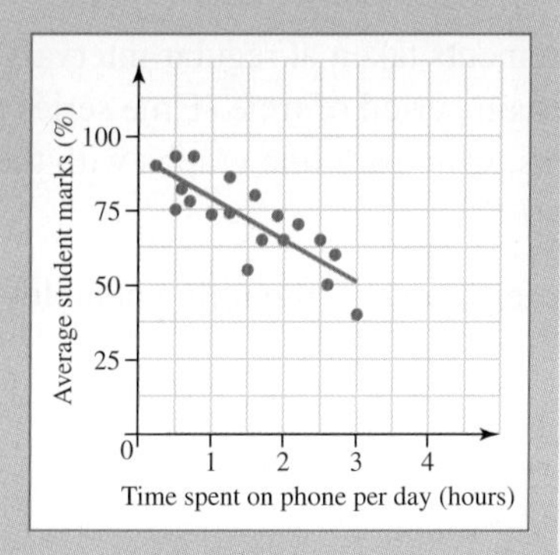

Correlations from scatter plots

- Correlation is a way of describing a connection between variables in a bivariate data set.
- Correlation between the two variables will have:
  - a type (linear or non-linear)
  - a direction (positive or negative)
  - a strength (strong, moderate or weak).
- Correlations can be used to make conclusions.
- There is no correlation if the data are spread out across the plot with no clear pattern.

Interpolation and extrapolation

- Interpolation and extrapolation can be used to make predictions.
- **Interpolation:**
  - is more reliable from a large number of data
  - used if the prediction sits within the given data.
- **Extrapolation:**
  - assumes the trend will continue
  - used if the prediction sits outside the given data.

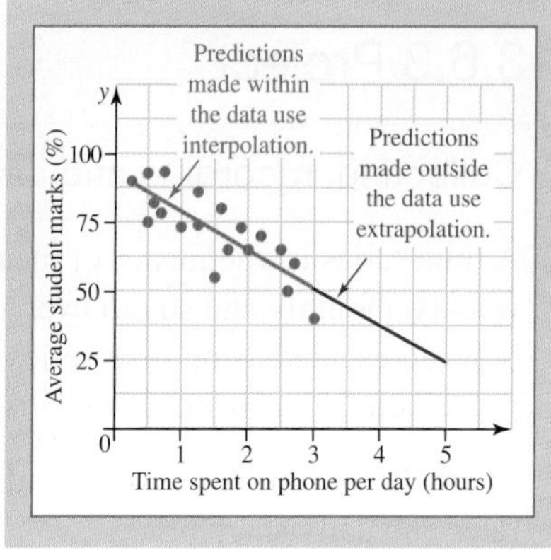

13.6.2 Success criteria

Tick the column to indicate that you have completed the subtopic and how well you have understood it using the traffic light system.

(**Green:** I understand; **Yellow:** I can do it with help; **Red:** I do not understand)

Subtopic	Success criteria	Green	Yellow	Red
13.2	I can recognise the independent and dependent variables in bivariate data.			
	I can represent bivariate data using a scatter plot.			
	I can describe the correlation between two variables in a bivariate data set.			
	I can draw conclusions about the correlation between two variables in a bivariate data set.			
13.3	I can draw a line of best fit by eye.			
	I can determine the equation of the line of best fit.			
	I can use the line of best fit to make predictions.			
	I can identify if a prediction is interpolation or extrapolation.			
13.4	I can display a scatter plot using technology.			
	I can determine the equation of the regression line using technology.			
	I can display a scatter plot with its regression line using technology.			
	I can use a regression line to make predictions.			
13.5	I can describe time series data using trends and patterns.			
	I can draw a line of best fit by eye and use it to make predictions for a time series.			

13.6.3 Project

Collecting, recording and analysing data over time

A time series is a sequence of measurements taken at regular intervals (daily, weekly, monthly and so on) over a certain period of time. Time series are best represented using time-series plots, which are line graphs with the time plotted on the horizontal axis.

Examples of time series include daily temperature, monthly unemployment rates and daily share prices.

When data are recorded on a regular basis, the value of the variable may go up and down in what seems to be an erratic pattern. These are called fluctuations. However, over a long period of time, the time series usually suggests a certain trend. These trends can be classified as being linear or non-linear, and upward, downward or stationary (no trend).

Time series are often used for forecasting, that is, making predictions about the future. The predictions made with the help of time series are always based on the assumption that the observed trend will continue in the future.

1. Choose a subject that is of interest to you and that can be observed and measured during one day or over the period of a week or more. (Suitable subjects are shown in the list below.)
2. Prepare a table for recording your results. Select appropriate regular time intervals. An example is shown below.

Time	8 am	9 am	10 am	11 am	12 pm	1 pm	2 pm	3 pm	4 pm	5 pm
Pulse rate										

3. Take your measurements at the selected time intervals and record them in the table.
4. Use your data to plot the time series. You can use software such as Excel or draw the scatterplot by hand.
5. Describe the graph and comment on its trend.
6. If appropriate, draw a line of best fit and predict the next few data values.
7. Take the actual measurements during the hours you have made predictions for. Compare the predictions with the actual measurements. Were your predictions good? Give reasons.

Here are some suitable subjects for data observation and recording:

- minimum and maximum temperatures each day for 2 weeks (use the TV news or online data as resources)
- the value of a stock on the share market (e.g. Telstra, Wesfarmers and Rio Tinto)
- your pulse over 12 hours (ask your teacher how to do this or check on the internet)
- the value of sales each day at the school canteen
- the number of students absent each day
- the position of a song in the Top 40 over a number of weeks
- petrol prices each day for 2 weeks
- other measurements (check with your teacher)
- world population statistics over time.

on Resources

eWorkbook Topic 13 Workbook (worksheets, code puzzle and project) (ewbk-2039)

Interactivities Crossword (int-2887)
Sudoku puzzle (int-3600)

Exercise 13.6 Review questions

learnON

To answer questions online and to receive **immediate corrective feedback** and **fully worked solutions** for all questions, go to your learnON title at www.jacplus.com.au.

Fluency

1. As preparation for a Mathematics test, a group of 20 students was given a revision sheet containing 60 questions. The table below shows the number of questions from the revision sheet successfully completed by each student and the mark, out of 100, of that student on the test.

Number of questions	9	12	37	60	55	40	10	25	50	48	60
Test result	18	21	52	95	100	67	15	50	97	85	89
Number of questions	50	48	35	29	19	44	49	20	16	58	52
Test result	97	85	62	54	30	70	82	37	28	99	80

a. State which of the variables is dependent and which is independent.
b. Construct a scatterplot of the data.
c. State the type of correlation between the two variables suggested by the scatterplot and draw a corresponding conclusion.
d. Suggest why the relationship is not perfectly linear.

2. Use the line of best fit shown on the graph to answer the following questions.

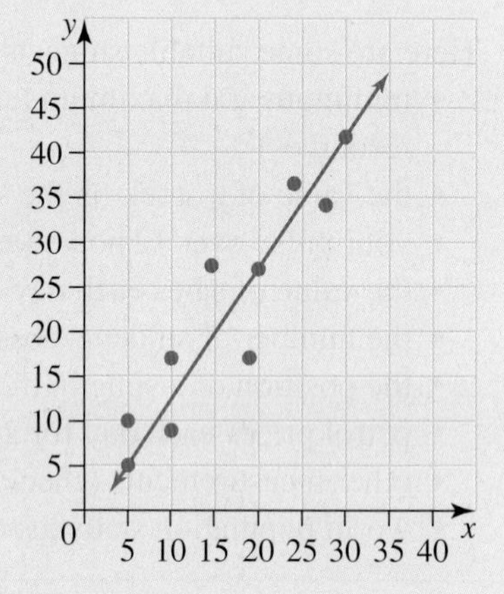

a. Predict the value of y, when the value of x is:
i. 10 ii. 35.
b. Predict the value of x, when the value of y is:
i. 15 ii. 30.
c. Determine the equation of a line of best fit if it is known that it passes through the points $(5, 5)$ and $(20, 27)$.
d. Use the equation of the line to algebraically verify the values obtained from the graph in parts **a** and **b**.

3. The graph shows the number of occupants of a large nursing home over the last 14 years.

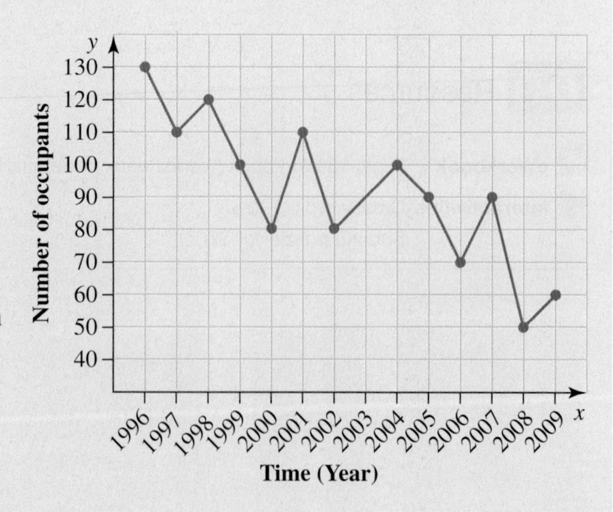

a. Comment on the type of trend displayed.
b. Explain why it is appropriate to draw in a line of best fit.
c. Draw a line of best fit and use it to predict the number of occupants in the nursing home in 3 years time.
d. State the assumption that have been made when predicting figures for part **c**.

4. The table below shows the advertised sale price ($'000) and the land size $\left(m^2\right)$ for ten vacant blocks of land.

Land size $\left(m^2\right)$	Sale price ($'000)
632	36
1560	58
800	40
1190	44
770	41
1250	52
1090	43
1780	75
1740	72
920	43

a. Construct a scatterplot and determine the equation of the line of best fit.
b. State what the gradient represents.
c. Using the line of best fit, predict the approximate sale price, to the nearest thousand dollars for a block of land with an area of $1600\,m^2$.
d. Using the line of best fit, predict the approximate land size, to the nearest 10 square metres, you could purchase with $50 000.

5. The table below shows, for fifteen students, the amount of pocket money they receive and spend at the school canteen in an average week.

Pocket money ($)	Canteen spending ($)
30	16
40	17
15	12
25	14
40	16
15	14
30	16
30	17
25	15
15	13
50	19
20	14
35	17
20	15
10	13

a. Construct a scatterplot and determine the equation of the line of best fit.
b. State what the gradient represents.
c. Using your line of best fit, predict the amount of money spent at the canteen for a student receiving $45 pocket money a week.
d. Using your line of best fit, predict the amount of money spent at the canteen by a student who receives $100 pocket money each week. Explain if this seems reasonable.

6. The table below shows, for 10 ballet students, the number of hours a week spent training and the number of pirouettes in a row they can complete.

Training (hours)	11	11	2	8	4	16	11	16	5	3
Number of pirouettes	15	13	3	12	7	17	13	16	8	5

a. Construct a scatterplot and determine the equation of the line of best fit.
b. State what the gradient represents.
c. Using your line of best fit, predict the number of pirouettes that could be completed if a student undertakes 14 hours of training.
d. Professional ballet dancers may undertake up to 30 hours of training a week. Using your line best fit, predict the number of pirouettes they should be able to do in a row. Comment on your findings.

7. Use the information in the data table to answer the following questions.

Age in years (x)	7	11	8	16	9	8	14	19	17	10	20	15
Hours of television watched in a week (y)	20	19	25	55	46	50	53	67	59	25	70	58

a. Use technology to determine the equation of the line of best fit for the following data.
b. Use technology to predict the value of the number of hours of television watched by a person aged 15.

Problem solving

8. Describe the trends present in the following time series data that shows the mean monthly daily hours of sunshine in Melbourne from January to December.

Month	1	2	3	4	5	6	7	8	9	10	11	12
Daily hours of sunshine	8.7	8.0	7.5	6.4	4.8	4.0	4.5	5.5	6.3	7.3	7.5	8.3

9. The existence of the following situations is often considered an obstacle to making estimates from data.
a. Outlier.
b. Extrapolation.
c. Small range of data.
d. Small number of data points.
Explain why each of these situations is considered an obstacle to making estimates of data and how each might be overcome.

10. The table shows the heights of 10 students and the distances along the ground between their feet as they attempt to do the splits.

Height (cm)	Distance stretched (cm)
134.5	150
156	160
133.5	147
145	160
160	162
135	149
163	163
138	149
152	158
159	160

Using the data, estimate the distance a person 1.8 m tall can achieve when attempting the splits. Write a detailed analysis of your result. Include:

- an explanation of the method(s) used
- any plots or formula generated
- comments on validity of the estimate
- any ways the validity of the estimate could be improved.

on To test your understanding and knowledge of this topic, go to your learnON title at www.jacplus.com.au and complete the **post-test**.

Online Resources

Below is a full list of **rich resources** available online for this topic. These resources are designed to bring ideas to life, to promote deep and lasting learning and to support the different learning needs of each individual.

eWorkbook

Download the workbook for this topic, which includes worksheets, a code puzzle and a project (ewbk-2039) ☐

Solutions

Download a copy of the fully worked solutions to every question in this topic (sol-0747) ☐

Digital documents

13.2 SkillSHEET Substitution into a linear rule (doc-5405) ☐
SkillSHEET Solving linear equations that arise when finding x- and y-intercepts (doc-5406) ☐
SkillSHEET Transposing linear equations to standard form (doc-5407) ☐
SkillSHEET Measuring the rise and the run (doc-5408) ☐
SkillSHEET Determining the gradient given two points (doc-5409) ☐
SkillSHEET Graphing linear equations using the x- and y-intercept method (doc-5410) ☐
SkillSHEET Determining independent and dependent variables (doc-5411) ☐
SkillSHEET Determining the type of relationship (doc-5413) ☐

Video eLessons

13.3 Bivariate data (eles-4965) ☐
Correlation (eles-4966) ☐
Drawing conclusions from correlation (eles-4967) ☐
13.4 Lines of best fit by eye (eles-4968) ☐
Predictions using lines of best fit (eles-4969) ☐
13.5 Scatter plots using technology (eles-4970) ☐
Regression lines using technology (eles-4971) ☐
Using regression lines to make predictions (eles-4972) ☐
13.6 Describing time series (eles-4973) ☐
Time series lines of best fit by eye (eles-4974) ☐
Fluctuations and cycles (eles-0181) ☐

Interactivities

13.2 Individual pathway interactivity: Bivariate data (int-4626) ☐
13.3 Individual pathway interactivity: Lines of best fit (int-4627) ☐
Lines of best fit (int-6180) ☐
Interpolation and extrapolation (int-6181) ☐
13.5 Individual pathway interactivity: Time series (int-4628) ☐
13.6 Crossword (int-2887) ☐
Sudoku puzzle (int-3600) ☐

Teacher resources

There are many resources available exclusively for teachers online.

To access these online resources, log on to **www.jacplus.com.au**.

Answers

Topic 13 Bivariate data

Exercise 13.1 Pre-test

1. B
2. C
3. Independent variable
4. a. B b. C c. A
5. D
6. B
7. Interpolation
8. The gradient of the line is $\frac{16}{11}$.
9. Independent variable
10. Explanatory variable
11. B
12. C
13. $x = 6$
14. a. See figure at the bottom of the page.*
    b. The number of COVID-19 cases started rising in March and peaked in April, then started to decline until June. There was an increase in cases in July and the cases reached peak again in August. Cases then started to decline again until December.
15. a. $y = 14$ b. $x = 12 \cdot 5$

Exercise 13.2 Bivariate data

1.

	Independent	Dependent
a.	Number of hours	Test results
b.	Rainfall	Attendance
c.	Hours in gym	Visits to the doctor
d.	Lengths of essay	Memory taken

2.

	Independent	Dependent
a.	Cost of care	Attendance
b.	Age of property	Cost of property
c.	Number of applicants	Cut-off OP score
d.	Running speed	Heart rate

3.

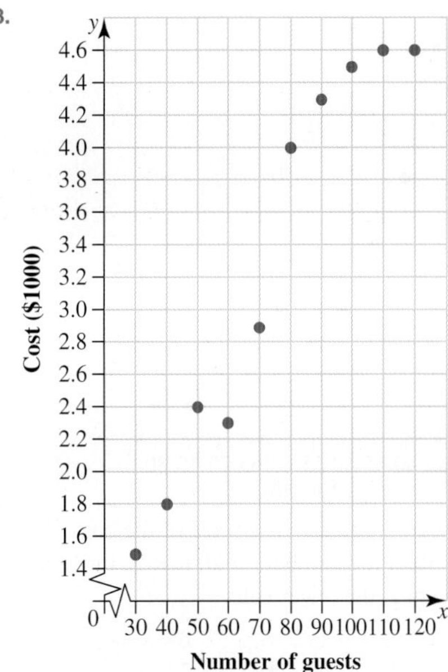

4. a. Perfectly linear, positive
   b. No correlation
   c. Non-linear, negative, moderate
   d. Strong, positive, linear
   e. No correlation
5. a. Non-linear, positive, strong
   b. Strong, negative, negative
   c. Non-linear, moderate, negative
   d. Weak, negative, linear
   e. Non-linear, moderate, positive
6. a. Positive, moderate, linear
   b. Non-linear, strong, negative
   c. Strong, negative, linear
   d. Weak, positive, linear
   e. Non-linear, moderate, positive

*14. a.

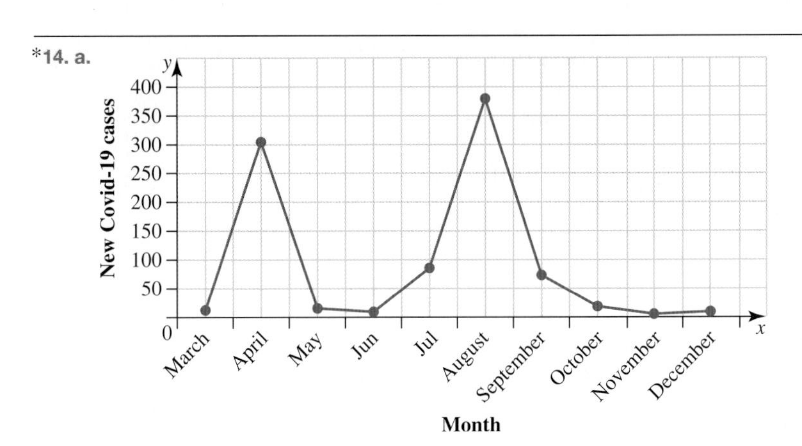

7. a. 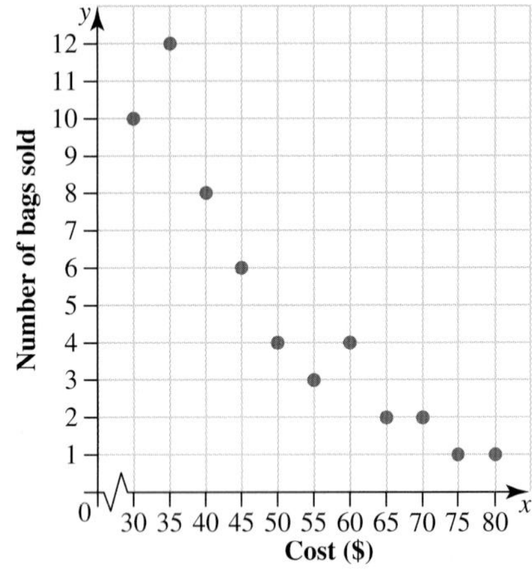

b. Negative, linear, moderate. The price of the bag appeared to affect the numbers sold; that is, the more expensive the bag, the fewer sold.

8. a. 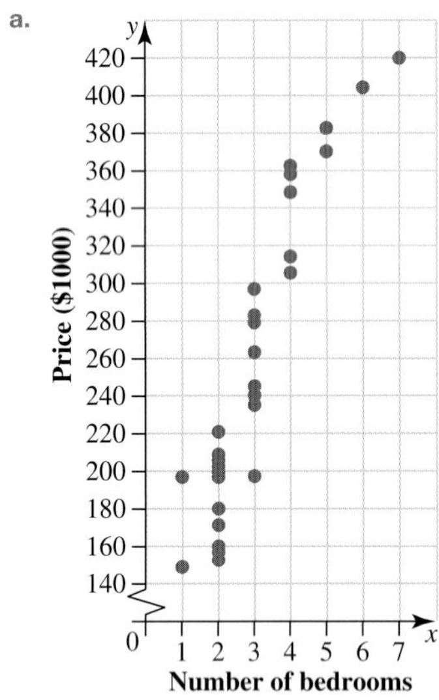

b. Moderate positive linear correlation. There is evidence to show that the larger the number of bedrooms, the higher the price of the house.

c. Various answers; location, age, number of people interested in the house, and so on.

9. a. 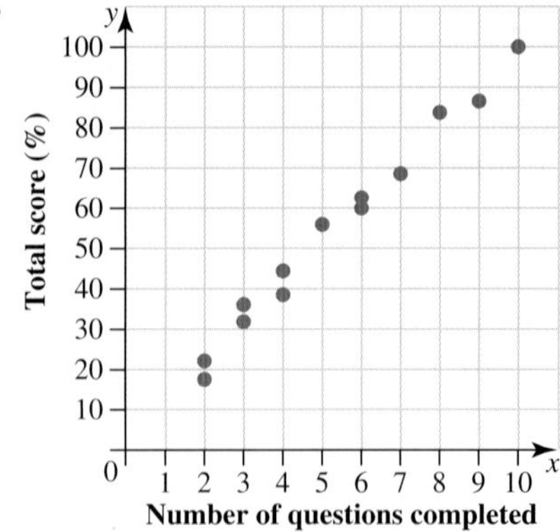

b. Strong, positive, linear correlation

c. Various answers; some students are of different ability levels and they may have attempted the questions but had incorrect answers.

10. a. 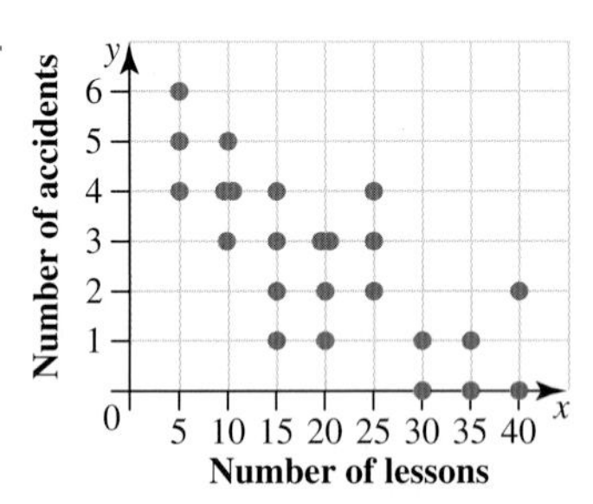

b. Weak, negative, linear relation

c. Various answers; some drivers are better than others, live in lower traffic areas, traffic conditions etc.

11. B

12. C

13. D

14. a. See figure at the bottom of the page.*

b. This scatterplot does not support the claim.

15. a. T b. F c. T d. F e. T

16. a. Mandy (iii) b. William (iv) c. Charlotte (viii)
d. Dario (vii) e. Edward (vi) f. Cindy (v)
g. Georgina (i) h. Harrison (ii)

*14. a. 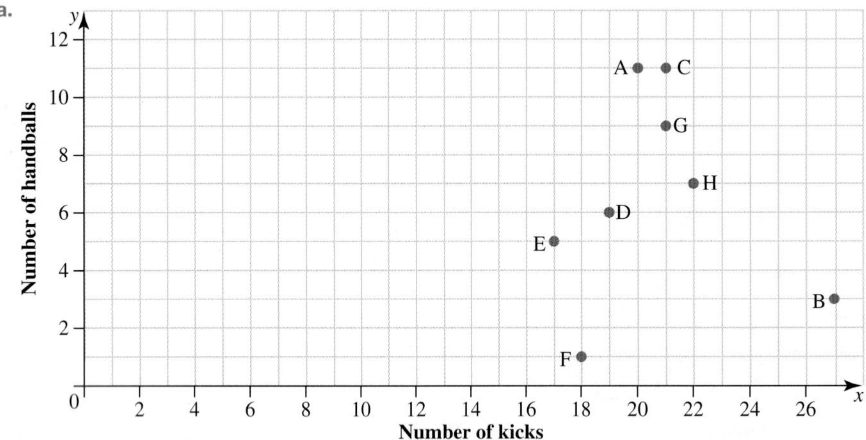

Exercise 13.3 Lines of best fit by eye

Note: Answers may vary slightly depending on the line of best fit drawn.

1. a

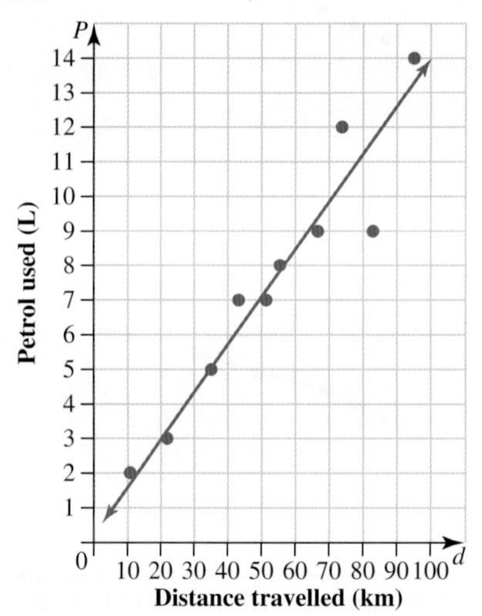

b. Using, (23, 3) and (56, 8), the equation is $P = \frac{5}{33}d - \frac{16}{33}$.

2. a. 38 b. 18

3. a. i. 510 ii. 315 iii. 125

b. i. 36.5 ii. 26 iii. 8

c. $y = -13x + 595$

d. y-values (a):

i. 594

ii. 309

iii. 127

x-values (b):

i. 36.54

ii. 25.55

iii. 5.86

4. a

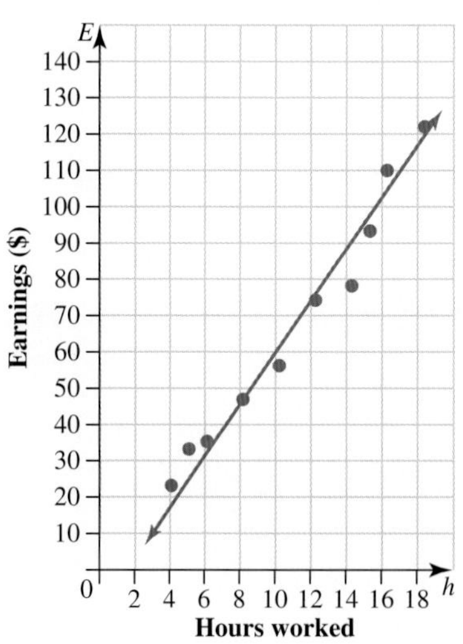

b. Using (8, 47) and (12, 74), the equation is $E = 6.75h - 7$.

c. On average, students were paid $6.75 per hour.

5. a.

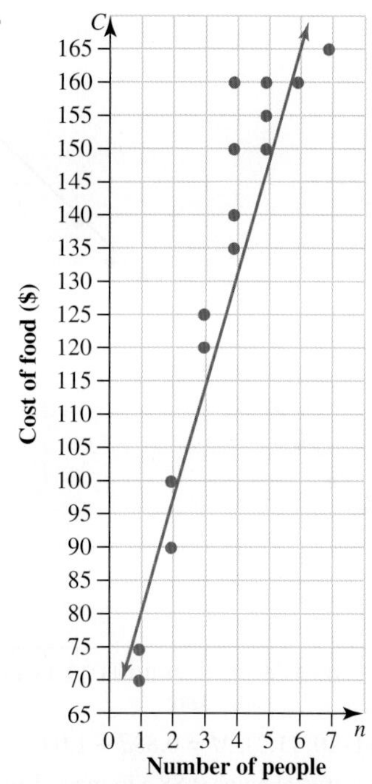

b. Using (1, 75) and (5, 150), the equation is $C = 18.75n + 56.25$

c. On average, weekly cost of food increases by $18.75 for every extra person.

d. i. $206.25 ii. $225.00 iii. $243.75

6. a. 35 to 90

b. 30 to 105

7. a.

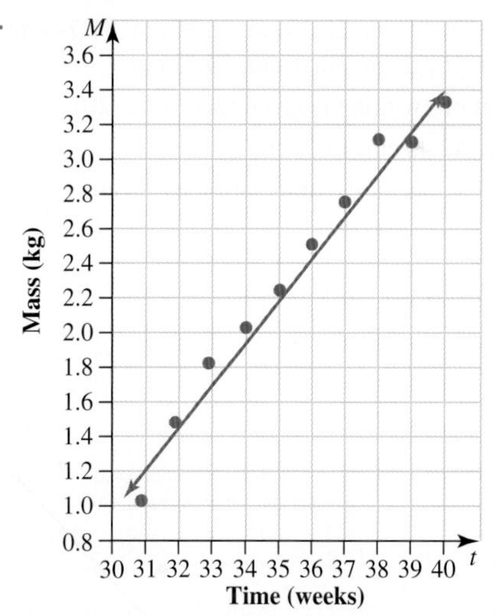

Positive, strong, linear correlation

b. Using (32, 1.470) and (35, 2.230), $M = 0.25t - 6.5$.

c. With every week of gestation the mass of the baby increases by approximately. 250 g.

d. 3.75 kg; 4 kg

e. Approximately 1 kg

f. Between 35 and 36 weeks

8. a. D b. C

9. E

10. a.

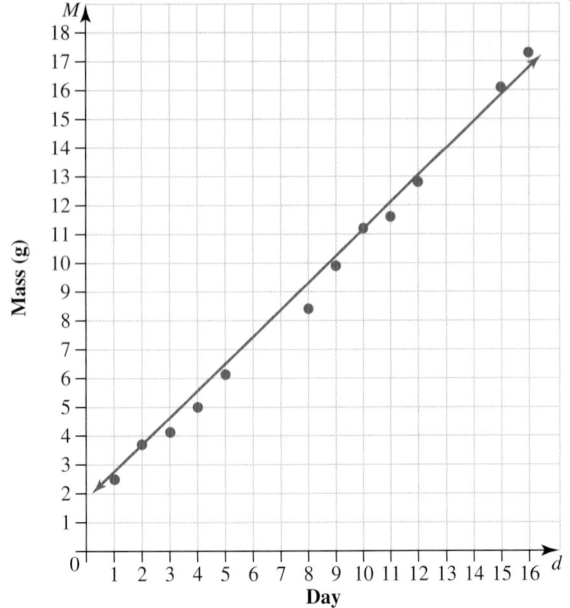

b. Using (2, 3.7) and (10, 11.2), $M = 0.88d + 1.94$.

c. Each day Rachel's crystal gains 0.88 *g* in mass. Line of best fit appears appropriate.

d. 7.22 g; 8.10 g; 13.38 g and 14.26 g; interpolation (within the given range of 1–16)

e. 16.9 g and 17.78 g; predictions are not reliable, since they were obtained using extrapolation.

11. a. See figure at the bottom of the page.*

b. $L = 1.07n + 18.9$

c. 24.25 cm; 26.39 cm; 28.53 cm; 30.67 cm; 31.74 cm; 34.95 cm; 36.02 cm; 38.16 cm; 39.23 cm

d. Interpolation (within the given range of 1–20)

e. 41.37 cm; 42.44 cm; 43.51 cm

f. Not reliable, because extrapolation has been used.

12. a.

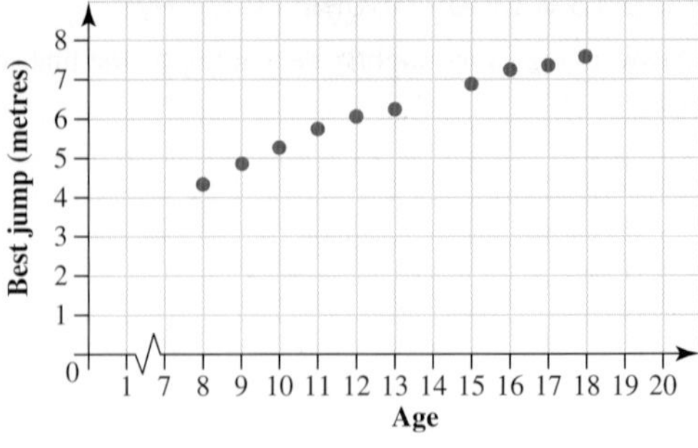

b.

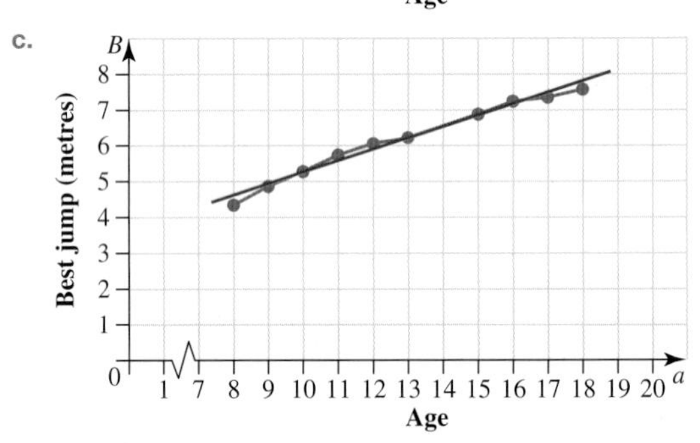

c.

d. Yes. Using points (9, 4.85) and (16, 7.24), $B = 0.34a + 1.8$; estimated best jump $= 8.6$ m.

e. No, trends work well over the short term but in the long term are affected by other variables.

***11. a.**

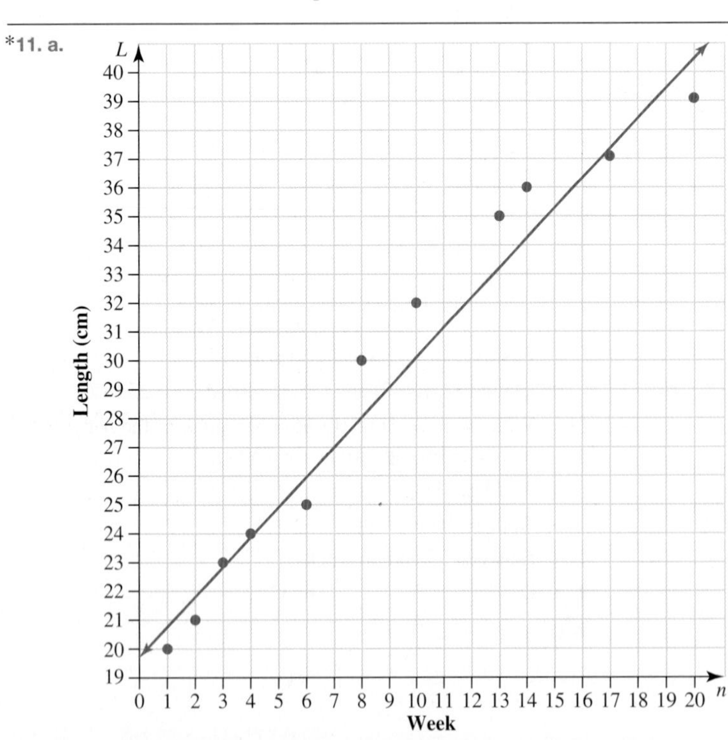

f. 24 years old: 9.97 m; 28 years old: 11.33 m. It is unrealistic to expect his jumping distance to increase indefinitely.

g. Equal first.

13. No. He would have to get 108% which would be impossible on a test.

Exercise 13.4 Linear regression using technology (10A)

1. Sample responses can be found in the worked solutions in the online resources.
2. Sample responses can be found in the worked solutions in the online resources.
3. a. \$174 b. \$418
   c. a. Reliable, but b. not reliable.
4. a. Sample responses can be found in the worked solutions in the online resources.
   b. Sample responses can be found in the worked solutions in the online resources.
   c. 3.8. hours d. 2.5 hours
   e. c. not reliable, d. reliable
5. a. Sample responses can be found in the worked solutions in the online resources.
   b. Sample responses can be found in the worked solutions in the online resources.
   c. 26 kg d. 19 kg
   e. c. not reliable, d. reliable
6. a. $y = 0.553x + 31$ b. 72.4°C
   c. Not possible. Extrapolation not reliable.
7. a. $y = 0.55x + 5.3$ b. 14
   c. No because no connection.
8. a. Sample responses can be found in the worked solutions in the online resources.
   b. Yard size independent, price dependent
   c. $y = 0.173x + 49.1$ d. \$50
9. A
10. D
11. a. Yousef: $y = 39x + 1$, Gavin: $y = 37x + 1$
    b. Yousef 39, Gavin 37
    c. Time to eat each apple.
    d. Gavin
12. a. $y = -0.094x + 3.35$ b. $y = -0.24x + 5.58$
    c. Sample responses can be found in the worked solutions in the online resources.
13. a. $y = 0.0508x + 2.96$
    b. No correlation
    c. $y = -0.913x + 4.32$
    d. Correlation is linear, negative and moderate/strong.
    e. Regression line from part c.
14. 128 cm
15. a. $y = 2.8x$ b. Gradient doubles.
    c. Add 5
16. a. Sample responses can be found in the worked solutions in the online resources.
    b. Sample responses can be found in the worked solutions in the online resources.

Exercise 13.5 Time series

1. a. Linear, downward
   b. Non-linear, upward
   c. Non-linear, stationary in the mean
   d. Linear, upward
2. a. Non-linear, downward
   b. Non-linear, downward
   c. Non-linear, downward
   d. Linear, upward
3. a. See figure at the bottom of the page.*
   b. Linear downward trend

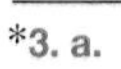
*3. a.

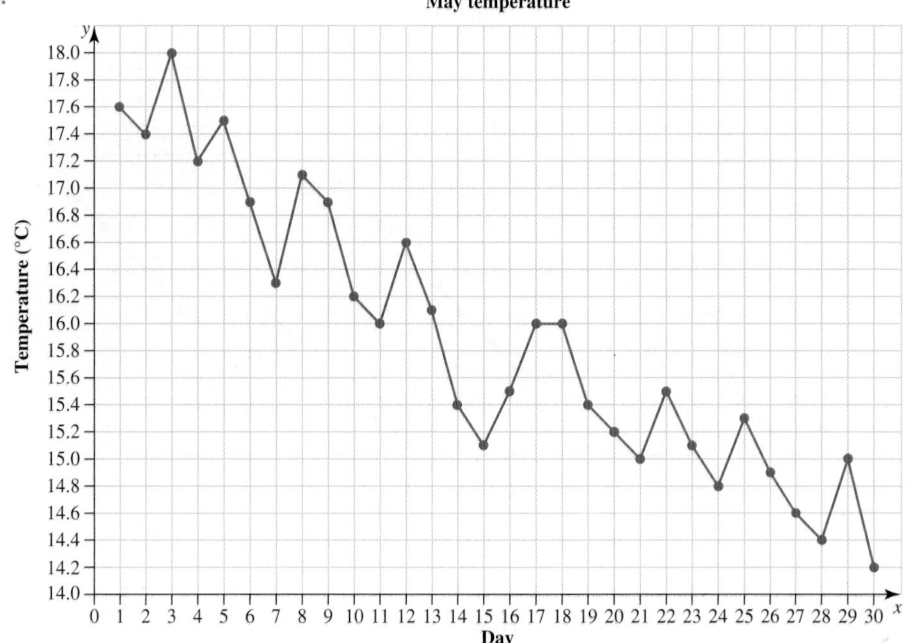

4. a.

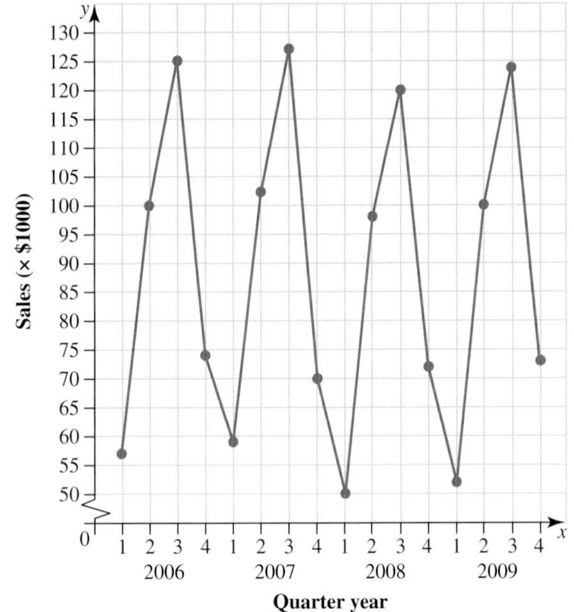

b. Sheepskin products more popular in the third quarter (presumably winter) — discount sales, increase in sales, and so on.

c. No trend.

5. a. See figure at the bottom of the page.*

b. General upward trend with peaks around December and troughs around April.

c. Peaks around Christmas where people have lots of parties, troughs around April where weather gets colder and people less inclined to go out.

d. Yes. Peaks in December, troughs in April.

6. a. Peaks around Christmas holidays and a minor peak at Easter. No camping in colder months.

b. Sample responses can be found in the worked solutions in the online resources.

7. a.

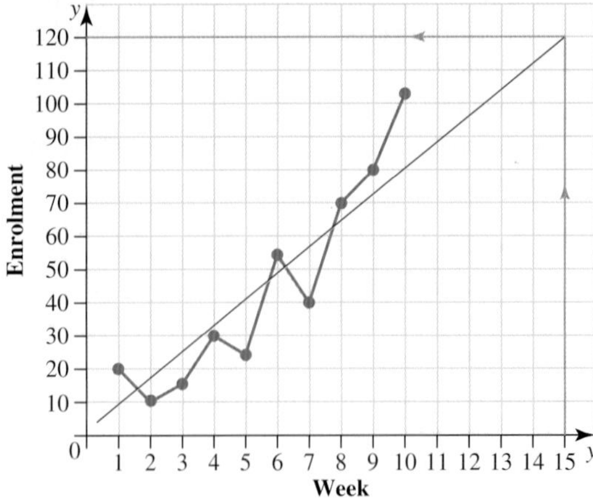

Upward linear.

b. In the 15th year the expected amount = 122.

8. a.

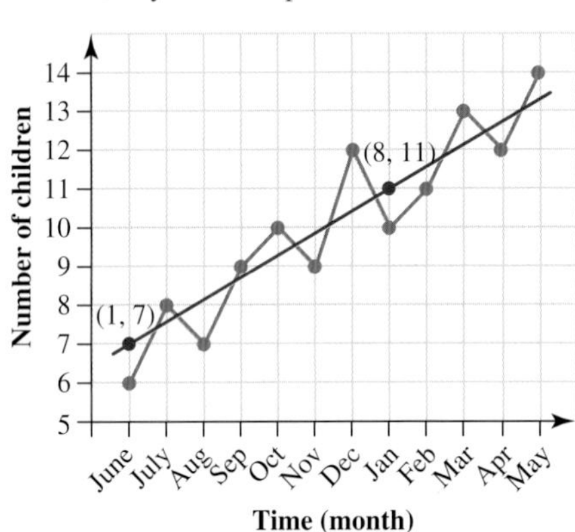

b. Yes, the graph shows an upward trend.

c. $y = \frac{4}{7}x + \frac{45}{7}$

d. i. 15 ii. 18

e. The assumption made was that business will continue on a linear upward trend.

9. The trend is non-linear, therefore unable to forecast future sales.

*5. a.

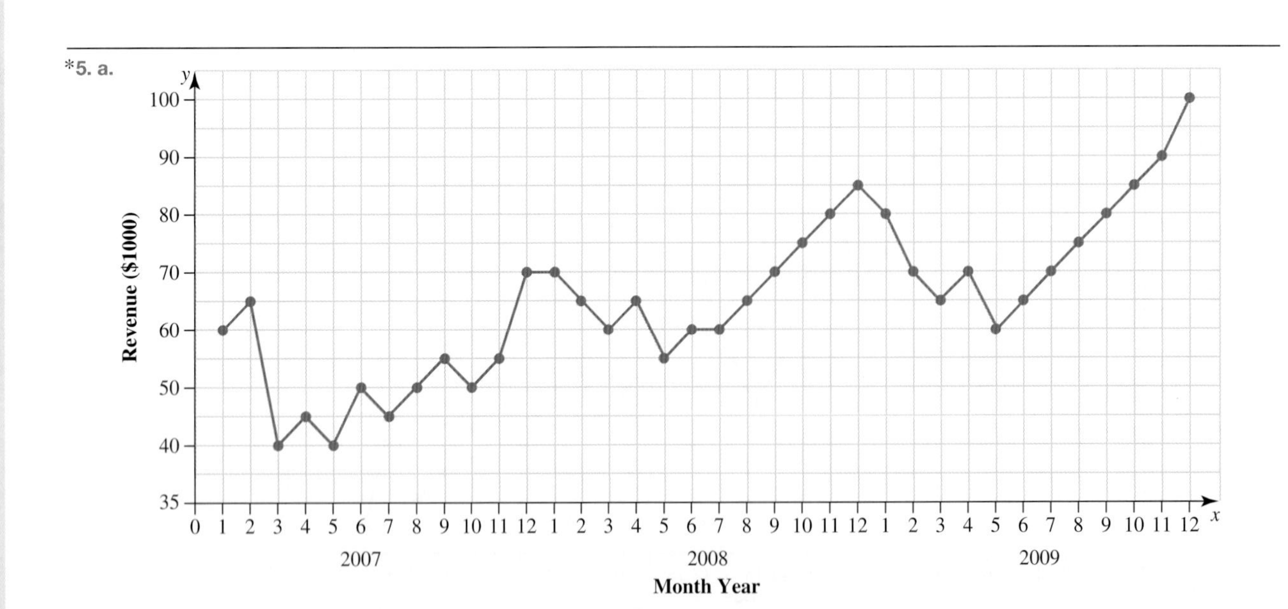

10. a. Sample responses can be found in the worked solutions in the online resources.
    b. Extrapolation is not reliable.
11. a. 90 seconds
    b. Sample responses can be found in the worked solutions in the online resources.
    c. Sample responses can be found in the worked solutions in the online resources.
    d. Approximately 920 seconds.
    e. Approximately 15.3 minutes.
12. a. See bottom of the page*
    b. Secondhand car sales per quarter have shown a general upward trend but with some major fluctuations.
    c. More cars are sold in the third and fourth quarters compared to the first and second quarters.
13. a. Sample responses can be found in the worked solutions in the online resources.
    b. Trend: non-linear, increasing; Pattern: seasonal

Project

1. Sample responses can be found in the worked solutions in the online resources. Students could choose any subject given in the list that can be observed and measure for one day or over the period of a week or more.
2. Sample responses can be found in the worked solutions in the online resources. Students need to create a data table for their recording. Students should use appropriate regular time intervals.
3. Sample responses can be found in the worked solutions in the online resources. For a selected subject, student's need to take their measurements at the selected time intervals and record them in the table.
4. Sample responses can be found in the worked solutions in the online resources. Students could use Excel or CAS to plot the time series.
5. Sample responses can be found in the worked solutions in the online resources. Students should describe their graph and comment on its trend.
6. Sample responses can be found in the worked solutions in the online resources. Students should draw a line of best fit and predict the next few data values.
7. Sample responses can be found in the worked solutions in the online resources. Students should take the actual measurements during the hours they have made predictions for and then compare the predictions with the actual measurements. Also comment on the accuracy of your predictions.

Exercise 13.6 Review questions

1. a. Number of questions: independent; test result: dependent
   b.

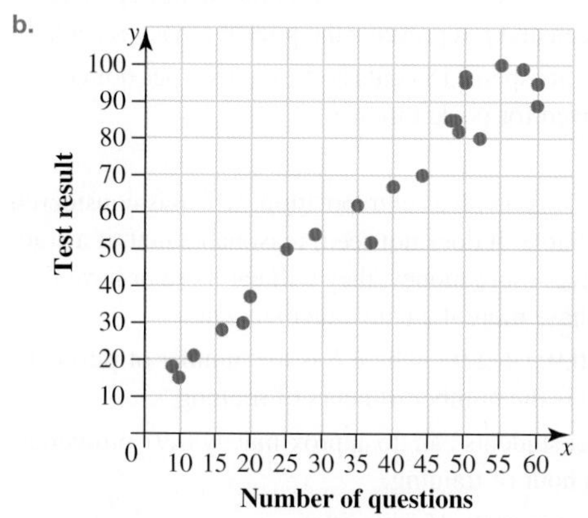

   c. Strong, positive, linear correlation; the larger the number of completed revision questions, the higher the mark on the test.
   d. Different abilities of the students
2. a. i. 12.5   ii. 49
   b. i. 12   ii. 22.5

*12. a.

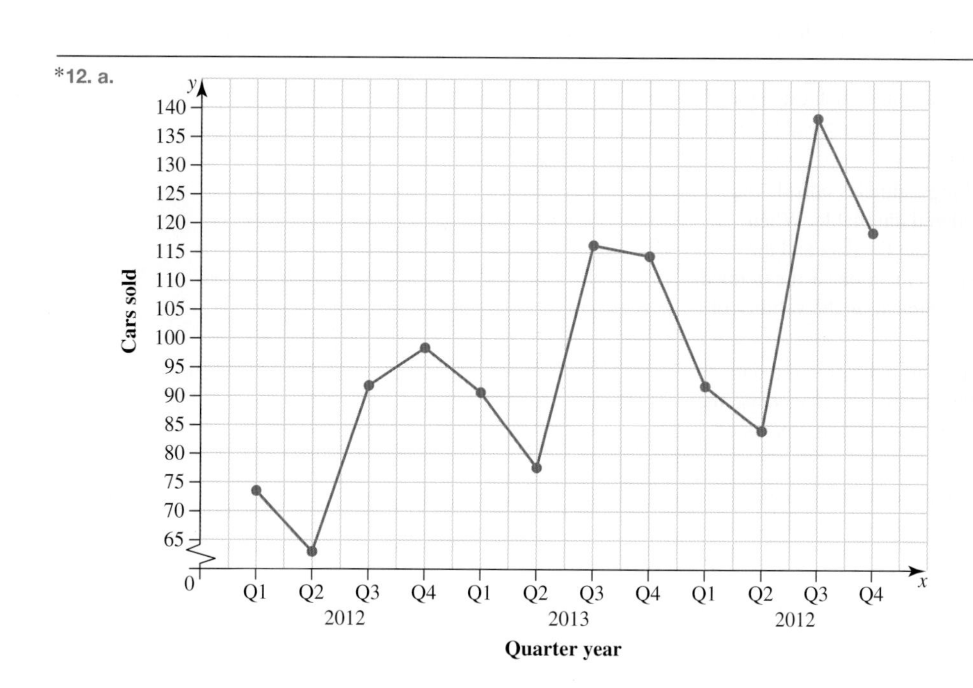

c. $y = \frac{22}{15}x - \frac{7}{3}$

d. i. 12.33 ii. 49

i. 11.82 ii. 22.05

3. a. Linear downwards

b. The trend is linear.

c. About 60–65 occupants

d. Assumes that the current trend will continue.

4. a. $P = 31.82a + 13070.4$, where P is the sale price and a is the land area.

b. The price of land is approximately \$31.82 per square metre.

c. \$64 000

d. $1160\,m^2$

5. a. $C = 0.15p + 11.09$, where C is the money spent at the canteen and p represents the pocket money received.

b. Students spend 15 cents at the canteen per dollar received for pocket money.

c. \$18

d. \$26. This involves extrapolation, which is considered unreliable. It does not seem reasonable that, if a student receives more money, they will eat more or have to purchase more than any other student.

6. a. $P = 0.91t + 2.95$, where P is the number of pirouettes and t is the number of hours of training.

b. Ballet students can do approximately 0.91 pirouettes for each hour of training.

c. Approximately 15 pirouettes.

d. Approximately 30 pirouettes. This estimate is based on extrapolation, which is considered unreliable. To model this data linearly as the number of hours of training becomes large is unrealistic.

7. a. $y = 3.31x + 3.05$

b. Approximately 53 hours.

8. Overall the data appears to be following a seasonal trend, with peaks at either end of the year and a trough in the middle.

9. a. Outliers can unfairly skew data and as such dramatically alter the line of best fit. Identify and remove any outliers from the data before determining the line of best fit.

b. Extrapolation involves making estimates outside the data range and this is considered unreliable. When extrapolation is required, consider the data and the likelihood that the data would remain linear if extended. When giving results, make comment on the validity of the estimation.

c. A small range may not give a fair indication if a data set shows a strong linear correlation. Try to increase the range of the data set by taking more measurements or undertaking more research.

d. A small number of data points may not be able to establish with confidence the existence of a strong linear correlation. Try to increase the number of data points by taking more measurements or undertaking more research.

10. About 170 cm; data was first plotted as a scatter plot. (145, 160) was identified as an outlier and removed from the data set. A line of best fit was then fitted to the remaining data and its equation determined as $d = 0.5h + 80$, where d is the distance stretched and h is the height. Substitution was used to obtain the estimate.
The estimation requires extrapolation and cannot be considered reliable. The presence of the outlier may indicate variation in flexibility rather than a strong linear correlation between the data. Estimate is based on a small set of data.

10A

14 Polynomials

LEARNING SEQUENCE

14.1 Overview

Why learn this?

Just as your knowledge of numbers is learned in stages, so too are graphs. You have been building your knowledge of graphs and functions over time. First, you encountered linear functions. You saw how straight lines are everywhere in our daily lives. Then you learned about quadratic functions, or parabolas. Again, you saw, in everyday situations, how bridges and arches can be based on quadratic or parabolic functions. Circles and hyperbolas are other functions that you have studied. A polynomial is an algebraic expression with integer powers that are greater than or equal to zero, such as a parabola. Polynomial functions are represented by smooth and continuous curves. They can be used to model situations in many different fields, including business, science, architecture, design and engineering. An engineer and designer would use polynomials to model the curves on a rollercoaster. Economists use polynomials to model changes and fluctuations in the stock market. Scientists and researchers use polynomials when looking at changes in the behaviour of objects in different circumstances. Designers and architects incorporate polynomial functions in many areas of their designs in buildings and in landscaping. This topic introduces the building blocks of polynomials.

Where to get help

Go to your learnON title at **www.jacplus.com.au** to access the following digital resources. The Online Resources Summary at the end of this topic provides a full list of what's available to help you learn the concepts covered in this topic.

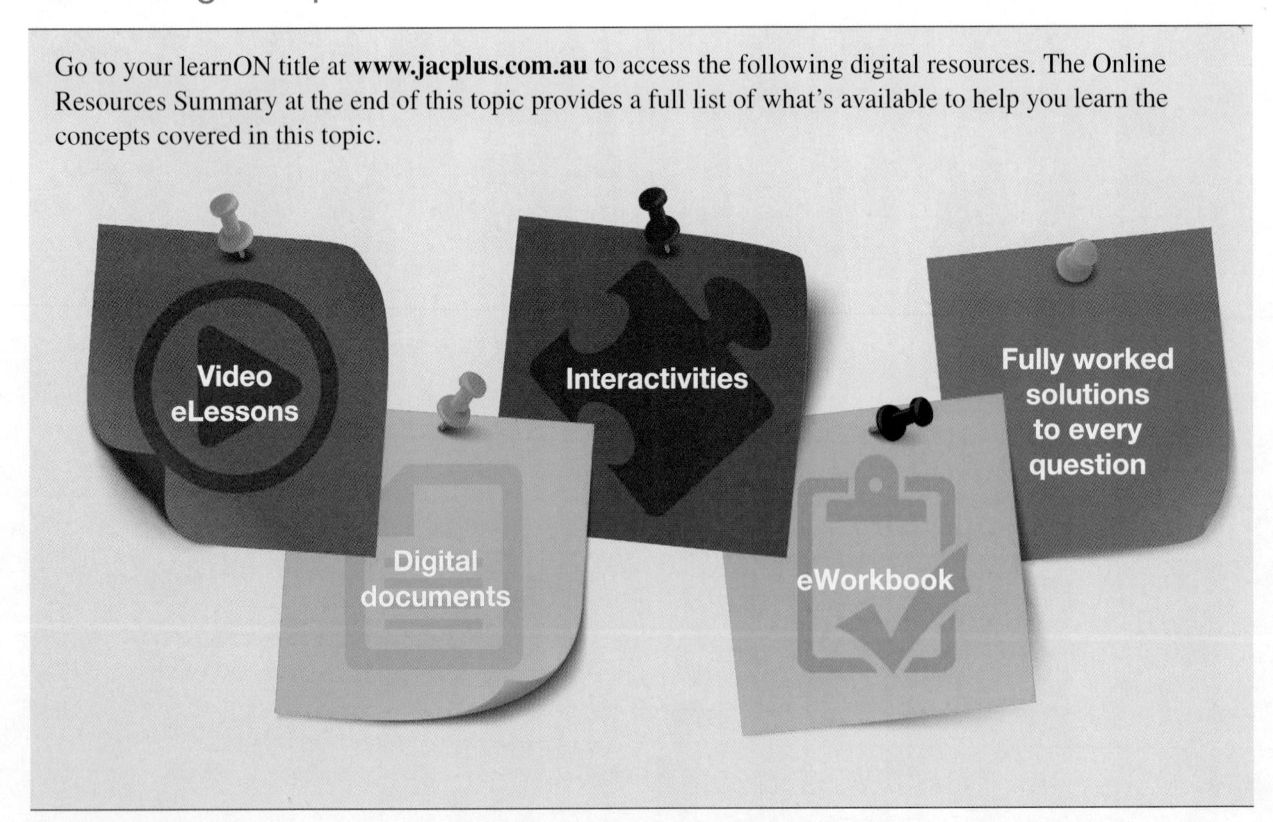

Exercise 14.1 Pre-test

learnon

Complete this pre-test in your learnON title at www.jacplus.com.au and receive **automatic marks, immediate corrective feedback** and **fully worked solutions**.

1. State the degree of the following polynomials.
   a. $3x^2 - 5x + 1$
   b. $2x^3 - 4x^2 + 3x$
   c. $\frac{1}{2}x + 7$
   d. 4

2. MC Choose which of the following is a polynomial.
   A. $\frac{x^3 + x}{2}$
   B. $x^{-2} - \sqrt{7}$
   C. $-x^2 + 3\sqrt{x}$
   D. $6x^4 - \frac{1}{x}$
   E. $2^x - 5x + 1$

3. If $x^2 + 4x - 2 \equiv (x + 1)^2 + a(x - 1) + b$, determine the values of a and b.

4. MC Select the expanded and simplified expression of $(2x)(-3x)(x + 1)$.
   A. $-6x^3 + 1$
   B. $-6x^2 + 1$
   C. $-6x^2 - 6x$
   D. $-6x^3 - 6x^2$
   E. $6x^3 + 1$

5. MC Select the expanded and simplified expression for $(1 - 2x)(3 - x)(4x + 1)(x - 5)$.
   A. $8x^4 - 66x^3 + 135x^2 - 22x - 15$
   B. $8x^4 - 66x^3 + 135x^2 - 12x - 10$
   C. $9x^4 - 66x^2 + 135x^2 - 8x - 10$
   D. $8x^3 - 66x^2 + 135x^2 - 22x - 15$
   E. $8x^5 - 66x^2 + 165x^2 - 22x - 15$

6. Consider the polynomial $-3x^3 + 2x^2 + 4x - 1$. State:
   a. the degree of the polynomial
   b. the leading coefficient
   c. the value of the constant term
   d. the coefficient of x^2.

7. MC Select the simplified expression for $(x^6 + 2x^3 + 3x + 1) - (x^5 - 2x^2 + 4x - 2)$.
   A. $x^6 - x^5 + 2x^3 + 2x^2 - x + 3$
   B. $2x^3 + 2x^2 - 2x + 3$
   C. $x^6 - x^5 + 4x^3 - x + 3$
   D. $x^6 - x^5 + 2x^3 + 2x^2 - 3x$
   E. $4x^5 - 2x + 3$

8. Calculate the quotient when $x^3 + 2x^2 + x - 1$ is divided by $x + 1$.

9. Determine the remainder when $x^3 - x^2 + 4x - 3$ is divided by $x - 2$.

10. If $P(x) = 3x^4 - 2x^3 + x^2 - 4x + 8$, calculate $P(-1)$.

11. MC If $P(x) = x^3 - x^2 + x + 1$, select the value of $P(x + 1)$:
    A. $x^3 + 2x^2 + 5x + 4$
    B. $2x + 3$
    C. $x^3 + 4x^2 + 5x + 4$
    D. $x^3 + 3x^2 + 3x + 1$
    E. $x^3 + 2x^2 + 2x + 2$

12. When $x^3 - 2x^2 + bx + 3$ is divided by $x - 1$ the remainder is 4. Calculate the value of b.

13. MC Select the correct factor for $x^3 - x^2 - 5x - 3$:
    A. x
    B. $x + 1$
    C. $x - 1$
    D. $x + 3$
    E. $x - 5$

14. MC Select the factorised from of $x^3 + 4x^2 + x - 6$.
    A. $(x + 1)(x - 2)(x + 3)$
    B. $(x + 1)(x - 2)(x - 3)$
    C. $(x - 1)(x + 2)(x + 3)$
    D. $(x - 1)(x - 2)(x - 3)$
    E. $(x - 1)(x + 2)(x - 3)$

15. Evaluate the correct value of x, for $2x^3 + 15x^2 + 19x + 6 = 0$.

14.2 Polynomials

LEARNING INTENTION

At the end of this subtopic you should be able to:
- identify polynomial expressions
- state the degree, leading term and leading coefficient of a polynomial.

14.2.1 Polynomials

eles-4975

A **polynomial** in x, sometimes denoted $P(x)$, is an expression containing only non-negative integer powers of x.

- The **degree** of a polynomial in x is the highest power of x in the expression.
  For example:

$3x+1$	is a polynomial of degree 1, or linear polynomial.
x^2+4x-7	is a polynomial of degree 2, or quadratic polynomial.
$-5x^3+\frac{x}{2}$	is a polynomial of degree 3, or cubic polynomial.
10	is a polynomial of degree 0 (think of 10 as $10x^0$).

- Expressions containing a term similar to any of the following terms are **not** polynomials:

$$\frac{1}{x}, \quad x^{-2}, \quad \sqrt{x}, \quad 2^x, \quad \sin x, \quad \text{and so on.}$$

For example, the following are not polynomials.

$$3x^2-4x+\frac{2}{x} \qquad -5x^4+x^3-2\sqrt{x} \qquad x^2+\sin x+1$$

- In the expression $6x^3+13x^2-x+1$
  x is the *variable*.
  6 is the *coefficient* of x^3.
  13 is the *coefficient* of x^2.
  -1 is the *coefficient* of x.
  $6x^3$, $13x^2$, $-x$ and $+1$ are all *terms*.
  The *constant* term is $+1$.
  The *degree* of the polynomial is 3.
- The **leading term** is $6x^3$ because it is the term that contains the highest power of x.
- The **leading coefficient** is 6.
- Any polynomial with a leading coefficient of 1 is called **monic**.

WORKED EXAMPLE 1 Identifying degrees and variables of polynomials

Answer the following questions.

i. **State the degree of each of the following polynomials.**

ii. **State the variable for each of the following polynomials.**

a. $x^3 + 3x^2 + x - 5$

b. $y^4 + 4y^3 - 8y^2 + 2y - 8$

c. $a^3 + 34a^6 - 12a - 72$

THINK	WRITE
a. i. Determine the highest power of x in the expression.	a. i. In the expression $x^3 + 3x^2 + x - 5$, the highest power of x is 3. Therefore, this polynomial is of degree 3.
ii. Determine the variable (unknown quantity) in the expression.	ii. x is the variable in this expression.
b. i. Determine the highest power of y in the expression.	b. i. In the expression $y^4 + 4y^3 - 8y^2 + 2y - 8$ the highest power of y is 4. Therefore, this polynomial is of degree 4.
ii. Determine the variable (unknown quantity) in the expression.	ii. y is the variable in this expression.
c. i. Determine the highest power of a in the expression.	c. i. In the expression $a^3 + 34a^6 - 12a - 72$, the highest power of a is 6. Therefore, this polynomial is of degree 6.
ii. Determine the variable (unknown quantity) in the expression.	ii. a is the variable in this expression.

WORKED EXAMPLE 2 Determining coefficients and terms of polynomials

Consider the polynomial $P(x) = 3x^4 - 5x^3 + 2x^2 + x - 12$

a. **State the degree and variable of the polynomial.**

b. **State the coefficient of x^3.**

c. **State the value of the constant term.**

d. **Determine the term that has a coefficient of 2.**

e. **Determine the leading term.**

THINK	WRITE
a. Determine the highest power of x and the variable (unknown quantity) in the expression.	a. The highest power of x is 4 and therefore, the degree is 4. x is the variable in this expression.
b. Determine the term with x^3 in the expression.	b. The coefficient in the term $-5x^3$ is -5.
c. Determine the term without variable in the expression.	c. The value of the constant term is -12.
d. Determine the term that has a coefficient of 2 in the expression.	d. The term that has a coefficient of 2 is $2x^2$.
e. Determine the term that contains the highest power of x in the expression.	e. The leading term is $3x^4$.

An example where polynomials are useful is in the construction of a greenhouse. The surface area, S, of a greenhouse of length L and height x can be approximated by the polynomial $S(x) = \pi x^2 + L\pi x - 4$.

on Resources

eWorkbook Topic 14 Workbook (worksheets, code puzzle and project) (ewbk-2040)

Interactivity Degrees of polynomials (int-6203)

Exercise 14.2 Polynomials

learnon

Individual pathways

■ PRACTISE	■ CONSOLIDATE	■ MASTER
1, 4, 5, 8, 11, 14	2, 6, 9, 12, 15	3, 7, 10, 13, 16

To answer questions online and to receive **immediate feedback** and **sample responses** for every question, go to your learnON title at www.jacplus.com.au.

Fluency

WE1 For questions 1 to 3, answer the following questions.

i. State the degree of each of the following polynomials.

ii. State the variable for each of the following polynomials.

1. a. $x^3 - 9x^2 + 19x + 7$ b. $65 + 2x^7$ c. $3x^2 - 8 + 2x$

2. a. $x^6 - 3x^5 + 2x^4 + 6x + 1$ b. $y^8 + 7y^3 - 5$ c. $\frac{1}{2}u^5 - \frac{u^4}{3} + 2u - 6$

3. a. $18 - \frac{e^5}{6}$ b. $2g - 3$ c. $1.5f^6 - 800f$

4. Identify the polynomials in questions 1 to 3 that are:

a. linear b. quadratic c. cubic d. monic.

For questions 5 to 7, state whether each of the following is a polynomial (P) or not (N).

5. a. $7x + 6x^2 + \frac{5}{x}$ b. $33 - 4p$ c. $\frac{x^2}{9} + x$

6. a. $3x^4 - 2x^3 - 3\sqrt{x} - 4$ b. $k^{-2} + k - 3k^3 + 7$ c. $5r - r^9 + \frac{1}{3}$

7. a. $\frac{4c^6 - 3c^3 + 1}{2}$ b. $2^x - 8x + 1$ c. $\sin x + x^2$

Understanding

8. **WE2** Consider the polynomial $P(x) = -2x^3 + 4x^2 + 3x + 5$.
   a. State the degree of the polynomial.
   b. State the variable.
   c. State the coefficient of x^2.
   d. State the value of the constant term.
   e. State the term that has a coefficient of 3.
   f. Determine the leading term.

9. Consider the polynomial $P(w) = 6w^7 + 7w^6 - 9$.
   a. State the degree of the polynomial.
   b. State the variable.
   c. State the coefficient of w^6.
   d. Determine the coefficient of w.
   e. State the value of the constant term.
   f. State the term that has a coefficient of 6.

10. Consider the polynomial $f(x) = 4 - x^2 + x^4$.
    a. State the degree of the polynomial.
    b. State the coefficient of x^4.
    c. Determine the leading term.
    d. State the leading coefficient.

Reasoning

11. Write the following polynomials as simply as possible, arranging terms in descending powers of x.
    a. $7x + 2x^2 - 8x + 15 + 4x^3 - 9x + 3$
    b. $x^2 - 8x^3 + 3x^4 - 2x^2 + 7x + 5x^3 - 7$
    c. $x^3 - 5x^2 - 11x - 1 + 4x^3 - 2x + x^2 - 5$

12. A sports scientist determines the following equation for the velocity of a breaststroke swimmer during one complete stroke:

$$v(t) = 63.876t^6 - 247.65t^5 + 360.39t^4 - 219.41t^3 + 53.816t^2 + 0.4746t.$$

    a. Determine the degree of the polynomial.
    b. State the variable.
    c. State the number of terms that are there.
    d. Use a graphics calculator or graphing software to draw the graph of this polynomial.
    e. Match what happens during one complete stroke with points on the graph.

13. The distance travelled by a body after t seconds is given by $d(t) = t^3 + 2t^2 - 4t + 5$. Using a graphing calculator or suitable computer software, draw a graph of the above motion for $0 \leq t \leq 3$.
Use the graph to help you answer the following:
    a. State what information the constant term gives.
    b. Evaluate the position of the body after 1 second.
    c. Describe in words the motion in the first 2 second.

Problem solving

14. If $x^2 - 3x + 5 = x^2 + (a + b)x + (a - b)$, determine the values of a and b.
15. If $x^2 + 2x - 1 \equiv (x - 1)^2 + a(x + 1) + b$, evaluate a and b.
16. If $x^3 + 9x^2 + 12x + 7 \equiv x^3 + (ax + b)^2 + 3$, evaluate a and b.

14.3 Adding, subtracting and multiplying polynomials

LEARNING INTENTION

At the end of this subtopic you should be able to:
- add, subtract and multiply polynomial expressions.

14.3.1 Operations with polynomials

eles-4976

- To add or subtract polynomials, simply add or subtract any like terms in the expressions.

WORKED EXAMPLE 3 Simplifying polynomial expressions

Simplify each of the following.

a. $(5x^3 + 3x^2 - 2x - 1) + (x^4 + 5x^2 - 4)$

b. $(5x^3 + 3x^2 - 2x - 1) - (x^4 + 5x^2 - 4)$

THINK	WRITE
a. 1. Write the expression.	**a.** $(5x^3 + 3x^2 - 2x - 1) + (x^4 + 5x^2 - 4)$
2. Remove any grouping symbols, watching any signs.	$= 5x^3 + 3x^2 - 2x - 1 + x^4 + 5x^2 - 4$
3. Re-order the terms with descending degrees of x.	$= x^4 + 5x^3 + 3x^2 + 5x^2 - 2x - 1 - 4$
4. Simplify by collecting like terms.	$= x^4 + 5x^3 + 8x^2 - 2x - 5$
b. 1. Write the expression.	**b.** $(5x^3 + 3x^2 - 2x - 1) - (x^4 + 5x^2 - 4)$
2. Remove any grouping symbols, watching any signs.	$= 5x^3 + 3x^2 - 2x - 1 - x^4 - 5x^2 + 4$
3. Re-order the terms with descending degrees of x.	$= -x^4 + 5x^3 + 3x^2 - 5x^2 - 2x - 1 + 4$
4. Simplify by collecting like terms.	$= -x^4 + 5x^3 - 2x^2 - 2x + 3$

TI | THINK	DISPLAY/WRITE	CASIO | THINK	DISPLAY/WRITE
a–b. In a new document, on a Calculator page, press: • MENU • 1: Actions • 1: Define Complete the entry lines as: Define $p1(x) = 5x^3 + 3x^2 - 2x - 1$ Define $p2(x) = x^4 + 5x^2 - 4$ $p1(x) + p2(x)$ $p1(x) - p2(x)$ Press ENTER after each entry.	**a–b.** 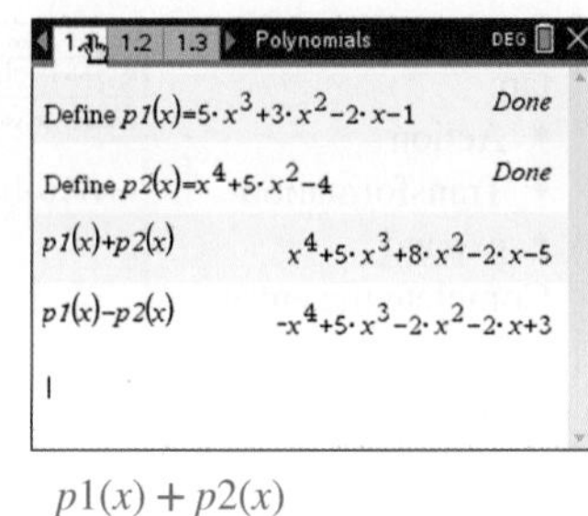 $p1(x) + p2(x)$ $= x^4 + 5x^3 + 8x^2 - 2x - 5$ $p1(x) - p2(x)$ $= -x^4 + 5x^3 - 2x^2 - 2x + 3$	**a–b.** On the Main screen, tap: • Action • Transformation • simplify Complete the entry lines as: simplify $((5x^3 + 3x^2 - 2x - 1)$ $+(x^4 + 5x^2 - 4))$ simplify $((5x^3 + 3x^2 - 2x - 1)$ $-(x^4 + 5x^2 - 4))$ Press EXE after each entry.	**a–b.** 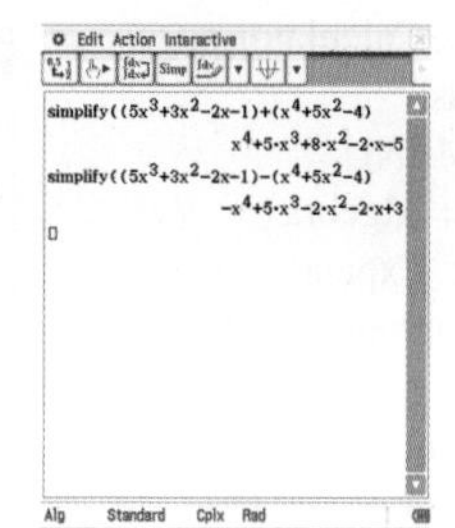 $x^4 + 5x^3 + 8x^2 - 2x - 5$ $-x^4 + 5x^3 - 2x^2 - 2x + 3$

- To expand linear factors, for example $(x + 1)(x + 2)(x - 7)$, use FOIL from quadratic expansions.

WORKED EXAMPLE 4 Expanding polynomial expressions

Expand and simplify:

a. $x(x + 2)(x - 3)$

b. $(x - 1)(x + 5)(x + 2)$.

THINK	WRITE
a. 1. Write the expression.	a. $x(x + 2)(x - 3)$
2. Expand the last two linear factors, using FOIL and simplify	$= x(x^2 - 3x + 2x - 6)$ $= x(x^2 - x - 6)$
3. Multiply the expression in brackets by x.	$= x^3 - x^2 - 6x$
b. 1. Write the expression.	b. $(x - 1)(x + 5)(x + 2)$
2. Expand the last two linear factors, using FOIL, and simplify	$= (x - 1)(x^2 + 2x + 5x + 10)$
3. Multiply the expression in the second bracket by x and then by -1.	$= (x - 1)\left(x^2 + 7x + 10\right)$ $= x^3 + 7x^2 + 10x - x^2 - 7x - 10$
4. Collect like terms.	$= x^3 + 6x^2 + 3x - 10$

TI | THINK

a–b.
On a Calculator page, press:
- MENU
- 3: Algebra
- 3: Expand

Complete the entry lines as:
expand
$(x \times (x+2) \times (x-3))$
expand
$(x \times (x-1) \times (x+5) \times (x+2))$
Press ENTER after each entry.
On the TI, the multiplication sign is displayed as a dot.

DISPLAY/WRITE

a–b.

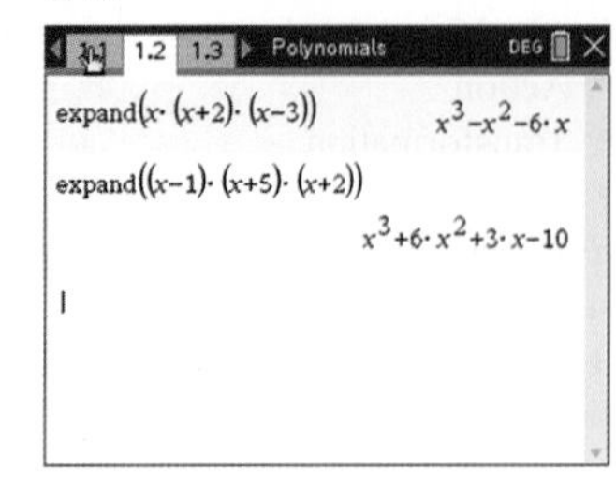

$x(x+2)(x-3) = x^3 - x^2 - 6x$

$(x-1)(x+5)(x+2) = x^3 + 6x^2 + 3x - 10$

CASIO | THINK

a–b.
On the Main screen, tap:
- Action
- Transformation
- expand

Complete the entry lines as:
expand
$(x \times (x+2) \times (x-3))$
expand
$((x-1) \times (x+5) \times (x+2))$
Press EXE after each entry.

DISPLAY/WRITE

a–b.

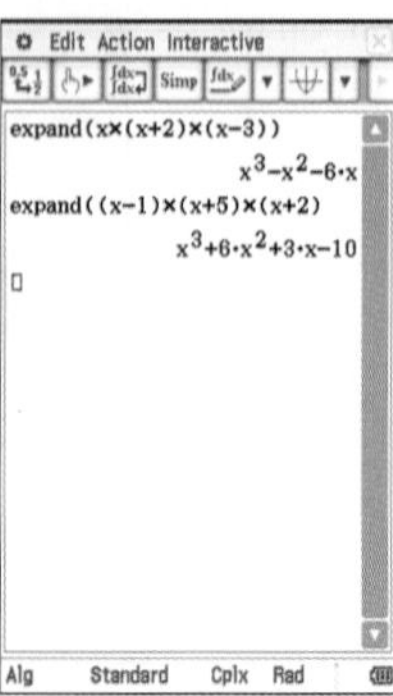

$x(x+2)(x-3) = x^3 - x^2 - 6x$

$(x-1)(x+5)(x+2) = x^3 + 6x^2 + 3x - 10$

on Resources

eWorkbook Topic 14 Workbook (worksheets, code puzzle and project) (ewbk-2040)

Digital document SkillSHEET Expanding the product of two linear factors (doc-5366)

Interactivity Adding and subtracting polynomials (int-6204)

Exercise 14.3 Adding, subtracting and multiplying polynomials learnON

Individual pathways

■ PRACTISE	■ CONSOLIDATE	■ MASTER
1, 3, 5, 7, 9, 13, 16	2, 4, 6, 10, 14, 17	8, 11, 12, 15, 18

To answer questions online and to receive **immediate feedback** and **sample responses** for every question, go to your learnON title at www.jacplus.com.au.

Fluency

1. **WE3a** Simplify each of the following.
   a. $(x^4 + x^3 - x^2 + 4) + (x^3 - 14)$
   b. $(x^6 + x^4 - 3x^3 + 6x^2) + (x^4 + 3x^2 + 5)$
   c. $(x^3 + x^2 + 2x - 4) + (4x^3 - 6x^2 + 5x - 9)$
   d. $(2x^4 - 3x^3 + 7x^2 + 9) + (6x^3 + 5x^2 - 4x + 5)$
   e. $(15x^4 - 3x^2 + 4x - 7) + (x^5 - 2x^4 + 3x^2 - 4x - 3)$
2. **WE3b** Simplify each of the following.
   a. $(x^4 + x^3 + 4x^2 + 5x + 5) - (x^3 + 2x^2 + 3x + 1)$
   b. $(x^6 + x^3 + 1) - (x^5 - x^2 - 1)$
   c. $(5x^7 + 6x^5 - 4x^3 + 8x^2 + 5x - 3) - (6x^5 + 8x^2 - 3)$
   d. $(10x^4 - 5x^2 + 16x + 11) - (2x^2 - 4x + 6)$
   e. $(6x^3 + 5x^2 - 7x + 12) - (4x^3 - x^2 + 3x - 3)$

3. WE4a Expand and simplify each of the following.

a. $x(x+6)(x+1)$
b. $x(x-9)(x+2)$
c. $x(x-3)(x+11)$
d. $2x(x+2)(x+3)$
e. $-3x(x-4)(x+4)$

4. Expand and simplify each of the following.

a. $5x(x+8)(x+2)$
b. $x^2(x+4)$
c. $-2x^2(7-x)$
d. $(5x)(-6x)(x+9)$
e. $-7x(x+4)^2$

WE4b For questions 5 to 10, expand and simplify each of the following.

5. a. $(x+7)(x+2)(x+3)$
b. $(x-2)(x+4)(x-5)$
c. $(x-1)(x-4)(x+8)$
d. $(x-1)(x-2)(x-3)$
e. $(x+6)(x-1)(x+1)$

6. a. $(x-7)(x+7)(x+5)$
b. $(x+11)(x+5)(x-12)$
c. $(x+5)(x-1)^2$
d. $(x+2)(x-7)^2$
e. $(x+1)(x-1)(x+1)$

7. a. $(x-2)(x+7)(x+8)$
b. $(x+5)(3x-1)(x+4)$
c. $(4x-1)(x+3)(x-3)(x+1)$
d. $(5x+3)(2x-3)(x-4)$
e. $(1-6x)(x+7)(x+5)$

8. a. $3x(7x-4)(x-4)(x+2)$
b. $-9x(1-2x)(3x+8)$
c. $(6x+5)(2x-7)^2$
d. $(3-4x)(2-x)(5x+9)(x-1)$
e. $2(7+2x)(x+3)(x+4)$

Understanding

9. a. $(x+2)^3$
b. $(x+5)^3$
c. $(x-1)^3$

10. a. $(x-3)^4$
b. $(2x-6)^3$
c. $(3x+4)^4$

11. Simplify the expression $2(ax+b)-5(c-bx)$.

12. Expand and simplify the expression $(x+a)(x-b)(x^2-3bx+2a)$.

Reasoning

13. If $(x-3)^4 = ax^4 + bx^3 + cx^2 + dx + e$, determine the values of a, b, c, d and e. Show your working.

14. Simplify the expression $(2x-3)^3-(4-3x)^2$.

15. Determine the difference in volume between a cube of side $\dfrac{3(x-1)}{2}$ and a cuboid whose sides are x, $(x+1)$ and $(2x+1)$. Show your working.

Problem solving

16. Determine the values of the pronumerals a and b if: $\dfrac{5x+1}{(x-1)(x+2)} \equiv \dfrac{a}{(x-1)} + \dfrac{b}{(x+2)}$

17. Evaluate the constants a, b and c if: $\dfrac{5x-7}{(x-1)(x+1)(x-2)} \equiv \dfrac{a}{(x-1)} + \dfrac{b}{(x+1)} + \dfrac{c}{(x-2)}$.

18. Write $\dfrac{3x-5}{(x^2+1)(x-1)}$ in the form $\dfrac{ax+b}{(x^2+1)} + \dfrac{c}{(x-1)}$ and hence determine the values of a, b and c.

14.4 Long division of polynomials

LEARNING INTENTION

At the end of this subtopic you should be able to:
- divide polynomials by linear expressions using long division
- determine the quotient and remainder when dividing polynomials by linear expressions.

14.4.1 Long division of polynomials

eles-4977

- The reverse of expanding is factorising (expressing a polynomial as a product of its linear factors).
- To factorise polynomials we need to use a form of long division.
- The following steps show how to divide a polynomial by a linear factor using long division.

Consider $\left(x^3+2x^2-13x+10\right) \div (x-3)$.		
Step 1	Write the division out using long division notation.	$x-3\overline{)x^3+2x^2-13x+10}$
Step 2	Consider the leading terms only. Determine how many times x goes into x^3.	$x-3\overline{)x^3+2x^2-13x+10}$
Step 3	x into x^3 goes x^2 times. Write x^2 above the x^2 term of the polynomial.	$\begin{array}{r} x^2 \qquad\qquad \\ x-3\overline{)x^3+2x^2-13x+10} \end{array}$
Step 4	Multiply the term at the top $\left(x^2\right)$ by the linear factor $(x-3)$: $x^2 \times (x-3)=x^3-3x^2$. Write the result beneath the first two terms of the polynomial.	$\begin{array}{r} x^2 \qquad\qquad \\ x-3\overline{)x^3+2x^2-13x+10} \\ x^3-3x^2 \qquad\qquad \end{array}$
Step 5	Subtract the first two terms of the polynomial by the terms written below them. $x^3-x^3=0$ and $2x^2-\left(-3x^2\right)=5x^2$	$\begin{array}{r} x^2 \qquad\qquad \\ x-3\overline{)x^3+2x^2-13x+10} \\ \underline{x^3-3x^2} \qquad\qquad \\ 5x^2 \qquad\qquad \end{array}$
Step 6	Bring the next term of the polynomial down to sit next to $5x^2$.	$\begin{array}{r} x^2 \qquad\qquad \\ x-3\overline{)x^3+2x^2-13x+10} \\ \underline{x^3-3x^2} \qquad\qquad \\ 5x^2-13x \qquad \end{array}$

- The process now restarts, looking at the newly created $5x^2-13x$ expression.

Step 7	Consider the leading terms only. Determine how many times x goes into $5x^2$.	$\begin{array}{r} x^2 \\ x-3\overline{)x^3+2x^2-13x+10} \\ \underline{x^3-3x^2} \\ 5x^2-13x \end{array}$
Step 8	x into $5x^2$ goes $5x$ times. Write $5x$ above the x term of the polynomial.	$\begin{array}{r} x^2+5x \\ x-3\overline{)x^3+2x^2-13x+10} \\ \underline{x^3-3x^2} \\ 5x^2-13x \end{array}$
Step 9	Multiply the term at the top ($5x$) by the linear factor $(x-3)$: $5x \times (x-3) = 5x^2 - 15x$. Write the result beneath the two terms written in step 6.	$\begin{array}{r} x^2+5x \\ x-3\overline{)x^3+2x^2-13x+10} \\ \underline{x^3-3x^2} \\ 5x^2-13x \\ 5x^2-15x \end{array}$
Step 10	Subtract the two terms of the polynomial by the terms written below them. $5x^2 - 5x^2 = 0$ and $(-13x) - (-15x) = 2x$	$\begin{array}{r} x^2+5x \\ x-3\overline{)x^3+2x^2-13x+10} \\ \underline{x^3-3x^2} \\ 5x^2-13x \\ \underline{5x^2-15x} \\ 2x \end{array}$
Step 11	Bring the next term of the polynomial down to sit next to $2x$.	$\begin{array}{r} x^2+5x \\ x-3\overline{)x^3+2x^2-13x+10} \\ \underline{x^3-3x^2} \\ 5x^2-13x \\ \underline{5x^2-15x} \downarrow \\ 2x+10 \end{array}$

- Once again the process restarts, looking at the newly created $2x + 10$ expression.

Step 12	Consider the leading terms only. Determine how many times x goes into $2x$.	$\begin{array}{r} x^2+5x \\ x-3\overline{)x^3+2x^2-13x+10} \\ \underline{x^3-3x^2} \\ 5x^2-13x \\ \underline{5x^2-15x}\downarrow \\ 2x+10 \end{array}$
Step 13	x into $2x$ goes 2 times. Write 2 above the constant term of the polynomial.	$\begin{array}{r} x^2+5x+2 \\ x-3\overline{)x^3+2x^2-13x+10} \\ \underline{x^3-3x^2} \\ 5x^2-13x \\ \underline{5x^2-15x} \\ 2x+10 \end{array}$
Step 14	Multiply the term at the top (2) by the linear factor $(x-3)$: $2\times(x-3)=2x-6$. Write the result beneath the two terms written in step 11.	$\begin{array}{r} x^2+5x+2 \\ x-3\overline{)x^3+2x^2-13x+10} \\ \underline{x^3-3x^2} \\ 5x^2-13x \\ \underline{5x^2-15x} \\ 2x+10 \\ 2x-6 \end{array}$
Step 15	Subtract the two terms of the polynomial by the terms written below them. $2x-2x=0$ and $10-(-6)=16$	$\begin{array}{r} x^2+5x+2 \\ x-3\overline{)x^3+2x^2-13x+10} \\ \underline{x^3-3x^2} \\ 5x^2-13x \\ \underline{5x^2-15x} \\ 2x+10 \\ \underline{2x-6} \\ 16 \end{array}$
Step 16	The division is now complete! The top line is the quotient (Q), and the bottom number is the remainder (R).	$\begin{array}{rl} x^2+5x+2 & \leftarrow \text{Quotient} \\ x-3\overline{)x^3+2x^2-13x+10} & \\ \underline{x^3-3x^2} & \\ 5x^2-13x & \\ \underline{5x^2-15x} & \\ 2x+10 & \\ \underline{2x-6} & \\ 16 & \leftarrow \text{Remainder} \end{array}$
Write the answer: $(x^3+2x^2-13x+10)\div(x-3)=x^2+5x+2$ remainder 16		

WORKED EXAMPLE 5 Performing long division of cubic polynomials

Perform the following long divisions and state the quotient and remainder.

a. $(x^3 + 3x^2 + x + 9) \div (x + 2)$

b. $(x^3 - 4x^2 - 7x - 5) \div (x - 1)$

c. $(2x^3 + 6x^2 - 3x + 2) \div (x - 6)$

THINK	WRITE
a. 1. Write the question in long division format. 2. Perform the long division process.	a. $$\begin{array}{r} x^2 + x - 1 \leftarrow Q \\ x+2\overline{)x^3 + 3x^2 + x + 9} \\ \underline{x^3 + 2x^2} \qquad\quad \\ x^2 + x \qquad \\ \underline{x^2 + 2x} \qquad \\ -x + 9 \\ \underline{-x - 2} \\ 11 \leftarrow R \end{array}$$
3. Write the quotient and remainder.	Quotient is $x^2 + x - 1$; remainder is 11.
b. 1. Write the question in long division format. 2. Perform the long division process.	b. $$\begin{array}{r} x^2 - 3x - 10 \leftarrow Q \\ x-1\overline{)x^3 - 4x^2 - 7x - 5} \\ \underline{x^3 - x^2} \qquad\qquad \\ -3x^2 - 7x \qquad \\ \underline{-3x^2 + 3x} \qquad \\ -10x - 5 \\ \underline{-10x + 10} \\ -15 \leftarrow R \end{array}$$
3. Write the quotient and remainder.	Quotient is $x^2 - 3x - 10$; remainder is -15.
c. 1. Write the question in long division format. 2. Perform the long division process.	c. $$\begin{array}{r} 2x^2 + 18x + 105 \leftarrow Q \\ x-6\overline{)2x^3 + 6x^2 - 3x + 2} \\ \underline{2x^3 - 12x^2} \qquad\qquad \\ 18x^2 - 3x \qquad \\ \underline{18x^2 - 108x} \qquad \\ 105x + 2 \\ \underline{105x - 630} \\ 632 \leftarrow R \end{array}$$
3. Write the quotient and remainder.	Quotient is $2x^2 + 18x + 105$; remainder is 632.

WORKED EXAMPLE 6 Determining the quotient and remainder of a degree 4 polynomial

Determine the quotient and the remainder when $x^4 - 3x^3 + 2x^2 - 8$ is divided by the linear expression $x + 2$.

THINK / **WRITE**

1. Set out the long division with each polynomial in descending powers of x. If one of the powers of x is missing, include it with 0 as the coefficient.

$$x+2\,\overline{)\,x^4-3x^3+2x^2+0x-8}$$

2. Divide x into x^4 and write the result above.

$$\begin{array}{r l} & x^3 \\ x+2 & \overline{)\,x^4-3x^3+2x^2+0x-8} \end{array}$$

3. Multiply the result x^3 by $x + 2$ and write the result underneath.

$$\begin{array}{r l} & x^3 \\ x+2 & \overline{)\,x^4-3x^3+2x^2+0x-8} \\ & x^4+2x^3 \end{array}$$

4. Subtract and bring down the remaining terms to complete the expression.

$$\begin{array}{r l} & x^3 \\ x+2 & \overline{)\,x^4-3x^3+2x^2+0x-8} \\ & \underline{x^4+2x^3} \\ & -5x^3+2x^2+0x-8 \end{array}$$

5. Divide x into $-5x^3$ and write the result above.
6. Continue this process to complete the long division.

$$\begin{array}{r l} & x^3-5x^2+12x-24 \\ x+2 & \overline{)\,x^4-3x^3+2x^2+0x-8} \\ & \underline{x^4+2x^3} \\ & -5x^3+2x^2+0x-8 \\ & \underline{-5x^3-10x^2} \\ & 12x^2+0x-8 \\ & \underline{12x^2+24x} \\ & -24x-8 \\ & \underline{-24x-48} \\ & 40 \end{array}$$

7. The polynomial $x^3 - 5x^2 + 12x - 24$, at the top, is the quotient.

The quotient is $x^3 - 5x^2 + 12x - 24$.

8. The result of the final subtraction, 40, is the remainder.

The remainder is 40.

on Resources

eWorkbook Topic 14 Workbook (worksheets, code puzzle and project) (ewbk-2040)

Interactivity Long division of polynomials (int-2793)

Exercise 14.4 Long division of polynomials

learnON

Individual pathways

PRACTISE	CONSOLIDATE	MASTER
1, 3, 5, 9, 14, 17	2, 4, 7, 10, 12, 15, 18	6, 8, 11, 13, 16, 19

To answer questions online and to receive **immediate feedback** and **sample responses** for every question, go to your learnON title at www.jacplus.com.au.

Fluency

1. WE5a Perform the following long divisions and state the quotient and remainder.
   a. $(x^3+4x^2+4x+9)\div(x+2)$
   b. $(x^3+2x^2+4x+1)\div(x+1)$
   c. $(x^3+6x^2+3x+1)\div(x+3)$
   d. $(x^3+3x^2+x+3)\div(x+4)$
2. Perform the following long divisions and state the quotient and remainder.
   a. $(x^3+6x^2+2x+2)\div(x+2)$
   b. $(x^3+x^2+x+3)\div(x+1)$
   c. $(x^3+8x^2+5x+4)\div(x+8)$
   d. $(x^3+x^2+4x+1)\div(x+2)$
3. WE5b State the quotient and remainder for each of the following.
   a. $(x^3+2x^2-5x-9)\div(x-2)$
   b. $(x^3+x^2+x+9)\div(x-3)$
   c. $(x^3+x^2-9x-5)\div(x-2)$
   d. $(x^3-4x^2+10x-2)\div(x-1)$
4. State the quotient and remainder for each of the following.
   a. $(x^3-5x^2+3x-8)\div(x-3)$
   b. $(x^3-7x^2+9x-7)\div(x-1)$
   c. $(x^3+9x^2+2x-1)\div(x-5)$
   d. $(x^3+4x^2-5x-4)\div(x-4)$

WE5c For questions **5** to **8**, divide the first polynomial by the second and state the quotient and remainder.

5. a. $3x^3-x^2+6x+5, x+2$
   b. $4x^3-4x^2+10x-4, x+1$
   c. $2x^3-7x^2+9x+1, x-2$
6. a. $2x^3+8x^2-9x-1, x+4$
   b. $4x^3-10x^2-9x+8, x-3$
   c. $3x^3+16x^2+4x-7, x+5$
7. a. $6x^3-7x^2+4x+4, 2x-1$
   b. $6x^3+23x^2+2x-31, 3x+4$
   c. $8x^3+6x^2-39x-13, 2x+5$
8. a. $2x^3-15x^2+34x-13,\ 2x-7$
   b. $3x^3+5x^2-16x-23,\ 3x+2$
   c. $9x^3-6x^2-5x+9,\ 3x-4$

Understanding

For questions **9** to **11**, state the quotient and remainder for each of the following.

9. a. $\dfrac{-x^3-6x^2-7x-16}{x+1}$
   b. $\dfrac{-3x^3+7x^2+10x-15}{x-3}$
   c. $\dfrac{-2x^3+9x^2+17x+15}{2x+1}$
   d. $\dfrac{4x^3-20x^2+23x-2}{-2x+3}$
10. a. $(x^3-3x+1)\div(x+1)$
    b. $(x^3+2x^2-7)\div(x+2)$
    c. $(x^3-5x^2+2x)\div(x-4)$
    d. $(-x^3-7x+8)\div(x-1)$

11. a. $(5x^2 + 13x + 1) \div (x + 3)$
b. $(2x^3 + 8x^2 - 4) \div (x + 5)$
c. $(-2x^3 - x + 2) \div (x - 2)$
d. $(-4x^3 + 6x^2 + 2x) \div (2x + 1)$

12. **WE6** Determine the quotient and the remainder when each polynomial is divided by the linear expression given.
a. $x^4 + x^3 + 3x^2 - 7x,\ x - 1$
b. $x^4 - 13x^2 + 36,\ x - 2$
c. $x^5 - 3x^3 + 4x + 3,\ x + 3$

13. Determine the quotient and the remainder when each polynomial is divided by the linear expression given.
a. $2x^6 - x^4 + x^3 + 6x^2 - 5x,\ x + 2$
b. $6x^4 - x^3 + 2x^2 - 4x,\ x - 3$
c. $3x^4 - 6x^3 + 12x,\ 3x + 1$

Reasoning

14. Determine the quotient and remainder when $3x^4 - 6x^3 + 12x + a$ is divided by $3x + 6$. Show your working.

15. Determine the quotient and remainder when $ax^2 + bx + c$ is divided by $(x - d)$. Show your working.

16. A birthday cake in the shape of a cube had side length $(x + p)$ cm. The cake was divided between $(x - p)$ guests. The left-over cake was used for lunch the next day. There were q^3 guests for lunch the next day and each received c^3 cm^3 of cake, which was then all finished.
Determine an expression for q in terms of p and c. Show your working.

Problem solving

17. When $x^3 - 2x^2 + 4x + a$ is divided by $x - 1$ the remainder is zero. Use long division to determine the value of a.

18. When $x^3 + 3x^2 + a$ is divided by $x + 1$, the remainder is 8. Use long division to determine the value of a.

19. When $2x^2 + ax + b$ is divided by $x - 1$ the remainder is zero but when $2x^2 + ax + b$ is divided by $x - 2$ the remainder is 9. Use long division to determine the value of the pronumerals a and b.

14.5 Polynomial values

LEARNING INTENTION

At the end of this subtopic you should be able to:
- determine the value of a polynomial for a given value.

14.5.1 Polynomial values

eles-4978

- Consider the polynomial $P(x) = x^3 - 5x^2 + x + 1$.
- The value of the polynomial when $x = 3$ is denoted by $P(3)$ and is found by substituting $x = 3$ into the equation in place of x, as shown.

$$P(3) = (3)^3 - 5(3)^2 + (3) + 1$$
$$P(3) = 27 - 5(9) + 3 + 1$$
$$P(3) = 27 - 45 + 4$$
$$P(3) = -14$$

WORKED EXAMPLE 7 Evaluating polynomials for values of x

If $P(x) = 2x^3 + x^2 - 3x - 4$, determine the value of:

a. $P(1)$ b. $P(-2)$ c. $P(a)$ d. $P(2b)$ e. $P(x+1)$.

THINK	WRITE
a. 1. Write the expression.	a. $P(x) = 2x^3 + x^2 - 3x - 4$
2. Replace x with 1.	$P(1) = 2(1)^3 + (1)^2 - 3(1) - 4$
3. Simplify and write the answer.	$= 2 + 1 - 3 - 4$ $= -4$
b. 1. Write the expression.	b. $P(x) = 2x^3 + x^2 - 3x - 4$
2. Replace x with −2.	$P(-2) = 2(-2)^3 + (-2)^2 - 3(-2) - 4$
3. Simplify and write the answer.	$= 2(-8) + (4) + 6 - 4$ $= -16 + 4 + 6 - 4$ $= -10$
c. 1. Write the expression.	c. $P(x) = 2x^3 + x^2 - 3x - 4$
2. Replace x with a.	$P(a) = 2a^3 + a^2 - 3a - 4$
3. No further simplification is possible.	
d. 1. Write the expression.	d. $P(x) = 2x^3 + x^2 - 3x - 4$
2. Replace x with $2b$.	$P(2b) = 2(2b)^3 + (2b)^2 - 3(2b) - 4$
3. Simplify and write the answer.	$= 2\left(8b^3\right) + 4b^2 - 6b + 4$ $= 16b^3 + 4b^2 - 6b + 4$
e. 1. Write the expression.	e. $P(x) = 2x^3 + x^2 - 3x - 4$
2. Replace x with $(x+1)$.	$P(x+1) = 2(x+1)^3 + (x+1)^2 - 3(x+1) - 4$
3. Expand the right-hand side and collect like terms.	$= 2(x+1)(x+1)(x+1) + (x+1)(x+1) - 3(x+1) - 4$ $= 2(x+1)\left(x^2 + 2x + 1\right) + x^2 + 2x + 1 - 3x - 3 - 4$ $= 2\left(x^3 + 2x^2 + x + x^2 + 2x + 1\right) + x^2 - x - 6$ $= 2\left(x^3 + 3x^2 + 3x + 1\right) + x^2 - x - 6$ $= 2x^3 + 6x^2 + 6x + 2 + x^2 - x - 6$
4. Write the answer.	$= 2x^3 + 7x^2 + 5x - 4$

TI | THINK	DISPLAY/WRITE	CASIO | THINK	DISPLAY/WRITE
a–e. On a Calculator page, press: • MENU • 1: Actions • 1: Define Complete the entry lines as: Define $p(x) = 2x^3 + x^2 - 3x - 4$ $p(1)$ $p(-2)$ $p(a)$ $p(2b)$ $p(x + 1)$ Press ENTER after each entry.	a–e. 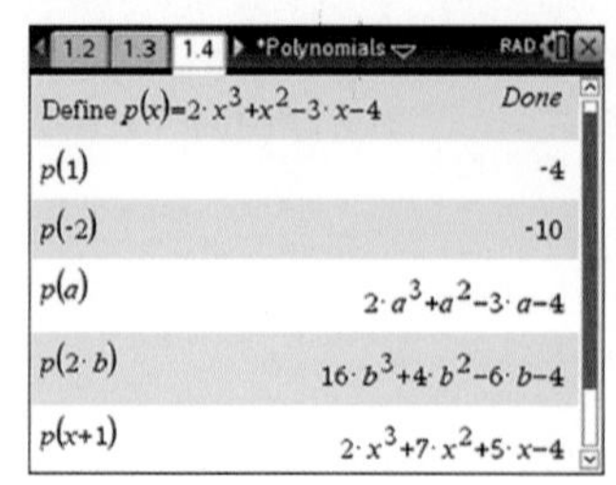 $P(1) = -4$ $P(-2) = -10$ $P(a) = 2a^3 + a^2 - 3a - 4$ $P(2b) = 16b^3 + 4b^2 - 6b - 4$ $P(x + 1) = 2x^3 + 7x^2 + 5x - 4$	a–e. On the Main screen, tap: • Action • Command • Define Complete the entry lines as: Define $p(x) = 2x^3 + x^2 - 3x - 4$ $p(1)$ $p(-2)$ $p(a)$ $p(2b)$ $p(x + 1)$ expand $(p(x + 1))$ Press EXE after each entry.	a–e. 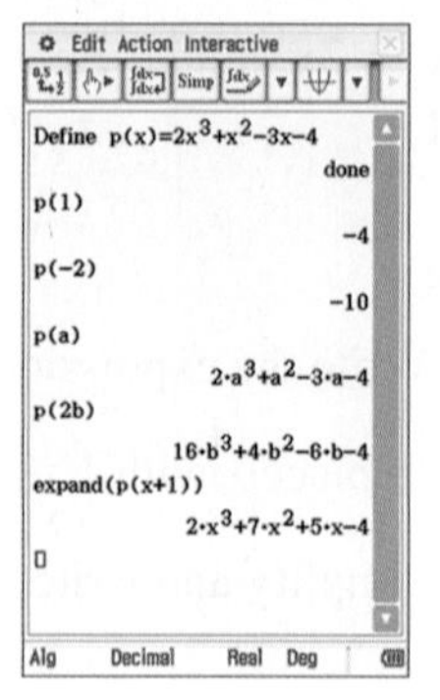 $P(1) = -4$ $P(-2) = -10$ $P(a) = 2a^3 + a^2 - 3a - 4$ $P(2b) = 16b^3 + 4b^2 - 6b - 4$ $P(x + 1) = 2x^3 + 7x^2 + 5x - 4$

on Resources

 **eWorkbook** Topic 14 Workbook (worksheets, code puzzle and project) (ewbk-2040)

 **Digital document** SkillSHEET Substitution into quadratic equations (doc-5367)

Exercise 14.5 Polynomial values

learnon

Individual pathways

■ PRACTISE	■ CONSOLIDATE	■ MASTER
1, 3, 7, 11, 12, 15	2, 5, 8, 9, 13, 16	4, 6, 10, 14, 17

To answer questions online and to receive **immediate feedback** and **sample responses** for every question, go to your learnON title at www.jacplus.com.au.

Fluency

WE7 For questions 1 to 6, $P(x) = 2x^3 - 3x^2 + 2x + 10$. Calculate the following.

1. a. $P(0)$ b. $P(1)$
2. a. $P(2)$ b. $P(3)$
3. a. $P(-1)$ b. $P(-2)$
4. a. $P(-3)$ b. $P(a)$
5. a. $P(2b)$ b. $P(x + 2)$
6. a. $P(x - 3)$ b. $P(-4y)$

Understanding

7. For the polynomial $P(x) = x^3 + x^2 + x + 1$, calculate the following showing your full working.
   a. $P(1)$
   b. $P(2)$
   c. $P(-1)$
   d. The remainder when $P(x)$ is divided by $(x - 1)$.
   e. The remainder when $P(x)$ is divided by $(x - 2)$.
   f. The remainder when $P(x)$ is divided by $(x + 1)$.

8. For the polynomial $P(x) = x^3 + 2x^2 + 5x + 2$, calculate the following showing your full working.
   a. $P(1)$
   b. $P(2)$
   c. $P(-2)$
   d. The remainder when $P(x)$ is divided by $(x - 1)$.
   e. The remainder when $P(x)$ is divided by $(x - 2)$.
   f. The remainder when $P(x)$ is divided by $(x + 2)$.

9. For the polynomial $P(x) = x^3 - x^2 + 4x - 1$, calculate the following showing your full working.
   a. $P(1)$
   b. $P(2)$
   c. $P(-2)$
   d. The remainder when $P(x)$ is divided by $(x - 1)$.
   e. The remainder when $P(x)$ is divided by $(x - 2)$.
   f. The remainder when $P(x)$ is divided by $(x + 2)$.

10. For the polynomial $P(x) = x^3 - 4x^2 - 7x + 3$, calculate the following showing your full working.
   a. $P(1)$
   b. $P(-1)$
   c. $P(-2)$
   d. The remainder when $P(x)$ is divided by $(x - 1)$.
   e. The remainder when $P(x)$ is divided by $(x + 1)$.
   f. The remainder when $P(x)$ is divided by $(x + 2)$.

Reasoning

11. Copy and complete:
   a. A quick way of determining the remainder when $P(x)$ is divided by $(x + 8)$ is to calculate ______.
   b. A quick way of determining the remainder when $P(x)$ is divided by $(x - 7)$ is to calculate ______.
   c. A quick way of determining the remainder when $P(x)$ is divided by $(x - a)$ is to calculate ______.

12. If $P(x) = 2(x - 3)^5 + 1$, determine:
   a. $P(2)$
   b. $P(-2)$
   c. $P(a)$

13. If $P(x) = -2x^3 - 3x^2 + x + 3$, evaluate:
   a. $P(a) + 1$
   b. $P(a + 1)$.

14. When $x^2 + bx + 2$ is divided by $(x - 1)$, the remainder is $b^2 - 4b + 7$. Determine the possible values of b.

Problem solving

15. If $P(x) = 2x^3 - 3x^2 + 4x + c$, determine the value of c if $P(2) = 20$.

16. If $P(x) = 3x^3 - 2x^2 - x + c$ and $P(2) = 8P(1)$, calculate the value of c.

17. If $P(x) = 5x^2 + bx + c$ and $P(-1) = 12$ while $P(2) = 21$, determine the values of b and c.

14.6 The remainder and factor theorems

LEARNING INTENTION

At the end of this subtopic you should be able to:
- identify factors of polynomials using the factor theorem.

14.6.1 The remainder theorem

eles-4979

- In the previous exercise, you may have noticed that:
  the remainder when $P(x)$ is divided by $(x-a)$ is equal to $P(a)$.
- This fact is summarised in the **remainder theorem**.

The remainder theorem

When $P(x)$ is divided by $(x-a)$, the remainder $R = P(a)$.

- If $P(x) = x^3 + x^2 + x + 1$ is divided by $(x-2)$, the quotient is $x^2 + 3x + 7$ and the remainder is $P(2)$, which equals 15. That is:

$$\left(x^3 + x^2 + x + 1\right) \div (x-2) = x^2 + 3x + 7 + \frac{15}{x-2}$$

$$\text{and } \left(x^3 + x^2 + x + 1\right) = \left(x^2 + 3x + 7\right)(x-2) + 15$$

Dividing a polynomial by a linear factor

If $P(x)$ is divided by $(x-a)$, the quotient is $Q(x)$ and the remainder is $R = P(a)$, we can write:

$$P(x) \div (x-a) = Q(x) + \frac{R}{(x-a)}$$

$$\Rightarrow P(x) = (x-a)\,Q(x) + R$$

WORKED EXAMPLE 8 Calculating remainders using the remainder theorem

Without actually dividing, determine the remainder when $x^3 - 7x^2 - 2x + 4$ is divided by:

a. $x - 3$

b. $x + 6$.

THINK	WRITE
a. 1. Name the polynomial.	**a.** Let $P(x) = x^3 - 7x^2 - 2x + 4$.
2. The remainder when $P(x)$ is divided by $(x-3)$ is equal to $P(3)$.	$R = P(3)$ $= 3^3 - 7(3)^2 - 2(3) + 4$ $= 27 - 7(9) - 6 + 4$ $= 27 - 63 - 6 + 4$
3. Write the remainder.	$= -38$

b. 1. The remainder when $P(x)$ is divided by $(x+6)$ is equal to $P(-6)$.

b. $R = P(-6)$
$= (-6)^3 - 7(-6)^2 - 2(-6) + 4$
$= -216 - 7(36) + 12 + 4$
$= 216 - 252 + 12 + 4$

2. Write the remainder.

$= -452$

14.6.2 The factor theorem

eles-4980

- The remainder when 12 is divided by 4 is zero, since 4 is a factor of 12.
- Similarly, if the remainder (R) when $P(x)$ is divided by $(x-a)$ is zero, then $(x-a)$ is a factor of $P(x)$.
- Since $R = P(a)$, determine the value of a that makes $P(a) = 0$, then $(x-a)$ is a factor.

This is summarised in the **factor theorem**.

The factor theorem

If $P(a) = 0$, then $(x-a)$ is a factor of $P(x)$.

- $P(x)$ could be factorised as follows:

$$P(x) = (x-a)Q(x), \text{ where } Q(x) \text{ is 'the other' factor of } P(x).$$

WORKED EXAMPLE 9 Applying the factor theorem to determine constants

$(x-2)$ is a factor of $x^3 + kx^2 + x - 2$. Determine the value of k.

THINK	WRITE
1. Name the polynomial.	Let $P(x) = x^3 + kx^2 + x - 2$.
2. The remainder when $P(x)$ is divided by $(x-2)$ is equal to $P(2) = 0$.	$0 = P(2)$ $= 2^3 + k(2)^2 + 2 - 2$ $0 = 8 + 4k$
3. Solve for k and write its value.	$4k = -8$ $k = -2$

Resources

eWorkbook Topic 14 Workbook (worksheets, code puzzle and project) (ewbk-2040)

Exercise 14.6 The remainder and factor theorems

learnON

Individual pathways

■ PRACTISE	■ CONSOLIDATE	■ MASTER
1, 3, 5, 8, 9, 16, 18, 21	2, 4, 6, 10, 11, 14, 17, 19, 22	7, 12, 13, 15, 20, 23, 24

To answer questions online and to receive **immediate feedback** and **sample responses** for every question, go to your learnON title at www.jacplus.com.au.

Fluency

WE6 For questions 1 and 2, without actually dividing, determine the remainder when $x^3 + 3x^2 - 10x - 24$ is divided by:

1. a. $x - 1$ b. $x + 2$ c. $x - 3$ d. $x + 5$
2. a. $x - 0$ b. $x - k$ c. $x + n$ d. $x + 3c$

For questions 3 to 7, determine the remainder when the first polynomial is divided by the second without performing long division.

3. a. $x^3 + 2x^2 + 3x + 4$, $(x - 3)$ b. $x^3 - 4x^2 + 2x - 1$, $(x + 1)$
4. a. $x^3 + 3x^2 - 3x + 1$, $(x + 2)$ b. $x^3 - x^2 - 4x - 5$, $(x - 1)$
5. a. $x^3 + x^2 + 8$, $(x - 5)$ b. $-3x^3 - 2x^2 + x + 6$, $(x + 1)$
6. a. $-x^3 + 8$, $(x + 3)$ b. $x^3 - 3x^2 - 2$, $(x - 2)$
7. a. $2x^3 + 3x^2 + 6x + 3$, $(x + 5)$ b. $x^3 + 2x^2$, $(x - 7)$

Understanding

8. WE9 The remainder when $x^3 + kx + 1$ is divided by $(x + 2)$ is -19. Calculate the value of k.
9. The remainder when $x^3 + 2x^2 + mx + 5$ is divided by $(x - 2)$ is 27. Determine the value of m.
10. The remainder when $x^3 - 3x^2 + 2x + n$ is divided by $(x - 1)$ is 1. Calculate the value of n.
11. The remainder when $ax^3 + 4x^2 - 2x + 1$ is divided by $(x - 3)$ is -23. Determine the value of a.
12. The remainder when $x^3 - bx^2 - 2x + 1$ is divided by $(x + 1)$ is 0. Calculate the value of b.
13. The remainder when $-4x^2 + 2x + 7$ is divided by $(x - c)$ is -5. Determine a possible whole number value of c.
14. The remainder when $x^2 - 3x + 1$ is divided by $(x + d)$ is 11. Calculate the possible values of d.
15. The remainder when $x^3 + ax^2 + bx + 1$ is divided by $(x - 5)$ is -14. When the cubic polynomial is divided by $(x + 1)$, the remainder is -2. Determine the values of a and b.
16. MC Answer the following. *Note:* There may be more than one correct answer.
    a. When $x^3 + 2x^2 - 5x - 5$ is divided by $(x + 2)$, the remainder is:
       A. -5 B. -2 C. 2 D. 5 E. 7
    b. Choose a factor of $2x^3 + 15x^2 + 22x - 15$ from the following.
       A. $(x - 1)$ B. $(x - 2)$ C. $(x + 3)$ D. $(x + 5)$ E. $(x - 3)$

c. When $x^3 - 13x^2 + 48x - 36$ is divided by $(x - 1)$, the remainder is:

A. -3 B. -2 C. -1 D. 0 E. 1

d. Select a factor of $x^3 - 5x^2 - 22x + 56$from the following.

A. $(x - 2)$ B. $(x + 2)$ C. $(x - 7)$ D. $(x + 4)$ E. $(x - 4)$

17. Determine one factor of each of the following cubic polynomials.

a. $x^3 - 3x^2 + 3x - 1$ b. $x^3 - 7x^2 + 16x - 12$ c. $x^3 + x^2 - 8x - 12$ d. $x^3 + 3x^2 - 34x - 120$

Reasoning

For questions 18 and 19, without actually dividing, show that the first polynomial is exactly divisible by the second (that is, the second polynomial is a factor of the first).

18. a. $x^3 + 5x^2 + 2x - 8$, $(x - 1)$
b. $x^3 - 7x^2 - x + 7$, $(x - 7)$
c. $x^3 - 7x^2 + 4x + 12$, $(x - 2)$
d. $x^3 + 2x^2 - 9x - 18$, $(x + 2)$

19. a. $x^3 + 3x^2 - 9x - 27$, $(x + 3)$
b. $-x^3 + x^2 + 9x - 9$, $(x - 1)$
c. $-2x^3 + 9x^2 - x - 12$, $(x - 4)$
d. $3x^3 + 22x^2 + 37x + 10$, $(x + 5)$

20. Prove that each of the following is a linear factor of $x^3 + 4x^2 - 11x - 30$ by substituting values into the cubic function: $(x + 2)$, $(x - 3)$, $(x + 5)$.

Problem solving

21. When $(x^3 + ax^2 - 4x + 1)$ and $\left(x^3 - ax^2 + 8x - 7\right)$ are each divided by $(x - 2)$, the remainders are equal. Determine the value of a.

22. When $x^4 + ax^3 - 4x^2 + b$ and $x^3 - ax^2 - 7x + b$ are each divided by $(x - 2)$, the remainders are 26 and 8 respectively. Calculate the values of a and b.

23. Both $(x - 1)$ and $(x - 2)$ are factors of $P(x) = x^4 + ax^3 - 7x^2 + bx - 30$. Determine the values of a and b and the remaining two linear factors.

24. The remainder when $2x - 1$ is divided into $6x^3 - x^2 + 3x + k$ is the same as when it is divided into $4x^3 - 8x^2 - 5x + 2$. Calculate the value of k.

14.7 Factorising polynomials

LEARNING INTENTION

At the end of this subtopic you should be able to:

- factorise polynomials using long division
- factorise polynomials using short division or inspection.

14.7.1 Using long division

eles-4981

- Once one factor of a polynomial has been found (using the factor theorem as in the previous section), long division may be used to find other factors.
- In the case of a cubic polynomial, one — possibly two — other factors may be found.

WORKED EXAMPLE 10 Factorising polynomials using long division

Use long division to factorise the following.

a. $x^3 - 5x^2 - 2x + 24$ b. $x^3 - 19x + 30$ c. $-2x^3 - 8x^2 + 6x + 4$

THINK	WRITE
a. 1. Name the polynomial.	a. $P(x) = x^3 - 5x^2 - 2x + 24$
2. Look for a value of x such that $P(x) = 0$. For cubics containing a single x^3, try a factor of the constant term (24 in this case). Try $P(1)$.	$P(1) = 1^3 - 5 \times 1^2 - 2 \times 1 + 24$ $= 1 - 5 - 2 + 24$ $= 18$ $\neq 0$
$P(1) \neq 0$, so $(x - 1)$ is not a factor. Try $P(2)$.	$P(2) = 2^3 - 5 \times 2^2 - 2 \times 2 + 24$ $= 8 - 20 - 4 + 24$ $\neq 0$
$P(2) \neq 0$, so $(x - 2)$ is not a factor. Try $P(-2)$.	$P(-2) = (-2)^3 - 5 \times (-2)^2 - 2 \times (-2) + 24$ $= -8 - 20 + 4 + 24$ $= -28 + 28$ $= 0$ So, $(x + 2)$ is a factor.
$P(-2)$ does equal 0, so $(x + 2)$ is a factor.	
3. Divide $(x + 2)$ into $P(x)$ using long division to determine a quadratic factor.	$x^2 - 7x + 12$ (quotient) $x + 2 \overline{)x^3 - 5x^2 - 2x + 24}$ $x^3 + 2x^2$ $-7x^2 - 2x$ $-7x^2 - 14x$ $12x + 24$ $12x + 24$ 0
4. Write $P(x)$ as a product of the two factors found so far.	$P(x) = (x + 2)(x^2 - 7x + 12)$
5. Factorise the quadratic factor if possible.	$P(x) = (x + 2)(x - 3)(x - 4)$
b. 1. Name the polynomial. *Note*: There is no x^2 term, so include $0x^2$.	b. $P(x) = x^3 - 19x + 30$ $P(x) = x^3 + 0x^2 - 19x + 30$
2. Look at the last term in $P(x)$, which is 30. This suggests it is worth trying $P(5)$ or $P(-5)$. Try $P(-5)$. $P(-5) = 0$ so $(x + 5)$ is a factor.	$P(-5) = (-5)^3 - 19 \times (-5) + 30$ $= -125 + 95 + 30$ $= 0$ So, $(x + 5)$ is a factor.
3. Divide $(x + 5)$ into $P(x)$ using long division to find a quadratic factor.	$x^2 - 5x + 6$ (quotient) $x + 5 \overline{)x^3 + 0x^2 - 19x + 30}$ $x^3 + 5x^2$ $-5x^2 - 19x$ $-5x^2 - 25x$ $6x + 30$ $6x + 30$ 0
4. Write $P(x)$ as a product of the two factors found so far.	$P(x) = (x + 5)(x^2 - 5x + 6)$

5. Factorise the quadratic factor if possible.	$P(x) = (x+5)(x-2)(x-3)$
c. 1. Write the given polynomial.	**c.** Let $P(x) = -2x^3 - 8x^2 + 6x + 4$
2. Take out a common factor of −2. (We could take out +2 as the common factor, but taking out −2 results in a positive leading term in the part still to be factorised.)	$= -2(x^3 + 4x^2 - 3x - 2)$
3. Let $Q(x) = x^3 + 4x^2 - 3x - 2$ (We have already used P earlier.)	Let $Q(x) = x^3 + 4x^2 - 3x - 2$
4. Evaluate $Q(1)$. $Q(1) = 0$, so $(x - 1)$ is a factor.	$Q(1) = 1 + 4 - 3 - 2$ $= 0$ So, $(x - 1)$ is a factor.
5. Divide $(x - 1)$ into $Q(x)$ using long division to determine a quadratic factor.	$$\begin{array}{r} x^2 + 5x + 2 \\ x - 1 \overline{) x^3 + 4x^2 - 3x - 2} \\ \underline{x^3 - x^2} \\ 5x^2 - 3x \\ \underline{5x^2 - 5x} \\ 2x - 2 \\ \underline{2x - 2} \\ 0 \end{array}$$
6. Write the original polynomial $P(x)$ as a product of the factors found so far. In this case, it is not possible to further factorise $P(x)$.	$P(x) = -2(x - 1)(x^2 + 5x + 2)$

- *Note:* In some of these examples, $P(x)$ may have been factorised without long division by finding all three values of x that make $P(x) = 0$ (and hence the three factors).

14.7.2 Using short division, or by inspection

eles-4982

- Short division, or factorising by inspection, is a quicker method than long division.
- Consider $P(x) = x^3 + 2x^2 - 13x + 10$
  Using the factor theorem:
  $P(1) = 0$ so $(x - 1)$ is a factor.
  $\therefore P(x) = (x - 1)\,Q(x)$, where $Q(x)$ is the quadratic quotient.
  so, $P(x) = (x - 1)\left(ax^2 + bx + c\right)$

Expanding and equating:

Term in x^3: $x \times ax^2$	$(x-1)\left(ax^2 + bx + c\right)$	**Equating with $P(x)$:** $ax^3 = x^3$ $\therefore a = 1$
Terms in x^2: $-1 \times ax^2 + bx^2$	$(x-1)\left(ax^2 + bx + c\right)$	**Equating with $P(x)$:** $-1 \times 1x^2 + bx^2 = 2x^2$ $-1 + b = 2$ $\therefore b = 3$
Constant term: $-1 \times c$	$(x-1)\left(ax^2 + bx + c\right)$	**Equating with $P(x)$:** $-c = 10$ $\therefore c = -10$

$$\therefore P(x) = (x-1)\left(x^2+3x-10\right)$$

Factorising the quadratic gives:

$$P(x) = (x-1)(x+5)(x-2)$$

- *Note*: In this example, the values of a and c can be seen simply by inspecting $P(x)$. Hence, $P(x) = (x-1)\left(x^2+bx-10\right)$, leaving only the value of b unknown.
- The following worked example is a repeat of a previous one, but explains the use of short, rather than long, division.

WORKED EXAMPLE 11 Factorising using short division

Use short division to factorise $x^3 - 5x^2 - 2x + 24$.

THINK	WRITE
1. Name the polynomial.	Let $P(x) = x^3 - 5x^2 - 2x + 24$
2. Look for a value of x such that $P(x) = 0$. Try $P(-2)$.	$P(-2) = (-2)^3 - 5\times(-2)^2 - 2\times(-2) + 24$ $= -8 - 20 + 4 + 24$ $= -28 + 28$ $= 0$
$P(-2)$ does equal 0, so $(x+2)$ is a factor.	So, $(x+2)$ is a factor.
3. Look again at the original and equate the factorised form to the expanded form. The values of a and c can be determined simply by inspection. Since the coefficient of the x^3 term is 1, $a = 1$. Since the constant term is 24, $c = 12$.	$x^3 - 5x^2 - 2x + 24 = (x+2)\left(ax^2+bx+c\right)$ $= (x+2)\left(x^2+bx+12\right)$
4. Expand the brackets and equate the coefficients of the x^2 terms. We can then solve for b.	$x^3 - 5x^2 - 2x + 24 = x^3 + bx^2 + 12x + 2x^2 + 2bx + 24$ $-5x^2 = bx^2 + 2x^2$ $-5x^2 = (b+2)x^2$ $b+2 = -5$ $b = -7$ $P(x) = (x+2)\left(x^2-7x+12\right)$
5. Factorise the expression in the second pair of brackets if possible.	$P(x) = (x+2)(x-3)(x-4)$

TI | THINK	DISPLAY/WRITE	CASIO | THINK	DISPLAY/WRITE
On a Calculator page, press: • MENU • 2: Number • 3: Factor Complete the entry lines as: factor $(x^3 - 5x^2 - 2x + 24)$ Then press ENTER.	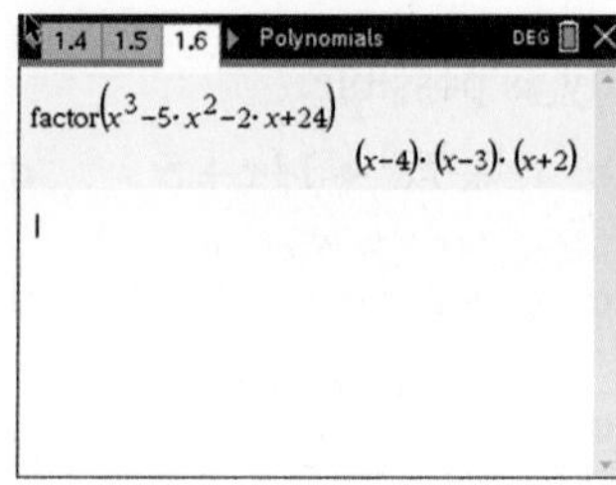 $x^3 - 5x^2 - 2x + 24$ $= (x - 4)(x - 3)(x + 2)$	On the Main screen, tap: • Action • Transformation • factor • factor Complete the entry line as: factor $(x^3 - 5x^2 - 2x + 24)$ Then press EXE.	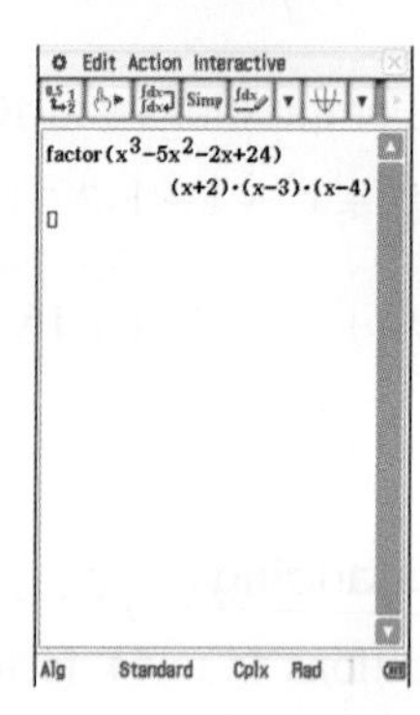 $x^3 - 5x^2 - 2x + 24$ $= (x + 2)(x - 3)(x - 4)$

on Resources

 **eWorkbook** Topic 14 Workbook (worksheets, code puzzle and project) (ewbk-2040)

 **Digital document** SkillSHEET Factorising quadratic trinomials (doc-5368)

Exercise 14.7 Factorising polynomials

learnon

Individual pathways

■ PRACTISE	■ CONSOLIDATE	■ MASTER
1, 2, 8, 11, 15, 18	3, 4, 5, 9, 12, 16, 19	6, 7, 10, 13, 14, 17, 20

To answer questions online and to receive **immediate feedback** and **sample responses** for every question, go to your learnON title at www.jacplus.com.au.

Fluency

WE10 For question 1 to 7 apply long division to factorise each dividend.

1. a. $x + 1\overline{)x^3 + 10x^2 + 27x + 18}$ b. $x + 2\overline{)x^3 + 8x^2 + 17x + 10}$

2. a. $x + 9\overline{)x^3 + 12x^2 + 29x + 18}$ b. $x + 1\overline{)x^3 + 8x^2 + 19x + 12}$

3. a. $x + 3\overline{)x^3 + 14x^2 + 61x + 84}$ b. $x + 7\overline{)x^3 + 12x^2 + 41x + 42}$

4. a. $x + 2\overline{)x^3 + 4x^2 + 5x + 2}$ b. $x + 3\overline{)x^3 + 7x^2 + 16x + 12}$

5. a. $x + 5\overline{)x^3 + 14x^2 + 65x + 100}$ b. $x\overline{)x^3 + 13x^2 + 40x}$

6. a. $x\overline{)x^3 + 7x^2 + 12x}$ b. $x + 5\overline{)x^3 + 10x^2 + 25x}$

7. a. $x+1\overline{)x^3+6x^2+5x}$ b. $x+6\overline{)x^3+6x^2}$

WE11 For questions 8 to 10, factorise the following as fully as possible.

8. a. x^3+x^2-x-1 b. x^3-2x^2-x+2 c. $x^3+7x^2+11x+5$ d. $x^3+x^2-8x-12$

9. a. $x^3+9x^2+24x+16$ b. $x^3-5x^2-4x+20$ c. x^3+2x^2-x-2 d. x^3-7x-6

10. a. $x^3+8x^2+17x+10$ b. x^3+x^2-9x-9 c. $x^3-x^2-8x+12$ d. $x^3+9x^2-12x-160$

Understanding

For questions 11 to 14, factorise as fully as possible.

11. a. $2x^3+5x^2-x-6$
    b. $3x^3+14x^2+7x-4$
    c. $3x^3+2x^2-12x-8$
    d. $4x^3+35x^2+84x+45$

12. a. x^3+x^2+x+1
    b. $4x^3+16x^2+21x+9$
    c. $6x^3-23x^2+26x-8$
    d. $7x^3+12x^2-60x+16$

13. a. $3x^3-x^2-10x$ b. $4x^3+2x^2-2x$ c. $3x^3-6x^2-24x$ d. $-2x^3-12x^2-18x$

14. a. $-x^3-7x^2-12x$ b. $-x^3-3x^2+x+3$ c. $-2x^3+10x^2-12x$ d. $-5x^3+24x^2-36x+16$

Reasoning

15. Factorise $x^4-9x^2-4x+12$.

16. Factorise $-x^5+6x^4+11x^3-84x^2-28x+240$.

17. Two of the factors of x^3+px^2+qx+r are $(x+a)$ and $(x+b)$. Determine the third factor.

Problem solving

18. Factorise $x^5-5x^4+5x^3+5x^2-6x$.

19. $(x-1)$ and $(x-2)$ are known to be factors of $x^5+ax^4-2x^3+bx^2+x-2$. Determine the values of a and b and hence fully factorise this fifth-degree polynomial.

20. The polynomial $x^4-6x^3+13x^2-12x-32$ has three factors, one of which is x^2-3x+8. Evaluate the other two factors.

14.8 Solving polynomial equations

LEARNING INTENTION

At the end of this subtopic you should be able to:

- solve polynomial equations by applying the Null Factor Law.

14.8.1 Solving polynomial equations

eles-4983

- To solve the polynomial equation of the form $P(x) = 0$:

  Step 1: factorise $P(x)$
  Step 2: apply the Null Factor Law
  Step 3: state the solutions.

- The Null Factor Law applies to polynomial equations just as it does for quadratics.
- If $P(x)$ is of degree n, then $P(x) = 0$ has up to n solutions.
  Solving each of these equations produces the solutions (roots).

$$x = a \qquad x = b \qquad x = c.$$

- If $P(x) = k(lx - a)(mx - b)(nx - c) = 0$, then the solutions can be found as follows. Let each factor $= 0$:

$$lx - a = 0 \qquad mx - b = 0 \qquad nx - c = 0$$

Solving each of these equations produces the solutions.
Note: The coefficient k used in this example does not produce a solution because $k \neq 0$.

$$x = \frac{a}{l} \qquad x = \frac{b}{m} \qquad x = \frac{c}{n}.$$

WORKED EXAMPLE 12 Solving polynomial equations

Solve:

a. $x^3 = 9x$ **b.** $-2x^3 + 4x^2 + 70x = 0$ **c.** $2x^3 - 11x^2 + 18x - 9 = 0$.

THINK	WRITE
a. 1. Write the equation.	**a.** $x^3 = 9x$
2. Rearrange so all terms are on the left.	$x^3 - 9x = 0$
3. Take out a common factor of x.	$x(x^2 - 9) = 0$
4. Factorise the quadratic expression using the difference of two squares.	$x(x + 3)(x - 3) = 0$
5. Use the Null Factor Law to solve.	$x = 0,\ x + 3 = 0$ or $x - 3 = 0$
6. Write the values of x.	$x = 0,\ x = -3$ or $x = 3$
b. 1. Write the equation.	**b.** $-2x^3 + 4x^2 + 70x = 0$
2. Take out a common factor of $-2x$.	$-2x(x^2 - 2x - 35) = 0$
3. Factorise the quadratic expression.	$-2x(x - 7)(x + 5) = 0$
4. Use the Null Factor Law to solve.	$-2x = 0,\ x - 7 = 0$ or $x + 5 = 0$
5. Write the values of x.	$x = 0,\ x = 7$ or $x = -5$

c. 1. Name the polynomial.

c. Let $P(x) = 2x^3 - 11x^2 + 18x - 9$.

2. Use the factor theorem to determine a factor (search for a value a such that $P(a) = 0$). Consider factors of the constant term (that is, factors of 9 such as 1 and 3). The simplest value to try is 1.

$$\begin{aligned} P(1) &= 2 - 11 + 18 - 9 \\ &= 0 \end{aligned}$$

So $(x - 1)$ is a factor.

3. Use long or short division to determine another factor of $P(x)$.

$$\begin{array}{r} 2x^2 - 9x + 9 \\ x - 1\overline{)2x^3 - 11x^2 + 18x - 9} \\ \underline{2x^3 - 2x^2} \qquad\qquad \\ -9x^2 + 18x \quad\; \\ \underline{-9x^2 + 9x} \quad\; \\ 9x - 9 \\ \underline{9x - 9} \\ 0 \end{array}$$

$P(x) = (x - 1)(2x^2 + 9x - 9)$

4. Factorise the quadratic factor.

$P(x) = (x - 1)(2x - 3)(x - 3)$

$(x - 1)(2x - 3)(x - 3) = 0$

5. Use the Null Factor Law to solve.

$x - 1 = 0,\ 2x - 3 = 0$ or $x - 3 = 0$

6. Write the values of x.

$x = 1,\ x = \dfrac{3}{2}$ or $x = 3$

Resources

eWorkbook Topic 14 Workbook (worksheets, code puzzle and project) (ewbk-2040)

Digital documents SkillSHEET Factorising difference of two squares expressions (doc-5369)
SkillSHEET Solving quadratic equations (doc-5370)

Exercise 14.8 Solving polynomial equations

learnON

Individual pathways

■ PRACTISE	■ CONSOLIDATE	■ MASTER
1, 5, 8, 10, 14, 17	2, 6, 9, 11, 12, 15, 18	3, 4, 7, 13, 16, 19

To answer questions online and to receive **immediate feedback** and **sample responses** for every question, go to your learnON title at www.jacplus.com.au.

Fluency

WE12a,b For questions 1 to 4, solve the following.

1. a. $x^3 - 4x = 0$ **b.** $x^3 - 16x = 0$ **c.** $2x^3 - 50x = 0$

2. a. $-3x^3 + 81 = 0$ **b.** $x^3 + 5x^2 = 0$ **c.** $x^3 - 2x^2 = 0$

3. a. $-4x^3 + 8x = 0$ **b.** $12x^3 + 3x^2 = 0$ **c.** $4x^2 - 20x^3 = 0$

4. a. $x^3 - 5x^2 + 6x = 0$ b. $x^3 - 8x^2 + 16x = 0$ c. $x^3 + 6x^2 = 7x$

WE12c For questions 5 to 7, apply the factor theorem to solve the following.

5. a. $x^3 - x^2 - 16x + 16 = 0$ b. $x^3 - 6x^2 - x + 30 = 0$
   c. $x^3 - x^2 - 25x + 25 = 0$ d. $x^3 + 4x^2 - 4x - 16 = 0$

6. a. $x^3 - 4x^2 + x + 6 = 0$ b. $x^3 - 4x^2 - 7x + 10 = 0$
   c. $x^3 + 6x^2 + 11x + 6 = 0$ d. $x^3 - 6x^2 - 15x + 100 = 0$

7. a. $x^3 - 3x^2 - 6x + 8 = 0$ b. $x^3 + 2x^2 - 29x + 42 = 0$
   c. $2x^3 + 15x^2 + 19x + 6 = 0$ d. $-4x^3 + 16x^2 - 9x - 9 = 0$

8. **MC** *Note:* There may be more than one correct answer.
   Select a solution to $x^3 - 7x^2 + 2x + 40 = 0$ from the following.
   A. 5 B. −4 C. −2 D. 1 E. 3

9. **MC** A solution of $x^3 - 9x^2 + 15x + 25 = 0$ is $x = 5$. Select the number of other (distinct) solutions there are.
   A. 0 B. 1 C. 2 D. 3 E. 4

Understanding

10. Solve $P(x) = 0$ for each of the following.
    a. $P(x) = x^3 + 4x^2 - 3x - 18$ b. $P(x) = 3x^3 - 13x^2 - 32x + 12$
    c. $P(x) = -x^3 + 12x - 16$ d. $P(x) = 8x^3 - 4x^2 - 32x - 20$

11. Solve $P(x) = 0$ for each of the following.
    a. $P(x) = x^4 + 2x^3 - 13x^2 - 14x + 24$ b. $P(x) = -72 - 42x + 19x^2 + 7x^3 - 2x^4$
    c. $P(x) = x^4 + 2x^3 - 7x^2 - 8x + 12$ d. $P(x) = 4x^4 + 12x^3 - 24x^2 - 32x$

12. Solve each of the following equations.
    a. $x^3 - 3x^2 - 6x + 8 = 0$ b. $x^3 + x^2 - 9x - 9 = 0$ c. $3x^4 + 3x^3 - 18x = 0$

13. Solve each of the following equations.
    a. $2x^4 + 10x^3 - 4x^2 - 48x = 0$ b. $2x^4 + x^3 - 14x^2 - 4x + 24 = 0$ c. $x^4 - 2x^3 + 1 = 0$

Reasoning

14. Solve for a if $x = 2$ is a solution of $ax^3 - 6x^2 + 3x - 4 = 0$.
15. Solve for p if $x = \frac{p}{2}$ is a solution of $x^3 - 5x^2 + 2x + 8 = 0$.
16. Show that it is possible for a cuboid of side lengths x cm, $(x - 1)$ cm and $(x + 2)$ cm to have a volume that is 4 cm^3 less than twice the volume of a cube of side length x cm. Comment on the shape of such a cuboid.

Problem solving

17. Solve the following equation for x.
$$x^3 + 8 = x(5x - 2)$$
18. Solve the following equation for x.
$$2\left(x^3 + 5\right) = 13x(x - 1)$$
19. Solve the following equation for z.
$$z(z - 1)^3 = -2(z^3 - 5z^2 + z + 3)$$

14.9 Review

14.9.1 Topic summary

POLYNOMIALS

Polynomials

- Polynomials are expressions with only non-negative integer powers.
- The degree of a polynomial is the highest power of the variable that it contains.
- The leading term is the term with highest power of the variable.
- Monic polynomials have a leading coefficient of 1.
- Polynomials are often denoted $P(x)$.
- The value of a polynomial can be determined by substituting the x-value into the expression.
  e.g. $P(x) = x^4 - 3x^2 + 8$ is a monic polynomial of degree 4. The coefficient of the x^2 term is -3 and the constant term is 8.

The remainder theorem

- When $P(x)$ is divided by $(x - a)$, the remainder, R, is given by:
  $$R = P(a)$$
  e.g.
  The remainder when $P(x) = 2x^3 + 3x^2 - 4x - 5$ is divided by $(x + 2)$ is $P(-2)$.
  $$\begin{aligned} P(-2) &= 2(-2)^3 + 3(-2)^2 - 4(-2) - 5 \\ &= 2(-8) + 3(4) + 8 - 5 \\ &= -16 + 12 + 8 - 5 \\ &= -1 \end{aligned}$$

Operations on polynomials

- To add or subtract polynomials simply add or subtract like terms.
  e.g.
  $$\begin{aligned} &(2x^3 - 5x + 1) + (-6x^3 + 8x^2 + 3x - 11) \\ &= (2x^3 - 6x^3) + 8x^2 + (-5x + 3x) + (1 - 11) \\ &= -4x^3 + 8x^2 - 2x - 10 \end{aligned}$$
- To multiply polynomials use the same methods as with quadratic expressions. Use FOIL and then simplify.
- For polynomials of degree 3 and higher you may need to use FOIL multiple times.
  e.g.
  $$\begin{aligned} (x + 3)(x - 1)(2x + 4) &= (x + 3)(2x^2 + 4x - 2x - 4) \\ &= (x + 3)(2x^2 + 2x - 4) \\ &= 2x^3 + 2x^2 - 4x + 6x^2 + 6x - 12 \\ &= 2x^3 + 8x^2 + 2x - 12 \end{aligned}$$

Factorising polynomials

- The factor theorem states the following:
  If $P(a) = 0$, then $(x - a)$ is a factor of $P(x)$.
- This can be used to find a factor, and then other factors can be found using the methods used to factorise quadratics.
  e.g.
  $$\begin{aligned} P(x) &= x^3 - 2x^2 - 5x + 6 \\ P(1) &= (1)^3 - 2(1)^2 - 5(1) + 6 \\ &= 1 - 2 - 5 + 6 \\ &= 0 \end{aligned}$$
  Therefore, $(x - 1)$ is a factor of $P(x)$.
  $$\begin{aligned} P(x) &= (x - 1)(x^2 - x - 6) \\ &= (x - 1)(x - 3)(x + 2) \end{aligned}$$

Long division

- Polynomials can be divided using long division.
- The example below shows the division of $P(x) = x^3 + 2x^2 - 13x + 10$ by $x - 3$.
  $$\begin{array}{rl} & x^3 + 5x + 2 \quad \leftarrow \text{Quotient} \\ x - 3\,\big) & x^3 + 2x^2 - 13x + 10 \\ & \underline{x^3 - 3x} \\ & 5x^2 - 13x \\ & \underline{5x^2 - 15x} \\ & 2x + 10 \\ & \underline{2x - 6} \\ & 16 \quad \leftarrow \text{Remainder} \end{array}$$
- The result is:
  $P(x) = (x - 3)(x^2 + 5x + 2) + 16$

Solving polynomial equations

- To solve a polynomial equation:
  1. express in the form $P(x) = 0$
  2. factorise $P(x)$
  3. solve using the Null Factor Law.

  e.g.
  $$\begin{aligned} 2x^3 + 7x^2 &= 9 \\ 2x^3 + 7x^2 - 9 &= 0 \end{aligned}$$
  Let $P(x) = 2x^3 + 7x^2 - 9$.
  $$\begin{aligned} P(1) &= 2(1)^3 + 7(1)^2 - 9 \\ &= 2 + 7 - 9 \\ &= 0 \end{aligned}$$
  Therefore, $(x - 1)$ is a factor of $P(x)$.
  $$\begin{aligned} P(x) &= (x - 1)(2x^2 + 9x + 9) \\ &= (x - 1)(2x - 3)(x + 3) \end{aligned}$$
  Using the Null Factor Law:
  $x = 1, x = -\dfrac{3}{2}$ or $x = -3$

14.9.2 Success criteria

Tick the column to indicate that you have completed the subtopic and how well you have understood it using the traffic light system.

(**Green:** I understand; **Yellow:** I can do it with help; **Red:** I do not understand)

Subtopic	Success criteria	Green	Yellow	Red
14.2	I can identify polynomial expressions.			
	I can state the degree, leading term and leading coefficient of a polynomial.			
14.3	I can add, subtract and multiply polynomial expressions.			
14.4	I can divide polynomials by linear expressions using long division.			
	I can determine the quotient and remainder when dividing polynomials by linear expressions.			
14.5	I can determine the value of a polynomial for a given value.			
14.6	I can identify factors of polynomials using the factor theorem.			
14.7	I can factorise polynomials using long division.			
	I can factorise polynomials using short division or inspection.			
14.8	I can solve polynomial equations by applying the Null Factor Law.			

14.9.3 Project

Investigating polynomials

A polynomial is a function involving the sum of integer powers of a variable (for example, $y = -4x^3 + 3x^2 - 4$). The highest power of the variable determines the degree of the polynomial. In the case of the given example, the degree is 3.

A polynomial of the first degree is a linear function (for example, $y = 3x - 8$), and a second-degree function is a quadratic (for example, $y = 5x^2 - 6x + 7$). Let us investigate how the degree of a polynomial affects the shape of its graph.

In order to simplify the graphing of these functions, the polynomials will be expressed in factor form. A graphics calculator or some other digital technology will make the graphing process less tedious.

It will be necessary to adjust the window of the calculator from time to time in order to capture the relevant features of the graph.

1. Consider the following polynomials.
   a. $y_1 = (x + 1)$
   b. $y_2 = (x + 1)(x - 2)$
   c. $y_3 = (x + 1)(x - 2)(x + 3)$
   d. $y_4 = (x + 1)(x - 2)(x + 3)(x - 4)$
   e. $y_5 = (x + 1)(x - 2)(x + 3)(x - 4)(x + 5)$
   f. $y_6 = (x + 1)(x - 2)(x + 3)(x - 4)(x + 5)(x - 6)$

   For each of the functions:
   i. give the degree of the polynomial
   ii. sketch the graph, marking in the x-intercepts
   iii. describe how the degree of the polynomial affects the shape of the graph.

   Complete question 1 on a separate sheet of paper.
2. Let us now look at the effect that the exponent of each factor has on the shape of the graph of the polynomial. Consider the following functions.
   a. $y_1 = (x + 1)(x - 2)(x + 3)$
   b. $y_2 = (x + 1)^2(x - 2)(x + 3)$
   c. $y_3 = (x + 1)^2(x - 2)^2(x + 3)$
   d. $y_4 = (x + 1)^2(x - 2)(x + 3)^3$
   e. $y_5 = (x + 1)^3(x - 2)(x + 3)^4$
   f. $y_6 = (x + 1)^5(x - 2)^3(x + 3)^2$

   i. On a separate sheet of paper, draw a sketch of each of the polynomials, marking in the x-intercepts.
   ii. Explain how the power of the factor affects the behaviour of the graph at the x-intercept.
3. Create and draw a sketch of polynomials with the following given characteristics. Complete your graphs on a separate sheet of paper.
   a. A first-degree polynomial that:
      i. crosses the x-axis
      ii. does not cross the x-axis.
   b. A second-degree polynomial that:
      i. crosses the x-axis twice
      ii. touches the x-axis at one and only one point.
   c. A third-degree polynomial that crosses the x-axis:
      i. three times
      ii. twice
      iii. once.
   d. A fourth-degree polynomial that crosses the x-axis:
      i. four times
      ii. three times
      iii. twice
      iv. once.
4. Considering the powers of factors of polynomials, write a general statement outlining the conditions under which the graph of a polynomial will pass through the x-axis or just touch the x-axis.

on Resources

eWorkbook Topic 14 Workbook (worksheets, code puzzle and project) (ewbk-2040)

Interactivities Crossword (int-2875)
Sudoku puzzle (int-3892)

To answer questions online and to receive **immediate corrective feedback** and **fully worked solutions** for all questions, go to your learnON title at www.jacplus.com.au.

Fluency

1. **MC** Select which of the following is *not* a polynomial.

A. $x^3 - \frac{x^2}{3} + 7x - 1$ **B.** $a^4 + 4a^3 + 2a + 2$ **C.** $\sqrt{x^2 + 3x + 2}$

D. 5 **E.** $2x^2 + 4x - 8$

2. Consider the polynomial $y = -\frac{1}{7}x^4 + x^5 + 3$.

a. State the degree of y.
b. State the coefficient of x^4.
c. State the constant term.
d. Determine the leading term.

3. **MC** The expansion of $(x + 5)(x + 1)(x - 6)$ is:

A. $x^3 - 30$ **B.** $x^3 + 12x^2 - 31x + 30$ **C.** $x^3 - 31x - 30$

D. $x^3 + 5x^2 - 36x - 30$ **E.** $x^3 - 31x^2 - 30$

4. **MC** $x^3 + 5x^2 + 3x - 9$ is the expansion of:

A. $(x + 3)^3$ **B.** $x(x + 3)(x - 3)$ **C.** $(x - 1)(x + 3)^2$

D. $(x - 1)(x + 1)(x + 3)$ **E.** $(x - 1)(x + 3)$

5. Expand each of the following.

a. $(x - 2)^2(x + 10)$ **b.** $(x + 6)(x - 1)(x + 5)$ **c.** $(x - 7)^3$ **d.** $(5 - 2x)(1 + x)(x + 2)$

6. **MC** Consider the following long division.

$$\begin{array}{r|l} & x^2 + x + 2 \\ x - 4 & x^3 + 5x^2 + 6x - 1 \\ & \underline{x^3 + 4x^2} \\ & \quad x^2 + 6x \\ & \quad \underline{x^2 + 4x} \\ & \qquad 2x - 1 \\ & \qquad \underline{2x + 8} \\ & \qquad\quad -9 \end{array}$$

a. The quotient is:

A. -9 **B.** 9 **C.** $x + 4$ **D.** $x^2 + x + 2$ **E.** $x^2 + 2$

b. The remainder is:

A. -9 **B.** 2 **C.** 4 **D.** $2x - 1$ **E.** 6

7. Determine the quotient and remainder when the first polynomial is divided by the second in each case.

a. $x^3 + 2x^2 - 16x - 3,\ (x + 2)$ **b.** $x^3 + 3x^2 - 13x - 7,\ (x - 3)$ **c.** $-x^3 + x^2 + 4x - 7,\ (x + 1)$

8. **MC** If $P(x) = x^3 - 3x^2 + 7x + 1$, then $P(-2)$ equals:

A. -34 **B.** -33 **C.** -9 **D.** 9 **E.** 33

9. If $P(x) = -3x^3 + 2x^2 + x - 4$, calculate:
   a. $P(1)$
   b. $P(-4)$
   c. $P(2a)$

10. Without dividing, determine the remainder when $x^3 + 3x^2 - 16x + 5$ is divided by $(x - 1)$.

11. Show that $(x + 3)$ is a factor of $x^3 - 2x^2 - 29x - 42$.

12. Factorise $x^3 + 4x^2 - 100x - 400$.

13. Solve:
   a. $(2x + 1)(x - 3)^2 = 0$
   b. $x^3 - 9x^2 + 26x - 24 = 0$
   c. $x^4 - 4x^3 - x^2 + 16x - 12 = 0$

Problem solving

14. Let $P(x) = a_n x^n + a_{n-1}x^{n-1} + ... + a_1 x + a_0$ be a polynomial where the coefficients are integers. Also let $P(w) = 0$ where w is an integer. Show that w is a factor of a_0.

15. Evaluate the area of a square whose sides are $(2x - 3)$ cm. Expand and simplify your answer. If the area is $16\,\text{cm}^2$, determine the value of x.

16. A window is in the shape of a semicircle above a rectangle. The height of the window is $(6x + 1)$ cm and its width is $(2x + 2)$ cm.
   a. Evaluate the total area of the window.
   b. Expand and simplify your answer.
   c. Determine the perimeter of the window.

17. Answer the following questions.
   a. Determine the volume of a cube of side $(x + 4)$ cm.
   b. Evaluate the surface area of the cube.
   c. Determine the value of x for which the volume and surface are numerically equal.
   d. Calculate the value of x if the numerical value of the volume is 5 less than the numerical value of the surface area.

18. Determine the quotient and remainder when $mx^2 + nx + q$ is divided by $(x - p)$.

19. When $P(x)$ is divided by $(x - n)$, the quotient is $x^2 - 2x + n$ and the remainder is $(n + 1)$. Evaluate the value of $P(x)$.

on To test your understanding and knowledge of this topic, go to your learnON title at www.jacplus.com.au and complete the **post-test**.

Online Resources

Below is a full list of **rich resources** available online for this topic. These resources are designed to bring ideas to life, to promote deep and lasting learning and to support the different learning needs of each individual.

eWorkbook

Download the workbook for this topic, which includes worksheets, a code puzzle and a project (ewbk-2040) ☐

Solutions

Download a copy of the fully worked solutions to every question in this topic (sol-0748) ☐

Digital documents

14.3 SkillSHEET Expanding the product of two linear factors (doc-5366) ☐
14.5 SkillSHEET Substitution into quadratic equations (doc-5367) ☐
14.7 SkillSHEET Factorising quadratic trinomials (doc-5368) ☐
14.8 SkillSHEET Factorising difference of two squares expressions (doc-5369) ☐
SkillSHEET Solving quadratic equations (doc-5370) ☐

Video eLessons

14.2 Polynomials (eles-4975) ☐
14.3 Operations with polynomials (eles-4976) ☐
14.4 Long division of polynomials (eles-4977) ☐
14.5 Polynomial values (eles-4978) ☐
14.6 The remainder theorem (eles-4979) ☐
The factor theorem (eles-4980) ☐
14.7 Using long division (eles-4981) ☐
Using short division, or by inspection (eles-4982) ☐
14.8 Solving polynomial equations (eles-4983) ☐

Interactivities

14.2 Degrees of polynomials (int-6203) ☐
14.3 Adding and subtracting polynomials (int-6204) ☐
14.4 Long division of polynomials (int-2793) ☐
14.9 Crossword (int-2875) ☐
Sudoku puzzle (int-3892) ☐

Teacher resources

There are many resources available exclusively for teachers online.

To access these online resources, log on to **www.jacplus.com.au**.

Answers

Topic 14 Polynomials

Exercise 14.1 Pre-test

1. a. 2 b. 3 c. 1 d. 0
2. A
3. $a = 2,\ b = -1$
4. D
5. A
6. a. 3 b. −3 c. −1 d. 2
7. A
8. $x^2 + x$
9. 9
10. 18
11. E
12. $b = 2$
13. B
14. C
15. $x = -1, -\frac{1}{2}$ or -6

Exercise 14.2 Polynomials

1. a. i. 3 ii. x
   b. i. 7 ii. x
   c. i. 2 ii. x
2. a. i. 6 ii. x
   b. i. 8 ii. y
   c. i. 5 ii. u
3. a. i. 5 ii. e
   b. i. 1 ii. g
   c. i. 6 ii. f
4. a. Polynomial 3b b. Polynomial 1c
   c. Polynomial 1a d. Polynomials 1a, 2a and 2b
5. a. N b. P c. P
6. a. N b. N c. P
7. a. P b. N c. N
8. a. 3 b. x c. 4
   d. 5 e. $3x$ f. $-2x^3$
9. a. 7 b. w c. 7
   d. 0 e. −9 f. $6w^7$
10. a. 4 b. 1 c. x^4 d. 1
11. a. $4x^3 + 2x^2 - 10x + 18$ b. $3x^4 - 3x^3 - x^2 + 7x - 7$
    c. $5x^3 - 4x^2 - 13x - 6$
12. a. 6
    b. t
    c. 6
    d. Sample responses can be found in the worked solutions in the online resources.
    e. Sample responses can be found in the worked solutions in the online resources.
13. a. 5 units to the right of the origin
    b. 4 units to the right of the origin
    c. The body moves towards the origin, then away.
14. $a = 1,\ b = -4$
15. $a = 4, b = -6$
16. $a = \pm 3,\ b = \pm 2$

Exercise 14.3 Adding, subtracting and multiplying polynomials

1. a. $x^4 + 2x^3 - x^2 - 10$
   b. $x^6 + 2x^4 - 3x^3 + 9x^2 + 5$
   c. $5x^3 - 5x^2 + 7x - 13$
   d. $2x^4 + 3x^3 + 12x^2 - 4x + 14$
   e. $x^5 + 13x^4 - 10$
2. a. $x^4 + 2x^2 + 2x + 4$ b. $x^6 - x^5 + x^3 + x^2 + 2$
   c. $5x^7 - 4x^3 + 5x$ d. $10x^4 - 7x^2 + 20x + 5$
   e. $2x^3 + 6x^2 - 10x + 15$
3. a. $x^3 + 7x^2 + 6x$ b. $x^3 - 7x^2 - 18x$
   c. $x^3 + 8x^2 - 33x$ d. $2x^3 + 10x^2 + 12x$
   e. $48x - 3x^3$
4. a. $5x^3 + 50x^2 + 80x$ b. $x^3 + 4x^2$
   c. $2x^3 - 14x^2$ d. $-30x^3 - 270x^2$
   e. $-7x^3 - 56x^2 - 112x$
5. a. $x^3 + 12x^2 + 41x + 42$ b. $x^3 - 3x^2 - 18x + 40$
   c. $x^3 + 3x^2 - 36x + 32$ d. $x^3 - 6x^2 + 11x - 6$
   e. $x^3 + 6x^2 - x - 6$
6. a. $x^3 + 5x^2 - 49x - 245$ b. $x^3 + 4x^2 - 137x - 660$
   c. $x^3 + 3x^2 - 9x + 5$ d. $x^3 - 12x^2 + 21x + 98$
   e. $x^3 + x^2 - x - 1$
7. a. $x^3 + 13x^2 + 26x - 112$
   b. $3x^3 + 26x^2 + 51x - 20$
   c. $4x^4 + 3x^3 - 37x^2 - 27x + 9$
   d. $10x^3 - 49x^2 + 27x + 36$
   e. $-6x^3 - 71x^2 - 198x + 35$
8. a. $21x^4 - 54x^3 - 144x^2 + 96x$
   b. $54x^3 + 117x^2 - 72x$
   c. $24x^3 - 148x^2 + 154x + 245$
   d. $20x^4 - 39x^3 - 50x^2 + 123x - 54$
   e. $4x^3 + 42x^2 + 146x + 168$
9. a. $x^3 + 6x^2 + 12x + 8$ b. $x^3 + 15x^2 + 75x + 125$
   c. $x^3 - 3x^2 + 3x - 1$
10. a. $x^4 - 12x^3 + 54x^2 - 108x + 81$
    b. $8x^3 - 72x^2 + 216x - 216$
    c. $81x^4 + 432x^3 + 864x^2 + 768x + 256$
11. $(2a + 5b)x + (2b - 5c)$
12. $x^4 + (a - 4b)x^3 + (2a - 4ab + 3b^2)x^2 +$
    $(2a^2 - 2ab + 3ab^2)x - 2a^2b$
13. $a = 1,\ b = -12,\ c = 54,\ d = -108,\ e = 81$
14. $8x^3 - 45x^2 + 78x - 43$
15. $\frac{1}{8}(11x^3 - 105x^2 + 73x - 27)$
16. $a = 2,\ b = 3$

17. $a = 1, b = -2$ and $c = 1$

18. $a = 1, b = 4$ and $c = -1$

Exercise 14.4 Long division of polynomials

1. a. $x^2 + 2x$, 9 b. $x^2 + x + 3$, -2
   c. $x^2 + 3x - 6$, 19 d. $x^2 - x + 5$, -17

2. a. $x^2 + 4x - 6$, 14 b. $x^2 + 1$, 2
   c. $x^2 + 5$, -36 d. $x^2 - x + 6$, -11

3. a. $x^2 + 4x + 3$, -3 b. $x^2 + 4x + 13$, 48
   c. $x^2 + 3x - 3$, -11 d. $x^2 - 3x + 7$, 5

4. a. $x^2 - 2x - 3$, -17 b. $x^2 - 6x + 3$, -4
   c. $x^2 + 14x + 72$, 359 d. $x^2 + 8x + 27$, 104

5. a. $3x^2 - 7x + 20$, -35 b. $4x^2 - 8x + 18$, -22
   c. $2x^2 - 3x + 3$, 7

6. a. $2x^2 - 9$, 35 b. $4x^2 + 2x - 3$, -1
   c. $3x^2 + x - 1$, -2

7. a. $3x^2 - 2x + 1$, 5 b. $2x^2 + 5x - 6$, -7
   c. $4x^2 - 7x - 2$, -3

8. a. $x^2 - 4x + 3$, 8 b. $x^2 + x - 6$, -11
   c. $3x^2 + 2x + 1$, 13

9. a. $-x^2 - 5x - 2$, -14 b. $-3x^2 - 2x + 4$, -3
   c. $-x^2 + 5x + 6$, 9 d. $-2x^2 + 7x - 1$, 1

10. a. $x^2 - x - 2$, 3 b. x^2, -7
    c. $x^2 - x - 2$, -8 d. $-x^2 - x - 8$, 0

11. a. $5x - 2$, 7 b. $2x^2 - 2x + 10$, -54
    c. $-2x^2 - 4x - 9$, -16 d. $-2x^2 + 4x - 1$, 1

12. a. $x^3 + 2x^2 + 5x - 2$, -2
    b. $x^3 + 2x^2 - 9x - 18$, 0
    c. $x^4 - 3x^3 + 6x^2 - 18x + 58$, -171

13. a. $2x^5 - 4x^4 + 7x^3 - 13x^2 + 32x - 69$, 138
    b. $6x^3 + 17x^2 + 53x + 155$, 465
    c. $x^3 - \frac{7}{3}x^2 + \frac{7}{9}x + 3\frac{20}{27}$, $-3\frac{20}{27}$

14. Quotient: $x^3 - 4x^2 + 8x - 12$
    Remainder: $(a + 72)$

15. Quotient $= ax + (b + ad)$
    Remainder $= Rc + d(b + ad)$

16. $q = \dfrac{2p}{c}$

17. $a = -3$

18. $a = 6$

19. $a = 3, b = -5$

Exercise 14.5 Polynomial values

1. a. 10 b. 11

2. a. 18 b. 43

3. a. 3 b. -22

4. a. -77 b. $2a^3 - 3a^2 + 2a + 10$

5. a. $16b^3 - 12b^2 + 4b + 10$
   b. $2x^3 + 9x^2 + 14x + 18$

6. a. $2x^3 - 21x^2 + 74x - 77$
   b. $-128y^3 - 48y^2 - 8y + 10$

7. a. 4 b. 15 c. 0
   d. 4 e. 15 f. 0

8. a. 10 b. 28 c. -8
   d. 10 e. 28 f. -8

9. a. 3 b. 11 c. -21
   d. 3 e. 11 f. -21

10. a. -7 b. 5 c. -7
    d. -7 e. 5 f. -7

11. a. $P(-8)$ b. $P(7)$ c. $P(a)$

12. a. -1 b. -6249 c. $2(a-3)^5 + 1$

13. a. $-2a^3 - 3a^2 + a + 4$ b. $-2a^3 - 9a^2 - 11a - 1$

14. $b = 1, 4$

15. $c = 8$

16. $c = 2$

17. $b = -2, c = 5$

Exercise 14.6 The remainder and factor theorems

1. a. -30 b. 0 c. 0 d. -24

2. a. -24 b. $k^3 + 3k^2 - 10k - 24$
   c. $-n^3 + 3n^2 + 10n - 24$ d. $-27c^3 + 27c^2 + 30c - 24$

3. a. 58 b. -8

4. a. 11 b. -9

5. a. 158 b. 6

6. a. 35 b. -6

7. a. -202 b. 441

8. 6

9. 3

10. 1

11. -2

12. 2

13. 2

14. $-5, 2$

15. $a = -5, b = -3$

16. a. D b. C, D c. D d. A, C, D

17. a. $(x - 1)$ b. $(x - 3)$ or $(x - 2)$
    c. $(x - 3)$ or $(x + 2)$ d. $(x - 6)$ or $(x + 4)$ or $(x + 5)$

18. a–d. Sample responses can be found in the worked solutions in the online resources.

19. a–d. Sample responses can be found in the worked solutions in the online resources.

20. Sample responses can be found in the worked solutions in the online resources.

21. $a = 2$

22. $a = 3, b = 2$

23. $a = -5, b = 41$, $(x + 3)$ and $(x - 5)$

24. $k = -4$

Exercise 14.7 Factorising polynomials

1. a. $(x + 1)(x + 3)(x + 6)$ b. $(x + 1)(x + 2)(x + 5)$

2. a. $(x + 1)(x + 2)(x + 9)$ b. $(x + 1)(x + 3)(x + 4)$

3. a. $(x+3)(x+4)(x+7)$ b. $(x+2)(x+3)(x+7)$

4. a. $(x+1)^2(x+2)$ b. $(x+2)^2(x+3)$

5. a. $(x+4)(x+5)^2$ b. $x(x+5)(x+8)$

6. a. $x(x+3)(x+4)$ b. $x(x+5)^2$

7. a. $x(x+1)(x+5)$ b. $x^2(x+6)$

8. a. $(x-1)(x+1)^2$ b. $(x-2)(x-1)(x+1)$
c. $(x+1)^2(x+5)$ d. $(x-3)(x+2)^2$

9. a. $(x+1)(x+4)^2$ b. $(x-5)(x-2)(x+2)$
c. $(x-1)(x+1)(x+2)$ d. $(x-3)(x+1)(x+2)$

10. a. $(x+1)(x+2)(x+5)$ b. $(x-3)(x+1)(x+3)$
c. $(x-2)^2(x+3)$ d. $(x-4)(x+5)(x+8)$

11. a. $(2x+3)(x-1)(x+2)$ b. $(3x-1)(x+1)(x+4)$
c. $(3x+2)(x-2)(x+2)$ d. $(4x+3)(x+3)(x+5)$

12. a. $(x+1)(x^2+1)$ b. $(x+1)(2x+3)^2$
c. $(x-2)(2x-1)(3x-4)$ d. $(7x-2)(x-2)(x+4)$

13. a. $x(x-2)(3x+5)$ b. $2x(x+1)(2x-1)$
c. $3x(x-4)(x+2)$ d. $-2x(x+3)^2$

14. a. $-x(x+4)(x+3)$ b. $-(x-1)(x+1)(x+3)$
c. $-2x(x-3)(x-2)$ d. $-(x-2)^2(5x-4)$

15. $(x-1)(x+2)(x+2)(x-3)$

16. $-(x-2)(x+2)(x+3)(x-4)(x-5)$

17. $(x-p+(a+b))$

18. $x(x-1)(x+1)(x-2)(x-3)$

19. $a=-2, b=4, (x-1)^2\,(x+1)^2\,(x-2)$

20. The other two factors are $(x-4)$ and $(x+1)$.

Exercise 14.8 Solving polynomial equations

1. a. $-2, 0, 2$ b. $-4, 0, 4$ c. $-5, 0, 5$

2. a. 3 b. $-5, 0$ c. $0, 2$

3. a. $-\sqrt{2}, 0, \sqrt{2}$ b. $-\frac{1}{4}, 0$ c. $0, \frac{1}{5}$

4. a. $0, 2, 3$ b. $0, 4$ c. $-7, 0, 1$

5. a. $-4, 1, 4$ b. $-2, 3, 5$
c. $-5, 1, 5$ d. $-4, -2, 2$

6. a. $-1, 2, 3$ b. $-2, 1, 5$
c. $-3, -2, -1$ d. $-4, 5$

7. a. $-2, 1, 4$ b. $-7, 2, 3$
c. $-6, -\frac{1}{2}, -1$ d. $-\frac{1}{2}, \frac{3}{2}, 3$

8. A, C

9. B

10. a. $-3, 2$ b. $-2, \frac{1}{3}, 6$
c. $-4, 2$ d. $-1, \frac{5}{2}$

11. a. $-4, -2, 1, 3$ b. $-2, -\frac{3}{2}, 3, 4$
c. $-3, -2, 1, 2$ d. $-4, -1, 0, 2$

12. a. $-2, 1, 4$ b. $-3, -1, 3$ c. $-3, 0, 2$

13. a. $-4, -3, 0, 2$ b. $-2, \frac{3}{2}, 2$ c. $-1, 1$

14. 2.75

15. $-2, 4, 8$

16. $x = 1.48$ (to 2 decimal places)

17. $x = -1, 4$ and 2

18. $x = \frac{-1}{2}, 2, 5$

19. $z = -1, 1, -2$ and 3

Project

1. a. i. 1

ii.

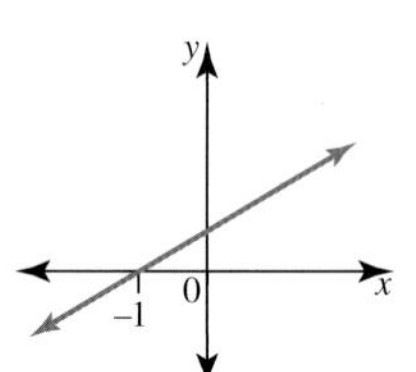

iii. The graph is linear and crosses the x-axis once (at $x = -1$).

b. i. 2

ii.

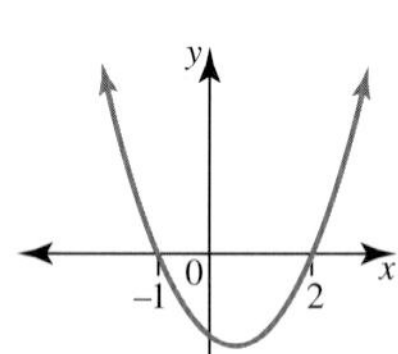

iii. The graph is quadratic and crosses the x-axis twice (at $x = -1$ and $x = 2$).

c. i. 3

ii.

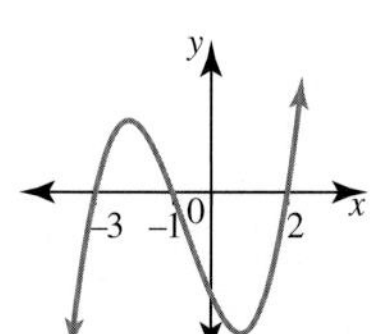

iii. The graph is a curve and crosses the x-axis 3 times (at $x = -1$, $x = 2$ and $x = -3$).

d. i. 4

ii.

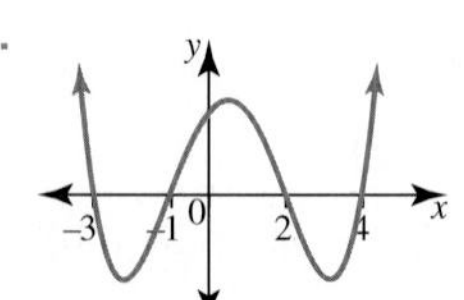

iii. The graph is a curve and crosses the x-axis 4 times (at $x = -1$, $x = 2$, $x = -3$ and $x = 4$).

e. i. 5

ii.

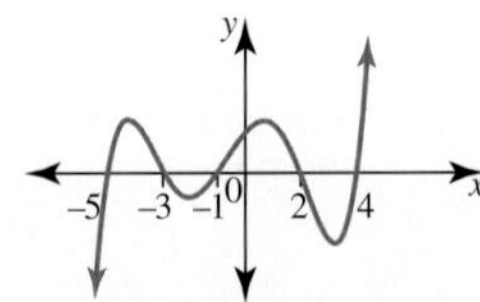

iii. The graph is a curve and crosses the x-axis 5 times (at $x = -1$, $x = 2$, $x = -3$, $x = 4$ and $x = -5$).

f. i. 6

ii.

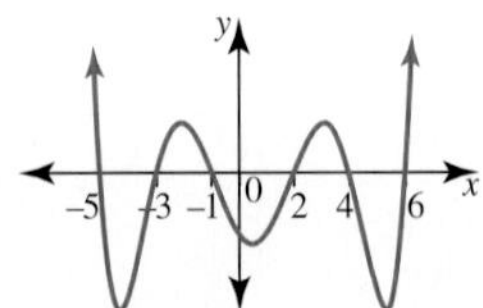

iii. The graph is a curve and crosses the x-axis 6 times (at $x = -1$, $x = 2$, $x = -3$, $x = 4$, $x = -5$ and $x = 6$).

2. a. i.

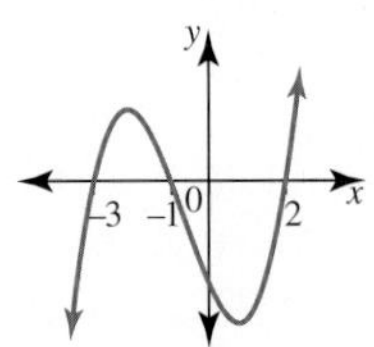

ii. Each factor is raised to the power 1. The polynomial is of degree 3 and the graph crosses the x-axis in 3 places(-3, -1 and 2).

b. i.

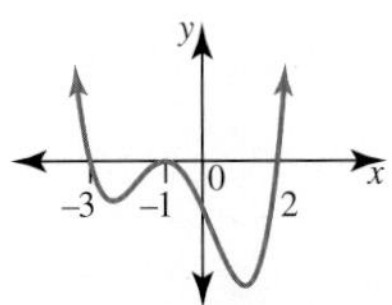

ii. The factor $(x + 1)$ is raised to the power 2 while the other two factors are raised to the power 1. The power 2 causes the curve not to cross the x-axis at $x = -1$ but to be curved back on itself.

c. i.

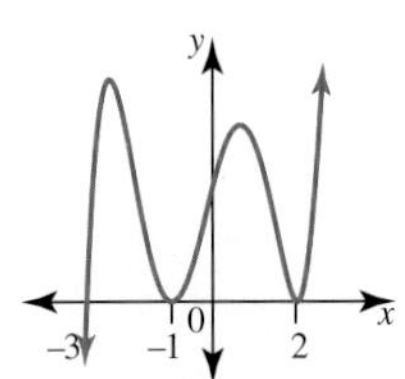

ii. The power 2 on the two factors $(x + 1)$ and $(x - 2)$ causes the curve to be directed back on itself and not to cross the x-axis at those two points ($x = -1$ and $x = 2$).

d. i.

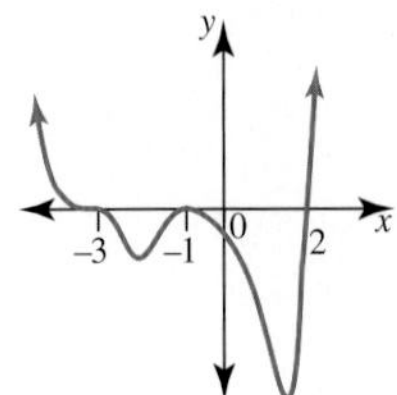

ii. The power 3 on the factor $(x + 3)$ causes the curve to run along the axis at that point then to cross the axis (at $x = -3$).

e. i.

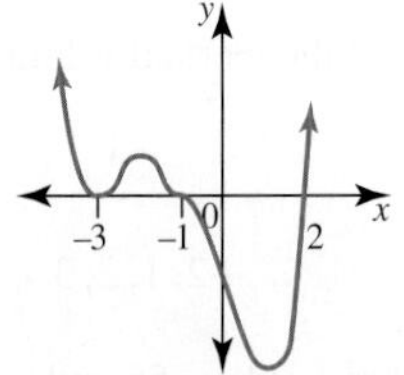

ii. The power 3 on the factor $(x + 1)$ causes the curve to run along the axis at $x = -1$, then cross the axis. The power 4 on the factor $(x + 3)$ causes the curve to be directed back on itself without crossing the axis at $x = -3$.

f. i.

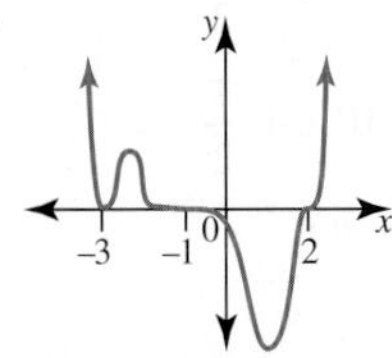

ii. The power 5 on the factor $(x + 1)$ causes the curve to run along the axis at $x = -1$, then cross the axis.

3. Answers will vary. Possible answers could be as follows.

a. i. $y = 3x + 2$

ii. $y = 4$

b. i. $y = (x + 1)(x + 2)$

ii. $y = (x + 1)^2$

c. i. $y = (x + 1)(x + 2)(x + 3)$

ii. Not possible

iii. $y = (x + 1)^2(x + 2)$

d. i. $y = (x + 1)(x + 2)(x + 3)(x + 4)$

ii. Not possible

iii. $y = (x + 1)^2(x + 2)(x + 3)$, $y = (x + 1)^3(x + 2)$

iv. Not possible

4. If the power of the factor of a polynomial is an odd integer, the curve will pass through the x-axis. If the power is 1, the curve passes straight through. If the power is 3, 5. . ., the curve will run along the x-axis before passing through it. On the other hand, an even power of a factor causes the curve to just touch the x-axis then move back on the same side of the x-axis.

Exercise 14.9 Review questions

1. C

2. a. 5 b. $-\dfrac{1}{7}$ c. 3 d. x^5

3. C

4. C

5. a. $x^3 + 6x^2 - 36x + 40$

b. $x^3 + 10x^2 + 19x - 30$

c. $x^3 - 21x^2 + 147x - 343$

d. $-2x^3 - x^2 + 11x + 10$

6. a. D b. A

7. a. $x^2 - 16, 29$ b. $x^2 + 6x + 5, 8$ c. $-x^2 + 2x + 2, -9$

8. B

9. a. -4

b. 216

c. $-24a^3 + 8a^2 + 2a - 4$

10. -7

11. Sample responses can be found in the worked solutions in the online resources.

12. $(x-10)(x+4)(x+10)$

13. a. $-\frac{1}{2}, 3$ b. 2, 3, 4 c. $-2, 1, 2, 3$

14. For example, given $P(x) = x^3 - x^2 - 34x - 56$ and $P(7) = 0 \Rightarrow (x-7)$ is a factor and 7 is a factor of 56.

15. $4x^2 - 12x + 9; x = -\frac{1}{2}, \frac{7}{2}$

16. a. Area $= (\frac{1}{2}\pi + 10)x^2 + (\pi + 10)x + \frac{\pi}{2}$

  b. Area $= (\frac{1}{2}\pi + 10)x^2 + (\pi + 10)x + \frac{\pi}{2}$

  c. Perimeter $= (12 + \pi)x + (2 + \pi)$

17. a. $(x+4)^3$

  b. $6(x+4)^2$

  c. $x = 2$

  d. $-3, \frac{-3+3\sqrt{5}}{2}$

18. $mx + (n + mp); q + p(n + mp)$

19. $x^3 - (2+n)x^2 + 3nx - (n^2 - n - 1)$

10A

15 Functions and relations

LEARNING SEQUENCE

15.1 Overview

Why learn this?

Functions and relations are broad and interesting topics of study. They are topics with many real-world applications and are very important topics to understand as you head towards higher studies in mathematics. You will have already seen some functions and relations in your maths classes; linear equations, quadratics and polynomials are all examples of functions, and circles are examples of relations.

In your previous study of quadratics you learned about graphs with an x^2 term, but have you wondered what a graph would look like if it had an x^3 term or an x^4 term? You will be learning about these and other graphs in this topic. Have you ever heard the phrases 'exponential growth' or 'exponential decay'? In this topic you will also learn exactly what these phrases mean and how to graph and interpret various exponential situations.

An understanding of how to apply and use functions and relations is relevant to many professionals. Medical teams working to map the spread of diseases, engineers designing complicated structures such as the Sydney Opera House, graphic designers creating a new logo, video game designers developing a new map for their game — all require the use and understanding of functions and relations.

This topic builds on what you already know and extends it into new areas of mathematics. By the end of this topic you will know all about different types of functions and relations, and how to graph them, interpret them and transform them.

Where to get help

Go to your learnON title at **www.jacplus.com.au** to access the following digital resources. The Online Resources Summary at the end of this topic provides a full list of what's available to help you learn the concepts covered in this topic.

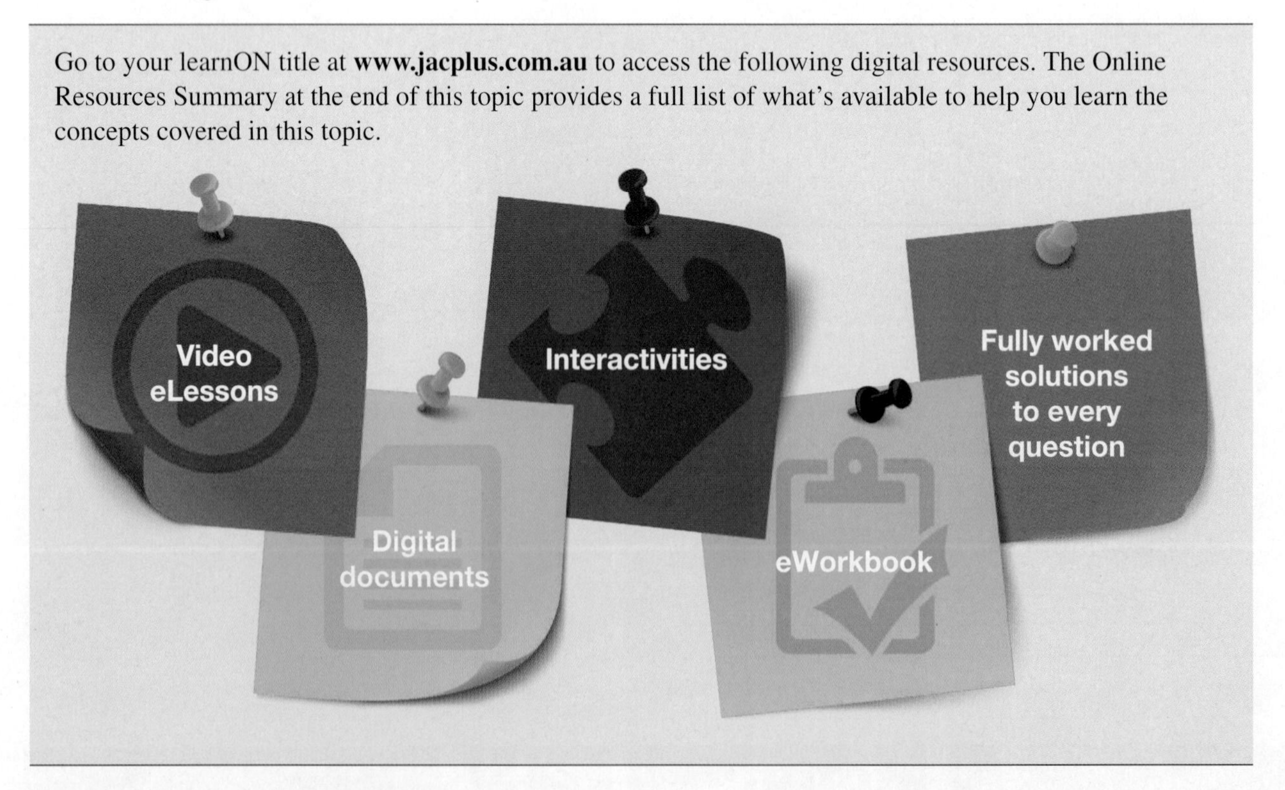

Exercise 15.1 Pre-test

Complete this pre-test in your learnON title at www.jacplus.com.au and receive **automatic marks**, **immediate corrective feedback** and **fully worked solutions**.

1. **MC** Choose the type of relation that the graph represents.
   - A. One-to-one relation
   - B. One-to-many relation
   - C. Many-to-one relation
   - D. Many-to-many relation
   - E. None of these

2. **MC** A function is a relation that is one-to-one or:
   - A. many-to-one
   - B. many-to-many
   - C. one-to-many
   - D. one-to-two
   - E. none of these

3. **MC** The graph below is a function.

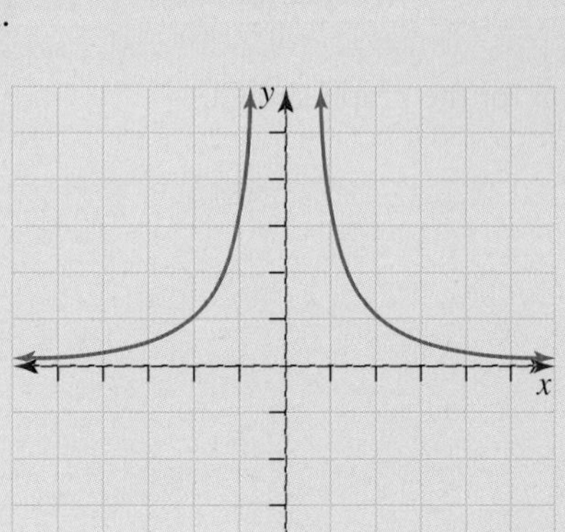

State whether this is true or false.

4. **MC** Select the correct domain of the relation shown in the graph.
   - A. $x \in R$
   - B. $x \in [-1, 2]$
   - C. $x \in [2, 8]$
   - D. $x \in [0, 8]$
   - E. $x \in [-1, 8]$

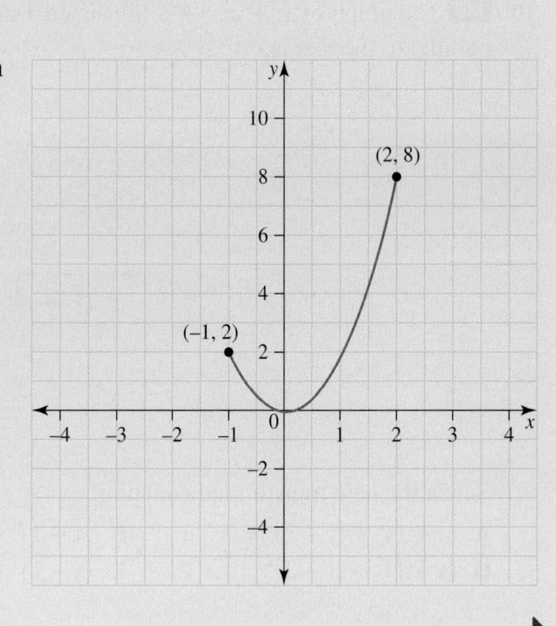

5. **MC** Select the correct range of the function $f(x) = \frac{1}{x-2} + 1$.

A. $y \in R$
B. $y \in R \backslash \{1\}$
C. $y \in R \backslash \{-1\}$
D. $y \in R \backslash \{2\}$
E. $y \in R \backslash \{-2\}$

6. **MC** Select the correct equation of the inverse function $f(x) = (x+1)^2 - 2, x \leq -1$.

A. $f^{-1}(x) = \sqrt{x+2} - 1, x \geq -1$
B. $f^{-1}(x) = \sqrt{x+2} - 1, x \leq -2$
C. $f^{-1}(x) = -\sqrt{x+2} - 1, x \geq -1$
D. $f^{-1}(x) = -\sqrt{x+2} - 1, x \geq -2$
E. $f^{-1}(x) = -\sqrt{x-1} + 2, x \geq -2$

7. **MC** For a function to have an inverse function, it must be:

A. one-to-one
B. one-to-many
C. many-to-one
D. many-to-many
E. all of these

8. **MC** Select the correct asymptote for the graph $y = 2^{x-1} + 3$.

A. $y = -1$
B. $x = 1$
C. $x = 3$
D. $y = 1$
E. $y = 3$

9. **MC** Select the correct equation for the graph shown.

A. $y = x(x-2)^2$
B. $y = x(x-2)$
C. $y = x(x+2)^2$
D. $y = x^2(x-2)$
E. $y = x^2(x+2)$

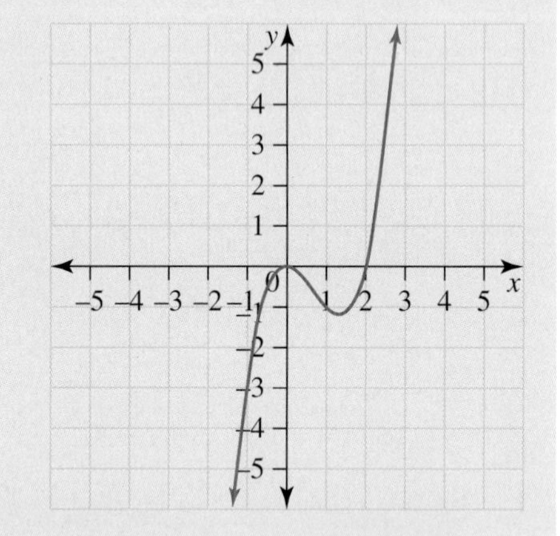

10. **MC** The graph of $x^2 + y^2 = 4$ is translated 1 unit to the left parallel to the x-axis and 2 units upwards, parallel to the y-axis.

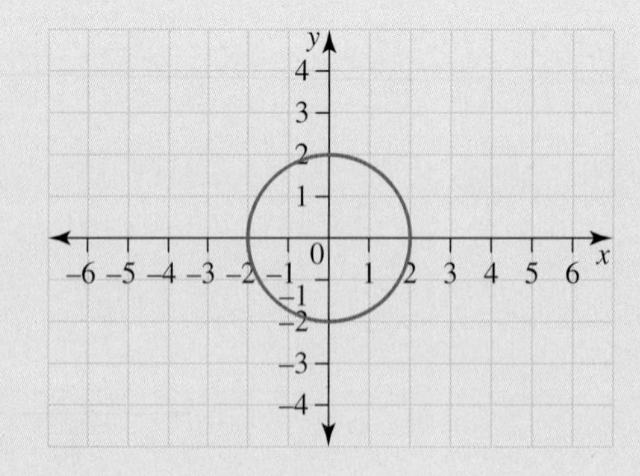

Select the new transformed equation.

A. $(x+1)^2 + (y+2)^2 = 7$
B. $(x+1)^2 + (y-2)^2 = 7$
C. $(x+1)^2 + (y+2)^2 = 4$
D. $(x+1)^2 + (y-2)^2 = 4$
E. $(x-1)^2 + (y+2)^2 = 1$

11. **MC** Consider the function $f(x) = \left(x^2 - 9\right)(x - 2)(1 - x)$.
The graph of $f(x)$ is best represented by:

A.
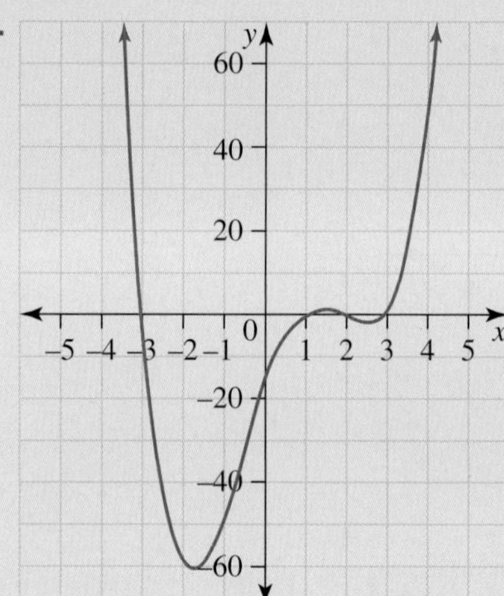

B.
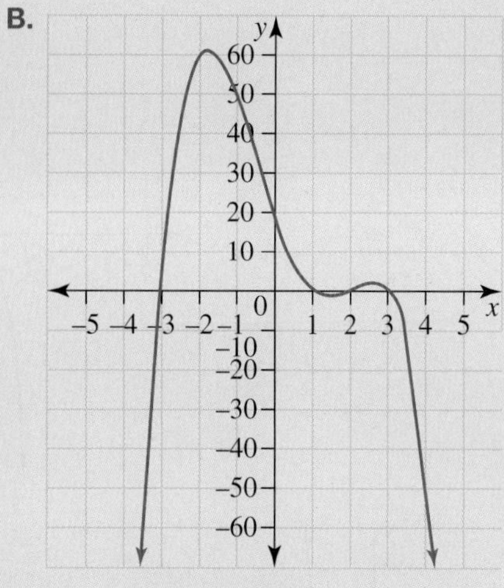

C.
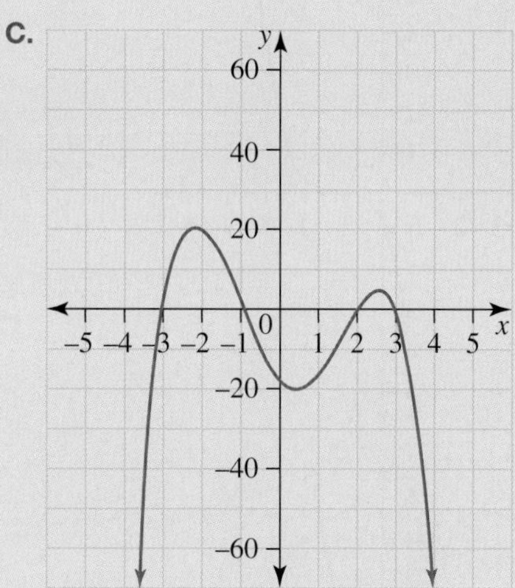

D.
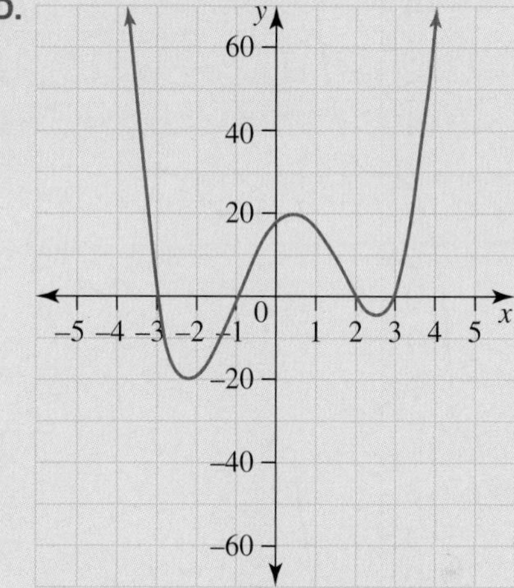

E.
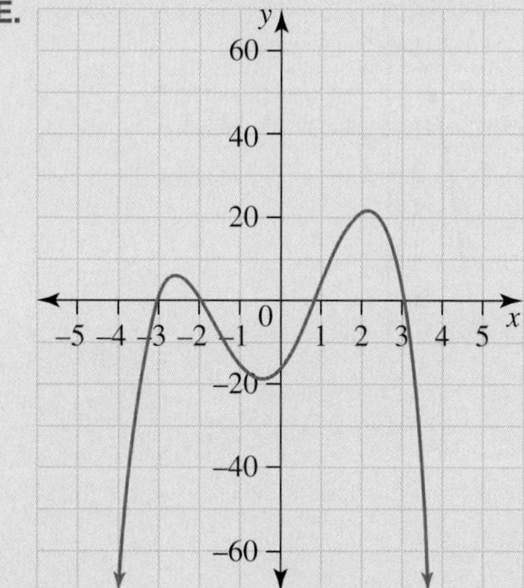

12. **MC** The quartic function has two x-intercepts at -1 and 4 and passes through the point $(0, -8)$. Select the equation that best represents the function.

A. $f(x) = -\frac{1}{2}(x+1)^2(x-4)^2$ **B.** $f(x) = \frac{1}{2}(x+1)^2(x-4)^2$ **C.** $f(x) = -\frac{1}{2}(x-1)^2(x+4)^2$
D. $f(x) = -2(x+1)^2(x-4)^2$ **E.** $f(x) = 2(x+1)^2(x-4)^2$

13. **MC** If the graph shown is represented by the equation $y = P(x)$, select the correct graph for the equation $y = P(x - 1)$.

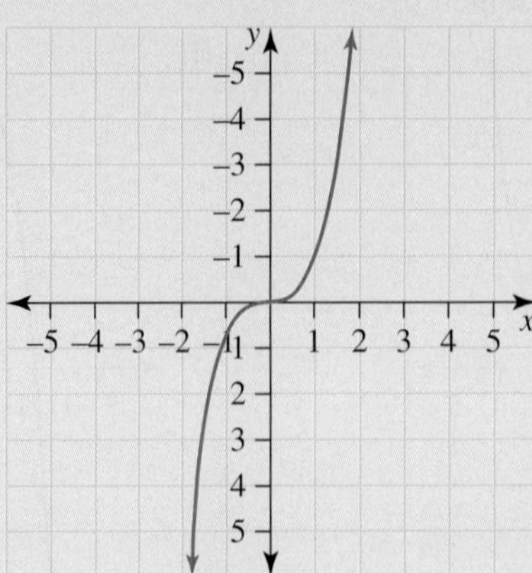

A.

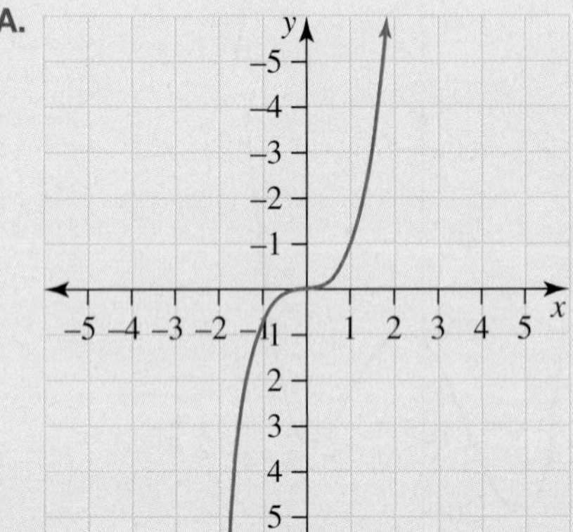

B.

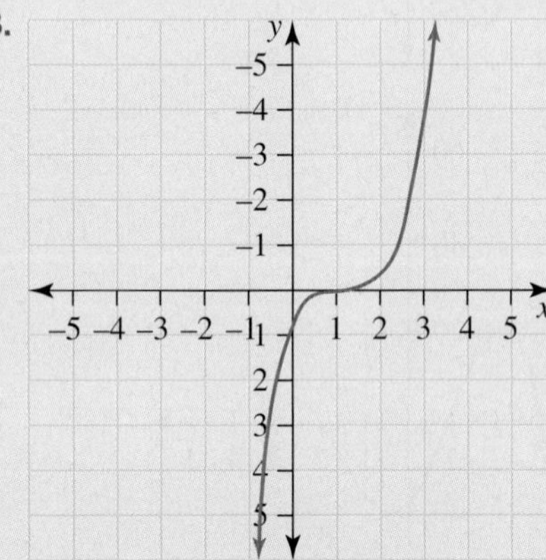

C.

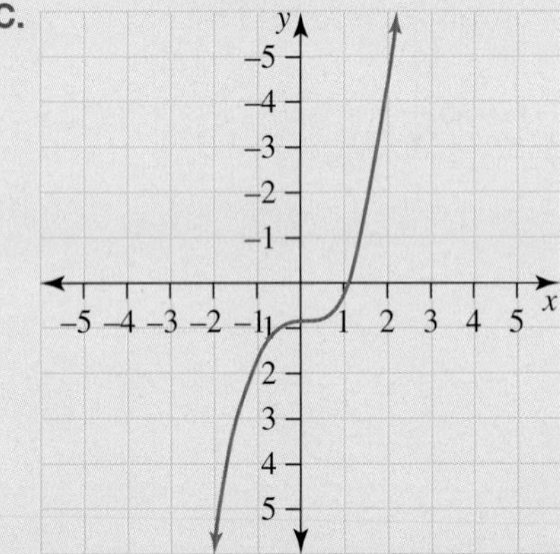

D.

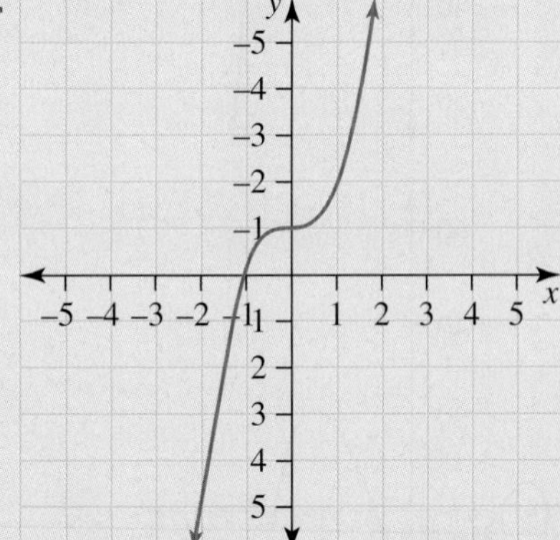

E.

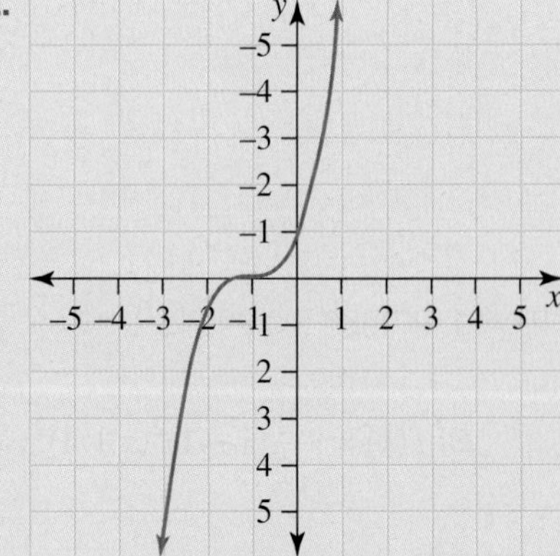

14. MC If the graph shown is represented by the equation $y = P(x)$, select the correct graph for the equation $y = -P(x) + 2$.

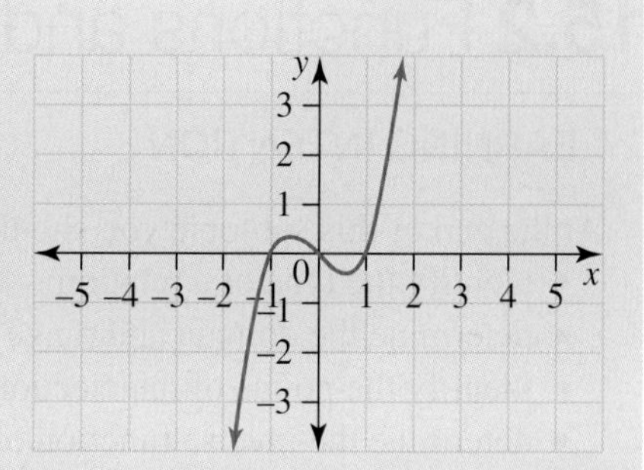

A.

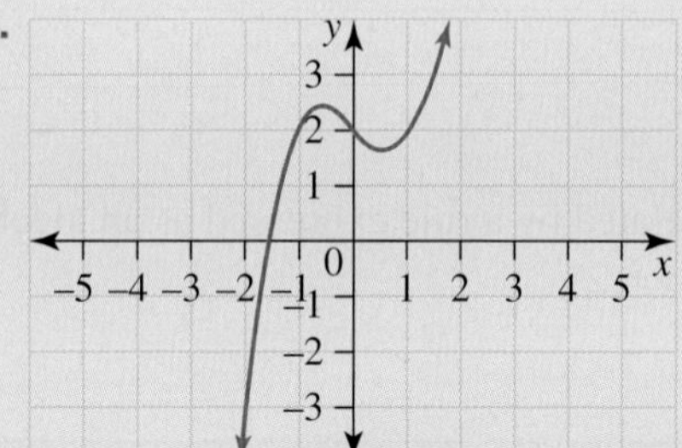

B.

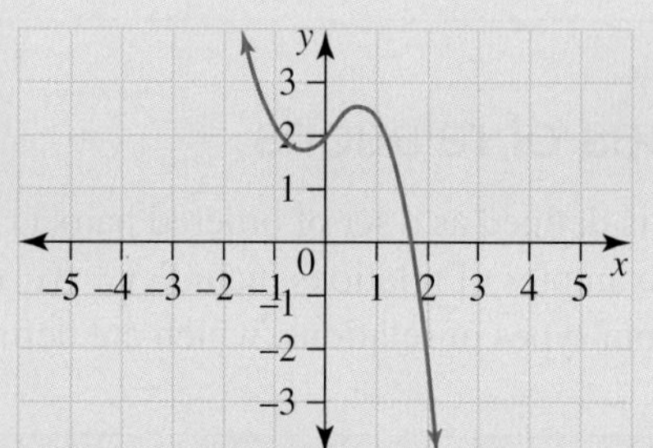

C.

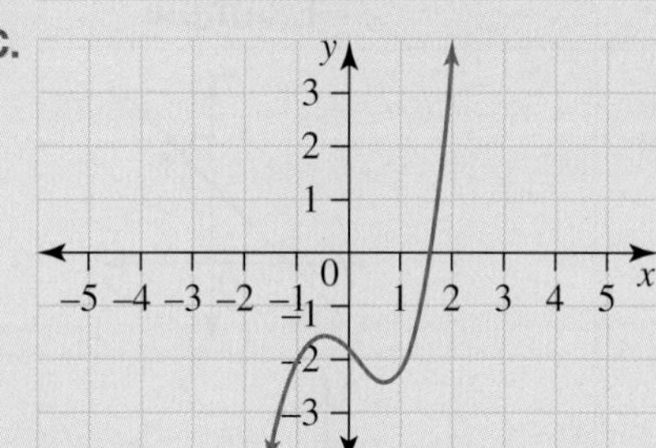

D.

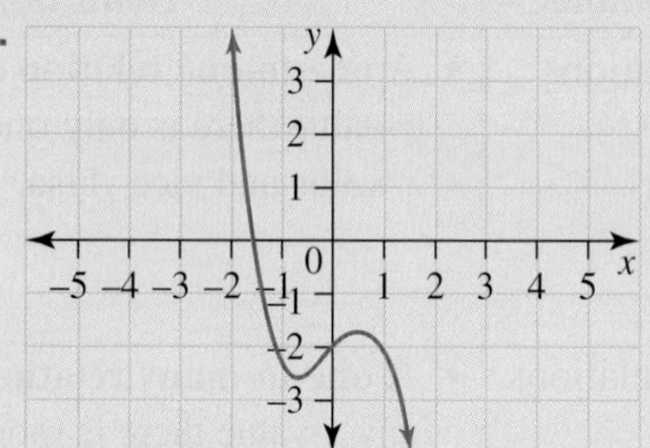

E.

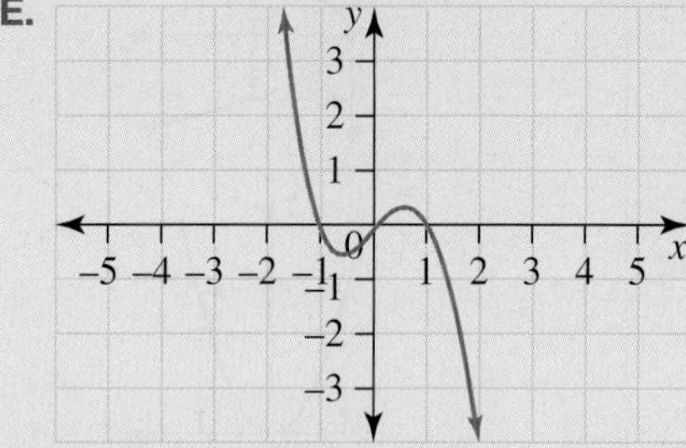

15. MC Consider the sketch of $y = P(x)$ and the graph of a transformation of $y = P(x)$.

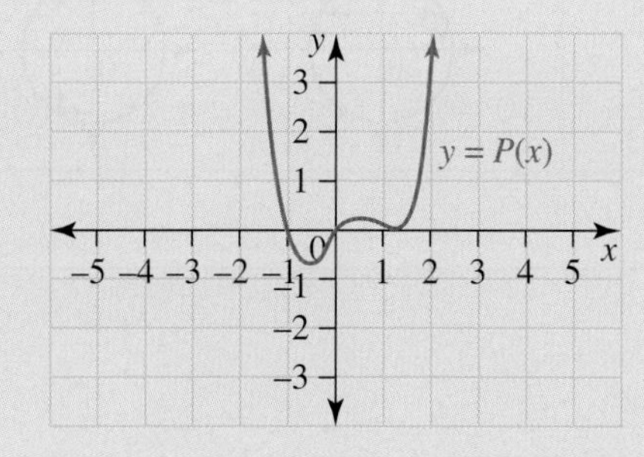

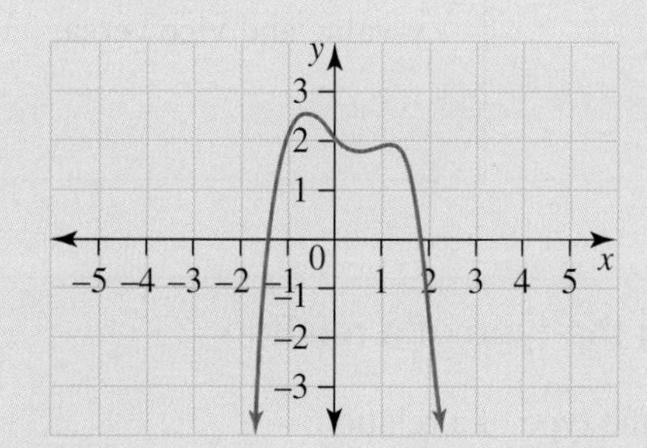

Select the possible equation in terms of $P(x)$ for the transformation of $y = P(x)$.

A. $y = P(x) + 2$ **B.** $y = P(x - 2)$ **C.** $y = P(x + 2)$ **D.** $y = -P(x) - 2$ **E.** $y = -P(x) + 2$

15.2 Functions and relations

LEARNING INTENTION

At the end of this subtopic you should be able to:
- identify the type of a relation
- determine the domain and range of a function or relation
- identify the points of intersection between two functions
- determine the inverse function of a one-to-one function.

15.2.1 Types of relations

eles-4984

- A **relation** is defined as a set of ordered pairs (x, y) which are related by a rule expressed as an algebraic equation. Examples of relations include $y = 3x, x^2 + y^2 = 4$ and $y = 2^x$.
- There are four types of relations, which are defined as follows.

Types of relations	Definition	Example
One-to-one relations	• A **one-to-one relation** exists if for any x-value there is only one corresponding y-value and vice versa.	
One-to-many relations	• A **one-to-many relation** exists if for any x-value there is more than one y-value, but for any y-value there is only one x-value.	
Many-to-one relations	• A **many-to-one relation** exists if there is more than one x-value for any y-value but for any x-value there is only one y-value.	
Many-to-many relations	• A **many-to-many relation** exists if there is more than one x-value for any y-value and vice versa.	

Determining the type of a relation

To determine the type of a relation:
- Draw a horizontal line through the graph so that it cuts the graph the maximum number of times. Determine whether the number of cuts is one or many.
- Draw a vertical line through the graph so that it cuts the graph the maximum number of times. Determine whether the number of cuts is one or many.

WORKED EXAMPLE 1 Identifying the types of relations

State the type of relation that each graph represents.

a.

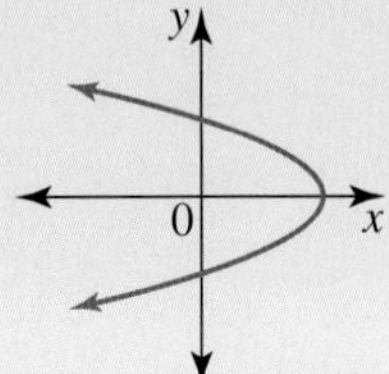

b.

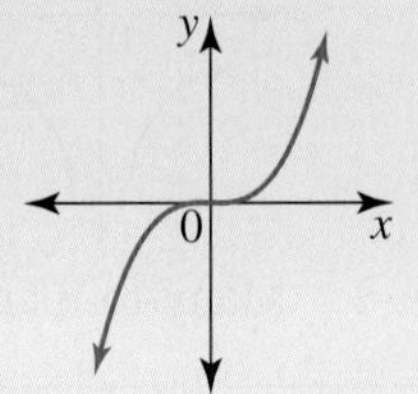

c.

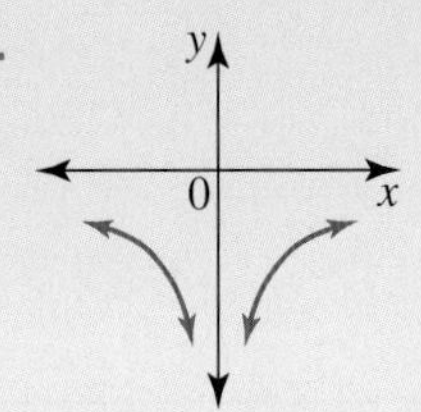

THINK	WRITE
a. 1. Draw a horizontal line through the graph. The line cuts the graph **one** time.	**a.** 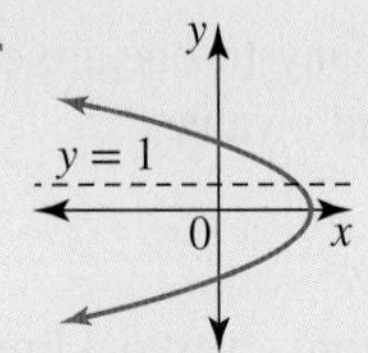 One-to-_____ relation
2. Draw a vertical line through the graph. The line cuts the graph **many** times.	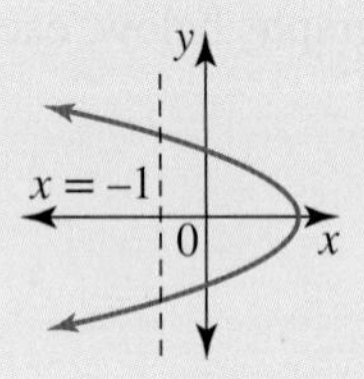 One-to-many relation
b. 1. Draw a horizontal line through the graph. The line cuts the graph **one** time.	**b.** 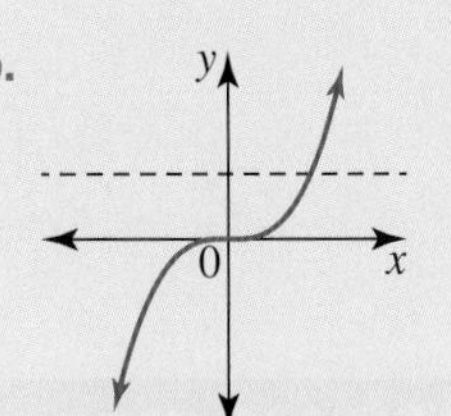 One-to-____ relation
2. Draw a vertical line through the graph. The line cuts the graph **one** time.	One-to-one relation
c. 1. Draw a horizontal line through the graph. The line cuts the graph **many** times.	**c.** 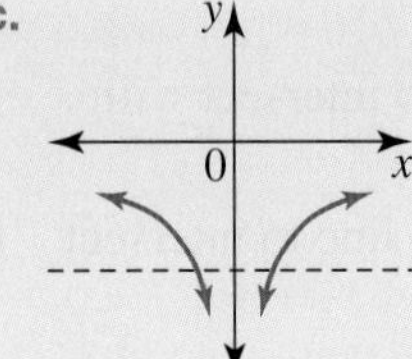 Many-to-____ relation

2. Draw a vertical line through the graph. The line cuts the graph **one** time.

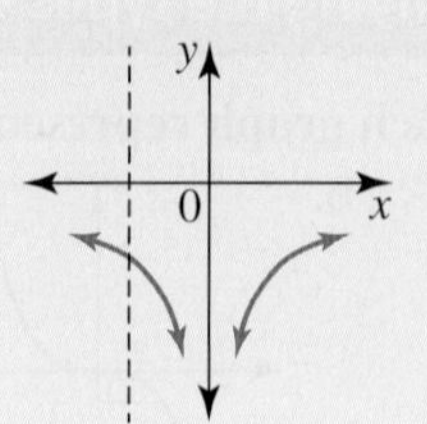

Many-to-many relation

15.2.2 Functions

eles-4985

- Relations that are one-to-one or many-to-one are called **functions**. That is, a function is a relation in which for any x-value there is at most one y-value.

Vertical line test

- To determine if a graph is a function, a vertical line is drawn anywhere on the graph. If it does not intersect with the curve more than once, then the graph is a function.
  For example, in each of the two graphs below, each vertical line intersects the graph only once.

1.

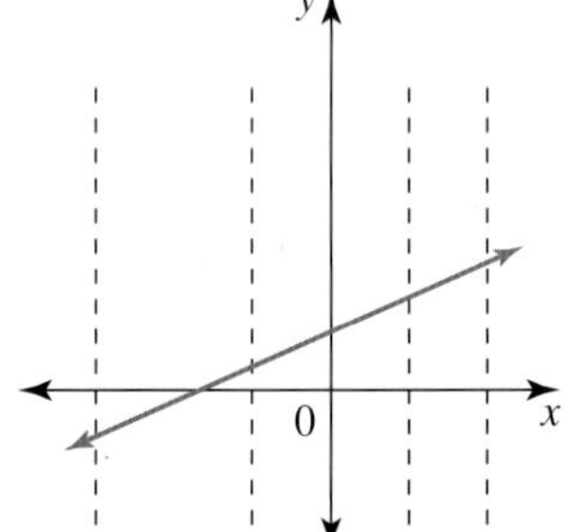

2.

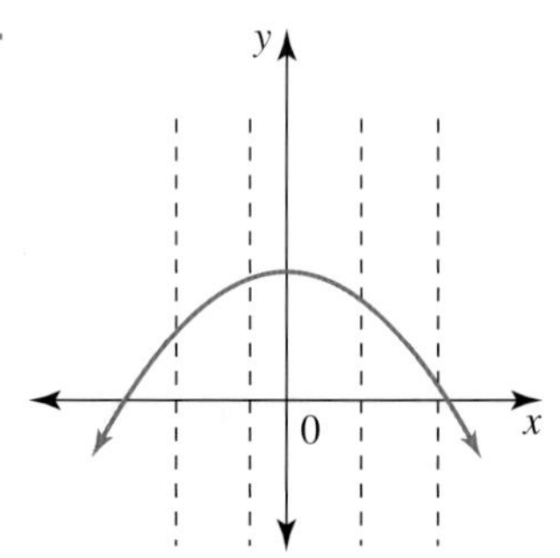

WORKED EXAMPLE 2 Identifying whether a relation is a function

State whether or not each of the following relations are functions.

a.

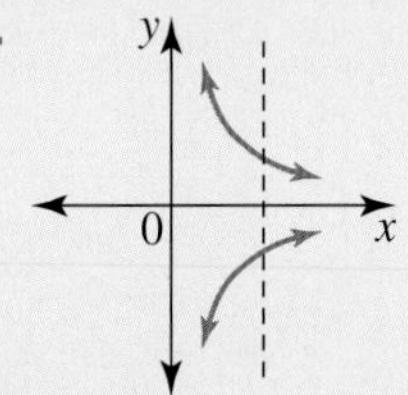

b.

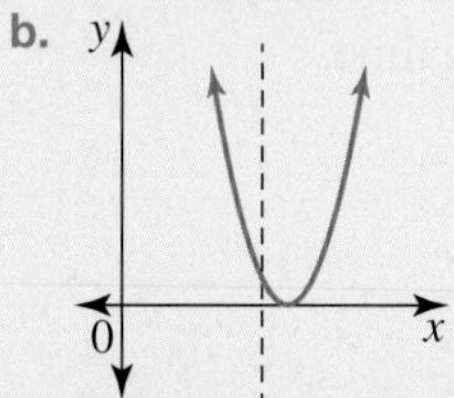

THINK	WRITE
a. It is possible for a vertical line to intersect with the curve more than once.	a. Not a function
b. It is not possible for any vertical line to intersect with the curve more than once.	b. Function

Function notation

- Consider the relation $y = 2x$, which is a function.
  The y-values are determined from the x-values, so we say 'y is a function of x', which is abbreviated to $y = f(x)$.
  So, the rule $y = 2x$ can also be written as $f(x) = 2x$.
- For a given function $y = f(x)$, the value of y when $x = 1$ is written as $f(1)$, the value of y when $x = 5$ is written as $f(5)$, the value of y when $x = a$ as $f(a)$, etc.
- For the function $f(x) = 2x$:

$$\begin{aligned}\text{when } x = 1, y &= f(1)\\ &= 2 \times 1\\ &= 2.\end{aligned}$$

$$\begin{aligned}\text{when } x = 2, y &= f(2)\\ &= 2 \times 2\\ &= 4, \text{ and so on.}\end{aligned}$$

Domain and range

- The **domain** of a function is the set of all allowable values of x. It is sometimes referred to as the **maximal domain**.
- The **range** of a function is the set of y-values produced by the function.
- The following examples show how to determine the domain and range of some graphs.

Graph	Domain	Range
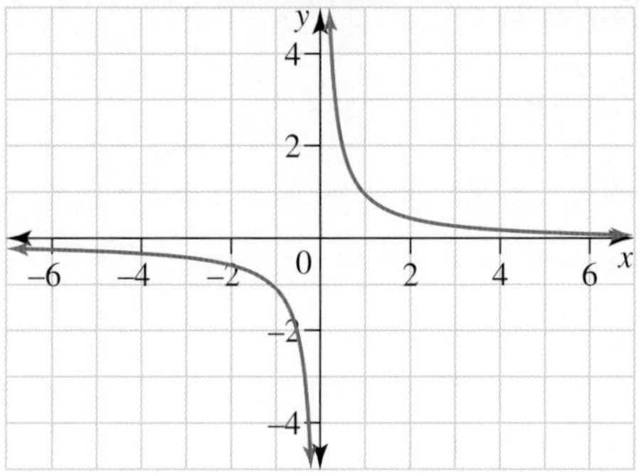	The domain is all x values except 0. Domain: $x \in R \setminus \{0\}$	The range is all y values except 0. Range: $y \in R \setminus \{0\}$
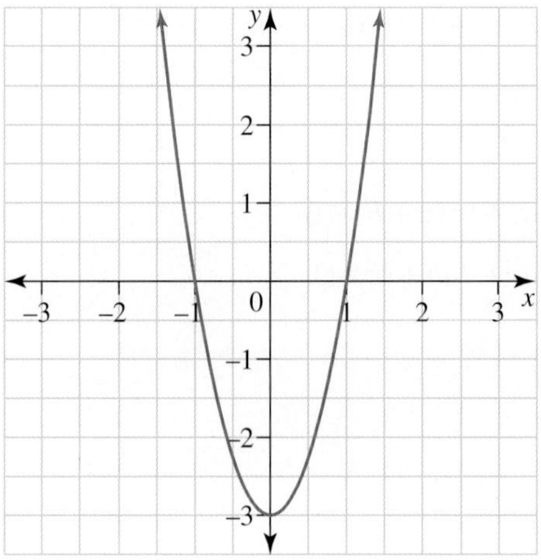	The domain is all x values. Domain: $x \in R$	The range is all y values that are greater than or equal to -3. Range: $y \geq -3$

WORKED EXAMPLE 3 Evaluating a function using function notation

If $f(x) = x^2 - 3$, calculate:

a. $f(1)$ b. $f(a)$ c. $3f(2a)$ d. $f(a) + f(b)$ e. $f(a + b)$.

THINK	WRITE
a. 1. Write the rule.	a. $f(x) = x^2 - 3$
2. Substitute $x = 1$ into the rule.	$f(1) = 1^2 - 3$
3. Simplify and write the answer.	$= 1 - 3$ $= -2$
b. 1. Write the rule.	b. $f(x) = x^2 - 3$
2. Substitute $x = a$ into the rule.	$f(a) = a^2 - 3$
c. 1. Write the rule.	c. $f(x) = x^2 - 3$
2. Substitute $x = 2a$ into the rule and simplify.	$f(2a) = (2a)^2 - 3$ $= 2^2a^2 - 3$ $= 4a^2 - 3$
3. Multiply the answer by 3 and simplify.	$3f(2a) = 3(4a^2 - 3)$
4. Write the answer.	$= 12a^2 - 9$
d. 1. Write the rule.	d. $f(x) = x^2 - 3$
2. Evaluate $f(a)$.	$f(a) = a^2 - 3$
3. Evaluate $f(b)$.	$f(b) = b^2 - 3$
4. Evaluate $f(a) + f(b)$.	$f(a) + f(b) = a^2 - 3 + b^2 - 3$
5. Write the answer.	$= a^2 + b^2 - 6$
e. 1. Write the rule	e. $f(x) = x^2 - 3$
2. Evaluate $f(a + b)$.	$f(a + b) = (a + b)^2 - 3$ $= (a + b)(a + b) - 3$
3. Write the answer.	$= a^2 + 2ab + b^2 - 3$

TI | THINK	DISPLAY/WRITE	CASIO | THINK	DISPLAY/WRITE
a-e. In a new document, on a Calculator page, press: • MENU • 1: Actions • 1: Define Complete the entry lines as: Define $f(x) = x^2 - 3$ $f(1)$ $f(a)$ $3f(2a)$ $f(a) + f(b)$ $f(a + b)$ Press ENTER after each entry.	a-e. 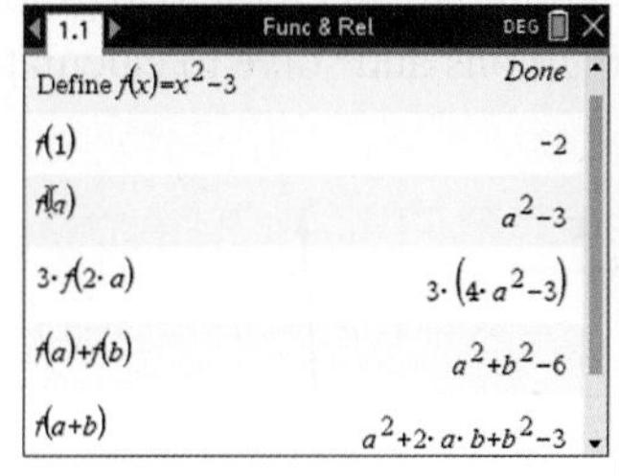 $f(1) = -2$ $f(a) = a^2 - 3$ $3f(2a) = 3\left(4a^2 - 3\right)$ $f(a) + f(b) = a^2 + b^2 - 6$ $f(a + b) = a^2 + 2ab + b^2 - 3$	a-e. On the Main screen, tap: • Action • Command • Define Complete the entry lines as: Define $f(x) = x^2 - 3$ $f(1)$ $f(a)$ $3f(2a)$ $f(a) + f(b)$ $f(a + b)$ expand $\left((a + b)^2 - 3\right)$ Press EXE after each entry.	a-e. 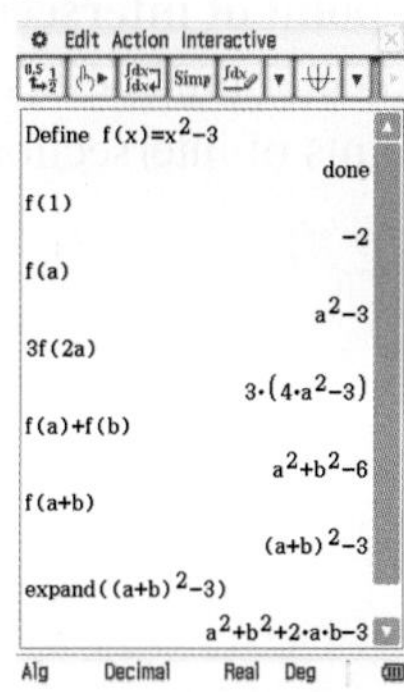 $f(1) = -2$ $f(a) = a^2 - 3$ $3f(2a) = \left(12a^2 - 9\right)$ $f(a) + f(b) = a^2 + b^2 - 6$ $f(a + b) = a^2 + 2ab + b^2 - 3$

15.2.3 Identifying features of functions

eles-4986

Behaviour of functions as they approach extreme values

- We can identify features of certain functions by observing what happens to the function value (y value) when x approaches a very small value such as 0 ($x \to 0$) or a very large value such as ∞ ($x \to \infty$).

WORKED EXAMPLE 4 Identifying end behaviour of a function

Describe what happens to these functions as the value of x increases, that is, as $x \to \infty$.

a. $y = x^2$ **b.** $f(x) = 2^{-x}$ **c.** $f(x) = \dfrac{1}{x} + 1$

THINK	WRITE
a. 1. Write the function.	**a.** $y = x^2$
2. Substitute large x values into the function, such as $x = 10\,000$ and $x = 1\,000\,000$.	$f(10\,000) = 100\,000\,000$ $f(1\,000\,000) = 1 \times 10^{12}$
3. Write a conclusion.	As $x \to \infty$, $f(x)$ also increases; that is, $f(x) \to \infty$.
b. 1. Write the function.	**b.** $f(x) = 2^{-x}$
2. Substitute large x values into the function, such as $x = 10\,000$ and $x = 1\,000\,000$.	$f(10\,000) \approx 0$ $f(1\,000\,000) \approx 0$
3. Write a conclusion.	As $x \to \infty$, $f(x) \to 0$.
c. 1. Write the function.	**c.** $f(x) = \dfrac{1}{x} + 1$
2. Substitute large x values into the function, such as $x = 10\,000$ and $x = 1\,000\,000$.	$f(10\,000) = 1.0001$ $f(1\,000\,000) = 1.000\,001$
3. Write a conclusion.	As $x \to \infty$, $f(x) \to 1$.

Points of intersection

- A **point of intersection** between two functions is a point at which the two graphs cross paths.
- To determine points of intersection, equate the two graphs and solve to calculate the coordinates of the points of intersection.

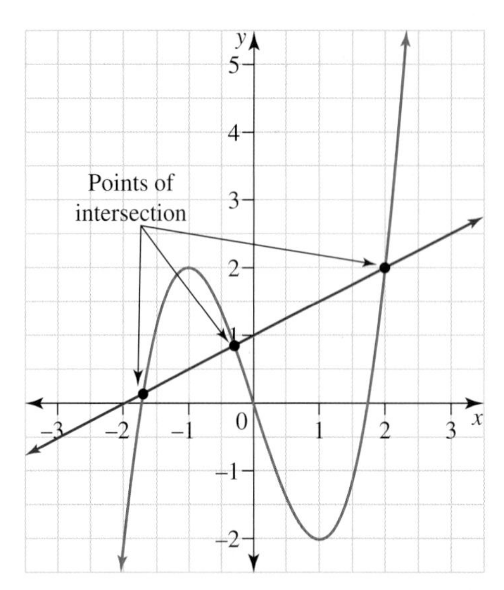

WORKED EXAMPLE 5 Determining points of intersection

Determine any points of intersection between $f(x) = 2x + 1$ and $g(x) = \frac{1}{x}$.

THINK	WRITE
1. Write the two equations.	$f(x) = 2x + 1$ $g(x) = \frac{1}{x}$
2. Points of intersection are common values between the two curves. To solve the equations simultaneously, equate both functions.	For points of intersection: $2x + 1 = \frac{1}{x}$
3. Rearrange the resulting equation and solve for x.	$2x^2 + x = 1$ $2x^2 + x - 1 = 0$ $(2x - 1)(x + 1) = 0$ $x = \frac{1}{2}$ or -1
4. Substitute the x values into either function to calculate the y values.	$f\left(\frac{1}{2}\right) = 2 \times \frac{1}{2} + 1 = 2$ $f(-1) = 2 \times -1 + 1 = -1$
5. Write the coordinates of the two points of intersection.	The points of intersection are $\left(\frac{1}{2}, 2\right)$ and $(-1, -1)$.

TI | THINK

1. In a new problem, on a Calculator page, press:
   - MENU
   - 1: Actions
   - 1: Define

   Complete the entry lines as:
   Define $f1(x) = 2x + 1$
   Define $f2(x) = \dfrac{1}{x}$

   Press ENTER after each entry. Then press:
   - MENU
   - 3: Algebra
   - 1: Solve

   Complete the entry lines as:
   solve $(f1(x) = f2(x), x)$
   $f1(-1)$
   $f2\left(\dfrac{1}{2}\right)$

   Press ENTER after each entry.

DISPLAY/WRITE

1.1 2.1 Func & Rel
Define $f1(x)=2\cdot x+1$ Done
Define $f2(x)=\frac{1}{x}$ Done
solve$(f1(x)=f2(x),x)$ $x=-1$ or $x=\frac{1}{2}$
$f1(-1)$ -1
$f2\left(\frac{1}{2}\right)$ 2

The points of intersection are $(-1, -1)$ and $\left(\dfrac{1}{2}, 2\right)$.

2. Alternatively, open a Graphs page in the current document. Since the functions have already been entered, just select the functions and press ENTER.
   The graphs will be displayed.

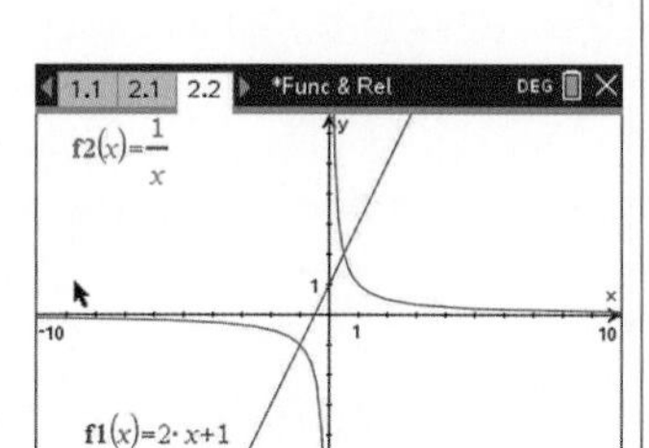

3. To locate the points of intersection between the two functions, press:
   - MENU
   - 6: Analyze Graph
   - 4: Intersection

   Move the cursor to the left of the intersection point and press ENTER. Then move the cursor to the right of the intersection point and press ENTER. The intersection point is displayed. Repeat for the other intersection point.

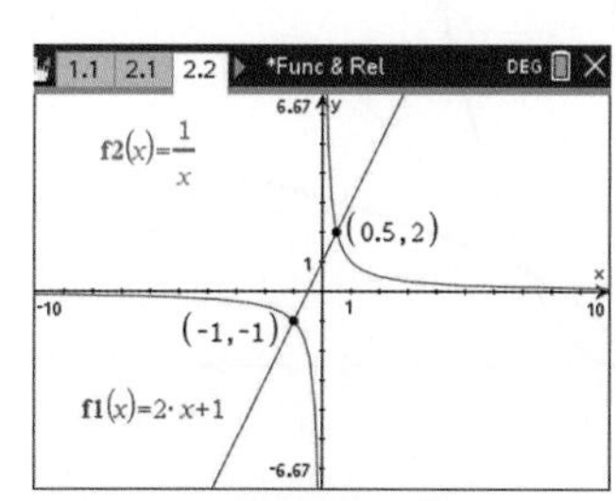

The points of intersection are $(-1, -1)$ and $(0.5, 2)$.

CASIO | THINK

1. On the Main screen, complete the entry line as:

   solve $\left(2x + 1 = \dfrac{1}{x}, x\right)$
   $2x + 1 \mid x = -1$
   $2x + 1 \mid x = \dfrac{1}{2}$

   Press EXE after each entry.

DISPLAY/WRITE

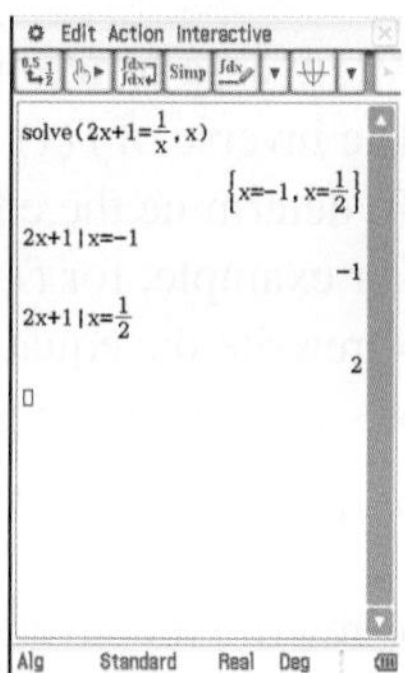

The points of intersection are $(-1, -1)$ and $\left(\dfrac{1}{2}, 2\right)$.

2. Alternatively, tap the graphing icon.
   Highlight each side of the equation separately and drag it down to the axis below.

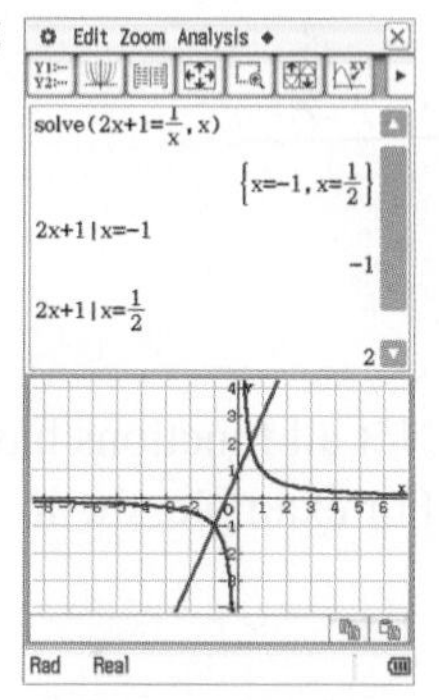

3. To locate the points of intersection, tap:
   - Analysis
   - G-Solve
   - Intersection

   To find the other intersection point, tap the right arrow.

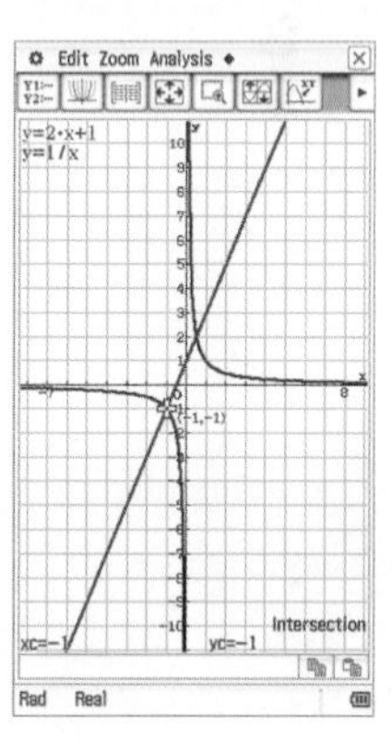

The points of intersection are $(-1, -1)$ and $(0.5, 2)$.

15.2.4 Inverse functions

eles-4987

- An **inverse function** is when a function is reflected across the line $y = x$.
- The inverse of $f(x)$ is written as $f^{-1}(x)$.
- To determine the equation of the inverse function, swap x and y and rearrange to make y the subject. For example, for $f(x) = 2x$:
  - rewrite the equation as $y = 2x$
  - swap x and y to get $x = 2y$
  - rearrange to make y the subject to get $y = \frac{1}{2}x$
  - rewrite the inverse in function notation $f^{-1}(x) = \frac{1}{2}x$.

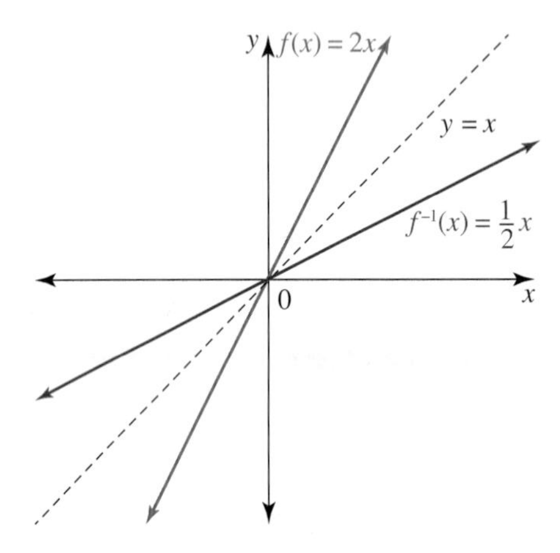

- Not all functions have inverses that are functions. For example, the inverse of the function $y = x^2$ is $x = y^2$ or $y = \pm\sqrt{x}$ which is a one-to-many relation.

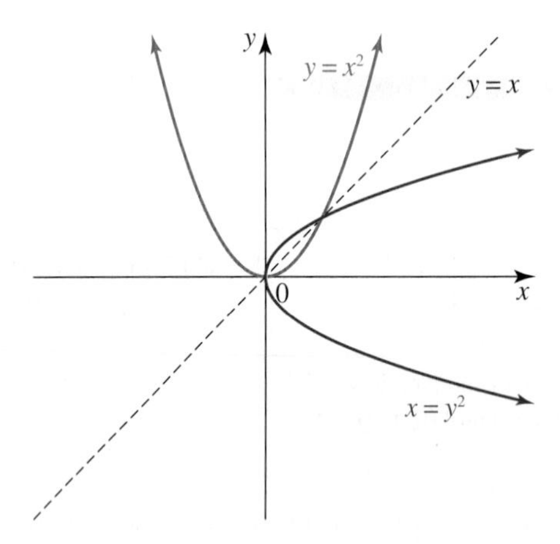

- For a function to have an inverse function it must be a one-to-one function.
- To determine whether a function has an inverse which is a function we can use the horizontal line test.
- Only when a function is one-to-one, will it have an inverse function. This can be tested by drawing a horizontal line to see if it cuts the graph once (has an inverse function) or more than once (does not have an inverse function).
- If a graph fails the horizontal line test its domain can be restricted so that it passes the test. The function with the restricted domain will then have an inverse function.

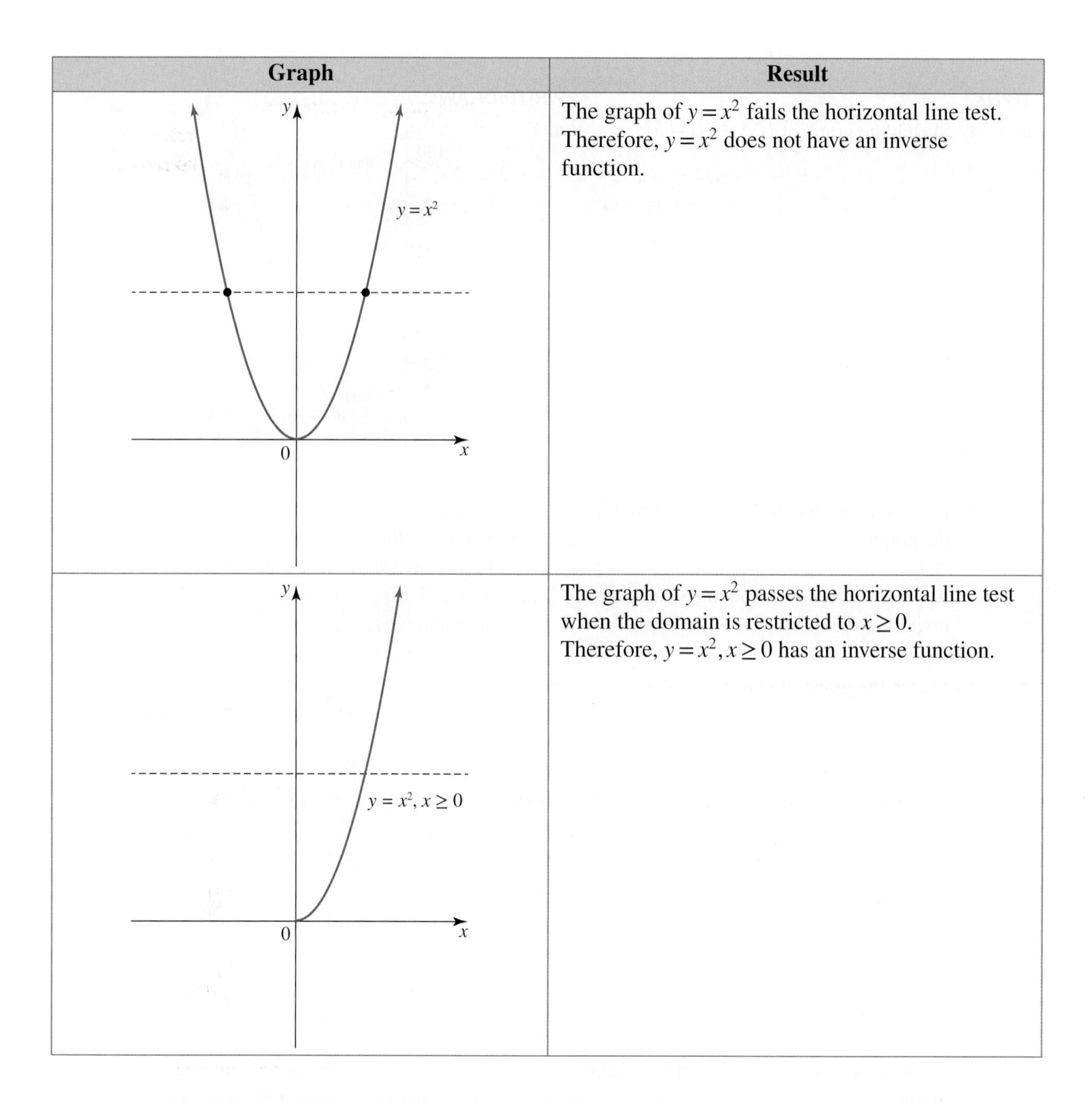

Graph	Result
	The graph of $y = x^2$ fails the horizontal line test. Therefore, $y = x^2$ does not have an inverse function.
	The graph of $y = x^2$ passes the horizontal line test when the domain is restricted to $x \geq 0$. Therefore, $y = x^2, x \geq 0$ has an inverse function.

WORKED EXAMPLE 6 Determining inverse functions of functions

Answer the following questions.

a. i. **Show that the function $f(x) = x(x - 5)$ will have not have an inverse function.**

ii. **Suggest a restriction that would result in an inverse function.**

b. i. **Show that the function $f(x) = x^2 + 4,\ x \geq 0$ will have an inverse function.**

ii. **Determine the equation of the inverse function.**

THINK | **WRITE/DRAW**

a. i. 1. Sketch the graph of $f(x) = x(x - 5)$.

a. i.

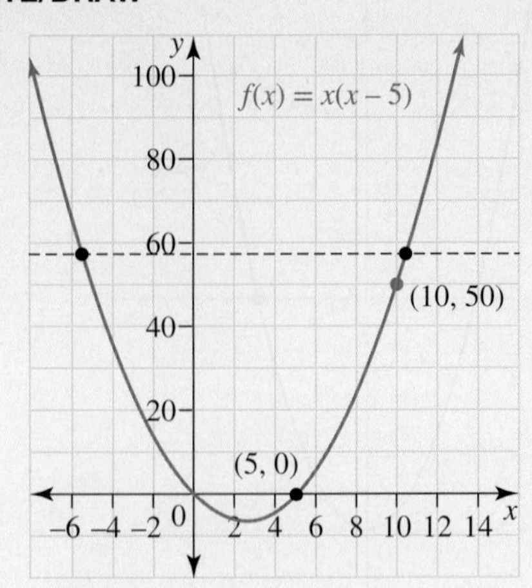

2. Draw a dotted horizontal line(s) through the graph.

The graph does not satisfy the horizontal line test, so the function $f(x) = x(x - 5)$ will not have an inverse function.

ii. Apply a restriction to the function so that it will have an inverse.

ii. An inverse function will exist if $f(x) = x(x - 5), x \leq 2.5$ or $f(x) = x(x - 5), x \geq 2.5$.

b. i. 1. Sketch the graph of $f(x) = x^2 + 4, x \geq 0$.

b. i.

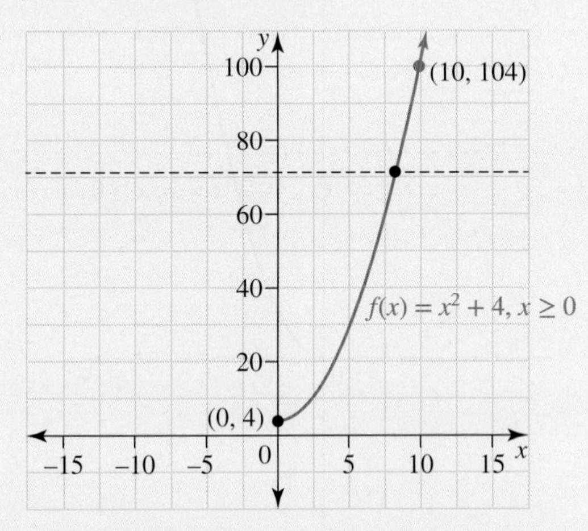

2. Draw a dotted horizontal line through the graph.

The graph satisfies the horizontal line test, so the function $f(x) = x^2 + 4, x \geq 0$ has an inverse function.

ii. 1. Determine the equation of the inverse function by swapping x and y.

ii. Let $y = x^2 + 4, x \geq 0$.

Swap x and y.

$x = y^2 + 4$

Make y the subject.

$$x = y^2 + 4$$
$$x - 4 = y^2$$
$$\sqrt{x - 4} = y$$
$$y = \sqrt{x - 4}$$

2. Write the answer in correct form, noting the domain.

The inverse of $f(x) = x^2 + 4$ is $f^{-1}(x) = \sqrt{x - 4}, x \geq 4$.

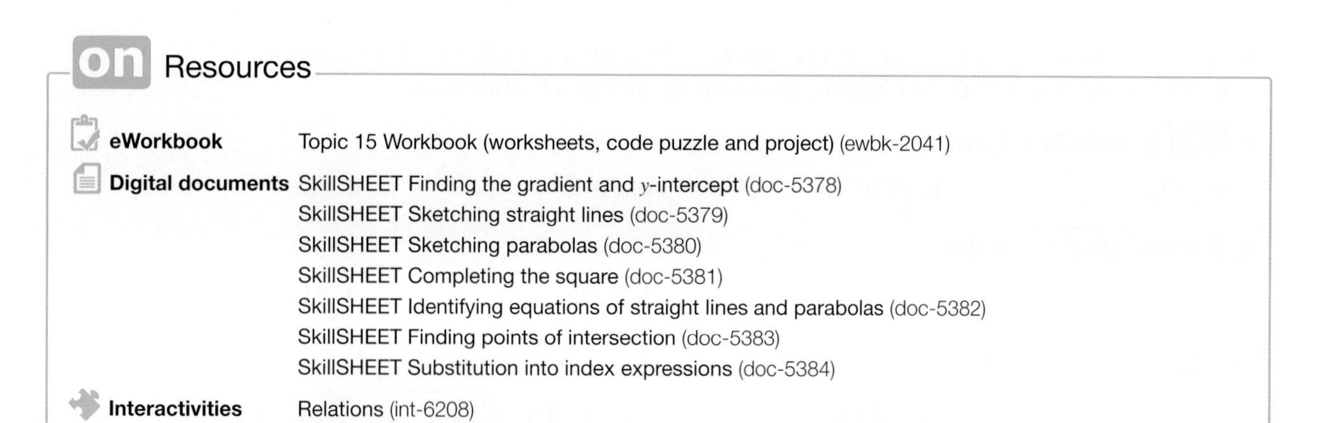
Resources

eWorkbook Topic 15 Workbook (worksheets, code puzzle and project) (ewbk-2041)

Digital documents SkillSHEET Finding the gradient and y-intercept (doc-5378)
SkillSHEET Sketching straight lines (doc-5379)
SkillSHEET Sketching parabolas (doc-5380)
SkillSHEET Completing the square (doc-5381)
SkillSHEET Identifying equations of straight lines and parabolas (doc-5382)
SkillSHEET Finding points of intersection (doc-5383)
SkillSHEET Substitution into index expressions (doc-5384)

Interactivities Relations (int-6208)
Evaluating functions (int-6209)

Exercise 15.2 Functions and relations

learnon

Individual pathways

■ PRACTISE	■ CONSOLIDATE	■ MASTER
1, 4, 7, 10, 12, 15, 17, 22	2, 5, 8, 11, 13, 18, 19, 23	3, 6, 9, 14, 16, 20, 21, 24

To answer questions online and to receive **immediate feedback** and **sample responses** for every question, go to your learnON title at www.jacplus.com.au.

Fluency

WE1 For questions 1 to 3, state the type of relation that each graph represents.

1. a. 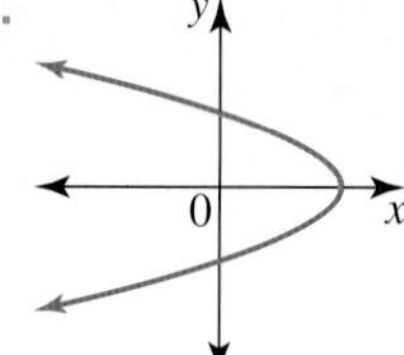
b. 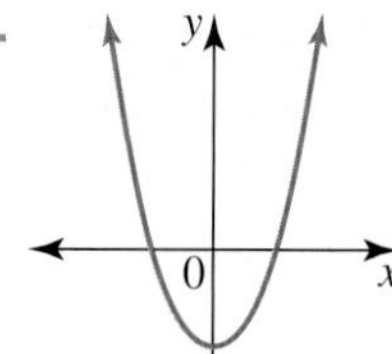
c. 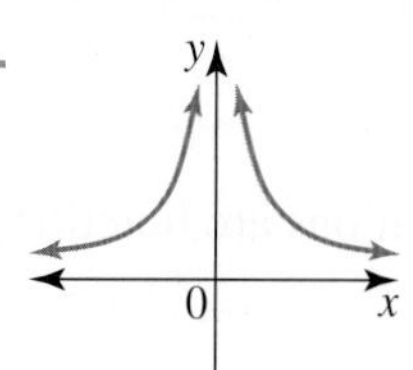
d. 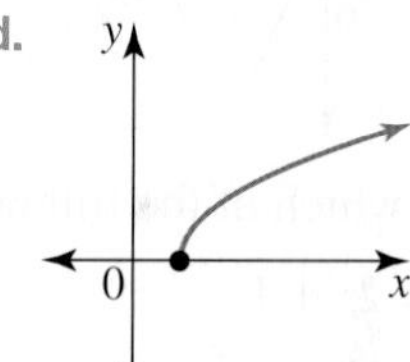

2. a. 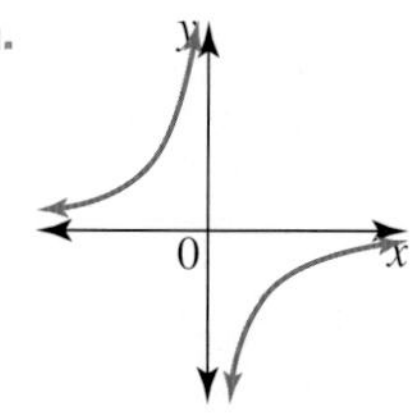
b. 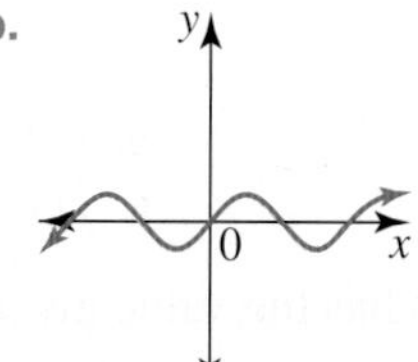
c. 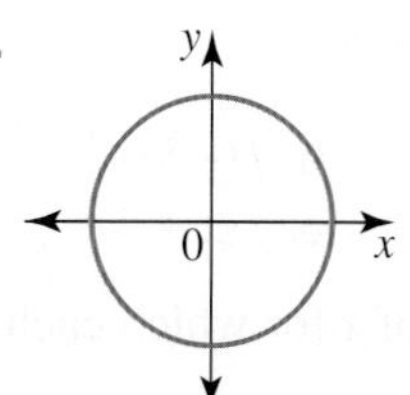
d. 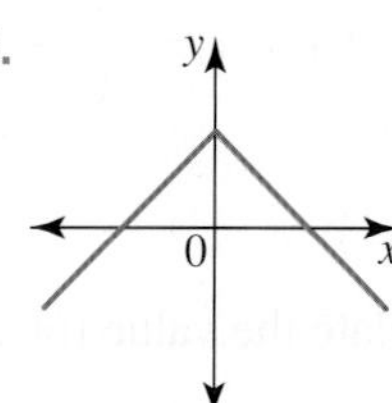

3. a. 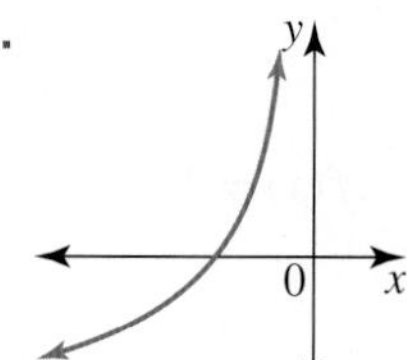
b. 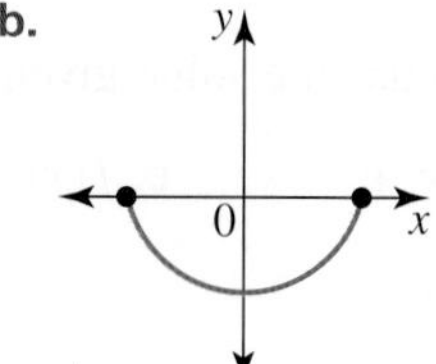
c. 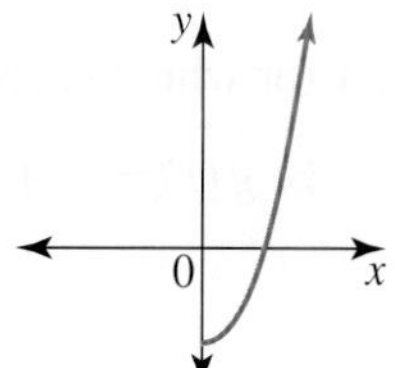
d. 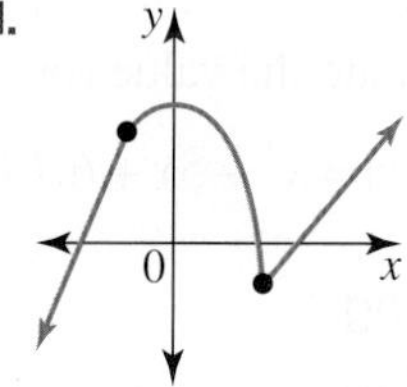

WE2 For questions 4 to 6, answer the following questions.

4. a. Use the vertical line test to determine which of the relations in question 1 are functions.
b. Determine which of these functions have inverses that are also functions.

5. a. Use the vertical line test to determine which of the relations in question 2 are functions.
b. Determine which of these functions have inverses that are also functions.

6. a. Use the vertical line test to determine which of the relations in question 3 are functions.
   b. Determine which of these functions have inverses that are also functions.

7. WE3 If $f(x) = 3x + 1$, calculate:

a. $f(0)$ b. $f(2)$ c. $f(-2)$ d. $f(5)$

8. If $g(x) = \sqrt{x+4}$, calculate:

a. $g(0)$ b. $g(-3)$ c. $g(5)$ d. $g(-4)$

9. If $g(x) = 4 - \frac{1}{x}$, calculate:

a. $g(1)$ b. $g\left(\frac{1}{2}\right)$ c. $g\left(-\frac{1}{2}\right)$ d. $g\left(-\frac{1}{5}\right)$

10. If $f(x) = (x+3)^2$, calculate:

a. $f(0)$ b. $f(-2)$ c. $f(1)$ d. $f(a)$

11. If $h(x) = \frac{24}{x}$, calculate:

a. $h(2)$ b. $h(4)$ c. $h(-6)$ d. $h(12)$

Understanding

12. State which of the following relations are functions.

a. 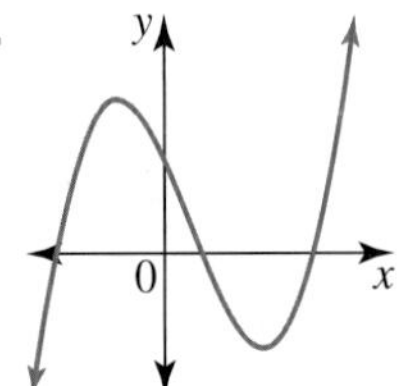

b. $x^2 + y^2 = 9$

c. $y = 8x - 3$

d. 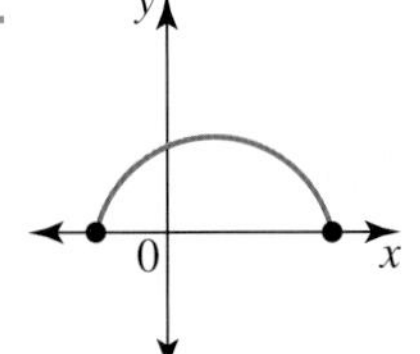

13. State which of the following relations are functions.

a. $y = 2x + 1$ b. $y = x^2 + 2$ c. $y = 2^x$
d. $x^2 + y^2 = 25$ e. $x^2 + 4x + y^2 + 6y = 14$ f. $y = -4x$

14. Given that $f(x) = \frac{10}{x} - x$, determine:

a. $f(2)$ b. $f(-5)$ c. $f(2x)$
d. $f\left(x^2\right)$ e. $f(x+3)$ f. $f(x-1)$

15. Calculate the value (or values) of x for which each function has the value given.

a. $f(x) = 3x - 4, f(x) = 5$ b. $g(x) = x^2 - 2, g(x) = 7$ c. $f(x) = \frac{1}{x}, f(x) = 3$

16. Calculate the value (or values) of x for which each function has the value given.

a. $h(x) = x^2 - 5x + 6, h(x) = 0$ b. $g(x) = x^2 + 3x, g(x) = 4$ c. $f(x) = \sqrt{8-x}, f(x) = 3$

Reasoning

17. WE4 Describe what happens to:

a. $f(x) = x^2 + 3$ as $x \to \infty$ b. $f(x) = 2^x$ as $x \to -\infty$ c. $f(x) = \frac{1}{x}$ as $x \to \infty$
d. $f(x) = x^3$ as $x \to -\infty$ e. $f(x) = -5^x$ as $x \to -\infty$

18. WE5 Determine any points of intersection between the following curves.

a. $f(x) = 2x - 4$ and $g(x) = x^2 - 4$

b. $f(x) = -3x + 1$ and $g(x) = -\frac{2}{x}$

c. $f(x) = x^2 - 4$ and $g(x) = 4 - x^2$

d. $f(x) = \frac{3}{4}x - 6\frac{1}{4}$ and $x^2 + y^2 = 25$

19. Determine the equation of the inverse function of each of the following, placing restrictions on the original values of x as required.

a. $f(x) = 2x - 1$

b. $f(x) = x^2 - 3$

c. $f(x) = (x-2)^2 + 4$

20. WE6 Answer the following questions with full working.

a. Show that the function $f(x) = x(x-2)$ will not have an inverse function.
b. Suggest a restriction that would result in an inverse function.

21. Answer the following questions with full working.

a. Show that the function $f(x) = -x^2 + 4, x \leq 0$ will have an inverse function.
b. Determine the equation of the inverse function.

Problem solving

22. Determine the value(s) of for which:

a. $f(x) = x^2 + 7$ and $f(x) = 16$

b. $g(x) = \frac{1}{x-2}$ and $g(x) = 3$

c. $h(x) = \sqrt{8+x}$ and $h(x) = 6$.

23. Compare the graphs of the inverse functions $y = a^x$ and $y = \log_a x$, choosing various values for a. Explain why these graphs are inverses.

24. Consider the function defined by the rule $f: R \rightarrow R, f(x) = (x-1)^2 + 2$.

a. State the range of the function.
b. Determine the type of mapping for the function.
c. Sketch the graph of the function stating where it cuts the y-axis and its turning point.
d. Select a domain where x is positive such that f is a one-to-one function.
e. Determine the inverse function. Give the domain and range of the inverse function.
f. Sketch the graph of the inverse function on the same set of axes used for part **c**.
g. Determine where f and the function $g(x) = x + 3$ intersect each other.

15.3 Exponential functions

LEARNING INTENTION

At the end of this subtopic you should be able to:
- sketch graphs of exponential functions
- determine the equation of an exponential function.

15.3.1 Exponential functions

eles-4988

- **Exponential functions** can be used to model many real situations involving natural growth and decay.

Exponential growth	Exponential decay
Exponential growth is when a quantity grows by a constant percentage in each fixed period of time. Examples of exponential growth include growth of investment at a certain rate of compound interest and growth in the number of cells in a bacterial colony.	**Exponential decay** is when a quantity decreases by a constant percentage in each fixed period of time. Examples of exponential decay include yearly loss of value of an item (called depreciation) and radioactive decay.
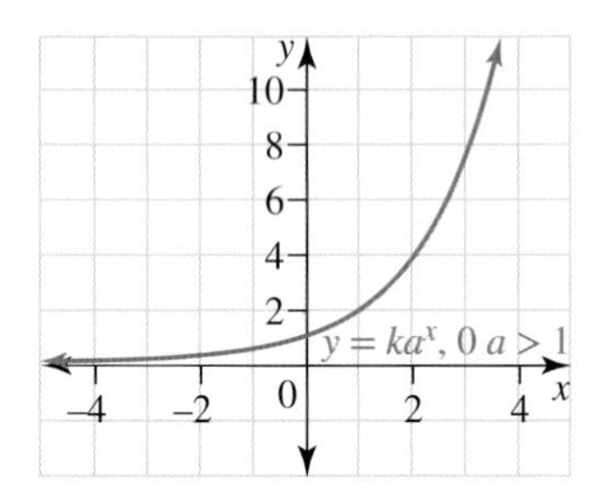	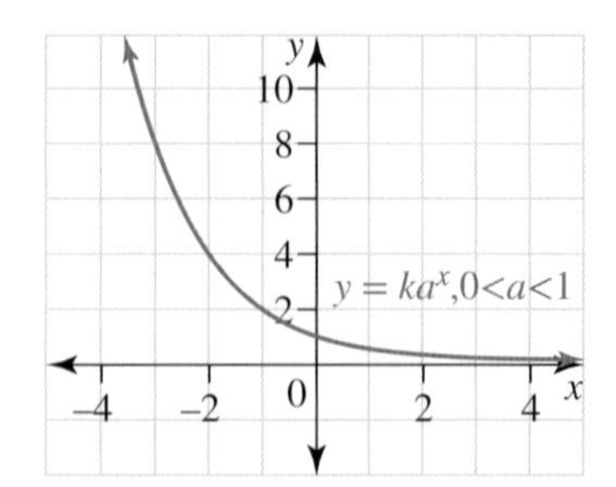

- Both exponential growth and decay can be modelled by exponential functions of the type $y = ka^x$ ($y = k \times a^x$). The difference is in the value of the base a. When $a > 1$, there is exponential growth and when $0 < a < 1$ there is exponential decay.

  When x is used to represent time, the value of k corresponds to the initial quantity that is growing or decaying.

WORKED EXAMPLE 7 Modelling bacterial growth using exponential functions

The number of bacteria, N, in a Petri dish after x hours is given by the equation $N = 50 \times 2^x$.

a. **Identify the initial number of bacteria in the Petri dish.**
b. **Determine the number of bacteria in the Petri dish after 3 hours.**
c. **Draw the graph of the function of N against x.**
d. **Use the graph to estimate the length of time it will take for the initial number of bacteria to treble.**

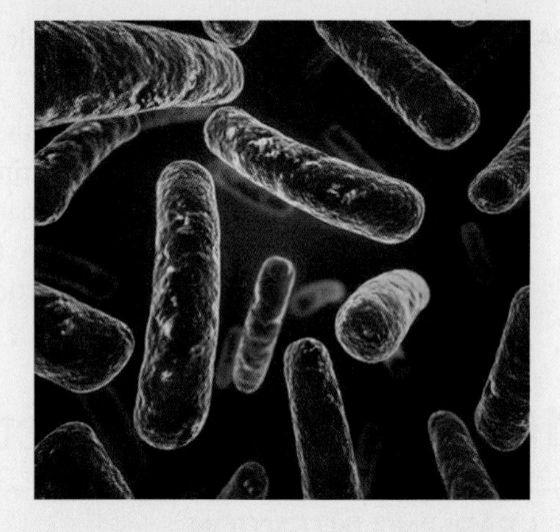

THINK	WRITE/DRAW
a. 1. Write the equation.	a. $N = 50 \times 2^x$
2. Substitute $x = 0$ into the given formula and evaluate. (Notice that this is the value of x for equations of the form $y = k \times a^x$.)	When $x = 0$, $N = 50 \times 2^0$ $= 50 \times 1$ $= 50$
3. Write the answer in a sentence.	The initial number of bacteria in the Petri dish is 50.
b. 1. Substitute $x = 3$ into the formula and evaluate.	b. When $x = 3$, $N = 50 \times 2^3$ $= 50 \times 8$ $= 400$
2. Write the answer in a sentence.	After 3 hours there are 400 bacteria in the Petri dish.

c. 1. Calculate the value of N when $x = 1$ and $x = 2$.

c. 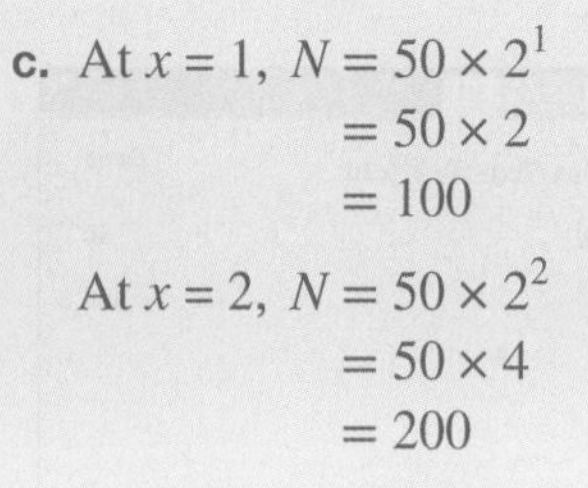

At $x = 1$, $N = 50 \times 2^1$
$= 50 \times 2$
$= 100$

At $x = 2$, $N = 50 \times 2^2$
$= 50 \times 4$
$= 200$

2. Draw a set of axes, labelling the horizontal axis as x and the vertical axis as N.

3. Plot the points generated by the answers to parts **a**, **b** and **c 1**.

4. Join the points plotted with a smooth curve.

5. Label the graph.

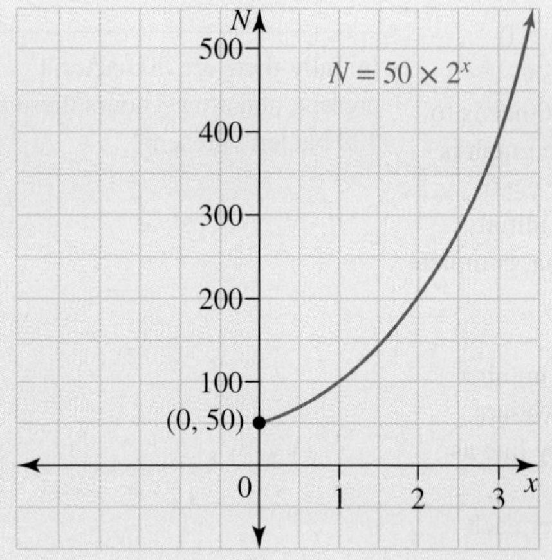

d. 1. Determine the number of bacteria required.

d. Number of bacteria $= 3 \times 50$
$= 150$

2. Draw a horizontal line from $N = 150$ to the curve and from this point draw a vertical line to the x-axis. This will help us to the estimate the time taken for the number of bacteria to treble.

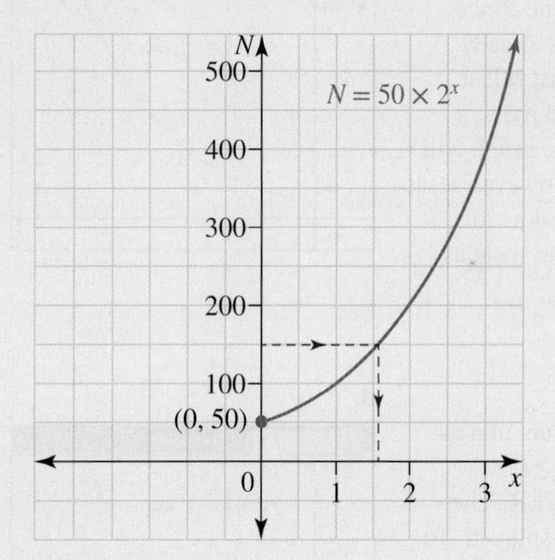

3. Write the answer in a sentence.

The time taken will be approximately 1.6 hours.

TI | THINK

DISPLAY/WRITE

a-b.

In a new problem, on a Calculator page, press:

- MENU
- 1: Actions
- 1: Define

Complete the function entry line as:
Define
$f1(x) = 50 \times 2^x | x \geq 0$
Then press ENTER.
Note that the $x \geq 0$ needs to be included as the graph is only sketched for $x \geq 0$.
To determine the initial number of bacteria, complete the entry line as:
$f1(0)$
To determine the number of bacteria after 3 hours, complete the entry line as:
$f1(3)$
Press ENTER after each entry.

a-b.

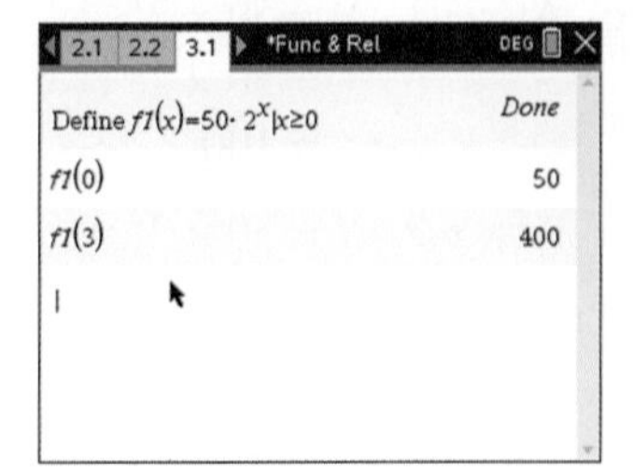

Initially there are 50 bacteria present, and after 3 hours there are 400 bacteria present.

c.

Open a Graphs page in the current document. Since the function has already been entered, just select the function and press ENTER, and the graph will be displayed. However, reset the viewing window to a more appropriate scale as shown.

c.

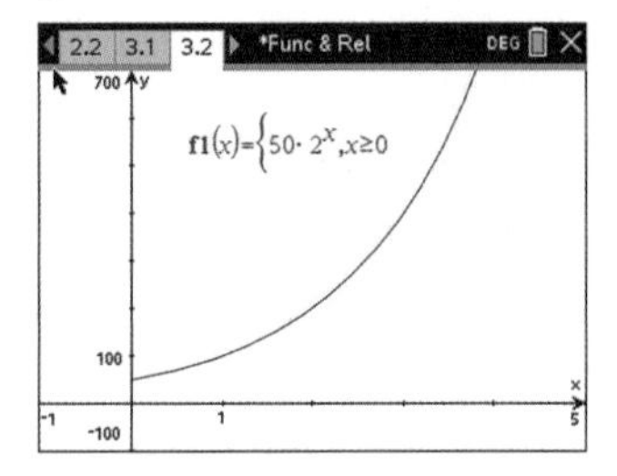

d.

Now enter the function as:
$f2(x) = 150 | x \geq 0$
Then press ENTER. The graph will be displayed. To find the point of intersection between the two graphs, press:

- MENU
- 6: Analyze Graph
- 4: Intersection

Move the cursor to the left of the intersection point and press
ENTER. Then move the cursor to the right of the intersection point and press ENTER. The intersection point is displayed.

d.

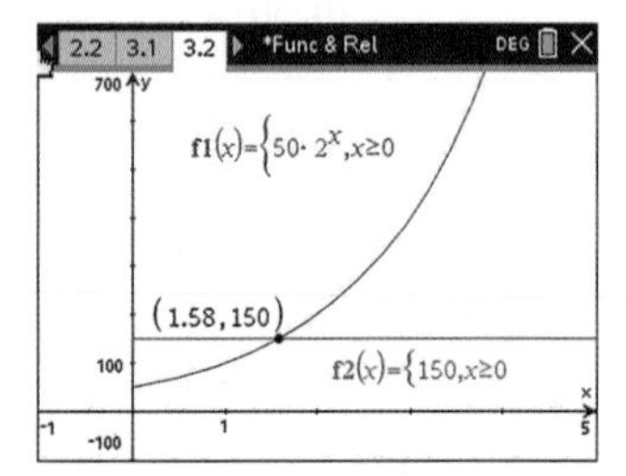

The point of intersection is at (1.58, 150). The initial number of bacteria will treble after 1.58 hours.

CASIO | THINK

DISPLAY/WRITE

a-b.

On the Main screen, complete the entry line as:
Define
$f1(x) = 50 \times 2^x | x \geq 0$
Then press EXE.
Note that we need to include the $x \geq 0$ as we want to sketch only the graph for $x \geq 0$.
To determine the initial number of bacteria, type:
$f1(0)$
To determine the number of bacteria after 3 hours, type:
$f1(3)$

a-b.

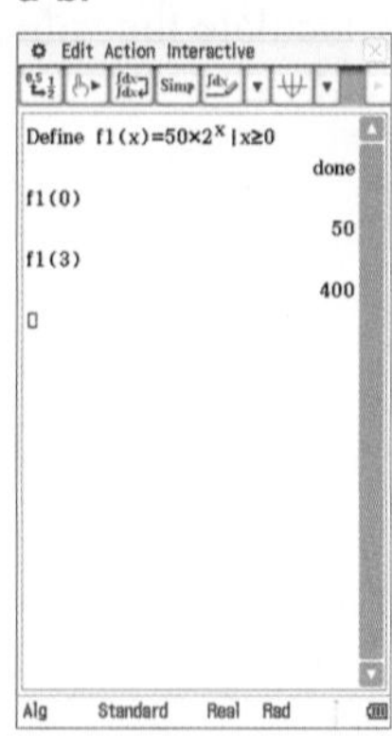

Initially there are 50 bacteria present, and after 3 hours there are 400 bacteria present.

c.

On the Graph & Table screen, complete the function entry line as:
$y1 = 50 \times 2^x | x \geq 0$
Tick the $y1$ box and tap the graphing icon.
Reset the viewing window to a more appropriate scale as shown.

c.

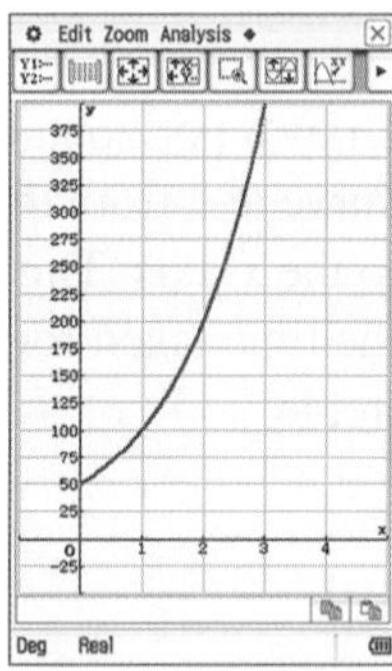

d.

Enter the equation by typing the entry line:
$y2 = 150 | x \geq 0$
Tick the $y2$ box and tap the graphing icon.
To find the point of intersection, tap:

- Analysis
- G-Solve
- Intersection

d.

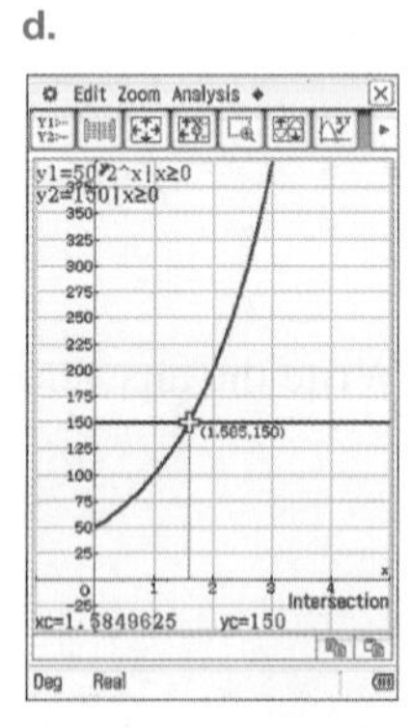

The point of intersection is at (1.58, 150). The initial number of bacteria will treble after 1.58 hours.

WORKED EXAMPLE 8 Determining the equation of an exponential function

A new computer costs $3000. It is estimated that each year it will be losing 12% of the previous year's value.

a. Calculate the value, $\$V$, of the computer after the first year.
b. Calculate the value of the computer after the second year.
c. Determine the equation that relates the value of the computer to the number of years, n, it has been used.
d. Use your equation to calculate the value of the computer in 10 years' time.

THINK	WRITE
a. 1. State the original value of the computer.	a. $V_0 = 3000$
2. Since 12% of the value is being lost each year, the value of the computer will be 88% or (100 − 12) % of the previous year's value. Therefore, the value after the first year (V_1) is 88% of the original cost.	$V_1 = 88\%$ of 3000 $= 0.88 \times 3000$ $= 2640$
3. Write the answer in a sentence.	The value of the computer after 1 year is $2640.
b. 1. The value of the computer after the second year, V_2, is 88% of the value after the first year.	b. $V_2 = 88\%$ of 2640 $= 0.88 \times 2640$ $= 2323.2$
2. Write the answer in a sentence.	The value of the computer after the second year is $2323.20.
c. 1. The original value is V_0.	c. $V_0 = 3000$
2. The value after the first year, V_1, is obtained by multiplying the original value by 0.88.	$V_1 = 3000 \times 0.88$
3. The value after the second year, V_2, is obtained by multiplying V_1 by 0.88, or by multiplying the original value, V_0, by $(0.88)^2$.	$V_2 = (3000 \times 0.88) \times 0.88$ $= 3000 \times (0.88)^2$
4. The value after the third year, V_3, is obtained by multiplying V_2 by 0.88, or V_0 by $(0.88)^3$.	$V_3 = (3000 \times 0.88)^2 \times 0.88$ $= 3000 \times (0.88)^3$
5. By observing the pattern we can generalise as follows: the value after the nth year, V_n, can be obtained by multiplying the original value, V_0, by 0.88 n times; that is, by $(0.88)^n$.	$V_n = 3000 \times (0.88)^n$
d. 1. Substitute $n = 10$ into the equation obtained in part **c** to calculate the value of the computer after 10 years.	d. When $n = 10$, $V_{10} = 3000 \times (0.88)^{10}$ $= 835.50$
2. Write the answer in a sentence.	The value of the computer after 10 years is $835.50.

15.3.2 Determining the equation of an exponential function using data

eles-5349

- If the initial value, k, is known, it is possible to substitute in a data point to determine the value of a and thereby find the equation of the function.
- Sometimes the relationship between the two variables closely resembles an exponential pattern, but cannot be described exactly by an exponential function. In such cases, part of the data are used to model the relationship with exponential growth or the decay function.

WORKED EXAMPLE 9 Determining the equation of an exponential function using data

The population of a certain city is shown in the table below.

Year	2000	2005	2010	2015	2020
Population (×1000)	128	170	232	316	412

Assume that the relationship between the population, P, and the year, x, can be modelled by the function $P = ka^x$, where x is the number of years after 2000.
Multiply the value of P by 1000 to determine the actual population.

a. State the value of k, which is the population, in thousands, at the start of the period.
b. Use a middle point in the data set to calculate the value of a, correct to 2 decimal places. Hence, write the formula, connecting the population, P, with the number of years, x, since 2000.
c. For the years given, calculate the size of the population using the formula obtained in part b. Compare it with the actual size of the population in those years.
d. Predict the population of the city in the years 2030 and 2035.

THINK	WRITE
a. From the given table, state the value of k that corresponds to the population of the city in the year 2000.	a. $k = 128$
b. 1. Write the given formula for the population of the city.	b. $P = ka^x$
2. Replace the value of k with the value found in a.	$P = 128 \times a^x$
3. Using a middle point of the data, replace x with the number of years since 1985 and P with the corresponding value	Middle point is (2010, 232). When $x = 10, P = 232$, so $232 = 128 \times a^{10}$.
4. Solve the equation for a.	$a^{10} = \frac{232}{128}$ $a^{10} = 1.8125$ $a = \sqrt[10]{1.8125}$ $a = 1.0613...$
5. Round the answer to 2 decimal places.	$a \approx 1.06$
6. Rewrite the formula with this value of a.	So, $P = 128 \times (1.06)^x$.
c. 1. Draw a table of values and enter the given years, the number of years since 2000, x, and the population for each year, P. Round values of P to the nearest whole number.	c. (see table below)
2. Comment on the closeness of the fit.	The values for the population obtained using the formula closely resemble the actual data.

Year	2000	2005	2010	2015	2020
x	0	5	10	15	20
P	128	171	229	307	411

d. 1. Determine the value of x, the number of years after 2000.	**d.** For the year, $x = 30$.
2. Substitute this value of x into the formula and evaluate.	$P = 128 \times (1.06)^{30}$ $= 735.166\,87...$
3. Round to the nearest whole number.	$P \approx 735$
4. Write the answer in a sentence.	The predicted population for 2030 is 735 000.
5. Repeat for the year 2035.	For the year 2035, $x = 35$. $P = 128 \times (1.06)^{35}$ $= 983.819...$ $P \approx 984$
6. Write the answer in a sentence.	The predicted population for 2035 is 984 000.

Resources

eWorkbook Topic 15 Workbook (worksheets, code puzzle and project) (ewbk-2041)

Digital documents SkillSHEET Converting a percentage to a decimal (doc-5386)
SkillSHEET Decreasing a quantity by a percentage (doc-5387)

Interactivity Exponential growth and decay (int-6211)

Exercise 15.3 Exponential functions

learnon

Individual pathways

PRACTISE	CONSOLIDATE	MASTER
1, 4, 7, 10, 13	2, 5, 8, 11, 14	3, 6, 9, 12, 15

To answer questions online and to receive **immediate feedback** and **sample responses** for every question, go to your learnON title at www.jacplus.com.au.

Fluency

1. WE7 The number of micro-organisms, N, in a culture dish after x hours is given by the equation $N = 2000 \times 3^x$.
   a. Identify the initial number of micro-organisms in the dish.
   b. Determine the number of micro-organisms in a dish after 5 hours.
   c. Draw the graph of N against x.
   d. Use the graph to estimate the number of hours needed for the initial number of micro-organisms to quadruple.

2. The value of an investment (in dollars) after n years is given by $A = 5000 \times (1.075)^n$.
   a. Identify the size of the initial investment.
   b. Determine the value of the investment (to the nearest dollar) after 6 years.
   c. Draw the graph of A against n.
   d. Use the graph to estimate the number of years needed for the initial investment to double.

3. **MC** a. The function $P = 300 \times (0.89)^n$ represents an:

A. exponential growth with the initial amount of 300
B. exponential growth with the initial amount of 0.89
C. exponential decay with the initial amount of 300
D. exponential decay with the initial amount of 0.89
E. exponential decay with the initial amount of 300×0.89

b. The relationship between two variables, A and t, is described by the function $A = 45 \times (1.095)^t$, where t is the time, in months, and A is the amount, in dollars. This function indicates:

A. a monthly growth of $45
B. a monthly growth of 9.5 cents
C. a monthly growth of 1.095%
D. a monthly growth of 9.5%
E. a yearly growth of 9.5%

Understanding

4. **WE8** A new washing machine costs $950. It is estimated that each year it will be losing 7% of the previous year's value.

a. Calculate the value of the machine after the first year.
b. Calculate the value of the machine after the second year.
c. Determine the equation that relates the value of the machine, $\$V$, to the number of years, n, that it has been used.
d. Use your equation to find the value of the machine in 12 years' time.

5. A certain radioactive element decays in such a way that every 50 years the amount present decreases by 15%. In 1900, 120 mg of the element was present.

a. Calculate the amount present in 1950.
b. Calculate the amount present in the year 2000.
c. Determine the rule that connects the amount of the element present, A, with the number of 50-year intervals, t, since 1900.
d. Calculate the amount present in the year 2010. Round your answer to 3 decimal places.
e. Graph the function of A against t.
f. Use the graph to estimate the half-life of this element (that is, the number of years needed for half the initial amount to decay).

6. When a shirt made of a certain fabric is washed, it loses 2% of its colour.

a. Determine the percentage of colour that remains after:

i. two washes
ii. five washes.

b. Write a function for the percentage of colour, C, remaining after w washings.
c. Draw the graph of C against w.
d. Use the graph to estimate the number of washes after which there is only 85% of the original colour left.

7. WE9 The population of a certain country is shown in the table.

Year	Population (in millions)
2000	118
2005	130
2010	144
2015	160
2020	178

Assume that the relationship between the population, P and the year, n can be modelled by the formula $P = ka^n$, where n is the number of years since 2000.

a. State the value of k.
b. Use the middle point of the data set to calculate the value of a rounded to 2 decimal places. Hence, write the formula that connects the two variables, P and n.
c. For the years given in the table, determine the size of the population, using your formula. Compare the numbers obtained with the actual size of the population.
d. Predict the population of the country in the year 2045.

8. The temperature in a room (in degrees Celsius), recorded at 10-minute intervals after the air conditioner was turned on, is shown in the table below.

Time (min)	0	10	20	30	40
Temperature (°C)	32	26	21	18	17

Assume that the relationship between the temperature, T, and the time, t, can be modelled by the formula $T = ca^t$, where t is the time, in minutes, since the air conditioner was turned on.

a. State the value of c.
b. Use the middle point in the data set to determine the value of a to 2 decimal places.
c. Write the rule connecting T and t.
d. Using the rule, calculate the temperature in the room, 10, 20, 30 and 40 minutes after the air conditioner was turned on and compare your numbers with the recorded temperature. Comment on your findings. (Give answers correct to 1 decimal place.)

9. The population of a species of dogs (D) increases exponentially and is described by the equation $D = 60\left(1 - 0.6^t\right) + 3$, where t represents the time in years.

a. Calculate the initial number of dogs.
b. Calculate the number of dogs after 1 year.
c. Determine the time taken for the population to reach 50 dogs.

Reasoning

10. Fiona is investing \$20 000 in a fixed term deposit earning 6% p.a. interest. When Fiona has \$30 000 she intends to put a deposit on a house.

a. Determine an exponential function that will model the growth of Fiona's investment.
b. Graph this function.
c. Determine the length of time (correct to the nearest year) that it will take for Fiona's investment to grow to \$30 000.
d. Suppose Fiona had been able to invest at 8% p.a. Explain how much quicker Fiona's investment would have grown to the \$30 000 she needs.
e. Alvin has \$15 000 to invest. Determine the interest rate at which Alvin must invest his money, if his investment is to grow to \$30 000 in less than 8 years.

11. A Petri dish containing a bacteria colony was exposed to an antiseptic. The number of bacteria within the colony, B, over time, t, in hours is shown in the graph.

   a. Using the graph, predict the number of bacteria in the Petri dish after 5 hours.
   b. Using the points from the graph, show that if B (in thousands) can be modelled by the function $B = ab^t$, then $a = 120$ and $b = 0.7$.
   c. After 8 hours, another type of antiseptic was added to the Petri dish. Within three hours, the number of bacteria in the Petri dish had decreased to 50.
   If the number of bacteria decreased at a constant rate, show that the total of number of bacteria that had decreased within two hours was approximately 6700.

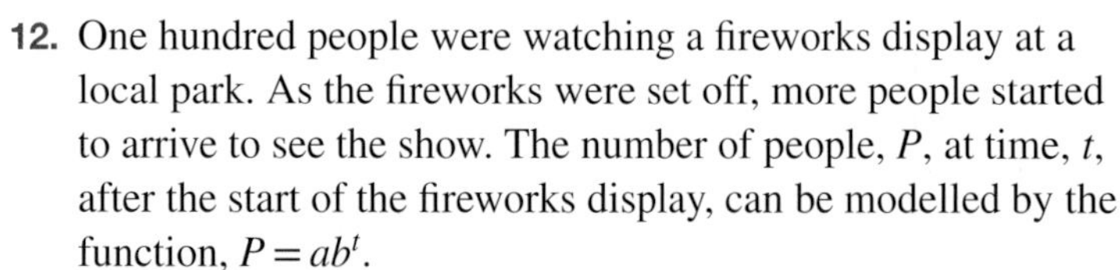

12. One hundred people were watching a fireworks display at a local park. As the fireworks were set off, more people started to arrive to see the show. The number of people, P, at time, t, after the start of the fireworks display, can be modelled by the function, $P = ab^t$.
   a. If after 5 minutes there were approximately 249 people, show that the number of people arriving at the park to watch the fireworks increased by 20% each minute.
   b. The fireworks display lasted for 40 minutes. After 40 minutes, people started to leave the park. The number of people leaving the park could be modelled by an exponential function.
   15 minutes after the fireworks ceased there were only 700 people in the park.
   Derive an exponential function that can determine the number of people, N, remaining in the park after the fireworks had finished at any time, m, in minutes.

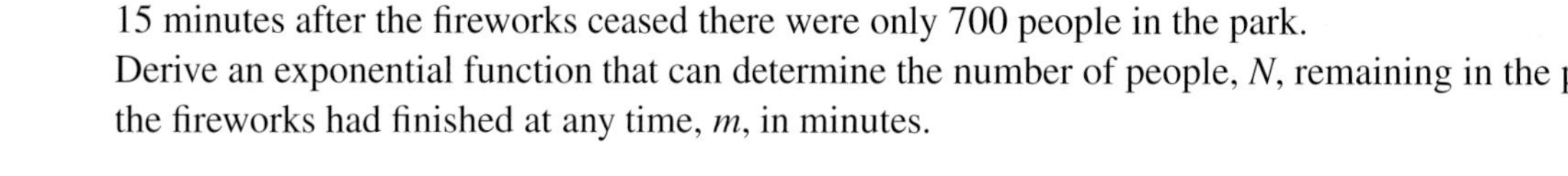

Problem solving

13. A hot plate used as a camping stove is cooling down. The formula that describes this cooling pattern is $T = 500 \times 0.5^t$ where T is the temperature in degrees Celsius and t is the time in hours.

   a. Identify the initial temperature of the stove.
   b. Determine the temperature of the stove after 2 hours.
   c. Decide when the stove will be cool enough to touch and give reasons.

14. The temperature in a greenhouse is monitored when the door is left open. The following measurements are taken.

Time (min)	0	5	10	15	20
Temperature (°C)	45	35	27	21	16

   a. State the initial temperature of the greenhouse.
   b. Determine an exponential equation to fit the collected data.
   c. Evaluate the temperature after 30 minutes.
   d. Explain whether the temperature will ever reach 0°C.

15. Carbon-14 decomposes in such a way that the amount present can be calculated using the equation $Q = Q_0(1 - 0.038)^t$, where Q is measured in milligrams and t in centuries.

   a. If there is 40 mg present initially, evaluate how much is present in:
      i. 10 years' time
      ii. 2000 years' time.
   b. Determine how many years will it take for the amount to be less than 10 mg.

15.4 Cubic functions

LEARNING INTENTION

At the end of this subtopic you should be able to:

- plot the graph of a cubic function using a table of values
- sketch the graph of a cubic function by calculating its intercepts
- determine the equation of a cubic function by inspection.

15.4.1 Cubic functions

- **Cubic functions** are **polynomials** where the highest power of the variable is three or the product of pronumeral makes up three.
- Some examples of cubic functions are $y = x^3$, $y = (x+1)(x-2)(x+3)$ and $y = 2x^2(4x-1)$.
- The following worked examples show how the graphs of cubic functions can be created by plotting points.

WORKED EXAMPLE 10 Plotting a cubic function using a table of values

Plot the graph of $y = x^3 - 1$ by completing a table of values.

THINK

1. Prepare a table of values, taking x-values from −3 to 3. Fill in the table by substituting each x-value into the given equation to determine the corresponding y-value.

WRITE/DRAW

x	−3	−2	−1	0	1	2	3
y	−28	−9	−2	−1	0	7	26

2. Draw a set of axes and plot the points from the table. Join them with a smooth curve.

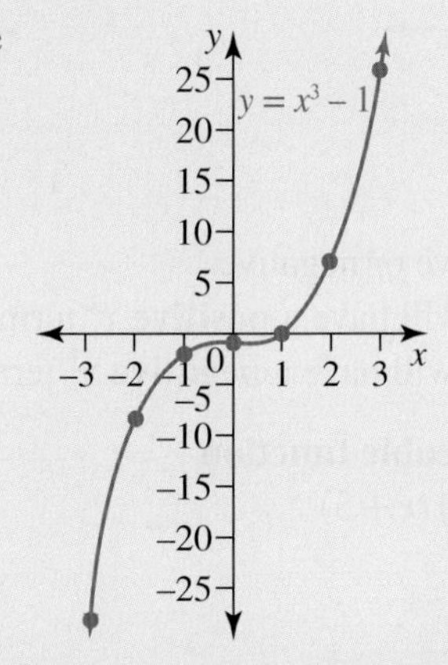

WORKED EXAMPLE 11 Plotting a cubic function using a table of values

Plot the curve of $y = x(x-2)(x+2)$ by completing a table of values.

THINK

1. Prepare a table of values, taking x-values from −3 to 3. Fill in the table by substituting each x-value into the given equation.

WRITE/DRAW

x	−3	−2	−1	0	1	2	3
y	−15	0	3	0	−3	0	15

2. Draw a set of axes and plot the points from the table. Join them with a smooth curve.

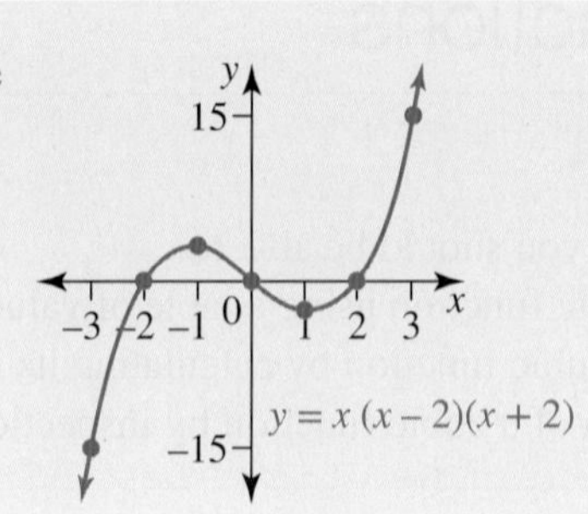

15.4.2 Sketching cubic functions

eles-4990

- Graphs of cubic functions have either *two turning points* or *one point of inflection.*
  - These two types of graphs are shown. Note that a point of inflection is a point where the gradient of the graph changes from decreasing to increasing or vice versa.
  - For the purposes of this topic, we will only consider the points of inflection where the graph momentarily flattens out.

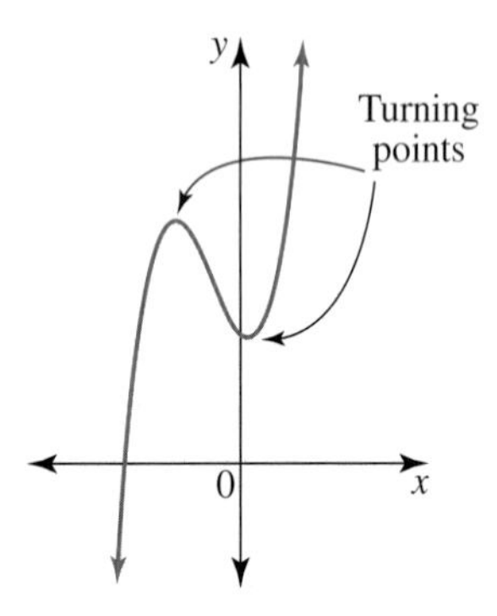

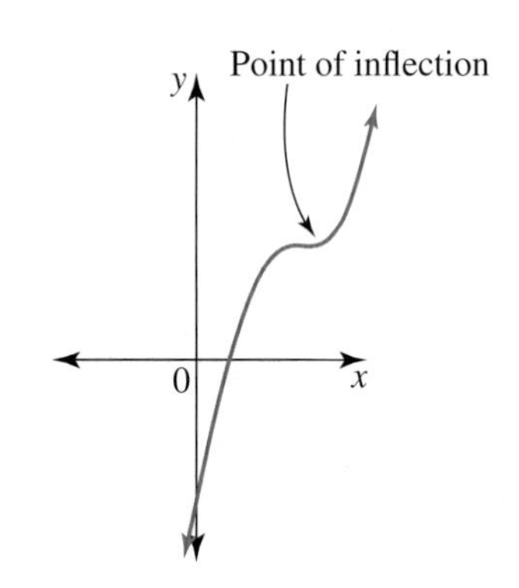

- Cubic functions can be positive or negative.
  - A positive cubic function will have a **positive** x^3 term and will have a general **upward** slope.
  - A negative cubic function will have a **negative** x^3 term and will have a general **downward** slope.

Example of a positive cubic function

$y = (x + 2)(x - 1)(x + 3)$

Example of a negative cubic function

$y = (x + 2)(1 - x)(x + 3)$

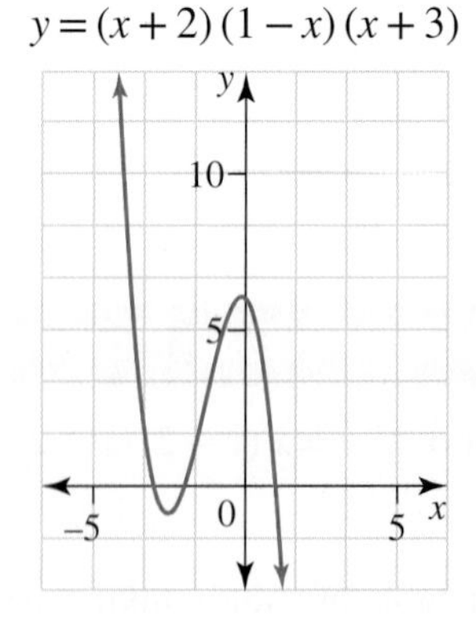

- To sketch the graph of a cubic function:
  1. determine the y-intercept by setting $x = 0$ and solving for y
  2. determine the x-intercepts by setting $y = 0$ and solving for x (you will need to use the Null Factor Law)
  3. draw the intercepts on the graph, then use those points to sketch the cubic graph.

- There are two special cases when sketching a cubic graph in factorised form:
  - Cubic functions of the form $y = (x - a)(x - b)^2$ will have a turning point on the x-axis at the point $(0, b)$.
  - Cubic functions of the form $y = (x - a)^3$ will have a point of inflection on the x-axis at the point $(0, a)$.

WORKED EXAMPLE 12 Sketching a cubic graph by determining intercepts

Sketch the following, showing all intercepts.

a. $y = (x - 2)(x - 3)(x + 5)$ b. $y = (x - 6)^2 (4 - x)$ c. $y = (x - 2)^3$

THINK	WRITE/DRAW
a. 1. Write the equation.	a. $y = (x - 2)(x - 3)(x + 5)$
2. The y-intercept occurs where $x = 0$. Substitute $x = 0$ into the equation.	y-intercept: if $x = 0$, $y = (-2)(-3)(5)$ $= 30$ Point: $(0, 30)$
3. Solve $y = 0$ to calculate the x-intercepts.	x-intercepts: if $y = 0$, $x - 2 = 0, x - 3 = 0$ or $x + 5 = 0$ $x = 2, x = 3$ or $x = -5$ Points: $(2, 0), (3, 0), (-5, 0)$
4. Combine the above steps to sketch.	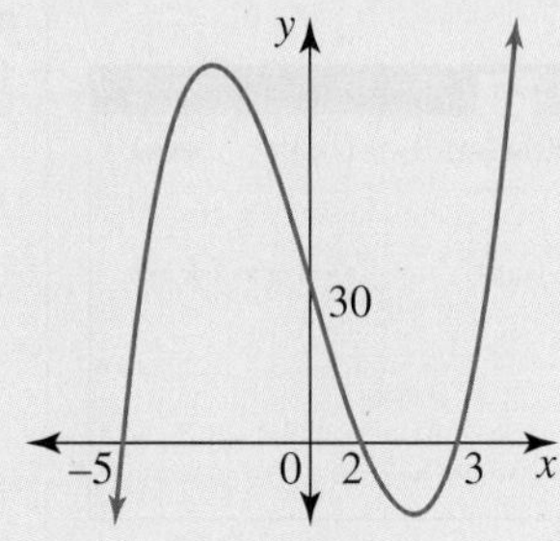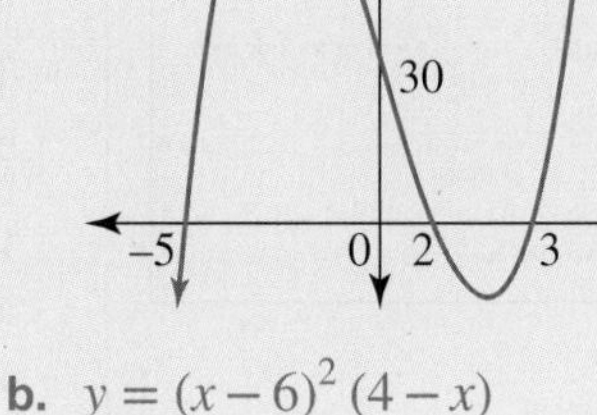
b. 1. Write the equation.	b. $y = (x - 6)^2 (4 - x)$
2. Substitute $x = 0$ to calculate the y-intercept.	y-intercept: if $x = 0$, $y = (-6)^2 (4)$ $= 144$ Point: $(0, 144)$
3. Solve $y = 0$ to calculate the x-intercepts.	x-intercept: if $y = 0$, $x - 6 = 0$ or $4 - x = 0$ $x = 6$ or $x = 4$ Point: $(6, 0), (4, 0)$
4. Combine all information and sketch the graph. *Note:* The curve just touches the x-axis at $x = 6$. This occurs with a double factor such as $(x - 6)^2$.	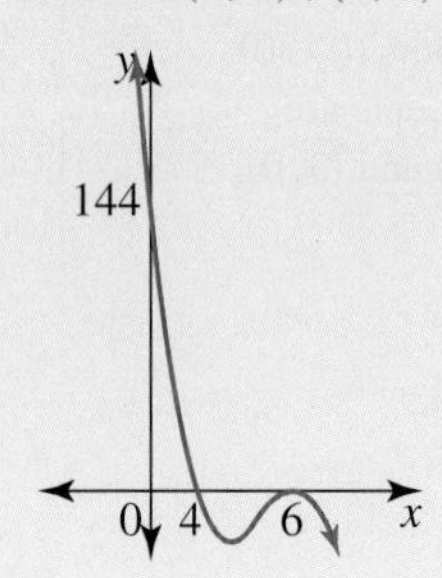

c. 1. Write the equation.

c. $y = (x - 2)^3$

2. Substitute $x = 0$ to calculate the y-intercept.

y-intercept: if $x = 0$,

$y = (-2)^3$

$= -8$

3. Solve $y = 0$ to calculate the x-intercepts.

x-intercept: if $x = 0$,

$x - 2 = 0$

$x = 2$

4. Combine all information and sketch the graph.
Note: The point of inflection is at $x = 2$. This occurs with a triple factor such as $(x - 2)^3$.

TI | THINK | **DISPLAY/WRITE**

a.

1. In a new problem, on a Calculator page, complete the entry lines as:
Define
$f1(x) = (x - 2)(x - 3)(x + 5)$
$f1(0)$
solve $\left(f1(x) = 0, x\right)$
Press ENTER after each entry.
Remember to include the implied multiplication sign between the brackets.

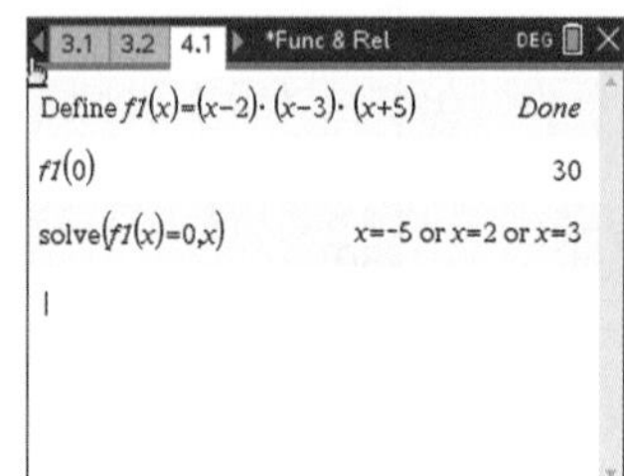

2. Open a Graphs page in the current document. Since the function has already been entered, just select the function and press ENTER and the graph will be displayed. Reset the viewing window to a more appropriate scale as shown. This graph does cross the x-axis at three distinct points. Find all the axial intercepts as described earlier.

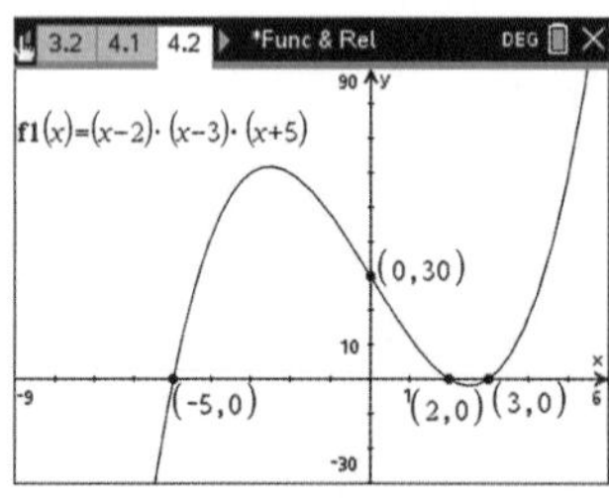

The y- intercept is (0, 30) and the x-intercepts are $(-5, 0)$, $(2, 0)$ and $(3, 0)$.

CASIO | THINK | **DISPLAY/WRITE**

a.

1. On the Graph & Table screen, complete the function entry line as:
$y1 = (x - 2)(x - 3)(x + 5)$
Touch the graphing icon and resize the graph. Touch the Y = 0 icon for the x-intercepts. Use the right arrow to move between the x-intercepts.

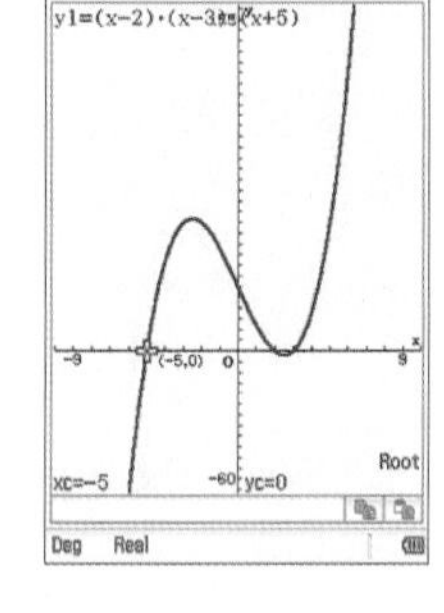

2. For the y-intercept, tap:
   - Analysis
   - G-Solve
   - y-intercept

The y-intercept is $(0, 30)$ and the x-intercepts are $(-5, 0)$, $(2, 0)$ and $(3, 0)$.

b.

1. On a Calculator page, complete the entry lines as:
   Define
   $f1(x) = (x - 6)^2 (4 - x)$
   $f1(0)$
   solve $\left(f1(x) = 0, x\right)$
   Then press ENTER.
   Remember to include the implied multiplication sign between the brackets.

2. Open a Graphs page in the current document. Since the function has already been entered, just select the function and press ENTER and the graph will be displayed. Reset the viewing window to a more appropriate scale as shown. This graph does cross the x-axis at two distinct points; however, this is not clear from the graph shown. Find all the axial intercepts as described earlier.

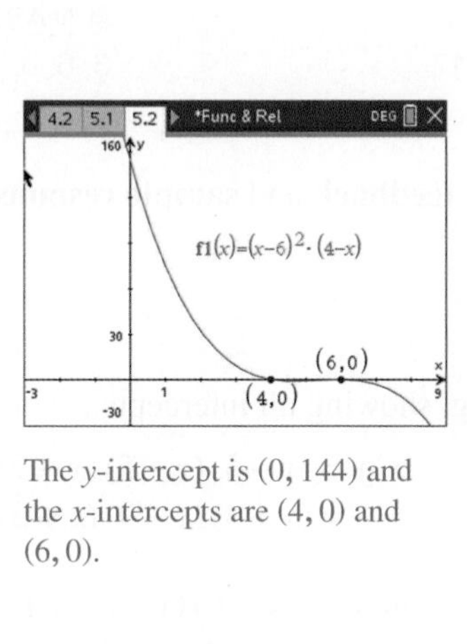

The y-intercept is $(0, 144)$ and the x-intercepts are $(4, 0)$ and $(6, 0)$.

b.

1. On the Graph & Table screen, complete the function entry line as:
   $y1 = (x - 6)^2 (4 - x)$
   Touch the graphing icon and resize the graph. Touch the Y = 0 icon for the x-intercepts. Use the right arrow to move between the x-intercepts.

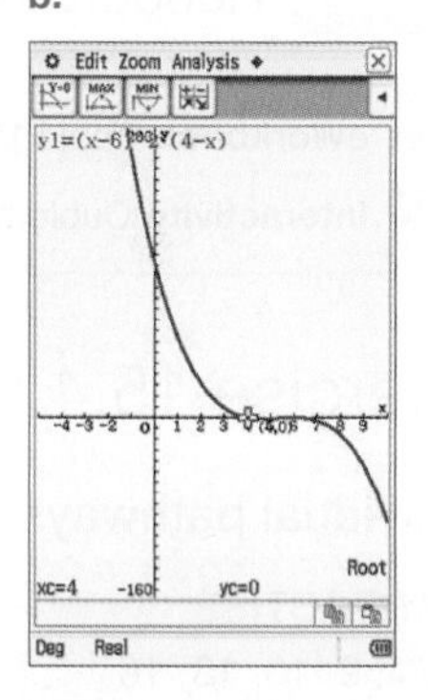

2. For the y-intercept, tap:
   - Analysis
   - G-Solve
   - y-intercept

The y-intercept is $(0, 144)$ and the x-intercepts are $(4, 0)$ and $(6, 0)$.

c.

1. In a new problem, on a Calculator page, complete the entry lines as:
   Define $f1\ (x) = (x - 2)^3$
   $f1\ (0)$
   solve $\left(f1(x) = 0, x\right)$
   Press ENTER after each entry.

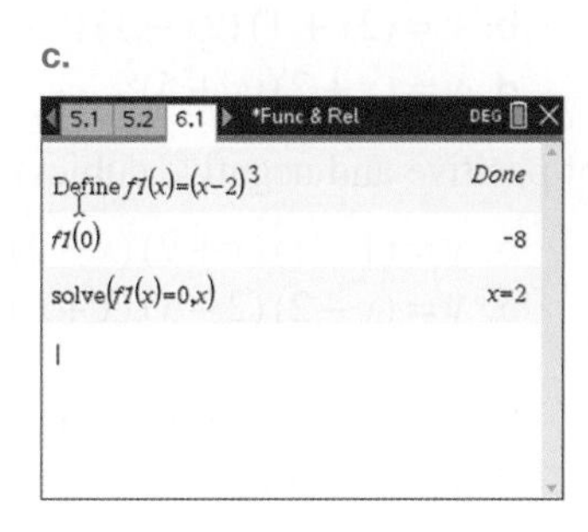

2. Open a Graphs page in the current document. Since the function has already been entered, just select the function and press ENTER and the graph will be displayed. Reset the viewing window to a more appropriate scale as shown. Find all the axial intercepts as described earlier.

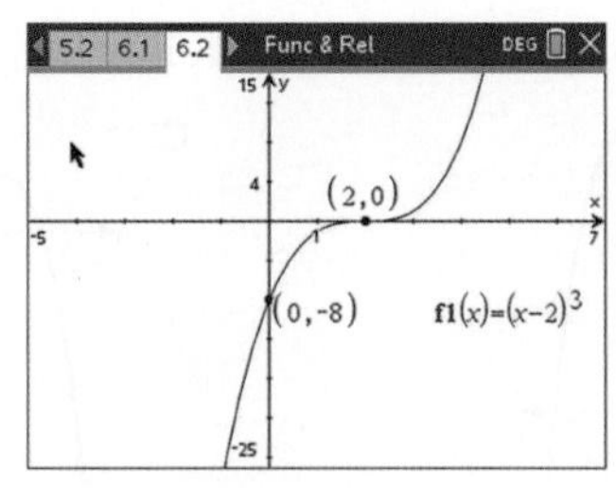

The y-intercept is $(0, -8)$ and the x-intercept is $(2, 0)$. For this example there is only one x-intercept as it is a triple factor; this point is called a point of inflection.

c.

1. On the Graph & Table screen, complete the function entry as:
   $y1 = (x - 2)^3$
   Touch the graphing icon and resize the graph. Touch the Y = 0 icon for the x-intercepts. Use the right arrow to move between the x-intercepts.

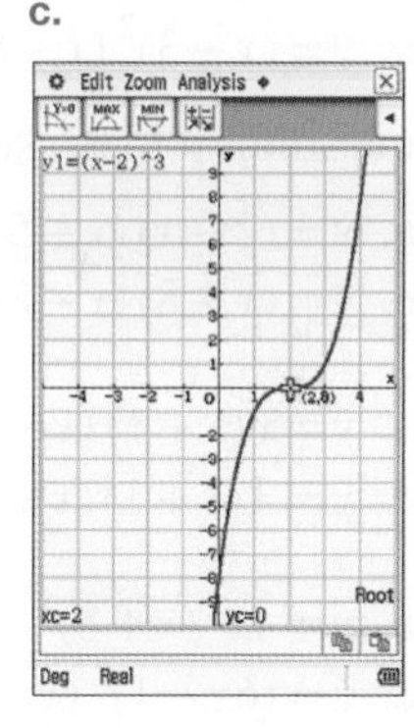

2. For the y-intercept, tap:
   - Analysis
   - G-Solve
   - y-intercept

The y-intercept is $(0, -8)$ and the x-intercept is $(2, 0)$. For this example there is only one x-intercept as it is a triple factor; this point is called a point of inflection.

Resources

eWorkbook Topic 15 Workbook (worksheets, code puzzle and project) (ewbk-2041)

Interactivity Cubic polynomials (int-2566)

Exercise 15.4 Cubic functions

learnon

Individual pathways

■ PRACTISE	■ CONSOLIDATE	■ MASTER
1, 4, 9, 10, 13, 16	2, 5, 7, 11, 14, 17	3, 6, 8, 12, 15, 18

To answer questions online and to receive **immediate feedback** and **sample responses** for every question, go to your learnON title at www.jacplus.com.au.

Fluency

WE10, 11, 12 For questions 1 to 3, sketch the following, showing all intercepts.

1. a. $y=(x-1)(x-2)(x-3)$
   b. $y=(x-3)(x-5)(x+2)$
   c. $y=(x+6)(x+1)(x-7)$
   d. $y=(x+4)(x+9)(x+3)$
2. a. $y=(x+8)(x-11)(x+1)$
   b. $y=(2x-6)(x-2)(x+1)$
   c. $y=(2x-5)(x+4)(x-3)$
   d. $y=(3x+7)(x-5)(x+6)$
3. a. $y=(4x-3)(2x+1)(x-4)$
   b. $y=(2x+1)(2x-1)(x+2)$
   c. $y=(x-3)^2(x-6)$
   d. $y=(x+2)(x+5)^2$

For questions 4 to 6, sketch the following (a mixture of positive and negative cubics).

4. a. $y=(2-x)(x+5)(x+3)$
   b. $y=(1-x)(x+7)(x-2)$
   c. $y=(x+8)(x-8)(2x+3)$
   d. $y=(x-2)(2-x)(x+6)$
5. a. $y=x(x+1)(x-2)$
   b. $y=-2(x+3)(x-1)(x+2)$
   c. $y=3(x+1)(x+10)(x+5)$
   d. $y=-3x(x-4)^2$
6. a. $y=4x^2(x+8)$
   b. $y=(5-3x)(x-1)(2x+9)$
   c. $y=(6x-1)^2(x+7)$
   d. $y=-2x^2(7x+3)$
7.  **MC** Select a reasonable sketch of $y=(x+2)(x-3)(2x+1)$ from the following.

A. 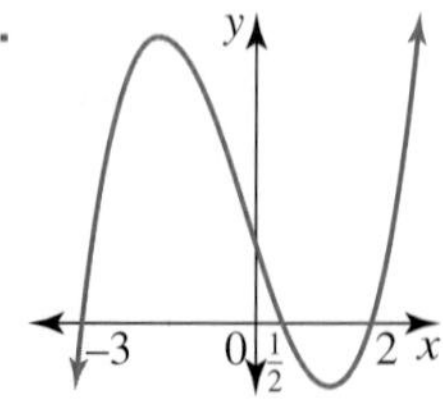

B. 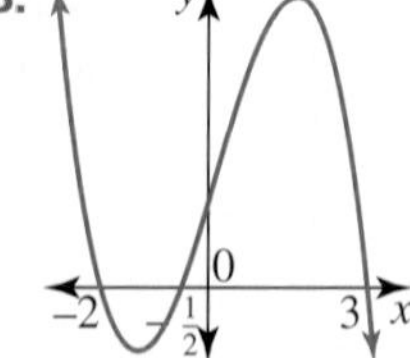

C. 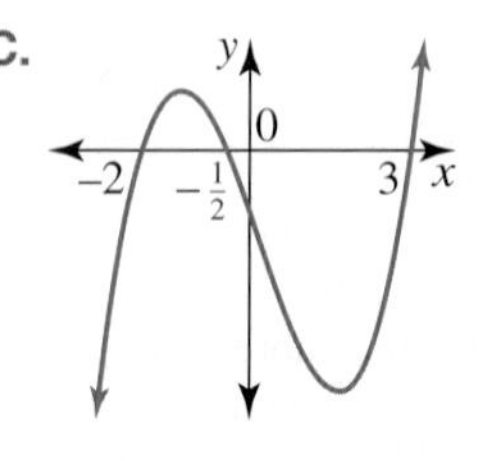

D. 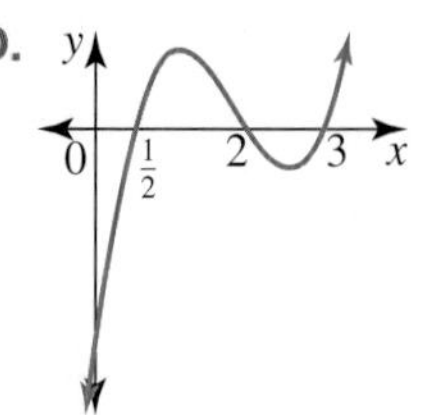

E. None of these.

8. **MC** The graph shown could be that of:

A. $y = x^2(x+2)$
B. $y = (x+2)^3$
C. $y = (x-2)(x+2)^2$
D. $y = (x-2)^2(x+2)$
E. $y = (x-2)(x-8)(x+2)$

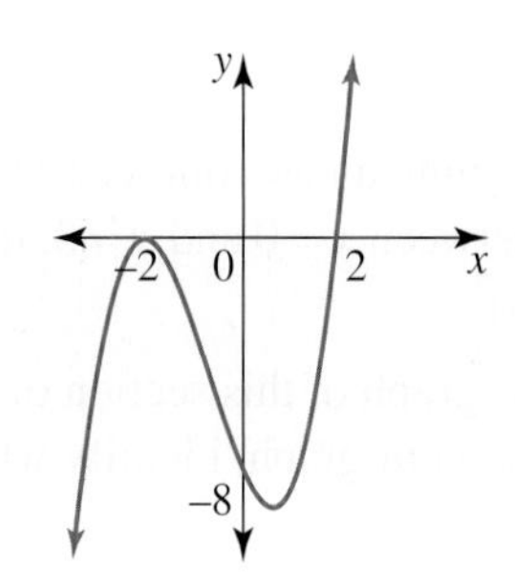

9. **MC** The graph shown has the equation:

A. $y = (x+1)(x+2)(x+3)$
B. $y = (x+1)(x-2)(x+3)$
C. $y = (x-1)(x+2)(x+3)$
D. $y = (x-1)(x+2)(x-3)$
E. $y = (x-3)(x-1)(x-6)$

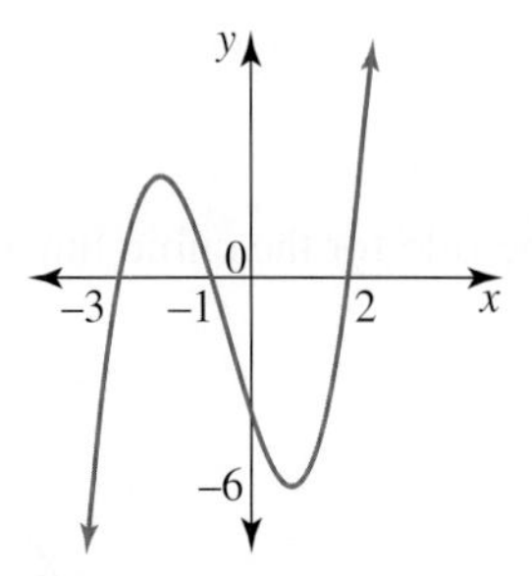

10. **MC** If a, b and c are positive numbers, the equation of the graph shown could be:

A. $y = (x-a)(x-b)(x-c)$
B. $y = (x+a)(x-b)(x+c)$
C. $y = (x+a)(x+b)(x-c)$
D. $y = (x-a)(x+b)(x-c)$
E. $y = x(x+b)(x-c)$

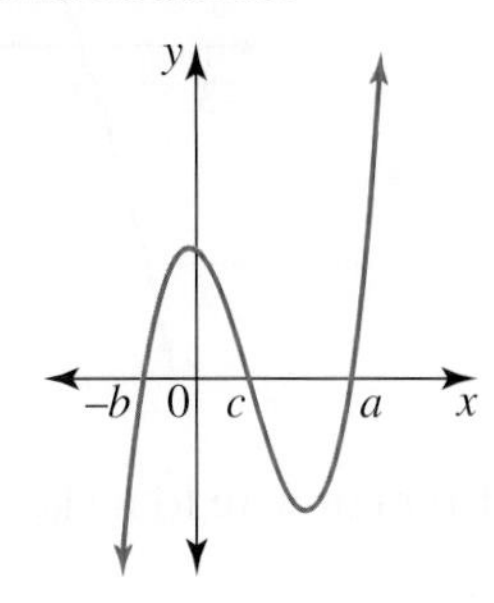

Understanding

11. Sketch the graph of each of the following.

a. $y = x(x-1)^2$
b. $y = -(x+1)^2(x-1)$

12. Sketch the graph of each of the following.

a. $y = (2-x)\left(x^2-9\right)$
b. $y = -x\left(1-x^2\right)$

Reasoning

13. For the graph shown, explain whether:

a. the gradient is positive, negative or zero to the left of the point of inflection.
b. the gradient is positive, negative or zero to the right of the point of inflection.
c. the gradient is positive, negative or zero at the point of inflection.
d. this is a positive or negative cubic graph.

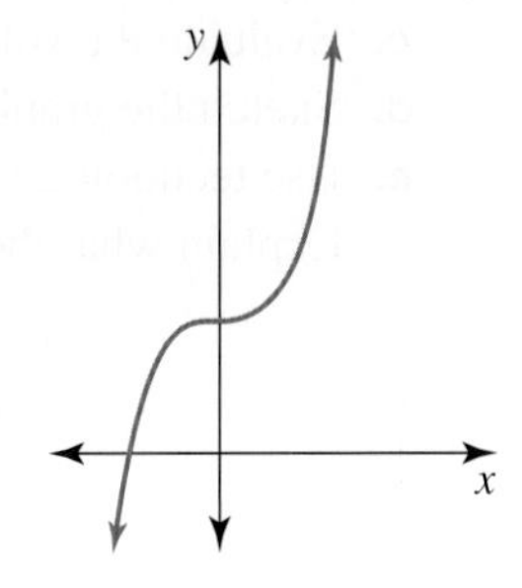

14. The function $f(x) = x^3 + ax^2 + bx + 4$ has x-intercepts at $(1, 0)$ and $(-4, 0)$. Determine the values of a and b. Show full working.

15. The graphs of the functions $f(x) = x^3 + (a+b)x^2 + 3x - 4$ and $g(x) = (x-3)^3 + 1$ touch. Express a in terms of b.

Problem solving

16. Susan is designing a new rollercoaster ride using maths. For the section between $x = 0$ and $x = 3$, the equation of the ride is $y = x(x - 3)^2$.

a. Sketch the graph of this section of the ride.
b. Looking at your graph, identify where the ride touches the ground.
c. The maximum height for this section is reached when $x = 1$. Use algebra to calculate the maximum height.

17. Determine the rule for the cubic function shown.

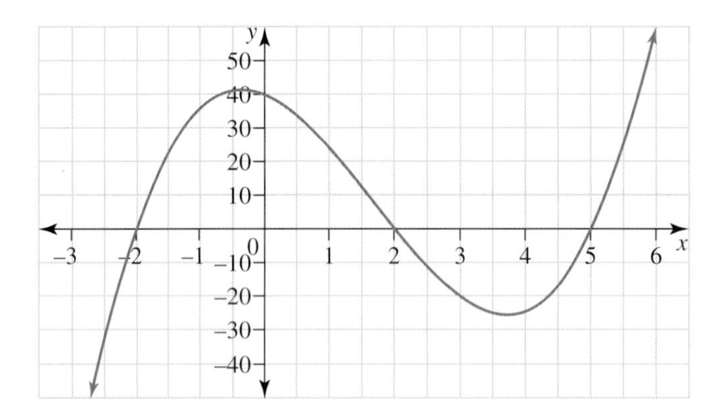

18. A girl uses 140 cm of wire to make a frame of a cuboid with a square base as shown.

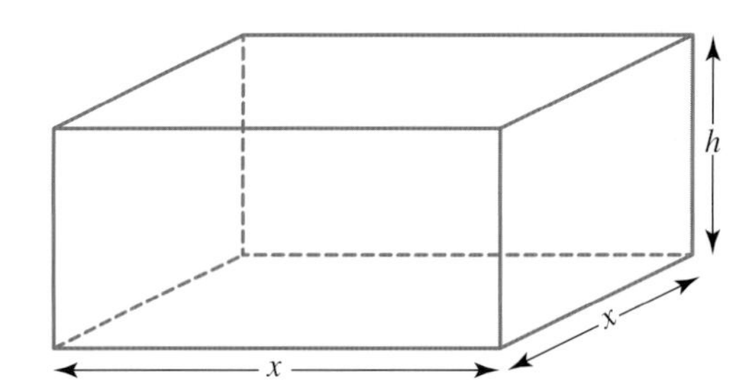

The base length of the cuboid is x cm and the height is h cm.

a. Explain why the volume cm^3 is given by $V = 35x^2 - 2x^3$.
b. Determine possible values that x can assume.
c. Evaluate the volume of the cuboid when the base area is 81 cm^2.
d. Sketch the graph of V versus x.
e. Use technology to determine the coordinates of the maximum turning point. Explain what these coordinates mean.

15.5 Quartic functions

LEARNING INTENTION

At the end of this subtopic you should be able to:
- factorise the equation of a quartic function
- sketch the graph of a quartic function from a factorised equation.

15.5.1 Quartic functions

eles-4991

- **Quartic function** are polynomials where the highest power of the variable is 4 or the product of pronumeral makes up four.
- Some examples of quartic functions are $y = x^4$ and $y = (x + 1)(x - 2)(x + 3)(x - 4)$.
- There are three types of quartic functions: those with *one turning point*, those with *three turning points* and those with *one turning point and one point of inflection*.
- The table below includes the standard types of positive quartic equations.

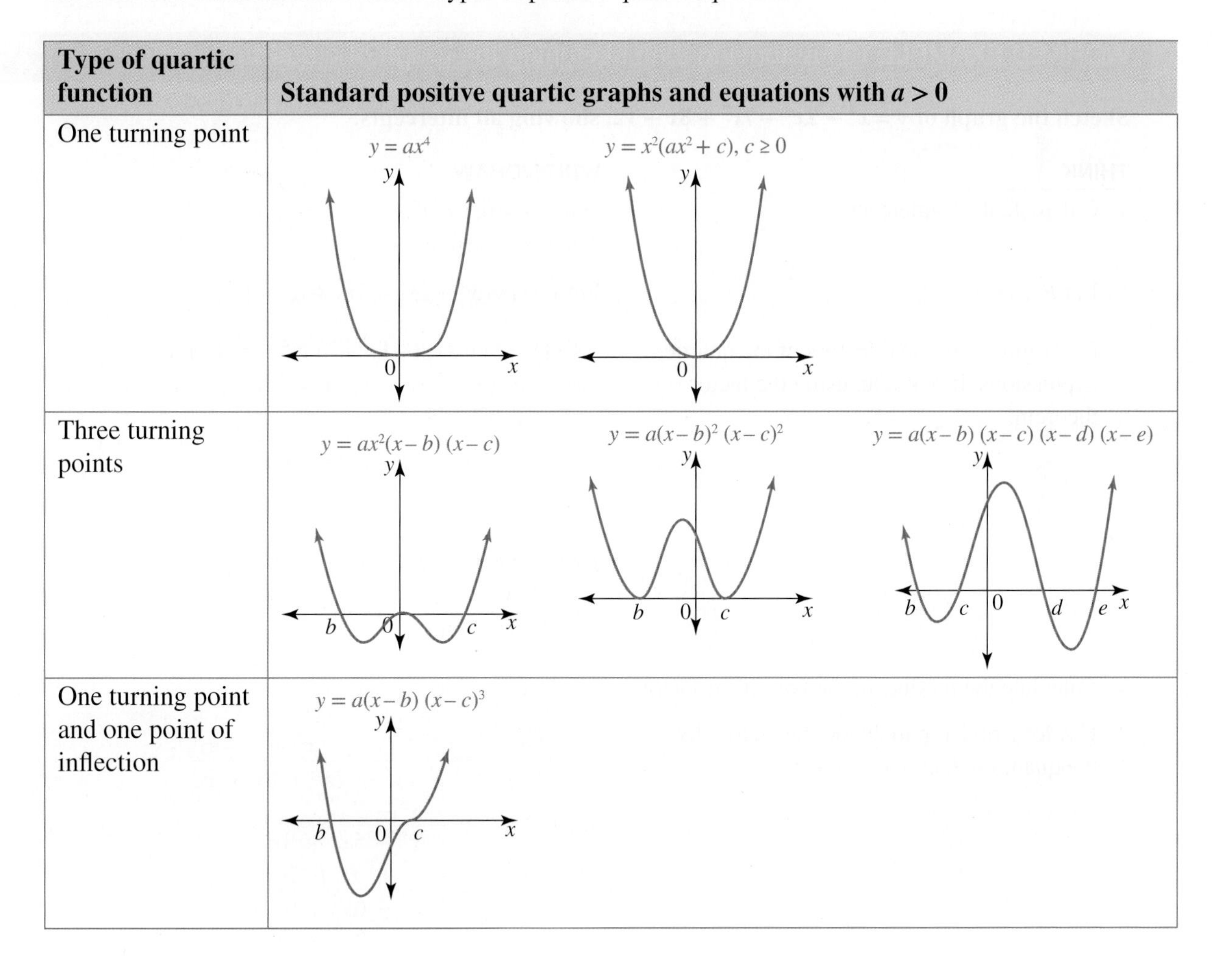

Type of quartic function	Standard positive quartic graphs and equations with $a > 0$
One turning point	$y = ax^4$; $y = x^2(ax^2 + c), c \geq 0$
Three turning points	$y = ax^2(x - b)(x - c)$; $y = a(x - b)^2(x - c)^2$; $y = a(x - b)(x - c)(x - d)(x - e)$
One turning point and one point of inflection	$y = a(x - b)(x - c)^3$

- There are also negative equivalents to all of the above graphs when $a < 0$. The negative graphs have the same shape, but are reflected across the x-axis. For example, the graph of $y = ax^4$ for $a < 0$ is shown.

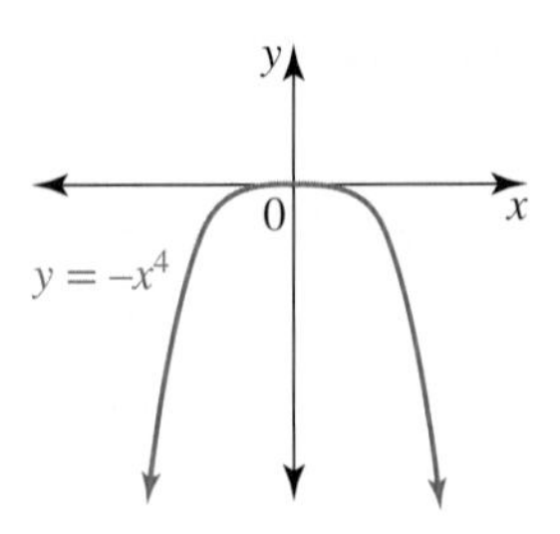

15.5.2 Sketching quartic functions

eles-4992

- To sketch a quartic function:
  1. factorise the equation so that the equation is in the form matching one of the standard quartics. To factorise, use the **factor theorem** and long division. (If the equation is already factorised, skip this step.)
  2. calculate all x- and y-intercepts
  3. draw the intercepts on the graph, then use those points to sketch the cubic graph.

WORKED EXAMPLE 13 Factorising then sketching the graph of a quartic function

Sketch the graph of $y = x^4 - 2x^3 - 7x^2 + 8x + 12$, showing all intercepts.

THINK	WRITE/DRAW
1. Calculate the y-intercept.	When $x = 0, y = 12$. The y-intercept is 12.
2. Let $P(x) = y$.	Let $P(x) = x^4 - 2x^3 - 7x^2 + 8x + 12$.
3. Determine two linear factors of the quartic expressions, if possible, using the factor theorem.	$P(1) = (1)^4 - 2(-1)^3 - 7(1)^2 + 8(1) + 12$ $= 12$ $\neq 0$ $P(-1) = (-1)^4 - 2(-1)^3 - 7(-1)^2 + 8(-1) + 12$ $= 0$ $(x+1)$ is a factor. $P(2) = (2)^4 - 2(2)^3 - 7(2)^2 + 8(2) + 12$ $= 0$ $(x-2)$ is a factor.
4. Calculate the product of the two linear factors	$(x+1)(x-2) = x^2 - x - 2$
5. Use long division to divide the quartic by the quadratic factor $x^2 - x - 2$.	$$\begin{array}{r} x^2 - x - 6 \\ x^2 - x - 2\overline{)x^4 - 2x^3 - 7x^2 + 8x + 12} \\ \underline{x^4 - x^3 - 2x^2} \\ -x^3 - 5x^2 + 8x \\ \underline{-x^3 + x^2 + 2x} \\ -6x^2 + 6x + 12 \\ \underline{-6x^2 + 6x + 12} \\ 0 \end{array}$$
6. Express the quartic in factorised form.	$y = (x+1)(x-2)(x^2 - x - 6)$ $= (x+1)(x-2)(x-3)(x+2)$

7. To calculate the x- intercepts, solve $y = 0$.

If $0 = (x + 1)(x - 2)(x - 3)(x + 2)$
$x = -1, 2, 3, -2$.

8. State the x-intercepts.

The x-intercepts are $-2, -1, 2, 3$.

9. Sketch the graph of the quartic.

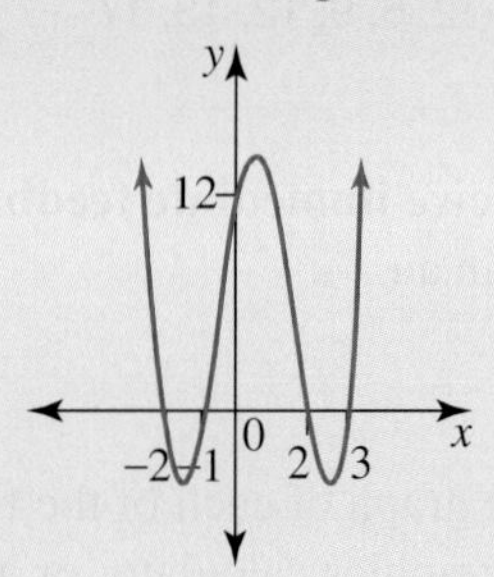

TI | THINK	DISPLAY/WRITE
1. In a new problem, on a Calculator page, complete the entry lines as: Define $f1(x) = x^4 - 2x^3 - 7x^2 + 8x + 12$ $f1(0)$ solve $(f1(x) = 0, x)$ Press ENTER after each entry.	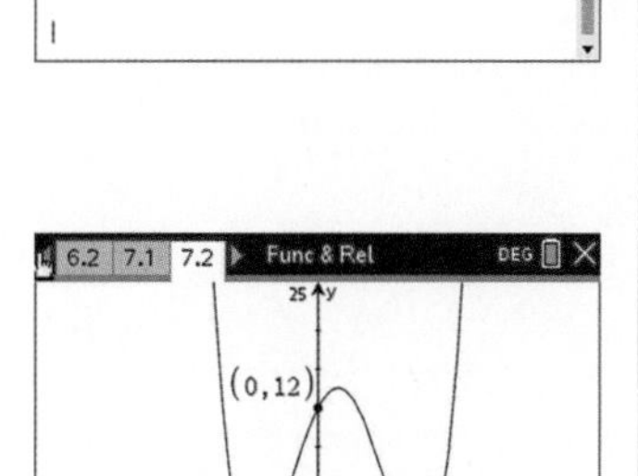
2. Open a Graphs page in the current document. Since the function has already been entered, just select the function and press ENTER and the graph will be displayed. Reset the viewing window to a more appropriate scale as shown. This graph does cross the x-axis at four distinct points. Find all the axial intercepts as described earlier.	The y-intercept is $(0, 12)$ and the x-intercepts are $(-2, 0)$, $(-1, 0)$, $(2, 0)$ and $(3, 0)$.

CASIO | THINK	DISPLAY/WRITE
1. On the Graph & Table screen, complete the function entry line as: $y1 = x^4 - 2x^3 - 7x^2 + 8x + 12$ Touch the graphing icon and resize the graph. Touch the Y = 0 icon for the x-intercepts. Use the right arrow to move between the x-intercepts.	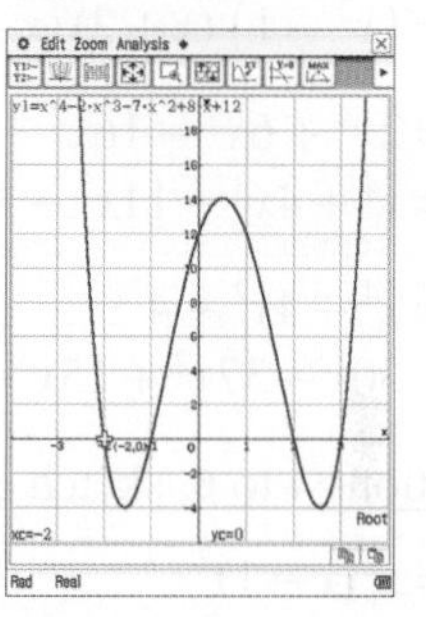
2. For the y-intercept, tap: • Analysis • G-Solve • y-intercept	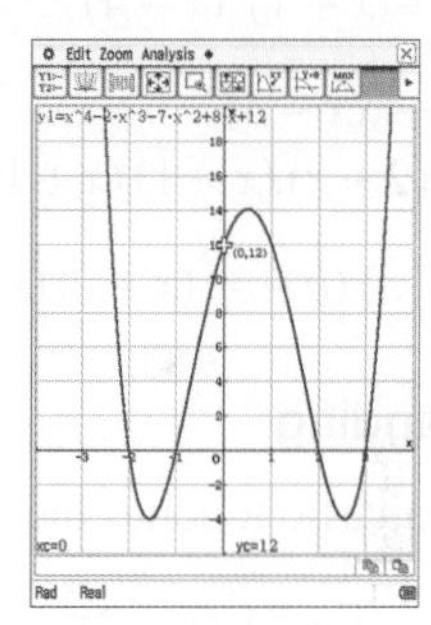 The y-intercept is $(0, 12)$ and the x-intercepts are $(-2, 0)$, $(-1, 0)$, $(2, 0)$ and $(3, 0)$.

on Resources

 **eWorkbook** Topic 15 Workbook (worksheets, code puzzle and project) (ewbk-2041)

 **Interactivity** Quartic functions (int-6213)

Exercise 15.5 Quartic functions

learnON

Individual pathways

■ PRACTISE	■ CONSOLIDATE	■ MASTER
1, 4, 7, 10, 11, 16	2, 5, 9, 12, 13, 17	3, 6, 8, 14, 15, 18

To answer questions online and to receive **immediate feedback** and **sample responses** for every question, go to your learnON title at www.jacplus.com.au.

Fluency

WE13 For questions 1 to 3, sketch the graph of each of the following showing all intercepts. You may like to verify the shape of the graph using a graphics calculator or another form of digital technology.

1. a. $y=(x-2)(x+3)(x-4)(x+1)$
   b. $y=\left(x^2-1\right)(x+2)(x-5)$

2. a. $y=2x^4+6x^3-16x^2-24x+32$
   b. $y=x^4+4x^3-11x^2-30x$

3. a. $y=x^4-4x^2+4$
   b. $y=30x-37x^2+15x^3-2x^4$

For questions 4 to 6, sketch each of the following.

4. a. $y=x^2(x-1)^2$
   b. $y=-(x+1)^2(x-4)^2$

5. a. $y=-x(x-3)^3$
   b. $y=(2-x)(x-1)(x+1)(x-4)$

6. $y=(x-a)(b-x)(x+c)(x+d)$, where $a, b, c, d>0$

Understanding

7. MC A quartic touches the x-axis at $x=-3$ and $x=2$. It crosses the y-axis at $y=-9$. A possible equation is:
   A. $y=\frac{1}{4}(x+3)^2(x-2)^2$
   B. $y=-\frac{1}{6}(x+3)^3(x-2)$
   C. $y=-\frac{3}{8}(x+3)(x-2)^3$
   D. $y=-\frac{1}{4}(x+3)^2(x-2)^2$
   E. $y=-\frac{1}{4}(x-3)^2(x+2)^2$

8. MC Consider the function $f(x)=x^4-8x^2-16$. When factorised, $f(x)$ is equal to:
   A. $(x+2)(x-2)(x-1)(x+4)$
   B. $(x+3)(x-2)(x-1)(x+1)$
   C. $(x-2)^3(x+2)$
   D. $(x-2)^2(x+2)^2$
   E. $(x-2)(x+2)^2$

9. **MC** Consider the function $f(x) = x^4 - 8x^2 - 16$.
The graph of $f(x)$ is best represented by:

A.
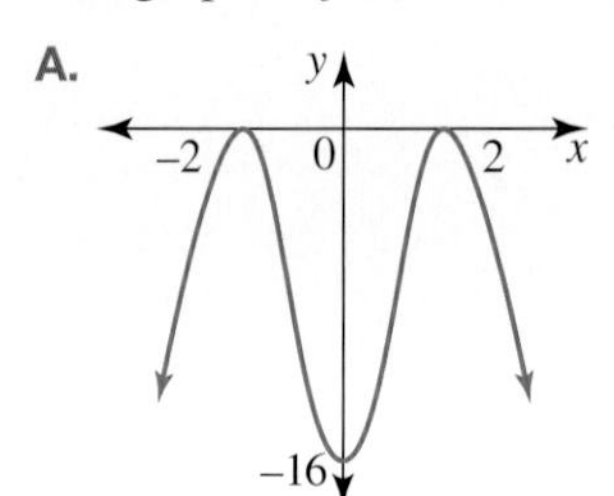

B.
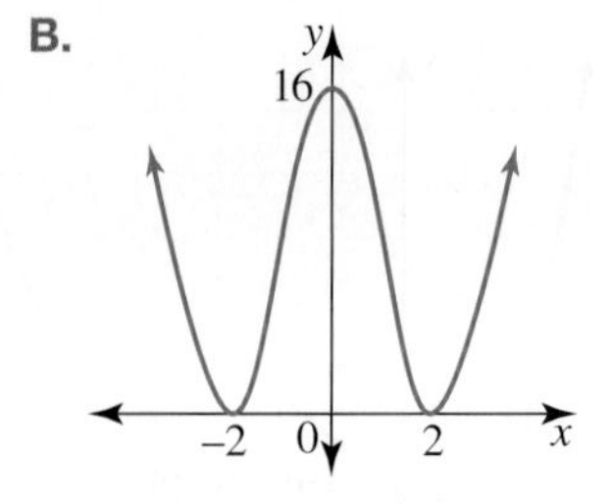

C.
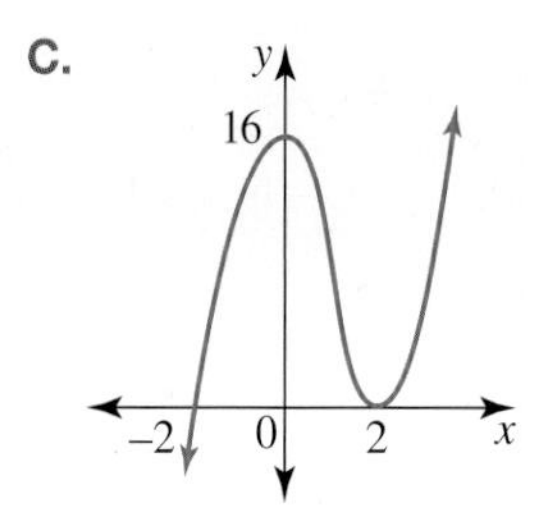

D.
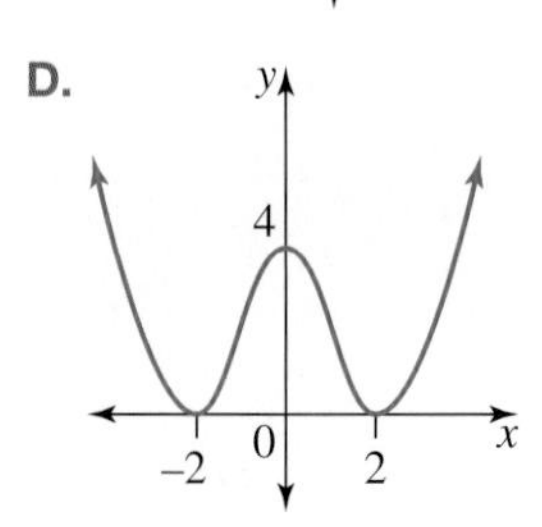

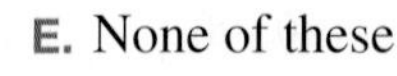
E. None of these

Reasoning

For questions 10 to 12, sketch the graph of each of the following functions. Verify your answers using a graphics calculator.

10. a. $y = x(x-1)^3$ b. $y = (2-x)\left(x^2-4\right)(x+3)$

11. a. $y = (x+2)^3(x-3)$ b. $y = 4x^2 - x^4$

12. a. $y = -(x-2)^2(x+1)^2$ b. $y = x^4 - 6x^2 - 27$

13. The functions $y = (a-2b)x^4 - 3x - 2$ and $y = x^4 - x^3 + (a+5b)x^2 - 5x + 7$ both have an x-intercept of 1. Determine the value of a and b. Show your working.

14. Sketch the graph of each of the following functions. Verify your answers using a graphics calculator.
a. $y = x^4 - x^2$ b. $y = 9x^4 - 30x^3 + 13x^2 + 20x + 4$

15. Patterns emerge when we graph polynomials with repeated factors, that is, polynomials of the form $P(x) = (x-a)^n, n > 1$. Discuss what happens if:
a. n is even b. n is odd.

Problem solving

16. The function $f(x) = x^4 + ax^3 - 4x^2 + bx + 6$ has x-intercepts $(2, 0)$ and $(-3, 0)$. Determine the values of a and b.

17. A carnival ride has a piece of the track modelled by the rule

$$h = -\frac{1}{300}x(x-12)^2(x-20) + 15, 0 \le x \le 20$$

where x metres is the horizontal displacement from the origin and h metres is the vertical displacement of the track above the horizontal ground.

a. Determine how high above the ground level the track is at the origin.
b. Use technology to sketch the function. Give the coordinates of any stationary points (that is, turning points or points of inflection).
c. Evaluate how high above ground level the track is when $x = 3$.

18. Determine the rule for the quartic function shown.

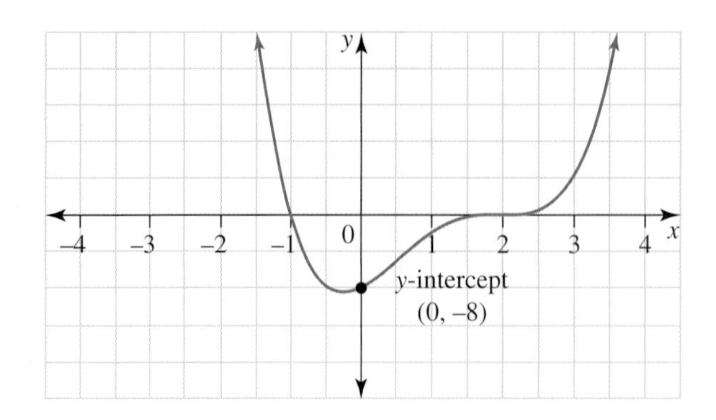

15.6 Transformations

LEARNING INTENTION

At the end of this subtopic you should be able to:
- sketch the graph of a function that has undergone some transformations
- describe a transformation using the words translation, dilation and reflection.

15.6.1 General transformations

eles-4993

- When the graph of a function has been moved, stretched and/or flipped, this is called a **transformation**.
- There are three types of transformations:
  - **translations** are movements of graphs left, right, up or down
  - **dilations** are stretches of graphs to make them thinner or wider
  - **reflections** are when a graph is flipped in the x- or y-axis.
- The following table summarises transformations of the general polynomial $P(x)$ with examples given for transformations of the basic quadratic function $y = x^2$ shown here:

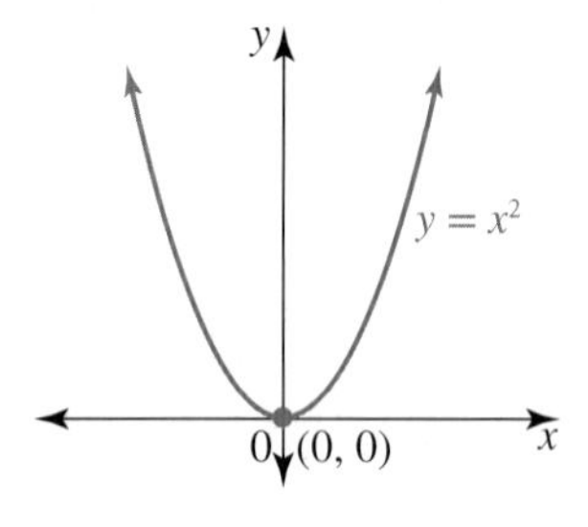

Transformation	Equation and explanation	Example(s)
Vertical translation	$y = P(x) + c$ This is a vertical translation of c units. If c is positive the translation is up, if c is negative the translation is down.	—— original function —— translated function $y = x^2 + 2$, $y = x^2$, $(0, 2)$ $y = x^2$, $y = x^2 - 3$, $(0, -3)$

Horizontal translation	$y=P(x-b)$ This is a horizontal translation of b units. If b is positive the translation is right, if b is negative the translation is left.	$y = x^2$, $y = (x-2)^2$, $(0, 4)$, 0, $(2, 0)$; $y = (x+1)^2$, $y = x^2$, $(0, 1)$, $(-1, 0)$, 0
Dilation	$y = aP(x)$ This is a dilation by a factor of a in the x direction. If $a > 1$ the graph becomes thinner, if $0 < a < 1$ the graph becomes wider.	$y = 2x^2$, $y = x^2$, $(0, 0)$; $y = \frac{1}{4}x^2$, $y = x^2$, $(0, 0)$
Reflection	$y=-P(x)$ This is a reflection in the x-axis. $y=P(-x)$ This is a reflection in the y-axis.	$y = x^2$, $(0, 0)$, $y = -x^2$

- Note that the graph $y=x^2$ does not change when reflected across the y-axis because it is symmetrical about the y-axis.
- With knowledge of the transformations discussed in this section, it is possible to generate many other graphs without knowing the equation of the original function.

WORKED EXAMPLE 14 Sketching transformations

Use the sketch of $y = P(x)$ shown to sketch:

a. $y=P(x)+1$ b. $y=P(x)-1$ c. $y=-P(x)$.

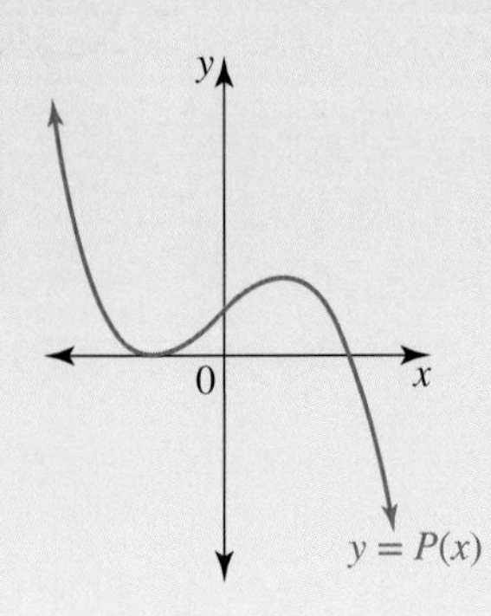

THINK | **WRITE/DRAW**

a. 1. Sketch the original $y = P(x)$.

2. Look at the equation $y = P(x) + 1$. This is a translation of one unit in the vertical direction — one unit up.

a.

$y = P(x)$

3. Sketch the graph of $y = P(x) + 1$ using a similar scale to the original.

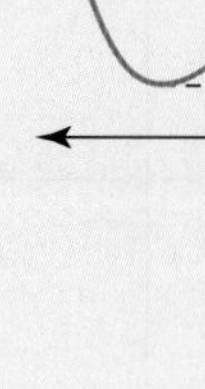

$y = P(x) + 1$

b. 1. Sketch the original $y = P(x)$.

2. Look at the equation $y = P(x) - 1$. This is a translation of negative one unit in the vertical direction; that is, one unit down.

b.

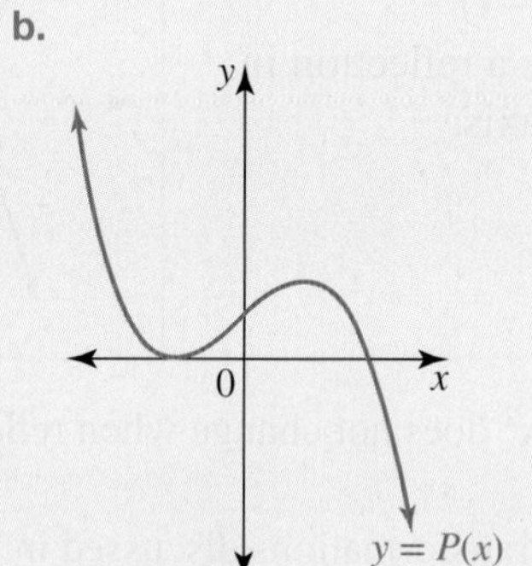

3. Sketch the graph of $y = P(x) - 1$ using a similar scale to the original.

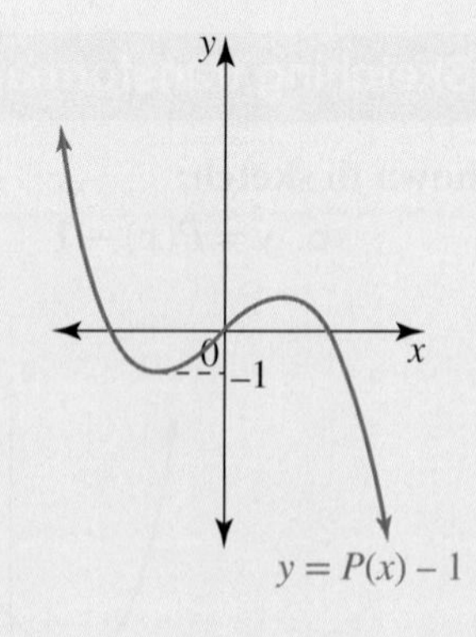

c. 1. Sketch the original $y = P(x)$.

2. Look at the equation $y = -P(x)$. This is a reflection across the x-axis.

c.

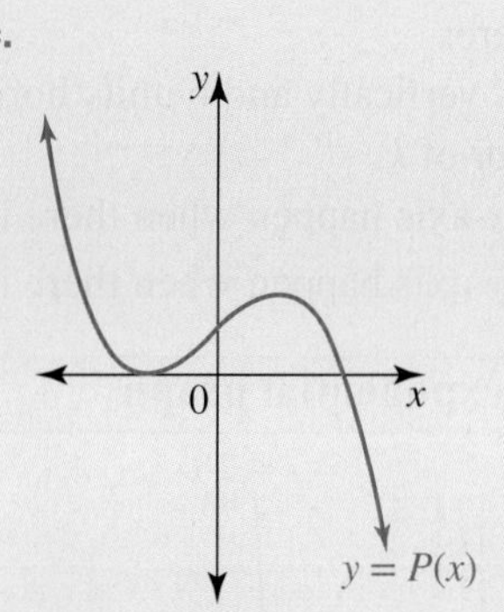

3. Sketch the graph of $y = -P(x)$ using a similar scale to the original.

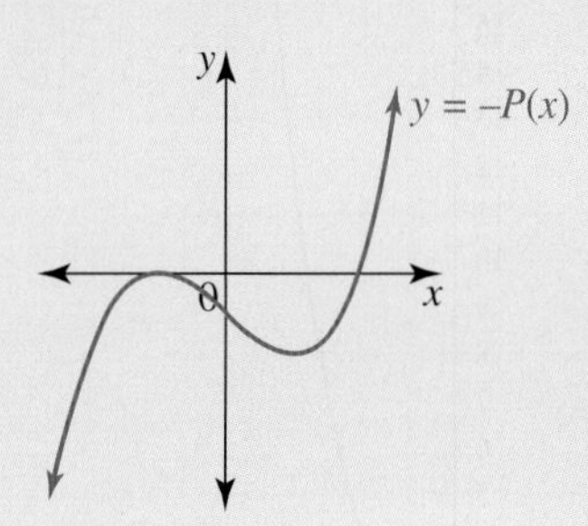

15.6.2 Transformations of hyperbolas, exponential functions and circles

eles-4994

- For transformations of hyperbolas, start with the standard hyperbola $y = \frac{1}{x}$ then transform using $y = \frac{a}{x-b} + c$ where:
  - translations are c units vertically and b units horizontally
  - dilations are by a factor of a
  - reflections across the x-axis happen when there is a negative at the front, for example $y = -\frac{a}{x-b} + c$; and reflections across the y-axis happen when there is x is negative, for example $y = \frac{a}{-x-b} + c$.

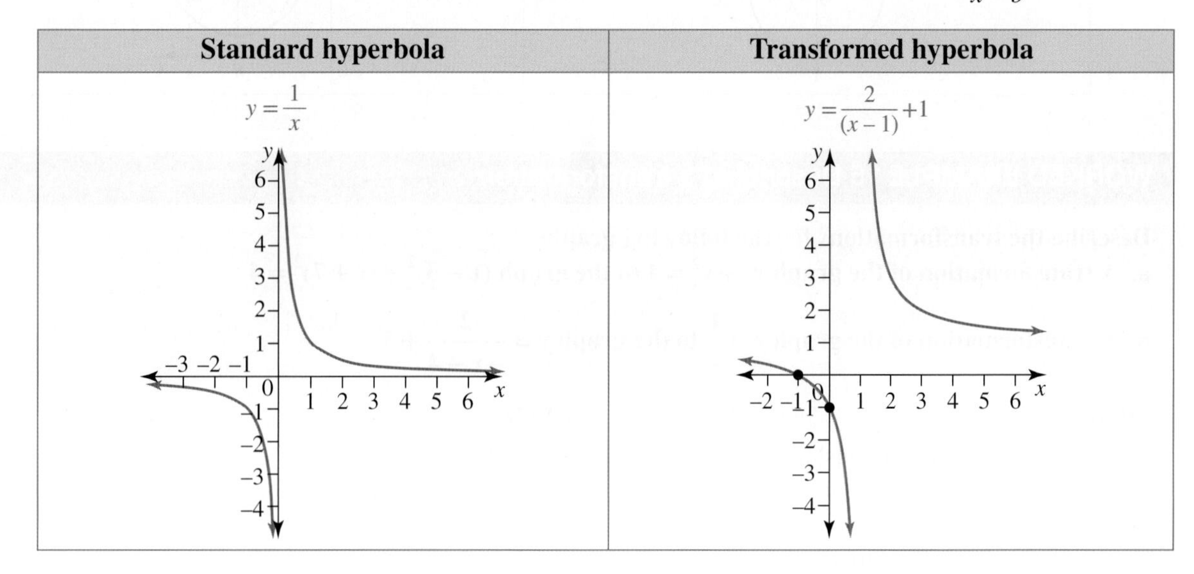

- For transformations of exponentials, start with the standard exponential $y = a^x$ where $a \neq 1$ then transform using $y = ka^{(x-b)} + c$ where:
  - translations are c units vertically and b units horizontally.
  - dilations are by a factor of k.
  - reflections across the x-axis happen when there is a negative out the front, $y = -\left(ka^{(x-b)} + c\right)$ and reflections across the y-axis happen when there is a negative on the x, $y = ka^{(-x-b)} + c$.

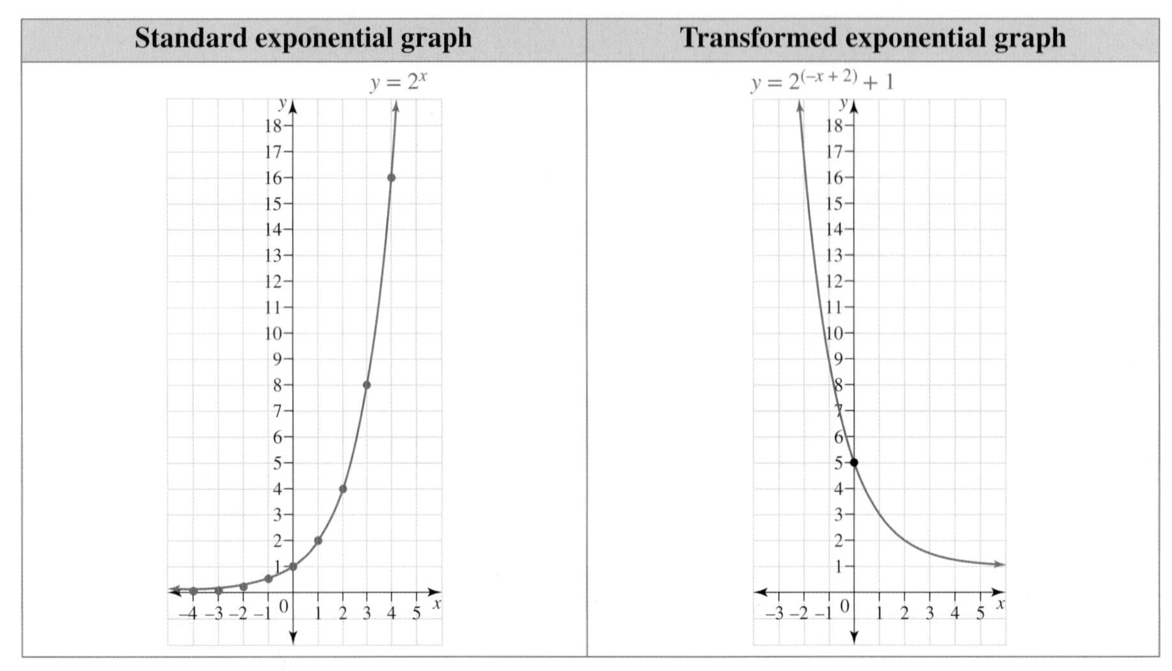

- For transformations of circles, start with the standard circle $x^2 + y^2 = r^2$ then transform using $(x-h)^2 + (y-k)^2 = r^2$ where:
  - translations are h units horizontally and k units vertically.

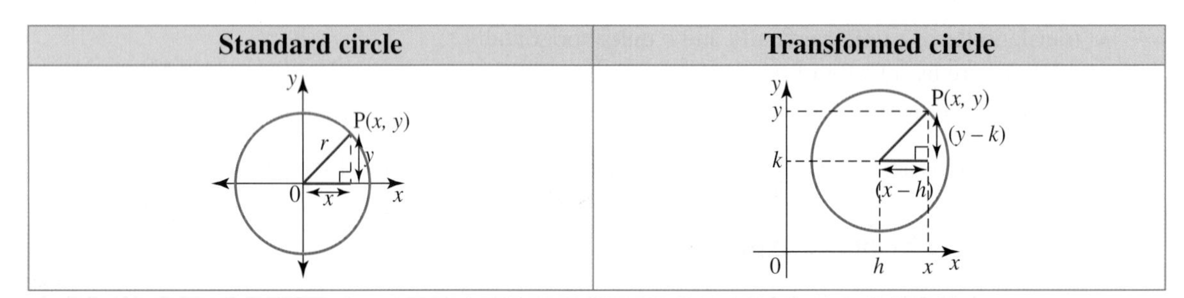

WORKED EXAMPLE 15 Describing a transformation

Describe the transformations for the following graphs.

a. A transformation of the graph $x^2 + y^2 = 4$ to the graph $(x-3)^2 + (y+7)^2 = 4$

b. A transformation of the graph $y = \dfrac{1}{x}$ to the graph $y = \dfrac{2}{-x-4} + 5$

THINK	WRITE
a. 1. Using the standard translated formula $(x-h)^2 + (y-k)^2 = r^2$ identify the value of h for the horizontal translation and the value of k for the vertical translation.	a. $h = 3, k = -7$. The graph is translated 3 units in the horizontal direction and –7 units in the vertical direction.

b. 1. Using the standard translated formula $y = \frac{a}{x-b} + c$ identify the value of c for the vertical translation and the value of b for the horizontal translation.

b. $b = 4, c = 5$.
The graph is translated 5 units vertically and 4 units horizontally.

2. Using the standard translated formula $y = \frac{a}{x-b} + c$ identify the value of a for the dilation factor.

$a = 2$.
The graph is dilated by a factor of 2.

3. Look for $y = -P(x)$ (reflection in the x-axis) and/or $y = P(-x)$ (reflection in the y-axis). This equation has a negative on the x-value so it is $y = P(-x)$.

The graph is reflected in the y-axis.

Resources

eWorkbook Topic 15 Workbook (worksheets, code puzzle and project) (ewbk-2041)

Interactivities Horizontal translations of parabolas (int-6054)
Vertical translations of parabolas (int-6055)
Dilation of parabolas (int-6096)
Exponential functions (int-5959)
Reflection of parabolas (int-6151)
Hyperbolas (int-6155)
Translations of circles (int-6214)
Transformations of exponentials (int-6216)
Transformations of cubics (int-6217)
The rectangular hyperbola (int-2573)

Exercise 15.6 Transformations

learnon

Individual pathways

■ PRACTISE	■ CONSOLIDATE	■ MASTER
1, 4, 7, 10, 13	2, 5, 8, 11, 14	3, 6, 9, 12, 15

To answer questions online and to receive **immediate feedback** and **sample responses** for every question, go to your learnON title at www.jacplus.com.au.

Fluency

1. **WE14** Use the sketch of $y = P(x)$ shown to sketch:

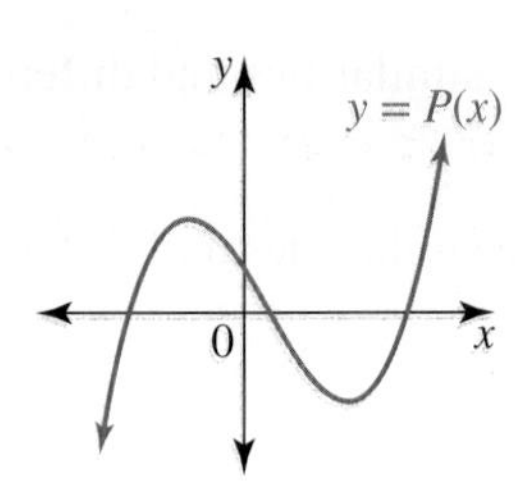

a. $y = P(x) + 1$

b. $y = -P(x)$

2. Use the sketch of $y = P(x)$ shown to sketch:

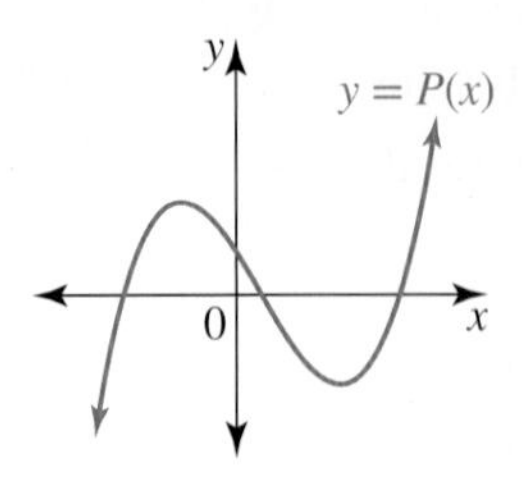

a. $y = P(x) - 2$

b. $y = 2P(x)$.

3. Consider the sketch of $y = P(x)$ shown. Sketch:

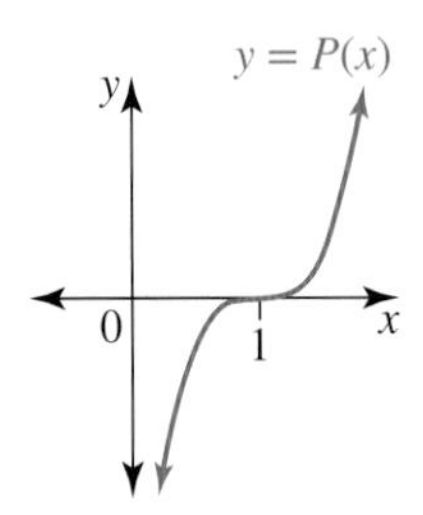

a. $y = P(x) + 1$

b. $y = -P(x)$

c. $y = P(x + 2)$.

4. **WE15** Describe the transformations for the following graphs.

a. A transformation of the graph $x^2 + y^2 = 9$ to the graph $(x - 2)^2 + (y - 1)^2 = 9$

b. A transformation of the graph $y = \frac{1}{x}$ to the graph $y = \frac{3}{x - 2} + 7$

5. Describe the transformations for the following graphs.

a. A transformation of the graph $y = \frac{1}{x}$ to the graph $y = \frac{5}{x - 1} + 2$

b. A transformation of the graph $y = 3^x$ to the graph $y = 3^{(-x+2)} + 4$

6. Describe the transformations for the following graphs.

a. A transformation of the graph $y = \frac{1}{x}$ to the graph $y = \frac{4}{-x + 3} + 10$

b. A transformation of the graph $y = 5^x$ to the graph $y = -\left(5^{(-x+7)} - 6\right)$

Understanding

7. Draw any polynomial $y = P(x)$. Discuss the similarities and differences between the graphs of $y = P(x)$ and $y = -P(x)$.

8. Draw any polynomial $y = P(x)$. Discuss the similarities and differences between the graphs of $y = P(x)$ and $y = 2P(x)$.

9. Draw any polynomial $y = P(x)$. Discuss the similarities and differences between the graphs of $y = P(x)$ and $y = P(x) - 2$.

Reasoning

10. Consider the sketch of $y = P(x)$ shown below.

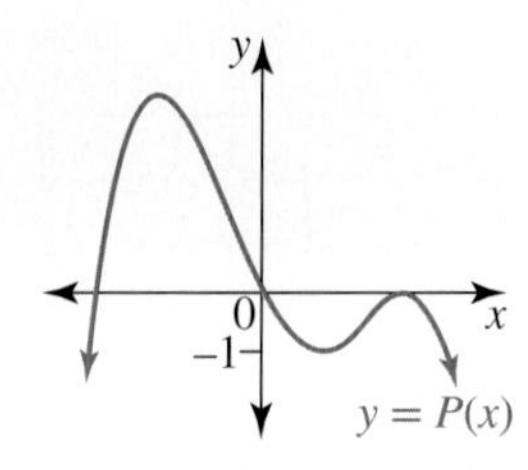

Give a possible equation for each of the following in terms of $P(x)$.

a.

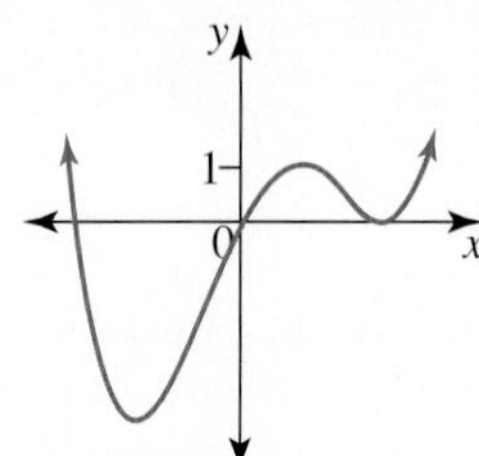

b.

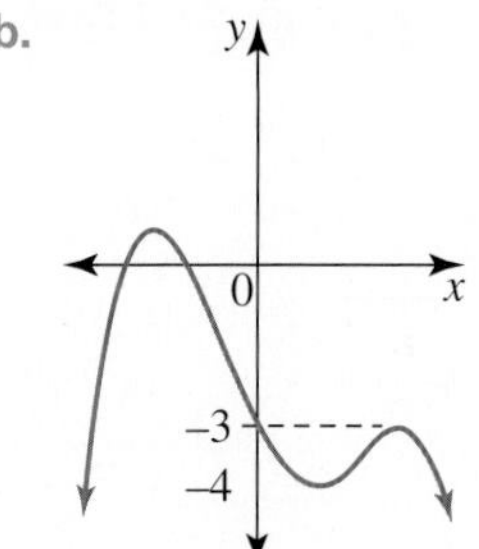

c.

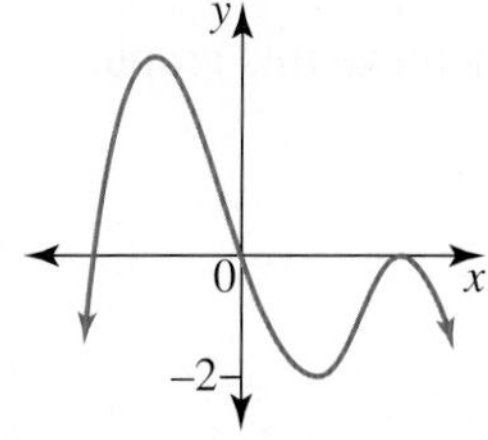

11. $y = x(x-2)(x-3)$ and $y = -2x(x-2)(x-3)$ are graphed on the same set of axes. Describe the relationship between the two graphs using the language of transformations.

12. If $y = -hr^{-(x+p)} - r$, explain what translations take place from the original graph, $y = r^x$.

Problem solving

13. a. Sketch the graph of $y = 3 \times 2^x$.
    b. If the graph of $y = 3 \times 2^x$ is transformed into the graph $y = 3 \times 2^x + 4$, describe the transformation.
    c. Sketch the graph of $y = 3 \times 2^x + 4$.
    d. Determine the coordinates of the y-intercept of $y = 3 \times 2^x + 4$.

14. The graph of $y = \dfrac{1}{x}$ is reflected in the y-axis, dilated by a factor of 2 parallel to the x-axis, translated 2 units to the left and up 1 unit. Determine the equation of the resultant curve. Give the equations of any asymptotes.

15. The graph of an exponential function is shown.

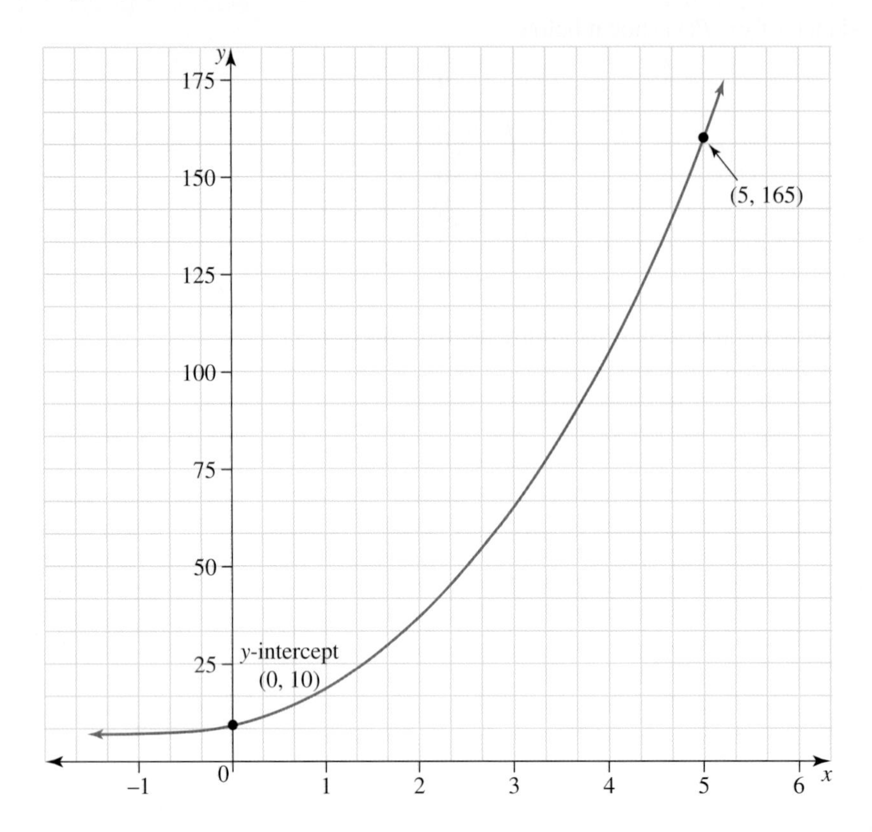

Its general rule is given by $y = a\left(2^x\right) + b$.

a. Determine the values of a and b.

b. Describe any transformations that had to be applied to the graph of $y = 2^x$ to achieve this graph.

15.7 Review

15.7.1 Topic summary

FUNCTIONS AND RELATIONS

Relations

- There are four types of relations: one-to-one, one-to-many, many-to-one and many-to-many.
- Relations that are one-to-one or many-to-one are called functions.

Domain and range of functions

- The domain of a function is the set of allowable values of x.
- The range of a function is the set of y-values it produces.
- Consider the function $y = x^2 - 1$. The domain of the function is $x \in R$. The range of the function is $y \geq -1$.

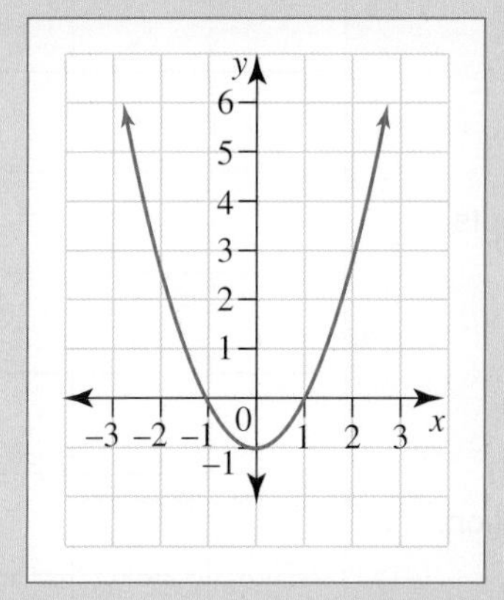

Inverse functions

- Functions have an inverse function if they are one-to-one.
- The inverse of a function $f(x)$ is denoted $f^{-1}(x)$.
- To determine the equation of the inverse function:
  1. Let $y = f(x)$.
  2. Switch x and y in the equation.
  3. Rearrange for y.

  e.g.

$$\begin{aligned} f(x) &= 2x^3 - 4 \\ y &= 2x^3 - 4 \\ x &= 2y^3 - 4 \\ x + 4 &= 2y^3 \\ \frac{x+4}{2} &= y^3 \\ y &= \sqrt[3]{\frac{x+4}{2}} \\ f^{-1}(x) &= \sqrt[3]{\frac{x+4}{2}} \end{aligned}$$

- Sometimes a function needs to have its domain restricted for the inverse to exist. For example, $f(x) = x^2$ is many-to-one, so does not have an inverse; but $f(x) = x^2, x \geq 0$ is one-to-one, and therefore does have an inverse function.

Function notation

- Functions are denoted $f(x)$.
- The value of a function at a point can be determined by substituting the x-value into the equation.

  e.g.

  $f(x) = 2x^3 - 3x + 5$

  The value of the function when $x = 2$ is:

$$\begin{aligned} f(2) &= 2(2)^3 - 3(2) + 5 \\ &= 2(8) - 6 + 5 \\ &= 15 \end{aligned}$$

Transformations

- There are 3 types of transformations that can be applied to functions and relations:
  - Translations
  - Dilations
  - Reflections
- The equation of a hyperbola $\left(f(x) = \frac{1}{x}\right)$ when transformed is $f(x) = \frac{a}{x-b} + c$.
- The equation of an exponential ($f(x) = a^x$) when transformed is $f(x) = ka^{(x-b)} + c$.
- The equation of a circle ($x^2 + y^2 = r^2$) when transformed is $(x - h)^2 + (y - k)^2 = r^2$.

Types of functions

- Cubic functions are functions where the highest power of x is 3. Cubic functions have 2 turning points, or 1 point of inflection.
- Quartic functions are functions where the highest power of x is 4. Quartic functions have 1 turning point, 3 turning points or 1 turning point and 1 point of inflection.
- Exponential functions are of the form $f(x) = k \times a^x$. The y-intercept is $y = k$.
- Examples of a cubic (top), a quartic (middle) and an exponential (bottom) function are shown below.

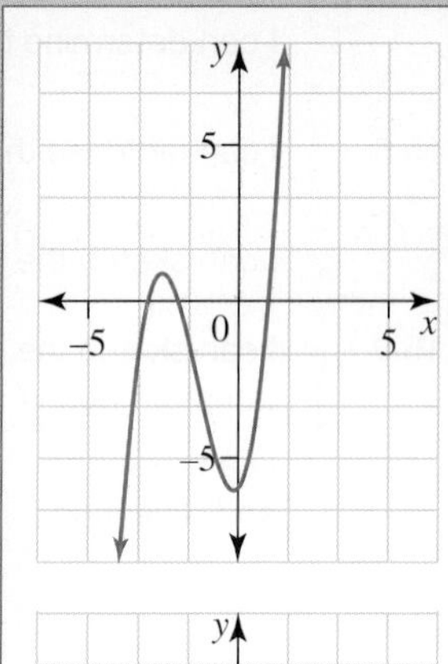

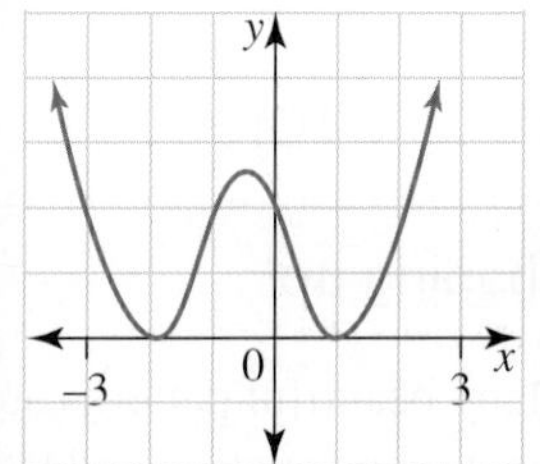

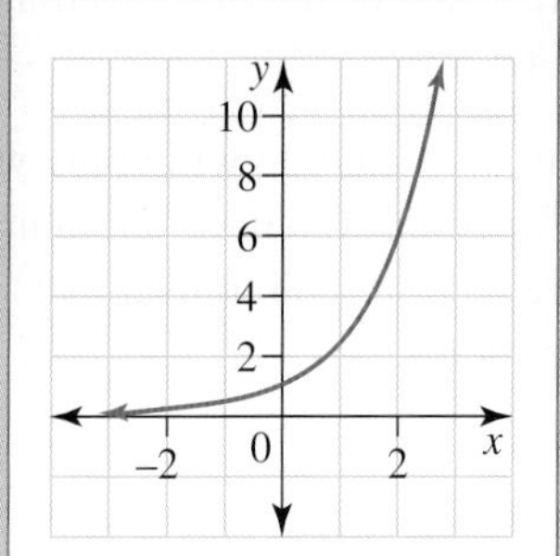

15.7.2 Success criteria

Tick the column to indicate that you have completed the subtopic and how well you have understood it using the traffic light system.

(**Green:** I understand; **Yellow:** I can do it with help; **Red:** I do not understand)

Subtopic	Success criteria	Green	Yellow	Red
15.2	I can identify the type of a relation.			
	I can determine the domain and range of a function or relation.			
	I can identify the points of intersection between two functions.			
	I can determine the inverse function of a one-to-one function.			
15.3	I can sketch graphs of exponential functions.			
	I can determine the equation of an exponential function.			
15.4	I can plot the graph of a cubic function using a table of values.			
	I can sketch the graph of a cubic function by calculating its intercepts			
	I can determine the equation of a cubic function by inspection.			
15.5	I can factorise the equation of a quartic function.			
	I can sketch the graph of a quartic function from a factorised equation.			
15.6	I can sketch the graph of a function that has undergone some transformations.			
	I can describe a transformation using the words translation, dilation and reflection.			

15.7.3 Project

Shaping up!

Many beautiful patterns are created by starting with a single function or relation and transforming and repeating it over and over.

In this task you will apply what you have learned about functions, relations and transformations (dilations, reflections and translations) to explore mathematical patterns.

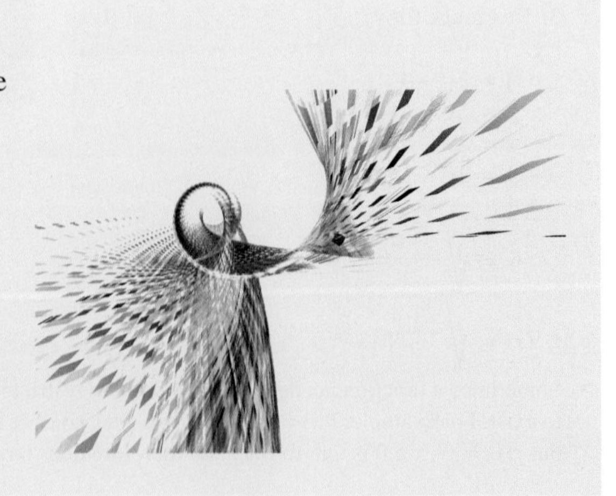

Exploring patterns using transformations

1. a. On the same set of axes, draw the graphs of:
   i. $y = x^2 - 4x + 1$
   ii. $y = x^2 - 3x + 1$
   iii. $y = x^2 - 2x + 1$
   iv. $y = x^2 + 2x + 1$
   v. $y = x^2 + 3x + 1$
   vi. $y = x^2 + 4x + 1$
   b. Describe the pattern formed by your graphs. Use mathematical terms such as intercepts, turning point, shape and transformations.

What you have drawn is referred to as a family of curves — curves in which the shape of the curve changes if the values of a, b and c in the general equation $y = ax^2 + bx + c$ change.

   c. Explore the family of parabolas formed by changing the values of a and c. Comment on your findings.
   d. Explore exponential functions belonging to the family of curves with equation $y = ka^x$, families of cubic functions with equations $y = ax^3$ or $y = ax^3 + bx^2 + cx + d$, and families of quartic functions with equations $y = ax^4$ or $y = ax^4 + bx^3 + cx^2 + dx + e$. Comment on your findings.
   e. Choose one of the designs shown below and recreate it (or a simplified version of it). Record the mathematical equations used to complete the design.

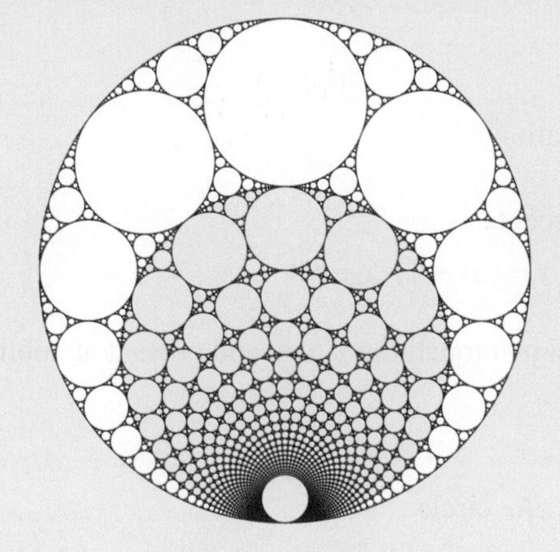

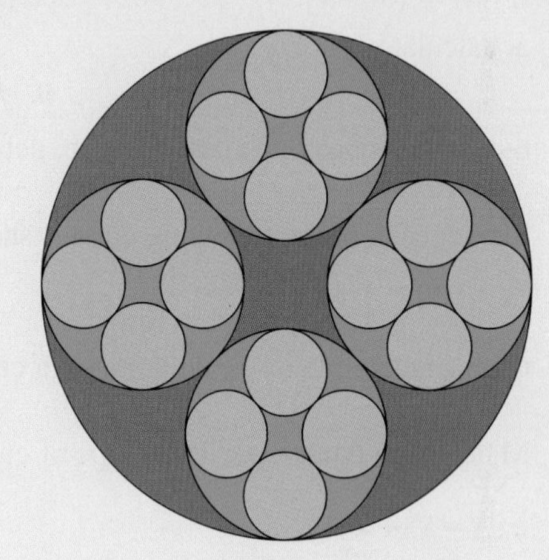

Coming up with your design

2. Use what you know about transformations to functions and relations to create your own design from a basic graph. You could begin with a circle, add some line segments and then repeat the pattern with some change. Record all the equations and restrictions you use.
   It may be helpful to apply your knowledge of inverse functions too.
   A digital technology will be very useful for this task.
   Create a poster of your design to share with the class.

on Resources

eWorkbook Topic 15 Workbook (worksheets, code puzzle and project) (ewbk-2041)

Interactivities Crossword (int-2878)
Sudoku puzzle (int-3893)

Exercise 15.7 Review questions

learnON

To answer questions online and to receive **immediate corrective feedback** and **fully worked solutions** for all questions, go to your learnON title at www.jacplus.com.au.

Fluency

1. State which of the following are functions.

a.

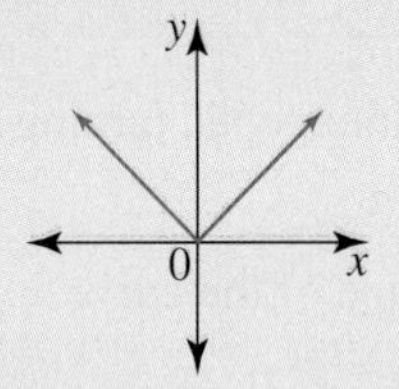

b.

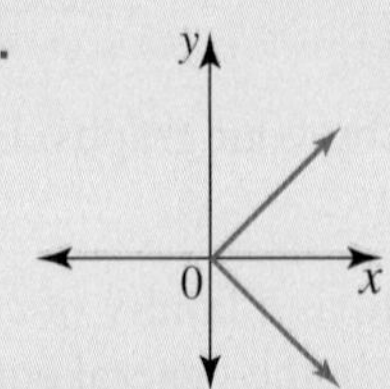

2. Identify which of the following are functions. For each identified as functions, state the equation of the inverse function, if it exists.

a. $y = 2x - 7$ b. $x^2 + y^2 = \sqrt{30}$ c. $y = 2^x$ d. $y = \dfrac{1}{x+1}$

3. If $f(x) = \sqrt{4 - x^2}$:

a. calculate:

i. $f(0)$ ii. $f(1)$ iii. $f(2)$

b. state whether $f^{-1}(x)$ exists. If so, determine its equation.

4. Sketch each of the following curves, showing all intercepts.

a. $y = (x-1)(x+2)(x-3)$ b. $y = (2x+1)(x+5)^2$

5. Give an example of the equation of a cubic that would just touch the x-axis and cross it at another point.

6. Match each equation with its type of curve.

a. $y = x^2 + 2$	A. circle
b. $x^2 + y^2 = 9$	B. cubic
c. $f(x) = \dfrac{2}{x+2}$	C. exponential
d. $g(x) = 6^{-x}$	D. parabola
e. $h(x) = (x+1)(x-3)(x+5)$	E. hyperbola

7. **MC** The equation for the graph shown could be:

A. $y = (x-5)(x+1)(x+3)$ B. $y = (x-3)(x-1)(x+5)$ C. $y = (x-3)(x+1)(x+5)$
D. $y = (5-x)(1+x)(3+x)$ E. $y = x(x-3)(x-1)$

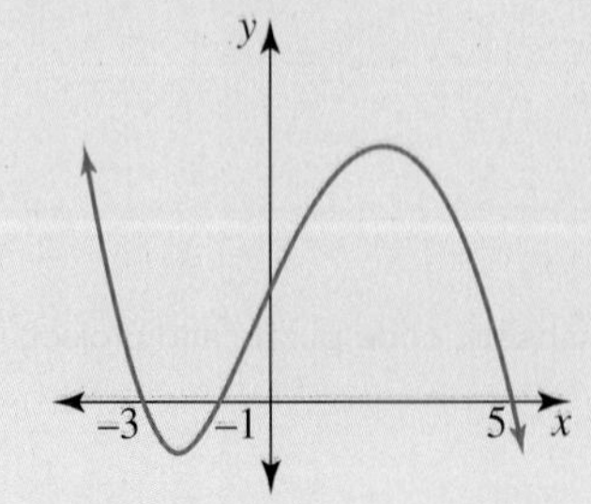

8. **MC** Select which of the following shows the graph of $y = -2(x+5)^3 - 12$.

A. (−5, −12)

B. 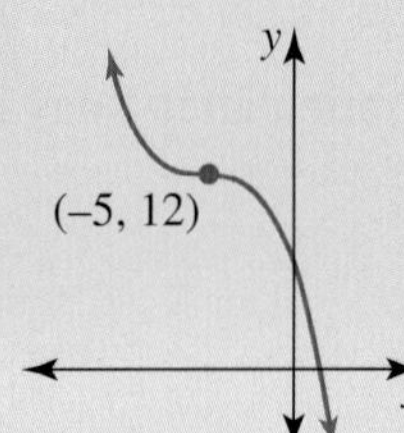

C. 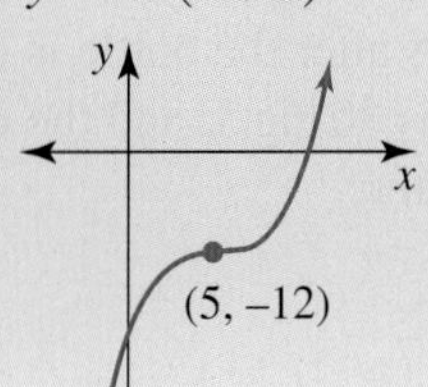

D. 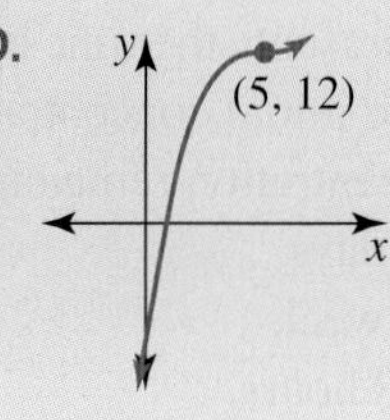

E. None of these

9. Sketch the following functions:

a. $y = x(x-2)(x+11)$
b. $y = x^3 + 6x^2 - 15x + 8$
c. $y = -2x^3 + x^2$

10. **MC** The rule for the graph shown could be:

A. $f(x) = x(x+2)^3$
B. $f(x) = -x(x-2)^2$
C. $f(x) = x^2(x-2)^2$
D. $f(x) = x(x-2)^3$
E. None of these

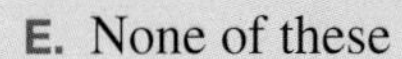

11. **MC** The graph of $y = (x+3)^2(x-1)(x-3)$ is best represented by:

A.

B. 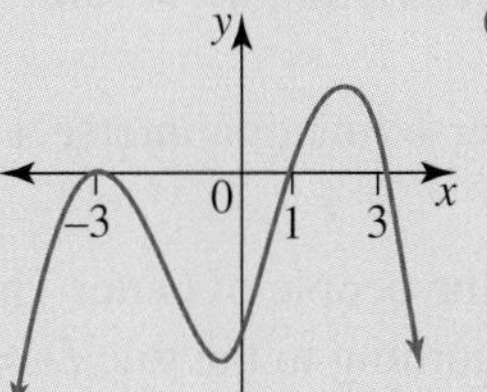

C. 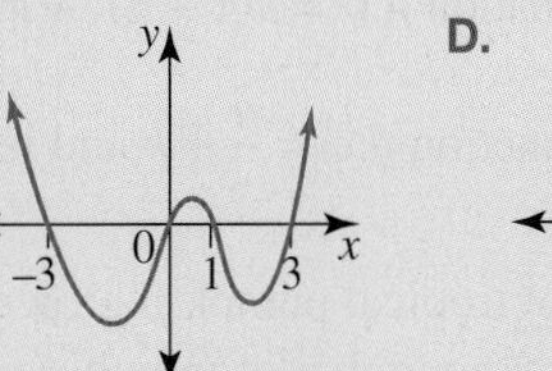

D. 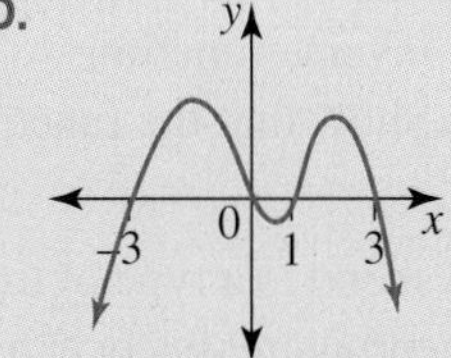

E. None of these

Problem solving

12. Sketch the graph of $y = x^4 - 7x^3 + 12x^2 + 4x - 16$, showing all intercepts.

13. Consider the sketch of $y = P(x)$ shown. Sketch $y = -P(x)$.

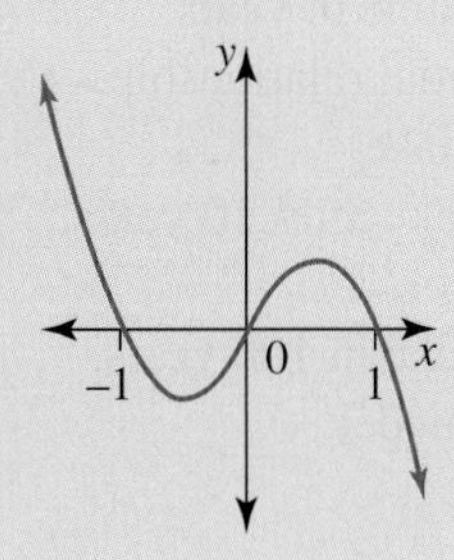

14. Draw any polynomial $y = P(x)$. Discuss the similarities and differences between the graphs of $y = P(x)$ and $y = P(x) + 3$.

15. Describe what happens to $f(x) = -2^x$ as $x \to \infty$ and $x \to -\infty$.

16. Determine any points of intersection between $f(x) = x^2 - 4$ and $g(x) = x^3 + x^2 - 12$.

17. The concentration of alcohol (mg/L) in a bottle of champagne is modelled by $C = C_0 \times 0.33^{kt}$ where t represents the time in days after the bottle is opened.
If the initial concentration is 80 mg/L and the concentration after 1 day is 70 mg/L, evaluate the concentration remaining after:
a. 3 days
b. 1 week
c. 18 hours.

18. The number of hyenas, H, in the zoo is given by $H = 20(10^{0.1t})$, where t is the number of years since counting started. At the same time, the number of dingoes, D, is given by $D = 25(10^{0.05t})$.
a. Calculate the number of:
i. hyenas
ii. dingoes when counting began.
b. Calculate the numbers of each after:
i. 1 year
ii. 18 months.
c. Which of the animals is the first to reach a population of 40 and by how long?
d. After how many months are the populations equal and what is this population?

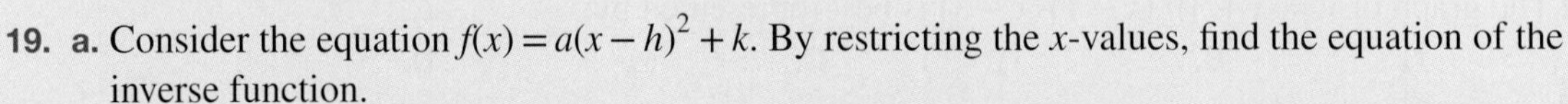

19. a. Consider the equation $f(x) = a(x - h)^2 + k$. By restricting the x-values, find the equation of the inverse function.
b. Show that the function $f(x) = \dfrac{a}{x} + b$ and its inverse function intersect on the line $y = x$.

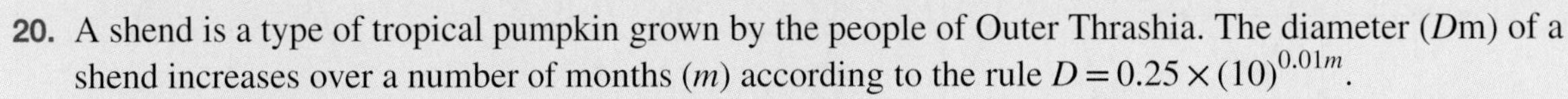

20. A shend is a type of tropical pumpkin grown by the people of Outer Thrashia. The diameter (Dm) of a shend increases over a number of months (m) according to the rule $D = 0.25 \times (10)^{0.01m}$.
a. Determine the diameter of the shend after 4 months.
b. If the shend is not harvested it will explode when it reaches a critical diameter of 0.5 metres. Show that it takes approximately 30 months for an unharvested shend to explode.

21. The surface area of a lake is evaporating at a rate of 5% per year due to climate change. To model this situation, the surface area of the lake (S km^2) over time is given by $S = 20\,000 \times 0.95^x$, where x is the time in years.
a. Explain whether this is an exponential relationship.
b. State the initial surface area of the lake.
c. Determine the surface area of the lake after 10 years.
d. Plot a graph for this relationship.
e. Determine the surface area of the lake in 100 years.
f. Explain whether this is a realistic model.

on To test your understanding and knowledge of this topic, go to your learnON title at www.jacplus.com.au and complete the **post-test**.

Online Resources

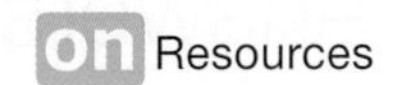

Below is a full list of **rich resources** available online for this topic. These resources are designed to bring ideas to life, to promote deep and lasting learning and to support the different learning needs of each individual.

eWorkbook

Download the workbook for this topic, which includes worksheets, a code puzzle and a project (ewbk-2041) ☐

Solutions

Download a copy of the fully worked solutions to every question in this topic (sol-0749) ☐

Digital documents

15.2 SkillSHEET Finding the gradient and y-intercept (doc-5378) ☐
SkillSHEET Sketching straight lines (doc-5379) ☐
SkillSHEET Sketching parabolas (doc-5380) ☐
SkillSHEET Completing the square (doc-5381) ☐
SkillSHEET Identifying equations of straight lines and parabolas (doc-5382) ☐
SkillSHEET Calculating points of intersection (doc-5383) ☐
SkillSHEET Substitution into index expressions (doc-5384) ☐
15.3 SkillSHEET Converting a percentage to a decimal (doc-5386) ☐
SkillSHEET Decreasing a quantity by a percentage (doc-5387) ☐

Video eLessons

15.2 Types of relations (eles-4984) ☐
Functions (eles-4985) ☐
Identifying features of functions (eles-4986) ☐
Inverse functions (eles-4987) ☐
15.3 Exponential functions (eles-4988) ☐
Determining the equation of an exponential function using data (eles-5349) ☐
15.4 Cubic functions (eles-4989) ☐
Sketching cubic functions (eles-4990) ☐
15.5 Quartic functions (eles-4991) ☐
Sketching quartic functions (eles-4992) ☐
15.6 General transformations (eles-4993) ☐
Transformations of hyperbolas, exponential functions and circles (eles-4994) ☐

Interactivities

15.2 Relations (int-6208) ☐
Evaluating functions (int-6209) ☐
15.3 Exponential growth and decay (int-6211) ☐
15.4 Cubic polynomials (int-2566) ☐
15.5 Quartic functions (int-6213) ☐
15.6 Horizontal translations of parabolas (int-6054) ☐
Vertical translations of parabolas (int-6055) ☐
Dilation of parabolas (int-6096) ☐
Exponential functions (int-5959) ☐
Reflection of parabolas (int-6151) ☐
Hyperbolas (int-6155) ☐
Translations of circles (int-6214) ☐
Transformations of exponentials (int-6216) ☐
Transformations of cubics (int-6217) ☐
The rectangular hyperbola (int-2573) ☐
15.7 Crossword (int-2878) ☐
Sudoku puzzle (int-3893) ☐

Teacher resources

There are many resources available exclusively for teachers online.

To access these online resources, log on to **www.jacplus.com.au**.

Answers

Topic 15 Functions and relations

Exercise 15.1 Pre-test

1. B
2. A.
3. True
4. B
5. B
6. D
7. A
8. E
9. D
10. D
11. B
12. A
13. B
14. B
15. E

Exercise 15.2 Functions and relations

1. a. One-to-many b. Many-to-one
   c. Many-to-one d. One-to-one
2. a. One-to-one b. Many-to-one
   c. Many-to-many d. Many-to-one
3. a. One-to-one b. Many-to-one
   c. One-to-one d. Many-to-one
4. a. b, c, d b. d
5. a. a, b, d b. a
6. a. a, b, c, d b. a, c
7. a. 1 b. 7 c. −5 d. 16
8. a. 2 b. 1 c. 3 d. 0
9. a. 3 b. 2 c. 6 d. 9
10. a. 9 b. 1
    c. 16 d. $a^2 + 6a + 9$
11. a. 12 b. 6 c. −4 d. 2
12. a, c, d
13. a, b, c, f
14. a. 3 b. 3 c. $\frac{5}{x} - 2x$
    d. $\frac{10}{x^2} - x^2$ e. $\frac{10}{x+3} - x - 3$ f. $\frac{10}{x-1} - x + 1$
15. a. 3 b. −3 or 3 c. $\frac{1}{3}$
16. a. 2 or 3 b. −4 or 1 c. −1
17. a. $f(x) \to \infty$ b. $f(x) \to 0$ c. $f(x) \to 0$
    d. $f(x) \to -\infty$ e. $f(x) \to 0$
18. a. $(0, -4), (2, 0)$ b. $(1, -2), \left(-\frac{2}{3}, 3\right)$
    c. $(2, 0), (-2, 0)$ d. $(3, -4)$
19. a. $f^{-1}(x) = \frac{x+1}{2}$
    b. $f^{-1}(x) = \sqrt{x+3}$ for original $x > 0$.
    c. $f^{-1}(x) = \sqrt{x-4} + 2$ for original $x > 2$
20. a. The horizontal line test fails.
    b. An inverse function will exist for $f(x) = x(x-2), x \le 1$ or $f(x) = x(x-2), x \ge 1$.
21. a. The horizontal line test is upheld.
    b. $f^{-1}(x) = -\sqrt{4-x}, x \le 4$.
22. a. $x = \pm 3$ b. $x = 2\frac{1}{3}$ c. $x = 28$
23. These graphs are inverse because they are the mirror images of each other through the line $y = x$.
24. a. Ran = [2, ∞)
    b. Many-to-one
    c. and f.

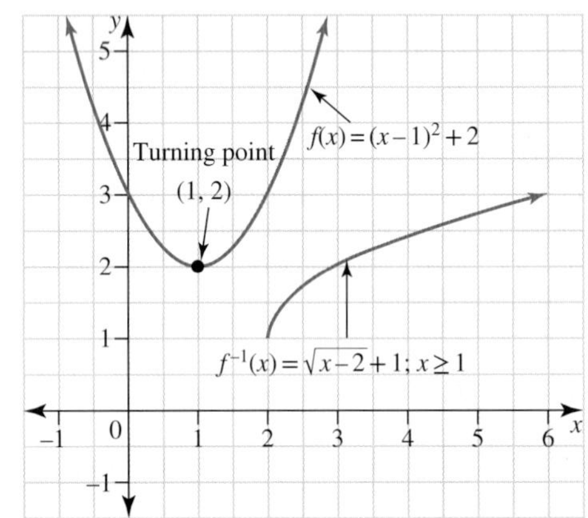

    d. Dom = [1, ∞)
    e. $f^{-1}(x) = \sqrt{x-2} + 1$, Dom = [2, ∞), Ran = [1, ∞)
    g. (0, 3) and (3, 6)

Exercise 15.3 Exponential functions

1. a. 2000
   b. 486 000
   c.

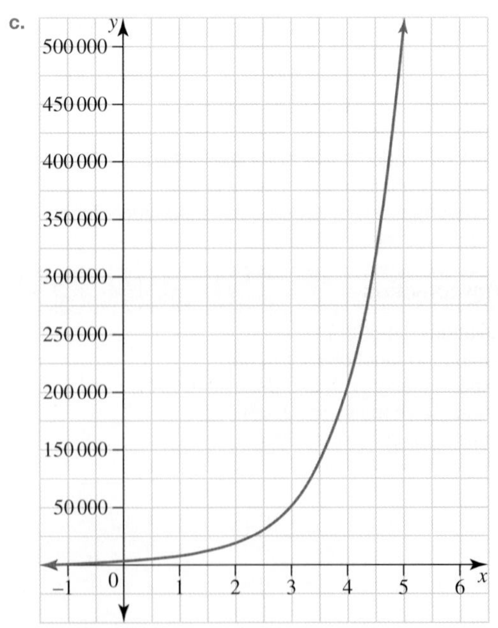

   d. 1.26 hours

2. a. \$5000

b. \$7717

c. 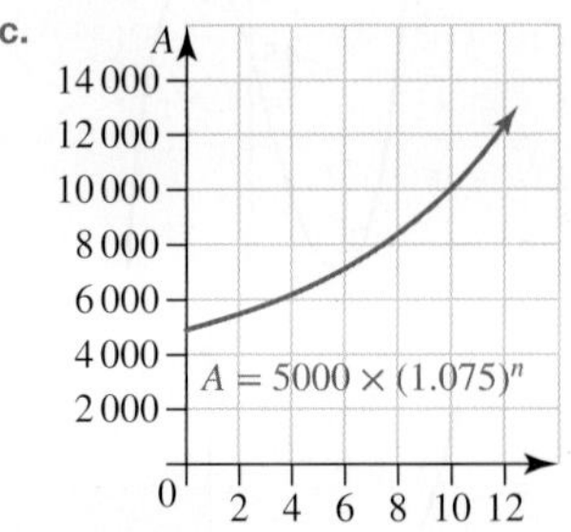

d. 10 years

3. a. C b. D

4. a. \$883.50 b. \$821.66

c. $V = 950 \times (0.93)^n$ d. \$397.67

5. a. 102 mg

b. 86.7 mg

c. $A = 120 \times (0.85)^t$

d. 83.927 mg

e. 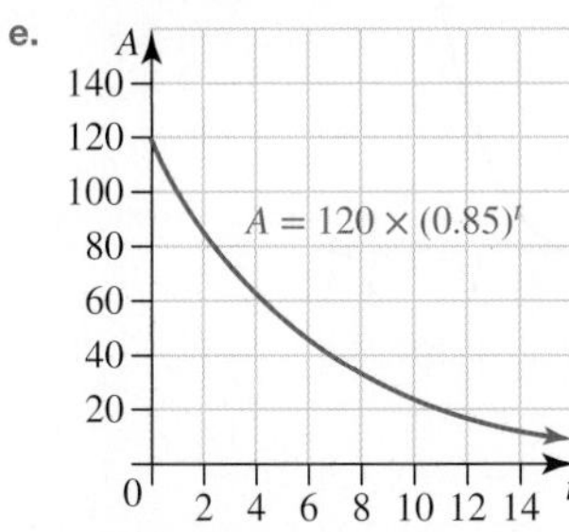

f. Approximately 213 years

6. a. i. 96.04% ii. 90.39%

b. $C = 100(0.98)^w$

c. 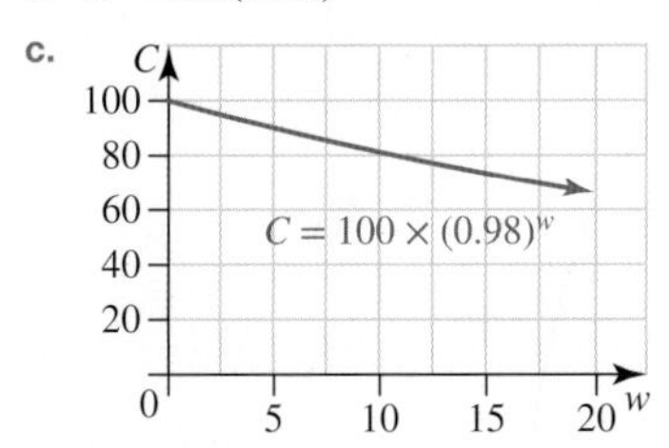

d. 8 washings

7. a. 118 (million)

b. $a = 1.02; P = 118 \times (1.02)^n$

c.

Year	2000	2005	2010	2015	2020
Population	118	130	144	159	175

Calculated population is less accurate after 10 years.

d. 288 (million)

8. a. 32

b. 0.98

c. $T = 32 \times (0.98)^t$

d. 26.1, 21.4, 17.5, 14.3; values are close except for $t = 40$.

9. a. 3 dogs b. 27 dogs c. 3 years

10. a. $A = 20\,000 \times 1.06^x$

b. Investment (\$) 30 000, 25 000, 20 000, 15 000, 10 000, 5 000; Years 0 1 2 3 4 5 6 7 8

c. 7 years

d. 6 years —1 year quicker

e. 9.05% p.a.

11. a. Approximately 20 000

b, c. Sample responses can be found in the worked solutions in the online resources.

12. a. $a = 100, b = 1.20$, increase = 20%/ min

b. $N = 146977 \times 0.70^m$

13. a. 500 °C

b. 125 °C.

c. Between 5 and 6 hours once it has cooled to below 15 °C.

14. a. 45 °C

b. $T = 45 \times 0.95^t$

c. 10 °C

d. No. The line $T = 0$ is an asymptote.

15. a. 1. 39.85 mg 2. 18.43 mg

b. More than 35.78 centuries.

Exercise 15.4 Cubic functions

1. a. 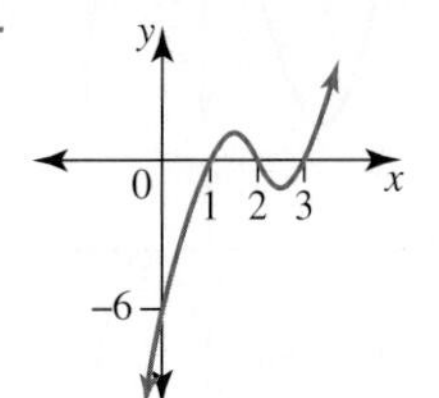

b. 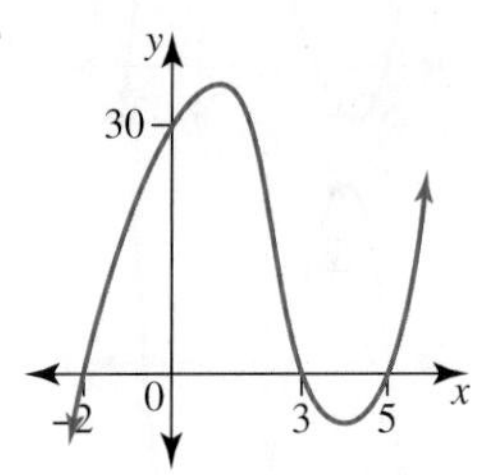

c. 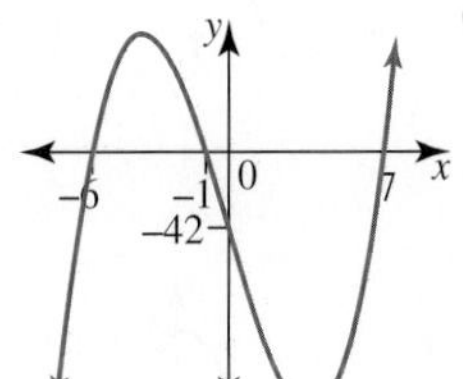

d. 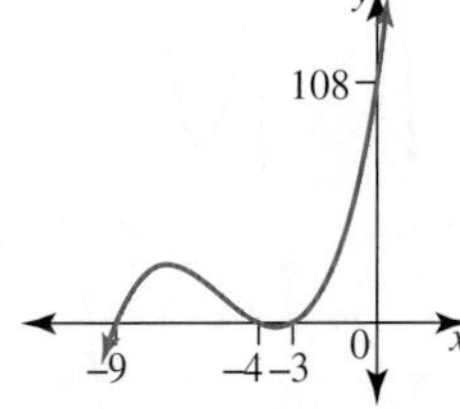

2. a. 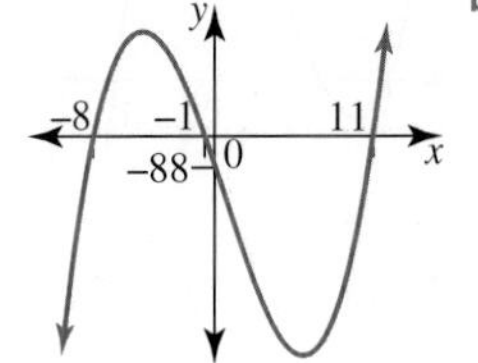
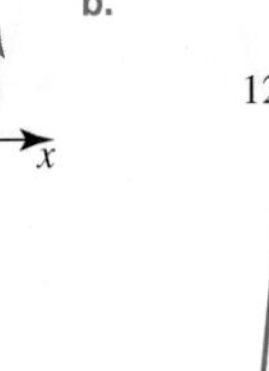

b. 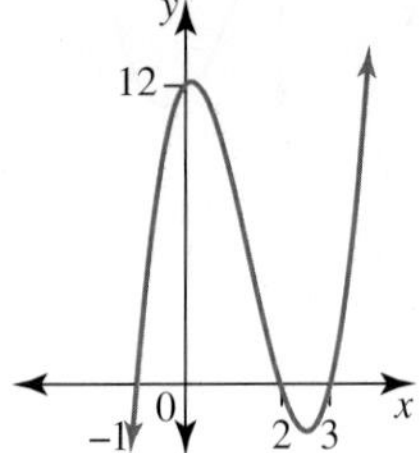

c.

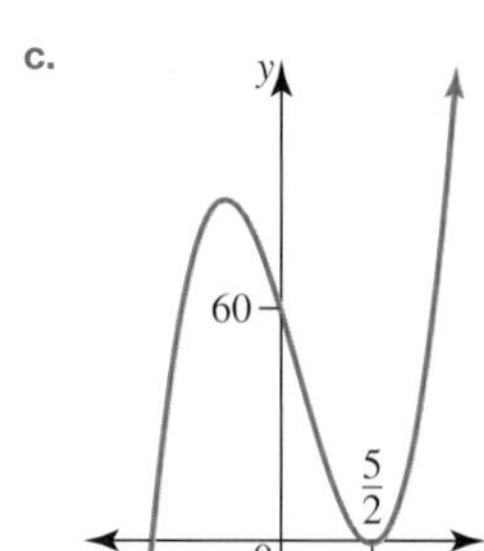

d.

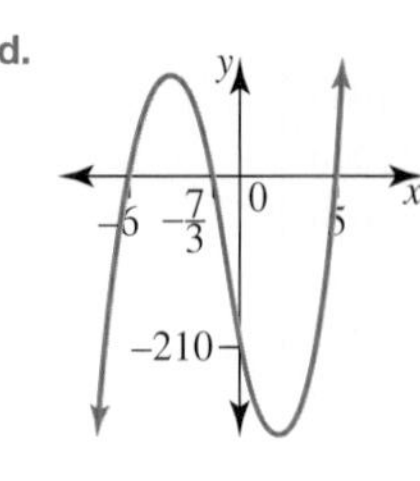

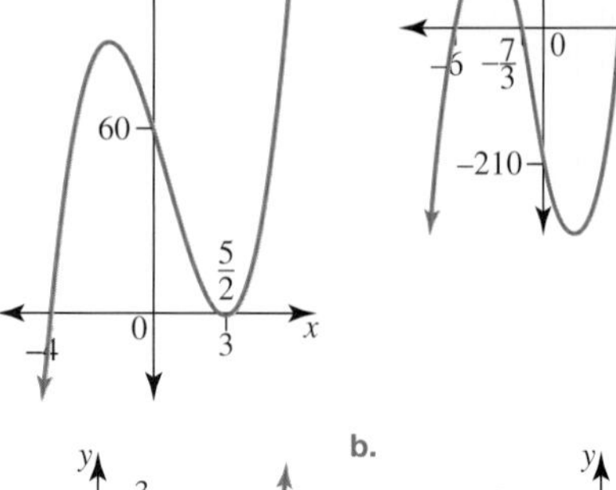

3. a.

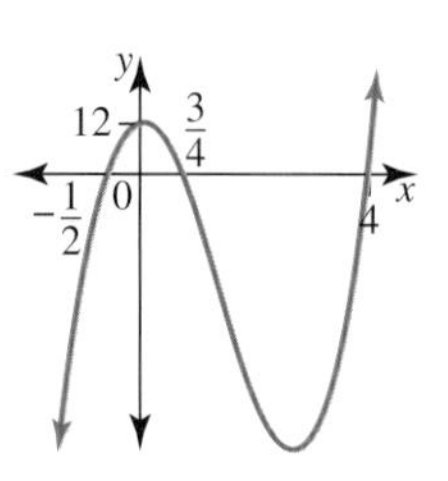

b.

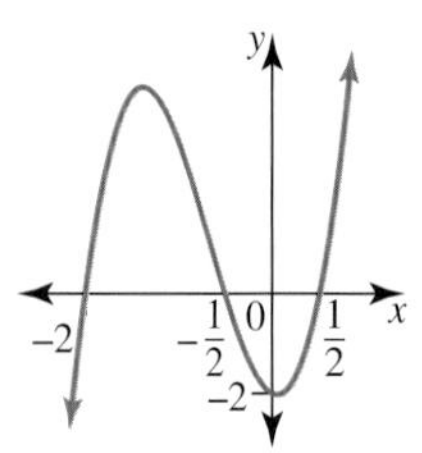

c.

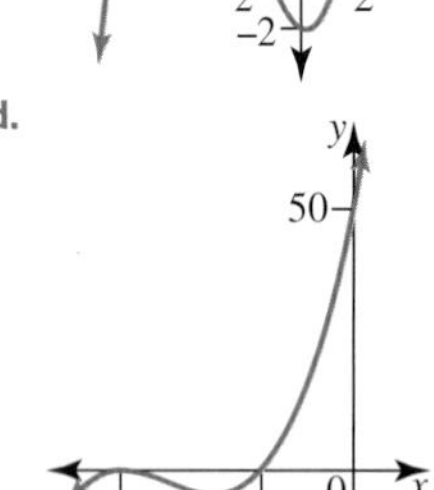

d.

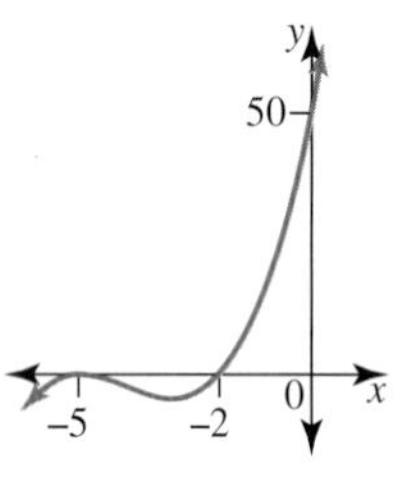

4. a.

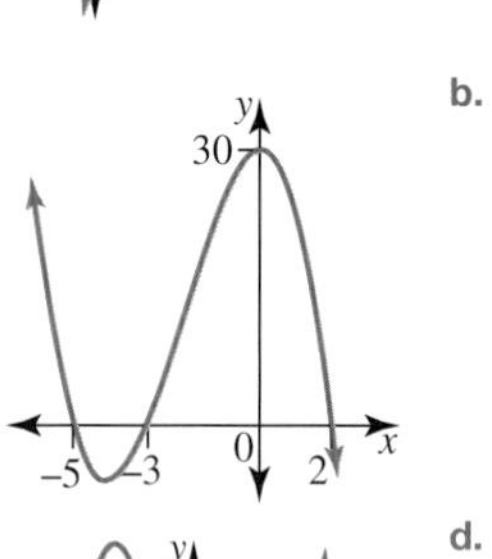

b.

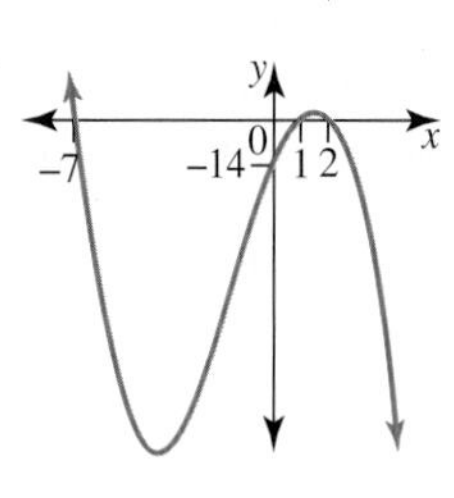

c.

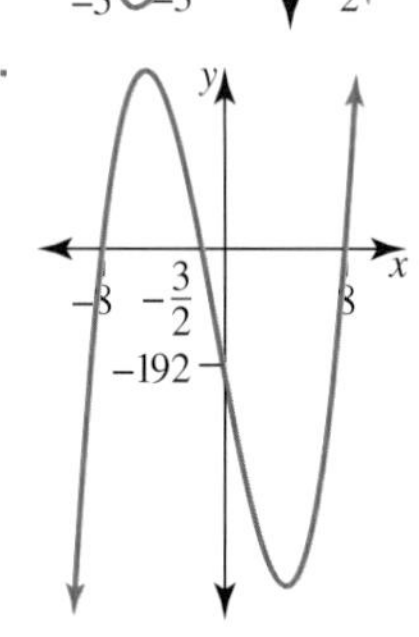

d.

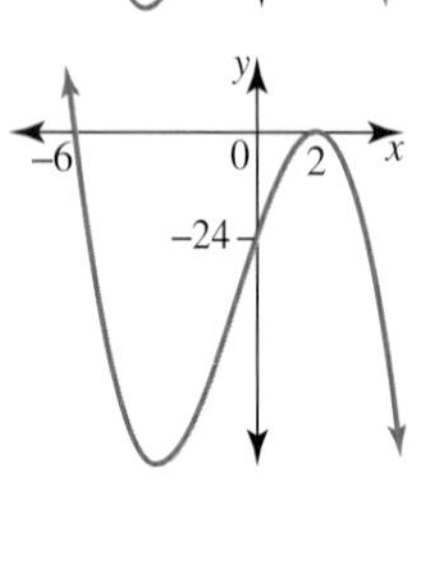

5. a.

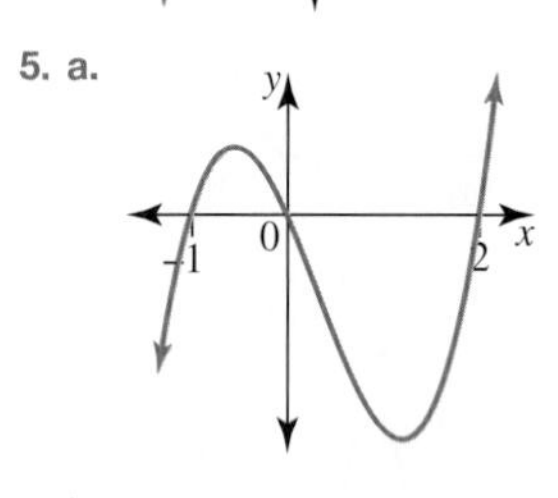

b.

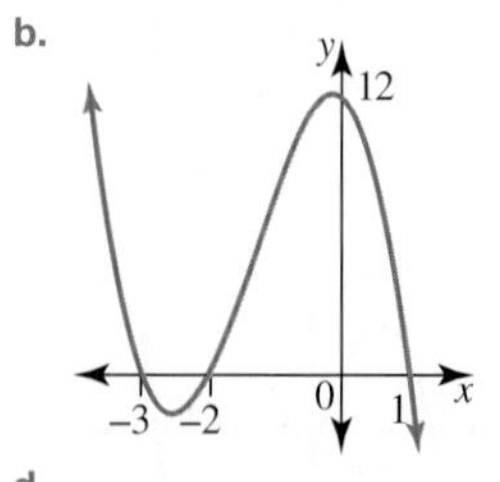

c.

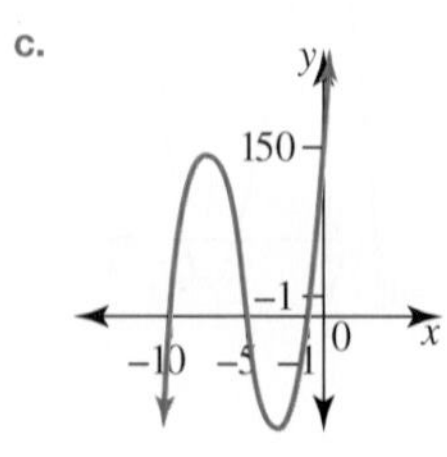

d.

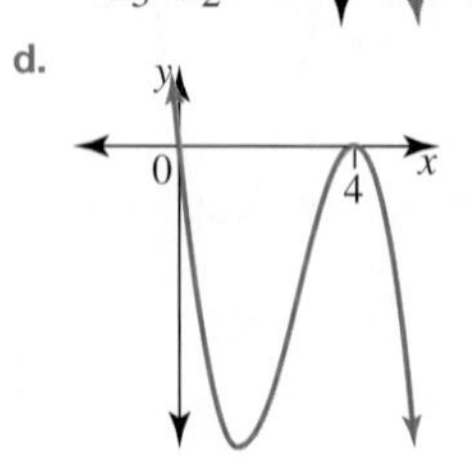

6. a.

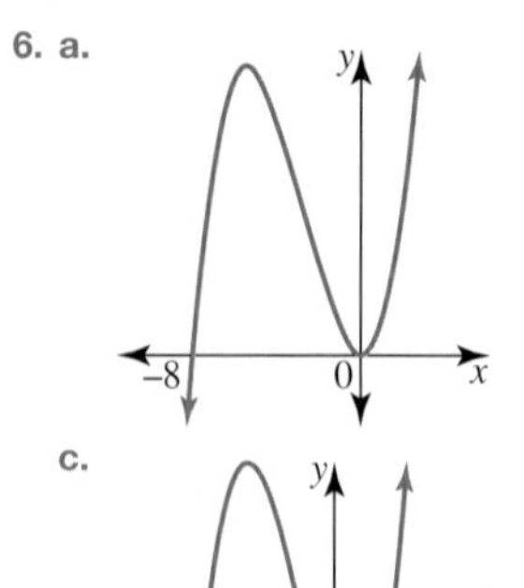

b.

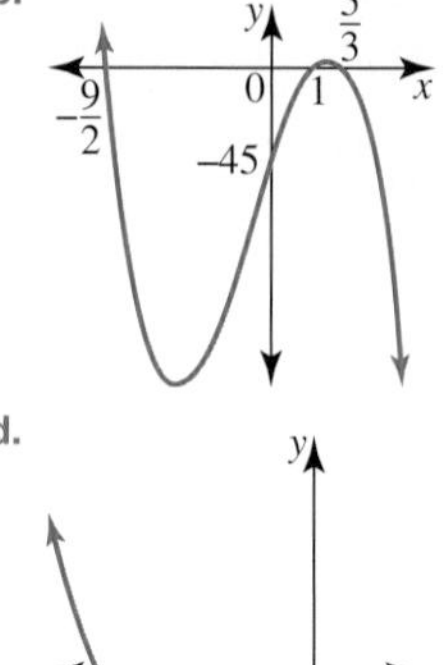

c.

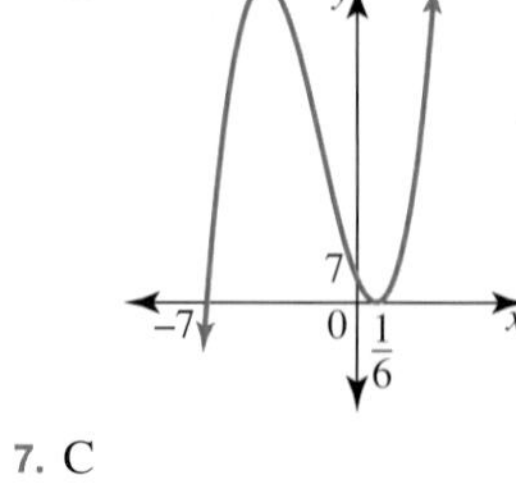

d.

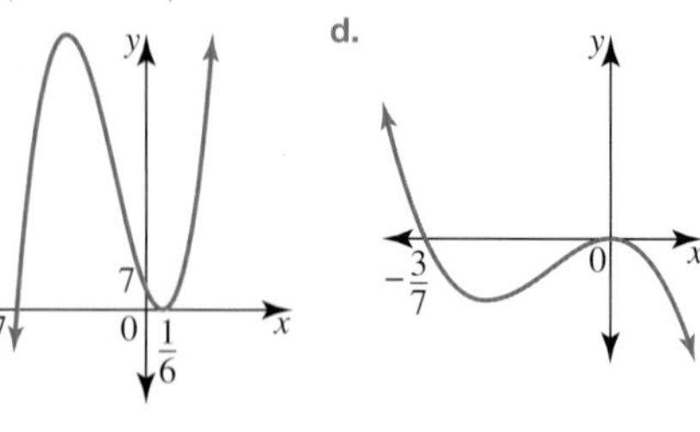

7. C

8. C

9. B

10. D

11. a.

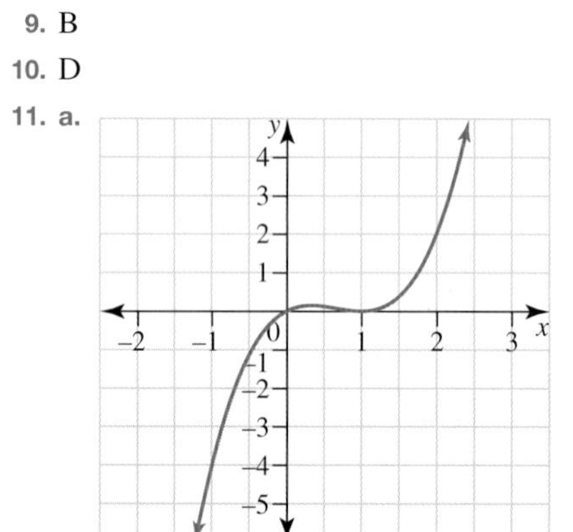

b.

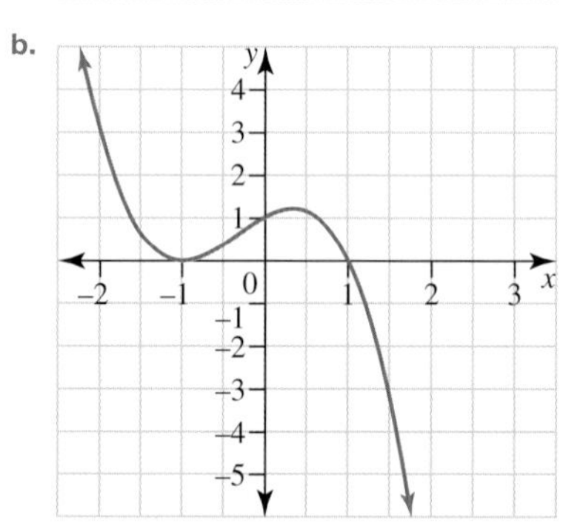

12. a.

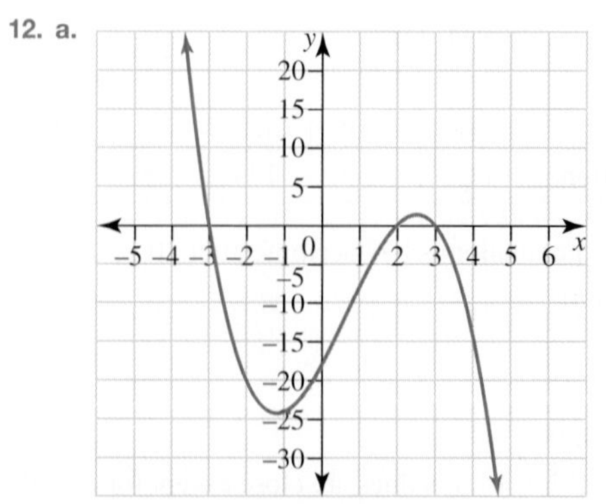

b.

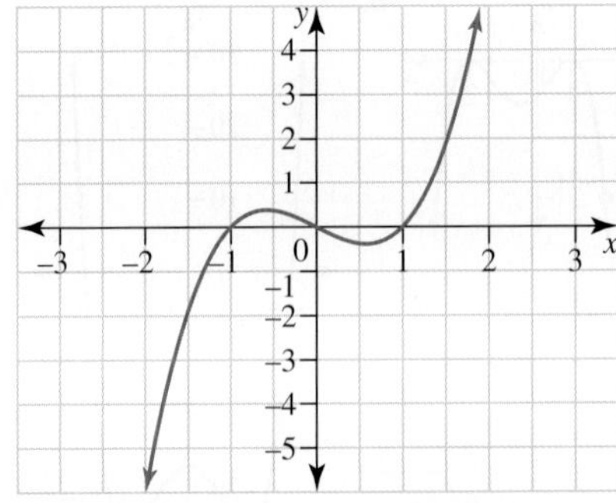

13. a. Positive
    b. Positive
    c. Zero
    d. Positive

14. $a = 2, b = -7$

15. $a = \dfrac{-(27 + 11b)}{11}$

16. a.

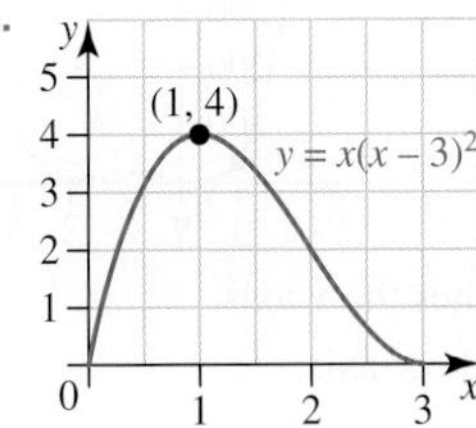

b. $x = 0$ and $x = 3$

c. 4 units

17. $y = 2(x + 2)(x - 2)(x - 5)$

18. a. Sample responses can be found in the worked solutions in the online resources.

b. $0 < x < 17.5$

c. 1377 cm^3

d.

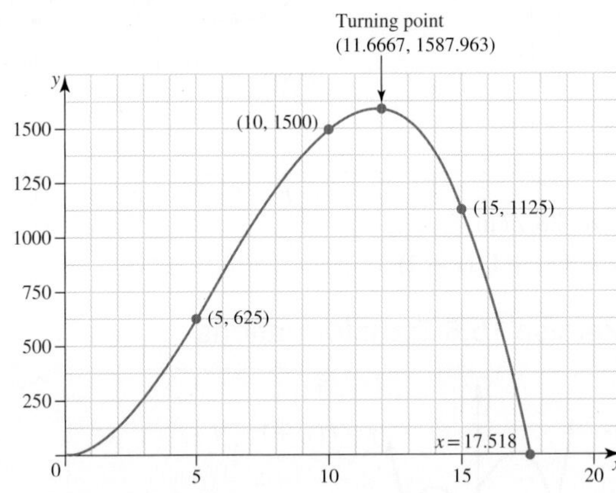

e. (11.6667, 1587.963); this is the value of x which creates the maximum volume.

Exercise 15.5 Quartic functions

1. a.

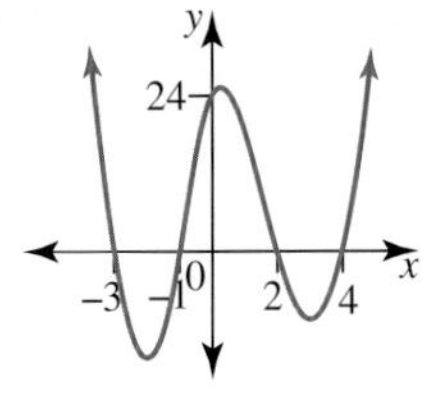

b.

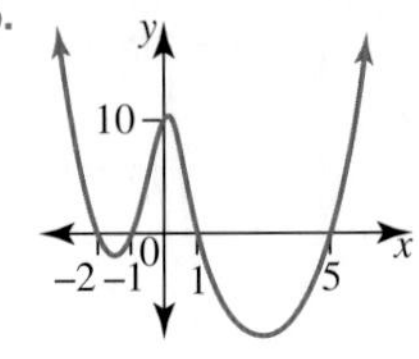

2. a.

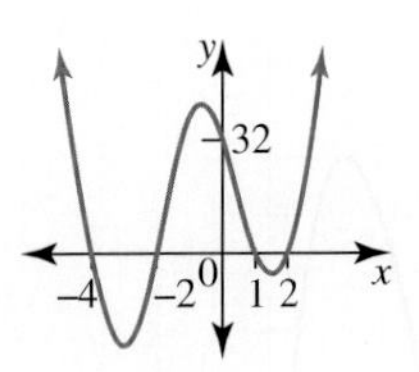

b.

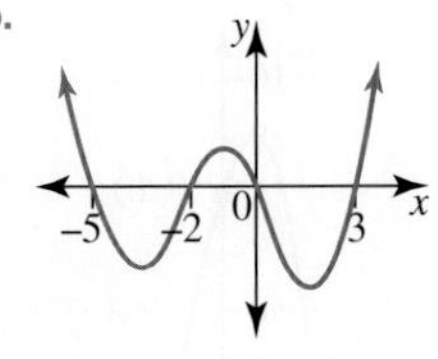

3. a.

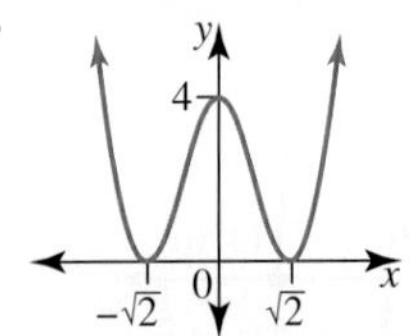

b.

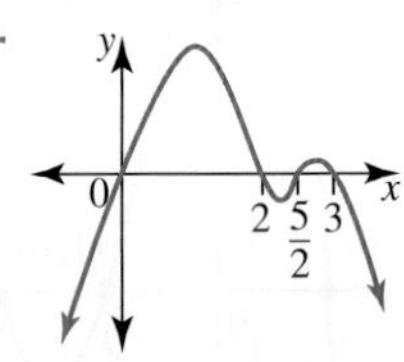

4. a.

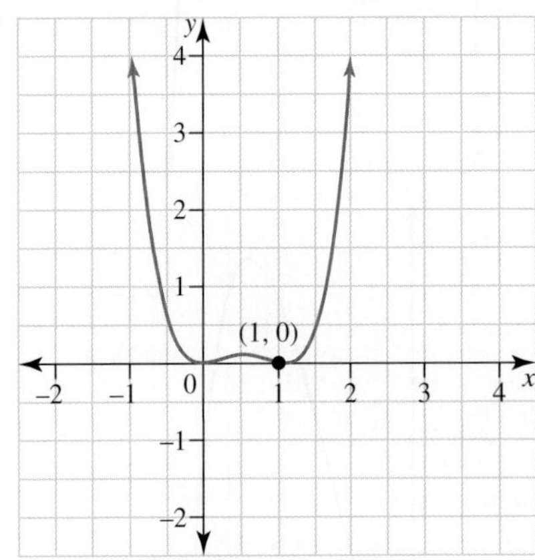

b.

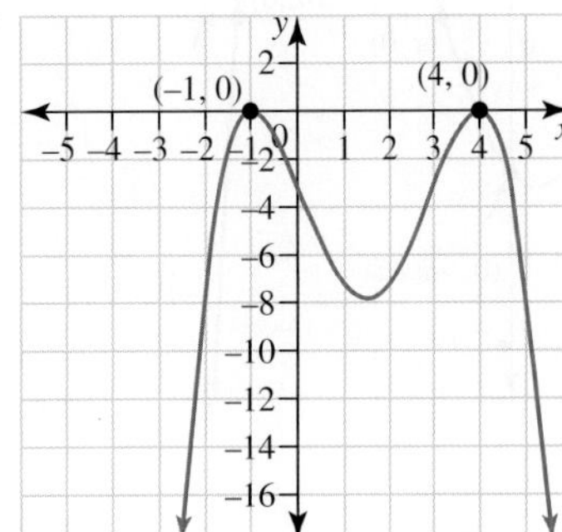

5. a.

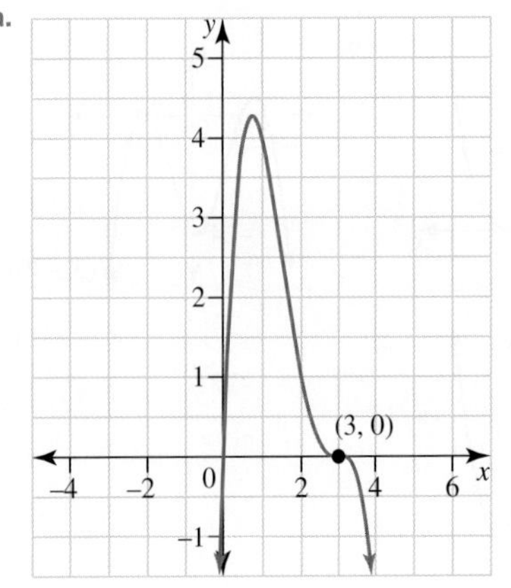

b.

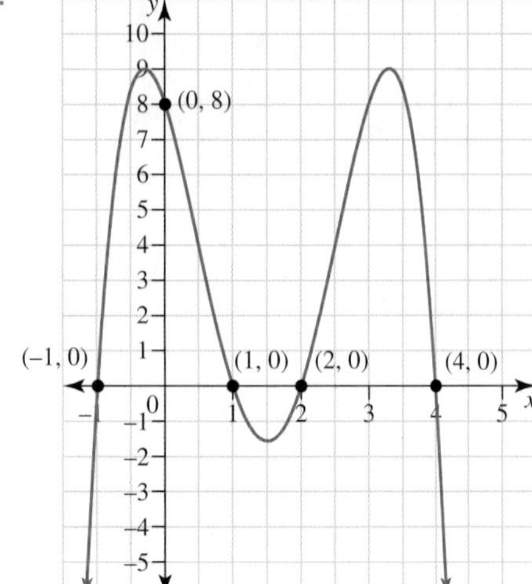

6.

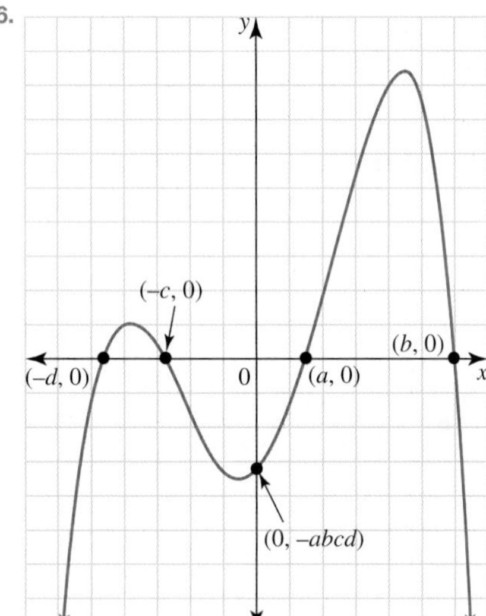

7. D

8. D

9. B

10. a.

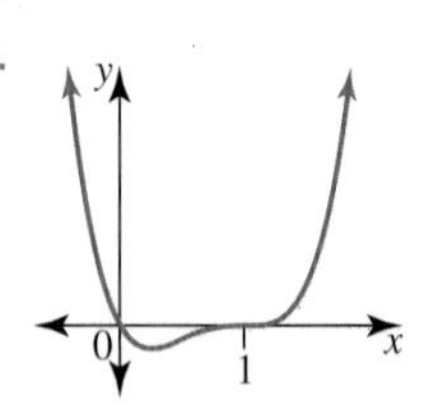

b.

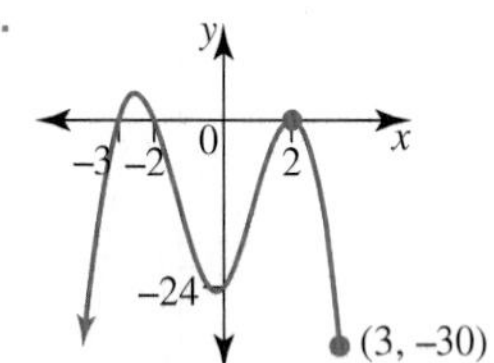

11. a.

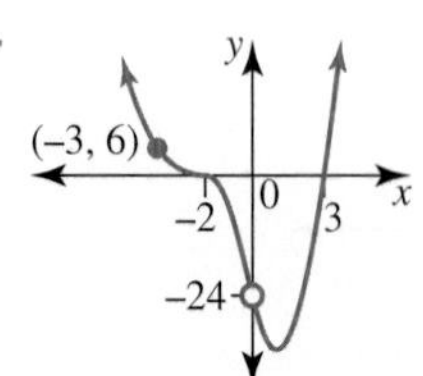

b.

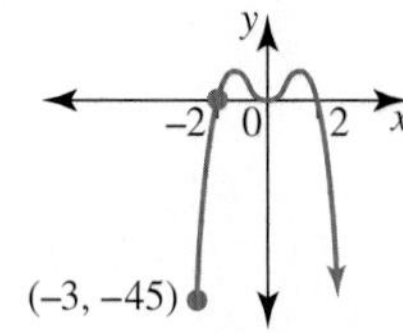

12. a. **b.**

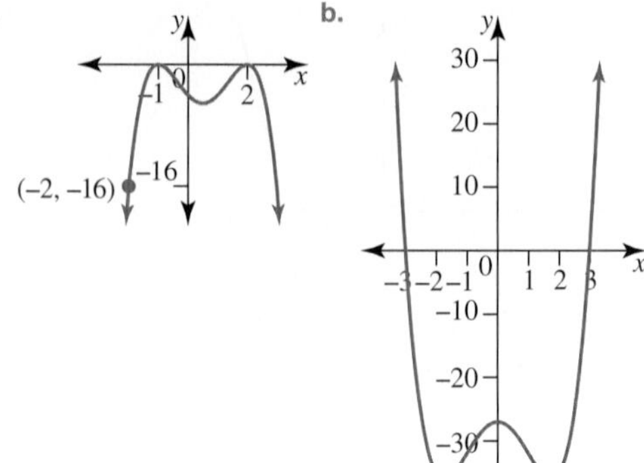

13. $a = 3, b = -1$

14. a. **b.**

15. a. If n is even, the graph touches the x-axis.

b. If n is odd, the graph cuts the x-axis.

16. $a = 4, b = -19$

17. a. 15 m

b.

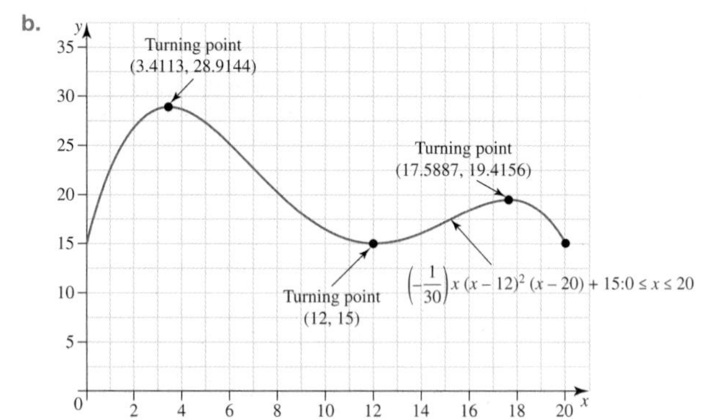

c. 28.77 m

18. $y = (x+1)(x-2)^3$

Exercise 15.6 Transformations

1.

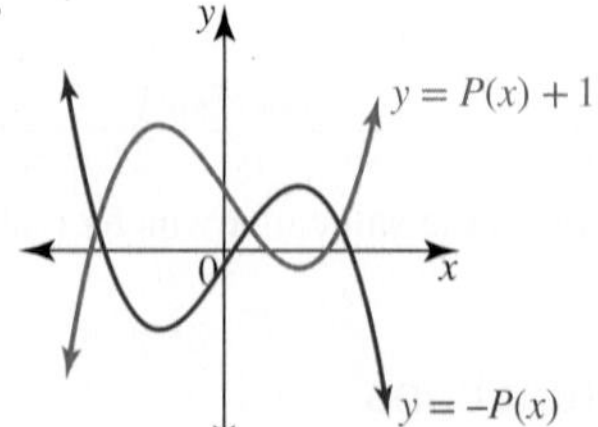

2.

3. $y = P(x) + 1$, $y = P(x + 2)$, $y = P(x)$, $y = -P(x)$

4. a. The transformation is a translation 2 units horizontally and a translation 1 unit vertically.
   b. The transformation is a dilation by a factor of 3 followed by a translation 2 units horizontally and 7 units vertically.
5. a. The transformation is a dilation by a factor of 5 followed by a translation 1 unit horizontally and 2 units vertically.
   b. The transformation is a reflection in the y-axis followed by a translation -2 units horizontally and 4 units vertically.
6. a. The transformation is a dilation by a factor of 4 followed by a reflection in the y-axis and then a translation -3 units horizontally and 10 units vertically.
   b. The transformation is a reflection in the y-axis followed by a translation -7 units horizontally and -6 units vertically and then a reflection in the x-axis.
7. They have the same x-intercepts, but $y = -P(x)$ is a reflection of $y = P(x)$ in the x-axis.
8. They have the same x-intercepts, but the y-values in $y = 2P(x)$ are all twice as large.
9. The entire graph is moved down 2 units. The shape is identical.
10. a. $y = -P(x)$ b. $y = P(x) - 3$ c. $y = 2P(x)$
11. The original graph has been reflected in the x-axis and dilated by a factor of 2. The location of the intercepts remains unchanged.
12. Dilation by a factor of h from the x-axis, reflection in the x-axis, dilation by a factor of $\frac{1}{q}$ from the y-axis, reflection in the y-axis, translation of p units left, translation of r units down.
13. a.

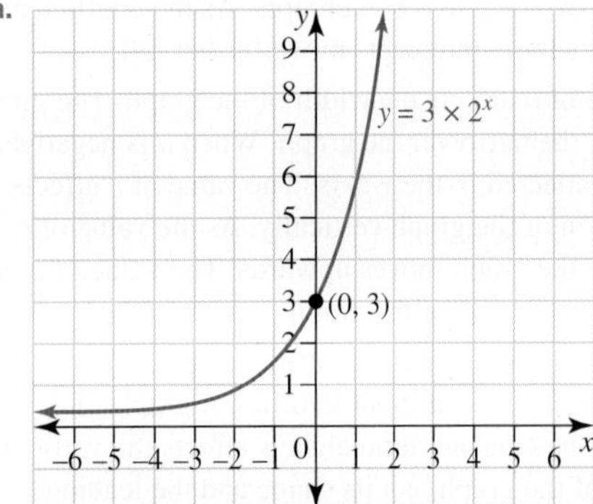

   b. Translation 4 units vertically.
   c.

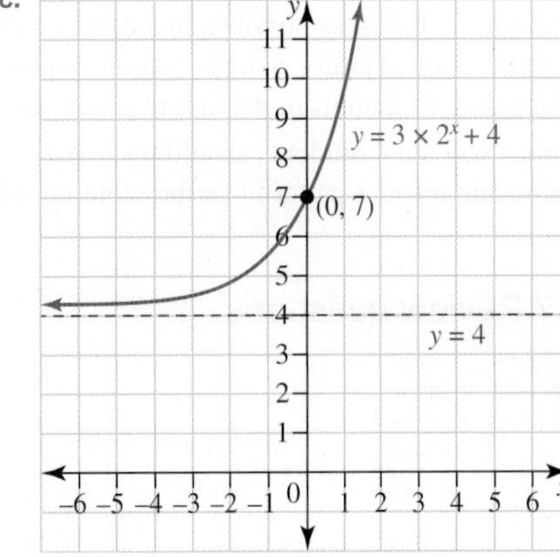

   d. $(0, 7)$
14. $y = -\dfrac{2}{(x+2)} + 1, x = -2, y = 1$
15. a. $y = 5(2^x) + 5, a = 5, b = 5$
   b. Dilation by a factor of 5 parallel to the y-axis and translation of 5 units up. Graph asymptotes to $y = 5$.

Project

1. a. See figure at the bottom of the page.*
   b. All of the graphs have a y-intercept of $y = 1$. The graphs which have a positive coefficient of x have two positive x-intercepts. The graphs which have a negative coefficient

*1. a.

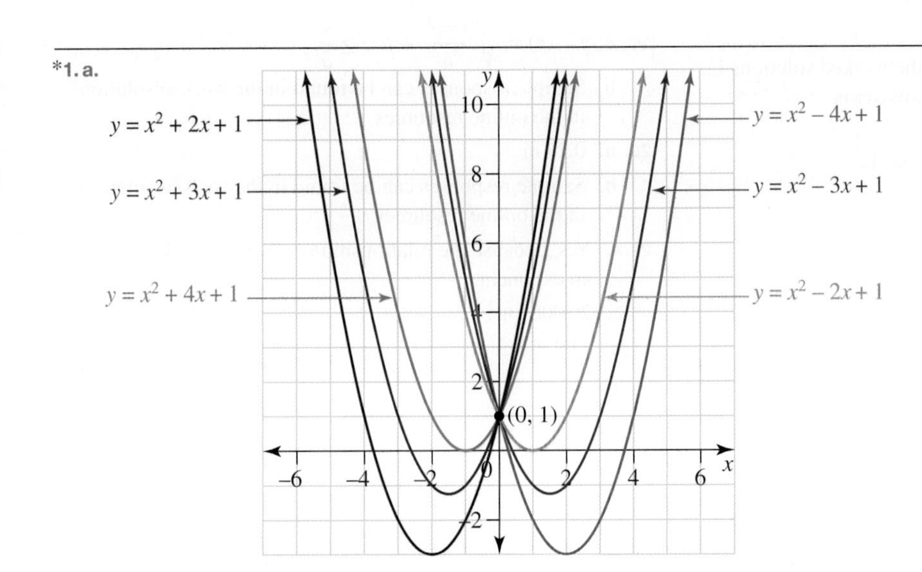

of x have two negative x-intercepts. As the coefficient of x decreases the x-intercepts move further left.

c. The value of a affects the width of the graph. The greater the value, the narrower the graph. When a is negative, the graph is reflected in the x-axis. The value of c affects the position of the graph vertically. As the value of c increases, the graph moves upwards. The value of c does not affect the shape of the graph.

d. Students may find some similarities in the effect of the coefficients on the different families of graphs. For example, the constant term always affects the vertical position of the graph, not its shape and the leading coefficient always affects the width of the graph.

e. Student responses will vary. Students should aim to recreate the pattern as best as possible, and describe the process in mathematical terms. Sample responses can be found in the worked solutions in the online resources.

2. Student responses will vary. Students should try to use their imagination and create a pattern that they find visually beautiful.

Exercise 15.7 Review questions

1. a

2. a, c, d, $\frac{x+7}{2}$, $\log_2 x$, $\frac{1}{x} - 1$

3. a. i. 2 ii. $\sqrt{3}$ iii. 0

b. No.

4. a.

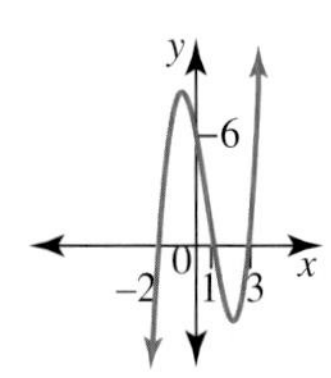

$y = (x - 1)(x + 2)(x - 3)$

b.

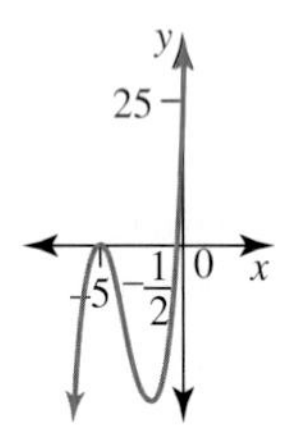

$y = (2x + 1)(x + 5)^2$

5. Sample responses can be found in the worked solutions in the online resources. One possible answer is $y = (x - 1)(x - 2)^2$.

6. a. D b. A c. E
d. C e. B

7. D

8. A

9. a.

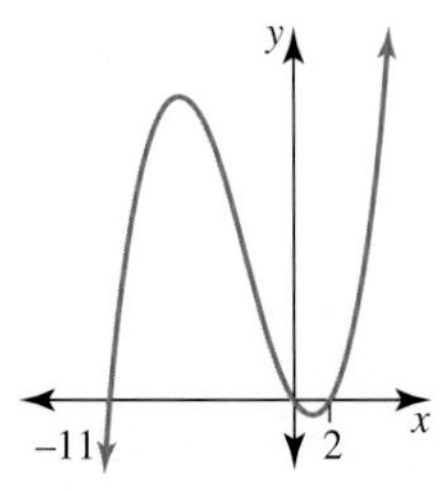

b.

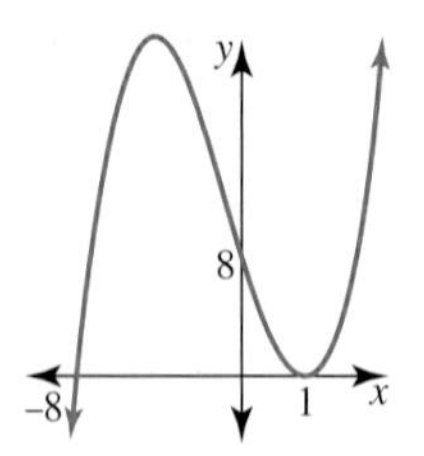

c.

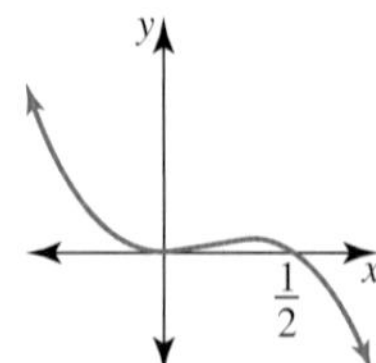

10. D

11. A

12.

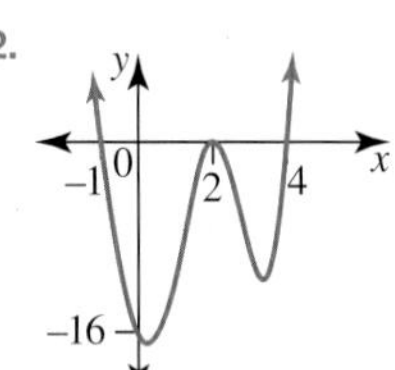

13.

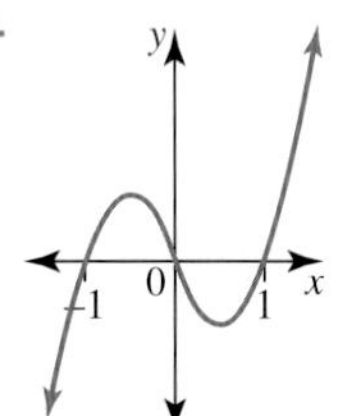

14. The entire graph is moved up 3 units. The shape is identical.

15. As $x \to \infty, f(x) \to -\infty$
As $x \to -\infty, f(x) \to 0$

16. $(2, 0)$

17. a. 53.59 mg/L b. 31.42 mg/L c. 72.38 mg/L

18. a. i. 20 ii. 25

b. i. $H = 25$; $D = 28$ ii. $H = 28$; $D = 30$

c. The hyenas are the first species to reach a population of 40 by 1 year.

d. After about 23 months; 31 animals

19. a. $f^{-1}(x) = \sqrt{\frac{x-k}{a}} + h, x \geq \frac{k}{a}$

b. Sample responses can be found in the worked solutions in the online resources.

20. a. 0.27 m

b. Sample responses can be found in the worked solutions in the online resources.

21. a. Yes, because the relationship involves a variable as an exponent.

b. $20\,000\ \text{km}^2$

c. $11\,975\ \text{km}^2$

d. 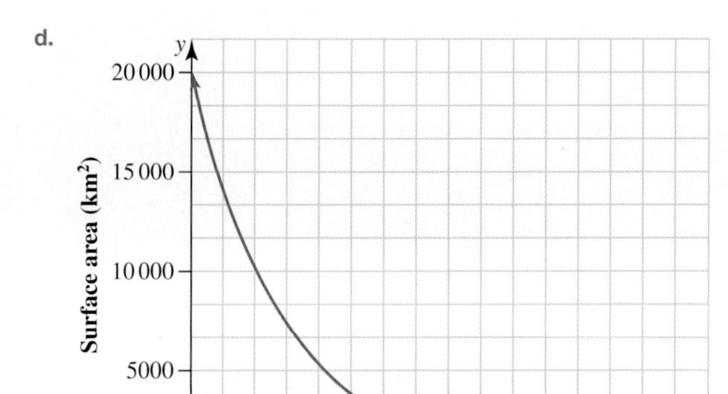

e. $118\,\text{km}^2$

f. No, this is not a realistic model as is it does not take into account changes to climate, rain, runoff from mountains, glaciers and so on.

10A

16 Circle geometry

LEARNING SEQUENCE

16.1 Overview

Why learn this?

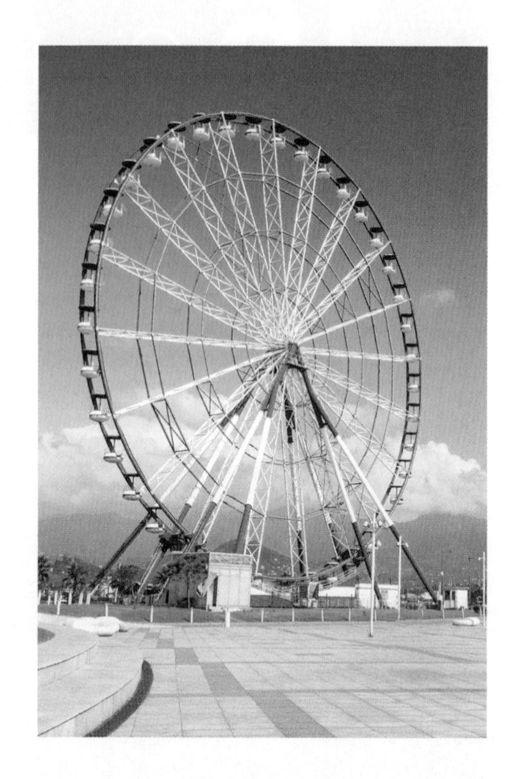

For thousands of years, humans have been fascinated by circles. Since they first looked upwards towards the sun and moon, which, from a distance at least, looked circular, artists have created circular monuments to nature. The most famous circular invention, one that has been credited as the most important invention of all, is the wheel. The potter's wheel can be traced back to around 3500 BC, approximately 300 years before wheels were used on chariots for transportation. Our whole transportation system revolves around wheels — bicycles, cars, trucks, trains and planes. Scholars as early as Socrates and Plato, Greek philosophers of the fourth century BCE, have been fascinated with the sheer beauty of the properties of circles. Many scholars made a life's work out of studying them, most famously Euclid, a Greek mathematician. It is in circle geometry that the concepts of congruence and similarity, studied earlier, have a powerful context. Today, we see circles in many different areas. Some buildings are now constructed based on circular designs. Engineers, designers and architects understand the various properties of circles. Road systems often have circular interchanges, and amusement parks usually include ferris wheels. As with the simple rectangle, circles are now part of our everyday life. Knowing the various properties of circles helps with our understanding and appreciation of this simple shape.

Where to get help

Go to your learnON title at **www.jacplus.com.au** to access the following digital resources. The Online Resources Summary at the end of this topic provides a full list of what's available to help you learn the concepts covered in this topic.

Video eLessons

Digital documents

Interactivities

eWorkbook

Fully worked solutions to every question

Exercise 16.1 Pre-test

learnON

Complete this pre-test in your learnON title at www.jacplus.com.au and receive **automatic marks**, **immediate corrective feedback** and **fully worked solutions**.

1. State the name of the area of the circle between a chord and the circumference.

2. MC Select the name of a line that touches the circumference of a circle at one point only.
   A. Secant B. Radius C. Chord D. Tangent E. Sector

3. Calculate the angle of x within the circle shown.

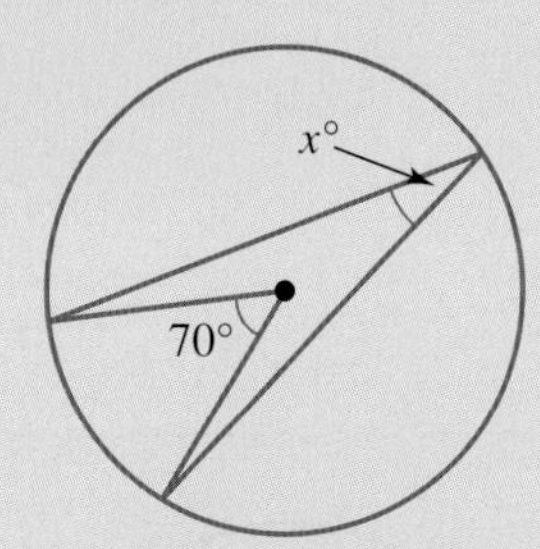

4. MC Select the correct angles for y and z in the circle shown.

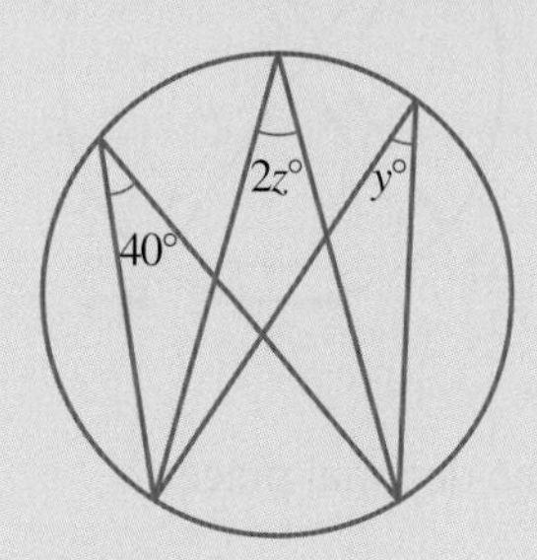

   A. $y = 40°$ and $z = 80°$
   B. $y = 20°$ and $z = 40°$
   C. $y = 40°$ and $z = 80°$
   D. $y = 40°$ and $z = 40°$
   E. $y = 40°$ and $z = 20°$

5. Determine the value of x in the circle shown.

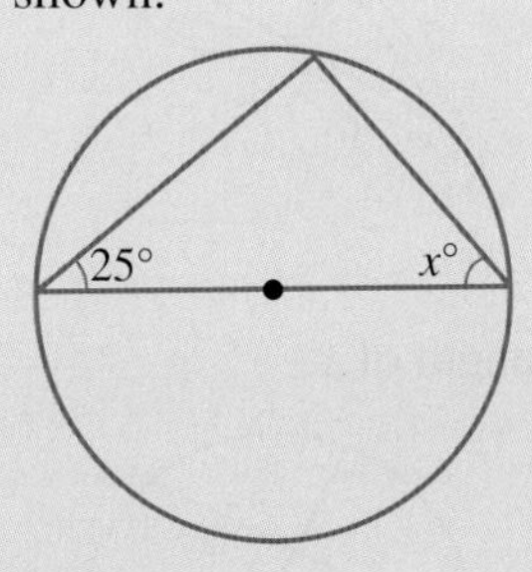

6. Determine the value of y in the shape shown.

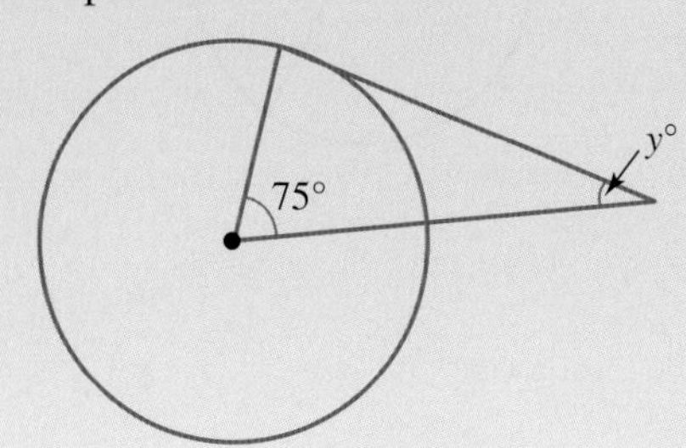

7. **MC** Select the correct values of x, y and z from the following list:

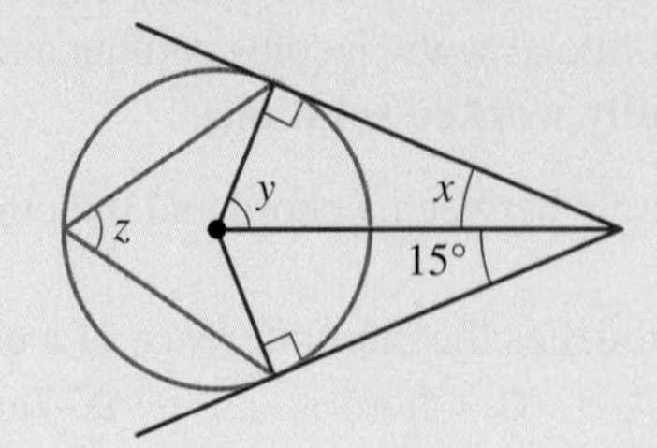

A. $x = 15°, y = 75°$ and $z = 150°$
B. $x = 15°, y = 30°$ and $z = 150°$
C. $x = 15°, y = 15°$ and $z = 37.5°$
D. $x = 15°, y = 30°$ and $z = 37.5°$
E. $x = 15°, y = 75°$ and $z = 75°$

8. Calculate the length of p in the circle.

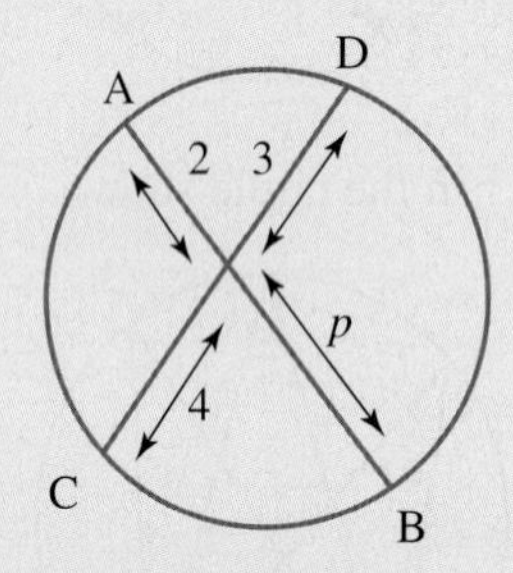

9. Calculate the length of m, correct to one decimal place.

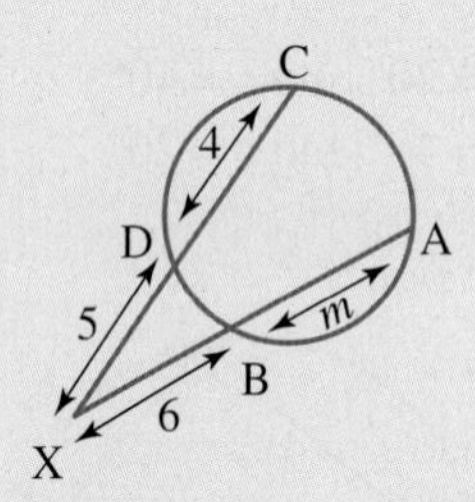

10. **MC** Choose the correct value for the length of x.

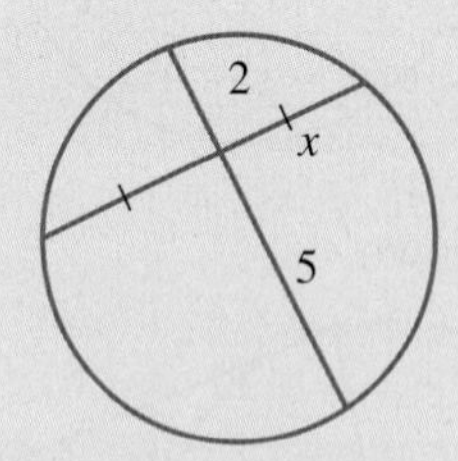

A. $\sqrt{10}$
B. $\sqrt{5}$
C. 2
D. 10
E. 5

11. Determine the values of m and p.

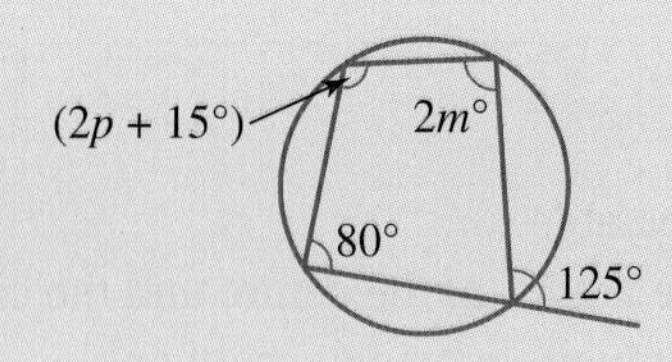

12. MC Select the correct values for x and y.

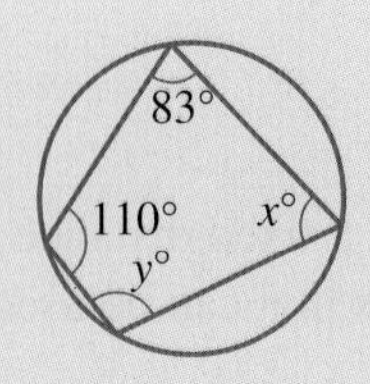

A. $x = 110°$ and $y = 83°$
B. $x = 83°$ and $y = 110°$
C. $x = 70°$ and $y = 97°$
D. $x = 97°$ and $y = 70°$
E. $x = 70°$ and $y = 107°$

13. MC Choose which of the following correctly states the relationships between x, y and z in the diagram.

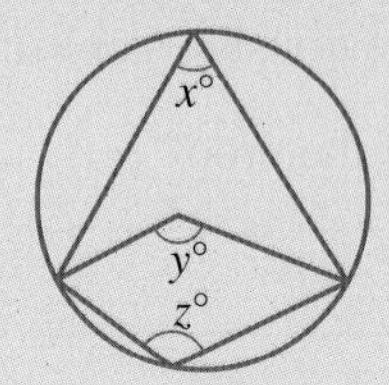

A. $x = 2y$ and $z + y = 180°$
B. $y = \frac{x}{2}$ and $z + y = 180°$
C. $x = 2y$ and $z + x = 180°$
D. $x = \frac{y}{2}$ and $z + y = 180°$
E. $y = 2x$ and $z + x = 180°$

14. MC Choose the correct value for x.

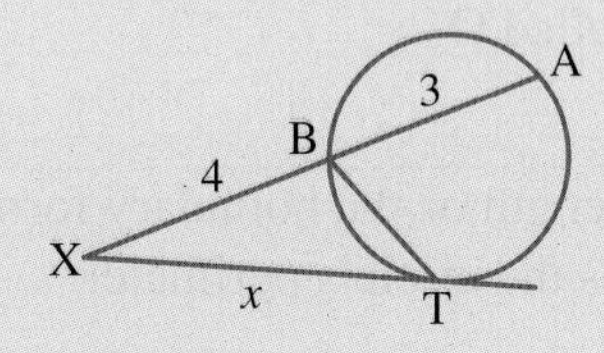

A. $2\sqrt{5}$
B. 28
C. 7
D. $2\sqrt{7}$
E. $\sqrt{5}$

15. MC AB = 9 and P divides AB in the ratio 2 : 3.
If PO = 3 where O is the centre of the circle, select the exact length of the diameter of the circle.

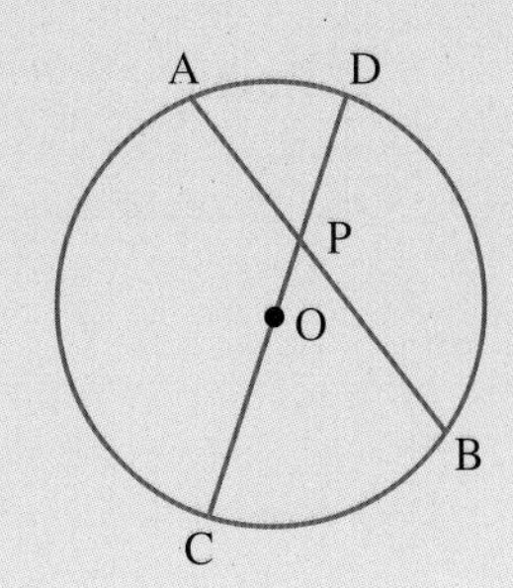

A. $\frac{6\sqrt{79}}{5}$
B. $\frac{4\sqrt{79}}{2}$
C. $\frac{3\sqrt{79}}{5}$
D. $\frac{6\sqrt{81}}{5}$
E. $\frac{3\sqrt{79}}{2}$

16.2 Angles in a circle

LEARNING INTENTION

At the end of this subtopic you should be able to:
- determine relationships between the angles at the centre and the circumference of a circle
- understand and use the angle in a semicircle being a right angle
- apply the relationships between tangents and radii of circles.

16.2.1 Circles

eles-4995

- A **circle** is a connected set of points that lie a fixed distance (the **radius**) from a fixed point (the **centre**).
- In circle geometry, there are many theorems that can be used to solve problems. It is important that we are also able to prove these theorems.

Steps to prove a theorem

Step 1. State the aim of the proof

Step 2. Use given information and previously established theorems to establish the result

Step 3. Give a reason for each step of the proof

Step 4. State a clear conclusion.

Parts of a circle

Part (name)	Description	Diagram
Centre	The middle point, equidistant from all points on the circumference. It is usually shown by a dot and labelled O.	O
Circumference	The outside length or the boundary forming the circle. It is the circle's perimeter.	O
Radius Radii	A straight line from the centre to any point on the circumference. Plural of radius.	O

Part (name)	Description	Diagram
Diameter	A straight line from one point on the circumference to another, passing through the centre	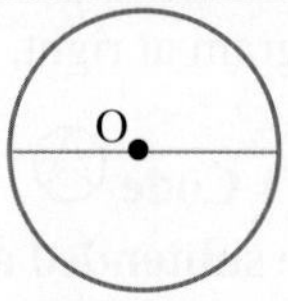
Chord	A straight line from one point on the circumference to another	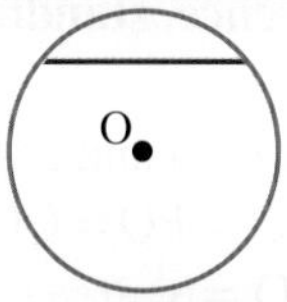
Segment	The area of the circle between a chord and the circumference. The smaller segment is called the minor segment and the larger segment is the major segment.	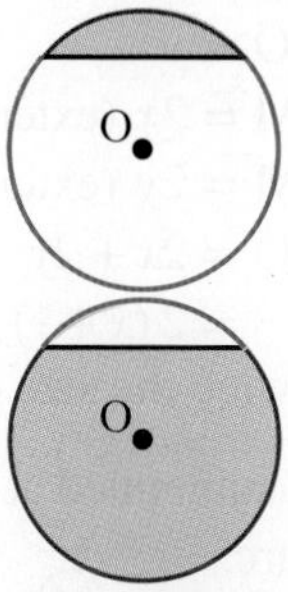
Sector	An area of a circle enclosed by 2 radii and the circumference	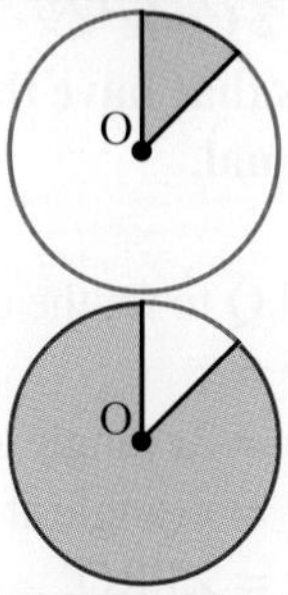
Arc	A portion of the circumference	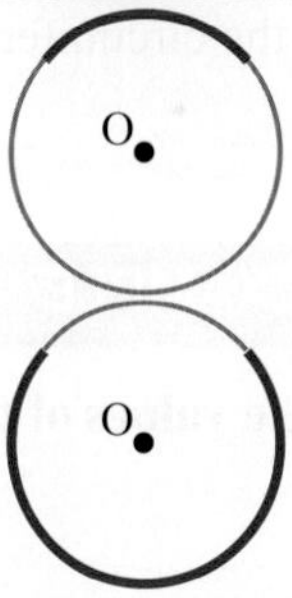
Tangent	A straight line that touches the circumference at one point only	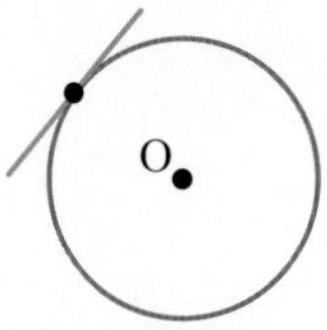
Secant	A chord extended beyond the circumference on one side	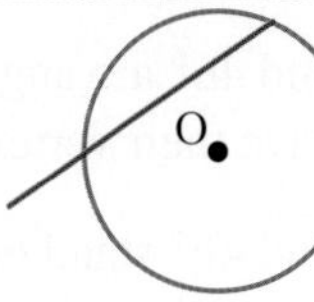

Angles in a circle

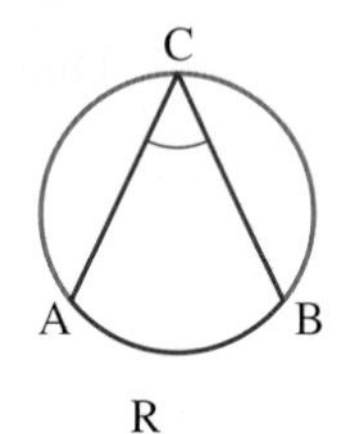

- In the diagram at right, chords AC and BC form the angle ACB. Arc AB has **subtended** angle ACB.
- **Theorem 1 Code**

  The angle subtended at the centre of a circle is twice the angle subtended at the circumference, standing on the same arc.

  Proof:

  Let $\angle PRO = x$ and $\angle QRO = y$

  $RO = PO = QO$ (radii of the same circle are equal)

  $\angle RPO = x$

  and $\angle RQO = y$

  $\angle POM = 2x$ (exterior angle of triangle)

  and $\angle QOM = 2y$ (exterior angle of triangle)

  $\angle POQ = 2x + 2y$

  $= 2(x + y)$

  which is twice the size of $\angle PRQ = x + y$.

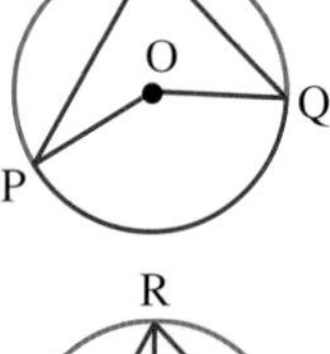

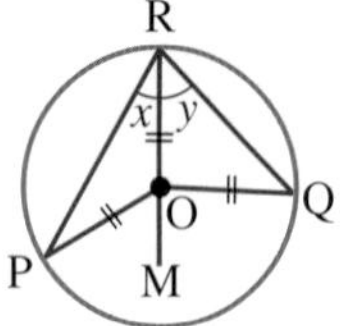

  The angle subtended at the centre of a circle is twice the angle subtended at the circumference, standing on the same arc.
- **Theorem 2 Code**

  All angles that have their vertex on the circumference and are subtended by the same arc are equal.

  Proof:

  Join P and Q to O, the centre of the circle.

  Let $\angle PSQ = x$

  $\angle POQ = 2x$ (angle at the centre is twice the angle at the circumference)

  $\angle PRQ = x$ (angle at the circumference is half the angle of the centre)

  $\angle PSQ = \angle PRQ$.

  Angles at the circumference subtended by the same arc are equal.

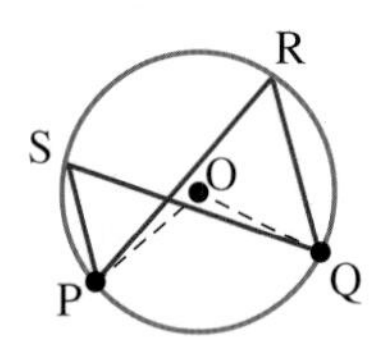

WORKED EXAMPLE 1 Determining angles in a circle

Determine the values of the pronumerals x and y in the diagram, giving reasons for your answers.

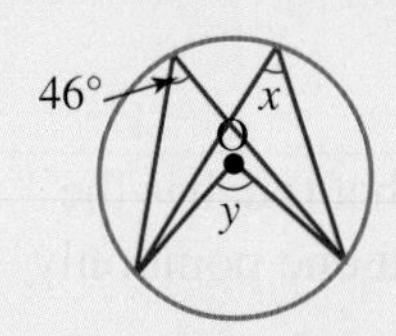

THINK	WRITE
1. Angles x and 46° are angles subtended by the same arc and both have their vertex on the circumference.	$x = 46°$
2. Angles y and 46° stand on the same arc. The 46° angle has its vertex on the circumference and y has its vertex at the centre. The angle at the centre is twice the angle at the circumference.	$y = 2 \times 46°$ $= 92°$ 

- **Theorem 3 Code**
  Angles subtended by the diameter, that is, angles in a semicircle, are right angles.
  In the diagram, PQ is the diameter. Angles a, b and c are right angles. This theorem is in fact a special case of Theorem 1.

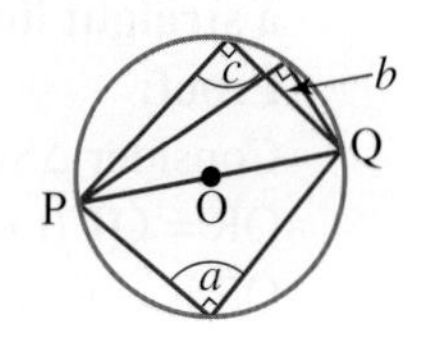

  Proof:
  $\angle POQ = 180°$ (straight line)
  Let S refer to the angle at the circumference subtended by the diameter. In the figure, S could be at the points where a, b and c are represented on the diagram.
  $\angle PSQ = 90°$(angle at the circumference is half the angle at the centre)
  Angles subtended by a diameter are right angles.

16.2.2 Tangents to a circle

eles-4996

- In the diagram, the tangent touches the circumference of the circle at the point of contact.

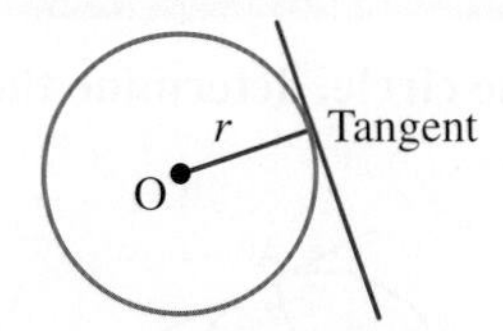

- **Theorem 4 Code**
  A tangent to a circle is perpendicular to the radius of the circle at the point of contact on the circumference.
  In the diagram, the radius is drawn to a point, P, on the circumference. The tangent to the circle is also drawn at P. The radius and the tangent meet at right angles, that is, the angle at P equals 90°.

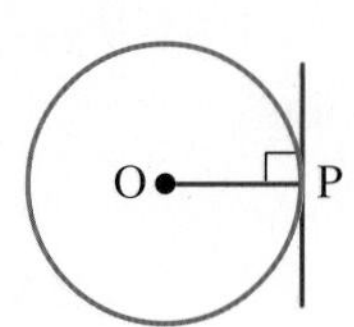

WORKED EXAMPLE 2 Determining angles when tangents are drawn to a circle

Determine the values of the pronumerals in the diagram, giving a reason for your answer.

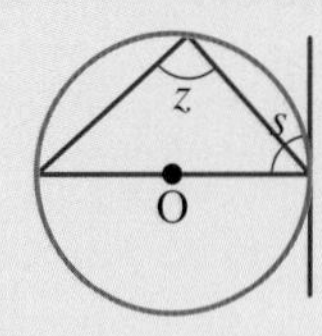

THINK	WRITE
1. Angle z is subtended by the diameter. Use an appropriate theorem to state the value of z.	$z = 90°$
2. Angle s is formed by a tangent and a radius, drawn to the point of contact. Apply the corresponding theorem to find the value of s.	$s = 90°$

- **Theorem 5 Code**

The angle formed by two tangents meeting at an external point is bisected by a straight line joining the centre of the circle to that external point.

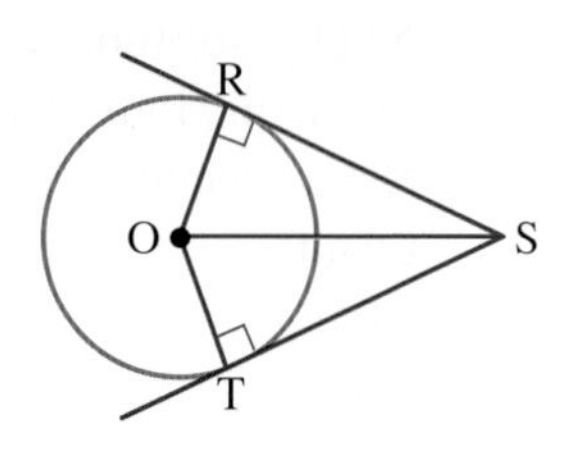

Proof:

Consider ΔSOR and ΔSOT.

OR = OT (radii of the same circle are equal)

OS is common.

∠ORS = ∠OTS = 90° (angle between a tangent and radii is 90°)

∴ ΔSOR ≅ ΔSOT (RHS)

So ∠ROS = ∠TOS and ∠OSR = ∠OST (corresponding angles in congruent triangles are equal).

The angle formed by two tangents meeting at an external point is bisected by a straight line joining the centre of the circle to the external point.

WORKED EXAMPLE 3 Determining angles using properties of tangents

Given that BA and BC are tangents to the circle, determine the values of the pronumerals in the diagram. Give reasons for your answers.

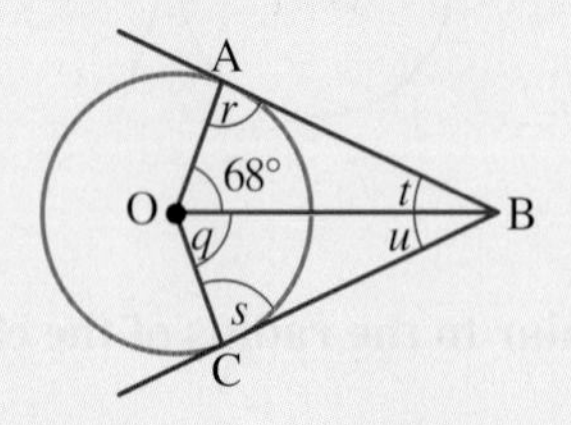

THINK	WRITE
1. Angles r and s are angles formed by the tangent and the radius, drawn to the same point on the circle. State their size.	$s = r = 90°$
2. In the triangle ABO, two angles are already known and so angle t can be found using our knowledge of the sum of the angles in a triangle.	ΔABO: $t + 90° + 68° = 180°$ $t + 158° = 180°$ $t = 22°$
3. ∠ABC is formed by the two tangents, so the line BO, joining the vertex B with the centre of the circle, bisects this angle. This means that angles t and u are equal.	∠ABO = ∠CBO ∠ABO = $t = 22°$, ∠CBO = u $u = 22°$
4. All angles in a triangle have a sum of 180°. ΔAOB and ΔCOB are congruent triangles using RHS, they are both right angled triangles, the hypotenuse is common and OA = OB, radii.	In ΔAOB and ΔCOB $r + t + 68° = 180°$ $s + u + q = 180°$ $r = s = 90°$ (proved previously) $t = u = 22°$ (proved previously $\therefore q = 68°$

DISCUSSION

What are the common steps in proving a theorem?

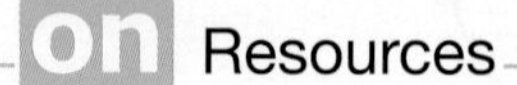

eWorkbook Topic 16 Workbook (worksheets, code puzzle and project) (ewbk-2042)

Digital documents SkillSHEET Using tests to prove congruent triangles (doc-5390)
SkillSHEET Corresponding sides and angles of congruent triangles (doc-5391)
SkillSHEET Using tests to prove similar triangles (doc-5392)
SkillSHEET Angles in a triangle (doc-5393)
SkillSHEET More angle relations (doc-5394)

Interactivities Circle theorem 1 (int-6218)
Circle theorem 2 (int-6219)
Circle theorem 3 (int-6220)
Circle theorem 4 (int-6221)
Circle theorem 5 (int-6222)

Exercise 16.2 Angles in a circle

learnon

Individual pathways

■ PRACTISE	■ CONSOLIDATE	■ MASTER
1, 4, 6, 8, 13, 17	2, 5, 9, 11, 14, 18	3, 7, 10, 12, 15, 16, 19

To answer questions online and to receive **immediate corrective feedback** and **fully worked solutions** for all questions, go to your learnON title at www.jacplus.com.au.

Fluency

WE1 For questions 1 to 3, calculate the values of the pronumerals in each of the following, giving reasons for your answers.

1. a. 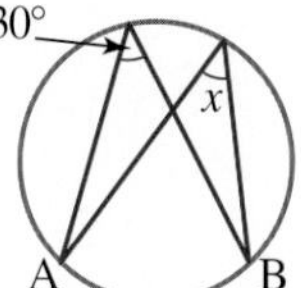

b. 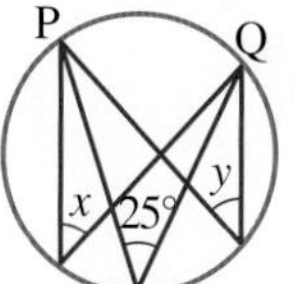

c. 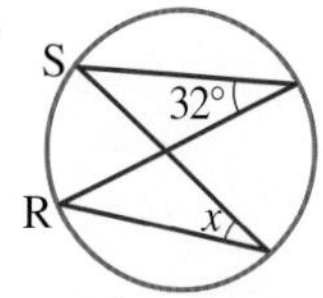

2. a. 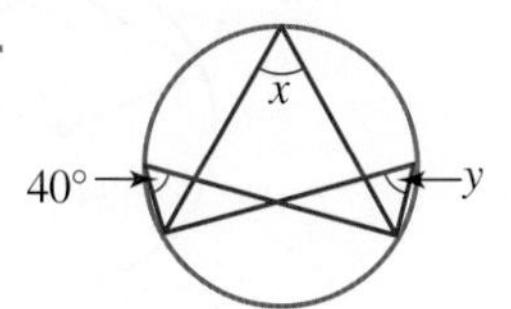

b. 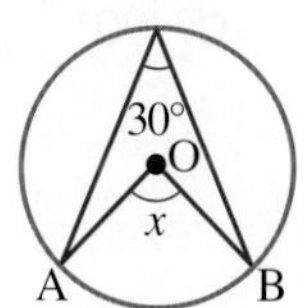

c. 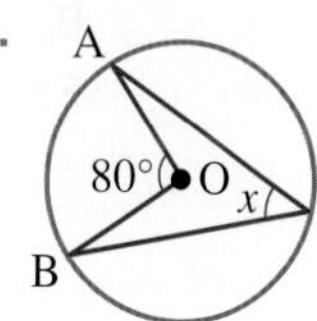

3. a. 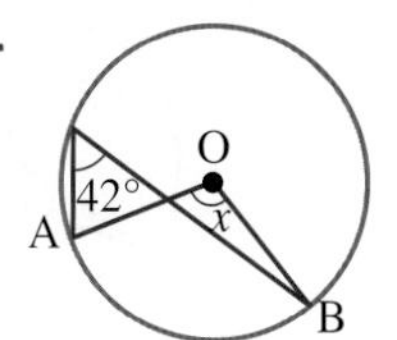

b. 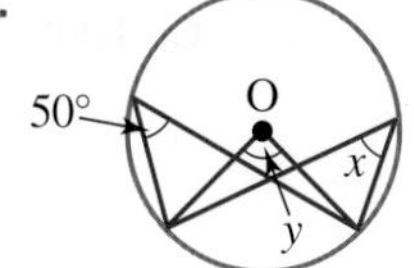

c. B
x
28°
A
O

4. **WE2** Determine the values of the pronumerals in each of the following figures, giving reasons for your answers.

a. 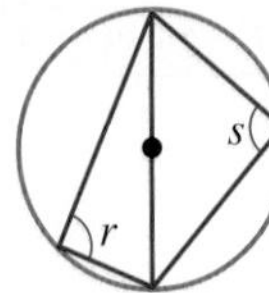

b. 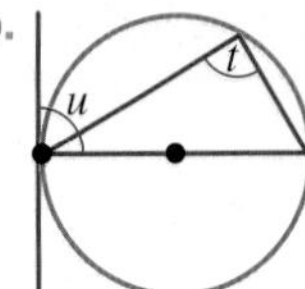

c. 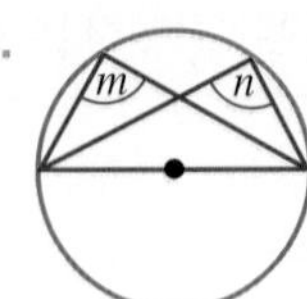

5. Calculate the values of the pronumerals in each of the following figures, giving reasons for your answers.

a. 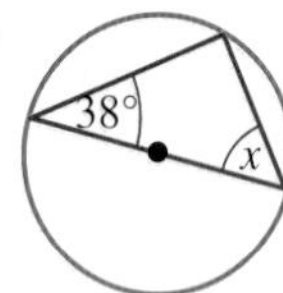

b. 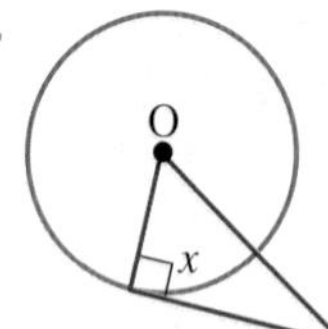

c. 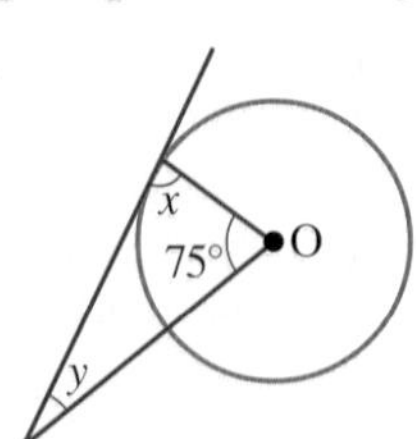

Understanding

6. **WE3** Given that AB and DB are tangents, determine the value of the pronumerals in each of the following, giving reasons for your answers.

a. 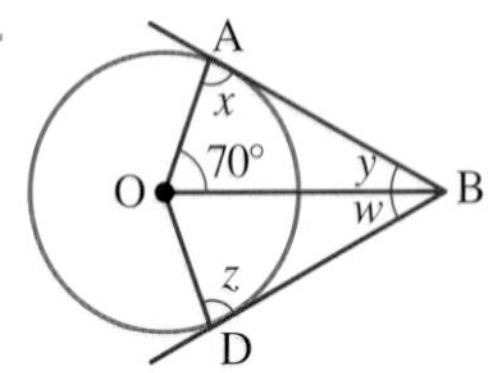

b. 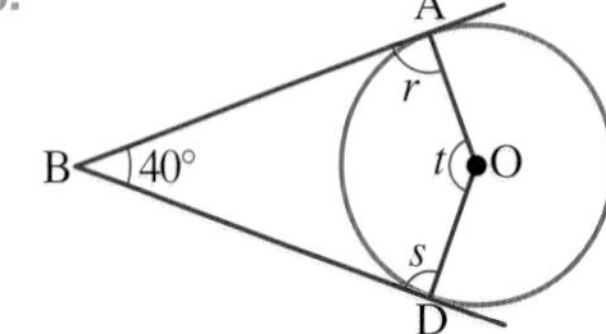

c. 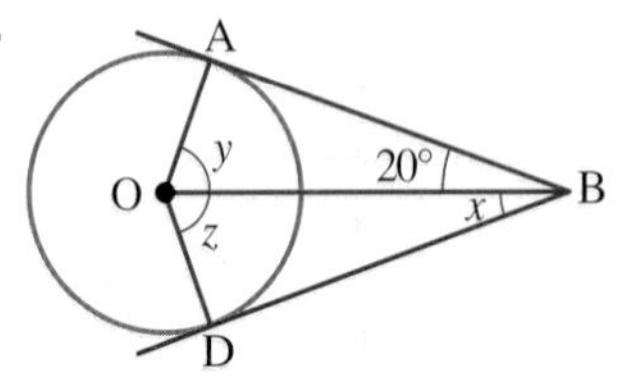

7. Given that AB and DB are tangents, determine the value of the pronumerals in each of the following, giving reasons for your answers.

a. 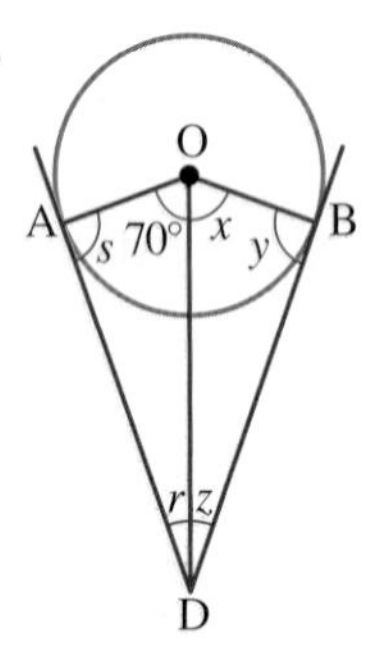

b. 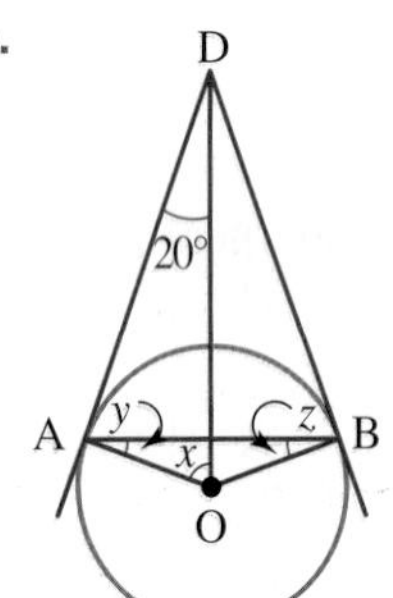

c. 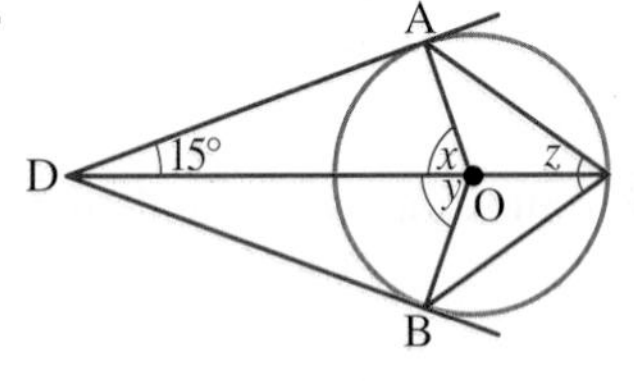

8. **MC** The value of x in the diagram is:

A. 240° B. 120° C. 90°
D. 60° E. 100°

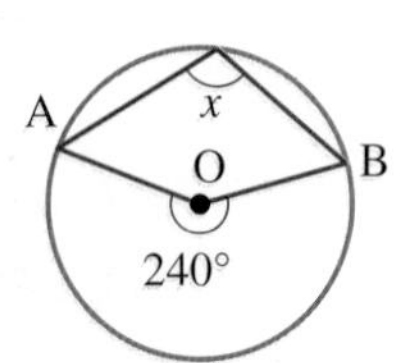

9. **MC** The value of x in the diagram is:

A. 50° B. 90° C. 100°
D. 80° E. 200°

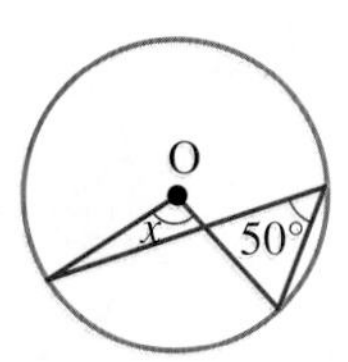

10. **MC** Choose which of the following statements is true for this diagram.

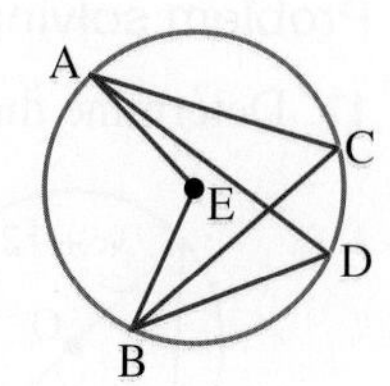

A. $\angle ACB = 2 \times \angle ADB$ **B.** $\angle AEB = \angle ACB$ **C.** $\angle ACB = \angle ADB$
D. $\angle AEB = \angle ADB$ **E.** $2 \times \angle ACB = \angle ADB$

11. **MC** In the diagram shown, determine which angle is subtended by the same arc as $\angle$APB.
Note: There may be more than one correct answer.

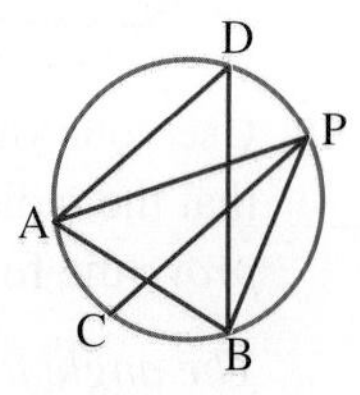

A. $\angle$APC **B.** $\angle$BPC **C.** $\angle$ABP
D. $\angle$ADB **E.** $\angle$BPD

12. **MC** For the diagram shown, determine which of the statements is true.
Note: There may be more than one correct answer.

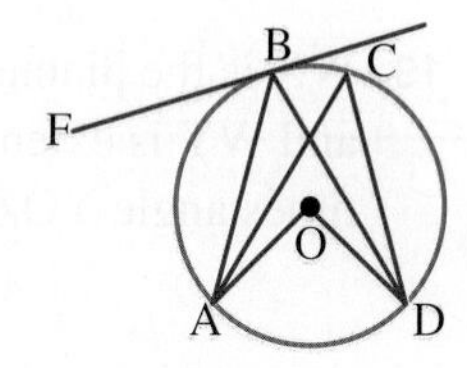

A. $2\angle AOD = \angle ABD$ **B.** $\angle AOD = 2\angle ACD$ **C.** $\angle ABF = \angle ABD$
D. $\angle ABD = \angle ACD$ **E.** $\angle AOD = \angle ABF$

Reasoning

13. Values are suggested for the pronumerals in the diagram shown. AB is a tangent to a circle and O is the centre.
In each case give reasons to justify suggested values.

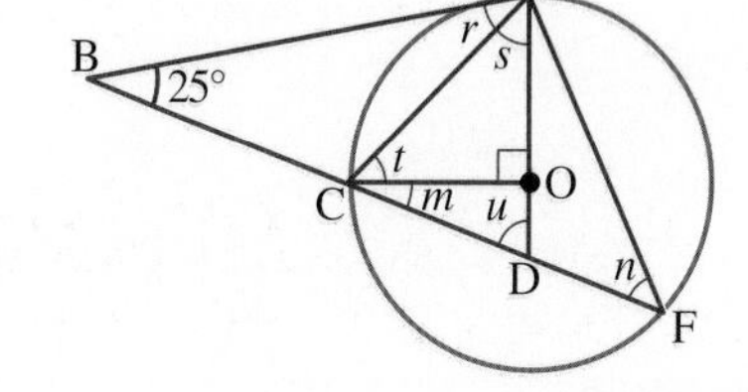

a. $s = t = 45°$ b. $r = 45°$
c. $u = 65°$ d. $m = 25°$
e. $n = 45°$

14. Set out below is the proof of this result: The angle at the centre of a circle is twice the angle at the circumference standing on the same arc.
Copy and complete the following to show that $\angle POQ = 2 \times \angle PRQ$.
Construct a diameter through R. Let the opposite end of the diameter be S.

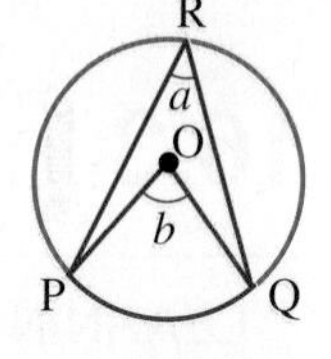

Let $\angle ORP = x$ and $\angle ORQ = y$.
OR = OP (____________________________)
$\angle OPR = x$ (____________________________)
$\angle SOP = 2x$ (exterior angle equals ____________________________)

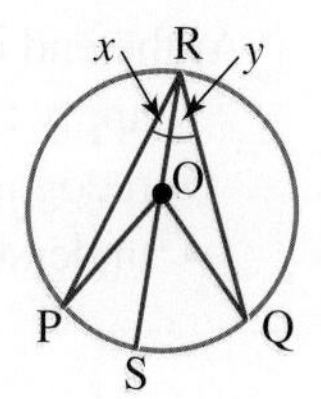

OR = OQ (____________________________)

$\angle OQR =$ ____________________________ (____________________________)
$\angle SOQ =$ ____________________________ (____________________________)
Now $\angle PRQ =$ ____________ and $\angle POQ =$ ____________.
Therefore $\angle POQ = 2 \times \angle PRQ$.

15. Prove that the segments formed by drawing tangents from an external point to a circle are equal in length.

16. Use the figure shown to prove that angles subtended by the same arc are equal.

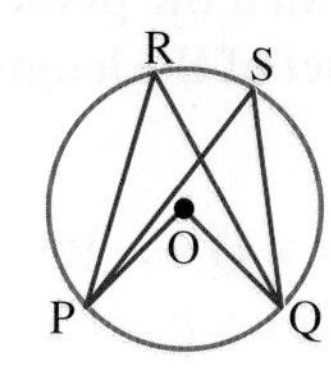

Problem solving

17. Determine the value of x in each of the following diagrams.

a.

$3x + 12$

O

$2x - 2$

b.

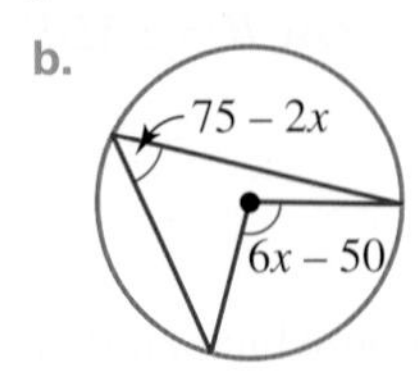

18. Use your knowledge of types of triangles, angles in triangles and the fact that the radius of a circle meets the tangent to the circle at right angles *to prove* the following theorem:

The angle formed between two tangents meeting at an external point is bisected by a line from the centre of the circle to the external point.

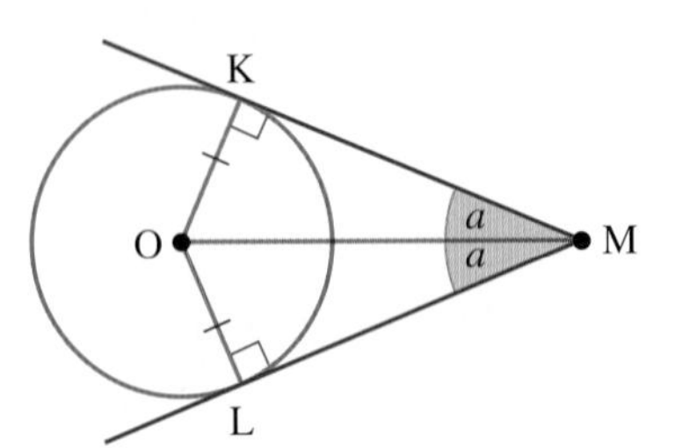

19. WX is the diameter of a circle with centre at O. Y is a point on the circle and WY is extended to Z so that OY = YZ. Prove that angle ZOX is three times angle YOZ.

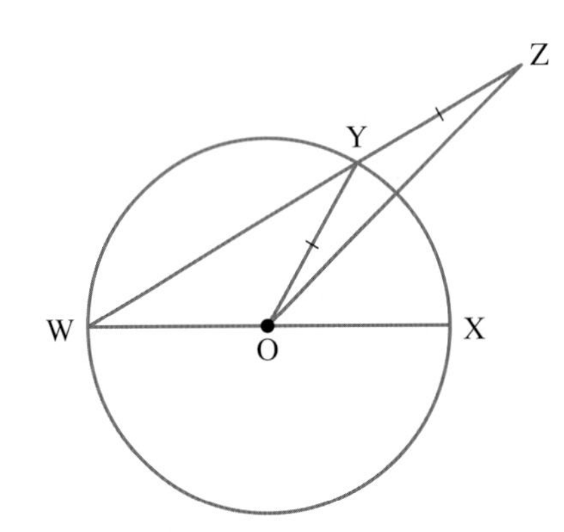

16.3 Intersecting chords, secants and tangents

LEARNING INTENTION

At the end of this subtopic you should be able to:

- apply the intersecting chords, secants and tangents theorems
- recognise that a radius will bisect a chord at right angles
- understand the concept of the circumcentre of a triangle.

16.3.1 Intersecting chords

eles-4997

- In the diagram at right, chords PQ and RS intersect at X.

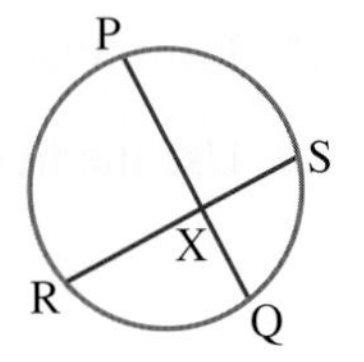

- **Theorem 6 Code**
  If the two chords intersect inside a circle, then the point of intersection divides each chord into two segments so that the product of the lengths of the segments for both chords is the same.

$$\mathbf{PX \times QX = RX \times SX}$$
$$\text{or } \boldsymbol{a \times b = c \times d}$$

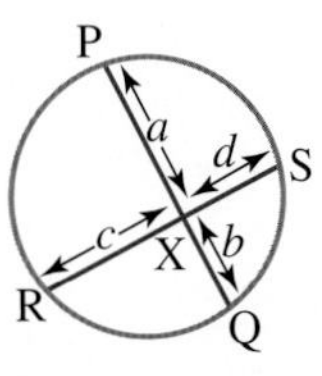

Proof:
Join PR and SQ.
Consider ΔPRX and ΔSQX.
$\angle PXR = \angle SXQ$ (vertically opposite angles are equal)
$\angle RSQ = \angle RPQ$ (angles at the circumference standing on the same arc are equal)
$\angle PRS = \angle PQS$ (angles at the circumference standing on the same arc are equal)
$\Delta PRX \sim \Delta SQX$ (**equiangular**)
$\frac{PX}{SX} = \frac{RX}{QX}$ (ratio of sides in similar triangles is equal)
or, $PX \times QX = RX \times SX$

WORKED EXAMPLE 4 Determining values using intersecting chords

Determine the value of the pronumeral *m*.

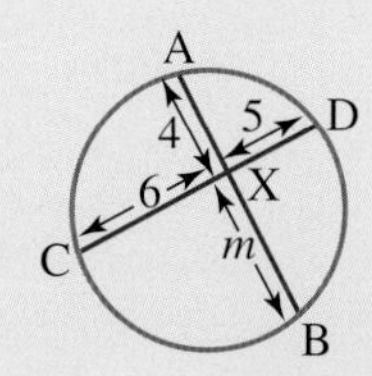

THINK	WRITE
1. Chords AB and CD intersect at X. Point X divides each chord into two parts so that the products of the lengths of these parts are equal. Write this as a mathematical statement.	$AX \times BX = CX \times DX$
2. Identify the lengths of the line segments.	$AX = 4$, $BX = m$, $CX = 6$, $DX = 5$
3. Substitute the given lengths into the formula and solve for *m*.	$4m = 6 \times 5$ $m = \frac{30}{4}$ $= 7.5$

16.3.2 Intersecting secants

eles-4998

- In the diagram at right, chords CD and AB are extended to form secants CX and AX respectively. They intersect at X.

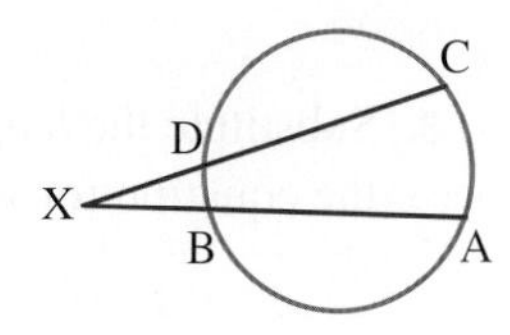

- **Theorem 7 Code**
If two secants intersect outside the circle as shown, then the following relationship is always true:

$$AX \times XB = XC \times DX$$
or $a \times b = c \times d$.

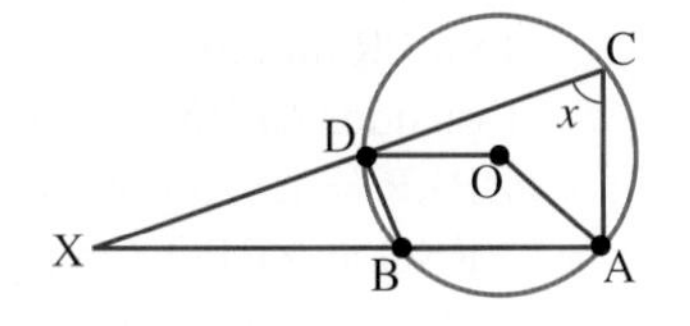

Proof:

Join D and A to O, the centre of the circle.

Let $\angle DCA = x$

$\angle DOA = 2x$ (angle at the centre is twice the angle at the circumference standing on the same arc)

Reflex $\angle DOA = 360° - 2x$ (angles in a revolution add to 360°)

$\angle DBA = 180° - x$ (angle at the centre is twice the angle at the circumference standing on the same arc)

$\angle DBX = x$ (angle sum of a straight line is 180°)

$\angle DCA = \angle DBX$

Consider ΔBXD and ΔCXA.

$\angle BXD$ is common.

$\angle DCA = \angle DBX$ (shown previously)

$\angle XAC = \angle XDB$ (angle sum of a triangle is 180°)

$\angle AXC \sim \Delta DXB$ (equiangular)

$$\frac{AX}{DX} = \frac{XC}{XB}$$

or, $AX \times XB = XC \times DX$

WORKED EXAMPLE 5 Determining pronumerals using intersecting secants.

Determine the value of the pronumeral *y*.

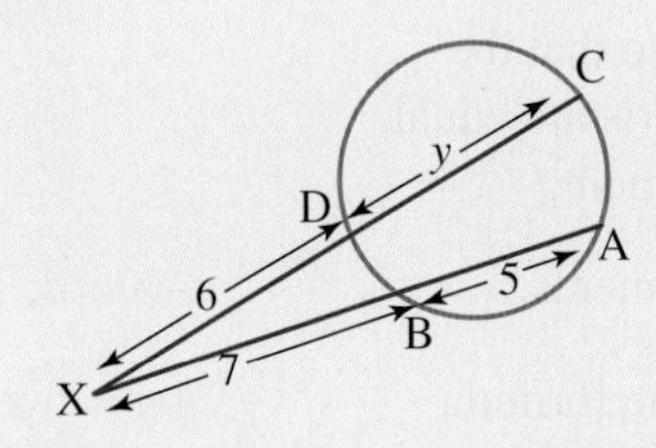

THINK	WRITE
1. Secants XC and AX intersect outside the circle at X. Write the rule connecting the lengths of XC, DX, AX and XB.	$XC \times DX = AX \times XB$
2. State the length of the required line segments.	$XC = y + 6 \quad DX = 6$ $AX = 7 + 5 \quad XB = 7$ $= 12$
3. Substitute the length of the line segments and solve the equation for *y*.	$(y + 6) \times 6 = 12 \times 7$ $6y + 36 = 84$ $6y = 48$ $y = 8$

16.3.3 Intersecting tangents

eles-4999

- In the diagram, the tangents AC and BC intersect at C.

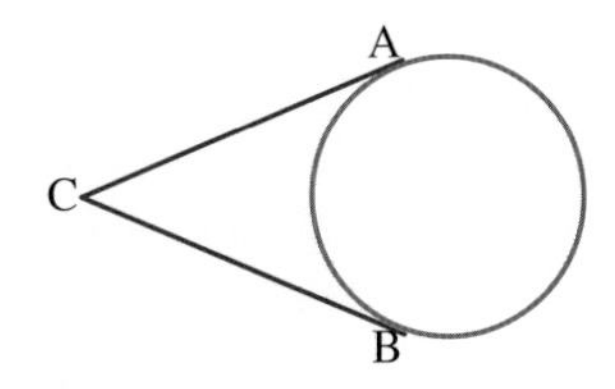

- **Theorem 8 Code**
If two tangents meet outside a circle, then the lengths from the external point to where they meet the circle are equal.
Proof:
Join A and B to O, the centre of the circle.
Consider ΔOCA and ΔOCB.
OC is common.
OA = OB (radii of the same circle are equal)
∠OAC = ∠OBC (radius is perpendicular to tangent through the point of contact)
ΔOCA ≅ ΔOCB (RHS)
AC = BC (corresponding sides of congruent triangles are equal)
If two tangents meet outside a circle, the lengths from the external point to the point of contact are equal.

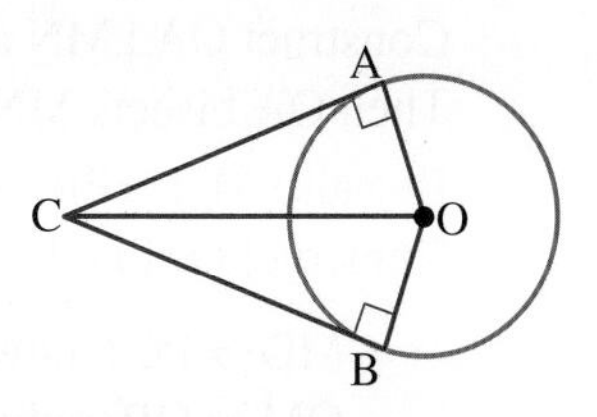

WORKED EXAMPLE 6 Determining the pronumeral using lengths of tangents.

Determine the value of the pronumeral *m*.

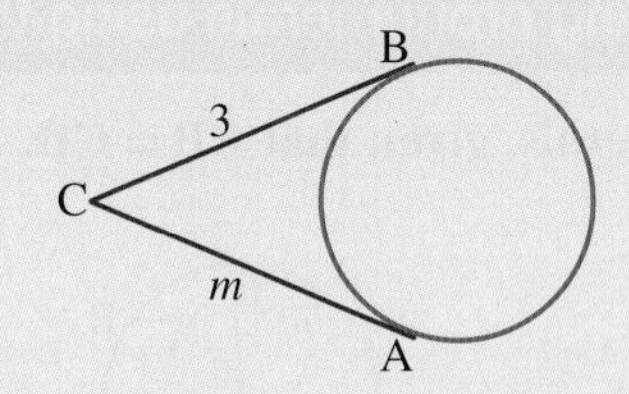

THINK	WRITE
1. BC and AC are tangents intersecting at C. State the rule that connects the lengths BC and AC.	AC = BC 
2. State the lengths of BC and AC.	AC = m, BC = 3
3. Substitute the required lengths into the equation to find the value of m.	$m = 3$

16.3.4 Chords and radii

eles-5000

- In the diagram at right, the chord AB and the radius OC intersect at X at 90°; that is, ∠OXB = 90°. OC bisects the chord AB; that is, AX = XB.

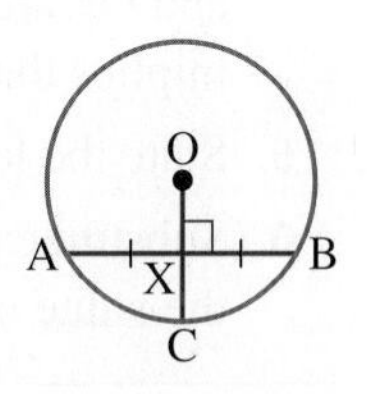

- **Theorem 9 Code** 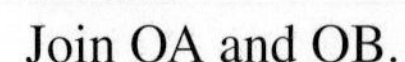
If a radius and a chord intersect at right angles, then the radius bisects the chord.
Proof:
Join OA and OB.
Consider ΔOAX and ΔOBX.
OA = OB (radii of the same circle are equal)
∠OXB = ∠OXA (given)
OX is common.
ΔOAX ≅ ΔOBX (RHS)
AX = BX (corresponding sides in congruent triangles are equal)
If a radius and a chord intersect at right angles, then the radius bisects the chord.

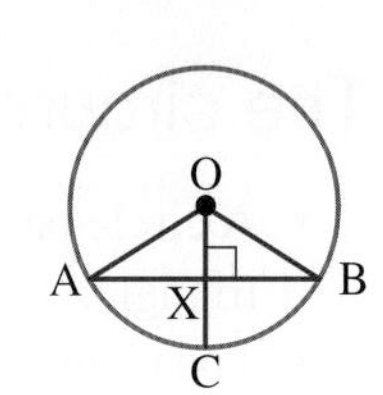

- The converse is also true:
If a radius bisects a chord, the radius and the chord meet at right angles.

- **Theorem 10 Code** 
  Chords equal in length are equidistant from the centre.
  This theorem states that if the chords MN and PR are of equal length, then OD = OC.
  Proof:
  Construct OA⊥MN and OB⊥PR.
  Then OA bisects MN and OB bisects PR (Theorem 9)
  Because MN = PR, MD = DN = PC = CR.
  Construct OM and OP, and consider ΔODM and ΔOCP.

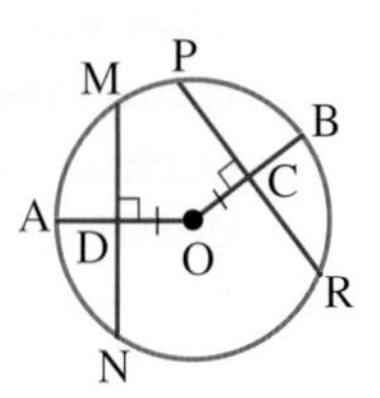

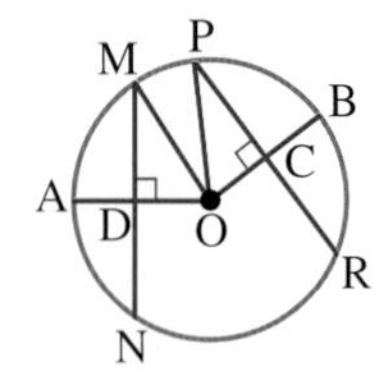

MD = PC (shown above)
OM = OP (radii of the same circle are equal)
$\angle$ODM = $\angle$OCP = 90° (by construction)
ΔODM ≅ ΔOCP (RHS)
So OD = OC (corresponding sides in congruent triangles are equal)

Chords equal in length are equidistant from the centre.

WORKED EXAMPLE 7 Determining pronumerals using theorems on chords

Determine the values of the pronumerals, given that AB = CD.

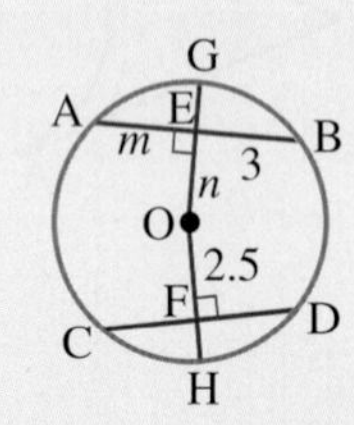

THINK	WRITE
1. Since the radius OG is perpendicular to the chord AB, the radius bisects the chord.	AE = EB
2. State the lengths of AE and EB.	AE = m, EB = 3
3. Substitute the lengths into the equation to find the value of m.	$m = 3$
4. AB and CD are chords of equal length and OE and OF are perpendicular to these chords. This implies that OE and OF are equal in length.	OE = OF
5. State the lengths of OE and OF.	OE = n, OF = 2.5
6. Substitute the lengths into the equation to find the value of n.	$n = 2.5$

The circumcentre of a triangle

- A circle passing through the three vertices of a triangle is called the **circumcircle** of the triangle.
- The centre of this circle is called the **circumcentre** of the triangle.

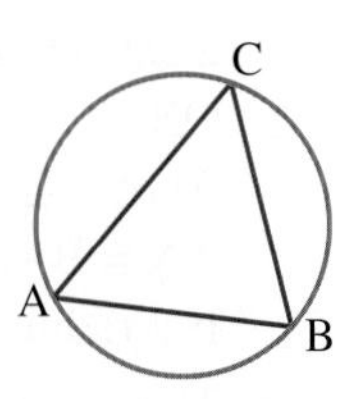

- The circumcentre is located by:
  Step 1: drawing any triangle ABC and label the vertices

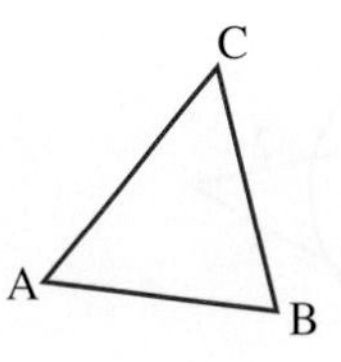

  Step 2: constructing the perpendicular bisectors of the three sides
  Step 3: let the bisectors intersect at O.
- This means OA = OB = OC, by congruent triangles.
- A circle, centre O, can be drawn through the vertices A, B and C.
- The point O is the circumcentre of the triangle.

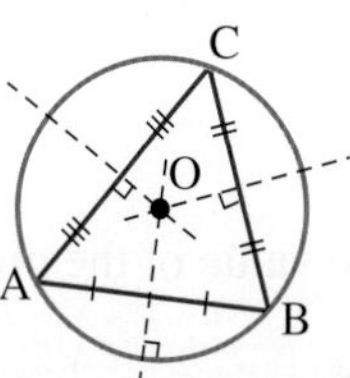

DISCUSSION

What techniques will you use to prove circle theorem?

on Resources

 **eWorkbook** Topic 16 Workbook (worksheets, code puzzle and project) (ewbk-2042)

 **Interactivities** Circle theorem 6 (int-6223)
Circle theorem 7 (int-6224)
Circle theorem 8 (int-6225)
Circle theorem 9 (int-6226)
Circle theorem 10 (int-6227)

Exercise 16.3 Intersecting chords, secants and tangents

learnon

Individual pathways

■ PRACTISE	■ CONSOLIDATE	■ MASTER
1, 4, 9, 12	2, 5, 7, 10, 13	3, 6, 8, 11, 14

To answer questions online and to receive **immediate corrective feedback** and **fully worked solutions** for all questions, go to your learnON title at www.jacplus.com.au.

Fluency

1. WE4 Determine the value of the pronumeral in each of the following.

a. 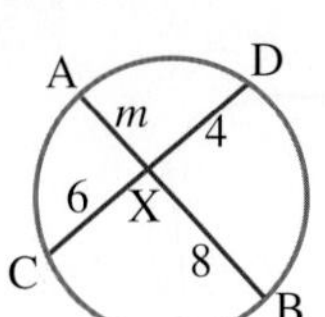

b. 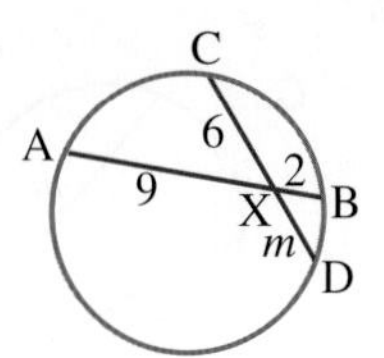

c.

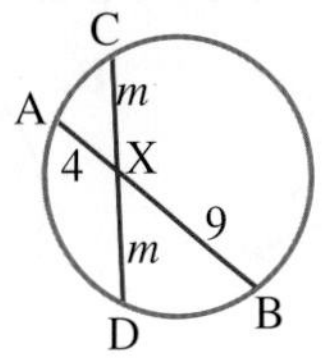

2. **WE5** Determine the value of the pronumeral in each of the following.

a. 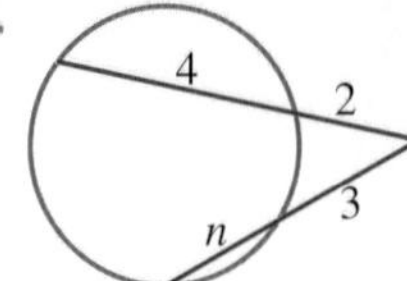

b. 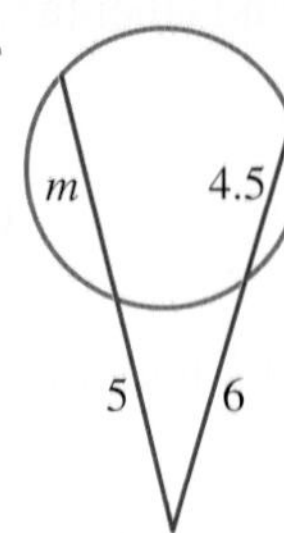

3. Determine the value of the pronumeral in each of the following.

a. 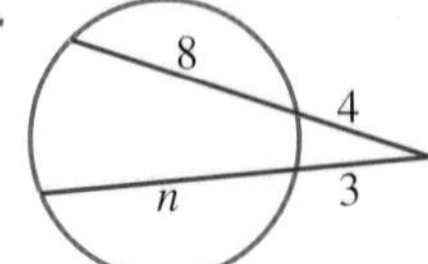

b. 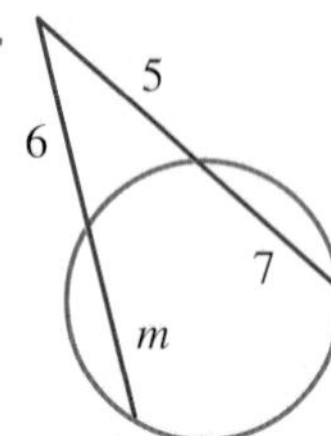

4. **WE6** Determine the value of the pronumerals in each of the following.

a. 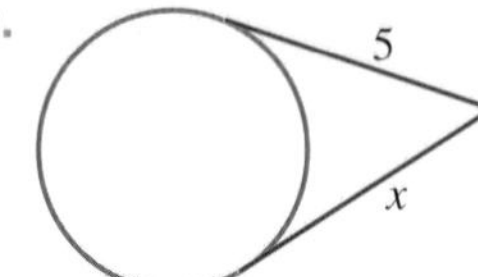

b. 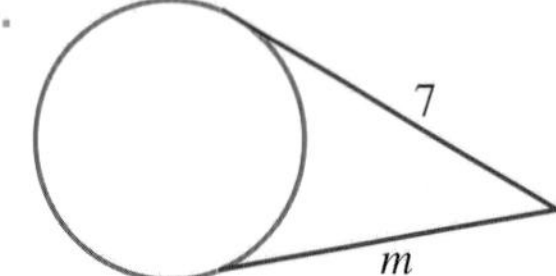

c. 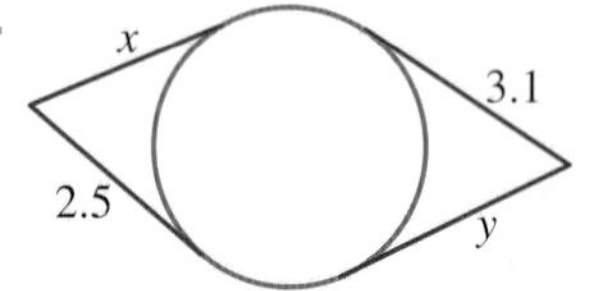

5. **WE7** Determine the value of the pronumeral in each of the following.

a. 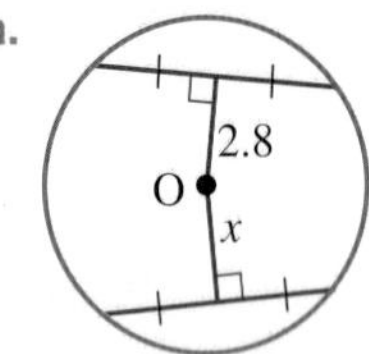

b. 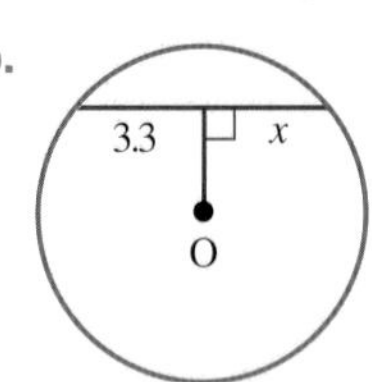

6. Determine the value of the pronumeral in each of the following.

a. 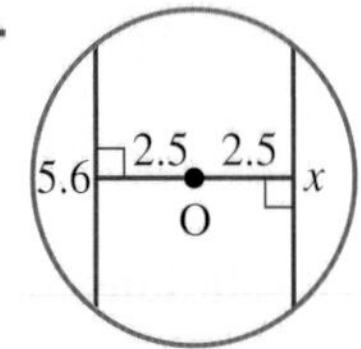

b. 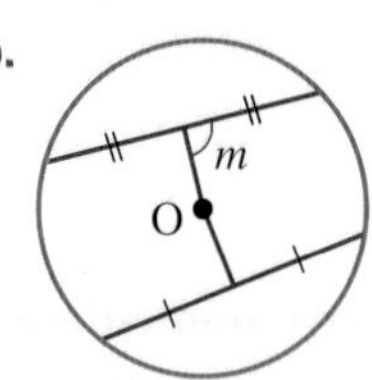

Understanding

7. **MC** Select which of the following figures allows the value of m to be determined by solving a linear equation.
Note: There may be more than one correct answer.

A. 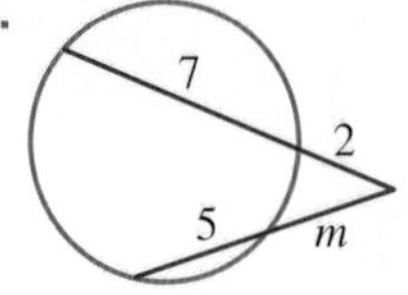

B. 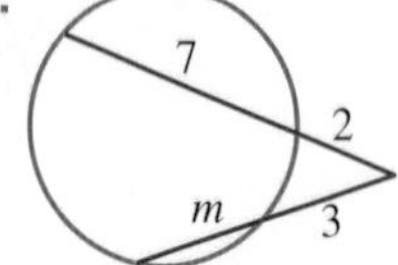

C. 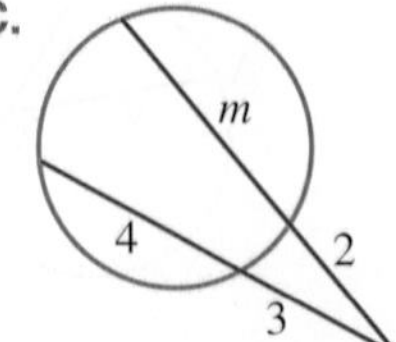

D. 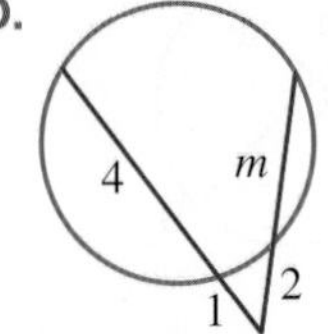

8. Calculate the length ST in the diagram.

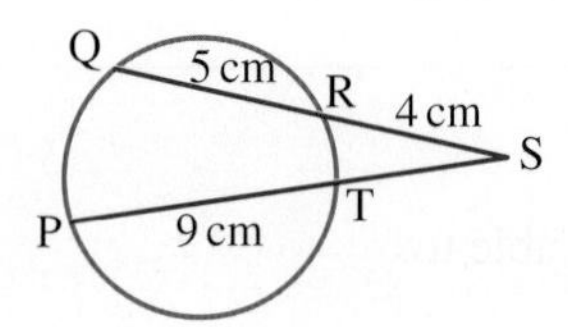

Reasoning

9. Prove the result: If a radius bisects a chord, then the radius meets the chord at right angles. Remember to provide reasons for your statements.
10. Prove the result: Chords that are an equal distance from the centre are equal in length. Provide reasons for your statements.
11. Prove that the line joining the centres of two intersecting circles bisects their common chord at right angles. Provide reasons for your statements.

Problem solving

12. Determine the value of the pronumeral in each of the following diagrams.

a.

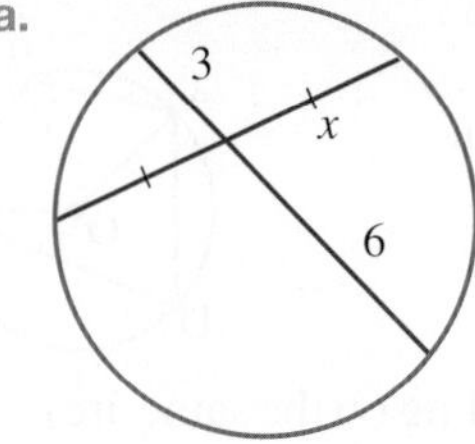

b.

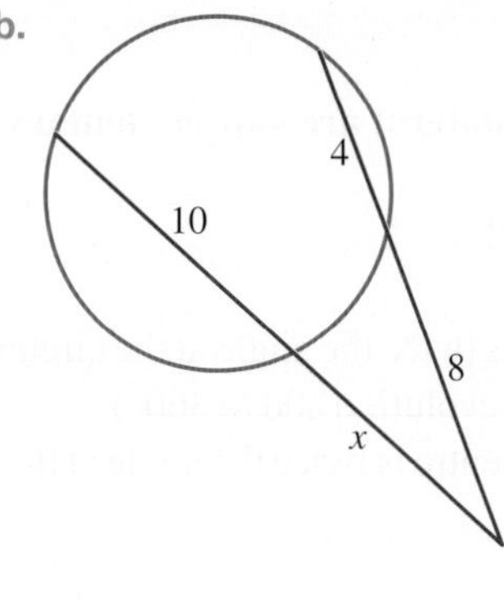

c.

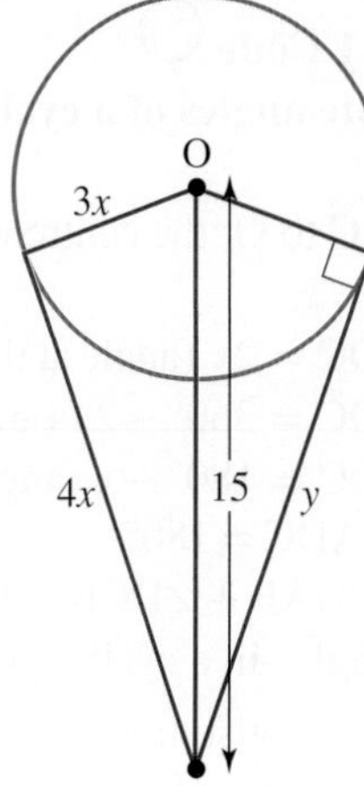

13. AOB is the diameter of the circle. CD is a chord perpendicular to AB and meeting AB at M.
    a. Explain why M is the midpoint of CD.
    b. If $CM = c$, $AM = a$ and $MB = b$, prove that $c^2 = ab$.
    c. Explain why the radius of the circle is equal to $\dfrac{a+b}{2}$.

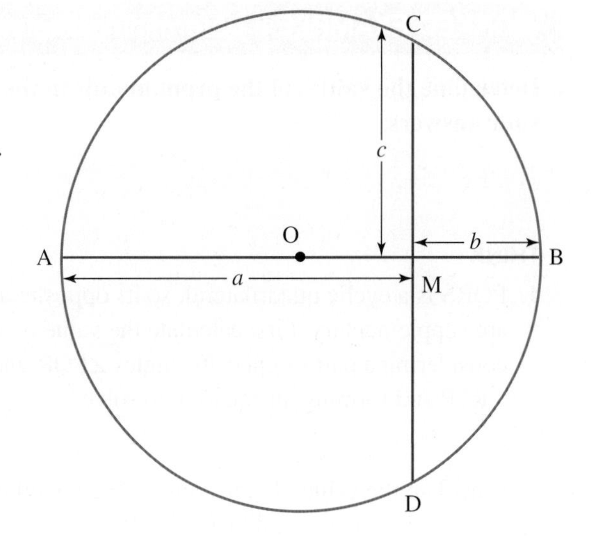

14. An astroid is the curve traced by a point on the circumference of a small circle as it rolls around the inside circumference of a circle that is four times larger than it. Draw the shape of an astroid.

16.4 Cyclic quadrilaterals

LEARNING INTENTION

At the end of this subtopic you should be able to:

- recognize a cyclic quadrilateral
- determine pronumerals knowing the opposite angles of a cyclic quadrilateral are supplementary
- determine pronumerals knowing the exterior angle of a cyclic quadrilateral equals the interior opposite angle.

16.4.1 Quadrilaterals in circles

eles-5001

- A **cyclic quadrilateral** has all four vertices on the circumference of a circle; that is, the quadrilateral is inscribed in the circle.
- In the diagram at right, points A, B, C and D lie on the circumference; hence, ABCD is a cyclic quadrilateral.
- It can also be said that points A, B, C and D are **concyclic**; that is, the circle passes through all the points.

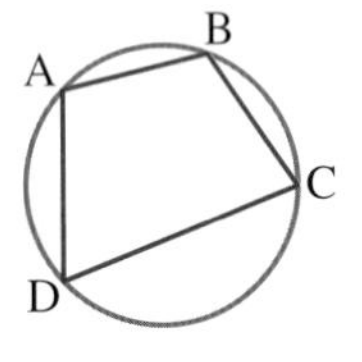

- **Theorem 11 Code**
  The opposite angles of a cyclic quadrilateral are supplementary (add to 180°).
  Proof:
  Join A and C to O, the centre of the circle.
  Let $\angle ABC = x$.
  Reflex $\angle AOC = 2x$ (angle at the centre is twice the angle at the circumference standing on the same arc)
  Reflex $\angle AOC = 360° - 2x$ (angles in a revolution add to 360°)
  $\angle ADC = 180° - x$ (angle at the centre is twice the angle at the circumference standing on the same arc)
  $\angle ABC + \angle ADC = 180°$
  Similarly, $\angle DAB + \angle DCB = 180°$.
  Opposite angles in a cyclic quadrilateral are supplementary.
- The converse is also true:
  If opposite angles of a quadrilateral are supplementary, then the quadrilateral is cyclic.

WORKED EXAMPLE 8 Determining angles in a cyclic quadrilateral

Determine the values of the pronumerals in the diagram below. Give reasons for your answers.

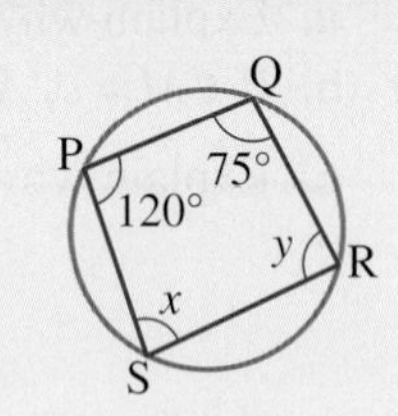

THINK	WRITE
1. PQRS is a cyclic quadrilateral, so its opposite angles are supplementary. First calculate the value of x by considering a pair of opposite angles $\angle PQR$ and $\angle RSP$ and forming an equation to solve.	$\angle PQR + \angle RSP = 180°$ (The opposite angles of a cyclic quadrilateral are supplementary.) $\angle PQR = 75°, \angle RSP = x$ $x + 75° = 180°$ $x = 105°$
2. Calculate the value of y by considering the other pair of opposite angles ($\angle SPQ$ and $\angle QRS$).	$\angle SPQ + \angle QRS = 180°$ $\angle SPQ = 120°, \angle QRS = y$ $y + 120° = 180°$ $y = 60°$

- **Theorem 12 Code**

  The exterior angle of a cyclic quadrilateral is equal to the interior opposite angle.

  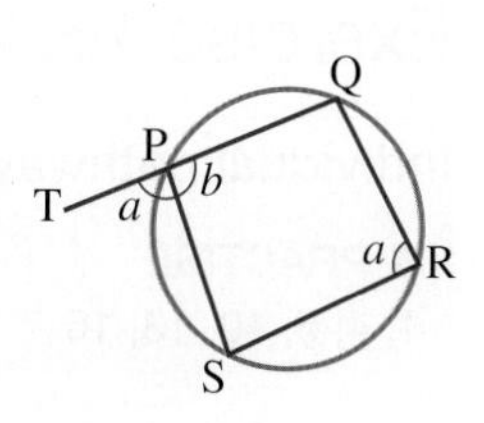

  Proof:

  $\angle QPS + \angle QRS = 180°$ (opposite angles of a cyclic quadrilateral)

  $\angle QPS + \angle SPT = 180°$ (adjacent angles on a straight line)

  Therefore $\angle SPT = \angle QRS$.

  The exterior angle of a cyclic quadrilateral is equal to the interior opposite angle.

WORKED EXAMPLE 9 Determining pronumerals in a cyclic quadrilateral

Determine the value of the pronumerals *x* and *y* in the diagram below.

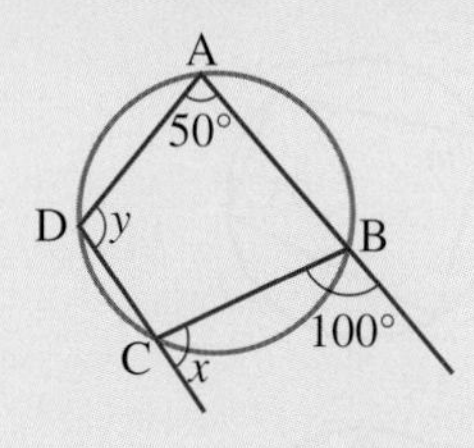

THINK	WRITE
1. ABCD is a cyclic quadrilateral. The exterior angle, *x*, is equal to its interior opposite angle, $\angle DAB$.	$x = \angle DAB$, $\angle DAB = 50°$ So $x = 50°$.
2. The exterior angle, 100°, is equal to its interior opposite angle, $\angle ADC$.	$\angle ADC = 100°$, $\angle ADC = y$ So $y = 100°$. 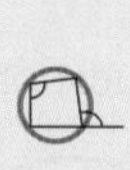

DISCUSSION

What is a cyclic quadrilateral?

 Resources

 **eWorkbook** Topic 16 Workbook (worksheets, code puzzle and project) (ewbk-2042)

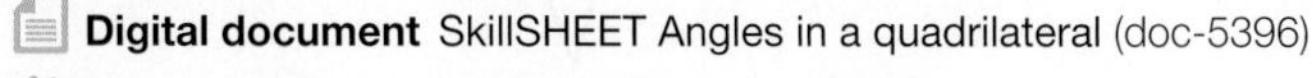

Digital document SkillSHEET Angles in a quadrilateral (doc-5396)

Interactivities Circle theorem 11 (int-6228)
Circle theorem 12 (int-6229)

Exercise 16.4 Cyclic quadrilaterals

learnON

Individual pathways

■ PRACTISE	■ CONSOLIDATE	■ MASTER
1, 4, 5, 10, 13, 16	2, 3, 6, 11, 14, 17	7, 8, 9, 12, 15, 18

To answer questions online and to receive **immediate corrective feedback** and **fully worked solutions** for all questions, go to your learnON title at www.jacplus.com.au.

Fluency

WE8 In questions 1 and 2, calculate the values of the pronumerals in each of the following.

1. a.

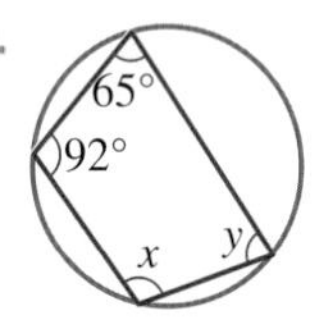

b.

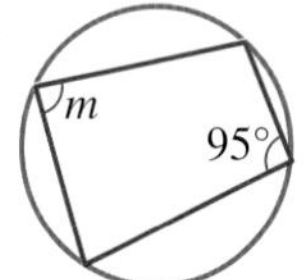

c.

2. a.

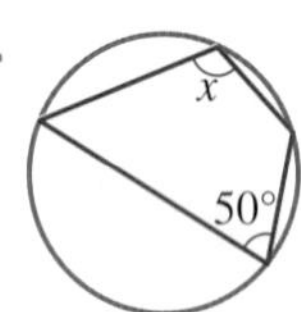

b.

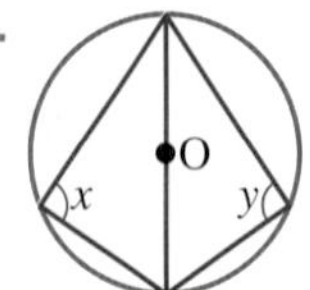

c.

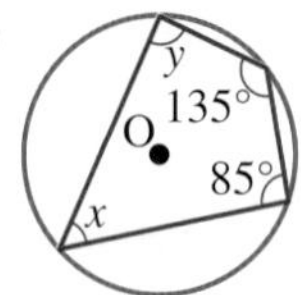

WE9 In questions 3 to 6, calculate the values of the pronumerals in each of the following.

3. a.

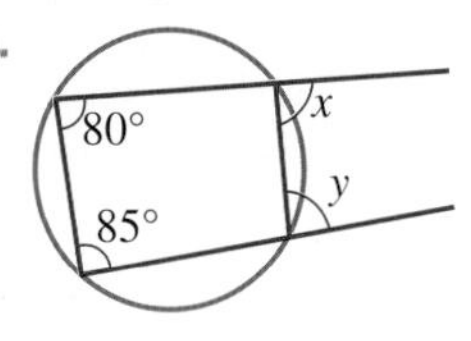

b.

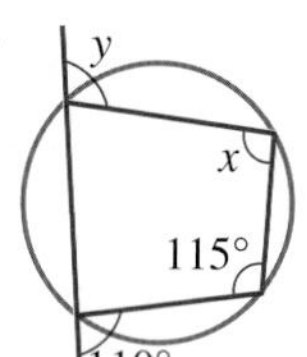

c.

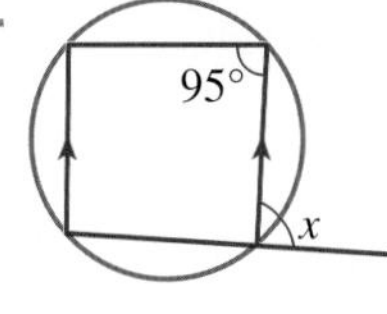

4. a.

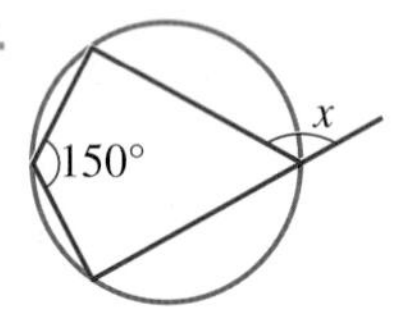

b.

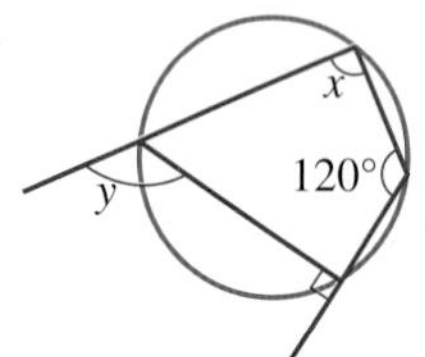

c.

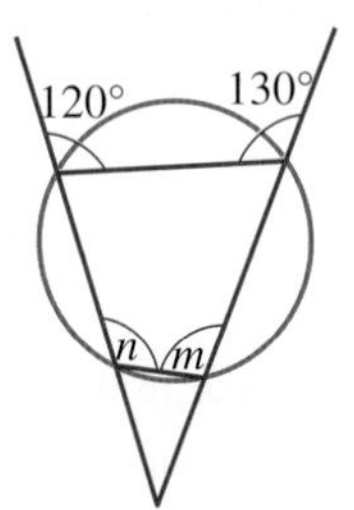

5. a.

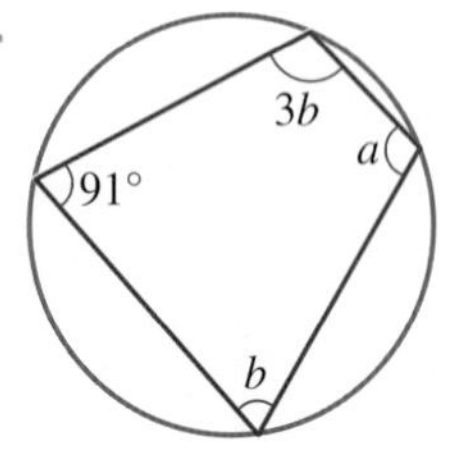

b.

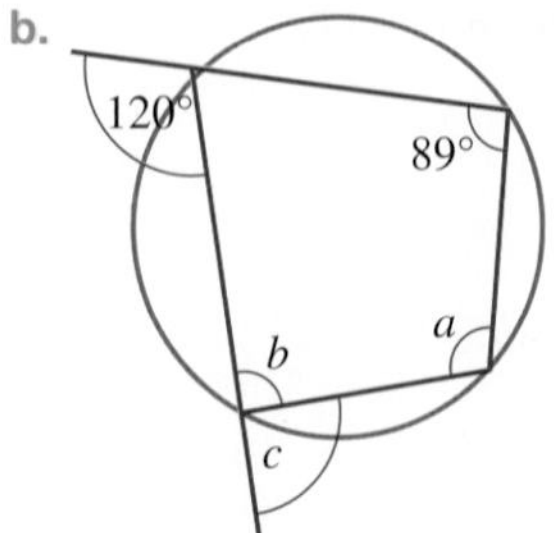

6. a.

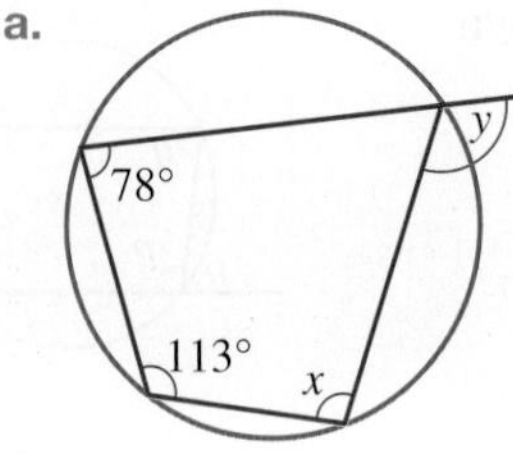

b.

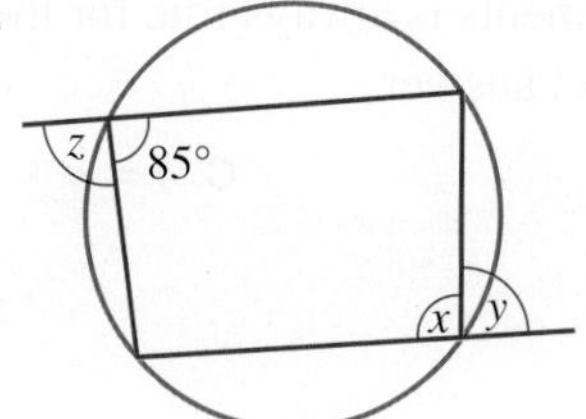

c.

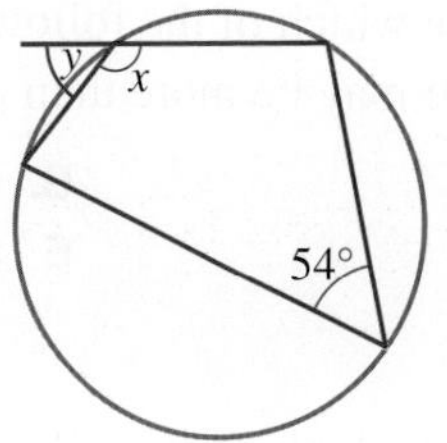

d.

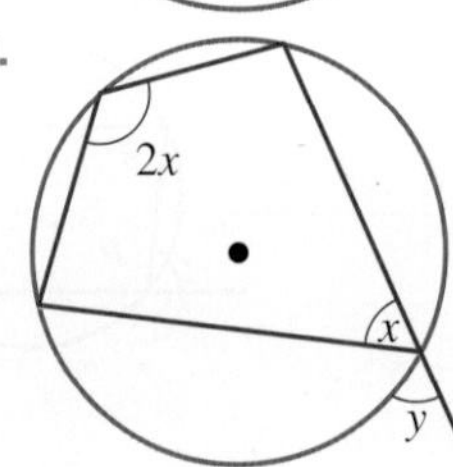

e.

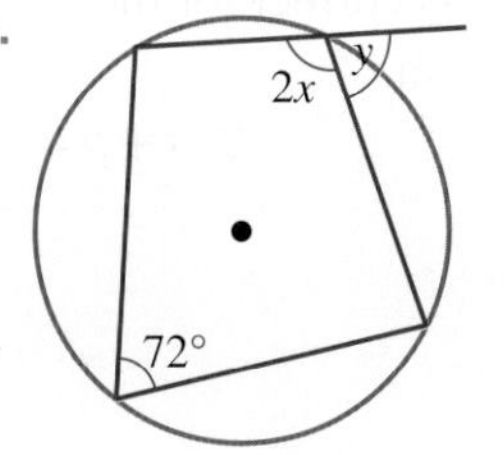

f.

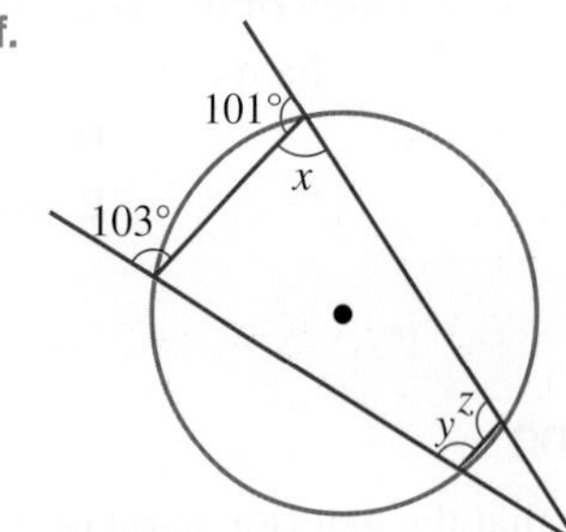

Understanding

7. **MC** Which of the following statements is true for this diagram?

A. $r = q$
B. $r + n = 180°$
C. $m + n = 180°$
D. $r + m = 180°$
E. $q = s$

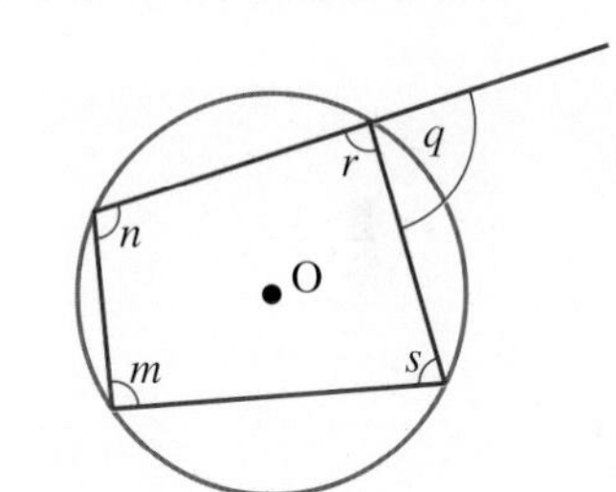

8. **MC** Which of the following statements is *not* true for this diagram?

A. $b + f = 180°$
B. $a = f$
C. $e = d$
D. $c + d = 180°$
E. $c + f = 180°$

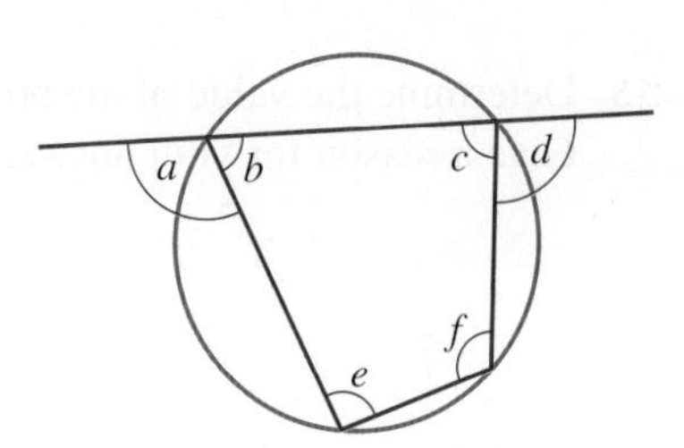

9. **MC** Choose which of the following correctly states the relationship between x, y and z in the diagram.

A. $x = y$ and $x = 2z$
B. $x = 2y$ and $y + z = 180°$
C. $z = 2x$ and $y = 2z$
D. $x + y = 180°$ and $z = 2x$
E. $x + y = 180°$ and $y + z = 180°$

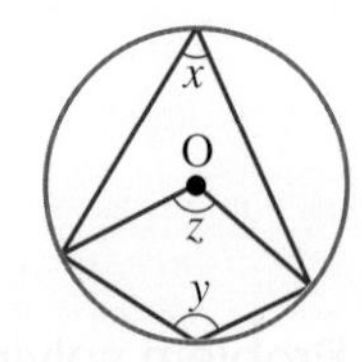

10. Follow the steps below to set out the proof that the opposite angles of a cyclic quadrilateral are equal.

a. Calculate the size of ∠DOB.
b. Calculate the size of the reflex angle DOB.
c. Calculate the size of ∠BCD.
d. Calculate ∠DAB + ∠BCD.

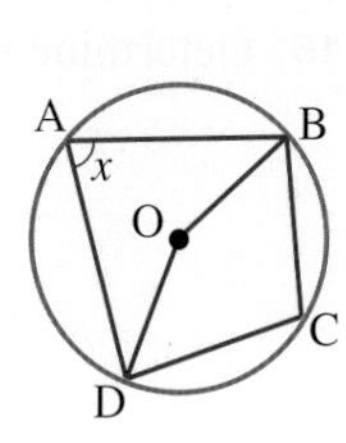

11. **MC** Choose which of the following statements is always true for the diagram shown.
Note: There may be more than one correct answer.

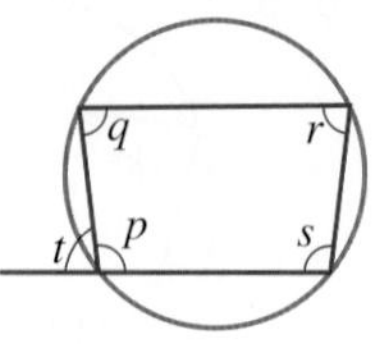

A. $r = t$ B. $r = p$ C. $r = q$
D. $r = s$ E. $r < 90°$

12. **MC** Choose which of the following statements is correct for the diagram shown.
Note: There may be more than one correct answer.

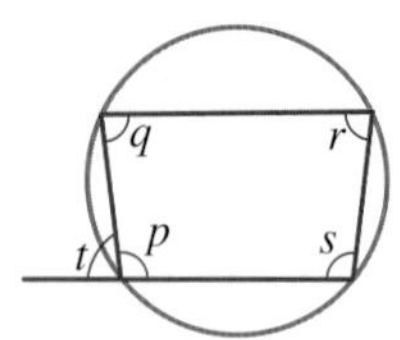

A. $r + p = 180°$ B. $q + s = 180°$ C. $t + p = 180°$
D. $t = r$ E. $t = q$

Reasoning

13. Prove that the exterior angle of a cyclic quadrilateral is equal to the interior opposite angle.

14. Determine the value of the pronumerals in the diagram.
Give a reason for your answer.

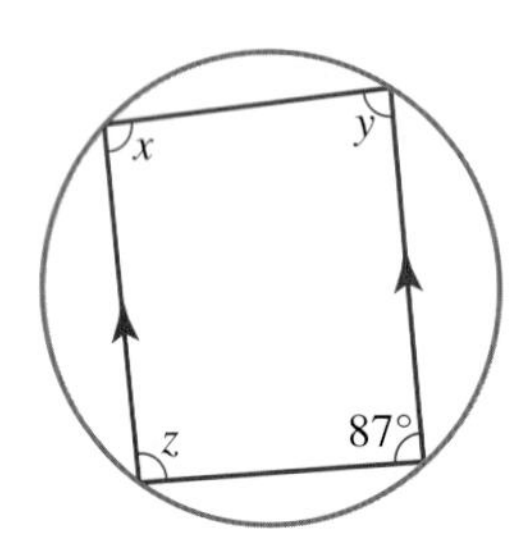

15. Determine the value of the pronumerals in the diagram.
Give a reason for your answer.

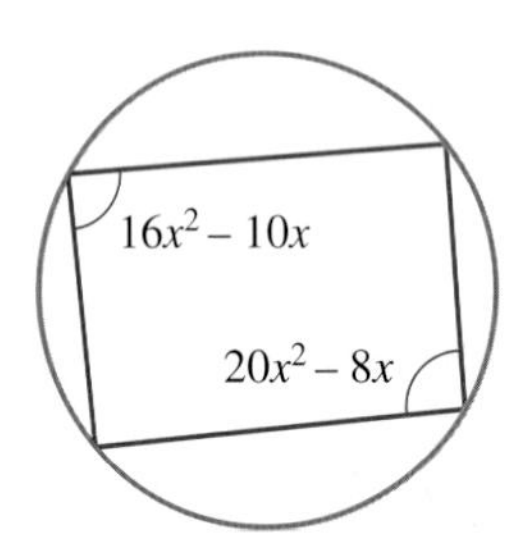

Problem solving

16. Determine the value of the pronumeral x in the diagram shown.

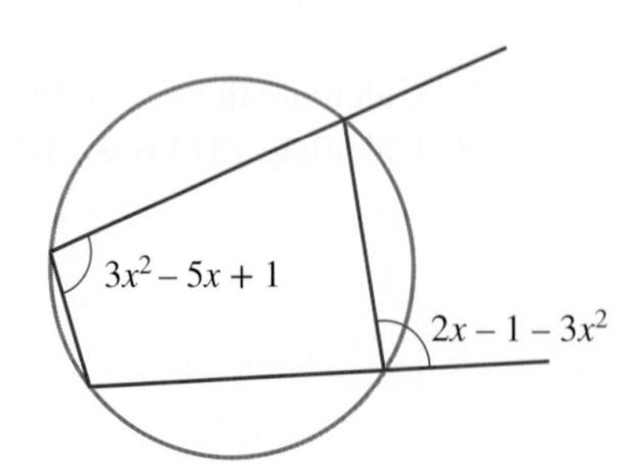

17. Determine the value of each pronumeral in the diagram shown.

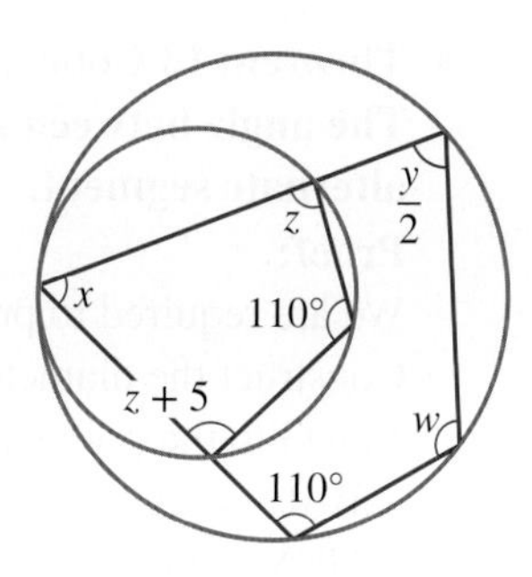

18. $\angle FAB = 70°$, $\angle BEF = a°$, $\angle BED = b°$ and $\angle BCD = c°$.

a. Calculate the values of a, b and c.
b. Prove that CD is parallel to AF.

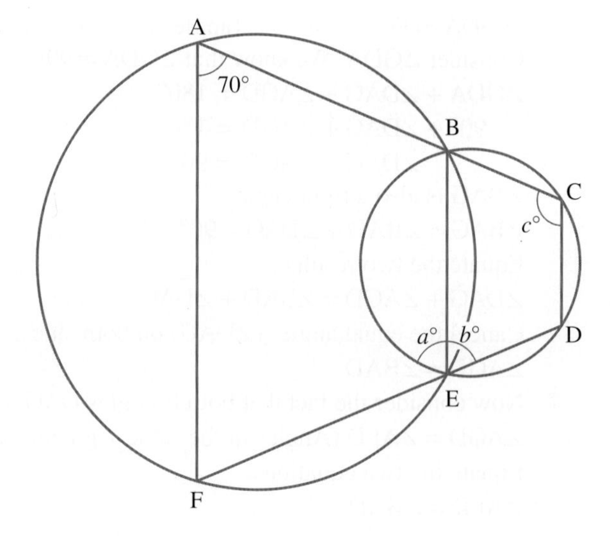

16.5 Tangents, secants and chords

LEARNING INTENTION

At the end of this subtopic you should be able to:
- apply the alternate segment theorem to evaluate pronumerals
- determine the lengths of tangents and secants when they intersect.

16.5.1 The alternate segment theorem

eles-5002

- Consider the figure shown. Line BC is a tangent to the circle at the point A.
- A line is drawn from A to anywhere on the circumference, point D.
  The angle $\angle BAD$ defines a segment (the shaded area).
  The unshaded part of the circle is called the **alternate segment** to $\angle BAD$.

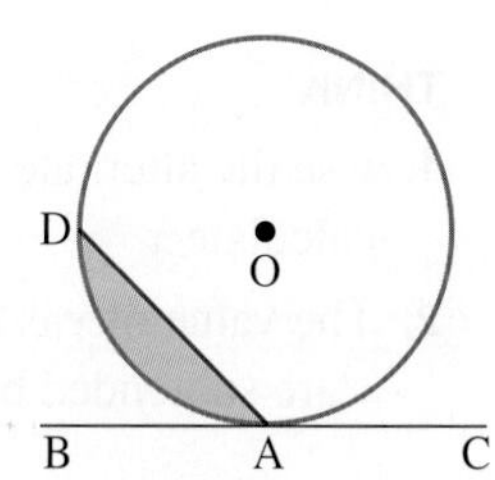

- Now consider angles subtended by the chord AD in the alternate segment, such as the angles marked in pink and blue.
- The alternate segment theorem states that these are equal to the angle that made the segment, namely:

$$\angle BAD = \angle AED \text{ and } \angle BAD = \angle AFD$$

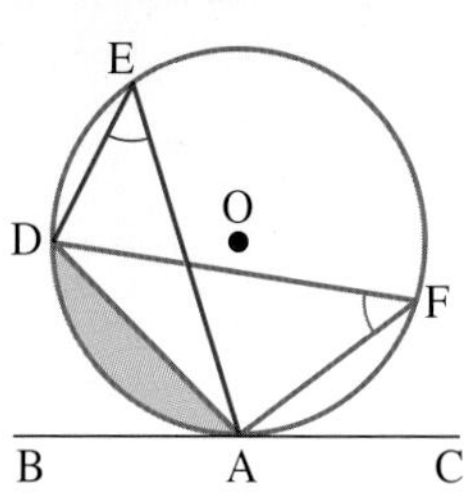

- **Theorem 13 Code**
  The angle between a tangent and a chord is equal to the angle in the alternate segment.

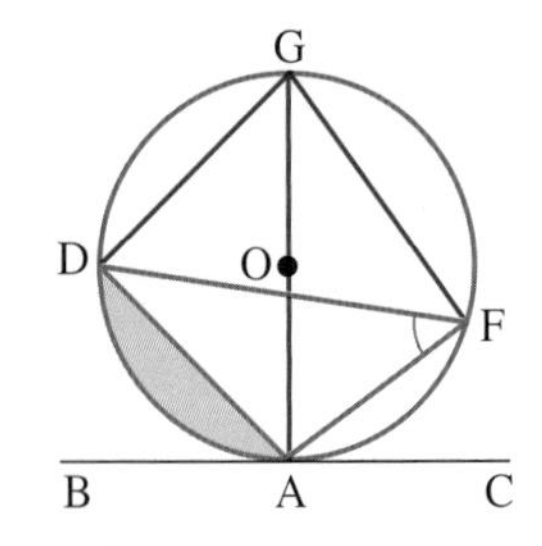

Proof:

We are required to prove that $\angle BAD = \angle AFD$.

Construct the diameter from A through O, meeting the circle at G.

Join G to the points D and F.

$\angle BAG = \angle CAG = 90°$ (radii $\perp$tangent at point of contact)

$\angle GFA = 90°$ (angle in a semicircle is 90°)

$\angle GDA = 90°$ (angle in a semicircle is 90°)

Consider ΔGDA. We know that $\angle GDA = 90°$.

$\angle GDA + \angle DAG + \angle AGD = 180°$

$90° + \angle DAG + \angle AGD = 180°$

$\angle DAG + \angle AGD = 90°$

$\angle BAG$ is also a right angle.

$\angle BAG = \angle BAD + \angle DAG = 90°$

Equate the two results.

$\angle DAG + \angle AGD = \angle BAD + \angle DAG$

Cancel the equal angles ($\angle DAG$) on both sides.

$\angle AGD = \angle BAD$

Now consider the fact that both triangles DAG and DAF are subtended from the same chord (DA).

$\angle AGD = \angle AFD$ (Angles in the same segment standing on the same arc are equal).

Equate the two equations.

$\angle AFD = \angle BAD$

WORKED EXAMPLE 10 Determining pronumerals using the alternate segment theorem

Determine the value of the pronumerals x and y, giving reasons.

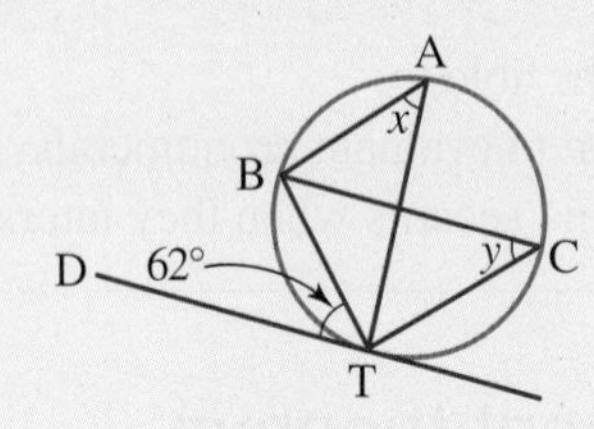

THINK	WRITE
1. Use the alternate segment theorem to calculate x.	$x = 62°$ (angle between a tangent and a chord is equal to the angle in the alternate segment.)
2. The value of y is the same as x because x and y are subtended by the same chord BT.	$y = 62°$ (angles in the same segment standing on the same arc are equal.)

16.5.2 Tangents and secants

eles-5003

- **Theorem 14 Code**

  If a tangent and a secant intersect as shown, the following relationship is always true:
  $XA \times XB = (XT)^2$ or $a \times b = c^2$.

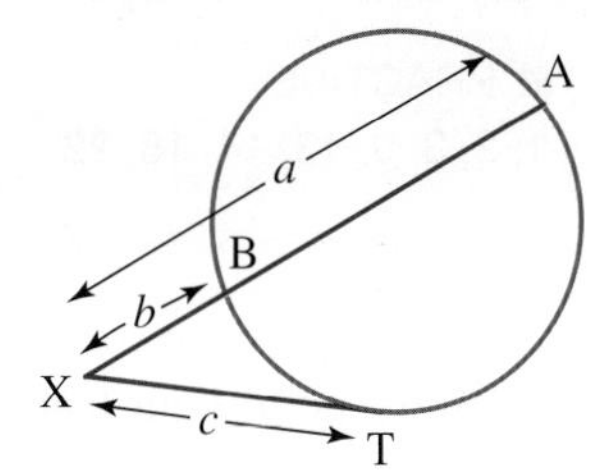

Proof:
Join BT and AT.
Consider ΔTXB and ΔAXT.
∠TXB is common.
∠XTB = ∠XAT (angle between a tangent and a chord is equal to the angle in the alternate segment)
∠XBT = ∠XTA (angle sum of a triangle is 180°)
ΔTXB~ΔAXT (equiangular)
So, $\frac{XB}{XT} = \frac{XT}{XA}$
or, $XA \times XB = (XT)^2$.

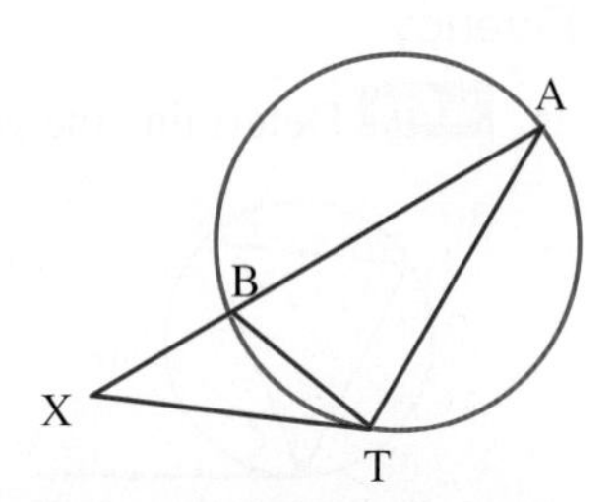

WORKED EXAMPLE 11 Determining pronumerals with intersecting tangents and secants

Determine the value of the pronumeral *m*.

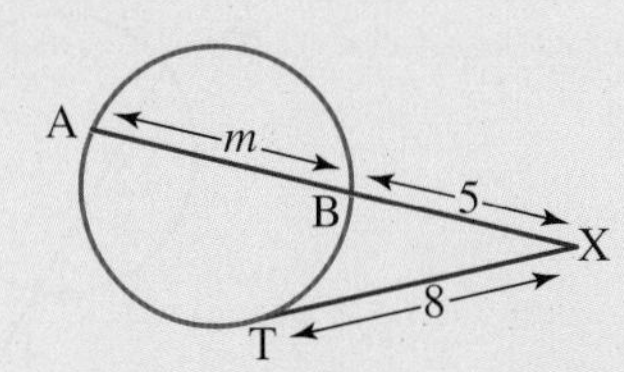

THINK	WRITE
1. Secant XA and tangent XT intersect at X. Write the rule connecting the lengths of XA, XB and XT.	$XA \times XB = (XT)^2$
2. State the values of XA, XB and XT.	$XA = m + 5,\ XB = 5,\ XT = 8$
3. Substitute the values of XA, XB and XT into the equation and solve for *m*.	$(m+5) \times 5 = 8^2$ $5m + 25 = 64$ $5m = 39$ $m = 7.8$

DISCUSSION

Describe the alternate segment of a circle.

on Resources

eWorkbook Topic 16 Workbook (worksheets, code puzzle and project) (ewbk-2042)

Interactivities Circle theorem 13 (int-6230)
Circle theorem 14 (int-6231)

Exercise 16.5 Tangents, secants and chords

learnON

Individual pathways

■ PRACTISE	■ CONSOLIDATE	■ MASTER
1, 2, 3, 9, 13, 14, 18, 22	4, 5, 6, 10, 15, 16, 17, 19, 23	7, 8, 11, 12, 20, 21, 24

To answer questions online and to receive **immediate corrective feedback** and **fully worked solutions** for all questions, go to your learnON title at www.jacplus.com.au.

Fluency

1. **WE10** Determine the value of the pronumerals in the following diagrams.

a.

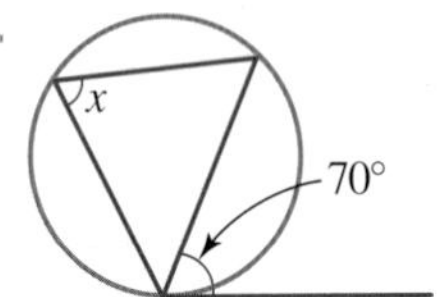

b.

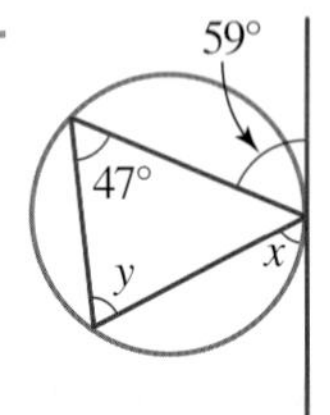

2. **WE11** Calculate the value of the pronumerals in the following diagrams.

a.

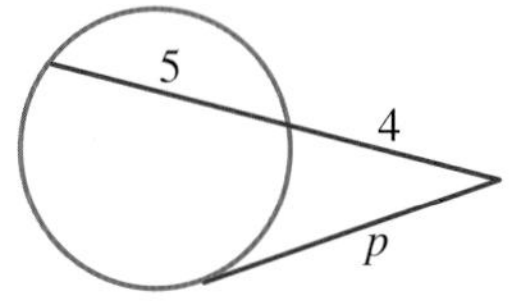

b.

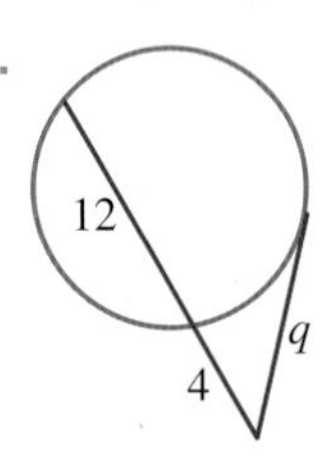

3. Line AB is a tangent to the circle as shown in the figure. Calculate the values of the angles labelled x and y.

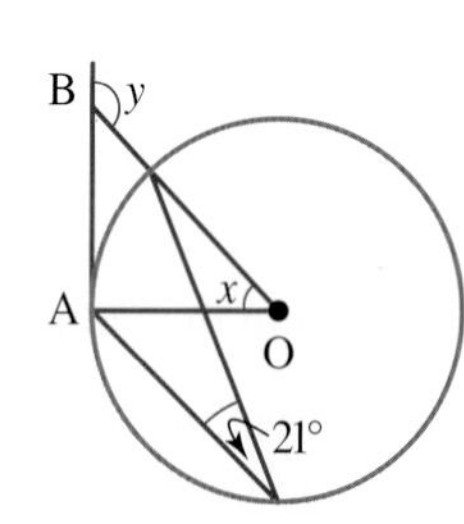

Questions 4 to 6 refer to the figure shown. The line MN is a tangent to the circle, and EA is a straight line. The circles have the same radius.

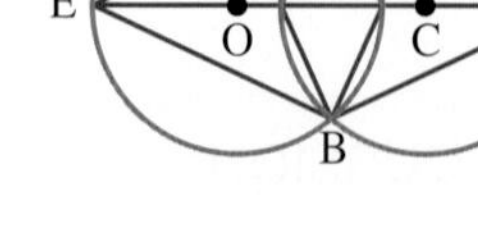

4. **MC** If $\angle DAC = 20°$, then $\angle CFD$ and $\angle FDG$ are respectively:

A. 70° and 50°
B. 70° and 40°
C. 40° and 70°
D. 70° and 70°
E. 40° and 50°

5. **MC** A triangle similar to FDA is:

A. FDG B. FGB C. EDA D. GDE E. ABD

6. State six different right angles.

7. Calculate the values of the angles x and y in the figure shown.

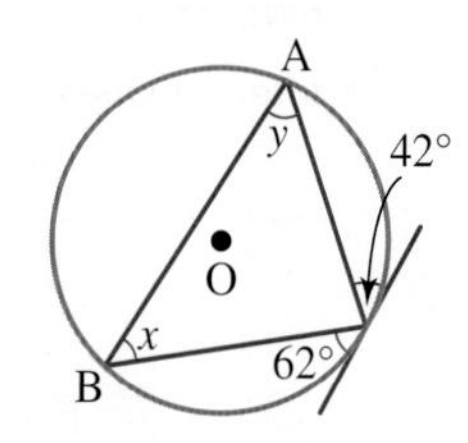

Understanding

8. Show that if the sum of the two given angles in question **7** is 90°, then the line AB must be a diameter.

9. Calculate the value of x in the figure shown, given that the line underneath the circle is a tangent.

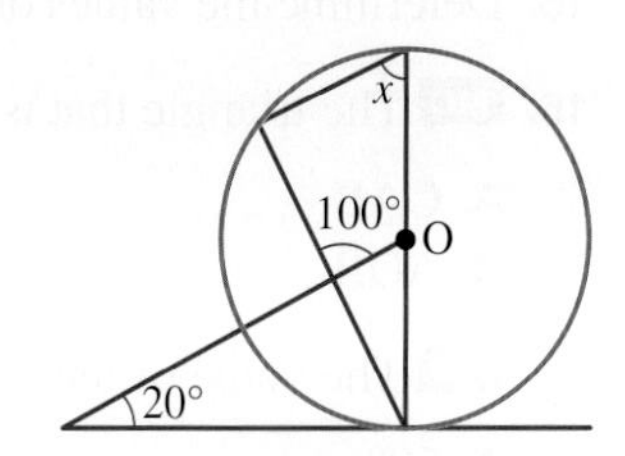

10. In the figure shown, express x in terms of a and b. This is the same diagram as in question 9.

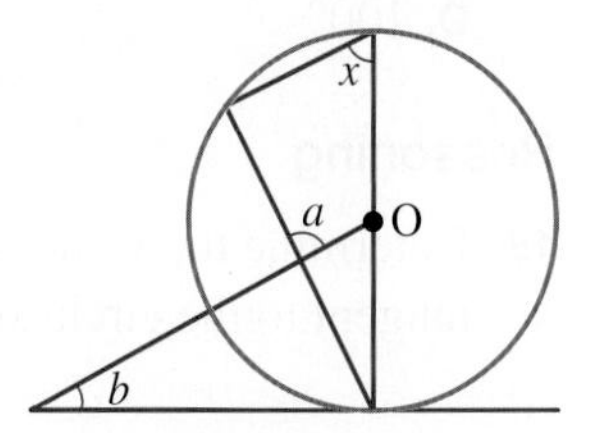

11. Two tangent lines to a circle meet at an angle y, as shown in the figure. Determine the values of the angles x, y and z.

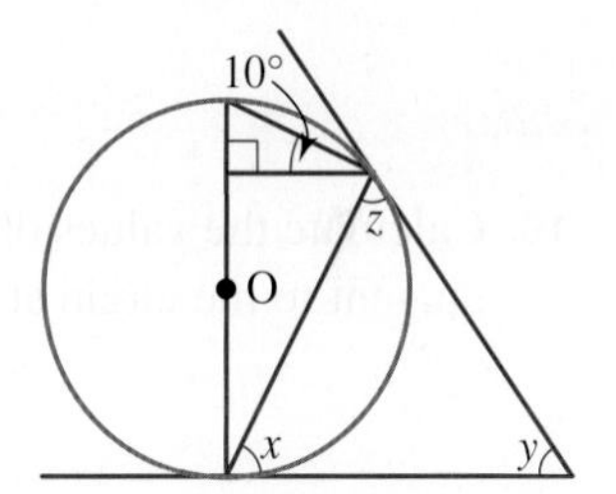

12. Solve question **11** in the general case (see the figure) and show that $y = 2a$. This result is important for space navigation (imagine the circle to be the Earth) in that an object at y can be seen by people at x and z at the same time.

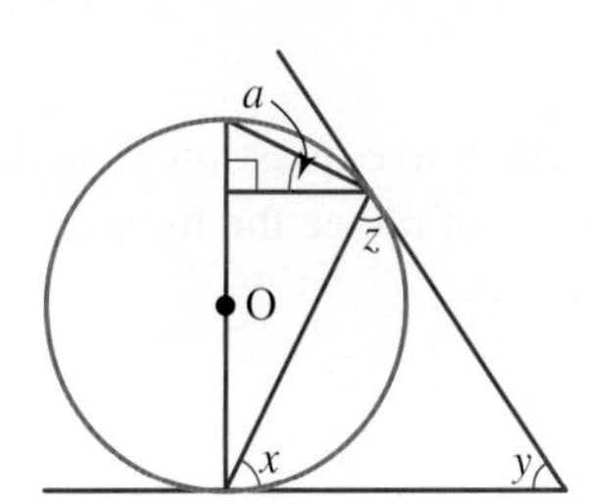

13. In the figure shown, determine the values of the angles x, y and z.

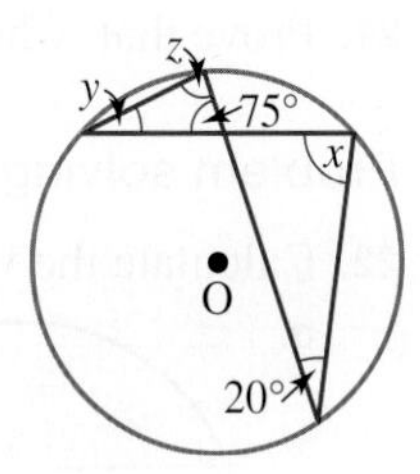

14. **MC** Examine the figure shown. The angles x and y (in degrees) are respectively:

A. 51 and 99
B. 51 and 129
C. 39 and 122
D. 51 and 122
E. 39 and 99

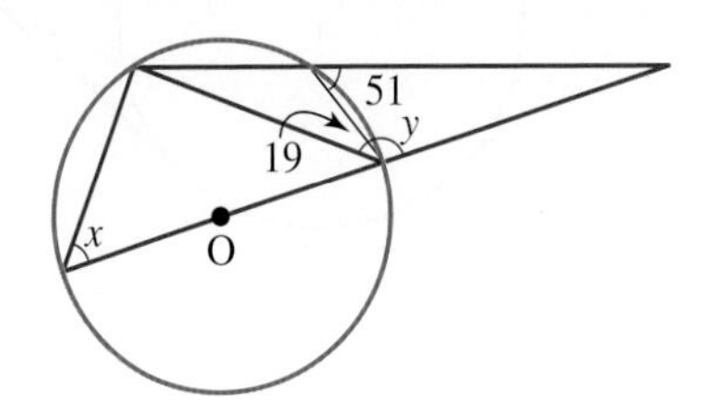

Questions **15** to **17** refer to the figure shown. The line BA is a tangent to the circle at point B. Line AC is a chord that meets the tangent at A.

15. Determine the values of the angles x and y.

16. **MC** The triangle that is similar to triangle BAD is:

A. CAB **B.** BCD **C.** BDC
D. AOB **E.** DOC

17. **MC** The value of the angle z is:

A. 50° **B.** 85° **C.** 95°
D. 100° **E.** 75°

Reasoning

18. Determine the values of the angles x, y and z in the figure shown. The line AB is tangent to the circle at B.

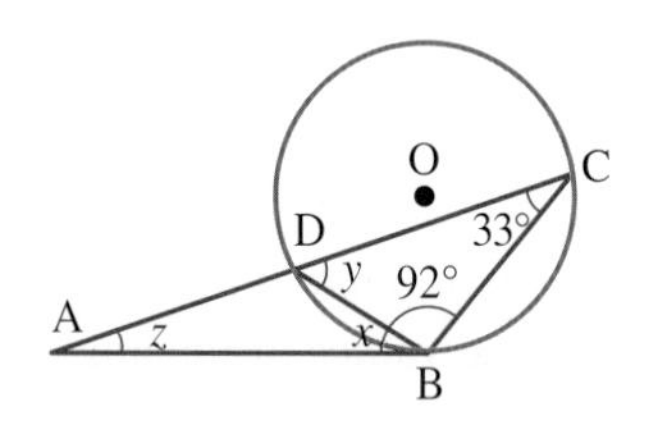

19. Calculate the values of the angles x, y and z in the figure shown. The line AB is tangent to the circle at B. The line CD is a diameter.

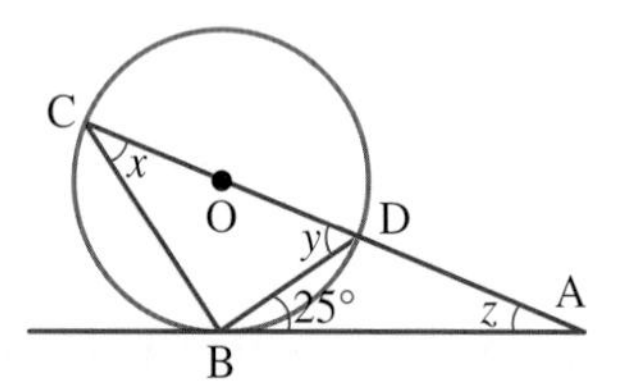

20. Solve question **19** in the general case; that is, express angles x, y and z in terms of a (see the figure).

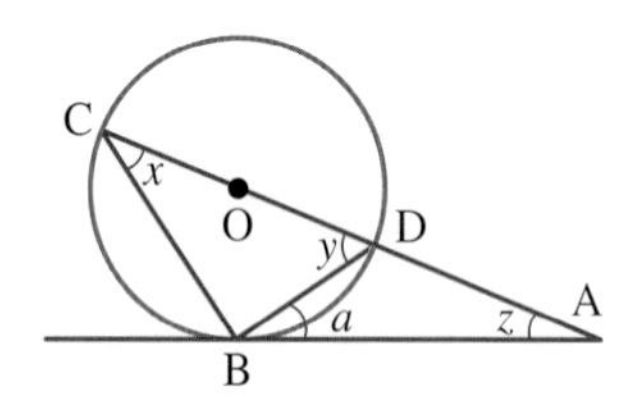

21. Prove that, when two circles touch, their centres and the point of contact are collinear.

Problem solving

22. Calculate the value of the pronumerals in the following.

a.

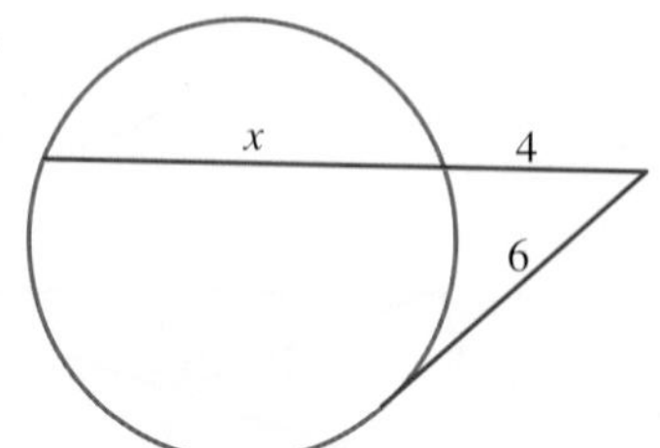

b.

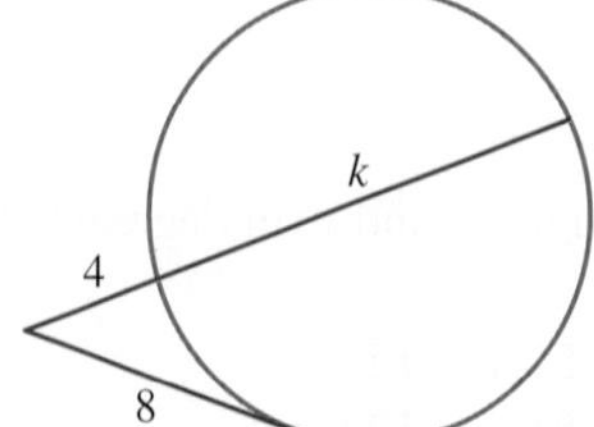

c.

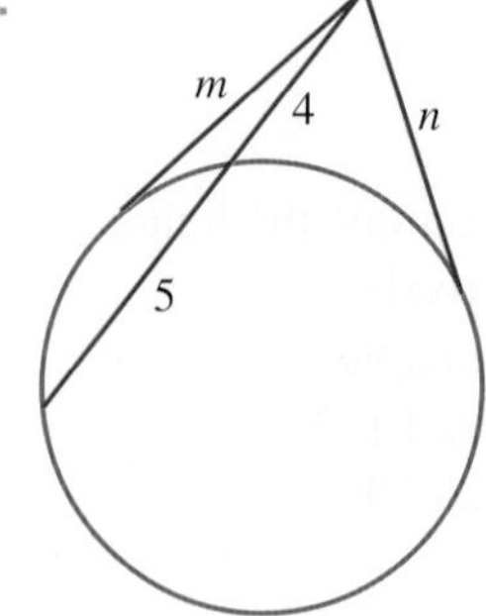

23. Determine the value of the pronumerals in the following.

a.

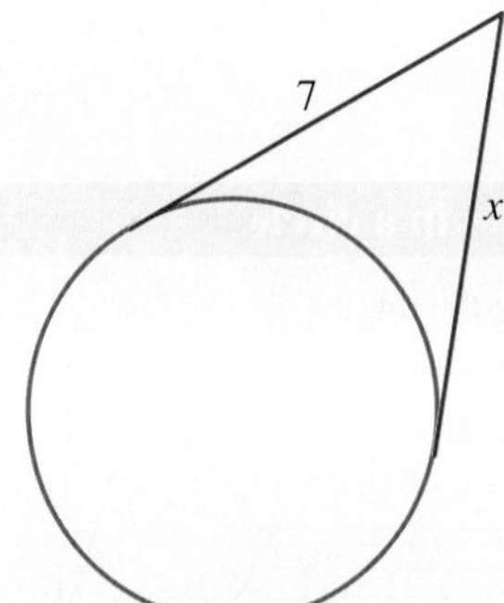

b.

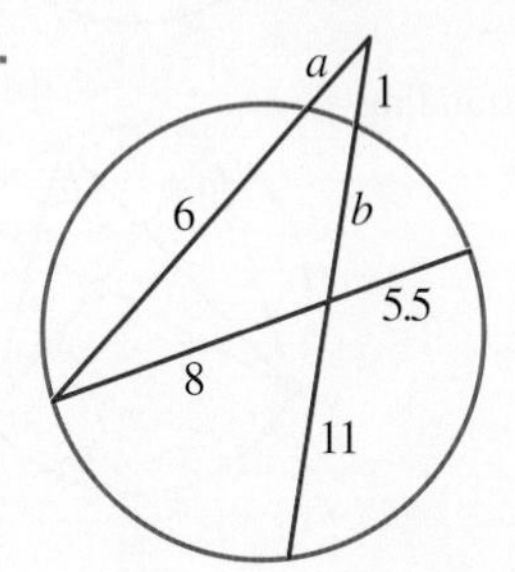

c.

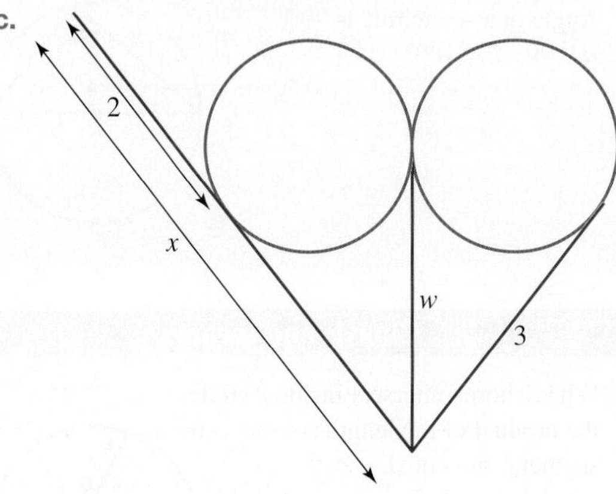

24. Calculate the values of a, b and c in each case.

a. $\angle BCE = 50°$ and $\angle ACE = c$

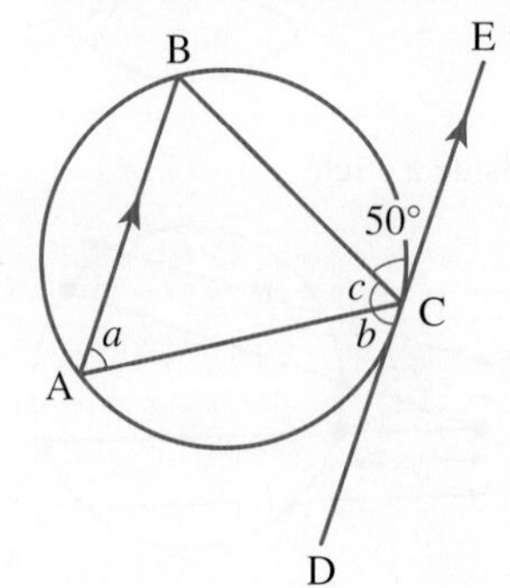

b.

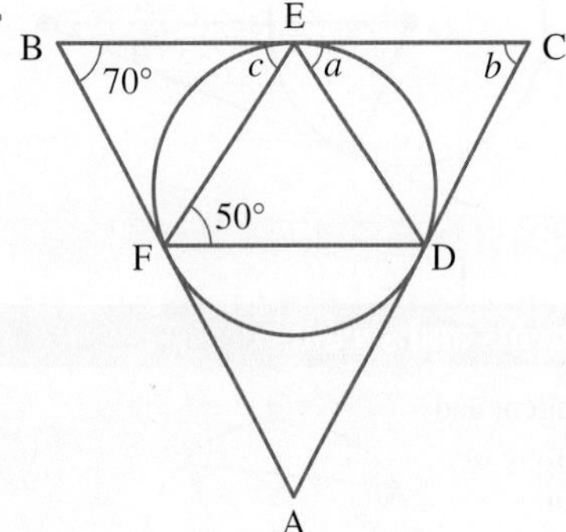

16.6 Review

16.6.1 Topic summary

CIRCLE GEOMETRY

Cyclic quadrilaterals

- **Opposite angles** of cyclic quadrilateral are supplementary.
  e.g. $a + b = 180^\circ$
- **Exterior angle** of a cyclic quadrilateral equals the interior opposite angle.

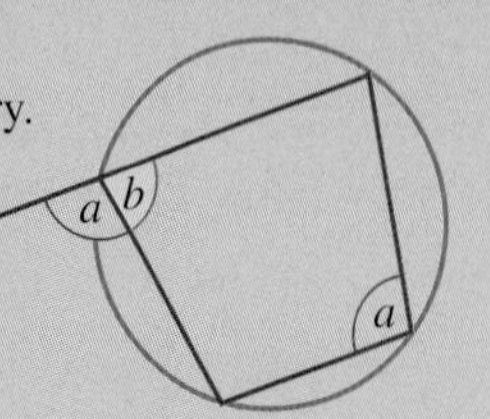

Angles in a circle

- Angle at centre is twice the angles at circumference standing on the same arc.
  e.g. $b = 2a$

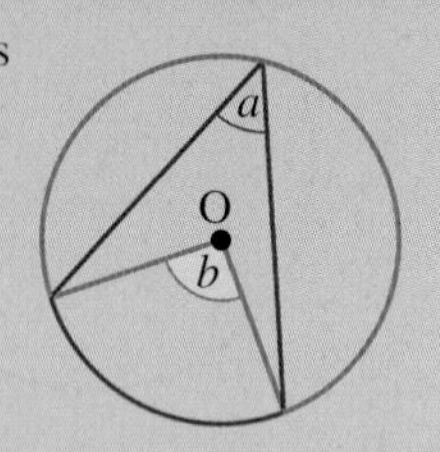

- Angles at circumference standing on the same arc are equal.
  $a = b = c$

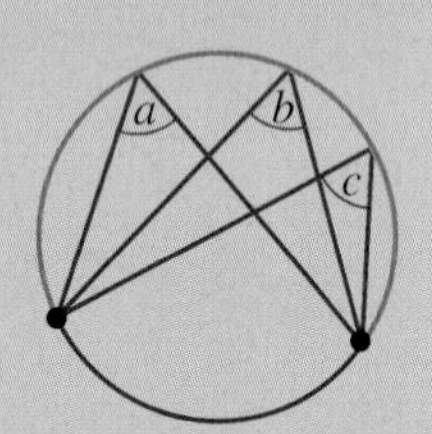

- Angle in a semicircle is 90°.
  $a = 90^\circ$, $b = 180^\circ$

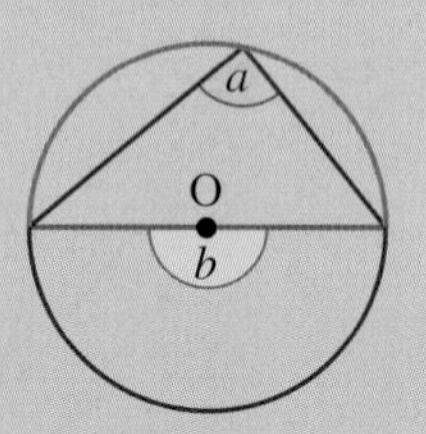

Tangents to a circle

- Tangent is perpendicular to the radius at the point of contact.
- Lengths of tangents from an external point are equal.
  $a = b$

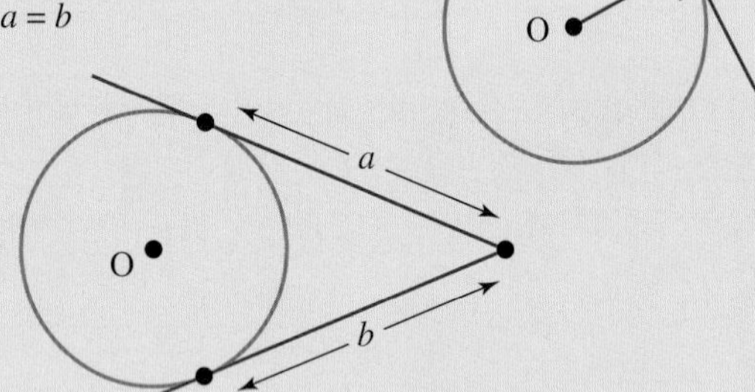

- Line from the centre bisects the external angle between two tangents.

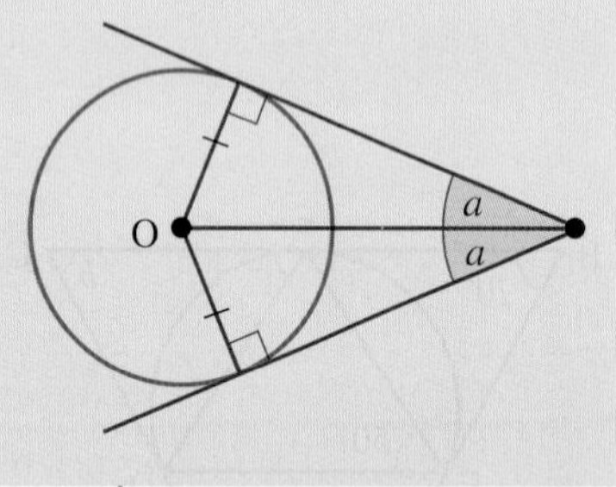

Chords and secants

- When chords intersect inside a circle, the product of the lengths of the segments are equal.
  $a \times b = c \times d$

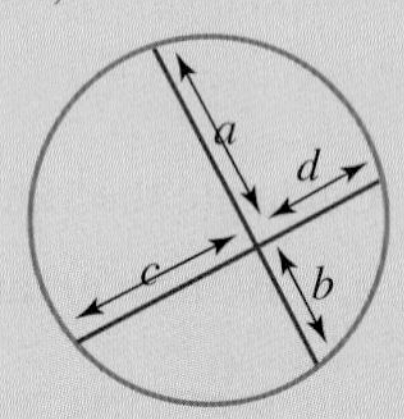

- When chords intersect outside a circle, then the following is true:
  $a \times b = c \times d$

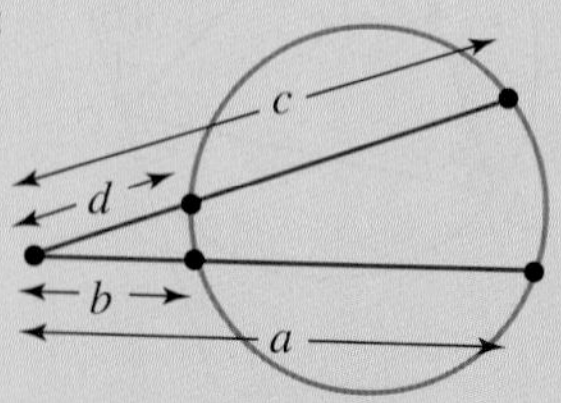

- Perpendicular bisector of a chord passes through the centre of the circle.

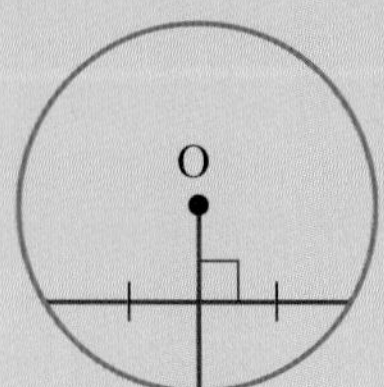

Tangents and secants

- Angle between a tangent and a chord equals the angle in the alternate segment.
- If tangent and segment intersect then: $a \times b = c^2$

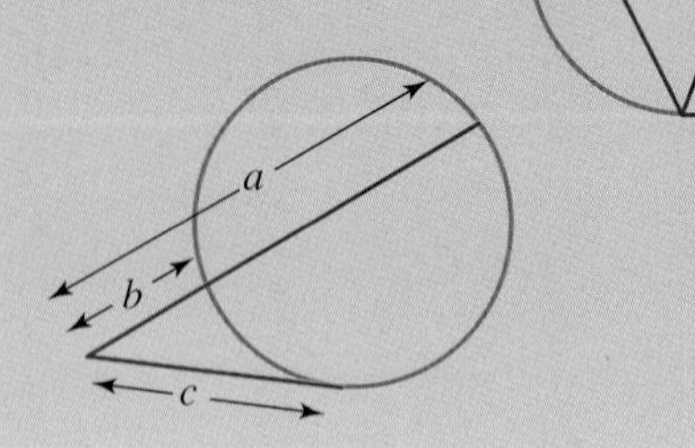

16.6.2 Success criteria

Tick the column to indicate that you have completed the subtopic and how well you have understood it using the traffic light system.

(**Green:** I understand; **Yellow:** I can do it with help; **Red:** I do not understand)

Subtopic	Success criteria	Green	Yellow	Red
16.2	I can determine relationships between the angles at the centre and the circumference of a circle.			
	I can understand and use the angle in a semicircle being a right angle.			
	I can apply the relationships between tangents and radii or circles.			
16.3	I can apply the intersecting chords, secants and tangent theorems.			
	I can recognise that a radius will bisect a chord at right angles.			
	I can understand the concept of the circumcentre of a triangle.			
16.4	I can recognise a cyclic quadrilateral.			
	I can determine pronumerals knowing the opposite angles of a cyclic quadrilateral are supplementary.			
	I can determine pronumerals knowing the exterior angle of a cyclic quadrilateral equals the interior opposite angle.			
16.5	I can apply the alternate segment theorem to evaluate pronumerals.			
	I can determine the lengths of tangents and secants when they intersect.			

16.6.3 Project

Variation of distance

The Earth approximates the shape of a sphere. Lines of longitude travel between the North and South poles, while lines of latitude travel east–west, parallel to the equator. While the lines of longitude are all approximately the same length, this is not the case with lines of latitude. The line of latitude at the equator is the maximum length and these lines decrease in length on approaching both the North and South poles.

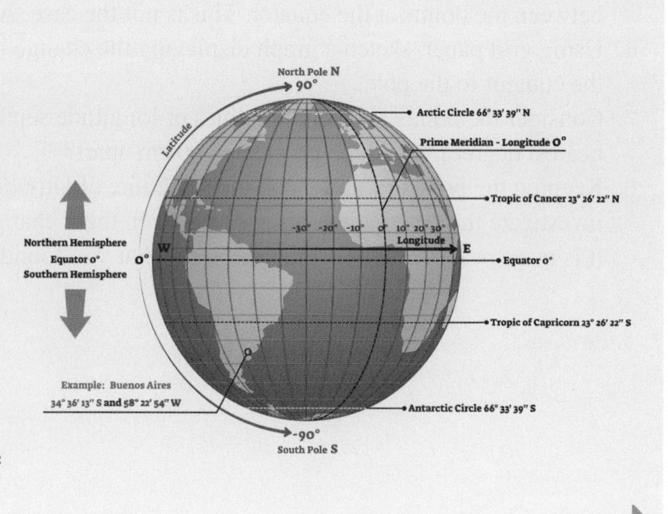

This investigation looks at how the distance between points on two given lines of longitude and the same line of latitude changes as we move from the equator to the pole.

Consider two lines of longitude, 0° and 100°E. Take two points, P_1 and P_2, lying on the equator on lines of longitude 0° and 100°E respectively.

The distance (in km) between two points on the same line of latitude is given by the formula:

$$\text{Distance} = \text{angle sector between the two points} \times 111 \times \cos(\text{degree of latitude})$$

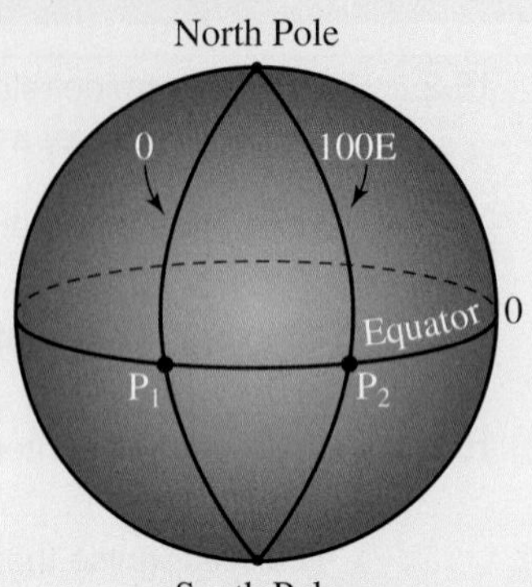

1. The size of the angle sector between P_1 and P_2 is 100° and these two points lie on 0° latitude. The distance between the points would be calculated as $100 \times 111 \times \cos(0°)$. Determine this distance.
2. Move the two points to the 10° line of latitude. Calculate the distance between P_1 and P_2 in this position. Round your answer to the nearest kilometre.
3. Complete the following table showing the distance (rounded to the nearest kilometre) between the points P_1 and P_2 as they move from the equator towards the pole.

Latitude	Distance between P_1 and P_2 (km)
0°	
10°	
20°	
30°	
40°	
50°	
60°	
70°	
80°	
90°	

4. Describe what happens to the distance between P_1 and P_2 as we move from the equator to the pole. Is there a constant change? Explain your answer.
5. You would perhaps assume that, at a latitude of 45°, the distance between P_1 and P_2 is half the distance between the points at the equator. This is not the case. At what latitude does this occur?
6. Using grid paper, sketch a graph displaying the change in distance between the points in moving from the equator to the pole.
7. Consider the points P_1 and P_2 on lines of longitude separated by 1°. On what line of latitude (to the nearest degree) would the points be 100 km apart?
8. Keeping the points P_1 and P_2 on the same line of latitude, and varying their lines of longitude, investigate the rate that the distance between them changes from the equator to the pole. Explain whether it is more or less rapid in comparison to what you found earlier.

Resources

eWorkbook Topic 16 Workbook (worksheets, code puzzle and project) (ewbk-2042)

Interactivities Crossword (int-2881)

Sudoku puzzle (int-3894)

Exercise 16.6 Review questions

learnON

To answer questions online and to receive **immediate corrective feedback** and **fully worked solutions** for all questions, go to your learnON title at www.jacplus.com.au.

Fluency

For questions **1** to **3**, determine the values of the pronumerals in each of the diagrams.

1. a.

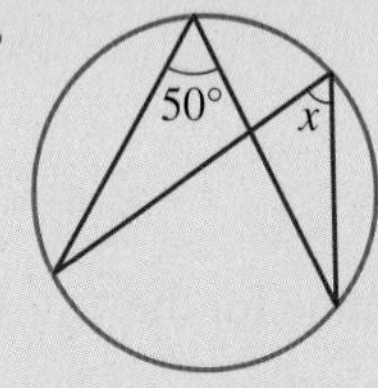

b.

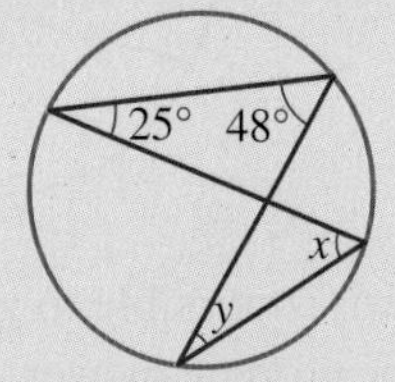

c.

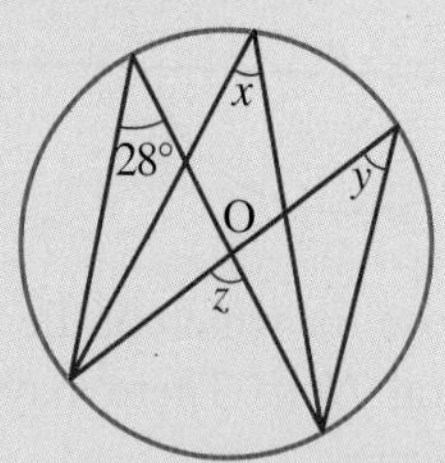

d.

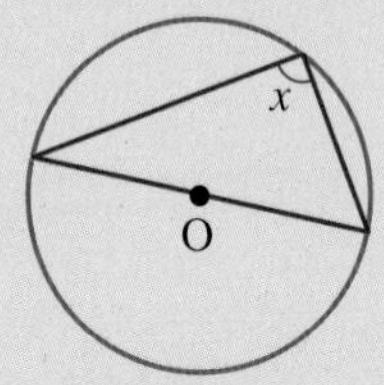

e.

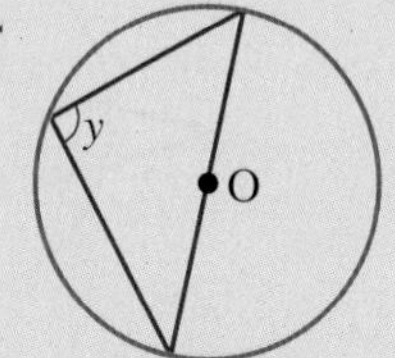

f.

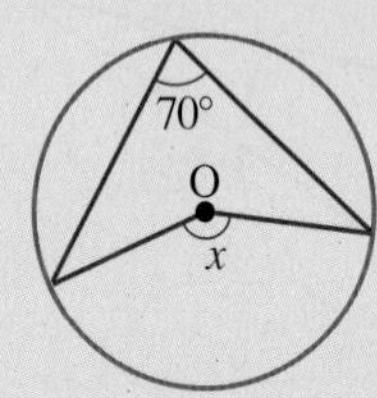

2. a.

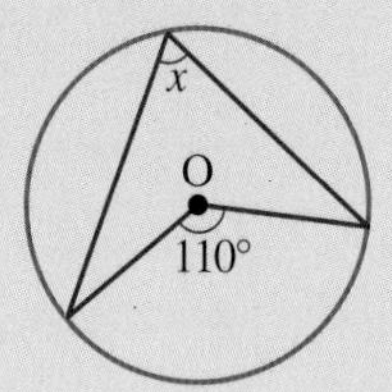

b.

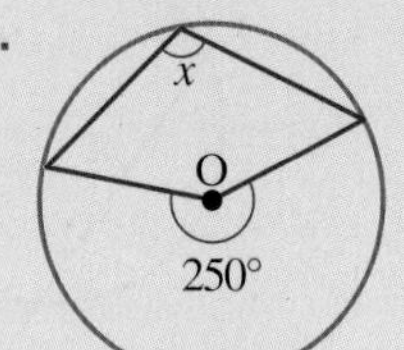

c.

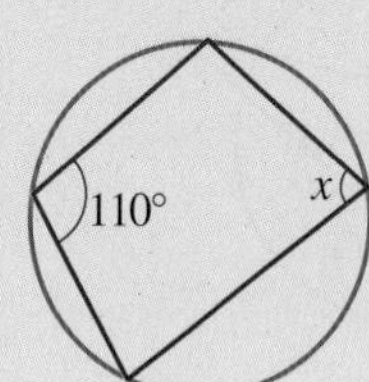

d.

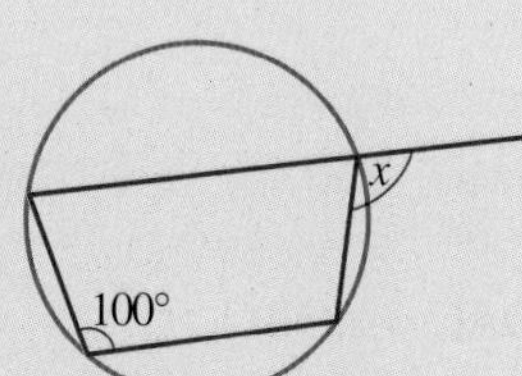

e.

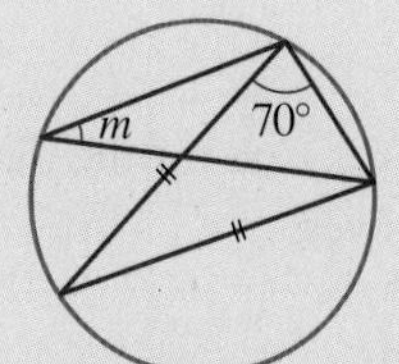

f.

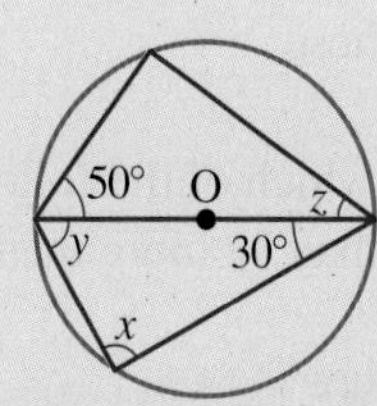

3. a.

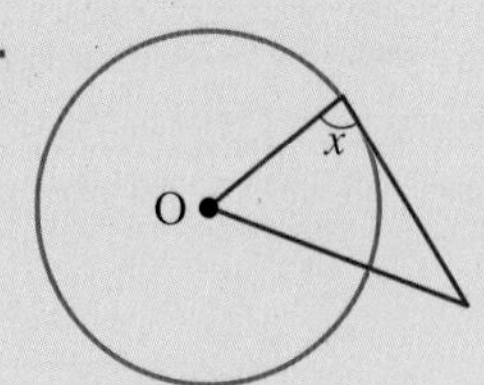

b.

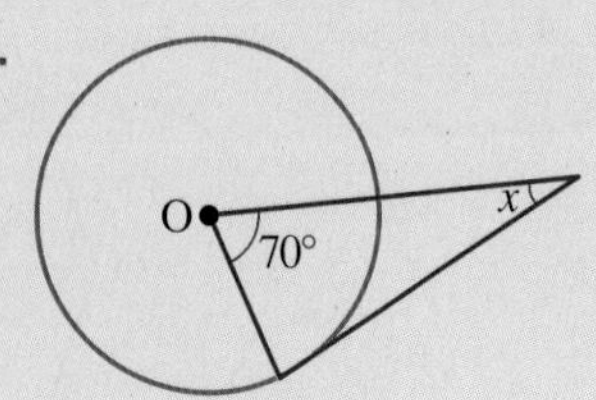

c.

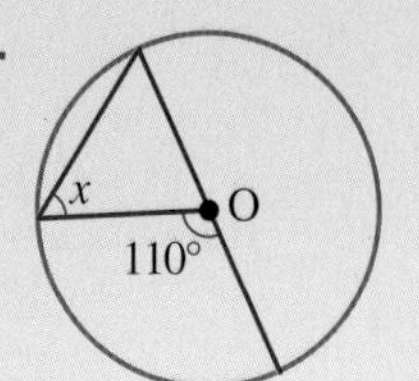

d.

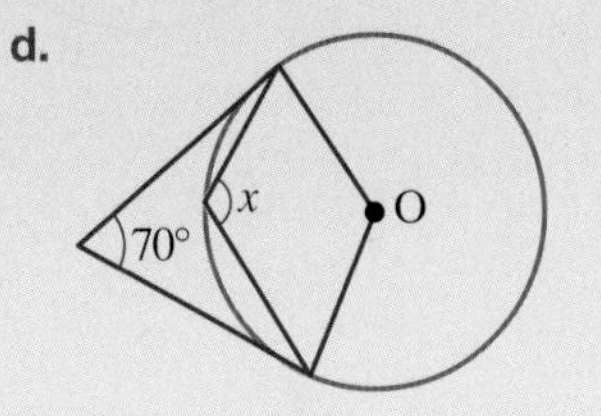

4. Determine the value of m in each of the following.

a.

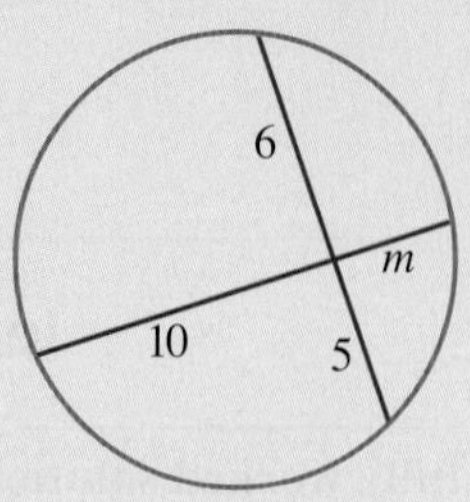

b.

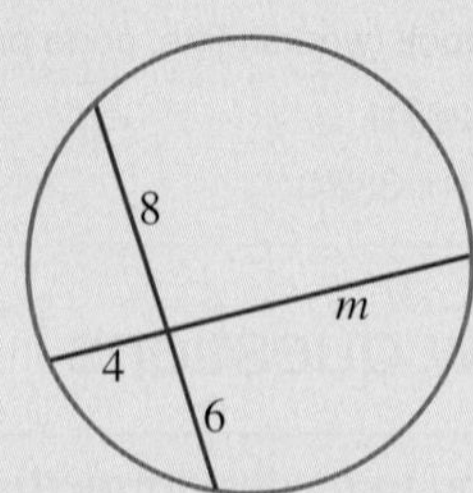

c.

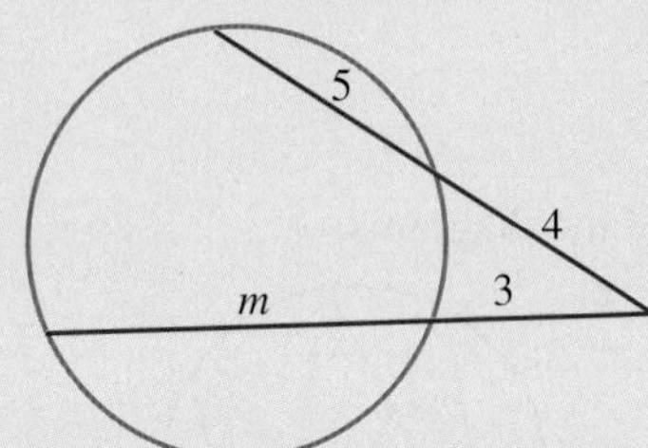

d.

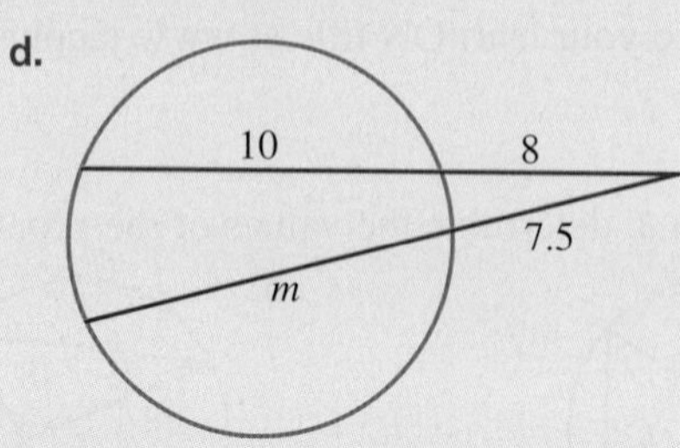

5. **MC** Choose for which of the following figures it is possible to get a reasonable value for the pronumeral. *Note*: There may be more than one correct answer.

A.

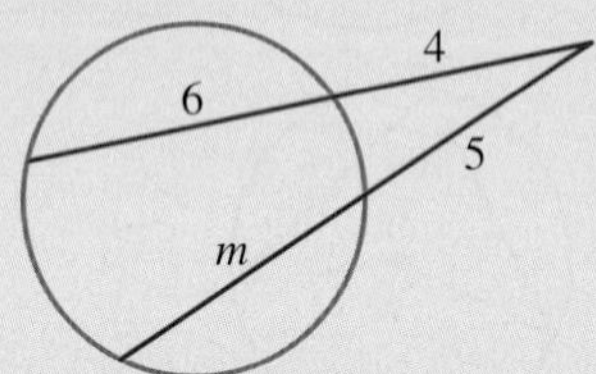

B.

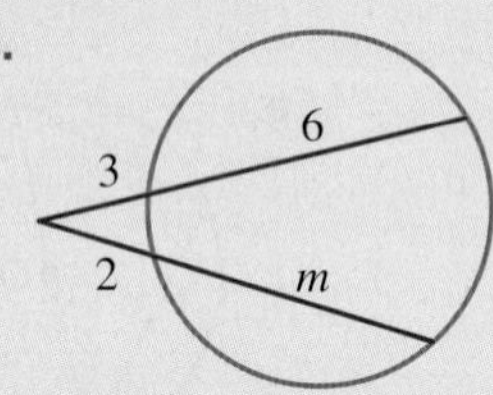

C.

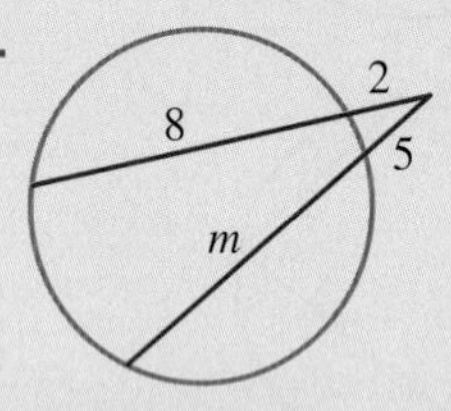

D.

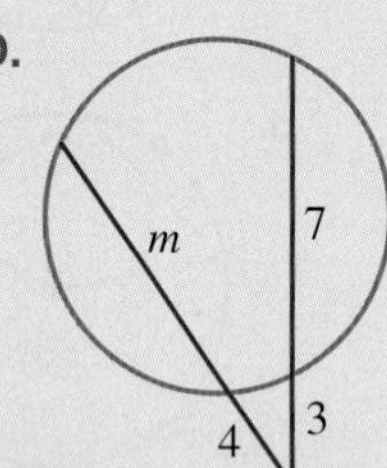

E. None of these

6. **MC** Choose which of the following statements is true for the diagram shown.
Note: There may be more than one correct answer.

A. AO = BO
B. AC = BC
C. ∠OAC = ∠OBC
D. ∠AOC = 90°
E. AC = OC

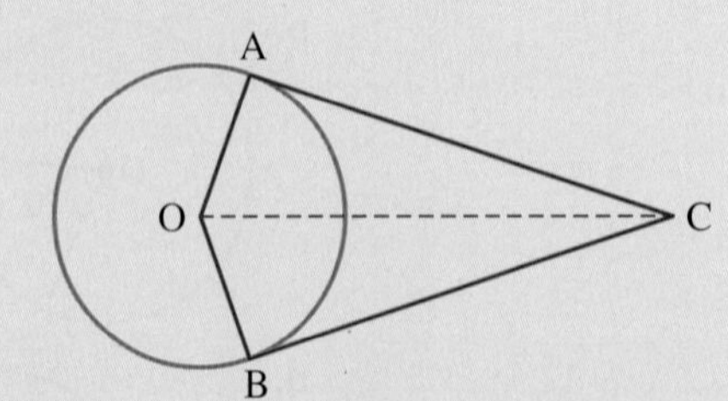

7. Determine the values of the pronumerals in the following figures.

a.

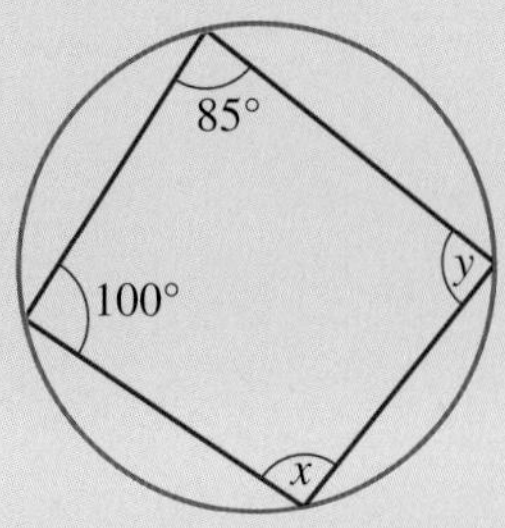

b.

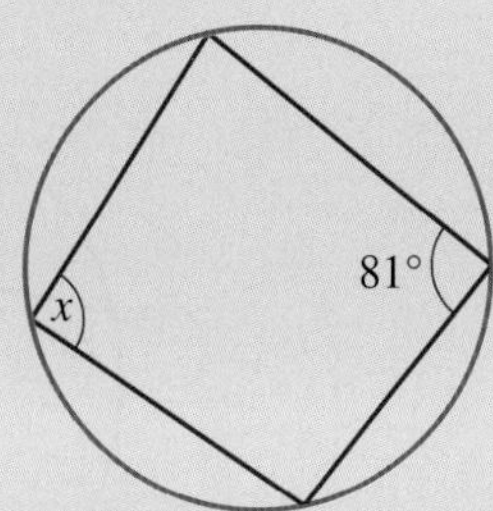

c.

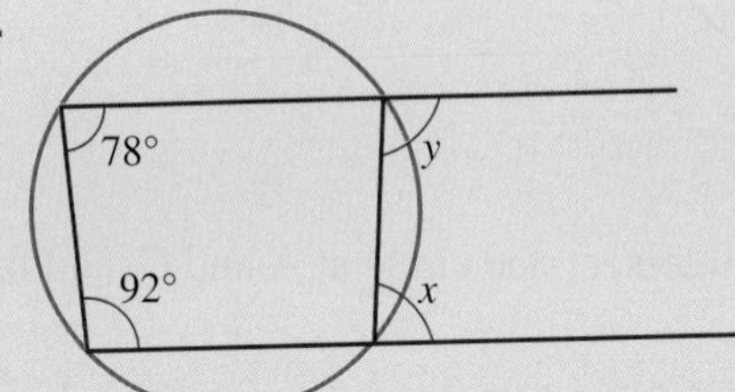

d.

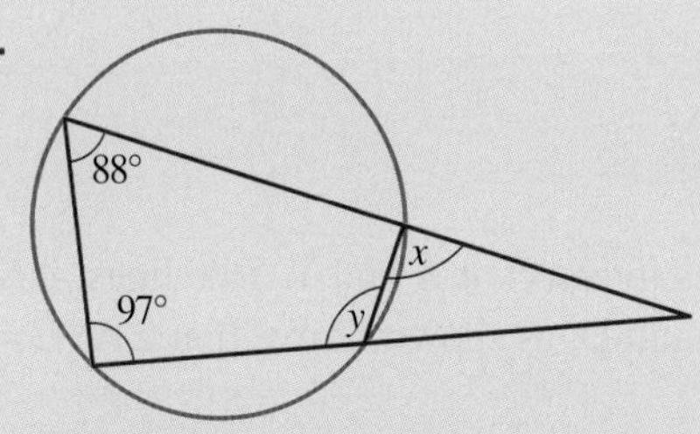

8. **MC** Choose which of the following statements is *not* always true for the diagram shown.

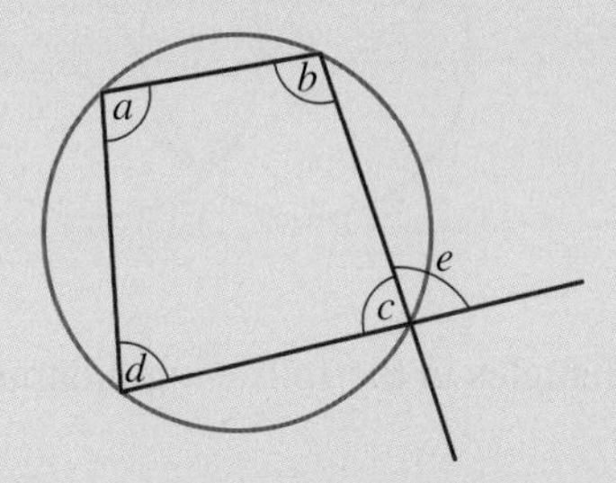

A. $\angle a + \angle c = 180°$

B. $\angle b + \angle d = 180°$

C. $\angle e + \angle c = 180°$

D. $\angle a + \angle e = 180°$

E. $\angle a + \angle b + \angle c + \angle d = 360°$

Problem solving

9. Determine the values of the pronumerals in the following figures.

a.

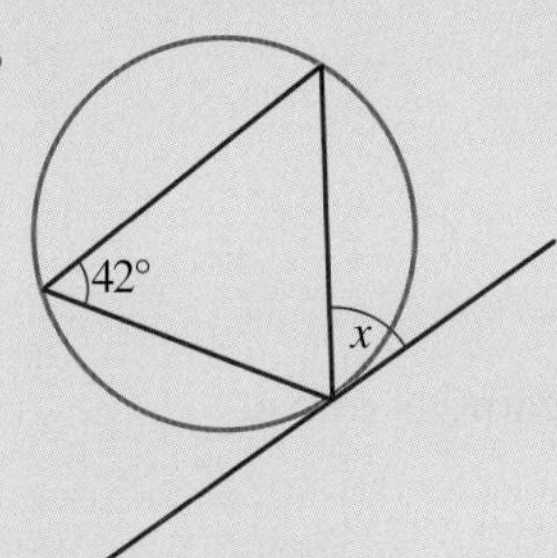

b.

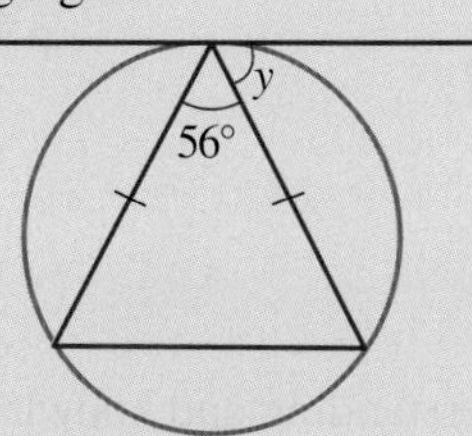

c.

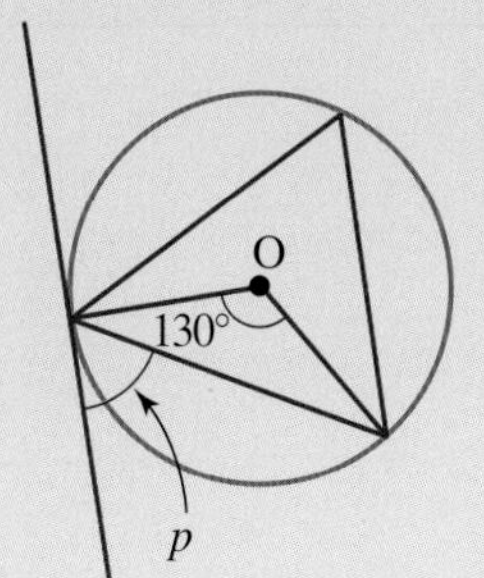

10. Two chords, AB and CD, intersect at E as shown. If AE = CE, prove that EB = ED.

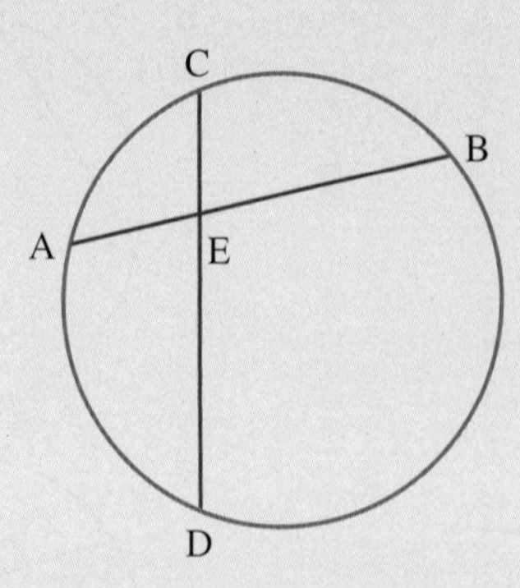

11. Two circles intersect at X and Y. Two lines, AXB and CXD, intersect one circle at A and C, and the other at B and D, as shown. Prove that ∠AYC = ∠BYD.

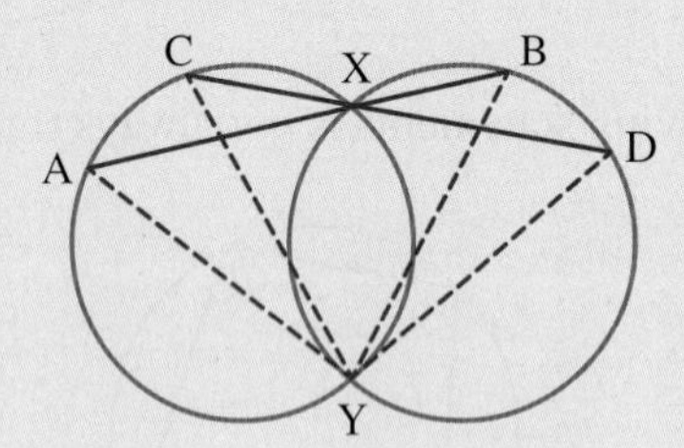

12. Name at least five pairs of equal angles in the following diagram.

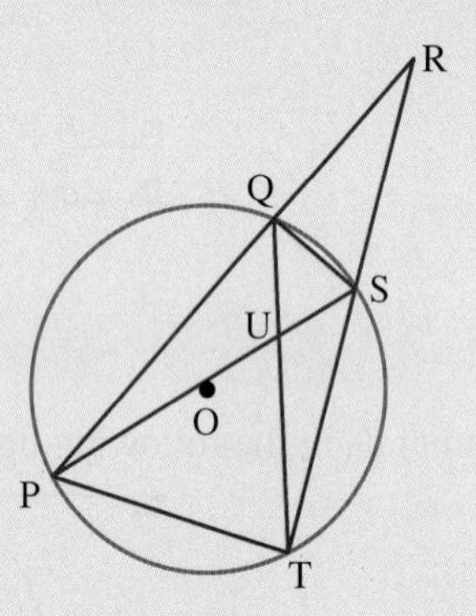

on To test your understanding and knowledge of this topic, go to your learnON title at www.jacplus.com.au and complete the **post-test**.

Online Resources

Below is a full list of **rich resources** available online for this topic. These resources are designed to bring ideas to life, to promote deep and lasting learning and to support the different learning needs of each individual.

eWorkbook

Download the workbook for this topic, which includes worksheets, a code puzzle and a project (ewbk-2042) ☐

Solutions

Download a copy of the fully worked solutions to every question in this topic (sol-0750) ☐

Digital documents

16.2 SkillSHEET Using tests to prove congruent triangles (doc-5390) ☐
SkillSHEET Corresponding sides and angles of congruent triangles (doc-5391) ☐
SkillSHEET Using tests to prove similar triangles (doc-5392) ☐
SkillSHEET Angles in a triangle (doc-5393) ☐
SkillSHEET More angle relations (doc-5394) ☐
16.4 SkillSHEET Angles in a quadrilateral (doc-5396) ☐

Video eLessons

16.2 Circle (eles-4995) ☐
Tangents to a circle (eles-4996) ☐
16.3 Intersecting chords (eles-4997) ☐
Intersecting secants (eles-4998) ☐
Intersecting tangents (eles-4999) ☐
Chords and radii (eles-5000) ☐
16.4 Quadrilaterals in circles (eles-5001) ☐
16.5 The alternate segment theorem (eles-5002) ☐
Tangents and secants (eles-5003) ☐

Interactivities

16.2 Circle theorem 1 (int-6218) ☐
Circle theorem 2 (int-6219) ☐
Circle theorem 3 (int-6220) ☐
Circle theorem 4 (int-6221) ☐
Circle theorem 5 (int-6222) ☐
16.3 Circle theorem 6 (int-6223) ☐
Circle theorem 7 (int-6224) ☐
Circle theorem 8 (int-6225) ☐
Circle theorem 9 (int-6226) ☐
Circle theorem 10 (int-6227) ☐
16.4 Circle theorem 11 (int-6228) ☐
Circle theorem 12 (int-6229) ☐
16.5 Circle theorem 13 (int-6230) ☐
Circle theorem 14 (int-6231) ☐
16.6 Crossword (int-2881) ☐
Sudoku puzzle (int-3894) ☐

Teacher resources

There are many resources available exclusively for teachers online.

To access these online resources, log on to **www.jacplus.com.au**.

Answers

Topic 16 Circle geometry

Exercise 16.1 Pre-test

1. Segment
2. D
3. 35°
4. E
5. 65°
6. 15°
7. E
8. 6
9. 1.5
10. A
11. $m = 55°$, $p = 50°$
12. C
13. E
14. D
15. A

Exercise 16.2 Angles in a circle

1. a. $x = 30°$ (theorem 2)
   b. $x = 25°$, $y = 25°$ (theorem 2 for both angles)
   c. $x = 32°$ (theorem 2)
2. a. $x = 40°$, $y = 40°$ (theorem 2 for both angles)
   b. $x = 60°$ (theorem 1)
   c. $x = 40°$ (theorem 1)
3. a. $x = 84°$(theorem 1)
   b. $x = 50°$(theorem 2); $y = 100°$ (theorem 1)
   c. $x = 56°$(theorem 1)
4. a. $s = 90°$, $r = 90°$ (theorem 3 for both angles)
   b. $u = 90°$ (theorem 4); $t = 90°$ (theorem 3)
   c. $m = 90°$, $n = 90°$(theorem 3 for both angles)
5. a. $x = 52°$ (theorem 3 and angle sum in a triangle = 180°)
   b. $x = 90°$(theorem 4)
   c. $x = 90°$ (theorem 4); $y = 15°$ (angle sum in a triangle = 180°)
6. a. $x = z = 90°$ (theorem 4); $y = w = 20°$ (theorem 5 and angle sum in a triangle = 180°)
   b. $s = r = 90°$ (theorem 4); $t = 140°$ (angle sum in a quadrilateral = 360°)
   c. $x = 20°$ (theorem 5); $y = z = 70°$ (theorem 4 and angle sum in a triangle = 180°)
7. a. $s = y = 90°$(theorem 4); $x = 70°$ (theorem 5); $r = z = 20°$ (angle sum in a triangle = 180°)
   b. $x = 70°$(theorem 4 and angle sum in a triangle = 180°); $y = z = 20°$ (angle sum in a triangle = 180°)
   c. $x = y = 75°$ (theorem 4 and angle sum in a triangle = 180°); $z = 75°$ (theorem 1)
8. B
9. C
10. C
11. D
12. B, D
13. a. Base angles of a right-angled isosceles triangle.
    b. $r + s = 90°$, $s = 45° \Rightarrow r = 45°$
    c. u is the third angle in Δ ABD, which is right-angled.
    d. m is the third angle in Δ OCD, which is right-angled.
    e. $\angle$AOC and $\angle$AFC stand on the same arc with $\angle$AOC at the centre and $\angle$AFC at the circumference.
14. OR = OP (radii of the circle)
    $\angle$OPR $= x$ (equal angles lie opposite equal sides)
    $\angle$SOP $= 2x$ (exterior angle equals the sum of the two interior opposite angles)
    OR = OQ (radii of the circle)
    $\angle$OQR $= y$ (equal angles lie opposite equal sides)
    $\angle$SOQ $= 2y$ (exterior angle equals the sum of the two interior opposite angles)
    Now $\angle$PRQ $= x + y$ and $\angle$POQ $= 2x + 2y = 2(x + y)$.
    Therefore $\angle$POQ $= 2 \times \angle$PRQ.
15, 16. Sample responses can be found in the worked solutions in the online resources.
17. a. 16° b. 20°
18, 19. Sample responses can be found in the worked solutions in the online resources.

Exercise 16.3 Intersecting chords, secants and tangents

1. a. $m = 3$ b. $m = 3$ c. $m = 6$
2. a. $n = 1$ b. $m = 7.6$
3. a. $n = 13$ b. $m = 4$
4. a. $x = 5$ b. $m = 7$ c. $x = 2.5$, $y = 3.1$
5. a. $x = 2.8$ b. $x = 3.3$
6. a. $x = 5.6$ b. $m = 90°$
7. B, C, D
8. ST = 3 cm
9–11. Sample responses can be found in the worked solutions in the online resources.
12. a. $x = 3\sqrt{2}$ b. $x = 6$ c. $x = 3$, $y = 12$
13. a. Line from centre perpendicular to the chord bisects the chord, giving M as the midpoint.
    b, c. Sample responses can be found in the worked solutions in the online resources.
14.

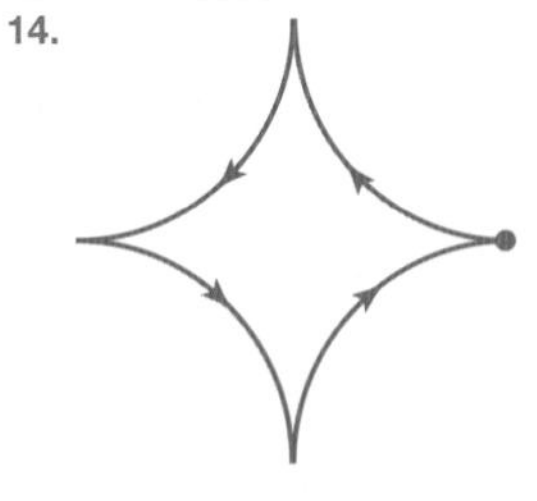

Exercise 16.4 Cyclic quadrilaterals

1. a. $x = 115°$, $y = 88°$ b. $m = 85°$
   c. $n = 25°$
2. a. $x = 130°$ b. $x = y = 90°$ c. $x = 45°$, $y = 95°$
3. a. $x = 85°$, $y = 80°$
   b. $x = 110°$, $y = 115°$
   c. $x = 85°$

4. a. $x = 150°$ b. $x = 90°$, $y = 120°$
   c. $m = 120°$, $n = 130°$
5. a. $a = 89°$, $b = 45°$
   b. $a = 120°$, $b = 91°$, $c = 89°$
6. a. $x = 102°$, $y = 113°$
   b. $x = 95°$, $y = 85°$, $z = 95°$
   c. $x = 126°$, $y = 54°$
   d. $x = 60°$, $y = 120°$
   e. $x = 54°$, $y = 72°$
   f. $x = 79°$, $y = 101°$, $z = 103°$
7. D
8. E
9. D
10. a. $2x$ b. $360° - 2x$
    c. $180° - x$ d. $180°$
11. A
12. A, B, C, D
13. Sample responses can be found in the worked solutions in the online resources.
14. $x = 93°$, $y = 87°$, $z = 93°$
15. $x = -2°$ or $\frac{5°}{2}$
16. $x = \frac{2}{3}^{\circ}$ or $\frac{1}{2}^{\circ}$
17. $w = 110°$, $x = 70°$, $y = 140°$, $z = 87.5°$
18. a. $a = 110°$, $b = 70°$ and $c = 110°$
    b. Sample responses can be found in the worked solutions in the online resources.

Exercise 16.5 Tangents, secants and chords

1. a. $x = 70°$ b. $x = 47°$, $y = 59°$
2. a. $p = 6$ b. $q = 8$
3. $x = 42°$, $y = 132°$
4. B
5. D
6. MAC, NAC, FDA, FBA, EDG, EBG.
7. $x = 42°$, $y = 62°$
8. Sample responses can be found in the worked solutions in the online resources.
9. 60°
10. $x = 180° - a - b$
11. $x = 80°$, $y = 20°$, $z = 80°$
12. Sample responses can be found in the worked solutions in the online resources.
13. $x = 85°$, $y = 20°$, $z = 85°$
14. D
15. $x = 50°$, $y = 95°$
16. A
17. C
18. $x = 33°$, $y = 55°$, $z = 22°$
19. $x = 25°$, $y = 65°$, $z = 40°$
20. $x = a$, $y = 90° - a$, $z = 90° - 2a$
21. Sample responses can be found in the worked solutions in the online resources.
22. a. $x = 5$ b. $k = 12$ c. $m = 6$, $n = 6$
23. a. $x = 7$ b. $b = 4$, $a = 2$ c. $w = 3$, $x = 5$
24. a. $a = 50°$, $b = 50°$ and $c = 80°$
    b. $a = 50°$, $b = 70°$ and $c = 70°$

Project

1. 11 100 km
2. 10931 km
3.

Latitude	Distance between P_1 and P_2 (km)
0°	11 100
10°	10 931
20°	10 431
30°	9613
40°	8503
50°	7135
60°	5550
70°	3796
80°	1927
90°	0

4. The distance between P_1 and P_2 decreases from 11 100 km at the equator to 0 km at the pole. The change is not constant. The distance between the points decreases more rapidly on moving towards the pole.
5. Latitude 60°
6.

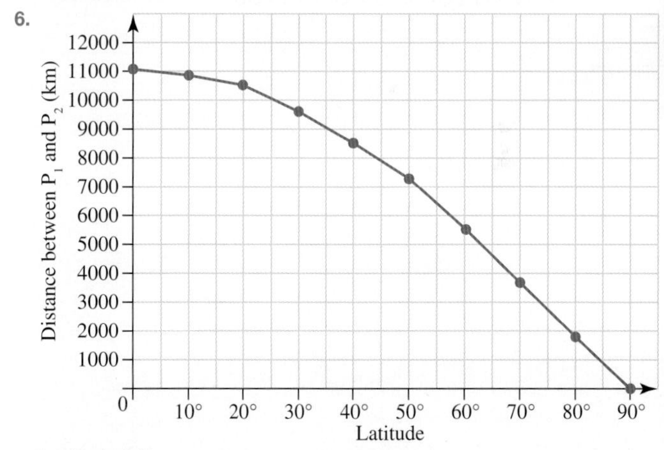

7. Latitude 26°
8. Sample responses can be found in the worked solutions in the online resources. Students need to investigate the rate that the distance between them changes from the equator to the pole and also comparison with the earlier values.

Exercise 16.6 Review questions

1. a. $x = 50°$ b. $x = 48°$, $y = 25°$
   c. $x = y = 28°$, $z = 56°$ d. $x = 90°$
   e. $y = 90°$ f. $y = 140°$
2. a. $x = 55°$ b. $x = 125°$
   c. $x = 70°$ d. $x = 100°$
   e. $m = 40°$ f. $x = 90°$, $y = 60°$, $z = 40°$

3. a. $x = 90°$ b. $x = 20°$
   c. $x = 55°$ d. $x = 125°$

4. a. $m = 3$ b. $m = 12$
   c. $m = 9$ d. $m = 11.7$

5. A, B, D

6. A, B, C

7. a. $x = 95°, y = 80°$
   b. $x = 99°$
   c. $x = 78°, y = 92°$
   d. $x = 97°, y = 92°$

8. D

9. a. $x = 42°$ b. $y = 62°$ c. $p = 65°$

10. CE × ED = AE × EB

    AE = CE (given)
    ∴ ED = EB

11. ∠AYC = ∠AXC

    ∠BXD = ∠BYD

    But ∠AXC = ∠BXD
    ⇒ ∠AYC = ∠BYD

12. ∠PQT and ∠PST, ∠PTS and ∠RQS, ∠TPQ and ∠QSR, ∠QPS and ∠QTS, ∠TPS and ∠TQS, ∠PQS and ∠PTS, ∠PUT and ∠QUS, ∠PUQ and ∠TUS

10A

17 Trigonometry II

LEARNING SEQUENCE

17.1 Overview

Why learn this?

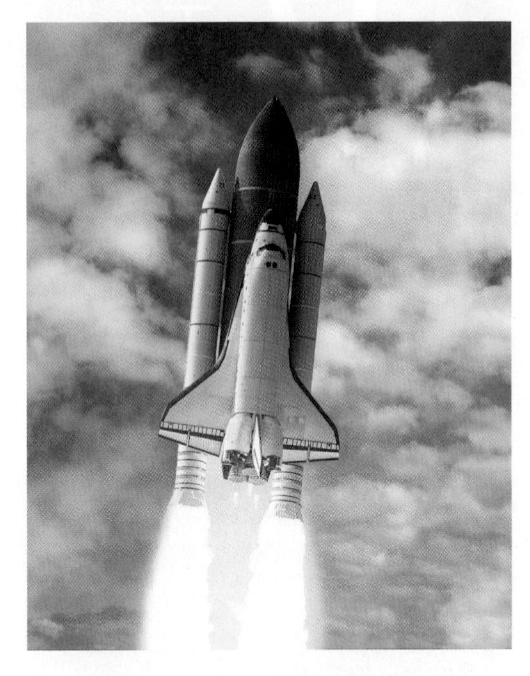

Trigonometry is the branch of mathematics that describes the relationship between the angles and side lengths in triangles. The ability to calculate distances using angles has long been critical. As early as the third century BCE, trigonometry was being used in the study of astronomy. Early explorers, using rudimentary calculations and the stars, were able to navigate their way around the world. They were even able to map coastlines along the way. Cartographers use trigonometry when they are making maps. It is essential to be able to calculate distances that can't be physically measured. Astronomers use trigonometry to calculate distances such as that from a particular planet to Earth. Our explorations have now turned towards the skies and outer space. Scientists design and launch space shuttles and rockets to explore our universe. By applying trigonometry, they can approximate the distances to other planets. As well as in astronomy and space exploration, trigonometry is widely used in many other areas. Surveyors use trigonometry in setting out a land subdivision. Builders, architects and engineers use angles, lengths and forces in the design and construction of all types of buildings, both domestic and industrial. In music, a single note is a sine wave. Sound engineers manipulate sine waves to create the desired effect. Trigonometry has many real-life applications.

Where to get help

Go to your learnON title at **www.jacplus.com.au** to access the following digital resources. The Online Resources Summary at the end of this topic provides a full list of what's available to help you learn the concepts covered in this topic.

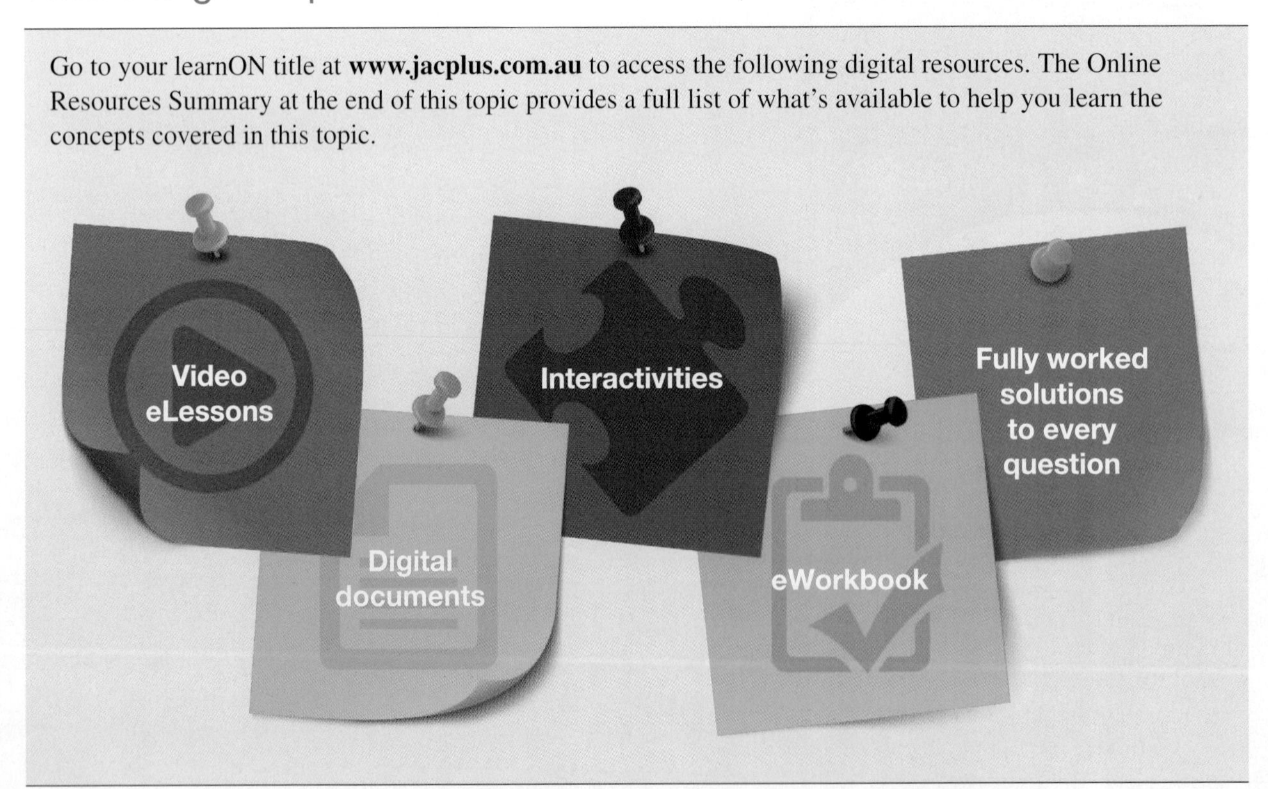

Exercise 17.1 Pre-test

learnON

Complete this pre-test in your learnON title at www.jacplus.com.au and receive **automatic marks**, **immediate corrective feedback** and **fully worked solutions**.

1. **MC** From the following options select the exact value of sin(30°).

   A. $\frac{1}{\sqrt{2}}$ B. $\frac{1}{2}$ C. $\frac{\sqrt{3}}{2}$

   D. 1 E. $\sqrt{3}$

2. Solve for x, correct to two decimal places.

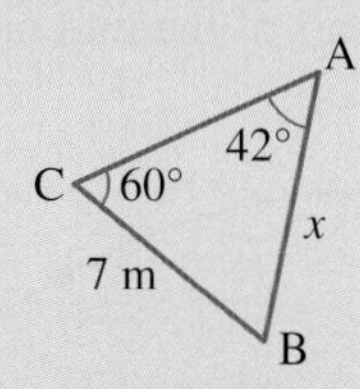

3. Solve for y, correct to two decimal places.

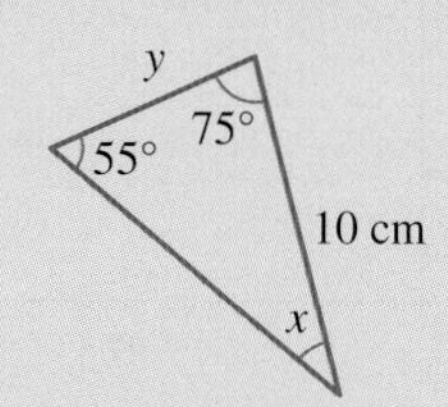

4. **MC** Choose the values of the angles B and B′ in the below triangle, to the nearest degree. (Assume $BC = B'C$.)

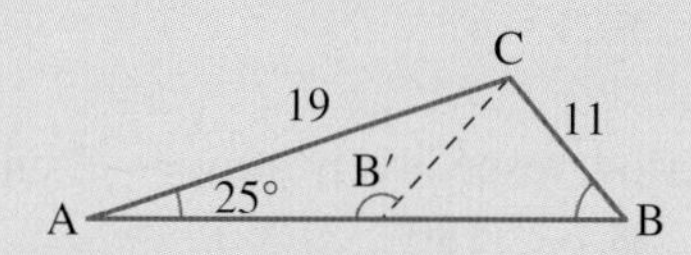

   A. 47° and 133° B. 46° and 134° C. 47° and 153°

   D. 25° and 155° E. 65° and 115°

5. Calculate the perimeter of the following triangle, correct to two decimal places.

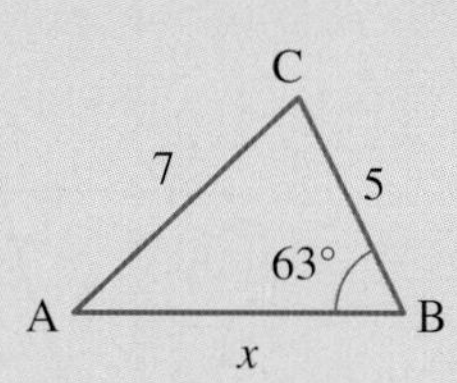

6. Solve for x, correct to one decimal place.

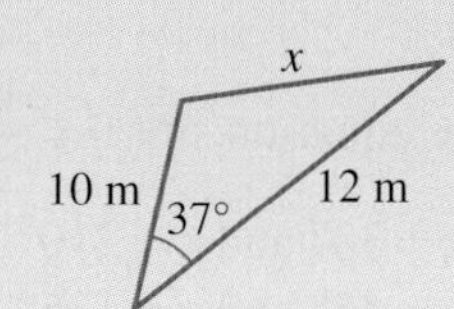

7. **MC** Calculate the area of the triangle shown.

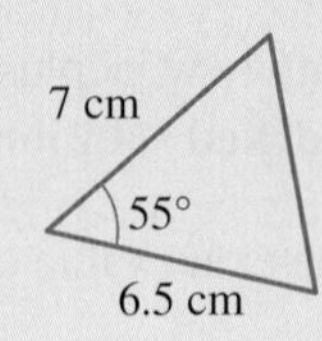

A. 13.05 cm^2
B. 18.64 cm^2
C. 22.75 cm^2
D. 26.1 cm^2
E. 37.27 cm^2

8. State in which quadrant of the unit circle is the angle 203° located.

9. Determine the value of $\cos(60°)$ using part of the unit circle.

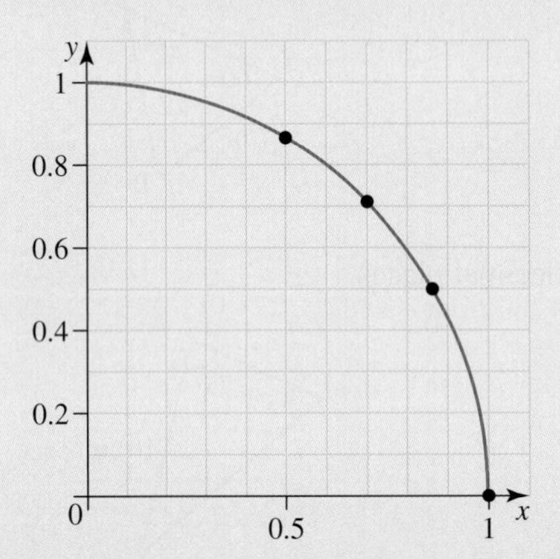

10. **MC** If $\cos(x°) = p$ for $0 \le x \le 90°$, then $\sin(180 - x°)$ in terms of p is:

A. p
B. $180 - p$
C. $1 - p$
D. $\sqrt{1 - p^2}$
E. $1 - p^2$

11. **MC** Select the amplitude and period, respectively, of $y = -2\sin(2x)$ from the following options.

A. $-2, 360°$
B. $-2, 180°$
C. $2, 2$
D. $2, 180°$
E. $-2, 2$

12. **MC** Select the correct equation for the graph shown.

A. $y = 4\cos(2x)$
B. $y = -4\cos(2x)$
C. $y = -4\sin(2x)$
D. $y = 4\sin(2x)$
E. $y = 4\cos\left(\frac{x}{2}\right)$

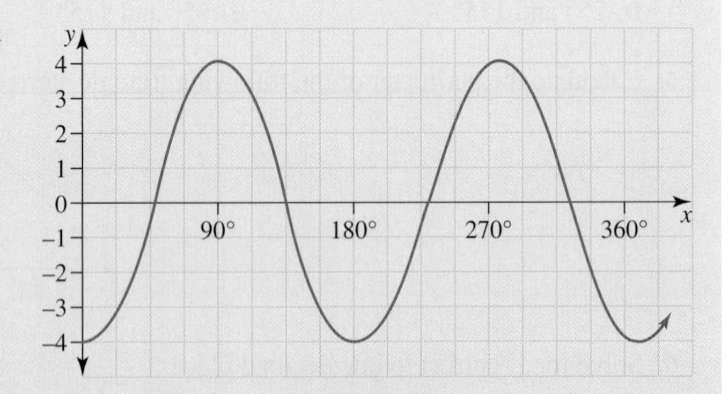

13. **MC** Select the correct solutions for the equation $\sin(x) = \frac{1}{2}$ for x over the domain $0 \le x \le 360°$.

A. $x = 30°$ and $x = 150°$
B. $x = 30°$ and $x = 210°$
C. $x = 60°$ and $x = 120°$
D. $x = 60°$ and $x = 240°$
E. $x = 45°$ and $x = 135°$

14. **MC** Select the correct solutions for the equation $\cos(2x) = -\frac{\sqrt{2}}{2}$ for x over the domain $0 \leq x \leq 360°$.

A. $x = 22.5°$ and $x = 337.5°$
B. $x = 45°$ and $x = 315°$
C. $x = 67.5°$ and $x = 112.5°$
D. $x = 157.5°$ and $x = 202.5°$
E. $x = 22.5°$ and $x = 157.5°$

15. Using the graph shown, solve the equation $7\sin(x) = -7$ for $0 \leq x \leq 360°$.

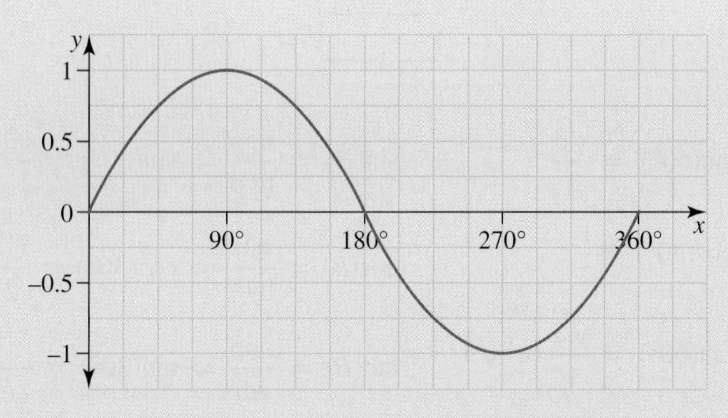

17.2 The sine rule

LEARNING INTENTION

At the end of this subtopic you should be able to:

- apply the exact values of sin, cos and tan for 30°, 45° and 60° angles
- apply the sine rule to evaluate angles and sides of triangles
- recognise when the ambiguous case of the sine rule exists.

17.2.1 Exact values of trigonometric functions and the sine rule

eles-5004

- Most of the trigonometric values that we will deal with in this topic are approximations.
- However, angles of 30° 45° and 60° have exact values of sine, cosine and tangent.
- Consider an equilateral triangle, ABC, of side length 2 cm.
  Let BD be the perpendicular bisector of AC, then:
  $\Delta ABD \cong \Delta CBD$ (using RHS)
  giving:
  $AD = CD = 1$ cm
  $\angle ABD = \angle CBD = 30°$
  and:
  $(AB)^2 = (AD)^2 + (BD)^2$ (using Pythagoras' theorem)
  $2^2 = 1^2 + (BD)^2$
  $BD = \sqrt{3}$

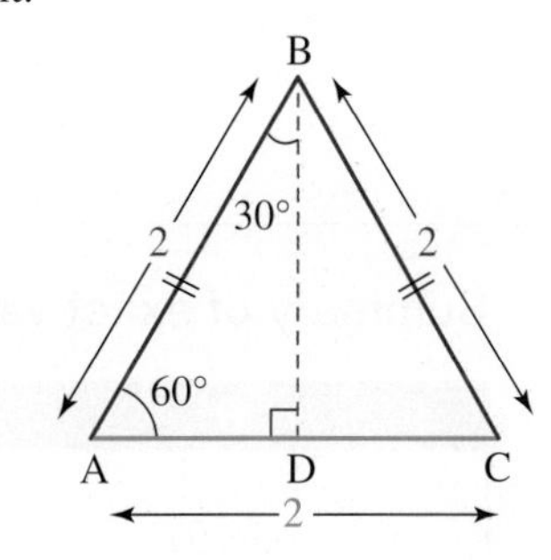

- Using ΔABD, the following exact values are obtained:

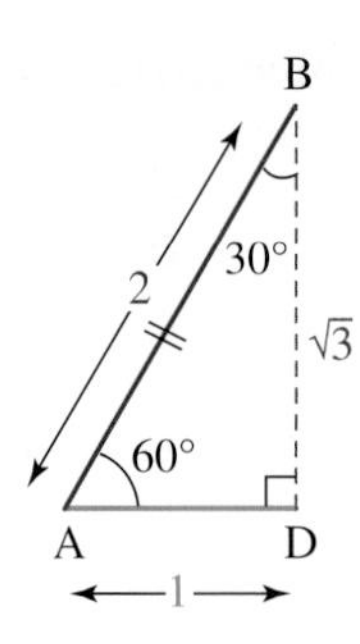

$$\sin(A) = \frac{\text{opp}}{\text{hyp}} \Rightarrow \sin(60°) = \frac{\sqrt{3}}{2}$$

$$\cos(A) = \frac{\text{adj}}{\text{hyp}} \Rightarrow \cos(60°) = \frac{1}{2}$$

$$\tan(A) = \frac{\text{opp}}{\text{adj}} \Rightarrow \tan(60°) = \frac{\sqrt{3}}{1} \text{ or } \sqrt{3}$$

$$\sin(B) = \frac{\text{opp}}{\text{hyp}} \Rightarrow \sin(30°) = \frac{1}{2}$$

$$\cos(B) = \frac{\text{adj}}{\text{hyp}} \Rightarrow \cos(30°) = \frac{\sqrt{3}}{2}$$

$$\tan(B) = \frac{\text{opp}}{\text{adj}} \Rightarrow \tan(30°) = \frac{1}{\sqrt{3}} \text{ or } \frac{\sqrt{3}}{3}$$

- Consider a right-angled isosceles ΔEFG with equal sides of 1 unit.

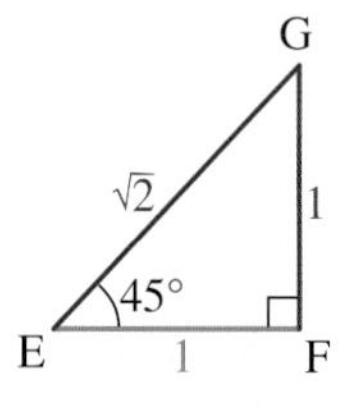

$$(\text{EG})^2 = (\text{EF})^2 + (\text{FG})^2 \quad \text{(using Pythagoras' theorem)}$$

$$(\text{EG})^2 = 1^2 + 1^2$$

$$\text{EG} = \sqrt{2}$$

- Using ΔEFG, the following exact values are obtained:

$$\sin(E) = \frac{\text{opp}}{\text{hyp}} \Rightarrow \sin(45°) = \frac{1}{\sqrt{2}} \text{ or } \frac{\sqrt{2}}{2}$$

$$\cos(E) = \frac{\text{adj}}{\text{hyp}} \Rightarrow \cos(45°) = \frac{1}{\sqrt{2}} \text{ or } \frac{\sqrt{2}}{2}$$

$$\tan(E) = \frac{\text{opp}}{\text{adj}} \Rightarrow \tan(45°) = \frac{1}{1} \text{ or } 1$$

Summary of exact values

θ	30º	45º	60º
$\sin(\theta)$	$\frac{1}{2}$	$\frac{1}{\sqrt{2}} = \frac{\sqrt{2}}{2}$	$\frac{\sqrt{3}}{2}$
$\cos(\theta)$	$\frac{\sqrt{3}}{2}$	$\frac{1}{\sqrt{2}} = \frac{\sqrt{2}}{2}$	$\frac{1}{2}$
$\tan(\theta)$	$\frac{1}{\sqrt{3}} = \frac{\sqrt{3}}{3}$	1	$\sqrt{3}$

The sine rule

- In any triangle, label the angles are named by the vertices A, B and C and the corresponding opposite sides as a, b and c as shown in the diagram at right.
- Let BD be the perpendicular line from B to AC, of length h, giving two right-angled triangles, ΔADB and ΔCDB.

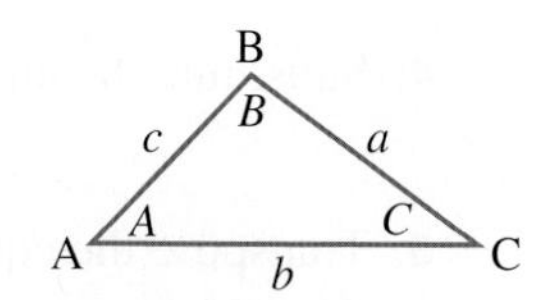

Using ΔADB:

$$\sin(A) = \frac{h}{c}$$

$$h = c\sin(A)$$

Using ΔCDB:

$$\sin(C) = \frac{h}{a}$$

$$h = a\sin(C)$$

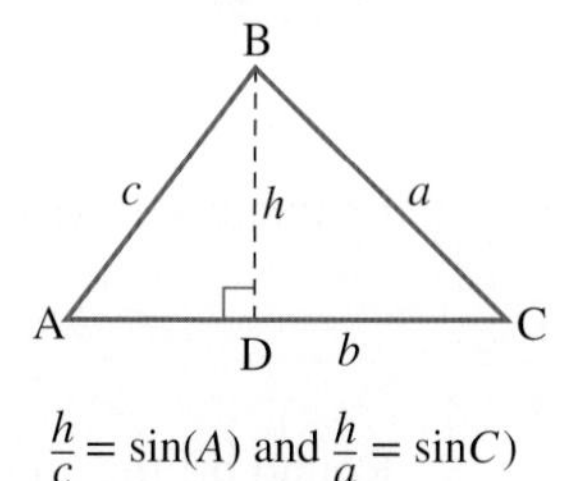

$\frac{h}{c} = \sin(A)$ and $\frac{h}{a} = \sin C)$

Equating the values of h:

$$c\sin(A) = a\sin(C)$$

giving:

$$\frac{c}{\sin(C)} = \frac{a}{\sin(A)}$$

- Similarly, if a perpendicular line is drawn from vertex A to BC, then:

$$\frac{c}{\sin(C)} = \frac{b}{\sin(B)}$$

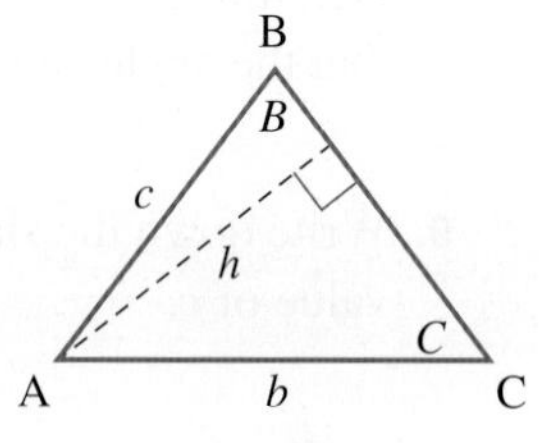

$h = c\sin(B)$ and $h = b\sin(C)$

Sine rule

- The **sine rule** for any triangle ABC is:

$$\frac{a}{\sin(A)} = \frac{b}{\sin(B)} = \frac{c}{\sin(C)}$$

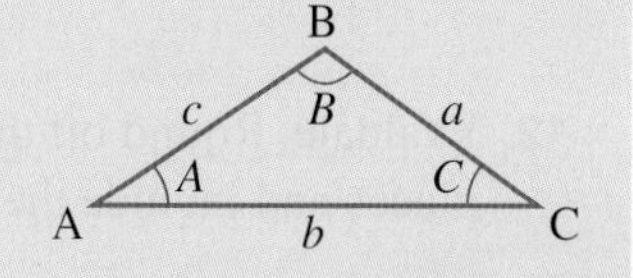

- The sine rule can be used to solve non-right-angled triangles if we are given:
  1. two angles and one side
  2. two sides and an angle opposite one of these sides.

WORKED EXAMPLE 1 Determining unknown angles and sides of a given triangle

In the triangle ABC, $a = 4$ m, $b = 7$ m and $B = 80°$. Calculate the values of A, C and c.

THINK	WRITE/DRAW
1. Draw a labelled diagram of the triangle ABC and fill in the given information.	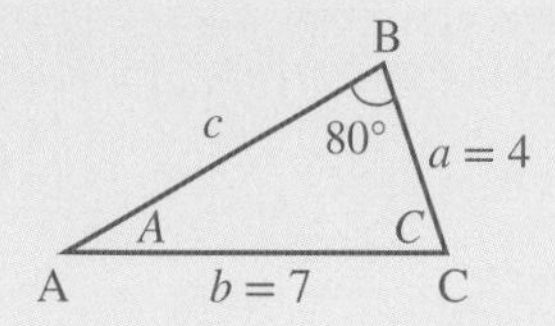
2. Check that one of the criteria for the sine rule has been satisfied.	The sine rule can be used since two side lengths and an angle opposite one of these side lengths have been given.
3. Write down the sine rule to calculate A.	To calculate angle A: $\frac{a}{\sin(A)} = \frac{b}{\sin(B)}$

THINK	WRITE
4. Substitute the known values into the rule.	$\dfrac{4}{\sin(A)} = \dfrac{7}{\sin(80°)}$
5. Transpose the equation to make $\sin(A)$ the subject.	$4\sin(80°) = 7\sin(A)$ $\sin(A) = \dfrac{4\sin(80°)}{7}$
6. Evaluate and write your answer.	$A = \sin^{-1}\left(\dfrac{4\sin(80°)}{7}\right)$ $\approx 34.246\,004\,71°$
7. Round off the answer to degrees and minutes.	$\approx 34°15'$
8. Determine the value of angle C using the fact that the angle sum of any triangle is 180°.	$C \approx 180° - (80° + 34°15')$ $= 65°45'$
9. Write down the sine rule to calculate the value of c.	To calculate side length c: $\dfrac{c}{\sin(C)} = \dfrac{b}{\sin(B)}$
10. Substitute the known values into the rule.	$\dfrac{c}{\sin(65°45')} = \dfrac{b}{\sin(80°)}$
11. Transpose the equation to make c the subject.	$c = \dfrac{7\sin(65°45')}{\sin(80°)}$
12. Evaluate. Round off the answer to 2 decimal places and include the appropriate unit.	≈ 6.48 m

TI | THINK

Open a new document and a Calculator page. Ensure your calculator is set to the degree and approximate mode, as shown for the next set of examples.
To do this, press:
- HOME
- 5: Settings
- 2: Document Settings

For Display Digits select 'Fix 2'.
TAB to Angle and select 'Degree'.
TAB to Calculation Mode and select 'Approximate'.
TAB to OK and press ENTER.

DISPLAY/WRITE

Document Settings
Display Digits: Fix 2
Angle: Degree
Exponential Format: Normal
Real or Complex: Rectangular
Calculation Mode: Approximate
CAS Mode: On
OK Cancel

CASIO | THINK

Ensure your calculator is set to the degree and decimal modes. To do this, at the bottom of the Main screen, tap on the mode options until you have Decimal and Deg.

DISPLAY/WRITE

1. In a new problem on the Calculator page, press TRIG to access and select the appropriate trigonometric ration $(\sin^{-1})$. Complete the entry line as:

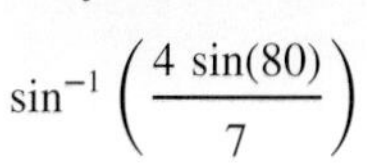

$\sin^{-1}\left(\dfrac{4\sin(80)}{7}\right)$

Then press ENTER.
To convert the decimal degree into degrees, minutes and seconds, press:
- CATALOG
- 1
- D

Scroll to and select ▶ DMS then press ENTER.
Complete the entry line as:
180 − (ans + 80)
Then press ENTER.
Repeat the above process to convert to degrees, minutes and seconds.

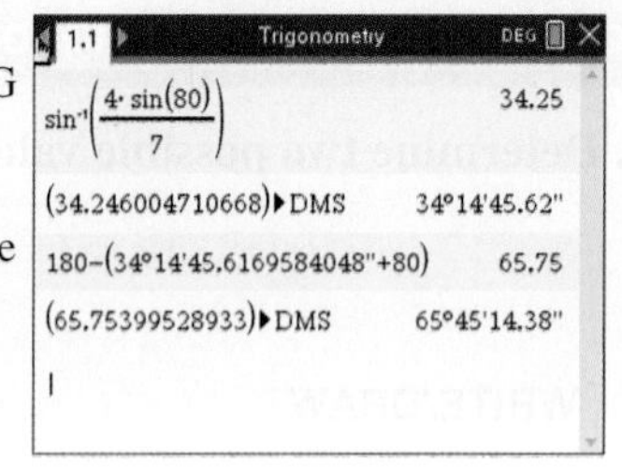

$A = 34°15'$ and $C = 65°45'$

2. To find the value of c, complete the entry line as:
$\dfrac{7\sin(\text{ans})}{\sin 80}$
Then press ENTER.

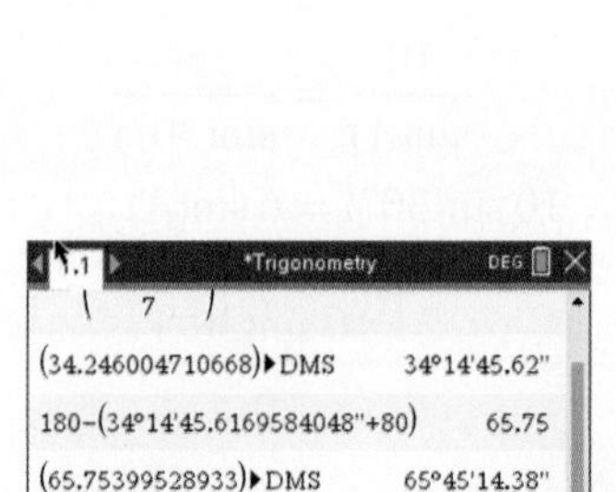

$c = 6.48$ m

1. On the Main screen, complete the entry line as:

$\sin^{-1}\left(\dfrac{4\sin(80)}{7}\right)$

$180 - \left(80 + \sin^{-1}\left(\dfrac{4\sin(80)}{7}\right)\right)$

Press EXE after each entry.
To convert the decimal degree into degrees, minutes and seconds, tap:
- Action
- Transformation
- DMS
- toDMS

Highlight and drag each of the decimal answers into the entry line and press EXE.

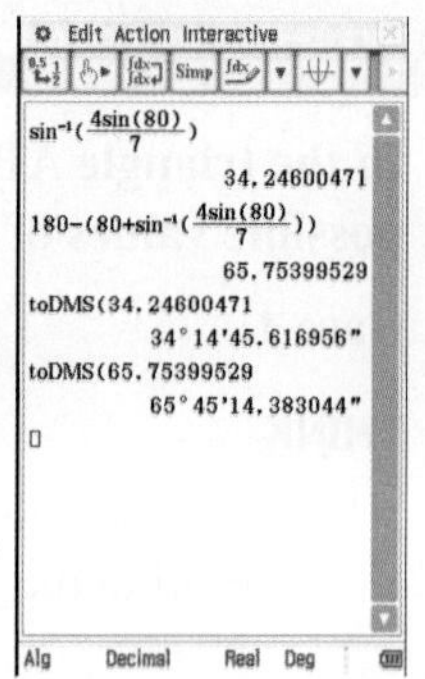

$A = 34°15'$ and $C = 65°45'$ rounded to the nearest minute.

2. To calculate the value of c, complete the entry lines as:
dms(65, 45)
$\dfrac{7\sin(65, 75)}{\sin(80)}$
Press EXE after each entry line.

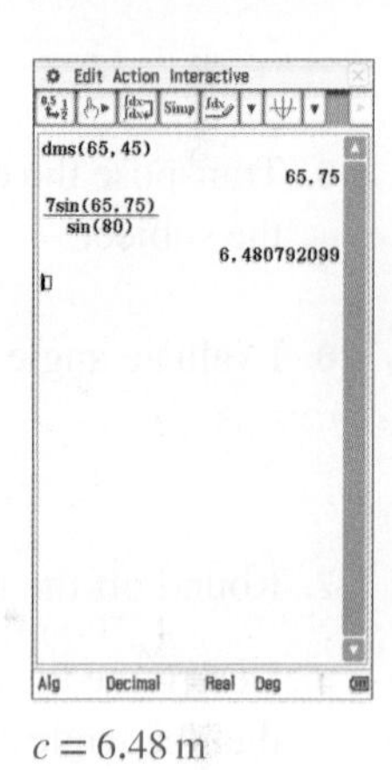

$c = 6.48$ m

17.2.2 The ambiguous case

eles-5005

- If two side lengths and an angle opposite one of these side lengths are given, then two different triangles may be drawn.
- For example, if $a = 10$, $c = 6$ and $C = 30°$, two possible triangles could be created.

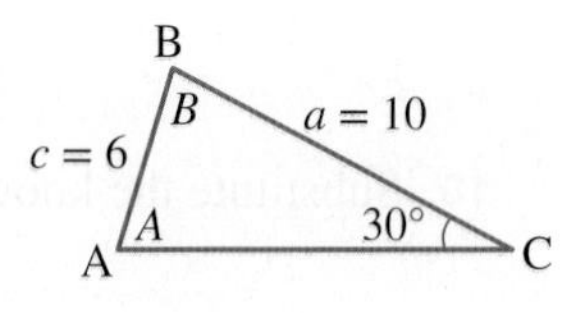

- In the first case angle A is an acute angle, while in the second case angle A is an obtuse angle.
- When using the sine rule to determine an angle, the inverse sine function is used.
- In subtopic 17.5, you will see that the sine of an angle between 0° and 90° has the same value as the sine of its supplement.
For example, $\sin 40° \approx 0.6427$ and $\sin 140° \approx 0.6427$.

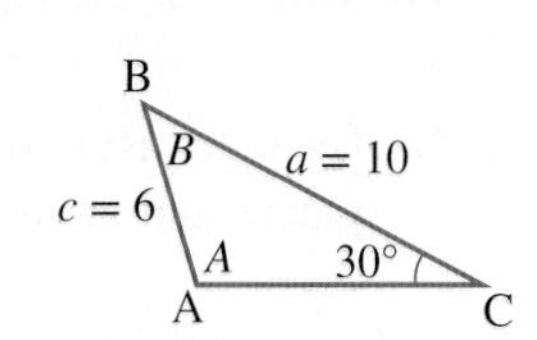

WORKED EXAMPLE 2 Solving triangles and checking for the ambiguous case

In the triangle ABC, $a = 10$ m, $c = 6$ m and $C = 30°$. Determine two possible values of A, and hence two possible values of B and b.

Case 1

THINK	WRITE/DRAW
1. Draw a labelled diagram of the triangle ABC and fill in the given information.	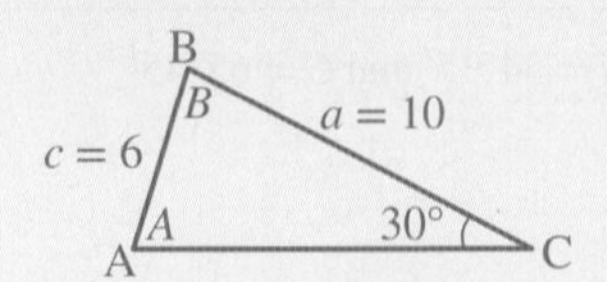
2. Check that one of the criteria for the sine rule has been satisfied.	The sine rule can be used since two side lengths and an angle opposite one of these side lengths have been given.
3. Write down the sine rule to determine A.	To determine angle A: $\dfrac{a}{\sin(A)} = \dfrac{c}{\sin(C)}$
4. Substitute the known values into the rule.	$\dfrac{10}{\sin(A)} = \dfrac{6}{\sin(30°)}$ $10\sin(30°) = 6\sin(A)$
5. Transpose the equation to make $\sin(A)$ the subject.	$\sin(A) = \dfrac{10\sin(30°)}{6}$
6. Evaluate angle A.	$A = \sin^{-1}\left(\dfrac{10\sin(30°)}{6}\right)$ $\approx 56.442\,690\,24°$
7. Round off the answer to degrees and minutes.	$A = 56°27'$
8. Determine the value of angle B, using the fact that the angle sum of any triangle is 180°.	$B \approx 180° - (30° + 56°27')$ $= 93°33'$
9. Write down the sine rule to calculate b.	To calculate side length b: $\dfrac{b}{\sin(B)} = \dfrac{c}{\sin(C)}$
10. Substitute the known values into the rule.	$\dfrac{b}{\sin(93°33')} = \dfrac{6}{\sin(30°)}$
11. Transpose the equation to make b the subject.	$b = \dfrac{6\sin(93°33')}{\sin(30°)}$
12. Evaluate. Round off the answer to 2 decimal places and include the appropriate unit.	≈ 11.98 m

Note: The values we have just obtained are only one set of possible answers for the given dimensions of the triangle ABC.
We are told that $a = 10$ m, $c = 6$ m and $C = 30°$. Since side a is larger than side c, it follows that angle A will be larger than angle C. Angle A must be larger than 30°; therefore it may be an acute angle or an obtuse angle.

Case 2

THINK	WRITE/DRAW
1. Draw a labelled diagram of the triangle ABC and fill in the given information.	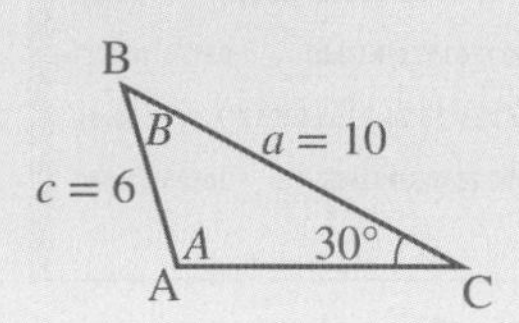
2. Write down the alternative value for angle A. Simply subtract the value obtained for A in case 1 from 180°.	To determine the alternative angle A: If $\sin A = 0.8333$, then A could also be: $A \approx 180° - 56°27'$ $= 123°33'$
3. Determine the alternative value of angle B, using the fact that the angle sum of any triangle is 180°.	$B \approx 180° - (30° + 123°33')$ $= 26°27'$
4. Write down the sine rule to determine the alternative b.	To calculate side length b: $\dfrac{b}{\sin(B)} = \dfrac{c}{\sin(C)}$
5. Substitute the known values into the rule.	$\dfrac{b}{\sin(26°27')} = \dfrac{6}{\sin(30°)}$
6. Transpose the equation to make b the subject.	$b = \dfrac{6\sin(26°27')}{\sin(30°)}$
7. Evaluate. Round off the answer to 2 decimal places and include the appropriate unit.	≈ 5.34 m

TI | THINK	DISPLAY/WRITE	CASIO | THINK	DISPLAY/WRITE
1. In a new problem, on a Calculator page, complete the entry line as: solve $\left(\sin(a) = \dfrac{10\sin(30)}{6}, a\right) \mid$ $0 < a < 180$ Press the up arrow and highlight the answer shown. Press ENTER to bring this answer down into the new line. Convert these angles to degrees and minutes as shown in Worked Example 1 by completing the entry as shown. Press ENTER after each entry.	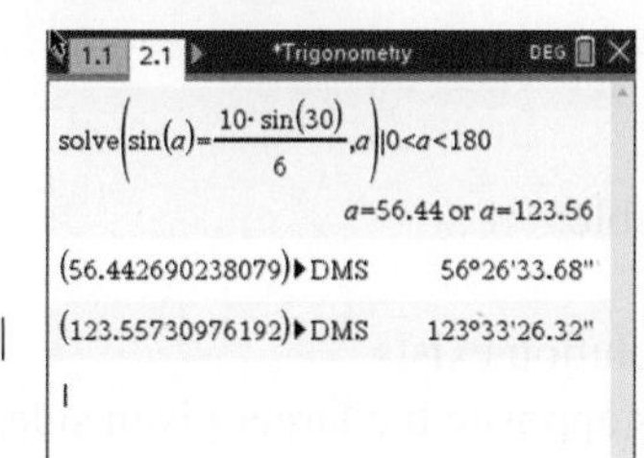 $A = 56°27'$ or $123°34'$	1. On the Main screen, complete the entry line as: solve $\left(\sin(a) = \dfrac{10\sin(30)}{6}, a\right) \mid$ $0 < a < 180$ Convert these angles to degrees and minutes as shown. Press EXE after each entry.	 $A = 56°27'$ or $123°33'$ rounded to the nearest minute.

2. Solve for the two values of B as shown in the screenshot. Instead of typing the angle manually, press the up arrow to highlight the previous answer you want, then press ENTER. Press ENTER after each entry. Convert these angles to degrees and minutes.

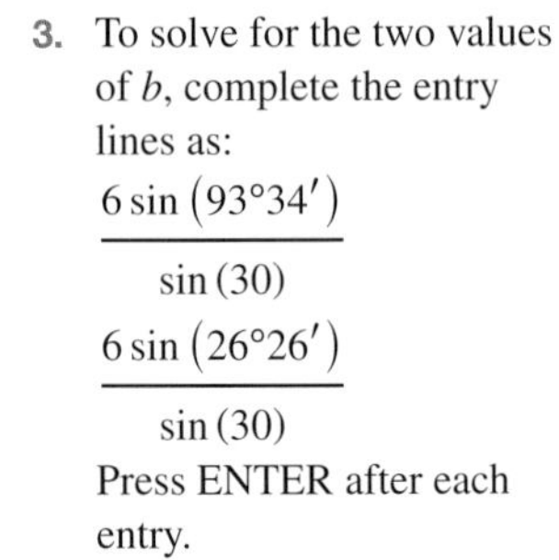

$B = 93°33'$ or $26°27'$

3. To solve for the two values of b, complete the entry lines as:
$\frac{6\sin(93°34')}{\sin(30)}$
$\frac{6\sin(26°26')}{\sin(30)}$
Press ENTER after each entry.

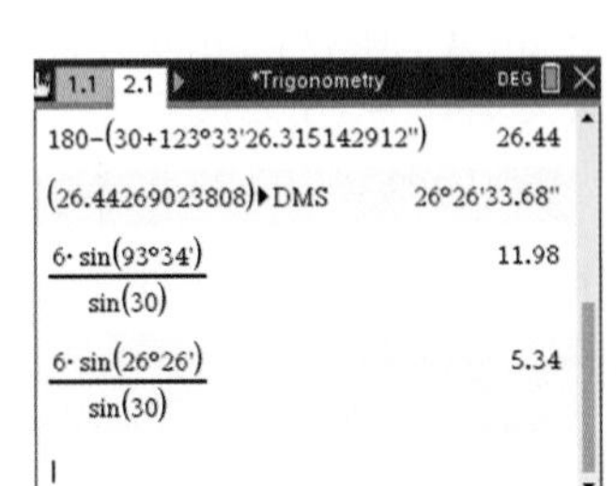

$b = 11.98$ m or 5.34 m

2. Solve for the two values of B as shown at right. [This uses $B = 180 - (30 + A)$.]
Press EXE after each entry, and convert these angles to degrees and minutes.

$B = 93°33'$ or $26°27'$

3. To solve for the two values of b, complete the entry lines as:
$\frac{6\sin(93.55730976)}{\sin(30)}$
$\frac{6\sin(26.4426902)}{\sin(30)}$
Press after each entry.

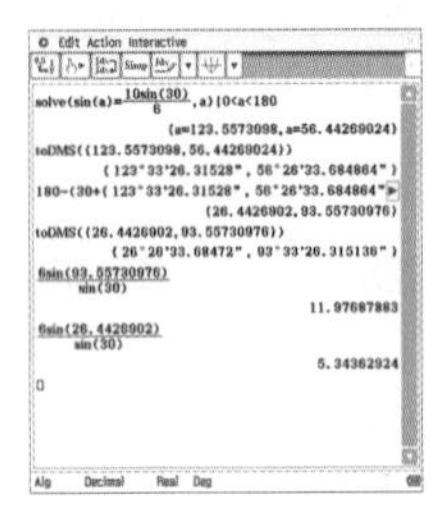

$b = 11.98$ m or 5.34 m

- In Worked example 2 there were two possible solutions, as shown by the diagrams below.

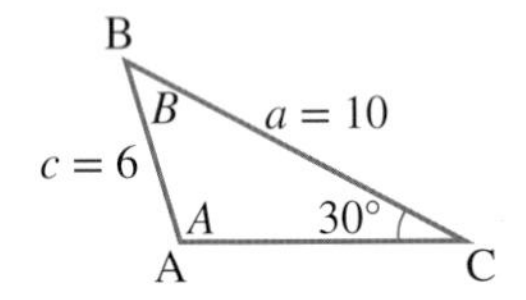

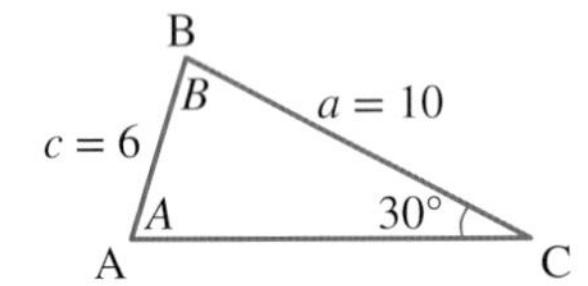

- The ambiguous case does not apply to every question.
Consider Worked example 1.
  - Since $\angle A = 34°15'$, then it could also have been $\angle A = 145°45'$, the supplementary angle.
  - If $\angle A = 34°15'$ and $\angle B = 80°$, then
  $\angle C = 65°45'$ (angle sum of triangle).
  - If $\angle A = 145°45'$ and $\angle B = 80°$, then
  $\angle C = 180° - (145°45' + 80°)$ which is not possible.
  $\angle C = -45°45'$

  Hence, for Worked example 1, only one possible solution exists.
- The ambiguous case may exist if the angle found is opposite the larger given side.

WORKED EXAMPLE 3 Calculating heights given angles of elevation

To calculate the height of a building, Kevin measures the angle of elevation to the top as 52°. He then walks 20 m closer to the building and measures the angle of elevation as 60°. Calculate the height of the building.

THINK

1. Draw a labelled diagram of the situation and fill in the given information.

WRITE/DRAW

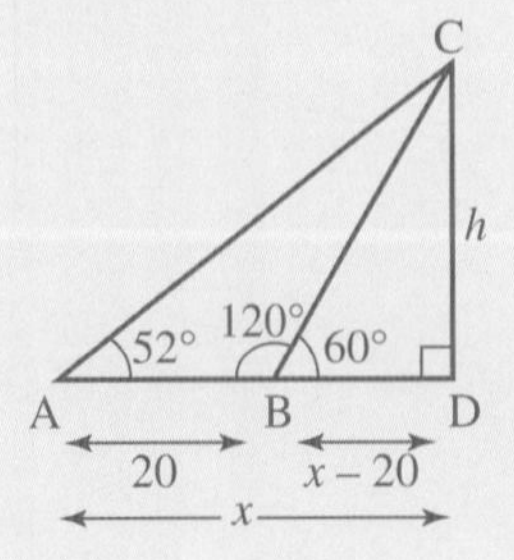

2. Check that one of the criteria for the sine rule has been satisfied for triangle ABC. | The sine rule can be used for triangle ABC since two angles and one side length have been given.

3. Calculate the value of angle ACB, using the fact that the angle sum of any triangle is 180°. | $\angle ACB = 180° - (52° + 120°)$
$= 8°$

4. Write down the sine rule to calculate b (or AC). | To calculate side length b of triangle ABC:
$$\frac{b}{\sin(B)} = \frac{c}{\sin(C)}$$

5. Substitute the known values into the rule. | $$\frac{b}{\sin(120°)} = \frac{20}{\sin(8°)}$$

6. Transpose the equation to make b the subject. | $$b = \frac{20\sin(120°)}{\sin(8°)}$$

7. Evaluate. Round off the answer to 2 decimal places and include the appropriate unit. | ≈ 124.45 m

8. Draw a diagram of the situation, that is, triangle ADC, labelling the required information. *Note*: There is no need to solve the rest of the triangle in this case as the values will not assist in calculating the height of the building.

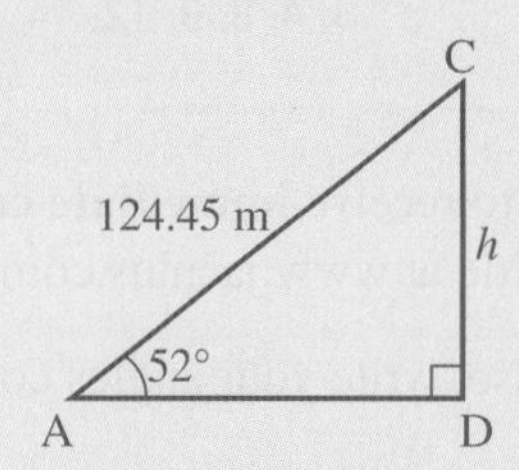

9. Write down what is given for the triangle. | Have: angle and hypotenuse

10. Write down what is needed for the triangle. | Need: opposite side

11. Determine which of the trigonometric ratios is required (SOH – CAH – TOA). | $\sin(\theta) = \frac{O}{H}$

12. Substitute the given values into the appropriate ratio. | $\sin(52°) = \frac{h}{124.45}$

13. Transpose the equation and solve for h. | $124.45\sin(52°) = h$
$h = 124.45\sin(52°)$

14. Round off the answer to 2 decimal places. | ≈ 98.07

15. Write the answer. | The height of the building is 98.07 m.

DISCUSSION

In what situations can the sine rule be used?

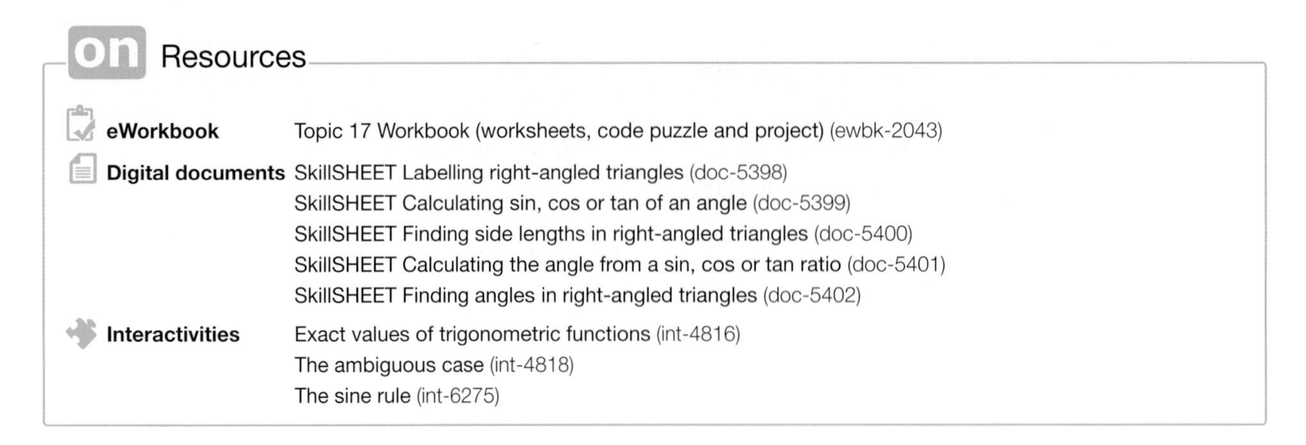

Resources

eWorkbook Topic 17 Workbook (worksheets, code puzzle and project) (ewbk-2043)

Digital documents SkillSHEET Labelling right-angled triangles (doc-5398)
SkillSHEET Calculating sin, cos or tan of an angle (doc-5399)
SkillSHEET Finding side lengths in right-angled triangles (doc-5400)
SkillSHEET Calculating the angle from a sin, cos or tan ratio (doc-5401)
SkillSHEET Finding angles in right-angled triangles (doc-5402)

Interactivities Exact values of trigonometric functions (int-4816)
The ambiguous case (int-4818)
The sine rule (int-6275)

Exercise 17.2 The sine rule

learnON

Individual pathways

■ PRACTISE	■ CONSOLIDATE	■ MASTER
1, 2, 6, 7, 11, 13, 17, 18, 21	3, 4, 8, 9, 12, 14, 16, 19, 22	5, 10, 15, 20, 23, 24, 25

To answer questions online and to receive **immediate corrective feedback** and **fully worked solutions** for all questions, go to your learnON title at www.jacplus.com.au.

Where appropriate in this exercise, write your angles correct to the nearest minute and side lengths correct to 2 decimal places.

Fluency

1. **WE1** In the triangle ABC, $a = 10$, $b = 12$ and $B = 58°$. Calculate A, C and c.
2. In the triangle ABC, $c = 17.35$, $a = 26.82$ and $A = 101°47'$. Calculate C, B and b.
3. In the triangle ABC, $a = 5$, $A = 30°$ and $B = 80°$. Calculate C, b and c.
4. In the triangle ABC, $c = 27$, $C = 42°$ and $A = 105°$. Calculate B, a and b.
5. In the triangle ABC, $a = 7$, $c = 5$ and $A = 68°$. Determine the perimeter of the triangle.
6. Calculate all unknown sides and angles for the triangle ABC, given $A = 57°$, $B = 72°$ and $a = 48.2$.
7. Calculate all unknown sides and angles for the triangle ABC, given $a = 105$, $B = 105°$ and $C = 15°$.
8. Calculate all unknown sides and angles for the triangle ABC, given $a = 32$, $b = 51$ and $A = 28°$.
9. Calculate the perimeter of the triangle ABC if $a = 7.8$, $b = 6.2$ and $A = 50°$.
10. **MC** In a triangle ABC, $B = 40°$, $b = 2.6$ and $c = 3$. Identify the approximate value of C.
    Note: There may be more than one correct answer.
    A. 47°
    B. 48°
    C. 132°
    D. 133°
    E. 139°

Understanding

11. **WE2** In the triangle ABC, $a = 10$, $c = 8$ and $C = 50°$. Determine two possible values of A, and hence two possible values of b.
12. In the triangle ABC, $a = 20$, $b = 12$ and $B = 35°$. Determine two possible values for the perimeter of the triangle.

13. Calculate all unknown sides and angles for the triangle ABC, given $A = 27°$, $B = 43°$ and $c = 6.4$.

14. Calculate all unknown sides and angles for the triangle ABC, given $A = 100°$, $b = 2.1$ and $C = 42°$.

15. Calculate all unknown sides and angles for the triangle ABC, given $A = 25°$, $b = 17$ and $a = 13$.

16. WE3 To calculate the height of a building, Kevin measures the angle of elevation to the top as 48°. He then walks 18 m closer to the building and measures the angle of elevation as 64°. Calculate the height of the building.

Reasoning

17. Calculate the value of h, correct to 1 decimal place. Show the full working.

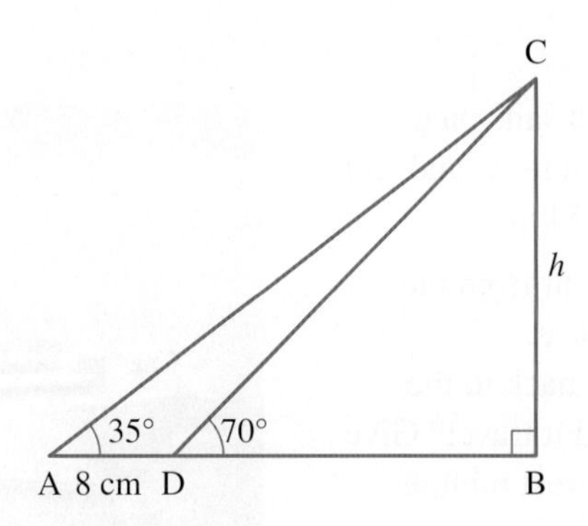

18. A boat sails on a bearing of N15°E for 10 km and then on a bearing of S85°E until it is due east of the starting point. Determine the distance from the starting point to the nearest kilometre. Show all your working.

19. A hill slopes at an angle of 30° to the horizontal. A tree that is 8 m tall and leaning downhill is growing at an angle of 10° m to the vertical and is part-way up the slope. Evaluate the vertical height of the top of the tree above the slope. Show all your working.

20. A cliff is 37 m high. The rock slopes outward at an angle of 50° to the horizontal and then cuts back at an angle of 25° to the vertical, meeting the ground directly below the top of the cliff.

Carol wishes to abseil from the top of the cliff to the ground as shown in the diagram. Her climbing rope is 45 m long, and she needs 2 m to secure it to a tree at the top of the cliff. Determine if the rope will be long enough to allow her to reach the ground.

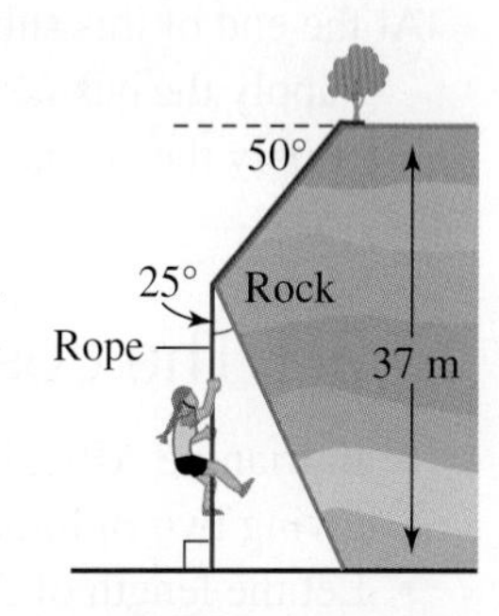

Problem solving

21. A river has parallel banks that run directly east–west. From the south bank, Kylie takes a bearing to a tree on the north side. The bearing is 047°T. She then walks 10 m due east, and takes a second bearing to the tree. This is 305°T. Determine:
    a. her distance from the second measuring point to the tree
    b. the width of the river, to the nearest metre.

22. A ship sails on a bearing of S20°W for 14 km; then it changes direction and sails for 20 km and drops anchor. Its bearing from the starting point is now N65°W.
    a. Determine the distance of the ship from the starting point of it.
    b. Calculate the bearing on which the ship sails for the 20 km leg.

23. A cross-country runner runs at 8 km/h on a bearing of 150°T for 45 mins; then she changes direction to a bearing of 053°T and runs for 80 mins at a different speed until she is due east of the starting point.

a. Calculate the distance of the second part of the run.
b. Calculate her speed for this section, correct to 2 decimal places.
c. Evaluate how far she needs to run to get back to the starting point.

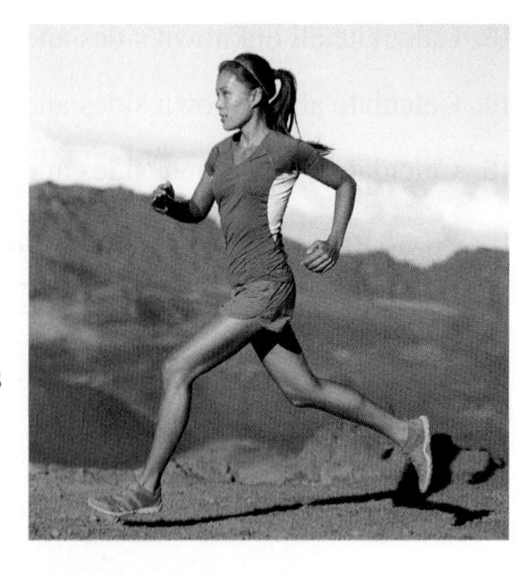

24. From a fire tower, A, a fire is spotted on a bearing of N42°E. From a second tower, B, the fire is on a bearing of N12°W. The two fire towers are 23 km apart, and A is N63°W of B. Determine how far the fire is from each tower.

25. A yacht sets sail from a marina and sails on a bearing of 065°T for 3.5 km. It then turns and sails on a bearing of 127°T for another 5 km.

a. Evaluate the distance of the yacht from the marina, correct to 1 decimal place.
b. If the yacht was to sail directly back to the marina, on what bearing should it travel? Give your answer rounded to the nearest minute.

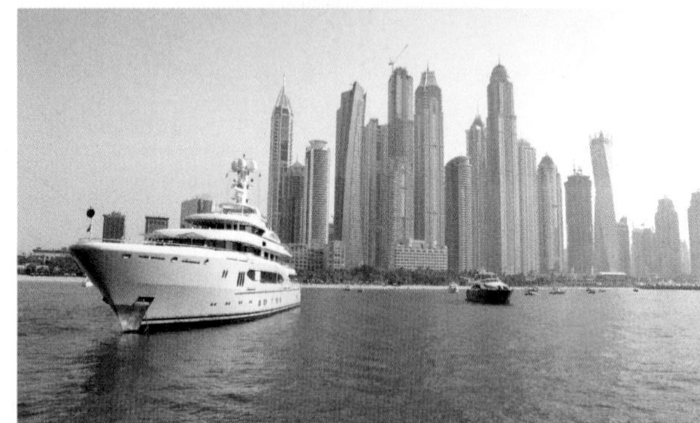

17.3 The cosine rule

LEARNING INTENTION

At the end of this subtopic you should be able to:
- apply the cosine rule to calculate a side of a triangle
- apply the cosine rule to calculate the angles of a triangle.

17.3.1 The cosine rule

eles-5006

- In triangle ABC, let BD be the perpendicular line from B to AC, of length h, giving two right-angled triangles, ΔADB and ΔCDB.
- Let the length of AD $= x$, then DC $= (b-x)$.
- Using triangle ADB and Pythagoras' theorem, we obtain:

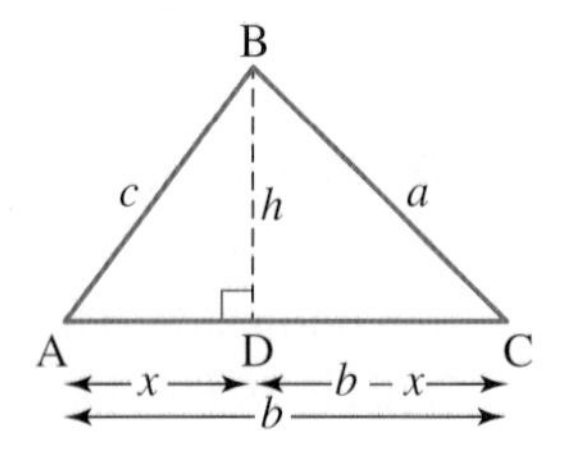

$$c^2 = h^2 + x^2 \qquad [1]$$

Using triangle CDB and Pythagoras' theorem, we obtain:

$$a^2 = h^2 + (b-x)^2 \qquad [2]$$

Expanding the brackets in equation [2]:

$$a^2 = h^2 + b^2 - 2bx + x^2$$

Rearranging equation [2] and using $c^2 = h^2 + x^2$ from equation [1]:

$$\begin{aligned} a^2 &= h^2 + x^2 + b^2 - 2bx \\ &= c^2 + b^2 - 2bx \\ &= b^2 + c^2 - 2bx \end{aligned}$$

From triangle ABD, $x = c\cos(A)$; therefore $a^2 = b^2 + c^2 - 2bx$ becomes

$$a^2 = b^2 + c^2 - 2bc\cos(A).$$

Cosine rule

- The **cosine rule** for any triangle ABC is:

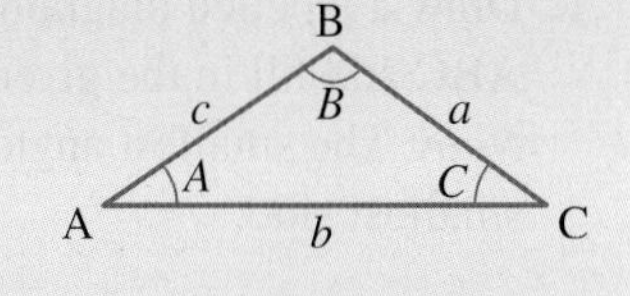

$$\mathbf{a^2 = b^2 + c^2 - 2bc\cos(A)}$$
$$\mathbf{b^2 = a^2 + c^2 - 2ac\cos(B)}$$
$$\mathbf{c^2 = a^2 + b^2 - 2ab\cos(C)}$$

- The cosine rule can be used to solve non-right-angled triangles if we are given:
  1. three sides
  2. two sides and the included angle.
- *Note:* Once the third side has been calculated, the sine rule could be used to determine other angles if necessary.
- If three sides of a triangle are known, an angle could be found by transposing the cosine rule to make $\cos(A)$, $\cos(B)$ or $\cos(C)$ the subject.

$$\mathbf{a^2 = b^2 + c^2 - 2bc\cos(A) \Rightarrow \cos(A) = \frac{b^2 + c^2 + a^2}{2bc}}$$

$$\mathbf{b^2 = a^2 + c^2 - 2ac\cos(B) \Rightarrow \cos(B) = \frac{a^2 + c^2 + b^2}{2ac}}$$

$$\mathbf{c^2 = a^2 + b^2 - 2ab\cos(C) \Rightarrow \cos(C) = \frac{a^2 + b^2 + c^2}{2ab}}$$

WORKED EXAMPLE 4 Calculating sides using the cosine rule

Calculate the third side of triangle ABC given $a = 6$, $c = 10$ and $B = 76°$.

THINK	WRITE/DRAW
1. Draw a labelled diagram of the triangle ABC and fill in the given information.	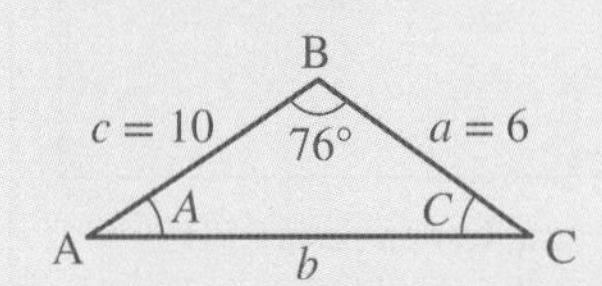
2. Check that one of the criteria for the cosine rule has been satisfied.	Yes, the cosine rule can be used since two side lengths and the included angle have been given.
3. Write down the appropriate cosine rule to calculate side b.	To calculate side b: $b^2 = a^2 + c^2 - 2ac\cos(B)$
4. Substitute the given values into the rule.	$= 6^2 + 10^2 - 2 \times 6 \times 10 \times \cos(76°)$
5. Evaluate.	$\approx 106.969\,372\,5$ $b \approx \sqrt{106.969\,372\,5}$
6. Round off the answer to 2 decimal places.	≈ 10.34

WORKED EXAMPLE 5 Calculating angles using the cosine rule

Calculate the smallest angle in the triangle with sides 4 cm, 7 cm and 9 cm.

THINK	WRITE/DRAW
1. Draw a labelled diagram of the triangle, call it ABC and fill in the given information. *Note:* The smallest angle will correspond to the smallest side.	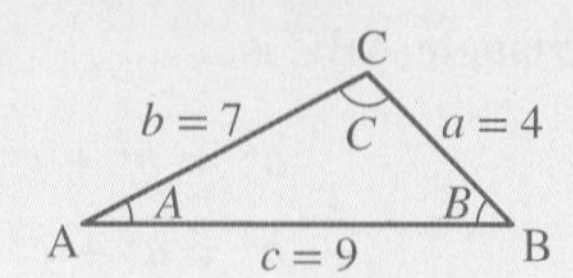 Let $a = 4, b = 7, c = 9$
2. Check that one of the criteria for the cosine rule has been satisfied.	The cosine rule can be used since three side lengths have been given.
3. Write down the appropriate cosine rule to calculate angle A.	$\cos(A) = \dfrac{b^2 + c^2 - a^2}{2bc}$
4. Substitute the given values into the rearranged rule.	$= \dfrac{7^2 + 9^2 - 4^2}{2 \times 7 \times 9}$
5. Evaluate.	$= \dfrac{114}{126}$
6. Transpose the equation to make A the subject by taking the inverse cos of both sides.	$A = \cos^{-1}\left(\dfrac{114}{126}\right)$ $\approx 25.208\,765\,3°$
7. Round off the answer to degrees and minutes.	$\approx 25°13'$

TI | THINK	DISPLAY/WRITE	CASIO | THINK	DISPLAY/WRITE
In a new problem, on a Calculator page, complete the entry lines as: $\dfrac{b^2 + c^2 - a^2}{2bc} \mid a = 4 \text{ and } b = 7 \text{ and } c = 9$ $\cos^{-1}$ (ans) Convert the angle to DMS as shown previously. Press ENTER after each entry.	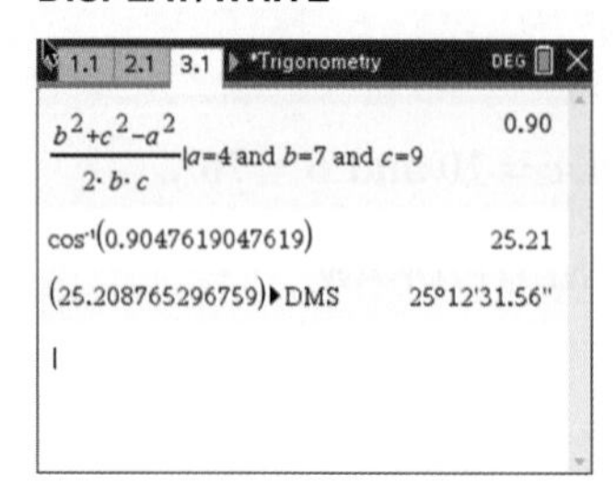 The smallest angle is 25°13′ rounded up to the nearest minute.	On the Main screen, complete the entry lines as: $\dfrac{b^2 + c^2 - a^2}{2b \times c} \mid a = 4 \mid b = 7 \mid c = 9$ $\cos^{-1}$ (0.9047619048) Convert the angle to DMS as shown previously. Press EXE after each entry.	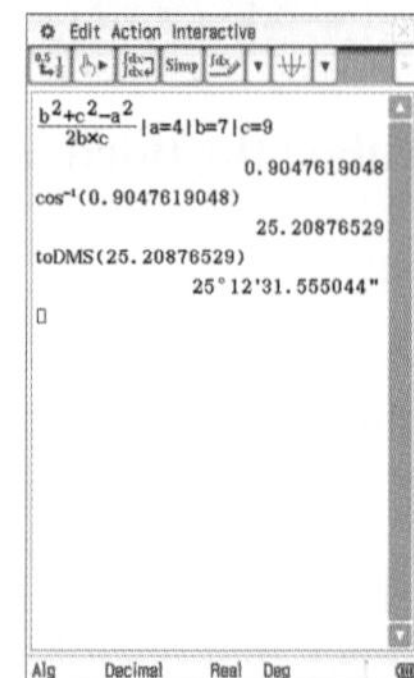 The smallest angle, rounded up to the nearest minute, is 25°13′.

WORKED EXAMPLE 6 Applying the cosine rule to solve problems

Two rowers, Harriet and Kate, set out from the same point. Harriet rows N70°E for 2000 m and Kate rows S15°W for 1800 m. Calculate the distance between the two rowers, correct to 2 decimal places.

THINK	WRITE/DRAW
1. Draw a labelled diagram of the triangle, call it ABC and fill in the given information.	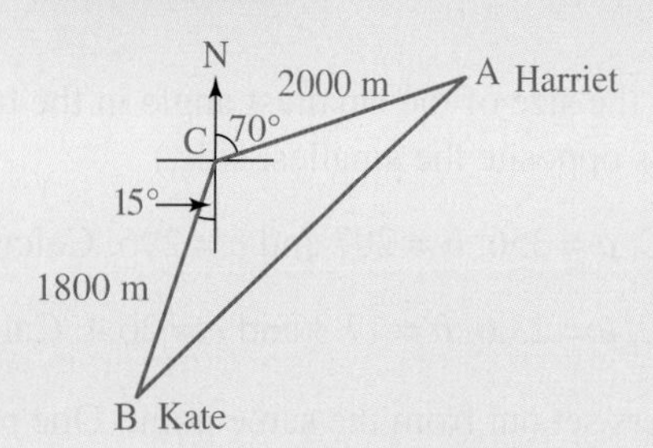
2. Check that one of the criteria for the cosine rule has been satisfied.	The cosine rule can be used since two side lengths and the included angle have been given.
3. Write down the appropriate cosine rule to calculate side c.	To calculate side c: $c^2 = a^2 + b^2 - 2ab\cos(C)$
4. Substitute the given values into the rule.	$= 2000^2 + 1800^2 - 2 \times 2000 \times 1800\cos(125°)$
5. Evaluate.	$\approx 11\,369\,750.342$ $c \approx \sqrt{11\,369\,750.342}$
6. Round off the answer to 2 decimal places.	≈ 3371.91
7. Write the answer.	The rowers are 3371.91 m apart.

DISCUSSION

In what situations would you use the sine rule rather than the cosine rule?

on Resources

 **eWorkbook** Topic 17 Workbook (worksheets, code puzzle and project) (ewbk-2043)

 **Interactivity** The cosine rule (int-6276)

Exercise 17.3 The cosine rule

learn on

Individual pathways

■ PRACTISE	■ CONSOLIDATE	■ MASTER
1, 2, 3, 4, 7, 9, 14, 17	5, 6, 8, 10, 15, 18	11, 12, 13, 16, 19

To answer questions online and to receive **immediate corrective feedback** and **fully worked solutions** for all questions, go to your learnON title at www.jacplus.com.au.

Where appropriate in this exercise, write your angles correct to the nearest minute and side lengths correct to 2 decimal places.

Fluency

1. WE4 Calculate the third side of triangle ABC given $a = 3.4$, $b = 7.8$ and $C = 80°$.
2. In triangle ABC, $b = 64.5$, $c = 38.1$ and $A = 58°34'$. Calculate the value of a.
3. In triangle ABC, $a = 17$, $c = 10$ and $B = 115°$. Calculate the value of b, and hence calculate the values of A and C.
4. WE5 Calculate the size of the smallest angle in the triangle with sides 6 cm, 4 cm and 8 cm. (*Hint:* The smallest angle is opposite the smallest side.)
5. In triangle ABC, $a = 356$, $b = 207$ and $c = 296$. Calculate the size of the largest angle.
6. In triangle ABC, $a = 23.6$, $b = 17.3$ and $c = 26.4$. Calculate the size of all the angles.
7. WE6 Two rowers set out from the same point. One rows N30°E for 1500 m and the other rows S40°E for 1200 m. Calculate the distance between the two rowers, correct to the nearest metre.
8. Maria cycles 12 km in a direction N68°W and then 7 km in a direction of N34°E.
   a. Calculate her distance from the starting point.
   b. Determine the bearing of the starting point from her finishing point.

Understanding

9. A garden bed is in the shape of a triangle, with sides of length 3 m, 4.5 m and 5.2 m.
   a. Calculate the size of the smallest angle.
   b. Hence, calculate the area of the garden, correct to 2 decimal places. (*Hint:* Draw a diagram, with the longest length as the base of the triangle.)
10. A hockey goal is 3 m wide. When Sophie is 7 m from one post and 5.2 m from the other, she shoots for goal. Determine within what angle, to the nearest degree, the shot must be made if it is to score a goal.
11. An advertising balloon is attached to two ropes 120 m and 100 m long. The ropes are anchored to level ground 35 m apart. Calculate the height of the balloon when both ropes are taut.

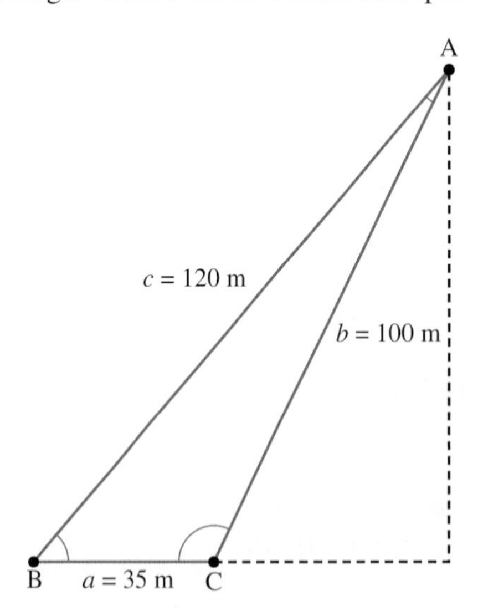

12. A plane flies in a direction of N70°E for 80 km and then on a bearing of S10°W for 150 km.
    a. Calculate the plane's distance from its starting point, correct to the nearest km.
    b. Calculate the plane's direction from its starting point.
13. Ship A is 16.2 km from port on a bearing of 053°T and ship B is 31.6 km from the same port on a bearing of 117°T. Calculate the distance between the two ships, in km correct to 1 decimal place.

Reasoning

14. A plane takes off at 10.00 am from an airfield and flies at 120 km/h on a bearing of N35°W. A second plane takes off at 10.05 am from the same airfield and flies on a bearing of S80°E at a speed of 90 km/h. Determine how far apart the planes are at 10.25 am, in km correct to 1 decimal place.

15. Three circles of radii 5 cm, 6 cm and 8 cm are positioned so that they just touch one another. Their centres form the vertices of a triangle. Determine the largest angle in the triangle. Show your working.

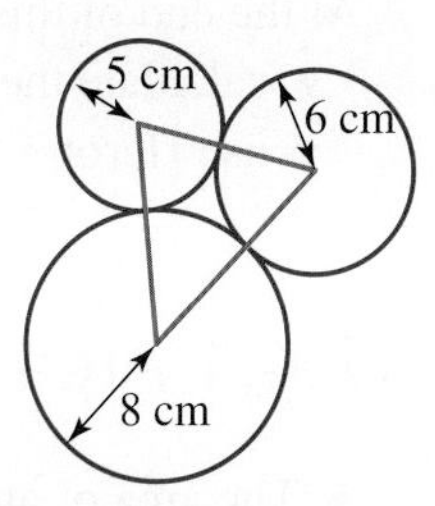

16. For the shape shown, determine:
    a. the length of the diagonal
    b. the magnitude (size) of angle B
    c. the length of x.

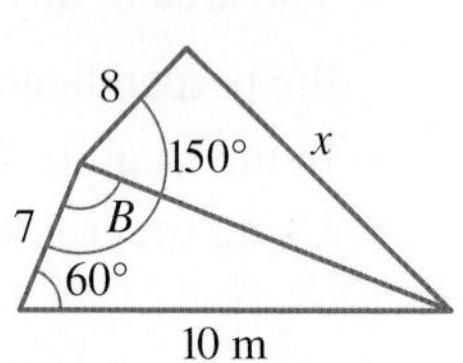

Problem solving

17. From the top of a vertical cliff 68 m high, an observer notices a yacht at sea. The angle of depression to the yacht is 47°. The yacht sails directly away from the cliff, and after 10 minutes the angle of depression is 15°. Determine the speed of the yacht, in km/h correct to 2 decimal places.

18. Determine the size of angles CAB, ABC and BCA.
    Give your answers in degrees correct to 2 decimal places.

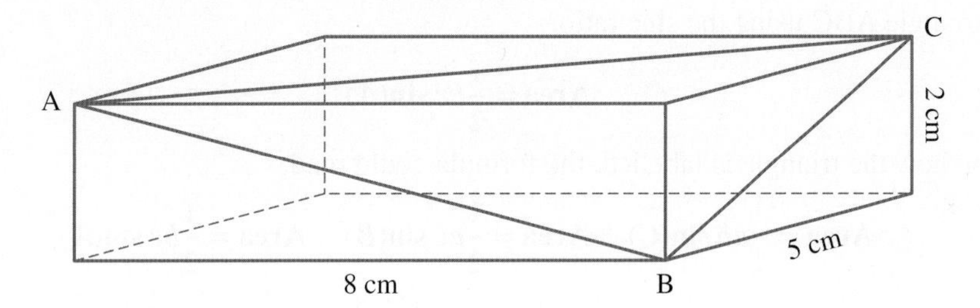

19. A vertical flag pole DB is supported by two wires AB and BC. AB is 5.2 metres long, BC is 4.7 metres long and B is 3.7 metres above ground level. Angle ADC is a right angle.
    a. Evaluate the distance from A to C, in metres correct to 4 decimal places.
    b. Determine the angle between AB and BC, in degrees correct to 2 decimal places.

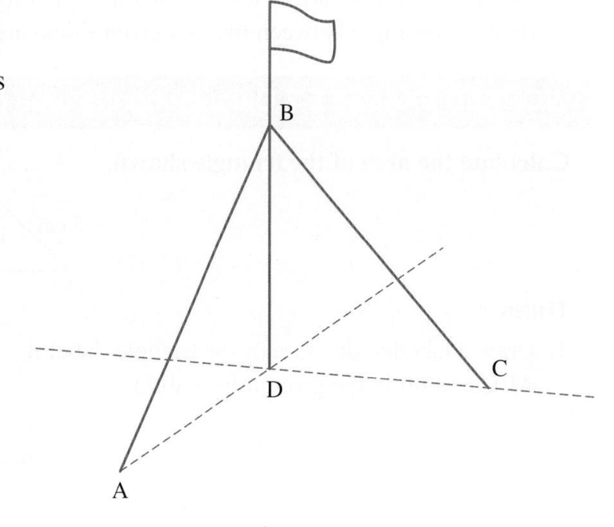

17.4 Area of triangles

LEARNING INTENTION

At the end of this subtopic you should be able to:
- calculate the area of a triangle, given two sides and the included angle
- use Heron's formula to calculate the area of a triangle, given the three sides.

17.4.1 Area of triangles

eles-5007

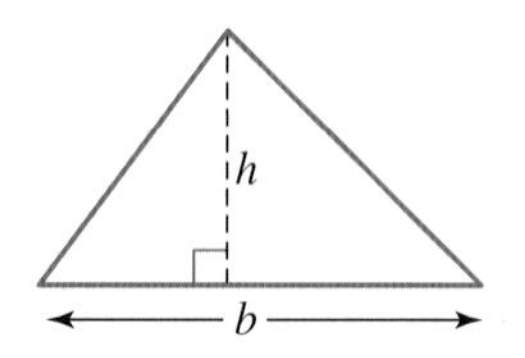

- The area of any triangle is given by the rule area $= \frac{1}{2}bh$, where b is the base and h is the perpendicular height of the triangle.
- In the triangle ABC, b is the base and h is the perpendicular height of the triangle.
- Using the trigonometric ratio for sine:

$$\sin(A) = \frac{h}{c}$$

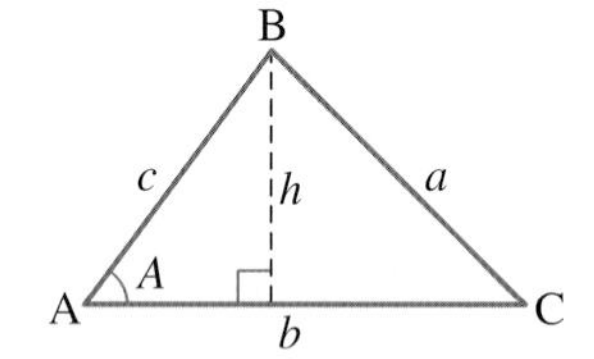

Transposing the equation to make h the subject, we obtain:

$$h = c\sin(A)$$

Area of triangle

- The area of triangle ABC using the sine ratio:

$$\textbf{Area} = \frac{1}{2}bc\sin(A)$$

- Depending on how the triangle is labelled, the formula could read:

$$\textbf{Area} = \frac{1}{2}ab\sin(C) \quad \textbf{Area} = \frac{1}{2}ac\sin(B) \quad \textbf{Area} = \frac{1}{2}bc\sin(A)$$

- The area formula may be used on any triangle provided that two sides of the triangle and the included angle (that is, the angle between the two given sides) are known.

WORKED EXAMPLE 7 Calculating the area of a triangle

Calculate the area of the triangle shown.

THINK	WRITE/DRAW
1. Draw a labelled diagram of the triangle, label it ABC and fill in the given information.	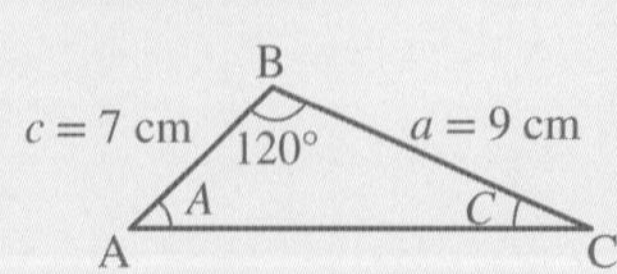 Let $a = 9$ cm, $c = 7$ cm, $B = 120°$.
2. Check that the criterion for the area rule has been satisfied.	The area rule can be used since two side lengths and the included angle have been given.
3. Write down the appropriate rule for the area.	Area $= \frac{1}{2}ac\sin(B)$

4. Substitute the known values into the rule. $= \frac{1}{2} \times 9 \times 7 \times \sin(120°)$

5. Evaluate. Round off the answer to 2 decimal places and include the appropriate unit. $\approx 27.28\text{ cm}^2$

WORKED EXAMPLE 8 Determining angles in a triangle and its area

A triangle has known dimensions of $a = 5$ cm, $b = 7$ cm and $B = 52°$. Determine A and C and hence the area.

THINK / **WRITE/DRAW**

1. Draw a labelled diagram of the triangle, label it ABC and fill in the given information.

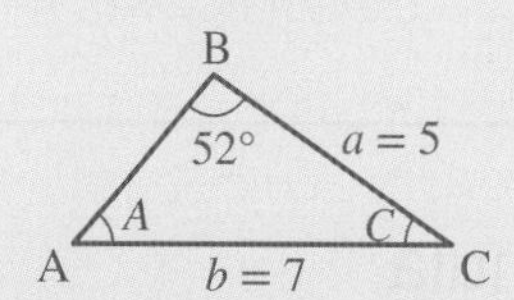

Let $a = 5, b = 7, B = 52°$.

2. Check whether the criterion for the area rule has been satisfied.

The area rule cannot be used since the included angle has not been given.

3. Write down the sine rule to calculate A.

To calculate angle A:

$$\frac{a}{\sin(A)} = \frac{b}{\sin(B)}$$

4. Substitute the known values into the rule.

$$\frac{5}{\sin(A)} = \frac{7}{\sin(52°)}$$

5. Transpose the equation to make $\sin A$ the subject.

$$5\sin(52°) = 7\sin(A)$$

$$\sin(A) = \frac{5\sin(52°)}{7}$$

6. Evaluate.

$$A = \sin^{-1}\left(\frac{5\sin(52°)}{7}\right)$$

$$\approx 34.254\,15187°$$

7. Round off the answer to degrees and minutes.

$\approx 34°15'$

8. Determine the value of the included angle, C, using the fact that the angle sum of any triangle is 180°.

$$C \approx 180° - (52° + 34°15')$$
$$= 93°45'$$

9. Write down the appropriate rule for the area.

$$\text{Area} = \frac{1}{2}ab\sin(C)$$

10. Substitute the known values into the rule.

$$\approx \frac{1}{2} \times 5 \times 7 \times \sin(93°45')$$

11. Evaluate. Round off the answer to 2 decimal places and include the appropriate unit.

$\approx 17.46\text{ cm}^2$

TI | THINK	DISPLAY/WRITE	CASIO | THINK	DISPLAY/WRITE
In a new problem, open a Calculator page. To calculate the angle A, complete the entry line as: $\sin^{-1}\left(\frac{5\sin(52)}{7}\right)$ Note that you can leave the angle in decimal degrees and work with this value. Determine the value of C as shown in the screenshot. Then calculate the area by completing the entry line as: $\frac{1}{2}\times 5\times 7\sin(93.75)$ Press ENTER after each entry.	2.1 3.1 4.1 *Trigonometry DEG $\sin^{-1}\left(\frac{5\cdot\sin(52)}{7}\right)$ 34.25 180−(52+34.254151866661) 93.75 $\frac{1}{2}\cdot 5\cdot 7\cdot\sin(93.75)$ 17.46 $A = 34.25°$ $C = 93.75°$ The area of the triangle is 17.46 cm^2.	To calculate the angle A, on the Main screen, complete the entry line as: $\sin^{-1}\left(\frac{5\sin(52)}{7}\right)$ Note that we can leave the angle in decimal degrees and work with this value. Determine the value of C as shown. Then calculate the area by completing the entry line as: $\frac{1}{2}\times 5\times 7\sin(93.74584813)$ Press EXE after each entry.	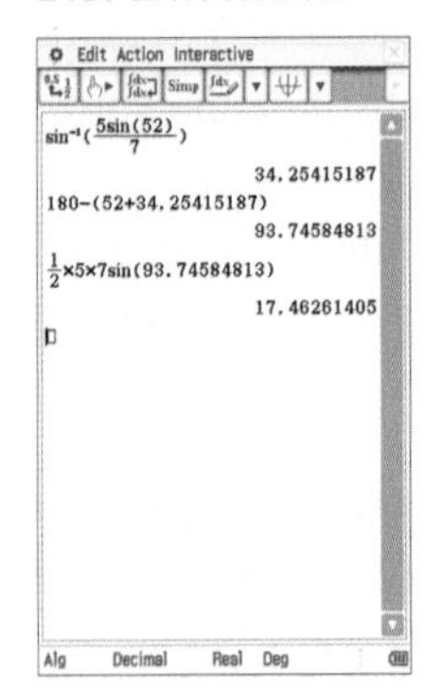 $A = 34.25°$ $C = 93.75°$ The area of the triangle is 17.46 cm^2.

17.4.2 Heron's formula

eles-5008

- If the lengths of all the sides of the triangle are known but none of the angles **Heron's formula** could be used to calculate the area.

Heron's formula

- The area of a triangle is given as:

$$\textbf{Area} = \sqrt{s(s-a)(s-b)(s-c)}$$

where s is the *semi-perimeter* of the triangle:

$$s = \frac{1}{2}(a+b+c)$$

Note: The proof of this formula is beyond the scope of this course.

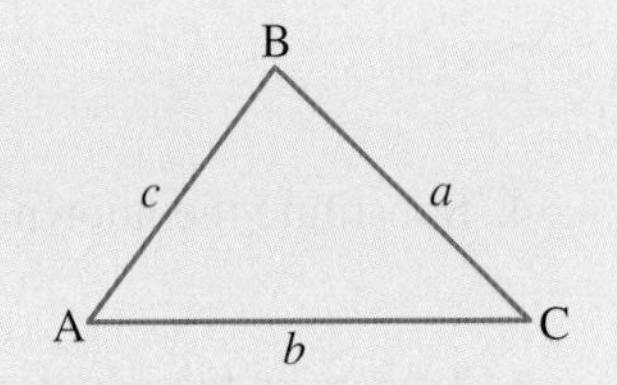

WORKED EXAMPLE 9 Calculating the area of a triangle using Heron's formula

Calculate the area of the triangle with sides of 4 cm, 6 cm and 8 cm.

THINK	WRITE/DRAW
1. Draw a labelled diagram of the triangle, call it ABC and fill in the given information.	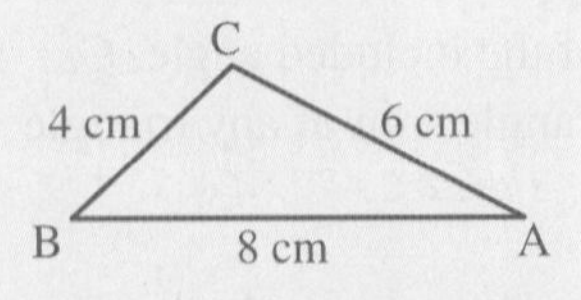 Let $a = 4$, $b = 6$, $c = 8$.
2. Determine which area rule will be used.	Since three side lengths have been given, use Heron's formula.
3. Write down the rule for Heron's formula.	Area $= \sqrt{s(s-a)(s-b)(s-c)}$
4. Write down the rule for s, the semi-perimeter of the triangle.	$s = \frac{1}{2}(a+b+c)$

5. Substitute the given values into the rule for the semi-perimeter.

$$= \frac{1}{2}(4+6+8)$$
$$= 9$$

6. Substitute all of the known values into Heron's formula.

$$\text{Area} = \sqrt{9(9-4)(9-6)(9-8)}$$

7. Evaluate.

$$= \sqrt{9 \times 5 \times 3 \times 1}$$
$$= \sqrt{135}$$
$$\approx 11.618\,950\,04$$

8. Round off the answer to 2 decimal places and include the appropriate unit.

$$\approx 11.62\,\text{cm}^2$$

DISCUSSION

List three formulas for calculating the area of a triangle.

on Resources

 **eWorkbook** Topic 17 Workbook (worksheets, code puzzle and project) (ewbk-2043)

 **Interactivities** Area of triangles (int-6483)
Using Heron's formula to calculate the area of a triangle (int-6475)

Exercise 17.4 Area of triangles

learnon

Individual pathways

■ PRACTISE	■ CONSOLIDATE	■ MASTER
1, 4, 7, 10, 12, 15, 18, 21	2, 5, 8, 11, 13, 16, 19, 22	3, 6, 9, 14, 17, 20, 23, 24

To answer questions online and to receive **immediate corrective feedback** and **fully worked solutions** for all questions, go to your learnON title at www.jacplus.com.au.

Where appropriate in this exercise, write your angles correct to the nearest minute and other measurements correct to 2 decimal places.

Fluency

1. **WE7** Calculate the area of the triangle ABC with $a = 7$, $b = 4$ and $C = 68°$.
2. Calculate the area of the triangle ABC with $a = 7.3$, $c = 10.8$ and $B = 104°40'$.
3. Calculate the area of the triangle ABC with $b = 23.1$, $c = 18.6$ and $A = 82°17'$.
4. **WE8** A triangle has $a = 10$ cm, $c = 14$ cm and $C = 48°$. Determine A and B and hence the area.
5. A triangle has $a = 17$ m, $c = 22$ m and $C = 56°$. Determine A and B and hence the area.
6. A triangle has $b = 32$ mm, $c = 15$ mm and $B = 38°$. Determine A and C and hence the area.

7. **MC** In a triangle, $a = 15$ m, $b = 20$ m and $B = 50°$. The area of the triangle is:

A. $86.2\ \text{m}^2$ B. $114.9\ \text{m}^2$ C. $149.4\ \text{m}^2$ D. $172.4\ \text{m}^2$ E. $183.2\ \text{m}^2$

8. **WE9** Calculate the area of the triangle with sides of 5 cm, 6 cm and 8 cm.

9. Calculate the area of the triangle with sides of 40 mm, 30 mm and 5.7 cm.

10. Calculate the area of the triangle with sides of 16 mm, 3 cm and 2.7 cm.

11. **MC** A triangle has sides of length 10 cm, 14 cm and 20 cm. The area of the triangle is:

A. $41\ \text{cm}^2$ B. $65\ \text{cm}^2$ C. $106\ \text{cm}^2$ D. $137\ \text{cm}^2$ E. $155\ \text{cm}^2$

Understanding

12. A piece of metal is in the shape of a triangle with sides of length 114 mm, 72 mm and 87 mm. Calculate its area using Heron's formula.

13. A triangle has the largest angle of 115°. The longest side is 62 cm and another side is 35 cm. Calculate the area of the triangle to the nearest whole number.

14. A triangle has two sides of 25 cm and 30 cm. The angle between the two sides is 30°. Determine:

a. its area b. the length of its third side c. its area using Heron's formula.

15. The surface of a fish pond has the shape shown in the diagram. Calculate how many goldfish can the pond support if each fish requires $0.3\ \text{m}^2$ surface area of water.

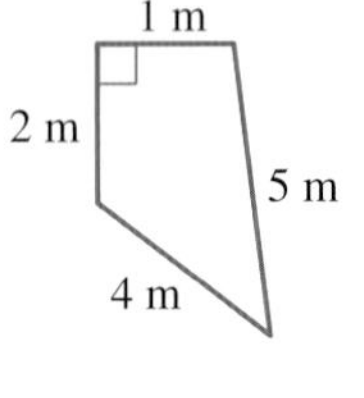

16. **MC** A parallelogram has sides of 14 cm and 18 cm and an angle between them of 72°. The area of the parallelogram is:

A. $118.4\ \text{cm}^2$ B. $172.4\ \text{cm}^2$ C. $239.7\ \text{cm}^2$ D. $252\ \text{cm}^2$ E. $388.1\ \text{cm}^2$

17. **MC** An advertising hoarding is in the shape of an isosceles triangle, with sides of length 15 m, 15 m and 18 m. It is to be painted with two coats of purple paint. If the paint covers $12\ \text{m}^2$ per litre, the amount of paint needed, to the nearest litre, would be:

A. 9 L B. 18 L C. 24 L D. 36 L E. 41 L

Reasoning

18. A parallelogram has diagonals of length 10 cm and 17 cm. An angle between them is 125°. Determine:

a. the area of the parallelogram
b. the dimensions of the parallelogram.

19. A lawn is to be made in the shape of a triangle, with sides of length 11 m, 15 m and 17.2 m. Determine how much grass seed, to the nearest kilogram, needs to be purchased if it is sown at the rate of 1 kg per $5\ \text{m}^2$.

20. A bushfire burns out an area of level grassland shown in the diagram. (*Note:* This is a sketch of the area and is not drawn to scale.) Evaluate the area, in hectares correct to 1 decimal place, of the land that is burned.

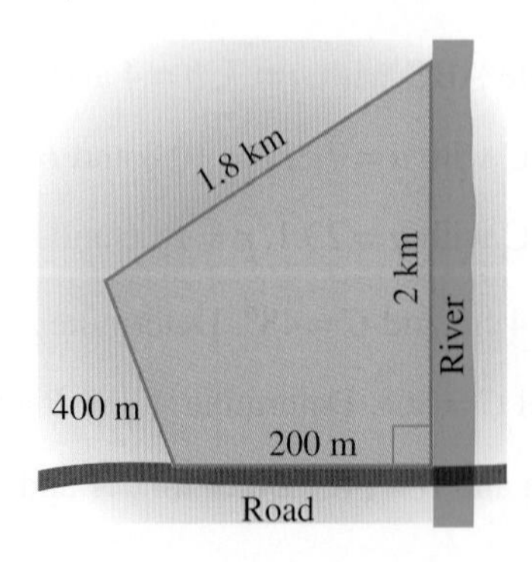

Problem solving

21. An earth embankment is 27 m long and has a vertical cross-section shown in the diagram. Determine the volume of earth needed to build the embankment, correct to the nearest cubic metre.

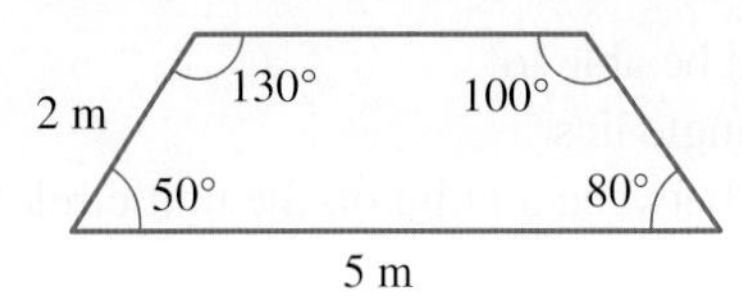

22. Evaluate the area of this quadrilateral.

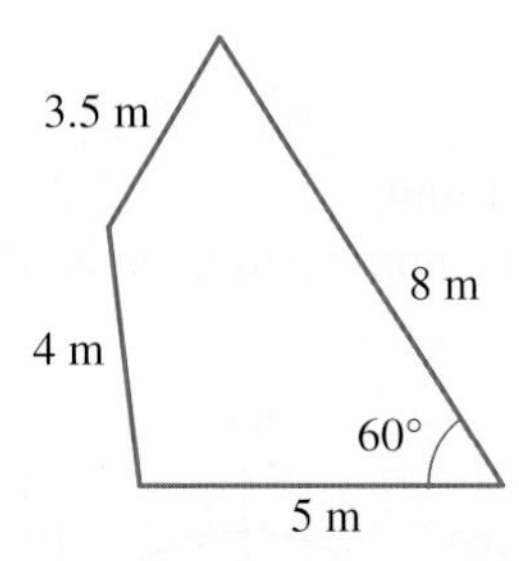

23. A surveyor measured the boundaries of a property as shown. The side AB could not be measured because it crossed through a marsh.
The owner of the property wanted to know the total area and the length of the side AB.
Give all lengths correct to 2 decimal places and angles to the nearest degree.
 a. Calculate the area of the triangle ACD.
 b. Calculate the distance AC.
 c. Calculate the angle CAB.
 d. Calculate the angle ACB.
 e. Calculate the length AB.
 f. Determine the area of the triangle ABC.
 g. Determine the area of the property.

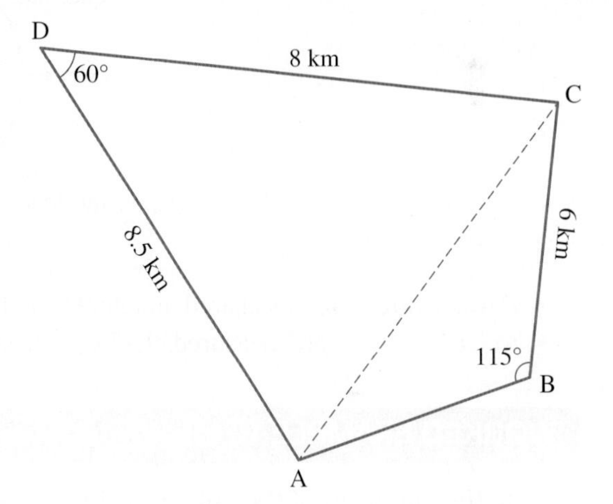

24. A regular hexagon has sides of length 12 centimetres. It is divided into six smaller equilateral triangles. Evaluate the area of the hexagon, giving your answer correct to 2 decimal places.

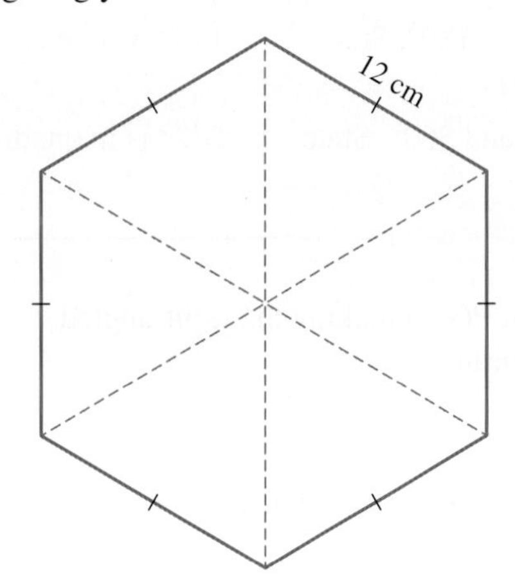

17.5 The unit circle

LEARNING INTENTION

At the end of this subtopic you should be able to:
- determine in which quadrant an angle lies
- use and interpret the relationship between a point on the unit circle and the angle made with the positive x-axis
- use the unit circle to determine approximate trigonometric ratios for angles greater than 90°.

17.5.1 The unit circle

eles-5009

- A **unit circle** is a circle with a radius of 1 unit.
- The unit circle is divided into 4 quadrants, numbered in an anticlockwise direction, as shown in the diagram.

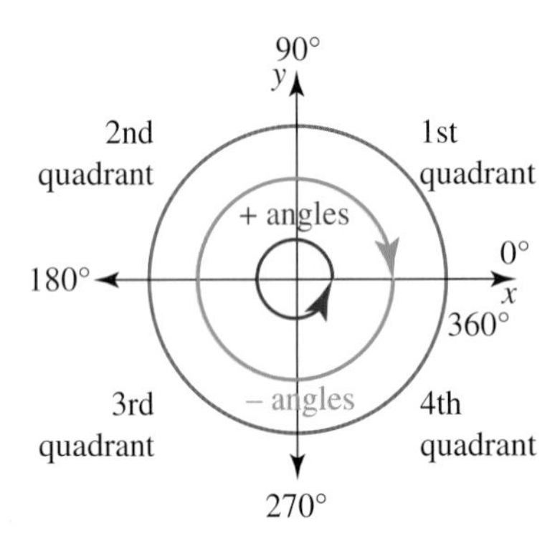

- *Positive angles* are measured anticlockwise from 0°.
- *Negative angles* are measured clockwise from 0°.

WORKED EXAMPLE 10 Identifying where an angle lies on the unit circle

State the quadrant of the unit circle in which each of the following angles is found.

a. 145° **b. 282°**

THINK	WRITE
a. The given angle is between 90° and 180°. State the appropriate quadrant.	a. 145° is in quadrant 2.
b. The given angle is between 270° and 360°. State the appropriate quadrant.	b. 282° is in quadrant 4.

- Consider the unit circle with point P(x, y) making the right-angled triangle OPN as shown in the diagram.
- Using the trigonometric ratios:

$$\frac{x}{1} = \cos(\theta), \ \frac{y}{1} = \sin(\theta), \ \frac{y}{x} = \tan(\theta)$$

where θ is measured anticlockwise from the positive x-axis.

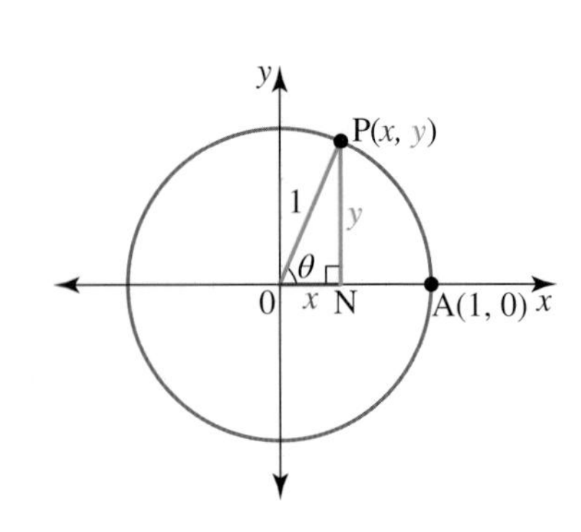

Calculate value of sine, cosine and tangent

To calculate the value of sine, cosine or tangent of any angle θ from the unit circle:

$$\cos(\theta) = x$$

$$\sin(\theta) = y$$

$$\tan(\theta) = \frac{y}{x} = \frac{\sin(\theta)}{\cos(\theta)}$$

17.5.2 The four quadrants of the unit circle

eles-5010

- Approximate values for sine, cosine and tangent of an angle can be found from the unit circle using the following steps, as shown in the diagram.
  Step 1: Draw a unit circle, label the x- and y-axes.
  Step 2: Mark the angles 0°, 90°, 180°, 270°, and 360°.
  Step 3: Draw the given angle θ.
  Step 4: Mark $x = \cos(\theta)$, $y = \sin(\theta)$.
  Step 5: Approximate the values of x and y and equate to give the values of $\cos(\theta)$ and $\sin(\theta)$.
- Where the angle lies in the unit circle determines whether the trigonometric ratio is positive or negative.

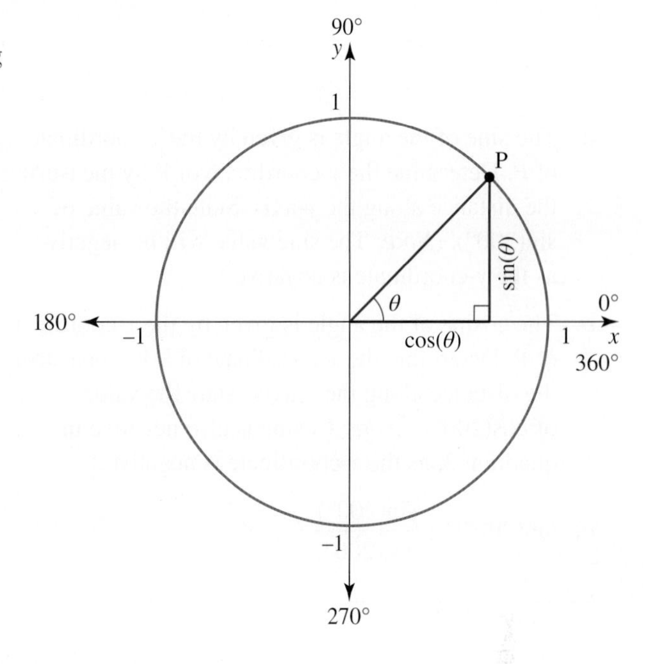

Sign of the trigonometric functions

- Consider the following.
  - In the first quadrant $x > 0$, $y > 0$; therefore **All** trig ratios are positive.
  - In the second quadrant $x < 0$, $y > 0$; therefore **Sine** (the y-value) is positive.
  - In the third quadrant $x < 0$, $y < 0$; therefore **Tangent** $\left(\frac{y}{x} \text{ values}\right)$ is positive.
  - In the fourth quadrant $x > 0$, $y < 0$; therefore **Cosine** (the x-value) is positive.

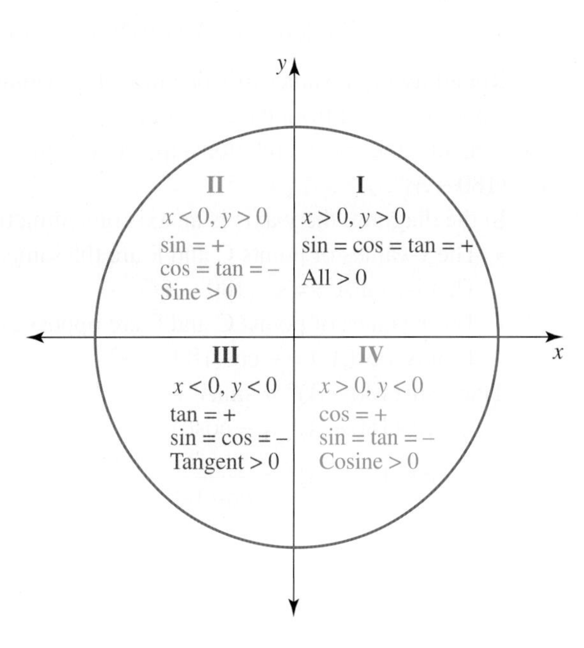

WORKED EXAMPLE 11 Using the unit circle to approximate trigonometric ratios of an angle

Determine the approximate value of each of the following using the unit circle.

a. $\sin(200°)$ **b.** $\cos(200°)$ **c.** $\tan(200°)$

THINK	WRITE/DRAW
Draw a unit circle and construct an angle of 200°. Label the point corresponding to the angle of 200° on the circle P. Highlight the lengths, representing the x- and y-coordinates of point P.	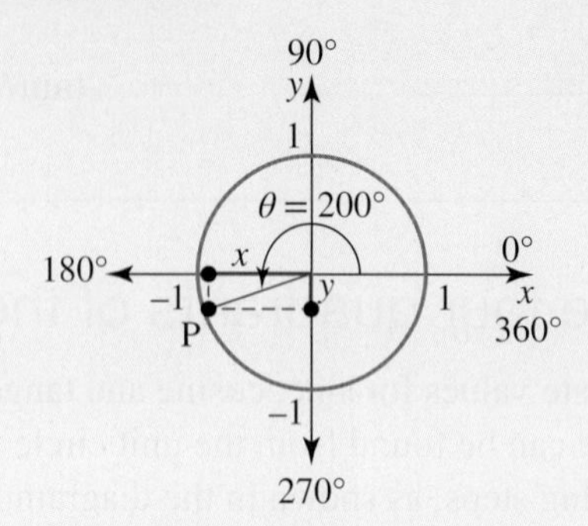
a. The sine of the angle is given by the y-coordinate of P. Determine the y-coordinate of P by measuring the distance along the y-axis. State the value of $\sin(200°)$. (*Note:* The sine value will be negative as the y-coordinate is negative.)	**a.** $\sin(200°) = -0.3$
b. The cosine of the angle is given by the x-coordinate of P. Determine the x-coordinate of P by measuring the distance along the x-axis. State the value of $\cos(200°)$. (*Note:* Cosine is also negative in quadrant 3, as the x-coordinate is negative.)	**b.** $\cos(200°) = -0.9$
c. $\tan(200°) = \dfrac{\sin(200°)}{\cos(200°)}$	**c.** $\dfrac{-0.3}{-0.9} = \dfrac{1}{3} = 0.3333$

- The approximate results obtained in Worked example 11 can be verified with the aid of a calculator:

$$\sin(200°) = -0.342\,020\,143, \cos(200°) = -0.939\,692\,62 \text{ and } \tan(200°) = 0.3640.$$

Rounding these values to 1 decimal place would give −0.3, −0.9 and 0.4 respectively, which match the values obtained from the unit circle.

- Consider the special relationship between the sine, cosine and tangent of supplementary angles, say $A°$ and $(180 - A)°$.

In the diagram, the y-axis is an axis of symmetry.

  - The y-values of points C and E are the same. That is, $\sin(A°) = \sin(180 - A)°$
  - The x-values of points C and E are opposites in value. That is, $\cos(A°) = -\cos(180 - A)°$

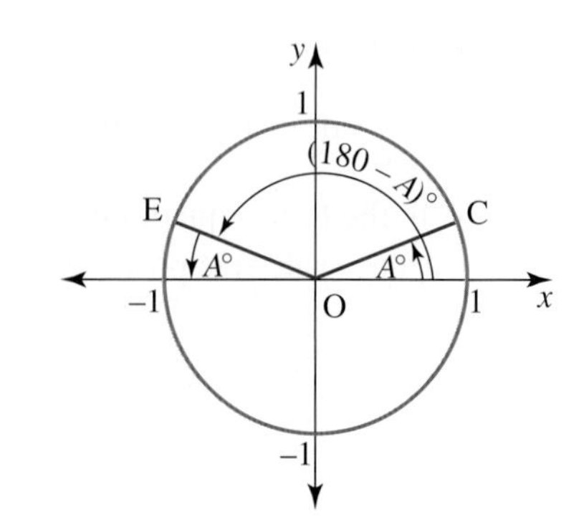

Thus: $\sin(180 - A)° = \sin(A°)$

$\cos(180 - A)° = -\cos(A°)$

$\cos(180 - A)° = \dfrac{\sin(180 - A)°}{\cos(180 - A)°} = \dfrac{\sin(A°)}{-\cos(A°)} = -\tan(A°)$

Resources

eWorkbook Topic 17 Workbook (worksheets, code puzzle and project) (ewbk-2043)

Exercise 17.5 The unit circle

learnON

Individual pathways

PRACTISE	CONSOLIDATE	MASTER
1, 2, 6, 8, 12, 17, 20	3, 4, 7, 9, 11, 13, 18, 21	5, 10, 14, 15, 16, 19, 22

To answer questions online and to receive **immediate corrective feedback** and **fully worked solutions** for all questions, go to your learnON title at www.jacplus.com.au.

Where appropriate in this exercise, give answers correct to 2 decimal places.

Fluency

1. **WE10** State which quadrant of the unit circle each of the following angles is in.
   a. 60° b. 130° c. 310°
   d. 260° e. 100° f. 185°
2. **MC** If $\theta = 43°$, the triangle drawn to show this would be in:
   A. quadrant 1 B. quadrant 2 C. quadrant 3 D. quadrant 4 E. none of these
3. **MC** If $\theta = 295°$, the triangle drawn to show this would be in:
   A. quadrant 1 B. quadrant 2 C. quadrant 3 D. quadrant 4 E. none of these
4. **WE11** Determine the approximate value of each of the following using the unit circle.
   a. sin(20°) b. cos(20°) c. cos(100°) d. sin(100°)
5. Determine the approximate value of each of the following using the unit circle.
   a. sin(320°) b. cos(320°) c. sin(215°) d. cos(215°)
6. Use the unit circle to determine the approximate value of each of the following.
   a. sin(90°) b. cos(90°) c. sin(180°) d. cos(180°)
7. Use the unit circle to determine the approximate value of each of the following.
   a. sin(270°) b. cos(270°) c. sin(360°) d. cos(360°)

Understanding

8. On the unit circle, use a protractor to measure an angle of 30° from the positive x-axis. Mark the point P on the circle. Use this point to construct a triangle in quadrant 1 as shown.
   a. Calculate the value of cos(30°). (Remember that the length of the adjacent side of the triangle is cos(30°).)
   b. Calculate the value of sin(30°). (This is the length of the opposite side of the triangle.)
   c. Check your answers in **a** and **b** by finding these values with a calculator.

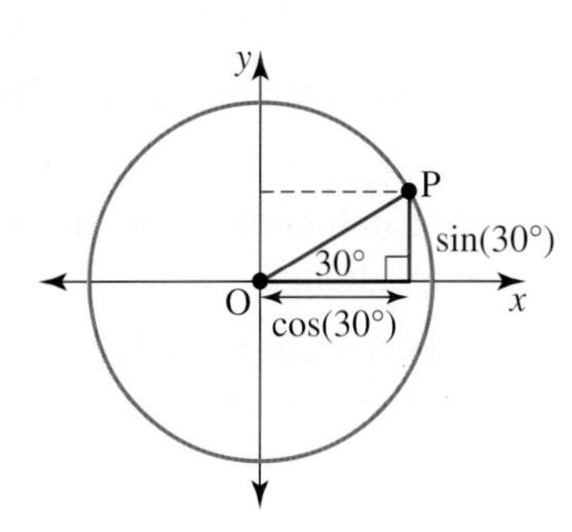

9. Using a graph of the unit circle, measure 150° with a protractor and mark the point P on the circle. Use this point to draw a triangle in quadrant 2 as shown.

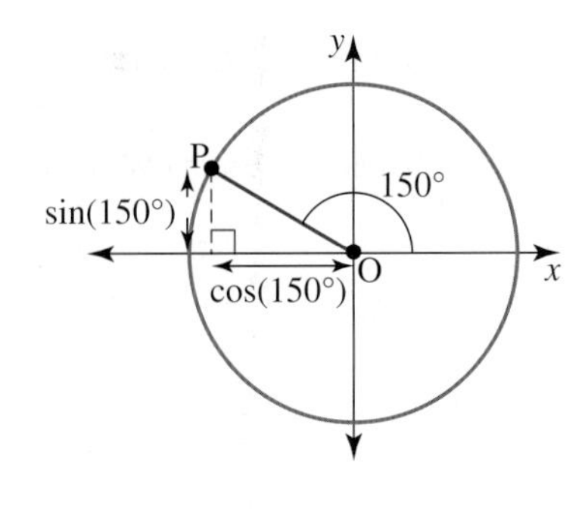

a. Determine the angle the radius OP makes with the negative x-axis.
b. Remembering that $x = \cos(\theta)$, use your circle to determine the value of $\cos(150°)$.
c. Comment on how $\cos(150°)$ compares to $\cos(30°)$.
d. Remembering that $y = \sin(\theta)$, use your circle to determine the value of $\sin(150°)$.
e. Comment on how $\sin(150°)$ compares with $\sin(30°)$.

10. On a unit circle, measure 210° with a protractor and mark the point P on the circle. Use this point to draw a triangle in quadrant 3 as shown.

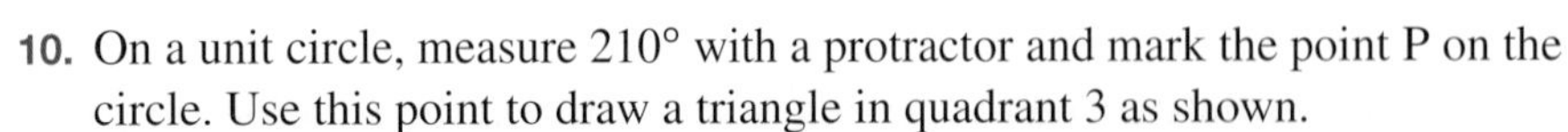

a. Determine the angle the radius OP makes with the negative x-axis.
b. Use your circle to determine the value of $\cos(210°)$.
c. Comment on how $\cos(210°)$ compares to $\cos(30°)$.
d. Use your circle to determine the value of $\sin(210°)$.
e. Comment on how $\sin(210°)$ compares with $\sin(30°)$.

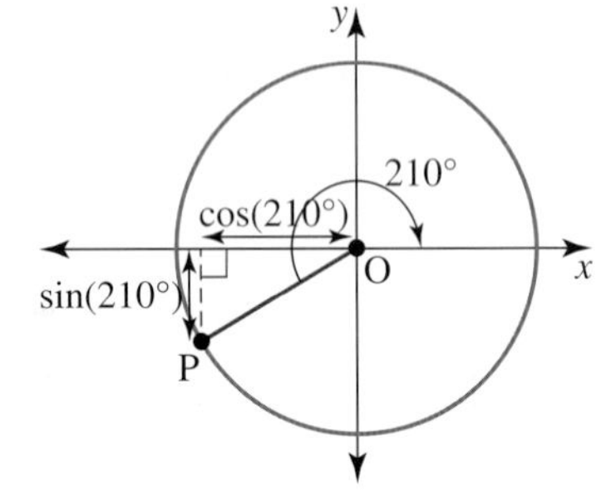

11. On a unit circle, measure 330° with a protractor and mark the point P on the circle. Use this point to draw a triangle in quadrant 4 as shown.

a. Determine the angle the radius OP makes with the positive x-axis.
b. Use your circle to determine the value of $\cos(330°)$.
c. Comment on how $\cos(330°)$ compares to $\cos(30°)$.
d. Use your circle to determine the value of $\sin(330°)$.
e. Comment on how $\sin(330°)$ compares with $\sin(30°)$.

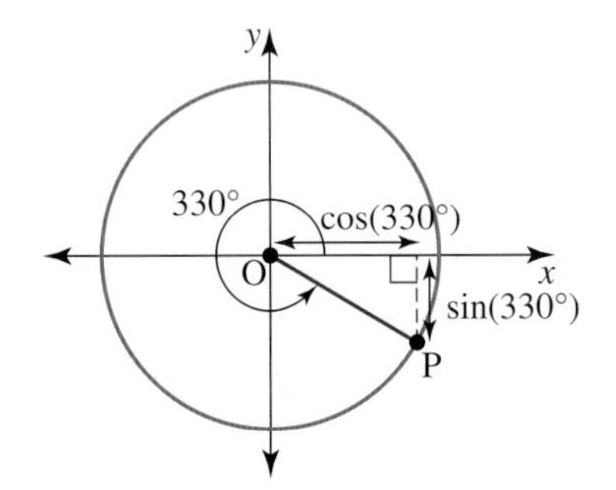

12. On a unit circle, draw an appropriate triangle for the angle of 20° in quadrant 1.

a. Determine the value of $\sin(20°)$.
b. Determine the value of $\cos(20°)$.
c. Draw a tangent line and extend the hypotenuse of the triangle to meet the tangent as shown.
Accurately measure the length of the tangent between the x-axis and the point where it meets the hypotenuse and, hence, state the value of $\tan(20°)$.
d. Determine the value of $\dfrac{\sin(20°)}{\cos(20°)}$.
e. Comment on how $\tan(20°)$ compares with $\dfrac{\sin(20°)}{\cos(20°)}$.

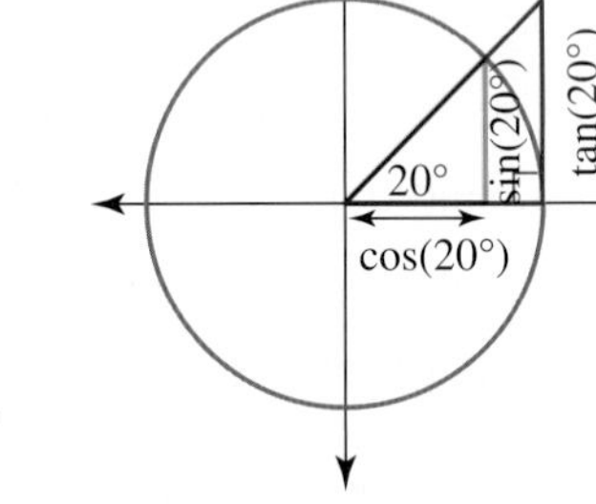

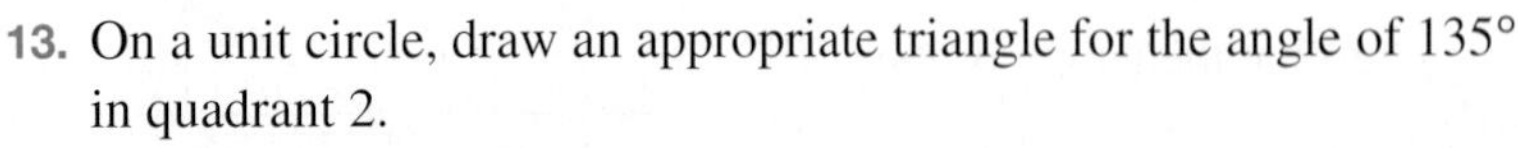

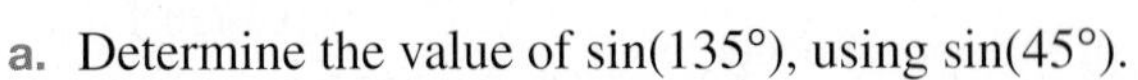

13. On a unit circle, draw an appropriate triangle for the angle of 135° in quadrant 2.

a. Determine the value of $\sin(135°)$, using $\sin(45°)$.
b. Determine the value of $\cos(135°)$, using $\cos(45°)$.
c. Draw a tangent line and extend the hypotenuse of the triangle to meet the tangent as shown.
Accurately measure the length of the tangent to where it meets the hypotenuse to calculate the value of $\tan(135°)$.

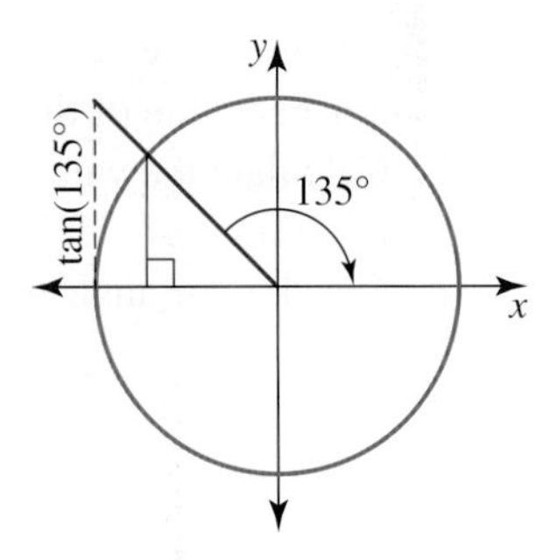

d. Determine the value of $\frac{\sin(135°)}{\cos(135°)}$.

e. Comment on how tan(135°) compares with $\frac{\sin(135°)}{\cos(135°)}$.

f. Comment on how tan(135°) compares with tan(45°).

14. On a unit circle, draw an appropriate triangle for the angle of 220° in quadrant 3.

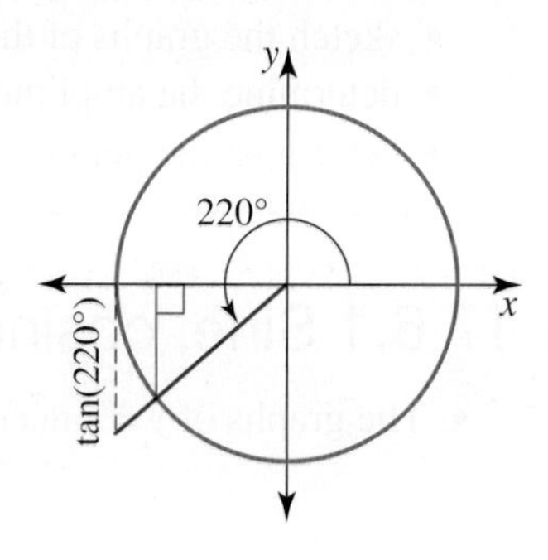

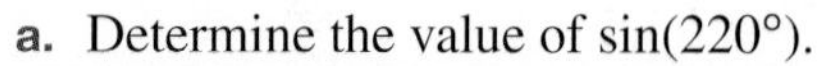

a. Determine the value of sin(220°).

b. Determine the value of cos(220°).

c. Draw a tangent line and extend the hypotenuse of the triangle to meet the tangent as shown.
Determine the value of tan(220°) by accurately measuring the length of the tangent to where it meets the hypotenuse.

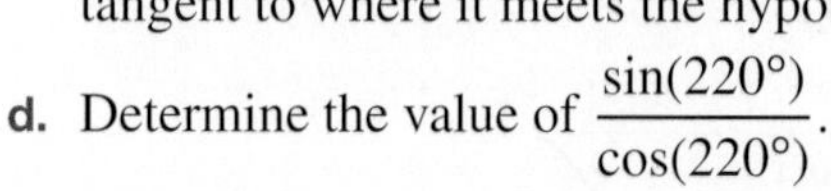

d. Determine the value of $\frac{\sin(220°)}{\cos(220°)}$.

e. Comment on how tan(220°) compares with $\frac{\sin(220°)}{\cos(220°)}$.

f. Comment on how tan(220°) compares with tan(40°). (Use a calculator.)

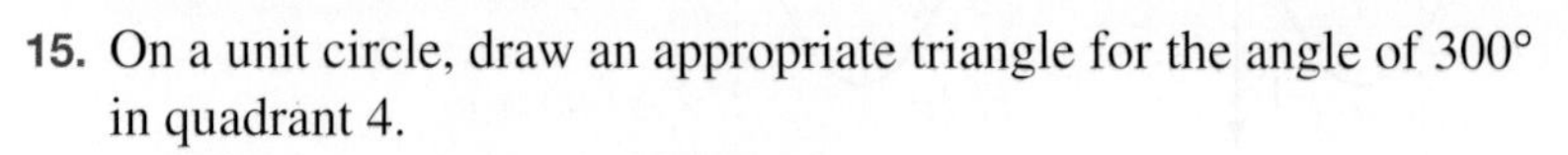

15. On a unit circle, draw an appropriate triangle for the angle of 300° in quadrant 4.

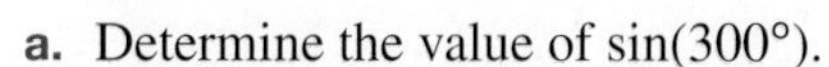

a. Determine the value of sin(300°).

b. Determine the value of cos(300°).

c. Draw a tangent line and extend the hypotenuse of the triangle to meet the tangent as shown.
Determine the value of tan(300°) by accurately measuring the length of the tangent to where it meets the hypotenuse.

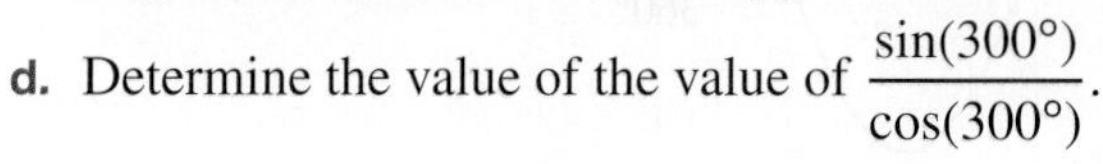

d. Determine the value of the value of $\frac{\sin(300°)}{\cos(300°)}$.

e. Comment on how tan(300°) compares with $\frac{\sin(300°)}{\cos(300°)}$.

f. Comment on how tan(300°) compares with tan(60°). (Use a calculator.)

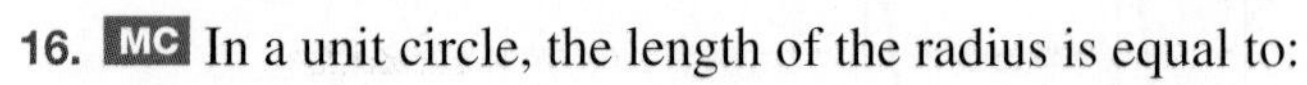

16. **MC** In a unit circle, the length of the radius is equal to:

A. $\sin(\theta)$ B. $\cos(\theta)$ C. $\tan(\theta)$ D. 1 E. none of these

Reasoning

17. Show that $\sin^2(\alpha°) + \cos^2(\alpha°) = 1$.

18. Show that $1 - \sin^2(180 - \alpha)° = \cos^2(180 - \alpha)°$.

19. Show that $1 + \tan^2(\alpha°) = \sec^2(\alpha°)$, where $\sec(\alpha°) = \frac{1}{\cos(\alpha°)}$.

Problem solving

20. If $\sin(x°) = p$, $0 \le x \le 90°$, write each of the following in terms of p.

a. $\cos(x°)$ b. $\sin(180 - x)°$ c. $\cos(180 - x)°$

21. Simplify $\sin(180 - x)° - \sin(x°)$.

22. Simplify $\cos(180 - x)° + \cos(x°)$, where $0 < x° < 90°$.

17.6 Trigonometric functions

LEARNING INTENTION

At the end of this subtopic you should be able to:
- sketch the graphs of the sine, cosine and tangent graphs
- determine the amplitude of a given trigonometric function
- determine the period of a given trigonometric function.

17.6.1 Sine, cosine and tangent graphs

eles-5011

- The graphs of $y = \sin(x)$, $y = \cos(x)$ and $y = \tan(x)$ are shown below.

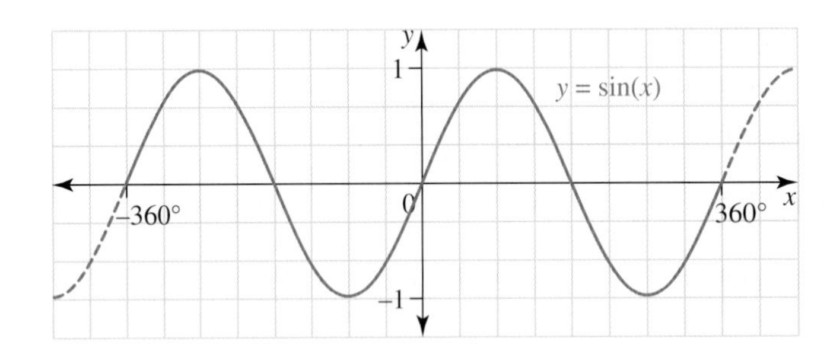

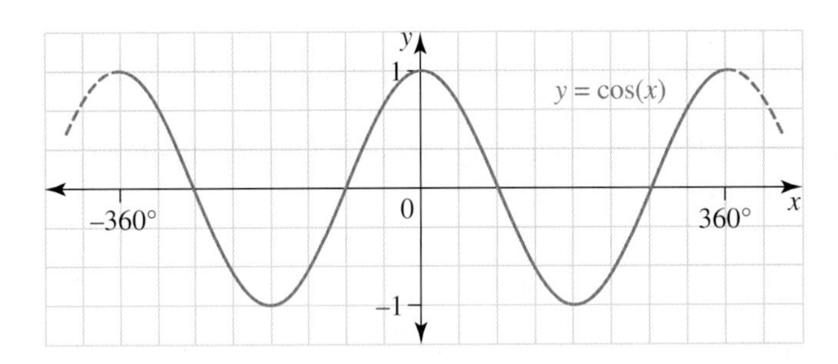

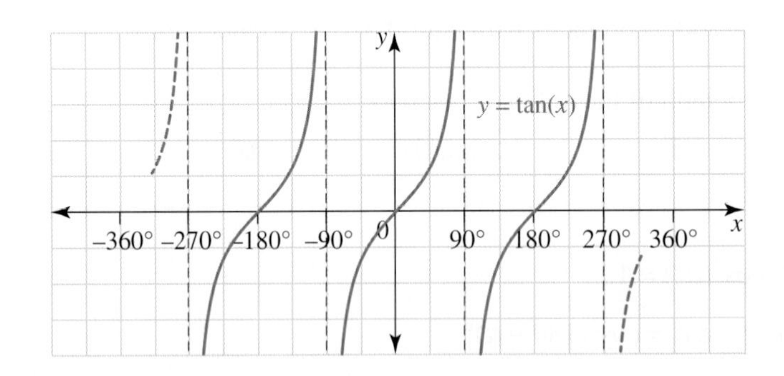

- Trigonometric graphs repeat themselves continuously in cycles, and hence they are called **periodic functions**.
- The **period** of the graph is the horizontal distance between repeating peaks or troughs. The period between the repeating peaks for $y = \sin(x)$ and $y = \cos(x)$ is 360°. The period of the graph $y = \tan(x)$ is 180°, and **asymptotes** occur at $x = 90°$ and intervals of 180°.
- The **amplitude** of a periodic graph is half the distance between the maximum and minimum values of the function. Amplitude can also be described as the amount by which the graph goes above and below its mean value, which is the x-axis for the graphs of $y = \sin(x)$, $y = \cos(x)$ and $y = \tan(x)$.

- The following can be summarised from the graphs of the trigonometric functions.

Graph	Period	Amplitude
$y = \sin(x)$	360°	1
$y = \cos(x)$	360°	1
$y = \tan(x)$	180°	Undefined

Translations of trigonometric graphs

- The sine, cosine and tangent graphs can be dilated, translated and reflected in the same way as other functions, studied earlier.
- These translations are summarised in the table below.

Graph	Period	Amplitude
$y = a\sin(nx)$	$\frac{360°}{n}$	a
$y = a\cos(nx)$	$\frac{360°}{n}$	a
$y = a\tan(nx)$	$\frac{180°}{n}$	Undefined

- If $a < 0$, the graph is reflected in the x-axis. The amplitude is always the positive value of a.

WORKED EXAMPLE 12 Sketching periodic functions

Sketch the graphs of a $y = 2\sin(x)$ and b $y = \cos(2x)$ for $0° \le x \le 360°$.

THINK

a. 1. The graph must be drawn from 0° to 360°.
2. Compared to the graph of $y = \sin(x)$ each value of $\sin(x)$ has been multiplied by 2, therefore the amplitude of the graph must be 2.
3. Label the graph $y = 2\sin(x)$.

WRITE/DRAW

a.

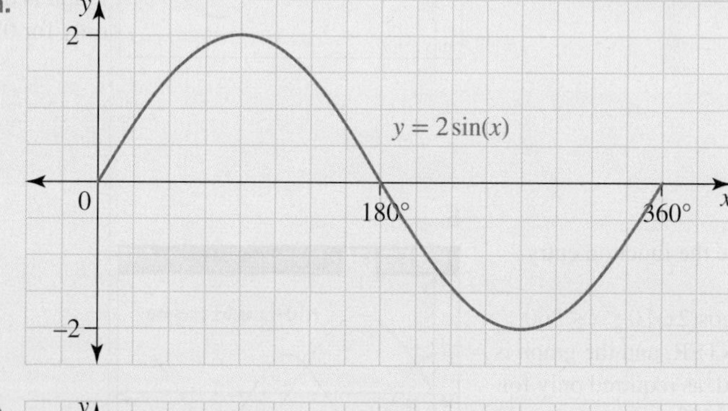

THINK

b. 1. The graph must be drawn from 0° to 360°.
2. Compared to the graph of $y = \cos(x)$, each value of x has been multiplied by 2, therefore the period of the graph must become 180°.
3. Label the graph $y = \cos(2x)$.

WRITE/DRAW

b.

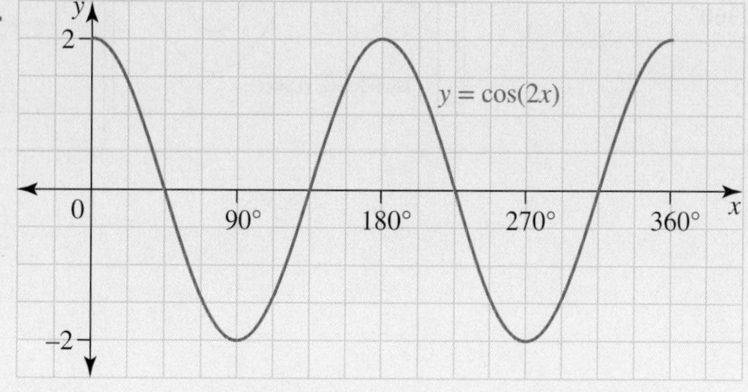

TI | THINK

DISPLAY/WRITE

a.

1. In a new problem, on a Graphs page, ensure the Graphs & Geometry Settings are set to the degrees mode, as shown in the screenshot.
   To do this, press:
   - MENU
   - 9: Settings

   For Display Digits select 'Fix 2'.
   TAB to Graphing Angle and select 'Degree'. TAB to OK and press ENTER.

a.

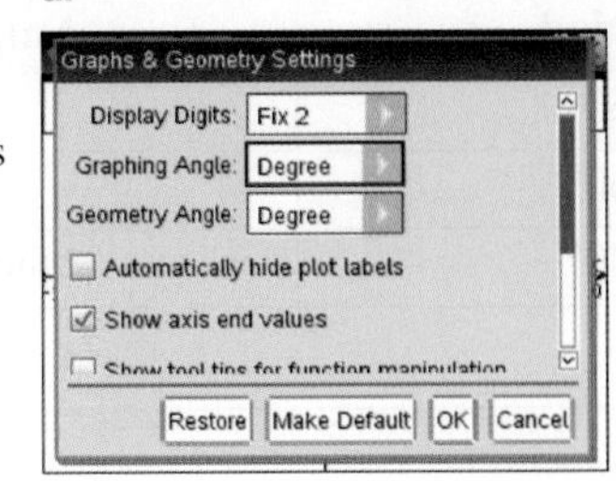

2. To set an appropriate viewing window, press:
   - MENU
   - 4: Window/Zoom
   - 1: Window Settings...

   Select the values as shown in the screenshot.
   TAB to OK and press ENTER.

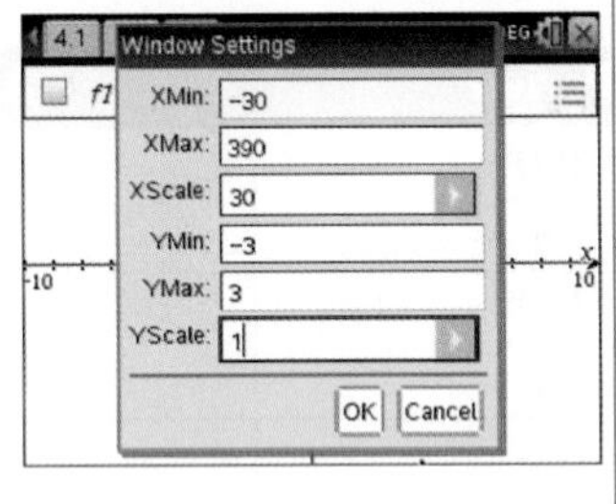

3. Complete the function entry line as:
   $f1(x) = 2\sin(x) \mid 0 \le x \le 360$
   Press ENTER. The graph is displayed as required for $0° \le x \le 360°$.

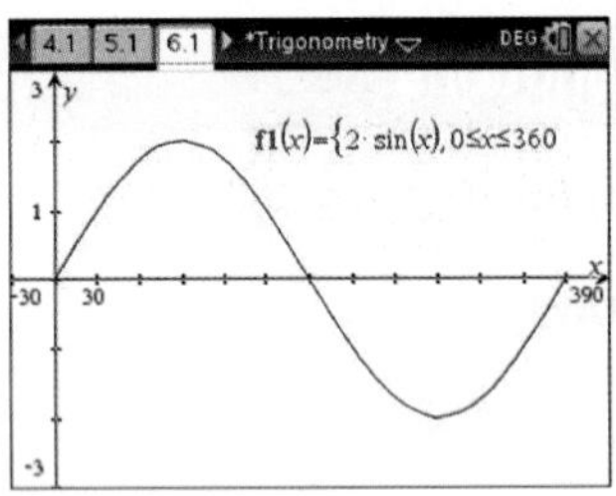

b.

Complete the function entry line as:
$f2(x) = \cos(2x) \mid 0 \le x \le 360$
Press ENTER, and the graph is displayed, as required only for $0° \le x \le 360°$.

b.

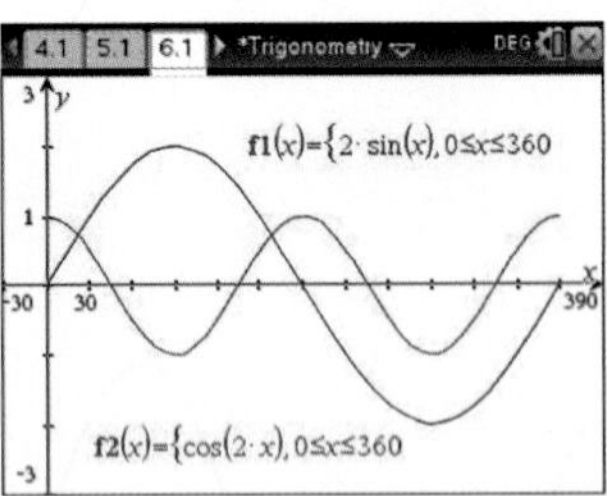

CASIO | THINK

DISPLAY/WRITE

On a Graph & Table screen, set an appropriate viewing window as shown.

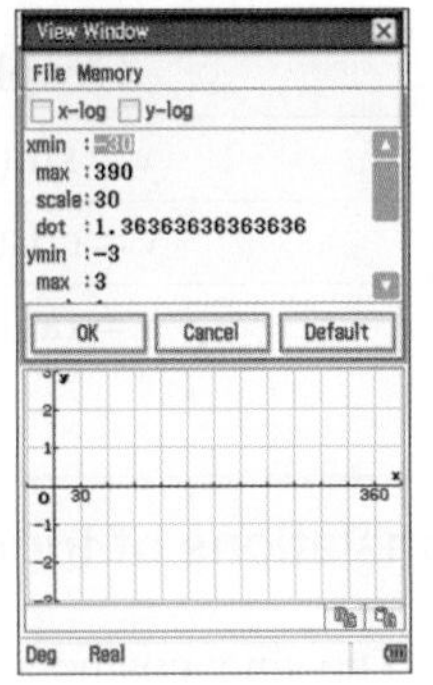

a.

Complete the function entry line as:
$y1 = 2\sin(x) \mid 0 \le x \le 360$
Press EXE.
Tap the graphing icon and the graph is displayed as required only for $0° \le x \le 360°$.

a.

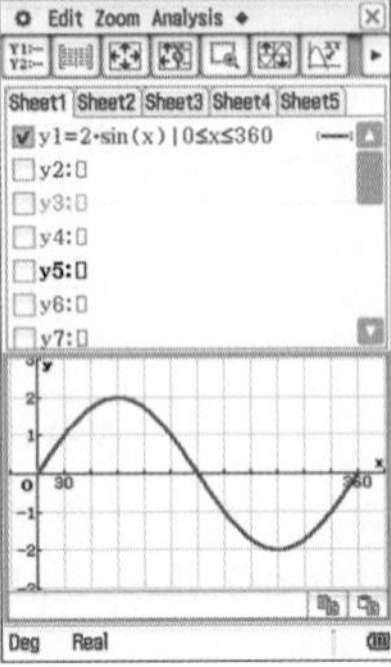

b.

Complete the function entry line as:
$y1 = \cos(2x) \mid 0 \le x \le 360$
Press EXE.
Tap the graphing icon and the graph is displayed as required only for $0° \le x \le 360°$.

b.

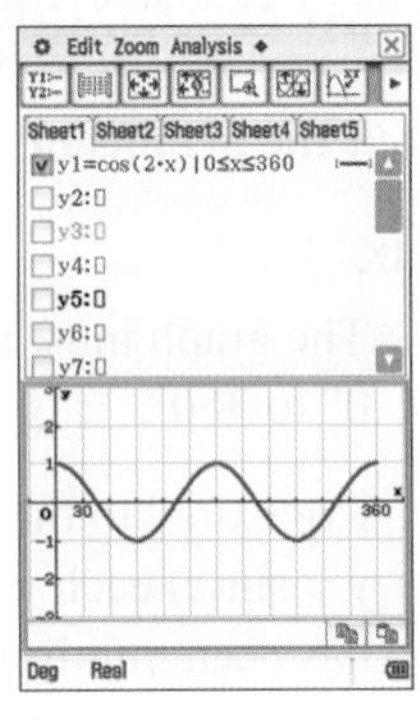

WORKED EXAMPLE 13 Stating the amplitude and period of given periodic functions

For each of the following graphs, state:

i. the amplitude **ii. the period.**

a. $y = 2\sin(3x)$ **b.** $y = \cos\left(\frac{x}{3}\right)$ **c.** $y = \tan(2x)$

THINK	WRITE
a. The value of a is 2.	a. i. Amplitude = 2
The periods is $\frac{360°}{n}$.	ii. Period = $\frac{360}{3} = 120°$
b. The value of a is 1.	b. i. Amplitude = 1
The period is $\frac{360°}{n}$.	ii. Period = $\frac{360}{\frac{1}{3}} = 1080°$
c. The tangent curve has an undefined amplitude.	c. i. Amplitude = undefined
The period is $\frac{180°}{2}$.	ii. Period = $\frac{180°}{2} = 90°$

DISCUSSION

For the graph of $y = a\tan(nx)$, what would be the period and amplitude?

on Resources

eWorkbook Topic 17 Workbook (worksheets, code puzzle and project) (ewbk-2043)

Interactivity Graphs of trigonometric functions (int-4821)

Exercise 17.6 Trigonometric functions

learnon

Individual pathways

■ PRACTISE	■ CONSOLIDATE	■ MASTER
1, 4, 8, 12, 16, 20, 25, 26, 28, 29, 32	3, 5, 7, 10, 13, 14, 17, 21, 23, 27, 30, 33	2, 6, 9, 11, 15, 18, 19, 22, 24, 31, 34

To answer questions online and to receive **immediate corrective feedback** and **fully worked solutions** for all questions, go to your learnON title at www.jacplus.com.au.

Fluency

1. Using your calculator (or the unit circle if you wish), complete the following table.

x	0°	30°	60°	90°	120°	150°	180°	210°	240°	270°	300°	330°	360°
$\sin(x)$													
x	390°	420°	450°	480°	510°	540°	570°	600°	630°	660°	690°	720°	
$\sin(x)$													

For questions **2** to **7**, using graph paper, rule x- and y-axes and carefully mark a scale along each axis.

2. Use 1 cm = 30° on the x-axis to show x-values from 0° to 720°.
Use 2 cm = 1 unit along the y-axis to show y-values from −1 to 1.
Carefully plot the graph of $y = \sin(x)$ using the values from the table in question 1.

3. State how long it takes for the graph of $y = \sin(x)$ to complete one full cycle.

4. From your graph of $y = \sin(x)$, estimate to 1 decimal place the value of y for each of the following.

a. $x = 42°$ b. $x = 130°$ c. $x = 160°$ d. $x = 200°$

5. From your graph of $y = \sin(x)$, estimate to 1 decimal place the value of y for each of the following.

a. $x = 180$ b. $x = 70°$ c. $x = 350°$ d. $x = 290°$

6. From your graph of $y = \sin(x)$, estimate to the nearest degree a value of x for each of the following.

a. $y = 0.9$ b. $y = -0.9$ c. $y = 0.7$

7. From your graph of $y = \sin(x)$, estimate to the nearest degree a value of x for each of the following.

a. $y = -0.5$ b. $y = -0.8$ c. $y = 0.4$

8. Using your calculator (or the unit circle if you wish), complete the following table.

x	0°	30°	60°	90°	120°	150°	180°	210°	240°	270°	300°	330°	360°
cos(x)													
x	390°	420°	450°	480°	510°	540°	570°	600°	630°	660°	690°	720°	
cos(x)													

For questions **9** to **14**, using graph paper, rule x- and y-axes and carefully mark a scale along each axis.

9. Use 1 cm = 30° on the x-axis to show x-values from 0° to 720°.
Use 2 cm = 1 unit along the y-axis to show y-values from −1 to 1.
Carefully plot the graph of $y = \cos(x)$ using the values from the table in question 8.

10. If you were to continue the graph of $y = \cos(x)$, state what shape you would expect it to take.

11. State whether the graph of $y = \cos(x)$ is the same as the graph of $y = \sin(x)$. Explain how it differs. State what features are the same.

12. Using the graph of $y = \cos(x)$, estimate to 1 decimal place the value of y for each of the following.

a. $x = 48°$ b. $x = 155°$ c. $x = 180°$ d. $x = 340°$

13. Using the graph of $y = \cos(x)$, estimate to 1 decimal place the value of y for each of the following.

a. $x = 240°$ b. $x = 140°$ c. $x = 40°$ d. $x = 200°$

14. Using the graph of $y = \cos(x)$, estimate to the nearest degree a value of x for each of the following.

a. $y = -0.5$ b. $y = 0.8$ c. $y = 0.7$

15. Using the graph of $y = \cos(x)$, estimate to the nearest degree a value of x for each of the following.

a. $y = -0.6$ b. $y = 0.9$ c. $y = -0.9$

16. Using your calculator (or the unit circle if you wish), complete the following table.

x	0°	30°	60°	90°	120°	150°	180°	210°	240°	270°	300°	330°	360°
tan(x)													
x	390°	420°	450°	480°	510°	540°	570°	600°	630°	660°	690°	720°	
tan(x)													

For questions 17 to 22, using graph paper, rule x- and y-axes and carefully mark a scale along each axis.

17. Use 1 cm = 30° on the x-axis to show x-values from 0° to 720°.
Use 2 cm = 1 unit along the y-axis to show y-values from −2 to 2.
Carefully plot the graph of $y = \tan(x)$ using the values from the table in question 16.

18. If you were to continue the graph of $y = \tan(x)$, state what shape would you expect it to take.

19. State whether the graph of $y = \tan(x)$ is the same as the graphs of $y = \sin(x)$ and $y = \cos(x)$. Explain how it differs. State what features are the same.

20. Using the graph of $y = \tan(x)$, estimate to 1 decimal place the value of y for each of the following.

a. $x = 60°$ b. $x = 135°$ c. $x = 310°$ d. $x = 220°$

21. Using the graph of $y = \tan(x)$, determine the value of y for each of the following.

a. $x = 500°$ b. $x = 590°$ c. $x = 710°$ d. $x = 585°$

22. Using the graph of $y = \tan(x)$, estimate to the nearest degree a value of x for each of the following.

a. $y = 1$ b. $y = 1.5$ c. $y = -0.4$
d. $y = -2$ e. $y = 0.2$ f. $y = -1$

WE12,13 For each of the graphs in questions 23 and 24:

i. state the period
ii. state the amplitude
iii. sketch the graph.

23. a. $y = \cos(x)$, for $x \in [-180°, 180°]$
b. $y = \sin(x)$, for $x \in [0°, 720°]$

24. a. $y = \sin(2x)$, for $x \in [0°, 360°]$
b. $y = 2\cos(x)$, for $x \in [-360°, 0°]$

25. For each of the following, state:

i. the period
ii. the amplitude.

a. $y = 3\cos(2x)$ b. $y = 4\sin(3x)$ c. $y = 2\cos\left(\frac{x}{2}\right)$

d. $y = \frac{1}{2}\sin\left(\frac{x}{4}\right)$ e. $y = -\sin(x)$ f. $y = -\cos(2x)$

Understanding

26. **MC** Use the graph shown to answer the following.

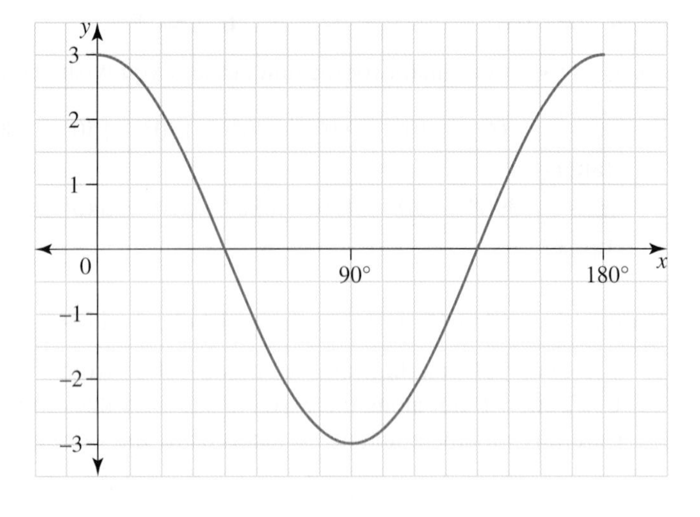

a. The amplitude of the graph is:

A. 180° B. 90°
C. 3 D. −3
E. 6

b. The period of the graph is:

A. 180° B. 360°
C. 90° D. 3
E. −3

c. The equation of the graph could be:

A. $y = \cos(x)$ B. $y = \sin(x)$
C. $y = 3\cos\left(\frac{x}{3}\right)$ D. $y = 3\cos(2x)$
E. $y = 3\sin(2x)$

27. Sketch each of the following graphs, stating the period and amplitude of each.

a. $y = 2\cos\left(\frac{x}{3}\right)$, for $x \in [0°, 1080°]$
b. $y = -3\sin(2x)$, for $x \in [0°, 360°]$
c. $y = 3\sin\left(\frac{x}{2}\right)$, for $x \in [-180°, 180°]$
d. $y = -\cos(3x)$, for $x \in [0°, 360°]$
e. $y = 5\cos(2x)$, for $x \in [0°, 180°]$
f. $y = -\sin(4x)$, for $x \in [0°, 180°]$

28. Use technology to sketch the graphs of each of the following for $0° \leq x \leq 360°$.

a. $y = \cos(x) + 1$
b. $y = \sin(2x) - 2$
c. $y = \cos\left(\frac{\pi}{180}(x - 60)\right)$
d. $y = 2\sin(4x) + 3$

Reasoning

29. a. Sketch the graph of $y = \cos(2x)$ for $x \in [0°, 360°]$.
i. State the minimum value of y for this graph. ii. State the maximum value of y for this graph.
b. Using the answers obtained in part **a** write down the maximum and minimum values of $y = \cos(2x) + 2$.
c. Determine what would be the maximum and minimum values of the graph of $y = 2\sin(x) + 3$. Explain how you obtained these values.

30. a. Complete the table below by filling in the exact values of $y = \tan(x)$

x	0°	30°	60°	90°	120°	150°	180°
$y = \tan(x)$							

b. Sketch the graph of $y = \tan(x)$ for $[0°, 180°]$.
c. Determine what happens at $x = 90°$.
d. For the graph of $y = \tan(x)$, $x = 90°$ is called an asymptote. Write down when the next asymptote would occur.
e. Determine the period and amplitude of $y = \tan(x)$.

31. a. Sketch the graph of $y = \tan(2x)$ for $[0°, 180°]$.
b. Determine when the asymptotes occur.
c. State the period and amplitude of $y = \tan(2x)$.

Problem solving

32. The height of the tide above the mean sea level on the first day of the month is given by the rule

$$h = 3\sin(30t°)$$

where t is the time in hours since midnight.

a. Sketch the graph of h versus t.
b. Determine the height of the high tide.
c. Calculate the height of the tide at 8 am.

33. The height, h metres, of the tide on the first day of January at Trig Cove is given by the rule

$$h = 6 + 4\sin(30t°)$$

where t is the time in hours since midnight.

a. Sketch the graph of h versus t, for $0 \le t \le 24$.
b. Determine the height of the high tide.
c. Determine the height of the low tide.
d. Calculate the height of the tide at 10 am, correct to the nearest centimetre.

34. The temperature, T, inside a house t hours after 3 am is given by the rule

$$T = 22 - 2\cos(15t°) \text{ for } 0 \le t \le 24$$

where T is the temperature in degrees Celsius.

a. Determine the temperature inside the house at 9 am.
b. Sketch the graph of T versus t.
c. Determine the warmest and coolest temperatures that it gets inside the house over the 24-hour period.

17.7 Solving trigonometric equations

LEARNING INTENTION

At the end of this subtopic you should be able to:
- solve trigonometric equations graphically for a given domain
- solve trigonometric equations algebraically, using exact values, for a given domain.

17.7.1 Solving trigonometric equations

eles-5012

Solving trigonometric equations graphically

- Because of the periodic nature of circular functions, there are infinitely many solutions to unrestricted trigonometric equations.
- Equations are usually solved within a particular domain (x-values), to restrict the number of solutions.
- The sine graph below shows the solutions between 0° and 360° for the equation $\sin(x) = 0.6$.

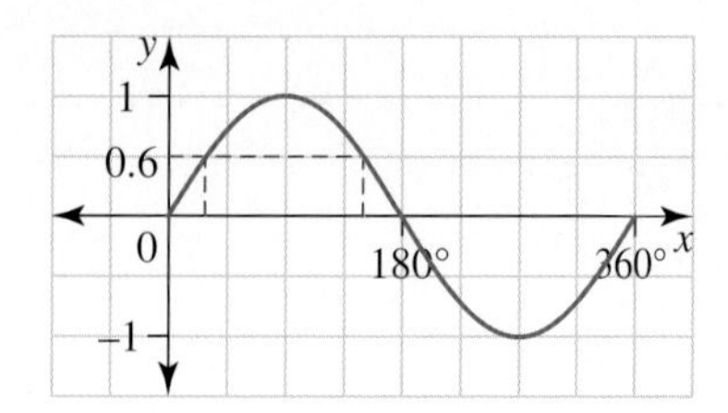

In the example above, it can clearly be seen that there are two solutions to this equation, which are approximately $x = 37°$ and $x = 143°$.

- It is difficult to obtain accurate answers from a graph. More accurate answers can be obtained using technology.

Solving trigonometric equations algebraically

Exact answers can be found for some trigonometric equations using the table in section 17.2.1.

WORKED EXAMPLE 14 Solving trigonometric equations using exact values

Solve the following equations.

a. $\sin(x) = \frac{\sqrt{3}}{2}, x \in [0°, 360°]$

b. $\cos(2x) = -\frac{1}{\sqrt{2}}, x \in [0°, 360°]$

THINK	WRITE
a. 1. The inverse operation of sine is $\sin^{-1}$.	**a.** $x = \sin^{-1}\left(\frac{\sqrt{3}}{2}\right)$
2. The first solution in the given domain from the table in subsection 17.2.1 is $x = 60°$.	
3. Since sine is positive in the first and second quadrants, another solution must be $x = 180° - 60° = 120°$.	There are two solutions in the given domain, $x = 60°$ and $x = 120°$.
b. 1. The inverse operation of cosine is $\cos^{-1}$.	**b.** $2x = \cos^{-1}\left(\frac{-1}{\sqrt{2}}\right)$
2. From the table of values, $\cos^{-1}\left(\frac{1}{\sqrt{2}}\right) = 45°$.	
3. Cosine is negative in the second and third quadrants, which gives the first two solutions to the equation as: $180° - 45°$ and $180° + 45°$.	$2x = 135°, 225°$
4. Solve for x by dividing by 2.	$x = 67.5°, 112.5°$
5. Since the domain in this case is [0°, 360°] and the period has been halved, there must be 4 solutions altogether. The other 2 solutions can be found by adding the period onto each solution.	The period $= \frac{360°}{2} = 180°$ $x = 67.5° + 180°, 112.5° + 180°$ $x = 67.5°, 112.5°, 247.5°, 292.5°$

TI | THINK

a.

In a new problem, on a Calculator page, complete the entry line as:

$$\text{solve}\left(\sin(x) = \frac{\sqrt{3}}{2}, x\right) |$$

$0 \le x \le 360$

Then press ENTER.

Note that the calculator is set to the degrees mode.

DISPLAY/WRITE

a.

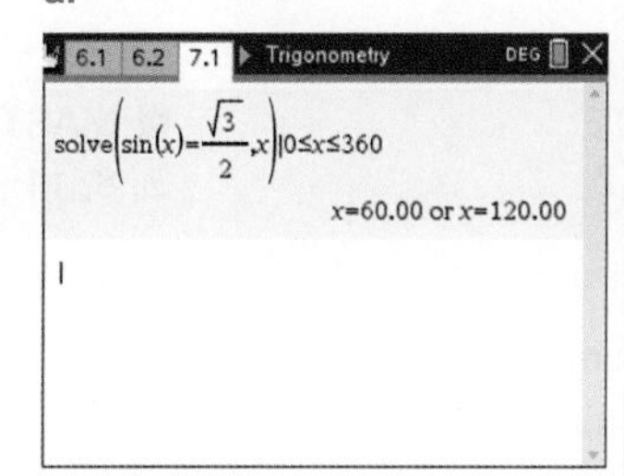

$\sin(x) = \frac{\sqrt{3}}{2}$ for $x \in [0°, 360°]$

$\Rightarrow x = 60°$ or $120°$

b.

On a Calculator page, complete the entry line as:

$$\text{solve}\left(\cos(2x) = -\frac{1}{\sqrt{2}}, x\right) |$$

$0 \le x \le 360$

Then press ENTER.

Note that the calculator is set to the degrees mode.

b.

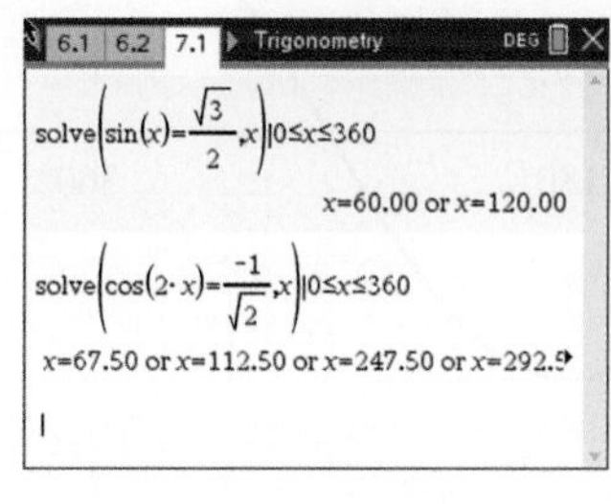

$\cos(2x) = -\frac{1}{\sqrt{2}}$ for $x \in [0°, 360°]$

$\Rightarrow x = 67.5°, 112.5°, 247.5°$ or $292.5°$

CASIO | THINK

a.

On the Main screen, complete the entry line as:

$$\text{solve}\left(\sin(x) = \frac{\sqrt{3}}{2}, x\right) |$$

$0 \le x \le 360$

Then press EXE.

Note that the calculator is set to the degrees mode.

DISPLAY/WRITE

a.

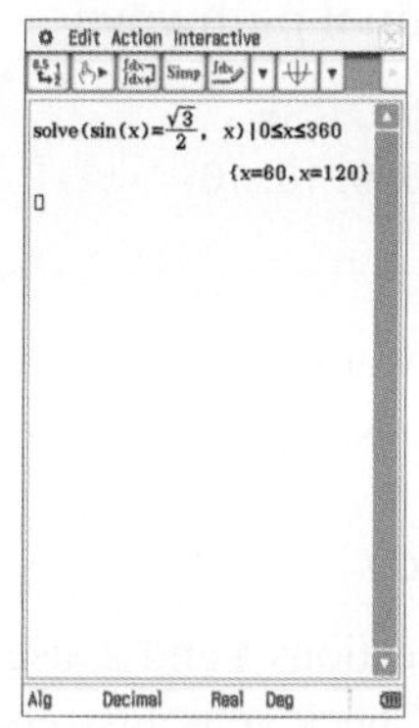

$\sin(x) = \frac{\sqrt{3}}{2}$ for $x \in [0°, 360°]$

$\Rightarrow x = 60°$ or $120°$

b.

On the Main screen, complete the entry line as:

$$\text{solve}\left(\cos(2x) = -\frac{1}{\sqrt{2}}, x\right) |$$

$0 \le x \le 360$

Then press EXE.

Note that the calculator is set to the degrees mode.

b.

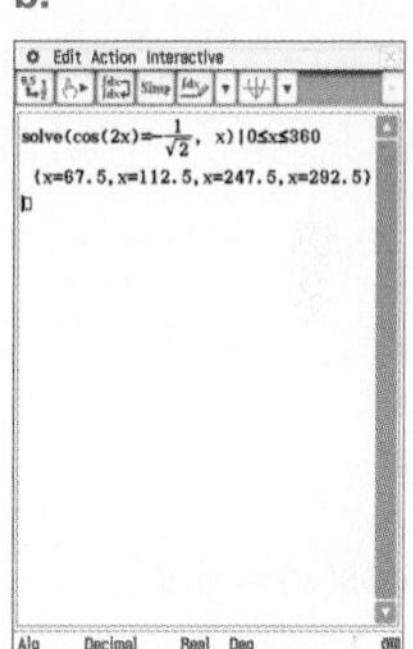

$\cos(2x) = -\frac{1}{\sqrt{2}}$ for $x \in [0°, 360°]$

$\Rightarrow x = 67.5°, 112.5°, 247.5°$ or $292.5°$

DISCUSSION

Explain why sine and cosine functions can be used to model situations that occur in nature such as tide heights and sound waves.

on Resources

 **eWorkbook** Topic 17 Workbook (worksheets, code puzzle and project) (ewbk-2043)

 **Interactivity** Solving trigonometric equations graphically (int-4822)

Exercise 17.7 Solving trigonometric equations

learnON

Individual pathways

PRACTISE	CONSOLIDATE	MASTER
1, 3, 6, 9, 13, 16	2, 4, 7, 10, 14, 17	5, 8, 11, 12, 15, 18

To answer questions online and to receive **immediate corrective feedback** and **fully worked solutions** for all questions, go to your learnON title at www.jacplus.com.au.

Fluency

For questions 1 and 2, use the graph to determine approximate answers to the equations for the domain $0 \leq x \leq 360°$. Check your answers using a calculator.

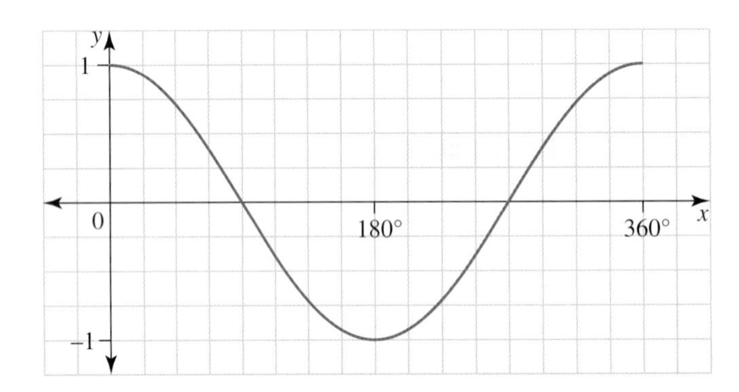

1. a. $\cos(x) = 0.9$
   b. $\cos(x) = 0.3$
2. a. $\cos(x) = -0.2$
   b. $\cos(x) = -0.6$

For questions 3 to 8, solve the equations for the domain $0° \leq x \leq 360°$.

3. a. $\sin(x) = \frac{1}{2}$
   b. $\sin(x) = \frac{\sqrt{3}}{2}$
4. a. $\cos(x) = -\frac{1}{2}$
   b. $\cos(x) = -\frac{1}{\sqrt{2}}$
5. a. $\sin(x) = 1$
   b. $\cos(x) = -1$
6. a. $\sin(x) = -\frac{1}{2}$
   b. $\sin(x) = -\frac{1}{\sqrt{2}}$
7. a. $\cos(x) = \frac{\sqrt{3}}{2}$
   b. $\cos(x) = -\frac{\sqrt{3}}{2}$
8. a. $\sin(x) = 1$
   b. $\cos(x) = 0$

Understanding

WE14 For questions 9 and 12, solve the following equations for the given values of x.

9. a. $\sin(2x)=\frac{\sqrt{3}}{2}, x\in[0°, 360°]$

b. $\cos(2x)=-\frac{\sqrt{3}}{2}, x\in[0°, 360°]$

10. a. $\tan(2x)=\frac{1}{\sqrt{3}}, x\in[0°, 360°]$

b. $\sin(3x)=-\frac{1}{2}, x\in[0°, 180°]$

11. a. $\sin(4x)=-\frac{1}{2}, x\in[0°, 180°]$

b. $\sin(3x)=-\frac{1}{\sqrt{2}}, x\in[-180°, 180°]$

12. a. $\tan(3x)=-1, x\in[0°, 90°]$

b. $\cos(3x)=0, x\in[0°, 360°]$

Reasoning

13. Solve the following equations for $x\in[0°, 360°]$.

a. $2\sin(x)-1=0$

b. $2\cos(x)=\sqrt{3}$

14. Solve the following equations for $x\in[0°, 360°]$.

a. $\sqrt{2}\cos(x)-1=0$

b. $\tan(x)+1=0$

15. Sam measured the depth of water at the end of the Intergate jetty at various times on Thursday 13 August 2020. The table below provides her results.

Time	6 am	7	8	9	10	11	12 pm	1	2	3	4	5	6	7	8	9
Depth	1.5	1.8	2.3	2.6	2.5	2.2	1.8	1.2	0.8	0.5	0.6	1.0	1.3	1.8	2.2	2.5

a. Plot the data.

b. Determine:

i. the period

ii. the amplitude.

c. Sam fishes from the jetty when the depth is a maximum. Specify these times for the next 3 days.

d. Sam's mother can moor her yacht when the depth is above 1.5 m. Determine during what periods she can moor the yacht on Sunday 16 January.

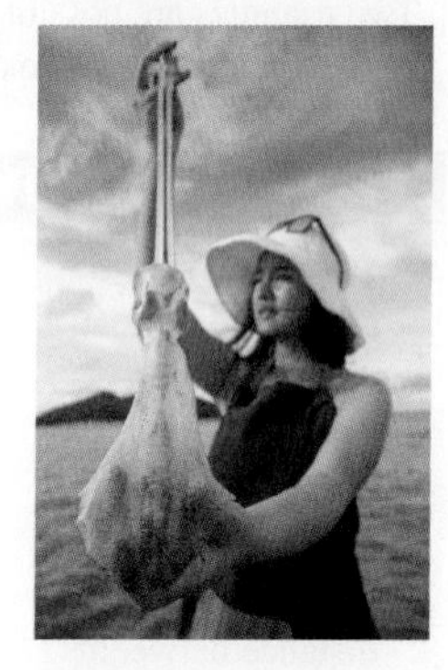

Problem solving

16. Solve:

a. $\sqrt{3}\sin(x°)=\cos(x°)$ for $0°\leq x\leq 360°$

b. $2\sin(x°)+\cos(x°)=0$ for $0°\leq x\leq 360°$.

17. Solve $2\sin^2(x°)+3\sin(x°)-2=0$ for $0°\leq x\leq 360°$.

18. The grad (^{g}) is another measurement used when measuring the size of angles. A grad is equivalent to $\frac{1}{400}$ of a full circle. Write each of the following as grads (1 grad is written as 1^g).

a. 90°

b. 180°

c. 270°

d. 360°

17.8 Review

17.8.1 Topic summary

TRIGONOMETRY II

Sine rule

- Connects two sides with the two opposite angles in a triangle.
- $\frac{a}{\sin A} = \frac{b}{\sin B} = \frac{c}{\sin C}$
- Used to solve triangles:
  - Two angles and one side.
  - Two sides and an angle opposite one of these sides.

Area of a triangle

- Given two sides and the included angle:
  Area $= \frac{1}{2} ab \sin(C)$
- Given three sides, Heron's formula:
  Area $= \sqrt{(s(s-a)(s-b)(s-c))}$
  where s is the semi-perimeter: $s = \frac{a+b+c}{2}$

Cosine rule

- Connects three sides and one angle of a triangle.
  $a^2 = b^2 + c^2 - 2bc\cos(A)$
  Used to solve triangles given:
  - three sides, or
  - two sides and the included angle.

Ambiguous case

- When using the sine rule to calculate an angle, there may be two answers.
- The ambiguous case may occur when determining the angle opposite the larger side.
- Always check that the three angles add to 180°.

e.g. In the triangle ABC, $a = 10$, $b = 6$ and $B = 30°$, using the sine rule, $A = 56°$ or $(180 - 56)°$.
Angles in triangle would be:
$A = 56°$, $B = 30°$ giving $C = 94°$
$A = 124°$, $B = 30°$ giving $C = 26°$
Two triangles are possible, so the ambiguous case exists.

Trigonometric equations

- Infinite number of solutions, so domain is normally restricted:
  e.g. $0° \leq x \leq 180°$
- Equations can be solved:
  - graphically — not very accurate
  - using technology
  - algebraically, using the exact values.

e.g. $\sin\alpha = -\frac{1}{2}$, $0° \leq x \leq 360°$
from exact values: $\sin 30° = \frac{1}{2}$
sine is negative in 3rd and 4th quadrants and the angle is from the x-axis
$\alpha = (180 + 30)°$ or $(360 - 30)°$
$\alpha = 210°$ or $330°$

Exact values

- Exact trig ratios can be found using triangles.
- For 30° and 60° use the equilateral triangle.

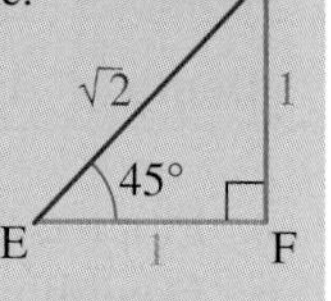

- For 45°, use a right-angled isosceles triangle.

e.g. $\sin 30° = \frac{1}{2}$

$\tan 45° = 1$

$\cos 60° = \frac{1}{2}$

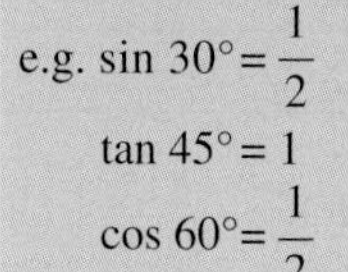

Trigonometric graphs

- Trigonometric graphs repeat themselves continuously in cycles.
- Period: horizontal distance between repeating peaks or troughs.
- Amplitude: half the distance between the maximum and minimum values.

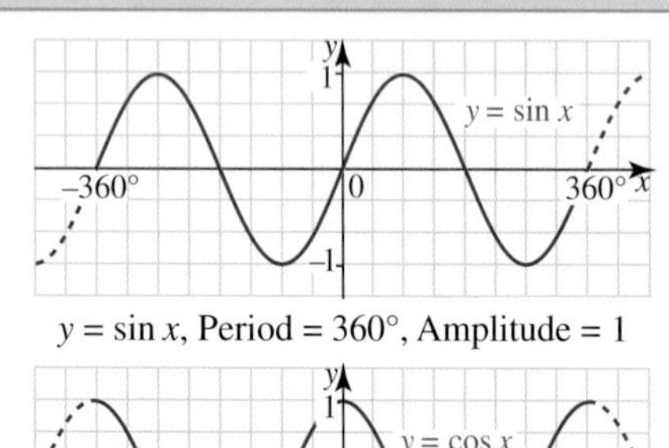

$y = \sin x$, Period = 360°, Amplitude = 1

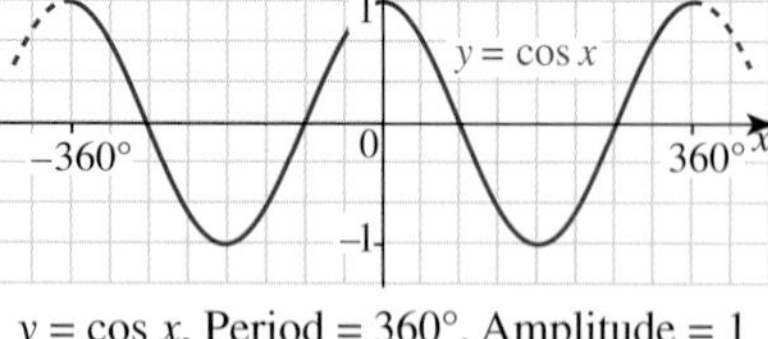

$y = \cos x$, Period = 360°, Amplitude = 1

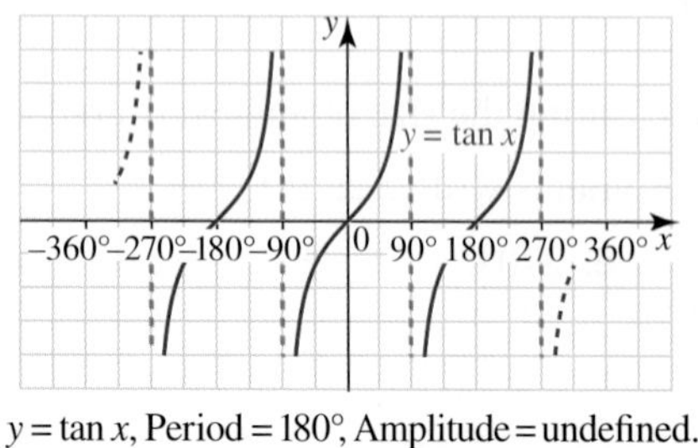

$y = \tan x$, Period = 180°, Amplitude = undefined.

Unit circle

- Equation of the unit circle: $x^2 + y^2 = 1$
- Radius of length 1 unit.
- For any point on circumference:
  $x = \cos\theta$
  $y = \sin\theta$
- $\tan\theta = \frac{y}{x} = \frac{\sin\theta}{\cos\theta}$
- $\sin(180 - A)° = \sin A°$
  $\cos(180 - A)° = -\cos A°$
  $\tan(180 - A)° = -\tan A°$
- Quadrants are positive for:

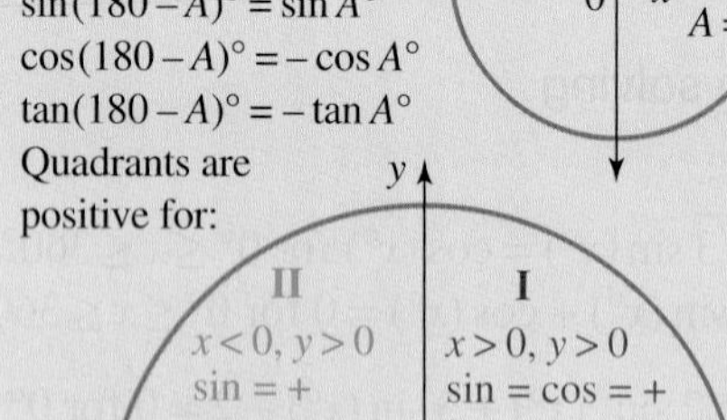

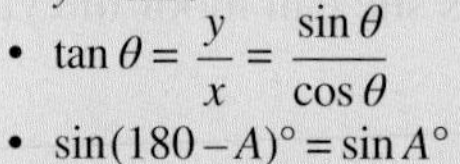

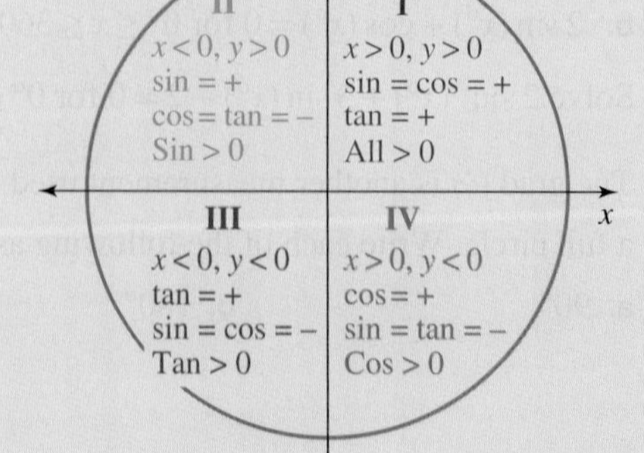

17.8.2 Success criteria

Tick the column to indicate that you have completed the subtopic and how well you have understood it using the traffic light system.

(**Green:** I understand; **Yellow:** I can do it with help; **Red:** I do not understand)

Subtopic	Success criteria	Green	Yellow	Red
17.2	I can apply the exact values of sin, cos and tan for 30°, 45° and 60° angles.			
	I can apply the sine rule to evaluate angles and sides of triangles.			
	I can recognise when the ambiguous case of the sine rule exists.			
17.3	I can apply the cosine rule to calculate a side of a triangle.			
	I can apply the cosine rule to calculate the angles of a triangle.			
17.4	I can calculate the area of a triangle, given two sides and the included angle.			
	I can use Heron's formula to calculate the area of a triangle, given the three sides.			
17.5	I can determine in which quadrant an angle lies.			
	I can use and interpret the relationship between a point on the unit circle and the angle made with the positive x-axis.			
	I can use the unit circle to determine approximate trigonometric ratios for angles greater than 90°.			
17.6	I can sketch the graphs of the sine, cosine and tangent graphs.			
	I can determine the amplitude of a given trigonometric function.			
	I can determine the period of a given trigonometric function.			
17.7	I can solve trigonometric equations graphically for a given domain.			
	I can solve trigonometric equations algebraically, using exact values, for a given domain.			

17.8.3 Project

What's an arbelos?

As an introduction to this task, you are required to complete the following construction. The questions that follow require the application of measurement formulas, and an understanding of semicircles related to this construction.

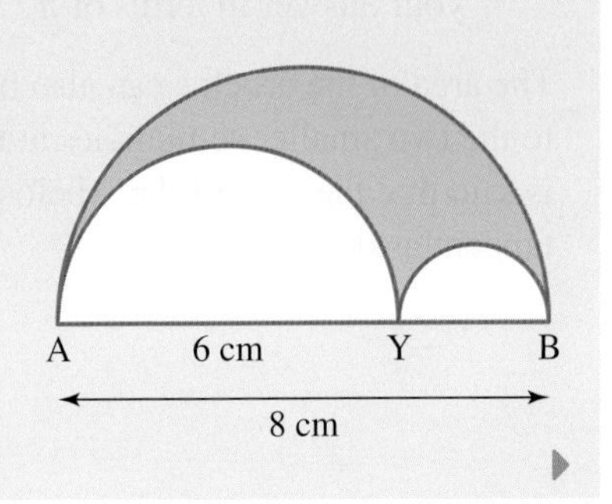

1. **Constructing an arbelos**
   - Rule a horizontal line AB 8 cm long.
   - Determine the midpoint of the line and construct a semicircle on top of the line with AB as the diameter.
   - Mark Y as a point on AB such that AY = 6 cm.
   - Determine the midpoint of AY and draw a small semicircle inside the larger semicircle with AY as the diameter.
   - Determine the midpoint of YB and construct a semicircle (also inside the larger semicircle) with a diameter YB.

The shape enclosed by the three semicircles is known as an arbelos. The word, in Greek, means *shoemaker's knife* as it resembles the blade of a knife used by cobblers. The point Y is not fixed and can be located anywhere along the diameter of the larger semicircle, which can also vary in size.

2. **Perimeter of an arbelos**
   The perimeter of an arbelos is the sum of the arc length of the three semicircles. Perform the following calculations, leaving each answer in terms of π.
   a. Calculate the arc length of the semicircle with diameter AB.
   b. Calculate the arc length of the semicircle with diameter AY.
   c. Calculate the arc length of the semicircle on diameter YB.
   d. Compare the largest arc length with the two smaller arc lengths. What do you conclude?
3. We can generalise the arc length of an arbelos. The point Y can be located anywhere on the line AB, which can also vary in length. Let the diameter AB be d cm, AY be d_1 cm and YB be d_2 cm. Prove that your conclusion from question 2d holds true for any value of d, where $d_1 + d_2 = d$.
4. **Area of an arbelos**
   The area of an arbelos may be treated as the area of a composite shape.
   a. Using your original measurements, calculate the area of the arbelos you drew in question 1. Leave your answer in terms of π.

The area of the arbelos can also be calculated using another method. We can draw the common tangent to the two smaller semicircles at their point of contact and extend this tangent to the larger semicircle. It is said that the area of the arbelos is the same as the area of the circle constructed on this common tangent as diameter.

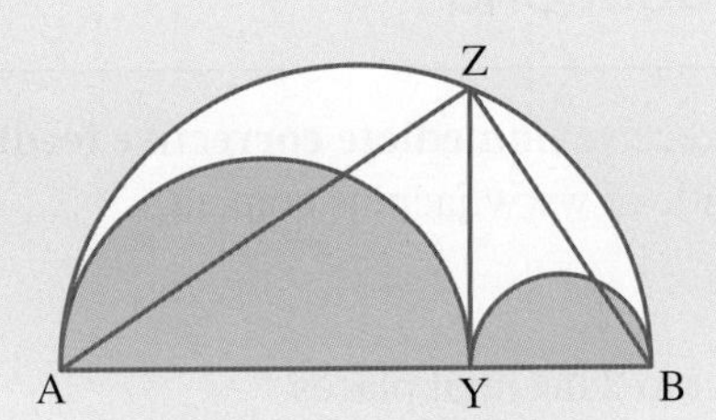

YZ is the common tangent.

Triangles AYZ, BYZ and AZB are all right-angled triangles. We can use Pythagoras' theorem, together with a set of simultaneous equations, to determine the length of the tangent YZ.

b. Complete the following.

In ΔAYZ, $\quad AZ^2 = AY^2 + YZ^2$

$= 6^2 + YZ^2$

In ΔBYZ, $\quad BZ^2 = BY^2 + YZ^2$

$=$ .... $+ YZ^2$

Adding these two equations,

$AZ^2 + BZ^2 =$ .... + ...

$AZ^2 + BZ^2 = AB^2$

But, in ΔAZB $\quad =$ ....

.... + ...

So, $\quad YZ =$ .... (Leave your answer in surd form.)

c. Now calculate the area of the circle with diameter YZ. Is your answer the same as that calculated in question 4a?

The area of an arbelos can be generalised.
Let the radii of the two smaller semicircles be r_1 and r_2.

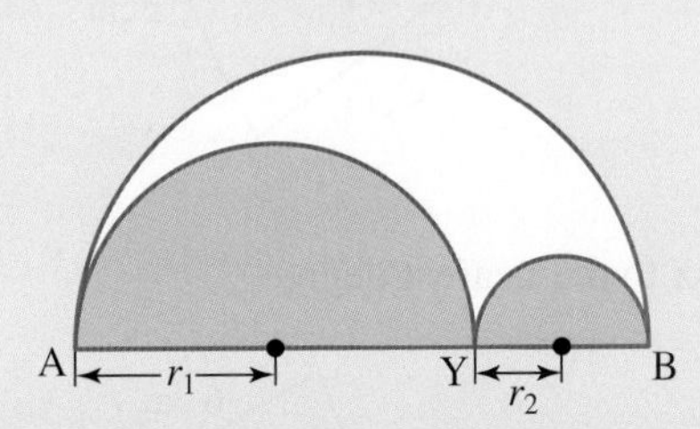

5. Develop a formula for the area of the arbelos in terms of r_1 and r_2. Demonstrate the use of your formula by checking your answer to question 4a.

on Resources

 **eWorkbook** Topic 17 Workbook (worksheets, code puzzle and project) (ewbk-2043)

 **Interactivities** Crossword (int-2884)
Sudoku puzzle (int-3895)

Exercise 17.8 Review questions

learnON

To answer questions online and to receive **immediate corrective feedback** and **fully worked solutions** for all questions, go to your learnON title at www.jacplus.com.au.

Fluency

1. Calculate the value of x, correct to 2 decimal places.

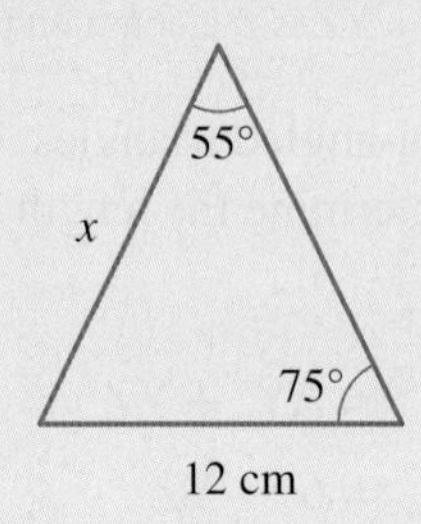

2. Calculate the value of θ, correct to the nearest minute.

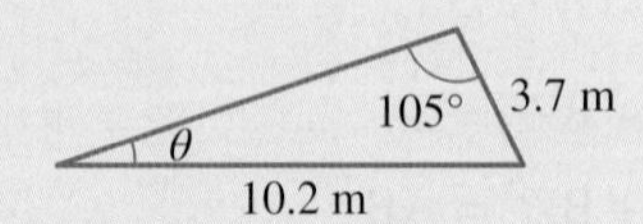

3. Determine all unknown sides (correct to 2 decimal places) and angles (correct to the nearest minute) of triangle ABC, given $a = 25$ m, $A = 120°$ and $B = 50°$.

4. Calculate the value of x, correct to 2 decimal places.

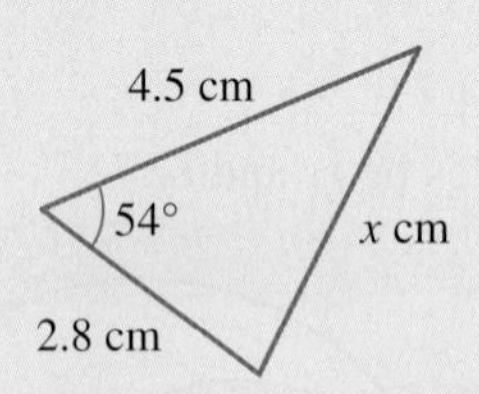

5. Calculate the value of θ, correct to the nearest degree.

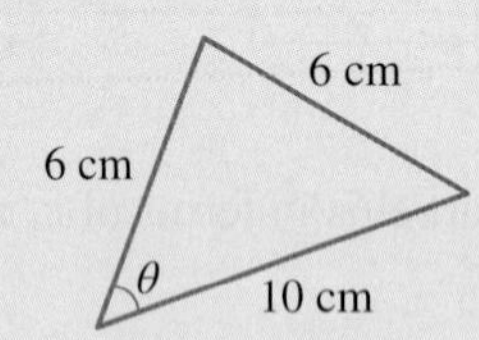

6. A triangle has sides of length 12 m, 15 m and 20 m. Calculate the magnitude (size) of the largest angle, correct to the nearest minute.

7. A triangle has two sides of 18 cm and 25 cm. The angle between the two sides is 45°. Calculate, correct to 2 decimal places:
   a. its area
   b. the length of its third side
   c. its area using Heron's formula.

8. If an angle of $\theta = 290°$ was represented on the unit circle, state which quadrant the triangle to show this would be drawn in.

9. On the unit circle, draw an appropriate triangle for the angle 110° in quadrant 2.
a. Determine the value of $\sin(110°)$ and $\cos(110°)$, correct to 2 decimal places.
b. Determine the value of $\tan(110°)$, correct to 2 decimal places.

10. **MC** The value of $\sin(53°)$ is equal to:
A. $\cos(53°)$ **B.** $\cos(37°)$ **C.** $\sin(37°)$ **D.** $\tan(53°)$ **E.** $\tan(37°)$

11. Simplify $\dfrac{\sin(53°)}{\sin(37°)}$.

12. Draw a sketch of $y = \sin(x)$ from $0° \le x \le 360°$.

13. Draw a sketch of $y = \cos(x)$ from $0° \le x \le 360°$.

14. Draw a sketch of $y = \tan(x)$ from $0° \le x \le 360°$.

15. Label this triangle so that $\dfrac{x}{\sin(46°)} = \dfrac{y}{\sin(68°)}$.

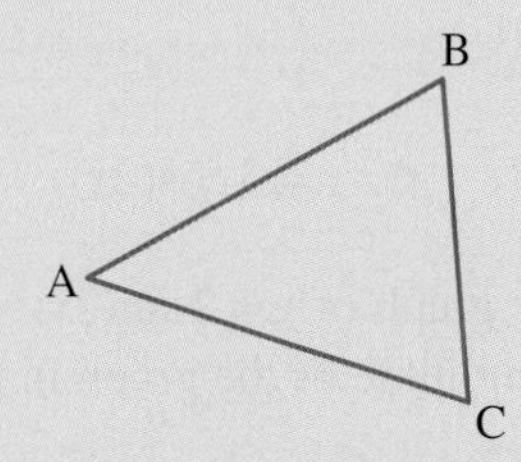

16. State the period and amplitude of each of the following graphs.
a. $y = 2\sin(3x)$
b. $y = -3\cos(2x)$
c.

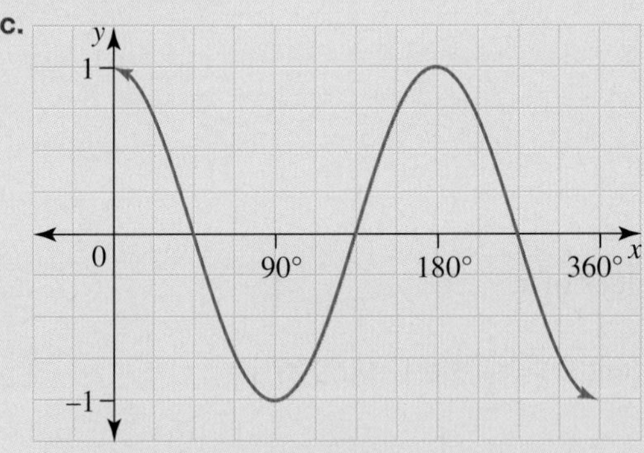

17. Sketch the following graphs.
a. $y = 2\sin(x), x \in [0°, 360°]$ **b.** $y = \cos(2x), x \in [-180°, 180°]$

18. Use technology to write down the solutions to the following equations for the domain $0° \le x \le 360°$ to 2 decimal places.
a. $\sin(x) = -0.2$ **b.** $\cos(2x) = 0.7$ **c.** $3\cos(x) = 0.1$ **d.** $2\tan(2x) = 0.5$

19. Solve each of the following equations.

a. $\sin(x) = \frac{1}{2}, x \in [0°, 360°]$

b. $\cos(x) = \frac{\sqrt{3}}{2}, x \in [0°, 360°]$

c. $\cos(x) = \frac{1}{\sqrt{2}}, x \in [0°, 360°]$

d. $\sin(x) = \frac{1}{\sqrt{2}}, x \in [0°, 360°]$

20. **MC** The equation that represents the graph shown could be:

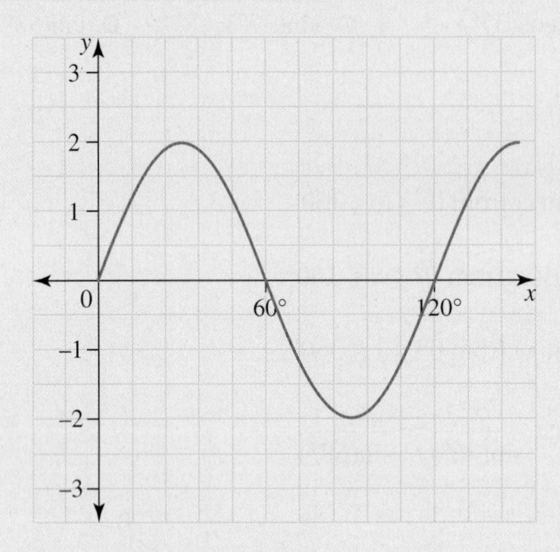

A. $y = 3\sin(2x)$ B. $y = 2\cos(3x)$ C. $y = 3\cos(2x)$ D. $y = 2\sin(2x)$ E. $y = 2\sin(3x)$

21. a. Use technology to help sketch the graph of $y = 2\sin(2x) - 3$.
b. Write down the period and the amplitude of the graph in part **a**.

Problem solving

22. Sketch the graphs of each of the following, stating:
i. the period
ii. the amplitude.
a. $y = 2\cos(2x), x \in [0°, 360°]$
b. $y = 3\sin(4x), x \in [0°, 180°]$
c. $y = -2\cos(3x), x \in [-60°, 60°]$
d. $y = 4\sin(2x), x \in [-90°, 90°]$

23. Solve each of the following equations for the given values of x.

a. $\cos(2x) = \frac{\sqrt{3}}{2}, x \in [0°, 360°]$

b. $\sin(3x) = \frac{1}{2}, x \in [-90°, 90°]$

c. $\sin(2x) = \frac{1}{\sqrt{2}}, x \in [0°, 360°]$

d. $\cos(3x) = -\frac{1}{\sqrt{2}}, x \in [0°, 360°]$

e. $\sin(4x) = 0, x \in [0°, 180°]$

f. $\tan(4x) = -1, x \in [0°, 180°]$

24. Solve the following for $x \in [0°, 360°]$.
a. $2\cos(x) - 1 = 0$ b. $2\sin(x) = -\sqrt{3}$ c. $-\sqrt{2}\cos(x) + 1 = 0$ d. $\sqrt{2}\sin(x) + 1 = 0$

25. Sketch the graph of $y = \tan(2x), x \in [0°, 180°]$. Write down the period, amplitude and the equations of any asymptotes.

26. A satellite dish is placed on top of an apartment building as shown in the diagram. Determine the height of the satellite dish, in metres correct to 2 decimal places.

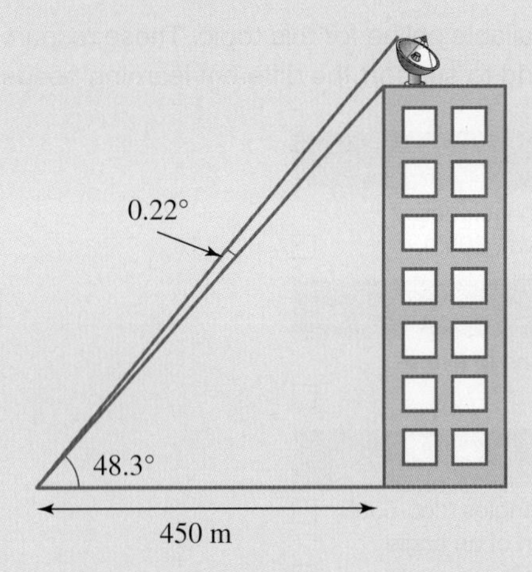

27. Australian power points supply voltage, V, in volts, where $V = 240\,(\sin 18\,000t)$ and t is measured in seconds.

t	V
0.000	
0.005	
0.010	
0.015	
0.020	
0.025	
0.030	
0.035	
0.040	

a. Copy and complete the table and sketch the graph, showing the fluctuations in voltage over time.
b. State the times at which the maximum voltage output occurs.
c. Determine how many seconds there are between times of maximum voltage output.
d. Determine how many periods (or cycles) are there per second.

on To test your understanding and knowledge of this topic, go to your learnON title at www.jacplus.com.au and complete the **post-test**.

Online Resources

Below is a full list of **rich resources** available online for this topic. These resources are designed to bring ideas to life, to promote deep and lasting learning and to support the different learning needs of each individual.

eWorkbook

Download the workbook for this topic, which includes worksheets, a code puzzle and a project (ewbk-2043) ☐

Solutions

Download a copy of the fully worked solutions to every question in this topic (sol-0751) ☐

Digital documents

17.2 SkillSHEET Labelling right-angled triangles (doc-5398) ☐
SkillSHEET Calculating sin, cos or tan of an angle (doc-5399) ☐
SkillSHEET Finding side lengths in right-angled triangles (doc-5400) ☐
SkillSHEET Calculating the angle from a sin, cos or tan ratio (doc-5401) ☐
SkillSHEET Finding angles in right-angled triangles (doc-5402) ☐

Video eLessons

17.2 Exact values of trigonometric functions and the sine rule (eles-5004) ☐
The ambiguous case (eles-5005) ☐
17.3 The cosine rule (eles-5006) ☐
17.4 Area of triangles (eles-5007) ☐
Heron's formula (eles-5008) ☐
17.5 The unit circle (eles-5009) ☐
The four quadrants of the unit circle (eles-5010) ☐
17.6 Sine, cosine and tangent graphs (eles-5011) ☐
17.7 Solving trigonometric equations (eles-5012) ☐

Interactivities

17.2 Exact values of trigonometric functions (int-4816) ☐
The ambiguous case (int-4818) ☐
The sine rule (int-6275) ☐
17.3 The cosine rule (int-6276) ☐
17.4 Area of triangles (int-6483) ☐
Using Heron's formula to calculate the area of a triangle (int-6475) ☐
17.6 Graphs of trigonometric functions (int-4821) ☐
17.7 Solving trigonometric equations graphically (int-4822) ☐
17.8 Crossword (int-2884) ☐
Sudoku puzzle (int-3895) ☐

Teacher resources

There are many resources available exclusively for teachers online.

To access these online resources, log on to **www.jacplus.com.au**.

Answers

Topic 17 Trigonometry II

Exercise 17.1 Pre-test

1. B
2. 9.06 m
3. 9.35 cm
4. A
5. 19.67
6. 7.2 m
7. B
8. 3rd quadrant
9. 0.5
10. D
11. D
12. B
13. A
14. C
15. 270°

Exercise 17.2 The sine rule

1. $44°58'$, $77°2'$, 13.79
2. $39°18'$, $38°55'$, 17.21
3. 70°, 9.85, 9.40
4. 33°, 38.98, 21.98
5. 19.12
6. $C = 51°, b = 54.66, c = 44.66$
7. $A = 60°, b = 117.11, c = 31.38$
8. $B = 48°26', C = 103°34', c = 66.26$; or $B = 131°34'$, $C = 20°26', c = 23.8$
9. 24.17
10. B, C
11. $A = 73°15', b = 8.73$; or $A = 106°45', b = 4.12$
12. 51.90 or 44.86
13. $C = 110°, a = 3.09, b = 4.64$
14. $B = 38°, a = 3.36, c = 2.28$
15. $B = 33°33', C = 121°27', c = 26.24$; or $B = 146°27'$, $C = 8°33', c = 4.57$
16. 43.62 m
17. $h = 7.5$ cm
18. 113 km
19. 8.68 m
20. Yes, she needs 43 m altogether.
21. a. 6.97 m b. 4 m
22. a. 13.11 km b. N20°47′W
23. a. 8.63 km b. 6.48 km/h c. 9.90 km
24. 22.09 km from A and 27.46 km from B.
25. a. 7.3 km b. 282°3′

Exercise 17.3 The cosine rule

1. 7.95
2. 55.22
3. 23.08, $41°53'$, $23°7'$
4. $28°57'$
5. $88°15'$
6. $A = 61°15', B = 40°, C = 78°45'$
7. 2218 m
8. a. 12.57 km b. S35°1′E
9. a. $35°6'$ b. 6.73 m^2
10. 23°
11. 89.12 m
12. a. 130 km b. S22°12′E
13. 28.5 km
14. 74.3 km
15. $70°49'$
16. a. 8.89 m b. $77°0'$ c. $x = 10.07$m
17. 1.14 km/h
18. $\angle CAB = 34.65°$, $\angle ABC = 84.83°$ and $\angle BCA = 60.52°$
19. a. 4.6637 m b. 55.93°

Exercise 17.4 Area of triangles

1. 12.98
2. 38.14
3. 212.88
4. $A = 32°4', B = 99°56'$, area = 68.95 cm^2
5. $A = 39°50', B = 84°10'$, area = 186.03 m^2
6. $A = 125°14', B = 16°46'$, area = 196.03 mm^2
7. C
8. 14.98 cm^2
9. 570.03 mm^2
10. 2.15 cm^2
11. B
12. 3131.41 mm^2
13. 610 cm^2
14. a. 187.5 cm^2 b. 15.03 cm^2 c. 187.47 cm^2
15. 17 goldfish
16. C
17. B
18. a. Area = 69.63 cm^2
    b. Dimensions are 12.08 cm and 6.96 cm.
19. 17 kg
20. 52.2 hectares
21. 175 m^3
22. 22.02 m^2
23. a. 29.44 km^2 b. 8.26 km c. 41°
    d. 24° e. 3.72 km f. 10.11 km^2
    g. 39.55 km^2
24. 374.12 cm^2

Exercise 17.5 The unit circle

1. a. 1st b. 2nd c. 4th
   d. 3rd e. 2nd f. 3rd
2. A
3. D

4. a. 0.34 b. 0.94 c. −0.17 d. 0.98

5. a. −0.64 b. 0.77 c. −0.57 d. −0.82

6. a. 1 b. 0 c. 0 d. −1

7. a. −1 b. 0 c. 0 d. 1

8. a. 0.87 b. 0.50

9. a. 30°
b. −0.87
c. $\cos(150°) = -\cos(30°)$
d. 0.5
e. $\sin(150°) = \sin(30°)$

10. a. 30°
b. −0.87
c. $\cos(210°) = -\cos(30°)$
d. −0.50
e. $\sin(210°) = -\sin(30°)$

11. a. 30°
b. 0.87
c. $\cos(330°) = -\cos(30°)$
d. −0.50
e. $\sin(330°) = -\sin(30°)$

12. a. 0.34 b. 0.94
c. 0.36 d. 0.36
e. They are equal.

13. a. 0.71 b. −0.71
c. −1 d. −1
e. They are equal. f. $\tan(135°) = -\tan(45°)$

14. a. −0.64
b. −0.77
c. 0.84
d. 0.83
e. They are approx. equal.
f. $\tan(220°) = \tan(40°)$

15. a. −0.87
b. 0.5
c. −1.73
d. −1.74
e. They are approx. equal.
f. $\tan(300°) = -\tan(60°)$

16. D

17, 18, 19. Sample responses can be found in the worked solutions in the online resources.

20. a. $\sqrt{1-p^2}$ b. p c. $-\sqrt{1-p^2}$

21. 0

22. 0

Exercise 17.6 Trigonometric functions

1. See table at the bottom of the page.*

2.

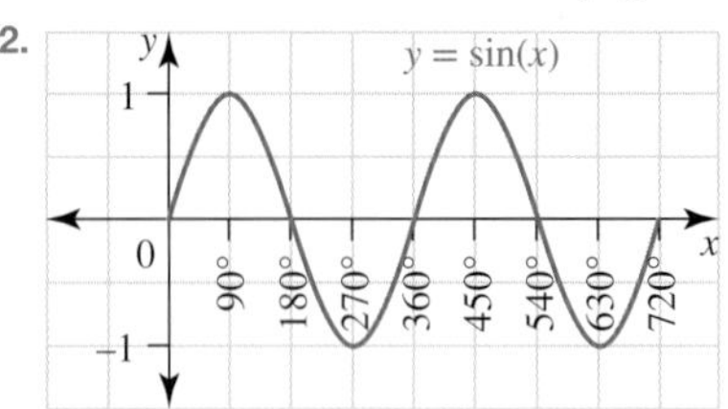

3. 360°

4. a. 0.7 b. 0.8 c. 0.3 d. −0.3

5. a. 0 b. 0.9 c. −0.2 d. −0.9

6. a. 64°, 116°, 424°, 476°
b. 244°, 296°, 604°, 656°
c. 44°, 136°, 404°, 496°

7. a. 210°, 330°, 570°, 690°
b. 233°, 307°, 593°, 667°
c. 24°, 156°, 384°, 516°

8. See table at the bottom of the page.*

9.

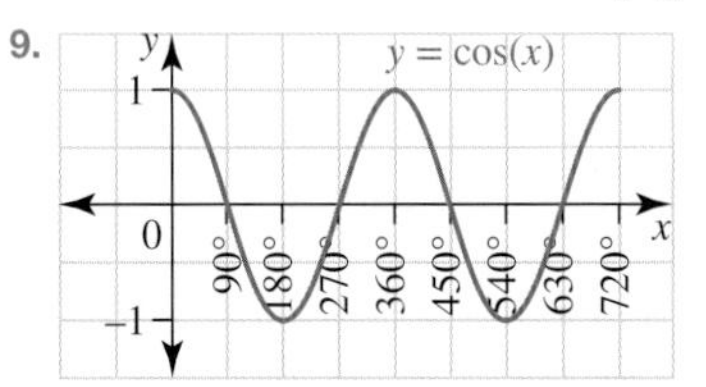

10. The graph would continue with the cycle.

11. It is a very similar graph with the same shape; however, the sine graph starts at (0, 0), whereas the cosine graph starts at (0, 1).

12. a. 0.7 b. −0.9 c. −1 d. 0.9

13. a. −0.5 b. −0.8 c. 0.8 d. −0.9

14. a. 120°, 240°, 480°, 600°
b. 37°, 323°, 397°, 683°
c. 46°, 314°, 406°, 674°

15. a. 127°, 233°, 487°, 593°
b. 26°, 334°, 386°, 694°
c. 154°, 206°, 514°, 566°

*1.

x	0°	30°	60°	90°	120°	150°	180°	210°	240°	270°	300°	330°	360°
sin(x)	0	0.5	0.87	1	0.87	0.5	0	−0.5°	−0.87	−1	−0.87	−0.5	0
x	390°	420°	450°	480°	510°	540°	570°	600°	630°	660°	690°	720°	
sin(x)	0.5	0.87	1	0.87	0.5	0	−0.5	−0.87	−1	−0.87	−0.5	0	

*8.

x	0°	30°	60°	90°	120°	150°	180°	210°	240°	270°	300°	330°	360°
cos(x)	1	0.87	0.5	0	−0.5	−0.87	−1	−0.87	−0.5	0	0.5	0.87	1
x	390°	420°	450°	480°	510°	540°	570°	600°	630°	660°	690°	720°	
cos(x)	0.87	0.5	0	−0.5	−0.87	−1	−0.87	−0.5	0	0.5	0.87	1	

16. See table at the bottom of the page.*

17. $y = \tan(x)$

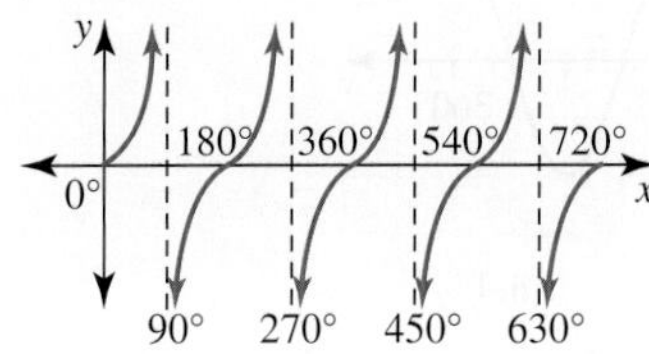

18. The graph would continue repeating every 180° as above.

19. Quite different. $y = \tan(x)$ has undefined values (asymptotes) and repeats every 180° rather than 360°. It also gives all y-values, rather than just values between -1 and 1.

20. a. 1.7 b. -1 c. -1.2 d. 0.8

21. a. -0.8 b. 1.2 c. -0.2 d. 1

22. a. 45°, 225°, 405°, 585°
b. 56°, 236°, 416°, 596°
c. 158°, 338°, 518°, 698°
d. 117°, 297°, 477°, 657°
e. 11°, 191°, 371°, 551°
f. 135°, 315°, 495°, 675°

23. a. i. 360°
ii. 1

iii.

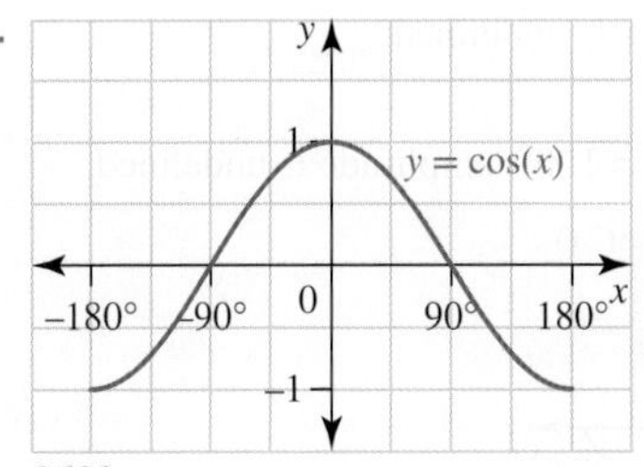

b. i. 360°
ii. 1

iii.

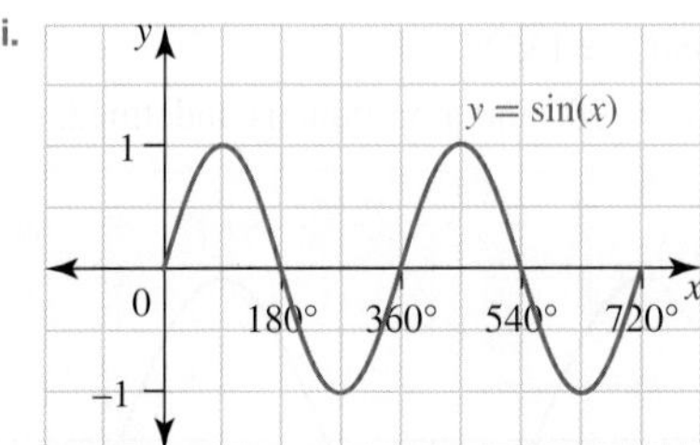

24. a. i. 180°
ii. 1
iii.

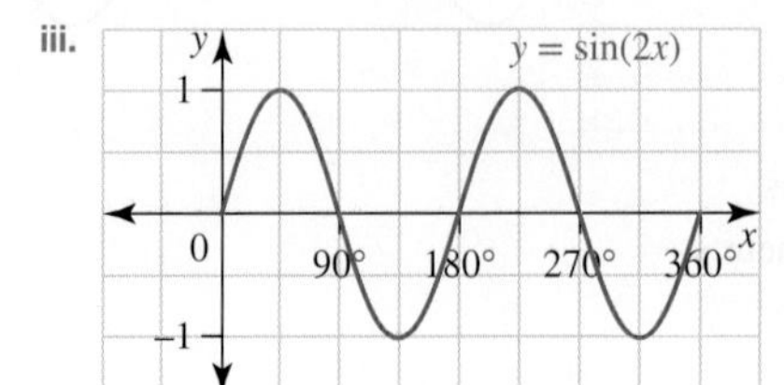

b. i. 360°
ii. 2
iii.

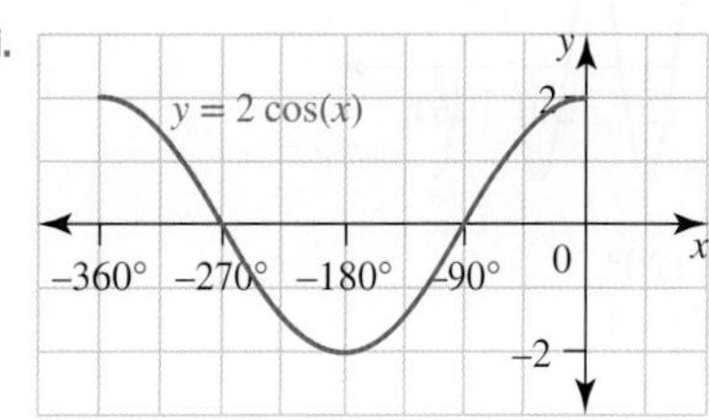

25. a. i. 180° ii. 3
b. i. 120° ii. 4
c. i. 720° ii. 2
d. i. 1440° ii. $\frac{1}{2}$
e. i. 360° ii. 1
f. i. 180° ii. 1

26. a. C b. A c. D

27. a.

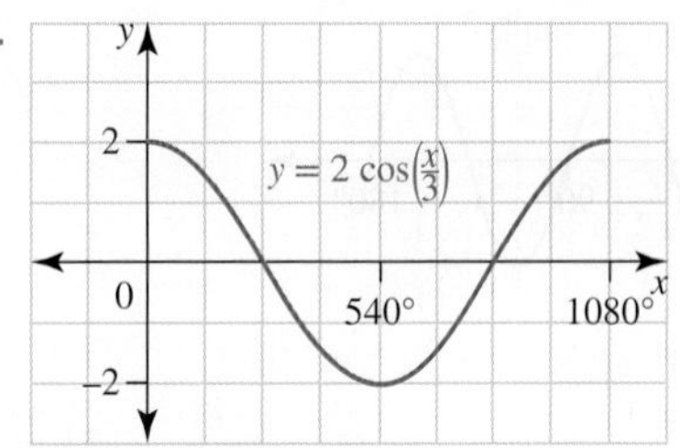

Period = 1080°
Amplitude = 2

b.

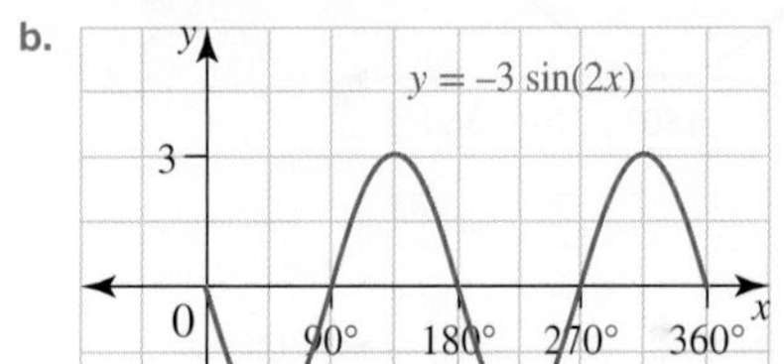

Period = 180°
Amplitude = 3

c.

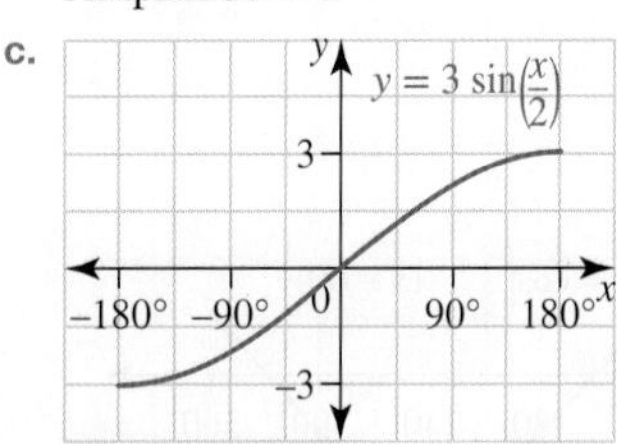

Period = 720°
Amplitude = 3

*16.

x	0°	30°	60°	90°	120°	150°	180°	210°	240°	270°	300°	330°	360°
tan(x)	0	0.58	1.73	undef.	-1.73	-0.58	0	0.58	1.73	undef.	-1.73	-0.58	0
x	390°	420°	450°	480°	510°	540°	570°	600°	630°	660°	690°	720°	
tan(x)	0.58	1.73	undef.	-1.73	-0.58	0	0.58	1.73	undef.	-1.73	-0.58	0	

d. 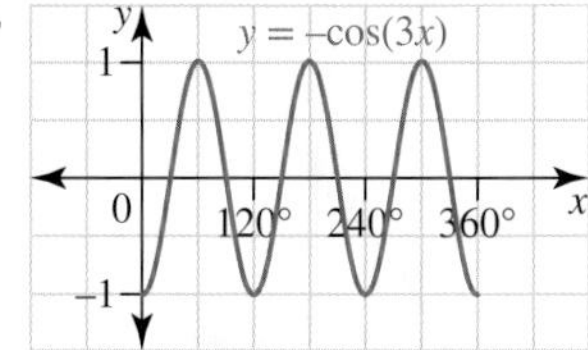

Period = 120°
Amplitude = 1

e. 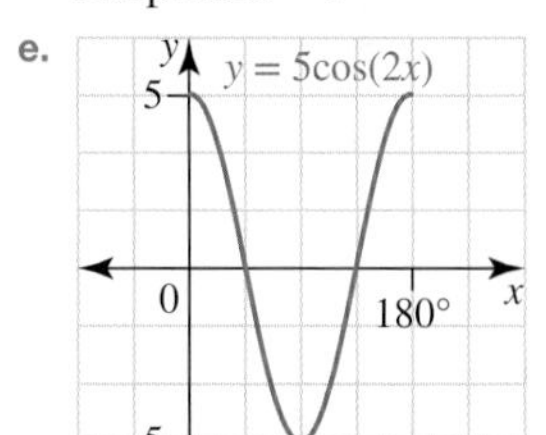

Period = 180°
Amplitude = 5

f. 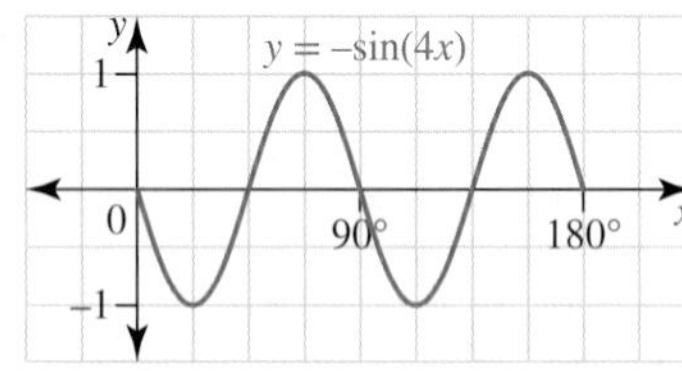

Period = 90°
Amplitude = 1

28. a. 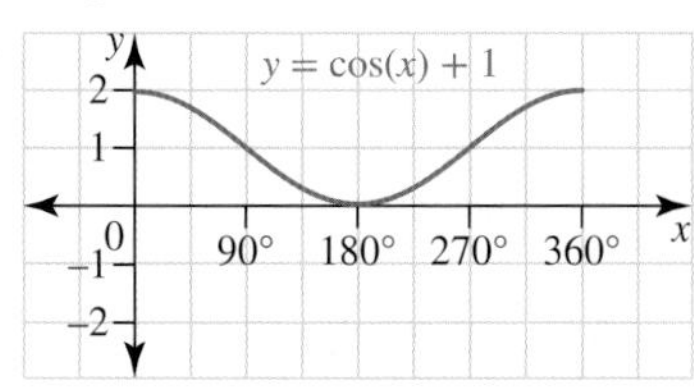

b. 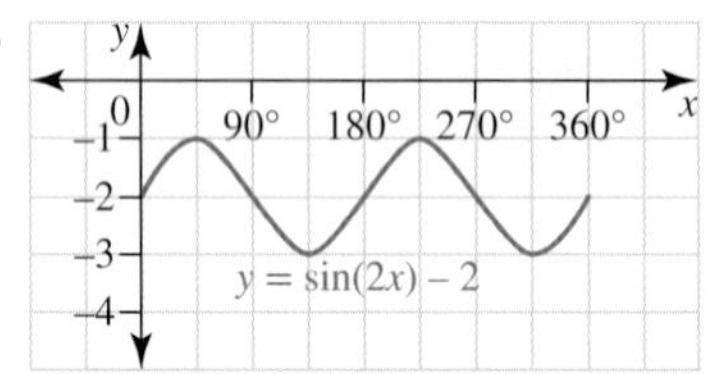

c. 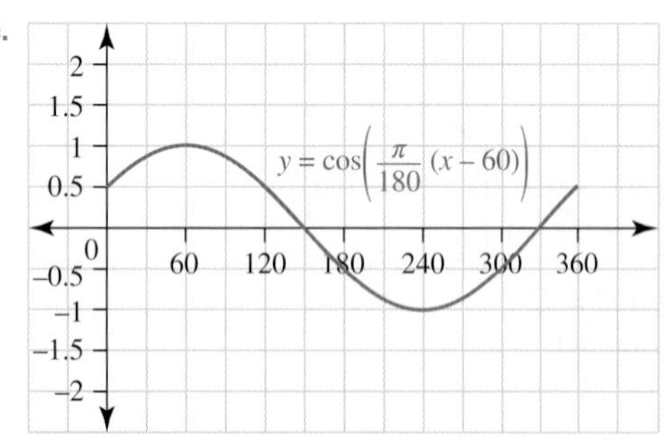

d. 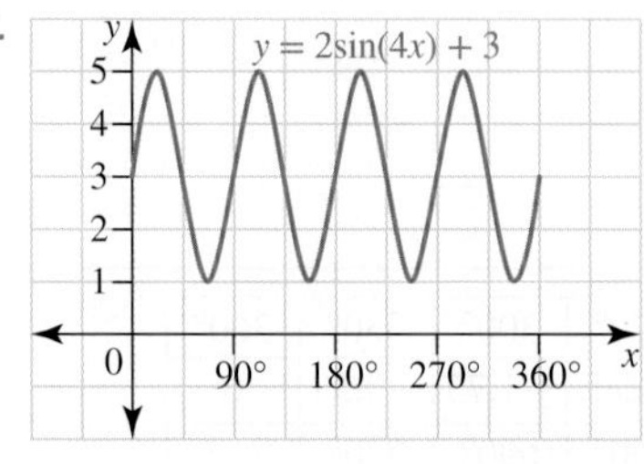

29. a. 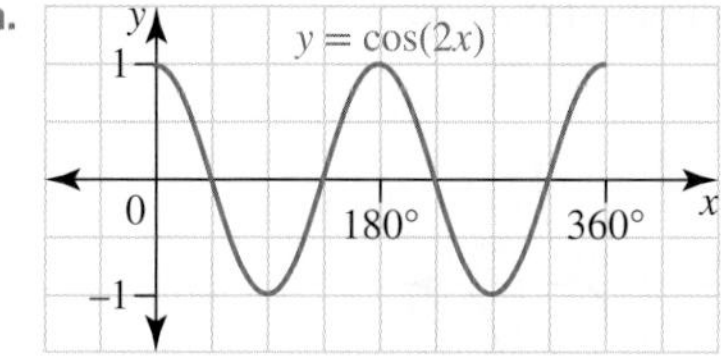

i. -1 ii. 1

b. Max value = 3, min value = 1

c. Max value of $\sin(x) = 1$, hence max value of $y = 2 \times 1 + 3 = 5$
Min value of $\sin(x) = -1$, hence min value of $y = 2 \times -1 + 3 = 1$

30. a.

x	0	30°	60°	90°	120°	150°	180°
y	0	$\frac{\sqrt{3}}{3}$	$\sqrt{3}$	undef	$-\sqrt{3}$	$-\frac{\sqrt{3}}{3}$	0

b. 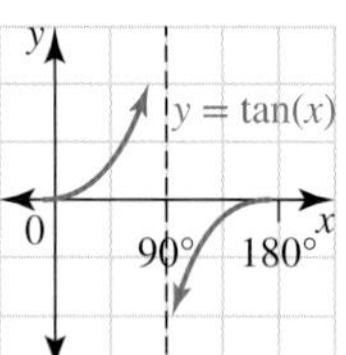

c. At $x = 90°$, y is undefined.

d. $x = 270°$

e. The period = 180°, amplitude is undefined.

31. a. 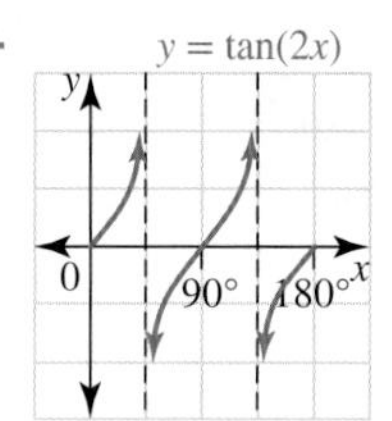

b. $x = 45°$ and $x = 135°$

c. The period = 90° and amplitude is undefined.

32. a. 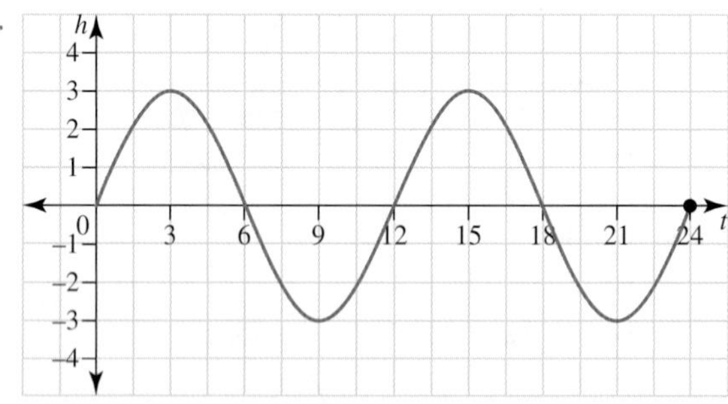

b. 3 metres

c. -2.6 metres

33. a.

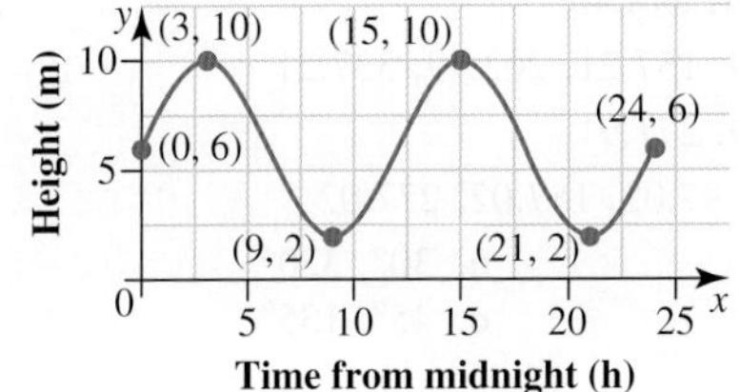

b. 10 metres

c. 2 metres

d. 2.54 metres

34. a. 22 °C

b.

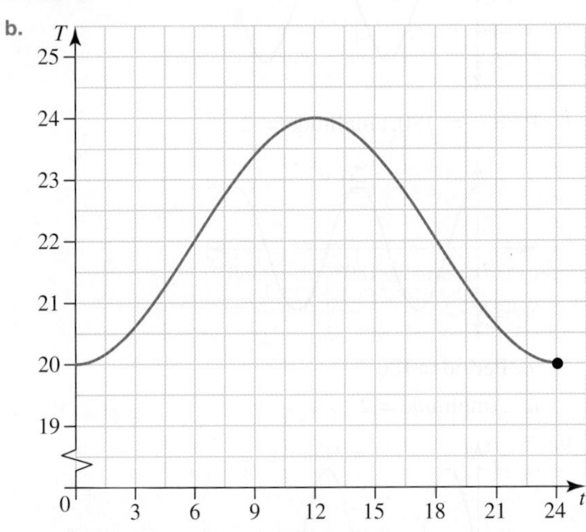

c. Coolest 20 °C, warmest 24 °C

Exercise 17.7 Solving trigonometric equations

1. Calculator answers
   a. 25.84°, 334.16° b. 72.54°, 287.46°
2. a. 101.54°, 258.46° b. 126.87°, 233.13°
3. a. 30°, 150° b. 60°, 120°
4. a. 120°, 240° b. 135°, 225°
5. a. 90° b. 180°
6. a. 210°, 330° b. 225°, 315°
7. a. 30°, 330° b. 150°, 210°
8. a. 90° b. 90°, 270°
9. a. 30°, 60°, 210°, 240° b. 75°, 105°, 255°, 285°
10. a. 15°, 105°, 195°, 285°
    b. 70°, 110°
11. a. 52.5°, 82.5°, 142.5°, 172.5°
    b. −165°, −135°, −45°, −15°, 75°, 105°
12. a. 45°
    b. 30°, 90°, 150°, 210°, 270°, 330°
13. a. 30°, 150° b. 30°, 330°
14. a. 45°, 315° b. 135°, 315°
15. a.

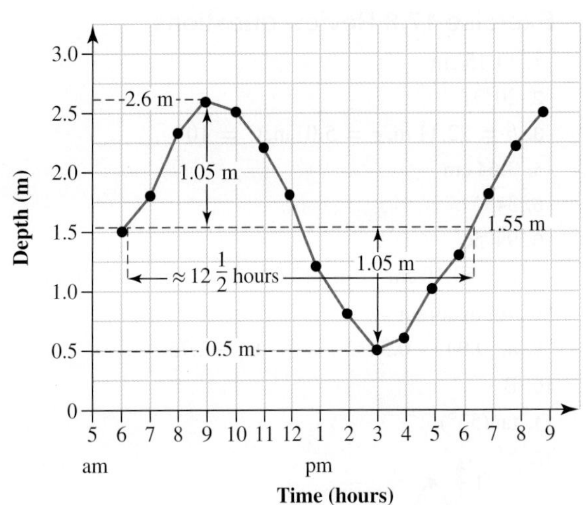

b. i. $12\frac{1}{2}$ hours ii. 1.05 m

c. 10.00 am, 10.30 pm, 11.00 am, 11.30 pm, noon.

d. Until 1.45 am Sunday, 8 am to 2.15 pm and after 8.30 pm.

16. a. $x = 30°, 210°$ b. $x = 153.43°, 333.43°$
17. $x = 30°, 150°$
18. a. 100^g b. 200^g c. 300^g d. 400^g

Project

1. Follow given instructions.
2. a. 4π cm
   b. 3π cm
   c. π cm
   d. The largest arc length equals the sum of the two smaller arc lengths.
3. Sample responses can be found in the worked solutions in the online resources.
4. a. 3π cm^2
   b. In ΔAYZ: $AZ^2 = AY^2 + YZ^2 = 6^2 + YZ^2$

   In ΔBYZ: $BZ^2 = BY^2 + YZ^2 = 2^2 + YZ^2$

   Adding these equations: $AZ^2 + BZ^2 = 6^2 + YZ^2 + 2^2 + YZ^2$

   But in ΔAZB: $AZ^2 + BZ^2 = AB^2$

   $$6^2 + YZ^2 + 2^2 + YZ^2 = 8^2$$
   $$2\,YZ^2 = 64–36–4$$
   $$YZ^2 = 12$$
   $$YZ = \pm\sqrt{12}$$

   But $YZ > 0$ as it is a length: $YZ = 2\sqrt{3}$

   c. 3π cm^2
   Yes, same area
5. Area of the arbelos $= \pi r_1 r_2$
   Sample responses can be found in the worked solutions in the online resources.

Exercise 17.8 Review questions

1. 14.15 cm
2. $20°31'$
3. $b = 22.11$ m, $c = 5.01$ m, $C = 10°$
4. 3.64 cm
5. 34°
6. $94°56'$
7. a. 159.10 cm^2 b. 17.68 cm c. 159.10 cm^2
8. 4th quadrant
9. a. 0.94, −0.34 b. −2.75
10. B
11. tan(53°)
12. 

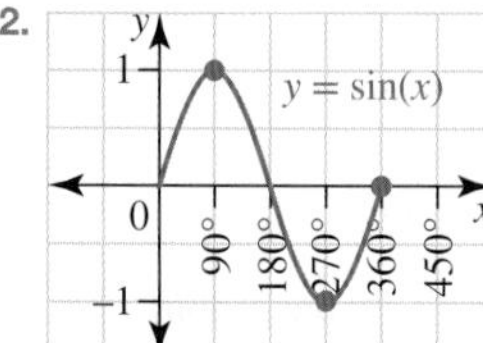

13. 

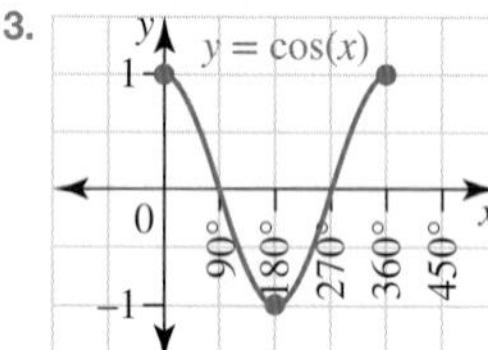

14. 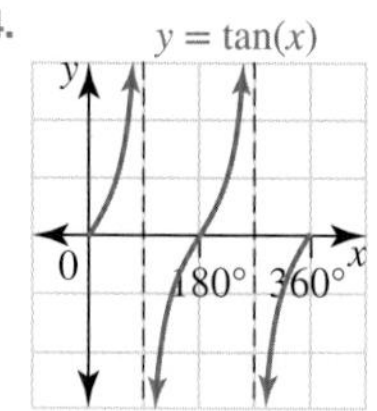

15. 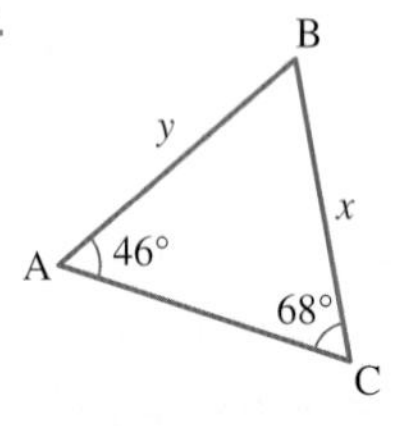

16. a. Period = 120°, amplitude = 2
    b. Period = 180°, amplitude = 3
    c. Period = 180°, amplitude = 1
17. a. 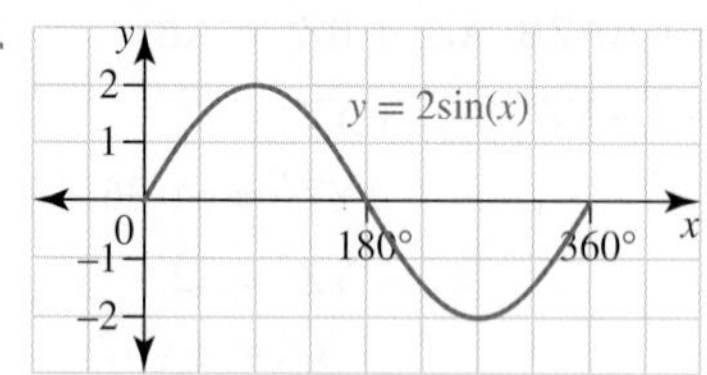

    b. 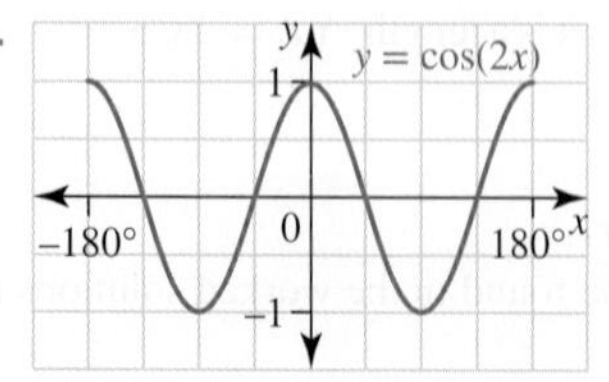

18. a. $x = 191.54, 348.46$
    b. $x = 22.79, 157.21, 202.79, 337.21$
    c. $x = 88.09, 271.91$
    d. $x = 7.02, 97.02, 187.02, 277.02$
19. a. 30°, 150° b. 30°, 330°
    c. 45°, 315° d. 45°, 135°
20. E
21. a. 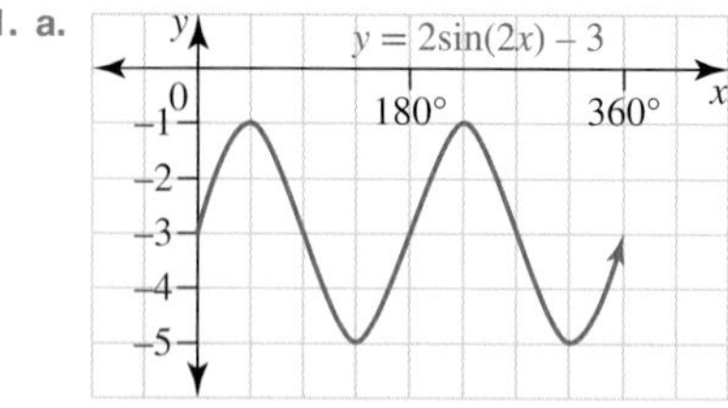

    b. Period = 180, amplitude = 2
22. a. 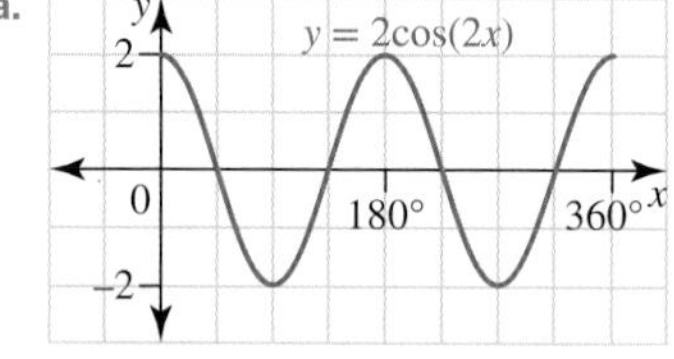

    i. Period = 180°
    ii. Amplitude = 2

    b. 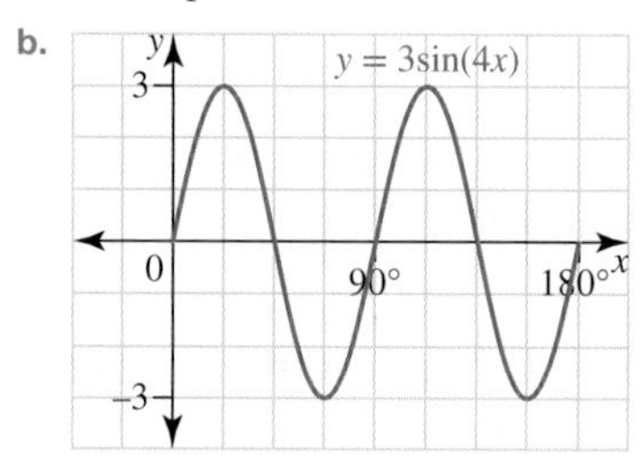

    i. Period = 90°
    ii. Amplitude = 3

    c. 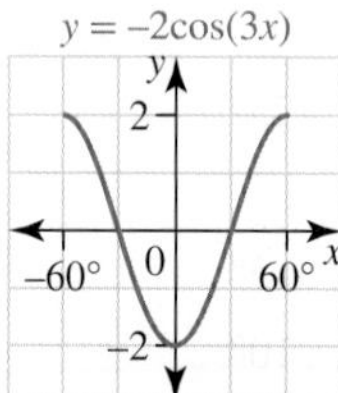

    i. Period = 120°
    ii. Amplitude = 2

    d. 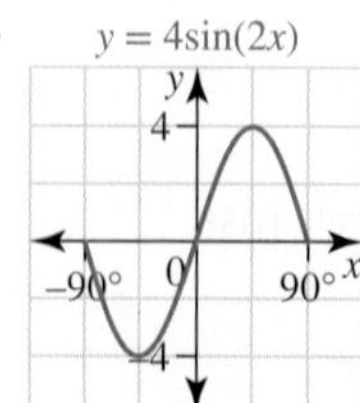

    i. Period = 180°
    ii. Amplitude= 4

23. a. 15°, 165°, 195°, 345°
b. −70°, 10°, 50°
c. 22.5°, 67.5°, 202.5°, 247.5°
d. 45°, 75°, 165°, 195°, 285°, 315°
e. 0°, 45°, 90°, 135°, 180°
f. 33.75°, 78.75°, 123.75°, 168.75°

24. a. 60°, 300°
b. 240°, 300°
c. 45°, 315°
d. 225°, 315°

25. 

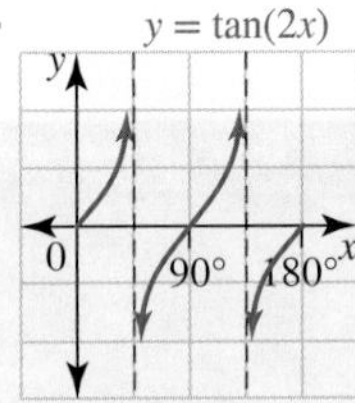

Period = 90°, amplitude is undefined.
Asymptotes are at $x = 45°$ and $x = 135°$.

26. 3.92 m

27. a.

t	V
0.000	0
0.005	240
0.010	0
0.015	−240
0.020	0
0.025	240
0.030	0
0.035	−240
0.040	0

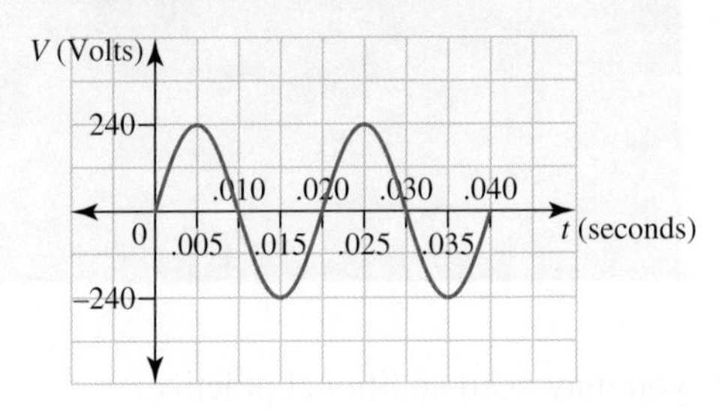

b. Maximum voltage occurs at $t = 0.005$ s, 0.025 s
c. 0.02 s
d. 50 cycles per second

Semester review 2

The learnON platform is a powerful tool that enables students to complete revision independently and allows teachers to set mixed and spaced practice with ease.

Student self-study

Review the **Course Content** to determine which topics and subtopics you studied throughout the year. Notice the green bubbles showing which elements were covered.

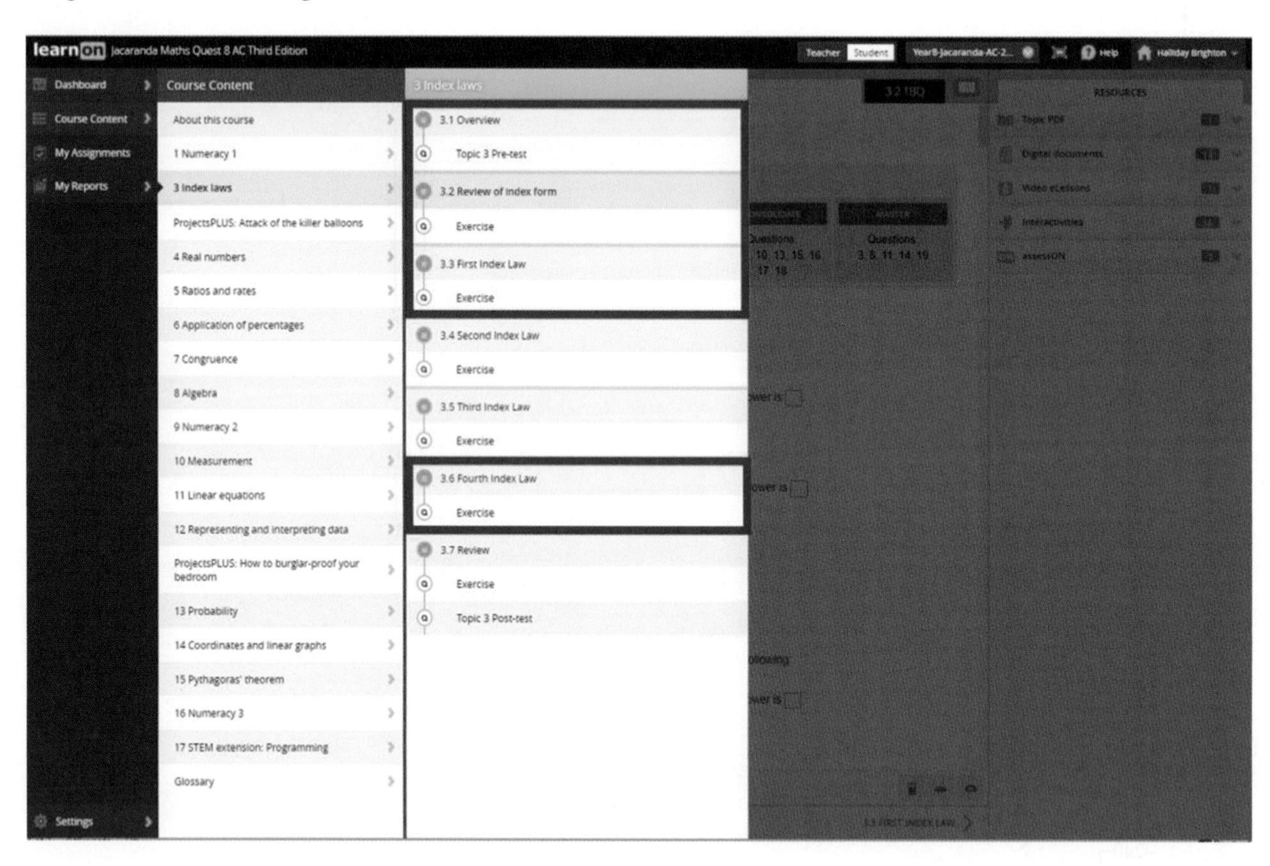

Review your results in **My Reports** and highlight the areas where you may need additional practice.

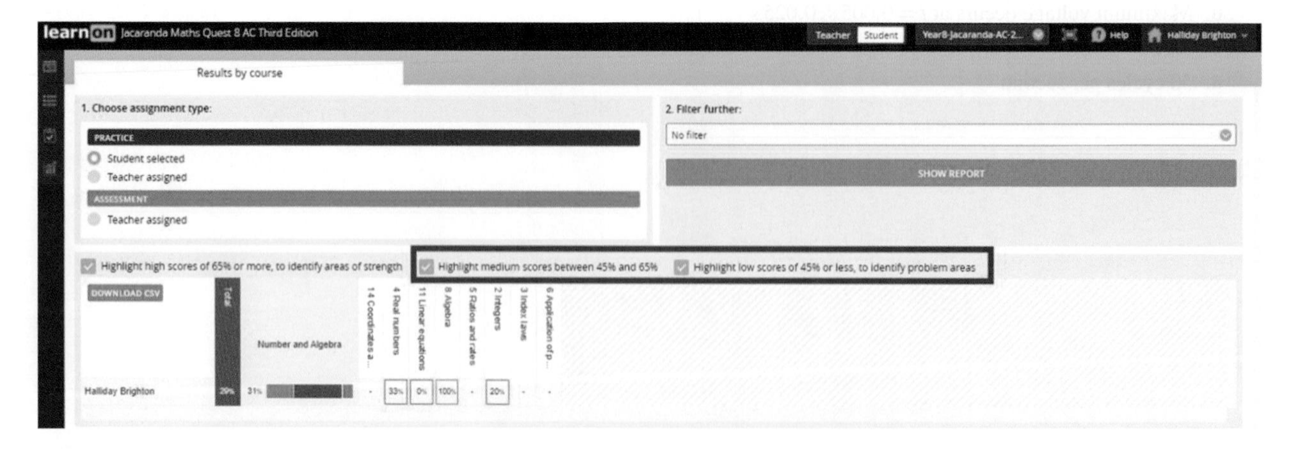

Use these and other tools to help identify areas of strengths and weakness and target those areas for improvement.

Teachers

It is possible to set questions that span multiple topics. These assignments can be given to individual students, to groups or to the whole class in a few easy steps.

Go to **Menu** and select **Assignments** and then **Create Assignment**. You can select questions from one or many topics simply by ticking the boxes as shown below.

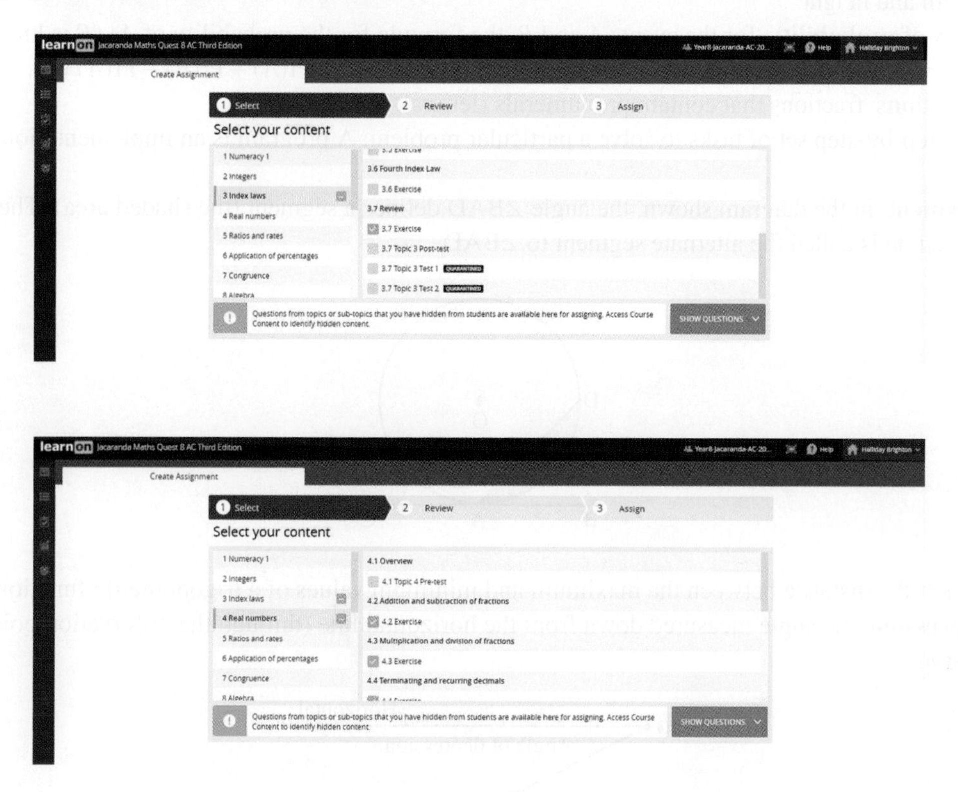

Once your selections are made, you can assign to your whole class or subsets of your class, with individualised start and finish times. You can also share with other teachers.

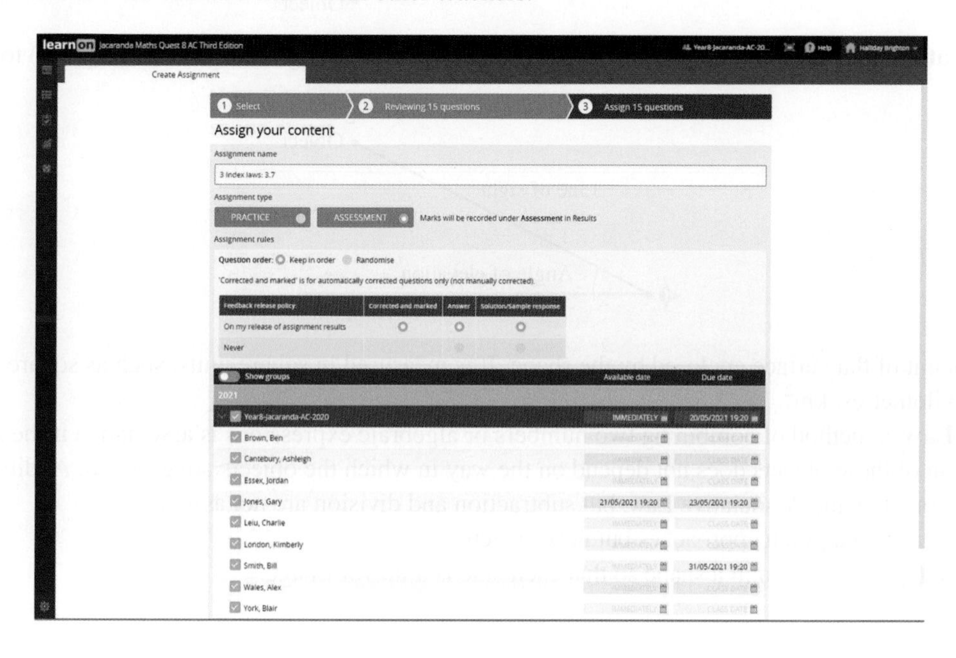

More instructions and helpful hints are available at www.jacplus.com.au.

GLOSSARY

3-dimensional a shape that occupies space (a solid); that is, one that has dimensions in three directions — length, width and height.

Addition Law of probability for the events A and B, the formula for the probability of $A \cup B$ is known as the Addition Law of probability and is given by the formula: $\Pr(A \cup B) = \Pr(A) + \Pr(B) - \Pr(A \cap B)$.

algebraic fractions fractions that contain pronumerals (letters).

algorithm a step-by-step set of tasks to solve a particular problem. A program is an implementation of an algorithm.

alternate segment in the diagram shown, the angle ∠BAD defines a segment (the shaded area). The unshaded part of the circle is called the alternate segment to ∠BAD.

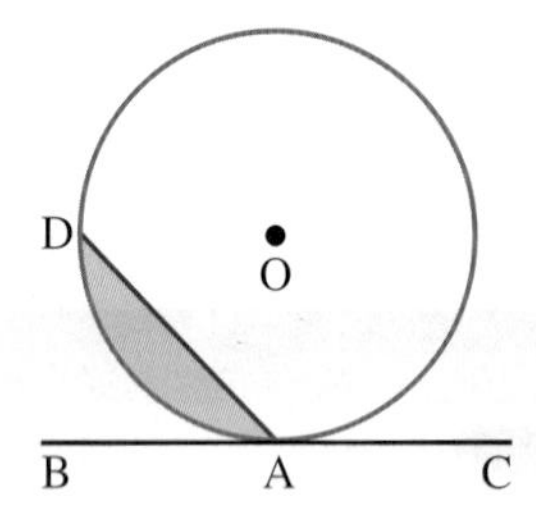

amplitude half the distance between the maximum and minimum values of a trigonometric function.

angle of depression the angle measured down from the horizontal line (through the observation point) to the line of vision.

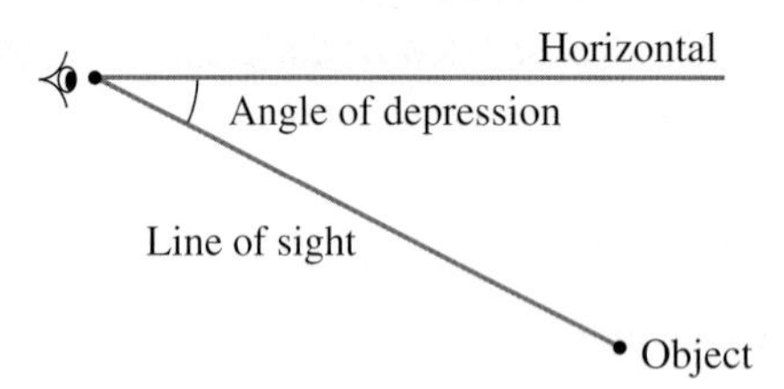

angle of elevation the angle measured up from the horizontal line (through the observation point) to the line of vision.

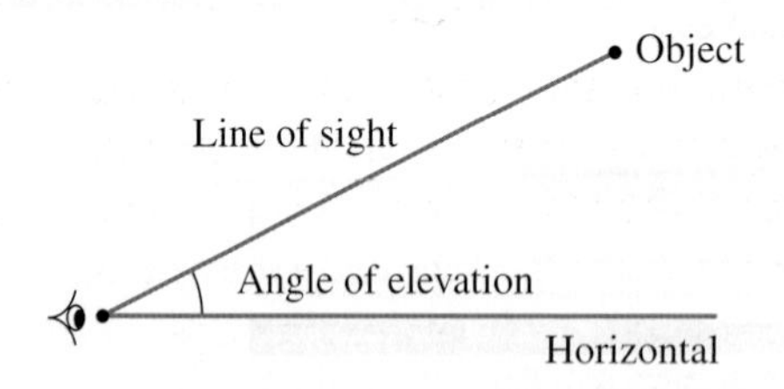

area the amount of flat surface enclosed by the shape. It is measured in square units, such as square metres, m^2, or square kilometres, km^2.

Associative Law a method of combining two numbers or algebraic expressions is associative if the result of the combination of these objects does not depend on the way in which the objects are grouped. Addition and multiplication obey the Associative Law, but subtraction and division are not associative.

asymptote a line that a graph approaches but never meets.

axis of symmetry a line through a shape so that each side is a mirror image.

back-to-back stem-and-leaf plot a method for comparing two data distributions by attaching two sets of 'leaves' to the same 'stem' in a stem-and-leaf plot; for example, comparing the pulse rate before and after exercise. A key should always be included.

Key: 8 | 6 = 86

Leaf *Before exercise*	*Stem*	*Leaf* *After exercise*
9 8 8 8	6	
8 6 6 4 1 1 0	7	
8 8 6 2	8	6 7 8 8
6 0	9	0 2 2 4 5 8 9 9
4	10	0 4 4
0	11	8
	12	4 4
	13	
	14	6

base the digit at the bottom of a number written in index form. For example, in 6^4, the base is 6. This tells us that 6 is multiplied by itself four times.
bias designing a questionnaire or choosing a method of sampling that would not be representative of the population as a whole.
bivariate data sets of data where each piece is represented by two variables.
Boolean a JavaScript data type with two possible values: true or false. JavaScript Booleans are used to make logical decisions.
boundary line indicates whether the points on a line satisfy the inequality.
box plot a graphical representation of the 5-number summary; that is, the lowest score, lower quartile, median, upper quartile and highest score, for a particular set of data. Two or more box plots can be drawn on the same scale to visually compare the five-number summaries of the data sets. These are called parallel box plots.

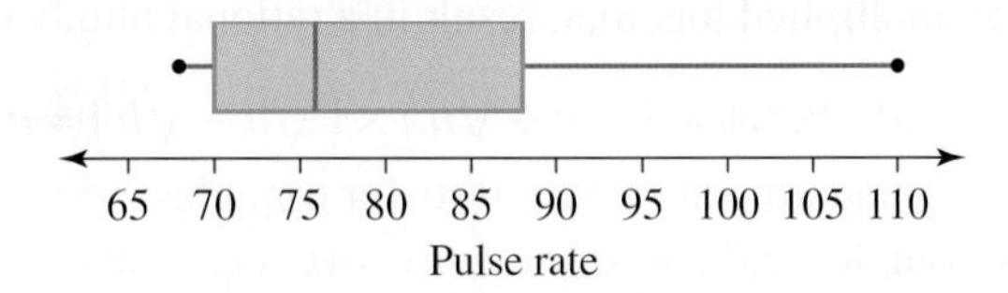

canvas a defined area on a web page where graphics can be drawn with JavaScript.
capacity the maximum amount of fluid that can be contained in an object. It is usually applied to the measurement of liquids and is measured in units such as millilitres (mL), litres (L) and kilolitres (kL).
Cartesian plane the area formed by a horizontal line with a scale (x-axis) joined to a vertical line with a scale (y-axis). The point of intersection of the lines is called the origin.
census collection of data from a population (e.g. all Year 10 students) rather than a sample.
centre middle point of a circle, equidistant (equal in distance) from all points on its circumference.
centre of enlargement the point from which the enlargement of an image is measured.
character in programming, a string of length 1. A JavaScript character is used to represent a letter, digit or symbol.
circle the general equation of a circle with centre (0, 0) and radius r is $x^2 + y^2 = r^2$.
circumcentre the centre of a circle drawn so that it touches all three vertices of a triangle.
circumcircle a circle drawn so that it touches all three vertices of a triangle.
Closure Law when an operation is performed on an element (or elements) of a set, the result produced must also be an element of that set.
coincident lines that lie on top of each other.
collinear points points that all lie on the same straight line.

Commutative Law a method of combining two numbers or algebraic expressions is commutative if the result of the combination does not depend on the order in which the objects are given. For example, the addition of 2 and 3 is commutative, since $2+3=3+2$. However, subtraction is not commutative, since $2-3 \neq 3-2$.

compass bearing directions measured in degrees from the north–south line in either a clockwise or anticlockwise direction. To write the compass bearing we need to state whether the angle is measured from the north or south, the size of the angle and whether the angle is measured in the direction of east or west; for example, N27°W, S32°E.

complement the complement of a set, A, written A', is the set of elements that are in ξ but not in A.

complementary angles two angles that add to 90°; for example, 24° and 66° are complementary angles.

complementary events events that have no common elements and together make up the sample space. If A and A' are complementary events, then $\Pr(A)+\Pr(A')=1$.

completing the square the process of writing a general quadratic expression in turning point form.

composite figure a figure made up of more than one basic shape.

compound interest the interest earned by investing a sum of money (the principal) when each successive interest payment is added to the principal for the purpose of calculating the next interest payment. The formula used for compound interest is: $A=P(1+i)^n$, where A is the amount to which the investment grows, P is the principal or initial amount invested, i is the interest rate per compounding period (as a decimal) and n is the number of compounding periods. The compound interest is calculated by subtracting the principal from the amount: $I=A-P$.

compounded value the value of the investment with accrued interest included.

compounding period the period of time over which interest is calculated.

concave polygon a polygon with at least one reflex interior angle.

concyclic points that lie on the circumference of a circle.

conditional probability where the probability of an event is conditional (depends) on another event occurring first. For two events A and B, the conditional probability of event B, given that event A occurs, is denoted by $\mathrm{P}(B|A)$ and can be calculated using the formula:

$$\mathrm{P}(B|A)=\frac{\mathrm{P}(A\cap B)}{\mathrm{P}(A)}, \mathrm{P}(A)\neq 0.$$

conjugate surds surds that, when multiplied together, result in a rational number. For example, $\left(\sqrt{a}+\sqrt{b}\right)$ and $\left(\sqrt{a}-\sqrt{b}\right)$ are conjugate surds, because $\left(\sqrt{a}+\sqrt{b}\right)\times\left(\sqrt{a}-\sqrt{b}\right)=a-b$.

congruent triangles there are five standard congruence tests for triangles: SSS (side, side, side), SAS (side, included angle, side), ASA (two angles and one side), AAS (two angles and a non-included side) and RHS (right angle, hypotenuse, side).

console a special region in a web browser for monitoring the running of JavaScript programs.

constant of proportionality used to prove that a proportionality relationship (direct or inverse) exists between two or more variables (or quantities).

convex polygon a polygon with no interior reflex angles.

coordinates a pair of values (typically x and y) that represent a point on the screen.

correlation a measure of the relationship between two variables. Correlation can be classified as linear, non-linear, positive, negative, weak, moderate or strong.

cosine ratio the ratio of the adjacent side to the hypotenuse in a right-angled triangle; $\cos\theta=\frac{\text{adjacent}}{\text{hypotenuse}}$.

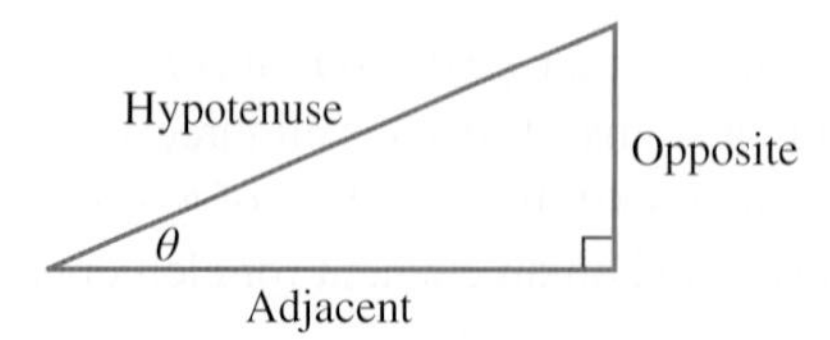

cosine rule in any triangle ABC, $a^2 = b^2 + c^2 - 2bc\cos(A)$.

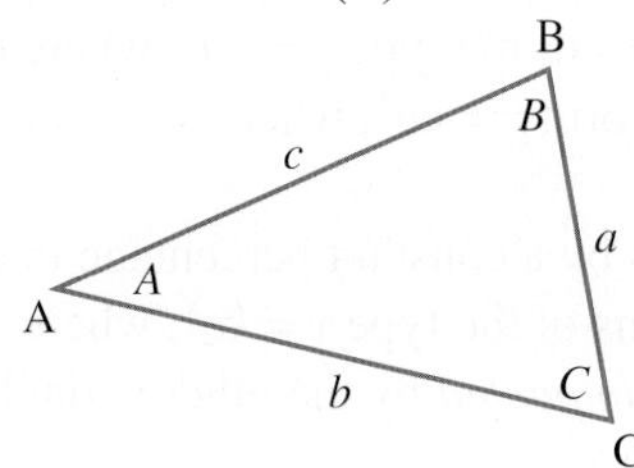

cubic function the basic form of a cubic function is $y = ax^3$. These functions can have 1, 2 or 3 roots.

cumulative frequency the total of all frequencies up to and including the frequency for a particular score in a frequency distribution.

cumulative frequency curve a line graph that is formed when the cumulative frequencies of a set of data are plotted against the end points of their respective class intervals and then joined up by straight-line segments. It is also called an ogive.

cyclic quadrilateral a quadrilateral that has all four vertices on the circumference of a circle. That is, the quadrilateral is inscribed in the circle.

data various forms of information.

degree the degree of a polynomial in x is the highest power of x in the expression.

degrees a unit used to measure the size of an angle.

denominator the lower number of a fraction that represents the number of equal fractional parts a whole has been divided into.

dependent events successive events in which one event affects the occurrence of the next.

dependent variable this variable is graphed on the y-axis.

deviation the difference between a data value and the mean.

dilation occurs when a graph is made thinner or wider.

discriminant referring to the quadratic equation $ax^2 + bx + c = 0$, the discriminant is given by $\Delta = b^2 - 4ac$. It is the expression under the square-root sign in the quadratic formula and can be used to determine the number and type of solutions of a quadratic equation.

domain the set of all allowable values of x.

dot plot this graphical representation uses one dot to represent a single observation. Dots are placed in columns or rows, so that each column or row corresponds to a single category or observation.

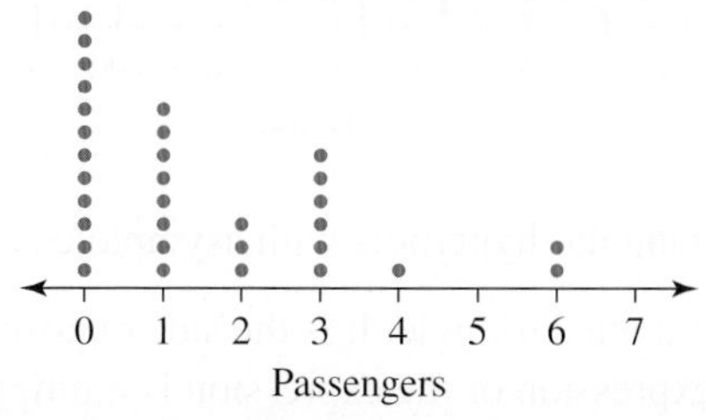

elimination method a method used to solve simultaneous equations. This method combines the two equations into a third equation involving only one of the variables.

enlargement a scaled-up (or down) version of a figure in which the transformed figure is in proportion to the original figure; that is, the two figures are similar.

equate the process of writing one expression as equal to another.

equation a statement that asserts that two expressions are equal in value. An equation must have an equal sign. For example, $x + 4 = 12$.

equiangular when two or more shapes have all corresponding angles equal.

equilateral triangle a triangle with all sides equal in length, and all angles equal to 60°.

event space a list of all the possible outcomes obtained from a probability experiment. It is written as ξ or S, and the list is enclosed in a pair of curled brackets $\{\}$. It is also called the sample space.

experimental probability the probability of an event based on the outcomes of experiments, simulations or surveys.

exponential decay a quantity that decreases by a constant percentage in each fixed period of time. This growth can be modelled by exponential functions of the type $y = ka^x$, where $0 < a < 1$.

exponential function relationship of the form $y = ka^x$, where $a \neq 1$, are called exponential functions with base a.

exponential growth a quantity that grows by a constant percentage in each fixed period of time. This growth can be modelled by exponential functions of the type $y = ka^x$, where $a > 1$.

explanatory variable this variable *is not impacted* by the other variable. This is the x-axis (or horizontal) variable. Also known as the independent variable.

extrapolation the process of predicting a value of a variable outside the range of the data.

factor theorem if $P(x)$ is a polynomial, and $P(a) = 0$ for some number a, then $P(x)$ is divisible by $(x - a)$.

FOIL a diagrammatic method of expanding a pair of brackets. The letters in FOIL represent the order of the expansion: First, Outer, Inner and Last.

frequency the number of times a particular score appears.

function a process that takes a set of x-values and produces a related set of y-values. For each distinct x-value, there is only one related y-value. They are usually defined by a formula for $f(x)$ in terms of x; for example, $f(x) = x^2$.

future value the future value of a loan or investment.

half plane a region that represents all the points that satisfy an inequality.

Heron's formula this formula is used to calculate the area of a triangle when all three sides are known. The formula is $A = \sqrt{s(s-a)(s-b)(s-c)}$, where a, b and c are the lengths of the sides and s is the semi-perimeter or $s = \dfrac{a+b+c}{2}$.

histogram graph that displays continuous numerical variables and does not retain all original data.

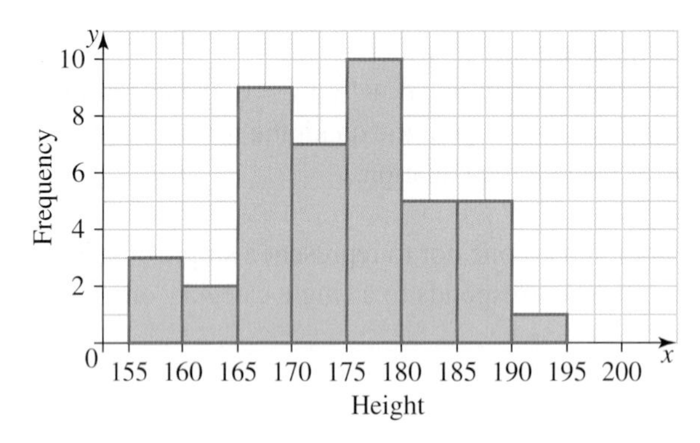

hyperbola the graph of $y = \dfrac{1}{x}$ is a rectangular hyperbola with asymptotes on the x- and y-axes.

hypotenuse the longest side of a right-angled triangle. It is the side opposite the right angle.

Identity Law when 0 is added to an expression or the expression is multiplied by 1, the value of the expression does not change. For example, $x + 0 = x$ and $x \times 1 = x$.

image the enlarged (or reduced) figure produced.

independent events successive events that have no effect on each other.

independent variable this variable *is not impacted* by the other variable. This is the x-axis (or horizontal) variable. Also known as the explanatory variable.

index (index notation) the number that indicates how many times the base is being multiplied by itself when an expression is written in index form, also known as an *exponent* or *power*; (algorithms) an integer that points to a particular item in an array.

inequality when one algebraic expression or one number is greater than or less than another.

inequation similar to equations, but contain an inequality sign instead of an equal sign. For example, $x = 3$ is an equation, but $x < 3$ is an inequation.

integers these include the positive and negative whole numbers, as well as zero; that is, $\ldots, -3, -2, -1, 0, 1, 2, \ldots$

interpolation the process of predicting a value of a variable from within the range of the data.

interquartile range (IQR) the difference between the upper (or third) quartile, Q_{upper} (or Q_3), and the lower (or first) quartile, Q_{lower} (or Q_1); that is, $IQR = Q_{upper} - Q_{lower} = Q_3 - Q_1$. It is the range of approximately the middle half of the data.

inverse function when a function is reflected across the line $y = x$.

Inverse Law when the additive inverse of a number or pronumeral is added to itself, the sum is equal to 0. When the multiplicative inverse of a number or pronumeral is multiplied by itself, the product is equal to 1. So, $x + (-x) = 0$ and $x \times \frac{1}{x} = 1$.

inverse proportion describes a particular relationship between two variables (or quantities); that is, as one variable increases, the other decreases. The rule used to relate the two variables is $y = \frac{k}{x}$.

irrational number numbers that cannot be written as fractions. Examples of irrational numbers include surds, π and non-terminating, non-recurring decimals.

isosceles triangle a triangle with exactly two sides equal in length.

leading coefficient the coefficient of the leading term in a polynomial.

leading term the term in a polynomial that contains the highest power of x.

line of best fit a straight line that best fits the data points of a scatterplot that appear to follow a linear trend. It is positioned on the scatterplot so that there is approximately an equal number of data points on either side of the line, and so that all the points are as close to the line as possible.

line segment a line segment or interval is a part of a line with end points.

linear graph consist of an infinite number of points that can be joined to form a straight line.

linked list a list of objects. Each object stores data and points to the next object in the list. The last object points to a terminator to indicate the end of the list.

literal equation an equation that includes two or more pronumerals or variables.

logarithm the power to which a given positive number b, called the base, must be raised in order to produce the number x. The logarithm of x, to the base b, is denoted by $\log_b(x)$. Algebraically: $\log_b(x) = y \leftrightarrow b^y = x$; for example, $\log_{10}(100) = 2$ because $10^2 = 100$.

logarithmic equation an equation that requires the application of the laws of indices and logarithms to solve.

loop in JavaScript, a process that executes the same code many times with different data each time.

many-to-many relation a relation in which one range value may yield more than one domain value and vice versa.

many-to-one relation a function or mapping that takes the same value for at least two different elements of its domain.

matrix a rectangular array of numbers arranged in rows and columns.

maximal domain the limit of the x-values that a function can have.

mean one measure of the centre of a set of data. It is given by mean $= \frac{\text{sum of all scores}}{\text{number of scores}}$ or $\bar{x} = \frac{\sum x}{n}$.

When data are presented in a frequency distribution table, $\bar{x} = \frac{\sum (f \times x)}{n}$.

measures of central tendency mean, median and mode.

measures of spread range, interquartile range, standard deviation.

median one measure of the centre of a set of data. It is the middle score for an odd number of scores arranged in numerical order. If there is an even number of scores, the median is the mean of the two middle scores when they are ordered. Its location is determined by the rule $\frac{n+1}{2}$.
For example, the median value of the set 1 3 3 4 5 6 8 9 9 is 5, while the median value for the set 1 3 3 4 5 6 8 9 9 10 is the mean of 5 and 6 (5.5).

midpoint the midpoint of a line segment is the point that divides the segment into two equal parts. The coordinates of the midpoint M between the two points $P(x_1, y_1)$ and $Q(x_2, y_2)$ is given by the formula $\left(\frac{x_1+x_2}{2}, \frac{y_1+y_2}{2}\right)$.

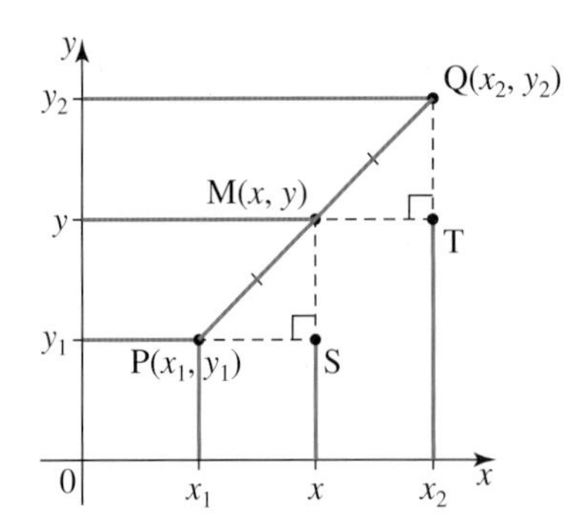

mode one measure of the centre of a set of data. It is the score that occurs most often. There may be no mode, one mode or more than one mode (two or more scores occur equally frequently).

monic any polynomial with a leading coefficient of 1.

Multiplication Law of probability if events A and B are independent, then: $\Pr(A \text{ and } B) = \Pr(A) \times \Pr(B)$ or $\Pr(A \cap B) = \Pr(A) \times \Pr(B)$.

mutually exclusive events that cannot occur together. On a Venn diagram, two mutually exclusive events will appear as disjoint sets.

natural numbers the set of positive integers, or counting numbers; that is, the set 1, 2, 3, …

negatively skewed showing larger amounts of data as the values of the data increase.

nested loop a loop within a loop. The outer loop contains an inner loop. The first iteration of the outer loop triggers a full cycle of the inner loop until the inner loop completes. This triggers the second iteration of the outer loop, which triggers a full cycle of the inner loop again. This process continues until the outer loop finishes a full cycle.

number a JavaScript data type that represents a numerical value.

object a general JavaScript data type that can have many properties. Each property is a name–value pair so that the property has a name to reference a value.

ogive a graph formed by joining the top right-hand corners of the columns of a cumulative frequency histogram.

one-dimensional array a simple array of values in which the values can be of any type except for another array.

one-to-many relation a relation in which there may be more than one range value for one domain value but only one domain value for each range value.

one-to-one relation refers to the relationship between two sets such that every element of the first set corresponds to one and only one element of the second set.

outlier a piece of data that is considerably different from the rest of the values in a set of data; for example, 24 is the outlier in the set of ages $\{12, 12, 13, 13, 13, 13, 13, 14, 14, 24\}$. Outliers sit $1.5 \times \text{IQR}$ or greater away from Q_1 or Q_3.

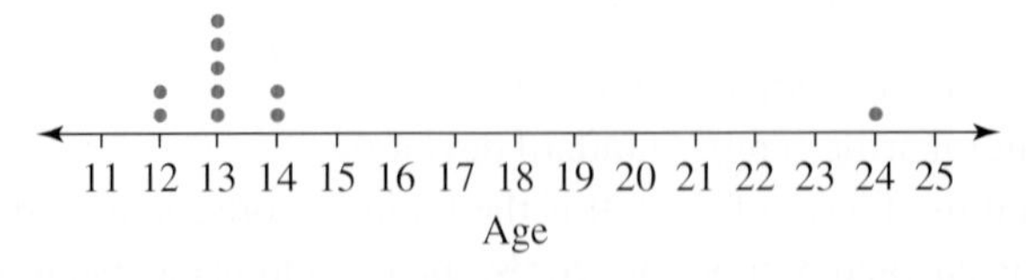

pi (π) the Greek letter π represents the ratio of the circumference of any circle to its diameter. The number π is irrational, with an approximate value of $\frac{22}{7}$, and a decimal value of $\pi = 3.141\,59\ldots.$

parabola the graph of a quadratic function has the shape of a parabola. For example, the typical shape is that of the graph of $y = x^2$.

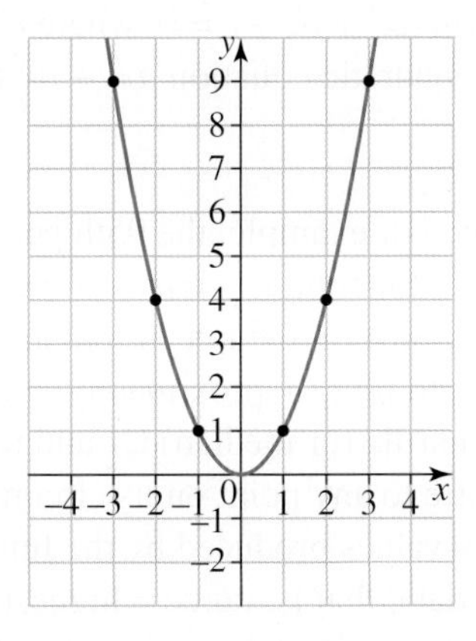

parallel parallel lines in a plane never meet, no matter how far they are extended. Parallel lines have the same gradient.

parallelogram a quadrilateral with both pairs of opposite sides parallel.

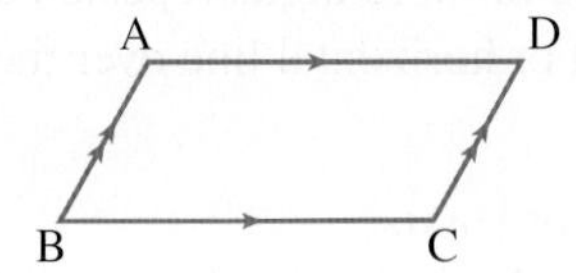

percentile the value below which a given percentage of all scores lie. For example, the 20th percentile is the value below which 20% of the scores in the set of data lie.

period the distance between repeating peaks or troughs of periodic functions.

periodic function a function that has a graph that repeats continuously in cycles, for example, graphs of $y = \sin(x)$ and $y = \cos(x)$.

perpendicular perpendicular lines are at right angles to each other. The product of the gradients of two perpendicular lines is -1.

point of intersection point of intersection a point where two or more lines intersect. The solution to a pair of simultaneous equations can be found by graphing the two equations and identifying the coordinates of the point of intersection.

pointer in JavaScript, a variable that points to a JavaScript object or array. Multiple pointers can point to the same object or array.

polygon a plane figure bounded by line segments.

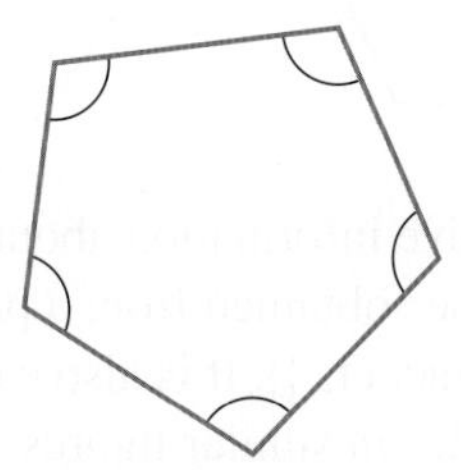

polynomial an expression containing only non-negative integer powers of a variable.

population the whole group from which a sample is drawn.

positively skewed showing smaller amounts of data as the values of the data decrease.

primary data data collected by the user.

probability the likelihood or chance of a particular event (result) occurring.

$$\Pr(\text{event}) = \frac{\text{number of favourable outcomes}}{\text{number of possible outcomes}}.$$

The probability of an event occurring ranges from 0 (impossible — will not occur) to 1 (certainty — will definitely occur).

proof an argument that shows why a statement is true.

property references a value on an object. A complex object may have many properties. Each property has a unique name on the object.
quadratic equation an equation in the form $ax^2 + bx + c = 0$, where a, b and c are numbers.
quadratic formula gives the roots of the quadratic equation $ax^2 + bx + c = 0$. It is expressed as $x = \frac{-b \pm \sqrt{b^2 - 4ac}}{2a}$.
quantile percentiles expressed as decimals. For example, the 95th percentile is the same as the 0.95 quantile.
quartic function the basic form of a quartic function is $y = ax^4$. If the value of a is positive, the curve is upright, whereas a negative value of a results in an inverted graph. A maximum of 4 roots can result.
quartile values that divide an ordered set into four (approximately) equal parts. There are three quartiles — the first (or lower) quartile Q_1, the second quartile (or median) Q_2 and the third (or upper) quartile Q_3.
radius the straight line from a circle's centre to any point on its circumference.
range (functions and relations) the set of y-values produced by the function; (statistics) the difference between the highest and lowest scores in a set of data; that is, range = highest score − lowest score.
rational number numbers that can be written as fractions, where the denominator is not zero.
real numbers rational and irrational numbers combine to form the set of real numbers.
reciprocal when a number is multiplied by its reciprocal, the result is 1.
recurring decimal a decimal which has one or more digits repeated continuously; for example, 0.999 .... They can be expressed exactly by placing a dot or horizontal line over the repeating digits; for example, $8.343\,434 = 8.\dot{3}\dot{4}$ or $8.\overline{34}$.
reflection when a graph is flipped in the x- or y-axis.
regression line a line of best fit that is created using technology.
regular polygon a polygon with sides of the same length and interior angles of the same size.
relation a set of ordered pairs.
relative frequency represents the frequency of a particular score divided by the total sum of the frequencies. It is given by the rule:

$$\text{relative frequency of a score} = \frac{\text{frequency of the score}}{\text{total sum of frequencies}}.$$

remainder theorem if a polynomial $P(x)$ is divided by $x - a$, where a is any real number, the remainder is $P(a)$.
required region the region that contains the points that satisfy an inequality.
response variable another way to reference the dependent variable; this variable is graphed on the y-axis.
rhombus a parallelogram with all sides equal.

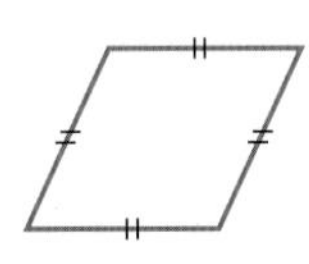
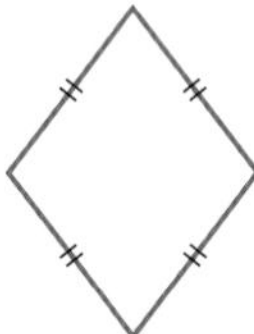

sample part of a population chosen so as to give information about the population as a whole.
sample space a list of all the possible outcomes obtained from a probability experiment. It is written as ξ or S, and the list is enclosed in a pair of curled brackets {}. It is also called the event space.
scale factor the ratio of the corresponding sides in similar figures, where the enlarged (or reduced) figure is referred to as the image and the original figure is called the object.

$$\text{scale factor} = \frac{\text{image length}}{\text{object length}}$$

scatter plot a graphical representation of bivariate data that displays the degree of correlation between two variables. Each piece of data on a scatterplot is shown by a point. The x-coordinate of this point is the value of the independent variable and the y-coordinate is the corresponding value of the dependent variable.
secondary data data collected by others.
similar triangles triangles that have similar shape but different size. There are four standard tests to determine whether two triangles are similar: AAA (angle, angle, angle), SAS (side, angle, side), SSS (side, side, side) and RHS (right angle, hypotenuse side).

simultaneous occurring at the same time.
simultaneous equations the equations of two (or more) linear graphs that have the same solution.
sine ratio the ratio of the opposite side to the hypotenuse in a right-angled triangle; $\sin\theta = \frac{\text{opposite}}{\text{hypotenuse}}$.
sine rule in any triangle ABC, $\frac{a}{\sin A} = \frac{b}{\sin B} = \frac{c}{\sin C}$.

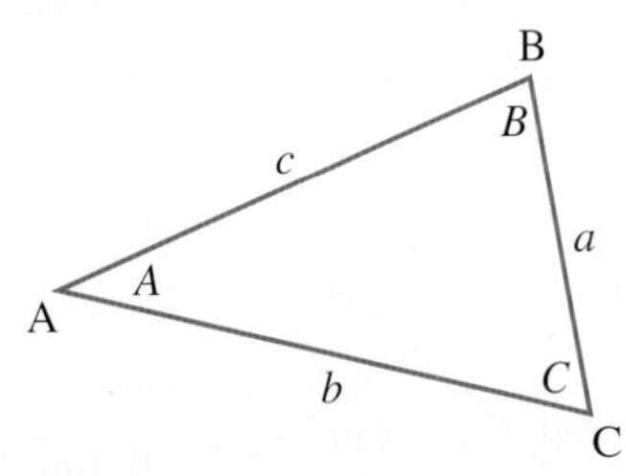

standard deviation a measure of the variability of spread of a data set. It gives an indication of the degree to which the individual data values are spread around the mean.
string a JavaScript data type that represents text.
subject of an equation the variable that is expressed in terms of the other variables. In the equation $y = 3x + 4$, the variable y is the subject.
substitution method a method used to solve simultaneous equations. It is useful when one (or both) of the equations has one of the variables as the subject.
subtended In the diagram shown, chords AC and BC form the angle ACB. Arc AB has subtended angle ACB.

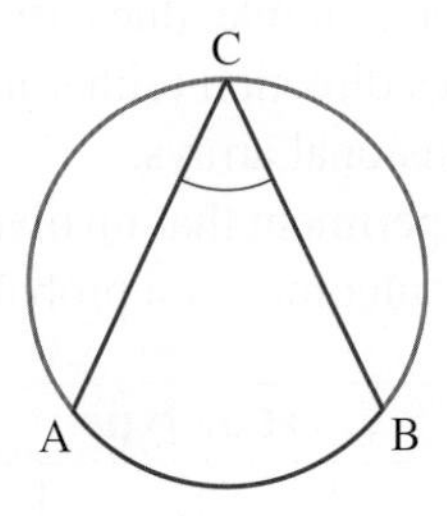

supplementary angles angles that add to 180°.
surd roots of numbers that do not have an exact answer, so they are irrational numbers. Surds themselves are exact numbers; for example, $\sqrt{6}$ or $\sqrt[3]{5}$.
symmetrical the identical size, shape and arrangement of parts of an object on opposite sides of a line or plane.
system of equations a set of two or more equations with the same variables.
tangent ratio the ratio of the opposite side to the adjacent side in a right-angled triangle; $\tan\theta = \frac{\text{opposite}}{\text{adjacent}}$.
terminating decimal a decimal which has a fixed number of places; for example, 0.6 and 2.54.
theorem rules or laws.
theoretical probability given by the rule $\Pr(\text{event}) = \frac{\text{number of favourable outcomes}}{\text{number of possible outcomes}}$ or $\Pr(E) = \frac{n(E)}{n(S)}$, where $n(E)$ = number of times or ways an event, E, can occur and $n(S)$ = number of elements in the sample space or number of ways all outcomes can occur, given all the outcomes are equally likely.
time series a sequence of measurements taken at regular intervals (daily, weekly, monthly and so on) over a certain period of time. They are used for analysing general trends and making predictions for the future.
total surface area (TSA) the area of the outside surface of a 3-dimensional figure.
transformation changes that occur to the basic parabola $y = x^2$ in order to obtain another graph. Examples of transformations are translations, reflections or dilations. Transformations can also be applied to non-quadratic functions.
translation movements of graphs left, right, up or down.

transversal a line that meets two or more other lines in a plane.

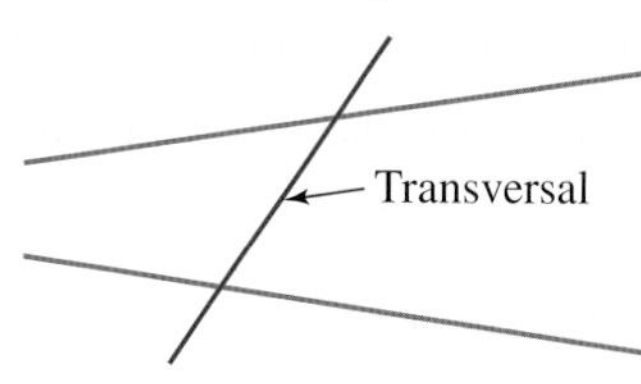

tree diagram branching diagram that lists all the possible outcomes of a probability experiment. This diagram shows the outcomes when a coin is tossed twice.

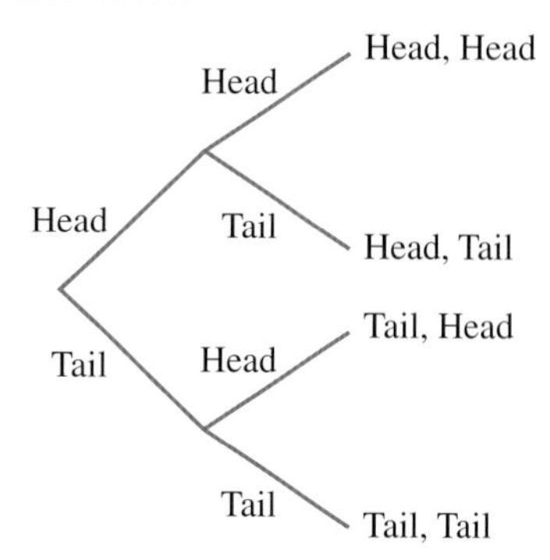

trial the number of times a probability experiment is conducted.

trigonometric ratios three different ratios of one side of a triangle to another. The three ratios are the sine, cosine and tangent.

true bearing a direction which is written as the number of degrees (3 digits) from north in a clockwise direction, followed by the word *true* or T; for example, due east would be 090° true or 090°T.

turning point a point at which a graph changes direction (either up or down).

two-dimensional array an array of one-dimensional arrays.

two-step chance experiment a probability experiment that involves two trials.

two-way table tables that list all the possible outcomes of a probability experiment in a logical manner.

Hair colour	Hair type		Total
Red	1	1	2
Brown	8	4	12
Blonde	1	3	4
Black	7	2	9
Total	17	10	27

unit circle a circle with its centre at the origin and having a radius of 1 unit.

univariate data data relating to a single variable.

variable a named container or memory location that holds a value.

vertex the point at which the graph of a quadratic function (parabola) changes direction (either up or down).

vertically opposite angles when two lines intersect, four angles are formed at the point of intersection, and two pairs of vertically opposite angles result. Vertically opposite angles are equal.

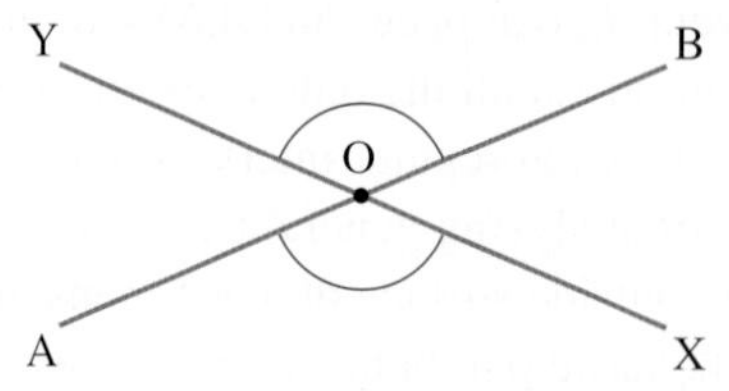

volume the amount of space a 3-dimensional object occupies. The units used are cubic units, such as cubic centimetres (cm^3) and cubic metres (m^3).

***x*-intercept** the point where a graph intersects the x-axis.

***y*-intercept** the point where a graph intersects the y-axis. In the equation of a straight line, $y = mx + c$, the constant term, c, represents the y-intercept of that line.

INDEX

D

E

F

G

H

I

L

M

T

U

V

W

X

Y